# HPLC METHODS
# FOR PHARMACEUTICAL
# ANALYSIS

# HPLC METHODS
# FOR PHARMACEUTICAL
# ANALYSIS

George Lunn and Norman R. Schmuff

A WILEY-INTERSCIENCE PUBLICATION

**JOHN WILEY & SONS, INC.**

New York / Chichester / Weinheim / Brisbane / Singapore / Toronto

Copyright © 1997 by John Wiley & Sons, Inc.

*Library of Congress Cataloging in Publication Data:*
Lunn, George.
    HPLC methods for pharmaceutical analysis / by George Lunn and
  Norman R. Schmuff.
      p.   cm.
    Includes bibliographical references and index.
    ISBN 0-471-18176-5 (cloth : alk. paper)
    1. High performance liquid chromatography.  2. Drugs—analysis.
I. Schmuff, Norman R.  II. Title.
    [DNLM: 1. Drugs—analysis.  2. Chromatography, High Pressure
Liquid—methods.  3. Chemistry, Pharmaceutical—methods.  QV 25
L936h 1996]
    RS189.5.H54L77   1996
    615′.1901—dc20          IOO 1130970
    DNLM/DLC
    for Library of Congress                      96-29417
                                             CIP

ISBN 0-471-18176-5 (Book only)
ISBN 0-471-18198-6 (CD-ROM only)
ISBN 0-471-13078-8 (Book/CD-ROM set)

# CONTENTS

# PREFACE

This book is a collection of procedures for the analysis of a number of pharmaceuticals using high-performance liquid chromatography (HPLC). For each compound various techniques are described in sufficient detail that the analyst can replicate the procedure without reference to the original publication. Since detailed procedures for the same drug are listed together, it is very easy for the researcher to combine features of different methods, e.g., the extraction procedure from one paper and the chromatographic procedure from another paper, to provide methods tailored to the researcher's requirements. In addition to the detailed procedures, bibliographies are provided listing other references. These references are annotated so that the reader can rapidly determine those procedures likely to be of the most utility. In the current volume, we have listed procedures for the analysis of the most commonly used drugs in the United States.[1,2] In future volumes we hope to cover the remaining drugs used for medical and veterinary purposes.

The impetus for writing this book was the realization that there was no single volume listing analytical procedures and that there was, in particular, no ready source of information on the analysis of drugs in biological fluids other than the original literature. Although a number of methods are in common use for the analysis of pharmaceuticals, HPLC may be regarded as the "gold standard"[3] and so we have decided to concentrate on this procedure. For example, HPLC assays of antibiotics have advantages such as specificity, better accuracy and precision, and wider availability of equipment, over other methods. Thus, HPLC procedures are gradually coming to replace other techniques.[4]

Although the universal penetration of computers has led to readily available laboratory-based searches of the literature, this resource is not exploited as much as it might be. An FDA inspector has stated[5] that many pharmaceutical firms, when questioned about deficiencies in this area, admit to never having performed a literature search for HPLC methods. One reason for this reluctance is, of course, that a computer search merely produces a listing of possibly relevant references. Tedious and time-consuming searches in the library are necessary to find the most relevant reference that can be turned into a practical analytical procedure in the searcher's own laboratory. The reference finally chosen will depend on the individual circumstances such as the matrix in which the drug is present and availability of equipment. This book circumvents this lengthy process by providing a number of abstracted and evaluated procedures for the analysis of each drug. The analyst can rapidly identify a relevant procedure and put it into practice without having to consult the original literature. For many compounds the number of analytical procedures is so large that it is not possible to fully abstract all of them. For this reason we have added annotated bibliographies so that the researcher can rapidly identify a relevant paper without having to locate and evaluate a large number of irrelevant procedures.

In addition to the analytical matrix, other factors may be important when choosing an analytical procedure. Accordingly, we have noted other features of analytical procedures such as sensitivity, mode of detection, other compounds that interfere with the analysis, and other drugs that may be determined at the same time.

We would like to thank the staff of the NCI-Frederick Cancer Research and Development Center Scientific Library for their help with this project. In particular, we would like to thank Pat Kuhns-Kelly for her diligent assistance with the computer searches and Ethel Armstrong for her indefatigable efforts in obtaining many of the more obscure references via Interlibrary Loan. The use of the NIH Library, National Institutes of Health, Bethesda, MD, is also greatly appreciated. We would also like to thank Perry King of John Wiley for help with the electronic version. Special thanks are due to our editor, Betty Sun, who brought the whole project together. The initial research which later became the nucleus of this book was supported by the Division of Safety, NIH, through NCI contract NO1-CO-74102 with Program Resources, Inc. Although many people have helped with the preparation of this work, the mistakes are our own. We would appreciate hearing from anyone who has corrections, comments, or suggestions. We can be reached at 71061.2731@compuserve.com or norman.schmuff@tcs.wap.org.

The content of this publication does not necessarily reflect the views, policies, or guidelines of the Department of Health and Human Services or the Food and Drug Administration, nor does mention of trade names, commercial products, or organizations imply endorsement by the U.S. Government.

## REFERENCES

1. *American Druggist* **1993**, *207(3)*, 49–53.
2. *Med Ad News*, **1995 (May)**, 22.
3. Jones, P.M.; Brune, K. Monitoring cyclosporine by HPLC with cyclosporin C as internal standard. *Clin. Chem.* **1993**, *39*, 168–169.
4. Wright, W.W. Use of liquid chromatography for the assay of antibiotics. *Pharmacopeial Forum* **1994**, *20*, 8155–8159.
5. Roos, R.W. Validation issues in high pressure liquid chromatographic methods for pharmaceuticals. Abstract #15, 27th Middle Atlantic Regional Meeting of the American Chemical Society, Hempstead, NY, June 2–4, 1993.

# ABOUT THIS BOOK

## SCOPE

Based on surveys of the top 200 drugs by number of prescriptions filled[1] and the top 100 drugs by dollar sales,[2] we selected the most commonly used drugs in the United States. Analytical procedures for these drugs are described in this book. In future volumes, we hope to cover other drugs used for medical and veterinary purposes.

After the target compounds were identified, a computer search was used to identify relevant references. In general, the computer search was conducted using Medline, 1980 to 1996, but for a few of the most common drugs the complete *Chemical Abstracts* file on Dialog was used. Tests showed that retrieval using Dialog and Medline was similar, except that references from the *Journal of Liquid Chromatography* are not included in Medline. All relevant references from the *Journal of Liquid Chromatography* (1980 to 1996) were manually added to the database.

In Medline the search strategy was:

HPLC (tw) **or** HPLC (mh) **or** liquid chromatography (mh)

**and** USAN drug name (tw or mh) where tw = text word and mh = MESH heading.

In addition to the Medline search some journals were routinely surveyed for relevant articles. These journals were:

*American Journal of Health-System Pharmacy* (formerly *American Journal of Hospital Pharmacy*)

*Analyst*

*Analytical Chemistry*

*Antimicrobial Agents and Chemotherapy*

*Arzneimittelforschung*

*Biochemical Pharmacology*

*Biological and Pharmaceutical Bulletin*

*Biomedical Chromatography*

*Biopharmaceutics and Drug Disposition*

*Chemical and Pharmaceutical Bulletin*

*Chromatographia*

*Clinical Chemistry*
*Clinical Pharmacology and Therapeutics*
*Drug Metabolism and Disposition*
*Drug Metabolism Reviews*
*Farmaco*
*Journal of Analytical Toxicology*
*Journal of AOAC International* (formerly *Journal of the Association of Official Analytical Chemists*)
*Journal of Chromatographic Science*
*Journal of Chromatography* (*Part A* and *Part B*)
*Journal of Clinical Pharmacology*
*Journal of Forensic Sciences*
*Journal of Liquid Chromatography & Related Technology* (formerly *Journal of Liquid Chromatography*)
*Journal of Medicinal Chemistry*
*Journal of Pharmaceutical and Biomedical Analysis*
*Journal of Pharmaceutical Sciences*
*Journal of Pharmacology and Experimental Therapeutics*
*Pharmaceutical Research*
*Pharmazie*
*Therapeutic Drug Monitoring*
*Xenobiotica*

Many other journals were consulted when relevant articles were identified by computer searches. In general, the literature is covered from 1980 to 1996, but a few earlier references are included.

Two other books list some HPLC procedures for drug analyses. The *United States Pharmacopeia*[3] lists procedures for the assay of drugs in formulations and *Official Methods of Analysis*[4] lists some procedures for determining drug residues in food, drink, formulations, and so on. However, these books do not list procedures for determining drugs in biological fluids, raw materials, etc. Additionally, not all the procedures are HPLC procedures. Since these books are widely available, the HPLC procedures they contain are not abstracted in this book.

This book uses non-English characters that are available in extended ASCII. Other non-English characters will be represented by the closest English equivalent or will be spelled out. This is particularly important with the names of authors. For example, Bronnum, where the o has a stroke through it, will be printed as shown because an o with a stroke through it is not available in extended ASCII. On the other hand, Carratù will be printed as shown because u with an accent is available in extended ASCII (#151). A similar situation applies with Greek characters. Thus $\alpha$, $\beta$, and so on, are available and will be printed as such, but a capital delta is not available and will be represented by "delta." The extended ASCII characters which are used include the following: $\alpha$, $\beta$, $\delta$, $\varepsilon$, $\Theta$, $\mu$, $\pi$, $\sigma$, $\Sigma$, $\tau$, á, à, â, ä, Ä, å, Å, ç, Ç, é, É, è, ê, ë, í, ì, î, ï, ñ, Ñ, ó, ò, ô, ö, Ö, ú, ù, û, ü, Ü, $\leq$, $\pm$, $\geq$, and °.

## MONOGRAPH STRUCTURE

Each monograph is headed by the name and structure of the target compound as well as the CAS Registry Number, molecular formula, and molecular weight. At the end of the book other names, such as trade names, that are used for this compound are given with references to the relevant monograph. Note that these names may be for formulations containing salts (e.g., the hydrochloride), mixtures of the compound with other drugs, or derivatives of the compound (e.g., esters). Names of salts (which would have identical chromatographic properties) are not further identified. Names of derivatives, such as esters, which would have different chromatographic properties, are identified by placing the derivative name in parentheses. In general, the United States Adopted Name (USAN) is used for the title of each monograph, although we have sometimes used a truncated version of this name, e.g., adiphenine for adiphenine hydrochloride. Mention of other names is only for the purpose of locating the drug under its US Adopted Name. The only exceptions are isotretinoin and tretinoin, which are listed in the monograph "Retinoic Acid," some of the steroids, which are listed for convenience in the monograph "Estrogens, conjugated," hyoscyamine, which is listed in the monograph for its racemate, atropine, levonorgestrel, which is listed in the "Norgestrel" monograph, and clavulanate potassium, which is listed in the "Clavulanic Acid" monograph. Separations that include estradiol and other conjugated estrogens are dealt with in the "Estrogens, conjugated" monograph. Separations that deal only with estradiol are dealt with in the estradiol monograph. Conjugated Estrogens, USP is defined[3] as ". . . a mixture of sodium estrone sulfate and sodium equilin sulfate, derived wholly or in part from equine urine or synthetically from estrone and equilin." This drug substance is further specified as containing a range of sodium estrone sulfate and sodium equilin sulfate as major components. Ranges of minor components, described as ". . . sodium sulfate conjugates . . . of 17α-dihydroequilin, . . . 17α-estradiol, . . . and 17β-dihydroequilin . . . ," are also specified. The current FDA Office of Generic Drugs guidance (August 21, 1991) on "Conjugated Estrogen Tablets—In Vivo Bioequivalence and In Vitro Drug release" (page 4) states that ". . . bioequivalence must be demonstrated with respect to the plasma concentrations of unconjugated estrone and equilin, as well as with respect to the plasma concentrations of conjugated estrone and equilin (the sulfate conjugates of total estrone and equilin)." The manufacturer states[5] that conjugated estrogens is a mixture containing estrone, equilin, 17α-dihydroequilin, 17α-estradiol, equilenin, and 17α-dihydroequilenin as salts of their sulfate esters. As the situation may change, readers are encouraged to consult the latest USP Supplement, and the latest Pharmacopeial Forum. The latest FDA guidance can be obtained from the FDA's Office of Generic Drugs, Division of Bioequivalence, HFD-650, MPN-2, 5600 Fishers Lane, Rockville, MD 20857.

In general, molecular formulae and molecular weights are given for the simplest form of the compound. Thus, although a particular drug might be used as the hydrochloride, the molecular formula and molecular weight will refer to the free base. We have listed the CAS Registry Numbers that refer to the various forms of each drug. For example, CAS Registry Numbers might be listed for free base, hydrochloride, and hydrochloride dihydrate.

At the end of the book cross references are given to the relevant abstracts in *The Merck Index*[6] and the relevant sections in the series *Organic Chemistry of Drug*

*Synthesis* by Lednicer and Mitscher.[7-11] Much useful information, such as melting point, solubility, optical rotation, and references to reviews, can be found in *The Merck Index*. The series by Lednicer and Mitscher gives valuable information about the syntheses of various drugs, and this may be helpful in determining impurities, understanding degradation reactions, and so on.

Each monograph comprises two sections: Procedures and Annotated Bibliography. The first section presents detailed procedures which should enable anyone to reproduce the analyses. The second section lists other relevant papers, but does not give any experimental details. However, some key words are given to indicate key features of the procedure that are not referred to in the title. The intent of this section is to allow readers to determine rapidly if a paper is relevant and worth looking up in the library.

## ABSTRACT STRUCTURE

The detailed procedures given in the first section of each monograph normally contain the following sections. Of course, not all papers give full details, so some sections may be missing.

Reference
Matrix
Sample Preparation
Guard Column
Column
Mobile Phase
Flow Rate
Injection Volume
Retention Time
Detector
Internal Standard
Limit of Detection
Limit of Quantitation
Drugs that Are Extracted under These Conditions
Drugs that Are Chromatographed Simultaneously under These Conditions
Drugs that Are Also Chromatographed under These Conditions
Drugs that Are Non-interfering
Drugs that Are Interfering
Key Words

## ABSTRACT CONVENTIONS

If not otherwise indicated, the detailed procedures describe procedures carried out at ambient temperature using stainless steel columns with biological fluids from humans. In some cases, these parameters may be specified. For example, if **both** human blood and rat blood are analyzed, **both** human and rat will be indicated in the key words section. Note that the noun is used instead of the adjective, e.g., cow **not** bovine.

Similarly, if not otherwise indicated, the procedures in the Annotated Bibliography describe conventional isocratic reverse-phase HPLC at ambient temperature using stainless steel columns with UV detection and biological fluids from humans. In some cases, these parameters may be specified. For example, if **both** normal-phase and reverse-phase techniques are used, **both** will be indicated. Notations are made only if the operating conditions were different from those specified above. For example, if a gradient or column heater was used, this fact will be noted. On the other hand, if the detector is a UV detector, no reference to the detector is made. Some annotation terms that may be used are as follows: chiral, column-switching, column temp [if other than ambient (in °C)], derivatization, gradient, matrix, metabolites (if method resolves compound and metabolites), normal phase, radiolabeled, species (if other than human), stability-indicating, and so on.

Note that the Injection Volume may be either the volume actually injected or the volume of the injection loop. If it is the volume actually injected, this value is also given in the Sample Preparation section. If the actual injection volume is not given in the Sample Preparation section, the Injection Volume given is that of the injection loop.

Gradients are linear and mobile phases are v/v unless otherwise noted. Times given when describing gradient elution, and other procedures such as column switching, are the times for each step, e.g., "MeOH:water 15:85 for 4 min, to 50:50 over 2 min, maintain at 50:50 for 4 min." If we were to include the cumulative times (t) in the example above it would read: "MeOH:water 15:85 for 4 min (t = 4), to 50:50 over 2 min (t = 6), maintain at 50:50 for 4 min (t = 10)."

For the sake of consistency, conditioning procedures for solid-phase extraction (SPE) cartridges are always described at the beginning of the sample preparation sections. Bear in mind, however, that the conditioning procedure should be carried out just prior to use. Thus, if sample preparation is a lengthy procedure, it may be necessary to delay SPE cartridge conditioning until the step requiring the cartridge.

Retention times are frequently estimated from reproduced chromatograms and so the accuracy may not be high. In particular, differences in retention times between adjacent peaks, e.g., enantiomers, may have a high margin of error.

## EXTRACTION FROM BIOLOGICAL MATRICES

In this book certain terms concerning extraction from biological matrices have highly specific meanings. These terms are as follows:

| | |
|---|---|
| Also | Compounds that can be analyzed under the same conditions as the target compound. It is not specified whether they interfere or can be extracted from the biological matrix in question. |
| Extracted | Compounds that can be extracted from the biological matrix in question, can be chromatographed under the same conditions, and do not interfere with the determination of the target compound. |

| | |
|---|---|
| Interfering | Compounds that interfere with the analysis of the target compound. Compounds that interfere with the chromatography of the internal standard are not listed in this category because another internal standard can always be selected or an external standard procedure can be used. |
| Non-interfering | Compounds that do not interfere with the analysis because no peaks appear on the chromatogram. |
| Simultaneous | Compounds that can be chromatographed at the same time as the target compound and do not interfere with the determination of the target compound. Note that the compound cannot necessarily be extracted from the biological matrix in question (although it may be). |

It is common practice, during the validation of a method for the analysis of a drug in biological fluids, to run a variety of other drugs so as to demonstrate interference or non-interference with the extracted drug. However, these other drugs are generally analyzed as solutions in solvents rather than in biological fluids. Methods in which the drugs are analyzed as solutions in solvents are generally not listed in the monographs. However, these procedures can be found using the "simultaneous" feature of the database. In these instances, the chromatographic procedure will be valid, but the extraction procedure from the biological fluid may not necessarily be valid for this drug.

## MATRICES

In an attempt to simplify searching procedures, we have made an effort to minimize the variety of terms used in the matrix heading. However, in a number of cases, the matrix is associated with various key words that can be used to narrow the search. For example, the term "formulations" has the key words tablets, creams, ointments, and injections associated with it. Thus, to find references applicable to tablets, search first for formulations under the matrix heading and then tablets under the key word heading. Note that the term "bulk" is used instead of "raw materials." Some of the more common matrix terms and their associated key words are given below.

| Matrix Term | Associated Key Word |
|---|---|
| bile | |
| blood | plasma, serum, whole blood |
| bulk | |
| CSF | |
| dialysate | |
| formulations | capsules, injections, tablets, creams, ointment, etc. |
| microsomal incubations | |
| milk | |
| perfusate | |
| reaction mixtures | |
| saliva | |
| solutions | |
| tissue | muscle, kidney, liver, heart, spleen, brain, etc. |
| urine | |

## DETECTORS

The following terms are used for HPLC detectors.

| | |
|---|---|
| Chemiluminescence | |
| Conductivity | |
| E | electrochemical |
| ELSD | evaporative light-scattering detector |
| F | fluorescence |
| MS | mass spectrometry |
| Radioactivity | |
| RI | refractive index |
| UV | ultraviolet |

## UNITS

The units used are as follows:

column dimensions in mm (length × internal diameter)

flow rates in mL/min

injection volume in μL

retention time in min

temperatures in °C

wavelengths in nm

## ABBREVIATIONS

| | |
|---|---|
| $\alpha$ | separation factor; defined by $k_2'/k_1'$ where $k_2'$ is the capacity factor of the second peak and $k_1'$ is the capacity factor of the first peak. |
| BHT | 2,6-di-tert-butyl-4-methylphenol, butylated hydroxytoluene |
| CE | capillary electrophoresis |
| DMF | dimethylformamide |
| DMSO | dimethyl sulfoxide |
| em | emission wavelength |
| EtOH | ethanol |
| ex | excitation wavelength |
| F | fluorescence detection |
| FW | formula weight |
| GPC | gel permeation chromatography |
| h | hour |
| IS | internal standard |
| $k'$ | capacity factor; defined by $(t_R - t_0)/t_0$ where $t_R$ is the retention time and $t_0$ is the column dead time |
| L | liter |
| LOD | limit of detection or some other description indicating that this is the smallest concentration or quantity that can be detected or analyzed for |

LOQ          lower limit of quantitation, either given as such in the paper or taken
             as the lower limit of the linear quantitation range
M            molar (i.e., moles/L)
MeCN         acetonitrile
MeOH         methanol
min          minutes
mL           milliliter
mM           milli-molar (i.e., milli-moles/L)
MTBE         methyl *tert*-butyl ether
nM           nano-molar (i.e., nano-moles/L)
RT           retention time
s            seconds
SEC          size exclusion chromatography
SFC          supercritical fluid chromatography
SFE          supercritical fluid extraction
SIM          selected-ion monitoring
SPE          solid phase extraction
Temp         temperature
U            units
UV           ultraviolet detection

## PIC REAGENTS

These reagents are offered by Waters as buffered solutions containing the following compounds:

   PIC A is tetrabutylammonium sulfate

   PIC B5 is pentanesulfonic acid

   PIC B6 is hexanesulfonic acid

   PIC B7 is heptanesulfonic acid

   PIC B8 is 1-octanesulfonic acid

   PIC D4 is dibutylamine phosphate

## WORKING PRACTICES

In general, good working practice, e.g., filtering and degassing mobile phases, using in-line filters, HPLC or analytical grade materials, and high-quality water is assumed. Solutions containing compounds should be protected from light and silanized glassware should be used, unless you have good reason to believe that these precautions are not necessary. A number of excellent texts[12-14] discuss good working practices and procedures in HPLC and these should be consulted.

Details of solution preparation are generally not given. It should be remembered that the preparation of a dilute aqueous solution of a relatively water-insoluble compound can frequently be made by dissolving the compound in a small volume of a water-miscible organic solvent and diluting this solution with water.

It is also assumed that safe working practices are observed. In particular, organic solvents should only be evaporated in a properly functioning chemical fume hood, correct protective equipment should be worn when dealing with potentially hazardous chemical or biological materials, and waste solutions should be disposed of in accordance with all applicable regulations.

A number of solvents used in HPLC are particularly hazardous. For example, benzene is a human carcinogen;[15] chloroform,[16] dichloromethane,[17] dioxane,[18] and carbon tetrachloride[19] are carcinogenic in experimental animals; and DMF[20] and MTBE[21,22] may be carcinogenic. Organic solvents are, in general, flammable and toxic by inhalation, ingestion, and skin absorption. Sodium azide is carcinogenic and toxic and liberates explosive, volatile, toxic hydrazoic acid with acid. Sodium azide can form explosive heavy metal azides, e.g., with plumbing fixtures, and so should not be discharged down the drain.[23] Disposal procedures have been described for a number of hazardous drugs and reagents[23] and recent papers describe a procedure for the hydrolysis of acetonitrile in waste solvent to the much less toxic acetic acid and ammonia.[24,25] Recent work has shown that n-hexane is surprisingly toxic.[26]

## SUPPLIERS

Suppliers of critical items such as columns are given in the abstracts but the suppliers for widely available items are not listed. These suppliers are as follows:

| Item | Supplier |
| --- | --- |
| Adsorbosphere | Alltech Associates |
| Asahipak | Asahi Chemical |
| Bakerbond | J.T. Baker |
| Bond Elut | Varian |
| µBondapak | Waters |
| Chiralcel | Daicel |
| Co:Pell | Whatman |
| Corasil | Waters |
| Cyclobond | Advanced Separation Technologies |
| Econosil | Alltech Associates |
| Econosphere | Alltech Associates |
| Extrelut | E. Merck |
| Hypersil | Shandon |
| Inertsil | MetaChem |
| LiChroprep | E. Merck |
| LiChrosorb | E. Merck |
| LiChrosphere | E. Merck |
| Micropak | Varian |
| Microsorb | Rainin |
| NewGuard | Applied Biosystems |
| Nova-Pak | Waters |
| Nucleosil | Macherey Nagel |
| Partisil | Whatman |

| Item | Supplier |
| --- | --- |
| Pecosphere | Perkin-Elmer |
| Porasil | Waters |
| Sep-Pak | Waters |
| Spheri-5 | Applied Biosystems |
| Spheri-10 | Applied Biosystems |
| Spherisorb | Phase Separations |
| SPICE | Analtech |
| Supelcosil | Supelco |
| Ultrasphere | Beckman |
| Ultremex | Phenomenex |
| Vydac | The Separations Group |
| Zorbax | Mac-Mod Analytical |

This list is not intended to be definitive. Many other companies supply these pieces of equipment.

## TRADEMARKS

The following trademarks are used:

| Trademark | Company |
| --- | --- |
| Adsorbosphere | Alltech Associates, Inc. |
| Asahipak | Asahi Chemical Industry Co. Ltd. |
| Bakerbond | J.T. Baker |
| Bond Elut | Varian Associates, Inc. |
| μBondapak | Waters Associates, Inc. |
| Chiralcel | Daicel Chemical Industries, Ltd. |
| Co:Pell | Whatman Chemical Separation Co. |
| Corasil | Waters Associates, Inc. |
| Cyclobond | Advanced Separation Technologies, Inc. |
| Econosil | Alltech Associates, Inc. |
| Econosphere | Alltech Associates, Inc. |
| Extrelut | E. Merck |
| Hypersil | Shandon Scientific, Ltd. |
| Inertsil | GL Sciences Inc. |
| LiChroprep | E. Merck |
| LiChrosorb | E. Merck |
| LiChrosphere | E. Merck |
| Micropak | Varian Associates, Inc. |
| Microsorb | Rainin Instrument Co. Inc. |
| NewGuard | Applied Biosystems |
| Nova-Pak | Waters Associates, Inc. |
| Nucleosil | Macherey Nagel |
| Partisil | Whatman Chemical Separation Co. |
| Pecosphere | Perkin-Elmer |
| PIC | Waters Associates, Inc. |
| Porasil | Waters Associates, Inc. |

| Trademark | Company |
|---|---|
| Resolve | Waters Associates, Inc. |
| Sep-Pak | Waters Associates, Inc. |
| Spheri-5 | Applied Biosystems |
| Spheri-10 | Applied Biosystems |
| Spherisorb | Phase Separations, Ltd. |
| SPICE | Analtech |
| Supelcosil | Supelco, Inc. |
| Ultrasphere | Beckman Instruments, Inc. |
| Ultremex | Phenomenex, Inc. |
| Vydac | The Separations Group |
| Zorbax | DuPont Company |

## REFERENCES

1. The top 200 drugs. *American Druggist* **Feb. 15, 1993**, 18–28.

2. *Med Ad News*, **1995 (May)**, 22.

3. *United States Pharmacopeia*, 23rd revision, Unites States Pharmacopeial Convention, Inc.: Rockville, MD, 1994.

4. Helrich, K., Ed., *Official Methods of Analysis*, 15th edition, Association of Official Analytical Chemists, Inc.: Arlington, VA, 1990.

5. *Physicians' Desk Reference*, 48th edition, Medical Economics Company Inc.: Oradell, NJ, 1994, p. 2594.

6. Budavari, S., Ed., *The Merck Index*, 12th edition, Merck & Co. Inc.: Whitehouse Station, NJ, 1996.

7. Lednicer, D.; Mitscher, L.A. *The Organic Chemistry of Drug Synthesis*, John Wiley & Sons, Inc.: New York, 1977.

8. Lednicer, D.; Mitscher, L.A. *The Organic Chemistry of Drug Synthesis*, Volume 2, John Wiley & Sons, Inc.: New York, 1980.

9. Lednicer, D.; Mitscher, L.A. *The Organic Chemistry of Drug Synthesis*, Volume 3, John Wiley & Sons, Inc.: New York, 1984.

10. Lednicer, D.; Mitscher, L.A.; Georg, G.I. *The Organic Chemistry of Drug Synthesis*, Volume 4, John Wiley & Sons, Inc.: New York, 1990.

11. Lednicer, D. *The Organic Chemistry of Drug Synthesis*, Volume 5, John Wiley & Sons, Inc.: New York, 1995.

12. Snyder, L.R.; Kirkland, J.J. *Introduction to Modern Liquid Chromatography*, 2nd edition, John Wiley & Sons, Inc.: New York, 1979.

13. Lawrence, J.F. *Organic Trace Analysis by Liquid Chromatography*, Academic Press: New York, 1981.

14. Snyder, L.R.; Glajch, J.L.; Kirkland, J.J. *Practical HPLC Method Development*, John Wiley & Sons, Inc.: New York, 1988.

15. Lewis, R.J., Sr. *Sax's Dangerous Properties of Industrial Materials,* 8th edition, van Nostrand-Reinhold: New York, 1992, pp. 356–358.

16. Reference 15, pp. 815–816.

17. Reference 15, pp. 2311–2312.

18. Reference 15, pp. 1449–1450.

19. Reference 15, pp. 701–702.

20. Reference 15, p. 1378.

21. Belpoggi, F.; Soffritti, M.; Maltoni, C. Methyl-tertiary-butyl ether (MTBE)—a gasoline additive—causes testicular and lympho-haematopoietic cancers in rats. *Toxicol. Ind. Health* **1995**, *11*, 119–149.

22. Mehlman, M.A. Dangerous and cancer-causing properties of products and chemicals in the oil refining and petrochemical industry: Part XV. Health hazards and health risks from oxygenated automobile fuels (MTBE): Lessons not heeded. *Int. J. Occup. Med. Toxicol.* **1995**, *4*, 219–236.

23. Lunn, G.; Sansone, E.B. *Destruction of Hazardous Chemicals in the Laboratory*, 2nd edition, John Wiley & Sons, Inc.: New York, 1994.

24. Gilomen, K.; Stauffer, H.P.; Meyer, V.R. Detoxification of acetonitrile—water wastes from liquid chromatography. *Chromatographia* **1995**, *41*, 488–491.

25. Gilomen, K.; Stauffer, H.P.; Meyer, V.R. Management and detoxification of acetonitrile wastes from liquid chromatography. *LC.GC* **1996**, *14*, 56–58.

26. Meyer, V. A safer solvent. *Anal. Chem.* **1997,** *69*, 18A.

# Acetaminophen

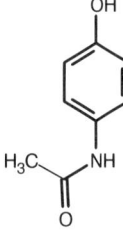

**Molecular formula:** $C_8H_9NO_2$
**Molecular weight:** 151.2
**CAS Registry No.:** 103-90-2

## SAMPLE
**Matrix:** bile, blood, tissue, urine
**Sample preparation:** Tissue. Macerate 2 g tissue with 5 mL water, add 15 mL MeCN, add 0.1 mL 1 mg/mL 2-acetamidophenol in ethanol, shake, centrifuge at 5200 g. Transfer supernatant to 50 mL tube containing 8 mL diethyl ether + 12 mL dichloromethane + 1 mL citrate buffer, vortex, proceed as in (A). Bile. 2 mL Bile + 3 mL water, add 15 mL MeCN, add 0.1 mL 1 mg/mL 2-acetamidophenol in ethanol, vortex, add 8 mL diethyl ether + 12 mL dichloromethane, vortex, centrifuge at 5200 g, proceed as in (A). Blood, urine. 5 mL Blood or urine + 15 mL acetone + 0.1 mL 1 mg/mL 2-acetamidophenol in ethanol, vortex for a few s, add 8 mL diethyl ether, vortex, centrifuge 5200 g. Transfer supernatant to 50 mL tube, add 12 mL dichloromethane, add 1 mL citrate buffer, vortex, proceed as in (A). (A). Discard lower, aqueous layer. Filter the organic layer through 3 g florisil + 8 g anhydrous sodium sulfate and wash through with 15 mL diethyl ether. Evaporate filtrate to dryness under a stream of air at 40°. Reconstitute in 5 mL MeCN:0.1 N $NaH_2PO_4$ 60:40 + 3 mL hexane, vortex. Remove and discard upper hexane layer, add 8 mL 20% (v/v) isopropanol in chloroform to the aqueous layer, vortex. Remove and discard the upper aqueous layer and evaporate lower layer. Reconstitute residue in MeCN:water 10:90, inject an aliquot. (Citrate buffer was saturated sodium citrate containing enough sodium tunstate (sic) to bring pH to 8. To each 1 L of diethyl ether 10 mL of water and 1 mL of citrate buffer are added.)

## HPLC VARIABLES
**Column:** 100 × 4.6 C18 microbore
**Mobile phase:** MeCN:dilute phosphoric acid (1 mL 85% phosphoric acid in 140 mL water) 7:93
**Column temperature:** 40
**Flow rate:** 0.3
**Detector:** UV 271

## CHROMATOGRAM
**Retention time:** 2.4
**Internal standard:** 2-acetamidophenol (4.9)
**Limit of detection:** 50 (blood, urine),200 (bile, tissue) ng/mL

## OTHER SUBSTANCES
**Extracted:** theophylline

## KEY WORDS
liver

## REFERENCE
Mathis, D.F.; Budd, R.D. Extraction of acetaminophen and theophylline from post-mortem tissues and urine for high-performance liquid chromatographic analysis. *J.Chromatogr.*, **1988**, *439*, 466–469

## SAMPLE
**Matrix:** blood
**Sample preparation:** 500 µL Plasma + 25 µL 40 µg/mL 3-acetamidophenol, extract with ether:dichloromethane:isopropanol 59:40:1. Remove the organic layer and evaporate it

to dryness under a stream of nitrogen, reconstitute the residue in 250 μL mobile phase, inject an aliquot.

## HPLC VARIABLES
**Column:** 100 × 4.6 microsphere C18 (Chrompack)
**Mobile phase:** MeOH:50 mM pH 7.8 sodium phosphate buffer 4:96
**Flow rate:** 1
**Detector:** UV 278

## CHROMATOGRAM
**Internal standard:** 3-acetamidophenol

## KEY WORDS
plasma; pig; pharmacokinetics

## REFERENCE
Monshouwer, M.; Witkamp, R.F.; Nijmeijer, S.M.; Pijpers, A.; Verheijden, J.H.M.; Van Miert, A.S.J.P.A.M. Selective effects of a bacterial infection (*Actinobacillus pleuropneumoniae*) on the hepatic clearance of caffeine, antipyrine, paracetamol, and indocyanine green in the pig. *Xenobiotica*, **1995**, *25*, 491–499

## SAMPLE
**Matrix:** blood
**Sample preparation:** 100 μL Plasma + 1 μg p-anisamide + 1 mL 500 mM pH 7.4 $Na_2HPO_4/KH_2PO_4$ buffer + 1 mL ethyl acetate, shake for 10 min, centrifuge at 1300 g for 5 min. Evaporate the supernatant to dryness under reduced pressure, reconstitute the residue in 200 μL mobile phase, inject a 100 μL aliquot.

## HPLC VARIABLES
**Column:** 150 × 4 Nucleosil 7C18
**Mobile phase:** MeCN:1% acetic acid 10:90
**Flow rate:** 1.5
**Injection volume:** 100
**Detector:** UV 254

## CHROMATOGRAM
**Internal standard:** p-anisamide
**Limit of detection:** 100 ng/mL

## OTHER SUBSTANCES
**Extracted:** sulfapyridine

## KEY WORDS
plasma; dog; pharmacokinetics

## REFERENCE
Sagara, K.; Mizuta, H.; Ohshiko, M.; Shibata, M.; Haga, K. Relationship between the phasic period of interdigestive migrating contraction and the systemic bioavailability of acetaminophen in dogs. *Pharm.Res.*, **1995**, *12*, 594–598

## SAMPLE
**Matrix:** blood
**Sample preparation:** 0.5 mL Plasma + 25 μL 40 μg/mL 3-acetamidophenol extracted with ether:dichloromethane:isopropanol 60:40:1. Evaporate organic layer under a stream of nitrogen and take up residue in 50 mM pH 7.8 phosphate buffer, inject an aliquot.

## HPLC VARIABLES
**Column:** 100 × 4.6 Chrompack microsphere C18
**Mobile phase:** Gradient. A was 50 mM pH 7.8 phosphate buffer. B was MeOH:50 mM pH 7.8 phosphate buffer 50:50. A:B from 100:0 to 50:50 over 38 min, maintain at 50:50 for 2 min.
**Flow rate:** 0.7
**Injection volume:** 250
**Detector:** UV 254

## CHROMATOGRAM
**Internal standard:** 3-acetamidophenol

## OTHER SUBSTANCES
**Extracted:** metabolites, antipyrine

## KEY WORDS
plasma; pig; pharmacokinetics

## REFERENCE
Monshouwer, M.; Witkamp, R.F.; Pijpers, A.; Verheijden, J.H.M.; Van Miert, A.S.J.P.A.M. Dose-dependent pharmacokinetic interaction between antipyrine and paracetamol *in vivo* and *in vitro* when administered as cocktail in pig. *Xenobiotica*, **1994**, *24*, 347–355

## SAMPLE
**Matrix:** blood
**Sample preparation:** 100 μL Serum + 100 μL buffer + 1.5 mL IS in 5% isopropanol in chloroform, vortex for 30 s, centrifuge. Remove the organic layer and evaporate it to dryness under a stream of air at room temperature, reconstitute the residue in 100 μL mobile phase, inject a 6-10 μL aliquot. (Buffer was 13.6 g $KH_2PO_4$ in 90 mL water, pH adjusted to 6.8 with about 3 mL 10 M NaOH, made up to 100 mL.)

## HPLC VARIABLES
**Guard column:** 20 × 4.6 Supelguard LC-1 (Supelco)
**Column:** 250 × 4.6 5 μm Supelcosil LC-1 (Supelco)
**Mobile phase:** MeOH:MeCN:buffer 17.5:17.5:65 (Buffer was 2.72 g $KH_2PO_4$ in 1.9 L water, pH adjusted to 6.3 with about 2 mL 1 M NaOH, made up to 2 L.)
**Flow rate:** 2
**Injection volume:** 6-10
**Detector:** UV 273

## CHROMATOGRAM
**Retention time:** 2.09
**Internal standard:** 3-isobutyl-1-methylxanthine (3.15)

## OTHER SUBSTANCES
**Extracted:** amobarbital, barbital, caffeine, carbamazepine, chloramphenicol, ethosuximide, mephobarbital, methsuximide, pentobarbital, phenobarbital, phenytoin, primidone, secobarbital, theophylline, thiopental
**Also analyzed:** acetanilide, butabarbital, butalbital, cimetidine, cyheptamide, diazoxide, diflunisal, disopyramide, ethchlorvynol, glutethimide, heptabarbital, hexobarbital, ibuprofen, indomethacin, ketoprofen, mefenamic acid, mephenytoin, methaqualone, methsuximide, methyl salicylate, methyprylon, naproxen, nirvanol, phenacetin, phensuximide, phenylbutazone, salicylamide, sulindac, tolmetin
**Noninterfering:** N-acetylcysteine, N-acetylprocainamide, amikacin, ampicillin, aspirin, chlorpropamide, codeine, diphylline, gentamicin, gentisic acid, meprobamate, morphine, netilmicin, procainamide, quinidine, salicylic acid, sulfamethoxazole, tetracycline, tobramycin, trimethoprim, valproic acid, vancomycin
**Interfering:** oxphenylbutazone

## KEY WORDS
serum

## REFERENCE
Meatherall, R.; Ford, D. Isocratic liquid chromatographic determination of theophylline, acetaminophen, chloramphenicol, caffeine, anticonvulsants, and barbiturates in serum. *Ther.Drug Monit.*, **1988**, *10*, 101–115

## SAMPLE
**Matrix:** blood
**Sample preparation:** Activate a 1 mL Bond-Elut C8 SPE cartridge with 2 mL MeOH then 1 mL 10 mM HCl, do not allow it to dry completely. Sonicate 1 mL whole blood for 20-30 min then apply to cartridge. Wash with 100 µL water, elute with three 500 µL portions of MeOH:MeCN:1% aqueous ammonium hydroxide 50:20:30, combine eluents and evaporate to dryness under a stream of nitrogen at 40°. Redissolve in 1 mL MeOH, inject a 20 µL aliquot.

## HPLC VARIABLES
**Column:** 250 × 4.5 5 µm Spherisorb ODS
**Mobile phase:** MeCN:MeOH:buffer 35:13:52 (Buffer was water adjusted to pH 3.2 with orthophosphoric acid.)
**Flow rate:** 1
**Injection volume:** 20
**Detector:** UV 250

## CHROMATOGRAM
**Retention time:** 2.5

## OTHER SUBSTANCES
**Simultaneous:** ketoprofen, salicylic acid, naproxen, fenoprofen, ibuprofen, indomethacin

## KEY WORDS
whole blood; SPE

## REFERENCE
Moore, C.M.; Tebbett, I.R. Rapid extraction of anti-inflammatory drugs in whole blood for HPLC analysis. *Forensic Sci.Int.*, **1987**, *34*, 155–158

## SAMPLE
**Matrix:** blood
**Sample preparation:** 500 µL Plasma + 75 µL water, mix, inject a 20 µL aliquot directly.

## HPLC VARIABLES
**Guard column:** 23 × 3.9 37-50 µm Bondapak C18/Corasil
**Column:** 300 × 3.9 10 µm µBondapak C18
**Mobile phase:** 50 mM pH 6.0 phosphate buffer
**Flow rate:** 2.5
**Injection volume:** 20
**Detector:** UV 254

## CHROMATOGRAM
**Retention time:** 18
**Internal standard:** acetaminophen

## OTHER SUBSTANCES
**Extracted:** uric acid, oxipurinol, allopurinol

## KEY WORDS
plasma; renew guard column after 50-70 injections; acetaminophen is IS

## REFERENCE
Nissen, P. Simultaneous determination of allopurinol, oxipurinol and uric acid in human plasma by high-performance liquid chromatography. *J.Chromatogr.*, **1982**, *228*, 382–386

## SAMPLE
**Matrix:** blood
**Sample preparation:** 50 μL Serum + 50 μL 100 mM HCl + 50 μL 15 μg/mL β-hydroxyethyltheophylline in MeCN + 3 mL ethyl acetate, mix for 30 s, centrifuge at 3000 g for 3 min. Remove the organic layer and evaporate it to dryness under a stream of nitrogen, reconstitute the residue in 50 μL MeOH, inject a 20 μL aliquot.

## HPLC VARIABLES
**Column:** μBondapak C18
**Mobile phase:** MeCN:buffer 9.75:90.25 (Buffer was 100 mM $KH_2PO_4$ adjusted to pH 4.0 with phosphoric acid.) (At the end of each day clean column with water for 20 min and MeOH for 30 min.)
**Flow rate:** 2
**Injection volume:** 20
**Detector:** UV 254

## CHROMATOGRAM
**Retention time:** 4.3
**Internal standard:** β-hydroxyethyltheophylline (5.8)
**Limit of detection:** 500 ng/mL

## OTHER SUBSTANCES
**Simultaneous:** dyphylline, theophylline, caffeine, aspirin, salicylic acid, procainamide, N-acetylprocainamide
**Noninterfering:** benzoic acid

## KEY WORDS
serum

## REFERENCE
Ou, C.-N.; Frawley, V.L. Theophylline, dyphylline, caffeine, acetaminophen, salicylate, acetylsalicylate, procainamide, and N-acetylprocainamide determined in serum with a single liquid-chromatographic assay. *Clin.Chem.*, **1982**, *28*, 2157–2160

## SAMPLE
**Matrix:** blood
**Sample preparation:** 200 μL Serum + 200 μL 50 μg/mL hexobarbital in MeCN + 25 μL glacial acetic acid, vortex for 10 s, centrifuge for 1 min, inject a 30-100 μL aliquot of the supernatant.

## HPLC VARIABLES
**Column:** μBondapak C18
**Mobile phase:** Gradient. MeCN:7.5 g/L $NaH_2PO_4$ adjusted to pH 3.2 with phosphoric acid 5:95 to 22:78 over 24 min, to 45:55 over 10 min, maintain at 45:55 for 5 min. Re-equilibrate with 5:95 for 5 min.
**Column temperature:** 50
**Flow rate:** 3

**Injection volume:** 30-100
**Detector:** UV 210

## CHROMATOGRAM
**Retention time:** 3.3
**Internal standard:** hexobarbital (20.6)
**Limit of detection:** 200-2000 ng/mL

## OTHER SUBSTANCES
**Extracted:** amobarbital, butabarbital, butalbital, chlordiazepoxide, diazepam, ethchlorvynol, flurazepam, glutethimide, methaqualone, methyprylon, nitrazepam, pentobarbital, phenobarbital, phenytoin, primidone, salicylic acid, secobarbital, theophylline
**Simultaneous:** amitriptyline, caffeine, clomipramine, codeine, desipramine, ethotoin, imipramine, lidocaine, mesantoin, methsuximide, nirvanol, nortriptyline, oxazepam, propranolol, quinidine
**Interfering:** procainamide, phenylpropanolamine

## KEY WORDS
serum

## REFERENCE
Kabra, P.M.; Stafford, B.E.; Marton, L.J. Rapid method for screening toxic drugs in serum with liquid chromatography. *J.Anal.Toxicol.*, **1981**, *5*, 177–182

## SAMPLE
**Matrix:** blood, CSF, gastric fluid, urine
**Sample preparation:** 200 μL Serum, urine, CSF, or gastric fluid + 300 μL reagent. Flush column A to waste with 500 μL 500 mM ammonium sulfate, inject sample onto column A, flush column A to waste with 500 μL 500 mM ammonium sulfate, backflush the contents of column A onto column B with mobile phase, monitor the effluent from column B. (Reagent was 8.05 M guanidine hydrochloride and 1.02 M ammonium sulfate in water.)

## HPLC VARIABLES
**Column:** A 40 μm preparative grade C18 (Analytichem); B 75 × 2.1 pellicular C18 (Whatman) + 250 × 4.6 5 μm C8 end-capped (Whatman)
**Mobile phase:** Gradient. A was 50 mM pH 4.5 $KH_2PO_4$. B was MeCN:isopropanol 80:20. A:B 90:10 for 1 min, to 30:70 over 20 min.
**Column temperature:** 50
**Flow rate:** 1.5
**Detector:** UV 220

## CHROMATOGRAM
**Retention time:** 4.93
**Internal standard:** heptanophenone (19)

## OTHER SUBSTANCES
**Extracted:** allobarbital, azinphos, barbital, brallobarbitone, bromazepam, butethal, caffeine, carbamazepine, carbaryl, cephaloridine, chloramphenicol, chlordiazepoxide, chlorothiazide, chlorvinphos, clothiapine, cocaine, coomassie blue, desipramine, diazepam, diphenhydramine, dipipanone, ethylbromphos, flufenamic acid, formothion, griseofulvin, indomethacin, lidocaine, lorazepam, malathion, medazepam, midazolam, oxazepam, paraoxon, penicillin G, pentobarbital, prazepam, propoxyphene, prothiophos, quinine, salicylic acid, secobarbital, strychnine, sulfamethoxazole, theophylline, thiopental, thioridazine, trimethoprim

## KEY WORDS
serum; column-switching

## REFERENCE

Kruger, P.B.; Albrecht, C.F.De V.; Jaarsveld, P.P. Use of guanidine hydrochloride and ammonium sulfate in comprehensive in-line sorption enrichment of xenobiotics in biological fluids by high-performance liquid chromatography. *J.Chromatogr.*, **1993**, *612*, 191–198

## SAMPLE

**Matrix:** blood, saliva

**Sample preparation:** 500 μL Saliva or plasma + 100 μL 100 μg/mL o-hydroxyacetanilide + 5 mL ethyl acetate, shake for 15 min, centrifuge at 2500 rpm for 5 min. Remove 4 mL of the organic layer and evaporate it to dryness under a stream of nitrogen at 50°, reconstitute the residue in 200 μL mobile phase, inject a 50 μL aliquot.

## HPLC VARIABLES

**Column:** 150 × 4 Inertsil ODS-2

**Mobile phase:** MeOH:1.5% acetic acid 15:85 (plasma) or MeOH:MeCN:50 mM pH 2.5 potassium phosphate buffer 8:7:85 (saliva)

**Column temperature:** 50

**Flow rate:** 1

**Injection volume:** 50

**Detector:** UV 240

## CHROMATOGRAM

**Retention time:** 4.7 (plasma), 4.0 (saliva)

**Internal standard:** o-hydroxyacetanilide (9.9 (plasma), 8.0 (saliva))

**Limit of quantitation:** 100 ng/mL

## KEY WORDS

plasma; human; dog; pharmacokinetics

## REFERENCE

Katori, N.; Aoyagi, N.; Terao, T. Estimation of agitation intensity in the GI tract in humans and dogs based on *in vitro/in vivo* correlation. *Pharm.Res.*, **1995**, *12*, 237–243

## SAMPLE

**Matrix:** blood, saliva, urine

**Sample preparation:** 900 μL Plasma or saliva + 10 μL 10 μg/mL 3-acetaminophen in 30% perchloric acid, mix, centrifuge at 3000 rpm for 15 min, inject an aliquot of the supernatant

## HPLC VARIABLES

**Guard column:** 50 × 4.6 37-53 μm pellicular ODS (Whatman)

**Column:** 250 × 4.6 5 μm Spherisorb ODS2

**Mobile phase:** MeCN:20 mM orthophosphoric acid 4:96, pH 3.5 (urine) or MeOH:MeCN: 20 mM orthophosphoric acid 2:4:94, adjusted to pH 3.2 (plasma) or MeCN:20 mM orthophosphoric acid 6:94 (saliva)

**Column temperature:** 25

**Detector:** UV (wavelength not specified)

## CHROMATOGRAM

**Retention time:** 8.3 (urine), 6.6 (plasma), 6.0 (saliva)

**Internal standard:** 3-acetaminophen (12.1 (urine), 9.3 (plasma), 8.3 (saliva))

**Limit of detection:** 200 ng/mL

## KEY WORDS

plasma; pharmacokinetics

## REFERENCE

Al-Obaidy, S.S.; Po, A.L.W.; McKiernan, P.J.; Glasgow, J.F.T.; Millership, J. Assay of paracetamol and its metabolites in urine, plasma and saliva of children with chronic liver disease. *J.Pharm. Biomed.Anal.*, **1995**, *13*, 1033–1039

## SAMPLE

**Matrix:** blood, urine

**Sample preparation:** 200 µL Plasma or urine spiked with β-hydroxyethyltheophylline + 200 µL 1 M tetrabutylammonium dihydrogen phosphate + 0.2 g potassium sulfate, mix 1 min, add 3 mL chloroform:isopropanol 1:1, mix 1 min, centrifuge 5 min 2500 g. Remove 2 mL of organic layer, evaporate to dryness at 35° under stream of nitrogen, take up in 200 µL mobile phase by mixing for 1 min, inject an aliquot.

## HPLC VARIABLES

**Column:** 150 × 4 5 µm C18 (FSA Laboratory Supplies, Loughborough)

**Mobile phase:** MeOH:1% acetic acid containing 0.5 mM tetrabutylammonium dihydrogenphosphate and 20 mM potassium sulfate 18:82

**Flow rate:** 1.2

**Detector:** UV 254

## CHROMATOGRAM

**Retention time:** 5

**Internal standard:** β-hydroxyethyltheophylline

## OTHER SUBSTANCES

**Extracted:** metabolites

## KEY WORDS

plasma

## REFERENCE

Kamali, F.; Herd, B. Liquid-liquid extraction and analysis of paracetamol (acetaminophen) and its major metabolites in biological fluids by reversed-phase ion-pair chromatography. *J.Chromatogr.*, **1990**, *530*, 222–225

## SAMPLE

**Matrix:** bulk

**Sample preparation:** Prepare a 750 µg/mL solution in 10 mM pH 2.5 orthophosphoric acid, sonicate for 10 min, filter (0.2 µm), inject a 15 µL aliquot.

## HPLC VARIABLES

**Guard column:** 4 × 4 5 µm LiChrospher 100

**Column:** 125 × 4 3 µm Spherisorb ODS-1

**Mobile phase:** Gradient. A was water containing 5 mL/L 85% orthophosphoric acid and 0.56 mL/L hexylamine. B was MeCN:water 90:10 containing 5 mL/L 85% orthophosphoric acid and 0.56 mL/L hexylamine. A:B from 91:9 to 86:14 over 4 min, maintain at 86:14 for 13 min, to 55:45 over 11 min, maintain at 55:45 for 8 min, re-equilibrate at initial conditions for 20 min.

**Flow rate:** 0.7

**Injection volume:** 15

**Detector:** UV 210

## CHROMATOGRAM

**Retention time:** 5.7

## OTHER SUBSTANCES
**Simultaneous:** acetylcodeine, benzocaine, caffeine, cocaine, codeine, diamorphine, lidocaine, 6-monoacetylmorphine, morphine, noscapine, papaverine, procaine

## REFERENCE
Grogg-Sulser, K.; Helmlin, H.-J.; Clerc, J.-T. Qualitative and quantitative determination of illicit heroin street samples by reversed-phase high-performance liquid chromatography: method development by CARTAGO-S. *J.Chromatogr.A*, **1995**, *692*, 121–129

## SAMPLE
**Matrix:** bulk, formulations
**Sample preparation:** Weigh out bulk drug, capsule contents, granules, or powders equivalent to 35 mg cephalexin, add 1 mL 2.52 mg/mL acetaminophen in MeOH:water 20:80, make up to 50 mL with water, inject a 20 μL aliquot.

## HPLC VARIABLES
**Column:** 300 × 3.9 10 μm μBondapak C18
**Mobile phase:** MeOH:1.25% acetic acid 25:75
**Flow rate:** 1
**Injection volume:** 20
**Detector:** UV 254

## CHROMATOGRAM
**Retention time:** 6
**Internal standard:** acetaminophen

## OTHER SUBSTANCES
**Simultaneous:** cephalexin

## KEY WORDS
capsules; granules; powder; acetaminophen is IS

## REFERENCE
Hsu, M.C.; Lin, Y.-S.; Chung, H.-C. High-performance liquid chromatographic method for potency determination of cephalexin in commercial preparations and for stability studies. *J.Chromatogr.A*, **1995**, *692*, 67–72

## SAMPLE
**Matrix:** dialysate
**Sample preparation:** Dialyze blood with Ringer's solution, inject a 0.5 μL aliquot of the dialysate.

## HPLC VARIABLES
**Column:** 14 × 1 3 μm BAS Sep-Stik ODS (Bioanalytical Systems)
**Mobile phase:** 50 mM pH 5.0 ammonium acetate buffer
**Flow rate:** 0.2
**Injection volume:** 0.5
**Detector:** UV 250

## CHROMATOGRAM
**Retention time:** 0.5
**Limit of quantitation:** 300 ng/mL

## OTHER SUBSTANCES
**Extracted:** metabolites

## KEY WORDS
rat; microbore

## REFERENCE
Chen, A.; Lunte, C.E. Microdialysis sampling coupled on-line to fast microbore liquid chromatography. *J.Chromatogr.A*, **1995**, *691*, 29–35

## SAMPLE
**Matrix:** dialysate
**Sample preparation:** Inject an aliquot directly.

## HPLC VARIABLES
**Column:** 100 × 1 5 μm ODS SepStik (Bioanalytical Systems)
**Mobile phase:** MeCN:50 mM pH 2.5 ammonium phosphate buffer 7:93
**Flow rate:** 0.1
**Injection volume:** 7
**Detector:** UV 250

## CHROMATOGRAM
**Retention time:** 4.5
**Limit of detection:** 100 ng/mL

## OTHER SUBSTANCES
**Extracted:** metabolites

## KEY WORDS
microbore; rat

## REFERENCE
Steele, K.M.; Lunte, C.E. Microdialysis sampling coupled to on-line microbore liquid chromatography for pharmacokinetic studies. *J.Pharm.Biomed.Anal.*, **1995**, *13*, 149–154

## SAMPLE
**Matrix:** formulations
**Sample preparation:** Weigh out powdered sample containing 51 mg acetaminophen, add 80 mL MeOH, sonicate for 10 min, dilute to 100 mL with MeOH, centrifuge. Remove a 5 mL aliquot of the supernatant and add it to 1 mL 2 mg/mL resorcinol, add 2 mL MeOH, make up to 20 mL with 50 mM pH 3.0 triethylamine phosphate, inject an aliquot.

## HPLC VARIABLES
**Column:** 150 × 3.2 5 μm Hypersil ODS
**Mobile phase:** THF:50 mM pH 3.0 triethylamine phosphate 12:88
**Flow rate:** 0.6
**Injection volume:** 20
**Detector:** UV 275

## CHROMATOGRAM
**Retention time:** 15
**Internal standard:** resorcinol (9)

## OTHER SUBSTANCES
**Simultaneous:** aspirin (post-column irradiation gives an increase in peak height), caffeine, propyphenazone

## REFERENCE

Di Pietra, A.M.; Gatti, R.; Andrisano, V.; Cavrini, V. Application of high-performance liquid chromatography with diode-array detection and on-line post-column photochemical derivatization to the determination of analgesics. *J.Chromatogr.A*, **1996**, *729*, 355–361

## SAMPLE

**Matrix:** formulations
**Sample preparation:** Finely powder half a tablet, add 9 mL mobile phase, sonicate for 20 min, make up to 10 mL with mobile phase, filter (Whatman type 40 and 0.2 μm Millipore), inject an aliquot of the filtrate.

## HPLC VARIABLES

**Column:** 250 × 4 5 μm LiChrospher 100 CN
**Mobile phase:** MeCN:THF:buffer 7:6:87 (Buffer was 0.8% acetic acid containing 5 mM sodium hexanesulfonate, 10 mM di-n-butylamine, and 0.12% phosphoric acid, pH 3.3.)
**Flow rate:** 1
**Injection volume:** 20
**Detector:** UV 310

## CHROMATOGRAM

**Retention time:** 4.25
**Limit of detection:** 1.9 μg/mL

## OTHER SUBSTANCES

**Simultaneous:** caffeine (UV 298), chlorpheniramine (UV 265), guaifenesin (glycerylguaicolate) (UV 284), phenylpropanolamine (UV 260)

## KEY WORDS

tablets

## REFERENCE

Indrayanto, G.; Sunarto, A.; Adriani, Y. Simultaneous assay of phenylpropanolamine hydrochloride, caffeine, paracetamol, glycerylguaiacolate and chlorpheniramine in Silabat™ tablet using HPLC with diode array detection. *J.Pharm.Biomed.Anal.*, **1995**, *13*, 1555–1559

## SAMPLE

**Matrix:** formulations
**Sample preparation:** 75 μL Sample + 6 mL mobile phase, vortex for 2 min, add 300 μL 300 μg/mL acetaminophen, make up to 10 mL with mobile phase, mix for 2 min, inject a 5 μL aliquot.

## HPLC VARIABLES

**Column:** 250 × 4.6 5 μm Spheri-5 ODS (Applied Biosystems)
**Mobile phase:** MeOH:buffer 30:70 (Buffer was 900 mL 50 mM $Na_2HPO_4$ + 18.75 mL tetrabutylammonium phosphate, pH adjusted to 6.8 with 1 N phosphoric acid.)
**Flow rate:** 1.2
**Injection volume:** 5
**Detector:** UV 254

## CHROMATOGRAM

**Retention time:** 3.8
**Internal standard:** acetaminophen

## OTHER SUBSTANCES

**Simultaneous:** 5-aminosalicylic acid

## KEY WORDS
acetaminophen is IS; rectal suspension; enema

## REFERENCE
Henderson, L.M.; Johnson, C.E.; Berardi, R.R. Stability of mesalamine in rectal suspension diluted with distilled water. *Am.J.Hosp.Pharm.*, **1994**, *51*, 2955–2957

## SAMPLE
**Matrix:** formulations
**Sample preparation:** Add one tablet to 10 mL MeOH and 80 mL dichloromethane, sonicate for 5 min, dilute to 100 mL with dichloromethane, dilute a 2 mL aliquot to 25 mL with mobile phase, inject a 10 μL aliquot.

## HPLC VARIABLES
**Guard column:** 20 × 4.6 5 μm Supelcosil
**Column:** 33 × 4.6 3 μm Supelcosil
**Mobile phase:** Dichloromethane:3.33% ammonium hydroxide in MeOH 98.5:1.5
**Flow rate:** 2
**Injection volume:** 10
**Detector:** UV 244

## CHROMATOGRAM
**Retention time:** 4
**Limit of detection:** 100 ng/mL

## OTHER SUBSTANCES
**Simultaneous:** propoxyphene
**Noninterfering:** impurities, p-nitrophenol, p-hydroxyacetophenone, 4-aminophenol

## KEY WORDS
tablets; normal phase

## REFERENCE
Ascah, T.L.; Hunter, B.T. Simultaneous high-performance liquid chromatographic determination of propoxyphene and acetaminophen in pharmaceutical preparations. *J.Chromatogr.*, **1988**, *455*, 279–289

## SAMPLE
**Matrix:** formulations
**Sample preparation:** Powder levodopa/carbidopa tablets or contents of capsules, weigh out an amount equivalent to about 100 ng levodopa, add 30 mL 0.1 M HCl, sonicate, make up to 50 mL with 0.1 M HCl, mix, filter (0.45μm), discard first 5 mL filtrate. 10 mL Filtrate + 50 mL 0.5 mg/mL acetaminophen in MeOH:mobile phase 75:175, make up to 100 mL with mobile phase, mix, inject a 20 μL aliquot.

## HPLC VARIABLES
**Column:** 300 × 3.9 10 μm μBondapak C18
**Mobile phase:** 3% aqueous acetic acid
**Flow rate:** 1.5
**Injection volume:** 20
**Detector:** UV 280

## CHROMATOGRAM
**Retention time:** 9
**Internal standard:** acetaminophen

## OTHER SUBSTANCES
**Simultaneous:** levodopa, carbidopa

## KEY WORDS
tablets; capsules; acetaminophen is IS

## REFERENCE
Ting, S. Liquid chromatographic determination of levodopa and levodopa-carbidopa in solid dosage forms: collaborative study. *J.Assoc.Off.Anal.Chem.*, **1987**, *70*, 987–990

## SAMPLE
**Matrix:** formulations

## HPLC VARIABLES
**Column:** 300 × 3.9 μBondapak C18
**Mobile phase:** MeOH:buffer 20:80 (Buffer was 15 mM 1-butanesulfonic acid + 15 mM $KH_2PO_4$ + 2 mL/L triethylamine, pH adjusted to 4.8 ± 0.1 with dilute phosphoric acid.)
**Column temperature:** 50
**Flow rate:** 2
**Injection volume:** 20
**Detector:** UV 214

## CHROMATOGRAM
**Retention time:** 2.8

## OTHER SUBSTANCES
**Simultaneous:** codeine, benzoic acid, p-aminophenol, codeine N-oxide, codeinone

## KEY WORDS
elixir; stability-indicating

## REFERENCE
Sisco, W.R.; Rittenhouse, C.T.; Everhart, L.A.; McLaughlin, A.M. Simultaneous high-performance liquid chromatographic stability-indicating analysis of acetaminophen, codeine phosphate, and sodium benzoate in elixirs. *J.Chromatogr.*, **1986**, *354*, 355–366

## SAMPLE
**Matrix:** formulations
**Sample preparation:** Dissolve capsules and tablets in MeOH:pH 4.0 water 1:1, shake for 1 (capsules) or 4 (tablets) h, dilute a 10 mL aliquot with 40 mL pH 3.2 water, filter (0.45 μm), collect last portion of filtrate, inject a 20 μL aliquot. (pH 3.2 and 4.0 water are prepared by adjusting pH of distilled water with phosphoric acid.)

## HPLC VARIABLES
**Column:** 300 × 3.9 μBondapak C18
**Mobile phase:** MeOH:buffer 7:93 (Buffer was 15 mM $KH_2PO_4$ + 2 mL triethylamine per liter. Adjusted to pH 2.35 ± 0.1 with concentrated phosphoric acid.)
**Column temperature:** 40
**Flow rate:** 3
**Injection volume:** 20
**Detector:** UV 214

## CHROMATOGRAM
**Retention time:** 2.7

## OTHER SUBSTANCES
**Simultaneous:** codeine, p-aminophenol, codeine-N-oxide, codeinone, degradation products

## KEY WORDS
capsules; tablets; rugged

## REFERENCE
Sisco, W.R.; Rittenhouse, C.T.; Everhart, L.A. Simultaneous high-performance liquid chromatographic stability-indicating analysis of acetaminophen and codeine phosphate in tablets and capsules. *J.Chromatogr.*, **1985**, *348*, 253–263

## SAMPLE
**Matrix:** formulations
**Sample preparation:** Pulverize tablets and weigh out 1 g, add 1 mL formic acid, add 25 mL MeOH, shake mechanically for 10 min, make up to 50 mL with methanol. Remove 10 mL and centrifuge. 5 mL Supernatant + 5 mL 0.0025% p-hydroxybenzoic acid in MeOH:water 20:80, make up to 25 mL with water, inject an aliquot. (Analyze within 1 h.)

## HPLC VARIABLES
**Column:** 250 × 4.6 LiChrosorb RP8
**Mobile phase:** MeOH:200 mM pH 3.5 phosphate buffer:water 20:10:70
**Flow rate:** 1
**Injection volume:** 10
**Detector:** UV 254

## CHROMATOGRAM
**Retention time:** 8.5
**Internal standard:** p-hydroxybenzoic acid (18)

## OTHER SUBSTANCES
**Simultaneous:** aspirin, p-aminophenol, 3-O-acetylascorbic acid, 2-O-acetylascorbic acid, Vitamin C, saccharin, O-acetyl-p-aminophenol, salicylic acid (UV 280), diacetyl-p-aminophenol (UV 280)

## KEY WORDS
tablets

## REFERENCE
Thomis, R.; Roets, E.; Hoogmartens, J. Analysis of tablets containing aspirin, acetaminophen, and ascorbic acid by high-performance liquid chromatography. *J.Pharm.Sci.*, **1984**, *73*, 1830–1833

## SAMPLE
**Matrix:** formulations
**Sample preparation:** Dilute with water, filter (0.45 μm), inject a 20 μL aliquot.

## HPLC VARIABLES
**Column:** 300 × 4 μBondapak C18
**Mobile phase:** MeOH:water:glacial acetic acid 45:55:2 containing 5 mM octanesulfonic acid
**Flow rate:** 2.5
**Injection volume:** 20
**Detector:** UV 280

## CHROMATOGRAM
**Retention time:** 1.5

## OTHER SUBSTANCES
**Simultaneous:** guaifenesin, pseudoephedrine, pholcodine, methyl paraben, ethyl paraben, propyl paraben, butyl paraben

## KEY WORDS
cough mixture

## REFERENCE
Carnevale, L. Simultaneous determination of acetaminophen, guaifenesin, pseudoephedrine, pholcodine, and paraben preservatives in cough mixture by high-performance liquid chromatography. *J.Pharm.Sci.*, **1983**, *72*, 196–198

## SAMPLE
**Matrix:** formulations
**Sample preparation:** Add 1 tablet to 95 mL water, place on a steam bath for 15 min, cool, mix for 15 min, sonicate, allow to stand, filter, inject a 13 μL aliquot.

## HPLC VARIABLES
**Column:** μBondapak C18
**Mobile phase:** MeOH:buffer 25:75 (Buffer was 0.01 N $KH_2PO_4$ + 50 mM $KNO_3$, adjusted to pH 4.5 with 3 N phosphoric acid.)
**Flow rate:** 1.1
**Injection volume:** 13
**Detector:** UV 283

## CHROMATOGRAM
**Retention time:** 5.9

## OTHER SUBSTANCES
**Simultaneous:** hydrocodone, p-aminophenol, hydromorphone, codeine, p-chloroacetanilide

## KEY WORDS
tablets; stability-indicating

## REFERENCE
Wallo, W.E.; D'Adamo, A. Simultaneous assay of hydrocodone bitartrate and acetaminophen in a tablet formulation. *J.Pharm.Sci.*, **1982**, *71*, 1115–1118

## SAMPLE
**Matrix:** formulations
**Sample preparation:** 3 mL Sample + 5 mL 200 mg/mL o-dinitrobenzene in 1:1 MeOH:water, dilute to 50 mL with 1:1 MeOH:water, inject a 15 μL aliquot.

## HPLC VARIABLES
**Column:** 300 × 4 μBondapak C18
**Mobile phase:** MeOH:water:ammonium formate buffer 45:54:1 (Prepare ammonium formate buffer by diluting 34 mL 28-30% ammonia with 30 mL water, add 30 mL 98% formic acid (Caution! Exothermic!). When cool dilute this mixture (pH 3.9) to 100 mL with water.)
**Flow rate:** 2
**Injection volume:** 15
**Detector:** UV 280

## CHROMATOGRAM
**Retention time:** 4
**Internal standard:** o-dinitrobenzene (11)

## OTHER SUBSTANCES
**Simultaneous:** guaifenesin, dextromethorphan, p-aminophenol

## KEY WORDS
cough syrup

## REFERENCE
McSharry, W.O.; Savage, I.V.E. Simultaneous high-pressure liquid chromatographic determination of acetaminophen, guaifenesin, and dextromethorphan hydrobromide in cough syrup. *J.Pharm.Sci.*, **1980**, *69*, 212–214

## SAMPLE
**Matrix:** microsomal incubations
**Sample preparation:** 250 μL Microsomal incubation + 1 mL ice-cold ethyl acetate + 100 μL 50 mM 11β-hydroxytestosterone, mix, add 1.5 mL ethyl acetate, vortex, centrifuge. Remove the organic layer and evaporate it to dryness under a stream of nitrogen, reconstitute the residue in 150 μL MeOH:water 20:80, inject an aliquot.

## HPLC VARIABLES
**Guard column:** 5 μm Supelco LC-18
**Column:** 150 × 4.6 5 μm Supelco LC-18
**Mobile phase:** Gradient. A was MeOH:water 30:70 adjusted to pH 4.5 with glacial acetic acid. B was MeCN:MeOH 10:90 adjusted to pH 4.5 with glacial acetic acid. A:B 87:13 for 34 min, to 50:50 over 10 min, return to initial conditions over 5 min.
**Flow rate:** 1.3
**Detector:** UV 240

## CHROMATOGRAM
**Retention time:** 1.8
**Internal standard:** 11β-hydroxytestosterone (26)

## OTHER SUBSTANCES
**Extracted:** 6β-hydroxytestosterone

## REFERENCE
Chiba, M.; Nishime, J.A.; Lin, J.H. Potent and selective inactivation of human liver microsomal cytochrome P-450 isoforms by L-754,394, an investigational human immune deficiency virus protease inhibitor. *J.Pharmacol.Exp.Ther.*, **1995**, *275*, 1527–1534

## SAMPLE
**Matrix:** solutions

## HPLC VARIABLES
**Column:** 250 × 4.6 8 μm Unisphere-PBD (polybutadiene on alumina) (Biotage, Charlottesville, VA)
**Mobile phase:** MeCN:water 20:80
**Flow rate:** 1
**Detector:** UV 254

## CHROMATOGRAM
**Retention time:** 3

## OTHER SUBSTANCES
**Simultaneous:** aspirin, phenacetin, salicylamide

## REFERENCE
Jedrejewski, P.T.; Taylor, L.T. Comparison of silica-, alumina-, and polymer-based stationary phases for reversed-phase liquid chromatography. *J.Chromatogr.Sci.*, **1995**, *33*, 438–445

## SAMPLE
**Matrix:** solutions

## HPLC VARIABLES
**Column:** 250 × 4.6 5 μm Supelcosil LC-DP (A) or 250 × 4 5 μm LiChrospher 100 RP-8 (B)
**Mobile phase:** MeCN:0.025% phosphoric acid:buffer 25:10:5 (A) or 60:25:15 (B) (Buffer was 9 mL concentrated phosphoric acid and 10 mL triethylamine in 900 mL water, adjust pH to 3.4 with dilute phosphoric acid, make up to 1 L.)
**Flow rate:** 0.6
**Injection volume:** 25
**Detector:** UV 229

## CHROMATOGRAM
**Retention time:** 4.75 (A), 3.68 (B)

## OTHER SUBSTANCES
**Also analyzed:** acebutolol, acepromazine, acetazolamide, acetophenazine, albuterol, alprazolam, amitriptyline, amobarbital, amoxapine, antipyrine, atenolol, atropine, azatadine, baclofen, benzocaine, bromocriptine, brompheniramine, brotizolam, bupivacaine, buspirone, butabarbital, butalbital, caffeine, carbamazepine, cetirizine, chlorcyclizine, chlordiazepoxide, chlormezanone, chloroquine, chlorpheniramine, chlorpromazine, chlorpropamide, chlorprothixene, chlorthalidone, chlorzoxazone, cimetidine, cisapride, clomipramine, clonazepam, clonidine, clozapine, cocaine, codeine, colchicine, cyclizine, cyclobenzaprine, dantrolene, desipramine, diazepam, diclofenac, diflunisal, diltiazem, diphenhydramine, diphenidol, diphenoxylate, dipyridamole, disopyramide, dobutamine, doxapram, doxepin, droperidol, encainide, ethidium bromide, ethopropazine, fenoprofen, fentanyl, flavoxate, fluoxetine, fluphenazine, flurazepam, flurbiprofen, fluvoxamine, furosemide, glutethimide, glyburide, guaifenesin, haloperidol, homatropine, hydralazine, hydrochlorothiazide, hydrocodone, hydromorphone, hydroxychloroquine, hydroxyzine, ibuprofen, imipramine, indomethacin, ketoconazole, ketoprofen, ketorolac, labetalol, levorphanol, lidocaine, loratadine, lorazepam, lovastatin, loxapine, mazindol, mefenamic acid, meperidine, mephenytoin, mepivacaine, mesoridazine, metaproterenol, methadone, methdilazine, methocarbamol, methotrexate, methotrimeprazine, methoxamine, methyldopa, methylphenidate, metoclopramide, metolazone, metoprolol, metronidazole, midazolam, moclobemide, morphine, nadolol, nalbuphine, naloxone, naphazoline, naproxen, nifedipine, nizatidine, norepinephrine, nortriptyline, oxazepam, oxycodone, oxymetazoline, paroxetine, pemoline, pentazocine, pentobarbital, pentoxifylline, perphenazine, pheniramine, phenobarbital, phenol, phenolphthalein, phentolamine, phenylbutazone, phenyltoloxamine, phenytoin, pimozide, pindolol, piroxicam, pramoxine, prazepam, prazosin, probenecid, procainamide, procaine, prochlorperazine, procyclidine, promazine, promethazine, propafenone, propantheline, propiomazine, propofol, propranolol, protriptyline, quazepam, quinidine, quinine, racemethorphan, ranitidine, remoxipride, risperidone, salicylic acid, scopolamine, secobarbital, sertraline, sotalol, spironolactone, sulfinpyrazone, sulindac, temazepam, terbutaline, terfenadine, tetracaine, theophylline, thiethylperazine, thiopental, thioridazine, thiothixene, timolol, tocainide, tolbutamide, tolmetin, trazodone, triamterene, triazolam, trifluoperazine, triflupromazine, trimeprazine, trimethoprim, trimipramine, verapamil, warfarin, xylometazoline, yohimbine, zopiclone

## KEY WORDS
also details of plasma extraction

## REFERENCE
Koves, E.M. Use of high-performance liquid chromatography-diode array detection in forensic toxicology. *J.Chromatogr.A*, **1995**, *692*, 103–119

## SAMPLE
**Matrix:** solutions
**Sample preparation:** Inject a 2 µL aliquot.

## HPLC VARIABLES
**Column:** 30 × 4.6 3 µm Hypersil ODS
**Mobile phase:** MeOH:50 mM pH 6.2 phosphate buffer 15:85
**Flow rate:** 2
**Injection volume:** 2
**Detector:** UV 280

## REFERENCE
Shah, K.P.; Chang, M.; Riley, C.M. Automated analytical systems for drug development studies. 3. Multivessel dissolution testing system based on microdialysis sampling. *J.Pharm.Biomed.Anal.*, **1995**, *13*, 1235–1241

## SAMPLE
**Matrix:** solutions
**Sample preparation:** Prepare a 0.5 mg/mL solution in water, inject a 5 µL aliquot.

## HPLC VARIABLES
**Column:** 250 × 4.6 Zorbax RX
**Mobile phase:** Gradient. A was 150 mM phosphoric acid and 50 mM triethylamine. B was MeCN:water 80:20 containing 150 mM phosphoric acid and 50 mM triethylamine. A:B 100:0 for 2.2 min then to 0:100 over 30 min.
**Column temperature:** 30
**Flow rate:** 2
**Injection volume:** 5
**Detector:** UV 210

## CHROMATOGRAM
**Retention time:** 8.3

## OTHER SUBSTANCES
**Simultaneous:** aprobarbital, butabarbital, chlordiazepoxide, chloroxylenol, chlorpromazine, clenbuterol, cortisone, danazol, diflunisal, doxapram, estrone, fluoxymesterone, mefenamic acid, methyltestosterone, nicotine, oxazepam, phentermine, phenylpropanolamine, progesterone, sulfamethazine, sulfanilamide, testosterone, testosterone propionate, tranylcypromine, tripelennamine
**Interfering:** N-acetylprocainamide

## KEY WORDS
details for purification of triethylamine in paper

## REFERENCE
Hill, D.W.; Kind, A.J. The effects of type B silica and triethylamine on the retention of drugs in silica based reverse phase high performance chromatography. *J.Liq.Chromatogr.*, **1993**, *16*, 3941–3964

## SAMPLE
**Matrix:** solutions
**Sample preparation:** Dissolve compounds in MeCN:water 80:20, inject a 1 µL aliquot.

## HPLC VARIABLES
**Column:** 150 × 1 3 μm Hitachi-Gel 3057 ODS silica (Hitachi)
**Mobile phase:** MeCN:water 25:75
**Flow rate:** 0.03
**Injection volume:** 1
**Detector:** UV 254

## CHROMATOGRAM
**Retention time:** 8

## OTHER SUBSTANCES
**Simultaneous:** caffeine, dipyrone (sulpyrin), guaifenesin (guaiacol glycerol ether), bucetin (3-hydroxy-p-butyrophenetidine), methyl p-hydroxybenzoate, phenacetin

## KEY WORDS
semi-micro

## REFERENCE
Matsushima, Y.; Nagata, Y.; Niyomura, M.; Takakusagi, K.; Takai, N. Analysis of antipyretics by semimicro liquid chromatography. *J.Chromatogr.*, **1985**, *332*, 269–273

## SAMPLE
**Matrix:** solutions
**Sample preparation:** Dissolve compounds in MeOH, inject a 1 μL aliquot.

## HPLC VARIABLES
**Column:** 150 × 1 3 μm Hitachi-Gel 3011 porous polymer (Hitachi)
**Mobile phase:** MeOH:ammonia 99:1
**Flow rate:** 0.03
**Injection volume:** 1
**Detector:** UV 254

## CHROMATOGRAM
**Retention time:** 3.26

## OTHER SUBSTANCES
**Also analyzed:** caffeine, dipyrone (sulpyrin), bucetin (3-hydroxy-p-butyrophenetidine), phenacetin, mefenamic acid, aspirin, salicylamide, salicylic acid, ethenzamide (o-ethoxybenzamide), theobromine, theophylline

## KEY WORDS
semi-micro

## REFERENCE
Matsushima, Y.; Nagata, Y.; Niyomura, M.; Takakusagi, K.; Takai, N. Analysis of antipyretics by semimicro liquid chromatography. *J.Chromatogr.*, **1985**, *332*, 269–273

## SAMPLE
**Matrix:** urine
**Sample preparation:** Dilute urine with water, filter (0.45 μm), inject a 20 μL aliquot of the filtrate.

## HPLC VARIABLES
**Guard column:** 5 μm Resolve C18 Guard-Pak (Waters)
**Column:** 100 × 8 5 μm Resolve C18 Radial Compression Module (Waters)

**Mobile phase:** MeOH:100 mM $KH_2PO_4$:glacial acetic acid 4:95:1
**Flow rate:** 2.5
**Injection volume:** 20
**Detector:** UV 248

## CHROMATOGRAM
**Retention time:** 4.47
**Limit of detection:** 2 ng/mL

## OTHER SUBSTANCES
**Extracted:** metabolites, acetaminophen glucuronide, acetaminophen sulfate

## KEY WORDS
pharmacokinetics

## REFERENCE
Goicoechea, A.G.; López de Alda, M.J.; Vila-Jato, J.L. A validated high-performance liquid chromatographic method for the determination of paracetamol and its major metabolites in urine. *J.Liq.Chromatogr.*, **1995**, *18*, 3257–3268

## SAMPLE
**Matrix:** urine
**Sample preparation:** 0.3 mL Urine + 1 mL 100 mM pH 4.5 acetate buffer containing 20 mg/mL sodium pyrosulfite + 20 mg Limpet Acetone Powder. Incubate 3 h 37°, add 100 μL 400 mg/L phenacetin, extract twice with 4 mL of ether:dichloromethane:isopropanol 60:40:1. Add 0.5 mL pH 6.5 50 mM phosphate buffer containing 20 mg/mL sodium pyrosulfite, shake, allow phases to separate, evaporate upper organic layer under a stream of nitrogen and inject aqueous layer.

## HPLC VARIABLES
**Column:** 100 × 4.6 Chrompack microsphere C18
**Mobile phase:** Gradient. A was 50 mM pH 6.6 phosphate buffer. B was MeCN:50 mM pH 6.5 phosphate buffer 25:75. A:B from 90:10 to 10:90 over 50 min, maintain at 10:90 for 2 min.
**Flow rate:** 1.2
**Injection volume:** 250
**Detector:** UV 254

## CHROMATOGRAM
**Internal standard:** phenacetin

## OTHER SUBSTANCES
**Simultaneous:** metabolites

## KEY WORDS
pig

## REFERENCE
Monshouwer, M.; Witkamp, R.F.; Pijpers, A.; Verheijden, J.H.M.; Van Miert, A.S.J.P.A.M. Dose-dependent pharmacokinetic interaction between antipyrine and paracetamol *in vivo* and *in vitro* when administered as cocktail in pig. *Xenobiotica*, **1994**, *24*, 347–355

## SAMPLE
**Matrix:** urine

**Sample preparation:** 1 mL Urine + 4 mL 2 M pH 5.0 acetate buffer (+ 50 µL β-gluco-ronidase-sulfatase at 37° overnight if required to hydrolyze conjugates). Centrifuge 200 µL aliquots for 5 min, inject a 10-20 µL aliquot.

## HPLC VARIABLES
**Column:** 300 × 4.6 10 µm µBondapak C18
**Mobile phase:** MeOH:HOAc:100 mM KH$_2$PO$_4$ 7:0.75:92.25
**Flow rate:** 2
**Injection volume:** 10-20
**Detector:** UV 248; E, Bioanalytical Systems LC-4, TL-3 glassy carbon electrode + 0.60 V, Ag/AgCl reference electrode, output 10-50 nA/V

## CHROMATOGRAM
**Retention time:** 8
**Limit of detection:** 70 ng/mL (UV), 15 ng/mL (E)

## OTHER SUBSTANCES
**Extracted:** metabolites

## REFERENCE
Wilson, J.M.; Slattery, J.T.; Forte, A.J.; Nelson, S.D. Analysis of acetaminophen metabolites in urine by high-performance liquid chromatography with UV and amperometric detection. *J.Chromatogr.*, **1982**, *227*, 453–462

◆———————————————◆———————————————◆

## ANNOTATED BIBLIOGRAPHY
Visentini, J.; Kwong, E.C.; Carrier, A.; Zidarov, D.; Bertrand, M.J. Comparison of softwares used for the detection of analytes present at low levels in liquid chromatographic-mass spectrometric experiments. *J.Chromatogr.A*, **1995**, *712*, 31–43 [LC-MS]

Akhtar, M.J.; Khan, S.; Hafiz, M. High-performance liquid chromatographic assay for the determination of paracetamol, pseudoephedrine hydrochloride and triprolidine hydrochloride. *J.Pharm. Biomed.Anal.*, **1994**, *12*, 379–382 [simultaneous pseudoephedrine, triprolidine; formulations]

Bogusz, M.; Erkens, M. Reversed-phase high-performance liquid chromatographic database of retention indices and UV spectra of toxicologically relevant substances and its interlaboratory use. *J.Chromatogr.A*, **1994**, *674*, 97–126 [gradient; also acebutolol, acecarbromal, acepromazine, acetazolamide, allobarbital, allopurinol, alprazolam, alprenolol, amiloride, amiodarone, amitriptyline, amobarbital, amoxapine, amphetamine, aprobarbital, vitamin C, aspirin, atenolol, atrazine, atropine, barbital, benzocaine, benztropine, betaxolol, bisacodyl, brallobarbital, bromazepam, brompheniramine, bupivacaine, bupranolol, buprenorphine, buspirone, butalbital, butaperazine, butobarbital, caffeine, carbamazepine, carbaryl, carbromal, chlordiazepoxide, chlormezanone, chloroquine, chlorpromazine, chlorprothixene, chlorthalidone, cimetidine, cinchocaine, clobazam, clomipramine, clonazepam, clonidine, clopamide, clopenthixol, clorazepate, clozapine, cocaine, codeine, colchicine, coumarin, cyclobarbital, cyclopentobarbital, demoxepam, desipramine, dextromethorphan, dextromoramide, dextropropoxyphene, diamorphine, diazepam, diazoxide, diclofenac, digoxin, diltiazem, dimenhydrinate, diphenhydramine, dipyridamole, disopyramide, dothiepin, doxepin, doxylamine, droperidol, ephedrine, estazolam, ethacrynic acid, ethosuximide, fenfluramine, fenoprofen, fentanyl, flecainide, flumazenil, flunarizine, flunitrazepam, fluoxetine, flupenthixol, flurazepam, fluvoxamine, furosemide, glyburide, glipizide, glutethimide, guaifenesin, haloperidol, hydrochlorothiazide, hydrocodone, hydromorphone, hydroxyzine, ibuprofen, imipramine, indomethacin, isoniazid, isosorbidedinitrate, ketamine, labetalol, methotrimeprazine, levorphanol, lidocaine, loprazolam, lorazepam, loxapine, matrotiline, meclozine, medazepam, mephenytoin, mepivacaine, metapramine, methadone, methamphetamine, methaqualone, methohexital, methylphenidate, metipranolol, metoclopramide, metoprolol, metronidazole, midazolam, monolinuron, morazone, morphine, nadolol, nalorphine, naloxone, naproxene, nicotinamide, nicotine, niflumic acid, nikethamide, nimodipine, nitrazepam, nitrendipine, nitrofurantoin, nitroglycerine, noscapine, orciprenaline, orphenadrine, oxazepam, oxazolam, oxprenolol, oxycodone, oxyphenbutazone, papaverine , pemoline, pentazocine, pentobarbital, pentoxifylline, perazine, perphenazine, meperidine, phenacemide, phenacetin, phenazone, phcencyclidine, phenel-

zine, pheniramine, phenobarbital, phentermine, phenylbutazone, phenytoin, pindolol, piroxicam, prazepam, primidone, probenecid, procainamide, procaine, prochlorperazine, promazine, promethazine,
propafenone, propoxur, propranolol, propyphenazone, protriptyline, quinidine, quinine, ranitidine,
reserpine, saccharin, salicylamide, salicylic acid, scopolamine, secobarbital, sotalol, spironolactone,
strychnine, sulphadiazine, sulpiride, suprofen, temazepam, terfenadine, tetracaine, tetrazepam, thebaine, theobromine, theophylline, thiabendazole, thiopental, thioridazine, tiaprofenic acid, timolol,
tocainide, tolbutamide, trancylpromine, trazodone, triamterene, triazolam, trichlormethiazide, trifluoperazine, trifluperidol, triflupromazine, trimethoprim, trimipramine, tripelenamine, tripolidine,
verapamil, vinylbarbital, warfarin, yohimbine, zolpidem]

Gurley, B.J.; Zermatten, S.; Skelton, D. Determination of antipyrine in human serum by direct injection
restricted access media liquid chromatography. *J.Pharm.Biomed.Anal.*, **1994**, *12*, 1591–1595 [ex-
tracted antipyrine; serum; column temp 37; acetaminophen is IS]

Hill, D.W.; Kind, A.J. Reversed-phase solvent gradient HPLC retention indexes of drugs. *J.Anal.Toxicol.*,
**1994**, *18*, 233–242 [also acepromazine, acetophenazine, albuterol, aminophylline, amitriptyline, amobarbital, amoxapine, amphetamine, amylocaine, antipyrine, aprobarbital, aspirin, atenolol, atropine,
avermectin, barbital, bebrisoquine, benzocaine, benzoic acid, benzotropine, benzphetamine, berberine, bibucaine, bromazepan, brompheniramine, buprenorphine, buspirone, butabarbital, butacaine,
butethal, caffeine, carbamazepine, carbromal, chloamphenicol, chlordiazepoxide, chloroquine, chlorothiazide, chloroxylenol, chlorphenesin, chlorpheniramine, chlorpromazine, chlorpropamide, chlortetracycline, cimetidine, cinchonidine, cinchonine, clenbuterol, clonazepam, clonixin, clorazepate, cocaine, codeine, colchicine, cortisone, coumarin, cyclazocine, cyclobenzaprine, cyclothiazide,
cyheptamide, cymarin, danazol, danthron, dapsone, desipramine, dexamethasone, dextromethorphan, dextropropoxyphene, diamorphine, diazepam, diclofenac, diethylpropion, diethylstilbestrol, diflunisal, digitoxin, digoxin, diltiazem, diphenhydramine, diphenoxylate, diprenorphine, dipyrone, disulfiram, dopamine, doxapram, doxepin, dronabinol, ephedine, epinephrine, epinine, estradiol,
estriol, estrone, ethacrynic acid, ethosuximide, etonitazene, etorphine, eugenol, famotidine, fenbendazole, fencamfamine, fenoprofen, fenproporex, fentanyl, flubendazole, flufenamic acid, flunitrazepam, 5-fluorouracil, fluoxymesterone, fluphenazine, furosemide, gentisic acid, gitoxigenin, glipizide,
glunixin, glutethimide, glybenclamide, guaiacol, halazepam, haloperidol, hydrochlorothiazide, hydrocodone, hydrocortisone, hydromorphone, hydroxyquinoline, ibogaine, ibuprofen, iminostilbene, imipramine, indomethacin, isocarbostyril, isocarboxazid, isoniazid, isoproterenol, isoxsuprine, ivermectin,
ketamine, ketoprofen, kynurenic acid, levorphanol, lidocaine, lorazepam, lormetazepam, loxapine,
mazindol, mebendazole, meclizine, meclofenamic acid, medazepam, mefenamic acid, megestrol, mepacrine, meperidine, mephentermine, mephenytoin, mephesin, mephobarbital, mepivacaine, mescaline, mesoridazine, methadone, methamphetamine, methapyrilene, methaqualone, methazolamide,
methocarbamol, methoxamine, methsuximide, methyl salicylate, methyldopa, methyldopamine,
methylphenidate, methylprednisolone, methyltestosterone, methyprylon, metoprolol, mibolerone,
morphine, nadolol, nalorphine, naloxone, naltrexone, naphazoline, naproxen, nefopam, niacinamide,
nicotine, nicotinic acid, nifedipine, niflumic acid, nitrazepam, norepinephrine, nortriptyline, noscapine, nylidrin, oxazepam, oxycodone, oxymorphone, oxyphenbutazone, oxytetracycline, papaverine,
pargyline, pemoline, pentazocine, pentobarbital, persantine, phenacetin, phenazocine, phenazopyridine, phencyclidine, phendimetrazine, phenelzine, pheniramine, phenobarbital]

Ismail, S.; Kokwaro, G.O.; Back, D.J.; Edwards, G. Effect of malaria infection on the pharmacokinetics
of paracetamol in rat. *Xenobiotica*, **1994**, *24*, 527–533 [pharmacokinetics; rat; plasma; urine; extracted metabolites; 3-acetamidophenol is IS]

Konishi, H.; Yamaji, A. Measurement of theophylline metabolites produced by reaction with hepatic
microsome by high performance liquid chromatography following solid phase extraction. *Bio-
med.Chromatogr.*, **1994**, *8*, 189–192 [extracted theophylline; microsomal incubations; SPE; acetaminophen is IS; mouse]

Lau, G.S.N.; Critchley, J.A.J.H. The estimation of paracetamol and its major metabolites in both plasma
and urine by a single high-performance liquid chromatography assay. *J.Pharm.Biomed.Anal.*, **1994**,
*12*, 1563–1572 [extracted metabolites; plasma; urine; blood; LOD 1 ng]

Nivaud-Guernet, E.; Guernet, M.; Ivanovic, D.; Medenica, M. Effect of eluent pH on the ionic and molecular forms of the non-steroidal anti-inflammatory agents in reversed-phase high-performance liquid chromatography. *J.Liq.Chromatogr.*, **1994**, *17*, 2343–2357 [also aspirin, flufenamic acid, keta-
zone, mefenamic acid, niflumic, niflumic acid, nixylic acid, oxyphenbutazone, phenacetin,
salicylamide, salicylic acid, sulfinpyrazone]

Oguro, T.; Gregus, Z.; Madhu, C.; Liu, L.; Klaassen, C.D. Molybdate depletes hepatic 3-phosphoadenosine 5-phosphosulfate and impairs the sulfation of acetaminophen in rats. *J.Pharmacol.Exp.Ther.*, **1994**, *270*, 1145–1151 [rat, extracted metabolites; serum; urine; bile]

Shah, K.P.; Chang, M.; Riley, C.M. Automated analytical systems for drug development studies. II-A system for dissolution testing. *J.Pharm.Biomed.Anal.*, **1994**, *12*, 1519–1527 [dissolution testing; formulations; tablets; also sulfamethoxazole, trimethoprim]

Thomas, B.R.; Fang, X.G.; Shen, P.; Ghodbane, S. Mixed ion pair liquid chromatography method for the simultaneous assay of ascorbic acid, caffeine, chlorpheniramine maleate, dextromethorphan HBr monohydrate and paracetamol in Frenadol sachets. *J.Pharm.Biomed.Anal.*, **1994**, *12*, 85–90 [simultaneous caffeine, chlorpheniramine, dextromethorphan, vitamin C; formulations]

van der Veen, J.; Eissens, A.C.; Lerk, C.F. Controlled release of paracetamol from amylodextrin tablets: In vitro and in vivo results. *Pharm.Res.*, **1994**, *11*, 384–387 [plasma; pharmacokinetics]

Abounassif, M.A.; Abdel-Moety, E.M.; Gad-Kariem, R.A. HPLC-quantification of diethylamine salicylate and methyl nicotinate in ointments. *J.Liq.Chromatogr.*, **1992**, *15*, 625–636 [formulations; ointments; simultabeous diethylamine salicylate, methyl nicotinate; acetaminophen is IS; simultaneous methyl paraben, propyl paraben

Bannwarth, B.; Netter, P.; Lapicque, F.; Gillet, P.; Pere, P.; Boccard, E.; Royer, R.J.; Gaucher, A. Plasma and cerebrospinal fluid concentrations of paracetamol after a single intravenous dose of propacetamol. *Br.J.Clin.Pharmacol.*, **1992**, *34*, 79–81 [plasma; CSF; LOD 2 ng/mL]

Curtis, M.A.; Pullen, R.H.; McKenna, K. HPLC determination of analgesics in human plasma and serum by direct injection on 80 Angstrom pore methyl bonded phase silica columns. *J.Liq.Chromatogr.*, **1991**, *14*, 165–178 [plasma; serum; extracted salicylic acid]

Aloba, O.T.; Adusumilli, P.S.; Nigalaye, A.G. High performance liquid chromatographic analysis of a multicomponent product using a silica stationary phase and an aqueous-organic mobile phase. *J.Pharm.Biomed.Anal.*, **1991**, *9*, 335–340 [simultaneous butalbital, caffeine]

Hays, P.A.; Lurie, I.S. Quantitative analysis of adulterants in illicit heroin samples via reversed phase HPLC. *J.Liq.Chromatogr.*, **1991**, *14*, 3513–3517 [also acetylcodeine, acetylmorphine, aspirin, benzocaine, caffeine, chloroquine, diamorphine, diazepam, diphenhydramine, dipyrone, lidocaine, methaqualone, monoacetylmorphine, morphine, nicotinamide, noscapine, papaverine, phenacetin, phenobarbital, phenolphthalein, N-phenyl-2-naphthylamine, salicylic acid, strychnine]

McCormick, C.P.; Shihabi, Z.K. HPLC of fluorescent products of acetaminophen reaction with peroxidase. *J.Liq.Chromatogr.*, **1990**, *13*, 1159–1171 [fluorescence detection; UV detection; simultaneous degradation products]

Molokhia, A.M.; Niazy, E.M.; El-Hoofy, S.A.; El-Dardari, M.E. Improved liquid chromatographic method for acyclovir determination in plasma. *J.Liq.Chromatogr.*, **1990**, *13*, 981–989 [extracted acyclovir; plasma; acetaminophen is IS]

Brinkman, U.A.T.; Frei, R.W.; Lingeman, H. Post-column reactors for sensitive and selective detection in high-performance liquid chromatography: Categorization and applications. *J.Chromatogr.*, **1989**, *492*, 251–298 [post-column reaction; review]

Siegers, C.P.; Möller-Hartmann, W. Cholestyramine as an antidote against paracetamol-induced hepato- and nephrotoxicity in the rat. *Toxicol.Lett.*, **1989**, *47*, 179–184 [rat; urine; extracted metabolites]

Lam, S.; Malikin, G. An improved micro-scale protein precipitation procedure for HPLC assay of therapeutic drugs in serum. *J.Liq.Chromatogr.*, **1989**, *12*, 1851–1872 [serum; also amiodarone, aspirin, caffeine, chloramphenicol, flecainide, pentobarbital, procainamide, pyrimethamine, quinidine, theophylline, tocainide, trazodone; fluorescence detection; UV detection]

Alvi, S.U.; Castro, F. A stability-indicating simultaneous analysis of acetaminophen and hydrocodone bitartrate in tablets formulation by HPLC. *J.Liq.Chromatogr.*, **1987**, *10*, 3413–3426 [stability-indicating; simultaneous codeine, hydrocodone; hydromorphone; tablets; column temp 30]

Kinney, C.D.; Kelly, J.G. Liquid chromatographic determination of paracetamol and dextropropoxyphene in plasma. *J.Chromatogr.*, **1987**, *419*, 433–437 [plasma; salicylamide is IS; LOD 100 ng/mL; pharmacokinetics]

Fatmi, A.A.; Williams, G.V. Simultaneous determination of acetaminophen and hydrocodone bitartrate in solid dosage forms by HPLC. *J.Liq.Chromatogr.*, **1987**, *10*, 2461–2472 [simultaneous hydrocodone; formulations]

Hill, D.W.; Langner, K.J. HPLC photodiode array UV detection for toxicological drug analysis. *J.Liq.Chromatogr.*, **1987**, *10*, 377–409 [also acepromazine, acetophenazine, acetophenetidine, acetylprocainamide, aflatoxins, allylcyclopentenylbarbituric acide, allylisobarbituric acid, alphaprodine, alphenal, aminoantipyrine, aminobenzamide, aminobenzoate, aminobenzoic, aminobenzoic acid, aminophylline, amitriptyline, amobarbital, amylocaine, anisic acid, anthranilamide, antipyrine, aprobarbital, aspirin, atropine, barbital, benzoate, benzocaine, benzoic, benzoic acid, benzoylecgonine, benzphetamine, brucine, butabarbital, butacaine, butethal, butyl paraben, caffeine, cannabichromene, cannabinol, carbostyril, carvacrole, chloramphenicol, chlordiazepoxide, chloroethyltheophylline, chlorophenol, chloroquine, chlorotheophylline, chlorothiazide, chlorphenesin, chlorpheniramine, chlorpromazine, chlorpropamide, cinchonidine, cinchonine, clonazepam, cocaine, codeine, colchicine, cortisone, coumarin, creatinine, cyclothiazide, cyheptamide, danazol, danthron, dapsone, DES, despropanylfentanyl, dexamethasone, diallylbarbituric acid, diazepam, dibucaine, dichlorophene, didrate, diethylstilbestrol, diethyltryptamine, dihydroxyphenethyl alcohol, dihydroxyphenylacetic acid, dihydroxyphenylglycol, dimethylbarbituric acid, diphenhydramine, diphenoxylate, diphenylhydantoin, dipropyltryptamine, dithiodinicotinic acid, doxapram, dronabinol, dyphylline, estradiol, estriol, estrone, ethonitazene, ethosuximide, ethylmorphine, ethylnornicotine, ethylphenylmalonamide, ethyltolylmalonamide, etonitazene, eugenol, fenfluramine, fenoprofen, fentanyl, 5-fluorouracil, fluoxymesterone, flurazepam, furosemide, gentisic, gentisic acid, gitoxigenin, glutethimide, guaiacol, hexabarbital, hexahydrocannabinol, hexylresorcinol, hippuric acid, homovanillic acid, hydrocortisone, hydromorphone, hydroquinone, hydroxyethyltheophylline, hydroxyindoleacetic acid, hydroxyisoquinoline, hydroxymethyltestosterone, hydroxynicotinic acid, hydroxyphenobarbital, hydroxyphenylpyruvic acid, hydroxyquinoline, ibuprofen, imipramine, indoleacetic acid, indolecarboxaldehyde, indomethacin, isobutylmethylxanthine, isocarbostyril, isoquinoline-N-oxide, lasix, levorphanol, lidocaine, LSD, meclizine, mefenamate, mefenamic, mefenamic acid, meperidine, mephenesin, mephobarbital, mepivacaine, mescaline, methocarbamal, methoxamine, methoxydichlorobenzoic acid, methyl salicylate, methyldopa, methyldopamine, methylparaben, methylphenidate, methylprimidone, methyltestosterone, methylxanthine, morphine, nalorphine, naloxone, naltrexol, naphthalene, naphthol, naphthoylacetic acid, naproxen, nicotine, nicotinic acid, nikethamide, nitrofurantoin, nitrophenol, normethsuximide, oxazepam, oxyphenbutazone, papaverine, paraxanthine, pemoline, pentazocine, pentobarbital, phencyclidine, phenetidine, phenobarbital, phentermine, phenylbutazone, phenytoin, physostigmine, piperonyl butoxide, prednisolone, prednisone, primidone, probenecid, procaine, progesterone, propiomazine, propyl paraben, pyrilamine, pyrithydione, pyrocatechol, quinoline-N-oxide, reserpine, resorcinol, saccharin, salicylamide, salicylate, salicylic, salicylic acid, secobarbital, stanozolol, sulfacetamide, sulfadimethoxine, sulfaethidole, sulfamerazine, sulfamethazine, sulfamethizole, sulfamethoxazole, sulfanilamide, sulfapyridine, sulfasoxazole, sulindac, testosterone, tetracaine, tetrahydocannabinol, thebaine, theobromine, theophylline, thiamylal, thienylcyclohexylpiperidine, thiobarbituric acid, thiobarbituric, thiosalicylate, thiosalicylic, thiosalicylic acid, tolbutamide, tolmetin, triamcinolone, trifluoromethylbenzoic acid]

Lurie, I.S.; McGuiness, K. The quantitation of heroin and selected basic impurities via reversed phase HPLC. II. The analysis of adulterated samples. *J.Liq.Chromatogr.*, **1987**, *10*, 2189–2204 [also impurities, acetaminophen, acetylcodeine, acetylmorphine, acetylprocaine, aminopyrene, amitriptyline, antipyrene, aspirin, barbital, benztropine, caffeine, cocaine, codeine, diamorphine, diazepam, diphenhydramine, dipyrone, ephedrine, ethylmorphine, lidocaine, meconin, methamphetamine, methapyrilene, methaqualone, monoacetylmorphine, morphine, nalorphine, niacinamide, nicotinamide, noscapine, papaverine, phenacetin, phenmetrazine, phenobarbital, phenolphthalein, procaine, propanophenone , propoxyphene, pyrilamine, quinidine, quinine, salicylamide, salicylate, salicylic, salicylic acid, secobarbital, strychnine, tetracaine, thebaine, tripelennamine, tropacocaine, vitamin B3, vitamin B5; electrochemical detection]

Wilson, T.D. Recent advances in HPLC analysis of analgesics. *J.Liq.Chromatogr.*, **1986**, *9*, 2309–2410 [also antipyrine, aspirin, buprenorphine, butorphanol, codeine, cyclazocine, diamorphine, dihydrocodeine, etorphine, fentanyl, heroin, hydrocodone, hydromorphone, ibuprofen, LAAM, levorphanol, meperidine, methadone, morphine, nalbuphine, naproxen, oxycodone, oxymorphone, pentazocine, phenacetin, phenazocine, phenylbutazone, propoxyphene, salicylamide, salicylic acid, zomepirac]

Ting, S. Liquid chromatographic determination of levodopa, levodopa-carbidopa, and related impurities in solid dosage forms. *J.Assoc.Off.Anal.Chem.*, **1986**, *69*, 169–173 [simultaneous carbidopa, levodopa; acetaminophen is IS; formulations]

Bouquet, S.; Regnier, B.; Quehen, S.; Brisson, A.M.; Courtois, P.; Fourtillan, J.B. Rapid determination of acyclovir in plasma by reversed phase high-performance liquid chromatography. *J.Liq. Chromatogr.*, **1985**, *8*, 1663–1675 [extracted acyclovir; plasma; acetaminophen is IS]

Wong, S.H.Y.; Marzouk, N.; McHugh, S.L.; Cazes, E. Simultaneous determination of theophylline and caffeine by reversed phase liquid chromatography using phenyl column. *J.Liq.Chromatogr.*, **1985**, *8*, 1797–1816 [also caffeine, cimetidine, codeine, dimethylxanthine, meperidine, pentobarbital, phenobarbital, secobarbital, theobromine, theophylline; hydroxyethyltheophylline is IS]

Meinsma, D.A.; Radzik, D.M.; Kissinger, P.T. Determination of common analgesics in serum and urine by liquid chromatography/electrochemistry. *J.Liq.Chromatogr.*, **1983**, *6*, 2311–2335 [serum; urine; electrochemical detection; simultaneous codeine, methyl salicylate, naproxen, phenacetin, salicylic acid]

Miner, D.J.; Skibic, M.J.; Bopp, R.J. Practical aspects of LC/EC determinations of pharmaceuticals in biological media. *J.Liq.Chromatogr.*, **1983**, *6*, 2209–2230 [plasma; column-switching; fluorescence detection; also diethylstilbestrol, enviradene, enviroxime, enviroxine, hexestrol, pergolide, zinviroxime]

Ferrell, W.J.; Goyette, G.W. Analysis of acetaminophen and salicylate by reverse phase HPLC. *J.Liq.Chromatogr.*, **1982**, *5*, 93–96 [also salicylic acid; serum]

van der Wal, S.; Bannister, S.J.; Snyder, L.R. Automated analysis of acetaminophen and caffeine in serum using the FAST-LC system: contributions to assay imprecision in procedures based on HPLC with sample pretreatment. *J.Chromatogr.Sci.*, **1982**, *20*, 260–265 [serum; extracted caffeine, carbamazepine, phenobarbital, phenylethylmalonamide, phenytoin, primidone, theophylline; cyclobarbital is IS]

Lurie, I.S.; Demchuk, S.M. Optimization of a reverse phase ion-pair chromatographic separation for drugs of forensic interest. II. Factors effecting selectivity. *J.Liq.Chromatogr.*, **1981**, *4*, 357–374 [also acetylcodeine, acetylmorphine, aminopyrene, aminopyrine, amobarbital, amphetamine, antipyrine, benzocaine, butabarbital, caffeine, cocaine, codeine, diamorphine, diazepam, diethylpropion, ephedrine, glutethimide, lidocaine, mecloqualone, mescaline, methamphetamine, methapyrilene, methaqualone, methpyrilene, methylphenidate, morphine, narcotine, papaverine, pentobarbital, phencyclidine, phendimetrazine phenmetrazine, phenobarbital, phentermine, phenylpropanolamine, procaine, quinidine, quinine, secobarbital, strychnine, tetracaine, thebaine, theophylline]

Teffera, Y.; Abramson, F. Application of high-performance liquid chromatography/chemical reaction interface mass spectrometry for the analysis of conjugated metabolites: a demonstration using deuterated acetaminophen. *Biol.Mass.Spectrom.*, **1994**, *23*, 776–783

Osterloh, J.; Yu, S. Simultaneous ion-pair and partition liquid chromatography of acetaminophen, theophylline and salicylate with application to 500 toxicologic specimens. *Clin.Chim.Acta*, **1988**, *175*, 239–248

Betowski, L.D.; Korfmacher, W.A.; Lay, J.O.J.; Potter, D.W.; Hinson, J.A. Direct analysis of rat bile for acetaminophen and two of its conjugated metabolites via thermospray liquid chromatography/mass spectrometry. *Biomed.Environ.Mass.Spectrom.*, **1987**, *14*, 705–709

Biemer, T.A. Simultaneous analysis of acetaminophen, pseudoephedrine hydrochloride and chlorpheniramine maleate in a cold tablet using an isocratic, mixed micellar high-performance liquid chromatographic mobile phase. *J.Chromatogr.*, **1987**, *410*, 206–210

Colin, P.; Sirois, G.; Chakrabarti, S. Rapid high-performance liquid chromatographic assay of acetaminophen in serum and tissue homogenates. *J.Chromatogr.*, **1987**, *413*, 151–160

Starkey, B.J.; Loscombe, S.M.; Smith, J.M. Paracetamol (acetaminophen) analysis by high performance liquid chromatography: interference studies and comparison with an enzymatic procedure. *Ther.Drug Monit.*, **1986**, *8*, 78–84

Jung, D.; Zafar, N.U. Micro high-performance liquid chromatographic assay of acetaminophen and its major metabolites in plasma and urine. *J.Chromatogr.*, **1985**, *339*, 198–202

Mamolo, M.G.; Vio, L.; Maurich, V. High-pressure liquid chromatographic analysis of paracetamol, caffeine and acetylsalicylic acid in tablets. Salicylic acid quantitation. *Farmaco.[Prat].*, **1985**, *40*, 111–123

To, E.C.; Wells, P.G. Repetitive microvolumetric sampling and analysis of acetaminophen and its toxicologically relevant metabolites in murine plasma and urine using high performance liquid chromatography. *J.Anal.Toxicol.*, **1985**, *9*, 217–221

Das Gupta, V.; Heble, A.R. Quantitation of acetaminophen, chlorpheniramine maleate, dextromethorphan hydrobromide, and phenylpropanolamine hydrochloride in combination using high-performance liquid chromatography. *J.Pharm.Sci.*, **1984**, *73*, 1553–1556

Krieger, D.J. Liquid chromatographic determination of acetaminophen in multicomponent analgesic tablets. *J.Assoc.Off.Anal.Chem.*, **1984**, *67*, 339–341

Quattrone, A.J.; Putnam, R.S. A single liquid-chromatographic procedure for therapeutic monitoring of theophylline, acetaminophen, or ethosuximide. *Clin.Chem.*, **1981**, *27*, 129–132

West, J.C. Rapid HPLC analysis of paracetamol (acetaminophen) in blood and postmortem viscera. *J.Anal.Toxicol.*, **1981**, *5*, 118–121

Das Gupta, V. Simultaneous quantitation of acetaminophen, aspirin, caffeine, codeine phosphate, phenacetin, and salicylamide by high-pressure liquid chromatography. *J.Pharm.Sci.*, **1980**, *69*, 110–113

# Acyclovir

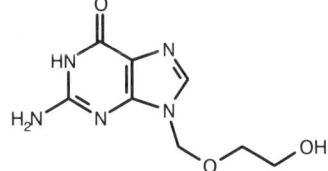

**Molecular formula:** $C_8H_{11}N_5O_3$
**Molecular weight:** 225.2
**CAS Registry No.:** 59277-89-3, 69657-51-8 (sodium salt)

## SAMPLE
**Matrix:** blood
**Sample preparation:** 100 µL Plasma + 10 µL water + 100 µL 200 mM pH 7 sodium phosphate buffer, mix well, inject a 5 µL aliquot.

## HPLC VARIABLES
**Guard column:** µBondapak CN guard-PAK
**Column:** 300 × 3.9 µBondapak C18
**Mobile phase:** MeCN:50 mM ammonium acetate 2:98
**Flow rate:** 1
**Injection volume:** 5
**Detector:** UV 254

## CHROMATOGRAM
**Retention time:** 18
**Internal standard:** acyclovir

## OTHER SUBSTANCES
**Extracted:** ceftibuten
**Noninterfering:** acetaminophen, amoxicillin, ampicillin, aspirin, aztreonam, caffeine, cefamandole, cefotiam, cefsulodin, ceftazidime, ceftriaxone, cefuroxime, cephaloridine, cephalothin, chlorpheniramine, gentamicin, moxolactam, nafcillin, piperacillin, pseudoephedrine, theophylline, ticarcillin, vancomycin

## KEY WORDS
plasma; acyclovir is IS; column-switching

## REFERENCE
Lim, J.M.; Kim, H.; Marco, A.; Mojaverian, P.; Lin, C.-C. Liquid chromatographic determination of ceftibuten, a new oral cephalosporin, in human plasma and urine. *J.Pharm.Biomed.Anal.*, **1994**, *12*, 699–703

## SAMPLE
**Matrix:** blood
**Sample preparation:** Condition a 1 mL 100 mg Sep-Pak Vac trifunctional C18 SPE cartridge with 1 mL MeOH and 1 mL buffer. 0.5 mL Plasma + 0.5 mL buffer, vortex, add to SPE cartridge, wash with 0.5 mL buffer, elute with 300 µL MeOH:5 mM sodium octanesulfonate 20:80 adjusted to pH 8.50 with 4 M NaOH, inject a 130 µL aliquot of the eluate. (Procedure was automated (ASPEC system). Buffer was 5 mM sodium octanesulfonate adjusted to pH 2.85 with concentrated orthophosphoric acid.)

## HPLC VARIABLES
**Guard column:** 20 × 2 30-40 µm Perisorb RP-18 (change each day)
**Column:** 150 × 3.9 4 µm Nova Pak C18
**Mobile phase:** MeOH:10 mM $Na_2HPO_4$ + 10 mM sodium octanesulfonate 7:93 with the final apparent pH adjusted to 2.80 with concentrated orthophosphoric acid
**Column temperature:** 40
**Flow rate:** 1

**Injection volume:** 130
**Detector:** UV 250

## CHROMATOGRAM
**Retention time:** 4
**Limit of quantitation:** 10 ng/mL

## KEY WORDS
plasma; SPE

## REFERENCE
Swart, K.J.; Hundt, H.K.L.; Groenewald, A.M. Automated high-performance liquid chromatographic method for the determination of acyclovir in plasma. *J.Chromatogr.A*, **1994**, *663*, 65–69

## SAMPLE
**Matrix:** blood
**Sample preparation:** Heat-inactivate serum at 56° for 1 h. 500 μL Serum + 50 μL 100 μg/mL guanosine, filter (Centrisart I $M_r$ 5000 cut-off), dilute ultrafiltrate 1:30 with mobile phase buffer, inject an aliquot.

## HPLC VARIABLES
**Column:** 125 × 4 5μm RP8e (Merck)
**Mobile phase:** MeOH:100 mM pH 3.0 phosphate buffer containing 50 mM 1-octanesulfonic acid 5:95
**Flow rate:** 1
**Detector:** UV 254

## CHROMATOGRAM
**Retention time:** 6.36
**Internal standard:** guanosine (7.35)
**Limit of detection:** 50 ng/mL

## OTHER SUBSTANCES
**Noninterfering:** gangiclovir, zidovudine

## KEY WORDS
serum; ultrafiltrate

## REFERENCE
Nebinger, P.; Koel, M. Determination of acyclovir by ultrafiltration and high-performance liquid chromatography. *J.Chromatogr.*, **1993**, *619*, 342–344

## SAMPLE
**Matrix:** blood
**Sample preparation:** 1 mL Plasma + 300 μL 3 M perchloric acid, vortex 15 s, centrifuge at 2000 g for 2 min, inject a 20 μL aliquot of the supernatant.

## HPLC VARIABLES
**Column:** 80 × 4 3 μm Nucleosil 120 3C18
**Mobile phase:** Gradient. A was 20 mM perchloric acid. B was MeCN:20 mM perchloric acid 45:55. A:B 100:0 for 3 min, 0:100 for 3 min (step gradient).
**Column temperature:** 30
**Flow rate:** 1.5
**Injection volume:** 20
**Detector:** F ex 260 em 375

## CHROMATOGRAM
**Retention time:** 2.3
**Limit of detection:** 6-10 ng/mL

## KEY WORDS
plasma

## REFERENCE
Mascher, H.; Kikuta, C.; Metz, R.; Vergin, H. New, high-sensitivity high-performance liquid chromato-graphic method for the determination of acyclovir in human plasma, using fluorometric detection. *J.Chromatogr.*, **1992**, *583*, 122–127

## SAMPLE
**Matrix:** blood
**Sample preparation:** 200 µL Blood + 600 µL 16% trichloroacetic acid, centrifuge, inject an aliquot of the supernatant.

## HPLC VARIABLES
**Guard column:** C18 Guard-Pak (Waters)
**Column:** 100 × 8 Nova-Pak C18 (in a Z module)
**Mobile phase:** Gradient. A was 50 mM sodium hydrogen phosphate. B was MeOH:water 80:20 containing 5 mM $NaH_2PO_4$. A:B 99:1 for 1.5 min, to 5:95 over 18.5 min, re-equilibrate at initial conditions for 10 min.
**Flow rate:** 1.6
**Detector:** UV 254

## CHROMATOGRAM
**Retention time:** 11.6
**Limit of detection:** 250 ng/mL

## OTHER SUBSTANCES
**Extracted:** famciclovir (as penciclovir, the active metabolite)

## KEY WORDS
mouse; pharmacokinetics

## REFERENCE
Boyd, M.R.; Bacon, T.H.; Sutton, D. Antiherpesvirus activity of 9-(4-hydroxy-3-hydroxymethylbut-1-yl) guanine (BRL 39123) in animals. *Antimicrob.Agents Chemother.*, **1988**, *32*, 358–363

## SAMPLE
**Matrix:** blood
**Sample preparation:** Inject an aliquot directly onto the column.

## HPLC VARIABLES
**Guard column:** 20 × 4.6 16 µm Spheron Micro 300
**Column:** 150 × 3.2 12.5 µm Spheron Micro 300 (glass column) (Lachema)
**Mobile phase:** pH 1.8 Buffer containing 100 mM phosphoric acid and 100 mM sodium sulfate
**Column temperature:** 45
**Flow rate:** 1
**Injection volume:** 10
**Detector:** F ex 285 em 370

## CHROMATOGRAM
**Retention time:** 2
**Limit of detection:** 100 ng/mL

## KEY WORDS
plasma; effluent cooled to 2° before entering detector; GPC; dog; pharmacokinetics; direct injection

## REFERENCE
Salamoun, J.; Sprta, V.; Sladek, T.; Smrz, M. Determination of acyclovir in plasma by column liquid chromatography with fluorescence detection. *J.Chromatogr.*, **1987**, *420*, 197–202

## SAMPLE
**Matrix:** blood, dialysate, urine
**Sample preparation:** 100 μL Plasma, urine, or dialysate + 100 μL 200 mM pH 7 phosphate buffer, inject a 5 μL aliquot.

## HPLC VARIABLES
**Column:** μBondapak C18
**Mobile phase:** MeCN:50 mM ammonium acetate 2:98
**Flow rate:** 1
**Injection volume:** 5
**Detector:** UV 254

## CHROMATOGRAM
**Internal standard:** acyclovir

## OTHER SUBSTANCES
**Extracted:** ceftibuten

## KEY WORDS
plasma; acyclovir is IS

## REFERENCE
Kelloway, J.S.; Awni, W.M.; Lin, C.C.; Lim, J.; Affrime, M.B.; Keane, W.F.; Matzke, G.R.; Halstenson, C.E. Pharmacokinetics of ceftibuten-*cis* and its *trans* metabolite in healthy volunteers and in patients with chronic renal insufficiency. *Antimicrob.Agents Chemother.*, **1991**, *35*, 2267–2274

## SAMPLE
**Matrix:** blood, urine
**Sample preparation:** 1 mL Plasma or urine + 300 μL 3 M perchloric acid, vortex for 15 s, centrifuge at 2000 g for 2 min, inject a 20 μL aliquot of the supernatant.

## HPLC VARIABLES
**Column:** 80 × 4 3 μm Nucleosil 3C-18
**Mobile phase:** Gradient. MeCN:20 mM perchloric acid 0:100 for 3 min, 45:55 for 3 min (step gradient).
**Column temperature:** 30
**Flow rate:** 1.5
**Injection volume:** 20
**Detector:** F ex 260 em 375

## CHROMATOGRAM
**Limit of detection:** 6-10 ng/mL (plasma); 25 ng/mL (urine)

## KEY WORDS
plasma; pharmacokinetics

## REFERENCE
Vergin, H.; Kikuta, C.; Mascher, H.; Metz, R. Pharmacokinetics and bioavailability of different formulations of aciclovir. *Arzneimittelforschung*, **1995**, *45*, 508–515

## SAMPLE
**Matrix:** blood, urine
**Sample preparation:** Plasma. 200 µL Plasma + 200 µL 200 mM pH 7.0 sodium phosphate, vortex, allow to sit for 15 min, add 800 µL MeCN, vortex for 20 s, centrifuge at 2500 g at 25° for 2 min. Remove the supernatant and add it to 1.6 mL dichloromethane, vortex for 20 s, centrifuge at 2500 g at 25° for 1 min, inject an aliquot of the organic layer. Urine. Dilute urine samples 10-20-fold with water, treat with Nonidet P-40 detergent, let stand for 5 min, inject an aliquot.

## HPLC VARIABLES
**Column:** 300 × 3.9 µBondapak C18
**Mobile phase:** MeCN:50 mM ammonium acetate 2:98
**Column temperature:** 30
**Flow rate:** 1
**Injection volume:** 20
**Detector:** UV 262 (plasma); UV 254 (urine)

## CHROMATOGRAM
**Internal standard:** acyclovir

## OTHER SUBSTANCES
**Extracted:** ceftibuten

## KEY WORDS
plasma; acyclovir is IS

## REFERENCE
Kearns, G.L.; Reed, M.D.; Jacobs, R.F.; Ardite, M.; Yogev, R.D.; Blumer, J.L. Single-dose pharmacokinetics of ceftibuten (SCH 39720) in infants and children. *Antimicrob.Agents Chemother.*, **1991**, *35*, 2078–2084

## SAMPLE
**Matrix:** perfusate

## HPLC VARIABLES
**Column:** Pecosphere 5C-C18
**Mobile phase:** MeCN:0.1% acetic acid 2:98
**Flow rate:** 1
**Detector:** UV 254

## REFERENCE
Volpato, N.M.; Santi, P.; Colombo, P. Iontophoresis enhances the transport of acyclovir through nude mouse skin by electrorepulsion and electroosmosis. *Pharm.Res.*, **1995**, *12*, 1623–1627

## ANNOTATED BIBLIOGRAPHY
Shao, Z.; Park, G.-B.; Krishnomoorthy, R.; Mitra, A.K. The physicochemical properties, plasma enzymatic hydrolysis, and nasal absorption of acyclovir and its 2'-ester prodrugs. *Pharm.Res.*, **1994**, *11*, 237–242 [nasal perfusate]

Shao, Z.; Mitra, A.K. Bile salt-fatty acid mixed micelles as nasal absorption promoters. III. Effects on nasal transport and enzymatic degradation of acyclovir prodrugs. *Pharm.Res.*, **1994**, *11*, 243–240 [nasal perfusate]

Macka, M.; Borák, J.; Seménková, L.; Popl, M.; Mikes, V. Determination of acyclovir in blood serum and plasma by micellar liquid chromatography with fluorimetric determination. *J.Liq.Chromatogr.*, **1993**, *16*, 2359–2386 [serum; plasma; fluorescence detection; LOD 80 ng/mL]

Molokhia, A.M.; Niazy, E.M.; El-Hoofy, S.A.; El-Dardari, M.E. Improved liquid chromatographic method for acyclovir determination in plasma. *J.Liq.Chromatogr.*, **1990**, *13*, 981–989 [plasma; acetaminophen is IS]

Cronqvist, J.; Nilsson-Ehle, I. Determination of acyclovir in human serum by high-performance liquid chromatography. *J.Liq.Chromatogr.*, **1988**, *11*, 2593–2601 [serum; non-interfering acetaminophen, allopurinol, baclofen, carbacholine, cefuroxime, chlorpropamide, cilastatin, cloxacillin, diazepam, dicumarol, digoxin, flucloxacillin, furosemide, fusidic acid, fusidic, glipizide, heparin, hydrochlorothiazide, imipenem, insulin, isoniazid, ketoprofen, metronidazole, naproxen, perphenazine, phenytoin, prednisolone, propranolol, pyrazinamide, pyridoxine, ranitidine, rifampicin, rifampin, spironolactone, streptomycin, sulfamethoxazole, trimethoprim, warfarin]

Bouquet, S.; Regnier, B.; Quehen, S.; Brisson, A.M.; Courtois, P.; Fourtillan, J.B. Rapid determination of acyclovir in plasma by reversed phase high-performance liquid chromatography. *J.Liq. Chromatogr.*, **1985**, *8*, 1663–1675 [plasma; acetaminophen is IS; LOD 250 ng/mL]

Smith, R.L.; Walker, D.D. High-performance liquid chromatographic determination of acyclovir in serum. *J.Chromatogr.*, **1985**, *343*, 203–207 [serum; LOQ 1.2 μg/mL]

Hoogewijs, G.; Massart, D.L. Development of a standardized analysis strategy for basic drugs, using ion-pair extraction and high-performance liquid chromatography. IV. Application to solid pharmaceutical dosage forms. *J.Liq.Chromatogr.*, **1983**, *6*, 2521–2541 [capsules; tablets; also benzocaine, caffeine, carbinoxamine, fenfluramine, flupentixol, lidocaine, melitracen, mepivacaine, phenylephrine, piperocaine, procaine, tetracaine; normal phase]

Nilsson-Ehle, I. High-performance liquid chromatography for analyses of antibiotics in biological fluids. *J.Liq.Chromatogr.*, **1983**, *6, Supp. 2*, 251–293 [review]

Zhang, C.; Dong, S.N. [Determination of acyclovir in human plasma by RP-HPLC]. *Yao. Hsueh.Hsueh.Pao.*, **1993**, *28*, 629–632

Marini, D.; Pollino, G.; Balestrieri, F. Liquid chromatographic determination of acyclovir. *Boll. Chim.Farm.*, **1991**, *130*, 101–104

# Adiphenine

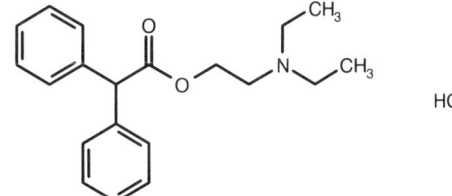

**Molecular formula:** $C_{20}H_{25}NO_2$
**Molecular weight:** 311.4
**CAS Registry No.:** 64-95-9 (adiphenine), 50-42-0
(adiphenine hydrochloride)

HCl

## SAMPLE
**Matrix:** blood, tissue
**Sample preparation:** Tissue. Homogenize brain with 4 volumes MeOH, stir for 15 min, centrifuge at 1300 g for 10 min, acidify supernatant with 300 µL 1 M HCl, evaporate to dryness under reduced pressure, reconstitute with mobile phase, inject an aliquot. Plasma. Extract 1 mL plasma with 15 mL MeOH, stir for 15 min, centrifuge at 1300 g for 10 min, acidify supernatant with 300 µL 1 M HCl, evaporate to dryness under reduced pressure, reconstitute with mobile phase, inject an aliquot.

## HPLC VARIABLES
**Column:** 300 × 4 10 µm Si 100 Lichrosorb
**Mobile phase:** Gradient. A was dichloromethane:diethylamine 100:0.2. B was dichloromethane:EtOH:diethylamine 80:20:0.2. A:B 100:0 for 4 min, to 50:50 over 13 min, maintain at 50:50 for 13 min.
**Flow rate:** 2
**Detector:** UV 254; Radioactivity

## CHROMATOGRAM
**Retention time:** 3.7

## KEY WORDS
normal phase; tritium labeled; $^{14}C$ labeled; plasma; brain; rat

## REFERENCE
Michelot, J.; Moreau, M.F.; Veyre, A.; Labarre, P.; Meyniel, G. Adiphenine plasma levels and blood-brain barrier crossing in the rat. *Eur.J.Drug Metab.Pharmacokinet.*, **1985**, *10*, 273–278

## SAMPLE
**Matrix:** formulations
**Sample preparation:** Tablets. Grind 5 tablets to a fine powder, dissolve in 100 mL MeOH:0.5% acetic acid 1:1, filter (paper), inject an aliquot. Suppositories. Cut up 3 suppositories, add to 100 mL MeOH:0.5% acetic acid 1:1, heat at 40° until all the fat melts, shake, filter (paper), inject a 25 µL aliquot. Liquid formulations. Dilute 10 mL formulation to 100 mL with MeOH:0.5% acetic acid 1:1, inject a 25 µL aliquot.

## HPLC VARIABLES
**Column:** 300 × 4 µBondapak phenyl
**Mobile phase:** Gradient. A was 10 mM heptanesulfonic acid in 1 mM acetic acid. B was 10 mM heptanesulfonic acid and 1 mM acetic acid in MeOH. A:B from 60:40 to 25:75 over 30 min
**Column temperature:** 35
**Flow rate:** 1.75
**Injection volume:** 25
**Detector:** UV 225

## CHROMATOGRAM
**Retention time:** 20

## OTHER SUBSTANCES
**Simultaneous:** dipyrone (metamizol), diphenhydramine, drofenine, ethyldiphenacetate
**Interfering:** promazine

## KEY WORDS
tablets; suppositories; liquid formulations

## REFERENCE
Facchini, G.; Zaccheo, F.; Nannetti, M. Simultaneous determination of hydrochloride salts of adiphenine,
diphenhydramine, ethyldiphenacetate, drofenine and promazine by ion-pair HPLC. *Boll.Chim.Farm.*,
**1983**, *122*, 405–411

## SAMPLE
**Matrix:** solutions

## HPLC VARIABLES
**Column:** 300 × 4 10 μm LiChrosorb RP-18
**Mobile phase:** Gradient. A was 27 mM ammonia in water. B was 27 mM ammonia in
MeOH. A:B from 100:0 to 70:30 over 6 min, maintain at 70:30 for 5 min, to 30:70 over
15 min, maintain at 30:70 for 15 min, to 0:100 over 2 min, maintain at 0:100 for 20
min.
**Flow rate:** 2
**Detector:** UV 254; Radioactivity

## CHROMATOGRAM
**Retention time:** 37.3

## OTHER SUBSTANCES
**Extracted:** metabolites

## KEY WORDS
tritium labeled; [14]C labeled

## REFERENCE
Michelot, J.; Madelmont, J.C.; Rousset, B.; Labarre, P.; Mornex, R.; Meyniel, G. Metabolism of adiphenine. II. Identification of major excretion metabolites in rats. *Xenobiotica*, **1982**, *12*, 457–462

## ANNOTATED BIBLIOGRAPHY
Michelot, J.; Moreau, M.F.; Madelmont, J.C.; Labarre, P.; Meyniel, G. Determination of adiphenine,
diphenylacetic acid and diethylamino-ethanol by high-performance liquid chromatography.
*J.Chromatogr.*, **1983**, *257*, 395–399 [same procedure as Xenobiotica 1982, 12, 457]

# Albuterol

**Molecular formula:** $C_{13}H_{21}NO_3$
**Molecular weight:** 239.3
**CAS Registry No.:** 18559-94-9, 51022-70-9 (sulfate)

## SAMPLE
**Matrix:** blood
**Sample preparation:** 2 mL Whole blood or plasma + 2 mL buffer + 5 mL chloroform: isopropanol:n-heptane 60:14:26, shake gently horizontally for 10 min, centrifuge at 2800 g for 10 min. Remove the lower organic layer and evaporate it to dryness under vacuum at 45°, reconstitute the residue in 100 μL mobile phase, centrifuge at 2800 g for 5 min, inject a 50 μL aliquot of the supernatant. (Buffer was saturated ammonium chloride solution 25% diluted with water, adjusted to pH 9.5 with 25% ammonia solution.)

## HPLC VARIABLES
**Column:** 300 × 3.9 4 μm NovaPack C18
**Mobile phase:** MeOH:THF:buffer 65:5:30 (Buffer was 0.68 g/L (10 mM (sic)) $KH_2PO_4$ adjusted to pH 2.6 with concentrated orthophosphoric acid.) (At the end of each session wash the column with water for 1 h and MeOH for 1 h, re-equilibrate for 30 min.)
**Column temperature:** 30
**Flow rate:** 0.8
**Injection volume:** 50
**Detector:** UV 226

## CHROMATOGRAM
**Retention time:** 3.37
**Limit of detection:** <120 ng/mL

## OTHER SUBSTANCES
**Extracted:** acebutolol, acenocoumarol, acepromazine, aceprometazine, acetaminophen, aconitine, ajmaline, alimemazine, alminoprofen, alpidem, alprazolam, alprenolol, amisulpride, amitriptyline, amodiaquine, amoxapine, astemizole, benazepril, benperidol, benzocaine, benzoylecgonine, bepridil, betaxolol, bisoprolol, bromazepam, brompheniramine, bumadizone, bupivacaine, buprenorphine, buspirone, caffeine, carbamazepine, carbinoxamine, carpipramine, carteolol, celiprolol, cetirizine, chlorambucil, chlordiazepoxide, chlorophenacinone, chloroquine, chlorpheniramine, chlorpromazine, chlorpropamide, cibenzoline, cicletanine, clemastine, clobazam, clomipramine, clonazepam, clonidine, clorazepate, clozapine, cocaine, colchicine, cyamemazine, cyclizine, cycloguanil, cyproheptadine, cytarabine, dacarbazine, daunorubicin, debrisoquine, demexiptiline, desipramine, dextromethorphan, dextromoramide, dextropropoxyphene, diazepam, diazoxide, diclofenac, dihydralazine, diltiazem, diphenhydramine, dipyridamole, disopyramide, dosulepine, doxepin, doxylamine, droperidol, ephedrine, estazolam, etodolac, fenfluramine, fenoprofen, fentiazac, flecainide, floctafenine, flumazenil, flunitrazepam, fluoxetine, fluphenazine, flurbiprofen, fluvoxamine, glibenclamide, glibornuride, glipizide, glutethimide, haloperidol, histapyrrodine, hydroxychloroquine, hydroxyzine, ibuprofen, imipramine, indomethacin, iproniazid, ketamine, ketoprofen, labetalol, levomepromazine, lidocaine, lidoflazine, lisinopril, loperamide, loprazolam, loratadine, lorazepam, loxapine, maprotiline, medazepam, medifoxamine, mefenamic acid, mefenidramine, mefloquine, melphalan, meperidine, mephenesin, mephentermine, mepivacaine, metapramine, methadone, methaqualone, methocarbamol, methotrexate, metipranolol, metoclopramide, metoprolol, mexiletine, mianserine, midazolam, minoxidil, moclobemide, moperone, nadolol, nalbuphine, nalorphine, naloxone, naltrexone, naproxen, nialamide, nicardipine, nifedipine, niflumic acid, nimodipine, nitrazepam, nitrendipine, nizatidine, nomifensine, nortriptyline, omeprazole, opipramol, oxazepam, oxprenolol, penbutolol, penfluridol, pentazocine, phencyclidine, phenylbutazone, pimozide, pindolol, pipamperone, piroxicam,

prazepam, prazosin, prilocaine, procainamide, procarbazine, proguanil, promethazine, propafenone, propranolol, protriptyline, pyrimethamine, quinidine, quinine, quinupramine, ramipril, reserpine, secobarbital, sotalol, strychnine, sulfinpyrazole, sulindac, sulpride, suriclone, temazepam, tenoxicam, terfenadine, tetracaine, tetrazepam, thiopental, thioproperazine, thioridazine, tianeptine, tiaprofenic acid, ticlopidine, timolol, tioclomarol, tofisopam, tolbutamide, trazodone, triazolam, trifluoperazine, trifluperidol, trimipramine, triprolidine, tropatenine, verapamil, viloxazine, vinblastine, vincristine, vindesine, warfarin, yohimbine, zolpidem, zopiclone, zorubicine

**Interfering:** aspirin, atenolol, chlormezanone, codeine, metformin, morphine, phenobarbital, phenol, ranitidine, ritodrine, sultopride, terbutaline, tiapride, toloxatone

## KEY WORDS
whole blood; plasma

## REFERENCE
Tracqui, A.; Kintz, P.; Mangin, P. Systematic toxicological analysis using HPLC/DAD. *J.Forensic Sci.*, **1995**, *40*, 254–262

## SAMPLE
**Matrix:** blood

**Sample preparation:** Condition a Sep-Pak SPE cartridge with 10 mL MeOH and 10 mL water. 500 μL Serum + 1 mL water + 15 μL 100 ng/mL fenoterol in water, mix, force slowly through SPE cartridge, wash twice with 2 mL water, elute with 2 mL MeOH (discard first 2 drops). Evaporate eluate to dryness under a stream of nitrogen at 40°, vortex for 1 min with 70 μL buffer + 300 μL 0.05% di(2-ethylhexyl) phosphate in ethyl acetate, centrifuge at 5000 g for 30 s, transfer organic phase to another tube and add 40 μL buffer to it, vortex for 1 min, centrifuge at 5000 g for 30 s, transfer ethyl acetate layer to another tube and add 70 μL 10 mM HCl to it, vortex 1 min, centrifuge at 5000 g for 30 s, remove acid layer and wash it with 150 μL chloroform, centrifuge at 5000 g for 30 s, inject a 40-60 μL aliquot of the aqueous phase. (Buffer was 70 mM $NaH_2PO_4$ + 1 mM chloride + 2 mM 1-heptanesulfonic acid, pH 6.8.)

## HPLC VARIABLES
**Column:** 75 × 4.6 3 μm Ultrasphere ODS

**Mobile phase:** MeOH:buffer 25:75 (Buffer was 70 mM $NaH_2PO_4$ + 1 mM chloride + 2 mM 1-heptanesulfonic acid, pH 6.8.)

**Flow rate:** 0.5

**Injection volume:** 40-60

**Detector:** E, BioAnalytical System Model LC-4, TL-5 glassy carbon working electrode, Ag/AgCl reference electrode, +0.80 V, 10 nA full scale

## CHROMATOGRAM
**Retention time:** 7

**Internal standard:** fenoterol (13)

**Limit of detection:** 0.4 ng/mL

## OTHER SUBSTANCES
**Noninterfering:** buphenine, carbamazepine, dobutamine, epinephrine, ethosuximide, gentamycin, isoproterenol, isoxsuprine, metaproterenol, metaraminol, oxymetazoline, phenobarbital, phentolamine, phenylephrine, phenytoin, primidone, terbutaline, theophylline, valproic acid

## KEY WORDS
serum; heparin interferes with IS; SPE; pharmacokinetics

## REFERENCE

Tan, Y.K.; Soldin, S.J. Determination of salbutamol in human serum by reversed-phase high-performance liquid chromatography with amperometric detection. *J.Chromatogr.*, **1984**, *311*, 311–317

## SAMPLE

**Matrix:** blood, urine

**Sample preparation:** Condition a 1 mL 100 mg (or 2.8 mL 500 mg) Extract-Clean silica SPE cartridge (Alltech) with 1 volume MeCN and 1 volume water. Add 1 mL urine or 3 mL plasma to the SPE cartridge, dry, wash with 1 volume water, wash with 1 volume MeCN, dry, elute with 2 volumes MeOH. Evaporate the eluate under reduced pressure at 45°, reconstitute with 1 (urine) or 0.3 (plasma) mL mobile phase, vortex for 1 min, inject an aliquot.

## HPLC VARIABLES

**Guard column:** 30 × 4 Chirex 3022 naphthyl urea (Phenomenex)

**Column:** 250 × 4 Chirex naphthyl urea (Phenomenex)

**Mobile phase:** Hexane:1,2-dichloromethane (sic):MeOH:trifluoroacetic acid 60.75:35:4.25:0.25

**Flow rate:** 1

**Injection volume:** 200

**Detector:** F ex 220 em 309

## CHROMATOGRAM

**Retention time:** 22 (S-(+)), 27 (R-(−))

**Limit of detection:** 0.25 ng/mL

## KEY WORDS

chiral; plasma; SPE; pharmacokinetics

## REFERENCE

Boulton, D.W.; Fawcett, J.P. Determination of salbutamol enantiomers in human plasma and urine by chiral high-performance liquid chromatography. *J.Chromatogr.B*, **1995**, *672*, 103–109

## SAMPLE

**Matrix:** blood, urine

**Sample preparation:** Condition a 100 mg Baxter C18 SPE cartridge with one volume MeOH and two volumes water. Dilute 10 μL urine to 500 μL with water. 500 μL Serum or diluted urine + 50 μL water, vortex for 30 s, add to SPE cartridge, wash with three 200 μL aliquots of water, elute with two 500 μL aliquots of MeOH. Evaporate the eluates to dryness under a stream of air at 40-45°, reconstitute the residue in 150 μL water, vortex for 30 s, centrifuge at 14000 g for 4 min, inject a 50 μL aliqu

## HPLC VARIABLES

**Guard column:** 5 μm Adsorbosphere C-18

**Column:** 250 × 4.6 5 μm Adsorbosphere C-18

**Mobile phase:** MeCN:buffer 7:93 adjusted to pH 3.0 with 85% phosphoric acid (Buffer was 25 mM $(NH_4)H_2PO_4$ and 1 mM N,N-dimethyloctylamine.)

**Flow rate:** 1.5

**Injection volume:** 50

**Detector:** UV 224; F ex 228 em 310

## CHROMATOGRAM

**Retention time:** 7.1

**Internal standard:** albuterol

## OTHER SUBSTANCES
**Extracted:** atenolol

## KEY WORDS
serum; SPE; albuterol is IS

## REFERENCE
Chatterjee, D.J.; Li, W.Y.; Hurst, A.K.; Koda, R.T. High-performance liquid chromatographic method for determination of atenolol from human plasma and urine: Simultaneous fluorescence and ultraviolet detection. *J.Liq.Chromatogr.*, **1995**, *18*, 791–806

## SAMPLE
**Matrix:** bulk
**Sample preparation:** Inject a 5 μL aliquot of a solution.

## HPLC VARIABLES
**Column:** 250 × 4.6 5 μm LiChrosphere Diol
**Mobile phase:** Gradient. Carbon dioxide:MeOH containing 0.5% n-propylamine 70:30 for 9.5 min, to 55:45 over 12 min.
**Column temperature:** 70
**Flow rate:** 1.5
**Injection volume:** 5
**Detector:** UV

## CHROMATOGRAM
**Retention time:** 6.2
**Limit of detection:** 1.3 μg/mL

## OTHER SUBSTANCES
**Simultaneous:** impurities

## KEY WORDS
300 bar; SFC

## REFERENCE
Bernal, J.L.; del Nozal, M.J.; Rivera, J.M.; Serna, M.L.; Toribo, L. Separation of salbutamol and six related impurities by packed column supercritical fluid chromatography. *Chromatographia*, **1996**, *42*, 89–94

## SAMPLE
**Matrix:** formulations
**Sample preparation:** Dilute with water to a concentration of 833 μg/mL, add a 300 μL aliquot to 250 μL 2 mg/mL mepivacaine hydrochloride in water, vortex, inject a 10 μL aliquot.

## HPLC VARIABLES
**Column:** 100 × 8 4 μm NovaPak C18 radial compression
**Mobile phase:** Gradient. A was 2.5 mM PIC B-8 Low UV (Waters) in THF:water 40:60. B was water. C was MeOH:water 50:50. A:B:C 50:50:0 for 7.7 min, to 60:15:25 over 5.3 min, re-equilibrate at initial conditions for 5 min.
**Flow rate:** 2
**Injection volume:** 10
**Detector:** UV 220

## CHROMATOGRAM
**Retention time:** 3.2
**Internal standard:** mepivacaine (8.2)

## OTHER SUBSTANCES
**Simultaneous:** fenoterol, ipratropium bromide, terbutaline

## KEY WORDS
nebulizer solutions; stability-indicating

## REFERENCE
Jacobson, G.A.; Peterson, G.M. High-performance liquid chromatographic assay for the simultaneous determination of ipratropium bromide, fenoterol, salbutamol and terbutaline in nebulizer solution. *J.Pharm.Biomed.Anal.*, **1994**, *12*, 825−832

## SAMPLE
**Matrix:** formulations
**Sample preparation:** Tablets, capsules. Mix tablets or capsules with 10 mL water, sonicate 30 min, centrifuge, inject an aliquot of the supernatant. Liquid formulations. Dilute liquid formulations with water, inject an aliquot.

## HPLC VARIABLES
**Column:** 125 × 4 5 μm LiChrospher 100 RP-18 endcapped
**Mobile phase:** MeOH:water 40:60 containing 2 mM KOH + 10 mM hexanoic acid
**Flow rate:** 0.4
**Injection volume:** 20
**Detector:** UV 214

## CHROMATOGRAM
**Retention time:** 3.5
**Limit of detection:** 1760 ng/mL

## OTHER SUBSTANCES
**Also analyzed:** terbutaline, fenoterol

## KEY WORDS
tablets; capsules; liquid formulations

## REFERENCE
Ackermans, M.T.; Beckers, J.L.; Everaerts, F.M.; Seelen, I.G. Comparison of isotachophoresis, capillary zone electrophoresis and high-performance liquid chromatography for the determination of salbutamol, terbutaline sulphate and fenoterol hydrobromide in pharmaceutical dosage forms. *J.Chromatogr.*, **1992**, *590*, 341−353

## SAMPLE
**Matrix:** blood
**Sample preparation:** Condition a 1 mL Bond Elut silica (not C18) SPE cartridge twice with 1 mL MeOH and with 1 mL water. 1 mL Plasma + 200 μL 50 μg/mL bamethan sulfate in 1% potassium bicarbonate, add to SPE cartridge without applying vacuum, after 2 min apply vacuum to move samples through at 1 mL/min then increase vacuum to remove all liquid. Wash twice with 1 mL water and with 1 mL MeCN (draining completely each time). Elute with 1 mL MeOH, expelling last drop with positive pressure. Evaporate eluate under vacuum without heating, dissolve residue in 40 μL mobile phase, vortex, centrifuge at 300 g for 20 s, inject whole sample.

## HPLC VARIABLES
**Guard column:** 15 × 3.2 Brownlee 7 μm C8
**Column:** 150 × 4.6 5 μm Ultrasphere Octylsilica

**Mobile phase:** MeCN:MeOH:water:phosphoric acid:KH$_2$PO$_4$:octanesulfonic acid 50:150: 900:0.25:0.75:0.05 (v/v/v/v/w/w)
**Flow rate:** 1.6
**Injection volume:** 40
**Detector:** F ex 275 em 310

## CHROMATOGRAM
**Retention time:** 7
**Internal standard:** bamethan sulfate (17.5)
**Limit of quantitation:** 0.2 ng/mL

## KEY WORDS
SPE; plasma

## REFERENCE
Gupta, R.N.; Fuller, H.D.; Dolovich, M.B. Optimization of a column liquid chromatographic procedure for the determination of plasma salbutamol concentrations. *J.Chromatogr.B*, **1994**, *654*, 205–211

## SAMPLE
**Matrix:** blood
**Sample preparation:** Condition a Bond Elut Si SPE cartridge by washing twice with 1 mL MeOH, twice with 1 mL water, and once with 1 mL 100 mM pH 9.2 K$_2$HPO$_4$. Add 1 mL plasma + 100 μL 500 ng/mL atenolol in water, wash twice with 1 mL water, centrifuge at 1000 g for 5 min, elute with 1 mL MeOH. Evaporate MeOH to dryness at 40° under a stream of air and dissolve residue in 200 μL mobile phase, inject an aliquot.

## HPLC VARIABLES
**Column:** 250 × 4.6 Spherisorb S5 SCX
**Mobile phase:** MeOH:MeCN:water 40:40:20 containing 0.2% perchloric acid (apparent pH 1.7)
**Flow rate:** 1.5
**Injection volume:** 100
**Detector:** F ex 200 no emission filter

## CHROMATOGRAM
**Retention time:** 8
**Internal standard:** atenolol (13)
**Limit of detection:** 1000 ng/mL

## OTHER SUBSTANCES
**Extracted:** terbutaline
**Noninterfering:** aminophylline, beclomethasone, cloprednol, dexamethasone, fenoterol, ipratropium bromide, methylprednisolone, orciprenaline, prednisolone, reproterol, rimiterol, salmeterol, sodium cromoglycate, theophylline

## KEY WORDS
SPE; plasma

## REFERENCE
McCarthy, P.T.; Atwal, S.; Sykes, A.P.; Ayres, J.G. Measurement of terbutaline and salbutamol in plasma by high performance liquid chromatography with fluorescence detection. *Biomed.Chromatogr.*, **1993**, 7, 25–28

## SAMPLE
**Matrix:** blood

**Sample preparation:** Condition a 1 mL Bond-Elut octadecylsilane SPE cartridge with 2 mL MeOH, 2 mL water, and 2 mL buffer (do not allow to dry). 1 mL Serum + 10 μL 100 μg/mL bamethan in water + 2 mL buffer, mix, add to SPE cartridge, wash with 1 mL buffer, 2 mL water, 2 mL MeOH:MeCN 15:85. Dry under full vacuum for 10 min and elute with 1 mL MeOH. Evaporate the eluate to dryness under a stream of nitrogen at room temperature. Dissolve the residue in 200 μL MeCN:triethylamine 199:1, heat at 45°for 20 min, add 10 μL 5 mg/mL TAGIT in MeCN, heat at 45°for 2 h. Evaporate at room temperature under a stream of nitrogen, reconstitute in 250 μL mobile phase, inject a 100 μL aliquot. (Buffer was 100 mM $Na_2HPO_4$ adjusted to pH 7.3 with concentrated phosphoric acid. The chiral derivatizing agent TAGIT was 2,3,4,6-tetra-O-acetyl-β-D-glucopyranosyl isothiocyanate. Prepare solutions of TAGIT in MeCN weekly.)

## HPLC VARIABLES
**Column:** 100 × 4.6 5 μm Brownlee octadecylsilyl
**Mobile phase:** MeCN:water 29:71 containing 0.1% triethylamine (pH adjusted to 4.0 with concentrated phosphoric acid)
**Flow rate:** 0.8
**Injection volume:** 100
**Detector:** F ex 223 no emission filter

## CHROMATOGRAM
**Retention time:** 5.77 (R(−)), 6.83 (S(+))
**Internal standard:** bamethan (9, 10)
**Limit of detection:** 1 ng/mL

## KEY WORDS
SPE; chiral; derivatization; serum

## REFERENCE
He, L.; Stewart, J.T. A high performance liquid chromatographic method for the determination of albuterol enantiomers in human serum using solid phase extraction and chemical derivatization. *Biomed.Chromatogr.*, **1992**, *6*, 291−294

## SAMPLE
**Matrix:** solutions
**Sample preparation:** Dilute 800 μL solution to 10 mL with water, filter, inject a 20 μL aliquot of the filtrate.

## HPLC VARIABLES
**Guard column:** RP-18
**Column:** 125 × 4 5 μm LiChrosorb RP-18
**Mobile phase:** Gradient. MeCN:buffer 4:96 for 6 min, to 9:91 (step gradient). (Buffer was 40 mM $NaH_2PO_4$ containing 5.74 mM triethylamine, adjusted to pH 3.0 with phosphoric acid.)
**Flow rate:** 1.5
**Injection volume:** 20
**Detector:** UV 265

## KEY WORDS
comparison with capillary electrophoresis

## REFERENCE
Mälkki-Laine, L.; Hartikainen, E. Electrokinetic behaviour of salbutamol and its decomposition products and determination of salbutamol by micellar electrokinetic capillary chromatography. *J.Chromatogr.A*, **1996**, *724*, 297−306

## SAMPLE
**Matrix:** solutions

## HPLC VARIABLES
**Column:** 250 × 4.6 CSP-4 (Prepare as follows. Add a solution of 1.07 g L-valyl-L-valyl-L-valine isopropylester (Bunseki Kagaku 1979, 28, 125) in 30 mL dry dioxane (Caution! Dioxane is a carcinogen!) dropwise to a mixture of 2.2 g 2,4,6-trichloro-1,3,5-triazine (cyanuric chloride) in 20 mL dry dioxane stirred at 0°, add 3 g anhydrous sodium carbonate at room temperature, stir, filter, evaporate to give a colorless solid. Dissolve 8.3 g of this solid in 30 mL dry dioxane, add 2 g N-(2-aminoethyl)-3-aminopropyltrimethoxysilane, add 1.5 g anhydrous sodium carbonate, reflux with stirring for 40 h, filter, add 3 g dried 10 μm LiChrosorb Si 100, reflux with slow stirring for 10 h, cool, filter. Wash the solid with dioxane, MeOH, and diethyl ether, dry under reduced pressure (J.Chromatogr. 1984, 292, 427).)
**Mobile phase:** Hexane:1,2-dichloroethane:MeOH:trifluoroacetic acid 60:37.5:3.75:0.25
**Detector:** UV

## CHROMATOGRAM
**Retention time:** k′ 5.84 (first enantiomer)

## KEY WORDS
chiral; α = 1.06

## REFERENCE
Oi, N.; Kitahara, H.; Matsushita, Y.; Kisu, N. Enantiomer separation by gas and high-performance liquid chromatography with tripeptide derivatives as chiral stationary phases. *J.Chromatogr.A*, **1996**, *722*, 229–232

## SAMPLE
**Matrix:** solutions
**Sample preparation:** Inject an aliquot of a 200 μM solution in MeOH.

## HPLC VARIABLES
**Column:** 100 × 4.7 7 μm Hypercarb (Shandon)
**Mobile phase:** MeOH containing 5 mM N-benzyloxycarbonylglycyl-L-proline and 4.5 mM NaOH
**Column temperature:** 17
**Injection volume:** 20
**Detector:** UV 270

## CHROMATOGRAM
**Retention time:** k′ 3.2 (first enantiomer)

## KEY WORDS
chiral; α = 1.09

## REFERENCE
Huynh, N.-H.; Karlsson, A.; Pettersson, C. Enantiomeric separation of basic drugs using N-benzyloxycarbonylglyclyl-L-proline as counter ion in methanol. *J.Chromatogr.A*, **1995**, *705*, 275–287

## SAMPLE
**Matrix:** solutions

## HPLC VARIABLES
**Column:** 250 × 4.6 5 μm Supelcosil LC-DP (A) or 250 × 4 5 μm LiChrospher 100 RP-8 (B)

**Mobile phase:** MeCN:0.025% phosphoric acid:buffer 25:10:5 (A) or 60:25:15 (B) (Buffer was 9 mL concentrated phosphoric acid and 10 mL triethylamine in 900 mL water, adjust pH to 3.4 with dilute phosphoric acid, make up to 1 L.)
**Flow rate:** 0.6
**Injection volume:** 25
**Detector:** UV 229

## CHROMATOGRAM
**Retention time:** 5.57 (A), 3.27 (B)

## OTHER SUBSTANCES
**Also analyzed:** acebutolol, acepromazine, acetaminophen, acetazolamide, acetophenazine, alprazolam, amitriptyline, amobarbital, amoxapine, antipyrine, atenolol, atropine, azatadine, baclofen, benzocaine, bromocriptine, brompheniramine, brotizolam, bupivacaine, buspirone, butabarbital, butalbital, caffeine, carbamazepine, cetirizine, chlorcyclizine, chlordiazepoxide, chlormezanone, chloroquine, chlorpheniramine, chlorpromazine, chlorpropamide, chlorprothixene, chlorthalidone, chlorzoxazone, cimetidine, cisapride, clomipramine, clonazepam, clonidine, clozapine, cocaine, codeine, colchicine, cyclizine, cyclobenzaprine, dantrolene, desipramine, diazepam, diclofenac, diflunisal, diltiazem, diphenhydramine, diphenidol, diphenoxylate, dipyridamole, disopyramide, dobutamine, doxapram, doxepin, droperidol, encainide, ethidium bromide, ethopropazine, fenoprofen, fentanyl, flavoxate, fluoxetine, fluphenazine, flurazepam, flurbiprofen, fluvoxamine, furosemide, glutethimide, glyburide, guaifenesin, haloperidol, homatropine, hydralazine, hydrochlorothiazide, hydrocodone, hydromorphone, hydroxychloroquine, hydroxyzine, ibuprofen, imipramine, indomethacin, ketoconazole, ketoprofen, ketorolac, labetalol, levorphanol, lidocaine, loratadine, lorazepam, lovastatin, loxapine, mazindol, mefenamic acid, meperidine, mephenytoin, mepivacaine, mesoridazine, metaproterenol, methadone, methdilazine, methocarbamol, methotrexate, methotrimeprazine, methoxamine, methyldopa, methylphenidate, metoclopramide, metolazone, metoprolol, metronidazole, midazolam, moclobemide, morphine, nadolol, nalbuphine, naloxone, naphazoline, naproxen, nifedipine, nizatidine, norepinephrine, nortriptyline, oxazepam, oxycodone, oxymetazoline, paroxetine, pemoline, pentazocine, pentobarbital, pentoxifylline, perphenazine, pheniramine, phenobarbital, phenol, phenolphthalein, phentolamine, phenylbutazone, phenyltoloxamine, phenytoin, pimozide, pindolol, piroxicam, pramoxine, prazepam, prazosin, probenecid, procainamide, procaine, prochlorperazine, procyclidine, promazine, promethazine, propafenone, propantheline, propiomazine, propofol, propranolol, protriptyline, quazepam, quinidine, quinine, racemethorphan, ranitidine, remoxipride, risperidone, salicylic acid, scopolamine, secobarbital, sertraline, sotalol, spironolactone, sulfinpyrazone, sulindac, temazepam, terbutaline, terfenadine, tetracaine, theophylline, thiethylperazine, thiopental, thioridazine, thiothixene, timolol, tocainide, tolbutamide, tolmetin, trazodone, triamterene, triazolam, trifluoperazine, triflupromazine, trimeprazine, trimethoprim, trimipramine, verapamil, warfarin, xylometazoline, yohimbine, zopiclone

## KEY WORDS
details of plasma extraction also in paper

## REFERENCE
Koves, E.M. Use of high-performance liquid chromatography-diode array detection in forensic toxicology. *J.Chromatogr.A*, **1995**, *692*, 103–119

## SAMPLE
**Matrix:** solutions

## HPLC VARIABLES
**Column:** 250 × 4.6 Sumchiral CSP 10 (Sumika Chemical Analysis Service)
**Mobile phase:** n-Hexane:1,2-dichloroethane:MeOH:trifluoroacetic acid 250:140:20:1

**Flow rate:** 1
**Detector:** UV 230-280

## CHROMATOGRAM
**Retention time:** k′ 5.41 (first enantiomer)

## KEY WORDS
chiral; α = 1.33

## REFERENCE
Oi, N.; Kitahara, H.; Aoki, F. Direct enantiomer separations by high-performance liquid chromatography with chiral urea derivatives as stationary phases. *J.Chromatogr.A*, **1995**, *694*, 129–134

## SAMPLE
**Matrix:** solutions

## HPLC VARIABLES
**Column:** 250 × 4.6 Zorbax RX
**Mobile phase:** Gradient. A was 10 mL concentrated orthophosphoric acid and 7 mL triethylamine in 1 L water. B was 10 mL concentrated orthophosphoric acid and 7 mL triethylamine in 200 mL water, make up to 1 L with MeCN. A:B from 100:0 to 0:100 over 30 min, maintain at 0:100 for 5 min.
**Column temperature:** 30
**Flow rate:** 2
**Detector:** UV 210

## OTHER SUBSTANCES
**Also analyzed:** 5-fluorouracil, acepromazine, acetaminophen, acetophenazine, aminophylline, amitriptyline, amobarbital, amoxapine, amphetamine, amylocaine, antipyrine, aprobarbital, aspirin, atenolol, atropine, avermectin, barbital, bebrisoquine, benzocaine, benzoic acid, benzotropine, benzphetamine, berberine, bibucaine, bromazepan, brompheniramine, buprenorphine, buspirone, butabarbital, butacaine, butethal, caffeine, carbamazepine, carbromal, chloramphenicol, chlordiazepoxide, chloroquine, chlorothiazide, chloroxylenol, chlorphenesin, chlorpheniramine, chlorpromazine, chlorpropamide, chlortetracycline, cimetidine, cinchonidine, cinchonine, clenbuterol, clonazepam, clonixin, clorazepate, cocaine, codeine, colchicine, cortisone, coumarin, cyclazocine, cyclobenzaprine, cyclothiazide, cyheptamide, cymarin, danazol, danthron, dapsone, desipramine, dexamethasone, dextromethorphan, dextropropoxyphene, diamorphine, diazepam, diclofenac, diethylpropion, diethylstilbestrol, diflunisal, digitoxin, digoxin, diltiazem, diphenhydramine, diphenoxylate, diprenorphine, dipyrone, disulfiram, dopamine, doxapram, doxepin, dronabinol, ephedrine, epinephrine, epinine, estradiol, estriol, estrone, ethacrynic acid, ethosuximide, etonitazene, etorphine, eugenol, famotidine, fenbendazole, fencamfamine, fenoprofen, fenproporex, fentanyl, flubendazole, flufenamic acid, flunitrazepam, fluoxymesterone, fluphenazine, furosemide, gentisic acid, gitoxigenin, glipizide, glunixin, glutethimide, glybenclamide, guaiacol, halazepam, haloperidol, hydrochlorothiazide, hydrocodone, hydrocortisone, hydromorphone, hydroxyquinoline, ibogaine, ibuprofen, iminostilbene, imipramine, indomethacin, isocarbostyril, isocarboxazid, isoniazid, isoproterenol, isoxsuprine, ivermectin, ketamine, ketoprofen, kynurenic acid, levorphanol, lidocaine, lorazepam, lormetazepam, loxapine, mazindol, mebendazole, meclizine, meclofenamic acid, medazepam, mefenamic acid, megestrol, mepacrine, meperidine, mephentermine, mephenytoin, mephesin, mephobarbital, mepivacaine, mescaline, mesoridazine, methadone, methamphetamine, methapyrilene, methaqualone, methazolamide, methocarbamol, methoxamine, methsuximide, methyl salicylate, methyldopa, methyldopamine, methylphenidate, methylprednisolone, methyltestosterone, methyprylon, metoprolol, mibolerone, morphine, nadolol, nalorphine, naloxone, naltrexone, naphazoline, naproxen, nefopam, niacinamide, nicotine, nicotinic acid, nifedipine, niflumic acid, nitrazepam, norepinephrine, nortriptyline, noscapine, nylidrin, oxazepam, oxycodone, oxymorphone,

oxyphenbutazone, oxytetracycline, papaverine, pargyline, pemoline, pentazocine, pentobarbital, persantine, phenacetin, phenazocine, phenazopyridine, phencyclidine, phendimetrazine, phenelzine, pheniramine, phenobarbital, phenothiazine, phensuximide, phentermine, phenylbutazone, phenylephrine, phenylpropanolamine, piperocaine, prazepam, prednisolone, primidone, probenecid, progesterone, propiomazine, propranolol, propylparaben, pseudoephedrine, puromycin, pyrilamine, pyrithyldione, quazepam, quinaldic acid, quinidine, quinine, ranitidine, recinnamine, reserpine, resorcinol, saccharin, albuterol, salicylamide, salicylic acid, scopolamine, scopoletin, secobarbital, strychnine, sulfacetamide, sufadiazine, sulfadimethoxine, sulfaethidole, sulfamerazine, sulfamethazine, sulfamethoxizole, sulfanilamide, sulfapyridine, sulfasoxizole, sulindac, tamoxifen, temazepam, testosterone, tetracaine, tetracycline, dronabinol, tetramisole, thebaine, theobromine, theophylline, thiabendazole, thiamine, thiamylal, thiobarbituric acid, thioridazine, thiosalicylic acid, thiothixene, thymol, tolazamide, tolazoline, tobutamide, tolmetin, tranylcypromine, triamcinolone, tribenzylamine, trichloromethiazide, trifluoperazine, trihexyphenidyl, trimethoprim, tripelennamine, triproilidine, tropacocaine, tyramine, verapamil, vincamine, warfarin, yohimbine, zoxazolamine

## REFERENCE

Hill, D.W.; Kind, A.J. Reversed-phase solvent gradient HPLC retention indexes of drugs. *J.Anal.Toxicol.*, **1994**, *18*, 233-242

## SAMPLE

**Matrix:** solutions
**Sample preparation:** Prepare a 10 μg/mL solution in MeOH, inject a 20 μL aliquot.

## HPLC VARIABLES

**Column:** 125 × 4.9 Spherisorb S5W silica
**Mobile phase:** MeOH containing 10 mM ammonium perchlorate and 1 mL/L 100 mM NaOH in MeOH, pH 6.7
**Flow rate:** 2
**Injection volume:** 20
**Detector:** E, LeCarbone, V25 glassy carbon electrode, + 1.2 V

## CHROMATOGRAM

**Retention time:** 1.76

## OTHER SUBSTANCES

**Also analyzed:** acebutolol, acepromazine, acetophenazine, N-acetylprocainamide, alprenolol, amethocaine, amiodarone, amitriptyline, antazoline, atenolol, azacyclonal, bamethane, benactyzine, benperidol, benzethidine, benzocaine, benzoctamine, benzphetamine, benzquinamide, bromhexine, bromodiphenhydramine, bromperidol, brompheniramine, brompromazine, buclizine, bufotenine, bupivacaine, buprenorphine, butacaine, butethamate, chlorcyclizine, chlorpheniramine, chlorphenoxamine, chlorprenaline, chlorpromazine, chlorprothixene, cimetidine, cinchonidine, cinnarizine, clemastine, clomipramine, clonidine, cocaine, cyclazocine, cyclizine, cyclopentamine, cyproheptadine, deserpidine, desipramine, dextromoramide, dextropropoxyphene, dicyclomine, diethylcarbamazine, diethylpropion, diethylthiambutene, dihydroergotamine, dimethindene, dimethothiazine, diphenhydramine, diphenoxylate, dipipanone, diprenorphine, dipyridamole, disopyramide, dothiepin, doxapram, doxepin, doxylamine, droperidol, ephedrine, ergocornine, ergocristine, ergocristinine, ergocryptine, ergometrine, ergosine, ergosinine, ergotamine, ethopropazine, etorphine, etoxeridine, fenethazine, fenfluramine, fenoterol, fentanyl, flavoxate, fluopromazine, flupenthixol, fluphenazine, flurazepam, haloperidol, hydroxyzine, hyoscine, ibogaine, imipramine, indapamine, iprindole, isothipendyl, isoxsuprine, ketanserin, laudanosine, lidocaine, lofepramine, loxapine, maprotiline, mecamylamine, meclophenoxate, meclozine, medazepam, mephentermine, mepivacaine, meptazinol, mepyramine, mesoridazine, metaraminol, methadone, methamphetamine, methapyrilene, methdilazene, methotrimeprazine, methoxamine, methoxyphenamine, methoxypromazine, meth-

ylephedrine, methylergonovine, methysergide, metoclopramide, metopimazine, metoprolol, mianserin, morazone, nadolol, nalorphine, naloxone, naphazoline, nicotine, nifedipine,
nomifensine, nortriptyline, noscapine, orphenadrine, oxeladin, oxprenolol, oxymetazolin,
papaverine, pargyline, pecazine, penbutolol, pentazocine, penthienate, pericyazine,
perphenazine, phenadoxone, phenampromide, phenazocine, phenbutrazate, phendimetrazine, phenelzine, phenglutarimide, phenindamine, pheniramine, phenmetrazine, phenomorphan, phenoperidine, phenothiazine, phenoxybenzamine, phentolamine, phenylephrine, phenyltoloxamine, physostigmine, piminodine, pimozide, pindolol, pipamazine,
pipazethate, piperacetazine, piperidolate, pipradol, pirenzepine, piritramide, pizotifen,
practolol, pramoxine, prazosin, prenylamine, prilocaine, primaquine, proadifen, procainamide, procaine, prochlorperazine, procyclidine, proheptazine, prolintane, promazine, promethazine, pronethalol, properidine, propiomazine, propranolol, prothipendyl, protriptyline, proxymetacaine, pseudoephedrine, pyrimethamine, quinidine, quinine, ranitidine,
rescinnamine, sotalol, tacrine, terazosin, terbutaline, terfenadine, thenyldiamine, theophylline, thiethylperazine, thiopropazate, thioproperazine, thioridazine, thiothixene, thonzylamine, timolol, tocainide, tolpropamine, tolycaine, tranylcypromine, trazodone, trifluoperazine, trifluperidol, trimeperidine, trimeprazine, trimethobenzamide, trimethoprim,
trimipramine, tripelennamine, triprolidine, tryptamine, verapamil, xylometazoline

## REFERENCE

Jane, I.; McKinnon, A.; Flanagan, R.J. High-performance liquid chromatographic analysis of basic drugs
on silica columns using non-aqueous ionic eluents. II. Application of UV, fluorescence and electrochemical oxidation detection. *J.Chromatogr.*, **1985**, *323*, 191–225

◆━━━━━━━━━━━━━━━━━━━━━◆━━━━━━━━━━━━━━━━━━━━━◆

## ANNOTATED BIBLIOGRAPHY

Tsai, C.E.; Kondo, F. Liquid chromatographic determination of salbutamol and clenbuterol residues in
swine serum and muscle. *Microbios*, **1994**, *80*, 251–258

Malkki, L.; Tammilehto, S. Optimization of the separation of salbutamol and its decomposition products
by liquid chromatography with diode-array detection. *J.Pharm.Biomed.Anal.*, **1993**, *11*, 79–84

Nasr, M.M. Single-puff particle-size analysis of albuterol metered-dose inhalers (MDIs) by high-pressure
liquid chromatography with electrochemical detection (HPLC-EC). *Pharm.Res.*, **1993**, *10*, 1381–1384

Sagar, K.A.; Kelly, M.T.; Smyth, M.R. Simultaneous determination of salbutamol and terbutaline at
overdose levels in human plasma by high performance liquid chromatography with electrochemical
detection. *Biomed.Chromatogr.*, **1993**, *7*, 29–33

Degroodt, J.M.; Wyhowski de Bukanski, B.; Srebrnik, S. Immunoaffinity-chromatography purification
of salbutamol in liver and HPLC-fluorometric detection at trace residue level. *Z.Lebensm.
Unters.Forsch.*, **1992**, *195*, 566–568

Tamisier-Karolak, L.; Delhotal-Landes, B.; Jolliet-Riant, P.; Milliez, J.; Jannet, D.; Barre, J.; Flouvat,
B. Plasma assay of salbutamol by means of high-performance liquid chromatography with amperometric determination using a loop column for injection of plasma extracts. Application to the evaluation of subcutaneous administration of salbutamol. *Ther.Drug Monit.*, **1992**, *14*, 243–248

Meyer, H.H.; Rinke, L.; Dursch, I. Residue screening for the beta-agonists clenbuterol, salbutamol and
cimaterol in urine using enzyme immunoassay and high-performance liquid chromatography.
*J.Chromatogr.*, **1991**, *564*, 551–556 [extracted cimaterol, clenbuterol; urine; cow; SPE; enzyme immunoassay detection; UV detection]

Wu, Y.Q.; Shi, R.; Williams, R.L.; Lin, E.T. High-performance liquid chromatographic assay for basic
amine drugs in human plasma with a silica gel column and an aqueous mobile phase. IV. Albuterol.
*J.Liq.Chromatogr.*, **1991**, *14*, 253–264 [plasma; fluorescence detection; LOD 0.2 ng/mL; metaproterenol (IS)]

Beaulieu, N.; Cyr, T.D.; Lovering, E.G. Liquid chromatographic methods for the determination of albuterol (salbutamol), albuterol sulphate and related compounds in drug raw materials, tablets and
inhalers. *J.Pharm.Biomed.Anal.*, **1990**, *8*, 583–589

Bland, R.E.; Tanner, R.J.; Chern, W.H.; Lang, J.R.; Powell, J.R. Determination of albuterol concentrations in human plasma using solid-phase extraction and high-performance liquid chromatography
with fluorescence detection. *J.Pharm.Biomed.Anal.*, **1990**, *8*, 591–596

Ong, H.; Adam, A.; Perreault, S.; Marleau, S.; Bellemare, M.; Du Souich, P.; Beaulieu, N. Analysis of albuterol in human plasma based on immunoaffinity chromatographic clean-up combined with high-performance liquid chromatography with fluorimetric detection. *J.Chromatogr.*, **1989**, *497*, 213–221

Emm, T.; Lesko, L.J.; Leslie, J.; Perkal, M.B. Determination of albuterol in human serum by reversed-phase high-performance liquid chromatography with electrochemical detection. *J.Chromatogr.*, **1988**, *427*, 188–194

Tan, Y.K.; Soldin, S.J. Analysis of salbutamol enantiomers in human urine by chiral high-performance liquid chromatography and preliminary studies related to the stereoselective disposition kinetics in man. *J.Chromatogr.*, **1987**, *422*, 187–195

Miller, L.G.; Greenblatt, D.J. Determination of albuterol in human plasma by high-performance liquid chromatography with fluorescence detection. *J.Chromatogr.*, **1986**, *381*, 205–208

Kucharczyk, N.; Segelman, F.H. Drug level monitoring of antiasthmatic drugs. *J.Chromatogr.*, **1985**, *340*, 243–271 [review]

Kurosawa, N.; Morishima, S.; Owada, E.; Ito, K. Reversed-phase high-performance liquid chromatographic determination of salbutamol in rabbit plasma. *J.Chromatogr.*, **1984**, *305*, 485–488 [rabbit; plasma; column temp 60; SPE; LOD 4 ng/mL; pharmacokinetics]

Eggers, N.J.; Saint-Joly, C.M. The effect of amine modifiers on the chromatographic behavior of salbutamol on reversed phase chemically bonded silica gel. *J.Liq.Chromatogr.*, **1983**, *6*, 1955–1967

Hutchings, M.J.; Paull, J.D.; Morgan, D.J. Determination of salbutamol in plasma by high-performance liquid chromatography with fluorescence detection. *J.Chromatogr.*, **1983**, *277*, 423–426 [plasma; fluorescence detection; LOD 1 ng/mL; pharmacokinetics]

Agostini, O.; Chiari, A.; Ciofi Baffoni, D. Simultaneous high-performance liquid chromatographic determination of salbutamol sulphate, theophylline, and saccharin in a hydroalcoholic formulation. *Boll.Chim.Farm.*, **1982**, *121*, 612–618

Oosterhuis, B.; van Boxtel, C.J. Determination of salbutamol in human plasma with bimodal high-performance liquid chromatography and a rotated disc amperometric detector. *J.Chromatogr.*, **1982**, *232*, 327–334 [SPE; electrochemical detection; plasma; LOD 0.5 ng/mL; pharmacokinetics]

# Alprazolam

**Molecular formula:** $C_{17}H_{13}ClN_4$
**Molecular weight:** 308.8
**CAS Registry No.:** 28981-97-7

## SAMPLE
**Matrix:** blood
**Sample preparation:** 2 mL Whole blood or plasma + 2 mL buffer + 5 mL chloroform: isopropanol:n-heptane 60:14:26, shake gently horizontally for 10 min, centrifuge at 2800 g for 10 min. Remove the lower organic layer and evaporate it to dryness under vacuum at 45°, reconstitute the residue in 100 μL mobile phase, centrifuge at 2800 g for 5 min, inject a 50 μL aliquot of the supernatant. (Buffer was saturated ammonium chloride solution 25% diluted with water, adjusted to pH 9.5 with 25% ammonia solution.)

## HPLC VARIABLES
**Column:** 300 × 3.9 4 μm NovaPack C18
**Mobile phase:** MeOH:THF:buffer 65:5:30 (Buffer was 0.68 g/L (10 mM (sic)) $KH_2PO_4$ adjusted to pH 2.6 with concentrated orthophosphoric acid.) (At the end of each session wash the column with water for 1 h and MeOH for 1 h, re-equilibrate for 30 min.)
**Column temperature:** 30
**Flow rate:** 0.8
**Injection volume:** 50
**Detector:** UV 222

## CHROMATOGRAM
**Retention time:** 4.17
**Limit of detection:** <120 ng/mL

## OTHER SUBSTANCES
**Extracted:** acebutolol, acenocoumarol, acepromazine, aceprometazine, acetaminophen, aconitine, ajmaline, albuterol, alimemazine, alminoprofen, alpidem, alprenolol, amisulpride, amitriptyline, amodiaquine, amoxapine, aspirin, astemizole, atenolol, benazepril, benperidol, benzocaine, benzoylecgonine, bepridil, betaxolol, bisoprolol, bromazepam, brompheniramine, bumadizone, bupivacaine, buprenorphine, buspirone, caffeine, carbamazepine, carbinoxamine, carpipramine, carteolol, celiprolol, cetirizine, chlorambucil, chlordiazepoxide, chlormezanone, chlorophenacinone, chloroquine, chlorpheniramine, chlorpromazine, chlorpropamide, cibenzoline, cicletanine, clemastine, clobazam, clomipramine, clonazepam, clonidine, clorazepate, clozapine, cocaine, codeine, colchicine, cyamemazine, cyclizine, cyproheptadine, cytarabine, dacarbazine, daunorubicin, demexiptiline, desipramine, dextromethorphan, dextromoramide, dextropropoxyphene, diazepam, diazoxide, diclofenac, dihydralazine, diltiazem, diphenhydramine, dipyridamole, disopyramide, dosulepine, doxepin, doxylamine, droperidol, ephedrine, estazolam, etodolac, fenfluramine, fenoprofen, fentiazac, flecainide, floctafenine, flumazenil, flunitrazepam, fluoxetine, fluphenazine, flurbiprofen, fluvoxamine, glibenclamide, glibornuride, glipizide, glutethimide, haloperidol, histapyrrodine, hydroxychloroquine, hydroxyzine, ibuprofen, imipramine, indomethacin, iproniazid, ketoprofen, labetalol, levomepromazine, lidocaine, lidoflazine, lisinopril, loperamide, loprazolam, loratadine, loxapine, maprotiline, medazepam, medifoxamine, mefenamic acid, mefenidramine, mefloquine, melphalan, meperidine, mephenesin, mephentermine, mepivacaine, metapramine, metformin, methadone, methocarbamol, methotrexate, metipranolol, metoclopramide, mexiletine, mianserine, midazolam, minoxidil, moclobemide, moperone, morphine, nadolol, nalbuphine, nalor-

phine, naloxone, naltrexone, naproxen, nialamide, nicardipine, niflumic acid, nimodipine, nitrazepam, nitrendipine, nizatidine, nomifensine, nortriptyline, omeprazole, opipramol, oxazepam, oxprenolol, penbutolol, penfluridol, pentazocine, phencyclidine, phenobarbital, phenol, phenylbutazone, pimozide, pindolol, pipamperone, prazepam, prazosin, prilocaine, procainamide, procarbazine, proguanil, promethazine, propafenone, propranolol, protriptyline, pyrimethamine, quinidine, quinine, quinupramine, ramipril, ranitidine, reserpine, ritodrine, secobarbital, sotalol, strychnine, sulfinpyrazole, sulpride, sultopride, suriclone, temazepam, tenoxicam, terbutaline, terfenadine, tetracaine, tetrazepam, thiopental, thioproperazine, thioridazine, tianeptine, tiapride, tiaprofenic acid, ticlopidine, timolol, tioclomarol, tofisopam, tolbutamide, toloxatone, trazodone, triazolam, trifluoperazine, trifluperidol, trimipramine, triprolidine, tropatenine, verapamil, viloxazine, vinblastine, vincristine, vindesine, warfarin, yohimbine, zolpidem, zopiclone, zorubicine

**Interfering:** cycloguanil, debrisoquine, ketamine, lorazepam, methaqualone, metoprolol, nifedipine, piroxicam, sulindac

## KEY WORDS
whole blood; plasma

## REFERENCE
Tracqui, A.; Kintz, P.; Mangin, P. Systematic toxicological analysis using HPLC/DAD. *J.Forensic Sci.*, **1995**, *40*, 254–262

## SAMPLE
**Matrix:** blood
**Sample preparation:** 50 µL Serum + 25 µL 0.5 µg/mL demoxepam in water + 100 µL 1 M pH 9.0 borate buffer, mix well, add 2 mL diethyl ether, vortex for 40 s, centrifuge at 1100 g for 5 min. Remove ether layer and evaporate it at 40° under nitrogen. Take up residue in 50 µL mobile phase, inject an aliquot.

## HPLC VARIABLES
**Column:** 150 × 2 5µm Ultrasphere C18
**Mobile phase:** MeCN:MeOH:43 mM pH 2.4 sodium acetate buffer 8:45:47
**Flow rate:** 0.3
**Injection volume:** 20
**Detector:** UV 230

## CHROMATOGRAM
**Retention time:** 7
**Internal standard:** demoxepam (4)
**Limit of detection:** 5 ng/mL

## OTHER SUBSTANCES
**Extracted:** metabolites, triazolam
**Simultaneous:** chlorpromazine, clonazepam, diazepam, flurazepam, hexobarbital, oxazepam, phenobarbital, temazepam
**Noninterfering:** amphetamine, buspirone, chlordiazepoxide, cocaine, cocathylene, flumazenil, midazolam, norcocaine

## KEY WORDS
rat; serum

## REFERENCE
Jin, L.; Lau, C.E. Determination of alprazolam and its major metabolites in serum microsamples by high-performance liquid chromatography and its application to pharmacokinetics in rats. *J.Chromatogr.B*, **1994**, *654*, 77–83

## SAMPLE
**Matrix:** blood
**Sample preparation:** Inject 100-200 μL plasma onto column A with mobile phase A and elute to waste, after 5 min backflush the contents of column A onto column B with mobile phase B, after 5 min remove column A from the circuit, elute column B with mobile phase B, monitor the effluent from column B. Wash column A with MeCN:water 60:40 at 1 mL/min for 6 min then re-equilibrate with pH 7.5 buffer for 10 min.

## HPLC VARIABLES
**Column:** A 45 × 4 12 μm TSK-gel G 3 PW (Tosohass); B 75 × 4.6 Ultrasphere ODS C18 3 μm
**Mobile phase:** A 50 mM pH 7.5 phosphate buffer; B Gradient. × was MeCN. Y was 65 mM $KH_2PO_4$ + 1% diethylamine adjusted to pH 5.4 with phosphoric acid. X:Y 22:78 for 5 min, to 25:75 over 10 min, to 60:40 over 15 min.
**Flow rate:** 1
**Injection volume:** 100-200
**Detector:** UV 230

## CHROMATOGRAM
**Retention time:** 23

## OTHER SUBSTANCES
**Extracted:** bromazepam, chlordiazepoxide, clobazam, clonazepam, clorazepate, clotiazepam, desmethylclobazam, desmethyldiazepam, diazepam, estazolam, flunitrazepam, loflazepate, lorazepam, medazepam, nitrazepam, oxazepam, prazepam, temazepam, tetrazepam, tofisopam, triazolam
**Noninterfering:** carbamazepine, phenytoin, ethosuximide, phenobarbital, primidone, valproic acid

## KEY WORDS
plasma; column-switching

## REFERENCE
Lacroix, C.; Wojciechowski, F.; Danger, P. Monitoring of benzodiazepines (clobazam, diazepam and their main active metabolites) in human plasma by column-switching high-performance liquid chromatography. *J.Chromatogr.*, **1993**, *617*, 285–290

## SAMPLE
**Matrix:** blood
**Sample preparation:** 3 mL Plasma + 30 μL 10 μg/mL triazolam in water, mix 1 min, allow to stand for 15 min at room temperature, add to 3 mL Extrelut SPE cartridge and allow to soak in for 10 min, elute with 20 mL dichloromethane. Evaporate eluant at 30° under reduced pressure, take up residue in 1 mL MeCN:water 5:95, stand for 15 min, centrifuge at 14000 g for 2 min, remove supernatant. Inject a 250 μL aliquot of the supernatant onto column A with mobile phase A and elute to waste. After 7 min forward flush the contents of column A onto column B with mobile phase B. After 0.47 min remove column A from the circuit and elute column B with mobile phase B, monitor the effluent from column B. When not in use, flush column A with mobile phase A. Between injections clean column A with two injections of 250 μL MeCN.

## HPLC VARIABLES
**Column:** A 30 × 2.1 10 μm MPLC cartridge PRP-1 (Kontron); B 100 × 4.6 MPLC cartridge 5 μm RP-8 Spheri-5 (Kontron)
**Mobile phase:** A 1 L water + 20 mL MeCN + 50 μL phosphoric acid (pH 3.2); B MeCN: buffer 40:60 (Buffer was 1 L water + 20 mL MeCN + 50 μL phosphoric acid (pH 3.2).)
**Flow rate:** A 0.3; B 1

**Injection volume:** 250
**Detector:** UV 230

## CHROMATOGRAM
**Retention time:** 7.6
**Internal standard:** triazolam (7.0)
**Limit of quantitation:** 1 ng/mL

## OTHER SUBSTANCES
**Simultaneous:** metabolites, bromazepam, clobazam, diazepam, lorazepam, oxazepam

## KEY WORDS
plasma; SPE; column-switching; pharmacokinetics

## REFERENCE
Rieck, W.; Platt, D. High-performance liquid chromatographic method for the determination of alprazolam in plasma using the column-switching technique. *J.Chromatogr.*, **1992**, *578*, 259–263

## SAMPLE
**Matrix:** blood
**Sample preparation:** 1 mL Serum + 25 µL 1 µg/mL triazolam in toluene + 75 µL 0.1% ammonium hydroxide, vortex 30 s, add 5 mL methylene chloride + 5 mL toluene, shake 15 min, centrifuge at 177 g for 10 min. Remove aqueous layer and freeze residual aqueous layer in dry ice-acetone for 30 s. Decant organic layer, dry under nitrogen at 50°, vortex residue with 200 µL mobile phase, inject a 125 µL aliquot.

## HPLC VARIABLES
**Column:** 150 × 4.6 5 µm C18 (Supelco)
**Mobile phase:** MeOH:buffer 40:60 (Buffer was 1 mM phosphate and 3 mM hexyltriethylammonium phosphate in water at pH 7.4.)
**Column temperature:** 35
**Flow rate:** 2
**Injection volume:** 125
**Detector:** UV 221

## CHROMATOGRAM
**Retention time:** 26
**Internal standard:** triazolam (30)
**Limit of detection:** 1 ng/mL

## OTHER SUBSTANCES
**Extracted:** metabolites

## KEY WORDS
serum

## REFERENCE
Schmith, V.D.; Cox, S.R.; Zemaitis, M.A.; Kroboth, P.D. New high-performance liquid chromatographic method for the determination of alprazolam and its metabolites in serum: instability of 4-hydroxyalprazolam. *J.Chromatogr.*, **1991**, *568*, 253–260

## SAMPLE
**Matrix:** blood
**Sample preparation:** 1 mL Plasma + 10 µL 4 µg/mL triazolam in methanol + 0.5 mL pH 9.12 saturated solution of sodium borate + 4.0 mL ethyl acetate:heptane 85:15, shake

10 min, centrifuge at 220 g for 10 min. Remove organic layer and evaporate it to dryness under a stream of nitrogen, reconstitute in 100 μL mobile phase, inject an aliquot.

## HPLC VARIABLES
**Column:** 100 × 4.6 5 μm C18 (IBM)
**Mobile phase:** MeCN:50 mM pH 6 potassium phosphate 30:70 (Between injections wash column with 10 mL MeCN:water 70:30 then 10 mL mobile phase.)
**Flow rate:** 1.5
**Detector:** UV 214

## CHROMATOGRAM
**Retention time:** 12
**Internal standard:** triazolam (13.2)
**Limit of quantitation:** 2 ng/mL

## OTHER SUBSTANCES
**Extracted:** metabolites
**Simultaneous:** nitrazepam

## KEY WORDS
plasma; pharmacokinetics

## REFERENCE
Miller, R.L.; DeVane, C.L. Alprazolam, α-hydroxy- and 4-hydroxyalprazolam analysis in plasma by high-performance liquid chromatography. *J.Chromatogr.*, **1988**, *430*, 180–186

## SAMPLE
**Matrix:** blood
**Sample preparation:** 2 mL Serum or plasma + 100 μL 1 μg/mL IS in water + 0.5 mL water, vortex, extract with 10 mL toluene:isoamyl alcohol 99:1 for 10 min on a rotator, centrifuge for 5 min. Remove upper organic layer, evaporate under a stream of nitrogen at 37°, take up in 150 μL mobile phase, vortex for 2 min, add 0.5 mL hexane, vortex briefly, centrifuge for 5 min, discard upper hexane layer, inject a 100 μL aliquot of the lower layer.

## HPLC VARIABLES
**Column:** 250 × 4 Bio-Sil ODS-10 (Bio-Rad)
**Mobile phase:** MeCN:pH 4.5 50 mM phosphate buffer 30:70 (Buffer was 6.9 g $KH_2PO_4$ in 1 L water adjusted to pH 4.5 with orthophosphoric acid.)
**Column temperature:** 45
**Flow rate:** 2.5
**Injection volume:** 100
**Detector:** UV 202

## CHROMATOGRAM
**Retention time:** 8.4
**Internal standard:** U-31485 (6.9)
**Limit of detection:** 1 ng/mL

## OTHER SUBSTANCES
**Extracted:** metabolites, desipramine, protriptyline
**Noninterfering:** N-acetylprocainamide, amitriptyline, caffeine, chlordiazepoxide, chlorpromazine, diazepam, flurazepam, lorazepam, oxazepam, prazepam, procainamide, propranolol, thioridazine
**Interfering:** imipramine, nortriptyline, triazolam

## KEY WORDS
serum; plasma

## REFERENCE
McCormick, S.R.; Nielsen, J.; Jatlow, P. Quantification of alprazolam in serum or plasma by liquid chromatography. *Clin.Chem.*, **1984**, *30*, 1652−1655

## SAMPLE
**Matrix:** solutions

## HPLC VARIABLES
**Column:** 250 × 4.6 5 μm Supelcosil LC-DP (A) or 250 × 4 5 μm LiChrospher 100 RP-8 (B)
**Mobile phase:** MeCN:0.025% phosphoric acid:buffer 25:10:5 (A) or 60:25:15 (B) (Buffer was 9 mL concentrated phosphoric acid and 10 mL triethylamine in 900 mL water, adjust pH to 3.4 with dilute phosphoric acid, make up to 1 L.)
**Flow rate:** 0.6
**Injection volume:** 25
**Detector:** UV 229

## CHROMATOGRAM
**Retention time:** 6.40 (A), 6.44 (B)

## OTHER SUBSTANCES
**Also analyzed:** acebutolol, acepromazine, acetaminophen, acetazolamide, acetophenazine, albuterol, amitriptyline, amobarbital, amoxapine, antipyrine, atenolol, atropine, azatadine, baclofen, benzocaine, bromocriptine, brompheniramine, brotizolam, bupivacaine, buspirone, butabarbital, butalbital, caffeine, carbamazepine, cetirizine, chlorcyclizine, chlordiazepoxide, chlormezanone, chloroquine, chlorpheniramine, chlorpromazine, chlorpropamide, chlorprothixene, chlorthalidone, chlorzoxazone, cimetidine, cisapride, clomipramine, clonazepam, clonidine, clozapine, cocaine, codeine, colchicine, cyclizine, cyclobenzaprine, dantrolene, desipramine, diazepam, diclofenac, diflunisal, diltiazem, diphenhydramine, diphenidol, diphenoxylate, dipyridamole, disopyramide, dobutamine, doxapram, doxepin, droperidol, encainide, ethidium bromide, ethopropazine, fenoprofen, fentanyl, flavoxate, fluoxetine, fluphenazine, flurazepam, flurbiprofen, fluvoxamine, furosemide, glutethimide, glyburide, guaifenesin, haloperidol, homatropine, hydralazine, hydrochlorothiazide, hydrocodone, hydromorphone, hydroxychloroquine, hydroxyzine, ibuprofen, imipramine, indomethacin, ketoconazole, ketoprofen, ketorolac, labetalol, levorphanol, lidocaine, loratadine, lorazepam, lovastatin, loxapine, mazindol, mefenamic acid, meperidine, mephenytoin, mepivacaine, mesoridazine, metaproterenol, methadone, methdilazine, methocarbamol, methotrexate, methotrimeprazine, methoxamine, methyldopa, methylphenidate, metoclopramide, metolazone, metoprolol, metronidazole, midazolam, moclobemide, morphine, nadolol, nalbuphine, naloxone, naphazoline, naproxen, nifedipine, nizatidine, norepinephrine, nortriptyline, oxazepam, oxycodone, oxymetazoline, paroxetine, pemoline, pentazocine, pentobarbital, pentoxifylline, perphenazine, pheniramine, phenobarbital, phenol, phenolphthalein, phentolamine, phenylbutazone, phenyltoloxamine, phenytoin, pimozide, pindolol, piroxicam, pramoxine, prazepam, prazosin, probenecid, procainamide, procaine, prochlorperazine, procyclidine, promazine, promethazine, propafenone, propantheline, propiomazine, propofol, propranolol, protriptyline, quazepam, quinidine, quinine, racemethorphan, ranitidine, remoxipride, risperidone, salicylic acid, scopolamine, secobarbital, sertraline, sotalol, spironolactone, sulfinpyrazone, sulindac, temazepam, terbutaline, terfenadine, tetracaine, theophylline, thiethylperazine, thiopental, thioridazine, thiothixene, timolol, tocainide, tolbutamide, tolmetin, trazodone, triamterene, triazolam, trifluoperazine, triflupromazine, trimeprazine, trimethoprim, trimipramine, verapamil, warfarin, xylometazoline, yohimbine, zopiclone

## KEY WORDS
some details of plasma extraction in paper

## REFERENCE
Koves, E.M. Use of high-performance liquid chromatography-diode array detection in forensic toxicology. *J.Chromatogr.A*, **1995**, *692*, 103–119

## SAMPLE
**Matrix:** urine
**Sample preparation:** Heat 5 mL urine + 1 mL temazepam in MeOH with 1 mL β-glucuronidase at 37° for 2.5 h, cool, adjust to pH 8.5 with saturated Na$_2$CO$_3$, extract with 10 mL dichloromethane. Evaporate, take up the residue in 200 µL mobile phase, inject an aliquot.

## HPLC VARIABLES
**Column:** 250 × 4.6 Brownlee 5 µm RP-8
**Mobile phase:** MeCN:10 mM KH$_2$PO$_4$:n-nonylamine 450:550:0.6 adjusted to pH 3.2 with phosphoric acid
**Flow rate:** 1.6
**Detector:** UV 225

## CHROMATOGRAM
**Retention time:** 8
**Internal standard:** temazepam (7)

## OTHER SUBSTANCES
**Extracted:** metabolites
**Interfering:** triazolam

## REFERENCE
Fraser, A.D. Urinary screening for alprazolam, triazolam, and their metabolites with the EMIT d.a.u. benzodiazepine metabolite assay. *J.Anal.Toxicol.*, **1987**, *11*, 263–266

## ANNOTATED BIBLIOGRAPHY
Hall, M.A.; Robinson, C.A.; Brissie, R.M. High-performance liquid chromatography of alprazolam in postmortem blood using solid-phase extraction. *J.Anal.Toxicol.*, **1995**, *19*, 511–513 [SPE; blood]

Goldnik, A.; Gajewska, M. Determination of estazolam and alprazolam in serum by HPLC. *Acta Pol.Pharm.*, **1994**, *51*, 311–312

Bogusz, M.; Erkens, M.; Franke, J.P.; Wijskbeek, J.; de Zeeuw, R.A. Interlaboratory applicability of a retention index library of drugs for screening by reversed phase HPLC in systematic toxicological analysis. *J.Liq.Chromatogr.*, **1993**, *16*, 1341–1354 [gradient; also acebutolol, acetazolamide, aminophenazone, amphetamine, aspirin, atenolol, caffeine, carbromal, clomipramine, codeine, cyclobarbital, diamorphine, diazepam, dibenzepine, diclofenac, droperidol, flunarizine, fluphenazine, flurazepam, hydrochlorothiazide, ibuprofen, lidocaine, lormetazepam, methadone, mianserin, mianserine, normethadone, oxycodone, perphenazine, phenacetin, phenazone, phenobarbital, phenylbutazone, promazine, propranolol, propyphenazone, salicylamide, strychnine, tetrazepam, theobromine, thiopental, tilidine, tolbutamide, trifluoperazine, trifluoropromazine, triflupromazine, vinylbital, warfarin]

Atta-Politou, J.; Parissi-Poulou, M.; Dona, A.; Koutselinis, A. A simple and rapid reversed phase high performance liquid chromatographic method for quantification of alprazolam and α- hydroxyalprazolam in plasma. *J.Liq.Chromatogr.*, **1991**, *14*, 3531–3546 [plasma; LOD 1 ng/mL; extracted hydroxyalprazolam; flunitrazepam (IS)]

Noggle, F.T., Jr.; Clark, C.R.; DeRuiter, J. Liquid chromatographic separation of some common benzodiazepines and their metabolites. *J.Liq.Chromatogr.*, **1990**, *13*, 4005–4021 [also bromazepam, chlordiazepoxide, clobazam, clonazepam, clorazepate, demoxepam, diazepam, estazolam, fludiazepam, flunitrazepam, flurazepam, halazepam, lorazepam, midazolam, nitrazepam, nordiazepam, oxazepam, prazepam, temazepam, triazolam]

Adams, W.J.; Bombardt, P.A.; Brewer, J.E. Normal-phase liquid chromatographic determination of alprazolam in human serum. *Anal.Chem.*, **1984**, *56*, 1590–1594

# Alteplase

**Molecular formula:** $C_{2736}H_{4174}N_{914}O_{824}S_{45}$
**Molecular weight:** 59050.0
**CAS Registry No.:** 105857-23-6

## SAMPLE
**Matrix:** formulations
**Sample preparation:** 100 μL Formulation + 300 μL 20 mM dithiothreitol in mobile phase, heat at 37° for 15 min, inject a 25 μL aliquot.

## HPLC VARIABLES
**Column:** 300 × 10 10 μm TSK3000SW (Hewlett-Packard)
**Mobile phase:** 200 mM pH 6.8 $NaH_2PO_4$ containing 0.1% sodium dodecyl sulfate
**Flow rate:** 0.35
**Injection volume:** 25
**Detector:** UV 214

## CHROMATOGRAM
**Retention time:** 16 (single-chain), 19 (two-chain)

## KEY WORDS
injections; water; saline; 5% dextrose

## REFERENCE
Lam, X.M.; Ward, C.A.; du Mée, C.P.R.C. Stability and activity of alteplase with injectable drugs commonly used in cardiac therapy. *Am.J.Health-Syst.Pharm.*, **1995**, *52*, 1904–1909

## SAMPLE
**Matrix:** solutions

## HPLC VARIABLES
**Column:** Bakerbond C4
**Mobile phase:** Gradient. MeCN:0.1% trifluoroacetic acid from 0:100 to 60:40 over 90 min
**Flow rate:** 1
**Injection volume:** 200
**Detector:** UV 210

## OTHER SUBSTANCES
**Also analyzed:** streptokinase, urokinase, anistreplase

## REFERENCE
Werner, R.G.; Bassarab, S.; Hoffmann, H.; Schlüter, M. Quality aspects of fibrinolytic agents based on biochemical characterization. *Arzneimittelforschung*, **1991**, *41*, 1196–1200

## SAMPLE
**Matrix:** solutions

## HPLC VARIABLES
**Column:** 600 × 7.5 Spherogel 3000 SW
**Mobile phase:** 230 mM pH 6.8 phosphate buffer containing 0.1% sodium dodecyl sulfate
**Flow rate:** 0.5
**Detector:** UV 280

## CHROMATOGRAM
**Retention time:** 27

## OTHER SUBSTANCES
**Simultaneous:** aggregates
**Also analyzed:** anistreplase, streptokinase, urokinase

## KEY WORDS
SEC; GPC

## REFERENCE
Werner, R.G.; Bassarab, S.; Hoffmann, H.; Schlüter, M. Quality aspects of fibrinolytic agents based on biochemical characterization. *Arzneimittelforschung*, **1991**, *41*, 1196–1200

# Amitriptyline

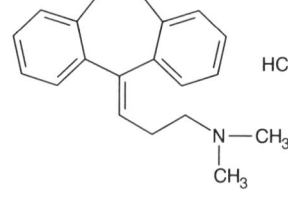

**Molecular formula:** $C_{20}H_{23}N$
**Molecular weight:** 277.4
**CAS Registry No.:** 50-48-6 (amitriptyline), 549-18-8 (amitriptyline hydrochloride)

## SAMPLE
**Matrix:** blood
**Sample preparation:** 1 mL Serum + 500 μL 750 mM pH 10 sodium bicarbonate/carbonate buffer + 50 μL IS in EtOH:water 50:50 + 8 mL heptane:isoamyl alcohol 98:2, shake at 250 cycles/min for 5 min, centrifuge at 1500 g for 10 min, freeze in dry ice/EtOH. Remove the organic layer and add it to 150 μL 22 mM pH 2.5 $KH_2PO_4$/phosphoric acid buffer, shake at 250 cycles/min for 5 min, centrifuge at 1500 g for 10 min, freeze in dry ice/EtOH. Discard the organic layer, inject a 65 μL aliquot of the aqueous layer.

## HPLC VARIABLES
**Column:** 250 × 4.6 Supelco C18
**Mobile phase:** MeCN:buffer 45:55 (Buffer was 44 mM $KH_2PO_4$ containing 1.5 mL/L triethylamine, adjusted to pH 2.5 with phosphoric acid.)
**Flow rate:** 1.5
**Injection volume:** 65
**Detector:** UV 240

## CHROMATOGRAM
**Retention time:** 11.5
**Internal standard:** 1-(3-(dimethylamino)propyl)-1-(p-chlorophenyl)-1,3-dihydroisobenzo-furan-5-carbonitrile (LU 10-202) (Lundbeck, Copenhagen) (8.33)

## OTHER SUBSTANCES
**Extracted:** citalopram, nortriptyline
**Simultaneous:** chlorprothixene, clomipramine, clozapine, flupenthixol, haloperidol, levomepromazine, perphenazine, zuclopenthixol
**Noninterfering:** benzodiazepines
**Interfering:** didesmethylclomipramine, levomepromazine

## KEY WORDS
serum

## REFERENCE
Olesen, O.V.; Linnet, K. Simplified high-performance liquid chromatographic method for the determination of citalopram and desmethylcitalopram in serum without interference from commonly used psychotropic drugs and their metabolites. *J.Chromatogr.B*, **1996**, *675*, 83–88

## SAMPLE
**Matrix:** blood
**Sample preparation:** 1 mL Plasma + 500 μL 1 M sodium carbonate, vortex for 30 s, add 5 mL diethyl ether, vortex for 2 min, centrifuge at 4000 rpm for 10 min. Remove the organic layer and evaporate it to dryness under a stream of nitrogen at 40°, reconstitute the residue in 250 μL mobile phase, vortex for 30 s, centrifuge at 12000 rpm for 5 min, inject an aliquot of the supernatant.

## HPLC VARIABLES
**Column:** 300 × 3.9 10 μm μBondapak C18

**Mobile phase:** MeOH:water 62:38 adjusted to pH 3.5 with phosphoric acid
**Column temperature:** 40
**Flow rate:** 1.2
**Detector:** UV 262

## CHROMATOGRAM
**Retention time:** 6.1
**Internal standard:** amitriptyline

## OTHER SUBSTANCES
**Extracted:** pheniramine
**Simultaneous:** chlorpheniramine, diazepam, diltiazem, flurbiprofen, ibuprofen, itraconazole, ketoprofen, mebeverine, metoclopramide, phenylbutazone

## KEY WORDS
dog; plasma; amitriptyline is IS

## REFERENCE
El-Sayed, Y.M.; Niazy, E.M.; Khidir, S.H. High-performance liquid chromatographic method for the quantitative determination of pheniramine in plasma. *J.Liq.Chromatogr.*, **1995**, *18*, 763–777

## SAMPLE
**Matrix:** blood
**Sample preparation:** 1 mL Plasma + doxepin + NaOH + hexane:isoamyl alcohol 98:2, extract. Remove the organic phase and add it to 0.03% phosphoric acid, extract, inject an aliquot of the aqueous phase.

## HPLC VARIABLES
**Guard column:** C18
**Column:** 100 × 8 10 μm Resolve C8 (Waters)
**Mobile phase:** MeCN:MeOH:56 mM ammonium acetate:1 M ammonium hydroxide 100:10:4.5:2.6
**Flow rate:** 2.5
**Detector:** UV 220

## CHROMATOGRAM
**Retention time:** 17.8
**Internal standard:** doxepin (11.6)

## OTHER SUBSTANCES
**Extracted:** fluoxetine, norfluoxetine, nortriptyline

## KEY WORDS
plasma

## REFERENCE
el-Yazigi, A.; Chaleby, K.; Gad, A.; Raines, D.A. Steady-state kinetics of fluoxetine and amitriptyline in patients treated with a combination of these drugs as compared with those treated with amitriptyline alone. *J.Clin.Pharmacol.*, **1995**, *35*, 17–21

## SAMPLE
**Matrix:** blood
**Sample preparation:** 2 mL Whole blood or plasma + 2 mL buffer + 5 mL chloroform:isopropanol:n-heptane 60:14:26, shake gently horizontally for 10 min, centrifuge at 2800 g for 10 min. Remove the lower organic layer and evaporate it to dryness under vacuum

at 45°, reconstitute the residue in 100 μL mobile phase, centrifuge at 2800 g for 5 min, inject a 50 μL aliquot of the supernatant. (Buffer was saturated ammonium chloride solution 25% diluted with water, adjusted to pH 9.5 with 25% ammonia solution.)

## HPLC VARIABLES
**Column:** 300 × 3.9 4 μm NovaPack C18
**Mobile phase:** MeOH:THF:buffer 65:5:30 (Buffer was 0.68 g/L (10 mM (sic)) $KH_2PO_4$ adjusted to pH 2.6 with concentrated orthophosphoric acid.) (At the end of each session wash the column with water for 1 h and MeOH for 1 h, re-equilibrate for 30 min.)
**Column temperature:** 30
**Flow rate:** 0.8
**Injection volume:** 50
**Detector:** UV 240

## CHROMATOGRAM
**Retention time:** 9.23
**Limit of detection:** <120 ng/mL

## OTHER SUBSTANCES
**Extracted:** acebutolol, acenocoumarol, acepromazine, aceprometazine, acetaminophen, aconitine, ajmaline, albuterol, alimemazine, alminoprofen, alpidem, alprazolam, alprenolol, amisulpride, amodiaquine, amoxapine, aspirin, astemizole, atenolol, benazepril, benperidol, benzocaine, benzoylecgonine, bepridil, betaxolol, bisoprolol, bromazepam, brompheniramine, bumadizone, bupivacaine, buprenorphine, buspirone, caffeine, carbamazepine, carbinoxamine, carpipramine, carteolol, celiprolol, cetirizine, chlorambucil, chlordiazepoxide, chlormezanone, chlorophenacinone, chloroquine, chlorpheniramine, chlorpromazine, chlorpropamide, cibenzoline, cicletanine, clemastine, clobazam, clomipramine, clonazepam, clonidine, clorazepate, clozapine, cocaine, codeine, colchicine, cyamemazine, cyclizine, cycloguanil, cyproheptadine, cytarabine, dacarbazine, daunorubicin, debrisoquine, demexiptiline, desipramine, dextromethorphan, dextromoramide, dextropropoxyphene, diazepam, diazoxide, diclofenac, dihydralazine, diltiazem, diphenhydramine, dipyridamole, disopyramide, dosulepine, doxepin, doxylamine, droperidol, ephedrine, estazolam, fenfluramine, fenoprofen, fentiazac, flecainide, floctafenine, flumazenil, flunitrazepam, fluphenazine, flurbiprofen, fluvoxamine, glibenclamide, glibornuride, glipizide, glutethimide, haloperidol, histapyrrodine, hydroxychloroquine, hydroxyzine, ibuprofen, imipramine, indomethacin, iproniazid, ketamine, ketoprofen, labetalol, levomepromazine, lidocaine, lidoflazine, lisinopril, loperamide, loprazolam, loratadine, lorazepam, loxapine, maprotiline, medazepam, medifoxamine, mefenamic acid, mefenidramine, mefloquine, melphalan, meperidine, mephenesin, mephentermine, mepivacaine, metapramine, metformin, methadone, methaqualone, methocarbamol, methotrexate, metipranolol, metoclopramide, metoprolol, mexiletine, mianserin, midazolam, minoxidil, moclobemide, moperone, morphine, nadolol, nalbuphine, nalorphine, naloxone, naltrexone, naproxen, nialamide, nicardipine, nifedipine, niflumic acid, nimodipine, nitrazepam, nitrendipine, nizatidine, nomifensine, omeprazole, opipramol, oxazepam, oxprenolol, penbutolol, penfluridol, pentazocine, phencyclidine, phenobarbital, phenol, phenylbutazone, pimozide, pindolol, pipamperone, piroxicam, prazepam, prazosin, prilocaine, procainamide, procarbazine, proguanil, promethazine, propafenone, propranolol, protriptyline, pyrimethamine, quinidine, quinine, quinupramine, ramipril, ranitidine, reserpine, ritodrine, secobarbital, sotalol, strychnine, sulfinpyrazole, sulindac, sulpride, sultopride, suriclone, temazepam, tenoxicam, terbutaline, terfenadine, tetracaine, tetrazepam, thiopental, thioproperazine, thioridazine, tianeptine, tiapride, tiaprofenic acid, ticlopidine, timolol, tofisopam, tolbutamide, toloxatone, trazodone, triazolam, trifluoperazine, trifluperidol, trimipramine, triprolidine, tropatenine, verapamil, viloxazine, vinblastine, vincristine, vindesine, warfarin, yohimbine, zolpidem, zopiclone, zorubicine
**Interfering:** etodolac, fluoxetine, nortriptyline, tioclomarol

## KEY WORDS
whole blood; plasma

## REFERENCE

Tracqui, A.; Kintz, P.; Mangin, P. Systematic toxicological analysis using HPLC/DAD. *J.Forensic Sci.*, **1995**, *40*, 254–262

## SAMPLE

**Matrix:** blood

**Sample preparation:** 1 mL Serum + 500 μL 250 μg/mL protriptyline hydrochloride + 1 mL 500 mM NaOH + 4 mL toluene:n-hexane:isoamyl alcohol 77:22:3, mix for 10 min, centrifuge at 3000 rpm for 5 min. Remove the upper organic layer and evaporate it to dryness under a stream of air at 40°, reconstitute the residue in 200 μL MeOH, inject a 50 μL aliquot.

## HPLC VARIABLES

**Column:** 150 × 4.7 5 μm Supelcosil LC-PCN cyanopropyl

**Mobile phase:** MeCN:MeOH:10 mM pH 7.2 potassium phosphate buffer 60:15:25 (Prepare buffer by mixing 194 mL 1.36 g/L $KH_2PO_4$ with 274 mL 1.74 g/L $K_2HPO_4$.)

**Flow rate:** 2

**Injection volume:** 50

**Detector:** UV 254

## CHROMATOGRAM

**Retention time:** 3.8

**Internal standard:** protriptyline (8.1)

## OTHER SUBSTANCES

**Extracted:** norcycyclobenzaprine, nortriptyline

**Interfering:** cyclobenzaprine

## KEY WORDS

serum

## REFERENCE

Wong, E.C.C.; Koenig, J.; Turk, J. Potential interference of cyclobenzaprine and norcyclobenzaprine with HPLC measurement of amitriptyline and nortriptyline: resolution by GC-MS analysis. *J.Anal.Toxicol.*, **1995**, *19*, 218–224

## SAMPLE

**Matrix:** blood

**Sample preparation:** Automated SPE by ASPEC system. Condition a C18 Clean-Up SPE cartridge (CEC 18111, Worldwide Monitoring) with 2 mL MeOH then 2 mL water. 1 mL Plasma + 1 mL 400 ng/mL protriptyline in water, vortex, add to SPE cartridge, wash with 3 mL water, wash with 3 mL 750 mL/L methanol. Elute with three aliquots of 300 μL 0.1 M ammonium acetate in MeOH. Add 0.5 mL 0.5 M NaOH and 4 mL 50 mL/L isopropanol in heptane to eluate, mix thoroughly. Allow 5 min for phase separation. Remove upper heptane phase and add it to 300 μL 0.1 M phosphoric acid (pH 2.5), mix, separate, inject a 100 μL aliquot of the aqueous phase.

## HPLC VARIABLES

**Guard column:** LC-8-DB (Supelco)

**Column:** 150 × 4.6 LC-8-DB (Supelco)

**Mobile phase:** MeCN:buffer 35:65 (Buffer was 10 mL/L triethylamine in water adjusted to pH 5.5 with glacial acetic acid.)

**Flow rate:** 2

**Injection volume:** 100

**Detector:** UV 228

## CHROMATOGRAM
**Retention time:** 5.6
**Internal standard:** protriptyline (4)

## OTHER SUBSTANCES
**Extracted:** chlordiazepoxide, chlorimipramine, chlorpromazine, desipramine, dextromethorphan, diazepam, diphenhydramine, doxepin, encainide, fentanyl, flecainide, fluoxetine, flurazepam, haloperidol, hydroxyethylflurazepam, ibuprofen, imipramine, lidocaine,
maprotiline, methaqualone, mexiletine, midazolam, norchlorimipramine, nordoxepin, nordiazepam, nortriptyline, norverapamil, pentazocine, promazine, propafenone, propoxyphene, propranolol, protriptyline, quinidine, trazodone, verapamil
**Noninterfering:** acetaminophen, acetylmorphine, amiodarone, amobarbital, amphetamine,
bendroflumethiazide, benzocaine, benzoylecgonine, benzthiazide, butalbital, carbamazepine, chlorothiazide, clonazepam, cocaine, codeine, cotinine, cyclosporine, cyclothiazide,
desalkylflurazepam, diamorphine, dicumerol, ephedrine, ethacrynic acid, ethanol, ethchlorvynol, ethosuximide, furosemide, glutethimide, hydrochlorothiazide, hydrocodone,
hydroflumethiazide, hydromorphone, lorazepam, mephentermine, meprobamate, methamphetamine, metharbital, methoxsalen, methoxyphenteramine, methsuximide, methylcyclothiazide, metoprolol, MHPG, monoacetylmorphine, morphine, normethsuximide,
oxazepam, oxycodone, oxymorphone, pentobarbital, phencyclidine, phenteramine, phenylephrine, phenytoin, polythiazide, primidone, prochlorperazine, salicylic acid, sulfanilamide, THC-COOH, theophylline, thiazolam, thiopental, thioridazine, tocainide, trichloromethiazide, trifluoperazine, valproic acid, warfarin
**Interfering:** acetazolamide, methadone, norfluoxetine, temazepam, trimipramine

## KEY WORDS
SPE; plasma

## REFERENCE
Nichols, J.H.; Charlson, J.R.; Lawson, G.M. Automated HPLC assay of fluoxetine and norfluoxetine in
serum. *Clin.Chem.*, **1994**, *40*, 1312–1316

## SAMPLE
**Matrix:** blood
**Sample preparation:** Condition a 1 mL BondElut C18 SPE cartridge with 1 mL 1 M HCl,
1 mL MeOH, 1 mL water, and 1 mL 1% potassium carbonate. 700 µL Serum + 50 µL 5
µg/mL trimipramine in 5% potassium bicarbonate + 700 µL MeCN, vortex, centrifuge at
1500 g for 5 min, add supernatant to SPE cartridge (at ca. 1 mL/min). Wash with 2 mL
water and 1 mL MeCN, elute with 250 µL MeOH:35% perchloric acid 20:1 by gravity
(10 min) then centrifuge for 20 s to remove rest of eluant, inject a 50 µL aliquot of the
elu

## HPLC VARIABLES
**Guard column:** 15 mm 7 µm Brownlee RP-8
**Column:** 150 × 4.6 5 µm Ultrasphere Octyl
**Mobile phase:** MeCN:water 37.5:62.5 containing 0.5 g/L tetramethylammonium perchlo-
rate and 0.5 mL/L 7% perchloric acid
**Flow rate:** 1.5
**Injection volume:** 50
**Detector:** UV 215

## CHROMATOGRAM
**Retention time:** 8.3
**Internal standard:** trimipramine (9.6)
**Limit of quantitation:** 5 ng/mL

## OTHER SUBSTANCES
**Extracted:** clomipramine, desipramine, doxepin, fluoxetine, fluvoxamine, imipramine, nortriptyline, protriptyline
**Interfering:** desmethyltrimipramine, maprotiline

## KEY WORDS
SPE; serum

## REFERENCE
Gupta, R.N. An improved solid phase extraction procedure for the determination of antidepressants in serum by column liquid chromatography. *J.Liq.Chromatogr.*, **1993**, *16*, 2751–2765

## SAMPLE
**Matrix:** blood
**Sample preparation:** 500 µL Serum + 250 µL 100 mM lauryl sulfate, centrifuge at 2500 g for 8 min, inject a 250 µL aliquot of the supernatant onto column A with mobile phase A, elute with mobile phase A for 6 min, backflush contents of column A onto column B with mobile phase B for 4 min. Elute column B with mobile phase B for 6 min and conduct analysis. When not in use flush column A with mobile phase A. Every eight injections backflush column A with MeCN:water 70:30.

## HPLC VARIABLES
**Column:** A Guard-Pak 10 µm Resolve CN (Waters); B 150 × 3 7 µm Separon SGX CN (Tessek)
**Mobile phase:** A MeCN:water 3:97; B MeCN:buffer 26:74 (Buffer was 50 mM phosphoric acid, 50 mM ammonium phosphate, and 28 mM diethylamine, pH 2.55.)
**Flow rate:** 1
**Injection volume:** 250
**Detector:** UV 210

## CHROMATOGRAM
**Retention time:** 12
**Limit of detection:** 20-25 ng/mL

## OTHER SUBSTANCES
**Extracted:** nortriptyline

## KEY WORDS
column-switching; serum

## REFERENCE
Dolezalová, M. On-line solid-phase extraction and high-performance liquid chromatographic determination of nortriptyline and amitriptyline in serum. *J.Chromatogr.*, **1992**, *579*, 291–297

## SAMPLE
**Matrix:** blood
**Sample preparation:** For each 1 mL plasma or serum add 10 µL 14 µg/mL trimipramine in MeOH. Inject serum or plasma directly onto column A with mobile phase A, elute with mobile phase A to waste. After 15 min elute column A onto column B (foreflush) with mobile phase B. After 2 min remove column A from the circuit, elute column B with mobile phase B, monitor the effluent from column B. Re-equilibrate column A with mobile phase A.

## HPLC VARIABLES
**Column:** A 20 × 4.6 10 µm Hypersil MOS C8; B 20 × 4.6 5 µm Hypersil CPS CN + 250 × 4.6 5 µm Nucleosil 100 CN

**Mobile phase:** A MeOH:water 5:95; B MeCN:MeOH:buffer 578:188:235 (Buffer was 10 mM $K_2HPO_4$ adjusted to pH 6.8 with 85% phosphoric acid.)
**Flow rate:** 1.5
**Injection volume:** 100
**Detector:** UV 214

## CHROMATOGRAM
**Retention time:** 8.54
**Internal standard:** trimipramine (6.5)
**Limit of detection:** 5-10 ng/mL (1 ng/mL with 3 injections)

## OTHER SUBSTANCES
**Extracted:** metabolites, desipramine, fluvoxamine, imipramine, maprotiline, nortriptyline
**Noninterfering:** chlordiazepoxide, clobazam, clozapine, diazepam, flurazepam, fluspirilene, haloperidol, nitrazepam, oxazepam, perazine, pimozide, spiroperidol, trifluperidol
**Interfering:** clomipramine, doxepin

## KEY WORDS
plasma; serum; column-switching

## REFERENCE
Härtter, S.; Hiemke, C. Column switching and high-performance liquid chromatography in the analysis of amitriptyline, nortriptyline and hydroxylated metabolites in human plasma or serum. *J.Chromatogr.*, **1992**, *578*, 273–282

## SAMPLE
**Matrix:** blood
**Sample preparation:** 1 mL Serum + 3 μL 20 ng/mL clobazam in methanol + 1 mL saturated sodium borate (pH adjusted to 11 with 6 M NaOH) + 5 mL n-hexane, mix for 2 min, centrifuge at 3000 g for 10 min. Separate organic phase, evaporate to dryness under a stream of helium at 30°, reconstitute in 20 μL mobile phase, inject a 10 μL aliquot.

## HPLC VARIABLES
**Guard column:** 20 mm 40 μm Pelliguard LC-8 (Supelco)
**Column:** 150 × 4.6 Supelco 5 μm C8
**Mobile phase:** MeCN:buffer 50:50 (Buffer was 10 mM $NaH_2PO_4$ + 1.2 mL/L butylamine, pH adjusted to 3 with phosphoric acid.)
**Flow rate:** 1
**Injection volume:** 10
**Detector:** UV 254

## CHROMATOGRAM
**Retention time:** 9
**Internal standard:** clobazam (5)
**Limit of detection:** 10 ng/mL

## OTHER SUBSTANCES
**Extracted:** clomipramine, desipramine, imipramine, nortriptyline
**Simultaneous:** alprazolam, clonazepam, diazepam, flunitrazepam, haloperidol, lorazepam, maprotiline, nitrazepam, triazolam

## KEY WORDS
serum

## REFERENCE

Segatti, M.P.; Nisi, G.; Grossi, F.; Mangiarotti, M.; Lucarelli, C. Rapid and simple high-performance liquid chromatographic determination of tricyclic antidepressants for routine and emergency serum analysis. *J.Chromatogr.*, **1991**, *536*, 319–325

## SAMPLE

**Matrix:** blood

**Sample preparation:** 1 mL Plasma + 20 µL 10 µg/mL clovoxamine + 120 µL 2 M NaOH + 4 mL heptane:isopropanol 98:2, shake for 30 min, centrifuge at 3000 g for 10 min. Remove the organic layer and add it to 100 µL 100 mM HCl, shake for 20 min, centrifuge at 3000 g for 10 min, inject a 20 µL aliquot of the aqueous layer.

## HPLC VARIABLES

**Column:** 120 × 4.6 5 µm Nucleosil C8

**Mobile phase:** MeCN:buffer 36:64 (Buffer was 16 mM $KH_2PO_4$ adjusted to pH 2.5 with concentrated phosphoric acid.)

**Flow rate:** 1

**Injection volume:** 20

**Detector:** UV 215

## CHROMATOGRAM

**Retention time:** 5.5

**Internal standard:** clovoxamine (3.3)

## OTHER SUBSTANCES

**Extracted:** chlorimipramine, desipramine, doxepin, fluvoxamine, imipramine, nortriptyline, trimipramine

## KEY WORDS

plasma

## REFERENCE

Foglia, J.P.; Birder, L.A.; Perel, J.M. Determination of fluvoxamine in human plasma by high-performance liquid chromatography with ultraviolet detection. *J.Chromatogr.*, **1989**, *495*, 295–302

## SAMPLE

**Matrix:** blood

**Sample preparation:** Inject 200 µL serum onto column A and elute with mobile phase A for 10 min then back-flush column A onto column B with mobile phase B for 4 min. Elute column B with mobile phase B and monitor the effluent. Remove column A from circuit and wash with MeCN:water 60:40 for 6 min then with mobile phase A for 10 min.

## HPLC VARIABLES

**Column:** A 40 × 4 TSKprecolumn PW (Tosoh); B 150 × 4 TSKgel ODS-80TM (Tosoh)

**Mobile phase:** A 50 mM pH 7.5 potassium phosphate; B MeCN:100 mM pH 2.7 potassium phosphate 32.5:67.5, containing 0.2 g/L sodium 1-heptanesulfonate

**Flow rate:** 1

**Injection volume:** 200

**Detector:** UV 210

## CHROMATOGRAM

**Retention time:** 17

**Limit of detection:** 10 ng/mL

## OTHER SUBSTANCES
**Simultaneous:** amoxapine, clomipramine, doxepin, desipramine, imipramine, maprotiline, nortriptyline, trimipramine

## KEY WORDS
column-switching; use gradient to determine metabolites; serum

## REFERENCE
Matsumoto, K.; Kanba, S.; Kubo, H.; Yagi, G.; Iri, H.; Yuki, H. Automated determination of drugs in serum by column-switching high-performance liquid chromatography. IV. Separation of tricyclic and tetracyclic antidepressants and their metabolites. *Clin.Chem.*, **1989**, *35*, 453–456

## SAMPLE
**Matrix:** blood
**Sample preparation:** 1 mL Plasma + 50 μL MeOH:water 50:50 + 50 μL 10 μg/mL desmethyldoxepin in MeOH:water 50:50, mix, inject a 250 μL aliquot of this mixture onto column A and elute to waste with mobile phase A. After 1.5 min backflush the contents of column A onto column B with mobile phase B, monitor the effluent from column B.

## HPLC VARIABLES
**Column:** A 37-50 μm 10 × 1.5 Corasil RP C18; B 100 × 4 Techsphere 3CN (HPLC Technology)
**Mobile phase:** A water; B MeCN:50 mM acetate buffer 60:40, pH 7
**Flow rate:** A 0.8; B 0.9
**Injection volume:** 250
**Detector:** UV 215

## CHROMATOGRAM
**Retention time:** 12.5
**Internal standard:** desmethyldoxepin (8.5)
**Limit of detection:** 5-10 ng/mL

## OTHER SUBSTANCES
**Extracted:** metabolites, nortriptyline
**Simultaneous:** cianopramine, chloripramine, clomipramine, desipramine, doxepin, protriptyline, trimipramine
**Noninterfering:** tranylcypromine
**Interfering:** imipramine

## KEY WORDS
column-switching; plasma

## REFERENCE
Dadgar, D.; Power, A. Applications of column-switching technique in biopharmaceutical analysis. I. High-performance liquid chromatographic determination of amitriptyline and its metabolites in human plasma. *J.Chromatogr.*, **1987**, *416*, 99–109

## SAMPLE
**Matrix:** blood
**Sample preparation:** Condition a Bond Elut C-18 SPE cartridge twice with MeOH and twice with water. 500 μL Serum + 50 μL 1 μg/mL N-propionylprocainamide in 2.5 mM HCl, add to SPE cartridge, wash with 2 volumes water, wash with 2 volumes 0.1 M acetic acid, wash with 1 volume MeOH:2.5 mM HCl 10:90. Add 200 μL 10 mM acetic acid and 5 mM diethylamine in MeOH to column, let stand 1 min, elute under vacuum, repeat, evaporate eluents to dryness under nitrogen at room temperature, reconstitute in 100 μL mobile phase, inject a 40 μL aliquot.

## HPLC VARIABLES
**Guard column:** Pelliguard LC-CN (Supelco)
**Column:** 150 × 4.6 5 μm Supelcosil LC-PCN
**Mobile phase:** MeCN:MeOH:10 mM pH 7.0 phosphate buffer 58:14:28
**Flow rate:** 1.2
**Injection volume:** 40
**Detector:** UV 254

## CHROMATOGRAM
**Retention time:** 9.1
**Internal standard:** N-propionylprocainamide (6)
**Limit of quantitation:** 25 ng/mL

## OTHER SUBSTANCES
**Extracted:** desipramine, doxepin, imipramine, nortriptyline, protriptyline, trimipramine
**Simultaneous:** atropine, butalbital, chlorpromazine, maprotiline, methadone, norpropoxy-phene, phenylpropanolamine, procainamide, prochlorperazine, promethazine, proprano-lol, quinidine, trifluoperazine, trimeprazine
**Noninterfering:** acetaminophen, allopurinol, amikacin, amoxapine, amytal, bretylium, caf-feine, carbamazepine, carisoprodol, chloramphenicol, chlordiazepoxide, chlorpropamide, clonazepam, codeine, diazepam, disopyramide, droperidol, ethinamate, ethinamate, etho-suximide, fluphenazine, flurazepam, furosemide, gentamicin, haloperidol, hydrochloro-thiazide, hydroxyzine, ibuprofen, kanamycin, lidocaine, loxapine, meperidine, mephobar-bital, meprobamate, methaqualone, methotrexate, morphine, nafcillin, naloxone, neomycin, perphenazine, phenacetin, phenobarbital, phenytoin, prazepam, primidone, procaine, propoxyphene, reserpine, salicylamide, salicylic acid, secobarbital, spironolac-tone, theophylline, thiopental, tobramycin, valproic acid, verapamil
**Interfering:** thioridazine

## KEY WORDS
serum; SPE

## REFERENCE
Lin, W.-N.; Frade, P.D. Simultaneous quantitation of eight tricyclic antidepressants in serum by high-performance liquid chromatography. *Ther.Drug Monit.*, **1987**, *9*, 448–455

## SAMPLE
**Matrix:** blood
**Sample preparation:** 500 μL Serum + 250 μL di-iso-propyl ether:n-butyl alcohol 7:3 con-taining 800 ng/mL minaprine, centrifuge 2 min, shake, centrifuge 5 min, inject 50 μL aliquot of top organic layer.

## HPLC VARIABLES
**Guard column:** 30 × 4.6 5 μm Brownlee cyano spheri-5
**Column:** 250 × 4.6 5 μm Altex Ultrasphere cyano
**Mobile phase:** MeCN:THF:water:2 M ammonium formate (pH 4.0) 700:100:195:5
**Column temperature:** 20
**Flow rate:** 1.5
**Injection volume:** 50
**Detector:** UV 240

## CHROMATOGRAM
**Retention time:** 7
**Internal standard:** minaprine (5.5)
**Limit of detection:** 20 ng/mL

## OTHER SUBSTANCES
**Simultaneous:** diltiazem, nortriptyline
**Also analyzed:** amiodarone, clomipramine, desipramine, haloperidol, imipramine, propafenone, verapamil

## KEY WORDS
serum

## REFERENCE
Mazzi, G. Simple and practical high-performance liquid chromatographic assay of some tricyclic drugs, haloperidol, diltiazem, verapamil, propafenone, and amiodarone. *Chromatographia*, **1987**, *24*, 313–316

## SAMPLE
**Matrix:** blood
**Sample preparation:** 1 mL Serum + 1 mL 450 mM NaOH + 5 mL hexane:isopropanol 95:5, shake for 5 min, centrifuge. Remove 4 mL of the organic layer and add it to 50 µL 200 mM HCl, shake for 2 min, centrifuge. Inject a 20 µL aliquot of the aqueous layer.

## HPLC VARIABLES
**Column:** 100 × 3 octyl CP-tm-Spher C8 glass column (Chrompack)
**Mobile phase:** MeCN:500 mM $NaH_2PO_4$ 35:65 adjusted to pH 2.2 with phosphoric acid
**Flow rate:** 0.8
**Injection volume:** 20
**Detector:** UV 210

## CHROMATOGRAM
**Retention time:** 5.8
**Limit of detection:** 10 ng/mL

## OTHER SUBSTANCES
**Simultaneous:** doxepin, nortriptyline

## KEY WORDS
serum

## REFERENCE
Van Damme, M.; Molle, L.; Abi Khalil, F. Useful sample handlings for reversed phase high performance liquid chromatography in emergency toxicology. *J.Toxicol.Clin.Toxicol.*, **1985**, *23*, 589–614

## SAMPLE
**Matrix:** blood
**Sample preparation:** 500 µL Plasma + 37 µL 2 µg/mL IS in MeOH + 500 µL pH 10 borate buffer + 1.5 mL hexane:isoamyl alcohol 95:5, shake for 10 min. Evaporate the organic layer to dryness under a stream of nitrogen, reconstitute in 100 µL MeOH, inject a 50 µL aliquot. (The borate buffer was prepared as follows. Prepare a solution of 61.8 g boric acid and 74.6 g KCl in 1 L water. Add 630 mL of this solution to 370 mL 106 g/L sodium carbonate solution. Adjust pH to 10.0 with 6 M NaOH and store at 35-37°.)

## HPLC VARIABLES
**Column:** 250 × 4.6 Zorbax Sil
**Mobile phase:** MeOH:ammonium hydroxide 998:2
**Flow rate:** 1.5
**Injection volume:** 50
**Detector:** UV 254

## CHROMATOGRAM
**Retention time:** 4
**Internal standard:** N-desmethylclomipramine hydrochloride (10)
**Limit of quantitation:** 20 ng/mL

## OTHER SUBSTANCES
**Extracted:** desipramine, 2-hydroxydesipramine, 2-hydroxyimipramine, imipramine, me-
tabolites, nortriptyline
**Also analyzed:** doxepin, desmethyldoxepin, desmethylclomipramine, clomipramine, ma-
protiline, protriptyline
**Noninterfering:** chlordiazepoxide, diazepam, flurazepam, oxazepam, thioridazine

## KEY WORDS
plasma

## REFERENCE
Sutfin, T.A.; D'Ambrosio, R.; Jusko, W.J. Liquid-chromatographic determination of eight tri- and tetracyclic antidepressants and their major active metabolites. *Clin.Chem.*, **1984**, *30*, 471–474

## SAMPLE
**Matrix:** blood
**Sample preparation:** Condition a Bond-Elut C18 column with 2 volumes MeOH then 2
volumes water. Add 1 mL serum then 200 μL 700 ng/mL promazine in MeOH:0.1 M HCl
13:87 to each column, wash with 2 volumes water, wash with 2 volumes 0.1 M acetic
acid, wash with MeOH/water, add 200 μL 10 mM ammonium acetate in MeOH, wait for
30 s, elute with vacuum, repeat elution process two more times. Combine eluates and
evaporate them to dryness at 56-8°under compressed air. Reconstitute with 200 μL mobile
phase, vortex 10 s, inject 75-100 μL aliquot. (MeOH/water was 500 mL MeOH:water
65:35 plus 25 μL concentrated HCl.)

## HPLC VARIABLES
**Column:** 250 × 4.6 5 μm Supelco
**Mobile phase:** 1 gal EtOH + 77 mL MeCN + 1.9 mL t-butylamine (EtOH:MeCN:t-buty-
lamine 98:2:0.05)
**Flow rate:** 2
**Injection volume:** 75-100
**Detector:** UV 254

## CHROMATOGRAM
**Retention time:** 3.2
**Internal standard:** promazine (5.2)
**Limit of detection:** 2 ng/mL

## OTHER SUBSTANCES
**Simultaneous:** N-acetylprocainamide, amoxapine, amphetamine, buprion, chlordiazepoxide, chlorpheniramine, chlorpromazine, cocaine, codeine, demoxepam, desipramine,
desmethylchlordiazepoxide, desmethyldisopyramide, desmethyldoxepin, dextropropoxyphene, diazepam, disopyramide, fluphenazine, 2-hydroxydesipramine, 2-hydroxyimipra-
mine, 10-hydroxynortriptyline, iminostilbene, imipramine, iprindole, loxepin, maprotiline,
methadone, mianserin, morphine, nortriptyline, norzimeldine, oxapam, oxaprotiline, procainamide, prochlorperazine, prolixin, promethazine, propoxyphene, protriptyline, pyrilamine, trifluoperazine, trimeprazine, trimipramine, zimeldine
**Noninterfering:** thiopropazine
**Interfering:** chlorimipramine, doxepin, hydroxyamoxapine, meperidine, perphenazine,
phenteramine, quinidine, thioridazine, trifluopromazine

## KEY WORDS
normal phase; serum

## REFERENCE
Beierle, F.A.; Hubbard, R.W. Liquid chromatographic separation of antidepressant drugs: I. Tricyclics. *Ther.Drug Monit.*, **1983**, *5*, 279–292

## SAMPLE
**Matrix:** blood
**Sample preparation:** 2 mL Plasma + 100 µL 1 µg/mL loxapine in isopropanol:diethylamine 99.9:0.1 + 250 µL 25% potassium carbonate containing 0.1% diethylamine + 5 mL hexane:isoamyl alcohol 97:3, vortex for 30 s, centrifuge at 500 g for 3 min. Remove the organic layer and add it to 100 µL 250 mM HCl, vortex for 30 s, inject a 50 µL aliquot of the aqueous phase.

## HPLC VARIABLES
**Guard column:** 50 × 4.6 40 µm C8 (Supelco)
**Column:** 250 × 4.6 5 µm Supelcosil C8
**Mobile phase:** MeCN:water:diethylamine:85% phosphoric acid 53.3:45.1:1:0.4, pH adjusted to 7.2 with NaOH or phosphoric acid
**Flow rate:** 2
**Injection volume:** 50
**Detector:** UV 254

## CHROMATOGRAM
**Retention time:** k' 6.41
**Internal standard:** loxapine (k' 7.18)
**Limit of detection:** 2.5 ng/mL

## OTHER SUBSTANCES
**Extracted:** chlordiazepoxide, chlorpromazine, desipramine, desmethldiazepam, desmethylchlordiazepoxide, desmethyldoxepin, diazepam, doxepin, fluphenazine, haloperidol, imipramine, nortriptyline, oxazepam, thiothixene
**Noninterfering:** molindone, perphenazine, trifluoperazine

## KEY WORDS
plasma

## REFERENCE
Kiel, J.S.; Abramson, R.K.; Morgan, S.L.; Voris, J.C. A rapid high performance liquid chromatographic method for the simultaneous measurement of six tricyclic antidepressants. *J.Liq.Chromatogr.*, **1983**, *6*, 2761–2773

## SAMPLE
**Matrix:** blood
**Sample preparation:** Plasma. 1-5 mL Plasma + 1 mL 1 M NaOH, extract with mixed hexanes for 30 min, centrifuge. Remove a 9 mL aliquot of the hexane layer and evaporate it to dryness under a stream of nitrogen at 30°, dissolve residue in 100 µL mobile phase, inject a 50 µL aliquot. Whole blood. 10 mL Whole blood + 1 mL 1 M NaOH, extract with 15 mL mixed hexanes for 1 h. Remove an aliquot of the hexane layer and evaporate it to dryness, reconstitute the residue in 1 mL 100 mM HCl, extract with 5 mL chloroform by vortexing for 1 min, centrifuge. Remove a 4.5 mL aliquot of the chloroform layer, evaporate to dryness, dissolve in 10 µL mobile phase, inject an aliquot.

## HPLC VARIABLES
**Column:** 10 µm Micropak CN

**Mobile phase:** MeCN:20 mM ammonium acetate 90:10 (vary ammonium acetate concentration to achieve best separation)
**Flow rate:** 2.5
**Injection volume:** 10-50
**Detector:** UV 254; E, Bioanalytical Systems LC-4A, glassy carbon electrode +0.9 V, Ag/AgCl reference electrode

## CHROMATOGRAM
**Retention time:** 8.8
**Limit of detection:** 10 ng/mL (UV); 0.1 ng/mL (E)

## OTHER SUBSTANCES
**Extracted:** acetophenazine, benztropine, butaperazine, carphenazine, chlorpromazine, fluphenazine, haloperidol, imipramine, mesoridazine, nortriptyline, orphenadrine, piperacetazine, promazine, promethazine, thioridazine, thiothixene, trifluoperazine, triflupromazine, trihexyphenidyl, trimeprazine

## KEY WORDS
plasma; whole blood

## REFERENCE
Curry, S.H.; Brown, E.A.; Hu, O.Y.-P.; Perrin, J.H. Liquid chromatographic assay of phenothiazine, thioxanthene and butyrophenone neuroleptics and antihistamines in blood and plasma with conventional and radial compression columns and UV and electrochemical detection. *J.Chromatogr.*, **1982**, *231*, 361–376

## SAMPLE
**Matrix:** blood
**Sample preparation:** 1 mL Serum + 200 µL 1.5 M NaOH, vortex for 5 s, add 6 mL hexane:isoamyl alcohol 99:1, add 200 µL 1 µg/mL perazine dimalonate in EtOH, rotate at 60 rpm for 30 min, centrifuge at 3000 g for 4 min. Remove the upper organic layer and add it to 20 µL 2.5 mg/mL maprotiline in MeOH, vortex for 5 s, evaporate to dryness under a stream of nitrogen at 40°, reconstitute the residue in 50 µL mobile phase, inject a 20 µL aliquot.

## HPLC VARIABLES
**Column:** 100 × 3 5 µm Lichrosorb SI60
**Mobile phase:** MeCN:MeOH:ammonium hydroxide 250:55:13
**Flow rate:** 1.2
**Injection volume:** 20
**Detector:** UV 240

## CHROMATOGRAM
**Retention time:** 3.5
**Internal standard:** perazine (5)
**Limit of detection:** 5 ng/mL

## OTHER SUBSTANCES
**Extracted:** metabolites
**Simultaneous:** N-acetylprocainamide, butaperazine, chlorimipramine, chlorpromazine, codeine, desipramine, dimethacrine, diphenhydramine, disopyramide, doxepin, hydroquinidine, maprotiline, melitracene, mesoridazine, nortriptyline, opipramol, perphenazine, procainamide, prochlorperazine, promazine, prothipendyl, protriptyline, quinidine, thioperazine, thioridazine, trifluoperazine
**Noninterfering:** acenocoumaron, acetaminophen, acetophenetidine, aspirin, benzodiazepines, bibenzepin, butriptyline, caffeine, chlorprothixene, clopenthixol, clothiapine, dixyrazine, droperidol, fluphenazine, haloperidol, hydroxyzine, isoniazid, methotrimeprazine,

metopimazine, moperone, noxiptyline, orphenadrine, pericyazine, phenprocoumon, pipothiazine, promethazine, salicylic acid, theophylline, thiopropazate, trimeprazine, trimipramine

**Interfering:** imipramine, pipamperone, thiethylperazine, thiothixene

## KEY WORDS
maprotiline prevents adsorption on glass; serum; pharmacokinetics

## REFERENCE
Edelbroek, P.M.; de Haas, E.J.M.; de Wolff, F.A. Liquid-chromatographic determination of amitryptyline and its metabolites in serum, with adsorption onto glass minimized. *Clin.Chem.*, **1982**, *28*, 2143–2148

## SAMPLE
**Matrix:** blood
**Sample preparation:** 1 mL Serum + 200 μL 10 μg/mL protriptyline in water + 200 μL 80 g/L NaHCO$_3$ + 5 mL hexane, vortex for 15 s, centrifuge for 5 min. Remove the hexane layer and evaporate it in a stream of nitrogen at 60°. Reconstitute in 100 μL mobile phase, vortex for 15 s, inject a 50 μL aliquot.

## HPLC VARIABLES
**Column:** 300 × 4 10 μm μBondapak CN
**Mobile phase:** MeCN:MeOH:5 mM phosphate buffer 60:15:25, adjusted to pH 7.0
**Flow rate:** 2
**Injection volume:** 50
**Detector:** UV 254

## CHROMATOGRAM
**Retention time:** 3.29
**Internal standard:** protriptyline (12.20)
**Limit of detection:** 6 ng/mL

## OTHER SUBSTANCES
**Simultaneous:** chlorpromazine, desipramine, desmethyldoxepin, disopyramide, doxepin, imipramine, maprotiline, nortriptyline, procainamide, propoxyphene, propranolol, thioridazine, trimipramine
**Noninterfering:** acetaminophen, caffeine, chlordiazepoxide, diazepam, methaqualone, salicylic acid, theophylline, trifluoperazine

## KEY WORDS
serum

## REFERENCE
Koteel, P.; Mullins, R.E.; Gadsden, R.H. Sample preparation and liquid-chromatographic analysis for tricyclic antidepressants in serum. *Clin.Chem.*, **1982**, *28*, 462–466

## SAMPLE
**Matrix:** blood
**Sample preparation:** 2 mL Plasma + 1600 ng clomipramine in MeOH + 2 mL 1 M NaOH + 5 mL hexane:isoamyl alcohol 99:1, shake mechanically for 15 min, centrifuge at 1686 g for 5 min. Remove the organic phase and add it to 200 μL 0.05% orthophosphoric acid, shake for 15 min, centrifuge for 5 min, inject a 50 μL aliquot of the aqueous phase.

## HPLC VARIABLES
**Guard column:** μBondapak/Porasil
**Column:** μBondapak C18

**Mobile phase:** MeCN:buffer 40:60 (Buffer was 13.68 g $KH_2PO_4$ in 2 L water, adjusted to pH 4.7 with dilute KOH.)
**Column temperature:** 50
**Flow rate:** 2
**Injection volume:** 50
**Detector:** UV 254

## CHROMATOGRAM
**Retention time:** 5.5
**Internal standard:** clomipramine (7.5)
**Limit of detection:** 0.7 ng

## OTHER SUBSTANCES
**Extracted:** desipramine, imipramine, nortriptyline
**Simultaneous:** chlordiazepoxide, chlorpromazine, cimetidine, clomipramine, diazepam, doxepin, flurazepam, lorazepam, oxazepam, pentobarbital, perphenazine, phenobarbital, phenytoin, prochlorperazine, secobarbital, thioridazine, trifluoperazine
**Noninterfering:** acetaminophen, codeine, meperidine
**Interfering:** propoxyphene

## KEY WORDS
plasma

## REFERENCE
Wong, S.H.Y.; McCauley, T. Reversed phase high-performance liquid chromatographic analysis of tricyclic antidepessants in plasma. *J.Liq.Chromatogr.*, **1981**, *4*, 849–862

## SAMPLE
**Matrix:** blood, dialysate
**Sample preparation:** Adjust pH of serum samples to 7.4 and dialyze 3 mL serum against 5 mL dialysis buffer at 37° in PTFE chambers for 4 h. Inject 1 mL mobile phase, 1 mL water, 2 mL dialysate, 1 mL water, and 1 mL MeCN:water 50:50 onto column A. Elute column A onto column B with mobile phase for 30 s then remove it from the circuit. Elute column B with mobile phase and monitor the effluent. (Dialysis buffer was 3.998 g $Na_2HPO_4.2H_2O$ + 0.775 g $NaH_2PO_4$ + 2.250 g NaCl + 0.055 g $Hg(NO_3)_2$ in 1 L water, pH was 7.4.)

## HPLC VARIABLES
**Column:** A 10 × 6 packed with 40 μm material from a Bond Elut cartridge (cat. no. 620303); B 100 × 4 3 μm Spherisorb ODS Superpac
**Mobile phase:** MeCN:85% phosphoric acid:triethylamine:water 49.55:0.225:0.225:50
**Flow rate:** 0.65
**Injection volume:** 1000-2000
**Detector:** UV 238

## CHROMATOGRAM
**Retention time:** 4.85
**Limit of detection:** 1 nM (5 mL sample)

## OTHER SUBSTANCES
**Extracted:** nortriptyline
**Simultaneous:** alprazolam, chlorpromazine, chlorprothixene, clomipramine, desmethylimipramine, diazepam, flunitrazepam, fluphenazine, haloperidol, imipramine, maprotiline, perphenazine, promethazine, protriptyline, thioridazine, thioridazine sulfone, thioridazine sulfoxide, zimeldine, zuclopenthixol
**Noninterfering:** carbamazepine, clonazepam, lorazepam, nitrazepam, oxazepam, phenytoin
**Interfering:** desclomipramine, levomepromazine, trimipramine

## KEY WORDS
column-switching; serum

## REFERENCE
Svensson, C.; Nyberg, G.; Mårtensson, E. High-performance liquid chromatographic quantitation of am-
itriptyline and nortriptyline in dialysate from plasma or serum using on-line solid-phase extraction.
*J.Chromatogr.*, **1988**, *432*, 363–369

## SAMPLE
**Matrix:** blood, gastric contents, tissue, urine
**Sample preparation:** 1 mL Blood, urine, or gastric contents or 1 g tissue homogenate +
500 µL buffer + 8 mL n-hexane:ethyl acetate 70:30, mix on a rotary mixer for 10 min,
centrifuge at 3000 g for 8 min. Remove the organic layer and evaporate it to dryness
under a stream of nitrogen, reconstitute the residue in 100 µL 12.5 mM NaOH in
MeOH:water 50:50, inject a 50 µL aliquot. (Buffer was 13.8 g potassium carbonate in
100 mL water, pH adjusted to 9.5 with concentrated HCl.)

## HPLC VARIABLES
**Guard column:** 4 × 4 30 µm LiChrocart Aluspher RP-select B (Merck)
**Column:** 125 × 4 5 µm Aluspher RP-select B (Merck)
**Mobile phase:** Gradient. A was 12.5 mM NaOH in MeOH. B was 12.5 mM NaOH in water.
A:B 10:90 for 5 min, to 90:10 over 15 min, maintain at 90:10 for 5 min, return to initial
conditions over 1 min, re-equilibrate for 5 min.
**Flow rate:** 1
**Injection volume:** 50
**Detector:** UV 230; UV 254

## CHROMATOGRAM
**Retention time:** 20

## OTHER SUBSTANCES
**Extracted:** alprenolol, bromazepam, carbamazepine, chlordiazepoxide, chlorpromazine,
clonazepam, desipramine, diazepam, flunitrazepam, haloperidol, nitrendipine, nordiazepam, nortriptyline, pindolol, zolpidem
**Also analyzed:** acebutolol, acetaminophen, alprazolam, amphetamine, atenolol, betaxolol,
brotizolam, caffeine, camazepam, captopril, chloroquine, clobazam, clomipramine, clothiapine, clotiazepam, cloxazolam, cocaine, codeine, diclofenac, dihydralazine, dihydrocodeine, dihydroergotamine, diphenhydramine, domperidone, doxepin, droperidol, ergotamine, ethyl loflazepate, fenethylline, fluoxetine, flupentixol, flurazepam, furosemide,
gliclazide, hydrochlorothiazide, hydroxyzine, ibuprofen, imipramine, ketazolam, loprazolam, lorazepam, lormetazepam, maprotiline, medazepam, mepyramine, methadone,
methaqualone, methyldopa, methylphenidate, metoclopramide, metoprolol, mexiletine,
mianserin, midazolam, minoxidil, morphine, nadolol, nitrazepam, oxprenolol, papaverine,
pentazocine, phenprocoumon, phenylbutazone, pipamperone, piritramide, practolol, prazepam, prazosin, promazine, promethazine, propoxyphene, propranolol, prothipendyl, quinine, sotalol, sulpride, thioridazine, trazodone, triazolam, trimipramine, tripelennamine,
tyramine, verapamil, yohimbine

## REFERENCE
Lambert, W.E.; Meyer, E.; De Leenheer, A.P. Systematic toxicological analysis of basic drugs by gradient
elution of an alumina-based HPLC packing material under alkaline conditions. *J.Anal.Toxicol.*, **1995**,
*19*, 73–78

## SAMPLE
**Matrix:** blood, tissue

**Sample preparation:** Blood or serum. 1 mL Blood or serum + 1 μg cianopramine + 1 mL water, vortex, add 1 mL 200 mM sodium carbonate, vortex, add 6 mL hexane:1-butanol 95:5, gently agitate for 30 min, centrifuge at 2500 g for 5 min. Remove the organic layer and add it to 100 μL 0.2% phosphoric acid, agitate gently for 30 min, centrifuge for 5 min. Remove the organic layer and inject a 30 μL aliquot of the aqueous layer. Liver homogenate. 0.5 mL Liver homogenate + 10 μg cianopramine + 500 μL 2% sodium tetraborate + 8 mL hexane:1-butanol 95:5, gently agitate for 30 min, centrifuge at 2500 g for 5 min. Remove the organic layer and add it to 400 μL 0.2% phosphoric acid, agitate gently for 30 min, centrifuge for 5 min. Remove the organic layer and inject a 30 μL aliquot of the aqueous layer.

## HPLC VARIABLES
**Guard column:** 15 × 3.2 7 μm RP-18 Newguard (Applied Biosystems)
**Column:** 100 × 4.6 5 μm Brownlee Spheri-5 RP-18
**Mobile phase:** MeCN:100 mM $NaH_2PO_4$:diethylamine 40:57.5:2.5
**Flow rate:** 2
**Injection volume:** 30
**Detector:** UV 220

## CHROMATOGRAM
**Retention time:** 22.5
**Internal standard:** cianopramine (8.93)
**Limit of detection:** 50 ng/mL

## OTHER SUBSTANCES
**Extracted:** amoxapine, brompheniramine, chlorpheniramine, chlorpromazine, clomipramine, cyproheptadine, dsipramine, diphenhydramine, dothiepin, doxepin, fluoxetine, haloperidol, imipramine, loxapine, maprotiline, meperidine, mesoridazine, methadone, metoclopramide, mianserin, moclobemide, nomifensine, nordoxepin, norfluoxetine, norpropoxyphene, northiaden, nortriptyline, pentobarbital, pheniramine, propoxyphene, propranolol, protriptyline, quinidine, quinine, sulforidazine, thioridazine, thiothixene, tranylcypromine, trazodone, trihexyphenidyl, triprolidine
**Noninterfering:** dextromethorphan, norphetidine, phenoxybenzamine, prochlorperazine, trifluoperazine
**Interfering:** benztropine, promethazine, trimipramine

## KEY WORDS
serum; liver

## REFERENCE
McIntyre, I.M.; King, C.V.; Skafidis, S.; Drummer, O.H. Dual ultraviolet wavelength high-performance liquid chromatographic method for the forensic or clinical analysis of seventeen antidepressants and some selected metabolites. *J.Chromatogr.*, **1993**, *621*, 215–223

## SAMPLE
**Matrix:** formulations
**Sample preparation:** Tablets, capsules. Crush tablet or capsule, weigh out amount corresponding to 2 mg amitriptyline, add 20 mL MeOH, shake for 30 min, centrifuge at 2000 rpm for 5 min, to 5 mL supernatant add 4 mL 1.25 mg/mL norephedrine.HCl in MeOH, dilute to 10 mL with MeOH, inject a 10 μL aliquot. Liquid formulations. Take a 5 mL aliquot of a 2 mg/mL solution, make up to 10 mL with 1% HCl, remove a 5 mL aliquot and add it to 40 mL 1.25 mg/mL norephedrine.HCl in MeOH, make up to 100 mL with MeOH, inject a 10 μL aliquot.

## HPLC VARIABLES
**Column:** 150 × 4.6 5 μm Zorbax CN
**Mobile phase:** MeCN:MeOH:25 mM pH 4.8 sodium acetate-acetic acid buffer 35:45:20

**Flow rate:** 2.5
**Injection volume:** 10
**Detector:** UV 254

## CHROMATOGRAM
**Retention time:** 4
**Internal standard:** norephedrine (2.7)

## OTHER SUBSTANCES
**Also analyzed:** chlorpromazine, imipramine, thioridazine, trifluoperazine

## KEY WORDS
tablets; capsules; liquid formulations

## REFERENCE
Lovering, E.G.; Beaulieu, N.; Lawrence, R.C.; Sears, R.W. Liquid chromatographic method for identity, assay, and content uniformity of five tricyclic drugs. *J.Assoc.Off.Anal.Chem.*, **1985**, *68*, 168–171

## SAMPLE
**Matrix:** hair
**Sample preparation:** Wash hair in water, rinse 3 times with MeOH, dry, weigh. 5-25 mg Washed hair + 1 mL 1 M NaOH, heat at 70° for 30 min, adjust pH to 9.5-10. 1 mL Extract + 1 μg protriptyline + 1 mL water + 1 mL 200 mM sodium carbonate buffer, mix, extract with hexane:butanol 95:5 for 20 min. Remove the organic layer and add it to 100 μL 0.2% orthophosphoric acid, mix for 20 min, inject a 30 μL aliquot of the aqueous layer.

## HPLC VARIABLES
**Guard column:** 15 × 3.2 7 μm Newguard RP-18
**Column:** 100 × 4.6 Spheri-5 RP-C18
**Mobile phase:** MeCN:buffer 40:60 (Buffer was 1.2 L 100 mM pH 7.0 $NaH_2PO_4$ + 30 mL diethylamine.)
**Flow rate:** 2
**Injection volume:** 30
**Detector:** UV 214

## CHROMATOGRAM
**Retention time:** 13
**Internal standard:** protriptyline (4)

## OTHER SUBSTANCES
**Extracted:** clomipramine, desipramine, dothiepin, doxepin, haloperidol, imipramine, mianserin, nortriptyline

## KEY WORDS
there may be interferences

## REFERENCE
Couper, F.J.; McIntyre, I.M.; Drummer, O.H. Extraction of psychotropic drugs from human scalp hair. *J.Forensic Sci.*, **1995**, *40*, 83–86

## SAMPLE
**Matrix:** microsomal incubations
**Sample preparation:** 250 μL Microsomal incubation + 50 μL 1 M HCl, cool on ice, add desipramine, centrifuge at 16000 g for 5 min, inject an aliquot of the supernatant.

## HPLC VARIABLES
**Column:** 300 × 3.9 μBondapak C18
**Mobile phase:** MeCN:50 mM pH 6 potassium phosphate buffer 35:65
**Flow rate:** 1.4
**Detector:** UV 220

## CHROMATOGRAM
**Retention time:** 12
**Internal standard:** desipramine (9)

## OTHER SUBSTANCES
**Extracted:** metabolites, nortriptyline
**Noninterfering:** α-naphthoflavone, ketoconazole, quinidine

## KEY WORDS
human; liver

## REFERENCE
Schmider, J.; Greenblatt, D.J.; von Moltke, L.L.; Harmatz, J.S.; Shader, R.I. N-Demethylation of amitriptyline *in vitro*: Role of cytochrome P- 450 3A (CYP3A) isoforms and effect of metabolic inhibitors. *J.Pharm.Exp.Ther.*, **1995**, *275*, 592–597

## SAMPLE
**Matrix:** solutions

## HPLC VARIABLES
**Column:** 250 × 4.6 5 μm Supelcosil LC-DP (A) or 250 × 4 5 μm LiChrospher 100 RP-8 (B)
**Mobile phase:** MeCN:0.025% phosphoric acid:buffer 25:10:5 (A) or 60:25:15 (B) (Buffer was 9 mL concentrated phosphoric acid and 10 mL triethylamine in 900 mL water, adjust pH to 3.4 with dilute phosphoric acid, make up to 1 L.)
**Flow rate:** 0.6
**Injection volume:** 25
**Detector:** UV 229

## CHROMATOGRAM
**Retention time:** 15.80 (A), 7.27 (B)

## OTHER SUBSTANCES
**Also analyzed:** acebutolol, acepromazine, acetaminophen, acetazolamide, acetophenazine, albuterol, alprazolam, amobarbital, amoxapine, antipyrine, atenolol, atropine, azatadine, baclofen, benzocaine, bromocriptine, brompheniramine, brotizolam, bupivacaine, buspirone, butabarbital, butalbital, caffeine, carbamazepine, cetirizine, chlorcyclizine, chlordiazepoxide, chlormezanone, chloroquine, chlorpheniramine, chlorpromazine, chlorpropamide, chlorprothixene, chlorthalidone, chlorzoxazone, cimetidine, cisapride, clomipramine, clonazepam, clonidine, clozapine, cocaine, codeine, colchicine, cyclizine, cyclobenzaprine, dantrolene, desipramine, diazepam, diclofenac, diflunisal, diltiazem, diphenhydramine, diphenidol, diphenoxylate, dipyridamole, disopyramide, dobutamine, doxapram, doxepin, droperidol, encainide, ethidium bromide, ethopropazine, fenoprofen, fentanyl, flavoxate, fluoxetine, fluphenazine, flurazepam, flurbiprofen, fluvoxamine, furosemide, glutethimide, glyburide, guaifenesin, haloperidol, homatropine, hydralazine, hydrochlorothiazide, hydrocodone, hydromorphone, hydroxychloroquine, hydroxyzine, ibuprofen, imipramine, indomethacin, ketoconazole, ketoprofen, ketorolac, labetalol, levorphanol, lidocaine, loratadine, lorazepam, lovastatin, loxapine, mazindol, mefenamic acid, meperidine, mephenytoin, mepivacaine, mesoridazine, metaproterenol, methadone, methdilazine, methocarbamol, methotrexate, methotrimeprazine, methoxamine, methyldopa, methylphenidate, metoclopramide, metolazone, metoprolol, metronidazole, midazolam, moclo-

bemide, morphine, nadolol, nalbuphine, naloxone, naphazoline, naproxen, nifedipine, nizatidine, norepinephrine, nortriptyline, oxazepam, oxycodone, oxymetazoline, paroxetine, pemoline, pentazocine, pentobarbital, pentoxifylline, perphenazine, pheniramine, phenobarbital, phenol, phenolphthalein, phentolamine, phenylbutazone, phenyltoloxamine, phenytoin, pimozide, pindolol, piroxicam, pramoxine, prazepam, prazosin, probenecid, procainamide, procaine, prochlorperazine, procyclidine, promazine, promethazine, propafenone, propantheline, propiomazine, propofol, propranolol, protriptyline, quazepam, quinidine, quinine, racemethorphan, ranitidine, remoxipride, risperidone, salicylic acid, scopolamine, secobarbital, sertraline, sotalol, spironolactone, sulfinpyrazone, sulindac, temazepam, terbutaline, terfenadine, tetracaine, theophylline, thiethylperazine, thiopental, thioridazine, thiothixene, timolol, tocainide, tolbutamide, tolmetin, trazodone, triamterene, triazolam, trifluoperazine, triflupromazine, trimeprazine, trimethoprim, trimipramine, verapamil, warfarin, xylometazoline, yohimbine, zopiclone

## KEY WORDS
some details of plasma extraction

## REFERENCE
Koves, E.M. Use of high-performance liquid chromatography-diode array detection in forensic toxicology. *J.Chromatogr.A*, **1995**, *692*, 103–119

## SAMPLE
**Matrix:** solutions

## HPLC VARIABLES
**Column:** 150 × 3.9 5 µm C8 Symmetry end-capped (prepared in the laboratory from Waters silica)
**Mobile phase:** MeOH:20 mM pH 7.00 potassium phosphate buffer 65:35
**Column temperature:** 23 ± 0.5
**Flow rate:** 1
**Detector:** UV

## CHROMATOGRAM
**Retention time:** 22

## OTHER SUBSTANCES
**Simultaneous:** propranolol

## REFERENCE
O'Gara, J.E.; Alden, B.A.; Walter, T.H.; Petersen, J.S.; Niederländer, C.L.; Neue, U.D. Simple preparation of a $C_8$ HPLC stationary phase with an internal polar functional group. *Anal.Chem.*, **1995**, *67*, 3809–3813

## SAMPLE
**Matrix:** solutions
**Sample preparation:** Prepare a 1-10 µg/mL solution in water, inject an aliquot.

## HPLC VARIABLES
**Column:** 250 × 4.6 5 µm Hypersil SCX/C18
**Mobile phase:** MeCN:25 mM pH 3 $Na_2HPO_4$ 50:50
**Injection volume:** 20
**Detector:** UV 254

## CHROMATOGRAM
**Retention time:** k' 3.79

## OTHER SUBSTANCES
**Also analyzed:** barbital, benzoic acid, butabarbital, clomipramine, clonazepam, desipramine, diazepam, flurazepam, furosemide, imipramine, nitrazepam, phenobarbital, phenol, phenolphthalein, pindolol, propranolol, resorcinol, salicylic acid, secobarbital, terbutaline, xylazine

## KEY WORDS
effect of mobile phase pH on capacity factor is discussed

## REFERENCE
Walshe, M.; Kelly, M.T.; Smyth, M.R.; Ritchie, H. Retention studies on mixed-mode columns in high-performance liquid chromatography. *J.Chromatogr.A*, **1995**, *708*, 31–40

## SAMPLE
**Matrix:** solutions
**Sample preparation:** Prepare a 1 mg/mL solution in MeOH, inject a 5 μL aliquot.

## HPLC VARIABLES
**Column:** 250 × 4.6 5 μm Lichrosphere cyanopropyl
**Mobile phase:** Carbon dioxide:MeOH:isopropylamine 90:10:0.05
**Column temperature:** 50
**Flow rate:** 3
**Injection volume:** 5
**Detector:** UV 220

## CHROMATOGRAM
**Retention time:** 2.75

## OTHER SUBSTANCES
**Simultaneous:** benactyzine, buclizine, hydroxyzine, perphenazine, thioridazine, desipramine, imipramine, nortriptyline, protriptyline

## KEY WORDS
SFC; pressure 200 bar

## REFERENCE
Berger, T.A.; Wilson, W.H. Separation of drugs by packed column supercritical fluid chromatography. 2. Antidepressants. *J.Pharm.Sci.*, **1994**, *83*, 287–290

## SAMPLE
**Matrix:** solutions

## HPLC VARIABLES
**Column:** 250 × 4.6 Zorbax RX
**Mobile phase:** Gradient. A was 10 mL concentrated orthophosphoric acid and 7 mL triethylamine in 1 L water. B was 10 mL concentrated orthophosphoric acid and 7 mL triethylamine in 200 mL water, make up to 1 L with MeCN. A:B from 100:0 to 0:100 over 30 min, maintain at 0:100 for 5 min.
**Column temperature:** 30
**Flow rate:** 2
**Detector:** UV 210

## OTHER SUBSTANCES
**Also analyzed:** acepromazine, acetaminophen, acetophenazine, albuterol, aminophylline, amobarbital, amoxapine, amphetamine, amylocaine, antipyrine, aprobarbital, aspirin,

atenolol, atropine, avermectin, barbital, bebrisoquine, benzocaine, benzoic acid, benzotropine, benzphetamine, berberine, bibucaine, bromazepan, brompheniramine, buprenorphine, buspirone, butabarbital, butacaine, butethal, caffeine, carbamazepine, carbromal, chloramphenicol, chlordiazepoxide, chloroquine, chlorothiazide, chloroxylenol, chlorphenesin, chlorpheniramine, chlorpromazine, chlorpropamide, chlortetracycline, cimetidine, cinchonidine, cinchonine, clenbuterol, clonazepam, clonixin, clorazepate, cocaine, codeine, colchicine, cortisone, coumarin, cyclazocine, cyclobenzaprine, cyclothiazide, cyheptamide, cymarin, danazol, danthron, dapsone, desipramine, dexamethasone, dextromethorphan, dextropropoxyphene, diamorphine, diazepam, diclofenac, diethylpropion, diethylstilbestrol, diflunisal, digitoxin, digoxin, diltiazem, diphenhydramine, diphenoxylate, diprenorphine, dipyrone, disulfiram, dopamine, doxapram, doxepin, dronabinol, ephedrine, epinephrine, epinine, estradiol, estriol, estrone, ethacrynic acid, ethosuximide, etonitazene, etorphine, eugenol, famotidine, fenbendazole, fencamfamine, fenoprofen, fenproporex, fentanyl, flubendazole, flufenamic acid, flunitrazepam, 5-fluorouracil, fluoxymesterone, fluphenazine, furosemide, gentisic acid, gitoxigenin, glipizide, glunixin, glutethimide, glybenclamide, guaiacol, halazepam, haloperidol, hydrochlorothiazide, hydrocodone, hydrocortisone, hydromorphone, hydroxyquinoline, ibogaine, ibuprofen, iminostilbene, imipramine, indomethacin, isocarbostyril, isocarboxazid, isoniazid, isoproterenol, isoxsuprine, ivermectin, ketamine, ketoprofen, kynurenic acid, levorphanol, lidocaine, lorazepam, lormetazepam, loxapine, mazindol, mebendazole, meclizine, meclofenamic acid, medazepam, mefenamic acid, megestrol, mepacrine, meperidine, mephentermine, mephenytoin, mephesin, mephobarbital, mepivacaine, mescaline, mesoridazine, methadone, methamphetamine, methapyrilene, methaqualone, methazolamide, methocarbamol, methoxamine, methsuximide, methyl salicylate, methyldopa, methyldopamine, methylphenidate, methylprednisolone, methyltestosterone, methyprylon, metoprolol, mibolerone, morphine, nadolol, nalorphine, naloxone, naltrexone, naphazoline, naproxen, nefopam, niacinamide, nicotine, nicotinic acid, nifedipine, niflumic acid, nitrazepam, norepinephrine, nortriptyline, noscapine, nylidrin, oxazepam, oxycodone, oxymorphone, oxyphenbutazone, oxytetracycline, papaverine, pargyline, pemoline, pentazocine, pentobarbital, persantine, phenacetin, phenazocine, phenazopyridine, phencyclidine, phendimetrazine, phenelzine, pheniramine, phenobarbital, phenothiazine, phensuximide, phentermine, phenylbutazone, phenylephrine, phenylpropanolamine, piperocaine, prazepam, prednisolone, primidone, probenecid, progesterone, propiomazine, propranolol, propylparaben, pseudoephedrine, puromycin, pyrilamine, pyrithyldione, quazepam, quinaldic acid, quinidine, quinine, ranitidine, recinnamine, reserpine, resorcinol, saccharin, albuterol, salicylamide, salicylic acid, scopolamine, scopoletin, secobarbital, strychnine, sulfacetamide, sufadiazine, sulfadimethoxine, sulfaethidole, sulfamerazine, sulfamethazine, sulfamethoxizole, sulfanilamide, sulfapyridine, sulfasoxizole, sulindac, tamoxifen, temazepam, testosterone, tetracaine, tetracycline, tetramisole, thebaine, theobromine, theophylline, thiabendazole, thiamine, thiamylal, thiobarbituric acid, thioridazine, thiosalicylic acid, thiothixene, thymol, tolazamide, tolazoline, tobutamide, tolmetin, tranylcypromine, triamcinolone, tribenzylamine, trichloromethiazide, trifluoperazine, trihexyphenidyl, trimethoprim, tripelennamine, triproilidine, tropacocaine, tyramine, verapamil, vincamine, warfarin, yohimbine, zoxazolamine

## REFERENCE

Hill, D.W.; Kind, A.J. Reversed-phase solvent gradient HPLC retention indexes of drugs. *J.Anal.Toxicol.*, **1994**, *18*, 233–242

## SAMPLE
**Matrix:** solutions

## HPLC VARIABLES
**Column:** 150 × 4.6 5 μm Adsorbosphere C18 (PEEK column) (retention times are longer and peaks broader with stainless steel column)
**Mobile phase:** MeCN:20 mM pH 3.2 $KH_2PO_4$ 23.4:76.6 containing 0.05% nonylamine
**Flow rate:** 1.2
**Detector:** UV 214

## CHROMATOGRAM
**Retention time:** 16.5

## OTHER SUBSTANCES
**Simultaneous:** desipramine, desmethyldoxepin, doxepin, imipramine, loxapine, maprotiline, nortriptyline, trazodone

## REFERENCE
*Alltech Chromatography Catalog 300,* **1993,** p. 440

## SAMPLE
**Matrix:** solutions

## HPLC VARIABLES
**Column:** 250 × 4.6 Econosil C8
**Mobile phase:** MeCN:buffer 30:70 (Buffer was 20 mM $KH_2PO_4$ and 14 mM triethylamine adjusted to pH 3.0 with phosphoric acid.)
**Injection volume:** 20
**Detector:** UV 210

## CHROMATOGRAM
**Retention time:** 11.0
**Limit of quantitation:** < 1000 ng/mL

## OTHER SUBSTANCES
**Simultaneous:** amoxapine, carbamazepine, imipramine, nortriptyline
**Also analyzed:** doxepin, desipramine, protriptyline, cyclobenzaprine, maprotiline

## KEY WORDS
UV spectra given

## REFERENCE
Ryan, T.W. Identification and quantification of tricyclic antidepressants by UV-photodiode array detection with multicomponent analysis. *J.Liq.Chromatogr.,* **1993,** *16,* 1545–1560

## SAMPLE
**Matrix:** solutions
**Sample preparation:** Dissolve in MeOH:water 1:1 at a concentration of 50 μg/mL, inject a 10 μL aliquot.

## HPLC VARIABLES
**Column:** 300 × 3.9 10 μm μBondapak C18
**Mobile phase:** MeOH:acetic acid:triethylamine:water 60:1.5:0.5:38
**Flow rate:** 1.5
**Injection volume:** 10
**Detector:** UV

## CHROMATOGRAM
**Retention time:** k′ 2.54

## REFERENCE
Roos, R.W.; Lau-Cam, C.A. General reversed-phase high-performance liquid chromatographic method for the separation of drugs using triethylamine as a competing base. *J.Chromatogr.,* **1986,** *370,* 403–418

## SAMPLE
**Matrix:** solutions
**Sample preparation:** Prepare a 10 μg/mL solution in MeOH, inject a 20 μL aliquot.

## HPLC VARIABLES
**Column:** 125 × 4.9 Spherisorb S5W silica
**Mobile phase:** MeOH containing 10 mM ammonium perchlorate and 1 mL/L 100 mM NaOH in MeOH, pH 6.7
**Flow rate:** 2
**Injection volume:** 20
**Detector:** E, LeCarbone, V25 glassy carbon electrode, + 1.2 V

## CHROMATOGRAM
**Retention time:** 3.74

## OTHER SUBSTANCES
**Also analyzed:** acebutolol, acepromazine, acetophenazine, N-acetylprocainamide, albuterol, alprenolol, amethocaine, amiodarone, amitriptyline, antazoline, atenolol, azacyclonal, bamethane, benactyzine, benperidol, benzethidine, benzocaine, benzoctamine, benzphetamine, benzquinamide, bromhexine, bromodiphenhydramine, bromperidol, brompheniramine, brompromazine, buclizine, bufotenine, bupivacaine, buprenorphine, butacaine, butethamate, chlorcyclizine, chlorpheniramine, chlorphenoxamine, chlorprenaline, chlorpromazine, chlorprothixene, cimetidine, cinchonidine, cinnarizine, clemastine, clomipramine, clonidine, cocaine, cyclazocine, cyclizine, cyclopentamine, cyproheptadine, deserpidine, desipramine, dextromoramide, dextropropoxyphene, dicyclomine, diethylcarbamazine, diethylpropion, diethylthiambutene, dihydroergotamine, dimethindene, dimethothiazine, diphenhydramine, diphenoxylate, dipipanone, diprenorphine, dipyridamole, disopyramide, dothiepin, doxapram, doxepin, doxylamine, droperidol, ephedrine, ergocornine, ergocristine, ergocristinine, ergocryptine, ergometrine, ergosine, ergosinine, ergotamine, ethopropazine, etorphine, etoxeridine, fenethazine, fenfluramine, fenoterol, fentanyl, flavoxate, fluopromazine, flupenthixol, fluphenazine, flurazepam, haloperidol, hydroxyzine, hyoscine, ibogaine, imipramine, indapamine, iprindole, isothipendyl, isoxsuprine, ketanserin, laudanosine, lidocaine, lofepramine, loxapine, maprotiline, mecamylamine, meclophenoxate, meclozine, medazepam, mephentermine, mepivacaine, meptazinol, mepyramine, mesoridazine, metaraminol, methadone, methamphetamine, methapyrilene, methdilazene, methotrimeprazine, methoxamine, methoxyphenamine, methoxypromazine, methylephedrine, methylergonovine, methysergide, metoclopramide, metopimazine, metoprolol, mianserin, morazone, nadolol, nalorphine, naloxone, naphazoline, nicotine, nifedipine, nomifensine, nortriptyline, noscapine, orphenadrine, oxeladin, oxprenolol, oxymetazolin, papaverine, pargyline, pecazine, penbutolol, pentazocine, penthienate, pericyazine, perphenazine, phenadoxone, phenampromide, phenazocine, phenbutrazate, phendimetrazine, phenelzine, phenglutarimide, phenindamine, pheniramine, phenmetrazine, phenomorphan, phenoperidine, phenothiazine, phenoxybenzamine, phentolamine, phenylephrine, phenyltoloxamine, physostigmine, piminodine, pimozide, pindolol, pipamazine, pipazethate, piperacetazine, piperidolate, pipradol, pirenzepine, piritramide, pizotifen, practolol, pramoxine, prazosin, prenylamine, prilocaine, primaquine, proadifen, procainamide, procaine, prochlorperazine, procyclidine, proheptazine, prolintane, promazine, promethazine, pronethalol, properidine, propiomazine, propranolol, prothipendyl, protriptyline, proxymetacaine, pseudoephedrine, pyrimethamine, quinidine, quinine, ranitidine, rescinnamine, sotalol, tacrine, terazosin, terbutaline, terfenadine, thenyldiamine, theophylline, thiethylperazine, thiopropazate, thioproperazine, thioridazine, thiothixene, thonzylamine, timolol, tocainide, tolpropamine, tolycaine, tranylcypromine, trazodone, trifluoperazine, trifluperidol, trimeperidine, trimeprazine, trimethobenzamide, trimethoprim, trimipramine, tripelennamine, triprolidine, tryptamine, verapamil, xylometazoline

## REFERENCE

Jane, I.; McKinnon, A.; Flanagan, R.J. High-performance liquid chromatographic analysis of basic drugs on silica columns using non-aqueous ionic eluents. II. Application of UV, fluorescence and electro-chemical oxidation detection. *J.Chromatogr.*, **1985**, *323*, 191–225

## SAMPLE

**Matrix:** urine
**Sample preparation:** 1 mL Urine + 0.5 mL 1% trichloroacetic acid, centrifuge at 5200 g for 10 min, filter (0.2 μm), inject 20 μL aliquot.

## HPLC VARIABLES

**Column:** 250 × 4 5 μm Lichrospher 60 RP-select B
**Mobile phase:** Gradient. MeCN:50 mM pH 3.2 potassium phosphate buffer from 10:90 to 75:25 over 7 min, hold at 75:25 for 3 min, return to 10:90 over 5 min, equilibrate at 10:90 for 5 min
**Flow rate:** 1.5
**Injection volume:** 20
**Detector:** UV 190-370

## CHROMATOGRAM

**Retention time:** 7.7

## OTHER SUBSTANCES

**Extracted:** amphetamine, benzoylecgonine, cocaine (different gradient), codeine, diphen-hydramine (different gradient), ephedrine (different gradient), lidocaine (different gradient), meperidine, morphine, nordiazepam, phenylpropanolamine (different gradient), nor-propoxyphene, nortriptyline (different gradient)

## REFERENCE

Li, S.; Gemperline, P.J.; Briley, K.; Kazmierczak, S. Identification and quantitation of drugs of abuse in urine using the generalized rank annihilation method of curve resolution. *J.Chromatogr.B*, **1994**, *655*, 213–223

## SAMPLE

**Matrix:** urine
**Sample preparation:** 500 μL Urine + N-ethylnordiazepam + chlorpheniramine + 100 μL buffer, centrifuge at 11000 g for 30 s, inject a 500 μL aliquot onto column A with mobile phase A, after 0.6 min backflush column A with mobile phase A to waste for 1.6 min, elute column A with 250 μL mobile phase B, with 200 μL mobile phase C, and with 1.15 mL mobile phase D. Elute column A to waste until drugs start to emerge then elute onto column B. Elute column B to waste until drugs started to emerge, then elute onto column C. When all the drugs have emerged from column B remove it from the circuit, elute column C with mobile phase D, monitor the effluent from column C. Flush column A with 7 mL mobile phase E, with mobile phase D, and mobile phase A. Flush column B with 5 mL mobile phase E then with mobile phase D. (Buffer was 6 M ammonium acetate adjusted to pH 8.0 with 2 M KOH.)

## HPLC VARIABLES

**Column:** A 10 × 2.1 12-20 μm PRP-1 spherical poly(styrene-divinylbenzene) (Hamilton); B 10 × 3.2 11 μm Aminex A-28 (Bio-Rad); C 25 × 3.2 5 μm C8 (Phenomenex) + 150 × 4.6 5 μm silica (Macherey-Nagel)
**Mobile phase:** A 0.1% pH 8.0 potassium borate buffer; B 6 mM $KH_2PO_4$ containing 5 mM tetramethylammonium hydroxide, and 2 mM dimethyloctylamine, pH adjusted to 6.50 with phosphoric acid; C MeCN:buffer 40:60 (Buffer was 6 mM $KH_2PO_4$ containing 5 mM tetramethylammonium hydroxide, and 2 mM dimethyloctylamine, pH adjusted to 6.50 with phosphoric acid.); D MeCN:buffer 33:67 (Buffer was 6 mM $KH_2PO_4$ containing 5

mM tetramethylammonium hydroxide, and 2 mM dimethyloctylamine, pH adjusted to 6.50 with phosphoric acid.); E MeCN:buffer 70:30 (Buffer was 6 mM $KH_2PO_4$ containing 5 mM tetramethylammonium hydroxide, and 2 mM dimethyloctylamine, pH adjusted to 6.50 with phosphoric acid.)

**Column temperature:** 40 (B, C only)
**Flow rate:** A 5; B-E 1
**Injection volume:** 500
**Detector:** UV 210; UV 235

## CHROMATOGRAM
**Retention time:** k′ 4.3
**Internal standard:** N-ethylnordiazepam (k′ 2.1), chlorpheniramine (k′ 5.9)
**Limit of detection:** 300 ng/mL

## OTHER SUBSTANCES
**Extracted:** amphetamine, benzoylecgonine, caffeine, codeine, cotinine, desipramine, diazepam, diphenhydramine, ephedrine, hydrocodone, hydromorphone, lidocaine, methadone, methamphetamine, morphine, nordiazepam, nortriptyline, oxazepam, pentazocine, phenmetrazine, phenobarbital, phentermine, phenylpropanolamine, secobarbital
**Interfering:** flurazepam, imipramine

## KEY WORDS
column-switching

## REFERENCE
Binder, S.R.; Regalia, M.; Biaggi-McEachern, M.; Mazhar, M. Automated liquid chromatographic analysis of drugs in urine by on-line sample cleanup and isocratic multi-column separation. *J.Chromatogr.*, **1989**, *473*, 325–341

## SAMPLE
**Matrix:** vitreous humor
**Sample preparation:** 600 µL Vitreous humor + 3 mL 0.1 M NaCl + 50 µL 4 µg/mL desmethylclomipramine in water, mix for a few s, add to a C18 SepPak SPE cartridge attached to a 5 mL syringe, allow to flow through (10-15 min). Wash with 1 mL 0.1 M NaCl, wash with 1 mL water, wash with 3 mL reagent by gravity. Elute with 3 mL MeOH and push air through to remove as much as possible. Evaporate under vacuum at 37°, vortex with 50 µL mobile phase for 1 min, inject a 25 µL aliquot. (Reagent was isopropanol:n-heptane:1 M sulfuric acid 40:320:1.)

## HPLC VARIABLES
**Guard column:** 50 × 4.6 30 µm Permaphase ETH (Du Pont)
**Column:** 250 × 4.6 5-6 µm Zorbax cyanopropyl
**Mobile phase:** MeCN:0.5 M acetic acid:n-butylamine 40:60:0.0022
**Flow rate:** 2.5
**Injection volume:** 25
**Detector:** UV 254

## CHROMATOGRAM
**Retention time:** 23
**Internal standard:** desmethylclomipramine (26)
**Limit of detection:** 16.7 ng/mL

## OTHER SUBSTANCES
**Simultaneous:** doxepin, imipramine, nortriptyline
**Noninterfering:** acetaminophen, N-acetylprocainamide, amikacin, caffeine, carbamazepine, chloramphenicol, clonazepam, cyclosporine, diazepam, digoxin, disopyramide, ethosuximide, flurazepam, gentamicin, haloperidol, kanamycin, lidocaine, meprobamate,

methapyriline, methaqualone, methotrexate, methyprylon, netilmicin, pentazocine, pentobarbital, phenobarbital, phenytoin, prazepam, primidone, procainamide, propranolol, quinidine, salicylic acid, secobarbital, streptomycin, theophylline, tobramycin, tocainide, valproic acid, vancomycin

## KEY WORDS
SPE

## REFERENCE
Evenson, M.A.; Engstrand, D.A. A SepPak HPLC method for tricyclic antidepressant drugs in human vitreous humor. *J.Anal.Toxicol.*, **1989**, *13*, 322–325

## ANNOTATED BIBLIOGRAPHY

McGuire, M.; Wong, S.H.; Skrinska, V.; Fogelman, K.; Miles, W. Totally automated sequential analysis of tricyclic antidepressants by SPE module system and reversed-phase HPLC analysis with a base-deactivated C18 column. *J.Anal.Toxicol.*, **1996**, *20*, 65 [SPE]

Atta-Politou, J.; Tsarpalis, K.; Koutselinis, A. A modified simple and rapid reversed phase high performance liquid chromatographic method for quantification of amitriptyline and nortriptyline in plasma. *J.Liq.Chromatogr.*, **1994**, *17*, 3969–3982 [plasma; extracted metabolites, clomipramine, doxepin, nortriptyline, protriptyline; LOD 5 ng/mL; LOQ 10 ng/mL; simultaneous chlordiazepoxide, chlorpheniramine, chlorpromazine, phenobarbital, propranolol; non-interfering alprazolam, artane, bromazepam, diazepam, haloperidol, lorazepam, nitrazepam, oxazepam, pseudoephedrine, theophylline, triazolam; interfering propoxyphene]

Joron, S.; Robert, H. Simultaneous determination of antidepressant drugs and metabolites by HPLC. Design and validation of a simple and reliable analytical procedure. *Biomed.Chromatogr.*, **1994**, *8*, 158–164 [LOQ 3-17 ng/mL; plasma; also amitriptyline, amineptine, amoxapine, clomipramine, demexiptiline, desipramine, dosulepine, doxepine, doxepin, fluoxetine, fluvoxamine, imipramine, maprotiline, medifloxamine, mianserine, opipramol, quinupramine, tianeptine, toloxatone, trazodone, trimipramine, viloxazine]

Kirkland, J.J. Trends in HPLC column design for improved method development. *Am.Lab.*, **1994**, *26(9)*, 28K–28R [simultaneous desipramine, doxepin, trimipramine]

Kirkland, J.J.; Boyes, B.E.; DeStefano, J.J. Changing band spacing in reversed-phase HPLC. A technique for varying column stationary-phase selectivity. *Am.Lab.*, **1994**, *26(14) (Sept.)*, 36–41 [simultaneous desipramine, doxepin, trimipramine]

Oshima, N.; Kotaki, H.; Sawada, Y.; Iga, T. Tissue distribution of amitriptyline after repeated administration in rats. *Drug Metab.Dispos.*, **1994**, *22*, 21–25 [rat; plasma; liver; kidney; lung; brain; muscle; heart; extracted nortriptyline; clomipramine is IS; column temp 35; LOQ 10 ng/mL; pharmacokinetics]

Piperaki, S.; Parissi-Poulou, M.; Koupparis, M. A separation study of tricyclic antidepressant drugs by HPLC with β-cyclodextrin bonded stationary phase. *J.Liq.Chromatogr.*, **1993**, *16*, 3487–3508 [simultaneous chloripramine, doxepin, imipramine, maprotiline, nortriptyline, protriptyline]

Bogusz, M.; Erkens, M.; Maier, R.D.; Schröder, I. Applicability of reversed-phase base-deactivated columns for systematic toxicological analysis. *J.Liq.Chromatogr.*, **1992**, *15*, 127–150 [simultaneous brallobarbital, diphenhydramine, fluphenazine, imipramine, pentobarbital, salicylamide, secobarbital, thiopental, thioridazine]

Coudore, F.; Ardid, D.; Eschalier, A.; Lavarenne, J.; Fialip, J. High-performance liquid chromatographic determination of amitriptyline and its main metabolites using a silica column with reversed-phase eluent. Application in mice. *J.Chromatogr.*, **1992**, *584*, 249–255 [mouse+asma;brain;simultaneous metabolites, nortriptyline;impiramine (IS)]

Härtter, S.; Wetzel, H.; Hiemke, C. Automated determination of fluvoxamine in plasma by column-switching high-performance liquid chromatography. *Clin.Chem.*, **1992**, *38*, 2082–2086 [column-switching; not validated for extraction from biological fluids; interfering clomipramine, doxepin, fluoxetine; simultaneous clozapine, demethyldoxepin, desipramine, desmethylclomipramine, desmethylmaprotiline, fluvoxamine, imipramine, maprotiline, norfluoxetine, nortriptyline, oxaprotiline]

Jalal, I.M.; Sa'sa', S.I.; Khalil, H.S. Simultaneous high performance liquid chromatographic determination of amitriptyline hydrochloride and perphenazine in tablet formulations. *J.Liq.Chromatogr.*, **1988**, *11*, 1531–1544 [tablets;simultaneous perphenazine;naphthalene (IS)]

Liang, M.Z.; Huang, Y.; Qin, Y.P.; Zeng, J.Z. [Reversed phase HPLC determination of amitriptyline and nortriptyline in plasma]. *Hua.Hsi.I.Ko.Ta.Hsueh.Hsueh.Pao.*, **1987**, *18*, 144–147

Kiel, J.S.; Abramson, R.K.; Smith, C.S.; Morgan, S.L. Development of a rapid extraction and high-performance liquid chromatographic separation for amitriptyline and six biological metabolites. *J.Chromatogr.*, **1986**, *383*, 119–127

Smith, C.S.; Abramson, R.K.; Morgan, S.L. An investigation of the metabolism of amitriptyline using high performance liquid chromatography. *J.Liq.Chromatogr.*, **1986**, *9*, 1727–1745 [rat; metabolites; tissue; liver]

Terlinden, R.; Borbe, H.O. Determination of amitriptylinoxide and its major metabolites amitriptyline and nortriptyline in plasma by high-performance liquid chromatography. *J.Chromatogr.*, **1986**, *382*, 372–376 [plasma; column temp 45; extracted amitriptylinoxide; desipramine (IS); LOQ 10 ng/mL; pharmacokinetics; dog]

Wong, S.H.Y.; McHugh, S.L.; Dolan, J.; Cohen, K.A. Tricyclic antidepressant analysis by reversed-phase liquid chromatography using phenyl columns. *J.Liq.Chromatogr.*, **1986**, *9*, 2511–2538 [also acetaminophen, amobarbital, amoxapine, barbital, chlordiazepoxide, chlorpromazine, cimetidine, clomipramine, codeine, desipramine, desmethyldoxepin, diazepam, doxepin, fluphenazine, flurazepam, glutethimide, hydroxyamoxapine, imipramine, lorazepam, maprotiline, meperidine, metabolites, nortriptyline, oxazepam, pentobarbital, perphenazine, phenobarbital, phenytoin, propoxyphene, protriptyline, secobarbital, thioridazine, trazodone]

Rop, P.P.; Viala, A.; Durand, A.; Conquy, T. Determination of citalopram, amitriptyline and clomipramine in plasma by reversed-phase high-performance liquid chromatography. *J.Chromatogr.*, **1985**, *338*, 171–178

Walker, S.T. Rapid high pressure liquid chromatographic determination of amitriptyline hydrochloride in tablets and injectables: collaborative study. *J.Assoc.Off.Anal.Chem.*, **1983**, *66*, 1196–1202

Preskorn, S.H.; Glotzbach, R.K. A liquid chromatographic method for quantitating amitriptyline in brain tissue. *Psychopharmacology (Berl)*, **1982**, *78*, 23–24

Smith, G.A.; Schulz, P.; Giacomini, K.M.; Blaschke, T.F. High-pressure liquid chromatographic determination of amitriptyline and its major metabolites in human whole blood. *J.Pharm.Sci.*, **1982**, *71*, 581–583

Suckow, R.F.; Cooper, T.B. Simultaneous determination of amitriptyline, nortriptyline and their respective isomeric 10-hydroxy metabolites in plasma by liquid chromatography. *J.Chromatogr.*, **1982**, *230*, 391–400

Kabra, P.M.; Mar, N.A.; Marton, L.J. Simultaneous liquid chromatographic analysis of amitriptyline, nortriptyline, imipramine, desipramine, doxepin, and nordoxepin. *Clin.Chim.Acta*, **1981**, *111*, 123–132

Burke, D.; Sokoloff, H. Simultaneous high-performance liquid chromatographic determination of chlordiazepoxide and amitriptyline hydrochloride in two-component tablet formulations. *J.Pharm.Sci.*, **1980**, *69*, 138–140

Jensen, K.M. Determination of amitriptyline-N-oxide, amitriptyline and nortriptyline in serum and plasma by high-performance liquid chromatography. *J.Chromatogr.*, **1980**, *183*, 321–329

Preskorn, S.H.; Leonard, K.; Hignite, C. Liquid chromatography of amitriptyline and related tricyclic compounds. *J.Chromatogr.*, **1980**, *197*, 246–250

# Amlodipine

**Molecular formula:** $C_{20}H_{25}ClN_2O_5$
**Molecular weight:** 408.9
**CAS Registry No.:** 88150-42-9 (amlodipine), 111470-99-6
(amlodipine besylate), 88150-47-4 (amlodipine maleate)

## SAMPLE
**Matrix:** blood
**Sample preparation:** Condition a 3 mL 100 mg Bond Elut C2 SPE cartridge with 2 mL MeCN, 1 mL water, and 1 mL buffer. Add 500 μL buffer, 50 μL 200 ng/mL IS in MeOH:water 50:50, 1 mL plasma, and 500 μL buffer sequentially to the SPE cartridge. Wash with 2 mL MeCN:water 20:80, wash with 1 mL MeCN, elute with 1 mL 2.5% ammonia in MeCN. Evaporate the eluate to dryness under reduced pressure, reconstitute with 50 μL MeOH:100 mM pH 4.0 acetate buffer 50:50, inject an aliquot. (Buffer was 25 mM pH 7.0 phosphate buffer.)

## HPLC VARIABLES
**Column:** 150 × 2.1 Zorbax SB-Phenyl
**Mobile phase:** MeOH:100 mM pH 4.0 acetate buffer containing 2 mM sodium dodecyl sulfate and 1 mg/L EDTA
**Column temperature:** 30
**Flow rate:** 0.3
**Detector:** E, Antec Decade, Antec VT-03 analytical cell with 50 μm spacer 0.95 V, ESA 5020 guard cell 0.5 V

## CHROMATOGRAM
**Retention time:** 9
**Internal standard:** 2-[(2-aminoethoxy)methyl]-4-(2,3-dichlorophenyl)-1,4-dihydro-6-methyl-3,5-pyridinedicarboxylic acid 3-ethyl 5-methyl ester (UK52.829) (13)
**Limit of quantitation:** 0.2 ng/mL

## KEY WORDS
SPE; narrow-bore; plasma

## REFERENCE
Josefsson, M.; Zackrisson, A.-L.; Norlander, B. Sensitive high-performance liquid chromatographic analysis of amlodipine in human plasma with amperometric detection and a single-step solid-phase sample preparation. *J.Chromatogr.B*, **1995**, *672*, 310−313

## SAMPLE
**Matrix:** blood
**Sample preparation:** 1 mL Plasma + 100 μL 2 μg/mL desipramine in 10 mM HCl + 200 μL 10% ammonium carbonate (final pH 8.7), vortex, extract with 5 mL MTBE for 20 min (Vibrax VXR2), centrifuge at 4° at 1720 g for 10 min. Remove the organic layer and evaporate it to dryness under a stream of nitrogen at 55°, reconstitute the residue in 100 μL 10 mM HCl. Add 2 mL MTBE, vortex for 1 min, remove the aqueous layer and dry at 68° under high vacuum, reconstitute the residue in 100 μL mobile phase, inject a 20-40 μL aliquot.

## HPLC VARIABLES
**Guard column:** 45 × 4.6 5 μm Ultrasphere-ODS
**Column:** 250 × 4.6 5 μm Ultrasphere-ODS

**Mobile phase:** MeOH:MeCN:40 mM ammonium acetate 38:24:38 containing 0.02% triethylamine, final pH adjusted to 7.1 with glacial acetic acid
**Flow rate:** 1.2
**Injection volume:** 20-40
**Detector:** UV 240

## CHROMATOGRAM
**Retention time:** 10.6
**Internal standard:** desipramine (12.9)
**Limit of quantitation:** 2.5 ng/mL

## KEY WORDS
plasma; rabbit; pharmacokinetics

## REFERENCE
Yeung, P.K.F.; Mosher, S.J.; Pollak, P.T. Liquid chromatography assay for amlodipine: chemical stability and pharmacokinetics in rabbits. *J.Pharm.Biomed.Anal.*, **1991**, *9*, 565–571

## SAMPLE
**Matrix:** blood
**Sample preparation:** Add IS to plasma, add pH 9 borate buffer, extract with diethyl ether. Remove the organic layer and add it to 100 µL 100 mM citric acid, extract, inject an 80 µL aliquot of the aqueous layer.

## HPLC VARIABLES
**Column:** 250 × 4.9 5 µm Spherisorb nitrile (dog) or 125 × 4.9 5 µm Spherisorb ODS (rat)
**Mobile phase:** MeOH:buffer 35:65 (dog) or 45:65 (rat) (Buffer was 100 mM pH 5 N, N, N', N'-tetramethylethylenediamine phosphate.)
**Flow rate:** 1
**Injection volume:** 80
**Detector:** F ex 230 em 370

## CHROMATOGRAM
**Internal standard:** 2-[(2-dimethylaminoethoxy)methyl]-4-(2-chlorophenyl)-3-ethoxycarbonyl-6-methyl-1,4-dihydropyridine

## KEY WORDS
plasma; dog; rat; pharmacokinetics

## REFERENCE
Stopher, D.A.; Beresford, A.P.; Macrae, P.V.; Humphrey, M.J. The metabolism and pharmacokinetics of amlodipine in humans and animals. *J.Cardiovasc.Pharmacol.*, **1988**, *12 Suppl 7*, S55–S59

## SAMPLE
**Matrix:** microsomal incubations, perfusate
**Sample preparation:** Basify 500 µL perfusate or 1 mL microsomal incubation with 1 mL 200 mM pH 9.0 sodium borate buffer, extract into MTBE, evaporate extract to dryness under a stream of nitrogen, reconstitute the residue in 100 µL mobile phase, inject an aliquot.

## HPLC VARIABLES
**Column:** 250 × 5 Hichrom HiRPB deactivated reverse-phase
**Mobile phase:** MeOH:25 mM pH 5.0 N, N, N', N'-tetramethylethylenediamine phosphate buffer 50:50
**Flow rate:** 1
**Detector:** UV 245

## CHROMATOGRAM
**Internal standard:** UK 46,129
**Limit of detection:** 10 ng/mL

## KEY WORDS
rat; liver

## REFERENCE
Walker, D.K.; Humphrey, M.J.; Smith, D.A. Importance of metabolic stability and hepatic distribution to the pharmacokinetic profile of amlodipine. *Xenobiotica*, **1994**, *24*, 243–250

## SAMPLE
**Matrix:** solutions
**Sample preparation:** Inject a 20 μL aliquot of a 100 μg/mL solution in MeOH or MeCN.

## HPLC VARIABLES
**Column:** 100 × 4.6 PGC Hypercarb-S
**Mobile phase:** MeOH:dichloromethane 75:25 containing 5 mM (1S)-(+)-10-camphorsulfonic acid
**Column temperature:** 30
**Flow rate:** 1.5
**Injection volume:** 20
**Detector:** UV 250

## CHROMATOGRAM
**Retention time:** 11 (S-(−)), 12 (R-(+))

## OTHER SUBSTANCES
**Also analyzed:** UK52.829

## KEY WORDS
recirculate mobile phase for at least 12 h before use; chiral

## REFERENCE
Josefsson, M.; Carlsson, M.; Norlander, B. Chiral ion-pair chromatographic separation of two dihydropyridines with camphorsulfonic acids on porous graphitic carbon. *J.Chromatogr.A*, **1994**, *684*, 23–27

## SAMPLE
**Matrix:** solutions
**Sample preparation:** 100 μL 0.1-1 mg/mL Amlodipine in MeCN:water 50:50 + 100 μL 1 M pH 6.8 sodium borate buffer, vortex, add 50 μL 100 mM (−)-(1R)-menthyl chloroformate in acetone, vortex, let stand for 5 min, add 1 mL water, add 2 mL dichloromethane, rotate for 10 min, centrifuge at 2000 rpm for 5 min. Remove the organic layer and evaporate it to dryness under a stream of nitrogen at 40°, reconstitute the residue in 200 μL MeCN, inject a 20 μL aliquot.

## HPLC VARIABLES
**Column:** 100 × 10 Hypercarb-S porous graphitic carbon
**Mobile phase:** MeCN:dichloromethane:formic acid 20:50:30
**Flow rate:** 1.5
**Injection volume:** 20
**Detector:** UV 265

## CHROMATOGRAM
**Retention time:** 6.5, 7.5 (enantiomers)

## OTHER SUBSTANCES
**Simultaneous:** mexiletine, UK52.829
**Interfering:** propranolol

## KEY WORDS
derivatization; chiral

## REFERENCE
Josefsson, M.; Carlsson, B.; Norlander, B. Fast chromatographic separation of (−)-menthyl chloroformate derivatives of some chiral drugs, with special reference to amlodipine, on porous graphitic carbon. *Chromatographia*, **1993**, *37*, 129–132

## SAMPLE
**Matrix:** urine
**Sample preparation:** Inject 1 mL urine.

## HPLC VARIABLES
**Column:** two 125 × 4.8 columns of Spherisorb 5 ODS in series
**Mobile phase:** Gradient. A was 100 mM ammonium acetate in water. B was 100 mM ammonium acetate in MeOH. A:B 100:0 for 15 min, to 90:10 (step gradient), to 10:90 over 1 h.
**Flow rate:** 1 for 15 min, then 0.8
**Injection volume:** 1000
**Detector:** MS, VG 12-250 quadrupole, thermospray. For the first 15 min the column effluent was diverted from the detector and the make-up flow to the detector was 1 mL/min. After 15 min the column effluent mixed with the make-up solvent pumped at 0.2 mL/min and the mixture flowed to the detector. The make-up solvent was 100 mM ammonium acetate in MeOH:water 50:50.

## KEY WORDS
rat

## REFERENCE
Beresford, A.P.; Macrae, P.V.; Alker, D.; Kobylecki, R.J. Biotransformation of amlodipine. Identification and synthesis of metabolites found in rat, dog and human urine/confirmation of structures by gas chromatography-mass spectrometry and liquid chromatography-mass spectrometry. *Arzneimittelforschung*, **1989**, *39*, 201–209

## ANNOTATED BIBLIOGRAPHY
Shimooka, K.; Sawada, Y.; Tatematsu, H. Analysis of amlodipine in serum by a sensitive high-performance liquid chromatographic method with amperometric detection. *J.Pharm.Biomed.Anal.*, **1989**, *7*, 1267–1272

# Amoxicillin

**Molecular formula:** $C_{16}H_{19}N_3O_5S$
**Molecular weight:** 365.4
**CAS Registry No.:** 26787-78-0 (anhydrous), 61336-70-7 (trihydrate)

## SAMPLE
**Matrix:** blood
**Sample preparation:** 100 μL Serum + 100 μL 10 M urea, filter (Amicon MPS-1 with YMT membrane) while centrifuging at 1500 g for 30 min. 80 μL Ultrafiltrate + 80 μL 100 mM pH 9.0 borate buffer + 8 μL 200 mM acetic anhydride in MeCN, let stand for 3 min, add 160 μL reagent, heat at 60° for 10 min, inject a 20 μL aliquot. (Reagent was 13.81 g 1,2,4-triazole in 60 mL water + 10 mL HgCl₂ solution (0.27 g HgCl₂ in 100 mL water), adjust pH to 9.0 ± 0.05 with 4 M NaOH, dilute to 100 mL (Analyst 1985, 110, 1277).)

## HPLC VARIABLES
**Guard column:** used but not specified
**Column:** 100 × 3 5 μm Chromospher Spherisorb ODS-2
**Mobile phase:** MeCN:buffer 15:85 (Buffer was 50 mM pH 4.6 sodium phosphate containing 10 mM thiosulfate.)
**Flow rate:** 0.8
**Injection volume:** 20
**Detector:** UV 328

## CHROMATOGRAM
**Limit of detection:** 1000 ng/mL

## KEY WORDS
derivatization; pharmacokinetics; ultrafiltrate; serum

## REFERENCE
Huisman-de Boer, J.J.; van den Anker, J.N.; Vogel, M.; Goessens, W.H.F.; Schoemaker, R.C.; de Groot, R. Amoxicillin pharmacokinetics in preterm infants with gestational ages of less than 32 weeks. *Antimicrob.Agents Chemother.*, **1995**, *39*, 431–434

## SAMPLE
**Matrix:** blood
**Sample preparation:** Condition an AASP C8 SPE cartridge (Varian) by washing with two 1 mL portions of MeCN and two 1 mL volumes of 67 mM pH 7.0 buffer. 150 μL Plasma + 1.5 mL 67 mM pH 7.0 phosphate buffer + 20 μL water, add a 1 mL aliquot to the SPE cartridge, wash with 200 μL 67 mM pH 7.0 phosphate buffer, wash three times with 200 mg/L sodium azide in 67 mM pH 7.6 buffer. Place SPE cartridge on-line so that it is eluted onto the analytical column.

## HPLC VARIABLES
**Guard column:** 4 × 6 10 μm μBondapak C18 Guard-Pak
**Column:** 250 × 4.6 5 μm Chromspher C18
**Mobile phase:** MeOH:buffer 20:80 (Buffer was 14.2 g NaH₂PO₄ in 900 mL water, add 12.5 mL 1 M tetrabutylammonium dihydrogenphosphate, adjust pH to 7.60 with dilute NaOH, make up to 1 L.) (100 × 3 Chrompack reversed-phase saturation column between pump and injector used to saturate mobile phase with stationary phase.)
**Column temperature:** 30

**Flow rate:** 1
**Detector:** UV 234

## CHROMATOGRAM
**Retention time:** 11.8
**Limit of detection:** 25 ng/mL

## KEY WORDS
pharmacokinetics; plasma; SPE

## REFERENCE
Krauwinkel, W.J.; Volkers-Kamermans, N.J.; van Zijtveld, J. Determination of amoxicillin in human
plasma by high-performance liquid chromatography and solid phase extraction. *J.Chromatogr.*, **1993**,
*617*, 334–338

## SAMPLE
**Matrix:** blood
**Sample preparation:** 1 mL Plasma + 150 μL 20% perchloric acid, centrifuge at 2000g for
4 min, inject a 20 μL aliquot of the supernatant within 15 min.

## HPLC VARIABLES
**Column:** 80 × 4 Nucleosil 120 3C18
**Mobile phase:** MeCN:20 mM methanesulfonic acid 7.5:92.5
**Injection volume:** 20
**Detector:** E, ESA Model 5010, 0.78 V then F ex 255 em 400 (Oxidation by the electrochem-
ical detector produces a fluorescent product which is then detected.)

## CHROMATOGRAM
**Retention time:** 2.1
**Limit of detection:** 50 ng/mL

## KEY WORDS
plasma

## REFERENCE
Mascher, H.; Kikuta, C. Determination of amoxicillin in plasma by high-performance liquid chromatography with fluorescence detection after on-line oxidation. *J.Chromatogr.*, **1990**, *506*, 417–421

## SAMPLE
**Matrix:** blood
**Sample preparation:** 500 μL Plasma + 4 mL water + 3 mL 10% trichloroacetic acid,
centrifuge at 800-1000 g for 5 min. Remove 3 mL of the supernatant and add it to 0.5
mL 2 M NaOH, let stand for 5 min, add 0.5 mL 2 M HCl, add 2 mL 0.002% mercury(II)
chloride in 500 mM $Na_2HPO_4$ (to adjust pH to 6.0), heat at 50° for 25 min, add 6 mL ethyl
acetate saturated with water, shake vigorously for 5 min, centrifuge. Remove 5 mL of the
organic layer and evaporate it to dryness under reduced pressure, reconstitute the residue
in 100 μL MeOH containing IS, inject a 20 μL aliquot.

## HPLC VARIABLES
**Column:** 250 × 4 5 μm Nucleosil C18
**Mobile phase:** MeOH:water 55:45
**Column temperature:** 55
**Injection volume:** 20
**Detector:** F ex 355 em 435

## CHROMATOGRAM
**Internal standard:** methyl anthranylate
**Limit of detection:** 10 ng/mL

## KEY WORDS
plasma; derivatization

## REFERENCE
Miyazaki, K.; Ohtani, K.; Sunada, K.; Arita, T. Determination of ampicillin, amoxicillin, cephalexin, and cephradine in plasma by high-performance liquid chromatography using fluorometric detection. *J.Chromatogr.*, **1983**, *276*, 478–482

## SAMPLE
**Matrix:** blood, broncho-alveolar lavage fluid
**Sample preparation:** 100 μL Plasma or 500 μL broncho-alveolar lavage fluid + 2 mL water + 1.5 mL 10% trichloroacetic acid, vortex for 30 s, centrifuge at 5000 rpm for 5 min. 1.5 mL Supernatant + 0.2 mL 2 M NaOH + 0.5 mL 0.1% (w/v) HgCl$_2$ in 67 mM pH 4.8 phosphate buffer (Sorensen), after 5 min bring to pH 6.2 with 1 mL 0.67 M Na$_2$HPO$_4$, keep at 40° for 25 min. Add 3 mL ethyl acetate, shake horizontally for 5 min, centrifuge at 3015 g for 5 min. Evaporate the organic phase to dryness under nitrogen; reconstitute in 100 μL mobile phase, inject an aliquot.

## HPLC VARIABLES
**Column:** 250 × 4 Spherisorb 5 ODS
**Mobile phase:** MeOH:5 mM 1-heptanesulfonic acid adjusted to pH 3.7 with acetic acid 60:40
**Flow rate:** 1
**Injection volume:** 20
**Detector:** F ex 345 em 425

## CHROMATOGRAM
**Retention time:** 4.67
**Internal standard:** amoxicillin
**Limit of detection:** 50 ng/mL (plasma), 10 ng/mL (lavage)

## OTHER SUBSTANCES
**Extracted:** ampicillin

## KEY WORDS
derivatization; amoxicillin is IS; plasma

## REFERENCE
Rosseel, M.T.; Bogaert, M.G.; Valcke, Y.J. High-performance liquid chromatographic assay of ampicillin in plasma and broncho-alveolar lavage fluid, using fluorescence detection. *Chromatographia*, **1989**, *27*, 243–246

## SAMPLE
**Matrix:** blood, middle ear fluid
**Sample preparation:** Condition a 2.8 mL 500 mg Bond Elut C18 SPE cartridge with 4 mL MeOH and 1 mL 50 mM pH 6.8 phosphate buffer. 200 μL Plasma or 50 μL middle ear fluid + 35 μL 60 μg/mL cefadroxil in MeOH:water 5:95 + 1 mL 50 mM pH 6.8 phosphate buffer, vortex, add to the SPE cartridge, wash with 1 mL 50 mM pH 6.8 phosphate buffer, wash with 1 mL buffer, dry under vacuum, elute with 1 mL MeOH:water 40:60. Evaporate the eluate to dryness under a stream of nitrogen at 40°, reconstitute with 35 μL MeOH:water 5:95, inject a 25 μL aliquot.

## HPLC VARIABLES
**Guard column:** 10 × 2 5 μm MOS Hypersil C8
**Column:** 150 × 2 5 μm MOS Hypersil C8
**Mobile phase:** MeCN:5 mM phosphate buffer containing 5 mM tetrabutylammonium 6:
94, adjusted to pH 6.5 (After 14 min wash column with MeCN:buffer 25:75 for 2 min,
re-equilibrate for 4 min.)
**Column temperature:** 40
**Flow rate:** 0.35
**Injection volume:** 25
**Detector:** UV 210

## CHROMATOGRAM
**Retention time:** 6.45
**Internal standard:** cefadroxil (12.30)
**Limit of quantitation:** 125 ng/mL (plasma); 500 ng/mL (fluid)

## KEY WORDS
plasma; SPE; pharmacokinetics

## REFERENCE
Yuan, Z.; Russlie, H.Q.; Canafax, D.M. Sensitive assay for measuring amoxicillin in human plasma and
middle ear fluid using solid-phase extraction and reversed-phase high-performance liquid chromatography. *J.Chromatogr.B*, **1995**, *674*, 93−99

## SAMPLE
**Matrix:** blood, middle-ear effusion
**Sample preparation:** 75 μL Plasma or middle-ear effusion + 50 μL 50 μg/mL hydroflumethiazide in water, mix, add 25 μL 10% perchloric acid, vortex, add 25 μL KCl solution.
Mix, centrifuge, remove supernatant, add 25 μL pH 10.4 800 mM Na$_2$HPO$_4$ to the supernatant, inject a 6 μL aliquot. (Hydroflumethiazide solution prepared by dissolving hydroflumethiazide in a few drops MeOH and then diluting with water.)

## HPLC VARIABLES
**Guard column:** 20 × 3.2 Brownlee C8 precolumn
**Column:** 150 × 4.6 5 μm Zorbax C8
**Mobile phase:** MeOH:MeCN:10 mM NaH$_2$PO$_4$ 10:2:88
**Column temperature:** 40
**Flow rate:** 1.4
**Injection volume:** 6
**Detector:** UV 230

## CHROMATOGRAM
**Retention time:** 3.9
**Internal standard:** hydroflumethiazide (6.4)
**Limit of quantitation:** 500 ng/mL

## KEY WORDS
chinchilla; plasma

## REFERENCE
Erdmann, G.R.; Walker, K.; Giebink, G.S.; Canafax, D.M. High performance liquid chromatographic
analysis of amoxicillin in microliter volumes of chinchilla middle ear effusion and plasma.
*J.Liq.Chromatogr.*, **1990**, *13*, 3339−3350

## SAMPLE
**Matrix:** blood, tissue

**Sample preparation:** Homogenize (Ultra-Turrax) 300 mg tissue at 4° for 45 s, centrifuge. 500 μL Serum or tissue homogenate supernatant + 500 μL 200 mM pH 7.0 ammonium acetate, vortex for 30 s, add 1 mL MeCN, mix for 15 s, centrifuge at 3000 rpm for 10 min. Add 3 mL dichloromethane to the supernatant, vortex for 30 s, centrifuge at 3000 rpm for 5 min, inject a 50 μL aliquot of the aqueous phase.

## HPLC VARIABLES
**Guard column:** 25 × 4 5 μm LiChrospher RP 18 E
**Column:** 125 × 4 5 μm LiChrospher RP 18 E
**Mobile phase:** MeCN:20 mM pH 6.8 Na$_2$HPO$_4$ 3:97
**Flow rate:** 1
**Injection volume:** 50
**Detector:** UV 225

## CHROMATOGRAM
**Limit of detection:** 100 ng/mL (serum), 200 ng/mL (tissue)

## KEY WORDS
fat; colon; serum

## REFERENCE
Martin, C.; Mallet, M.-N.; Sastre, B.; Viviand, X.; Martin, A.; De Micco, P.; Gouin, F. Comparison of concentrations of two doses of clavulanic acid (200 and 400 milligrams) administered with amoxicillin (2,000 milligrams) in tissues of patients undergoing colorectal surgery. *Antimicrob.Agents Chemother.*, **1995**, *39*, 94–98.

## SAMPLE
**Matrix:** blood, urine
**Sample preparation:** Serum. 200 μL Serum + 100 μL 10 M urea, ultrafilter with Amicon YMT membrane at 1500 g for 10 min. 200 μL Ultrafiltrate + 200 μL 0.1 M pH 9 borate buffer + 20 μL 0.2 M acetic anhydride solution in MeCN. Stand at RT for 3 min, add 400 μL reagent, heat at 60° in a water bath for 10 min, cool, inject a 40-80 μL aliquot. Urine. Dilute urine 10 fold with water and filter through 0.45 μm acrylate copolymer membrane. 200 μL Filtrate + 200 μL 0.1 M pH 9 borate buffer + 20 μL 0.2 M acetic anhydride solution in MeCN. Stand at RT 3 min, add 400 μL reagent, heat at 60° in a water bath for 10 min, cool, inject a 20-80 μL aliquot. (Reagent was 13.81 g 1,2,4-triazole in 60 mL water + 10 mL HgCl$_2$ solution (0.27 g HgCl$_2$ in 100 mL water), adjust pH to 9.0 ± 0.05 with 4 M NaOH, dilute to 100 mL.)

## HPLC VARIABLES
**Column:** 150 × 4.6 7 μm Zorbax ODS-7
**Mobile phase:** MeCN:20 mM NaH$_2$PO$_4$:20 mM sodium thiosulfate:MeCN 24:38:38
**Flow rate:** 1
**Injection volume:** 20-80
**Detector:** UV 328

## CHROMATOGRAM
**Retention time:** 6
**Limit of quantitation:** 50 ng/mL

## OTHER SUBSTANCES
**Also analyzed:** ampicillin, cyclacillin (ciclacillin)

## KEY WORDS
serum; derivatization

## REFERENCE

Haginaka, J.; Wakai, J. High-performance liquid chromatographic assay of ampicillin, amoxicillin and ciclacillin in serum and urine using a pre-column reaction with 1,2,4-triazole and mercury(II) chloride. *Analyst*, **1985**, *110*, 1277–1281

## SAMPLE

**Matrix:** blood, urine

**Sample preparation:** Serum. Wash Amicon YMB filter membrane by stirring gently in 200 mL 100 mM pH 7.0 sodium phosphate buffer for 30 min, blot dry with filter paper. Dilute serum with an equal volume of 100 mM pH 7.0 sodium phosphate buffer, filter (Amicon YMB) while centrifuging at 5° at 1000 g for 15 min, inject a 25-50 μL aliquot. Urine. Dilute 10-fold with 100 mM pH 7.0 sodium phosphate buffer. Remove a 600 μL aliquot and add it to 200 μL buffer, inject a 25 μL aliquot.

## HPLC VARIABLES

**Guard column:** CO:PEL ODS C18

**Column:** 250 × 4.6 μBondapak C18

**Mobile phase:** MeOH:buffer 6:94 (serum) or 4:96 (urine) (Buffer was 100 mM $KH_2PO_4$ adjusted to pH 3.2 with phosphoric acid.)

**Flow rate:** 2.5

**Injection volume:** 25-50

**Detector:** UV 227

## CHROMATOGRAM

**Retention time:** 4 (serum), 7 (urine)

**Limit of detection:** 500 ng/mL

## KEY WORDS

derivatization; ultrafiltrate; serum; pharmacokinetics

## REFERENCE

Foulstone, M.; Reading, C. Assay of amoxicillin and clavulanic acid, the components of Augmentin, in biological fluids with high-performance liquid chromatography. *Antimicrob.Agents Chemother.*, **1982**, *22*, 753–762

## SAMPLE

**Matrix:** broncho-alveolar lavage fluid

**Sample preparation:** Filter 1 mL broncho-alveolar lavage fluid (Tosoh Ultracent-30 with a molecular mass cut-off at 30000) while centrifuging at 1500 g at 5° for 30 min, inject a 100 μL aliquot of the ultrafiltrate.

## HPLC VARIABLES

**Column:** 150 × 4.6 5 μm Shodex C18 5A (Showa Denko)

**Mobile phase:** MeCN:50 mM pH 3.0 potassium hydrogen phosphate containing 20 mM sodium 1-heptanesulfonate and 5 mg/L EDTA 10:100

**Column temperature:** 40

**Flow rate:** 1.2

**Injection volume:** 100

**Detector:** E, Irica Σ875, glassy carbon electrode 800 mV, Ag/AgCl reference electrode, following post-column reaction. The column effluent passed through a 10 m × 0.3 mm coil of PTFE tubing irradiated by a GL-10 10 W mercury lamp and flowed to the detector.

## CHROMATOGRAM

**Retention time:** 17

**Internal standard:** amoxicillin

**Limit of detection:** 1 ng/mL

## OTHER SUBSTANCES
**Extracted:** aspoxicillin

## KEY WORDS
ultrafiltrate; post-column reaction; amoxicillin is IS

## REFERENCE
Yamazaki, T.; Ishikawa, T.; Nakai, H.; Miyai, M.; Tsubota, T.; Asano, K. Determination of aspoxicillin in broncho-alveolar lavage fluid by high-performance liquid chromatography with photolysis and electrochemical detection. *J.Chromatogr.*, **1993**, *615*, 180−185

## SAMPLE
**Matrix:** bulk
**Sample preparation:** Dissolve 200 mg amoxicillin trihydrate in 8 mL 200 mM pH 11.0 phosphate buffer, add 10 mL 70 mg/L sulfamethazine in solvent, make up to 20 mL with solvent, inject an aliquot within 1 min. Dissolve 200 mg amoxicillin sodium salt in 8 mL solvent, add 10 mL 70 mg/L sulfamethazine in solvent, make up to 20 mL with solvent, inject an aliquot within 1 min. (Solvent was MeOH:200 mM pH 7.0 potassium phosphate buffer:water 5:5:90.)

## HPLC VARIABLES
**Guard column:** 40 × 4.6 10 μm LiChrosorb RP-2
**Column:** 250 × 4.6 7μm Zorbax C8
**Mobile phase:** Gradient. A is MeOH:200 mM pH 7.0 potassium phosphate buffer:water 5:5:90. B is MeOH:200 mM pH 7.0 potassium phosphate buffer:water 50:5:45. A:B 95:5 for 5 min, to 35:65 over 30 min, to 95:5 over 7.5 min.
**Column temperature:** 30
**Flow rate:** 1
**Injection volume:** 25
**Detector:** UV 274

## CHROMATOGRAM
**Retention time:** 11
**Internal standard:** sulfamethazine (35)

## OTHER SUBSTANCES
**Simultaneous:** impurities, amoxicillin dimer, amoxicillin trimer, amoxicillin piperazine-2,5-dione, amoxicilloates

## REFERENCE
De Pourcq, P.; Hoebus, J.; Roets, E.; Hoogmartens, J.; Vanderhaeghe, H. Quantitative determination of amoxicillin and its decomposition products by high-performance liquid chromatography. *J.Chromatogr.*, **1985**, *321*, 441−449

## SAMPLE
**Matrix:** formulations
**Sample preparation:** Capsules. Open five capsules, dissolve contents and shells in water, make up to 1 L with water, shake thoroughly, filter (0.2 μm) an aliquot, dilute the filtrate 10-20-fold, inject an aliquot. Oral suspensions. Dilute 200-400-fold, filter, inject an aliquot of the filtrate.

## HPLC VARIABLES
**Column:** 100 × 2 5 μm Hypersil ODS
**Mobile phase:** MeOH:50 mM phosphate buffer 3:97, adjusted to pH 7.0 with orthophosphoric acid

**Flow rate:** 0.4
**Injection volume:** 20
**Detector:** UV 230

## CHROMATOGRAM
**Retention time:** 6.5

## OTHER SUBSTANCES
**Simultaneous:** degradation products

## KEY WORDS
capsules; oral suspensions

## REFERENCE
Shakoor, O.; Taylor, R.B.; Moody, R.R. Analysis of amoxycillin in capsules and oral suspensions by high-performance liquid chromatography. *Analyst*, **1995**, *120*, 2191–2194

## SAMPLE
**Matrix:** formulations
**Sample preparation:** Weigh out contents of amoxicillin/dicloxacillin capsules equivalent to 100 mg amoxicillin, add 10 mL water, stir magnetically for 10 min, filter, discard first 5 mL of the filtrate. 5 mL filtrate + 10 mL 1 mg/mL albuterol sulfate in water, make up to 100 mL with water, filter (0.45 μm), inject a 10 μL aliquot of the filtrate.

## HPLC VARIABLES
**Column:** 200 × 4.6 10 μm LiChrosorb RP-8
**Mobile phase:** MeOH:20 mM ammonium acetate 50:50, pH adjusted to 5 with acetic acid
**Flow rate:** 1
**Injection volume:** 10
**Detector:** UV 230

## CHROMATOGRAM
**Retention time:** 2.555
**Internal standard:** albuterol sulfate (3.388)

## OTHER SUBSTANCES
**Simultaneous:** dicloxacillin

## KEY WORDS
capsules

## REFERENCE
el Walily, A.F.M.; el-Anwar, F.; Eid, M.A.; Awaad, H. High-performance liquid chromatographic and derivative ultraviolet spectrophotometric determination of amoxicillin and dicloxacillin mixtures in capsules. *Analyst*, **1992**, *117*, 981–984

## SAMPLE
**Matrix:** formulations
**Sample preparation:** Dilute with water and filter (reject first few mL of filtrate)

## HPLC VARIABLES
**Column:** 300 × 4 10 μm μBondapak C18
**Mobile phase:** MeOH:buffer:water 15:1:84 (Buffer was 50 mL 200 mM $KH_2PO_4$ + 5.7 mL 200 mM NaOH made up to 200 mL, pH 6.)
**Flow rate:** 1

**Injection volume:** 20
**Detector:** UV 235

## CHROMATOGRAM
**Retention time:** 5.3

## OTHER SUBSTANCES
**Simultaneous:** clavulanic acid

## KEY WORDS
tablets; suspensions

## REFERENCE
Abounassif, M.A.; Abdel-Moety, E.M.; Mohamed, M.E.; Gad-Kariem, R.A. Liquid chromatographic determination of amoxycillin and clavulanic acid in pharmaceutical preparations. *J.Pharm. Biomed.Anal.*, **1991**, *9*, 731–735

## SAMPLE
**Matrix:** formulations
**Sample preparation:** Dilute and filter

## HPLC VARIABLES
**Column:** 150 × 3.9 5μm Nova Pak C18
**Mobile phase:** MeOH:50 mM pH 4.0 phosphate buffer 3:97
**Flow rate:** 0.8
**Detector:** UV 254

## CHROMATOGRAM
**Retention time:** 4.5

## OTHER SUBSTANCES
**Simultaneous:** clavulanic acid
**Noninterfering:** degradation products, mannitol, saccharin

## KEY WORDS
oral suspensions; stability-indicating

## REFERENCE
Tu, Y.H.; Stiles, M.L.; Allen, L.V.J.; Olsen, K.M.; Barton, C.I.; Greenwood, R.B. Stability of amoxicillin trihydrate-potassium clavulanate in original containers and unit dose oral syringes. *Am.J. Hosp.Pharm.*, **1988**, *45*, 1092–1099

## SAMPLE
**Matrix:** formulations
**Sample preparation:** Blend tablets and capsules with water in a high-speed blender for 5 min, filter, dilute with mobile phase, inject a 20 μL aliquot. Dilute oral suspensions and injections with mobile phase, inject a 20 μL aliquot.

## HPLC VARIABLES
**Guard column:** 70 mm long Co:Pell ODS
**Column:** 300 × 4.6 10 μm Chromegabond C18 (E.S. Industries)
**Mobile phase:** MeCN:MeOH:10 mM $KH_2PO_4$ 19:11:70
**Flow rate:** 1
**Injection volume:** 20
**Detector:** UV 225

## CHROMATOGRAM
**Retention time:** 3.6
**Limit of detection:** 342 ng/mL

## OTHER SUBSTANCES
**Simultaneous:** ampicillin, cloxacillin, dicloxacillin, methicillin, nafcillin, oxacillin, penicillin G, penicillin V

## KEY WORDS
tablets; capsules; oral suspensions; injections

## REFERENCE
Briguglio, G.T.; Lau-Cam, C.A. Separation and identification of nine penicillins by reverse phase liquid chromatography. *J.Assoc.Off.Anal.Chem.*, **1984**, *67*, 228–231

## SAMPLE
**Matrix:** formulations
**Sample preparation:** Grind capsule to a fine powder. Weigh 500 mg amoxicillin, make up to 500 mL with water, sonicate ca. 30 min keeping temp below 30°, filter (0.45 μm), inject a 50 μL aliquot of the filtrate within 3 h.

## HPLC VARIABLES
**Column:** Two 150 × 4.6 5 μm Spherisorb-ODS columns in series
**Mobile phase:** Gradient. A was buffer. B was MeOH:MeCN 75:25. A:B 100:0 for 5 min, to 60:40 over 5 min, maintain at 60:40 for 5 min, to 100:0 over 5 min. (Buffer was 10 mL 0.5 M $K_2HPO_4$ + 90 mL 0.5 M $KH_2PO_4$ made up to 1 L, pH ca. 5.9.)
**Flow rate:** 1
**Injection volume:** 50
**Detector:** UV 220

## CHROMATOGRAM
**Retention time:** 16

## OTHER SUBSTANCES
**Simultaneous:** impurities, 6-aminopenicillanic acid, amoxicillin penicilloic acids, p-hydroxyphenylglycine

## KEY WORDS
capsules

## REFERENCE
Fong, G.W.K.; Martin, D.T.; Johnson, R.N.; Kho, B.T. Determination of degradation products and impurities of amoxicillin capsules using ternary gradient elution high-performance liquid chromatography. *J.Chromatogr.*, **1984**, *298*, 459–472

## SAMPLE
**Matrix:** milk
**Sample preparation:** 10 mL Milk + 2 mL 200 mM tetraethylammonium chloride, stir, slowly add 38 mL MeCN over 30 s, let stand for 5 min, decant the supernatant through a plug of glass wool. 40 mL Filtrate + 1 mL water, evaporate under reduced pressure to 1-2 mL, make up to 4 mL with water, filter (0.45 μm polyvinylidene difluoride). Inject 2 mL into an LC system (150 × 4.6 5 μm Supelcosil LC-18≥CN:10 mM $KH_2PO_4$ 0:100 for 3 min, to 40:60 over 27 min, to 0:100 over 1 min; 1 mL/min;UV 210 and 295), collect a 2 mL fraction at retention time for amoxicillin (11.5 min), add 100 μL reagent, evaporate

to 1 mL, inject a 200 μL aliquot. (Reagent was 10 mM phosphoric acid, 10 mM KH$_2$PO$_4$, and 10 mM sodium decanesulfonate.)

## HPLC VARIABLES
**Column:** 150 × 4.6 5 μm Supelcosil LC-18
**Mobile phase:** MeCN:buffer 33:67 (Buffer was 15 mM phosphoric acid and 7.5 mM sodium dodecyl sulfate.)
**Flow rate:** 1
**Injection volume:** 200
**Detector:** UV 214

## CHROMATOGRAM
**Limit of quantitation:** 2-5 ppb

## OTHER SUBSTANCES
**Also analyzed:** ampicillin, cephapirin, penicillin G, ceftiofur, penicillin V, cloxacillin

## KEY WORDS
cow

## REFERENCE
Moats, W.A.; Harik-Khan, R. Liquid chromatographic determination of β-lactam antibiotics in milk: A multiresidue approach. *J.AOAC Int.*, **1995**, *78*, 49–54

## SAMPLE
**Matrix:** milk
**Sample preparation:** Condition a Bond Elut C8 SPE cartridge with 5 mL MeOH and 5 mL water. 20 mL Milk + 1 mL 1 M oxalic acid, heat at 60° for 10 min, centrifuge for 10 min, remove the supernatant and add it to 20 mL water and 400 μL tributylamine, shake well, add to the SPE cartridge, wash with two 2.5 mL portions of water, elute with 2.5 mL MeOH. Evaporate the eluate to dryness under a stream of nitrogen, extract the residue with three 100 μL portions of 50 mM pH 6.0 potassium phosphate buffer, filter (0.2 μm), inject an aliquot of the filtrate. (Buffer was 545 mL 100 mM citric acid, 455 mL 200 mM Na$_2$HPO$_4$, and 74.4 g EDTA, adjust to pH 4.5 with ammonium hydroxide, make up to 2 L with water.)

## HPLC VARIABLES
**Column:** 250 × 4.6 10 μm Lichrosorb RP-8
**Mobile phase:** MeOH:50 mM pH 6.0 potassium phosphate buffer 35:65
**Flow rate:** 1
**Injection volume:** 200
**Detector:** UV 210; Charm II assay

## CHROMATOGRAM
**Retention time:** 3.91

## OTHER SUBSTANCES
**Extracted:** cefadroxil, ticarcillin
**Simultaneous:** ampicillin, ceftiofur, cephapirin, cloxacillin, dicloxacillin, nafcillin, oxacillin, penicillin G

## KEY WORDS
SPE

## REFERENCE

Zomer, E.; Quintana, J.; Saul, S.; Charm, S.E. LC-Receptograms: A method for identification and quantitation of β-lactams in milk by liquid chromatography with microbial receptor assay. *J.AOAC Int.*, **1995**, *78*, 1165–1172

## SAMPLE

**Matrix:** milk
**Sample preparation:** 500 µL Milk + 500 µL MeCN:MeOH:water 40:20:40, vortex for 10-15 s, filter (Centricon-10, molecular mass cut-off filter 10000 daltons) with centrifuging at 2677 g for 30 min, inject a 10-100 µL aliquot of the ultrafiltrate.

## HPLC VARIABLES

**Column:** 220 × 2.1 5 µm Spheri-5 phenyl microbore (UV detection) or 220 × 4.6 5 µm Spheri-5 phenyl microbore (MS detection)
**Mobile phase:** MeCN:MeOH:triethylamine:85% phosphoric acid:water 15:5:0.4:0.4:79.2 containing 2 mM octanesulfonate and 2 mM dodecanesulfonate (UV) or isopropanol:acetic acid in 200 mM ammonium acetate:water 1.5:5:93.5 (MS)
**Column temperature:** 40
**Flow rate:** 0.2-0.45 (UV) or 0.8-1.2 (MS)
**Injection volume:** 10-100
**Detector:** UV 220; MS, Finnigan MAT 4800 quadrupole, thermospray, source 320°, vaporizer 120°, pulsed positive ion negative ion

## CHROMATOGRAM

**Retention time:** 8.2 (UV), 4 (MS)
**Limit of detection:** 100 ng/mL (UV); 200 ng/mL (MS)

## OTHER SUBSTANCES

**Also analyzed:** ampicillin, cloxacillin

## KEY WORDS

ultrafiltrate

## REFERENCE

Voyksner, R.D.; Tyczkowska, K.L.; Aronson, A.L. Development of analytical methods for some penicillins in bovine milk by ion-paired chromatography and confirmation by thermospray mass spectrometry. *J.Chromatogr.*, **1991**, *567*, 389–404

## SAMPLE

**Matrix:** reaction mixtures
**Sample preparation:** If necessary, remove oxidizing power of solution by adding sodium metabisulfite, inject a 20 µL aliquot.

## HPLC VARIABLES

**Guard column:** 15 × 4.6 5 µm Microsorb C8
**Column:** 250 × 4.6 5 µm Microsorb C8
**Mobile phase:** MeCN:5.5 mM sodium octanesulfonate + 20 mM trisodium citrate dihydrate adjusted to pH 3 with concentrated HCl 18:82
**Flow rate:** 1
**Injection volume:** 20
**Detector:** UV 236

## CHROMATOGRAM

**Retention time:** 6.2
**Limit of detection:** 500 ng/mL

## REFERENCE

Lunn, G.; Rhodes, S.W.; Sansone, E.B.; Schmuff, N.R. Photolytic destruction and polymeric resin decontamination of aqueous solutions of pharmaceuticals. *J.Pharm.Sci.*, **1994**, *83*, 1289–1293

## SAMPLE

**Matrix:** solutions

## HPLC VARIABLES

**Column:** 300 × 3.9 10 μm μBondapak phenyl
**Mobile phase:** MeOH:10 mM phosphate buffer 27:73, pH 3.6
**Column temperature:** 27
**Flow rate:** 1
**Detector:** UV 254

## CHROMATOGRAM

**Retention time:** 3.5

## OTHER SUBSTANCES

**Simultaneous:** ampicillin, cefaclor, cephalexin, cephradine

## REFERENCE

Huang, H.-S.; Wu, J.-R.; Chen, M.-L. Reversed-phase high-performance liquid chromatography of amphoteric β-lactam antibiotics: effects of columns, ion-pairing reagents and mobile phase pH on their retention times. *J.Chromatogr.*, **1991**, *564*, 195–203

## SAMPLE

**Matrix:** solutions
**Sample preparation:** Prepare an aqueous solution, inject a 200 μL aliquot.

## HPLC VARIABLES

**Guard column:** present but not specified
**Column:** 150 × 4.6 4 μm Micropak SPC-18 C18
**Mobile phase:** Gradient. MeCN:10 mM orthophosphoric acid containing 10 mM tetramethylammonium chloride from 15:85 to 60:40 over 20 min
**Flow rate:** 1
**Injection volume:** 200
**Detector:** UV 220

## CHROMATOGRAM

**Retention time:** 3.5

## OTHER SUBSTANCES

**Simultaneous:** ampicillin, cephapirin

## REFERENCE

Moats, W.A. Effect of the silica support of bonded reversed-phase columns on chromatography of some antibiotic compounds. *J.Chromatogr.*, **1986**, *366*, 69–78

## SAMPLE

**Matrix:** tissue
**Sample preparation:** Condition a 3 mL 500 mg Sep-Pak C18 SPE cartridge with 5 mL MeOH, 2 mL water, and 2 mL 2% trichloroacetic acid. Homogenize (Ultra-Turrax T25) 5 g blended tissue with 20 mL 10 mM pH 4.5 phosphate buffer at 10000 rpm for 1.5 min, centrifuge at 4500 rpm for 10 min, decant supernatant, homogenize residue with another

20 mL buffer, centrifuge. Combine the supernatants and filter them through glass wool, add 1 mL 75% trichloroacetic acid to the filtrate, vortex for 30 s, centrifuge at 4500 rpm for 20 min, filter the supernatant through glass wool. Add the filtrate to the SPE cartridge at 1-2 mL/min, wash with 2 mL 2% trichloroacetic acid, wash with 2 mL water, elute with 1.5 mL MeCN at 0.7 mL/min. Add the eluate to 500 μL water and 3 mL ethyl ether, vortex gently for 30 s, centrifuge at 2000 rpm for 3 min, discard the organic layer. Add 200 μL 20% trichloroacetic acid solution to the aqueous phase, vortex for 15 s, add 200 μL 7% formaldehyde in 400 mM citric acid, vortex for 30 s, heat in boiling water bath for 30 min, cool to room temperature, add 500 mg NaCl, mix briefly, add 3 mL ethyl ether, vortex for 1 min, centrifuge at 2000 rpm for 3 min, repeat extraction twice more. Combine the organic layers and evaporate them to dryness under a stream of nitrogen at 40°, reconstitute the residue in 500 μL mobile phase, vortex thoroughly, inject a 50 μL aliquot.

## HPLC VARIABLES
**Column:** 250 × 4.6 S5 ODS2
**Mobile phase:** MeCN:buffer 20:80 (Buffer was 50 mM $KH_2PO_4$ adjusted to pH 5.6 with KOH.)
**Flow rate:** 1 for 10 min then 2
**Injection volume:** 50
**Detector:** F ex 358 em 440

## CHROMATOGRAM
**Retention time:** 6
**Limit of detection:** 0.5 ppb (catfish); 0.8 ppb (salmon)
**Limit of quantitation:** 1.2 ppb (catfish); 2.0 ppb (salmon)

## KEY WORDS
derivatization; fish; catfish; salmon; SPE

## REFERENCE
Ang, C.Y.W.; Luo, W.; Hansen, E.B., Jr.; Freeman, J.P.; Thompson, H.C., Jr. Determination of amoxicillin in catfish and salmon tissues by liquid chromatography with precolumn formaldehyde derivatization. *J.AOAC Int.*, **1996**, *79*, 389–396

## SAMPLE
**Matrix:** urine
**Sample preparation:** Condition a Sep-Pak C18 SPE cartridge with 5 mL MeOH then 5 mL water. 1 mL Urine + 100 μL 0.5 M tetrabutylammonium bromide in water, vortex 30 s, add to SPE cartridge, elute with 9 mL MeCN:buffer 3:97, make up to 10 mL with mobile phase, inject an aliquot. (Buffer was 100 mL 0.5 M disodium hydrogen orthophosphate + 350 mL water adjusted to pH 4.85 with 1 M citric acid then made up to 500 mL.)

## HPLC VARIABLES
**Column:** 250 × 4.6 5 μm Ultrasphere C18
**Mobile phase:** MeCN:pH 7.1 100 mM disodium hydrogen phosphate:water 32.5:100:900
**Flow rate:** 1.2
**Injection volume:** 30
**Detector:** UV 229

## CHROMATOGRAM
**Retention time:** 20
**Limit of quantitation:** 7.5 μg/mL

## OTHER SUBSTANCES
**Simultaneous:** metabolites, amoxicillin piperazine-2′,5′-dione, amoxicilloic acid

## KEY WORDS
SPE

## REFERENCE
Chulavatnatol, S.; Charles, B.G. High-performance liquid chromatographic determination of amoxicillin in urine using solid-phase, ion-pair extraction and ultraviolet detection. *J.Chromatogr.*, **1993**, *615*, 91−96

◆━━━━━━━━━━━━━━━━━━━━◆━━━━━━━━━━━━━━━━━━━━◆

## ANNOTATED BIBLIOGRAPHY

Harik-Khan, R.; Moats, W.A. Identification and measurement of β-lactam antibiotic residues in milk: Integration of screening kits with liquid chromatography. *J.AOAC Int.*, **1995**, *78*, 978−986 [simultaneous ampicillin, ceftiofur, cephapirin, cloxacilin, penicillin G; milk; gradient]

Chesa-Jiménez, J.; Peris, J.E.; Torres-Molina, F.; Granero, L. Low bioavailability of amoxicillin in rats as a consequence of presystemic degradation in the intestine. *Antimicrob.Agents Chemother.*, **1994**, *38*, 842−847 [tissue homogenate; rat]

Moats, W.A. Determination of ampicillin and amoxicillin in milk with an automated liquid chromatographic cleanup. *J.AOAC Int.*, **1994**, 77, 41−45 [milk; extracted ampicillin, cephapirin; gradient; LOQ 10 ppb]

Snippe, N.; Van de Merbel, N.C.; Ruiter, F.P.M.; Steijer, O.M.; Lingeman, H.; Brinkman, U.A.T. Automated column liquid chromatographic determination of amoxicillin and cefadroxil in bovine serum and muscle tissue using on-line dialysis for sample preparation. *J.Chromatogr.B*, **1994**, *662*, 61−70 [serum; muscle; cow; post-column reaction; LOD 50 ng/mL; LOD 200 ng/g; SPE; extracted; cefadroxil; dialysis]

Straub, R.; Linder, M.; Voyksner, R.D. Determination of β-lactam residues in milk using perfusive-particle liquid chromatography combined with ultrasonic nebulization electrospray mass spectrometry. *Anal.Chem.*, **1994**, *66*, 3651−3658 [milk; LC-MS; electrospray; extracted ampicillin, ceftiofur, cephapirin, cloxacillin, penicillin G; LOD 10 ppb]

Tyczkowska, K.L.; Voyksner, R.D.; Straub, R.F.; Aronson, A.L. Simultaneous multiresidue analysis of β-lactam antibiotics in bovine milk by liquid chromatography with ultraviolet detection and confirmation by electrospray mass spectrometry. *J.AOAC Int.*, **1994**, 77, 1122−1131 [milk; cow; LC-MS; electrospray; UV detection; LOD 10 ppb; extracted ampicillin, ceftiofur, cephapirin, cloxacillin, penicillin G; column temp 40]

Kaniou, I.P.; Zachariadis, G.A.; Stratis, J.A. Separation and determination of five penicillins by reversed phase HPLC. *J.Liq.Chromatogr.*, **1993**, *16*, 2891−2897 [simultaneous ampicillin, cloxacillin, dicloxacillin, penicillin G; LOD 30-50 pb]

Parker, C.E.; Perkins, J.R.; Tomer, K.B.; Shida, Y.; O'Hara, K. Nanoscale packed capillary liquid chromatography-electrospray ionization mass spectrometry: analysis of penicillins and cephems. *J.Chromatogr.*, **1993**, *616*, 45−57 [capillary HPLC; electrospray; LC-MS; serum; also ampicillin, carbenicillin, cefalothin, cefazolin, cefmenoxime, cefmetazole, cefoperazone, cefotaxime, cefotiam, cefoxitin, cephalexin, cloxacillin, dicloxacillin, penicillin G, pipericillin, sulbenicillin]

Straub, R.F.; Voyksner, R.D. Determination of penicillin G, ampicillin, amoxicillin, cloxacillin and cephapirin by high-performance liquid chromatography-electrospray mass spectrometry. *J.Chromatogr.*, **1993**, *647*, 167−181

Hsu, M.C.; Hsu, P.W. High-performance liquid chromatographic method for potency determination of amoxicillin in commercial preparations and for stability studies. *Antimicrob.Agents Chemother.*, **1992**, *36*, 1276−1279

Leroy, P.; Gavriloff, C.; Nicolas, A.; Archimbault, P.; Ambroggi, G. Comparative assay of amoxicillin by high-performance liquid chromatography and microbiological methods for pharmacokinetic studies in calves. *Int.J.Pharm.*, **1992**, *82*, 157−164

Nelis, H.J.; Vandenbranden, J.; De Kruif, A.; De Leenheer, A.P. Liquid chromatographic determination of amoxicillin concentrations in bovine plasma by using a tandem solid-phase extraction method. *Antimicrob.Agents Chemother.*, **1992**, *36*, 1859−1863

Doadrio, A.L.; Sotelo, J. Determination of hydrolysis constants for amoxicillin by liquid chromatography. *An.R.Acad.Farm.*, **1989**, *55*, 203−212

Mendez, R.; Alemany, M.T.; Jurado, C.; Martin, J. Study on the rate of decomposition of amoxicillin in solid state using high-performance liquid chromatography. *Drug Dev.Ind.Pharm.*, **1989**, *15*, 1263–1274

Martín, J.; Méndez, R.; Negro, A. Effect of temperature on HPLC separations of penicillins. *J.Liq.Chromatogr.*, **1988**, *11*, 1707–1716 [simultaneous ampicillin, cloxacillin, penicillin G, penicillin V, piperacillin; column temperature 15-55°]

Haginaka, J.; Wakai, J. Liquid chromatographic determination of amoxicillin and its metabolites in human urine by postcolumn degradation with sodium hypochlorite. *J.Chromatogr.*, **1987**, *413*, 219–226 [extracted metabolites; post-column reaction; urine; LOD 1 μg/mL]

Fong, G.W.K.; Johnson, R.N.; Kho, B.T. Study on the rate of epimerization of amoxicillin β-penicilloic acid to its α-form in aqueous solutions using high-performance liquid chromatography. *J.Chromatogr.*, **1983**, *255*, 199–207

Nakagawa, T.; Shibukawa, A.; Uno, T. Liquid chromatography with crown ether-containing mobile phases. II. Retention behavior of β-lactam antibiotics in reversed- phase high-performance liquid chromatography. *J.Chromatogr.*, **1982**, *239*, 695–706 [ampicillin, carbenicillin, cefradine, cephalexin, cephaloglycin, cephaloridin, ciclacillin, cloxacillin, dicloxacillin, oxacillin, penicillin G]

Brooks, M.A.; Hackman, M.R.; Mazzo, D.J. Determination of amoxicillin by high-performance liquid chromatography with amperometric detection. *J.Chromatogr.*, **1981**, *210*, 531–535 [electrochemical detection; UV detection]

# Ampicillin

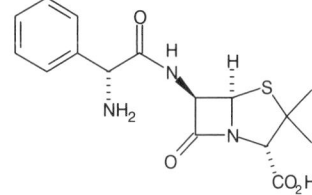

**Molecular formula:** $C_{16}H_{19}N_3O_4S$
**Molecular weight:** 349.4
**CAS Registry No.:** 69-53-4 (anhydrous), 32388-53-7 (monohydrate),
23277-71-6 (K salt), 7177-48-2 (trihydrate), 69-52-3 (Na salt)

## SAMPLE
**Matrix:** bile, blood, urine
**Sample preparation:** Dilute bile and urine with 50 mM pH 7.0 phosphate buffer. 50 μL Plasma + 100 μL cephalexin in 50 mM pH 7.0 phosphate buffer, mix. Inject 100 μL onto column A and elute to waste with mobile phase A, after 5 min backflush the contents of column A onto column B with mobile phase B, elute column B with mobile phase B and monitor the effluent. Re-equilibrate for 4 min before the next injection.

## HPLC VARIABLES
**Column:** A 20 × 3.9 25-40 μm LiChrosorb RP-8; B 10 × 4 Nova-Pak C8 guard column + 250 × 4.6 5 μm Ultracarb 5 ODS-30 (Phenomenex)
**Mobile phase:** A 50 mM pH 7.0 phosphate buffer; B Gradient. X was MeCN:20 mM pH 7.0 phosphate buffer 4:96. Y was MeCN:20 mM pH 7.0 phosphate buffer 30:70. X:Y 55:45 for 10 min, to 0:100 over 8 min, maintain at 0:100 for 12 min.
**Flow rate:** 1
**Injection volume:** 100
**Detector:** UV 230

## CHROMATOGRAM
**Retention time:** 9.0
**Internal standard:** cephalexin (7.4)
**Limit of detection:** 100 ng/mL

## OTHER SUBSTANCES
**Extracted:** metampicillin
**Noninterfering:** acetaminophen, caffeine, ibuprofen, phenobarbital, sulbactam

## KEY WORDS
column-switching; rat; pharmacokinetics; plasma

## REFERENCE
Lee, H.; Lee, J.S.; Lee, H.S. Simultaneous determination of ampicillin and metampicillin in biological fluids using high-performance liquid chromatography with column switching. *J.Chromatogr.B*, **1995**, *664*, 335–340

## SAMPLE
**Matrix:** bile, blood, urine
**Sample preparation:** Serum. 0.5 mL Serum + 0.5 mL MeCN mix in 7 mL tube on vortex mixer, shake by rotation (20 rpm) for 10 min, centrifuge at 1000 g for 10 min, transfer supernatant to another tube, add 7 volumes dichloromethane, equilibrate for 10 min, shake by rotation (20 rpm) for 10 min, centrifuge at 1000 g for 10 min, inject aliquot of the upper aqueous layer. Urine. Centrifuge urine and dilute 1:20, inject an aliquot. Bile. Centrifuge bile and dilute 1:10, inject an aliquot.

## HPLC VARIABLES
**Column:** 150 × 4.6 5 μm Ultrasphere ODS

**Mobile phase:** MeCN:20 mM ammonium acetate adjusted to pH 5 with glacial acetic acid
8:92
**Flow rate:** 1
**Injection volume:** 20
**Detector:** UV 214

## CHROMATOGRAM
**Retention time:** 6.3
**Limit of detection:** 500 ng/mL

## OTHER SUBSTANCES
**Also analyzed:** azlocillin, aztreonam, cefmenoxime, cefoperazone, cefotaxime, cefsulodin,
ceftazidime, ceftriaxone, cloxacillin, desacetylcefotaxime, mezlocillin, penicillin G, piperacillin, ticarcillin

## KEY WORDS
serum

## REFERENCE
Jehl, F.; Birckel, P.; Monteil, H. Hospital routine analysis of penicillins, third-generation cephalosporins
and aztreonam by conventional and high-speed high-performance liquid chromatography.
*J.Chromatogr.*, **1987**, *413*, 109–119

## SAMPLE
**Matrix:** blood
**Sample preparation:** 500 μL Serum + 1 mL 20% trichloroacetic acid, vortex for 15 s,
centrifuge at 1000 g for 20 min. 1 mL Supernatant + 500 μL 7% formaldehyde in 400
mM citric acid, vortex for 15 s, heat at 90° for 2 h, cool to room temperature. Either inject
an aliquot of this solution directly or extract it twice with 3 mL portions of diethyl ether.
Evaporate the extracts to dryness under reduced pressure, reconstitute with 100 μL mobile phase, inject an aliquot.

## HPLC VARIABLES
**Guard column:** 30 × 4.6 10 μm RP-18 (Pierce)
**Column:** 100 × 4.6 10 μm RP-18 (Pierce)
**Mobile phase:** MeCN:100 mM pH 5.6 $KH_2PO_4$ 23:77
**Flow rate:** 1
**Injection volume:** 50
**Detector:** F ex 346 em 422

## CHROMATOGRAM
**Retention time:** 7
**Limit of detection:** 2 ng/mL (with extraction)

## KEY WORDS
serum; derivatization

## REFERENCE
Lal, J.; Paliwal, J.K.; Grover, P.K.; Gupta, R.C. Determination of ampicillin in serum by high-performance liquid chromatography with precolumn derivatization. *J.Chromatogr.B*, **1994**, *655*, 142–146

## SAMPLE
**Matrix:** blood
**Sample preparation:** 0.5 mL Plasma + 1 mL MeOH, stir for 5 min, centrifuge at 2400 g
for 10 min. Remove 1 mL supernatant, add 2 μg cefazolin, inject an aliquot.

## HPLC VARIABLES
**Column:** 300 × 3.9 5 μm μBondapak C18
**Mobile phase:** MeOH:67 mM KH$_2$PO$_4$ 20:80
**Flow rate:** 1.5
**Injection volume:** 50
**Detector:** UV 225

## CHROMATOGRAM
**Retention time:** 9
**Internal standard:** cefazolin (14)
**Limit of detection:** 500 ng/mL

## OTHER SUBSTANCES
**Extracted:** bacampicillin (detected as ampicillin), lenampicillin (detected as ampicillin)

## KEY WORDS
pharmacokinetics; plasma

## REFERENCE
Marzo, A.; Monti, N.; Ripamonti, M.; Arrigoni Martelli, E.; Picari, M. High-performance liquid chromatographic assay of ampicillin and its prodrug lenampicillin. *J.Chromatogr.*, **1990**, *507*, 235–239

## SAMPLE
**Matrix:** blood
**Sample preparation:** 500 μL Plasma + 4 mL water + 3 mL 10% trichloroacetic acid, centrifuge at 800-1000 g for 5 min. Remove 3 mL of the supernatant and add it to 500 μL 2 M NaOH, let stand for 5 min, add 500 μL 2 M HCl, add 1 mL 0.1% mercury(II) chloride in buffer, let stand for 5 min, add 2 mL 0.67 M Na$_2$HPO$_4$ warmed to 40° to adjust pH to 6.2, heat mixture at 40° for 25 min, add 6 mL ethyl acetate saturated with water, shake vigorously for 5 min, centrifuge. Remove 5 mL of the organic layer and evaporate it to dryness under reduced pressure, reconstitute the residue in 100 μL MeOH containing IS, inject a 20 μL aliquot. (Prepare buffer by dissolving 21 g citric acid in 200 mL 1 M NaOH, make up to 1 L with water, adjust pH to 2.5 with 100 mM HCl.)

## HPLC VARIABLES
**Column:** 250 × 4 5 μm Nucleosil C18
**Mobile phase:** MeOH:water 60:40
**Column temperature:** 55
**Injection volume:** 20
**Detector:** F ex 345 em 420

## CHROMATOGRAM
**Retention time:** 6 (?)
**Internal standard:** methyl anthranylate (9 (?))
**Limit of detection:** 0.5 ng/mL

## OTHER SUBSTANCES
**Interfering:** cephalexin, cephradine

## KEY WORDS
plasma; derivatization

## REFERENCE
Miyazaki, K.; Ohtani, K.; Sunada, K.; Arita, T. Determination of ampicillin, amoxicillin, cephalexin, and cephradine in plasma by high-performance liquid chromatography using fluorometric detection. *J.Chromatogr.*, **1983**, *276*, 478–482

## SAMPLE

**Matrix:** blood, broncho-alveolar lavage fluid

**Sample preparation:** 100 µL Plasma or 500 µL broncho-alveolar lavage fluid + 40 µL 1 mg/mL (plasma) or 75 µL 100 µg/mL (broncho-alveolar lavage fluid) amoxicillin in 67 mM pH 4.8 phosphate buffer + 2 mL water + 1.5 mL 10% trichloroacetic acid, vortex for 30 s, centrifuge at 5000 rpm for 5 min. 1.5 mL Supernatant + 0.2 mL 2 M NaOH + 0.5 mL 0.1% (w/v) $HgCl_2$ in 67 mM pH 4.8 phosphate buffer (Sorensen), after 5 min bring to pH 6.2 with 1 mL 0.67 M $Na_2HPO_4$, keep at 40° for 25 min. Add 3 mL ethyl acetate, shake horizontally for 5 min, centrifuge at 3015 g for 5 min. Evaporate the organic phase to dryness under nitrogen; reconstitute in 100 µL mobile phase, inject an aliquot.

## HPLC VARIABLES

**Column:** 250 × 4 Spherisorb 5 ODS

**Mobile phase:** MeOH:5 mM 1-heptanesulfonic acid adjusted to pH 3.7 with acetic acid 60:40

**Flow rate:** 1

**Injection volume:** 20

**Detector:** F ex 345 em 425

## CHROMATOGRAM

**Retention time:** 3.45

**Internal standard:** amoxicillin (4.67)

**Limit of detection:** 50 ng/mL (plasma); 10 ng/mL (lavage)

## KEY WORDS

derivatization; plasma

## REFERENCE

Rosseel, M.T.; Bogaert, M.G.; Valcke, Y.J. High-performance liquid chromatographic assay of ampicillin in plasma and broncho-alveolar lavage fluid, using fluorescence detection. *Chromatographia*, **1989**, 27, 243–246

## SAMPLE

**Matrix:** blood, CSF

**Sample preparation:** 200 µL Serum, plasma, or CSF + 300 µL reagent. Flush column A to waste with 500 µL 500 mM ammonium sulfate, inject sample onto column A, flush column A to waste with 500 µL 500 mM ammonium sulfate, elute the contents of column A onto column B with mobile phase, monitor the effluent from column B. (Reagent was 8.05 M guanidine hydrochloride and 1.02 M ammonium sulfate in water.)

## HPLC VARIABLES

**Column:** A 30 × 2.1 40 µm preparative grade C18 (Analytichem); B 250 × 4.6 10 µm Partisil C8

**Mobile phase:** Gradient. A was 50 mM pH 4.5 $KH_2PO_4$. B was MeCN:isopropanol 80:20. A:B 90:10 for 1 min, to 30:70 over 15 min, maintain at 30:70 for 4 min.

**Column temperature:** 50

**Flow rate:** 1.5

**Detector:** UV 280 for 5 min then UV 254

## CHROMATOGRAM

**Retention time:** 4.98

**Internal standard:** heptanophenone (19.2)

## OTHER SUBSTANCES

**Extracted:** bromazepam, caffeine, carbamazepine, chloramphenicol, chlorothiazide, diazepam, droperidol, ethionamide, furosemide, isoniazid, methadone, penicillin G, phenobarbital, phenytoin, prazepam, propoxyphene, pyrazinamide, rifampin, trimeprazine, trimethoprim

**Interfering:** acetazolamide

## KEY WORDS

column-switching; serum; plasma

## REFERENCE

Seifart, H.I.; Kruger, P.B.; Parkin, D.P.; van Jaarsveld, P.P.; Donald, P.R. Therapeutic monitoring of antituberculosis drugs by direct in-line extraction on a high-performance liquid chromatography system. *J.Chromatogr.*, **1993**, *619*, 285–290

## SAMPLE

**Matrix:** blood, saliva, urine

**Sample preparation:** Plasma, saliva. 200 $\mu$L Plasma or saliva + 10 $\mu$L 60% perchloric acid + 200 $\mu$L dichloromethane, vortex, centrifuge, inject a 50 $\mu$L aliquot of the aqueous phase. Urine. 10 $\mu$L Urine + 500 $\mu$L 330 mM perchloric acid, mix, inject a 100 $\mu$L aliquot.

## HPLC VARIABLES

**Column:** 300 $\times$ 4.6 Magnusphere C18 (Magnus Scientific, Sandbach, England)

**Mobile phase:** MeOH:67 mM pH 4.6 $KH_2PO_4$ 15:85

**Flow rate:** 1

**Injection volume:** 50-100

**Detector:** UV 225

## KEY WORDS

plasma; pharmacokinetics

## REFERENCE

Rogers, H.J.; Bradbrook, I.D.; Morrison, P.J.; Spector, R.G.; Cox, D.A.; Lees, L.J. Pharmacokinetics and bioavailability of sultamicillin estimated by high performance liquid chromatography. *J.Antimicrob.Chemother.*, **1983**, *11*, 435–445

## SAMPLE

**Matrix:** blood, urine

**Sample preparation:** Plasma. 100 $\mu$L Plasma + 200 $\mu$L water + 100 $\mu$L 70% perchloric acid:pH 5.4 buffer 25:75, vortex for 1 min, centrifuge at ca. 2400 g for 10 min. Remove 300 $\mu$L supernatant and add it to 75 $\mu$L 1 M NaOH, mix, inject a 100 $\mu$L aliquot onto column A and elute to waste with mobile phase A, after 2 min collect the effluent from column A in a 1 mL sample loop, after 1 min inject the contents of the sample loop onto column B with mobile phase B, elute column B with mobile phase B and monitor the effluent from column B. Urine. Dilute urine if necessary. 80 $\mu$L Urine + 720 $\mu$L pH 4.85 buffer, vortex, inject a 20 $\mu$L aliquot onto column A and elute to waste with mobile phase A, after 2 min collect the effluent from column A in a 1 mL sample loop, after 1 min inject the contents of the sample loop onto column B with mobile phase B, elute column B with mobile phase B and monitor the effluent from column B. (Prepare pH 5.4 buffer by dissolving 19.9 g $Na_2HPO_4$ and 8.4 g citric acid monohydrate in 250 mL water. Prepare pH 4.85 buffer by mixing 10 mL 500 mM $Na_2HPO_4$ and 350 mL water, adjust pH to 4.85 with 1 M citric acid, make up to 500 mL with water.)

## HPLC VARIABLES

**Column:** A 15 $\times$ 3.2 7 $\mu$m New Guard RP18 + 3 $\times$ 3 3 $\mu$m Perkin-Elmer; B 100 $\times$ 4.6 3 $\mu$m Microspher C18 (Chrompack)

**Mobile phase:** A MeOH:buffer 17:83 containing 1 mM sodium hexylsulfate (Buffer was pH 7.4 phosphate buffer, ionic strength 0.05.); B MeOH:buffer 35:65 (plasma) or 30:70 (urine) (Buffer was pH 7.4 phosphate buffer, ionic strength 0.05.)
**Flow rate:** 1
**Injection volume:** 20-100
**Detector:** F ex 372 em 470 following post-column reaction. The effluent from column B mixed with 160 µg/mL fluorescamine in MeCN pumped at 0.2 mL/min and the mixture flowed through a 5 m × 0.4 mm i.d. knitted PTFE tube to the detector.

## CHROMATOGRAM
**Retention time:** 8
**Limit of detection:** 570 nM (urine); 14 nM (plasma)

## KEY WORDS
derivatization; plasma; post-column reaction; column-switching; heart-cut

## REFERENCE
Lanbeck-Vallén, K. Carlqvist, J.; Nordgren, T. Determination of ampicillin in biological fluids by coupled-column liquid chromatography and post-column derivatization. *J.Chromatogr.*, **1991**, *567*, 121–128

## SAMPLE
**Matrix:** blood, urine
**Sample preparation:** Serum. Filter using Molcut II (Millipore), inject a 50 µL aliquot of the ultrafiltrate. Urine. Dilute ten-fold with water, filter (Gelman acrylate copolymer 0.45 µm), inject a 20 µL aliquot of the filtrate.

## HPLC VARIABLES
**Guard column:** 4 × 4 5 µm LiChrospher RP-18(e)
**Column:** 250 × 4 5 µm LiChrospher RP-18(e)
**Mobile phase:** MeOH:20 mM tetrabutylammonium bromide + 5 mM Na₂HPO₄ + 5 mM NaH₂PO₄ 1:1.75
**Flow rate:** 0.8
**Injection volume:** 20-50
**Detector:** UV 270 following post-column reaction. The column effluent mixed with 2 M NaOH and 0.05% sodium hypochlorite solution pumped at 0.1 mL/min in a 400 × 0.5 mm hollow fiber membrane reactor at 40° and this mixture flowed through a 1400 × 0.3 mm knitted open tubular reactor at 50° to the detector.

## CHROMATOGRAM
**Retention time:** 30
**Limit of detection:** 20 ng

## OTHER SUBSTANCES
**Extracted:** sulbactam

## KEY WORDS
post-column reactiom; serum

## REFERENCE
Haginaka, J.; Nishimura, Y. Simultaneous determination of ampicillin and sulbactam by liquid chromatography: post-column reaction with sodium hydroxide and sodium hypochlorite using an active hollow-fiber membrane reactor. *J.Chromatogr.*, **1990**, *532*, 87–94

## SAMPLE
**Matrix:** blood, urine

**Sample preparation:** Serum. 200 μL Serum + 100 μL 10 M urea, ultrafilter with Amicon YMT membrane at 1500 g for 10 min. 200 μL Ultrafiltrate + 200 μL 0.1 M pH 9 borate buffer + 20 μL 0.2 M acetic anhydride solution in MeCN. Stand at RT for 3 min, add 400 μL reagent, heat at 60° in a water bath for 10 min, cool, inject a 40-80 μL aliquot. Urine. Dilute urine 10 fold with water and filter through 0.45 μm acrylate copolymer membrane. 200 μL Filtrate + 200 μL 0.1 M pH 9 borate buffer + 20 μL 0.2 M acetic anhydride solution in MeCN. Stand at RT 3 min, add 400 μL reagent, heat at 60° in a water bath for 10 min, cool, inject a 20-80 μL aliquot. (Reagent was 13.81 g 1,2,4-triazole in 60 mL water + 10 mL HgCl$_2$ solution (0.27 g HgCl$_2$ in 100 mL water), adjust pH to 9.0 ± 0.05 with 4 M NaOH, dilute to 100 mL.)

## HPLC VARIABLES
**Column:** 150 × 4.6 5 μm Develosil ODS-5 (Nomura Chemicals)
**Mobile phase:** MeCN:20 mM NaH$_2$PO$_4$:20 mM sodium thiosulfate:MeCN 25:37.5:37.5
**Flow rate:** 0.8
**Injection volume:** 20-80
**Detector:** UV 328

## CHROMATOGRAM
**Retention time:** 6
**Limit of quantitation:** 100 ng/mL

## OTHER SUBSTANCES
**Also analyzed:** amoxicillin, cyclacillin (ciclacillin)

## KEY WORDS
derivatization; serum

## REFERENCE
Haginaka, J.; Wakai, J. High-performance liquid chromatographic assay of ampicillin, amoxicillin and ciclacillin in serum and urine using a pre-column reaction with 1,2,4-triazole and mercury(II) chloride. *Analyst*, **1985**, *110*, 1277–1281

## SAMPLE
**Matrix:** bulk, formulations
**Sample preparation:** Prepare a solution of capsule contents or bulk drug in the mobile phase, inject a 20 μL aliquot.

## HPLC VARIABLES
**Column:** 125 × 4 5 μm LiChrospher RP-18
**Mobile phase:** MeCN:1% acetic acid 39:61
**Flow rate:** 2
**Injection volume:** 20
**Detector:** UV 240

## CHROMATOGRAM
**Retention time:** 2

## OTHER SUBSTANCES
**Simultaneous:** degradation products, dicloxacillin

## KEY WORDS
capsules; stability-indicating

## REFERENCE
Al-Rashood, K. Simultaneous determination of ampicillin and dicloxacillin in pharmaceutical formulations by high-performance liquid chromatography. *J.Liq.Chromatogr.*, **1995**, *18*, 2457–2465

## SAMPLE
**Matrix:** formulations

## HPLC VARIABLES
**Column:** 250 × 4.6 5 μm Bakerbond C18
**Mobile phase:** MeCN:water:1 M KH$_2$PO$_4$:1 M acetic acid 80:909:10:1
**Flow rate:** 0.8
**Detector:** UV 254

## CHROMATOGRAM
**Retention time:** 3.29

## KEY WORDS
injections; saline; water; stability-indicating

## REFERENCE
Stiles, M.L.; Allen, L.V., Jr.; Prince, S.J. Stability of various antibiotics kept in an insulated pouch during administration via portable infusion pump. *Am.J.Health-Syst.Pharm.*, **1995**, *52*, 70–74

## SAMPLE
**Matrix:** formulations
**Sample preparation:** Dilute 1:8 with water, combine a 100 μL aliquot of the diluted solution with 100 μL cimetidine solution and 200 μL water, inject a 20 μL aliquot.

## HPLC VARIABLES
**Column:** 3.9 × 300 μBondapak C18
**Mobile phase:** MeCN:MeOH:10 mM pH 2.6-2.7 phosphate buffer 7:14:79 containing 5 mM tetrabutylammonium hydrogen sulfate
**Flow rate:** 1
**Injection volume:** 20
**Detector:** UV 225

## CHROMATOGRAM
**Retention time:** 4.35
**Internal standard:** cimetidine (3.27)

## OTHER SUBSTANCES
**Simultaneous:** aztreonam, sulbactam

## KEY WORDS
stability-indicating; saline; injections

## REFERENCE
Belliveau, P.P.; Nightingale, C.H.; Quintiliani, R. Stability of aztreonam and ampicillin sodium-sulbactam sodium in 0.9% sodium chloride injection. *Am.J.Hosp.Pharm.*, **1994**, *51*, 901–904

## SAMPLE
**Matrix:** formulations
**Sample preparation:** Dilute injection with mobile phase, inject an aliquot.

## HPLC VARIABLES
**Column:** μBondapak C18
**Mobile phase:** MeCN:buffer 17.5:82.5 (Buffer was 5 mM tetrabutylammonium hydroxide adjusted to pH 5.0 with concentrated phosphoric acid.)

**Column temperature:** 25
**Flow rate:** 2
**Injection volume:** 10
**Detector:** UV 230

## CHROMATOGRAM
**Retention time:** 3.3

## OTHER SUBSTANCES
**Simultaneous:** sulbactam

## KEY WORDS
injections

## REFERENCE
Mushinsky, R.F.; Reynolds, M.L.; Nicholson, C.A.; Crider, L.L.; Forcier, G.A. Stability of sulbactam/ampicillin in diluents for parenteral administration. *Rev.Infect.Dis.*, **1986**, *8 Suppl 5*, S523–S527

## SAMPLE
**Matrix:** formulations
**Sample preparation:** Blend tablets and capsules with water in a high-speed blender for 5 min, filter, dilute with mobile phase, inject a 20 μL aliquot. Dilute oral suspensions and injections with mobile phase, inject a 20 μL aliquot.

## HPLC VARIABLES
**Guard column:** 70 mm long Co:Pell ODS
**Column:** 300 × 4.6 10 μm Chromegabond C18 (E.S. Industries)
**Mobile phase:** MeCN:MeOH:10 mM $KH_2PO_4$ 19:11:70
**Flow rate:** 1
**Injection volume:** 20
**Detector:** UV 225

## CHROMATOGRAM
**Retention time:** 4.0
**Limit of detection:** 699 ng/mL

## OTHER SUBSTANCES
**Simultaneous:** amoxicillin, cloxacillin, dicloxacillin, methicillin, nafcillin, oxacillin, penicillin G, penicillin V

## KEY WORDS
tablets; capsules; oral suspensions; injections

## REFERENCE
Briguglio, G.T.; Lau-Cam, C.A. Separation and identification of nine penicillins by reverse phase liquid chromatography. *J.Assoc.Off.Anal.Chem.*, **1984**, *67*, 228–231

## SAMPLE
**Matrix:** milk
**Sample preparation:** 10 mL Milk + 2 mL 200 mM tetraethylammonium chloride, stir, slowly add 38 mL MeCN over 30 s, let stand for 5 min, decant the supernatant through a plug of glass wool. 40 mL Filtrate + 1 mL water, evaporate under reduced pressure to 1-2 mL, make up to 4 mL with water, filter (0.45 μm polyvinylidene difluoride). Inject 2 mL into an LC system (150 × 4.6 5 μm Supelcosil LC-18; MeCN:10 mM $KH_2PO_4$ 0:100

for 3 min, to 40:60 over 27 min, to 0:100 over 1 min; 1 mL/min; UV 210 and 295), collect a 1.5 mL fraction at retention time for ampicillin (16.5 min), add 100 µL reagent, evaporate to 1 mL, inject a 200 µL aliquot. (Reagent was 10 mM phosphoric acid, 10 mM $KH_2PO_4$, and 10 mM sodium decanesulfonate.)

## HPLC VARIABLES
**Column:** 150 × 4.6 5 µm Supelcosil LC-18
**Mobile phase:** MeCN:buffer 33:67 (Buffer was 67 mM phosphoric acid, 3.3 mM $KH_2PO_4$, and 5 mM sodium dodecyl sulfate.)
**Flow rate:** 1
**Injection volume:** 200
**Detector:** UV 210

## CHROMATOGRAM
**Limit of quantitation:** 2-5 ppb

## OTHER SUBSTANCES
**Also analyzed:** amoxicillin, ceftiofur, cephapirin, cloxacillin, penicillin G, penicillin V

## KEY WORDS
cow

## REFERENCE
Moats, W.A.; Harik-Khan, R. Liquid chromatographic determination of β-lactam antibiotics in milk: A multiresidue approach. *J.AOAC Int.*, **1995**, *78*, 49–54

## SAMPLE
**Matrix:** milk
**Sample preparation:** Condition a Bond Elut C8 SPE cartridge with 5 mL MeOH and 5 mL water. 20 mL Milk + 20 mL buffer, heat at 60° for 20 min or until milk curdles, centrifuge for 10 min, add the supernatant to the SPE cartridge, wash with two 2.5 mL portions of water, elute with 2.5 mL MeOH. Evaporate the eluate to dryness under a stream of nitrogen, extract the residue with three 100 µL portions of 50 mM pH 6.0 potassium phosphate buffer, filter (0.2 µm), inject an aliquot of the filtrate. (Buffer was 545 mL 100 mM citric acid, 455 mL 200 mM $Na_2HPO_4$, and 74.4 g EDTA, adjust to pH 4.5 with ammonium hydroxide, make up to 2 L with water.)

## HPLC VARIABLES
**Column:** 250 × 4.6 10 µm Lichrosorb RP-8
**Mobile phase:** MeOH:50 mM pH 6.0 potassium phosphate buffer 35:65
**Flow rate:** 1
**Injection volume:** 200
**Detector:** UV 210; Charm II assay

## CHROMATOGRAM
**Retention time:** 7.19

## OTHER SUBSTANCES
**Extracted:** ceftiofur, cephapirin, cloxacillin, dicloxacillin, nafcillin, oxacillin, penicillin G
**Simultaneous:** amoxicillin

## KEY WORDS
SPE

## REFERENCE
Zomer, E.; Quintana, J.; Saul, S.; Charm, S.E. LC-Receptograms: A method for identification and quantitation of β-lactams in milk by liquid chromatography with microbial receptor assay. *J.AOAC Int.*, **1995**, *78*, 1165–1172

## SAMPLE
**Matrix:** milk
**Sample preparation:** Add 2 volumes MeCN to milk, stand 5 min, decant aqueous portion, suction filter, extract with an equal volume of 1:1 methylene chloride:hexane, centrifuge aqueous phase at 3000 rpm for 10 min. Dilute 3:1 with 20 mM sodium acetate buffer and filter (0.2 μm nylon). Inject 50 μL onto column with mobile phase A, run mobile phase A for 30 min and elute to waste. After 30 min switch to mobile phase B and elute through detector.

## HPLC VARIABLES
**Column:** 100 × 8 Radial-Pak 10μm μBondapak C18
**Mobile phase:** A 20 mM sodium acetate buffer; B Gradient. MeCN:MeOH:20 mM sodium acetate buffer from 15:10:75 to 30:0:70 over 15 min and hold at 30:0:70
**Flow rate:** A 3; B 2
**Injection volume:** 50
**Detector:** E, Waters 464 pulsed electrochemical detector, thin layer cell, Ag/AgCl reference electrode, E1 = 1300 mV for 0.166 s, E2 = 1500 mV for 0.166 s, E3 = -200 mV for 0.333 s.

## CHROMATOGRAM
**Retention time:** 3.2
**Limit of detection:** 0.3 ppm

## OTHER SUBSTANCES
**Simultaneous:** cloxacillin, dicloxacillin, methicillin, nafcillin, oxacillin, penicillin G, penicillin V

## REFERENCE
Kirchmann, E.; Earley, R.L.; Welch, L.E. The electrochemical detection of penicillins in milk. *J.Liq.Chromatogr.*, **1994**, *17*, 1755–1772

## SAMPLE
**Matrix:** milk
**Sample preparation:** 500 μL Milk + 500 μL MeCN:MeOH:water 40:20:40, vortex for 10-15 s, filter (Centricon-10, molecular mass cut-off filter 10000 daltons) with centrifuging at 2677 g for 30 min, inject a 10-100 μL aliquot of the ultrafiltrate.

## HPLC VARIABLES
**Column:** 220 × 2.1 5 μm Spheri-5 phenyl microbore (UV detection) or 220 × 4.6 5 μm Spheri-5 phenyl microbore (MS detection)
**Mobile phase:** MeCN:85% phosphoric acid:triethylamine:water 20:0.4:0.4:79.2 containing 5 mM dodecanesulfonate (UV) or isopropanol:acetic acid in 200 mM ammonium acetate:water 10:2:88 (MS)
**Column temperature:** 50
**Flow rate:** 0.2-0.45 (UV) or 0.8-1.2 (MS)
**Injection volume:** 10-100
**Detector:** UV 220; MS, Finnigan MAT 4800 quadrupole, thermospray, source 320°, vaporizer 120°, pulsed positive ion negative ion

## CHROMATOGRAM
**Retention time:** 8.3 (UV), 14.5 (MS)
**Limit of detection:** 200 ng/mL (MS)

## OTHER SUBSTANCES
**Also analyzed:** amoxicillin, cloxacillin

## KEY WORDS
ultrafiltrate; LC-MS

## REFERENCE
Voyksner, R.D.; Tyczkowska, K.L.; Aronson, A.L. Development of analytical methods for some penicillins in bovine milk by ion-paired chromatography and confirmation by thermospray mass spectrometry. *J.Chromatogr.*, **1991**, *567*, 389–404

## SAMPLE
**Matrix:** milk
**Sample preparation:** Condition a Sep-Pak C18 SPE cartridge with 20 mL MeOH, 20 mL water, and 2 mL 2% NaCl. Pass through 30 g filtered (glass-wool plug) milk at 2 mL/min, wash with 5 mL water, wash with 10 mL MeOH:water:20% NaCl 10:80:10 containing 20 mM 18-crown-6, elute with 10 mL 15% (v/v) MeOH, inject a 100 μL aliquot.

## HPLC VARIABLES
**Guard column:** 50 × 2.1 Permaphase ETH (Du Pont)
**Column:** 150 × 4.3 LiChrosorb RP-18
**Mobile phase:** MeOH:water:0.2 M pH 4.0 phosphate buffer 25:65:10 containing 11 mM sodium 1-heptanesulfonate
**Column temperature:** 45
**Flow rate:** 1
**Injection volume:** 100
**Detector:** UV 210

## CHROMATOGRAM
**Retention time:** 11
**Limit of detection:** 30 ng/g

## OTHER SUBSTANCES
**Extracted:** penicillin G, penicillin V

## KEY WORDS
cow; SPE

## REFERENCE
Terada, H.; Sakabe, Y. Studies on residual antibacterials in foods. IV. Simultaneous determination of penicillin G, penicillin V and ampicillin in milk by high-performance liquid chromatography. *J.Chromatogr.*, **1985**, *348*, 379–387

## SAMPLE
**Matrix:** reaction mixtures
**Sample preparation:** If necessary, remove oxidizing power of solution by adding sodium metabisulfite, inject a 20 μL aliquot.

## HPLC VARIABLES
**Guard column:** 15 × 4.6 5 μm Microsorb C8
**Column:** 250 × 4.6 5 μm Microsorb C8
**Mobile phase:** MeCN:5.5 mM sodium octanesulfonate + 20 mM trisodium citrate dihydrate adjusted to pH 3 with concentrated HCl 23:77
**Flow rate:** 1

**Injection volume:** 20
**Detector:** UV 230

## CHROMATOGRAM
**Retention time:** 6.8
**Limit of detection:** 1 μg/mL

## REFERENCE
Lunn, G.; Rhodes, S.W.; Sansone, E.B.; Schmuff, N.R. Photolytic destruction and polymeric resin decontamination of aqueous solutions of pharmaceuticals. *J.Pharm.Sci.*, **1994**, *83*, 1289–1293

## SAMPLE
**Matrix:** solutions

## HPLC VARIABLES
**Column:** 300 × 3.9 10 μm μBondapak phenyl
**Mobile phase:** MeOH:10 mM phosphate buffer 27:73, pH 3.6
**Column temperature:** 27
**Flow rate:** 1
**Detector:** UV 254

## CHROMATOGRAM
**Retention time:** 7.5

## OTHER SUBSTANCES
**Simultaneous:** amoxicillin, cefaclor, cephalexin, cephradine

## REFERENCE
Huang, H.-S.; Wu, J.-R.; Chen, M.-L. Reversed-phase high-performance liquid chromatography of amphoteric β-lactam antibiotics: effects of columns, ion-pairing reagents and mobile phase pH on their retention times. *J.Chromatogr.*, **1991**, *564*, 195–203

## SAMPLE
**Matrix:** solutions
**Sample preparation:** Prepare an aqueous solution, inject a 200 μL aliquot.

## HPLC VARIABLES
**Guard column:** present but not specified
**Column:** 150 × 4.6 4 μm Micropak SPC-18 C18
**Mobile phase:** Gradient. MeCN:10 mM orthophosphoric acid and 10 mM tetramethylammonium chloride from 15:85 to 60:40 over 20 min.
**Flow rate:** 1
**Injection volume:** 200
**Detector:** UV 220

## CHROMATOGRAM
**Retention time:** 4.5

## OTHER SUBSTANCES
**Simultaneous:** amoxicillin, cephapirin

## REFERENCE
Moats, W.A. Effect of the silica support of bonded reversed-phase columns on chromatography of some antibiotic compounds. *J.Chromatogr.*, **1986**, *366*, 69–78

◆━━━━━━━━━━━━━━━━━━━━━◆━━━━━━━━━━━━━━━━━━━━━◆

## ANNOTATED BIBLIOGRAPHY

Moats, W.A. Determination of ampicillin and amoxicillin in milk with an automated liquid chromatographic cleanup. *J.AOAC Int.*, **1994**, *77*, 41–45 [milk; gradient; extracted amoxicillin, cephapirin; LOQ 10 ppb]

Straub, R.; Linder, M.; Voyksner, R.D. Determination of β-lactam residues in milk using perfusive-particle liquid chromatography combined with ultrasonic nebulization electrospray mass spectrometry. *Anal.Chem.*, **1994**, *66*, 3651–3658 [milk; electrospray; LC-MS; microbore; LOD 10 ppb; extracted amoxicillin, ceftiofur, cephapirin, cloxacillin, penicillin G]

Tyczkowska, K.L.; Voyksner, R.D.; Straub, R.F.; Aronson, A.L. Simultaneous multiresidue analysis of β-lactam antibiotics in bovine milk by liquid chromatography with ultraviolet detection and confirmation by electrospray mass spectrometry. *J.AOAC Int.*, **1994**, *77*, 1122–1131 [cow; milk; LC-MS; UV detection; LOD 10 ppb; column temp 40; extracted amoxicillin, ceftiofur, cephapirin, cloxacillin, penicillin G]

Akhtar, M.J.; Khan, S.; Khan, M.A. Determination of ampicillin in human plasma by high-performance liquid chromatography using ultraviolet detection. *J.Pharm.Biomed.Anal.*, **1993**, *11*, 375–378

Kaniou, I.P.; Zachariadis, G.A.; Stratis, J.A. Separation and determination of five penicillins by reversed phase HPLC. *J.Liq.Chromatogr.*, **1993**, *16*, 2891–2897 [simultaneous amoxicillin, cloxacillin, dicloxacillin, penicillin G]

Straub, R.F.; Voyksner, R.D. Determination of penicillin G, ampicillin, amoxicillin, cloxacillin and cephapirin by high-performance liquid chromatography-electrospray mass spectrometry. *J. Chromatogr.*, **1993**, *647*, 167–181 [milk; extracted amoxicillin, cephapirin, cloxacillin, penicillin G; LC-MS; electrospray; UV detection; LOD 100 ppb]

Nelis, H.J.; Vandenbranden, J.; Verhaeghe, B.; De Kruif, A.; Mattheeuws, D.; De Leenheer, A.P. Liquid chromatographic determination of ampicillin in bovine and dog plasma by using a tandem solid-phase extraction method. *Antimicrob.Agents Chemother.*, **1992**, *36*, 1606–1610

Burns, D.T.; O'Callaghan, M.; Smyth, W.F.; Ayling, C.J. High-performance liquid chromatographic analysis of ampicillin and cloxacillin and its application to an intramammary veterinary preparation. *Fresenius' J.Anal.Chem.*, **1991**, *340*, 53–56

Komarova, N.I.; Krylova, N.S.; Lushanova, G.I.; Chimitova, T..A.; Chernyshev, V.V. Determination of antibacterial drugs in blood serum and urine by microcolumn high-performance liquid chromatography. I. Ampicillin. *Farmakol.Toksikol.(Moscow)*, **1991**, *54*, 65–67

Zhao, C.; He, C.; Zhao, H.; Xie, J.; Wu, Q. HPLC determination of ampicillin in urine. *Shenyang Yaoxueyuan Xuebao*, **1990**, *7*, 1–4

He, S.; Zhu, X.; Ge, L.; Yang, J.; Tian, X.; Zhao, H. Determination of ampicillin in human plasma and its pharmacokinetics by high-pressure liquid chromatography. *Zhongguo Yaoxue Zazhi*, **1989**, *24*, 736–738

Suwanrumpha, S.; Freas, R.B. Identification of metabolites of ampicillin using liquid chromatography/thermospray mass spectrometry and fast atom bombardment tandem mass spectrometry. *Biomed.Environ.Mass.Spectrom.*, **1989**, *18*, 983–994

Abuirjeie, M.A.; Abdel-Hamid, M.E. Simultaneous high-pressure liquid chromatographic analysis of ampicillin and cloxacillin in serum and urine. *J.Clin.Pharm.Ther.*, **1988**, *13*, 101–108

Hikida, K.; Ishii, N.; Inoue, Y.; Ohkura, Y. Determination of ampicillin in serum by automated column-switching HPLC. *Bunseki Kagaku*, **1988**, *37*, 566–569

Ibrahim, E.S.A.; Abdel-Hamid, M.E.; Abuirjeie, M.A.; Hurani, A.M. Rapid high performance liquid chromatographic determination of ampicillin in human urine. *Anal.Lett.*, **1988**, *21*, 423–434

Martín, J.; Méndez, R.; Negro, A. Effect of temperature on HPLC separations of penicillins. *J.Liq.Chromatogr.*, **1988**, *11*, 1707–1716 [simultaneous amoxicillin, cloxacillin, penicillin G, penicillin V, piperacillin%lumn temp 15-55°]

Saesmaa, T. Quantitative high-performance liquid chromatographic determination of ampicillin embonate and amoxycillin embonate. *J.Chromatogr.*, **1988**, *455*, 415–419 [column temp 40; simultaneous amoxicillin, embonic acid; penicillin V]

Haginaka, J.; Wakai, J.; Yasuda, H.; Uno, T.; Takahashi, K.; Katagi, T. High-performance liquid chromatographic determination of ampicillin and its metabolites in rat plasma, bile and urine by post-column degradation with sodium hypochlorite. *J.Chromatogr.*, **1987**, *400*, 101–111

Haginaka, J.; Wakai, J. Liquid chromatographic determination of ampicillin and its metabolites in human urine by postcolumn alkaline degradation. *J.Pharm.Pharmacol.*, **1987**, *39*, 5–8

Margosis, M. Quantitative liquid chromatography of ampicillin: collaborative study. *J.Assoc.Off. Anal.Chem.*, **1987**, *70*, 206–212

Salem, M.A.S.; Alkaysi, H.N. High performance liquid chromatographic analysis and dissolution of ampicillin and cloxacillin in capsule formulation. *Drug Dev.Ind.Pharm.*, **1987**, *13*, 2771–2787

Hutchins, J.E.; Tyczkowska, K.; Aronson, A.L. Determination of ampicillin in serum by using simple ultrafiltration technique and liquid chromatographic analysis. *J.Assoc.Off.Anal.Chem.*, **1986**, *69*, 757–759

Nagata, T.; Saeki, M. Determination of ampicillin residues in fish tissues by liquid chromatography. *J.Assoc.Off.Anal.Chem.*, **1986**, *69*, 448–450

Hikal, A.H.; Jones, A.B. Determination of ampicillin in plasma by paired ion high performance liquid chromatography. *J.Liq.Chromatogr.*, **1985**, *8*, 1455–1464 [plasma; propiophenone (IS)]

Nakagawa, H.; Nishiyama, K.; Higashitani, T.; Ishikawa, S.; Fukui, Y. [Application of high performance liquid chromatography with fluorescence detection to determination of ampicillin in plasma deproteinized by the phenol method]. *Yakugaku Zasshi*, **1985**, *105*, 1096–1099

Lauback, R.G.; Rice, J.J.; Bleiberg, B.; Muhammad, H.; Hanna, S.A. Specific high-performance liquid chromatographic determination of ampicillin in bulks, injectables, capsules, and oral suspensions by reverse-phase ion-pair chromatography. *J.Liq.Chromatogr.*, **1984**, *7*, 1243–1265 [capsules; oral suspensions; injections; bulk; simultaneous degradation products, penicillin V, probenecid]

Sjövall, J.; Westerlund, D.; Alván, G.; Magni, L.; Nord, C.E.; Sörstad, J. Rectal bioavailability of bacampicillin hydrochloride in man as determined by reversed-phase liquid chromatography. *Chemotherapy*, **1984**, *30*, 137–147 [plasma; urine; post-column reaction; LOD 100 ng/mL; pharmacokinetics]

# Aspirin

**Molecular formula:** C₉H₈O₄
**Molecular weight:** 180.2
**CAS Registry No.:** 50-78-2

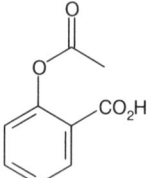

## SAMPLE
**Matrix:** blood
**Sample preparation:** Plasma. 200 μL Plasma + 200 μL 5 μg/mL IS in 200 mM HCL:200 mM orthophosphoric acid 50:50, vortex for 1-2 s, add 400 μL MeCN, vortex, let stand at 4° for 15 min, centrifuge at 10500 g for 1 min. Remove the supernatant and add it to 100-120 mg NaCl, vortex briefly, let stand at 4° for 10 min, vortex, centrifuge at 10500 g for 1 min, inject a 10 μL aliquot of the upper organic layer. Whole blood. 200 μL Lysed whole blood + 400 μL 5 μg/mL IS in 200 mM HCL:200 mM orthophosphoric acid 50:50, vortex for 1-2 s, add 600 μL MeCN, vortex, let stand at 4° for 15 min, centrifuge at 10500 g for 1 min. Remove the supernatant and add it to 200 mg NaCl, vortex briefly, let stand at 4° for 10 min, vortex, centrifuge at 10500 g for 1 min, inject a 10 μL aliquot of the upper organic layer.

## HPLC VARIABLES
**Column:** 150 × 3.9 4 μm Novapak C18
**Mobile phase:** MeCN:water:85% orthophosphoric acid 18:74:0.09 (Before use prime column by recycling 200 mL mobile phase + 400 μL di-n-butylamine overnight at 0.3 mL/min.)
**Column temperature:** 30
**Flow rate:** 1
**Injection volume:** 10
**Detector:** UV 237

## CHROMATOGRAM
**Retention time:** 4.2
**Internal standard:** 2-methylbenzoic acid (8.9)
**Limit of quantitation:** 100 ng/mL

## OTHER SUBSTANCES
**Extracted:** metabolites, gentisic acid, salicylic acid, salicyluric acid

## KEY WORDS
plasma; whole blood; pharmacokinetics

## REFERENCE
Kees, F.; Jehnich, D.; Grobecker, H. Simultaneous determination of acetylsalicylic acid and salicylic acid in human plasma by high-performance liquid chromatography. *J.Chromatogr.B*, **1996**, *677*, 172–177

## SAMPLE
**Matrix:** blood
**Sample preparation:** Add o-anisic acid to 1 mL plasma, acidify with HCl, extract with diethyl ether. Remove the organic layer and evaporate it to dryness, reconstitute the residue, inject an aliquot.

## HPLC VARIABLES
**Column:** 150 × 4.6 5 μm Supelcosil LC-8
**Mobile phase:** MeCN:20 mM phosphoric acid 15:85
**Flow rate:** 1.6
**Detector:** UV 237

## CHROMATOGRAM
**Internal standard:** o-anisic acid
**Limit of quantitation:** 20 ng/mL

## OTHER SUBSTANCES
**Extracted:** salicylic acid

## KEY WORDS
pharmacokinetics; plasma

## REFERENCE
Benedek, I.H.; Joshi, A.S.; Pieniaszek, H.J.; King, S.-Y.P.; Kornhauser, D.M. Variability in the pharmacokinetics and pharmacodynamics of low dose aspirin in healthy male volunteers. *J.Clin. Pharmacol.*, **1995**, *35*, 1181–1186

## SAMPLE
**Matrix:** blood
**Sample preparation:** Filter (0.22 μm), inject a 2 μL aliquot of the filtrate.

## HPLC VARIABLES
**Column:** 250 × 4 5 μm LiChrospher 100 Diol
**Mobile phase:** MeCN:50 mM pH 3.0 phosphate buffer 1.5:98.5
**Flow rate:** 0.5
**Injection volume:** 2
**Detector:** UV 254

## CHROMATOGRAM
**Retention time:** 8

## OTHER SUBSTANCES
**Extracted:** salicylic acid

## KEY WORDS
direct injection; serum

## REFERENCE
Nimura, N.; Itoh, H.; Kinoshita, T. Diol-bonded silica gel as a restricted access packing forming a binary-layered phase for direct injection of serum for the determination of drugs. *J.Chromatogr.A*, **1995**, *689*, 203–210

## SAMPLE
**Matrix:** blood
**Sample preparation:** 2 mL Whole blood or plasma + 2 mL buffer + 5 mL chloroform: isopropanol:n-heptane 60:14:26, shake gently horizontally for 10 min, centrifuge at 2800 g for 10 min. Remove the lower organic layer and evaporate it to dryness under vacuum at 45°, reconstitute the residue in 100 μL mobile phase, centrifuge at 2800 g for 5 min, inject a 50 μL aliquot of the supernatant. (Buffer was saturated ammonium chloride solution 25% diluted with water, adjusted to pH 9.5 with 25% ammonia solution.)

## HPLC VARIABLES
**Column:** 300 × 3.9 4 μm NovaPack C18
**Mobile phase:** MeOH:THF:buffer 65:5:30 (Buffer was 0.68 g/L (10 mM (sic)) $KH_2PO_4$ adjusted to pH 2.6 with concentrated orthophosphoric acid.) (At the end of each session wash the column with water for 1 h and MeOH for 1 h, re-equilibrate for 30 min.)
**Column temperature:** 30

**Flow rate:** 0.8
**Injection volume:** 50
**Detector:** UV 233

## CHROMATOGRAM
**Retention time:** 3.40
**Limit of detection:** <120 ng/mL

## OTHER SUBSTANCES
**Extracted:** acebutolol, acenocoumarol, acepromazine, aceprometazine, acetaminophen, aconitine, ajmaline, alimemazine, alminoprofen, alpidem, alprazolam, alprenolol, amitriptyline, amodiaquine, amoxapine, astemizole, benazepril, benperidol, benzocaine, benzoylecgonine, bepridil, betaxolol, bisoprolol, bromazepam, brompheniramine, bumadizone, bupivacaine, buprenorphine, buspirone, caffeine, carbamazepine, carbinoxamine, carpipramine, carteolol, celiprolol, cetirizine, chlorambucil, chlordiazepoxide, chlorophenacinone, chloroquine, chlorpheniramine, chlorpromazine, chlorpropamide, cibenzoline, cicletanine, clemastine, clobazam, clomipramine, clonazepam, clonidine, clorazepate, clozapine, cocaine, colchicine, cyamemazine, cyclizine, cycloguanil, cyproheptadine, cytarabine, dacarbazine, daunorubicin, debrisoquine, demexiptiline, desipramine, dextromethorphan, dextromoramide, dextropropoxyphene, diazepam, diazoxide, diclofenac, dihydralazine, diltiazem, diphenhydramine, dipyridamole, disopyramide, dosulepine, doxepin, doxylamine, droperidol, ephedrine, estazolam, etodolac, fenfluramine, fenoprofen, fentiazac, flecainide, floctafenine, flumazenil, flunitrazepam, fluoxetine, fluphenazine, flurbiprofen, fluvoxamine, glibenclamide, glibortnuride, glipizide, glutethimide, haloperidol, histapyrrodine, hydroxychloroquine, hydroxyzine, ibuprofen, imipramine, indomethacin, iproniazid, ketamine, ketoprofen, labetalol, levomepromazine, lidocaine, lidoflazine, loperamide, loprazolam, loratadine, lorazepam, loxapine, maprotiline, medazepam, medifoxamine, mefenamic acid, mefenidramine, mefloquine, melphalan, meperidine, mephenesin, mephentermine, mepivacaine, metapramine, methadone, methaqualone, methocarbamol, methotrexate, metipranolol, metoclopramide, metoprolol, mexiletine, mianserine, midazolam, minoxidil, moclobemide, moperone, nadolol, nalbuphine, nalorphine, naloxone, naproxen, nialamide, nicardipine, nifedipine, niflumic acid, nimodipine, nitrazepam, nitrendipine, nizatidine, nomifensine, nortriptyline, omeprazole, opipramol, oxazepam, oxprenolol, penbutolol, penfluridol, pentazocine, phencyclidine, phenylbutazone, pimozide, pindolol, pipamperone, piroxicam, prazepam, prazosin, prilocaine, procainamide, procarbazine, proguanil, promethazine, propafenone, propranolol, protriptyline, pyrimethamine, quinidine, quinine, quinupramine, ramipril, reserpine, secobarbital, sotalol, strychnine, sulfinpyrazole, sulindac, sulpride, suriclone, temazepam, tenoxicam, terfenadine, tetracaine, tetrazepam, thiopental, thioproperazine, thioridazine, tianeptine, tiaprofenic acid, ticlopidine, timolol, tioclomarol, tofisopam, tolbutamide, trazodone, triazolam, trifluoperazine, trifluperidol, trimipramine, triprolidine, tropatenine, verapamil, viloxazine, vinblastine, vincristine, vindesine, warfarin, yohimbine, zolpidem, zopiclone, zorubicine

**Interfering:** albuterol, amisulpride, atenolol, chlormezanone, codeine, lisinopril, metformin, naltrexone, phenobarbital, phenol, ranitidine, ritodrine, sultopride, terbutaline, tiapride, toloxatone

## KEY WORDS
whole blood; plasma

## REFERENCE
Tracqui, A.; Kintz, P.; Mangin, P. Systematic toxicological analysis using HPLC/DAD. *J.Forensic Sci.*, **1995**, *40*, 254–262

## SAMPLE
**Matrix:** blood
**Sample preparation:** 1 mL Plasma + 0.2 mL 1 M HCl + 10 mL diethyl ether, gently mix for 10 min, centrifuge at 1500 rpm for 4 min. Remove the organic phase, evaporate it to dryness at 0° under a stream of nitrogen, add 200 µL mobile phase, vortex 90 s, inject a 5-100 µL aliquot.

## HPLC VARIABLES
**Guard column:** 23 × 3.9 μBondapak C18/Porasil B
**Column:** 300 × 3.9 10 μm μBondapak C18
**Mobile phase:** MeOH:water:1-butanol:orthophosphoric acid 270:720:10:0.13
**Column temperature:** 47
**Flow rate:** 1.8
**Injection volume:** 5-100
**Detector:** UV 234

## CHROMATOGRAM
**Retention time:** 5.6
**Internal standard:** m-anisic acid (9.6)
**Limit of quantitation:** 10-15 ng/mL

## OTHER SUBSTANCES
**Simultaneous:** salicylic acid
**Noninterfering:** acetaminophen, albuterol, aminophylline, amitriptyline, atenolol, beclomethasone, bromazepam, caffeine, carbamazepine, chloral hydrate, chlordiazepoxide, cimetidine, clonazepam, codeine, desipramine, dexamethasone, dextropropoxyphene, diazepam, dicyclomine, digoxin, disopyramide, doxycycline, ergotamine, ethosuximide,
furosemide, gentisic acid, haloperidol, hydrocortisone, imipramine, indomethacin, levodopa, lignocaine, lithium carbonate, meperidine, methdilazine, methylphenobarbitone,
methylprednisolone, methysergide, metoclopramide, metoprolol, mexiletine, midazolam,
naphthoxyacetic acid, nitrazepam, nitroglycerin, nortripyline, oxazepam, oxpranolol, pentobarbitone, pethidine, phenytoin, prednisolone, prednisone, primidone, procainamide,
prochlorperazine, propranolol, quinidine, salicyluric acid, spironolactone, sulfamethoxazole, theophylline, trimethoprim, valproic acid, verapamil, warfarin
**Interfering:** methyclothiazide

## KEY WORDS
plasma; pharmacokinetics

## REFERENCE
Brandon, R.A.; Eadie, M.J.; Smith, M.T. A sensitive liquid chromatographic assay for plasma aspirin
and salicylate concentrations after low doses of aspirin. *Ther.Drug Monit.*, **1985**, *7*, 216–221

## SAMPLE
**Matrix:** blood
**Sample preparation:** 50 μL Serum + 50 μL 15 μg/mL β-hydroxyethyltheophylline in
MeCN, mix for 30 s, centrifuge at 13000 g for 5 min. Inject the supernatant (about 20
μL).

## HPLC VARIABLES
**Column:** μBondapak C18
**Mobile phase:** MeCN:buffer 9.75:90.25 (Buffer was 100 mM $KH_2PO_4$ adjusted to pH 4.0
with phosphoric acid.) (At the end of each day clean column with water for 20 min and
MeOH for 30 min.)
**Flow rate:** 2
**Injection volume:** 20
**Detector:** UV 254

## CHROMATOGRAM
**Retention time:** 12
**Internal standard:** β-hydroxyethyltheophylline (5.8)
**Limit of detection:** 500 ng/mL

## OTHER SUBSTANCES
**Extracted:** salicylic acid
**Simultaneous:** acetaminophen, acetylprocainamide, caffeine, dyphylline, procainamide, theophylline
**Noninterfering:** benzoic acid

## KEY WORDS
serum

## REFERENCE
Ou, C.-N.; Frawley, V.L. Theophylline, dyphylline, caffeine, acetaminophen, salicylate, acetylsalicylate, procainamide, and N-acetylprocainamide determined in serum with a single liquid-chromatographic assay. *Clin.Chem.*, **1982**, *28*, 2157–2160

## SAMPLE
**Matrix:** blood
**Sample preparation:** Mix plasma with an equal volume of MeCN, vortex for 30 s, centrifuge at 900 g for 5 min, inject a 100 μL aliquot of the supernatant.

## HPLC VARIABLES
**Column:** 300 × 3.9 10 μm μBondapak C18
**Mobile phase:** MeOH:acetic acid:water 22:5:73
**Flow rate:** 2.6
**Injection volume:** 100
**Detector:** UV 280

## CHROMATOGRAM
**Retention time:** 4.75
**Limit of quantitation:** 2 μg/mL

## OTHER SUBSTANCES
**Extracted:** salicylic acid
**Noninterfering:** acetaminophen, albuterol, aminophylline, amitriptyline, amoxicillin, ampicillin, amobarbital, beclomethasone, carbamazepine, carbenicillin, chlordiazepoxide, cimetidine, clonazepam, cyproheptadine, debrisoquine, dextropropoxyphene, diazepam, digoxin, dihydroxyanthraquinone, ergotamine, ethosuximide, fluphenazine, furosemide, gentamicin, gentisic acid, guaifenesin, haloperidol, heparin, hydrocortisone, indomethacin, methdilazine, methyclothiazide, methylphenobarbitone, methysergide, metoclopramide, naproxen, nitrazepam, nystatin, penicillin, pentobarbitone, phenytoin, phenytoin, pizotifen, prazosin, prednisone, prochlorperazine, propranolol, spironolactone, sulfamethoxazole, theophylline, trifluoperazine, trimethoprim, valproic acid

## KEY WORDS
plasma

## REFERENCE
Cham, B.E.; Ross-Lee, L.; Bochner, F.; Imhoff, D.M. Measurement and pharmacokinetics of acetylsalicylic acid by a novel high performance liquid chromatographic assay. *Ther.Drug Monit.*, **1980**, *2*, 365–372

## SAMPLE
**Matrix:** blood, urine
**Sample preparation:** Plasma. 500 μL Plasma + 900 μL 270 mM HCl + 100 μL 100 μg/mL α-phenylcinnamic acid in MeOH + 10 mL dichloromethane, shake at 125 cycles/min for 15 min, centrifuge at 750 g for 5 min. Remove the organic layer and evaporate it to

dryness, reconstitute the residue in 500 μL MeOH, inject a 25 μL aliquot. Urine. 2 mL Urine + 900 μL 270 mM HCl + 100 μL 100 μg/mL α-phenylcinnamic acid in MeOH + 10 mL hexane, shake at 125 cycles/min for 15 min, centrifuge at 750 g for 5 min. Remove the organic layer and evaporate it to dryness, reconstitute the residue in 500 μL MeOH, inject a 25 μL aliquot.

## HPLC VARIABLES
**Column:** 300 × 4 μBondapak C18
**Mobile phase:** MeOH:1% acetic acid 60:40
**Flow rate:** 2
**Injection volume:** 25
**Detector:** UV 280

## CHROMATOGRAM
**Retention time:** 2.5
**Internal standard:** α-phenylcinnamic acid (8.0)
**Limit of detection:** 2 μg/mL

## OTHER SUBSTANCES
**Extracted:** salicylic acid (UV 300), salsalate (UV 300)

## KEY WORDS
plasma; pharmacokinetics

## REFERENCE
Harrison, L.I.; Funk, M.L.; Ober, R.E. High-pressure liquid chromatographic determination of salicyl-salicylic acid, aspirin, and salicylic acid in human plasma and urine. *J.Pharm.Sci.*, **1980**, *69*, 1268–1271

## SAMPLE
**Matrix:** bulk, formulations
**Sample preparation:** Bulk. Prepare a 20 mg/mL solution of bulk aspirin in dichloromethane, inject a 10 μL aliquot as soon as dissolution is complete. Tablets. Prepare a 20 mg/mL solution of ground aspirin tablets in dichloromethane, filter (0.45 μm) immediately, immediately inject a 10 μL aliquot of the filtrate.

## HPLC VARIABLES
**Column:** 150 × 4.6 6 μm Zorbax SIL
**Mobile phase:** Hexane:chloroform:acetic acid 80:19:3 (Before first use pump 10 column volumes of dichloromethane:acetic acid:2,3-dimethoxypropane 96:2:2 through column at 3 mL/min.)
**Flow rate:** 3
**Injection volume:** 10
**Detector:** UV 254

## CHROMATOGRAM
**Retention time:** 3.6

## OTHER SUBSTANCES
**Simultaneous:** impurities, salicylic acid, salsalate

## KEY WORDS
normal phase; tablets

## REFERENCE
Pfeiffer, C.D.; Pankey, J.W. Determination of related compounds in aspirin by liquid chromatography. *J.Pharm.Sci.*, **1982**, *71*, 511–514

## SAMPLE
**Matrix:** formulations
**Sample preparation:** Weigh out powdered sample containing 68 mg aspirin, add 80 mL MeOH, sonicate for 10 min, dilute to 100 mL with MeOH, centrifuge. Remove a 5 mL aliquot of the supernatant and add it to 1 mL 2 mg/mL resorcinol, add 2 mL MeOH, make up to 20 mL with 50 mM pH 3.0 triethylamine phosphate, inject an aliquot.

## HPLC VARIABLES
**Column:** 150 × 3.2 5 µm Hypersil ODS
**Mobile phase:** THF:50 mM pH 3.0 triethylamine phosphate 12:88
**Flow rate:** 0.6
**Injection volume:** 20
**Detector:** UV 275 following post-column reaction. The column effluent flowed through a 10 m × 0.3 mm ID crocheted PTFE coil irradiated with an 8 W low-pressure mercury lamp at 254 nm to the detector.

## CHROMATOGRAM
**Retention time:** 15
**Internal standard:** resorcinol (9)

## OTHER SUBSTANCES
**Simultaneous:** acetaminophen (post-column irradiation gives little increase in peak height), caffeine (post-column irradiation gives little increase in peak height), propyphenazone (post-column irradiation gives a decrease in peak height)

## KEY WORDS
post-column reaction

## REFERENCE
Di Pietra, A.M.; Gatti, R.; Andrisano, V.; Cavrini, V. Application of high-performance liquid chromatography with diode-array detection and on-line post-column photochemical derivatization to the determination of analgesics. *J.Chromatogr.A*, **1996**, *729*, 355–361

## SAMPLE
**Matrix:** formulations
**Sample preparation:** Condition a 500 mg Extract Clean silica SPE cartridge (Alltech stock no. 209250) with 2 mL hexane. Allow a solution of aspirin in 10 mM sorbitan trioleate in CFC-11 to evaporate, dissolve the residue in 5 mL hexane. Add 1 mL to the SPE cartridge, elute with two 2 mL portions of mobile phase, make up eluate to 5 mL with mobile phase, inject a 20 µL aliquot.

## HPLC VARIABLES
**Guard column:** 20 × 2 30-40 µm Perisorb RP-8 Pellicular (Upchurch)
**Column:** 250 × 4.6 5 µm Econosphere C8
**Mobile phase:** MeOH:THF:1 M phosphoric acid:water 44:5:5:46
**Flow rate:** 1
**Injection volume:** 20
**Detector:** UV 275

## CHROMATOGRAM
**Retention time:** 5.6

## OTHER SUBSTANCES
**Extracted:** degradation products, salicylic acid

## KEY WORDS
aerosols; SPE

## REFERENCE
Blondino, F.E.; Byron, P.R. The quantitative determination of aspirin and its degradation products in a model solution aerosol. *J.Pharm.Biomed.Anal.*, **1995**, *13*, 111–119

## SAMPLE
**Matrix:** formulations
**Sample preparation:** Powder tablets, add 40 mg pyrimethamine, dissolve in 20 mL MeCN, add 40 mL mobile phase, filter (paper), wash filter with mobile phase, make up filtrate to 100 mL with mobile phase. Dilute a 5 mL aliquot to 50 mL with mobile phase, inject a 20 µL aliquot.

## HPLC VARIABLES
**Column:** 250 × 4 10 µm Nucleosil C18
**Mobile phase:** MeOH:MeCN:water:triethylamine 55:5:40:0.1, pH adjusted to 4.0 with phosphoric acid
**Flow rate:** 1.5
**Injection volume:** 20
**Detector:** UV 254

## CHROMATOGRAM
**Retention time:** 2.5
**Internal standard:** pyrimethamine (3.5)
**Limit of quantitation:** 6 µg/mL

## OTHER SUBSTANCES
**Simultaneous:** dipyridamole

## KEY WORDS
tablets

## REFERENCE
Sane, R.T.; Ghadge, J.K.; Jani, A.B.; Vaidya, A.J.; Kotwal, S.S. Simultaneous high-performance liquid chromatographic determination of haloperidol with propantheline bromide, nalidixic acid with phenazopyridine hydrochloride, and dipyridamole with aspirin in combined dosage (forms). *Indian Drugs*, **1992**, *29*, 240–244

## SAMPLE
**Matrix:** formulations
**Sample preparation:** Oils. 1 mL Sample + 25 mL MeOH:water 90:10, shake vigorously for 5 min, centrifuge, inject a 10 µL aliquot of the supernatant. Tablets. Grind a tablet to a fine powder, add 25 mL MeOH, sonicate for 5-10 min, filter (0.45 µm), discard first 5 mL of filtrate, inject a 10 µL aliquot of the remaining filtrate. Suspensions (aqueous). Make up 5 mL to 50 mL with MeOH, filter (0.45 µm), discard first 5 mL of filtrate, inject a 10 µL aliquot of the remaining filtrate.

## HPLC VARIABLES
**Column:** 250 × 4.6 5 µm Zorbax ODS
**Mobile phase:** MeOH:water 75:25
**Flow rate:** 1.5
**Injection volume:** 10
**Detector:** UV 240

## CHROMATOGRAM
**Retention time:** 2.5
**Limit of detection:** 5 µg/mL

## OTHER SUBSTANCES
**Simultaneous:** benzyl alcohol, benzyl benzoate, boldenone, calusterone, cortisone, dehydroepiandrosterone (UV 210), ethisterone, fluoxymesterone, mesterolone (UV 210), methandriol (UV 210), methandrostenolone, methenolone acetate, methyltestosterone, mibolerone, nandrolone, nandrolone acetate, nandrolone propionate, norethandrolone, norethindrone, norethindrone acetate, norgestrel, oxandrolone (UV 210), oxymetholone, stanozolol, testosterone, testosterone acetate, testosterone propionate, trenbolone acetate
**Interfering:** caffeine, formebolone, testolactone

## KEY WORDS
oils; tablets; suspensions

## REFERENCE
Walters, M.J.; Ayers, R.J.; Brown, D.J. Analysis of illegally distributed anabolic steroid products by liquid chromatography with identity confirmation by mass spectrometry or infrared spectrophotometry. *J.Assoc.Off.Anal.Chem.*, **1990**, *73*, 904–926

## SAMPLE
**Matrix:** formulations
**Sample preparation:** Place 5 tablets in MeCN:MeOH:85% phosphoric acid 92:8:0.5, sonicate 15 min, shake 15 min, dilute to 250 mL. Centrifuge an aliquot in 50 mL tube at 2000 rpm for 15 min and filter supernatant (0.45 µm).

## HPLC VARIABLES
**Column:** 150 × 3.9 Resolve (Waters)
**Mobile phase:** MeCN:water:85% phosphoric acid 24:76:0.5
**Column temperature:** 35
**Flow rate:** 2
**Injection volume:** 10
**Detector:** UV 295

## CHROMATOGRAM
**Retention time:** 2.2

## OTHER SUBSTANCES
**Simultaneous:** acetaminophen, caffeine, salicylic acid

## KEY WORDS
film coated tablets; tablets

## REFERENCE
Fogel, J.; Epstein, P.; Chen, P. Simultaneous high-performance liquid chromatography assay of acetylsalicylic acid and salicylic acid in film-coated aspirin tablets. *J.Chromatogr.*, **1984**, *317*, 507–511

## SAMPLE
**Matrix:** formulations
**Sample preparation:** Grind tablets to a fine powder and add about 250 mg aspirin to 100 mL chloroform saturated with citric acid containing 500 µL formic acid, add 500 mg solid citric acid, sonicate for 2 min, centrifuge or filter, inject an aliquot. (If buffers or antacid are present, add ground tablets equivalent to about 500 mg aspirin to 3 g acid-washed siliceous earth, mix, add 2 mL 6 M HCl, mix, add to a 200 × 25 column, dry wash

container with siliceous earth, add to column, elute column with chloroform saturated with citric acid at 10 mL/min. Collect 150 mL eluent, add 1 mL formic acid, make up to 200 mL with chloroform saturated with citric acid, add 500 mg citric acid, shake, inject an aliquot.)

## HPLC VARIABLES
**Column:** 250 × 4.6 5μm Zorbax-Sil
**Mobile phase:** Chloroform:dichloromethane:acetonitrile:formic acid 700:300:30:4 (At the end of the day wash the column with 200 mL MeOH.)
**Flow rate:** 2
**Injection volume:** 20
**Detector:** UV 300

## CHROMATOGRAM
**Retention time:** 8

## OTHER SUBSTANCES
**Simultaneous:** acetylsalicylic acid anhydride, acetylsalicylsalicylic acid, salicylic acid, excipients

## KEY WORDS
normal phase; tablets; SPE

## REFERENCE
Galante, R.N.; Visalli, A.J.; Grim, W.M. Stabilized normal-phase high-performance liquid chromatographic analysis of aspirin and salicylic acid in solid pharmaceutical dosage forms. *J.Pharm.Sci.*, **1984**, *73*, 195–197

## SAMPLE
**Matrix:** formulations
**Sample preparation:** Pulverize tablets and weigh out 1 g, add 1 mL formic acid, add 25 mL MeOH, shake mechanically for 10 min, make up to 50 mL with methanol. Remove 10 mL and centrifuge. 5 mL Supernatant + 5 mL 0.0025% p-hydroxybenzoic acid in MeOH:water 20:80, make up to 25 mL with water, inject an aliquot. (Analyze within 1 h.)

## HPLC VARIABLES
**Column:** 250 × 4.6 LiChrosorb RP8
**Mobile phase:** MeOH:200 mM pH 3.5 phosphate buffer:water 20:10:70
**Flow rate:** 1
**Injection volume:** 10
**Detector:** UV 254

## CHROMATOGRAM
**Retention time:** 24
**Internal standard:** p-hydroxybenzoic acid (18)

## OTHER SUBSTANCES
**Simultaneous:** acetaminophen, O-acetyl-p-aminophenol, 2-O-acetylascorbic acid, 3-O-acetylascorbic acid, p-aminophenol, diacetyl-p-aminophenol (UV 280), saccharin, salicylic acid (UV 280), vitamin C

## KEY WORDS
tablets

## REFERENCE

Thomis, R.; Roets, E.; Hoogmartens, J. Analysis of tablets containing aspirin, acetaminophen, and ascorbic acid by high-performance liquid chromatography. *J.Pharm.Sci.*, **1984**, *73*, 1830–1833

## SAMPLE

**Matrix:** solutions

## HPLC VARIABLES

**Column:** 250 × 4.1 6 μm PolyEncap ODS (n-octadecylacrylate copolymerized with vinyl silica in heptane, carrier Ultrasep ES 100; preparation described in paper)
**Mobile phase:** MeCN:pH 2.2 phosphate buffer 32:68
**Flow rate:** 1
**Detector:** UV 220

## CHROMATOGRAM

**Retention time:** 4

## OTHER SUBSTANCES

**Simultaneous:** diazepam, diphenhydramine, o-hydroxyhippuric acid, MPPH, niacin, toluene

## REFERENCE

Engelhardt, H.; Cuñat-Walter, M.A. Polymer encapsulated stationary phases with improved efficiency. *Chromatographia*, **1995**, *40*, 657–661

## SAMPLE

**Matrix:** solutions

## HPLC VARIABLES

**Column:** 250 × 4.6 8 μm Unisphere-PBD (polybutadiene on alumina) (Biotage, Charlottesville, VA)
**Mobile phase:** MeCN:water 20:80
**Flow rate:** 1
**Detector:** UV 254

## CHROMATOGRAM

**Retention time:** 5

## OTHER SUBSTANCES

**Simultaneous:** acetaminophen, phenacetin, salicylamide

## REFERENCE

Jedrejewski, P.T.; Taylor, L.T. Comparison of silica-, alumina-, and polymer-based stationary phases for reversed-phase liquid chromatography. *J.Chromatogr.Sci.*, **1995**, *33*, 438–445

## SAMPLE

**Matrix:** solutions

## HPLC VARIABLES

**Guard column:** 30 × 3.2 7 μm SI 100 ODS (not commercially available)
**Column:** 150 × 3.2 7 μm SI 100 ODS (not commercially available)
**Mobile phase:** MeCN:buffer 31.2:68.8 (Buffer was 6.66 g $KH_2PO_4$ and 4.8 g 85% phosphoric acid in 1 L water, pH 2.3.)
**Flow rate:** 0.5-1
**Detector:** UV 231

## CHROMATOGRAM
**Retention time:** 3.4
**Internal standard:** 5-(4-methylphenyl)-5-phenylhydantoin (7.3)

## OTHER SUBSTANCES
**Also analyzed:** carbamazepine, chlordiazepoxide, chlorprothixene, clonazepam, caffeine, diazepam, doxylamine, ethosuximide, furosemide, haloperidol, hydrochlorothiazide, methocarbamol, methotrimeprazine, nicotine, oxazepam, procaine, promazine, propafenone, propranolol, salicylamide, temazepam, tetracaine, thiopental, triamterene, verapamil, zolpidem, zopiclone

## REFERENCE
Below, E.; Burrmann, M. Application of HPLC equipment with rapid scan detection to the identification of drugs in toxicological analysis. *J.Liq.Chromatogr.*, **1994**, *17*, 4131–4144

## SAMPLE
**Matrix:** solutions

## HPLC VARIABLES
**Column:** 250 × 4.6 Zorbax RX
**Mobile phase:** Gradient. A was 10 mL concentrated orthophosphoric acid and 7 mL triethylamine in 1 L water. B was 10 mL concentrated orthophosphoric acid and 7 mL triethylamine in 200 mL water, make up to 1 L with MeCN. A:B from 100:0 to 0:100 over 30 min, maintain at 0:100 for 5 min.
**Column temperature:** 30
**Flow rate:** 2
**Detector:** UV 210

## OTHER SUBSTANCES
**Also analyzed:** acepromazine, acetaminophen, acetophenazine, albuterol, aminophylline, amitriptyline, amobarbital, amoxapine, amphetamine, amylocaine, antipyrine, aprobarbital, atenolol, atropine, avermectin, barbital, benzocaine, benzoic acid, benzotropine, benzphetamine, berberine, bibucaine, bromazepan, brompheniramine, buprenorphine, buspirone, butabarbital, butacaine, butethal, caffeine, carbamazepine, carbromal, chloramphenicol, chlordiazepoxide, chloroquine, chlorothiazide, chloroxylenol, chlorphenesin, chlorpheniramine, chlorpromazine, chlorpropamide, chlortetracycline, cimetidine, cinchonidine, cinchonine, clenbuterol, clonazepam, clonixin, clorazepate, cocaine, codeine, colchicine, cortisone, coumarin, cyclazocine, cyclobenzaprine, cyclothiazide, cyheptamide, cymarin, danazol, danthron, dapsone, debrisoquine, desipramine, dexamethasone, dextromethorphan, dextropropoxyphene, diamorphine, diazepam, diclofenac, diethylpropion, diethylstilbestrol, diflunisal, digitoxin, digoxin, diltiazem, diphenhydramine, diphenoxylate, diprenorphine, dipyrone, disulfiram, dopamine, doxapram, doxepin, dronabinol, ephedrine, epinephrine, epinine, estradiol, estriol, estrone, ethacrynic acid, ethosuximide, etonitazene, etorphine, eugenol, famotidine, fenbendazole, fencamfamine, fenoprofen, fenproporex, fentanyl, flubendazole, flufenamic acid, flunitrazepam, 5-fluorouracil, fluoxymesterone, fluphenazine, furosemide, gentisic acid, gitoxigenin, glipizide, glunixin, glutethimide, glybenclamide, guaiacol, halazepam, haloperidol, hydrochlorothiazide, hydrocodone, hydrocortisone, hydromorphone, hydroxyquinoline, ibogaine, ibuprofen, iminostilbene, imipramine, indomethacin, isocarbostyril, isocarboxazid, isoniazid, isoproterenol, isoxsuprine, ivermectin, ketamine, ketoprofen, kynurenic acid, levorphanol, lidocaine, lorazepam, lormetazepam, loxapine, mazindol, mebendazole, meclizine, meclofenamic acid, medazepam, mefenamic acid, megestrol, mepacrine, meperidine, mephentermine, mephenytoin, mephesin, mephobarbital, mepivacaine, mescaline, mesoridazine, methadone, methamphetamine, methapyrilene, methaqualone, methazolamide, methocarbamol, methoxamine, methsuximide, methyl salicylate, methyldopa, methyldopamine, methylphenidate, methylprednisolone, methyltestosterone, methyprylon, metoprolol, mibolerone, morphine, nadolol, nalorphine, naloxone, naltrexone, naphazoline, na-

proxen, nefopam, niacinamide, nicotine, nicotinic acid, nifedipine, niflumic acid, nitrazepam, norepinephrine, nortriptyline, noscapine, nylidrin, oxazepam, oxycodone, oxymorphone, oxyphenbutazone, oxytetracycline, papaverine, pargyline, pemoline, pentazocine, pentobarbital, persantine, phenacetin, phenazocine, phenazopyridine, phencyclidine, phendimetrazine, phenelzine, pheniramine, phenobarbital, phenothiazine, phensuximide, phentermine, phenylbutazone, phenylephrine, phenylpropanolamine, piperocaine, prazepam, prednisolone, primidone, probenecid, progesterone, propiomazine, propranolol, propylparaben, pseudoephedrine, puromycin, pyrilamine, pyrithyldione, quazepam, quinaldic acid, quinidine, quinine, ranitidine, recinnamine, reserpine, resorcinol, saccharin, albuterol, salicylamide, salicylic acid, scopolamine, scopoletin, secobarbital, strychnine, sulfacetamide, sufadiazine, sulfadimethoxine, sulfaethidole, sulfamerazine, sulfamethazine, sulfamethoxizole, sulfanilamide, sulfapyridine, sulfasoxizole, sulindac, tamoxifen, temazepam, testosterone, tetracaine, tetracycline, tetramisole, thebaine, theobromine, theophylline, thiabendazole, thiamine, thiamylal, thiobarbituric acid, thioridazine, thiosalicylic acid, thiothixene, thymol, tolazamide, tolazoline, tobutamide, tolmetin, tranylcypromine, triamcinolone, tribenzylamine, trichloromethiazide, trifluoperazine, trihexyphenidyl, trimethoprim, tripelennamine, triproilidine, tropacocaine, tyramine, verapamil, vincamine, warfarin, yohimbine, zoxazolamine

## REFERENCE
Hill, D.W.; Kind, A.J. Reversed-phase solvent gradient HPLC retention indexes of drugs. *J.Anal.Toxicol.*, **1994**, *18*, 233–242

## SAMPLE
**Matrix:** solutions
**Sample preparation:** Prepare a 50 µg/mL solution in the mobile phase, inject a 10 µL aliquot.

## HPLC VARIABLES
**Column:** 250 × 4.6 7 µm Lichrosorb RP 18
**Mobile phase:** MeOH:water 45:55 containing 1% acetic acid
**Flow rate:** 1
**Injection volume:** 10
**Detector:** UV 230

## CHROMATOGRAM
**Retention time:** 5.09

## OTHER SUBSTANCES
**Simultaneous:** acetaminophen, phenacetin, salicylamide, salicylic acid

## REFERENCE
Nivaud-Guernet, E.; Guernet, M.; Ivanovic, D.; Medenica, M. Effect of eluent pH on the ionic and molecular forms of the non-steroidal anti-inflammatory agents in reversed-phase high-performance liquid chromatography. *J.Liq.Chromatogr.*, **1994**, *17*, 2343–2357

## SAMPLE
**Matrix:** solutions

## HPLC VARIABLES
**Column:** 250 × 4 OmniPac PAX-500 (Dionex)
**Mobile phase:** Gradient. A was MeCN:10 mM sodium carbonate 18:82. B was MeCN:50 mM sodium carbonate 33:67. A:B from 100:0 to 0:100 over 10 min.
**Flow rate:** 1
**Detector:** UV 254

## CHROMATOGRAM
**Retention time:** 3

## OTHER SUBSTANCES
**Simultaneous:** carprofen, diflunisal, fenbufen, ibuprofen, indomethacin, naproxen, tolmetin

## REFERENCE
Slingsby, R.W.; Rey, M. Determination of pharmaceuticals by multi-phase chromatography: Combined reversed phase and ion exchange in one column. *J.Liq.Chromatogr.*, **1990**, *13*, 107−134

## SAMPLE
**Matrix:** solutions
**Sample preparation:** Dissolve compounds in MeOH, inject a 1 μL aliquot.

## HPLC VARIABLES
**Column:** 150 × 1 3 μm Hitachi-Gel 3011 porous polymer (Hitachi)
**Mobile phase:** MeOH:ammonia 99:1
**Flow rate:** 0.03
**Injection volume:** 1
**Detector:** UV 254

## CHROMATOGRAM
**Retention time:** 3.52

## OTHER SUBSTANCES
**Also analyzed:** acetaminophen, bucetin (3-hydroxy-p-butyrophenetidine), caffeine, dipyrone (sulpyrin), ethenzamide (o-ethoxybenzamide), mefenamic acid, phenacetin, salicylamide, salicylic acid, theobromine, theophylline

## KEY WORDS
semi-micro; porous polymer

## REFERENCE
Matsushima, Y.; Nagata, Y.; Niyomura, M.; Takakusagi, K.; Takai, N. Analysis of antipyretics by semimicro liquid chromatography. *J.Chromatogr.*, **1985**, *332*, 269−273

## SAMPLE
**Matrix:** urine
**Sample preparation:** Acidify 5 mL urine to pH 1 with 40% phosphoric acid, shake with two 5 mL aliquots of diethyl ether, centrifuge. Remove the organic layer and evaporate it to dryness under a stream of nitrogen at 40°, reconstitute the residue in 2 mL MeOH, inject a 25 μL aliquot.

## HPLC VARIABLES
**Column:** μBondapak C18 radial compression
**Mobile phase:** MeCN:0.085% phosphoric acid 20:80
**Flow rate:** 1.5
**Injection volume:** 25
**Detector:** UV 237

## CHROMATOGRAM
**Limit of detection:** 500 ng/mL

## KEY WORDS
horse

## REFERENCE
Beaumier, P.M.; Fenwick, J.D.; Stevenson, A.J.; Weber, M.P.; Young, L.M. Presence of salicylic acid in standardbred horse urine and plasma after various feed and drug administrations. *Equine.Vet.J.*, **1987**, *19*, 207–213

◆─────────────────────────◆─────────────────────────◆

## ANNOTATED BIBLIOGRAPHY

Smigol, V.; Svec, F.; Fréchet, J.M.J. Novel uniformly sized polymeric stationary phase with hydrophilized large pores for direct injection HPLC determination of drugs in biological fluids. *J.Liq.Chromatogr.*, **1994**, *17*, 891–911 [plasma; cow; extracted salicylic acid]

Wongyai, S. Synthesis and characterization of phenylpropanolamine bonded silica for multimode liquid chromatography of small molecules. *Chromatographia*, **1994**, *38*, 485–490 [simultaneous benzoic acid, salicylic acid]

Hays, P.A.; Lurie, I.S. Quantitative analysis of adulterants in illicit heroin samples via reversed phase HPLC. *J.Liq.Chromatogr.*, **1991**, *14*, 3513–3517 [simultaneous acetaminophen, acetylcodeine, acetylmorphine, benzocaine, caffeine, chloroquine, diamorphine, diazepam, diphenhydramine, dipyrone, lidocaine, methaqualone, monoacetylmorphine, morphine, nicotinamide, noscapine, papaverine, phenacetin, phenobarbital, phenolphthalein, N-phenyl-2-naphthylamine, salicylic acid, strychnine]

Shen, J.; Wanwimolruk, S.; Clark, C.R.; Roberts, M.S. A sensitive assay for aspirin and its metabolites using reversed-phase ion-pair high-performance liquid chromatography. *J.Liq.Chromatogr.*, **1990**, *13*, 751–761 [simultaneous metabolites gentisic acid, salicyluric acid; LOD 50 ng/mL]

Lam, S.; Malikin, G. An improved micro-scale protein precipitation procedure for HPLC assay of therapeutic drugs in serum. *J.Liq.Chromatogr.*, **1989**, *12*, 1851–1872 [serum; also acetaminophen, amiodarone, caffeine, chloramphenicol, flecainide, pentobarbital, procainamide, pyrimethamine, quinidine, theophylline, tocainide, trazodone; fluorescence detection; UV detection]

Lurie, I.S.; McGuiness, K. The quantitation of heroin and selected basic impurities via reversed phase HPLC. II. The analysis of adulterated samples. *J.Liq.Chromatogr.*, **1987**, *10*, 2189–2204 [UV detection; electrochemical detection; simultaneous acetaminophen, acetylcodeine, acetylmorphine, acetylprocaine, aminopyrene, amitriptyline, antipyrene, barbital, benztropine, caffeine, cocaine, codeine, diamorphine, diazepam, diphenhydramine, dipyrone, ephedrine, ethylmorphine, lidocaine, meconin, methamphetamine, methapyrilene, methaqualone, monoacetylmorphine, morphine, nalorphine (IS), niacinamide, noscapine, papaverine, phenacetin, phenmetrazine, phenobarbital, phenolphthalein, procaine, propanophenone, propoxyphene, pyrilamine, quinidine, quinine, salicylamide, salicylic acid, secobarbital, strychnine, tartaric acid, tetracaine, thebaine, tripelennamine, tropacocaine, vitamin $B_5$, vitamin $B_3$]

Lau, A.H.; Chang, C.W.; Schlesinger, P.K. Evaluation of a potential drug interaction between sucralfate and aspirin. *Clin.Pharmacol.Ther.*, **1986**, *39*, 151–155 [plasma; pharmacokinetics; extracted metabolites, salicylic acid, salicyluric acid; α-phenylcinnamic acid (IS); gradient; column temp 30]

Mamolo, M.G.; Vio, L.; Maurich, V. High-pressure liquid chromatographic analysis of paracetamol, caffeine and acetylsalicylic acid in tablets. Salicylic acid quantitation. *Farmaco.[Prat].*, **1985**, *40*, 111–123 [tablets; simultaneous acetaminophen, caffeine, phenazone, salicylic acid]

Pedersen, A.K.; FitzGerald, G.A. Preparation and analysis of deuterium-labeled aspirin: application to pharmacokinetic studies. *J.Pharm.Sci.*, **1985**, *74*, 188–192 [stability; simultaneous salicylic acid]

Bevitt, R.N.; Mather, J.R.; Sharman, D.C. Minimization of salicylic acid formation during preparation of aspirin products for analysis by high-performance liquid chromatography. *Analyst*, **1984**, *109*, 1327–1329

Sreenivasan, K.; Nair, P.D.; Rathinam, K. A GPC method for analysis of low molecular weight drugs. *J.Liq.Chromatogr.*, **1984**, *7*, 2297–2305 [GPC; SEC; simultaneous salicylic acid]

Das Gupta, V. Simultaneous quantitation of acetaminophen, aspirin, caffeine, codeine phosphate, phenacetin, and salicylamide by high-pressure liquid chromatography. *J.Pharm.Sci.*, **1980**, *69*, 110–113

Kirchhoefer, R.D. Simultaneous determination of aspirin and salicylic acid in bulk aspirin and in plain, buffered, and enteric-coated tablets by high-pressure liquid chromatography with UV and fluorescence detectors. *J.Pharm.Sci.*, **1980**, *69*, 1188–1191

# Astemizole

**Molecular formula:** $C_{28}H_{31}FN_4O$
**Molecular weight:** 458.6
**CAS Registry No.:** 68844-77-9

## SAMPLE
**Matrix:** blood
**Sample preparation:** 2 mL Whole blood or plasma + 2 mL buffer + 5 mL chloroform: isopropanol:n-heptane 60:14:26, shake gently horizontally for 10 min, centrifuge at 2800 g for 10 min. Remove the lower organic layer and evaporate it to dryness under vacuum at 45°, reconstitute the residue in 100 μL mobile phase, centrifuge at 2800 g for 5 min, inject a 50 μL aliquot of the supernatant. (Buffer was saturated ammonium chloride solution 25% diluted with water, adjusted to pH 9.5 with 25% ammonia solution.)

## HPLC VARIABLES
**Column:** 300 × 3.9 4 μm NovaPack C18
**Mobile phase:** MeOH:THF:buffer 65:5:30 (Buffer was 0.68 g/L (10 mM (sic)) $KH_2PO_4$ adjusted to pH 2.6 with concentrated orthophosphoric acid.) (At the end of each session wash the column with water for 1 h and MeOH for 1 h, re-equilibrate for 30 min.)
**Column temperature:** 30
**Flow rate:** 0.8
**Injection volume:** 50
**Detector:** UV 278

## CHROMATOGRAM
**Retention time:** 5.73
**Limit of detection:** <120 ng/mL

## OTHER SUBSTANCES
**Extracted:** acebutolol, acenocoumarol, acepromazine, aceprometazine, acetaminophen, aconitine, ajmaline, albuterol, alimemazine, alminoprofen, alpidem, alprazolam, alprenolol, amisulpride, amitriptyline, amodiaquine, amoxapine, aspirin, atenolol, benazepril, benperidol, benzocaine, benzoylecgonine, bepridil, betaxolol, bromazepam, brompheniramine, bumadizone, bupivacaine, buprenorphine, buspirone, caffeine, carbamazepine, carbinoxamine, carpipramine, carteolol, celiprolol, chlorambucil, chlordiazepoxide, chlormezanone, chlorophenacinone, chloroquine, chlorpromazine, chlorpropamide, cicletanine, clemastine, clobazam, clomipramine, clonazepam, clonidine, clorazepate, clozapine, cocaine, codeine, colchicine, cyamemazine, cyclizine, cycloguanil, cyproheptadine, cytarabine, dacarbazine, daunorubicin, debrisoquine, demexiptiline, desipramine, dextromethorphan, dextromoramide, dextropropoxyphene, diazepam, diazoxide, diclofenac, dihydralazine, diphenhydramine, dipyridamole, disopyramide, dosulepine, doxepin, doxylamine, droperidol, ephedrine, estazolam, etodolac, fenfluramine, fenoprofen, fentiazac, flecainide, floctafenine, flumazenil, flunitrazepam, fluoxetine, fluphenazine, flurbiprofen, fluvoxamine, glibenclamide, glipizide, glutethimide, haloperidol, histapyrrodine, hydroxychloroquine, hydroxyzine, ibuprofen, imipramine, indomethacin, iproniazid, ketamine, ketoprofen, labetalol, levomepromazine, lidocaine, lidoflazine, lisinopril, loperamide, loratadine, lorazepam, loxapine, maprotiline, medazepam, mefenamic acid, mefenidramine, mefloquine, melphalan, meperidine, mephenesin, mephentermine, mepivacaine, metapramine, metformin, methadone, methaqualone, methocarbamol, methotrexate, metipranolol, metoclopramide, metoprolol, mexiletine, mianserine, midazolam, minoxidil, moclobemide, morphine, nadolol, nalbuphine, nalorphine, naloxone, naltrexone, naproxen, nialamide, nifedipine, niflumic acid, nimodipine, nitrazepam, nitrendipine, nizatidine, nomifensine, nortriptyline, omeprazole, opipramol, oxazepam, oxprenolol, penbutolol, pen-

fluridol, pentazocine, phencyclidine, phenobarbital, phenol, phenylbutazone, pimozide, pindolol, pipamperone, piroxicam, prazepam, prazosin, prilocaine, procainamide, procarbazine, proguanil, promethazine, propafenone, propranolol, protriptyline, pyrimethamine, quinidine, quinine, quinupramine, ramipril, ranitidine, reserpine, ritodrine, secobarbital, sotalol, strychnine, sulfinpyrazole, sulindac, sulpride, sultopride, suriclone, temazepam, tenoxicam, terbutaline, terfenadine, tetracaine, tetrazepam, thiopental, thioproperazine, thioridazine, tianeptine, tiapride, tiaprofenic acid, ticlopidine, timolol, tioclomarol, tofisopam, tolbutamide, toloxatone, trazodone, triazolam, trifluoperazine, trifluperidol, trimipramine, triprolidine, tropatenine, verapamil, viloxazine, vincristine, vindesine, warfarin, yohimbine, zolpidem, zopiclone, zorubicine

**Interfering:** bisoprolol, cetirizine, chlorpheniramine, cibenzoline, diltiazem, glibornuride, loprazolam, medifoxamine, moperone, nicardipine, vinblastine

## KEY WORDS
whole blood; plasma

## REFERENCE
Tracqui, A.; Kintz, P.; Mangin, P. Systematic toxicological analysis using HPLC/DAD. *J.Forensic Sci.*, **1995**, *40*, 254–262

## SAMPLE
**Matrix:** blood, tissue

**Sample preparation:** Plasma. 2 mL Plasma + 100 μL 1 μg/mL IS in MeOH + 2 mL 50 mM borax solution + 4 mL heptane:isoamyl alcohol 95:5, rotate at 10 rpm for 10 min, centrifuge at 1000 g for 5 min. Remove upper organic layer and repeat extraction of aqueous layer. Combine extracts, add 3 mL 50 mM sulfuric acid, extract. Remove acidic aqueous phase and make it alkaline with concentrated ammonia, extract twice with 2 mL heptane:isoamyl alcohol 95:5, combine organic layers and evaporate them to dryness under a stream of nitrogen at 55°, take up in 50 μL MeOH, inject the whole sample. Tissue. Grind tissue in a Waring blender, homogenize 1:4 in water. 2 mL homogenate + 100 μL 1 μg/mL IS in MeOH + 2 mL 50 mM borax solution + 4 mL heptane:isoamyl alcohol 95:5, rotate at 10 rpm for 10 min, centrifuge at 1000 g for 5 min. Remove upper organic layer and repeat extraction of aqueous layer. Combine extracts, add 3 mL 50 mM sulfuric acid, extract. Remove acidic aqueous phase and make it alkaline with concentrated ammonia, extract twice with 2 mL heptane:isoamyl alcohol 95:5, combine organic layers and evaporate them to dryness under a stream of nitrogen at 55°, take up in 50 μL MeOH, inject the whole sample.

## HPLC VARIABLES
**Column:** 150 × 2.1 5 μm Alltech RSiL C18HL
**Mobile phase:** MeCN:water 50:50 + 0.05% diethylamine
**Flow rate:** 0.6
**Injection volume:** 50
**Detector:** UV 254

## CHROMATOGRAM
**Retention time:** 5.8
**Internal standard:** 1-[(4-fluorophenyl)methyl]-N-{1-[2-(4-ethoxyphenyl)ethyl]-4-piperidinyl}-1H-benzimidazol-2-amine (R 44 180) (8.3)
**Limit of detection:** 1 ng/mL (plasma); 5 ng/g (tissue)

## OTHER SUBSTANCES
**Extracted:** metabolites, desmethylastemizole

## KEY WORDS
plasma; human; dog

## REFERENCE

Woestenborghs, R.; Embrechts, L.; Heykants, J. Simultaneous determination of astemizole and its demethylated metabolite in animal plasma and tissues by high-performance liquid chromatography. *J.Chromatogr.*, **1983**, *278*, 359–366

## SAMPLE
**Matrix:** solutions
**Sample preparation:** Inject an aliquot of a solution in MeOH.

## HPLC VARIABLES
**Column:** 150 × 4.6 5 μm MicroPack MCH-5
**Mobile phase:** MeOH:30 mM pH 3.0 phosphate buffer 85:15
**Flow rate:** 1.5
**Injection volume:** 20
**Detector:** UV 276

## REFERENCE

Fernández Otero, G.C.; Lucangioli, S.E.; Carducci, C.N. Adsorption of drugs in high-performance liquid chromatography injector loops. *J.Chromatogr.A*, **1993**, *654*, 87–91

# Atenolol

**Molecular formula:** $C_{14}H_{22}N_2O_3$
**Molecular weight:** 266.3
**CAS Registry No.:** 29122-68-7

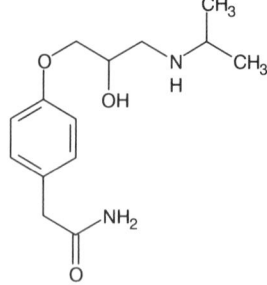

## SAMPLE
**Matrix:** blood

## HPLC VARIABLES
**Column:** 83 × 4.6 3 μm Pecosphere C18
**Mobile phase:** MeCN:MeOH:10 mM pH 4.0 sodium acetate containing 10 mM octanesulfonic acid 17:2:80.1
**Detector:** F ex 228 em 310

## KEY WORDS
pharmacokinetics; plasma

## REFERENCE
Hartmann, D.; Stief, G.; Lingenfelder, M.; Güzelhan, C.; Horsch, A.K. Study on the possible interaction between tenoxicam and atenolol in hypertensive patients. *Arzneimittelforschung*, **1995**, *45*, 494–498

## SAMPLE
**Matrix:** blood
**Sample preparation:** Condition a 3 mL cyano SPE cartridge (J.T.Baker) with 2 mL water. 400 μL Serum + 20 μL practolol solution, vortex, add to the SPE cartridge, elute with 1 mL MeOH:triethylamine 99:1. Evaporate the eluate to dryness under a stream of nitrogen at 37°, reconstitute the residue in 80 μL mobile phase, inject a 50 μL aliquot.

## HPLC VARIABLES
**Column:** 250 × 4.6 5 μm Supelcosil LC-8-OB
**Mobile phase:** MeCN:20 mM ammonium dihydrogen phosphate 6:94 containing 2% triethylamine, pH adjusted to 5.0
**Flow rate:** 1
**Injection volume:** 50
**Detector:** UV 220

## CHROMATOGRAM
**Retention time:** 10
**Internal standard:** practolol (14)
**Limit of detection:** 50 ng/mL

## KEY WORDS
serum; SPE; pharmacokinetics

## REFERENCE
Phelps, S.J.; Alpert, B.S.; Ward, J.L.; Pieper, J.A.; Lima, J.J. Absorption pharmacokinetics of atenolol in patients with the Marfan syndrome. *J.Clin.Pharmacol.*, **1995**, *35*, 268–274

## SAMPLE

**Matrix:** blood

**Sample preparation:** 1 mL Plasma + 50 µL 2 M KOH + 100 µL pH 12.2 glycine buffer + 5 mg NaCl, vortex, add 8 mL dichloromethane:1-butanol 95:5, extract. Remove 7 mL of the organic layer and evaporate it to dryness under a stream of nitrogen, reconstitute the residue in 1 mL 1 mg/mL 4-dimethylaminopyridine in dioxane, add 500 µL 20% phosgene in toluene, vortex, heat at 70° for 1 h, heat at 40° overnight. Evaporate to dryness under a stream of nitrogen, reconstitute the residue in 50 µL mobile phase, inject a 20 µL aliquot.

## HPLC VARIABLES

**Column:** 250 × 4 LiChrosorb Si 100 modified with (R, R)-DACH-DNB (see J. Chromatogr. 1991, 539, 25)

**Mobile phase:** Dichloromethane:MeOH 98:2

**Flow rate:** 1

**Injection volume:** 20

**Detector:** F ex 230 em 300

## CHROMATOGRAM

**Internal standard:** (R, S)-n-pentyl propranolol hydrochloride

**Limit of detection:** 0.5-0.6 ng/mL

## KEY WORDS

plasma; chiral; derivatization

## REFERENCE

Stoschitzky, K.; Kahr, S.; Donnerer, J.; Schumacher, M.; Luha, O.; Maier, R.; Klein, W.; Lindner, W. Stereoselective increase of plasma concentrations of the enantiomers of propranolol and atenolol during exercise. *Clin.Pharmacol.Ther.*, **1995**, *57*, 543–551

## SAMPLE

**Matrix:** blood

**Sample preparation:** 2 mL Whole blood or plasma + 2 mL buffer + 5 mL chloroform: isopropanol:n-heptane 60:14:26, shake gently horizontally for 10 min, centrifuge at 2800 g for 10 min. Remove the lower organic layer and evaporate it to dryness under vacuum at 45°, reconstitute the residue in 100 µL mobile phase, centrifuge at 2800 g for 5 min, inject a 50 µL aliquot of the supernatant. (Buffer was saturated ammonium chloride solution 25% diluted with water, adjusted to pH 9.5 with 25% ammonia solution.)

## HPLC VARIABLES

**Column:** 300 × 3.9 4 µm NovaPack C18

**Mobile phase:** MeOH:THF:buffer 65:5:30 (Buffer was 0.68 g/L (10 mM (sic)) $KH_2PO_4$ adjusted to pH 2.6 with concentrated orthophosphoric acid.) (At the end of each session wash the column with water for 1 h and MeOH for 1 h, re-equilibrate for 30 min.)

**Column temperature:** 30

**Flow rate:** 0.8

**Injection volume:** 50

**Detector:** UV 225

## CHROMATOGRAM

**Retention time:** 3.33

**Limit of detection:** <120 ng/mL

## OTHER SUBSTANCES

**Extracted:** acebutolol, acenocoumarol, acepromazine, aceprometazine, acetaminophen, aconitine, ajmaline, alimemazine, alminoprofen, alpidem, alprazolam, alprenolol, amisulpride, amitriptyline, amodiaquine, amoxapine, astemizole, benazepril, benperidol, benzocaine, benzoylecgonine, bepridil, betaxolol, bisoprolol, bromazepam, brompheniramine, bumadizone, bupivacaine, buprenorphine, buspirone, caffeine, carbamazepine, carbinoxamine, carpipramine, carteolol, celiprolol, cetirizine, chlorambucil, chlordiazepoxide, chlorophenacinone, chloroquine, chlorpheniramine, chlorpromazine, chlorpropamide, cibenzoline, cicletanine, clemastine, clobazam, clomipramine, clonazepam, clonidine, clorazepate, clozapine, cocaine, codeine, colchicine, cyamemazine, cyclizine, cycloguanil, cyproheptadine, cytarabine, dacarbazine, daunorubicin, debrisoquine, demexiptiline, desipramine, dextromethorphan, dextromoramide, dextropropoxyphene, diazepam, diazoxide, diclofenac, dihydralazine, diltiazem, diphenhydramine, dipyridamole, disopyramide, dosulepine, doxepin, doxylamine, droperidol, ephedrine, estazolam, etodolac, fenfluramine, fenoprofen, fentiazac, flecainide, floctafenine, flunitrazepam, fluoxetine, fluphenazine, flurbiprofen, fluvoxamine, glibenclamide, glibornuride, glipizide, glutethimide, haloperidol, histapyrrodine, hydroxychloroquine, hydroxyzine, ibuprofen, imipramine, indomethacin, iproniazid, ketamine, ketoprofen, labetalol, levomepromazine, lidocaine, lidoflazine, lisinopril, loperamide, loprazolam, loratadine, lorazepam, loxapine, maprotiline, medazepam, medifoxamine, mefenamic acid, mefenidramine, mefloquine, melphalan, meperidine, mephenesin, mephentermine, mepivacaine, metapramine, methadone, methaqualone, methocarbamol, methotrexate, metipranolol, metoclopramide, metoprolol, mexiletine, mianserine, midazolam, minoxidil, moclobemide, moperone, nadolol, nalbuphine, nalorphine, naloxone, naltrexone, naproxen, nialamide, nicardipine, nifedipine, niflumic acid, nimodipine, nitrazepam, nitrendipine, nizatidine, nomifensine, nortriptyline, omeprazole, opipramol, oxazepam, oxprenolol, penbutolol, penfluridol, pentazocine, phencyclidine, phenylbutazone, pimozide, pindolol, pipamperone, piroxicam, prazepam, prazosin, prilocaine, procainamide, procarbazine, proguanil, promethazine, propafenone, propranolol, protriptyline, pyrimethamine, quinidine, quinine, quinupramine, ramipril, reserpine, secobarbital, sotalol, strychnine, sulindac, sultopride, suriclone, temazepam, tenoxicam, terfenadine, tetracaine, tetrazepam, thiopental, thioproperazine, thioridazine, tianeptine, tiaprofenic acid, ticlopidine, timolol, tioclomarol, tofisopam, tolbutamide, trazodone, triazolam, trifluoperazine, trifluperidol, trimipramine, triprolidine, tropatenine, verapamil, viloxazine, vinblastine, vincristine, vindesine, warfarin, yohimbine, zolpidem, zopiclone, zorubicine

**Interfering:** albuterol, aspirin, chlormezanone, flumazenil, metformin, morphine, phenobarbital, phenol, ranitidine, ritodrine, sulfinpyrazole, sulpride, terbutaline, tiapride, toloxatone

## KEY WORDS
whole blood; plasma

## REFERENCE
Tracqui, A.; Kintz, P.; Mangin, P. Systematic toxicological analysis using HPLC/DAD. *J.Forensic Sci.*, **1995**, *40*, 254–262

## SAMPLE
**Matrix:** blood

**Sample preparation:** Inject 150 μL plasma onto column A with mobile phase A and elute to waste, after 11 min switch 10 mL of eluate containing atenolol onto column B. After 5 min elute column A to waste again and elute column B with mobile phase C onto column C, after 1.25 min, remove column B from the circuit, elute column C with mobile phase C, monitor the effluent from column C. When not in use column B is washed with mobile phase B.

## HPLC VARIABLES
**Column:** A 100 × 7.6 Asahipak GS220M size exclusion (Asahi Chemical); B 50 × 4.6 Chemcosorb 7-ODS-H (Chemco Scientific); C 150 × 6 Shinwa β-perphenylcarbamate bonded silica (Shinwa Chemical)

**Mobile phase:** A pH 7.5 phosphate buffer (I = 0.01); B water; C EtOH:20 mM pH 4.6 $NaH_2PO_4$ 10:90

**Flow rate:** A 2; B 1; C 1
**Injection volume:** 150
**Detector:** F ex 270 em 305

## CHROMATOGRAM
**Retention time:** 19.8 ((S)-atenolol), 24.3 ((R)-atenolol)
**Limit of quantitation:** 10 ng/mL

## KEY WORDS
plasma; chiral; direct injection; column-switching; heart-cut

## REFERENCE
He, J.; Shibukawa, A.; Nakagawa, T.; Wada, H.; Fujima, H.; Imai, E.; Go-oh, Y. Direct injection analysis of atenolol enantiomers in plasma using an achiral/chiral coupled column HPLC system. *Chem.Pharm.Bull.*, **1993**, *41*, 544–548

## SAMPLE
**Matrix:** blood
**Sample preparation:** Condition a Bond Elut Si SPE cartridge by washing twice with 1 mL MeOH, twice with 1 mL water, and once with 1 mL 100 mM pH 9.2 $K_2HPO_4$. Add 1 mL plasma + 100 μL water, wash twice with 1 mL water, centrifuge at 1000 g for 5 min, elute with 1 mL MeOH. Evaporate MeOH to dryness at 40° under a stream of air and dissolve residue in 200 μL mobile phase, inject an aliquot.

## HPLC VARIABLES
**Column:** 250 × 4.6 Spherisorb S5 SCX
**Mobile phase:** MeOH:MeCN:water 40:40:20 containing 0.2% perchloric acid (apparent pH 1.7)
**Flow rate:** 1.5
**Injection volume:** 100
**Detector:** F ex 200 no emission filter

## CHROMATOGRAM
**Retention time:** 13
**Internal standard:** atenolol

## OTHER SUBSTANCES
**Extracted:** albuterol, terbutaline
**Noninterfering:** aminophylline, beclomethasone, cloprednol, dexamethasone, fenoterol, ipratropium bromide, methylprednisolone, orciprenaline, prednisolone, reproterol, rimiterol, salmeterol, sodium cromoglycate, theophylline

## KEY WORDS
plasma; SPE; atenolol is IS

## REFERENCE
McCarthy, P.T.; Atwal, S.; Sykes, A.P.; Ayres, J.G. Measurement of terbutaline and salbutamol in plasma by high performance liquid chromatography with fluorescence detection. *Biomed.Chromatogr.*, **1993**, *7*, 25–28

## SAMPLE
**Matrix:** blood
**Sample preparation:** 1 mL Whole blood + 1 mL 2.5 μg/mL methoxamine in 100 mM NaOH, vortex, add 5 mL ethyl acetate, shake at 230 oscillations/min on a reciprocating shaker for 15 min, centrifuge at 2000 g for 15 min, repeat the extraction with 3 mL ethyl acetate. Combine the organic layers and evaporate them to dryness under a stream of

nitrogen at 37°, reconstitute the residue in 50 μL 100 mM NaOH, vortex briefly, add 200 μL 200 mM (−)-menthyl chloroformate in MeCN, vortex for 30 s, let stand at room temperature for 10 min, inject a 50 μL aliquot.

## HPLC VARIABLES
**Column:** 150 × 4.6 5 μm Hypersil ODS
**Mobile phase:** MeCN:MeOH:water 43:25:32
**Flow rate:** 1.2
**Injection volume:** 50
**Detector:** F ex 230 em 305

## CHROMATOGRAM
**Retention time:** 9.8 (S-(−)), 11.0 (R-(+))
**Internal standard:** methoxamine (14.0, 14.8 (enantiomers))
**Limit of quantitation:** 12.5 ng/mL

## KEY WORDS
derivatization; chiral; whole blood; pharmacokinetics

## REFERENCE
Miller, R.B.; Guertin, Y. High-performance liquid chromatographic assay for the derivatized enantiomers of atenolol in whole blood. *J.Liq.Chromatogr.*, **1992**, *15*, 1289–1302

## SAMPLE
**Matrix:** blood
**Sample preparation:** 500 μL Serum + 500 μL 10 M NaOH + 300 mg NaCl + 5 mL diethyl ether, shake for 10 min, centrifuge at 1500 g for 5 min. Remove 4 mL of the organic layer and evaporate it to dryness under a stream of nitrogen, dissolve the residue in 200 μL mobile phase, inject a 50 μL aliquot.

## HPLC VARIABLES
**Column:** 150 × 4.6 5 μm STR ODS-M (Shimadzu)
**Mobile phase:** MeCN:50 mM ammonium acetate adjusted to pH 4.5 with acetic acid 15:85
**Column temperature:** 35
**Flow rate:** 0.8
**Injection volume:** 50
**Detector:** F ex 230 em 300

## CHROMATOGRAM
**Retention time:** 4
**Internal standard:** atenolol

## OTHER SUBSTANCES
**Extracted:** nadolol

## KEY WORDS
atenolol is IS; serum

## REFERENCE
Noguchi, H.; Yoshida, K.; Murano, M.; Naruto, S. Determination of nadolol in serum by high-performance liquid chromatography with fluorimetric detection. *J.Chromatogr.*, **1992**, *573*, 336–338

## SAMPLE
**Matrix:** blood

**Sample preparation:** 100 μL Plasma + 5 μL 100 μg/mL racemic practolol in MeOH + 50 μL 0.5 M pH 12 glycine buffer + 50 μL 2 M NaOH + 1 mL saturated NaCl + 4 mL dichloromethane containing 3% (v/v) heptafluoro-1-butanol, extract for 10 min, centrifuge at 3015 g at 4° for 5 min, remove organic phase and evaporate to dryness at room temperature under a stream of nitrogen. Dissolve the residue in 40 μL 1 M pH 8.5 borate buffer and 50 μL 1 mM (+)-1-(9-fluorenyl)ethyl chloroformate (Flec) in acetone, let stand 30 min at room temperature, add 100 μL 30 mM hydroxyproline, after 2 min vortex mix with 300 μL n-pentane for 15 s, centrifuge at 3015 g at 4° for 10 min, discard n-pentane. Shake aqueous layer with 3 mL dichloromethane for 10 min, centrifuge at 3015 g at 4° for 10 min, remove organic phase and evaporate to dryness at room temperature under a stream of nitrogen. Dissolve the residue in 50 μL mobile phase, inject an aliquot.

## HPLC VARIABLES
**Column:** 100 × 4.6 3μm Microspher C18 (Chrompack)
**Mobile phase:** MeCN:10 mM pH 7 sodium acetate buffer 50:50
**Flow rate:** 0.8
**Injection volume:** 10
**Detector:** F ex 227 em 310

## CHROMATOGRAM
**Retention time:** 5.95 (S-(−)), 6.55 (R-(+))
**Internal standard:** practolol (8, 9)
**Limit of quantitation:** 10 ng/mL

## KEY WORDS
plasma; chiral; rat; derivatization

## REFERENCE
Rosseel, M.T.; Vermeulen, A.M.; Belpaire, F.M. Reversed-phase high-performance liquid chromatographic analysis of atenolol enantiomers in plasma after chiral derivatization with (+)-1-(9-fluorenyl)ethyl chloroformate. *J.Chromatogr.*, **1991**, *568*, 239–245

## SAMPLE
**Matrix:** blood
**Sample preparation:** 500 μL Serum or plasma + 50 μL water + 100 μL buffer + 5 mL dichloromethane:isopropanol, rotate at 40 rpm for 5 min, centrifuge at 800 g for 3 min. Remove the lower organic layer and evaporate it to dryness under a stream of nitrogen at 55°, reconstitute the residue in 100 μL 50 mM sulfuric acid, vortex for 15 s, inject a 30-50 μL aliquot. (Buffer was prepared by adjusting the pH of a saturated solution of disodium tetraborate to 9 with 6 M HCl.)

## HPLC VARIABLES
**Column:** 150 × 4.5 5 μm ODS Hypersil
**Mobile phase:** MeCN:10 mM pH 3.2 phosphate buffer 20:80 containing 3 mM sodium 1-octanesulfonate
**Flow rate:** 1
**Injection volume:** 30-50
**Detector:** UV 226

## CHROMATOGRAM
**Retention time:** 4.0
**Internal standard:** atenolol

## OTHER SUBSTANCES
**Extracted:** sotalol

## KEY WORDS
serum; plasma; atenolol is IS

## REFERENCE
Urech, R.; Chan, L.; Duffy, P. High-performance liquid chromatographic assay of sotalol: improved procedure and investigation of peak broadening. *J.Chromatogr.*, **1990**, *534*, 271–278

## SAMPLE
**Matrix:** blood
**Sample preparation:** Condition a 1 mL Bond-Elut CN SPE cartridge with 2 mL MeOH then 2 mL water. 400 µL Plasma + 200 ng practolol, add to SPE cartridge, wash twice with 1 mL water, wash with 1 mL acetone, allow to go dry. Elute with three 200 µL aliquots of eluting solvent, combine the fractions, evaporate under nitrogen, suspend in 80 µL mobile phase, inject a 50 µL aliquot. (Eluting solvent was 10 mM acetic acid and 50 mM triethylamine in MeOH.)

## HPLC VARIABLES
**Guard column:** Brownlee cyano
**Column:** 250 × 4.6 6 µm Zorbax CN
**Mobile phase:** MeCN:12.5 mM $(NH_4)H_2PO_4$:triethylamine 4:96:0.25, pH adjusted to 5.5 with 1.0 M phosphoric acid
**Flow rate:** 1.5
**Injection volume:** 50
**Detector:** UV 224

## CHROMATOGRAM
**Retention time:** 6
**Internal standard:** practolol (9)
**Limit of detection:** < 10 ng/mL
**Limit of quantitation:** 25 ng/mL

## OTHER SUBSTANCES
**Simultaneous:** disopyramide, lidocaine, metoprolol, nadolol, procainamide, quinidine, timolol, verapamil

## KEY WORDS
plasma; SPE

## REFERENCE
Verghese, C.; McLeod, A.; Shand, D. Rapid high-performance liquid chromatographic method for the measurement of atenolol in plasma using UV detection. *J.Chromatogr.*, **1983**, *275*, 367–375

## SAMPLE
**Matrix:** blood, urine
**Sample preparation:** Condition a 100 mg Baxter C18 SPE cartridge with one volume MeOH and two volumes water. Dilute 10 µL urine to 500 µL with water. 500 µL Serum or diluted urine + 50 µL 5 µg/mL albuterol in water, vortex for 30 s, add to SPE cartridge, wash with three 200 µL aliquots of water, elute with two 500 µL aliquots of MeOH. Evaporate the eluates to dryness under a stream of air at 40-45°, reconstitute the residue in 150 µL water, vortex for 30 s, centrifuge at 14000 g for 4 min, inject a 50 µL aliquot.

## HPLC VARIABLES
**Guard column:** 5 µm Adsorbosphere C-18
**Column:** 250 × 4.6 5 µm Adsorbosphere C-18
**Mobile phase:** MeCN:buffer 7:93 adjusted to pH 3.0 with 85% phosphoric acid (Buffer was 25 mM $(NH_4)H_2PO_4$ and 1 mM N,N-dimethyloctylamine.)

**Flow rate:** 1.5
**Injection volume:** 50
**Detector:** UV 224; F ex 228 em 310

## CHROMATOGRAM
**Retention time:** 10.4
**Internal standard:** albuterol (7.1)
**Limit of quantitation:** 50 ng/mL (F, UV)

## KEY WORDS
pharmacokinetics; SPE; serum

## REFERENCE
Chatterjee, D.J.; Li, W.Y.; Hurst, A.K.; Koda, R.T. High-performance liquid chromatographic method for determination of atenolol from human plasma and urine: Simultaneous fluorescence and ultraviolet detection. *J.Liq.Chromatogr.*, **1995**, *18*, 791–806

## SAMPLE
**Matrix:** blood, urine
**Sample preparation:** Serum. Condition a 3 mL Supelclean LC-18 SPE cartridge (Supelco) with MeOH and water. Hydrolyze 900 μL serum with β-glucuronidase (EC 3.2.1.31 type H-1 from Helix pomatia) at 60° for 1 h, add 500 μL (?) MeOH, centrifuge at 2000 g, add the supernatant to the SPE cartridge, wash with 1 mL water, dry under vacuum, elute with 2 mL MeOH:water 90:10, filter, inject an aliquot. Urine. 900 μL Urine + 500 μL MeOH, filter, inject an aliquot of the filtrate.

## HPLC VARIABLES
**Guard column:** 10 × 4.6 5 μm HP C18
**Column:** 150 × 4.6 5 μm C8P-50 (Asahipak)
**Mobile phase:** Gradient. MeOH:buffer 30:70 for 4 min, to 45:55 over 6 min, to 50:50 over 2 min, to 60:40 over 2 min, re-equilibrate at initial conditions for 10 min. (Prepare buffer by mixing 100 mM $NaH_2PO_4$ and 100 mM $Na_2HPO_4$ to achieve a pH of 7.0 and adding 10 mM N-cetyl-N, N, N-trimethylammonium bromide.)
**Injection volume:** 20
**Detector:** UV 260

## CHROMATOGRAM
**Retention time:** 2.5

## OTHER SUBSTANCES
**Extracted:** acebutolol, alprenolol, metoprolol, oxprenolol, propranolol

## KEY WORDS
serum; comparison with CE; SPE

## REFERENCE
Lukkari, P.; Sirén, H. Ion-pair chromatography and micellar electrokinetic capillary chromatography in analyzing beta-adrenergic blocking agents from human biological fluids. *J.Chromatogr.A*, **1995**, *717*, 211–217

## SAMPLE
**Matrix:** blood, urine
**Sample preparation:** 1 mL Plasma or 0.1 mL urine + 100 μL 1 M pH 12 sodium phosphate buffer + 100 μL 1 M NaOH + 10 mL dichloromethane:heptafluorobutanol 97:3, extract. Remove 8 mL of the organic layer and evaporate it to dryness under a stream of nitrogen at 30°, reconstitute the residue in mobile phase, inject an aliquot.

## HPLC VARIABLES

**Column:** Partisil C8

**Mobile phase:** MeCN:MeOH:buffer 3:3:94 (Buffer was 10 mM ammonium phosphate buffer containing 11.6 mM phosphoric acid, pH 2.9.)

**Detector:** F ex 220 em 200 (no cut-off filter)

## CHROMATOGRAM

**Limit of quantitation:** 5 ng/mL (plasma); 1000 ng/mL (urine)

## KEY WORDS

plasma; pharmacokinetics

## REFERENCE

Sowinski, K.M.; Forrest, A.; Wilton, J.H.; Taylor, A.M., II; Wilson, M.F.; Kazierad, D.J. Effect of aging on atenolol pharmacokinetics and pharmacodynamics. *J.Clin.Pharmacol.*, **1995**, *35*, 807–814

## SAMPLE

**Matrix:** blood, urine

**Sample preparation:** Plasma. 1 mL Plasma + 50 μL 10 μg/mL methoxamine in water + 100 μL 1 M NaOH + 4 mL ethyl acetate, vortex for 30 s, centrifuge at 3000 rpm for 10 min. Remove the organic layer and evaporate it to dryness under a stream of nitrogen at room temperature, reconstitute the residue in 200 μL saturated sodium carbonate and 200 μL 187 mM (−)-menthyl chloroformate in MeCN, vortex for 30 s, add 1 mL water, add 2 mL chloroform, vortex for 30 s, centrifuge for 3 min. Remove the organic layer and evaporate it to dryness under a stream of nitrogen at room temperature, reconstitute the residue in 200 μL mobile phase, vortex for 5 s, centrifuge for 5 min, inject a 20-60 μL aliquot of the supernatant. Urine. Dilute urine 10 times with water. 100 μL Diluted urine + 50 μL 10 μg/mL methoxamine in water + 100 μL saturated sodium carbonate + 200 μL 187 mM (−)-menthyl chloroformate in MeCN, vortex for 30 s, add 1 mL water, add 2 mL chloroform, vortex for 30 s, centrifuge for 3 min. Remove the organic layer and evaporate it to dryness under a stream of nitrogen at room temperature, reconstitute the residue in 200 μL mobile phase, vortex for 5 s, centrifuge for 5 min, inject a 20-60 μL aliquot of the supernatant.

## HPLC VARIABLES

**Guard column:** 50 mm long pellicular ODS (Whatman)

**Column:** 100 × 4.6 5 μm Partisil 5 ODS3

**Mobile phase:** MeCN:MeOH:water 35:22:43

**Flow rate:** 1.2

**Injection volume:** 20-60

**Detector:** F ex 195 em no emission filter

## CHROMATOGRAM

**Retention time:** 13, 15 (enantiomers)

**Internal standard:** methoxamine (18 (−), 20 (+))

**Limit of detection:** 2.5 ng/mL

## KEY WORDS

plasma; derivatization; pharmacokinetics

## REFERENCE

Mehvar, R. Liquid chromatographic analysis of atenolol enantiomers in human plasma and urine. *J.Pharm.Sci.*, **1989**, *78*, 1035–1039

## SAMPLE

**Matrix:** bulk

**Sample preparation:** Dissolve 10 μmole compound (as free base or hydrochloride) in 500 μL MeCN, add 250 μL 5% sodium carbonate (for hydrochlorides only), add 500 μL 100 mM reagent in MeCN, vortex for 1 min, heat at 60°for 2 h, add 100 μmole L-proline, heat at 60°for 30 min. Remove a 100 μL aliquot and dilute it with mobile phase, neutralize with acetic acid, inject a 10 μL aliquot. (Prepare the reagent ((R, R)-N-(3,5-dinitrobenzoyl)-2-aminocyclohexylisothiocyanate) as follows. Add 0.7 mL carbon disulfide to 6 mL (1R,2R)-(−)-1,2-diaminocyclohexane, 12 mL water, and 12 mL EtOH, heat the oil bath to 80°, add 2.8 mL carbon disulfide dropwise (making sure that the product does not start to precipitate), when addition is complete reflux for 1 h, acidify with 500 μL 5 M HCl, reflux for 12 h, cool, filter, wash the solid with a little cold EtOH to give trans-4,5-tetramethyleneimidazolidine-2-thione as a white fluffy solid (mp 148-150°) (Tetrahedron 1993, 49, 4419). Stir 7.97 g 3,5-dinitrobenzoyl chloride in 30 mL dichloroethane at 50°, add a solution of 6 g trans-4,5-tetramethyleneimidazolidine-2-thione in 120 mL dichloroethane containing a catalytic amount of 4-(dimethylamino)pyridine over 15 min, reflux for 2 h, remove the crystals of (R, R)-N-(3,5-dinitrobenzoyl)-2-aminocyclohexylisothiocyanate by filtration, evaporate the filtrate to dryness and dissolve the residue in 60 mL dichloroethane, reflux for 16 h to obtain more (R, R)-N-(3,5-dinitrobenzoyl)-2-aminocyclohexylisothiocyanate (mp >250°, $[\alpha]_{546}$ = -133° (c=1) in MeCN).)

## HPLC VARIABLES
**Column:** 125 × 4 5 μm Lichrospher 60 RP Select B
**Mobile phase:** MeCN:20 mM ammonium acetate 55:45
**Flow rate:** 1
**Injection volume:** 10
**Detector:** UV 254

## CHROMATOGRAM
**Retention time:** k′ 1.63, k′ 2.24 (enantiomers)

## OTHER SUBSTANCES
**Also analyzed:** acebutolol, alprenolol, carazolol, carvedilol, formoterol, methamphetamine, metipranolol, metoprolol, nifenanol, nitrilo atenolol, oxprenolol, pindolol, propranolol, xamoterol

## KEY WORDS
derivatization; chiral

## REFERENCE
Kleidernigg, O.P.; Posch, K.; Lindner, W. Synthesis and application of a new isothiocyanate as a chiral derivatizing agent for the indirect resolution of chiral amino alcohols and amines. *J.Chromatogr.A*, **1996**, *729*, 33–42

## SAMPLE
**Matrix:** formulations
**Sample preparation:** Dissolve 2 mg tablet or capsule in 10 mL pH 10 solution, extract twice with 2 mL ether, combine extracts, filter, inject a 20 μL aliquot.

## HPLC VARIABLES
**Column:** 250 × 4.6 5 μm β-cyclodextrin bonded C18 (Advanced Separation Technologies)
**Mobile phase:** MeCN:MeOH:acetic acid:triethylamine 95:5:0.3:0.2
**Flow rate:** 1
**Injection volume:** 20
**Detector:** UV 254

## CHROMATOGRAM
**Retention time:** 33, 37 (enantiomers)

## OTHER SUBSTANCES
**Simultaneous:** metoprolol, propranolol

## KEY WORDS
capsules; tablets; chiral

## REFERENCE
Tran, C.D.; Dotlich, M. Enantiomeric separation of beta-blockers by high performance liquid chromatography. *J.Chem.Educ.*, **1995**, *72*, 71–73

## SAMPLE
**Matrix:** formulations
**Sample preparation:** Dilute 500 μL to 10 mL with water. 100 μL Diluted sample + 100 μL 2.5 μg/mL sotalol + 100 μL saturated sodium tetraborate adjusted to pH 9 with HCl + 500 μL water + 5 mL dichloromethane:isopropanol 3:1, agitate on mechanical shaker for 5 min, centrifuge at 800 g for 3 min. Evaporate organic layer to dryness at 45° under a stream of nitrogen. Dissolve residue in 200 μL 50 mM sulfuric acid, mix for 30 s, inject a 30 μL aliquot.

## HPLC VARIABLES
**Guard column:** Novapak C18 guard insert
**Column:** 100 × 5 Novapak C18
**Mobile phase:** MeCN:10 mM potassium phosphate buffer adjusted to pH 3.2 with 0.2 M phosphoric acid 20:80 containing 3 mM 1-octanesulfonic acid
**Flow rate:** 1
**Injection volume:** 30
**Detector:** UV 226

## CHROMATOGRAM
**Retention time:** 3.5
**Internal standard:** sotalol (5)

## KEY WORDS
stability indicating; oral liquid

## REFERENCE
Garner, S.S.; Wiest, D.B.; Reynolds, E.R., Jr. Stability of atenolol in an extemporaneously compounded oral liquid. *Am.J.Hosp.Pharm.*, **1994**, *51*, 508–511

## SAMPLE
**Matrix:** perfusate
**Sample preparation:** 50 μL Perfusate + 50 μL pH 7.4 phosphate-buffered saline or 100 mM HCl + 100 μL 30 μg/mL salicylic acid in MeOH, centrifuge at 12000 g for 10 min, inject a 50 μL aliquot of the supernatant.

## HPLC VARIABLES
**Column:** 150 × 4.6 Cosmosil 5C18-P (Nacalai Tesque)
**Mobile phase:** MeOH:50 mM $NaH_2PO_4$ 20:80
**Flow rate:** 1
**Injection volume:** 50
**Detector:** UV 220

## CHROMATOGRAM
**Internal standard:** salicylic acid

## KEY WORDS
rabbit

## REFERENCE
Sasaki, H.; Igarashi, Y.; Nagano, T.; Nishida, K.; Nakamura, J. Different effects of absorption promoters on corneal and conjunctival penetration of ophthalmic beta-blockers. *Pharm.Res.*, **1995**, *12*, 1146–1150

## SAMPLE
**Matrix:** perfusate
**Sample preparation:** Add perfusate to an equal volume of MeOH, vortex, centrifuge, inject an aliquot.

## HPLC VARIABLES
**Column:** 250 × 4.6 Nucleosil 5C18
**Mobile phase:** MeCN:1% phosphoric acid 60:40 containing 2.5 mM sodium dodecyl sulfate
**Detector:** F ex 280 em 333

## REFERENCE
Kobayashi, D.; Matsuzawa, T.; Sugibayashi, K.; Morimoto, Y.; Kobayashi, M.; Kimura, M. Feasibility of use of several cardiovascular agents in transdermal therapeutic systems with *l*-menthol-ethanol system on hairless rat and human skin. *Biol.Pharm.Bull.*, **1993**, *16*, 254–258

## SAMPLE
**Matrix:** saliva
**Sample preparation:** Condition a 100 mg 1 mL Bond-Elut C2 SPE cartridge with 1 mL MeOH, 1 mL water, and 1 mL pH 9.0 borate buffer. Centrifuge a cotton roll soaked with saliva at 1000 g for 5 min, remove the liquid supernatant. 1 mL Supernatant + 50 μL 100 μg/mL tertatolol, add to the SPE cartridge, wash with 500 μL water, wash with 500 μL MeCN, elute with two 500 μL portions of acidified MeOH. Evaporate the eluate to dryness under a stream of nitrogen at 60°, reconstitute the residue in 50 μL mobile phase, mix for 15 s, inject a 40 μL aliquot. (Acidified MeOH was 50 mL MeOH + 300 μL 96% acetic acid.)

## HPLC VARIABLES
**Guard column:** RCSS silica guard-pack (Waters)
**Column:** 250 × 4.6 Chiralcel OD-H
**Mobile phase:** n-Hexane:EtOH:diethylamine 50:50:1
**Flow rate:** 1
**Injection volume:** 40
**Detector:** F ex 225 em 290 cut-off filter

## CHROMATOGRAM
**Internal standard:** (R, S)-tertatolol

## KEY WORDS
SPE; chiral

## REFERENCE
Höld, K.M.; de Boer, D.; Zuidema, J.; Maes, R.A.A. Evaluation of the Salivette as sampling device for monitoring β-adrenoceptor blocking drugs in saliva. *J.Chromatogr.B*, **1995**, *663*, 103–110

## SAMPLE
**Matrix:** solutions

**Sample preparation:** Mix 300 μL of a 30 μM solution in dichloromethane with 10 μL 20 mM 1-(6-methoxy-2-naphthyl)ethyl isothiocyanate in anhydrous dichloromethane and 50 μL 0.1% triethylamine in dichloromethane, vortex thoroughly, heat at 50° for 1.5 h, inject an aliquot. (Synthesize 1-(6-methoxy-2-naphthyl)ethyl isothiocyanate as follows (protect from light). Dissolve 500 mg (S)-(+)-naproxen in 50 mL dry toluene, slowly add 5 mL freshly distilled thionyl chloride, reflux for 1 h, evaporate to dryness under vacuum, dry the acyl chloride (mp 87.5°) under vacuum over KOH for 2 days. Dissolve 0.5 mmoles acyl chloride in 5 mL acetone, stir at 0°, add 0.6 mmoles sodium azide dissolved in ice water, stir at 0° for 30 min, add 10 mL ice-cold water, filter, dry solid in a desiccator under vacuum. Dissolve the solid in 1 mL toluene or dichloromethane (dried over 3 Å molecular sieve), reflux for 10 min, evaporate, store resulting isocyanate (mp 51°) under vacuum over a desiccant. Dissolve 0.5 mmole isocyanate in 5 mL acetone, add 20 mL 8.5% phosphoric acid, heat to 80° for 1.5 h, adjust to pH 13, extract with diethyl ether:dichloromethane 4:1. Wash the organic layer twice with water, dry over anhydrous sodium sulfate, evaporate to dryness, dissolve in 1 mL toluene, evaporate to give the amine from naproxen as crystals (mp 53°) (Pharm.Res. 1990, 7, 1262). Dissolve 1 mmole 1,1-thiocarbonyldiimidazole in 15 mL ice-cold chloroform, stir at 0°, add dropwise 1 mmole of the amine dissolved in 10 mL chloroform, stir at room temperature for 1.5 h, evaporate to dryness, reconstitute with carbon tetrachloride (Caution! Carbon tetrachloride is a carcinogen!), filter, evaporate the filtrate to dryness, store the resulting oil in a desiccator, purify on a short silica gel column with dichloromethane:light petroleum 50:50 to give 1-(6-methoxy-2-naphthyl)ethyl isothiocyanate as a slightly yellow liquid (store in the freezer under argon).)

## HPLC VARIABLES
**Column:** 250 × 4 5 μm Zorbax ODS
**Mobile phase:** MeCN:water 50:50
**Flow rate:** 1
**Injection volume:** 100
**Detector:** UV 230; F ex 270 em 350

## CHROMATOGRAM
**Retention time:** k' 5.2 (S-(−)), 6.1 (R-(+))

## OTHER SUBSTANCES
**Simultaneous:** diacetolol

## KEY WORDS
derivatization; chiral; F not much more sensitive than UV; α = 1.17

## REFERENCE
Büschges, R.; Linde, H.; Mutschler, E.; Spahn-Langguth, H. Chloroformates and isothiocyanates derived from 2-arylpropionic acids as chiral reagents: synthetic routes and chromatographic behaviour of the derivatives. *J.Chromatogr.A*, **1996**, *725*, 323−334

## SAMPLE
**Matrix:** solutions

## HPLC VARIABLES
**Column:** 250 × 4.6 10 μm Partisil ODS1
**Mobile phase:** MeOH:50 mM pH 3.0 phosphoric acid 10:90
**Column temperature:** 30
**Flow rate:** 1.5
**Detector:** Radioactivity

## OTHER SUBSTANCES
**Also analyzed:** cimetidine, hydrochlorothiazide, ranitidine

## KEY WORDS
tritium labeled

## REFERENCE
Collett, A.; Sims, E.; Walker, D.; He, Y.-L.; Ayrton, J.; Rowland, M.; Warhurst, G. Comparison of HT29-18-C$_1$ and Caco-2 cell lines as models for studying intestinal paracellular drug absorption. *Pharm.Res.*, **1996**, *13*, 216–221

## SAMPLE
**Matrix:** solutions

## HPLC VARIABLES
**Guard column:** 10 × 3.2 5 μm Partisil ODS3
**Column:** 100 × 4.6 5 μm Partisil ODS3
**Mobile phase:** MeCN:buffer 10:90 (Buffer was 60 mM KH$_2$PO$_4$ adjusted to pH 3.0 with phosphoric acid.)
**Flow rate:** 0.6-1
**Injection volume:** 10-100
**Detector:** UV 270

## OTHER SUBSTANCES
**Also analyzed:** practolol

## REFERENCE
Palm, K.; Luthman, K.; Ungell, A.-L.; Strandlund, G.; Artursson, P. Correlation of drug absorption with molecular surface properties. *J.Pharm.Sci.*, **1996**, *85*, 32–39

## SAMPLE
**Matrix:** solutions

## HPLC VARIABLES
**Column:** 250 × 4.6 Vydac C18
**Mobile phase:** Gradient. A was 0.1% trifluoroacetic acid in water. B was 0.1% trifluoroacetic acid in MeCN. A:B from 95:5 to 65:35 over 9 min.
**Column temperature:** 40
**Flow rate:** 1
**Detector:** UV (wavelength not given)

## OTHER SUBSTANCES
**Simultaneous:** dexamethasone

## REFERENCE
Rubas, W.; Cromwell, M.E.M.; Shahrokh, Z.; Villagran, J.; Nguyen, T.-N.; Wellton, M.; Nguyen, T.-H.; Mrsny, R.J. Flux measurements across Caco-2 monolayers may predict transport in human large intestinal tissue. *J.Pharm.Sci.*, **1996**, *85*, 165–169

## SAMPLE
**Matrix:** solutions
**Sample preparation:** Inject a 40 μL aliquot.

## HPLC VARIABLES
**Column:** 250 × 4.6 5 μm Hypersil MOS C-8
**Mobile phase:** MeOH:water 70:30 containing 0.02% dimethyloctylamine, 25 mM sodium hexanesulfonate, and 20 mM acetic acid

**Flow rate:** 1
**Injection volume:** 40
**Detector:** F ex 275 em 305

## CHROMATOGRAM
**Retention time:** 3.3

## OTHER SUBSTANCES
**Simultaneous:** alprenolol, pindolol, propranolol (UV 288)

## REFERENCE
Adson, A.; Burton, P.S.; Raub, T.J.; Barsuhn, C.L.; Audus, K.L.; Ho, N.F.H. Passive diffusion of weak
    organic electrolytes across Caco-2 cell monolayers: Uncoupling the contributions of hydrodynamic,
    transcellular, and paracellular barriers. *J.Pharm.Sci.*, **1995**, *84*, 1197–1204

## SAMPLE
**Matrix:** solutions
**Sample preparation:** Inject a 20 µL aliquot of a 1 mg/mL solution.

## HPLC VARIABLES
**Column:** 250 × 4.6 10 µm Chiralcel OD
**Mobile phase:** Hexane:isopropanol:diethylamine 60:40:0.1
**Flow rate:** 0.5
**Injection volume:** 20
**Detector:** UV 275

## CHROMATOGRAM
**Retention time:** $k'$ 0.94, 1.75 (enantiomers)

## KEY WORDS
chiral

## REFERENCE
Ekelund, J.; van Arkens, A.; Bronnum-Hansen, K.; Fich, K.; Olsen, L.; Petersen, P.V. Chiral separations
    of β-blocking drug substances using chiral stationary phases. *J.Chromatogr.A*, **1995**, *708*, 253–261

## SAMPLE
**Matrix:** solutions
**Sample preparation:** Inject an aliquot of a 200 µM solution in MeOH.

## HPLC VARIABLES
**Column:** 100 × 4.7 7 µm Hypercarb (Shandon)
**Mobile phase:** MeOH containing 5 mM N-benzyloxycarbonylglycyl-L-proline and 4.5 mM
    NaOH
**Column temperature:** 17
**Injection volume:** 20
**Detector:** UV 270

## CHROMATOGRAM
**Retention time:** $k'$ 13 (first enantiomer)

## KEY WORDS
chiral; $\alpha = 1.09$

## REFERENCE

Huynh, N.-H.; Karlsson, A.; Pettersson, C. Enantiomeric separation of basic drugs using N-benzyloxy-carbonylglyclyl-L-proline as counter ion in methanol. *J.Chromatogr.A*, **1995**, *705*, 275–287

## SAMPLE

**Matrix:** solutions

## HPLC VARIABLES

**Column:** 150 × 4.6 12 μm 1-myristoyl-2-[(13-carboxyl)-tridecoyl]-sn-3-glycerophosphocholine chemically bonded to silica (Regis)
**Mobile phase:** MeCN:100 mM pH 7.0 phosphate buffer 20:80
**Flow rate:** 1
**Detector:** UV 254

## CHROMATOGRAM

**Retention time:** k′ 0.71

## OTHER SUBSTANCES

**Also analyzed:** acebutolol, alprenolol, antazoline, betaxolol, bisoprolol, bopindolol, bupranolol, carteolol, celiprolol, chloropyramine, chlorpheniramine, cicloprolol, cimetidine, cinnarizine, cirazoline, clonidine, dilevalol, dimethindene, diphenhydramine, doxazosin, esmolol, famotidine, isothipendyl, ketotifen, metiamide, metoprolol, moxonidine, nadolol, naphazoline, nifenalol, nizatidine, oxprenolol, pheniramine, phentolamine, pindolol, pizotyline (pizotifen), practolol, prazosin, promethazine, propranolol, pyrilamine (mepyramine), ranitidine, roxatidine, sotalol, tiamenidine, timolol, tramazoline, tripelennamine, triprolidine, tymazoline, UK-14,304

## REFERENCE

Kaliszan, R.; Nasal, A.; Turowski, M. Binding site for basic drugs on $\alpha_1$-acid glycoprotein as revealed by chemometric analysis of biochromatographic data. *Biomed.Chromatogr.*, **1995**, *9*, 211–215

## SAMPLE

**Matrix:** solutions

## HPLC VARIABLES

**Column:** 250 × 4.6 5 μm Supelcosil LC-DP (A) or 250 × 4 5 μm LiChrospher 100 RP-8 (B)
**Mobile phase:** MeCN:0.025% phosphoric acid:buffer 25:10:5 (A) or 60:25:15 (B) (Buffer was 9 mL concentrated phosphoric acid and 10 mL triethylamine in 900 mL water, adjust pH to 3.4 with dilute phosphoric acid, make up to 1 L.)
**Flow rate:** 0.6
**Injection volume:** 25
**Detector:** UV 229

## CHROMATOGRAM

**Retention time:** 5.33 (A), 3.11 (B)

## OTHER SUBSTANCES

**Also analyzed:** acebutolol, acepromazine, acetaminophen, acetazolamide, acetophenazine, albuterol, alprazolam, amitriptyline, amobarbital, amoxapine, antipyrine, atropine, azatadine, baclofen, benzocaine, bromocriptine, brompheniramine, brotizolam, bupivacaine, buspirone, butabarbital, butalbital, caffeine, carbamazepine, cetirizine, chlorcyclizine, chlordiazepoxide, chlormezanone, chloroquine, chlorpheniramine, chlorpromazine, chlorpropamide, chlorprothixene, chlorthalidone, chlorzoxazone, cimetidine, cisapride, clomipramine, clonazepam, clonidine, clozapine, cocaine, codeine, colchicine, cyclizine, cyclobenzaprine, dantrolene, desipramine, diazepam, diclofenac, diflunisal, diltiazem, diphenhydramine, diphenidol, diphenoxylate, dipyridamole, disopyramide, dobutamine, doxapram, doxepin, droperidol, encainide, ethidium bromide, ethopropazine, fenoprofen, fentanyl, flavoxate, fluoxetine, fluphenazine, flurazepam, flurbiprofen, fluvoxamine, furosemide, glutethimide, glyburide, guaifenesin, haloperidol, homatropine, hydralazine, hydrochlorothiazide, hydrocodone, hydromorphone, hydroxychloroquine, hydroxyzine, ibuprofen, imipramine, indomethacin, ketoconazole, ketoprofen, ketorolac, labetalol, levorphanol, lidocaine, loratadine, lorazepam, lovastatin, loxapine, mazindol, mefenamic acid, meperidine, mephenytoin, mepivacaine, mesoridazine, metaproterenol, methadone, methdilazine, methocarbamol, methotrexate, methotrimeprazine, methoxamine, methyldopa, methylphenidate, metoclopramide, metolazone, metoprolol, metronidazole, midazolam, moclobemide, morphine, nadolol, nalbuphine, naloxone, naphazoline, naproxen, nifedipine, nizatidine, norepinephrine, nortriptyline, oxazepam, oxycodone, oxymetazoline, paroxetine, pemoline, pentazocine, pentobarbital, pentoxifylline, perphenazine, pheniramine, phenobarbital, phenol, phenolphthalein, phentolamine, phenylbutazone, phenyltoloxamine, phenytoin, pimozide, pindolol, piroxicam, pramoxine, prazepam, prazosin, probenecid, procainamide, procaine, prochlorperazine, procyclidine, promazine, promethazine, propafenone, propantheline, propiomazine, propofol, propranolol, protriptyline, quazepam, quinidine, quinine, racemethorphan, ranitidine, remoxipride, risperidone, salicylic acid, scopolamine, secobarbital, sertraline, sotalol, spironolactone, sulfinpyrazone, sulindac, temazepam, terbutaline, terfenadine, tetracaine, theophylline, thiethylperazine, thiopental, thioridazine, thiothixene, timolol, tocainide, tolbutamide, tolmetin, trazodone, triamterene, triazolam, trifluoperazine, triflupromazine, trimeprazine, trimethoprim, trimipramine, verapamil, warfarin, xylometazoline, yohimbine, zopiclone

## KEY WORDS
some details of plasma extraction

## REFERENCE
Koves, E.M. Use of high-performance liquid chromatography-diode array detection in forensic toxicology. *J.Chromatogr.A*, **1995**, *692*, 103–119

## SAMPLE
**Matrix:** solutions

## HPLC VARIABLES
**Column:** 150 × 4.6 cellulose 3,5-dimethylphenylcarbamate/10-undecenoate bonded to allylsilica
**Mobile phase:** Heptane:isopropanol:diethylamine 80:20:0.1
**Flow rate:** 1
**Injection volume:** 1000
**Detector:** UV 254

## CHROMATOGRAM
**Retention time:** k' 9.78

## KEY WORDS
chiral; α 1.14

## REFERENCE
Oliveros, L.; Lopez, P.; Minguillon, C.; Franco, P. Chiral chromatographic discrimination ability of a cellulose 3,5-dimethylphenylcarbamate/10-undecenoate mixed derivative fixed on several chromatographic matrices. *J.Liq.Chromatogr.*, **1995**, *18*, 1521–1532

## SAMPLE
**Matrix:** solutions

## HPLC VARIABLES
**Column:** 250 × 4.6 Zorbax RX
**Mobile phase:** Gradient. A was 10 mL concentrated orthophosphoric acid and 7 mL triethylamine in 1 L water. B was 10 mL concentrated orthophosphoric acid and 7 mL triethylamine in 200 mL water, make up to 1 L with MeCN. A:B from 100:0 to 0:100 over 30 min, maintain at 0:100 for 5 min.
**Column temperature:** 30
**Flow rate:** 2
**Detector:** UV 210

## OTHER SUBSTANCES
**Also analyzed:** acepromazine, acetaminophen, acetophenazine, albuterol, aminophylline, amitriptyline, amobarbital, amoxapine, amphetamine, amylocaine, antipyrine, aprobarbital, aspirin, atropine, avermectin, barbital, benzocaine, benzoic acid, benzotropine, benzphetamine, berberine, bibucaine, bromazepan, brompheniramine, buprenorphine, buspirone, butabarbital, butacaine, butethal, caffeine, carbamazepine, carbromal, chloramphenicol, chlordiazepoxide, chloroquine, chlorothiazide, chloroxylenol, chlorphenesin, chlorpheniramine, chlorpromazine, chlorpropamide, chlortetracycline, cimetidine, cinchonidine, cinchonine, clenbuterol, clonazepam, clonixin, clorazepate, cocaine, codeine, colchicine, cortisone, coumarin, cyclazocine, cyclobenzaprine, cyclothiazide, cyheptamide, cymarin, danazol, danthron, dapsone, debrisoquine, desipramine, dexamethasone, dextromethorphan, dextropropoxyphene, diamorphine, diazepam, diclofenac, diethylpropion, diethylstilbestrol, diflunisal, digitoxin, digoxin, diltiazem, diphenhydramine, diphenoxylate, diprenorphine, dipyrone, disulfiram, dopamine, doxapram, doxepin, dronabinol, ephedrine, epinephrine, epinine, estradiol, estriol, estrone, ethacrynic acid, ethosuximide, etonitazene, etorphine, eugenol, famotidine, fenbendazole, fencamfamine, fenoprofen, fenproporex, fentanyl, flubendazole, flufenamic acid, flunitrazepam, 5-fluorouracil, fluoxymesterone, fluphenazine, furosemide, gentisic acid, gitoxigenin, glipizide, glunixin, glutethimide, glybenclamide, guaiacol, halazepam, haloperidol, hydrochlorothiazide, hydrocodone, hydrocortisone, hydromorphone, hydroxyquinoline, ibogaine, ibuprofen, iminostilbene, imipramine, indomethacin, isocarbostyril, isocarboxazid, isoniazid, isoproterenol, isoxsuprine, ivermectin, ketamine, ketoprofen, kynurenic acid, levorphanol, lidocaine, lorazepam, lormetazepam, loxapine, mazindol, mebendazole, meclizine, meclofenamic acid, medazepam, mefenamic acid, megestrol, mepacrine, meperidine, mephentermine, mephenytoin, mephesin, mephobarbital, mepivacaine, mescaline, mesoridazine, methadone, methamphetamine, methapyrilene, methaqualone, methazolamide, methocarbamol, methoxamine, methsuximide, methyl salicylate, methyldopa, methyldopamine, methylphenidate, methylprednisolone, methyltestosterone, methyprylon, metoprolol, mibolerone, morphine, nadolol, nalorphine, naloxone, naltrexone, naphazoline, naproxen, nefopam, niacinamide, nicotine, nicotinic acid, nifedipine, niflumic acid, nitrazepam, norepinephrine, nortriptyline, noscapine, nylidrin, oxazepam, oxycodone, oxymorphone, oxyphenbutazone, oxytetracycline, papaverine, pargyline, pemoline, pentazocine, pentobarbital, persantine, phenacetin, phenazocine, phenazopyridine, phencyclidine, phendimetrazine, phenelzine, pheniramine, phenobarbital, phenothiazine, phensuximide, phentermine, phenylbutazone, phenylephrine, phenylpropanolamine, piperocaine, prazepam, prednisolone, primidone, probenecid, progesterone, propiomazine, propranolol, propylparaben, pseudoephedrine, puromycin, pyrilamine, pyrithyldione, quazepam, quinaldic acid, quinidine, quinine, ranitidine, recinnamine, reserpine, resorcinol, saccharin, albuterol, salicylamide, salicylic acid, scopolamine, scopoletin, secobarbital, strychnine, sulfacetamide, sufadiazine, sulfadimethoxine, sulfaethidole, sulfamerazine, sulfamethazine, sulfamethoxizole, sulfanilamide, sulfapyridine, sulfasoxizole, sulindac, tamoxifen, temazepam, testosterone, tetracaine, tetracycline, tetramisole, thebaine, theobromine, theophylline, thiabendazole, thiamine, thiamylal, thiobarbituric acid, thioridazine, thiosalicylic acid, thiothixene, thymol, tolazamide, tolazoline, tobutamide, tolmetin, tranylcypromine, triamcinolone, tribenzylamine, trichloromethiazide, trifluoperazine, trihexyphenidyl, trimethoprim, tripelennamine, triproilidine, tropacocaine, tyramine, verapamil, vincamine, warfarin, yohimbine, zoxazolamine

## REFERENCE

Hill, D.W.; Kind, A.J. Reversed-phase solvent gradient HPLC retention indexes of drugs. *J.Anal.Toxicol.*, **1994**, *18*, 233–242

## SAMPLE

**Matrix:** solutions
**Sample preparation:** 50 μL Solution + 50 μL pH 7.4 PBS + 100 μL 30 μg/mL salicylic acid in MeOH, centrifuge at 12000 g for 10 min, inject a 50 μL aliquot.

## HPLC VARIABLES

**Column:** 150 × 4.6 Cosmosil 5C18-P (Nacalai Tesque)
**Mobile phase:** MeOH:50 mM $NaH_2PO_4$ 20:80
**Flow rate:** 1
**Injection volume:** 50
**Detector:** UV 220

## CHROMATOGRAM

**Internal standard:** salicylic acid

## KEY WORDS

buffer; Earle's balanced salt solution

## REFERENCE

Sasaki, H.; Igarishi, Y.; Nishida, K.; Nakamura, J. Intestinal permeability of ophthalmic β-blockers for predicting ocular permeability. *J.Pharm.Sci.*, **1994**, *83*, 1335–1338

## SAMPLE

**Matrix:** solutions
**Sample preparation:** Prepare a 25 μg/mL solution in MeCN:water 40:60, inject an aliquot.

## HPLC VARIABLES

**Column:** 250 × 4.6 3 μm silica (Phenomenex)
**Mobile phase:** MeCN:6.25 mM pH 3.0 phosphate buffer 40:60
**Flow rate:** 1
**Injection volume:** 50
**Detector:** UV 254

## CHROMATOGRAM

**Retention time:** 6.55

## OTHER SUBSTANCES

**Also analyzed:** clonidine, diltiazem, metoprolol, nifedipine, prazosin, propranolol, verapamil

## REFERENCE

Simmons, B.R.; Stewart, J.T. HPLC separation of selected cardiovascular agents on underivatized silica using an aqueous organic mobile phase. *J.Liq.Chromatogr.*, **1994**, *17*, 2675–2690

## SAMPLE

**Matrix:** solutions
**Sample preparation:** Filter (0.22 μm), inject a 10 μL aliquot.

## HPLC VARIABLES

**Column:** 250 × 4.6 internal surface reversed-phase silica (Pinkerton) (Regis Chemical)

**Mobile phase:** Isopropanol:100 mM pH 6.8 KH$_2$PO$_4$ 10:90
**Flow rate:** 1
**Injection volume:** 10
**Detector:** UV 232-274 (wavelength of maximum absorption used)

## CHROMATOGRAM
**Retention time:** 13.6

## OTHER SUBSTANCES
**Simultaneous:** acebutolol, alprenolol, carteolol, metoprolol, oxprenolol, pindolol

## REFERENCE
Ohshima, T.; Takagi, K.; Miyamoto, K.-I. High performance liquid chromatographic retention time of β-blockers as an index of pharmacological activity. *J.Liq.Chromatogr.*, **1993**, *16*, 3933–3939

## SAMPLE
**Matrix:** solutions

## HPLC VARIABLES
**Column:** 250 × 4.6 cellulose tris(3,5-dimethylphenylcarbamate)
**Mobile phase:** Hexane:isopropanol:diethylamine 80:20:0.1
**Flow rate:** 0.5
**Detector:** UV

## CHROMATOGRAM
**Retention time:** k′ 3.54 (of first (+) enantiomer)

## KEY WORDS
chiral; α 1.58

## REFERENCE
Okamoto, Y.; Aburatani, R.; Hatano, K.; Hatada, K. Optical resolution of racemic drugs by chiral HPLC on cellulose and amylose tris(phenylcarbamate) derivatives. *J.Liq.Chromatogr.*, **1988**, *11*, 2147–2163

## SAMPLE
**Matrix:** solutions
**Sample preparation:** Prepare a 10 μg/mL solution in MeOH, inject a 20 μL aliquot.

## HPLC VARIABLES
**Column:** 125 × 4.9 Spherisorb S5W silica
**Mobile phase:** MeOH containing 10 mM ammonium perchlorate and 1 mL/L 100 mM NaOH in MeOH, pH 6.7
**Flow rate:** 2
**Injection volume:** 20
**Detector:** E, LeCarbone, V25 glassy carbon electrode, + 1.2 V

## CHROMATOGRAM
**Retention time:** 2.03

## OTHER SUBSTANCES
**Also analyzed:** acebutolol, acepromazine, acetophenazine, N-acetylprocainamide, albuterol, alprenolol, amethocaine, amiodarone, amitriptyline, antazoline, azacyclonal, amethane, benactyzine, benperidol, benzethidine, benzocaine, benzoctamine, benzphetamine, benzquinamide, bromhexine, bromodiphenhydramine, bromperidol, brompheniramine, brompromazine, buclizine, bufotenine, bupivacaine, buprenorphine, butacaine, butetha-

mate, chlorcyclizine, chlorpheniramine, chlorphenoxamine, chlorprenaline, chlorpromazine, chlorprothixene, cimetidine, cinchonidine, cinnarizine, clemastine, clomipramine, clonidine, cocaine, cyclazocine, cyclizine, cyclopentamine, cyproheptadine, deserpidine, desipramine, dextromoramide, dextropropoxyphene, dicyclomine, diethylcarbamazine, diethylpropion, diethylthiambutene, dihydroergotamine, dimethindene, dimethothiazine, diphenhydramine, diphenoxylate, dipipanone, diprenorphine, dipyridamole, disopyramide, dothiepin, doxapram, doxepin, doxylamine, droperidol, ephedrine, ergocornine, ergocristine, ergocristinine, ergocryptine, ergometrine, ergosine, ergosinine, ergotamine, ethopropazine, etorphine, etoxeridine, fenethazine, fenfluramine, fenoterol, fentanyl, flavoxate, fluopromazine, flupenthixol, fluphenazine, flurazepam, haloperidol, hydroxyzine, hyoscine, ibogaine, imipramine, indapamine, iprindole, isothipendyl, isoxsuprine, ketanserin, laudanosine, lidocaine, lofepramine, loxapine, maprotiline, mecamylamine, meclophenoxate, meclozine, medazepam, mephentermine, mepivacaine, meptazinol, mepyramine, mesoridazine, metaraminol, methadone, methamphetamine, methapyrilene, methdilazene, methotrimeprazine, methoxamine, methoxyphenamine, methoxypromazine, methylephedrine, methylergonovine, methysergide, metoclopramide, metopimazine, metoprolol, mianserin, morazone, nadolol, nalorphine, naloxone, naphazoline, nicotine, nifedipine, nomifensine, nortriptyline, noscapine, orphenadrine, oxeladin, oxprenolol, oxymetazolin, papaverine, pargyline, pecazine, penbutolol, pentazocine, penthienate, pericyazine, perphenazine, phenadoxone, phenampromide, phenazocine, phenbutrazate, phendimetrazine, phenelzine, phenglutarimide, phenindamine, pheniramine, phenmetrazine, phenomorphan, phenoperidine, phenothiazine, phenoxybenzamine, phentolamine, phenylephrine, phenyltoloxamine, physostigmine, piminodine, pimozide, pindolol, pipamazine, pipazethate, piperacetazine, piperidolate, pipradol, pirenzepine, piritramide, pizotifen, practolol, pramoxine, prazosin, prenylamine, prilocaine, primaquine, proadifen, procainamide, procaine, prochlorperazine, procyclidine, proheptazine, prolintane, promazine, promethazine, pronethalol, properidine, propiomazine, propranolol, prothipendyl, protriptyline, proxymetacaine, pseudoephedrine, pyrimethamine, quinidine, quinine, ranitidine, rescinnamine, sotalol, tacrine, terazosin, terbutaline, terfenadine, thenyldiamine, theophylline, thiethylperazine, thiopropazate, thioproperazine, thioridazine, thiothixene, thonzylamine, timolol, tocainide, tolpropamine, tolycaine, tranylcypromine, trazodone, trifluoperazine, trifluperidol, trimeperidine, trimeprazine, trimethobenzamide, trimethoprim, trimipramine, tripelennamine, triprolidine, tryptamine, verapamil, xylometazoline

## REFERENCE
Jane, I.; McKinnon, A.; Flanagan, R.J. High-performance liquid chromatographic analysis of basic drugs on silica columns using non-aqueous ionic eluents. II. Application of UV, fluorescence and electrochemical oxidation detection. *J.Chromatogr.*, **1985**, *323*, 191–225

## SAMPLE
**Matrix:** tablets
**Sample preparation:** Grind tablet equivalent to about 50 mg nadolol, add 200 mL mobile phase, sonicate for 15 min, make up to 250 mL with mobile phase, filter or centrifuge, to 20 mL solution add 5 mL 1.2 mg/mL atenolol in mobile phase, mix, inject an aliquot.

## HPLC VARIABLES
**Column:** 250 × 4.6 10 μm LiChrosorb C2
**Mobile phase:** MeCN:buffer 35:65 (1 mL 100 mM HCl + 1200 mL water + 5.84 g NaCl, mix to dissolve, add 700 mL MeOH, make up to 2 L, apparent pH 4.5.)
**Flow rate:** 1.2
**Injection volume:** 20
**Detector:** UV 254

## CHROMATOGRAM
**Retention time:** 4.5
**Internal standard:** atenolol

## OTHER SUBSTANCES
**Simultaneous:** acebutolol, alprenolol, metoprolol, nadolol, oxprenolol, pindolol, practolol, propranolol, sotalol, timolol

## KEY WORDS
stability-indicating; atenolol is IS

## REFERENCE
Patel, B.R.; Kirschbaum, J.J.; Poet, R.B. High-pressure liquid chromatography of nadolol and other beta-adrenergic blocking drugs. *J.Pharm.Sci.*, **1981**, *70*, 336–338

## SAMPLE
**Matrix:** urine
**Sample preparation:** 1 mL Urine + 10 mg β-glucuronidase/arylsulfatase (Helix pomatia, Sigma), heat at 37° overnight, add an equal volume of buffer, centrifuge at 2000 g for 5 min, inject an aliquot of the supernatant onto column A with mobile phase A and elute to waste. After 2.5 min backflush the contents of column A onto column B with mobile phase B, monitor the effluent from column B. Re-equilibrate both columns for 12.5 min before the next injection. (Buffer was 200 mM boric acid adjusted to pH 9.5 with 5 M NaOH.)

## HPLC VARIABLES
**Column:** A 10 × 4.6 5 μm Spherisorb cyanopropyl; B 250 × 4.6 Capcell Pak C18 UG-120 (Shiseido)
**Mobile phase:** A water; B Gradient. MeCN:buffer from 3:97 to 30:70 over 30 min, to 40:60 over 8 min (Buffer was 3.4 mL/L phosphoric acid adjusted to pH 3.0 with 5 M NaOH.)
**Flow rate:** A 1.25; B 1
**Injection volume:** 100
**Detector:** UV 220

## CHROMATOGRAM
**Retention time:** 7
**Limit of detection:** 250 ng/mL

## OTHER SUBSTANCES
**Extracted:** acebutolol, alprenolol, amphetamine, bopindolol, codeine, ephedrine, labetalol, metoprolol, morphine, nadolol, oxprenolol, pindolol, propranolol, timolol

## KEY WORDS
column-switching

## REFERENCE
Saarinen, M.T.; Sirén, H.; Riekkola, M.-L. Screening and determination of β-blockers, narcotic analgesics and stimulants in urine by high-performance liquid chromatography with column switching. *J.Chromatogr.B*, **1995**, *664*, 341–346

## ANNOTATED BIBLIOGRAPHY
Nakamura, K.; Fujima, H.; Kitagawa, H.; Wada, H.; Makino, K. Preparation and chromatographic characteristics of a chiral-recognizing perphenylated cyclodextrin column. *J.Chromatogr.A*, **1995**, *694*, 111–118 [chiral; also acetylpheneturide, alprenolol, arotinolol, benzoin, biperiden, bunitrolol, chlormezanone, chlorphenesin, chlorpheniramine, eperisone, flavanone, ibuprofen, oxprenolol, phenylethyl alcohol, phenylethylamine, pindolol, proglumide, propranolol, trihexyphenidyl]

Bailey, C.J.; Ruane, R.J.; Wilson, I.D. Packed-column supercritical fluid chromatography of β-blockers. *J.Chromatogr.Sci.*, **1994**, *32*, 426–429 [SFC; also alprenolol, labetalol, metoprolol, oxprenolol, pindolol, practolol, propranolol, toliprolol, xamoterol]

Hamoir, T.; Massart, D.L. Retention prediction for β-adrenergic blocking drugs in normal-phase liquid chromatography. *J.Chromatogr.A*, **1994**, *673*, 1–10 [column temp 30; cyanopropyl column; also acebutolol, alprenolol, bunitrolol, bupranolol, carazolol, mepindolol, metipranolol, metoprolol, nadolol, oxprenolol, penbutolol, pindolol, practolol, prenalterol, propranolol, tertatolol]

Hermansson, J.; Grahn, A. Resolution of racemic drugs on a new chiral column based on silica-immobilized cellobiohydrolase. Characterization of the basic properties of the column. *J.Chromatogr.*, **1994**, *687*, 45–59 [chiral; also acebutolol, betaxolol, bisoprolol, carbuterol, cathinone, cimetidine, dobutamine, dopropizine, epanolol, epinephrine, laudanosine, metanephrine, metoprolol, moprolol, norepinephrine, normetanephrine, octopamine, oxybutynine, pamatolol, practolol, prilocaine, propafenone, proxyphylline, sotalol, talinolol, tetrahydropapaveroline, tetramisole, timolol, tolamolol, toliprolol]

Kobayashi, D.; Matsuzawa, T.; Sugibayashi, K.; Morimoto, Y.; Kimura, M. Analysis of the combined effect of 1-menthol and ethanol as skin permeation enhancers based on a two-layer skin model. *Pharm.Res.*, **1994**, *11*, 96–103 [also morphine, naloxone, nifedipine, nitrendipine, vinpocetine]

Egginger, G.; Lindner, W.; Kahr, S.; Stoschitzky, K. Stereoselective HPLC bioanalysis of atenolol enantiomers in plasma: application to a comparative human pharmacokinetic study. *Chirality*, **1993**, *5*, 505–512

Josefsson, M.; Carlsson, B.; Norlander, B. Fast chromatographic separation of (−)-menthyl chloroformate derivatives of some chiral drugs, with special reference to amlodipine, on porous graphitic carbon. *Chromatographia*, **1993**, *37*, 129–132 [derivatization; chiral; also amlodipine, mexiletine, propranolol, sotalol]

Armstrong, D.W.; Chen, S.; Chang, C.; Chang, S. A new approach for the direct resolution of racemic beta adrenergic blocking agents by HPLC. *J.Liq.Chromatogr.*, **1992**, *15*, 545–556 [chiral; also alprenolol, carteolol, labetolol, metoprolol, nadolol, oxprenolol, pindolol, propranolol, timolol]

Sallustio, B.C.; Morris, R.G.; Horowitz, J.D. High-performance liquid chromatographic determination of sotalol in plasma. I. Application to the disposition of sotalol enantiomers in humans. *J.Chromatogr.*, **1992**, *576*, 321–327 [atenolol is IS; extracted sotalol; derivatization; chiral; achiral; fluorescence detection; UV detection; plasma; SPE]

Miller, R.B. A validated high-performance liquid chromatographic method for the determination of atenolol in whole blood. *J.Pharm.Biomed.Anal.*, **1991**, *9*, 849–853

Morris, R.G.; Saccoia, N.C.; Sallustio, B.C.; Zacest, R. Improved high-performance liquid chromatography assay for atenolol in plasma and urine using fluorescence detection. *Ther.Drug Monit.*, **1991**, *13*, 345–349

Owino, E.; Clark, B.J.; Fell, A.F. Diode array detection and simultaneous quantitation of the coeluting atenolol-related synthetic route impurities, PPA-Diol. *J.Chromatogr.Sci.*, **1991**, *29*, 450–456 [bulk; simultaneous impurities]

Shen, J.; Wanwimolruk, S.; Hung, C.T.; Zoest, A.R. Quantitative analysis of β-blockers in human plasma by reversed-phase ion-pair high-performance liquid chromatography using a microbore column. *J.Liq.Chromatogr.*, **1991**, *14*, 777–793 [plasma; microbore; fluorescence detection; UV detection; LOD 1-10 ng/mL; also labetalol, metoprolol, pindolol, propranolol;oxprenolol (IS)]

Teitelbaum, Z.; Ben-Dom, N.; Terry, S. A liquid chromatographic method for the determination of atenolol in human plasma. *J.Liq.Chromatogr.*, **1991**, *14*, 3735–3744 [plasma; metoprolol (IS); LOD 12.6 ng/mL; fluorescence detection]

Alebic-Kolbah, T.; Plavsic, F.; Wolf-Coporda, A. Determination of serum atenolol using HPLC with fluorescence detection following isolation with activated charcoal. *J.Pharm.Biomed.Anal.*, **1989**, *7*, 1777–1781

Bui, K.H.; French, S.B. Direct serum injection and analysis of drugs with aqueous mobile phases containing triethylammonium acetate. *J.Liq.Chromatogr.*, **1989**, *12*, 861–873 [serum; plasma; dog; rat; fluorescence detection; UV detection; also antipyrine, hexaphenone, metoprolol, naproxen, propranolol]

Chin, S.K.; Hui, A.C.; Giacomini, K.M. High-performance liquid chromatographic determination of the enantiomers of beta-adrenoceptor blocking agents in biological fluids. II. Studies with atenolol. *J.Chromatogr.*, **1989**, *489*, 438–445

Johannsson, M.; Forsmo-Bruce, H. Determination of atenolol in plasma by dual-column liquid chromatography and fluorimetric detection. *J.Chromatogr.*, **1988**, *432*, 265–272

Keech, A.C.; Harrison, P.M.; McLean, A.J. Simple extraction of atenolol from urine and its determination by high-performance liquid chromatography. *J.Chromatogr.*, **1988**, *426*, 234–236

Sa'sa', S.I.; Jalal, I.M.; Khalil, H.S. Determination of atenolol combinations with hydrochlorothiazide and chlorthalidone in tablet formulations by reverse-phase HPLC. *J.Liq.Chromatogr.*, **1988**, *11*, 1673–1696 [tablets; stability-indicating; simultaneous chlorthalidone, hydrochlorothiazide; methyl p-hydroxybenzoate (IS)]

Sa'sa', S.I. Determination of atenolol and its related compounds by ion pair high performance liquid chromatography. *J.Liq.Chromatogr.*, **1988**, *11*, 929–942 [stability-indicating]

Wilson, M.J.; Ballard, K.D.; Walle, T. Preparative resolution of the enantiomers of the beta-blocking drug atenolol by chiral derivatization and high performance liquid chromatography. *J.Chromatogr.*, **1988**, *431*, 222–227

Buhring, K.U.; Garbe, A. Determination of the new beta-blocker bisoprolol and of metoprolol, atenolol and propranolol in plasma and urine by high-performance liquid chromatography. *J.Chromatogr.*, **1986**, *382*, 215–224

Miller, L.G.; Greenblatt, D.J. Determination of atenolol in plasma by high-performance liquid chromatography with application to single-dose pharmacokinetics. *J.Chromatogr.*, **1986**, *381*, 201–204

Ficarra, R.; Ficarra, P.; Tommasini, A.; Calabro, M.L.; Guarniera Fenech, C. [HPLC determination of atenolol and chlorthalidone associated in pharmaceutical preparations]. *Farmaco [Prat]*, **1985**, *40*, 307–312

Harrison, P.M.; Tonkin, A.M.; McLean, A.J. Simple and rapid analysis of atenolol and metoprolol in plasma using solid-phase extraction and high-performance liquid chromatography. *J.Chromatogr.*, **1985**, *339*, 429–433 [plasma; SPE; fluorescence detection; LOD 10 ng/mL; non-interfering chlorothiazide, disopyramide, furosemide, hydralazine, lidocaine, methydopa, prazosin, verapamil; simultaneous alprenolol, oxprenolol, pindolol, practolol, propranolol, timolol]

Bhamra, R.K.; Thorley, K.J.; Vale, J.A.; Holt, D.W. High-performance liquid chromatographic measurement of atenolol: methodology and clinical applications. *Ther.Drug Monit.*, **1983**, *5*, 313–318

Winkler, H.; Ried, W.; Lemmer, B. High-performance liquid chromatographic method for the quantitative analysis of the aryloxypropanolamines propranolol, metoprolol and atenolol in plasma and tissue. *J.Chromatogr.*, **1982**, *228*, 223–234

Lefebvre, M.A.; Girault, J.; Fourtillan, J.B. β-Blocking agents: Determination of biological levels using high performance liquid chromatography. *J.Liq.Chromatogr.*, **1981**, *4*, 483–500 [plasma; fluorescence detection; also acebutolol, metoprolol, oxprenolol, pindolol, propranolol, sotalol, timolol]

# Atropine

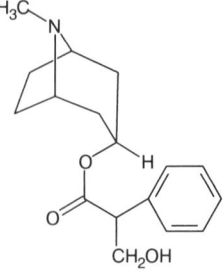

**Molecular formula:** $C_{17}H_{23}NO_3$
**Molecular weight:** 289.4
**CAS Registry No.:** 51-55-8 (atropine), 52-88-0 (atropine methylnitrate),
101-31-5 (hyoscyamine), 306-03-6 (hyoscyamine hydrobromide), 6835-16-1
(hyoscyamine sulfate dihydrate), 620-61-1 (hyoscyamine sulfate)

## SAMPLE
**Matrix:** blood
**Sample preparation:** 1 mL Plasma + 100 μL 5 ng/mL scopolamine in MeOH, vortex
briefly, add 50 μL 1 M ammonium hydroxide, mix, add 5 mL dichloromethane, shake
horizontally for 5 min, centrifuge at 2500 rpm for 5 min. Remove the organic layer and
evaporate it to dryness under a stream of nitrogen at 40°, reconstitute the residue in 100
μL mobile phase, inject a 20 μL aliquot.

## HPLC VARIABLES
**Guard column:** 10 × 2 5 μm BDS C18 (Keystone)
**Column:** 50 × 3 3 μm BDS C18 (Keystone)
**Mobile phase:** MeCN:MeOH:10 mM ammonium acetate 62.5:37.5:15
**Flow rate:** 0.5
**Injection volume:** 20
**Detector:** MS, Perkin Elmer Sciex API III-Plus triple quadrupole, APCI, nebulizer 400°
and 80 psi, auxiliary nitrogen 1.2 L/min, curtain gas 1.2 L/min, interface 55°, collision
gas argon, electron multiplier 3000 V, declustering potential 35 V, collision energy 35 eV

## CHROMATOGRAM
**Retention time:** 1.2
**Internal standard:** scopolamine (0.8)
**Limit of quantitation:** 20 pg/mL

## KEY WORDS
plasma; protect from light

## REFERENCE
Xu, A.; Havel, J.; Linderholm, K.; Hulse, J. Development and validation of an LC/MS/MS method for
the determination of L-hyoscyamine in human plasma. *J.Pharm.Biomed.Anal.*, **1996**, *14*, 33–42

## SAMPLE
**Matrix:** blood
**Sample preparation:** Condition a Sep-Pak C18 SPE cartridge with water, MeOH, and 100
mM ammonium acetate. Add 200 μL plasma to the SPE cartridge, wash with 100 mM
ammonium acetate, elute with MeOH:100 mM ammonium acetate 3:1. Evaporate the
eluate to dryness under reduced pressure, dissolve the residue in 200 μL mobile phase,
inject a 20 μL aliquot.

## HPLC VARIABLES
**Column:** 150 × 4.6 Hitachi gel 3056 octadecylsilica
**Mobile phase:** MeOH:100 mM ammonium acetate 60:40
**Flow rate:** 1
**Injection volume:** 20
**Detector:** MS, Hitachi M1000, APCI, nebulizer 260°, vaporizer 399°

## CHROMATOGRAM
**Retention time:** 4.3
**Limit of detection:** 0.5-2.5 ng/mL

## OTHER SUBSTANCES
**Simultaneous:** atipamezole, butorphanol, flumazenil, ketamine, medetomidine, midazolam, xylazine

## KEY WORDS
plasma; SPE; dog

## REFERENCE
Kanazawa, H.; Nagata, Y.; Matsushima, Y.; Takai, N.; Uchiyama, H.; Nishimura, R.; Takeuchi, A. Liquid chromatography-mass spectrometry for the determination of medetomidine and other anaesthetics in plasma. *J.Chromatogr.*, **1993**, *631*, 215–220

## SAMPLE
**Matrix:** blood
**Sample preparation:** 2 mL Plasma + 100 μL 1 M NaOH + 10 mL chloroform, shake for 30 s, remove a 9 mL aliquot of the organic phase and add it to 1 mL 100 mM HCl, extract. Remove a 900 μL aliquot of the aqueous layer and add it to 100 μL 5 M NaOH, heat at 38° for 3 h, acidify with 5 M HCl, add 8 mL dichloromethane, vortex for 2 min, centrifuge. Remove the organic layer and evaporate it to dryness under a stream of nitrogen at 40-45°, add 3 mg solid potassium bicarbonate, add 100 μL 10 μg/mL mandelic acid in MeCN, add 300 μL 50 μg/mL 4-bromomethyl-7-methoxycoumarin in MeCN, add 100 μL 15 μg/mL 18-crown-6 in MeCN, vortex for 15 s, heat at 70° for 45 min, inject a 25 μL aliquot.

## HPLC VARIABLES
**Column:** 150 × 4.6 5 μm ODS (Dupont)
**Mobile phase:** MeCN:buffer 33:67 (Buffer was 10 mM $(NH_4)H_2PO_4$ adjusted to pH 5.0.)
**Column temperature:** 40
**Flow rate:** 2
**Injection volume:** 25
**Detector:** F ex 328 em 389 (cutoff filter)

## CHROMATOGRAM
**Retention time:** 10
**Internal standard:** mandelic acid (7)
**Limit of detection:** 108 ng/mL
**Limit of quantitation:** 125 ng/mL

## KEY WORDS
plasma; derivatization; atropine determined after hydrolysis to tropic acid

## REFERENCE
Li, S.; Wahba Khalil, S.K. An HPLC method for determination of atropine in human plasma. *J.Liq.Chromatogr.*, **1990**, *13*, 1339–1350

## SAMPLE
**Matrix:** bulk, plants
**Sample preparation:** Place 0.5 g powdered crude drug in 25 mL mobile phase, reflux 30 min, cool, centrifuge at 1600 g, decant wash residue twice with 10 mL portions of mobile phase, combine extracts and washings, make up to 50 mL with mobile phase, inject 10 μL aliquot.

## HPLC VARIABLES
**Column:** 150 × 4 5 μm TSK gel 120A ODS
**Mobile phase:** MeCN:67 mM pH 2.5 phosphate buffer 35:65, containing 17.5 mM sodium dodecylsulfate
**Column temperature:** 35
**Flow rate:** 1.5
**Injection volume:** 10
**Detector:** UV 210

## CHROMATOGRAM
**Retention time:** 15

## OTHER SUBSTANCES
**Simultaneous:** scopolamine

## REFERENCE
Oshima, T.; Sagara, K.; Tong, Y.Y.; Zhang, G.; Chen, Y.H. Application of ion-pair high performance liquid chromatography for analysis of hyoscyamine and scopolamine in solanaceous crude drugs. *Chem.Pharm.Bull.*, **1989**, *37*, 2456–2458

## SAMPLE
**Matrix:** formulations
**Sample preparation:** Injections and ophthalmic solutions. Dilute with water to an atropine concentration of 80 μg/mL, inject a 20 μL aliquot. Ointment. Weigh out ointment equivalent to about 4 mg atropine sulfate, add 10 mL THF:water 80:20, sonicate and swirl until the ointment is completely dispersed, make up to 50 mL with water, filter (0.45 μm), inject a 20 μL aliquot

## HPLC VARIABLES
**Column:** 250 × 4.6 5 μm Spherisorb CN
**Mobile phase:** MeCN:50 mM $NaH_2PO_4$ 10:90, pH adjusted to 4.0 with 10% phosphoric acid
**Flow rate:** 1
**Injection volume:** 20
**Detector:** UV 220

## CHROMATOGRAM
**Retention time:** 8

## OTHER SUBSTANCES
**Simultaneous:** phenol, tropic acid
**Noninterfering:** benzalkonium chloride, benzyl alcohol, chlorobutanol, methylparaben

## KEY WORDS
injections; ophthalmic solutions; ointments

## REFERENCE
Lehr, G.J.; Yuen, S.M.; Lawrence, G.D. Liquid chromatographic determination of atropine in nerve gas antidotes and other dosage forms. *J.AOAC Int.*, **1995**, *78*, 339–343

## SAMPLE
**Matrix:** formulations
**Sample preparation:** Inject a 20 μL aliquot.

## HPLC VARIABLES
**Column:** 100 × 4.6 Spheri-5 RP-8

**Mobile phase:** MeCN:buffer 40:60 (Buffer was 10 mM $KH_2PO_4$ adjusted to pH 4.0 with 1 M KOH.)
**Flow rate:** 1
**Injection volume:** 20
**Detector:** UV 254

## CHROMATOGRAM
**Retention time:** 7.8
**Limit of detection:** 8.6 µg/mL

## OTHER SUBSTANCES
**Simultaneous:** ondansetron
**Noninterfering:** degradation products

## KEY WORDS
injections; saline

## REFERENCE
Venkateshwaran, T.G.; King, D.T.; Stewart, J.T. HPLC determination of ondansetron-atropine and ondansetron-glycopyrrolate mixtures in 0.9% sodium chloride injection. *J.Liq.Chromatogr.*, **1995**, *18*, 2647–2659

## SAMPLE
**Matrix:** formulations
**Sample preparation:** Dilute with water, inject a 20 µL aliquot.

## HPLC VARIABLES
**Guard column:** 12.5 × 4 5 µm Zorbax RX-C18
**Column:** 250 × 4.6 5 µm Zorbax RX-C18
**Mobile phase:** MeCN:buffer 20:80 (Buffer was 50 mM $NaH_2PO_4$ + 1 mM tetramethylammonium chloride + 0.5 mM 1-octanesulfonic acid adjusted to pH 3.5 with concentrated orthophosphoric acid.)
**Column temperature:** 25
**Flow rate:** 1
**Injection volume:** 20
**Detector:** UV 203

## CHROMATOGRAM
**Retention time:** 9

## OTHER SUBSTANCES
**Simultaneous:** HI-6, obidoxime, phenol, tropic acid
**Also analyzed:** pralidoxime chloride

## KEY WORDS
nerve agent antidote mixtures

## REFERENCE
Paddle, B.M.; Dowling, M.H. Simple high-performance liquid chromatographic method for assessing the deterioration of atropine-oxime mixtures employed as antidotes in the treatment of nerve agent poisoning. *J.Chromatogr.*, **1993**, *648*, 373–380

## SAMPLE
**Matrix:** formulations
**Sample preparation:** Dilute with water, inject an aliquot.

## HPLC VARIABLES
**Column:** 250 × 4.6 10 μm Whatman PXS ODS-3 C18
**Mobile phase:** MeCN:Buffer 25:75 (Buffer was 1.08 g sodium octanesulfonate in 900 mL of water adjusted to pH 3.5 with glacial acetic acid and diluted to 1 L with water.)
**Flow rate:** 2
**Injection volume:** 20
**Detector:** UV 229

## CHROMATOGRAM
**Retention time:** 5

## OTHER SUBSTANCES
**Simultaneous:** milrinone

## KEY WORDS
injections; 10% calcium chloride; 7.5% sodium bicarbonate; stability-indicating

## REFERENCE
Wilson, T.D.; Forde, M.D. Stability of milrinone and epinephrine, atropine sulfate, lidocaine hydrochloride, or morphine sulfate injection. *Am.J.Hosp.Pharm.*, **1990**, *47*, 2504–2507

## SAMPLE
**Matrix:** formulations
**Sample preparation:** Tablets. Crush tablet, add 2 mL water, add 500 μL 0.1 M NaOH, if not basic to litmus add 0.1 M NaOH dropwise until it is. Extract with 10 mL chloroform then with three 5 mL portions of chloroform. Pass all extracts through a sodium sulfate column and then wash column with 2 mL chloroform. Evaporate all organic extracts with heating under a stream of air. Take up residue in 1 mL MeOH, inject a 100 μL aliquot. Injections. Evaporate 2 mL to dryness on a steam bath, take up residue in 5 mL water, add 500 μL 0.1 M NaOH, if not basic to litmus add 0.1 M NaOH dropwise until it is. Extract with 10 mL chloroform then with three 5 mL portions of chloroform. Pass all extracts through a sodium sulfate column and then wash column with 2 mL chloroform. Evaporate all organic extracts with heating under a stream of air. Take up residue in 2 mL MeOH, inject a 100 μL aliquot. (Sodium sulfate column was a 300 × 20 glass chromatography column containing 10 g anhydrous sodium sulfate washed with 10 mL chloroform before use.)

## HPLC VARIABLES
**Column:** 250 × 4.6 Alltech C18
**Mobile phase:** MeOH:20 g/L sodium pentanesulfonate 95:5
**Flow rate:** 1
**Injection volume:** 100
**Detector:** F ex 255 em 285

## CHROMATOGRAM
**Retention time:** 3.5

## OTHER SUBSTANCES
**Simultaneous:** scopolamine
**Noninterfering:** ergotamine, phenobarbital

## KEY WORDS
tablets; injections

## REFERENCE
Cieri, U.R. Determination of small quantities of atropine in commercial preparations by liquid chromatography with fluorescence detection. *J.Assoc.Off.Anal.Chem.*, **1985**, *68*, 1042–1045

## SAMPLE

**Matrix:** formulations
**Sample preparation:** Dilute 10 mL eye drops to 100 mL with water, inject a 20 μL aliquot.

## HPLC VARIABLES

**Column:** 150 × 3.9 5 μm Perkin-Elmer C18
**Mobile phase:** MeOH:water:heptanesulfonic acid solution 500:500:25 (Heptanesulfonic acid solution was 5 g heptanesulfonic acid + 20 mL glacial acetic acid diluted to 150 mL with water.)
**Flow rate:** 1.8
**Injection volume:** 20
**Detector:** UV 225

## CHROMATOGRAM

**Retention time:** 4
**Limit of detection:** 5 μg/mL

## OTHER SUBSTANCES

**Simultaneous:** tropic acid

## KEY WORDS

eye drops

## REFERENCE

Richard, A.; Andermann, G. Simultaneous determination of atropine sulphate and tropic acid by reversed phase high-pressure liquid chromatography. *Pharmazie*, **1984**, *39*, 866–867

## SAMPLE

**Matrix:** formulations
**Sample preparation:** Tablets, capsules. Powder tablets or remove contents of capsules, weigh out amount equivalent to about 600 μg hyoscyamine sulfate-atropine sulfate, add 25 mL 25 mM sulfuric acid, shake for 15 min, centrifuge at 3000 rpm for 5 min. Remove 5 mL of the supernatant and extract it twice with 30 mL portions of dichloromethane, discard the organic phase, add 2 mL buffer to the aqueous phase, extract with four 30 mL portions of dichloromethane, filter extracts through dichloromethane-rinsed glass wool, add 3 mL 2.25 μg/mL theophylline in dichloromethane, distil off the dichloromethane through a Snyder column by using a steam bath, when the volume reaches 10 mL rinse the column with 1-2 mL dichloromethane, continue distillation to 0.5-1 mL, remove the column and rinse the concentrator tube-column junction with 1 mL dichloromethane, evaporate to 1 mL with a stream of air at 40°, add 100 μL 1% concentrated HCl in MeOH, mix, evaporate to dryness with a stream of air at 40°, rinse the sides of the concentrator tube with 500 μL MeOH, evaporate to dryness with a stream of air at 40°, dissolve the residue in 300 μL water, inject a 20 μL aliquot. Elixirs. Add an amount equivalent to about 600 μg hyoscyamine sulfate-atropine sulfate to a 150 mL beaker, warm at 40° with a current of air for 30 min to remove alcohol, cool, make up to 25 mL with water, remove 5 mL of this solution, add 2 mL 100 mM sulfuric acid, extract twice with 30 mL portions of dichloromethane, discard the organic phase, add 2 mL buffer to the aqueous phase, extract with four 30 mL portions of dichloromethane, filter extracts through dichloromethane-rinsed glass wool, add 3 mL 2.25 μg/mL theophylline in dichloromethane, distil off the dichloromethane through a Snyder column by using a steam bath, when the volume reaches 10 mL rinse the column with 1-2 mL dichloromethane, continue distillation to 0.5-1 mL, remove the column and rinse the concentrator tube-column junction with 1 mL dichloromethane, evaporate to 1 mL with a stream of air at 40°, add 100 μL 1% concentrated HCl in MeOH, mix, evaporate to dryness with a stream of air at 40°, rinse the sides of the concentrator tube with 500 μL MeOH, evaporate to dryness with a stream of air at 40°, dissolve the residue in 300 μL water, inject a 20 μL aliquot. (Buffer was 5.3 g

anhydrous sodium carbonate and 4.2 g sodium bicarbonate in 100 mL water, pH 9.4. Pass dichloromethane through 75 g basic aluminum oxide, Brockmann Activity Grade 1, store over 25 g alumina/4 L.)

## HPLC VARIABLES
**Column:** 250 × 4 5 μm Spherisorb ODS
**Mobile phase:** MeOH:buffer 250:525 (The 50 mM tetramethylammonium phosphate buffer was prepared from 500 mL water + 23 mL 20% tetramethylammonium hydroxide in MeOH + 10 mL concentrated phosphoric acid, adjust to pH 2.0 with concentrated phosphoric acid, make up to 1 L with water.
**Flow rate:** 0.8
**Injection volume:** 20
**Detector:** UV 220

## CHROMATOGRAM
**Retention time:** 7.5
**Internal standard:** theophylline (6.5)

## OTHER SUBSTANCES
**Simultaneous:** hyoscyamine, phenobarbital, scopolamine

## KEY WORDS
tablets; capsules; elixirs

## REFERENCE
Pennington, L.J.; Schmidt, W.F. Belladonna alkaloids and phenobarbital combination pharmaceuticals analysis I: High-performance liquid chromatographic determinations of hyoscyamine-atropine and scopolamine. *J.Pharm.Sci.*, **1982**, *71*, 951–953

## SAMPLE
**Matrix:** plants
**Sample preparation:** 100 mg Freeze-dried powdered plant leaves + 10 mL mobile phase, heat at 40° for 15 min, filter, inject a 20 μL aliquot.

## HPLC VARIABLES
**Column:** 150 × 4 4 μm Novapack C18
**Mobile phase:** MeCN:water 12.5:87.5 with 0.3% phosphoric acid adjusted to pH 2.2 with triethylamine
**Flow rate:** 0.8
**Injection volume:** 20
**Detector:** UV 204

## CHROMATOGRAM
**Retention time:** 11.7
**Limit of detection:** 50 μg/g

## OTHER SUBSTANCES
**Simultaneous:** tropic acid, scopolamine

## KEY WORDS
freeze-dried plant leaves

## REFERENCE
Fliniaux, M.-A.; Manceau, F.; Jacquin-Dubreuil, A. Simultaneous analysis of l-hyoscyamine, l-scopolamine and dl-tropic acid in plant material by reversed phase high-performance liquid chromatography. *J.Chromatogr.*, **1993**, *644*, 193–197

## SAMPLE
**Matrix:** plants
**Sample preparation:** Extract 0.1 g dry plant material with 10 mL MeOH for 10 min under reflux, filter, inject aliquot.

## HPLC VARIABLES
**Guard column:** 40 × 4 10 μm Hypersil ODS
**Column:** 250 × 4 10 μm Hypersil ODS
**Mobile phase:** MeOH:water 45:55 containing 0.1% phosphoric acid adjusted to pH 7 with triethylamine
**Flow rate:** 1
**Detector:** UV 229

## CHROMATOGRAM
**Retention time:** 20.7

## OTHER SUBSTANCES
**Simultaneous:** scopolamine

## REFERENCE
Hagemann, K.; Piek, K.; Stöckigt, J.; Weiler, E.W. Monoclonal antibody-based enzyme immunoassay for the quantitative determination of the tropane alkaloid, scopolamine. *Planta Med.*, **1992**, *58*, 68–72

## SAMPLE
**Matrix:** plants
**Sample preparation:** Dissolve plant extract in 1 mL MeOH, add 40 ng homatropine, inject aliquot.

## HPLC VARIABLES
**Column:** 150 × 4.1 5 μm Hamilton PRP-1
**Mobile phase:** MeCN:100 mM pH 10.4 ammonium acetate 30:70
**Flow rate:** 1
**Injection volume:** 20
**Detector:** MS thermospray, VG Trio-2, ion source 150°, vaporizer tip 170°, repeller electrode 150 V, m/z 290

## CHROMATOGRAM
**Internal standard:** homatropine (m/z 276)
**Limit of detection:** 2.5 ng/mL

## OTHER SUBSTANCES
**Simultaneous:** scopolamine

## KEY WORDS
total run time 6 min

## REFERENCE
Auriola, S.; Martinsen, A.; Oksman-Caldentey, K.M.; Naaranlahti, T. Analysis of tropane alkaloids with thermospray high-performance liquid chromatography-mass spectrometry. *J.Chromatogr.*, **1991**, *562*, 737–744

## SAMPLE
**Matrix:** solutions

## HPLC VARIABLES
**Column:** 250 × 4.1 6 μm PolyEncap ODS (n-octadecylacrylate copolymerized with vinyl silica in heptane, carrier Ultrasep ES 100; preparation described in paper)
**Mobile phase:** MeCN:pH 2.2 phosphate buffer 20:80
**Flow rate:** 1
**Detector:** UV 220

## CHROMATOGRAM
**Retention time:** 2.7

## OTHER SUBSTANCES
**Simultaneous:** barbituric acid, codeine, diphenhydramine, noscapine, papaverine

## REFERENCE
Engelhardt, H.; Cuñat-Walter, M.A. Polymer encapsulated stationary phases with improved efficiency. *Chromatographia*, **1995**, *40*, 657–661

## SAMPLE
**Matrix:** solutions

## HPLC VARIABLES
**Column:** 150 × 2 PRP-1 (Keystone)
**Mobile phase:** MeCN:2-butanone:100 mM pH 7.5 phosphate buffer 40:20:40
**Flow rate:** 0.15
**Injection volume:** 1
**Detector:** Chemiluminescence following post-column reaction. A 1 mM solution of Ru(2,2'-bipyridine)$_3^{2+}$ in 50 mM sodium sulfate (continuously sparged with helium) was oxidized to Ru(2,2'-bipyridine)$_3^{3+}$ using a Princeton Applied Research Model 174A polarographic analyzer with a platinum gauze working electrode, a platinum wire auxiliary electrode, and a silver wire reference electrode. The Ru solution pumped at 0.3 mL/min mixed with the column effluent in the flow cell of the detector, a fluorescence detector with the light source removed.

## CHROMATOGRAM
**Retention time:** 3
**Limit of detection:** 0.1-1 μg/mL

## OTHER SUBSTANCES
**Simultaneous:** cyclobenzaprine, cyclopentolate, dicyclomine, procyclidine

## KEY WORDS
post-column reaction

## REFERENCE
Holeman, J.A.; Danielson, N.D. Microbore liquid chromatography of tertiary amine anticholinergic pharmaceuticals with tris(2,2'-bipyridine)ruthenium(III) chemiluminescence detection. *J.Chromatogr. Sci.*, **1995**, *33*, 297–302

## SAMPLE
**Matrix:** solutions

## HPLC VARIABLES
**Column:** 250 × 4.6 5 μm Supelcosil LC-DP (A) or 250 × 4 5 μm LiChrospher 100 RP-8 (B)
**Mobile phase:** MeCN:0.025% phosphoric acid:buffer 25:10:5 (A) or 60:25:15 (B) (Buffer was 9 mL concentrated phosphoric acid and 10 mL triethylamine in 900 mL water, adjust pH to 3.4 with dilute phosphoric acid, make up to 1 L.)

**Flow rate:** 0.6
**Injection volume:** 25
**Detector:** UV 229

## CHROMATOGRAM
**Retention time:** 6.05 (A), 3.75 (B)

## OTHER SUBSTANCES
**Also analyzed:** acebutolol, acepromazine, acetaminophen, acetazolamide, acetophenazine, albuterol, alprazolam, amitriptyline, amobarbital, amoxapine, antipyrine, atenolol, azatadine, baclofen, benzocaine, bromocriptine, brompheniramine, brotizolam, bupivacaine, buspirone, butabarbital, butalbital, caffeine, carbamazepine, cetirizine, chlorcyclizine, chlordiazepoxide, chlormezanone, chloroquine, chlorpheniramine, chlorpromazine, chlorpropamide, chlorprothixene, chlorthalidone, chlorzoxazone, cimetidine, cisapride, clomipramine, clonazepam, clonidine, clozapine, cocaine, codeine, colchicine, cyclizine, cyclobenzaprine, dantrolene, desipramine, diazepam, diclofenac, diflunisal, diltiazem, diphenhydramine, diphenidol, diphenoxylate, dipyridamole, disopyramide, dobutamine, doxapram, doxepin, droperidol, encainide, ethidium bromide, ethopropazine, fenoprofen, fentanyl, flavoxate, fluoxetine, fluphenazine, flurazepam, flurbiprofen, fluvoxamine, furosemide, glutethimide, glyburide, guaifenesin, haloperidol, homatropine, hydralazine, hydrochlorothiazide, hydrocodone, hydromorphone, hydroxychloroquine, hydroxyzine, ibuprofen, imipramine, indomethacin, ketoconazole, ketoprofen, ketorolac, labetalol, levorphanol, lidocaine, loratadine, lorazepam, lovastatin, loxapine, mazindol, mefenamic acid, meperidine, mephenytoin, mepivacaine, mesoridazine, metaproterenol, methadone, methdilazine, methocarbamol, methotrexate, methotrimeprazine, methoxamine, methyldopa, methylphenidate, metoclopramide, metolazone, metoprolol, metronidazole, midazolam, moclobemide, morphine, nadolol, nalbuphine, naloxone, naphazoline, naproxen, nifedipine, nizatidine, norepinephrine, nortriptyline, oxazepam, oxycodone, oxymetazoline, paroxetine, pemoline, pentazocine, pentobarbital, pentoxifylline, perphenazine, pheniramine, phenobarbital, phenol, phenolphthalein, phentolamine, phenylbutazone, phenyltoloxamine, phenytoin, pimozide, pindolol, piroxicam, pramoxine, prazepam, prazosin, probenecid, procainamide, procaine, prochlorperazine, procyclidine, promazine, promethazine, propafenone, propantheline, propiomazine, propofol, propranolol, protriptyline, quazepam, quinidine, quinine, racemethorphan, ranitidine, remoxipride, risperidone, salicylic acid, scopolamine, secobarbital, sertraline, sotalol, spironolactone, sulfinpyrazone, sulindac, temazepam, terbutaline, terfenadine, tetracaine, theophylline, thiethylperazine, thiopental, thioridazine, thiothixene, timolol, tocainide, tolbutamide, tolmetin, trazodone, triamterene, triazolam, trifluoperazine, triflupromazine, trimeprazine, trimethoprim, trimipramine, verapamil, warfarin, xylometazoline, yohimbine, zopiclone

## KEY WORDS
some details of plasma extraction

## REFERENCE
Koves, E.M. Use of high-performance liquid chromatography-diode array detection in forensic toxicology. *J.Chromatogr.A*, **1995**, *692*, 103–119

## SAMPLE
**Matrix:** solutions

## HPLC VARIABLES
**Column:** 250 × 4.6 Zorbax RX
**Mobile phase:** Gradient. A was 10 mL concentrated orthophosphoric acid and 7 mL triethylamine in 1 L water. B was 10 mL concentrated orthophosphoric acid and 7 mL triethylamine in 200 mL water, make up to 1 L with MeCN. A:B from 100:0 to 0:100 over 30 min, maintain at 0:100 for 5 min.
**Column temperature:** 30

**Flow rate:** 2
**Detector:** UV 210

## OTHER SUBSTANCES
**Also analyzed:** acepromazine, acetaminophen, acetophenazine, albuterol, aminophylline, amitriptyline, amobarbital, amoxapine, amphetamine, amylocaine, antipyrine, aprobarbital, aspirin, atenolol, avermectin, barbital, benzocaine, benzoic acid, benzotropine, benzphetamine, berberine, bibucaine, bromazepan, brompheniramine, buprenorphine, buspirone, butabarbital, butacaine, butethal, caffeine, carbamazepine, carbromal, chloramphenicol, chlordiazepoxide, chloroquine, chlorothiazide, chloroxylenol, chlorphenesin, chlorpheniramine, chlorpromazine, chlorpropamide, chlortetracycline, cimetidine, cinchonidine, cinchonine, clenbuterol, clonazepam, clonixin, clorazepate, cocaine, codeine, colchicine, cortisone, coumarin, cyclazocine, cyclobenzaprine, cyclothiazide, cyheptamide, cymarin, danazol, danthron, dapsone, debrisoquine, desipramine, dexamethasone, dextromethorphan, dextropropoxyphene, diamorphine, diazepam, diclofenac, diethylpropion, diethylstilbestrol, diflunisal, digitoxin, digoxin, diltiazem, diphenhydramine, diphenoxylate, diprenorphine, dipyrone, disulfiram, dopamine, doxapram, doxepin, dronabinol, ephedrine, epinephrine, epinine, estradiol, estriol, estrone, ethacrynic acid, ethosuximide, etonitazene, etorphine, eugenol, famotidine, fenbendazole, fencamfamine, fenoprofen, fenproporex, fentanyl, flubendazole, flufenamic acid, flunitrazepam, 5-fluorouracil, fluoxymesterone, fluphenazine, furosemide, gentisic acid, gitoxigenin, glipizide, glunixin, glutethimide, glybenclamide, guaiacol, halazepam, haloperidol, hydrochlorothiazide, hydrocodone, hydrocortisone, hydromorphone, hydroxyquinoline, ibogaine, ibuprofen, iminostilbene, imipramine, indomethacin, isocarbostyril, isocarboxazid, isoniazid, isoproterenol, isoxsuprine, ivermectin, ketamine, ketoprofen, kynurenic acid, levorphanol, lidocaine, lorazepam, lormetazepam, loxapine, mazindol, mebendazole, meclizine, meclofenamic acid, medazepam, mefenamic acid, megestrol, mepacrine, meperidine, mephentermine, mephenytoin, mephesin, mephobarbital, mepivacaine, mescaline, mesoridazine, methadone, methamphetamine, methapyrilene, methaqualone, methazolamide, methocarbamol, methoxamine, methsuximide, methyl salicylate, methyldopa, methyldopamine, methylphenidate, methylprednisolone, methyltestosterone, methyprylon, metoprolol, mibolerone, morphine, nadolol, nalorphine, naloxone, naltrexone, naphazoline, naproxen, nefopam, niacinamide, nicotine, nicotinic acid, nifedipine, niflumic acid, nitrazepam, norepinephrine, nortriptyline, noscapine, nylidrin, oxazepam, oxycodone, oxymorphone, oxyphenbutazone, oxytetracycline, papaverine, pargyline, pemoline, pentazocine, pentobarbital, persantine, phenacetin, phenazocine, phenazopyridine, phencyclidine, phendimetrazine, phenelzine, pheniramine, phenobarbital, phenothiazine, phensuximide, phentermine, phenylbutazone, phenylephrine, phenylpropanolamine, piperocaine, prazepam, prednisolone, primidone, probenecid, progesterone, propiomazine, propranolol, propylparaben, pseudoephedrine, puromycin, pyrilamine, pyrithyldione, quazepam, quinaldic acid, quinidine, quinine, ranitidine, recinnamine, reserpine, resorcinol, saccharin, albuterol, salicylamide, salicylic acid, scopolamine, scopoletin, secobarbital, strychnine, sulfacetamide, sufadiazine, sulfadimethoxine, sulfaethidole, sulfamerazine, sulfamethazine, sulfamethoxizole, sulfanilamide, sulfapyridine, sulfasoxizole, sulindac, tamoxifen, temazepam, testosterone, tetracaine, tetracycline, tetramisole, thebaine, theobromine, theophylline, thiabendazole, thiamine, thiamylal, thiobarbituric acid, thioridazine, thiosalicylic acid, thiothixene, thymol, tolazamide, tolazoline, tobutamide, tolmetin, tranylcypromine, triamcinolone, tribenzylamine, trichloromethiazide, trifluoperazine, trihexyphenidyl, trimethoprim, tripelennamine, triproilidine, tropacocaine, tyramine, verapamil, vincamine, warfarin, yohimbine, zoxazolamine

## REFERENCE
Hill, D.W.; Kind, A.J. Reversed-phase solvent gradient HPLC retention indexes of drugs. *J.Anal.Toxicol.*, **1994**, *18*, 233–242

## SAMPLE
**Matrix:** solutions

## HPLC VARIABLES
**Column:** 250 × 4.6 cellulose tris(3,5-dimethylphenylcarbamate)
**Mobile phase:** Hexane:isopropanol:diethylamine 80:20:0.1
**Flow rate:** 0.5
**Detector:** UV

## CHROMATOGRAM
**Retention time:** k' 0.72 (of first (+) enantiomer)

## KEY WORDS
chiral; α 1.62

## REFERENCE
Okamoto, Y.; Aburatani, R.; Hatano, K.; Hatada, K. Optical resolution of racemic drugs by chiral HPLC on cellulose and amylose tris(phenylcarbamate) derivatives. *J.Liq.Chromatogr.*, **1988**, *11*, 2147–2163

## SAMPLE
**Matrix:** solutions
**Sample preparation:** Dissolve in MeOH or water to 0.1%, inject an aliquot

## HPLC VARIABLES
**Column:** two 250 mm β-cyclodextrin bonded phase columns in series (Advanced Separation Technologies)
**Mobile phase:** MeCN:1% pH 4.1 aqueous triethylammonium acetate 2:98
**Flow rate:** 0.5
**Injection volume:** 1
**Detector:** UV

## CHROMATOGRAM
**Retention time:** k' 6.83 (d-isomer)

## KEY WORDS
chiral; optical isomers are separated

## REFERENCE
Armstrong, D.W.; Han, S.M.; Han, Y.I. Separation of optical isomers of scopolamine, cocaine, homatropine, and atropine. *Anal.Biochem.*, **1987**, *167*, 261–264

## SAMPLE
**Matrix:** solutions

## HPLC VARIABLES
**Column:** 300 × 3.9 10 μm LiChrosorb Si-60
**Mobile phase:** MeOH:water 60:40 containing 4 mM disodium citrate and 4 mM tetrabutylammonium bromide, pH 5.9
**Flow rate:** 1
**Injection volume:** 10
**Detector:** UV 254

## CHROMATOGRAM
**Retention time:** 13

## OTHER SUBSTANCES
**Simultaneous:** codeine, dansylamide, dansylcadaverine, doxorubicin, methylatropine, naphazoline, noscapine, xylometazoline

## REFERENCE

Lingeman, H.; van Munster, H.A.; Beynen, J.H.; Underberg, W.J.; Hulshoff, A. High-performance liquid chromatographic analysis of basic compounds on non-modified silica gel and aluminium oxide with aqueous solvent mixtures. *J.Chromatogr.*, **1986**, *352*, 261–274

## SAMPLE

**Matrix:** solutions
**Sample preparation:** Dissolve in MeOH:water 1:1 at a concentration of 50 μg/mL, inject a 10 μL aliquot.

## HPLC VARIABLES

**Column:** 300 × 3.9 10 μm μBondapak C18
**Mobile phase:** MeOH:acetic acid:triethylamine:water 15:1.5:0.5:83
**Flow rate:** 1.5
**Injection volume:** 10
**Detector:** UV 230

## CHROMATOGRAM

**Retention time:** 17

## OTHER SUBSTANCES

**Simultaneous:** homatropine, methscopolamine, tropic acid, atropine methyl, scopolamine

## REFERENCE

Roos, R.W.; Lau-Cam, C.A. General reversed-phase high-performance liquid chromatographic method for the separation of drugs using triethylamine as a competing base. *J.Chromatogr.*, **1986**, *370*, 403 –418

## ANNOTATED BIBLIOGRAPHY

Theodoridis, G.; Papadoyannis, I.; Vasilikiotis, G.; Tsoukali-Papadopoulou, H. Reversed-phase high-performance liquid chromatography–photodiode-array analysis of alkaloid drugs of forensic interest. *J.Chromatogr.B*, **1995**, *668*, 253–263 [also amphetamine, bamifylline, caffeine, cocaine, codeine, diamorphine, ethylmorphine, flufenamic acid, methadone, morphine, nalorphine, norcodeine, papaverine, quinine, scopolamine, strychnine, theobromine, theophylline, tolfenamic acid]

Buch, U.; Isenberg, E.; Buch, H.P. HPLC assay for atropine in serum and protein solutions after in vitro addition of the tropane alkaloid. *Methods Find.Exp.Clin.Pharmacol.*, **1994**, *16*, 361–365

Pohjola, J.; Harpf, M. Determination of atropine and obidoxime in automatic injection devices used as antidotes against nerve agent intoxication. *J.Chromatogr.*, **1994**, *686*, 350–354 [simultaneous obidoxime; phenol (IS)]

Schill, G.; Wainer, I.W.; Barkan, S.A. Chiral separation of cationic drugs on an α1-acid glycoprotein bonded stationary phase. *J.Liq.Chromatogr.*, **1986**, *9*, 641–666 [chiral; also bromdiphenhydramine, brompheniramine, bupivacaine, butorphanol, carbinoxamine, chlorpheniramine, clidinium, cocaine, cyclopentolate, dimethindene, diperidone, disopyramide, doxylamine, ephedrine, homatropine, labetalol B, labetalol, labetalol A, mepensolate, mepivacaine, methadone, methorphan, methylatropine, methylhomatropine, methylphenidate, metoprolol, nadolol, nadolol A, nadolol B, oxprenolol, oxyphencyclimine, phenmetrazine, phenoxybenzamine, promethazine, pronethalol, propoxyphene, propranolol, pseudoephedrine, terbutaline, tocainide, tridihexethyl]

Achari, R.G.; Jacob, J.T. A study of the retention behavior of some basic drug substances by ion-pair HPLC. *J.Liq.Chromatogr.*, **1980**, *3*, 81–92 [also N-acetylprocainamide, antazoline, caffeine, chlorpheniramine, codeine, ephedrine, epinephrine, naphazoline, papaverine, pheniramine, phenylephrine, phenylpropanolamine, procainamide, quinidine, scopolamine, xylocaine]

# Azathioprine

**Molecular formula:** C$_9$H$_7$N$_7$O$_2$S
**Molecular weight:** 277.3
**CAS Registry No.:** 446-86-6

## SAMPLE
**Matrix:** blood
**Sample preparation:** 500 µL Serum + 25 ng 9-methylazathioprine + 4.5 mL ethyl acetate, vortex for 1 min, centrifuge for 1 min, repeat extraction. Combine the organic layers and evaporate them to dryness under reduced pressure at 35°, reconstitute the residue in 250 µL mobile phase, vortex for 10 s, sonicate for 10 min, inject a 200 µL aliquot.

## HPLC VARIABLES
**Guard column:** 4 × 4.6 5 µm LiChrospher 100 RP 18
**Column:** 250 × 4.6 5 µm LiChrospher 60 Rp-select B
**Mobile phase:** MeCN:10 mM pH 2.3 potassium phosphate buffer 12:88 (Flush with MeCN:buffer 50:50 for 2 min after each run.)
**Column temperature:** 22
**Flow rate:** 1
**Injection volume:** 200
**Detector:** UV 285

## CHROMATOGRAM
**Retention time:** 16
**Internal standard:** 9-methylazathioprine (Add 400 mg anhydrous potassium carbonate and 200 µL methyl iodide to a solution of 220 mg azathioprine in 7 mL DMF at 0-5°, stir under nitrogen for 24 h, add 14 mL water, neutralize with 1 M HCl and sodium bicarbonate solution, filter, wash the solid with water, dry under vacuum to give 9-methylazathioprine (mp 174-5°). Purify by precipitating from DMF solution with water.) (29)
**Limit of quantitation:** 2.5 ng/mL

## OTHER SUBSTANCES
**Noninterfering:** 6-mercaptopurine

## KEY WORDS
serum; pharmacokinetics

## REFERENCE
Binscheck, T.; Meyer, H.; Wellhömer, H.H. High-performance liquid chromatographic assay for the measurement of azathioprine in human serum samples. *J.Chromatogr.B*, **1996**, *675*, 287–294

## SAMPLE
**Matrix:** blood
**Sample preparation:** Condition a 100 mg Isolute C8 SPE cartridge (International Sorbent Technology) with 2.5 mL MeOH and 3.5 mL 10 mM pH 7.0 phosphate buffer. 1 mL Plasma + 50 µL 1.5 µg/mL guaneran + 2 mL 10 mM pH 7.0 phosphate buffer, mix, add to the SPE cartridge at 2.5 mL/min, wash with 3 mL MeCN:pH 7.0 phosphate buffer 1:99, air dry for 2 min, elute with 2 mL MeOH:ethyl acetate 5:95. Evaporate the eluate to dryness under a stream of nitrogen at 40°, reconstitute the residue in 50 µL mobile phase, inject a 20 µL aliquot.

## HPLC VARIABLES
**Column:** 80 × 4.6 3 μm HSpecosphere 3CR C8 (Perkin-Elmer)
**Mobile phase:** MeCN:10 mM pH 6.2 sodium phosphate buffer 9:91
**Flow rate:** 1.2
**Injection volume:** 20
**Detector:** UV 280

## CHROMATOGRAM
**Retention time:** 8.9
**Internal standard:** guaneran (6-[(1-methyl-4-nitro-5-imidazolyl)thio]-2-aminopurine, Wellcome Foundation) (7.7)
**Limit of detection:** 0.5 ng/mL

## OTHER SUBSTANCES
**Simultaneous:** caffeine
**Noninterfering:** aspirin, chloroquine, cyclosporin, diltiazem, nifedipine, prednisolone

## KEY WORDS
plasma; pharmacokinetics; SPE

## REFERENCE
Albertioni, F.; Pettersson, B.; Ohlman, S.; Peterson, C. Analysis of azathioprine and 6-mercaptopurine in plasma in renal transplant recipients after administration with oral azathioprine. *J.Liq. Chromatogr.*, **1995**, *18*, 3991–4005

## SAMPLE
**Matrix:** blood
**Sample preparation:** Condition a 1 mL Sep-Pak silica SPE cartridge with 3 mL ethyl acetate and vacuum dry for 1 min. 1 mL Plasma + 2.8 μg antipyrine, add to SPE cartridge, wash with 5 mL benzene (Caution! Benzene is a carcinogen!) or hexane over 1 min, vacuum dry for 1 min, elute with 5 mL ethyl acetate. Evaporate the eluate to dryness under a stream of nitrogen, reconstitute the residue in 200 μL mobile phase, sonicate, inject all of the sample.

## HPLC VARIABLES
**Guard column:** C18 Guard-Pak (Waters)
**Column:** 100 × 8 4 μm 8 NV C18 Radial-Pak (Waters)
**Mobile phase:** MeOH:10 mM pH 4.5 sodium phosphate 13:87
**Flow rate:** 3
**Injection volume:** 200
**Detector:** UV 280

## CHROMATOGRAM
**Retention time:** 7
**Internal standard:** antipyrine (17)
**Limit of quantitation:** 5 ng/mL

## OTHER SUBSTANCES
**Simultaneous:** acetaminophen, aspirin, carmustine, chlorambucil, cytarabine, dacarbazine, diclofenac, etoposide, 5-fluorouracil, ifosfamide, indomethacin, lomustine, methotrexate, naproxen, salicylic acid, tegafur, teniposide, thioguanine
**Noninterfering:** betamethasone, carboplatin, cyclophosphamide, cyclosporine A, ibuprofen, thiotepa

## KEY WORDS
plasma; rabbit; SPE; human; pharmacokinetics

## REFERENCE
el-Yazigi, A.; Abdel Wahab, F. Expedient liquid chromatographic analysis of azathioprine in plasma by use of silica solid phase extraction. *Ther.Drug Monit.*, **1992**, *14*, 312–316

## SAMPLE
**Matrix:** blood
**Sample preparation:** 200 μL Serum + 5 μL 50 μg/mL 2-ethyl-4-oxoquinazoline in EtOH + 100 μL reagent, let stand at room temperature for 1 h, add 1.8 mL ethyl acetate, mix, centrifuge at 1800 g for 5 min, remove 1.5 mL of the supernatant, repeat the extraction. Combine the organic layers and evaporate them to dryness under reduced pressure below 30°, reconstitute the residue in 100 μL initial mobile phase, inject a 90 μL aliquot. (Reagent was 30 mg N-ethylmaleimide in 2 mL 50 mM pH 7.0 phosphate buffer, prepare fresh daily.)

## HPLC VARIABLES
**Column:** 10 μm μBondapak C18
**Mobile phase:** Gradient. MeCN:10 mM $KH_2PO_4$ 9:91 for 26 min, then 50:50 for 1 min (step gradient).
**Flow rate:** 1.5
**Injection volume:** 90
**Detector:** UV 280

## CHROMATOGRAM
**Retention time:** 13.4
**Internal standard:** 2-ethyl-4-oxoquinazoline (28)
**Limit of quantitation:** 10 ng/mL

## OTHER SUBSTANCES
**Extracted:** 6-mercaptopurine

## KEY WORDS
serum

## REFERENCE
Tsutsumi, K.; Otsuki, Y.; Kinoshita, T. Simultaneous determination of azathioprine and 6-mercaptopurine in serum by reversed-phase high-performance liquid chromatography. *J.Chromatogr.*, **1982**, *231*, 393–399

## SAMPLE
**Matrix:** formulations
**Sample preparation:** Dissolve crushed tablets or the freeze-dried compound for injection in 20 mM NaOH. Add a 10 mL aliquot of this solution (or a saline injection) to 10 mL 3 mg/mL theophylline in 20 mM NaOH, inject a 1.5 μL aliquot.

## HPLC VARIABLES
**Column:** 100 × 5 5 μm ODS-Hypersil
**Mobile phase:** MeOH:25 mM $KH_2PO_4$:glacial acetic acid 20:79:5 adjusted to pH 4.50 (Flush column with MeOH:water 60:40 at the end of each day.)
**Flow rate:** 1.5
**Injection volume:** 1.5
**Detector:** UV 240

## CHROMATOGRAM
**Retention time:** 4.2
**Internal standard:** theophylline (3.5)

## OTHER SUBSTANCES
**Simultaneous:** impurities, degradation products, 6-mercaptopurine

## KEY WORDS
stability-indicating; injections; tablets

## REFERENCE
Fell, A.F.; Plag, S.M.; Neil, J.M. Stability-indicating assay for azathioprine and 6-mercaptopurine by reversed-phase high-performance liquid chromatography. *J.Chromatogr.*, **1979**, *186*, 691–704

## ANNOTATED BIBLIOGRAPHY
Ding, T.L.; Benet, L.Z. Determination of 6-mercaptopurine and azathioprine in plasma by high-performance liquid chromatography. *J.Chromatogr.*, **1979**, *163*, 281–288

# Azithromycin

**Molecular formula:** $C_{38}H_{72}N_2O_{12}$
**Molecular weight:** 749.0
**CAS Registry No.:** 83905-01-5

## SAMPLE
**Matrix:** blood
**Sample preparation:** 50 μL Serum + 50 μL 60 mM potassium carbonate + 50 μL 50 ng/mL IS in MeCN water 50:50, mix, add 200 μL water, vortex for several s, add 3 mL MTBE, vortex for 20 s, centrifuge at 3300 rpm for 3 min. Remove the organic layer and evaporate it to dryness in a vortex evaporator at 40°, reconstitute the residue in 100 μL mobile phase, sonicate, vortex, inject a 50 μL aliquot.

## HPLC VARIABLES
**Column:** 30 × 4.6 3 μm Hypersil ODS
**Mobile phase:** MeCN:MeOH:THF:50 mM ammonium acetate 44:19:3:34
**Flow rate:** 1
**Injection volume:** 50
**Detector:** MS, SCIEX API III, atmospheric pressure ionization, nebulizer probe 450°, 2.5 μÅ Corona discharge needle, quadrupole mass filter, 0.002 inch pinhole aperture, SIM m/z 749 and 752

## CHROMATOGRAM
**Retention time:** 1.9
**Internal standard:** trideuteroazithromycin
**Limit of quantitation:** 10 ng/mL

## KEY WORDS
serum

## REFERENCE
Fouda, H.G.; Schneider, R.P. Quantitative determination of the antibiotic azithromycin in human serum by high-performance liquid chromatography (HPLC)-atmospheric pressure chemical ionization mass spectrometry: Correlation with a standard HPLC-electrochemical method. *Ther.Drug Monit.*, **1995**, *17*, 179–183

## SAMPLE
**Matrix:** blood
**Sample preparation:** 1 mL Serum + 25 μL 10 μg/mL IS in MeCN + 1 mL 30 mM potassium carbonate, vortex, add 5 mL MTBE, vortex for 30 s, centrifuge at 2000 g for 10 min. Remove the organic layer and evaporate it to dryness under vacuum at 37°, reconstitute the residue in 300 μL mobile phase, vortex for 30 s, add 1 volume hexane, vortex, centrifuge, remove the aqueous layer, inject a 50 μL aliquot of the aqueous layer (J. Chromatogr. 1991, 565, 321).

## HPLC VARIABLES
**Guard column:** 20 × 4.6 5 μm Nucleosil C18
**Column:** 125 × 4.6 5 μm Nucleosil C18

**Mobile phase:** MeCN:40 mM $Na_2HPO_4$:5 mM tetrabutylammonium phosphate 33:50:50, adjusted to pH 7.0 with 25% phosphoric acid
**Flow rate:** 1
**Injection volume:** 50
**Detector:** E, ESA 5100A Coulochem, guard cell +1 V, dual electrode analytical cell +0.7 and +0.8 V

## CHROMATOGRAM
**Retention time:** 5.8
**Internal standard:** n-propyl analog of azithromycin (7.8)
**Limit of quantitation:** 8 ng/mL

## OTHER SUBSTANCES
**Extracted:** metabolites

## KEY WORDS
serum; pharmacokinetics

## REFERENCE
Riedel, K.-D.; Wildfeuer, A.; Laufen, H.; Zimmermann, T. Equivalence of a high-performance liquid chromatographic assay and a bioassay of azithromycin in human serum samples. *J.Chromatogr.*, **1992**, *576*, 358–362

## SAMPLE
**Matrix:** blood
**Sample preparation:** 1 mL Serum + 25 μL 10 μg/mL IS in MeCN + 1 mL 30 mM potassium carbonate, vortex, add 5 mL MTBE, vortex for 30 s, centrifuge at 2000 g for 10 min. Remove the organic layer and evaporate it to dryness under vacuum at 37°, reconstitute the residue in 300 μL mobile phase, vortex for 30 s, add 1 volume hexane, vortex, centrifuge, remove the aqueous layer, inject a 100 μL aliquot of the aqueous layer.

## HPLC VARIABLES
**Guard column:** 21 × 3 40 μm glass bead column (Waters)
**Column:** 50 × 4.6 5 μm Chromegabond alkylphenyl (ES Industries)
**Mobile phase:** MeCN:MeOH:20 mM ammonium acetate:20 mM sodium perchlorate 45:10:22:23, adjust apparent pH to 6.8-7.2 with glacial acetic acid
**Flow rate:** 1
**Injection volume:** 100
**Detector:** E, ESA 5100A Coulochem, ESA 5020 guard cell +1 V, ESA 5010 dual electrode analytical cell, screen electrode +0.7 V, detector electrode +0.8 V, porous carbon electrodes

## CHROMATOGRAM
**Retention time:** 9
**Internal standard:** 9a-N-propyl analog of azithromycin (13)
**Limit of quantitation:** 10 ng/mL

## OTHER SUBSTANCES
**Extracted:** metabolites

## KEY WORDS
serum; human; mouse; rat; dog; rabbit

## REFERENCE
Shepard, R.M.; Duthu, G.S.; Ferraina, R.A.; Mullins, M.A. High-performance liquid chromatographic assay with electrochemical detection for azithromycin in serum and tissues. *J.Chromatogr.*, **1991**, *565*, 321–337

## SAMPLE
**Matrix:** blood, tissue
**Sample preparation:** Serum. 500 μL Serum + 25 μL 2 μg/mL IS in MeCN + 500 μL 60 mM potassium carbonate, vortex, add 5 mL MTBE, vortex for 30 s, centrifuge at 2000 g for 10 min. Remove the organic layer and add it to 500 μL 15 mM pH 3.1 citric acid, vortex for 30 s, centrifuge at 2000 g for 5 min. Remove the aqueous layer and add it to 1 mL 60 mM potassium carbonate, vortex for 1 min, add 5 mL MTBE, vortex for 30 s, centrifuge at 2000 g for 10 min. Remove the organic layer and evaporate it to dryness under vacuum at 37°, reconstitute the residue in 300 μL MeCN:water 50:50, vortex for 30 s, add 1 volume hexane, vortex, centrifuge, remove the aqueous layer, inject a 60 μL aliquot of the aqueous layer. (For high azithromycin concentrations extraction into citric acid and back extraction is not necessary.) Tissue. 1 g Tissue + 9 volumes MeCN + 50 μL 20 μg/mL IS in MeOH, homogenize (Polytron PT 10/35) for 10 s, centrifuge at 2000 g for 10 min. Remove a 500 μL aliquot of the supernatant and evaporate it to dryness under vacuum at 50°, reconstitute the residue 500 μL 60 mM potassium carbonate, add 5 mL MTBE, vortex for 30 s, centrifuge at 2000 g for 10 min. Remove the organic layer and evaporate it to dryness under vacuum at 37°, reconstitute the residue in 300 μL MeCN:water 50:50, vortex for 30 s, add 1 volume hexane, vortex, centrifuge, remove the aqueous layer, inject a 20-60 μL aliquot of the aqueous layer.

## HPLC VARIABLES
**Guard column:** 21 × 3 40 μm glass bead column (Waters)
**Column:** 150 × 4.6 5 μm Chromegabond γ-RP-1 alumina (ES Industries)
**Mobile phase:** MeCN:50 mM potassium phosphate 30:70, adjusted to an apparent pH of 11.0 with 1 M KOH
**Flow rate:** 1
**Injection volume:** 60
**Detector:** E, BAS LC-4B (Bioanalytical Systems), glassy carbon electrode +0.8 V, Ag/AgCl reference electrode

## CHROMATOGRAM
**Retention time:** 8
**Internal standard:** 9a-N-propargyl analog of azithromycin (10)
**Limit of detection:** 100 ng/g (tissue)
**Limit of quantitation:** 10 ng/mL (serum)

## OTHER SUBSTANCES
**Extracted:** metabolites

## KEY WORDS
serum; human; mouse; rat; dog; rabbit; brain; muscle; liver; kidney

## REFERENCE
Shepard, R.M.; Duthu, G.S.; Ferraina, R.A.; Mullins, M.A. High-performance liquid chromatographic assay with electrochemical detection for azithromycin in serum and tissues. *J.Chromatogr.*, **1991**, *565*, 321–337

# Beclomethasone Dipropionate

**Molecular formula:** $C_{28}H_{37}ClO_7$

**Molecular weight:** 521.1

**CAS Registry No.:** 5334-09-8 (beclomethasone dipropionate), 4419-39-0 (beclomethasone)

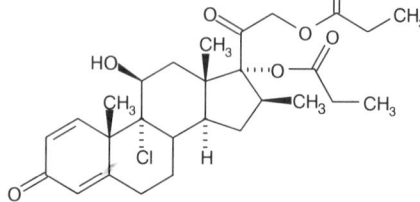

## SAMPLE
**Matrix:** blood

**Sample preparation:** 50 μL Plasma + 100 μL 20 μg/mL cloprednol + 3 mL ether, shake 10 min, centrifuge at 3000 g, remove the organic phase and evaporate it to dryness under nitrogen. Take up the residue in 200 μL mobile phase, inject a 50 μL aliquot.

## HPLC VARIABLES
**Column:** Nucleosil R 10 C 18

**Mobile phase:** MeOH:MeCN:water:acetic acid 400:100:200:1

**Injection volume:** 50

**Detector:** UV 254

## CHROMATOGRAM
**Internal standard:** cloprednol

**Limit of detection:** 500 ng/mL

## KEY WORDS
plasma

## REFERENCE
Würthwein, G.; Rohdewald, P. Activation of beclomethasone dipropionate by hydrolysis to beclomethasone-17-monopropionate. *Biopharm.Drug Dispos.*, **1990**, *11*, 381–394

## SAMPLE
**Matrix:** blood, tissue

**Sample preparation:** Acidify plasma or lung tissue homogenate to pH 2 with 500 mM HCl, add 100 μL 20 μg/mL IS, extract with 8 mL dichloromethane. Evaporate the organic layer to dryness under vacuum, reconstitute in 120 μL MeOH:5% acetic acid 50:50, inject an 80 μL aliquot.

## HPLC VARIABLES
**Column:** 250 × 4.6 5 μm Zorbax ODS C18

**Mobile phase:** MeCN:MeOH:water 44:11:45

**Flow rate:** 1

**Injection volume:** 80

**Detector:** UV 242; Radioactivity

## CHROMATOGRAM
**Internal standard:** hydrocortisone 21-S-propionate (JO 498)

## OTHER SUBSTANCES
**Extracted:** metabolites, budesonide

## KEY WORDS
rat; lung; radiolabeled; pharmacokinetics; plasma

## REFERENCE

Chanoine, F.; Grenot, C.; Heidmann, P.; Junien, J.L. Pharmacokinetics of butixocort 21-propionate, budesonide, and beclomethasone dipropionate in the rat after intratracheal, intravenous, and oral treat-
ments. *Drug Metab.Dispos.*, **1991**, *19*, 546–553

## SAMPLE

**Matrix:** formulations
**Sample preparation:** Ointment. Add pentane:EtOH 75:25 to ointment, sonicate for 20
min, dilute an aliquot to 100 mL with MeOH, allow to settle. Centrifuge and filter an
aliquot of the supernatant, inject an aliquot of the filtrate. Cream, lotion. Stir cream or
lotion in EtOH:THF:water 25:25:50 at 40° for 15 min, cool in an ice bath. Centrifuge
and filter an aliquot of the supernatant, inject an aliquot of the filtrate. Gel. Dissolve gel
in EtOH, sonicate, filter, inject an aliquot.

## HPLC VARIABLES

**Guard column:** present but not specified
**Column:** 250 × 2.1 10 μm Bondapak C18
**Mobile phase:** MeCN:water 48:52 containing 0.65% acetic acid, pH 3.18 (At the end of
each day flush guard column only with MeOH:THF 75:25 for 30 min.)
**Flow rate:** 1
**Injection volume:** 20
**Detector:** UV 251

## CHROMATOGRAM

**Retention time:** 8.78

## OTHER SUBSTANCES

**Simultaneous:** bamipine lactate, betamethasone-17-valerate, dexamethasone, hydrocorti-
sone-21-acetate

## KEY WORDS

ointment; creams; lotions; gels

## REFERENCE

Kountourellis, J.E.; Markopoulou, C.K.; Ebete, K.O.; Stratis, J.A. Separation and determination of some
corticosteroids combined with bamipine in pharmaceutical formulations by high performance liquid
chromatography. *J.Liq.Chromatogr.*, **1995**, *18*, 3507–3517

## SAMPLE

**Matrix:** ileostomy effluent
**Sample preparation:** Dilute ileostomy effluent 1:2 by weight with water and mix with
100 μL 11 μg/mL 17-hydroxyprogesterone. Extract 3 g aliquot three times with 10 mL
dichloromethane by shaking for 1 min and centrifuging at 2000 rpm for 2 min. Wash
combined extracts successively with 2 mL 0.1 M NaOH and 4 mL water by shaking for
30 s and centrifuging for 1 min then dry the organic layer under air at 40°. Take up the
extract in 1 mL MeOH, add 1.1 mL water and apply to C18 Bond Elut SPE cartridge.
Wash with 10 mL water, wash with 5 mL MeOH:water 45:55, elute with 2 mL MeOH.
Add 50 μL 20 μg/mL progesterone to the eluate, dry at 40°, take up in 100 μL MeOH,
inject 10 μL aliquot.

## HPLC VARIABLES

**Guard column:** Bondapak C18/Corasil
**Column:** 300 × 3.9 μBondapak C18
**Mobile phase:** MeOH:50 mM pH 3.0 sodium phosphate buffer 55:45
**Flow rate:** 3

**Injection volume:** 10
**Detector:** UV 254; UV 238

## CHROMATOGRAM
**Retention time:** 21.3
**Internal standard:** 17-hydroxyprogesterone (6.0), progesterone (11.6)

## OTHER SUBSTANCES
**Extracted:** beclomethasone alcohol, beclomethasone 17-monopropionate

## KEY WORDS
SPE

## REFERENCE
Levine, D.S.; Raisys, V.A.; Ainardi, V. Coating of oral beclomethasone dipropionate capsules with cellulose acetate phthalate enhances delivery of topically active antiinflammatory drug to the terminal ileum. *Gastroenterology*, **1987**, *92*, 1037–1044

## SAMPLE
**Matrix:** solutions

## HPLC VARIABLES
**Column:** 250 × 4.6 Zorbax SAX
**Mobile phase:** MeOH:buffer 50:50 (Buffer was 180 mM $Na_2HPO_4$ adjusted to pH 3.00 ± 0.05 with 180 mM orthophosphoric acid. Pass mobile phase through a 250 × 4.6 25-40 μm silica (HPLC Technology) column to saturate it with silica.)
**Flow rate:** 1
**Detector:** UV 253

## CHROMATOGRAM
**Retention time:** 2.7

## OTHER SUBSTANCES
**Simultaneous:** cromolyn, minocromil, nedocromil, quinoline yellow, saccharin, salicylic acid
**Interfering:** acetaminophen, albuterol, aspartame, aspirin, caffeine, isoproterenol, menthol, reproterol, riboflavin, sorbitan trioleate, terbutaline, theophylline

## REFERENCE
Baker, P.R.; Gardner, J.J.; Wilkinson, D. Automated high-performance liquid chromatographic method for the determination of nedocromil sodium in human urine using bimodal column switching. *J.Chromatogr.B*, **1995**, *668*, 59–65

## SAMPLE
**Matrix:** tissue
**Sample preparation:** 100 mg Tissue + 2 mL Ringer's pH 6.8 phosphate buffer + 2 mL EtOH, centrifuge, wash residue twice. Pool supernatant and washings and evaporate to dryness, take up in 400 μL EtOH, inject an aliquot.

## HPLC VARIABLES
**Guard column:** present but not specified
**Column:** μBondapak C18
**Mobile phase:** Gradient. MeOH:water 40:60 to 80:20, time not specified
**Flow rate:** 1.5
**Detector:** UV 254

## OTHER SUBSTANCES
**Extracted:** beclomethasone monopropionate, beclomethasone, cyclomethasone

## KEY WORDS
lung

## REFERENCE
Ronca-Testoni, S. Hydrolysis of cyclomethasone by the human lung. *Int.J.Clin.Pharmacol.Res.*, **1983**, *3*, 17–20

## ANNOTATED BIBLIOGRAPHY
Valvo, L.; Paris, A.; Savella, A.L.; Gallinella, B.; Ciranni Signoretti, E. General high-performance liquid chromatographic procedures for the rapid screening of natural and synthetic corticosteroids. *J.Pharm.Biomed.Anal.*, **1994**, *12*, 805–810 [gradient; reverse phase; normal phase; also betamethasone, betamethasone 21-acetate, betamethasone 17,21-dipropionate, betamethasone 21-disodium phosphate, betamethasone 17-valerate, cortisone, cortisone 21-acetate, 11-deoxycorticosterone 21-acetate, dexamethasone, dexamethasone 21-acetate, dexamethasone 21-disodium phosphate, fluocinolone, fluocinolone acetonide, 9α-fluorohydrocortisone, 9α-fluorohydrocortisone 21-acetate, 9α-fluoroprednisolone, 9α-fluoroprednisolone 21-acetate, hydrocortisone, hydrocortisone 21-acetate, hydrocortisone 21-hemisuccinate, 6α-methylprednisolone, 6α-methylprednisolone 21-acetate, 6α-methylprednisolone 21-sodium succinate, prednisolone, prednisolone 21-acetate, prednisolone 21-disodium phosphate, prednisolone 21-pivalate, prednisolone 21-sodium succinate, prednisone, triamcinolone, triamcinolone acetonide]

Girault, J.; Istin, B.; Malgouyat, J.M.; Brisson, A.M.; Fourtillan, J.B. Simultaneous determination of beclomethasone, beclomethasone monopropionate and beclomethasone dipropionate in biological fluids using a particle beam interface for combining liquid chromatography with negative-ion chemical ionization mass spectrometry. *J.Chromatogr.*, **1991**, *564*, 43–53 [plasma; urine; LC-MS; LOQ 1 ng/mL]

# Benazepril

**Molecular formula:** $C_{24}H_{28}N_2O_5$
**Molecular weight:** 424.5
**CAS Registry No.:** 86541-75-5 (benazepril),
86541-74-4 (benazepril hydrochloride)

## SAMPLE
**Matrix:** blood
**Sample preparation:** 2 mL Whole blood or plasma + 2 mL buffer + 5 mL chloroform: isopropanol:n-heptane 60:14:26, shake gently horizontally for 10 min, centrifuge at 2800 g for 10 min. Remove the lower organic layer and evaporate it to dryness under vacuum at 45°, reconstitute the residue in 100 μL mobile phase, centrifuge at 2800 g for 5 min, inject a 50 μL aliquot of the supernatant. (Buffer was saturated ammonium chloride solution 25% diluted with water, adjusted to pH 9.5 with 25% ammonia solution.)

## HPLC VARIABLES
**Column:** 300 × 3.9 4 μm NovaPack C18
**Mobile phase:** MeOH:THF:buffer 65:5:30 (Buffer was 0.68 g/L (10 mM (sic)) $KH_2PO_4$ adjusted to pH 2.6 with concentrated orthophosphoric acid.) (At the end of each session wash the column with water for 1 h and MeOH for 1 h, re-equilibrate for 30 min.)
**Column temperature:** 30
**Flow rate:** 0.8
**Injection volume:** 50
**Detector:** UV 239

## CHROMATOGRAM
**Retention time:** 5.01
**Limit of detection:** <120 ng/mL

## OTHER SUBSTANCES
**Extracted:** acebutolol, acepromazine, aceprometazine, acetaminophen, aconitine, ajmaline, albuterol, alimemazine, alminoprofen, alpidem, alprazolam, alprenolol, amisulpride, amitriptyline, amodiaquine, amoxapine, aspirin, astemizole, atenolol, benperidol, benzocaine, benzoylecgonine, bepridil, betaxolol, bisoprolol, bromazepam, brompheniramine, bumadizone, bupivacaine, buprenorphine, buspirone, caffeine, carbamazepine, carbinoxamine, carpipramine, carteolol, celiprolol, cetirizine, chlorambucil, chlormezanone, chlorophenacinone, chloroquine, chlorpheniramine, chlorpromazine, chlorpropamide, cibenzoline, cicletanine, clemastine, clobazam, clomipramine, clonazepam, clonidine, clozapine, cocaine, codeine, colchicine, cyamemazine, cyclizine, cycloguanil, cyproheptadine, cytarabine, dacarbazine, daunorubicin, debrisoquine, demexiptiline, desipramine, dextromethorphan, dextromoramide, dextropropoxyphene, diazepam, diazoxide, diclofenac, dihydralazine, diltiazem, diphenhydramine, disopyramide, dosulepine, doxepin, doxylamine, droperidol, ephedrine, estazolam, etodolac, fenfluramine, fenoprofen, fentiazac, flecainide, floctafenine, flumazenil, flunitrazepam, fluoxetine, fluphenazine, flurbiprofen, fluvoxamine, glibenclamide, glibornuride, glipizide, glutethimide, haloperidol, histapyrrodine, hydroxychloroquine, hydroxyzine, ibuprofen, imipramine, indomethacin, iproniazid, ketamine, ketoprofen, labetalol, levomepromazine, lidocaine, lidoflazine, lisinopril, loperamide, loprazolam, loratadine, lorazepam, loxapine, maprotiline, medazepam, medifoxamine, mefenamic acid, mefenidramine, mefloquine, melphalan, meperidine, mephenesin, mephentermine, mepivacaine, metformin, methadone, methaqualone, methocarbamol, methotrexate, metipranolol, metoclopramide, metoprolol, mianserine, midazolam, minoxidil, moclobemide, moperone, morphine, nadolol, nalbuphine, nalorphine, naloxone, naltrexone, naproxen, nialamide, nicardipine, nifedipine, niflumic acid, nimodipine, nitrazepam,

nitrendipine, nizatidine, nortriptyline, omeprazole, opipramol, oxazepam, penbutolol, penfluridol, pentazocine, phencyclidine, phenobarbital, phenol, phenylbutazone, pimozide, pindolol, piroxicam, prazepam, prazosin, prilocaine, procainamide, procarbazine, proguanil, promethazine, propafenone, propranolol, protriptyline, quinidine, quinine, quinupramine, ramipril, ranitidine, reserpine, ritodrine, secobarbital, sotalol, strychnine, sulfinpyrazole, sulindac, sulpride, sultopride, suriclone, temazepam, tenoxicam, terbutaline, terfenadine, tetracaine, tetrazepam, thiopental, thioproperazine, thioridazine, tianeptine, tiapride, tiaprofenic acid, timolol, tioclomarol, tofisopam, tolbutamide, toloxatone, triazolam, trifluoperazine, trifluperidol, trimipramine, triprolidine, tropatenine, verapamil, viloxazine, vinblastine, yohimbine, zolpidem, zopiclone, zorubicine

**Interfering:** acenocoumarol, benazepril, chlordiazepoxide, clorazepate, dipyridamole, metapramine, mexiletine, nomifensine, oxprenolol, pipamperone, pyrimethamine, ticlopidine, trazodone, vincristine, vindesine, warfarin

## KEY WORDS
whole blood; plasma

## REFERENCE
Tracqui, A.; Kintz, P.; Mangin, P. Systematic toxicological analysis using HPLC/DAD. *J.Forensic Sci.*, **1995**, *40*, 254−262

## SAMPLE
**Matrix:** blood
**Sample preparation:** Condition a Bond-Elut C2 SPE cartridge with 2 mL MeOH and 2 mL water. Adjust pH of plasma to 1.0 with dilute phosphoric acid, add 1 mL to the SPE cartridge, wash with pH 1.0 dilute phosphoric acid, wash with water, wash with EtOH, wash with MeCN, air dry, elute with 200 μL mobile phase, inject an aliquot.

## HPLC VARIABLES
**Guard column:** Newguard-Phenyl
**Column:** 150 × 3.9 μBondapak phenyl
**Mobile phase:** MeOH:water 49:51 with 1 vial PIC-A reagent/L, adjusted to pH 7.0 with phosphoric acid
**Flow rate:** 1
**Detector:** UV 210

## KEY WORDS
plasma; dog; SPE; pharmacokinetics

## REFERENCE
Kim, J.S.; Oberle, R.L.; Krummel, D.A.; Dressman, J.B.; Fleisher, D. Absorption of ACE inhibitors from small intestine and colon. *J.Pharm.Sci.*, **1994**, *83*, 1350−1356

## SAMPLE
**Matrix:** solutions

## HPLC VARIABLES
**Column:** 250 × 4.6 Spherisorb 5 ODS-2
**Mobile phase:** n-Propanol:buffer 20:80 (Buffer was pH 3.0 phosphate buffer containing 0.4% triethylamine.)
**Flow rate:** 1
**Detector:** UV 240

## CHROMATOGRAM
**Retention time:** 14

## OTHER SUBSTANCES
**Simultaneous:** captopril, cilazapril, enalapril, quinapril, ramipril

## REFERENCE
Barbato, F.; Morrica, P.; Quaglia, F. Analysis of ACE inhibitor drugs by high performance liquid chromatography. *Farmaco*, **1994**, *49*, 457–460

# Betamethasone

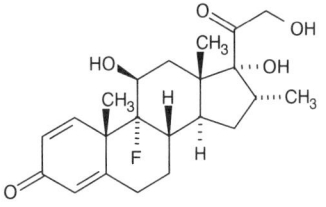

**Molecular formula:** $C_{22}H_{29}FO_5$

**Molecular weight:** 392.5

**CAS Registry No.:** 378-44-9, 987-24-6 (acetate), 22298-29-9 (benzoate), 5593-20-4 (dipropionate), 151-73-5 (sodium phosphate), 2152-44-5 (17-valerate), 5534-05-4 (acibutate), 360-63-4 (dihydrogen phosphate)

## SAMPLE

**Matrix:** blood

**Sample preparation:** 1 mL Serum + 100 μL water containing 5 μg/mL 2,3-diaminonaphthalene and 3.5 μg/mL 18-hydroxy-11-deoxycorticosterone + 1 mL 250 mM NaOH + 7 mL diethyl ether, shake on a rotary shaker for 15 min, repeat extraction. Combine the organic layers and evaporate them to dryness under a stream of nitrogen at 30-40°, reconstitute the residue in 70 μL MeOH:100 mM perchloric acid 50:50, inject a 20 μL aliquot.

## HPLC VARIABLES

**Column:** 150 × 3.9 4 μm Nova-Pak C18

**Mobile phase:** Gradient. A was 58 mM $NaH_2PO_4$ containing 6 mM sodium heptanesulfonate, adjusted to pH 3.1 with concentrated phosphoric acid. B was MeCN:MeOH 85:15. A:B from 100:0 to 78:22 over 5 min, to 70:30 over 12 min, maintain at 70:30 for 4 min, to 65:35 over 9 min.

**Flow rate:** 1

**Injection volume:** 20

**Detector:** UV 245; UV 256; UV 343

## CHROMATOGRAM

**Retention time:** 19.45

**Internal standard:** 2,3-diaminonaphthalene (10.71), 18-hydroxy-11-deoxycorticosterone (15.85)

**Limit of detection:** 1-10 ng/mL (245 nm)

## OTHER SUBSTANCES

**Extracted:** chloroquine, corticosterone, cortisone, dexamethasone, fluocinolone acetonide, fluendrenolide, fluorometholone, fluprednisolone, hydrocortisone, hydroxychloroquine, 17δ-hydroxyprogesterone, meprednisone, methylprednisolone, methylprednisolone acetate, paramethasone, prednisolone, prednisone, progesterone, triamcinolone

**Noninterfering:** aspirin, ibuprofen, indomethacin, phenylbutazone, pregnenolone

## KEY WORDS

serum

## REFERENCE

Volin, P. Simple and specific reversed-phase liquid chromatographic method with diode-array detection for simultaneous determination of serum hydroxychloroquine, chloroquine and some corticosteroids. *J.Chromatogr.B,* **1995**, *666*, 347–353

## SAMPLE

**Matrix:** blood

**Sample preparation:** Prepare a Sep-Pak Plus Environmental C18 cartridge by washing with 15 mL MeOH then 15 mL water. 1 mL Serum + 200 μL isopropanol:acetonitrile 1:1, mix, add to SPE cartridge, wash with 10 mL water, elute with 3 mL MeOH. Evaporate the eluate at 50° under a stream of nitrogen, reconstitute in 200 µL mobile phase A, inject a 20 µL aliquot.

## HPLC VARIABLES
**Guard column:** µBondapak C18 guard column
**Column:** 250 × 4.6 5 µm Hypersil ODS
**Mobile phase:** Gradient. A was isopropanol:50 mM pH 4.5 acetate buffer 10:90. B was isopropanol:50 mM pH 4.5 acetate buffer 30:70. A:B from 90:10 to 30:70 over 25 min, hold at 30:70 for 5 min, to 90:10 over 5 min, hold at 90:10 for 15 min before next injection.
**Column temperature:** 40
**Flow rate:** 1
**Injection volume:** 20
**Detector:** UV 254

## CHROMATOGRAM
**Retention time:** 33
**Internal standard:** betamethasone

## OTHER SUBSTANCES
**Simultaneous:** metabolites, cortisone, hydrocortisone, prednisolone, prednisone

## KEY WORDS
serum; SPE; betamethasone is IS

## REFERENCE
Hirata, H.; Kasama, T.; Sawai, Y.; Fike, R.R. Simultaneous determination of deflazacort metabolites II and III, cortisol, cortisone, prednisolone and prednisone in human serum by reversed-phase high-performance liquid chromatography. *J.Chromatogr.B*, **1994**, *658*, 55–61

## SAMPLE
**Matrix:** blood
**Sample preparation:** 1 mL Plasma + 15 mL dichloromethane, shake horizontally for 15 min, centrifuge at 1500 g for 15 min. Remove the organic layer and wash it with 100 µL 100 mM NaOH then 1 mL water. Remove the aqueous phase and dry the organic phase over 1 g of anhydrous sodium sulfate. Evaporate the organic phase to dryness under a stream of nitrogen at not more than 37°, reconstitute in 200 µL mobile phase, inject a 175 µL aliquot.

## HPLC VARIABLES
**Guard column:** 20 × 2 30-38 µm HC Pellosil
**Column:** 250 × 4.6 5-6 µm Zorbax SIL
**Mobile phase:** Heptane:dichloromethane:glacial acetic acid:ethanol 350:600:10:35
**Flow rate:** 2
**Injection volume:** 175
**Detector:** UV 254

## CHROMATOGRAM
**Retention time:** 12
**Internal standard:** betamethasone

## OTHER SUBSTANCES
**Extracted:** hydrocortisone, prednisolone, prednisone
**Noninterfering:** cyclosporin, ethinyl estradiol, ketoconazole, levonorgestrel, rapamycin, tacrolimus, tenidap, tetrahydrocortisone

## KEY WORDS
plasma; normal phase; betamethasone is IS

## REFERENCE
Jusko, W.J.; Pyszczynski, N.A.; Bushway, M.S.; D'Ambrosio, R.; Mis, S.M. Fifteen years of operation of a high-performance liquid chromatographic assay for prednisolone, cortisol and prednisone in plasma. *J.Chromatogr.B*, **1994**, *658*, 47–54

## SAMPLE
**Matrix:** blood
**Sample preparation:** Add 1 mL serum to a Sep Pak C18 SPE cartridge, wash with 4 mL water, elute with 4 mL MeOH, evaporate to dryness under vacuum, reconstitute in 50 μL MeCN:water 30:70, inject whole sample.

## HPLC VARIABLES
**Column:** 250 × 4.6 5 μm Ultrasphere ODS
**Mobile phase:** MeCN:water 30:70
**Flow rate:** 1
**Injection volume:** 50
**Detector:** enzyme immunoassay of fractions

## CHROMATOGRAM
**Retention time:** 16
**Limit of detection:** 0.3 pg

## OTHER SUBSTANCES
**Extracted:** dexamethasone, flumethasone, triamcinolone
**Noninterfering:** endogenous steroids

## KEY WORDS
serum; SPE; horse

## REFERENCE
Friedrich, A.; Schulz, R.; Meyer, H.H. Use of enzyme immunoassay and reverse-phase high-performance liquid chromatography to detect and confirm identity of dexamethasone in equine blood. *Am.J.Vet.Res.*, **1992**, *53*, 2213–2220

## SAMPLE
**Matrix:** blood
**Sample preparation:** 1 mL Plasma + 5 mL water + 1 mL 2 μg/mL equilenin in MeOH + 50 μL 0.1 M NaOH to adjust pH to 10, vortex briefly after each addition, shake with 10 mL dichloromethane for 10 min, centrifuge at 2000 g for 10 min. Wash organic layer twice with 2 mL water, centrifuge 5 min, evaporate 8 mL of organic phase to dryness at 40° under a stream of nitrogen, reconstitute residue in 150 μL mobile phase, inject 25 μL aliquot.

## HPLC VARIABLES
**Column:** 300 × 4 10 μm μBondapak C18
**Mobile phase:** MeOH:buffer 60:40 (Buffer was 10 mL 200 mM acetic acid + 15 mL 200 mM sodium acetate made up to 1 L, pH 4.8.)
**Flow rate:** 1
**Injection volume:** 25
**Detector:** UV 254

## CHROMATOGRAM
**Retention time:** 5.5

**Internal standard:** equilenin (7.5)
**Limit of detection:** 5 ng/mL

## OTHER SUBSTANCES
**Simultaneous:** deoxycortisol, hydrocortisone, prednisolone, prednisone, triamcinolone
**Interfering:** dexamethasone

## KEY WORDS
Anal. Abs. 1982, 43, 4D182; plasma

## REFERENCE
Bouquet, S.; Brisson, A.M.; Gombert, J. Dosage du cortisol et du 11-désoxycortisol plasmatiques par chromatographie liquide haute performance [Cortisol and 11-desoxycortisol determination in blood by high performance liquid chromatography]. *Ann.Biol.Clin.(Paris)*, **1981**, *39*, 189–191

## SAMPLE
**Matrix:** formulations
**Sample preparation:** Ointment. Add pentane:EtOH 75:25 to ointment, sonicate for 20 min, dilute an aliquot to 100 mL with MeOH, allow to settle. Centrifuge and filter an aliquot of the supernatant, inject an aliquot of the filtrate. Cream, lotion. Stir cream or lotion in EtOH:THF:water 25:25:50 at 40° for 15 min, cool in an ice bath. Centrifuge and filter an aliquot of the supernatant, inject an aliquot of the filtrate. Gel. Dissolve gel in EtOH, sonicate, filter, inject an aliquot.

## HPLC VARIABLES
**Guard column:** present but not specified
**Column:** 250 × 2.1 10 μm Bondapak C18
**Mobile phase:** MeCN:water 48:52 containing 0.65% acetic acid, pH 3.18 (At the end of each day flush guard column only with MeOH:THF 75:25 for 30 min.)
**Flow rate:** 1
**Injection volume:** 20
**Detector:** UV 251

## CHROMATOGRAM
**Retention time:** 5.17 (betamethasone-17-valerate)

## OTHER SUBSTANCES
**Simultaneous:** bamipine lactate, beclomethasone dipropionate, dexamethasone, hydrocortisone-21-acetate

## KEY WORDS
ointment; creams; lotions; gels

## REFERENCE
Kountourellis, J.E.; Markopoulou, C.K.; Ebete, K.O.; Stratis, J.A. Separation and determination of some corticosteroids combined with bamipine in pharmaceutical formulations by high performance liquid chromatography. *J.Liq.Chromatogr.*, **1995**, *18*, 3507–3517

## SAMPLE
**Matrix:** formulations
**Sample preparation:** Pulverize tablets, weigh out amount equivalent to about 500 μg betamethasone, add 10 mL water, sonicate for 15 min, extract three times with 15 mL chloroform:n-butanol 95:5. Combine extracts and filter them through 1 g anhydrous sodium sulfate moistened with chloroform:n-butanol 95:5. Collect filtrate and dilute it to 50 mL with chloroform:n-butanol 95:5. Remove a 1 mL aliquot, add 0.5 mL 40 μM cortisone in mobile phase, mix, inject a 5 μL aliquot.

## HPLC VARIABLES
**Guard column:** 5 μm Guard-Pak Resolve Si (dead volume 60-75 μL)
**Column:** 75 × 3.9 4 μm Nova-Pak silica
**Mobile phase:** Dichloromethane:EtOH 34:1
**Flow rate:** 0.7
**Injection volume:** 5
**Detector:** UV 240

## CHROMATOGRAM
**Retention time:** 6
**Internal standard:** cortisone (3)

## OTHER SUBSTANCES
**Simultaneous:** dexamethasone, hydrocortisone, 6α-methylprednisolone, prednisolone, prednisone

## KEY WORDS
tablets; normal phase

## REFERENCE
Liu, K.-R.; Chen, S.-H.; Wu, S.-M.; Kou, H.-S.; Wu, H.-L. High-performance liquid chromatographic determination of betamethasone and dexamethasone. *J.Chromatogr.A*, **1994**, *676*, 455–460

## SAMPLE
**Matrix:** formulations
**Sample preparation:** Triturate 1 tablet with a glass rod with 5 mL water, sonicate for 20 min, extract with 9 mL dichloromethane then three times with 5 mL dichloromethane, filter (paper), make up to 25 mL with dichloromethane. Remove a 500 μL aliquot, add 200 μL 1.2 mM phenacetin in dichloromethane + 100 μL 0.5 mM 4-dimethylaminopyridine + 100 μL 100 mM N-CBZ-Phe in dichloromethane + 100 μL 100 mM N, N'-dicyclohexylcarbodiimide in dichloromethane, shake mechanically at 30° for 1 h, inject a 10 μL aliquot.

## HPLC VARIABLES
**Column:** 75 × 3.9 4 μm Nova-Pak silica
**Mobile phase:** n-Hexane:dichloromethane:isopropanol 100:100:4
**Flow rate:** 1
**Injection volume:** 10
**Detector:** UV 240

## CHROMATOGRAM
**Retention time:** 7
**Internal standard:** phenacetin (10)
**Limit of detection:** 4.2 pmol

## OTHER SUBSTANCES
**Simultaneous:** dexamethasone

## KEY WORDS
tablets; normal phase; derivatization

## REFERENCE
Chen, S.-H.; Wu, S.-M.; Wu, H.-L. Stereochemical analysis of betamethasone and dexamethasone by derivatization and high-performance liquid chromatography. *J.Chromatogr.*, **1992**, *595*, 203–208

## SAMPLE
**Matrix:** solutions

## HPLC VARIABLES
**Column:** 250 × 4.6 10 µm Partisil 10 ODS
**Mobile phase:** MeOH:water 75:25
**Flow rate:** 1.2
**Detector:** UV 242

## CHROMATOGRAM
**Retention time:** 5 (betamethasone 17-valerate)

## REFERENCE
Mithani, S.D.; Bakatselou, V.; TenHoor, C.N.; Dressman, J.B. Estimation of the increase in solubility of drugs as a function of bile salt concentration. *Pharm.Res.*, **1996**, *13*, 163–167

## SAMPLE
**Matrix:** solutions
**Sample preparation:** Condition a Bond Elut C18 SPE cartridge with 4 mL water then 3 mL MeOH. Add aqueous steroid solution to cartridge, elute with MeOH, inject a 20 µL aliquot.

## HPLC VARIABLES
**Column:** 150 × 4 5 µm Nucleosil C18
**Mobile phase:** MeCN:water 70:30
**Flow rate:** 1
**Injection volume:** 20
**Detector:** UV 239

## CHROMATOGRAM
**Limit of detection:** 120 ng/mL

## OTHER SUBSTANCES
**Also analyzed:** dexamethasone, flumethasone 21-acetate

## KEY WORDS
SPE; for betamethasone 17-valerate

## REFERENCE
Valenta, C.; Janout, H. Corticosteroid analysis by HPLC with increased sensitivity by use of precolumn concentration. *J.Liq.Chromatogr.*, **1994**, *17*, 1141–1146

## SAMPLE
**Matrix:** solutions

## HPLC VARIABLES
**Column:** 500 × 1 C18 (Alltech)
**Mobile phase:** MeOH:water 65:35
**Flow rate:** 0.04
**Injection volume:** 0.5
**Detector:** UV 254; MS, Hewlett Packard 5985, home-made interface (details in paper)

## CHROMATOGRAM
**Retention time:** 21

## OTHER SUBSTANCES
**Simultaneous:** metabolites, hydrocortisone

## KEY WORDS
microbore

## REFERENCE
Eckers, C.; Skrabalak, D.S.; Henion, J. On-line direct liquid introduction interface for micro-liquid chromatography/mass spectrometry: application to drug analysis. *Clin.Chem.*, **1982**, *28*, 1882−1886

## SAMPLE
**Matrix:** solutions, tissue
**Sample preparation:** Buffer solutions. Condition a 3 mL Baker C18 SPE cartridge with 400 μL MeOH. 100-500 μL 0.1 M pH 4.5 acetate buffer containing steroids + 100-1000 ng prednisone in MeOH, add to SPE cartridge, wash with 1 mL water, elute with 1 mL MeOH. Evaporate the eluate to dryness under a stream of air, reconstitute in 40-200 μL dichloromethane:MeOH 98:2, inject a 40 μL aliquot. Skin tissue. Crush tissue with 3 mL MeOH:100 mM pH 4.5 acetate buffer 20:80 using a Polytron tissue homogenizer, wash homogenizer twice with the same solution, combine all solutions, add 400-4000 ng prednisone in MeOH, add 10 mL dichloromethane, vortex for 1 min, centrifuge at 3000 rpm for 5 min. Filter the organic phase through a column of Celite 545, evaporate to dryness under a stream of air, reconstitute in 40-200 μL dichloromethane:MeOH 98:2, inject a 40 μL aliquot.

## HPLC VARIABLES
**Column:** 250 × 4 5 μm LiChrosorb Si-60
**Mobile phase:** n-Hexane:dichloromethane:MeOH:water 63.9:30:6:0.1
**Flow rate:** 1.2
**Injection volume:** 40-200
**Detector:** UV 240

## CHROMATOGRAM
**Retention time:** 6.6
**Internal standard:** prednisone (5.3)
**Limit of detection:** 4 ng

## KEY WORDS
SPE; normal phase

## REFERENCE
Kubota, K.; Maibach, H.I. In vitro percutaneous permeation of betamethasone and betamethasone 17-valerate. *J.Pharm.Sci.*, **1993**, *82*, 1039−1045

## SAMPLE
**Matrix:** urine
**Sample preparation:** 3 mL Urine + 0.25 g NaCl, adjust pH to 9.0 with 0.5 g $Na_2HPO_4$, add 4 mL dichloromethane, vortex 1 min, centrifuge at 3700 g for 3 min. Remove organic phase and dry it over anhydrous sodium sulfate. Evaporate a 3 mL aliquot to dryness under vacuum, reconstitute residue with 200 μL 5 μg/mL IS in MeOH, inject 20 μL aliquot.

## HPLC VARIABLES
**Column:** 250 × 4.6 Hypersil ODS
**Mobile phase:** MeCN:water 32:68
**Column temperature:** 30

**Flow rate:** 1
**Injection volume:** 20
**Detector:** UV 245

## CHROMATOGRAM
**Retention time:** 10
**Internal standard:** methylprednisolone (9)

## OTHER SUBSTANCES
**Extracted:** corticosterone, cortisone, fluorocortisone, fluorocortisone acetate, hydrocortisone, hydroxyprogesterone, prednisolone, prednisone, triamcinolone, triamcinolone acetonide
**Interfering:** dexamethasone

## KEY WORDS
SPE also discussed

## REFERENCE
Santos-Montes, A.; Gonzalo-Lumbreras, R.; Gasco-Lopez, A.I.; Izquierdo-Hornillos, R. Solvent and solid-phase extraction of natural and synthetic corticoids in human urine. *J.Chromatogr.B*, **1994**, *652*, 83–89

## SAMPLE
**Matrix:** urine
**Sample preparation:** Dilute, if necessary, 100 μL-1 mL urine to 1 mL with water, add to a Chem Elut high surface-area diatomaceous earth extraction column, after 5 min elute with two 6 mL portions of ethyl acetate, combine the eluates and wash them twice with 1 mL 200 mM NaOH. Dry the organic layer over 1 g anhydrous sodium sulfate, evaporate to dryness at 30° under a stream of nitrogen, reconstitute the residue in 250 μL mobile phase, inject a 50 μL aliquot.

## HPLC VARIABLES
**Guard column:** 70 × 6 37-53 μm HC-Pellocil
**Column:** 250 × 4.6 5-6 μm Zorbax SIL
**Mobile phase:** Dichloromethane:glacial acetic acid:MeOH 91.3:7.5:1.2
**Flow rate:** 2
**Injection volume:** 50
**Detector:** UV 254

## CHROMATOGRAM
**Retention time:** 8.5
**Internal standard:** betamethasone

## OTHER SUBSTANCES
**Extracted:** metabolites, hydrocortisone, 6β-hydroxycortisol, 20β-hydroxyprednisone, 6β-hydroxyprednisolone, 20α-hydroxyprednisolone, 20β-hydroxyprednisolone, prednisolone, prednisone

## KEY WORDS
betamethasone is IS; normal phase

## REFERENCE
Garg, V.; Jusko, W.J. Simultaneous analysis of prednisone, prednisolone and their major hydroxylated metabolites in urine by high-performance liquid chromatography. *J.Chromatogr.*, **1991**, *567*, 39–47

## SAMPLE
**Matrix:** urine
**Sample preparation:** 3 mL Urine + 100 mg K$_2$HPO$_4$ + 500 mg anhydrous sodium sulfate + 5 mL diethyl ether, shake mechanically for 10 min, centrifuge at 2500 g for 5 min. Remove the organic layer and evaporate it to dryness under vacuum, reconstitute the residue in 200 μL MeOH, filter (0.45 μm), inject a 15 μL aliquot.

## HPLC VARIABLES
**Column:** 60 × 4.6 3 μm Hypersil ODS
**Mobile phase:** Gradient. MeOH:150 mM ammonium acetate from 40:60 to 50:50 over 6 min, maintain at 50:50 for 1 min, to 60:40 over 3 min, maintain at 60:40 for 5 min
**Flow rate:** 0.8
**Injection volume:** 15
**Detector:** MS, Hewlett-Packard HP 5988A, vaporizer probe 92° decreased to 89° over 6 min, decreased to 86° over 3 min, maintain at 86° for 5 min, ion source 276°, emission current 150 μA, electron energy 955 eV, positive ion mode, filament on

## CHROMATOGRAM
**Retention time:** 7
**Internal standard:** betamethasone

## OTHER SUBSTANCES
**Extracted:** corticosterone, cortisone, deoxycorticosterone, hydrocortisone, 11α-hydroxyprogesterone, prednisolone, prednisone, triamcinolone, triamcinolone acetonide

## KEY WORDS
betamethasone is IS

## REFERENCE
Park, S.-J.; Kim, Y.-J.; Pyo, H.-S.; Park, J. Analysis of corticosteroids in urine by HPLC and thermospray LC/MS. *J.Anal.Toxicol.*, **1990**, *14*, 102–108

## ANNOTATED BIBLIOGRAPHY
Jönsson, G.; Åström, A.; Andersson, P. Budesonide is metabolized by cytochrome P450 3A (CYP3A) enzymes in human liver. *Drug Metab.Dispos.*, **1995**, *23*, 137–142 [human; liver; microsomal incubations; extracted budesonide; betamethasone is IS; SPE; gradient]

Garg, V.; Jusko, W.J. Effects of indomethacin on the pharmacokinetics and pharmacodynamics of prednisolone in rats. *J.Pharm.Sci.*, **1994**, *83*, 747–750 [rat; plasma; extracted corticosterone, prednisolone, prednisone; betamethasone is IS]

Santos-Montes, A.; Gonzalo-Lumbreras, R.; Gasco-Lopez, A.I.; Izquierdo-Hornillos, R. Extraction and high-performance liquid chromatographic separation of deflazacort and its metabolite 21-hydroxydeflazacort. Application to urine samples. *J.Chromatogr.B*, **1994**, *657*, 248–253 [interfering triamcinolone acetonide; simultaneous corticosterone, cortisone, deflazacort, deoxycorticosterone, dexamethasone, fludrocortisone, fludrocortisone acetate, hydrocortisone, 21-hydroxydeflazacort, 11α-hydroxyprogesterone, methylprednisolone, prednisolone, prednisone, triamcinolone]

Santos-Montes, A.; Gasco-López, A.I.; Izquierdo-Hornillos, R. Simultaneous determination of dexamethasone and betamethasone in pharmaceuticals by reversed-phase HPLC. *Chromatographia*, **1994**, *39*, 539–542 [simultaneous dexamethasone; methylprednisolone (IS); tablets; column temp 30; LOD 6 ng/mL; non-interfering corticosterone, cortisone, deflazacort, fludrocortisone, fludrocortisone acetate, hydrocortisone, hydroxyprogesterone, methylprednisolone, prednisolone, prednisone, triamcinolone; interfering triamcinolone acetonide]

Valvo, L.; Paris, A.; Savella, A.L.; Gallinella, B.; Ciranni Signoretti, E. General high-performance liquid chromatographic procedures for the rapid screening of natural and synthetic corticosteroids. *J.Pharm.Biomed.Anal.*, **1994**, *12*, 805–810 [for betamethasone, betamethasone 21-acetate, betamethasone 17,21-dipropionate, betamethasone 21-disodium phosphate, betamethasone 17-valerate; gradient; reverse phase; normal phase; also beclomethasone, beclomethasone 17,21-dipropionate, cortisone, cortisone 21-acetate, 11-deoxycorticosterone 21-acetate, dexamethasone, dexamethasone 21-acetate, dexamethasone 21-disodium phosphate, fluocinolone, fluocinolone acetonide, 9α-fluorohydrocortisone 21-acetate, 9α-fluorohydrocortisone, 9α-fluoroprednisolone, 9α-fluoroprednisolone 21-acetate, hydrocortisone, hydrocortisone 21-acetate, hydrocortisone 21-hemisuccinate, 6α-methylprednisolone, 6α-methylprednisolone 21-acetate, 6α-methylprednisolone 21-sodium succinate, prednisolone, prednisolone 21-acetate, prednisolone 21-disodium phosphate, prednisolone 21-pivalate, prednisolone 21-sodium succinate, prednisone, triamcinolone, triamcinolone acetonide]

Santos-Montes, A.; Gasco-Lopez, A.I.; Izquierdo-Hornillos, R. Optimization of the high-performance liquid chromatographic separation of a mixture of natural and synthetic corticosteroids. *J.Chromatogr.*, **1993**, *620*, 15−23 [simultaneous corticosterone, cortisone, deoxycorticosterone, dexamethasone, fluorocortisone, hydrocortisone, hydroxyprogesterone, methylprednisolone, prednisolone, prednisone, triamcinolone]

Smith, E.W.; Haigh, J.M. In vitro diffusion cell design and validation. I. A stability-indicating high-performance liquid chromatographic assay for betamethasone 17-valerate in purified isopropyl myristate receptor phase. *Pharm.Res.*, **1989**, *6*, 431−435

Maron, N.; Cristi, E.A.; Ramos, A.A. Determination of betamethasone 17-benzoate in lipophilic vehicles by reversed-phase high-performance liquid chromatography. *J.Pharm.Sci.*, **1988**, *77*, 638−639

Skrabalak, D.S.; Cuddy, K.K.; Henion, J.D. Quantitative determination of betamethasone and its major metabolite in equine urine by micro-liquid chromatography-mass spectrometry. *J.Chromatogr.*, **1985**, *341*, 261−269

Tokunaga, H.; Kimura, T.; Kawamura, J. Determination of glucocorticoids by liquid chromatography. III. Application to ointments and a cream containing cortisone acetate, dexamethasone acetate, fluorometholone, and betamethasone valerate. *Chem.Pharm.Bull.*, **1984**, *32*, 4012−4016

Cairns, T.; Siegmund, E.G.; Stamp, J.J.; Skelly, J.P. Liquid chromatography mass spectrometry of dexamethasone and betamethasone. *Biomed.Mass.Spectrom.*, **1983**, *10*, 203−208

Okumura, T. Application of thin-layer chromatography to high-performance liquid chromatographic separation of steroidal hormones and cephalosporin antibiotics. *J.Liq.Chromatogr.*, **1981**, *4*, 1035−1064 [normal phase; also cephalexin, cephaloglycine, cephaloridine, cephalothin, cortisone, dexamethasone, hydrocortisone]

Petersen, M.C.; Nation, R.L.; Ashley, J.J. Simultaneous determination of betamethasone, betamethasone acetate and hydrocortisone in biological fluids using high-performance liquid chromatography. *J.Chromatogr.*, **1980**, *183*, 131−139

# Bromocriptine

**Molecular formula:** $C_{32}H_{40}BrN_5O_5$
**Molecular weight:** 654.6
**CAS Registry No.:** 25614-03-3, 22260-51-1 (mesylate)

## SAMPLE
**Matrix:** blood, tissue
**Sample preparation:** Plasma. 1 mL Plasma + 10 μL 50 μg/mL ergocriptine in MeCN:
MeOH:water 4:1:5 + 1 mL 2.5 M potassium carbonate, vortex, slowly add 4 mL MeCN
with vortexing for 30 s, centrifuge at 2200 g at -10° for 15 min, maintain at -10° for 30
min. Remove the organic layer and evaporate it to dryness, dissolve residue in 100 μL
MeCN:MeOH:water 4:1:5, filter (0.45 μm), inject a 50 μL aliquot. Tissue. Sonicate 300
mg rat brain tissue in 1 mL water (Heat Systems-Ultrasonics), add 10 μL 10 μg/mL
ergocriptine in MeCN:MeOH:water 4:1:5, add 1 mL 2.5 M aqueous potassium carbonate,
extract with 4 mL MeCN, centrifuge at 2200 g at -10° for 15 min, maintain at -10° for 30
min. Remove the organic layer and evaporate it to dryness, dissolve residue in 100 μL
MeCN:MeOH:water 4:1:5, filter (0.45 μm), inject a 50 μL aliquot. (All glassware should
be silanized.)

## HPLC VARIABLES
**Column:** 70 × 4.6 3 μm Ultrasphere XL C8
**Mobile phase:** MeCN:isopropanol:25.3 mM ammonium carbonate 40:6:54 (After each run
elute column with MeCN:water 80:20 for 15 min.)
**Flow rate:** 1.2
**Injection volume:** 50
**Detector:** UV 310

## CHROMATOGRAM
**Retention time:** 10.5
**Internal standard:** ergocriptine (6)
**Limit of detection:** 19.5 ng/mL (plasma); 65 ng/g (tissue)

## KEY WORDS
plasma; rat; brain

## REFERENCE
Phelan, D.G.; Greig, N.H.; Rapoport, S.I.; Soncrant, T.T. High-performance liquid chromatographic assay
of bromocriptine in rat plasma and brain. *J.Chromatogr.*, **1990**, *533*, 264–270

## SAMPLE
**Matrix:** solutions

## HPLC VARIABLES
**Column:** 250 × 4.6 5 μm Supelcosil LC-DP (A) or 250 × 4 5 μm LiChrospher 100 RP-8 (B)

**Mobile phase:** MeCN:0.025% phosphoric acid:buffer 25:10:5 (A) or 60:25:15 (B) (Buffer was 9 mL concentrated phosphoric acid and 10 mL triethylamine in 900 mL water, adjust pH to 3.4 with dilute phosphoric acid, make up to 1 L.)
**Flow rate:** 0.6
**Injection volume:** 25
**Detector:** UV 229

## CHROMATOGRAM
**Retention time:** 10.80 (A), 6.21 (B)

## OTHER SUBSTANCES
**Also analyzed:** acebutolol, acepromazine, acetaminophen, acetazolamide, acetophenazine, albuterol, alprazolam, amitriptyline, amobarbital, amoxapine, antipyrine, atenolol, atropine, azatadine, baclofen, benzocaine, brompheniramine, brotizolam, bupivacaine, buspirone, butabarbital, butalbital, caffeine, carbamazepine, cetirizine, chlorcyclizine, chlordiazepoxide, chlormezanone, chloroquine, chlorpheniramine, chlorpromazine, chlorpropamide, chlorprothixene, chlorthalidone, chlorzoxazone, cimetidine, cisapride, clomipramine, clonazepam, clonidine, clozapine, cocaine, codeine, colchicine, cyclizine, cyclobenzaprine, dantrolene, desipramine, diazepam, diclofenac, diflunisal, diltiazem, diphenhydramine, diphenidol, diphenoxylate, dipyridamole, disopyramide, dobutamine, doxapram, doxepin, droperidol, encainide, ethidium bromide, ethopropazine, fenoprofen, fentanyl, flavoxate, fluoxetine, fluphenazine, flurazepam, flurbiprofen, fluvoxamine, furosemide, glutethimide, glyburide, guaifenesin, haloperidol, homatropine, hydralazine, hydrochlorothiazide, hydrocodone, hydromorphone, hydroxychloroquine, hydroxyzine, ibuprofen, imipramine, indomethacin, ketoconazole, ketoprofen, ketorolac, labetalol, levorphanol, lidocaine, loratadine, lorazepam, lovastatin, loxapine, mazindol, mefenamic acid, meperidine, mephenytoin, mepivacaine, mesoridazine, metaproterenol, methadone, methdilazine, methocarbamol, methotrexate, methotrimeprazine, methoxamine, methyldopa, methylphenidate, metoclopramide, metolazone, metoprolol, metronidazole, midazolam, moclobemide, morphine, nadolol, nalbuphine, naloxone, naphazoline, naproxen, nifedipine, nizatidine, norepinephrine, nortriptyline, oxazepam, oxycodone, oxymetazoline, paroxetine, pemoline, pentazocine, pentobarbital, pentoxifylline, perphenazine, pheniramine, phenobarbital, phenol, phenolphthalein, phentolamine, phenylbutazone, phenyltoloxamine, phenytoin, pimozide, pindolol, piroxicam, pramoxine, prazepam, prazosin, probenecid, procainamide, procaine, prochlorperazine, procyclidine, promazine, promethazine, propafenone, propantheline, propiomazine, propofol, propranolol, protriptyline, quazepam, quinidine, quinine, racemethorphan, ranitidine, remoxipride, risperidone, salicylic acid, scopolamine, secobarbital, sertraline, sotalol, spironolactone, sulfinpyrazone, sulindac, temazepam, terbutaline, terfenadine, tetracaine, theophylline, thiethylperazine, thiopental, thioridazine, thiothixene, timolol, tocainide, tolbutamide, tolmetin, trazodone, triamterene, triazolam, trifluoperazine, triflupromazine, trimeprazine, trimethoprim, trimipramine, verapamil, warfarin, xylometazoline, yohimbine, zopiclone

## KEY WORDS
some details of plasma extraction

## REFERENCE
Koves, E.M. Use of high-performance liquid chromatography-diode array detection in forensic toxicology. *J.Chromatogr.A*, **1995**, *692*, 103–119

## SAMPLE
**Matrix:** solutions

## HPLC VARIABLES
**Column:** 150 × 3.9 5 μm Spherisorb C8
**Mobile phase:** MeCN:buffer 60:40 (Buffer was 1.5 mL triethylamine in 1 L water adjusted to pH 3.0 with 85% phosphoric acid.)

**Flow rate:** 1.5
**Injection volume:** 200
**Detector:** UV 199

## CHROMATOGRAM
**Retention time:** 5.5

## OTHER SUBSTANCES
**Simultaneous:** benztropine mesylate, biperiden, desipramine, hyoscyamine, orphenadrine
**Noninterfering:** amantadine, carbidopa, levodopa

## REFERENCE
Selinger, K.; Lebel, G.; Hill, H.M.; Discenza, C. High-performance liquid chromatographic method for the analysis of benztropine in human plasma. *J.Chromatogr.*, **1989**, *491*, 248–252

# Buspirone

**Molecular formula:** $C_{21}H_{31}N_5O_2$
**Molecular weight:** 385.5
**CAS Registry No.:** 36505-84-7 (buspirone),
33386-08-2 (buspirone hydrochloride)

HCl

## SAMPLE
**Matrix:** blood
**Sample preparation:** 2 mL Whole blood or plasma + 2 mL buffer + 5 mL chloroform: isopropanol:n-heptane 60:14:26, shake gently horizontally for 10 min, centrifuge at 2800 g for 10 min. Remove the lower organic layer and evaporate it to dryness under vacuum at 45°, reconstitute the residue in 100 μL mobile phase, centrifuge at 2800 g for 5 min, inject a 50 μL aliquot of the supernatant. (Buffer was saturated ammonium chloride solution 25% diluted with water, adjusted to pH 9.5 with 25% ammonia solution.)

## HPLC VARIABLES
**Column:** 300 × 3.9 4 μm NovaPack C18
**Mobile phase:** MeOH:THF:buffer 65:5:30 (Buffer was 0.68 g/L (10 mM (sic)) $KH_2PO_4$ adjusted to pH 2.6 with concentrated orthophosphoric acid.) (At the end of each session wash the column with water for 1 h and MeOH for 1 h, re-equilibrate for 30 min.)
**Column temperature:** 30
**Flow rate:** 0.8
**Injection volume:** 50
**Detector:** UV 236

## CHROMATOGRAM
**Retention time:** 5.39
**Limit of detection:** <120 ng/mL

## OTHER SUBSTANCES
**Also analyzed:** acebutolol, acenocoumarol, acepromazine, aceprometazine, acetaminophen, aconitine, ajmaline, albuterol, alimemazine, alminoprofen, alpidem, alprazolam, alprenolol, amisulpride, amitriptyline, amodiaquine, amoxapine, aspirin, astemizole, atenolol, benazepril, benperidol, benzocaine, benzoylecgonine, bepridil, betaxolol, bisoprolol, bromazepam, brompheniramine, bumadizone, bupivacaine, buprenorphine, buspirone, caffeine, carbamazepine, carbinoxamine, carpipramine, carteolol, celiprolol, cetirizine, chlorambucil, chlordiazepoxide, chlormezanone, chlorophenacinone, chloroquine, chlorpheniramine, chlorpromazine, chlorpropamide, cibenzoline, cicletanine, clemastine, clobazam, clomipramine, clonazepam, clonidine, clorazepate, clozapine, cocaine, codeine, colchicine, cyamemazine, cyclizine, cycloguanil, cyproheptadine, cytarabine, dacarbazine, daunorubicin, debrisoquine, demexiptiline, desipramine, dextromethorphan, dextromoramide, dextropropoxyphene, diazepam, diazoxide, diclofenac, dihydralazine, diltiazem, diphenhydramine, dipyridamole, disopyramide, dosulepine, doxepin, doxylamine, droperidol, ephedrine, estazolam, etodolac, fenfluramine, fenoprofen, fentiazac, flecainide, floctafenine, flumazenil, flunitrazepam, fluoxetine, fluphenazine, flurbiprofen, fluvoxamine, glibenclamide, glibornuride, glipizide, glutethimide, haloperidol, histapyrrodine, hydroxychloroquine, hydroxyzine, ibuprofen, imipramine, indomethacin, iproniazid, ketamine, ketoprofen, labetalol, levomepromazine, lidocaine, lidoflazine, lisinopril, loperamide, loprazolam, loratadine, lorazepam, loxapine, maprotiline, medazepam, medifoxamine, mefenamic acid, mefenidramine, mefloquine, melphalan, meperidine, mephenesin, mephentermine, mepivacaine, metapramine, metformin, methadone, methaqualone, methocarbamol, methotrexate, metipranolol, metoclopramide, metoprolol, mexiletine, mianserine, midazolam, minoxidil, moclobemide, moperone, morphine, nadolol, nalbuphine, nalorphine, naloxone, naltrexone, naproxen, nialamide, nicardipine, nifedipine, ni-

flumic acid, nimodipine, nitrazepam, nitrendipine, nizatidine, nomifensine, nortriptyline, omeprazole, opipramol, oxazepam, oxprenolol, penbutolol, penfluridol, pentazocine, phencyclidine, phenobarbital, phenol, phenylbutazone, pimozide, pindolol, pipamperone, piroxicam, prazepam, prazosin, prilocaine, procainamide, procarbazine, proguanil, promethazine, propafenone, propranolol, protriptyline, pyrimethamine, quinidine, quinine, quinupramine, ramipril, ranitidine, reserpine, ritodrine, secobarbital, sotalol, strychnine, sulfinpyrazole, sulindac, sulpride, sultopride, suriclone, temazepam, tenoxicam, terbutaline, terfenadine, tetracaine, tetrazepam, thiopental, thioproperazine, thioridazine, tianeptine, tiapride, tiaprofenic acid, ticlopidine, timolol, tioclomarol, tofisopam, tolbutamide, toloxatone, trazodone, triazolam, trifluoperazine, trifluperidol, trimipramine, triprolidine, tropatenine, verapamil, viloxazine, vinblastine, vincristine, vindesine, warfarin, yohimbine, zolpidem, zopiclone, zorubicine

## KEY WORDS
whole blood; plasma

## REFERENCE
Tracqui, A.; Kintz, P.; Mangin, P. Systematic toxicological analysis using HPLC/DAD. *J.Forensic Sci.*, **1995**, *40*, 254–262

## SAMPLE
**Matrix:** blood
**Sample preparation:** Condition a 3 mL Baker carboxylic acid SPE cartridge with 5 mL 2 M HCl and 10 mL water. Mix 500 μL plasma + 5 mL water and add to column. Wash with 5 mL water and elute with 2 mL 1 M formic acid. Evaporate eluate to dryness under vacuum and dissolve residue in 200 μL mobile phase, inject a 20 μL aliquot.

## HPLC VARIABLES
**Guard column:** 5 μm Cyanonitrile
**Column:** 33 × 4.6 Supelcosil LC-CN
**Mobile phase:** MeCN:20 mM pH 7 potassium phosphate buffer 43:57 adjusted to pH 7.34 with 0.5 M KOH
**Flow rate:** 0.4
**Injection volume:** 20
**Detector:** E, ESA Coulochem model 5100 A, two porous graphite working electrodes and associated palladium reference electrodes, +0.55 V for first electrode, +0.70 V for second (monitoring) electrode

## CHROMATOGRAM
**Retention time:** 10
**Limit of detection:** 1 ng/mL

## OTHER SUBSTANCES
**Extracted:** metabolites, 1-(2-pyrimidinyl)piperazine

## KEY WORDS
plasma; mouse; SPE

## REFERENCE
Betto, P.; Meneguz, A.; Ricciarello, G.; Pichini, S. Simultaneous high-performance liquid chromatographic analysis of buspirone and its metabolite 1-(2-pyrimidinyl)-piperazine in plasma using electrochemical detection. *J.Chromatogr.*, **1992**, *575*, 117–121

## SAMPLE
**Matrix:** blood

**Sample preparation:** Condition a 1 mL C18 Supelco SPE cartridge with two 1 mL portions of MeOH and two 1 mL portions of 50 mM $KH_2PO_4$ adjusted to pH 7.2 with 2 M NaOH, add 2 mL serum and flush through 2-5 mL air. Wash with two 1 mL portions of 50 mM $KH_2PO_4$ adjusted to pH 7.2 with 2 M NaOH, wash with two 1 mL and one 0.5 mL portions of MeOH:water 1:1 and flush through with 2-5 mL of air. Elute with 1 mL MeCN: triethylamine 99:1. Evaporate solvent at 37° under a stream of air and dissolve residue in 100 μL of MeCN:5 mM $KH_2PO_4$ + 0.1% triethylamine adjusted to pH 2.5 with ortho-phosphoric acid 45:55, centrifuge at 5000 g for 10 min, inject a 70 μL aliquot onto column A with mobile phase A and elute to waste, switch a 0.7 min fraction onto column B (retention time is ca. 3.6 min) and chromatograph on column B with mobile phase B, monitor the effluent from column B.

## HPLC VARIABLES
**Column:** A 150 × 4.6 5 μm Spherisorb ODS2; B 150 × 4.6 5 μm Spherisorb ODS2
**Mobile phase:** A MeCN:5 mM $KH_2PO_4$ + 0.1% triethylamine adjusted to pH 2.5 with or-thophosphoric acid 45:55; B MeCN:5 mM $KH_2PO_4$ + 0.2% triethylamine adjusted to pH 2.5 with orthophosphoric acid 55:45, containing 5 mM sodium lauryl sulfate
**Flow rate:** 1.2
**Injection volume:** 70
**Detector:** UV 235

## CHROMATOGRAM
**Retention time:** 7 (on column B)
**Limit of detection:** 0.2 ng/mL

## KEY WORDS
serum; column-switching; heart-cut; SPE; pharmacokinetics

## REFERENCE
Kristjánsson, F. Sensitive determination of buspirone in serum by solid-phase extraction and two-dimensional high-performance liquid chromatography. *J.Chromatogr.*, **1991**, *566*, 250–256

## SAMPLE
**Matrix:** blood
**Sample preparation:** 2 mL Plasma + 50 μL 400 ng/mL gepirone in EtOH + 0.5 mL 1 M NaOH + 4 mL ethyl acetate, shake slowly mechanically for 10 min, centrifuge at 400 g for 10 min. Freeze lower plasma layer in dry ice/acetone and decant ethyl acetate into a centrifuge tube containing 2 mL 50 mM HCl, shake 10 min, centrifuge 400 g 10 min, remove, discard organic phase. Add 500 μL 4 M ammonia solution and 400 μL butyl acetate to the aqueous phase, mix for 15-20 sec, centrifuge at 400 g for 10 min. Remove organic layer and evaporate it to dryness at 30° under vacuum. Reconstitute in 90 μL mobile phase, inject an aliquot.

## HPLC VARIABLES
**Guard column:** 5 μm Cyanonitrile (Anachem)
**Column:** 10 × 4.6 5 μm cyanonitrile (Brownlee)
**Mobile phase:** MeCN:40 mM potassium phosphate buffer adjusted to pH 6.6 with 2 M KOH 34:66 (Columns were initially conditioned with 50 mL water; then 5 mM pH 4.8 sodium acetate buffer; then MeCN:5 mM pH 4.8 sodium acetate buffer 40:60; then 5 mM pH 4.8 potassium acetate; then mobile phase.)
**Flow rate:** 1.5
**Injection volume:** 50
**Detector:** E, ESA Coulochem model 5100 A, Model 5020 guard cell, guard cell 0.3 V, detector 0.55 V for first electrode and 0.70 V for second electrode

## CHROMATOGRAM
**Retention time:** 7.3

**Internal standard:** gepirone (5.4)
**Limit of detection:** 0.5 ng/mL

## OTHER SUBSTANCES
**Simultaneous:** 1-(2-pyrimidinyl)piperazine, metabolites, amitriptyline, chlorpromazine, clomipramine, fluphenazine, imipramine
**Noninterfering:** caffeine, diazepam, desipramine, mianserin, zimeldine
**Interfering:** haloperidol

## KEY WORDS
plasma

## REFERENCE
Franklin, M. Determination of plasma buspirone by high-performance liquid chromatography with coulometric detection. *J.Chromatogr.*, **1990**, *526*, 590–596

## SAMPLE
**Matrix:** blood
**Sample preparation:** 2 mL Plasma + 100 μL 50 μg/mL 1-phenylpiperazine + 1 mL pH 10 borate buffer + 5 mL chloroform:MeCN 8:2, agitate, centrifuge, repeat extraction. Evaporate organic layers under reduced pressure below 40°, dissolve residue in 100 μL mobile phase, inject a 20 μL aliquot.

## HPLC VARIABLES
**Guard column:** 23 × 3.6 37-50 μm CN/Corasil
**Column:** 250 × 4.6 5 μm Spherisorb CN
**Mobile phase:** MeOH:5 mM $KH_2PO_4$ pH 7.4 35:65
**Flow rate:** 1.7
**Injection volume:** 20
**Detector:** UV 254

## CHROMATOGRAM
**Retention time:** 10
**Internal standard:** 1-phenylpiperazine (13)
**Limit of detection:** 5 ng/mL

## OTHER SUBSTANCES
**Extracted:** 1-(2-pyrimidinyl)piperazine, metabolites

## KEY WORDS
plasma; rat

## REFERENCE
Diaz-Marot, A.; Puigdellivol, E.; Salvatella, C.; Comellas, L.; Gassiot, M. Determination of buspirone and 1-(2-pyrimidinyl)piperazine in plasma samples by high-performance liquid chromatography. *J.Chromatogr.*, **1989**, *490*, 470–473

## SAMPLE
**Matrix:** solutions

## HPLC VARIABLES
**Column:** 250 × 4.6 5 μm Supelcosil LC-DP (A) or 250 × 4 5 μm LiChrospher 100 RP-8 (B)
**Mobile phase:** MeCN:0.025% phosphoric acid:buffer 25:10:5 (A) or 60:25:15 (B) (Buffer was 9 mL concentrated phosphoric acid and 10 mL triethylamine in 900 mL water, adjust pH to 3.4 with dilute phosphoric acid, make up to 1 L.)

**Flow rate:** 0.6
**Injection volume:** 25
**Detector:** UV 229

## CHROMATOGRAM
**Retention time:** 9.07 (A), 4.98 (B)

## OTHER SUBSTANCES
**Also analyzed:** acebutolol, acepromazine, acetaminophen, acetazolamide, acetophenazine, albuterol, alprazolam, amitriptyline, amobarbital, amoxapine, antipyrine, atenolol, atropine, azatadine, baclofen, benzocaine, bromocriptine, brompheniramine, brotizolam, bupivacaine, butabarbital, butalbital, caffeine, carbamazepine, cetirizine, chlorcyclizine, chlordiazepoxide, chlormezanone, chloroquine, chlorpheniramine, chlorpromazine, chlorpropamide, chlorprothixene, chlorthalidone, chlorzoxazone, cimetidine, cisapride, clomipramine, clonazepam, clonidine, clozapine, cocaine, codeine, colchicine, cyclizine, cyclobenzaprine, dantrolene, desipramine, diazepam, diclofenac, diflunisal, diltiazem, diphenhydramine, diphenidol, diphenoxylate, dipyridamole, disopyramide, dobutamine, doxapram, doxepin, droperidol, encainide, ethidium bromide, ethopropazine, fenoprofen, fentanyl, flavoxate, fluoxetine, fluphenazine, flurazepam, flurbiprofen, fluvoxamine, furosemide, glutethimide, glyburide, guaifenesin, haloperidol, homatropine, hydralazine, hydrochlorothiazide, hydrocodone, hydromorphone, hydroxychloroquine, hydroxyzine, ibuprofen, imipramine, indomethacin, ketoconazole, ketoprofen, ketorolac, labetalol, levorphanol, lidocaine, loratadine, lorazepam, lovastatin, loxapine, mazindol, mefenamic acid, meperidine, mephenytoin, mepivacaine, mesoridazine, metaproterenol, methadone, methdilazine, methocarbamol, methotrexate, methotrimeprazine, methoxamine, methyldopa, methylphenidate, metoclopramide, metolazone, metoprolol, metronidazole, midazolam, moclobemide, morphine, nadolol, nalbuphine, naloxone, naphazoline, naproxen, nifedipine, nizatidine, norepinephrine, nortriptyline, oxazepam, oxycodone, oxymetazoline, paroxetine, pemoline, pentazocine, pentobarbital, pentoxifylline, perphenazine, pheniramine, phenobarbital, phenol, phenolphthalein, phentolamine, phenylbutazone, phenyltoloxamine, phenytoin, pimozide, pindolol, piroxicam, pramoxine, prazepam, prazosin, probenecid, procainamide, procaine, prochlorperazine, procyclidine, promazine, promethazine, propafenone, propantheline, propiomazine, propofol, propranolol, protriptyline, quazepam, quinidine, quinine, racemethorphan, ranitidine, remoxipride, risperidone, salicylic acid, scopolamine, secobarbital, sertraline, sotalol, spironolactone, sulfinpyrazone, sulindac, temazepam, terbutaline, terfenadine, tetracaine, theophylline, thiethylperazine, thiopental, thioridazine, thiothixene, timolol, tocainide, tolbutamide, tolmetin, trazodone, triamterene, triazolam, trifluoperazine, triflupromazine, trimeprazine, trimethoprim, trimipramine, verapamil, warfarin, xylometazoline, yohimbine, zopiclone

## KEY WORDS
some details of plasma extraction

## REFERENCE
Koves, E.M. Use of high-performance liquid chromatography-diode array detection in forensic toxicology. *J.Chromatogr.A*, **1995**, *692*, 103–119

## SAMPLE
**Matrix:** solutions

## HPLC VARIABLES
**Column:** 250 × 4.6 Zorbax RX
**Mobile phase:** Gradient. A was 10 mL concentrated orthophosphoric acid and 7 mL triethylamine in 1 L water. B was 10 mL concentrated orthophosphoric acid and 7 mL triethylamine in 200 mL water, make up to 1 L with MeCN. A:B from 100:0 to 0:100 over 30 min, maintain at 0:100 for 5 min.

**Column temperature:** 30
**Flow rate:** 2
**Detector:** UV 210

## OTHER SUBSTANCES

**Also analyzed:** acepromazine, acetaminophen, acetophenazine, albuterol, aminophylline, amitriptyline, amobarbital, amoxapine, amphetamine, amylocaine, antipyrine, aprobarbital, aspirin, atenolol, atropine, avermectin, barbital, benzocaine, benzoic acid, benzotropine, benzphetamine, berberine, bibucaine, bromazepan, brompheniramine, buprenorphine, butabarbital, butacaine, butethal, caffeine, carbamazepine, carbromal, chloramphenicol, chlordiazepoxide, chloroquine, chlorothiazide, chloroxylenol, chlorphenesin, chlorpheniramine, chlorpromazine, chlorpropamide, chlortetracycline, cimetidine, cinchonidine, cinchonine, clenbuterol, clonazepam, clonixin, clorazepate, cocaine, codeine, colchicine, cortisone, coumarin, cyclazocine, cyclobenzaprine, cyclothiazide, cyheptamide, cymarin, danazol, danthron, dapsone, debrisoquine, desipramine, dexamethasone, dextromethorphan, dextropropoxyphene, diamorphine, diazepam, diclofenac, diethylpropion, diethylstilbestrol, diflunisal, digitoxin, digoxin, diltiazem, diphenhydramine, diphenoxylate, diprenorphine, dipyrone, disulfiram, dopamine, doxapram, doxepin, dronabinol, ephedrine, epinephrine, epinine, estradiol, estriol, estrone, ethacrynic acid, ethosuximide, etonitazene, etorphine, eugenol, famotidine, fenbendazole, fencamfamine, fenoprofen, fenproporex, fentanyl, flubendazole, flufenamic acid, flunitrazepam, 5-fluorouracil, fluoxymesterone, fluphenazine, furosemide, gentisic acid, gitoxigenin, glipizide, glunixin, glutethimide, glybenclamide, guaiacol, halazepam, haloperidol, hydrochlorothiazide, hydrocodone, hydrocortisone, hydromorphone, hydroxyquinoline, ibogaine, ibuprofen, iminostilbene, imipramine, indomethacin, isocarbostyril, isocarboxazid, isoniazid, isoproterenol, isoxsuprine, ivermectin, ketamine, ketoprofen, kynurenic acid, levorphanol, lidocaine, lorazepam, lormetazepam, loxapine, mazindol, mebendazole, meclizine, meclofenamic acid, medazepam, mefenamic acid, megestrol, mepacrine, meperidine, mephentermine, mephenytoin, mephesin, mephobarbital, mepivacaine, mescaline, mesoridazine, methadone, methamphetamine, methapyrilene, methaqualone, methazolamide, methocarbamol, methoxamine, methsuximide, methyl salicylate, methyldopa, methyldopamine, methylphenidate, methylprednisolone, methyltestosterone, methyprylon, metoprolol, mibolerone, morphine, nadolol, nalorphine, naloxone, naltrexone, naphazoline, naproxen, nefopam, niacinamide, nicotine, nicotinic acid, nifedipine, niflumic acid, nitrazepam, norepinephrine, nortriptyline, noscapine, nylidrin, oxazepam, oxycodone, oxymorphone, oxyphenbutazone, oxytetracycline, papaverine, pargyline, pemoline, pentazocine, pentobarbital, persantine, phenacetin, phenazocine, phenazopyridine, phencyclidine, phendimetrazine, phenelzine, pheniramine, phenobarbital, phenothiazine, phensuximide, phentermine, phenylbutazone, phenylephrine, phenylpropanolamine, piperocaine, prazepam, prednisolone, primidone, probenecid, progesterone, propiomazine, propranolol, propylparaben, pseudoephedrine, puromycin, pyrilamine, pyrithyldione, quazepam, quinaldic acid, quinidine, quinine, ranitidine, recinnamine, reserpine, resorcinol, saccharin, albuterol, salicylamide, salicylic acid, scopolamine, scopoletin, secobarbital, strychnine, sulfacetamide, sufadiazine, sulfadimethoxine, sulfaethidole, sulfamerazine, sulfamethazine, sulfamethoxizole, sulfanilamide, sulfapyridine, sulfasoxizole, sulindac, tamoxifen, temazepam, testosterone, tetracaine, tetracycline, tetramisole, thebaine, theobromine, theophylline, thiabendazole, thiamine, thiamylal, thiobarbituric acid, thioridazine, thiosalicylic acid, thiothixene, thymol, tolazamide, tolazoline, tobutamide, tolmetin, tranylcypromine, triamcinolone, tribenzylamine, trichloromethiazide, trifluoperazine, trihexyphenidyl, trimethoprim, tripelennamine, triproilidine, tropacocaine, tyramine, verapamil, vincamine, warfarin, yohimbine, zoxazolamine

## REFERENCE

Hill, D.W.; Kind, A.J. Reversed-phase solvent gradient HPLC retention indexes of drugs. *J.Anal.Toxicol.*, **1994**, *18*, 233–242

**ANNOTATED BIBLIOGRAPHY**

Gil, M.S.; Ochoa, C.; Vega, S. High performance liquid chromatography of new potential anxiolytic drugs and related benzodiazepines: A comparative study of hydrophobicity. *J.Liq.Chromatogr.*, **1991**, *14*, 2141–2156 [also chlordiazepoxide, diazepam]

Sarati, S.; Guiso, G.; Spinelli, R.; Caccia, S. Determination of piribedil and its basic metabolites in plasma by high-performance liquid chromatography. *J.Chromatogr.*, **1991**, *563*, 323–332 [plasma; rat; extracted piribedil; buspirone is IS; gradient]

# Butalbital

**Molecular formula:** $C_{11}H_{16}N_2O_3$
**Molecular weight:** 224.3
**CAS Registry No.:** 77-26-9

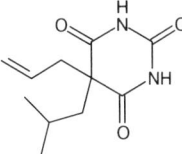

## SAMPLE
**Matrix:** blood
**Sample preparation:** 1 mL Blood + 1 mL water + 50 µL 76 mg/L allobarbital in EtOH:
water 10:90 + 5 mL ethyl acetate, shake by hand, add 2 mL of 0.1 M HCl. Mix by
inversion with a mechanical shaker for 5 min. Centrifuge at 2700 rpm for 5-10 min.
Remove ethyl acetate and evaporate to dryness under a stream of nitrogen at room temperature. Take up in 200 µL MeOH, filter (0.45 µm), inject a 20 µL aliquot.

## HPLC VARIABLES
**Guard column:** Guard-Pak precolumn insert
**Column:** 200 × 4.6 5 µm Hypersil octadecylsilane
**Mobile phase:** Gradient. MeCN:1 mM pH 3.2 phosphate buffer from 20:80 to 40:60 over
10 min, stay at 40:60 for 6 min, to 20:80 over 4 min.
**Column temperature:** 60
**Flow rate:** 3
**Injection volume:** 20
**Detector:** UV 202

## CHROMATOGRAM
**Retention time:** 5.0
**Internal standard:** allobarbital (2.9)
**Limit of quantitation:** 500 ng/mL

## OTHER SUBSTANCES
**Extracted:** salicylic acid
**Simultaneous:** aspirin, caffeine

## KEY WORDS
pharmacokinetics

## REFERENCE
Drost, M.L.; Walter, L. Blood and plasma concentrations of butalbital following single oral doses in man.
*J.Anal.Toxicol.*, **1988**, *12*, 322−324

## SAMPLE
**Matrix:** blood
**Sample preparation:** 200 µL Serum + 200 µL 50 µg/mL hexobarbital in MeCN + 25 µL
glacial acetic acid, vortex for 10 s, centrifuge for 1 min, inject a 30-100 µL aliquot of the
supernatant.

## HPLC VARIABLES
**Column:** µBondapak C18
**Mobile phase:** Gradient. MeCN:7.5 g/L $NaH_2PO_4$ adjusted to pH 3.2 with phosphoric acid
5:95 to 22:78 over 24 min, to 45:55 over 10 min, maintain at 45:55 for 5 min. Re-
equilibrate with 5:95 for 5 min.
**Column temperature:** 50
**Flow rate:** 3

**Injection volume:** 30-100
**Detector:** UV 210

## CHROMATOGRAM
**Retention time:** 17.1
**Internal standard:** hexobarbital (20.6)
**Limit of detection:** 200-2000 ng/mL

## OTHER SUBSTANCES
**Extracted:** acetaminophen, amobarbital, butabarbital, chlordiazepoxide, diazepam, ethchlorvynol, flurazepam, glutethimide, methaqualone, methyprylon, nitrazepam, pentobarbital, phenobarbital, phenytoin, primidone, salicylic acid, secobarbital, theophylline
**Simultaneous:** amitriptyline, caffeine, clomipramine, codeine, desipramine, ethotoin, imipramine, lidocaine, mesantoin, methsuximide, nirvanol, nortriptyline, oxazepam, procainamide, phenylpropanolamine, propranolol, quinidine

## KEY WORDS
serum

## REFERENCE
Kabra, P.M.; Stafford, B.E.; Marton, L.J. Rapid method for screening toxic drugs in serum with liquid chromatography. *J.Anal.Toxicol.*, **1981**, *5*, 177–182

## SAMPLE
**Matrix:** solutions

## HPLC VARIABLES
**Column:** 250 × 4.6 5 μm Supelcosil LC-DP (A) or 250 × 4 5 μm LiChrospher 100 RP-8 (B)
**Mobile phase:** MeCN:0.025% phosphoric acid:buffer 25:10:5 (A) or 60:25:15 (B) (Buffer was 9 mL concentrated phosphoric acid and 10 mL triethylamine in 900 mL water, adjust pH to 3.4 with dilute phosphoric acid, make up to 1 L.)
**Flow rate:** 0.6
**Injection volume:** 25
**Detector:** UV 229

## CHROMATOGRAM
**Retention time:** 5.77 (A), 5.12 (B)

## OTHER SUBSTANCES
**Also analyzed:** acebutolol, acepromazine, acetaminophen, acetazolamide, acetophenazine, albuterol, alprazolam, amitriptyline, amobarbital, amoxapine, antipyrine, atenolol, atropine, azatadine, baclofen, benzocaine, bromocriptine, brompheniramine, brotizolam, bupivacaine, buspirone, butabarbital, caffeine, carbamazepine, cetirizine, chlorcyclizine, chlordiazepoxide, chlormezanone, chloroquine, chlorpheniramine, chlorpromazine, chlorpropamide, chlorprothixene, chlorthalidone, chlorzoxazone, cimetidine, cisapride, clomipramine, clonazepam, clonidine, clozapine, cocaine, codeine, colchicine, cyclizine, cyclobenzaprine, dantrolene, desipramine, diazepam, diclofenac, diflunisal, diltiazem, diphenhydramine, diphenidol, diphenoxylate, dipyridamole, disopyramide, dobutamine, doxapram, doxepin, droperidol, encainide, ethidium bromide, ethopropazine, fenoprofen, fentanyl, flavoxate, fluoxetine, fluphenazine, flurazepam, flurbiprofen, fluvoxamine, furosemide, glutethimide, glyburide, guaifenesin, haloperidol, homatropine, hydralazine, hydrochlorothiazide, hydrocodone, hydromorphone, hydroxychloroquine, hydroxyzine, ibuprofen, imipramine, indomethacin, ketoconazole, ketoprofen, ketorolac, labetalol, levorphanol, lidocaine, loratadine, lorazepam, lovastatin, loxapine, mazindol, mefenamic acid, meperidine, mephenytoin, mepivacaine, mesoridazine, metaproterenol, methadone, methdilazine, methocarbamol, methotrexate, methotrimeprazine, methoxamine, methyldopa, methylphenidate, metoclopramide, metolazone, metoprolol, metronidazole, mida-

zolam, moclobemide, morphine, nadolol, nalbuphine, naloxone, naphazoline, naproxen, nifedipine, nizatidine, norepinephrine, nortriptyline, oxazepam, oxycodone, oxymetazo-line, paroxetine, pemoline, pentazocine, pentobarbital, pentoxifylline, perphenazine, pheniramine, phenobarbital, phenol, phenolphthalein, phentolamine, phenylbutazone, phenyltoloxamine, phenytoin, pimozide, pindolol, piroxicam, pramoxine, prazepam, pra-zosin, probenecid, procainamide, procaine, prochlorperazine, procyclidine, promazine, pro-methazine, propafenone, propantheline, propiomazine, propofol, propranolol, protripty-line, quazepam, quinidine, quinine, racemethorphan, ranitidine, remoxipride, risperidone, salicylic acid, scopolamine, secobarbital, sertraline, sotalol, spironolactone, sulfinpyra-zone, sulindac, temazepam, terbutaline, terfenadine, tetracaine, theophylline, thiethyl-perazine, thiopental, thioridazine, thiothixene, timolol, tocainide, tolbutamide, tolmetin, trazodone, triamterene, triazolam, trifluoperazine, triflupromazine, trimeprazine, tri-methoprim, trimipramine, verapamil, warfarin, xylometazoline, yohimbine, zopiclone

## KEY WORDS
some details of plasma extraction

## REFERENCE
Koves, E.M. Use of high-performance liquid chromatography-diode array detection in forensic toxicology. *J.Chromatogr.A*, **1995**, *692*, 103–119

## SAMPLE
**Matrix:** solutions
**Sample preparation:** Dissolve in mobile phase to a concentration of 50 µg/mL.

## HPLC VARIABLES
**Column:** 250 × 4 β-cyclodextrin polymer-coated silica (Chromatographia 1993, 36, 373)
**Mobile phase:** MeOH:water 50:50
**Flow rate:** 0.6
**Injection volume:** 20
**Detector:** UV 240

## CHROMATOGRAM
**Retention time:** k' 1.15

## OTHER SUBSTANCES
**Simultaneous:** amobarbital, aprobarbital, butabarbital, pentobarbital, phenobarbital, sec-obarbital, thiopental

## REFERENCE
Forgács, E.; Cserháti, T. Retention behaviour of barbituric acid derivatives on a β-cyclodextrin polymer-coated silicon column. *J.Chromatogr.A*, **1994**, *668*, 395–402

## SAMPLE
**Matrix:** solutions
**Sample preparation:** Dissolve in mobile phase at a concentration of 100 µg/mL, inject a 5 µL aliquot.

## HPLC VARIABLES
**Column:** 300 × 2 µBondapak C18
**Mobile phase:** MeCN:water 30:70 adjusted to pH 3.0 with formic acid
**Flow rate:** 0.27
**Injection volume:** 5
**Detector:** MS, VG TRIO 2000 single quadrupole MS with EI or CI; UV 270

## CHROMATOGRAM
**Retention time:** 11.25

## OTHER SUBSTANCES
**Extracted:** amobarbital, butabarbital, butethal, pentobarbital, talbutal

## KEY WORDS
mass spectra given

## REFERENCE
Ryan, T.W. Identification of barbiturates using high performance liquid chromatography-particle beam
   EI/CI mass spectrometry. *J.Liq.Chromatogr.*, **1994**, *17*, 867–881

## SAMPLE
**Matrix:** solutions

## HPLC VARIABLES
**Column:** 300 × 3.9 μBondapak C18
**Mobile phase:** MeCN:10 mm $KH_2PO_4$ + 5 mM 1-decanesulfonic acid 30:70, adjusted to
   pH 3.2 with 85% phosphoric acid
**Flow rate:** 1
**Injection volume:** 10
**Detector:** UV 214

## CHROMATOGRAM
**Retention time:** 7.9
**Internal standard:** methyl paraben (7.0)
**Limit of detection:** 100 ng/mL

## OTHER SUBSTANCES
**Simultaneous:** allobarbital, aprobarbital, barbital, mephobarbital, pentobarbital, phenobarbital, secobarbital, talbutal, vinbarbital

## KEY WORDS
stability-indicating

## REFERENCE
Ibrahim, F.B. Simultaneous determination and separation of several barbiturates and analgesic products
   by ion-pair high-performance liquid chromatography. *J.Liq.Chromatogr.*, **1993**, *16*, 2835–2851

◆――――――――――――――――――◆――――――――――――――――――◆

## ANNOTATED BIBLIOGRAPHY
Ryan, T.W. Resolution of the non-specific spectra of barbiturates by UV-photodiode array detection.
   *J.Liq.Chromatogr.*, **1993**, *16*, 315–329 [allobarbital, amobarbital, barbital, mephobarbital, pentobar-
   bital, phenobarbital, secobarbital, talbutal, vinbarbital]

# Captopril

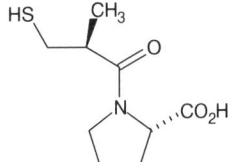

**Molecular formula:** C$_9$H$_{15}$NO$_3$S
**Molecular weight:** 217.3
**CAS Registry No.:** 62571-86-2

## SAMPLE
**Matrix:** blood
**Sample preparation:** 500 µL Plasma + 30 µL 1 mg/mL p-bromophenacyl bromide in MeCN + 50 µL 100 mM NaOH, shake for 15 min, add 75 µL 1 M HCl, add 500 ng nitrazepam for each mL, add 150 µL 200 mM pH 4.0 acetate buffer. Extract with 4 mL benzene (Caution! Benzene is a carcinogen!), centrifuge. Remove the organic layer and evaporate it to dryness under a stream of nitrogen, reconstitute the residue in 200 µL mobile phase, inject a 20 µL aliquot.

## HPLC VARIABLES
**Column:** 250 × 4.6 Kontron Analytical S5 ODS2
**Mobile phase:** MeCN:1% acetic acid 60:40
**Flow rate:** 1.3
**Injection volume:** 20
**Detector:** UV 260

## CHROMATOGRAM
**Retention time:** 4
**Internal standard:** nitrazepam (4.5)
**Limit of detection:** 15 ng/mL
**Limit of quantitation:** 30 ng/mL

## KEY WORDS
plasma; derivatization; pharmacokinetics

## REFERENCE
Jankowski, A.; Skorek, A.; Krzysko, K.; Zarzyxki, P.K.; Ochocka, R.J.; Lamparczyk, H. Captopril: determination in blood and pharmacokinetics after single oral dose. *J.Pharm.Biomed.Anal.*, **1995**, *13*, 655–660

## SAMPLE
**Matrix:** blood
**Sample preparation:** 1 mL Blood + 50 µL solution containing 100 mM EDTA and 100 mM ascorbic acid, centrifuge at 13000 g for 2 min. 500 µL supernatant + 2 mL 100 mM pH 7 phosphate buffer + 200 ng IS + 200 µL 1.5 mg/mL N-(3-pyrenyl)maleimide in MeCN, shake for 15 min, acidify with 100 µL 11 M HCl, add 6 mL ethyl acetate, vortex for 20 min, centrifuge at 2500 g for 5 min. Remove the organic layer, dry it under nitrogen, dissolve in 50 or 200 µL MeCN, inject a 5-15 µL aliquot.

## HPLC VARIABLES
**Column:** 100 × 4.6 5 µm Partisil ODS-3 C18
**Mobile phase:** MeCN:1% acetic acid 37:63
**Flow rate:** 1.5
**Injection volume:** 5-15
**Detector:** F ex 340 em 389

## CHROMATOGRAM
**Retention time:** 8

**Internal    standard:** (4R)-2-(2-Hydroxyphenyl)-3-(3-mercaptopropionyl)-4-thiazolidinecar-
boxylic acid (SA 446) (14)
**Limit of detection:** 10 ng/mL

## OTHER SUBSTANCES
**Noninterfering:** chloral hydrate, chlorpromazine, furosemide, digoxin, promethazine

## KEY WORDS
with modifications can be used to determine captopril disulfide; derivatization; pharmacokinetics

## REFERENCE
Pereira, C.M.; Tam, Y.K.; Collins-Nakai, R.L.; Ng, P. Simplified determination of captopril in plasma by
high-performance liquid chromatography. *J.Chromatogr.*, **1988**, *425*, 208–213

## SAMPLE
**Matrix:** blood
**Sample preparation:** 500 μL Whole blood + 20 μL 1% N-(4-dimethylaminophenyl)
maleimide in acetone + 300 μL 33.3 mM pH 6.85 phosphate buffer, add 100 ng IS, vortex,
let stand at 0° for 30 min, freeze in dry ice/acetone, thaw, wash twice with 2 mL portions
of ether. Add 500 μg glutathione to the aqueous layer, let stand at 0° for 20 min, add 3
mL acetone, centrifuge at 1580 g for 5 min. Remove the supernatant and wash the precipitate with 3 mL acetone. Combine the supernatant and the wash and evaporate them
to about 1 mL under reduced pressure at room temperature, dilute the residue with 6
mL water, add to a Sep-Pak C18 SPE cartridge, wash with 2 mL water, elute with 8 mL
MeCN. Evaporate the eluate to dryness under reduced pressure below 40°, reconstitute
the residue in 200 μL MeOH, inject an aliquot. (Prepare N-(4-dimethylamino-
phenyl)maleimide as follows. Mix equimolar amounts of maleic anhydride and N, N-di-
methyl-1,4-phenylenediamine in ether or THF with cooling in ice and stirring, allow to
stand overnight, remove the maleamic acid by filtration, wash with THF. Heat 1 mmole
of the maleamic acid with 3 mmole acetic anhydride and 0.3 mmole sodium acetate at
100° for 5-10 min (until the solution goes clear), cool, add ice water, neutralize with sodium bicarbonate, extract with ethyl acetate. Wash the organic layer with saturated sodium chloride and dry over anhydrous sodium sulfate (Chem. Pharm. Bull. 1976, 24, 3045;
1977, 25, 2739). Evaporate to dryness and recrystallize from acetone to give
N-(4-dimethylaminophenyl)maleimide as reddish-brown crystals (mp 153-154° (J. Org.
Chem. 1963, 28, 2018)).)

## HPLC VARIABLES
**Column:** 300 × 3.9 8-10 μm μBondapak C18
**Mobile phase:** MeCN:0.8% pH 3.0 (NH$_4$)H$_2$PO$_4$ 1:2
**Flow rate:** 1
**Detector:** E, Yanagimoto VMD 101, +0.9 V, Ag/AgCl reference electrode

## CHROMATOGRAM
**Retention time:** 10
**Internal    standard:** (4R)-2-(2-hydroxyphenyl)-3-(3-mercaptopropionyl)-4-thiazolidinecar-
boxylic acid (Sankyo SA 446) (19)
**Limit of detection:** 10 ng/mL

## KEY WORDS
derivatization; whole blood; SPE; pharmacokinetics

## REFERENCE
Shimada, K.; Tanaka, M.; Nambara, T.; Imai, Y.; Abe, K.; Yoshinaga, K. Determination of captopril in
human blood by high-performance liquid chromatography with electrochemical detection.
*J.Chromatogr.*, **1982**, *227*, 445–451

## SAMPLE

**Matrix:** blood

**Sample preparation:** 3 mL Blood + 1.5 mL 0.5% p-bromophenacyl bromide in acetone, vortex for 30 s, let stand for 5 min, acidify with 300 mM HCl, store below -15°, extract with 16 mL benzene (Caution! Benzene is a carcinogen!), extract with 8 mL benzene. Combine the organic layers and evaporate them to dryness under reduced pressure, reconstitute the residue in 4 mL 50 mM pH 7 phosphate buffer, add 20 mL hexane, sonicate, wash with 6 mL hexane, discard the hexane layer, acidify the aqueous layer with 100 μL 2 M HCl, extract with 6 mL benzene, extract with 2 mL benzene. Combine the organic layers and add 500 ng IS, evaporate to dryness under reduced pressure, reconstitute the residue in 200 μL MeCN, inject a 5-25 μL aliquot.

## HPLC VARIABLES

**Column:** μBondapak C18
**Mobile phase:** MeCN:water:acetic acid 48:51.5:0.5
**Injection volume:** 5-25
**Detector:** UV 254

## CHROMATOGRAM

**Internal standard:** thiosalicylic acid-p-bromophenacyl bromide adduct (Prepare by dissolving 2.4 mmoles thiosalicylic acid and 2.4 mmoles p-bromophenacyl bromide in 40 mL MeOH, adjust to pH 7 by the dropwise addition of 1 M NaOH, allow to stand at room temperature for 10 min, evaporate to dryness under reduced pressure, reconstitute with 40 mL 50 mM pH 7.0 phosphate buffer, wash twice with 20 mL portions of hexane, adjust pH to 2 with dilute HCl, extract with 40 mL ethyl acetate, evaporate to dryness under reduced pressure, recrystallize the residue from benzene to give the adduct as pale yellow plates.)

**Limit of quantitation:** 5 ng/mL

## KEY WORDS

derivatization; whole blood

## REFERENCE

Kawahara, Y.; Hisaoka, M.; Yamazaki, Y.; Inage, A.; Morioka, T. Determination of captopril in blood and urine by high-performance liquid chromatography. *Chem.Pharm.Bull.*, **1981**, *29*, 150–157

## SAMPLE

**Matrix:** blood, CSF

**Sample preparation:** 1 mL blood + 50 μL of a solution containing 100 mM EDTA and 100 mM ascorbic acid, centrifuge at 13000 g for 2 min . Proceed immediately. 500 μL Plasma or CSF + 50 μL 1 mg/mL p-bromophenacyl bromide in acetonitrile, vortex for 15 s, leave at room temperature for 30 min, add 100 μL 1 M HCl, add 50 μL 0.02% phenylacetic acid in methanol, add 5 mL ethyl acetate:benzene 50:50 (Caution! Benzene is a carcinogen!), vortex for 3 min, shake gently for 15 min. Saturate the aqueous phase with NaCl, centrifuge. Remove the organic layer and evaporate it to dryness under a stream of nitrogen at 50°, reconstitute in 250 μL MeCN, inject a 20 μL aliquot.

## HPLC VARIABLES

**Column:** 300 × 4 5 μm Lichrosorb RP-18
**Mobile phase:** MeCN:water:acetic acid 220:180:2.5
**Flow rate:** 1; increase to 3 after captopril elutes
**Injection volume:** 20
**Detector:** UV 260

## CHROMATOGRAM

**Retention time:** 4.7
**Internal standard:** phenylacetic acid (3.2)
**Limit of detection:** 5 ng/mL

## KEY WORDS
plasma; with modification can be used to determine captopril disulfides; derivatization

## REFERENCE
Colin, P.; Scherer, E. Simple high-performance liquid chromatography determination of captopril in human plasma and cerebrospinal fluid. *J.Liq.Chromatogr.*, **1989**, *12*, 629–643

## SAMPLE
**Matrix:** blood, urine
**Sample preparation:** Whole blood. 3 mL Whole blood + 100 µL 100 mM EDTA + 100 µL 200 mM ascorbic acid + 2 mL 1 M pH 8.2 Tris buffer + 200 µL 3 µg/mL IS + 100 µL 20 µg/mL 1-benzyl-2-chloropyridinium bromide, vortex for 15 min, centrifuge at 3000 g for 10 min. Remove a 1 mL aliquot of the supernatant and add it to 400 µL 3 M perchloric acid, centrifuge for 15 min, rinse the precipitate with 500 µL portions of water. Combine the supernatant and the rinses and adjust the pH to 2.5-3.0 (indicator paper) with 100 mM NaOH, add to a conditioned Bakerbond C18 SPE cartridge, wash with 1 mL water, dry under vacuum suction for 10 min, elute with 200 µL MeOH:acetic acid 80:20, elute with two 200 µL portions of MeOH:water 80:20. Combine the eluates and evaporate them to dryness at 60°, reconstitute with 50 µL water, inject a 20 µL aliquot. Urine. 500 µL Urine + 100 µL 200 mM EDTA + 100 µL 200 mM ascorbic acid + 3 mL 1 M pH 8.2 Tris buffer + 200 µL 3 µg/mL IS + 100 µL 20 µg/mL 1-benzyl-2-chloropyridinium bromide, vortex for 15 min, adjust the pH to 2.5-3.0 with 4 M phosphoric acid, add to the SPE cartridge, wash with 1 mL water, dry under vacuum suction for 10 min, elute with 200 µL MeOH:acetic acid 80:20, elute with two 200 µL portions of MeOH:water 80:20. Combine the eluates and evaporate them to dryness at 60°, reconstitute with 50 µL water, inject a 20 µL aliquot.

## HPLC VARIABLES
**Guard column:** 20 × 2.1 5 µm Hypersil
**Column:** 150 × 3.3 7 µm Separon SGX (Struzeni, Prague)
**Mobile phase:** Gradient. A was MeCN:100 mM pH 2.5 citric acid buffer containing 20 mM sodium octanesulfonate 25:75. B was MeCN:MeOH 50:50. A:B 100:0 for 10 min, to 80:20 over 10 min, maintain at 80:20 for 5 min, to 60:40 over 5 min, return to initial conditions over 7 min.
**Column temperature:** 50
**Flow rate:** 0.5
**Injection volume:** 20
**Detector:** UV 314

## CHROMATOGRAM
**Retention time:** 23
**Internal standard:** 1-benzyl-2-chloro-4-methylpyridinium bromide-captopril adduct (27)
**Limit of detection:** 0.3 ng/mL
**Limit of quantitation:** 10 ng/mL

## KEY WORDS
derivatization; whole blood; SPE

## REFERENCE
Sypniewski, S.; Bald, E. Determination of captopril and its disulphides in whole human blood and urine by high-performance liquid chromatography with ultraviolet detection and precolumn derivatization. *J.Chromatogr.A*, **1996**, *729*, 335–340

## SAMPLE
**Matrix:** blood, urine

**Sample preparation:** Plasma. 1.5 mL Blood + 50 μL 100 mM ascorbic acid and 100 mM disodium ethylenediaminetetraacetate, centrifuge. Remove 0.5 mL plasma and immediately add it to 50 μL 1 mg/mL p-bromophenacyl bromide, vortex 30 s, let stand at room temperature for 20 min, add 300 μL 6% perchloric acid, vortex for 30 s, centrifuge at 10000 g for 10 min, inject a 500 μL aliquot onto column A with mobile phase A and elute to waste, after 3 min elute the contents of column A onto column B with mobile phase B, after 2 min remove column A, elute column B with mobile phase B, monitor the effluent from column B. Urine. 50 μL Urine + 50 μL 1 mg/mL p-bromophenacyl bromide + 600 μL water, vortex 30 s, let stand at room temperature for 20 min, inject a 500 μL aliquot onto column A with mobile phase A and elute to waste, after 3 min elute the contents of column A onto column B with mobile phase B, after 2 min remove column A, elute column B with mobile phase B, monitor the effluent from column B. (Column A should be washed with MeOH for 2 min then re-equilibrated with mobile phase A for 2 min.)

## HPLC VARIABLES
**Column:** A 50 × 5 37-50 μm μBondapak C18; B 150 × 5 5 μm Tianjing silica gel YWG-C18
**Mobile phase:** A 0.2% Acetic acid in water; B MeCN:water:acetic acid 35:65:0.4
**Flow rate:** A 3; B 2
**Injection volume:** 500
**Detector:** UV 260

## CHROMATOGRAM
**Retention time:** 9
**Limit of detection:** 10 ng/mL

## KEY WORDS
plasma; column-switching; pharmacokinetics

## REFERENCE
Gao, S.; Tian, W.; Wang, S. Simple high-performance liquid chromatographic method for the determination of captopril in biological fluids. *J.Chromatogr.*, **1992**, *582*, 258–262

## SAMPLE
**Matrix:** blood, urine
**Sample preparation:** Plasma. 1 mL Plasma + 2 mL 100 mM pH 6.0 phosphate buffer + 500 μL 0.5% N-(4-benzoylphenyl)maleimide in acetone, vortex for 15 s, let stand at room temperature for 10 min, add 2 mL 500 mM pH 7.0 phosphate buffer, add 100 μL 40 μg/mL IS1 in acetone, wash twice with 4 mL portions of ether, acidify the aqueous phase with 500 μL 6 M HCl, extract with 7 mL chloroform. Remove the organic layer and evaporate it to dryness, reconstitute the residue in 100 μL MeOH, inject a 20 μL aliquot. Urine. 200 μL Urine + 200 μL 0.5% N-(4-benzoylphenyl)maleimide in acetone + 200 μL 100 mM pH 6.5 phosphate buffer, mix, let stand at room temperature for 15 min, add 2.5 mL 500 mM pH 7.0 phosphate buffer, wash with 4 mL diethyl ether, add 100 μL 10 μg/mL IS2 in acetone to the aqueous phase, acidify with 500 μL 6 M HCl, extract with 6 mL chloroform. Remove the organic layer and evaporate it to dryness, reconstitute the residue in 200 μL MeCN, inject a 20 μL aliquot. (Prepare N-(4-benzoylphenyl)maleimide by adding 5.3 g maleic anhydride to 9.6 g 4-aminobenzophenone in dioxane (Caution! Dioxane is a carcinogen!), stir at room temperature (Japan Pat. 59,204,171 (19 Nov. 1984); Chem. Abstr. 1985, 102, 113288t).)

## HPLC VARIABLES
**Column:** 300 × 4 10 μm μBondapak C18
**Mobile phase:** MeCN:MeOH:1% acetic acid 45:11:75 (plasma) or 42.5:8.2:47.3 (urine)
**Flow rate:** 1
**Injection volume:** 20
**Detector:** UV 254

## CHROMATOGRAM
**Retention time:** 21 (plasma), 10 (urine)
**Internal standard:** adduct of N-(4-benzoylphenyl)maleimide with (4R)-2-(2-hydroxy-phenyl)-3-(3-mercaptopropionyl)-4-thiazolidinecarboxylic acid (IS1) 30 min), adduct of N-(4-benzoylphenyl)maleimide with thiosalicylic acid (IS2) (14 min) (Prepare adducts as follows. Add 150 mg N-(4-benzoylphenyl)maleimide in 2 mL acetone to 500 µmoles compound in 2 mL water, add 1 drop triethylamine, let stand at room temperature for 15 min, evaporate to dryness under reduced pressure.)
**Limit of detection:** 50 ng/mL

## KEY WORDS
derivatization; plasma; pharmacokinetics

## REFERENCE
Hayashi, K.; Miyamoto, M.; Sekine, Y. Determination of captopril and its mixed disulfides in plasma and urine by high-performance liquid chromatography. *J.Chromatogr.*, **1985**, *338*, 161–169

## SAMPLE
**Matrix:** formulations
**Sample preparation:** Dilute 5 fold with mobile phase. Mix the diluted formulation with an equal volume of 50 µg/mL hydrochlorothiazide, inject a 10 µL aliquot.

## HPLC VARIABLES
**Column:** 300 × 3.9 10 µm µBondapak C18
**Mobile phase:** MeOH:0.1% phosphoric acid 55:45
**Flow rate:** 1
**Injection volume:** 10
**Detector:** UV 260

## CHROMATOGRAM
**Retention time:** 7.1
**Internal standard:** hydrochlorothiazide (5.6)

## OTHER SUBSTANCES
**Simultaneous:** degradation products

## KEY WORDS
stability-indicating; syrup; water

## REFERENCE
Nahata, M.C.; Morosco, R.S.; Hipple, T.F. Stability of captopril in three liquid dosage forms. *Am.J.Hosp.Pharm.*, **1994**, *51*, 95–96

## SAMPLE
**Matrix:** intestinal mucosal homogenate
**Sample preparation:** 400 µL Homogenate mixture + 400 µL 1 M HCl, mix, centrifuge at 4° at 34000 g for 10 min, filter (0.45 µm) the supernatant, inject an aliquot of the filtrate.

## HPLC VARIABLES
**Guard column:** 20 mm long Supelguard LC-18S (Supelco)
**Column:** 250 × 4.6 Suplecosil LC-18S
**Mobile phase:** MeOH:water:85% phosphoric acid 54.97:44.98:0.05
**Flow rate:** 1
**Detector:** UV 210

## KEY WORDS
rat

## REFERENCE
Sinko, P.J.; Hu, P. Determining intestinal metabolism and permeability for several compounds in rats. Implications on regional bioavailability in humans. *Pharm.Res.*, **1996**, *13*, 108−113

## SAMPLE
**Matrix:** perfusate
**Sample preparation:** Centrifuge perfusate, add 100 μL supernatant to 50 μL 2 mM N-(1-pyrenyl)maleimide in acetone, add mixture to 2 mL pH 7 phosphate buffer, stir for 15 min at room temperature, inject an aliquot.

## HPLC VARIABLES
**Column:** 250 × 4.6 Nucleosil 5C18
**Mobile phase:** MeCN:0.1% phosphoric acid 47:53
**Detector:** F ex 340 em 390

## KEY WORDS
derivatization

## REFERENCE
Kobayashi, D.; Matsuzawa, T.; Sugibayashi, K.; Morimoto, Y.; Kobayashi, M.; Kimura, M. Feasibility of use of several cardiovascular agents in transdermal therapeutic systems with *l*-menthol-ethanol system on hairless rat and human skin. *Biol.Pharm.Bull.*, **1993**, *16*, 254−258

## SAMPLE
**Matrix:** solutions

## HPLC VARIABLES
**Column:** 250 × 4.6 Spherisorb 5 ODS-2
**Mobile phase:** n-Propanol:buffer 20:80 (Buffer was pH 3.0 phosphate buffer containing 0.4% triethylamine.)
**Flow rate:** 1
**Detector:** UV 240

## CHROMATOGRAM
**Retention time:** 4

## OTHER SUBSTANCES
**Simultaneous:** benzepril, cilazapril, enalapril, quinapril, ramipril

## REFERENCE
Barbato, F.; Morrica, P.; Quaglia, F. Analysis of ACE inhibitor drugs by high performance liquid chromatography. *Farmaco*, **1994**, *49*, 457−460

## SAMPLE
**Matrix:** solutions

## HPLC VARIABLES
**Column:** 150 × 3.9 μBondapak phenyl
**Mobile phase:** MeOH:water:85% phosphoric acid 45:55:0.05
**Column temperature:** 30-40
**Detector:** UV 215-220

## REFERENCE

Ranadive, S.A.; Chen, A.X.; Serajuddin, A.T. Relative lipophilicities and structural-pharmacological considerations of various angiotensin-converting enzyme (ACE) inhibitors. *Pharm.Res.*, **1992**, *9*, 1480–1486

## SAMPLE

**Matrix:** urine
**Sample preparation:** 5 mL Urine + 2 mL 500 mM pH 7.0 phosphate buffer + 0.5 mL 20 mg/mL p-bromophenacyl bromide in MeOH, shake vigorously. Remove a 1 mL aliquot and add it to 1.5 mL water and 6 mL hexane, mix, discard the hexane layer. Remove a 2 mL aliquot of the aqueous layer and add it to 100 μL 2% tributylphosphine in MeOH, heat at 50°for 30 min, wash with 6 mL hexane, add 200 μL 0.2% N-(4-dimethylamino-3,5-dinitrophenyl)maleimide in acetone to the aqueous layer, mix, let stand at room temperature for 5 min, wash with 6 mL hexane, discard the hexane layer, acidify the aqueous layer with about 200 μL 2 M HCl, extract twice with 6 mL portions of benzene (Caution! Benzene is a carcinogen!). Combine the organic layers and add 10 μg IS, evaporate to dryness under reduced pressure, reconstitute the residue in 200 μL MeOH, inject a 5-20 μL aliquot. (Free captopril is derivatized as its p-bromophenacyl bromide adduct then oxidized captopril is reduced and derivatized as its N-(4-dimethylamino-3,5-dinitrophenyl)maleimide adduct.)

## HPLC VARIABLES

**Column:** μBondapak C18
**Mobile phase:** MeCN:water:acetic acid 46.5:53:0.5
**Injection volume:** 5-20
**Detector:** UV 254

## CHROMATOGRAM

**Retention time:** 7 (free captopril (as p-bromophenacyl bromide adduct)), 8 (oxidized captopril (as N-(4-dimethylamino-3,5-dinitrophenyl)maleimide adduct))
**Internal standard:** thiosalicylic acid-p-bromophenacyl bromide adduct (Prepare by dissolving 2.4 mmoles thiosalicylic acid and 2.4 mmoles p-bromophenacyl bromide in 40 mL MeOH, adjust to pH 7 by the dropwise addition of 1 M NaOH, allow to stand at room temperature for 10 min, evaporate to dryness under reduced pressure, reconstitute with 40 mL 50 mM pH 7.0 phosphate buffer, wash twice with 20 mL portions of hexane, adjust pH to 2 with dilute HCl, extract with 40 mL ethyl acetate, evaporate to dryness under reduced pressure, recrystallize the residue from benzene to give the adduct as pale yellow plates.) (12.5)
**Limit of quantitation:** 100 ng/mL

## KEY WORDS

derivatization

## REFERENCE

Kawahara, Y.; Hisaoka, M.; Yamazaki, Y.; Inage, A.; Morioka, T. Determination of captopril in blood and urine by high-performance liquid chromatography. *Chem.Pharm.Bull.*, **1981**, *29*, 150–157

◆━━━━━━━━━━━━━━━━◆━━━━━━━━━━━━━━━━◆

## ANNOTATED BIBLIOGRAPHY

Chan, D.S.; Sato, A.K.; Claybaugh, J.R. Degradation of captopril in solutions compounded from tablets and standard powder. *Am.J.Hosp.Pharm.*, **1994**, *51*, 1205–1207 [stability-indicating; tablets; powder]

Wakabayashi, H.; Yamato, S.; Nakajima, M.; Shimada, K. Application of an electrochemical detector with a graphite electrode to liquid chromatographic determination of penicillamine and captopril in biological samples. *J.Pharm.Biomed.Anal.*, **1994**, *12*, 1147–1152 [rat; serum; liver; kidney; SPE; electrochemical detection; LOD 20-300 pg; extracted penicillamine; homocysteine (IS)]

Liu, C.; Chen, G. Quantitation determination of captopril intravenous injection by HPLC. *Zhongguo Yiyuan Yaoxue Zazhi*, **1992**, *12*, 170–171

Tian, W.R.; Gao, S.; Wang, S.X. Determination of captopril in plasma and urine by high-performance liquid chromatography with column switching. *Yaoxue Xuebao*, **1992**, *27*, 613–617

Jain, R.; Jain, C.L. Simultaneous quantification of captopril and hydrochlorothiazide using high performance liquid chromatography. *Indian Drugs*, **1991**, *28*, 380–382

Tan, H.; Zhu, D.; Tang, S. HPLC determination in dissolution of captopril tablets. *Zhongguo Yaoxue Zazhi*, **1991**, *26*, 546–548

Klein, J.; Colin, P.; Scherer, E.; Levy, M.; Koren, G. Simple measurement of captopril in plasma by high-performance liquid chromatography with ultraviolet detection. *Ther.Drug Monit.*, **1990**, *12*, 105–110

Arzamastsev, A.P.; Volchenok, V.I.; Ordabaeva, S.K.; Ryzhenkova, A..P.; Nasyrov, S.N.; Shvarts, G.Y. HPLC determination of captopril in biological fluids. *Khim.-Farm.Zh.*, **1989**, *23*, 1404–1406

Hu, M.; Amidon, G.L. Passive and carrier-mediated intestinal absorption components of captopril. *J.Pharm.Sci.*, **1988**, *77*, 1007–1011 [rat; perfusate; simultaneous captopril disulfide, cephradine]

Xu, Z. Determination of captopril by HPLC. *Yiyao Gongye*, **1985**, *16*, 536–538

Kirschbaum, J.; Perlman, S. Analysis of captopril and hydrochlorothiazide combination tablet formulations by liquid chromatography. *J.Pharm.Sci.*, **1984**, *73*, 686–687

Perrett, D.; Rudge, S.R.; Drury, P.L. Determination of captopril by an improved high-performance liquid chromatography-electrochemical assay. *Biochem.Soc.Trans.*, **1984**, *12*, 1059–1060

Toyooka, T.; Imai, K.; Kawahara, Y. Determination of total captopril in dog plasma by HPLC after prelabeling with ammonium 7-fluorobenzo-2-oxa-1,3-diazole-4-sulfonate (SBD-F). *J.Pharm. Biomed.Anal.*, **1984**, *2*, 473–479 [derivatization]

Perrett, D.; Drury, P.L. The determination of captopril in physiological fluids using high-performance liquid chromatography with electrochemical detection. *J.Liq.Chromatogr.*, **1982**, *5*, 97–110 [LOD 1 pmole; electrochemical detection; plasma; urine]

# Carbamazepine

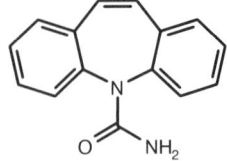

**Molecular formula:** $C_{15}H_{12}N_2O$
**Molecular weight:** 236.3
**CAS Registry No.:** 298-46-4

## SAMPLE
**Matrix:** blood
**Sample preparation:** Add two volumes of MeCN to the mouse serum, mix, centrifuge at 1500 g for 5 min, inject a 5 μL aliquot of the supernatant.

## HPLC VARIABLES
**Guard column:** Sentry (Waters)
**Column:** 150 × 4.6 Nova-Pak C18
**Mobile phase:** MeCN:MeOH:10 mM pH 7.0 phosphate buffer 50:30:110
**Column temperature:** 40
**Flow rate:** 0.5
**Injection volume:** 5
**Detector:** UV 214

## CHROMATOGRAM
**Retention time:** 11

## OTHER SUBSTANCES
**Extracted:** carbamazepine-10,11-epoxide, phenobarbital, phenylethyl malonamide, phenytoin, primidone

## KEY WORDS
serum; mouse

## REFERENCE
Capparella, M.; Foster, W., III; Larrousse, M.; Phillips, D.J.; Pomfret, A.; Tuvim, Y. Characteristics and applications of a new high-performance liquid chromatography guard column. *J.Chromatogr.A*, **1995**, *691*, 141–150

## SAMPLE
**Matrix:** blood
**Sample preparation:** 200 μL Plasma + 750 ng carbamazepine-10-hydroxide + 200 μL 1.5 M NaOH + 2 mL ethyl acetate:chloroform 50:50, extract. Remove the organic layer and evaporate it to dryness under a stream of air, reconstitute the residue in 100 μL mobile phase and 100 μL hexane, inject a 20 μL aliquot of the lower aqueous phase.

## HPLC VARIABLES
**Guard column:** 50 × 4 5 μm Lichrospher 100 RP-18
**Column:** 125 × 4 5 μm Lichrocart C18
**Mobile phase:** MeCN:water 30:70
**Flow rate:** 1
**Injection volume:** 20
**Detector:** UV 220

## CHROMATOGRAM
**Internal standard:** carbamazepine-10-hydroxide

## OTHER SUBSTANCES
**Extracted:** carbamazepine-10,11-epoxide, metabolites

## KEY WORDS
plasma

## REFERENCE
Lanchote, V.L.; Bonato, P.S.; Campos, G.M.; Rodrigues, I. Factors influencing plasma concentrations of carbamazepine and carbamazepine-10,11-epoxide in epileptic children and adults. *Ther.Drug Monit.*, **1995**, *17*, 47−52

## SAMPLE
**Matrix:** blood
**Sample preparation:** Filter (0.22 μm), inject a 5 μL aliquot of the filtrate.

## HPLC VARIABLES
**Column:** 250 × 4 5 μm LiChrospher 100 Diol
**Mobile phase:** MeCN:50 mM pH 6.9 phosphate buffer 12:88
**Flow rate:** 0.6
**Injection volume:** 5
**Detector:** UV 254

## CHROMATOGRAM
**Retention time:** 9.5

## OTHER SUBSTANCES
**Extracted:** phenobarbital, phenytoin

## KEY WORDS
serum; direct injection

## REFERENCE
Nimura, N.; Itoh, H.; Kinoshita, T. Diol-bonded silica gel as a restricted access packing forming a binary-layered phase for direct injection of serum for the determination of drugs. *J.Chromatogr.A*, **1995**, *689*, 203−210

## SAMPLE
**Matrix:** blood
**Sample preparation:** Condition an Extrelut-1 glass SPE cartridge with 5 mL dichloromethane:isopropanol 90:10, dry under nitrogen. 1 mL Serum + 100 μL EtOH, vortex for 30 s, add to the SPE cartridge, let stand for 10 min, elute with 5 mL dichloromethane:isopropanol 90:10. Evaporate the eluate to dryness under a stream of nitrogen, reconstitute the residue in 100 μL mobile phase, inject a 20 μL aliquot.

## HPLC VARIABLES
**Column:** 250 × 4.6 10 μm Chiralcel OD + 250 × 4.6 10 μm Chiralcel ODH
**Mobile phase:** n-Hexane:EtOH 70:30
**Column temperature:** 40 (2nd column)
**Flow rate:** 0.9
**Injection volume:** 20
**Detector:** UV 220

## CHROMATOGRAM
**Retention time:** 21.9
**Internal standard:** carbamazepine

## OTHER SUBSTANCES
**Extracted:** oxcarbazepine
**Simultaneous:** phenobarbital
**Noninterfering:** phenytoin, valproic acid

## KEY WORDS
serum; SPE; chiral (for oxcarbazepine metabolites); carbamazepine is IS

## REFERENCE
Pichini, S.; Altieri, I.; Passa, A.R.; Zuccaro, P.; Pacifici, R. Stereoselective bioanalysis of oxcarbazepine and the enantiomers of its metabolites by high-performance liquid chromatography. *J.Liq. Chromatogr.*, **1995**, *18*, 1533–1541

## SAMPLE
**Matrix:** blood
**Sample preparation:** 500 μL Serum + 6 mL MTBE, vortex for 30 s, shake for 5 min, centrifuge at 800 g for 5 min. Remove the organic layer and evaporate it to dryness under a stream of nitrogen in a warm water bath, reconstitute the residue in 40 μL MeOH: water 5:2, inject a 20 μL aliquot.

## HPLC VARIABLES
**Guard column:** 4 × 4 5 μm LiChrospher 100 RP-18
**Column:** 125 × 4 4 μm Superspher 60 RP-select B (Merck)
**Mobile phase:** MeCN:20 mM $KH_2PO_4$ 20:80 containing 0.05% triethylamine, pH 6.30
**Flow rate:** 1
**Injection volume:** 20
**Detector:** UV 212

## CHROMATOGRAM
**Retention time:** 27.43
**Limit of detection:** 10 ng/mL
**Limit of quantitation:** 39 ng/mL

## OTHER SUBSTANCES
**Extracted:** metabolites, oxcarbazepine

## KEY WORDS
serum

## REFERENCE
Pienimäki, P.; Fuchs, S.; Isojärvi, J.; Vähäkangas, K. Improved detection and determination of carbamazepine and oxcarbazepine and their metabolites by high-performance liquid chromatography. *J.Chromatogr.B*, **1995**, *673*, 97–105

## SAMPLE
**Matrix:** blood
**Sample preparation:** 50 μL Plasma + 250 μL 3 μg/mL methyl p-hydroxybenzoate in MeCN, mix, centrifuge at 10000 rpm for 5 min, inject a 10 μL aliquot of the supernatant.

## HPLC VARIABLES
**Column:** 150 × 16 5 μm Cosmosil 5C8 (Shimadzu)
**Mobile phase:** MeCN:5 mM $KH_2PO_4$ 31:69
**Flow rate:** 1
**Injection volume:** 10
**Detector:** UV 210

## CHROMATOGRAM
**Internal standard:** methyl p-hydroxybenzoate
**Limit of detection:** 500 ng/mL

## OTHER SUBSTANCES
**Extracted:** metabolites

## KEY WORDS
plasma; rat; pharmacokinetics

## REFERENCE
Shinoda, M.; Akita, M.; Hasegawa, M.; Nadai, M.; Hasegawa, T.; Nabeshima, T. Pharmaceutical evaluation of carbamazepine suppositories in rats. *Biol.Pharm.Bull.*, **1995**, *18*, 1289–1291

## SAMPLE
**Matrix:** blood
**Sample preparation:** 50 μL Plasma + 100 μL MeCN, centrifuge, inject a 5 μL aliquot of the supernatant.

## HPLC VARIABLES
**Column:** 100 × 4.6 3.5 μm Zorbax SB
**Mobile phase:** MeCN:MeOH:10 mM pH 7.1 phosphate buffer 7:34:59
**Flow rate:** 1.5
**Injection volume:** 5
**Detector:** UV 220

## CHROMATOGRAM
**Retention time:** 7.4
**Limit of detection:** <1 μM

## OTHER SUBSTANCES
**Extracted:** metabolites, carbamazepine epoxide, hydroxycarbamazepine, lamotrigine (UV 310), oxcarbazepine, phenobarbital, phenytoin
**Also analyzed:** ibuprofen, naproxen, trimethoprim

## KEY WORDS
plasma

## REFERENCE
Svensson, J.O. Simple HPLC method for determination of antiepileptic drugs in plasma (Abstract 102). *Ther.Drug Monit.*, **1995**, *17*, 408

## SAMPLE
**Matrix:** blood
**Sample preparation:** 2 mL Whole blood or plasma + 2 mL buffer + 5 mL chloroform: isopropanol:n-heptane 60:14:26, shake gently horizontally for 10 min, centrifuge at 2800 g for 10 min. Remove the lower organic layer and evaporate it to dryness under vacuum at 45°, reconstitute the residue in 100 μL mobile phase, centrifuge at 2800 g for 5 min, inject a 50 μL aliquot of the supernatant. (Buffer was saturated ammonium chloride solution 25% diluted with water, adjusted to pH 9.5 with 25% ammonia solution.)

## HPLC VARIABLES
**Column:** 300 × 3.9 4 μm NovaPack C18
**Mobile phase:** MeOH:THF:buffer 65:5:30 (Buffer was 0.68 g/L (10 mM (sic)) $KH_2PO_4$ adjusted to pH 2.6 with concentrated orthophosphoric acid.) (At the end of each session wash the column with water for 1 h and MeOH for 1 h, re-equilibrate for 30 min.)

**Column temperature:** 30
**Flow rate:** 0.8
**Injection volume:** 50
**Detector:** UV 283

## CHROMATOGRAM
**Retention time:** 3.67
**Limit of detection:** <120 ng/mL

## OTHER SUBSTANCES
**Extracted:** acebutolol, acenocoumarol, acepromazine, aceprometazine, acetaminophen, aconitine, ajmaline, albuterol, alimemazine, alminoprofen, alpidem, alprazolam, alprenolol, amisulpride, amitriptyline, amodiaquine, amoxapine, aspirin, astemizole, atenolol, benazepril, benperidol, benzocaine, benzoylecgonine, bepridil, betaxolol, bisoprolol, brompheniramine, bumadizone, bupivacaine, buprenorphine, buspirone, caffeine, carbinoxamine, carpipramine, celiprolol, cetirizine, chlorambucil, chlordiazepoxide, chlormezanone, chlorophenacinone, chloroquine, chlorpheniramine, chlorpromazine, chlorpropamide, cibenzoline, cicletanine, clemastine, clobazam, clomipramine, clonazepam, clonidine, clorazepate, clozapine, cocaine, codeine, colchicine, cyamemazine, cyclizine, cycloguanil, cyproheptadine, cytarabine, dacarbazine, daunorubicin, debrisoquine, demexiptiline, desipramine, dextromethorphan, dextromoramide, dextropropoxyphene, diazepam, diazoxide, diclofenac, diltiazem, diphenhydramine, dipyridamole, disopyramide, dosulepine, doxepin, doxylamine, droperidol, ephedrine, estazolam, etodolac, fenfluramine, fenoprofen, fentiazac, flecainide, floctafenine, flumazenil, flunitrazepam, fluoxetine, fluphenazine, flurbiprofen, fluvoxamine, glibenclamide, glibornuride, glipizide, glutethimide, haloperidol, histapyrrodine, hydroxychloroquine, hydroxyzine, ibuprofen, imipramine, indomethacin, iproniazid, ketamine, ketoprofen, labetalol, levomepromazine, lidocaine, lidoflazine, lisinopril, loperamide, loprazolam, loratadine, lorazepam, loxapine, maprotiline, medazepam, medifoxamine, mefenamic acid, mefenidramine, mefloquine, melphalan, meperidine, mephenesin, mephentermine, mepivacaine, metapramine, metformin, methadone, methaqualone, methocarbamol, methotrexate, metipranolol, metoclopramide, metoprolol, mexiletine, mianserine, midazolam, minoxidil, moclobemide, moperone, morphine, nalorphine, naloxone, naltrexone, naproxen, nialamide, nicardipine, nifedipine, niflumic acid, nimodipine, nitrazepam, nitrendipine, nizatidine, nomifensine, nortriptyline, opipramol, oxazepam, oxprenolol, penbutolol, penfluridol, pentazocine, phencyclidine, phenobarbital, phenol, phenylbutazone, pimozide, pindolol, pipamperone, piroxicam, prazepam, prazosin, prilocaine, proguanil, promethazine, propafenone, propranolol, protriptyline, pyrimethamine, quinidine, quinine, quinupramine, ramipril, ranitidine, reserpine, ritodrine, secobarbital, sulfinpyrazole, sulindac, sulpride, sultopride, suriclone, temazepam, tenoxicam, terbutaline, terfenadine, tetracaine, tetrazepam, thiopental, thioproperazine, thioridazine, tianeptine, tiapride, tiaprofenic acid, ticlopidine, timolol, tioclomarol, tofisopam, tolbutamide, toloxatone, trazodone, triazolam, trifluoperazine, trifluperidol, trimipramine, triprolidine, tropatenine, verapamil, viloxazine, vinblastine, vincristine, vindesine, warfarin, yohimbine, zolpidem, zopiclone, zorubicine
**Interfering:** bromazepam, carbamazepine, carteolol, dihydralazine, nadolol, nalbuphine, omeprazole, procainamide, procarbazine, sotalol, strychnine

## KEY WORDS
whole blood; plasma

## REFERENCE
Tracqui, A.; Kintz, P.; Mangin, P. Systematic toxicological analysis using HPLC/DAD. *J.Forensic Sci.*, **1995**, *40*, 254–262

## SAMPLE
**Matrix:** blood
**Sample preparation:** 500 μL Serum + 600 μL allobarbital in 75 mM pH 6.8 phosphate buffer, add 200 units β-glucuronidase, heat at 37° for 30 min, add 1 mL of this solution

to an Extrelut-1 SPE cartridge, let stand for 10 min, elute with 2.5 mL MTBE. Evaporate the eluate to dryness under a stream of nitrogen, reconstitute the residue in 50 μL MeOH:water 50:50, inject a 10 μL aliquot onto columns A and B in series with mobile phase A. After 12 min elute column A with mobile phase B, continue to elute column B with mobile phase A. Carbamazepine diol, carbamazepine epoxide, phenytoin, and carbamazepine elute from column A and the enantiomers of 5-(p-hydroxyphenyl)-5-phenylhydantoin and mephobarbital, phenobarbital, zonisamide, and allobarbital elute from column B. Re-equilibrate columns A and B with mobile phase A for 5 min before the next injection.

## HPLC VARIABLES
**Column:** A 250 × 4 4 μm Superspher RP-18e (E. Merck); B 250 × 4 4 μm Superspher RP-18e (E. Merck)
**Mobile phase:** A MeOH:11.2 mM β-cyclodextrin in 20 mM $KH_2PO_4$ 5:95; B MeCN:20 mM $KH_2PO_4$ 16:84
**Flow rate:** 0.8
**Injection volume:** 10
**Detector:** UV 210 (A); UV 210 (B)

## CHROMATOGRAM
**Retention time:** 24 (column A)
**Internal standard:** allobarbital
**Limit of detection:** 2.2 ng/mL

## OTHER SUBSTANCES
**Extracted:** metabolites, carbamazepine diol, carbamazepine epoxide, 5-(p-hydroxyphenyl)-5-phenylhydantoin, mephobarbital, phenobarbital, phenytoin, zonisamide

## KEY WORDS
serum; column-switching; SPE; chiral

## REFERENCE
Eto, S.; Noda, H.; Noda, A. Simultaneous determination of antiepileptic drugs and their metabolites, including chiral compounds, via β-cyclodextrin inclusion complexes by a column-switching chromatographic technique. *J.Chromatogr.B*, **1994**, *658*, 385–390

## SAMPLE
**Matrix:** blood
**Sample preparation:** Inject sample onto column A with mobile phase A and elute for 3 min. Backflush contents of column A onto column B with mobile phase B for 6 min and elute column B with mobile phase B and monitor eluant.

## HPLC VARIABLES
**Column:** A 10 × 3 BioTrap Acid C18 (ChromTech); B 10 × 3 CT-sil C18 guard column + 150 × 4.6 5 μm CT-sil C18 (ChromTech)
**Mobile phase:** A 82 mM pH 6.0 phosphate buffer; B MeCN:82 mM pH 6.0 phosphate buffer 50:50
**Flow rate:** A 0.55; B 1
**Injection volume:** 50
**Detector:** UV 254

## CHROMATOGRAM
**Retention time:** 13
**Limit of quantitation:** 2050 ng/mL

## OTHER SUBSTANCES
**Simultaneous:** phenytoin

## KEY WORDS
plasma; column-switching; direct injection

## REFERENCE
Hermansson, J.; Grahn, A. Determination of drugs by direct injection of plasma into a biocompatible extraction column based on a protein-entrapped hydrophobic phase. *J.Chromatogr.A*, **1994**, *660*, 119–129

## SAMPLE
**Matrix:** blood
**Sample preparation:** Condition an Extrashot-Silica (diatomaceous earth) SPE cartridge (Kusano Scientific) with 200 μL EtOH and 200 μL dichloromethane, force out the remaining solvent with 500 μL air. Add 5 μL serum to the surface of the cartridge and pass 130 μL dichloromethane gently through the cartridge into the 100 μL sample loop.

## HPLC VARIABLES
**Column:** 125 × 4 5 μm LiChrosorb Si60
**Mobile phase:** n-Hexane:dichloromethane:EtOH:acetic acid 82.8:15:2:0.2
**Flow rate:** 1
**Injection volume:** 100
**Detector:** UV 240

## CHROMATOGRAM
**Retention time:** 14.7
**Limit of quantitation:** 1 μg/mL

## OTHER SUBSTANCES
**Extracted:** phenobarbital, phenytoin

## KEY WORDS
serum; normal phase; SPE

## REFERENCE
Kouno, Y.; Ishikura, C.; Homma, M.; Oka, K. Simple and accurate high-performance liquid chromatographic method for the measurement of three antiepileptics in therapeutic drug monitoring. *J.Chromatogr.*, **1993**, *622*, 47–52

## SAMPLE
**Matrix:** blood
**Sample preparation:** 400 μL Serum + 50 μL 100 μg/mL IS in water, mix, add 400 μL 1.5 M NaOH, add 100 mg NaCl, add 4 mL ethyl acetate:chloroform 50:50, shake for 15 min, centrifuge. Remove the organic layer and evaporate it to dryness, reconstitute the residue in 200 μL hexane and 200 μL mobile phase, mix, inject a 20 μL aliquot of the mobile phase layer.

## HPLC VARIABLES
**Guard column:** 10 × 4 5 μm Spherisorb ODS-2
**Column:** 250 × 4 5 μm Spherisorb ODS-2
**Mobile phase:** MeCN:water 30:70
**Flow rate:** 1.1
**Injection volume:** 20
**Detector:** UV 210

## CHROMATOGRAM
**Retention time:** 10.9

**Internal standard:** 1,3-dimethyl-7-benzylxanthine (8.2) (synthesized from theophylline, bromobenzene, and potassium carbonate in boiling benzene)
**Limit of detection:** 80 ng/mL
**Limit of quantitation:** 270 ng/mL

## OTHER SUBSTANCES
**Extracted:** metabolites, carbamazepine epoxide, phenytoin

## KEY WORDS
serum

## REFERENCE
Martens, J.; Banditt, P. Validation of the analysis of carbamazepine and its 10,11-epoxide metabolite by high-performance liquid chromatography from plasma: comparison with gas chromatography and the enzyme-multiplied immunoassay technique. *J.Chromatogr.*, **1993**, *620*, 169–173

## SAMPLE
**Matrix:** blood
**Sample preparation:** 100 μL Serum + 200 μL 20 μg/mL butalbital in MeCN, vortex, centrifuge 5 min, inject supernatant

## HPLC VARIABLES
**Column:** 125 × 4 LiChroSpher RP-8 5 μm
**Mobile phase:** MeCN:water:100 mM pH 7.0 phosphate buffer 20:75:5
**Column temperature:** 45
**Flow rate:** 2
**Injection volume:** 50
**Detector:** UV 212

## CHROMATOGRAM
**Retention time:** 9.4
**Internal standard:** butalbital (3.8)

## OTHER SUBSTANCES
**Simultaneous:** phenobarbital, phenytoin

## KEY WORDS
serum

## REFERENCE
Hannak, D.; Haux, P.; Scharbert, F.; Kattermann, R. Liquid chromatographic analysis of phenobarbital, phenytoin, and theophylline. *Wien.Klin.Wochenschr.Suppl.*, **1992**, *191*, 27–31

## SAMPLE
**Matrix:** blood
**Sample preparation:** Inject 10 μL serum onto column A with mobile phase A, after 2 min backflush contents of column A onto column B with mobile phase B, monitor the effluent from column B.

## HPLC VARIABLES
**Column:** A μBondapak phenyl Guard Pak; B 150 × 3.9 μBondapak phenyl
**Mobile phase:** A 5 mM sodium dodecyl sulfate in water; B MeOH:water 30:70 containing 50 mM sodium dodecyl sulfate
**Flow rate:** 2

**Injection volume:** 10
**Detector:** UV 214

---

## CHROMATOGRAM
**Retention time:** 9

---

## OTHER SUBSTANCES
**Extracted:** metabolites, carbamazepine epoxide

---

## KEY WORDS
serum; column-switching; cow

---

## REFERENCE
Bentrop, D.; Warren, F.V., Jr.; Schmitz, S.; Bidlingmeyer, B.A. Analysis of carbamazepine in serum by liquid chromatography with direct sample injection and surfactant-containing eluents. *J.Chromatogr.*, **1990**, *535*, 293–304

---

## SAMPLE
**Matrix:** blood
**Sample preparation:** 250 µL Plasma + 2 µg 10-methoxycarbamazepine + 25 µL 1 M NaOH + 1.2 mL dichloromethane, mix for 15 min, centrifuge. Remove the organic layer and evaporate it to dryness under a stream of nitrogen at 40°, reconstitute the residue in 20 µL MeCN, inject a 10 µL aliquot.

---

## HPLC VARIABLES
**Column:** 250 × 4.9 10 µm LiChrosorb RP8
**Mobile phase:** MeCN:water 32:68
**Flow rate:** 1.8
**Injection volume:** 10
**Detector:** UV 215

---

## CHROMATOGRAM
**Retention time:** 7.0
**Internal standard:** 10-methoxycarbamazepine (9.3)

---

## OTHER SUBSTANCES
**Extracted:** oxcarbazepine, phenobarbital, primidone
**Noninterfering:** clobazam, clonazepam, diazepam, ethosuximide, phenytoin, valproic acid

---

## KEY WORDS
plasma

---

## REFERENCE
Elyas, A.A.; Goldberg, V.D.; Patsalos, P.N. Simple and rapid micro-analytical high-performance liquid chromatographic technique for the assay of oxcarbazepine and its primary active metabolite 10-hydroxycarbazepine. *J.Chromatogr.*, **1990**, *528*, 473–479

---

## SAMPLE
**Matrix:** blood
**Sample preparation:** Inject 20 µL serum onto column A with mobile phase A and elute to waste, after 1.5 min backflush the contents of column A onto column B with mobile phase B, after 1 min remove column A from the circuit, elute column B with mobile phase B, monitor the effluent from column B. Re-equilibrate column A with mobile phase A.

## HPLC VARIABLES
**Column:** A 30 × 4.6 IRSP silica (for preparation see Anal. Chem. 1989, 61, 2445); B 150 × 4.6 5 µm Nucleosil C18
**Mobile phase:** A 14 mM $NaH_2PO_4$ containing 6 mM $Na_2HPO_4$; B MeCN:MeOH:14 mM $NaH_2PO_4$ containing 6 mM $Na_2HPO_4$ 15:20:65
**Flow rate:** 0.8
**Injection volume:** 20
**Detector:** UV 230

## CHROMATOGRAM
**Retention time:** 30
**Limit of detection:** 100 ng/mL

## OTHER SUBSTANCES
**Extracted:** phenobarbital, phenytoin, primidone

## KEY WORDS
serum; column-switching

## REFERENCE
Haginaka, J.; Wakai, J.; Yasuda, H.; Kimura, Y. Determination of anticonvulsant drugs and methyl xanthine derivatives in serum by liquid chromatography with direct injection: column-switching method using a new internal-surface reversed-phase silica support as a precolumn. *J.Chromatogr.*, **1990**, *529*, 455–461

## SAMPLE
**Matrix:** blood
**Sample preparation:** 500 µL Plasma + 500 µL 1 M pH 5.0 sodium acetate buffer + 50 µL 50 µg/mL cyheptamide in MeOH, vortex for 15 s, add 4 mL dichloromethane:ethyl acetate 2:1, shake for 5 min, repeat extraction. Combine the organic layers and evaporate them to dryness under a stream of nitrogen at 40°. Reconstitute the residue in 200 µL mobile phase, inject a 20 µL aliquot.

## HPLC VARIABLES
**Column:** 250 × 4.6 5 µm Spherisorb Octyl C8
**Mobile phase:** MeOH:MeCN:THF:10 mM pH 6.5 ammonium phosphate buffer 16:11:7:66
**Column temperature:** 40
**Flow rate:** 1.5
**Injection volume:** 20
**Detector:** UV 215

## CHROMATOGRAM
**Retention time:** 8.3
**Internal standard:** cyheptamide (12.6)
**Limit of detection:** 200 ng/mL

## OTHER SUBSTANCES
**Simultaneous:** metabolites, carbamazepine-10,11-epoxide, carbamazepinediol, felbamate, 5-(p-hydroxyphenyl)-5-phenylhydantoin, phenytoin
**Also analyzed:** ethosuximide, ethotoin, lorazepam, phenobarbital, phenylethylmalonamide, primidone

## KEY WORDS
plasma

## REFERENCE

Remmel, R.P.; Miller, S.A.; Graves, N.M. Simultaneous assay of felbamate plus carbamazepine, phenytoin, and their metabolites by liquid chromatography with mobile phase optimization. *Ther.Drug Monit.*, **1990**, *12*, 90–96

## SAMPLE

**Matrix:** blood

**Sample preparation:** Prepare an SPE cartridge by plugging the end of a 1 mL disposable pipette tip with glass wool and adding about 100 mg Chromosorb P/NAW. Add 50 μL plasma then 50 μL 10 μg/mL tolylphenobarbital in 200 mM HCl to the SPE cartridge, let stand for 2 min, elute with 1 mL chloroform:isopropanol 6:1. Evaporate the eluate to dryness under a stream of nitrogen at 30°, reconstitute the residue in 100 μL mobile phase, inject a 15 μL aliquot.

## HPLC VARIABLES

**Column:** 150 × 4.6 5 μm Supelcosil-LC-8
**Mobile phase:** MeCN:water 20:80
**Flow rate:** 3.3
**Injection volume:** 15
**Detector:** UV 208

## CHROMATOGRAM

**Retention time:** 11.37
**Internal standard:** tolylphenobarbital (7.57)
**Limit of detection:** 50-100 ng/mL

## OTHER SUBSTANCES

**Extracted:** amobarbital, barbital, butabarbital, caffeine, carbamazepinediol, carbamazepine epoxide, chloramphenicol, ethosuximide, glutethimide, mephenytoin, methaqualone, methyprylon, nirvanol, pentobarbital, phenacemide, phenobarbital, phenytoin, primidone, secobarbital, theophylline

**Noninterfering:** acetaminophen, N-acetylprocainamide, amikacin, amitriptyline, clonazepam, cyclosporine, desipramine, diazepam, digoxin, disopyramide, gentamicin, p-hydroxyphenobarbital, imipramine, lidocaine, methotrexate, netilmicin, nortriptyline, procainamide, quinidine, salicylic acid, sulfamethoxazole, tobramycin, trimethoprim, valproic acid, vancomycin

## KEY WORDS

SPE

## REFERENCE

Svinarov, D.A.; Dotchev, D.C. Simultaneous liquid-chromatographic determination of some bronchodilators, anticonvulsants, chloramphenicol, and hypnotic agents, with Chromosorb P columns used for sample preparation. *Clin.Chem.*, **1989**, *35*, 1615–1618

## SAMPLE

**Matrix:** blood

**Sample preparation:** 100 μL Serum + 100 μL buffer + 1.5 mL IS in 5% isopropanol in chloroform, vortex for 30 s, centrifuge. Remove the organic layer and evaporate it to dryness under a stream of air at room temperature, reconstitute the residue in 100 μL mobile phase, inject a 6-10 μL aliquot. (Buffer was 13.6 g $KH_2PO_4$ in 90 mL water, pH adjusted to 6.8 with about 3 mL 10 M NaOH, made up to 100 mL.)

## HPLC VARIABLES

**Guard column:** 20 × 4.6 Supelguard LC-1 (Supelco)
**Column:** 250 × 4.6 5 μm Supelcosil LC-1 (Supelco)

**Mobile phase:** MeOH:MeCN:buffer 17.5:17.5:65 (Buffer was 2.72 g $KH_2PO_4$ in 1.9 L water, pH adjusted to 6.3 with about 2 mL 1 M NaOH, made up to 2 L.)
**Flow rate:** 2
**Injection volume:** 6-10
**Detector:** UV 204

## CHROMATOGRAM
**Retention time:** 6.68
**Internal standard:** 5-ethyl-5-p-tolybarbituric acid (tolybarb) (4.80)

## OTHER SUBSTANCES
**Extracted:** acetaminophen, amobarbital, barbital, caffeine, chloramphenicol, ethosuximide, mephobarbital, methsuximide, pentobarbital, phenobarbital, phenytoin, primidone, secobarbital, theophylline, thiopental
**Simultaneous:** acetanilide, N-acetylcysteine, N-acetylprocainamide, ampicillin, aspirin, butabarbital, butalbital, chlorpropamide, cimetidine, codeine, cyheptamide, diazoxide, diflunisal, diphylline, disopyramide, ethchlorvynol, gentisic acid, glutethimide, heptabarbital, hexobarbital, ibuprofen, indomethacin, ketoprofen, mefenamic acid, mephenytoin, methaqualone, methsuximide, methyl salicylate, methyprylon, morphine, naproxen, nirvanol, oxphenylbutazone, phenacetin, phensuximide, phenylbutazone, procainamide, salicylamide, salicylic acid, sulfamethoxazole, sulindac, tolmetin, trimethoprim, vancomycin
**Noninterfering:** amikacin, gentamicin, meprobamate, netilmicin, quinidine, tetracycline, tobramycin, valproic acid

## KEY WORDS
serum

## REFERENCE
Meatherall, R.; Ford, D. Isocratic liquid chromatographic determination of theophylline, acetaminophen, chloramphenicol, caffeine, anticonvulsants, and barbiturates in serum. *Ther.Drug Monit.*, **1988**, *10*, 101–115

## SAMPLE
**Matrix:** blood
**Sample preparation:** Add 20 μL 30 μg/mL cyheptamide in MeOH to a tube and evaporate the MeOH, add 1 mL plasma, add 2 mL buffer, add 20 mL chloroform:t-butyl alcohol 95:5, shake horizontally at 180 cycles/min, centrifuge at 750 g for 5 min. Remove the organic layer and evaporate it to dryness under reduced pressure at 65°, reconstitute the residue in 50 μL mobile phase, vortex, inject a 10 μL aliquot. (Buffer was 50 mL 200 mM $Na_2HPO_4.7H_2O$ + 6.3 mL 400 mM NaOH, pH 11.2.)

## HPLC VARIABLES
**Column:** 200 × 2.1 5 μm ODS C18 (Hewlett-Packard)
**Mobile phase:** MeOH:water 45:55
**Column temperature:** 36
**Flow rate:** 0.5
**Injection volume:** 10
**Detector:** UV 212

## CHROMATOGRAM
**Retention time:** 5.78
**Internal standard:** cyheptamide (8.92)
**Limit of quantitation:** 25 ng/mL

## OTHER SUBSTANCES
**Extracted:** metabolites, carbamazepine diol, carbamazepine epoxide
**Noninterfering:** clonazepam, phenobarbital, phenytoin

## KEY WORDS
plasma; microbore

## REFERENCE
Riad, L.E.; Sawchuk, R.J. Simultaneous determination of carbamazepine and its epoxide and transdiol metabolites in plasma by microbore liquid chromatography. *Clin.Chem.*, **1988**, *34*, 1863–1866

## SAMPLE
**Matrix:** blood
**Sample preparation:** 100 μL Plasma + 200 μL 1 M HCl saturated with ammonium sulfate, vortex for 20 s, add 60 μL 10 μg/mL 4-methylprimidone in MeCN, vortex for 20 s, centrifuge at 2700 g for 5 min, inject a 5-10 μL aliquot of the MeCN layer.

## HPLC VARIABLES
**Column:** 250 × 4 5 μm LiChrosorb RP-18
**Mobile phase:** MeOH:THF:50 mM pH 5.9 phosphate buffer 44:1:55
**Column temperature:** 50
**Flow rate:** 1.1
**Injection volume:** 5-10
**Detector:** UV 210

## CHROMATOGRAM
**Retention time:** 10
**Internal standard:** 4-methylprimidone (5)
**Limit of detection:** 50 ng/mL

## OTHER SUBSTANCES
**Extracted:** phenobarbital, phenytoin, primidone, valproic acid
**Simultaneous:** metabolites, acetaminophen, caffeine, chloramphenicol, diazepam, ethosuximide, ethylphenylmalonamide, glutethimide, lidocaine, methylphenobarbital, pentobarbital, salicylic acid, theophylline

## KEY WORDS
plasma

## REFERENCE
Kushida, K.; Ishizaki, T. Concurrent determination of valproic acid with other antiepileptic drugs by high-performance liquid chromatography. *J.Chromatogr.*, **1985**, *338*, 131–139

## SAMPLE
**Matrix:** blood
**Sample preparation:** 50 μL Serum + 50 μL 10 μg/mL IS in MeCN, vortex for 10 s, centrifuge at 3000 g for 1 min, remove the supernatant and place it in another tube, centrifuge for 1 min, inject a 20 μL aliquot of the supernatant.

## HPLC VARIABLES
**Column:** 100 × 8 5 μm Nova Pak C18 Radial pak
**Mobile phase:** MeCN:MeOH:acetone:buffer 8:21:10:61 adjusted to pH 7.95 ± 0.02 with NaOH (Buffer was 1.36 g/L $KH_2PO_4$.)
**Flow rate:** 2.8
**Injection volume:** 20
**Detector:** UV 200

## CHROMATOGRAM
**Retention time:** 6.44

**Internal standard:** tolybarb (5-ethyl-5-(p-methylphenyl)barbituric acid) (4.89)
**Limit of detection:** 500 ng/mL

## OTHER SUBSTANCES
**Extracted:** metabolites, ethosuximide, primidone, phenobarbital, phenytoin
**Simultaneous:** acetaminophen, N-acetylprocainamide, aspirin, ampicillin, caffeine, cephapirin, chloramphenicol, digoxin, disopyramide, hexobarbital, indomethacin, lidocaine, mephobarbital, methsuximide, nafcillin, pentobarbital, phenylethylmalonamide, procainamide, quinidine, salicylic acid, secobarbital, sulfamerazine, sulfamethazine, terbutaline, tetracycline, theobromine, theophylline
**Noninterfering:** acetazolamide, amikacin, cephalosporin C, gentamicin, propranolol, sulfadiazine, sulfamethoxazole, sulfisoxazole, tobramycin, valproic acid, verapamil

## KEY WORDS
serum

## REFERENCE
Ou, C.-N.; Rognerud, C.L. Simultaneous measurement of ethosuximide, primidone, phenobarbital, phenytoin, carbamazepine, and their bioactive metabolites by liquid chromatography. *Clin.Chem.*, **1984**, *30*, 1667–1670

## SAMPLE
**Matrix:** blood
**Sample preparation:** 500 μL Serum + 50 μL 7 μg/mL IS in water + 1 mL buffer, vortex for 10 s, add 5 mL n-hexane:ether:n-propanol 49:49:2, shake gently for 20 min, centrifuge at 1000 g for 5 min, repeat the extraction. Combine the organic layers and evaporate them to dryness under a stream of nitrogen at 50°, reconstitute the residue in 300 μL mobile phase, inject a 50-100 μL aliquot. (Buffer was 10 mM sodium acetate:10 mM acetic acid 88.5:11.5, pH 5.5.)

## HPLC VARIABLES
**Column:** 250 × 4.6 5 μm Partisil 5 ODS-3
**Mobile phase:** MeCN:buffer 28:72 (Buffer was 300 μL 1 M $KH_2PO_4$ and 50 μL 900 mM phosphoric acid in 1.8 L water, pH 4.4.)
**Column temperature:** 50
**Flow rate:** 2.8
**Injection volume:** 50-100
**Detector:** UV 195

## CHROMATOGRAM
**Retention time:** 7.8
**Internal standard:** 5-(4-methylphenyl)-5-phenylhydantoin (11.5)

## OTHER SUBSTANCES
**Extracted:** ethosuximide, secobarbital
**Simultaneous:** mephobarbital, paramethadione, phenobarbital, primidone
**Noninterfering:** chlorazepate, clonazepam, diazepam, thioridazine, valproic acid
**Interfering:** phenytoin

## KEY WORDS
serum

## REFERENCE
Levine, H.L.; Cohen, M.E.; Duffner, P.K.; Kustas, K.A.; Shen, D.D. An improved high-pressure liquid chromatographic assay for secobarbital in serum. *J.Pharm.Sci.*, **1982**, *71*, 1281–1283

## SAMPLE

**Matrix:** blood
**Sample preparation:** 400 µL Serum or plasma + 400 µL 10 µg/mL IS in acetone, vortex for 10 s, centrifuge at 4500-5000 g for 1 min, remove the supernatant to another tube, centrifuge for 30 s, inject a 5-7.5 µL aliquot.

## HPLC VARIABLES

**Column:** 300 × 3.9 µBondapak C18
**Mobile phase:** MeCN:MeOH:buffer 17:28:55, final pH 6.8-7.0 (Buffer was 400 µL 1 M KH$_2$PO$_4$ in 1 L water, pH adjusted to 6.0 with 900 mM phosphoric acid.)
**Column temperature:** 30
**Flow rate:** 0.7
**Injection volume:** 5-7.5
**Detector:** UV 195

## CHROMATOGRAM

**Retention time:** 18.6
**Internal standard:** tolybarb (5-ethyl-5-(p-methylphenyl)barbituric acid) (13.8)

## OTHER SUBSTANCES

**Extracted:** N-desmethylmethsuximide, ethosuximide, phenobarbital, phenytoin, primidone
**Simultaneous:** acetaminophen, butalbital, caffeine, hexobarbital, methsuximide, phenacetin, phenylethylmalonamide, salicylic acid

## KEY WORDS

serum; plasma

## REFERENCE

Szabo, G.K.; Browne, T.R. Improved isocratic liquid-chromatographic simultaneous measurement of phenytoin, phenobarbital, primidone, carbamazepine, ethosuximide, and N-desmethylmethsuximide in serum. *Clin.Chem.*, **1982**, *28*, 100−104

## SAMPLE

**Matrix:** blood
**Sample preparation:** 500 µL Plasma + 100 µL heptabarbital in MeOH + 500 µL 400 mM pH 7.0 sodium phosphate buffer + 10 mL ethyl acetate, extract. Evaporate the extract to dryness at 50°, reconstitute the residue in 20 µL MeOH, inject a 3 µL aliquot.

## HPLC VARIABLES

**Guard column:** 50 × 2.1 Whatman Co:Pell ODS
**Column:** 125 × 4.5 5 µm SAS Hypersil
**Mobile phase:** MeCN:buffer 20:80 (Buffer was 5 mM tetrabutylammonium hydroxide adjusted to pH 7.5 with phosphoric acid.)
**Flow rate:** 1.6
**Injection volume:** 3
**Detector:** UV 200

## CHROMATOGRAM

**Retention time:** 16.4
**Internal standard:** heptabarbital (9.8)
**Limit of quantitation:** 3.8 µM

## OTHER SUBSTANCES

**Extracted:** ethosuximide, primidone, pheneturide, phenobarbital, phenytoin

**Simultaneous:** amobarbital, barbital, butabarbital, cyclobarbital, ethotoin, ethylphenacemide, glutethimide, methsuximide, pentobarbital, phenylethylmalonamide, sulfamethoxazole, sulthiame
**Interfering:** secobarbital

## KEY WORDS
plasma; horse

## REFERENCE
Christofides, J.A.; Fry, D.E. Measurement of anticonvulsants in serum by reversed-phase ion-pair liquid
chromatography. *Clin.Chem.*, **1980**, *26*, 499–501

## SAMPLE
**Matrix:** blood
**Sample preparation:** 200 µL Serum or plasma + 200 µL 20 µg/mL IS in MeOH:water
10:90 + 75 µL glacial acetic acid, vortex for 30 s, add 5 mL chloroform, shake for 5 min,
centrifuge at 2000 rpm for 5 min. Remove the organic layer and evaporate it to dryness
under a stream of nitrogen, reconstitute the residue in 200 µL mobile phase, inject a 40
µL aliquot.

## HPLC VARIABLES
**Guard column:** 30 × 2.1 Permaphase ETH (DuPont)
**Column:** 250 × 4.6 CLC 1 C8 (DuPont)
**Mobile phase:** MeCN:buffer 35:65 (Buffer was 20 mM $KH_2PO_4$ and 1 mM $K_2HPO_4$ ad-
justed to pH 5.6.)
**Column temperature:** 25
**Flow rate:** 2
**Injection volume:** 40
**Detector:** UV 220

## CHROMATOGRAM
**Retention time:** 5.9
**Internal standard:** alphenal (5-allyl-5-phenylbarbituric acid) (4.4)
**Limit of quantitation:** 200 ng/mL

## OTHER SUBSTANCES
**Extracted:** ethosuximide, phenytoin, primidone, phenobarbital
**Simultaneous:** amobarbital, barbital, chlordiazepoxide, codeine, cortisol, ethotoin, glutethimide, hexobarbital, mephenytoin, mephobarbital, metharbital, methsuximide, nitrazepam, pentobarbital, phenacetin, secobarbital
**Noninterfering:** acetaminophen, acetazolamide, amphetamine, bilirubin, caffeine, diazepam, dimenhydrinate, meperidine, meprobamate, methamphetamine, methaqualone,
methylphenidate, nicotine, propoxyphene, theophylline, valproate
**Interfering:** phensuximide

## KEY WORDS
serum; plasma

## REFERENCE
Rydzewski, R.S.; Gadsden, R.H.; Phelps, C.A. Simultaneous rapid HPLC determination of anticonvulsant drugs in plasma and correlation with EMIT. *Ann.Clin.Lab.Sci.*, **1980**, *10*, 89–94

## SAMPLE
**Matrix:** blood, CSF
**Sample preparation:** 200 µL Serum, plasma, or CSF + 300 µL reagent. Flush column A
to waste with 500 µL 500 mM ammonium sulfate, inject sample onto column A, flush

column A to waste with 500 μL 500 mM ammonium sulfate, elute the contents of column A onto column B with mobile phase, monitor the effluent from column B. (Reagent was 8.05 M guanidine hydrochloride and 1.02 M ammonium sulfate in water.)

## HPLC VARIABLES
**Column:** A 30 × 2.1 40 μm preparative grade C18 (Analytichem); B 250 × 4.6 10 μm Partisil C8
**Mobile phase:** Gradient. A was 50 mM pH 4.5 $KH_2PO_4$. B was MeCN:isopropanol 80:20. A:B 90:10 for 1 min, to 30:70 over 15 min, maintain at 30:70 for 4 min.
**Column temperature:** 50
**Flow rate:** 1.5
**Detector:** UV 280 for 5 min then UV 254

## CHROMATOGRAM
**Retention time:** 11.66
**Internal standard:** heptanophenone (19.2)

## OTHER SUBSTANCES
**Extracted:** acetazolamide, ampicillin, bromazepam, caffeine, chloramphenicol, chlorothiazide, diazepam, droperidol, ethionamide, furosemide, isoniazid, methadone, penicillin G, phenobarbital, phenytoin, prazepam, propoxyphene, pyrazinamide, rifampin, trimeprazine, trimethoprim

## KEY WORDS
serum; plasma; column-switching

## REFERENCE
Seifart, H.I.; Kruger, P.B.; Parkin, D.P.; van Jaarsveld, P.P.; Donald, P.R. Therapeutic monitoring of antituberculosis drugs by direct in-line extraction on a high-performance liquid chromatography system. *J.Chromatogr.*, **1993**, *619*, 285–290

## SAMPLE
**Matrix:** blood, CSF, gastric fluid, urine
**Sample preparation:** 200 μL Serum, urine, CSF, or gastric fluid + 300 μL reagent. Flush column A to waste with 500 μL 500 mM ammonium sulfate, inject sample onto column A, flush column A to waste with 500 μL 500 mM ammonium sulfate, backflush the contents of column A onto column B with mobile phase, monitor the effluent from column B. (Reagent was 8.05 M guanidine hydrochloride and 1.02 M ammonium sulfate in water.)

## HPLC VARIABLES
**Column:** A 40 μm preparative grade C18 (Analytichem); B 75 × 2.1 pellicular C18 (Whatman) + 250 × 4.6 5 μm C8 end-capped (Whatman)
**Mobile phase:** Gradient. A was 50 mM pH 4.5 $KH_2PO_4$. B was MeCN:isopropanol 80:20. A:B 90:10 for 1 min, to 30:70 over 20 min.
**Column temperature:** 50
**Flow rate:** 1.5
**Detector:** UV 220

## CHROMATOGRAM
**Retention time:** 11.7
**Internal standard:** heptanophenone (19)

## OTHER SUBSTANCES
**Extracted:** acetaminophen, allobarbital, azinphos, barbital, brallobarbitone, bromazepam, butethal, caffeine, carbaryl, cephaloridine, chloramphenicol, chlordiazepoxide, chlorothiazide, chlorvinphos, clothiapine, cocaine, coomassie blue, desipramine, diazepam, diphenhydramine, dipipanone, ethylbromphos, flufenamic acid, formothion, griseofulvin, indomethacin, lidocaine, lorazepam, malathion, medazepam, midazolam, oxazepam, paraoxon, penicillin G, pentobarbital, prazepam, propoxyphene, prothiophos, quinine, salicylic acid, secobarbital, strychnine, sulfamethoxazole, theophylline, thiopental, thioridazine, trimethoprim

## KEY WORDS
serum; column-switching

## REFERENCE
Kruger, P.B.; Albrecht, C.F.De V.; Jaarsveld, P.P. Use of guanidine hydrochloride and ammonium sulfate in comprehensive in-line sorption enrichment of xenobiotics in biological fluids by high-performance liquid chromatography. *J.Chromatogr.*, **1993**, *612*, 191–198

## SAMPLE
**Matrix:** blood, dialysate
**Sample preparation:** Dialyze 400 μL plasma against 175 μL acceptor solution through a Cuprophane membrane (15 kDa cut-off) at 37° for 10 min, inject 500 μL acceptor solution (including the portion used for dialysis) onto column A at 0.71 mL/min, elute the contents of column A onto column B with mobile phase, remove column A from circuit and condition it with 1 mL acceptor solution, elute column B with mobile phase and monitor the effluent. Flush acceptor channel with 5 mL acceptor solution and plasma channel with 8 mL acceptor solution containing 25 μg/mL Triton X-100. (Acceptor solution contained 5.9 g NaCl, 4.1 g sodium acetate, 0.3 g KCl, and 1.65 g sodium citrate in 1 L water, adjusted to pH 7.4 with citric acid.)

## HPLC VARIABLES
**Column:** A 5 × 1.6 Hypersil ODS-2; B 100 × 3 5 μm Spherisorb ODS-2
**Mobile phase:** MeCN:THF:20 mM pH 6.0 phosphate buffer 22:6.5:71.5
**Column temperature:** 37
**Flow rate:** 0.6
**Detector:** UV 240

## CHROMATOGRAM
**Retention time:** 4
**Limit of detection:** 100 ng/mL

## OTHER SUBSTANCES
**Extracted:** phenobarbital, phenytoin

## KEY WORDS
plasma; column-switching; dialysis

## REFERENCE
Johansen, K.; Krogh, M.; Andresen, A.T.; Christophersen, A.S.; Lehne, G.; Rasmussen, K.E. Automated analysis of free and total concentrations of three antiepileptic drugs in plasma with on-line dialysis and high-performance liquid chromatography. *J.Chromatogr.B*, **1995**, *669*, 281–288

## SAMPLE
**Matrix:** blood, gastric contents, tissue, urine
**Sample preparation:** 1 mL Blood, urine, or gastric contents or 1 g tissue homogenate + 500 μL buffer + 8 mL n-hexane:ethyl acetate 70:30, mix on a rotary mixer for 10 min,

centrifuge at 3000 g for 8 min. Remove the organic layer and evaporate it to dryness under a stream of nitrogen, reconstitute the residue in 100 μL 12.5 mM NaOH in MeOH:water 50:50, inject a 50 μL aliquot. (Buffer was 13.8 g potassium carbonate in 100 mL water, pH adjusted to 9.5 with concentrated HCl.)

## HPLC VARIABLES
**Guard column:** 4 × 4 30 μm LiChrocart Aluspher RP-select B (Merck)
**Column:** 125 × 4 5 μm Aluspher RP-select B (Merck)
**Mobile phase:** Gradient. A was 12.5 mM NaOH in MeOH. B was 12.5 mM NaOH in water. A:B 10:90 for 5 min, to 90:10 over 15 min, maintain at 90:10 for 5 min, return to initial conditions over 1 min, re-equilibrate for 5 min.
**Flow rate:** 1
**Injection volume:** 50
**Detector:** UV 230; UV 254

## CHROMATOGRAM
**Retention time:** 12.5

## OTHER SUBSTANCES
**Extracted:** alprenolol, amitriptyline, bromazepam, chlordiazepoxide, chlorpromazine, clonazepam, desipramine, diazepam, flunitrazepam, haloperidol, nitrendipine, nordiazepam, nortriptyline, pindolol, zolpidem

**Also analyzed:** acebutolol, acetaminophen, alprazolam, amphetamine, atenolol, betaxolol, brotizolam, caffeine, camazepam, captopril, chloroquine, clobazam, clomipramine, clothiapine, clotiazepam, cloxazolam, cocaine, codeine, diclofenac, dihydralazine, dihydrocodeine, dihydroergotamine, diphenhydramine, domperidone, doxepin, droperidol, ergotamine, ethyl loflazepate, fenethylline, fluoxetine, flupentixol, flurazepam, furosemide, gliclazide, hydrochlorothiazide, hydroxyzine, ibuprofen, imipramine, ketazolam, loprazolam, lorazepam, lormetazepam, maprotiline, medazepam, mepyramine, methadone, methaqualone, methyldopa, methylphenidate, metoclopramide, metoprolol, mexiletine, mianserin, midazolam, minoxidil, morphine, nadolol, nitrazepam, oxprenolol, papaverine, pentazocine, phenprocoumon, phenylbutazone, pipamperone, piritramide, practolol, prazepam, prazosin, promazine, promethazine, propoxyphene, propranolol, prothipendyl, quinine, sotalol, sulpride, thioridazine, trazodone, triazolam, trimipramine, tripelennamine, tyramine, verapamil, yohimbine

## REFERENCE
Lambert, W.E.; Meyer, E.; De Leenheer, A.P. Systematic toxicological analysis of basic drugs by gradient elution of an alumina-based HPLC packing material under alkaline conditions. *J.Anal.Toxicol.*, **1995**, *19*, 73–78

## SAMPLE
**Matrix:** blood, saliva, urine
**Sample preparation:** Serum. 100 μL Serum + 200 μL MeCN, vortex for 10 s, centrifuge at 1500 g for 5 min, inject a 2 μL aliquot of the supernatant. Saliva. 250 μL Saliva + 50 μL MeCN, centrifuge at 1500 g for 5 min, inject a 2 μL aliquot of the supernatant. Urine. Condition a Sep-Pak SPE cartridge with 5 mL MeCN then 20 mL water. Add 2 mL urine to the cartridge, wash with 20 mL water, elute with 500 μL MeCN, inject 2 μL of the eluent.

## HPLC VARIABLES
**Guard column:** 20 × 2 3 μm ODS-Hypersil
**Column:** 250 × 2 3 μm ODS-Hypersil
**Mobile phase:** MeCN:MeOH:10 mM pH 7.0 phosphate buffer 50:30:110
**Column temperature:** 40
**Flow rate:** 0.2

**Injection volume:** 2
**Detector:** UV 200

## CHROMATOGRAM
**Retention time:** 13.1
**Limit of quantitation:** 780 ng/mL

## OTHER SUBSTANCES
**Simultaneous:** carbamazepine-10,11-epoxide, clonazepam, dihydrodihydroxycarbamazepine, hexobarbital, p-hydroxyphenobarbital, 5-(m-hydroxyphenyl)-5-phenylhydantoin, 5-
   (p-hydroxyphenyl)-5-phenylhydantoin, nitrazepam, phenobarbital, phenylethylmaleimide,
   phenytoin, primidone
**Noninterfering:** chlordiazepoxide, cyheptamide, diazepam, lorazepam, nordiazepam, oxazepam, prazepam, temazepam

## KEY WORDS
serum; SPE

## REFERENCE
Liu, H.; Delgado, M.; Forman, L.J.; Eggers, C.M.; Montoya, J.L. Simultaneous determination of carbamazepine, phenytoin, phenobarbital, primidone and their principal metabolites by high-performance
   liquid chromatography with photodiode-array detection. *J.Chromatogr.*, **1993**, *616*, 105−115

## SAMPLE
**Matrix:** blood, tissue
**Sample preparation:** Tissue. Homogenize 20-200 mg brain tissue with 1 mL 1.5 μg/mL
   IS in 1% ammonium acetate + 1% sodium azide buffer:MeCN 99:1, flush apparatus with
   1 mL extraction buffer, add 1 mL acetone, shake for 5 min, centrifuge for 10 min. Add
   sample to an Extrelut-3 SPE cartridge (Kieselguhr), add 1 mL extraction buffer, wait for
   10 min, elute with 15 mL extraction solvent. Evaporate the eluate, take up residue in 50
   μL MeOH, add 50 μL water, inject a 10-25 μL aliquot. Serum. 100 μL Serum + 1 mL
   1.5 μg/mL IS in 1% ammonium acetate + 1% sodium azide buffer:MeCN 99:1, mix, add
   1 mL extraction buffer, mix, add 1 mL acetone, shake for 5 min, centrifuge for 10 min.
   Add sample to an Extrelut-3 SPE cartridge (Kieselguhr), add 1 mL extraction buffer, wait
   for 10 min, elute with 15 mL extraction solvent. Evaporate the eluate, take up residue
   in 50 μL MeOH, add 50 μL water, inject a 10-25 μL aliquot. (Extraction buffer was 20 g
   $NaH_2PO_4.2H_2O$ + 4.5 g $Na_2HPO_4.2H_2O$ + 1.5 g $NaN_3$ in 1 L water, pH 6. Extraction
   solvent was dichloromethane:isopropanol 97:3.)

## HPLC VARIABLES
**Column:** 200 × 2.1 5 μm Hypersil ODS
**Mobile phase:** Gradient. A was MeCN:50 mM $(NH_4)H_2PO_4$ (pH 4.4) 10:90. B was MeCN:
   50 mM $(NH_4)H_2PO_4$ (pH 4.4) 60:40. A:B from 85:15 to 55:45 over 9.5 min, keep at 55:
   45 for 0.5 min, return to 85:15 over 0.5 min.
**Column temperature:** 65
**Flow rate:** 0.3
**Injection volume:** 10-25
**Detector:** UV 207

## CHROMATOGRAM
**Retention time:** 10.63
**Internal standard:** 5-ethyl-5-(p-tolyl)barbituric acid (9.07)
**Limit of quantitation:** 150 ng/mL

## OTHER SUBSTANCES
**Simultaneous:** carbamazepine-10,11-epoxide, N-desmethylmethsuximide, phenobarbital,
   phenytoin, primidone

## KEY WORDS
serum; SPE; brain

## REFERENCE
Juergens, U.; Rambeck, B. Sensitive analysis of antiepileptic drugs in very small portions of human brain by microbore HPLC. *J.Liq.Chromatogr.*, **1987**, *10*, 1847–1863

## SAMPLE
**Matrix:** dialysate
**Sample preparation:** Inject a 25 µL aliquot of dialysate containing 1 µg/mL IS.

## HPLC VARIABLES
**Guard column:** 30 × 2.1 5 µm Spheri-5 ODS
**Column:** 220 × 2.1 5 µm Spheri-5 ODS
**Mobile phase:** MeCN:buffer 27:73 (After 8 min increase flow to 0.5 mL/min over 2 min, maintain at 0.5 mL/min for 12 min, return to 0.2 mL/min over 3 min. Buffer was 50 mM $KH_2PO_4$ adjusted to pH 6.5 with 5 M NaOH.)
**Flow rate:** 0.2
**Injection volume:** 25
**Detector:** UV 212

## CHROMATOGRAM
**Internal standard:** 2-methyl-5H-dibenz(b,f)azepine-5-carboxamide
**Limit of detection:** 2.5 ng/mL

## OTHER SUBSTANCES
**Extracted:** metabolites

## KEY WORDS
pharmacokinetics; narrow bore; rat

## REFERENCE
Van Belle, K.; Sarre, S.; Ebinger, G.; Michotte, Y. Brain, liver and blood distribution kinetics of carbamazepine and its metabolic interaction with clomipramine in rats: A quantitative microdialysis study. *J.Pharmacol.Exp.Ther.*, **1995**, *272*, 1217–1222

## SAMPLE
**Matrix:** dialysate
**Sample preparation:** Inject a 10 µL aliquot of the dialysate.

## HPLC VARIABLES
**Column:** 100 × 4.6 3µm Econosphere C18
**Mobile phase:** MeOH:MeCN:water 17:23:60
**Column temperature:** 40
**Flow rate:** 0.6
**Injection volume:** 10
**Detector:** UV 210

## CHROMATOGRAM
**Retention time:** 7.55
**Limit of quantitation:** 20 ng/mL

## OTHER SUBSTANCES
**Extracted:** metabolites, carbamazepine-10,11-epoxide, carbamazepine-10,11-trans-diol

## KEY WORDS
brain

## REFERENCE
Scheyer, R.D.; During, M.J.; Cramer, J.A.; Toftness, B.R.; Hochholzer, J.M.; Mattson, R.H. Simultaneous HPLC analysis of carbamazepine and carbamazepine epoxide in human brain microdialysate. *J.Liq.Chromatogr.*, **1994**, *17*, 1567–1576

## SAMPLE
**Matrix:** formulations
**Sample preparation:** 100 μL Solution + 100 μL IS solution, make up to 10 mL with mobile phase, inject an aliquot.

## HPLC VARIABLES
**Column:** 250 × 4.6 Spherisorb ODS-II
**Mobile phase:** MeOH:water 60:40
**Flow rate:** 1
**Detector:** UV 212

## CHROMATOGRAM
**Retention time:** 4.3
**Internal standard:** carbamazepine 10,11-epoxide (6.6)

## KEY WORDS
suspensions; saline; 5% dextrose

## REFERENCE
Clark-Schmidt, A.L.; Garnett, W.R.; Lowe, D.R.; Karnes, H.T. Loss of carbamazepine suspension through nasogastric feeding tubes. *Am.J.Hosp.Pharm.*, **1990**, *47*, 2034–2037

## SAMPLE
**Matrix:** formulations
**Sample preparation:** Finely powder tablets, weigh out amount equivalent to 30 mg guanabenz acetate, add 190 mL MeCN, sonicate for a few minutes, make up to 200 mL with MeCN, filter. Remove a 14-59 mL aliquot and add it to 9 mL 1 mg/mL carbamazepine in MeCN, make up to 100 mL with MeCN, inject a 20 μL aliquot.

## HPLC VARIABLES
**Column:** 250 × 4.6 10 μm octadecylsilane (Perkin-Elmer)
**Mobile phase:** MeCN:buffer 35:65 (Buffer was 4 mM $KH_2PO_4$ adjusted to pH 3.25 with phosphoric acid.)
**Flow rate:** 2.5
**Injection volume:** 20
**Detector:** UV 265

## CHROMATOGRAM
**Retention time:** 3
**Internal standard:** carbamazepine

## OTHER SUBSTANCES
**Simultaneous:** guanabenz, mefruside

## KEY WORDS
tablets; carbamazepine is IS

## REFERENCE

Vio, L.; Mamolo, M.G.; Furlan, G. Quantitative high pressure liquid chromatographic determination of guanabenz and mephruside in pharmaceutical formulations. *Farmaco.[Prat].*, **1988**, *43*, 27–36

## SAMPLE
**Matrix:** solutions

## HPLC VARIABLES
**Column:** 250 × 4.6 5 μm Supelcosil LC-DP (A) or 250 × 4 5 μm LiChrospher 100 RP-8 (B)
**Mobile phase:** MeCN:0.025% phosphoric acid:buffer 25:10:5 (A) or 60:25:15 (B) (Buffer was 9 mL concentrated phosphoric acid and 10 mL triethylamine in 900 mL water, adjust pH to 3.4 with dilute phosphoric acid, make up to 1 L.)
**Flow rate:** 0.6
**Injection volume:** 25
**Detector:** UV 229

## CHROMATOGRAM
**Retention time:** 6.15 (A), 5.49 (B)

## OTHER SUBSTANCES
**Also analyzed:** acebutolol, acepromazine, acetaminophen, acetazolamide, acetophenazine, albuterol, alprazolam, amitriptyline, amobarbital, amoxapine, antipyrine, atenolol, atropine, azatadine, baclofen, benzocaine, bromocriptine, brompheniramine, brotizolam, bupivacaine, buspirone, butabarbital, butalbital, caffeine, cetirizine, chlorcyclizine, chlordiazepoxide, chlormezanone, chloroquine, chlorpheniramine, chlorpromazine, chlorpropamide, chlorprothixene, chlorthalidone, chlorzoxazone, cimetidine, cisapride, clomipramine, clonazepam, clonidine, clozapine, cocaine, codeine, colchicine, cyclizine, cyclobenzaprine, dantrolene, desipramine, diazepam, diclofenac, diflunisal, diltiazem, diphenhydramine, diphenidol, diphenoxylate, dipyridamole, disopyramide, dobutamine, doxapram, doxepin, droperidol, encainide, ethidium bromide, ethopropazine, fenoprofen, fentanyl, flavoxate, fluoxetine, fluphenazine, flurazepam, flurbiprofen, fluvoxamine, furosemide, glutethimide, glyburide, guaifenesin, haloperidol, homatropine, hydralazine, hydrochlorothiazide, hydrocodone, hydromorphone, hydroxychloroquine, hydroxyzine, ibuprofen, imipramine, indomethacin, ketoconazole, ketoprofen, ketorolac, labetalol, levorphanol, lidocaine, loratadine, lorazepam, lovastatin, loxapine, mazindol, mefenamic acid, meperidine, mephenytoin, mepivacaine, mesoridazine, metaproterenol, methadone, methdilazine, methocarbamol, methotrexate, methotrimeprazine, methoxamine, methyldopa, methylphenidate, metoclopramide, metolazone, metoprolol, metronidazole, midazolam, moclobemide, morphine, nadolol, nalbuphine, naloxone, naphazoline, naproxen, nifedipine, nizatidine, norepinephrine, nortriptyline, oxazepam, oxycodone, oxymetazoline, paroxetine, pemoline, pentazocine, pentobarbital, pentoxifylline, perphenazine, pheniramine, phenobarbital, phenol, phenolphthalein, phentolamine, phenylbutazone, phenyltoloxamine, phenytoin, pimozide, pindolol, piroxicam, pramoxine, prazepam, prazosin, probenecid, procainamide, procaine, prochlorperazine, procyclidine, promazine, promethazine, propafenone, propantheline, propiomazine, propofol, propranolol, protriptyline, quazepam, quinidine, quinine, racemethorphan, ranitidine, remoxipride, risperidone, salicylic acid, scopolamine, secobarbital, sertraline, sotalol, spironolactone, sulfinpyrazone, sulindac, temazepam, terbutaline, terfenadine, tetracaine, theophylline, thiethylperazine, thiopental, thioridazine, thiothixene, timolol, tocainide, tolbutamide, tolmetin, trazodone, triamterene, triazolam, trifluoperazine, triflupromazine, trimeprazine, trimethoprim, trimipramine, verapamil, warfarin, xylometazoline, yohimbine, zopiclone

## KEY WORDS
details of plasma extraction

## REFERENCE
Koves, E.M. Use of high-performance liquid chromatography-diode array detection in forensic toxicology. *J.Chromatogr.A*, **1995**, *692*, 103−119

## SAMPLE
**Matrix:** solutions

## HPLC VARIABLES
**Guard column:** 30 × 3.2 7 μm SI 100 ODS (not commercially available)
**Column:** 150 × 3.2 7 μm SI 100 ODS (not commercially available)
**Mobile phase:** MeCN:buffer 31.2:68.8 (Buffer was 6.66 g $KH_2PO_4$ and 4.8 g 85% phosphoric acid in 1 L water, pH 2.3.)
**Flow rate:** 0.5-1
**Detector:** UV 208; UV 233

## CHROMATOGRAM
**Retention time:** 4.2
**Internal standard:** 5-(4-methylphenyl)-5-phenylhydantoin (7.3)

## OTHER SUBSTANCES
**Also analyzed:** aspirin, chlordiazepoxide, chlorprothixene, clonazepam, caffeine, diazepam, doxylamine, ethosuximide, furosemide, haloperidol, hydrochlorothiazide, methocarbamol, methotrimeprazine, nicotine, oxazepam, procaine, promazine, propafenone, propranolol, salicylamide, temazepam, tetracaine, thiopental, triamterene, verapamil, zolpidem, zopiclone

## REFERENCE
Below, E.; Burrmann, M. Application of HPLC equipment with rapid scan detection to the identification of drugs in toxicological analysis. *J.Liq.Chromatogr.*, **1994**, *17*, 4131−4144

## SAMPLE
**Matrix:** solutions

## HPLC VARIABLES
**Column:** 250 × 4.6 Zorbax RX
**Mobile phase:** Gradient. A was 10 mL concentrated orthophosphoric acid and 7 mL triethylamine in 1 L water. B was 10 mL concentrated orthophosphoric acid and 7 mL triethylamine in 200 mL water, make up to 1 L with MeCN. A:B from 100:0 to 0:100 over 30 min, maintain at 0:100 for 5 min.
**Column temperature:** 30
**Flow rate:** 2
**Detector:** UV 210

## OTHER SUBSTANCES
**Also analyzed:** acepromazine, acetaminophen, acetophenazine, albuterol, aminophylline, amitriptyline, amobarbital, amoxapine, amphetamine, amylocaine, antipyrine, aprobarbital, aspirin, atenolol, atropine, avermectin, barbital, benzocaine, benzoic acid, benzotropine, benzphetamine, berberine, bibucaine, bromazepan, brompheniramine, buprenorphine, buspirone, butabarbital, butacaine, butethal, caffeine, carbromal, chloramphenicol, chlordiazepoxide, chloroquine, chlorothiazide, chloroxylenol, chlorphenesin, chlorpheniramine, chlorpromazine, chlorpropamide, chlortetracycline, cimetidine, cinchonidine, cinchonine, clenbuterol, clonazepam, clonixin, clorazepate, cocaine, codeine, colchicine, cortisone, coumarin, cyclazocine, cyclobenzaprine, cyclothiazide, cyheptamide, cymarin, danazol, danthron, dapsone, debrisoquine, desipramine, dexamethasone, dextromethorphan, dextropropoxyphene, diamorphine, diazepam, diclofenac, diethylpropion, diethylstilbestrol, diflunisal, digitoxin, digoxin, diltiazem, diphenhydramine, diphenoxylate, di-

prenorphine, dipyrone, disulfiram, dopamine, doxapram, doxepin, dronabinol, ephedrine, epinephrine, epinine, estradiol, estriol, estrone, ethacrynic acid, ethosuximide, etonitazene, etorphine, eugenol, famotidine, fenbendazole, fencamfamine, fenoprofen, fenproporex, fentanyl, flubendazole, flufenamic acid, flunitrazepam, 5-fluorouracil, fluoxymesterone, fluphenazine, furosemide, gentisic acid, gitoxigenin, glipizide, glunixin, glutethimide, glybenclamide, guaiacol, halazepam, haloperidol, hydrochlorothiazide, hydrocodone, hydrocortisone, hydromorphone, hydroxyquinoline, ibogaine, ibuprofen, iminostilbene, imipramine, indomethacin, isocarbostyril, isocarboxazid, isoniazid, isoproterenol, isoxsuprine, ivermectin, ketamine, ketoprofen, kynurenic acid, levorphanol, lidocaine, lorazepam, lormetazepam, loxapine, mazindol, mebendazole, meclizine, meclofenamic acid, medazepam, mefenamic acid, megestrol, mepacrine, meperidine, mephentermine, mephenytoin, mephesin, mephobarbital, mepivacaine, mescaline, mesoridazine, methadone, methamphetamine, methapyrilene, methaqualone, methazolamide, methocarbamol, methoxamine, methsuximide, methyl salicylate, methyldopa, methyldopamine, methylphenidate, methylprednisolone, methyltestosterone, methyprylon, metoprolol, mibolerone, morphine, nadolol, nalorphine, naloxone, naltrexone, naphazoline, naproxen, nefopam, niacinamide, nicotine, nicotinic acid, nifedipine, niflumic acid, nitrazepam, norepinephrine, nortriptyline, noscapine, nylidrin, oxazepam, oxycodone, oxymorphone, oxyphenbutazone, oxytetracycline, papaverine, pargyline, pemoline, pentazocine, pentobarbital, persantine, phenacetin, phenazocine, phenazopyridine, phencyclidine, phendimetrazine, phenelzine, pheniramine, phenobarbital, phenothiazine, phensuximide, phentermine, phenylbutazone, phenylephrine, phenylpropanolamine, piperocaine, prazepam, prednisolone, primidone, probenecid, progesterone, propiomazine, propranolol, propylparaben, pseudoephedrine, puromycin, pyrilamine, pyrithyldione, quazepam, quinaldic acid, quinidine, quinine, ranitidine, recinnamine, reserpine, resorcinol, saccharin, albuterol, salicylamide, salicylic acid, scopolamine, scopoletin, secobarbital, strychnine, sulfacetamide, sufadiazine, sulfadimethoxine, sulfaethidole, sulfamerazine, sulfamethazine, sulfamethoxizole, sulfanilamide, sulfapyridine, sulfasoxizole, sulindac, tamoxifen, temazepam, testosterone, tetracaine, tetracycline, tetramisole, thebaine, theobromine, theophylline, thiabendazole, thiamine, thiamylal, thiobarbituric acid, thioridazine, thiosalicylic acid, thiothixene, thymol, tolazamide, tolazoline, tobutamide, tolmetin, tranylcypromine, triamcinolone, tribenzylamine, trichloromethiazide, trifluoperazine, trihexyphenidyl, trimethoprim, tripelennamine, triproilidine, tropacocaine, tyramine, verapamil, vincamine, warfarin, yohimbine, zoxazolamine

## REFERENCE

Hill, D.W.; Kind, A.J. Reversed-phase solvent gradient HPLC retention indexes of drugs. *J.Anal.Toxicol.*, **1994**, *18*, 233–242

## SAMPLE
**Matrix:** solutions

## HPLC VARIABLES
**Column:** 250 × 4.6 Econosil C8
**Mobile phase:** MeCN:buffer 30:70 (Buffer was 20 mM $KH_2PO_4$ and 14 mM triethylamine adjusted to pH 3.0 with phosphoric acid.)
**Injection volume:** 20
**Detector:** UV 210

## CHROMATOGRAM
**Retention time:** 7.3
**Limit of quantitation:** < 1000 ng/mL

## OTHER SUBSTANCES
**Simultaneous:** amitriptyline, amoxapine, imipramine, nortriptyline
**Also analyzed:** cyclobenzaprine, desipramine, doxepin, maprotiline, protriptyline

## KEY WORDS
UV spectra given

## REFERENCE
Ryan, T.W. Identification and quantification of tricyclic antidepressants by UV-photodiode array detection with multicomponent analysis. *J.Liq.Chromatogr.*, **1993**, *16*, 1545–1560

## SAMPLE
**Matrix:** urine
**Sample preparation:** Buffer urine to 4.9 by mixing with an equal volume of pH 4.9 200 mM sodium phosphate buffer. Inject a 40 µL aliquot onto column A with mobile phase A, after 3 min backflush the contents of column A onto column B with mobile phase B and start the gradient. At the end of the run re-equilibrate for 10 min.

## HPLC VARIABLES
**Column:** A 20 × 4 5 µm Hypersil octadecylsilica ODS; B 200 × 4.6 5 µm Shiseido SG-120 polymer-based C18
**Mobile phase:** A water; B Gradient. MeCN:buffer from 7:93 to 15:85 over 3.5 min, to 50:50 over 8.5 min, maintain at 50:50 for 11 min (Buffer was 6.9 g $NaH_2PO_4.H_2O$ in 1 L water, pH adjusted to 3.1 with phosphoric acid.)
**Flow rate:** 1
**Injection volume:** 40
**Detector:** UV 230

## CHROMATOGRAM
**Retention time:** 16.2

## OTHER SUBSTANCES
**Extracted:** acetazolamide, amiloride, bendroflumethiazide, benzthiazide, bumetanide, caffeine, chlorothiazide, chlorthalidone, clopamide, dichlorfenamide, ethacrynic acid, furosemide, hydrochlorothiazide, metyrapone, probenecid, spironolactone, triamterene, trichlormethiazide

## KEY WORDS
column-switching; optimum detection wavelengths vary for each drug

## REFERENCE
Saarinen, M.; Sirén, H.; Riekkola, M.-L. A column switching technique for the screening of diuretics in urine by high performance liquid chromatography. *J.Liq.Chromatogr.*, **1993**, *16*, 4063–4078

## ANNOTATED BIBLIOGRAPHY

Kanda, T.; Kutsuna, H.; Ohtsu, Y.; Yamaguchi, M. Synthesis of polymer-coated mixed-functional packing materials for direct analysis of drug-containing serum and plasma by high-performance liquid chromatography. *J.Chromatogr.A*, **1994**, *672*, 51–57 [serum;column temp 40;also chloramphenicol, indomethacin, phenobarbital, phenytoin, theophylline, trimethoprim]

Liu, H.; Delgado, N.R. Improved therapeutic monitoring of drug interactions in epileptic children using carbamazepine polytherapy. *Ther.Drug Monit.*, **1994**, *16*, 132–138 [column temp 40; simultaneous metabolites]

Ramachandran, S.; Underhill, S.; Jones, S.R. Measurement of lamotrigine under conditions measuring phenobarbitone, phenytoin, and carbamazepine using reversed-phase high-performance liquid chromatography at dual wavelengths. *Ther.Drug Monit.*, **1994**, *16*, 75–82 [serum;LOD 200 ng/mL;extracted lamotrigine, phenobarbital, phenytoin;hexabarbital (IS);non-interfering ethosuximide, oxcarbazepine, primidone, valproic acid]

Rambeck, B.; May, T.W.; Jurgens, M.U.; Blankenhorn, V.; Jurges, U.; Korn-Merker, E.; Salke-Kellermann, A. Comparison of phenytoin and carbamazepine serum concentrations measured by high-performance liquid chromatography, the standard TDx assay, the enzyme multiplied immunoassay technique, and a new patient-side immunoassay cartridge system. *Ther.Drug Monit.*, **1994**, *16*, 608−612

Romanyshyn, L.A.; Wichmann, J.K.; Kucharczyk, N.; Shumaker, R.C.; Ward, D.; Sofia, R.D. Simultaneous determination of felbamate, primidone, phenobarbital, carbamazepine, two carbamazepine metabolites, phenytoin, and one phenytoin metabolite in human plasma by high-performance liquid chromatography. *Ther.Drug Monit.*, **1994**, *16*, 90−99 [plasma; extracted metabolites, felbamate, phenobarbital, primidone; column temp 40-50; LOQ 195-391 ng/mL; simultaneous acetaminophen, aspirin, brompheniramine, caffeine, chlorpheniramine, dextromethorphan, dimethadione, ethosuximide, ethotoin, ibuprofen, iministilbene, mephenytoin, mephobarbital, metharbital, methsuximide, paramethadione, phenacemide, phensuximide, phenylpropanolamine, theophylline, trimethadione; non-interfering clonazepam; valproic acid]

Smigol, V.; Svec, F.; Fréchet, J.M.J. Novel uniformly sized polymeric stationary phase with hydrophilized large pores for direct injection HPLC determination of drugs in biological fluids. *J.Liq.Chromatogr.*, **1994**, *17*, 891−911 [direct injection; cow; plasma; also aspirin, caffeine, lidocaine, phenytoin, salicylic acid, theobromine, theophylline]

Spigset, O.; Carleborg, L.; Mjörndal, T.; Norström, Å.; Sundgren, M. Carbamazepine interference in a high-performance liquid chromatography analysis for perphenazine. *Ther.Drug Monit.*, **1994**, *16*, 332−333 [interfering perphenazine; serum]

Sudo, Y.; Akiba, M.; Sakaki, T.; Takahata, Y. Glycerylalkylsilylated silica gels for direct injection analysis of drugs in serum by high-performance liquid chromatography. *J.Liq.Chromatogr.*, **1994**, *17*, 1743−1754 [direct injection; serum; extracted phenobarbital, phenytoin]

Betlach, C.J.; Gonzalez, M.A.; McKiernan, B.C.; Neff-Davis, C.; Bodor, N. Oral phrmacokinetics of carbamazepine in dogs from commercial tablets and a cyclodextrin complex. *J.Pharm.Sci.*, **1993**, *82*, 1058−1060 [dog; plasma; SPE; tolybarb (IS); LOQ 330 ng/mL; pharmacokinetics]

Bonato, P.S.; Lanchote, V.L. A rapid procedure for the purification of biological samples to be analysed by high performance liquid chromatography. *J.Liq.Chromatogr.*, **1993**, *16*, 2299−2308 [also albendazole, clonazepam, desalkylflurazepam, mebendazole; methaqualone (IS)]

Liu, H.; Delgado, M.; Iannaccone, S.T.; Forman, L.J.; Eggers, C.M. Determination of total and free carbamazepine and the principal metabolites in serum by high-performance liquid chromatography with photodiode-array detection. *Ther.Drug Monit.*, **1993**, *15*, 317−327

Miller, R.B.; Vranderick, M. A validated HPLC method for the determination of carbamazepine and carbamazepine-10,11-epoxide in human plasma. *J.Liq.Chromatogr.*, **1993**, *16*, 1249−1261 [plasma; extracted metabolites; lorazepam (IS); non-interfering acetaminophen, aspirin, caffeine, ethosuximide, ibuprofen, nicotine, phenytoin, theophylline, valproic acid]

Vatassery, G.T.; Holden, L.A.; Dysken, M.W. Resolution of the interference from carbamazepine and diphenhydramine during reversed-phase liquid chromatographic determination of haloperidol and reduced haloperidol. *J.Anal.Toxicol.*, **1993**, *17*, 304−306

Bonato, P.S.; Lanchote, V.L.; de Carvalho, D.; Ache, P. Measurement of carbamazepine and its main biotransformation products in plasma by HPLC. *J.Anal.Toxicol.*, **1992**, *16*, 88−92

He, J.; Shibukawa, A.; Nakagawa, T. Direct injection analysis of carbamazepine and its active 10,11-epoxide metabolite in plasma by use of a semipermeable surface (SPS) silica column in LC. *J.Pharm.Biomed.Anal.*, **1992**, *10*, 289−294

Schmitz, S.; Warren, F.V.; Bidlingmeyer, B.A. Analysis of carbamazepine and its 10,11-epoxide in serum by direct sample injection using surfactant containing eluents and column switching. *J.Pharm.Biomed.Anal.*, **1991**, *9*, 985−994

Schramm, W.; Annesley, T.M.; Siegel, G.J.; Sackellares, J.C.; Smith, R.H. Measurement of phenytoin and carbamazepine in an ultrafiltrate of saliva. *Ther.Drug Monit.*, **1991**, *13*, 452−460

Tsaprounis, C.K.; Kajbaf, M.; Gorrod, J.W. Simultaneous determination of carbamazepine and its major metabolites in human plasma and urine by HPLC. *J.Clin.Pharm.Ther.*, **1991**, *16*, 257−262

Moor, M.J.; Rashed, M.S.; Kalhorn, T.F.; Levy, R.H.; Howald, W.N. Application of thermospray liquid chromatography-mass spectrometry to the simultaneous quantification of tracer concentrations of isotopically labelled carbamazepine epoxide and steady-state levels of carbamazepine and carbamazepine epoxide. *J.Chromatogr.*, **1989**, *474*, 223−230

Asberg, A.; Haffner, F. Analysis of serum concentration of phenobarbital, phenytoin, carbamazepine and carbamazepine 10,11-epoxide by solvent-recycled liquid chromatography. *Scand.J.Clin.Lab.Invest.*, **1987**, *47*, 389–392

Cyr, T.D.; Matsui, F.; Sears, R.W.; Curran, N.M.; Lovering, E.G. Liquid chromatographic methods for assay of carbamazepine, 10,11-dihydrocarbamazepine, and related compounds in carbamazepine drug substance and tablets. *J.Assoc.Off.Anal.Chem.*, **1987**, *70*, 836–840

Hartley, R.; Lucock, M.; Forsythe, W.I.; Smithells, R.W. Solid-phase extraction of carbamazepine and two major metabolites from plasma for analysis by HPLC. *J.Liq.Chromatogr.*, **1987**, *10*, 2393–2409 [extracted metabolites; plasma; SPE; LOD 50 ng/mL; nitrazepam (IS)]

Menge, G.P.; Dubois, J.P.; Bauer, G. Simultaneous determination of carbamazepine, oxcarbazepine and their main metabolites in plasma by liquid chromatography. *J.Chromatogr.*, **1987**, *414*, 477–483

Regnaud, L.; Sirois, G.; Colin, P.; Chakrabarti, S. Simultaneous ion-pairing chromatography of the major metabolites of styrene and carbamazepine, and of unchanged carbamazepine in urine. *J.Liq.Chromatogr.*, **1987**, *10*, 2369–2382 [simultaneous metabolites, styrene metabolites; urine; rat]

Shihabi, Z.K.; Dyer, R.D. Serum injection of the HPLC column for carbamazepine assay. *J.Liq.Chromatogr.*, **1987**, *10*, 2383–2391

Messiha, F.S. Determination of carbamazepine by HPLC electrochemical detection and application for estimation of imipramine, desipramine, doxepin and nordoxepin. *Alcohol*, **1986**, *3*, 135–138

Soto-Otero, R.; Méndez-Alvarez, E.; Sierra-Marcuño, G. Simultaneous determination of ethosuximide, phenobarbital, phenytoin, and carbamazepine in brain tissue by HPLC. *J.Liq.Chromatogr.*, **1985**, *8*, 753–763

Gerson, B.; Bell, F.; Chan, S. Antiepileptic agents–primidone, phenobarbital, phenytoin, and carbamazepine by reversed-phase liquid chromatography. *Clin.Chem.*, **1984**, *30*, 105–108

Kapetanovic, I.M.; Kupferberg, H.J. Nafimidone, an imidazole anticonvulsant, and its metabolite as potent inhibitors of microsomal metabolism of phenytoin and carbamazepine. *Drug Metab.Dispos.*, **1984**, *12*, 560–564 [microsomal incubations; rat; liver; column temp 50; extracted metabolites; 2-methylcarbamazepine (IS)]

Kumps, A. Simultaneous HPLC determination of oxcarbazepine, carbamazepine and their metabolites in serum. *J.Liq.Chromatogr.*, **1984**, *7*, 1235–1241

Kabra, P.M.; Nelson, M.A.; Marton, L.J. Simultaneous very fast liquid-chromatographic analysis of ethosuximide, primidone, phenobarbital, phenytoin, and carbamazepine in serum. *Clin.Chem.*, **1983**, *29*, 473–476

Neels, H.M.; Totte, J.A.; Verkerk, R.M.; Vlietinck, A.J.; Scharpe, S.L. Simultaneous high performance liquid-chromatographic determination of carbamazepine, carbamazepine-10,11-epoxide, ethosuximide, phenobarbital, phenytoin, primidone and phenylethylmalonamide in plasma. *J.Clin. Chem.Clin.Biochem.*, **1983**, *21*, 295–299

Turnell, D.C.; Trevor, S.C.; Cooper, J.D. A rapid procedure for the simultaneous estimation of the anticonvulsant drugs, ethosuximide, phenobarbitone, phenytoin, and carbamazepine in serum using high-pressure liquid chromatography. *Ann.Clin.Biochem.*, **1983**, *20 Pt 1*, 37–40

Kinberger, B.; Holmen, A. Analysis for carbamazepine and phenytoin in serum with a high-speed liquid chromatography system (Perkin-Elmer). *Clin.Chem.*, **1982**, *28*, 718–719

MacKichan, J.J. Simultaneous liquid chromatographic analysis for carbamazepine and carbamazepine 10,11-epoxide in plasma and saliva by use of double internal standardization. *J.Chromatogr.*, **1980**, *181*, 373–383

# Carboplatin

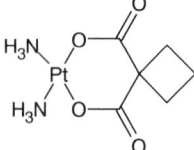

**Molecular formula:** $C_6H_{12}N_2O_4Pt$
**Molecular weight:** 371.3
**CAS Registry No.:** 41575-94-4

## SAMPLE
**Matrix:** blood
**Sample preparation:** 500 μL Plasma + 500 μL MeCN:MeOH:water 10:40:50, vortex, filter through a Centricon-10 10000 molecular mass cut-off filter (Amicon) with centrifuging at 2677 g for 30 min, inject a 10-30 μL aliquot of the ultrafiltrate.

## HPLC VARIABLES
**Column:** 100 × 4.6 3 μm Spherisorb phenyl
**Mobile phase:** MeOH:water 5:95 for 5 min then MeCN:MeOH:isopropanol:water 45:45:5:5 at 2.5 mL/min, re-equilibrate with MeOH:water 5:95 for 10 min
**Column temperature:** 50
**Flow rate:** 0.5
**Injection volume:** 10-30
**Detector:** UV 210

## CHROMATOGRAM
**Retention time:** 3.6
**Limit of detection:** 10 ppb (100 μL injection)

## KEY WORDS
plasma; dog; ultrafiltrate

## REFERENCE
Tyczkowska, K.; Page, R.L.; Riviere, J.E. Determination of carboplatin in canine plasma by liquid chromatography with ultraviolet-visible detection and confirmation by atomic absorption spectroscopy. *J.Chromatogr.*, **1990**, *527*, 447–453

## SAMPLE
**Matrix:** blood, urine
**Sample preparation:** Plasma. Filter (Amicon MPS-1 with a YMT membrane) while centrifuging at 3000 g at 4° for 15 min, inject an aliquot of the ultrafiltrate. Urine. Inject an aliquot directly.

## HPLC VARIABLES
**Column:** 250 × 4.6 Inertsil ODS-2
**Mobile phase:** MeCN:10 mM pH 5.5 buffer 5:95
**Column temperature:** 40
**Flow rate:** 1
**Injection volume:** 100
**Detector:** UV 290 following post-column derivatization. The column effluent mixed with the reagent pumped at 0.3 mL/min and the mixture flowed through a 10 m × 0.5 mm i.d. coil of PTFE tubing held at 60° to the detector. (The reagent was 40 mM sodium bisulfite and 10 mM acetate buffer, pH 5.5.)

## CHROMATOGRAM
**Retention time:** 6.5
**Limit of detection:** 60 nM

## OTHER SUBSTANCES
**Simultaneous:** oxaliplatin, tetraplatin

## KEY WORDS
plasma; ultrafiltrate; rabbit; human; post-column reaction

## REFERENCE
Kizu, R.; Yamamoto, T.; Yokoyama, T.; Tanaka, M.; Miyazaki, M. A sensitive postcolumn derivatization/UV detection system for HPLC determination of antitumor divalent and quadrivalent platinum complexes. *Chem.Pharm.Bull.*, **1995**, *43*, 108–114

## SAMPLE
**Matrix:** formulations
**Sample preparation:** Dilute with mobile phase, inject an aliquot.

## HPLC VARIABLES
**Column:** 300 × 4.6 5 µm C18
**Mobile phase:** water
**Flow rate:** 2.5
**Injection volume:** 20
**Detector:** UV 228

## CHROMATOGRAM
**Retention time:** 3.10

## KEY WORDS
stability-indicating; injections; saline

## REFERENCE
Mayron, D.; Gennaro, A.R. Stability and compatibility of granisetron hydrochloride in i.v. solutions and oral liquids and during simulated Y-site injection with selected drugs. *Am.J.Health-Syst.Pharm.*, **1996**, *53*, 294–304

## SAMPLE
**Matrix:** formulations
**Sample preparation:** Emulsion. 500 µL Emulsion + 10 mL 400 µg/mL hydroquinone in MeOH + 40 mL 0.1% Tween 80, shake until homogeneous, inject a 10 µL aliquot. Drug release medium. 1 mL Drug release medium + 200 µL 100 µg/mL hydroquinone, mix, inject a 50 µL aliquot.

## HPLC VARIABLES
**Column:** 250 × 4.6 10 µm Cosmosil 10 C18 (Nacalai Tesque)
**Mobile phase:** Gradient. MeCN:10 mM pH 3.0 phosphate buffer 2:98 for 1 min, to 45:55 over 5.5 min, maintain at 45:55 for 2 min, return to initial conditions over 1 min.
**Flow rate:** 2
**Injection volume:** 10-50
**Detector:** UV 220

## CHROMATOGRAM
**Retention time:** 2.2
**Internal standard:** hydroquinone (4.2)
**Limit of detection:** 500 ng/mL

## OTHER SUBSTANCES
**Simultaneous:** epirubicin, mitomycin C, iomeprol

## KEY WORDS
emulsions; drug release medium; injections

## REFERENCE
Yamazoe, K.; Horiuchi, T.; Sugiyama, T.; Katagiri, Y. Simultaneous high-performance liquid chromatographic determination of carboplatin, epirubicin hydrochloride and mitomycin C in a Lipiodol emulsion. *J.Chromatogr.A*, **1996**, *726*, 241–245

## SAMPLE
**Matrix:** formulations
**Sample preparation:** Adjust pH to 7.0, dilute if necessary, inject an aliquot.

## HPLC VARIABLES
**Column:** 150 × 4.2 5 μm Nucleosil C18
**Mobile phase:** 10 mM pH 7.0 phosphate buffer containing 0.55 mM hexadecyltrimethylammonium bromide (Condition column before use with 0.5% hexadecyltrimethylammonium bromide.)
**Flow rate:** 1
**Detector:** UV 216

## CHROMATOGRAM
**Retention time:** 3
**Limit of detection:** 1 μg/mL
**Limit of quantitation:** 5 μg/mL

## OTHER SUBSTANCES
**Simultaneous:** cisplatin

## KEY WORDS
infusion fluids; stability-indicating

## REFERENCE
Rochard, E.; Boutelet, H.; Griesemann, E.; Barthes, D.; Courtois, P. Simultaneous high performance liquid chromatographic analysis of carboplatin and cisplatin in infusion fluids. *J.Liq.Chromatogr.*, **1993**, *16*, 1505–1516

◆————————————————◆————————————————◆

## ANNOTATED BIBLIOGRAPHY
Prat, J.; Pujol, M.; Girona, V.; Munoz, M.; Sole, L.A. Stability of carboplatin in 5% glucose solution in glass, polyethylene and polypropylene containers. *J.Pharm.Biomed.Anal.*, **1994**, *12*, 81–84 [stability-indicating; 5% dextrose]

Hadfield, J.A.; McGown, A.T.; Dawson, M.J.; Thatcher, N.; Fox, B.W. The suitability of carboplatin solutions for 14-day continuous infusion by ambulatory pump: an HPLC-dynamic FAB study. *J.Pharm.Biomed.Anal.*, **1993**, *11*, 723–727

Allsopp, M.A.; Sewell, G.J.; Rowland, C.G. A column-switching liquid chromatography assay for the analysis of carboplatin in plasma ultrafiltrate. *J.Pharm.Biomed.Anal.*, **1992**, *10*, 375–381

Duncan, G.F.; Faulkner, H.C.; Farmen, R.H.; Pittman, K.A. Liquid chromatographic procedure for the quantitative analysis of carboplatin in beagle dog plasma ultrafiltrate. *J.Pharm.Sci.*, **1988**, *77*, 273–276

Elferink, F.; van der Vijgh, W.J.; Pinedo, H.M. On-line differential pulse polarographic detection of carboplatin in biological samples after chromatographic separation. *Anal.Chem.*, **1986**, *58*, 2293–2296

Gaver, R.C.; Deeb, G. High-performance liquid chromatographic procedures for the analysis of carboplatin in human plasma and urine. *Cancer Chemother.Pharmacol.*, **1986**, *16*, 201–206

Ding, X.-D.; Krull, I.S. Dual electrode liquid chromatography-electrochemical detection (LCEC) for platinum-derived cancer chemotherapy agents. *J.Liq.Chromatogr.*, **1983**, *6*, 2173–2194

# Carisoprodol

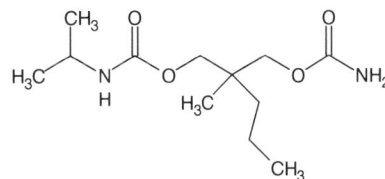

**Molecular formula:** $C_{12}H_{24}N_2O_4$
**Molecular weight:** 260.3
**CAS Registry No.:** 78-44-4

## SAMPLE
**Matrix:** blood
**Sample preparation:** 500 μL Serum + 100 μL 200 mM HCl + 1 mL chloroform, shake for 2 min, sonicate for 1 min, centrifuge for 1 min. Remove the organic layer and add it to 200 mM NaOH, shake for 1 min, sonicate for 1 min, centrifuge for 1 min. Remove 750 μL of the organic layer and evaporate it to dryness, dissolve the residue in 50 μL mobile phase, inject a 20 μL aliquot.

## HPLC VARIABLES
**Column:** 100 × 3 octyl CP-tm-Spher C8 glass column (Chrompack)
**Mobile phase:** MeCN:50 mM $NaH_2PO_4$ 35:65, adjusted to pH 2.2 with phosphoric acid
**Flow rate:** 0.8
**Injection volume:** 20
**Detector:** UV 190

## CHROMATOGRAM
**Retention time:** 5.5
**Internal standard:** carisoprodol

## OTHER SUBSTANCES
**Simultaneous:** meprobamate

## KEY WORDS
serum; carisoprodol is IS

## REFERENCE
Van Damme, M.; Molle, L.; Abi Khalil, F. Useful sample handlings for reversed phase high performance liquid chromatography in emergency toxicology. *J.Toxicol.Clin.Toxicol.*, **1985**, *23*, 589–614

## SAMPLE
**Matrix:** solutions
**Sample preparation:** Inject a 40 μL aliquot of a solution in mobile phase.

## HPLC VARIABLES
**Column:** 300 × 4 10 μm μPorasil
**Mobile phase:** THF:toluene 50:50
**Flow rate:** 2
**Injection volume:** 40
**Detector:** RI

## CHROMATOGRAM
**Retention time:** 2
**Internal standard:** acetaminophen (3)

## OTHER SUBSTANCES
**Simultaneous:** caffeine, phenacetin

## KEY WORDS
normal phase

## REFERENCE

Honigberg, I.L.; Stewart, J.T.; Smith, M. Liquid chromatography in pharmaceutical analysis IX: Determination of muscle relaxant-analgesic mixtures using normal phase chromatography. *J.Pharm.Sci.*, **1978**, *67*, 675–679

# Cefaclor

**Molecular formula:** $C_{15}H_{14}ClN_3O_4S$
**Molecular weight:** 367.8
**CAS Registry No.:** 53994-73-3, 70356-03-5 (monohydrate)

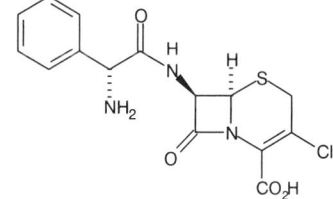

## SAMPLE
**Matrix:** blood
**Sample preparation:** 100 µL Serum + 100 µL 6 µg/mL cefotaxime + 1 mL MeCN, vortex, centrifuge at 2000 g for 10 min. Remove the aqueous phase and add it to 2.5 mL dichloromethane, vortex, centrifuge, inject a 25 µL aliquot of the upper aqueous layer.

## HPLC VARIABLES
**Column:** 150 × 3.9 5 µm C18 (Waters)
**Mobile phase:** MeCN:100 mM phosphate 8:92, pH 5.6
**Flow rate:** 1.2
**Injection volume:** 25
**Detector:** UV 254

## CHROMATOGRAM
**Internal standard:** cefotaxime
**Limit of detection:** 200 ng/mL
**Limit of quantitation:** 500 ng/mL

## KEY WORDS
serum; mouse; pharmacokinetics

## REFERENCE
Onyeji, C.O.; Nicolau, D.P.; Nightingale, C.H.; Quintiliani, R. Optimal times above MICs of ceftibuten and cefaclor in experimental intra-abdominal infections. *Antimicrob.Agents Chemother.*, **1994**, *38*, 1112–1117

## SAMPLE
**Matrix:** blood
**Sample preparation:** Condition a 1 mL Bond-Elut C18 SPE cartridge with 2 mL MeOH and 2 mL 8.5% phosphoric acid. Condition an NH2 SPE cartridge with 1 mL hexane. 500 µL Plasma + 25 µL 8.5% phosphoric acid + 250 µL 1 mg/mL coumarin-3-carboxylic acid in water, add to the C18 SPE cartridge, wash with 500 µL water, wash with 1 mL 8.5% phosphoric acid, wash with 5% MeOH:8.5% phosphoric acid 20:1, elute with 1 mL MeOH:8.5% phosphoric acid 60:40 into the NH2 SPE cartridge. Wash the NH2 SPE cartridge with 1 mL hexane, wash with 1 mL MeCN, elute with 1 mL water:10% ammonium sulfate 95:5, inject a 20 µL aliquot of the eluate.

## HPLC VARIABLES
**Column:** 250 × 4.6 C18
**Mobile phase:** Water:2 mM tetramethylammonium hydroxide in MeOH:acetic acid 60:40:0.5
**Flow rate:** 0.8
**Injection volume:** 20
**Detector:** UV 262

## CHROMATOGRAM
**Internal standard:** coumarin-3-carboxylic acid (13)

## OTHER SUBSTANCES
**Extracted:** cefazolin, ceftizoxime, cephalexin

## KEY WORDS
plasma; SPE

## REFERENCE
Moore, C.M.; Sato, K.; Hattori, H.; Katsumata, Y. Improved HPLC method for the determination of cephalosporins in human plasma and a new solid-phase extraction procedure for cefazolin and ceftizoxime. *Clin.Chim.Acta*, **1990**, *190*, 121–123

## SAMPLE
**Matrix:** blood
**Sample preparation:** 100 μL Serum + 10 μL 100 μg/mL cephradine in water + 100 μL MeCN, vortex, centrifuge at 9000 g for 10 min. Remove 100 μL supernatant, evaporate to dryness at room temperature under reduced pressure, dissolve residue in 100 μL 20 mM $NaH_2PO_4$ adjusted to pH 3.5 with phosphoric acid, centrifuge at 9000 g for 5 min, inject 20 μL supernatant.

## HPLC VARIABLES
**Guard column:** Guard Pak C18 (Waters)
**Column:** 200 × 4.6 5 μm Nucleosil C18
**Mobile phase:** MeCN:buffer 30:70, pH adjusted to 7.0 with NaOH (Buffer was 20 mM sodium phosphate and 5 mM tetrabutylammonium hydrogen sulfate.)
**Flow rate:** 1
**Injection volume:** 20
**Detector:** UV 265

## CHROMATOGRAM
**Retention time:** 5.8
**Internal standard:** cephradine
**Limit of detection:** 1 μg/mL

## KEY WORDS
serum

## REFERENCE
Lindgren, K. Determination of cefaclor and cephradine in serum by ion-pair reversed-phase chromatography. *J.Chromatogr.*, **1987**, *413*, 351–354

## SAMPLE
**Matrix:** blood
**Sample preparation:** 100 μL Serum + 10 μL 5 μg/mL cefixime in MeOH + 100 μL MeCN, vortex for 15 s, centrifuge at 14000 g for 2 min. Remove the supernatant and evaporate it under a stream of nitrogen, reconstitute in 100 μL mobile phase, inject a 50-80 μL aliquot.

## HPLC VARIABLES
**Guard column:** RCSS Silica Guard Pak (Waters)
**Column:** 150 × 4.6 5μm Ultrasphere Octyl C8
**Mobile phase:** MeOH:12.5 mM pH 2.6 $NaH_2PO_4$ (pH adjusted with concentrated phosphoric acid) 20:80
**Flow rate:** 2
**Injection volume:** 50-80
**Detector:** UV 240

## CHROMATOGRAM
**Retention time:** 6
**Internal standard:** cefixime (11)
**Limit of detection:** 1 μg/mL

## OTHER SUBSTANCES
**Extracted:** cefadroxil, cephalexin, cephradine
**Noninterfering:** acetaminophen, cimetidine, diazepam, digoxin, ibuprofen, phenytoin, propranolol, salicylic acid, warfarin

## KEY WORDS
serum

## REFERENCE
McAteer, J.A.; Hiltke, M.F.; Silber, B.M.; Faulkner, R.D. Liquid-chromatographic determination of five orally active cephalosporins−cefixime, cefaclor, cefadroxil, cephalexin, and cephradine−in human serum. *Clin.Chem.*, **1987**, *33*, 1788−1790

## SAMPLE
**Matrix:** blood
**Sample preparation:** Filter plasma (0.22 μm), inject a 10 μL aliquot.

## HPLC VARIABLES
**Column:** 250 × 4.6 5 μm GFF-S5-80 internal-surface reversed phase "Pinkerton" (Regis)
**Mobile phase:** 100 mM pH 4.38 sodium phosphate buffer containing 20 mM sodium dodecyl sulfate
**Flow rate:** 0.8
**Injection volume:** 10
**Detector:** UV 254

## CHROMATOGRAM
**Retention time:** 13

## KEY WORDS
plasma; direct injection

## REFERENCE
Nakagawa, T.; Shibukawa, A.; Shimono, N.; Kawashima, T.; Tanaka, H.; Haginaka, J. Retention properties of internal-surface reversed-phase silica packing and recovery of drugs from human plasma. *J.Chromatogr.*, **1987**, *420*, 297−311

## SAMPLE
**Matrix:** blood
**Sample preparation:** 300 μL Plasma + 300 μL IS in ice-cold MeOH:100 mM pH 5.2 sodium acetate 70:30, vortex for 30 s, let stand at -20° for 10 min, centrifuge at 1500 g for 10 min, inject a 10 μL aliquot.

## HPLC VARIABLES
**Guard column:** 4 × 4 10 μm C18
**Column:** 300 × 4 10 μm μBondapak C18
**Mobile phase:** MeCN:MeOH:100 mM sodium acetate 8.64:0.36:91, pH 5.2
**Flow rate:** 2.5
**Injection volume:** 10
**Detector:** UV 254

## CHROMATOGRAM
**Retention time:** 6
**Internal standard:** 8-chlorotheophylline (8.5)
**Limit of detection:** 500 ng/mL

## KEY WORDS
plasma

## REFERENCE
Signs, S.A.; File, T.M.; Tan, J.S. High-pressure liquid chromatographic method for analysis of cephalo-sporins. *Antimicrob.Agents Chemother.*, **1984**, *26*, 652–655

## SAMPLE
**Matrix:** blood, sinus mucosa
**Sample preparation:** Plasma. Condition a 3 mL 500 mg Bond Elut C8 SPE cartridge with 3 mL MeOH and 2 mL 1% phosphoric acid. 500 μL Plasma + 1 mL 1% phosphoric acid, mix, add to the SPE cartridge, wash with 3 mL 1% perchloric acid, elute with 750 μL MeOH, inject a 50 μL aliquot of the eluate. Sinus mucosa. Condition a 1 mL Bond Elut C8 SPE cartridge with 1 mL MeOH and 1 mL 1% phosphoric acid. Chop sample with a scalpel, weigh out 20 mg and add it to 500 μL 10 mM pH 7.0 phosphate buffer, rotate at 4° for 12 h, centrifuge at 800 g for 10 min. 400 μL Supernatant + 1 mL 1% phosphoric acid (?), mix, add to the SPE cartridge, wash with 1% perchloric acid, elute with 150 μL MeOH, inject a 75 μL aliquot of the eluate.

## HPLC VARIABLES
**Guard column:** 20 × 4.6 5 μm C18 (Shandon)
**Column:** 250 × 4.6 5 μm Supelcosil LC 18
**Mobile phase:** MeOH:MeCN:50 mM pH 3.8 acetate buffer 10:3:87 (plasma) or 12:2:86 (sinus mucosa)
**Flow rate:** 1
**Injection volume:** 50-75
**Detector:** UV 235

## CHROMATOGRAM
**Retention time:** 18.2
**Internal standard:** cefaclor

## OTHER SUBSTANCES
**Extracted:** cefpodoxime

## KEY WORDS
plasma; cefaclor is IS; SPE

## REFERENCE
Camus, F.; Deslandes, A.; Harcouet, L.; Farinotti, R. High-performance liquid chromatographic method for the determination of cefpodoxime levels in plasma and sinus mucosa. *J.Chromatogr.B*, **1994**, *656*, 383–388

## SAMPLE
**Matrix:** blood, urine
**Sample preparation:** Plasma or serum. Condition a Sep-Pak C18 SPE cartridge with 2 mL MeOH then 2 mL water, do not allow to go dry. 500 μL Plasma or serum + 100 μL 100 μg/mL cephalexin in water + 50 μL 25% acetic acid, mix, add to SPE cartridge, wash with two 1 mL portions of water, elute with 3 mL MeOH. Evaporate eluate under nitrogen, add 200 μL mobile phase, vortex, inject a 25 μL aliquot. Urine. Dilute 100:1 (ratio may vary depending on concentration) with water, inject a 50 μL aliquot.

## HPLC VARIABLES
**Column:** 150 × 4.6 5 μm Supelcosil LC-18-DB
**Mobile phase:** MeOH:THF:buffer 16:4:80 (Buffer was 1 g sodium 1-heptanesulfonate + 15 mL triethylamine in 1 L water with the pH adjusted to 2.3 with concentrated phosphoric acid.)
**Column temperature:** 30
**Flow rate:** 1.4
**Injection volume:** 25-50
**Detector:** UV 265

## CHROMATOGRAM
**Retention time:** 6
**Internal standard:** cephalexin (9.2)
**Limit of quantitation:** 500 ng/mL

## OTHER SUBSTANCES
**Extracted:** hydroxyloracarbef, loracarbef
**Noninterfering:** acetaminophen, caffeine

## KEY WORDS
plasma; serum; SPE; pharmacokinetics

## REFERENCE
Kovach, P.M.; Lantz, R.J.; Brier, G. High-performance liquid chromatographic determination of loracarbef, a potential metabolite, cefaclor and cephalexin in human plasma, serum and urine. *J.Chromatogr.*, **1991**, *567*, 129–139

## SAMPLE
**Matrix:** blood, urine
**Sample preparation:** 1 mL Plasma + 1 mL 6% trichloroacetic acid, mix, centrifuge at 4000 rpm for 10 min, inject an aliquot of the supernatant. Inject an aliquot of urine directly.

## HPLC VARIABLES
**Guard column:** 10 × 4 7 μm Lichrosorb RP 18
**Column:** 250 × 4 7 μm Lichrosorb RP 18
**Mobile phase:** MeCN:25 mM pH 7 phosphate buffer 10:90
**Flow rate:** 1
**Injection volume:** 10
**Detector:** F ex 385 em 485 following post-column reaction. The column effluent mixed with 200 μg/mL fluorescamine in MeCN pumped at 0.25 mL/min and the mixture flowed through a 4.5 m × 0.25 mm ID coil of PTFE tubing to the detector.;UV 260

## CHROMATOGRAM
**Retention time:** 14
**Limit of detection:** 1.3 ng/mL (F); 1.6 ng/mL (UV)

## OTHER SUBSTANCES
**Also analyzed:** cefroxadine, cephalexin, cephradine
**Noninterfering:** amidopyrin, aspirin, barbital, caffeine, cefmenoxime, cefotaxime, ceftizoxime, ceftriaxone, cetazidime, diazepam, dibekacin, gentamycin,      nycin, lidocaine, netilmicin, tetracaine, theophylline, tobramycin

## KEY WORDS
post-column reaction; plasma; F detection may be less susceptible to interferences

## REFERENCE
Blanchine, M.D.; Fabre, H.; Mandrou, B. Fluorescamine post-column derivatization for the HPLC determination of cephalosporins in plasma and urine. *J.Liq.Chromatogr.*, **1988**, *11*, 2993–3010

## SAMPLE
**Matrix:** bulk
**Sample preparation:** Dissolve in water at a concentration of 5 mg/mL, inject a 20 μL aliquot.

## HPLC VARIABLES
**Column:** 250 × 4.6 5 μm YMC-ODS (YMC)
**Mobile phase:** Gradient. A was 50 mM sodium dihydrogen phosphate adjusted to pH 4.0 with phosphoric acid. B was MeCN:50 mM sodium dihydrogen phosphate adjusted to pH 4.0 with phosphoric acid 45:55. A:B from 95:5 to 75:25 over 30 min, to 0:100 over 15 min, maintain at 0:100 for 5 min, return to 95:5 and equilibrate for 14 min.
**Flow rate:** 1
**Injection volume:** 20
**Detector:** UV 220

## CHROMATOGRAM
**Retention time:** 28

## OTHER SUBSTANCES
**Simultaneous:** impurities, excipients
**Also analyzed:** cephalexin

## REFERENCE
Olsen, B.A.; Baertschi, S.W.; Riggin, R.M. Multidimensional evaluation of impurity profiles for generic cephalexin and cefaclor antibiotics. *J.Chromatogr.*, **1993**, *648*, 165–173

## SAMPLE
**Matrix:** bulk, formulations
**Sample preparation:** Dissolve in water, inject a 20 μL aliquot.

## HPLC VARIABLES
**Column:** 300 × 3.9 10 μm μBondapak C18
**Mobile phase:** MeOH:water:acetic acid 30:70:0.1
**Flow rate:** 1
**Injection volume:** 20
**Detector:** UV 254

## CHROMATOGRAM
**Retention time:** 7.5
**Limit of quantitation:** 2 μg/mL

## OTHER SUBSTANCES
**Simultaneous:** impurities, cefadroxil, cefamandole, cefamandole nafate, cefazolin, cefoperazone, cefotaxime, cefoxitin, ceftizoxime, cephalexin, cephalothin, cephapirin, cephradine

## REFERENCE
Ting, S. Reverse-phase liquid chromatographic analysis of cephalosporins. *J.Assoc.Off.Anal.Chem.*, **1988**, *71*, 1123–1130

## SAMPLE
**Matrix:** solutions

## HPLC VARIABLES
**Column:** 300 × 3.9 10 μm μBondapak phenyl
**Mobile phase:** MeOH:10 mM phosphate buffer 27:73, pH 3.6
**Column temperature:** 27
**Flow rate:** 1
**Detector:** UV 254

## CHROMATOGRAM
**Retention time:** 6.7

## OTHER SUBSTANCES
**Simultaneous:** amoxicillin, ampicillin, cephalexin, cephradine

## REFERENCE
Huang, H.-S.; Wu, J.-R.; Chen, M.-L. Reversed-phase high-performance liquid chromatography of amphoteric β-lactam antibiotics: effects of columns, ion-pairing reagents and mobile phase pH on their retention times. *J.Chromatogr.*, **1991**, *564*, 195–203

## SAMPLE
**Matrix:** solutions
**Sample preparation:** Inject 100 μL onto column A with mobile phase A, after 3 min backflush the contents of column A onto column B with mobile phase B, elute column B with mobile phase B, monitor the effluent from column B.

## HPLC VARIABLES
**Column:** A 30 × 0.3 5 μm ODS C18 (Nomura); B 150 × 0.3 5 μm ODS C18 (Nomura)
**Mobile phase:** A 10 mM ammonium acetate adjusted to pH 5 with acetic acid; B MeOH: water:acetic acid 40:60:0.5
**Flow rate:** A 0.1; B 0.004
**Injection volume:** 100
**Detector:** UV 262

## CHROMATOGRAM
**Retention time:** 6.71
**Limit of detection:** 10 ng/mL

## OTHER SUBSTANCES
**Simultaneous:** cefaloridine, cefazolin, ceftizoxime

## KEY WORDS
microbore; column-switching

## REFERENCE
Moore, C.M.; Sato, K.; Katsumata, Y. High-performance liquid chromatographic determination of cephalosporin antibiotics using 0.3 mm I.D. columns. *J.Chromatogr.*, **1991**, *539*, 215–220

## ANNOTATED BIBLIOGRAPHY
Muranushi, N.; Horie, K.; Masuda, K.; Hirano, K. Characteristics of ceftibuten uptake into Caco-2 cells. *Pharm.Res.*, **1994**, *11*, 1761–1765 [also cefadroxil, cefazolin, cephalexin, cephradine, cyclacillin, latamoxef]

Lorenz, L.J.; Bashore, F.N.; Olsen, B.A. Determination of process-related impurities and degradation products in cefaclor by high-performance liquid chromatography. *J.Chromatogr.Sci.*, **1992**, *30*, 211–216 determination, [simultaneous impurities, degradation products; bulk; gradient; column temp 25]

Nahata, M.C.; Jackson, D.S. Liquid chromatographic method for the determination of cefadroxil in its suspension and in serum. *J.Liq.Chromatogr.*, **1990**, *13*, 1651–1656 [simultaneous cefadroxil; suspensions; cefaclor is IS]

Nahata, M.C. Determination of cefaclor by high-performance liquid chromatography. *J.Chromatogr.*, **1982**, *228*, 429–433

Rotschafer, J.C.; Crossley, K.B.; Lesar, T.S.; Zaske, D.; Miller, K. Cefaclor pharmacokinetic parameters: serum concentrations determined by a new high-performance liquid chromatographic technique. *Antimicrob.Agents Chemother.*, **1982**, *21*, 170–172

Ullmann, U. High-pressure liquid chromatography and microbiological assay in the determination of serum levels using cefradine and cefaclor. *Zentralbl.Bakteriol.[A].*, **1980**, *248*, 414–421

# Cefadroxil

**Molecular formula:** $C_{16}H_{17}N_3O_5S$
**Molecular weight:** 363.4
**CAS Registry No.:** 66592-87-8

## SAMPLE
**Matrix:** blood
**Sample preparation:** 100 µL Plasma + 8 mL dichloromethane, shake for 20 min, centrifuge at 2500 rpm for 20 min. Remove 7 mL of the organic layer and evaporate it to dryness under nitrogen or at 60°. Dissolve residue in 200 µL mobile phase, inject a 20 µL aliquot.

## HPLC VARIABLES
**Column:** 150 × 6 Shimpack CLS-ODS (Shimadzu)
**Mobile phase:** MeCN:0.5 mM phosphoric acid 12:88
**Column temperature:** 40
**Flow rate:** 1.5
**Injection volume:** 20
**Detector:** UV 254

## CHROMATOGRAM
**Internal standard:** cefadroxil

## OTHER SUBSTANCES
**Simultaneous:** antipyrine

## KEY WORDS
plasma; rat; cefadroxil is IS

## REFERENCE
Lee, C.K.; Uchida, T.; Kitagawa, K.; Yagi, A.; Kim, N.-S.; Goto, S. Skin permeability of various drugs with different lipophilicity. *J.Pharm.Sci.*, **1994**, *83*, 562–565

## SAMPLE
**Matrix:** blood
**Sample preparation:** 100 µL Serum + 1 mL MeCN, vortex, centrifuge at 2000 g for 10 min. Remove the aqueous phase and add it to 2.5 mL dichloromethane, vortex, centrifuge, inject a 25 µL aliquot of the upper aqueous layer.

## HPLC VARIABLES
**Column:** 150 × 3.9 5 µm C18 (Waters)
**Mobile phase:** MeCN:150 mM ammonium acetate 0.7:99.3, pH 7.0
**Flow rate:** 1.1
**Injection volume:** 25
**Detector:** UV 254

## CHROMATOGRAM
**Internal standard:** cefadroxil

## OTHER SUBSTANCES
**Extracted:** ceftibuten

## KEY WORDS
serum; mouse; cefadroxil is IS

## REFERENCE
Onyeji, C.O.; Nicolau, D.P.; Nightingale, C.H.; Quintiliani, R. Optimal times above MICs of ceftibuten and cefaclor in experimental intra-abdominal infections. *Antimicrob.Agents Chemother.*, **1994**, *38*, 1112–1117

## SAMPLE
**Matrix:** blood
**Sample preparation:** 500 μL Plasma + 100 μL 100 μg/mL cefadroxil + 300 μL 5% trichloroacetic acid + 500 μL MeCN + 1.5 mL dichloromethane, vortex for 10 s, centrifuge at 500-600 g at 5° for 10 min, inject a 25 μL aliquot of the aqueous supernatant.

## HPLC VARIABLES
**Guard column:** 23 × 4 37-50 μm Corasil C18
**Column:** 150 × 4 Nova-Pak
**Mobile phase:** MeCN:5 mM 1-octanesulfonic acid 12:88
**Flow rate:** 1
**Injection volume:** 25
**Detector:** UV 280

## CHROMATOGRAM
**Retention time:** 10
**Internal standard:** cefadroxil

## OTHER SUBSTANCES
**Extracted:** cefepime

## KEY WORDS
plasma; rat; cefadroxil is IS

## REFERENCE
Barbhaiya, R.H.; Forgue, S.T.; Shyu, W.C.; Papp, E.A.; Pittman, K.A. High-pressure liquid chromatographic analysis of BMY-28142 in plasma and urine. *Antimicrob.Agents Chemother.*, **1987**, *31*, 55–59

## SAMPLE
**Matrix:** blood
**Sample preparation:** 100 μL Serum + 10 μL 200 μg/mL cephradine in water + 100 μL 6% trichloroacetic acid, vortex, centrifuge at 9000 g for 10 min, inject 25 μL supernatant.

## HPLC VARIABLES
**Guard column:** Waters Guard-Pak C18
**Column:** 200 × 4.6 5 μm Nucleosil SA
**Mobile phase:** 20 mM Ammonium dihydrogen phosphate to final concentration of 20 mM in water:MeOH:MeCN 30:35:35. The pH was adjusted to 3.0 with concentrated phosphoric acid.
**Flow rate:** 1.5
**Injection volume:** 25
**Detector:** UV 240

## CHROMATOGRAM
**Retention time:** 7.7
**Internal standard:** cephradine
**Limit of quantitation:** 1 μg/mL

## KEY WORDS
serum

## REFERENCE
Lindgren, K. Determination of cefadroxil in serum by high-performance liquid chromatography with cephradine as internal standard. *J.Chromatogr.*, **1987**, *413*, 347–350

## SAMPLE
**Matrix:** blood
**Sample preparation:** 100 μL Serum + 10 μL 5 μg/mL cefixime in MeOH + 100 μL MeCN, vortex for 15 s, centrifuge at 14000 g for 2 min. Remove the supernatant and evaporate it under a stream of nitrogen, reconstitute in 100 μL mobile phase, inject a 50-80 μL aliquot.

## HPLC VARIABLES
**Guard column:** RCSS Silica Guard Pak (Waters)
**Column:** 150 × 4.6 5μm Ultrasphere Octyl C8
**Mobile phase:** MeOH:12.5 mM pH 2.6 $NaH_2PO_4$ (pH adjusted with concentrated phosphoric acid) 20:80
**Flow rate:** 2
**Injection volume:** 50-80
**Detector:** UV 240

## CHROMATOGRAM
**Retention time:** 3
**Internal standard:** cefixime (11)
**Limit of detection:** 1 μg/ mL

## OTHER SUBSTANCES
**Extracted:** cefaclor, cephalexin, cephradine
**Noninterfering:** acetaminophen, cimetidine, diazepam, digoxin, ibuprofen, phenytoin, propranolol, salicylic acid, warfarin

## KEY WORDS
serum

## REFERENCE
McAteer, J.A.; Hiltke, M.F.; Silber, B.M.; Faulkner, R.D. Liquid-chromatographic determination of five orally active cephalosporins–cefixime, cefaclor, cefadroxil, cephalexin, and cephradine–in human serum. *Clin.Chem.*, **1987**, *33*, 1788–1790

## SAMPLE
**Matrix:** blood
**Sample preparation:** 1 mL Plasma + 1 mL MeCN, shake 30 s, centrifuge 5 min. Transfer upper layer phase, add 6 mL dichloromethane, shake 5 min, centrifuge 5 min, inject 10-100 μL upper aqueous phase.

## HPLC VARIABLES
**Column:** 300 × 3.9 μBondapak C18
**Mobile phase:** MeOH:10 mM pH 4.8 buffer 5:95
**Flow rate:** 1.5
**Injection volume:** 10-100
**Detector:** UV 240

## CHROMATOGRAM
**Retention time:** 6.4
**Limit of detection:** 150 ng/mL

## KEY WORDS
plasma

## REFERENCE
Brisson, A.M.; Fourtillan, J.B. Pharmacokinetic study of cefadroxil following single and repeated doses. *J.Antimicrob.Chemother.*, **1982**, *10 Suppl B*, 11–15

## SAMPLE
**Matrix:** blood, middle ear fluid
**Sample preparation:** Condition a 2.8 mL 500 mg Bond Elut C18 SPE cartridge with 4 mL MeOH and 1 mL 50 mM pH 6.8 phosphate buffer. 200 µL Plasma or 50 µL middle ear fluid + 1 mL 50 mM pH 6.8 phosphate buffer, vortex, add to the SPE cartridge, wash with 1 mL 50 mM pH 6.8 phosphate buffer, wash with 1 mL buffer, dry under vacuum, elute with 1 mL MeOH:water 40:60. Evaporate the eluate to dryness under a stream of nitrogen at 40°, reconstitute with 35 µL MeOH:water 5:95, inject a 25 µL aliquot.

## HPLC VARIABLES
**Guard column:** 10 × 2 5 µm MOS Hypersil C8
**Column:** 150 × 2 5 µm MOS Hypersil C8
**Mobile phase:** MeCN:5 mM phosphate buffer containing 5 mM tetrabutylammonium 6:94, adjusted to pH 6.5 (After 14 min wash column with MeCN:buffer 25:75 for 2 min, re-equilibrate for 4 min.)
**Column temperature:** 40
**Flow rate:** 0.35
**Injection volume:** 25
**Detector:** UV 210

## CHROMATOGRAM
**Retention time:** 12.30
**Internal standard:** cefadroxil

## OTHER SUBSTANCES
**Extracted:** amoxicillin

## KEY WORDS
plasma; SPE; cefadroxil is IS

## REFERENCE
Yuan, Z.; Russlie, H.Q.; Canafax, D.M. Sensitive assay for measuring amoxicillin in human plasma and middle ear fluid using solid-phase extraction and reversed-phase high-performance liquid chromatography. *J.Chromatogr.B*, **1995**, *674*, 93–99

## SAMPLE
**Matrix:** blood, urine
**Sample preparation:** Plasma. 150 µL Plasma + 150 µL MeCN, vortex, rotate at 20 rpm for 10 min; centrifuge at 1000 g for 10 min. Transfer supernatant to another tube and add 7 volumes dichloromethane, equilibrate for 10 min; rotate at 20 rpm for 10 min; centrifuge at 1000 g for 10 min, inject an aliquot of the upper aqueous layer (J.Chromatogr. 1987, 413, 109). Urine. Dilute with water, inject an aliquot.

## HPLC VARIABLES
**Guard column:** C18
**Column:** 150 × 1.6 Spherisorb S5-ODS2 C18
**Mobile phase:** MeOH:100 mM pH 3 acetate buffer 13:87
**Flow rate:** 1
**Detector:** UV 254

## CHROMATOGRAM
**Limit of detection:** 300 ng/mL

## KEY WORDS
plasma; rat; pharmacokinetics

## REFERENCE
Gimeno, M.J.; Martínez, M.; Granero, L.; Torres-Molina, F.; Peris, J.-E. Influence of probenecid on the renal excretion mechanisms of cefadroxil. *Drug Metab.Dispos.*, **1996**, *24*, 270–272

## SAMPLE
**Matrix:** blood, urine
**Sample preparation:** 1 mL Plasma + 1 mL 6% trichloroacetic acid, mix, centrifuge at 4000 rpm for 10 min, inject an aliquot of the supernatant. Inject an aliquot of urine directly.

## HPLC VARIABLES
**Guard column:** 10 × 4 7 μm Lichrosorb RP 18
**Column:** 250 × 4 7 μm Lichrosorb RP 18
**Mobile phase:** MeCN:25 mM pH 7 phosphate buffer 5:95
**Flow rate:** 1
**Injection volume:** 10
**Detector:** F ex 385 em 485 following post-column reaction. The column effluent mixed with 200 μg/mL fluorescamine in MeCN pumped at 0.25 mL/min and the mixture flowed through a 4.5 m × 0.25 mm ID coil of PTFE tubing to the detector.; UV 260

## CHROMATOGRAM
**Limit of detection:** 0.3 ng/mL (F); 0.6 ng/mL (UV)

## OTHER SUBSTANCES
**Noninterfering:** amidopyrin, aspirin, barbital, caffeine, cefmenoxime, cefotaxime, ceftizoxime, ceftriaxone, cetazidime, diazepam, dibekacin, gentamycin, kanamycin, lidocaine, netilmicin, tetracaine, theophylline, tobramycin

## KEY WORDS
post-column reaction; plasma; F detection may be less susceptible to interferences

## REFERENCE
Blanchine, M.D.; Fabre, H.; Mandrou, B. Fluorescamine post-column derivatization for the HPLC determination of cephalosporins in plasma and urine. *J.Liq.Chromatogr.*, **1988**, *11*, 2993–3010

## SAMPLE
**Matrix:** bulk, formulations
**Sample preparation:** Homogenize sample, weigh out sample equivalent to 50 mg cefadroxil, make up to 50 mL with 100 mM pH 4.5 phosphate buffer. Take 5 mL of this solution, add 0.5 mL 30 mg/mL dimethyl phthalate in MeCN:water 1:1, make up to 50 mL with 100 mM pH 4.5 phosphate buffer.

## HPLC VARIABLES
**Column:** 300 × 3.9 10 μm μBondapak C18
**Mobile phase:** MeCN:10 mM pH 4.5 phosphate buffer 60:40
**Flow rate:** 1
**Injection volume:** 10
**Detector:** UV 254

## CHROMATOGRAM
**Retention time:** 2
**Internal standard:** dimethyl phthalate
**Limit of quantitation:** 20 μg/mL

## KEY WORDS
capsules; powders

## REFERENCE
Hsu, M.-C.; Chang, Y.-W.; Lee, Y.-T. Column liquid chromatography and microbiological assay compared for determination of cefadroxil preparations. *J.Chromatogr.*, **1992**, *609*, 181–186

## SAMPLE
**Matrix:** bulk, formulations
**Sample preparation:** Dissolve in water to a concentration of 40 μg/mL, inject a 20 μL aliquot.

## HPLC VARIABLES
**Column:** 300 × 3.9 10 μm μBondapak C18
**Mobile phase:** MeOH:water:acetic acid 30:70:0.1
**Flow rate:** 1
**Injection volume:** 20
**Detector:** UV 254

## CHROMATOGRAM
**Retention time:** 5
**Limit of quantitation:** 1.8 μg/mL

## OTHER SUBSTANCES
**Simultaneous:** impurities, cefaclor, cefamandole, cefamandole nafate, cefazolin, cefoperazone, cefotaxime, cefoxitin, ceftizoxime, cephalexin, cephalothin, cephapirin, cephradine

## REFERENCE
Ting, S. Reverse-phase liquid chromatographic analysis of cephalosporins. *J.Assoc.Off.Anal.Chem.*, **1988**, *71*, 1123–1130

## SAMPLE
**Matrix:** milk
**Sample preparation:** Condition a Bond Elut C8 SPE cartridge with 5 mL MeOH and 5 mL water. 20 mL Milk + 1 mL 1 M oxalic acid, heat at 60° for 10 min, centrifuge for 10 min, remove the supernatant and add it to 20 mL water and 400 μL tributylamine, shake well, add to the SPE cartridge, wash with two 2.5 mL portions of water, elute with 2.5 mL MeOH. Evaporate the eluate to dryness under a stream of nitrogen, extract the residue with three 100 μL portions of 50 mM pH 6.0 potassium phosphate buffer, filter (0.2 μm), inject an aliquot of the filtrate. (Buffer was 545 mL 100 mM citric acid, 455 mL 200 mM $Na_2HPO_4$, and 74.4 g EDTA, adjust to pH 4.5 with ammonium hydroxide, make up to 2 L with water.)

## HPLC VARIABLES
**Column:** 250 × 4.6 10 μm Lichrosorb RP-8
**Mobile phase:** MeOH:50 mM pH 6.0 potassium phosphate buffer 35:65
**Flow rate:** 1
**Injection volume:** 200
**Detector:** UV 210; Charm II assay

## OTHER SUBSTANCES
**Extracted:** amoxicillin, ticarcillin
**Simultaneous:** ampicillin, ceftiofur, cephapirin, cloxacillin, dicloxacillin, nafcillin, oxacillin, penicillin G

## KEY WORDS
SPE

## REFERENCE
Zomer, E.; Quintana, J.; Saul, S.; Charm, S.E. LC-Receptograms: A method for identification and quantitation of β-lactams in milk by liquid chromatography with microbial receptor assay. *J.AOAC Int.*, **1995**, *78*, 1165–1172

## SAMPLE
**Matrix:** perfusate

## HPLC VARIABLES
**Guard column:** 40 μm C18 (Teknochroma)
**Column:** 150 × 3.9 4 μm Novapak C18 (Teknochroma)
**Mobile phase:** MeCN:buffer 16:84, adjusted to pH 3.9 with dilute NaOH (Buffer was 5.8 mL glacial acetic acid and 2.456 g sodium laurylsulfate in 1 L water.)
**Flow rate:** 1
**Detector:** UV 280

## KEY WORDS
rat

## REFERENCE
Sancho-Chust, V.; Fabra-Campos, S.; Gómez-Meseguer, V.; Bengochea, M.; Martín-Villodre, A. Experimental studies on the influence of surfactants on intestinal absorption of drugs. Cefadroxil as model drug and sodium lauryl sulfate as model surfactant: Studies in rat colon. *Arzneimittelforschung*, **1995**, *45*, 595–601

## SAMPLE
**Matrix:** solutions

## HPLC VARIABLES
**Column:** 250 × 4 OmniPac PCX-500 (Dionex)
**Mobile phase:** Gradient. A was MeCN:90 mM perchloric acid 13.5:86.5. B was MeCN:300 mM perchloric acid 45:55. A:B from 100:0 to 0:100 over 7 min, maintain at 0:100.
**Flow rate:** 1
**Detector:** UV 254

## CHROMATOGRAM
**Retention time:** 6.5

## OTHER SUBSTANCES
**Simultaneous:** 7-aminocephalosproranic acid, cefazolin, cefotaxime, cephalexin, cephaloridine, cephalosporin C, cephalothin, cephapirin, D-hydroxyphenylglycine

## REFERENCE
Slingsby, R.W.; Rey, M. Determination of pharmaceuticals by multi-phase chromatography: Combined reversed phase and ion exchange in one column. *J.Liq.Chromatogr.*, **1990**, *13*, 107–134

## SAMPLE
**Matrix:** surface wipes
**Sample preparation:** Swab 100 × 100 mm surface with water (total volume 10 mL), remove excess liquid with a second swab, vortex swabs for 45 s, filter (0.45 μm polycarbonate), inject a 50 μL aliquot.

## HPLC VARIABLES
**Column:** 150 × 4.6 5 μm Nucleosil C18
**Mobile phase:** MeCN:70 mM $KH_2PO_4$ 4:96
**Flow rate:** 1
**Injection volume:** 50
**Detector:** UV 254

## CHROMATOGRAM
**Retention time:** 5.6
**Limit of quantitation:** 100 ng/mL

## REFERENCE
Gorski, R.J.; Plasz, A.C.; Elrod, L.J.; Yoder, J.; White, L.B. Determination of cefsulodin, cefmenoxime, and cefadroxil as residues on surfaces. *Pharm.Res.*, **1991**, *8*, 1525–1527

## SAMPLE
**Matrix:** tissue
**Sample preparation:** Homogenize muscle with three volumes phosphate-buffered saline (Polytron, level 3) for 2 min, centrifuge at 1300 g for 10 min. 125 μL Supernatant + 100 μL water + 800 μL MeCN, vortex for 30 s, centrifuge at 1600 g for 5 min. Remove the supernatant and evaporate it to dryness under a stream of nitrogen, reconstitute the residue in 125 μL mobile phase, inject a 50 μL aliquot.

## HPLC VARIABLES
**Column:** 4 μm Novapak C18
**Mobile phase:** MeCN:5 mm sodium heptanesulfonic acid 9:91, adjust pH to 3.33 with glacial acetic acid
**Flow rate:** 2
**Injection volume:** 50
**Detector:** UV 280

## CHROMATOGRAM
**Retention time:** 6.7
**Internal standard:** cefadroxil

## OTHER SUBSTANCES
**Extracted:** cefepime

## KEY WORDS
mouse; muscle; cefadroxil is IS

## REFERENCE
Darouiche, R.; Musher, D.; Hamill, R.; Ou, C.; Rognerud, C. Cephalosporin penetration into soft tissue of paralyzed limbs. *Antimicrob.Agents Chemother.*, **1989**, *33*, 1326–1328

## ANNOTATED BIBLIOGRAPHY
Changqin, H.; Shaohong, S.; Kaimin, W. The chromatographic behavior of cephalosporins in gel filtration chromatography, a novel method to separate high molecular weight impurities. *J.Pharm.*

*Biomed.Anal.*, **1994**, *12*, 533–541 [also cefamandole, cefmenoxime, cefoperazone, cefotaxime, cefta-zidime, ceftriaxone, cephalexin, cephaloridine, cephalothin, cephradine]

Muranushi, N.; Horie, K.; Masuda, K.; Hirano, K. Characteristics of ceftibuten uptake into Caco-2 cells. *Pharm.Res.*, **1994**, *11*, 1761–1765 [also cefaclor, cefazolin, ceftibuten, cephalexin, cephradine, cycla-cillin, latamoxef]

Snippe, N.; Van de Merbel, N.C.; Ruiter, F.P.M.; Steijer, O.M.; Lingeman, H.; Brinkman, U.A.T. Auto-mated column liquid chromatographic determination of amoxicillin and cefadroxil in bovine serum and muscle tissue using on-line dialysis for sample preparation. *J.Chromatogr.B*, **1994**, *662*, 61–70 [extracted amoxicillin; cow; serum; muscle; on-line dialysis; SPE; post-column reaction; LOD 50 ng/mL; LOD 200 ng/g]

Nahata, M.C.; Jackson, D.S. Liquid chromatographic method for the determination of cefadroxil in its suspension and in serum. *J.Liq.Chromatogr.*, **1990**, *13*, 1651–1656 [suspensions; serum; cefaclor (IS)]

# Cefixime

**Molecular formula:** $C_{16}H_{15}N_5O_7S_2$
**Molecular weight:** 453.4
**CAS Registry No.:** 79350-37-1

## SAMPLE
**Matrix:** bile, blood
**Sample preparation:** Plasma. 25 µL Plasma + 25 µL buffer + 450 µL 5% trichloroacetic acid, mix, centrifuge, inject a 10 µL aliquot of the supernatant. Bile. 100 µL Bile + 900 µL buffer, mix, centrifuge at 2000 rpm for 5 min, inject a 10 µL aliquot of the supernatant. (Buffer was 2.54% $NaH_2PO_4.2H_2O$ and 4.41% $Na_2HPO_4.12H_2O$, pH 7.4.)

## HPLC VARIABLES
**Column:** $150 \times 4.6$ Chemcosorb 7C18 (Chemco)
**Mobile phase:** MeCN:buffer 11:89 (plasma) or 13:87 (bile) (Buffer was 20 mM $(NH_4)H_2PO_4$ adjusted to pH 3.2 with phosphoric acid.)
**Column temperature:** 40
**Flow rate:** 2 (plasma), 1 (bile)
**Injection volume:** 10
**Detector:** UV 290

## KEY WORDS
rat; plasma

## REFERENCE
Yasui, H.; Yamaoka, K.; Nakagawa, T. Alternative continuous infusion method for analysis of entero-hepatic circulation and biliary excretion of cefixime in the rat. *J.Pharm.Sci.*, **1994**, *83*, 819–823

## SAMPLE
**Matrix:** blood
**Sample preparation:** 100 µL Serum + 10 µL 5 µg/mL cephalexin in MeOH + 100 µL MeCN, vortex for 15 s, centrifuge at 14000 g for 2 min. Remove the supernatant and evaporate it under a stream of nitrogen, reconstitute in 100 µL mobile phase, inject a 50-80 µL aliquot.

## HPLC VARIABLES
**Guard column:** RCSS Silica Guard Pak (Waters)
**Column:** $150 \times 4.6$ 5µm Ultrasphere Octyl C8
**Mobile phase:** MeOH:12.5 mM pH 2.6 $NaH_2PO_4$ (pH adjusted with concentrated phosphoric acid) 20:80
**Flow rate:** 2
**Injection volume:** 50-80
**Detector:** UV 240

## CHROMATOGRAM
**Retention time:** 11
**Internal standard:** cephalexin (15)
**Limit of detection:** 100 ng/mL

## OTHER SUBSTANCES
**Extracted:** cefaclor, cefadroxil, cephradine

**Noninterfering:** acetaminophen, cimetidine, diazepam, digoxin, ibuprofen, phenytoin, propranolol, salicylic acid, warfarin

## KEY WORDS
serum

## REFERENCE
McAteer, J.A.; Hiltke, M.F.; Silber, B.M.; Faulkner, R.D. Liquid-chromatographic determination of five orally active cephalosporins—cefixime, cefaclor, cefadroxil, cephalexin, and cephradine—in human serum. *Clin.Chem.*, **1987**, *33*, 1788–1790

## SAMPLE
**Matrix:** blood, CSF
**Sample preparation:** 150 µL Serum or CSF + 150 µL 6% trichloroacetic acid, centrifuge for 3 min, inject 75 µL supernatant.

## HPLC VARIABLES
**Column:** 150 × 4.6 3 µm Techsphere C18
**Mobile phase:** 170 mL MeCN + 1.36 g $NaH_2PO_4$ + 2 mL phosphoric acid + 828 mL water adjusted to pH 2.7
**Flow rate:** 2
**Injection volume:** 75
**Detector:** UV 313

## KEY WORDS
serum; method has advantages over that of Falkowski et al. (J. Chromatogr. 1987; 422; 145-152)

## REFERENCE
White, L.O.; Reeves, D.S.; Lovering, A.M.; MacGowan, A.P. HPLC assay of cefixime in serum and CSF. *J.Antimicrob.Chemother.*, **1993**, *31*, 450–451

## SAMPLE
**Matrix:** blood, tissue, urine
**Sample preparation:** Serum, plasma. Dilute serum or plasma 1:2 to 1:10 with buffer, centrifuge, inject a 20 µL aliquot of supernatant. Urine. Dilute urine 1:10 to 1:100 with buffer, centrifuge, inject a 20 µL aliquot of supernatant. Tissue (lung, gut). Cut tissue with a scalpel, homogenize with 1-3 mL buffer, centrifuge at 9600 g for 5 min three times, inject a 20 µL aliquot. Tissue (chondral). Cut tissue with a scalpel, homogenize with 3-6 mL buffer in an ice bath for 2-3 min, centrifuge at 9600 g for 5 min four or five times, inject a 100 µL aliquot. Pleural. Dilute human pleural samples with buffer, centrifuge, inject a 20 µL aliquot. (Buffer was 66.6 mM $K_2HPO_4$ adjusted to pH 7.40 with $KH_2PO_4$.)

## HPLC VARIABLES
**Column:** 200 × 4 5 µm Nucleosil C18
**Mobile phase:** MeOH:buffer 15:85, adjusted to pH 5.2 with phosphoric acid (Buffer was 57.4 mM $K_2HPO_4$ adjusted to pH 5.2 with phosphoric acid.)
**Flow rate:** 1
**Injection volume:** 20-100
**Detector:** UV 230

## CHROMATOGRAM
**Retention time:** 9
**Limit of detection:** 100 ng/mL

## KEY WORDS
serum; plasma; lung; gut; pleural; chondral

## REFERENCE
Knöller, J.; König, W.; Schönfeld, W.; Bremm, K.D.; Köller, M. Application of high-performance liquid chromatography of some antibiotics in clinical microbiology. *J.Chromatogr.*, **1988**, *427*, 257–267

## SAMPLE
**Matrix:** blood, urine
**Sample preparation:** Serum. 250 µL Serum + 250 µL 4 µg/mL IS in 6% trichloroacetic acid in water, vortex at high speed for 15 s, centrifuge at 30000 g for 2 min, inject a 75 µL aliquot of the supernatant. Urine. 100 µL Urine + 2 mL 0.25 µg/mL IS in 6% trichloroacetic acid in water, vortex at high speed for 15 s, centrifuge at 30000 g for 2 min, inject a 75 µL aliquot of the supernatant.

## HPLC VARIABLES
**Column:** 100 × 8 5 µm Nova-Pak C18 RCM-100
**Mobile phase:** MeCN:water:85% phosphoric acid 17:82.8:0.2 containing 1.36 g/L $KH_2PO_4$, pH 2.7 (serum) or MeCN:water:85% phosphoric acid 20:79.8:0.2 containing 1.36 g/L $KH_2PO_4$, pH 2.7 (urine)
**Flow rate:** 2
**Injection volume:** 75
**Detector:** UV 280 (serum); UV 313 (urine)

## CHROMATOGRAM
**Retention time:** 3.5 (serum), 4 (urine)
**Internal standard:** 7-hydroxycoumarin (5.6 (serum), 6.8 (urine))
**Limit of quantitation:** 50 ng/mL (serum); 1000 ng/mL (urine)

## OTHER SUBSTANCES
**Noninterfering:** acetaminophen, caffeine, salicylic acid, theophylline

## KEY WORDS
serum; human; rat; dog; monkey

## REFERENCE
Falkowski, A.J.; Look, Z.M.; Noguchi, H.; Silber, B.M. Determination of cefixime in biological samples by reversed- phase high-performance liquid chromatography. *J.Chromatogr.*, **1987**, *422*, 145–152

## SAMPLE
**Matrix:** blood, urine
**Sample preparation:** Dilute serum 1:10 with buffer, dilute urine 1:10-100 with buffer, centrifuge at 9300 g for 4 min, inject 20 µL of supernatant. (Buffer was 60 mmol $K_2HPO_4$ adjusted to pH 7.40 with 50 mmol $KH_2PO_4$ (Soerensen buffer).)

## HPLC VARIABLES
**Column:** 200 × 4 Nucleosil 5 C18
**Mobile phase:** MeOH:43 mM $K_2HPO_4$ 15:85 adjusted to pH 5.20 with phosphoric acid
**Flow rate:** 1
**Injection volume:** 20
**Detector:** UV 230

## CHROMATOGRAM
**Retention time:** 12
**Limit of detection:** 100 ng/mL

## KEY WORDS
serum

## REFERENCE
Knöller, J.; Schönfeld, W.; Bremm, K.D.; König, W. *In vitro* stability of cefixime (FK-027) in serum, urine and buffer. *J.Chromatogr.*, **1987**, *389*, 312–316

## ANNOTATED BIBLIOGRAPHY
Liu, G.L.; Sha, R.G.; Gao, S.; Shen, Y.X.; Wang, S.X. [Determination of cefixime in human plasma and urine using high performance liquid chromatography column switching technique]. *Yao Hsueh Hsueh Pao*, **1993**, *28*, 216–221

Nahata, M.C. Measurement of cefixime in serum and cerebrospinal fluid by high-performance liquid chromatography. *J.Liq.Chromatogr.*, **1991**, *14*, 3755–3759 [serum; CSF; 7-hydroxycoumarin (IS); LOD 30 ng/mL]

# Cefotaxime

**Molecular formula:** $C_{16}H_{17}N_5O_7S_2$
**Molecular weight:** 455.5
**CAS Registry No.:** 63527-52-6 (cefotaxime),
64485-93-4 (cefotaxime sodium)

## SAMPLE
**Matrix:** bile, blood, urine
**Sample preparation:** Serum. 0.5 mL serum + 0.5 mL MeCN mix in 7 mL tube on vortex mixer, shake by rotation (20 rpm) 10 min, centrifuge 10 min 1000 g, transfer supernatant to another tube, add 7 aliquots dichloromethane, equilibrate 10 min, shake by rotation (20 rpm) 10 min, centrifuge 10 min 1000 g, inject aliquot of upper aqueous layer. Urine. Centrifuge urine and dilute 1:20. Bile. Centrifuge bile and dilute 1:10.

## HPLC VARIABLES
**Column:** 250 × 4.6 5 μm Ultrasphere ODS
**Mobile phase:** 8:92 MeCN:20 mM ammonium acetate adjusted to pH 5 with glacial acetic acid
**Flow rate:** 1
**Injection volume:** 50
**Detector:** UV 254

## CHROMATOGRAM
**Retention time:** 8.4
**Limit of detection:** 200 ng/mL

## OTHER SUBSTANCES
**Also analyzed:** ampicillin, azlocillin, aztreonam, cefmenoxime, cefoperazone, cefsulodin, ceftazidime, ceftriaxone, cloxacillin, desacetylcefotaxime, mezlocillin, penicillin G, piperacillin, ticarcillin

## KEY WORDS
serum

## REFERENCE
Jehl, F.; Birckel, P.; Monteil, H. Hospital routine analysis of penicillins, third-generation cephalosporins and aztreonam by conventional and high-speed high-performance liquid chromatography. *J.Chromatogr.*, **1987**, *413*, 109–119

## SAMPLE
**Matrix:** blood
**Sample preparation:** 100 μL Serum + 1 mL MeCN, vortex, centrifuge at 2000 g for 10 min. Remove the aqueous phase and add it to 2.5 mL dichloromethane, vortex, centrifuge, inject a 25 μL aliquot of the upper aqueous layer.

## HPLC VARIABLES
**Column:** 150 × 3.9 5 μm C18 (Waters)
**Mobile phase:** MeCN:100 mM phosphate 8:92, pH 5.6
**Flow rate:** 1.2
**Injection volume:** 25
**Detector:** UV 254

## CHROMATOGRAM
**Internal standard:** cefotaxime

## OTHER SUBSTANCES
**Extracted:** cefaclor

## KEY WORDS
serum; mouse; cefotaxime is IS

## REFERENCE
Onyeji, C.O.; Nicolau, D.P.; Nightingale, C.H.; Quintiliani, R. Optimal times above MICs of ceftibuten and cefaclor in experimental intra-abdominal infections. *Antimicrob.Agents Chemother.*, **1994**, *38*, 1112–1117

## SAMPLE
**Matrix:** blood
**Sample preparation:** Condition a C8 SPE cartridge with 1 mL MeOH:DMF 90:10 and 1 mL 1% phosphoric acid, do not allow to go dry. 200 μL Plasma + 1 mL 1 μg/mL cefaclor in 1% phosphoric acid + 200 μL MeCN:1% phosphoric acid 1:99, add to the SPE cartridge, wash with 1 mL MeOH:1% phosphoric acid 5:95, wash with 500 μL 1% phosphoric acid, elute the contents of the SPE cartridge onto the analytical column with the mobile phase.

## HPLC VARIABLES
**Guard column:** 12 × 4.6 7 μm Newguard C8
**Column:** 250 × 4.6 5 μm IB-SIL C18 (Phenomenex)
**Mobile phase:** MeCN:MeOH:50 mM pH 6.0 sodium acetate buffer 4:4:92 (After elution of IS inject 1 mL MeCN:water 90:10 to remove late eluting peaks.)
**Flow rate:** 2
**Detector:** UV 254

## CHROMATOGRAM
**Retention time:** 18.2
**Internal standard:** cefaclor (14.1)

## OTHER SUBSTANCES
**Extracted:** caffeine, cefpodoxime
**Noninterfering:** acetaminophen, amikacin, ceftazidime, ceftriaxone, gentamicin, nafcillin, phenytoin, ticarcillin, tobramycin, vancomycin
**Interfering:** theophylline

## KEY WORDS
SPE; plasma

## REFERENCE
Steenwyk, R.C.; Brewer, J.E.; Royer, M.E.; Cathcart, K.S. Reversed-phase liquid chromatographic determination of cefpodoxime in human plasma. *J.Liq.Chromatogr.*, **1991**, *14*, 3641–3656

## SAMPLE
**Matrix:** blood
**Sample preparation:** 500 μL Serum + 500 μL ice-cold 100 μg/mL cefoperazone in MeOH:100 mM pH 5.2 sodium acetate 70:30, vortex for 30 s, hold at -20° for 10 min, centrifuge at 1500 g for 10 min, inject 15 μL of supernatant.

## HPLC VARIABLES
**Guard column:** 10 μm C18 Guard-PAK

**Column:** 300 × 3.9 10 μm μBondapak C18
**Mobile phase:** MeCN:10 mM pH 7.5 phosphate buffer containing 10 mM hexadecyltri-methylammonium bromide 18:82
**Flow rate:** 2
**Injection volume:** 15
**Detector:** UV 254

## CHROMATOGRAM
**Internal standard:** cefoperazone
**Limit of detection:** 800 ng/mL

## KEY WORDS
serum

## REFERENCE
Deeter, R.G.; Weinstein, M.P.; Swanson, K.A.; Gross, J.S.; Bailey, L.C. Crossover assessment of serum bactericidal activity and pharmacokinetics of five broad-spectrum cephalosporins in the elderly. *Antimicrob.Agents Chemother.*, **1990**, *34*, 1007−1013

## SAMPLE
**Matrix:** blood
**Sample preparation:** Mix 100 μL plasma + 300 μL 5 μg/mL cefotaxime in pH 3.5 10 mM acetate buffer and keep at 4°. Inject 100 μL onto column A with mobile phase A. After 5 min backflush column A with mobile phase B onto column B for 3 min. Re-equilibrate column A with mobile phase A for 16 min.

## HPLC VARIABLES
**Column:** A 40 × 2 37-50 μm Corasil RP C18; B 20 × 4 25-40 μm Lichrosorb RP-8 + 250 × 4 Partisil ODS-3
**Mobile phase:** A 10 mM pH 3.5 acetate buffer; B MeCN:20 mM pH 4.3 acetate buffer 15:85
**Flow rate:** 1
**Injection volume:** 100
**Detector:** UV 254

## CHROMATOGRAM
**Retention time:** 10.3
**Internal standard:** cefotaxime
**Limit of detection:** 500 ng/mL

## OTHER SUBSTANCES
**Simultaneous:** cefoxitin, cefuroxime, cephalexin, cephaloridine
**Noninterfering:** alclofenac, aspirin, caffeine, cefadroxil, cefamandole, cefazolin, cefopera-zone, cefotiam, cephalothin, diclofenac, ibuprofen, indomethacin, ketoprofen, lonazolac, mefenamic acid, naproxen, phenylbutazone, piroxicam

## KEY WORDS
plasma; column-switching; cefotaxime is IS; rat; human

## REFERENCE
Lee, Y.J.; Lee, H.S. Simultaneous determination of cefoxitin, cefuroxime, cephalexin and cephaloridine in plasma using HPLC and a column-switching technique. *Chromatographia*, **1990**, *30*, 80−84

## SAMPLE
**Matrix:** blood

**Sample preparation:** 500 µL Plasma + 1 mL MeCN, vortex for 3 s, centrifuge for 5 min. Remove the upper layer and add it to 3 mL dichloromethane, shake for 5 min, centrifuge, inject a 20 µL aliquot of the aqueous phase.

## HPLC VARIABLES
**Column:** 250 × 4.6 7 µm Eicompak MA-ODS (Eicom Corp.)
**Mobile phase:** MeCN:10 mM pH 4.2 acetate buffer 15:85
**Injection volume:** 20
**Detector:** UV 254

## CHROMATOGRAM
**Retention time:** 4.1
**Internal standard:** cefotaxime

## OTHER SUBSTANCES
**Extracted:** cefotiam

## KEY WORDS
plasma; cefotaxime is IS

## REFERENCE
Chiba, K.; Tsuchiya, M.; Kato, J.; Ochi, K.; Kawa, Z.; Ishizaki, T. Cefotiam disposition in markedly obese athlete patients, Japanese sumo wrestlers. *Antimicrob.Agents Chemother.*, **1989**, *33*, 1188–1192

## SAMPLE
**Matrix:** blood
**Sample preparation:** 200 µL Serum + 100 µL water + 1 mL MeCN, vortex for 5 s, centrifuge at 30 g for 5 min. Remove the supernatant and add it to 1.5 mL dichloromethane, vortex for 5 s, centrifuge for 5 min, inject a 10-20 µL aliquot of the upper aqueous layer.

## HPLC VARIABLES
**Guard column:** 50 mm long CO:PELL ODS
**Column:** 300 × 3.9 µBondapak C18
**Mobile phase:** MeCN:water:glacial acetic acid 13:84.2:2.8
**Flow rate:** 1.5
**Injection volume:** 10-20
**Detector:** UV 310

## CHROMATOGRAM
**Retention time:** 9
**Internal standard:** cefotaxime
**Limit of detection:** 250 ng/mL

## OTHER SUBSTANCES
**Extracted:** ceftizoxime
**Simultaneous:** cefamandole, cefazolin, cefoxitin, ceftriaxone, cephalexin, cephaloridine, cephapirin, moxalactam
**Noninterfering:** amikacin, apalcillin, carbenicillin, cefoperazone, clindamycin, erythromycin, gentamicin, penicillin, piperacillin, ticarcillin, tobramycin, vancomycin

## KEY WORDS
serum; cefotaxime is IS; this assay can be used for cefotaxime-see Antimicrob. Agents Chemother. 1987; 31; 1177

## REFERENCE
McCormick, E.M.; Echols, R.M.; Rosano, T.G. Liquid chromatographic assay of ceftizoxime in sera of normal and uremic patients. *Antimicrob.Agents Chemother.*, **1984**, *25*, 336–338

## SAMPLE
**Matrix:** blood
**Sample preparation:** 300 μL Plasma + 300 μL IS in ice-cold MeOH:100 mM pH 5.2 sodium acetate 70:30, vortex for 30 s, let stand at -20° for 10 min, centrifuge at 1500 g for 10 min, inject a 10 μL aliquot.

## HPLC VARIABLES
**Guard column:** 4 × 4 10 μm C18
**Column:** 300 × 4 10 μm μBondapak C18
**Mobile phase:** MeCN:MeOH:100 mM sodium acetate 11.52:0.48:88, pH 5.2
**Flow rate:** 2.5
**Injection volume:** 10
**Detector:** UV 254

## CHROMATOGRAM
**Retention time:** 6
**Internal standard:** 8-chlorotheophylline (4)
**Limit of detection:** 500 ng/mL

## OTHER SUBSTANCES
**Extracted:** cefoperazone, cefoxitin, cephalexin, cephaloridine

## KEY WORDS
plasma

## REFERENCE
Signs, S.A.; File, T.M.; Tan, J.S. High-pressure liquid chromatographic method for analysis of cephalosporins. *Antimicrob.Agents Chemother.*, **1984**, *26*, 652–655

## SAMPLE
**Matrix:** blood
**Sample preparation:** Mix serum with an equal volume of 250 μg/mL 4'-nitroacetanilide in MeCN:MeOH 90:10, mix, let stand at room temperature for 10 min, mix, centrifuge at 12800 g for 2 min, inject a 25 μL aliquot of the supernatant.

## HPLC VARIABLES
**Guard column:** RCSS Guard-Pak (Waters)
**Column:** 100 × 8 C18 Radial Pak (Waters)
**Mobile phase:** MeOH:0.75% acetic acid 30:70, pH adjusted to 5.5 with triethylamine
**Flow rate:** 3
**Injection volume:** 25
**Detector:** UV 254

## CHROMATOGRAM
**Retention time:** 3.0
**Internal standard:** 4'-nitroacetanilide (12.4)
**Limit of detection:** 3 μg/mL

## OTHER SUBSTANCES
**Extracted:** cefamandole, cefazolin, cefoxitin, cephapirin, chloramphenicol
**Simultaneous:** acetaminophen, N-acetylprocainamide, cefaclor, cephalexin, cephalothin, cimetidine, miconazole, moxalactam, procainamide, sulfamethoxazole, theophylline, tobramycin, vancomycin

## KEY WORDS
serum

## REFERENCE

Danzer, L.A. Liquid-chromatographic determination of cephalosporins and chloramphenicol in serum. *Clin.Chem.*, **1983**, *29*, 856–858

## SAMPLE

**Matrix:** blood

**Sample preparation:** Prepare an anion-exchange SPE cartridge in a 6 mL syringe barrel with a filter paper disc in the bottom. Pack with DEAE-A-25 Sephadex in PBS to a bed volume of 3 mL, wash with PBS, place filter paper on top. Add 500 μL serum to SPE cartridge, add 500 μL PBS to SPE cartridge, wash with 4 mL PBS, elute with 5 mL 1 M NaCl, inject a 100 μL aliquot of the eluate. (PBS was 8 g NaCl, 1.15 g $Na_2HPO_4$, 0.2 g KCl, and 0.2 g $KH_2PO_4$ in 1 L water, pH 7.2.)

## HPLC VARIABLES

**Column:** 300 × 4 10 μm octadecylsilane

**Mobile phase:** MeCN:buffer 13:87 (Buffer was water adjusted to pH 2.8 with acetic acid, about 1.5 mL/L.)

**Flow rate:** 1.5

**Injection volume:** 100

**Detector:** UV 270

## CHROMATOGRAM

**Retention time:** 7.7

**Limit of quantitation:** 1 μg/mL

## OTHER SUBSTANCES

**Extracted:** metabolites, cephapirin

**Noninterfering:** amikacin, amphotericin B, azathioprine, carbenicillin, chloral hydrate, cimetidine, dopamine, fluphenazine, furosemide, hydrochlorothiazide, insulin, levothyroxine, methylprednisolone, nitroglycerin, oxacillin, prednisone, procainamide, sulfamethoxazole, tolazamide, tolbutamide, triamterene, trimethoprim

**Interfering:** cefoxitin

## KEY WORDS

serum; SPE

## REFERENCE

Fasching, C.E.; Peterson, L.R. Anion-exchange extraction of cephapirin, cefotaxime, and cefoxitin from serum for liquid chromatography. *Antimicrob.Agents Chemother.*, **1982**, *21*, 628–633

## SAMPLE

**Matrix:** blood, urine

**Sample preparation:** Plasma. 200 μL Plasma + 20 μL 0.45 N phosphoric acid + 100 μL methanol + 20 μL 270 μmol/L cephalexin, vortex 15 s, centrifuge for 3 min, remove 100 μL supernatant, inject 20 μL. Urine. 10 μL Urine + 0.5 mL water + 20 μL 270 μmol/L cephalexin, vortex 15 s, remove 100 μL supernatant, inject 20 μL.

## HPLC VARIABLES

**Guard column:** 100 × 4.7 Co:Pell ODS

**Column:** 120 × 4.7 LiChrosorb RP-18

**Mobile phase:** MeOH: 10 mM tetrabutylammonium hydrogen sulfate and 20 mM $K_3PO_4$ and 20 mM $KH_2PO_4$ 20:80

**Flow rate:** 1

**Injection volume:** 20

**Detector:** UV 254

## CHROMATOGRAM
**Retention time:** 11
**Internal standard:** cephalexin
**Limit of detection:** 1 nmol/mL (plasma); 50 nmol/mL (urine)

## OTHER SUBSTANCES
**Simultaneous:** cefotiam

## KEY WORDS
plasma

## REFERENCE
Lecaillon, J.B.; Rouan, M.C.; Souppart, C.; Febvre, N.; Juge, F. Determination of cefsulodin, cefotiam, cephalexin, cefotaxime, deacetylcefotaxime, cefuroxime and cefroxadin in plasma and urine by high-performance liquid chromatography. *J.Chromatogr.*, **1982**, *228*, 257–267

## SAMPLE
**Matrix:** bulk, formulations
**Sample preparation:** Dissolve in water, inject a 20 µL aliquot.

## HPLC VARIABLES
**Column:** 300 × 3.9 10 µm µBondapak C18
**Mobile phase:** MeOH:water:acetic acid 30:70:0.1
**Flow rate:** 1
**Injection volume:** 20
**Detector:** UV 254

## CHROMATOGRAM
**Retention time:** 8.5
**Limit of quantitation:** 760 ng/mL

## OTHER SUBSTANCES
**Simultaneous:** impurities, cefaclor, cefadroxil, cefamandole, cefamandole nafate, cefazolin, cefoperazone, cefoxitin, ceftizoxime, cephalexin, cephalothin, cephapirin, cephradine

## REFERENCE
Ting, S. Reverse-phase liquid chromatographic analysis of cephalosporins. *J.Assoc.Off.Anal.Chem.*, **1988**, *71*, 1123–1130

## SAMPLE
**Matrix:** cell suspensions
**Sample preparation:** Filter (0.45 µm).

## HPLC VARIABLES
**Column:** 150 × 4.6 5 µm Ultrasphere IP ion pair
**Mobile phase:** MeOH:100 mM sodium perchlorate adjusted to pH 2.5 with concentrated sulfuric acid 35:65
**Flow rate:** 1
**Injection volume:** 20
**Detector:** UV 254

## CHROMATOGRAM
**Retention time:** 4.6

## OTHER SUBSTANCES
**Extracted:** cefpirome
**Interfering:** carumonam (UV 295), ceftriaxone

## REFERENCE
Bellido, F.; Pechère, J.-C.; Hancock, R.E.W. Novel method for measurement of outer membrane permeability to new β-lactams in intact *Enterobacter cloacae* cells. *Antimicrob.Agents Chemother.*, **1991**, *35*, 68–72

## SAMPLE
**Matrix:** formulations
**Sample preparation:** Mix an aliquot with an equal volume of 5 mg/mL cefoxitin, dilute with water, inject a 20 μL aliquot.

## HPLC VARIABLES
**Column:** 150 × 3.9 5 μm Resolve (Waters)
**Mobile phase:** MeCN:buffer 18:86 (Buffer was 2.46 g anhydrous sodium acetate, 8 mL glacial acetic acid, and 200 mg tetrabutylammonium hydrogen sulfate in 1 L water, pH 3.0.)
**Flow rate:** 1.2
**Injection volume:** 20
**Detector:** UV 254

## CHROMATOGRAM
**Retention time:** 2.3
**Internal standard:** cefoxitin (3.0)

## OTHER SUBSTANCES
**Simultaneous:** metronidazole

## KEY WORDS
stability-indicating; injections; saline

## REFERENCE
Belliveau, P.P.; Nightingale, C.H.; Quintiliani, R. Stability of cefotaxime sodium and metronidazole in 0.9% sodium chloride injection or in ready-to-use metronidazole bags. *Am.J.Health-Syst.Pharm.*, **1995**, *52*, 1561–1563

## SAMPLE
**Matrix:** formulations
**Sample preparation:** Dilute 1:5 with water, inject a 20 μL aliquot.

## HPLC VARIABLES
**Column:** 150 × 3.9 4 μm μBondapak C18
**Mobile phase:** MeCN:buffer 7:93 (Buffer was 20 mM $KH_2PO_4$ and 5 mM triethylamine adjusted to pH 4.8 with NaOH.)
**Flow rate:** 1.5
**Injection volume:** 20
**Detector:** UV 270

## CHROMATOGRAM
**Retention time:** 8.44

## OTHER SUBSTANCES
**Simultaneous:** desacetylcefotaxime, metronidazole

## KEY WORDS
stability-indicating; injections; water

## REFERENCE
Rivers, T.E.; McBride, H.A.; Trang, J.M. Stability of cefotaxime sodium and metronidazole in an i.v. admixture at 8°C. *Am.J.Hosp.Pharm.*, **1991**, *48*, 2638–2640

## SAMPLE
**Matrix:** solutions
**Sample preparation:** Inject an aliquot of a 50 µg/mL solution in water.

## HPLC VARIABLES
**Guard column:** 4 × 4 5 µm Lichrospher 100 C18
**Column:** 250 × 4 5 µm Lichrospher 100 C18
**Mobile phase:** MeOH:buffer 20:80 (Buffer was 3.5 g $KH_2PO_4$ and 11.6 g $Na_2HPO_4.12H_2O$ in 1 L water.)
**Flow rate:** 1
**Injection volume:** 20
**Detector:** UV 254

## KEY WORDS
comparison with CE

## REFERENCE
Fabre, H.; Castaneda Penalvo, G. Capillary electrophoresis as an alternative method for the determination of cefotaxime. *J.Liq.Chromatogr.*, **1995**, *18*, 3877–3887

## SAMPLE
**Matrix:** solutions

## HPLC VARIABLES
**Column:** 250 × 4 5 µm Spherisorb ODS II
**Mobile phase:** Gradient. A was MeOH:buffer 4:20. B was MeOH:buffer 7:20. A:B 100:0 for 8 min, to 0:100 over 21 min, maintain at 0:100 for 4 min. Re-equilibrate at initial conditions for 2 min. (Buffer was $KH_2PO_4/Na_2HPO_4$.)
**Column temperature:** 25
**Flow rate:** 1
**Detector:** UV 235

## OTHER SUBSTANCES
**Simultaneous:** impurities

## REFERENCE
Wetterich, U.; Mutschler, E. Quality of cefotaxime sodium preparations. *Arzneimittelforschung*, **1995**, *45*, 74–80

## SAMPLE
**Matrix:** solutions
**Sample preparation:** Prepare a 2.5-5 µg/mL solution, inject a 20 µL aliquot.

## HPLC VARIABLES
**Column:** 80 × 4.6 3.65 µm Zorbax Rx-SIL (similar to Zorbax SB-C8 (Mac-Mod Analytical))
**Mobile phase:** MeCN:0.1% trifluoroacetic acid 18:82
**Flow rate:** 1

**Injection volume:** 20
**Detector:** UV 254

## CHROMATOGRAM
**Retention time:** k′ 2.6

## REFERENCE
Kirkland, K.M.; McCombs, D.A.; Kirkland, J.J. Rapid, high-resolution high-performance liquid chromatographic analysis of antibiotics. *J.Chromatogr.A*, **1994**, *660*, 327–337

## SAMPLE
**Matrix:** solutions
**Sample preparation:** Dissolve in 100 mM pH 10.5 carbonate buffer at a concentration of 0.113 mM, inject 50 μL aliquots.

## HPLC VARIABLES
**Column:** 250 × 4.6 10 μm Spherisorb ODS-2
**Mobile phase:** MeCN:10 mM ammonium acetate 6:94, pH 6.50
**Flow rate:** 1.1
**Injection volume:** 50
**Detector:** UV 235

## CHROMATOGRAM
**Retention time:** 9.28

## OTHER SUBSTANCES
**Simultaneous:** degradation products

## REFERENCE
Vilanova, B.; Muñoz, F.; Donoso, J.; Frau, J.; Garcia Blanco, F. Alkaline hydrolysis of cefotaxime. A HPLC and $^1$H NMR study. *J.Pharm.Sci.*, **1994**, *83*, 322–327

## SAMPLE
**Matrix:** solutions
**Sample preparation:** Separate buffer containing drug from human serum albumin by centrifuging at 37° at 700 g for 3 min using a Micropartition System MPS-1 (Amicon) unit, inject a 10-20 μL aliquot of the ultrafiltrate.

## HPLC VARIABLES
**Guard column:** C18/Corasil (Waters)
**Column:** 300 × 3.9 μBondapak C18
**Mobile phase:** MeCN:10 mM ammonium acetate 10:90
**Flow rate:** 1.5
**Injection volume:** 10-20
**Detector:** UV 240

## OTHER SUBSTANCES
**Also analyzed:** cefuroxime, cephacetrile

## REFERENCE
Terasaki, T.; Nouda, H.; Tsuji, A. Relationship between lipophilicity and binding affinity with human serum albumin for penicillin and cephem antibiotics. *J.Pharmacobiodyn.*, **1992**, *15*, 99–106

## SAMPLE
**Matrix:** solutions

## HPLC VARIABLES
**Column:** 250 × 4 OmniPac PCX-500 (Dionex)
**Mobile phase:** Gradient. A was MeCN:90 mM perchloric acid 13.5:86.5. B was MeCN:300 mM perchloric acid 45:55. A:B from 100:0 to 0:100 over 7 min, maintain at 0:100.
**Flow rate:** 1
**Detector:** UV 254

## CHROMATOGRAM
**Retention time:** 9.5

## OTHER SUBSTANCES
**Simultaneous:** 7-aminocephalosproranic acid, cefadroxil, cefazolin, cephalexin, cephaloridine, cephalosporin C, cephalothin, cephapirin, D-hydroxyphenylglycine

## REFERENCE
Slingsby, R.W.; Rey, M. Determination of pharmaceuticals by multi-phase chromatography: Combined reversed phase and ion exchange in one column. *J.Liq.Chromatogr.*, **1990**, *13*, 107–134

## ANNOTATED BIBLIOGRAPHY
Changqin, H.; Shaohong, S.; Kaimin, W. The chromatographic behavior of cephalosporins in gel filtration chromatography, a novel method to separate high molecular weight impurities. *J.Pharm. Biomed.Anal.*, **1994**, *12*, 533–541 [also cefadroxil, cefamandole, cefmenoxime, cefoperazone, ceftazidime, ceftriaxone, cephalexin, cephaloridine, cephalothin, cephradine]

Fabre, H.; Fell, A.F. Comparison of techniques for peak purity testing of cephalosporins. *J.Liq.Chromatogr.*, **1992**, *15*, 3031–3043 [also theophylline]

Haginaka, J.; Yasuda, N.; Wakai, J.; Matsunaga, H.; Yasuda, H.; Kimura, Y. Internal-surface reversed-phase silica support for direct injection determination of drugs in biological fluids by liquid chromatography. *Anal.Chem.*, **1989**, *61*, 2445–2448 [direct injection; ISRP; serum; extracted cefamandole, cefmenoxime]

Paap, C.M.; Nahata, M.C. A novel micromethod for the simultaneous analysis of cefotaxime and desacetylcefotaxime from plasma using ion pair high performance liquid chromatography. *J.Liq.Chromatogr.*, **1989**, *12*, 2385–2395 [plasma; cefoxitin (IS); also acetaminophen, caffeine, cefazolin, methicillin, theophylline, vancomycin; non-interfering ampicillin, gentamicin, ibuprofen, phenobarbital]

Hakim, L.; Bourne, D.W.; Triggs, E.J. High-performance liquid chromatographic assay of cefotaxime, desacetylcefotaxime and ceftriaxone in rat plasma. *J.Chromatogr.*, **1988**, *424*, 111–117

Hary, L.; Andrejak, M. [Analysis of serum cefotaxime and desacetylcefotaxime by high performance liquid ion exchange chromatography]. *J.Chromatogr.*, **1987**, *419*, 396–400

Yost, R.L.; Derendorf, H. Rapid chromatographic determination of cefotaxime and its metabolite in biological fluids. *J.Chromatogr.*, **1985**, *341*, 131–138

Bawdon, R.E.; Novick, W.J.; Hemsell, D.L.; Welch, W.D. High-pressure liquid chromatographic assay of cefotaxime and desacetylcefotaxime in human myometrium. *J.Liq.Chromatogr.*, **1984**, *7*, 2483–2491 [SPE]

Demotes-Mainard, F.; Vinçon, G.; Bouchet, J.L.; Jarry, C.; Albin, H. [Assay of cefotaxime and desacetylcefotaxime in plasma and urine by high performance liquid chromatography]. *Ann Biol.Clin.(Paris)*, **1984**, *42*, 301–305

LeBel, M.; Ericson, J.F.; Pitkin, D.H. Improved high-performance liquid chromatographic (HPLC) assay method for ceftizoxime. *J.Liq.Chromatogr.*, **1984**, *7*, 961–968 [simultaneous ceftizoxime; cefotaxime is IS]

Nygard, G.; Wahba Kahlil, S.K. An isocratic HPLC method for the determination of cephalosporins in plasma. *J.Liq.Chromatogr.*, **1984**, *7*, 1461–1475 [plasma; cephapirin (IS); column temp 45; extracted cefamandole, cefazolin, cefonicid, cefoperazone, cefoxitin, cephalothin]

# Cefprozil

**Molecular formula:** $C_{18}H_{19}N_3O_5S$
**Molecular weight:** 389.4
**CAS Registry No.:** 92665-29-7, 121123-17-9 (monohydrate)

## SAMPLE

**Matrix:** blood
**Sample preparation:** 100 μL Plasma + 150 μL 300 μg/mL cephalexin + 20 μL 5% trichloroacetic acid + 100 μL MeCN, vortex, centrifuge at 13000 g for 3 min, inject a 40 μL aliquot of the supernatant.

## HPLC VARIABLES

**Guard column:** 37 μm Corasil C18
**Column:** 100 × 9.4 Partisil-5-CCS-C8RAC C8
**Mobile phase:** MeCN:3 mM pH 3.8 sodium acetate buffer 13:87
**Flow rate:** 0.9
**Injection volume:** 40
**Detector:** UV 280

## CHROMATOGRAM

**Retention time:** 10.2
**Internal standard:** cephalexin (12.4)
**Limit of quantitation:** 500 ng/mL

## KEY WORDS

plasma; pharmacokinetics

## REFERENCE

Barbhaiya, R.H.; Shukla, U.A.; Gleason, C.R.; Shyu, W.C.; Wilber, R.B.; Martin, R.R.; Pittman, K.A. Phase I study of multiple-dose cefprozil and comparison with cefaclor. *Antimicrob.Agents Chemother.*, **1990**, *34*, 1198–1203

## SAMPLE

**Matrix:** blood, urine
**Sample preparation:** Plasma. 500 μL Plasma + 100 μL 1 mg/mL cephalexin + 150 μL 10% trichloroacetic acid + 500 μL MeCN, mix, add 1.5 mL dichloromethane, vortex, centrifuge, inject a 60 μL aliquot of the aqueous supernatant. Urine. 5 mL Urine + 5 mL 20 mM pH 3.8 sodium acetate buffer, mix. 500 μL Buffered urine + 100 μL 1.5 mg/mL cephalexin + 150 μL 5% trichloroacetic acid, vortex, inject a 10 μL aliquot.

## HPLC VARIABLES

**Guard column:** 45 × 4 37-50 μm C18/Corasil
**Column:** Zorbax C8 (plasma) or Partisil 5 ODS-3 RAC C18 (urine)
**Mobile phase:** MeCN:glacial acetic acid:water 17:2:81 (plasma) or MeCN:MeOH:THF: trichloroacetic acid:glacial acetic acid:sodium acetate trihydrate:sodium dodecyl sulfate: water 25:3:0.925:0.075:0.25:0.077:0.134:70.75 (v/v/v/w/v/w/w/v) (urine)
**Flow rate:** 1.5 (plasma) or 2 (urine)
**Injection volume:** 10-60
**Detector:** UV 280

## CHROMATOGRAM

**Retention time:** 14 (cis, plasma), 15 (cis, urine), 18 (trans, urine), 20 (trans, plasma)

**Internal standard:** cephalexin (17 (plasma), 23 (urine))
**Limit of quantitation:** 100 ng/mL (plasma); 5000 ng/mL (urine)

## KEY WORDS
plasma; human; rat; pharmacokinetics

## REFERENCE
Shyu, W.C.; Shukla, U.A.; Shah, V.R.; Papp, E.A.; Barbhaiya, R.H. Simultaneous high-performance liquid chromatographic analysis of cefprozil diastereomers in a pharmacokinetic study. *Pharm.Res.*, **1991**, *8*, 992–996

## SAMPLE
**Matrix:** urine
**Sample preparation:** Mix urine with an equal volume of 20 mM pH 3.57 sodium acetate buffer. 500 µL Buffered urine + 100 µL 1.5 µg/mL cephalexin + 150 µL 5% trichloroacetic acid, vortex for 30 s, inject a 10 µL aliquot.

## HPLC VARIABLES
**Guard column:** 37 µm Corasil C18
**Column:** 100 × 9.4 Partisil-5-ODS-3RAC C18
**Mobile phase:** MeCN:MeOH:THF:buffer 25:3:0.925:71.075 (Prepare buffer by dissolving 1.54 g sodium acetate trihydrate and 2.67 g sodium dodecyl sulfate in 1 L water and adding 5 mL glacial acetic acid, 30 mL 5% trichloroacetic acid, 500 mL MeCN, 60 mL MeOH, and 18.5 mL THF, make up to 2 L with water.)
**Flow rate:** 2
**Injection volume:** 10
**Detector:** UV 280

## CHROMATOGRAM
**Retention time:** 12
**Internal standard:** cephalexin (17)
**Limit of quantitation:** 5 µg/mL

## KEY WORDS
pharmacokinetics

## REFERENCE
Barbhaiya, R.H.; Shukla, U.A.; Gleason, C.R.; Shyu, W.C.; Wilber, R.B.; Martin, R.R.; Pittman, K.A. Phase I study of multiple-dose cefprozil and comparison with cefaclor. *Antimicrob.Agents Chemother.*, **1990**, *34*, 1198–1203

# Ceftazidime

**Molecular formula:** $C_{22}H_{22}N_6O_7S_2$
**Molecular weight:** 546.6
**CAS Registry No.:** 72558-82-8, 78439-06-2 (pentahydrate)

## SAMPLE
**Matrix:** bile, blood, urine
**Sample preparation:** Serum. 0.5 mL serum + 0.5 mL MeCN, mix in 7 mL tube on vortex mixer, shake by rotation (20 rpm) 10 min, centrifuge 10 min 1000 g, transfer supernatant to another tube, add 7 aliquots dichloromethane, equilibrate 10 min, shake by rotation (20 rpm) 10 min, centrifuge 10 min 1000 g, inject aliquot of upper aqueous layer. Urine. Centrifuge urine and dilute 1:20. Bile. Centrifuge bile and dilute 1:10.

## HPLC VARIABLES
**Column:** 150 × 4.6 5 μm Ultrasphere ODS
**Mobile phase:** MeCN:20 mM ammonium acetate adjusted to pH 5 with glacial acetic acid 9:91
**Flow rate:** 1
**Injection volume:** 20
**Detector:** UV 254

## CHROMATOGRAM
**Retention time:** 6.3
**Limit of detection:** LOD 200 ng/mL

## OTHER SUBSTANCES
**Also analyzed:** ampicillin, azlocillin, aztreonam, cefmenoxime, cefoperazone, cefsulodin, cefotaxime, ceftriaxone, cloxacillin, desacetylcefotaxime, mezlocillin, penicillin G, piperacillin, ticarcillin

## KEY WORDS
serum

## REFERENCE
Jehl, F.; Birckel, P.; Monteil, H. Hospital routine analysis of penicillins, third-generation cephalosporins and aztreonam by conventional and high-speed high-performance liquid chromatography. *J.Chromatogr.*, **1987**, *413*, 109–119

## SAMPLE
**Matrix:** blister fluid, blood
**Sample preparation:** Serum. 0.5 mL Serum + 2.5 mL MeCN, vortex, centrifuge at 3000 g for 5 min. Remove the supernatant and add it to 5 mL dichloromethane, vortex, inject a 20 μL aliquot of the aqueous layer. Blister fluid. Inject directly.

## HPLC VARIABLES
**Column:** μBondapak C18
**Mobile phase:** MeCN:100 mM phosphate buffer 2.5:97.5
**Flow rate:** 2
**Injection volume:** 20
**Detector:** UV 229

## CHROMATOGRAM
**Retention time:** 14

## KEY WORDS
serum; pharmacokinetics

## REFERENCE
Kalman, D.; Barriere, S.L.; Johnson, B.L., Jr. Pharmacokinetic disposition and bactericidal activities of cefepime, ceftazidime, and cefoperazone in serum and blister fluid. *Antimicrob.Agents Chemother.*, **1992**, *36*, 453–457

## SAMPLE
**Matrix:** blood
**Sample preparation:** Dilute serum with 3 volumes 10 mM $NaH_2PO_4$, inject a 50 µL aliquot onto column A and elute to waste with mobile phase A, after 0.9 min elute the contents of column A onto column B with mobile phase A, after 2.6 min remove column A from the circuit and elute column B with mobile phase B, monitor the effluent from column B.

## HPLC VARIABLES
**Column:** A 50 × 4 40 µm C8 (Backflush each day with MeCN:water 50:50 for 20 min.); B 150 × 4 5 µm HP ODS (Hewlett-Packard)
**Mobile phase:** A MeCN:10 mM $NaH_2PO_4$ 4:96, pH 5.0; B MeCN:10 mM $NaH_2PO_4$ 2:92, pH 5.0 (sic, 8:92 ?)
**Flow rate:** 1
**Injection volume:** 50
**Detector:** UV 258

## CHROMATOGRAM
**Retention time:** 8.9
**Limit of detection:** 500 ng/mL

## KEY WORDS
serum; column-switching; pharmacokinetics

## REFERENCE
Bompadre, S.; Ferrante, L.; Alò, F.P.; Leone, L. On-line solid-phase extraction of ceftazidime in serum and determination by high-performance liquid chromatography. *J.Chromatogr.B*, **1995**, *669*, 265–269

## SAMPLE
**Matrix:** blood
**Sample preparation:** 200 µL Serum + 50 µL cefpirome sulfate solution + 100 µL 10% trichloroacetic acid, inject a 20 µL aliquot of the supernatant.

## HPLC VARIABLES
**Column:** C18
**Mobile phase:** MeCN:50 mM pH 3.5-4.0 ammonium phosphate buffer 8:92
**Flow rate:** 1.2
**Injection volume:** 20
**Detector:** UV 257

## CHROMATOGRAM
**Internal standard:** cefpirome
**Limit of quantitation:** 500 ng/mL

## KEY WORDS
pharmacokinetics; serum

## REFERENCE

Klepser, M.E.; Patel, K.B.; Nicolau, D.P.; Quintiliani, R.; Nightingale, C.H. Comparison of the bacteri-cidal activities of ofloxacin and ciprofloxacin alone and in combination with ceftazidime and pipera-cillin against clinical strains of *Pseudomonas aeruginosa. Antimicrob.Agents Chemother.*, **1995**, *39*, 2503–2510

## SAMPLE

**Matrix:** blood
**Sample preparation:** 50 μL Serum + 50 μL 50 μg/mL cephaloridine in 6% perchloric acid, mix, centrifuge at 1500 g for 5 min, inject a 25 μL aliquot of the supernatant.

## HPLC VARIABLES

**Guard column:** present but not specified
**Column:** 100 × 8 Resolve C18 Radial Pak glass column (Waters)
**Mobile phase:** MeCN:MeOH:20 mM pH 3.6 sodium acetate buffer 4.8:13.5:81.7
**Flow rate:** 2
**Injection volume:** 25
**Detector:** UV 254; UV 265

## CHROMATOGRAM

**Internal standard:** cephaloridine
**Limit of detection:** 500 ng/mL

## KEY WORDS

pharmacokinetics; serum

## REFERENCE

van den Anker, J.N.; Schoemaker, R.C.; Hop, W.C.J.; van der Heijden, B.J.; Weber, A.; Sauer, P.J.J.; Neijens, H.J.; de Groot, R. Ceftazidime pharmacokinetics in preterm infants: Effects of renal func-tion and gestational age. *Clin.Pharmacol.Ther.*, **1995**, *58*, 650–659

## SAMPLE

**Matrix:** blood
**Sample preparation:** 500 μL Plasma + 500 μL MeCN, vortex for 10 s, centrifuge at 3000 g for 10 min. Remove 800 μL of the supernatant and add it to 5 mL dichloromethane, vortex for 10 s, centrifuge, inject a 50 μL aliquot of the aqueous layer.

## HPLC VARIABLES

**Guard column:** 20 × 4.6 10 μm Spherisorb C8
**Column:** 250 × 4.6 5 μm Spherisorb ODS
**Mobile phase:** MeCN:buffer 4.5:95.5 (Buffer was 3.85 g/L ammonium acetate + 2 mL triethylamine, adjusted to pH 4 with formic acid.)
**Column temperature:** 50
**Flow rate:** 1.75
**Injection volume:** 50
**Detector:** UV 254

## CHROMATOGRAM

**Retention time:** 19.8
**Internal standard:** ceftazidime

## OTHER SUBSTANCES

**Extracted:** ceftibuten
**Noninterfering:** acyclovir, amikacin, norfloxacin, ofloxacin, pefloxacin, tobramycin

## KEY WORDS
plasma; column-switching; ceftazidime is IS

## REFERENCE
Kinowswki, J.M.; Bressolle, F.; Fabre, D.; Goncalves, F.; Rouzier-Panis, R.; Galtier, M. High-performance liquid chromatographic determination of ceftibuten and its metabolite in biological fluids: Applications in pharmacokinetic studies. *J.Pharm.Sci.*, **1994**, *83*, 736–741

## SAMPLE
**Matrix:** blood
**Sample preparation:** 500 μL Serum + 500 μL MeCN:EtOH:water (40:40:20), vortex for 10-15 s, centrifuge through a Centricon-10 filter unit with a 10000 dalton cut-off (Amicon) at 4000 g for 30 min, inject a 10-60 μL aliquot of the colorless ultrafiltrate.

## HPLC VARIABLES
**Column:** 150 × 4.6 3 μm Ultremex phenyl (Phenomenex)
**Mobile phase:** MeCN:water 25:75 containing 5 mM sodium dodecanesulfonate and 0.1% phosphoric acid
**Column temperature:** 40
**Flow rate:** 0.8
**Injection volume:** 10-60
**Detector:** UV 259

## CHROMATOGRAM
**Retention time:** 7-8
**Limit of detection:** 50 ng/mL

## KEY WORDS
serum; dolphin; ultrafiltrate

## REFERENCE
Tyczkowska, K.L.; Seay, S.S.; Stoskopf, M.K.; Aucoin, D.P. Determination of ceftazidime in dolphin serum by liquid chromatography with ultraviolet-visible detection and confirmation by thermospray liquid chromatography-mass spectrometry. *J.Chromatogr.*, **1992**, *576*, 305–313

## SAMPLE
**Matrix:** blood
**Sample preparation:** 500 μL Serum + 500 μL ice-cold 50 μg/mL cefotaxime in MeOH: 100 mM pH 5.2 sodium acetate 70:30, vortex for 30 s, hold at -20° for 10 min, centrifuge at 1500 g for 10 min, inject 15 μL of supernatant.

## HPLC VARIABLES
**Guard column:** 10 μm C18 Guard-PAK
**Column:** 300 × 3.9 10 μm μBondapak C18
**Mobile phase:** MeCN:water:glacial acetic acid 100:876:24
**Flow rate:** 1.5
**Injection volume:** 15
**Detector:** UV 254

## CHROMATOGRAM
**Internal standard:** cefotaxime
**Limit of detection:** 800 ng/mL

## OTHER SUBSTANCES
**Also analyzed:** ceftizoxime

## KEY WORDS
serum

## REFERENCE
Deeter, R.G.; Weinstein, M.P.; Swanson, K.A.; Gross, J.S.; Bailey, L.C. Crossover assessment of serum bactericidal activity and pharmacokinetics of five broad-spectrum cephalosporins in the elderly. *Antimicrob.Agents Chemother.*, **1990**, *34*, 1007–1013

## SAMPLE
**Matrix:** blood
**Sample preparation:** 500 µL Plasma + 500 µL MeCN, mix vigorously on a Whirlmixer for 30 s, centrifuge at 1200 g for 5 min. Remove 400 µL of the supernatant and add it to 3 mL dichloromethane, centrifuge at 1200 g for 5 min, inject a 20 µL aliquot of the upper aqueous layer.

## HPLC VARIABLES
**Column:** 100 × 3 5 µm Hypersil ODS
**Mobile phase:** MeCN:5 mM pH 5.5 acetate buffer 0.7:97.3
**Flow rate:** 1
**Injection volume:** 20
**Detector:** UV 254

## CHROMATOGRAM
**Limit of detection:** 500 ng/mL

## OTHER SUBSTANCES
**Also analyzed:** cefepime, ceftriaxone

## KEY WORDS
plasma; mouse

## REFERENCE
van Ogtrop, M.L.; Mattie, H.; Guiot, H.F.L.; van Strijen, E.; Hazekamp-van Dokkum, A.-M.; van Furth, R. Comparative study of the effects of four cephalosporins against *Escherichia coli* in vitro and in vivo. *Antimicrob.Agents Chemother.*, **1990**, *34*, 1932–1937

## SAMPLE
**Matrix:** blood, CSF
**Sample preparation:** Deproteinize serum or CSF with MeCN, centrifuge, add the supernatant to dichloromethane, inject a 100 µL aliquot of the aqueous layer.

## HPLC VARIABLES
**Column:** reversed-phase
**Mobile phase:** MeCN:100 mM pH 5.0 $NaH_2PO_4$ buffer 8:92, containing 5 mM pentane-sulfonic acid
**Injection volume:** 100
**Detector:** UV 254

## CHROMATOGRAM
**Limit of quantitation:** 94 ng/mL (serum); 67 ng/mL (CSF)

## KEY WORDS
serum; pharmacokinetics

# REFERENCE
Nau, R.; Prange, H.W.; Kinzig, M.; Frank, A.; Dressel, A.; Scholz, P.; Kolenda, H.; Sörgel, F. Cerebro-spinal fluid ceftazidime kinetics in patients with external ventriculostomies. *Antimicrob.Agents Chemother.*, **1996**, *40*, 763–766

# SAMPLE
**Matrix:** blood, dialysate
**Sample preparation:** Plasma. 0.5 mL Plasma + 0.5 mL MeCN, mix in 7 mL tube on vortex mixer, shake by rotation (20 rpm) 10 min, centrifuge 10 min 1000 g, transfer supernatant to another tube, add 7 aliquots dichloromethane, equilibrate 10 min, shake by rotation (20 rpm) 10 min, centrifuge 10 min 1000 g, inject aliquot of upper aqueous layer (J.Chromatogr. 1987, 413, 109). Dialysate. Inject a 30 µL aliquot directly.

# HPLC VARIABLES
**Mobile phase:** MeCN:100 nM (sic) pH 3.5 acetate buffer:40 mM tetradecyltrimethylam-monium bromide 20:78:2
**Flow rate:** 1
**Injection volume:** 30
**Detector:** UV 270

# CHROMATOGRAM
**Limit of detection:** 50 ng/mL

# KEY WORDS
rat; plasma; pharmacokinetics

# REFERENCE
Granero, L.; Santiago, M.; Cano, J.; Machado, A.; Peris, J.-E. Analysis of ceftriaxone and ceftazidime distribution in cerebrospinal fluid of and cerebral extracellular space in awake rats by in vivo mi-crodialysis. *Antimicrob.Agents Chemother.*, **1995**, *39*, 2728–2731

# SAMPLE
**Matrix:** blood, urine
**Sample preparation:** Add aminophylline, precipitate proteins with 0.8 M perchloric acid, inject an aliquot.

# HPLC VARIABLES
**Column:** µBondapak C18
**Mobile phase:** MeOH:90 mM sodium acetate 14:86, pH adjusted to 4.2 with acetic acid
**Flow rate:** 1
**Detector:** UV 254

# CHROMATOGRAM
**Retention time:** 7.2
**Internal standard:** aminophylline
**Limit of detection:** 150 ng/mL

# OTHER SUBSTANCES
**Also analyzed:** cefpirome

# KEY WORDS
plasma

## REFERENCE

Paradis, D.; Vallée, F.; Allard, S.; Bisson, C.; Daviau, N.; Drapeau, C.; Auger, F.; LeBel, M. Comparative study of pharmacokinetics and serum bactericidal activities of cefpirome, ceftazidime, ceftriaxone, imipenem, and ciprofloxacin. *Antimicrob.Agents Chemother.*, **1992**, *36*, 2085–2092

## SAMPLE

**Matrix:** cecal contents
**Sample preparation:** Dilute cecal contents in 2 mL phosphate-buffered saline, centrifuge at 1500 g for 10 min. 500 μL Sample + 500 μL MeCN, vortex for 30 s, centrifuge at 1200 g for 5 min. Remove 400 μL of the supernatant and add it to 3 mL dichloromethane, mix for 30 s, centrifuge at 1200 g for 5 min, inject a 20 μL aliquot of the upper aqueous phase.

## HPLC VARIABLES

**Column:** 100 × 3 5 μm Hypersil ODS
**Mobile phase:** MeCN:5 mM pH 5.5 acetate buffer 0.7:99.3
**Flow rate:** 1
**Injection volume:** 20
**Detector:** UV 254

## OTHER SUBSTANCES

**Also analyzed:** cefoperazone, ceftriaxone

## KEY WORDS

mouse; pharmacokinetics

## REFERENCE

van Ogtrop, M.L.; Guiot, H.F.L.; Mattie, H.; van Furth, R. Modulation of the intestinal flora of mice by parenteral treatment with broad-spectrum cephalosporins. *Antimicrob.Agents Chemother.*, **1991**, *35*, 976–982

## SAMPLE

**Matrix:** formulations
**Sample preparation:** Dilute with mobile phase, inject an aliquot.

## HPLC VARIABLES

**Column:** 250 × 4.6 5 μm cyano
**Mobile phase:** MeCN:100 mM $NaH_2PO_4$ 20:80, adjusted to pH 4.2 with phosphoric acid
**Flow rate:** 2
**Injection volume:** 20
**Detector:** UV 210

## CHROMATOGRAM

**Retention time:** 1.73

## OTHER SUBSTANCES

**Simultaneous:** granisetron (UV 300)

## KEY WORDS

stability-indicating; injections; saline

## REFERENCE

Mayron, D.; Gennaro, A.R. Stability and compatibility of granisetron hydrochloride in i.v. solutions and oral liquids and during simulated Y-site injection with selected drugs. *Am.J.Health-Syst.Pharm.*, **1996**, *53*, 294–304

## SAMPLE
**Matrix:** formulations
**Sample preparation:** Dilute 100 μL eye drops to 25 mL with water, inject a 10 μL aliquot.

## HPLC VARIABLES
**Guard column:** 10 × 4.6 5 μm Spherisorb hexyl
**Column:** 100 × 4.6 5 μm Spherisorb hexyl
**Mobile phase:** MeCN:50 mM pH 7.0 ammonium acetate 7:93
**Flow rate:** 2
**Injection volume:** 10
**Detector:** UV 254

## CHROMATOGRAM
**Retention time:** 1.3

## OTHER SUBSTANCES
**Simultaneous:** degradation products, pyridine

## KEY WORDS
stability-indicating; eye drops

## REFERENCE
Barnes, A.R. Determination of ceftazidime and pyridine by HPLC: Application to a viscous eye drop formulation. *J.Liq.Chromatogr.*, **1995**, *18*, 3117–3128

## SAMPLE
**Matrix:** formulations
**Sample preparation:** Dilute 1000-fold, mix a 200 μL aliquot with 200 μL 100 μg/mL hydrochlorothiazide, inject a 20 μL aliquot.

## HPLC VARIABLES
**Column:** Microsorb MV C-18
**Mobile phase:** MeCN:water:acetic acid 6:93:1, adjusted to pH 4.0 with 6 M NaOH
**Flow rate:** 2
**Injection volume:** 20
**Detector:** UV 254

## CHROMATOGRAM
**Internal standard:** hydrochlorothiazide

## KEY WORDS
injections; saline; stability-indicating

## REFERENCE
Bednar, D.A.; Klutman, N.E.; Henry, D.W.; Fox, J.L.; Strayer, A.H. Stability of ceftazidime (with arginine) in an elastomeric infusion device. *Am.J.Health-Syst.Pharm.*, **1995**, *52*, 1912–1914

## SAMPLE
**Matrix:** formulations
**Sample preparation:** Dilute 50-fold with water, inject an aliquot.

## HPLC VARIABLES
**Column:** 150 × 3.9 Nova-Pak C18

**Mobile phase:** MeCN:20 mM KH$_2$PO$_4$ 7:93 containing 10 mM triethylamine, adjusted to pH 4.8 with HCl
**Flow rate:** 1.5
**Injection volume:** 20
**Detector:** UV 270

## CHROMATOGRAM
**Retention time:** 1.7

## OTHER SUBSTANCES
**Simultaneous:** ceftizoxime, ceftriaxone, metronidazole
**Noninterfering:** degradation products

## KEY WORDS
saline; injections

## REFERENCE
Rivers, T.E.; Webster, A.A. Stability of ceftizoxime sodium, ceftriaxone sodium, and ceftazidime with metronidazole in ready-to-use metronidazole bags. *Am.J.Health-Syst.Pharm.*, **1995**, *52*, 2568–2570

## SAMPLE
**Matrix:** formulations
**Sample preparation:** 100 µL Solution + 4.9 mL MeOH:water 20:80, inject a 50 µL aliquot.

## HPLC VARIABLES
**Guard column:** 5 µm Adsorbosphere C18
**Column:** 250 × 4.6 5 µm Adsorbosphere C18
**Mobile phase:** MeCN:50 mM pH 7 phosphate buffer 7.5:92.5
**Flow rate:** 1
**Injection volume:** 50
**Detector:** UV 254

## CHROMATOGRAM
**Retention time:** 6.8

## KEY WORDS
stability-indicating; injections; 5% dextrose

## REFERENCE
Inagaki, K.; Gill, M.A.; Okamoto, M.P.; Takagi, J. Stability of ranitidine hydrochloride with aztreonam, ceftazidime, or piperacillin sodium during simulated Y-site administration. *Am.J.Hosp.Pharm.*, **1992**, *49*, 2769–2772

## SAMPLE
**Matrix:** formulations
**Sample preparation:** Dilute with water, inject an aliquot.

## HPLC VARIABLES
**Column:** 150 × 3.9 5 µm Nova Pak C18
**Mobile phase:** MeOH:5 mM pH 7.5 phosphate buffer 10:90
**Flow rate:** 0.6
**Detector:** UV 254

## CHROMATOGRAM
**Retention time:** 2.3

## OTHER SUBSTANCES
**Simultaneous:** degradation products

## KEY WORDS
injections; water; stability-indicating

## REFERENCE
Stiles, M.L.; Tu, Y.H.; Allen, L.V., Jr. Stability of cefazolin sodium, cefoxitin sodium, ceftazidime, and penicillin G sodium in portable pump reservoirs. *Am.J.Hosp.Pharm.*, **1989**, *46*, 1408–1412

## SAMPLE
**Matrix:** formulations
**Sample preparation:** Add theophylline (100 μg/mL), inject a 10 μL aliquot.

## HPLC VARIABLES
**Column:** 300 × 3.9 μBondapak C18
**Mobile phase:** MeCN:acetic acid:water 10:1:89, adjusted to pH 4 with 5 M NaOH
**Flow rate:** 1.5
**Injection volume:** 10
**Detector:** UV 293 (?)

## CHROMATOGRAM
**Retention time:** 4.6
**Internal standard:** theophylline (3.3)

## OTHER SUBSTANCES
**Simultaneous:** cefazolin

## KEY WORDS
stability-indicating; 5% dextrose; injections

## REFERENCE
Bosso, J.A.; Prince, R.A.; Fox, J.L. Compatibility of ondansetron hydrochloride with fluconazole, ceftazidime, aztreonam, and cefazolin sodium under simulated Y- site conditions. *Am.J.Hosp.Pharm.*, **1994**, *51*, 389–391

## SAMPLE
**Matrix:** solutions
**Sample preparation:** Dilute with an equal volume of water or buffer, inject an aliquot.

## HPLC VARIABLES
**Column:** 150 × 4.6 5 μm Spherisorb C6
**Mobile phase:** MeCN:buffer 4:96 (Buffer was 100 mM acetic acid containing 25 mM sodium acetate.)
**Flow rate:** 1.5
**Injection volume:** 20
**Detector:** UV 254

## CHROMATOGRAM
**Retention time:** 8.9

## OTHER SUBSTANCES
**Simultaneous:** degradation products

## KEY WORDS
stability-indicating

## REFERENCE
Zhou, M.; Notari, R.E. Influence of pH, temperature, and buffers on the kinetics of ceftazidime degradation in aqueous solution. *J.Pharm.Sci.*, **1995**, *84*, 534–538

## SAMPLE
**Matrix:** solutions
**Sample preparation:** Prepare a 2.5-5 µg/mL solution, inject a 20 µL aliquot.

## HPLC VARIABLES
**Column:** 80 × 4.6 3.65 µm Zorbax Rx-SIL (similar to Zorbax SB-C8 (Mac-Mod Analytical))
**Mobile phase:** MeCN:0.1% trifluoroacetic acid 12:88
**Flow rate:** 1
**Injection volume:** 20
**Detector:** UV 254

## CHROMATOGRAM
**Retention time:** k′ 2.4

## REFERENCE
Kirkland, K.M.; McCombs, D.A.; Kirkland, J.J. Rapid, high-resolution high-performance liquid chromatographic analysis of antibiotics. *J.Chromatogr.A*, **1994**, *660*, 327–337

## SAMPLE
**Matrix:** solutions
**Sample preparation:** Dilute 1:100 with water.

## HPLC VARIABLES
**Column:** Novapak C18 (model no. PN86344)
**Mobile phase:** MeCN:glacial acetic acid:water 6:1:93, adjusted to a pH of 4.0
**Flow rate:** 1
**Injection volume:** 20
**Detector:** UV 254

## CHROMATOGRAM
**Retention time:** 5.3

## OTHER SUBSTANCES
**Interfering:** cefuroxime

## KEY WORDS
water; stability-indicating

## REFERENCE
Stiles, M.L.; Allen, L.V., Jr.; Fox, J.L. Stability of ceftazidime (with arginine) and of cefuroxime sodium in infusion-pump reservoirs. *Am.J.Hosp.Pharm.*, **1992**, *49*, 2761–2764

## SAMPLE
**Matrix:** urine
**Sample preparation:** Dilute urine three-fold with 200 mM pH 4.5 sodium acetate buffer, vortex for 30 s, inject a 10 µL aliquot.

## HPLC VARIABLES
**Guard column:** 23 × 4 37-50 μm Corasil C18
**Column:** 100 × 9.4 Partisil 5 ODS-3RAC C18
**Mobile phase:** MeOH:10 mM sodium dodecyl sulfate adjusted to pH 3.0 with glacial acetic
   acid:5% trichloroacetic acid:850 mM phosphoric acid:THF 49.7:40.4:3.9:0.7:5.3
**Flow rate:** 2.8
**Injection volume:** 10
**Detector:** UV 280

## CHROMATOGRAM
**Retention time:** 10
**Internal standard:** ceftazidime

## OTHER SUBSTANCES
**Extracted:** cefepime

## KEY WORDS
rat; ceftazidime is IS

## REFERENCE
Barbhaiya, R.H.; Forgue, S.T.; Shyu, W.C.; Papp, E.A.; Pittman, K.A. High-pressure liquid chromatographic analysis of BMY-28142 in plasma and urine. *Antimicrob.Agents Chemother.*, **1987**, *31*, 55–
   59

◆————————————————◆————————————————◆

## ANNOTATED BIBLIOGRAPHY
Changqin, H.; Shaohong, S.; Kaimin, W. The chromatographic behavior of cephalosporins in gel filtration
   chromatography, a novel method to separate high molecular weight impurities. *J.Pharm.
   Biomed.Anal.*, **1994**, *12*, 533–541 [also cefadroxil, cefamandole, cefmenoxime, cefoperazone, cefotaxime, ceftriaxone, cephalexin, cephaloridine, cephalothin, cephradine]

Faouzi, M.A.; Dine, T.; Luyckx, M.; Gressier, B.; Brunet, C.; Goudaliez, F.; Mallevais, M.L.; Cazin, M.;
   Cazin, J.C. Stability and compatibility studies of cefaloridine, cefuroxime and ceftazidime with PVC
   infusion bags. *Pharmazie*, **1994**, *49*, 425–427 [also cefaloridine, cefuroxime; formulations; saline; 5%
   dextrose]

Kinowski, J.M.; Bressolle, F.; Fabre, D.; Goncalves, F.; Rouzier-Panis, R.; Galtier, M. High-performance
   liquid chromatographic determination of ceftibuten and its metabolite in biological fluids: Application in pharmacokinetic studies. *J.Pharm.Sci.*, **1994**, *83*, 736–741 [plasma; extracted ceftibuten; ceftazidime is IS]

Vinks, A.A.T.M.M.; Touw, D.J.; Heijerman, H.G.M.; Danhof, M.; de Leede, G.P.J.; Bakker, W. Pharmacokinetics of ceftazidime in adult cystic fibrosis patients during continuous infusion and ambulatory
   treatment at home. *Ther.Drug Monit.*, **1994**, *16*, 341–348 [pharmacokinetics; serum; urine; sputum;
   LOD 500 ng/mL; 8-chlorotheophylline (IS)]

Nahata, M.C.; Morosco, R.S. Measurement of ceftazidime arginine in aqueous solution by HPLC.
   *J.Liq.Chromatogr.*, **1992**, *15*, 1507–1511

Fasching, C.E.; Peterson, L.R.; Gerding, D.N. High pressure liquid chromatographic analysis for the
   quantitation of BMY-28142 and ceftazidime in human and rabbit serum. *J.Liq.Chromatogr.*, **1986**,
   *9*, 1803–1814 [human; rabbit; serum; extracted BMY-28142; LOD 1.5 μg/mL]

Hwang, P.T.R.; Drexler, P.G.; Meyer, M.C. High-performance liquid chromatographic determination of
   ceftazidime in serum, urine, CSF, and peritoneal dialysis fluid. *J.Liq.Chromatogr.*, **1984**, *7*, 979–987
   [serum; urine; CSF; peritoneal dialysis fluid; hydrochlorothiazide (IS); also amikacin, ampicillin, caffeine, chloramphenicol]

# Ceftriaxone

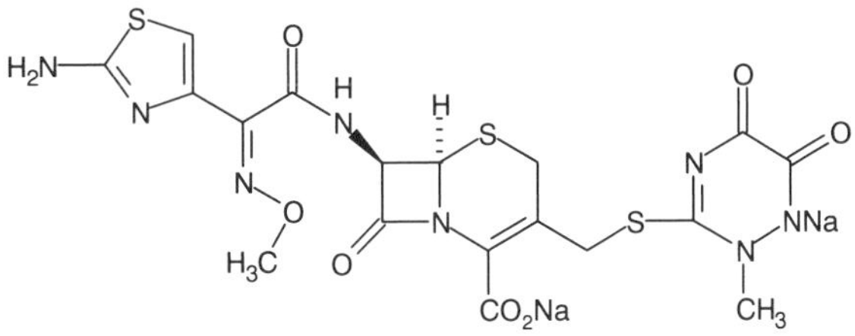

**Molecular formula:** $C_{18}H_{18}N_8O_7S_3$

**Molecular weight:** 554.6

**CAS Registry No.:** 73384-59-5 (ceftriaxone),

104376-79-6 (ceftriaxone sodium)

## SAMPLE

**Matrix:** bile

**Sample preparation:** 100 μL Bile + 200 μL water, vortex for 20 s, add 3 mL 45.2 μg/mL o-phthalic acid in mobile phase, vortex, centrifuge, inject a 20-50 μL aliquot of the supernatant.

## HPLC VARIABLES

**Column:** 150 × 3.2 5 μm LiChrosorb RP-18

**Mobile phase:** MeCN:pH 7.0 Titrisol phosphate buffer (Merck):water 440:60:500, containing 3.5 g/L tetraoctylammonium bromide

**Flow rate:** 1.5

**Injection volume:** 20-50

**Detector:** UV 274

## CHROMATOGRAM

**Retention time:** 7.5

**Internal standard:** o-phthalic acid (10.1)

**Limit of detection:** 5 μg/mL

## KEY WORDS

dog; human

## REFERENCE

Trautmann, K.H.; Haefelfinger, P. Determination of the cephalosporin Ro 13-9904 in plasma, urine, and bile by means of ion-pair reversed phase chromatography. *J.High Res.Chromatogr.*, **1981**, *4*, 54–59

## SAMPLE

**Matrix:** bile, blood, urine

**Sample preparation:** Serum. 0.5 mL serum + 0.5 mL MeCN, mix in 7 mL tube on vortex mixer, shake by rotation (20 rpm) 10 min, centrifuge 10 min 1000 g, transfer supernatant to another tube, add 7 aliquots dichloromethane, equilibrate 10 min, shake by rotation (20 rpm) 10 min, centrifuge 10 min 1000 g, inject aliquot of upper aqueous layer. Urine. Centrifuge urine and dilute 1:20. Bile. Centrifuge bile and dilute 1:10.

## HPLC VARIABLES

**Column:** 250 × 4.6 5 μm Ultrasphere ODS

**Mobile phase:** MeCN:10 mM phosphate buffered saline containing 11 mM hexadecyltrimethylammonium bromide 50:50, adjusted to pH 8 with glacial acetic acid

**Flow rate:** 2

**Injection volume:** 50

**Detector:** UV 254

## CHROMATOGRAM

**Retention time:** 17.5

**Limit of detection:** 200 ng/mL

## OTHER SUBSTANCES

**Also analyzed:** ampicillin, azlocillin, aztreonam, cefmenoxime, cefoperazone, cefsulodin, cefotaxime, ceftazidime, cloxacillin, desacetylcefotaxime, mezlocillin, penicillin G, piperacillin, ticarcillin

## KEY WORDS

serum

## REFERENCE

Jehl, F.; Birckel, P.; Monteil, H. Hospital routine analysis of penicillins, third-generation cephalosporins and aztreonam by conventional and high-speed high-performance liquid chromatography. *J.Chromatogr.*, **1987**, *413*, 109–119

## SAMPLE

**Matrix:** blood
**Sample preparation:** Dilute serum 1:10 with cold MeOH or filter (Millipore Ultraspec-MC, molecular weight limit 10000), inject an aliquot.

## HPLC VARIABLES

**Column:** 25 × 4.6 5 μm C18
**Mobile phase:** MeCN:1 M pH 7 phosphate buffer:water 50:1:49 containing 3 g/L hexadecyltrimethylammonium bromide
**Flow rate:** 1
**Detector:** UV 280

## CHROMATOGRAM

**Retention time:** 6-7
**Limit of quantitation:** 500 ng/mL (filtered sample)

## KEY WORDS

serum; ultrafiltrate; pharmacokinetics

## REFERENCE

Hayward, C.J.; Nafziger, A.N.; Kohlhepp, S.J.; Bertino, J.S., Jr. Investigation of bioequivalence and tolerability of intramuscular ceftriaxone injections using 1% lidocaine, buffered lidocaine, and sterile water diluents. *Antimicrob.Agents Chemother.*, **1996**, *40*, 485–487

## SAMPLE

**Matrix:** blood
**Sample preparation:** 500 μL Serum + 500 μL ice-cold 100 μg/mL cefoperazone in MeOH:100 mM pH 5.2 sodium acetate 70:30, vortex for 30 s, hold at -20° for 10 min, centrifuge at 1500 g for 10 min, inject 15 μL of supernatant.

## HPLC VARIABLES

**Guard column:** 10 μm C18 Guard-PAK
**Column:** 300 × 3.9 10 μm μBondapak C18
**Mobile phase:** MeCN:10 mM pH 7.5 phosphate buffer containing 10 mM hexadecyltrimethylammonium bromide 35:65
**Flow rate:** 1.2
**Injection volume:** 15
**Detector:** UV 254

## CHROMATOGRAM

**Internal standard:** cefoperazone
**Limit of detection:** 800 ng/mL

## KEY WORDS
serum

## REFERENCE
Deeter, R.G.; Weinstein, M.P.; Swanson, K.A.; Gross, J.S.; Bailey, L.C. Crossover assessment of serum bactericidal activity and pharmacokinetics of five broad-spectrum cephalosporins in the elderly. *Antimicrob.Agents Chemother.*, **1990**, *34*, 1007–1013

## SAMPLE
**Matrix:** blood
**Sample preparation:** 500 μL Plasma + 500 μL MeCN, mix vigorously on a Whirlmixer for 30 s, centrifuge at 1200 g for 5 min. Remove 400 μL of the supernatant and add it to 3 mL dichloromethane, centrifuge at 1200 g for 5 min, inject a 20 μL aliquot of the upper aqueous layer.

## HPLC VARIABLES
**Column:** 100 × 3 5 μm Hypersil ODS
**Mobile phase:** MeCN:5 mM pH 5.5 acetate buffer 0.7:97.3
**Flow rate:** 1
**Injection volume:** 20
**Detector:** UV 274

## CHROMATOGRAM
**Limit of detection:** 400 ng/mL

## OTHER SUBSTANCES
**Also analyzed:** cefepime, cefoperazone, ceftazidime

## KEY WORDS
plasma; mouse

## REFERENCE
van Ogtrop, M.L.; Mattie, H.; Guiot, H.F.L.; van Strijen, E.; Hazekamp-van Dokkum, A.-M.; van Furth, R. Comparative study of the effects of four cephalosporins against *Escherichia coli* in vitro and in vivo. *Antimicrob.Agents Chemother.*, **1990**, *34*, 1932–1937

## SAMPLE
**Matrix:** blood
**Sample preparation:** 100 μL Plasma + 300 μL water, vortex for 20 s, add 2 mL 25 μg/mL probenecid in EtOH, shake 20-30 times by hand, rotate at 20 rpm for 5 min, centrifuge at 1000 g for 2 min, inject a 60 μL aliquot.

## HPLC VARIABLES
**Column:** 150 × 3.2 7 μm LiChrosorb RP-18
**Mobile phase:** MeCN:pH 7.0 Titrisol phosphate buffer (Merck):water 500:50:450, containing 4 g/L hexadecyltrimethylammonium bromide
**Flow rate:** 1.5
**Injection volume:** 60
**Detector:** UV 274

## CHROMATOGRAM
**Retention time:** 3.8
**Internal standard:** probenecid (6.7)
**Limit of detection:** 500 ng/mL

## KEY WORDS
plasma; dog; human

## REFERENCE
Trautmann, K.H.; Haefelfinger, P. Determination of the cephalosporin Ro 13-9904 in plasma, urine, and bile by means of ion-pair reversed phase chromatography. *J.High Res.Chromatogr.*, **1981**, *4*, 54–59

## SAMPLE
**Matrix:** blood, CSF, urine
**Sample preparation:** Dilute urine 1:10 with normal saline. 50 μL Serum, CSF, or diluted urine + 50 μL 200 μg/mL moxalactam in MeCN, vortex for 30 s, centrifuge at 13000 g for 3 min, inject a 10 μL aliquot.

## HPLC VARIABLES
**Guard column:** 30 × 4 18 μm spherical silica gel
**Column:** 300 × 4 μBondapak C18
**Mobile phase:** MeCN:10 mM pH 9.0 potassium phosphate buffer containing 10 mM hexadecyltrimethylammonium bromide 46:54
**Flow rate:** 1.5
**Injection volume:** 10
**Detector:** UV 274

## CHROMATOGRAM
**Retention time:** 4
**Internal standard:** moxalactam (7)
**Limit of detection:** 1 μg/mL

## OTHER SUBSTANCES
**Noninterfering:** acetaminophen, N-acetylprocainamide, amikacin, amitriptyline, ampicillin, carbamazepine, cefamandole, cefazolin, cefoperazone, cefotaxime, cefoxitin, ceftazidime, ceftizoxime, cephalexin, cephalothin, cephapirin, chloramphenicol, ciprofloxacin, clonazepam, cyclosporine, desipramine, digoxin, disopyramide, ethosuximide, gentamicin, haloperidol, imipramine, kanamycin, lidocaine, mezlocillin, netilmicin, nortriptyline, phenobarbital, phenytoin, primidone, procainamide, propranolol, quinidine, salicylic acid, streptomycin, sulfamethoxazole, theophylline, thiamphenicol, ticarcillin, tobramycin, trimethoprim, vancomycin

## KEY WORDS
serum

## REFERENCE
Granich, G.G.; Krogstad, D.J. Ion pair high-performance liquid chromatographic assay for ceftriaxone. *Antimicrob.Agents Chemother.*, **1987**, *31*, 385–388

## SAMPLE
**Matrix:** blood, dialysate
**Sample preparation:** Plasma. 0.5 mL Plasma + 0.5 mL MeCN, mix in 7 mL tube on vortex mixer, shake by rotation (20 rpm) 10 min, centrifuge 10 min 1000 g, transfer supernatant to another tube, add 7 aliquots dichloromethane, equilibrate 10 min, shake by rotation (20 rpm) 10 min, centrifuge 10 min 1000 g, inject aliquot of upper aqueous layer (J.Chromatogr. 1987, 413, 109). Dialysate. Inject a 30 μL aliquot directly.

## HPLC VARIABLES
**Mobile phase:** MeCN:MeOH:50 mM pH 6.5 phosphate buffer:40 mM tetradecyltrimethylammonium bromide 30:10:45:15

**Flow rate:** 1
**Injection volume:** 30
**Detector:** UV 270

## CHROMATOGRAM
**Limit of detection:** 50 ng/mL

## KEY WORDS
rat; plasma; pharmacokinetics

## REFERENCE
Granero, L.; Santiago, M.; Cano, J.; Machado, A.; Peris, J.-E. Analysis of ceftriaxone and ceftazidime distribution in cerebrospinal fluid of and cerebral extracellular space in awake rats by in vivo microdialysis. *Antimicrob.Agents Chemother.*, **1995**, *39*, 2728–2731

## SAMPLE
**Matrix:** blood, urine
**Sample preparation:** Precipitate proteins with MeCN containing cefotetan.

## HPLC VARIABLES
**Column:** μBondapak C18
**Mobile phase:** MeCN:70 mM sodium acetate 44:56, pH adjusted to 5.7 with acetic acid
**Flow rate:** 1
**Detector:** UV 270

## CHROMATOGRAM
**Retention time:** 7.4
**Internal standard:** cefotetan
**Limit of detection:** 600 ng/mL

## KEY WORDS
plasma

## REFERENCE
Paradis, D.; Vallée, F.; Allard, S.; Bisson, C.; Daviau, N.; Drapeau, C.; Auger, F.; LeBel, M. Comparative study of pharmacokinetics and serum bactericidal activities of cefpirome, ceftazidime, ceftriaxone, imipenem, and ciprofloxacin. *Antimicrob.Agents Chemother.*, **1992**, *36*, 2085–2092

## SAMPLE
**Matrix:** blood, urine
**Sample preparation:** 500 (?) μL Plasma or urine + 1500 (?) μL MeCN, vortex, centrifuge, inject a 20 μL aliquot of the supernatant.

## HPLC VARIABLES
**Column:** Brownlee 10 μm RP-8
**Mobile phase:** MeCN:MeOH:20 mM pH 5 phosphate buffer 5:25:70
**Flow rate:** 2
**Injection volume:** 20
**Detector:** UV 254

## CHROMATOGRAM
**Retention time:** 4

## OTHER SUBSTANCES
**Also analyzed:** cefaronide, ceftriaxone

## KEY WORDS
plasma; sheep; pharmacokinetics

## REFERENCE
Guerrini, V.H.; Filippich, L.J.; Cao, G.R.; English, P.B.; Bourne, D.W.A. Pharmacokinetics of cefaronide, ceftriaxone and cefoperazone in sheep. *J.Vet.Pharmacol.Ther.*, **1985**, *8*, 120–127

## SAMPLE
**Matrix:** cecal contents
**Sample preparation:** Weigh contents of cecum, dilute with 2 mL PBS, centrifuge at 1500 g for 10 min. Add a 500 μL aliquot of supernatant to 500 μL MeCN, mix on a whirlmixer for 30 s, centrifuge at 1200 g for 5 min. Remove 400 μL of the supernatant and add it to 3 mL dichloromethane, mix for 30 s, centrifuge at 1200 g for 5 min, inject a 20 μL aliquot of the upper aqueous phase.

## HPLC VARIABLES
**Column:** 100 × 3 5 μm Hypersil ODS
**Mobile phase:** MeCN:5 mM pH 5.5 acetate buffer 0.7:99.3
**Flow rate:** 1
**Injection volume:** 20
**Detector:** UV 274

## CHROMATOGRAM
**Limit of quantitation:** 500 ng/mL

## OTHER SUBSTANCES
**Also analyzed:** cefepime, cefoperazone, ceftazidime

## KEY WORDS
mouse

## REFERENCE
van Ogtrop, M.L.; Guiot, H.F.L.; Mattie, H.; van Furth, R. Modulation of the intestinal flora of mice by parenteral treatment with broad-spectrum cephalosporins. *Antimicrob.Agents Chemother.*, **1991**, *35*, 976–982

## SAMPLE
**Matrix:** formulations
**Sample preparation:** Dilute 50-fold with water, inject an aliquot.

## HPLC VARIABLES
**Column:** 150 × 3.9 Nova-Pak C18
**Mobile phase:** MeCN:20 mM $KH_2PO_4$ 7:93 containing 10 mM triethylamine, adjusted to pH 4.8 with HCl
**Flow rate:** 1.5
**Injection volume:** 20
**Detector:** UV 270

## CHROMATOGRAM
**Retention time:** 2.1

## OTHER SUBSTANCES
**Simultaneous:** ceftazidime, ceftizoxime, metronidazole
**Noninterfering:** degradation products

## KEY WORDS
saline; injections

## REFERENCE
Rivers, T.E.; Webster, A.A. Stability of ceftizoxime sodium, ceftriaxone sodium, and ceftazidime with metronidazole in ready-to-use metronidazole bags. *Am.J.Health-Syst.Pharm.*, **1995**, *52*, 2568–2570

## SAMPLE
**Matrix:** formulations
**Sample preparation:** Inject a 20 μL aliquot.

## HPLC VARIABLES
**Guard column:** 10 μm Dynamax C18
**Column:** 250 × 4.6 10 μm Dynamax C18
**Mobile phase:** MeCN:6.5 mM tetraheptylammonium bromide:100 mM pH 7.0 phosphate buffer:100 mM pH 5.0 citrate buffer 40:55.2:4.4:0.4
**Flow rate:** 1
**Injection volume:** 20
**Detector:** UV 271

## CHROMATOGRAM
**Retention time:** 14.9

## OTHER SUBSTANCES
**Simultaneous:** degradation products, theophylline

## KEY WORDS
injections; stability-indicating; saline; use low actinic glassware; 5% dextrose

## REFERENCE
Parrish, M.A.; Bailey, L.C.; Medwick, T. Stability of ceftriaxone sodium and aminophylline or theophylline in intravenous mixtures. *Am.J.Hosp.Pharm.*, **1994**, *51*, 92–94

## SAMPLE
**Matrix:** urine
**Sample preparation:** 100 μL Urine + 200 μL water, vortex for 20 s, add 3 mL 9 μg/mL 4-nitrobenzoic acid in mobile phase, vortex, centrifuge, inject a 20-50 μL aliquot of the supernatant.

## HPLC VARIABLES
**Column:** 150 × 3.2 5 μm LiChrosorb RP-18
**Mobile phase:** MeCN:buffer:water 200:38:762, containing 3.8 g/L tetrapentylammonium bromide (Buffer was 500 mL of 0.5 M $Na_2HPO_4$ adjusted to pH 7.8 with about 50 mL 0.5 M $NaH_2PO_4$.)
**Flow rate:** 1.5
**Injection volume:** 20-50
**Detector:** UV 300

## CHROMATOGRAM
**Retention time:** 6.4
**Internal standard:** 4-nitrobenzoic acid (20.4)
**Limit of detection:** 5 μg/mL

## KEY WORDS
dog; human

## REFERENCE

Trautmann, K.H.; Haefelfinger, P. Determination of the cephalosporin Ro 13-9904 in plasma, urine, and bile by means of ion-pair reversed phase chromatography. *J.High Res.Chromatogr.*, **1981**, *4*, 54–59

## ANNOTATED BIBLIOGRAPHY

Changqin, H.; Shaohong, S.; Kaimin, W. The chromatographic behavior of cephalosporins in gel filtration chromatography, a novel method to separate high molecular weight impurities. *J.Pharm. Biomed.Anal.*, **1994**, *12*, 533–541 [also cefadroxil, cefamandole, cefmenoxime, cefoperazone, cefotaxime, ceftazidime, cephalexin, cephaloridine, cephalothin, cephradine]

Bailey, L.C.; Tang, K.T.; Medwick, T. Stability of ceftriaxone sodium in infusion-pump syringes. *Am.J.Hosp.Pharm.*, **1993**, *50*, 2092–2094

Bellido, F.; Pechère, J.-C.; Hancock, R.E.W. Novel method for measurement of outer membrame permeability to new β-lactams in intact *Enterobacter cloacae* cells. *Antimicrob.Agents Chemother.*, **1991**, *35*, 68–72

Nahata, M.C. Measurement of ceftriaxone in peritoneal dialysis solutions by high-performance liquid chromatography. *J.Liq.Chromatogr.*, **1991**, *14*, 179–185

Mohler, J.; Meulemans, A.; Vulpillat, M. High-performance liquid chromatographic determination of the binding of ceftriaxone to human serum albumin solution and albumin from diluted human serum. *J.Chromatogr.*, **1990**, *528*, 415–423

Jungbluth, G.L.; Jusko, W.J. Ion-paired reversed-phase high-performance liquid chromatography assay for determination of ceftriaxone in human plasma and urine. *J.Pharm.Sci.*, **1989**, *78*, 968–970

Hakim, L.; Bourne, D.W.; Triggs, E.J. High-performance liquid chromatographic assay of cefotaxime, desacetylcefotaxime and ceftriaxone in rat plasma. *J.Chromatogr.*, **1988**, *424*, 111–117

Marini, D.; Balestrieri, F. Determination of ceftriaxone by HPLC. *Farmaco.[Prat].*, **1986**, *41*, 172–176

Bawdon, R.E.; Hemsell, D.L.; Hemsell, P.G. Serum and pelvic tissue concentrations of ceftriaxone and cefazolin at hysterectomy. *J.Liq.Chromatogr.*, **1984**, *7*, 2011–2020

Bowman, D.B.; Aravind, M.K.; Miceli, J.N.; Kauffman, R.E. Reversed-phase high-performance liquid chromatographic method to determine ceftriaxone in biological fluids. *J.Chromatogr.*, **1984**, *309*, 209–213

Ascalone, V.; Dal, B. Determination of ceftriaxone, a novel cephalosporin, in plasma, urine and saliva by high-performance liquid chromatography on an NH2 bonded-phase column. *J.Chromatogr.*, **1983**, *273*, 357–366

# Cefuroxime

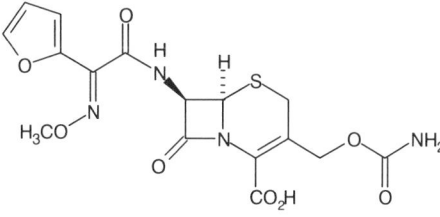

**Molecular formula:** $C_{16}H_{16}N_4O_8S$

**Molecular weight:** 424.4

**CAS Registry No.:** 55268-75-2, 56238-63-2 (Na salt), 64544-07-6 (cefuroxime axetil), 100680-33-9 (cefuroxime pivoxetil)

## SAMPLE

**Matrix:** blood

**Sample preparation:** Deproteinize with 12.5% trichloroacetic acid.

## HPLC VARIABLES

**Column:** $150 \times 4.6$ 5 μm Beckman Ultrasphere ODS

**Mobile phase:** MeCN:50 mM phosphate buffer 10:90

**Flow rate:** 1.5

**Detector:** UV 280

## CHROMATOGRAM

**Retention time:** 6.3

**Internal standard:** cephaloridine

**Limit of quantitation:** 500 ng/mL

## KEY WORDS

serum

## REFERENCE

Donn, K.H.; James, N.C.; Powell, J.R. Bioavailability of cefuroxime axetil formulations. *J.Pharm.Sci.*, **1994**, *83*, 842–844

## SAMPLE

**Matrix:** blood

**Sample preparation:** 100 μL Serum + 400 μL A, vortex, centrifuge, inject 10 μL supernatant. (A was 1 mg/mL cephalexin in water:20% perchloric acid:water 3:2:7.)

## HPLC VARIABLES

**Guard column:** Used but not specified

**Column:** $100 \times 3$ 5 μm ChromSpher C18 in a glass column

**Mobile phase:** MeOH:Sorensen buffer (67 mM $KH_2PO_4$) 15:85

**Flow rate:** 0.4

**Injection volume:** 10

**Detector:** UV 275

## CHROMATOGRAM

**Retention time:** 7.7

**Internal standard:** cephalexin

**Limit of quantitation:** 700 ng/mL

## OTHER SUBSTANCES

**Noninterfering:** diazepam, etomidate, fentanyl, heparin, pancuronium, papaverine, polygeline, procaine, protamine, sodium nitroprusside

## KEY WORDS

serum

## REFERENCE

Koot, M.J.; IJdenberg, F.N.; Stuurman, R.M.; Poell, J.; Bras, L.J.; Langemeijer, J.J.; Lie-A-Huen, L. High pressure liquid chromatographic analysis of the serum concentration of cefuroxime after an intravenous bolus injection of cefuroxime in patients with a coronary artery bypass grafting. *Pharm.Weekbl.[Sci].*, **1992**, *14*, 360–364

## SAMPLE

**Matrix:** blood
**Sample preparation:** Mix 100 μL plasma + 300 μL 5 μg/mL cefotaxime in pH 3.5 10 mM acetate buffer and keep at 4°. Inject 100 μL onto column A with mobile phase A. After 5 min backflush column A with mobile phase B onto column B for 3 min. Re-equilibrate column A with mobile phase A for 16 min.

## HPLC VARIABLES

**Column:** A 40 × 2 37-50 μm Corasil RP C18; B 20 × 4 25-40 μm Lichrosorb RP-8 + 250 × 4 Partisil ODS-3
**Mobile phase:** A 10 mM pH 3.5 acetate buffer; B MeCN:20 mM pH 4.3 acetate buffer 15:85
**Flow rate:** 1
**Injection volume:** 100
**Detector:** UV 254

## CHROMATOGRAM

**Retention time:** 10.3
**Internal standard:** cefotaxime
**Limit of detection:** 500 ng/mL

## OTHER SUBSTANCES

**Simultaneous:** cefoxitin, cephalexin, cephaloridine
**Noninterfering:** alclofenac, aspirin, caffeine, cefadroxil, cefamandole, cefazolin, cefoperazone, cefotiam, cephalothin, diclofenac, ibuprofen, indomethacin, ketoprofen, lonazolac, mefenamic acid, naproxen, phenylbutazone, piroxicam

## KEY WORDS

plasma; column-switching; rat; human

## REFERENCE

Lee, Y.J.; Lee, H.S. Simultaneous determination of cefoxitin, cefuroxime, cephalexin and cephaloridine in plasma using HPLC and a column-switching technique. *Chromatographia*, **1990**, *30*, 80–84

## SAMPLE

**Matrix:** blood, ear fluid
**Sample preparation:** 50 μL Plasma or ear effusion + 50 μL water + 2 mL MeCN, vortex briefly, centrifuge at 1500 g for 10 min. Remove the organic layer and evaporate it to dryness under a stream of nitrogen at 50°, reconstitute the residue in 75 μL mobile phase, inject a 25 μL aliquot.

## HPLC VARIABLES

**Guard column:** 10 × 2.1 5 μm Hypersil C18
**Column:** 250 × 2.1 5 μm Hypersil C18
**Mobile phase:** MeCN:buffer 7.5:92.5 After elution of IS increase ratio to 50:50 to clean column. (Buffer was 25 mM acetate and 15 mM triethylamine adjusted to pH 4.3 with NaOH.)
**Column temperature:** 40
**Flow rate:** 0.35

**Injection volume:** 25
**Detector:** UV 254

## CHROMATOGRAM
**Retention time:** 5.9
**Internal standard:** cefuroxime

## OTHER SUBSTANCES
**Extracted:** cefpodoxime

## KEY WORDS
plasma; chinchilla; middle ear effusion; cefuroxime is IS

## REFERENCE
Lovdahl, M.J.; Reher, K.E.; Russlie, H.Q.; Canafax, D.M. Determination of cefpodoxime levels in chinchilla middle ear fluid and plasma by high-performance liquid chromatography. *J.Chromatogr.B*, **1994**, *653*, 227–232

## SAMPLE
**Matrix:** blood, urine
**Sample preparation:** Plasma. 200 µL Plasma + 20 µL 0.45 N phosphoric acid + 100 µL methanol + 20 µL 270 µmol/L cephalexin, vortex 15 s, centrifuge for 3 min, remove 100 µL supernatant, inject 20 µL. Urine. 10 µL Urine + 0.5 mL water + 20 µL 270 µmol/L cephalexin, vortex 15 s, remove 100 µL supernatant, inject 20 µL.

## HPLC VARIABLES
**Guard column:** 100 × 4.7 Co:Pell ODS
**Column:** 150 × 4.7 LiChrosorb RP-18
**Mobile phase:** MeOH:20 mM tetrabutylammonium hydrogen sulfate and 24 mM $K_3PO_4$ and 16 mM $KH_2PO_4$ 23:77
**Flow rate:** 1
**Injection volume:** 20
**Detector:** UV 254

## CHROMATOGRAM
**Retention time:** 13
**Internal standard:** cephalexin
**Limit of detection:** 1 nmol/mL (plasma); 50 nmol/mL (urine)

## OTHER SUBSTANCES
**Simultaneous:** cefotiam

## KEY WORDS
plasma

## REFERENCE
Lecaillon, J.B.; Rouan, M.C.; Souppart, C.; Febvre, N.; Juge, F. Determination of cefsulodin, cefotiam, cephalexin, cefotaxime, deacetylcefotaxime, cefuroxime and cefroxadin in plasma and urine by high-performance liquid chromatography. *J.Chromatogr.*, **1982**, *228*, 257–267

## SAMPLE
**Matrix:** cell suspensions
**Sample preparation:** 100 µL Cell suspension + 100 µL cefoperazone solution + 100 µL Hanks balanced salt solution, sonicate 30 min, add 800 µL MeCN, centrifuge at 13000 g

for 5 min, remove supernatant. Dry supernatant under air, dissolve in 100 µL mobile phase, inject 75 µL.

## HPLC VARIABLES
**Column:** µBondapak C18
**Mobile phase:** MeCN:10 mM pH 5.2 ammonium acetate 15:85
**Flow rate:** 1
**Injection volume:** 75
**Detector:** UV 254

## CHROMATOGRAM
**Retention time:** 6.8
**Internal standard:** cefoperazone
**Limit of detection:** 100-1000 ng/mL

## REFERENCE
Darouiche, R.O.; Hamill, R.J. Antibiotic penetration of and bactericidal activity within endothelial cells. *Antimicrob.Agents Chemother.*, **1994**, *38*, 1059–1064

## SAMPLE
**Matrix:** formulations
**Sample preparation:** Dilute 1 mL to 10 mL with water, add 3-6 mL to 20 mL 1.5 mg/mL 5-methylresorcinol in water, make up to 100 mL with water, inject a 20 µL aliquot.

## HPLC VARIABLES
**Column:** 150 × 4.6 5 µm Ultrasphere C18
**Mobile phase:** MeCN:100 mM pH 3.4 acetate buffer 1:10 (Buffer was 50 mL 100 mM sodium acetate diluted to 1 L with 100 mM acetic acid.)
**Flow rate:** 2
**Injection volume:** 20
**Detector:** UV 254

## CHROMATOGRAM
**Internal standard:** 5-methylresorcinol

## OTHER SUBSTANCES
**Simultaneous:** aminophylline, theophylline

## KEY WORDS
injections; 5% dextrose; saline; stability-indicating

## REFERENCE
Stewart, J.T.; Warren, F.W.; Johnson, S.M. Stability of cefuroxime sodium and aminophylline or theophylline. *Am.J.Hosp.Pharm.*, **1994**, *51*, 809–811

## SAMPLE
**Matrix:** solutions
**Sample preparation:** Dilute 1:100 with water.

## HPLC VARIABLES
**Column:** Novapak C18 (model no. PN86344)
**Mobile phase:** MeCN:glacial acetic acid:water 6:1:93, adjusted to a pH of 4.0
**Flow rate:** 1
**Injection volume:** 20
**Detector:** UV 254

## CHROMATOGRAM
**Retention time:** 5.4

## OTHER SUBSTANCES
**Interfering:** ceftazidime

## KEY WORDS
water; stability-indicating

## REFERENCE
Stiles, M.L.; Allen, L.V., Jr.; Fox, J.L. Stability of ceftazidime (with arginine) and of cefuroxime sodium in infusion-pump reservoirs. *Am.J.Hosp.Pharm.*, **1992**, *49*, 2761–2764

## SAMPLE
**Matrix:** solutions
**Sample preparation:** Separate the buffer containing the drug from human serum albumin by centrifuging at 37° at 700 g for 3 min using a Micropartition System MPS-1 (Amicon) unit, inject a 10-20 µL aliquot of the ultrafiltrate.

## HPLC VARIABLES
**Guard column:** C18/Corasil (Waters)
**Column:** 300 × 3.9 µBondapak C18
**Mobile phase:** MeCN:10 mM ammonium acetate 10:90
**Flow rate:** 1.5
**Injection volume:** 10-20
**Detector:** UV 270

## OTHER SUBSTANCES
**Also analyzed:** cefotaxime, cephacetrile

## REFERENCE
Terasaki, T.; Nouda, H.; Tsuji, A. Relationship between lipophilicity and binding affinity with human serum albumin for penicillin and cephem antibiotics. *J.Pharmacobiodyn.*, **1992**, *15*, 99–106

## SAMPLE
**Matrix:** solutions
**Sample preparation:** Add 100 µL solution to 1 mL 1 mg/mL cefsulodin in water, vortex for 15 s, inject a 10 µL aliquot.

## HPLC VARIABLES
**Column:** 250 × 4.6 10 µm Alltech C8
**Mobile phase:** MeOH:10 mM sodium acetate 30:70 containing 1.7 g/L tetrabutylammonium hydrogen sulfate, pH adjusted to 6.5 with 5 M NaOH
**Flow rate:** 1.5
**Injection volume:** 10
**Detector:** UV 254

## CHROMATOGRAM
**Retention time:** 10.0
**Internal standard:** cefsulodin (4.0)

## OTHER SUBSTANCES
**Simultaneous:** degradation products

## KEY WORDS
5% dextrose; saline

## REFERENCE
Marble, D.A.; Bosso, J.A.; Townsend, R.J. Compatibility of clindamycin phosphate with aztreonam in polypropylene syringes and with cefoperazone sodium, cefonicid sodium, and cefuroxime sodium in partial-fill glass bottles. *Drug Intell.Clin.Pharm.*, **1988**, *22*, 54–57

◆────────────────────────────◆────────────────────────────◆

## ANNOTATED BIBLIOGRAPHY
Fabre, H.; Ibork, H.; Lerner, D.A. Photoisomerization kinetics of cefuroxime and related compounds. *J.Pharm.Sci.*, **1994**, *83*, 553–558 [simultaneous degradation products; cefuroxime axetil determined]

Faouzi, M.A.; Dine, T.; Luyckx, M.; Gressier, B.; Brunet, C.; Goudaliez, F.; Mallevais, M.L.; Cazin, M.; Cazin, J.C. Stability and compatibility studies of cefaloridine, cefuroxime and ceftazidime with PVC infusion bags. *Pharmazie*, **1994**, *49*, 425–427 [simultaneous cefaloridine, ceftazidime; saline; 5% dextrose]

Wang, D.; Notari, R.E. Cefuroxime hydrolysis kinetics and stability predictions in aqueous solution. *J.Pharm.Sci.*, **1994**, *83*, 577–581 [simultaneous degradation products]

Zhang, H.; Stewart, J.T. Determination of a cefuroxime and aminophylline/theophylline mixture by high-performance liquid chromatography. *J.Liq.Chromatogr.*, **1994**, *17*, 1327–1335 [orcinol (IS)]

Kim, M.; Stewart, J.T. HPLC post-column ion-pair extraction of acidic drugs using a substituted α-phenylcinnamonitrile quaternary ammonium salt as a new fluorescent ion-pair reagent. *J.Liq.Chromatogr.*, **1990**, *13*, 213–237 [derivatization; fluorescence detection; also benzoic acid, flufenamic acid, ibuprofen, ketoprofen, mefenamic acid, probenecid, salicylic acid, valproic acid]

Das Gupta, V.; Stewart, K.R. Stability of cefuroxime sodium in some aqueous buffered solutions and intravenous admixtures. *J.Clin.Hosp.Pharm.*, **1986**, *11*, 47–54 [5% dextrose; saline; stability-indicating; simultaneous degradation products; cefazolin (IS)]

Sanders, C.A.; Moore, E.S. Liquid-chromatographic assay of cefuroxime in plasma. *Clin.Chem.*, **1986**, *32*, 2109

Campbell, C.J.; Langley, C. Measurement of rat-intestinal cefuroxime axetil esterase activity: comparison of an h.p.l.c. and coupled-enzyme assay. *Xenobiotica*, **1985**, *15*, 1011–1019 [enzyme incubations; cefuroxime determined]

Coomber, P.A.; Jefferies, J.P.; Woodford, J.D. High-performance liquid chromatographic determination of cefuroxime. *Analyst*, **1982**, *107*, 1451–1456

Bundtzen, R.W.; Toothaker, R.D.; Nielson, O.S.; Madsen, P.O.; Welling, P.G.; Craig, W.A. Pharmacokinetics of cefuroxime in normal and impaired renal function: comparison of high-pressure liquid chromatography and microbiological assays. *Antimicrob.Agents Chemother.*, **1981**, *19*, 443–449

Hekster, Y.A.; Baars, A.M.; Vree, T.B.; Van Klingeren, B.; Rutgers, A. Comparison of high performance liquid chromatography and microbiological assay in the determination of plasma cefuroxime concentrations in rabbits. *J.Antimicrob.Chemother.*, **1980**, *6*, 65–71

# Cephalexin

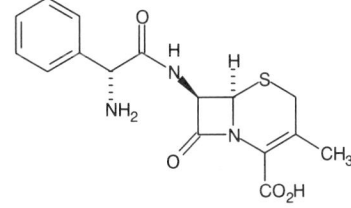

**Molecular formula:** $C_{16}H_{17}N_3O_4S$
**Molecular weight:** 347.4
**CAS Registry No.:** 15686-71-2, 23325-78-2 (monohydrate),
105879-42-3 (HCl)

## SAMPLE
**Matrix:** blood
**Sample preparation:** 50 µL Plasma + 100 µL 50 mM pH 7.0 phosphate buffer, mix. Inject 100 µL onto column A and elute to waste with mobile phase A, after 5 min backflush the contents of column A onto column B with mobile phase B, elute column B with mobile phase B and monitor the effluent. Re-equilibrate for 4 min before the next injection.

## HPLC VARIABLES
**Column:** A 20 × 3.9 25-40 µm LiChrosorb RP-8; B 10 × 4 Nova-Pak C8 guard column + 250 × 4.6 5 µm Ultracarb 5 ODS-30 (Phenomenex)
**Mobile phase:** A 50 mM pH 7.0 phosphate buffer; B Gradient. A was MeCN:20 mM pH 7.0 phosphate buffer 4:96. B was MeCN:20 mM pH 7.0 phosphate buffer 30:70. A:B 55:45 for 10 min, to 0:100 over 8 min, maintain at 0:100 for 12 min.
**Flow rate:** 1
**Injection volume:** 100
**Detector:** UV 230

## CHROMATOGRAM
**Retention time:** 7.4
**Internal standard:** cephalexin

## OTHER SUBSTANCES
**Extracted:** ampicillin, metampicillin
**Noninterfering:** acetaminophen, caffeine, ibuprofen, phenobarbital, sulbactam

## KEY WORDS
plasma; rat; column-switching; cephalexin is IS

## REFERENCE
Lee, H.; Lee, J.S.; Lee, H.S. Simultaneous determination of ampicillin and metampicillin in biological fluids using high-performance liquid chromatography with column switching. *J.Chromatogr.B*, **1995**, *664*, 335–340

## SAMPLE
**Matrix:** blood
**Sample preparation:** Mix 100 µL plasma + 300 µL 5 µg/mL cefotaxime in pH 3.5 10 mM acetate buffer and keep at 4°. Inject 100 µL onto column A with mobile phase A. After 5 min backflush column A with mobile phase B onto column B for 3 min. Re-equilibrate column A with mobile phase A for 16 min.

## HPLC VARIABLES
**Column:** A 40 × 2 37-50 µm Corasil RP C18; B 20 × 4 25-40 µm Lichrosorb RP-8 + 250 × 4 Partisil ODS-3
**Mobile phase:** A 10 mM pH 3.5 acetate buffer; B MeCN:20 mM pH 4.3 acetate buffer 15:85
**Flow rate:** 1

**Injection volume:** 100
**Detector:** UV 254

## CHROMATOGRAM
**Retention time:** 6.2
**Internal standard:** cefotaxime
**Limit of detection:** 500 ng/mL

## OTHER SUBSTANCES
**Simultaneous:** cefoxitin, cefuroxime, cephaloridine
**Noninterfering:** alclofenac, aspirin, caffeine, cefadroxil, cefamandole, cefazolin, cefoperazone, cefotiam, cephalothin, diclofenac, ibuprofen, indomethacin, ketoprofen, lonazolac,
mefenamic acid, naproxen, phenylbutazone, piroxicam

## KEY WORDS
plasma; column-switching; rat; human

## REFERENCE
Lee, Y.J.; Lee, H.S. Simultaneous determination of cefoxitin, cefuroxime, cephalexin and cephaloridine
in plasma using HPLC and a column-switching technique. *Chromatographia*, **1990**, *30*, 80–84

## SAMPLE
**Matrix:** blood
**Sample preparation:** Condition a 1 mL Bond-Elut C18 SPE cartridge with 2 mL MeOH
and 2 mL 8.5% phosphoric acid. Condition an NH2 SPE cartridge with 1 mL hexane. 500
μL Plasma + 25 μL 8.5% phosphoric acid + 250 μL 1 mg/mL coumarin-3-carboxylic acid
in water, add to the C18 SPE cartridge, wash with 500 μL water, wash with 1 mL 8.5%
phosphoric acid, wash with 5% MeOH:8.5% phosphoric acid 20:1, elute with 1 mL
MeOH:8.5% phosphoric acid 60:40 into the NH2 SPE cartridge. Wash the NH2 SPE
cartridge with 1 mL hexane, wash with 1 mL MeCN, elute with 1 mL water:10% ammonium sulfate 95:5, inject a 20 μL aliquot of the eluate.

## HPLC VARIABLES
**Column:** 250 × 4.6 C18
**Mobile phase:** Water:2 mM tetramethylammonium hydroxide in MeOH:acetic acid 60:
40:0.5
**Flow rate:** 0.8
**Injection volume:** 20
**Detector:** UV 262

## CHROMATOGRAM
**Internal standard:** coumarin-3-carboxylic acid (13)

## OTHER SUBSTANCES
**Extracted:** cefaclor, cefazolin, ceftizoxime

## KEY WORDS
plasma; SPE

## REFERENCE
Moore, C.M.; Sato, K.; Hattori, H.; Katsumata, Y. Improved HPLC method for the determination of
cephalosporins in human plasma and a new solid-phase extraction procedure for cefazolin and ceftizoxime. *Clin.Chim.Acta*, **1990**, *190*, 121–123

## SAMPLE
**Matrix:** blood

**Sample preparation:** 200 µL Plasma + 70 µL 1.2 M perchloric acid + 70 µL 500 mM sodium heptanesulfonate, vortex 15 s, centrifuge for 3 min, inject a 5 µL aliquot of the supernatant.

## HPLC VARIABLES
**Guard column:** 40 × 1 Co:Pell ODS
**Column:** 150 × 1 5 µm Nucleosil C18 in a glass-lined stainless steel column
**Mobile phase:** MeOH:2 mM phosphoric acid 26:74
**Flow rate:** 0.05
**Injection volume:** 5
**Detector:** UV 254 (2.4 µL flow cell)

## CHROMATOGRAM
**Retention time:** 20
**Limit of quantitation:** 200 ng/mL

## OTHER SUBSTANCES
**Extracted:** cefroxadin

## KEY WORDS
plasma; microbore

## REFERENCE
Rouan, M.C. Microbore liquid chromatographic determination of cadralazine and cephalexin in plasma with large-volume injection. *J.Chromatogr.*, **1988**, *426*, 335–344

## SAMPLE
**Matrix:** blood
**Sample preparation:** 100 µL Serum + 10 µL 5 µg/mL cefixime in MeOH + 100 µL MeCN, vortex for 15 s, centrifuge at 14000 g for 2 min. Remove the supernatant and evaporate it under a stream of nitrogen, reconstitute in 100 µL mobile phase, inject a 50-80 µL aliquot.

## HPLC VARIABLES
**Guard column:** RCSS Silica Guard Pak (Waters)
**Column:** 150 × 4.6 5µm Ultrasphere Octyl C8
**Mobile phase:** MeOH:12.5 mM pH 2.6 $NaH_2PO_4$ (pH adjusted with concentrated phosphoric acid) 20:80
**Flow rate:** 2
**Injection volume:** 50-80
**Detector:** UV 240

## CHROMATOGRAM
**Retention time:** 15
**Internal standard:** cefixime (11)
**Limit of detection:** 1 µg/ mL

## OTHER SUBSTANCES
**Extracted:** cefaclor, cefadroxil, cephradine
**Noninterfering:** acetaminophen, cimetidine, diazepam, digoxin, ibuprofen, phenytoin, propranolol, salicylic acid, warfarin

## KEY WORDS
serum

## REFERENCE

McAteer, J.A.; Hiltke, M.F.; Silber, B.M.; Faulkner, R.D. Liquid-chromatographic determination of five orally active cephalosporins–cefixime, cefaclor, cefadroxil, cephalexin, and cephradine–in human serum. *Clin.Chem.*, **1987**, *33*, 1788–1790

## SAMPLE

**Matrix:** blood
**Sample preparation:** 300 μL Plasma + 300 μL IS in ice-cold MeOH:100 mM pH 5.2 sodium acetate 70:30, vortex for 30 s, let stand at -20° for 10 min, centrifuge at 1500 g for 10 min, inject a 10 μL aliquot.

## HPLC VARIABLES

**Guard column:** 4 × 4 10 μm C18
**Column:** 300 × 4 10 μm μBondapak C18
**Mobile phase:** MeCN:MeOH:100 mM sodium acetate 11.52:0.48:88, pH 5.2
**Flow rate:** 2.5
**Injection volume:** 10
**Detector:** UV 254

## CHROMATOGRAM

**Retention time:** 5
**Internal standard:** 8-chlorotheophylline (4)
**Limit of detection:** 500 ng/mL

## OTHER SUBSTANCES

**Extracted:** cefoperazone, cefotaxime
**Interfering:** cefoxitin, cephaloridine

## KEY WORDS

plasma

## REFERENCE

Signs, S.A.; File, T.M.; Tan, J.S. High-pressure liquid chromatographic method for analysis of cephalosporins. *Antimicrob.Agents Chemother.*, **1984**, *26*, 652–655

## SAMPLE

**Matrix:** blood
**Sample preparation:** 500 μL Plasma + 4 mL water + 3 mL 10% trichloroacetic acid, centrifuge at 800-1000 g for 5 min. Remove 3 mL of the supernatant and add it to 2 mL buffer, add 1 mL 0.5% hydrogen peroxide in buffer, heat in a boiling water bath for 70 min, cool to room temperature, add 2 mL 500 mM $Na_2HPO_4$, add 7 mL acetone:chloroform 40:60, shake vigorously for 5 min, centrifuge. Remove 5 mL of the organic layer and evaporate it to dryness under reduced pressure, reconstitute the residue in 100 μL MeOH containing IS, inject a 20 μL aliquot. (Prepare buffer by dissolving 21 g citric acid in 200 mL 1 M NaOH, make up to 1 L with water, adjust pH to 2.5 with 100 mM HCl.)

## HPLC VARIABLES

**Column:** 250 × 4 5 μm Nucleosil C18
**Mobile phase:** MeOH:water 60:40
**Column temperature:** 55
**Injection volume:** 20
**Detector:** F ex 345 em 420

## CHROMATOGRAM

**Retention time:** 6 (?)

**Internal standard:** methyl anthranylate (9 (?))
**Limit of detection:** 2 ng/mL

## OTHER SUBSTANCES
**Interfering:** ampicillin, cephradine

## KEY WORDS
plasma; derivatization; rat; pharmacokinetics

## REFERENCE
Miyazaki, K.; Ohtani, K.; Sunada, K.; Arita, T. Determination of ampicillin, amoxicillin, cephalexin, and cephradine in plasma by high-performance liquid chromatography using fluorometric detection. *J.Chromatogr.*, **1983**, *276*, 478–482

## SAMPLE
**Matrix:** blood
**Sample preparation:** 200 μL Plasma + 70 μL 10% trichloroacetic acid, vortex 15 s, centrifuge for 3 min, remove 100 μL supernatant, inject 20 μL.

## HPLC VARIABLES
**Guard column:** 100 × 4.7 Co:Pell ODS
**Column:** 100 × 7.5 LiChrosorb RP-8
**Mobile phase:** MeOH:2 mM phosphoric acid 28:72
**Flow rate:** 1
**Injection volume:** 20
**Detector:** UV 254

## CHROMATOGRAM
**Retention time:** 19
**Limit of detection:** 1 nmol/mL

## KEY WORDS
plasma

## REFERENCE
Lecaillon, J.B.; Rouan, M.C.; Souppart, C.; Febvre, N.; Juge, F. Determination of cefsulodin, cefotiam, cephalexin, cefotaxime, deacetylcefotaxime, cefuroxime and cefroxadin in plasma and urine by high-performance liquid chromatography. *J.Chromatogr.*, **1982**, *228*, 257–267

## SAMPLE
**Matrix:** blood, urine
**Sample preparation:** Plasma or serum. Condition a Sep-Pak C18 SPE cartridge with 2 mL MeOH then 2 mL water, do not allow to go dry. 500 μL Plasma or serum + 100 μL water + 50 μL 25% acetic acid, mix, add to SPE cartridge, wash with two 1 mL portions of water, elute with 3 mL MeOH. Evaporate eluate under nitrogen, add 200 μL mobile phase, vortex, inject a 25 μL aliquot. Urine. Dilute 100:1 (ratio may vary depending on concentration) with water, inject a 50 μL aliquot.

## HPLC VARIABLES
**Column:** 150 × 4.6 5 μm Supelcosil LC-18-DB
**Mobile phase:** MeOH:THF:buffer 16:4:80 (Buffer was 1 g sodium 1-heptanesulfonate + 15 mL triethylamine in 1 L water with the pH adjusted to 2.3 with concentrated phosphoric acid.)
**Column temperature:** 30
**Flow rate:** 1.4

**Injection volume:** 25-50
**Detector:** UV 265

## CHROMATOGRAM
**Retention time:** 9.2
**Internal standard:** cephalexin

## OTHER SUBSTANCES
**Extracted:** cefaclor, hydroxyloracarbef, loracarbef
**Noninterfering:** acetaminophen, caffeine

## KEY WORDS
plasma; serum; SPE; pharmacokinetics; cephalexin is IS

## REFERENCE
Kovach, P.M.; Lantz, R.J.; Brier, G. High-performance liquid chromatographic determination of loracarbef, a potential metabolite, cefaclor and cephalexin in human plasma, serum and urine. *J.Chromatogr.*, **1991**, *567*, 129–139

## SAMPLE
**Matrix:** blood, urine
**Sample preparation:** Plasma. 500 μL Plasma + 150 μL 10% trichloroacetic acid + 500 μL MeCN, mix, add 1.5 mL dichloromethane, vortex, centrifuge, inject a 60 μL aliquot of the aqueous supernatant. Urine. 5 mL Urine + 5 mL 20 mM pH 3.8 sodium acetate buffer, mix. 500 μL Buffered urine + 150 μL 5% trichloroacetic acid, vortex, inject a 10 μL aliquot.

## HPLC VARIABLES
**Guard column:** 45 × 4 37-50 μm C18/Corasil
**Column:** Zorbax C8 (plasma) or Partisil 5 ODS-3 RAC C18 (urine)
**Mobile phase:** MeCN:glacial acetic acid:water 17:2:81 (plasma) or MeCN:MeOH:THF: trichloroacetic acid:glacial acetic acid:sodium acetate trihydrate:sodium dodecyl sulfate: water 25:3:0.925:0.075:0.25:0.077:0.134:70.75 (v/v/v/w/v/w/w/v)
**Flow rate:** 1.5 (plasma) or 2 (urine)
**Injection volume:** 10-60
**Detector:** UV 280

## CHROMATOGRAM
**Retention time:** 17 (plasma), 23 (urine)
**Internal standard:** cephalexin

## OTHER SUBSTANCES
**Extracted:** cefprozil

## KEY WORDS
plasma; cephalexin is IS; human; rat; pharmacokinetics

## REFERENCE
Shyu, W.C.; Shukla, U.A.; Shah, V.R.; Papp, E.A.; Barbhaiya, R.H. Simultaneous high-performance liquid chromatographic analysis of cefprozil diastereomers in a pharmacokinetic study. *Pharm.Res.*, **1991**, *8*, 992–996

## SAMPLE
**Matrix:** blood, urine

**Sample preparation:** 1 mL Plasma + 1 mL 6% trichloroacetic acid, mix, centrifuge at 4000 rpm for 10 min, inject an aliquot of the supernatant. Inject an aliquot of urine directly.

## HPLC VARIABLES
**Guard column:** 10 × 4 7 μm Lichrosorb RP 18
**Column:** 250 × 4 7 μm Lichrosorb RP 18
**Mobile phase:** MeCN:25 mM pH 7 phosphate buffer 10:90
**Flow rate:** 1
**Injection volume:** 10
**Detector:** F ex 385 em 485 following post-column reaction. The column effluent mixed with 200 μg/mL fluorescamine in MeCN pumped at 0.25 mL/min and the mixture flowed through a 4.5 m × 0.25 mm ID coil of PTFE tubing to the detector.; UV 260

## CHROMATOGRAM
**Limit of detection:** 1.4 ng/mL (F); 1.9 ng/mL (UV)

## OTHER SUBSTANCES
**Also analyzed:** cefaclor, cefroxadine, cephradine
**Noninterfering:** amidopyrin, aspirin, barbital, caffeine, cefmenoxime, cefotaxime, ceftizoxime, ceftriaxone, cetazidime, diazepam, dibekacin, gentamycin, kanamycin, lidocaine, netilmicin, tetracaine, theophylline, tobramycin

## KEY WORDS
post-column reaction; plasma; F detection may be less susceptible to interferences

## REFERENCE
Blanchine, M.D.; Fabre, H.; Mandrou, B. Fluorescamine post-column derivatization for the HPLC determination of cephalosporins in plasma and urine. *J.Liq.Chromatogr.*, **1988**, *11*, 2993–3010

## SAMPLE
**Matrix:** blood, urine
**Sample preparation:** Serum. 200 μL Serum + 250 μL 4 μg/mL ceftazidime in 6% perchloric acid, vortex 15 s, centrifuge at 1500 g for 15 min, inject 25 μL supernatant. Urine. 250 μL Urine + 1 mL 4 μg/mL ceftazidime in 6% perchloric acid, mix, inject 10 μL supernatant.

## HPLC VARIABLES
**Guard column:** 30 × 2 30-38 μm CO:Pell ODS
**Column:** 150 × 4.6 5 μm Pecosphere C18
**Mobile phase:** 100 mL MeCN + 90 mL 0.5 M $(NH_4)H_2PO_4$ adjusted to pH 3.0 with 10% phosphoric acid. This mixture was made up to 1 L with water.
**Flow rate:** 1.7
**Injection volume:** 10-25
**Detector:** UV 254

## CHROMATOGRAM
**Retention time:** 8
**Internal standard:** ceftazidime

## OTHER SUBSTANCES
**Noninterfering:** amikacin, carbenicillin, cefazolin, cefoperazine, cefoxitin, chloramphenicol, ciprofloxacin, gentamicin, imipenem, moxalactam, oxacillin, ticarcillin, tobramycin, vancomycin
**Interfering:** cefatoxime, cefonicid, methicillin, penicillin G

## KEY WORDS
serum

## REFERENCE
Emm, T.A.; Leslie, J.; Chai, M.; Lesko, L.J.; Perkal, M.B. High-performance liquid chromatographic assay of cephalexin in serum and urine. *J.Chromatogr.*, **1988**, *427*, 162–165

## SAMPLE
**Matrix:** bulk
**Sample preparation:** Dissolve in water at a concentration of 5 mg/mL, inject a 20 μL aliquot.

## HPLC VARIABLES
**Column:** 250 × 4.6 5 μm YMC-ODS (YMC)
**Mobile phase:** Gradient. A was 50 mM sodium dihydrogen phosphate adjusted to pH 4.0 with phosphoric acid. B was MeCN:50 mM sodium dihydrogen phosphate adjusted to pH 4.0 with phosphoric acid 45:55. A:B from 95:5 to 75:25 over 30 min, to 0:100 over 15 min, maintain at 0:100 for 5 min, return to 95:5 and equilibrate for 14 min.
**Flow rate:** 1
**Injection volume:** 20
**Detector:** UV 220

## CHROMATOGRAM
**Retention time:** 28

## OTHER SUBSTANCES
**Simultaneous:** impurities, excipients
**Also analyzed:** cefaclor

## REFERENCE
Olsen, B.A.; Baertschi, S.W.; Riggin, R.M. Multidimensional evaluation of impurity profiles for generic cephalexin and cefaclor antibiotics. *J.Chromatogr.*, **1993**, *648*, 165–173

## SAMPLE
**Matrix:** bulk, formulations
**Sample preparation:** Weigh out bulk drug, capsule contents, granules, or powders equivalent to 35 mg cephalexin, add 1 mL 2.52 mg/mL acetaminophen in MeOH:water 20:80, make up to 50 mL with water, inject a 20 μL aliquot.

## HPLC VARIABLES
**Column:** 300 × 3.9 10 μm μBondapak C18
**Mobile phase:** MeOH:1.25% acetic acid 25:75
**Flow rate:** 1
**Injection volume:** 20
**Detector:** UV 254

## CHROMATOGRAM
**Retention time:** 11.3
**Internal standard:** acetaminophen (6)

## OTHER SUBSTANCES
**Simultaneous:** degradation products

## KEY WORDS
capsules; granules; powder

## REFERENCE
Hsu, M.C.; Lin, Y.-S.; Chung, H.-C. High-performance liquid chromatographic method for potency determination of cephalexin in commercial preparations and for stability studies. *J.Chromatogr.A*, **1995**, *692*, 67–72

## SAMPLE
**Matrix:** bulk, formulations
**Sample preparation:** Dissolve in water to a concentration of 70 μg/mL, inject a 20 μL aliquot.

## HPLC VARIABLES
**Column:** 300 × 3.9 10 μm μBondapak C18
**Mobile phase:** MeOH:water:acetic acid 30:70:0.1
**Flow rate:** 1
**Injection volume:** 20
**Detector:** UV 254

## CHROMATOGRAM
**Retention time:** 10
**Limit of quantitation:** 5.6 μg/mL

## OTHER SUBSTANCES
**Simultaneous:** impurities, cefaclor, cefadroxil, cefamandole, cefamandole nafate, cefazolin, cefoperazone, cefotaxime, cefoxitin, ceftizoxime, cephalothin, cephapirin, cephradine

## REFERENCE
Ting, S. Reverse-phase liquid chromatographic analysis of cephalosporins. *J.Assoc.Off.Anal.Chem.*, **1988**, *71*, 1123–1130

## SAMPLE
**Matrix:** perfusate
**Sample preparation:** Filter, inject an aliquot.

## HPLC VARIABLES
**Column:** 5 μm Ultrasphere C18
**Mobile phase:** MeOH:50 mM pH 5 sodium phosphate buffer 20:80
**Detector:** UV 220

## CHROMATOGRAM
**Retention time:** 9

## REFERENCE
Hu, M.; Zheng, L.; Chen, J.; Liu, L.; Zhu, Y.; Dantzig, A.H.; Stratford, R.E., Jr. Mechanisms of transport of quinapril in Caco-2 cell monolayers: Comparison with cephalexin. *Pharm.Res.*, **1995**, *12*, 1120–1125

## SAMPLE
**Matrix:** solutions
**Sample preparation:** Inject a 10 μL aliquot.

## HPLC VARIABLES
**Column:** 220 × 4.6 Spheri 5 ODS-224
**Mobile phase:** 100 mM sodium dodecyl sulfate, pH 6.72
**Flow rate:** 1

**Injection volume:** 10
**Detector:** UV 260

## CHROMATOGRAM
**Retention time:** 4.5

## OTHER SUBSTANCES
**Simultaneous:** 7-aminocephalorospanic acid, 7-aminodesacetoxycephalosporanic acid, cefazolin, cephaloridine, cephalothin, cephradine

## REFERENCE
Garcia Pinto, C.; Pérez Pavón, J.L.; Moreno Cordero, B. Micellar liquid chromatography of zwitterions: Retention mechanism of cephalosporins. *Analyst*, **1995**, *120*, 53–62

## SAMPLE
**Matrix:** solutions

## HPLC VARIABLES
**Column:** 300 × 3.9 10 μm μBondapak phenyl
**Mobile phase:** MeOH:10 mM phosphate buffer 27:73, pH 3.6
**Column temperature:** 27
**Flow rate:** 1
**Detector:** UV 254

## CHROMATOGRAM
**Retention time:** 9

## OTHER SUBSTANCES
**Simultaneous:** amoxicillin, ampicillin, cefaclor, cephradine

## REFERENCE
Huang, H.-S.; Wu, J.-R.; Chen, M.-L. Reversed-phase high-performance liquid chromatography of amphoteric β-lactam antibiotics: effects of columns, ion-pairing reagents and mobile phase pH on their retention times. *J.Chromatogr.*, **1991**, *564*, 195–203

## SAMPLE
**Matrix:** solutions

## HPLC VARIABLES
**Column:** 250 × 4 OmniPac PCX-500 (Dionex)
**Mobile phase:** Gradient. A was MeCN:90 mM perchloric acid 13.5:86.5. B was MeCN:300 mM perchloric acid 45:55. A:B from 100:0 to 0:100 over 7 min, maintain at 0:100.
**Flow rate:** 1
**Detector:** UV 254

## CHROMATOGRAM
**Retention time:** 8.5

## OTHER SUBSTANCES
**Simultaneous:** 7-aminocephalosproranic acid, cefadroxil, cefazolin, cefotaxime, cephaloridine, cephalosporin C, cephalothin, cephapirin, D-hydroxyphenylglycine

## REFERENCE
Slingsby, R.W.; Rey, M. Determination of pharmaceuticals by multi-phase chromatography: Combined reversed phase and ion exchange in one column. *J.Liq.Chromatogr.*, **1990**, *13*, 107–134

## SAMPLE
**Matrix:** tissue
**Sample preparation:** Condition a 100 mg Sep-Pak SPE cartridge with 5 mL MeOH and 5 mL water. Homogenize tissue with 4 (liver, lung) or 29 (spleen) volumes of water (Thomas tissue grinder series 3431-D70). 1 mL Homogenate + 50 μL 8.5% phosphoric acid, vortex for 30 s, centrifuge at 2000 g for 5 min, add to the SPE cartridge, wash with 3 mL water, elute with 2 mL MeOH:water 60:40, inject a 10 μL aliquot of the eluate.

## HPLC VARIABLES
**Guard column:** Nova-Pak C18 guard column
**Column:** 250 × 4.6 5 μm Econosphere C18
**Mobile phase:** MeOH:20 mM $NaH_2PO_4$ 23:77, pH 5.0
**Flow rate:** 1
**Injection volume:** 100
**Detector:** UV 270

## CHROMATOGRAM
**Retention time:** 11.5
**Internal standard:** cephalexin

## OTHER SUBSTANCES
**Extracted:** cefazolin

## KEY WORDS
rat; liver; spleen; lung; SPE; cephalexin is IS

## REFERENCE
Liang, D.; Chow, D.; White, C. High-performance liquid chromatographic assay of cefazolin in rat tissues. *J.Chromatogr.B*, **1994**, *656*, 460–465

## SAMPLE
**Matrix:** tissue
**Sample preparation:** Muscle, fat. 10 g Minced tissue + 1 mL cephradine in 10 mM pH 3.0 phosphate buffer, let stand for 30 min, add 19 mL 5% trichloroacetic acid, chill to 5°, homogenize (Virtis model 45), centrifuge at 1000 g for 5 min, filter (0.2 μm), inject 2 mL of the filtrate onto column A with mobile phase A, elute to waste with mobile phase A for 1.5 min then flush contents of column A onto column B with mobile phase B, elute column B with mobile phase B and monitor the effluent. Liver, kidney. 10 g Minced tissue + 1 mL cephradine in 10 mM pH 3.0 phosphate buffer, let stand for 30 min, add 19 mL 5% trichloroacetic acid, chill to 5°, homogenize (Virtis model 45), centrifuge at 1000 g for 5 min, filter (0.2 μm). 10 mL Filtrate + 20 mL dichloromethane:isopropanol 95:5, stir for 2 min, centrifuge at 1000 g for 5 min. Discard the organic layer, add 250 μL concentrated ammonia solution to the aqueous layer, add 20 mL dichloromethane:isopropanol 95:5, stir for 2 min, centrifuge at 1000 g for 5 min. Discard the organic layer, restore the initial pH of the aqueous layer with concentrated HCl. Inject 2 mL of this solution onto column A with mobile phase A, elute to waste with mobile phase A for 1.5 min then flush contents of column A onto column B with mobile phase B, elute column B with mobile phase B and monitor the effluent.

## HPLC VARIABLES
**Column:** A 25 × 4 25-40 μm LiChroprep RP 18; B 4 × 4 5 μm LiChrospher 100 CH-18 +250 × 4 5 μm LiChrospher 100 CH-18
**Mobile phase:** A MeOH:10 mM pH 3.0 phosphate buffer 15:85;B MeOH:10 mM pH 3.0 phosphate buffer 30:70 (Every 30 injections change column A and column B guard column, flush system with 60 mL MeOH:water 30:70 and 30 mL MeOH.)
**Flow rate:** 1

**Injection volume:** 2000
**Detector:** UV 260

## CHROMATOGRAM
**Retention time:** 11
**Internal standard:** cephradine (15)
**Limit of quantitation:** 45 ng/g

## KEY WORDS
cow; muscle; fat; liver; kidney

## REFERENCE
Leroy, P.; Decolin, D.; Nicolas, S.; Archimbault, P.; Nicolas, A. Residue determination of two coadministered antibacterial agents–cephalexin and colistin–in calf tissues using high-performance liquid chromatography and microbiological methods. *J.Pharm.Biomed.Anal.*, **1989**, *7*, 1837–1846

◆─────────────────────◆─────────────────────◆

## ANNOTATED BIBLIOGRAPHY

Changqin, H.; Shaohong, S.; Kaimin, W. The chromatographic behavior of cephalosporins in gel filtration chromatography, a novel method to separate high molecular weight impurities. *J.Pharm. Biomed.Anal.*, **1994**, *12*, 533–541 [also cefadroxil, cefamandole, cefmenoxime, cefoperazone, cefotaxime, ceftazidime, ceftriaxone, cephaloridine, cephalothin, cephradine]

Granero, L.; Gimeno, M.J.; Torres-Molina, F.; Chesa-Jiménez, J.; Peris, J.E. Studies on the renal excretion mechanisms of cefadroxil. *Drug Metab.Dispos.*, **1994**, *22*, 447–450 [rat; plasma; urine; gradient; extracted cefadroxil; LOD 300 ng/mL]

Muranushi, N.; Horie, K.; Masuda, K.; Hirano, K. Characteristics of ceftibuten uptake into Caco-2 cells. *Pharm.Res.*, **1994**, *11*, 1761–1765 [also cefaclor, cefadroxil, cefazolin, cephalexin, cephradine, cyclacillin, latamoxef]

Yongxin, Z.; Hendrix, C.; Busson, R.; Janssen, G.; Roets, E.; Hoogmartens, J. Isolation and structural elucidation of an impurity of cefradine. *J.Pharm.Biomed.Anal.*, **1994**, *12*, 1137–1140 [also cefradine; column temp 55]

Parker, C.E.; Perkins, J.R.; Tomer, K.B.; Shida, Y.; O'Hara, K. Nanoscale packed capillary liquid chromatography-electrospray ionization mass spectrometry: analysis of penicillins and cephems. *J.Chromatogr.*, **1993**, *616*, 45–57 [serum; also amoxicillin, ampicillin, carbenicillin, cefalothin, cefazolin, cefmenoxime, cefmetazole, cefoperazone, cefotaxime, cefotiam, cefoxitin, cloxacillin, dicloxacillin, penicillin G, pipericillin, sulbenicillin]

Kelly, J.W.; Stewart, J.T. Separation of selected beta lactam antibiotic epimers on gamma cyclodextrin, ion exchange ethylvinylbenzene/divinylbenzene/copolymer and poly(styrene- divinylbenzene) copolymer stationary phases. *J.Liq.Chromatogr.*, **1991**, *14*, 2235–2250 [also carbemicillin, moxalactam, ticarcillin]

Wang, D.P.; Yeh, M.K. Stability-indicating method for cephalexin in capsules by high-performance liquid chromatography. *Chung-hua Yao Hsueh Tsa Chih*, **1990**, *42*, 349–353

Marincel, J.; Bosnjak, N.; Lamut, M. Comparison between HPLC and microbiological methods in assays of cephalexin in samples. *Acta Pharm.Jugosl.*, **1988**, *38*, 35–45

Das Gupta, V.; Parasrampuria, J. Quantitation of cephalexin in pharmaceutical dosage forms using high-performance liquid chromatography. *Drug Dev.Ind.Pharm.*, **1987**, *13*, 2231–2238

Kovacic-Bosnjak, N.; Mandic, Z.; Kovacevic, M. Reversed-phase HPLC separation of delta$^2$ and delta$^3$ isomers of 7-ADCA and cephalexin monohydrate. *Chromatographia*, **1987**, *23*, 350–354

Najib, N.M.; Suleiman, M.S.; El-Sayed, Y.M.; Abdulhameed, M.E. High performance liquid chromatographic analysis of cephalexin in serum and urine. *J.Clin.Pharm.Ther.*, **1987**, *12*, 419–426

Nakagawa, T.; Shibukawa, A.; Uno, T. Liquid chromatography with crown ether-containing mobile phases. II. Retention behavior of β-lactam antibiotics in reversed-phase high-performance liquid chromatography. *J.Chromatogr.*, **1982**, *239*, 695–706 [also amoxicillin, ampicillin, benzylpenicillin, carbenicillin, cefradine, cephalexin, cephaloglycin, cephaloridin, ciclacillin, cloxacillin, dicloxacillin, oxacillin, penicillin G]

Brunetta, A.; Mosconi, L.; Pongiluppi, S.; Scagnolari, U.; Zambonin, G. [Chemical and microbiological determination of cephalexin and sodium flucloxacillin in combination, following separation by HPLC]. *Boll.Chim.Farm.*, **1981**, *120*, 335–342

Mason, B.; Tranter, J. Use of high-performance liquid chromatography for the determination of cephalexin. *Anal.Proc.*, **1981**, *18*, 310–313

Okumura, T. Application of thin-layer chromatography to high-performance liquid chromatographic separation of steroidal hormones and cephalosporin antibiotics. *J.Liq.Chromatogr.*, **1981**, *4*, 1035–1064 [normal phase; also betamethasone, cefaloridine, cephaloglycine, cephaloridine, cephalothin, cortisone, dexamethasone, hydrocortisone]

Tsutsumi, K.; Kubo, H.; Kinoshita, T. Determination of serum cephalexin by high performance liquid chromatography. *Anal.Lett.*, **1981**, *14*, 1735–1743

Fabregas, J.L.; Beneyto, J.E. Simultaneous determination of cephalexin and lysine in their salt using high-performance liquid chromatography of derivatives. *J.Pharm.Sci.*, **1980**, *69*, 1378–1380

# Cilastatin

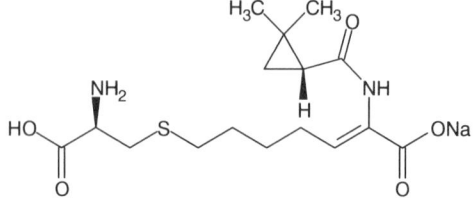

**Molecular formula:** $C_{16}H_{26}N_2O_5S$
**Molecular weight:** 358.5
**CAS Registry No.:** 82009-34-5 (cilastatin),
81129-83-1 (cilastatin sodium)

## SAMPLE
**Matrix:** bile, blood, urine
**Sample preparation:** Plasma, urine. Condition a 3 mL 200 mg C8 Bond Elut SPE cartridge with 1 mL MeOH and 2 mL water. 100-400 μL Plasma or 10 μL urine + 10 μL 1 mg/mL N-propionylcilastatin in MeCN + 1 mL water, acidify with 1 drop concentrated HCl, vortex, add to SPE cartridge, elute with 1 mL MeOH:water 80:20. Evaporate the eluate to dryness under a stream of nitrogen, reconstitute the residue in 1 mL mobile phase, inject a 50-100 μL aliquot. Bile. 10 μL Bile + 1 mL mobile phase + 10 μL 1 mg/mL N-propionylcilastatin in MeCN, vortex for 30 s, centrifuge at 1500 g for 5 min, inject a 50-100 μL aliquot.

## HPLC VARIABLES
**Guard column:** 30 × 4.6 5 μm Brownlee RP-18
**Column:** 100 × 4.6 5 μm Partisil ODS-3 RAC 2
**Mobile phase:** MeCN:50 mM $NaH_2PO_4$ containing 5 mM PIC B-8 (Waters), pH adjusted to 4.0 with 85% phosphoric acid 12:88
**Flow rate:** 1.2
**Injection volume:** 50-100
**Detector:** UV 210

## CHROMATOGRAM
**Retention time:** 4.6
**Internal standard:** N-propionylcilastatin (15.1)
**Limit of detection:** 500 ng/mL

## OTHER SUBSTANCES
**Simultaneous:** N-acetylcilastatin

## KEY WORDS
plasma; rat; SPE

## REFERENCE
Chen, I.-W.; Hsieh, J.Y.-K.; Lin, J.H.; Duggan, D.E. High-performance liquid chromatographic determination of cilastatin and its major metabolite N-acetylcilastatin in rat plasma, urine and bile. *J.Chromatogr.*, **1990**, *534*, 119–126

## SAMPLE
**Matrix:** blood
**Sample preparation:** Stabilize plasma with an equal volume of ethylene glycol:1 M pH 6.0 morpholineethanesulfonate buffer 50:50.

## HPLC VARIABLES
**Column:** 150 × 4.6 5 μm Adsorbosphere C8
**Mobile phase:** MeOH:buffer:water 4:10:86 (Buffer was 20.9 g 3-(N-morpholino) propanesulfonic acid in 1 L water, pH 7.0.)
**Column temperature:** 50
**Flow rate:** 4

**Injection volume:** 10
**Detector:** UV 245

## KEY WORDS
plasma; dog; pharmacokinetics

## REFERENCE
Wong, J.; Kuu, W.-Y.; Burke, R.; Johnson, R.; Wood, R.W. Comparison of simulated and in-vivo plasma levels of cilastatin following intravenous in-line drug administration. *Pharm.Res.*, **1995**, *12*, 144–148

## SAMPLE
**Matrix:** blood
**Sample preparation:** Dilute serum with an equal volume of 500 mM pH 6.0 2-[N-morpholine]ethanesulfonic acid (MES) buffer, filter (Amicon Centrifree with YMT membrane) a 700 μL aliquot while centrifuging at 1000-2000 g for 30 min, vortex filtrate, inject a 50 μL aliquot.

## HPLC VARIABLES
**Guard column:** 40 × 4 40 μm pellicular reverse-phase (Vydac)
**Column:** 200 × 4 Micro Pak MCH 10 reverse-phase
**Mobile phase:** MeOH:100 mM pH 2.5 potassium phosphate buffer 24:76 (Wash with 50:50 at the end of each run.)
**Column temperature:** 30
**Flow rate:** 1
**Injection volume:** 50
**Detector:** UV 220

## CHROMATOGRAM
**Retention time:** 20
**Limit of detection:** 500 ng/mL

## OTHER SUBSTANCES
**Simultaneous:** caffeine, carbenicillin, ceftazidime, chloramphenicol, phenobarbital, phenytoin, salicylic acid, sulfamethodoxazole, theophylline, ticarcillin
**Noninterfering:** acetaminophen, cimetidine, imipenem, moxalactam, theobromine

## KEY WORDS
serum; pharmacokinetics

## REFERENCE
Myers, C.M.; Blumer, J.L. Determination of imipenem and cilastatin in serum by high-pressure liquid chromatography. *Antimicrob.Agents Chemother.*, **1984**, *26*, 78–81

## SAMPLE
**Matrix:** blood, tissue
**Sample preparation:** Serum. 500 μL Serum + 500 μL MeOH, vortex vigorously, heat at 35° for 15 min, centrifuge at 4° at 4000 rpm for 10 min, inject a 20 μL aliquot of the supernatant. Tissue. Homogenize tissue with pH 7.2 sodium borate buffer, centrifuge at 4° at 4000 rpm for 10 min, remove a 500 μL aliquot of the supernatant and add it to 500 μL MeOH, vortex vigorously, heat at 35° for 15 min, centrifuge at 4° at 4000 rpm for 10 min, inject a 20 μL aliquot of the supernatant.

## HPLC VARIABLES
**Guard column:** C18 Corasil
**Column:** 300 mm long μBondapak C18

**Mobile phase:** Isopropanol:water 7:93, adjusted to pH 3 with phosphoric acid
**Flow rate:** 1.5
**Injection volume:** 20
**Detector:** UV 210

## CHROMATOGRAM
**Retention time:** 11.50
**Limit of detection:** 500 ng/mL

## KEY WORDS
serum

## REFERENCE
Krausse, R.; Ullmann, U. Determination of imipenem and cilastatin in serum and tissue by high-pressure liquid chromatography. *Infection*, **1986**, *14*, 243–245

## SAMPLE
**Matrix:** blood, urine
**Sample preparation:** Plasma. 500 μL Plasma + 75 μL 30 μg/mL IS + 2 mL 500 mM pH 3 $KH_2PO_4$, vortex, add to an activated Sep-Pak C18 SPE cartridge, wash with 20 mL 1 mM orthophosphoric acid, elute with 1.5 mL MeOH, add the eluate to 1 mL water, vortex, inject a 50-200 μL aliquot. Urine. Stabilize urine by mixing with an equal volume of 1 M pH 6.8 MOPS buffer:ethylene glycol 50:50. 1 mL Stabilized urine + 50 μL 100 μg/mL IS + 2.5 mL 20 mM orthophosphoric acid, vortex, add to an activated Sep-Pak C18 SPE cartridge, wash with 20 mL 1 mM orthophosphoric acid, elute with 1.5 mL MeOH, add the eluate to 1 mL water, vortex, inject a 25-75 μL aliquot.

## HPLC VARIABLES
**Guard column:** 40 × 4.6 ODS-10 (Bio-Rad)
**Column:** 250 × 4.6 Bio-Sil ODS-10 (Bio-Rad)
**Mobile phase:** Isopropanol:0.2% orthophosphoric acid 10.9:89.1, pH 3 (plasma) or isopropanol:0.2% orthophosphoric acid 6:94, pH 3 (urine)
**Flow rate:** 2
**Injection volume:** 25-200
**Detector:** F ex 335 em 455 following post-column reaction with o-phthalaldehyde reagent solution (Pierce) pumped at 1 mL/min. The mixture flowed through a 250 × 4.6 column packed with 40 μm glass beads (Whatman) to the detector.

## CHROMATOGRAM
**Retention time:** 5.04 (plasma), 7.95 (urine)
**Internal standard:** S-(p-methylbenzyl)-L-cysteine (8.04 (plasma), 12.01 (urine))
**Limit of detection:** 750 ng/mL (plasma); 2500 ng/mL (urine)

## OTHER SUBSTANCES
**Noninterfering:** metabolites, imipenem

## KEY WORDS
plasma; derivatization; post-column reaction; SPE

## REFERENCE
Demetriades, J.L.; Souder, P.R.; Entwistle, L.A.; Vincek, W.C.; Musson, D.G.; Bayne, W.F. High-performance liquid chromatographic determination of cilastatin in biological fluids. *J.Chromatogr.*, **1986**, *382*, 225–231

## SAMPLE
**Matrix:** formulations

## HPLC VARIABLES
**Column:** 150 × 4.6 5 μm Adsorbosphere C8
**Mobile phase:** MeOH:water:100 mM pH 7 3-[N-morpholino]propanesulfonic acid buffer 4:86:10
**Detector:** UV 245

## OTHER SUBSTANCES
**Simultaneous:** imipenem

## KEY WORDS
injections; saline; 5% dextrose

## REFERENCE
Jenke, D.R. Drug binding by reservoirs in elastomeric infusion devices. *Pharm.Res.*, **1994**, *11*, 984–989

## SAMPLE
**Matrix:** formulations
**Sample preparation:** Dilute 1:100 with mobile phase, inject a 30 μL aliquot.

## HPLC VARIABLES
**Column:** 200 × 4.6 RP-8 (Hewlett-Packard)
**Mobile phase:** MeCN:MeOH:buffer 0.4:0.5:99.1, adjusted to pH 7.00 with NaOH (Buffer was 4 mM 3-[N-morpholino]propanesulfonic acid (MOPS) containing 2 g/L sodium hexane sulfate.)
**Flow rate:** 1.8
**Injection volume:** 30
**Detector:** UV 250

## CHROMATOGRAM
**Retention time:** 8.8

## OTHER SUBSTANCES
**Extracted:** imipenem

## KEY WORDS
injections; total parenteral nutrition; stability-indicating

## REFERENCE
Zaccardelli, D.S.; Krcmarik, C.S.; Wolk, R.; Khalidi, N. Stability of imipenem and cilastatin sodium in total parenteral nutrient solution. *J.Parenter.Enteral.Nutr.*, **1990**, *14*, 306–309

# Cimetidine

**Molecular formula:** C$_{10}$H$_{16}$N$_6$S
**Molecular weight:** 252.3
**CAS Registry No.:** 51481-61-9, 70059-30-2 (HCl)

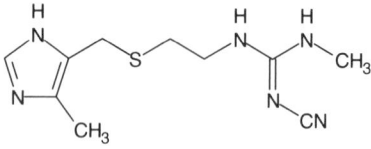

## SAMPLE
**Matrix:** blood
**Sample preparation:** 250 µL Plasma + 25 µL 2.5 µg/mL IS in MeOH:water 50:50 + 20 µL 2.5 M NaOH + 100 µL saturated potassium carbonate, vortex for 1 min, add 1 mL ethyl acetate, vortex for 90 s, centrifuge at 9500 g. Remove the organic layer and evaporate it to dryness under a stream of nitrogen at 37°, reconstitute the residue in 500 µL water, inject a 50 µL aliquot.

## HPLC VARIABLES
**Column:** 250 × 4.6 5 µm Nucleosil C18
**Mobile phase:** MeCN:10 mM pH 6.2 phosphate buffer 25:75, containing 2.5 g/L heptane-sulfonic acid
**Flow rate:** 0.9
**Injection volume:** 50
**Detector:** UV 228

## CHROMATOGRAM
**Retention time:** 6.3
**Internal standard:** BN CK249 (Glaxo) (7.1)
**Limit of detection:** 15 ng/mL
**Limit of quantitation:** 50 ng/mL

## KEY WORDS
plasma; pharmacokinetics

## REFERENCE
Kelly, M.T.; McGuirk, D.; Bloomfield, F.J. Determination of cimetidine in human plasma by high-performance liquid chromatography following liquid-liquid extraction. *J.Chromatogr.B*, **1995**, *668*, 117–123

## SAMPLE
**Matrix:** blood
**Sample preparation:** 100 µL Plasma + 50 µL 2 M NaOH + 5 mL dichloromethane, shake for 10 min, centrifuge at 850 g for 5 min. Remove 4 mL of the organic layer and evaporate it to dryness under a stream of nitrogen at 40°, reconstitute the residue in 100 µL 2.5 µg/mL 1,3-dimethyluric acid in MeOH, inject a 20 µL aliquot.

## HPLC VARIABLES
**Column:** reverse-phase
**Mobile phase:** MeOH:10 mM pH 5.2 sodium acetate 25:75
**Flow rate:** 1.2
**Injection volume:** 20
**Detector:** UV 228

## CHROMATOGRAM
**Internal standard:** 1,3-dimethyluric acid

## KEY WORDS
plasma; rat

## REFERENCE

Nagai, N.; Furuhata, M.; Ogata, H. Drug interactions between theophylline and H₂-antagonists, roxatidine acetate hydrochloride and cimetidine: Pharmacokinetic analysis in rats in vivo. *Biol.Pharm.Bull.*, **1995**, *18*, 1610–1613

## SAMPLE

**Matrix:** blood
**Sample preparation:** 300 μL Plasma + 150 μL water + 300 μL 2.5 M NaOH + 2 μg codeine (dissolved in water), extract with 5 mL dichloromethane. Remove the organic layer and evaporate it to dryness at 37°, reconstitute the residue in 200 μL mobile phase, inject an 80 μL aliquot.

## HPLC VARIABLES

**Guard column:** 30 × 4.6 10 μm Spheri-10 RP-18 (Brownlee)
**Column:** 250 × 4.6 10 μm Partisil ODS/10
**Mobile phase:** MeOH:buffer 15:85 (Buffer was 50 mM $KH_2PO_4$ adjusted to pH 2.7 with 1 M HCl.)
**Flow rate:** 2
**Injection volume:** 80
**Detector:** UV 228

## CHROMATOGRAM

**Retention time:** 6-7
**Internal standard:** codeine (10-11)
**Limit of detection:** 100 ng/mL
**Limit of quantitation:** 250 ng/mL

## KEY WORDS

dog; plasma; pharmacokinetics

## REFERENCE

Langguth, P.; Lee, K.M.; Spahn-Langguth, H.; Amidon, G.L. Variable gastric emptying and discontinuities in drug absorption profiles: Dependence of rates and extent of cimetidine absorption on motility phase and pH. *Biopharm.Drug Dispos.*, **1994**, *15*, 719–746

## SAMPLE

**Matrix:** blood
**Sample preparation:** Mix 1 mL plasma + 0.1 mL 20 μg/mL IS in water, add to 1 mL Baker C-18 column, rinse twice with 1 mL water, elute with 0.5 mL MeOH. In each case passage of fluid through the column is helped by centrifugation.

## HPLC VARIABLES

**Column:** 250 × 4 5 μm LiChrosorb RP-18
**Mobile phase:** MeOH:10 mM pH 8.9 ammonium carbamate 40:60
**Column temperature:** 45
**Flow rate:** 1.2
**Injection volume:** 25
**Detector:** UV 220

## CHROMATOGRAM

**Retention time:** 4
**Internal standard:** N-cyano-N'-{2-[(5-methyl-1H-imidazol-4-yl)-methylthio]ethyl}-S-methyl-isothourea

## KEY WORDS

plasma

## REFERENCE

Nitsche, V.; Mascher, H. New rapid assay of cimetidine in human plasma by reversed-phase high-performance liquid chromatography. *J.Chromatogr.*, **1983**, *273*, 449–452

## SAMPLE

**Matrix:** blood, gastric fluid

**Sample preparation:** Plasma (human). 250 μL Plasma + 150 μL 8 μg/mL IS in MeOH + 25 μL 5 M NaOH, vortex, add 2 mL MeCN, shake 5 min, centrifuge at 4000 g for 10 min. Remove organic layer and add to 1 mL 20 mM HCl saturated with NaCL (300 g/L). Shake 10 min, centrifuge at 4000 g for 10 min. Remove aqueous layer and add it to 100 μL 5 M NaOH + 2 mL MeCN. Shake 5 min, remove organic layer and evaporate under dry nitrogen in a water bath at 40°. Take up residue in 250 μL 1 mM HCl in MeOH, inject 15 μL. Gastric Fluid (monkey). 250 μL Gastric fluid + 150 μL 43 μg/mL IS in MeOH + 2 mL 20 mM HCl saturated with NaCL (300 g/L), vortex, add 2 mL MeCN, shake 10 min, centrifuge at 4000 g for 10 min. Discard organic layer and extract again with 2 mL MeCN. Remove aqueous layer and add it to 100 μL 5 M NaOH + 2 mL MeCN. Shake 10 min, remove organic layer and evaporate under dry nitrogen in a water bath at 40°. Take up residue in 250 μL 1 mM HCl in MeOH, inject 15 μL.

## HPLC VARIABLES

**Column:** Waters RCM-100 radial compression cartridge

**Mobile phase:** MeOH:8.7 mM $KH_2PO_4$ + 3.04 mM $Na_2HPO_4$ (pH 7.41) 34:66 (w/w)

**Flow rate:** 3

**Injection volume:** 15

**Detector:** UV 228

## CHROMATOGRAM

**Retention time:** 2.78

**Internal standard:** N-cyano-N'-methyl-N''-[3-(4-imidazolyl)propyl]-guanidine

**Limit of detection:** 200 ng/mL

## KEY WORDS

plasma

## REFERENCE

Abdel-Rahim, M.; Ezra, D.; Peck, C.; Lazar, J. Liquid-chromatographic assay of cimetidine in plasma and gastric fluid. *Clin.Chem.*, **1985**, *31*, 621–623

## SAMPLE

**Matrix:** blood, milk

**Sample preparation:** 100 μL Serum or whole milk + 25 μL 2.5 μg/mL ranitidine in MeOH + 4 mL dichloromethane, vortex, centrifuge. Remove the organic layer and evaporate it to dryness under a stream of nitrogen, reconstitute the residue in 200 μL mobile phase, inject a 50 μL aliquot.

## HPLC VARIABLES

**Guard column:** 15 × 3.2 RP-18 (Applied Biosystems)

**Column:** 300 × 3.9 μBondapak C18

**Mobile phase:** MeCN:water:glacial acetic acid:triethylamine 15:85:0.15:0.02

**Flow rate:** 1

**Injection volume:** 50

**Detector:** UV 228

## CHROMATOGRAM

**Retention time:** 7.8

**Internal standard:** ranitidine (10.7)
**Limit of detection:** 50 ng/mL

## KEY WORDS
serum; whole milk; pharmacokinetics

## REFERENCE
Oo, C.Y.; Kuhn, R.J.; Desai, N.; McNamara, P.J. Active transport of cimetidine into human milk. *Clin.Pharmacol.Ther.*, **1995**, *58*, 548–555

## SAMPLE
**Matrix:** blood, CSF, tissue
**Sample preparation:** Blood. Hemolyze 25 μL whole blood with 50 μL water. 25 μL Plasma or hemolyzed blood + 100 μL 100 μg/mL ranitidine + 100 μL 5 M NaOH + 5 mL dichloromethane, mix, shake for 10 min, centrifuge at 1650 g for 10 min. Remove 4 mL of the organic layer and evaporate it to dryness, reconstitute the residue in 100 μL mobile phase, inject a 25 μL aliquot. (To measure unbound cimetidine in plasma inject 25 μL ultrafiltrate (Amicon MPS-3 centrifree).) Tissue. Brain tissue + 100 μL 50 μg/mL ranitidine + 1 mL saline, homogenize in an ice bath for 1 min, add 100 μL 1 M NaOH, add 5 mL dichloromethane, extract. Remove 3 mL of the organic layer and evaporate it to dryness, reconstitute the residue in 100 μL mobile phase, centrifuge at 10000 g, inject a 25 μL aliquot. CSF. Inject an aliquot directly.

## HPLC VARIABLES
**Column:** 250 × 4 Senshu gel 7C18H (Senshu)
**Mobile phase:** MeCN:5 mM NaH$_2$PO$_4$ containing 5 mM tetramethylammonium chloride 5:95
**Column temperature:** 40
**Flow rate:** 2
**Injection volume:** 25
**Detector:** UV 225

## CHROMATOGRAM
**Internal standard:** ranitidine
**Limit of detection:** 50 (CSF), 500 (brain), 1000 (blood) ng/g

## KEY WORDS
rat; plasma; brain; ultrafiltrate; whole blood; pharmacokinetics

## REFERENCE
Nakada, Y.; Yamamoto, K.; Kawakami, J.; Sawada, Y.; Iga, T. Effect of acute renal failure on neurotoxicity of cimetidine in rats. *Pharm.Res.*, **1995**, *12*, 1953–1957

## SAMPLE
**Matrix:** blood, urine
**Sample preparation:** 200 μL Serum or urine + 100 μL 0.072 μg/mL procaine hydrochloride (+ 100 μL 6.25% NaHCO$_3$ solution for urine samples) + 5 mL dichloromethane, vortex 210 s. Evaporate organic phase to dryness under nitrogen, take up in 100 μL mobile phase, inject a 20 μL aliquot.

## HPLC VARIABLES
**Column:** 250 × 4.6 Partisil-10 ODS-3
**Mobile phase:** MeCN:10 mM pH 4.8 potassium phosphate buffer 7:93
**Flow rate:** 2
**Injection volume:** 20
**Detector:** UV 228

## CHROMATOGRAM
**Retention time:** 6.2
**Internal standard:** procaine
**Limit of detection:** 100 ng/mL

## OTHER SUBSTANCES
**Noninterfering:** acetaminophen, caffeine, cimetidine sulfoxide, diazepam, digoxin, flurazepam, furosemide, methyldopa, minoxidil, propranolol, quinidine, sulfinpyrazone
**Interfering:** procainamide, tolazamide

## KEY WORDS
serum

## REFERENCE
Guay, D.R.; Bockbrader, H.N.; Matzke, G.R. High-performance liquid chromatographic analysis of cimetidine in serum and urine. *J.Chromatogr.*, **1982**, *228*, 398–403

## SAMPLE
**Matrix:** formulations
**Sample preparation:** Dilute with mobile phase, inject an aliquot.

## HPLC VARIABLES
**Column:** 300 × 4.6 5 μm C18
**Mobile phase:** MeCN:100 mM $NaH_2PO_4$ 20:80, adjusted to pH 4.2 with phosphoric acid
**Flow rate:** 1.5
**Injection volume:** 20
**Detector:** UV 228

## CHROMATOGRAM
**Retention time:** 2.71

## OTHER SUBSTANCES
**Simultaneous:** cisplatin (UV 198), dacarbazine (UV 300), granisetron (UV 300)

## KEY WORDS
stability-indicating; injections; saline

## REFERENCE
Mayron, D.; Gennaro, A.R. Stability and compatibility of granisetron hydrochloride in i.v. solutions and oral liquids and during simulated Y-site injection with selected drugs. *Am.J.Health-Syst.Pharm.*, **1996**, *53*, 294–304

## SAMPLE
**Matrix:** formulations
**Sample preparation:** Inject a 20 μL aliquot.

## HPLC VARIABLES
**Column:** 250 × 4.6 10 μm Econosil C18
**Mobile phase:** MeOH:water:phosphoric acid:sodium 1-hexanesulfonate 24:76:0.03:0.094
**Flow rate:** 2
**Injection volume:** 20
**Detector:** UV 220

## CHROMATOGRAM
**Retention time:** 16.8

## KEY WORDS
injections; saline; stability-indicating

## REFERENCE
Ku, Y.-M.; Min, D.I.; Kumar, V.; Noormohamed, S.E. Compatibility of tacrolimus injection with cimetidine hydrochloride injection in 0.9% sodium chloride injection. *Am.J.Health-Syst.Pharm.*, **1995**, *52*, 2024–2025

## SAMPLE
**Matrix:** formulations
**Sample preparation:** Dilute 1:8 with water, combine a 100 μL aliquot of the diluted solution with 100 μL cimetidine solution and 200 μL water, inject a 20 μL aliquot.

## HPLC VARIABLES
**Column:** 3.9 × 300 μBondapak C18
**Mobile phase:** MeCN:MeOH:10 mM pH 2.6-2.7 phosphate buffer 7:14:79, containing 5 mM tetrabutylammonium hydrogen sulfate
**Flow rate:** 1
**Injection volume:** 20
**Detector:** UV 225

## CHROMATOGRAM
**Retention time:** 3.27
**Internal standard:** cimetidine

## OTHER SUBSTANCES
**Simultaneous:** ampicillin, aztreonam, sulbactam

## KEY WORDS
saline; injections; cimetidine is IS

## REFERENCE
Belliveau, P.P.; Nightingale, C.H.; Quintiliani, R. Stability of aztreonam and ampicillin sodium-sulbactam sodium in 0.9% sodium chloride injection. *Am.J.Hosp.Pharm.*, **1994**, *51*, 901–904

## SAMPLE
**Matrix:** injections
**Sample preparation:** Inject a 10 μL aliquot.

## HPLC VARIABLES
**Column:** 250 × 4.8 Spherisorb S5CN
**Mobile phase:** MeCN:buffer 50:50 (Buffer was 20 mM $KH_2PO_4$ adjusted to pH 5.4 with 1 M NaOH.)
**Flow rate:** 1.5
**Injection volume:** 10
**Detector:** UV 216

## CHROMATOGRAM
**Retention time:** 4.5
**Internal standard:** cimetidine

## OTHER SUBSTANCES
**Simultaneous:** ondansetron

## KEY WORDS
5% dextrose; cimetidine is IS

## REFERENCE
Bosso, J.A.; Prince, R.A.; Fox, J.L. Compatibility of ondansetron hydrochloride with fluconazole, cefta-zidime, aztreonam, and cefazolin sodium under simulated Y-site conditions. *Am.J.Hosp.Pharm.*, **1994**, *51*, 389–391

## SAMPLE
**Matrix:** perfusate
**Sample preparation:** 1 mL Perfusate + 100 μL 20 μg/mL procainamide + 1 mL 2 M NaOH + 6 mL ethyl acetate, vortex for 3 min, centrifuge, repeat the extraction. Combine the organic layers and evaporate them to dryness under a stream of nitrogen at 40°, reconstitute the residue in 110 μL mobile phase, inject a 40 μL aliquot.

## HPLC VARIABLES
**Column:** 100 mm long 5 μm Radpak C18
**Mobile phase:** MeCN:water:triethylamine 6:94:1, adjusted to pH 3 with concentrated phosphoric acid
**Injection volume:** 40
**Detector:** UV 228

## CHROMATOGRAM
**Internal standard:** procainamide
**Limit of quantitation:** 100 ng/mL

## REFERENCE
Bassily, M.; Ghabrial, H.; Smallwood, R.A.; Morgan, D.J. Determinants of placental drug transfer: Studies in the isolated perfused human placenta. *J.Pharm.Sci.*, **1995**, *84*, 1054–1060

## SAMPLE
**Matrix:** perfusate, urine
**Sample preparation:** Urine. Add 10 μL urine diluted 10 times with 10 mM pH 7.5 $Na_2HPO_4$ buffer to 1-5 μg nizatidine, make up volume to 300 μL with 10 mM pH 7.5 $Na_2HPO_4$ buffer. Place solution on YM-10 ultrafiltration membrane with a cut-off of 10000, centrifuge at 4000 g for 20 min. Mix 180 μL filtrate with 20 μL MeOH, inject 50 μL. Perfusate. Add 10-100 μL perfusate to 1-5 μg nizatidine, make up volume to 300 μL with 10 mM pH 7.5 $Na_2HPO_4$ buffer. Place solution on YM-10 ultrafiltration membrane with a cut-off of 10 000, centrifuge at 4000 g for 20 min. Mix 180 μL filtrate with 20 μL MeOH, inject 50 μL.

## HPLC VARIABLES
**Guard column:** 75 × 2.1 5 μm LiChrosorb RP-18
**Column:** 150 × 4.6 5 μm LiChrosorb RP-18
**Mobile phase:** MeOH:10 mM pH 7.5 $Na_2HPO_4$ 30:70
**Column temperature:** 40
**Flow rate:** 1
**Injection volume:** 50
**Detector:** UV 228

## CHROMATOGRAM
**Retention time:** 7
**Internal standard:** nizatidine
**Limit of detection:** 50 ng/mL

## REFERENCE
Boom, S.P.A.; Moons, M.M.; Russel, F.G.M. Renal tubular transport of cimetidine in the isolated perfused kidney of the rat. *Drug Metab.Dispos.*, **1994**, *22*, 148–153

## SAMPLE
**Matrix:** solutions

## HPLC VARIABLES
**Column:** 250 × 4.6 10 μm Partisil ODS1
**Mobile phase:** MeOH:50 mM pH 3.0 phosphoric acid 10:90
**Column temperature:** 30
**Flow rate:** 1.5
**Detector:** Radioactivity

## OTHER SUBSTANCES
**Also analyzed:** atenolol, hydrochlorothiazide, ranitidine

## KEY WORDS
tritium labeled

## REFERENCE
Collett, A.; Sims, E.; Walker, D.; He, Y.-L.; Ayrton, J.; Rowland, M.; Warhurst, G. Comparison of HT29-18-C$_1$ and Caco-2 cell lines as models for studying intestinal paracellular drug absorption. *Pharm.Res.*, **1996**, *13*, 216–221

## SAMPLE
**Matrix:** solutions

## HPLC VARIABLES
**Column:** 150 × 4.6 12 μm 1-myristoyl-2-[(13-carboxyl)-tridecoyl]-sn-3-glycerophosphocholine chemically bonded to silica (Regis)
**Mobile phase:** MeCN:100 mM pH 7.0 phosphate buffer 20:80
**Flow rate:** 1
**Detector:** UV 254

## CHROMATOGRAM
**Retention time:** k' 0.54

## OTHER SUBSTANCES
**Also analyzed:** acebutolol, alprenolol, antazoline, atenolol, betaxolol, bisoprolol, bopindolol, bupranolol, carteolol, celiprolol, chloropyramine, chlorpheniramine, cicloprolol, cinnarizine, cirazoline, clonidine, dilevalol, dimethindene, diphenhydramine, doxazosin, esmolol, famotidine, isothipendyl, ketotifen, metiamide, metoprolol, moxonidine, nadolol, naphazoline, nifenalol, nizatidine, oxprenolol, pheniramine, phentolamine, pindolol, pizotyline (pizotifen), practolol, prazosin, promethazine, propranolol, pyrilamine (mepyramine), ranitidine, roxatidine, sotalol, tiamenidine, timolol, tramazoline, tripelennamine, triprolidine, tymazoline, UK-14,304

## REFERENCE
Kaliszan, R.; Nasal, A.; Turowski, M. Binding site for basic drugs on $\alpha_1$-acid glycoprotein as revealed by chemometric analysis of biochromatographic data. *Biomed.Chromatogr.*, **1995**, *9*, 211–215

## SAMPLE
**Matrix:** solutions

## HPLC VARIABLES
**Column:** 250 × 4.6 5 μm Supelcosil LC-DP (A) or 250 × 4 5 μm LiChrospher 100 RP-8 (B)
**Mobile phase:** MeCN:0.025% phosphoric acid:buffer 25:10:5 (A) or 60:25:15 (B) (Buffer was 9 mL concentrated phosphoric acid and 10 mL triethylamine in 900 mL water, adjust pH to 3.4 with dilute phosphoric acid, make up to 1 L.)
**Flow rate:** 0.6
**Injection volume:** 25
**Detector:** UV 229

## CHROMATOGRAM
**Retention time:** 5.47 (A), 3.16 (B)

## OTHER SUBSTANCES
**Also analyzed:** acebutolol, acepromazine, acetaminophen, acetazolamide, acetophenazine, albuterol, alprazolam, amitriptyline, amobarbital, amoxapine, antipyrine, atenolol, atropine, azatadine, baclofen, benzocaine, bromocriptine, brompheniramine, brotizolam, bupivacaine, buspirone, butabarbital, butalbital, caffeine, carbamazepine, cetirizine, chlorcyclizine, chlordiazepoxide, chlormezanone, chloroquine, chlorpheniramine, chlorpromazine, chlorpropamide, chlorprothixene, chlorthalidone, chlorzoxazone, cisapride, clomipramine, clonazepam, clonidine, clozapine, cocaine, codeine, colchicine, cyclizine, cyclobenzaprine, dantrolene, desipramine, diazepam, diclofenac, diflunisal, diltiazem, diphenhydramine, diphenidol, diphenoxylate, dipyridamole, disopyramide, dobutamine, doxapram, doxepin, droperidol, encainide, ethidium bromide, ethopropazine, fenoprofen, fentanyl, flavoxate, fluoxetine, fluphenazine, flurazepam, flurbiprofen, fluvoxamine, furosemide, glutethimide, glyburide, guaifenesin, haloperidol, homatropine, hydralazine, hydrochlorothiazide, hydrocodone, hydromorphone, hydroxychloroquine, hydroxyzine, ibuprofen, imipramine, indomethacin, ketoconazole, ketoprofen, ketorolac, labetalol, levorphanol, lidocaine, loratadine, lorazepam, lovastatin, loxapine, mazindol, mefenamic acid, meperidine, mephenytoin, mepivacaine, mesoridazine, metaproterenol, methadone, methdilazine, methocarbamol, methotrexate, methotrimeprazine, methoxamine, methyldopa, methylphenidate, metoclopramide, metolazone, metoprolol, metronidazole, midazolam, moclobemide, morphine, nadolol, nalbuphine, naloxone, naphazoline, naproxen, nifedipine, nizatidine, norepinephrine, nortriptyline, oxazepam, oxycodone, oxymetazoline, paroxetine, pemoline, pentazocine, pentobarbital, pentoxifylline, perphenazine, pheniramine, phenobarbital, phenol, phenolphthalein, phentolamine, phenylbutazone, phenyltoloxamine, phenytoin, pimozide, pindolol, piroxicam, pramoxine, prazepam, prazosin, probenecid, procainamide, procaine, prochlorperazine, procyclidine, promazine, promethazine, propafenone, propantheline, propiomazine, propofol, propranolol, protriptyline, quazepam, quinidine, quinine, racemethorphan, ranitidine, remoxipride, risperidone, salicylic acid, scopolamine, secobarbital, sertraline, sotalol, spironolactone, sulfinpyrazone, sulindac, temazepam, terbutaline, terfenadine, tetracaine, theophylline, thiethylperazine, thiopental, thioridazine, thiothixene, timolol, tocainide, tolbutamide, tolmetin, trazodone, triamterene, triazolam, trifluoperazine, triflupromazine, trimeprazine, trimethoprim, trimipramine, verapamil, warfarin, xylometazoline, yohimbine, zopiclone

## KEY WORDS
some details of plasma extraction

## REFERENCE
Koves, E.M. Use of high-performance liquid chromatography-diode array detection in forensic toxicology. *J.Chromatogr.A*, **1995**, *692*, 103–119

## SAMPLE
**Matrix:** solutions

## HPLC VARIABLES
**Column:** 250 × 4.6 Partisil-10 ODS-3
**Mobile phase:** MeOH:50 mM $KH_2PO_4$ 15:85, adjusted to pH 2.7
**Flow rate:** 2
**Injection volume:** 100
**Detector:** UV 228

## CHROMATOGRAM
**Retention time:** 6-8
**Internal standard:** codeine (10-12)

## REFERENCE
Mummaneni, V.; Amidon, G.L.; Dressman, J.B. Gastric pH influences the appearance of double peaks in the plasma concentration-time profiles of cimetidine after oral administration in dogs. *Pharm.Res.*, **1995**, *12*, 780–786

## SAMPLE
**Matrix:** solutions

## HPLC VARIABLES
**Column:** 250 × 4.6 Zorbax RX
**Mobile phase:** Gradient. A was 10 mL concentrated orthophosphoric acid and 7 mL triethylamine in 1 L water. B was 10 mL concentrated orthophosphoric acid and 7 mL triethylamine in 200 mL water, make up to 1 L with MeCN. A:B from 100:0 to 0:100 over 30 min, maintain at 0:100 for 5 min.
**Column temperature:** 30
**Flow rate:** 2
**Detector:** UV 210

## OTHER SUBSTANCES
**Also analyzed:** acepromazine, acetaminophen, acetophenazine, albuterol, aminophylline, amitriptyline, amobarbital, amoxapine, amphetamine, amylocaine, antipyrine, aprobarbital, aspirin, atenolol, atropine, avermectin, barbital, benzocaine, benzoic acid, benzotropine, benzphetamine, berberine, bibucaine, bromazepan, brompheniramine, buprenorphine, buspirone, butabarbital, butacaine, butethal, caffeine, carbamazepine, carbromal, chloramphenicol, chlordiazepoxide, chloroquine, chlorothiazide, chloroxylenol, chlorphenesin, chlorpheniramine, chlorpromazine, chlorpropamide, chlortetracycline, cinchonidine, cinchonine, clenbuterol, clonazepam, clonixin, clorazepate, cocaine, codeine, colchicine, cortisone, coumarin, cyclazocine, cyclobenzaprine, cyclothiazide, cyheptamide, cymarin, danazol, danthron, dapsone, debrisoquine, desipramine, dexamethasone, dextromethorphan, dextropropoxyphene, diamorphine, diazepam, diclofenac, diethylpropion, diethylstilbestrol, diflunisal, digitoxin, digoxin, diltiazem, diphenhydramine, diphenoxylate, diprenorphine, dipyrone, disulfiram, dopamine, doxapram, doxepin, dronabinol, ephedrine, epinephrine, epinine, estradiol, estriol, estrone, ethacrynic acid, ethosuximide, etonitazene, etorphine, eugenol, famotidine, fenbendazole, fencamfamine, fenoprofen, fenproporex, fentanyl, flubendazole, flufenamic acid, flunitrazepam, 5-fluorouracil, fluoxymesterone, fluphenazine, furosemide, gentisic acid, gitoxigenin, glipizide, glunixin, glutethimide, glybenclamide, guaiacol, halazepam, haloperidol, hydrochlorothiazide, hydrocodone, hydrocortisone, hydromorphone, hydroxyquinoline, ibogaine, ibuprofen, iminostilbene, imipramine, indomethacin, isocarbostyril, isocarboxazid, isoniazid, isoproterenol, isoxsuprine, ivermectin, ketamine, ketoprofen, kynurenic acid, levorphanol, lidocaine, lorazepam, lormetazepam, loxapine, mazindol, mebendazole, meclizine, meclofenamic acid, medazepam, mefenamic acid, megestrol, mepacrine, meperidine, mephentermine, mephenytoin, mephesin, mephobarbital, mepivacaine, mescaline, mesoridazine, methadone, methamphetamine, methapyrilene, methaqualone, methazolamide, methocarbamol, methoxamine, methsuximide, methyl salicylate, methyldopa, methyldopamine, methylphenidate, methylprednisolone, methyltestosterone, methyprylon, meto-

prolol, mibolerone, morphine, nadolol, nalorphine, naloxone, naltrexone, naphazoline, naproxen, nefopam, niacinamide, nicotine, nicotinic acid, nifedipine, niflumic acid, nitrazepam, norepinephrine, nortriptyline, noscapine, nylidrin, oxazepam, oxycodone, oxymorphone, oxyphenbutazone, oxytetracycline, papaverine, pargyline, pemoline, pentazocine, pentobarbital, persantine, phenacetin, phenazocine, phenazopyridine, phencyclidine, phendimetrazine, phenelzine, pheniramine, phenobarbital, phenothiazine, phensuximide, phentermine, phenylbutazone, phenylephrine, phenylpropanolamine, piperocaine, prazepam, prednisolone, primidone, probenecid, progesterone, propiomazine, propranolol, propylparaben, pseudoephedrine, puromycin, pyrilamine, pyrithyldione, quazepam, quinaldic acid, quinidine, quinine, ranitidine, recinnamine, reserpine, resorcinol, saccharin, albuterol, salicylamide, salicylic acid, scopolamine, scopoletin, secobarbital, strychnine, sulfacetamide, sufadiazine, sulfadimethoxine, sulfaethidole, sulfamerazine, sulfamethazine, sulfamethoxizole, sulfanilamide, sulfapyridine, sulfasoxizole, sulindac, tamoxifen, temazepam, testosterone, tetracaine, tetracycline, tetramisole, thebaine, theobromine, theophylline, thiabendazole, thiamine, thiamylal, thiobarbituric acid, thioridazine, thiosalicylic acid, thiothixene, thymol, tolazamide, tolazoline, tobutamide, tolmetin, tranylcypromine, triamcinolone, tribenzylamine, trichloromethiazide, trifluoperazine, trihexyphenidyl, trimethoprim, tripelennamine, triproilidine, tropacocaine, tyramine, verapamil, vincamine, warfarin, yohimbine, zoxazolamine

## REFERENCE

Hill, D.W.; Kind, A.J. Reversed-phase solvent gradient HPLC retention indexes of drugs. *J.Anal.Toxicol.*, **1994**, *18*, 233–242

## SAMPLE
**Matrix:** solutions
**Sample preparation:** Dilute solution 10-fold with mobile phase. Mix 100 μL with 100 μL 300 μg/mL hydrocortisone, inject 10 μL.

## HPLC VARIABLES
**Column:** 300 × 3.9 10 μm μBondapak C18
**Mobile phase:** MeCN:50 mM sodium acetate 33:67
**Flow rate:** 1
**Injection volume:** 10
**Detector:** UV 248

## CHROMATOGRAM
**Retention time:** 7.1
**Internal standard:** hydrocortisone

## KEY WORDS
water; stability-indicating

## REFERENCE

Mahata, M.C.; Morosco, R.S.; Hipple, T.F. Stability of cimetidine hydrochloride and of clindamycin phosphate in water for injection stored in glass vials at two temperatures. *Am.J.Hosp.Pharm.*, **1993**, *50*, 2559–2561

## SAMPLE
**Matrix:** solutions
**Sample preparation:** Prepare a 10 μg/mL solution in MeOH, inject a 20 μL aliquot.

## HPLC VARIABLES
**Column:** 125 × 4.9 Spherisorb S5W silica
**Mobile phase:** MeOH containing 10 mM ammonium perchlorate and 1 mL/L 100 mM NaOH in MeOH, pH 6.7

**Flow rate:** 2
**Injection volume:** 20
**Detector:** E, LeCarbone, V25 glassy carbon electrode, + 1.2 V

## CHROMATOGRAM
**Retention time:** 1.22

## OTHER SUBSTANCES
**Simultaneous:** acebutolol, acepromazine, acetophenazine, N-acetylprocainamide, albuterol, alprenolol, amethocaine, amiodarone, amitriptyline, antazoline, atenolol, azacyclonal, bamethane, benactyzine, benperidol, benzethidine, benzocaine, benzoctamine, benzphetamine, benzquinamide, bromhexine, bromodiphenhydramine, bromperidol, brompheniramine, brompromazine, buclizine, bufotenine, bupivacaine, buprenorphine, butacaine, butethamate, chlorcyclizine, chlorpheniramine, chlorphenoxamine, chlorprenaline, chlorpromazine, chlorprothixene, cinchonidine, cinnarizine, clemastine, clomipramine, clonidine, cocaine, cyclazocine, cyclizine, cyclopentamine, cyproheptadine, deserpidine, desipramine, dextromoramide, dextropropoxyphene, dicyclomine, diethylcarbamazine, diethylpropion, diethylthiambutene, dihydroergotamine, dimethindene, dimethothiazine, diphenhydramine, diphenoxylate, dipipanone, diprenorphine, dipyridamole, disopyramide, dothiepin, doxapram, doxepin, doxylamine, droperidol, ephedrine, ergocornine, ergocristine, ergocristinine, ergocryptine, ergometrine, ergosine, ergosinine, ergotamine, ethopropazine, etorphine, etoxeridine, fenethazine, fenfluramine, fenoterol, fentanyl, flavoxate, fluopromazine, flupenthixol, fluphenazine, flurazepam, haloperidol, hydroxyzine, hyoscine, ibogaine, imipramine, indapamine, iprindole, isothipendyl, isoxsuprine, ketanserin, laudanosine, lidocaine, lofepramine, loxapine, maprotiline, mecamylamine, meclophenoxate, meclozine, medazepam, mephentermine, mepivacaine, meptazinol, mepyramine, mesoridazine, metaraminol, methadone, methamphetamine, methapyrilene, methdilazene, methotrimeprazine, methoxamine, methoxyphenamine, methoxypromazine, methylephedrine, methylergonovine, methysergide, metoclopramide, metopimazine, metoprolol, mianserin, morazone, nadolol, nalorphine, naloxone, naphazoline, nicotine, nifedipine, nomifensine, nortriptyline, noscapine, orphenadrine, oxeladin, oxprenolol, oxymetazolin, papaverine, pargyline, pecazine, penbutolol, pentazocine, penthienate, pericyazine, perphenazine, phenadoxone, phenampromide, phenazocine, phenbutrazate, phendimetrazine, phenelzine, phenglutarimide, phenindamine, pheniramine, phenmetrazine, phenomorphan, phenoperidine, phenothiazine, phenoxybenzamine, phentolamine, phenylephrine, phenyltoloxamine, physostigmine, piminodine, pimozide, pindolol, pipamazine, pipazethate, piperacetazine, piperidolate, pipradol, pirenzepine, piritramide, pizotifen, practolol, pramoxine, prazosin, prenylamine, prilocaine, primaquine, proadifen, procainamide, procaine, prochlorperazine, procyclidine, proheptazine, prolintane, promazine, promethazine, pronethalol, properidine, propiomazine, propranolol, prothipendyl, protriptyline, proxymetacaine, pseudoephedrine, pyrimethamine, quinidine, quinine, ranitidine, rescinnamine, sotalol, tacrine, terazosin, terbutaline, terfenadine, thenyldiamine, theophylline, thiethylperazine, thiopropazate, thioproperazine, thioridazine, thiothixene, thonzylamine, timolol, tocainide, tolpropamine, tolycaine, tranylcypromine, trazodone, trifluoperazine, trifluperidol, trimeperidine, trimeprazine, trimethobenzamide, trimethoprim, trimipramine, tripelennamine, triprolidine, tryptamine, verapamil, xylometazoline

## REFERENCE
Jane, I.; McKinnon, A.; Flanagan, R.J. High-performance liquid chromatographic analysis of basic drugs on silica columns using non-aqueous ionic eluents. II. Application of UV, fluorescence and electrochemical oxidation detection. *J.Chromatogr.*, **1985**, *323*, 191–225

◆          ◆          ◆

## ANNOTATED BIBLIOGRAPHY
Busch, U.; Heinzel, G.; Narjes, H.; Nehmiz, G. Interaction of meloxicam with cimetidine, Maalox, or aspirin. *J.Clin.Pharmacol.*, **1996**, *36*, 79–84 [also aspirin, meloxicam+asma; pharmacokinetics]

Hermansson, J.; Grahn, A. Resolution of racemic drugs on a new chiral column based on silica-immobilized cellobiohydrolase. Characterization of the basic properties of the column. *J.Chromatogr.*, **1994**, *687*, 45–59 [chiral; also acebutolol, atenolol, betaxolol, bisoprolol, carbuterol, cathinone, dobutamine, dopropizine, epanolol, epinephrine, laudanosine, metanephrine, metoprolol, moprolol, norepinephrine, normetanephrine, octopamine, oxybutynine, pamatolol, practolol, prilocaine, propafenone, proxyphylline, sotalol, talinolol, tetrahydropapaveroline, tetramisole, timolol, tolamolol, toliprolol]

Russel, F.G.; Creemers, M.C.; Tan, Y.; van Riel, P.L.; Gribnau, F.W. Ion-pair solid-phase extraction of cimetidine from plasma and subsequent analysis by high-performance liquid chromatography. *J.Chromatogr.B*, **1994**, *661*, 173–177 [plasma; serum; SPE; ranitidine (IS); LOD 5 ng/mL; LOQ 25 ng/mL; column temp 40]

Kaka, J.S. Rapid method for cimetidine and ranitidine determination in human and rat plasma by HPLC. *J.Liq.Chromatogr.*, **1988**, *11*, 3447–3456 [LOD 50-100 ng/mL]

Tracqui, A.; Kintz, P.; Mangin, P.; Lugnier, A.A.; Chaumont, A.J. A new rapid HPLC assay for the simultaneous determination of two histamine H2-receptor antagonists, cimetidine and ranitidine, in human plasma. *J.Toxicol.Clin.Exp.*, **1988**, *8*, 387–394 [LOD 25 ng/mL; plasma]

Wong, S.H.Y.; McHugh, S.L.; Dolan, J.; Cohen, K.A. Tricyclic antidepressant analysis by reversed-phase liquid chromatography using phenyl columns. *J.Liq.Chromatogr.*, **1986**, *9*, 2511–2538 [also acetaminophen, amitriptyline, amobarbital, amoxapine, barbital, chlordiazepoxide, chlorpromazine, clomipramine, codeine, desipramine, desmethyldoxepin, diazepam, doxepin, fluphenazine, flurazepam, glutethimide, hydroxyamoxapine, imipramine, lorazepam, maprotiline, meperidine, metabolites, nortriptyline, oxazepam, pentobarbital, perphenazine, phenobarbital, phenytoin, propoxyphene, protriptyline, secobarbital, thioridazine, trazodone]

Wong, S.H.Y.; Marzouk, N.; McHugh, S.L.; Cazes, E. Simultaneous determination of theophylline and caffeine by reversed phase liquid chromatography using phenyl column. *J.Liq.Chromatogr.*, **1985**, *8*, 1797–1816 [also acetaminophen, caffeine, cimetidine, codeine, dimethylxanthine, meperidine, pentobarbital, phenobarbital, secobarbital, theobromine, theophylline; hydroxyethyltheophylline(IS)]

Boutagy, J.; More, D.G.; Munro, I.A.; Shenfield, G.M. Simultaneous analysis of cimetidine and ranitidine in human plasma by HPLC. *J.Liq.Chromatogr.*, **1984**, *7*, 1651–1664 [also metabolites, N-acetylprocainamide, procainamide]

Elliott, G.T.; McKenzie, M.W.; Curry, S.H.; Pieper, J.A.; Quinn, S.L. Stability of cimetidine hydrochloride in admixtures after microwave thawing. *Am.J.Hosp.Pharm.*, **1983**, *40*, 1002–1006 [stability-indicating; 5% dextrose; saline]

Apffel, J.A.; Brinkman, U.A.T.; Frei, R.W. Analysis of cimetidine in biological fluids by high performance liquid chromatography. *J.Liq.Chromatogr.*, **1982**, *5*, 2413–2422 [urine; procaine (IS); blood; SPE]

Fleitman, J.; Torosian, G.; Perrin, J.H. Improved high-performance liquid chromatographic assay for cimetidine using ranitidine as an internal standard. *J.Chromatogr.*, **1982**, *229*, 255–258 [plasma; ranitidine (IS); LOD 50 ng/mL]

Mihaly, G.W.; Cockbain, S.; Jones, D.B.; Hanson, R.G.; Smallwood, R.A. High-pressure liquid chromatographic determination of cimetidine in plasma and urine. *J.Pharm.Sci.*, **1982**, *71*, 590–592 [plasma; urine; LOD 25 ng/mL; pharmacokinetics; procainamide can be used as IS (J.Pharm.Sci. 1984, 73, 1015)]

Kunitani, M.G.; Johnson, D.A.; Upton, R.A.; Riegelman, S. Convenient and sensitive high-performance liquid chromatography assay for cimetidine in plasma or urine. *J.Chromatogr.*, **1981**, *224*, 156–161 [plasma; urine; non-interfering acebutolol, caffeine, ketoprofen, naproxen, theophylline; pharmacokinetics; LOQ 100 ng/mL]

Lorenzo, B.; Drayer, D.E. Improved method for the measurement of cimetidine in human serum by reverse-phase high-pressure liquid chromatography. *J.Lab.Clin.Med.*, **1981**, *97*, 545–550 [serum; plasma; LOD 50 ng/mL]

Ziemniak, J.A.; Chiarmonte, D.A.; Schentag, J.J. Liquid-chromatographic determination of cimetidine, its known metabolites, and creatinine in serum and urine. *Clin.Chem.*, **1981**, *27*, 272–275 [serum; plasma; urine; extracted metabolites, creatinine; pharmacokinetics; LOD 50 ng/mL]

La Rotonda, M.I.; Cozzolino, S.; Schettino, O. [Analysis of active principles in pharmaceutical dosage forms by high pressure liquid chromatography–cimetidine and zolimidine]. *Boll.Soc.Ital.Biol.Sper.*, **1980**, *56*, 1394–1398

# Ciprofloxacin

**Molecular formula:** $C_{17}H_{18}FN_3O_3$
**Molecular weight:** 331.4
**CAS Registry No.:** 85721-33-1, 86393-32-0
(hydrochloride monohydrate)

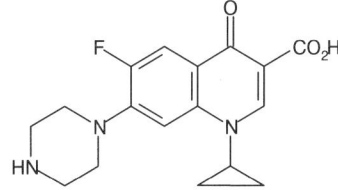

## SAMPLE
**Matrix:** bile, perfusate
**Sample preparation:** Add an equal volume of MeCN to perfusate or bile, centrifuge, inject a 50 μL aliquot of the supernatant.

## HPLC VARIABLES
**Column:** 5 μm RP-18 (Merck)
**Mobile phase:** MeCN:buffer 22:78 (Buffer was 25 mM pH 3 phosphate containing 5 mM tetraethylammonium bromide.)
**Flow rate:** 1.5
**Injection volume:** 50
**Detector:** F ex 270 em 440

## CHROMATOGRAM
**Retention time:** 3.5
**Limit of detection:** 100 ng/mL

## KEY WORDS
rat; liver; pharmacokinetics

## REFERENCE
Abadìa, A.R.; De Francesco, L.; Guaitani, A. Disposition of ciprofloxacin in the isolated perfused rat liver. *Drug Metab.Dispos.*, **1995**, *23*, 197–200

## SAMPLE
**Matrix:** blood
**Sample preparation:** 200 μL Serum + 50 μL 20 μg/mL pipemidic acid + 200 μL 25% sodium sulfate + 3.5 mL dichloromethane, extract. Extract the organic phase with 200 μL 100 mM NaOH, inject a 20 μL aliquot.

## HPLC VARIABLES
**Column:** C18
**Mobile phase:** MeCN:buffer 15:85 (Buffer was 10 mM $NaH_2PO_4$ and 5 mM tetrabutylammonium hydrogen sulfate, pH 2.7.)
**Flow rate:** 2
**Injection volume:** 20
**Detector:** F ex 278

## CHROMATOGRAM
**Internal standard:** pipemidic acid
**Limit of quantitation:** 100 ng/mL

## KEY WORDS
serum; pharmacokinetics

## REFERENCE
Klepser, M.E.; Patel, K.B.; Nicolau, D.P.; Quintiliani, R.; Nightingale, C.H. Comparison of the bactericidal activities of ofloxacin and ciprofloxacin alone and in combination with ceftazidime and piperacillin against clinical strains of *Pseudomonas aeruginosa*. *Antimicrob.Agents Chemother.*, **1995**, *39*,
2503–2510

## SAMPLE
**Matrix:** blood
**Sample preparation:** Precipitate proteins with MeCN and perchloric acid, inject an
aliquot.

## HPLC VARIABLES
**Guard column:** Alltech C18
**Column:** 150 × 3.9 10 μm μBondapak C18
**Mobile phase:** MeCN:MeOH:100 mM citric acid 4:21:75 containing 0.54 g/L ammonium
perchlorate and 0.65 mL/L tetrabutylammonium hydroxide
**Flow rate:** 1
**Detector:** F ex 270 em 440

## CHROMATOGRAM
**Limit of quantitation:** 50 ng/mL

## KEY WORDS
plasma; pharmacokinetics

## REFERENCE
Shah, A.; Lettieri, J.; Nix, D.; Wilton, J.; Heller, A.H. Pharmacokinetics of high-dose intravenous ciprofloxacin in young and elderly and in male and female subjects. *Antimicrob.Agents Chemother.*, **1995**,
*39*, 1003–1006

## SAMPLE
**Matrix:** blood
**Sample preparation:** 100 μL Plasma + 100 μL pH 7.4 phosphate buffer + 50 μL 20 μg/mL
β-hydroxypropyltheophylline in pH 7.4 phosphate buffer + 5 mL chloroform:isopropanol
95:5, shake on a rotary mixer for 15 min, centrifuge at 800 g for 5 min. Evaporate organic
layer under nitrogen at 45°, sonicate residue with 100 μL mobile phase, inject 25 μL
aliquot.

## HPLC VARIABLES
**Guard column:** 10 × 4.9 Spherisorb ODS
**Column:** 250 × 4.9 Spherisorb S5 ODS2
**Mobile phase:** MeCN:buffer 15:85 adjusted, to pH 3.0 with 85% phosphoric acid imme-
diately before use. (Buffer was 4.54 g $KH_2PO_4$ + 5.94 g $Na_2HPO_4.2H_2O$ + 1.49 g tetra-
butylammonium hydrogen sulfate per L.)
**Flow rate:** 1.3
**Injection volume:** 25
**Detector:** UV 280

## CHROMATOGRAM
**Retention time:** 6.2
**Internal standard:** β-hydroxypropyltheophylline
**Limit of detection:** 500 ng/mL

## OTHER SUBSTANCES
**Simultaneous:** enoxacin, norfloxacin, theophylline

## KEY WORDS
plasma; rat

## REFERENCE
Davis, J.D.; Aarons, L.; Houston, J.B. Simultaneous assay of fluoroquinolones and theophylline in plasma by high-performance liquid chromatography. *J.Chromatogr.*, **1993**, *621*, 105–109

## SAMPLE
**Matrix:** blood
**Sample preparation:** 200 μL Serum + 50 μL 40 μg/mL quinine hemisulfate in water, vortex 30 s, add 400 μL MeCN, vortex 1 min, centrifuge at 4000 rpm for 10 min. Remove organic layer and evaporate to 200 μL under a stream of dry nitrogen at 45°. Inject 10-20 μL aliquot.

## HPLC VARIABLES
**Column:** 10 μm μBondapak C18
**Mobile phase:** MeCN:100 mM $NaH_2PO_4$ adjusted to pH 3.9 with phosphoric acid 20:80
**Flow rate:** 2.5
**Injection volume:** 10-20
**Detector:** F ex 280 em 455

## CHROMATOGRAM
**Retention time:** 4.0
**Internal standard:** quinine
**Limit of detection:** 25 ng/mL

## OTHER SUBSTANCES
**Simultaneous:** acebutolol, lomefloxacin, norfloxacin, ofloxacin, pefloxacin
**Noninterfering:** atenolol, deoxyprenaline, digoxin, gentamicin, hyoscine, metoclopramide, metronidazole, midodrine, nadolol, netilmicin, prednisolone, ranitidine, verapamil, vitamin B1

## KEY WORDS
serum

## REFERENCE
Jim, L.K.; el-Sayed, N.; al-Khamis, K.I. A simple high-performance liquid chromatographic assay for ciprofloxacin in human serum. *J.Clin.Pharm.Ther.*, **1992**, *17*, 111–115

## SAMPLE
**Matrix:** blood
**Sample preparation:** 50 μL Plasma + 1 mL 100 mM pH 7.0 $K_2HPO_4$ adjusted to pH 7.0 with 85% orthophosphoric acid + 100 μL 300 μg/mL nalidixic acid in water + 3 mL dichloromethane:isoamyl alcohol 9:1, shake vigorously for 10 min, centrifuge at 2270 g for 10 min. Remove 2 mL of the organic phase and evaporate it to dryness under a stream of nitrogen at 40°. Reconstitute residue in 100 μL MeOH:50 mM NaOH 2:1, vortex, inject a 10 μL aliquot.

## HPLC VARIABLES
**Column:** 150 × 4.6 5 μm Chemcosorb 5-ODS-H
**Mobile phase:** MeOH:5 mM sodium lauryl sulfate 2:1, adjusted to pH 2.35 with 85% phosphoric acid
**Column temperature:** 40
**Flow rate:** 0.6

**Injection volume:** 10
**Detector:** UV 275

## CHROMATOGRAM
**Retention time:** 6
**Internal standard:** nalidixic acid (5)
**Limit of quantitation:** 200 ng/mL

## OTHER SUBSTANCES
**Extracted:** felbinac, fenbufen

## KEY WORDS
plasma; rat; pharmacokinetics

## REFERENCE
Naora, K.; Katagiri, Y.; Ichikawa, N.; Hayashibara, M.; Iwamoto, K. Simultaneous high-performance liquid chromatographic determination of ciprofloxacin, fenbufen and felbinac in rat plasma. *J.Chromatogr.*, **1990**, *530*, 186–191

## SAMPLE
**Matrix:** blood
**Sample preparation:** 500 μL Serum + 500 μL 7% perchloric acid, vortex for 10 s, centrifuge at >700 g for 10 min, inject a 20 μL aliquot of the supernatant.

## HPLC VARIABLES
**Column:** 150 × 3.9 Nova-Pak C18
**Mobile phase:** MeOH:18 mM $KH_2PO_4$ containing 0.13 mM heptanesulfonic acid:concentrated phosphoric acid 30:70:0.1
**Injection volume:** 20
**Detector:** F ex 278 em 475

## CHROMATOGRAM
**Retention time:** 8.5
**Limit of detection:** 200 ng/mL

## OTHER SUBSTANCES
**Extracted:** metabolites

## KEY WORDS
serum

## REFERENCE
Griggs, D.J.; Wise, R. A simple isocratic high-pressure liquid chromatographic assay of quinolones in serum. *J.Antimicrob.Chemother.*, **1989**, *24*, 437–445

## SAMPLE
**Matrix:** blood, intestinal efflux
**Sample preparation:** Intestinal efflux. Freeze intestinal efflux at -80°, lyophilize, reconstitute with 1 mL ofloxacin in MeOH:100 mM phosphoric acid 50:50, centrifuge at 3000 rpm for 10 min, inject a 20 μL aliquot. Serum. Deproteinize serum with MeOH containing ofloxacin.

## HPLC VARIABLES
**Column:** 150 × 3.9 Novapack C18

**Mobile phase:** MeOH:buffer 25:75 (Buffer was 10 mM pH 3.0 potassium phosphate buffer containing 25 mM sodium heptanesulfonate (PIC B7) and 20 mM triethylamine.)
**Flow rate:** 1.5
**Injection volume:** 20
**Detector:** F ex 330 em 440

## CHROMATOGRAM
**Internal standard:** ofloxacin
**Limit of detection:** 10 ng/mL

## KEY WORDS
serum; rat

## REFERENCE
Rubinstein, E.; Dautrey, S.; Farinoti, R.; St.Julien, L.; Ramon, J.; Carbon, C. Intestinal elimination of sparfloxacin, fleroxacin, and ciprofloxacin in rats. *Antimicrob.Agents Chemother.*, **1995**, *39*, 99–102

## SAMPLE
**Matrix:** blood, middle-ear fluid
**Sample preparation:** 50 µL Plasma or middle-ear fluid + 20 µL 10 µg/mL difloxacin in water + 2 mL MeCN, vortex, centrifuge at 1500 g for 10 min. Evaporate organic layer to dryness under nitrogen at 50°, reconstitute in 75 µL mobile phase, inject 5 µL.

## HPLC VARIABLES
**Guard column:** 10 × 2.1 5 µm C18 Hypersil
**Column:** 100 × 2.1 5 µm C18 Hypersil
**Mobile phase:** MeCN:buffer 40:60 (Buffer was 30 mM $NaH_2PO_4$, 20 mM triethylamine, 20 mM sodium dodecyl sulfate adjusted to pH 3.0 with phosphoric acid.)
**Column temperature:** 45
**Flow rate:** 0.35
**Injection volume:** 5
**Detector:** F ex 278 em 456

## CHROMATOGRAM
**Retention time:** 3.0
**Internal standard:** difloxacin
**Limit of detection:** 5 ng/mL

## KEY WORDS
plasma; chinchilla

## REFERENCE
Lovdahl, M.; Steury, J.; Russlie, H.; Canafax, D.M. Determination of ciprofloxacin levels in chinchilla middle ear effusion and plasma by high-performance liquid chromatography with fluorescence detection. *J.Chromatogr.*, **1993**, *617*, 329–333

## SAMPLE
**Matrix:** blood, milk
**Sample preparation:** 500 µL Milk or plasma + 500 µL MeCN:100 mM NaOH, vortex for 10-15 s, filter (Centricon-3, 3000 Dalton cut-off) while centrifuging at 4000 g for 30 min, inject a 50-150 µL aliquot of the ultrafiltrate.

## HPLC VARIABLES
**Column:** 250 × 4.6 3 µm Spherisorb phenyl
**Mobile phase:** MeCN:MeOH:triethylamine:85% phosphoric acid:water 9:9:0.45:0.4:81.15 containing 5 mM dodecanesulfonate

**Column temperature:** 50
**Flow rate:** 1
**Injection volume:** 50-150
**Detector:** UV 278

## CHROMATOGRAM
**Retention time:** 10.7
**Limit of detection:** 5 ng/mL

## OTHER SUBSTANCES
**Extracted:** enrofloxacin

## KEY WORDS
plasma; cow; ultrafiltrate

## REFERENCE
Tyczkowska, K.L.; Voyksner, R.D.; Anderson, K.L.; Papich, M.G. Simultaneous determination of enrofloxacin and its primary metabolite ciprofloxacin in bovine milk and plasma by ion-pairing liquid chromatography. *J.Chromatogr.B*, **1994**, *658*, 341–348

## SAMPLE
**Matrix:** blood, saliva
**Sample preparation:** 500 μL Plasma or saliva + 50 μL 50 ng/mL difloxacin, vortex briefly, add 500 μL 100 mM pH 7.4 phosphate buffer, add 4 mL dichloromethane, add 1 mL isopropanol, vortex for 30 s, shake gently for 30 min, centrifuge at 1500 g for 20 min. Remove the lower organic layer and evaporate it to dryness under a stream of nitrogen at 45°, reconstitute the residue in 500 μL mobile phase, inject a 50-200 μL aliquot.

## HPLC VARIABLES
**Guard column:** μBondapak C18 Guard-Pak
**Column:** 300 × 3.9 10 μm μBondapak C18
**Mobile phase:** MeOH:buffer 35:100 (Buffer was 5.44 g $KH_2PO_4$ and 4 mL tetrabutylammonium hydroxide in 1 L water, adjust pH to 2.5 with 85% phosphoric acid.)
**Flow rate:** 2
**Injection volume:** 50-200
**Detector:** UV 268

## CHROMATOGRAM
**Retention time:** 7.1
**Internal standard:** difloxacin (8.8)
**Limit of detection:** 50 ng/mL

## OTHER SUBSTANCES
**Extracted:** enoxacin, theophylline
**Simultaneous:** caffeine, 1,7-dimethylxanthine
**Noninterfering:** 1,3-dimethyluric acid, hypoxanthine, 1-methyluric acid, 1-methylxanthine, 3-methylxanthine, 7-methylxanthine, theobromine

## KEY WORDS
plasma; pharmacokinetics

## REFERENCE
Zhai, S.; Korrapati, M.R.; Wei, X.; Muppalla, S.; Vestal, R.E. Simultaneous determination of theophylline, enoxacin and ciprofloxacin in human plasma and saliva by high-performance liquid chromatography. *J.Chromatogr.B*, **1995**, *669*, 372–376

## SAMPLE

**Matrix:** blood, tissue

**Sample preparation:** Lung. Homogenize (Ultra-Turrax T25) mouse lung in 1-3 mL pH 6.8 Soerensen phosphate buffer, centrifuge. Mix an aliquot of the supernatant with an equal volume of MeCN, centrifuge, inject a 5-20 μL aliquot of the supernatant. Serum. Mix serum with an equal volume of MeCN, centrifuge, inject a 5-20 μL aliquot of the supernatant.

## HPLC VARIABLES

**Column:** 150 × 4.6 5 μm Ultrabase C8 (SFCC, Neuilly Plaisance, France)

**Mobile phase:** MeCN:MeOH:5% acetic acid 5:3:92

**Flow rate:** 1

**Injection volume:** 5-20

**Detector:** F ex 278 em 418

## CHROMATOGRAM

**Retention time:** 5

**Limit of detection:** 15 ng

## KEY WORDS

serum; lung; mouse; pharmacokinetics

## REFERENCE

Vallée, E.; Azoulay-Dupuis, E.; Bauchet, J.; Pocidalo, J.-J. Kinetic disposition of temafloxacin and ciprofloxacin in a murine model of pneumococcal pneumonia. Relevance for drug efficacy. *J.Pharmacol.Exp.Ther.*, **1992**, *262*, 1203–1208

## SAMPLE

**Matrix:** blood, tissue

**Sample preparation:** Serum. 500 μL Serum + 500 μL MeCN:100 mM NaOH 50:50, vortex for 10-15 s, filter (Amicon Centricon-10, 10000 Daltons) while centrifuging at 2677 g for 30 min, inject a 30-120 μL aliquot of the ultrafiltrate. Tissue. Cut up prostate tissue with a scalpel. Weigh out 100-130 mg tissue, make up to 500 μL with MeCN:100 mM NaOH 50:50, sonicate for 30 min, filter (Amicon Centricon-10, 10000 Daltons) while centrifuging at 2677 g for 30 min, inject a 80-120 μL aliquot of the ultrafiltrate.

## HPLC VARIABLES

**Column:** 100 × 4.6 3 μm Spherisorb phenyl

**Mobile phase:** MeCN:MeOH:water 15:2:83 containing 3 mM dodecanesulfonate, 1.5 mM octanesulfonate, 0.4% phosphoric acid, and 0.4% triethylamine

**Column temperature:** 40

**Injection volume:** 30-120

**Detector:** UV 278.6

## CHROMATOGRAM

**Retention time:** 7.75

**Limit of detection:** 2 ng/mL

## OTHER SUBSTANCES

**Extracted:** enrofloxacin

## KEY WORDS

serum; dog; prostate; ultrafiltrate

## REFERENCE

Tyczkowska, K.; Hedeen, K.M.; Aucoin, D.P.; Aronson, A.L. High-performance liquid chromatographic method for the simultaneous determination of enrofloxacin and its primary metabolite ciprofloxacin in canine serum and prostatic tissue. *J.Chromatogr.*, **1989**, *493*, 337–346

## SAMPLE

**Matrix:** blood, tissue, urine

**Sample preparation:** Serum, plasma. Dilute serum or plasma 1:2 to 1:10 with 30 mM phosphoric acid, centrifuge, inject a 20 μL aliquot of supernatant. Urine. Dilute urine 1: 10 to 1:100 with 30 mM phosphoric acid, centrifuge, inject a 20 μL aliquot of supernatant. Tissue (lung, gut). Cut tissue with a scalpel, homogenize with 1-3 mL buffer, centrifuge at 9600 g for 5 min three times, inject a 20 μL aliquot. Tissue (chondral). Cut tissue with a scalpel, homogenize with 3-6 mL buffer in an ice bath for 2-3 min, centrifuge at 9600 g for 5 min four or five times, inject a 100 μL aliquot. Pleural. Dilute human pleural samples with buffer, centrifuge, inject a 20 μL aliquot. (Buffer was 66.6 mM $K_2HPO_4$ adjusted to pH 7.40 with $KH_2PO_4$.)

## HPLC VARIABLES

**Column:** 200 × 4 5 μm Nucleosil C18

**Mobile phase:** MeOH:MeCN:buffer 13:7:80, adjusted to pH 3.0 with phosphoric acid (Buffer was 15 mM phosphoric acid adjusted to pH 3.0 with tetrabutylammonium hydroxide.)

**Flow rate:** 1

**Injection volume:** 20-100

**Detector:** F ex 278 em 446

## CHROMATOGRAM

**Retention time:** 6

**Limit of detection:** 2.5 ng/mL

## OTHER SUBSTANCES

**Simultaneous:** norfloxacin, ofloxacin

## KEY WORDS

serum; plasma; lung; gut; pleural; chondral

## REFERENCE

Knöller, J.; König, W.; Schönfeld, W.; Bremm, K.D.; Köller, M. Application of high-performance liquid chromatography of some antibiotics in clinical microbiology. *J.Chromatogr.*, **1988**, *427*, 257–267

## SAMPLE

**Matrix:** blood, urine

**Sample preparation:** Blood. 500 μL Serum or plasma + 100 μL 20 μg/mL IS in 100 mM phosphoric acid + 300 μL MeCN:5 M trichloroacetic acid 50:50, vortex, add 100 μL MeCN, add 300 μL water, vortex, centrifuge at 1500 g for 15 min, inject a 10 μL aliquot of the supernatant. Urine. Dilute urine 1:20 (or more) with 50 mM pH 3.0 $KH_2PO_4$, remove a 500 μL aliquot and add it to 100 μL 20 μg/mL IS in 100 mM phosphoric acid, add 700 μL 100 mM trichloroacetic acid, vortex, inject a 10 μL aliquot.

## HPLC VARIABLES

**Guard column:** 5 × 3 PLRP-S (Polymer Laboratories)

**Column:** 150 × 4.6 PLRP-S (Polymer Laboratories)

**Mobile phase:** MeCN:MeOH:20 mM pH 3.0 trichloroacetic acid 22:4:74

**Column temperature:** 30

**Flow rate:** 0.7

**Injection volume:** 10

**Detector:** F ex 277 em 418 following post-column photolysis. The column effluent flowed through a 10 m × 0.25 mm knitted PTFE coil irradiated with a UV 254 low pressure lamp and flowed to the detector.

## CHROMATOGRAM
**Retention time:** 8
**Internal standard:** 1-isopropyl-6-fluoro-1,4-dihydro-4-oxo-7-(1-piperazinyl)-3-quinolinecar-boxylic acid (13)
**Limit of quantitation:** 50 ng/mL

## OTHER SUBSTANCES
**Extracted:** metabolites

## KEY WORDS
serum; plasma; post-column reaction

## REFERENCE
Krol, G.J.; Beck, G.W.; Benham, T. HPLC analysis of ciprofloxacin and ciprofloxacin metabolites in body fluids. *J.Pharm.Biomed.Anal.*, **1996**, *14*, 181–190

## SAMPLE
**Matrix:** blood, urine
**Sample preparation:** Extract with chloroform:isopropanol 95:5.

## HPLC VARIABLES
**Column:** μBondapak C18
**Mobile phase:** MeOH:40 mM potassium phosphate + 20 mM sodium phosphate 65:35, containing 2 g/L hexadecyltrimethylammonium bromide, pH adjusted to 7.4 with HCl
**Flow rate:** 1
**Detector:** F ex 280 em 425

## CHROMATOGRAM
**Retention time:** 4.5
**Internal standard:** fleroxacin
**Limit of detection:** 30 ng/mL

## KEY WORDS
plasma

## REFERENCE
Paradis, D.; Vallée, F.; Allard, S.; Bisson, C.; Daviau, N.; Drapeau, C.; Auger, F.; LeBel, M. Comparative study of pharmacokinetics and serum bactericidal activities of cefpirome, ceftazidime, ceftriaxone, imipenem, and ciprofloxacin. *Antimicrob.Agents Chemother.*, **1992**, *36*, 2085–2092

## SAMPLE
**Matrix:** blood, urine
**Sample preparation:** Dilute with one or more volumes of water, filter (0.6 μm)

## HPLC VARIABLES
**Column:** 200 × 4 5 μm Nucleosil C18
**Mobile phase:** MeCN:25 mM orthophosphoric acid adjusted to pH 3.0 with tetrabutylam-monium hydroxide 11:89
**Flow rate:** 1.5
**Injection volume:** 10-20
**Detector:** F ex 278 em 445

## CHROMATOGRAM
**Retention time:** 3.1
**Limit of detection:** 10 ng/mL

## OTHER SUBSTANCES
**Noninterfering:** acetaminophen, cefuroxime, cloxacillin, dextropropoxyphene, digoxin, doxycycline, erythromycin, furosemide, metronidazole, netilmicin, penicillin G, prednisolone, salicylic acid, sulfamethoxazole, trimethoprim, warfarin
**Interfering:** norfloxacin

## KEY WORDS
serum

## REFERENCE
Nilsson-Ehle, I. Assay of ciprofloxacin and norfloxacin in serum and urine by high-performance liquid chromatography. *J.Chromatogr.*, **1987**, *416*, 207–211

## SAMPLE
**Matrix:** blood, urine
**Sample preparation:** Serum. 300 µL Serum + 300 µL 6% aqueous trichloroacetic acid, centrifuge at 5000 rpm for 5 min, inject 10 µL of supernatant. Urine. Dilute 1:1000 or 1:100 with mobile phase, inject 10 µL directly.

## HPLC VARIABLES
**Column:** 250 × 4.6 Spherisorb ODS
**Mobile phase:** MeCN:25 mM phosphoric acid adjusted to pH 3.0 with tetrabutylammonium hydroxide 11:89
**Flow rate:** 2
**Injection volume:** 10
**Detector:** F ex 278 em 456

## CHROMATOGRAM
**Retention time:** 4

## KEY WORDS
serum; a similar analysis is stability-indicating (Am.J.Hosp.Pharm. 1994, 51, 373-7)

## REFERENCE
Joos, B.; Ledergerber, B.; Flepp, M.; Bettex, J.-D.; Lüthy, R.; Siegenthaler, W. Comparison of high-pressure liquid chromatography and bioassay for determination of ciprofloxacin in serum and urine. *Antimicrob.Agents Chemother.*, **1985**, *27*, 353–356

## SAMPLE
**Matrix:** blood, vitreous humor
**Sample preparation:** Serum. 20 µL Serum + 130 µL mobile phase, mix, filter, inject a 100 µL aliquot. Vitreous humor. 15 µL Vitreous humor + 135 µL mobile phase, mix, filter, inject a 100 µL aliquot.

## HPLC VARIABLES
**Column:** 220 × 2.1 5 µm Nucleosil C18
**Mobile phase:** MeCN:MeOH:50 mM $KH_2PO_4$:100 mM tetrabutylammonium hydroxide 10:7:73:10
**Column temperature:** 25
**Flow rate:** 0.2

**Injection volume:** 100
**Detector:** UV 240; UV 280; F ex 280 em 445

## CHROMATOGRAM
**Limit of detection:** 2 ng/mL

## KEY WORDS
serum; rabbit; pharmacokinetics

## REFERENCE
Drusano, G.L.; Liu, W.; Perkins, R.; Madu, A.; Madu, C.; Mayers, M.; Miller, M.H. Determination of robust ocular pharmacokinetic parameters in serum and vitreous humor of albino rabbits following systemic administration of ciprofloxacin from sparse data sets by using IT2S, a population pharmacokinetic modeling program. *Antimicrob.Agents Chemother.*, **1995**, *39*, 1683–1687

## SAMPLE
**Matrix:** bulk
**Sample preparation:** Prepare a 100 μg/mL solution in mobile phase, inject a 20 μL aliquot.

## HPLC VARIABLES
**Column:** 300 × 3.5 10 μm C18 (Flexit, Pune, India)
**Mobile phase:** MeOH:water:acetic acid 84:15.9:0.1
**Flow rate:** 1
**Injection volume:** 20
**Detector:** UV 254

## CHROMATOGRAM
**Retention time:** 2.25
**Limit of detection:** 5 ng

## OTHER SUBSTANCES
**Simultaneous:** impurities

## REFERENCE
Husain, S.; Khalid, S.; Nagaraju, V.; Rao, R.N. High-performance liquid chromatographic separation and determination of small amounts of process impurities of ciprofloxacin in bulk drugs and formulations. *J.Chromatogr.A*, **1995**, *705*, 380–384

## SAMPLE
**Matrix:** cell suspensions
**Sample preparation:** 100 μL Cell suspension + 100 μL cefoperazone solution + 100 μL Hanks balanced salt solution, sonicate 30 min, add 800 μL MeCN, centrifuge at 13000 g for 5 min, remove supernatant. Dry the supernatant under a stream of air, dissolve the residue in 100 μL mobile phase, inject a 75 μL aliquot.

## HPLC VARIABLES
**Column:** μBondapak C18
**Mobile phase:** MeCN:5 mM pH 2.0 tetrabutylammonium hydrogen sulfate 10:90
**Flow rate:** 1
**Injection volume:** 75
**Detector:** UV 280

## CHROMATOGRAM
**Retention time:** 14

**Internal standard:** ofloxacin
**Limit of detection:** 100-1000 ng/mL

## REFERENCE

Darouiche, R.O.; Hamill, R.J. Antibiotic penetration of and bactericidal activity within endothelial cells. *Antimicrob.Agents Chemother.*, **1994**, *38*, 1059–1064

## SAMPLE

**Matrix:** cells
**Sample preparation:** Incubate cells in 2 mL 100 mM pH 3.0 glycine-HCl buffer for 2 h at room temperature, centrifuge at 5600 g for 5 min, inject an aliquot.

## HPLC VARIABLES

**Column:** Bondapak C18
**Mobile phase:** MeCN:25 mM phosphoric acid adjusted to pH 3.0 with tetrabutylammonium hydroxide 25:75
**Flow rate:** 1.5
**Detector:** F ex 340 em 425

## OTHER SUBSTANCES

**Also analyzed:** fleroxacin, lomefloxacin, norfloxacin, ofloxacin, temafloxacin

## REFERENCE

Pascual, A.; Garcia, I.; Conejo, M.C.; Perea, E.J. Fluorometric and high-performance liquid chromatographic measurement of quinolone uptake by human neutrophils. *Eur.J.Clin.Microbiol.Infect.Dis.*, **1991**, *10*, 969–971

## SAMPLE

**Matrix:** hair
**Sample preparation:** Wash hair successively with 0.1% sodium dodecyl sulfate and water for 30 min, repeat twice, blot between 2 sheets of paper towel, allow to dry at room temperature. Take a 1 cm fragment of hair, add 500 μL 1 M NaOH, heat at 80° for 30 min, cool, add 500 μL 1 M HCl, add 1 mL 100 mM pH 4.6 potassium hydrogen citrate buffer, add 50 μL 1 μg/mL IS in water. Add the mixture to a Bond-Elut C8 SPE cartridge, elute with 2 mL THF:25 mM orthophosphoric acid 20:80, evaporate eluate to dryness in vacuum, dissolve residue in 150 μL mobile phase, vortex, inject a 60 μL aliquot.

## HPLC VARIABLES

**Column:** 150 × 4.6 Tosoh 5 μm TSKgel ODS-80Ts
**Mobile phase:** MeCN:25 mM orthophosphoric acid adjusted to pH 3.0 with 0.5 M tetra-n-butylammonium hydroxide 5:95
**Column temperature:** 40
**Flow rate:** 1
**Injection volume:** 60
**Detector:** F ex 280 em 445

## CHROMATOGRAM

**Retention time:** 13.5
**Internal standard:** (R)-9-fluoro-2,3-dihydro-3-methyl-10-(4-ethyl-1-piperazinyl)-7-oxo-7H-pyrido[1,2,3-de][1,4]benzoxazine-6-carboxylic acid (DS-4632) (10.2)
**Limit of detection:** 0.3 ng/mL

## OTHER SUBSTANCES

**Simultaneous:** norfloxacin, ofloxacin (F ex 295 em 490)

## KEY WORDS
SPE

## REFERENCE
Mizuno, A.; Uematsu, T.; Nakashima, M. Simultaneous determination of ofloxacin, norfloxacin and ciprofloxacin in human hair by high-performance liquid chromatography and fluorescence detection. *J.Chromatogr.B*, **1994**, *653*, 187–193

## SAMPLE
**Matrix:** solutions
**Sample preparation:** Filter (0.45 μm) a solution in MeCN:water 10:90, inject an aliquot of the filtrate.

## HPLC VARIABLES
**Column:** 250 × 4 5 μm LiChrospher 100 RP-18
**Mobile phase:** MeCN:buffer 7:93 (Buffer was 25 mM phosphoric acid adjusted to pH 3.89 with 100 mM tetrabutylammonium hydroxide.)
**Flow rate:** 1
**Injection volume:** 10
**Detector:** UV 280

## CHROMATOGRAM
**Retention time:** 10

## OTHER SUBSTANCES
**Simultaneous:** enoxacin, fleroxacin, norfloxacin, ofloxacin (UV 295), pipemidic acid

## REFERENCE
Barbosa, J.; Bergés, R.; Sanz-Nebot, V. Solvatochromic parameter values and pH in aqueous-organic mixtures used in liquid chromatography. Prediction of retention of a series of quinolones. *J.Chromatogr.A*, **1996**, *719*, 27–36

## SAMPLE
**Matrix:** solutions
**Sample preparation:** Prepare a 20 μg/mL solution in MeCN:water 10:90, filter (0.45 μm), inject an aliquot.

## HPLC VARIABLES
**Column:** 250 × 4 5 μm LiChrospher 100 RP-18
**Mobile phase:** MeCN:25 mM phosphoric acid 7:93, adjusted to pH 3.09 with 100 mM tetrabutylammonium hydroxide
**Flow rate:** 1
**Injection volume:** 10
**Detector:** UV 280

## CHROMATOGRAM
**Retention time:** 9.8

## OTHER SUBSTANCES
**Simultaneous:** norfloxacin, ofloxacin (UV 295), pipemidic acid

## REFERENCE
Barbosa, J.; Bergés, R.; Sanz-Nebot, V. Linear solvation energy relationships in reversed-phase liquid chromatography. Prediction of retention of several quinolones. *J.Liq.Chromatogr.*, **1995**, *18*, 3445–3463

## SAMPLE
**Matrix:** solutions

## HPLC VARIABLES
**Column:** 150 × 3.9 4 μm Nova-Pak C18
**Mobile phase:** MeCN:20 mM pH 2.3 phosphoric acid 15:85 containing 2.5 mM sodium 1-heptanesulfonate
**Flow rate:** 1.5
**Detector:** UV 278

## CHROMATOGRAM
**Retention time:** 25

## OTHER SUBSTANCES
**Simultaneous:** degradation products

## REFERENCE
Torniainen, K.; Mäki, E. Development of an isocratic high-performance liquid chromatographic method for monitoring of ciprofloxacin photodegradation. *J.Chromatogr.A*, **1995**, *697*, 397–405

## SAMPLE
**Matrix:** solutions
**Sample preparation:** Prepare a 2.5-5 μg/mL solution, inject a 20 μL aliquot.

## HPLC VARIABLES
**Column:** 80 × 4.6 3.65 μm Zorbax Rx-SIL (similar to Zorbax SB-C8 (Mac-Mod Analytical))
**Mobile phase:** MeCN:0.1% trifluoroacetic acid 20:80
**Flow rate:** 1
**Injection volume:** 20
**Detector:** UV 277

## CHROMATOGRAM
**Retention time:** k' 3.1

## REFERENCE
Kirkland, K.M.; McCombs, D.A.; Kirkland, J.J. Rapid, high-resolution high-performance liquid chromatographic analysis of antibiotics. *J.Chromatogr.A*, **1994**, *660*, 327–337

## SAMPLE
**Matrix:** solutions

## HPLC VARIABLES
**Column:** 150 × 4.6 5 μm BDS-Hypersil C18
**Mobile phase:** MeOH:THF:670 mM pH 3.0 phosphate buffer 20:0.8:79.2 plus 2 g/L tetrabutylammonium hydrogen sulfate and 2 mL/L 85% phosphoric acid
**Flow rate:** 1
**Injection volume:** 20
**Detector:** UV 275

## CHROMATOGRAM
**Retention time:** 5.83

## OTHER SUBSTANCES
**Simultaneous:** photodegradation products
**Interfering:** fleroxacin, ofloxacin

## KEY WORDS
water

## REFERENCE
Tiefenbacher, E.-M.; Haen, E.; Przybilla, B.; Kurz, H. Photodegradation of some quinolones used as antimicrobial therapeutics. *J.Pharm.Sci.*, **1994**, *83*, 463−467

## SAMPLE
**Matrix:** solutions
**Sample preparation:** Prepare a 450 μg/mL solution in MeCN:water 50:50. 5 mL Solution + 5 mL THF + 200 molar excess of acetic anhydride + 3 molar excess of 1 M NaOH, sonicate for 15 min, add 15 mL mobile phase, sonicate for 15 min, cool to room temperature, make up to 50 mL with mobile phase, inject a 50 μL aliquot.

## HPLC VARIABLES
**Column:** 150 × 4.6 5 μm Nucleosil C18
**Mobile phase:** MeCN:buffer 35:65 (Buffer was prepared by mixing equal volumes of 20 mM citric acid and 20 mM sodium citrate, pH adjusted to 2.4 with perchloric acid.)
**Flow rate:** 1
**Injection volume:** 50
**Detector:** UV 280

## CHROMATOGRAM
**Retention time:** 6.4

## OTHER SUBSTANCES
**Simultaneous:** norfloxacin, sarafloxacin, temafloxacin

## KEY WORDS
derivatization

## REFERENCE
Morley, J.A.; Elrod, L., Jr. Determination of fluoroquinolone antibacterials as N-Acyl derivatives. *Chromatographia*, **1993**, *37*, 295−299

## SAMPLE
**Matrix:** tissue
**Sample preparation:** Wash a 500 mg 2.8 mL Bond-Elut SCX cartridge with 1% acetic acid in EtOH. Homogenize 2 g muscle tissue in 20 mL 1% acetic acid in EtOH, sonicate 3 min, centrifuge at 4200 g for 5 min, decant supernatant. Repeat extraction, combine supernatants, centrifuge at 4200 g for 5 min. Pass supernatants through SPE cartridge, wash cartridge with 5 mL MeOH, 10 mL water, 5 mL MeOH, and elute with 25% aqueous ammonia (specific gravity 0.88) in MeOH. Evaporate eluate to dryness under a stream of nitrogen at 50°, evaporation of the final portion is aided by the addition of 1 mL MeCN. Add 1 mL mobile phase, vortex 15 s, sonicate 3 min, centrifuge at 1860 g for 5 min, filter (0.45 μm), inject 20 μL.

## HPLC VARIABLES
**Column:** 250 × 4.6 Zorbax RXC8
**Mobile phase:** MeCN:buffer 20:80 (Buffer was 0.68 mL orthophosphoric acid in 900 mL water, taken to pH 3.0 with triethylamine, made up to 1 L.)

**Flow rate:** 0.5
**Injection volume:** 20
**Detector:** F ex 278 em 445

## CHROMATOGRAM
**Retention time:** 10
**Limit of quantitation:** <10 ng/g

## OTHER SUBSTANCES
**Simultaneous:** enrofloxacxin

## KEY WORDS
muscle; pig; cow; SPE

## REFERENCE
Tarbin, J.A.; Tyler, D.J.; Shearer, G. Analysis of enrofloxacin and its metabolite ciprofloxacin in bovine and porcine muscle by high-performance liquid chromatography following cation exchange clean-up. *Food Addit.Contam.*, **1992**, *9*, 345–350

## SAMPLE
**Matrix:** urine
**Sample preparation:** 50 μL Urine + 100 μL IS + 3.85 mL water, mix, inject a 10 μL aliquot.

## HPLC VARIABLES
**Column:** 250 × 4.6 10 μm Nucleosil C18
**Mobile phase:** MeCN:50 mM citric acid:1 M ammonium acetate 22:77:1
**Flow rate:** 1.5
**Injection volume:** 10
**Detector:** F ex 280 em 418

## CHROMATOGRAM
**Retention time:** 3.8
**Internal standard:** KK-123 (G.D.Searle) (6.3)
**Limit of quantitation:** 2 μg/mL

## OTHER SUBSTANCES
**Extracted:** lomefloxacin

## KEY WORDS
plasma

## REFERENCE
Stuht, H.; Lode, H.; Koeppe, P.; Rost, K.L.; Schaberg, T. Interaction study of lomefloxacin and ciprofloxacin with omeprazole and comparative pharmacokinetics. *Antimicrob.Agents Chemother.*, **1995**, *39*, 1045–1049

◆─────────────────────◆─────────────────────◆

## ANNOTATED BIBLIOGRAPHY
Delon, A.; Favreliere, S.; Couet, W.; Courtois, P.; Bouquet, S. Rapid and sensitive determination of thalidomide in human plasma by high-performance liquid chromatography. *J.Liq.Chromatogr.*, **1995**, *18*, 297–309 [SPE; ciprofloxacin is IS; simultaneous acyclovir, azathioprine, cefotaxime, ceftazidime, flucytosine, metronidazole; non-interfering amphotericin, clobazam, clonazepam, cyclophosphamide, cyclosporin, diazepam, diltiazem, hydroxyzine, nifedipine, prednisolone]

Barbato, F.; Morrica, P.; Seccia, S.; Ventriglia, M. High performance liquid chromatographic analysis of quinolone antibacterial agents. *Farmaco*, **1994**, *49*, 407–410 [simultaneous cinoxacin, ciprofloxacin, flumequine, nalidixic acid, norfloxacin, ofloxacin, oxolinic acid, pefloxacin, piromidic acid]

Davis, J.D.; Aarons, L.; Houston, J.B. Relationship between enoxacin and ciprofloxacin plasma concentrations and theophylline disposition. *Pharm.Res.*, **1994**, *11*, 1424–1428 [extracted enoxacin, theophylline; plasma; hydroxypropyltheophylline (IS); LOD 500 ng/mL]

Kane, M.P.; Bailie, G.R.; Moon, D.G.; Siu, I. Stability of ciprofloxacin injection in peritoneal dialysis solutions. *Am.J.Hosp.Pharm.*, **1994**, *51*, 373–377 [stability-indicating]

Mueller, B.A.; Brierton, D.G.; Abel, S.R.; Bowman, L. Effect of feeding with Ensure on oral bioavailabilities of ofloxacin and ciprofloxacin. *Antimicrob.Agents Chemother.*, **1994**, *38*, 2101–2105 [extracted ofloxacin; plasma; ultrafiltrate; fluorescence detection; A-57084 (IS); LOQ 9.4 ng/mL; pharmacokinetics]

Pei, Y.Y.; Meng, X.; Nightingale, C.H. An improved HPLC assay for ciprofloxacin in biological samples. *Chung Kuo Yao Li Hsueh Pao*, **1994**, *15*, 197–201

Ramon, J.; Dautrey, S.; Farinoti, R.; Carbon, C.; Rubinstein, E. Intestinal elimination of ciprofloxacin in rabbits. *Antimicrob.Agents Chemother.*, **1994**, *38*, 757–760 [rabbit; ofloxacin (IS); serum; intestinal efflux; fluorescence detection; LOD 50 ng/mL; pharmacokinetics]

Reid, G.; Sharma, S.; Advikolanu, K.; Tieszer, C.; Martin, R.A.; Bruce, A.W. Effects of ciprofloxacin, norfloxacin, and ofloxacin on in vitro adhesion and survival of pseudomonas aeruginosa AK1 on urinary catheters. *Antimicrob.Agents Chemother.*, **1994**, *38*, 1490–1495 [column temp 40]

Saux, P.; Martin, C.; Mallet, M.-N.; Papazian, L.; Bruguerolle, B.; De Mico, P.; Gouin, F. Penetration of ciprofloxacin into bronchial secretions from mechanically ventilated patients with nosocomial bronchopneumonia. *Antimicrob.Agents Chemother.*, **1994**, *38*, 901–904 [serum; pharmacokinetics; column-switching; fluorescence detection; LOD 15 ng/mL]

Israel, D.; Gillum, G.; Turik, M.; Harvey, K.; Ford, J.; Dalton, H.; Towle, M.; Echols, R.; Heller, A.H.; Polk, R. Pharmacokinetics and serum bactericidal titers of ciprofloxacin and ofloxacin following multiple oral doses in healthy volunteers. *Antimicrob.Agents Chemother.*, **1993**, *37*, 2193–2199 [serum; urine; extracted ofloxacin; fluorescence detection; LOD 50 ng/mL]

Budvári-Bárány, Z.; Szász, G.; Takács-Novák, K.; Hermecz, I.; Lore, A. The pH influence on the HPLC-retention of chemotherapeutic fluoroquinolone derivatives. *J.Liq.Chromatogr.*, **1991**, *14*, 3411–3424 [also amifloxacin, lomefloxacin, nalidixic acid, norfloxacin, ofloxacin, oxolinic acid, pefloxacin]

Van Slooten, A.D.; Nix, D.E.; Wilton, J.H.; Love, J.H.; Spivey, J.M.; Goldstein, H.R. Combined use of ciprofloxacin and sucralfate. *DICP*, **1991**, *25*, 578–582 [plasma; urine; fluorescence detection; LOQ 50 nM; SPE; pharmacokinetics]

Katagiri, Y.; Naora, K.; Ichikawa, N.; Hayashibara, M.; Iwamoto, K. High-performance liquid chromatographic determination of ciprofloxacin in rat brain and cerebrospinal fluid. *Chem.Pharm.Bull.*, **1990**, *38*, 2884–2886 [derivatization; rat; brain; CSF]

Scholl, H.; Schmidt, K.; Weber, B. Sensitive and selective determination of picogram amounts of ciprofloxacin and its metabolites in biological samples using high-performance liquid chromatography and photothermal post-column derivatization. *J.Chromatogr.*, **1987**, *416*, 321–330 [post-column reaction]

Groeneveld, A.J.; Brouwers, J.R. Quantitative determination of ofloxacin, ciprofloxacin, norfloxacin and pefloxacin in serum by high pressure liquid chromatography. *Pharm.Weekbl.[Sci].*, **1986**, *8*, 79–84

Krol, G.J.; Noe, A.J.; Beermann, D. Liquid chromatographic analysis of ciprofloxacin and ciprofloxacin metabolites in body fluids. *J.Liq.Chromatogr.*, **1986**, *9*, 2897–2919 [bile; saliva; urine; serum; plasma; SPE; fluorescence detection; UV detection; column temp 30-40; extracted metabolites; LOD 10 ng/mL]

Fasching, C.E.; Peterson, L.R. High pressure liquid chromatography of (BAY o 9867) ciprofloxacin in serum samples. *J.Liq.Chromatogr.*, **1985**, *8*, 555–562 [LOD 50 ng/mL; fluorescence detection]

Gau, W.; Ploschke, H.J.; Schmidt, K.; Weber, B. Determination of ciprofloxacin (BAY o 9867) in biological fluids by high-performance liquid chromatography. *J.Liq.Chromatogr.*, **1985**, *8*, 485–497 [serum; plasma; urine; fluorescence detection; LOD 8 ng/mL (plasma, serum); LOD 50 ng/mL (urine); pharmacokinetics]

Aronoff, G.E.; Kenner, C.H.; Sloan, R.S.; Pottratz, S.T. Multiple-dose ciprofloxacin kinetics in normal subjects. *Clin.Pharmacol.Ther.*, **1984**, *36*, 384–388 [plasma; urine; fluorescence detection; column temp 58; LOD 30 ng/mL; pharmacokinetics]

# Cisapride

**Molecular formula:** $C_{23}H_{29}ClFN_3O_4$
**Molecular weight:** 466.0
**CAS Registry No.:** 81098-60-4

## SAMPLE
**Matrix:** blood
**Sample preparation:** 100 µL Plasma + 70 µL 1 M NaOH + 1 mL 30 ng/mL IS in chloroform:isopropanol 90:10, vortex for 2 min, centrifuge at 1800-1900 g for 1 min. Remove the organic layer and evaporate it to dryness under a stream of air at 60°, reconstitute the residue in 100 µL mobile phase, inject an aliquot.

## HPLC VARIABLES
**Column:** 150 × 3.9 octyl Symmetry (Waters)
**Mobile phase:** MeCN:20 mM pH 5.2 phosphate buffer 37:63
**Flow rate:** 1
**Detector:** F ex 295 em 350

## CHROMATOGRAM
**Retention time:** 5
**Internal standard:** cis-4-amino-5-chloro-N-[1-[5-(4-fluorophenoxy)pentyl]-3-methoxy-4-piperidinyl]-2-methoxybenzamide monohydrate (R 54 680, Jansen-Cilag) (8)
**Limit of detection:** 5 ng/mL
**Limit of quantitation:** 8 ng/mL

## OTHER SUBSTANCES
**Extracted:** metabolites, norcisapride
**Simultaneous:** furosemide
**Noninterfering:** amoxicillin, caffeine, dexamethasone, gentamicin, hydrocortisone, indomethacin, metoclopramide, midazolam, theobromine, theophylline, tolazoline

## KEY WORDS
plasma

## REFERENCE
Preechagoon, Y.; Charles, B.G. Analysis of cisapride in neonatal plasma using high-performance liquid chromatography with a base-stable column and fluorescence detection. *J.Chromatogr.B*, **1995**, *670*, 139-143

## SAMPLE
**Matrix:** blood, tissue
**Sample preparation:** Plasma. 2 mL Plasma + 100 ng IS + 500 µL 1 M NaOH, mix, add 6 mL heptane:isoamyl alcohol 95:5, rotate at 10 rpm for 10 min, centrifuge at 1000 g for 5 min. Remove the organic layer and add it to 3 mL 50 mM sulfuric acid, extract, centrifuge. Remove the aqueous layer and make it alkaline with 150 µL concentrated ammonia, add 4 mL heptane:isoamyl alcohol 95:5, extract, centrifuge. Remove the organic layer and evaporate it to dryness under a stream of nitrogen at 55°, reconstitute the residue in 120 µL mobile phase, inject a 40 µL aliquot. Tissue. Grind tissue (Waring blender), homogenize (Ultra-Turrax) with three volumes 10 mM pH 7.4 phosphate buffer containing 1.15% KCl. 1 mL Homogenate + 200 ng IS + 500 µL 1 M NaOH, mix, add 6 mL heptane:isoamyl alcohol 95:5, rotate at 10 rpm for 10 min, centrifuge at 1000 g for 5 min. Remove the organic layer and add it to 3 mL 50 mM sulfuric acid, extract, centrifuge. Remove the aqueous layer and make it alkaline with 150 µL concentrated am-

monia, add 4 mL heptane:isoamyl alcohol 95:5, extract, centrifuge. Remove the organic layer and evaporate it to dryness under a stream of nitrogen at 55°, reconstitute the residue in 120 μL mobile phase, inject a 40 μL aliquot.

## HPLC VARIABLES
**Column:** 150 × 2.1 5 μm ODS-Hypersil
**Mobile phase:** MeCN:water:diethylamine 44:56:0.02
**Flow rate:** 0.8
**Injection volume:** 40
**Detector:** UV 276

## CHROMATOGRAM
**Retention time:** 3.08
**Internal standard:** cis-4-amino-5-chloro-N-[1-[5-(4-fluorophenoxy)pentyl]-3-methoxy-4-piperidinyl]-2-methoxybenzamide monohydrate (R 54 680) (5.96)
**Limit of detection:** 1 ng/mL

## KEY WORDS
plasma; method can also be used for urine (Drug Metab. Dispos. 1988, 16, 403); method can also be used for human breast milk (Eur.J.Clin.Pharmacol. 1986, 30, 735); human; rat; liver

## REFERENCE
Woestenborghs, R.; Lorreyne, W.; Van Rompaey, F.; Heykants, J. Determination of cisapride in plasma and animal tissues by high-performance liquid chromatography. *J.Chromatogr.*, **1988**, *424*, 195–200

## SAMPLE
**Matrix:** feces, urine
**Sample preparation:** Urine. Inject a 275 μL aliquot of urine directly. Feces. Extract feces with MeOH.

## HPLC VARIABLES
**Column:** 300 × 4.6 5 μm Lichrosorb RP-8
**Mobile phase:** Gradient. A was 100 mM ammonium acetate containing 40 mM diisopropylamine, adjusted to pH 8.0 with ammonia. B was MeCN:MeOH:1 M pH 8.0 ammonium acetate containing 400 mM diisopropylamine 45:45:10. A:B from 90:10 to 30:70 over 40 min.
**Flow rate:** 1
**Injection volume:** 275
**Detector:** Radioactivity; UV 230; UV 306

## CHROMATOGRAM
**Retention time:** 52

## OTHER SUBSTANCES
**Extracted:** metabolites

## KEY WORDS
rat; tritium labeled

## REFERENCE
Meuldermans, W.; Hendrickx, J.; Lauwers, W.; Hurkmans, R.; Mostmans, E.; Swysen, E.; Bracke, J.; Knaeps, A.; Heykants, J. Excretion and biotransformation of cisapride in rats after oral administration. *Drug Metab.Dispos.*, **1988**, *16*, 410–419

## SAMPLE
**Matrix:** formulations
**Sample preparation:** Dilute syrup 10-fold with MeOH, remove a 200 μL aliquot and add it to 20 μL IS solution, inject a 10 μL aliquot.

## HPLC VARIABLES
**Column:** 160 × 4 5 μm Zorbax Rx-C8
**Mobile phase:** MeCN:water:triethylamine 65:35:0.02
**Flow rate:** 1
**Injection volume:** 10
**Detector:** UV 276

## CHROMATOGRAM
**Retention time:** 2.7
**Internal standard:** cis-4-amino-5-chloro-n-{1-[5-(4-fluorophenoxy)pentyl]-3-methoxy-4-piperidinyl}-2-methoxybenzamide monohydrate (3.8)

## KEY WORDS
syrup; stability-indicating

## REFERENCE
Nahata, M.C.; Morosco, R.S.; Hipple, T.F. Stability of cisapride in a liquid dosage form at two temperatures. *Ann.Pharmacother.*, **1995**, *29*, 125–126

## SAMPLE
**Matrix:** solutions

## HPLC VARIABLES
**Column:** 250 × 4.6 5 μm Supelcosil LC-DP (A) or 250 × 4 5 μm LiChrospher 100 RP-8 (B)
**Mobile phase:** MeCN:0.025% phosphoric acid:buffer 25:10:5 (A) or 60:25:15 (B) (Buffer was 9 mL concentrated phosphoric acid and 10 mL triethylamine in 900 mL water, adjust pH to 3.4 with dilute phosphoric acid, make up to 1 L.)
**Flow rate:** 0.6
**Injection volume:** 25
**Detector:** UV 229

## CHROMATOGRAM
**Retention time:** 10.90 (A), 5.81 (B)

## OTHER SUBSTANCES
**Also analyzed:** acebutolol, acepromazine, acetaminophen, acetazolamide, acetophenazine, albuterol, alprazolam, amitriptyline, amobarbital, amoxapine, antipyrine, atenolol, atropine, azatadine, baclofen, benzocaine, bromocriptine, brompheniramine, brotizolam, bupivacaine, buspirone, butabarbital, butalbital, caffeine, carbamazepine, cetirizine, chlorcyclizine, chlordiazepoxide, chlormezanone, chloroquine, chlorpheniramine, chlorpromazine, chlorpropamide, chlorprothixene, chlorthalidone, chlorzoxazone, cimetidine, clomipramine, clonazepam, clonidine, clozapine, cocaine, codeine, colchicine, cyclizine, cyclobenzaprine, dantrolene, desipramine, diazepam, diclofenac, diflunisal, diltiazem, diphenhydramine, diphenidol, diphenoxylate, dipyridamole, disopyramide, dobutamine, doxapram, doxepin, droperidol, encainide, ethidium bromide, ethopropazine, fenoprofen, fentanyl, flavoxate, fluoxetine, fluphenazine, flurazepam, flurbiprofen, fluvoxamine, furosemide, glutethimide, glyburide, guaifenesin, haloperidol, homatropine, hydralazine, hydrochlorothiazide, hydrocodone, hydromorphone, hydroxychloroquine, hydroxyzine, ibuprofen, imipramine, indomethacin, ketoconazole, ketoprofen, ketorolac, labetalol, levorphanol, lidocaine, loratadine, lorazepam, lovastatin, loxapine, mazindol, mefenamic acid, meperidine, mephenytoin, mepivacaine, mesoridazine, metaproterenol,

methadone, methdilazine, methocarbamol, methotrexate, methotrimeprazine, methoxamine, methyldopa, methylphenidate, metoclopramide, metolazone, metoprolol, metronidazole, midazolam, moclobemide, morphine, nadolol, nalbuphine, naloxone, naphazoline, naproxen, nifedipine, nizatidine, norepinephrine, nortriptyline, oxazepam, oxycodone, oxymetazoline, paroxetine, pemoline, pentazocine, pentobarbital, pentoxifylline, perphenazine, pheniramine, phenobarbital, phenol, phenolphthalein, phentolamine, phenylbutazone, phenyltoloxamine, phenytoin, pimozide, pindolol, piroxicam, pramoxine, prazepam, prazosin, probenecid, procainamide, procaine, prochlorperazine, procyclidine, promazine, promethazine, propafenone, propantheline, propiomazine, propofol, propranolol, protriptyline, quazepam, quinidine, quinine, racemethorphan, ranitidine, remoxipride, risperidone, salicylic acid, scopolamine, secobarbital, sertraline, sotalol, spironolactone, sulfinpyrazone, sulindac, temazepam, terbutaline, terfenadine, tetracaine, theophylline, thiethylperazine, thiopental, thioridazine, thiothixene, timolol, tocainide, tolbutamide, tolmetin, trazodone, triamterene, triazolam, trifluoperazine, triflupromazine, trimeprazine, trimethoprim, trimipramine, verapamil, warfarin, xylometazoline, yohimbine, zopiclone

## KEY WORDS
some details of plasma extraction

## REFERENCE
Koves, E.M. Use of high-performance liquid chromatography-diode array detection in forensic toxicology. *J.Chromatogr.A*, **1995**, *692*, 103–119

# Clarithromycin

**Molecular formula:** $C_{38}H_{69}NO_{13}$
**Molecular weight:** 748.0
**CAS Registry No.:** 81103-11-9

## SAMPLE
**Matrix:** alveolar cells, blood, bronchoalveolar lavage fluid
**Sample preparation:** Freeze dry bronchoalveolar lavage fluid and reconstitute in water to yield a 10-fold concentration. Suspend alveolar cells in pH 8.0 potassium phosphate buffer, sonicate (Fisher Model 50 Sonic Dismembrator) at 50% cycle for 2 min. 500 μL Alveolar cell suspension, bronchoalveolar lavage fluid concentrate, or plasma 60 μL 10 μg/mL IS + 200 μL 100 mM sodium carbonate + 3 mL hexane:ethyl acetate 50:50, vortex , centrifuge at 800 g for 5 min. Remove the organic layer and evaporate it to dryness under a stream of nitrogen, reconstitute the residue in 300 μL MeCN:water 50:50, inject a 50 μL aliquot.

## HPLC VARIABLES
**Column:** 100 × 8 4 μm Nova-Pak phenyl radial compression
**Mobile phase:** MeCN:MeOH:1 M NaH₂PO₄:water 35:4:4:57, pH 6.85
**Flow rate:** 1.6
**Injection volume:** 50
**Detector:** E, Environmental Sciences Associates Model 5100A, screen electrode +0.5 V, analytical cell +0.79 V, Model 5020 pre-column guard cell +0.85 V

## CHROMATOGRAM
**Internal standard:** erythromycin A 9-O-methyloxime
**Limit of quantitation:** 50 ng/mL

## OTHER SUBSTANCES
**Extracted:** metabolites

## KEY WORDS
plasma

## REFERENCE
Conte, J.E., Jr.; Golden, J.A.; Duncan, S.; McKenna, E.; Zurlinden, E. Intrapulmonary pharmacokinetics of clarithromycin and of erythromycin. *Antimicrob.Agents Chemother.*, **1995**, *39*, 334–338

## SAMPLE
**Matrix:** blood
**Sample preparation:** Condition a 10 × 2 20 mg 30-40 μm Baker CN SPE cartridge with 2 mL MeOH, 2 mL MeOH:water 10:90, and 4 mL MeCN:pH 10.5 phosphate buffer (I = 0.10) 10:90 at 2 mL/min. Centrifuge plasma at 1300 g for 5 min, 100 μL plasma + 100 μL roxithromycin in MeCN:pH 10.5 phosphate buffer (I = 0.10) 10:90, mix, add a 20-100 μL aliquot to the SPE cartridge, wash SPE cartridge with MeCN:pH 10.5 phosphate buffer (I = 0.10) 10:90 at 0.5 mL/min, after 5 min backflush the contents of the SPE cartridge onto the column with the mobile phase, elute the column with the mobile phase and monitor the effluent.

## HPLC VARIABLES
**Column:** 100 × 4.6 3 μm Hypersil BDS C18
**Mobile phase:** MeCN:water 54:46 containing 4.5 mM $NaH_2PO_4$ and 6.8 mM $Na_2HPO_4$, pH 7
**Column temperature:** 55
**Flow rate:** 1
**Detector:** E, ESA Coulochem II, Model 5011 dual analytical cell, upstream +0.65 V, downstream +0.85 V (monitored), analytical cell protected by an ESA carbon in-line filter

## CHROMATOGRAM
**Retention time:** 6
**Internal standard:** roxithromycin (7)
**Limit of quantitation:** 500 nM

## KEY WORDS
plasma; SPE

## REFERENCE
Hedenmo, M.; Eriksson, B.-M. Liquid chromatographic determination of the macrolide antibiotics roxithromycin and clarithromycin in plasma by automated solid-phase extraction and electrochemical detection. *J.Chromatogr.A*, **1995**, *692*, 161–166

## SAMPLE
**Matrix:** blood
**Sample preparation:** 500 μL Serum + 50 μL 15 μg/mL IS in MeCN:water 50:50 + 0.2 g (?) sodium carbonate + 3 mL hexane:ethyl acetate 50:50, vortex for 1 min, centrifuge at 800 g for 5 min. Remove the organic layer and evaporate it to dryness under a stream of nitrogen at 40°, reconstitute the residue in 150 μL mobile phase, sonicate for 1 min, inject a 75 μL aliquot of the supernatant.

## HPLC VARIABLES
**Guard column:** Supelguard C8 (Supelco)
**Column:** 150 × 4.6 5 μm Supelcosil C8
**Mobile phase:** MeCN:MeOH:25 mM acetic acid 46:10:44 adjusted to pH 6.8 with NaOH
**Flow rate:** 1
**Injection volume:** 50
**Detector:** E, ESA Coulochem 5100A, 5010 analytical cell, +0.78 V

## CHROMATOGRAM
**Retention time:** 17
**Internal standard:** erythromycin A-6-O-methyloxime (25)
**Limit of detection:** 50 ng/mL

## OTHER SUBSTANCES
**Extracted:** metabolites

## KEY WORDS
serum; pharmacokinetics

## REFERENCE
Nilsen, O.G.; Aamo, T.; Zahlsen, K.; Svarva, P. Macrolide pharmacokinetics and dose scheduling of roxithromycin. *Diagn.Microbiol.Infect.Dis.*, **1992**, *15*, 71S–76S

## SAMPLE
**Matrix:** blood, urine

**Sample preparation:** Plasma. 500 μL Plasma + 75 μL 10 μg/mL IS in 1:1 MeCN:water + 200 μL 100 mM sodium carbonate + 3 mL ethyl acetate:hexane 1:1, stir vigorously for 1 min, centrifuge at 800 g for 5 min. Evaporate organic layer to dryness at 45° under a stream of air, dissolve residue in 200-400 μL 1:1 MeCN:water, inject 20-80 μL aliquot. Urine. 200 μL Urine + 300 μL 10 μg/mL IS in 1:1 MeCN:water + 100 μL 100 mM sodium carbonate + 3-4 mL ethyl acetate:hexane 1:1, stir vigorously for 1 min on a vortex mixer, centrifuge at 800 g for 5 min. Evaporate organic layer to dryness at 45° under a stream of air, dissolve residue in 800-1200 μL 1:1 MeCN:water, inject aliquot.

## HPLC VARIABLES
**Column:** 250 × 4.6 5 μm Nucleosil C8
**Mobile phase:** MeCN:MeOH:water 39:9:52, containing 0.04 M $NaH_2PO_4$ and NaOH to bring the pH to 6.8
**Flow rate:** 1.2-1.4
**Injection volume:** 20-80
**Detector:** E, Environmental Sciences Assoc. Model 5100A, screening electrode +0.5 V, working electrode +0.78 ± 0.04 V

## CHROMATOGRAM
**Retention time:** 21
**Internal standard:** erythromycin A 9-O-methyloxime
**Limit of quantitation:** 30 ng/mL

## OTHER SUBSTANCES
**Simultaneous:** metabolites, 14-hydroxyclarithromycin

## KEY WORDS
plasma

## REFERENCE
Chu, S.-Y.; Sennello, L.T.; Sonders, R.C. Simultaneous determination of clarithromycin and (14*R*)-hydroxyclarithromycin in plasma and urine using high-performance liquid chromatography with electrochemical detection. *J.Chromatogr.*, **1991**, *571*, 199–208

## SAMPLE
**Matrix:** solutions

## HPLC VARIABLES
**Column:** 150 × 4.6 ODS-80TM (Tosoh)
**Mobile phase:** MeCN:67 mM $KH_2PO_4$ 35:65
**Column temperature:** 50
**Flow rate:** 1
**Detector:** UV 210

## REFERENCE
Ishii, K.; Katayama, Y.; Itai, S.; Ito, Y.; Hayashi, H. *In vitro* dissolution tests corresponding to the *in vivo* dissolution of clarithromycin tablets in the stomach and intestine. *Chem.Pharm.Bull.*, **1995**, *43*, 1943–1948

## ANNOTATED BIBLIOGRAPHY
Morgan, D.K.; Brown, D.M.; Rotsch, T.D.; Plasz, A.C. A reversed-phase high-performance liquid chromatographic method for the determination and identification of clarithromycin as the drug substance and in various dosage forms. *J.Pharm.Biomed.Anal.*, **1991**, *9*, 261–269

# Clavulanic Acid

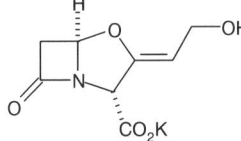

**Molecular formula:** $C_8H_9NO_5$
**Molecular weight:** 199.2
**CAS Registry No.:** 58001-44-8 (clavulanic acid), 61177-45-5 (clavulanate potassium)

## SAMPLE
**Matrix:** blood
**Sample preparation:** Filter (Amicon MPS-1 with YMT membrane) while centrifuging at 1500 g for 10 min, inject a 20 μL aliquot of the ultrafiltrate.

## HPLC VARIABLES
**Guard column:** 30 × 4.6 10 μm Develosil ODS-10
**Column:** 7 μm Zorbax ODS-7
**Mobile phase:** MeOH:buffer 1:2.7 (1:3.5 for concentrations <100 ng/mL) (Prepare buffer by dissolving 1.791 g $Na_2HPO_4.12H_2O$ and 0.780 g $NaH_2PO_4.2H_2O$ in 1 L water, add tetrabutylammonium bromide to a final concentration of 5 mM.)
**Flow rate:** 0.8
**Injection volume:** 20
**Detector:** UV 272 following post-column reaction. The column effluent mixed with MeOH: 500 mM NaOH 1:2.7 (1:3.5 for concentrations <100 ng/mL) pumped at 0.2 mL/min and this mixture flowed through a 1 m × 0.5 mm ID coil to the detector.

## CHROMATOGRAM
**Retention time:** 5
**Limit of detection:** 25 ng/mL
**Limit of quantitation:** 50 ng/mL

## OTHER SUBSTANCES
**Noninterfering:** ampicillin, cefoperazone, ticarcillin

## KEY WORDS
plasma; post-column reaction; ultrafiltrate

## REFERENCE
Haginaka, J.; Wakai, J.; Yasuda, H.; Uno, T.; Nakagawa, T. Improved high-performance liquid chromatographic assay of clavulanic acid and sulbactam by postcolumn alkaline degradation. *J.Liq.Chromatogr.*, **1985**, *8*, 2521–2534

## SAMPLE
**Matrix:** blood, tissue
**Sample preparation:** Homogenize (Ultra-Turrax) 300 mg tissue at 4° for 45 s, centrifuge. 400 μL Serum or tissue homogenate supernatant + 100 mM pH 6.8 ammonium citrate containing 3 M imidazole, vortex for 30 s, add 1 mL MeCN, mix for 15 s, centrifuge at 3000 rpm for 10 min. Add 3 mL dichloromethane to the supernatant, vortex for 30 s, centrifuge at 3000 rpm for 5 min, inject a 50 μL aliquot of the aqueous phase.

## HPLC VARIABLES
**Guard column:** 25 × 4 5 μm LiChrospher RP 18 E
**Column:** 125 × 4 5 μm LiChrospher RP 18 E
**Mobile phase:** MeCN:10 mM pH 3.2 $KH_2PO_4$ 4:96
**Flow rate:** 1.3

**Injection volume:** 50
**Detector:** UV 311

## CHROMATOGRAM
**Limit of detection:** 100 ng/mL

## KEY WORDS
serum; fat; colon

## REFERENCE
Martin, C.; Mallet, M.-N.; Sastre, B.; Viviand, X.; Martin, A.; De Micco, P.; Gouin, F. Comparison of
concentrations of two doses of clavulanic acid (200 and 400 milligrams) administered with amoxicillin
(2,000 milligrams) in tissues of patients undergoing colorectal surgery. *Antimicrob.Agents Chemother.*, **1995**, *39*, 94−98

## SAMPLE
**Matrix:** blood, urine
**Sample preparation:** Serum. 1 mL Serum + 1 mL MeCN, shake at 0° or 15 min, centrifuge
at 8000 g for 10 min. Add supernatant to 10 mL dichloromethane, shake at 0° or 15 min,
centrifuge at 8000 g for 10 min, discard organic layer. 200 µL Aqueous layer + 400 µL
reagent, after 5 min inject 40 µL aliquot. Urine. Dilute 10-fold, take 200 µL + 400 µL
reagent, after 5 min inject 40 µL aliquot. (Reagent was 3.45 g 1,2,4-triazole dissolved in
15 mL water, adjust pH to 7.0 with 4 M NaOH, make up to 25 mL.)

## HPLC VARIABLES
**Guard column:** 50 × 4.6 5 µm Spherisorb C-18
**Column:** 250 × 4.6 5 µm Spherisorb C-18
**Mobile phase:** Gradient. MeCN:20 mM pH 7.0 phosphate buffer from 2:98 to 25:75 over
25 min
**Flow rate:** 0.5
**Injection volume:** 40
**Detector:** UV 315

## CHROMATOGRAM
**Retention time:** 19
**Limit of detection:** 50 ng/mL

## KEY WORDS
serum; derivatization

## REFERENCE
Shah, A.J.; Adlard, M.W.; Stride, J.D. A sensitive assay for clavulanic acid and sulbactam in biological
fluids by high-performance liquid chromatography and precolumn derivatization. *J.Pharm.
Biomed.Anal.*, **1990**, *8*, 437−443

## SAMPLE
**Matrix:** blood, urine
**Sample preparation:** Serum. 500 µL Serum + 200 µL 200 mM pH 7.0 phosphate buffer,
vortex for 10 s, filter (Amicon YMT membrane) while centrifuging at 5° at 1500 g for 15
min. Mix 250 µL ultrafiltrate with 250 µL reagent, heat at 30° for 5 min, inject a 50 µL
aliquot. 500 µL Urine + 4.5 mL water, mix vigorously for 20 s, filter (0.45 µm) an aliquot.
Mix 250 µL filtrate with 250 µL reagent, heat at 30° for 5 min, inject a 50 µL aliquot.
(Prepare reagent by dissolving 13.81 g 1,2,4-triazole in 70 mL water and adjusting the
pH to 9.00 ± 0.05 with 4 M NaOH, make up to 100 mL.)

## HPLC VARIABLES
**Guard column:** 30 × 4.6 μBondapak C18
**Column:** 300 × 3.9 μBondapak C18
**Mobile phase:** MeOH:30 mM pH 7.0 phosphate buffer 20:80
**Flow rate:** 2
**Injection volume:** 50
**Detector:** UV 313

## CHROMATOGRAM
**Retention time:** 7
**Limit of detection:** 100 ng/mL

## OTHER SUBSTANCES
**Noninterfering:** degradation products, penicillins, penicilloic acids

## KEY WORDS
serum; derivatization

## REFERENCE
Martín, J.; Méndez, R. High-performance liquid chromatographic determination of clavulanic acid in human serum and urine using a pre-column reaction with 1,2,4-triazole. *J.Liq.Chromatogr.*, **1988**, *11*, 1697−1705

## SAMPLE
**Matrix:** blood, urine
**Sample preparation:** Plasma. 150-200 μL Plasma ultrafiltered (Amicon MPS-1, YMT membrane) at 1500 g for 10 min. 50 μL Ultrafiltrate + 150 μL 1 M pH 3.8 phosphate buffer + 20 μL 2% benzaldehyde in MeOH, heat at 100° for 20 min, cool to room temperature, inject 20-50 μL aliquot. Urine. Dilute 10-fold with water, filter (0.45 μm). 100 μL Filtrate + 300 μL 1 M pH 3.8 phosphate buffer + 40 μL 2% benzaldehyde in MeOH, heat at 100° for 20 min, cool to room temperature, inject 20-50 μL aliquot.

## HPLC VARIABLES
**Guard column:** 30 × 4.6 5 μm Develosil ODS-5
**Column:** 150 × 4.6 5 μm Develosil ODS-5
**Mobile phase:** MeOH:water 1:1
**Flow rate:** 0.8
**Injection volume:** 20-50
**Detector:** F ex 386 em 460

## CHROMATOGRAM
**Retention time:** 9
**Limit of detection:** 10 ng/mL

## OTHER SUBSTANCES
**Simultaneous:** amoxicillin
**Noninterfering:** ticarcillin

## KEY WORDS
plasma

## REFERENCE
Haginaka, J.; Yasuda, H.; Uno, T.; Nakagawa, T. High-performance liquid chromatographic assay of clavulanate in human plasma and urine by fluorimetric detection. *J.Chromatogr.*, **1986**, *377*, 269−277

## SAMPLE
**Matrix:** blood, urine
**Sample preparation:** Serum. 500 μL Serum + 500 μL 100 mM pH 7.0 phosphate buffer, mix, filter (Amicon MPS-1 ultrafiltration) with centrifugation at 4° and 1500 g for 20 min. Remove 100 μL ultrafiltrate, add 100 μL reagent, mix, inject a 75 μL aliquot taken from the top 5 mm. Urine. 100 μL urine + 100 μL reagent, mix, inject a 75 μL aliquot taken from the top 5 mm. (Reagent prepared by dissolving 8.25 g imidazole in 24 mL water + 2 mL 5 M HCl, adjust pH to 6.8 with 5 M HCl, make up to 40 mL with water.)

## HPLC VARIABLES
**Column:** 250 × 4.6 5 μm Spherisorb ODS
**Mobile phase:** MeOH:100 mM KH$_2$PO$_4$ + 50 mM pentanesulfonic acid + 100 mM ethanolamine 10:90
**Flow rate:** 1.5
**Injection volume:** 75
**Detector:** UV 313

## CHROMATOGRAM
**Retention time:** 7
**Limit of detection:** 100 ng/mL

## KEY WORDS
serum; derivatization

## REFERENCE
Watson, I.D. Clavulanate-potentiated ticarcillin: high-performance liquid chromatographic assays for clavulanic acid and ticarcillin isomers in serum and urine. *J.Chromatogr.*, **1985**, *337*, 301–309

## SAMPLE
**Matrix:** blood, urine
**Sample preparation:** Plasma. Vortex plasma with 2 volumes of MeCN for 30 s, centrifuge at 3600 rpm for 5 min, inject a 50 μL aliquot of the supernatant. Urine. Filter (0.45 μm) urine, inject a 25 μL aliquot of the filtrate.

## HPLC VARIABLES
**Guard column:** 50 × 4.6 LiChrosorb RP-2
**Column:** 250 × 4.6 Develosil ODS-10 (Nomura Chemicals)
**Mobile phase:** MeOH:buffer A 25:75 (plasma) or MeOH:buffer B 20:100 (urine) (Buffer A was 0.1 mM Na$_2$HPO$_4$ containing 0.1 mM NaH$_2$PO$_4$ and 5 mM tetrabutylammonium bromide. Buffer B was 1 mM Na$_2$HPO$_4$ containing 1 mM NaH$_2$PO$_4$ and 5 mM tetrabutylammonium bromide.)
**Flow rate:** 1.2
**Injection volume:** 25-50
**Detector:** UV 270 following post-column reaction. The column effluent mixed with 500 mM NaOH pumped at 0.6 mL/min and the mixture flowed through a 2 m × 0.25 mm ID coil to the detector.)

## CHROMATOGRAM
**Retention time:** 13 (plasma), 20 (urine)
**Limit of detection:** 100 ng/mL

## KEY WORDS
post-column reaction; plasma

## REFERENCE

Haginaka, J.; Yasuda, H.; Uno, T.; Nakagawa, T. Alkaline degradation of clavulanic acid and high performance liquid chromatographic determination by post-column alkaline degradation. *Chem. Pharm.Bull.*, **1983**, *31*, 4436−4447

## SAMPLE
**Matrix:** blood, urine
**Sample preparation:** Serum. Wash Amicon YMB filter membrane by stirring gently in 200 mL 100 mM pH 7.0 sodium phosphate buffer for 30 min, blot dry with filter paper. Dilute serum with an equal volume of 100 mM pH 7.0 sodium phosphate buffer, filter (Amicon YMB) while centrifuging at 5° at 1000 g for 15 min. Add 1 part reagent to 4 parts ultrafiltrate, let stand for 10 min, inject a 25-50 µL aliquot. Urine. Dilute 10-fold with 100 mM pH 7.0 sodium phosphate buffer. Add 1 part reagent to 4 parts diluted urine, let stand for 10 min, inject a 25-50 µL aliquot. (Reagent was 8.25 g imidazole, 24 mL water, and 2 mL 5 M HCl made up to 40 mL with water.)

## HPLC VARIABLES
**Guard column:** CO:PEL ODS C18
**Column:** 250 × 4.6 µBondapak C18
**Mobile phase:** MeOH:buffer 6:94 (Use 4:96 for clavulanic acid concentrations of <2 µg/mL in urine.) (Buffer was 100 mM $KH_2PO_4$ adjusted to pH 3.2 with phosphoric acid.)
**Flow rate:** 2.5
**Injection volume:** 25-50
**Detector:** UV 311

## CHROMATOGRAM
**Retention time:** 4
**Limit of detection:** 100 ng/mL

## OTHER SUBSTANCES
**Noninterfering:** degradation products, amoxicillin

## KEY WORDS
derivatization; ultrafiltrate; serum; pharmacokinetics

## REFERENCE

Foulstone, M.; Reading, C. Assay of amoxicillin and clavulanic acid, the components of Augmentin, in biological fluids with high-performance liquid chromatography. *Antimicrob.Agents Chemother.*, **1982**, *22*, 753−762

## SAMPLE
**Matrix:** formulations
**Sample preparation:** Dilute with mobile phase, inject an aliquot.

## HPLC VARIABLES
**Column:** 250 × 4.6 5 µm cyano
**Mobile phase:** MeCN:100 mM $NaH_2PO_4$ 20:80 adjusted to pH 4.2 with phosphoric acid
**Flow rate:** 0.8
**Injection volume:** 20
**Detector:** UV 195

## CHROMATOGRAM
**Retention time:** 2.71

## OTHER SUBSTANCES
**Simultaneous:** granisetron (UV 300), ticarcillin

## KEY WORDS
stability-indicating; injections; saline

## REFERENCE
Mayron, D.; Gennaro, A.R. Stability and compatibility of granisetron hydrochloride in i.v. solutions and oral liquids and during simulated Y-site injection with selected drugs. *Am.J.Health-Syst.Pharm.*, **1996**, *53*, 294–304

## SAMPLE
**Matrix:** formulations
**Sample preparation:** Dilute with water and filter (reject first few mL of filtrate)

## HPLC VARIABLES
**Column:** 300 × 4 μ-Bondapak 10 μm C18
**Mobile phase:** MeOH:buffer:water 15:1:84 (Buffer was 50 mL 200 mM $KH_2PO_4$ + 5.7 mL 200 mM NaOH made up to 200 mL, pH 6.)
**Flow rate:** 1
**Injection volume:** 20
**Detector:** UV 235

## CHROMATOGRAM
**Retention time:** 2.3

## OTHER SUBSTANCES
**Simultaneous:** amoxicillin

## KEY WORDS
tablets; suspensions

## REFERENCE
Abounassif, M.A.; Abdel-Moety, E.M.; Mohamed, M.E.; Gad-Kariem, R.A. Liquid chromatographic determination of amoxycillin and clavulanic acid in pharmaceutical preparations. *J.Pharm. Biomed.Anal.*, **1991**, *9*, 731–735

## SAMPLE
**Matrix:** formulations
**Sample preparation:** Dilute, filter, inject an aliquot of the filtrate.

## HPLC VARIABLES
**Column:** 150 × 3.9 5μm Nova Pak C18
**Mobile phase:** MeOH:50 mM pH 4.0 phosphate buffer 3:97
**Flow rate:** 0.8
**Detector:** UV 254

## CHROMATOGRAM
**Retention time:** 2.5

## OTHER SUBSTANCES
**Simultaneous:** amoxicillin
**Noninterfering:** degradation products, mannitol, saccharin

## KEY WORDS
oral suspensions; stability-indicating

## REFERENCE
Tu, Y.H.; Stiles, M.L.; Allen, L.V., Jr.; Olsen, K.M.; Barton, C.I.; Greenwood, R.B. Stability of amoxicillin trihydrate-potassium clavulanate in original containers and unit dose oral syringes. *Am.J. Hosp.Pharm.*, **1988**, *45*, 1092–1099

## SAMPLE
**Matrix:** urine
**Sample preparation:** Dilute 10-fold with water, filter (0.45 µm acrylate copolymer), inject a 20 µL aliquot.

## HPLC VARIABLES
**Guard column:** 30 × 4.6 10 µm Develosil ODS-10
**Column:** 150 × 4.6 5 µm Develosil ODS-5 (Nomura Chemicals)
**Mobile phase:** MeOH:buffer 1:2.5 (Prepare buffer by dissolving 1.791 g $Na_2HPO_4.12H_2O$ and 0.780 g $NaH_2PO_4.2H_2O$ in 5 L water, add tetrabutylammonium bromide to a final concentration of 5 mM.)
**Flow rate:** 0.8
**Injection volume:** 20
**Detector:** UV 272 following post-column reaction. The column effluent mixed with MeOH: 500 mM NaOH 1:2.5 pumped at 0.2 mL/min and this mixture flowed through a 1 m × 0.5 mm ID coil to the detector.

## CHROMATOGRAM
**Retention time:** 7
**Limit of detection:** 500 ng/mL
**Limit of quantitation:** 1 µg/mL

## OTHER SUBSTANCES
**Noninterfering:** ampicillin, cefoperazone, ticarcillin

## KEY WORDS
post-column reaction

## REFERENCE
Haginaka, J.; Wakai, J.; Yasuda, H.; Uno, T.; Nakagawa, T. Improved high-performance liquid chromatographic assay of clavulanic acid and sulbactam by postcolumn alkaline degradation. *J.Liq.Chromatogr.*, **1985**, *8*, 2521–2534

## ANNOTATED BIBLIOGRAPHY
Eckers, C.; Hutton, K.A.; de Biasi, V.; East, P.B.; Haskins, N.J.; Jacewicz, V.W. Determination of clavam-2-carboxylate in clavulanate potassium and tablet material by liquid chromatography-tandem mass spectrometry. *J.Chromatogr.*, **1994**, *686*, 213–218 [LC-MS; simultaneous impurities; bulk]

Jenke, D.R. Drug binding by reservoirs in elastomeric infusion devices. *Pharm.Res.*, **1994**, *11*, 984–989 [saline; 5% dextrose; also cilastatin, fluconazole, foscarnet, gentamicin, imipenem, lidocaine, penicillin G, tobramycin, vancomycin]

Low, A.S.; Taylor, R.B.; Gould, I.M. Determination of clavulanic acid by a sensitive HPLC method. *J.Antimicrob.Chemother.*, **1989**, *24 Suppl B*, 83–86

Haginaka, J.; Wakai, J.; Yasuda, H. Liquid chromatographic assay of β-lactamase inhibitors in human serum and urine using a hollow-fiber postcolumn reactor. *Anal.Chem.*, **1987**, *59*, 324–327 [serum; urine; post-column reaction]

Jehl, F.; Monteil, H.; Brogard, J.M. [Direct determination of clavulanic acid in biological fluids using HPLC]. *Pathol.Biol.(Paris)*, **1987**, *35*, 702–706

Bawdon, R.E.; Leveno, K.L.; Cunningham, F.G.; Nobles, B.; Nelson, S. Intrapartum pharmacokinetics of ticarcillin and clavulanic acid in serum as determined by high-pressure liquid chromatography. *Adv.Therapy*, **1984**, *1*, 419–426 [derivatization; SPE; serum; pharmacokinetics]

Haginaka, J.; Nakagawa, T.; Nishino, Y.; Uno, T. High performance liquid chromatographic determination of clavulanic acid in human urine. *J.Antibiot.(Tokyo)*, **1981**, *34*, 1189–1194

# Clindamycin

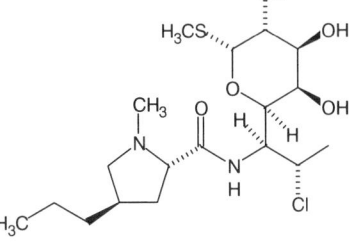

**Molecular formula:** $C_{18}H_{33}ClN_2O_5S$

**Molecular weight:** 425.0

**CAS Registry No.:** 18323-44-9, 58207-19-5 (HCl monohydrate),
36688-78-5 (palmitate), 25507-04-4
(palmitate HCl), 24729-96-2 (phosphate)

## SAMPLE
**Matrix:** blood
**Sample preparation:** 200 µL Plasma or serum + 500 µL 0.44 µg/L triazolam in MeCN,
vortex 20 s, centrifuge at 3000 g for 10 min, remove supernatant, evaporate under nitrogen to a volume of 250 µL, inject 15-30 µL aliquots

## HPLC VARIABLES
**Column:** 150 × 3.8 5 µm Nova-Pak C18
**Mobile phase:** MeCN:water:85% phosphoric acid:7.6 mM tetramethylammonium chloride
30:70:0.2:0.075, final apparent pH 6.7 (adjusted with 1 M NaOH)
**Flow rate:** 1
**Injection volume:** 15-30
**Detector:** UV 198 (nitrogen purged)

## CHROMATOGRAM
**Retention time:** 8.4
**Internal standard:** triazolam

## OTHER SUBSTANCES
**Simultaneous:** clindamycin B, diazepam, mezlocillin, oxazepam, phenobarbital
**Noninterfering:** cefoperazone, cefotaxime, cephalothin, ticarcillin

## KEY WORDS
plasma; serum; human; rabbit; dog; pig

## REFERENCE
La Follette, G.; Gambertoglio, J.; White, J.A.; Knuth, D.W.; Lin, E.T. Determination of clindamycin in
plasma or serum by high-performance liquid chromatography with ultraviolet detection.
*J.Chromatogr.*, **1988**, *431*, 379–388

## SAMPLE
**Matrix:** bulk, formulations
**Sample preparation:** Capsules. Extract capsule contents with 0.5% phenethyl alcohol in
mobile phase for 30 min, filter, inject an aliquot. Syrup. Measure out an amount of syrup
containing 50 mg clindamycin, mix with 5 mL 0.5% phenethyl alcohol in mobile phase,
inject an aliquot. Bulk. Dissolve 15 mg drug in 1 mL 0.5% phenethyl alcohol in mobile
phase, inject a 25 µL aliquot.

## HPLC VARIABLES
**Column:** 300 × 3.9 µBondapak C18
**Mobile phase:** MeOH:water:glacial acetic acid 60:40:0.2 containing 5 mM D, L-10-sodium
camphor sulfonate, adjusted to pH 6.0 (RI detection) or MeOH:water 60:40 containing
10 mM phosphate buffer and 5 mM sodium pentanesulfonate, pH 6 (UV detection)
**Flow rate:** 1

**Injection volume:** 25
**Detector:** RI; UV 214

## CHROMATOGRAM
**Retention time:** 11.3
**Internal standard:** phenethyl alcohol (5.9)

## OTHER SUBSTANCES
**Simultaneous:** clindamycin B, 7-epiclindamycin
**Noninterfering:** lincomycin

## KEY WORDS
capsules; syrup

## REFERENCE
Landis, J.B.; Grant, M.E.; Nelson, S.A. Determination of clindamycin in pharmaceuticals by high-performance liquid chromatography using ion-pair formation. *J.Chromatogr.*, **1980**, *202*, 99–106

## SAMPLE
**Matrix:** formulations
**Sample preparation:** Dilute 1:10 with 5% dextrose, remove a 900 μL aliquot and add it to 100 μL 1 mg/mL IS in MeOH, mix, inject a 20 μL aliquot.

## HPLC VARIABLES
**Column:** 250 × 4.6 Nucleosil C18
**Mobile phase:** MeOH:10 mM pH 6.3 phosphate buffer 40:60
**Flow rate:** 1.5
**Injection volume:** 20
**Detector:** UV 214

## CHROMATOGRAM
**Retention time:** 7.5
**Internal standard:** 2-nitrobenzenesulfonamide

## OTHER SUBSTANCES
**Simultaneous:** degradation products

## KEY WORDS
injections; 5% dextrose; stability-indicating

## REFERENCE
Sarkar, M.A.; Rogers, E.; Reinhard, M.; Wells, B.; Karnes, H.T. Stability of clindamycin phosphate, ranitidine hydrochloride, and piperacillin sodium in polyolefin containers. *Am.J.Hosp.Pharm.*, **1991**, *48*, 2184–2186

## SAMPLE
**Matrix:** formulations
**Sample preparation:** 100 μL solution + 2 mL 20 μg/mL propyl paraben in water, mix 15 s, inject 10 μL aliquot

## HPLC VARIABLES
**Column:** 300 × 4.1 10 μm Versapak C18
**Mobile phase:** MeOH:10 mM phosphate buffer + 5 mM pentanesulfonic acid 50:50, final pH 6.3
**Flow rate:** 1.2

**Injection volume:** 10
**Detector:** UV 214

## CHROMATOGRAM
**Retention time:** 9
**Internal standard:** propyl paraben

## OTHER SUBSTANCES
**Noninterfering:** degradation products, aztreonam, ceftazidime, ceftriaxone, piperacillin

## KEY WORDS
injections; 5% dextrose; saline

## REFERENCE
Marble, D.A.; Bosso, J.A.; Townsend, R.J. Stability of clindamycin phosphate with aztreonam, ceftazidime sodium, ceftriaxone sodium, or piperacillin sodium in two intravenous solutions. *Am.J.Hosp.Pharm.*, **1986**, *43*, 1732–1736

## SAMPLE
**Matrix:** formulations
**Sample preparation:** Dilute 100 μL with 2 mL 20 μg/mL propyl paraben in water, vortex for 15 s, inject a 10 μL aliquot.

## HPLC VARIABLES
**Column:** 250 × 4.1 10 μm Versapak C18 (Alltech)
**Mobile phase:** MeOH:10 mM phosphate buffer containing 5 mM sodium pentanesulfonate 50:50, pH 6.3
**Flow rate:** 1.2
**Injection volume:** 10
**Detector:** UV 214

## CHROMATOGRAM
**Retention time:** 9
**Internal standard:** propyl paraben (15)

## KEY WORDS
injections; saline; 5% dextrose

## REFERENCE
Bosso, J.A.; Townsend, R.J. Stability of clindamycin phosphate and ceftizoxime sodium, cefoxitin sodium, cefamandole nafate, or cefazolin sodium in two intravenous solutions. *Am.J.Hosp.Pharm.*, **1985**, *42*, 2211–2214

## SAMPLE
**Matrix:** solutions
**Sample preparation:** Dilute solution 10-fold with mobile phase. Mix 100 μL + 100 μL 50 μg/mL propyl paraben, inject 10 μL.

## HPLC VARIABLES
**Column:** 300 × 3.9 10 μm μBondapak C18
**Mobile phase:** MeOH:10 mM $KH_2PO_4$ + 5 mM 1-pentanesulfonic acid 55:45
**Flow rate:** 1.2
**Injection volume:** 10
**Detector:** UV 214

## CHROMATOGRAM
**Retention time:** 8.1
**Internal standard:** propyl paraben

## KEY WORDS
water; stability-indicating

## REFERENCE
Mahata, M.C.; Morosco, R.S.; Hipple, T.F. Stability of cimetidine hydrochloride and of clindamycin phosphate in water for injection stored in glass vials at two temperatures. *Am.J.Hosp.Pharm.*, **1993**, *50*, 2559–2561

## SAMPLE
**Matrix:** solutions
**Sample preparation:** Centrifuge and filter cell solutions (0.22 µm), inject an aliquot.

## HPLC VARIABLES
**Guard column:** Guard-PAK C18 (Waters)
**Column:** 150 × 3.9 5 µm NOVA PAK C18
**Mobile phase:** MeCN:50 mM pH 6.0 $KH_2PO_4$ 35:65
**Flow rate:** 1
**Detector:** UV 214

## CHROMATOGRAM
**Retention time:** 3.8

## REFERENCE
Koga, H. High-performance liquid chromatography measurement of antimicrobial concentrations in polymorphonuclear leukocytes. *Antimicrob.Agents Chemother.*, **1987**, *31*, 1904–1908

## SAMPLE
**Matrix:** solutions
**Sample preparation:** Prepare a 400 µg/mL solution of clindamycin in mobile phase, inject a 25 µL aliquot.

## HPLC VARIABLES
**Column:** 250 × 4.6 Zorbax C8
**Mobile phase:** MeCN:water 12:88 containing 0.25 g/L tetrabutylammonium perchlorate and 2 mL/L 70% perchloric acid, apparent pH adjusted to 2.5 with 50% NaOH
**Flow rate:** 1.5
**Injection volume:** 25
**Detector:** UV 214

## CHROMATOGRAM
**Retention time:** 22.5

## OTHER SUBSTANCES
**Simultaneous:** benzyl alcohol, lincomycin, pirlimycin

## REFERENCE
Theis, D.L. Ion-pairing liquid chromatographic method for the determination of pirlimycin hydrochloride. *J.Chromatogr.*, **1987**, *402*, 335–343

# Clonazepam

**Molecular formula:** $C_{15}H_{10}ClN_3O_3$
**Molecular weight:** 315.7
**CAS Registry No.:** 1622-61-3

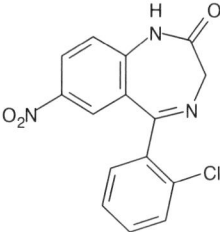

## SAMPLE
**Matrix:** blood
**Sample preparation:** 500 µL Whole blood + 15 µL 10 µg/mL demoxepam, mix, add 500 µL 200 mM pH 11.5 carbonate buffer, mix, add 6 mL butyl chloride, rotate gently for 20 min, centrifuge at 2500 g for 10 min. Remove the organic layer and evaporate it to dryness, reconstitute the residue in 200 µL mobile phase, inject a 50 µL aliquot.

## HPLC VARIABLES
**Column:** 150 × 3.9 4 µm Nova-Pak phenyl
**Mobile phase:** MeCN:40 mM $KH_2PO_4$ 28:72, pH 3.75
**Flow rate:** 0.8
**Injection volume:** 50
**Detector:** UV 240

## CHROMATOGRAM
**Retention time:** 16
**Internal standard:** demoxepam (8)
**Limit of detection:** 6 ng/mL;
**Limit of quantitation:** 10 ng/mL

## OTHER SUBSTANCES
**Extracted:** 7-aminoclonazepam, flunitrazepam, nitrazepam
**Simultaneous:** alprazolam, bromazepam, carbamazepine, carbamazepine epoxide, desalkylflurazepam, desmethyldiazepam, diazepam, flurazepam, lorazepam, moclobemide, naproxen, oxazepam, phenobarbital, phenytoin, quinidine, quinine, salicylic acid, temazepam, triazolam
**Noninterfering:** acetaminophen, theophylline

## KEY WORDS
whole blood

## REFERENCE
Robertson, M.D.; Drummer, O.H. High-performance liquid chromatographic procedure for the measurement of nitrobenzodiazepines and their 7-amino metabolites in blood. *J.Chromatogr.B*, **1995**, *667*, 179–184

## SAMPLE
**Matrix:** blood
**Sample preparation:** 2 mL Whole blood or plasma + 2 mL buffer + 5 mL chloroform:isopropanol:n-heptane 60:14:26, shake gently horizontally for 10 min, centrifuge at 2800 g for 10 min. Remove the lower organic layer and evaporate it to dryness under vacuum at 45°, reconstitute the residue in 100 µL mobile phase, centrifuge at 2800 g for 5 min, inject a 50 µL aliquot of the supernatant. (Buffer was saturated ammonium chloride solution 25% diluted with water, adjusted to pH 9.5 with 25% ammonia solution.)

## HPLC VARIABLES
**Column:** 300 × 3.9 4 μm NovaPack C18
**Mobile phase:** MeOH:THF:buffer 65:5:30 (Buffer was 0.68 g/L (10 mM (sic)) $KH_2PO_4$ adjusted to pH 2.6 with concentrated orthophosphoric acid.) (At the end of each session wash the column with water for 1 h and MeOH for 1 h, re-equilibrate for 30 min.)
**Column temperature:** 30
**Flow rate:** 0.8
**Injection volume:** 50
**Detector:** UV 246

## CHROMATOGRAM
**Retention time:** 3.92
**Limit of detection:** <120 ng/mL

## OTHER SUBSTANCES
**Extracted:** acebutolol, acenocoumarol, acepromazine, aceprometazine, acetaminophen, aconitine, ajmaline, albuterol, alimemazine, alminoprofen, alpidem, alprazolam, alprenolol, amisulpride, amitriptyline, amodiaquine, amoxapine, aspirin, astemizole, atenolol, benazepril, benperidol, benzocaine, benzoylecgonine, bepridil, betaxolol, bisoprolol, bromazepam, brompheniramine, bumadizone, bupivacaine, buprenorphine, buspirone, caffeine, carbamazepine, carbinoxamine, carpipramine, carteolol, celiprolol, cetirizine, chlorambucil, chlordiazepoxide, chlormezanone, chlorophenacinone, chloroquine, chlorpheniramine, chlorpromazine, cibenzoline, cicletanine, clemastine, clomipramine, clorazepate, clozapine, cocaine, codeine, colchicine, cyamemazine, cyclizine, cycloguanil, cyproheptadine, cytarabine, dacarbazine, daunorubicin, debrisoquine, demexiptiline, desipramine, dextromethorphan, dextromoramide, dextropropoxyphene, diazepam, diazoxide, diclofenac, dihydralazine, diltiazem, diphenhydramine, dipyridamole, dosulepine, doxepin, doxylamine, droperidol, etodolac, fenfluramine, fenoprofen, fentiazac, flecainide, floctafenine, flumazenil, fluoxetine, fluphenazine, flurbiprofen, fluvoxamine, glibenclamide, glibornuride, haloperidol, histapyrrodine, hydroxychloroquine, hydroxyzine, ibuprofen, imipramine, indomethacin, iproniazid, ketamine, ketoprofen, labetalol, levomepromazine, lidocaine, lidoflazine, lisinopril, loperamide, loprazolam, loratadine, lorazepam, loxapine, maprotiline, medazepam, medifoxamine, mefenamic acid, mefenidramine, mefloquine, meperidine, mephenesin, mephentermine, mepivacaine, metapramine, metformin, methadone, methaqualone, methocarbamol, methotrexate, metipranolol, metoprolol, mexiletine, mianserine, midazolam, moclobemide, moperone, morphine, nadolol, nalbuphine, nalorphine, naloxone, naltrexone, naproxen, nialamide, nicardipine, nifedipine, niflumic acid, nimodipine, nitrazepam, nitrendipine, nizatidine, nomifensine, nortriptyline, omeprazole, opipramol, oxazepam, oxprenolol, penbutolol, penfluridol, pentazocine, phencyclidine, phenobarbital, phenol, phenylbutazone, pimozide, pipamperone, piroxicam, prazepam, prilocaine, procainamide, procarbazine, proguanil, promethazine, propafenone, propranolol, protriptyline, pyrimethamine, quinidine, quinine, quinupramine, ramipril, ranitidine, reserpine, ritodrine, secobarbital, sotalol, strychnine, sulfinpyrazone, sulindac, sulpride, sultopride, suriclone, temazepam, tenoxicam, terbutaline, terfenadine, tetracaine, tetrazepam, thiopental, thioproperazine, thioridazine, tianeptine, tiapride, tiaprofenic acid, ticlopidine, timolol, tioclomarol, tofisopam, toloxatone, trazodone, trifluoperazine, trifluperidol, trimipramine, triprolidine, tropatenine, verapamil, viloxazine, vinblastine, vincristine, vindesine, warfarin, yohimbine, zolpidem, zopiclone, zorubicine
**Interfering:** chlorpropamide, clobazam, clonidine, disopyramide, ephedrine, estazolam, flunitrazepam, glipizide, melphalan, metoclopramide, minoxidil, pindolol, prazosin, tolbutamide, triazolam

## KEY WORDS
whole blood; plasma

## REFERENCE
Tracqui, A.; Kintz, P.; Mangin, P. Systematic toxicological analysis using HPLC/DAD. *J.Forensic Sci.*, **1995**, *40*, 254−262

## SAMPLE
**Matrix:** blood
**Sample preparation:** Rock 5 mL whole blood + 10 mL water + 8.5 mL Na$_2$WO$_4$ in a 50 mL stoppered tube for 1 min, add 6 mL NiCl$_2$, rock for 5 min, add 15 mL 1-chlorobutane:isobutyl alcohol:THF 40:40:20, centrifuge at 2500 g for 15 min. Remove organic phase and repeat the process. Filter all organic phases through a 40-90 µm filter and evaporate to dryness in a 100 mL porcelain dish at a moderate temperature in a sand bath. Take up residue in 500 µL MeCN:water 80:20, inject a 20 µL aliquot. (Na$_2$WO$_4$ prepared by mixing 10 g Na$_2$WO$_4$.2H$_2$O in 38 mL of 2 M NaOH and 2.5 g of NaHCO$_3$ and making up to 100 mL. NiCl$_2$ was 17% w/v NiCl$_2$ in water.)

## HPLC VARIABLES
**Column:** 200 × 4.6 5 µm Hypersil C8
**Mobile phase:** Gradient. A was MeCN. B was 20 mM n-hexylamine adjusted to pH 4 with 85% phosphoric acid. A:B from 25:75 to 40:60 over 25 min to 50:50 over another 5 min
**Injection volume:** 20
**Detector:** UV 230

## CHROMATOGRAM
**Retention time:** 14
**Limit of detection:** 0.30 ppm

## OTHER SUBSTANCES
**Extracted:** bromazepam, diazepam, flunitrazepam, flurazepam, medazepam, nitrazepam, oxazepam
**Also analyzed:** buprenorphine, caffeine, cocaine, codeine, diamorphine, ethylmorphine, lidocaine, methaqualone, morphine, naloxone, noscapine, papaverine, pentazocine, procaine

## KEY WORDS
whole blood

## REFERENCE
Bernal, J.L.; Del Nozal, M.J.; Rosas, V.; Villarino, A. Extraction of basic drugs from whole blood and determination by high performance liquid chromatography. *Chromatographia*, **1994**, *38*, 617–623

## SAMPLE
**Matrix:** blood
**Sample preparation:** Condition a 12 mL 500 mg PrepSep C1 SPE cartridge with 3 mL MeOH and 3 mL water. Add 1 mL plasma to SPE cartridge, wash with two 3 mL portions of water, wash with two 1 mL portions of MeOH:water 30:70, elute with two 1 mL portions of MeOH:50 mM pH 9.0 (NH$_4$)H$_2$PO$_4$ 90:10, evaporate the eluents under vacuum, dissolve the residue in 200 µL mobile phase, inject a 100 µL aliquot.

## HPLC VARIABLES
**Column:** 100 × 4.6 5 µm Spherisorb C8
**Mobile phase:** MeCN:MeOH:20 mM (NH$_4$)H$_2$PO$_4$ 5:35:60 containing 2 mL/L 200 mM tetrabutylammonium bromide, final pH adjusted to 4.10
**Column temperature:** 30
**Flow rate:** 1.5
**Injection volume:** 100
**Detector:** UV 254

## CHROMATOGRAM
**Retention time:** 12.4
**Internal standard:** clonazepam

## OTHER SUBSTANCES
**Extracted:** midazolam

## KEY WORDS
plasma; clonazepam is IS; SPE

## REFERENCE
Mastey, V.; Panneton, A.-C.; Donati, F.; Varin, F. Determination of midazolam and two of its metabolites in human plasma by high-performance liquid chromatography. *J.Chromatogr.B*, **1994**, *655*, 305–310

## SAMPLE
**Matrix:** blood
**Sample preparation:** Inject 100-200 μL plasma onto column A with mobile phase A and elute to waste, after 5 min backflush the contents of column A onto column B with mobile phase B, after 5 min remove column A from the circuit, elute column B with mobile phase B, monitor the effluent from column B. Wash column A with MeCN:water 60:40 at 1 mL/min for 6 min then re-equilibrate with pH 7.5 buffer for 10 min.

## HPLC VARIABLES
**Column:** A 45 × 4 12 μm TSK-gel G 3 PW (Tosohass); B 75 × 4.6 Ultrasphere ODS C18 3 μm
**Mobile phase:** A 50 mM pH 7.5 phosphate buffer; B Gradient. A was MeCN. B was 65 mM $KH_2PO_4$ + 1% diethylamine adjusted to pH 5.4 with phosphoric acid. A:B 22:78 for 5 min, to 25:75 over 10 min, to 60:40 over 15 min.
**Flow rate:** 1
**Injection volume:** 100-200
**Detector:** UV 230

## CHROMATOGRAM
**Retention time:** 22

## OTHER SUBSTANCES
**Extracted:** alprazolam, bromazepam, chlordiazepoxide, clobazam, clorazepate, clotiazepam, desmethylclobazam, desmethyldiazepam, diazepam, estazolam, flunitrazepam, loflazepate, medazepam, nitrazepam, oxazepam, prazepam, temazepam, tetrazepam, tofisopam, triazolam
**Noninterfering:** carbamazepine, ethosuximide, phenobarbital, phenytoin, primidone, valproic acid
**Interfering:** lorazepam

## KEY WORDS
plasma; column-switching

## REFERENCE
Lacroix, C.; Wojciechowski, F.; Danger, P. Monitoring of benzodiazepines (clobazam, diazepam and their main active metabolites) in human plasma by column-switching high- performance liquid chromatography. *J.Chromatogr.*, **1993**, *617*, 285–290

## SAMPLE
**Matrix:** blood
**Sample preparation:** 1 mL Plasma + 30 μL 0.5 μg/mL flunitrazepam in MeOH + 50 μL 1 M ammonium hydroxide, vortex, add 8 mL diethyl ether, mix 2 min, remove organic phase and dry with 0.5 g anhydrous sodium sulfate. Add a few grains of sodium chloride to the ether to prevent bumping and evaporate to dryness in a water bath at 55°. Dissolve residue in 25 μL MeOH and inject a 20 μL aliquot.

## HPLC VARIABLES
**Column:** 30 × 4.6 3 μm Perkin-Elmer C18
**Mobile phase:** MeCN:water 25:75
**Flow rate:** 2.5
**Injection volume:** 20
**Detector:** UV 306

## CHROMATOGRAM
**Retention time:** 2.4
**Internal standard:** flunitrazepam
**Limit of detection:** 3 ng/mL

## OTHER SUBSTANCES
**Simultaneous:** clobazam, diazepam
**Noninterfering:** caffeine, carbamazepine, esobarbital, ethosuximide, flurazepam, medazepam, nitrazepam, oxazepam, phenobarbital, phenytoin, primidone, theobromine, theophylline
**Interfering:** desmethylclobazam, lorazepam

## KEY WORDS
plasma

## REFERENCE
Valenza, T.; Rosselli, P. Rapid and specific high-performance liquid chromatographic determination of clonazepam in plasma. *J.Chromatogr.*, **1987**, *386*, 363–366

## SAMPLE
**Matrix:** blood
**Sample preparation:** 1 mL Serum + 1 mL saturated sodium tetraborate + 100 ng flunitrazepam + 7.5 mL n-hexane:ethyl acetate 9:1, vortex 5 min, centrifuge 1000 g. Repeat extraction. Combine extracts, evaporate to dryness, dissolve in 120 μL mobile phase, inject 100 μL.

## HPLC VARIABLES
**Guard column:** 20 mm Supelco 40 μm Pelliguard LC-8
**Column:** 150 × 4.6 5 μm Supelco reversed phase C8
**Mobile phase:** MeCN:1.75 mM HCl:50 mM sodium acetate 36:10:54
**Flow rate:** 1.5
**Injection volume:** 100
**Detector:** UV 220

## CHROMATOGRAM
**Retention time:** 5
**Internal standard:** flunitrazepam
**Limit of detection:** 1 ng/mL

## OTHER SUBSTANCES
**Simultaneous:** clobazam, nitrazepam

## KEY WORDS
serum

## REFERENCE
Zilli, M.A.; Nisi, G. Simple and sensitive method for the determination of clobazam, clonazepam and nitrazepam in human serum by high-performance liquid chromatography. *J.Chromatogr.*, **1986**, *378*, 492–497

## SAMPLE
**Matrix:** blood
**Sample preparation:** Prepare a C18 Bond-Elut column by rinsing with 2 column volumes of MeOH and 2 column volumes of water using vacuum. Place 100 μL IS solution then 1 mL serum on column, wash column with 2 column volumes of water, wash with 50 μL MeOH, elute with two 200 μL portions of MeOH. Combine eluents, evaporate to dryness under a stream of air at 45°, reconstitute with 40 μL MeOH, inject whole sample. (IS was 50 ng/mL methylclonazepam in 1 M glycine buffer (pH adjusted to 10.5 with NaOH).)

## HPLC VARIABLES
**Column:** 150 × 4.6 5 μm Ultrasphere ODS C18
**Mobile phase:** MeCN:20 mM pH 3.8 phosphate buffer 30:70
**Column temperature:** 50
**Flow rate:** 2
**Injection volume:** 40
**Detector:** UV 254

## CHROMATOGRAM
**Retention time:** 6
**Internal standard:** methylclonazepam
**Limit of detection:** 2 ng/mL
**Limit of quantitation:** 5 ng/mL

## OTHER SUBSTANCES
**Simultaneous:** amitriptyline, amobarbital, butalbital, carbamazepine, chlordiazepoxide, chlorpromazine, clonazepam, demoxepam, desipramine, diazepam, ethinamate, flurazepam, glutethimide, heptobarbital, hexobarbital, lidocaine, medazepam, meperidine, mephobarbital, mesoridazine, methaqualone, pentathal, nortriptyline, oxazepam, pentobarbital, perphenazine, phenytoin, promazine, propranolol, protriptyline, quinidine, secobarbital
**Noninterfering:** aprobarbital, barbituric acid, ethinamate, ethosuximide, gentamicin, lidocaine, mebutamate, meprobamate, methadone, methyprylon, nirvanol, phenobarbital, procainamide, propoxyphene, thioamyl, thioridazine, trifluoperazine, triflupromazine, tybamate, vinbarbital

## KEY WORDS
serum

## REFERENCE
Kabra, P.M.; Nzekwe, E.U. Liquid chromatographic analysis of clonazepam in human serum with solid-phase (Bond-Elut) extraction. *J.Chromatogr.*, **1985**, *341*, 383–390

## SAMPLE
**Matrix:** blood, gastric contents, tissue, urine
**Sample preparation:** 1 mL Blood, urine, or gastric contents or 1 g tissue homogenate + 500 μL buffer + 8 mL n-hexane:ethyl acetate 70:30, mix on a rotary mixer for 10 min, centrifuge at 3000 g for 8 min. Remove the organic layer and evaporate it to dryness under a stream of nitrogen, reconstitute the residue in 100 μL 12.5 mM NaOH in MeOH:water 50:50, inject a 50 μL aliquot. (Buffer was 13.8 g potassium carbonate in 100 mL water, pH adjusted to 9.5 with concentrated HCl.)

## HPLC VARIABLES
**Guard column:** 4 × 4 30 μm LiChrocart Aluspher RP-select B (Merck)
**Column:** 125 × 4 5 μm Aluspher RP-select B (Merck)
**Mobile phase:** Gradient. A was 12.5 mM NaOH in MeOH. B was 12.5 mM NaOH in water. A:B 10:90 for 5 min, to 90:10 over 15 min, maintain at 90:10 for 5 min, return to initial conditions over 1 min, re-equilibrate for 5 min.

**Flow rate:** 1
**Injection volume:** 50
**Detector:** UV 230; UV 254

## CHROMATOGRAM
**Retention time:** 7

## OTHER SUBSTANCES
**Extracted:** alprenolol, amitriptyline, bromazepam, carbamazepine, chlordiazepoxide, chlorpromazine, desipramine, diazepam, flunitrazepam, haloperidol, nitrendipine, nordiazepam, nortriptyline, pindolol, zolpidem
**Also analyzed:** acebutolol, acetaminophen, alprazolam, amphetamine, atenolol, betaxolol, brotizolam, caffeine, camazepam, captopril, chloroquine, clobazam, clomipramine, clothiapine, clotiazepam, cloxazolam, cocaine, codeine, diclofenac, dihydralazine, dihydrocodeine, dihydroergotamine, diphenhydramine, domperidone, doxepin, droperidol, ergotamine, ethyl loflazepate, fenethylline, fluoxetine, flupentixol, flurazepam, furosemide, gliclazide, hydrochlorothiazide, hydroxyzine, ibuprofen, imipramine, ketazolam, loprazolam, lorazepam, lormetazepam, maprotiline, medazepam, mepyramine, methadone, methaqualone, methyldopa, methylphenidate, metoclopramide, metoprolol, mexiletine, mianserin, midazolam, minoxidil, morphine, nadolol, nitrazepam, oxprenolol, papaverine, pentazocine, phenprocoumon, phenylbutazone, pipamperone, piritramide, practolol, prazepam, prazosin, promazine, promethazine, propoxyphene, propranolol, prothipendyl, quinine, sotalol, sulpride, thioridazine, trazodone, triazolam, trimipramine, tripelennamine, tyramine, verapamil, yohimbine

## REFERENCE
Lambert, W.E.; Meyer, E.; De Leenheer, A.P. Systematic toxicological analysis of basic drugs by gradient elution of an alumina-based HPLC packing material under alkaline conditions. *J.Anal.Toxicol.*, **1995**, *19*, 73–78

## SAMPLE
**Matrix:** blood, urine
**Sample preparation:** Prepare urine samples by adjusting pH to 5.0 with 2 M HCl then hydrolyzing with 20 µL 20 000 U/mL β-D-glucuronidase at 37° for 5 h. 500 µL Plasma or urine + 15 µL 2.5 mg/L desmethylflunitrazepam in MeOH + 50 µL 0.5 M NaOH + 5 mL diethyl ether, agitate, centrifuge, evaporate to dryness at 45° under vacuum, take up residue in 100 µL MeOH, inject 25 µL aliquot.

## HPLC VARIABLES
**Column:** $150 \times 4.6$ 5 µm Nova Pak C18
**Mobile phase:** MeCN:MeOH:buffer 30:10:60 adjusted to pH 5.7 with 0.1 M HCl (Buffer was 94 mL 0.2 M $NaH_2PO_4$ + 6 mL 0.2 M $Na_2HPO_4$.)
**Flow rate:** 1.3
**Injection volume:** 25
**Detector:** UV 242

## CHROMATOGRAM
**Retention time:** 4.43
**Internal standard:** desmethylflunitrazepam (3.65)
**Limit of detection:** 3 ng/mL

## OTHER SUBSTANCES
**Simultaneous:** alprazoplam, bromazepam, chlordiazepoxide, clobazam, diazepam, estazolam, flunitrazepam, lorazepam, medazepam, triazolam
**Noninterfering:** acepromazine, aceprometazine, aprobarbital, barbital, butabarbital, clothiazepam, heptabarbital, hexobarbital, loprazolam, prazepam
**Interfering:** nitrazepam, oxazepam

## KEY WORDS
plasma

## REFERENCE
Boukhabza, A.; Lugnier, A.A.; Kintz, P.; Mangin, P.; Chaumont, A.J. Simple and sensitive method for
monitoring clonazepam in human plasma and urine by high-performance liquid chromatography.
*J.Chromatogr.*, **1990**, *529*, 210–216

## SAMPLE
**Matrix:** solutions
**Sample preparation:** Inject a 30 µL aliquot of a solution in MeOH.

## HPLC VARIABLES
**Column:** 250 × 4.6 Spherisorb S5SCX in a PEEK column
**Mobile phase:** MeOH:water:60% perchloric acid 97.5:1.75:0.75
**Flow rate:** 1
**Injection volume:** 30
**Detector:** UV 220

## CHROMATOGRAM
**Retention time:** 6

## OTHER SUBSTANCES
**Simultaneous:** diazepam, dothiepin, dothiepin sulfoxide, nordiazepam, nordothiepin, nordothiepin sulfoxide

## REFERENCE
Croes, K.; McCarthy, P.T.; Flanagan, R.J. HPLC of basic drugs and quaternary ammonium compounds
on microparticulate strong cation-exchange materials using methanolic or aqueous methanol eluents
containing an ionic modifier. *J.Chromatogr.A*, **1995**, *693*, 289–306

## SAMPLE
**Matrix:** solutions

## HPLC VARIABLES
**Column:** 250 × 4.6 5 µm Supelcosil LC-DP (A) or 250 × 4 5 µm LiChrospher 100 RP-8 (B)
**Mobile phase:** MeCN:0.025% phosphoric acid:buffer 25:10:5 (A) or 60:25:15 (B) (Buffer
was 9 mL concentrated phosphoric acid and 10 mL triethylamine in 900 mL water, adjust
pH to 3.4 with dilute phosphoric acid, make up to 1 L.)
**Flow rate:** 0.6
**Injection volume:** 25
**Detector:** UV 229

## CHROMATOGRAM
**Retention time:** 6.45 (A), 6.02 (B)

## OTHER SUBSTANCES
**Also analyzed:** acebutolol, acepromazine, acetaminophen, acetazolamide, acetophenazine,
albuterol, alprazolam, amitriptyline, amobarbital, amoxapine, antipyrine, atenolol, atropine, azatadine, baclofen, benzocaine, bromocriptine, brompheniramine, brotizolam,
bupivacaine, buspirone, butabarbital, butalbital, caffeine, carbamazepine, cetirizine,
chlorcyclizine, chlordiazepoxide, chlormezanone, chloroquine, chlorpheniramine,
chlorpromazine, chlorpropamide, chlorprothixene, chlorthalidone, chlorzoxazone, cimetidine, cisapride, clomipramine, clonidine, clozapine, cocaine, codeine, colchicine, cyclizine,
cyclobenzaprine, dantrolene, desipramine, diazepam, diclofenac, diflunisal, diltiazem, di-

phenhydramine, diphenidol, diphenoxylate, dipyridamole, disopyramide, dobutamine, doxapram, doxepin, droperidol, encainide, ethidium bromide, ethopropazine, fenoprofen, fentanyl, flavoxate, fluoxetine, fluphenazine, flurazepam, flurbiprofen, fluvoxamine, furosemide, glutethimide, glyburide, guaifenesin, haloperidol, homatropine, hydralazine, hydrochlorothiazide, hydrocodone, hydromorphone, hydroxychloroquine, hydroxyzine, ibuprofen, imipramine, indomethacin, ketoconazole, ketoprofen, ketorolac, labetalol, levorphanol, lidocaine, loratadine, lorazepam, lovastatin, loxapine, mazindol, mefenamic acid, meperidine, mephenytoin, mepivacaine, mesoridazine, metaproterenol, methadone, methdilazine, methocarbamol, methotrexate, methotrimeprazine, methoxamine, methyldopa, methylphenidate, metoclopramide, metolazone, metoprolol, metronidazole, midazolam, moclobemide, morphine, nadolol, nalbuphine, naloxone, naphazoline, naproxen, nifedipine, nizatidine, norepinephrine, nortriptyline, oxazepam, oxycodone, oxymetazoline, paroxetine, pemoline, pentazocine, pentobarbital, pentoxifylline, perphenazine, pheniramine, phenobarbital, phenol, phenolphthalein, phentolamine, phenylbutazone, phenyltoloxamine, phenytoin, pimozide, pindolol, piroxicam, pramoxine, prazepam, prazosin, probenecid, procainamide, procaine, prochlorperazine, procyclidine, promazine, promethazine, propafenone, propantheline, propiomazine, propofol, propranolol, protriptyline, quazepam, quinidine, quinine, racemethorphan, ranitidine, remoxipride, risperidone, salicylic acid, scopolamine, secobarbital, sertraline, sotalol, spironolactone, sulfinpyrazone, sulindac, temazepam, terbutaline, terfenadine, tetracaine, theophylline, thiethylperazine, thiopental, thioridazine, thiothixene, timolol, tocainide, tolbutamide, tolmetin, trazodone, triamterene, triazolam, trifluoperazine, triflupromazine, trimeprazine, trimethoprim, trimipramine, verapamil, warfarin, xylometazoline, yohimbine, zopiclone

## KEY WORDS
some details of plasma extraction

## REFERENCE
Koves, E.M. Use of high-performance liquid chromatography-diode array detection in forensic toxicology. *J.Chromatogr.A*, **1995**, *692*, 103–119

## SAMPLE
**Matrix:** solutions
**Sample preparation:** Prepare a 1-10 μg/mL solution in water, inject an aliquot.

## HPLC VARIABLES
**Column:** 250 × 4.6 5 μm Hypersil SCX/C18
**Mobile phase:** MeCN : 25 mM pH 3 $Na_2HPO_4$ 50 : 50
**Injection volume:** 20
**Detector:** UV 254

## CHROMATOGRAM
**Retention time:** k' 1.35

## OTHER SUBSTANCES
**Also analyzed:** amitriptyline, barbital, benzoic acid, butabarbital, clomipramine, desipramine, diazepam, flurazepam, furosemide, imipramine, nitrazepam, phenobarbital, phenol, phenolphthalein, pindolol, propranolol, resorcinol, salicylic acid, secobarbital, terbutaline, xylazine

## KEY WORDS
effect of mobile phase pH on capacity factor is discussed

## REFERENCE
Walshe, M.; Kelly, M.T.; Smyth, M.R.; Ritchie, H. Retention studies on mixed-mode columns in high-performance liquid chromatography. *J.Chromatogr.A*, **1995**, *708*, 31–40

## SAMPLE
**Matrix:** solutions

## HPLC VARIABLES
**Guard column:** 30 × 3.2 7 μm SI 100 ODS (not commercially available)
**Column:** 150 × 3.2 7 μm SI 100 ODS (not commercially available)
**Mobile phase:** MeCN:buffer 31.2:68.8 (Buffer was 6.66 g $KH_2PO_4$ and 4.8 g 85% phosphoric acid in 1 L water, pH 2.3.)
**Flow rate:** 0.5-1
**Detector:** UV 210; UV 245

## CHROMATOGRAM
**Retention time:** 6.6
**Internal standard:** 5-(4-methylphenyl)-5-phenylhydantoin (7.3)

## OTHER SUBSTANCES
**Also analyzed:** aspirin, carbamazepine, chlordiazepoxide, chlorprothixene, caffeine, diazepam, doxylamine, ethosuximide, furosemide, haloperidol, hydrochlorothiazide, methocarbamol, methotrimeprazine, nicotine, oxazepam, procaine, promazine, propafenone, propranolol, salicylamide, temazepam, tetracaine, thiopental, triamterene, verapamil, zolpidem, zopiclone

## REFERENCE
Below, E.; Burrmann, M. Application of HPLC equipment with rapid scan detection to the identification of drugs in toxicological analysis. *J.Liq.Chromatogr.*, **1994**, *17*, 4131–4144

## SAMPLE
**Matrix:** solutions

## HPLC VARIABLES
**Column:** 250 × 4.6 Zorbax RX
**Mobile phase:** Gradient. A was 10 mL concentrated orthophosphoric acid and 7 mL triethylamine in 1 L water. B was 10 mL concentrated orthophosphoric acid and 7 mL triethylamine in 200 mL water, make up to 1 L with MeCN. A:B from 100:0 to 0:100 over 30 min, maintain at 0:100 for 5 min.
**Column temperature:** 30
**Flow rate:** 2
**Detector:** UV 210

## OTHER SUBSTANCES
**Also analyzed:** acepromazine, acetaminophen, acetophenazine, albuterol, aminophylline, amitriptyline, amobarbital, amoxapine, amphetamine, amylocaine, antipyrine, aprobarbital, aspirin, atenolol, atropine, avermectin, barbital, benzocaine, benzoic acid, benzotropine, benzphetamine, berberine, bibucaine, bromazepan, brompheniramine, buprenorphine, buspirone, butabarbital, butacaine, butethal, caffeine, carbamazepine, carbromal, chloramphenicol, chlordiazepoxide, chloroquine, chlorothiazide, chloroxylenol, chlorphenesin, chlorpheniramine, chlorpromazine, chlorpropamide, chlortetracycline, cimetidine, cinchonidine, cinchonine, clenbuterol, clonixin, clorazepate, cocaine, codeine, colchicine, cortisone, coumarin, cyclazocine, cyclobenzaprine, cyclothiazide, cyheptamide, cymarin, danazol, danthron, dapsone, debrisoquine, desipramine, dexamethasone, dextromethorphan, dextropropoxyphene, diamorphine, diazepam, diclofenac, diethylpropion, diethylstilbestrol, diflunisal, digitoxin, digoxin, diltiazem, diphenhydramine, diphenoxylate, diprenorphine, dipyrone, disulfiram, dopamine, doxapram, doxepin, dronabinol, ephedrine, epinephrine, epinine, estradiol, estriol, estrone, ethacrynic acid, ethosuximide, etonitazene, etorphine, eugenol, famotidine, fenbendazole, fencamfamine, fenoprofen, fenproporex, fentanyl, flubendazole, flufenamic acid, flunitrazepam, 5-fluorouracil, fluoxy-

mesterone, fluphenazine, furosemide, gentisic acid, gitoxigenin, glipizide, glunixin, glutethimide, glybenclamide, guaiacol, halazepam, haloperidol, hydrochlorothiazide, hydrocodone, hydrocortisone, hydromorphone, hydroxyquinoline, ibogaine, ibuprofen, iminostilbene, imipramine, indomethacin, isocarbostyril, isocarboxazid, isoniazid, isoproterenol, isoxsuprine, ivermectin, ketamine, ketoprofen, kynurenic acid, levorphanol, lidocaine, lorazepam, lormetazepam, loxapine, mazindol, mebendazole, meclizine, meclofenamic acid, medazepam, mefenamic acid, megestrol, mepacrine, meperidine, mephentermine, mephenytoin, mephesin, mephobarbital, mepivacaine, mescaline, mesoridazine, methadone, methamphetamine, methapyrilene, methaqualone, methazolamide, methocarbamol, methoxamine, methsuximide, methyl salicylate, methyldopa, methyldopamine, methylphenidate, methylprednisolone, methyltestosterone, methyprylon, metoprolol, mibolerone, morphine, nadolol, nalorphine, naloxone, naltrexone, naphazoline, naproxen, nefopam, niacinamide, nicotine, nicotinic acid, nifedipine, niflumic acid, nitrazepam, norepinephrine, nortriptyline, noscapine, nylidrin, oxazepam, oxycodone, oxymorphone, oxyphenbutazone, oxytetracycline, papaverine, pargyline, pemoline, pentazocine, pentobarbital, persantine, phenacetin, phenazocine, phenazopyridine, phencyclidine, phendimetrazine, phenelzine, pheniramine, phenobarbital, phenothiazine, phensuximide, phentermine, phenylbutazone, phenylephrine, phenylpropanolamine, piperocaine, prazepam, prednisolone, primidone, probenecid, progesterone, propiomazine, propranolol, propylparaben, pseudoephedrine, puromycin, pyrilamine, pyrithyldione, quazepam, quinaldic acid, quinidine, quinine, ranitidine, recinnamine, reserpine, resorcinol, saccharin, albuterol, salicylamide, salicylic acid, scopolamine, scopoletin, secobarbital, strychnine, sulfacetamide, sufadiazine, sulfadimethoxine, sulfaethidole, sulfamerazine, sulfamethazine, sulfamethoxizole, sulfanilamide, sulfapyridine, sulfasoxizole, sulindac, tamoxifen, temazepam, testosterone, tetracaine, tetracycline, tetramisole, thebaine, theobromine, theophylline, thiabendazole, thiamine, thiamylal, thiobarbituric acid, thioridazine, thiosalicylic acid, thiothixene, thymol, tolazamide, tolazoline, tobutamide, tolmetin, tranylcypromine, triamcinolone, tribenzylamine, trichloromethiazide, trifluoperazine, trihexyphenidyl, trimethoprim, tripelennamine, triproilidine, tropacocaine, tyramine, verapamil, vincamine, warfarin, yohimbine, zoxazolamine

## REFERENCE
Hill, D.W.; Kind, A.J. Reversed-phase solvent gradient HPLC retention indexes of drugs. *J.Anal.Toxicol.*, **1994**, *18*, 233–242

## ANNOTATED BIBLIOGRAPHY
Sallustio, B.C.; Kassapidis, C.; Morris, R.G. High-performance liquid chromatography determination of clonazepam in plasma using solid-phase extraction. *Ther.Drug Monit.*, **1994**, *16*, 174–178 [SPE; plasma; methylclonazepam (IS); column temp 40 ; LOQ 5 ng/mL; LOD 2 ng/mL; non-interfering bromazepam, medazepam; simultaneous alprazolam, clobazam, chlordiazepoxide, diazepam, flunitrazepam, lorazepam, midazolam, nitrazepam, oxazepam, temazepam]

Bonato, P.S.; Lanchote, V.L. A rapid procedure for the purification of biological samples to be analysed by high performance liquid chromatography. *J.Liq.Chromatogr.*, **1993**, *16*, 2299–2308 [also albendazole, carbamazepine, clonazepam, desalkylflurazepam, mebendazole, methaqualone]

Furuno, K.; Gomita, Y.; Araki, Y.; Fukuda, T. Clonazepam serum levels in epileptic patients determined simply and rapidly by high-performance liquid chromatography using a solid-phase extraction column. *Acta Med.Okayama.*, **1991**, *45*, 123–127

Noggle, F.T., Jr.; Clark, C.R.; DeRuiter, J. Liquid chromatographic separation of some common benzodiazepines and their metabolites. *J.Liq.Chromatogr.*, **1990**, *13*, 4005–4021 [also alprazolam, bromazepam, chlordiazepoxide, clobazam, clorazepate, demoxepam, diazepam, estazolam, fludiazepam, flunitrazepam, flurazepam, halazepam, lorazepam, midazolam, nitrazepam, nordiazepam, oxazepam, prazepam, temazepam, triazolam]

Doran, T.C. Liquid chromatographic assay for serum clonazepam. *Ther.Drug Monit.*, **1988**, *10*, 474–479

Dusci, L.J.; Hackett, L.P. Simultaneous determination of clobazam, N-desmethyl clobazam and clonazepam in plasma by high performance liquid chromatography. *Ther.Drug Monit.*, **1987**, *9*, 113–116

Lin, W.N. Determination of clonazepam in serum by high performance liquid chromatography. *Ther.Drug Monit.*, **1987**, *9*, 337–342

Haver, V.M.; Porter, W.H.; Dorie, L.D.; Lea, J.R. Simplified high performance liquid chromatographic method for the determination of clonazepam and other benzodiazepines in serum. *Ther.Drug Monit.*, **1986**, *8*, 352–357

Wad, N. Degradation of clonazepam in serum by light confirmed by means of a high performance liquid chromatographic method. *Ther.Drug Monit.*, **1986**, *8*, 358–360

Taylor, E.H.; Sloniewsky, D.; Gadsden, R.H. Automated extraction and high-performance liquid chromatographic determination of serum clonazepam. *Ther.Drug Monit.*, **1984**, *6*, 474–477

Bouquet, S.; Aucourtier, P.; Brisson, A.M.; Courtois, P.; Fourtillan, J.B. High-performance liquid chromatographic determination in human plasma of an anticonvulsant benzodiazepine: Clonazepam. *J.Liq.Chromatogr.*, **1983**, *6*, 301–310 [chlordiazepoxide (IS)]

Remmel, R.P.; Elmer, G.W. Separation of clonazepam and five metabolites by reverse phase HPLC and quantitation from rat liver microsomal incubations. *J.Liq.Chromatogr.*, **1983**, *6*, 585–598 [flunitrazepam (IS)]

Wittwer, J.D., Jr. Application of high pressure liquid chromatography to the forensic analysis of several benzodiazepines. *J.Liq.Chromatogr.*, **1980**, *3*, 1713–1724 [simultaneous chlordiazepoxide, clorazepate, cyprazepam, demoxapam, desmethyldiazepam, diazepam, flurazepam, medazepam, nitrazepam, oxazepam, prazepam; normal phase]

# Clotrimazole

**Molecular formula:** $C_{22}H_{17}ClN_2$
**Molecular weight:** 344.8
**CAS Registry No.:** 23593-75-1

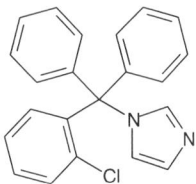

## SAMPLE
**Matrix:** blood
**Sample preparation:** Condition a 100 mg Bond-Elut C18 SPE cartridge with 2 mL 95% EtOH and 2 mL MeCN:water 15:85. 200 μL Plasma or whole blood + 50 μL 100 μM testosterone propionate in MeOH + 3 mL MeCN:water 15:85, vortex for 30 s, add to the SPE cartridge, wash with 9 mL MeCN:water 30:70, dry, elute with 200 μL 95% EtOH, inject a 10 μL aliquot of the eluate.

## HPLC VARIABLES
**Column:** 50 × 4.6 5 μm Supelcosil LC-8DB
**Mobile phase:** MeOH:buffer 72.5:27.5 (Buffer was 25 mM $K_2HPO_4$ adjusted to pH 3 with 670 mM phosphoric acid.)
**Flow rate:** 1
**Injection volume:** 10
**Detector:** UV 210

## CHROMATOGRAM
**Retention time:** 2.60
**Internal standard:** testosterone propionate (3.60)

## OTHER SUBSTANCES
**Extracted:** metabolites, doxepin
**Noninterfering:** acetaminophen, N-acetylprocainamide, amitriptyline, aspirin, barbituric acid, brompheniramine, caffeine, carbamazepine, chloramphenicol, chlorpheniramine, clonazepam, desipramine, desmethyldoxepin, digitoxin, digoxin, disopyramide, ethosuximide, felbamate, gentamicin, ibuprofen, imipramine, lidocaine, maprotiline, mephenytoin, mephobarbital, metharbital, methsuximide, methylsuccinimide, nortriptyline, paramethadione, phenacemide, phenobarbital, phensuximide, phenylpropanolamine, phenytoin, primidone, procainamide, protriptyline, quinidine, theophylline, tobramycin, trimethadione, valproic acid, vancomycin
**Interfering:** itraconazole

## KEY WORDS
plasma; SPE; whole blood; pharmacokinetics

## REFERENCE
Rifai, N.; Sakamoto, M.; Law, T.; Platt, O.; Mikati, M.; Armsby, C.C.; Brugnara, C. HPLC measurement, blood distribution, and pharmacokinetics of oral clotrimazole, potentially useful antisickling agent. *Clin.Chem.*, **1995**, *41*, 387–391

## SAMPLE
**Matrix:** blood
**Sample preparation:** Prepare a C18 SPE cartridge (Analytichem part 607303) by washing with 2 mL MeOH then 5 mL water (J. Chromatogr. 1986, 377, 287). 1 mL Serum + 200 μL ammonium hydroxide, add to cartridge, wash with 6 mL water, elute with 3 mL MeOH. Evaporate the residue to dryness under a stream of nitrogen at 45°, reconstitute in 1 mL mobile phase, inject 50 μL aliquot.

## HPLC VARIABLES
**Column:** 150 × 3.9 Novopak C18
**Mobile phase:** MeOH:MeCN:20 mM $KH_2PO_4$ 30:30:35, adjusted to pH 6.8
**Flow rate:** 2
**Injection volume:** 50
**Detector:** UV 254

## CHROMATOGRAM
**Retention time:** 7.0
**Internal standard:** clotrimazole

## OTHER SUBSTANCES
**Extracted:** ketoconazole

## KEY WORDS
serum; SPE; cotrimazole is IS

## REFERENCE
Piscitelli, S.C.; Goss, T.F.; Wilton, J.H.; D'Andrea, D.T.; Goldstein, H.; Schentag, J.J. Effects of ranitidine and sucralfate on ketoconazole bioavailability. *Antimicrob.Agents Chemother.*, **1991**, *35*, 1765–1771

## SAMPLE
**Matrix:** blood, tissue
**Sample preparation:** Prepare a SPICE reversed-phase SPE cartridge by washing with 2 mL MeOH then 5 mL water. Tissue. 0.5 g Tissue + 4 mL MeCN, homogenize for 2 min using a PTFE pestle in a tissue grinder, centrifuge at 1500 g for 15 min. Remove supernatant and evaporate it under a stream of nitrogen at 45°. Reconstitute residue in 1 mL 10 mM HCl, add 200 µL ammonium hydroxide to adjust pH to about 10.5, add to cartridge, wash with 6 mL water, elute with 3 mL MeOH. Evaporate the residue to dryness under a stream of nitrogen at 45°, reconstitute in 200 µL mobile phase, inject 50 µL aliquot. Plasma. 1 mL Plasma + 200 µL ammonium hydroxide, add to cartridge, wash with 6 mL water, elute with 3 mL MeOH. Evaporate the residue to dryness under a stream of nitrogen at 45°, reconstitute in 1 mL mobile phase, inject 50 µL aliquot.

## HPLC VARIABLES
**Column:** 150 × 3.9 5 µm Novopak C18
**Mobile phase:** MeCN:MeOH:20 mM pH 6.8 $KH_2PO_4$/NaOH 30:35:35
**Flow rate:** 2
**Injection volume:** 50
**Detector:** UV 254

## CHROMATOGRAM
**Retention time:** 6.8
**Internal standard:** clotrimazole

## OTHER SUBSTANCES
**Extracted:** ketoconazole

## KEY WORDS
plasma; SPE; clotrimazole is IS; lung; liver; adrenal

## REFERENCE
Riley, C.M.; James, M.O. Determination of ketoconazole in the plasma, liver, lung and adrenal of the rat by high-performance liquid chromatography. *J.Chromatogr.*, **1986**, *377*, 287–294

## SAMPLE

**Matrix:** formulations
**Sample preparation:** Dissolve in MeCN:THF 94.3:5.7, inject an aliquot.

## HPLC VARIABLES

**Column:** 250 × 4.6 5 μm Suplex pKb-100 (Supelco)
**Mobile phase:** MeCN:THF:15 mM pH 4.1 acetate buffer 41.5:2.5:56
**Flow rate:** 2
**Detector:** UV 254

## CHROMATOGRAM

**Retention time:** 8

## OTHER SUBSTANCES

**Simultaneous:** degradation products, impurities, benzaldehyde, benzyl alcohol, mometasone furoate, orthochlorophenyl-diphenylmethanol

## KEY WORDS

creams

## REFERENCE

Spangler, M. Isocratic reversed phase HPLC analysis of a pharmaceutical cream. *Supelco Reporter*, **1994**, *13(2)*, 12–13

## SAMPLE

**Matrix:** formulations
**Sample preparation:** Tablets. Powder tablets, weigh out amount equivalent to about 30 mg, add 100 mL MeOH, sonicate for 5 min, filter. Add a 2 mL aliquot of filtrate to 5 mL of 100 μg/mL ketoconazole in MeOH, make up to 25 mL with MeOH, inject 20 μL aliquot. Cream. Condition a 500 mg Bond-Elut diol cartridge with 6 mL dichloromethane. Weigh out cream equivalent to about 5 mg of drug, add 30 mL dichloromethane, sonicate for 3 min, make up to 100 mL with dichloromethane, filter. Add a 2 mL aliquot to the cartridge, wash with 2 mL dichloromethane:methanol 4:1, wash with 2 mL dichloromethane, elute with 3 mL MeOH:buffer 85:15. Add eluate to 0.5 mL 100 μg/mL ketoconazole in MeOH, make up to 5 mL with MeOH, inject 20 μL aliquot. (Buffer was 50 mM triethylamine adjusted to pH 7.0 with phosphoric acid.)

## HPLC VARIABLES

**Column:** 250 × 4.6 5 μm Spherisorb CN
**Mobile phase:** THF:buffer 30:70 (Buffer was 50 mM triethylamine adjusted to pH 3.0 with phosphoric acid.)
**Flow rate:** 1
**Injection volume:** 20
**Detector:** UV 230 [Enhanced sensitivity with photoreactor (Beam Boost model C6808 with 10 m × 0.3 mm reaction coil) followed by UV detection at 270 nm.]

## CHROMATOGRAM

**Retention time:** 9.5
**Internal standard:** ketoconazole (7)

## OTHER SUBSTANCES

**Simultaneous:** bifonazole, econazole, fenticonazole, isoconazole, miconazole, tioconazole

## KEY WORDS

tablets; creams

## REFERENCE
Di Pietra, A.M.; Cavrini, V.; Andrisano, V.; Gatti, R. HPLC analysis of imidazole antimycotic drugs in pharmaceutical formulations. *J.Pharm.Biomed.Anal.*, **1992**, *10*, 873–879

## SAMPLE
**Matrix:** formulations
**Sample preparation:** 5 g Ointment containing 1% clotrimazole + 70 mL MeOH, sonicate for 15 min, make up to 100 mL with MeOH, filter (0.22 μm PTFE), mix an aliquot of the filtrate with an equal volume of 6 μg/mL chrysene in MeOH, inject a 20 μL aliquot.

## HPLC VARIABLES
**Column:** 150 × 4 5 μm Nucleosil C18
**Mobile phase:** MeOH:water 9:1
**Flow rate:** 1
**Injection volume:** 20
**Detector:** UV 258

## CHROMATOGRAM
**Retention time:** 5.23
**Internal standard:** chrysene (9.54)

## OTHER SUBSTANCES
**Interfering:** octanol

## KEY WORDS
ointment

## REFERENCE
Valenta, C.; Lexer, A.; Spiegl, P. Analysis of clotrimazole in ointments by high-performance liquid chromatography. *Pharmazie*, **1992**, *47*, 641–642

## SAMPLE
**Matrix:** formulations
**Sample preparation:** 0.5 g Ointment containing 1% clotrimazole + 7 mL octanol, sonicate for 15 min, make up to 10 mL with octanol, filter (0.22 μm PTFE), inject a 20 μL aliquot.

## HPLC VARIABLES
**Column:** 150 × 4 5 μm Nucleosil C18
**Mobile phase:** MeOH:4.25% phosphoric acid 9:1
**Flow rate:** 1
**Injection volume:** 20
**Detector:** UV 258

## CHROMATOGRAM
**Retention time:** 2.33

## KEY WORDS
ointment

## REFERENCE
Valenta, C.; Lexer, A.; Spiegl, P. Analysis of clotrimazole in ointments by high-performance liquid chromatography. *Pharmazie*, **1992**, *47*, 641–642

## ANNOTATED BIBLIOGRAPHY

Stuber, B.; Muller, K.H. [High pressure liquid chromatography of combined clotrimazole-containing hydrocreams and hydrocream pastes]. *Pharm.Acta Helv.*, **1984**, *59*, 210–212

Hoogerheide, J.G.; Strusiak, S.H.; Taddei, C.R.; Townley, E.R.; Wyka, B.E. High performance liquid chromatographic determination of clotrimazole in pharmaceutical formulations. *J.Assoc. Off.Anal.Chem.*, **1981**, *64*, 864–869

# Clozapine

**Molecular formula:** $C_{18}H_{19}ClN_4$
**Molecular weight:** 326.8
**CAS Registry No.:** 5786-21-0

## SAMPLE
**Matrix:** bile, blood, urine
**Sample preparation:** 1 mL Bile, plasma, or urine + 20 µL 40 µg/mL clozapine in EtOH + 200 µL 2.5 mM pH 6.0 sodium pentanesulfonate, mix for 3 s, add 6 mL diethyl ether, shake vigorously for 2 min, centrifuge at 2500 g for 5 min. Remove the organic layer and evaporate it to dryness under a stream of nitrogen at 40°, reconstitute the residue in 100 µL mobile phase, vortex for 30 s, inject a 25 µL aliquot.

## HPLC VARIABLES
**Guard column:** 25 × 4.6 30 µm C18 (Merck)
**Column:** 100 × 8 10 µm µBondapak C18
**Mobile phase:** MeOH:water:triethylamine 65:35:0.5, adjust pH to 6-7 with glacial acetic acid
**Flow rate:** 2
**Injection volume:** 25
**Detector:** UV 254

## CHROMATOGRAM
**Retention time:** 9.7
**Internal standard:** clozapine

## OTHER SUBSTANCES
**Extracted:** moricizine

## KEY WORDS
plasma; rat; clozapine is IS

## REFERENCE
Yang, J.M.; Chan, K. Simultaneous determination of moricizine and its sulphoxidation metabolites in biological fluids by high-performance liquid chromatography. *J.Chromatogr.B*, **1995**, *663*, 172–176

## SAMPLE
**Matrix:** blood
**Sample preparation:** Condition a Bond Elut C18 SPE cartridge with 1 mL MeOH and 1 mL buffer. Mix 100 (rat) or 250 (human) µL plasma or serum with 100 µL 5 µg/mL N-methylspiperone, add to SPE cartridge, wash twice with 1 mL MeCN:buffer 28:72, elute with 1 mL MeOH:triethylamine 99.3:0.7. Evaporate the eluate to dryness under a stream of nitrogen at 50°, reconstitute the residue in 100 µL mobile phase, inject a 55 µL aliquot. (Buffer was 7.8 g $K_2HPO_4$ in 1 L water, pH 9.0.)

## HPLC VARIABLES
**Guard column:** 50 mm long C18 (Waters)
**Column:** 300 × 3.9 µBondapak C18

**Mobile phase:** MeCN:buffer 36:64, pH adjusted to 3.7 with 10% phosphoric acid (Buffer was 7.8 g $K_2HPO_4$ in 500 mL water, adjust pH to 4.0 with phosphoric acid, make up to 1 L with water.)
**Flow rate:** 1
**Injection volume:** 55
**Detector:** UV 254

## CHROMATOGRAM
**Retention time:** 6.0
**Internal standard:** N-methylspiperone (11.2)
**Limit of quantitation:** 60 ng/mL (rat); 15 ng/mL (human)

## OTHER SUBSTANCES
**Extracted:** metabolites
**Simultaneous:** amitriptyline, diazepam, nitrazepam, oxazepam, protriptyline

## KEY WORDS
plasma; serum; human; rat; pharmacokinetics; SPE

## REFERENCE
Fadiran, E.O.; Leslie, J.; Fossler, M.; Young, D. Determination of clozapine and its major metabolites in human serum and rat plasma by liquid chromatography using solid-phase extraction and ultraviolet detection. *J.Pharm.Biomed.Anal.*, **1995**, *13*, 185–190

## SAMPLE
**Matrix:** blood
**Sample preparation:** Condition a 1 mL BondElut C18 SPE cartridge with 1 volume of 1 M HCl, 2 volumes of MeOH, and 1 volume of water. 500 μL Serum + 50 μL amoxapine in 1% potassium bicarbonate + 500 μL MeCN, mix, centrifuge at 1500 g for 3 min, add the supernatant to the SPE cartridge, wash with 2 volumes of water, wash with 1 volume of MeCN, elute with 250 μL MeOH:35% perchloric acid 100:1, remove all eluate by centrifuging at 1000 g for 30 s, inject a 15 μL aliquot.

## HPLC VARIABLES
**Guard column:** 15 × 3.2 7 μm C8 (Applied Biosystems)
**Column:** 150 × 4.6 5 μm Ultrasphere Octyl
**Mobile phase:** MeCN:0.1% tetramethylammonium perchlorate 27:73, adjusted to pH 4.2 with 10% perchloric acid
**Flow rate:** 2
**Injection volume:** 15
**Detector:** UV 245

## CHROMATOGRAM
**Retention time:** 7.3
**Internal standard:** amoxapine (15.8)
**Limit of quantitation:** 15 ng/mL

## OTHER SUBSTANCES
**Extracted:** metabolites

## KEY WORDS
serum; SPE; protect from light

## REFERENCE
Gupta, R.N. Column liquid chromatographic determination of clozapine and N-desmethylclozapine in human serum using solid-phase extraction. *J.Chromatogr.B*, **1995**, *673*, 311–315

## SAMPLE
**Matrix:** blood
**Sample preparation:** 200 μL Plasma or serum + 50 μL 4 μg/mL nortriptyline hydrochloride in 100 mM HCl + 100 μL 2 M pH 10.6 Tris buffer + 200 μL MTBE, vortex for 30 s, centrifuge at 9950 g for 3 min, inject a 50 μL aliquot of the organic layer.

## HPLC VARIABLES
**Guard column:** 10 × 4.6 SCX-10C5 (Hichrom)
**Column:** 150 × 4.6 Spherisorb S5 SCX (sulfopropyl-bonded silica) cation exchange
**Mobile phase:** 35 mM ammonium perchlorate in MeOH adjusted to an apparent pH of 6.7 with 100 mM NaOH in MeOH
**Flow rate:** 1.5
**Injection volume:** 50
**Detector:** UV 215

## CHROMATOGRAM
**Retention time:** 6
**Internal standard:** nortriptyline hydrochloride (9)
**Limit of detection:** 50 ng/mL

## OTHER SUBSTANCES
**Extracted:** metabolites
**Simultaneous:** amitriptyline, chlorpromazine, clomipramine, dothiepin, doxepin, fluoxetine, fluphenazine, haloperidol, imipramine, maprotiline, mianserin, norclomipramine, nordothiepin, nordoxepin, norfluoxetine, nortriptyline, paroxetine, remoxipride, sertraline, sulpride, thioridazine, trazodone
**Noninterfering:** carbamazepine, clonazepam, diazepam, flunitrazepam, lorazepam, nordiazepam, theophylline
**Interfering:** fluvoxamine

## KEY WORDS
plasma; serum

## REFERENCE
McCarthy, P.T.; Hughes, S.; Paton, C. Measurement of clozapine and norclozapine in plasma/serum by high-performance liquid chromatography with ultraviolet detection. *Biomed.Chromatogr.*, **1995**, *9*, 36−41

## SAMPLE
**Matrix:** blood
**Sample preparation:** 2 mL Whole blood or plasma + 2 mL buffer + 5 mL chloroform: isopropanol:n-heptane 60:14:26, shake gently horizontally for 10 min, centrifuge at 2800 g for 10 min. Remove the lower organic layer and evaporate it to dryness under vacuum at 45°, reconstitute the residue in 100 μL mobile phase, centrifuge at 2800 g for 5 min, inject a 50 μL aliquot of the supernatant. (Buffer was saturated ammonium chloride solution 25% diluted with water, adjusted to pH 9.5 with 25% ammonia solution.)

## HPLC VARIABLES
**Column:** 300 × 3.9 4 μm NovaPack C18
**Mobile phase:** MeOH:THF:buffer 65:5:30 (Buffer was 0.68 g/L (10 mM (sic)) $KH_2PO_4$ adjusted to pH 2.6 with concentrated orthophosphoric acid.) (At the end of each session wash the column with water for 1 h and MeOH for 1 h, re-equilibrate for 30 min.)
**Column temperature:** 30
**Flow rate:** 0.8
**Injection volume:** 50
**Detector:** UV 269

## CHROMATOGRAM
**Retention time:** 6.53
**Limit of detection:** <120 ng/mL

## OTHER SUBSTANCES
**Extracted:** acebutolol, acenocoumarol, acepromazine, aceprometazine, acetaminophen, aconitine, ajmaline, albuterol, alimemazine, alminoprofen, alpidem, alprazolam, alprenolol, amisulpride, amitriptyline, amodiaquine, amoxapine, aspirin, astemizole, atenolol, benazepril, benperidol, benzocaine, benzoylecgonine, bepridil, betaxolol, bisoprolol, bromazepam, brompheniramine, bupivacaine, buprenorphine, buspirone, caffeine, carbamazepine, carbinoxamine, carpipramine, carteolol, celiprolol, cetirizine, chlorambucil, chlordiazepoxide, chlormezanone, chlorophenacinone, chloroquine, chlorpheniramine, chlorpromazine, chlorpropamide, cibenzoline, cicletanine, clemastine, clobazam, clomipramine, clonazepam, clonidine, clorazepate, cocaine, codeine, colchicine, cycloguanil, cyproheptadine, cytarabine, dacarbazine, daunorubicin, debrisoquine, desipramine, dextromethorphan, dextromoramide, dextropropoxyphene, diazepam, diazoxide, diclofenac, dihydralazine, diltiazem, dipyridamole, disopyramide, dosulepine, doxepin, doxylamine, droperidol, ephedrine, estazolam, etodolac, fenfluramine, fenoprofen, fentiazac, flecainide, floctafenine, flumazenil, flunitrazepam, fluoxetine, fluphenazine, flurbiprofen, fluvoxamine, glibenclamide, glibornuride, glipizide, glutethimide, haloperidol, hydroxychloroquine, hydroxyzine, ibuprofen, imipramine, indomethacin, iproniazid, ketamine, ketoprofen, labetalol, levomepromazine, lidocaine, lidoflazine, lisinopril, loperamide, loprazolam, loratadine, lorazepam, loxapine, maprotiline, medifoxamine, mefenamic acid, mefenidramine, mefloquine, melphalan, meperidine, mephenesin, mephentermine, mepivacaine, metapramine, metformin, methadone, methaqualone, methocarbamol, methotrexate, metipranolol, metoclopramide, metoprolol, mexiletine, mianserine, midazolam, minoxidil, moclobemide, moperone, morphine, nadolol, nalbuphine, nalorphine, naloxone, naltrexone, naproxen, nialamide, nicardipine, nifedipine, niflumic acid, nitrazepam, nitrendipine, nizatidine, nomifensine, nortriptyline, omeprazole, opipramol, oxazepam, oxprenolol, penbutolol, penfluridol, pentazocine, phencyclidine, phenobarbital, phenol, pimozide, pindolol, pipamperone, piroxicam, prazepam, prazosin, prilocaine, procainamide, procarbazine, promethazine, propafenone, propranolol, protriptyline, pyrimethamine, quinidine, quinine, quinupramine, ramipril, ranitidine, reserpine, ritodrine, secobarbital, sotalol, strychnine, sulfinpyrazole, sulindac, sulpride, sultopride, suriclone, temazepam, tenoxicam, terbutaline, terfenadine, tetracaine, tetrazepam, thiopental, thioproperazine, thioridazine, tianeptine, tiapride, tiaprofenic acid, ticlopidine, timolol, tioclomarol, tofisopam, tolbutamide, toloxatone, trazodone, triazolam, trifluoperazine, trimipramine, triprolidine, tropatenine, verapamil, viloxazine, vinblastine, vincristine, vindesine, warfarin, yohimbine, zolpidem, zopiclone, zorubicine

**Interfering:** bumadizone, cyamemazine, cyclizine, demexiptiline, diphenhydramine, histapyrrodine, medazepam, nimodipine, phenylbutazone, proguanil, trifluperidol

## KEY WORDS
whole blood; plasma

## REFERENCE
Tracqui, A.; Kintz, P.; Mangin, P. Systematic toxicological analysis using HPLC/DAD. *J.Forensic Sci.*, **1995**, *40*, 254–262

## SAMPLE
**Matrix:** blood
**Sample preparation:** Inject plasma onto column A and elute to waste with mobile phase A, elute the contents of column A onto column B with mobile phase B, monitor the effluent from column B.

## HPLC VARIABLES
**Column:** A 20 × 4.6 10 μm CN; B 250 × 4.6 5 μm Hypersil C18 ODS

**Mobile phase:** A water; B MeCN:water 40:60 containing 0.4% TEMED adjusted to pH 6.5 with glacial acetic acid
**Detector:** UV 254

## CHROMATOGRAM
**Limit of detection:** 10 ng/mL

## OTHER SUBSTANCES
**Extracted:** metabolites

## KEY WORDS
column-switching; plasma

## REFERENCE
Weigmann, H.; Hiemke, C. Automated determination of clozapine, N-desmethylclozapine and clozapine-N-oxide by column switching and on-line high-performance liquid chromatography (Abstract 110). *Ther.Drug Monit.*, **1995**, *17*, 410

## SAMPLE
**Matrix:** blood
**Sample preparation:** 1 mL Plasma + 100 μL 4 μg/mL protriptyline + 200 μL 2 M NaOH, vortex, allow to stand for 5 min, add 5 mL n-hexane:isoamyl alcohol 98.5:1.5, shake for 20 min, centrifuge at 3000 g for 5 min. Remove the organic phase and add it to 1 mL 100 mM HCl, shake for 5 min, centrifuge at 3000 g for 2 min. Remove the aqueous layer and add it to 200 μL 2 M NaOH, add n-hexane:isoamyl alcohol 98.5:1.5, extract, repeat extraction. Combine the organic layers and evaporate them at 45° under a stream of nitrogen, reconstitute in 100 μL 100 mM HCl, inject a 80 μL aliquot.

## HPLC VARIABLES
**Guard column:** 70 mm Whatman column survival kit 10 μm ODS
**Column:** 250 × 4.6 10 μm Partisil 10 ODS-3
**Mobile phase:** MeCN:MeOH:buffer 24:12:64 (Buffer was 5 g $Na_2HPO_4$ (per L?) adjusted to pH 4.0 with phosphoric acid.)
**Flow rate:** 2
**Injection volume:** 80
**Detector:** UV 230

## CHROMATOGRAM
**Retention time:** 9
**Internal standard:** protriptyline (15)
**Limit of detection:** 2 ng/mL

## OTHER SUBSTANCES
**Extracted:** metabolites

## KEY WORDS
plasma

## REFERENCE
Chung, M.-C.; Lin, S.-K.; Chang, W.-H.; Jann, M.W. Determination of clozapine and desmethylclozapine in human plasma by high-performance liquid chromatography with ultraviolet detection. *J.Chromatogr.*, **1993**, *613*, 168–173

## SAMPLE
**Matrix:** blood

**Sample preparation:** 1 mL Serum + 1 mL 100 mM NaOH + 50 μL 4.098 μg/mL trifluoperazine and 2.801 μg/mL imipramine in EtOH:water 1:1, mix, add 5 mL heptane:isoamyl alcohol 99:1, shake at 250 shakings/min for 5 min, centrifuge at 1500 g for 10 min, freeze in dry ice/ethanol. Remove the organic layer and evaporate it to dryness at 50° under a stream of nitrogen, dissolve the residue in 75 μL mobile phase, inject a 65 μL aliquot.

## HPLC VARIABLES
**Column:** 150 × 4.6 5 μm Spherisorb S5W
**Mobile phase:** MeOH:50 mM pH 9.9 ammonium acetate buffer 85:15
**Flow rate:** 1.4
**Injection volume:** 65
**Detector:** UV 261

## CHROMATOGRAM
**Retention time:** 2.3
**Internal standard:** trifluoperazine (3.5), imipramine (7.7)
**Limit of detection:** 15 ng/mL
**Limit of quantitation:** 30 ng/mL

## OTHER SUBSTANCES
**Simultaneous:** metabolites, amitriptyline, carbamazepine, chlorprothixene, clomipramine, levomepromazine, nortriptyline
**Interfering:** mianserine, perphenazine, zuclopenthixol

## KEY WORDS
serum

## REFERENCE
Olesen, O.V.; Poulsen, B. On-line fully automated determination of clozapine and desmethylclozapine in human serum by solid-phase extraction on exchangeable cartridges and liquid chromatography using a methanol buffer mobile phase on unmodified silica. *J.Chromatogr.*, **1993**, *622*, 39–46

## SAMPLE
**Matrix:** blood
**Sample preparation:** Condition a 10 × 2 SPE cartridge packed with 40 μm cyanopropyl bonded phase (J.T. Baker) with MeOH at 1.5 mL/min for 1 min, with water at 4.5 mL/min for 2 min, with 100 mM pH 8.0 ammonium acetate buffer at 1.5 mL/min for 2 min. 600 μL Serum + 300 μL trifluoperazine and imipramine in water, mix, add a 200 μL aliquot to the SPE cartridge, wash with 100 mM pH 8.0 ammonium acetate buffer at 1.5 mL/min for 0.5 min, wash with MeOH:water 9:91 at 1.5 mL/min for 1 min, elute the contents of the SPE cartridge directly onto the analytical column with mobile phase for 1 min.

## HPLC VARIABLES
**Column:** 150 × 4.6 5 μm Spherisorb S5W
**Mobile phase:** MeOH:50 mM pH 9.9 ammonium acetate buffer 85:15
**Flow rate:** 1.4
**Detector:** UV 261

## CHROMATOGRAM
**Retention time:** 2.3
**Internal standard:** trifluoperazine (3.5), imipramine (7.7)
**Limit of detection:** 15 ng/mL
**Limit of quantitation:** 30 ng/mL

## OTHER SUBSTANCES
**Simultaneous:** metabolites, amitriptyline, carbamazepine, chlorprothixene, clomipramine, levomepromazine, nortriptyline
**Interfering:** mianserine, perphenazine, zuclopenthixol

## KEY WORDS
serum; SPE

## REFERENCE
Olesen, O.V.; Poulsen, B. On-line fully automated determination of clozapine and desmethylclozapine in human serum by solid-phase extraction on exchangeable cartridges and liquid chromatography using a methanol buffer mobile phase on unmodified silica. *J.Chromatogr.*, **1993**, *622*, 39–46

## SAMPLE
**Matrix:** blood
**Sample preparation:** 600 μL Serum + 100 μL 300 μg/mL triprolidine in 400 mM NaOH + 7 mL ethyl acetate, vortex for 1 min, centrifuge at 1000 g for 10 min. Remove the organic layer and add it to 150 μL 100 mM HCl, vortex for 1 min, centrifuge, inject a 50 μL aliquot of the aqueous phase.

## HPLC VARIABLES
**Column:** 250 × 4.6 5 μm Spherisorb-C6
**Mobile phase:** MeCN:30 mM $KH_2PO_4$ 45:55 containing 2 g sodium hexanesulfonate, pH adjusted to 2.7 with phosphoric acid
**Column temperature:** 30
**Flow rate:** 1.5
**Injection volume:** 50
**Detector:** UV 254

## CHROMATOGRAM
**Retention time:** 4.5
**Internal standard:** triprolidine (5.5)
**Limit of detection:** 3-4 ng/mL

## OTHER SUBSTANCES
**Extracted:** metabolites
**Interfering:** carbinoxamine, chlorpheniramine, cocaine, dipyridamole, oxazolam, trazodone

## KEY WORDS
serum

## REFERENCE
Volpicelli, S.A.; Centorrino, F.; Puopolo, P.R.; Kando, J.; Frankenburg, F.R.; Baldessarini, R.J.; Flood, J.G. Determination of clozapine, norclozapine, and clozapine-N-oxide in serum by liquid chromatography. *Clin.Chem.*, **1993**, *39*, 1656–1659

## SAMPLE
**Matrix:** blood
**Sample preparation:** Add 10 μL 20 μg/mL oxaprotiline in MeOH to 990 μL plasma or serum. Inject 100 μL plasma or serum onto column A with mobile phase A and elute to waste, after 15 min elute column A onto column B with mobile phase B for 2 min. Remove column A from circuit and re-equilibrate it with mobile phase A for 5 min. Chromatograph on column B with mobile phase B.

## HPLC VARIABLES
**Column:** A 20 × 4.6 10 μm Hypersil MOS C8; B 20 × 4.6 5 μm Hypersil CPS CN + 250 × 4.6 5 μm Nucleosil 100 CN
**Mobile phase:** A MeOH:water 5:95; B MeOH:MeCN:10 mM pH 6.8 potassium phosphate buffer 188:578:235
**Flow rate:** 1.5
**Injection volume:** 100
**Detector:** UV 214

## CHROMATOGRAM
**Retention time:** 6.2
**Internal standard:** oxaprotiline (9.5)
**Limit of detection:** 20 ng/mL

## OTHER SUBSTANCES
**Simultaneous:** amitriptyline, clomipramine, desipramine, doxepin, fluoxetine, imipramine, maprotiline, metoclopramide, norfluoxetine, nortriptyline
**Noninterfering:** carbamazepine, chlordiazepoxide, clobazam, diazepam, flurazepam, fluspirilene, haloperidol, lorazepam, nitrazepam, nordiazepam, oxazepam, perazine, pimozide, spiroperidol, trifluperidol
**Interfering:** fluvoxamine

## KEY WORDS
plasma; serum; column-switching

## REFERENCE
Härtter, S.; Wetzel, H.; Hiemke, C. Automated determination of fluvoxamine in plasma by column-switching high-performance liquid chromatography. *Clin.Chem.*, **1992**, *38*, 2082–2086

## SAMPLE
**Matrix:** blood, saliva, tissue, urine
**Sample preparation:** Homogenize (Polytron) tissue with 4 (whole brain) or 8 (brain striata) volumes of 100 mM pH 4.5 NaH$_2$PO$_4$ containing 0.5% NaF. Add 500 μL brain homogenate or 500 μL plasma, saliva, or urine containing 15 μL saturated NaF solution to 75 μL 150 μg/mL IS, add 50 μL 50% perchloric acid, mix vigorously for 10 s, let stand at room temperature for 10 min, add 1 mL water, mix briefly, centrifuge at 10° at 2500 (?) for 30 min. Remove the supernatant and add it to 750 μL saturated sodium carbonate solution, mix briefly, add 7.5 mL pentane:chloroform 95:5, rock gently for 10 min, centrifuge in a desk-top centrifuge for 2 min, freeze in dry ice/acetone for 2 min. Remove the organic layer and add it to 250 μL 100 mM HCl, mix vigorously for 10 s, centrifuge in a desk-top centrifuge for 1-2 min, freeze in dry ice/acetone for 3-5 min, discard the organic layer. Allow the aqueous layer to thaw, remove any trace of organic solvent with a stream of nitrogen, inject a 75 μL aliquot of the aqueous layer.

## HPLC VARIABLES
**Guard column:** 15 × 3.2 7 μm Brownlee RP-8
**Column:** 250 × 4.6 5 μm Zorbax RX-C18
**Mobile phase:** MeCN:buffer 18:82 (Buffer was 100 mM K$_2$HPO$_4$ containing 0.5% triethylamine, adjusted to pH 2.7 with phosphoric acid.)
**Flow rate:** 2
**Injection volume:** 75
**Detector:** UV 235

## CHROMATOGRAM
**Retention time:** 5.6
**Internal standard:** 2β-carbomethoxy-3β-(4-chlorophenyl)tropane (RTI-31) (Research Biochemical International, Natick MA) (11.4)

## OTHER SUBSTANCES
**Extracted:** chlordiazepoxide, cocaine, gepirone, methylphenidate, pentazocine, pseudococaine

**Simultaneous:** acetaminophen, acetophenazine, amoxapine, amphetamine, atropine, benperidol, buspirone, caffeine, carbamazepine, chlorpheniramine, codeine, dextromethorphan, diazepam, diphenhydramine, flupenthixol, flurazepam, haloperidol, hydergine, hydrocodone, hydromorphone, lidocaine, loxapine, mepazine, meperidine, mesoridazine, methaquaalude, 3,4-methylenedioxyamphetamine, 3,4-methylenedioxyethylamphetamine, 3,4-methylenedioxymethamphetamine, morphine, norcocaine, oxazepam, pentobarbital, phenylpropanolamine, procainamide, procaine, propyl benzoylecgonine, quinidine, quinine, salicylic acid, secobarbital, theophylline, trazodone, 3-tropanyl-3,5-dichlorobenzoate, vancomycin, WIN 35428

**Noninterfering:** amitriptyline, benztropine methanesulfonate, butaperazine, butriptyline, carphenazine, chlorpromazine, clomipramine, cyclobenzaprine, dextropropoxyphene, dronabinol, ephedrine, ethchlorvynol, fluoxetine, fluphenazine, imipramine, meprobamate, methadone, methamphetamine, nicotine, norfluoxetine, nortriptyline, PCP, phenothiazine, pseudoephedrine

## KEY WORDS
rat; cow; plasma; brain

## REFERENCE
Bonate, P.L.; Davis, C.M.; Silverman, P.B.; Swann, A. Determination of cocaine in biological matrices using reversed phase HPLC: Application to plasma and brain tissue. *J.Liq.Chromatogr.*, **1995**, *18*, 3473–3494

## SAMPLE
**Matrix:** enzyme incubations
**Sample preparation:** Inject a 20 μL aliquot.

## HPLC VARIABLES
**Guard column:** 30 × 2 5 μm Ultracarb ODS 30 (Phenomenex)
**Column:** 100 × 2 5 μm Ultracarb ODS 30 (Phenomenex)
**Mobile phase:** MeCN:water:acetic acid 75:25:1 containing 1 mM ammonium acetate
**Flow rate:** 0.2
**Injection volume:** 20
**Detector:** UV 254; MS, Sciex API III triple quadrupole, IonSpray interface, ionizing voltage 5 kV, nebulizing gas air at 40 psi, orifice voltage 60 V, target gas argon 26 eV, post-column splitter to decrease flow to 20 μL/min

## OTHER SUBSTANCES
**Extracted:** metabolites, degradation products

## REFERENCE
Liu, Z.C.; Uetrecht, J.P. Clozapine is oxidized by activated human neutrophils to a reactive nitrenium ion that irreversibly binds to the cells. *J.Pharmacol.Exp.Ther.*, **1995**, *275*, 1476–1483

## SAMPLE
**Matrix:** microsomal incubations
**Sample preparation:** Microsomal incubation + 2 mL ice-cold MeOH, let stand at 4° for 16 h, centrifuge. Remove the supernatant and evaporate it to dryness under a stream of nitrogen at 37°, reconstitute the residue in 150 μL MeOH:water 50:50, inject a 10 μL aliquot.

## HPLC VARIABLES
**Column:** 250 × 3.2 5 μm Nucleosil C8

**Mobile phase:** Gradient. MeCN:6 mM pH 3.5 ammonium formate from 10:90 to 25:75 over 15 min, to 55:45 over 20 min.
**Flow rate:** 0.7
**Injection volume:** 10
**Detector:** UV 254; MS, Fisons Quattro II quadrupole, column effluent passed through a splitter and 40 µL/min passed through a 150 cm × 75 µm fused-silica capillary to the electrospray probe, nebulizing gas nitrogen at 12 L/h, drying gas nitrogen at 280 L/h, interface 60°, capillary voltage 4 kV, con (orifice) voltage 30-80 V, photomultiplier 550 V, m/z 327

## CHROMATOGRAM
**Retention time:** 27

## OTHER SUBSTANCES
**Extracted:** metabolites

## KEY WORDS
human; mouse; rat; liver

## REFERENCE
Maggs, J.L.; Williams, D.; Pirmohamed, M.; Park, B.K. The metabolic formation of reactive intermediates from clozapine, a drug associated with agranulocytosis in man. *J.Pharmacol.Exp.Ther.*, **1995**, *275*, 1463−1475

## SAMPLE
**Matrix:** microsomal incubations
**Sample preparation:** 1 mL Microsomal incubation + 2 mL ice-cold MeOH, let stand at 4° overnight, centrifuge. Remove the supernatant and evaporate it to dryness under a stream of nitrogen, reconstitute the residue in EtOH:water 50:50, inject a 50 µL aliquot.

## HPLC VARIABLES
**Column:** 250 × 4.6 Nucleosil 5 C8
**Mobile phase:** Gradient. MeCN:100 mM pH 3.8 ammonium acetate buffer from 20:80 to 40:60 over 15 min, return to 20:80 over 5 min.
**Flow rate:** 1.5
**Injection volume:** 50
**Detector:** UV 254; Radioactivity

## CHROMATOGRAM
**Retention time:** 13

## OTHER SUBSTANCES
**Extracted:** metabolites

## KEY WORDS
liver

## REFERENCE
Pirmohamed, M.; Williams, D.; Madden, S.; Templeton, E.; Park, B.K. Metabolism and bioactivation of clozapine by human liver *in vitro*. *J.Pharmacol.Exp.Ther.*, **1995**, *272*, 984−990

## SAMPLE
**Matrix:** solutions

## HPLC VARIABLES
**Column:** 250 × 4.6 5 µm Supelcosil LC-DP (A) or 250 × 4 5 µm LiChrospher 100 RP-8 (B)

**Mobile phase:** MeCN:0.025% phosphoric acid:buffer 25:10:5 (A) or 60:25:15 (B) (Buffer was 9 mL concentrated phosphoric acid and 10 mL triethylamine in 900 mL water, adjust pH to 3.4 with dilute phosphoric acid, make up to 1 L.)
**Flow rate:** 0.6
**Injection volume:** 25
**Detector:** UV 229

## CHROMATOGRAM
**Retention time:** 10.90 (A), 5.15 (B)

## OTHER SUBSTANCES
**Also analyzed:** acebutolol, acepromazine, acetaminophen, acetazolamide, acetophenazine, albuterol, alprazolam, amitriptyline, amobarbital, amoxapine, antipyrine, atenolol, atropine, azatadine, baclofen, benzocaine, bromocriptine, brompheniramine, brotizolam, bupivacaine, buspirone, butabarbital, butalbital, caffeine, carbamazepine, cetirizine, chlorcyclizine, chlordiazepoxide, chlormezanone, chloroquine, chlorpheniramine, chlorpromazine, chlorpropamide, chlorprothixene, chlorthalidone, chlorzoxazone, cimetidine, cisapride, clomipramine, clonazepam, clonidine, cocaine, codeine, colchicine, cyclizine, cyclobenzaprine, dantrolene, desipramine, diazepam, diclofenac, diflunisal, diltiazem, diphenhydramine, diphenidol, diphenoxylate, dipyridamole, disopyramide, dobutamine, doxapram, doxepin, droperidol, encainide, ethidium bromide, ethopropazine, fenoprofen, fentanyl, flavoxate, fluoxetine, fluphenazine, flurazepam, flurbiprofen, fluvoxamine, furosemide, glutethimide, glyburide, guaifenesin, haloperidol, homatropine, hydralazine, hydrochlorothiazide, hydrocodone, hydromorphone, hydroxychloroquine, hydroxyzine, ibuprofen, imipramine, indomethacin, ketoconazole, ketoprofen, ketorolac, labetalol, levorphanol, lidocaine, loratadine, lorazepam, lovastatin, loxapine, mazindol, mefenamic acid, meperidine, mephenytoin, mepivacaine, mesoridazine, metaproterenol, methadone, methdilazine, methocarbamol, methotrexate, methotrimeprazine, methoxamine, methyldopa, methylphenidate, metoclopramide, metolazone, metoprolol, metronidazole, midazolam, moclobemide, morphine, nadolol, nalbuphine, naloxone, naphazoline, naproxen, nifedipine, nizatidine, norepinephrine, nortriptyline, oxazepam, oxycodone, oxymetazoline, paroxetine, pemoline, pentazocine, pentobarbital, pentoxifylline, perphenazine, pheniramine, phenobarbital, phenol, phenolphthalein, phentolamine, phenylbutazone, phenyltoloxamine, phenytoin, pimozide, pindolol, piroxicam, pramoxine, prazepam, prazosin, probenecid, procainamide, procaine, prochlorperazine, procyclidine, promazine, promethazine, propafenone, propantheline, propiomazine, propofol, propranolol, protriptyline, quazepam, quinidine, quinine, racemethorphan, ranitidine, remoxipride, risperidone, salicylic acid, scopolamine, secobarbital, sertraline, sotalol, spironolactone, sulfinpyrazone, sulindac, temazepam, terbutaline, terfenadine, tetracaine, theophylline, thiethylperazine, thiopental, thioridazine, thiothixene, timolol, tocainide, tolbutamide, tolmetin, trazodone, triamterene, triazolam, trifluoperazine, triflupromazine, trimeprazine, trimethoprim, trimipramine, verapamil, warfarin, xylometazoline, yohimbine, zopiclone

## KEY WORDS
some details of plasma extraction

## REFERENCE
Koves, E.M. Use of high-performance liquid chromatography-diode array detection in forensic toxicology. *J.Chromatogr.A*, **1995**, *692*, 103–119

◆━━━━━━━━━━━━━━━━◆━━━━━━━━━━━━━━━━◆

## ANNOTATED BIBLIOGRAPHY
Hubmann, M.R.; Waschgler, R.; Moll, W.; Conca, A.; König, P. Simultaneous drug monitoring of citalopram, clozapine, fluoxetine, maprotiline, and trazodone by HPLC analysis (Abstract 41). *Ther.Drug Monit.*, **1995**, *17*, 393 [simultaneous citalopram, clozapine, fluoxetine, maprotiline, trazodone; LOQ 50 ng/mL]

Schulz, E.; Fleischhaker, C.; Remschmidt, H. Determination of clozapine and its major metabolites in serum samples of adolescent schizophrenic patients by high-performance liquid chromatography. Data from a prospective clinical trial. *Pharmacopsychiatry*, **1995**, *28*, 20–25

Lin, S.-K.; Chang, W.-H.; Chung, M.-C.; Lam.Y.W.F.; Jann, M.W. Disposition of clozapine and desmethylclozapine in schizophrenic patients. *J.Clin.Pharmacol.*, **1994**, *34*, 318–324 [plasma; protriptyline (IS); LOD 2 ng/mL]

Weigmann, H.; Hiemke, C. Determination of clozapine and its major metabolites in human serum using automated solid-phase extraction and subsequent isocratic high-performance liquid chromatography with ultraviolet detection. *J.Chromatogr.*, **1992**, *583*, 209–216

Lovdahl, M.J.; Perry, P.J.; Miller, D.D. The assay of clozapine and N-desmethylclozapine in human plasma by high-performance liquid chromatography. *Ther.Drug Monit.*, **1991**, *13*, 69–72

Wilhelm, D.; Kemper, A. High-performance liquid chromatographic procedure for the determination of clozapine, haloperidol, droperidol and several benzodiazepines in plasma. *J.Chromatogr.*, **1990**, *525*, 218–224

Humpel, C.; Haring, C.; Saria, A. Rapid and sensitive determination of clozapine in human plasma using high-performance liquid chromatography and amperometric detection. *J.Chromatogr.*, **1989**, *491*, 235–239

Haring, C.; Humpel, C.; Auer, B.; Saria, A.; Barnas, C.; Fleischhacker, W.; Hinterhuber, H. Clozapine plasma levels determined by high-performance liquid chromatography with ultraviolet detection. *J.Chromatogr.*, **1988**, *428*, 160–166

Yin, J.L. [HPLC determination of clozapine in plasma]. *Chung Hua Shen Ching Ching Shen Ko Tsa Chih*, **1987**, *20*, 78–80

Wang, Z.R.; Lu, M.L.; Xu, P.P.; Zeng, Y.L. Determination of clozapine and its metabolites in serum and urine by reversed phase HPLC. *Biomed.Chromatogr.*, **1986**, *1*, 53–57

# Codeine

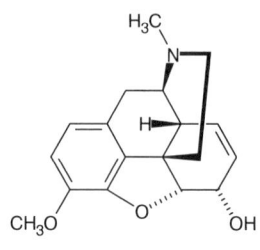

**Molecular formula:** $C_{18}H_{21}NO_3$
**Molecular weight:** 299.4
**CAS Registry No.:** 76-57-3, 6069-47-8 (monohydrate), 5913-71-3
(acetate), 125-25-7 (HBr), 1422-07-7 (HCl), 6020-73-1
(salicylate), 125-27-9 (methyl bromide), 52-28-8 (phosphate),
41444-62-6 (phosphate hemihydrate), 1420-53-7 (sulfate), 6854-40-6
(sulfate trihydrate)

## SAMPLE
**Matrix:** bile, blood
**Sample preparation:** 0.5 mL Blood or bile + 10 (blood) or 15 (bile) μL 100 μg/mL nalorphine in MeOH + 300 μL 1.1 M pH 5.0 sodium acetate buffer + 3000-3500 U of Patella
vulgata glucuronidase, incubate at 55° overnight, add 0.5 mL borate buffer to achieve a
pH of 8.3-8.5. Add 8 mL chloroform:isopropanol 90:10, gently rotate for 30 min, centrifuge at 3500 rpm for 10 min, remove aqueous layer. Wash organic layer (twice for blood,
three times for bile) with 3 mL 100 mM pH 9.9 sodium phosphate buffer with gentle
rotation for 10 min and centrifugation each time. Add organic layer to 200 (blood) or 400
(bile) μL 0.2% phosphoric acid, gently rotate for 30 min, discard organic layer, inject 50
μL of the acid layer. (Borate buffer was 50 mM boric acid and 43 mM sodium tetraborate,
adjusted to pH 9.8.)

## HPLC VARIABLES
**Guard column:** Nova-Pak phenyl guard column
**Column:** 150 × 3.9 5 μm Nova-Pak phenyl
**Mobile phase:** MeCN:10 mM pH 6.6 $NaH_2PO_4$ 10:90
**Flow rate:** 1.2
**Injection volume:** 50
**Detector:** UV 210; F ex 220 em 370 (cut-off)

## CHROMATOGRAM
**Retention time:** 19.2
**Internal standard:** nalorphine (23.5)
**Limit of detection:** 60 ng/mL (blood), 200 ng/mL (bile)

## OTHER SUBSTANCES
**Simultaneous:** dihydrocodeine, hydrocodone, 6-monoacetylmorphine, morphine, oxycodone
**Noninterfering:** acetylcodeine, amitriptyline, amphetamine, diamorphine, diazepam, dothiepin, doxepin, ephedrine, ephedrine, hydromorphone, mesoridazine, methadone, methamphetamine, 3-monoacetylmorphine, nordiazepam, norpropoxyphene, nortriptyline,
oxazepam, propoxyphene, pseudoephedrine, quinidine, quinine, sulfamethoxazole, sulforidazine, thioridazine

## KEY WORDS
UV and F detection used together

## REFERENCE
Crump, K.L.; McIntyre, I.M.; Drummer, O.H. Simultaneous determination of morphine and codeine in
blood and bile using dual ultraviolet and fluorescence high-performance liquid chromatography.
*J.Anal.Toxicol.*, **1994**, *18*, 208–212

## SAMPLE
**Matrix:** bile, blood, tissue

**Sample preparation:** 250 μL Bile, 3 mL blood, or 5 mL tissue homogenate + 1 mL 200 μg/mL nalorphine in water + 2 mL 200 mM pH 8.9 sodium borate buffer + 5 (bile) or 10 (blood, tissue) mL chloroform:isopropanol 90:10, rotate gently for 20 min, centrifuge at 2000 rpm for 10 min. Remove the organic layer and add it to 2 mL 500 mM HCl, rotate for 20 min, centrifuge for 5 min. Remove 1.8 mL of the upper aqueous phase, adjust to pH 8.6 ± 0.2 by very carefully adding powdered ammonium carbonate until the solution is saturated, add 5 mL ethyl acetate:isopropanol 90:10, rotate for 20 min, centrifuge for 5 min. Remove 4.8 mL of the upper organic layer and evaporate it to dryness under a stream of nitrogen at 40°, reconstitute the residue in 50 μL MeOH, vortex for 30 s, inject a 20 μL aliquot.

## HPLC VARIABLES
**Guard column:** 30 × 4.6 5 μm RP-18 Spheri-5
**Column:** 250 × 4.6 5 μm Ultrasphere ODS
**Mobile phase:** MeOH:50 mM pH 7 phosphate buffer 40:60 (Place a 70 × 2 30-38 μm Co-Pell ODS column before the injection valve.)
**Column temperature:** 50
**Flow rate:** 2
**Injection volume:** 20
**Detector:** E, Environmental Sciences Associates Model 5100, porous graphite electrode W1 900 mV W2 400 mV, difference in electrolysis current monitored

## CHROMATOGRAM
**Retention time:** 8.95
**Internal standard:** nalorphine (14.72)
**Limit of detection:** 5 ng/mL

## OTHER SUBSTANCES
**Extracted:** hydromorphone, morphine, norcodeine, normorphine
**Simultaneous:** acetaminophen, atropine, epinephrine, ethylmorphine, hydrocodone, hydroxyzine, naloxone, oxycodone, pentazocine, phenylpropanolamine, pseudomorphine, scopolamine, secobarbital
**Noninterfering:** brompheniramine, chloroprocaine, dextromethorphan, diazepam, diphenhydramine, fentanyl, flurazepam, meperidine, methadone, neostigmine, propoxyphene

## REFERENCE
Hepler, B.R.; Sutheimer, C.; Sunshine, I.; Sebrosky, G.F. Combined enzyme immunoassay-LCEC method for the identification, confirmation, and quantitation of opiates in biological fluids. *J.Anal.Toxicol.*, **1984**, *8*, 78–90

## SAMPLE
**Matrix:** blood
**Sample preparation:** Condition a 1 mL 100 mg ethyl SPE cartridge (J.T.Baker) with 2 volumes of MeOH, 1 volume of water, and 2 volumes of 10 mM pH 9.3 ammonium hydrogen carbonate buffer. 1 mL Serum + 100 μL water, add to the SPE cartridge, wash with 1 volume of 10 mM ammonium hydrogen carbonate buffer, elute with 1 volume of MeOH. Evaporate the eluate to dryness under a stream of nitrogen, reconstitute the residue in 100 μL mobile phase, inject a 20 μL aliquot.

## HPLC VARIABLES
**Column:** 250 × 4.6 5 μm Supelcosil ABZ
**Mobile phase:** Gradient. MeOH:water from 15:85 to 60:40 over 10 min. (Convex gradient where MeOH% = -0.46exp(-x/1.18) + 0.6 where x = time in min.)
**Flow rate:** 0.8 (0.018 mL/min entered MS)
**Injection volume:** 20
**Detector:** MS, Fisons TRIO 2, electrospray, capillary tip 2.97 kV, counter electrode 390 V, sampling cone voltages 66 V, -106 V, -17 V, source 60°, SIM m/z 300

## CHROMATOGRAM
**Retention time:** 7.03
**Internal standard:** codeine
**Limit of quantitation:** 10 ng/mL

## OTHER SUBSTANCES
**Extracted:** morphine (m/z 286)

## KEY WORDS
serum; human; mouse; SPE, codeine is IS

## REFERENCE
Saarinen, M.T.; Sirén, H.; Riekkola, M.-L. Screening and determination of β-blockers, narcotic analgesics and stimulants in urine by high-performance liquid chromatography with column switching. *J.Chromatogr.B*, **1995**, *664*, 341–346

## SAMPLE
**Matrix:** blood
**Sample preparation:** 2 mL Whole blood or plasma + 2 mL buffer + 5 mL chloroform: isopropanol:n-heptane 60:14:26, shake gently horizontally for 10 min, centrifuge at 2800 g for 10 min. Remove the lower organic layer and evaporate it to dryness under vacuum at 45°, reconstitute the residue in 100 µL mobile phase, centrifuge at 2800 g for 5 min, inject a 50 µL aliquot of the supernatant. (Buffer was saturated ammonium chloride solution 25% diluted with water, adjusted to pH 9.5 with 25% ammonia solution.)

## HPLC VARIABLES
**Column:** 300 × 3.9 4 µm NovaPack C18
**Mobile phase:** MeOH:THF:buffer 65:5:30 (Buffer was 0.68 g/L (10 mM (sic)) $KH_2PO_4$ adjusted to pH 2.6 with concentrated orthophosphoric acid.) (At the end of each session wash the column with water for 1 h and MeOH for 1 h, re-equilibrate for 30 min.)
**Column temperature:** 30
**Flow rate:** 0.8
**Injection volume:** 50
**Detector:** UV 286

## CHROMATOGRAM
**Retention time:** 3.44
**Limit of detection:** <120 ng/mL

## OTHER SUBSTANCES
**Extracted:** acebutolol, acenocoumarol, acepromazine, aceprometazine, acetaminophen, aconitine, ajmaline, alimemazine, alminoprofen, alpidem, alprazolam, alprenolol, amitriptyline, amodiaquine, amoxapine, astemizole, atenolol, benazepril, benperidol, benzoylecgonine, bepridil, betaxolol, bisoprolol, bromazepam, brompheniramine, bumadizone, bupivacaine, buprenorphine, buspirone, caffeine, carbamazepine, carbinoxamine, carpipramine, carteolol, celiprolol, cetirizine, chlorambucil, chlordiazepoxide, chlorophenacinone, chloroquine, chlorpheniramine, chlorpromazine, chlorpropamide, cibenzoline, cicletanine, clemastine, clobazam, clomipramine, clonazepam, clonidine, clorazepate, clozapine, cocaine, colchicine, cyamemazine, cyclizine, cycloguanil, cyproheptadine, cytarabine, dacarbazine, daunorubicin, debrisoquine, demexiptiline, desipramine, dextromethorphan, dextromoramide, dextropropoxyphene, diazepam, diazoxide, diclofenac, dihydralazine, diltiazem, diphenhydramine, dipyridamole, disopyramide, dosulepine, doxepin, doxylamine, droperidol, ephedrine, estazolam, etodolac, fenfluramine, fenoprofen, fentiazac, flecainide, floctafenine, flumazenil, flunitrazepam, fluoxetine, fluphenazine, flurbiprofen, fluvoxamine, glibenclamide, glibornuride, glipizide, glutethimide, haloperidol, histapyrrodine, hydroxychloroquine, hydroxyzine, ibuprofen, imipramine, indomethacin,

iproniazid, ketamine, ketoprofen, labetalol, levomepromazine, lidocaine, lidoflazine, loperamide, loprazolam, loratadine, lorazepam, loxapine, maprotiline, medazepam, medifoxamine, mefenamic acid, mefenidramine, mefloquine, melphalan, meperidine, mephentermine, mepivacaine, metapramine, methadone, methaqualone, methocarbamol, methotrexate, metipranolol, metoclopramide, metoprolol, mexiletine, mianserine, midazolam, minoxidil, moclobemide, moperone, morphine, nadolol, nalbuphine, naloxone, naproxen, nialamide, nicardipine, nifedipine, niflumic acid, nimodipine, nitrazepam, nitrendipine, nomifensine, nortriptyline, omeprazole, opipramol, oxazepam, oxprenolol, penbutolol, penfluridol, pentazocine, phencyclidine, phenylbutazone, pimozide, pindolol, pipamperone, piroxicam, prazepam, prazosin, prilocaine, procainamide, procarbazine, proguanil, promethazine, propafenone, propranolol, protriptyline, pyrimethamine, quinidine, quinine, quinupramine, ramipril, reserpine, secobarbital, sotalol, strychnine, sulfinpyrazole, sulindac, sulpride, suriclone, temazepam, tenoxicam, terfenadine, tetracaine, tetrazepam, thiopental, thioproperazine, thioridazine, tianeptine, tiaprofenic acid, ticlopidine, timolol, tioclomarol, tofisopam, tolbutamide, trazodone, triazolam, trifluoperazine, trifluperidol, trimipramine, triprolidine, tropatenine, verapamil, viloxazine, vinblastine, vincristine, vindesine, warfarin, yohimbine, zolpidem, zopiclone, zorubicine

**Interfering:** albuterol, amisulpride, aspirin, benzocaine, chlormezanone, codeine, lisinopril, mephenesin, metformin, nalorphine, naltrexone, nizatidine, phenobarbital, phenol, ranitidine, ritodrine, sultopride, terbutaline, tiapride, toloxatone

## KEY WORDS
whole blood; plasma

## REFERENCE
Tracqui, A.; Kintz, P.; Mangin, P. Systematic toxicological analysis using HPLC/DAD. *J.Forensic Sci.*, **1995**, *40*, 254–262

## SAMPLE
**Matrix:** blood
**Sample preparation:** Condition a 3 mL Bond Elut Certify SPE cartridge with 2 mL MeOH and 2 mL water. 1 mL Plasma + 100 μL 1000 ng/mL nalorphine + 1 mL water, add to the SPE cartridge at 2 mL/min, wash with 2 mL water, wash with 2 mL MeCN, dry under vacuum for 1 min, elute with 2 mL dichloromethane:isopropanol:ammonium hydroxide 80:20:2. Evaporate the eluate to dryness under a stream of nitrogen at 40°, reconstitute in 100 μL mobile phase, inject a 60 μL aliquot.

## HPLC VARIABLES
**Column:** 150 × 4.6 3 μm Basic C8 (YMC)
**Mobile phase:** MeCN:5 mM $(NH_4)_2HPO_4$ 8:92 adjusted to pH 5.8 with phosphoric acid
**Flow rate:** 1
**Injection volume:** 60
**Detector:** F ex 214 em 345

## CHROMATOGRAM
**Retention time:** 15
**Internal standard:** nalorphine (18)
**Limit of detection:** 5 ng/mL
**Limit of quantitation:** 10 ng/mL

## OTHER SUBSTANCES
**Simultaneous:** morphine, norcodeine

## KEY WORDS
plasma; SPE; pharmacokinetics

## REFERENCE

Weingarten, B.; Wang, H.-Y.; Roberts, D.M. Determination of codeine in human plasma by high-performance liquid chromatography with fluorescence detection. *J.Chromatogr.A*, **1995**, *696*, 83−92

## SAMPLE

**Matrix:** blood

**Sample preparation:** Rock 5 mL whole blood + 10 mL water + 8.5 mL $Na_2WO_4$ in a 50 mL stoppered tube for 1 min, add 6 mL $NiCl_2$, rock for 5 min, add 15 mL dichloromethane:isobutyl alcohol:THF 30:45:25, centrifuge at 2500 g for 15 min. Remove organic phase and repeat the process. Filter all organic phases through a 40-90 μm filter and evaporate to dryness in a 100 mL porcelain dish at a moderate temperature in a sand bath. Take up residue in 500 μL MeCN:water 80:20, inject a 20 μL aliquot. ($Na_2WO_4$ prepared by mixing 10 g $Na_2WO_4.2H_2O$ in 38 mL of 2 M NaOH and 2.5 g of $NaHCO_3$ and making up to 100 mL. $NiCl_2$ was 17% w/v $NiCl_2$ in water.)

## HPLC VARIABLES

**Column:** 200 × 4.6 5 μm Hypersil C8

**Mobile phase:** Gradient. A was MeCN. B was 20 mM n-propylamine adjusted to pH 5 with 85% phosphoric acid. A:B from 15:85 to 20:80 over 5 min to 45:55 over another 15 min to 65:35 over another 5 min.

**Injection volume:** 20

**Detector:** UV 230

## CHROMATOGRAM

**Retention time:** 11

**Limit of detection:** 0.20 ppm

## OTHER SUBSTANCES

**Extracted:** buprenorphine, caffeine, cocaine, diamorphine, ethylmorphine, lidocaine, methaqualone, morphine, naloxone, noscapine, papaverine, pentazocine, procaine

**Also analyzed:** bromazepam, clonazepam, diazepam, flunitrazepam, flurazepam, medazepam, nitrazepam, oxazepam

## KEY WORDS

whole blood

## REFERENCE

Bernal, J.L.; Del Nozal, M.J.; Rosas, V.; Villarino, A. Extraction of basic drugs from whole blood and determination by high performance liquid chromatography. *Chromatographia*, **1994**, *38*, 617−623

## SAMPLE

**Matrix:** blood

**Sample preparation:** 1 mL plasma + IS, vortex 30 s, add 1 mL 50 mM pH 8 phosphate buffer, vortex, add 6 mL hexane:dichloromethane 2:1, shake 5 min, centrifuge, repeat extraction. Combine organic phases and extract with 1 mL 50 mM pH 3 acetate buffer. Make aqueous phase alkaline with 1 mL 0.1 M NaOH and extract with 6 mL hexane:dichloromethane 2:1. Evaporate organic phase to dryness, dissolve residue in 100 μL mobile phase, vortex vigorously, inject 50 μL aliquot.

## HPLC VARIABLES

**Guard column:** pellicular cyano

**Column:** 150 × 4.6 5 μm Zorbax CN

**Mobile phase:** MeCN:50 mM $KH_2PO_4$ 17:83 containing 5 mM sodium octanesulfonate, pH 4.9

**Flow rate:** 1.2
**Injection volume:** 50
**Detector:** F ex 285 em 345

## CHROMATOGRAM
**Retention time:** 5.7
**Internal standard:** isopropylnorcodeine (9)
**Limit of detection:** 5 ng/mL

## KEY WORDS
plasma

## REFERENCE
Mohammed, S.S.; Butschkau, M.; Derendorf, H. A reversed phase liquid chromatographic method for the determination of codeine in biological fluids with applications. *J.Liq.Chromatogr.*, **1993**, *16*, 2325–2334

## SAMPLE
**Matrix:** blood
**Sample preparation:** Condition a 3 mL 40 μm bonded silica Clean Screen SPE cartridge (Worldwide Monitoring) with 3 mL MeOH, 3 mL water, and 1 mL pH 3 phosphate buffer. 1 mL Plasma + 2 mL 10 mM phosphoric acid, mix, add to the SPE cartridge, air dry for 30 s, wash with 1 mL pH 3 phosphate buffer, wash with 1 mL MeOH, air dry for 30 s, elute with 3 mL 2% ammoniacal MeOH. Evaporate the eluate to dryness under a stream of nitrogen at 45°, reconstitute the residue in 50 μL mobile phase, inject an aliquot.

## HPLC VARIABLES
**Column:** 200 × 4.5 5 μm LiChrosphere diol
**Mobile phase:** MeCN:50 mM $NaH_2PO_4$ 80:20 pH, adjusted to 3 with orthophosphoric acid
**Flow rate:** 1
**Injection volume:** 20
**Detector:** UV 230

## CHROMATOGRAM
**Retention time:** 6
**Limit of detection:** 1 ng/mL

## OTHER SUBSTANCES
**Extracted:** metabolites, glucuronides, morphine, normorphine

## KEY WORDS
plasma; SPE

## REFERENCE
Wielbo, D.; Bhat, R.; Chari, G.; Vidyasagar, D.; Tebbett, I.R.; Gulati, A. High-performance liquid chromatographic determination of morphine and its metabolites in plasma using diode-array detection. *J.Chromatogr.*, **1993**, *615*, 164–168

## SAMPLE
**Matrix:** blood
**Sample preparation:** 1 mL Plasma + 100 μL 1 M phosphoric acid + 5 mL butyl chloride, mix for 1.5 min, centrifuge at 1500 g for 3 min, discard upper organic layer. To the aqueous layer add 500 μL pH 10 1 M carbonate buffer and 5 mL butyl chloride, mix for 1.5 min, centrifuge at 1500 g for 3 min, remove organic layer and repeat extraction. Combine butyl chloride layers and evaporate them to dryness under a stream of air at 40°. Reconstitute the residue in 200 μL mobile phase, inject a 150 μL aliquot.

## HPLC VARIABLES
**Guard column:** Novapak C18 guard column
**Column:** 4 μm Novapak C18 in a Waters RCM 8 × 10 radial compression unit
**Mobile phase:** MeOH:MeCN:10 mM pH 7 phosphate buffer 230:20:1000, containing 40 mg/L cetyltrimethylammonium bromide (cetavlon)
**Flow rate:** 2
**Injection volume:** 150
**Detector:** E, Waters Model 460, working electrode 1.10 V

## CHROMATOGRAM
**Retention time:** 13.5
**Internal standard:** codeine

## OTHER SUBSTANCES
**Simultaneous:** oxycodone
**Noninterfering:** acetaminophen, amitriptyline, aspirin, atenolol, camazepam, carbamazepine, chlorimipramine, chlorthalidone, clonazepam, cortisone, desipramine, diazepam, halazepam, hydrochlorothiazide, hydrocortisone, imipramine, lorazepam, maprotiline, meperidine, methylphenobarbital, methylprednisolone, metoclopramide, midazolam, morphine, nalorphine, naloxone, nitrazepam, nortriptyline, oxazepam, oxprenolol, phenobarbital, phenytoin, pindolol, prazepam, prednisolone, prednisone, primidone, prochlorperazine, propranolol, salicylic acid, temazepam

## KEY WORDS
plasma; codeine is IS

## REFERENCE
Smith, M.T.; Watt, J.A.; Mapp, G.P.; Cramond, T. Quantitation of oxycodone in human plasma using high-performance liquid chromatography with electrochemical detection. *Ther.Drug Monit.*, **1991**, *13*, 126–130

## SAMPLE
**Matrix:** blood, CSF
**Sample preparation:** Prepare 500 mg 3 mL Bond Elut C2 SPE cartridges by rinsing with 2 mL MeOH then 2 mL 50 mM pH 7.5 Tris-HCl buffer. Apply 1 mL serum or CSF + 1 mL 50 mM pH 7.5 Tris-HCl buffer to the cartridge and wash with 10 mL 50 mM pH 7.5 Tris-HCl buffer. Elute with 2 mL 50% MeCN containing 0.1% trifluoroacetic acid. Freeze dry eluent or dry an aliquot at 40° under a stream of nitrogen, dissolve residue in 2560 μL mobile phase, inject 20-200 μL aliquot.

## HPLC VARIABLES
**Guard column:** hexyl
**Column:** 150 × 4.6 Spherisorb S5 C6
**Mobile phase:** Gradient. A 0.1% trifluoroacetic acid in water, B 0.1% trifluoroacetic acid in 40% MeCN. 16% B for 2 min then to 50% B over 10 min then to 100% B over 2 min, after 7 min return to original conditions over 2 min.
**Flow rate:** 1
**Injection volume:** 20-200
**Detector:** F ex 280 em 335

## CHROMATOGRAM
**Retention time:** 13
**Limit of detection:** 1.11 ng/mL

## OTHER SUBSTANCES
**Extracted:** metabolites, dihydrocodeine, morphine, normorphine
**Simultaneous:** diamorphine, dihydrocodeine

## KEY WORDS
serum; SPE

## REFERENCE
Venn, R.F.; Michalkiewicz, A. Fast reliable assay for morphine and its metabolites using high-performance liquid chromatography and native fluorescence detection. *J.Chromatogr.*, **1990**, *525*, 379–388

## SAMPLE
**Matrix:** blood, urine
**Sample preparation:** Condition two 130 mg Sep-Pak Light C18 SPE cartridges with 1 mL MeOH and 1 mL water. Dilute urine, if necessary, 20-fold with water. 1 mL Plasma, urine, or diluted urine + 1 mL 500 mM pH 9.3 ammonium sulfate buffer, mix, add 1.9 mL of this mixture to a SPE cartridge at 0.75 mL/min, wash with 4 mL 5 mM pH 9.1 ammonium sulfate buffer at 1.5 mL/min, wash with 200 μL MeCN:30 mM pH 2.1 potassium phosphate buffer 15:85 at 0.75 mL/min, elute with 600 μL MeCN:30 mM pH 2.1 potassium phosphate buffer 15:85 at 0.75 mL/min. Mix the eluate with 1 mL 500 mM pH 9.3 ammonium sulfate buffer, add to a second SPE cartridge at 0.75 mL/min, wash with 4 mL 5 mM pH 9.1 ammonium sulfate buffer at 1.5 mL/min, wash with 200 μL MeCN:30 mM pH 2.1 potassium phosphate buffer 15:85 at 0.75 mL/min, elute with 600 μL MeCN:30 mM pH 2.1 potassium phosphate buffer 15:85 at 0.75 mL/min, inject a 400 μL aliquot of the eluate.

## HPLC VARIABLES
**Column:** 100 × 4 3 μm Spherisorb S3 ODS2
**Mobile phase:** MeCN:buffer 22:78 (Buffer was 30 mM $KH_2PO_4$ containing 3 mM dodecyl sulfate, adjusted to pH 2.1 with phosphoric acid.)
**Flow rate:** 1.5
**Injection volume:** 400
**Detector:** UV 214

## CHROMATOGRAM
**Retention time:** 22
**Limit of detection:** 20 nM

## OTHER SUBSTANCES
**Extracted:** metabolites, morphine (electrochemical detection), norcodeine, normorphine (electrochemical detection)

## KEY WORDS
plasma; SPE

## REFERENCE
Svensson, J.O.; Yue, Q.Y.; Säwe, J. Determination of codeine and metabolites in plasma and urine using ion-pair high-performance liquid chromatography. *J.Chromatogr.B*, **1995**, *674*, 49–55

## SAMPLE
**Matrix:** blood, urine
**Sample preparation:** Plasma. Condition a Toxiclean SPE cartridge (Alltech) with 3 mL MeOH, two 3 mL portions of water, and 2 mL buffer. 100 μL Plasma or serum + 100 μL MeOH + 200 μL MeCN + 100 μL buffer, vortex for 1 min, centrifuge at 4000 rpm for 15 min, add the supernatant to the SPE cartridge, wash with two 3 mL portions of water, dry under vacuum for 10 min, elute with 2 mL MeOH. Evaporate the eluate to dryness under a stream of nitrogen at 45°, reconstitute the residue in 100 μL 2.5 μg/mL flufenamic acid in MeOH (?), inject an aliquot. Urine. Condition a Bond Elut C8 SPE cartridge with 3 mL MeOH, two 3 mL portions of water, and 2 mL buffer. 100 μL Urine + 100 μL MeOH + 200 μL MeCN + 500 μL buffer, vortex for 1 min, centrifuge at 2000 rpm for 5 min,

add the supernatant to the SPE cartridge, wash with two 3 mL portions of water, dry under vacuum for 10 min, elute with 2 mL MeOH. Evaporate the eluate to dryness under a stream of nitrogen at 45°, reconstitute the residue in 100 μL 2.5 μg/mL flufenamic acid in MeOH (?), inject an aliquot. (Buffer was 250 mL 25 mM sodium borate and 18 mL 100 mM NaOH, pH 9.2.)

## HPLC VARIABLES
**Column:** 250 × 4.6 5 μm Adsorbosphere HS C18
**Mobile phase:** MeCN:MeOH:1.2% ammonium acetate 15:40:45
**Flow rate:** 0.8
**Detector:** UV 239

## CHROMATOGRAM
**Retention time:** 9.17
**Internal standard:** flufenamic acid (24.39)
**Limit of quantitation:** 100 ng/mL (blood); 300 ng/mL (urine)

## OTHER SUBSTANCES
**Extracted:** monoacetylmorphine, morphine, papaverine

## KEY WORDS
SPE; plasma; serum

## REFERENCE
Theodoridis, G.; Papadoyannis, I.; Tsoukali-Papadopoulou, H.; Vasilikiotis, G. A comparative study of different solid phase extraction procedures for the analysis of alkaloids of forensic interest in biological fluids by RP-HPLC/diode array. *J.Liq.Chromatogr.*, **1995**, *18*, 1973–1975

## SAMPLE
**Matrix:** bulk
**Sample preparation:** Prepare a 750 μg/mL solution in 10 mM pH 2.5 orthophosphoric acid, sonicate for 10 min, filter (0.2 μm), inject a 15 μL aliquot.

## HPLC VARIABLES
**Guard column:** 4 × 4 5 μm LiChrospher 100
**Column:** 125 × 4 3 μm Spherisorb ODS-1
**Mobile phase:** Gradient. A was water containing 5 mL/L 85% orthophosphoric acid and 0.56 mL/L hexylamine. B was MeCN:water 90:10 containing 5 mL/L 85% orthophosphoric acid and 0.56 mL/L hexylamine. A:B from 91:9 to 86:14 over 4 min, maintain at 86:14 for 13 min, to 55:45 over 11 min, maintain at 55:45 for 8 min, re-equilibrate at initial conditions for 20 min.
**Flow rate:** 0.7
**Injection volume:** 15
**Detector:** UV 210

## CHROMATOGRAM
**Retention time:** 6.7

## OTHER SUBSTANCES
**Simultaneous:** acetaminophen, acetylcodeine, benzocaine, caffeine, cocaine, diamorphine, lidocaine, 6-monoacetylmorphine, morphine, noscapine, papaverine, procaine

## REFERENCE
Grogg-Sulser, K.; Helmlin, H.-J.; Clerc, J.-T. Qualitative and quantitative determination of illicit heroin street samples by reversed-phase high-performance liquid chromatography: method development by CARTAGO-S. *J.Chromatogr.A*, **1995**, *692*, 121–129

## SAMPLE
**Matrix:** bulk
**Sample preparation:** Dissolve in mobile phase, filter (0.45 μm), inject an aliquot.

## HPLC VARIABLES
**Column:** 300 × 3.9 10 μm μBondapak phenyl
**Mobile phase:** MeOH:7 mM pH 3.1 triethylammonium phosphate buffer 20:80
**Flow rate:** 1
**Injection volume:** 25
**Detector:** UV 254

## CHROMATOGRAM
**Retention time:** 7

## OTHER SUBSTANCES
**Simultaneous:** α-codeimethine, $O^6$-codeine methyl ether, meconic acid, morphine

## REFERENCE
Ayyangar, N.R.; Bhide, S.R.; Kalkote, U.R. Assay of semi-synthetic codeine base with simultaneous determination of alpha-codeimethine and O6-codeine methyl ether as by-product impurities by high-performance liquid chromatography. *J.Chromatogr.*, **1990**, *519*, 250–255

## SAMPLE
**Matrix:** formulations
**Sample preparation:** Dilute syrup with mobile phase to a concentration of 5-100 μg/mL, shake, filter, inject an aliquot.

## HPLC VARIABLES
**Column:** 250 × 4.6 5 μm 80 Å Ultrasphere CN
**Mobile phase:** MeCN:water:EtOH 60:38:2 containing 1 mM perchloric acid
**Column temperature:** 30
**Flow rate:** 1
**Injection volume:** 20
**Detector:** Conductivity, zero suppression 2, range 1 or 10

## CHROMATOGRAM
**Retention time:** 9.0

## OTHER SUBSTANCES
**Simultaneous:** bromhexine, chlorpheniramine, dextromethorphan, diphenhydramine, ephedrine, papaverine, phenylephrine

## KEY WORDS
syrup; indirect conductometric detection; presence of compound causes a decrease in mobile phase conductivity

## REFERENCE
Lau, O.-W.; Mok, C.-S. High-performance liquid chromatographic determination of active ingredients in cough-cold syrups with indirect conductometric detection. *J.Chromatogr.A*, **1995**, *693*, 45–54

## SAMPLE
**Matrix:** formulations

## HPLC VARIABLES
**Column:** 300 × 3.9 μBondapak C18

**Mobile phase:** MeOH:buffer 20:80 (Buffer was 15 mM 1-butanesulfonic acid + 15 mM KH$_2$PO$_4$ + 2 mL/L triethylamine, pH adjusted to 4.8 ± 0.1 with dilute phosphoric acid.)
**Column temperature:** 50
**Flow rate:** 2
**Injection volume:** 20
**Detector:** UV 214

## CHROMATOGRAM
**Retention time:** 4.9

## OTHER SUBSTANCES
**Simultaneous:** acetaminophen, p-aminophenol, benzoic acid, codeine N-oxide, codeinone

## KEY WORDS
elixir; stability-indicating

## REFERENCE
Sisco, W.R.; Rittenhouse, C.T.; Everhart, L.A.; McLaughlin, A.M. Simultaneous high-performance liquid chromatographic stability-indicating analysis of acetaminophen, codeine phosphate, and sodium benzoate in elixirs. *J.Chromatogr.*, **1986**, *354*, 355–366

## SAMPLE
**Matrix:** formulations
**Sample preparation:** Dissolve capsules and tablets in MeOH:pH 4.0 water 1:1, shake for 1 (capsules) or 4 (tablets) h, dilute a 10 mL aliquot with 40 mL pH 3.2 water, filter (0.45 μm), collect last portion of filtrate, inject a 20 μL aliquot. (pH 3.2 and 4.0 water are prepared by adjusting pH of distilled water with phosphoric acid.)

## HPLC VARIABLES
**Column:** 300 × 3.9 μBondapak C18
**Mobile phase:** MeOH:buffer 7:93 (Buffer was 15 mM KH$_2$PO$_4$ + 2 mL triethylamine per liter. Adjusted to pH 2.35 ± 0.1 with concentrated phosphoric acid.)
**Column temperature:** 40
**Flow rate:** 3
**Injection volume:** 20
**Detector:** UV 214

## CHROMATOGRAM
**Retention time:** 5.3

## OTHER SUBSTANCES
**Simultaneous:** degradation products, acetaminophen, p-aminophenol, codeine-N-oxide, codeinone

## KEY WORDS
capsules; tablets; rugged

## REFERENCE
Sisco, W.R.; Rittenhouse, C.T.; Everhart, L.A. Simultaneous high-performance liquid chromatographic stability-indicating analysis of acetaminophen and codeine phosphate in tablets and capsules. *J.Chromatogr.*, **1985**, *348*, 253–263

## SAMPLE
**Matrix:** formulations

**Sample preparation:** Add 1 tablet to 95 mL water, place on a steam bath for 15 min, cool, mix for 15 min, sonicate, allow to stand, filter, inject 13 μL aliquot.

## HPLC VARIABLES
**Column:** μBondapak C18
**Mobile phase:** MeOH:buffer 25:75 (Buffer was 0.01 N $KH_2PO_4$ + 50 mM $KNO_3$, adjusted to pH 4.5 with 3 N phosphoric acid.)
**Flow rate:** 1.1
**Injection volume:** 13
**Detector:** UV 283

## CHROMATOGRAM
**Retention time:** 7.3

## OTHER SUBSTANCES
**Simultaneous:** acetaminophen, p-aminophenol, p-chloroacetanilide, hydrocodone, hydromorphone

## KEY WORDS
tablets

## REFERENCE
Wallo, W.E.; D'Adamo, A. Simultaneous assay of hydrocodone bitartrate and acetaminophen in a tablet formulation. *J.Pharm.Sci.*, **1982**, *71*, 1115–1118

## SAMPLE
**Matrix:** microsomal incubations
**Sample preparation:** Prepare a 1 mL 100 mg C18 Bond Elut SPE cartridge by washing with 1 mL MeOH, 1 mL water, 1 mL 5 mM pH 9.0 carbonate buffer. Mix 100 μL microsomal incubation, 20 μL 25 μg/mL 10,11-dihydrocarbamazepine in MeCN:water 25:75, 600 μL 200 mM pH 10.2 carbonate buffer, 80 μL 20 mM tetrabutylammonium hydrogen sulfate in water with vortex mixing after each addition. Add to the SPE cartridge, wash with 1 mL 5 mM pH 9.0 carbonate buffer, elute with 0.5 mL MeCN:mobile phase buffer 40:60.

## HPLC VARIABLES
**Guard column:** 20 × 2 Phase Separations pellicular ODS
**Column:** 250 × 4.6 5μm Hypersil CPS (cyanopropyl)
**Mobile phase:** MeCN:buffer 24:76 (Buffer was 50 mM potassium hydrogen phosphate containing 1 mM sodium dodecyl sulfate adjusted to pH 2.5 with orthophosphoric acid.)
**Flow rate:** 1
**Injection volume:** 20
**Detector:** UV 210; E, ESA Coulochem II with a 5020 guard cell (+0.60 V) and a 5011 analytical cell (cell 1 +0.22 V, cell 2 +0.45 V)

## CHROMATOGRAM
**Retention time:** 14
**Internal standard:** 10,11-dihydrocarbamazepine
**Limit of detection:** 5 ng/mL

## OTHER SUBSTANCES
**Simultaneous:** metabolites, morphine, norcodeine

## KEY WORDS
SPE

## REFERENCE

Pawula, M.; Shaw, P.N.; Barrett, D.A. Determination of codeine and its metabolites in microsomal incubates by high-performance liquid chromatography. *J.Chromatogr.B*, **1994**, *653*, 106–111

## SAMPLE
**Matrix:** solutions

## HPLC VARIABLES
**Column:** 250 × 4.1 6 µm PolyEncap ODS (n-octadecylacrylate copolymerized with vinyl silica in heptane, carrier Ultrasep ES 100; preparation described in paper)
**Mobile phase:** MeCN:pH 2.2 phosphate buffer 20:80
**Flow rate:** 1
**Detector:** UV 220

## CHROMATOGRAM
**Retention time:** 2

## OTHER SUBSTANCES
**Simultaneous:** atropine, barbituric acid, diphenhydramine, noscapine, papaverine

## REFERENCE

Engelhardt, H.; Cuñat-Walter, M.A. Polymer encapsulated stationary phases with improved efficiency. *Chromatographia*, **1995**, *40*, 657–661

## SAMPLE
**Matrix:** solutions

## HPLC VARIABLES
**Column:** 250 × 4.6 5 µm Supelcosil LC-DP (A) or 250 × 4 5 µm LiChrospher 100 RP-8 (B)
**Mobile phase:** MeCN:0.025% phosphoric acid:buffer 25:10:5 (A) or 60:25:15 (B) (Buffer was 9 mL concentrated phosphoric acid and 10 mL triethylamine in 900 mL water, adjust pH to 3.4 with dilute phosphoric acid, make up to 1 L.)
**Flow rate:** 0.6
**Injection volume:** 25
**Detector:** UV 229

## CHROMATOGRAM
**Retention time:** 6.09 (A), 3.42 (B)

## OTHER SUBSTANCES
**Also analyzed:** acebutolol, acepromazine, acetaminophen, acetazolamide, acetophenazine, albuterol, alprazolam, amitriptyline, amobarbital, amoxapine, antipyrine, atenolol, atropine, azatadine, baclofen, benzocaine, bromocriptine, brompheniramine, brotizolam, bupivacaine, buspirone, butabarbital, butalbital, caffeine, carbamazepine, cetirizine, chlorcyclizine, chlordiazepoxide, chlormezanone, chloroquine, chlorpheniramine, chlorpromazine, chlorpropamide, chlorprothixene, chlorthalidone, chlorzoxazone, cimetidine, cisapride, clomipramine, clonazepam, clonidine, clozapine, cocaine, colchicine, cyclizine, cyclobenzaprine, dantrolene, desipramine, diazepam, diclofenac, diflunisal, diltiazem, diphenhydramine, diphenidol, diphenoxylate, dipyridamole, disopyramide, dobutamine, doxapram, doxepin, droperidol, encainide, ethidium bromide, ethopropazine, fenoprofen, fentanyl, flavoxate, fluoxetine, fluphenazine, flurazepam, flurbiprofen, fluvoxamine, furosemide, glutethimide, glyburide, guaifenesin, haloperidol, homatropine, hydralazine, hydrochlorothiazide, hydrocodone, hydromorphone, hydroxychloroquine, hydroxyzine, ibuprofen, imipramine, indomethacin, ketoconazole, ketoprofen, ketorolac, labetalol, levorphanol, lidocaine, loratadine, lorazepam, lovastatin, loxapine, mazindol,

mefenamic acid, meperidine, mephenytoin, mepivacaine, mesoridazine, metaproterenol, methadone, methdilazine, methocarbamol, methotrexate, methotrimeprazine, methoxamine, methyldopa, methylphenidate, metoclopramide, metolazone, metoprolol, metronidazole, midazolam, moclobemide, morphine, nadolol, nalbuphine, naloxone, naphazoline, naproxen, nifedipine, nizatidine, norepinephrine, nortriptyline, oxazepam, oxycodone, oxymetazoline, paroxetine, pemoline, pentazocine, pentobarbital, pentoxifylline, perphenazine, pheniramine, phenobarbital, phenol, phenolphthalein, phentolamine, phenylbutazone, phenyltoloxamine, phenytoin, pimozide, pindolol, piroxicam, pramoxine, prazepam, prazosin, probenecid, procainamide, procaine, prochlorperazine, procyclidine, promazine, promethazine, propafenone, propantheline, propiomazine, propofol, propranolol, protriptyline, quazepam, quinidine, quinine, racemethorphan, ranitidine, remoxipride, risperidone, salicylic acid, scopolamine, secobarbital, sertraline, sotalol, spironolactone, sulfinpyrazone, sulindac, temazepam, terbutaline, terfenadine, tetracaine, theophylline, thiethylperazine, thiopental, thioridazine, thiothixene, timolol, tocainide, tolbutamide, tolmetin, trazodone, triamterene, triazolam, trifluoperazine, triflupromazine, trimeprazine, trimethoprim, trimipramine, verapamil, warfarin, xylometazoline, yohimbine, zopiclone

## KEY WORDS
some details of plasma extraction

## REFERENCE
Koves, E.M. Use of high-performance liquid chromatography-diode array detection in forensic toxicology. *J.Chromatogr.A*, **1995**, *692*, 103–119

## SAMPLE
**Matrix:** solutions

## HPLC VARIABLES
**Column:** 250 × 4.6 Zorbax RX

**Mobile phase:** Gradient. A was 10 mL concentrated orthophosphoric acid and 7 mL triethylamine in 1 L water. B was 10 mL concentrated orthophosphoric acid and 7 mL triethylamine in 200 mL water, make up to 1 L with MeCN. A:B from 100:0 to 0:100 over 30 min, maintain at 0:100 for 5 min.

**Column temperature:** 30

**Flow rate:** 2

**Detector:** UV 210

## OTHER SUBSTANCES
**Also analyzed:** acepromazine, acetaminophen, acetophenazine, albuterol, aminophylline, amitriptyline, amobarbital, amoxapine, amphetamine, amylocaine, antipyrine, aprobarbital, aspirin, atenolol, atropine, avermectin, barbital, benzocaine, benzoic acid, benzotropine, benzphetamine, berberine, bibucaine, bromazepan, brompheniramine, buprenorphine, buspirone, butabarbital, butacaine, butethal, caffeine, carbamazepine, carbromal, chloramphenicol, chlordiazepoxide, chloroquine, chlorothiazide, chloroxylenol, chlorphenesin, chlorpheniramine, chlorpromazine, chlorpropamide, chlortetracycline, cimetidine, cinchonidine, cinchonine, clenbuterol, clonazepam, clonixin, clorazepate, cocaine, colchicine, cortisone, coumarin, cyclazocine, cyclobenzaprine, cyclothiazide, cyheptamide, cymarin, danazol, danthron, dapsone, debrisoquine, desipramine, dexamethasone, dextromethorphan, dextropropoxyphene, diamorphine, diazepam, diclofenac, diethylpropion, diethylstilbestrol, diflunisal, digitoxin, digoxin, diltiazem, diphenhydramine, diphenoxylate, diprenorphine, dipyrone, disulfiram, dopamine, doxapram, doxepin, dronabinol, ephedrine, epinephrine, epinine, estradiol, estriol, estrone, ethacrynic acid, ethosuximide, etonitazene, etorphine, eugenol, famotidine, fenbendazole, fencamfamine, fenoprofen, fenproporex, fentanyl, flubendazole, flufenamic acid, flunitrazepam, 5-fluorouracil, fluoxymesterone, fluphenazine, furosemide, gentisic acid, gitoxigenin, glipizide, glunixin, glutethimide, glybenclamide, guaiacol, halazepam, haloperidol, hydrochlorothiazide, hydrocodone, hydrocortisone, hydromorphone, hydroxyquinoline, ibogaine, ibuprofen, im-

inostilbene, imipramine, indomethacin, isocarbostyril, isocarboxazid, isoniazid, isoproterenol, isoxsuprine, ivermectin, ketamine, ketoprofen, kynurenic acid, levorphanol, lidocaine, lorazepam, lormetazepam, loxapine, mazindol, mebendazole, meclizine, meclofenamic acid, medazepam, mefenamic acid, megestrol, mepacrine, meperidine, mephentermine, mephenytoin, mephesin, mephobarbital, mepivacaine, mescaline, mesoridazine, methadone, methamphetamine, methapyrilene, methaqualone, methazolamide, methocarbamol, methoxamine, methsuximide, methyl salicylate, methyldopa, methyldopamine, methylphenidate, methylprednisolone, methyltestosterone, methyprylon, metoprolol, mibolerone, morphine, nadolol, nalorphine, naloxone, naltrexone, naphazoline, naproxen, nefopam, niacinamide, nicotine, nicotinic acid, nifedipine, niflumic acid, nitrazepam, norepinephrine, nortriptyline, noscapine, nylidrin, oxazepam, oxycodone, oxymorphone, oxyphenbutazone, oxytetracycline, papaverine, pargyline, pemoline, pentazocine, pentobarbital, persantine, phenacetin, phenazocine, phenazopyridine, phencyclidine, phendimetrazine, phenelzine, pheniramine, phenobarbital, phenothiazine, phensuximide, phentermine, phenylbutazone, phenylephrine, phenylpropanolamine, piperocaine, prazepam, prednisolone, primidone, probenecid, progesterone, propiomazine, propranolol, propylparaben, pseudoephedrine, puromycin, pyrilamine, pyrithyldione, quazepam, quinaldic acid, quinidine, quinine, ranitidine, recinnamine, reserpine, resorcinol, saccharin, albuterol, salicylamide, salicylic acid, scopolamine, scopoletin, secobarbital, strychnine, sulfacetamide, sufadiazine, sulfadimethoxine, sulfaethidole, sulfamerazine, sulfamethazine, sulfamethoxizole, sulfanilamide, sulfapyridine, sulfasoxizole, sulindac, tamoxifen, temazepam, testosterone, tetracaine, tetracycline, tetramisole, thebaine, theobromine, theophylline, thiabendazole, thiamine, thiamylal, thiobarbituric acid, thioridazine, thiosalicylic acid, thiothixene, thymol, tolazamide, tolazoline, tobutamide, tolmetin, tranylcypromine, triamcinolone, tribenzylamine, trichloromethiazide, trifluoperazine, trihexyphenidyl, trimethoprim, tripelennamine, triproilidine, tropacocaine, tyramine, verapamil, vincamine, warfarin, yohimbine, zoxazolamine

## REFERENCE
Hill, D.W.; Kind, A.J. Reversed-phase solvent gradient HPLC retention indexes of drugs. *J.Anal.Toxicol.*, **1994**, *18*, 233–242

## SAMPLE
**Matrix:** solutions

## HPLC VARIABLES
**Column:** 150 × 4.6 10 μm PRP-1 (Hamilton)
**Mobile phase:** Gradient. MeCN:20 mM ammonium hydroxide from 15:85 to 100:0 over 17 min
**Flow rate:** 1
**Detector:** UV 220

## CHROMATOGRAM
**Retention time:** 5

## OTHER SUBSTANCES
**Simultaneous:** cocaine, methadone, reserpine, thebaine, yohimbine

## REFERENCE
Keystone Scientific Catalog, **1993-4**, p. 22

## SAMPLE
**Matrix:** solutions

## HPLC VARIABLES
**Column:** 150 × 4.6 Supelcosil LC-ABZ
**Mobile phase:** MeCN:25 mM pH 6.9 potassium phosphate buffer 35:65

**Flow rate:** 1.5
**Injection volume:** 25
**Detector:** UV 254

## CHROMATOGRAM
**Retention time:** 2.016

## OTHER SUBSTANCES
**Also analyzed:** 6-acetylmorphine, amiloride, amphetamine, benzocaine, benzoylecgonine, caffeine, cocaine, doxylamine, fluoxetine, glutethimide, hexobarbital, hypoxanthine, levorphanol, LSD, meperidine, mephobarbital, methadone, methylphenidate, methyprylon, N-norcodeine, oxazepam, oxycodone, phenylpropanolamine, prilocaine, procaine, terfenadine

## REFERENCE
Ascah, T.L. Improved separations of alkaloid drugs and other substances of abuse using Supelcosil LC-ABZ column. *Supelco Reporter*, **1993**, *12(3)*, 18–21

## SAMPLE
**Matrix:** solutions
**Sample preparation:** Inject a 20 µL aliquot.

## HPLC VARIABLES
**Column:** 250 × 4.6 5 µm Supelco C18
**Mobile phase:** MeCN:buffer 70:30 (Buffer contained 2.88% sodium lauryl sulfate and 1.248% $NaH_2PO_4$, adjusted to pH 3 with orthophosphoric acid.)
**Flow rate:** 2
**Injection volume:** 20
**Detector:** UV 254

## CHROMATOGRAM
**Retention time:** 2.3

## OTHER SUBSTANCES
**Simultaneous:** drotaverine, ethaverine, moxaverine, papaverine

## REFERENCE
Girgis, E.H. Ion-pair reversed-phase liquid chromatographic identification and quantitation of papaverine congeners. *J.Pharm.Sci.*, **1993**, *82*, 503–505

## SAMPLE
**Matrix:** solutions
**Sample preparation:** Dissolve in mobile phase, inject an aliquot.

## HPLC VARIABLES
**Guard column:** 15 × 3.2 7 µm Applied Biosystems pre-column
**Column:** 100 × 2 10 µm µPorasil
**Mobile phase:** MeCN:5 mM pH 3.75 sodium acetate 80:20
**Flow rate:** 1
**Injection volume:** 200
**Detector:** UV 214

## CHROMATOGRAM
**Retention time:** 16.0
**Limit of detection:** 2.9 ng/mL

## OTHER SUBSTANCES
**Simultaneous:** buprenorphine, butorphanol, ethylmorphine, fentanyl, meperidine, morphine, nalbuphine, nalorphine, tramadol
**Noninterfering:** atropine, diazepam, neostigmine, pancuronium, succinylcholine, thiopental

## REFERENCE
Ho, S.-T.; Wang, J.-J.; Ho, W.; Hu, O.Y.-P. Determination of buprenorphine by high-performance liquid chromatography with fluorescence detection: application to human and rabbit pharmacokinetic studies. *J.Chromatogr.*, **1991**, *570*, 339–350

## SAMPLE
**Matrix:** solutions

## HPLC VARIABLES
**Column:** 300 × 3.9 10 μm LiChrosorb Si-60
**Mobile phase:** MeOH:water 60:40 containing 4 mM disodium citrate and 4 mM tetrabutylammonium bromide, pH 5.9
**Flow rate:** 1
**Injection volume:** 10
**Detector:** UV 254

## CHROMATOGRAM
**Retention time:** 15

## OTHER SUBSTANCES
**Simultaneous:** atropine, dansylamide, dansylcadaverine, doxorubicin, methylatropine, naphazoline, noscapine, xylometazoline

## REFERENCE
Lingeman, H.; van Munster, H.A.; Beynen, J.H.; Underberg, W.J.; Hulshoff, A. High-performance liquid chromatographic analysis of basic compounds on non-modified silica gel and aluminium oxide with aqueous solvent mixtures. *J.Chromatogr.*, **1986**, *352*, 261–274

## SAMPLE
**Matrix:** solutions

## HPLC VARIABLES
**Column:** Supelco LC-8
**Mobile phase:** MeOH:water:acetic acid 40:59:1 containing 100 mM potassium nitrate, 10 mM tetramethylammonium bromide, and 2.5 mM heptanesulfonic acid
**Flow rate:** 1
**Detector:** E, Metrohm 1096/2, platinum working electrode +0.4 V, Ag/AgCl reference electrode following post-column reaction. The column effluent passed through an electrochemical cell (construction details in paper) and the bromide was oxidized to bromine at 3 μA. The mixture flowed through a 20 s reaction coil (3.9 m (?) × 0.33 mm ID) to the detector.

## CHROMATOGRAM
**Retention time:** 5
**Limit of detection:** 0.5 ng

## OTHER SUBSTANCES
**Simultaneous:** morphine, noscapine, papaverine

## KEY WORDS
post-column reaction

## REFERENCE
Kok, W.T.; Brinkman, U.A.T.; Frei, R.W. On-line electrochemical reagent production for detection in liquid chromatography and continuous flow systems. *Anal.Chim.Acta*, **1984**, *162*, 19−32

## SAMPLE
**Matrix:** solutions
**Sample preparation:** Dissolve in MeOH at a concentration of 1 mg/mL, inject a 20 μL aliquot.

## HPLC VARIABLES
**Column:** 250 × 5 Spherisorb S5W
**Mobile phase:** MeOH:buffer 90:10 (Buffer was 94 mL 35% ammonia and 21.5 mL 70% nitric acid in 884 mL water, adjust the pH to 10.1 with ammonia.)
**Flow rate:** 2
**Injection volume:** 20
**Detector:** UV 254

## CHROMATOGRAM
**Retention time:** 2.92

## OTHER SUBSTANCES
**Simultaneous:** acetylcodeine, amphetamine, benzphetamine, benzylmorphine, buprenorphine, caffeine, chlorphentermine, dextromoramide, dextropropoxyphene, diamorphine, diethylpropion, dihydrocodeine, dihydromorphine, dimethylamphetamine, dipipanone, ephedrine, epinephrine, ethoheptazine, etorphine, fencamfamin, fenethyline, fenfluramine, fentanyl, hydrocodone, hydroxypethidine, levallorphan, levorphanol, mazindol, meperidine, mephentermine, mescaline, methadone, methamphetamine, methylenedioxyamphetamine, methylephedrine, methylphenidate, monoacetylmorphine, morphine-3-glucuronide, nalorphine, naloxone, norcodeine, norlevorphanol, normethadone, normorphine, norpethidine, norpipanone, norpseudoephedrine, noscapine, oxycodone, papaverine, pemoline, pentazocine, phenazocine, phendimetrazine, phenelzine, phenoperidine, phentermine, phenylephrine, phenylpropanolamine, pholcodine, pipradol, piritramide, pseudoephedrine, thebacon, thebaine, tranylcypromine, trimethoxyamphetamine, tyramine
**Noninterfering:** dopamine, levodopa, methyldopa, methyldopate, norepinephrine
**Interfering:** bromo-STP, codeine-N-oxide, ethylmorphine, 4-hydroxyamphetamine, morphine, morphine-N-oxide, normetanephrine, 2-phenethylamine, prolintane, STP

## REFERENCE
Law, B.; Gill, R.; Moffat, A.C. High-performance liquid chromatography retention data for 84 basic drugs of forensic interest on a silica column using an aqueous methanol eluent. *J.Chromatogr.*, **1984**, *301*, 165−172

## SAMPLE
**Matrix:** urine
**Sample preparation:** Condition a 300 mg Bond Elut Certify SPE cartridge with 2 mL MeOH and 2 mL water. 5 mL Urine + 1 mL concentrated HCl, vortex, heat at 120° for 30 min, cool, adjust pH to between 7.0 and 8.0 with 10 M KOH. 5 mL Urine or hydrolyzed urine + nalorphine, add to the SPE cartridge, wash with 2 mL water, wash with 1 mL pH 4 acetate buffer, wash with 2 mL MeOH, elute with 2 mL dichloromethane:isopropanol 80:20 containing 2% ammonia. Evaporate the eluate to dryness under a stream of nitrogen, reconstitute the residue in 0.5-1 mL pentane:dichloromethane 90:10. (Use unhydrolyzed urine to determine diamorphine and unconjugated compounds.)

## HPLC VARIABLES
**Column:** 200 × 2 3 μm Hypersil

**Mobile phase:** Pentane:dichloromethane:MeOH containing 0.5% diethylamine 65:29.8:
5.2
**Flow rate:** 0.4
**Injection volume:** 50
**Detector:** UV 280

## CHROMATOGRAM
**Retention time:** 8
**Internal standard:** nalorphine (5)
**Limit of detection:** <20 ng/mL

## OTHER SUBSTANCES
**Extracted:** diamorphine, 6-monoacetylmorphine, pholcodine, dihydrocodeine, morphine
**Simultaneous:** diphenhydramine, ephedrine, hydrocodone
**Noninterfering:** aspirin, caffeine, chlordiazepoxide, dextropropoxyphene, diazepam, lig-
nocaine, naloxone, norcodeine, normorphine, papaverine, procaine, quinine, theobromine,
theophylline

## KEY WORDS
normal phase; SPE

## REFERENCE
Low, A.S.; Taylor, R.B. Analysis of common opiates and heroin metabolites in urine by high-performance
liquid chromatography. *J.Chromatogr.B*, **1995**, *663*, 225–233

## SAMPLE
**Matrix:** urine
**Sample preparation:** 1 mL Urine + 0.5 mL 1% trichloroacetic acid, centrifuge at 5200 g
for 10 min, filter (0.2 μm), inject 20 μL aliquot.

## HPLC VARIABLES
**Column:** 250 × 4 Lichrospher 5μm 60 RP-select B
**Mobile phase:** Gradient. MeCN:50 mM pH 3.2 potassium phosphate buffer from 10:90 to
75:25 over 7 min, hold at 75:25 for 3 min, return to 10:90 over 5 min, equilibrate at
10:90 for 5 min.
**Flow rate:** 1.5
**Injection volume:** 20
**Detector:** UV 190-370

## CHROMATOGRAM
**Retention time:** 4.5

## OTHER SUBSTANCES
**Extracted:** amitriptyline, amphetamine, benzoylecgonine, meperidine, morphine, nordiaze-
pam, norpropoxyphene
**Also analyzed:** cocaine (different gradient), diphenhydramine, lidocaine, nortriptyline,
phenylpropanolamine
**Interfering:** ephedrine

## REFERENCE
Li, S.; Gemperline, P.J.; Briley, K.; Kazmierczak, S. Identification and quantitation of drugs of abuse in
urine using the generalized rank annihilation method of curve resolution. *J.Chromatogr.B*, **1994**,
*655*, 213–223

## SAMPLE
**Matrix:** urine

**Sample preparation:** 1 mL Urine + 10 µL nalorphine solution + 3000-3500 U glucuron-idase (Patella vulgata, Sigma) + 300 µL 1.1 M pH 5 sodium acetate buffer, heat overnight at 55°, add 500 µL buffer, add 8 mL chloroform:isopropanol 90:10, rotate gently for 30 min, centrifuge at 2500 g for 10 min. Remove the organic layer and add it to 3 mL pH 9.9 NaH$_2$PO$_4$ buffer, rotate gently for 10 min, centrifuge, discard the aqueous layer, repeat the wash. Remove the organic layer and add it to 200 µL 0.2% phosphoric acid, rotate gently for 30 min, inject a 50 µL aliquot of the aqueous layer. (Buffer was 50 mM boric acid and 43 mM sodium tetraborate, pH adjusted to 9.9.)

## HPLC VARIABLES
**Guard column:** Nova-Pak phenyl
**Column:** 150 × 3.9 5 µm Nova-Pak phenyl
**Mobile phase:** MeCN:10 mM pH 6.6 NaH$_2$PO$_4$ 10:90
**Flow rate:** 1.2
**Injection volume:** 50
**Detector:** E, ESA Coulochem, Model 5010 analytical cell, detector cell 1 +0.20 V, detector cell 2 + 0.55 V, model 5020 guard cell + 0.75 V; UV 210

## CHROMATOGRAM
**Retention time:** 19.2
**Internal standard:** nalorphine (25.2)
**Limit of detection:** 40 ng/mL

## OTHER SUBSTANCES
**Extracted:** 6-monoacetylmorphine, morphine
**Simultaneous:** dihydrocodone, hydrocodone, oxycodone
**Noninterfering:** 7-aminoclonazepam, 7-aminoflunitrazepam, amitriptyline, amphetamine, diazepam, dothiepin, doxepin, ephedrine, mesoridazine, methadone, methamphetamine, nordiazepam, norpropoxyphene, nortriptyline, oxazepam, propoxyphene, quinidine, quinine, sulfamethoxazole, sulforidazine, thioridazine, trimethoprim

## REFERENCE
Gerostamoulos, J.; Crump, K.; McIntyre, I.M.; Drummer, O.H. Simultaneous determination of 6-monoacetylmorphine, morphine and codeine in urine using high-performance liquid chromatography with combined ultraviolet and electrochemical detection. *J.Chromatogr.*, **1993**, *617*, 152–156

## SAMPLE
**Matrix:** urine
**Sample preparation:** 10 mL Urine + 500 µL 100 µg/mL nalorphine hydrobromide in MeOH + 1 mL concentrated HCl, heat at 100° for 1 h, cool, add 500 µL saturated ammonium sulfate solution, adjust pH to 9 with 25% NaOH, dilute to 20 mL with water, add mixture to an Extrelut 20 column, let stand for 10 min, elute with 40 mL dichloromethane:isopropanol 85:15. Add the eluate to 3 mL 200 mM HCl, extract, repeat extraction. Combine the aqueous phases and add them to 500 µL saturated ammonium sulfate solution, adjust pH to 9.2 with 25% NaOH, dilute to 20 mL with water, add to another Extrelut 20 column, let stand for 10 min, elute with 40 mL dichloromethane:isopropanol 85:15. Evaporate the eluate to dryness under a stream of nitrogen at 40°, reconstitute the residue in 100 µL mobile phase, inject a 20 µL aliquot.

## HPLC VARIABLES
**Guard column:** 4 × 4 5 µm Lichrosorb
**Column:** 250 × 4 5 µm Lichrospher Si 100
**Mobile phase:** n-Hexane:dichloromethane:MeOH containing 0.75% diethylamine 72.5:20:7.5
**Flow rate:** 1.35
**Injection volume:** 20
**Detector:** UV 225

## CHROMATOGRAM
**Retention time:** 8.3
**Internal standard:** nalorphine (6)

## OTHER SUBSTANCES
**Extracted:** morphine
**Noninterfering:** acetaminophen, aspirin, amitriptyline, buprenorphine, caffeine, carbamazepine, chlorpromazine, desipramine, dextromethorphan, doxepin, ephedrine, fenfluramine, imipramine, lidocaine, loxapine, meperidine, methadone, methaqualone, naloxone, naltrexone, nicotine, orphenadrine, oxycodone, papaverine, pentazocine, phendimetrazine, phenmetrazine, phentermine, phenylpropanolamine, phenytoin, primidone, procaine, promethazine, propoxyphene, propyphenazone, theobromine, theophylline, trazodone, triflupromazine, trimethoprim, trimipramine

## KEY WORDS
SPE; normal phase

## REFERENCE
Ferrara, S.D.; Tedeschi, L.; Frison, G.; Castagna, F. Solid-phase extraction and HPLC-UV confirmation of drugs of abuse in urine. *J.Anal.Toxicol.*, **1992**, *16*, 217–222

## SAMPLE
**Matrix:** urine
**Sample preparation:** 500 µL Urine + N-ethylnordiazepam + 100 µL buffer, centrifuge at 11000 g for 30 s, inject a 500 µL aliquot onto column A with mobile phase A, after 0.6 min backflush column A with mobile phase A to waste for 1.6 min, elute column A with 250 µL mobile phase B, with 200 µL mobile phase C, and with 1.15 mL mobile phase D. Elute column A to waste until drugs start to emerge then elute onto column B. Elute column B to waste until drugs start to emerge, then elute onto column C. When all the drugs have emerged from column B remove it from the circuit, elute column C with mobile phase D, monitor the effluent from column C. Flush column A with 7 mL mobile phase E, with mobile phase D, and mobile phase A. Flush column B with 5 mL mobile phase E then with mobile phase D. (Buffer was 6 M ammonium acetate adjusted to pH 8.0 with 2 M KOH.)

## HPLC VARIABLES
**Column:** A 10 × 2.1 12-20 µm PRP-1 spherical poly(styrene-divinylbenzene) (Hamilton); B 10 × 3.2 11 µm Aminex A-28 (Bio-Rad); C 25 × 3.2 5 µm C8 (Phenomenex) + 150 × 4.6 5 µm silica (Macherey-Nagel)
**Mobile phase:** A 0.1% pH 8.0 potassium borate buffer; B 6 mM $KH_2PO_4$ containing 5 mM tetramethylammonium hydroxide, and 2 mM dimethyloctylamine, pH adjusted to 6.50 with phosphoric acid; C MeCN:buffer 40:60 (Buffer was 6 mM $KH_2PO_4$ containing 5 mM tetramethylammonium hydroxide, and 2 mM dimethyloctylamine, pH adjusted to 6.50 with phosphoric acid.); D MeCN:buffer 33:67 (Buffer was 6 mM $KH_2PO_4$ containing 5 mM tetramethylammonium hydroxide, and 2 mM dimethyloctylamine, pH adjusted to 6.50 with phosphoric acid.); E MeCN:buffer 70:30 (Buffer was 6 mM $KH_2PO_4$ containing 5 mM tetramethylammonium hydroxide, and 2 mM dimethyloctylamine, pH adjusted to 6.50 with phosphoric acid.)
**Column temperature:** 40 (B, C only)
**Flow rate:** A 5; B-E 1
**Injection volume:** 500
**Detector:** UV 210; UV 235

## CHROMATOGRAM
**Retention time:** k' 5.7
**Internal standard:** N-ethylnordiazepam (k' 2.1)
**Limit of detection:** 300 ng/mL

## OTHER SUBSTANCES
**Extracted:** amitriptyline, amphetamine, benzoylecgonine, caffeine, cotinine, desipramine, diazepam, diphenhydramine, ephedrine, flurazepam, hydrocodone, hydromorphone, imipramine, lidocaine, methadone, methamphetamine, morphine, nordiazepam, nortriptyline, oxazepam, pentazocine, phenmetrazine, phenobarbital, phentermine, phenylpropanolamine, secobarbital
**Interfering:** chlorpheniramine

## KEY WORDS
column-switching

## REFERENCE
Binder, S.R.; Regalia, M.; Biaggi-McEachern, M.; Mazhar, M. Automated liquid chromatographic analysis of drugs in urine by on-line sample cleanup and isocratic multi-column separation. *J.Chromatogr.*, **1989**, *473*, 325–341

## ANNOTATED BIBLIOGRAPHY
Theodoridis, G.; Papadoyannis, I.; Vasilikiotis, G.; Tsoukali-Papadopoulou, H. Reversed-phase high-performance liquid chromatography–photodiode-array analysis of alkaloid drugs of forensic interest. *J.Chromatogr.B*, **1995**, *668*, 253–263 [also amphetamine, bamifylline, caffeine, cocaine, diamorphine, ethylmorphine, flufenamic acid, hyoscyamine, methadone, morphine, nalorphine, norcodeine, papaverine, quinine, scopolamine, strychnine, theobromine, theophylline, tolfenamic acid]

Band, C.J.; Band, P.R.; Deschamps, M.; Besner, J.-G.; Coldman, A.J. Human pharmacokinetic study of immediate-release (codeine phosphate) and sustained-release (codeine contin) codeine. *J.Clin. Pharmacol.*, **1994**, *34*, 938–943 [electrochemical detection; SPE; plasma; ethylmorphine (IS)]

Papadoyannis, I.; Zotou, A.; Samanidou, V.; Theodoridis, G.; Zougrou, F. Comparative study of different solid-phase extraction cartridges in the simultaneous RP-HPLC analysis of morphine and codeine in biological fluids. *J.Liq.Chromatogr.*, **1993**, *16*, 3017–3040 [simultaneous caffeine, morphine, quinine, strychnine; SPE; urine; plasma; LOD 10-20 ng/mL]

Berthod, A.; Laserna, J.J.; Carretero, I. Oil-in-water microemulsions as mobile phases for rapid screening of illegal drugs in sports. *J.Liq.Chromatogr.*, **1992**, *15*, 3115–3127 [also acebutolol, chlorthalidone, hydrochlorothiazide, methoxamine, methyltestosterone, nadolol, norcodeine, oxprenolol, phenylephrine, probenecid]

Heybroek, W.M.; Caulfield, M.; Johnston, A.; Turner, P. Automatic on-line extraction coupled with electrochemical detection as an improved method for the HPLC co-analysis of codeine and morphine in plasma and gastric juice. *J.Pharm.Biomed.Anal.*, **1990**, *8*, 1021–1027

Chen, Z.R.; Bochner, F.; Somogyi, A. Simultaneous determination of codeine, norcodeine and morphine in biological fluids by high-performance liquid chromatography with fluorescence detection. *J.Chromatogr.*, **1989**, *491*, 367–378

Persson, K.; Lindstrom, B.; Spalding, D.; Wahlstrom, A.; Rane, A. Determination of codeine and its metabolites in human blood plasma and in microsomal incubates by high-performance liquid chromatography with ultraviolet detection. *J.Chromatogr.*, **1989**, *491*, 473–480

Harris, S.C.; Miller, M.A.; Wallace, J.E. Determination of codeine and morphine in human plasma by high performance liquid chromatography with serial electrochemical detection. *Ann.Clin.Lab.Sci.*, **1988**, *18*, 297–305 [also ethylmorphine, morphine, nalorphine; electrochemical detection]

Janicot, J.L.; Caude, M.; Rosset, R. Separation of opium alkaloids by carbon dioxide sub- and supercritical fluid chromatography with packed columns. Application to the quantitative analysis of poppy straw extracts. *J.Chromatogr.*, **1988**, *437*, 351–364 [SFC; simultaneous cryptopine, morphine, narcotine, papaverine, thebaine]

Alvi, S.U.; Castro, F. A stability-indicating simultaneous analysis of acetaminophen and hydrocodone bitartrate in tablets formulation by HPLC. *J.Liq.Chromatogr.*, **1987**, *10*, 3413–3426 [stability-indicating; simultaneous acetaminophen, hydrocodone, hydromorphone; tablets; column temp 30]

Gibson, M.; Jefferies, T.M.; Soper, C.J. Isolation of codeine and norcodeine from microbial transformation liquors by preparative high-performance liquid chromatography. *Analyst*, **1987**, *112*, 1667–1670

Lurie, I.S.; McGuiness, K. The quantitation of heroin and selected basic impurities via reversed phase HPLC. II. The analysis of adulterated samples. *J.Liq.Chromatogr.*, **1987**, *10*, 2189–2204 [UV detection; electrochemical detection also acetaminophen, acetylcodeine, acetylmorphine, acetylprocaine, aminopyrene, amitriptyline, antipyrine, aspirin, barbital, benztropine, caffeine, cocaine, diamorphine, diazepam, diphenhydramine, dipyrone, ephedrine, ethylmorphine, lidocaine, meconin, methamphetamine, methapyrilene, methaqualone, morphine, nalorphine, niacinamide, noscapine, papaverine, phenacetin, phenmetrazine, phenobarbital, phenolphthalein, procaine, propanophenone, propoxyphene, pyrilamine, quinidine, quinine, salicylamide, salicylic acid, secobarbital, strychnine, tartaric acid, tetracaine, thebaine, tripelennamine, tropacocaine, vitamin B3, vitamin B5]

Huttner, A.; Eigendorf, H.G. [Simultaneous determination of propyphenazone, caffeine and codeine in drug mixtures by reverse phase HPLC]. *Pharmazie*, **1986**, *41*, 59

Stubbs, R.J.; Chiou, R.; Bayne, W.F. Determination of codeine in plasma and urine by reversed-phase high-performance liquid chromatography. *J.Chromatogr.*, **1986**, *377*, 447–453

Bedford, K.R.; White, P.C. Improved method for the simultaneous determination of morphine, codeine and dihydrocodeine in blood by high-performance liquid chromatography with electrochemical detection. *J.Chromatogr.*, **1985**, *347*, 398–404

Nitsche, V.; Mascher, H. Determination of codeine in human plasma by reverse-phase high-performance liquid chromatography. *J.Pharm.Sci.*, **1984**, *73*, 1556–1558

Posey, B.L.; Kimble, S.N. High-performance liquid chromatographic study of codeine, norcodeine, and morphine as indicators of codeine ingestion. *J.Anal.Toxicol.*, **1984**, *8*, 68–74

Meinsma, D.A.; Radzik, D.M.; Kissinger, P.T. Determination of common analgesics in serum and urine by liquid chromatography/electrochemistry. *J.Liq.Chromatogr.*, **1983**, *6*, 2311–2335 [serum; urine; electrochemical detection; extracted methyl salicylate, naproxen, phenacetin, salicylic acid]

Posey, B.L.; Kimble, S.N. Simultaneous determination of codeine and morphine in urine and blood by HPLC. *J.Anal.Toxicol.*, **1983**, *7*, 241–245

Visser, J.; Grasmeijer, G.; Moolenaar, F. Determination of therapeutic concentrations of codeine by high-performance liquid chromatography. *J.Chromatogr.*, **1983**, *274*, 372–375

Stuber, B.; Muller, K.H. [High pressure liquid chromatography of paracetamol, acetylsalicylic acid and codeine phosphate]. *Pharm.Acta Helv.*, **1982**, *57*, 181

Tsina, I.W.; Fass, M.; Debban, J.A.; Matin, S.B. Liquid chromatography of codeine in plasma with fluorescence detection. *Clin.Chem.*, **1982**, *28*, 1137–1139

Lurie, I.S.; Demchuk, S.M. Optimization of a reverse phase ion-pair chromatographic separation for drugs of forensic interest. II. Factors effecting selectivity. *J.Liq.Chromatogr.*, **1981**, *4*, 357–374 [also acetaminophen, acetylcodeine, acetylmorphine, aminopyrene, aminopyrine, amobarbital, amphetamine, antipyrine, benzocaine, butabarbital, caffeine, cocaine, diamorphine, diazepam, diethylpropion, DMT, ephedrine, glutethimide, Lampa, lidocaine, LSD, MDA, mecloqualone, mescaline, methamphetamine, methapyrilene, methaqualone, methpyrilene, methylphenidate, morphine, narcotine, papaverine, PCP, pentobarbital, phencyclidine, phendimetrazine, phenmetrazine, phenobarbital, phentermine, phenylpropanolamine, procaine, quinidine, quinine, secobarbital, strychnine, TCP, tetracaine, thebaine, theophylline]

Lurie, I.S.; Demchuk, S.M. Optimization of a reverse phase ion-pair chromatographic separation for drugs of forensic interest. I. Variables affecting capacity factors. *J.Liq.Chromatogr.*, **1981**, *4*, 337–355 [acetylcodeine, acetylmorphine, aminopyrene, aminopyrine, amobarbital, antipyrine, butabarbital, diamorphine, methapyrilene, morphine, narcotine, papaverine, pentobarbital, phenobarbital, quinidine, quinine, secobarbital, strychnine, thebaine]

Achari, R.G.; Jacob, J.T. A study of the retention behavior of some basic drug substances by ion-pair HPLC. *J.Liq.Chromatogr.*, **1980**, *3*, 81–92 [also N-acetylprocainamide, antazoline, atropine, caffeine, chlorpheniramine, ephedrine, epinephrine, naphazoline, papaverine, pheniramine, phenylephrine, phenylpropanolamine, procainamide, quinidine, scopolamine, xylocaine]

Das Gupta, V. Simultaneous quantitation of acetaminophen, aspirin, caffeine, codeine phosphate, phenacetin, and salicylamide by high-pressure liquid chromatography. *J.Pharm.Sci.*, **1980**, *69*, 110–113

Harbin, D.N.; Lott, P.F. The identification of drugs of abuse in urine using reverse phase high pressure liquid chromatography. *J.Liq.Chromatogr.*, **1980**, *3*, 243–256 [urine; also amobarbital, amphetamine, caffeine, chlordiazepoxide, diazepam, glutethimide, indole, meperidine, methamphetamine, methaqualone, morphine, pentobarbital, phenobarbital, secobarbital]

Ko, C.Y.; Marziani, F.C.; Janicki, C.A. High-performance liquid chromatographic assay of codeine in acetaminophen with codeine dosage forms. *J.Pharm.Sci.*, **1980**, *69*, 1081–1084

Kubiak, E.J.; Munson, J.W. High-performance liquid chromatographic analysis of codeine in syrups using ion-pair formation. *J.Pharm.Sci.*, **1980**, *69*, 152–156 [syrup; simultaneous ethylmorphine, morphine]

Muhammad, N.; Bodnar, J.A. Quantitative determination of guaifenesin, phenylpropanolamine hydro-chloride, sodium benzoate & codeine phosphate in cough syrups by high-pressure liquid chromatography. *J.Liq.Chromatogr.*, **1980**, *3*, 113–122

Ulrich, L.; Ruegsegger, P. [Determination of morphine and codeine in urine by high-pressure liquid chromatography]. *Arch.Toxicol.*, **1980**, *45*, 241–248

# Cromolyn

**Molecular formula:** $C_{23}H_{16}O_{11}$
**Molecular weight:** 468.4
**CAS Registry No.:** 16110-51-3 (cromolyn),
15826-37-6 (cromolyn sodium)

## SAMPLE
**Matrix:** bile, blood, urine
**Sample preparation:** Plasma. 500 µL Plasma + 1 mL water + 200 µL concentrated HCl + 3 mL ethyl acetate, shake vigorously for 10 min, centrifuge at 3000 rpm for 10 min. Remove a 2.5 mL aliquot of the organic layer and add it to 100 µL 100 mM pH 7.0 phosphate buffer, shake vigorously for 10 min, centrifuge at 3000 rpm for 10 min, inject a 30 µL aliquot of the aqueous layer. Urine, bile. 5 mL Urine or bile + 500 µL concentrated HCl, mix, add 5 mL ethyl acetate, shake vigorously for 10 min, centrifuge at 3000 rpm for 10 min. Remove a 4 mL aliquot of the organic layer and add it to 500 µL 100 mM pH 7.0 phosphate buffer, shake vigorously for 10 min, centrifuge at 3000 rpm for 10 min, inject a 15 µL aliquot of the aqueous layer.

## HPLC VARIABLES
**Column:** 200 × 4 10 µm Nucleosil NH2
**Mobile phase:** MeCN:68 mM pH 3.0 $KH_2PO_4$/phosphoric acid 35:65 (rat) or MeCN:78 mM pH 2.5 $KH_2PO_4$/phosphoric acid 35:65 (rabbit)
**Flow rate:** 2
**Injection volume:** 15-30
**Detector:** UV 240

## KEY WORDS
plasma; rat; rabbit; pharmacokinetics

## REFERENCE
Yoshimi, A.; Hashizume, H.; Kitagawa, M.; Nishimura, K.; Kakeya, N. Characteristics of 1,3-bis-(2-ethoxycarbonylchromon-5-yloxy)-2-((S)-lysyloxy)propane dihydrochloride (N-556), a prodrug for the oral delivery of disodium cromoglycate, in absorption and excretion in rats and rabbits. *J.Pharmacobiodyn.*, **1992**, *15*, 681–686

## SAMPLE
**Matrix:** bulk
**Sample preparation:** Inject a 10 µL aliquot of a solution in MeCN:water 50:50.

## HPLC VARIABLES
**Column:** 150 × 3.9 5 µm Resolve octadecylsilane (Waters)
**Mobile phase:** MeCN:water 50:50 containing 0.5 g/L cetyltrimethylammonium bromide (Use a 100 × 9.4 column of 8 µm CSC-S silica between pump and injector.)
**Flow rate:** 1
**Injection volume:** 10
**Detector:** UV 254

## CHROMATOGRAM
**Retention time:** 11.79

## OTHER SUBSTANCES
**Simultaneous:** impurities

## REFERENCE

Duhaime, R.M.; Rollins, L.K.; Gorecki, D.J.K.; Lovering, E.G. Liquid chromatographic determination of cromolyn sodium and related compounds in raw materials. *J.AOAC Int.*, **1994**, 77, 1439–1442

## SAMPLE

**Matrix:** formulations

**Sample preparation:** Solutions, capsules. Dilute solutions and capsule contents with water so as to achieve a cromolyn concentration of 40 μg/mL, inject a 20 μL aliquot. Aerosols. Direct aerosol into a flask, dissolve collected sample in water so as to achieve a cromolyn concentration of 40 μg/mL, filter (0.45 μm), inject a 20 μL aliquot of the filtrate

## HPLC VARIABLES

**Column:** 250 × 4.6 5 μm Nucleosil octadecylsilane
**Mobile phase:** MeCN:1% phosphoric acid 25:75
**Flow rate:** 1.5
**Injection volume:** 20
**Detector:** UV 238

## CHROMATOGRAM

**Retention time:** 5.15

## KEY WORDS

inhalation solution; nasal solution; capsules; aerosols

## REFERENCE

Ng, L.L. Reversed-phase liquid chromatographic determination of cromolyn sodium in drug substances and select dosage forms. *J.AOAC Int.*, **1994**, 77, 1689–1694

## SAMPLE

**Matrix:** formulations

**Sample preparation:** Extract capsule contents or gels with water, filter (0.45 μm), inject a 10 μL aliquot. Dilute drops to 0.39 mM, filter (0.45 μm), inject a 10 μL aliquot.

## HPLC VARIABLES

**Column:** 150 × 3.9 4 μm C18 Nova-Pak
**Mobile phase:** MeOH:10 mM ammonium dihydrogen phosphate 50:50, pH adjusted to 2.3 with orthophosphoric acid
**Flow rate:** 1
**Injection volume:** 10
**Detector:** UV 326

## CHROMATOGRAM

**Retention time:** 2.65

## KEY WORDS

capsules; gels; drops

## REFERENCE

Radulovic, D.; Kocic-Pesic, V.; Pecanac, D.; Zivanovic, L. HPLC determination of sodium cromoglycate in pharmaceutical dosage forms. *Farmaco*, **1994**, 49, 375–376

## SAMPLE

**Matrix:** solutions

## HPLC VARIABLES
**Column:** 250 × 4.6 Zorbax SAX
**Mobile phase:** MeOH:buffer 50:50 (Buffer was 180 mM $Na_2HPO_4$ adjusted to pH 3.00 ± 0.05 with 180 mM orthophosphoric acid. Pass mobile phase through a 250 × 4.6 25-40 μm silica (HPLC Technology) column to saturate it with silica.)
**Flow rate:** 1
**Detector:** UV 253

## CHROMATOGRAM
**Retention time:** 9.5

## OTHER SUBSTANCES
**Simultaneous:** acetaminophen, albuterol, aspartame, aspirin, beclomethasone dipropionate, caffeine, isoproterenol, menthol, minocromil, nedocromil, quinoline yellow, reproterol, riboflavin, saccharin, salicylic acid, sorbitan trioleate, terbutaline, theophylline

## REFERENCE
Baker, P.R.; Gardner, J.J.; Wilkinson, D. Automated high-performance liquid chromatographic method for the determination of nedocromil sodium in human urine using bimodal column switching. *J.Chromatogr.B*, **1995**, *668*, 59–65

## SAMPLE
**Matrix:** urine
**Sample preparation:** 10 mL Urine + 5 g NaCl + 1 mL water + 1 mL concentrated HCl + 10 mL diethyl ether, shake for 10 min at 200 oscillations/min, centrifuge at 1540 g for 10 min. Repeat extraction. Combine extracts, add 1 mL 1 M pH 3.5 glycine HCl buffer, shake, centrifuge, inject aliquot of lower aqueous phase

## HPLC VARIABLES
**Column:** 250 × 4.6 10 μm Partisil SAX
**Mobile phase:** 0.9 M Orthophosphoric acid adjusted to pH 2.30 ± 0.01 with 5 M NaOH
**Flow rate:** 3.6
**Injection volume:** 120
**Detector:** UV 325

## CHROMATOGRAM
**Retention time:** 4.5
**Limit of detection:** 50 ng/mL

## OTHER SUBSTANCES
**Noninterfering:** acetaminophen, aspirin, hydrocortisone, phenylbutazone, prednisolone, salicylic acid, terbutaline, theophylline

## REFERENCE
Gardner, J.J. Determination of sodium cromoglycate in human urine by high-performance liquid chromatography on an anion-exchange column. *J.Chromatogr.*, **1984**, *305*, 228–232

# Cyclobenzaprine

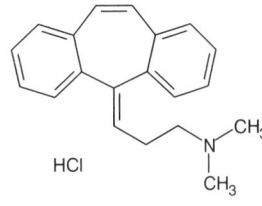

**Molecular formula:** C$_{20}$H$_{21}$N
**Molecular weight:** 275.4
**CAS Registry No.:** 303-53-7 (cyclobenzaprine), 6202-23-9
(cyclobenzaprine hydrochloride)

## SAMPLE
**Matrix:** blood
**Sample preparation:** 1 mL Serum + 500 μL 250 μg/mL protriptyline hydrochloride + 1 mL 500 mM NaOH + 4 mL toluene:n-hexane:isoamyl alcohol 77:22:3, mix for 10 min, centrifuge at 3000 rpm for 5 min. Remove the upper organic layer and evaporate it to dryness under a stream of air at 40°, reconstitute the residue in 200 μL MeOH, inject a 50 μL aliquot.

## HPLC VARIABLES
**Column:** 150 × 4.7 5 μm Supelcosil LC-PCN cyanopropyl
**Mobile phase:** MeCN:MeOH:10 mM pH 7.2 potassium phosphate buffer 60:15:25 (Prepare buffer by mixing 194 mL 1.36 g/L KH$_2$PO$_4$ with 274 mL 1.74 g/L K$_2$HPO$_4$.)
**Flow rate:** 2
**Injection volume:** 50
**Detector:** UV 254

## CHROMATOGRAM
**Retention time:** 3.7
**Internal standard:** protriptyline (8.1)

## OTHER SUBSTANCES
**Extracted:** norcyclobenzaprine, nortriptyline
**Interfering:** amitriptyline

## KEY WORDS
serum

## REFERENCE
Wong, E.C.C.; Koenig, J.; Turk, J. Potential interference of cyclobenzaprine and norcyclobenzaprine with HPLC measurement of amitriptyline and nortriptyline: resolution by GC-MS analysis. *J.Anal.Toxicol.*, **1995**, *19*, 218−224

## SAMPLE
**Matrix:** blood
**Sample preparation:** 2 mL Plasma + 500 μL 10 μg/mL nortriptyline hydrochloride in water + 500 μL 100 mM HCl, mix, add 500 μL saturated sodium bicarbonate, vortex briefly, when effervescence ceases add 10 mL dichloromethane:pentane 30:70, shake for 25 min, centrifuge at 4° at 1400 g for 25 min. Remove the organic layer and evaporate it to dryness under a stream of nitrogen at room temperature, reconstitute the residue in 50 μL mobile phase, inject the whole amount.

## HPLC VARIABLES
**Guard column:** Resolve Si Guard-PAK (Waters)
**Column:** 300 × 3.9 10 μm μPorasil
**Mobile phase:** MeCN:EtOH:tert-butylamine 10:90:0.025
**Flow rate:** 1

**Injection volume:** 50
**Detector:** UV 230

## CHROMATOGRAM
**Retention time:** 7.5
**Internal standard:** nortriptyline (13.7)
**Limit of quantitation:** 1 ng/mL

## OTHER SUBSTANCES
**Noninterfering:** acetaminophen, aspirin, caffeine, ibuprofen

## KEY WORDS
plasma; normal phase; use silanized glassware

## REFERENCE
Hwang, P.T.R.; Young, D.A.; Straughn, A.B.; Meyer, M.C. Quantitative determination of cyclobenzaprine in human plasma by high pressure liquid chromatography. *J.Liq.Chromatogr.*, **1993**, *16*, 1163–1171

## SAMPLE
**Matrix:** blood, urine
**Sample preparation:** Plasma. 1 mL Plasma + 100 μL 500 ng/mL trimipramine in MeOH + 1 mL 200 mM pH 9.8 carbonate buffer + 5 mL hexane, rotate for 15 min, centrifuge at 3000 g for 5 min. Remove the organic layer and evaporate it to dryness under a stream of nitrogen at 50°, reconstitute the residue in 300 μL mobile phase, inject a 150 μL aliquot. Urine. 1 mL Urine + 1 mL β-glucuronidase (7200 Fishman Units) in 20 mM pH 6.5 phosphate buffer, heat at 37° for 24 h, add 100 μL 10 M NaOH, add 5 mL hexane, rotate for 15 min, centrifuge at 3000 g for 5 min. Remove the organic layer and evaporate it to dryness under a stream of nitrogen at 50°, reconstitute the residue in 300 μL mobile phase, inject a 150 μL aliquot.

## HPLC VARIABLES
**Guard column:** 20 mm long C18 base-deactivated silica (BDS) (Keystone)
**Column:** 250 × 4.6 5 μm C18 base-deactivated silica (BDS) (Keystone)
**Mobile phase:** MeCN:buffer 50:50 (plasma) or 43:57 (urine) (Buffer was 0.085% phosphoric acid adjusted to pH 6.5 with triethylamine.)
**Flow rate:** 1
**Injection volume:** 150
**Detector:** UV 229

## CHROMATOGRAM
**Retention time:** 7.8 (plasma), 9.6 (urine)
**Internal standard:** trimipramine (10.5 (plasma), 12.8 (urine))
**Limit of quantitation:** 0.5 ng/mL (plasma); 10 ng/mL (urine)

## KEY WORDS
plasma

## REFERENCE
Constanzer, M.; Chavez, C.; Matuszewski, B. Development and comparison of high-performance liquid chromatographic methods with tandem mass spectrometric and ultraviolet absorbance detection for the determination of cyclobenzaprine in human plasma and urine. *J.Chromatogr.B*, **1995**, *666*, 117–126

## SAMPLE
**Matrix:** blood, urine

**Sample preparation:** Plasma. 1 mL Plasma + 100 μL 500 ng/mL trimipramine in MeOH + 1 mL 200 mM pH 9.8 carbonate buffer + 5 mL hexane, rotate for 15 min, centrifuge at 3000 g for 5 min. Remove the organic layer and evaporate it to dryness under a stream of nitrogen at 50°, reconstitute the residue in 150 μL mobile phase, inject a 75 μL aliquot. Urine. 1 mL Urine + 1 mL β-glucuronidase (7200 Fishman Units) in 20 mM pH 6.5 phosphate buffer, heat at 37° for 24 h, add 100 μL 10 M NaOH, add 5 mL hexane, rotate for 15 min, centrifuge at 3000 g for 5 min. Remove the organic layer and evaporate it to dryness under a stream of nitrogen at 50°, reconstitute the residue in 1 mL mobile phase, inject a 50 μL aliquot.

## HPLC VARIABLES
**Guard column:** 20 × 4.6 C18 base-deactivated silica (BDS) (Keystone)
**Column:** 50 × 4.6 5 μm C18 base-deactivated silica (BDS) (Keystone)
**Mobile phase:** MeCN:water 90:10 containing 0.1% formic acid and 10 mM ammonium acetate
**Flow rate:** 1
**Injection volume:** 50-75
**Detector:** MS, PE Sciex API III, heated nebulized interface, corona discharge needle +4 μA, nebulizer probe 500°, nebulizing gas was air at 2 L/min and 80 psi, curtain gas flow was nitrogen at 0.9 L/min, sampling orifice +45 V, dwell time 400 ms, interface heater 60°, electron multiplier−3.7 kV, collision gas was argon $355 \times 10^{12}$ atoms/cm$^2$, first quadrupole filter admits m/z 276 (cyclobenzaprine) and 295 (trimipramine, collisional fragmentation at second filter, monitor m/z 215 (cyclobenzaprine) and 208 (trimipramine) at third quadrupole filter

## CHROMATOGRAM
**Retention time:** 1.9
**Internal standard:** trimipramine (2.2)
**Limit of quantitation:** 0.1 ng/mL (plasma); 10 ng/mL (urine)

## KEY WORDS
plasma; pharmacokinetics

## REFERENCE
Constanzer, M.; Chavez, C.; Matuszewski, B. Development and comparison of high-performance liquid chromatographic methods with tandem mass spectrometric and ultraviolet absorbance detection for the determination of cyclobenzaprine in human plasma and urine. *J.Chromatogr.B*, **1995**, *666*, 117−126

## SAMPLE
**Matrix:** formulations
**Sample preparation:** Powder tablets (60 mesh), shake with 25 mL 25 mM sulfuric acid for 1 h, add MeOH to 85 mL, swirl, allow to cool to room temperature, make up to 100 mL. Remove a 10 mL aliquot, add 5 mL 1 mg/mL naphazoline hydrochloride in MeOH, make up to 100 mL, filter (0.45 μm), inject 20 μL aliquot.

## HPLC VARIABLES
**Column:** 250 × 4.6 Zorbax C8
**Mobile phase:** MeCN:buffer 75:25 (Buffer was 12 g KH$_2$PO$_4$ in 1800 mL water, adjust pH to 3.0 with 1:3 phosphoric acid, make up to 2 L.)
**Flow rate:** 1.5
**Injection volume:** 20
**Detector:** UV 254

## CHROMATOGRAM
**Retention time:** 6.8
**Internal standard:** naphazoline hydrochloride

## OTHER SUBSTANCES
**Simultaneous:** desipramine
**Interfering:** amitriptyline

## KEY WORDS
tablets

## REFERENCE
Heinitz, M.L. Determination of cyclobenzaprine in tablets by high-performance liquid chromatography. *J.Pharm.Sci.*, **1982**, *71*, 656–658

## SAMPLE
**Matrix:** solutions

## HPLC VARIABLES
**Column:** 150 × 2 Deltabond C8 (Keystone)
**Mobile phase:** MeCN:2-butanone:50 mM pH 7.0 phosphate buffer 27:13:60
**Flow rate:** 0.15
**Injection volume:** 1
**Detector:** Chemiluminescence following post-column reaction. A 1 mM solution of Ru(2,2'-bipyridine)$_3^{2+}$ in 50 mM sodium sulfate (continuously sparged with helium) was oxidized to Ru(2,2'-bipyridine)$_3^{3+}$ using a Princeton Applied Research Model 174A polarographic analyzer with a platinum gauze working electrode, a platinum wire auxiliary electrode, and a silver wire reference electrode. The Ru solution was pumped at 0.3 mL/min and mixed with the column effluent in the flow cell of the detector, a fluorescence detector with the light source removed.

## CHROMATOGRAM
**Retention time:** 6
**Limit of detection:** 0.1-1 µg/mL

## OTHER SUBSTANCES
**Simultaneous:** dicyclomine

## KEY WORDS
post-column reaction

## REFERENCE
Holeman, J.A.; Danielson, N.D. Microbore liquid chromatography of tertiary amine anticholinergic pharmaceuticals with tris(2,2'-bipyridine)ruthenium(III) chemiluminescence detection. *J.Chromatogr. Sci.*, **1995**, *33*, 297–302

## SAMPLE
**Matrix:** solutions

## HPLC VARIABLES
**Column:** 250 × 4.6 5 µm Supelcosil LC-DP (A) or 250 × 4 5 µm LiChrospher 100 RP-8 (B)
**Mobile phase:** MeCN:0.025% phosphoric acid:buffer 25:10:5 (A) or 60:25:15 (B) (Buffer was 9 mL concentrated phosphoric acid and 10 mL triethylamine in 900 mL water, adjust pH to 3.4 with dilute phosphoric acid, make up to 1 L.)
**Flow rate:** 0.6
**Injection volume:** 25
**Detector:** UV 229

## CHROMATOGRAM
**Retention time:** 15.50 (A), 7.05 (B)

## OTHER SUBSTANCES
**Also analyzed:** acebutolol, acepromazine, acetaminophen, acetazolamide, acetophenazine, albuterol, alprazolam, amitriptyline, amobarbital, amoxapine, antipyrine, atenolol, atropine, azatadine, baclofen, benzocaine, bromocriptine, brompheniramine, brotizolam, bupivacaine, buspirone, butabarbital, butalbital, caffeine, carbamazepine, cetirizine, chlorcyclizine, chlordiazepoxide, chlormezanone, chloroquine, chlorpheniramine, chlorpromazine, chlorpropamide, chlorprothixene, chlorthalidone, chlorzoxazone, cimetidine, cisapride, clomipramine, clonazepam, clonidine, clozapine, cocaine, codeine, colchicine, cyclizine, dantrolene, desipramine, diazepam, diclofenac, diflunisal, diltiazem, diphenhydramine, diphenidol, diphenoxylate, dipyridamole, disopyramide, dobutamine, doxapram, doxepin, droperidol, encainide, ethidium bromide, ethopropazine, fenoprofen, fentanyl, flavoxate, fluoxetine, fluphenazine, flurazepam, flurbiprofen, fluvoxamine, furosemide, glutethimide, glyburide, guaifenesin, haloperidol, homatropine, hydralazine, hydrochlorothiazide, hydrocodone, hydromorphone, hydroxychloroquine, hydroxyzine, ibuprofen, imipramine, indomethacin, ketoconazole, ketoprofen, ketorolac, labetalol, levorphanol, lidocaine, loratadine, lorazepam, lovastatin, loxapine, mazindol, mefenamic acid, meperidine, mephenytoin, mepivacaine, mesoridazine, metaproterenol, methadone, methdilazine, methocarbamol, methotrexate, methotrimeprazine, methoxamine, methyldopa, methylphenidate, metoclopramide, metolazone, metoprolol, metronidazole, midazolam, moclobemide, morphine, nadolol, nalbuphine, naloxone, naphazoline, naproxen, nifedipine, nizatidine, norepinephrine, nortriptyline, oxazepam, oxycodone, oxymetazoline, paroxetine, pemoline, pentazocine, pentobarbital, pentoxifylline, perphenazine, pheniramine, phenobarbital, phenol, phenolphthalein, phentolamine, phenylbutazone, phenyltoloxamine, phenytoin, pimozide, pindolol, piroxicam, pramoxine, prazepam, prazosin, probenecid, procainamide, procaine, prochlorperazine, procyclidine, promazine, promethazine, propafenone, propantheline, propiomazine, propofol, propranolol, protriptyline, quazepam, quinidine, quinine, racemethorphan, ranitidine, remoxipride, risperidone, salicylic acid, scopolamine, secobarbital, sertraline, sotalol, spironolactone, sulfinpyrazone, sulindac, temazepam, terbutaline, terfenadine, tetracaine, theophylline, thiethylperazine, thiopental, thioridazine, thiothixene, timolol, tocainide, tolbutamide, tolmetin, trazodone, triamterene, triazolam, trifluoperazine, triflupromazine, trimeprazine, trimethoprim, trimipramine, verapamil, warfarin, xylometazoline, yohimbine, zopiclone

## KEY WORDS
some details of plasma extraction

## REFERENCE
Koves, E.M. Use of high-performance liquid chromatography-diode array detection in forensic toxicology. *J.Chromatogr.A*, **1995**, *692*, 103–119

## SAMPLE
**Matrix:** solutions

## HPLC VARIABLES
**Column:** 250 × 4.6 Zorbax RX
**Mobile phase:** Gradient. A was 10 mL concentrated orthophosphoric acid and 7 mL triethylamine in 1 L water. B was 10 mL concentrated orthophosphoric acid and 7 mL triethylamine in 200 mL water, make up to 1 L with MeCN. A:B from 100:0 to 0:100 over 30 min, maintain at 0:100 for 5 min.
**Column temperature:** 30
**Flow rate:** 2
**Detector:** UV 210

## OTHER SUBSTANCES
**Also analyzed:** acepromazine, acetaminophen, acetophenazine, albuterol, aminophylline, amitriptyline, amobarbital, amoxapine, amphetamine, amylocaine, antipyrine, aprobarbital, aspirin, atenolol, atropine, avermectin, barbital, benzocaine, benzoic acid, benzotro-

pine, benzphetamine, berberine, bibucaine, bromazepan, brompheniramine, buprenorphine, buspirone, butabarbital, butacaine, butethal, caffeine, carbamazepine, carbromal, chloramphenicol, chlordiazepoxide, chloroquine, chlorothiazide, chloroxylenol, chlorphenesin, chlorpheniramine, chlorpromazine, chlorpropamide, chlortetracycline, cimetidine, cinchonidine, cinchonine, clenbuterol, clonazepam, clonixin, clorazepate, cocaine, codeine, colchicine, cortisone, coumarin, cyclazocine, cyclothiazide, cyheptamide, cymarin, danazol, danthron, dapsone, debrisoquine, desipramine, dexamethasone, dextromethorphan, dextropropoxyphene, diamorphine, diazepam, diclofenac, diethylpropion, diethylstilbestrol, diflunisal, digitoxin, digoxin, diltiazem, diphenhydramine, diphenoxylate, diprenorphine, dipyrone, disulfiram, dopamine, doxapram, doxepin, dronabinol, ephedrine, epinephrine, epinine, estradiol, estriol, estrone, ethacrynic acid, ethosuximide, etonitazene, etorphine, eugenol, famotidine, fenbendazole, fencamfamine, fenoprofen, fenproporex, fentanyl, flubendazole, flufenamic acid, flunitrazepam, 5-fluorouracil, fluoxymesterone, fluphenazine, furosemide, gentisic acid, gitoxigenin, glipizide, glunixin, glutethimide, glybenclamide, guaiacol, halazepam, haloperidol, hydrochlorothiazide, hydrocodone, hydrocortisone, hydromorphone, hydroxyquinoline, ibogaine, ibuprofen, iminostilbene, imipramine, indomethacin, isocarbostyril, isocarboxazid, isoniazid, isoproterenol, isoxsuprine, ivermectin, ketamine, ketoprofen, kynurenic acid, levorphanol, lidocaine, lorazepam, lormetazepam, loxapine, mazindol, mebendazole, meclizine, meclofenamic acid, medazepam, mefenamic acid, megestrol, mepacrine, meperidine, mephentermine, mephenytoin, mephesin, mephobarbital, mepivacaine, mescaline, mesoridazine, methadone, methamphetamine, methapyrilene, methaqualone, methazolamide, methocarbamol, methoxamine, methsuximide, methyl salicylate, methyldopa, methyldopamine, methylphenidate, methylprednisolone, methyltestosterone, methyprylon, metoprolol, mibolerone, morphine, nadolol, nalorphine, naloxone, naltrexone, naphazoline, naproxen, nefopam, niacinamide, nicotine, nicotinic acid, nifedipine, niflumic acid, nitrazepam, norepinephrine, nortriptyline, noscapine, nylidrin, oxazepam, oxycodone, oxymorphone, oxyphenbutazone, oxytetracycline, papaverine, pargyline, pemoline, pentazocine, pentobarbital, persantine, phenacetin, phenazocine, phenazopyridine, phencyclidine, phendimetrazine, phenelzine, pheniramine, phenobarbital, phenothiazine, phensuximide, phentermine, phenylbutazone, phenylephrine, phenylpropanolamine, piperocaine, prazepam, prednisolone, primidone, probenecid, progesterone, propiomazine, propranolol, propylparaben, pseudoephedrine, puromycin, pyrilamine, pyrithyldione, quazepam, quinaldic acid, quinidine, quinine, ranitidine, recinnamine, reserpine, resorcinol, saccharin, albuterol, salicylamide, salicylic acid, scopolamine, scopoletin, secobarbital, strychnine, sulfacetamide, sufadiazine, sulfadimethoxine, sulfaethidole, sulfamerazine, sulfamethazine, sulfamethoxizole, sulfanilamide, sulfapyridine, sulfasoxizole, sulindac, tamoxifen, temazepam, testosterone, tetracaine, tetracycline, tetramisole, thebaine, theobromine, theophylline, thiabendazole, thiamine, thiamylal, thiobarbituric acid, thioridazine, thiosalicylic acid, thiothixene, thymol, tolazamide, tolazoline, tobutamide, tolmetin, tranylcypromine, triamcinolone, tribenzylamine, trichloromethiazide, trifluoperazine, trihexyphenidyl, trimethoprim, tripelennamine, triproilidine, tropacocaine, tyramine, verapamil, vincamine, warfarin, yohimbine, zoxazolamine

## REFERENCE

Hill, D.W.; Kind, A.J. Reversed-phase solvent gradient HPLC retention indexes of drugs. *J.Anal.Toxicol.*, **1994**, *18*, 233–242

## SAMPLE

**Matrix:** solutions
**Sample preparation:** Inject an aliquot of a solution in mobile phase.

## HPLC VARIABLES

**Column:** 250 × 4.6 Econosil C8
**Mobile phase:** MeCN:buffer 30:70 (Buffer was 20 mM $KH_2PO_4$ and 14 mM triethylamine adjusted to pH 3.0 with phosphoric acid.)
**Injection volume:** 20
**Detector:** UV 210

## CHROMATOGRAM
**Retention time:** 8.6
**Limit of quantitation:** <1 µg/mL

## OTHER SUBSTANCES
**Simultaneous:** doxepin, desipramine, protriptyline, maprotiline
**Also analyzed:** amitriptyline, amoxapine, carbamazepine, imipramine, nortriptyline

## KEY WORDS
UV spectra given

## REFERENCE
Ryan, T.W. Identification and quantification of tricyclic antidepressants by UV-photodiode array detection with multicomponent analysis. *J.Liq.Chromatogr.*, **1993**, *16*, 1545–1560

## ANNOTATED BIBLIOGRAPHY
Puopolo, P.R.; Flood, J.G. Detection of interference by cyclobenzaprine in liquid- chromatographic assays of tricyclic antidepressants. *Clin.Chem.*, **1987**, *33*, 819–820 [LOD 10 ng/mL]

Schneider, M.; Giardina, E.-G.V. Inteference by Flexeril, a tricyclic muscle relaxant, with liquid-chromatographic determination of imipramine. *Clin.Chem.*, **1986**, *32*, 1599 [plasma; interfering imipramine]

# Cyclosporine

**Molecular formula:** $C_{62}H_{111}N_{11}O_{12}$
**Molecular weight:** 1202.6
**CAS Registry No.:** 59865-13-3

## SAMPLE
**Matrix:** aqueous humor
**Sample preparation:** 10 Volumes aqueous humor + 1 volume 20 mg/mL cyclosporine D, mix, inject an aliquot.

## HPLC VARIABLES
**Guard column:** Supelco guard column
**Column:** 250 × 4.6 5 μm C18 (Supelco)
**Mobile phase:** MeCN:water 70:30
**Column temperature:** 70
**Flow rate:** 1.5
**Detector:** UV 204

## CHROMATOGRAM
**Internal standard:** cyclosporine D
**Limit of quantitation:** 250 ng/mL

## KEY WORDS
rabbit; eye; pharmacokinetics

## REFERENCE
Oh, C.; Saville, B.A.; Cheng, Y.-L.; Rootman, D.S. A compartmental model for the ocular pharmacokinetics of cyclosporine in rabbits. *Pharm.Res.*, **1995**, *12*, 433–437

## SAMPLE
**Matrix:** bile, blood, feces, urine
**Sample preparation:** Whole blood. Add whole blood to 7-15 volumes MeOH, sonicate for 15 min, centrifuge at 6000 g for 20 min. Remove the supernatant and concentrate it under reduced pressure at 35°, inject an aliquot of the concentrate. Feces. Extract feces with 2 volumes MeOH, repeat extraction 3-4 times, combine the extracts, filter (5 μm), inject an aliquot. Urine, bile. Inject urine and bile directly.

## HPLC VARIABLES
**Column:** two 150 × 4.6 Supelcosil LC-18 columns in series
**Mobile phase:** Gradient. A was MeOH:water 10:90. B was MeOH:MeCN 10:90. A:B from 100:0 to 55:45 over 29 min, to 45:55 over 60 min, to 10:90 over 32 min, maintain at 10:90 for 10 min.
**Column temperature:** 70

**Flow rate:** 1.5
**Detector:** UV 210

## CHROMATOGRAM
**Retention time:** 109.7 (cyclosporin G)

## OTHER SUBSTANCES
**Extracted:** metabolites

## KEY WORDS
mouse; rat; dog; whole blood

## REFERENCE
Mangold, J.B.; Rodriguez, L.C.; Wang, Y.K. Metabolism of cyclosporin G in the mouse, rat, and dog. *Drug Metab.Dispos.*, **1995**, *23*, 615–621

## SAMPLE
**Matrix:** blood
**Sample preparation:** Serum + cyclosporine D, extract with 6 mL ethyl ether/petroleum ether. Remove the organic layer and evaporate it to dryness, reconstitute the residue in 1 mL MeOH, wash the MeOH layer with 1.5 mL hexane/heptane. Evaporate the MeOH layer at 40°, reconstitute with mobile phase, inject a 20 µL aliquot.

## HPLC VARIABLES
**Column:** 300 × 4.6 10 µm YWG
**Mobile phase:** MeOH:isopropanol:water 65:20:20
**Column temperature:** 65
**Flow rate:** 1
**Injection volume:** 20
**Detector:** UV 215

## CHROMATOGRAM
**Internal standard:** cyclosporine D
**Limit of detection:** 25 ng/mL
**Limit of quantitation:** 50 ng/mL

## KEY WORDS
serum

## REFERENCE
Liang, M.Z.; Zou, Y.G.; Yu, Q.; Qing, Y.P. RPHPLC for monitoring blood concentration of cyclosporine A in kidney transplantation patient and clinical application (Abstract 65). *Ther.Drug Monit.*, **1995**, *17*, 399

## SAMPLE
**Matrix:** blood

## HPLC VARIABLES
**Column:** YWG-C18
**Mobile phase:** MeCN:water 80:22.5
**Column temperature:** 60
**Flow rate:** 1
**Detector:** UV 214

## CHROMATOGRAM
**Retention time:** 6.8 (cyclosporine A)

**Internal standard:** cyclosporine D (9)
**Limit of detection:** 10 ng/mL

---

## REFERENCE

Lin, S.G.; Yu, X.Y.; Yang, M. A modified HPLC method for measuring cyclosporine A and monitoring its blood concentration (Abstract 66). *Ther.Drug Monit.*, **1995**, *17*, 399

---

## SAMPLE

**Matrix:** blood
**Sample preparation:** Condition a Varian C18 SPE cartridge with two 2 mL portions of MeCN and two 2 mL portions of MeCN:water 32:68. Condition a Varian silica SPE cartridge with two 2 mL portions of MeCN. 1 mL Whole blood + 400 μL water + 3 mL cyclosporine C in MeCN:MeOH 90:10, vortex for 30 s, centrifuge at 3000 rpm for 10 min. Dilute the supernatant with 4.5 mL water, add to the C18 SPE cartridge, wash with 4 mL MeCN:water 32:68, wash with two 1.4 mL portions of MeOH:water 62:38, wash with 1.4 mL MeOH:300 mM acetic acid 60:40, wash with 3 mL hexane:acetone 99:1, elute with two 1.4 mL portions of MeCN:MeOH 75:25 through the silica SPE cartridge. Evaporate the eluate to dryness under a stream of nitrogen at 40°, reconstitute the residue in 100 μL mobile phase, add 300 μL heptane, vortex for 30 s, centrifuge at 3000 rpm for 5 min, discard the heptane layer, repeat the heptane wash, inject an aliquot of the MeCN layer.

---

## HPLC VARIABLES

**Column:** C8 (Beckman)
**Mobile phase:** MeCN:MeOH:water 51:20:9
**Column temperature:** 70
**Flow rate:** 0.6
**Detector:** UV 214

---

## CHROMATOGRAM

**Retention time:** 38
**Internal standard:** cyclosporine C
**Limit of detection:** 10 ng/mL

---

## OTHER SUBSTANCES

**Extracted:** metabolites

---

## KEY WORDS

whole blood; SPE

---

## REFERENCE

Liu, W.T.; Levy, G.A.; Wong, P.Y. Measurement of AM19 and other cyclosporine metabolites in the blood of liver transplant patients with stable liver function. *Ther.Drug Monit.*, **1995**, *17*, 479–486

---

## SAMPLE

**Matrix:** blood
**Sample preparation:** 1 mL Whole blood + cyclosporine D in MeCN, centrifuge. Remove the supernatant and dilute it with water, add to a solid-phase extraction disk (15 mg SPEC RP 18 AR), wash with MeCN:water 20:80, wash with n-hexane, elute with MeCN. Evaporate the eluate to dryness, reconstitute, inject an aliquot.

---

## HPLC VARIABLES

**Column:** 250 × 4.6 3 μm nitrile
**Mobile phase:** n-Heptane:isopropanol 86:14
**Column temperature:** 54

**Flow rate:** 0.85
**Detector:** UV 210

## CHROMATOGRAM
**Retention time:** 11.8 (cyclosporine A)
**Internal standard:** cyclosporine D (9.8)
**Limit of detection:** 25 ng/mL

## OTHER SUBSTANCES
**Extracted:** metabolites

## KEY WORDS
whole blood; SPE

## REFERENCE
Wenk, M.; Haefeli, W.E. Improved determination of cyclosporine and its metabolites using solid phase
extraction disks and normal phase liquid chromatography (Abstract 72). *Ther.Drug Monit.*, **1995**, *17*,
401

## SAMPLE
**Matrix:** blood
**Sample preparation:** 2 mL Whole blood + 4 mL MeCN:MeOH 9:1, mix, centrifuge, add
the supernatant to an Analytichem 6 mL C18 SPE cartridge, wash with 3 mL MeOH:
water 70:30, wash with 3 mL hexane:acetone 99:1, elute with 3 mL ethyl acetate:iso-
propanol 3:1. Pass the eluate through an Analytichem 3 mL silica SPE cartridge, collect
the eluate and evaporate it to dryness, reconstitute the residue in 100 μL mobile phase,
inject an aliquot.

## HPLC VARIABLES
**Column:** 50 mm long LC-1 (Supelco)
**Mobile phase:** MeCN:water 46:54
**Flow rate:** 1.2
**Detector:** UV 214

## CHROMATOGRAM
**Retention time:** 11.17
**Internal standard:** cyclosporin C (9.39)
**Limit of detection:** 25 ng/mL

## KEY WORDS
whole blood; cyclosporin D, previously used as IS, is now unavailable; SPE

## REFERENCE
Jones, P.M.; Brune, K. Monitoring cyclosporine by HPLC with cyclosporin C as internal standard.
*Clin.Chem.*, **1993**, *39*, 168–169

## SAMPLE
**Matrix:** blood
**Sample preparation:** Condition an AASP C8 SPE cartridge (Analytichem) with 1.5 mL
isopropanol then with 1.5 mL MeCN:water 2:3, re-wet with a few drops of MeCN:water
2:3. 200 μL Whole blood + 20 μL 5 μg/mL cyclosporin D in MeOH + 1.5 mL MeCN:
water 2:3, vortex for 10 s, let stand for 20 min, centrifuge at 1300 g for 10 min. Add the
supernatant to the SPE cartridge, wash with 1.5 mL MeCN:water 2:3, purge with nitrogen at 1.4 bar for 5 min. Place SPE cartridge in a vacuum desiccator for 5 min, wash
with 1 mL hexane, purge with nitrogen for 5 min. Purge with nitrogen for 3 min imme-

diately before injection, elute contents of SPE cartridge onto analytical column with mobile phase for 0.6 min.

## HPLC VARIABLES
**Column:** 150 × 4.6 5 μm CPS Hypersil (cyanopropyl)
**Mobile phase:** Hexane:EtOH 91:9 (Place a 100 × 4.6 mm column packed with 37-53μm silica gel (Whatman Pre Column Gel) held at 53° between pump and SPE cartridge.)
**Column temperature:** 53
**Flow rate:** 0.7
**Detector:** UV 210

## CHROMATOGRAM
**Retention time:** 7.2
**Internal standard:** cyclosporin D (8.2)
**Limit of detection:** 12.5 ng/mL

## KEY WORDS
whole blood; SPE

## REFERENCE
Lachno, D.R.; Patel, N.; Rose, M.L.; Yacoub, M.H. Improved high-performance liquid chromatographic method for analysis of cyclosporin A using an automated sample processor. *J.Chromatogr.*, **1990**, *525*, 123–132

## SAMPLE
**Matrix:** blood
**Sample preparation:** 0.5 mL Whole blood + 50 μL 5 μg/mL cyclosporin D, mix, add 2 mL 90 mM HCl, add 5 mL MTBE, rotate for 9 min, centrifuge at 500 g for 5 min. Remove the organic layer, add 2 mL 90 mM NaOH, shake vigorously for 3 min, centrifuge at 500 g for 5 min. Remove the organic layer and evaporate it to dryness under a stream of air, reconstitute in 60 μL mobile phase and 100 μL heptane, vortex, centrifuge, inject a 20 μL aliquot of the lower layer.

## HPLC VARIABLES
**Guard column:** 15 mm long RP-8 guard column (Brownlee)
**Column:** 250 × 2 5 μm Astec microbore octyl (Advanced Separation Technologies)
**Mobile phase:** MeCN:MeOH:water 52:19:29
**Column temperature:** 70
**Flow rate:** 0.25
**Injection volume:** 20
**Detector:** UV 214

## CHROMATOGRAM
**Retention time:** 8.5
**Internal standard:** cyclosporin D (10)
**Limit of detection:** 3 ng/mL

## OTHER SUBSTANCES
**Noninterfering:** acetaminophen, N-acetylprocainamide, caffeine, carbamazepine, digoxin, ethosuximide, gentamicin, lidocaine, phenobarbital, phenytoin, primidone, procainamide, salicylic acid, theophylline, tobramycin, valproic acid

## KEY WORDS
microbore; whole blood

## REFERENCE

Annesley, T.; Matz, K.; Balogh, L.; Clayton, L.; Giacherio, D. Liquid-chromatographic analysis for cyclosporine with use of a microbore column and small sample volume. *Clin.Chem.*, **1986**, *32*, 1407-1409

## SAMPLE

**Matrix:** blood

**Sample preparation:** 2 mL Whole blood or plasma + 25 μL 25 μg/mL cyclosporin D in MeOH, rinse pipette used for blood or plasma with 2 mL water and add the rinse to the mixture, add 14 mL ether, shake horizontally at 180 cycles/min for 15 min, centrifuge at 750 g for 5 min. Remove 11.5 mL of the organic phase and evaporate it to dryness, add 1 mL 25 mM HCl, add 2 mL MeOH, add 7 mL n-hexane, shake horizontally at 180 cycles/min for 5 min, centrifuge at 750 g for 5 min. Discard the n-hexane and wash the aqueous layer with another 7 mL n-hexane. Remove the aqueous phase and add it to 1 mL 25 mM NaOH and 7 mL ether, shake horizontally at 180 cycles/min for 10 min, centrifuge at 750 g for 5 min. Remove the ether layer and evaporate it to dryness, reconstitute with 100 μL mobile phase, inject a 90 μL aliquot.

## HPLC VARIABLES

**Column:** 150 × 4.6 5 μm LC-18 (Supelco)

**Mobile phase:** MeCN:water 68.5:31.5

**Column temperature:** 75

**Flow rate:** 1.4

**Injection volume:** 90

**Detector:** UV 202

## CHROMATOGRAM

**Retention time:** 5.77

**Internal standard:** cyclosporin D (7.82)

**Limit of quantitation:** 25 ng/mL

## KEY WORDS

whole blood; plasma

## REFERENCE

Sawchuk, R.J.; Cartier, L.L. Liquid-chromatographic determination of cyclosporin A in blood and plasma. *Clin.Chem.*, **1981**, *27*, 1368–1371

## SAMPLE

**Matrix:** blood, tissue

**Sample preparation:** Whole blood. 1 mL Whole blood + 300 ng cyclosporine C + 3 mL MeCN:water 50:50 saturated with zinc sulfate, vortex, centrifuge at 6000 rpm for 10 min, wash the supernatant twice with 2 mL portions of hexane, add the supernatant to a 1 mL 100 mg Bond Elut LRC C18 SPE cartridge, wash with 3 mL portions of MeCN:water 35:65, elute with 1 mL MeCN, evaporate the eluate to dryness under a stream of nitrogen, reconstitute in 350 μL MeCN:water 50:50, inject a 150 μL aliquot. Tissue. Homogenize 50-100 mg liver or spleen with 300 ng cyclosporine C in MeCN:water 35:65 containing 5% zinc sulfate, centrifuge, wash the supernatant twice with 2 mL portions of hexane, add the supernatant to a 1 mL 100 mg Bond Elut LRC C18 SPE cartridge, wash with 3 mL portions of MeCN:water 35:65, elute with 1 mL MeCN, evaporate the eluate to dryness under a stream of nitrogen, reconstitute in 350 μL MeCN:water 50:50, inject a 150 μL aliquot.

## HPLC VARIABLES

**Guard column:** Adsorbosphere Direct-Connect guard column

**Column:** 250 × 4.6 5 μm Ultrasphere RP-18

**Mobile phase:** Gradient. MeCN:water 50:50 for 20 min, to 65:35 over 5 min, maintain at
   65:35 over 15 min, to 70:30 over 10 min, maintain at 70:30 for 7 min, wash column with
   85:15 for 15 min, re-equilibrate for 10 min.
**Column temperature:** 70
**Injection volume:** 150
**Detector:** UV 214

## CHROMATOGRAM
**Internal standard:** cyclosporine C
**Limit of detection:** 15 ng/mL

## OTHER SUBSTANCES
**Extracted:** metabolites

## KEY WORDS
rat; whole blood; SPE; pharmacokinetics; liver; spleen

## REFERENCE
Wacher, V.J.; Liu, T.; Roberts, J.P.; Ascher, N.L.; Benet, L.Z. Time course of cyclosporine and its metabolites in blood, liver and spleen of naive Lewis rats: Comparison with preliminary data obtained in
   transplanted animals. *Biopharm.Drug Dispos.*, **1995**, *16*, 303–312

## SAMPLE
**Matrix:** formulations
**Sample preparation:** 50 mg Paste + 10 mL acetone, mix by inverting at 30 rpm for 15
   min, centrifuge at 2500 rpm for 10 min. Remove 4 mL of the upper layer and evaporate
   it to dryness under a stream of nitrogen, reconstitute the residue in 400 µL ethyl acetate,
   inject a 5 µL aliquot.

## HPLC VARIABLES
**Column:** 300 × 3.9 10 µm µBondapak C18
**Mobile phase:** MeCN:MeOH:40 mM pH 5 KH$_2$PO$_4$ 60:10:30
**Column temperature:** 70
**Flow rate:** 2
**Injection volume:** 5
**Detector:** UV 214

## CHROMATOGRAM
**Retention time:** 8.1

## KEY WORDS
gel; paste; stability-indicating

## REFERENCE
Ghnassia, L.T.; Yau, D.F.; Kaye, K.I.; Duggin, G.G. Stability of cyclosporine in an extemporaneously
   compounded paste. *Am.J.Health-Syst.Pharm.*, **1995**, *52*, 2204–2207

## SAMPLE
**Matrix:** microsomal incubations
**Sample preparation:** 500 µL Microsomal incubation + 500 µL MeCN, mix, centrifuge,
   inject a 100 µL aliquot of the supernatant.

## HPLC VARIABLES
**Guard column:** ultra sepharose ODS (Beckman)
**Column:** 250 × 4.6 ultra sepharose ODS (Beckman)

**Mobile phase:** Gradient. MeCN:water from 55:45 to 60:40 over 15 min, to 70:30 over 10 min, to 90:10 over 15 min, return to initial conditions over 5 min.
**Column temperature:** 70
**Injection volume:** 100

## CHROMATOGRAM
**Retention time:** 34.4 (cyclosporin G)

## OTHER SUBSTANCES
**Extracted:** metabolites

## KEY WORDS
human; liver

## REFERENCE
Pichard, L.; Domergue, J.; Fourtanier, G.; Koch, P.; Schran, H.F.; Maurel, P. Metabolism of the new immunosuppressor cyclosporin G by human liver cytochromes P450. *Biochem.Pharmacol.*, **1996**, *51*, 591–598

## SAMPLE
**Matrix:** microsomal incubations
**Sample preparation:** 1 mL Microsomal incubation + 3 mL MeCN:MeOH:5% (?) zinc sulfate 20:30:50, add benzo[a]pyrene 9,10-diol, mix, centrifuge at 600 g. Add the supernatant to a Sep-Pak C18 SPE cartridge, wash with 2 mL water, elute with 4 mL MeOH. Evaporate the eluate to dryness under a stream of nitrogen, reconstitute with MeOH: water 50:50, inject an aliquot.

## HPLC VARIABLES
**Column:** 250 × 4.6 5 μm Ultrasphere octyl
**Mobile phase:** Gradient. MeCN:THF:pH 3 phosphoric acid 30:5:65 for 5 min, to 35:5:60 over 23 min, to 57:5:38 over 12 min.
**Flow rate:** 1
**Detector:** UV 230

## CHROMATOGRAM
**Internal standard:** benzo[a]pyrene 9,10-diol

## OTHER SUBSTANCES
**Extracted:** metabolites

## KEY WORDS
rat; human; liver; SPE

## REFERENCE
Gelboin, H.V.; Krausz, K.W.; Goldfarb, I.; Buters, J.T.M.; Yang, S.K.; Gonzalez, F.J.; Korzekwa, K.R.; Shou, M. Inhibitory and non-inhibitory monoclonal antibodies to human cytochrome P450 3A3/4. *Biochem.Pharmacol.*, **1995**, *50*, 1841–1850

## SAMPLE
**Matrix:** solutions

## HPLC VARIABLES
**Column:** 250 × 4.6 5 μm Vydac C18
**Mobile phase:** Gradient. MeCN:water from 30:70 to 70:30 over 40 min, maintain at 70:30 for 10 min, to 90:10 over 2.5 min, to 70:30 over 7.5 min.

**Column temperature:** 70
**Flow rate:** 1
**Detector:** Radioactivity

## CHROMATOGRAM
**Retention time:** 39.2

## OTHER SUBSTANCES
**Extracted:** metabolites

## KEY WORDS
tritium labeled

## REFERENCE
Gan, L.-S.L.; Moseley, M.A.; Khosla, B.; Augustijns, P.F.; Bradshaw, T.P.; Hendren, R.W. CYP3A-like cytochrome P450-mediated metabolism and polarized efflux of cyclosporin A in Caco-2 cells. Interaction between the two biochemical barriers to intestinal transport. *Drug Metab.Dispos.*, **1996**, *24*, 344–349

## SAMPLE
**Matrix:** solutions

## HPLC VARIABLES
**Column:** 250 × 1 5 μm Vydac C18 peptide/protein
**Mobile phase:** Gradient. MeCN:1% acetic acid from 30:70 to 70:30 over 40 min, maintain at 70:30 for 10 min, to 90:10 over 2.5 min, to 70:30 over 7.5 min.
**Column temperature:** 70
**Flow rate:** 0.05
**Detector:** MS, Sciex API-III, IonSpray (pneumatically assisted electrospray)

## CHROMATOGRAM
**Retention time:** 39

## OTHER SUBSTANCES
**Extracted:** metabolites

## KEY WORDS
tritium labeled; microbore

## REFERENCE
Gan, L.-S.L.; Moseley, M.A.; Khosla, B.; Augustijns, P.F.; Bradshaw, T.P.; Hendren, R.W. CYP3A-like cytochrome P450-mediated metabolism and polarized efflux of cyclosporin A in Caco-2 cells. Interaction between the two biochemical barriers to intestinal transport. *Drug Metab.Dispos.*, **1996**, *24*, 344–349

## SAMPLE
**Matrix:** solutions

## HPLC VARIABLES
**Column:** 250 × 4.6 10 μm Spherisorb ODS-2
**Mobile phase:** MeCN:water 70:30
**Flow rate:** 2
**Detector:** UV 215

## CHROMATOGRAM
**Retention time:** 10

## REFERENCE

Mithani, S.D.; Bakatselou, V.; TenHoor, C.N.; Dressman, J.B. Estimation of the increase in solubility of drugs as a function of bile salt concentration. *Pharm.Res.*, **1996**, *13*, 163–167

## SAMPLE
**Matrix:** solutions

## HPLC VARIABLES
**Column:** 150 mm long 5 μm Ultrasphere octyl
**Mobile phase:** MeCN:water 67:33
**Flow rate:** 1
**Detector:** UV 250

## CHROMATOGRAM
**Retention time:** 7.5

## REFERENCE

Choi, H.-K.; Flynn, G.L.; Amidon, G.L. Percutaneous absorption and dermal delivery of cyclosporin A. *J.Pharm.Sci.*, **1995**, *84*, 581–583

## ANNOTATED BIBLIOGRAPHY

Pham-Huy, C.; Sadeg, N.; Becue, T.; Martin, C.; Mahuzier, G.; Warnet, J.M.; Hamon, M.; Claude, J.R. In vitro metabolism of cyclosporin A with rabbit renal or hepatic microsomes: analysis by HPLC-FPIA and HPLC-MS. *Arch.Toxicol.*, **1995**, *69*, 346–349

Mangold, J.B.; Schran, H.F.; Tse, F.L.S. Pharmacokinetics and metabolism of cyclosporin G in humans. *Drug Metab.Dispos.*, **1994**, *22*, 873–879 [cyclosporine G; plasma; urine; feces; column temp 70; gradient]

Nishikawa, T.; Hasumi, H.; Susuki, S.; Kubo, H.; Ohtani, H. Interconversion of cyclosporin molecular form inducing peak broadening, tailing and splitting during reversed-phase liquid chromatography. *Chromatographia*, **1994**, *38*, 359–364 [cyclosporin A; column temp 0-60]

Poirier, J.-M.; Lebot, M.; Cheymol, G. Cyclosporine in whole blood: Drug monitoring difficulties and presentation of a reliable normal-phase liquid chromatographic assay. *Ther.Drug Monit.*, **1994**, *16*, 388–394 [whole blood; normal-phase; column temp 60; SPE]

Maguire, S.; Kyne, F. A rapid selective high-performance liquid chromatography assay for cyclosporine. *Ann.Clin.Biochem.*, **1993**, *30*, 488–489

Sukhanov, A.V.; Shoikhet, I.N. [The determination of cyclosporin A concentration in whole blood by high-pressure liquid chromatography on the Milikhrom-1 microcolumn chromatograph]. *Klin.Lab.Diagn.*, **1993**, , 7–9

Annesley, T.M.; Matz, K.; Leichtman, A.B. High-performance liquid chromatographic analysis of cyclosporin G (Nva2-cyclosporine) in human blood. *Ther.Drug Monit.*, **1992**, *14*, 397–401

Raghuveeran, C.D.; Gopalan, N.; Dangi, R.S.; Kaushik, M.P.; Venkateswaran, K.S. Preparative scale high-performance liquid chromatography and identification of cyclosporine-A from an indigenous fungal isolate. *J.Liq.Chromatogr.*, **1992**, *15*, 2407–2416 [cyclosporine A; fungus; fermentation broth; preparative; gradient; column temp 75]

Svinarov, D.A.; Dimova, M.N. Liquid chromatographic determination of cyclosporine-A in blood, with Chromosorb P columns used for sample purification. *J.Liq.Chromatogr.*, **1991**, *14*, 1683–1690 [cyclosporine A; whole blood; cyclosporine D (IS); column temp 60]

Bowers, L.D. Cyclosporine analysis by high-performance liquid chromatography: precision, accuracy, and minimum detectable quantity. *Transplant.Proc.*, **1990**, *22*, 1150–1154

Gupta, S.K.; Benet, L.Z. HPLC measurement of cyclosporine in blood plasma and urine and simultaneous measurement of its four metabolites in blood. *J.Liq.Chromatogr.*, **1989**, *12*, 1451–1462 [cyclosporine A; plasma; urine; simultaneous metabolites; LOD 30 ng/mL; cyclosporine D (IS); column temp 70]

Kabra, P.M.; Wall, J.H. Evaluation of polymeric reverse phase extraction columns for liquid chromatographic analysis of cyclosporine in whole blood and comparison with silica based bonded reversed phase extraction columns. *J.Liq.Chromatogr.*, **1989**, *12*, 1819–1834 [whole blood; cyclosporin D (IS); column temp 70; LOD 10 ng/mL]

Awni, W.M.; Maloney, J.A. Optimized high-performance liquid chromatographic method for the analysis of cyclosporine and three of its metabolites in blood and urine. *J.Chromatogr.*, **1988**, *425*, 233–236

Yee, G.C.; Gmur, D.J.; Meier, P. Measurement of blood cyclosporine metabolite concentrations with a new column-switching high-performance liquid chromatographic assay. *Transplant.Proc.*, **1988**, *20*, 585–590

Bowers, L.D.; Singh, J. A gradient HPLC method for quantitation of cyclosporine and its metabolites in blood and bile. *J.Liq.Chromatogr.*, **1987**, *10*, 411–420 [bile; plasma; whole blood]

Buice, R.G.; Stentz, F.B.; Gurley, B.J. Analytical methodologies for cyclosporine pharmacokinetics: A comparison of radioimmunoassay with high performance liquid chromatography. *J.Liq.Chromatogr.*, **1987**, *10*, 421–438 [pharmacokinetics; bile; plasma; serum; dog]

Kabra, P.M.; Wall, J.H. Improved liquid chromatographic analysis of cyclosporine in whole blood with solid phase (Bond-Elut™) extraction. *J.Liq.Chromatogr.*, **1987**, *10*, 477–490 [whole blood; SPE; LOD 10 ng/mL]

Kabra, P.M.; Wall, J.H.; Dimson, P. Automated solid-phase extraction and liquid chromatography for assay of cyclosporine in whole blood. *Clin.Chem.*, **1987**, *33*, 2272–2274

Sangalli, L.; Bonati, M. Reversed-phase high-performance liquid chromatography determination of cyclosporin in human blood. *Ther.Drug Monit.*, **1987**, *9*, 353–357

Shibata, N.; Minouchi, T.; Hayashi, Y.; Ono, T.; Shimakawa, H. Quantitative determination of cyclosporin A in whole blood and plasma by high performance liquid chromatography. *Res.Commun. Chem.Pathol.Pharmacol.*, **1987**, *57*, 261–271

Moyer, T.P.; Johnson, P.; Faynor, S.M.; Sterioff, S. Cyclosporine: a review of drug monitoring problems and presentation of a simple, accurate liquid chromatographic procedure that solves these problems. *Clin.Biochem.*, **1986**, *19*, 83–89

Moyer, T.P.; Charlson, J.R.; Ebnet, L.E. Improved chromatography of cyclosporine. *Ther.Drug Monit.*, **1986**, *8*, 466–468 [SPE; LOD 25 ng/mL; whole blood; column temp 60; cyclosporine A; cyclosporine D (IS)]

Bowers, L.D.; Mathews, S.E. Investigation of the mechanism of peak broadening observed in the high-performance liquid chromatographic analysis of cyclosporine. *J.Chromatogr.*, **1985**, *333*, 231–238

Hoffman, N.E.; Rustum, A.M.; Quebbeman, E.J.; Hamid, A.A.R.; Ausman, R.K. HPLC determination of cyclosporin in whole blood. *J.Liq.Chromatogr.*, **1985**, *8*, 2511–2520 [whole blood; column temp 70]

Kabra, P.M.; Wall, J.H.; Blanckaert, N. Solid-phase extraction and liquid chromatography for improved assay of cyclosporine in whole blood or plasma. *Clin.Chem.*, **1985**, *31*, 1717–1720

Shihabi, Z.K.; Scaro, J.; David, R.M. A rapid method for cyclosporine A determination by HPLC. *J.Liq.Chromatogr.*, **1985**, *8*, 2641–2648

Kates, R.E.; Latini, R. Simple and rapid high-performance liquid chromatographic analysis of cyclosporine in human blood and serum. *J.Chromatogr.*, **1984**, *309*, 441–447

Smith, H.T.; Robinson, W.T. Semi-automated high-performance liquid chromatographic method for the determination of cyclosporine in plasma and blood using column switching. *J.Chromatogr.*, **1984**, *305*, 353–362

Carruthers, S.G.; Freeman, D.J.; Koegler, J.C.; Howson, W.; Keown, P.A.; Laupacis, A.; Stiller, C.R. Simplified liquid-chromotographic analysis for cyclosporin A, and comparison with radioimmunoassay. *Clin.Chem.*, **1983**, *29*, 180–183

Yee, G.C.; Gmur, D.J.; Kennedy, M.S. Liquid-chromatographic determination of cyclosporine in serum with use of a rapid extraction procedure. *Clin.Chem.*, **1982**, *28*, 2269–2271

Allwood, M.C.; Lawrance, R. High pressure liquid chromatographic determination of cyclosporin A in plasma. *J.Clin.Hosp.Pharm.*, **1981**, *6*, 195–199

Niederberger, W.; Schaub, P.; Beveridge, T. High-performance liquid chromatographic determination of cyclosporin A in human plasma and urine. *J.Chromatogr.*, **1980**, *182*, 454–458

# Desogestrel

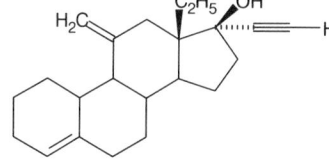

**Molecular formula:** $C_{22}H_{30}O$
**Molecular weight:** 310.5
**CAS Registry No.:** 54024-22-5

## SAMPLE
**Matrix:** mucosal fluid
**Sample preparation:** Extract 1 mL mucosal fluid twice with 5 mL diethyl ether, evaporate to dryness, reconstitute in 100 µL MeOH, inject an aliquot. (Aqueous layer can be incubated with β-glucuronidase/sulfatase at 37° for 3 h before extraction.)

## HPLC VARIABLES
**Guard column:** present but not specified
**Column:** 100 × 8 µBondapak C18
**Mobile phase:** Gradient. MeCN:water from 55:45 to 100:0 over 15 min, maintain at 100:0 for 30 min
**Detector:** UV 214

## OTHER SUBSTANCES
**Extracted:** metabolites

## KEY WORDS
also for microsomal incubations (J. Steroid Biochem. 1990; 35; 281)

## REFERENCE
Madden, S.; Back, D.J.; Martin, C.A.; Orme, M.L. Metabolism of the contraceptive steroid desogestrel by the intestinal mucosa. *Br.J.Clin.Pharmacol.*, **1989**, *27*, 295–299

## SAMPLE
**Matrix:** solutions

## HPLC VARIABLES
**Column:** 200 × 4 5 µm Nucleosil RP-8
**Mobile phase:** MeCN:MeOH:water 3:76:21
**Flow rate:** 1
**Injection volume:** 20
**Detector:** UV 205

## CHROMATOGRAM
**Retention time:** 7

## OTHER SUBSTANCES
**Simultaneous:** metabolites, ethinyl estradiol, 3-hydroxydesogestrel, 3-ketodesogestrel, 6-ketoethinyl estradiol

## REFERENCE
Smilde, A.K.; Bruins, C.H.P.; Doornbos, D.A.; Vink, J. Optimization of the reversed-phase high-performance liquid chromatographic separation of synthetic estrogenic and progestogenic steroids using the multi-criteria decision making method. *J.Chromatogr.*, **1987**, *410*, 1–12

# Diazepam

**Molecular formula:** $C_{16}H_{13}ClN_2O$
**Molecular weight:** 284.8
**CAS Registry No.:** 439-14-5

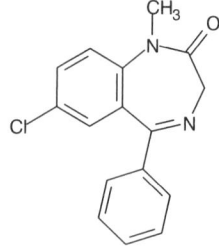

## SAMPLE
**Matrix:** blood
**Sample preparation:** 1-2 mL Plasma + 200 ng lorazepam + 1 mL 1 M pH 10 bicarbonate buffer + 8 mL n-hexane:ethyl acetate 70:30, shake for 10 min, centrifuge. Remove the organic layer and evaporate it to dryness under a stream of nitrogen, reconstitute the residue in 200 μL mobile phase, inject an aliquot.

## HPLC VARIABLES
**Column:** 250 × 4.6 Ultrasphere ODS
**Mobile phase:** Gradient. A was MeCN:buffer 30:70. B was MeCN:buffer 70:30. A:B from 85:15 to 40:60 over 10 min (Waters curve no. 5), to 0:100 over 5 min (Waters curve no. 1), return to initial conditions over 10 min (Waters curve no. 1). (Buffer was 10 mM pH 3.35 $NaH_2PO_4$.)
**Flow rate:** 1
**Detector:** UV 229

## CHROMATOGRAM
**Internal standard:** lorazepam
**Limit of detection:** 2 ng/mL

## OTHER SUBSTANCES
**Extracted:** metabolites, desmethyldiazepam

## KEY WORDS
plasma; pharmacokinetics

## REFERENCE
Caraco, Y.; Tateishi, T.; Wood, A.J.J. Interethnic difference in omeprazole's inhibition of diazepam metabolism. *Clin.Pharmacol.Ther.*, **1995**, *58*, 62–72

## SAMPLE
**Matrix:** blood
**Sample preparation:** 1 mL Plasma + 100 μL 2 μg/mL prazepam in MeOH + 1 mL saturated trisodium phosphate + 3 mL dichloromethane, shake, centrifuge. Remove the organic layer and evaporate it to dryness under reduced pressure, reconstitute the residue in 250 μL mobile phase, inject a 50 μL aliquot.

## HPLC VARIABLES
**Column:** 250 × 4.6 5 μm CAPCELL PAK C18 (Shiseido)
**Mobile phase:** MeOH:water 65:35, adjusted to pH 3.4 with phosphoric acid
**Column temperature:** 30
**Flow rate:** 0.7
**Injection volume:** 50
**Detector:** UV 240

## CHROMATOGRAM
**Retention time:** 11.8
**Internal standard:** prazepam (20.4)
**Limit of detection:** 2 ng/mL

## OTHER SUBSTANCES
**Extracted:** metabolites, demethyldiazepam

## KEY WORDS
plasma; pharmacokinetics

## REFERENCE
Ishizaki, T.; Chiba, K.; Manabe, K.; Koyama, E.; Hayashi, M.; Yasuda, S.; Horai, Y.; Tomono, Y.; Yamato, C.; Toyoki, T. Comparison of the interaction potential of a new proton pump inhibitor, E3810, versus omeprazole with diazepam in extensive and poor metabolizers of *S*-mephenytoin 4'-hydroxylation. *Clin.Pharmacol.Ther.*, **1995**, *58*, 155–164

## SAMPLE
**Matrix:** blood
**Sample preparation:** 2 mL Whole blood or plasma + 2 mL buffer + 5 mL chloroform : isopropanol : n-heptane 60:14:26, shake gently horizontally for 10 min, centrifuge at 2800 g for 10 min. Remove the lower organic layer and evaporate it to dryness under vacuum at 45°, reconstitute the residue in 100 μL mobile phase, centrifuge at 2800 g for 5 min, inject a 50 μL aliquot of the supernatant. (Buffer was saturated ammonium chloride solution 25% diluted with water, adjusted to pH 9.5 with 25% ammonia solution.)

## HPLC VARIABLES
**Column:** $300 \times 3.9$ 4 μm NovaPack C18
**Mobile phase:** MeOH:THF:buffer 65:5:30 (Buffer was 0.68 g/L (10 mM (sic)) $KH_2PO_4$ adjusted to pH 2.6 with concentrated orthophosphoric acid.) (At the end of each session wash the column with water for 1 h and MeOH for 1 h, re-equilibrate for 30 min.)
**Column temperature:** 30
**Flow rate:** 0.8
**Injection volume:** 50
**Detector:** UV 229

## CHROMATOGRAM
**Retention time:** 6.01
**Limit of detection:** <120 ng/mL

## OTHER SUBSTANCES
**Extracted:** acebutolol, acenocoumarol, acepromazine, acetaminophen, ajmaline, albuterol, alimemazine, alminoprofen, alpidem, alprazolam, amisulpride, amitriptyline, amodiaquine, amoxapine, aspirin, astemizole, atenolol, benazepril, benperidol, benzocaine, benzoylecgonine, bepridil, betaxolol, bromazepam, brompheniramine, bumadizone, bupivacaine, buprenorphine, buspirone, caffeine, carbamazepine, carbinoxamine, carpipramine, carteolol, celiprolol, cetirizine, chlorambucil, chlordiazepoxide, chlormezanone, chlorophenacinone, chloroquine, chlorpheniramine, chlorpromazine, chlorpropamide, cibenzoline, cicletanine, clemastine, clobazam, clomipramine, clonazepam, clonidine, clorazepate, clozapine, cocaine, codeine, colchicine, cyamemazine, cyclizine, cycloguanil, cyproheptadine, cytarabine, dacarbazine, daunorubicin, debrisoquine, demexiptiline, desipramine, dextromethorphan, dextromoramide, dextropropoxyphene, diazoxide, diclofenac, dihydralazine, diphenhydramine, dipyridamole, disopyramide, dosulepine, doxepin, doxylamine, droperidol, ephedrine, estazolam, etodolac, fenfluramine, fenoprofen, fentiazac, flecainide, floctafenine, flumazenil, flunitrazepam, fluoxetine, fluphenazine, flurbiprofen, fluvoxamine, glibenclamide, glipizide, glutethimide, histapyrrodine, hydroxychloroquine,

hydroxyzine, ibuprofen, imipramine, indomethacin, iproniazid, ketamine, ketoprofen, labetalol, levomepromazine, lidocaine, lidoflazine, lisinopril, loperamide, loprazolam, loratadine, lorazepam, loxapine, maprotiline, medazepam, medifoxamine, mefenamic acid, mefenidramine, mefloquine, melphalan, meperidine, mephenesin, mephentermine, mepivacaine, metapramine, metformin, methadone, methaqualone, methocarbamol, methotrexate, metipranolol, metoclopramide, metoprolol, mexiletine, midazolam, minoxidil, moclobemide, moperone, morphine, nadolol, nalbuphine, nalorphine, naloxone, naltrexone, naproxen, nialamide, nifedipine, niflumic acid, nimodipine, nitrazepam, nizatidine, nomifensine, nortriptyline, omeprazole, opipramol, oxazepam, oxprenolol, penbutolol, penfluridol, pentazocine, phencyclidine, phenobarbital, phenol, phenylbutazone, pimozide, pindolol, pipamperone, piroxicam, prazepam, prazosin, prilocaine, procainamide, procarbazine, proguanil, promethazine, propafenone, propranolol, protriptyline, pyrimethamine, quinidine, quinine, quinupramine, ranitidine, ritodrine, secobarbital, sotalol, strychnine, sulfinpyrazole, sulindac, sulpride, sultopride, suriclone, temazepam, tenoxicam, terbutaline, terfenadine, tetrazepam, thiopental, thioproperazine, thioridazine, tianeptine, tiapride, tiaprofenic acid, ticlopidine, timolol, tioclomarol, tofisopam, tolbutamide, toloxatone, trazodone, triazolam, trifluoperazine, trifluperidol, trimipramine, triprolidine, tropatenine, verapamil, viloxazine, vinblastine, vincristine, vindesine, warfarin, yohimbine, zolpidem, zopiclone, zorubicine
**Interfering:** aceprometazine, aconitine, alprenolol, bisoprolol, diazepam, diltiazem, glibornuride, haloperidol, mianserine, nicardipine, nitrendipine, ramipril, reserpine, tetracaine

## KEY WORDS
whole blood; plasma

## REFERENCE
Tracqui, A.; Kintz, P.; Mangin, P. Systematic toxicological analysis using HPLC/DAD. *J.Forensic Sci.*, **1995**, *40*, 254–262

## SAMPLE
**Matrix:** blood
**Sample preparation:** Make 200 µL serum alkaline with borate buffer, extract with cyclohexane:dichloromethane 60:40. Remove the organic layer and evaporate it to dryness, reconstitute the residue in mobile phase, inject an aliquot

## HPLC VARIABLES
**Column:** C18 DB (Supelco)
**Mobile phase:** MeCN:pH 2.5 phosphate buffer 37:63
**Detector:** UV 254

## OTHER SUBSTANCES
**Extracted:** bromazepam, clobazam, fluvoxamine, lorazepam, oxazepam

## KEY WORDS
serum

## REFERENCE
Vandenberghe, H.; MacDonald, J.C. Analysis of fluvoxamine, clobazam and other benzodiazepines on the same HPLC system (Abstract 40). *Ther.Drug Monit.*, **1995**, *17*, 393

## SAMPLE
**Matrix:** blood
**Sample preparation:** Wash PCPure SPE cartridge (Moritex) containing 0.4 g hydroxyapatite with 10 mL MeCN and remove MeCN by evaporation. 110 µL Plasma + 10 µL of 100 µg/mL IS in 5% aqueous MeCN, inject onto PCPure cartridge, elute with MeCN:water 1:1. Use first 600 µL of eluate, inject a 20 µL aliquot.

## HPLC VARIABLES
**Column:** 150 × 4.6 5 µm Inertsil ODS-2
**Mobile phase:** MeCN : water 50 : 50
**Column temperature:** 40
**Flow rate:** 1
**Injection volume:** 20
**Detector:** UV 240

## CHROMATOGRAM
**Retention time:** 8
**Internal standard:** 4,4'-difluorobenzophenone (15)
**Limit of detection:** 0.25 ng

## KEY WORDS
plasma; SPE

## REFERENCE
Iwase, H.; Gondo, K.; Koike, T.; Ono, I. Novel precolumn deproteinization method using a hydroxyapatite cartridge for the determination of theophylline and diazepam in human plasma by high-performance liquid chromatography with ultraviolet detection. *J.Chromatogr.B*, **1994**, *655*, 73−81

## SAMPLE
**Matrix:** blood
**Sample preparation:** Automated SPE by ASPEC system. Condition a C18 Clean-Up SPE cartridge (CEC 18111, Worldwide Monitoring) with 2 mL MeOH then 2 mL water. 1 mL Plasma + 1 mL 400 ng/mL protriptyline in water, vortex, add to column, wash with 3 mL water, wash with 3 mL 750 mL/L methanol. Elute with three aliquots of 300 µL 0.1 M ammonium acetate in MeOH. Add 0.5 mL 0.5 M NaOH and 4 mL 50 mL/L isopropanol in heptane to eluate, mix thoroughly. Allow 5 min for phase separation. Remove upper heptane phase and add it to 300 µL 0.1 M phosphoric acid (pH 2.5), mix, separate, inject a 100 µL aliquot of the aqueous phase.

## HPLC VARIABLES
**Guard column:** LC-8-DB (Supelco)
**Column:** 150 × 4.6 LC-8-DB (Supelco)
**Mobile phase:** MeCN : buffer 35 : 65 (Buffer was 10 mL/L triethylamine in water adjusted to pH 5.5 with glacial acetic acid.)
**Flow rate:** 2
**Injection volume:** 100
**Detector:** UV 228

## CHROMATOGRAM
**Retention time:** 8.3
**Internal standard:** protriptyline (4)

## OTHER SUBSTANCES
**Extracted:** acetazolamide, amitriptyline, chlordiazepoxide, chlorpromazine, desipramine, dextromethorphan, diphenhydramine, doxepin, encainide, fentanyl, flecainide, fluoxetine, flurazepam, haloperidol, hydroxyethylflurazepam, ibuprofen, imipramine, lidocaine, maprotiline, methadone, methaqualone, mexiletine, norchlorimipramine, nordoxepin, norfluoxetine, nordiazepam, nortriptyline, norverapamil, pentazocine, promazine, propafenone, propoxyphene, propranolol, protriptyline, quinidine, temazepam, trazodone, trimipramine, verapamil
**Noninterfering:** acetaminophen, acetylmorphine, amiodarone, amobarbital, amphetamine, bendroflumethiazide, benzocaine, benzoylecgonine, benzthiazide, butalbital, carbamazepine, chlorothiazide, clonazepam, cocaine, codeine, cotinine, cyclosporine, cyclothiazide,

desalkylflurazepam, diamorphine, dicumerol, ephedrine, ethacrynic acid, ethanol, ethchlorvynol, ethosuximide, furosemide, glutethimide, hydrochlorothiazide, hydrocodone,
hydroflumethiazide, hydromorphone, lorazepam, mephentermine, meprobamate, methamphetamine, metharbital, methoxsalen, methoxyphenteramine, methsuximide, methylcyclothiazide, metoprolol, MHPG, monoacetylmorphine, morphine, normethsuximide,
oxazepam, oxycodone, oxymorphone, pentobarbital, phencyclidine, phenteramine, phenylephrine, phenytoin, polythiazide, primidone, prochlorperazine, salicylic acid, sulfanilamide, THC-COOH, theophylline, thiazolam, thiopental, thioridazine, tocainide, trichloromethiazide, trifluoperazine, valproic acid, warfarin
**Interfering:** chlorimipramine, midazolam

## KEY WORDS
plasma; SPE

## REFERENCE
Nichols, J.H.; Charlson, J.R.; Lawson, G.M. Automated HPLC assay of fluoxetine and norfluoxetine in
serum. *Clin.Chem.*, **1994**, *40*, 1312–1316

## SAMPLE
**Matrix:** blood
**Sample preparation:** Inject 100-200 μL plasma onto column A with mobile phase A and
elute to waste, after 5 min backflush the contents of column A onto column B with mobile
phase B, after 5 min remove column A from the circuit, elute column B with mobile phase
B, monitor the effluent from column B. Wash column A with MeCN:water 60:40 at 1
mL/min for 6 min then re-equilibrate with pH 7.5 buffer for 10 min.

## HPLC VARIABLES
**Column:** A 45 × 4 12 μm TSK-gel G 3 PW (Tosohass); B 75 × 4.6 Ultrasphere ODS C18 3
μm
**Mobile phase:** A 50 mM pH 7.5 phosphate buffer; B Gradient. A was MeCN. B was 65 mM
$KH_2PO_4$ + 1% diethylamine adjusted to pH 5.4 with phosphoric acid. A:B 22:78 for 5
min, to 25:75 over 10 min, to 60:40 over 15 min.
**Flow rate:** 1
**Injection volume:** 100-200
**Detector:** UV 230

## CHROMATOGRAM
**Retention time:** 26
**Limit of detection:** 25 ng/mL

## OTHER SUBSTANCES
**Extracted:** alprazolam, bromazepam, chlordiazepoxide, clobazam, clonazepam, clorazepate,
clotiazepam, desmethylclobazam, desmethyldiazepam, estazolam, flunitrazepam, loflazepate, lorazepam, medazepam, nitrazepam, oxazepam, prazepam, temazepam, tetrazepam,
tofisopam, triazolam
**Noninterfering:** carbamazepine, ethosuximide, phenobarbital, phenytoin, primidone, valproic acid

## KEY WORDS
plasma; column-switching

## REFERENCE
Lacroix, C.; Wojciechowski, F.; Danger, P. Monitoring of benzodiazepines (clobazam, diazepam and their
main active metabolites) in human plasma by column-switching high-performance liquid chromatography. *J.Chromatogr.*, **1993**, *617*, 285–290

## SAMPLE

**Matrix:** blood

**Sample preparation:** 1 mL Whole blood or plasma + 1 mL water + 50 µL 10 µg/mL IS in MeOH + 1 mL 60% aqueous KOH, vortex for 1 min, heat for 3 min on a boiling water bath, pass onto a 3 mL Extrelut cartridge. Elute with diethyl ether:dichloromethane 70:30, evaporate eluate to dryness under a stream of air at 40°, vortex in 100 µL initial mobile phase, inject a 40 µL aliquot.

## HPLC VARIABLES

**Column:** 300 × 3.9 10 µm µBondapak C18

**Mobile phase:** Gradient. A was 12.5 mM $KH_2PO_4$:1 N phosphoric acid 999:1. B was MeCN. A:B 65:35 for 12 min at 1 mL/min and UV 343 nm then 55:45 at 2 mL/min and UV 242 nm (step gradient).

**Flow rate:** 1-2

**Injection volume:** 40

**Detector:** UV 343; UV 242

## CHROMATOGRAM

**Retention time:** 19.5

**Internal standard:** papaverine hydrochloride (10.7)

**Limit of detection:** 4 ng/mL

## OTHER SUBSTANCES

**Extracted:** metabolites, chloroquine, monodesethylchloroquine, nordiazepam

**Simultaneous:** epinephrine, mefloquine, pyrimethamine, quinine, sulfadoxine

## KEY WORDS

whole blood; plasma

## REFERENCE

Estadieu, M.; Durand, A.; Viala, A.; Rop, P.P.; Fornaris, M.; Quicke, J. A rapid HPLC procedure for the simultaneous determination of chloroquine, monodesethylchloroquine, diazepam, and nordiazepam in blood. *J.Anal.Toxicol.*, **1989**, *13*, 89–93

## SAMPLE

**Matrix:** blood

**Sample preparation:** Filter (0.5 µm) serum, inject 200 µL directly onto column A with mobile phase A, run with mobile phase A for 1.5 min then change to mobile phase B over 0.1 min, wash column A with mobile phase B for 10.5 min, backflush column A onto column B with mobile phase C for 7.5 min then switch column B out of circuit, elute column B with mobile phase C and monitor the eluant, re-equilibrate column A with mobile phase A for at least 5 min.

## HPLC VARIABLES

**Column:** A 15 × 3.2 5 µm Brownlee ODS; B 250 × 1 5 µm Adsorbosphere ODS

**Mobile phase:** A 10 mM sodium dodecyl sulfate; B water; C MeOH:water 65:35

**Flow rate:** A 1; B 1; C 0.06

**Injection volume:** 200

**Detector:** UV 242

## CHROMATOGRAM

**Retention time:** 30

**Limit of detection:** 30 ng/mL

## OTHER SUBSTANCES

**Simultaneous:** nordiazepam, oxazepam, temazepam

## KEY WORDS
serum; column-switching; microbore

## REFERENCE
Koenigbauer, M.J.; Curtis, M.A. Use of micellar mobile phases and microbore column switching for the assay of drugs in physiological fluids. *J.Chromatogr.*, **1988**, *427*, 277–285

## SAMPLE
**Matrix:** blood
**Sample preparation:** 1 mL Plasma + 100 μL 1 μg/mL diazepam in MeOH + 2 mL phosphate buffer, vortex for 30 s, add 8 mL hexane:isoamyl alcohol 95:5, shake gently by hand for 10 min, vortex for 1 min, centrifuge at 400 g at 4° for 10 min. Remove organic layer and add it to 2 mL 6 M HCl, shake for 10 min, vortex for 1 min, centrifuge at 400 g for 10 min. Remove the aqueous layer and slowly add about 2 mL 6 M NaOH to it to achieve a pH greater than 7.0, add 2 mL phosphate buffer, vortex for 10 s, add 8 mL hexane:isoamyl alcohol 95:5, shake for 10 min, vortex for 1 min, centrifuge at 400 g for 10 min. Remove the organic layer and evaporate it to dryness at 40° with nitrogen, rinse residue from sides with hexane:isoamyl alcohol 95:5, again dry with nitrogen, take up residue in 40 μL MeOH, inject a 20-40 μL aliquot. (Phosphate buffer was 136.1 g $KH_2PO_4$ in 1 L water, adjusted to pH 7 with 1 M $K_2HPO_4$.)

## HPLC VARIABLES
**Column:** 300 × 3.9 10 μm μBondapak C18
**Mobile phase:** MeOH:MeCN:10 mM sodium acetate 40:12.5:47.5, at pH 4.6
**Flow rate:** 2.2
**Injection volume:** 20-40
**Detector:** UV 254

## CHROMATOGRAM
**Retention time:** 9.2
**Internal standard:** diazepam

## OTHER SUBSTANCES
**Extracted:** lorazepam
**Simultaneous:** clonazepam, flunitrazepam, flurazepam, midazolam, nitrazepam, oxazepam, temazepam

## KEY WORDS
plasma; diazepam is IS

## REFERENCE
Egan, J.M.; Abernethy, D.R. Lorazepam analysis using liquid chromatography: improved sensitivity for single-dose pharmacokinetic studies. *J.Chromatogr.*, **1986**, *380*, 196–201

## SAMPLE
**Matrix:** blood
**Sample preparation:** 1 mL Serum + 2 mL water + 50 μL 3.2 μg/mL estazolam in MeOH + 2 mL 100 mM NaOH, mix gently, add 8 mL diethyl ether, shake for 15 min, centrifuge at 2500 rpm for 5 min. Remove 4 mL of the organic layer and evaporate it to dryness under a stream of nitrogen at 40°, reconstitute the residue in 100 μL mobile phase, vortex for 30 s, inject a 50 μL aliquot.

## HPLC VARIABLES
**Column:** 50 × 4.6 Shim-pack FLC-C8 (Shimadzu)
**Mobile phase:** MeOH:buffer 53:47 (Buffer was 5 mM $Na_2HPO_4$ adjusted to pH 6.0 with phosphoric acid.)

**Flow rate:** 0.6
**Injection volume:** 50
**Detector:** UV 254

## CHROMATOGRAM
**Retention time:** 8.5
**Internal standard:** estazolam (4)
**Limit of detection:** 8 ng/mL

## OTHER SUBSTANCES
**Extracted:** clorazepate, nordiazepam, oxazepam, temazepam, triazolam
**Simultaneous:** bromazepam, flunitrazepam, nitrazepam, sulpride
**Noninterfering:** haloperidol, trihexyphenidyl

## KEY WORDS
serum; pharmacokinetics

## REFERENCE
Tada, K.; Moroji, T.; Sekiguchi, R.; Motomura, H.; Noguchi, T. Liquid-chromatographic assay of diazepam and its major metabolites in serum, and application to pharmacokinetic study of high doses of diazepam in schizophrenics. *Clin.Chem.*, **1985**, *31*, 1712–1715

## SAMPLE
**Matrix:** blood
**Sample preparation:** 2 mL Plasma + 100 μL 1 μg/mL loxapine in isopropanol:diethylamine 99.9:0.1 + 250 μL 25% potassium carbonate containing 0.1% diethylamine + 5 mL hexane:isoamyl alcohol 97:3, vortex for 30 s, centrifuge at 500 g for 3 min. Remove the organic layer and add it to 100 μL 250 mM HCl, vortex for 30 s, inject a 50 μL aliquot of the aqueous phase.

## HPLC VARIABLES
**Guard column:** 50 × 4.6 40 μm C8 (Supelco)
**Column:** 250 × 4.6 5 μm Supelcosil C8
**Mobile phase:** MeCN:water:diethylamine:85% phosphoric acid 53.3:45.1:1:0.4, pH adjusted to 7.2 with NaOH or phosphoric acid
**Flow rate:** 2
**Injection volume:** 50
**Detector:** UV 254

## CHROMATOGRAM
**Retention time:** k' 5.20
**Internal standard:** loxapine (k' 7.18)

## OTHER SUBSTANCES
**Extracted:** amitriptyline, chlordiazepoxide, chlorpromazine, desipramine, desmethldiazepam, desmethylchlordiazepoxide, desmethyldoxepin, doxepin, fluphenazine, haloperidol, nortriptyline, oxazepam, thiothixene
**Noninterfering:** molindone, perphenazine, trifluoperazine
**Interfering:** imipramine

## KEY WORDS
plasma

## REFERENCE
Kiel, J.S.; Abramson, R.K.; Morgan, S.L.; Voris, J.C. A rapid high performance liquid chromatographic method for the simultaneous measurement of six tricyclic antidepressants. *J.Liq.Chromatogr.*, **1983**, *6*, 2761–2773

## SAMPLE
**Matrix:** blood
**Sample preparation:** 200 μL Serum + 200 μL 50 μg/mL hexobarbital in MeCN + 25 μL glacial acetic acid, vortex for 10 s, centrifuge for 1 min, inject a 30-100 μL aliquot of the supernatant.

## HPLC VARIABLES
**Column:** μBondapak C18
**Mobile phase:** Gradient. MeCN:7.5 g/L NaH$_2$PO$_4$ adjusted to pH 3.2 with phosphoric acid 5:95 to 22:78 over 24 min, to 45:55 over 10 min, maintain at 45:55 for 5 min. Re-equilibrate with 5:95 for 5 min.
**Column temperature:** 50
**Flow rate:** 3
**Injection volume:** 30-100
**Detector:** UV 210

## CHROMATOGRAM
**Retention time:** 32.5
**Internal standard:** hexobarbital (20.6)
**Limit of detection:** 200-2000 ng/mL

## OTHER SUBSTANCES
**Extracted:** acetaminophen, amobarbital, butabarbital, butalbital, chlordiazepoxide, ethchlorvynol, flurazepam, glutethimide, methaqualone, methyprylon, nitrazepam, pentobarbital, phenobarbital, phenytoin, primidone, salicylic acid, secobarbital, theophylline
**Simultaneous:** amitriptyline, caffeine, clomipramine, codeine, desipramine, ethotoin, imipramine, lidocaine, mesantoin, methsuximide, nirvanol, nortriptyline, oxazepam, procainamide, phenylpropanolamine, propranolol, quinidine

## KEY WORDS
serum

## REFERENCE
Kabra, P.M.; Stafford, B.E.; Marton, L.J. Rapid method for screening toxic drugs in serum with liquid chromatography. *J.Anal.Toxicol.*, **1981**, *5*, 177–182

## SAMPLE
**Matrix:** blood, CSF
**Sample preparation:** 200 μL Serum, plasma, or CSF + 300 μL reagent. Flush column A to waste with 500 μL 500 mM ammonium sulfate, inject sample onto column A, flush column A to waste with 500 μL 500 mM ammonium sulfate, elute the contents of column A onto column B with mobile phase, monitor the effluent from column B. (Reagent was 8.05 M guanidine hydrochloride and 1.02 M ammonium sulfate in water.)

## HPLC VARIABLES
**Column:** A 30 × 2.1 40 μm preparative grade C18 (Analytichem); B 250 × 4.6 10 μm Partisil C8
**Mobile phase:** Gradient. A was 50 mM pH 4.5 KH$_2$PO$_4$. B was MeCN:isopropanol 80:20. A:B 90:10 for 1 min, to 30:70 over 15 min, maintain at 30:70 for 4 min.
**Column temperature:** 50
**Flow rate:** 1.5
**Detector:** UV 280 for 5 min then UV 254

## CHROMATOGRAM
**Retention time:** 15.21
**Internal standard:** heptanophenone (19.2)

## OTHER SUBSTANCES
**Extracted:** acetazolamide, ampicillin, bromazepam, caffeine, carbamazepine, chloramphenicol, chlorothiazide, droperidol, ethionamide, furosemide, isoniazid, methadone, penicillin
G, phenobarbital, phenytoin, prazepam, propoxyphene, pyrazinamide, rifampin, trimeprazine, trimethoprim

## KEY WORDS
serum; plasma; column-switching

## REFERENCE
Seifart, H.I.; Kruger, P.B.; Parkin, D.P.; van Jaarsveld, P.P.; Donald, P.R. Therapeutic monitoring of
antituberculosis drugs by direct in-line extraction on a high-performance liquid chromatography system. *J.Chromatogr.*, **1993**, *619*, 285–290

## SAMPLE
**Matrix:** blood, CSF, gastric fluid, urine
**Sample preparation:** 200 μL Serum, urine, CSF, or gastric fluid + 300 μL reagent. Flush
column A to waste with 500 μL 500 mM ammonium sulfate, inject sample onto column
A, flush column A to waste with 500 μL 500 mM ammonium sulfate, backflush the con-
tents of column A onto column B with mobile phase, monitor the effluent from column B.
(Reagent was 8.05 M guanidine hydrochloride and 1.02 M ammonium sulfate in water.)

## HPLC VARIABLES
**Column:** A 40 μm preparative grade C18 (Analytichem); B 75 × 2.1 pellicular C18 (What-
man) + 250 × 4.6 5 μm C8 end-capped (Whatman)
**Mobile phase:** Gradient. A was 50 mM pH 4.5 $KH_2PO_4$. B was MeCN:isopropanol 80:20.
A:B 90:10 for 1 min, to 30:70 over 20 min.
**Column temperature:** 50
**Flow rate:** 1.5
**Detector:** UV 220

## CHROMATOGRAM
**Retention time:** 14.95
**Internal standard:** heptanophenone (19)

## OTHER SUBSTANCES
**Extracted:** acetaminophen, allobarbital, azinphos, barbital, brallobarbitone, bromazepam,
butethal, caffeine, carbamazepine, carbaryl, cephaloridine, chloramphenicol, chlordiazepoxide, chlorothiazide, chlorvinphos, clothiapine, cocaine, coomassie blue, desipramine,
diphenhydramine, dipipanone, ethylbromphos, flufenamic acid, formothion, griseofulvin,
indomethacin, lidocaine, lorazepam, malathion, medazepam, midazolam, oxazepam, paraoxon, penicillin G, pentobarbital, prazepam, propoxyphene, prothiophos, quinine, salicylic acid, secobarbital, strychnine, sulfamethoxazole, theophylline, thiopental, thioridazine, trimethoprim

## KEY WORDS
serum; column-switching

## REFERENCE
Kruger, P.B.; Albrecht, C.F.De V.; Jaarsveld, P.P. Use of guanidine hydrochloride and ammonium sulfate
in comprehensive in-line sorption enrichment of xenobiotics in biological fluids by high-performance
liquid chromatography. *J.Chromatogr.*, **1993**, *612*, 191–198

## SAMPLE
**Matrix:** blood, gastric contents, tissue, urine

**Sample preparation:** 1 mL Blood, urine, or gastric contents or 1 g tissue homogenate + 500 μL buffer + 8 mL n-hexane:ethyl acetate 70:30, mix on a rotary mixer for 10 min, centrifuge at 3000 g for 8 min. Remove the organic layer and evaporate it to dryness under a stream of nitrogen, reconstitute the residue in 100 μL 12.5 mM NaOH in MeOH:water 50:50, inject a 50 μL aliquot. (Buffer was 13.8 g potassium carbonate in 100 mL water, pH adjusted to 9.5 with concentrated HCl.)

## HPLC VARIABLES
**Guard column:** 4 × 4 30 μm LiChrocart Aluspher RP-select B (Merck)
**Column:** 125 × 4 5 μm Aluspher RP-select B (Merck)
**Mobile phase:** Gradient. A was 12.5 mM NaOH in MeOH. B was 12.5 mM NaOH in water. A:B 10:90 for 5 min, to 90:10 over 15 min, maintain at 90:10 for 5 min, return to initial conditions over 1 min, re-equilibrate for 5 min.
**Flow rate:** 1
**Injection volume:** 50
**Detector:** UV 230; UV 254

## CHROMATOGRAM
**Retention time:** 21

## OTHER SUBSTANCES
**Extracted:** alprenolol, amitriptyline, bromazepam, carbamazepine, chlordiazepoxide, chlorpromazine, clonazepam, desipramine, flunitrazepam, haloperidol, nitrendipine, nordiazepam, nortriptyline, pindolol, zolpidem
**Also analyzed:** acebutolol, acetaminophen, alprazolam, amphetamine, atenolol, betaxolol, brotizolam, caffeine, camazepam, captopril, chloroquine, clobazam, clomipramine, clothiapine, clotiazepam, cloxazolam, cocaine, codeine, diclofenac, dihydralazine, dihydrocodeine, dihydroergotamine, diphenhydramine, domperidone, doxepin, droperidol, ergotamine, ethyl loflazepate, fenethylline, fluoxetine, flupentixol, flurazepam, furosemide, gliclazide, hydrochlorothiazide, hydroxyzine, ibuprofen, imipramine, ketazolam, loprazolam, lorazepam, lormetazepam, maprotiline, medazepam, mepyramine, methadone, methaqualone, methyldopa, methylphenidate, metoclopramide, metoprolol, mexiletine, mianserin, midazolam, minoxidil, morphine, nadolol, nitrazepam, oxprenolol, papaverine, pentazocine, phenprocoumon, phenylbutazone, pipamperone, piritramide, practolol, prazepam, prazosin, promazine, promethazine, propoxyphene, propranolol, prothipendyl, quinine, sotalol, sulpride, thioridazine, trazodone, triazolam, trimipramine, tripelennamine, tyramine, verapamil, yohimbine

## REFERENCE
Lambert, W.E.; Meyer, E.; De Leenheer, A.P. Systematic toxicological analysis of basic drugs by gradient elution of an alumina-based HPLC packing material under alkaline conditions. *J.Anal.Toxicol.*, **1995**, *19*, 73–78

## SAMPLE
**Matrix:** blood, gastric contents, tissue, urine
**Sample preparation:** Tissue homogenates were 1:2 in water. 1 mL Sample + 1 mL saturated sodium borate buffer + 100 μL 20 μg/mL methyl clonazepam in water + 5 mL n-butyl chloride, rotate at 40 rpm for 30 min, centrifuge at 2500 rcf for 5 min. Remove the organic phase and evaporate it to dryness at 70° under a stream of air, reconstitute the residue in 300 μL mobile phase, vortex for 30 s, inject a 20 μL aliquot.

## HPLC VARIABLES
**Column:** 250 × 2.1 Analytichem ODS with an integral guard column
**Mobile phase:** MeCN:100 mM KH$_2$PO$_4$ 300:700, adjust pH to 3.00 with concentrated phosphoric acid
**Column temperature:** 60
**Flow rate:** 1.5

**Injection volume:** 20
**Detector:** UV 242

## CHROMATOGRAM
**Retention time:** 7.46
**Internal standard:** methyl clonazepam (5.36)
**Limit of detection:** 50 ng/mL

## OTHER SUBSTANCES
**Extracted:** temazepam, trazodone
**Also analyzed:** acetaminophen, alprazolam, amitriptyline, amoxapine, carbamazepine, chlordiazepoxide, chlorpromazine, chlorprothixene, clonazepam, demoxepam, desipramine, diphenhydramine, disopyramide, doxepin, ethotoin, flurazepam, glutethimide, haloperidol, haloperidol, imipramine, lidocaine, lorazepam, loxapine, maprotiline, mesantoin, mesoridazine, methaqualone, methotrimeprazine, nordiazepam, nortriptyline, oxazepam, pentazocine, perphenazine, phenacetin, phenobarbital, phenytoin, promazine, promethazine, propranolol, protriptyline, salicylic acid, thiothixene, trifluoperazine, triflupromazine, trimipramine
**Noninterfering:** chloral hydrate, codeine, ketamine, meperidine, methadone, methamphetamine, methypyrlon, thioridazine

## KEY WORDS
serum; plasma; whole blood

## REFERENCE
Root, I.; Ohlson, G.B. Trazodone overdose: report of two cases. *J.Anal.Toxicol.*, **1984**, *8*, 91–94

## SAMPLE
**Matrix:** blood, milk
**Sample preparation:** 500 μL Plasma or milk + 25 μL 5 μg/mL flurazepam in water:MeCN 2.5:97.5 + 500 μL 67 mM pH 7.4 phosphate buffer + 7 mL diethyl ether, extract for 15 min (A). Remove ether layer and add it to 1 mL 1.5 M HCl, shake for 15 min. Freeze and discard ether phase. Basify aqueous phase with 1 mL 2 M NaOH, extract with 7 mL diethyl ether for 15 min. Evaporate ether at 37° under a stream of nitrogen and take up residue in mobile phase, inject an aliquot. (For plasma **only** ether at (A) can be evaporated at 37° under a stream of nitrogen, take up residue in mobile phase, inject an aliquot.)

## HPLC VARIABLES
**Guard column:** 25 × 4 5 μm LiChrospher 60 RP-select B
**Column:** 125 × 4 5 μm LiChrospher 60 RP-select B
**Mobile phase:** MeCN: 10 mM $KH_2PO_4$ 31:69, adjusted to pH 2.80 with phosphoric acid
**Column temperature:** 45
**Flow rate:** 2
**Injection volume:** 50
**Detector:** UV 241

## CHROMATOGRAM
**Retention time:** 8.6
**Internal standard:** flurazepam (3.0)
**Limit of detection:** 20 ng/mL

## OTHER SUBSTANCES
**Extracted:** nordiazepam, oxazepam, temazepam

## KEY WORDS
plasma; human; rabbit

## REFERENCE

Stebler, T.; Guentert, T.W. Determination of diazepam and nordazepam in milk and plasma in the presence of oxazepam and temazepam. *J.Chromatogr.*, **1991**, *564*, 330–337

## SAMPLE

**Matrix:** blood, stomach contents, tissue, urine

**Sample preparation:** Whole blood, stomach contents. 1 mL Whole blood or stomach contents + IS + 500 μL 1 M potassium carbonate + 8 mL n-hexane:ethyl acetate 70:30, vortex for 1 min, centrifuge at 2000 rpm for 5 min. Remove the organic layer and evaporate it to dryness under a stream of nitrogen, reconstitute the residue in 100 μL initial mobile phase, vortex, centrifuge, inject a 50 μL aliquot. Tissue. Cut 1 g tissue into small pieces, make up to 5 mL with water, homogenize (Ultraturrax). 1 mL Homogenate + IS + 500 μL 1 M potassium carbonate + 8 mL n-hexane:ethyl acetate 70:30, vortex for 1 min, centrifuge at 2000 rpm for 5 min. Remove the organic layer and evaporate it to dryness under a stream of nitrogen, reconstitute the residue in 100 μL initial mobile phase, vortex, centrifuge, inject a 50 μL aliquot. Urine. 1 mL Urine + IS + 250 μL concentrated HCl, heat at 100° for 1 h, cool, adjust pH to 9 with NaOH pellets and 1 M potassium carbonate. Add 8 mL n-hexane:ethyl acetate 70:30, vortex for 1 min, centrifuge at 2000 rpm for 5 min. Remove the organic layer and evaporate it to dryness under a stream of nitrogen, reconstitute the residue in 100 μL initial mobile phase, vortex, centrifuge, inject a 50 μL aliquot.

## HPLC VARIABLES

**Guard column:** 10 × 2.1 pellicular reverse phase (Chrompack)

**Column:** 100 × 3 5 μm Chromspher C8 (Chrompack)

**Mobile phase:** Gradient. MeOH containing 0.03% isopropylamine:water containing 0.03% isopropylamine 20:80 for 2 min, to 30:70 over 0.2 min, maintain at 30:70 for 1.8 min, to 40:60 over 0.2 min, maintain at 40:60 for 0.3 min, to 43:57 over 0.5 min, to 45:55 over 1 min, to 52:48 over 1 min, to 58:42 over 2.5 min, to 75:25 over 1 min, maintain at 75:25 for 4.5 min, return to initial conditions over 0.3 min, re-equilibrate for 3.7 min before next injection.

**Flow rate:** 0.7

**Injection volume:** 50

**Detector:** UV 230

## CHROMATOGRAM

**Retention time:** 10

**Internal standard:** camazepam (10.5), clotiazepam (11)

**Limit of detection:** 10 ng/mL

**Limit of quantitation:** 30 ng/mL

## OTHER SUBSTANCES

**Extracted:** bromazepam, flunitrazepam, nitrazepam, nordiazepam, oxazepam

**Simultaneous:** alprazolam, brotizolam, chlordiazepoxide, clonazepam, cloxazolam, flurazepam, loprazolam, lormetazepam, medazepam, prazepam, triazolam

## KEY WORDS

whole blood; liver; kidney

## REFERENCE

Lambert, W.E.; Meyer, E.; Xue-Ping, Y.; De Leenheer, A.P. Screening, identification, and quantitation of benzodiazepines in postmortem samples by HPLC with photodiode array detection. *J.Anal.Toxicol.*, **1995**, *19*, 35–40

## SAMPLE

**Matrix:** blood, urine

**Sample preparation:** Condition a 3 mL 200 mg Bond Elut C18 SPE cartridge with 2 mL MeOH and 2 mL water. 1 mL Serum or urine + 50 ng diazepam-d$_5$ + 50 ng N-desmethyldiazepam-d$_5$, add to the SPE cartridge, wash with 2 mL water, elute with 500 μL MeOH. Evaporate the eluate to dryness under a stream of nitrogen at 60°, reconstitute the residue in 100 μL mobile phase, inject a 5 μL aliquot. Alternatively, condition a 6 mL narc-2 Bakerbond SPE cartridge with 2 mL MeOH and 2 mL 100 mM pH 6.0 KH$_2$PO$_4$. 1 mL Serum or urine + 50 ng diazepam-d$_5$ + 50 ng N-desmethyldiazepam-d$_5$, add to the SPE cartridge at 1 mL/min, wash with 1 mL MeOH:100 mM pH 6.0 KH$_2$PO$_4$ 20:80, wash with 1 mL 1 M acetic acid, dry under vacuum for 5 min, wash with 1 mL hexane, dry under vacuum for 1 min, elute with two 2 mL portions of dichloromethane:ammonium hydroxide 96:4 (pH 11). Evaporate the eluate to dryness under a stream of nitrogen at 50°, reconstitute the residue in 100 μL mobile phase, inject a 5 μL aliquot.

## HPLC VARIABLES
**Column:** 100 × 2 5 μm LiChrospher 60-RP select B
**Mobile phase:** MeCN:MeOH:water 1:1:1, pH 6
**Flow rate:** 0.1
**Injection volume:** 5
**Detector:** MS, Finnigan MAT TSQ 7000 triple stage quadrupole, selected reaction monitoring mode (m/z 285/257, offset -30 eV), collision gas argon at 0.27 Pa, heating capillary 250°, repeller 20 V, electrospray capillary 6 kV, electron multiplier 2.3 kV, sheath gas nitrogen at 344.7 kPa, auxiliary gas nitrogen at 5 L/min, octapole offset 5 eV

## CHROMATOGRAM
**Retention time:** 6.4
**Internal standard:** diazepam-d$_5$, N-desmethyldiazepam-d$_5$
**Limit of quantitation:** 2 ng/mL

## OTHER SUBSTANCES
**Extracted:** flunitrazepam, medazepam, nitrazepam

## KEY WORDS
SPE; serum

## REFERENCE
Kleinschnitz, M.; Herderich, M.; Schreier, P. Determination of 1,4-benzodiazepines by high-performance liquid chromatography-electrospray tandem mass spectrometry. *J.Chromatogr.B*, **1996**, *676*, 61–67

## SAMPLE
**Matrix:** blood, urine
**Sample preparation:** Condition a C2 Bond-Elut SPE cartridge with 1 column volume methanol and 1 column volume buffer. Add 1 mL of urine buffered with pH 6 100 mM phosphate buffer or plasma buffered with pH 8 100 mM phosphate buffer to the SPE cartridge, wash with 3 column volumes of water, wash with 1 mL of MeOH:water 30:70, elute with 1 mL of MeOH:water 60:40. Evaporate the eluate to dryness and take up the residue in 200 μL mobile phase, inject an aliquot.

## HPLC VARIABLES
**Column:** 35 × 4.6 5 μm ultrabase C18 (Scharlau)
**Mobile phase:** MeOH:water 60:40
**Flow rate:** 1.3
**Injection volume:** 20
**Detector:** UV 229

## CHROMATOGRAM
**Internal standard:** prazepam
**Limit of detection:** 63 ng/mL

## OTHER SUBSTANCES
**Also analyzed:** adinazolam, brotizolam, midazolam, nordazepam, oxazepam, temazepam

## KEY WORDS
plasma; SPE

## REFERENCE
Casas, M.; Berrueta, L.A.; Gallo, B.; Vicente, F. Solid-phase extraction of 1,4-benzodiazepines from biological fluids. *J.Pharm.Biomed.Anal.*, **1993**, *11*, 277-284

## SAMPLE
**Matrix:** diffusate, tissue
**Sample preparation:** Homogenize (Polytron PCU-2) 150-200 mg skin with 4 mL chloroform, repeat homogenization, filter (phase-separating paper) extracts. Make the residue alkaline with 2 mL 10% NaOH, extract twice with 4 mL portions of chloroform, wash the extracts twice with 2 mL portions of water, filter (phase-separating paper) the organic layer. Combine all the chloroform layers and evaporate them to dryness under a stream of air, reconstitute the residue in 1 mL mobile phase, filter (microfilter), inject an aliquot.

## HPLC VARIABLES
**Guard column:** 20 × 4 40 μm ODS (Valco)
**Column:** 150 × 4.6 5 μm Spherisorb ODS-I
**Mobile phase:** MeCN:water 52:48 containing 10 mM octanesulfonic acid and 1% acetic acid, pH 3.5
**Flow rate:** 1
**Injection volume:** 50
**Detector:** UV 254

## CHROMATOGRAM
**Retention time:** 6.0
**Internal standard:** diazepam

## OTHER SUBSTANCES
**Extracted:** physostigmine, tacrine

## KEY WORDS
skin; diazepam is IS

## REFERENCE
Lau, S.W.J.; Chow, D.; Feldman, S. Simultaneous determination of physostigmine and tetrahydroaminoacridine in a transdermal permeation study by high-performance liquid chromatography. *J.Chromatogr.*, **1990**, *526*, 87-95

## SAMPLE
**Matrix:** formulations
**Sample preparation:** Stir weighed tablet with 5 mL water until coating is dissolved; add 5 mL 1 mg/mL IS in MeCN; make up to 50 mL with MeCN. Centrifuge an aliquot at 2472 g for 5 min, inject a 10 μL aliquot.

## HPLC VARIABLES
**Guard column:** 4 × 4 5 μm spherical octyl derivative silica (Merck)
**Column:** 250 × 4 5 μm spherical octyl derivative silica (Merck)
**Mobile phase:** MeOH:buffer 70:30 adjusted to pH 6.0 with glacial acetic acid (Buffer was 500 mM sodium acetate and 5 mM sodium 1-heptanesulfonate.)
**Column temperature:** 50

**Flow rate:** 1
**Injection volume:** 10
**Detector:** UV 230

## CHROMATOGRAM
**Retention time:** 4.8
**Internal standard:** n-butyl p-hydroxybenzoate (UV 254)

## OTHER SUBSTANCES
**Simultaneous:** otilonium bromide (UV 290)

## KEY WORDS
tablets; stability-indicating

## REFERENCE
Mannucci, C.; Bertini, J.; Cocchini, A.; Perico, A.; Salvagnini, F.; Triolo, A. High-performance liquid chromatographic method for assay of otilonium bromide, diazepam, and related compounds in finished pharmaceutical forms. *J.Pharm.Sci.*, **1993**, *82*, 367–370

## SAMPLE
**Matrix:** formulations
**Sample preparation:** Dilute with saline, inject a 10 μL aliquot.

## HPLC VARIABLES
**Column:** 250 × 4.6 Lichrosorb 10 RP 8
**Mobile phase:** MeOH:THF:water 50:5:50
**Flow rate:** 3
**Injection volume:** 10
**Detector:** UV 254

## CHROMATOGRAM
**Retention time:** 10

## OTHER SUBSTANCES
**Simultaneous:** lorazepam, thiopental

## KEY WORDS
injections; saline

## REFERENCE
Martens, H.J.; de Goede, P.N.; van Loenen, A.C. Sorption of various drugs in polyvinyl chloride, glass, and polyethylene-lined infusion containers. *Am.J.Hosp.Pharm.*, **1990**, *47*, 369–373

## SAMPLE
**Matrix:** formulations
**Sample preparation:** Sonicate 2 mg tablet in MeOH, make up to 20 mL with MeOH, filter (0.45 μm), dilute 5 mL filtrate to 100 mL with MeOH, inject a 20 μL aliquot.

## HPLC VARIABLES
**Column:** 250 × 4.6 LiChrosorb 10 RP-18
**Mobile phase:** MeOH:water 60:40
**Flow rate:** 3.5
**Injection volume:** 20
**Detector:** UV 258

## CHROMATOGRAM
**Retention time:** 8

## OTHER SUBSTANCES
**Simultaneous:** degradation products, temazepam

## KEY WORDS
tablets

## REFERENCE
Gordon, S.M.; Freeston, L.K.; Collins, A.J. Determination of temazepam and its major degradation products in soft gelatin capsules by isocratic reversed-phase high-performance liquid chromatography. *J.Chromatogr.*, **1986**, *368*, 180–183

## SAMPLE
**Matrix:** hair
**Sample preparation:** Wash hair in water, rinse 3 times with MeOH, dry, weigh. 5-25 mg Washed hair + 1 mL MeOH, heat at 55° for 18 h, adjust pH to 9.5-10. 1 mL Extract + 1 μg protriptyline + 1 mL water + 1 mL 200 mM sodium carbonate buffer, mix, extract with hexane:butanol 95:5 for 20 min. Remove the organic layer and add it to 100 μL 0.2% orthophosphoric acid, mix for 20 min, inject a 30 μL aliquot of the aqueous layer.

## HPLC VARIABLES
**Guard column:** 15 × 3.2 7 μm Newguard RP-18
**Column:** 100 × 4.6 Spheri-5 RP-C18
**Mobile phase:** MeCN:buffer 40:60 (Buffer was 1.2 L 100 mM pH 7.0 $NaH_2PO_4$ + 30 mL diethylamine.)
**Flow rate:** 2
**Injection volume:** 30
**Detector:** UV 214

## CHROMATOGRAM
**Internal standard:** protriptyline (4)

## OTHER SUBSTANCES
**Extracted:** amitriptyline, desipramine, dothiepin, flunitrazepam, haloperidol, imipramine, imipramine, nitrazepam, nortriptyline, oxazepam, temazepam

## KEY WORDS
may be interferences

## REFERENCE
Couper, F.J.; McIntyre, I.M.; Drummer, O.H. Extraction of psychotropic drugs from human scalp hair. *J.Forensic Sci.*, **1995**, *40*, 83–86

## SAMPLE
**Matrix:** microsomal incubations
**Sample preparation:** 1 mL Microsomal incubation + 5 mL dichloromethane, extract, evaporate to dryness, reconstitute with MeOH, inject an aliquot.

## HPLC VARIABLES
**Column:** 150 × 4.6 Zorbax SB-C18
**Mobile phase:** MeCN:MeOH:water 10:40:50
**Flow rate:** 1
**Detector:** UV 232

## CHROMATOGRAM
**Internal standard:** 6β-hydroxyprogesterone

## OTHER SUBSTANCES
**Extracted:** metabolites, temazepam

## KEY WORDS
rat; human; liver

## REFERENCE
Gelboin, H.V.; Krausz, K.W.; Goldfarb, I.; Buters, J.T.M.; Yang, S.K.; Gonzalez, F.J.; Korzekwa, K.R.; Shou, M. Inhibitory and non-inhibitory monoclonal antibodies to human cytochrome P450 3A3/4. *Biochem.Pharmacol.*, **1995**, *50*, 1841–1850

## SAMPLE
**Matrix:** microsomal incubations
**Sample preparation:** 1 mL Microsomal incubation + 20 μL 10 M NaOH + 100 μL 70 μM prazepam in MeOH + 1 mL 100 mM pH 10 sodium carbonate buffer + 5 mL ethyl acetate, rotate for 25 min, centrifuge at 900 g for 10 min. Remove the organic layer and evaporate it to dryness under a stream of nitrogen at 50°, reconstitute the residue in mobile phase, inject an aliquot.

## HPLC VARIABLES
**Column:** 250 × 5 Spherisorb S5 ODS2
**Mobile phase:** MeOH:water 65:35 containing 0.02% triethylamine, adjusted to pH 7.0 with phosphoric acid
**Flow rate:** 1
**Detector:** UV 236

## CHROMATOGRAM
**Internal standard:** prazepam

## OTHER SUBSTANCES
**Extracted:** metabolites

## KEY WORDS
rat; liver

## REFERENCE
Zomorodi, K.; Carlile, D.J.; Houston, J.B. Kinetics of diazepam metabolism in rat hepatic microsomes and hepatocytes and their use in predicting *in vivo* hepatic clearance. *Xenobiotica*, **1995**, *25*, 907–916

## SAMPLE
**Matrix:** solutions

## HPLC VARIABLES
**Column:** 250 × 4.6 10 μm Spherisorb ODS-2
**Mobile phase:** MeCN:MeOH:water 5:45:50
**Flow rate:** 2
**Detector:** UV 228

## CHROMATOGRAM
**Retention time:** 12

## REFERENCE

Mithani, S.D.; Bakatselou, V.; TenHoor, C.N.; Dressman, J.B. Estimation of the increase in solubility of drugs as a function of bile salt concentration. *Pharm.Res.*, **1996**, *13*, 163–167

## SAMPLE

**Matrix:** solutions
**Sample preparation:** Inject a 30 μL aliquot of a solution in MeOH.

## HPLC VARIABLES

**Column:** 250 × 4.6 Spherisorb S5SCX in a PEEK column
**Mobile phase:** MeOH:water:60% perchloric acid 97.5:1.75:0.75
**Flow rate:** 1
**Injection volume:** 30
**Detector:** UV 220

## CHROMATOGRAM

**Retention time:** 9.5

## OTHER SUBSTANCES

**Simultaneous:** clonazepam, dothiepin, dothiepin sulfoxide, nordiazepam, nordothiepin, nordothiepin sulfoxide

## REFERENCE

Croes, K.; McCarthy, P.T.; Flanagan, R.J. HPLC of basic drugs and quaternary ammonium compounds on microparticulate strong cation-exchange materials using methanolic or aqueous methanol eluents containing an ionic modifier. *J.Chromatogr.A*, **1995**, *693*, 289–306

## SAMPLE

**Matrix:** solutions

## HPLC VARIABLES

**Column:** 250 × 4.1 6 μm PolyEncap ODS (n-octadecylacrylate copolymerized with vinyl silica in heptane, carrier Ultrasep ES 100; preparation described in paper)
**Mobile phase:** MeCN:pH 2.2 phosphate buffer 32:68
**Flow rate:** 1
**Detector:** UV 220

## CHROMATOGRAM

**Retention time:** 14

## OTHER SUBSTANCES

**Simultaneous:** aspirin, diphenhydramine, o-hydroxyhippuric acid, MPPH, niacin, toluene

## REFERENCE

Engelhardt, H.; Cuñat-Walter, M.A. Polymer encapsulated stationary phases with improved efficiency. *Chromatographia*, **1995**, *40*, 657–661

## SAMPLE

**Matrix:** solutions

## HPLC VARIABLES

**Column:** 150 × 3.9 4 μm Nova pack C18
**Mobile phase:** MeOH:water 52:48
**Column temperature:** 48

**Flow rate:** 0.8
**Injection volume:** 20
**Detector:** UV 254

## CHROMATOGRAM
**Retention time:** 16

## OTHER SUBSTANCES
**Simultaneous:** bromazepam, chlordiazepoxide, clobazam, clorazepate, flunitrazepam, lorazepam, nitrazepam, oxazepam, tofisopam

## REFERENCE
Guillaume, Y.; Guinchard, C. Thermodynamic behavior of mixed benzodiazepines by a new liquid chromatographic method. *Chromatographia*, **1995**, *40*, 193–196

## SAMPLE
**Matrix:** solutions

## HPLC VARIABLES
**Column:** 150 × 3.9 4 μm Nova pak C18
**Mobile phase:** MeCN:water 57:43
**Column temperature:** 44
**Flow rate:** 1.1
**Injection volume:** 20
**Detector:** UV 254

## CHROMATOGRAM
**Retention time:** 12.2

## OTHER SUBSTANCES
**Simultaneous:** bromazepam, chlordiazepoxide, clobazam, clorazepate, flunitrazepam, lorazepam, nitrazepam, oxazepam, tofisopam

## REFERENCE
Guillaume, Y.; Guinchard, C. Marked difference between acetonitrile/water and methanol/water mobile phase systems on the thermodynamic behavior of benzodiazepines in reversed phase liquid chromatography. *Chromatographia*, **1995**, *41*, 84–87

## SAMPLE
**Matrix:** solutions

## HPLC VARIABLES
**Column:** 250 × 4.6 5 μm Supelcosil LC-DP (A) or 250 × 4 5 μm LiChrospher 100 RP-8 (B)
**Mobile phase:** MeCN:0.025% phosphoric acid:buffer 25:10:5 (A) or 60:25:15 (B) (Buffer was 9 mL concentrated phosphoric acid and 10 mL triethylamine in 900 mL water, adjust pH to 3.4 with dilute phosphoric acid, make up to 1 L.)
**Flow rate:** 0.6
**Injection volume:** 25
**Detector:** UV 229

## CHROMATOGRAM
**Retention time:** 7.71 (A), 8.83 (B)

## OTHER SUBSTANCES
**Also analyzed:** acebutolol, acepromazine, acetaminophen, acetazolamide, acetophenazine, albuterol, alprazolam, amitriptyline, amobarbital, amoxapine, antipyrine, atenolol, atropine, azatadine, baclofen, benzocaine, bromocriptine, brompheniramine, brotizolam, bupivacaine, buspirone, butabarbital, butalbital, caffeine, carbamazepine, cetirizine, chlorcyclizine, chlordiazepoxide, chlormezanone, chloroquine, chlorpheniramine, chlorpromazine, chlorpropamide, chlorprothixene, chlorthalidone, chlorzoxazone, cimetidine, cisapride, clomipramine, clonazepam, clonidine, clozapine, cocaine, codeine, colchicine, cyclizine, cyclobenzaprine, dantrolene, desipramine, diclofenac, diflunisal, diltiazem, diphenhydramine, diphenidol, diphenoxylate, dipyridamole, disopyramide, dobutamine, doxapram, doxepin, droperidol, encainide, ethidium bromide, ethopropazine, fenoprofen, fentanyl, flavoxate, fluoxetine, fluphenazine, flurazepam, flurbiprofen, fluvoxamine, furosemide, glutethimide, glyburide, guaifenesin, haloperidol, homatropine, hydralazine, hydrochlorothiazide, hydrocodone, hydromorphone, hydroxychloroquine, hydroxyzine, ibuprofen, imipramine, indomethacin, ketoconazole, ketoprofen, ketorolac, labetalol, levorphanol, lidocaine, loratadine, lorazepam, lovastatin, loxapine, mazindol, mefenamic acid, meperidine, mephenytoin, mepivacaine, mesoridazine, metaproterenol, methadone, methdilazine, methocarbamol, methotrexate, methotrimeprazine, methoxamine, methyldopa, methylphenidate, metoclopramide, metolazone, metoprolol, metronidazole, midazolam, moclobemide, morphine, nadolol, nalbuphine, naloxone, naphazoline, naproxen, nifedipine, nizatidine, norepinephrine, nortriptyline, oxazepam, oxycodone, oxymetazoline, paroxetine, pemoline, pentazocine, pentobarbital, pentoxifylline, perphenazine, pheniramine, phenobarbital, phenol, phenolphthalein, phentolamine, phenylbutazone, phenyltoloxamine, phenytoin, pimozide, pindolol, piroxicam, pramoxine, prazepam, prazosin, probenecid, procainamide, procaine, prochlorperazine, procyclidine, promazine, promethazine, propafenone, propantheline, propiomazine, propofol, propranolol, protriptyline, quazepam, quinidine, quinine, racemethorphan, ranitidine, remoxipride, risperidone, salicylic acid, scopolamine, secobarbital, sertraline, sotalol, spironolactone, sulfinpyrazone, sulindac, temazepam, terbutaline, terfenadine, tetracaine, theophylline, thiethylperazine, thiopental, thioridazine, thiothixene, timolol, tocainide, tolbutamide, tolmetin, trazodone, triamterene, triazolam, trifluoperazine, triflupromazine, trimeprazine, trimethoprim, trimipramine, verapamil, warfarin, xylometazoline, yohimbine, zopiclone

## KEY WORDS
some details of plasma extraction

## REFERENCE
Koves, E.M. Use of high-performance liquid chromatography-diode array detection in forensic toxicology. *J.Chromatogr.A*, **1995**, *692*, 103–119

## SAMPLE
**Matrix:** solutions
**Sample preparation:** Prepare a 1-10 µg/mL solution in water, inject an aliquot.

## HPLC VARIABLES
**Column:** 250 × 4.6 5 µm Hypersil SCX/C18
**Mobile phase:** MeCN:25 mM pH 3 Na$_2$HPO$_4$ 50:50
**Injection volume:** 20
**Detector:** UV 254

## CHROMATOGRAM
**Retention time:** k′ 2.32

## OTHER SUBSTANCES
**Also analyzed:** amitriptyline, barbital, benzoic acid, butabarbital, clomipramine, clonazepam, desipramine, flurazepam, furosemide, imipramine, nitrazepam, phenobarbital, phenol, phenolphthalein, pindolol, propranolol, resorcinol, salicylic acid, secobarbital, terbutaline, xylazine

## KEY WORDS
effect of mobile phase pH on capacity factor is discussed

## REFERENCE
Walshe, M.; Kelly, M.T.; Smyth, M.R.; Ritchie, H. Retention studies on mixed-mode columns in high-performance liquid chromatography. *J.Chromatogr.A*, **1995**, *708*, 31–40

## SAMPLE
**Matrix:** solutions

## HPLC VARIABLES
**Column:** 100 × 4.6 3 μm 208HS3410 (Vydac)
**Mobile phase:** Gradient. MeCN:water from 15:85 to 60:40 over 10 min.
**Flow rate:** 1.5
**Detector:** UV 210 (?)

## CHROMATOGRAM
**Retention time:** 9.4

## OTHER SUBSTANCES
**Simultaneous:** barbital, carbamazepine, ethotoin, mephenytoin, methsuximide, phenacemide, phenobarbital, phensuximide

## REFERENCE
*Vydac HPLC Catalog*, **1994-5**, p. 26

## SAMPLE
**Matrix:** solutions

## HPLC VARIABLES
**Guard column:** 30 × 2.1 Spheri-5 RP-8
**Column:** 220 × 2.1 Spheri-5 RP-8
**Mobile phase:** Gradient. A was 0.08% diethylamine and 0.09% phosphoric acid in water, pH 2.3. B was MeCN:water 90:10 containing 0.08% diethylamine and 0.09% phosphoric acid. A:B 95:5 for 2 min, to 0:100 over 15 min (?), maintain at 0:100 for 5 min.
**Column temperature:** 50
**Flow rate:** 0.5
**Detector:** UV 200

## CHROMATOGRAM
**Retention time:** 13

## OTHER SUBSTANCES
**Simultaneous:** chlordiazepoxide, desalkylflurazepam, flurazepam, norchlordiazepoxide, nordiazepam, oxazepam, prazepam
**Also analyzed:** amitriptyline, amphetamine, chlorpromazine, desipramine, desmethyldoxepin, diethylpropion, doxepin, ephedrine, fenfluramine, imipramine, mesoridazine, methamphetamine, nortriptyline, phentermine, phenylpropanolamine, promazine, thioridazine, thiothixene, trifluoperazine

## REFERENCE
*Rainin Catalog Ci-94*, **1994-5**, p. 7.24

## SAMPLE
**Matrix:** solutions
**Sample preparation:** Dilute in MeOH to a concentration of 10-80 mg/mL, inject an aliquot.

## HPLC VARIABLES
**Column:** 150 × 3.9 5 μm Nova pak RP 18
**Mobile phase:** MeOH:water 50:50
**Column temperature:** 50
**Flow rate:** 0.82
**Injection volume:** 20
**Detector:** UV 254

## CHROMATOGRAM
**Retention time:** 16

## OTHER SUBSTANCES
**Simultaneous:** bromazepam, chlorazepate, chlordiazepoxide, clobazam, flunitrazepam, lorazepam, nitrazepam, oxazepam, tofisopam

## KEY WORDS
conditions are optimized

## REFERENCE
Guillaume, Y.; Guinchard, C. Study and optimization of column efficiency in HPLC: Comparison of two methods for separating ten benzodiazepines. *J.Liq.Chromatogr.*, **1994**, *17*, 1443–1459

## SAMPLE
**Matrix:** solutions

## HPLC VARIABLES
**Column:** 250 × 4.6 Zorbax RX
**Mobile phase:** Gradient. A was 10 mL concentrated orthophosphoric acid and 7 mL triethylamine in 1 L water. B was 10 mL concentrated orthophosphoric acid and 7 mL triethylamine in 200 mL water, make up to 1 L with MeCN. A:B from 100:0 to 0:100 over 30 min, maintain at 0:100 for 5 min.
**Column temperature:** 30
**Flow rate:** 2
**Detector:** UV 210

## OTHER SUBSTANCES
**Also analyzed:** acepromazine, acetaminophen, acetophenazine, albuterol, aminophylline, amitriptyline, amobarbital, amoxapine, amphetamine, amylocaine, antipyrine, aprobarbital, aspirin, atenolol, atropine, avermectin, barbital, benzocaine, benzoic acid, benzotropine, benzphetamine, berberine, bibucaine, bromazepan, brompheniramine, buprenorphine, buspirone, butabarbital, butacaine, butethal, caffeine, carbamazepine, carbromal, chloramphenicol, chlordiazepoxide, chloroquine, chlorothiazide, chloroxylenol, chlorphenesin, chlorpheniramine, chlorpromazine, chlorpropamide, chlortetracycline, cimetidine, cinchonidine, cinchonine, clenbuterol, clonazepam, clonixin, clorazepate, cocaine, codeine, colchicine, cortisone, coumarin, cyclazocine, cyclobenzaprine, cyclothiazide, cyheptamide, cymarin, danazol, danthron, dapsone, debrisoquine, desipramine, dexamethasone, dextromethorphan, dextropropoxyphene, diamorphine, diclofenac, diethylpropion, diethylstilbestrol, diflunisal, digitoxin, digoxin, diltiazem, diphenhydramine, diphenoxylate, diprenorphine, dipyrone, disulfiram, dopamine, doxapram, doxepin, dronabinol, ephedrine, epinephrine, epinine, estradiol, estriol, estrone, ethacrynic acid, ethosuximide, etonitazene, etorphine, eugenol, famotidine, fenbendazole, fencamfamine, fenoprofen, fenproporex, fentanyl, flubendazole, flufenamic acid, flunitrazepam, 5-fluorouracil, fluoxymesterone, fluphenazine, furosemide, gentisic acid, gitoxigenin, glipizide, glunixin, glutethimide, glybenclamide, guaiacol, halazepam, haloperidol, hydrochlorothiazide, hydrocodone, hydrocortisone, hydromorphone, hydroxyquinoline, ibogaine, ibuprofen, iminostilbene, imipramine, indomethacin, isocarbostyril, isocarboxazid, isoniazid, isoproter-

enol, isoxsuprine, ivermectin, ketamine, ketoprofen, kynurenic acid, levorphanol, lidocaine, lorazepam, lormetazepam, loxapine, mazindol, mebendazole, meclizine, meclofenamic acid, medazepam, mefenamic acid, megestrol, mepacrine, meperidine, mephentermine, mephenytoin, mephesin, mephobarbital, mepivacaine, mescaline, mesoridazine, methadone, methamphetamine, methapyrilene, methaqualone, methazolamide, methocarbamol, methoxamine, methsuximide, methyl salicylate, methyldopa, methyldopamine, methylphenidate, methylprednisolone, methyltestosterone, methyprylon, metoprolol, mibolerone, morphine, nadolol, nalorphine, naloxone, naltrexone, naphazoline, naproxen, nefopam, niacinamide, nicotine, nicotinic acid, nifedipine, niflumic acid, nitrazepam, norepinephrine, nortriptyline, noscapine, nylidrin, oxazepam, oxycodone, oxymorphone, oxyphenbutazone, oxytetracycline, papaverine, pargyline, pemoline, pentazocine, pentobarbital, persantine, phenacetin, phenazocine, phenazopyridine, phencyclidine, phendimetrazine, phenelzine, pheniramine, phenobarbital, phenothiazine, phensuximide, phentermine, phenylbutazone, phenylephrine, phenylpropanolamine, piperocaine, prazepam, prednisolone, primidone, probenecid, progesterone, propiomazine, propranolol, propylparaben, pseudoephedrine, puromycin, pyrilamine, pyrithyldione, quazepam, quinaldic acid, quinidine, quinine, ranitidine, recinnamine, reserpine, resorcinol, saccharin, albuterol, salicylamide, salicylic acid, scopolamine, scopoletin, secobarbital, strychnine, sulfacetamide, sufadiazine, sulfadimethoxine, sulfaethidole, sulfamerazine, sulfamethazine, sulfamethoxizole, sulfanilamide, sulfapyridine, sulfasoxizole, sulindac, tamoxifen, temazepam, testosterone, tetracaine, tetracycline, tetramisole, thebaine, theobromine, theophylline, thiabendazole, thiamine, thiamylal, thiobarbituric acid, thioridazine, thiosalicylic acid, thiothixene, thymol, tolazamide, tolazoline, tobutamide, tolmetin, tranylcypromine, triamcinolone, tribenzylamine, trichloromethiazide, trifluoperazine, trihexyphenidyl, trimethoprim, tripelennamine, triproilidine, tropacocaine, tyramine, verapamil, vincamine, warfarin, yohimbine, zoxazolamine

## REFERENCE

Hill, D.W.; Kind, A.J. Reversed-phase solvent gradient HPLC retention indexes of drugs. *J.Anal.Toxicol.*, **1994**, *18*, 233–242

## SAMPLE
**Matrix:** solutions

## HPLC VARIABLES
**Guard column:** 30 × 3.2 7 µm SI 100 ODS (not commercially available)
**Column:** 150 × 3.2 7 µm SI 100 ODS (not commercially available)
**Mobile phase:** MeCN:buffer 31.2:68.8 (Buffer was 6.66 g $KH_2PO_4$ and 4.8 g 85% phosphoric acid in 1 L water, pH 2.3.)
**Flow rate:** 0.5-1
**Detector:** UV 227; UV 279

## CHROMATOGRAM
**Retention time:** 11.5
**Internal standard:** 5-(4-methylphenyl)-5-phenylhydantoin (7.3)

## OTHER SUBSTANCES
**Also analyzed:** aspirin, caffeine, carbamazepine, chlordiazepoxide, chlorprothixene, clonazepam, doxylamine, ethosuximide, furosemide, haloperidol, hydrochlorothiazide, methocarbamol, methotrimeprazine, nicotine, oxazepam, procaine, promazine, propafenone, propranolol, salicylamide, temazepam, tetracaine, thiopental, triamterene, verapamil, zolpidem, zopiclone

## REFERENCE

Below, E.; Burrmann, M. Application of HPLC equipment with rapid scan detection to the identification of drugs in toxicological analysis. *J.Liq.Chromatogr.*, **1994**, *17*, 4131-4144

## SAMPLE

**Matrix:** urine

**Sample preparation:** 1 mL Urine + 100 μL 5 mM pH 5.5 acetate buffer + 25 μL β-glucuronidase/arylsulfatase (0.235/0.065 U, Calbiochem), mix, heat at 37°for 16 h, add 50 μL 5-50 μg/mL prazepam in MeOH, add 1 mL saturated trisodium phosphate, add 3 mL dichloromethane, vortex for 2 min, centrifuge at 1610 g for 5 min. Remove a 2 mL aliquot of the organic layer and add it to 2 mL hexane and 2 mL 6 M HCl, vortex for 2 min, centrifuge at 1610 g for 5 min. Remove 1 mL of the aqueous phase and adjust pH to 6 with 1 mL 6 M NaOH and 1 mL saturated trisodium phosphate, add 3 mL dichloromethane, vortex for 2 min, centrifuge at 1610 g for 5 min. Remove the organic layer and evaporate it to dryness under a stream of nitrogen at 50°, reconstitute the residue in 150 μL mobile phase, inject a 60 μL aliquot.

## HPLC VARIABLES

**Column:** 250 × 4 5 μm LiChrospher 100 RP-18(e)

**Mobile phase:** MeOH:water:triethylamine 30:70:0.1 adjusted to pH 5.5 with phosphoric acid

**Flow rate:** 0.7

**Injection volume:** 60

**Detector:** UV 240

## CHROMATOGRAM

**Retention time:** 10.3

**Internal standard:** prazepam (17.0)

**Limit of detection:** 2 ng/mL

## OTHER SUBSTANCES

**Extracted:** metabolites, desmethyldiazepam, oxazepam, temazepam

**Simultaneous:** amitriptyline, caffeine, carbamazepine, chlordiazepoxide, chlorpromazine, clonazepam, desipramine, flunitrazepam, flurazepam, haloperidol, imipramine, levomepromazine, maprotiline, mianserin, nitrazepam, nortriptyline, perphenazine, phenobarbital, phenytoin, sulpride, thioridazine, triazolam

## REFERENCE

Chiba, K.; Horii, H.; Chiba, T.; Kato, Y.; Hirano, T.; Ishizaki, T. Development and preliminary application of high-performance liquid chromatographic assay of urinary metabolites of diazepam in humans. *J.Chromatogr.B*, **1995**, *668*, 77–84

## SAMPLE

**Matrix:** urine

**Sample preparation:** 500 μL Urine + N-ethylnordiazepam + chlorpheniramine + 100 μL buffer, centrifuge at 11000 g for 30 s, inject a 500 μL aliquot onto column A with mobile phase A, after 0.6 min backflush column A with mobile phase A to waste for 1.6 min, elute column A with 250 μL mobile phase B, with 200 μL mobile phase C, and with 1.15 mL mobile phase D. Elute column A to waste until drugs start to emerge then elute onto column B. Elute column B to waste until drugs started to emerge, then elute onto column C. When all the drugs have emerged from column B remove it from the circuit, elute column C with mobile phase D, monitor the effluent from column C. Flush column A with 7 mL mobile phase E, with mobile phase D, and mobile phase A. Flush column B with 5 mL mobile phase E then with mobile phase D. (Buffer was 6 M ammonium acetate adjusted to pH 8.0 with 2 M KOH.)

## HPLC VARIABLES

**Column:** A 10 × 2.1 12-20 μm PRP-1 spherical poly(styrene-divinylbenzene) (Hamilton); B 10 × 3.2 11 μm Aminex A-28 (Bio-Rad); C 25 × 3.2 5 μm C8 (Phenomenex) + 150 × 4.6 5 μm silica (Macherey-Nagel)

**Mobile phase:** A 0.1% pH 8.0 potassium borate buffer; B 6 mM KH$_2$PO$_4$ containing 5 mM tetramethylammonium hydroxide, and 2 mM dimethyloctylamine, pH adjusted to 6.50 with phosphoric acid; C MeCN:buffer 40:60 (Buffer was 6 mM KH$_2$PO$_4$ containing 5 mM tetramethylammonium hydroxide, and 2 mM dimethyloctylamine, pH adjusted to 6.50 with phosphoric acid.); D MeCN:buffer 33:67 (Buffer was 6 mM KH$_2$PO$_4$ containing 5 mM tetramethylammonium hydroxide, and 2 mM dimethyloctylamine, pH adjusted to 6.50 with phosphoric acid.); E MeCN:buffer 70:30 (Buffer was 6 mM KH$_2$PO$_4$ containing 5 mM tetramethylammonium hydroxide, and 2 mM dimethyloctylamine, pH adjusted to 6.50 with phosphoric acid.)

**Column temperature:** 40 (B, C only)

**Flow rate:** A 5; B-E 1

**Injection volume:** 500

**Detector:** UV 210; UV 235

## CHROMATOGRAM

**Retention time:** k$'$ 1.6

**Internal standard:** N-ethylnordiazepam (k$'$ 2.1), chlorpheniramine (k$'$ 5.9)

**Limit of detection:** 300 ng/mL

## OTHER SUBSTANCES

**Extracted:** amitriptyline, amphetamine, benzoylecgonine, caffeine, codeine, cotinine, desipramine, diphenhydramine, ephedrine, flurazepam, hydrocodone, hydromorphone, imipramine, lidocaine, methadone, methamphetamine, morphine, nortriptyline, oxazepam, pentazocine, phenmetrazine, phenobarbital, phentermine, phenylpropanolamine, secobarbital

**Interfering:** nordiazepam

## KEY WORDS

column-switching

## REFERENCE

Binder, S.R.; Regalia, M.; Biaggi-McEachern, M.; Mazhar, M. Automated liquid chromatographic analysis of drugs in urine by on-line sample cleanup and isocratic multi-column separation. *J.Chromatogr.*, **1989**, *473*, 325–341

◆ ━━━━━━━━━━━━━━━━━━ ◆ ━━━━━━━━━━━━━━━━━━ ◆

## ANNOTATED BIBLIOGRAPHY

Goldnik, A.; Gajewska, M.; Jaworska, M. Determination of oxazepam and diazepam in body fluids by HPLC. *Acta Pol.Pharm.*, **1993**, *50*, 421–422

Kamali, F. Determination of plasma diazepam and desmethyldiazepam by solid-phase extraction and reversed-phase high-performance liquid chromatography. *J.Pharm.Biomed.Anal.*, **1993**, *11*, 625–627

Fernández, P.; Hermida, I.; Bermejo, A.M.; López-Rivadulla, M.; Cruz, A.; Concheiro, L. Simultaneous determination of diazepam and its metabolites in plasma by high-performance liquid chromatography. *J.Liq.Chromatogr.*, **1991**, *14*, 2587–2599 [also nordiazepam, oxazepam, temazepam; carbamazepine (IS); SPE]

Gil, M.S.; Ochoa, C.; Vega, S. High performance liquid chromatography of new potential anxiolytic drugs and related benzodiazepines: A comparative study of hydrophobicity. *J.Liq.Chromatogr.*, **1991**, *14*, 2141–2156 [also buspirone, chlordiazepoxide]

Hays, P.A.; Lurie, I.S. Quantitative analysis of adulterants in illicit heroin samples via reversed phase HPLC. *J.Liq.Chromatogr.*, **1991**, *14*, 3513–3517 [simultaneous acetaminophen, acetylcodeine, acetylmorphine, aspirin, benzocaine, caffeine, chloroquine, diamorphine, diphenhydramine, dipyrone, lidocaine, methaqualone, monoacetylmorphine, morphine, nicotinamide, noscapine, papaverine, phenacetin, phenobarbital, phenolphthalein, N-phenyl-2-naphthylamine, salicylic acid, strychnine]

Moro, M.E.; Novillo-Fertrell, J.; Velazquez, M.M.; Rodriguez, L.J. Kinetics of the acid hydrolysis of diazepam, bromazepam, and flunitrazepam in aqueous and micellar systems. *J.Pharm.Sci.*, **1991**, *80*, 459–468

Brazeau, G.A.; Fung, H.L. Effect of organic cosolvent-induced skeletal muscle damage on the bioavailability of intramuscular [$^{14}$C]diazepam. *J.Pharm.Sci.*, **1990**, *79*, 773–777 [rabbit; plasma]

Noggle, F.T., Jr.; Clark, C.R.; DeRuiter, J. Liquid chromatographic separation of some common benzodiazepines and their metabolites. *J.Liq.Chromatogr.*, **1990**, *13*, 4005–4021 [also metabolites, alprazolam, bromazepam, chlordiazepoxide, clobazam, clonazepam, clorazepate, demoxepam, estazolam, fludiazepam, flunitrazepam, flurazepam, halazepam, lorazepam, midazolam, nitrazepam, nordiazepam, oxazepam, prazepam, temazepam, triazolam]

Bastos, M.L.A. Improvement of HPLC conditions for separation of diazepam and its metabolites in biological extracts. *J.Liq.Chromatogr.*, **1989**, *12*, 1919–1934 [also metabolites, desmethyldiazepam, oxazepam, temazepam; prazepam (IS); liver; brain; bile; kidney; gradient]

Abdel-Hamid, M.E.; Abuirjeie, M.A. Determination of diazepam and oxazepam using high-performance liquid chromatography and fourth-derivative spectrophotometric techniques. *Analyst*, **1988**, *113*, 1443–1446 [powders; tablets; simultaneous degradation products, oxazepam]

Pietrogrande, M.C.; Dondi, F.; Blo, G.; Borea, P.A.; Bighi, C. Retention behavior of benzodiazepines in normal-phase HPLC. Silica, cyano, and amino phases comparison. *J.Liq.Chromatogr.*, **1988**, *11*, 1313–1333 [normal phase; also lorazepam, medazepam, methyllorazepam, oxazepam, prazepam, temazepam]

Wildmann, J. Increase of natural benzodiazepines in wheat and potato during germination. *Biochem.Biophys.Res.Commun.*, **1988**, *157*, 1436–1443 [plants]

Klockowski, P.M.; Levy, G. Simultaneous determination of diazepam and its active metabolites in rat serum, brain and cerebrospinal fluid by high-performance liquid chromatography. *J.Chromatogr.*, **1987**, *422*, 334–339

Koenigbauer, M.J.; Assenza, S.P.; Willoughby, R.C.; Curtis, M.A. Trace analysis of diazepam in serum using microbore high-performance liquid chromatography and on-line preconcentration. *J.Chromatogr.*, **1987**, *413*, 161–169 [serum; column-switching; extracted metabolites, nordiazepam, oxazepam, temazepam; non-interfering caffeine; LOQ 4 ng/mL]

Lau, C.E.; Dolan, S.; Tang, M. Microsample determination of diazepam and its three metabolites in serum by reversed-phase high-performance liquid chromatography. *J.Chromatogr.*, **1987**, *416*, 212–218

Lurie, I.S.; McGuiness, K. The quamtitation of heroin and selected basic impurities via reversed phase HPLC. II. The analysis of adulterated samples. *J.Liq.Chromatogr.*, **1987**, *10*, 2189–2204 [electrochemical detection; UV detection; also acetaminophen, acetylcodeine, acetylmorphine, acetylprocaine, aminopyrene, amitriptyline, antipyrene, aspirin, barbital, benztropine, caffeine, cocaine, codeine, diamorphine, diphenhydramine, dipyrone, ephedrine, ethylmorphine, lidocaine, meconin, methamphetamine, methapyrilene, methaqualone, morphine, nalorphine, niacinamide, nicotinamide, noscapine, papaverine, phenacetin, phenmetrazine, phenobarbital, phenolphthalein, procaine, propanophenone, propoxyphene, pyrilamine, quinidine, quinine, salicylamide, salicylic acid, secobarbital, strychnine, tartaric acid, tetracaine, thebaine, tripelennamine, tropacocaine, vitamin B3, vitamin B5]

St-Pierre, M.V.; Pang, K.S. Determination of diazepam and its metabolites by high-performance liquid chromatography and thin-layer chromatography. *J.Chromatogr.*, **1987**, *421*, 291–307

Faibushevich, A.A.; Kuramshin, R.K.; Iushkevich, A.M.; Kolesnikov, S.I. [Determination of diazepam and its metabolites in the blood by microcolumn high-performance liquid chromatography]. *Farmakol.Toksikol.(Moscow)*, **1986**, *49*, 20–22

Wong, S.H.Y.; McHugh, S.L.; Dolan, J.; Cohen, K.A. Tricyclic antidepressant analysis by reversed-phase liquid chromatography using phenyl columns. *J.Liq.Chromatogr.*, **1986**, *9*, 2511–2538 [also acetaminophen, amitriptyline, amobarbital, amoxapine, barbital, chlordiazepoxide, chlorpromazine, cimetidine, clomipramine, codeine, desipramine, desmethyldoxepin, doxepin, fluphenazine, flurazepam, glutethimide, hydroxyamoxapine, imipramine, lorazepam, maprotiline, meperidine, metabolites, nortriptyline, oxazepam, pentobarbital, perphenazine, phenobarbital, phenytoin, propoxyphene, protriptyline, secobarbital, thioridazine, trazodone]

Pietrogrande, M.C.; Bighi, C.; Borea, P.A.; Barbaro, A.M.; Guerra, M.C.; Biagi, G.L. Relationship between log k′ values of benzodiazepines and composition of the mobile phase. *J.Liq.Chromatogr.*, **1985**, *8*, 1711–1729 [also carbenicillin, chlordiazepoxide, dicloxacillin, flurazepam, lorazepam, medazepam, methyllorazepam, nitrazepam, oxazepam, prazepam, temazepam, testosterone]

Pakuts, A.P.; Downie, R.H.; Matula, T.I. A rapid HPLC analysis of diazepam in animal feed. *J.Liq.Chromatogr.*, **1983**, *6*, 2557–2564 [SPE]

Rao, S.N.; Dhar, A.K.; Kutt, H.; Okamoto, M. Determination of diazepam and its pharmacologically active metabolites in blood by Bond Elut column extraction and reversed-phase high-performance liquid chromatography. *J.Chromatogr.*, **1982**, *231*, 341–348 [SPE]

Cotler, S.; Puglisi, C.V.; Gustafson, J.H. Determination of diazepam and its major metabolites in man and in the cat by high-performance liquid chromatography. *J.Chromatogr.*, **1981**, *222*, 95–106

Lurie, I.S.; Demchuk, S.M. Optimization of a reverse phase ion-pair chromatographic separation for drugs of forensic interest. II. Factors effecting selectivity. *J.Liq.Chromatogr.*, **1981**, *4*, 357–374 [also acetaminophen, acetylcodeine, acetylmorphine, aminopyrine, amobarbital, amphetamine, antipyrine, benzocaine, butabarbital, caffeine, cocaine, codeine, diamorphine, diethylpropion, DMT, ephedrine, glutethimide, Lampa, lidocaine, LSD, MDA, mecloqualone, mescaline, methamphetamine, methapyrilene, methaqualone, methpyrilene, methylphenidate, morphine, narcotine, papaverine, PCP, pentobarbital, phencyclidine, phendimetrazine, phenmetrazine, phenobarbital, phentermine, phenylpropanolamine, procaine, quinidine, quinine, secobarbital, strychnine, TCP, tetracaine, thebaine, theophylline]

Ratnaraj, N.; Goldberg, V.D.; Elyas, A.; Lascelles, P.T. Determination of diazepam and its major metabolites using high-performance liquid chromatography. *Analyst*, **1981**, *106*, 1001–1004

Foreman, J.M.; Griffiths, W.C.; Dextraze, P.G.; Diamond, I. Simultaneous assay of diazepam, chlordiazepoxide, N-desmethyldiazepam, N-desmethylchloridazepoxide, and demoxepam in serum by high performance, liquid chromatography. *Clin.Biochem.*, **1980**, *13*, 122–125

Harbin, D.N.; Lott, P.F. The identification of drugs of abuse in urine using reverse phase high pressure liquid chromatography. *J.Liq.Chromatogr.*, **1980**, *3*, 243–256 [urine; also amobarbital, amphetamine, caffeine, chlordiazepoxide, codeine, glutethimide, indole, meperidine, methamphetamine, methaqualone, morphine, pentobarbital, phenobarbital, secobarbital]

Raisys, V.A.; Friel, P.N.; Graaff, P.R.; Opheim, K.E.; Wilensky, A.J. High-performance liquid chromatographic and gas-liquid chromatographic determination of diazepam and nordiazepam in plasma. *J.Chromatogr.*, **1980**, *183*, 441–448

Wittwer, J.D., Jr. Application of high pressure liquid chromatography to the forensic analysis of several benzodiazepines. *J.Liq.Chromatogr.*, **1980**, *3*, 1713–1724 [simultaneous chlordiazepoxide, clonazepam, clorazepate, cyprazepam, demoxapam, desmethyldiazepam, flurazepam, medazepam, nitrazepam, oxazepam, prazepam; normal phase]

# Diclofenac

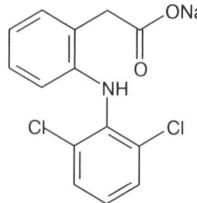

**Molecular formula:** $C_{14}H_{11}Cl_2NO_2$
**Molecular weight:** 295.1
**CAS Registry No.:** 15307-79-6 (diclofenac sodium)

## SAMPLE
**Matrix:** aqueous humor
**Sample preparation:** 100 µL Aqueous humor + 500 µL MeCN + 30 µL 400 ng/mL (+)-naproxen in MeOH, mix mechanically for 90 s, centrifuge at 3000 g for 20 min. Remove the supernatant and dry it under nitrogen at room temperature, dissolve the residue in 50 µL mobile phase by swirl-mixing for 1 min, centrifuge at 3000 g for 20 s, reduce volume to 20-30 µL, inject an aliquot.

## HPLC VARIABLES
**Column:** 150 × 4.5 5 µm Ultrasphere octyl
**Mobile phase:** MeCN:triethylamine:1.65% glacial acetic acid 505:0.65:495, pH 4.35
**Column temperature:** 30
**Flow rate:** 1
**Injection volume:** 20
**Detector:** UV 280

## CHROMATOGRAM
**Retention time:** 7.14
**Internal standard:** naproxen (3.89)
**Limit of detection:** 0.3 ng

## OTHER SUBSTANCES
**Extracted:** flurbiprofen, indomethacin, meclofenamic acid
**Simultaneous:** bacitracin, cortisone acetate, diazepam, fluorometholone, hydrocortisone acetate, imipramine, ketoprofen, ketorolac tromethamine, levobunolol, metipranolol, neomycin, prednisolone acetate, proparacaine, propranolol, salicylic acid, sulfacetamide, suprofen
**Noninterfering:** acebutolol, acetaminophen, acetazolamide, alprenolol, apraclonidine, atenolol, atropine, betamethasone, betaxolol, bupivacaine, caffeine, cyclopentolate, dexamethasone, diphenhydramine, erythromycin, haloperidol, lidocaine, phenylephrine, polymyxin B, procaine, scopolamine, timolol, tropicamide

## KEY WORDS
human; rabbit

## REFERENCE
Riegel, M.; Ellis, P.P. High-performance liquid chromatography assay for antiinflammatory agents diclofenac and flurbiprofen in ocular fluids. *J.Chromatogr.B*, **1994**, *654*, 140–145

## SAMPLE
**Matrix:** blood
**Sample preparation:** 500 µL Plasma + 1 µg naproxen + 500 µL 500 mM HCl, vortex for 1 min, add 5 mL ethyl acetate, extract for 20 min, centrifuge at 2500 rpm for 10 min. Remove the organic layer and evaporate it to dryness, reconstitute the residue in 200 µL MeCN, inject a 100 µL aliquot.

## HPLC VARIABLES
**Column:** 250 × 5 5 µm C18 (Machery & Nagel)

**Mobile phase:** MeCN:water:acetic acid 50:50:0.1
**Flow rate:** 1
**Injection volume:** 100
**Detector:** UV 280

## CHROMATOGRAM
**Internal standard:** naproxen
**Limit of quantitation:** 50 ng/mL

## KEY WORDS
plasma; pharmacokinetics

## REFERENCE
Ramakrishna, S.; Fadnavis, N.W.; Diwan, P.V. Comparative pharmacokinetic evaluation of compressed suppositories of diclofenac sodium in humans. *Arzneimittelforschung*, **1996**, *46*, 175–177

## SAMPLE
**Matrix:** blood
**Sample preparation:** 1 mL Plasma + 50 μL 10 μg/mL flufenamic acid in MeCN + 4 mL MeCN, vortex for 1 min, centrifuge at 2500 rpm for 10 min. Remove the supernatant and evaporate it to dryness under a stream of nitrogen at 45°, reconstitute the residue in 200 μL mobile phase, vortex for 30 s, centrifuge at 11500 rpm for 2 min, inject a 140 μL aliquot of the supernatant.

## HPLC VARIABLES
**Column:** 250 × 4.6 5 μm Supelcosil LC-8
**Mobile phase:** MeCN:water 50:50 adjusted to pH 3.3 with glacial acetic acid (After 13 min increase flow rate to 2.7 mL/min over 4 min, maintain at 2.7 mL/min for 11 min, return to initial conditions.)
**Flow rate:** 2
**Injection volume:** 140
**Detector:** UV 280

## CHROMATOGRAM
**Retention time:** 6
**Internal standard:** flufenamic acid (10)
**Limit of detection:** 25 ng/mL

## OTHER SUBSTANCES
**Extracted:** nitrofenac (UV 275), metabolites

## KEY WORDS
plasma; rat; pharmacokinetics

## REFERENCE
Benoni, G.; Terzi, M.; Adami, A.; Grigolini, L.; Del Soldato, P.; Cuzzolin, L. Plasma concentrations and pharmacokinetic parameters of nitrofenac using a simple and sensitive HPLC method. *J.Pharm.Sci.*, **1995**, *84*, 93–95

## SAMPLE
**Matrix:** blood
**Sample preparation:** 500 μL Plasma + 600 μL 1 M phosphoric acid, vortex for 10 s, add 5 mL 30 ng/mL diphenylamine in hexane:isopropanol 95:5, vortex for 1 min, centrifuge at 1000 g for 10 min. Remove 4 mL of the organic layer and evaporate it to dryness under a stream of air at 40°, reconstitute the residue in 150 μL mobile phase, vortex for 30 s, inject a 25 μL aliquot.

## HPLC VARIABLES
**Column:** 200 × 4.6 10 μm Spherisorb ODS
**Mobile phase:** MeOH:buffer 68:32 (Buffer was 6.8 g/L sodium acetate adjusted to pH 4.2 with HCl.)
**Flow rate:** 1.4
**Injection volume:** 25
**Detector:** UV 274

## CHROMATOGRAM
**Retention time:** 4.8
**Internal standard:** diphenylamine (6.4)
**Limit of detection:** 30 ng/mL
**Limit of quantitation:** 100 ng/mL

## OTHER SUBSTANCES
**Noninterfering:** aspirin, chlorphenpyridamine, ciprofloxacin, ibuprofen, lomefloxacin, norfloxacin, ofloxacin

## KEY WORDS
plasma; pharmacokinetics

## REFERENCE
Li, K.; Zhao, F.-L.; Yuan, Y.-S.; Tan, L. Determination of diclofenac sodium in human plasma by reversed-phase liquid chromatography. *J.Liq.Chromatogr.*, **1995**, *18*, 2205–2216

## SAMPLE
**Matrix:** blood
**Sample preparation:** 1 mL Plasma + 1 μg naproxen + 100 μL 5% zinc sulfate in water, vortex for 2 min, add 3 mL MeOH, vortex for 2 min, add 440 μL buffer, vortex for 1 min, centrifuge at 27° at 2000 g for 10 min, inject a 100 μL aliquot of the supernatant. (Buffer was 100 mM $NaH_2PO_4$ containing 10 mM sodium lauryl sulfate, adjust pH to 2.8 with orthophosphoric acid, filter (0.45 μm).)

## HPLC VARIABLES
**Column:** 250 × 4.6 5 μm Nucleosil C18
**Mobile phase:** MeCN:water 35:65 containing 1 mM sodium lauryl sulfate and 10 mM $NaH_2PO_4$, pH adjusted to 2.8 with orthophosphoric acid
**Flow rate:** 1.5
**Injection volume:** 100
**Detector:** UV 280

## CHROMATOGRAM
**Retention time:** 10.6
**Internal standard:** naproxen (5.8)
**Limit of detection:** 30 ng/mL

## KEY WORDS
plasma; pharmacokinetics

## REFERENCE
Mason, J.L.; Hobbs, G.J. A rapid high performance liquid chromatographic assay for the measurement of diclofenac in human plasma. *J.Liq.Chromatogr.*, **1995**, *18*, 2045–2058

## SAMPLE
**Matrix:** blood

**Sample preparation:** 2 mL Whole blood or plasma + 2 mL buffer + 5 mL chloroform: isopropanol:n-heptane 60:14:26, shake gently horizontally for 10 min, centrifuge at 2800 g for 10 min. Remove the lower organic layer and evaporate it to dryness under vacuum at 45°, reconstitute the residue in 100 μL mobile phase, centrifuge at 2800 g for 5 min, inject a 50 μL aliquot of the supernatant. (Buffer was saturated ammonium chloride solution 25% diluted with water, adjusted to pH 9.5 with 25% ammonia solution.)

## HPLC VARIABLES
**Column:** 300 × 3.9 4 μm NovaPack C18
**Mobile phase:** MeOH:THF:buffer 65:5:30 (Buffer was 0.68 g/L (10 mM (sic)) $KH_2PO_4$ adjusted to pH 2.6 with concentrated orthophosphoric acid.) (At the end of each session wash the column with water for 1 h and MeOH for 1 h, re-equilibrate for 30 min.)
**Column temperature:** 30
**Flow rate:** 0.8
**Injection volume:** 50
**Detector:** UV 275

## CHROMATOGRAM
**Retention time:** 9.51
**Limit of detection:** <120 ng/mL

## OTHER SUBSTANCES
**Extracted:** acebutolol, acenocoumarol, acepromazine, aceprometazine, acetaminophen, aconitine, ajmaline, albuterol, alimemazine, alminoprofen, alpidem, alprazolam, alprenolol, amisulpride, amodiaquine, amoxapine, aspirin, astemizole, atenolol, benazepril, benperidol, benzocaine, benzoylecgonine, bepridil, betaxolol, bisoprolol, bromazepam, brompheniramine, bumadizone, bupivacaine, buprenorphine, buspirone, caffeine, carbamazepine, carbinoxamine, carpipramine, carteolol, celiprolol, cetirizine, chlorambucil, chlordiazepoxide, chlormezanone, chlorophenacinone, chloroquine, chlorpheniramine, chlorpromazine, chlorpropamide, cibenzoline, cicletanine, clemastine, clobazam, clomipramine, clonazepam, clonidine, clorazepate, clozapine, cocaine, codeine, colchicine, cyamemazine, cyclizine, cycloguanil, cyproheptadine, cytarabine, dacarbazine, daunorubicin, debrisoquine, demexiptiline, desipramine, dextromethorphan, dextromoramide, dextropropoxyphene, diazepam, diazoxide, dihydralazine, diltiazem, diphenhydramine, dipyridamole, disopyramide, dosulepine, doxepin, doxylamine, droperidol, ephedrine, estazolam, etodolac, fenfluramine, fenoprofen, fentiazac, flecainide, floctafenine, flumazenil, flunitrazepam, fluoxetine, fluphenazine, flurbiprofen, fluvoxamine, glibenclamide, glibornuride, glipizide, glutethimide, haloperidol, histapyrrodine, hydroxychloroquine, hydroxyzine, ibuprofen, imipramine, indomethacin, iproniazid, ketamine, ketoprofen, labetalol, levomepromazine, lidocaine, lidoflazine, lisinopril, loperamide, loprazolam, loratadine, lorazepam, loxapine, maprotiline, medazepam, medifoxamine, mefenamic acid, mefenidramine, melphalan, meperidine, mephenesin, mephentermine, mepivacaine, metapramine, metformin, methadone, methaqualone, methocarbamol, methotrexate, metipranolol, metoclopramide, metoprolol, mexiletine, mianserin, midazolam, minoxidil, moclobemide, moperone, morphine, nadolol, nalbuphine, nalorphine, naloxone, naltrexone, naproxen, nialamide, nicardipine, nifedipine, niflumic acid, nimodipine, nitrazepam, nitrendipine, nizatidine, nomifensine, omeprazole, opipramol, oxazepam, oxprenolol, penbutolol, penfluridol, pentazocine, phencyclidine, phenobarbital, phenol, phenylbutazone, pimozide, pindolol, pipamperone, piroxicam, prazepam, prazosin, prilocaine, procainamide, procarbazine, proguanil, promethazine, propafenone, propranolol, protriptyline, pyrimethamine, quinidine, quinine, quinupramine, ramipril, ranitidine, reserpine, ritodrine, secobarbital, sotalol, strychnine, sulfinpyrazone, sulindac, sulpride, sultopride, suriclone, temazepam, tenoxicam, terbutaline, terfenadine, tetracaine, tetrazepam, thiopental, thioproperazine, thioridazine, tianeptine, tiapride, tiaprofenic acid, ticlopidine, timolol, tofisopam, tolbutamide, toloxatone, trazodone, triazolam, trifluoperazine, trifluperidol, triprolidine, tropatenine, verapamil, viloxazine, vinblastine, vincristine, vindesine, warfarin, yohimbine, zolpidem, zopiclone, zorubicine
**Interfering:** amitriptyline, diclofenac, mefloquine, nortriptyline, tioclomarol, trimipramine

## KEY WORDS
whole blood; plasma

## REFERENCE
Tracqui, A.; Kintz, P.; Mangin, P. Systematic toxicological analysis using HPLC/DAD. *J.Forensic Sci.*, **1995**, *40*, 254–262

## SAMPLE
**Matrix:** blood
**Sample preparation:** 0.5 mL Plasma + 10 μL 50 μg/mL mefenamic acid + 50 μL 85% phosphoric acid, vortex 10 sec, add 3 mL chloroform, vortex 1 min, centrifuge at 6000 rpm for 5 min. Remove organic layer and evaporate it to dryness at 45° under a stream of nitrogen. Vortex residue with 200 μL mobile phase for 10 s, inject 50 μL aliquot.

## HPLC VARIABLES
**Guard column:** 30-40 μm C18 pellicular
**Column:** 150 × 3.9 Novapak C18
**Mobile phase:** MeCN:water 50:50 adjusted to pH 3.5 with glacial acetic acid
**Flow rate:** 1.5
**Injection volume:** 50
**Detector:** UV 278

## CHROMATOGRAM
**Retention time:** 3.8
**Internal standard:** mefenamic acid (6.3)
**Limit of quantitation:** 10 ng/mL

## KEY WORDS
plasma; dog

## REFERENCE
Mohamed, F.A.; Jun, H.W.; Elfaham, T.H.; Sayed, H.A.; Hafez, E. An improved HPLC procedure for the quantitation of diclofenac in plasma. *J.Liq.Chromatogr.*, **1994**, *17*, 1065–1088

## SAMPLE
**Matrix:** blood
**Sample preparation:** Deproteinize plasma with HCl and extract with dichloromethane. Remove the organic layer and evaporate it to dryness, reconstitute the residue in 100 μL mobile phase, inject a 40 μL aliquot.

## HPLC VARIABLES
**Column:** 150 × 3.6 ODS Hypersil
**Mobile phase:** MeCN:isopropanol:pH 7 sodium acetate buffer:water 21:6:20:53
**Flow rate:** 0.4
**Injection volume:** 40
**Detector:** UV 280

## CHROMATOGRAM
**Limit of detection:** 20 ng/mL

## OTHER SUBSTANCES
**Noninterfering:** ranitidine

## KEY WORDS
plasma; bioequivalence

## REFERENCE

Van Gelderen, M.E.M.; Olling, M.; Barends, D.M.; Meulenbelt, J.; Salomons, P.; Rauws, A.G. The bio-availability of diclofenac from enteric coated products in healthy volunteers with normal and artificially decreased gastric acidity. *Biopharm.Drug Dispos.*, **1994**, *15*, 775–788

## SAMPLE

**Matrix:** blood

**Sample preparation:** 1 mL Plasma + 200 μL 1.5 μg/mL IS in water + 4 mL 2 M phosphoric acid + 6 mL hexane:isopropanol 9:1, shake at 150 oscillations/min for 15 min, centrifuge at 1500 g for 10 min. Remove organic layer and evaporate it to dryness at 37° under a gentle stream of nitrogen. Reconstitute in 250 μL mobile phase, inject 100 μL aliquot onto column A, after 2 min switch eluent from column A onto column B, after another 2 min switch column A out of circuit and continue to flush it to waste with mobile phase. Monitor eluent from column B.

## HPLC VARIABLES

**Column:** A 35 × 4.6 10 μm Nucleosil C18; B 150 × 4.6 10 μm Nucleosil C18

**Mobile phase:** MeCN:MeOH:22 mM pH 7.1 sodium acetate 23:25:52 (both columns)

**Flow rate:** 1.5

**Injection volume:** 100

**Detector:** UV 280

## CHROMATOGRAM

**Retention time:** 6.4

**Internal standard:** CGP-4287 (7.6)

**Limit of detection:** 2.5 ng/mL

## KEY WORDS

plasma; column-switching

## REFERENCE

Miller, R.B. High-performance liquid chromatographic determination of diclofenac in human plasma using automated column switching. *J.Chromatogr.*, **1993**, *616*, 283–290

## SAMPLE

**Matrix:** blood

**Sample preparation:** 50 μL Plasma + 50 μL MeCN, mix, centrifuge at 12000 g for 2 min, inject a 50 μL aliquot of the supernatant.

## HPLC VARIABLES

**Guard column:** 15 × 3.2 5 μm NewGuard ODS

**Column:** 150 × 4 5 μm Nucleosil C18

**Mobile phase:** MeCN:buffer 32:68 (Buffer was 40 mL 1 M $NaH_2PO_4$ and 40 mL 500 mM $Na_2HPO_4$ made up to 680 mL with water, pH 6.6.)

**Flow rate:** 0.7

**Injection volume:** 50

**Detector:** F ex 288 em 360 following post-column reaction. The column effluent flowed through a 1.3 m × 1 mm ID PTFE tube irradiated by a UV 254 lamp (Philips TUV 6W, TYP 103314) to the detector.

## CHROMATOGRAM

**Retention time:** 10

**Limit of detection:** 6 ng/mL

## KEY WORDS

post-column reaction; plasma; human; rat

## REFERENCE
Wiese, B.; Hermansson, J. Bioanalysis of diclofenac as its fluorescent carbazole acetic acid derivative by a post-column photoderivatization high performance liquid chromatographic method. *J.Chromatogr.*, **1991**, *567*, 175–183

## SAMPLE
**Matrix:** blood, CSF
**Sample preparation:** Condition a Baker SPE-Octadecyl (C18) SPE cartridge with 2 mL MeOH and 1 mL 1 M phosphoric acid. 0.5 mL Plasma or CSF + 2.5 ng (CSF) or 50 ng (plasma) pirprofen, shake, add 1 mL of 1 M phosphoric acid, add to the cartridge, wash twice with 1 mL 1 M phosphoric acid, wash twice with 1 mL water, elute with two 250 µL aliquots of MeOH. Evaporate MeOH at room temperature under a stream of nitrogen, dissolve residue in 100 µL mobile phase, inject 10-20 µL.

## HPLC VARIABLES
**Column:** 80 × 4.6 3µm Perkin-Elmer C8
**Mobile phase:** MeCN:buffer 35:65 (Buffer was 50 mM sodium acetate adjusted to pH 3.00 with phosphoric acid.)
**Flow rate:** 1.2
**Injection volume:** 10-20
**Detector:** E, BAS LC-4B/17AT glassy carbon electrode, Ag/AgCl reference electrode, +0.95 V

## CHROMATOGRAM
**Retention time:** 23
**Internal standard:** pirprofen (12)
**Limit of detection:** 0.7 ng/mL

## OTHER SUBSTANCES
**Extracted:** metbolites

## KEY WORDS
plasma; SPE

## REFERENCE
Zecca, L.; Ferrario, P.; Costi, P. Determination of diclofenac and its metabolites in plasma and cerebro-spinal fluid by high-performance liquid chromatography with electrochemical detection. *J.Chromatogr.*, **1991**, *567*, 425–432

## SAMPLE
**Matrix:** blood, exudate
**Sample preparation:** 100 µL Plasma or exudate + 50 µL 50 µg/mL IS + 250 µL 900 mM phosphoric acid, vortex, add 2 mL hexane:isopropanol 90:10, rotate for 10 min, centrifuge, freeze. Remove the organic layer and evaporate it to dryness under a stream of nitrogen at 45°, reconstitute the residue in 120 µL mobile phase, inject a 100 µL aliquot.

## HPLC VARIABLES
**Column:** 200 × 4.6 Hypersil ODS
**Mobile phase:** MeOH:MeCN:1% acetic acid 55:16:29
**Flow rate:** 1
**Injection volume:** 100
**Detector:** UV 282

## CHROMATOGRAM
**Internal standard:** 2-(p-cyclohexen-1'-ylphenyl)propionic acid
**Limit of detection:** 100 ng/mL

## KEY WORDS
plasma; rat; pharmacokinetics

## REFERENCE
Stevens, A.J.; Martin, S.W.; Brennan, B.S.; McLachlan, A.; Gifford, L.A.; Rowland, M.; Houston, J.B. Regional drug delivery II: Relationship between drug targeting index and pharmacokinetic parameters for three non-steroidal anti-inflammatory drugs using the rat air pouch model of inflammation. *Pharm.Res.*, **1995**, *12*, 1987–1996

## SAMPLE
**Matrix:** blood, synovial fluid
**Sample preparation:** 0.5 mL Plasma or synovial fluid + 50 µL 24 µg/mL flurbiprofen + 200 µL 2 M HCl + 5 mL hexane, tumble 10 min on a rotary mixer, centrifuge at 10 000 g for 5 min. Remove organic layer and evaporate it to dryness under vacuum centrifugation. Reconstitute residue in 150 µL MeOH + 100 µL water, vortex mix, inject aliquot.

## HPLC VARIABLES
**Guard column:** 20 × 2 Perisorb RP18 30-40 µm pellicular
**Column:** 125 × 4.6 5 µm Spherisorb ODS 1
**Mobile phase:** MeOH:water 63:37 adjusted to pH 3.3 with phosphoric acid
**Flow rate:** 1
**Injection volume:** 20
**Detector:** UV 280

## CHROMATOGRAM
**Retention time:** 7.5
**Internal standard:** flurbiprofen (9)
**Limit of detection:** <100 ng/mL

## KEY WORDS
plasma

## REFERENCE
Blagbrough, I.S.; Daykin, M.M.; Doherty, M.; Pattrick, M.; Shaw, P.N. High-performance liquid chromatographic determination of naproxen, ibuprofen and diclofenac in plasma and synovial fluid in man. *J.Chromatogr.*, **1992**, *578*, 251–257

## SAMPLE
**Matrix:** blood, urine
**Sample preparation:** Plasma. 500 µL Plasma + 50 µL 20 µM mefenamic acid + 200 µL 2 M HCl + 5 mL dichloromethane, rotate for 10 min, centrifuge at 5000 rpm for 8 min. Remove 4 mL of the organic layer and evaporate it to dryness under a stream of nitrogen, reconstitute the residue in 250 µL plasma mobile phase, inject a 100 µL aliquot. Urine. 100 µL Urine + 100 µL 400 µg/mL ascorbic acid + 100 µL 5 M NaOH, vortex for 30 s, heat at 75°for 1 h, add 50 µL 20 µM mefenamic acid, add 500 µL 2 M HCl, add 5 mL dichloromethane, rotate for 10 min, centrifuge at 5000 rpm for 8 min. Remove 4 mL of the organic layer and evaporate it to dryness under a stream of nitrogen, reconstitute the residue in 500 µL urine mobile phase, inject a 100 µL aliquot.

## HPLC VARIABLES
**Guard column:** 4 × 4 4 µm Lichrocart C18 (Merck)
**Column:** 50 × 4 4 µm Lichrocart C18 (Merck)
**Mobile phase:** MeCN:100 mM pH 7.4 phosphate buffer:triethylamine 25:75:0.02 (plasma) or 20:80:0 (urine)
**Flow rate:** 1

**Injection volume:** 100
**Detector:** UV 282

## CHROMATOGRAM
**Internal standard:** mefenamic acid

## OTHER SUBSTANCES
**Extracted:** metabolites
**Noninterfering:** fluvastatin

## KEY WORDS
plasma; pharmacokinetics

## REFERENCE
Transon, C.; Leemann, T.; Vogt, N.; Dayer, P. In vivo inhibition profile of cytochrome P450TB (CYP2C9) by (±)-fluvastatin. *Clin.Pharmacol.Ther.*, **1995**, *58*, 412–417

## SAMPLE
**Matrix:** bulk, formulations
**Sample preparation:** Powder tablets, shake (ca. 100 mg diclofenac) with 25 mL MeCN: water 25:75 for 30 min, centrifuge at 3500 rpm for 10 min.

## HPLC VARIABLES
**Column:** 250 × 4.6 Chromatography Sciences Co. octadecylsilane bonded phase
**Mobile phase:** MeCN:THF:buffer 300:75:625 (Buffer was 50 mM $(NH_4)H_2PO_4$ adjusted to pH 5.0 with 50 mM ammonium hydroxide.)
**Flow rate:** 1.5
**Injection volume:** 10
**Detector:** UV 229

## CHROMATOGRAM
**Retention time:** 17
**Limit of quantitation:** 400 ng/mL

## KEY WORDS
tablets

## REFERENCE
Beaulieu, N.; Lovering, E.G.; Lefrançois, J.; Ong, H. Determination of diclofenac sodium and related compounds in raw materials and formulations. *J.Assoc.Off.Anal.Chem.*, **1990**, *73*, 698–701

## SAMPLE
**Matrix:** perfusate
**Sample preparation:** 250 µL Perfusate + 500 µL MeCN, vortex for 1 min, centrifuge for 10 min, inject a 150 µL aliquot of the supernatant.

## HPLC VARIABLES
**Column:** 100 mm long 5 µm Radpak C18
**Mobile phase:** MeOH:water:triethylamine 70:30:0.2 adjusted to pH 7 with concentrated phosphoric acid
**Injection volume:** 150
**Detector:** UV 273

## CHROMATOGRAM
**Limit of quantitation:** 100 ng/mL

## REFERENCE

Bassily, M.; Ghabrial, H.; Smallwood, R.A.; Morgan, D.J. Determinants of placental drug transfer: Studies in the isolated perfused human placenta. *J.Pharm.Sci.*, **1995**, *84*, 1054–1060

## SAMPLE
**Matrix:** perfusate

## HPLC VARIABLES
**Column:** 250 × 4.6 Nucleosil 100-5C18
**Mobile phase:** MeOH:18.7 mM phosphoric acid 80:20
**Flow rate:** 1
**Detector:** UV 282

## REFERENCE

Takahashi, K.; Suzuki, T.; Sakano, H.; Mizuno, N. Effect of vehicles on diclofenac permeation across excised rat skin. *Biol.Pharm.Bull.*, **1995**, *18*, 571–575

## SAMPLE
**Matrix:** solutions

## HPLC VARIABLES
**Column:** 125 × 3 Ecocart LiChrospher 100 RP-18
**Mobile phase:** Isopropanol:100 mM $KH_2PO_4$:formic acid 54:100:0.1
**Flow rate:** 0.6
**Detector:** UV 254

## CHROMATOGRAM
**Retention time:** 14.7
**Limit of quantitation:** 200-500 ng/mL

## OTHER SUBSTANCES
**Simultaneous:** acemetacin, flurbiprofen, indomethacin, lonazolac, ketoprofen, naproxen, piroxicam, sulindac, tenoxicam

## REFERENCE

Baeyens, W.R.G.; Van Der Weken, G.; Van Overbeke, A.; Zhang, Z.D. Preliminary results on the LC-separation of non-steroidal anti-inflammatory agents in conventional and narrow-bore RP set-ups applying columns with different internal diameters. *Biomed.Chromatogr.*, **1995**, *9*, 261–262

## SAMPLE
**Matrix:** solutions

## HPLC VARIABLES
**Column:** 250 × 4.6 5 μm Supelcosil LC-DP (A) or 250 × 4 5 μm LiChrospher 100 RP-8 (B)
**Mobile phase:** MeCN:0.025% phosphoric acid:buffer 25:10:5 (A) or 60:25:15 (B) (Buffer was 9 mL concentrated phosphoric acid and 10 mL triethylamine in 900 mL water, adjust pH to 3.4 with dilute phosphoric acid, make up to 1 L.)
**Flow rate:** 0.6
**Injection volume:** 25
**Detector:** UV 229

## CHROMATOGRAM
**Retention time:** 8.68 (A), 9.97 (B)

## OTHER SUBSTANCES

**Also analyzed:** acebutolol, acepromazine, acetaminophen, acetazolamide, acetophenazine, albuterol, alprazolam, amitriptyline, amobarbital, amoxapine, antipyrine, atenolol, atropine, azatadine, baclofen, benzocaine, bromocriptine, brompheniramine, brotizolam, bupivacaine, buspirone, butabarbital, butalbital, caffeine, carbamazepine, cetirizine, chlorcyclizine, chlordiazepoxide, chlormezanone, chloroquine, chlorpheniramine, chlorpromazine, chlorpropamide, chlorprothixene, chlorthalidone, chlorzoxazone, cimetidine, cisapride, clomipramine, clonazepam, clonidine, clozapine, cocaine, codeine, colchicine, cyclizine, cyclobenzaprine, dantrolene, desipramine, diazepam, diflunisal, diltiazem, diphenhydramine, diphenidol, diphenoxylate, dipyridamole, disopyramide, dobutamine, doxapram, doxepin, droperidol, encainide, ethidium bromide, ethopropazine, fenoprofen, fentanyl, flavoxate, fluoxetine, fluphenazine, flurazepam, flurbiprofen, fluvoxamine, furosemide, glutethimide, glyburide, guaifenesin, haloperidol, homatropine, hydralazine, hydrochlorothiazide, hydrocodone, hydromorphone, hydroxychloroquine, hydroxyzine, ibuprofen, imipramine, indomethacin, ketoconazole, ketoprofen, ketorolac, labetalol, levorphanol, lidocaine, loratadine, lorazepam, lovastatin, loxapine, mazindol, mefenamic acid, meperidine, mephenytoin, mepivacaine, mesoridazine, metaproterenol, methadone, methdilazine, methocarbamol, methotrexate, methotrimeprazine, methoxamine, methyldopa, methylphenidate, metoclopramide, metolazone, metoprolol, metronidazole, midazolam, moclobemide, morphine, nadolol, nalbuphine, naloxone, naphazoline, naproxen, nifedipine, nizatidine, norepinephrine, nortriptyline, oxazepam, oxycodone, oxymetazoline, paroxetine, pemoline, pentazocine, pentobarbital, pentoxifylline, perphenazine, pheniramine, phenobarbital, phenol, phenolphthalein, phentolamine, phenylbutazone, phenyltoloxamine, phenytoin, pimozide, pindolol, piroxicam, pramoxine, prazepam, prazosin, probenecid, procainamide, procaine, prochlorperazine, procyclidine, promazine, promethazine, propafenone, propantheline, propiomazine, propofol, propranolol, protriptyline, quazepam, quinidine, quinine, racemethorphan, ranitidine, remoxipride, risperidone, salicylic acid, scopolamine, secobarbital, sertraline, sotalol, spironolactone, sulfinpyrazone, sulindac, temazepam, terbutaline, terfenadine, tetracaine, theophylline, thiethylperazine, thiopental, thioridazine, thiothixene, timolol, tocainide, tolbutamide, tolmetin, trazodone, triamterene, triazolam, trifluoperazine, triflupromazine, trimeprazine, trimethoprim, trimipramine, verapamil, warfarin, xylometazoline, yohimbine, zopiclone

## KEY WORDS

some details of plasma extraction

## REFERENCE

Koves, E.M. Use of high-performance liquid chromatography-diode array detection in forensic toxicology. *J.Chromatogr.A*, **1995**, *692*, 103–119

## SAMPLE

**Matrix:** solutions

## HPLC VARIABLES

**Column:** 250 × 4.6 Zorbax RX
**Mobile phase:** Gradient. A was 10 mL concentrated orthophosphoric acid and 7 mL triethylamine in 1 L water. B was 10 mL concentrated orthophosphoric acid and 7 mL triethylamine in 200 mL water, make up to 1 L with MeCN. A:B from 100:0 to 0:100 over 30 min, maintain at 0:100 for 5 min.
**Column temperature:** 30
**Flow rate:** 2
**Detector:** UV 210

## OTHER SUBSTANCES

**Also analyzed:** acepromazine, acetaminophen, acetophenazine, albuterol, aminophylline, amitriptyline, amobarbital, amoxapine, amphetamine, amylocaine, antipyrine, aprobarbital, aspirin, atenolol, atropine, avermectin, barbital, benzocaine, benzoic acid, benzotropine, benzphetamine, berberine, bibucaine, bromazepan, brompheniramine, buprenorphine, buspirone, butabarbital, butacaine, butethal, caffeine, carbamazepine, carbromal, chloramphenicol, chlordiazepoxide, chloroquine, chlorothiazide, chloroxylenol, chlorphenesin, chlorpheniramine, chlorpromazine, chlorpropamide, chlortetracycline, cimetidine, cinchonidine, cinchonine, clenbuterol, clonazepam, clonixin, clorazepate, cocaine, codeine, colchicine, cortisone, coumarin, cyclazocine, cyclobenzaprine, cyclothiazide, cyheptamide, cymarin, danazol, danthron, dapsone, debrisoquine, desipramine, dexamethasone, dextromethorphan, dextropropoxyphene, diamorphine, diazepam, diethylpropion, diethylstilbestrol, diflunisal, digitoxin, digoxin, diltiazem, diphenhydramine, diphenoxylate, diprenorphine, dipyrone, disulfiram, dopamine, doxapram, doxepin, dronabinol, ephedrine, epinephrine, epinine, estradiol, estriol, estrone, ethacrynic acid, ethosuximide, etonitazene, etorphine, eugenol, famotidine, fenbendazole, fencamfamine, fenoprofen, fenproporex, fentanyl, flubendazole, flufenamic acid, flunitrazepam, 5-fluorouracil, fluoxymesterone, fluphenazine, furosemide, gentisic acid, gitoxigenin, glipizide, glunixin, glutethimide, glybenclamide, guaiacol, halazepam, haloperidol, hydrochlorothiazide, hydrocodone, hydrocortisone, hydromorphone, hydroxyquinoline, ibogaine, ibuprofen, iminostilbene, imipramine, indomethacin, isocarbostyril, isocarboxazid, isoniazid, isoproterenol, isoxsuprine, ivermectin, ketamine, ketoprofen, kynurenic acid, levorphanol, lidocaine, lorazepam, lormetazepam, loxapine, mazindol, mebendazole, meclizine, meclofenamic acid, medazepam, mefenamic acid, megestrol, mepacrine, meperidine, mephentermine, mephenytoin, mephesin, mephobarbital, mepivacaine, mescaline, mesoridazine, methadone, methamphetamine, methapyrilene, methaqualone, methazolamide, methocarbamol, methoxamine, methsuximide, methyl salicylate, methyldopa, methyldopamine, methylphenidate, methylprednisolone, methyltestosterone, methyprylon, metoprolol, mibolerone, morphine, nadolol, nalorphine, naloxone, naltrexone, naphazoline, naproxen, nefopam, niacinamide, nicotine, nicotinic acid, nifedipine, niflumic acid, nitrazepam, norepinephrine, nortriptyline, noscapine, nylidrin, oxazepam, oxycodone, oxymorphone, oxyphenbutazone, oxytetracycline, papaverine, pargyline, pemoline, pentazocine, pentobarbital, persantine, phenacetin, phenazocine, phenazopyridine, phencyclidine, phendimetrazine, phenelzine, pheniramine, phenobarbital, phenothiazine, phensuximide, phentermine, phenylbutazone, phenylephrine, phenylpropanolamine, piperocaine, prazepam, prednisolone, primidone, probenecid, progesterone, propiomazine, propranolol, propylparaben, pseudoephedrine, puromycin, pyrilamine, pyrithyldione, quazepam, quinaldic acid, quinidine, quinine, ranitidine, recinnamine, reserpine, resorcinol, saccharin, albuterol, salicylamide, salicylic acid, scopolamine, scopoletin, secobarbital, strychnine, sulfacetamide, sufadiazine, sulfadimethoxine, sulfaethidole, sulfamerazine, sulfamethazine, sulfamethoxizole, sulfanilamide, sulfapyridine, sulfasoxizole, sulindac, tamoxifen, temazepam, testosterone, tetracaine, tetracycline, tetramisole, thebaine, theobromine, theophylline, thiabendazole, thiamine, thiamylal, thiobarbituric acid, thioridazine, thiosalicylic acid, thiothixene, thymol, tolazamide, tolazoline, tobutamide, tolmetin, tranylcypromine, triamcinolone, tribenzylamine, trichloromethiazide, trifluoperazine, trihexyphenidyl, trimethoprim, tripelennamine, triproilidine, tropacocaine, tyramine, verapamil, vincamine, warfarin, yohimbine, zoxazolamine

## REFERENCE

Hill, D.W.; Kind, A.J. Reversed-phase solvent gradient HPLC retention indexes of drugs. *J.Anal.Toxicol.*, **1994**, *18*, 233–242

## SAMPLE

**Matrix:** urine

**Sample preparation:** 250 µL Urine + 50 µL 12.5 µg/mL 4'-hydroxy-5-chlorodiclofenac in MeOH + 150 µL 5 M NaOH, vortex at medium speed for 5-10 s, heat at 70° for 1 h, cool to room temperature, neutralize with 1 M HCl, add 750 µL buffer, vortex, add 7 mL dichloromethane:isopropanol 95:5, shake horizontally at 180 cycles/min for 10 min, centrifuge at 750 g for 5 min. Remove 5 mL of the organic layer and evaporate it to dryness under a vacuum at 30°, reconstitute the residue in 500 µL mobile phase containing 0.1% ascorbic acid, inject a 20 µL aliquot. (Buffer was 877 mL 1 M $KH_2PO_4$ and 123 mL 1 M $Na_2HPO_4$, pH 6.0.)

## HPLC VARIABLES
**Column:** 150 mm long 5 μm ODS (Supelco)
**Mobile phase:** Gradient. MeOH:MeCN:buffer 57.5:0.3:42.2 for 12 min, to 57.5:1.5:41.0 over 2 min, maintain at 57.5:1.5:41.0 for 12 min, re-equilibrate at initial conditions for 4 min. After 12 min increase flow rate to 1.6 mL/min over 2 min, maintain at 1.6 mL/min for 12 min, return to initial conditions. (Buffer was 1.156 g $(NH_4)H_2PO_4$ in 1 L water adjusted to pH 2.66 with concentrated phosphoric acid.)
**Flow rate:** 1
**Injection volume:** 20
**Detector:** UV 270

## CHROMATOGRAM
**Retention time:** 21.7
**Internal standard:** 4'-hydroxy-5-chlorodiclofenac (18.66)
**Limit of quantitation:** 400 ng/mL

## OTHER SUBSTANCES
**Extracted:** metabolites

## KEY WORDS
pharmacokinetics

## REFERENCE
Sawchuk, R.J.; Maloney, J.A.; Cartier, L.L.; Rackley, R.J.; Chan, K.K.H.; Lau, H.S.L. Analysis of diclofenac and four of its metabolites in human urine by HPLC. *Pharm.Res.*, **1995**, *12*, 756–762

◆ ─────────────────── ◆ ─────────────────── ◆

## ANNOTATED BIBLIOGRAPHY
Hanses, A.; Spahn-Langguth, H.; Mutschler, E. A new rapid and sensitive high-performance liquid chromatographic assay for diclofenac in human plasma. *Arch.Pharm.(Weinheim).*, **1995**, *328*, 257–260

Fukuyama, T.; Yamaoka, K.; Ohata, Y.; Nakagawa, T. A new analysis method for disposition kinetics of enterohepatic circulation of diclofenac in rats. *Drug Metab.Dispos.*, **1994**, *22*, 479–485 [rat; plasma; bile; column temp 40]

Maitani, Y.; Kugo, M.; Nagai, T. Permeation of diclofenac salts through silicone membrane: A mechanistic study of percutaneous absorption of ionizable drugs. *Chem.Pharm.Bull.*, **1994**, *42*, 1297–1301

Oberle, R.L.; Das, H.; Wong, S.L.; Chan, K.K.; Sawchuk, R.J. Pharmacokinetics and metabolism of diclofenac sodium in Yucatan miniature pigs. *Pharm.Res.*, **1994**, *11*, 698–703 [pharmacokinetics; pig; metabolites; plasma; indomethacin (IS); LOQ 50 ng/mL]

Reer, O.; Bock, T.K.; Müller, B.W. In vitro corneal permeability of diclofenac sodium in formulations containing cyclodextrins compared to the commercial product Voltaren ophtha. *J.Pharm.Sci.*, **1994**, *83*, 1345–1349 [perfusate]

Zhang, S.Y.; Zou, H.Q.; Zhang, Z.Y.; Peng, W.L.; Liu, L.Q. [High-performance liquid chromatographic method for the determination of diclofenac in serum and its pharmacokinetics in healthy volunteers]. *Yao Hsueh Hsueh Pao*, **1994**, *29*, 228–231

Avgerinos, A.; Karidas, T.; Malamataris, S. Extractionless high-performance liquid chromatographic method for the determination of diclofenac in human plasma and urine. *J.Chromatogr.*, **1993**, *619*, 324–329 [plasma; urine; indomethacin (IS); column temp 40; LOD 200 ng/mL; pharmacokinetics]

De Bernardi di Valserra, M.; Feletti, F.; Tripodi, A.S.; Contos, S.; Carabelli, A.; Maggi, L.; Germogli, R. Pharmacokinetic studies in healthy volunteers on a new gastroprotective pharmaceutic form of diclofenac. *Arzneimittelforschung*, **1993**, *43*, 373–377 [plasma; column temp 30; LOD 20 ng/mL; pharmacokinetics]

Schmitz, G.; Lepper, H.; Estler, C.-J. High-performance liquid chromatographic method for the routine determination of diclofenac and its hydroxy and methoxy metabolites from in vitro systems. *J.Chromatogr.*, **1993**, *620*, 158–163 [gradient; cell suspensions; extracted metabolites; LOD 5 ng/mL]

Szász, G.; Budvári-Bárany, Z.; Löre, A.; Radeczky, G.; Shalaby, A. HPLC of antiphlogistic acids on silica dynamically modified with cetylpyridinium chloride. *J.Liq.Chromatogr.*, **1993**, *16*, 2335–2345 [fenoprofen, ibuprofen, ketoprofen, naproxen, nicotinic acid, nifluminic acid, salicylic acid]

Moncrieff, J. Extractionless determination of diclofenac sodium in serum using reversed-phase high-performance liquid chromatography with fluorimetric detection. *J.Chromatogr.*, **1992**, *577*, 185–189 [fluorescence detection; harmol (IS); serum; column temp 40; LOD 20 ng/mL]

Santos, S.R.; Donzella, H.; Bertoline, M.A.; Pereira, M.D.; Omosako, C.E.; Porta, V. Simplified micromethod for the HPLC measurement of diclofenac in plasma. *Braz.J.Med.Biol.Res.*, **1992**, *25*, 125–128

Brunner, L.A.; Luders, R.C. An automated method for the determination of diclofenac sodium in human plasma. *J.Chromatogr.Sci.*, **1991**, *29*, 287–291 [plasma; LOQ 5 ng/mL; LOD 2.5 ng/mL]

Sioufi, A.; Richard, J.; Mangoni, P.; Godbillon, J. Determination of diclofenac in plasma using a fully automated analytical system combining liquid-solid extraction with liquid chromatography. *J.Chromatogr.*, **1991**, *565*, 401–407 [plasma; SPE; pharmacokinetics; LOQ 31 nM]

Lansdorp, D.; Janssen, T.J.; Guelen, P.J.; Vree, T.B. High-performance liquid chromatographic method for the determination of diclofenac and its hydroxy metabolites in human plasma and urine. *J.Chromatogr.*, **1990**, *528*, 487–494 [plasma; urine; column temp 30; extracted metabolites; pharmacokinetics; LOD 20 ng/mL (plasma); LOD 2.5 μg/mL (urine)]

Lee, H.S.; Kim, E.J.; Zee, O.P.; Lee, Y.J. High performance liquid chromatographic determination of diclofenac sodium in plasma using column-switching technique for sample clean-up. *Arch.Pharm. (Weinheim).*, **1989**, *322*, 801–806

Zecca, L.; Ferrario, P. Determination of diclofenac in plasma and synovial fluid by high-performance liquid chromatography with electrochemical detection. *J.Chromatogr.*, **1989**, *495*, 303–308 [plasma; electrochemical detection]

El-Sayed, Y.M.; Abdel-Hameed, M.E.; Suleiman, M.S.; Najib, N.M. A rapid and sensitive high-performance liquid chromatographic method for the determination of diclofenac sodium in serum and its use in pharmacokinetic studies. *J.Pharm.Pharmacol.*, **1988**, *40*, 727–729

Godbillon, J.; Gauron, S.; Metayer, J.P. High-performance liquid chromatographic determination of diclofenac and its monohydroxylated metabolites in biological fluids. *J.Chromatogr.*, **1985**, *338*, 151–159

Plavsic, F.; Culig, J. Determination of serum diclofenac by high-performance liquid chromatography by electromechanical detection. *Hum.Toxicol.*, **1985**, *4*, 317–322

Said, S.A.; Sharaf, A.A. Pharmacokinetics of diclofenac sodium using a developed HPLC method. *Arzneimittelforschung*, **1981**, *31*, 2089–2092

Tammara, V.K.; Narurkar, M.M.; Crider, A.M.; Khan, M.A. Morpholinoalkyl ester prodrugs of diclofenac: Synthesis, *in vitro* and *in vivo* evaluation. *J.Pharm.Sci.*, **1994**, *83*, 644–648 [rat; plasma; mefenamic acid (IS)]

# Dicyclomine

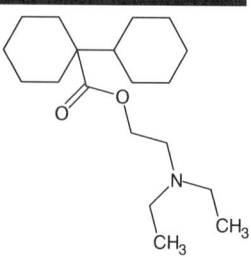

**Molecular formula:** $C_{19}H_{35}NO_2$
**Molecular weight:** 309.5
**CAS Registry No.:** 77-19-0 (dicyclomine), 67-92-5 (dicyclomine hydrochloride)

## SAMPLE
**Matrix:** blood
**Sample preparation:** 2 mL Plasma + 100 μL 1 μg/mL dicyclomine in water + 1 mL MeCN, vortex, allow to stand for 10 min, add 200 μL 1 M pH 9.4 tris(hydroxymethyl)methylamine (TRIS), add 5 mL hexane, shake horizontally for 10 min, centrifuge at 2000 g for 5 min. Remove the aqueous layer and add it to 3 mL hexane, shake horizontally for 10 min, centrifuge at 2000 g for 5 min. Combine the hexane layers and add them to 1 mL 100 mM HCl, shake for 10 min, centrifuge. Remove the aqueous layer and evaporate it to dryness under vacuum, reconstitute in 250 μL mobile phase, inject a 100 μL aliquot.

## HPLC VARIABLES
**Column:** 100 mm long 5 μm Techsil CN
**Mobile phase:** MeOH:20 mM pH 6.2 orthophosphoric acid buffer 40:60
**Column temperature:** 30
**Flow rate:** 0.6
**Injection volume:** 100
**Detector:** E, ESA Coulochem 5100-A, guard cell 1.0 V, dual porous graphite electrode 0.85 and 0.95 V

## CHROMATOGRAM
**Retention time:** 23
**Internal standard:** dicyclomine

## OTHER SUBSTANCES
**Extracted:** oxybutynin

## KEY WORDS
plasma; dicyclomine is IS

## REFERENCE
Hughes, K.M.; Lang, J.C.T.; Lazare, R.; Gordon, D.; Stanton, S.L.; Malone-Lee, J.; Geraint, M. Measurement of oxybutynin and its N-desethyl metabolite in plasma, and its application to pharmacokinetic studies in young, elderly and frail elderly volunteers. *Xenobiotica*, **1992**, *22*, 859–869

## SAMPLE
**Matrix:** solutions

## HPLC VARIABLES
**Column:** 150 × 2 Deltabond C8 (Keystone)
**Mobile phase:** MeCN:2-butanone:50 mM pH 7.0 phosphate buffer 27:13:60
**Flow rate:** 0.15
**Injection volume:** 1
**Detector:** Chemiluminescence following post-column reaction. A 1 mM solution of Ru(2,2'-bipyridine)$_3^{2+}$ in 50 mM sodium sulfate (continuously sparged with helium) was oxidized to Ru(2,2'-bipyridine)$_3^{3+}$ using a Princeton Applied Research Model 174A polarographic

analyzer with a platinum gauze working electrode, a platinum wire auxiliary electrode, and a silver wire reference electrode. The Ru solution was pumped at 0.3 mL/min and mixed with the column effluent in the flow cell of the detector, a fluorescence detector with the light source removed.

## CHROMATOGRAM
**Retention time:** 14
**Limit of detection:** 0.1-1 μg/mL

## OTHER SUBSTANCES
**Simultaneous:** cyclobenzaprine

## KEY WORDS
post-column reaction

## REFERENCE
Holeman, J.A.; Danielson, N.D. Microbore liquid chromatography of tertiary amine anticholinergic pharmaceuticals with tris(2,2'-bipyridine)ruthenium(III) chemiluminescence detection. *J.Chromatogr. Sci.*, **1995**, *33*, 297–302

## SAMPLE
**Matrix:** solutions
**Sample preparation:** Prepare a 10 μg/mL solution in MeOH, inject a 20 μL aliquot.

## HPLC VARIABLES
**Column:** 125 × 4.9 Spherisorb S5W silica
**Mobile phase:** MeOH containing 10 mM ammonium perchlorate and 1 mL/L 100 mM NaOH in MeOH, pH 6.7
**Flow rate:** 2
**Injection volume:** 20
**Detector:** E, LeCarbone, V25 glassy carbon electrode, + 1.2 V

## CHROMATOGRAM
**Retention time:** 1.8

## OTHER SUBSTANCES
**Also analyzed:** acebutolol, acepromazine, acetophenazine, N-acetylprocainamide, albuterol, alprenolol, amethocaine, amiodarone, amitriptyline, antazoline, atenolol, azacyclonal, bamethane, benactyzine, benperidol, benzethidine, benzocaine, benzoctamine, benzphetamine, benzquinamide, bromhexine, bromodiphenhydramine, bromperidol, brompheniramine, brompromazine, buclizine, bufotenine, bupivacaine, buprenorphine, butacaine, butethamate, chlorcyclizine, chlorpheniramine, chlorphenoxamine, chlorprenaline, chlorpromazine, chlorprothixene, cimetidine, cinchonidine, cinnarizine, clemastine, clomipramine, clonidine, cocaine, cyclazocine, cyclizine, cyclopentamine, cyproheptadine, deserpidine, desipramine, dextromoramide, dextropropoxyphene, diethylcarbamazine, diethylpropion, diethylthiambutene, dihydroergotamine, dimethindene, dimethothiazine, diphenhydramine, diphenoxylate, dipipanone, diprenorphine, dipyridamole, disopyramide, dothiepin, doxapram, doxepin, doxylamine, droperidol, ephedrine, ergocornine, ergocristine, ergocristinine, ergocryptine, ergometrine, ergosine, ergosinine, ergotamine, ethopropazine, etorphine, etoxeridine, fenethazine, fenfluramine, fenoterol, fentanyl, flavoxate, fluopromazine, flupenthixol, fluphenazine, flurazepam, haloperidol, hydroxyzine, hyoscine, ibogaine, imipramine, indapamine, iprindole, isothipendyl, isoxsuprine, ketanserin, laudanosine, lidocaine, lofepramine, loxapine, maprotiline, mecamylamine, meclophenoxate, meclozine, medazepam, mephentermine, mepivacaine, meptazinol, mepyramine, mesoridazine, metaraminol, methadone, methamphetamine, methapyrilene, methdilazine, methotrimeprazine, methoxamine, methoxyphenamine, methoxypromazine, methylephedrine, methylergonovine, methysergide, metoclopramide, metopimazine,

metoprolol, mianserin, morazone, nadolol, nalorphine, naloxone, naphazoline, nicotine, nifedipine, nomifensine, nortriptyline, noscapine, orphenadrine, oxeladin, oxprenolol, oxymetazolin, papaverine, pargyline, pecazine, penbutolol, pentazocine, penthienate, pericyazine, perphenazine, phenadoxone, phenampromide, phenazocine, phenbutrazate, phendimetrazine, phenelzine, phenglutarimide, phenindamine, pheniramine, phenmetrazine, phenomorphan, phenoperidine, phenothiazine, phenoxybenzamine, phentolamine, phenylephrine, phenyltoloxamine, physostigmine, piminodine, pimozide, pindolol, pipamazine, pipazethate, piperacetazine, piperidolate, pipradol, pirenzepine, piritramide, pizotifen, practolol, pramoxine, prazosin, prenylamine, prilocaine, primaquine, proadifen, procainamide, procaine, prochlorperazine, procyclidine, proheptazine, prolintane, promazine, promethazine, pronethalol, properidine, propiomazine, propranolol, prothipendyl, protriptyline, proxymetacaine, pseudoephedrine, pyrimethamine, quinidine, quinine, ranitidine, rescinnamine, sotalol, tacrine, terazosin, terbutaline, terfenadine, thenyldiamine, theophylline, thiethylperazine, thiopropazate, thioproperazine, thioridazine, thiothixene, thonzylamine, timolol, tocainide, tolpropamine, tolycaine, tranylcypromine, trazodone, trifluoperazine, trifluperidol, trimeperidine, trimeprazine, trimethobenzamide, trimethoprim, trimipramine, tripelennamine, triprolidine, tryptamine, verapamil, xylometazoline

## REFERENCE

Jane, I.; McKinnon, A.; Flanagan, R.J. High-performance liquid chromatographic analysis of basic drugs on silica columns using non-aqueous ionic eluents. II. Application of UV, fluorescence and electrochemical oxidation detection. *J.Chromatogr.*, **1985**, *323*, 191–225

# Digoxin

**Molecular formula:** $C_{41}H_{64}O_{14}$
**Molecular weight:** 781.0
**CAS Registry No.:** 20830-75-5

## SAMPLE
**Matrix:** blood
**Sample preparation:** Condition a 1 mL Cyclobond I β-cyclodextrin SPE cartridge (Astec) with 2 mL MeOH, 2 mL MeCN, 2 mL isopropanol, and 2 mL water (SPE cartridge A). Condition a 1 mL Cyclobond I β-cyclodextrin SPE cartridge (Astec) with 2 mL MeOH, 2 mL MeCN, and 2 mL dichloromethane (SPE cartridge B). Condition a 1 mL Bond Elut C1 SPE cartridge with 2 mL MeOH and 2 mL MeCN (SPE cartridge C). 1 mL Serum + 10 ng digoxin + 1 mL water, add to SPE cartridge A, wash with 2 mL water, wash with 1 mL MeOH:7.5 mM pH 7.0 potassium phosphate buffer 20:80, wash with 3 mL water, wash with 1 mL isopropanol:water 10:90, dry under vacuum for 5 min, wash with ten 100 μL aliquots of dichloromethane, dry under vacuum for 5 min, elute with 1 mL isopropanol. Evaporate the eluate to dryness under a stream of nitrogen at room temperature, add 50 μL 10% 4-dimethylaminopyridine in MeCN then 50 μL 4% 1-naphthoyl chloride in MeCN under nitrogen in a glove box (relative humidity <26%), mix thoroughly, heat at 50°for 1 h, centrifuge briefly, evaporate under a stream of nitrogen, add 2 mL 5% pH 10.0 sodium bicarbonate solution, shake for 1 min, add 2 mL chloroform, shake, centrifuge. Remove the organic layer and wash it with 2 mL 5% sodium bicarbonate, wash twice with 2 mL portions of 50 mM HCl, evaporate to dryness under a stream of nitrogen at room temperature, reconstitute with 200 μL dichloromethane, add to SPE cartridge B, wash with eight 100 μL aliquots of dichloromethane, elute with 1 mL MeOH. Evaporate the eluate to dryness under a stream of nitrogen, reconstitute with 250 μL MeCN, add to SPE cartridge C, rinse container with 250 μL MeCN, add rinse to the SPE cartridge, add 500 μL MeCN to the SPE cartridge. Collect all the eluates and evaporate them to dryness under a stream of nitrogen, reconstitute with mobile phase, inject an aliquot. (Purify 4-dimethylaminopyridine by passing a 30% solution in MeCN through a layer of silica gel covered with a layer of activated charcoal, evaporate the filtrate under reduced pressure, store the residue in a desiccator. Immerse glassware in sulfuric acid:nitric acid 80:20 for 24 h, wash with water, treat with 1% Surfasil (Pierce) in toluene, rinse with water, dry in an oven.)

## HPLC VARIABLES
**Guard column:** 15 × 3.2 7 μm silica (Applied Biosystems)
**Column:** 150 × 4.6 3 μm Spherisorb silica
**Mobile phase:** Hexane:dichloromethane:MeCN:MeOH 36:6.3:5.4:0.2
**Flow rate:** 1.6
**Injection volume:** 20
**Detector:** F ex 217 em 340

## CHROMATOGRAM
**Retention time:** 10
**Internal standard:** digitoxin (9.5)
**Limit of detection:** 0.25 ng/mL

## OTHER SUBSTANCES
**Extracted:** metabolites, digoxigenin, digoxigenin bisdigitoxoside, digoxigenin monodigitoxoside, dihydrodigoxin
**Noninterfering:** acetaminophen, acetazolamide, acyclovir, albuterol, allopurinol, amiodarone, amitriptyline, amoxicillin, ampicillin, aspirin, atenolol, atropine, azathioprine, bumetanide, calcitriol, captopril, carbamazepine, cefazolin, cefoperazone, ceftazidime, ceftizoxime, cephalexin, chlordiazepoxide, ciprofloxacin, clavulanic acid, clindamycin, clonidine, clotrimazole, codeine, conjugated estrogens, cyclophosphamide, diazepam, diphenhydramine, dipyridamole, dobutamine, docusate sodium, dopamine, enalapril, erythromycin, famotidine, fluconazole, furosemide, gemfibrozil, gentamicin, glyburide, heparin, hydralazine, hydrochlorothiazide, ibuprofen, ipratropium bromide, isosorbide dinitrate, isradipine, labetalol, lidocaine, lorazepam, lovastatin, medroxyprogesterone acetate, meperidine, metoclopramide, metolazone, metoprolol, midazolam, minoxidil, morphine, nicotine, nifedipine, nitroglycerin, norepinephrine, nystatin, oxybutynin, oxycodone, pentoxiphylline, phenytoin, piroxicam, prednisone, procainamide, procaine, promethazine, propoxyphene, ranitidine, sotalol, spironolactone, sulbactam, sulfamethoxazole, sulfisoxazole, temazepam, tetracycline, timolol, tobramycin, triamcinolone acetonide, triamterene, trimethoprim, vancomycin, verapamil, warfarin

## KEY WORDS
normal phase; derivatization; serum; SPE; pharmacokinetics

## REFERENCE
Tzou, M.-C.; Sams, R.A.; Reuning, R.H. Specific and sensitive determination of digoxin and metabolites in human serum by high-performance liquid chromatography with cyclodextrin solid-phase extraction and precolumn fluorescence derivatization. *J.Pharm.Biomed.Anal.*, **1995**, *13*, 1531–1540

## SAMPLE
**Matrix:** blood
**Sample preparation:** 3 mL Serum + 20 µL 8 µg/mL IS in EtOH + 3 mL acetone, vortex for 20 s, centrifuge at 1000 g for 5 min, remove the supernatant and add it to 2 mL isooctane:dichloromethane 80:20, vortex for 1 min, centrifuge at 1000 g for 5 min. Remove the acetone/water layer and evaporate it to 3 mL under a stream of nitrogen at 37°, add 10 mL dichloromethane:n-propanol 98:2, rotate for 10 min, centrifuge at 1000 g for 5 min, repeat extraction, filter the organic layers and evaporate them to dryness under a stream of nitrogen at 37°, reconstitute the residue in 100 µL MeOH:water 50:50, inject the whole amount.

## HPLC VARIABLES
**Guard column:** 15 × 3.2 ODS (Brownlee)
**Column:** 150 × 4.6 3 µm Spherisorb ODS II
**Mobile phase:** MeOH:EtOH:isopropanol:buffer 52:3:1:45 (Prepare buffer by mixing 12.5 mL 0.15% hydrogen peroxide in water with 500 mL 500 µg/mL L-ascorbic acid in water, stir for 2 h. Prepare fresh each week.)
**Flow rate:** 0.4
**Injection volume:** 100
**Detector:** F ex 360 (filter) em 425 (filter) following post-column reaction. The column effluent mixed with concentrated HCl pumped at 0.5 mL/min and flowed through a 20 m × 0.3 mm i.d. PTFE coil at 79 ± 1° to the detector. (The flow of concentrated HCl was generated by displacing concentrated HCl from a pressure vessel with hexane. The hexane was pumped into the pressure vessel by an HPLC pump.)

## CHROMATOGRAM
**Retention time:** 18.5
**Internal standard:** digitoxigenin (25.5)
**Limit of quantitation:** 0.5 ng/mL

## OTHER SUBSTANCES
**Simultaneous:** digoxigenin, digoxigenin bisdigitoxoside, digoxigenin monodigitoxoside, dihydrodigoxigenin, dihydrodigoxin, furosemide, spironolactone
**Noninterfering:** mexiletine, captopril, dipyridamole, disopyramide, procainamide, propafenone, quinidine, sulfamethoxazole, trimethoprim, verapamil

## KEY WORDS
serum; post-column reaction

## REFERENCE
Embree, L.; McErlane, K.M. Development of a high-performance liquid chromatographic-post-column fluorogenic assay for digoxin in serum. *J.Chromatogr.*, **1989**, *496*, 321–334

## SAMPLE
**Matrix:** blood
**Sample preparation:** Wash a C18 Sep-Pak with 24 mL MeOH then 24 mL water. Wash a Diol Sep-Pak with 6 mL MeOH. 300 µL Serum + 25 µL 23.86 nmol/L deslanoside + 25 µL 28.68 µmol/L gitoxigenin, vortex, add 300 µL to the C18 Sep-Pak, wash with 1 mL water, 1 mL ice-cold 100 g/L $ZnSO_4$, 1 mL 20 mL/L MeCN, 3 mL water, remove excess water by applying vacuum for several min. Elute the C18 Sep-Pak with 3 mL MeOH through the Diol Sep-Pak and collect the eluate. Dry the eluate under nitrogen at 37°, reconstitute in 200 µL mobile phase, vortex 30 s, centrifuge at 1100 g for 15 min, inject 185-195 µL. After each run clean column by injecting 200 µL MeOH + 1 mL THF. Immunoassay detection. Gitoxigenin elutes at 16 min as a marker. Collect deslanoside (10-12 min), digoxin (24-30 min), and 1 min fractions on either side of digoxin (4 tubes). Dry under air at 25°, reconstitute residue with 230 µL digoxin-free serum, vortex 20 s, centrifuge at 1100 g for 5 min, use 200 µL. Determine digoxin by fluorescence polarization immunoassay in accordance with manufacturer's instructions.

## HPLC VARIABLES
**Column:** 100 × 4.6 3 µm ODS2 (Chromatography Sciences Co.)
**Mobile phase:** THF:water 20.5:79.5
**Flow rate:** 0.6
**Injection volume:** 185-195
**Detector:** UV 218; Immunoassay

## CHROMATOGRAM
**Retention time:** 25
**Internal standard:** deslanoside

## OTHER SUBSTANCES
**Noninterfering:** dexamethasone, hydroxyprogesterone, methylprednisolone, progesterone
**Interfering:** cortisone (with UV assay), fludrocortisone (with UV assay), prednisone (with UV assay)

## KEY WORDS
serum

## REFERENCE
Stone, J.A.; Soldin, S.J. Improved liquid chromatographic/immunoassay of digoxin in serum. *Clin.Chem.*, **1988**, *34*, 2547–2551

## SAMPLE
**Matrix:** blood
**Sample preparation:** 3 mL Plasma + 3 mL acetone, vortex, centrifuge at 1000 g for 5 min, remove the supernatant and add it to 2 mL isooctane, vortex, centrifuge at 1000 g

for 5 min. Remove the acetone/water layer and evaporate it to 3 mL under a stream of nitrogen at 37°, add 10 mL dichloromethane:n-propanol 98:2, rotate for 10 min, centrifuge at 1000 g for 5 min, filter the organic layer and evaporate it to dryness under a stream of nitrogen at 37°, reconstitute the residue in 100 μL MeOH:water 50:50, inject the whole amount.

## HPLC VARIABLES

**Guard column:** 37 μm ODS
**Column:** 150 × 4.6 3 μm Spherisorb ODS II
**Mobile phase:** MeOH:EtOH:isopropanol:water 52:3:1:45
**Flow rate:** 0.3
**Injection volume:** 100
**Detector:** F ex 360 (filter) em 425 (filter) following post-column reaction. The column effluent mixed with the reagent and flowed through a 10 m × 0.3 mm i.d. knitted PTFE coil at 79 ± 1° to the detector. The reagent was generated by mixing 1.1 mM hydrogen peroxide in 0.1% ascorbic acid pumped at 0.038 mL/min and concentrated HCl pumped at 0.192 mL/min and allowing this mixture to flow through a 2 m × 0.8 mm i.d. PTFE coil to the point where it mixed with the column effluent (J.Chromatogr. 1986, 377, 233).

## CHROMATOGRAM

**Retention time:** 33
**Internal standard:** digitoxigenin (42)
**Limit of detection:** 0.5 ng/mL

## OTHER SUBSTANCES

**Simultaneous:** digoxigenin, digoxigenin bisdigitoxoside, digoxigenin monodigitoxoside, dihydrodigoxigenin, dihydrodigoxin, furosemide, spironolactone
**Noninterfering:** captopril, dipyridamole, disopyramide, procainamide, propafenone, quinidine, sulfamethoxazole, trimethoprim, verapamil

## KEY WORDS

plasma; post-column reaction; comparison with RIA

## REFERENCE

Kwong, E.; McErlane, K.M. Analysis of digoxin at therapeutic concentrations using high-performance liquid chromatography with post-column derivatization. *J.Chromatogr.*, **1986**, *381*, 357–363

## SAMPLE

**Matrix:** blood, perfusate
**Sample preparation:** 200 μL Plasma or perfusate + 20 μL ethinyl estradiol solution + 5 mL dichloromethane, vortex, centrifuge for 10 min. Remove a 4.5 mL aliquot of the lower organic layer and evaporate it to dryness under a stream of nitrogen, reconstitute the residue in 200 μL mobile phase, inject a 50 μL aliquot.

## HPLC VARIABLES

**Column:** LiChrocart 100 RP-18
**Mobile phase:** MeOH:isopropanol:dichloromethane:water 40:9:4:47
**Flow rate:** 1
**Injection volume:** 50
**Detector:** UV 220

## CHROMATOGRAM

**Internal standard:** ethinyl estradiol (17α-ethynylestradiol)

## KEY WORDS

plasma; rat; pharmacokinetics

## REFERENCE

Su, S.-F.; Huang, J.-D. Inhibition of the intestinal digoxin absorption and exsorption by quinidine. *Drug Metab.Dispos.*, **1996**, *24*, 142–147

## SAMPLE

**Matrix:** blood, urine

**Sample preparation:** Urine. Extract 20 mL urine with 20 mL dichloromethane containing 3% heptafluorobutanol for 15 min, centrifuge at 1000 g for 10 min. Remove 15 mL aqueous phase, extract with 15 mL dichloromethane containing 3% heptafluorobutanol for 15 min, centrifuge at 1000 g for 10 min. Combine 10 mL volumes of each organic phase, evaporate under nitrogen to about 0.5 mL, add 20 μL 1-pentanol, evaporate to 20 μL, dissolve residue in 250 μL mobile phase, inject a 100 μL aliquot. Plasma. 10 mL Plasma + 3 g sodium chloride, extract with 15 mL dichloromethane containing 3% heptafluorobutanol for 15 min, centrifuge at 1000 g for 10 min. Evaporate 10 mL organic phase under nitrogen to about 0.5 mL, add 25 μL phosphate buffer, evaporate to 25 μL, dissolve residue in mobile phase, inject whole amount. (Plasma extraction in Acta Pharmacol. Toxicol. (Copenh.) 1986, 59 (Suppl. 4), 1-62.)

## HPLC VARIABLES

**Column:** 150 × 4.5 7 μm LiChrosorb RP-8

**Mobile phase:** Isopropanol:pH 6.3 phosphate buffer (I=0.1) 16.5:83.5

**Flow rate:** 1

**Injection volume:** 100

**Detector:** UV 220

## CHROMATOGRAM

**Retention time:** 23

**Limit of detection:** 2 ng/mL

## KEY WORDS

plasma

## REFERENCE

Eriksson, B.-M.; Tekensbergs, L.; Magnusson, J.-O.; Molin, L. Determination of tritiated digoxin and metabolites in urine by liquid chromatography. *J.Chromatogr.*, **1981**, *223*, 401–408

## SAMPLE

**Matrix:** bulk

**Sample preparation:** Dissolve a small amount in 200 μL dry pyridine, add 15 mg 3,5-dinitrobenzoyl chloride, shake for 2 h, evaporate to dryness under nitrogen under reduced pressure. Reconstitute with 1.5 mL ethyl acetate, wash 4 times with 1 mL portions of 5% sodium bicarbonate containing 2.5 mg/mL 4-dimethylaminopyridine, wash 4 times with 1 mL portions of 1% HCl, wash 4 times with 1 mL portions of water, evaporate to dryness under a stream of nitrogen, reconstitute with mobile phase, inject an aliquot (J.Chromatogr.Sci. 1983, 21, 495).

## HPLC VARIABLES

**Guard column:** 15 × 3.2 Brownlee ODS

**Column:** 150 × 4.6 3 μm Spherisorb ODS II

**Mobile phase:** MeOH:EtOH:MeCN:isopropanol:100 mM pH 4.6 sodium acetate buffer 40:3:60:2:22

**Flow rate:** 1

**Detector:** UV 254; E, ESA Coulochem Model 5100A, Model 5020 guard cell -0.8 V (placed before the injector), Model 5010 dual-electrode analytical cell with glassy-carbon electrodes (-0.8 V first electrode, +0.8 V second electrode)

## CHROMATOGRAM
**Retention time:** 13
**Limit of detection:** 0.39 ng (electrochemical detection)

## OTHER SUBSTANCES
**Simultaneous:** digoxigenin bisdigitoxoside, digoxigenin monodigitoxoside, digoxigenin, dihydrodigoxigenin, dihydrodigoxin

## KEY WORDS
derivatization

## REFERENCE
Embree, L.; McErlane, K.M. Electrochemical detection of 3,5-dinitrobenzoyl derivatives of digoxin by high-performance liquid chromatography. *J.Chromatogr.*, **1990**, *526*, 439–446

## SAMPLE
**Matrix:** bulk, formulations
**Sample preparation:** Ampoules. Add the contents of 1 ampoule (2 mL) to 15 mL 2% sodium bicarbonate solution, extract 5 times with 10 mL portions of chloroform:isopropanol 60:40, wash each extract with the same 10 mL portion of water, wash with another 10 mL portion of water. Combine the organic layers and evaporate them to dryness, transfer the residue to another tube with two 1 mL portions of chloroform:pyridine 10:1, evaporate to dryness under reduced pressure at 50°, add 200 μL reagent, shake well, let stand at room temperature for 10 min, evaporate to dryness under reduced pressure at 50°, flush the tube with a stream of air or nitrogen, add 2 mL 5% sodium carbonate solution containing 2.5 mg/mL 4-dimethylaminopyridine, shake or sonicate for 5 min, extract with 2 mL chloroform. Wash the extract with 2 mL 5% sodium bicarbonate solution, wash twice with 3 mL portions of 50 mM HCl containing 5% NaCl, inject a 20 μL aliquot. Bulk. Prepare a solution in pyridine containing not more than 10 mg/mL. Add 150 μL reagent to 50 μL solution, shake well, let stand at room temperature for 10 min, evaporate to dryness under reduced pressure at 50°, flush the tube with a stream of air or nitrogen, add 2 mL 5% sodium carbonate solution containing 2.5 mg/mL 4-dimethylaminopyridine, shake or sonicate for 5 min, extract with 2 mL chloroform. Wash the extract with 2 mL 5% sodium bicarbonate solution, wash twice with 3 mL portions of 50 mM HCl containing 5% NaCl, inject a 20 μL aliquot. (Prepare reagent fresh each day by dissolving 100 mg 4-nitrobenzoyl chloride in 1 mL pyridine with gentle warming.)

## HPLC VARIABLES
**Column:** 200 × 3 5 μm Merckosorb SI 60
**Mobile phase:** n-Hexane:chloroform:MeCN 30:10:9
**Flow rate:** 1.5
**Injection volume:** 20
**Detector:** UV 254

## CHROMATOGRAM
**Retention time:** 6
**Limit of detection:** 11 ng/mL (100 μL injection)

## OTHER SUBSTANCES
**Simultaneous:** diginatigenin, diginatin, digitoxigenin, digitoxin, digoxigenin, gitaloxigenin, gitaloxin, gitoxigenin, gitoxin, lanatoside A, lanatoside B, lanatoside C, lanatoside D, lanatoside E

## KEY WORDS
ampoules; normal phase; derivatization

## REFERENCE

Nachtmann, F.; Spitzy, H.; Frei, R.W. Rapid and sensitive high-resolution procedure for digitalis glyco-side analysis by derivatization liquid chromatography. *J.Chromatogr.*, **1976**, *122*, 293–303

## SAMPLE

**Matrix:** feces, urine

**Sample preparation:** Urine. Place 1 mL 100 ng/mL digitoxin in isopropanol in a tube and evaporate. Add 1 mL urine + 2 mL dichloromethane, shake by hand 4 times, centrifuge 1650 g. Remove organic layer and wash it twice with 2 mL 5% sodium bicarbonate solution, evaporate under nitrogen at 50°. Add 25 mg 4-dimethylaminopyridine and 10 µL 1-naphthoyl chloride, add 100 µL MeCN, vortex thoroughly, place in water bath at 50° for 1 h, centrifuge, evaporate at 50° under nitrogen. Add 2 mL 5% sodium bicarbonate solution, shake mechanically for 5 min, add 2 mL chloroform, shake by hand. Remove organic layer and wash it twice with 2 mL 5% sodium bicarbonate solution, wash three times with 0.05 M HCl containing 5% NaCl, evaporate chloroform, dissolve residue in mobile phase. Feces. Dilute 5:1 (v/w) with 5 µg/mL clindamycin in water to stop bacterial metabolism, homogenize with mechanical shaking for 15 min. Evaporate 1 mL 100 ng/mL digitoxin in isopropanol into a tube, weigh ca. 1 g homogenate into the tube, add 1 mL water, vortex 30 s, shake 15 min, centrifuge 1 h. Pour off supernatant and extract it with 2 mL dichloromethane. Wash the extract twice with 2 mL 5% sodium bicarbonate solution, evaporate under nitrogen at 50°. Add 25 mg 4-dimethylaminopyridine and 10 µL 1-na-phthoyl chloride, add 100 µL MeCN, vortex thoroughly, place in water bath at 50° for 1 h, centrifuge, evaporate at 50° under nitrogen. Add 2 mL 5% sodium bicarbonate solution, shake mechanically for 5 min, add 2 mL chloroform, shake by hand. Remove organic layer and wash it twice with 2 mL 5% sodium bicarbonate solution, wash three times with 0.05 M HCl containing 5% NaCl, evaporate chloroform, dissolve residue in mobile phase.

## HPLC VARIABLES

**Column:** 150 × 4.6 3 µm Adsorbosphere SI
**Mobile phase:** Hexane:dichloromethane:MeCN 6:1:1
**Flow rate:** 1.8-2
**Injection volume:** 20-175
**Detector:** F ex 217 em 340 cut-off filter (372 nm max)

## CHROMATOGRAM

**Retention time:** 9.4
**Internal standard:** digitoxin (8.1)
**Limit of detection:** 5 ng/mL (urine); 50 ng/g (feces)

## OTHER SUBSTANCES

**Simultaneous:** metabolites

## KEY WORDS

normal phase

## REFERENCE

Shepard, T.A.; Hui, J.; Chandrasekaran, A.; Sams, R.A.; Reuning, R.H.; Robertson, L.W.; Caldwell, J.H.; Donnerberg, R.L. Digoxin and metabolites in urine and feces: a fluorescence derivatization–high-performance liquid chromatographic technique. *J.Chromatogr.*, **1986**, *380*, 89–98

## SAMPLE

**Matrix:** formulations
**Sample preparation:** Dilute a 1 mL sample to 10 mL with mobile phase, inject an aliquot.

## HPLC VARIABLES

**Column:** 250 × 4.6 10 µm Partisil ODS III C18

**Mobile phase:** MeCN:water:phosphoric acid 35:65:0.1
**Flow rate:** 2
**Injection volume:** 20
**Detector:** UV 220

## CHROMATOGRAM
**Retention time:** 7

## OTHER SUBSTANCES
**Noninterfering:** amrinone

## KEY WORDS
injections; stability-indicating; 5% dextrose; 0.45% NaCl

## REFERENCE
Riley, C.M.; Junkin, P. Stability of amrinone and digoxin, procainamide hydrochloride, propranolol hydrochloride, sodium bicarbonate, potassium chloride, or verapamil hydrochloride in intravenous admixtures. *Am.J.Hosp.Pharm.*, **1991**, *48*, 1245–1252

## SAMPLE
**Matrix:** formulations
**Sample preparation:** One tablet (0.25 mg digoxin) + 5 mL acetone:ethanol 9:1 containing 0.11826 mg dexamethasone, sonicate 5 min, centrifuge at 1400 g for 5 min, evaporate supernatant under vacuum, dissolve residue in 100 µL MeOH, inject 0.2 µL aliquots.

## HPLC VARIABLES
**Column:** 95 × 0.5 Japan Spectroscopic SC-01 (5 µm octadecylsilyl silica in a PTFE tube)
**Mobile phase:** MeCN:water 28:72
**Flow rate:** 0.008
**Injection volume:** 0.2
**Detector:** UV 220

## CHROMATOGRAM
**Retention time:** 10
**Internal standard:** dexamethasone (14)

## OTHER SUBSTANCES
**Simultaneous:** digoxigenin, digoxigenin bisdigitoxoside, digoxigenin monodigitoxoside, dimethyldigoxin, β-methyldigoxin

## KEY WORDS
tablets; microbore

## REFERENCE
Fujii, Y.; Ikeda, Y.; Yamazaki, M. High-performance liquid chromatographic determination of secondary cardiac glycosides in Digitalis purpurea leaves. *J.Chromatogr.*, **1989**, *479*, 319–325

## SAMPLE
**Matrix:** formulations
**Sample preparation:** 1 mL Sample + 9 mL mobile phase, mix, inject an aliquot.

## HPLC VARIABLES
**Column:** 150 × 4.6 5 µm Zorbax ODS
**Mobile phase:** MeCN:phosphoric acid:water 26:0.1:58, adjusted to pH 6.5 with 10 M NaOH

**Flow rate:** 2
**Injection volume:** 20
**Detector:** UV 218

## CHROMATOGRAM
**Retention time:** 6

## OTHER SUBSTANCES
**Simultaneous:** digoxigenin, digoxigenin didigitoxoside, digoxigenin monodigitoxoside
**Noninterfering:** milrinone

## KEY WORDS
stability-indicating; 5% dextrose; injections

## REFERENCE
Riley, C.M. Stability of milrinone and digoxin, furosemide, procainamide hydrochloride, propranolol hydrochloride, quinidine gluconate, or verapamil hydrochloride in 5% dextrose injection. *Am.J.Hosp.Pharm.*, **1988**, *45*, 2079–2091

## SAMPLE
**Matrix:** formulations
**Sample preparation:** Directly inject a 20 μL aliquot of a 250 μg/mL digoxin injection.

## HPLC VARIABLES
**Column:** 300 × 3.9 10 μm μBondapak C18
**Mobile phase:** MeCN:water 29:71
**Flow rate:** 2
**Injection volume:** 20
**Detector:** UV 218

## CHROMATOGRAM
**Retention time:** 9

## OTHER SUBSTANCES
**Simultaneous:** mercaptobenzothiazole

## KEY WORDS
injections

## REFERENCE
Reepmeyer, J.C.; Juhl, Y.H. Contamination of injectable solutions with 2-mercaptobenzothiazole leached from rubber closures. *J.Pharm.Sci.*, **1983**, *72*, 1302–1305

## SAMPLE
**Matrix:** solutions
**Sample preparation:** Inject a 10 μL aliquot of a solution in mobile phase.

## HPLC VARIABLES
**Column:** 250 × 4.6 Deltabond C18 (Keystone)
**Mobile phase:** Gradient. MeCN:water from 10:90 to 45:55 over 8 min.
**Flow rate:** 1.3
**Injection volume:** 10
**Detector:** E, Dionex pulsed electrochemical detector, integrated amperometry mode, 1.4 mm gold working electrode with 0.005 inch gasket, E1 +0.07 V, t1 400 ms, E2 +0.70 V, t2 120 ms, E3 1.00 V, t3 300 ms, stainless steel counter electrode, Ag/AgCl reference

electrode, following post-column reaction. The column effluent mixed with 1 M NaOH pumped at 0.5 mL/min and the mixture flowed through a 500 μL reaction coil (Dionex) to the detector.

## CHROMATOGRAM
**Retention time:** 9
**Limit of detection:** 230 ng/mL

## OTHER SUBSTANCES
**Simultaneous:** digitoxigenin, digitoxigenin bisdigitoxoside, digitoxigenin monodigitoxoside, digitoxin, digoxigenin, digoxigenin bisdigitoxoside, digoxigenin monodigitoxoside

## KEY WORDS
post-column reaction

## REFERENCE
Kelly, K.L.; Kimball, B.A.; Johnston, J.J. Quantitation of digitoxin, digoxin, and their metabolites by high-performance liquid chromatography using pulsed amperometric detection. *J.Chromatogr.A*, **1995**, *711*, 289–295

## SAMPLE
**Matrix:** solutions

## HPLC VARIABLES
**Column:** 250 × 4.6 Zorbax RX
**Mobile phase:** Gradient. A was 10 mL concentrated orthophosphoric acid and 7 mL triethylamine in 1 L water. B was 10 mL concentrated orthophosphoric acid and 7 mL triethylamine in 200 mL water, make up to 1 L with MeCN. A:B from 100:0 to 0:100 over 30 min, maintain at 0:100 for 5 min.
**Column temperature:** 30
**Flow rate:** 2
**Detector:** UV 210

## OTHER SUBSTANCES
**Also analyzed:** acepromazine, acetaminophen, acetophenazine, albuterol, aminophylline, amitriptyline, amobarbital, amoxapine, amphetamine, amylocaine, antipyrine, aprobarbital, aspirin, atenolol, atropine, avermectin, barbital, benzocaine, benzoic acid, benzotropine, benzphetamine, berberine, bibucaine, bromazepan, brompheniramine, buprenorphine, buspirone, butabarbital, butacaine, butethal, caffeine, carbamazepine, carbromal, chloramphenicol, chlordiazepoxide, chloroquine, chlorothiazide, chloroxylenol, chlorphenesin, chlorpheniramine, chlorpromazine, chlorpropamide, chlortetracycline, cimetidine, cinchonidine, cinchonine, clenbuterol, clonazepam, clonixin, clorazepate, cocaine, codeine, colchicine, cortisone, coumarin, cyclazocine, cyclobenzaprine, cyclothiazide, cyheptamide, cymarin, danazol, danthron, dapsone, debrisoquine, desipramine, dexamethasone, dextromethorphan, dextropropoxyphene, diamorphine, diazepam, diclofenac, diethylpropion, diethylstilbestrol, diflunisal, digitoxin, diltiazem, diphenhydramine, diphenoxylate, diprenorphine, dipyrone, disulfiram, dopamine, doxapram, doxepin, dronabinol, ephedrine, epinephrine, epinine, estradiol, estriol, estrone, ethacrynic acid, ethosuximide, etonitazene, etorphine, eugenol, famotidine, fenbendazole, fencamfamine, fenoprofen, fenproporex, fentanyl, flubendazole, flufenamic acid, flunitrazepam, 5-fluorouracil, fluoxymesterone, fluphenazine, furosemide, gentisic acid, gitoxigenin, glipizide, glunixin, glutethimide, glybenclamide, guaiacol, halazepam, haloperidol, hydrochlorothiazide, hydrocodone, hydrocortisone, hydromorphone, hydroxyquinoline, ibogaine, ibuprofen, iminostilbene, imipramine, indomethacin, isocarbostyril, isocarboxazid, isoniazid, isoproterenol, isoxsuprine, ivermectin, ketamine, ketoprofen, kynurenic acid, levorphanol, lidocaine, lorazepam, lormetazepam, loxapine, mazindol, mebendazole, meclizine, meclofenamic acid, medazepam, mefenamic acid, megestrol, mepacrine, meperidine, mephen-

termine, mephenytoin, mephesin, mephobarbital, mepivacaine, mescaline, mesoridazine, methadone, methamphetamine, methapyrilene, methaqualone, methazolamide, methocarbamol, methoxamine, methsuximide, methyl salicylate, methyldopa, methyldopamine, methylphenidate, methylprednisolone, methyltestosterone, methyprylon, metoprolol, mibolerone, morphine, nadolol, nalorphine, naloxone, naltrexone, naphazoline, naproxen, nefopam, niacinamide, nicotine, nicotinic acid, nifedipine, niflumic acid, nitrazepam, norepinephrine, nortriptyline, noscapine, nylidrin, oxazepam, oxycodone, oxymorphone, oxyphenbutazone, oxytetracycline, papaverine, pargyline, pemoline, pentazocine, pentobarbital, persantine, phenacetin, phenazocine, phenazopyridine, phencyclidine, phendimetrazine, phenelzine, pheniramine, phenobarbital, phenothiazine, phensuximide, phentermine, phenylbutazone, phenylephrine, phenylpropanolamine, piperocaine, prazepam, prednisolone, primidone, probenecid, progesterone, propiomazine, propranolol, propylparaben, pseudoephedrine, puromycin, pyrilamine, pyrithyldione, quazepam, quinaldic acid, quinidine, quinine, ranitidine, recinnamine, reserpine, resorcinol, saccharin, albuterol, salicylamide, salicylic acid, scopolamine, scopoletin, secobarbital, strychnine, sulfacetamide, sufadiazine, sulfadimethoxine, sulfaethidole, sulfamerazine, sulfamethazine, sulfamethoxizole, sulfanilamide, sulfapyridine, sulfasoxizole, sulindac, tamoxifen, temazepam, testosterone, tetracaine, tetracycline, tetramisole, thebaine, theobromine, theophylline, thiabendazole, thiamine, thiamylal, thiobarbituric acid, thioridazine, thiosalicylic acid, thiothixene, thymol, tolazamide, tolazoline, tobutamide, tolmetin, tranylcypromine, triamcinolone, tribenzylamine, trichloromethiazide, trifluoperazine, trihexyphenidyl, trimethoprim, tripelennamine, triproilidine, tropacocaine, tyramine, verapamil, vincamine, warfarin, yohimbine, zoxazolamine

## REFERENCE

Hill, D.W.; Kind, A.J. Reversed-phase solvent gradient HPLC retention indexes of drugs. *J.Anal.Toxicol.*, **1994**, *18*, 233–242

## SAMPLE

**Matrix:** solutions

## HPLC VARIABLES

**Guard column:** 37 μm ODS
**Column:** 150 × 4.6 3 μm Spherisorb ODS II
**Mobile phase:** MeOH:EtOH:isopropanol:water 52:3:1:45
**Flow rate:** 0.3
**Detector:** F ex 360 (filter) em 425 (filter) following post-column reaction. The column effluent mixed with the reagent and flowed through a 10 m × 0.3 mm i.d. knitted PTFE coil at 79 ± 1° to the detector. The reagent was generated by mixing 1.1 mM hydrogen peroxide in 0.1% ascorbic acid pumped at 0.038 mL/min and concentrated HCl pumped at 0.192 mL/min and allowing this mixture to flow through a 2 m × 0.8 mm i.d. PTFE coil to the point where it mixed with the column effluent.

## CHROMATOGRAM

**Retention time:** 33

## OTHER SUBSTANCES

**Simultaneous:** digitoxigenin, digoxigenin, digoxigenin bisdigitoxoside, digoxigenin monodigitoxoside, dihydrodigoxigenin, dihydrodigoxin, furosemide, spironolactone

## KEY WORDS

post-column reaction

## REFERENCE

Kwong, E.; McErlane, K.M. Development of a high-performance liquid chromatographic assay for digoxin using post-column fluorogenic derivatization. *J.Chromatogr.*, **1986**, *377*, 233–242

## SAMPLE
**Matrix:** urine
**Sample preparation:** 10 mL Urine + 2 mL 1 M HCl (check pH is 1-2), heat 37° for 3 h, add 5 mL pH 6.5 phosphate buffer, add 2 mL 1 M NaOH (check pH is 6.5-7.0). Add to a 20 cm Extrelut SPE column, rinse flask with 3 mL water, add rinsings to column, dry for 15 min, elute with 40 mL dichloromethane, evaporate eluent to dryness, dry over concentrated sulfuric acid. Prepare a 100 mg/mL solution of 4-nitrobenzoyl chloride (4-NBP) in dry pyridine with gentle heating. Use immediately. Dissolve residue from column in 30 μL dry pyridine, add 20 μL 2 mg/mL digitoxigenin in pyridine, add 300 μL 4-NBP solution, shake well. Heat at 70° for 1 h, add 2 mL 5% sodium bicarbonate, shake until precipitate has dissolved, add 2 mL chloroform, shake, centrifuge, repeat extraction twice. Combine chloroform layers, wash three times with 2 mL 1 M HCl, inject an aliquot of chloroform solution directly.

## HPLC VARIABLES
**Column:** 200 × 4 Hibar 5 μm Lichrosorb Si 60
**Mobile phase:** n-Hexane:dichloromethane:methanol 82.9:14.2:2.9
**Flow rate:** 1.2
**Injection volume:** 20
**Detector:** UV 258

## CHROMATOGRAM
**Retention time:** 12
**Internal standard:** digitoxigenin (8)
**Limit of detection:** 1 μg/mL

## KEY WORDS
normal phase; derivatization; SPE; digoxin is hydrolysed to digoxigenin and determined as its 4-NBP derivative

## REFERENCE
Jakobsen, P.; Waldorff, S. Determination of digoxin, digoxigenin and dihydrodigoxigenin in urine by extraction, derivatization and high-performance liquid chromatography. *J.Chromatogr.*, **1986**, *382*, 349–354

## SAMPLE
**Matrix:** urine
**Sample preparation:** 10 mL Urine + 0.5 mL 20 μg/mL digitoxigenin in dichloromethane + 20 mL dichloromethane, shake for 15 min, centrifuge for 20 min. Remove the organic phase and add it to 15 mL 5% sodium bicarbonate, shake for 15 min, centrifuge for 20 min. Remove the organic phase and evaporate it to dryness at 50° under a stream of nitrogen, add 200 μL derivatizing solution to the residue, shake gently at room temperature for 10 min, evaporate to dryness under a stream of nitrogen at 50°, add 2 mL 2 mg/mL 4-dimethylaminopyridine in 5% sodium bicarbonate, shake for 5 min, add 1 mL chloroform, rock on an Aliquot Mixer. Remove the organic phase and add it to 2 mL 5% sodium bicarbonate solution, mix for 2 min. Remove the organic phase and add it to 3 mL 50 mM HCl containing 5% NaCl, mix for 2 min. Remove the organic phase and repeat the acid wash 3 more times, inject a 100 μL aliquot of the organic phase. (The derivatizing solution was 85 mg/mL 3,5-dinitrobenzoyl chloride in pyridine, prepared with gentle warming to help the solid dissolve.)

## HPLC VARIABLES
**Column:** 250 × 4.6 10 μm Partisil 10
**Mobile phase:** Hexane:dichloromethane:MeCN 60:20:20
**Flow rate:** 1.8

**Injection volume:** 100
**Detector:** UV 254

## CHROMATOGRAM
**Retention time:** 54
**Internal standard:** digitoxigenin (17)
**Limit of detection:** 100 ng/mL

## OTHER SUBSTANCES
**Simultaneous:** digoxigenin, digoxigenin bisdigitoxoside, digoxigenin monodigitoxoside

## KEY WORDS
normal phase

## REFERENCE
Bockbrader, H.N.; Reuning, R.H. Digoxin and metabolites in urine: A derivatization-high-performance liquid chromatographic method capable of quantitating individual epimers of dihydrodigoxin. *J.Chromatogr.*, **1984**, *310*, 85–95

## SAMPLE
**Matrix:** urine
**Sample preparation:** Extract 20 mL urine with 20 mL dichloromethane containing 3% heptafluorobutanol for 15 min, centrifuge at 1000 g for 10 min. Remove 15 mL aqueous phase, extract with 15 mL dichloromethane containing 3% heptafluorobutanol for 15 min, centrifuge at 1000 g for 10 min. Combine 10 mL volumes of each organic phase, evaporate under nitrogen to about 0.5 mL, add 20 μL 1-pentanol, evaporate to 20 μL, dissolve residue in 250 μL mobile phase, inject a 100 μL aliquot.

## HPLC VARIABLES
**Column:** 150 × 4.5 5 μm LiChrosorb SI 60
**Mobile phase:** n-Heptane:1-pentanol:MeCN:water 64:26:9:1
**Flow rate:** 1.5
**Injection volume:** 100
**Detector:** UV 220

## CHROMATOGRAM
**Retention time:** 15
**Limit of detection:** 10 ng/mL

## KEY WORDS
normal phase

## REFERENCE
Eriksson, B.-M.; Tekensbergs, L.; Magnusson, J.-O.; Molin, L. Determination of tritiated digoxin and metabolites in urine by liquid chromatography. *J.Chromatogr.*, **1981**, *223*, 401–408

## ANNOTATED BIBLIOGRAPHY
Ikeda, Y.; Fujii, Y.; Nakaya, I.; Yamazaki, M. Quantitative HPLC analysis of cardiac glycosides in Digitalis purpurea leaves. *J.Nat.Prod.*, **1995**, *58*, 897–901

Hui, J.; Geraets, D.R.; Chandrasekaran, A.; Wang, Y.-M.C.; Caldwell, J.H.; Robertson, L.W.; Donnerberg, R.L.; Reuning, R.H. Digoxin disposition in elderly humans with hypochlorhydria. *J.Clin.Pharmacol.*, **1994**, *34*, 734–741 [urine; feces; derivatization; fluorescence detection; normal phase; digitoxin (IS); extracted metabolites; LOD 5 ng/mL]

Oosterkamp, A.J.; Irth, H.; Beth, M.; Unger, K.K.; Tjaden, U.R.; van de Greef, J. Bioanalysis of digoxin and its metabolites using direct serum injection combined with liquid chromatography and on-line immunochemical detection. *J.Chromatogr.B*, **1994**, *653*, 55−61 [LOD 160 pg/mL; serum; column-switching; post-column reaction; fluorescence detection; extracted metabolites; pharmacokinetics]

Ikeda, Y.; Fujii, Y.; Yamazaki, M. Determination of lanatoside C and digoxin in Digitalis lanata by HPLC and its application to analysis of the fermented leaf powder. *J.Nat.Prod.*, **1992**, *55*, 748−752

Nakashima, H.; Tsutsumi, K.; Hashiguchi, M.; Kumagai, Y.; Ebihara, A. Determination of β-methyldigoxin and its metabolites by high-performance liquid chromatography and fluorescence polarization immunoassay. *J.Chromatogr.*, **1989**, *489*, 425−431

Fujii, Y.; Ikeda, Y.; Yamazaki, M. Micro high-performance liquid chromatographic determination of cardiac glycosides in β-methyldigoxin and digoxin tablets. *J.Chromatogr.*, **1988**, *448*, 157−164

Reh, E. Determination of digoxin in serum by on-line immunoadsorptive clean-up high-performance liquid chromatographic separation and fluorescence-reaction detection. *J.Chromatogr.*, **1988**, *433*, 119−130 [serum; column-switching; LOD 300 pg/mL; post-column reaction; fluorescence detection]

Desta, B. Separation of digoxin from dihydrodigoxin and the other metabolites by high-performance liquid chromatography with post-column derivatization. *J.Chromatogr.*, **1987**, *421*, 381−386 [post-column reaction; fluorescence detection; simultaneous metabolites]

Plum, J.; Daldrup, T. Detection of digoxin, digitoxin, their cardioactive metabolites and derivatives by high-performance liquid chromatography and high-performance liquid chromatography-radioimmunoassay. *J.Chromatogr.*, **1986**, *377*, 221−231 [gradient; tissue; extracted metabolites; SPE]

Vetticaden, S.J.; Barr, W.H.; Beightol, L.A. Improved method for assaying digoxin in serum using high-performance liquid chromatography-radioimmunoassay. *J.Chromatogr.*, **1986**, *383*, 187−193 [serum; dog; RIA detection]

de Jong, H.C.; Voogt, W.H.; Bos, P.; Frei, R.W. Tensammetric detection in high performance liquid chromatography. Application to lynestrenol and some cardiac glycosides. *J.Liq.Chromatogr.*, **1983**, *6*, 1745−1758 [also lynestrenol; electrochemical detection]

Gandelman, M.S.; Birks, J.W. Liquid chromatographic detection of cardiac glycosides, saccharides and hydrocortisone based on the photoreduction of 2-tert-butylanthraquinone. *Anal.Chim.Acta*, **1983**, *155*, 159−171 [post-column reaction; simultaneous diginatin; also hydrocortisone; LOD 2 ng]

Wagner, J.G.; Dick, M.; Behrendt, D.M.; Lockwood, G.F.; Sakmar, E.; Hees, P. Determination of myocardial and serum digoxin concentrations in children by specific and nonspecific assay methods. *Clin.Pharmacol.Ther.*, **1983**, *33*, 577−584 [heart; tissue; serum; ethoxzolamide (IS); RIA detection]

Desta, B.; Kwong, E.; McErlane, K.M. Separation of digoxin, digitoxin and their potential metabolites, impurities or degradation products by high-performance liquid chromatography. *J.Chromatogr.*, **1982**, *240*, 137−143

Gault, H.; Kalra, J.; Ahmed, M.; Kepkay, D.; Longerich, L.; Barrowman, J. Influence of gastric pH on digoxin biotransformation. II. Extractable urinary metabolites. *Clin.Pharmacol.Ther.*, **1981**, *29*, 181−190 [urine; radioactivity detection; tritium labeled]

Loo, J.C.K.; McGilveray, I.J.; Jordan, N. The estimation of serum digoxin by combined HPLC separation and radioimmunological assay. *J.Liq.Chromatogr.*, **1981**, *4*, 879−886 [serum; RIA detection; extracted metabolites]

# Diltiazem

**Molecular formula:** $C_{22}H_{26}N_2O_4S$
**Molecular weight:** 414.5
**CAS Registry No.:** 42399-41-7 (diltiazem), 33286-22-5
(diltiazem hydrochloride)

## SAMPLE
**Matrix:** blood
**Sample preparation:** 1 mL Plasma + 200 µL 500 ng/mL 4-methylpropranolol in solvent + 1 mL buffer, mix briefly, add 5 mL MTBE, shake vigorously for 10 min, centrifuge at 2500 g for 10 min. Remove 4 mL of the organic layer and evaporate it to dryness using a vortex evaporator at 30°, reconstitute the residue in 2 mL hexane and 200 µL solvent, vortex for 2 min, discard the upper hexane layer, wash again with 2 mL hexane, inject a 100 µL aliquot of the aqueous phase. (Buffer was 200 mM $K_2HPO_4$ adjusted to pH 10 with 5 M KOH. Solvent was MeCN:MeOH:10 mM pH 2 sulfuric acid 10:45:45 containing 56 mM sodium octanesulfonate. MTBE was stored over activated charcoal and filtered (Whatman No. 2v) immediately before use. Hexane was purified by stirring 4 volumes hexane with 1 volume concentrated sulfuric acid overnight then washing twice with 1 volume water.)

## HPLC VARIABLES
**Guard column:** 20 × 4.6 5 µm Suplex pKb-100 (Supelco)
**Column:** 150 × 4.6 5 µm Suplex pKb-100 (Supelco)
**Mobile phase:** MeCN:MeOH:10 mM pH 2 sulfuric acid 10:45:45 containing 10 mM sodium octanesulfonate
**Flow rate:** 1
**Injection volume:** 200
**Detector:** UV 237

## CHROMATOGRAM
**Retention time:** 10.3
**Internal standard:** 4-methylpropranolol (Wyeth-Ayerst) (17.1)
**Limit of quantitation:** 2 ng/mL

## OTHER SUBSTANCES
**Extracted:** metabolites, quinidine (F ex 247 em 270)

## KEY WORDS
plasma

## REFERENCE
Carignan, G.; Carrier, K.; Laganière, S.; Lessard, M. Simultaneous determination of diltiazem and quinidine in human plasma by liquid chromatography. *J.Chromatogr.B*, **1995**, *672*, 261–269

## SAMPLE
**Matrix:** blood
**Sample preparation:** Make serum alkaline with 10% sodium carbonate, extract with diisopropyl ether (Caution! Diisopropyl ether readily forms explosive peroxides!). Remove the organic layer and extract it with 10 mM HCl, inject an aliquot of the aqueous layer.

## HPLC VARIABLES
**Column:** 150 × 4.6 5 μm Supelcosil LC-CN
**Mobile phase:** MeCN:water:500 mM KH$_2$PO$_4$ 36:62:2
**Flow rate:** 1.8
**Detector:** UV 210

## CHROMATOGRAM
**Limit of detection:** 10 ng/mL

## OTHER SUBSTANCES
**Extracted:** mexiletine, propafenone

## KEY WORDS
serum

## REFERENCE
Kunicki, P.K.; Sitkiewicz, D. High-performance liquid chromatographic determination of some antiar-rhythmic drugs using cyanopropyl derivatized silica phase (Abstract 43). *Ther.Drug Monit.*, **1995**, *17*, 394

## SAMPLE
**Matrix:** blood
**Sample preparation:** 2 mL Whole blood or plasma + 2 mL buffer + 5 mL chloroform: isopropanol:n-heptane 60:14:26, shake gently horizontally for 10 min, centrifuge at 2800 g for 10 min. Remove the lower organic layer and evaporate it to dryness under vacuum at 45°, reconstitute the residue in 100 μL mobile phase, centrifuge at 2800 g for 5 min, inject a 50 μL aliquot of the supernatant. (Buffer was saturated ammonium chloride solution 25% diluted with water, adjusted to pH 9.5 with 25% ammonia solution.)

## HPLC VARIABLES
**Column:** 300 × 3.9 4 μm NovaPack C18
**Mobile phase:** MeOH:THF:buffer 65:5:30 (Buffer was 0.68 g/L (10 mM (sic)) KH$_2$PO$_4$ adjusted to pH 2.6 with concentrated orthophosphoric acid.) (At the end of each session wash the column with water for 1 h and MeOH for 1 h, re-equilibrate for 30 min.)
**Column temperature:** 30
**Flow rate:** 0.8
**Injection volume:** 50
**Detector:** UV 238

## CHROMATOGRAM
**Retention time:** 5.90
**Limit of detection:** <120 ng/mL

## OTHER SUBSTANCES
**Extracted:** acebutolol, acenocoumarol, acepromazine, aceprometazine, acetaminophen, ajmaline, albuterol, alimemazine, alminoprofen, alpidem, alprazolam, alprenolol, amisul-pride, amitriptyline, amodiaquine, amoxapine, aspirin, atenolol, benazepril, benperidol, benzocaine, benzoylecgonine, bepridil, betaxolol, bromazepam, brompheniramine, bumad-izone, bupivacaine, buprenorphine, buspirone, caffeine, carbamazepine, carbinoxamine, carpipramine, carteolol, celiprolol, cetirizine, chlorambucil, chlordiazepoxide, chlormeza-none, chlorophenacinone, chloroquine, chlorpheniramine, chlorpromazine, chlorpropam-ide, cibenzoline, cicletanine, clemastine, clobazam, clomipramine, clonazepam, clonidine, clorazepate, clozapine, cocaine, codeine, colchicine, cyamemazine, cyclizine, cycloguanil, cyproheptadine, cytarabine, dacarbazine, daunorubicin, debrisoquine, demexiptiline, de-sipramine, dextromethorphan, dextromoramide, dextropropoxyphene, diazoxide, diclo-fenac, dihydralazine, diphenhydramine, dipyridamole, disopyramide, dosulepine, doxepin,

doxylamine, droperidol, ephedrine, estazolam, etodolac, fenfluramine, fenoprofen, fentiazac, flecainide, floctafenine, flumazenil, flunitrazepam, fluoxetine, fluphenazine, flurbiprofen, fluvoxamine, glibenclamide, glipizide, glutethimide, histapyrrodine, hydroxychloroquine, hydroxyzine, ibuprofen, imipramine, indomethacin, iproniazid, ketamine,
ketoprofen, labetalol, levomepromazine, lidocaine, lidoflazine, lisinopril, loperamide, loprazolam, loratadine, lorazepam, loxapine, maprotiline, medazepam, mefenamic acid,
mefenidramine, mefloquine, melphalan, meperidine, mephenesin, mephentermine, mepivacaine, metapramine, metformin, methadone, methaqualone, methocarbamol, methotrexate, metipranolol, metoclopramide, metoprolol, mexiletine, midazolam, minoxidil, moclobemide, moperone, morphine, nadolol, nalbuphine, nalorphine, naloxone, naltrexone,
naproxen, nialamide, nifedipine, niflumic acid, nimodipine, nitrazepam, nizatidine, nomifensine, nortriptyline, omeprazole, opipramol, oxazepam, oxprenolol, penbutolol, penfluridol, pentazocine, phencyclidine, phenobarbital, phenol, phenylbutazone, pimozide,
pindolol, pipamperone, piroxicam, prazepam, prazosin, prilocaine, procainamide, procarbazine, proguanil, promethazine, propafenone, propranolol, protriptyline, pyrimethamine,
quinidine, quinine, quinupramine, ranitidine, ritodrine, secobarbital, sotalol, strychnine,
sulfinpyrazole, sulindac, sulpride, sultopride, suriclone, temazepam, tenoxicam, terbutaline, terfenadine, tetracaine, tetrazepam, thiopental, thioproperazine, thioridazine, tianeptine, tiapride, tiaprofenic acid, ticlopidine, timolol, tioclomarol, tofisopam, tolbutamide, toloxatone, trazodone, triazolam, trifluoperazine, trifluperidol, trimipramine,
triprolidine, tropatenine, verapamil, viloxazine, vincristine, vindesine, warfarin, yohimbine, zolpidem, zopiclone, zorubicine
**Interfering:** aconitine, astemizole, bisoprolol, diazepam, diltiazem, glibornuride, haloperidol, medifoxamine, mianserine, nicardipine, nitrendipine, ramipril, reserpine, vinblastine

## KEY WORDS
whole blood; plasma

## REFERENCE
Tracqui, A.; Kintz, P.; Mangin, P. Systematic toxicological analysis using HPLC/DAD. *J.Forensic Sci.*,
**1995**, *40*, 254–262

## SAMPLE
**Matrix:** blood
**Sample preparation:** Wash a Lida 100 mg C18 SPE column with 3 mL MeCN then 3 mL
100 mM ammonium dihydrogen phosphate. 1 mL Plasma + 20 μL 7.5 μg/mL propyldiltiazem in MeOH + 500 μL 100 mM ammonium dihydrogen phosphate, vortex, add to
SPE column, wash with 2 mL MeCN:water 20:80, wash with 1 mL MeCN:water 40:60,
air dry column for 30 s, elute with 500 μL MeCN:100 mM ammonium dihydrogen phosphate 80:20 containing 0.06% triethylamine (final pH 6.8). Evaporate eluate to dryness
under nitrogen at 40-45°, dissolve residue in 250 μL MeCN:50 mM $KH_2PO_4$ pH 2.9 20:
80, inject 100 μL aliquot.

## HPLC VARIABLES
**Guard column:** 20 × 4.6 40 μm Pelliguard LC8
**Column:** 150 × 4.6 5 μm Hypersil C8 BDS
**Mobile phase:** MeCN:50 mM $KH_2PO_4$ adjusted to pH 2.9 with phosphoric acid:triethylam-
ine 398:600:2
**Flow rate:** 1
**Injection volume:** 100
**Detector:** UV 238

## CHROMATOGRAM
**Retention time:** 6
**Internal standard:** propyldiltiazem (8)
**Limit of quantitation:** 2.5 ng/mL

## OTHER SUBSTANCES
**Simultaneous:** metabolites

## KEY WORDS
plasma; SPE

## REFERENCE
Ascalone, V.; Locatelli, M.; Malavasi, B. Determination of diltiazem and its main metabolites in human plasma by automated solid-phase extraction and high-performance liquid chromatography: a new method overcoming instability of the compounds and interference problems. *J.Chromatogr.B*, **1994**, *657*, 133–140

## SAMPLE
**Matrix:** blood
**Sample preparation:** 1 mL Plasma + 5 mL MTBE + 40 μL 10 μg/mL verapamil in MeOH, vortex, centrifuge at 3000 g for 15 min. Remove the supernatant and add it to 80 μL 50 mM sulfuric acid, vortex, centrifuge, inject a 25 μL aliquot of the lower aqueous phase.

## HPLC VARIABLES
**Guard column:** 42 × 3 30-35 μm CO-PELL ODS
**Column:** 250 × 4.6 10 μm Econosil-CN
**Mobile phase:** MeOH:50 mM ammonium dihydrogen phosphate:triethylamine 45:55: 0.25, pH adjusted to 5.0 with 1 M phosphoric acid
**Injection volume:** 25
**Detector:** UV 237

## CHROMATOGRAM
**Internal standard:** verapamil

## KEY WORDS
plasma; pharmacokinetics

## REFERENCE
Bialer, M.; Hadad, S.; Golomb, G.; Barel, S.; Samara, E.; Abu Salach, O.; Berkman, N.; Danenberg, H.D.; Ben David, J.; Caron, D. Pharmacokinetic analysis of two new sustained-release products of diltiazem designed for twice- and once-daily treatment. *Biopharm.Drug Dispos.*, **1994**, *15*, 45–52

## SAMPLE
**Matrix:** blood
**Sample preparation:** 1 mL Serum + 200 μL 1 M NaOH, vortex 30 s, dilute to 5 mL with MeCN, vortex 2 min, centrifuge at 2200 g for 5 min, inject a 100 μL aliquot of supernatant.

## HPLC VARIABLES
**Guard column:** 23 × 3.6 10 μm Waters C8
**Column:** 300 × 4.1 10 μm LiChrosorb RP-8
**Mobile phase:** MeCN:10 mM $Na_2HPO_4$ + 0.1% triethanolamine 40:60, pH adjusted to 3.0 ± 0.1 with 85% phosphoric acid
**Flow rate:** 1.2
**Injection volume:** 100
**Detector:** UV 237

## CHROMATOGRAM
**Retention time:** 7

**Limit of detection:** 2.5 ng/mL
**Limit of quantitation:** 10 ng/mL

## KEY WORDS
serum

## REFERENCE
Chaudhary, R.S.; Gangwal, S.S.; Avachat, M.K.; Shah, Y.N.; Jindal, K.C. Determination of diltiazem hydrochloride in human serum by high-performance liquid chromatography. *J.Chromatogr.*, **1993**, *614*, 261–266

## SAMPLE
**Matrix:** blood
**Sample preparation:** 0.5 mL Plasma + hexane:2-propanol 98:2, stir 15 min, centrifuge 1500 g 10 min. Remove organic layer and evaporate it to dryness under a stream of nitrogen at 50°. Reconstitute residue in 200 μL mobile phase, vortex 1 min, inject 20 μL aliquot.

## HPLC VARIABLES
**Guard column:** 18 × 0.4 5 μm Ultron ES-OVM
**Column:** 150 × 4.6 5 μ m Ultron ES-OVM (ovomucoid chemically bonded to aminopropyl-silica gel)
**Mobile phase:** EtOH:20 mM $KH_2PO_4$ 3:97, pH adjusted to 4.5 with phosphoric acid or KOH
**Flow rate:** 1
**Injection volume:** 20
**Detector:** UV 237

## CHROMATOGRAM
**Retention time:** 3.5 (cis-(+)-diltiazem), 7.5 (cis-(−)-diltiazem)
**Limit of detection:** 64 ng/mL

## KEY WORDS
plasma; chiral; effect of organic modifier and pH on retention time and separation is discussed

## REFERENCE
Rosell, G.; Camacho, A.; Parra, P. Direct enantiomeric separation of *cis*-(±)diltiazem in plasma by high-performance liquid chromatography with ovomucoid column. *J.Chromatogr.*, **1993**, *619*, 87–92

## SAMPLE
**Matrix:** blood
**Sample preparation:** 2 mL Plasma + 100 μL 3 μg/mL propranolol hydrochloride in water + 6 mL MTBE, shake 15 min, centrifuge at 1500 g for 15 min. Remove organic layer and add it to 100 μL 0.05 M sulfuric acid, shake 15 min, centrifuge at 1500 g at 4° for 10 min, discard organic layer, inject 50 μL aliquots of aqueous layer.

## HPLC VARIABLES
**Column:** 100 × 2 5 μm ODS Hypersil
**Mobile phase:** MeCN:10 mM $Na_2HPO_4$ 40:60 containing 40 mM sodium dodecyl sulfate and 3 mM tetrabutylammonium bromide, adjusted to pH 2 with orthophosphoric acid
**Flow rate:** 0.5
**Injection volume:** 50
**Detector:** UV 240

## CHROMATOGRAM
**Retention time:** 6
**Internal standard:** propranolol (10.5)
**Limit of detection:** 1 ng/mL

## OTHER SUBSTANCES
**Simultaneous:** metabolites

## KEY WORDS
plasma; microbore

## REFERENCE
Zoest, A.R.; Hung, C.T.; Wanwimolruk, S. Diltiazem: a sensitive HPLC assay and application to pharmacokinetic study. *J.Liq.Chromatogr.*, **1992**, *15*, 1277−1287

## SAMPLE
**Matrix:** blood
**Sample preparation:** 2 mL Plasma + 100 μL 6 μg/mL imipramine in 10 mM HCl + 200 μL 10% ammonium carbonate (final pH 8.7), vortex gently, add 5 mL MTBE, extract (Vibrax VXR2) for 20 min, centrifuge at 4° at 1720 g for 10 min, remove the organic layer. Add 5 mL dichloromethane to the aqueous layer, shake for 20 min on a reciprocating shaker, centrifuge at 0° at 1720 g for 10 min. Combine the organic layers and evaporate them to dryness under a stream of nitrogen at 50°, reconstitute the residue in 100 μL 10 mM HCl, wash with 2 mL MTBE, wash with 2 mL hexane, inject a 3-20 μL aliquot of the aqueous layer.

## HPLC VARIABLES
**Column:** 250 × 4.6 5 μm Ultrasphere-ODS C18
**Mobile phase:** MeCN:MeOH:40 mM ammonium acetate 24:40:36 containing 0.04% triethylamine, pH adjusted to 7.3 with glacial acetic acid
**Flow rate:** 1.2
**Injection volume:** 3-20
**Detector:** UV 237

## CHROMATOGRAM
**Retention time:** 13.0
**Internal standard:** imipramine (17.9)

## OTHER SUBSTANCES
**Extracted:** metabolites
**Simultaneous:** alprazolam, amitriptyline, desipramine, loxapine, nortriptyline
**Noninterfering:** clomipramine

## KEY WORDS
plasma; pharmacokinetics

## REFERENCE
Yeung, P.K.F.; Montague, T.J.; Tsui, B.; McGregor, C. High-performance liquid chromatographic assay of diltiazem and six of its metabolites in plasma: application to a pharmacokinetic study in healthy volunteers. *J.Pharm.Sci.*, **1989**, *78*, 592−597

## SAMPLE
**Matrix:** blood
**Sample preparation:** 500 μL Serum + 250 μL di-iso-propyl ether:n-butyl alcohol 7:3 containing 800 ng/mL minaprine, centrifuge 2 min, shake, centrifuge 5 min, inject 50 μL aliquot of top organic layer.

## HPLC VARIABLES
**Guard column:** 30 × 4.6 5 µm Brownlee cyano spheri-5
**Column:** 250 × 4.6 5 µm Altex ultrasphere cyano
**Mobile phase:** MeCN:THF:water:2 M ammonium formate (pH 4.0) 700:100:195:5
**Column temperature:** 20
**Flow rate:** 1.5
**Injection volume:** 50
**Detector:** UV 240

## CHROMATOGRAM
**Retention time:** 6
**Internal standard:** minaprine (5.5)
**Limit of detection:** 20 ng/mL

## OTHER SUBSTANCES
**Simultaneous:** amitriptyline, nortriptyline
**Also analyzed:** amiodarone, clomipramine, desipramine, haloperidol, imipramine, propa-
fenone, verapamil

## KEY WORDS
serum

## REFERENCE
Mazzi, G. Simple and practical high-performance liquid chromatographic assay of some tricyclic drugs,
haloperidol, diltiazem, verapamil, propafenone, and amiodarone. *Chromatographia*, **1987**, *24*, 313–
316

## SAMPLE
**Matrix:** blood, tissue
**Sample preparation:** Homogenize liver at 20 mg/mL in 50 mM pH 7.4 Tris-HCl buffer.
200 µL Plasma or 250 µL liver homogenate + 250 ng IS + pH 7.3 ammonium phosphate
buffer + MTBE, extract. Remove the organic layer and add it to 250 µL 50 mM phosphoric
acid, extract. Remove the aqueous layer and add it to 100 µL MeCN, inject an aliquot.

## HPLC VARIABLES
**Column:** 100 × 4.6 3 µm Chromegabond (ES Industries)
**Mobile phase:** MeCN:100 mM sodium perchlorate containing 50 mM phosphoric acid 34:
66
**Column temperature:** 30
**Flow rate:** 1
**Detector:** UV 237

## CHROMATOGRAM
**Retention time:** 4.5
**Internal standard:** N-methyl-N-ethyldiltiazem (9.5)
**Limit of quantitation:** 10 ng/mL

## OTHER SUBSTANCES
**Extracted:** metabolites

## KEY WORDS
rat; plasma; liver; pharmacokinetics

## REFERENCE
Los, L.E.; Welsh, D.A.; Herold, E.G.; Bagdon, W.J.; Zacchei, A.G. Gender differences in toxicokinetics,
liver metabolism, and plasma esterase activity: Observations from a chronic (27-week) toxicity study
of enalapril/diltiazem combinations in rats. *Drug Metab.Dispos.*, **1996**, *24*, 28–33

## SAMPLE
**Matrix:** bulk
**Sample preparation:** Inject a 10 µL aliquot of a 1 mg/mL diltiazem solution containing 1 mg/mL IS.

## HPLC VARIABLES
**Column:** 250 × 4.6 10 µm Chiralcel OF cellulose tris(4-chlorophenylcarbamate)
**Mobile phase:** Hexane:isopropanol 50:50 containing 0.1% diethylamine
**Column temperature:** 30
**Flow rate:** 1
**Injection volume:** 20
**Detector:** UV 254

## CHROMATOGRAM
**Retention time:** 9 ((−)-trans), 11 ((−)-cis), 15 ((+)-trans), 19 ((+)-cis)
**Internal standard:** (+)-cis-5-acetyl-2,3-dihydro-3-hydroxy-2-(p-methoxyphenyl)-1,5-benzo-thiazepin-4(5H)-one (acetylthiazepin) (24)

## KEY WORDS
chiral

## REFERENCE
Ishii, K.; Minato, K.; Nakai, H.; Sato, T. Simultaneous assay of four stereoisomers of diltiazem hydrochloride. Application to in vitro chiral inversion studies. *Chromatographia*, **1995**, *41*, 450–454

## SAMPLE
**Matrix:** formulations
**Sample preparation:** Powder tablets, weigh out powder equivalent to about 50 mg diltiazem, dissolve in 100 mL MeOH, filter. Remove 5 mL aliquot, add 5 mL 0.5 mg/mL cyproheptadine hydrochloride in MeOH, make up to 50 mL with water, inject 50 µL aliquot.

## HPLC VARIABLES
**Column:** 250 × 4.6 Rexchrome ODS
**Mobile phase:** MeCN:MeOH:50 mM $KH_2PO_4$ 25:20:55
**Flow rate:** 2
**Injection volume:** 50
**Detector:** UV 240

## CHROMATOGRAM
**Retention time:** 6
**Internal standard:** cyproheptadine (8)

## KEY WORDS
tablets; stability-indicating

## REFERENCE
Shivram, K.; Shah, A.C.; Newalkar, B.L.; Kamath, B.V. Stability indicating high-performance liquid chromatographic method for the assay of diltiazem hydrochloride in tablets. *J.Liq.Chromatogr.*, **1992**, *15*, 2417–2422

## SAMPLE
**Matrix:** perfusate

## HPLC VARIABLES
**Column:** 100 × 8 4 μm Novapak C18
**Mobile phase:** MeCN:0.092% phosphoric acid containing 0.2% triethylamine 26:74
**Flow rate:** 2
**Detector:** UV 214

## CHROMATOGRAM
**Internal standard:** lidocaine
**Limit of quantitation:** 10 ng/mL

## OTHER SUBSTANCES
**Simultaneous:** metabolites, diphenhydramine
**Also analyzed:** bupivacaine

## KEY WORDS
rat; liver

## REFERENCE
Hussain, M.D.; Tam, Y.K.; Gray, M.R.; Coutts, K.T. Kinetic interactions of lidocaine, diphenhydramine, and verapamil with diltiazem: A study using isolated perfused rat liver. *Drug Metab.Dispos.*, **1994**, *22*, 530–536

## SAMPLE
**Matrix:** solutions

## HPLC VARIABLES
**Column:** 250 × 4.6 5 μm Supelcosil LC-DP (A) or 250 × 4 5 μm LiChrospher 100 RP-8 (B)
**Mobile phase:** MeCN:0.025% phosphoric acid:buffer 25:10:5 (A) or 60:25:15 (B) (Buffer was 9 mL concentrated phosphoric acid and 10 mL triethylamine in 900 mL water, adjust pH to 3.4 with dilute phosphoric acid, make up to 1 L.)
**Flow rate:** 0.6
**Injection volume:** 25
**Detector:** UV 229

## CHROMATOGRAM
**Retention time:** 11.10 (A), 5.42 (B)

## OTHER SUBSTANCES
**Also analyzed:** acebutolol, acepromazine, acetaminophen, acetazolamide, acetophenazine, albuterol, alprazolam, amitriptyline, amobarbital, amoxapine, antipyrine, atenolol, atropine, azatadine, baclofen, benzocaine, bromocriptine, brompheniramine, brotizolam, bupivacaine, buspirone, butabarbital, butalbital, caffeine, carbamazepine, cetirizine, chlorcyclizine, chlordiazepoxide, chlormezanone, chloroquine, chlorpheniramine, chlorpromazine, chlorpropamide, chlorprothixene, chlorthalidone, chlorzoxazone, cimetidine, cisapride, clomipramine, clonazepam, clonidine, clozapine, cocaine, codeine, colchicine, cyclizine, cyclobenzaprine, dantrolene, desipramine, diazepam, diclofenac, diflunisal, diphenhydramine, diphenidol, diphenoxylate, dipyridamole, disopyramide, dobutamine, doxapram, doxepin, droperidol, encainide, ethidium bromide, ethopropazine, fenoprofen, fentanyl, flavoxate, fluoxetine, fluphenazine, flurazepam, flurbiprofen, fluvoxamine, furosemide, glutethimide, glyburide, guaifenesin, haloperidol, homatropine, hydralazine, hydrochlorothiazide, hydrocodone, hydromorphone, hydroxychloroquine, hydroxyzine, ibuprofen, imipramine, indomethacin, ketoconazole, ketoprofen, ketorolac, labetalol, levorphanol, lidocaine, loratadine, lorazepam, lovastatin, loxapine, mazindol, mefenamic acid, meperidine, mephenytoin, mepivacaine, mesoridazine, metaproterenol, methadone, methdilazine, methocarbamol, methotrexate, methotrimeprazine, methoxamine, methyldopa, methylphenidate, metoclopramide, metolazone, metoprolol, metronidazole, mida-

zolam, moclobemide, morphine, nadolol, nalbuphine, naloxone, naphazoline, naproxen, nifedipine, nizatidine, norepinephrine, nortriptyline, oxazepam, oxycodone, oxymetazoline, paroxetine, pemoline, pentazocine, pentobarbital, pentoxifylline, perphenazine, pheniramine, phenobarbital, phenol, phenolphthalein, phentolamine, phenylbutazone, phenyltoloxamine, phenytoin, pimozide, pindolol, piroxicam, pramoxine, prazepam, prazosin, probenecid, procainamide, procaine, prochlorperazine, procyclidine, promazine, promethazine, propafenone, propantheline, propiomazine, propofol, propranolol, protriptyline, quazepam, quinidine, quinine, racemethorphan, ranitidine, remoxipride, risperidone, salicylic acid, scopolamine, secobarbital, sertraline, sotalol, spironolactone, sulfinpyrazone, sulindac, temazepam, terbutaline, terfenadine, tetracaine, theophylline, thiethylperazine, thiopental, thioridazine, thiothixene, timolol, tocainide, tolbutamide, tolmetin, trazodone, triamterene, triazolam, trifluoperazine, triflupromazine, trimeprazine, trimethoprim, trimipramine, verapamil, warfarin, xylometazoline, yohimbine, zopiclone

## KEY WORDS
some details of plasma extraction

## REFERENCE
Koves, E.M. Use of high-performance liquid chromatography-diode array detection in forensic toxicology. *J.Chromatogr.A*, **1995**, *692*, 103–119

## SAMPLE
**Matrix:** solutions

## HPLC VARIABLES
**Column:** 250 × 4.6 Zorbax RX
**Mobile phase:** Gradient. A was 10 mL concentrated orthophosphoric acid and 7 mL triethylamine in 1 L water. B was 10 mL concentrated orthophosphoric acid and 7 mL triethylamine in 200 mL water, make up to 1 L with MeCN. A:B from 100:0 to 0:100 over 30 min, maintain at 0:100 for 5 min.
**Column temperature:** 30
**Flow rate:** 2
**Detector:** UV 210

## OTHER SUBSTANCES
**Also analyzed:** acepromazine, acetaminophen, acetophenazine, albuterol, aminophylline, amitriptyline, amobarbital, amoxapine, amphetamine, amylocaine, antipyrine, aprobarbital, aspirin, atenolol, atropine, avermectin, barbital, benzocaine, benzoic acid, benzotropine, benzphetamine, berberine, bibucaine, bromazepan, brompheniramine, buprenorphine, buspirone, butabarbital, butacaine, butethal, caffeine, carbamazepine, carbromal, chloramphenicol, chlordiazepoxide, chloroquine, chlorothiazide, chloroxylenol, chlorphenesin, chlorpheniramine, chlorpromazine, chlorpropamide, chlortetracycline, cimetidine, cinchonidine, cinchonine, clenbuterol, clonazepam, clonixin, clorazepate, cocaine, codeine, colchicine, cortisone, coumarin, cyclazocine, cyclobenzaprine, cyclothiazide, cyheptamide, cymarin, danazol, danthron, dapsone, debrisoquine, desipramine, dexamethasone, dextromethorphan, dextropropoxyphene, diamorphine, diazepam, diclofenac, diethylpropion, diethylstilbestrol, diflunisal, digitoxin, digoxin, diphenhydramine, diphenoxylate, diprenorphine, dipyrone, disulfiram, dopamine, doxapram, doxepin, dronabinol, ephedrine, epinephrine, epinine, estradiol, estriol, estrone, ethacrynic acid, ethosuximide, etonitazene, etorphine, eugenol, famotidine, fenbendazole, fencamfamine, fenoprofen, fenproporex, fentanyl, flubendazole, flufenamic acid, flunitrazepam, 5-fluorouracil, fluoxymesterone, fluphenazine, furosemide, gentisic acid, gitoxigenin, glipizide, glunixin, glutethimide, glybenclamide, guaiacol, halazepam, haloperidol, hydrochlorothiazide, hydrocodone, hydrocortisone, hydromorphone, hydroxyquinoline, ibogaine, ibuprofen, iminostilbene, imipramine, indomethacin, isocarbostyril, isocarboxazid, isoniazid, isoproterenol, isoxsuprine, ivermectin, ketamine, ketoprofen, kynurenic acid, levorphanol, lidocaine, lorazepam, lormetazepam, loxapine, mazindol, mebendazole, meclizine, meclofenamic acid,

medazepam, mefenamic acid, megestrol, mepacrine, meperidine, mephentermine, mephenytoin, mephesin, mephobarbital, mepivacaine, mescaline, mesoridazine, methadone, methamphetamine, methapyrilene, methaqualone, methazolamide, methocarbamol, methoxamine, methsuximide, methyl salicylate, methyldopa, methyldopamine, methylphenidate, methylprednisolone, methyltestosterone, methyprylon, metoprolol, mibolerone, morphine, nadolol, nalorphine, naloxone, naltrexone, naphazoline, naproxen, nefopam, niacinamide, nicotine, nicotinic acid, nifedipine, niflumic acid, nitrazepam, norepinephrine, nortriptyline, noscapine, nylidrin, oxazepam, oxycodone, oxymorphone, oxyphenbutazone, oxytetracycline, papaverine, pargyline, pemoline, pentazocine, pentobarbital, persantine, phenacetin, phenazocine, phenazopyridine, phencyclidine, phendimetrazine, phenelzine, pheniramine, phenobarbital, phenothiazine, phensuximide, phentermine, phenylbutazone, phenylephrine, phenylpropanolamine, piperocaine, prazepam, prednisolone, primidone, probenecid, progesterone, propiomazine, propranolol, propylparaben, pseudoephedrine, puromycin, pyrilamine, pyrithyldione, quazepam, quinaldic acid, quinidine, quinine, ranitidine, recinnamine, reserpine, resorcinol, saccharin, albuterol, salicylamide, salicylic acid, scopolamine, scopoletin, secobarbital, strychnine, sulfacetamide, sufadiazine, sulfadimethoxine, sulfaethidole, sulfamerazine, sulfamethazine, sulfamethoxizole, sulfanilamide, sulfapyridine, sulfasoxizole, sulindac, tamoxifen, temazepam, testosterone, tetracaine, tetracycline, tetramisole, thebaine, theobromine, theophylline, thiabendazole, thiamine, thiamylal, thiobarbituric acid, thioridazine, thiosalicylic acid, thiothixene, thymol, tolazamide, tolazoline, tobutamide, tolmetin, tranylcypromine, triamcinolone, tribenzylamine, trichloromethiazide, trifluoperazine, trihexyphenidyl, trimethoprim, tripelennamine, triproilidine, tropacocaine, tyramine, verapamil, vincamine, warfarin, yohimbine, zoxazolamine

## REFERENCE

Hill, D.W.; Kind, A.J. Reversed-phase solvent gradient HPLC retention indexes of drugs. *J.Anal.Toxicol.*, **1994**, *18*, 233–242

## SAMPLE

**Matrix:** solutions
**Sample preparation:** Prepare a 50 µg/mL solution in MeCN:water 40:60, inject an aliquot.

## HPLC VARIABLES

**Column:** 250 × 4.6 3 µm silica (Phenomenex)
**Mobile phase:** MeCN:6.25 mM pH 3.0 phosphate buffer 40:60
**Flow rate:** 1
**Injection volume:** 50
**Detector:** UV 254

## CHROMATOGRAM

**Retention time:** 9.08

## OTHER SUBSTANCES

**Also analyzed:** atenolol, clonidine, metoprolol, nifedipine, prazosin, propranolol, verapamil

## REFERENCE

Simmons, B.R.; Stewart, J.T. HPLC separation of selected cardiovascular agents on underivatized silica using an aqueous organic mobile phase. *J.Liq.Chromatogr.*, **1994**, *17*, 2675–2690

## SAMPLE

**Matrix:** solutions

## HPLC VARIABLES

**Column:** 250 × 4.6 cellulose tris(3,5-dimethylphenylcarbamate)

**Mobile phase:** Hexane:isopropanol 90:10
**Flow rate:** 0.5
**Detector:** UV

## CHROMATOGRAM
**Retention time:** 21 (−), 28 (+)

## KEY WORDS
chiral

## REFERENCE
Okamoto, Y.; Aburatani, R.; Hatano, K.; Hatada, K. Optical resolution of racemic drugs by chiral HPLC on cellulose and amylose tris(phenylcarbamate) derivatives. *J.Liq.Chromatogr.*, **1988**, *11*, 2147–2163

◆ ━━━━━━━━━━━━━━━━━ ◆ ━━━━━━━━━━━━━━━━━ ◆

## ANNOTATED BIBLIOGRAPHY

Higashidate, S.; Imai, K.; Prados, P.; Adachi-Akahane, S.; Nagao, T. Relations between blood pressure and plasma norepinephrine concentrations after administration of diltiazem to rats: HPLC-peroxyoxalate chemiluminescence determination on an individual basis. *Biomed.Chromatogr.*, **1994**, *8*, 19–21 [plasma; extracted diltiazem, dopamine, epinephrine, norepinephrine; rat; chemiluminescence detection;SPE]

Hussain, M.D.; Tam, Y.K.; Gray, M.R.; Coutts, R.T. Mechanisms of time-dependent kinetics of diltiazem in the isolated perfused rat liver. *Drug Metab.Dispos.*, **1994**, *22*, 36–42 [perfusate; extracted metabolites; LOQ 30 nM; bupivacaine (IS)]

Ishii, K.; Minato, K.; Nishimura, N.; Miyamoto, T.; Sato, T. Direct chromatographic resolution of four optical isomers of diltiazem hydrochloride on a Chiralcel OF column. *J.Chromatogr.*, **1994**, *686*, 93–100 [chiral; column temp 10-40]

Rutledge, D.R.; Abadi, A.H.; Lopez, L.M. Liquid chromatographic determination of celiprolol, diltiazem, desmethyldiltiazem and deacetyldiltiazem in plasma using a short alkyl chain silanol deactivated column. *J.Pharm.Biomed.Anal.*, **1994**, *12*, 135–140 [extracted metabolites, celiprolol; LOD 3 ng/mL]

Sigusch, H.; Henschel, L.; Kraul, H.; Merkel, U.; Hoffmann, A. Lack of effect of grapefruit juice on diltiazem bioavailability in normal subjects. *Pharmazie*, **1994**, *49*, 675–679 [pharmacokinetics; SPE; extracted metabolites; flurazepam (IS)]

Rutledge, D.R.; Abadi, A.H.; Lopez, L.M.; Beaudreau, C.A. High-performance liquid chromatographic determination of diltiazem and two of its metabolites in plasma using a short alkyl chain silanol deactivated column. *J.Chromatogr.*, **1993**, *615*, 111–116 [plasma; extracted metabolites; imipramine IS; LOD 4 ng/mL; pharmacokinetics; interfering theophylline; simultaneous desipramine, propranolol, verapamil; non-interfering aspirin, atenolol, caffeine, ibuprofen, lidocaine, metoprolol, nifedipine]

Hubert, P.; Chiap, P.; Crommen, J. Automatic determination of diltiazem and deacetyldiltiazem in human plasma using liquid-solid extraction on disposable cartridges coupled to HPLC–Part I: optimization of the HPLC system and method validation. *J.Pharm.Biomed.Anal.*, **1991**, *9*, 877–882

Hubert, P.; Chiap, P.; Crommen, J. Automatic determination of diltiazem and deacetyldiltiazem in human plasma using liquid-solid extraction on disposable cartridges coupled to HPLC–Part II: optimization of liquid- solid extraction. *J.Pharm.Biomed.Anal.*, **1991**, *9*, 883–887

Ishii, K.; Banno, K.; Miyamoto, T.; Kakimoto, T. Determination of diltiazem hydrochloride enantiomers in dog plasma using chiral stationary-phase liquid chromatography. *J.Chromatogr.*, **1991**, *564*, 338–345 [dog; plasma; chiral; achiral; extracted metabolites; column temp 40]

Jensen, B.H.; Larsen, C. Quantitation of diltiazem in human plasma by HPLC using an end-capped reversed-phase column. *Acta Pharm.Nord.*, **1991**, *3*, 179–180

Leneveu, A.; Stheneur, A.; Bousquet, A.; Roux, A. Automated high-performance liquid chromatographic technique for determining diltiazem and its three main metabolites in serum. *J.Liq.Chromatogr.*, **1991**, *14*, 3519–3530 [serum; extracted metabolites; SPE; pharmacokinetics; LOD 2.5 ng/mL]

Bonnefous, J.L.; Boulieu, R. Comparison of solid-phase extraction and liquid-liquid extraction methods for liquid chromatographic determination of diltiazem and its metabolites in plasma.

*J.Liq.Chromatogr.*, **1990**, *13*, 3799–3807 [extracted metabolites; plasma; SPE; propionyldeacetyldiltiazem (IS); LOD 5 ng/mL]

Boulieu, R.; Bonnefous, J.L.; Ferry, S. Determination of diltiazem and its metabolites in plasma by high performance liquid chromatography. *J.Liq.Chromatogr.*, **1990**, *13*, 291–301 [plasma; extracted metabolites; propionyldeacetyldiltiazem (IS); LOD 5 ng/mL]

Boulieu, R.; Bonnefous, J.L.; Ferry, S. Solid-phase extraction of diltiazem and its metabolites from plasma prior to high-performance liquid chromatography. *J.Chromatogr.*, **1990**, *528*, 542–546 [plasma; propionyldeacetyldiltiazem (IS); SPE; extracted metabolites; LOD 0.3 ng; non-interfering diazepam, flunitrazepam, midazolam, nifedipine, pancuronium bromide, procainamide, propranolol, quinidine, verapamil]

Johnson, K.E.; Pieper, J.A. An HPLC method for the determination of diltiazem and three of its metabolites in serum. *J.Liq.Chromatogr.*, **1990**, *13*, 951–960 [extracted metabolites; serum; doxepin (IS); LOD 3 ng/mL; non-interfering carbamazepine, chlorpromazine, gallopamil, imipramine, lidocaine, prochlorperazine, quinidine, thioridazine, trimeprazine; pharmacokinetics]

Parissi-Poulou, M.; Ismailos, G.; Macheras, P. Modified HPLC analysis of diltiazem in plasma for pharmacokinetic studies. *Int.J.Pharm.*, **1990**, *62*, R13–R16

Shah, Y.; Khanna, S.; Dighe, V.S.; Jindal, K.C. High-performance liquid chromatographic determination of diltiazem hydrochloride in tablets. *Indian Drugs*, **1990**, *27*, 363–364

Ververs, F.F.T.; Schaefer, H.G.; Lefevre, J.F.; Lopez, L.M.; Derendorf, H. Simultaneous assay of propranolol, diltiazem and metabolites of diltiazem in human plasma by liquid chromatography. *J.Pharm.Biomed.Anal.*, **1990**, *8*, 535–539

Yamahara, H.; Suzuki, T.; Mizobe, M.; Noda, K.; Samejima, M. In situ perfusion system for oral mucosal absorption in dogs. *J.Pharm.Sci.*, **1990**, *79*, 963–967 [perfusate]

Ascalone, V.; Flaminio, L. Automated high-performance liquid chromatography with column switching for on-line clean-up and analysis of diltiazem and metabolites in human plasma. *J.Chromatogr.*, **1989**, *495*, 358–360 [plasma; column-switching; extracted metabolites; LOD 2 ng/mL; improved version of method in J.Chromatogr. 1987, 423, 239]

Boucher, S.; Varin, F.; Theoret, Y.; Du Souich, P.; Caille, G. High-performance liquid chromatographic method for the determination of diltiazem and two of its metabolites in human plasma: application to a new sustained release formulation. *J.Pharm.Biomed.Anal.*, **1989**, *7*, 1925–1930

Caille, G.; Dube, L.M.; Theoret, Y.; Varin, F.; Mousseau, N.; McGilveray, I.J. Stability study of diltiazem and two of its metabolites using a high performance liquid chromatographic method. *Biopharm.Drug Dispos.*, **1989**, *10*, 107–114

Lacroix, P.M.; Beaulieu, N.; Cyr, T.D.; Lovering, E.G. High-performance liquid chromatography method for assay of diltiazem hydrochloride and its related compounds in bulk drug and finished tablets. *J.Pharm.Sci.*, **1989**, *78*, 243–246

Rustum, A.M. Determination of diltiazem in human whole blood and plasma by high-performance liquid chromatography using a polymeric reversed-phase column and utilizing a salting-out extraction procedure. *J.Chromatogr.*, **1989**, *490*:, 365–375

Zhao, H.; Chow, M.S.S. Analysis of diltiazem and desacetyldiltiazem in plasma using modified high-performance liquid chromatography: improved sensitivity and reproducibility. *Pharm.Res.*, **1989**, *6*, 428–430

Dube, L.M.; Mousseau, N.; McGilveray, I.J. High-performance liquid chromatographic determination of diltiazem and four of its metabolites in plasma: evaluation of their stability. *J.Chromatogr.*, **1988**, *430*, 103–111

Ascalone, V.; Dal Bo', L. Automated high-performance liquid chromatographic and column-switching technique for on-line clean-up and analysis of diltiazem in human plasma. *J.Chromatogr.*, **1987**, *423*, 239–249

Bhamra, R.K.; Ward, A.E.; Holt, D.W. HPLC measurement of diltiazem and desacetyldiltiazem in serum or plasma. *Biomed.Chromatogr.*, **1987**, *2*, 180–182

Hoglund, P.; Nilsson, L.G. Liquid chromatographic determination of diltiazem and its metabolites using trans isomers as internal standards, with dynamic modification of the solid phase by addition of an amine to the mobile phase. *J.Chromatogr.*, **1987**, *414*, 109–120

Johnson, S.M.; Wahba Khalil, S.K. An HPLC method for the determination of diltiazem and desacetyldiltiazem in human plasma. *J.Liq.Chromatogr.*, **1987**, *10*, 673–685 [plasma; extracted desacetyldiltiazem; diazepam (IS); LOD 2 ng/mL; non-interfering atenolol, chlorthalidone, furosemide, hydralazine, methyldopa, pentoxifylline; also captopril, chlorothiazide, dipyridamole, disopyramide, isosorbide dinitrate, labetalol, metoprolol, pindolol, procainamide, propranolol, quinidine, warfarin]

Montamat, S.C.; Abernethy, D.R.; Mitchell, J.R. High-performance liquid chromatographic determination of diltiazem and its major metabolites, N-monodemethyldiltiazem and desacetyldiltiazem, in plasma. *J.Chromatogr.*, **1987**, *415*, 203–207

Kinney, C.D.; Kelly, J.G. Estimation of concentrations of diltiazem in plasma using normal-phase column liquid chromatography with ultraviolet detection. *J.Chromatogr.*, **1986**, *382*, 377–381

Abernethy, D.R.; Schwartz, J.B.; Todd, E.L. Diltiazem and desacetyldiltiazem analysis in human plasma using high-performance liquid chromatography: improved sensitivity without derivation. *J.Chromatogr.*, **1985**, *342*, 216–220

Goebel, K.J.; Kolle, E.U. High-performance liquid chromatographic determination of diltiazem and four of its metabolites in plasma. Application to pharmacokinetics. *J.Chromatogr.*, **1985**, *345*, 355–363

Clozel, J.P.; Caille, G.; Taeymans, Y.; Theroux, P.; Biron, P.; Trudel, F. High-performance liquid chromatographic determination of diltiazem and six of its metabolites in human urine. *J.Pharm.Sci.*, **1984**, *73*, 771–773

Wiens, R.E.; Runser, D.J.; Lacz, J.P.; Dimmitt, D.C. Quantitation of diltiazem and desacetyldiltiazem in dog plasma by high-performance liquid chromatography. *J.Pharm.Sci.*, **1984**, *73*, 688–689

Verghese, C.; Smith, M.S.; Aanonsen, L.; Pritchett, E.L.; Shand, D.G. High-performance liquid chromatographic analysis of diltiazem and its metabolite in plasma. *J.Chromatogr.*, **1983**, *272*, 149–155

Hussain, M.D.; Tam, Y.K.; Finegan, B.A.; Coutts, R.T. Simple and sensitive high-performance liquid chromatographic method for the determination of diltiazem and six of its metabolites in human plasma. *J.Chromatogr.*, **1992**, *582*, 203–209 [plasma; benzylamphetamine (IS); extracted metabolites; pharmacokinetics; LOQ 5 ng/mL; non-interfering bupivacaine, diphenhydramine, lidocaine, metoprolol]

# Doxazosin

**Molecular formula:** $C_{23}H_{25}N_5O_5$
**Molecular weight:** 451.5
**CAS Registry No.:** 74191-85-8
(doxazosin),
77883-43-3 (doxazosin mesylate)

## SAMPLE
**Matrix:** blood
**Sample preparation:** Condition a Bond Elut C18 SPE cartridge with 1 mL MeOH and 1 mL water. 1 mL Plasma + 50 μL 1 μg/mL propranolol in MeOH + 250 μL MeOH, agitate briefly, let stand for 10 min, centrifuge at 3000 g for 10 min. Add the supernatant to the SPE cartridge, wash with 1 mL MeOH:water 30:70, wash with 1 mL water, elute with 1 mL MeOH:acetic acid 99.5:0.5. Evaporate the eluate to dryness under a stream of air, reconstitute the residue in 150 μL mobile phase, inject an aliquot.

## HPLC VARIABLES
**Column:** 150 × 4.6 5 μm Zorbax CN
**Mobile phase:** MeOH:buffer 50:50 (Buffer was 10 mM perchloric acid and 1.8 mM sodium heptanesulfonate.)
**Flow rate:** 1
**Detector:** F ex 245 em bandpass 320-390 (Corning 7-60 filter)

## CHROMATOGRAM
**Retention time:** 5
**Internal standard:** propranolol (3.5)
**Limit of detection:** 1 ng/mL

## KEY WORDS
plasma; pharmacokinetics; SPE

## REFERENCE
Jackman, G.P.; Colagrande, F.; Louis, W.J. Validation of a solid-phase extraction high-performance liquid chromatographic assay for doxazosin. *J.Chromatogr.*, **1991**, *566*, 234–238

## SAMPLE
**Matrix:** blood
**Sample preparation:** 250 μL Serum + 250 μL 20 ng/mL prazosin in MeOH:water 30:70, vortex for 30 s, add 1.5 mL ethyl acetate, vortex for 15 s, centrifuge at 3000 rpm for 2 min. Remove the organic layer and evaporate it to dryness in a vortex evaporator at 40° for 10 min, reconstitute the residue in 200 μL MeOH:water 50:50, inject a 50 μL aliquot.

## HPLC VARIABLES
**Guard column:** 50 × 3.9 40 μm glass beads
**Column:** 150 × 4.6 alumina-based reversed-phase gamma RP-1 (ES Industries)
**Mobile phase:** MeCN:MeOH:25 mM pH 10.9 sodium carbonate 15:15:70 (At the end of each day flush with about 75 mL MeOH:water 50:50 until pH of effluent is neutral.)
**Flow rate:** 2
**Injection volume:** 50
**Detector:** F ex 246 em 389 (cutoff filter)

## CHROMATOGRAM
**Retention time:** 4.9

**Internal standard:** prazosin (1.8)
**Limit of quantitation:** 0.5 ng/mL

## KEY WORDS
serum

## REFERENCE
Fouda, H.G.; Twomey, T.M.; Schneider, R.P. Liquid chromatographic analysis of doxazosin in human serum with manual and robotic sample preparation. *J.Chromatogr.Sci.*, **1988**, *26*, 570–573

## SAMPLE
**Matrix:** blood
**Sample preparation:** 1 mL Whole blood + 5 mL diethyl ether, shake for 10 min, centrifuge at 2000 rpm for 5 min, freeze in acetone/dry ice. Remove the organic layer and add it to 100 µL 50 mM sulfuric acid, shake for 10 min, centrifuge at 2000 rpm for 5 min, inject a 20 µL aliquot of the aqueous layer.

## HPLC VARIABLES
**Column:** 250 × 4 5 µm Spherisorb ODS
**Mobile phase:** MeOH:water 55:45 containing 10 mM pentane sodium sulfate and 9 mM tetramethylammonium chloride, adjusted to pH 3.4 with glacial acetic acid
**Flow rate:** 1.8
**Injection volume:** 20
**Detector:** F ex 254 em 400 (cut-off filter)

## CHROMATOGRAM
**Retention time:** 9
**Internal standard:** doxazosin

## OTHER SUBSTANCES
**Extracted:** trimazosin

## KEY WORDS
whole blood; doxazosin is IS

## REFERENCE
Hughes, M.A.; Meredith, P.A.; Elliott, H.L. The determination of trimazosin and its metabolite CP23445 in whole blood by high performance liquid chromatography using fluorescence detection. *J.Pharmacol.Methods*, **1984**, *12*, 29–34

## SAMPLE
**Matrix:** solutions

## HPLC VARIABLES
**Column:** 150 × 4.6 12 µm 1-myristoyl-2-[(13-carboxyl)-tridecoyl]-sn-3-glycerophosphocholine chemically bonded to silica (Regis)
**Mobile phase:** MeCN:100 mM pH 7.0 phosphate buffer 20:80
**Flow rate:** 1
**Detector:** UV 254

## CHROMATOGRAM
**Retention time:** k' 96.16

## OTHER SUBSTANCES
**Also analyzed:** acebutolol, alprenolol, antazoline, atenolol, betaxolol, bisoprolol, bopindolol, bupranolol, carteolol, celiprolol, chloropyramine, chlorpheniramine, cicloprolol, cimetidine, cinnarizine, cirazoline, clonidine, dilevalol, dimethindene, diphenhydramine, esmolol, famotidine, isothipendyl, ketotifen, metiamide, metoprolol, moxonidine, nadolol, naphazoline, nifenalol, nizatidine, oxprenolol, pheniramine, phentolamine, pindolol, pizotyline (pizotifen), practolol, prazosin, promethazine, propranolol, pyrilamine (mepyramine), ranitidine, roxatidine, sotalol, tiamenidine, timolol, tramazoline, tripelennamine, triprolidine, tymazoline, UK-14,304

## REFERENCE
Kaliszan, R.; Nasal, A.; Turowski, M. Binding site for basic drugs on $\alpha_1$-acid glycoprotein as revealed by chemometric analysis of biochromatographic data. *Biomed.Chromatogr.*, **1995**, *9*, 211–215

## SAMPLE
**Matrix:** solutions

## HPLC VARIABLES
**Column:** 125 × 5 5 μm Spherisorb C8
**Mobile phase:** MeCN:water 25:45 containing 5 mM dibutylamine
**Flow rate:** 2
**Detector:** F ex 346 em 340 (filter)

## REFERENCE
Ferry, D.G.; Caplan, N.B.; Cubeddu, L.X. Interaction between antidepressants and alpha 1-adrenergic receptor antagonists on the binding to alpha 1-acid glycoprotein. *J.Pharm.Sci.*, **1986**, *75*, 146–149

# Doxycycline

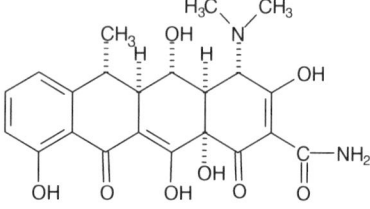

**Molecular formula:** $C_{22}H_{24}N_2O_8$
**Molecular weight:** 444.4
**CAS Registry No.:** 564-25-0, 17086-28-1 (monohydrate),
24390-14-5 (HCl monohydrate), 83038-87-3 (fosfatex),
24390-14-5 (hyclate)

## SAMPLE
**Matrix:** blood
**Sample preparation:** 100 μL Serum + 50 μL 6% aqueous ascorbic acid + 50 ng demeclo-cycline in MeOH + 400 μL buffer, vortex 30 s, add 3 mL ethyl acetate, vortex 5 min, centrifuge at 3000 rpm for 6 min. Remove organic layer and add it to 100 μL 0.2% ascorbic acid in MeOH. Evaporate to dryness at 20° in a vortex evaporator, dissolve residue in 100 μL mobile phase, inject entire amount. (Buffer was 2 M $NaH_2PO_4$ and 2 M $Na_2SO_3$, pH 6.1.)

## HPLC VARIABLES
**Guard column:** 4 μm Nova-Pak C18 Guard-Pak
**Column:** 150 × 4.6 5 μm Ultrabase C18
**Mobile phase:** MeCN:water adjusted to pH 2.5 with phosphoric acid 28:72
**Flow rate:** 1
**Injection volume:** 100
**Detector:** UV 350

## CHROMATOGRAM
**Retention time:** 4.2
**Internal standard:** demeclocycline (2.7)
**Limit of quantitation:** 20 ng/mL

## KEY WORDS
serum

## REFERENCE
Gastearena, I.; Dios-Viéitez, M.C.; Segura, E.; Goñi, M.M.; Renedo, M.J.; Fos, D. Determination of dox-ycycline in small serum samples by liquid chromatography. Application to pharmacokinetical studies on small laboratory animals. *Chromatographia*, **1993**, *35*, 524–526

## SAMPLE
**Matrix:** blood
**Sample preparation:** 100 μL Plasma + 20 μL trifluoroacetic acid, mix 30 s in a whirl mixer, centrifuge at 5400 g for 5 min, inject supernatant (80 μL).

## HPLC VARIABLES
**Guard column:** 10 μm Waters RP phenyl
**Column:** 125 × 4.6 10 μm Waters RP phenyl
**Mobile phase:** MeCN:10 mM phosphoric acid 30:70
**Flow rate:** 2
**Injection volume:** 80
**Detector:** UV 270

## CHROMATOGRAM
**Retention time:** 2.2
**Limit of detection:** 15 ng/mL

## KEY WORDS
plasma

## REFERENCE
Krämer-Horaczynska, F. High-performance liquid chromatographic procedures for the quantitative analysis of 15 tetracycline derivatives in small blood samples. *J.Chromatogr.Sci.*, **1991**, *29*, 107–113

## SAMPLE
**Matrix:** blood, urine
**Sample preparation:** Serum. 0.5 mL Serum + 0.5 mL MeCN:85% phosphoric acid:water 20:2:78, vortex, filter (10 000 or 30 000 Da cutoff) by centrifuging at 2200 g for 30 min, inject 10 μL aliquot of filtrate. Urine. Dilute urine 5 to 10 times with MeCN:85% phosphoric acid:water 20:1.7:78.3, vortex, filter (10 000 or 30 000 Da cutoff) by centrifuging at 2200 g for 30 min, inject 10 μL aliquot of filtrate.

## HPLC VARIABLES
**Column:** 220 × 4.6 phenyl
**Mobile phase:** MeOH:MeCN:triethylamine:phosphoric acid:80 mM pH 2.4 sodium phosphate buffer 10:1.5:0.5:1.7:86.3
**Column temperature:** 50
**Flow rate:** 0.6-0.8
**Injection volume:** 10
**Detector:** UV 268; UV 345

## CHROMATOGRAM
**Retention time:** 8
**Limit of detection:** <10 ng/mL

## KEY WORDS
serum; cow

## REFERENCE
Riond, J.L.; Hedeen, K.M.; Tyczkowska, K.; Riviere, J.E. Determination of doxycycline in bovine tissues and body fluids by high-performance liquid chromatography using photodiode array ultraviolet-visible detection. *J.Pharm.Sci.*, **1989**, *78*, 44–47

## SAMPLE
**Matrix:** blood, urine
**Sample preparation:** Serum. Condition a Bond-Elut C18 SPE cartridge with 1 volume MeOH and 2 volumes water. 1 mL Serum + 5 mL buffer, add to SPE cartridge, wash with 10 mL water, elute with 10 mL 10 mM phosphoric acid in MeCN. Evaporate eluate to dryness at 50° under a stream of nitrogen and resuspend residue in 1 mL water. Centrifuge at 10000 g for 1 min, inject 100 μL aliquot. Urine. Activate a Bond-Elut C18 cartridge with 1 volume MeOH and 2 volumes water. 1 mL Urine + 5 mL buffer, add to cartridge, wash with 10 mL MeCN, elute with 10 mL 10 mM phosphoric acid in MeCN. Evaporate eluate to dryness at 50° under a stream of nitrogen and resuspend residue in 1 mL water. Centrifuge at 10 000 g for 1 min, inject 100 μL aliquot. (Buffer was 0.1 M citric acid:0.2 M $Na_2HPO_4$ 61.4:38.6 (McIlvaines buffer) containing 0.1 M disodium EDTA.)

## HPLC VARIABLES
**Guard column:** LiChrosorb RP-18
**Column:** 150 × 3.9 4 μm Nova-Pak C18
**Mobile phase:** MeCN:acetic acid:100 mM $KH_2PO_4$ 75:150:125 (serum) or 65:150:125 (urine)
**Flow rate:** 1

**Injection volume:** 100
**Detector:** UV 340

## CHROMATOGRAM
**Retention time:** 4
**Limit of detection:** 25 ng/mL

## KEY WORDS
serum; SPE; protect from light with amber glassware

## REFERENCE
Sheridan, M.E.; Clarke, G.S. Improved high-performance liquid chromatographic determination of doxycycline in serum and urine using solid-phase extraction columns. *J.Chromatogr.*, **1988**, *434*, 253–258

## SAMPLE
**Matrix:** blood, urine
**Sample preparation:** Serum. 500 μL Serum + 50 μL 6% ascorbic acid in water + 50 μL demeclocycline in MeOH/100 mM HCl + 1 mL buffer, mix for 30 s, add 6 mL ethyl acetate, rotate for 10 min, centrifuge at 3000 rpm for 6 min. Remove the organic layer and add it to 100 μL 0.2% ascorbic acid in MeOH, evaporate to dryness under vacuum while vortexing, reconstitute the residue in 200 μL mobile phase, mix, filter, keep in ice, inject a 20 μL aliquot. Urine. 100 μL Urine + 50 μL 6% ascorbic acid in water + 50 μL demeclocycline in MeOH/100 mM HCl + 400 μL buffer, mix for 30 s, add 3 mL ethyl acetate, rotate for 10 min, centrifuge at 3000 rpm for 6 min. Remove the organic layer and add it to 100 μL 0.2% ascorbic acid in MeOH, evaporate to dryness under vacuum while vortexing, reconstitute the residue in 200 μL mobile phase, mix, filter, keep in ice, inject a 20 μL aliquot. (Buffer was 27.6 g $NaH_2PO_4$ + 25.2 g sodium sulfite in 100 mL water, pH 6.1.)

## HPLC VARIABLES
**Column:** 100 × 2 5 μm Lichrosorb RP8
**Mobile phase:** MeCN:100 mM citric acid 24:76
**Flow rate:** 0.5
**Injection volume:** 20
**Detector:** UV 350

## CHROMATOGRAM
**Retention time:** 9
**Internal standard:** demeclocycline (4)
**Limit of detection:** 50 ng/mL

## OTHER SUBSTANCES
**Extracted:** chlortetracycline, methacycline, oxytetracycline, tetracycline

## KEY WORDS
serum

## REFERENCE
De Leenheer, A.P.; Nelis, H.J.C.F. Doxycycline determination in human serum and urine by high-performance liquid chromatography. *J.Pharm.Sci.*, **1979**, *68*, 999–1002

## SAMPLE
**Matrix:** bulk, formulations
**Sample preparation:** Bulk. Prepare a 10-100 μg/mL solution in buffer, inject an aliquot. Capsules, tablets. Prepare a 1 mg/mL solution of capsule contents or crushed tablets in

buffer, sonicate for 10 min, filter (0.45 μm), dilute with buffer, inject an aliquot. (Buffer was 20 mM sodium perchlorate adjusted to pH 2.0 with perchloric acid.)

## HPLC VARIABLES
**Column:** 250 × 4.6 5 μm 100 Å PLRP-S polystyrene-divinylbenzene (Polymer Laboratories)
**Mobile phase:** MeCN:buffer 25:75 (Buffer was 20 mM sodium perchlorate adjusted to pH 2.0 with perchloric acid.)
**Flow rate:** 1
**Detector:** UV 280

## CHROMATOGRAM
**Retention time:** 40

## OTHER SUBSTANCES
**Simultaneous:** impurities

## KEY WORDS
capsules; tablets

## REFERENCE
Bryan, P.D.; Stewart, J.T. Chromatographic analysis of selected tetracyclines from dosage forms and bulk drug substance using polymeric columns with acidic mobile phases. *J.Pharm.Biomed.Anal.*, **1994**, *12*, 675–692

## SAMPLE
**Matrix:** cell suspensions
**Sample preparation:** 300 μL Cell suspension + 300 μL MeCN, vortex, centrifuge, inject a 10 μL aliquot.

## HPLC VARIABLES
**Column:** 125 × 4 Nucleosil 100 5CN
**Mobile phase:** MeCN:THF:phosphate/citrate buffer 10:10:80
**Injection volume:** 10
**Detector:** UV 350

## CHROMATOGRAM
**Retention time:** 2.1

## REFERENCE
Kersten, A.; Poitschek, C.; Rauch, S.; Aberer, E. Effects of penicillin, ceftriaxone, and doxycycline on morphology of *Borrelia burgdorferi*. *Antimicrob.Agents Chemother.*, **1995**, *39*, 1127–1133

## SAMPLE
**Matrix:** food
**Sample preparation:** Condition a 100 mg Baker 10 C18 SPE cartridge by washing with MeOH, water, and 10 mL saturated aqueous Na₂EDTA. Dissolve 5 g honey in 20 mL 100 mM pH 4.0 Na₂EDTA-McIlvaine buffer, filter, apply to the SPE cartridge, wash with 20 mL water, air dry under vacuum for 5 min. Condition a Baker 10 COOH cartridge with ethyl acetate. Elute contents of C18 cartridge onto COOH cartridge with 50 mL ethyl acetate. Wash COOH cartridge with 10 mL MeOH, elute with 10 mL mobile phase, inject 100 μL aliquot.

## HPLC VARIABLES
**Column:** 250 × 4.6 5 μm Bakerbond C8
**Mobile phase:** MeOH:MeCN:10 mM aqueous oxalic acid 1:1.5:3

**Flow rate:** 1
**Injection volume:** 100
**Detector:** UV 350

## CHROMATOGRAM
**Retention time:** 6
**Limit of detection:** 0.05 ppm

## OTHER SUBSTANCES
**Extracted:** chlortetracycline, oxytetracycline, tetracycline

## KEY WORDS
honey; SPE

## REFERENCE
Oka, H.; Ikai, Y.; Kawamura, N.; Uno, K.; Yamada, M.; Harada, K.; Uchiyama, M.; Asukabe, H.; Mori, Y.; Suzuki, M. Improvement of chemical analysis of antibiotics. IX. A simple method for residual tetracyclines analysis in honey using a tandem cartridge clean-up system. *J.Chromatogr.*, **1987**, *389*, 417–426

## SAMPLE
**Matrix:** food
**Sample preparation:** Condition a 500 mg Baker-10 C18 SPE cartridge with 10 mL MeOH, 10 mL water, and 10 mL saturated aqueous disodium EDTA. Condition a 500 mg Baker-10 COOH cartridge with MeOH:ethyl acetate 10:90. Dissolve 25 g honey in 50 mL 100 mM pH 4.0 disodium EDTA-McIlvaine buffer, filter. Add the filtrate to the C18 SPE cartridge, wash with 20 mL water, wash with 400 µL ethyl acetate, air dry under vacuum for 5 min, elute with 50 mL MeOH:ethyl acetate 10:90. Add a 5 mL aliquot to the COOH SPE cartridge, wash with 5 mL MeOH (?), elute with 10 mL mobile phase, inject a 100 µL aliquot.

## HPLC VARIABLES
**Column:** 75 × 4.6 3 µm Chemcosorb 3C8 (Chemco)
**Mobile phase:** MeCN:MeOH:10 mM aqueous oxalic acid 3:2:16, pH 3.0
**Flow rate:** 1
**Injection volume:** 100
**Detector:** UV 350

## CHROMATOGRAM
**Retention time:** 9
**Limit of detection:** 0.1 ppm

## OTHER SUBSTANCES
**Extracted:** chlortetracycline, demeclocycline (demethylchlortetracycline), methacycline, minocycline, oxytetracycline, tetracycline

## KEY WORDS
honey; SPE

## REFERENCE
Oka, H.; Ikai, Y.; Kawamura, N.; Uno, K.; Yamada, M.; Harada, K.; Suzuki, M. Improvement of chemical analysis of antibiotics. XII. Simultaneous analysis of seven tetracyclines in honey. *J.Chromatogr.*, **1987**, *400*, 253–261

## SAMPLE
**Matrix:** formulations

## HPLC VARIABLES
**Column:** 250 × 4.6 5 μm Bakerbond phenylethyl
**Mobile phase:** MeOH:100 mM NaH$_2$PO$_4$ 70:30
**Flow rate:** 0.8
**Detector:** UV 280

## CHROMATOGRAM
**Retention time:** 5.25 (doxycline hyclate)

## KEY WORDS
injections; saline; water; stability-indicating

## REFERENCE
Stiles, M.L.; Allen, L.V., Jr.; Prince, S.J. Stability of various antibiotics kept in an insulated pouch during administration via portable infusion pump. *Am.J.Health-Syst.Pharm.*, **1995**, *52*, 70–74

## SAMPLE
**Matrix:** formulations
**Sample preparation:** Dissolve ointment in petroleum ether, add an equal volume of EtOH:water 70:30, dilute with MeOH to 100 μg/mL, inject a 10 μL aliquot.

## HPLC VARIABLES
**Column:** 300 × 3.9 10 μm LiChrosorb Si-60
**Mobile phase:** MeOH:water 5:95 containing 1.3 mM disodium citrate, 1 mM tetrabutylammonium bromide, 1.1 mM citric acid, and 8 mM EDTA.
**Flow rate:** 1
**Injection volume:** 10
**Detector:** UV 254

## CHROMATOGRAM
**Retention time:** k′ 0.56

## OTHER SUBSTANCES
**Simultaneous:** anhydrotetracycline, chlortetracycline, demeclocycline, epianhydrotetracycline, oxytetracycline, quatrimycin, rolitetracycline, tetracycline

## KEY WORDS
ointment

## REFERENCE
Lingeman, H.; van Munster, H.A.; Beynen, J.H.; Underberg, W.J.; Hulshoff, A. High-performance liquid chromatographic analysis of basic compounds on non-modified silica gel and aluminium oxide with aqueous solvent mixtures. *J.Chromatogr.*, **1986**, *352*, 261–274

## SAMPLE
**Matrix:** milk
**Sample preparation:** Prepare a column as follows. Swirl Chelating Sepharose Fast Flow resin (Pharmacia) in its bottle, add it to a polypropylene column to give a bed volume of 1.0-1.2 mL, wash 3 times with 2 mL portions of water, wash with 2 mL 10 mM copper sulfate, wash with two 2 mL portions of water. Centrifuge 5 mL milk at 10° at 1500 g for 15 min, remove the lower layer and add it to 10 mL succinate buffer, mix, centrifuge at 1500 g for 30 min, add the supernatant to the column. Wash with 2 mL succinate buffer, wash with 2 mL water, wash with 2 mL MeOH, wash with 2 mL water, wash with 700 μL citrate/phosphate buffer (be careful not to disturb bed), elute with 2.5 mL citrate/phosphate buffer (column is white and eluate is blue). Filter (Amicon Centricon 30,

MW 30000 cut-off; pre-washed by centrifuging with 2 mL water) while centrifuging at 5000 g for 30-90 min, inject a 600 μL aliquot of the ultrafiltrate. (Prepare succinate buffer by dissolving 11.8 g succinic acid in 980 mL water, adjust pH to 4.0 with 10 M NaOH, make up to 1 L. Prepare the citrate/phosphate buffer by dissolving 12.9 g citric acid monohydrate, 10.9 g $Na_2HPO_4$, 37.2 g disodium EDTA dihydrate, and 29.2 g NaCl in 1 L water.)

## HPLC VARIABLES
**Column:** 150 × 4.6 5 μm PLRP-S (Polymer Labs)
**Mobile phase:** Gradient. MeCN:MeOH:10 mM oxalic acid 0:0:100 for 1 min, to 22:8:70 over 5 min, maintain at 22:8:70 for 11 min, return to initial conditions.
**Flow rate:** 1
**Injection volume:** 600
**Detector:** UV 355

## CHROMATOGRAM
**Retention time:** 16.6
**Limit of detection:** 1.15 ng/mL
**Limit of quantitation:** 2.22 ng/mL

## OTHER SUBSTANCES
**Extracted:** chlortetracycline, demeclocycline, methacycline, minocycline, oxytetracycline, tetracycline
**Noninterfering:** chloramphenicol, gentian violet, hydromycin B, ivermectin, spectinomycin, sulfa drugs

## KEY WORDS
cow; SPE; ultrafiltrate

## REFERENCE
Carson, M.C. Simultaneous determination of multiple tetracycline residues in milk using metal chelate affinity chromatography. *J.AOAC Int.*, **1993**, *76*, 329–334

## SAMPLE
**Matrix:** tissue
**Sample preparation:** Prepare an affinity column by filling a 10 mL column with 5 mL chelating Sepharose, allow to settle, wash with 20 mL 0.5% copper(II) sulfate solution, eliminate air bubbles by agitation, wash with 15 mL 50 mM pH 4 succinate buffer, do not allow to dry. Condition an Analytichem Bond Elut C18 SPE cartridge with 10 mL MeOH and 10 mL water, do not allow to dry. Homogenize 4 g minced kidney with 40 mL 50 mM pH 4 succinate buffer, sonicate for 10 min, centrifuge at 9000 rpm for 10 min, filter the supernatant through paper, repeat the extraction. Combine the supernatants and pass them through the affinity column at 5-7 mL/min, wash with 10 mL water, wash with 30 mL MeOH, wash with 20 mL water, elute with 50 mL 50 mM pH 4 succinate buffer containing 3.7% Titriplex III (ethylenedinitrilotetracacetic acid, disodium salt dihydrate). Add the eluate to the SPE cartridge at 5-7 mL/min, wash with 10 mL water, dry with air aspiration for 10 min, elute with 5 mL MeOH:MeCN 1:1, evaporate the eluate at 40° under a stream of nitrogen, dissolve the residue in 500 μL mobile phase, inject an aliquot. Protect from light through process. (The affinity columns may be re-used up to 15 times by washing with 20 mL water then 20 mL EtOH:water 20:80 then conditioning as described above.)

## HPLC VARIABLES
**Guard column:** Perisorb RP-8
**Column:** two 300 × 100 5 μm Chromspher C8 columns (cat. no. 28262) in series
**Mobile phase:** MeCN:10 mM pH 2 oxalic acid 20:80
**Flow rate:** 0.8
**Detector:** UV 365

## CHROMATOGRAM
**Retention time:** 26
**Limit of quantitation:** 30 ng/g

## OTHER SUBSTANCES
**Simultaneous:** chlortetracycline, demethylchlortetracycline, methacycline, oxytetracycline, tetracycline

## KEY WORDS
kidney; SPE

## REFERENCE
Degroodt, J.M.; Wyhowski de Bukanski, B.; Srebrnik, S. Multiresidue analysis of tetracyclines in kidney by HPLC and photodiode array detection. *J.Liq.Chromatogr.*, **1993**, *16*, 3515–3529

## SAMPLE
**Matrix:** tissue
**Sample preparation:** Mince 0.1-0.3 g tissue with a scalpel and incubate at 37° with 0.5 mL water for 1 h. Add MeCN:85% phosphoric acid:water 20:2:78 (muscle, renal medulla, lung) or MeOH:MeCN:85% phosphoric acid:water 30:10:2:58 (renal cortex, liver) to a total volume of 1 mL, sonicate 30 min, filter (10 000 or 30 000 Da cutoff) by centrifuging at 2200 g for 30 min, inject 10-30 μL aliquot of filtrate.

## HPLC VARIABLES
**Column:** 220 × 2.1 Brownlee phenyl Spheri-5 MPLC cartridge
**Mobile phase:** MeOH:MeCN:triethylamine:phosphoric acid:80 mM pH 2.4 sodium phosphate buffer 22.5:2.5:0.5:1.7:72.8
**Column temperature:** 60
**Flow rate:** 0.3-0.4
**Injection volume:** 10-30
**Detector:** UV 268; UV 345

## CHROMATOGRAM
**Retention time:** 6
**Limit of detection:** <5-10 ng/g

## KEY WORDS
cow; muscle; renal cortex; renal medulla; liver; lung

## REFERENCE
Riond, J.L.; Hedeen, K.M.; Tyczkowska, K.; Riviere, J.E. Determination of doxycycline in bovine tissues and body fluids by high-performance liquid chromatography using photodiode array ultraviolet-visible detection. *J.Pharm.Sci.*, **1989**, *78*, 44–47

## ANNOTATED BIBLIOGRAPHY

Prevosto, J.M.; Beraud, B.; Cheminel, V.; Gaillard, Y.; Mounier, C.; Chaulet, J.F. Determination of doxycycline in human plasma and urine samples by high performance liquid chromatography. Application for drug monitoring in malaria chemoprophylaxis. *Ann.Biol.Clin.(Paris)*, **1995**, *53*, 29–32

Colmenero, J.D.; Fernández-Gallardo, L.C.; Agúndez, J.A.G.; Sedeño, J.; Benítez, J.; Valverde, E. Possible implications of doxycycline-rifampin interaction for treatment of brucellosis. *Antimicrob.Agents Chemother.*, **1994**, *38*, 2798–2802 [extracted rifampin; plasma; serum; papaverine (IS); LOQ 200 ng/mL]

Hoogmartens, J.; Khan, N.H.; Vanderhaeghe, H.; Van der Leeden, A..L.; Oosterbaan, M.; Veld-Tulp, G.L.; Plugge, W.; Van der Vlies, C.; Mialanne, D.; et al. A collaborative study of the analysis of doxycycline hyclate by high-performance liquid chromatography on polystyrene-divinylbenzene packing materials. *J.Pharm.Biomed.Anal.*, **1989**, *7*, 601–610

Nieder, M.; Jaeger, H. Selective quantification of doxycycline in human plasma and urine with optimized chromatography. *Chromatographia*, **1988**, *25*, 526–530 [column temp 30; plasma; urine; SPE; demeclocycline (IS); pharmacokinetics; non-interfering other tetracyclines, caffeine, nicotine, salicylic acid; LOQ 125 ng/mL]

Dihuidi, K.; Kucharski, M.J.; Roets, E.; Hoogmartens, J.; Vanderhaeghe, H. Quantitative analysis of doxycycline and related substances by high-performance liquid chromatography. *J.Chromatogr.*, **1985**, *325*, 413–424 [column temp 60; bulk; tablets; capsules; simultaneous impurities, methacycline, oxytetracycline]

Böcker, R. Rapid analysis of doxycycline from biological samples by high-performance liquid chromatography. *J.Chromatogr.*, **1980**, *187*, 439–441 [whole blood; serum; tissue; mouse; liver]

De Leenheer, A.P.; Nelis, H.J.C.F. Reversed-phase high-performance liquid chromatography of doxycycline. *J.Chromatogr.*, **1977**, *140*, 293–299

# Enalapril

**Molecular formula:** $C_{20}H_{28}N_2O_5$
**Molecular weight:** 376.5
**CAS Registry No.:** 75847-73-3 (enalapril), 76095-16-4
(enalapril maleate)

## SAMPLE
**Matrix:** formulations
**Sample preparation:** Finely powder tablets, weigh out amount equivalent to 20 mg enalapril maleate, suspend in 100 mL mobile phase, filter, inject a 5 μL aliquot.

## HPLC VARIABLES
**Column:** $250 \times 4.6$ 12 μm Hypersil C18
**Mobile phase:** MeCN:water 20:80 adjusted to pH 3.8 with acetic acid
**Flow rate:** 1
**Injection volume:** 5
**Detector:** UV 215 for 3.5 min, then UV 275

## CHROMATOGRAM
**Retention time:** 1.9
**Internal standard:** caffeine (4.8)

## OTHER SUBSTANCES
**Simultaneous:** hydrochlorothiazide

## KEY WORDS
tablets

## REFERENCE
el Walily, A.F.M.; Belal, S.F.; Heaba, E.A.; El Kersh, A. Simultaneous determination of enalapril maleate and hydrochlorothiazide by first-derivative ultraviolet spectrophotometry and high-performance liquid chromatography. *J.Pharm.Biomed.Anal.*, **1995**, *13*, 851–856

## SAMPLE
**Matrix:** formulations
**Sample preparation:** Dissolve tablets in MeCN:1 mM pH 2 $KH_2PO_4$ 50:50, centrifuge, inject a 50 μL aliquot of the supernatant.

## HPLC VARIABLES
**Column:** $250 \times 4.6$ 5 μm Spherisorb C8
**Mobile phase:** MeCN:buffer 35:65 (Buffer was 1 mM $KH_2PO_4$ adjusted to pH 2 with phosphoric acid.)
**Column temperature:** 40
**Flow rate:** 2.5
**Injection volume:** 50
**Detector:** UV 215

## CHROMATOGRAM
**Retention time:** 11

## OTHER SUBSTANCES
**Simultaneous:** degradation products, enalaprilat, felodipine

## KEY WORDS
tablets

## REFERENCE

Qin, X.-Z.; DeMarco, J.; Ip, D.P. Simultaneous determination of enalapril, felodipine and their degradation products in the dosage formulation by reversed-phase high-performance liquid chromatography using a Spherisorb $C_8$ column. *J.Chromatogr.A*, **1995**, *707*, 245–254

## SAMPLE

**Matrix:** formulations

**Sample preparation:** Crush tablets, weigh out powder equivalent to 50 mg enalapril maleate, dissolve in 25 mL mobile phase, filter, dilute filtrate with an equal volume 1 mg/mL lisinopril in mobile phase, inject an aliquot.

## HPLC VARIABLES

**Column:** 300 × 4 ODS

**Mobile phase:** MeOH:water:phosphoric acid 75:25:0.1

**Injection volume:** 20

**Detector:** UV 215

## CHROMATOGRAM

**Retention time:** 8.0

**Internal standard:** lisinopril (13.2)

## KEY WORDS

tablets

## REFERENCE

Sane, R.T.; Vaidya, A.J.; Ghadge, J.K.; Jani, A.B.; Kotwal, S.K. Estimation of enalapril maleate in pharmaceutical dosage by HPLC. *Indian Drugs*, **1992**, *29*, 244–245

## SAMPLE

**Matrix:** formulations

**Sample preparation:** Crush tablet, mix with 10 mL mobile phase, filter, dilute with mobile phase to ca. 25 µg/mL, inject an aliquot.

## HPLC VARIABLES

**Column:** 250 × 4.6 Nucleosil C18

**Mobile phase:** MeCN:MeOH:water 50:25:25

**Flow rate:** 1

**Injection volume:** 100

**Detector:** UV 230

## CHROMATOGRAM

**Retention time:** 2.3

## KEY WORDS

tablets

## REFERENCE

Rau, H.L.; Udupa, N.; Aroor, A.R. A new HPLC method for the estimation of enalapril maleate in tablets. *Indian Drugs*, **1991**, *29*, 46–48

## SAMPLE

**Matrix:** perfusate

**Sample preparation:** Hydrolyze with NaOH to enalaprilat.

## HPLC VARIABLES
**Column:** 5 μm Ultrasphere ODS
**Mobile phase:** MeCN:50 mM phosphate buffer 12:88, pH 3.2
**Flow rate:** 1
**Detector:** UV

## CHROMATOGRAM
**Limit of detection:** 50 nM

## KEY WORDS
rat

## REFERENCE
Friedman, D.I.; Amidon, G.L. Passive and carrier-mediated intestinal absorption components of two angiotensin converting enzyme (ACE) inhibitor prodrugs in rats: enalapril and fosinopril. *Pharm.Res.*, **1989**, *6*, 1043–1047

## SAMPLE
**Matrix:** solutions

## HPLC VARIABLES
**Column:** 250 × 4.6 Spherisorb 5 ODS-2
**Mobile phase:** n-Propanol:buffer 20:80 (Buffer was pH 3.0 phosphate buffer containing 0.4% triethylamine.)
**Flow rate:** 1
**Detector:** UV 240

## CHROMATOGRAM
**Retention time:** 6

## OTHER SUBSTANCES
**Simultaneous:** benzepril, captopril, cilazapril, quinapril, ramipril

## REFERENCE
Barbato, F.; Morrica, P.; Quaglia, F. Analysis of ACE inhibitor drugs by high performance liquid chromatography. *Farmaco*, **1994**, *49*, 457–460

## SAMPLE
**Matrix:** solutions

## HPLC VARIABLES
**Column:** 300 × 3.9 μBondapak phenyl
**Mobile phase:** MeOH:water:85% phosphoric acid 60:40:0.05
**Column temperature:** 30-40
**Detector:** UV 215-220

## OTHER SUBSTANCES
**Also analyzed:** lisinopril

## REFERENCE
Ranadive, S.A.; Chen, A.X.; Serajuddin, A.T. Relative lipophilicities and structural-pharmacological considerations of various angiotensin-converting enzyme (ACE) inhibitors. *Pharm.Res.*, **1992**, *9*, 1480–1486

# Epoetin

**Molecular formula:** $C_{809}H_{1301}N_{229}O_{240}S_5$
**Molecular weight:** $30400 \pm 400$
**CAS Registry No.:** 113427-24-0 ($\alpha$), 122312-54-3 ($\beta$)

## SAMPLE
**Matrix:** blood
**Sample preparation:** Acidify serum or plasma with 8 volumes of ice-cold 0.1% trifluoroacetic acid, centrifuge at 10000 g for 10 min, inject an aliquot of the supernatant.

## HPLC VARIABLES
**Guard column:** Whatman silica precolumn (Whatman No. 6561-403)
**Column:** two µBondapak C18 columns in series
**Mobile phase:** Gradient. MeCN:water:trifluoroacetic acid from 280:720:1 to 600:400:1 over 40 min, reequilibrate at 280:720:1 for 15 min before next injection.
**Flow rate:** 1.5
**Detector:** UV 280; collect 1.5 mL fractions and use bioassay

## CHROMATOGRAM
**Retention time:** 32
**Internal standard:** coproporphyrin I (20)

## KEY WORDS
serum; plasma; sheep; cow

## REFERENCE
Congote, L.F. High-performance liquid chromatographic separation of serum erythrotropin and erythropoietin. *J.Chromatogr.*, **1984**, *310*, 396–400

## SAMPLE
**Matrix:** formulations
**Sample preparation:** 300 µL Liposome suspension + 90 µL chloroform, centrifuge at 3000 rpm for 10 min, inject a 200 µL aliquot of the aqueous phase.

## HPLC VARIABLES
**Column:** $250 \times 4$ 5 µm Vydac C4
**Mobile phase:** Gradient. A was MeCN:water:trifluoroacetic acid 100:400:1. B was MeCN:water:trifluoroacetic acid 400:100:1. A:B 65:35 for 5 min, to 0:100 over 15 min, maintain at 0:100 for 2 min.
**Flow rate:** 1
**Injection volume:** 200
**Detector:** UV 214

## CHROMATOGRAM
**Retention time:** 20

## KEY WORDS
liposome suspensions

## REFERENCE
Qi, X.-R.; Maitani, Y.; Shimoda, N.; Sakaguchi, K.; Nagai, T. Evaluation of liposomal erythropoietin prepared with reverse-phase evaporation vesicle method by subcutaneous administration in rats. *Chem.Pharm.Bull.*, **1995**, *43*, 295–299

## SAMPLE
**Matrix:** solutions

## HPLC VARIABLES
**Column:** 600 × 7.5 10 μm TSK gel G3000SW (Toyosoda)
**Mobile phase:** 20 mM pH 7.0 sodium citrate containing 100 mM NaCl
**Flow rate:** 1
**Detector:** UV 240

## CHROMATOGRAM
**Retention time:** 15

## OTHER SUBSTANCES
**Simultaneous:** degradation products

## KEY WORDS
SEC

## REFERENCE
Depaolis, A.M.; Advani, J.V.; Sharma, B.G. Characterization of erythropoietin dimerization. *J.Pharm.Sci.*, **1995**, *84*, 1280–1284

## SAMPLE
**Matrix:** solutions

## HPLC VARIABLES
**Column:** 50 × 4.6 5 μm C4 214TP5405 (Vydac)
**Mobile phase:** Gradient. MeCN:0.06% trifluoroacetic acid 35:65 for 5 min, to 38:62 over 10 min, to 50:50 over 20 min, return to initial conditions over 5 min.
**Flow rate:** 1.5
**Detector:** UV 230

## CHROMATOGRAM
**Retention time:** 23

## OTHER SUBSTANCES
**Simultaneous:** degradation products

## REFERENCE
Depaolis, A.M.; Advani, J.V.; Sharma, B.G. Characterization of erythropoietin dimerization. *J.Pharm.Sci.*, **1995**, *84*, 1280–1284

## SAMPLE
**Matrix:** solutions

## HPLC VARIABLES
**Column:** 50 × 4.6 YMC AP-800 C4 (Yamamura)
**Mobile phase:** Gradient. EtOH:buffer from 50:50 to 90:10 over 1 h (Buffer was 10 mM pH 7.0 Tris-HCl.)
**Flow rate:** 0.5
**Detector:** UV 280

## CHROMATOGRAM
**Retention time:** 35

## REFERENCE

Inoue, N.; Wada, M.; Takeuchi, M. An improved method for the purification of human erythropoietin with high *in vivo* activity from the urine of anemic patients. *Biol.Pharm.Bull.*, **1994**, *17*, 180–184

## SAMPLE

**Matrix:** urine
**Sample preparation:** Concentrate urine by ultrafiltration-dialysis then chromatograph on Phenyl-Sepharose CL4B.

## HPLC VARIABLES

**Column:** 75 × 7.5 Waters DEAE 5PW
**Mobile phase:** Gradient. A was 15 mM Tris adjusted to pH 8.6 with acetic acid. B was 15 mM Tris + 500 mM NaCl adjusted to pH 8.6 with acetic acid. A:B from 100:0 to 0:100 over 40 min.
**Flow rate:** 0.5
**Detector:** UV 280; bioassay

## CHROMATOGRAM

**Retention time:** about 15

## REFERENCE

Lange, R.D.; Andrews, R.B.; Trent, D.J.; Reyniers, J.P.; Draganac, P.S.; Farkas, W.R. Preparation of purified erythropoietin by high performance liquid chromatography. *Blood Cells*, **1984**, *10*, 305–314

# Erythromycin

**Molecular formula:** $C_{37}H_{67}NO_{13}$

**Molecular weight:** 733.9

**CAS Registry No.:** 114-07-8, 41342-53-4 (ethylsuccinate), 96128-89-1 (acistrate), 3521-62-8 (estolate), 304-63-2 (gluheptonate), 23067-13-2 (gluheptonate), 3847-29-8 (lactobionate), 134-36-1 (propionate), 643-22-1 (stearate), 84252-03-9 (stinoprate)

## SAMPLE

**Matrix:** blood

**Sample preparation:** 200 µL Plasma or whole blood + 10 µL 10 µg/mL oleandomycin + 20 µL saturated sodium carbonate + 1 mL diethyl ether, mix vigorously for 30 s, centrifuge at 6000 g for 2 min. Remove 750 µL ether, evaporate to dryness under a stream of nitrogen at room temperature for 10 min, reconstitute residue in 50 µL mobile phase, inject 20 µL aliquot.

## HPLC VARIABLES

**Guard column:** 10 × 4.6 5 µm Asahi ODP-50G

**Column:** 150 × 4.6 5 µm Asahipak octadecyl polymer

**Mobile phase:** MeCN:50 mM pH 10.5 $KH_2PO_4$ 37:63

**Flow rate:** 1

**Injection volume:** 20

**Detector:** E, Shimadzu L-ECD-6A, glassy carbon electrode +0.72 V versus an Ag/AgCl reference electrode

## CHROMATOGRAM

**Retention time:** 12.1

**Internal standard:** oleandomycin (5.7)

**Limit of detection:** 100 ng/mL

## OTHER SUBSTANCES

**Noninterfering:** clenbuterol, diltiazem, dipyridamole, ketotifen, methacholine, orciprenaline, theophylline

## KEY WORDS

plasma; whole blood

## REFERENCE

Kato, Y.; Yokoyama, T.; Shimokawa, M.; Kudo, K.; Kabe, J.; Mohri, K. Determination of erythromycin in human plasma and whole blood by high-performance liquid chromatography. *J.Liq.Chromatogr.*, **1993**, *16*, 661–680

## SAMPLE

**Matrix:** blood

**Sample preparation:** 1 mL Plasma + 20 µL of 100 µg/mL oleandomycin phosphate in ethanol + 60 µL saturated potassium carbonate (final pH 10), mix briefly, add 5 mL t-butyl methyl ether, shake in a reciprocating shaker for 15 min, centrifuge at 800 g for 5 min. Remove 4 mL of the upper ether layer and evaporate it to dryness under a stream of nitrogen at room temperature. Wash down tube with 200 µL t-butyl methyl ether. Evaporate to dryness again, take up residue in 125 µL mobile phase, centrifuge 30 s, inject aliquot.

## HPLC VARIABLES
**Guard column:** Alltech Direct Connect packed with μBondapak C18
**Column:** 250 × 4.6 5 μm Ultrasphere C18
**Mobile phase:** MeCN:MeOH:buffer 42:10:48, final pH adjusted to 6.30-6.35 (Buffer was 100 mM sodium acetate adjusted to pH 5.0 with 100 mM acetic acid.)
**Flow rate:** 1.20
**Injection volume:** 20
**Detector:** E, ESA model 5100A, guard cell +0.95 V, detector 1 +0.65 V, detector 2 +0.85 V

## CHROMATOGRAM
**Retention time:** 6
**Internal standard:** oleandomycin phosphate (4)
**Limit of detection:** 250 ng/mL

## OTHER SUBSTANCES
**Extracted:** anhydroerythromycin, 2'-acetylerythromycin

## KEY WORDS
plasma; mobile phase recirculated

## REFERENCE
Laakso, S.; Scheinin, M.; Anttila, M. Determination of erythromycin base and 2'-acetylerythromycin in human plasma using high-performance liquid chromatography with electrochemical detection. *J.Chromatogr.*, **1990**, *526*, 475−486

## SAMPLE
**Matrix:** blood
**Sample preparation:** 1 mL Plasma + 400 μL 0.1 M NaOH to adjust pH to 10, mix 30 s, add 5 mL methyl t-butyl ether, vortex 1 min, centrifuge at 3000 rpm for 5 min, evaporate organic layer to dryness under a stream of nitrogen at 40°, dissolve residue in 150 μL mobile phase, vortex 2 min, inject 10 μL aliquot.

## HPLC VARIABLES
**Column:** 150 × 4 10 μm Techopak T-15 C18
**Mobile phase:** MeCN:MeOH:THF:buffer 86:3:3:8 (Buffer was 75 mM sodium acetate adjusted to pH 4.1 with glacial acetic acid.)
**Flow rate:** 1.5
**Injection volume:** 10
**Detector:** E, ESA Model 5100A with ESA model 5020 guard cell, analytical cell + 0.70 V (I) +0.85 V (II), guard cell +0.90 V, 0.5 μm carbon filters

## CHROMATOGRAM
**Retention time:** 10
**Limit of quantitation:** 50 ng/mL

## KEY WORDS
plasma

## REFERENCE
Kokkonen, P.; Haataja, H.; Välttilä, S. Determination of 2'-acetylerythromycin and erythromycin in plasma by HPLC using manual and robotic sample preparation. *Chromatographia*, **1987**, *24*, 680−682

## SAMPLE
**Matrix:** blood, gastric juice

**Sample preparation:** Centrifuge plasma or gastric juice at 1200 g for 5 min. 500 µL Plasma or gastric juice + 500 µL pH 11 phosphate buffer (ionic strength I=1.0) + 50 µL 100 µM oleandomycin in MeCN + 5 mL hexane:2-butanol 80:20, shake for 15 min, centrifuge at 1200 g for 5 min, freeze in dry ice/acetone. Remove the organic layer and evaporate it to dryness under a stream of nitrogen, reconstitute the residue in 300 µL mobile phase, vortex three times for 1 min, inject a 40 µL aliquot.

## HPLC VARIABLES
**Guard column:** 15 × 3.2 7 µm Brownlee CN

**Column:** 100 × 4.6 Hypersil C18 BDS base-deactivated

**Mobile phase:** MeCN:2.1 mM $NaH_2PO_4$:27.1 mM $Na_2HPO_4$ 40:30:30

**Column temperature:** 65

**Flow rate:** 1.2

**Injection volume:** 40

**Detector:** E, ESA Coulochem Model 5100A, Model 5020 guard cell before injector +1.0 V, model 5011 dual electrode analytical cell, screen electrode (detector 1) +0.65 V, sample electrode (detector 2) +0.85 V, ESA carbon filters before guard and analytical cells

## CHROMATOGRAM
**Retention time:** 10.5

**Internal standard:** oleandomycin (5)

**Limit of quantitation:** 20 nM (plasma); 100 nM (gastric juice)

## KEY WORDS
plasma; rugged; pharmacokinetics

## REFERENCE
Toreson, H.; Eriksson, B.M. Determination of erythromycin in gastric juice and blood plasma by liquid chromatography and electrochemical detection. *J.Chromatogr.B*, **1995**, *673*, 81–89

## SAMPLE
**Matrix:** blood, saliva, urine

**Sample preparation:** Plasma. 2 mL Plasma + 20 µL 750 µg/mL roxithromycin in MeCN + 5 mL diethyl ether, shake vigorously for 3 min, centrifuge at 900 g at 4° for 5 min. Remove upper layer and evaporate it to dryness under a stream of nitrogen at 45°. Reconstitute residue with 100 µL MeCN, vortex 5 s, inject 40 µL aliquot. Urine. 1.5 mL Urine + 100 µL 750 µg/mL roxithromycin in saturated $K_2HPO_4$ + 4 mL diethyl ether, shake vigorously for 3 min, centrifuge at 900 g at 4° for 5 min. Remove upper layer and evaporate it to dryness under a stream of nitrogen at 45°. Reconstitute residue with 100 µL MeCN, vortex 5 s, inject 40 µL aliquot. Saliva. 1.5 mL Saliva + 100 µL 750 µg/mL roxithromycin in saturated $K_2HPO_4$ + 4 mL diethyl ether, shake vigorously for 3 min, centrifuge at 900 g at 4° for 15 min. Remove upper layer and evaporate it to dryness under a stream of nitrogen at 45°. Reconstitute residue with 100 µL MeCN, vortex 5 s, inject 40 µL aliquot.

## HPLC VARIABLES
**Column:** Nova-Pak C18

**Mobile phase:** MeCN:MeOH:56 mM sodium acetate buffer 50:4:56, final pH adjusted to 7.0 with glacial acetic acid

**Flow rate:** 1.1

**Injection volume:** 40

**Detector:** E, Waters M460, +0.9 V versus Ag/AgCl reference electrode

## CHROMATOGRAM
**Retention time:** 6.0 (erythromycin base), 7.1 (erythromycin B), 34.5 (erythromycin esto-late), 35.5 (erythromycin ethylsuccinate)
**Internal standard:** roxithromycin (14.7)
**Limit of detection:** 12.5 ng/mL

## OTHER SUBSTANCES
**Simultaneous:** 4″-acetylerythromycin, 6-O-methylerythromycin

## KEY WORDS
plasma

## REFERENCE
Croteau, D.; Vallée, F.; Bergeron, M.G.; LeBel, M. High-performance liquid chromatographic assay of erythromycin and its esters using electrochemical detection. *J.Chromatogr.*, **1987**, *419*, 205–212

## SAMPLE
**Matrix:** blood, urine
**Sample preparation:** Urine. Centrifuge urine at 2500 g for 5 min, inject a 100 μL aliquot. Whole blood. 200 μL Whole blood + 100 μL 10% EDTA + 50 μL 20 μg/mL josamycin in MeOH:water 50:50, centrifuge to separate plasma. 200 μL Plasma + 20 μL saturated potassium carbonate + 1 mL MTBE, mix, centrifuge. Remove 800 μL of the MTBE layer and evaporate it to dryness, reconstitute the residue in 200 μL MeOH:water 50:50, inject a 100 μL aliquot.

## HPLC VARIABLES
**Column:** 250 × 4.6 8 μm PLRP-S 1000 Å (Polymer Labs)
**Mobile phase:** MeCN:t-butanol:200 mM pH 9.0 phosphate buffer:water 3:19:5:73
**Column temperature:** 70
**Flow rate:** 1.5
**Injection volume:** 100
**Detector:** F ex 365 em 450 following post-column extraction. The column effluent mixed with reagent pumped at 0.7 mL/min and this mixture flowed through a 1.5 m × 0.5 mm ID stainless steel coil. The effluent from the coil mixed with chloroform pumped at 1.5 mL/min and this mixture flowed through a 1.5 m × 0.5 mm ID stainless steel coil to a sandwich-type phase separator with a 40 μL groove volume (Vrije Universiteit, Amster-dam). Part of the organic layer was separated and flowed through the detector at 0.5 mL/min. (Reagent was 5 μM sodium 9,10-dimethoxyanthracene-2-sulfonate in 100 mM citric acid.)

## CHROMATOGRAM
**Retention time:** 11
**Internal standard:** josamycin (28)
**Limit of detection:** 12.5 ng/mL (plasma); 50 ng/mL (urine)

## OTHER SUBSTANCES
**Extracted:** metabolites
**Simultaneous:** midecamycin, troleandomycin

## KEY WORDS
plasma; whole blood; post-column extraction

## REFERENCE
Khan, K.; Paesen, J.; Roets, E.; Hoogmartens, J. Analysis of erythromycin A and its metabolites in biological samples by liquid chromatography with post-column ion-pair extraction. *J.Liq. Chromatogr.*, **1994**, *17*, 4195–4213

## SAMPLE
**Matrix:** blood, urine
**Sample preparation:** Dilute urine 1:2 with isotonic NaCl. 200 μL Plasma or diluted urine + 100 μL water + 600 μL pH 9 phosphate buffer + 3 mL dichloromethane, shake for 10 min, centrifuge at 2000 g for 5 min. Remove 2.5 mL of the organic layer and evaporate it to dryness under a stream of nitrogen, reconstitute the residue in 50 μL MeOH, vortex for 10 s, inject a 15 μL aliquot.

## HPLC VARIABLES
**Column:** 300 × 3.9 10 μm μBondapak C18
**Mobile phase:** MeCN:MeOH:83 mM ammonium acetate 55:22:23, pH adjusted to 7.5 with acetic acid
**Flow rate:** 1
**Injection volume:** 15
**Detector:** E, ESA Coulochem Model 5100A, Model 5020 guard cell 1.0 V (before injector), Model 5010 dual-electrode cell, screen electrode E1 + 0.7 V, sample electrode E2 +0.9 V, 0.5 μm ESA carbon filters placed before guard and analytical cells

## CHROMATOGRAM
**Retention time:** 7.0
**Internal standard:** erythromycin

## OTHER SUBSTANCES
**Extracted:** roxithromycin
**Simultaneous:** amitriptyline, clomipramine, disopyramide, erythromycin estolate, erythromycin ethylsuccinate, erythromycin stearate, imipramine, josamycin, lidocaine, spiramycin

## KEY WORDS
plasma; erythromycin is IS

## REFERENCE
Demotes-Mainaird, F.M.; Vinçon, G.A.; Jarry, C.H.; Albin, H.C. Micro-method for the determination of roxithromycin in human plasma and urine by high-performance liquid chromatography using electrochemical detection. *J.Chromatogr.*, **1989**, *490*, 115–123

## SAMPLE
**Matrix:** blood, urine
**Sample preparation:** Prewash a 1 mL C18 Bondelut C18 SPE cartridge with 3 mL MeCN and 3 mL water. 1 mL Serum or urine + 0.25 mL (0.50 mL for urine) 6-12 μg/mL oleandomycin phosphate in water, add 1 mL MeCN, vortex 1 min, centrifuge at 1600 g for 5 min, add to 8 mL water, load onto the SPE cartridge, wash with 5 mL water, wash with 5 mL MeCN:water 1:1, suck dry, elute with two 0.5 mL aliquots of MeCN:50 mM pH 6.30 phosphate buffer. Dry under vacuum in a rotary vacuum centrifuge, reconstitute in 20 μL water, vortex 1 min, add 25 μL MeCN, vortex 1 min, centrifuge at 1600 g for 1 min, inject 3-5 μL aliquot of upper layer.

## HPLC VARIABLES
**Guard column:** Waters Guard-Pak with Anatech 40-60 μm glass beads
**Column:** 150 × 3.9 Novapak C18
**Mobile phase:** MeCN:50 mM pH 6.30 phosphate buffer 30:70
**Column temperature:** 35
**Flow rate:** 1
**Injection volume:** 3-5
**Detector:** E, Metrohm 656 with a glassy carbon electrode, 1.15 V versus Ag/AgCl reference electrode; also UV at 200 nm with LOD 250-1000 ng/mL (J.Pharm.Sci. 1985, 74, 1126-1128)

## CHROMATOGRAM
**Retention time:** 6.3
**Internal standard:** oleandomycin phosphate (4.4)
**Limit of detection:** 100 ng/mL

## OTHER SUBSTANCES
**Simultaneous:** anhydroerythromycin

## KEY WORDS
serum; SPE; stability-indicating

## REFERENCE
Stubbs, C.; Haigh, J.M.; Kanfer, I. A stability-indicating liquid chromatographic method for the analysis of erythromycin in stored biological fluids using amperometric detection. *J.Liq.Chromatogr.*, **1987**, *10*, 2547–2557

## SAMPLE
**Matrix:** formulations

## HPLC VARIABLES
**Column:** 250 × 4.6 5 μm Bakerbond C18
**Mobile phase:** MeCN:MeOH:200 mM ammonium acetate:water 45:10:10:25, pH 6.25
**Flow rate:** 1
**Detector:** UV 215

## CHROMATOGRAM
**Retention time:** 7.46 (erythromycin lactobionate)

## KEY WORDS
injections; saline; water; stability-indicating

## REFERENCE
Stiles, M.L.; Allen, L.V., Jr.; Prince, S.J. Stability of various antibiotics kept in an insulated pouch during administration via portable infusion pump. *Am.J.Health-Syst.Pharm.*, **1995**, *52*, 70–74

## SAMPLE
**Matrix:** formulations
**Sample preparation:** Weigh out material corresponding to ca. 250 mg erythromycin ethylsuccinate, add 10 mL acetone, sonicate 5 min, centrifuge at 2500 g for 5 min, dilute a 6 mL aliquot of supernatant to 10 mL with 200 mM pH 6.5 tetrabutylammonium hydrogen sulfate:200 mM pH 6.5 phosphate buffer:water 12.5:7.5:80.

## HPLC VARIABLES
**Column:** 250 × 4.6 10 μm RSil LL C18 (RSL-Bio-Rad)
**Mobile phase:** MeCN:200 mM pH 6.5 tetrabutylammonium hydrogen sulfate (adjust pH with NaOH):200 mM pH 6.5 phosphate buffer:water 42.5:5:5:47.5
**Column temperature:** 35
**Flow rate:** 1.5
**Injection volume:** 20
**Detector:** UV 215

## CHROMATOGRAM
**Retention time:** 24 (erythromycin A ethylsuccinate), 8 (erythromycin A)

## KEY WORDS
powders; tablets

## REFERENCE
Cachet, T.; Lannoo, P.; Paesen, J.; Janssen, G.; Hoogmartens, J. Determination of erythromycin ethyl succinate by liquid chromatography. *J.Chromatogr.*, **1992**, *600*, 99–108

## SAMPLE
**Matrix:** solutions

## HPLC VARIABLES
**Column:** 250 × 4.6 Zorbax C8
**Mobile phase:** MeCN:200 mM pH 6.5 tetramethylammonium phosphate:200 mM pH 6.5 ammonium phosphate:water 35:20:5:40
**Column temperature:** 35
**Flow rate:** 1.5
**Injection volume:** 20
**Detector:** UV 215

## CHROMATOGRAM
**Retention time:** 12

## KEY WORDS
better results with aged columns

## REFERENCE
Cachet, T.; Quintens, I.; Roets, E.; Hoogmartens, J. Improved separation of erythromycin on aged reversed-phase columns. *J.Liq.Chromatogr.*, **1989**, *12*, 2171–2201

## ANNOTATED BIBLIOGRAPHY

Zierfels, G.; Petz, M. [Fluorimetric determination of erythromycin residues in foods of animal origin after derivatization with FMOC and HPLC separation]. *Z.Lebensm.Unters.Forsch.*, **1994**, *198*, 307–312

Janecek, M.; Quilliam, M.A.; Bailey, M.R.; North, D.H. Determination of erythromycin A by liquid chromatography and electrochemical detection, with application to salmon tissue. *J.Chromatogr.*, **1993**, *619*, 63–69 [electrochemical detection; fish; tissue; SPE; column temp 40; LOD 100 ng/g]

Paesen, J.; Calam, D.H.; Miller, J.H.McB.; Raiola, G.; Rozanski, A.; Silver, B.; Hoogmartens, J. Collaborative study of the analysis of erythromycin by liquid chromatography on wide-pore poly(styrene-divinylbenzene). *J.Liq.Chromatogr.*, **1993**, *16*, 1529–1544 [column temp 70; simultaneous impurities]

Cachet, T.; Quintens, I.; Paesen, J.; Roets, E.; Hoogmartens, J. Improved separation of erythromycin on aged reversed-phase columns. II. *J.Liq.Chromatogr.*, **1991**, *14*, 1203–1218

Paesen, J.; Roets, E.; Hoogmartens, J. Liquid chromatography of erythromycin A and related substances on poly(styrene-divinylbenzene). *Chromatographia*, **1991**, *32*, 162–166

Stubbs, C.; Kanfer, I. A stability-indicating high-performance liquid chromatographic assay of erythromycin estolate in pharmaceutical dosage forms. *Int.J.Pharm.*, **1990**, *63*, 113–119

Araman, A.; Temiz, D.; Guven, K.C. Stability studies of erythromycin in simulated gastric medium studied by high-performance liquid chromatography. *Acta Pharm.Turc.*, **1988**, *30*, 37–42

Croteau, D.; Bergeron, M.G.; LeBel, M. Pharmacokinetic advantages of erythromycin estolate over ethylsuccinate as determined by high-pressure liquid chromatography. *Antimicrob.Agents Chemother.*, **1988**, *32*, 561–565

Grgurinovich, N.; Matthews, A. Analysis of erythromycin and roxithromycin in plasma or serum by high-performance liquid chromatography using electrochemical detection. *J.Chromatogr.*, **1988**, *433*, 298–304

Haataja, H.; Kokkonen, P. Determination of 2'-acetylerythromycin and erythromycin in human tonsil tissue by HPLC with coulometric detection. *J.Antimicrob.Chemother.*, **1988**, *21*, 67–72

Stubbs, C.; Kanfer, I. High-performance liquid chromatography of erythromycin propionyl ester and erythromycin base in biological fluids. *J.Chromatogr.*, **1988**, *427*, 93–101 [extracted erythromycin, erythromycin propionate; oleandomycin (IS); electrochemical detection; column temp 35; serum; urine; SPE; pharmacokinetics; LOQ 250 ng/mL]

Cachet, T.; Kibwage, I.O.; Roets, E.; Hoogmartens, J.; Vanderhaeghe, H. Optimization of the separation of erythromycin and related substances by high-performance liquid chromatography. *J.Chromatogr.*, **1987**, *409*, 91–100 [column temp 35; simultaneous impurities, erythromycin A, erythromycin B, erythromycin C]

Geria, T.; Hong, W.H.; Daly, R.E. Improved high-performance liquid chromatographic assay of erythromycin in pharmaceutical solid dosage forms. *J.Chromatogr.*, **1987**, *396*, 191–198

Nilsson, L.G.; Walldorf, B.; Paulsen, O. Determination of erythromycin in human plasma, using column liquid chromatography with a polymeric packing material, alkaline mobile phase and amperometric detection. *J.Chromatogr.*, **1987**, *423*, 189–197

Kibwage, I.O.; Roets, E.; Hoogmartens, J.; Vanderhaeghe, H. Separation of erythromycin and related substances by high-performance liquid chromatography on poly(styrene-divinylbenzene) packing materials. *J.Chromatogr.*, **1985**, *330*, 275–286

Stubbs, C.; Haigh, J.M.; Kanfer, I. Determination of erythromycin in serum and urine by high-performance liquid chromatography with ultraviolet detection. *J.Pharm.Sci.*, **1985**, *74*, 1126–1128

Duthu, G.S. Assay of erythromycin from human serum by high performance liquid chromatography with electrochemical detection. *J.Liq.Chromatogr.*, **1984**, *7*, 1023–1032 [serum; electrochemical detection; also josamycin, oleandomycin, tylosin]

Chen, M.L.; Chiou, W.L. Analysis of erythromycin in biological fluids by high-performance liquid chromatography with electrochemical detection. *J.Chromatogr.*, **1983**, *278*, 91–100

Tsuji, K.; Kane, M.P. Improved high-pressure liquid chromatographic method for the analysis of erythromycin in solid dosage forms. *J.Pharm.Sci.*, **1982**, *71*, 1160–1164

# Estradiol

**Molecular formula:** $C_{18}H_{24}O_2$
**Molecular weight:** 272.4
**CAS Registry No.:** 50-28-2, 113-38-2 (dipropionate),
979-32-8 (valerate), 57-91-0 (α-estradiol),
50-50-0 (benzoate), 313-06-4 (cypionate),
4956-37-0 (enanthate), 3571-53-7 (undecylenate)

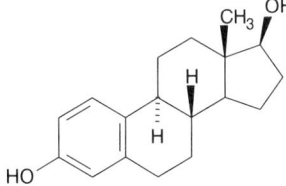

## SAMPLE
**Matrix:** blood
**Sample preparation:** Inject 10 µL plasma into MeCN pumped at 0.2 mL/min so that the precipitated proteins are removed by 0.5 and 0.2 µm filters in series. Swith the MeCN containing sample into the mobile phase and allow it to pass onto the analytical column, elute the analytical column in the usual way with mobile phase. Remove the filter unit from the circuit and back-flush it to waste with 100 mM sodium dodecyl sulfate at 2 mL/min, equilibrate filters with MeCN for 5 min before next injection.

## HPLC VARIABLES
**Guard column:** 20 mm Brownlee C18
**Column:** 250 × 4.6 5 µm Ultrasphere C18
**Mobile phase:** MeCN:water 33:67
**Flow rate:** 1
**Injection volume:** 10
**Detector:** UV 280

## CHROMATOGRAM
**Retention time:** 25.8

## OTHER SUBSTANCES
**Simultaneous:** equilin, estrone

## KEY WORDS
plasma; dog

## REFERENCE
Asafu-Adjaye, E.B.; Su, S.Y.; Shiu, G.K. Switching-valve-filter technique for the direct injection and analysis of drugs in plasma using high-performance liquid chromatography. *J.Chromatogr.B*, **1994**, *652*, 35−42

## SAMPLE
**Matrix:** blood
**Sample preparation:** 1 mL Plasma + 500 µL 10 M NaOH, shake on a slow rotatory mixer for 5 min, add 5 mL diethyl ether, rotomix 10 min, centrifuge at 700 g for 5 min, repeat extraction. Combine organic layers, evaporate to dryness under a stream of nitrogen at 37°, dissolve in 250 µL mobile phase, inject aliquot.

## HPLC VARIABLES
**Column:** 150 × 3.9 4 µm Novapack C18
**Mobile phase:** MeCN:MeOH:buffer 35:15:50 (Buffer was 50 mM $KH_2PO_4$ adjusted to pH 3.6 with phosphoric acid.)
**Flow rate:** 1.6

**Injection volume:** 50
**Detector:** E, Waters Model 464 pulsed electrochemical detector, + 1 V versus Ag/AgCl

## CHROMATOGRAM
**Retention time:** 2.44
**Limit of detection:** 50 pg/mL

## OTHER SUBSTANCES
**Simultaneous:** estriol, estrone, ethinylestradiol, heparin
**Noninterfering:** pentobarbital

## KEY WORDS
plasma; rabbit

## REFERENCE
Fernández, N.; Garcia, J.J.; Diez, M.J.; Terán, M.T.; Sierra, M. Rapid high-performance liquid chromatographic assay of ethynyloestradiol in rabbit plasma. *J.Chromatogr.*, **1993**, *619*, 143–147

## SAMPLE
**Matrix:** blood
**Sample preparation:** 100 µL Serum + 500 µL water + 100 µL 10 µg/mL 3,7-dimethoxyflavone in EtOH + 8 mL diethyl ether, shake, centrifuge at 4° at 1000 g for 5 min, freeze in acetone/dry ice. Remove the organic layer and dry it over anhydrous sodium sulfate, evaporate to dryness under a stream of nitrogen, reconstitute the residue in 100 µL MeOH:water 40:60, inject a 50 µL aliquot.

## HPLC VARIABLES
**Column:** 250 × 4.6 3 µm NS-Gel C18
**Mobile phase:** Gradient. MeOH:water from 40:60 to 55:45, maintain at 55:45 for 24 min, to 80:20 over 25 min.
**Column temperature:** 50
**Flow rate:** 1
**Injection volume:** 50
**Detector:** UV 210; UV 240

## CHROMATOGRAM
**Retention time:** 30.74
**Internal standard:** 3,7-dimethoxyflavone (47)

## OTHER SUBSTANCES
**Extracted:** aldosterone, androstenedione, dehydroepiandrosterone, deoxycorticosterone, 11-deoxycortisol, estrone, hydrocortisone, 17-hydroxyprogesterone, pregnenolone, progesterone

## KEY WORDS
serum

## REFERENCE
Ueshiba, H.; Segawa, M.; Hayashi, T.; Miyachi, Y.; Irie, M. Serum profiles of steroid hormones in patients with Cushing's syndrome determined by a new HPLC/RIA method. *Clin.Chem.*, **1991**, *37*, 1329–1333

## SAMPLE
**Matrix:** blood

**Sample preparation:** Extract 1 mL serum twice with 5 volumes ether by vortexing for 2 min, evaporate extracts to dryness under a stream of nitrogen at 35°, reconstitute in 100 μL MeOH.

## HPLC VARIABLES
**Column:** 240 × 4.5 Bio-Rad ODS-5S
**Mobile phase:** Gradient. MeOH:MeCN:water at 20:60:20 for 3 min then to 5:85:10 over 26 min.
**Flow rate:** 1
**Injection volume:** 50
**Detector:** UV 230

## OTHER SUBSTANCES
**Simultaneous:** androstenedione, progesterone, testosterone

## KEY WORDS
serum

## REFERENCE
Yu, F.H.; Yun, Y.W.; Yuen, B.H.; Moon, Y.S. Effects of hydroxyflutamide on rats treated with a superovulatory dose of pregnant mare serum gonadotropin. *Can.J.Physiol.Pharmacol.*, **1991**, *69*, 185–190

## SAMPLE
**Matrix:** blood
**Sample preparation:** 0.5-1 mL Plasma + 1 mL 500 mM pH 7 phosphate buffer + 12 mL hexane:ethyl acetate 70:30, extract. Remove a 10 mL aliquot of the organic layer and evaporate it to dryness under a stream of nitrogen at 50°, reconstitute the residue in 100 μL mobile phase, inject a 20-50 μL aliquot. (Hydrolyze 500 μL plasma by adding 500 μL 200 mM pH 5 acetate buffer and 100 μL beef liver β-glucuronidase (Sigma) or 10 μL β-glucuronidase/sulfatase (Glusulase), heat at 37° overnight, add 1 mL 500 mM pH 7 phosphate buffer + 12 mL hexane:ethyl acetate 70:30, extract. Remove a 10 mL aliquot of the organic layer and evaporate it to dryness under a stream of nitrogen at 50°, reconstitute the residue in 100 μL mobile phase, inject a 20-50 μL aliquot.)

## HPLC VARIABLES
**Column:** 250 × 4.6 5 μm Partisil 5/25 silica gel
**Mobile phase:** Hexane:EtOH 92.5:7.5
**Flow rate:** 1.5
**Injection volume:** 20-50
**Detector:** F ex 195 em 250 (cut-off filter)

## CHROMATOGRAM
**Retention time:** 8.9
**Limit of detection:** 3 ng/mL

## OTHER SUBSTANCES
**Extracted:** metabolites, estramustine, estromustine, estrone

## KEY WORDS
plasma; rat; dog; human; pharmacokinetics; normal phase

## REFERENCE
Dixon, R.; Brooks, M.; Gill, G. Estramustine phosphate: Plasma concentrations of its metabolites following oral administration to man, rat and dog. *Res.Commun.Chem.Pathol.Pharmacol.*, **1980**, *27*, 17–29

## SAMPLE
**Matrix:** blood, tissue
**Sample preparation:** 1 mL Blood or brain + 1 mL 50 mM ammonium acetate buffer, homogenize (Polytron PT-1200C), add 4 mL MeCN, vortex, add 1 mL concentrated brine, allow to stand at -5° for 1 h. Remove the organic phase and centrifuge it at 3000 g, filter, inject an aliquot.

## HPLC VARIABLES
**Column:** 250 × 4.6 5 μm Spherisorb C8
**Mobile phase:** MeCN:50 mM pH 6.8 ammonium acetate 52:48 containing 10 mM tetraethylammonium perchlorate
**Flow rate:** 1
**Detector:** UV 280

## CHROMATOGRAM
**Retention time:** 3.8
**Limit of detection:** 133 ng/g

## KEY WORDS
whole blood; rat; brain

## REFERENCE
Brewster, M.E.; Druzgala, P.J.; Anderson, W.R.; Huang, M.-J.; Bodor, N.; Pop, E. Efficacy of a 3-substituted versus 17-substituted chemical delivery system for estradiol brain targeting. *J.Pharm.Sci.*, **1995**, *84*, 38–43

## SAMPLE
**Matrix:** culture medium
**Sample preparation:** Extract culture medium twice with 2 volumes of ether, combine the extracts and evaporate them to dryness, reconstitute with MeOH, inject an aliquot.

## HPLC VARIABLES
**Column:** 300 × 3.9 Techopak 10 C18 (HPLC Technology)
**Mobile phase:** MeOH:0.5% pH 3.0 $(NH_4)H_2PO_4$ 62:39
**Flow rate:** 0.7
**Detector:** UV 280; Radioactivity

## CHROMATOGRAM
**Retention time:** 20

## OTHER SUBSTANCES
**Extracted:** estrone

## KEY WORDS
tritium labeled

## REFERENCE
Wild, M.J.; Rudland, P.S.; Back, D.J. Metabolism of the oral contraceptive steroids ethynylestradiol and norgestimate by normal (Huma 7) and malignant (MCF-7 and ZR-75-1) human breast cells in culture. *J.Steroid Biochem.Mol.Biol.*, **1991**, *39*, 535–543

## SAMPLE
**Matrix:** microsomal incubations
**Sample preparation:** 1 mL Microsomal incubation + 5 mL ethyl acetate, vortex, centrifuge at 2000 g for 8 min, remove organic phase, repeat extraction. Combine the organic layers

and evaporate them to dryness under a stream of nitrogen, reconstitute the residue in 100 μL MeOH:water 50:50, inject a 50 μL aliquot.

## HPLC VARIABLES
**Column:** 150 × 4.6 5 μm Ultracarb 30 ODS (Phenomenex)
**Mobile phase:** Gradient. MeCN:0.1% acetic acid in MeOH:0.1% acetic acid in water 16: 12:72 for 3 min, to 20:21:59 over 25 min (Waters no. 3 convex gradient), to 24:23:53 over 10 min (linear), to 55:24:21 over 10 min (linear), to 92:5:3 over 1 min, maintain at 92:5:3 for 7 min, return to initial conditions over 15 min.
**Flow rate:** 1.2
**Injection volume:** 50
**Detector:** UV 280

## CHROMATOGRAM
**Retention time:** 46

## OTHER SUBSTANCES
**Extracted:** metabolites, estrone

## KEY WORDS
rat

## REFERENCE
Suchar, L.A.; Chang, R.L.; Rosen, R.T.; Lech, J.; Conney, A.H. High-performance liquid chromatography separation of hydroxylated estradiol metabolites: Formation of estradiol metabolites by liver microsomes from male and female rats. *J.Pharmacol.Exp.Ther.*, **1995**, *272*, 197–206

## SAMPLE
**Matrix:** microsomal incubations
**Sample preparation:** 1 mL Human placental microsome suspension + 1 mL dichloromethane, extract, centrifuge, remove organic layer and evaporate it under vacuum, dissolve residue in 30 μL MeCN:water 50:50, centrifuge for 3 min, inject supernatant. After each run wash column with MeCN for 1 min, re-equilibrate for 1 min.

## HPLC VARIABLES
**Column:** 50 × 4.6 3 μm Spherisorb ODS-2
**Mobile phase:** MeCN:water 50:50
**Column temperature:** 60
**Flow rate:** 2
**Injection volume:** 30
**Detector:** UV 200

## CHROMATOGRAM
**Retention time:** 0.6
**Limit of detection:** <0.1 nmol/mL

## OTHER SUBSTANCES
**Simultaneous:** androstenedione, estrone, testosterone

## KEY WORDS
human; placenta

## REFERENCE
Taniguchi, H.; Feldmann, H.R.; Kaufmann, M.; Pyerin, W. Fast liquid chromatographic assay of androgen aromatase activity. *Anal.Biochem.*, **1989**, *181*, 167–171

## SAMPLE
**Matrix:** solutions

## HPLC VARIABLES
**Column:** 150 × 4.6 5 μm Ultrasphere
**Mobile phase:** MeCN:EtOH:water 54:1:45
**Flow rate:** 1.5
**Detector:** UV 270

## CHROMATOGRAM
**Retention time:** 2.4 (17β-estradiol)

## REFERENCE
Fridriksdottir, H.; Loftsson, T.; Gudmundsson, J.A.; Bjarnason, G.J.; Kjeld, M.; Thorsteinsson, T. Design and in vivo testing of 17β-estradiol-HPβCD sublingual tablets. *Pharmazie*, **1996**, *51*, 39–42

## SAMPLE
**Matrix:** solutions

## HPLC VARIABLES
**Column:** 250 × 4.6 5 μm Nucleosil phenyl
**Mobile phase:** Gradient. Carbon dioxide:MeOH from 98:2 to 78:22 over 40 min.
**Column temperature:** 50
**Flow rate:** 2
**Detector:** UV

## CHROMATOGRAM
**Retention time:** 10.3

## OTHER SUBSTANCES
**Simultaneous:** other steroids, estriol, hydrocortisone, hydroxyprogesterone, norethisterone, testosterone

## KEY WORDS
SFC; 200 bar

## REFERENCE
Hanson, M. Aspects of retention behaviour of steroids in packed column supercritical fluid chromatography. *Chromatographia*, **1995**, *40*, 58–68

## SAMPLE
**Matrix:** solutions
**Sample preparation:** Prepare an aqueous solution, inject a 20 μL aliquot.

## HPLC VARIABLES
**Column:** 150 × 4.6 3.5 μm Zorbax SB C18
**Mobile phase:** MeCN:MeOH:buffer 15:45:40 (Buffer was 10 mM $KH_2PO_4$ and 50 mM tetrabutylammonium chloride, pH adjusted to 3.0 with 1 M HCl.)
**Flow rate:** 0.9
**Injection volume:** 20
**Detector:** UV 220

## CHROMATOGRAM
**Retention time:** 9.6 (17β-estradiol), 6.9 (17β-estradiol-3-phosphate)

## OTHER SUBSTANCES
**Simultaneous:** estriol, estrone, estrone-3-phosphate

## KEY WORDS
stability-indicating (for 17β-estradiol-3-phosphate)

## REFERENCE
Miller, R.B.; Chen, C. A stability-indicating HPLC method for the determination of 17β-estradiol-3-phosphate in an ophthalmic solution. *Chromatographia*, **1995**, *40*, 204–206

## SAMPLE
**Matrix:** solutions
**Sample preparation:** Prepare a solution in n-propanol:water 80:20 or DMF:water 80:20, inject an aliquot.

## HPLC VARIABLES
**Column:** 250 × 4 5 μm LiChrospher 100 Diol
**Mobile phase:** Gradient. A was hexane. B was ethyl acetate. C was 0.1% formic acid in MeCN. D was 0.1% formic acid in water. A:B:C:D 100:0:0:0 for 5 min, to 0:100:0:0 over 15 min, maintain at 0:100:0:0 for 5 min, to 0:0:100:0 over 5 min, maintain at 0:0:100:0 for 5 min; to 0:0:0:100 over 25 min, maintain at 0:0:0:100 for 5 min.
**Flow rate:** 0.9
**Detector:** ELSD (Sédex 55, Sédéré)

## CHROMATOGRAM
**Retention time:** 18.22

## OTHER SUBSTANCES
**Simultaneous:** acetylcholine, cholesterol, choline, cortisone, dextrose, glycine, phenylalanine, testosterone

## REFERENCE
Treiber, L.R. Normal-phase high-performance liquid chromatography with relay gradient elution. I. Description of the method. *J.Chromatogr.A*, **1995**, *696*, 193–199

## SAMPLE
**Matrix:** solutions

## HPLC VARIABLES
**Column:** 250 × 4.6 Zorbax RX
**Mobile phase:** Gradient. A was 10 mL concentrated orthophosphoric acid and 7 mL triethylamine in 1 L water. B was 10 mL concentrated orthophosphoric acid and 7 mL triethylamine in 200 mL water, make up to 1 L with MeCN. A:B from 100:0 to 0:100 over 30 min, maintain at 0:100 for 5 min.
**Column temperature:** 30
**Flow rate:** 2
**Detector:** UV 210

## OTHER SUBSTANCES
**Also analyzed:** acepromazine, acetaminophen, acetophenazine, albuterol, aminophylline, amitriptyline, amobarbital, amoxapine, amphetamine, amylocaine, antipyrine, aprobarbital, aspirin, atenolol, atropine, avermectin, barbital, benzocaine, benzoic acid, benzotropine, benzphetamine, berberine, bibucaine, bromazepan, brompheniramine, buprenorphine, buspirone, butabarbital, butacaine, butethal, caffeine, carbamazepine, carbromal, chloramphenicol, chlordiazepoxide, chloroquine, chlorothiazide, chloroxylenol, chlorphe-

nesin, chlorpheniramine, chlorpromazine, chlorpropamide, chlortetracycline, cimetidine, cinchonidine, cinchonine, clenbuterol, clonazepam, clonixin, clorazepate, cocaine, codeine, colchicine, cortisone, coumarin, cyclazocine, cyclobenzaprine, cyclothiazide, cyheptamide, cymarin, danazol, danthron, dapsone, debrisoquine, desipramine, dexamethasone, dextromethorphan, dextropropoxyphene, diamorphine, diazepam, diclofenac, diethylpropion, diethylstilbestrol, diflunisal, digitoxin, digoxin, diltiazem, diphenhydramine, diphenoxylate, diprenorphine, dipyrone, disulfiram, dopamine, doxapram, doxepin, dronabinol, ephedrine, epinephrine, epinine, estriol, estrone, ethacrynic acid, ethosuximide, etonitazene, etorphine, eugenol, famotidine, fenbendazole, fencamfamine, fenoprofen, fenproporex, fentanyl, flubendazole, flufenamic acid, flunitrazepam, 5-fluorouracil, fluoxymesterone, fluphenazine, furosemide, gentisic acid, gitoxigenin, glipizide, glunixin, glutethimide, glybenclamide, guaiacol, halazepam, haloperidol, hydrochlorothiazide, hydrocodone, hydrocortisone, hydromorphone, hydroxyquinoline, ibogaine, ibuprofen, iminostilbene, imipramine, indomethacin, isocarbostyril, isocarboxazid, isoniazid, isoproterenol, isoxsuprine, ivermectin, ketamine, ketoprofen, kynurenic acid, levorphanol, lidocaine, lorazepam, lormetazepam, loxapine, mazindol, mebendazole, meclizine, meclofenamic acid, medazepam, mefenamic acid, megestrol, mepacrine, meperidine, mephentermine, mephenytoin, mephesin, mephobarbital, mepivacaine, mescaline, mesoridazine, methadone, methamphetamine, methapyrilene, methaqualone, methazolamide, methocarbamol, methoxamine, methsuximide, methyl salicylate, methyldopa, methyldopamine, methylphenidate, methylprednisolone, methyltestosterone, methyprylon, metoprolol, mibolerone, morphine, nadolol, nalorphine, naloxone, naltrexone, naphazoline, naproxen, nefopam, niacinamide, nicotine, nicotinic acid, nifedipine, niflumic acid, nitrazepam, norepinephrine, nortriptyline, noscapine, nylidrin, oxazepam, oxycodone, oxymorphone, oxyphenbutazone, oxytetracycline, papaverine, pargyline, pemoline, pentazocine, pentobarbital, persantine, phenacetin, phenazocine, phenazopyridine, phencyclidine, phendimetrazine, phenelzine, pheniramine, phenobarbital, phenothiazine, phensuximide, phentermine, phenylbutazone, phenylephrine, phenylpropanolamine, piperocaine, prazepam, prednisolone, primidone, probenecid, progesterone, propiomazine, propranolol, propylparaben, pseudoephedrine, puromycin, pyrilamine, pyrithyldione, quazepam, quinaldic acid, quinidine, quinine, ranitidine, recinnamine, reserpine, resorcinol, saccharin, albuterol, salicylamide, salicylic acid, scopolamine, scopoletin, secobarbital, strychnine, sulfacetamide, sufadiazine, sulfadimethoxine, sulfaethidole, sulfamerazine, sulfamethazine, sulfamethoxizole, sulfanilamide, sulfapyridine, sulfasoxizole, sulindac, tamoxifen, temazepam, testosterone, tetracaine, tetracycline, tetramisole, thebaine, theobromine, theophylline, thiabendazole, thiamine, thiamylal, thiobarbituric acid, thioridazine, thiosalicylic acid, thiothixene, thymol, tolazamide, tolazoline, tobutamide, tolmetin, tranylcypromine, triamcinolone, tribenzylamine, trichloromethiazide, trifluoperazine, trihexyphenidyl, trimethoprim, tripelennamine, triproilidine, tropacocaine, tyramine, verapamil, vincamine, warfarin, yohimbine, zoxazolamine

## REFERENCE

Hill, D.W.; Kind, A.J. Reversed-phase solvent gradient HPLC retention indexes of drugs. *J.Anal.Toxicol.*, **1994**, *18*, 233–242

## SAMPLE
**Matrix:** solutions
**Sample preparation:** Prepare a 25 µg/mL solution in mobile phase, inject an aliquot.

## HPLC VARIABLES
**Column:** 250 × 4.6 Partisil 10 ODS-1
**Mobile phase:** MeOH:water 55:45
**Column temperature:** 40
**Flow rate:** 1.5
**Detector:** UV 280

## CHROMATOGRAM
**Retention time:** k′ 3.462

## OTHER SUBSTANCES
**Also analyzed:** androsterone (UV 210), cortexolone (UV 240), cortisone (UV 240), estrone (UV 280), ethinyl estradiol (UV 280), ethisterone (UV 240), hydrocortisone (UV 240), hydroxyprogesterone (UV 240), lynestrenol (UV 210), medroxyprogesterone (UV 240), medroxyprogesterone acetate (UV 240), methandienone (UV 240), methylandrostenediol (UV 210), methylprednisolone (UV 240), methylprednisolone acetate (UV 240), methyltestosterone (UV 240), nandrolone (UV 240), norethisterone (UV 240), prednisolone (UV 240), prednisolone acetate (UV 240), prednisone (UV 240), pregnenolone (UV 210), progesterone (UV 240), testosterone (UV 240)

## REFERENCE
Sadlej-Sosnowska, N. Structure retention relationship for steroid hormones. Functional groups as structural descriptors. *J.Liq.Chromatogr.*, **1994**, *17*, 2319–2330

## SAMPLE
**Matrix:** solutions
**Sample preparation:** Inject an aliquot of a solution in MeOH.

## HPLC VARIABLES
**Column:** Radial-PAK μBondapak C18
**Mobile phase:** MeCN:water 50:50
**Flow rate:** 2
**Injection volume:** 100
**Detector:** UV 254; UV 214

## CHROMATOGRAM
**Retention time:** 5.6

## OTHER SUBSTANCES
**Simultaneous:** estriol, estrone, progesterone
**Interfering:** estradiol, testosterone

## REFERENCE
Erkoc, F.U.; Özsar, S.; Güven, B.; Kalkandelen, G.; Ugrar, E. High-performance liquid chromatographic analysis of steroid hormones. *J.Chromatogr.Sci.*, **1989**, *27*, 86–90

## SAMPLE
**Matrix:** solutions
**Sample preparation:** Dissolve in MeOH:water 1:1 at a concentration of 50 μg/mL, inject a 10 μL aliquot.

## HPLC VARIABLES
**Column:** 300 × 3.9 10 μm μBondapak C18
**Mobile phase:** MeOH:acetic acid:triethylamine:water 80:1.5:0.5:18
**Flow rate:** 1.5
**Injection volume:** 10
**Detector:** UV

## CHROMATOGRAM
**Retention time:** k′ 0.41 (estradiol), k′ 3.53 (estradiol benzoate), k′ 7.45 (estradiol cypionate), k′ 3.49 (estradiol valerate)

## REFERENCE
Roos, R.W.; Lau-Cam, C.A. General reversed-phase high-performance liquid chromatographic method for the separation of drugs using triethylamine as a competing base. *J.Chromatogr.*, **1986**, *370*, 403–418

## SAMPLE
**Matrix:** solutions
**Sample preparation:** Prepare a solution in EtOH, inject an aliquot.

## HPLC VARIABLES
**Column:** 150 × 4.6 5 μm Spherisorb S5-ODS
**Mobile phase:** Gradient. MeOH:20 mM ammonium sulfate from 30:70 to 100:0 over 35 min
**Column temperature:** 45
**Flow rate:** 1
**Injection volume:** 50
**Detector:** F ex 214 em 340 (cut-off); UV 280

## CHROMATOGRAM
**Retention time:** 13 (17β-estradiol-3-sulfate), 23 (17β-estradiol)

## OTHER SUBSTANCES
**Simultaneous:** estriol, estriol-3-sulfate, estrone, estrone-3-sulfate

## REFERENCE
Simonian, M.H.; Capp, M.W. Reversed-phase high-performance liquid chromatography of steroid 3-sulfates and the corresponding unconjugated steroids. *J.Chromatogr.*, **1984**, *287*, 97–104

## SAMPLE
**Matrix:** tissue
**Sample preparation:** Incubate endometrial tissue with buffer, remove tissue, extract medium twice with 2 volumes of diethyl ether, evaporate to dryness, reconstitute in a small volume of MeOH, inject an aliquot.

## HPLC VARIABLES
**Column:** 300 × 3.9 Technopak 10 C18
**Mobile phase:** MeOH:0.5% pH 3.0 $(NH_4)H_2PO_4$ 62:38
**Flow rate:** 0.7
**Detector:** UV 280

## CHROMATOGRAM
**Retention time:** 25

## OTHER SUBSTANCES
**Simultaneous:** estrone

## KEY WORDS
endometrial tissue

## REFERENCE
Wild, M.J.; Rudland, P.S.; Back, D.J. Metabolism of the oral contraceptive steroids ethynylestradiol, norgestimate and 3-ketodesogestrel by a human endometrial cancer cell line (HEC-1A) and endometrial tissue *in vitro*. *J.Steroid Biochem.Mol.Biol.*, **1993**, *45*, 407–420

## SAMPLE
**Matrix:** tissue
**Sample preparation:** Dry pack 60 × 8 mm glass columns with 250 mg Carbopack B (200-400 mesh) and 60 × 4 mm glass columns with 50 mg Amberlite CG-400 I (100-200 mesh). Wash Carbopack column with 5 mL MeOH, 15 mL dichloromethane:MeOH 70:30, and MeOH:water 85:15. Wash Amberlite column with 3 mL 0.5 M NaOH, 8 mL

dichloromethane: MeOH 70:30, 1 mL water, and 3 mL 1 M HCl. Repeat this cycle 4 times. Finally pass through 20 mL 50 mM NaOH then 1 mL water. Keep column in water. (Process converts Amberlite to OH form.) Homogenize 1 g of tissue in 5 mL MeOH, sonicate 5 min, centrifuge at 6000 rpm for 10 min. Add another 5 mL MeOH to pellet and repeat. Combine supernatants, make up to 6.8 mL with MeOH, add 1.2 mL water. Pass through Carbopack column, wash column with 2 mL MeOH:water 85:15 then 2 mL MeOH, elute column with 8 mL dichloromethane: MeOH 70:30. Pass eluate onto Amberlite column, wash with 1 mL MeOH, 1 mL 1 M HCl, elute with 2 mL 30 mM HCl in MeCN:MeOH 20:80. Evaporate eluate to dryness with nitrogen at 40°, take up in 100 µL MeCN:MeOH:THF:10 mM $KH_2PO_4$ adjusted to pH 3.0 with phosphoric acid 22:8: 13:57, inject 40 µL aliquot

## HPLC VARIABLES
**Guard column:** 20 × 4.6 5 µm Supelguard LC-18
**Column:** 250 × 4.6 5 µm Supelco C18
**Mobile phase:** MeCN:10 mM $KH_2PO_4$ adjusted to pH 3.0 with phosphoric acid 46:54
**Flow rate:** 1.2
**Injection volume:** 40
**Detector:** F ex 280 em 308

## CHROMATOGRAM
**Retention time:** 7
**Limit of detection:** 1 ng/g

## KEY WORDS
muscle; liver; chicken; ox; cow

## REFERENCE
Laganà, A.; Marino, A. General and selective isolation procedure for high-performance liquid chromatographic determination of anabolic steroids in tissues. *J.Chromatogr.*, **1991**, *588*, 89–98

## SAMPLE
**Matrix:** urine
**Sample preparation:** 50 mL Urine + 7 mL concentrated HCl, heat at 90° for 1 h, add 10 µL 1 mg/mL 4-phenylphenol in MeOH, extract 3 times with 10 mL diethyl ether, combine organic phases, wash twice with 20 mL portions of pH 10.5 $NaHCO_3$/NaOH buffer, wash with 20 mL water, dry over 5 g anhydrous sodium sulfate. Filter, evaporate under reduced pressure almost to dryness, take up residue in 1 mL mobile phase, inject 20 µL aliquot.

## HPLC VARIABLES
**Column:** 250 × 4.5 Beckman ODS
**Mobile phase:** MeCN:water 25:75 containing 14 mM β-cyclodextrin
**Column temperature:** 40
**Flow rate:** 1
**Injection volume:** 20
**Detector:** UV 280; F ex 280 em 312

## CHROMATOGRAM
**Retention time:** 8.1
**Internal standard:** 4-phenylphenol
**Limit of detection:** 1-3 ng/mL

## OTHER SUBSTANCES
**Simultaneous:** estriol, estrone

## REFERENCE

Lamparczyk, H.; Zarzycki, P.K.; Nowakowska, J.; Ochocka, R.J. Application of β-cyclodextrin for the analysis of estrogenic steroids in human urine by high-performance liquid chromatography. *Chromatographia*, **1994**, *38*, 168–172

◆————————————————◆————————————————◆

## ANNOTATED BIBLIOGRAPHY

Liu, P.; Higuchi, W.I.; Ghanem, A.-H.; Good, W.R. Transport of β-estradiol in freshly excised human skin in vitro: Diffusion and metabolism in each skin layer. *Pharm.Res.*, **1994**, *11*, 1777–1784 [perfusate]

Patel, J.U.; Prankerd, R.J.; Sloan, K.B. A prodrug approach to inceasing the oral potency of a phenolic drug. 1. Synthesis, characterization, and stability of an O-(imidomethyl) derivative of 17β-estradiol. *J.Pharm.Sci.*, **1994**, *83*, 1477–1481

Sheikh, S.U.; Touchstone, J.C. Determination of free estriol in amniotic fluid by high performance liquid chromatography. *J.Liq.Chromatogr.*, **1994**, *17*, 3813–3820 [amniotic fluid; plasma; column temp 20; also estriol; SPE]

Chong, K.Y.; Khoo, T.H.; Koo, F.S.; Ong, C.P.; Li, S.F.Y.; Lee, H.K.; Venkatesh, B.; Tan, C.H. Optimization of the high-performance liquid chromatographic separation of steroids by the overlapping resolution mapping procedure. *J.Liq.Chromatogr.*, **1991**, *14*, 2445–2455 [simultaneous androstenedione, 5α-dihydrotestosterone, 17α,20β-dihydroxyprogesterone, 17α-hydroxyprogesterone, 11β-hydroxytestosterone, 11-ketotestosterone, progesterone, testosterone]

Hines, G.A.; Watts, S.A.; Sower, S.A.; Walker, C.W. Sex steroid extraction from echinoderm tissues. *J.Liq.Chromatogr.*, **1990**, *13*, 2489–2498 [testis; ovary; pyloric caecal; also progesterone, testosterone; SPE; sea star; sea urchin; radiolabeled compounds]

Jiang, L.-X.; Wang, Z.-J.; Matlin, S.A. HPLC analysis of injectable contraceptive preparation containing norethisterone enanthate and estradiol valerate. *J.Liq.Chromatogr.*, **1990**, *13*, 3473–3479 [simultaneous benzyl benzoate, estradiol valerate, norethisterone enanthate; column temp 15]

Formento, J.L.; Moll, J.L.; Francoual, M.; Krebs, B.P.; Milano, G.; Renee, N.; Khater, R.; Frenay, M.; Namer, M. HPLC micromethod for simultaneous measurement of estradiol, progesterone, androgen and glucocorticoid receptor levels. Application to breast cancer biopsies. *Eur.J.Cancer Clin.Oncol.*, **1987**, *23*, 1307–1314

Sheikh, S.U.; Touchstone, J. HPLC of steroids in non-aqueous mobile phase at subambient temperature. *J.Liq.Chromatogr.*, **1987**, *10*, 2489–2496 [column temp -50; also cortisone, desoxycorticosterone, estrone; hydrocortisone]

Carignan, G.; Lodge, B.A.; Skakum, W. Simultaneous analysis of estradiol dienanthate, estradiol 3-benzoate and testosterone enanthate benzilic acid hydrazone in oily formulations by gradient HPLC. *J.Liq.Chromatogr.*, **1985**, *8*, 2567–2577 [formulations; oils; gradient; simultaneous estradiol dienanthate, estradiol 3-benzoate, testosterone enanthate benzilic acid hydrazone; 1,2,4,5-tetrachlorobenzene (IS)]

Hayashi, N.; Hayata, K.; Sekiba, K. Rapid and simultaneous measurement of estrone, estradiol, estriol and estetrol in serum by high performance liquid chromatography with electrochemical detection. *Acta Med.Okayama.*, **1985**, *39*, 143–153

Batra, S.K.; Saumande, J. High performance liquid chromatographic separation of estradiol- 17α and–17β in biological fluids; Application to plasma, milk and urine of cows. *J.Liq.Chromatogr.*, **1984**, *7*, 2431–2446 [plasma; milk; urine; cow; radioimmunoassay detection; LOD 100 pg/mL]

Carignan, G.; Lodge, B.A.; Skakum, W. High-performance liquid chromatographic analysis of estradiol valerate-testosterone enanthate in oily formulations. *J.Chromatogr.*, **1984**, *301*, 292–296

Fast, D.M.; Culbreth, P.H.; Sampson, E.J. Multivariate and univariate optimization studies of liquid-chromatographic separation of steroid mixtures. *Clin.Chem.*, **1982**, *28*, 444–448 [also estriol, hydrocortisone, progesterone, testosterone]

Kessler, M.J. A rapid high performance liquid chromatography system for the separation of gonadal steroids. *J.Liq.Chromatogr.*, **1982**, *5*, 125–139 [simultaneous androstandiol, androstenedione, androstenone, dehydroepiandrostenone, dihydrotestosterone, estriol, hydroxypregnenolone, hydroxyprogesterone, hydroxytestosterone, pregnenolone, progesterone, testosterone]

# Estrogens, Conjugated

**Molecular formula:** $C_{18}H_{18}O_2$ (equilenin), $C_{18}H_{22}O_2$ (17α-dihydroequilin), $C_{18}H_{20}O_2$ (equilin), $C_{18}H_{22}O_2$ (estrone), $C_{18}H_{24}O_2$ (estradiol)

**Molecular weight:** 266.3 (equilenin), 272.4 (estradiol), 268.3 (equilin), 270.4 (17α-dihydroequilin), 270.4 (estrone)

**CAS Registry No.:** 474-86-2 (equilin), 50-28-2 (estradiol), 57-91-0 (α-estradiol), 517-09-9 (equilenin), 53-16-7 (estrone), 338-67-5 (estrone sodium sulfate), 481-97-0 (estrone hydrogen sulfate)

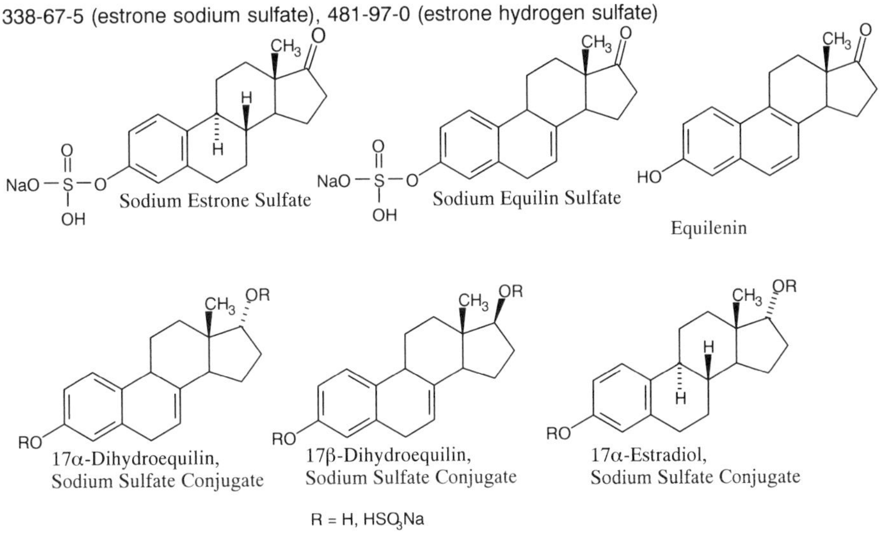

Sodium Estrone Sulfate

Sodium Equilin Sulfate

Equilenin

17α-Dihydroequilin, Sodium Sulfate Conjugate

17β-Dihydroequilin, Sodium Sulfate Conjugate

17α-Estradiol, Sodium Sulfate Conjugate

R = H, HSO₃Na

Conjugated Estrogens
(see Preface)

## SAMPLE
**Matrix:** blood
**Sample preparation:** Add 0.1 (rabbit) or 1 (rat, monkey) plasma to 1 mL 100 mM pH 5.0 acetate buffer and 50 μL Glusulase (from Helix Pomatia, contains 10000 U/mL sulfatase and 90000 U/mL β-glucuronidase, DuPont), heat at 37° for 1 h, cool to room temperature, add 15 mL diethyl ether, shake mechanically at high speed for 10 min, centrifuge at 3033 g for 10 min, freeze in dry ice/acetone. Remove the organic layer and evaporate it to dryness under a stream of nitrogen, reconstitute the residue in 500 μL mobile phase, filter (0.45 μm), inject a 150 μL aliquot of the filtrate.

## HPLC VARIABLES
**Column:** 150 × 3 5 μm C6 (Column Engineering, Ontario CA)
**Mobile phase:** MeCN:MeOH:50 mM pH 3.5 ammonium acetate 27:8:65
**Flow rate:** 0.35
**Injection volume:** 150
**Detector:** F ex 210 em 370

## CHROMATOGRAM
**Retention time:** 19.5 (17α-dihydroequilenin)
**Internal standard:** 14β-equilenin (24)
**Limit of quantitation:** 2.5 ng/mL (rat), 5 ng/mL (rabbit, monkey)

## OTHER SUBSTANCES
**Noninterfering:** equilenin, 17α-dihydroequilenin

## KEY WORDS
plasma; rat; rabbit; monkey; pharmacokinetics

## REFERENCE
Chandrasekaran, A.; Osman, M.; Adelman, S.J.; Warsheski, J.; Scatina, J.; Sisenwine, S.F. Determination of 17α-dihydroequilenin in rat, rabbit and monkey plasma by high-performance liquid chromatography with fluorimetric detection. *J.Chromatogr.B*, **1996**, *676*, 69–75

## SAMPLE
**Matrix:** blood
**Sample preparation:** 10 μL Plasma is injected into MeCN pumped at 0.2 mL/min, precipitated proteins are removed by 0.5 and 0.2 μm filters in series, MeCN containing sample is switched into mobile phase allowed to pass onto analytical column. Next filter unit is switched out of circuit and back-flushed to waste with 100 mM sodium dodecyl sulfate at 2 mL/min, analytical column is eluted in normal fashion with mobile phase. Equilibrate filters with MeCN for 5 min before next injection.

## HPLC VARIABLES
**Guard column:** 20 mm Brownlee C18
**Column:** 250 × 4.6 5 μm Ultrasphere C18
**Flow rate:** 1
**Injection volume:** 10
**Detector:** UV 280

## CHROMATOGRAM
**Retention time:** 25.8 (estradiol), 36.6 (equilin), 42.6 (estrone)

## KEY WORDS
plasma; dog

## REFERENCE
Asafu-Adjaye, E.B.; Su, S.Y.; Shiu, G.K. Switching-valve-filter technique for the direct injection and analysis of drugs in plasma using high-performance liquid chromatography. *J.Chromatogr.B*, **1994**, *652*, 35–42

## SAMPLE
**Matrix:** blood
**Sample preparation:** 100 μL Plasma + 10 μL IS in water, extract twice by shaking for 1 min with 1.2 mL dichloromethane, evaporate organic layer below 40° under reduced pressure, dissolve residue in 100 μL MeCN. Add 10 μL reagent 1, add 10 μL reagent 2, heat at 50° for 15 min, cool to room temperature, add 100 μL water, add 200 μL MeOH:water 1:1, add to Sep-Pak C18 cartridge, wash vial with 2 mL MeOH:water 1:1 and add washings to cartridge, wash cartridge with 40 mL MeOH:water 1:1, elute with 5 mL MeOH. Concentrate eluent to 500 μL by evaporation at 40° under reduced pressure, inject 20 μL aliquot. (Reagent 1 was 30 mg 2-(4-carboxyphenyl)-5,6-dimethylbenzimidazole in 3 mL pyridine, add 700 mg 4-piperidinopyridine, dilute to 10 mL with MeCN. Reagent 2 was 700 mg 1-isopropyl-3-(3-dimethylaminopropyl)carbodiimide perchlorate in 10 mL MeCN.)

## HPLC VARIABLES
**Guard column:** 50 × 4 5 μm Wakosil 5C18
**Column:** 300 × 4 5 μm Wakosil 5C18
**Mobile phase:** MeOH:water 90:10
**Flow rate:** 0.7
**Injection volume:** 20
**Detector:** F ex 336 em 440

## CHROMATOGRAM
**Retention time:** 10.5 (estriol), 15.4 (ethynylestradiol), 16.5 (equilin), 16.5 (equilenin), 17.2 (estrone), 18.2 (estradiol), 19.2 (estetrol), 28.1 (4-hydroxyestradiol), 35.5 (2-hydroxyestradiol)
**Internal standard:** sec-butyl p-hydroxybenzoate (14.3)
**Limit of detection:** 1-2 pg/mL

## KEY WORDS
plasma; equilin and equilenin not resolved

## REFERENCE
Katayama, M.; Taniguchi, H. Determination of estrogens in plasma by high-performance liquid chromatography after pre-column derivatization with 2-(4-carboxyphenyl)-5,6-dimethylbenzimidazole. *J.Chromatogr.*, **1993**, *616*, 317−322

## SAMPLE
**Matrix:** blood
**Sample preparation:** 0.5 mL Serum + 0.5 mL MeCN:water 1:1, vortex 15 s, add 3 mL MeCN, shake 1 min, centrifuge at 1800 rpm for 10 min. Remove supernatant and dry it under a stream of nitrogen at 55°, add 2 mL MeCN:MeOH 1:1, vortex 15 s, centrifuge at 1800 rpm for 10 min. Remove supernatant and dry it under a stream of nitrogen at 55°. Reconstitute in 200 μL MeCN:water 1:1.

## HPLC VARIABLES
**Column:** 250 × 4.6 5 μm Beckman ODS
**Mobile phase:** A 2% tetrabutylammonium hydroxide adjusted to pH 3 with phosphoric acid. B MeCN:water 33:67 A:B was 6.5:93.5
**Flow rate:** 0.8
**Injection volume:** 20
**Detector:** UV 210; F ex 280 em 312

## CHROMATOGRAM
**Retention time:** 53 (estrone), 48 (equilin), 41 (estrone sulfate), 38 (equilin sulfate), 32 (estradiol), 26 (17-α-dihydroequilin sulfate)
**Limit of detection:** 10-100 ng/mL

## KEY WORDS
serum

## REFERENCE
Su, S.Y.; Shiu, G.K.; Simmons, J.; Viswanathan, C.T.; Skelly, J.P. High performance liquid chromatographic analysis of six conjugated and unconjugated estrogens in serum. *Biomed.Chromatogr.*, **1992**, *6*, 265−268

## SAMPLE
**Matrix:** blood
**Sample preparation:** 1 mL Plasma + 5 mL water + 1 mL 2 μg/mL equilenin in MeOH + 50 μL 0.1 M NaOH to adjust pH to 10, vortex briefly after each addition, shake with 10 mL dichloromethane for 10 min, centrifuge at 2000 g for 10 min. Wash organic layer twice with 2 mL water, centrifuge 5 min, evaporate 8 mL of organic phase to dryness at 40°under a stream of nitrogen, reconstitute residue in 150 μL mobile phase, inject 25 μL aliquot.

## HPLC VARIABLES
**Column:** 300 × 4 10 μm μBondapak C18
**Mobile phase:** MeOH:buffer 65:35 (Buffer was 10 mL 200 mM acetic acid + 15 mL 200 mM sodium acetate made up to 1 L, pH 4.8.)

**Flow rate:** 1
**Injection volume:** 25
**Detector:** UV 254

## CHROMATOGRAM
**Retention time:** 5.3
**Internal standard:** equilenin
**Limit of detection:** 5 ng/mL

## OTHER SUBSTANCES
**Simultaneous:** betamethasone, deoxycortisol, dexamethasone, hydrocortisone, prednisone, triamcinolone

## KEY WORDS
Anal.Abs. 1982, 43, 4D182; plasma=uilenin is IS

## REFERENCE
Bouquet, S.; Brisson, A.M.; Gombert, J. Dosage du cortisol et du 11-désoxycortisol plasmatiques par chromatographie liquide haute performance [Cortisol and 11-desoxycortisol determination in blood by high performance liquid chromatography]. *Ann.Biol.Clin.(Paris)*, **1981**, *39*, 189–191

## SAMPLE
**Matrix:** cells
**Sample preparation:** Homogenize (glass-glass homogenizer) cells in medium, centrifuge at 1000 g for 5 min. Add 3 mL homogenate to 10 mL diethyl ether:acetone 90:10, mix thoroughly for a few s, let stand at 4° for 5 min. Remove the organic layer and evaporate it to dryness under a stream of nitrogen, reconstitute the residue in three 500 μL portions of acetone, evaporate to dryness under a stream of nitrogen, reconstitute with 30 μL MeCN, inject a 20 μL aliquot.

## HPLC VARIABLES
**Column:** Ultrasphere ODS
**Mobile phase:** MeCN:10 mM citric acid 40:60
**Column temperature:** 20 ± 1
**Flow rate:** 1
**Injection volume:** 20
**Detector:** UV 280; Radioactivity

## CHROMATOGRAM
**Retention time:** 14 (17β-estradiol), 16 (equilin), 19 (estrone)

## KEY WORDS
tritium labeled; $^{14}$C labeled

## REFERENCE
Castagnetta, L.A.; Granata, O.M.; Lo Casto, M.; Calabro, M.; Arcuri, F.; Carruba, G. Simple approach to measure metabolic pathways of steroids in living cells. *J.Chromatogr.*, **1991**, *572*, 25–39

## SAMPLE
**Matrix:** formulations
**Sample preparation:** Dissolve a quantity equivalent to about 25 mg of conjugated estrogens in 20 mL MeOH, add 20 mL water, add 4 mL concentrated HCl, add several boiling chips, boil for 5 min, cool to room temperature, add 2.5 mL 0.5 mg/mL estriol in MeOH, extract twice with 10 mL and once with 5 mL portions of chloroform, combine extracts, wash with 5 mL water, pass through 1 g anhydrous sodium sulfate, evaporate to dryness under nitrogen, dissolve residue in 25 mL MeOH:water 1:1, inject aliquot.

## HPLC VARIABLES
**Column:** 250 × 4.6 3 μm Nucleosil C18
**Mobile phase:** MeOH:water:isopropanol:dichloromethane 45:42.5:7.5:5
**Flow rate:** 0.7
**Injection volume:** 20
**Detector:** UV 280; E, Laboratorni Pristroje ADLC 2 detector, carbon fiber working electrode, stainless steel counter electrode, 1.1 V vs Ag/AgCl reference electrode, 0.6% $Na_2HPO_4.12H_2O$ added to mobile phase which was adjusted to pH 6.0-6.05 with acetic acid

## CHROMATOGRAM
**Retention time:** 23.18 (estrone), 20.59 (equilin), 18.59 (equilenin), 15.14 (17α-estradiol), 13.36 (17α-dihydroequilin), 11.85 (17α-dihydroequilenin)
**Internal standard:** estriol (6.44)
**Limit of detection:** 2-4 μg/mL

## KEY WORDS
tablets; capsules

## REFERENCE
Novakovic, J.; Tvrzická, E.; Pacáková, V. High-performance liquid chromatographic determination of equine estrogens with ultraviolet absorbance and electrochemical detection. *J.Chromatogr.A*, **1994**, *678*, 359–363

## SAMPLE
**Matrix:** formulations
**Sample preparation:** Dissolve in mobile phase, inject an aliquot.

## HPLC VARIABLES
**Column:** 150 × 3.9 4 μm NovaPak C18
**Mobile phase:** MeOH:25 mM $KH_2PO_4$ 40:60
**Flow rate:** 0.8
**Injection volume:** 50
**Detector:** UV 200

## CHROMATOGRAM
**Retention time:** 19.01 (estrone sulfate), 16.54 (equilin sulfate), 17.87 (17α-dihydroequilin sulfate), 16.16 (17β-dihydroequilin sulfate), 25.28 (17α-estradiol sulfate), 19.01 (17β-estradiol sulfate), 13.88 (equilenin sulfate), 14.26 (17α-dihydroequilenin sulfate)

## KEY WORDS
injections; tablets

## REFERENCE
Flann, B.; Lodge, B. Analysis of estrogen sulphate mixtures in pharmaceutical formulations by reversed-phase chromatography. *J.Chromatogr.*, **1987**, *402*, 273–282

## SAMPLE
**Matrix:** formulations
**Sample preparation:** Dissolve in mobile phase, inject an aliquot.

## HPLC VARIABLES
**Column:** 250 × 4.6 5 μm Ultrasphere Octyl
**Mobile phase:** MeOH:water:trichloroethanol 23:75:2, all 0.1 M in silver nitrate
**Column temperature:** 45
**Flow rate:** 1.0

**Injection volume:** 10
**Detector:** UV 270

## CHROMATOGRAM
**Retention time:** 50.61 (estrone sulfate), 33.40 (equilin sulfate), 19.74 (17α-dihydroequilin sulfate), 14.17 (17β-dihydroequilin sulfate), 47.07 (17α-estradiol sulfate), 33.43 (17β-estradiol sulfate), 30.37 (equilenin sulfate), 23.28 (17α-dihydroequilenin sulfate)

## KEY WORDS
tablets; injections

## REFERENCE
Flann, B.; Lodge, B. Analysis of estrogen sulphate mixtures in pharmaceutical formulations by reversed-phase chromatography. *J.Chromatogr.*, **1987**, *402*, 273–282

## SAMPLE
**Matrix:** formulations
**Sample preparation:** Dissolve in mobile phase, inject an aliquot.

## HPLC VARIABLES
**Column:** 250 × 4.6 5 μm Ultrasphere Octyl
**Mobile phase:** MeCN:MeOH:buffer 32:18:50 (Buffer was 1.7 mM cetyltrimethylammonium phosphate and 25 mM $KH_2PO_4$.)
**Flow rate:** 0.9
**Injection volume:** 50
**Detector:** UV 200

## CHROMATOGRAM
**Retention time:** 58.33 (estrone sulfate), 53.08 (equilin sulfate), 45.50 (17α-dihydroequilin sulfate), 34.41 (17β-dihydroequilin sulfate), 50.75 (17α-estradiol sulfate), 39.66 (17β-estradiol sulfate), 48.41 (equilenin sulfate), 37.91 (17α-dihydroequilenin sulfate), 17.50 (estrone), 15.75 (equilin), 14.00 (17α-dihydroequilin), 11.08 (17β-dihydroequilin), 15.75 (17α-estradiol), 13.42 (17β-estradiol), 14.58 (equilenin sulfate), 9.92 (17β-dihydroequilenin)

## KEY WORDS
tablets; injections

## REFERENCE
Flann, B.; Lodge, B. Analysis of estrogen sulphate mixtures in pharmaceutical formulations by reversed-phase chromatography. *J.Chromatogr.*, **1987**, *402*, 273–282

## SAMPLE
**Matrix:** formulations
**Sample preparation:** Powder tablets, weigh out amount corresponding to 6.9 mg conjugated estrogens, add 6 g Celite 545, add 4 mL water, mix, add to a mixture of 2 g Celite and 1 mL water in a 150 × 25 tube, dry rinse container with 1 g Celite and add this to the tube, elute with 100 mL water-saturated ether, collect this eluate (A), elute with 5 mL 20 mg/mL dicyclohexylamine acetate in chloroform, elute with 145 mL chloroform (B). Combine the chloroform eluates and evaporate them to dryness under a stream of air on a steam bath, reconstitute with 20 mL MeOH, add 6 mL 5% HCl, reflux for 12 min, cool in an ice bath, add 5 mL 400 μg/mL ethinyl estradiol in MeOH, add 70 mL water, add 50 mL benzene (Caution! Benzene is a carcinogen!), shake for 1 min. Remove the organic layer and wash it with 10 mL water, three 15 mL portions of 2% sodium carbonate solution, and two 10 mL portions of water. Pass the organic layer through 30 g anhydrous sodium sulfate in a column to give C. Evaporate a 2 mL aliquot of the solution (or an aliquot of eluate A) to dryness under a stream of air, reconstitute with 10 mL 200 μg/mL

dansyl chloride in acetone, add 15 mL buffer, let stand in the dark for 30 min, add 50 mL water, add 50 mL ether, shake for several min, extract the aqueous layer with 25 mL ether. Combine the ether layers and wash them with two 25 mL portions of water, pass the organic layer through a 150 × 25 column containing 50 g anhydrous sodium sulfate, wash the column with 25 mL ether. Combine the eluates and evaporate them to dryness under a stream of air on a steam bath, reconstitute with 10 mL chloroform, inject an aliquot. (Under these conditions estrone, equilin, and equilenin co-elute. They can be reduced to β-estradiol, β-dihydroequilin, and β-dihydroequilenin, respectively, as follows. Evaporate a 10 mL aliquot of eluate (A) or solution (C) to dryness under a stream of air, reconstitute with 20 mL MeOH, add 150 mg sodium borohydride (Caution! Flammable hydrogen gas is evolved!), let stand for 45 min, add 70 mL water, add 50 mL benzene, shake for 1 min. Remove the organic layer and wash it with four 20 mL portions of water, pass through 30 g of anhydrous sodium sulfate in a 150 × 25 tube, evaporate a 10 mL aliquot to dryness and proceed with the derivatization as described above. (Prepare the buffer by dissolving 366.7 mg anhydrous sodium carbonate in 300 mL water and adding 150 mL acetone. Note that the initial elution with ether (A) gives free estrogens and the elution with chloroform (B) gives 3-sulfate derivatives which are then hydrolyzed.)

## HPLC VARIABLES
**Column:** 250 × 4.6 5 μm Zorbax-Sil
**Mobile phase:** n-Heptane:chloroform:EtOH 50:49.5:0.5
**Flow rate:** 2
**Injection volume:** 10
**Detector:** F ex 240-420 (filter) em 440 (cutoff filter)

## CHROMATOGRAM
**Retention time:** 4 (estrone), 4 (equilin), 4 (equilenin), 12.5 (α-estradiol), 15 (α-dihydroequilin), 16 (α-dihydroequilenin), 18 (β-estradiol), 19.5 (β-dihydroequilin), 22 (β-dihydroequilenin)
**Internal standard:** ethinyl estradiol (9)

## KEY WORDS
normal phase; tablets; derivatization

## REFERENCE
Roos, R.W.; Lau-Cam, C.A. Liquid chromatographic analysis of conjugated and esterified estrogens in tablets. *J.Pharm.Sci.*, **1985**, *74*, 201–204

## SAMPLE
**Matrix:** formulations
**Sample preparation:** Finely powder tablets. Weigh out an amount equivalent to 3 mg piperazine estrone sulfate, add 10 mL mobile phase containing 100 μg/mL biphenyl, shake 30 min, inject 10 μL aliquot.

## HPLC VARIABLES
**Column:** 250 × 4.8 Brownlee RP-18
**Mobile phase:** MeCN:20 mM pH 5.0 phosphate buffer 55:45 containing 3 mM cetyltrimethylammonium bromide
**Flow rate:** 2
**Injection volume:** 10
**Detector:** UV 225

## CHROMATOGRAM
**Retention time:** $k'$ 8.06 (estrone sulfate), $k'$ 7.63 (equilin sulfate), $k'$ 6.45 (α-estradiol sulfate), $k'$ 5.30 (β-estradiol sulfate), $k'$ 2.78 (estrone), $k'$ 2.16 (α-estradiol), $k'$ 1.95 (β-estradiol)
**Internal standard:** biphenyl ($k'$ 11.04)
**Limit of detection:** 1 μg/mL

## OTHER SUBSTANCES
**Simultaneous:** methylparaben, propylparaben

## KEY WORDS
tablets

## REFERENCE
Carignan, G.; Lodge, B.A.; Skakum, W. Analysis of piperazine estrone sulfate in tablets by ion-pair high-performance liquid chromatography. *J.Chromatogr.*, **1982**, *234*, 240–243

## SAMPLE
**Matrix:** formulations
**Sample preparation:** Powder tablets (60 mesh), take powder equivalent to about 3.2 mg conjugated estrogens, add 50 mL MeOH, shake 30 min, dilute to 100 mL with MeOH, mix, filter, discard first 20 mL filtrate, collect the rest of the filtrate (A). Take a 25 mL aliquot, add 1 mL HCl, add boiling chips, heat on a steam bath for 5 min, cool, add 70 mL water, extract with 75 mL benzene (Caution! Benzene is a carcinogen!). Wash the benzene layer with 15 mL water, four times with 15 mL 2% sodium carbonate in water, and twice with 10 mL water. Pass the benzene through a tube containing 30 g anhydrous sodium sulfate, wash the tube with 25 mL benzene, evaporate to dryness. Add 10 mL 200 μg/mL dansyl chloride in acetone, swirl to dissolve, add 15 mL base solution, mix, stopper, allow to stand in the dark for 30 min. Extract twice with 50 mL ether, wash each extract twice with 25 mL water, pass the ether through a 150 × 25 mm tube containing 50 g anhydrous sodium sulfate, wash the column with 25 mL ether, evaporate the ether layers to dryness, dissolve residue in 5 mL chloroform, inject a 10 μL aliquot. (Prepare base solution by dissolving 366.7 mg anhydrous sodium carbonate in 300 mL water and adding 150 mL acetone. Estrone, equilin, and equilenin co-elute. They can be reduced to β-estradiol, β-dihydroequilin, and β-dihydroequilenin, respectively as follows. Add 150 mg sodium borohydride to 25 mL filtrate (A) (Caution! Flammable hydrogen gas is evolved!), let stand for 45 min, add 1 mL HCl, let stand for 15 min, add 25 mL water, add 25 mL benzene, shake, wash the organic layer with four 20 mL portions of water, pass the organic layer through 25 g anhydrous sodium sulfate in a 150 × 25 tube. Evaporate a 1 mL aliquot to dryness under a stream of air and proceed with the derivatization as described above.)

## HPLC VARIABLES
**Column:** 250 × 3.2 5 μm LiChrosorb Si-60
**Mobile phase:** n-Heptane:chloroform 50:50
**Flow rate:** 0.98
**Injection volume:** 10
**Detector:** F ex 240-420 em 440 (cut-off)

## CHROMATOGRAM
**Retention time:** 5 (estrone), 5 (equilin), 5 (equilenin), 14 (α-estradiol), 16 (α-dihydroequilin), 18 (α-dihydroequilenin), 20 (β-estradiol), 21 (β-dihydroequilin), 24 (β-dihydroequilenin)

## KEY WORDS
tablets; normal phase; derivatization

## REFERENCE
Roos, R.W.; Medwick, T. Application of dansyl derivatization to the high pressure liquid chromatographic identification of equine estrogens. *J.Chromatogr.Sci.*, **1980**, *18*, 626–630

## SAMPLE
**Matrix:** solutions

## HPLC VARIABLES
**Column:** 120 × 4 5 μm ODS-2 (Knauer)
**Mobile phase:** MeCN:water 30:70 containing 16 mM β-cyclodextrin
**Column temperature:** 40
**Flow rate:** 1
**Injection volume:** 20
**Detector:** UV 280

## CHROMATOGRAM
**Retention time:** 3 (17β-estradiol), 5 (17α-estradiol), 6 (equilin)

## REFERENCE
Lamparczyk, H.; Zarzycki, P.K. Effect of temperature on separation of estradiol stereoisomers and equilin by liquid chromatography using mobile phases modified with β-cyclodextrin. *J.Pharm. Biomed.Anal.*, **1995**, *13*, 543–549

## SAMPLE
**Matrix:** solutions
**Sample preparation:** Inject a 2 μL aliquot of a 1 mg/mL solution in MeOH.

## HPLC VARIABLES
**Column:** 150 × 4.6 5 μm Zorbax ODS
**Mobile phase:** MeOH:50 mM $KH_2PO_4$ 45:55 containing 5 mg/mL heptakis(2,6-di-O-methyl)-β-cyclodextrin
**Injection volume:** 2
**Detector:** UV 200

## CHROMATOGRAM
**Retention time:** 14 (equilin), 17 (estrone)

## OTHER SUBSTANCES
**Simultaneous:** 2-hydroxyestrone, 4-hydroxyestrone, 16α-hydroxyestrone

## REFERENCE
Spencer, B.J.; Purdy, W.C. High-performance liquid chromatographic separation of equilin, estrone, and estrone derivatives with cyclodextrins as mobile phase additives. *J.Liq.Chromatogr.*, **1995**, *18*, 4063–4080

## SAMPLE
**Matrix:** solutions
**Sample preparation:** Inject an aliquot of a solution in EtOH.

## HPLC VARIABLES
**Column:** 150 × 4.6 5 μm Spherisorb S5-ODS
**Mobile phase:** Gradient. MeOH:20 mM ammonium sulfate from 30:70 to 100:0 over 35 min.
**Column temperature:** 45
**Flow rate:** 1
**Injection volume:** 50
**Detector:** F ex 214 em 340 (cut-off); UV 280

## CHROMATOGRAM
**Retention time:** 5 (estriol-3-sulfate), 12 (estrone-3-sulfate), 13 (17β-estradiol-3-sulfate), 16 (estriol), 22 (estrone), 23 (17β-estradiol)

## REFERENCE

Wei, J.Q.; Wei, J.L.; Zhou, X.T. Optimization of an isocratic reversed phase liquid chromatographic system for the separation of fourteen steroids using factorial design and computer simulation. *Biomed.Chromatogr.*, **1990**, *4*, 34–38

## SAMPLE

**Matrix:** solutions
**Sample preparation:** Inject an aliquot of a solution in mobile phase.

## HPLC VARIABLES

**Column:** 250 × 4.6 7-8 μm Zorbax BP-ODS
**Mobile phase:** MeCN:water 35:65
**Flow rate:** 2
**Injection volume:** 50
**Detector:** UV 280

## CHROMATOGRAM

**Retention time:** 32 (estrone), 28.5 (equilin), 25.5 (equilenin), 23 (17β-estradiol), 18.5 (17α-dihydroequilin), 16 (17α-dihydroequilenin)

## KEY WORDS

also details of normal phase procedure

## REFERENCE

Lin, J.-T.; Heftmann, E. High-performance liquid chromatography of naturally occurring estrogens. *J.Chromatogr.*, **1981**, *212*, 239–244

## SAMPLE

**Matrix:** urine
**Sample preparation:** Condition a Sep-Pak C18 SPE cartridge with 10 mL water, 5 mL MeOH, and 10 mL water. 1 mL Urine + 2 nmoles equilin + 100 μL 1.5 M pH 3 acetate buffer, add to the SPE cartridge, wash with 10 mL 150 mM pH 3 acetate buffer, elute with 3 mL MeOH. Add HCl to the eluate so that the concentration of HCl is 500 mM, heat at 100° for 1.5 h, neutralize with sodium bicarbonate, extract with 2 mL chloroform. Evaporate the organic layer to dryness, reconstitute with 1 mL 5 μM 4-chloro-7-nitro-benzo-2-oxa-1,3-diazole (NBD-Cl) in MeCN containing 25 nM 18-crown-6 and 15 mM potassium carbonate, heat at 80° for 30 min, filter, inject a 10-15 μL aliquot.

## HPLC VARIABLES

**Column:** 5 μm Hypersil ODS
**Mobile phase:** MeOH:water 75:25
**Flow rate:** 1
**Injection volume:** 10-15
**Detector:** UV 380

## CHROMATOGRAM

**Retention time:** 6.50 (estrone), 8.39 (estradiol), 3.23 (estriol)
**Internal standard:** equilin (4.95)
**Limit of detection:** 30-50 nM

## KEY WORDS

derivatization; SPE; derivatives are not fluorescent

## REFERENCE

Tirendi, S.; Lancetta, T.; Bousquet, E. Estrogens determination in urine by RP-HPLC with UV detection. *Farmaco*, **1994**, *49*, 427–430

# Ethinyl Estradiol

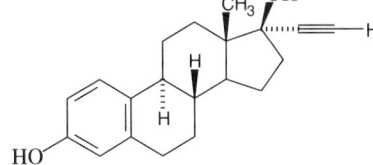

**Molecular formula:** $C_{20}H_{24}O_2$
**Molecular weight:** 296.4
**CAS Registry No.:** 57-63-6

## SAMPLE
**Matrix:** blood
**Sample preparation:** 1 mL Plasma + 500 µL 10 M NaOH, shake on a slow rotatory mixer for 5 min, add 5 mL diethyl ether, rotomix 10 min, centrifuge at 700 g for 5 min, repeat extraction. Combine organic layers, evaporate to dryness under a stream of nitrogen at 37°, dissolve in 250 µL mobile phase, inject aliquot.

## HPLC VARIABLES
**Column:** 150 × 3.9 4 µm Novapack C18
**Mobile phase:** MeCN:MeOH:buffer 35:15:50 (Buffer was 50 mM $KH_2PO_4$ adjusted to pH 3.6 with phosphoric acid.)
**Flow rate:** 1.6
**Injection volume:** 50
**Detector:** E, Waters Model 464 pulsed electrochemical detector, + 1 V versus Ag/AgCl

## CHROMATOGRAM
**Retention time:** 2.94
**Limit of detection:** 50 pg/mL

## OTHER SUBSTANCES
**Simultaneous:** estradiol, estriol, estrone, heparin
**Noninterfering:** pentobarbital

## KEY WORDS
plasma; rabbit

## REFERENCE
Fernández, N.; Garcia, J.J.; Diez, M.J.; Terán, M.T.; Sierra, M. Rapid high-performance liquid chromatographic assay of ethynyloestradiol in rabbit plasma. *J.Chromatogr.*, **1993**, *619*, 143–147

## SAMPLE
**Matrix:** blood
**Sample preparation:** 100 µL Plasma + 10 µL IS in water, extract twice by shaking for 1 min with 1.2 mL dichloromethane, evaporate organic layer below 40° under reduced pressure, dissolve residue in 100 µL MeCN. Add 10 µL reagent 1, add 10 µL reagent 2, heat at 50° for 15 min, cool to room temperature, add 100 µL water, add 200 µL MeOH:water 1:1, add to Sep-Pak C18 cartridge, wash vial with 2 mL MeOH:water 1:1 and add washings to cartridge, wash cartridge with 40 mL MeOH:water 1:1, elute with 5 mL MeOH. Concentrate eluent to 500 µL by evaporation at 40° under reduced pressure, inject 20 µL aliquot. (Reagent 1 was 30 mg 2-(4-carboxyphenyl)-5,6-dimethylbenzimidazole in 3 mL pyridine, add 700 mg 4-piperidinopyridine, dilute to 10 mL with MeCN. Reagent 2 was 700 mg 1-isopropyl-3-(3-dimethylaminopropyl)carbodiimide perchlorate in 10 mL MeCN.)

## HPLC VARIABLES
**Guard column:** 50 × 4 5 µm Wakosil 5C18
**Column:** 300 × 4 5 µm Wakosil 5C18
**Mobile phase:** MeOH:water 90:10

**Flow rate:** 0.7
**Injection volume:** 20
**Detector:** F ex 336 em 440

## CHROMATOGRAM
**Retention time:** 15.4
**Internal standard:** sec-butyl p-hydroxybenzoate (14.3)
**Limit of detection:** 1-2 pg/mL

## OTHER SUBSTANCES
**Simultaneous:** equilenin, equilin, estetrol, estradiol, estriol, estrone, 2-hydroxyestradiol, 4-hydroxyestradiol

## KEY WORDS
plasma; equilin and equilenin not resolved

## REFERENCE
Katayama, M.; Taniguchi, H. Determination of estrogens in plasma by high-performance liquid chromatography after pre-column derivatization with 2-(4-carboxyphenyl)-5,6-dimethylbenzimidazole. *J.Chromatogr.*, **1993**, *616*, 317–322

## SAMPLE
**Matrix:** blood, perfusate
**Sample preparation:** 200 μL Plasma or perfusate + 5 mL dichloromethane, vortex, centrifuge for 10 min. Remove a 4.5 mL aliquot of the lower organic layer and evaporate it to dryness under a stream of nitrogen, reconstitute the residue in 200 μL mobile phase, inject a 50 μL aliquot.

## HPLC VARIABLES
**Column:** LiChrocart 100 RP-18
**Mobile phase:** MeOH:isopropanol:dichloromethane:water 40:9:4:47
**Flow rate:** 1
**Injection volume:** 50
**Detector:** UV 220

## CHROMATOGRAM
**Internal standard:** ethinyl estradiol (17α-ethynylestradiol)

## OTHER SUBSTANCES
**Extracted:** digoxin

## KEY WORDS
plasma; rat; ethinyl estradiol is IS

## REFERENCE
Su, S.-F.; Huang, J.-D. Inhibition of the intestinal digoxin absorption and exsorption by quinidine. *Drug Metab.Dispos.*, **1996**, *24*, 142–147

## SAMPLE
**Matrix:** bulk

## HPLC VARIABLES
**Column:** 250 × 4 10 μm LiChrosorb RP-18
**Mobile phase:** MeOH:water 70:30
**Flow rate:** 1

**Injection volume:** 25
**Detector:** UV 280

## CHROMATOGRAM
**Retention time:** 9

## OTHER SUBSTANCES
**Simultaneous:** impurities, norethindrone
**Interfering:** norgestrel

## REFERENCE
Görög, S.; Herényi, B. Analysis of steroids. XXXVIII. The use of high-performance liquid chromatography
with diode-array UV detection for estimating impurity profiles of steroid drugs. *J.Chromatogr.*, **1987**,
*400*, 177–186

## SAMPLE
**Matrix:** formulations
**Sample preparation:** Dissolve 12 tablets in 600 mL water with stirring at 75 rpm, remove
3 mL sample, centrifuge at 3000 rpm for 15 min, inject a 250 µL aliquot.

## HPLC VARIABLES
**Column:** 100 × 4.6 3 µm Phenomenex IB-Sil 3 C18
**Mobile phase:** MeCN:water 40:60, pH 5.6
**Flow rate:** 1.2
**Injection volume:** 250
**Detector:** UV 200

## CHROMATOGRAM
**Retention time:** 9.5

## OTHER SUBSTANCES
**Simultaneous:** norethindrone

## KEY WORDS
tablets; modification of USP method

## REFERENCE
Dorantes, A.; Stavchansky, S. Modification of the U.S.P. dissolution method for the analysis of norethindrone and ethinyl estradiol tablets. *J.Pharm.Sci.*, **1994**, *83*, 379–381

## SAMPLE
**Matrix:** formulations
**Sample preparation:** Dissolve 6 tablets in 600 mL dissolution medium (100 mM HCl +
0.02% sodium lauryl sulfate), remove 5 mL samples, centrifuge at 1500 rpm for 10 min,
inject a 50-200 µL aliquot.

## HPLC VARIABLES
**Column:** 220 × 4.6 5 µm Spheri-5 C18
**Mobile phase:** MeCN:20 mM pH 6.0 phosphate buffer 35:65
**Flow rate:** 1.5
**Injection volume:** 50-200
**Detector:** UV 200

## CHROMATOGRAM
**Retention time:** 22.73

## OTHER SUBSTANCES
**Simultaneous:** norethindrone

## KEY WORDS
tablets; modified USP method

## REFERENCE
Nguyen, H.T.; Shiu, G.K.; Worsley, W.N.; Skelly, J.P. Dissolution testing of norethindrone : ethinyl estradiol, norethindrone : mestranol, and norethindrone acetate : ethinyl estradiol combination tablets. *J.Pharm.Sci.*, **1990**, *79*, 163–167

## SAMPLE
**Matrix:** formulations
**Sample preparation:** 5 Tablets + 2 glass beads + 25 mL 50 μg/mL dibutyl phthalate in MeOH, vortex 15 min or until tablets have completely disintegrated, sonicate 5 min, filter (2 μm), inject 25 μL aliquot.

## HPLC VARIABLES
**Column:** 50 × 4.5 5μm IBM C18
**Mobile phase:** MeOH : THF : water 10 : 25 : 65
**Flow rate:** 2.1
**Injection volume:** 25
**Detector:** UV 230

## CHROMATOGRAM
**Retention time:** 3.5
**Internal standard:** dibutyl phthalate

## OTHER SUBSTANCES
**Simultaneous:** degradation products, norgestimate

## KEY WORDS
tablets; stability-indicating

## REFERENCE
Lane, P.A.; Mayberry, D.O.; Young, R.W. Determination of norgestimate and ethinyl estradiol in tablets by high-performance liquid chromatography. *J.Pharm.Sci.*, **1987**, *76*, 44–47

## SAMPLE
**Matrix:** formulations
**Sample preparation:** Powder tablets (60 mesh), weigh out amount equivalent to one tablet, add 2 mL 50 μg/mL BHT in MeCN : water 80 : 20, shake 30 min, centrifuge.

## HPLC VARIABLES
**Column:** 250 × 3.2 Altex RP-2 express series
**Mobile phase:** MeCN : water 38 : 62
**Flow rate:** 1.75
**Injection volume:** 20
**Detector:** UV 210; UV 280

## CHROMATOGRAM
**Retention time:** k′ 3.85
**Internal standard:** BHT (butylated hydroxytoluene) (k′ 16.54)

## OTHER SUBSTANCES
**Simultaneous:** degradation products, ethynodiol diacetate, mestranol

## KEY WORDS
tablets

## REFERENCE
Carignan, G.; Lodge, B.A.; Skakum, W. Quantitative analysis of ethynodiol diacetate and ethinyl estradiol/mestranol in oral contraceptive tablets by high-performance liquid chromatography. *J.Pharm.Sci.*, **1982**, *71*, 264–266

## SAMPLE
**Matrix:** formulations
**Sample preparation:** 1 Tablet + 4 mL 50 mM $KH_2PO_4$, rotate 15 min, add 2 mL 1 μg/mL o-phenylphenol in mobile phase, add 4 mL MeOH, rotate 15 min, centrifuge. Remove supernatant, extract residue twice with 5 mL mobile phase (10 min rotation), combine supernatants, inject 50 μL aliquot.

## HPLC VARIABLES
**Column:** 250 × 4.6 10 μm LiChrosorb RP8
**Mobile phase:** MeOH:50 mM $KH_2PO_4$ 3:2
**Flow rate:** 2
**Injection volume:** 50
**Detector:** F ex 280 em 330

## CHROMATOGRAM
**Retention time:** 9
**Internal standard:** o-phenylphenol (6)

## OTHER SUBSTANCES
**Interfering:** norethindrone

## KEY WORDS
tablets; stability-indicating

## REFERENCE
Strusiak, S.H.; Hoogerheide, J.G.; Gardner, M.S. Determination of ethinyl estradiol in solid dosage forms by high-performance liquid chromatography. *J.Pharm.Sci.*, **1982**, *71*, 636–640

## SAMPLE
**Matrix:** media
**Sample preparation:** Extract culture medium twice with 2 volumes of ether, combine the extracts and evaporate them to dryness, reconstitute with MeOH, inject an aliquot.

## HPLC VARIABLES
**Column:** 300 × 3.9 Techopak 10 C18 (HPLC Technology)
**Mobile phase:** MeOH:0.5% pH 3.0 $(NH_4)H_2PO_4$ 62:38
**Flow rate:** 0.7
**Detector:** UV 280; Radioactivity

## CHROMATOGRAM
**Retention time:** 19

## OTHER SUBSTANCES
**Extracted:** estrone
**Interfering:** estradiol

## KEY WORDS
culture medium; tritium labeled

## REFERENCE
Wild, M.J.; Rudland, P.S.; Back, D.J. Metabolism of the oral contraceptive steroids ethynylestradiol and
norgestimate by normal (Huma 7) and malignant (MCF-7 and ZR-75-1) human breast cells in culture.
*J.Steroid Biochem.Mol.Biol.*, **1991**, *39*, 535–543

## SAMPLE
**Matrix:** solutions

## HPLC VARIABLES
**Column:** 50 × 4.6 5 μm Supelcosil LC-18
**Mobile phase:** MeOH:THF:water 10:20:70
**Flow rate:** 2
**Injection volume:** 20
**Detector:** UV 220

## CHROMATOGRAM
**Retention time:** 8.8

## OTHER SUBSTANCES
**Simultaneous:** norethindrone, norethindrone acetate, norethynodrel acetate, norgestrel

## REFERENCE
*Supelco Catalog*, **1994**, p. 779

## SAMPLE
**Matrix:** solutions
**Sample preparation:** Prepare a 25 μg/mL solution in mobile phase, inject an aliquot.

## HPLC VARIABLES
**Column:** 250 × 4.6 Partisil 10 ODS-1
**Mobile phase:** MeOH:water 55:45
**Column temperature:** 40
**Flow rate:** 1.5
**Detector:** UV 280

## CHROMATOGRAM
**Retention time:** k′ 3.204

## OTHER SUBSTANCES
**Also analyzed:** androsterone (UV 210), cortexolone (UV 240), cortisone (UV 240), estradiol
(UV 280), estrone (UV 280), ethisterone (UV 240), hydrocortisone (UV 240), hydroxyprogesterone (UV 240), lynestrenol (UV 210), medroxyprogesterone (UV 240), medroxyprogesterone acetate (UV 240), methandienone (UV 240), methylandrostenediol (UV 210),
methylprednisolone (UV 240), methylprednisolone acetate (UV 240), methyltestosterone
(UV 240), nandrolone (UV 240), norethisterone (UV 240), prednisolone (UV 240), prednisolone acetate (UV 240), prednisone (UV 240), pregnenolone (UV 210), progesterone
(UV 240), testosterone (UV 240)

## REFERENCE

Sadlej-Sosnowska, N. Structure retention relationship for steroid hormones. Functional groups as structural descriptors. *J.Liq.Chromatogr.*, **1994**, *17*, 2319–2330

## SAMPLE

**Matrix:** solutions

## HPLC VARIABLES

**Column:** 250 × 4.6 10 μm Nucleosil C18
**Mobile phase:** MeCN:THF:water 12.9:22.4:64.7
**Flow rate:** 1
**Detector:** UV 240

## CHROMATOGRAM

**Retention time:** 14

## OTHER SUBSTANCES

**Simultaneous:** estrone, mestranol, norethindrone, norethindrone acetate, norgestrel

## REFERENCE

Gazdag, M.; Szepesi, G.; Szeleczki, E. Selection of high-performance liquid chromatographic methods in pharmaceutical analysis. I. Optimization for selectivity in reversed-phase chromatography. *J.Chromatogr.*, **1988**, *454*, 83–94

## SAMPLE

**Matrix:** solutions

## HPLC VARIABLES

**Column:** 250 × 4.6 5 μm LiChrosorb Si 60
**Mobile phase:** Hexane:dioxane:isopropanol 95:3:2 (Caution! Dioxane is a carcinogen!)
**Flow rate:** 1
**Detector:** UV 254

## CHROMATOGRAM

**Retention time:** 17

## OTHER SUBSTANCES

**Simultaneous:** estrone, mestranol, norethindrone, norethindrone acetate, norgestrel

## KEY WORDS

normal phase

## REFERENCE

Gazdag, M.; Szepesi, G.; Fábián-Varga, K. Selection of high-performance liquid chromatographic methods in pharmaceutical analysis. II. Optimization for selectivity in normal-phase systems. *J.Chromatogr.*, **1988**, *454*, 95–107

## SAMPLE

**Matrix:** solutions

## HPLC VARIABLES

**Column:** 200 × 4 5 μm Nucleosil RP-8
**Mobile phase:** MeCN:MeOH:water 3:76:21
**Flow rate:** 1

**Injection volume:** 20
**Detector:** UV 205

## CHROMATOGRAM
**Retention time:** 2.8

## OTHER SUBSTANCES
**Simultaneous:** metabolites, desogestrel, 3-hydroxydesogestrel, 3-ketodesogestrel, 6-ketoethinyl estradiol

## REFERENCE
Smilde, A.K.; Bruins, C.H.P.; Doornbos, D.A.; Vink, J. Optimization of the reversed-phase high-performance liquid chromatographic separation of synthetic estrogenic and progestogenic steroids using the multi-criteria decision making method. *J.Chromatogr.*, **1987**, *410*, 1–12

## SAMPLE
**Matrix:** solutions
**Sample preparation:** Extract 15 mL water with dichloromethane, evaporate organic layer, take up residue in 3 mL mobile phase, inject 50 µL aliquot.

## HPLC VARIABLES
**Column:** reverse phase
**Mobile phase:** MeOH:water 82:18
**Injection volume:** 50
**Detector:** F ex 200 em 300

## CHROMATOGRAM
**Internal standard:** mestranol
**Limit of quantitation:** 10 ng/mL

## REFERENCE
de Leede, L.G.J.; Govers, C.P.M.; de Nijs, H. A multi-compartment vaginal ring system for independently adjustable release of contraceptive steroids. *Contraception*, **1986**, *34*, 589–602

## SAMPLE
**Matrix:** solutions
**Sample preparation:** Dissolve in MeOH:water 1:1 at a concentration of 50 µg/mL, inject a 10 µL aliquot.

## HPLC VARIABLES
**Column:** 300 × 3.9 10 µm µBondapak C18
**Mobile phase:** MeOH:acetic acid:triethylamine:water 70:1.5:0.5:28
**Flow rate:** 1.5
**Injection volume:** 10
**Detector:** UV

## CHROMATOGRAM
**Retention time:** k' 1.45

## REFERENCE
Roos, R.W.; Lau-Cam, C.A. General reversed-phase high-performance liquid chromatographic method for the separation of drugs using triethylamine as a competing base. *J.Chromatogr.*, **1986**, *370*, 403–418

## SAMPLE
**Matrix:** tissue
**Sample preparation:** Incubate endometrial tissue with buffer, remove tissue, extract medium twice with 2 volumes of diethyl ether, evaporate to dryness, reconstitute in a small volume of MeOH, inject an aliquot.

## HPLC VARIABLES
**Column:** 300 × 3.9 Technopak 10 C18
**Mobile phase:** MeOH:0.5% pH 3.0 $(NH_4)H_2PO_4$ 62:38
**Flow rate:** 0.7
**Detector:** UV 280

## CHROMATOGRAM
**Retention time:** 24

## OTHER SUBSTANCES
**Extracted:** metabolites

## KEY WORDS
endometrial tissue

## REFERENCE
Wild, M.J.; Rudland, P.S.; Back, D.J. Metabolism of the oral contraceptive steroids ethynylestradiol, norgestimate and 3-ketodesogestrel by a human endometrial cancer cell line (HEC-1A) and endometrial tissue *in vitro*. *J.Steroid Biochem.Mol.Biol.*, **1993**, *45*, 407–420

## ANNOTATED BIBLIOGRAPHY

Tacey, R.L.; Harman, W.J.; Kelly, L.L. Development of a highly sensitive and specific assay for plasma ethinylestradiol using combined extraction, liquid chromatography and radioimmunoassay. *J.Pharm.Biomed.Anal.*, **1994**, *12*, 1303–1310 [plasma; RIA detection; SPE; LOD 2 pg/mL]

Standeven, A.M.; Shi, Y.E.; Sinclair, J.F.; Sinclair, P.R.; Yager, J.D. Metabolism of the liver tumor promoter ethinyl estradiol by primary cultures of rat hepatocytes. *Toxicol.Appl.Pharmacol.*, **1990**, *102*, 486–496 [gradient; UV detection; radioactivity detection; extracted metabolites]

Backe, W. [Determination of ethinyl estradiol in feces of calves and cattle with high pressure liquid chromatography]. *Arch.Pharm.(Weinheim).*, **1988**, *321*, 431–432

Lee, G.J.-L.; Oyang, M.-H.; Bautista, J.; Kushinsky, S. Determination of ethinylestradiol and norethindrone in a single specimen of plasma by automated high-performance liquid chromatography and subsequent radioimmunoassay. *J.Liq.Chromatogr.*, **1987**, *10*, 2305–2318 [extracted norethindrone; plasma; RIA detection; LOD 20 pg/mL]

Nielen, M.W.F.; van Soest, R.E.J.; van Ingen, H.E.; Farjam, A.; Frei, R.W.; Brinkman, U.A.T. Selective on-line trace enrichment for the determination of ethynyl steroids in urine by liquid chromatography with precolumn technology. *J.Chromatogr.*, **1987**, *417*, 159–167 [column-switching; urine]

Reif, V.D.; Eickhoff, W.M.; Jackman, J.K.; DeAngelis, N.J. Automated stability-indicating high-performance liquid chromatographic assay for ethinyl estradiol and (levo)norgestrel tablets. *Pharm.Res.*, **1987**, *4*, 54–58

Roos, R.W.; Lau-Cam, C.A. Liquid chromatographic analysis of conjugated and esterified estrogens in tablets. *J.Pharm.Sci.*, **1985**, *74*, 201–204 [ethinyl estradiol is IS; derivatization; fluorescence detection]

Swynnerton, N.F.; Fischer, J.B. Determination of ethynylestradiol and norethindrone in synthetic intestinal fluid and in timed-release oral formulations. *J.Liq.Chromatogr.*, **1980**, *3*, 1195–1204

# Ethynodiol Diacetate

**Molecular formula:** $C_{24}H_{32}O_4$
**Molecular weight:** 384.5
**CAS Registry No.:** 297-76-7 (ethynodiol diacetate),
1231-93-2 (ethynodiol)

## SAMPLE
**Matrix:** formulations
**Sample preparation:** Powder tablets (60 mesh), weigh out amount equivalent to one tablet, add 2 mL 50 µg/mL BHT in MeCN:water 80:20, shake 30 min, centrifuge

## HPLC VARIABLES
**Column:** 250 × 3.2 Altex RP-2 express series
**Mobile phase:** MeCN:water 38:62
**Flow rate:** 1.75
**Injection volume:** 20
**Detector:** UV 210; UV 280

## CHROMATOGRAM
**Retention time:** k′ 22.65
**Internal standard:** BHT (butylated hydroxytoluene) (k′ 16.54)

## OTHER SUBSTANCES
**Simultaneous:** degradation products, ethinyl estradiol, mestranol

## KEY WORDS
tablets

## REFERENCE
Carignan, G.; Lodge, B.A.; Skakum, W. Quantitative analysis of ethynodiol diacetate and ethinyl estradiol/mestranol in oral contraceptive tablets by high-performance liquid chromatography. *J.Pharm.Sci.*, **1982**, *71*, 264–266

## SAMPLE
**Matrix:** solutions
**Sample preparation:** Inject a 20 µL aliquot of a 30 µg/mL solution.

## HPLC VARIABLES
**Column:** 50 × 4.6 5 µm Supelcosil LC-DP
**Mobile phase:** MeCN:THF:water 26:14:60
**Flow rate:** 1
**Injection volume:** 20
**Detector:** UV 220

## CHROMATOGRAM
**Retention time:** 15

## OTHER SUBSTANCES
**Simultaneous:** ethinyl estradiol, mestranol, norethindrone, norethindrone acetate, norethynodrel acetate, norgestrel

## REFERENCE
*Supelco Chromatography Products*, **1996**, p. A130

# Etodolac

**Molecular formula:** $C_{17}H_{21}NO_3$
**Molecular weight:** 287.4
**CAS Registry No.:** 41340-25-4

## SAMPLE
**Matrix:** blood
**Sample preparation:** 2 mL Whole blood or plasma + 2 mL buffer + 5 mL chloroform: isopropanol:n-heptane 60:14:26, shake gently horizontally for 10 min, centrifuge at 2800 g for 10 min. Remove the lower organic layer and evaporate it to dryness under vacuum at 45°, reconstitute the residue in 100 μL mobile phase, centrifuge at 2800 g for 5 min, inject a 50 μL aliquot of the supernatant. (Buffer was saturated ammonium chloride solution 25% diluted with water, adjusted to pH 9.5 with 25% ammonia solution.)

## HPLC VARIABLES
**Column:** 300 × 3.9 4 μm NovaPack C18
**Mobile phase:** MeOH:THF:buffer 65:5:30 (Buffer was 0.68 g/L (10 mM (sic)) $KH_2PO_4$ adjusted to pH 2.6 with concentrated orthophosphoric acid.) (At the end of each session wash the column with water for 1 h and MeOH for 1 h, re-equilibrate for 30 min.)
**Column temperature:** 30
**Flow rate:** 0.8
**Injection volume:** 50
**Detector:** UV 273

## CHROMATOGRAM
**Retention time:** 9.13
**Limit of detection:** <120 ng/mL

## OTHER SUBSTANCES
**Extracted:** acebutolol, acenocoumarol, acepromazine, aceprometazine, acetaminophen, aconitine, ajmaline, albuterol, alimemazine, alminoprofen, alpidem, alprazolam, alprenolol, amisulpride, amodiaquine, amoxapine, aspirin, astemizole, atenolol, benazepril, benperidol, benzocaine, benzoylecgonine, bepridil, betaxolol, bisoprolol, bromazepam, brompheniramine, bumadizone, bupivacaine, buprenorphine, buspirone, caffeine, carbamazepine, carbinoxamine, carpipramine, carteolol, celiprolol, cetirizine, chlorambucil, chlordiazepoxide, chlormezanone, chlorophenacinone, chloroquine, chlorpheniramine, chlorpromazine, chlorpropamide, cibenzoline, cicletanine, clemastine, clobazam, clomipramine, clonazepam, clonidine, clorazepate, clozapine, cocaine, codeine, colchicine, cyamemazine, cyclizine, cycloguanil, cyproheptadine, cytarabine, dacarbazine, daunorubicin, debrisoquine, demexiptiline, desipramine, dextromethorphan, dextromoramide, dextropropoxyphene, diazepam, diazoxide, diclofenac, dihydralazine, diltiazem, diphenhydramine, dipyridamole, disopyramide, dosulepine, doxepin, doxylamine, droperidol, ephedrine, estazolam, fenfluramine, fenoprofen, fentiazac, flecainide, floctafenine, flumazenil, flunitrazepam, fluphenazine, flurbiprofen, fluvoxamine, glibenclamide, glibornuride, glipizide, glutethimide, haloperidol, histapyrrodine, hydroxychloroquine, hydroxyzine, ibuprofen, imipramine, indomethacin, iproniazid, ketamine, ketoprofen, labetalol, levomepromazine, lidocaine, lidoflazine, lisinopril, loperamide, loprazolam, loratadine, lorazepam, loxapine, medazepam, medifoxamine, mefenamic acid, mefenidramine, mefloquine, melphalan, meperidine, mephenesin, mephentermine, mepivacaine, metapramine, metformin, methadone, methaqualone, methocarbamol, methotrexate, metipranolol, metoclopramide, metoprolol, mexiletine, mianserin, midazolam, minoxidil, moclobemide, moperone, morphine, nadolol, nalbuphine, nalorphine, naloxone, naltrexone, naproxen, nialamide, nicardipine, nifedipine, niflumic acid, nimodipine, nitrazepam, nitrendipine, nizatidine, nomifensine, omeprazole, opipramol, oxazepam, oxprenolol, penbutolol, penfluridol, pentazocine, phencyclidine, phenobarbital, phenol, phenylbutazone, pimozide,

pindolol, pipamperone, piroxicam, prazepam, prazosin, prilocaine, procainamide, procarbazine, proguanil, promethazine, propafenone, propranolol, protriptyline, pyrimethamine, quinidine, quinine, quinupramine, ramipril, ranitidine, reserpine, ritodrine, secobarbital, sotalol, strychnine, sulfinpyrazole, sulindac, sulpride, sultopride, suriclone, temazepam, tenoxicam, terbutaline, terfenadine, tetracaine, tetrazepam, thiopental, thioproperazine, thioridazine, tianeptine, tiapride, tiaprofenic acid, ticlopidine, timolol, tioclomarol, tofisopam, tolbutamide, toloxatone, trazodone, triazolam, trifluoperazine, trifluperidol, trimipramine, triprolidine, verapamil, viloxazine, vinblastine, vincristine, vindesine, warfarin, yohimbine, zolpidem, zopiclone, zorubicine
**Interfering:** amitriptyline, fluoxetine, maprotiline, nortriptyline, tropatenine

## KEY WORDS
whole blood; plasma

## REFERENCE
Tracqui, A.; Kintz, P.; Mangin, P. Systematic toxicological analysis using HPLC/DAD. *J.Forensic Sci.*, **1995**, *40*, 254–262

## SAMPLE
**Matrix:** blood
**Sample preparation:** 1 mL Serum or plasma + 4 mL 1 M HCl + 5 mL hexane:isopentyl alcohol 95:5, shake mechanically for 15 min, centrifuge at 1000 rpm for 5 min. Remove a 4 mL aliquot of the organic layer and add it to 1 mL 100 mM pH 11.0 Tris buffer, shake mechanically for 15 min. Remove an 800 µL aliquot of the aqueous layer and add it to 20 µL 2.5 M phosphoric acid, inject a 50-150 µL aliquot.

## HPLC VARIABLES
**Column:** 250 × 4.6 5 µm Spherisorb ODS
**Mobile phase:** MeCN:100 mM pH 6.0 potassium phosphate 30:70
**Column temperature:** 50
**Flow rate:** 1.8
**Injection volume:** 50-150
**Detector:** UV 226

## CHROMATOGRAM
**Retention time:** 5.0
**Limit of detection:** 200 ng/mL

## OTHER SUBSTANCES
**Simultaneous:** acetaminophen, indomethacin, salicylic acid
**Noninterfering:** dicumarol, phenylbutazone, ethacrynic acid, glyburide, hydrochlorothiazide, niacin, phenobarbital, propoxyphene, diazepam

## KEY WORDS
rat; dog; plasma; human; serum

## REFERENCE
Cosyns, L.; Spain, M.; Kraml, M. Sensitive high-performance liquid chromatographic method for the determination of etodolac in serum. *J.Pharm.Sci.*, **1983**, *72*, 275–277

## SAMPLE
**Matrix:** blood, urine
**Sample preparation:** Plasma. 500 µL Plasma + 50 µL 100 µg/mL IS in 10 mM NaOH + 100 µL 0.6 M sulfuric acid + 3 mL isooctane:isopropanol 95:5, vortex 30 s, centrifuge at 2500 rpm for 5 min. Remove organic layer and evaporate it to dryness. Reconstitute in 200 µL MeCN:water 1:1, inject aliquot

## HPLC VARIABLES
**Column:** $100 \times 4.6$ Partisil 5 ODS-3
**Mobile phase:** MeCN:triethylamine:70 mM $KH_2PO_4$ 35:0.02:65
**Flow rate:** 1
**Injection volume:** 5-50
**Detector:** UV 280

## CHROMATOGRAM
**Retention time:** 6.8
**Internal standard:** $(\pm)$-2-(4-benzoylphenyl)butyric acid

## KEY WORDS
plasma

## REFERENCE
Jamali, F.; Mehvar, R.; Lemko, C.; Eradiri, O. Application of a stereospecific high-performance liquid chromatography assay to a pharmacokinetic study of etodolac enantiomers in humans. *J.Pharm.Sci.*, **1988**, *77*, 963–966

## SAMPLE
**Matrix:** blood, urine
**Sample preparation:** Plasma. 500 μL Plasma + 50 μL 100 μg/mL IS in 10 mM NaOH + 100 μL 0.6 M sulfuric acid + 3 mL isooctane:isopropanol 95:5, vortex 30 s, centrifuge at 2500 rpm for 5 min. Remove organic layer and evaporate it to dryness. Reconstitute in 200 μL 50 mM triethylamine in MeCN, add 50 μL 6 mM ethyl chloroformate in MeCN, after 30 s add 50 μL 0.5 M S-(−)-α-methylbenzylamine in MeCN:triethylamine 80:20, after 2 min add 500 μL 0.25 M HCl, add 3 mL chloroform, vortex 15 s, centrifuge at 2500 rpm for 2 min. Remove organic layer and evaporate it. Dissolve residue in 200 μL mobile phase, inject 5-50 μL aliquot. Urine. 500 μL Urine + 250 μL 1 M NaOH, add 300 μL sulfuric acid, proceed as for plasma (above).

## HPLC VARIABLES
**Guard column:** 5 cm 30-38 μm HC Pellosil
**Column:** $250 \times 4.6$ 5 μm Partisil 5
**Mobile phase:** Hexane:ethyl acetate:isopropanol 85:15:0.2
**Flow rate:** 2
**Injection volume:** 5-50
**Detector:** UV 280

## CHROMATOGRAM
**Retention time:** 21.9 (R), 29.5 (S)
**Internal standard:** $(\pm)$-2-(4-benzoylphenyl)butyric acid (10, 17)
**Limit of detection:** <100 ng/mL

## KEY WORDS
plasma; normal phase; chiral; derivatization

## REFERENCE
Jamali, F.; Mehvar, R.; Lemko, C.; Eradiri, O. Application of a stereospecific high-performance liquid chromatography assay to a pharmacokinetic study of etodolac enantiomers in humans. *J.Pharm.Sci.*, **1988**, *77*, 963–966

## SAMPLE
**Matrix:** bulk
**Sample preparation:** 10 mg Etodolac + 10 mg 1-(3-dimethylaminopropyl)-3-ethylcarbo-diimide hydrochloride + 2 drops 3,5-dimethylaniline + 1.5 mL dichloromethane, mix,

after 30 min add 1 mL 1 M HCl, shake vigorously. Remove the lower organic layer and dry it over anhydrous magnesium sulfate, inject an aliquot.

## HPLC VARIABLES
**Column:** 250 × 4.6 5 μm D N-(3,5-dinitrobenzoyl)phenylglycine (Regis)
**Mobile phase:** Hexane:isopropanol 80:20
**Flow rate:** 2
**Injection volume:** 20
**Detector:** UV 254; UV 280

## CHROMATOGRAM
**Retention time:** k′ 1.33 (for first enantiomer)

## OTHER SUBSTANCES
**Also analyzed:** carprofen, cicloprofen, fenoprofen, flurbiprofen, ibuprofen, ketoprofen, naproxen, pirprofen, tiaprofenic acid

## KEY WORDS
derivatization; α = 1.35; chiral

## REFERENCE
Pirkle, W.H.; Murray, P.G. The separation of the enantiomers of a variety of non-steroidal anti-inflammatory drugs (NSAIDS) as their anilide derivatives using a chiral stationary phase. *J.Liq.Chromatogr.*, **1990**, *13*, 2123–2134

## SAMPLE
**Matrix:** solutions
**Sample preparation:** Inject an aliquot of a 500 μg/mL solution in MeOH.

## HPLC VARIABLES
**Column:** 250 × 4.6 5 μm CSP 2 polymeric chiral stationary phase (preparation details in paper)
**Mobile phase:** n-Hexane:EtOH:acetic acid 99:1:0.005
**Column temperature:** 25
**Flow rate:** 2
**Detector:** UV 220

## CHROMATOGRAM
**Retention time:** k′ 0.33 (for first enantiomer)

## OTHER SUBSTANCES
**Also analyzed:** carprofen, cicloprofen, fenoprofen, flurbiprofen, ibuprofen, naproxen, pirprofen

## KEY WORDS
α = 1.76; chiral

## REFERENCE
Terfloth, G.J.; Pirkle, W.H.; Lynam, K.G.; Nicolas, E.C. Broadly applicable polysiloxane-based chiral stationary phase for high-performance liquid chromatography and supercritical fluid chromatography. *J.Chromatogr.A*, **1995**, *705*, 185–194

## SAMPLE
**Matrix:** solutions

**Sample preparation:** Dilute 1 mL buffer solution with 4 mL mobile phase, inject a 40 μL aliquot.

## HPLC VARIABLES
**Column:** 150 × 4.1 5 μm Spherisorb ODS
**Mobile phase:** MeCN:50 mM KH$_2$PO$_4$ 45:55
**Flow rate:** 2
**Injection volume:** 40
**Detector:** UV 230

## CHROMATOGRAM
**Retention time:** 3

## OTHER SUBSTANCES
**Simultaneous:** degradation products

## REFERENCE
Lee, Y.J.; Padula, J.; Lee, H.K. Kinetics and mechanisms of etodolac degradation in aqueous solutions. *J.Pharm.Sci.*, **1988**, 77, 81−86

## SAMPLE
**Matrix:** solutions
**Sample preparation:** Inject a 10 μL aliquot of a 300 μg/mL solution in hexane:isopropanol 80:20.

## HPLC VARIABLES
**Column:** 250 × 4.6 Bakerbond-DNBPG 07651-2-20
**Mobile phase:** Hexane:isopropanol 99:1
**Flow rate:** 1
**Injection volume:** 10
**Detector:** UV 274

## KEY WORDS
chiral

## REFERENCE
Demerson, C.A.; Humber, L.G.; Abraham, N.A.; Schilling, G.; Martel, R.R.; Pace-Asciak, C. Resolution of etodolac and antiinflammatory and prostaglandin synthetase inhibiting properties of the enanti-omers. *J.Med.Chem.*, **1983**, 26, 1778−1780

## SAMPLE
**Matrix:** urine
**Sample preparation:** Dilute a 100-200 μL aliquot of urine to 1.1 mL with water, add 200 μL 1 M HCl, extract with 3 mL cyclohexane:ethyl acetate 95:5. Remove the organic layer and evaporate it to dryness under a stream of nitrogen, reconstitute the residue in 50 μL 100 μg/mL ibuprofen in MeCN, inject a 20 μL aliquot. (Hydrolyze conjugates by adding 100 μL 1 M NaOH and vortexing twice for 15 s periods, proceed as above.)

## HPLC VARIABLES
**Guard column:** 30 × 4 10 μm LiChrospher 60 CN
**Column:** 250 × 4 5 μm LiChrospher 100 RP-18
**Mobile phase:** MeCN:50 mM pH 4.0 phosphate buffer 45:55
**Flow rate:** 1.3
**Injection volume:** 20
**Detector:** UV 220

## CHROMATOGRAM
**Retention time:** 12.5
**Internal standard:** ibuprofen (16.9)
**Limit of quantitation:** 125 ng/mL

## KEY WORDS
pharmacokinetics

## REFERENCE
Becker-Scharfenkamp, U.; Blaschke, G. Evaluation of the stereoselective metabolism of the chiral analgesic drug etodolac by high-performance liquid chromatography. *J.Chromatogr.*, **1993**, *621*, 199–207

## SAMPLE
**Matrix:** urine
**Sample preparation:** Dilute a 100-200 μL aliquot of urine to 1.1 mL with water, add 200 μL 1 M HCl, extract with 3 mL cyclohexane:ethyl acetate 95:5. Remove the organic layer and evaporate it to dryness under a stream of nitrogen, reconstitute the residue in 50 μL isopropanol, inject a 20 μL aliquot. (Hydrolyze conjugates by adding 100 μL 1 M NaOH and vortexing twice for 15 s periods, proceed as above.)

## HPLC VARIABLES
**Guard column:** 30 × 4 10 μm LiChrosorb NH2
**Column:** 125 × 4 bovine serum albumin on silica, cross-linked with formaldehyde
**Mobile phase:** Isopropanol:50 mM pH 7.0 phosphate buffer 7:93
**Flow rate:** 1
**Injection volume:** 20
**Detector:** UV 220

## CHROMATOGRAM
**Retention time:** 14 (S-(+)), 22 (R-(−))
**Limit of quantitation:** 125 ng/mL

## KEY WORDS
pharmacokinetics; chiral

## REFERENCE
Becker-Scharfenkamp, U.; Blaschke, G. Evaluation of the stereoselective metabolism of the chiral analgesic drug etodolac by high-performance liquid chromatography. *J.Chromatogr.*, **1993**, *621*, 199–207

## SAMPLE
**Matrix:** urine
**Sample preparation:** Hydrolyze 1 mL urine in 200 mM pH 4.6 sodium acetate buffer with 40000 U Glusulase (DuPont) at 37° for 20 h, extract with 20 g neutral Amberlite XAD-2 resin. Elute the resin with 40 mL MeOH and methylate with excess (trimethylsilyl)diazomethane to give N-methyl etodolac methyl ester.

## HPLC VARIABLES
**Column:** 250 × 4.6 Microsorb C18
**Mobile phase:** MeCN:water 75:25
**Flow rate:** 1.5
**Detector:** UV 226

## REFERENCE

Humber, L.G.; Ferdinandi, E.; Demerson, C.A.; Ahmed, S.; Shah, U.; Mobilio, D.; Sabatucci, J.; De Lange, B.; Labbadia, F.; Hughes, P.; DeVirgilio, J.; Neuman, G.; Chau, T.T.; Weichman, B.M. Etodolac, a novel antiinflammatory agent. The syntheses and biological evaluation of its metabolites. *J.Med.Chem.*, **1988**, *31*, 1712–1719

---

## ANNOTATED BIBLIOGRAPHY

Caccamese, S. Direct high-performance liquid chromatography (HPLC) separation of etodolac enantiomers using chiral stationary phases. *Chirality*, **1993**, *5*, 164–167

# Etoposide

**Molecular formula:** $C_{29}H_{32}O_{13}$
**Molecular weight:** 588.6
**CAS Registry No.:** 33419-42-0, 117091-64-2 (phosphate)

## SAMPLE
**Matrix:** blood
**Sample preparation:** 500 μL Plasma + 50 μL 1 μg/mL 17β-estradiol + 200 μL 200 mM pH 8.0 $Na_2HPO_4$, extract into ethylene dichloride, centrifuge for 5 min. Remove the organic layer and evaporate it to dryness under a stream of nitrogen, reconstitute the residue in 200 μL MeCN:MeOH:water 30:15:55, inject a 50 μL aliquot.

## HPLC VARIABLES
**Column:** 5 μm Zorbax phenyl
**Mobile phase:** MeCN:MeOH:water:acetic acid 30:15:54.5:0.5, containing 10 mM tetramethylammonium hydroxide
**Flow rate:** 1
**Injection volume:** 50
**Detector:** E, +0.5 V

## CHROMATOGRAM
**Internal standard:** 17β-estradiol
**Limit of quantitation:** 10 ng/mL

## KEY WORDS
plasma; dog; pharmacokinetics

## REFERENCE
Igwemezie, L.N.; Kaul, S.; Barbhaiya, R.H. Assessment of toxicokinetics and toxicodynamics following intravenous administration of etoposide phosphate in beagle dogs. *Pharm.Res.*, **1995**, *12*, 117–123

## SAMPLE
**Matrix:** blood
**Sample preparation:** Sonicate 50 million leukemic cells in 1 mL phosphate buffered saline. 500 μL Plasma or 1 mL sonicated cells + 0.5 (cells) or 2.5 (plasma) μg teniposide + 2 mL chloroform, mix. Remove the organic layer and evaporate it to dryness under a stream of nitrogen, reconstitute the residue in 200 μL MeOH:water 50:50, sonicate for 5 min, inject a 100 μL aliquot. To measure non-protein-bound etoposide filter (Amicon Centrifree) while centrifuging at 20°, inject a 100-200 μL aliquot of the ultrafiltrate.

## HPLC VARIABLES
**Column:** 250 × 4.6 5 μm Spherisorb Phenyl
**Mobile phase:** MeOH:water:acetic acid 45:54:1
**Flow rate:** 1

**Injection volume:** 100-200
**Detector:** F ex 220 em 330

## CHROMATOGRAM
**Retention time:** 6.1
**Internal standard:** teniposide (9.3)
**Limit of detection:** 10 (blood), 25 (ultrafiltrate) ng/mL

## OTHER SUBSTANCES
**Extracted:** cis-etoposide

## KEY WORDS
plasma; cells; ultrafiltrate; pharmacokinetics

## REFERENCE
Liliemark, E.; Petterson, B.; Peterson, C.; Liliemark, J. High-performance liquid chromatography with fluorometric detection for monitoring of etoposide and its *cis*-isomer in plasma and leukaemic cells. *J.Chromatogr.B*, **1995**, *669*, 311–317

## SAMPLE
**Matrix:** blood
**Sample preparation:** 200 µL Blood + 1 mL 150 ng/mL teniposide in dichloromethane: hexane 1:1, vortex for 1 min, centrifuge at 15000 g for 3 min. Remove the supernatant and evaporate it under reduced pressure at 40° for 30 min, reconstitute with 60 µL MeCN:water:650 mM pH 4.0 sodium citrate 40:60:0.4, inject a 30 µL aliquot.

## HPLC VARIABLES
**Guard column:** 10 × 3.9 10 µm µBondapak phenyl
**Column:** 150 × 3.9 10 µm µBondapak phenyl
**Mobile phase:** MeCN:water:650 mM pH 4.0 sodium citrate 35:57.3:7.7. When run is over wash with MeCN:water:650 mM pH 4.0 sodium citrate 70:27.3:7.7 for 4 min, re-equilibrate with initial mobile phase for 9 min.
**Flow rate:** 2
**Injection volume:** 30
**Detector:** E, ESA 5100A detector, Model 5020 guard cell between pump and autosampler +0.7 V, Model 5011 dual electrode analytical cell, upstream (screening) electrode +0.2 V, downstream electrode +0.45 V against Ag/AgCl.

## CHROMATOGRAM
**Retention time:** 2.4
**Internal standard:** teniposide (5.8)
**Limit of detection:** 2.4 ng/mL
**Limit of quantitation:** 7.5 ng/mL

## KEY WORDS
dog; rat

## REFERENCE
Eisenberg, E.J.; Eickhoff, W.M. Determination of etoposide in blood by liquid chromatography with electrochemical detection. *J.Chromatogr.*, **1993**, *621*, 110–114

## SAMPLE
**Matrix:** blood
**Sample preparation:** 1 mL Plasma + 50 µL MeOH, vortex, add 2 mL dichloroethane, shake thoroughly for 1 min, centrifuge at 3000 g for 5 min. Remove 1. 5 mL of the organic

layer and evaporate it to dryness under a stream of nitrogen, reconstitute the residue in 150 μL MeOH:water 70:30, sonicate for 6 min, inject a 10 μL aliquot.

## HPLC VARIABLES
**Guard column:** 10 × 4.6 10 μm LiChrosorb C18
**Column:** 100 × 4.6 10 μm Novapak phenyl
**Mobile phase:** MeOH:10 mM pH 7.0 phosphate buffer 55:45
**Flow rate:** 0.7
**Injection volume:** 10
**Detector:** E, Metrohm Model 641 VA, EA 286/1 glassy carbon electrode + 500 mV, stainless-steel auxiliary electrode, Ag/AgCl reference electrode

## CHROMATOGRAM
**Retention time:** 4.5
**Internal standard:** etoposide

## OTHER SUBSTANCES
**Extracted:** teniposide

## KEY WORDS
plasma; etoposide is IS

## REFERENCE
van der Horst, F.A.L.; van Opstal, M.A.J.; Teeuwsen, J.; Post, M.H.; Holthuis, J.J.M.; Brinkman, U.A.T. Comparative study on the determination of the anti-neoplastic drug teniposide in plasma using micellar liquid chromatography and surfactant-mediated plasma clean-up. *J.Chromatogr.*, **1991**, *567*, 161–174

## SAMPLE
**Matrix:** blood
**Sample preparation:** 450 μL Plasma + 50 μL 380 mM sodium dodecyl sulfate in 59 mM pH 7 sodium phosphate buffer, sonicate for 5 min. Inject a 100 μL aliquot onto column A with mobile phase A, elute with mobile phase A for 7.5 min, elute the contents of column A onto column B with mobile phase B for 1 min. After 1 min remove column A from the circuit and re-equilibrate it with mobile phase A for 1.5 min, monitor the effluent from column B.

## HPLC VARIABLES
**Column:** A 10 × 2.1 40 μm Chromsep C18 (Chrompack); B 300 × 4.6 10 μm μBondapak phenyl
**Mobile phase:** A 10 mM pH 7.0 sodium phosphate; B MeOH:10 mM pH 7.0 sodium phosphate buffer 55:45
**Flow rate:** A 0.4; B 1
**Injection volume:** 100
**Detector:** UV 254; E, +500 mV vs Ag/AgCl

## CHROMATOGRAM
**Retention time:** 4.2
**Internal standard:** etoposide
**Limit of detection:** 100 ng/mL (UV)

## OTHER SUBSTANCES
**Extracted:** teniposide

## KEY WORDS
plasma;column-switching;etoposide is IS

## REFERENCE
van Opstal, M.A.J.; van der Horst, F.A.L.; Holthuis, J.J.M.; Van Bennekom, W.P.; Bult, A. Automated reversed-phase chromatographic analysis of etoposide and teniposide in plasma by using on-line surfactant-mediated sample clean-up and column-switching. *J.Chromatogr.*, **1989**, *495*, 139–151

## SAMPLE
**Matrix:** blood
**Sample preparation:** 1 mL Plasma + 3 mL chloroform, shake for 10 min, centrifuge at 1500 g for 10 min, repeat extraction. Combine the organic layers and evaporate a 5 mL aliquot to dryness under a stream of nitrogen at 50°, reconstitute the residue in 200 µL mobile phase, inject a 20 µL aliquot.

## HPLC VARIABLES
**Column:** 300 × 3.9 10 µm µBondapak C18
**Mobile phase:** MeOH:250 mM ammonium acetate:acetic acid 54:45:1
**Flow rate:** 1.5
**Injection volume:** 20
**Detector:** E, Bioanalytical Systems LC4, TL5 glassy carbon electrode, +900 mV, Ag/AgCl reference electrode

## CHROMATOGRAM
**Retention time:** 4.1
**Internal standard:** etoposide

## OTHER SUBSTANCES
**Extracted:** teniposide

## KEY WORDS
plasma; etoposide is IS

## REFERENCE
Canal, P.; Michel, C.; Bugat, R.; Soula, G.; Carton, M. Quantification of teniposide in human serum by high-performance liquid chromatography with electrochemical detection. *J.Chromatogr.*, **1986**, *375*, 451–456

## SAMPLE
**Matrix:** blood
**Sample preparation:** Mix plasma or serum with an equal volume of proteinase K, let stand for 10 min. (Alternatively, heat serum or plasma with an equal volume of 1 mg/mL subtilisin A at 50° for 15 min.) Inject 1.6 mL hydrolyzed blood or filtered serum onto column A with mobile phase A at 1 mL/min, backflush column A with mobile phase A to waste for 2 min at 2 mL/min, backflush the contents of column A onto column B with mobile phase B, after 30 s remove column A from the circuit, elute column B with mobile phase B, monitor the effluent from column B. (Clean column A by backflushing with MeOH at 2 mL/min for 3 min then forward flush with water at 1 mL/min for 2 min.)

## HPLC VARIABLES
**Column:** A 2 × 4.6 10 µm PRP-1 divinylbenzene-styrene copolymer (Hamilton); B 125 × 4 10 µm LiChrosorb RP-18
**Mobile phase:** A water; B MeOH:water 55:45
**Flow rate:** A 1-2; B 1
**Injection volume:** 1600
**Detector:** F ex 230 em 328 following post-column extraction. The column effluent mixed with dichloroethane pumped at 0.6 mL/min, the mixture flowed through a 2 mm i.d. glass reactor (Technicon) to a phase separator (Technicon (J.Chromatogr. 1979, 185, 473)) with a PTFE insert and 0.3 mL/min of the organic phase flowed through the detector.

## CHROMATOGRAM
**Retention time:** 5
**Limit of detection:** 8 ng (blood); 30 ng (urine)

## OTHER SUBSTANCES
**Extracted:** teniposide

## KEY WORDS
column-switching; post-column extraction; plasma; serum

## REFERENCE
Werkhoven-Goewie, C.E.; Brinkman, U.A.T.; Frei, R.W.; de Ruiter, C.; de Vries, J. Automated liquid chromatographic analysis of the anti-tumorigenic drugs etoposide (VP 16-213) and teniposide (VM 26). *J.Chromatogr.*, **1983**, *276*, 349–357

## SAMPLE
**Matrix:** blood
**Sample preparation:** 1 mL Plasma + 10 μL 100 μg/mL teniposide in MeOH + 5 mL chloroform, rock gently for 15 min, centrifuge. Remove 4.5 mL of the organic layer and evaporate it to dryness under a stream of nitrogen at 40°, dissolve residue in 50 μL MeOH, vortex, centrifuge for 5-10 min, inject a 25 μL aliquot.

## HPLC VARIABLES
**Guard column:** 70 × 2.1 30 μm Co:Pell (Whatman)
**Column:** 300 × 3.9 10 μm μBondapak C18
**Mobile phase:** MeOH:water 60:40
**Flow rate:** 1
**Injection volume:** 25
**Detector:** F ex 215 em 328

## CHROMATOGRAM
**Retention time:** 5.5
**Internal standard:** teniposide (8)
**Limit of detection:** 25 ng/mL
**Limit of quantitation:** 50 ng/mL

## KEY WORDS
plasma

## REFERENCE
Strife, R.J.; Jardine, I.; Colvin, M. Analysis of the anticancer drugs etoposide (VP 16-213) and teniposide (VM 26) by high-performance liquid chromatography with fluorescence detection. *J.Chromatogr.*, **1981**, *224*, 168–174

## SAMPLE
**Matrix:** blood
**Sample preparation:** 1 mL Plasma + 10 μL 1 mg/mL teniposide in MeOH + 5 mL chloroform, rock gently for 15 min, centrifuge. Remove 4.5 mL of the organic layer and evaporate it to dryness under a stream of nitrogen at 40°, dissolve residue in 50 μL MeOH, vortex, centrifuge for 5-10 min, inject a 20 μL aliquot.

## HPLC VARIABLES
**Column:** 300 × 3.9 10 μm μBondapak C18
**Mobile phase:** MeOH:water 60:40
**Flow rate:** 1-1.2

**Injection volume:** 20
**Detector:** UV 254

## CHROMATOGRAM
**Retention time:** 5
**Internal standard:** teniposide (7.5)
**Limit of quantitation:** 500 ng/mL

## KEY WORDS
plasma

## REFERENCE
Strife, R.J.; Jardine, I.; Colvin, M. Analysis of the anticancer drugs VP 16-213 and VM 26 and their metabolites by high-performance liquid chromatography. *J.Chromatogr.*, **1980**, *182*, 211–220

## SAMPLE
**Matrix:** blood, CSF, urine
**Sample preparation:** 500 μL Plasma, urine, or CSF + 500 μL saturated ammonium sulfate + 4 mL ethyl acetate + 5 μL 2.5 μg/mL IS, vortex for 5 min, centrifuge at 3000 rpm for 15 min, repeat the extraction. Combine the organic layers and evaporate them to dryness under a stream of nitrogen, reconstitute the residue in 200 μL MeOH, inject an aliquot.

## HPLC VARIABLES
**Guard column:** 10 μm μBondapak phenyl
**Column:** 250 × 4.6 10 μm μBondapak phenyl
**Mobile phase:** MeCN:water:acetic acid 25:74:1
**Flow rate:** 1
**Injection volume:** 50
**Detector:** UV 284; E, Bioanalytical Systems LC-4A, 0.85 V

## CHROMATOGRAM
**Retention time:** 12
**Internal standard:** trans/cis-hydroxy acid of teniposide (26)
**Limit of detection:** 20 ng/mL
**Limit of quantitation:** 50 ng/mL

## OTHER SUBSTANCES
**Extracted:** metabolites

## KEY WORDS
plasma; pharmacokinetics

## REFERENCE
Sinkule, J.A.; Evans, W.E. High-performance liquid chromatographic analysis of the semisynthetic epipodophyllotoxins teniposide and etoposide using electrochemical detection. *J.Pharm.Sci.*, **1984**, *73*, 164–168

## SAMPLE
**Matrix:** blood, urine
**Sample preparation:** Plasma. 1 mL Plasma + 50 μL 200 μg/mL phenytoin or methylphenytoin + 5 mL chloroform, rotate for 10 min, centrifuge at 400 g for 10 min. Remove the organic layer, filter it, and evaporate it to dryness at 50°. Reconstitute in 200 μL mobile phase, inject an aliquot. Urine. 200 μL Urine + 1 mL pH 7.3 phosphate-buffered saline + 50 μL 200 μg/mL phenytoin or methylphenytoin + 5 mL chloroform, rotate for

10 min, centrifuge at 400 g for 10 min. Remove the organic layer, filter it, and evaporate it to dryness at 50°. Reconstitute in 200 μL mobile phase, inject an aliquot.

## HPLC VARIABLES
**Column:** 100 × 5 5 μm ODS Hypersil
**Mobile phase:** MeOH:water 51:49
**Flow rate:** 2
**Injection volume:** 50
**Detector:** UV 229

## CHROMATOGRAM
**Retention time:** 2.1
**Internal standard:** phenytoin (2.9), methylphenytoin (4.5)
**Limit of detection:** 100 ng/mL

## OTHER SUBSTANCES
**Noninterfering:** metabolites, acetaminophen, aspirin, cyclophosphamide, dextropropoxyphene, diamorphine, dihydrocodeine, doxorubicin, methotrexate, metoclopramide, morphine, phenobarbital, prednisone, procarbazine, prochlorperazine, vincristine

## KEY WORDS
plasma

## REFERENCE
Harvey, V.J.; Joel, S.P.; Johnston, A.; Slevin, M.L. High-performance liquid chromatography of etoposide in plasma and urine. *J.Chromatogr.*, **1985**, *339*, 419–423

## SAMPLE
**Matrix:** formulations
**Sample preparation:** Dilute with mobile phase, inject an aliquot.

## HPLC VARIABLES
**Column:** 250 × 4.6 5 μm cyano
**Mobile phase:** MeCN:20 mM sodium acetate 26:74, pH adjusted to 4.0 with acetic acid
**Flow rate:** 1
**Injection volume:** 20
**Detector:** UV 290

## CHROMATOGRAM
**Retention time:** 6.19

## OTHER SUBSTANCES
**Simultaneous:** granisetron (UV 300)

## KEY WORDS
stability-indicating; injections; saline

## REFERENCE
Mayron, D.; Gennaro, A.R. Stability and compatibility of granisetron hydrochloride in i.v. solutions and oral liquids and during simulated Y-site injection with selected drugs. *Am.J.Health-Syst.Pharm.*, **1996**, *53*, 294–304

## SAMPLE
**Matrix:** formulations

**Sample preparation:** Dilute 5 mL of the injection to 50 mL with mobile phase, add a 5 mL aliquot of this solution to 5 mL 40 µg/mL methyl p-aminobenzoate in MeCN and make up to 50 mL with mobile phase, inject a 20 µL aliquot.

## HPLC VARIABLES
**Column:** 300 × 3.9 10 µm µBondapak phenyl
**Mobile phase:** MeCN:20 mM pH 4.0 sodium acetate 26:74
**Flow rate:** 1
**Injection volume:** 20
**Detector:** UV 254

## CHROMATOGRAM
**Retention time:** 16.3
**Internal standard:** methyl p-aminobenzoate (9)

## OTHER SUBSTANCES
**Simultaneous:** impurities, degradation products, benzaldehyde, benzyl alcohol

## KEY WORDS
injections; stability-indicating

## REFERENCE
Floor, B.J.; Klein, A.E.; Muhammad, N.; Ross, D. Stability-indicating liquid chromatographic determination of etoposide and benzyl alcohol in injectable formulations. *J.Pharm.Sci.*, **1985**, *74*, 197–200

◆━━━━━━━━━━━━━◆━━━━━━━━━━━━━◆

## ANNOTATED BIBLIOGRAPHY

Barthes, D.M.C.; Rochard, E.B.; Pouliquen, I.J.; Rabouan, S.M.; Courtois, P.Y. Stability and compatibility of etoposide in 0.9% sodium chloride injection in three containers. *Am.J.Hosp.Pharm.*, **1994**, *51*, 2706–2709

Stiff, D.D.; Schwinghammer, T.L.; Corey, S.E. High-performance liquid chromatographic analysis of etoposide in plasma using fluorescence detection. *J.Liq.Chromatogr.*, **1992**, *15*, 863–873 [plasma; fluorescence detection; teniposide (IS); LOQ 50 ng/mL]

Fleming, R.A.; Stewart, C.F. High-performance liquid chromatographic determination of etoposide in plasma. *J.Liq.Chromatogr.*, **1991**, *14*, 1275–1283 [plasma; phenacetin (IS)]

Saita, T.; Fujiwara, K.; Kitagawa, T.; Mori, M.; Takata, K. A highly sensitive enzyme-linked immunosorbent assay for etoposide using beta-D-galactosidase as a label. *Cancer Chemother.Pharmacol.*, **1990**, *27*, 115–120 [rat; serum; fluorescence detection]

van Opstal, M.A.; Krabbenborg, P.; Holthuis, J.J.; Van Bennekom, W.P.; Bult, A. Comparison of flow-injection analysis with high-performance liquid chromatography for the determination of etoposide in plasma. *J.Chromatogr.*, **1988**, *432*, 385–400

el-Yazigi, A.; Martin, C.R. Improved assay for etoposide in plasma by radial-compression liquid chromatography with electrochemical detection. *Clin.Chem.*, **1987**, *33*, 803–805

Ploegmakers, H.H.; Mertens, M.J.; van Oort, W.J. Improved HPLC-ECD analysis of mitomycin C, porfiromycin, VP 16-213 and VM 26 by implantation of software filters. *Anticancer Res.*, **1987**, *7*, 1315–1319

Hersh, M.R.; Ludden, T.M. High-performance liquid chromatographic assay for etoposide in human plasma. *J.Pharm.Sci.*, **1986**, *75*, 815–817

Danigel, H.; Pfluger, K.H.; Jungclas, H.; Schmidt, L.; Dellbrugge, J. Drug monitoring of etoposide (VP16-213). I. A combined method of liquid chromatography and mass spectrometry. *Cancer Chemother.Pharmacol.*, **1985**, *15*, 121–124

Littlewood, T.J.; Hutchings, A.L.; Bentley, D.P.; Spragg, B.P. High-performance liquid chromatographic determination of etoposide in plasma using electrochemical detection. *J.Chromatogr.*, **1984**, *336*, 434–437 [plasma; electrochemical detection; teniposide (IS); LOD 5 ng/mL]

# Famotidine

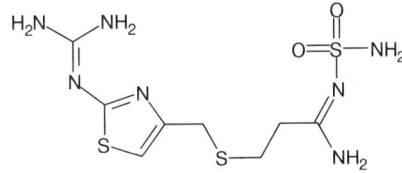

**Molecular formula:** $C_8H_{15}N_7O_2S_3$
**Molecular weight:** 337.4
**CAS Registry No.:** 76824-35-6

## SAMPLE
**Matrix:** blood
**Sample preparation:** Condition an SPE cartridge with 1 mL MeOH and 1 mL water. Add 1 mL plasma to the SPE cartridge, wash with 5 mL water, elute with 2 mL MeCN. Evaporate the eluate, reconstitute in acetic acid/phosphate buffer, centrifuge, inject an aliquot.

## HPLC VARIABLES
**Guard column:** Newguard RP-8 (Applied Biosystems)
**Column:** Spherisorb C8
**Mobile phase:** MeCN:30 mM pH 2.6 $NaH_2PO_4$ 7:93 containing 7.2 mM triethylamine
**Flow rate:** 1
**Detector:** UV 267

## CHROMATOGRAM
**Retention time:** 9.6
**Limit of quantitation:** 2 ng/mL

## KEY WORDS
plasma; SPE; pharmacokinetics

## REFERENCE
Schwartz, J.I.; Yeh, K.C.; Berger, M.L.; Tomasko, L.; Hoover, M.E.; Ebel, D.L.; Stauffer, L.A.; Han, R.; Bjornsson, T.D. Novel oral medication delivery system for famotidine. *J.Clin.Pharmacol.*, **1995**, *35*, 362–367

## SAMPLE
**Matrix:** blood
**Sample preparation:** 1.5 mL Plasma + 100 μL 4 M HCl, shake, add 8 mL diethyl ether, shake for 15 min, centrifuge at 1000 g at 4° for 5 min, discard ether layer. Add 0.5 mL saturated $Na_2CO_3$ solution, 0.5 mL saturated $NaHCO_3$ solution, 100 μL 5 μg/mL clopamide in water, and 7 mL ethyl acetate to aqueous layer, shake 15 min, centrifuge at 2000 g at 4° for 5 min, repeat extraction, combine organic layers and evaporate them to dryness under a stream of nitrogen at 37°. Reconstitute with 150 μL MeCN:water 12:88, vortex, inject 90 μL aliquot.

## HPLC VARIABLES
**Column:** 100 × 2 5 μm ODS Hypersil C18
**Mobile phase:** MeCN:water 12:88 containing 20 mM $Na_2HPO_4$ and 50 mM sodium dodecyl sulfate adjusted to pH 3 with orthophosphoric acid
**Flow rate:** 0.5
**Injection volume:** 90
**Detector:** UV 267

## CHROMATOGRAM
**Retention time:** 13
**Internal standard:** clopamide (9)
**Limit of detection:** 5 ng/mL

## OTHER SUBSTANCES
**Noninterfering:** acetaminophen, amiloride, aspirin, atenolol, chlorothiazide, cimetidine, cyclopenthiazide, diazepam, furosemide, indapamide, labetalol, lorazepam, metoprolol, phenytoin, propranolol, ranitidine, salicylic acid, theophylline

## KEY WORDS
plasma; microbore

## REFERENCE
Wanwimolruk, S.; Zoest, A.R.; Wanwimolruk, S.Z.; Hung, C.T. Sensitive high-performance liquid chromatographic determination of famotidine in plasma. Application to pharmacokinetic study. *J.Chromatogr.*, **1991**, *572*, 227–238

## SAMPLE
**Matrix:** blood
**Sample preparation:** 1 mL Plasma + 1 mL saturated potassium carbonate + 5 mL ethyl acetate, mix, centrifuge. Remove the organic layer and add it to 1 mL saturated NaCl solution and 1 mL 1 M HCl, centrifuge. Remove the aqueous phase and add it to 1 mL saturated potassium carbonate solution and 3 mL ethyl acetate, extract. Remove a 2.5 mL aliquot of the organic layer and evaporate it to dryness under reduced pressure, reconstitute the residue with 400 μL 250 μM phenanthrenequinone in MeOH, evaporate to dryness under reduced pressure, reconstitute with 40 μL DMF and 40 μL 2 M NaOH, heat at 60° for 15 min, cool on ice, neutralize with 5 M acetic acid, inject a 20 μL aliquot.

## HPLC VARIABLES
**Column:** reversed phase
**Mobile phase:** MeCN:10 mM pH 4 citrate buffer 40:50
**Flow rate:** 1
**Injection volume:** 20
**Detector:** F ex 296 em 411

## CHROMATOGRAM
**Limit of detection:** 5 ng/mL

## KEY WORDS
plasma; pharmacokinetics; derivatization

## REFERENCE
Echizen, H.; Shoda, R.; Umeda, N.; Ishizaki, T. Plasma famotidine concentration versus intragastric pH in patients with upper gastrointestinal bleeding and in healthy subjects. *Clin.Pharmacol.Ther.*, **1988**, *44*, 690–698

## SAMPLE
**Matrix:** blood, urine
**Sample preparation:** 50 μL Plasma or urine + 350 μL 0.2 μg/mL 3-butylxanthine in MeOH, centrifuge at 6000 g for 5 min. Evaporate supernatant under a stream of nitrogen at 40°, reconstitute in 250 μL mobile phase, inject an aliquot.

## HPLC VARIABLES
**Column:** 250 × 4.2 Cosmosil 5C18
**Mobile phase:** MeCN:30 mM pH 7.0 phosphate buffer 10:90
**Flow rate:** 1.5
**Detector:** UV 267

## CHROMATOGRAM
**Internal standard:** 3-butylxanthine
**Limit of quantitation:** 100 ng/mL

## KEY WORDS
plasma

## REFERENCE
Hasegawa, T.; Nadai, M.; Wang, L.; Takayama, Y.-I.; Kato, K.; Nabeshima, T.; Kato, N. Renal excretion of famotidine and role of adenosine in renal failure induced by bacterial lipopolysaccharide in rats. *Drug Metab.Dispos.*, **1994**, *22*, 8–13

## SAMPLE
**Matrix:** formulations
**Sample preparation:** Inject a 20 μL aliquot.

## HPLC VARIABLES
**Column:** Nova Pak C18
**Mobile phase:** MeCN:0.1% acetic acid:10 mM pH 7.8 $(NH_4)H_2PO_4$ 10:23:74
**Flow rate:** 1
**Injection volume:** 20
**Detector:** UV 300

## CHROMATOGRAM
**Retention time:** 13.9

## OTHER SUBSTANCES
**Simultaneous:** cefmetazole
**Noninterfering:** degradation products

## KEY WORDS
stability-indicating; injections; 5% dextrose

## REFERENCE
Lee, D.K.T.; Wong, C.-Y.; Wang, D.-P.; Chang, L.-C.; Wu, K.-H. Stability of cefmetazole sodium and famotidine. *Am.J.Health-Syst.Pharm.*, **1996**, *53*, 432–442

## SAMPLE
**Matrix:** formulations
**Sample preparation:** Dilute with 5% dextrose (if necessary), inject a 20 μL aliquot.

## HPLC VARIABLES
**Column:** Nova Pak C18
**Mobile phase:** MeOH:100 mM $(NH_4)_2HPO_4$ 20:80, pH 7.80
**Flow rate:** 1
**Injection volume:** 20
**Detector:** UV 322

## CHROMATOGRAM
**Retention time:** 4.6

## OTHER SUBSTANCES
**Simultaneous:** cefazolin

## KEY WORDS
injections; 5% dextrose; stability-indicating

## REFERENCE
Wang, D.-P.; Chang, L.-C.; Wong, C.-Y.; Lee, D.K.T. Stability of cefazolin sodium-famotidine admixture. *Am.J.Hosp.Pharm.*, **1994**, *51*, 2205–2209

## SAMPLE
**Matrix:** formulations
**Sample preparation:** Shake, remove 2 mL of oral suspension, dilute to 40 mL with water, vortex 1 min, centrifuge at 2000 rpm for 10 min. Dilute a 100 μL aliquot of supernatant with 100 μL of 100 μg/mL theophylline and add 800 μL water, inject 20 μL aliquot.

## HPLC VARIABLES
**Column:** 300 mm 10 μm Waters reversed-phase C18
**Mobile phase:** MeCN:50 mM sodium acetate buffer 8:92, adjusted to pH 6.5
**Flow rate:** 1.5
**Injection volume:** 20
**Detector:** UV 254

## CHROMATOGRAM
**Retention time:** 5.6
**Internal standard:** theophylline (4.3)

## KEY WORDS
stability-indicating

## REFERENCE
Quercia, R.A.; Jay, G.T.; Fan, C.; Chow, M.S. Stability of famotidine in an extemporaneously prepared oral liquid. *Am.J.Hosp.Pharm.*, **1993**, *50*, 691–693

## SAMPLE
**Matrix:** solutions

## HPLC VARIABLES
**Column:** 250 × 4.6 5 μm octadecylsilane (Beckman)
**Mobile phase:** MeOH:MeCN:glacial acetic acid:10 mM $KH_2PO_4$ 12:3:0.1:84.9, pH 5.0
**Flow rate:** 1
**Detector:** UV 266

## REFERENCE
Junnarkar, G.H.; Stavchansky, S. Isothermal and nonisothermal decomposition of famotidine in aqueous solution. *Pharm.Res.*, **1995**, *12*, 599–604

## SAMPLE
**Matrix:** solutions

## HPLC VARIABLES
**Column:** 150 × 4.6 12 μm 1-myristoyl-2-[(13-carboxyl)-tridecoyl]-sn-3-glycerophosphocholine chemically bonded to silica (Regis)
**Mobile phase:** MeCN:100 mM pH 7.0 phosphate buffer 20:80
**Flow rate:** 1
**Detector:** UV 254

## CHROMATOGRAM
**Retention time:** k′ 0.54

## OTHER SUBSTANCES
**Also analyzed:** acebutolol, alprenolol, antazoline, atenolol, betaxolol, bisoprolol, bopindolol, bupranolol, carteolol, celiprolol, chloropyramine, chlorpheniramine, cicloprolol, cimetidine, cinnarizine, cirazoline, clonidine, dilevalol, dimethindene, diphenhydramine, doxazosin, esmolol, isothipendyl, ketotifen, metiamide, metoprolol, moxonidine, nadolol, naphazoline, nifenalol, nizatidine, oxprenolol, pheniramine, phentolamine, pindolol, pizotyline (pizotifen), practolol, prazosin, promethazine, propranolol, pyrilamine (mepyramine), ranitidine, roxatidine, sotalol, tiamenidine, timolol, tramazoline, tripelennamine, triprolidine, tymazoline, UK-14,304

## REFERENCE
Kaliszan, R.; Nasal, A.; Turowski, M. Binding site for basic drugs on $\alpha_1$-acid glycoprotein as revealed by chemometric analysis of biochromatographic data. *Biomed.Chromatogr.*, **1995**, *9*, 211–215

## SAMPLE
**Matrix:** solutions

## HPLC VARIABLES
**Column:** 250 × 4.6 Zorbax RX
**Mobile phase:** Gradient. A was 10 mL concentrated orthophosphoric acid and 7 mL triethylamine in 1 L water. B was 10 mL concentrated orthophosphoric acid and 7 mL triethylamine in 200 mL water, make up to 1 L with MeCN. A:B from 100:0 to 0:100 over 30 min, maintain at 0:100 for 5 min.
**Column temperature:** 30
**Flow rate:** 2
**Detector:** UV 210

## OTHER SUBSTANCES
**Also analyzed:** acepromazine, acetaminophen, acetophenazine, albuterol, aminophylline, amitriptyline, amobarbital, amoxapine, amphetamine, amylocaine, antipyrine, aprobarbital, aspirin, atenolol, atropine, avermectin, barbital, benzocaine, benzoic acid, benzotropine, benzphetamine, berberine, bibucaine, bromazepan, brompheniramine, buprenorphine, buspirone, butabarbital, butacaine, butethal, caffeine, carbamazepine, carbromal, chloramphenicol, chlordiazepoxide, chloroquine, chlorothiazide, chloroxylenol, chlorphenesin, chlorpheniramine, chlorpromazine, chlorpropamide, chlortetracycline, cimetidine, cinchonidine, cinchonine, clenbuterol, clonazepam, clonixin, clorazepate, cocaine, codeine, colchicine, cortisone, coumarin, cyclazocine, cyclobenzaprine, cyclothiazide, cyheptamide, cymarin, danazol, danthron, dapsone, debrisoquine, desipramine, dexamethasone, dextromethorphan, dextropropoxyphene, diamorphine, diazepam, diclofenac, diethylpropion, diethylstilbestrol, diflunisal, digitoxin, digoxin, diltiazem, diphenhydramine, diphenoxylate, diprenorphine, dipyrone, disulfiram, dopamine, doxapram, doxepin, dronabinol, ephedrine, epinephrine, epinine, estradiol, estriol, estrone, ethacrynic acid, ethosuximide, etonitazene, etorphine, eugenol, fenbendazole, fencamfamine, fenoprofen, fenproporex, fentanyl, flubendazole, flufenamic acid, flunitrazepam, 5-fluorouracil, fluoxymesterone, fluphenazine, furosemide, gentisic acid, gitoxigenin, glipizide, glunixin, glutethimide, glybenclamide, guaiacol, halazepam, haloperidol, hydrochlorothiazide, hydrocodone, hydrocortisone, hydromorphone, hydroxyquinoline, ibogaine, ibuprofen, iminostilbene, imipramine, indomethacin, isocarbostyril, isocarboxazid, isoniazid, isoproterenol, isoxsuprine, ivermectin, ketamine, ketoprofen, kynurenic acid, levorphanol, lidocaine, lorazepam, lormetazepam, loxapine, mazindol, mebendazole, meclizine, meclofenamic acid, medazepam, mefenamic acid, megestrol, mepacrine, meperidine, mephentermine, mephenytoin, mephesin, mephobarbital, mepivacaine, mescaline, mesoridazine, methadone, methamphetamine, methapyrilene, methaqualone, methazolamide, methocarbamol, methoxamine, methsuximide, methyl salicylate, methyldopa, methyldopamine, methylphenidate, meth-

ylprednisolone, methyltestosterone, methyprylon, metoprolol, mibolerone, morphine, nadolol, nalorphine, naloxone, naltrexone, naphazoline, naproxen, nefopam, niacinamide,
nicotine, nicotinic acid, nifedipine, niflumic acid, nitrazepam, norepinephrine, nortriptyline, noscapine, nylidrin, oxazepam, oxycodone, oxymorphone, oxyphenbutazone, oxytetracycline, papaverine, pargyline, pemoline, pentazocine, pentobarbital, persantine, phenacetin, phenazocine, phenazopyridine, phencyclidine, phendimetrazine, phenelzine,
pheniramine, phenobarbital, phenothiazine, phensuximide, phentermine, phenylbutazone, phenylephrine, phenylpropanolamine, piperocaine, prazepam, prednisolone, primidone, probenecid, progesterone, propiomazine, propranolol, propylparaben, pseudoephedrine, puromycin, pyrilamine, pyrithyldione, quazepam, quinaldic acid, quinidine, quinine,
ranitidine, recinnamine, reserpine, resorcinol, saccharin, albuterol, salicylamide, salicylic
acid, scopolamine, scopoletin, secobarbital, strychnine, sulfacetamide, sufadiazine, sulfadimethoxine, sulfaethidole, sulfamerazine, sulfamethazine, sulfamethoxizole, sulfanilamide, sulfapyridine, sulfasoxizole, sulindac, tamoxifen, temazepam, testosterone, tetracaine, tetracycline, tetramisole, thebaine, theobromine, theophylline, thiabendazole,
thiamine, thiamylal, thiobarbituric acid, thioridazine, thiosalicylic acid, thiothixene, thymol, tolazamide, tolazoline, tobutamide, tolmetin, tranylcypromine, triamcinolone, tribenzylamine, trichloromethiazide, trifluoperazine, trihexyphenidyl, trimethoprim, tripelennamine, triproilidine, tropacocaine, tyramine, verapamil, vincamine, warfarin, yohimbine,
zoxazolamine

## REFERENCE

Hill, D.W.; Kind, A.J. Reversed-phase solvent gradient HPLC retention indexes of drugs. *J.Anal.Toxicol.*,
**1994**, *18*, 233–242

◆───────────────────────◆───────────────────────◆

## ANNOTATED BIBLIOGRAPHY

Qin, X.Z.; Ip, D.P.; Chang, K.H.; Dradransky, P.M.; Brooks, M.A.; Sakuma, T. Pharmaceutical application
of LC-MS. 1–Characterization of a famotidine degradate in a package screening study by LC-APCI
MS. *J.Pharm.Biomed.Anal.*, **1994**, *12*, 221–233

Kamath, B.V.; Shivram, K.; Newalkar, B.L.; Shah, A.C. Liquid chromatographic analysis and degradation kinetics of famotidine. *J.Liq.Chromatogr.*, **1993**, *16*, 1007–1014 [tablets; stability indicating]

Imai, Y.; Kobayashi, S. A simple method for the quantification of famotidine in human plasma and urine
by paired-ion high performance liquid chromatography. *Biomed.Chromatogr.*, **1992**, *6*, 222–223

Cvitkovic, L.; Zupancic-Kralj, L.; Marsel, J. Determination of famotidine in human plasma and urine
by high-performance liquid chromatography. *J.Pharm.Biomed.Anal.*, **1991**, *9*, 207–210

Guo, P.; Ye, L.M.; Lu, B.; He, Y.J.; Li, Z.W. [Direct injection of plasma to determine famotidine in plasma
using HPLC column switching technique]. *Yao Hsueh Hsueh Pao*, **1990**, *25*, 622–625

Bullock, L.; Fitzgerald, J.F.; Glick, M.R. Stability of famotidine 20 and 50 mg/L in total nutrient admixtures. *Am.J.Hosp.Pharm.*, **1989**, *46*, 2326–2329 [theophylline (IS)]

Bullock, L.; Fitzgerald, J.F.; Glick, M.R.; Parks, R.B.; Schnabel, J.G.; Hancock, B.G. Stability of famotidine 20 and 40 mg/L and amino acids in total parenteral nutrient solutions. *Am.J.Hosp.Pharm.*,
**1989**, *46*, 2321–2325 [stability-mdicating; theophylline (IS)]

DiStefano, J.E.; Mitrano, F.P.; Baptista, R.J.; Der, M.M.; Silvestri, A.P.; Palombo, J.D.; Bistrian, B.R.
Long-term stability of famotidine 20 mg/L in a total parenteral nutrient solution. *Am.J.Hosp.Pharm.*,
**1989**, *46*, 2333–2335 [non-interfering degradation products]

# Finasteride

**Molecular formula:** $C_{23}H_{36}N_2O_2$
**Molecular weight:** 372.6
**CAS Registry No.:** 98319-26-7

## SAMPLE
**Matrix:** blood
**Sample preparation:** 150 µL Plasma + 150 µL ethylene glycol:water 40:60, mix, filter (5 µm), filter (0.22 µm) while centrifuging at 4° at 1000 g for 5 min, inject a 150 µL aliquot of the filtrate onto column A and elute to waste with mobile phase A, after 5 min elute column A to waste with mobile phase B, after 8 min direct the effluent from column A onto column B, after 1.5 min remove column A from the circuit, elute column B with mobile phase D, monitor the effluent from column B. Clean column A by eluting to waste with mobile phase C for 19.5 min, re-equilibrate column A with mobile phase A for 4 min.

## HPLC VARIABLES
**Column:** A 35 × 4.6 5 µm Capcell Pak CN SG-120 (Shiseido); B 250 × 4.6 5 µm Inertsil ODS-2
**Mobile phase:** A MeCN:water 10:90; B MeCN:water 25:75; C MeCN:water 70:30; D MeCN:water 45:55 (Pass mobile phase A and mobile phase B through 35 × 4.6 5 µm Capcell Pak C18 SG-120 (Shiseido) columns before use. Clean these columns with mobile phase C each day.)
**Column temperature:** 40
**Flow rate:** C 1; D 1.1
**Injection volume:** 150
**Detector:** UV 210

## CHROMATOGRAM
**Retention time:** 24.4
**Limit of quantitation:** 1 ng/mL

## KEY WORDS
plasma; column-switching; heart-cut

## REFERENCE
Takano, T.; Hata, S. High-performance liquid chromatographic determination of finasteride in human plasma using direct injection with column switching. *J.Chromatogr.B*, **1996**, *676*, 141–146

## SAMPLE
**Matrix:** blood
**Sample preparation:** Condition a 1 mL Baker nitrile SPE cartridge with 1 mL MeOH and 1 mL water. 1 mL Plasma + 50 µL 2 µg/mL IS, vortex for 10 s, add to the SPE cartridge, wash with 2 mL acetone:water 10:90, wash with 2 mL water, elute with 250 µL MeOH, add 10 µL water to the eluate, inject a 200 µL aliquot.

## HPLC VARIABLES
**Guard column:** 25 × 4.6 5 µm RP-8 (Brownlee)
**Column:** 150 × 4.6 5 µm RP-8 (Altex) + 50 × 4.6 3 µm RP-18 (Analytichem)

**Mobile phase:** MeOH:MeCN:water 6:5:7
**Flow rate:** 1
**Injection volume:** 200
**Detector:** UV 210

## CHROMATOGRAM
**Retention time:** 13.1
**Internal standard:** 4-methylfinasteride (20.8)
**Limit of quantitation:** 1 ng/mL

## OTHER SUBSTANCES
**Extracted:** metabolites

## KEY WORDS
plasma; SPE

## REFERENCE
Constanzer, M.L.; Matuszewski, B.K.; Bayne, W.F. High-performance liquid chromatographic method for the determination of finasteride in human plasma at therapeutic doses. *J.Chromatogr.*, **1991**, *566*, 127–134

## SAMPLE
**Matrix:** blood, semen
**Sample preparation:** Condition a Baker 1 mL nitrile SPE cartridge with 1 mL MeOH and 1 mL water. 1 mL Plasma or semen + 100 µL 100 ng/mL IS, vortex for 10 s, add to the SPE cartridge, wash with 2 mL water, elute with 300 µL MeCN, inject a 100 µL aliquot of the eluate.

## HPLC VARIABLES
**Guard column:** 20 × 4.6 5 µm base deactivated C18 (Keystone)
**Column:** 33 × 4.6 3 µm C18 (Perkin-Elmer)
**Mobile phase:** MeCN:water 70:30 containing 0.1% formic acid
**Column temperature:** 70
**Flow rate:** 1
**Injection volume:** 100
**Detector:** MS, PE-SCIEX API III triple quadrupole, heated nebulizer, corona discharge (+5 µA), positive ion APCI, nebulizer probe 500°, collision gas argon at $350 \times 10^{12}$ molecules/cm$^2$, nebulizing gas nitrogen at 80 psi and 2 L/min, curtain gas nitrogen at 0.9 L/min, orifice +50 V, electron multiplier -3.8 kV, dwell time 400 ms, interface heater 60°, m/z 317

## CHROMATOGRAM
**Retention time:** 1
**Internal standard:** N-(1,1,3,3-tetramethylbutyl)-3-oxo-4-aza-5α-androst-1-ene-17β-carboxamide (1.5)
**Limit of detection:** 0.2 ng/mL

## KEY WORDS
plasma; SPE

## REFERENCE
Constanzer, M.L.; Chavez, C.M.; Matuszewski, B.K. Picogram determination of finasteride in human plasma and semen by high-performance liquid chromatography with atmospheric-pressure chemical-ionization tandem mass spectrometry. *J.Chromatogr.B*, **1994**, *658*, 281–287

## SAMPLE
**Matrix:** microsomal incubations

**Sample preparation:** 500 μL Microsomal incubation + 2 mL ice-cold water, add to a Millipore C18 SPE cartridge, elute with 5 mL MeOH. Evaporate the eluate to dryness, reconstitute with mobile phase, inject an aliquot.

## HPLC VARIABLES
**Guard column:** 30 × 4.6 7 μm Aquapore RP-300 (Applied Biosystems)
**Column:** 250 × 4.6 5 μm Zorbax ODS C18
**Mobile phase:** Gradient. A was 20 mM ammonium acetate buffer containing 0.1% trifluoroacetic acid. B was MeCN containing 0.1% trifluoroacetic acid. A:B 60:40 for 5 min, to 40:60 over 35 min.
**Flow rate:** 1
**Detector:** UV 210

## CHROMATOGRAM
**Retention time:** 28.0

## OTHER SUBSTANCES
**Extracted:** metabolites

## KEY WORDS
human; liver; SPE

## REFERENCE
Huskey, S.-E.W.; Dean, D.C.; Miller, R.R.; Rasmusson, G.H.; Chiu, S.-H.L. Identification of human cytochrome P450 isozymes responsible for the in vitro oxidative metabolism of finasteride. *Drug Metab.Dispos.*, **1995**, *23*, 1126–1135

## SAMPLE
**Matrix:** microsomal incubations
**Sample preparation:** Microsomal incubation + 200 μL 5 M HCl, extract with ethyl acetate. Evaporate the ethyl acetate, take up the residue in MeOH, inject an aliquot.

## HPLC VARIABLES
**Guard column:** LiChrospher 100 RP-8
**Column:** 250 × 4.6 Sepralyte C8 (Analytichem)
**Mobile phase:** MeCN:MeOH:water 35:10:55
**Flow rate:** 0.8-1.5
**Detector:** UV 210

## CHROMATOGRAM
**Internal standard:** 17-methyltestosterone

## OTHER SUBSTANCES
**Extracted:** metabolites

## KEY WORDS
rat; liver

## REFERENCE
Ishii, Y.; Mukoyama, H.; Ohtawa, M. *In vitro* biotransformation of finasteride in rat hepatic microsomes. Isolation and characterization of metabolites. *Drug Metab.Dispos.*, **1994**, *22*, 79–84

## ANNOTATED BIBLIOGRAPHY

Ishii, Y.; Mukoyama, H.; Hata, S. Metabolism of finasteride in rat hepatic microsomes: age and sex differences and effects of P450 inducers. *Xenobiotica*, **1994**, *24*, 863–872 [rat; liver; microsomal incubations; extracted metabolites; 17-methyltestosterone (IS)]

# Fluconazole

**Molecular formula:** $C_{13}H_{12}F_2N_6O$
**Molecular weight:** 306.3
**CAS Registry No.:** 86386-73-4

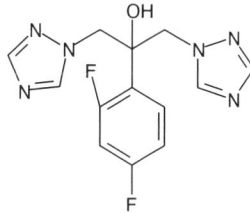

## SAMPLE
**Matrix:** blood
**Sample preparation:** 500 µL Plasma + 500 µL 10 µg/mL IS in MeOH:water 10:90 + 250 µL 1 M ammonium hydroxide, mix, add 5 mL ethyl acetate, mix, centrifuge. Remove the organic layer and add it to 1 mL 1 M HCl, mix, centrifuge. Remove the aqueous layer and add it to 1.5 mL 6 M ammonium hydroxide and 5 mL ethyl acetate, extract. Remove the organic layer and evaporate it to dryness under a stream of nitrogen at 40°, reconstitute the residue in mobile phase, inject an aliquot.

## HPLC VARIABLES
**Column:** Adsorbosphere C18
**Mobile phase:** MeOH:10 mM pH 7 phosphate buffer 50:50
**Detector:** UV 260

## CHROMATOGRAM
**Internal standard:** UK54373
**Limit of detection:** 1 ng/mL

## KEY WORDS
plasma; pharmacokinetics

## REFERENCE
Schwartz, E.L.; Hallam, S.; Gallagher, R.E.; Wiernik, P.H. Inhibition of all-*trans*-retinoic acid metabolism by fluconazole *in vitro* and in patients with acute promyelocytic leukemia. *Biochem.Pharmacol.*, **1995**, *50*, 923–928

## SAMPLE
**Matrix:** blood
**Sample preparation:** Condition a Bond-Elut C18 SPE cartridge with 2 column volumes MeOH and 2 column volumes 100 mM pH 6.0 sodium phosphate buffer. 1 mL Serum + 100 µL 100 µg/mL IS in MeOH:water 10:90 + 2 mL 100 mM pH 6.0 phosphate buffer, mix, add to the SPE cartridge, wash with 1 mL phosphate buffer, wash with 1 mL MeOH:phosphate buffer 15:85, dry under vacuum, elute with 500 µL MeOH. Evaporate the eluate to dryness under a stream of nitrogen at 40°, reconstitute with 200 µL mobile phase, filter, inject a 50 µL aliquot of the filtrate.

## HPLC VARIABLES
**Guard column:** 5 µm Adsorbosphere C18
**Column:** 250 × 4.6 5 µm Adsorbosphere C18
**Mobile phase:** MeCN:25 mM pH 7.0 Tris-phosphate buffer 25:75
**Flow rate:** 1
**Injection volume:** 50
**Detector:** UV 210

## CHROMATOGRAM
**Retention time:** 6.6

**Internal standard:** 2-(4-chlorophenyl)-1,3-bis(1H-1,2,4-triazol-1-yl)propan-2-ol (UK-48,134, Pfizer) (9.0)
**Limit of quantitation:** 100 ng/mL

## OTHER SUBSTANCES
**Simultaneous:** acyclovir, cefazolin, ceftazidime, clindamycin, metronidazole, piperacillin, sulfamethoxazole, trimethoprim
**Noninterfering:** amphotericin B

## KEY WORDS
serum; SPE

## REFERENCE
Inagaki, K.; Takagi, J.; Lor, E.; Okamoto, M.P.; Gill, M.A. Determination of fluconazole in human serum by solid-phase extraction and reversed-phase high-performance liquid chromatography. *Ther.Drug Monit.*, **1992**, *14*, 306–311

## SAMPLE
**Matrix:** blood, CSF
**Sample preparation:** 500 μL Serum, plasma, or CSF + 500 μL water + 50 μL 100 μg/mL IS in MeOH:water 10:90 + 250 μL 1 M ammonium hydroxide, mix, add 5 mL ethyl acetate, vortex for 30 s, centrifuge. Remove organic layer and add it to 1 mL 1 M HCl, vortex, centrifuge. Remove the aqueous layer, add it to 1.5 mL 6 M ammonium hydroxide, mix, add 5 mL ethyl acetate, mix, centrifuge. Remove the organic layer and evaporate it to dryness at 40°under a stream of nitrogen, dissolve the residue in 200 μL mobile phase, filter (0.20 μm), inject a 30 μL aliquot.

## HPLC VARIABLES
**Guard column:** Corasil C18 guard column
**Column:** 250 × 4.6 5 μm Microsorb ODS
**Mobile phase:** MeOH:10 mM pH 7.0 phosphate buffer 50:50
**Flow rate:** 1
**Injection volume:** 30
**Detector:** UV 260

## CHROMATOGRAM
**Retention time:** 5.0
**Internal standard:** 2-(4-chlorophenyl)-1,3-bis(1H-1,2,4-triazol-1-yl)propan-2-ol (UK-48,134) (6.2)
**Limit of quantitation:** 200 ng/mL

## KEY WORDS
serum; plasma; also for urine

## REFERENCE
Foulds, G.; Brennan, D.R.; Wajszczuk, C.; Catanzaro, A.; Garg, D.C.; Knopf, W.; Rinaldi, M.; Weidler, D.J. Fluconazole penetration into cerebrospinal fluid in humans. *J.Clin.Pharmacol.*, **1988**, *28*, 363–366

## SAMPLE
**Matrix:** blood, saliva
**Sample preparation:** Plasma. Add 100 μL 20 μg/mL phenacetin in MeOH to a tube, evaporate to dryness under a stream of nitrogen at 40°, add 500 μL plasma, vortex for 10 s, add 50 μL 5 M NaOH, vortex for 10 s, add 5 mL chloroform:isopropanol 80:20, shake for 15 min, centrifuge at 2500 g for 15 min. Remove the organic layer and evaporate it to dryness under a stream of nitrogen at 50°, reconstitute the residue in 60 μL MeOH, vortex

for 30 s, add 140 μL 10 mM pH 5.0 sodium acetate buffer, vortex for 30 s, centrifuge at 1000 g for 5 min, inject a 100 μL aliquot of the supernatant. Saliva. 250 μL Saliva + 50 μL 5 M NaOH, vortex for 10 s, add 5 mL chloroform:isopropanol 80:20, shake for 15 min, centrifuge at 2500 g for 15 min. Remove 4 mL of the organic layer and evaporate it to dryness under a stream of nitrogen at 50°, reconstitute the residue in 60 μL MeOH, vortex for 30 s, add 140 μL 10 mM pH 5.0 sodium acetate buffer, vortex for 30 s, centrifuge at 1000 g for 5 min, inject a 100 μL aliquot of the supernatant.

## HPLC VARIABLES

**Guard column:** 4 × 4 5 μm LiChrospher 100 RP-8
**Column:** 125 × 4 5 μm Lichrosorb RP-18
**Mobile phase:** MeOH:10 mM sodium acetate adjusted to pH 5.0 with concentrated HCl 30:70
**Flow rate:** 1
**Injection volume:** 100
**Detector:** UV 261

## CHROMATOGRAM

**Retention time:** 8
**Internal standard:** phenacetin (12.5)
**Limit of quantitation:** 100 ng/mL (plasma), 1000 ng/mL (saliva)

## OTHER SUBSTANCES

**Noninterfering:** acetaminophen, amphotericin B, brotizolam, ceftazidime, ciprofloxacin, codeine, diclofenac, didanosine, domperidone, fluoxetine, foscarnet, ganciclovir, methadone, metoclopramide, mianserin, nystatin, pyrimethamine, ranitidine, sulfamethoxazole, temazepam, trimethoprim, zidovudine

## KEY WORDS

plasma; pharmacokinetics

## REFERENCE

Koks, C.H.W.; Rosing, H.; Meenhorst, P.L.; Bult, A.; Beijnen, J.H. High-performance liquid chromatographic determination of the antifungal drug fluconazole in plasma and saliva of human immunodeficiency virus-infected patients. *J.Chromatogr.B*, **1995**, *663*, 345–351

## SAMPLE

**Matrix:** feed
**Sample preparation:** Stir 5 g feed with 20 mL dichloromethane at room temperature for 1.5 h, filter, add filtrate to two 2.8 mL 500 mg Bond Elut cyanopropyl SPE cartridges in series, add the eluate to the cartridges for a second pass, dry under vacuum, elute with 20 mL MeOH:water 35:65, make up the eluate to 25 mL with MeOH:water 35:65, inject an aliquot.

## HPLC VARIABLES

**Column:** 100 × 2 5 μm Spherisorb C8
**Mobile phase:** MeOH:water 25:75
**Column temperature:** 35
**Flow rate:** 0.4
**Detector:** UV 210

## KEY WORDS

SPE

## REFERENCE

Khundker, S.; Dean, J.R.; Jones, P. A comparison between solid phase extraction and supercritical fluid extraction for the determination of fluconazole from animal feed. *J.Pharm.Biomed.Anal.*, **1995**, *13*, 1441–1447

## SAMPLE

**Matrix:** injections
**Sample preparation:** 1 mL Sample + 50 μL 150 ng/mL cimetidine, inject a 10 μL aliquot.

## HPLC VARIABLES

**Column:** 250 × 4.8 Spherisorb S5CN
**Mobile phase:** MeCN:buffer 50:50 (Buffer was 20 mM $KH_2PO_4$ adjusted to pH 5.4 with 1 M NaOH.)
**Flow rate:** 2.2
**Injection volume:** 10
**Detector:** UV 254

## CHROMATOGRAM

**Retention time:** 2.2
**Internal standard:** cimetidine (3.6)

## KEY WORDS

stability-indicating; 5% dextrose

## REFERENCE

Bosso, J.A.; Prince, R.A.; Fox, J.L. Compatibility of ondansetron hydrochloride with fluconazole, ceftazidime, aztreonam, and cefazolin sodium under simulated Y-site conditions. *Am.J.Hosp.Pharm.*, **1994**, *51*, 389–391

## SAMPLE

**Matrix:** injections
**Sample preparation:** 100 μL Injection + 700 μL water + 100 μL 200 μg/mL IS in MeOH:water 10:90 + 100 μL 1 M ammonium hydroxide, mix, add 2.5 mL ethyl acetate, vortex for 1 min, centrifuge. Remove organic layer and add it to 1 mL 1 M HCl, mix, centrifuge. Remove the aqueous layer, add 1 mL 6 M ammonium hydroxide, mix, add 2.5 mL ethyl acetate, centrifuge. Remove the organic layer and evaporate it to dryness at 40° under a stream of nitrogen, dissolve the residue in 200 μL mobile phase, filter (0.45 μm), inject a 50 μL aliquot.

## HPLC VARIABLES

**Guard column:** 5 μm Adsorbosphere C18 guard column
**Column:** 250 × 4.6 5 μm Adsorbosphere C18
**Mobile phase:** MeCN:water 26:74
**Flow rate:** 1
**Injection volume:** 50
**Detector:** UV 260

## CHROMATOGRAM

**Retention time:** 6.8
**Internal standard:** 2-(4-chlorophenyl)-1,3-bis(1H-1,2,4-triazol-1-yl)propan-2-ol (9.5)

## OTHER SUBSTANCES

**Noninterfering:** acyclovir, amikacin, amphotericin B, cefazolin, ceftazidime, clindamycin, gentamicin, metronidazole, piperacillin, trimethoprim, sulfamethoxazole

## KEY WORDS
stability-indicating; water; saline; 5% dextrose

## REFERENCE
Inagaki, K.; Tagaki, J.; Lor, E.; Lee, K.-J.; Nii, L.; Gill, M.A. Stability of fluconazole in commonly used intravenous antibiotic solutions. *Am.J.Hosp.Pharm.*, **1993**, *50*, 1206–1208

## SAMPLE
**Matrix:** solutions

## HPLC VARIABLES
**Column:** C18
**Mobile phase:** MeOH:25 mM sodium phosphate buffer 45:55, pH adjusted to 7.0 with phosphoric acid
**Flow rate:** 1
**Injection volume:** 20
**Detector:** UV 260

## CHROMATOGRAM
**Retention time:** 6.1

## OTHER SUBSTANCES
**Simultaneous:** cefpirome

## KEY WORDS
stability-indicating

## REFERENCE
Allen, L.V., Jr.; Stiles, M.L.; Prince, S.J.; Sylvestri, M.F. Stability of cefpirome sulfate in the presence of commonly used intensive care drugs during simulated Y-site injection. *Am.J.Health-Syst.Pharm.*, **1995**, *52*, 2427–2433

## ANNOTATED BIBLIOGRAPHY
Thaler, F.; Bernard, B.; Tod, M.; Jednyak, C.P.; Petitjean, O.; Derome, P.; Loirat, P. Fluconazole penetration in cerebral parenchyma in humans at steady state. *Antimicrob.Agents Chemother.*, **1995**, *39*, 1154–1156 [brain; tissue; LOQ 150 ng/mL]

Burm, J.-P.; Choi, J.-S.; Jhee, S.S.; Chin, A.; Ulrich, R.W.; Gill, M.A. Stability of paclitaxel and fluconazole during simulated Y-site administration. *Am.J.Hosp.Pharm.*, **1994**, *51*, 2704–2706 [5% dextrose]

Flores-Murrieta, F.J.; Granados-Soto, V.; Hong, E. A simple and rapid method for determination of fluconazole in human plasma samples by high-performance liquid chromatography. *J.Liq. Chromatogr.*, **1994**, *17*, 3803–3811 [plasma; LOD 20 ng/mL]

Jenke, D.R. Drug binding by reservoirs in elastomeric infusion devices. *Pharm.Res.*, **1994**, *11*, 984–989 [saline; 5% dextrose]

Li, Z.W.; Guo, P.; Ye, L.M.; Hong, Z.; Wang, Y.S. Determination of fluconazole by direct injection of plasma and high performance liquid chromatography with column switching. *Yao Hsueh Hsueh Pao*, **1994**, *29*, 773–777

Madu, A.; Cioffe, C.; Mian, U.; Burroughs, M.; Tuomanen, E.; Mayers, M.; Schwartz, E.; Miller, M. Pharmacokinetics of fluconazole in cerebrospinal fluid and serum of rabbits: Validation of an animal model used to measure drug concentrations in cerebrospinal fluid. *Antimicrob.Agents Chemother.*, **1994**, *38*, 2111–2115 [pharmacokinetics; CSF; serum; plasma; rabbit; UK54373 (IS)]

Pompilio, F.M.; Fox, J.L.; Inagaki, K.; Burm, J.-P.; Jhee, S.; Gill, M.A. Stability of ranitidine hydrochloride with ondansetron hydrochloride or fluconazole during simulated Y-site administration. *Am.J.Hosp.Pharm.*, **1994**, *51*, 391–394 [saline; stability-indicating]

Johnson, C.E.; Jacobson, P.A.; Pillen, H.A.; Woycik, C.L. Stability and compatibility of fluconazole and aminophylline in intravenous admixtures. *Am.J.Hosp.Pharm.*, **1993**, *50*, 703–706 [stability-indicating; UK-48-134 (IS); saline; 5% dextrose; non-interfering aminophylline]

Yamreudeewong, W.; Lopez-Anaya, A.; Rappaport, H. Stability of fluconazole in an extemporaneously prepared oral liquid. *Am.J.Hosp.Pharm.*, **1993**, *50*, 2366–2367 [methyl p-hydroxybenzoate (IS); stability-indicating]

Wallace, J.E.; Harris, S.C.; Gallegos, J.; Foulds, G.; Chen, T.J.; Rinaldi, M.G. Assay of fluconazole by high-performance liquid chromatography with a mixed-phase column. *Antimicrob.Agents Chemother.*, **1992**, *36*, 603–606

Hosotsubo, K.K.; Hosotsubo, H.; Nishijima, M.K.; Okada, T.; Taenaka, N.; Yoshiya, I. Rapid determination of serum levels of a new antifungal agent, fluconazole, by high-performance liquid chromatography. *J.Chromatogr.*, **1990**, *529*, 223–228

# Fluoxetine

**Molecular formula:** C₁₇H₁₈F₃NO
**Molecular weight:** 309.3
**CAS Registry No.:** 54910-89-3, 59333-67-4 (HCl)

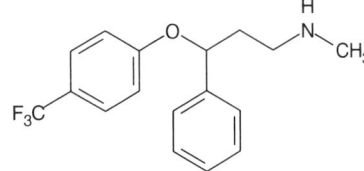

## SAMPLE
**Matrix:** blood
**Sample preparation:** 1 mL Plasma + doxepin + NaOH + hexane:isoamyl alcohol 98:2, extract. Remove the organic phase and add it to 0.03% phosphoric acid, extract, inject an aliquot of the aqueous phase.

## HPLC VARIABLES
**Guard column:** C18
**Column:** 100 × 8 10 μm Resolve C8 (Waters)
**Mobile phase:** MeCN:MeOH:56 mM ammonium acetate:1 M ammonium hydroxide 100:10:4.5:2.6
**Flow rate:** 2.5
**Detector:** UV 220

## CHROMATOGRAM
**Retention time:** 16
**Internal standard:** doxepin (11.6)

## OTHER SUBSTANCES
**Extracted:** amitriptyline, norfluoxetine, nortriptyline

## KEY WORDS
plasma

## REFERENCE
el-Yazigi, A.; Chaleby, K.; Gad, A.; Raines, D.A. Steady-state kinetics of fluoxetine and amitriptyline in patients treated with a combination of these drugs as compared with those treated with amitriptyline alone. *J.Clin.Pharmacol.*, **1995**, *35*, 17–21

## SAMPLE
**Matrix:** blood
**Sample preparation:** 2 mL Whole blood or plasma + 2 mL buffer + 5 mL chloroform:isopropanol:n-heptane 60:14:26, shake gently horizontally for 10 min, centrifuge at 2800 g for 10 min. Remove the lower organic layer and evaporate it to dryness under vacuum at 45°, reconstitute the residue in 100 μL mobile phase, centrifuge at 2800 g for 5 min, inject a 50 μL aliquot of the supernatant. (Buffer was saturated ammonium chloride solution 25% diluted with water, adjusted to pH 9.5 with 25% ammonia solution.)

## HPLC VARIABLES
**Column:** 300 × 3.9 4 μm NovaPack C18
**Mobile phase:** MeOH:THF:buffer 65:5:30 (Buffer was 0.68 g/L (10 mM (sic)) KH₂PO₄ adjusted to pH 2.6 with concentrated orthophosphoric acid.) (At the end of each session wash the column with water for 1 h and MeOH for 1 h, re-equilibrate for 30 min.)
**Column temperature:** 30
**Flow rate:** 0.8
**Injection volume:** 50
**Detector:** UV 226

## CHROMATOGRAM
**Retention time:** 9.14
**Limit of detection:** <120 ng/mL

## OTHER SUBSTANCES
**Extracted:** acebutolol, acenocoumarol, acepromazine, aceprometazine, acetaminophen, aconitine, ajmaline, albuterol, alimemazine, alminoprofen, alpidem, alprazolam, alprenolol, amisulpride, amodiaquine, amoxapine, aspirin, astemizole, atenolol, benazepril, benperidol, benzocaine, benzoylecgonine, bepridil, betaxolol, bisoprolol, bromazepam, brompheniramine, bumadizone, bupivacaine, buprenorphine, buspirone, caffeine, carbamazepine, carbinoxamine, carpipramine, carteolol, celiprolol, cetirizine, chlorambucil, chlordiazepoxide, chlormezanone, chlorophenacinone, chloroquine, chlorpheniramine, chlorpromazine, chlorpropamide, cibenzoline, cicletanine, clemastine, clobazam, clomipramine, clonazepam, clonidine, clorazepate, clozapine, cocaine, codeine, colchicine, cyamemazine, cyclizine, cycloguanil, cyproheptadine, cytarabine, dacarbazine, debrisoquine, demexiptiline, desipramine, dextromethorphan, dextromoramide, dextropropoxyphene, diazepam, diazoxide, diclofenac, dihydralazine, diltiazem, diphenhydramine, dipyridamole, disopyramide, dosulepine, doxepin, doxylamine, droperidol, ephedrine, estazolam, fenfluramine, fenoprofen, fentiazac, flecainide, floctafenine, flumazenil, flunitrazepam, fluphenazine, flurbiprofen, fluvoxamine, glibenclamide, glibornuride, glipizide, glutethimide, haloperidol, histapyrrodine, hydroxychloroquine, hydroxyzine, ibuprofen, imipramine, iproniazid, ketamine, ketoprofen, labetalol, levomepromazine, lidocaine, lidoflazine, lisinopril, loperamide, loprazolam, loratadine, lorazepam, loxapine, medazepam, medifoxamine, mefenamic acid, mefenidramine, mefloquine, melphalan, meperidine, mephenesin, mephentermine, mepivacaine, metapramine, metformin, methadone, methaqualone, methocarbamol, methotrexate, metipranolol, metoclopramide, metoprolol, mexiletine, mianserine, midazolam, minoxidil, moclobemide, moperone, morphine, nadolol, nalbuphine, nalorphine, naloxone, naltrexone, naproxen, nialamide, nicardipine, nifedipine, niflumic acid, nimodipine, nitrazepam, nitrendipine, nizatidine, nomifensine, omeprazole, opipramol, oxazepam, oxprenolol, penbutolol, penfluridol, pentazocine, phencyclidine, phenobarbital, phenol, phenylbutazone, pimozide, pindolol, pipamperone, piroxicam, prazepam, prazosin, prilocaine, procainamide, procarbazine, proguanil, promethazine, propafenone, propranolol, protriptyline, pyrimethamine, quinidine, quinine, quinupramine, ramipril, ranitidine, reserpine, ritodrine, secobarbital, sotalol, strychnine, sulfinpyrazone, sulindac, sulpride, sultopride, suriclone, temazepam, tenoxicam, terbutaline, terfenadine, tetracaine, tetrazepam, thiopental, thioproperazine, thioridazine, tianeptine, tiapride, tiaprofenic acid, ticlopidine, timolol, tofisopam, tolbutamide, toloxatone, trazodone, triazolam, trifluoperazine, trifluperidol, trimipramine, triprolidine, verapamil, viloxazine, vinblastine, vincristine, vindesine, warfarin, yohimbine, zolpidem, zopiclone, zorubicine
**Interfering:** amitriptyline, daunorubicin, etodolac, indomethacin, maprotiline, nortriptyline, tioclomarol, tropatenine

## KEY WORDS
whole blood; plasma

## REFERENCE
Tracqui, A.; Kintz, P.; Mangin, P. Systematic toxicological analysis using HPLC/DAD. *J.Forensic Sci.*, **1995**, *40*, 254–262

## SAMPLE
**Matrix:** blood
**Sample preparation:** Automated SPE by ASPEC system. Condition a C18 Clean-Up SPE cartridge (CEC 18111, Worldwide Monitoring) with 2 mL MeOH then 2 mL water. 1 mL Plasma + 1 mL 400 ng/mL protriptyline in water, vortex, add to the SPE cartridge, wash with 3 mL water, wash with 3 mL 750 mL/L methanol. Elute with three aliquots of 300 µL 0.1 M ammonium acetate in MeOH. Add 0.5 mL 0.5 M NaOH and 4 mL 50 mL/L

isopropanol in heptane to eluate, mix thoroughly. Allow 5 min for phase separation. Remove upper heptane phase and add it to 300 μL 0.1 M phosphoric acid (pH 2.5), mix, separate, inject a 100 μL aliquot of the aqueous phase.

## HPLC VARIABLES
**Guard column:** LC-8-DB (Supelco)
**Column:** 150 × 4.6 LC-8-DB (Supelco)
**Mobile phase:** MeCN:buffer 35:65 (Buffer was 10 mL/L triethylamine in water adjusted to pH 5.5 with glacial acetic acid.)
**Flow rate:** 2
**Injection volume:** 100
**Detector:** UV 228

## CHROMATOGRAM
**Retention time:** 6.5
**Internal standard:** protriptyline (4)

## OTHER SUBSTANCES
**Extracted:** acetazolamide, amitriptyline, chlordiazepoxide, chlorimipramine, chlorpromazine, desipramine, dextromethorphan, diazepam, diphenhydramine, doxepin, encainide, fentanyl, flecainide, haloperidol, ibuprofen, imipramine, lidocaine, maprotiline, methadone, methaqualone, mexiletine, midazolam, nordiazepam, nordoxepin, norfluoxetine, nortriptyline, norverapamil, pentazocine, promazine, propafenone, propoxyphene, propranolol, protriptyline, quinidine, temazepam, trazodone, trimipramine, verapamil
**Noninterfering:** acetaminophen, acetylmorphine, amiodarone, amobarbital, amphetamine, bendroflumethiazide, benzocaine, benzoylecgonine, benzthiazide, butalbital, carbamazepine, chlorothiazide, clonazepam, cocaine, codeine, cotinine, cyclosporine, cyclothiazide, desalkylflurazepam, diamorphine, dicumerol, ephedrine, ethacrynic acid, ethanol, ethchlorvynol, ethosuximide, furosemide, glutethimide, hydrochlorothiazide, hydrocodone, hydroflumethiazide, hydromorphone, lorazepam, mephentermine, meprobamate, methamphetamine, metharbital, methoxsalen, methoxyphenteramine, methsuximide, methylcyclothiazide, metoprolol, MHPG, monoacetylmorphine, morphine, normethsuximide, oxazepam, oxycodone, oxymorphone, pentobarbital, phencyclidine, phenteramine, phenylephrine, phenytoin, polythiazide, primidone, prochlorperazine, salicylic acid, sulfanilamide, THC-COOH, theophylline, thiazolam, thiopental, thioridazine, tocainide, trichloromethiazide, trifluoperazine, valproic acid, warfarin
**Interfering:** flurazepam, hydroxyethylflurazepam, norchlorimipramine

## KEY WORDS
plasma; SPE

## REFERENCE
Nichols, J.H.; Charlson, J.R.; Lawson, G.M. Automated HPLC assay of fluoxetine and norfluoxetine in serum. *Clin.Chem.*, **1994**, *40*, 1312–1316

## SAMPLE
**Matrix:** blood
**Sample preparation:** 1 mL Serum or plasma + 30 μL trimipramine in MeOH + 200 μL 0.33 M NaOH, shake 5 s, add 7 mL n-hexane:iso-amyl alcohol 985:15, shake 20 min, centrifuge at 2100 g for 5 min. Remove organic phase and add 200 μL 0.1 M HCl to it, shake for 1 min, discard organic phase, inject 30 μL of aqueous phase.

## HPLC VARIABLES
**Guard column:** 10 mm 10 μm Bischoff C18
**Column:** 125 × 4 5 μm Ecotube Nucleosil C8
**Mobile phase:** MeCN:water:diethylamine:PicB5 370:630:0.4:25 (PicB5 is water-MeOH-1-pentanesulfonic acid.)

**Column temperature:** 55
**Flow rate:** 1.7
**Injection volume:** 30
**Detector:** UV 230

## CHROMATOGRAM
**Retention time:** 5.6
**Internal standard:** trimipramine (8.4)
**Limit of detection:** 10 ng/mL

## OTHER SUBSTANCES
**Simultaneous:** norfluoxetine
**Noninterfering:** alprazolam, amitriptyline, bromazepam, clomipramine, clorazepate, desipramine, diazepam, flunitrazepam, fluvoxamine, imipramine, lorazepam, nortriptyline, oxazepam, triazolam

## KEY WORDS
serum; plasma

## REFERENCE
el Maanni, A.; Combourieu, I.; Bonini, M.; Creppy, E.E. Fluoxetine, an antidepressant, and norfluoxetine, its metabolite, determined by HPLC with a $C_8$ column and ultraviolet detection. *Clin.Chem.*, **1993**, *39*, 1749–1750

## SAMPLE
**Matrix:** blood
**Sample preparation:** Condition a 1 mL BondElut C18 SPE cartridge with 1 mL 1 M HCl, 1 mL MeOH, 1 mL water, and 1 mL 1% potassium carbonate. 700 μL Serum + 50 μL 5 μg/mL protriptyline in 5% potassium bicarbonate + 700 μL MeCN, vortex, centrifuge at 1500 g for 5 min, add supernatant to SPE cartridge (at ca. 1 mL/min). Wash with 2 mL water and 1 mL MeCN, elute with 250 μL MeOH:35% perchloric acid 20:1 by gravity (10 min) then centrifuge for 20 s to remove rest of eluate, inject a 50 μL aliquot of the eluate.

## HPLC VARIABLES
**Guard column:** 15 mm 7 μm Brownlee RP-8
**Column:** 150 × 4.6 5 μm Ultrasphere Octyl
**Mobile phase:** MeCN:water 37.5:62.5 containing 0.5 g/L tetramethylammonium perchlorate and 0.5 mL/L 7% perchloric acid
**Flow rate:** 1.5
**Injection volume:** 50
**Detector:** UV 226

## CHROMATOGRAM
**Retention time:** 10.9
**Internal standard:** protriptyline (6.6)
**Limit of quantitation:** 5 ng/mL

## OTHER SUBSTANCES
**Extracted:** amitriptyline, clomipramine, desipramine, doxepin, fluvoxamine, imipramine, maprotiline, nortriptyline, trimipramine
**Interfering:** desmethylclomipramine

## KEY WORDS
serum; SPE

## REFERENCE

Gupta, R.N. An improved solid phase extraction procedure for the determination of antidepressants in serum by column liquid chromatography. *J.Liq.Chromatogr.*, **1993**, *16*, 2751–2765

## SAMPLE

**Matrix:** blood

**Sample preparation:** Add 10 μL 20 μg/mL oxaprotiline in MeOH to 990 μL plasma or serum. Inject 100 μL plasma or serum onto column A with mobile phase A and elute to waste, after 15 min elute column A onto column B with mobile phase B for 2 min. Remove column A from circuit and re-equilibrate it with mobile phase A for 5 min. Chromatograph on column B with mobile phase B.

## HPLC VARIABLES

**Column:** A 20 × 4.6 10 μm Hypersil MOS C8; B 20 × 4.6 5 μm Hypersil CPS CN + 250 × 4.6 5 μm Nucleosil 100 CN

**Mobile phase:** A MeOH:water 5:95; B MeOH:MeCN:10 mM pH 6.8 potassium phosphate buffer 188:578:235

**Flow rate:** 1.5

**Injection volume:** 100

**Detector:** UV 214

## CHROMATOGRAM

**Retention time:** 8.7

**Internal standard:** oxaprotiline (9.5)

**Limit of detection:** 20 ng/mL

## OTHER SUBSTANCES

**Simultaneous:** clozapine, desipramine, doxepin, fluvoxamine, imipramine, maprotiline, metoclopramide, norfluoxetine, nortriptyline

**Noninterfering:** carbamazepine, chlordiazepoxide, clobazam, diazepam, flurazepam, fluspirilene, haloperidol, lorazepam, nitrazepam, nordiazepam, oxazepam, perazine, pimozide, spiroperidol, trifluperidol

**Interfering:** amitriptyline, clomipramine

## KEY WORDS

plasma; serum; column-switching

## REFERENCE

Härtter, S.; Wetzel, H.; Hiemke, C. Automated determination of fluvoxamine in plasma by column-switching high-performance liquid chromatography. *Clin.Chem.*, **1992**, *38*, 2082–2086

## SAMPLE

**Matrix:** blood

**Sample preparation:** 1 mL Plasma + 1 mL 0.6 M pH 9.8 carbonate buffer + 40 μL 5 μg/mL maprotiline in 10 mM HCl + 5 mL 200 g/L ethyl acetate in n-heptane, mix by rocking for 10 min, centrifuge at 1500 g for 10 min. Remove organic layer and add it to 150 μL 100 mM HCl, mix 10 min, centrifuge at 1500 g for 10 min. Discard organic layer and evaporate aqueous layer at 45° in a vacuum centrifuge for 1 h. Take up residue in 50 μL 1 M pH 10.3 carbonate buffer and 25 μL 10 mg/mL dansyl chloride in MeCN, vortex, allow to react at room temperature for 45 min, evaporate at 45° in a vacuum centrifuge for 20 min, reconstitute in 125 μL MeCN:water 75:25, vortex, centrifuge for 3-5 min, inject a 25-40 μL aliquot.

## HPLC VARIABLES

**Column:** 250 × 4.6 5 μm Supelcosil LC-18

**Mobile phase:** MeCN:25 mM $KH_2PO_4$ 75:25 containing 500 μL/L orthophosphoric acid and 600 μL/L n-butylamine
**Flow rate:** 2
**Injection volume:** 25-40
**Detector:** F ex 235 em 470 (cut-off)

## CHROMATOGRAM
**Retention time:** 10.4
**Internal standard:** maprotiline (12.8)
**Limit of quantitation:** 3 ng/mL

## OTHER SUBSTANCES
**Simultaneous:** amoxapine, clovoxamine, desipramine, fenfluramine, fluvoxamine, norfluoxetine, nortriptyline, propranolol, protriptyline, sertraline
**Noninterfering:** amitriptyline, atenolol, bupropion, carbamazepine, chlordiazepoxide, citalopram, clomipramine, clozapine, cyclobenzaprine, doxepin, imipramine, loxapine, metoprolol, mianserin, moclobemide, nomifensine, pindolol, thioridazine, tranylcypromine, trazodone, trimipramine

## KEY WORDS
plasma

## REFERENCE
Suckow, R.F.; Zhang, M.F.; Cooper, T.B. Sensitive and selective liquid-chromatographic assay of fluoxetine and norfluoxetine in plasma with fluorescence detection after precolumn derivatization. *Clin.Chem.*, **1992**, *38*, 1756–1761

## SAMPLE
**Matrix:** blood
**Sample preparation:** 2 mL Plasma + 160 μL 10 μg/mL clomipramine in MeOH, vortex, add 2 mL 1 M NaOH, vortex, add 5 mL n-hexane:isoamyl alcohol 99:1, rotate for 5 min, centrifuge for 10 min. Remove organic layer and add it to 200 μL 0.05% phosphoric acid, rotate for 5 min, centrifuge for 10 min, remove lower aqueous layer and inject a 25-50 μL aliquot of it.

## HPLC VARIABLES
**Guard column:** Bondapak/Corasil C18
**Column:** 300 × 4.6 μBondapak C18
**Mobile phase:** MeCN:50 mM $KH_2PO_4$ adjusted to pH 4.7 with KOH 40:60
**Column temperature:** 50
**Flow rate:** 2
**Injection volume:** 25-50
**Detector:** UV 214

## CHROMATOGRAM
**Retention time:** 7
**Internal standard:** clomipramine (9)
**Limit of detection:** 6 ng/mL

## OTHER SUBSTANCES
**Simultaneous:** amoxapine, chlordiazepoxide, chlorpromazine, cimetidine, desipramine, diazepam, doxepin, flurazepam, imipramine, lorazepam, norfluoxetine, nortriptyline, oxazepam, pentobarbital, perphenazine, phenobarbital, phenytoin, prochlorperazine, secobarbital, thioridazine, trifluoperazine
**Noninterfering:** acetaminophen, codeine, meperidine
**Interfering:** amitriptyline, propoxyphene

## KEY WORDS
plasma

## REFERENCE
Wong, S.H.; Dellafera, S.S.; Fernandes, R.; Kranzler, H. Determination of fluoxetine and norfluoxetine by high-performance liquid chromatography. *J.Chromatogr.*, **1990**, *499*, 601–608

## SAMPLE
**Matrix:** blood
**Sample preparation:** 1 mL Serum + 50 µL 1 mg/mL reduced haloperidol in water + 2 mL 0.5 M NaH$_2$PO$_4$ adjusted to pH 10 with 10 M NaOH, vortex 3-5 s, add 5 mL hexane:isoamyl alcohol 97:3, shake 20 min, centrifuge at 1000 g for 10 min. Remove organic layer and add 1 mL 0.1 M HCl to it, shake for 20 min, centrifuge for 10 min, discard organic layer. Add 1 mL 0.2 M NaOH to aqueous layer, add 5 mL hexane:isoamyl alcohol 97:3, shake for 20 min, centrifuge for 10 min. Remove organic layer, add 1 drop of 0.3 M HCl in MeOH, evaporate under nitrogen at 40°, reconstitute with 250 µL mobile phase, vortex, inject 100 µL aliquot.

## HPLC VARIABLES
**Guard column:** C18 Guard-Pak (Waters no. 88070)
**Column:** Nova-Pak phenyl (Waters no. 10656)
**Mobile phase:** MeCN:buffer 40:60 (Buffer was 600 mL water + 1 mL triethylamine, adjusted to pH 5.5 with acetic acid.)
**Flow rate:** 1.7
**Injection volume:** 100
**Detector:** UV 226

## CHROMATOGRAM
**Retention time:** 7.3
**Internal standard:** reduced haloperidol (3.9)
**Limit of detection:** 15 ng/mL

## OTHER SUBSTANCES
**Simultaneous:** alprazolam, amitriptyline, amoxapine, chlordiazepoxide, chlorimipramine, clonazepam, demoxepam, diazepam, doxepin, halazepam, haloperidol, lorazepam, maprotiline, norfluoxetine, nortriptyline, oxazepam, temazepam, trazodone, trimipramine
**Interfering:** desipramine, imipramine, loxapine, protriptyline

## KEY WORDS
serum

## REFERENCE
Orsulak, P.J.; Kenney, J.T.; Debus, J.R.; Crowley, G.; Wittman, P.D. Determination of the antidepressant fluoxetine and its metabolite norfluoxetine in serum by reversed-phase HPLC with ultraviolet detection. *Clin.Chem.*, **1988**, *34*, 1875–1878

## SAMPLE
**Matrix:** blood, tissue
**Sample preparation:** Blood or serum. 1 mL Blood or serum + 1 µg cianopramine + 1 mL water, vortex, add 1 mL 200 mM sodium carbonate, vortex, add 6 mL hexane:1-butanol 95:5, gently agitate for 30 min, centrifuge at 2500 g for 5 min. Remove the organic layer and add it to 100 µL 0.2% phosphoric acid, agitate gently for 30 min, centrifuge for 5 min. Remove the organic layer and inject a 30 µL aliquot of the aqueous layer. Liver homogenate. 0.5 mL Liver homogenate + 10 µg cianopramine + 500 µL 2% sodium tetraborate + 8 mL hexane:1-butanol 95:5, gently agitate for 30 min, centrifuge at 2500 g for 5 min. Remove the organic layer and add it to 400 µL 0.2% phosphoric acid, agitate

gently for 30 min, centrifuge for 5 min. Remove the organic layer and inject a 30 μL aliquot of the aqueous layer.

## HPLC VARIABLES
**Guard column:** 15 × 3.2 7 μm RP-18 Newguard (Applied Biosystems)
**Column:** 100 × 4.6 5 μm Brownlee Spheri-5 RP-18
**Mobile phase:** MeCN:100 mM $NaH_2PO_4$:diethylamine 40:57.5:2.5
**Flow rate:** 2
**Injection volume:** 30
**Detector:** UV 220

## CHROMATOGRAM
**Retention time:** 6.70
**Internal standard:** cianopramine (8.93)
**Limit of detection:** 100 ng/mL

## OTHER SUBSTANCES
**Extracted:** amitriptyline, benztropine, brompheniramine, chlorpheniramine, chlorpromazine, clomipramine, cyproheptadine, desipramine, diphenhydramine, dothiepin, doxepin, haloperidol, imipramine, loxapine, maprotiline, meperidine, mesoridazine, methadone, metoclopramide, mianserin, moclobemide, nomifensine, nordoxepin, norfluoxetine, norpropoxyphene, northiaden, nortriptyline, pentobarbital, pheniramine, promethazine, propoxyphene, propranolol, protriptyline, quinidine, quinine, sulforidazine, thioridazine, thiothixene, tranylcypromine, trazodone, trihexyphenidyl, trimipramine, triprolidine
**Noninterfering:** dextromethorphan, norphetidine, phenoxybenzamine, prochlorperazine, trifluoperazine
**Interfering:** amoxapine

## KEY WORDS
serum; whole blood; liver

## REFERENCE
McIntyre, I.M.; King, C.V.; Skafidis, S.; Drummer, O.H. Dual ultraviolet wavelength high-performance liquid chromatographic method for the forensic or clinical analysis of seventeen antidepressants and some selected metabolites. *J.Chromatogr.*, **1993**, *621*, 215–223

## SAMPLE
**Matrix:** blood, tissue
**Sample preparation:** Homogenize (Brinkman Polytron) tissue with 5-10 volumes of water. 500 μL Plasma or tissue homogenate + 500 μL water + 50 μL 2 μg/mL IS in water + 100 μL 1 M NaOH, vortex gently, add 5 mL hexane:butanol 99.7:0.3, shake mechanically at 125-150 cycles/min for 30 min, centrifuge at 2000 g for 15 min. Remove the organic layer and mix it with 100 μL 200 μM R-(−)-1-(1-naphthyl)ethyl isocyanate in hexane, evaporate to dryness at 50-55° over 20-30 min, dry more vigorously when all the hexane is gone, reconstitute with 200 μL mobile phase, inject a 75 μL aliquot.

## HPLC VARIABLES
**Column:** 250 × 4.6 5 μm Apex silica (Jones Chromatography)
**Mobile phase:** Isooctane:THF 70:30
**Column temperature:** 35
**Flow rate:** 1
**Injection volume:** 75
**Detector:** F ex 218 em 333

## CHROMATOGRAM
**Retention time:** 8.3 (S), 9.3 (R)

**Internal standard:** S-nornisoxetine (15)
**Limit of detection:** 5 ng/mL (plasma); 25 ng/g (tissue)

## OTHER SUBSTANCES
**Extracted:** metabolites, norfluoxetine

## KEY WORDS
chiral; derivatization; normal phase; plasma; silylate all glassware

## REFERENCE
Potts, B.D.; Parli, C.J. Analysis of the enantiomers of fluoxetine and norfluoxetine in plasma and tissue using chiral derivatization and normal-phase liquid chromatography. *J.Liq.Chromatogr.*, **1992**, *15*, 665–681

## SAMPLE
**Matrix:** bulk
**Sample preparation:** Prepare a 2 mg/mL solution of fluoxetine hydrochloride in mobile phase, inject a 10 μL solution.

## HPLC VARIABLES
**Column:** 150 × 4.6 3 μm Exsil 100Å/ODS-B octadecylsilane (Keystone)
**Mobile phase:** MeCN:THF:buffer 15:10:75 (Buffer was 50 mM ammonium acetate adjusted to pH 5.75 with 50 mM acetic acid.)
**Flow rate:** 1
**Injection volume:** 10
**Detector:** UV 214

## CHROMATOGRAM
**Retention time:** 31.9

## OTHER SUBSTANCES
**Simultaneous:** impurities, meta isomer

## REFERENCE
Lacroix, P.M.; Yat, P.N.; Lovering, E.G. Liquid chromatographic methods for fluoxetine hydrochloride, its *meta* isomer, and related compounds in raw materials. *J.AOAC Int.*, **1995**, *78*, 334–339

## SAMPLE
**Matrix:** bulk
**Sample preparation:** Reflux 1.23 g fluoxetine with 788 mg (R)-(−)-1-(1-naphthyl)ethyl isocyanate in 25 mL toluene for 2 h, evaporate to dryness under reduced pressure, reconstitute, inject an aliquot.

## HPLC VARIABLES
**Column:** 250 × 4.6 silica (IBM)
**Mobile phase:** Dichloromethane:MeOH 99.75:0.25
**Flow rate:** 2
**Detector:** UV 254

## CHROMATOGRAM
**Retention time:** 6.69 (S), 7.52 (R)

## KEY WORDS
derivatization; chiral; normal phase

## REFERENCE

Robertson, D.W.; Krushinski, J.H.; Fuller, R.W.; Leander, J.D. Absolute configurations and pharmacological activities of the optical isomers of fluoxetine, a selective serotonin-uptake inhibitor. *J.Med.Chem.*, **1988**, *31*, 1412–1417

## SAMPLE

**Matrix:** formulations
**Sample preparation:** Weigh out 4 mL of 1 mg/mL solution and make up to 100 mL with mobile phase, inject 20 μL aliquot.

## HPLC VARIABLES

**Column:** 250 × 4.6 Zorbax cyano special
**Mobile phase:** MeCN:buffer 50:50 (Buffer was 1% triethylamine, pH adjusted to 6 with concentrated phosphoric acid.)
**Flow rate:** 1
**Injection volume:** 20
**Detector:** UV 215

## CHROMATOGRAM

**Retention time:** 11

## OTHER SUBSTANCES

**Simultaneous:** benzoic acid

## KEY WORDS

syrup; elixir; stability-indicating

## REFERENCE

Peterson, J.A.; Risley, D.S.; Anderson, P.N.; Hostettler, K.F. Stability of fluoxetine hydrochloride in fluoxetine solution diluted with common pharmaceutical diluents. *Am.J.Hosp.Pharm.*, **1994**, *51*, 1342–1345

## SAMPLE

**Matrix:** solutions

## HPLC VARIABLES

**Column:** 250 × 4.6 5 μm Supelcosil LC-DP (A) or 250 × 4 5 μm LiChrospher 100 RP-8 (B)
**Mobile phase:** MeCN:0.025% phosphoric acid:buffer 25:10:5 (A) or 60:25:15 (B) (Buffer was 9 mL concentrated phosphoric acid and 10 mL triethylamine in 900 mL water, adjust pH to 3.4 with dilute phosphoric acid, make up to 1 L.)
**Flow rate:** 0.6
**Injection volume:** 25
**Detector:** UV 229

## CHROMATOGRAM

**Retention time:** 12.20 (A), 7.07 (B)

## OTHER SUBSTANCES

**Also analyzed:** acebutolol, acepromazine, acetaminophen, acetazolamide, acetophenazine, albuterol, alprazolam, amitriptyline, amobarbital, amoxapine, antipyrine, atenolol, atropine, azatadine, baclofen, benzocaine, bromocriptine, brompheniramine, brotizolam, bupivacaine, buspirone, butabarbital, butalbital, caffeine, carbamazepine, cetirizine, chlorcyclizine, chlordiazepoxide, chlormezanone, chloroquine, chlorpheniramine, chlorpromazine, chlorpropamide, chlorprothixene, chlorthalidone, chlorzoxazone, cimetidine, cisapride, clomipramine, clonazepam, clonidine, clozapine, cocaine, codeine, colchicine, cyclizine, cyclobenzaprine, dantrolene, desipramine, diazepam, diclofenac, diflunisal, diltiazem, diphenhydramine, diphenidol, diphenoxylate, dipyridamole, disopyramide, dobutamine, doxapram, doxepin, droperidol, encainide, ethidium bromide, ethopropazine, fenoprofen, fentanyl, flavoxate, fluphenazine, flurazepam, flurbiprofen, fluvoxamine, furosemide, glutethimide, glyburide, guaifenesin, haloperidol, homatropine, hydralazine, hydrochlorothiazide, hydrocodone, hydromorphone, hydroxychloroquine, hydroxyzine, ibuprofen, imipramine, indomethacin, ketoconazole, ketoprofen, ketorolac, labetalol, levorphanol, lidocaine, loratadine, lorazepam, lovastatin, loxapine, mazindol, mefenamic acid, meperidine, mephenytoin, mepivacaine, mesoridazine, metaproterenol, methadone, methdilazine, methocarbamol, methotrexate, methotrimeprazine, methoxamine, methyldopa, methylphenidate, metoclopramide, metolazone, metoprolol, metronidazole, midazolam, moclobemide, morphine, nadolol, nalbuphine, naloxone, naphazoline, naproxen, nifedipine, nizatidine, norepinephrine, nortriptyline, oxazepam, oxycodone, oxymetazoline, paroxetine, pemoline, pentazocine, pentobarbital, pentoxifylline, perphenazine, pheniramine, phenobarbital, phenol, phenolphthalein, phentolamine, phenylbutazone, phenyltoloxamine, phenytoin, pimozide, pindolol, piroxicam, pramoxine, prazepam, prazosin, probenecid, procainamide, procaine, prochlorperazine, procyclidine, promazine, promethazine, propafenone, propantheline, propiomazine, propofol, propranolol, protriptyline, quazepam, quinidine, quinine, racemethorphan, ranitidine, remoxipride, risperidone, salicylic acid, scopolamine, secobarbital, sertraline, sotalol, spironolactone, sulfinpyrazone, sulindac, temazepam, terbutaline, terfenadine, tetracaine, theophylline, thiethylperazine, thiopental, thioridazine, thiothixene, timolol, tocainide, tolbutamide, tolmetin, trazodone, triamterene, triazolam, trifluoperazine, triflupromazine, trimeprazine, trimethoprim, trimipramine, verapamil, warfarin, xylometazoline, yohimbine, zopiclone

## KEY WORDS
some details of plasma extraction

## REFERENCE
Koves, E.M. Use of high-performance liquid chromatography-diode array detection in forensic toxicology. *J.Chromatogr.A*, **1995**, *692*, 103–119

## SAMPLE
**Matrix:** solutions

## HPLC VARIABLES
**Column:** 150 × 4.6 Supelcosil LC-ABZ
**Mobile phase:** MeCN:25 mM pH 6.9 potassium phosphate buffer 35:65
**Flow rate:** 1.5
**Injection volume:** 25
**Detector:** UV 254

## CHROMATOGRAM
**Retention time:** 11.460

## OTHER SUBSTANCES
**Also analyzed:** 6-acetylmorphine, amiloride, amphetamine, benzocaine, benzoylecgonine, caffeine, cocaine, codeine, doxylamine, glutethimide, hexobarbital, hypoxanthine, levorphanol, LSD, meperidine, mephobarbital, methadone, methylphenidate, methyprylon, N-norcodeine, oxazepam, oxycodone, phenylpropanolamine, prilocaine, procaine, terfenadine

## REFERENCE
Ascah, T.L. Improved separations of alkaloid drugs and other substances of abuse using Supelcosil LC-ABZ column. *Supelco Reporter*, **1993**, *12(3)*, 18–21

## SAMPLE
**Matrix:** tissue

**Sample preparation:** Homogenize 1 mL liver in 5 mL water. Centrifuge 1 mL homogenate, add the supernatant to 1.2 µg clomipramine, add 75 µL MeOH, add 75 µL MeCN, add 100 µL 1 M HCl, vortex, centrifuge, inject an aliquot of the supernatant.

## HPLC VARIABLES
**Column:** 300 × 3.9 µBondapak C18
**Mobile phase:** MeCN:50 mM potassium phosphate buffer 35:65
**Flow rate:** 1.3
**Detector:** UV 226

## CHROMATOGRAM
**Internal standard:** clomipramine

## OTHER SUBSTANCES
**Extracted:** metabolites, norfluoxetine
**Also analyzed:** sertraline

## KEY WORDS
mouse; liver

## REFERENCE
von Moltke, L.L.; Greenblatt, D.J.; Cotreau-Bibbo, M.M.; Duan, S.X.; Harmatz, J.S.; Shader, R.I. Inhibition of desipramine hydroxylation in vitro by serotonin-reuptake-inhibitor antidepressants and by quinidine and ketoconazole: A model system to predict drug interactions in vivo. *J.Pharmacol.Exp.Ther.*, **1994**, *268*, 1278–1283

◆————————————————————◆————————————————————◆

## ANNOTATED BIBLIOGRAPHY
Hubmann, M.R.; Waschgler, R.; Moll, W.; Conca, A.; König, P. Simultaneous drug monitoring of citalopram, clozapine, fluoxetine, maprotiline, and trazodone by HPLC analysis (Abstract 41). *Ther.Drug Monit.*, **1995**, *17*, 393 [simultaneous citalopram, clozapine, maprotiline, trazodone; LOQ 50 ng/mL]

Joron, S.; Robert, H. Simultaneous determination of antidepressant drugs and metabolites by HPLC. Design and validation of a simple and reliable analytical procedure. *Biomed.Chromatogr.*, **1994**, *8*, 158–164 [simultaneous amineptine, amitriptyline, amoxapine, clomipramine, demexiptiline, desipramine, dosulepine, doxepin, doxepine, fluvoxamine, imipramine, maprotiline, medifloxamine, mianserine, opipramol, quinupramine, tianeptine, toloxatone, trazodone, trimipramine, viloxazine; LOQ 3-17 ng/mL; plasma]

Thomare, P.; Wang, K.; Van Der Meersch-Mougeot, V.; Diquet, B. Sensitive micromethod for column liquid chromatographic determination of fluoxetine and norfluoxetine in human plasma. *J.Chromatogr.*, **1992**, *583*, 217–221 [plasma; extracted metabolites; LOD 2 ng/mL; simultaneous amineptine, amitriptyline, chlordiazepoxide, chlorpromazine, clomipramine, clonazepam, clorazepate, desipramine, diazepam, doxepin, flunitrazepam, fluvoxamine, imipramine, levomepromazine, lorazepam, loxapine, maprotiline, mefloquine, nortriptyline, oxazepam, thioridazine]

Peyton, A.L.; Carpenter, R.; Rutkowski, K. The stereospecific determination of fluoxetine and norfluoxetine enantiomers in human plasma by high-pressure liquid chromatography (HPLC) with fluorescence detection. *Pharm.Res.*, **1991**, *8*, 1528–1532

Gupta, R.N.; Steiner, M. Determination of fluoxetine and norfluoxetine in serum by liquid chromatography with fluorescence detection. *J.Liq.Chromatogr.*, **1990**, *13*, 3785–3798 [extracted norfluoxetine; serum; fluorescence detection; SPE; protriptyline (IS); simultaneous amitriptyline, nortriptyline]

# Flutamide

**Molecular formula:** $C_{11}H_{11}F_3N_2O_3$
**Molecular weight:** 276.2
**CAS Registry No.:** 13311-84-7

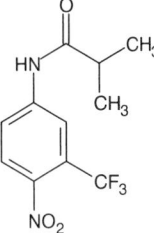

## SAMPLE
**Matrix:** blood
**Sample preparation:** 150 μL Plasma + 150 μL MeCN, vortex for 15 s, centrifuge at 13000 g for 10 min, inject a 50 μL aliquot of the supernatant

## HPLC VARIABLES
**Guard column:** 30 mm long 40-50 μm pellicular C18 (Replace guard column and prefilter before each run.)
**Column:** 150 × 3.2 3 μm Sphere 3 ODS C18 (Phenomenex)
**Mobile phase:** MeCN:1% acetic acid 50:50, pH 2.9 (dog) or MeCN:MeOH:water 30:20:50 (human)
**Flow rate:** 0.5
**Injection volume:** 50
**Detector:** UV 300

## CHROMATOGRAM
**Retention time:** 7.4
**Limit of detection:** 11.27 ng/mL
**Limit of quantitation:** 50 ng/mL

## OTHER SUBSTANCES
**Extracted:** metabolites

## KEY WORDS
dog; plasma; pharmacokinetics; human

## REFERENCE
Farthing, D.; Sica, D.; Fakhry, I.; Walters, D.L.; Cefali, E.A.; Allan, G. Determination of flutamide and hydroxyflutamide in dog plasma by a sensitive high performance liquid chromatography method utilizing mid-bore chromatography. *Biomed.Chromatogr.*, **1994**, *8*, 251–254

# Fluvastatin

**Molecular formula:** $C_{24}H_{26}FNO_4$
**Molecular weight:** 410.5
**CAS Registry No.:** 93957-55-2 (fluvastatin sodium)

## SAMPLE
**Matrix:** blood
**Sample preparation:** Dilute with an equal volume of water, precipitate with two 10 mL portions of acetone:MeOH 5:2, collect the supernatant, reduce it in volume, inject an aliquot.

## HPLC VARIABLES
**Column:** 300 × 3.9 10 μm fatty acid (Waters)
**Mobile phase:** Gradient. Buffer:MeOH 100:0 for 2 min, to 72:18 over 25 min, maintain at 72:28 for 18 min, to 62:38 over 19 min, maintain at 62:38 for 6 min, to 59.7:40.3 over 4 min, to 59.2:40.8 over 2 min, maintain at 59.2:40.8 for 2 min, to 58:42 over 2 min, maintain at 58:42 for 2 min, to 54:46 over 4 min, to 45:55 over 4 min, maintain at 45:55 for 2 min, to 20:80 over 6 min, maintain at 20:80 over 6 min, return to initial conditions over 4 min.
**Flow rate:** 1
**Detector:** UV 254

## CHROMATOGRAM
**Retention time:** 85

## OTHER SUBSTANCES
**Extracted:** metabolites

## KEY WORDS
plasma; human; rat; hamster

## REFERENCE
Dain, J.G.; Fu, E.; Gorski, J.; Nicoletti, J.; Scallen, T.J. Biotransformation of fluvastatin sodium in humans. *Drug Metab.Dispos.*, **1993**, *21*, 567–572

## SAMPLE
**Matrix:** blood
**Sample preparation:** 1 mL Plasma + 1 mL MeCN, mix on a Maxi-Mix for 5 s, add 1 mL 300 ng/mL IS in water, add 2 mL phosphate buffer, add 10 mL MTBE, shake horizontally on a platform shaker at 200 cycles/min for 15 min, centrifuge at 700 g for 5 min. Remove the upper organic layer and evaporate it to dryness under vacuum, reconstitute the residue in 400 μL mobile phase, inject a 200 μL aliquot.

## HPLC VARIABLES
**Column:** 150 × 4.6 5 μm Supelcosil LC-18
**Mobile phase:** MeOH:13 mM tetrabutylammonium fluoride 60:40
**Column temperature:** 50
**Injection volume:** 200

**Detector:** F ex 305 em 380

## CHROMATOGRAM
**Retention time:** 9.2
**Internal standard:** ([R*, S*)-(E)-]($\pm$)-7-[3-(4-fluorophenyl)-1-(1-methylethyl)-1H-indol-2-yl]-3,5-dihydroxy-6-methyl-6-heptenoic acid, monosodium salt (Sandoz 63-267, 6-methyl-fluvastatin) (12.8)
**Limit of quantitation:** 1 ng/mL

## KEY WORDS
plasma; protect from light; pharmacokinetics

## REFERENCE
Kalafsky, G.; Smith, H.T.; Choc, M.G. High-performance liquid chromatographic method for the determination of fluvastatin in human plasma. *J.Chromatogr.*, **1993**, *614*, 307–313

## SAMPLE
**Matrix:** blood
**Sample preparation:** Adjust pH of 500 μL blood to 7, extract with MTBE. Remove the organic layer and evaporate it to dryness, reconstitute the residue in mobile phase, inject an aliquot.

## HPLC VARIABLES
**Column:** 150 × 4.6 5 μm Supelcosil LC-18
**Mobile phase:** MeOH:water 60:40 containing 5 mL/L tetrabutylammonium fluoride
**Column temperature:** 50
**Detector:** F ex 305 em 380

## CHROMATOGRAM
**Internal standard:** sodium 3,5-dihydroxy-7-[3-(4-fluorophenyl)-1-(1-methylethyl)-1H-indole-2-yl]-6-methylhept-6-enoate
**Limit of detection:** 1 ng/mL

## KEY WORDS
pharmacokinetics; rabbit

## REFERENCE
Tse, F.L.; Labbadia, D. Absorption and disposition of fluvastatin, an inhibitor of HMG-CoA reductase, in the rabbit. *Biopharm.Drug Dispos.*, **1992**, *13*, 285–294

## SAMPLE
**Matrix:** blood
**Sample preparation:** Adjust pH of 500 μL blood to 7, extract with MTBE. Remove the organic layer and evaporate it to dryness, reconstitute the residue in MeCN:5 mM pH 6.5 hexyltriethylammonium phosphate 5:95, inject an aliquot.

## HPLC VARIABLES
**Column:** C8
**Mobile phase:** MeCN:5 mM pH 6.5 hexyltriethylammonium phosphate 40:60
**Column temperature:** 50
**Detector:** F ex 305 em 380

## CHROMATOGRAM
**Limit of detection:** 2 ng/mL

## KEY WORDS
pharmacokinetics; mouse; dog; monkey

## REFERENCE
Tse, F.L.; Smith, H.T.; Ballard, F.H.; Nicoletti, J. Disposition of fluvastatin, an inhibitor of HMG-CoA reductase, in mouse, rat, dog, and monkey. *Biopharm.Drug Dispos.*, **1990**, *11*, 519–531

# Furosemide

**Molecular formula:** $C_{12}H_{11}ClN_2O_5S$
**Molecular weight:** 330.7
**CAS Registry No.:** 54-31-9

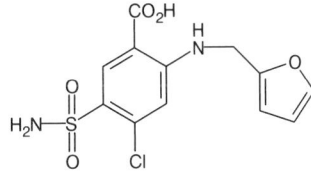

## SAMPLE
**Matrix:** bile
**Sample preparation:** 2 mL Bile + 1 mL pH 5.0 phosphate buffer, filter (0.5 μm). Remove a 1 mL aliquot and add it to 500 μL 20 μg/mL piretanide, inject a 10-50 μL aliquot. (Hydrolyze glucuronide by heating 2 mL bile with 1 mL 1000 U/mL β-glucuronidase in 100 mM pH 5.0 acetate buffer at 37° for 2 h, proceed as above.)

## HPLC VARIABLES
**Column:** 150 × 6 5 μm Shim-pack CLC-ODS (Shimadzu)
**Mobile phase:** Gradient. A was MeCN:water 20:80 containing 0.3% acetic acid. B was MeCN:water 80:20 containing 0.3% acetic acid. A:B 90:10 for 3 min, to 60:40 over 7 min, maintain at 60:40 for 5 min, to 40:60 over 3 min, to 60:40 over 2 min, to 90:10 over 10 min.
**Column temperature:** 40
**Injection volume:** 10-50
**Detector:** F ex 345 em 415

## CHROMATOGRAM
**Retention time:** 15
**Internal standard:** piretanide (21)
**Limit of detection:** 5 ng/mL

## KEY WORDS
pharmacokinetics

## REFERENCE
Sekikawa, H.; Yagi, N.; Oda, K.; Kenmotsu, H.; Takada, M.; Chen, H.-f.; Lin, E.T.; Benet, L.Z. Biliary excretion of furosemide glucuronide in rabbits. *Biol.Pharm.Bull.*, **1995**, *18*, 447–453

## SAMPLE
**Matrix:** blood
**Sample preparation:** Inject 50 μL plasma onto column A with mobile phase A, after 6 min the contents of column A were back-flushed onto column B with mobile phase B, after 3 min column A was removed from the circuit and column B was eluted with mobile phase B. Column A was washed with mobile phase C for 5 min then equilibrated with mobile phase A (1.5 mL/min) for 11 min until next injection.

## HPLC VARIABLES
**Column:** A 35 × 4.6 20 μm TSK BSA-ODS; B 150 × 4.6 5 μm Nucleosil 5C18
**Mobile phase:** A MeOH:2.5 mM pH 5 ammonium phosphate 1:50; B MeCN:2.5 mM pH 2.5 ammonium phosphate 31:69; C MeCN:water 50:50
**Flow rate:** A 1.2; B 1; C 1.5
**Injection volume:** 50
**Detector:** UV 254

## CHROMATOGRAM
**Retention time:** 19.7
**Limit of detection:** 100 ng/mL

## KEY WORDS
plasma; dog; beagle; column-switching

## REFERENCE
Matsuura, A.; Nagayama, T.; Kitagawa, T. Automated high-performance liquid chromatographic method for determination of furosemide in dog plasma. *J.Chromatogr.*, **1993**, *617*, 339–343

## SAMPLE
**Matrix:** blood
**Sample preparation:** 25 µL Plasma + 100 µL 10 µg/mL naproxen in MeCN, vortex 30 s, centrifuge at 11000-12300 g for 7 min. Remove supernatant and evaporate it under air at 55°. Dissolve residue in 50 µL mobile phase and inject a 20 µL aliquot.

## HPLC VARIABLES
**Column:** 300 × 3.9 10 µm µBondapak C18
**Mobile phase:** MeCN:80 mM pH 2.0 orthophosphoric acid 46:54
**Flow rate:** 1.1
**Injection volume:** 20
**Detector:** F ex 270 em 410

## CHROMATOGRAM
**Retention time:** 6.5
**Internal standard:** naproxen (11.5)
**Limit of detection:** 10 ng/mL
**Limit of quantitation:** 20 ng/mL

## OTHER SUBSTANCES
**Simultaneous:** degradation products
**Noninterfering:** amikacin, amoxicillin, dexamethasone, gentamicin, indomethacin, morphine, phenobarbital, theophylline, vitamins

## KEY WORDS
plasma; microscale; neonatal

## REFERENCE
Sidhu, J.S.; Charles, B.G. Simple microscale high-performance liquid chromatographic method for determination of furosemide in neonatal plasma. *J.Chromatogr.*, **1993**, *612*, 161–165

## SAMPLE
**Matrix:** blood
**Sample preparation:** Condition an Analytichem C2 ethyl sorbent SPE cartridge with 1 mL MeCN and 1 mL buffer. 25 µL Plasma + 1 mL buffer, add to the SPE cartridge, wash with 1 mL buffer, blow dry with nitrogen for 1 min, elute cartridge directly onto column (Varian AASP system). (Buffer was 10 mM $KH_2PO_4$ adjusted to pH 3.0 with concentrated phosphoric acid.)

## HPLC VARIABLES
**Guard column:** 3 × 4.6 30 µm C18 Alltech pellicular packing
**Column:** 150 × 4.6 5 µm Nucleosil C18
**Mobile phase:** MeCN:10 mM $KH_2PO_4$ adjusted to pH 3.0 with concentrated phosphoric acid 30:70
**Flow rate:** 1.5
**Detector:** F ex 272 em 410

## CHROMATOGRAM
**Retention time:** 9.1

**Internal standard:** metolazone (8.3)
**Limit of detection:** 1.8 ng/mL

## OTHER SUBSTANCES
**Noninterfering:** acetaminophen, bumetanide, chlorothiazide, chlorthalidone, hydrochlorothiazide, ibuprofen, salicylic acid

## KEY WORDS
plasma; SPE; better results with external standard

## REFERENCE
Farthing, D.; Karnes, T.; Gehr, T.W.; March, C.; Fakhry, I.; Sica, D.A. External-standard high-performance liquid chromatographic method for quantitative determination of furosemide in plasma by
using solid-phase extraction and on-line elution. *J.Pharm.Sci.*, **1992**, *81*, 569–571

## SAMPLE
**Matrix:** blood
**Sample preparation:** 1 mL Plasma + 250 μL 8.5 M acetic acid, mix, add 250 μL 125 mM
sodium dodecylsulfate, mix 5 s, add 100 μL 40 μg/mL naproxen in MeOH, add 7 mL ethyl
acetate saturated with water, mix by rotation at 60 rpm for 30 min, centrifuge at 5200 g
for 10 min. Remove organic phase and evaporate on a vortex evaporator at 35°. Dissolve
in 250 μL mobile phase, inject 100 μL aliquot.

## HPLC VARIABLES
**Guard column:** 10 × 2 5 μm Nucleosil 100 C18
**Column:** 100 × 3 5 μm Nucleosil 100 C18
**Mobile phase:** MeCN:125 mM sodium dodecylsulfate:10 mM pH 2.0 perchloric acid
234.6:35:665
**Flow rate:** 0.6
**Injection volume:** 100
**Detector:** F ex 360 em 413

## CHROMATOGRAM
**Retention time:** 4.5
**Internal standard:** naproxen (12)
**Limit of detection:** 0.3 ng/mL

## OTHER SUBSTANCES
**Simultaneous:** amiloride

## KEY WORDS
plasma

## REFERENCE
Reeuwijk, H.J.; Tjaden, U.R.; van der Greef, J. Simultaneous determination of furosemide and amiloride
in plasma using high-performance liquid chromatography with fluorescence detection. *J.Chromatogr.*,
**1992**, *575*, 269–274

## SAMPLE
**Matrix:** blood
**Sample preparation:** 1 mL Plasma + 500 ng naproxen + 1 mL 100 mM HCl + 10 mL
dichloromethane, extract. Dry organic layer at 50° under nitrogen, dissolve in 1 mL mobile
phase, inject a 20 μL aliquot.

## HPLC VARIABLES
**Column:** 150 × 4.6 3 μm Alltech C8

**Mobile phase:** MeCN:80 mM phosphoric acid 35:65
**Flow rate:** 1
**Injection volume:** 20
**Detector:** F ex 235 em 405

## CHROMATOGRAM
**Retention time:** 5.05
**Internal standard:** naproxen (3.5)
**Limit of detection:** 10 ng/mL

## OTHER SUBSTANCES
**Extracted:** bumetanide

## KEY WORDS
plasma; horse; pharmacokinetics

## REFERENCE
Singh, A.K.; McArdle, C.; Gordon, B.; Ashraf, M.; Granley, K. Simultaneous analysis of furosemide and bumetanide in horse plasma using high performance liquid chromatography. *Biomed.Chromatogr.*, **1989**, *3*, 262–265

## SAMPLE
**Matrix:** blood, CSF
**Sample preparation:** 200 µL Serum, plasma, or CSF + 300 µL reagent. Flush column A to waste with 500 µL 500 mM ammonium sulfate, inject sample onto column A, flush column A to waste with 500 µL 500 mM ammonium sulfate, elute the contents of column A onto column B with mobile phase, monitor the effluent from column B. (Reagent was 8.05 M guanidine hydrochloride and 1.02 M ammonium sulfate in water.)

## HPLC VARIABLES
**Column:** A 30 × 2.1 40 µm preparative grade C18 (Analytichem); B 250 × 4.6 10 µm Partisil C8
**Mobile phase:** Gradient. A was 50 mM pH 4.5 $KH_2PO_4$. B was MeCN:isopropanol 80:20. A:B 90:10 for 1 min, to 30:70 over 15 min, maintain at 30:70 for 4 min.
**Column temperature:** 50
**Flow rate:** 1.5
**Detector:** UV 280 for 5 min then UV 254

## CHROMATOGRAM
**Retention time:** 9.12
**Internal standard:** heptanophenone (19.2)

## OTHER SUBSTANCES
**Extracted:** acetazolamide, ampicillin, bromazepam, caffeine, carbamazepine, chloramphenicol, chlorothiazide, diazepam, droperidol, ethionamide, isoniazid, methadone, penicillin G, phenobarbital, phenytoin, prazepam, propoxyphene, pyrazinamide, rifampin, trimeprazine, trimethoprim

## KEY WORDS
serum; plasma; column-switching

## REFERENCE
Seifart, H.I.; Kruger, P.B.; Parkin, D.P.; van Jaarsveld, P.P.; Donald, P.R. Therapeutic monitoring of antituberculosis drugs by direct in-line extraction on a high-performance liquid chromatography system. *J.Chromatogr.*, **1993**, *619*, 285–290

## SAMPLE
**Matrix:** blood, urine
**Sample preparation:** Filter (0.45 μm) urine or plasma. Mix plasma filtrate with an equal volume of 50 mM pH 8.0 Tris-sulfuric acid buffer containing 0.1 mM zinc acetate and 40 mM sodium dodecyl sulfate. Inject a 50 μL aliquot of the urine filtrate or the diluted plasma onto column A with mobile phase A, elute to waste with mobile phase A, after 5 min backflush the contents of column A onto column B with mobile phase B, after 3 min remove column A from the circuit, elute column B with mobile phase B, monitor the effluent from column B. Re-equilibrate column A with mobile phase A for 5 min before next injection.

## HPLC VARIABLES
**Column:** A 10 × 4.6 carbonic anhydrase (Prepare by adding 3 g aminopropyl silica from a Sep-Pak NH2 SPE cartridge to 30 mL 100 mg/mL N, N′-disuccinimidyl carbonate in MeCN in portions over 30 min with gentle mixing, mix for 3 h, filter (G-5 glass), wash the solid 5 times with 50 mL portions of MeCN. Add 100 mg activated gel to 2 mL 200 mM pH 8.0 phosphate buffer containing 1 M NaCl, degas by sonicating under aspirator vacuum, add 2 mL 2.5 mg/mL carbonic anhydrase in water, shake at room temperature for 4 h, centrifuge, discard the supernatant, suspend the gel in 50 mM pH 8.0 Tris-sulfuric acid buffer, slurry pack into column. Store in 50 mM pH 8.0 Tris-sulfuric acid buffer containing 0.1 mM zinc acetate when not in use.); B 150 × 4.6 Cosmosil 5C18-AR (Nakarai Tesque)
**Mobile phase:** A 50 mM pH 8.0 Tris-sulfuric acid buffer containing 0.1 mM zinc acetate; B Gradient. MeCN:100 mM pH 5.2 acetate buffer 10:90 containing 500 mM NaCl for 3 min then MeCN:100 mM pH 5.2 acetate buffer 23:77 containing 500 mM NaCl (step gradient). (Increase in gradient occurs at the same time as column A is removed from the circuit.)
**Flow rate:** 1
**Injection volume:** 50
**Detector:** UV 270

## CHROMATOGRAM
**Retention time:** 13

## OTHER SUBSTANCES
**Also analyzed:** acetazolamide, chlorothiazide, chlorthalidone, hydrochlorothiazide
**Noninterfering:** acetaminophen, bumetanide, caffeine, phenylbutazone, salicylic acid, sulfamerazine, sulfamethiazole, sulfamethoxazole, sulfamonomethoxine, sulfisoxazole, sulfisomidine, theophylline, tolbutamide, warfarin

## KEY WORDS
plasma; column-switching

## REFERENCE
Ohta, T.; Takamiya, I.; Takitani, S. Carbonic anhydrase-immobilized precolumn for selective on-line sample pretreatment in high-performance liquid chromatographic determination of certain sulphonamide drugs. *Biomed.Chromatogr.*, **1994**, *8*, 184−188

## SAMPLE
**Matrix:** blood, urine
**Sample preparation:** Plasma. 100 μL Plasma + 100 μL MeCN, centrifuge at 3000 g for 5 min, inject 20 μL aliquot of supernatant. Urine. Dilute urine 1:1 with water and inject 20 μL.

## HPLC VARIABLES
**Guard column:** 75 × 2.1 pellicular reversed phase (Chrompack cat. no. 28653)
**Column:** 250 × 4.6 5 μm Cp Spherisorb ODS

**Mobile phase:** Gradient. MeCN:0.5% pH 2.1 orthophosphoric acid (98%), from 5:95 to 41:59 over 30 min, stay at 41:59 for 5 min, return to 5:95 over 5 min, equilibrate for 2 min before next injection.
**Flow rate:** 1.2
**Injection volume:** 20
**Detector:** F ex 345 em 405

## CHROMATOGRAM
**Retention time:** 28.77
**Limit of detection:** 5 ng/mL
**Limit of quantitation:** 7 ng/mL (plasma); 100 ng/mL (urine)

## OTHER SUBSTANCES
**Simultaneous:** metabolites, glucuronides

## KEY WORDS
plasma

## REFERENCE
Vree, T.B.; Van den Biggelaar-Martea, M.; Verwey-van Wissen, C.P.W.G.M. Determination of furosemide with its acyl glucoronide in human plasma and urine by means of direct gradient high-performance liquid chromatographic analysis with fluorescence detection. Preliminary pharmacokinetics and effect of probenecid. *J.Chromatogr.B*, **1994**, *655*, 53–62

## SAMPLE
**Matrix:** formulations
**Sample preparation:** Dilute with mobile phase, inject an aliquot.

## HPLC VARIABLES
**Column:** 300 × 4.6 5 μm C18
**Mobile phase:** MeCN:100 mM $NaH_2PO_4$ 20:80 adjusted to pH 4.2 with phosphoric acid
**Flow rate:** 1.2
**Injection volume:** 20
**Detector:** UV 228

## CHROMATOGRAM
**Retention time:** 4.52

## OTHER SUBSTANCES
**Simultaneous:** granisetron (UV 300)

## KEY WORDS
stability-indicating; injections; saline

## REFERENCE
Mayron, D.; Gennaro, A.R. Stability and compatibility of granisetron hydrochloride in i.v. solutions and oral liquids and during simulated Y-site injection with selected drugs. *Am.J.Health-Syst.Pharm.*, **1996**, *53*, 294–304

## SAMPLE
**Matrix:** formulations, urine
**Sample preparation:** Tablets. Pulverize tablets, add MeOH, shake for 30 min, sonicate for 5 min, filter (Albet 242 paper), wash solid with MeOH, make up filtrate to 50 mL with MeOH, inject a 20 μL aliquot. Urine. Adjust pH of 2 mL urine to 10.0 with 2 M KOH, add 1.5 mg NaCl, add 4 mL ethyl acetate, shake for 10 min, centrifuge at 2500 rpm for

5 min. Remove the organic layer and evaporate it to dryness under a stream of nitrogen at 40°, reconstitute the residue in 2 mL mobile phase, sonicate, inject a 20 μL aliquot.

## HPLC VARIABLES
**Guard column:** μBondapak C18
**Column:** 300 × 3.9 10 μm μBondapak C18
**Mobile phase:** MeCN:water 30:70 containing 5 mM $KH_2PO_4/K_2HPO_4$, pH adjusted to 5.5
**Flow rate:** 1
**Injection volume:** 20
**Detector:** E, EG&G Princeton Applied Research PAR Model 400, glassy carbon working electrode +1300 mV, Ag/AgCl reference electrode (At the end of each day clean electrode with mobile phase of MeOH at 1.5 mL/min, -800 mV for 2 min then +1600 mV for 5 min.)

## CHROMATOGRAM
**Retention time:** 6.70
**Limit of detection:** 15 ng/mL

## OTHER SUBSTANCES
**Extracted:** triamterene

## KEY WORDS
tablets; pharmacokinetics

## REFERENCE
Barroso, M.B.; Alonso, R.M.; Jiménez, R.M. Simultaneous determination of the diuretics triamterene and furosemide in pharmaceutical formulations and urine by HPLC-EC. *J.Liq.Chrom.Rel.Technol.*, **1996**, *19*, 231–246

## SAMPLE
**Matrix:** formulations, urine
**Sample preparation:** Tablets. Pulverize tablets, add MeOH, shake for 20 min, filter, wash solid with MeOH, dilute filtrate with mobile phase, inject a 20 μL aliquot. Urine. 2 mL Urine + 2 mL 1 M pH 3.25 $KH_2PO_4$ + 4 mL ethyl acetate, vortex for 20 min, centrifuge at 734 g for 5 min. Remove the organic layer and evaporate it to dryness under a stream of nitrogen at 40°, reconstitute the residue in 2 mL mobile phase, inject a 20 μL aliquot.

## HPLC VARIABLES
**Guard column:** μBondapak C18
**Column:** 300 × 3.9 10 μm μBondapak C18
**Mobile phase:** MeCN:water 40:60 containing 5 mM $KH_2PO_4/K_2HPO_4$, pH adjusted to 4.25
**Column temperature:** 30
**Flow rate:** 1
**Injection volume:** 20
**Detector:** E, EG&G Princeton Applied Research PAR Model 400, glassy carbon working electrode +1200 mV, Ag/AgCl reference electrode (At the end of each day clean electrode with mobile phase of MeOH at 1.5 mL/min, -800 mV for 2 min then +1600 mV for 15 min.)

## CHROMATOGRAM
**Retention time:** 7.7
**Limit of quantitation:** 15 ng/mL

## OTHER SUBSTANCES
**Extracted:** piretanide

## KEY WORDS
tablets; pharmacokinetics

## REFERENCE

Barroso, M.B.; Jiménez, R.M.; Alonso, R.M.; Ortiz, E. Determination of piretanide and furosemide in pharmaceuticals and human urine by high-performance liquid chromatography. *J.Chromatogr.B*, **1996**, *675*, 303–312

## SAMPLE

**Matrix:** perfusate
**Sample preparation:** Dilute perfusate with an equal volume of 15 mM pH 8 HEPES buffer, centrifuge at 2000 g for 2 min, inject an aliquot of the supernatant.

## HPLC VARIABLES

**Guard column:** 20 mm long Supelguard LC-18S (Supelco)
**Column:** 250 × 4.6 Supelcosil LC-18S
**Mobile phase:** MeOH:water 40:60 containing 10 mM $KH_2PO_4$
**Flow rate:** 1
**Detector:** UV 264

## KEY WORDS

rat; rabbit; pharmacokinetics

## REFERENCE

Sinko, P.J.; Hu, P.; Waclawski, A.P.; Patel, N.R. Oral absorption of anti-AIDS nucleoside analogues. 1. Intestinal transport of didanosine in rat and rabbit preparations. *J.Pharm.Sci.*, **1995**, *84*, 959–965

## SAMPLE

**Matrix:** solutions

## HPLC VARIABLES

**Column:** 250 × 4.6 5 μm Supelcosil LC-DP (A) or 250 × 4 5 μm LiChrospher 100 RP-8 (B)
**Mobile phase:** MeCN:0.025% phosphoric acid:buffer 25:10:5 (A) or 60:25:15 (B) (Buffer was 9 mL concentrated phosphoric acid and 10 mL triethylamine in 900 mL water, adjust pH to 3.4 with dilute phosphoric acid, make up to 1 L.)
**Flow rate:** 0.6
**Injection volume:** 25
**Detector:** UV 229

## CHROMATOGRAM

**Retention time:** 5.78 (A), 5.03 (B)

## OTHER SUBSTANCES

**Also analyzed:** acebutolol, acepromazine, acetaminophen, acetazolamide, acetophenazine, albuterol, alprazolam, amitriptyline, amobarbital, amoxapine, antipyrine, atenolol, atropine, azatadine, baclofen, benzocaine, bromocriptine, brompheniramine, brotizolam, bupivacaine, buspirone, butabarbital, butalbital, caffeine, carbamazepine, cetirizine, chlorcyclizine, chlordiazepoxide, chlormezanone, chloroquine, chlorpheniramine, chlorpromazine, chlorpropamide, chlorprothixene, chlorthalidone, chlorzoxazone, cimetidine, cisapride, clomipramine, clonazepam, clonidine, clozapine, cocaine, codeine, colchicine, cyclizine, cyclobenzaprine, dantrolene, desipramine, diazepam, diclofenac, diflunisal, diltiazem, diphenhydramine, diphenidol, diphenoxylate, dipyridamole, disopyramide, dobutamine, doxapram, doxepin, droperidol, encainide, ethidium bromide, ethopropazine, fenoprofen, fentanyl, flavoxate, fluoxetine, fluphenazine, flurazepam, flurbiprofen, fluvoxamine, glutethimide, glyburide, guaifenesin, haloperidol, homatropine, hydralazine, hydrochlorothiazide, hydrocodone, hydromorphone, hydroxychloroquine, hydroxyzine, ibuprofen, imipramine, indomethacin, ketoconazole, ketoprofen, ketorolac, labetalol, levorphanol, lidocaine, loratadine, lorazepam, lovastatin, loxapine, mazindol, mefenamic acid, meperidine, mephenytoin, mepivacaine, mesoridazine, metaproterenol, methadone, methdilazine, methocarbamol, methotrexate, methotrimeprazine, methoxamine, methyldopa, methylphenidate, metoclopramide, metolazone, metoprolol, metronidazole, midazolam, moclobemide, morphine, nadolol, nalbuphine, naloxone, naphazoline, naproxen, nifedipine, nizatidine, norepinephrine, nortriptyline, oxazepam, oxycodone, oxymetazoline, paroxetine, pemoline, pentazocine, pentobarbital, pentoxifylline, perphenazine, pheniramine, phenobarbital, phenol, phenolphthalein, phentolamine, phenylbutazone, phenyltoloxamine, phenytoin, pimozide, pindolol, piroxicam, pramoxine, prazepam, prazosin, probenecid, procainamide, procaine, prochlorperazine, procyclidine, promazine, promethazine, propafenone, propantheline, propiomazine, propofol, propranolol, protriptyline, quazepam, quinidine, quinine, racemethorphan, ranitidine, remoxipride, risperidone, salicylic acid, scopolamine, secobarbital, sertraline, sotalol, spironolactone, sulfinpyrazone, sulindac, temazepam, terbutaline, terfenadine, tetracaine, theophylline, thiethylperazine, thiopental, thioridazine, thiothixene, timolol, tocainide, tolbutamide, tolmetin, trazodone, triamterene, triazolam, trifluoperazine, triflupromazine, trimeprazine, trimethoprim, trimipramine, verapamil, warfarin, xylometazoline, yohimbine, zopiclone

## KEY WORDS
some details of plasma extraction

## REFERENCE
Koves, E.M. Use of high-performance liquid chromatography-diode array detection in forensic toxicology. *J.Chromatogr.A*, **1995**, *692*, 103–119

## SAMPLE
**Matrix:** solutions
**Sample preparation:** Prepare a solution in MeOH:water 80:20, inject a 6 μL aliquot.

## HPLC VARIABLES
**Guard column:** 5 × 4 10 μm LiChrosorb RP-8
**Column:** 100 × 4.6 5 μm Spheri RP-18 (Brownlee)
**Mobile phase:** MeOH:water 80:20 containing 2 g/L lithium perchlorate
**Flow rate:** 0.5
**Injection volume:** 6
**Detector:** E, ESA Model 5100A Coulochem, model 5020 guard cell +950 mV, Model 5010 analytical cell + 400 mV, palladium reference electrode, following post-column photolysis. The effluent from the column flowed through a 20 m × 0.3 mm coil of PTFE tubing irradiated at 254 nm with a Sylvania GTE 8 W low-pressure lamp to the detector.

## OTHER SUBSTANCES
**Also analyzed:** bendroflumethiazide, butizide, chlorthalidone, ethacrynic acid, hydrochlorothiazide

## KEY WORDS
post-column reaction

## REFERENCE
Macher, M.; Wintersteiger, R. Improved electrochemical detection of diuretics in high-performance liquid chromatographic analysis by postcolumn on-line photolysis. *J.Chromatogr.A*, **1995**, *709*, 257–264

## SAMPLE
**Matrix:** solutions
**Sample preparation:** Prepare a 1-10 μg/mL solution in water, inject an aliquot.

## HPLC VARIABLES
**Column:** 250 × 4.6 5 μm Hypersil SCX/C18
**Mobile phase:** MeCN:25 mM pH 3 $Na_2HPO_4$ 50:50

**Injection volume:** 20
**Detector:** UV 254

## CHROMATOGRAM
**Retention time:** k' 0.42

## OTHER SUBSTANCES
**Also analyzed:** amitriptyline, barbital, benzoic acid, butabarbital, clomipramine, clonazepam, desipramine, diazepam, flurazepam, imipramine, nitrazepam, phenobarbital, phenol, phenolphthalein, pindolol, propranolol, resorcinol, salicylic acid, secobarbital, terbutaline, xylazine

## KEY WORDS
effect of mobile phase pH on capacity factor is discussed

## REFERENCE
Walshe, M.; Kelly, M.T.; Smyth, M.R.; Ritchie, H. Retention studies on mixed-mode columns in high-performance liquid chromatography. *J.Chromatogr.A*, **1995**, *708*, 31–40

## SAMPLE
**Matrix:** solutions

## HPLC VARIABLES
**Guard column:** 30 × 3.2 7 μm SI 100 ODS (not commercially available)
**Column:** 150 × 3.2 7 μm SI 100 ODS (not commercially available)
**Mobile phase:** MeCN:buffer 31.2:68.8 (Buffer was 6.66 g $KH_2PO_4$ and 4.8 g 85% phosphoric acid in 1 L water, pH 2.3.)
**Flow rate:** 0.5-1
**Detector:** UV 227, 266

## CHROMATOGRAM
**Retention time:** 4.7
**Internal standard:** 5-(4-methylphenyl)-5-phenylhydantoin (7.3)

## OTHER SUBSTANCES
**Also analyzed:** aspirin, caffeine, carbamazepine, chlordiazepoxide, chlorprothixene, clonazepam, diazepam, doxylamine, ethosuximide, haloperidol, hydrochlorothiazide, methocarbamol, methotrimeprazine, nicotine, oxazepam, procaine, promazine, propafenone, propranolol, salicylamide, temazepam, tetracaine, thiopental, triamterene, verapamil, zolpidem, zopiclone

## REFERENCE
Below, E.; Burrmann, M. Application of HPLC equipment with rapid scan detection to the identification of drugs in toxicological analysis. *J.Liq.Chromatogr.*, **1994**, *17*, 4131–4144

## SAMPLE
**Matrix:** solutions

## HPLC VARIABLES
**Column:** 250 × 4.6 Zorbax RX
**Mobile phase:** Gradient. A was 10 mL concentrated orthophosphoric acid and 7 mL triethylamine in 1 L water. B was 10 mL concentrated orthophosphoric acid and 7 mL triethylamine in 200 mL water, make up to 1 L with MeCN. A:B from 100:0 to 0:100 over 30 min, maintain at 0:100 for 5 min.
**Column temperature:** 30

**Flow rate:** 2
**Detector:** UV 210

## OTHER SUBSTANCES
**Also analyzed:** acepromazine, acetaminophen, acetophenazine, albuterol, aminophylline, amitriptyline, amobarbital, amoxapine, amphetamine, amylocaine, antipyrine, aprobarbital, aspirin, atenolol, atropine, avermectin, barbital, benzocaine, benzoic acid, benzotropine, benzphetamine, berberine, bibucaine, bromazepan, brompheniramine, buprenorphine, buspirone, butabarbital, butacaine, butethal, caffeine, carbamazepine, carbromal, chloramphenicol, chlordiazepoxide, chloroquine, chlorothiazide, chloroxylenol, chlorphenesin, chlorpheniramine, chlorpromazine, chlorpropamide, chlortetracycline, cimetidine, cinchonidine, cinchonine, clenbuterol, clonazepam, clonixin, clorazepate, cocaine, codeine, colchicine, cortisone, coumarin, cyclazocine, cyclobenzaprine, cyclothiazide, cyheptamide, cymarin, danazol, danthron, dapsone, debrisoquine, desipramine, dexamethasone, dextromethorphan, dextropropoxyphene, diamorphine, diazepam, diclofenac, diethylpropion, diethylstilbestrol, diflunisal, digitoxin, digoxin, diltiazem, diphenhydramine, diphenoxylate, diprenorphine, dipyrone, disulfiram, dopamine, doxapram, doxepin, dronabinol, ephedrine, epinephrine, epinine, estradiol, estriol, estrone, ethacrynic acid, ethosuximide, etonitazene, etorphine, eugenol, famotidine, fenbendazole, fencamfamine, fenoprofen, fenproporex, fentanyl, flubendazole, flufenamic acid, flunitrazepam, 5-fluorouracil, fluoxymesterone, fluphenazine, gentisic acid, gitoxigenin, glipizide, glunixin, glutethimide, glybenclamide, guaiacol, halazepam, haloperidol, hydrochlorothiazide, hydrocodone, hydrocortisone, hydromorphone, hydroxyquinoline, ibogaine, ibuprofen, iminostilbene, imipramine, indomethacin, isocarbostyril, isocarboxazid, isoniazid, isoproterenol, isoxsuprine, ivermectin, ketamine, ketoprofen, kynurenic acid, levorphanol, lidocaine, lorazepam, lormetazepam, loxapine, mazindol, mebendazole, meclizine, meclofenamic acid, medazepam, mefenamic acid, megestrol, mepacrine, meperidine, mephentermine, mephenytoin, mephesin, mephobarbital, mepivacaine, mescaline, mesoridazine, methadone, methamphetamine, methapyrilene, methaqualone, methazolamide, methocarbamol, methoxamine, methsuximide, methyl salicylate, methyldopa, methyldopamine, methylphenidate, methylprednisolone, methyltestosterone, methyprylon, metoprolol, mibolerone, morphine, nadolol, nalorphine, naloxone, naltrexone, naphazoline, naproxen, nefopam, niacinamide, nicotine, nicotinic acid, nifedipine, niflumic acid, nitrazepam, norepinephrine, nortriptyline, noscapine, nylidrin, oxazepam, oxycodone, oxymorphone, oxyphenbutazone, oxytetracycline, papaverine, pargyline, pemoline, pentazocine, pentobarbital, persantine, phenacetin, phenazocine, phenazopyridine, phencyclidine, phendimetrazine, phenelzine, pheniramine, phenobarbital, phenothiazine, phensuximide, phentermine, phenylbutazone, phenylephrine, phenylpropanolamine, piperocaine, prazepam, prednisolone, primidone, probenecid, progesterone, propiomazine, propranolol, propylparaben, pseudoephedrine, puromycin, pyrilamine, pyrithyldione, quazepam, quinaldic acid, quinidine, quinine, ranitidine, recinnamine, reserpine, resorcinol, saccharin, albuterol, salicylamide, salicylic acid, scopolamine, scopoletin, secobarbital, strychnine, sulfacetamide, sufadiazine, sulfadimethoxine, sulfaethidole, sulfamerazine, sulfamethazine, sulfamethoxizole, sulfanilamide, sulfapyridine, sulfasoxizole, sulindac, tamoxifen, temazepam, testosterone, tetracaine, tetracycline, tetramisole, thebaine, theobromine, theophylline, thiabendazole, thiamine, thiamylal, thiobarbituric acid, thioridazine, thiosalicylic acid, thiothixene, thymol, tolazamide, tolazoline, tobutamide, tolmetin, tranylcypromine, triamcinolone, tribenzylamine, trichloromethiazide, trifluoperazine, trihexyphenidyl, trimethoprim, tripelennamine, triprolidine, tropacocaine, tyramine, verapamil, vincamine, warfarin, yohimbine, zoxazolamine

## REFERENCE
Hill, D.W.; Kind, A.J. Reversed-phase solvent gradient HPLC retention indexes of drugs. *J.Anal.Toxicol.*, **1994**, *18*, 233–242

## SAMPLE
**Matrix:** urine

**Sample preparation:** Direct injection into column A with mobile phase A for 1 min then back flush onto column B with mobile phase B.

## HPLC VARIABLES
**Column:** A 20 × 2.1 30 μm Hypersil ODS-C18; B 250 × 4 5 μm Hypersil ODS-C18
**Mobile phase:** A Water; B Gradient. MeCN:buffer 15:85 for 1.5 min then to 80:20 over 8 min. Keep at 80:20 for 2.5 min then re-equilibrate with 15:85. (Buffer was 50 mM NaH$_2$PO$_4$ + 1.4 mL propylamine hydrochloride per liter adjusted to pH 3 with concentrated phosphoric acid.)
**Flow rate:** 1
**Injection volume:** 50
**Detector:** UV 230

## CHROMATOGRAM
**Retention time:** 8.7
**Limit of detection:** 2 ng/mL

## OTHER SUBSTANCES
**Simultaneous:** acetazolamide, amiloride, bendroflumethiazide, bumetanide, chlorthalidone, cyclothiazide, ethacrynic acid, hydrochlorothiazide, probenecid, spironolactone, triamterene

## REFERENCE
Campíns-Falco, P.; Herráez-Hernández, R.; Sevillano-Cabeza, A. Column-switching techniques for screening of diuretics and probenecid in urine samples. *Anal.Chem.*, **1994**, *66*, 244–248

## SAMPLE
**Matrix:** urine
**Sample preparation:** Buffer urine to 4.9 by mixing with an equal volume of pH 4.9 200 mM sodium phosphate buffer. Inject a 40 μL aliquot onto column A with mobile phase A, after 3 min backflush the contents of column A onto column B with mobile phase B and start the gradient. At the end of the run re-equilibrate for 10 min.

## HPLC VARIABLES
**Column:** A 20 × 4 5 μm Hypersil octadecylsilica ODS; B 200 × 4.6 5 μm Shiseido SG-120 polymer-based C18
**Mobile phase:** A water; B Gradient. MeCN:buffer from 7:93 to 15:85 over 3.5 min, to 50:50 over 8.5 min, maintain at 50:50 for 11 min (Buffer was 6.9 g NaH$_2$PO$_4$.H$_2$O in 1 L water, pH adjusted to 3.1 with phosphoric acid.)
**Flow rate:** 1
**Injection volume:** 40
**Detector:** UV 230

## CHROMATOGRAM
**Retention time:** 16.4
**Limit of detection:** 1 μg/mL

## OTHER SUBSTANCES
**Extracted:** acetazolamide, amiloride, bendroflumethiazide, benzthiazide, bumetanide, caffeine, carbamazepine, chlorothiazide, chlorthalidone, clopamide, dichlorfenamide, ethacrynic acid, hydrochlorothiazide, metyrapone, probenecid, spironolactone, triamterene, trichlormethiazide

## KEY WORDS
column-switching; optimum detection wavelengths vary for each drug

## REFERENCE

Saarinen, M.; Sirén, H.; Riekkola, M.-L. A column switching technique for the screening of diuretics in urine by high performance liquid chromatography. *J.Liq.Chromatogr.*, **1993**, *16*, 4063–4078

## SAMPLE

**Matrix:** urine
**Sample preparation:** 5 mL Urine + 50 μL 100 μg/mL 7-propyltheophylline in MeOH + 200 μL ammonium chloride buffer + 2 g NaCl, extract with 6 mL ethyl acetate by rocking at 40 movements/min for 20 min and centrifuging at 800 g for 5 min, repeat extraction, combine organic layers, evaporate to dryness at 40° under a stream of nitrogen. Reconstitute in 200 μL MeCN:water 15:85 and inject 20 μL aliquots. (Ammonium chloride buffer was 28 g ammonium chloride in 100 mL water with the pH adjusted to 9.5 with concentrated ammonia solution.)

## HPLC VARIABLES

**Column:** 75 × 4.6 3 μm Ultrasphere ODS
**Mobile phase:** Gradient. MeCN:100 mM ammonium acetate adjusted to pH 3 with concentrated phosphoric acid. From 10:90 to 15:85 over 2 min, to 55:45 over 3 min, to 60:40 over 3 min. Kept at 60:40 for 1 min, decreased to 10:90 over 1 min and equilibrated at 10:90 for 2 min.
**Flow rate:** 1
**Injection volume:** 20
**Detector:** UV 270

## CHROMATOGRAM

**Retention time:** 5.9
**Internal standard:** 7-propyltheophylline (4.5)
**Limit of detection:** 50 ng/mL

## OTHER SUBSTANCES

**Simultaneous:** acetazolamide, amiloride, bendroflumethiazide, benzthiazide, bumetanide, buthiazide, caffeine, canrenone, chlorthalidone, clopamide, cyclothiazide, diclofenamide, ethacrynic acid, hydrochlorothiazide, mesocarb, morazone, piretanide, polythiazide, probenecid, spironolactone, torsemide, triamterene, xipamide

## REFERENCE

Ventura, R.; Nadal, T.; Alcalde, P.; Pascual, J.A.; Segura, J. Fast screening method for diuretics, probenecid and other compounds of doping interest. *J.Chromatogr.A*, **1993**, *655*, 233–242

## SAMPLE

**Matrix:** urine
**Sample preparation:** Make 5 mL urine alkaline (pH 9-10), add 2 g NaCl, extract twice with 6 mL ethyl acetate. Combine the organic layers and evaporate them to dryness under a stream of nitrogen, reconstitute the residue in 200 μL MeCN/water, inject a 10-20 μL aliquot.

## HPLC VARIABLES

**Column:** 100 × 4 5 μm SGE 100 GL-4 C18P (Scientific Glass Engineering)
**Mobile phase:** MeCN:MeOH:water:trifluoroacetic acid 4.5:10.5:85:0.5
**Flow rate:** 0.8 or 1
**Injection volume:** 10-20
**Detector:** MS, ZAB2-SEQ (VG), PSP source coupled to LC, source 250°, probe 240-260°, scan m/z 200-550; UV 270

## CHROMATOGRAM

**Retention time:** 6.3
**Limit of detection:** 500 ng (by MS)

## OTHER SUBSTANCES
**Extracted:** amiloride, bendroflumethiazide, benzthiazide, chlorthalidone, triamterene

## REFERENCE
Ventura, R.; Fraisse, D.; Becchi, M.; Paisse, O.; Segura, J. Approach to the analysis of diuretics and masking agents by high-performance liquid chromatography-mass spectrometry in doping control. *J.Chromatogr.*, **1991**, *562*, 723–736

## SAMPLE
**Matrix:** urine
**Sample preparation:** 2 mL Urine + 1 mL 10 mM HCl + 2000 ng bendroflumethiazide, extract with 5 mL ethyl acetate, centrifuge at 3000 rpm for 5 min. Remove the organic layer and dry it under a stream of nitrogen at 40°. Reconstitute with 100 μL MeOH, inject a 2 μL aliquot.

## HPLC VARIABLES
**Column:** 100 × 2.1 5 μm Hypersil ODS
**Mobile phase:** Gradient. MeOH: 50 mM ammonium acetate from 10:90 to 60:40 over 10 min, maintain at 60:40 for 10 min.
**Column temperature:** 40
**Flow rate:** 0.3
**Injection volume:** 2
**Detector:** UV 230

## CHROMATOGRAM
**Retention time:** 6.2
**Internal standard:** bendroflumethiazide (8.6)

## OTHER SUBSTANCES
**Extracted:** bumetanide, canrenone, cyclopenthiazide, etozolin, piretanide

## REFERENCE
Gradeen, C.Y.; Billay, D.M.; Chan, S.C. Analysis of bumetanide in human urine by high-performance liquid chromatography with fluorescence detection and gas chromatography/mass spectrometry. *J.Anal.Toxicol.*, **1990**, *14*, 123–126

## SAMPLE
**Matrix:** urine
**Sample preparation:** 2 mL Urine + 0.5 g solid buffer I (pH 5-5.5), vortex 15 s, add 4 mL ethyl acetate, agitate for 10 min, centrifuge at 600 g for 5 min. Remove organic layer and vortex it with 2 mL 5% aqueous lead acetate for 10 s, centrifuge at 600 g for 5 min, remove and keep organic phase. 2 mL Urine + 0.5 g solid buffer II (pH 9-9.5), vortex 15 s, add 4 mL ethyl acetate, agitate for 10 min, centrifuge at 600 g for 5 min. Remove organic layer and combine it with previous organic layer. Evaporate to dryness at 50°under a stream of nitrogen, reconstitute in 300 μL 50 μg/mL β-hydroxyethyltheophylline in MeOH, inject 5 μL aliquot. (Solid buffer I was $KH_2PO_4:Na_2HPO_4$ 99:1, solid buffer II was $NaHCO_3:K_2CO_3$ 3:2.)

## HPLC VARIABLES
**Column:** 250 × 4.6 5 μm HP Hypersil ODS (A) or HP LiChrosorb RP-18 (B)
**Mobile phase:** Gradient. MeCN:buffer from 15:85 at 2 min to 80:20 at 20 min (Buffer was 50 mM $NaH_2PO_4$ containing 16 mM propylamine hydrochloride, adjusted to pH 3 with concentrated phosphoric acid.)
**Flow rate:** 1
**Injection volume:** 5
**Detector:** UV 230;UV 275

## CHROMATOGRAM
**Retention time:** 12.16 (A); 12.9 (B)
**Internal standard:** β-hydroxyethyltheophylline (3.7 (A), 4.4 (B))
**Limit of detection:** 500 ng/mL

## OTHER SUBSTANCES
**Extracted:** acetazolamide, amiloride, bendroflumethiazide, benzthiazide, bumetanide, canrenone, chlorothiazide, chlorthalidone, cyclothiazide, dichlorphenamide, ethacrynic acid, flumethiazide, hydrochlorothiazide, hydroflumethiazide, methyclothiazide, polythiazide, probenecid, quinethazone, spironolactone, triamterene, trichloromethiazide
**Noninterfering:** acetaminophen, aspirin, caffeine, diflunisal, fenoprofen, ibuprofen, indomethacin, methocarbamol, naproxen, phenylbutazone, sulindac, tetracycline, theobromine, theophylline, tolmetin, trimethoprim, verapamil
**Interfering:** metolazone

## REFERENCE
Cooper, S.F.; Massé, R.; Dugal, R. Comprehensive screening procedure for diuretics in urine by high-performance liquid chromatography. *J.Chromatogr.*, **1989**, *489*, 65–88

## SAMPLE
**Matrix:** urine
**Sample preparation:** 2 mL Urine + 2 mL 1 M pH 4.1 $NaH_2PO_4$ + 4 mL ethyl acetate, vortex for 2 min, centrifuge at 1500 g for 5 min. Remove the organic phase and add it to 5 mL 100 mM pH 7.5 $Na_2HPO_4$, vortex for 2 min, centrifuge at 1500 g for 5 min. Remove the organic layer and evaporate it to dryness under a stream of nitrogen at 60°, reconstitute the residue in 100 μL MeCN:10 mM pH 3.0 phosphate buffer, inject a 5 μL aliquot.

## HPLC VARIABLES
**Column:** 125 × 4 5 μm LiChrosorb RP-18
**Mobile phase:** Gradient. MeCN:10 mM pH 3.0 phosphate buffer 10:90 for 1.5 min then to 35:65 over 2 min
**Column temperature:** 50
**Flow rate:** 1.5
**Injection volume:** 5
**Detector:** UV 271

## CHROMATOGRAM
**Retention time:** 5.0
**Limit of quantitation:** 1 μg/mL

## OTHER SUBSTANCES
**Extracted:** bendroflumethiazide, bumetanide, chlorothiazide, chlorthalidone, clopamide, cyclopenthiazide, hydrochlorothiazide, mefruside, methyclothiazide, metolazone, quinethazone
**Simultaneous:** clorexolone, ethacrynic acid, indapamide
**Noninterfering:** albuterol, allopurinol, alprenolol, aspirin, atenolol, captopril, carbimazole, clonidine, coloxyl, danthron, diazepam, digoxin, doxepin, glibenclamide, hydralazine, indomethacin, labetalol, metformin, methyldopa, metoprolol, mianserin, minoxidil, nifedipine, nitrazepam, oxazepam, oxprenolol, pindolol, prazosin, propranolol, senokot, theophylline, trifluoperazine

## REFERENCE
Fullinfaw, R.O.; Bury, R.W.; Moulds, R.F.W. Liquid chromatographic screening of diuretics in urine. *J.Chromatogr.*, **1987**, *415*, 347–356

◆━━━━━━━━━━━━━━━━━━━━━━━━━◆━━━━━━━━━━━━━━━━━━━━━━━━━◆

## ANNOTATED BIBLIOGRAPHY

Carretero, I.; Vadillo, J.M.; Laserna, J.J. Determination of antipyrine metabolites in human plasma by solid-phase extraction and micellar liquid chromatography. *Analyst*, **1995**, *120*, 1729–1732 [plasma; SPE; furosemide is IS]

Vree, T.B.; Van den Biggelaar-Martea, M.; Verwey-van Wissen, C.P. Determination of furosemide with its acyl glucuronide in human plasma and urine by means of direct gradient high-performance liquid chromatographic analysis with fluorescence detection. Preliminary pharmacokinetics and effect of probenecid. *J.Chromatogr.B*, **1994**, *655*, 53–62

Herráez-Hernández, R.; Campíns-Falcó, P.; Sevillano-Cabeza, A. Improved screening procedure for diuretics. *J.Liq.Chromatogr.*, **1992**, *15*, 2205–2224 [LOD 10-1000 ng/mL; gradient; urine; hydroxy-methyltheophylline (IS); extracted acetazolamide, amiloride, bendroflumethiazide, bumetanide, chlorthalidone, cyclothiazide, ethacrynic acid, hydrochlorothiazide, probenecid, spironolactone, triamterene]

Campíns-Falcó, P.; Herráez-Hernández, R.; Sevillano-Cabeza, A. Solid-phase extraction techniques for assay of diuretics in human urine samples. *J.Liq.Chromatogr.*, **1991**, *14*, 3575–3590 [urine; SPE; hydroxyethyltheophylline (IS); extracted acetazolamide, amiloride, bendroflumethiazide, bumetanide, chlorthalidone, ciclothiazide, ethacrynic acid, hydrochlorothiazide, probenecid, spironolactone, triamterene]

Li, H.Z.; Kubo, H.; Kobayashi, Y.; Kinoshita, T. [Quantitative determination of furosemide in serum and urine by high-performance liquid chromatography with electrochemical detector]. *Yao Hsueh Hsueh Pao*, **1991**, *26*, 923–927

Saugy, M.; Meuwly, P.; Munafo, A.; Rivier, L. Rapid high-performance liquid chromatographic determination with fluorescence detection of furosemide in human body fluids and its confirmation by gas chromatography-mass spectrometry. *J.Chromatogr.*, **1991**, *564*, 567–578 [serum; urine; fluorescence detection; warfarin (IS); pharmacokinetics; gradient; LOD 10 ng/mL]

Santasania, C.T. Direct injection analysis of diuretic and anti-inflammatory drugs on a shielded hydrophobic phase column. *J.Liq.Chromatogr.*, **1990**, *13*, 2605–2631 [direct injection; serum; gradient; horse; extracted hydrochlorothiazide, oxyphenbutazone, phenylbutazone]

Berthod, A.; Asensio, J.M.; Laserna, J.J. Micellar liquid chromatography for rapid screening of illegal drugs in sport. *J.Liq.Chromatogr.*, **1989**, *12*, 2621–2634 [fluorescence detection; UV detection; also bumetanide, caffeine, chlorthalidone, dihydrochlorothiazide, ephedrine, methyltestosterone, oxandrolone, propranolol, spironolactone, testosterone]

Radeck, W.; Heller, M. Improved method for the determination of furosemide in plasma by high-performance liquid chromatography. *J.Chromatogr.*, **1989**, *497*, 367–370

Russel, F.G.; Tan, Y.; Van Meijel, J.J.; Gribnau, F.W.; Van Ginneken, C.A. Solid-phase extraction of furosemide from plasma and urine and subsequent analysis by high-performance liquid chromatography. *J.Chromatogr.*, **1989**, *496*, 234–241

Miwa, Y.; Yamaji, A.; Nakahama, H.; Orita, Y.; Fukuhara, Y.; Kamada, T.; Ishibashi, M.; Ichikawa, Y.; Takahara, S.; Sonoda, T. [Determination of furosemide and its metabolic products in plasma and urine by high performance liquid chromatography and clinical application]. *Yakugaku Zasshi*, **1988**, *108*, 1087–1092

Pinkerton, T.C.; Perry, J.A.; Rateike, J.D. Separation of furosemide, phenylbutazone and oxyphenbutazone in plasma by direct injection onto internal surface reversed-phase columns with systematic optimization of selectivity. *J.Chromatogr.*, **1986**, *367*, 412–418

Lovett, L.J.; Nygard, G.; Dura, P.; Khalil, S.K.W. An improved HPLC method for the determination of furosemide in plasma and urine. *J.Liq.Chromatogr.*, **1985**, *8*, 1611–1628 [plasma; urine; desmethyl-naproxen (IS)]

Uchino, K.; Isozaki, S.; Saitoh, Y.; Nakagawa, F.; Tamura, Z.; Tanaka, N. Quantitative determination of furosemide in plasma, plasma water, urine and ascites fluid by high-performance liquid chromatography. *J.Chromatogr.*, **1984**, *308*, 241–249

Kerremans, A.L.; Tan, Y.; Van Ginneken, C.A.; Gribnau, F.W. Specimen handling and high-performance liquid chromatographic determination of furosemide. *J.Chromatogr.*, **1982**, *229*, 129–139

Rapaka, R.S.; Roth, J.; Viswanathan, C.; Goehl, T.J.; Prasad, V.K.; Cabana, B.E. Improved method for the analysis of furosemide in plasma by high-performance liquid chromatography. *J.Chromatogr.*, **1982**, *227*, 463–469

Snedden, W.; Sharma, J.N.; Fernandez, P.G. A sensitive assay method of furosemide in plasma and urine by high-performance liquid chromotography. *Ther.Drug Monit.*, **1982**, *4*, 381–383

Yoshitomi, H.; Ikeda, K.; Goto, S. [Analyses for furosemide and its metabolite in body fluids and urine of rabbit by high performance liquid chromatography]. *Yakugaku Zasshi*, **1982**, *102*, 1171–1176

Nation, R.L.; Peng, G.W.; Chiou, W.L. Quantitative analysis of furosemide in micro plasma volumes by high-performance liquid column chromatography. *J.Chromatogr.*, **1979**, *162*, 88–93 [plasma; fluorescence detection; non-interfering metabolites, acetaminophen, ampicillin, caffeine, digoxin, ephedrine, phenacetin, phenobarbital, phenytoin, salicylic acid, tetracycline, theobromine, theophylline; LOD 100 ng/mL]

# Gemfibrozil

**Molecular formula:** $C_{15}H_{22}O_3$
**Molecular weight:** 250.3
**CAS Registry No.:** 25812-30-0

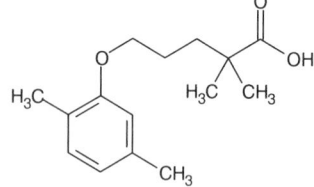

## SAMPLE
**Matrix:** blood
**Sample preparation:** Adjust pH of plasma to 4-5 with phosphoric acid (about 30 µL phosphoric acid per 4 mL of plasma). 500 µL Acidified plasma + 100 µL 2 µg/mL flurbiprofen in MeCN:water 30:70 + 1.3 mL MeCN, vortex, centrifuge at 1000 g for 15 min. Remove the supernatant and add it to 800 µL 1 M pH 3.0 glycine buffer and 4 mL ethyl acetate, shake horizontally at 70 rpm for 15 min, centrifuge at 1000 g for 15 min. Remove the organic layer and evaporate it to dryness in an evacuated centrifuge, reconstitute the residue in 250 µL mobile phase, inject a 10-100 µL aliquot.

## HPLC VARIABLES
**Column:** 250 × 4.6 5 µm Ultrasphere cyano
**Mobile phase:** MeCN:10 mM tetrabutylammonium sulfate 28:72, final apparent pH adjusted to 3.5
**Flow rate:** 1
**Injection volume:** 10-100
**Detector:** F ex 284 em 316

## CHROMATOGRAM
**Retention time:** 20.9
**Internal standard:** flurbiprofen (17.6)
**Limit of detection:** 100 ng/mL

## OTHER SUBSTANCES
**Extracted:** metabolites, glucuronide

## KEY WORDS
plasma

## REFERENCE
Sallustio, B.C.; Fairchild, B.A. Biosynthesis, characterization and direct high-performance liquid chromatographic analysis of gemfibrozil 1-O-β-acylglucuronide. *J.Chromatogr.B*, **1995**, *665*, 345–353

## SAMPLE
**Matrix:** blood
**Sample preparation:** 0.5 mL Plasma + 0.5 mL 10 µg/mL IS in MeCN:pH 7.4 phosphate buffered saline 1:99 + 20 µL formic acid + 5 mL cyclohexane:ethyl acetate 8:2, extract. Remove the organic layer and evaporate it, reconstitute the residue with 500 µL mobile phase, inject a 10 µL aliquot.

## HPLC VARIABLES
**Column:** 250 × 4.6 $C_6H_5$-1252N (Senshu Sci.)
**Mobile phase:** MeCN:10 mM pH 3.3 tartrate buffer:PIC-A 52:48:0.5
**Flow rate:** 1
**Injection volume:** 10
**Detector:** F ex 293 em 325

## CHROMATOGRAM
**Retention time:** 7.5

**Internal standard:** 4-(2,5-dimethylphenoxy)-2,2-dimethylbutanoic acid
**Limit of detection:** 100 ng/mL

## OTHER SUBSTANCES
**Simultaneous:** metabolites

## KEY WORDS
plasma

## REFERENCE
Nakagawa, A.; Shigeta, A.; Iwabuchi, H.; Horiguchi, M.; Nakamura, K.; Takahagi, H. Simultaneous determination of gemfibrozil and its metabolites in plasma and urine by a fully automated high performance liquid chromatographic system. *Biomed.Chromatogr.*, **1991**, *5*, 68–73

## SAMPLE
**Matrix:** blood
**Sample preparation:** 1 mL Serum + 2 mL MeCN:glacial acetic acid 75:10, vortex thoroughly, centrifuge. Remove the supernatant and add it to 500 mg NaCl, vortex, centrifuge. Remove the organic layer and evaporate it to dryness under a stream of nitrogen at room temperature, reconstitute the residue in 200 μL mobile phase, inject a 20 μL aliquot.

## HPLC VARIABLES
**Column:** 5 mm i.d. μBondapak C18 radial compression
**Mobile phase:** MeOH:water:glacial acetic acid 75:24:1
**Flow rate:** 0.8-1.2
**Injection volume:** 20
**Detector:** UV 276

## CHROMATOGRAM
**Retention time:** 9-10
**Limit of quantitation:** 1 μg/mL

## KEY WORDS
serum

## REFERENCE
Forland, S.C.; Chaplin, L.; Cutler, R.E. Assay of gemfibrozil in plasma by "high-performance" liquid chromatography. *Clin.Chem.*, **1987**, *33*, 1938

## SAMPLE
**Matrix:** blood
**Sample preparation:** 500 μL Plasma + 20 μL 100 μg/mL ibuprofen in MeCN:water 50:50, acidify with 3 drops 1 M HCl, add 5 mL cyclohexane, shake mechanically for 20 min, centrifuge. Remove the organic layer and evaporate it to dryness under reduced pressure, reconstitute the residue in 50-200 μL mobile phase, inject a 10-20 μL aliquot.

## HPLC VARIABLES
**Column:** 125 × 4.6 5 μm Spherisorb ODS II
**Mobile phase:** MeCN:water:phosphoric acid 50:50:0.2
**Flow rate:** 2
**Injection volume:** 10-20
**Detector:** UV 225

## CHROMATOGRAM
**Retention time:** 8.8

**Internal standard:** ibuprofen (5.8)
**Limit of detection:** 50 ng/mL

## OTHER SUBSTANCES
**Noninterfering:** metabolites

## KEY WORDS
plasma; pharmacokinetics

## REFERENCE
Hengy, H.; Kolle, E.U. Determination of gemfibrozil in plasma by high performance liquid chromatography. *Arzneimittelforschung*, **1985**, *35*, 1637–1639

## SAMPLE
**Matrix:** blood, tissue
**Sample preparation:** Acidify plasma with 10 μL 5 M orthophosphoric acid. Homogenize 1 g tissue with 2 mL ice-cold 10 mM pH 5.0 phosphate buffer containing 1.15% KCl. 1 mL Plasma or tissue homogenate + 5 mL MeCN:acetic acid 96:4, centrifuge at 1000 g for 5 min, discard the supernatant, wash the pellet with 5 mL diethyl ether, wash nine times with 5 mL 4% acetic acid in MeCN:10 mM pH 5.0 phosphate buffer 2:1. Add 1 mL 1 M KOH and 50 μL 5 μg/mL flurbiprofen in 0.06% MeCN to the pellet, heat at 80° overnight, cool, add 500 μL 4 M HCl, add 5 mL diethyl ether, shake horizontally at 80 oscillations/min for 15 min, centrifuge. Remove the organic layer and evaporate it to dryness under a stream of nitrogen at 40°, reconstitute the residue in 250 μL mobile phase, inject a 50 μL aliquot.

## HPLC VARIABLES
**Column:** 125 × 4 5 μm LiChrosphere 60 C8 RP-select B
**Mobile phase:** MeCN:water:acetic acid 51:48.5:0.5
**Flow rate:** 1
**Injection volume:** 50
**Detector:** F ex 284 em 316

## CHROMATOGRAM
**Retention time:** 9
**Internal standard:** flurbiprofen (5)
**Limit of quantitation:** 25 ng/mL

## OTHER SUBSTANCES
**Extracted:** metabolites, glucuronides

## KEY WORDS
rat; plasma; liver; kidney; heart; only gemfibrozil bound to protein is determined by this method

## REFERENCE
Sallustio, B.C.; Foster, D.J.R. Reactivity of gemfibrozil 1-O-β-acyl glucuronide. Pharmacokinetics of covalently bound gemfibrozil-protein adducts in rats. *Drug Metab.Dispos.*, **1995**, *23*, 892–899

## SAMPLE
**Matrix:** blood, urine
**Sample preparation:** 1 mL Plasma + 250 μL 200 μg/mL IS in MeCN:water 5:95 + 1 mL 1 M HCl + 10 mL diethyl ether, shake on a reciprocating shaker for 15 min, centrifuge at 700 g. Remove the organic layer and evaporate it to dryness under a stream of air at 55°, reconstitute the residue in 500 μL mobile phase, mix, inject a 40 μL aliquot. (To hydrolyze conjugates mix 1 mL plasma or urine with 25 μL glucuronidase/sulfatase (Glu-

sulase, DuPont), 1 mL water, and 1 mL 2 M pH 5.2 acetate buffer, heat at 37° overnight, add 2 mL 1 M HCl, proceed as above.)

## HPLC VARIABLES
**Column:** 100 × 4.6 5 μm Partisil ODS-3 RAC II
**Mobile phase:** MeCN:7 mM phosphoric acid 45:55
**Flow rate:** 2
**Injection volume:** 40
**Detector:** UV 276

## CHROMATOGRAM
**Retention time:** 10.4
**Internal standard:** 2,2'-dimethyl-5-(2,6-xylyloxy)valeric acid (7.4)
**Limit of quantitation:** 500 ng/mL

## OTHER SUBSTANCES
**Extracted:** metabolites

## KEY WORDS
plasma; pharmacokinetics; also for whole blood

## REFERENCE
Randinitis, E.J.; Parker, T.D.; Kinkel, A.W. Liquid chromatographic determination of gemfibrozil and its metabolite in plasma. *J.Chromatogr.*, **1986**, *383*, 444–448

## SAMPLE
**Matrix:** solutions
**Sample preparation:** Inject a 10 μL aliquot of a 200 μg/mL solution in mobile phase.

## HPLC VARIABLES
**Column:** 300 × 4 5 μm Suplecosil LC-18
**Mobile phase:** MeOH:water:glacial acetic acid 80:20:1
**Flow rate:** 0.8
**Injection volume:** 10
**Detector:** UV 276

## CHROMATOGRAM
**Retention time:** 14
**Internal standard:** 2,5-xylenol (5)

## REFERENCE
*Supelco Chromatography Products, Supleco, Inc., Bellefonte PA,* **1996**, p. 155

## SAMPLE
**Matrix:** solutions
**Sample preparation:** Inject a 10-100 μL aliquot.

## HPLC VARIABLES
**Column:** 50 × 4.6 5 μm Supelcosil LC-18-DB or 100 × 4.6 5 μm Supelcosil LC-18-DB
**Mobile phase:** MeOH:water:acetic acid 80:19:1
**Flow rate:** 1
**Injection volume:** 10-100
**Detector:** UV 276

## REFERENCE

Luner, P.E.; Babu, S.R.; Radebaugh, G.W. The effects of bile salts and lipids on the physicochemical behavior of gemfibrozil. *Pharm.Res.*, **1994**, *11*, 1755–1760

## SAMPLE

**Matrix:** urine
**Sample preparation:** 0.5 mL Urine + 0.5 mL 10 μg/mL IS in MeCN, vortex, centrifuge at 2000 g, inject a 10 μL aliquot of the upper layer.

## HPLC VARIABLES

**Column:** 150 × 4.6 YMC-A312 ODS (Yamamura Chemicals)
**Mobile phase:** MeCN:10 mM pH 4.7 acetate buffer 45:55 for 10.5 min then 80:20
**Flow rate:** 1 for 10.5 min then 2 mL/min
**Injection volume:** 10
**Detector:** F ex 283 em 315 (for gemfibrozil) and ex 300 em 340 (for metabolites)

## CHROMATOGRAM

**Retention time:** 16
**Internal standard:** 7-(2,5-dimethylphenoxy)-2,2-dimethylheptanoic acid
**Limit of detection:** 100 ng/mL

## OTHER SUBSTANCES

**Simultaneous:** metabolites

## REFERENCE

Nakagawa, A.; Shigeta, A.; Iwabuchi, H.; Horiguchi, M.; Nakamura, K.; Takahagi, H. Simultaneous determination of gemfibrozil and its metabolites in plasma and urine by a fully automated high performance liquid chromatographic system. *Biomed.Chromatogr.*, **1991**, *5*, 68–73

# Glipizide

**Molecular formula:** $C_{21}H_{27}N_5O_4S$
**Molecular weight:** 445.5
**CAS Registry No.:** 29094-61-9

## SAMPLE
**Matrix:** blood
**Sample preparation:** 2 mL Whole blood or plasma + 2 mL buffer + 5 mL chloroform: isopropanol:n-heptane 60:14:26, shake gently horizontally for 10 min, centrifuge at 2800 g for 10 min. Remove the lower organic layer and evaporate it to dryness under vacuum at 45°, reconstitute the residue in 100 μL mobile phase, centrifuge at 2800 g for 5 min, inject a 50 μL aliquot of the supernatant. (Buffer was saturated ammonium chloride solution 25% diluted with water, adjusted to pH 9.5 with 25% ammonia solution.)

## HPLC VARIABLES
**Column:** 300 × 3.9 4 μm NovaPack C18
**Mobile phase:** MeOH:THF:buffer 65:5:30 (Buffer was 0.68 g/L (10 mM (sic)) $KH_2PO_4$ adjusted to pH 2.6 with concentrated orthophosphoric acid.) (At the end of each session wash the column with water for 1 h and MeOH for 1 h, re-equilibrate for 30 min.)
**Column temperature:** 30
**Flow rate:** 0.8
**Injection volume:** 50
**Detector:** UV 226

## CHROMATOGRAM
**Retention time:** 3.86
**Limit of detection:** <120 ng/mL

## OTHER SUBSTANCES
**Extracted:** acenocoumarol, acepromazine, aceprometazine, acetaminophen, aconitine, ajmaline, albuterol, alimemazine, alminoprofen, alpidem, alprazolam, alprenolol, amisulpride, amitriptyline, amodiaquine, amoxapine, aspirin, astemizole, atenolol, benazepril, benperidol, benzocaine, benzoylecgonine, bepridil, betaxolol, bisoprolol, bromazepam, brompheniramine, bumadizone, bupivacaine, buprenorphine, buspirone, caffeine, carbamazepine, carbinoxamine, carpipramine, carteolol, celiprolol, cetirizine, chlorambucil, chlordiazepoxide, chlormezanone, chlorophenacinone, chloroquine, chlorpheniramine, chlorpromazine, cibenzoline, cicletanine, clemastine, clobazam, clomipramine, clorazepate, clozapine, cocaine, codeine, colchicine, cyamemazine, cyclizine, cycloguanil, cyproheptadine, cytarabine, dacarbazine, daunorubicin, debrisoquine, demexiptiline, desipramine, dextromethorphan, dextromoramide, dextropropoxyphene, diazepam, diazoxide, diclofenac, dihydralazine, diltiazem, diphenhydramine, dipyridamole, disopyramide, dosulepine, doxepin, doxylamine, droperidol, etodolac, fenfluramine, fenoprofen, fentiazac, flecainide, floctafenine, flumazenil, fluoxetine, fluphenazine, flurbiprofen, fluvoxamine, glibenclamide, glibornuride, haloperidol, histapyrrodine, hydroxychloroquine, hydroxyzine, ibuprofen, imipramine, indomethacin, iproniazid, ketamine, ketoprofen, labetalol, levomepromazine, lidocaine, lidoflazine, lisinopril, loperamide, loprazolam, loratadine, lorazepam, loxapine, maprotiline, medazepam, medifoxamine, mefenamic acid, mefenidramine, mefloquine, meperidine, mephenesin, mephentermine, mepivacaine, metapramine, metformin, methadone, methaqualone, methocarbamol, methotrexate, metipranolol, metoprolol, mexiletine, mianserine, midazolam, minoxidil, moclobemide, moperone, morphine, nadolol, nalbuphine, nalorphine, naloxone, naltrexone, naproxen, nialamide, nicardipine, nifedipine, niflumic acid, nimodipine, nitrazepam, nitrendipine, nizatidine, nomifensine, nortriptyline, omeprazole, opipramol, oxazepam, oxprenolol, penbutolol, penfluridol, pentazocine, phencyclidine, phenobarbital, phenol, phenylbutazone, pimozide,

pindolol, pipamperone, piroxicam, prazepam, prilocaine, procainamide, procarbazine, proguanil, promethazine, propafenone, propranolol, protriptyline, pyrimethamine, quinidine,
quinine, quinupramine, ramipril, ranitidine, reserpine, ritodrine, secobarbital, sotalol,
sulfinpyrazone, sulindac, sulpride, sultopride, suriclone, temazepam, tenoxicam, terbutaline, terfenadine, tetracaine, tetrazepam, thiopental, thioproperazine, thioridazine, tianeptine, tiapride, tiaprofenic acid, ticlopidine, timolol, tioclomarol, tofisopam, toloxatone,
trazodone, trifluoperazine, trifluperidol, trimipramine, triprolidine, tropatenine, verapamil, viloxazine, vinblastine, vincristine, vindesine, warfarin, yohimbine, zolpidem, zopiclone, zorubicine
**Interfering:** acebutolol, chlorpropamide, clonazepam, clonidine, ephedrine, estazolam, flunitrazepam, glipizide, glutethimide, melphalan, metoclopramide, prazosin, strychnine, tolbutamide, triazolam

## KEY WORDS
whole blood; plasma

## REFERENCE
Tracqui, A.; Kintz, P.; Mangin, P. Systematic toxicological analysis using HPLC/DAD. *J.Forensic Sci.*,
**1995**, *40*, 254–262

## SAMPLE
**Matrix:** blood
**Sample preparation:** 500 μL Plasma + 200 μL 2 M HCl + 2 mL diethyl ether, vortex for
30 s, centrifuge at 1500 g for 5 min, freeze in dry ice for 5 min. Remove the organic layer
and evaporate it to dryness under a stream of nitrogen at 35-40°, reconstitute the residue
in 100 μL mobile phase, vortex for 30 s, inject a 25-50 μL aliquot.

## HPLC VARIABLES
**Column:** 250 × 4.1 10 μm Versapack C18 (Alltech)
**Mobile phase:** MeCN:10 mM orthophosphoric acid 50:50
**Flow rate:** 1
**Injection volume:** 25-50
**Detector:** UV 230

## CHROMATOGRAM
**Retention time:** 6.49
**Limit of detection:** 100 ng/mL

## OTHER SUBSTANCES
**Extracted:** chlorpropamide, gliclazide, tolazamide, tolbutamide
**Simultaneous:** N-acetylsulfamethoxazole, sulfamethoxazole
**Noninterfering:** trimethoprim

## KEY WORDS
plasma

## REFERENCE
Shenfield, G.M.; Boutagy, J.S.; Webb, C. A screening test for detecting sulfonylureas in plasma.
*Ther.Drug Monit.*, **1990**, *12*, 393–397

## SAMPLE
**Matrix:** blood
**Sample preparation:** 2 mL Serum + 2 mL water + 200 μL 1 (?) M HCl + 200 μL 2.5
μg/mL glibornuride in MeOH + 7 mL diethyl ether, mix, centrifuge at 2000 rpm for 5
min. Remove 6.5 mL of the organic layer and evaporate it to dryness under a stream of
nitrogen, reconstitute the residue in 500 μL 2 mg/mL dinitrofluorobenzene in butyl ace-

tate, heat at 120° for 1 h, cool, evaporate to dryness under a stream of nitrogen, reconstitute the residue in 150 μL mobile phase, inject a 120 μL aliquot.

## HPLC VARIABLES
**Column:** 250 × 4.6 5 μm Spherisorb ODS 2
**Mobile phase:** MeCN:0.4% aqueous phosphoric acid 75:25
**Column temperature:** 40
**Flow rate:** 1.2
**Injection volume:** 120
**Detector:** UV 360

## CHROMATOGRAM
**Retention time:** 6.5
**Internal standard:** glibornuride (5.8)
**Limit of detection:** 40 ng/mL

## OTHER SUBSTANCES
**Extracted:** chlorpropamide, tolazamide, tolbutamide
**Interfering:** glyburide (glibenclamide) (forms same derivative)

## KEY WORDS
serum; derivatization

## REFERENCE
Starkey, B.J.; Mould, G.P.; Teale, J.D. The determination of sulphonylurea drugs by HPLC and its clinical application. *J.Liq.Chromatogr.*, **1989**, *12*, 1889–1896

## SAMPLE
**Matrix:** blood
**Sample preparation:** 500 μL Serum + 800 ng glibornuride + 1 mL 50 mM HCl + 3 mL benzene (Caution! Benzene is a carcinogen!), shake gently for 10 min, centrifuge. Remove organic phase and evaporate it to dryness at 45° under a stream of air. Dissolve residue in 50 μL MeOH and inject a 20 μL aliquot.

## HPLC VARIABLES
**Column:** 300 × 3.9 10 μm μBondapak C18
**Mobile phase:** MeOH:10 mM pH 3.5 phosphate buffer 60:40
**Flow rate:** 1
**Injection volume:** 20
**Detector:** UV 225

## CHROMATOGRAM
**Retention time:** 9
**Internal standard:** glibornuride (18)
**Limit of detection:** 10 ng/mL

## KEY WORDS
serum

## REFERENCE
Wåhlin-Boll, E.; Melander, A. High-performance liquid chromatographic determination of glipizide and some other sulfonylurea drugs in serum. *J.Chromatogr.*, **1979**, *164*, 541–546

## SAMPLE
**Matrix:** blood, urine

**Sample preparation:** 500 μL Plasma or urine + 300 μL 10 μg/mL tolbutamide in MeOH + 1 mL 50 mM HCl + 3 mL benzene (Caution! Benzene is a carcinogen!), shake gently for 15 min, centrifuge at 3250 g for 5 min. Remove organic layer and evaporate it to dryness under a stream of air. Dissolve residue in 50 μL mobile phase, vortex, inject 20 μL aliquot.

## HPLC VARIABLES
**Column:** 250 × 4.5 5 μm Spherisorb ODS C18
**Mobile phase:** MeCN:10 mM pH 3.5 phosphate buffer 35:65
**Flow rate:** 1.5
**Injection volume:** 20
**Detector:** UV 275

## CHROMATOGRAM
**Retention time:** 10.5
**Internal standard:** tolbutamide (8.2)
**Limit of detection:** 5 ng/mL

## OTHER SUBSTANCES
**Noninterfering:** metabolites

## KEY WORDS
plasma

## REFERENCE
Emilsson, H. High-performance liquid chromatographic determination of glipizide in human plasma and urine. *J.Chromatogr.*, **1987**, *421*, 319–326

## SAMPLE
**Matrix:** solutions

## HPLC VARIABLES
**Column:** 250 × 4.6 Zorbax RX
**Mobile phase:** Gradient. A was 10 mL concentrated orthophosphoric acid and 7 mL triethylamine in 1 L water. B was 10 mL concentrated orthophosphoric acid and 7 mL triethylamine in 200 mL water, make up to 1 L with MeCN. A:B from 100:0 to 0:100 over 30 min, maintain at 0:100 for 5 min.
**Column temperature:** 30
**Flow rate:** 2
**Detector:** UV 210

## OTHER SUBSTANCES
**Also analyzed:** acepromazine, acetaminophen, acetophenazine, albuterol, aminophylline, amitriptyline, amobarbital, amoxapine, amphetamine, amylocaine, antipyrine, aprobarbital, aspirin, atenolol, atropine, avermectin, barbital, benzocaine, benzoic acid, benzotropine, benzphetamine, berberine, bibucaine, bromazepan, brompheniramine, buprenorphine, buspirone, butabarbital, butacaine, butethal, caffeine, carbamazepine, carbromal, chloramphenicol, chlordiazepoxide, chloroquine, chlorothiazide, chloroxylenol, chlorphenesin, chlorpheniramine, chlorpromazine, chlorpropamide, chlortetracycline, cimetidine, cinchonidine, cinchonine, clenbuterol, clonazepam, clonixin, clorazepate, cocaine, codeine, colchicine, cortisone, coumarin, cyclazocine, cyclobenzaprine, cyclothiazide, cyheptamide, cymarin, danazol, danthron, dapsone, debrisoquine, desipramine, dexamethasone, dextromethorphan, dextropropoxyphene, diamorphine, diazepam, diclofenac, diethylpropion, diethylstilbestrol, diflunisal, digitoxin, digoxin, diltiazem, diphenhydramine, diphenoxylate, diprenorphine, dipyrone, disulfiram, dopamine, doxapram, doxepin, dronabinol, ephedrine, epinephrine, epinine, estradiol, estriol, estrone, ethacrynic acid, ethosuximide, etonitazene, etorphine, eugenol, famotidine, fenbendazole, fencamfamine, fenoprofen, fen-

proporex, fentanyl, flubendazole, flufenamic acid, flunitrazepam, 5-fluorouracil, fluoxymesterone, fluphenazine, furosemide, gentisic acid, gitoxigenin, glunixin, glutethimide, glybenclamide, guaiacol, halazepam, haloperidol, hydrochlorothiazide, hydrocodone, hydrocortisone, hydromorphone, hydroxyquinoline, ibogaine, ibuprofen, iminostilbene, imipramine, indomethacin, isocarbostyril, isocarboxazid, isoniazid, isoproterenol, isoxsuprine, ivermectin, ketamine, ketoprofen, kynurenic acid, levorphanol, lidocaine, lorazepam, lormetazepam, loxapine, mazindol, mebendazole, meclizine, meclofenamic acid, medazepam, mefenamic acid, megestrol, mepacrine, meperidine, mephentermine, mephenytoin, mephesin, mephobarbital, mepivacaine, mescaline, mesoridazine, methadone, methamphetamine, methapyrilene, methaqualone, methazolamide, methocarbamol, methoxamine, methsuximide, methyl salicylate, methyldopa, methyldopamine, methylphenidate, methylprednisolone, methyltestosterone, methyprylon, metoprolol, mibolerone, morphine, nadolol, nalorphine, naloxone, naltrexone, naphazoline, naproxen, nefopam, niacinamide, nicotine, nicotinic acid, nifedipine, niflumic acid, nitrazepam, norepinephrine, nortriptyline, noscapine, nylidrin, oxazepam, oxycodone, oxymorphone, oxyphenbutazone, oxytetracycline, papaverine, pargyline, pemoline, pentazocine, pentobarbital, persantine, phenacetin, phenazocine, phenazopyridine, phencyclidine, phendimetrazine, phenelzine, pheniramine, phenobarbital, phenothiazine, phensuximide, phentermine, phenylbutazone, phenylephrine, phenylpropanolamine, piperocaine, prazepam, prednisolone, primidone, probenecid, progesterone, propiomazine, propranolol, propylparaben, pseudoephedrine, puromycin, pyrilamine, pyrithyldione, quazepam, quinaldic acid, quinidine, quinine, ranitidine, recinnamine, reserpine, resorcinol, saccharin, albuterol, salicylamide, salicylic acid, scopolamine, scopoletin, secobarbital, strychnine, sulfacetamide, sufadiazine, sulfadimethoxine, sulfaethidole, sulfamerazine, sulfamethazine, sulfamethoxizole, sulfanilamide, sulfapyridine, sulfasoxizole, sulindac, tamoxifen, temazepam, testosterone, tetracaine, tetracycline, tetramisole, thebaine, theobromine, theophylline, thiabendazole, thiamine, thiamylal, thiobarbituric acid, thioridazine, thiosalicylic acid, thiothixene, thymol, tolazamide, tolazoline, tolbutamide, tolmetin, tranylcypromine, triamcinolone, tribenzylamine, trichloromethiazide, trifluoperazine, trihexyphenidyl, trimethoprim, tripelennamine, triproilidine, tropacocaine, tyramine, verapamil, vincamine, warfarin, yohimbine, zoxazolamine

## REFERENCE

Hill, D.W.; Kind, A.J. Reversed-phase solvent gradient HPLC retention indexes of drugs. *J.Anal.Toxicol.*, **1994**, *18*, 233–242

## SAMPLE

**Matrix:** whole blood
**Sample preparation:** 1 mL Whole blood + 1 mL 50 mM pH 6.6 KH$_2$PO$_4$, vortex, add 6 mL diethyl ether, shake at 150 ± 20 oscillations/min on a reciprocating shaker for 15 min, centrifuge at 1500 g for 10 min. Remove the organic layer and evaporate it to dryness at 37° under a stream of nitrogen. Dissolve residue in 200 µL mobile phase, centrifuge at 1000 g for 3 min, inject a 40 µL aliquot of the supernatant.

## HPLC VARIABLES

**Column:** 150 × 4.6 5 µm Nucleosil C18
**Mobile phase:** MeCN:isopropanol:buffer 30:5:65 (Buffer was 80 mM ammonium acetate adjusted to pH 3.5 with concentrated HCl.)
**Flow rate:** 1
**Injection volume:** 40
**Detector:** UV 241

## CHROMATOGRAM

**Retention time:** 5.9
**Internal standard:** glipizide

## OTHER SUBSTANCES

**Simultaneous:** indapamide

**Noninterfering:** acetaminophen, aspirin, caffeine, dextromethorphan, ibuprofen, nicotine, phenylpropanolamine, theophylline

## KEY WORDS
glipizide is IS

## REFERENCE
Miller, R.B.; Dadgar, D.; Lalande, M. High-performance liquid chromatographic method for the determination of indapamide in human whole blood. *J.Chromatogr.*, **1993**, *614*, 293–298

## ANNOTATED BIBLIOGRAPHY
Das Gupta, V. Quantitation of glipizide and glyburide in tablets using high performance liquid chromatography. *J.Liq.Chromatogr.*, **1986**, *9*, 3607–3615 [simultaneous glyburide; tablets; hydrocortisone (IS)]

# Glyburide

**Molecular formula:** $C_{23}H_{28}ClN_3O_5S$
**Molecular weight:** 494.0
**CAS Registry No.:** 10238-21-8

## SAMPLE
**Matrix:** blood
**Sample preparation:** 200 μL Plasma + 15 μL 5 μg/mL warfarin + 3 mL dichloromethane, vortex for 1 min, shake on a rotary mixer for 5 min, centrifuge at 1000 g for 15 min. Remove the organic layer and evaporate it to dryness under a stream of nitrogen, reconstitute the residue in 100 μL mobile phase, inject an 80 μL aliquot.

## HPLC VARIABLES
**Guard column:** Novapak C18
**Column:** 150 × 3.9 4 μm Novapak C18
**Mobile phase:** MeOH:50 mM $(NH_4)H_2PO_4$ 61:39, pH adjusted to 4.0
**Flow rate:** 1.2
**Injection volume:** 80
**Detector:** F ex 308 em 360

## CHROMATOGRAM
**Retention time:** 12.3
**Internal standard:** warfarin (6.7)
**Limit of detection:** 20 ng/mL

## KEY WORDS
plasma; rat; human; pharmacokinetics

## REFERENCE
al-Dhawailie, A.A.; Abdulaziz, M.A.; Tekle, A.; Matar, K.M. A simple, specific, and rapid high-performance liquid chromatographic assay for glibenclamide in plasma. *J.Liq.Chromatogr.*, **1995**, *18*, 3981–3990

## SAMPLE
**Matrix:** blood
**Sample preparation:** 1 mL Serum + 200 μL 1 M HCl + 1 μg tolbutamide + 5 mL toluene, shake gently for 15 min, centrifuge at 1500 g for 3 min. Remove the organic layer and evaporate it to dryness, reconstitute the residue in 25 μL 6 mg/mL dinitrofluorobenzene in n-butyl acetate (prepare fresh each week, store at 4° in the dark), heat at 120° for 30 min, evaporate to dryness, reconstitute with 50 μL mobile phase, inject a 25-50 μL aliquot. Alternatively, filter (Amicon YMT membrane, 30000 MW cutoff) 200 μL 100 mM NaOH while centrifuging at 4°, rinse filter with 500 μL water, filter 1 mL serum in the same unit while centrifuging at 4° at 2500 g for 1.5 h. Remove a 700 μL aliquot of the ultrafiltrate, add 200 μL 1 M HCl, add 1 μg tolbutamide, add 5 mL toluene, shake gently for 15 min, centrifuge at 1500 g for 3 min. Remove the organic layer and evaporate it to dryness, reconstitute the residue in 25 μL 6 mg/mL dinitrofluorobenzene in n-butyl acetate (prepare fresh each week, store at 4° in the dark), heat at 120° for 30 min, evaporate to dryness, reconstitute with 50 μL mobile phase, inject a 25-50 μL aliquot.

## HPLC VARIABLES
**Column:** 250 × 4.6 7 μm LiChrosorb RP18

**Mobile phase:** MeCN : 10 mM pH 3.5 phosphate buffer 80 : 20
**Flow rate:** 1
**Injection volume:** 25-50
**Detector:** UV 360

## CHROMATOGRAM
**Retention time:** 7
**Internal standard:** tolbutamide (5)
**Limit of detection:** 2 ng/mL

## KEY WORDS
derivatization; serum; ultrafiltrate; pharmacokinetics

## REFERENCE
Arcelloni, C.; Fermo, I.; Calderara, A.; Pacchioni, M.; Pontiroli, A.E.; Paroni, R. Glibenclamide and tolbutamide in human serum: Rapid measurement of the free fraction. *J.Liq.Chromatogr.*, **1990**, *13*, 175–189

## SAMPLE
**Matrix:** blood
**Sample preparation:** 500 µL Plasma + 200 µL 2 M HCl + 2 mL diethyl ether, vortex for 30 s, centrifuge at 1500 g for 5 min, freeze in dry ice for 5 min. Remove the organic layer and evaporate it to dryness under a stream of nitrogen at 35-40°, reconstitute the residue in 100 µL mobile phase, vortex for 30 s, inject a 25-50 µL aliquot.

## HPLC VARIABLES
**Column:** 250 × 4.1 10 µm Versapack C18 (Alltech)
**Mobile phase:** MeCN : 10 mM orthophosphoric acid 50 : 50
**Flow rate:** 1
**Injection volume:** 25-50
**Detector:** UV 230

## CHROMATOGRAM
**Retention time:** 18.52
**Limit of detection:** 100 ng/mL

## OTHER SUBSTANCES
**Extracted:** chlorpropamide, gliclazide, glipizide, tolazamide, tolbutamide
**Simultaneous:** N-acetylsulfamethoxazole, sulfamethoxazole
**Noninterfering:** trimethoprim

## KEY WORDS
plasma

## REFERENCE
Shenfield, G.M.; Boutagy, J.S.; Webb, C. A screening test for detecting sulfonylureas in plasma. *Ther.Drug Monit.*, **1990**, *12*, 393–397

## SAMPLE
**Matrix:** blood
**Sample preparation:** 2 mL Serum + 2 mL water + 200 µL 1 (?) M HCl + 200 µL 2.5 µg/mL glibornuride in MeOH + 7 mL diethyl ether, mix, centrifuge at 2000 rpm for 5 min. Remove 6.5 mL of the organic layer and evaporate it to dryness under a stream of nitrogen, reconstitute the residue in 500 µL 2 mg/mL dinitrofluorobenzene in butyl acetate, heat at 120° for 1 h, cool, evaporate to dryness under a stream of nitrogen, reconstitute the residue in 150 µL mobile phase, inject a 120 µL aliquot.

## HPLC VARIABLES

**Column:** 250 × 4.6 5 μm Spherisorb ODS 2
**Mobile phase:** MeCN:0.4% aqueous phosphoric acid 75:25
**Column temperature:** 40
**Flow rate:** 1.2
**Injection volume:** 120
**Detector:** UV 360

## CHROMATOGRAM

**Retention time:** 6.5
**Internal standard:** glibornuride (5.8)
**Limit of detection:** 40 ng/mL

## OTHER SUBSTANCES

**Extracted:** chlorpropamide, tolazamide, tolbutamide
**Interfering:** glipizide (forms same derivative)

## KEY WORDS

serum; derivatization

## REFERENCE

Starkey, B.J.; Mould, G.P.; Teale, J.D. The determination of sulphonylurea drugs by HPLC and its clinical application. *J.Liq.Chromatogr.*, **1989**, *12*, 1889–1896

## SAMPLE

**Matrix:** blood
**Sample preparation:** 1 mL Plasma + 100 μL 500 ng/mL IS in MeOH:water 1:1 + 2 mL 100 mM $KH_2PO_4$ adjusted to pH 4.0 with 1 M orthophosphoric acid, shake for 10 s, add dichloromethane, extract in Rollamix for 30 min, centrifuge at 2000 rpm for 5 min. Remove organic phase and evaporate it to dryness under a stream of air at 50°. Reconstitute residue in 100 μL mobile phase, vortex 30 s, inject 50 μL aliquot.

## HPLC VARIABLES

**Column:** 150 × 4.6 5 μm Microsorb C-18
**Mobile phase:** MeCN:50 mM ammonium sulfate adjusted to pH 3.0 with 1 M sulfuric acid 42:58
**Flow rate:** 1.4
**Injection volume:** 50
**Detector:** UV 229

## CHROMATOGRAM

**Retention time:** 17.8
**Internal standard:** N-(4-[2-(5-chloro-2-methoxybenzamido)ethyl]benzenesulfonyl)-N'-cyclopentylurea (11.4)
**Limit of detection:** 5 ng/mL

## KEY WORDS

plasma

## REFERENCE

Othman, S.; Shaheen, O.; Jalal, I.; Awidi, A.; Al-Turk, W. Liquid chromatographic determination of glibenclamide in human plasma. *J.Assoc.Off.Anal.Chem.*, **1988**, *71*, 942–944

## SAMPLE

**Matrix:** blood

**Sample preparation:** 2 mL Plasma + 100 ng tolbutamide + 500 μL 1 M HCl + 8 mL chloroform, shake on a reciprocal shaker, shake for 10 min in a reciprocal shaker, centrifuge at 2000 g for 15 min. Remove 7 mL of the lower organic layer and evaporate it to dryness under a stream of nitrogen at 45°, reconstitute the residue in 100 μL 3 mg/mL dinitrofluorobenzene in n-butyl acetate, heat at 120° for 30 min, evaporate to dryness under a stream of nitrogen at 60°, dissolve the residue in 100 μL mobile phase, inject a 30-70 μL aliquot. (Recrystallize dinitrofluorobenzene from diethyl ether. Prepare solutions weekly, store at 4° in the dark.)

## HPLC VARIABLES
**Column:** 125 × 4.6 5 μm C8 (Perkin-Elmer)
**Mobile phase:** MeCN:water 50:50 containing 0.15% phosphoric acid
**Flow rate:** 1.5
**Injection volume:** 30-70
**Detector:** UV 350

## CHROMATOGRAM
**Retention time:** 3.4
**Internal standard:** tolbutamide (4.5)
**Limit of detection:** 5 ng/mL

## OTHER SUBSTANCES
**Extracted:** chlorpropamide
**Noninterfering:** acetaminophen, aspirin, chlordiazepoxide, diazepam, phenobarbital, phenytoin, quinidine, theophylline

## KEY WORDS
plasma; derivatization

## REFERENCE
Zecca, L.; Trivulzio, S.; Pinelli, A.; Colombo, R.; Tofanetti, O. Determination of glibenclamide, chlorpropamide and tolbutamide in plasma by high-performance liquid chromatography with ultraviolet detection. *J.Chromatogr.*, **1985**, *339*, 203–209

## SAMPLE
**Matrix:** blood, urine
**Sample preparation:** Dilute serum 5-fold with water before analysis. 1 mL Serum or diluted urine + 25 μL 15 μg/mL glibornuride in MeOH + 100 μL 2 M HCl + 6 mL n-hexane: dichloromethane 1:1, rotate for 10 min, centrifuge at 700 g for 10 min. Remove 5 mL of the organic phase and evaporate it to dryness at 37° under a stream of air. Dissolve residue in 50 μL mobile phase, inject 10-25 μL aliquots. (If necessary to remove interferences, the organic layer can be washed with 5 mL 25 mM tetrabutylammonium hydrogen sulfate in 50 mM pH 12.2 phosphate buffer by rotation for 10 min.)

## HPLC VARIABLES
**Guard column:** Chrompack reversed-phase
**Column:** 100 × 4.6 3 μm Chrompack Chromsep microsphere C18
**Mobile phase:** MeCN:38 mM pH 7.49 phosphate buffer 28:72
**Flow rate:** 0.7
**Injection volume:** 10-25
**Detector:** UV 203

## CHROMATOGRAM
**Retention time:** 12.7
**Internal standard:** glibornuride (6)
**Limit of detection:** 1 ng/mL

## OTHER SUBSTANCES
**Extracted:** metabolites

## KEY WORDS
serum

## REFERENCE
Rydberg, T.; Wåhlin-Boll, E.; Melander, A. Determination of glibenclamide and its two major metabolites in human serum and urine by column liquid chromatography. *J.Chromatogr.*, **1991**, *564*, 223–233

## SAMPLE
**Matrix:** blood, urine
**Sample preparation:** 500 µL Plasma or urine + 1 mL 50 mM HCl +25 µL 10 µg/mL glibornuride in MeOH + 3 mL benzene (Caution! Benzene is a carcinogen!), shake gently for 15 min, centrifuge at 3250 g for 5 min. Remove the organic layer and evaporate it to dryness under a stream of air. Dissolve the residue in 50 µL mobile phase with vortexing, inject a 20 µL aliquot.

## HPLC VARIABLES
**Column:** 250 × 4.6 5 µm Spherisorb ODS
**Mobile phase:** MeCN:10 mM pH 3.5 phosphate buffer 50:50
**Flow rate:** 1.6
**Injection volume:** 20
**Detector:** UV 225

## CHROMATOGRAM
**Retention time:** 8
**Internal standard:** glibornuride (5.8)
**Limit of detection:** 5-10 ng/mL

## OTHER SUBSTANCES
**Noninterfering:** chlorpropamide, flunitrazepam, furosemide, glipizide, naproxen, sulfamethoxazole, theophylline, thioridazine, tolbutamide, trimethoprim

## KEY WORDS
plasma

## REFERENCE
Emilsson, H.; Sjöberg, S.; Svedner, M.; Christenson, I. High-performance liquid chromatographic determination of glibenclamide in human plasma and urine. *J.Chromatogr.*, **1986**, *383*, 93–102

## SAMPLE
**Matrix:** bulk
**Sample preparation:** Prepare a 2 mg/mL solution in MeOH, inject an aliquot.

## HPLC VARIABLES
**Column:** 300 × 3.9 10 µm Waters phenyl
**Mobile phase:** MeOH:MeCN:50 mM $(NH_4)H_2PO_4$ 50:10:40
**Flow rate:** 1.5
**Injection volume:** 10
**Detector:** UV 229

## CHROMATOGRAM
**Retention time:** 9.5
**Limit of quantitation:** 0.02%

## OTHER SUBSTANCES
**Simultaneous:** impurities

## REFERENCE
Beaulieu, N.; Graham, S.J.; Lovering, E.G. Liquid chromatographic determination of glyburide (gliben-clamide) and its related compounds in raw materials. *J.AOAC Int.*, **1993**, *76*, 962–965

## SAMPLE
**Matrix:** solutions

## HPLC VARIABLES
**Column:** 250 × 4.6 5 μm Supelcosil LC-DP (A) or 250 × 4 5 μm LiChrospher 100 RP-8 (B)
**Mobile phase:** MeCN:0.025% phosphoric acid:buffer 25:10:5 (A) or 60:25:15 (B) (Buffer was 9 mL concentrated phosphoric acid and 10 mL triethylamine in 900 mL water, adjust pH to 3.4 with dilute phosphoric acid, make up to 1 L.)
**Flow rate:** 0.6
**Injection volume:** 25
**Detector:** UV 229

## CHROMATOGRAM
**Retention time:** 8.51 (A), 9.83 (B)

## OTHER SUBSTANCES
**Also analyzed:** acebutolol, acepromazine, acetaminophen, acetazolamide, acetophenazine, albuterol, alprazolam, amitriptyline, amobarbital, amoxapine, antipyrine, atenolol, atropine, azatadine, baclofen, benzocaine, bromocriptine, brompheniramine, brotizolam, bupivacaine, buspirone, butabarbital, butalbital, caffeine, carbamazepine, cetirizine, chlorcyclizine, chlordiazepoxide, chlormezanone, chloroquine, chlorpheniramine, chlorpromazine, chlorpropamide, chlorprothixene, chlorthalidone, chlorzoxazone, cimetidine, cisapride, clomipramine, clonazepam, clonidine, clozapine, cocaine, codeine, colchicine, cyclizine, cyclobenzaprine, dantrolene, desipramine, diazepam, diclofenac, diflunisal, diltiazem, diphenhydramine, diphenidol, diphenoxylate, dipyridamole, disopyramide, dobutamine, doxapram, doxepin, droperidol, encainide, ethidium bromide, ethopropazine, fenoprofen, fentanyl, flavoxate, fluoxetine, fluphenazine, flurazepam, flurbiprofen, fluvoxamine, furosemide, glutethimide, guaifenesin, haloperidol, homatropine, hydralazine, hydrochlorothiazide, hydrocodone, hydromorphone, hydroxychloroquine, hydroxyzine, ibuprofen, imipramine, indomethacin, ketoconazole, ketoprofen, ketorolac, labetalol, levorphanol, lidocaine, loratadine, lorazepam, lovastatin, loxapine, mazindol, mefenamic acid, meperidine, mephenytoin, mepivacaine, mesoridazine, metaproterenol, methadone, methdilazine, methocarbamol, methotrexate, methotrimeprazine, methoxamine, methyldopa, methylphenidate, metoclopramide, metolazone, metoprolol, metronidazole, midazolam, moclobemide, morphine, nadolol, nalbuphine, naloxone, naphazoline, naproxen, nifedipine, nizatidine, norepinephrine, nortriptyline, oxazepam, oxycodone, oxymetazoline, paroxetine, pemoline, pentazocine, pentobarbital, pentoxifylline, perphenazine, pheniramine, phenobarbital, phenol, phenolphthalein, phentolamine, phenylbutazone, phenyltoloxamine, phenytoin, pimozide, pindolol, piroxicam, pramoxine, prazepam, prazosin, probenecid, procainamide, procaine, prochlorperazine, procyclidine, promazine, promethazine, propafenone, propantheline, propiomazine, propofol, propranolol, protriptyline, quazepam, quinidine, quinine, racemethorphan, ranitidine, remoxipride, risperidone, salicylic acid, scopolamine, secobarbital, sertraline, sotalol, spironolactone, sulfinpyrazone, sulindac, temazepam, terbutaline, terfenadine, tetracaine, theophylline, thiethylperazine, thiopental, thioridazine, thiothixene, timolol, tocainide, tolbutamide, tolmetin, trazodone, triamterene, triazolam, trifluoperazine, triflupromazine, trimeprazine, trimethoprim, trimipramine, verapamil, warfarin, xylometazoline, yohimbine, zopiclone

## KEY WORDS
some details of plasma extraction

## REFERENCE

Koves, E.M. Use of high-performance liquid chromatography-diode array detection in forensic toxicology. *J.Chromatogr.A*, **1995**, *692*, 103–119

## ANNOTATED BIBLIOGRAPHY

Coppack, S.W.; Lant, A.F.; McIntosh, C.S.; Rodgers, A.V. Pharmacokinetic and pharmacodynamic studies of glibenclamide in non-insulin dependent diabetes mellitus. *Br.J.Clin.Pharmacol.*, **1990**, *29*, 673–684 [plasma; gliburnide (IS); LOD 10 ng/mL; pharmacokinetics]

Abdel-Hamid, M.E.; Suleiman, M.S.; El-Sayed, Y.M.; Najib, N.M.; Hasan, M.M. A rapid high-performance liquid chromatography assay of glibenclamide in serum. *J.Clin.Pharm.Ther.*, **1989**, *14*, 181–188

Gupta, R.N. Determination of glyburide in human plasma by liquid chromatography with fluorescence detection. *J.Liq.Chromatogr.*, **1989**, *12*, 1741–1758 [plasma; fluorescence detection; SPE; chlorowarfarin (IS)]

Das Gupta, V. Quantitation of glipizide and glyburide in tablets using high performance liquid chromatography. *J.Liq.Chromatogr.*, **1986**, *9*, 3607–3615 [simultaneous glipizide; tablets; hydrocortisone (IS)]

Potter, H.; Hulm, M. [Determination of glibenclamide in blood using high performance liquid chromatography]. *J.Chromatogr.*, **1983**, *273*, 217–222

Adams, W.J.; Skinner, G.S.; Bombardt, P.A.; Courtney, M.; Brewer, J.E. Determination of glyburide in human serum by liquid chromatography with fluorescence detection. *Anal.Chem.*, **1982**, *54*, 1287–1291

Uihlein, M.; Sistovaris, N. High-performance liquid column and thin-layer chromatographic determination of human serum glibenclamide at therapeutic levels. *J.Chromatogr.*, **1982**, *227*, 93–101

# Goserelin

**Molecular formula:** $C_{59}H_{84}N_{18}O_{14}$
**Molecular weight:** 1269.4
**CAS Registry No.:** 65807-02-5

## SAMPLE
**Matrix:** solutions

## HPLC VARIABLES
**Column:** 200 × 3 Spherisorb S5ODS-2
**Mobile phase:** Gradient. A was 0.05% phosphoric acid containing 0.5% $(NH_4)_2SO_4$. B was MeCN. A:B from 82:18 to 64:36 over 25 min, maintain at 64:36 for 2.5 min, return to initial conditions over 1 min, re-equilibrate for 6.5 min; or isocratic 76:24
**Flow rate:** 0.5
**Detector:** UV 210

## CHROMATOGRAM
**Retention time:** 14.5 (gradient), 11 (isocratic)

## OTHER SUBSTANCES
**Simultaneous:** buserelin, deslorelin, gonadorelin, leuprolide, nafarelin

## KEY WORDS
comparison with capillary electrophoresis

## REFERENCE
Corran, P.H.; Sutcliffe, N. Identification of gonadorelin (LHRH) derivatives: comparison of reversed-phase high-performance liquid chromatography and micellar electrokinetic chromatography. *J.Chromatogr.*, **1993**, *636*, 87–94

# Guaifenesin

**Molecular formula:** $C_{10}H_{14}O_4$
**Molecular weight:** 198.2
**CAS Registry No.:** 93-14-1

## SAMPLE
**Matrix:** blood
**Sample preparation:** 1 mL Plasma + 100 µL 2.58 µg/mL laudanosine in MeCN + 500 µL saturated sodium carbonate solution, vortex for 10 s, add 5 mL chloroform, vortex for 10 s, mix on a rocking mixer for 40 min, centrifuge at 2000 g for 25 min. Remove the organic layer and evaporate it to dryness under a stream of nitrogen, reconstitute the residue in 400 µL mobile phase, inject a 300 µL aliquot. (Hydrolyze conjugates by heating 1 mL plasma with 1 mL 3000 U/mL β-glucuronidase (Helix pomatia type H-1 (Sigma)) in 100 mM pH 5.0 sodium citrate at 37° for 2 h, proceed as above.)

## HPLC VARIABLES
**Column:** 150 × 4.6 Spherisorb 5-CN
**Mobile phase:** MeCN:water:triethylamine 10:89:1 adjusted to pH 6 with orthophosphoric acid
**Flow rate:** 1
**Injection volume:** 300
**Detector:** F ex 280 em 315

## CHROMATOGRAM
**Retention time:** 1.275
**Internal standard:** laudanosine (8.603)
**Limit of detection:** 30 ng/mL

## OTHER SUBSTANCES
**Extracted:** levorphanol (dextrorphan)

## KEY WORDS
plasma

## REFERENCE
Stavchansky, S.; Demirbas, S.; Reyderman, L.; Chai, C.-K. Simultaneous determination of dextrorphan and guaifenesin in human plasma by liquid chromatography with fluorescence detection. *J.Pharm.Biomed.Anal.*, **1995**, *13*, 919–925

## SAMPLE
**Matrix:** blood
**Sample preparation:** 200 µL Plasma + 100 µL 200 µg/mL mephenesin in water, mix, add 5 mL ethyl acetate, shake for 15 min. Remove 3 mL of the organic layer and evaporate it to dryness under a stream of nitrogen. Reconstitute in 500 µL of water, inject a 50 µL aliquot.

## HPLC VARIABLES
**Column:** 100 × 4.6 3 µm Microsorb C18
**Mobile phase:** MeOH:100 mM $KH_2PO_4$:water 35:10:55
**Column temperature:** 40
**Flow rate:** 0.8
**Injection volume:** 50
**Detector:** UV 272

## CHROMATOGRAM
**Retention time:** 4.8
**Internal standard:** mephenesin (10.5)

## OTHER SUBSTANCES
**Simultaneous:** methocarbamol
**Noninterfering:** acetaminophen, ibuprofen

## KEY WORDS
plasma

## REFERENCE
Naidong, W.; Lee, J.W.; Hulse, J.D. Development and validation of a high-performance liquid chromatographic method for the determination of methocarbamol in human plasma. *J.Chromatogr.B*, **1994**, *654*, 287–292

## SAMPLE
**Matrix:** blood
**Sample preparation:** 500 μL Plasma + 100 μL 2.5 mg/mL O-desmethylnaproxen in MeOH + 400 μL acetone, homogenize for 10 min, centrifuge at 1000 g for 15 min. Remove supernatant and evaporate it to dryness under a stream of air at 35°. Take up residue in 500 μL mobile phase, inject 10 μL aliquot.

## HPLC VARIABLES
**Column:** 150 × 4.6 5 μm LiChrosorb RP-18
**Mobile phase:** MeOH:10 mM pH 6.5 citrate buffer 10:90
**Column temperature:** 35
**Flow rate:** 2
**Injection volume:** 10
**Detector:** F ex 230 em 306

## CHROMATOGRAM
**Retention time:** 37
**Internal standard:** O-desmethylnaproxen (26)

## OTHER SUBSTANCES
**Simultaneous:** metabolites

## KEY WORDS
plasma; horse

## REFERENCE
Ketelaars, H.C.; Peters, J.G.; Anzion, R.B.; Van Ginneken, C.A. Isolation, partial identification and quantitative determination of four guaiphenesin glucuronides in plasma and urine of the horse by high-performance liquid chromatography. *J.Chromatogr.*, **1984**, *288*, 423–429

## SAMPLE
**Matrix:** blood
**Sample preparation:** 500 μL Plasma + 50 μL 1 mg/mL mephenesin in water + 60 μL 1 M HCl, homogenize, add 5 mL diethyl ether, shake for 30 min, centrifuge at 1000 g for 15 min. Remove ether layer and evaporate it to dryness at 30° under a stream of air. Take up residue in 500 μL mobile phase and inject an aliquot.

## HPLC VARIABLES
**Column:** 150 × 4.6 5 μm LiChrosorb RP-8

**Mobile phase:** MeOH:10 mM pH 6.5 citrate buffer 40:60
**Column temperature:** 30
**Flow rate:** 1
**Detector:** UV 275

---

## CHROMATOGRAM
**Retention time:** 2.8
**Internal standard:** mephenesin (8.8)

---

## OTHER SUBSTANCES
**Simultaneous:** β-(2-methoxyphenoxy)lactic acid

---

## KEY WORDS
plasma

---

## REFERENCE
Ketelaars, H.C.J.; Peters, J.G.P. Determination of guaiphenesin and its metabolite, beta-(2-methoxy-phenoxy)-lactic acid, in plasma by high-performance liquid chromatography. *J.Chromatogr.*, **1981**, *224*, 144–148

---

## SAMPLE
**Matrix:** formulations
**Sample preparation:** Finely powder half a tablet, add 9 mL mobile phase, sonicate for 20 min, make up to 10 mL with mobile phase, filter (Whatman type 40 and 0.2 μm Millipore), inject an aliquot of the filtrate.

---

## HPLC VARIABLES
**Column:** 250 × 4 5 μm LiChrospher 100 CN
**Mobile phase:** MeCN:THF:buffer 7:6:87 (Buffer was 0.8% acetic acid containing 5 mM sodium hexanesulfonate, 10 mM di-n-butylamine, and 0.12% phosphoric acid, pH 3.3.)
**Flow rate:** 1
**Injection volume:** 20
**Detector:** UV 284

---

## CHROMATOGRAM
**Retention time:** 5.2
**Limit of detection:** 2.8 μg/mL

---

## OTHER SUBSTANCES
**Simultaneous:** acetaminophen (UV 310), caffeine (UV 298), chlorpheniramine (UV 265), phenylpropanolamine (UV 260)

---

## KEY WORDS
tablets

---

## REFERENCE
Indrayanto, G.; Sunarto, A.; Adriani, Y. Simultaneous assay of phenylpropanolamine hydrochloride, caffeine, paracetamol, glycerylguaiacolate and chlorpheniramine in Silabat™ tablet using HPLC with diode array detection. *J.Pharm.Biomed.Anal.*, **1995**, *13*, 1555–1559

---

## SAMPLE
**Matrix:** formulations
**Sample preparation:** Dilute 10 mL to 1 L with water.

## HPLC VARIABLES
**Column:** 300 × 3.9 10 μm μBondapak C18
**Mobile phase:** MeCN:water:diethylamine:glacial acetic acid 250:739:1:10, apparent pH 4.1
**Column temperature:** 35
**Flow rate:** 1.3
**Injection volume:** 25
**Detector:** UV 273

## CHROMATOGRAM
**Retention time:** 3.35

## OTHER SUBSTANCES
**Simultaneous:** benzoic acid, dextromethorphan

## KEY WORDS
liquid formulation; stability-indicating

## REFERENCE
Wilson, T.D.; Jump, W.G.; Neumann, W.C.; San Martin, T. Validation of improved methods for high performance liquid chromatographic determination of phenylpropanolamine, dextromethorphan, guaifenesin and sodium benzoate in a cough-cold formulation. *J.Chromatogr.*, **1993**, *641*, 241–248

## SAMPLE
**Matrix:** formulations
**Sample preparation:** Dilute 1 mL syrup to 50 mL with mobile phase, filter (0.45 μm) inject 20 μL aliquot.

## HPLC VARIABLES
**Column:** 250 × 4.6 Zorbax CN
**Mobile phase:** MeCN:water:formic acid:methanesulfonic acid 500:500:1:1, pH adjusted to 3.5 with 10% NaOH
**Flow rate:** 1
**Injection volume:** 20
**Detector:** UV 290

## CHROMATOGRAM
**Retention time:** 3.7

## OTHER SUBSTANCES
**Simultaneous:** benzoic acid, dextromethorphan, saccharin

## KEY WORDS
syrup

## REFERENCE
Chen, T.M.; Pacifico, J.R.; Daly, R.E. High-pressure liquid chromatographic assay of dextromethorphan hydrobromide, guaifenesin, and sodium benzoate in an expectorant syrup. *J.Chromatogr.Sci.*, **1988**, *26*, 636–639

## SAMPLE
**Matrix:** formulations
**Sample preparation:** Leach 200 or 300 mg ground capsule or tablet with water or mobile phase and dilute to 50 mL, sonicate for 5 min, centrifuge at 2500 rpm for 5 min, inject an aliquot. Dilute 4-25 mL of liquid formulations to 250 mL with water, inject an aliquot.

## HPLC VARIABLES
**Column:** Partisil-10 C8
**Mobile phase:** MeOH:MeCN:water:PIC-B5 50:170:755:25 (PIC-B5 (Waters) is 200 mM sodium pentanesulfonate in glacial acetic acid.)
**Flow rate:** 2
**Injection volume:** 20
**Detector:** UV 254

## CHROMATOGRAM
**Retention time:** 7.5

## OTHER SUBSTANCES
**Simultaneous:** impurities, degradation products, benzoic acid, phenylephrine, phenylpropanolamine

## KEY WORDS
tablets; capsules; liquid formulations; stability-indicating

## REFERENCE
Schieffer, G.W.; Smith, W.O.; Lubey, G.S.; Newby, D.G. Determination of the structure of a synthetic impurity in guaifenesin: modification of a high-performance liquid chromatographic method for phenylephrine hydrochloride, phenylpropanolamine hydrochloride, guaifenesin, and sodium benzoate in dosage forms. *J.Pharm.Sci.*, **1984**, *73*, 1856–1858

## SAMPLE
**Matrix:** formulations
**Sample preparation:** Dilute with water, filter (0.45 μm), inject a 20 μL aliquot.

## HPLC VARIABLES
**Column:** 300 × 4 μBondapak C18
**Mobile phase:** MeOH:water:glacial acetic acid 45:55:2 containing 5 mM octanesulfonic acid
**Flow rate:** 2.5
**Injection volume:** 20
**Detector:** UV 280

## CHROMATOGRAM
**Retention time:** 2.2

## OTHER SUBSTANCES
**Simultaneous:** acetaminophen, butyl paraben, ethyl paraben, methyl paraben, pholcodine, propyl paraben, pseudoephedrine

## KEY WORDS
cough mixture

## REFERENCE
Carnevale, L. Simultaneous determination of acetaminophen, guaifenesin, pseudoephedrine, pholcodine, and paraben preservatives in cough mixture by high-performance liquid chromatography. *J.Pharm.Sci.*, **1983**, *72*, 196–198

## SAMPLE
**Matrix:** formulations
**Sample preparation:** Capsules and Tablets. Leach 1 g of ground capsule or tablet with 250 mL 0.4 mg/mL 2,5-dihydroxybenzoic acid in water, sonicate for 10 min, centrifuge at

2500 rpm for 5 min, inject an aliquot. Liquid formulations. Dilute 4-25 mL of the formulation to 250 mL with 0.4 mg/mL 2,5-dihydroxybenzoic acid in water, inject an aliquot.

## HPLC VARIABLES
**Column:** 250 × 4.6 Partisil 10 C8
**Mobile phase:** MeOH:water:PIC-B5 300:675:25 (PIC-B5 (Waters) is 200 mM sodium pentanesulfonate in glacial acetic acid.)
**Flow rate:** 2
**Injection volume:** 20
**Detector:** UV 254

## CHROMATOGRAM
**Retention time:** 8
**Internal standard:** 2,5-dihydroxybenzoic acid (4.5)

## OTHER SUBSTANCES
**Simultaneous:** impurities, degradation products, phenylephrine, phenylpropanolamine

## KEY WORDS
capsules; tablets; liquid formulations; stability-indicating

## REFERENCE
Schieffer, G.W.; Hughes, D.E. Simultaneous stability-indicating determination of phenylephrine hydrochloride, phenylpropanolamine hydrochloride, and guaifenesin in dosage forms by reversed-phase paired-ion high-performance liquid chromatography. *J.Pharm.Sci.*, **1983**, *72*, 55–59

---

## SAMPLE
**Matrix:** formulations
**Sample preparation:** 3 mL Sample + 5 mL 200 mg/mL o-dinitrobenzene in 1:1 MeOH: water, dilute to 50 mL with 1:1 MeOH:water, inject a 15 μL aliquot.

## HPLC VARIABLES
**Column:** 300 × 4 μBondapak C18
**Mobile phase:** MeOH:water:ammonium formate buffer 45:54:1 (Prepare ammonium formate buffer by diluting 34 mL 28-30% ammonia with 30 mL water, add 30 mL 98% formic acid (Caution! Exothermic!), cool, dilute this mixture (pH 3.9) to 100 mL with water.)
**Flow rate:** 2
**Injection volume:** 15
**Detector:** UV 280

## CHROMATOGRAM
**Retention time:** 6
**Internal standard:** o-dinitrobenzene (11)

## OTHER SUBSTANCES
**Simultaneous:** acetaminophen, p-aminophenol, dextromethorphan

## KEY WORDS
cough syrup

## REFERENCE
McSharry, W.O.; Savage, I.V.E. Simultaneous high-pressure liquid chromatographic determination of acetaminophen, guaifenesin, and dextromethorphan hydrobromide in cough syrup. *J.Pharm.Sci.*, **1980**, *69*, 212–214

## SAMPLE
**Matrix:** solutions

## HPLC VARIABLES
**Column:** 250 × 4.6 5 μm Supelcosil LC-DP (A) or 250 × 4 5 μm LiChrospher 100 RP-8 (B)
**Mobile phase:** MeCN:0.025% phosphoric acid:buffer 25:10:5 (A) or 60:25:15 (B) (Buffer
  was 9 mL concentrated phosphoric acid and 10 mL triethylamine in 900 mL water, adjust
  pH to 3.4 with dilute phosphoric acid, make up to 1 L.)
**Flow rate:** 0.6
**Injection volume:** 25
**Detector:** UV 229

## CHROMATOGRAM
**Retention time:** 4.94 (A), 3.93 (B)

## OTHER SUBSTANCES
**Also analyzed:** acebutolol, acepromazine, acetaminophen, acetazolamide, acetophenazine,
  albuterol, alprazolam, amitriptyline, amobarbital, amoxapine, antipyrine, atenolol, atropine, azatadine, baclofen, benzocaine, bromocriptine, brompheniramine, brotizolam,
  bupivacaine, buspirone, butabarbital, butalbital, caffeine, carbamazepine, cetirizine,
  chlorcyclizine, chlordiazepoxide, chlormezanone, chloroquine, chlorpheniramine,
  chlorpromazine, chlorpropamide, chlorprothixene, chlorthalidone, chlorzoxazone, cimetidine, cisapride, clomipramine, clonazepam, clonidine, clozapine, cocaine, codeine, colchicine, cyclizine, cyclobenzaprine, dantrolene, desipramine, diazepam, diclofenac, diflunisal,
  diltiazem, diphenhydramine, diphenidol, diphenoxylate, dipyridamole, disopyramide, dobutamine, doxapram, doxepin, droperidol, encainide, ethidium bromide, ethopropazine,
  fenoprofen, fentanyl, flavoxate, fluoxetine, fluphenazine, flurazepam, flurbiprofen, fluvoxamine, furosemide, glutethimide, glyburide, haloperidol, homatropine, hydralazine, hydrochlorothiazide, hydrocodone, hydromorphone, hydroxychloroquine, hydroxyzine, ibuprofen, imipramine, indomethacin, ketoconazole, ketoprofen, ketorolac, labetalol,
  levorphanol, lidocaine, loratadine, lorazepam, lovastatin, loxapine, mazindol, mefenamic
  acid, meperidine, mephenytoin, mepivacaine, mesoridazine, metaproterenol, methadone,
  methdilazine, methocarbamol, methotrexate, methotrimeprazine, methoxamine, methyldopa, methylphenidate, metoclopramide, metolazone, metoprolol, metronidazole, midazolam, moclobemide, morphine, nadolol, nalbuphine, naloxone, naphazoline, naproxen,
  nifedipine, nizatidine, norepinephrine, nortriptyline, oxazepam, oxycodone, oxymetazoline, paroxetine, pemoline, pentazocine, pentobarbital, pentoxifylline, perphenazine,
  pheniramine, phenobarbital, phenol, phenolphthalein, phentolamine, phenylbutazone,
  phenyltoloxamine, phenytoin, pimozide, pindolol, piroxicam, pramoxine, prazepam, prazosin, probenecid, procainamide, procaine, prochlorperazine, procyclidine, promazine, promethazine, propafenone, propantheline, propiomazine, propofol, propranolol, protriptyline, quazepam, quinidine, quinine, racemethorphan, ranitidine, remoxipride, risperidone,
  salicylic acid, scopolamine, secobarbital, sertraline, sotalol, spironolactone, sulfinpyrazone, sulindac, temazepam, terbutaline, terfenadine, tetracaine, theophylline, thiethylperazine, thiopental, thioridazine, thiothixene, timolol, tocainide, tolbutamide, tolmetin,
  trazodone, triamterene, triazolam, trifluoperazine, triflupromazine, trimeprazine, trimethoprim, trimipramine, verapamil, warfarin, xylometazoline, yohimbine, zopiclone

## KEY WORDS
some details of plasma extraction

## REFERENCE
Koves, E.M. Use of high-performance liquid chromatography-diode array detection in forensic toxicology.
  *J.Chromatogr.A*, **1995**, *692*, 103–119

## SAMPLE
**Matrix:** solutions

**Sample preparation:** 450 μL Buffer solution + 50 μL 4 mg/mL acetanilide, cool in ice, inject a 10 μL aliquot.

## HPLC VARIABLES
**Column:** Perkin-Elmer 3 × 3 CR C-18
**Mobile phase:** MeCN:water 10:90 containing 1% acetic acid
**Flow rate:** 2.5
**Injection volume:** 10
**Detector:** UV 274

## CHROMATOGRAM
**Retention time:** 3
**Internal standard:** acetanilide (2.1)

## OTHER SUBSTANCES
**Simultaneous:** methocarbamol

## KEY WORDS
buffers

## REFERENCE
Pouli, N.; Antoniadou-Vyzas, A.; Foscolos, G.B. Methocarbamol degradation in aqueous solution. *J.Pharm.Sci.*, **1994**, *83*, 499–501

## SAMPLE
**Matrix:** solutions
**Sample preparation:** Dissolve compounds in MeCN:water 80:20, inject a 1 μL aliquot.

## HPLC VARIABLES
**Column:** 150 × 1 3 μm Hitachi-Gel 3057 ODS silica (Hitachi)
**Mobile phase:** MeCN:water 25:75
**Flow rate:** 0.03
**Injection volume:** 1
**Detector:** UV 254

## CHROMATOGRAM
**Retention time:** 8

## OTHER SUBSTANCES
**Simultaneous:** acetaminophen, bucetin (3-hydroxy-p-butyrophenetidine), caffeine, dipyrone (sulpyrin), methyl p-hydroxybenzoate, phenacetin

## KEY WORDS
semi-micro

## REFERENCE
Matsushima, Y.; Nagata, Y.; Niyomura, M.; Takakusagi, K.; Takai, N. Analysis of antipyretics by semimicro liquid chromatography. *J.Chromatogr.*, **1985**, *332*, 269–273

◆━━━━━━━━━━━━━━━━━━━◆━━━━━━━━━━━━━━━━━━━◆

## ANNOTATED BIBLIOGRAPHY
Alvi, S.U.; Castro, F. A simultaneous assay of theophylline, ephedrine hydrochloride, and phenobarbital in suspensions and tablets formulations by high performance liquid chromatography. *J.Liq. Chromatogr.*, **1986**, *9*, 2269–2279 [simultaneous ephedrine, phenobarbital, theophylline; suspensions; tablets; guaifenesin is IS]

Muhammad, N.; Bodnar, J.A. Quantitative determination of guaifenesin, phenylpropanolamine hydrochloride, sodium benzoate & codeine phosphate in cough syrups by high-pressure liquid chromatography. *J.Liq.Chromatogr.*, **1980**, *3*, 113–122 [simultaneous benzoic acid, codeine, phenylpropanolamine; syrup]

# Hydrochlorothiazide

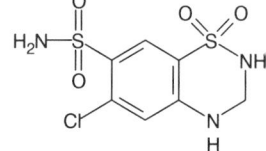

**Molecular formula:** $C_7H_8ClN_3O_4S_2$
**Molecular weight:** 297.7
**CAS Registry No.:** 58-93-5

## SAMPLE
**Matrix:** blood
**Sample preparation:** 1 mL Plasma + 100 μL 20 μg/mL hydroflumethiazide in MeOH + 1 mL buffer + 200 μL water + 6 mL ethyl acetate, shake for 5 min, centrifuge at 900 g for 5 min. Remove 5 mL organic layer and evaporate at 37°under a stream of nitrogen. Reconstitute with 100 μL MeOH, sonicate twice at 37°for 1 min, cool at 2-8°for 2 h to obtain a clear solution, inject a 20 μL aliquot. (Buffer was 0.38 g ammonium acetate in 500 mL water, acidified to pH 5.0 with glacial acetic acid.)

## HPLC VARIABLES
**Guard column:** 40 × 4 35-50 μm C18 Corasil
**Column:** 125 × 4 5 μm Nucleosil 100-5 C18
**Mobile phase:** Gradient. A was MeCN:acetic acid:water 25:1:975. B was MeCN:acetic acid:water 500:1:500. A:B from 100:0 to 36:64 over 16 min, re-equilibrate at 100:0 for 24 min before next injection
**Flow rate:** 1
**Injection volume:** 20
**Detector:** UV 280

## CHROMATOGRAM
**Retention time:** 9.0
**Internal standard:** hydroflumethiazide (12.0)
**Limit of detection:** 10 ng/mL

## OTHER SUBSTANCES
**Noninterfering:** acebutolol, acenocoumarol, acetaminophen, aspirin, allopurinol, ambroxol, amoxicillin, atenolol, bendroflumethiazide, benzbromarone, bezafibrate, biperiden, bisacodyl, bromazepam, butizide, caffeine, captopril, cimetidine, ciprofloxacin, clobutinol, clonidine, cotinine, diazepam, diclofenac, digitoxin, digoxin, dihydrocodeine, dihydroergotamine, diltiazem, doxepin, doxycycline, enalapril, erythromycin, fenoterol, furosemide, glibenclamide, heparin, hypoxanthine, ibuprofen, indomethacin, isosorbide mononitrate, lisinopril, lovastatin, maprotiline, methyldigoxin, methyldopa, metoclopramide, metoprolol, metronidazole, midazolam, naloxone, nifedipine, nicotine, norfloxacin, ofloxacin, oxazepam, oxipurinol, penicillin V, pentoxyfylline, phenacetin, phenazone, propyphenazone, phenprocoumon, ranitidine, salicylic acid, sotalol, sulfamethoxazole, trimethoprim, terbutaline, theophylline, tilidine, timolol, triamterene, uric acid, verapamil, vitamin C, warfarin, xanthine, purine and pyrimidine bases, nucleosides, nucleotides

## KEY WORDS
plasma; amiloride interferes with IS

## REFERENCE
de Vries, J.X.; Voss, A. Simple determination of hydrochlorothiazide in human plasma and urine by high performance liquid chromatography. *Biomed.Chromatogr.*, **1993**, 7, 12–14

## SAMPLE
**Matrix:** blood
**Sample preparation:** 500 μL Plasma + 25 μL 20 μg/mL procainamide.HCl in MeOH, vortex, add 5 mL MTBE, place on a reciprocating shaker at low speed for 15 min, centrifuge at 1250 g for 10 min. Remove organic layer and evaporate under a stream of nitrogen. Reconstitute in 200 μL mobile phase, inject 100 μL aliquot.

## HPLC VARIABLES
**Guard column:** 20 × 4.6 Supelco Pelliguard C18
**Column:** 150 × 4.6 3 μm Hypersil ODS
**Mobile phase:** MeCN:7 mM sodium heptanesulfonate 18:82, containing 1% glacial acetic acid and 0.035% triethylamine
**Flow rate:** 0.8
**Injection volume:** 100
**Detector:** UV 272

## CHROMATOGRAM
**Retention time:** 5.1
**Internal standard:** procainamide.HCl (10.1)
**Limit of quantitation:** 1 ng/mL

## OTHER SUBSTANCES
**Noninterfering:** acetaminophen, aspirin, ibuprofen

## KEY WORDS
plasma

## REFERENCE
Azumaya, C.T. Sensitive liquid chromatographic method for the determination of hydrochlorothiazide in human plasma. *J.Chromatogr.*, **1990**, *532*, 168–174

## SAMPLE
**Matrix:** blood
**Sample preparation:** 1 mL Plasma + 25 μL 10 mg/mL hydroflumethiazide in water + 1 mL 1 M pH 10 sodium carbonate-bicarbonate buffer + 5 mL ethyl acetate, vortex 1 min, centrifuge at 1250 g for 5 min. Remove the ethyl acetate layer and evaporate at 45° under nitrogen. Dissolve in 100 μL mobile phase, inject 50 μL aliquot.

## HPLC VARIABLES
**Column:** 125 × 4.6 5 μm Spherisorb ODSII
**Mobile phase:** MeCN:MeOH:buffer 10:9:100 (Buffer was 15.54 g tetraethylammonium hydroxide and 2.9 g 89% orthophosphoric acid in 500 mL water, pH was 2.8.)
**Flow rate:** 1.2
**Injection volume:** 50
**Detector:** UV 271

## CHROMATOGRAM
**Retention time:** 4.80
**Internal standard:** hydroflumethiazide (7.94)
**Limit of detection:** 10 ng/mL

## OTHER SUBSTANCES
**Simultaneous:** amiloride (detection by F)

## KEY WORDS
plasma

## REFERENCE
Van der Meer, M.J.; Brown, L.W. Simultaneous determination of amiloride and hydrochlorothiazide in plasma by reversed-phase high-performance liquid chromatography. *J.Chromatogr.*, **1987**, *423*, 351–357

## SAMPLE
**Matrix:** blood, urine
**Sample preparation:** Filter (0.45 μm) urine or plasma. Mix plasma filtrate with an equal volume of 50 mM pH 8.0 Tris-sulfuric acid buffer containing 0.1 mM zinc acetate and 40 mM sodium dodecyl sulfate. Inject a 50 μL aliquot of the urine filtrate or the diluted plasma onto column A with mobile phase A, elute to waste with mobile phase A, after 5 min backflush the contents of column A onto column B with mobile phase B, elute with mobile phase B, monitor the effluent from column B. Re-equilibrate column A with mobile phase A for 5 min before next injection.

## HPLC VARIABLES
**Column:** A 10 × 4.6 carbonic anhydrase (Prepare by adding 3 g aminopropyl silica from a Sep-Pak NH2 SPE cartridge to 30 mL 100 mg/mL N, N′-disuccinimidyl carbonate in MeCN in portions over 30 min with gentle mixing, mix for 3 h, filter (G-5 glass), wash the solid 5 times with 50 mL portions of MeCN. Add 100 mg activated gel to 2 mL 200 mM pH 8.0 phosphate buffer containing 1 M NaCl, degas by sonicating under aspirator vacuum, add 2 mL 2.5 mg/mL carbonic anhydrase in water, shake at room temperature for 4 h, centrifuge, discard the supernatant, suspend the gel in 50 mM pH 8.0 Tris-sulfuric acid buffer, slurry pack into column. Store in 50 mM pH 8.0 Tris-sulfuric acid buffer containing 0.1 mM zinc acetate when not in use.); B 150 × 4.6 Cosmosil 5C18-AR (Nakarai Tesque)
**Mobile phase:** A 50 mM pH 8.0 Tris-sulfuric acid buffer containing 0.1 mM zinc acetate; B MeCN:100 mM pH 5.2 acetate buffer 10:90 containing 500 mM NaCl
**Flow rate:** 1
**Injection volume:** 50
**Detector:** UV 270

## CHROMATOGRAM
**Retention time:** 7.5
**Limit of detection:** 100 nM
**Limit of quantitation:** 1 μM

## OTHER SUBSTANCES
**Also analyzed:** acetazolamide, chlorothiazide, chlorthalidone, furosemide
**Noninterfering:** acetaminophen, bumetanide, caffeine, phenylbutazone, salicylic acid, sulfamerazine, sulfamethiazole, sulfamethoxazole, sulfamonomethoxine, sulfisomidine, sulfisoxazole, theophylline, tolbutamide, warfarin

## KEY WORDS
plasma; column-switching

## REFERENCE
Ohta, T.; Takamiya, I.; Takitani, S. Carbonic anhydrase-immobilized precolumn for selective on-line sample pretreatment in high-performance liquid chromatographic determination of certain sulphonamide drugs. *Biomed.Chromatogr.*, **1994**, *8*, 184–188

## SAMPLE
**Matrix:** feed
**Sample preparation:** 2 g Feed + 20 mL MeCN, rotate at 20 rpm for 1 h, centrifuge at 1300 rpm for 15 min, inject an aliquot.

## HPLC VARIABLES
**Guard column:** 100 × 6.3 30-38 μm Co:Pell ODS (Whatman)
**Column:** 250 × 4.6 10 μm Lichrosorb RP-2
**Mobile phase:** MeOH:water: 5:95 (flush column with MeCN:water 50:50 after use)
**Flow rate:** 2

**Injection volume:** 50
**Detector:** UV 265

## CHROMATOGRAM
**Retention time:** 7

## REFERENCE
Spurlock, C.H.; Schneider, H.G. Liquid chromatographic and ultraviolet spectrophotometric determination of bevantolol and hydrochlorothiazide in feeds. *J.Assoc.Off.Anal.Chem.*, **1984**, *67*, 321–324

## SAMPLE
**Matrix:** formulations
**Sample preparation:** Dilute 1000-fold, mix a 200 μL aliquot with 200 μL 100 μg/mL hydrochlorothiazide, inject a 20 μL aliquot.

## HPLC VARIABLES
**Column:** Microsorb MV C-18
**Mobile phase:** MeCN:water:acetic acid 6:93:1, adjusted to pH 4.0 with 6 M NaOH
**Flow rate:** 2
**Injection volume:** 20
**Detector:** UV 254

## CHROMATOGRAM
**Internal standard:** hydrochlorothiazide

## OTHER SUBSTANCES
**Simultaneous:** ceftazidime

## KEY WORDS
injections; saline; stability-indicating; hydrochlorothiazide is IS

## REFERENCE
Bednar, D.A.; Klutman, N.E.; Henry, D.W.; Fox, J.L.; Strayer, A.H. Stability of ceftazidime (with arginine) in an elastomeric infusion device. *Am.J.Health-Syst.Pharm.*, **1995**, *52*, 1912–1914

## SAMPLE
**Matrix:** formulations
**Sample preparation:** Finely powder tablets, weigh out amount equivalent to 20 mg enalapril maleate, suspend in 100 mL mobile phase, filter, inject a 5 μL aliquot.

## HPLC VARIABLES
**Column:** 250 × 4.6 12 μm Hypersil C18
**Mobile phase:** MeCN:water 20:80 adjusted to pH 3.8 with acetic acid
**Flow rate:** 1
**Injection volume:** 5
**Detector:** UV 215 for 3.5 min, then UV 275

## CHROMATOGRAM
**Retention time:** 6.7
**Internal standard:** caffeine (4.8)

## OTHER SUBSTANCES
**Simultaneous:** enalapril

## KEY WORDS
tablets

## REFERENCE
el Walily, A.F.M.; Belal, S.F.; Heaba, E.A.; El Kersh, A. Simultaneous determination of enalapril maleate and hydrochlorothiazide by first-derivative ultraviolet spectrophotometry and high-performance liquid chromatography. *J.Pharm.Biomed.Anal.*, **1995**, *13*, 851–856

## SAMPLE
**Matrix:** formulations
**Sample preparation:** Tablets. Grind tablets, weigh out a portion, dissolve in 50 mL mobile phase, sonicate, filter (No. 4 sintered glass plate), dilute, inject an aliquot. Capsules. Dissolve 10 capsules (without opening) in 100 mL mobile phase, sonicate, inject an aliquot. Injections, ampules, sprays. Dilute, inject an aliquot.

## HPLC VARIABLES
**Column:** 120 × 4.6 Spherisorb C18 ODS-2
**Mobile phase:** Isopropanol:buffer 5:95 (Buffer was 100 mM sodium dodecyl sulfate containing 25 mM $Na_2HPO_4$, pH adjusted to 3.0 with HCl.)
**Flow rate:** 1
**Injection volume:** 20
**Detector:** UV 280

## CHROMATOGRAM
**Retention time:** 1.6
**Limit of detection:** 2 ng/mL

## OTHER SUBSTANCES
**Simultaneous:** carbidopa, dopamine, epinephrine, isoproterenol, levodopa, methyldopa, norepinephrine, phenylephrine

## KEY WORDS
tablets; capsules; injections; ampules; sprays

## REFERENCE
Villanueva Camañas, R.M.; Sanchis Mallols, J.M.; Torres Lapasió, J.R.; Ramis-Ramos, G. Analysis of pharmaceutical preparations containing catecholamines by micellar liquid chromatography with spectrophotometric detection. *Analyst*, **1995**, *120*, 1767–1772

## SAMPLE
**Matrix:** formulations
**Sample preparation:** Dilute 5 fold with mobile phase. Mix the diluted formulation with an equal volume of 50 μg/mL hydrochlorothiazide, inject a 10 μL aliquot.

## HPLC VARIABLES
**Column:** 300 × 3.9 10 μm μBondapak C18
**Mobile phase:** MeOH:0.1% phosphoric acid 55:45
**Flow rate:** 1
**Injection volume:** 10
**Detector:** UV 260

## CHROMATOGRAM
**Retention time:** 5.6
**Internal standard:** hydrochlorothiazide

## OTHER SUBSTANCES
**Simultaneous:** captopril

## KEY WORDS
syrup; hydrochlorothiazide is IS

## REFERENCE
Nahata, M.C.; Morosco, R.S.; Hipple, T.F. Stability of captopril in three liquid dosage forms. *Am.J.Hosp.Pharm.*, **1994**, *51*, 95−96

## SAMPLE
**Matrix:** formulations
**Sample preparation:** Crush tablets, weigh out amount equivalent to one tablet, add 40 mL MeOH, warm on a steam bath for 5 min, cool to room temperature, make up to 100 mL with MeOH, filter through paper, inject 100 µL aliquot.

## HPLC VARIABLES
**Column:** 300 × 3.9 µPorasil
**Mobile phase:** MeOH
**Flow rate:** 1.5
**Injection volume:** 100
**Detector:** F ex 280 em 360; UV 360

## CHROMATOGRAM
**Retention time:** 2

## OTHER SUBSTANCES
**Simultaneous:** reserpine (by F only)

## KEY WORDS
tablets; normal phase

## REFERENCE
Cieri, U.R. Determination of reserpine and hydrochlorothiazide in commercial tablets by liquid chromatography with fluorescence and UV absorption detectors in series. *J.Assoc.Off.Anal.Chem.*, **1988**, *71*, 515−518

## SAMPLE
**Matrix:** formulations
**Sample preparation:** Grind tablets, add 3-20 mL MeCN:water 15:85, sonicate for 10 min, filter, make up to 100 mL with MeCN:water 15:85. Remove a 500 µL aliquot and add it to 300 µL 250 µg/mL procaine hydrochloride in water, make up to 10 mL with mobile phase, inject a 50 µL aliquot.

## HPLC VARIABLES
**Column:** 250 × 4.6 5 µm ASI chromosphere 3869 octadecylsilane (Analytical Sciences, Inc.)
**Mobile phase:** MeCN:50 mM $NaH_2PO_4$ 30:70 containing sodium pentanesulfonate, pH adjusted to 2.5 with concentrated phosphoric acid
**Flow rate:** 1
**Injection volume:** 50
**Detector:** E, Metrohm model E-611, Bioanalytical Systems Kel F cell, glassy carbon electrode +1300 mV, auxiliary platinum electrode, Ag/AgCl reference electrode

## CHROMATOGRAM
**Retention time:** 4.5
**Internal standard:** procaine hydrochloride (5.7)
**Limit of quantitation:** 1.25 µg/mL

## OTHER SUBSTANCES
**Simultaneous:** guanethidine

## KEY WORDS
tablets; not stability-indicating

## REFERENCE
Stewart, J.T.; Clark, S.S. Liquid chromatographic determination of guanethidine salts and hydrochlorothiazide using electrochemical detection and ion-pair techniques. *J.Pharm.Sci.*, **1986**, *75*, 413–415

## SAMPLE
**Matrix:** formulations
**Sample preparation:** Injections. Dilute 1.5 mL of a 20 mg/mL injection to 100 mL with water, remove a 10 mL aliquot and add it to 3 mL 0.2% hydrochlorothiazide, make up to 100 mL with water, inject a 20 μL aliquot. Tablets. Grind tablets to a fine powder, weigh out amount equivalent to about 10 mg hydralazine, mix thoroughly with 2 mL 500 mM HCl, make up to 100 mL with water, shake for 2-3 min, filter, discard first 15 mL. 15 mL Filtrate + 1.5 mL 0.2% hydrochlorothiazide, make up to 50 mL with water, inject a 20 μL aliquot.

## HPLC VARIABLES
**Column:** 300 × 3.9 μBondapak phenyl
**Mobile phase:** MeOH:15 mM $KH_2PO_4$:glacial acetic acid 0.5:99.4:0.1
**Flow rate:** 3
**Injection volume:** 20
**Detector:** UV 256

## CHROMATOGRAM
**Retention time:** 8
**Internal standard:** hydrochlorothiazide

## OTHER SUBSTANCES
**Simultaneous:** hydralazine, phenylpropanolamine

## KEY WORDS
injections; tablets; hydrochlorothiazide is IS

## REFERENCE
Das Gupta, V. Quantitation of hydralazine hydrochloride in pharmaceutical dosage forms using high-performance liquid chromatography. *J.Liq.Chromatogr.*, **1985**, *8*, 2497–2509

## SAMPLE
**Matrix:** solutions

## HPLC VARIABLES
**Column:** 250 × 4.6 10 μm Partisil ODS1
**Mobile phase:** MeOH:50 mM pH 3.0 phosphoric acid 10:90
**Column temperature:** 30
**Flow rate:** 1.5
**Detector:** Radioactivity

## OTHER SUBSTANCES
**Also analyzed:** atenolol, cimetidine, ranitidine

## KEY WORDS
$^{14}$C labeled

## REFERENCE
Collett, A.; Sims, E.; Walker, D.; He, Y.-L.; Ayrton, J.; Rowland, M.; Warhurst, G. Comparison of HT29-18-C$_1$ and Caco-2 cell lines as models for studying intestinal paracellular drug absorption. *Pharm.Res.*, **1996**, *13*, 216–221

## SAMPLE
**Matrix:** solutions

## HPLC VARIABLES
**Column:** 250 × 4.6 5 μm Supelcosil LC-DP (A) or 250 × 4 5 μm LiChrospher 100 RP-8 (B)
**Mobile phase:** MeCN:0.025% phosphoric acid:buffer 25:10:5 (A) or 60:25:15 (B) (Buffer was 9 mL concentrated phosphoric acid and 10 mL triethylamine in 900 mL water, adjust pH to 3.4 with dilute phosphoric acid, make up to 1 L.)
**Flow rate:** 0.6
**Injection volume:** 25
**Detector:** UV 229

## CHROMATOGRAM
**Retention time:** 5.09 (A), 3.98 (B)

## OTHER SUBSTANCES
**Also analyzed:** acebutolol, acepromazine, acetaminophen, acetazolamide, acetophenazine, albuterol, alprazolam, amitriptyline, amobarbital, amoxapine, antipyrine, atenolol, atropine, azatadine, baclofen, benzocaine, bromocriptine, brompheniramine, brotizolam, bupivacaine, buspirone, butabarbital, butalbital, caffeine, carbamazepine, cetirizine, chlorcyclizine, chlordiazepoxide, chlormezanone, chloroquine, chlorpheniramine, chlorpromazine, chlorpropamide, chlorprothixene, chlorthalidone, chlorzoxazone, cimetidine, cisapride, clomipramine, clonazepam, clonidine, clozapine, cocaine, codeine, colchicine, cyclizine, cyclobenzaprine, dantrolene, desipramine, diazepam, diclofenac, diflunisal, diltiazem, diphenhydramine, diphenidol, diphenoxylate, dipyridamole, disopyramide, dobutamine, doxapram, doxepin, droperidol, encainide, ethidium bromide, ethopropazine, fenoprofen, fentanyl, flavoxate, fluoxetine, fluphenazine, flurazepam, flurbiprofen, fluvoxamine, furosemide, glutethimide, glyburide, guaifenesin, haloperidol, homatropine, hydralazine, hydrocodone, hydromorphone, hydroxychloroquine, hydroxyzine, ibuprofen, imipramine, indomethacin, ketoconazole, ketoprofen, ketorolac, labetalol, levorphanol, lidocaine, loratadine, lorazepam, lovastatin, loxapine, mazindol, mefenamic acid, meperidine, mephenytoin, mepivacaine, mesoridazine, metaproterenol, methadone, methdilazine, methocarbamol, methotrexate, methotrimeprazine, methoxamine, methyldopa, methylphenidate, metoclopramide, metolazone, metoprolol, metronidazole, midazolam, moclobemide, morphine, nadolol, nalbuphine, naloxone, naphazoline, naproxen, nifedipine, nizatidine, norepinephrine, nortriptyline, oxazepam, oxycodone, oxymetazoline, paroxetine, pemoline, pentazocine, pentobarbital, pentoxifylline, perphenazine, pheniramine, phenobarbital, phenol, phenolphthalein, phentolamine, phenylbutazone, phenyltoloxamine, phenytoin, pimozide, pindolol, piroxicam, pramoxine, prazepam, prazosin, probenecid, procainamide, procaine, prochlorperazine, procyclidine, promazine, promethazine, propafenone, propantheline, propiomazine, propofol, propranolol, protriptyline, quazepam, quinidine, quinine, racemethorphan, ranitidine, remoxipride, risperidone, salicylic acid, scopolamine, secobarbital, sertraline, sotalol, spironolactone, sulfinpyrazone, sulindac, temazepam, terbutaline, terfenadine, tetracaine, theophylline, thiethylperazine, thiopental, thioridazine, thiothixene, timolol, tocainide, tolbutamide, tolmetin, trazodone, triamterene, triazolam, trifluoperazine, triflupromazine, trimeprazine, trimethoprim, trimipramine, verapamil, warfarin, xylometazoline, yohimbine, zopiclone

## KEY WORDS
some details of plasma extraction

## REFERENCE
Koves, E.M. Use of high-performance liquid chromatography-diode array detection in forensic toxicology. *J.Chromatogr.A*, **1995**, *692*, 103–119

## SAMPLE
**Matrix:** solutions
**Sample preparation:** Prepare a solution in MeOH:water 80:20, inject a 6 μL aliquot.

## HPLC VARIABLES
**Guard column:** 5 × 4 10 μm LiChrosorb RP-8
**Column:** 100 × 4.6 5 μm Spheri RP-18 (Brownlee)
**Mobile phase:** MeOH:water 80:20 containing 2 g/L lithium perchlorate
**Flow rate:** 0.5
**Injection volume:** 6
**Detector:** E, ESA Model 5100A Coulochem, model 5020 guard cell +950 mV, Model 5010 analytical cell + 400 mV, palladium reference electrode, following post-column photolysis. The effluent from the column flowed through a 20 m × 0.3 mm coil of PTFE tubing irradiated at 254 nm with a Sylvania GTE 8 W low-pressure lamp to the detector.

## CHROMATOGRAM
**Limit of detection:** 133 ng/mL

## OTHER SUBSTANCES
**Also analyzed:** bendroflumethiazide, butizide, chlorthalidone, ethacrynic acid, furosemide

## KEY WORDS
post-column reaction

## REFERENCE
Macher, M.; Wintersteiger, R. Improved electrochemical detection of diuretics in high-performance liquid chromatographic analysis by postcolumn on-line photolysis. *J.Chromatogr.A*, **1995**, *709*, 257–264

## SAMPLE
**Matrix:** solutions

## HPLC VARIABLES
**Guard column:** 30 × 3.2 7 μm SI 100 ODS (not commercially available)
**Column:** 150 × 3.2 7 μm SI 100 ODS (not commercially available)
**Mobile phase:** MeCN:buffer 31.2:68.8 (Buffer was 6.66 g $KH_2PO_4$ and 4.8 g 85% phosphoric acid in 1 L water, pH 2.3.)
**Flow rate:** 0.5-1
**Detector:** UV 211, 268

## CHROMATOGRAM
**Retention time:** 1.6
**Internal standard:** 5-(4-methylphenyl)-5-phenylhydantoin (7.3)

## OTHER SUBSTANCES
**Also analyzed:** aspirin, caffeine, carbamazepine, chlordiazepoxide, chlorprothixene, clonazepam, diazepam, doxylamine, ethosuximide, furosemide, haloperidol, methocarbamol, methotrimeprazine, nicotine, oxazepam, procaine, promazine, propafenone, propranolol, salicylamide, temazepam, tetracaine, thiopental, triamterene, verapamil, zolpidem, zopiclone

## REFERENCE

Below, E.; Burrmann, M. Application of HPLC equipment with rapid scan detection to the identification of drugs in toxicological analysis. *J.Liq.Chromatogr.*, **1994**, *17*, 4131–4144

## SAMPLE

**Matrix:** solutions

## HPLC VARIABLES

**Column:** 250 × 4.6 Zorbax RX
**Mobile phase:** Gradient. A was 10 mL concentrated orthophosphoric acid and 7 mL triethylamine in 1 L water. B was 10 mL concentrated orthophosphoric acid and 7 mL triethylamine in 200 mL water, make up to 1 L with MeCN. A:B from 100:0 to 0:100 over 30 min, maintain at 0:100 for 5 min.
**Column temperature:** 30
**Flow rate:** 2
**Detector:** UV 210

## OTHER SUBSTANCES

**Also analyzed:** acepromazine, acetaminophen, acetophenazine, albuterol, aminophylline, amitriptyline, amobarbital, amoxapine, amphetamine, amylocaine, antipyrine, aprobarbital, aspirin, atenolol, atropine, avermectin, barbital, benzocaine, benzoic acid, benzotropine, benzphetamine, berberine, bibucaine, bromazepan, brompheniramine, buprenorphine, buspirone, butabarbital, butacaine, butethal, caffeine, carbamazepine, carbromal, chloramphenicol, chlordiazepoxide, chloroquine, chlorothiazide, chloroxylenol, chlorphenesin, chlorpheniramine, chlorpromazine, chlorpropamide, chlortetracycline, cimetidine, cinchonidine, cinchonine, clenbuterol, clonazepam, clonixin, clorazepate, cocaine, codeine, colchicine, cortisone, coumarin, cyclazocine, cyclobenzaprine, cyclothiazide, cyheptamide, cymarin, danazol, danthron, dapsone, debrisoquine, desipramine, dexamethasone, dextromethorphan, dextropropoxyphene, diamorphine, diazepam, diclofenac, diethylpropion, diethylstilbestrol, diflunisal, digitoxin, digoxin, diltiazem, diphenhydramine, diphenoxylate, diprenorphine, dipyrone, disulfiram, dopamine, doxapram, doxepin, dronabinol, ephedrine, epinephrine, epinine, estradiol, estriol, estrone, ethacrynic acid, ethosuximide, etonitazene, etorphine, eugenol, famotidine, fenbendazole, fencamfamine, fenoprofen, fenproporex, fentanyl, flubendazole, flufenamic acid, flunitrazepam, 5-fluorouracil, fluoxymesterone, fluphenazine, furosemide, gentisic acid, gitoxigenin, glipizide, glunixin, glutethimide, glybenclamide, guaiacol, halazepam, haloperidol, hydrochlorothiazide, hydrocodone, hydrocortisone, hydromorphone, hydroxyquinoline, ibogaine, ibuprofen, iminostilbene, imipramine, indomethacin, isocarbostyril, isocarboxazid, isoniazid, isoproterenol, isoxsuprine, ivermectin, ketamine, ketoprofen, kynurenic acid, levorphanol, lidocaine, lorazepam, lormetazepam, loxapine, mazindol, mebendazole, meclizine, meclofenamic acid, medazepam, mefenamic acid, megestrol, mepacrine, meperidine, mephentermine, mephenytoin, mephesin, mephobarbital, mepivacaine, mescaline, mesoridazine, methadone, methamphetamine, methapyrilene, methaqualone, methazolamide, methocarbamol, methoxamine, methsuximide, methyl salicylate, methyldopa, methyldopamine, methylphenidate, methylprednisolone, methyltestosterone, methyprylon, metoprolol, mibolerone, morphine, nadolol, nalorphine, naloxone, naltrexone, naphazoline, naproxen, nefopam, niacinamide, nicotine, nicotinic acid, nifedipine, niflumic acid, nitrazepam, norepinephrine, nortriptyline, noscapine, nylidrin, oxazepam, oxycodone, oxymorphone, oxyphenbutazone, oxytetracycline, papaverine, pargyline, pemoline, pentazocine, pentobarbital, persantine, phenacetin, phenazocine, phenazopyridine, phencyclidine, phendimetrazine, phenelzine, pheniramine, phenobarbital, phenothiazine, phensuximide, phentermine, phenylbutazone, phenylephrine, phenylpropanolamine, piperocaine, prazepam, prednisolone, primidone, probenecid, progesterone, propiomazine, propranolol, propylparaben, pseudoephedrine, puromycin, pyrilamine, pyrithyldione, quazepam, quinaldic acid, quinidine, quinine, ranitidine, recinnamine, reserpine, resorcinol, saccharin, albuterol, salicylamide, salicylic acid, scopolamine, scopoletin, secobarbital, strychnine, sulfacetamide, sufadiazine, sulfadimethoxine, sulfaethidole, sulfamerazine, sulfamethazine, sulfamethoxizole, sulfanilamide, sulfapyridine, sulfasoxizole, sulindac, tamoxifen, temazepam, testosterone, tetracaine, tetracycline, tetramisole, thebaine, theobromine, theophylline, thiabendazole, thiamine, thiamylal, thiobarbituric acid, thioridazine, thiosalicylic acid, thiothixene, thymol, tolazamide, tolazoline, tobutamide, tolmetin, tranylcypromine, triamcinolone, tribenzylamine, trichloromethiazide, trifluoperazine, trihexyphenidyl, trimethoprim, tripelennamine, triproilidine, tropacocaine, tyramine, verapamil, vincamine, warfarin, yohimbine, zoxazolamine

## REFERENCE

Hill, D.W.; Kind, A.J. Reversed-phase solvent gradient HPLC retention indexes of drugs. *J.Anal.Toxicol.*, **1994**, *18*, 233–242

## SAMPLE

**Matrix:** solutions
**Sample preparation:** Dissolve in MeOH:water 1:1 at a concentration of 50 μg/mL, inject a 10 μL aliquot.

## HPLC VARIABLES

**Column:** 300 × 3.9 10 μm μBondapak C18
**Mobile phase:** MeOH:acetic acid:triethylamine:water 10:1.5:0.5:88
**Flow rate:** 1.5
**Injection volume:** 10
**Detector:** UV

## CHROMATOGRAM

**Retention time:** k' 2.64

## REFERENCE

Roos, R.W.; Lau-Cam, C.A. General reversed-phase high-performance liquid chromatographic method for the separation of drugs using triethylamine as a competing base. *J.Chromatogr.*, **1986**, *370*, 403–418

## SAMPLE

**Matrix:** solutions

## HPLC VARIABLES

**Column:** 250 × 2 C18 glass lined (Whatman)
**Mobile phase:** MeCN:water 60:40
**Flow rate:** 0.04
**Injection volume:** 0.5
**Detector:** UV 254; MS, Hewlett Packard 5985, home-made interface (details in paper)

## CHROMATOGRAM

**Retention time:** 8.5

## OTHER SUBSTANCES

**Simultaneous:** chlorothiazide, trichlormethiazide

## KEY WORDS

microbore

## REFERENCE

Eckers, C.; Skrabalak, D.S.; Henion, J. On-line direct liquid introduction interface for micro-liquid chromatography/mass spectrometry: application to drug analysis. *Clin.Chem.*, **1982**, *28*, 1882–1886

## SAMPLE
**Matrix:** urine
**Sample preparation:** Inject an aliquot onto column A and elute to waste with mobile phase A, after 1 min backflush the contents of column A onto column B with mobile phase B, elute with mobile phase B, monitor the effluent from column B.

## HPLC VARIABLES
**Column:** A 30 μm 20 × 2.1 Hypersil ODS-C18; B 250 × 4 5 μm Hypersil ODS-C18
**Mobile phase:** A Water; B Gradient. MeCN:buffer 15:85 for 1.5 min then to 80:20 over 8 min. Keep at 80:20 for 2.5 min then re-equilibrate with 15:85. (Buffer was 50 mM $NaH_2PO_4$ + 1.4 mL propylamine hydrochloride per liter adjusted to pH 3 with concentrated phosphoric acid.)
**Flow rate:** 1
**Injection volume:** 50
**Detector:** UV 230

## CHROMATOGRAM
**Retention time:** 6.3
**Limit of detection:** 7 ng/mL

## OTHER SUBSTANCES
**Simultaneous:** acetazolamide, amiloride, bendroflumethiazide, bumetanide, chlorthalidone, cyclothiazide, ethacrynic acid, furosemide, probenecid, spironolactone, triamterene

## REFERENCE
Campíns-Falco, P.; Herráez-Hernández, R.; Sevillano-Cabeza, A. Column-switching techniques for screening of diuretics and probenecid in urine samples. *Anal.Chem.*, **1994**, *66*, 244–248

## SAMPLE
**Matrix:** urine
**Sample preparation:** 1 mL Urine + 100 μL 20 μg/mL hydroflumethiazide in MeOH + 1 mL buffer + 200 μL water + 6 mL ethyl acetate, shake for 5 min, centrifuge at 900 g for 5 min. Remove 5 mL organic layer and evaporate at 37° under a stream of nitrogen. Reconstitute with 100 μL mobile phase, inject a 20 μL aliquot. (Buffer was 0.38 g ammonium acetate in 500 mL water, acidified to pH 5.0 with glacial acetic acid.)

## HPLC VARIABLES
**Guard column:** 40 × 4 35-50 μm C18 Corasil
**Column:** 125 × 4 5 μm Nucleosil 100-5 C18
**Mobile phase:** MeCN:acetic acid:water 120:1:880
**Flow rate:** 1
**Injection volume:** 20
**Detector:** UV 280

## CHROMATOGRAM
**Retention time:** 5.0
**Internal standard:** hydroflumethiazide (10.0)
**Limit of detection:** 10 ng/mL

## OTHER SUBSTANCES
**Noninterfering:** amiloride, acebutolol, acenocoumarol, acetaminophen, aspirin, allopurinol, ambroxol, amoxicillin, atenolol, bendroflumethiazide, benzbromarone, bezafibrate, biperiden, bisacodyl, bromazepam, butizide, captopril, cimetidine, ciprofloxacin, clobutinol, clonidine, cotinine, diazepam, diclofenac, digitoxin, digoxin, dihydrocodeine, dihydroergotamine, diltiazem, doxepin, doxycycline, enalapril, erythromycin, fenoterol, furosemide, glibenclamide, heparin, hypoxanthine, ibuprofen, indomethacin, isosorbide

mononitrate, lisinopril, lovastatin, maprotiline, methyldigoxin, methyldopa, metoclopramide, metoprolol, metronidazole, midazolam, naloxone, nifedipine, nicotine, oxazepam, oxipurinol, penicillin V, pentoxyfylline, phenacetin, phenazone, propyphenazone, phenprocoumon, ranitidine, salicylic acid, sotalol, sulfamethoxazole, trimethoprim, terbutaline, theophylline, tilidine, timolol, triamterene, uric acid, verapamil, vitamin C, warfarin, xanthine, purine and pyrimidine bases, nucleosides, nucleotides
**Interfering:** caffeine

## KEY WORDS
norfloxacin and ofloxacin interfere with IS

## REFERENCE
de Vries, J.X.; Voss, A. Simple determination of hydrochlorothiazide in human plasma and urine by high performance liquid chromatography. *Biomed.Chromatogr.*, **1993**, *7*, 12–14

## SAMPLE
**Matrix:** urine
**Sample preparation:** Buffer urine to 4.9 by mixing with an equal volume of pH 4.9 200 mM sodium phosphate buffer. Inject a 40 μL aliquot onto column A and elute to waste with mobile phase A, after 3 min backflush the contents of column A onto column B with mobile phase B and start the gradient. At the end of the run re-equilibrate for 10 min.

## HPLC VARIABLES
**Column:** A 20 × 4 5 μm Hypersil octadecylsilica ODS; B 200 × 4.6 5 μm Shiseido SG-120 polymer-based C18
**Mobile phase:** A water; B Gradient. MeCN:buffer from 7:93 to 15:85 over 3.5 min, to 50:50 over 8.5 min, maintain at 50:50 for 11 min (Buffer was 6.9 g $NaH_2PO_4.H_2O$ in 1 L water, pH adjusted to 3.1 with phosphoric acid.)
**Flow rate:** 1
**Injection volume:** 40
**Detector:** UV 270

## CHROMATOGRAM
**Retention time:** 11.3
**Limit of detection:** 1 μg/mL

## OTHER SUBSTANCES
**Extracted:** acetazolamide, amiloride, bendroflumethiazide, benzthiazide, bumetanide, caffeine, carbamazepine, chlorothiazide, chlorthalidone, clopamide, dichlorfenamide, ethacrynic acid, furosemide, metyrapone, probenecid, spironolactone, triamterene, trichlormethiazide

## KEY WORDS
column-switching; optimum detection wavelengths vary for each drug

## REFERENCE
Saarinen, M.; Sirén, H.; Riekkola, M.-L. A column switching technique for the screening of diuretics in urine by high performance liquid chromatography. *J.Liq.Chromatogr.*, **1993**, *16*, 4063–4078

## SAMPLE
**Matrix:** urine
**Sample preparation:** 5 mL Urine + 50 μL 100 μg/mL 7-propyltheophylline in MeOH + 200 μL ammonium chloride buffer + 2 g NaCl, extract with 6 mL ethyl acetate by rocking at 40 movements/min for 20 min and centrifuging at 800 g for 5 min, repeat extraction, combine organic layers, evaporate to dryness at 40° under a stream of nitrogen. Reconstitute in 200 μL MeCN:water 15:85 and inject 20 μL aliquots. (Ammonium chloride

buffer was 28 g ammonium chloride in 100 mL water with the pH adjusted to 9.5 with concentrated ammonia solution.)

## HPLC VARIABLES
**Column:** 75 × 4.6 3 µm Ultrasphere ODS
**Mobile phase:** Gradient. MeCN:100 mM ammonium acetate adjusted to pH 3 with concentrated phosphoric acid. From 10:90 to 15:85 over 2 min to 55:45 over 3 min to 60: 40 over 3 min. Kept at 60:40 for 1 min, decreased to 10:90 over 1 min and equilibrated at 10:90 for 2 min.
**Flow rate:** 1
**Injection volume:** 20
**Detector:** UV 270

## CHROMATOGRAM
**Retention time:** 3.0
**Internal standard:** 7-propyltheophylline (4.5)
**Limit of detection:** 100 ng/mL

## OTHER SUBSTANCES
**Simultaneous:** acetazolamide, amiloride, bendroflumethiazide, benzthiazide, bumetanide, buthiazide, caffeine, canrenone, chlorthalidone, clopamide, cyclothiazide, diclofenamide, ethacrynic acid, furosemide, hydrochlorothiazide, mesocarb, morazone, piretanide, polythiazide, probenecid, spironolactone, torsemide, triamterene, xipamide

## REFERENCE
Ventura, R.; Nadal, T.; Alcalde, P.; Pascual, J.A.; Segura, J. Fast screening method for diuretics, probenecid and other compounds of doping interest. *J.Chromatogr.A*, **1993**, *655*, 233–242

## SAMPLE
**Matrix:** urine
**Sample preparation:** Make 5 mL urine alkaline (pH 9-10), add 2 g NaCl, extract twice with 6 mL ethyl acetate. Combine the organic layers and evaporate them to dryness under a stream of nitrogen, reconstitute the residue in 200 µL MeCN/water, inject a 10-20 µL aliquot.

## HPLC VARIABLES
**Column:** 100 × 4 5 µm SGE 100 GL-4 C18P (Scientific Glass Engineering)
**Mobile phase:** MeCN:MeOH:water:trifluoroacetic acid 0.3:0.7:99:0.5
**Flow rate:** 0.8 or 1
**Injection volume:** 10-20
**Detector:** MS, ZAB2-SEQ (VG), PSP source coupled to LC, source 250°, probe 240-260°, scan m/z 200-550; UV 270

## CHROMATOGRAM
**Retention time:** 4.3
**Limit of detection:** 150 ng (by MS)

## OTHER SUBSTANCES
**Extracted:** acetazolamide

## REFERENCE
Ventura, R.; Fraisse, D.; Becchi, M.; Paisse, O.; Segura, J. Approach to the analysis of diuretics and masking agents by high-performance liquid chromatography-mass spectrometry in doping control. *J.Chromatogr.*, **1991**, *562*, 723–736

## SAMPLE
**Matrix:** urine

**Sample preparation:** 2 mL Urine + 0.5 g solid buffer ɪ (pH 5-5.5), vortex 15 s, add 4 mL ethyl acetate, agitate for 10 min, centrifuge at 600 g for 5 min. Remove organic layer and vortex it with 2 mL 5% aqueous lead acetate for 10 s, centrifuge at 600 g for 5 min, remove and keep organic phase. 2 mL Urine + 0.5 g solid buffer II (pH 9-9.5), vortex 15 s, add 4 mL ethyl acetate, agitate for 10 min, centrifuge at 600 g for 5 min. Remove organic layer and combine it with previous organic layer. Evaporate to dryness at 50° under a stream of nitrogen, reconstitute in 300 μL 50 μg/mL β-hydroxyethyltheophylline in MeOH, inject 5 μL aliquot. (Solid buffer I was $KH_2PO_4:Na_2HPO_4$ 99:1, solid buffer II was $NaHCO_3:K_2CO_3$ 3:2.)

## HPLC VARIABLES
**Column:** 250 × 4.6 5 μm HP Hypersil ODS (A) or HP LiChrosorb RP-18 (B)
**Mobile phase:** Gradient. MeCN:buffer from 15:85 at 2 min to 80:20 at 20 min (Buffer was 50 mM $NaH_2PO_4$ containing 16 mM propylamine hydrochloride, adjusted to pH 3 with concentrated phosphoric acid.)
**Flow rate:** 1
**Injection volume:** 5
**Detector:** UV 230; UV 275

## CHROMATOGRAM
**Retention time:** 6.03 (A), 7.07 (B)
**Internal standard:** β-hydroxyethyltheophylline (3.7 (A), 4.4 ʿB))
**Limit of detection:** 500 ng/mL

## OTHER SUBSTANCES
**Extracted:** acetazolamide, amiloride, bendroflumethiazide, benthiazide, bumetanide, canrenone, chlorothiazide, chlorethalidone, cyclothiazide, dichlorphenamide, ethacrynic acid, flumethiazide, furosemide, hydroflumethiazide, methyclothiazide, metolazone, polythiazide, probenecid, quinethazone, spironolactone, triamterene, trichloromethiazide
**Noninterfering:** acetaminophen, aspirin, caffeine, diflunisal, fenoprofen, ibuprofen, indomethacin, methocarbamol, naproxen, phenylbutazone, sulindac, tetracycline, theobromine, theophylline, tolmetin, trimethoprim, verapamil

## REFERENCE
Cooper, S.F.; Massé, R.; Dugal, R. Comprehensive screening procedure for diuretics in urine by high-performance liquid chromatography. *J.Chromatogr.*, **1989**, *489*, 65–88

## SAMPLE
**Matrix:** urine
**Sample preparation:** 2 mL Urine + 2 mL 1 M pH 4.1 $NaH_2PO_4$ + 4 mL ethyl acetate, vortex for 2 min, centrifuge at 1500 g for 5 min. Remove the organic phase and add it to 5 mL 100 mM pH 7.5 $Na_2HPO_4$, vortex for 2 min, centrifuge at 1500 g for 5 min. Remove the organic layer and evaporate it to dryness under a stream of nitrogen at 60°, reconstitute the residue in 100 μL MeCN:10 mM pH 3.0 phosphate buffer, inject a 5 μL aliquot.

## HPLC VARIABLES
**Column:** 125 × 4 5 μm LiCHrosorb RP-18
**Mobile phase:** Gradient. MeCN:10 mM pH 3.0 phosphate buffer 10:90 for 1.5 min then to 35:65 over 2 min.
**Column temperature:** 50
**Flow rate:** 1.5
**Injection volume:** 5
**Detector:** UV 271

## CHROMATOGRAM
**Retention time:** 2.2
**Limit of quantitation:** 500 ng/mL

## OTHER SUBSTANCES

**Extracted:** bendroflumethiazide, bumetanide, chlorothiazide, chlorthalidone, clopamide, cyclopenthiazide, furosemide, mefruside, methyclothiazide, metolazone, quinethazone
**Simultaneous:** clorexolone, ethacrynic acid, indapamide
**Noninterfering:** aspirin, albuterol, allopurinol, alprenolol, atenolol, captopril, carbimazole, clonidine, coloxyl, danthron, diazepam, digoxin, doxepin, glibenclamide, hydralazine, indomethacin, labetalol, metformin, methyldopa, metoprolol, mianserin, minoxidil, nifedipine, nitrazepam, oxazepam, oxprenolol, pindolol, prazosin, propranolol, senokot, theophylline, trifluoperazine

## REFERENCE

Fullinfaw, R.O.; Bury, R.W.; Moulds, R.F.W. Liquid chromatographic screening of diuretics in urine. *J.Chromatogr.*, **1987**, *415*, 347–356

## ANNOTATED BIBLIOGRAPHY

Cieri, U.R. Determination of reserpine, hydralazine HCl, and hydrochlorothiazide in tablets by liquid chromatography on a short, normal-phase column. *J.AOAC Int.*, **1994**, *77*, 1104–1108 [simultaneous hydralazine, reserpine; tablets; normal-phase; fluorescence detection; UV detection]

Hsieh, J.Y.-K.; Lin, C.; Matuszewski, B.K.; Dobrinska, M.R. Fully automated methods for the determination of hydrochlorothiazide in human plasma and urine. *J.Pharm.Biomed.Anal.*, **1994**, *12*, 1555–1562 [plasma; urine; SPE; LOQ 2 ng/mL]

Ulvi, V.; Keski-Hynnilä, H. First-derivative UV spectrophotometric and high-performance liquid chromatographic analysis of some thiazide diuretics in the presence of their photodecomposition products. *J.Pharm.Biomed.Anal.*, **1994**, *12*, 917–922 [simultaneous degradation products; also chlorothiazide, trichlormethiazide]

Abdelhameed, M.H.; Chen, T.M.; Chi   W.L. Intrahepatic distribution of hydrochlorothiazide and quinidine in rats: Implications in ph  macokinetics. *J.Pharm.Sci.*, **1993**, *82*, 992–996 [whole blood; plasma; liver; chlorothiazide (IS); rat; pharmacokinetics]

Berthod, A.; Laserna, J.J.; Carretero,  . Oil-in-water microemulsions as mobile phases for rapid screening of illegal drugs in sports. *J.Liq.Chromatogr.*, **1992**, *15*, 3115–3127 [simultaneous acebutolol, chlorthalidone, codeine, hydrochloro thiazide, methoxamine, methyltestosterone, nadolol, norcodeine, oxprenolol, phenylephrine, probenecid]

Chen, T.M.; Abdelhameed, M.H.; Chiou, W.L. Erythrocytes as a total barrier for renal excretion of hydrochlorothiazide: slow influx and efflux across erythrocyte membranes. *J.Pharm.Sci.*, **1992**, *81*, 212–218 [whole blood; plasma; chlorothiazide (IS); human; rat; pharmacokinetics]

Herráez-Hernández, R.; Campíns-Falcó, P.; Sevillano-Cabeza, A. Improved screening procedure for diuretics. *J.Liq.Chromatogr.*, **1992**, *15*, 2205–2224 [LOD 10-1000 ng/mL; gradient; urine; hydroxymethyltheophylline (IS); extracted acetazolamide, amiloride, bendroflumethiazide, bumetanide, chlorthalidone, cyclothiazide, ethacrynic acid, furosemide, probenecid, spironolactone, triamterene]

Miller, R.B.; Amestoy, C. A liquid chromatographic method for the determination of hydrochlorothiazide in human plasma. *J.Pharm.Biomed.Anal.*, **1992**, *10*, 541–545

Campíns-Falcó, P.; Herráez-Hernández, R.; Sevillano-Cabeza, A. Solid-phase extraction techniques for assay of diuretics in human urine samples. *J.Liq.Chromatogr.*, **1991**, *14*, 3575–3590 [urine; SPE; hydroxyethyltheophylline (IS); extracted acetazolamide, amiloride, bendroflumethiazide, bumetanide, chlorthalidone, ciclothiazide, ethacrynic acid, furosemide, probenecid, spironolactone, triamterene]

Bachman, W.J.; Stewart, J.T. HPLC-photolysis-electrochemical detection in pharmaceutical analysis: application to the determination of spironolactone and hydrochlorothiazide in tablets. *J.Chromatogr.Sci.*, **1990**, *28*, 123–128 [post-column reaction; electrochemical detection; tablets; simultaneous spironolactone]

Kuo, B.S.; Mandagere, A.; Osborne, D.R.; Hwang, K.K. Column-switching high-performance liquid chromatographic (HPLC) determination of hydrochlorothiazide in rat, dog, and human plasma. *Pharm.Res.*, **1990**, *7*, 1257–1261

Santasania, C.T. Direct injection analysis of diuretic and anti-inflammatory drugs on a shielded hydrophobic phase column. *J.Liq.Chromatogr.*, **1990**, *13*, 2605–2631 [serum; direct injection; gradient; horse; extracted furosemide, oxyphenbutazone, phenylbutazone]

Sa'sa', S.I.; Jalal, I.M.; Khalil, H.S. Determination of atenolol combinations with hydrochlorothiazide and chlorthalidone in tablet formulations by reverse-phase HPLC. *J.Liq.Chromatogr.*, **1988**, *11*, 1673–1696 [simultaneous atenolol, chlorthalidone; tablets; methyl p-hydroxybenzoate (IS); stability-indicating]

Hitscherich, M.E.; Rydberg, E.M.; Tsilifonis, D.C.; Daly, R.E. Simultaneous determination of hydrochlorothiazide and propranolol hydrochloride in tablets by high-performance liquid chromatography. *J.Liq.Chromatogr.*, **1987**, *10*, 1011–1021 [simultaneous propranolol; tablets; stability indicating]

Valkó, K. RP-HPLC retention data for measuring structural similarity of compounds for QSAR studies. *J.Liq.Chromatogr.*, **1987**, *10*, 1663–1686 [also acetanilide, acetylazidomorphine, aspirin, azidocodeine, azidoethylmorphine, azidomorphine, barbital, benzaldehyde, benzoic acid, bromocyanonitrophenol, caffeine, chloramphenicol, chlorocyanonitrophenol, chloronitroaniline, cortexolone, cortisone, cyanodinitrophenol, cyanofluoronitrophenol, cyclopropylazidoethylmorphine, cyclopropylmethylazidomorphine, 11-deoxycorticosterone, dexamethasone, dichloronitroaniline, dinitroaniline, dinitrophenol, hydrocortisone, isoniazid, methyl salicylate, morphine, niacinamide, nicotinamide, nitroaniline, nitrophenol, norazidoethylmorphine, norazidomorphine, normorphine, phenacetin, phenobarbital, phenylethylazidoethylmorphine, phenylethylazidomorphine, prednisolone, progesterone, salicylamide, salicylic acid, sulfadimidine, sulfaguanidine, sulfamethazine, sulfamethoxypyridazine, testosterone, triamcinolone, trinitroaniline, trinitrophenol, vanillin, vitamin B3, vitamin B5]

Alton, K.B.; Desrivieres, D.; Patrick, J.E. High-performance liquid chromatographic assay for hydrochlorothiazide in human urine. *J.Chromatogr.*, **1986**, *374*, 103–110

Shiu, G.K.; Prasad, V.K.; Lin, J.; Worsley, W. Simple and selective high-performance liquid chromatographic method for the determination of hydrochlorothiazide in urine. *J.Chromatogr.*, **1986**, *377*, 430–435

Kirschbaum, J.; Perlman, S. Analysis of captopril and hydrochlorothiazide combination tablet formulations by liquid chromatography. *J.Pharm.Sci.*, **1984**, *73*, 686–687

Koopmans, P.P.; Tan, Y.; Van Ginneken, C.A.; Gribnau, F.W. High-performance liquid chromatographic determination of hydrochlorothiazide in plasma and urine. *J.Chromatogr.*, **1984**, *307*, 445–450

Yamazaki, M.; Ito, Y.; Suzuka, T.; Yaginuma, H.; Itoh, S.; Kamada, A.; Orita, Y.; Nakahama, H.; Nakanishi, T.; Ando, A. Biopharmaceutical studies of thiazide diuretics. II. High-performance liquid chromatographic method for determination of hydrochlorothiazide in plasma, urine, blood cells and bile. *Chem.Pharm.Bull.*, **1984**, *32*, 2387–2394

Shah, V.P.; Walker, M.A.; Prasad, V.K. Application of flow programming in the analysis of drugs and their metabolites in biological fluids. *J.Liq.Chromatogr.*, **1983**, *6*, 1949–1954 [extracted metabolites, chlorothiazide, triamterene; flow programming; urine; plasma]

Barbhaiya, R.H.; Phillips, T.A.; Welling, P.G. High-pressure liquid chromatographic determination of chlorothiazide and hydrochlorothiazide in plasma and urine: preliminary results of clinical studies. *J.Pharm.Sci.*, **1981**, *70*, 291–295

Daniels, S.L.; Vanderwielen, A.J. Stability-indicating assay for hydrochlorothiazide. *J.Pharm.Sci.*, **1981**, *70*, 211–215

Menon, G.N.; White, L.B. Simultaneous determination of hydrochlorothiazide and triamterene in capsule formulations by high-performance liquid chromatography. *J.Pharm.Sci.*, **1981**, *70*, 1083–1085 [capsules; m-hydroxyacetophenone (IS); simultaneous triamterene]

Henion, J.D.; Maylin, G.A. Qualitative and quantitative analysis of hydrochlorothiazide in equine plasma and urine by high-performance liquid chromatography. *J.Anal.Toxicol.*, **1980**, *4*, 185–191

# Hydrocodone

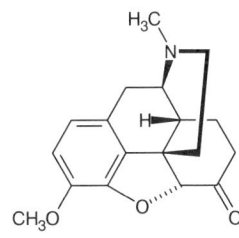

**Molecular formula:** $C_{18}H_{21}NO_3$
**Molecular weight:** 299.4
**CAS Registry No.:** 125-29-1 (hydrocodone),
34195-34-1 (hydrocodone bitartrate
hydrate), 143-71-5 (hydrocodone bitartrate)

## SAMPLE
**Matrix:** bile, blood
**Sample preparation:** 0.5 mL Blood or bile + 10 (blood) or 15 (bile) μL 100 μg/mL nalorphine in MeOH + 300 μL 1.1 M pH 5.0 sodium acetate buffer + 3000-3500 U of Patella
vulgata glucuronidase, incubate at 55° overnight, add 0.5 mL borate buffer to achieve a
pH of 8.3-8.5. Add 8 mL chloroform:isopropanol 90:10, gently rotate for 30 min, centrifuge at 3500 rpm for 10 min, remove aqueous layer. Wash organic layer (twice for blood,
three times for bile) with 3 mL 100 mM pH 9.9 sodium phosphate buffer with gentle
rotation for 10 min and centrifugation each time. Add organic layer to 200 (blood) or 400
(bile) μL 0.2% phosphoric acid, gently rotate for 30 min, discard organic layer, inject 50
μL of the acid layer. (Borate buffer was 50 mM boric acid and 43 mM sodium tetraborate,
adjusted to pH 9.8.)

## HPLC VARIABLES
**Guard column:** Nova-Pak phenyl guard column
**Column:** 150 × 3.9 5 μm Nova-Pak phenyl
**Mobile phase:** MeCN:10 mM pH 6.6 $NaH_2PO_4$ 10:90
**Flow rate:** 1.2
**Injection volume:** 50
**Detector:** UV 210; F ex 220 em 370 (cut-off)

## CHROMATOGRAM
**Retention time:** 38.1
**Internal standard:** nalorphine (23.5)

## OTHER SUBSTANCES
**Simultaneous:** codeine, dihydrocodeine, 6-monoacetylmorphine, morphine, oxycodone
**Noninterfering:** acetylcodeine, amitriptyline, amphetamine, diamorphine, diazepam, dothiepin, doxepin, ephedrine, ephedrine, hydromorphone, mesoridazine, methadone, methamphetamine, 3-monoacetylmorphine, nordiazepam, norpropoxyphene, nortriptyline,
oxazepam, propoxyphene, pseudoephedrine, quinidine, quinine, sulfamethoxazole, sulforidazine, thioridazine

## KEY WORDS
UV and F detection used together

## REFERENCE
Crump, K.L.; McIntyre, I.M.; Drummer, O.H. Simultaneous determination of morphine and codeine in
blood and bile using dual ultraviolet and fluorescence high-performance liquid chromatography.
*J.Anal.Toxicol.*, **1994**, *18*, 208–212

## SAMPLE
**Matrix:** formulations
**Sample preparation:** Measure out syrup equivalent to about 5 mg hydrocodone bitartrate,
add 5 mL water, add 1 mL 1.8 M sulfuric acid, wash twice with 40 mL portions of chloroform. Make the aqueous layer alkaline with 5 mL 1 M NaOH, extract twice with 25

mL portions of chloroform. Combine the organic layers and evaporate them to dryness, reconstitute the residue in 500 µL EtOH, inject the whole amount.

## HPLC VARIABLES

**Column:** 305 × 7 PRP-1 (Hamilton)
**Mobile phase:** Gradient. A was water:triethylamine 99.9:0.1. B was MeCN:triethylamine 99.9:0.1. A:B 60:40 for 7 min, to 20:80 over 5 min, maintain at 20:80 for 5 min, to 60:40 over 6 min, re-equilibrate at 60:40 for 2 min.
**Column temperature:** 40
**Flow rate:** 3.5
**Injection volume:** 500
**Detector:** UV 254

## CHROMATOGRAM

**Retention time:** 6.9

## OTHER SUBSTANCES

**Simultaneous:** diphenylpyraline, doxylamine, etafedrine, guaifenesin, pheniramine, phenylephrine, phenylpropanolamine, pyrilamine

## KEY WORDS

syrup

## REFERENCE

Black, D.B.; By, A.W.; Lodge, B.A. Isolation and identification of hydrocodone in narcotic cough syrups by high-performance liquid chromatography with infrared spectrometric identification. *J.Chromatogr.*, **1986**, *358*, 438–443

## SAMPLE

**Matrix:** formulations
**Sample preparation:** Add 1 tablet to 95 mL water, place on a steam bath for 15 min, cool, mix for 15 min, sonicate, allow to stand, filter, inject 13 µL aliquot

## HPLC VARIABLES

**Column:** µBondapak C18
**Mobile phase:** MeOH:buffer 25:75 (Buffer was 0.01 N $KH_2PO_4$ + 50 mM $KNO_3$, adjusted to pH 4.5 with 3 N phosphoric acid.)
**Flow rate:** 1.1
**Injection volume:** 13
**Detector:** UV 283

## CHROMATOGRAM

**Retention time:** 10

## OTHER SUBSTANCES

**Simultaneous:** acetaminophen, p-aminophenol, p-chloroacetanilide, codeine, hydromorphone

## KEY WORDS

tablets; stability-indicating

## REFERENCE

Wallo, W.E.; D'Adamo, A. Simultaneous assay of hydrocodone bitartrate and acetaminophen in a tablet formulation. *J.Pharm.Sci.*, **1982**, *71*, 1115–1118

## SAMPLE
**Matrix:** solutions

## HPLC VARIABLES
**Column:** 250 × 4.6 5 μm Supelcosil LC-DP (A) or 250 × 4 5 μm LiChrospher 100 RP-8 (B)
**Mobile phase:** MeCN : 0.025% phosphoric acid : buffer 25 : 10 : 5 (A) or 60 : 25 : 15 (B) (Buffer was 9 mL concentrated phosphoric acid and 10 mL triethylamine in 900 mL water, adjust pH to 3.4 with dilute phosphoric acid, make up to 1 L.)
**Flow rate:** 0.6
**Injection volume:** 25
**Detector:** UV 229

## CHROMATOGRAM
**Retention time:** 6.93 (A), 3.71 (B)

## OTHER SUBSTANCES
**Also analyzed:** acebutolol, acepromazine, acetaminophen, acetazolamide, acetophenazine, albuterol, alprazolam, amitriptyline, amobarbital, amoxapine, antipyrine, atenolol, atropine, azatadine, baclofen, benzocaine, bromocriptine, brompheniramine, brotizolam, bupivacaine, buspirone, butabarbital, butalbital, caffeine, carbamazepine, cetirizine, chlorcyclizine, chlordiazepoxide, chlormezanone, chloroquine, chlorpheniramine, chlorpromazine, chlorpropamide, chlorprothixene, chlorthalidone, chlorzoxazone, cimetidine, cisapride, clomipramine, clonazepam, clonidine, clozapine, cocaine, codeine, colchicine, cyclizine, cyclobenzaprine, dantrolene, desipramine, diazepam, diclofenac, diflunisal, diltiazem, diphenhydramine, diphenidol, diphenoxylate, dipyridamole, disopyramide, dobutamine, doxapram, doxepin, droperidol, encainide, ethidium bromide, ethopropazine, fenoprofen, fentanyl, flavoxate, fluoxetine, fluphenazine, flurazepam, flurbiprofen, fluvoxamine, furosemide, glutethimide, glyburide, guaifenesin, haloperidol, homatropine, hydralazine, hydrochlorothiazide, hydromorphone, hydroxychloroquine, hydroxyzine, ibuprofen, imipramine, indomethacin, ketoconazole, ketoprofen, ketorolac, labetalol, levorphanol, lidocaine, loratadine, lorazepam, lovastatin, loxapine, mazindol, mefenamic acid, meperidine, mephenytoin, mepivacaine, mesoridazine, metaproterenol, methadone, methdilazine, methocarbamol, methotrexate, methotrimeprazine, methoxamine, methyldopa, methylphenidate, metoclopramide, metolazone, metoprolol, metronidazole, midazolam, moclobemide, morphine, nadolol, nalbuphine, naloxone, naphazoline, naproxen, nifedipine, nizatidine, norepinephrine, nortriptyline, oxazepam, oxycodone, oxymetazoline, paroxetine, pemoline, pentazocine, pentobarbital, pentoxifylline, perphenazine, pheniramine, phenobarbital, phenol, phenolphthalein, phentolamine, phenylbutazone, phenyltoloxamine, phenytoin, pimozide, pindolol, piroxicam, pramoxine, prazepam, prazosin, probenecid, procainamide, procaine, prochlorperazine, procyclidine, promazine, promethazine, propafenone, propantheline, propiomazine, propofol, propranolol, protriptyline, quazepam, quinidine, quinine, racemethorphan, ranitidine, remoxipride, risperidone, salicylic acid, scopolamine, secobarbital, sertraline, sotalol, spironolactone, sulfinpyrazone, sulindac, temazepam, terbutaline, terfenadine, tetracaine, theophylline, thiethylperazine, thiopental, thioridazine, thiothixene, timolol, tocainide, tolbutamide, tolmetin, trazodone, triamterene, triazolam, trifluoperazine, triflupromazine, trimeprazine, trimethoprim, trimipramine, verapamil, warfarin, xylometazoline, yohimbine, zopiclone

## KEY WORDS
some details of plasma extraction

## REFERENCE
Koves, E.M. Use of high-performance liquid chromatography-diode array detection in forensic toxicology. *J.Chromatogr.A*, **1995**, *692*, 103−119

## SAMPLE
**Matrix:** solutions

## HPLC VARIABLES
**Column:** 250 × 4.6 Zorbax RX
**Mobile phase:** Gradient. A was 10 mL concentrated orthophosphoric acid and 7 mL triethylamine in 1 L water. B was 10 mL concentrated orthophosphoric acid and 7 mL triethylamine in 200 mL water, make up to 1 L with MeCN. A:B from 100:0 to 0:100 over 30 min, maintain at 0:100 for 5 min.
**Column temperature:** 30
**Flow rate:** 2
**Detector:** UV 210

## OTHER SUBSTANCES
**Also analyzed:** acepromazine, acetaminophen, acetophenazine, albuterol, aminophylline, amitriptyline, amobarbital, amoxapine, amphetamine, amylocaine, antipyrine, aprobarbital, aspirin, atenolol, atropine, avermectin, barbital, benzocaine, benzoic acid, benzotropine, benzphetamine, berberine, bibucaine, bromazepan, brompheniramine, buprenorphine, buspirone, butabarbital, butacaine, butethal, caffeine, carbamazepine, carbromal, chloramphenicol, chlordiazepoxide, chloroquine, chlorothiazide, chloroxylenol, chlorphenesin, chlorpheniramine, chlorpromazine, chlorpropamide, chlortetracycline, cimetidine, cinchonidine, cinchonine, clenbuterol, clonazepam, clonixin, clorazepate, cocaine, codeine, colchicine, cortisone, coumarin, cyclazocine, cyclobenzaprine, cyclothiazide, cyheptamide, cymarin, danazol, danthron, dapsone, debrisoquine, desipramine, dexamethasone, dextromethorphan, dextropropoxyphene, diamorphine, diazepam, diclofenac, diethylpropion, diethylstilbestrol, diflunisal, digitoxin, digoxin, diltiazem, diphenhydramine, diphenoxylate, diprenorphine, dipyrone, disulfiram, dopamine, doxapram, doxepin, dronabinol, ephedrine, epinephrine, epinine, estradiol, estriol, estrone, ethacrynic acid, ethosuximide, etonitazene, etorphine, eugenol, famotidine, fenbendazole, fencamfamine, fenoprofen, fenproporex, fentanyl, flubendazole, flufenamic acid, flunitrazepam, 5-fluorouracil, fluoxymesterone, fluphenazine, furosemide, gentisic acid, gitoxigenin, glipizide, glunixin, glutethimide, glybenclamide, guaiacol, halazepam, haloperidol, hydrochlorothiazide, hydrocortisone, hydromorphone, hydroxyquinoline, ibogaine, ibuprofen, iminostilbene, imipramine, indomethacin, isocarbostyril, isocarboxazid, isoniazid, isoproterenol, isoxsuprine, ivermectin, ketamine, ketoprofen, kynurenic acid, levorphanol, lidocaine, lorazepam, lormetazepam, loxapine, mazindol, mebendazole, meclizine, meclofenamic acid, medazepam, mefenamic acid, megestrol, mepacrine, meperidine, mephentermine, mephenytoin, mephesin, mephobarbital, mepivacaine, mescaline, mesoridazine, methadone, methamphetamine, methapyrilene, methaqualone, methazolamide, methocarbamol, methoxamine, methsuximide, methyl salicylate, methyldopa, methyldopamine, methylphenidate, methylprednisolone, methyltestosterone, methyprylon, metoprolol, mibolerone, morphine, nadolol, nalorphine, naloxone, naltrexone, naphazoline, naproxen, nefopam, niacinamide, nicotine, nicotinic acid, nifedipine, niflumic acid, nitrazepam, norepinephrine, nortriptyline, noscapine, nylidrin, oxazepam, oxycodone, oxymorphone, oxyphenbutazone, oxytetracycline, papaverine, pargyline, pemoline, pentazocine, pentobarbital, persantine, phenacetin, phenazocine, phenazopyridine, phencyclidine, phendimetrazine, phenelzine, pheniramine, phenobarbital, phenothiazine, phensuximide, phentermine, phenylbutazone, phenylephrine, phenylpropanolamine, piperocaine, prazepam, prednisolone, primidone, probenecid, progesterone, propiomazine, propranolol, propylparaben, pseudoephedrine, puromycin, pyrilamine, pyrithyldione, quazepam, quinaldic acid, quinidine, quinine, ranitidine, recinnamine, reserpine, resorcinol, saccharin, albuterol, salicylamide, salicylic acid, scopolamine, scopoletin, secobarbital, strychnine, sulfacetamide, sufadiazine, sulfadimethoxine, sulfaethidole, sulfamerazine, sulfamethazine, sulfamethoxizole, sulfanilamide, sulfapyridine, sulfasoxizole, sulindac, tamoxifen, temazepam, testosterone, tetracaine, tetracycline, tetramisole, thebaine, theobromine, theophylline, thiabendazole, thiamine, thiamylal, thiobarbituric acid, thioridazine, thiosalicylic acid, thiothixene, thymol, tolazamide, tolazoline, tobutamide, tolmetin, tranylcypromine, triamcinolone, tribenzylamine, trichloromethiazide, trifluoperazine, trihexyphenidyl, trimethoprim, tripelennamine, triproilidine, tropacocaine, tyramine, verapamil, vincamine, warfarin, yohimbine, zoxazolamine

## REFERENCE
Hill, D.W.; Kind, A.J. Reversed-phase solvent gradient HPLC retention indexes of drugs. *J.Anal.Toxicol.*, **1994**, *18*, 233–242

## SAMPLE
**Matrix:** solutions
**Sample preparation:** Dissolve in MeOH at a concentration of 1 mg/mL, inject a 20 μL aliquot.

## HPLC VARIABLES
**Column:** 250 × 5 Spherisorb S5W
**Mobile phase:** MeOH:buffer 90:10 (Buffer was 94 mL 35% ammonia and 21.5 mL 70% nitric acid in 884 mL water, adjust the pH to 10.1 with ammonia.)
**Flow rate:** 2
**Injection volume:** 20
**Detector:** UV 254

## CHROMATOGRAM
**Retention time:** 4.18

## OTHER SUBSTANCES
**Simultaneous:** acetylcodeine, amphetamine, benzphetamine, benzylmorphine, bromo-STP, buprenorphine, caffeine, chlorphentermine, codeine, codeine-N-oxide, dextromoramide, dextropropoxyphene, diamorphine, diethylpropion, dihydrocodeine, dihydromorphine, dimethylamphetamine, dipipanone, ephedrine, epinephrine, ethoheptazine, ethylmorphine, etorphine, fencamfamin, fenethyline, fenfluramine, fentanyl, 4-hydroxyamphetamine, hydroxypethidine, levallorphan, levorphanol, mazindol, meperidine, mephentermine, methadone, methylenedioxyamphetamine, methylephedrine, methylphenidate, monoacetylmorphine, morphine, morphine-3-glucuronide, morphine-N-oxide, nalorphine, naloxone, norcodeine, norlevorphanol, normetanephrine, normethadone, normorphine, norpipanone, norpseudoephedrine, noscapine, oxycodone, papaverine, pemoline, pentazocine, phenazocine, phendimetrazine, phenelzine, 2-phenethylamine, phenoperidine, phentermine, phenylephrine, phenylpropanolamine, pholcodeine, pipradol, piritramide, prolintane, pseudoephedrine, STP, thebacon, thebaine, tranylcypromine, trimethoxyamphetamine, tyramine
**Noninterfering:** dopamine, levodopa, methyldopa, methyldopate, norepinephrine
**Interfering:** mescaline, methamphetamine, norpethidine

## REFERENCE
Law, B.; Gill, R.; Moffat, A.C. High-performance liquid chromatography retention data for 84 basic drugs of forensic interest on a silica column using an aqueous methanol eluent. *J.Chromatogr.*, **1984**, *301*, 165–172

## SAMPLE
**Matrix:** urine
**Sample preparation:** 500 μL Urine + N-ethylnordiazepam + chlorpheniramine + 100 μL buffer, centrifuge at 11000 g for 30 s, inject a 500 μL aliquot onto column A with mobile phase A, after 0.6 min backflush column A with mobile phase A to waste for 1.6 min, elute column A with 250 μL mobile phase B, with 200 μL mobile phase C, and with 1.15 mL mobile phase D. Elute column A to waste until drugs start to emerge then elute onto column B. Elute column B to waste until drugs started to emerge, then elute onto column C. When all the drugs have emerged from column B remove it from the circuit, elute column C with mobile phase D, monitor the effluent from column C. Flush column A with 7 mL mobile phase E, with mobile phase D, and mobile phase A. Flush column B with 5 mL mobile phase E then with mobile phase D. (Buffer was 6 M ammonium acetate adjusted to pH 8.0 with 2 M KOH.)

## HPLC VARIABLES
**Column:** A 10 × 2.1 12-20 μm PRP-1 spherical poly(styrene-divinylbenzene) (Hamilton); B 10 × 3.2 11 μm Aminex A-28 (Bio-Rad); C 25 × 3.2 5 μm C8 (Phenomenex) + 150 × 4.6 5 μm silica (Macherey-Nagel)
**Mobile phase:** A 0.1% pH 8.0 potassium borate buffer; B 6 mM $KH_2PO_4$ containing 5 mM tetramethylammonium hydroxide, and 2 mM dimethyloctylamine, pH adjusted to 6.50 with phosphoric acid; C MeCN:buffer 40:60 (Buffer was 6 mM $KH_2PO_4$ containing 5 mM tetramethylammonium hydroxide, and 2 mM dimethyloctylamine, pH adjusted to 6.50 with phosphoric acid.); D MeCN:buffer 33:67 (Buffer was 6 mM $KH_2PO_4$ containing 5 mM tetramethylammonium hydroxide, and 2 mM dimethyloctylamine, pH adjusted to 6.50 with phosphoric acid.); E MeCN:buffer 70:30 (Buffer was 6 mM $KH_2PO_4$ containing 5 mM tetramethylammonium hydroxide, and 2 mM dimethyloctylamine, pH adjusted to 6.50 with phosphoric acid.)
**Column temperature:** 40 (B, C only)
**Flow rate:** A 5; B-E 1
**Injection volume:** 500
**Detector:** UV 210; UV 235

## CHROMATOGRAM
**Retention time:** k' 8.0
**Internal standard:** N-ethylnordiazepam (k' 2.1), chlorpheniramine (k' 5.9)
**Limit of detection:** 300 ng/mL

## OTHER SUBSTANCES
**Extracted:** amitriptyline, amphetamine, benzoylecgonine, caffeine, codeine, cotinine, desipramine, diazepam, diphenhydramine, ephedrine, flurazepam, hydromorphone, imipramine, lidocaine, methadone, methamphetamine, morphine, nordiazepam, nortriptyline, oxazepam, pentazocine, phenmetrazine, phenobarbital, phentermine, phenylpropanolamine, secobarbital

## KEY WORDS
column-switching

## REFERENCE
Binder, S.R.; Regalia, M.; Biaggi-McEachern, M.; Mazhar, M. Automated liquid chromatographic analysis of drugs in urine by on-line sample cleanup and isocratic multi-column separation. *J.Chromatogr.*, **1989**, *473*, 325–341

◆          ◆          ◆

## ANNOTATED BIBLIOGRAPHY
Alvi, S.U.; Castro, F. A stability-indicating simultaneous analysis of acetaminophen and hydrocodone bitartrate in tablets formulation by HPLC. *J.Liq.Chromatogr.*, **1987**, *10*, 3413–3426 [stability-indicating; simultaneous impurities, acetaminophen, codeine, hydromorphone; tablets; column temp 30]

Fatmi, A.A.; Williams, G.V. Simultaneous determination of acetaminophen and hydrocodone bitartrate in solid dosage forms by HPLC. *J.Liq.Chromatogr.*, **1987**, *10*, 2461–2472

# Hydrocortisone

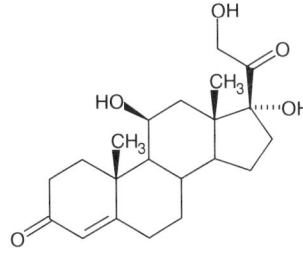

**Molecular formula:** $C_{21}H_{30}O_5$
**Molecular weight:** 362.5
**CAS Registry No.:** 50-23-7, 13609-67-1 (butyrate), 57524-89-7
(valerate), 50-03-3 (acetate), 3863-59-0 (phosphate),
6000-74-4 (sodium phosphate), 125-04-2
(21-sodium succinate), 508-96-3 (tebutate), 74050-20-7
(aceponate), 72590-77-3 (buteprate), 508-99-6 (cypionate),
83784-20-7 (hemisuccinate monohydrate), 2203-97-6 (hemisuccinate)

## SAMPLE
**Matrix:** amniotic fluid, blood
**Sample preparation:** Centrifuge serum or amniotic fluid for 10 min. 0.5-1 mL Serum or
  amniotic fluid + 500 μL MeOH:water 5:95, mix, inject 750 μL onto column A with mobile
  phase A, after 5 min elute contents of column A onto column B with mobile phase B,
  monitor effluent from column B.

## HPLC VARIABLES
**Column:** A Serumont-25 (Sekisui); B 260 × 4.6 5 μm Medipola-ODS C18 (Sekisui)
**Mobile phase:** A water; B MeCN:MeOH:buffer 2:7:20 (Buffer was 6.8 g/L $KH_2PO_4$, pH
  adjusted to 3.1 with concentrated phosphoric acid.)
**Column temperature:** 40
**Flow rate:** A 0.8; B 1
**Injection volume:** 750
**Detector:** UV 245

## CHROMATOGRAM
**Retention time:** 66
**Limit of detection:** 7.8 ng

## OTHER SUBSTANCES
**Extracted:** cortisone, estetrol, estriol
**Noninterfering:** androstenedione, corticosterone, hydroxyprogesterone, progesterone, tes-
  tosterone

## KEY WORDS
serum; column-switching

## REFERENCE
Noma, J.; Hayashi, N.; Sekiba, K. Automated direct high-performance liquid chromatographic assay for
  estetrol, estriol, cortisone and cortisol in serum and amniotic fluid. *J.Chromatogr.*, **1991**, *568*, 35–
  44

## SAMPLE
**Matrix:** blood
**Sample preparation:** 750 μL Serum + 75 μL MeOH + 100 μL 1.5 μg/mL dexamethasone
  in MeOH + 2 mL ethyl acetate, shake for 10 min, centrifuge at 2500 g for 10 min. Remove
  1.9 mL of the organic layer and evaporate it to dryness under a stream of nitrogen at
  45°, reconstitute the residue in 100 μL ethyl acetate, inject a 17 μL aliquot.

## HPLC VARIABLES
**Guard column:** 10 × 4 5 μm LiChrosorb Si 60

**Column:** 250 × 4 5 μm LiChrosorb Si 60
**Mobile phase:** n-Hexane:dichloromethane:MeOH:acetic acid 266:120:26:0.8 (Prepare by mixing an aliquot of mobile phase with an aliquot of mobile phase saturated with water.)
**Flow rate:** 2
**Injection volume:** 17
**Detector:** UV 242

## CHROMATOGRAM
**Retention time:** 12.79
**Internal standard:** dexamethasone (11.43)
**Limit of quantitation:** 5 ng/mL

## OTHER SUBSTANCES
**Extracted:** prednisolone, prednisolone acetate

## KEY WORDS
serum; normal phase

## REFERENCE
Döppenschmitt, S.A.; Scheidel, B.; Harrison, F.; Surmann, J.P. Simultaneous determination of prednisolone, prednisolone acetate and hydrocortisone in human serum by high-performance liquid chromatography. *J.Chromatogr.B*, **1995**, *674*, 237–246

## SAMPLE
**Matrix:** blood
**Sample preparation:** Centrifuge plasma at 2500 g for 10 min, mix the supernatant with an equal volume of 1 M pH 3.0 glycine buffer containing 0.2% Tween 20, centrifuge at 2500 g for 10 min, inject an aliquot of the supernatant onto column A and elute to waste with mobile phase, after 3 min divert the effluent from column A onto column B, after 3 min remove column A from the circuit, elute column B with mobile phase, monitor the effluent from column B. Backflush column A with mobile phase for 28 min.

## HPLC VARIABLES
**Column:** A 30 × 2.1 Spherisorb C1 pH stable; B 150 × 2.1 Spherisorb C1 pH stable
**Mobile phase:** 5 mM pH 7.3 Tris-nitric acid buffer containing 0.1% Tween 20 and 150 mM sodium nitrate
**Column temperature:** 40
**Flow rate:** 0.2
**Injection volume:** 50
**Detector:** UV

## CHROMATOGRAM
**Retention time:** 23

## OTHER SUBSTANCES
**Extracted:** cortisone, prednisolone

## KEY WORDS
plasma; column-switching; heart-cut

## REFERENCE
Lövgren, U.; Johansson, M.; Kronkvist, K.; Edholm, L.-E. Biocompatible sample pretreatment for immunochemical techniques using micellar liquid chromatography for separation of corticosteroids. *J.Chromatogr.B*, **1995**, *672*, 33–44

## SAMPLE

**Matrix:** blood

**Sample preparation:** Condition an Empore C8 extraction disc (3M Co.) by adding 500 µL MeOH and forcing through three drops, discard the remaining liquid, add water, force through three drops, discard the water. 300 µL Serum + 150 µL IS solution, let stand at room temperature for 10 min, add 800 µL saturated sodium borate solution, mix, centrifuge at 12400 g for 3 min (if necessary), add to the extraction disc, centrifuge at 100-120 g for 5 min, force through 200 µL water, force through 500 µL MeOH:water 18:82, elute with 50 µL MeCN then 150 µL water, mix the eluates, inject a 20 µL aliquot. (IS solution contained 0.5 mg/L fludrocortisone and 0.75 mg/L methylprednisolone in 400 mM HCl.) (The extraction disc permits use of lower volumes of eluate than a conventional SPE cartridge.)

## HPLC VARIABLES

**Guard column:** 20 × 2 30 µm Permaphase ETH (Du Pont)

**Column:** 250 × 2 Ultrasphere C18 or 250 × 4.6 Ultrasphere C18

**Mobile phase:** THF:water 20:80 (Use a 150 × 4.6 37-53 µm silica gel (Whatman) saturating column (held at 55°) between the pump and the injector.)

**Column temperature:** 55

**Flow rate:** 0.18 (250 × 2) or 0.8 (250 × 4.6)

**Injection volume:** 20

**Detector:** UV 254

## CHROMATOGRAM

**Retention time:** 13

**Internal standard:** fludrocortisone (15), methylprednisolone (20)

**Limit of detection:** 4 ng/mL

## OTHER SUBSTANCES

**Extracted:** corticosterone, cortisone, prednisolone, prednisone

**Simultaneous:** aldosterone, androsteindione, beclomethasone, 11-deoxycorticosterone, 11-deoxycortisol, 21-deoxycortisone, dexamethasone, 17-hydroxyprogesterone, metyrapone, pregnenolone, progesterone, testosterone, triamcinolone

## KEY WORDS

serum; SPE; extraction disc

## REFERENCE

Lensmeyer, G.L.; Onsager, C.; Carlson, I.H.; Wiebe, D.A. Use of particle-loaded membranes to extract steroids for high-performance liquid chromatographic analyses. Improved analyte stability and detection. *J.Chromatogr.A*, **1995**, *691*, 239–246

## SAMPLE

**Matrix:** blood

**Sample preparation:** Extract 1 mL plasma containing dexamethasone with 12 mL dichloromethane. Remove the organic phase and wash it with 2 mL 100 mM NaOH, wash with 1 mL water, dry over 1 g anhydrous sodium sulfate. Evaporate to dryness under a stream of nitrogen, reconstitute with mobile phase, inject an aliquot.

## HPLC VARIABLES

**Column:** 150 × 4.6 3 µm Spherisorb silica

**Mobile phase:** Hexane:dichloromethane:EtOH:glacial acetic acid 26:69:3.4:2

**Flow rate:** 0.75

**Detector:** UV 254

## CHROMATOGRAM

**Retention time:** 6.3

**Internal standard:** dexamethasone (5.1)
**Limit of quantitation:** 10 ng/mL

## OTHER SUBSTANCES
**Extracted:** methylprednisolone, prednisolone

## KEY WORDS
plasma; pharmacokinetics

## REFERENCE
Möllmann, H.; Hochhaus, G.; Rohatagi, S.; Barth, J.; Derendorf, H. Pharmacokinetic/pharmacodynamic
evaluation of deflazacort in comparison to methylprednisolone and prednisolone. *Pharm.Res.*, **1995**,
*12*, 1096–1100

## SAMPLE
**Matrix:** blood
**Sample preparation:** Extract plasma with 12 mL dichloromethane, wash the organic layer
with 2 mL 100 mM NaOH and 1 mL water, dry the organic layer over 1 g anhydrous
sodium sulfate. Evaporate the organic layer to dryness under a stream of nitrogen, reconstitute with mobile phase, inject an aliquot.

## HPLC VARIABLES
**Column:** 150 × 4.6 3 μm Spherisorb silica
**Mobile phase:** Hexane:dichloromethane:EtOH:glacial acetic acid 26:69:3.4:2
**Flow rate:** 0.75
**Detector:** UV 254

## CHROMATOGRAM
**Retention time:** 11.6
**Internal standard:** methylprednisolone (13.4)
**Limit of quantitation:** 5 ng/mL

## OTHER SUBSTANCES
**Extracted:** triamcinolone acetonide
**Noninterfering:** cortisone

## KEY WORDS
plasma; normal phase

## REFERENCE
Rohatagi, S.; Hochhaus, G.; Möllmann, J.; Barth, J.; Galia, E.; Erdmann, M.; Sourgens, H.; Derendorf,
H. Pharmacokinetic and pharmacodynamic evaluation of triamcinolone acetonide after intravenous,
oral, and inhaled administration. *J.Clin.Pharmacol.*, **1995**, *35*, 1187–1193

## SAMPLE
**Matrix:** blood
**Sample preparation:** 1 mL Serum + 100 μL water containing 5 μg/mL 2,3-diaminona-
phthalene and 3.5 μg/mL 18-hydroxy-11-deoxycorticosterone + 1 mL 250 mM NaOH + 7
mL diethyl ether, shake on a rotary shaker for 15 min, repeat extraction. Combine the
organic layers and evaporate them to dryness under a stream of nitrogen at 30-40°, reconstitute the residue in 70 μL MeOH:100 mM perchloric acid 50:50, inject a 20 μL
aliquot.

## HPLC VARIABLES
**Column:** 150 × 3.9 4 μm Nova-Pak C18

**Mobile phase:** Gradient. A was 58 mM $NaH_2PO_4$ containing 6 mM sodium heptanesulfonate, adjusted to pH 3.1 with concentrated phosphoric acid. B was MeCN:MeOH 85:15. A:B from 100:0 to 78:22 over 5 min, to 70:30 over 12 min, maintain at 70:30 for 4 min, to 65:35 over 9 min.
**Flow rate:** 1
**Injection volume:** 20
**Detector:** UV 245; UV 256; UV 343

## CHROMATOGRAM
**Retention time:** 14.37
**Internal standard:** 2,3-diaminonaphthalene (10.71), 18-hydroxy-11-deoxycorticosterone (15.85)
**Limit of detection:** 1-10 ng/mL (245 nm)

## OTHER SUBSTANCES
**Extracted:** betamethasone
**Noninterfering:** aspirin, ibuprofen, indomethacin, phenylbutazone, pregnenolone, chloroquine, corticosterone, cortisone, dexamethasone, fluendrenolide, fluocinolone acetonide, fluorometholone, fluprednisolone, hydroxychloroquine, 17δ-hydroxyprogesterone, meprednisone, methylprednisolone acetate, methylprednisolone, paramethasone, prednisolone, prednisone, progesterone, triamcinolone

## KEY WORDS
serum

## REFERENCE
Volin, P. Simple and specific reversed-phase liquid chromatographic method with diode-array detection for simultaneous determination of serum hydroxychloroquine, chloroquine and some corticosteroids. *J.Chromatogr.B*, **1995**, *666*, 347–353

## SAMPLE
**Matrix:** blood
**Sample preparation:** Prepare a Sep-Pak Plus Environmental C18 SPE cartridge by washing with 15 mL MeOH then 15 mL water. 1 mL Serum + 100 µL 3 µg/mL betamethasone in isopropanol:MeCN 1:1 + 100 µL isopropanol:acetonitrile 1:1, mix, add to SPE cartridge, wash with 10 mL water, elute with 3 mL MeOH. Evaporate the eluate at 50° under a stream of nitrogen, reconstitute in 200 µL mobile phase A, inject a 20 µL aliquot.

## HPLC VARIABLES
**Guard column:** µBondapak C18 guard column
**Column:** 250 × 4.6 5 µm Hypersil ODS
**Mobile phase:** Gradient. A was isopropanol:50 mM pH 4.5 acetate buffer 10:90. B was isopropanol:50 mM pH 4.5 acetate buffer 30:70. A:B from 90:10 to 30:70 over 25 min, hold at 30:70 for 5 min, to 90:10 over 5 min, hold at 90:10 for 15 min before next injection.
**Column temperature:** 40
**Flow rate:** 1
**Injection volume:** 20
**Detector:** UV 254

## CHROMATOGRAM
**Retention time:** 26
**Internal standard:** betamethasone (33)
**Limit of quantitation:** 10 ng/mL

## OTHER SUBSTANCES
**Simultaneous:** metabolites, cortisone, prednisolone, prednisone

## KEY WORDS
serum; SPE

## REFERENCE
Hirata, H.; Kasama, T.; Sawai, Y.; Fike, R.R. Simultaneous determination of deflazacort metabolites II and III, cortisol, cortisone, prednisolone and prednisone in human serum by reversed-phase high-performance liquid chromatography. *J.Chromatogr.B*, **1994**, *658*, 55–61

## SAMPLE
**Matrix:** blood
**Sample preparation:** 1 mL Plasma + 50 μL 4 μg/mL betamethasone in EtOH + 15 mL dichloromethane, shake horizontally for 15 min, centrifuge at 1500 g for 15 min. Remove the organic layer and wash it with 100 μL 100 mM NaOH then 1 mL water. Remove the aqueous phase and dry the organic phase over 1 g of anhydrous sodium sulfate. Evaporate the organic phase to dryness under a stream of nitrogen at not more than 37°, reconstitute in 200 μL mobile phase, inject a 175 μL aliquot.

## HPLC VARIABLES
**Guard column:** 20 × 2 30-38 μm HC Pellosil
**Column:** 250 × 4.6 5-6 μm Zorbax SIL
**Mobile phase:** Heptane:dichloromethane:glacial acetic acid:ethanol 350:600:10:35
**Flow rate:** 2
**Injection volume:** 175
**Detector:** UV 254

## CHROMATOGRAM
**Retention time:** 15
**Internal standard:** betamethasone (12)
**Limit of detection:** 5 ng/mL
**Limit of quantitation:** 10 ng/mL

## OTHER SUBSTANCES
**Simultaneous:** prednisolone, prednisone
**Noninterfering:** cyclosporin, ethinyl estradiol, ketoconazole, levonorgestrel, rapamycin, tacrolimus, tenidap, tetrahydrocortisone

## KEY WORDS
plasma; normal phase

## REFERENCE
Jusko, W.J.; Pyszczynski, N.A.; Bushway, M.S.; D'Ambrosio, R.; Mis, S.M. Fifteen years of operation of a high-performance liquid chromatographic assay for prednisolone, cortisol and prednisone in plasma. *J.Chromatogr.B*, **1994**, *658*, 47–54

## SAMPLE
**Matrix:** blood
**Sample preparation:** 100 μL Plasma + 1 mL 50 ng/mL beclomethasone in ethyl acetate, vortex, centrifuge at 11000-12300 g for 5 min. Evaporate the supernatant under a stream of nitrogen at 50-60°, reconstitute in 100 μL mobile phase, inject a 20-80 μL aliquot.

## HPLC VARIABLES
**Column:** 250 × 4.6 5 μm Ultrasphere ODS
**Mobile phase:** MeCN:10 mM pH 7.0 phosphate buffer 45:55
**Flow rate:** 1

**Injection volume:** 20-80
**Detector:** UV 240

## CHROMATOGRAM
**Retention time:** 9
**Internal standard:** beclomethasone (22)
**Limit of quantitation:** 15 ng/mL

## OTHER SUBSTANCES
**Extracted:** dexamethasone
**Noninterfering:** albuterol, amoxicillin, ceftriaxone, erythromycin, furosemide, gentamicin, indomethacin, midazolam, morphine, nystatin, theophylline, vancomycin

## KEY WORDS
plasma; pharmacokinetics

## REFERENCE
Schild, P.N.; Charles, B.G. Determination of dexamethasone in plasma of premature neonates using high-performance liquid chromatography. *J.Chromatogr.B*, **1994**, *658*, 189–192

## SAMPLE
**Matrix:** blood
**Sample preparation:** Condition a 2 mL 200 mg Tef Elutor C18 SPE cartridge (Versa Prep) with 3 mL MeOH and two 3 mL portions of water. 1 mL Plasma + 50 μL 400 ng/mL flumethasone in MeOH:water 5:95, heat at 50° for 10 min, add to the SPE cartridge, wash with 2 mL water, wash with 1 mL MeOH:water 10:90, wash with 4 mL acetone:water 20:80, air-dry for 10 min, elute with 1 mL MeOH. Evaporate the eluate to dryness under a stream of nitrogen at 45°, reconstitute the residue in 50 μL mobile phase, inject a 25 μL aliquot.

## HPLC VARIABLES
**Column:** 100 × 2 3 μm Hypersil
**Mobile phase:** MeCN:THF:water 8:10:82 containing 5 mL/L triethylamine, pH adjusted to 6.5 with citric acid
**Flow rate:** 0.6
**Injection volume:** 25
**Detector:** UV 242

## CHROMATOGRAM
**Retention time:** 4.5
**Internal standard:** flumethasone (13)
**Limit of detection:** 0.3 ng/mL

## OTHER SUBSTANCES
**Extracted:** corticosterone, cortisone
**Simultaneous:** acebutolol, acetazolamide, acetophenetidin, adrenosterone, aldosterone, amitriptyline, androsten-3,17-dione, aspirin, carbamazepine, cephalothin, chlorothiazide, dehydrocorticosterone, deoxycorticosterone, deoxycortisol, desipramine, dexamethasone, diazepam, equilenin, estradiol, estriol, estrone, fluoromethatone, furosemide, hydrochlorothiazide, hydroxycorticosterone, hydroxyprogesterone, hydroxyprogesterone, imipramine, indomethacin, methylhydroxyprogesterone, methylprednisolone, nandrolone, nordiazepam, nortriptyline, pheniramine, phenobarbital, phenytoin, prednisolone, prednisone, primidone, probenecid, progesterone, quinine, spironolactone, testosterone, theophylline, triamcinolone, tripelennamine
**Noninterfering:** allopurinol, caffeine, cotinine, ephedrine, nicotine, phenylephrine
**Interfering:** chlordiazepoxide, diphenhydramine, propranolol

## KEY WORDS
serum; SPE

## REFERENCE
Hariharan, M.; Naga, S.; VanNoord, T.; Kindt, E.K. Assay of human plasma cortisone by liquid
chromatography: normal plasma concentrations (between 8 and 10 a.m.) of cortisone and corticosterone. *J.Chromatogr.*, **1993**, *613*, 195–201

## SAMPLE
**Matrix:** blood
**Sample preparation:** 100 μL Plasma + 10 μL IS in water, extract twice by shaking for 1
min with 1.2 mL dichloromethane, evaporate organic layer below 40° under reduced pressure, dissolve residue in 100 μL MeCN. Add 10 μL reagent 1, add 10 μL reagent 2, heat
at 70° for 20 min, cool to room temperature, add 100 μL water, add 200 μL MeOH:water
1:1, add to Sep-Pak C18 cartridge, wash vial with 2 mL MeOH:water 1:1 and add washings to cartridge, wash cartridge with 40 mL MeOH:water 1:1, elute with 5 mL MeOH.
Concentrate eluent to 500 μL by evaporation at 40° under reduced pressure, inject 20 μL
aliquot. (Reagent 1 was 30 mg 2-(4-carboxyphenyl)-5,6-dimethylbenzimidazole in 3 mL
pyridine, add 700 mg 4-piperidinopyridine, dilute to 10 mL with MeCN. Reagent 2 was
700 mg 1-isopropyl-3-(3-dimethylaminopropyl)carbodiimide perchlorate in 10 mL MeCN.)

## HPLC VARIABLES
**Guard column:** 50 × 4.6 7 μm Zorbax ODS
**Column:** 250 × 4.6 7 μm Zorbax ODS
**Mobile phase:** MeOH:water 75:25 containing 5 mM tetramethylammonium hydrogen
sulfate
**Flow rate:** 0.4
**Injection volume:** 20
**Detector:** F ex 334 em 418

## CHROMATOGRAM
**Retention time:** 26.5
**Internal standard:** fluocinolone acetonide (40.7)
**Limit of detection:** 0.6-3 pg/mL

## OTHER SUBSTANCES
**Simultaneous:** aldosterone, corticosterone, cortisone, dexamethasone, triamcinolone

## KEY WORDS
plasma; derivatization

## REFERENCE
Katayama, M.; Masuda, Y.; Taniguchi, H. Determination of corticosteroids in plasma by high-performance liquid chromatography after pre-column derivatization with 2-(4-carboxyphenyl)-5,6-dime-
thylbenzimidazole. *J.Chromatogr.*, **1993**, *612*, 33–39

## SAMPLE
**Matrix:** blood
**Sample preparation:** Condition a Tef Elutor C18 SPE cartridge with two 3 mL portions
of MeOH then two 3 mL portions of water. 1 mL Plasma + 50 μL 400 ng/mL flumethasone
in 5:95 MeOH:water, heat at 50° for 10 min, add to the SPE cartridge, wash with 2 mL
water, 1 mL MeOH:water 10:90, 4 mL acetone:water 20:80, apply suction to cartridge
for 10 min to air dry. Elute with 1 mL MeOH, evaporate eluent at 45° under nitrogen,
reconstitute with 50 μL mobile phase, inject 25 μL aliquot.

## HPLC VARIABLES
**Column:** 100 × 2 3 μm C18 Hypersil
**Mobile phase:** MeCN:THF:water 8:10:82, containing 5 mL/L triethylamine, pH adjusted to 6.5 with citric acid
**Flow rate:** 0.6
**Injection volume:** 25
**Detector:** UV 242

## CHROMATOGRAM
**Retention time:** 3.98
**Internal standard:** flumethasone (11.50)
**Limit of detection:** 300 pg/mL

## OTHER SUBSTANCES
**Simultaneous:** adrenosterone, amitriptyline, aspirin, carbamazepine, corticosterone, cortisone, deoxycorticosterone, desipramine, dexamethasone, diazepam, diphenhydramine, equilenin, estradiol, estriol, estrone, fluorometholone, hydroxyprogesterone, imipramine, indomethacin, methylprednisolone, nordiazepam, nortriptyline, phenobarbital, prednisolone, prednisone, probenecid, progesterone, propranolol, spironolactone, testosterone, theophylline, tripelennamine
**Noninterfering:** acebutolol, acetazolamide, acetophenetidin, aldosterone, allopurinol, caffeine, cephalothin, chlorothiazide, cotinine, ephedrine, furosemide, hydrochlorothiazide, nicotine, pheniramine, phenylephrine, phenytoin, primidone, quinine, triamcinolone
**Interfering:** chlordiazepoxide

## KEY WORDS
plasma; SPE

## REFERENCE
Hariharan, M.; Naga, S.; VanNoord, T.; Kindt, E.K. Simultaneous assay of corticosterone and cortisol in plasma by reversed-phase liquid chromatography. *Clin.Chem.*, **1992**, *38*, 346–352

## SAMPLE
**Matrix:** blood
**Sample preparation:** 100 μL Serum + 500 μL water + 100 μL 10 μg/mL 3,7-dimethoxyflavone in EtOH + 8 mL diethyl ether, shake, centrifuge at 4° at 1000 g for 5 min, freeze in acetone/dry ice. Remove the organic layer and dry it over anhydrous sodium sulfate, evaporate to dryness under a stream of nitrogen, reconstitute the residue in 100 μL MeOH:water 40:60, inject a 50 μL aliquot.

## HPLC VARIABLES
**Column:** 250 × 4.6 3 μm NS-Gel C18
**Mobile phase:** Gradient. MeOH:water from 40:60 to 55:45, maintain at 55:45 for 24 min, to 80:20 over 25 min.
**Column temperature:** 50
**Flow rate:** 1
**Injection volume:** 50
**Detector:** UV 210; UV 240

## CHROMATOGRAM
**Retention time:** 17.00
**Internal standard:** 3,7-dimethoxyflavone (47)

## OTHER SUBSTANCES
**Extracted:** aldosterone, androstenedione, dehydroepiandrosterone, 11-deoxycortisol, deoxycorticosterone, estradiol, estrone, 17-hydroxyprogesterone, progesterone, pregnenolone

## KEY WORDS
serum

## REFERENCE
Ueshiba, H.; Segawa, M.; Hayashi, T.; Miyachi, Y.; Irie, M. Serum profiles of steroid hormones in patients with Cushing's syndrome determined by a new HPLC/RIA method. *Clin.Chem.*, **1991**, *37*, 1329–1333

## SAMPLE
**Matrix:** blood
**Sample preparation:** 1 mL Plasma + 100 μL 3 M sulfuric acid + 50 μL 6 μg/mL prednisone in MeCN:MeOH 50:50, mix, add 15 mL hexane:ethyl acetate 50:50, shake for 20 min, centrifuge, freeze at -70°. Remove the organic layer and add it to 1 mL 1 M nitric acid, shake, freeze. Remove the organic layer and dry it over anhydrous sodium sulfate, evaporate to dryness under a stream of nitrogen at 30°, reconstitute the residue in 100 μL mobile phase, inject an aliquot.

## HPLC VARIABLES
**Guard column:** 12.5 × 4 5 μm Zorbax SIL
**Column:** three 80 × 4 5 μm Zorbax SIL Reliance 5 columns in series
**Mobile phase:** Dichloromethane:hexane:EtOH:glacial acetic acid 69:26:2.3:1 (Pass the mobile phase through a 70 × 6 37-53 μm HC-Pellocil (Whatman) column.)
**Flow rate:** 2
**Detector:** UV 254

## CHROMATOGRAM
**Retention time:** 15
**Internal standard:** prednisone (9)
**Limit of detection:** 15 ng/mL

## OTHER SUBSTANCES
**Extracted:** methylprednisolone, methylprednisolone hemisuccinate

## KEY WORDS
plasma; pharmacokinetics

## REFERENCE
Kong, A.-N.; Slaughter, R.L.; Jusko, W.J. Simultaneous analysis of methylprednisolone hemisuccinate, cortisol and methylprednisolone by normal-phase high-performance liquid chromatography in human plasma. *J.Chromatogr.*, **1988**, *432*, 308–314

## SAMPLE
**Matrix:** blood
**Sample preparation:** 100 μL Plasma + 200 μL IS solution, vortex 1 min, centrifuge at 1500 g for 10 min, inject 50 μL of supernatant. (Prepare IS solution by dissolving 200 μg n-propyl p-hydroxybenzoate in 10 mL MeOH, add 2 mL glacial acetic acid, dilute 1 mL of this solution with 9 mL MeOH.)

## HPLC VARIABLES
**Column:** 300 × 4 μBondapak C18
**Mobile phase:** MeCN:buffer 23:77 (Buffer was 50 mM sodium acetate + 100 mM NaCl, adjusted to pH 2.8 with acetic acid.)
**Flow rate:** 2.5
**Injection volume:** 50
**Detector:** UV 254

## CHROMATOGRAM
**Retention time:** 4.8 (hydrocortisone), 8.4 (hydrocortisone succinate)
**Internal standard:** n-propyl p-hydroxybenzoate (7.2)
**Limit of detection:** 200 ng/mL; 500 ng/mL (succinate)

## OTHER SUBSTANCES
**Simultaneous:** bromhexine, noscapine, tipepidine
**Noninterfering:** albuterol, orciprenaline, terbutaline, theophylline

## KEY WORDS
plasma

## REFERENCE
Iwasaki, E. Hydrocortisone succinate and hydrocortisone simultaneously determined in plasma by reversed-phase liquid chromatography, and their pharmacokinetics in asthmatic children. *Clin.Chem.*, **1987**, *33*, 1412–1415

## SAMPLE
**Matrix:** blood
**Sample preparation:** Condition a Bond-Elut C18 SPE cartridge by washing with 2 mL MeCN, 2 mL acetone:water 2:98, and 4 mL water. Do not allow cartridge to run dry. 2 mL Plasma + 40 μL 5 μg/mL dexamethasone in MeOH, add to the SPE cartridge, allow to sit for 15 min, wash twice with 2 mL water, wash twice with 2 mL acetone:water 2:98, pull a vacuum on the column for 15 min, elute with 1 mL MeCN under vacuum. Evaporate the eluate to dryness under a stream of nitrogen at 40°, dissolve the residue in 150 μL dichloromethane, inject a 100 μL aliquot.

## HPLC VARIABLES
**Column:** 250 × 4.6 5 μm LiChrosorb Si-60
**Mobile phase:** Dichloromethane:water-saturated dichloromethane:THF:MeOH:glacial acetic acid 664.5:300:10:25:0.5
**Flow rate:** 0.8
**Injection volume:** 100
**Detector:** UV 254

## CHROMATOGRAM
**Retention time:** 27
**Internal standard:** dexamethasone (23.5)
**Limit of detection:** 10 ng/mL

## OTHER SUBSTANCES
**Simultaneous:** cortisone, prednisolone, prednisolone acetate, prednisone

## KEY WORDS
plasma; normal phase; pig; SPE

## REFERENCE
Prasad, V.K.; Ho, B.; Haneke, C. Simultaneous determination of prednisolone acetate, prednisolone, prednisone, cortisone and hydrocortisone in swine plasma using solid-phase and liquid-liquid extraction techniques. *J.Chromatogr.*, **1986**, *378*, 305–316

## SAMPLE
**Matrix:** blood
**Sample preparation:** 2 mL Plasma + 40 μL 5 μg/mL dexamethasone in MeOH, vortex 30 s, add 5 mL dichloromethane:diethyl ether 50:50, vortex for 15 s, repeat extraction, com-

bine organic layers and wash them with 4 mL 100 mM NaOH, centrifuge. Remove the organic layer and dry it over anhydrous sodium sulfate, evaporate to dryness under a stream of nitrogen at 40°, dissolve the residue in 150 μL dichloromethane, inject a 100 μL aliquot.

## HPLC VARIABLES
**Column:** 250 × 4.6 5 μm LiChrosorb Si-60
**Mobile phase:** Dichloromethane : water-saturated dichloromethane : THF : MeOH : glacial acetic acid 664.5 : 300 : 10 : 25 : 0.5
**Flow rate:** 0.8
**Injection volume:** 100
**Detector:** UV 254

## CHROMATOGRAM
**Retention time:** 27
**Internal standard:** dexamethasone (23.5)
**Limit of detection:** 5 ng/mL

## OTHER SUBSTANCES
**Simultaneous:** cortisone, prednisolone, prednisolone acetate, prednisone

## KEY WORDS
plasma; normal phase; pig

## REFERENCE
Prasad, V.K.; Ho, B.; Haneke, C. Simultaneous determination of prednisolone acetate, prednisolone, prednisone, cortisone and hydrocortisone in swine plasma using solid-phase and liquid-liquid extraction techniques. *J.Chromatogr.*, **1986**, *378*, 305–316

## SAMPLE
**Matrix:** blood
**Sample preparation:** 1 mL Plasma + 80 μL 3.125 μg/mL dexamethasone in MeOH, mix, add 15 mL dichloromethane, shake for 20 min, centrifuge. Remove organic phase and wash it with 1 mL 100 mM NaOH then with 1 mL water. Remove organic phase and dry it with 1 g anhydrous sodium sulfate. Evaporate to dryness at 45° under a stream of nitrogen, reconstitute in 200 μL mobile phase, inject.

## HPLC VARIABLES
**Guard column:** 70 × 6 37-53 μm Whatman HC-Pellocil
**Column:** 250 × 4.6 5-6 μm Zorbax SIL
**Mobile phase:** Hexane:dichloromethane:ethanol:acetic acid 26:69:3.4:1
**Flow rate:** 2
**Injection volume:** 200
**Detector:** UV 254

## CHROMATOGRAM
**Retention time:** 10
**Internal standard:** dexamethasone (8)
**Limit of detection:** 2 ng/mL
**Limit of quantitation:** 10 ng/mL

## OTHER SUBSTANCES
**Simultaneous:** beclomethasone, betamethasone, corticosterone, cortisone, fluocinonide, methylprednisolone, methylprednisone, prednisolone, prednisone

## KEY WORDS
plasma; normal phase

## REFERENCE
Ebling, W.F.; Szefler, S.J.; Jusko, W.J. Analysis of cortisol, methylprednisolone, and methylprednisolone hemisuccinate. Absence of effects of troleandomycin on ester hydrolysis. *J.Chromatogr.*, **1984**, *305*, 271–280

## SAMPLE
**Matrix:** blood
**Sample preparation:** 1 mL Plasma + 10 μL 10 μg/mL prednisolone in MeOH, add 1 mL 0.1 M NaOH, add 10 mL dichloromethane, shake for 10 min, centrifuge at 8400 g at 4° for 10 min. Remove organic layer and evaporate it at 40° under a stream of nitrogen. Dissolve residue in 100 μL mobile phase and inject.

## HPLC VARIABLES
**Column:** 100 × 8 radial compression 10 μm Radialpack B
**Mobile phase:** Dichloromethane:MeOH:acetic acid 96:4:0.4
**Flow rate:** 1.5
**Injection volume:** 100
**Detector:** UV 254

## CHROMATOGRAM
**Retention time:** 6.0
**Internal standard:** prednisolone (7.5)
**Limit of quantitation:** 2 ng/mL

## OTHER SUBSTANCES
**Simultaneous:** corticosterone, dexamethasone

## KEY WORDS
plasma; dog; normal phase

## REFERENCE
Alvinerie, M.; Toutain, P.L. Simultaneous determination of corticosterone, hydrocortisone, and dexamethasone in dog plasma using high-performance liquid chromatography. *J.Pharm.Sci.*, **1982**, *71*, 816–818

## SAMPLE
**Matrix:** blood
**Sample preparation:** 1 mL Plasma + 150 ng dexamethasone + 1 mL 100 mM NaOH + 10 mL ether:dichloromethane 60:40, shake for 10 min, centrifuge at 300 g for 5 min. Remove the organic layer and add it to 1 mL 100 mM HCl, shake for 5 min, centrifuge at 300 g for 5 min. Remove the organic layer and evaporate it to dryness under a stream of nitrogen, reconstitute the residue in 100 μL mobile phase, inject a 50 μL aliquot.

## HPLC VARIABLES
**Guard column:** μBondapak/Corasil (Waters)
**Column:** 300 × 3.9 10 μm μPorasil (Waters)
**Mobile phase:** Dichloromethane:glacial acetic acid 99:1 (Prepare dichloromethane as follows. Stir 500 mL dichloromethane, 30 mL EtOH, and 30 mL water for 1 h, use the lower organic layer.)
**Flow rate:** 2
**Injection volume:** 50
**Detector:** UV 254

## CHROMATOGRAM
**Retention time:** 6.5
**Internal standard:** dexamethasone (5)

## OTHER SUBSTANCES
**Extracted:** prednisolone, prednisone

## KEY WORDS
plasma; normal phase

## REFERENCE
Hartley, R.; Brocklebank, J.T. Determination of prednisolone in plasma by high-performance liquid chromatography. *J.Chromatogr.*, **1982**, *232*, 406–412

## SAMPLE
**Matrix:** blood
**Sample preparation:** 1 mL Plasma + 5 mL water + 1 mL 2 μg/mL equilenin in MeOH + 50 μL 0.1 M NaOH to adjust pH to 10, vortex briefly after each addition, shake with 10 mL dichloromethane for 10 min, centrifuge at 2000 g for 10 min. Wash organic layer twice with 2 mL water, centrifuge 5 min, evaporate 8 mL of organic phase to dryness at 40° under a stream of nitrogen, reconstitute residue in 150 μL mobile phase, inject 25 μL aliquot

## HPLC VARIABLES
**Column:** 300 × 4 10 μm μBondapak C18
**Mobile phase:** MeOH:buffer 60:40 (Buffer was 10 mL 200 mM acetic acid + 15 mL 200 mM sodium acetate made up to 1 L, pH 4.8.)
**Flow rate:** 1
**Injection volume:** 25
**Detector:** UV 254

## CHROMATOGRAM
**Retention time:** 4.5
**Internal standard:** equilenin (7.5)
**Limit of detection:** 5 ng/mL

## OTHER SUBSTANCES
**Simultaneous:** betamethasone, deoxycortisol, dexamethasone, prednisone, triamcinolone
**Interfering:** prednisolone

## KEY WORDS
Anal.Abs. 1982, 43, 4D182; plasma

## REFERENCE
Bouquet, S.; Brisson, A.M.; Gombert, J. Dosage du cortisol et du 11-désoxycortisol plasmatiques par chromatographie liquide haute performance [Cortisol and 11-desoxycortisol determination in blood by high performance liquid chromatography]. *Ann.Biol.Clin.(Paris)*, **1981**, *39*, 189–191

## SAMPLE
**Matrix:** blood
**Sample preparation:** 1 mL Plasma + 4 mL 59.5 ng/mL triamcinolone acetonide in dichloromethane, shake at high speed for 15 min, centrifuge at 2000 rpm for 10 min, remove aqueous layer, add 5 mL saturated sodium bicarbonate solution to the organic layer, shake at high speed for 5 min, centrifuge at 2000 rpm for 10 min, remove aqueous layer. Place

organic layer in a pointed tube and evaporate to dryness at 45° under a stream of nitrogen. Reconstitute with 50 µL mobile phase, inject 20 µL aliquot.

## HPLC VARIABLES
**Column:** 10 µm Porasil
**Mobile phase:** Hexane:dichloromethane:ethanol:acetic acid 68.8:25:6:0.2
**Flow rate:** 2.5
**Injection volume:** 20
**Detector:** UV 254

## CHROMATOGRAM
**Retention time:** 7
**Internal standard:** triamcinolone acetonide (3)

## OTHER SUBSTANCES
**Simultaneous:** prednisolone, prednisone

## KEY WORDS
plasma; normal phase

## REFERENCE
Agabeyoglu, I.T.; Wagner, J.G.; Kay, D.R. A sensitive high-pressure liquid chromatographic method for the determination of prednisone, prednisolone and hydrocortisone in plasma. *Res.Commun.Chem. Pathol.Pharmacol.*, **1980**, *28*, 163–176

## SAMPLE
**Matrix:** blood
**Sample preparation:** 1 mL Plasma + 100 µL 2 µg/mL dexamethasone in EtOH:water 10:90 + 100 µL 250 mM NaOH + 7 mL ether:dichloromethane 60:40, vortex for 30 s, centrifuge at 2000 rpm for 5 min. Remove 6 mL of the organic layer and evaporate it to dryness under a stream of air at 40°, reconstitute the residue in 100 µL dichloromethane:EtOH:water 95:4:1, inject a 50 µL aliquot.

## HPLC VARIABLES
**Column:** 250 × 4.5 5 µm Partisil silica
**Mobile phase:** Dichloromethane:EtOH:water 95:4:1
**Flow rate:** 1.5
**Injection volume:** 50
**Detector:** UV 239

## CHROMATOGRAM
**Retention time:** 14
**Internal standard:** dexamethasone (11.5)
**Limit of quantitation:** 25 ng/mL

## OTHER SUBSTANCES
**Extracted:** corticosterone, 11-deoxycortisol, 17-hydroxyprogesterone, 6α-methylprednisolone, prednisolone, prednisone, progesterone

## KEY WORDS
plasma; normal phase

## REFERENCE
Scott, N.R.; Chakraborty, J.; Marks, V. Determination of prednisolone, prednisone, and cortisol in human plasma by high-performance liquid chromatography. *Anal.Biochem.*, **1980**, *108*, 266–268

## SAMPLE
**Matrix:** blood, tissue, urine
**Sample preparation:** Urine. 1 mL Urine + 1 mL MeOH:EtOH 50:50, centrifuge at 4000 g for 10 min. Remove the supernatant and evaporate to about 200 μL under a stream of nitrogen at 37°, inject a 5-20 μL aliquot. Plasma. Mix plasma with an equal volume of MeOH:EtOH 50:50, let stand at -20°for 30 min or overnight. Remove supernatant and wash precipitate twice with equal volumes of MeOH:EtOH 50:50. Combine the organic layers and evaporate them to dryness under a stream of nitrogen at 37°, reconstitute the residue in 200 μL MeOH:water 65:35, inject a 5-20 μL aliquot. Tissue. Homogenize (Polytron) fetal tissue in 10-15 mL MeOH:dimethoxymethane 50:50 for 1 min or until breakup is complete, shake at 37° overnight, centrifuge at 4000 g for 5 min. Filter (Whatman No. 1 filter paper) supernatant. Resuspend precipitate in MeOH:dimethoxymethane 50:50, filter, wash precipitate with MeOH. Combine filtrates, evaporate to dryness under nitrogen, resuspend residue in up to 500 μL MeOH:water 65:35, centrifuge, inject a 5-20 μL aliquot of the supernatant.

## HPLC VARIABLES
**Guard column:** 70 × 6 35-50 μm Bondapak C18 Corasil
**Column:** 250 × 10 5 μm LiChrosorb RP-18
**Mobile phase:** Gradient. MeOH:10 mM pH 6.9 ammonium acetate from 10:90 to 100:0 over 50 min (Waters No. 5 convex gradient).
**Flow rate:** 1.5
**Injection volume:** 5-20
**Detector:** UV 254

## CHROMATOGRAM
**Retention time:** 31.07

## OTHER SUBSTANCES
**Extracted:** metabolites, cortexolone, cortisol glucuronide, cortisone, 6β-hydroxycortisol, triamcinolone, triamcinolone acetonide

## KEY WORDS
plasma; monkey

## REFERENCE
Althaus, Z.R.; Rowland, J.M.; Freeman, J.P.; Slikker, W., Jr. Separation of some natural and synthetic corticosteroids in biological fluids and tissues by high-performance liquid chromatography. *J.Chromatogr.*, **1982**, *227*, 11-23

## SAMPLE
**Matrix:** formulations
**Sample preparation:** Ointment. Add pentane:EtOH 75:25 to ointment, sonicate for 20 min, dilute an aliquot to 100 mL with MeOH, allow to settle. Centrifuge and filter an aliquot of the supernatant, inject an aliquot of the filtrate. Cream, lotion. Stir cream or lotion in EtOH:THF:water 25:25:50 at 40° for 15 min, cool in an ice bath. Centrifuge and filter an aliquot of the supernatant, inject an aliquot of the filtrate. Gel. Dissolve gel in EtOH, sonicate, filter, inject an aliquot.

## HPLC VARIABLES
**Guard column:** present but not specified
**Column:** 250 × 2.1 10 μm Bondapak C18
**Mobile phase:** MeCN:water 48:52 containing 0.65% acetic acid, pH 3.18 (At the end of each day flush guard column only with MeOH:THF 75:25 for 30 min.)
**Flow rate:** 1

**Injection volume:** 20
**Detector:** UV 251

## CHROMATOGRAM
**Retention time:** 2.54 (hydrocortisone-21-acetate)

## OTHER SUBSTANCES
**Simultaneous:** bamipine lactate, beclomethasone dipropionate, betamethasone-17-valerate, dexamethasone

## KEY WORDS
ointment; creams; lotions; gels

## REFERENCE
Kountourellis, J.E.; Markopoulou, C.K.; Ebete, K.O.; Stratis, J.A. Separation and determination of some corticosteroids combined with bamipine in pharmaceutical formulations by high performance liquid chromatography. *J.Liq.Chromatogr.*, **1995**, *18*, 3507–3517

## SAMPLE
**Matrix:** formulations
**Sample preparation:** Weigh out ointment corresponding to 50-300 µg hydrocortisone, add mobile phase, warm until a fine dispersion formed, make up to 100 mL with mobile phase, filter (0.45 µm), inject a 100 µL aliquot.

## HPLC VARIABLES
**Column:** 250 mm long 5 µm Hypersil ODS
**Mobile phase:** MeOH:water 65:35 adjusted to pH 3 with 85% phosphoric acid
**Flow rate:** 1
**Injection volume:** 100
**Detector:** UV (wavelength not specified)

## CHROMATOGRAM
**Limit of detection:** 20 ng/mL

## KEY WORDS
ointment

## REFERENCE
Preiss, A.; Mehnert, W.; Frömming, K.-H. Penetration of hydrocortisone into excised human skin under the influence of cyclodextrins. *Pharmazie*, **1995**, *50*, 121–126

## SAMPLE
**Matrix:** formulations
**Sample preparation:** Dissolve tablet in 10 mM HCl containing 90 mM KCl (pH 2.0), inject an aliquot.

## HPLC VARIABLES
**Column:** 50 mm long ODS Hypersil C18
**Mobile phase:** MeCN:50 mM pH 7.0 phosphate buffer 30:70
**Flow rate:** 1
**Detector:** UV 257

## CHROMATOGRAM
**Retention time:** 2.3

## OTHER SUBSTANCES
**Simultaneous:** nitrofurantoin

## KEY WORDS
tablets

## REFERENCE
Neervannan, S.; Dias, L.S.; Southard, M.Z.; Stella, V.J. A convective-diffusion model for dissolution of two non-interacting drug mixtures from co-compressed slabs under laminar hydrodynamic conditions. *Pharm.Res.*, **1994**, *11*, 1288–1295

## SAMPLE
**Matrix:** formulations
**Sample preparation:** Ointment. 50 mg Ointment + 10 mL ether, vortex until dissolved. Remove a 200 μL aliquot and add phenyl salicylate in mobile phase, evaporate to dryness under a stream of nitrogen at 40°, reconstitute the residue in 10 mL mobile phase, warm for 1 min on a steam bath, vortex for 1 min, cool. Remove an aliquot, dilute with mobile phase, inject an aliquot. Cream. Suspend 50 mg cream in 10 mL mobile phase by vortexing. Remove an aliquot and add phenyl salicylate in mobile phase, evaporate to dryness under a stream of nitrogen at 40°, suspend the residue in 10 mL mobile phase, warm for 1 min on a steam bath, vortex for 1 min, cool. Remove an aliquot, dilute with mobile phase, inject an aliquot.

## HPLC VARIABLES
**Guard column:** 40 × 5 RP-18-MPLC (Brownlee)
**Column:** 250 × 2.6 ODS-HC-SIL-X (Perkin-Elmer)
**Mobile phase:** MeOH:50 mM phosphoric acid 70:30 (Flush column with MeOH at the end of each day.)
**Column temperature:** 40
**Flow rate:** 1
**Injection volume:** 20
**Detector:** UV 256

## CHROMATOGRAM
**Retention time:** 3
**Internal standard:** phenyl salicylate (5.25)

## OTHER SUBSTANCES
**Simultaneous:** iodochlorhydroxyquin

## KEY WORDS
ointment; cream

## REFERENCE
Ezzedeen, F.W.; Stohs, S.J.; Masoud, A.N. High-performance liquid chromatographic analysis of iodochlorhydroxyquin and hydrocortisone in ointments and creams. *J.Pharm.Sci.*, **1983**, *72*, 1036–1039

## SAMPLE
**Matrix:** formulations
**Sample preparation:** 1 g Ointment + 30 mL trimethylpentane, warm on a water bath until ointment melts, add 10 mL 4% bromobenzene in MeOH:water 80:20, extract with 30 mL methanol:50 mM phosphoric acid 80:20 then twice with 20 mL methanol:50 mM phosphoric acid 80:20, combine extracts, cool, make up to 100 mL with methanol:50 mM phosphoric acid 80:20, inject 20 μL aliquot.

## HPLC VARIABLES
**Column:** 225 × 4 Hypersil-ODS
**Mobile phase:** MeOH:50 mM phosphoric acid 80:20
**Flow rate:** 2
**Injection volume:** 20
**Detector:** UV 240

## CHROMATOGRAM
**Retention time:** 2
**Internal standard:** bromobenzene (3.5)
**Limit of detection:** 500 ng/mL

## OTHER SUBSTANCES
**Simultaneous:** clioquinol

## KEY WORDS
ointment

## REFERENCE
Phoon, K.W.; Stubley, C. Rapid method for the simultaneous analysis of hydrocortisone and clioquinol in topical preparations by high-performance liquid chromatography. *J.Chromatogr.*, **1982**, *246*, 297–303

## SAMPLE
**Matrix:** formulations
**Sample preparation:** Weigh out a sample equivalent to about 10 mg active ingredient, add 20 mL warm MeOH:water 4:1, shake vigorously, add 20 mL n-hexane, extract. Remove the hexane and extract it twice with 10 mL MeOH:water 4:1. Combine all aqueous layers and make up to 50 mL with MeOH:water 4:1, inject 20 μL aliquot.

## HPLC VARIABLES
**Column:** 300 × 3.9 μBondapak C18
**Mobile phase:** MeOH:water 7:3
**Flow rate:** 1
**Injection volume:** 20
**Detector:** UV 240

## CHROMATOGRAM
**Retention time:** 6 (hydrocortisone acetate)

## KEY WORDS
ointment; cream

## REFERENCE
Lea, A.R.; Kennedy, J.M.; Low, G.K.C. Analysis of hydrocortisone acetate ointments and creams by high-performance liquid chromatography. *J.Chromatogr.*, **1980**, *198*, 41–47

## SAMPLE
**Matrix:** formulations
**Sample preparation:** Ointment. Dissolve 0.5 g ointment in 10 mL chloroform, make up to 25 mL with chloroform, inject 20 μL aliquot. Cream. Heat 0.5 g cream in a vacuum desiccator at 60° for 2-4 h, cool, dissolve residue in 10 mL chloroform, make up to 25 mL with chloroform, inject 20 μL aliquot.

## HPLC VARIABLES
**Column:** 250 × 2 Varian SI-10

**Mobile phase:** Cyclohexane:isopropanol 90:10
**Flow rate:** 1
**Injection volume:** 20
**Detector:** UV 240

## CHROMATOGRAM
**Retention time:** 4 (hydrocortisone acetate)

## KEY WORDS
ointment; cream; normal phase

## REFERENCE
Lea, A.R.; Kennedy, J.M.; Low, G.K.C. Analysis of hydrocortisone acetate ointments and creams by high-performance liquid chromatography. *J.Chromatogr.*, **1980**, *198*, 41–47

## SAMPLE
**Matrix:** formulations, solutions
**Sample preparation:** Ointment. 1 g Ointment + 5 mL MeOH + 5 mL water + 800 μL 1 mg/mL hydrocortisone in EtOH, stir until a clear solution forms, make up to 25 mL with water, inject a 20 μL aliquot. Solutions. 8 mL Solution + 800 μL 1 mg/mL hydrocortisone in EtOH + 5 mL MeOH, make up to 25 mL with water, inject a 20 μL aliquot.

## HPLC VARIABLES
**Column:** 300 × 4 μBondapak C18
**Mobile phase:** MeCN:200 mM $KH_2PO_4$ 32:68, pH 4.2
**Flow rate:** 3
**Injection volume:** 20
**Detector:** UV 254

## CHROMATOGRAM
**Retention time:** 4
**Internal standard:** hydrocortisone

## OTHER SUBSTANCES
**Simultaneous:** triamcinolone acetonide

## KEY WORDS
ointment; hydrocortisone is IS

## REFERENCE
Das Gupta, V. Stability of triamcinolone acetonide solutions as determined by high-performance liquid chromatography. *J.Pharm.Sci.*, **1983**, *72*, 1453–1456

## SAMPLE
**Matrix:** perfusate
**Sample preparation:** Condition a Sep-Pak C18 SPE cartridge with 10 mL MeOH and 10 mL water. 3 mL Perfusate + 500 ng 6α-methylprednisolone, add to the SPE cartridge, wash three times with 10 mL aliquots of water, elute with 5 mL MeOH. Evaporate the eluate to dryness under a stream of nitrogen at 35°, reconstitute the residue in 100 μL mobile phase, inject a 50 μL aliquot.

## HPLC VARIABLES
**Guard column:** 15 × 3.2 Newguard RP-18
**Column:** two 250 × 4.6 Spheri-5 RP-18 columns in series
**Mobile phase:** MeOH:water 53:47

**Column temperature:** 40
**Flow rate:** 1.1
**Injection volume:** 50
**Detector:** UV 242

## CHROMATOGRAM
**Retention time:** 19
**Internal standard:** 6α-methylprednisolone (30)
**Limit of detection:** 5 nM

## OTHER SUBSTANCES
**Extracted:** metabolites, cortisone, dihydrocortisol, dihydrocortisone
**Noninterfering:** acetaminophen, albuterol, betamethasone, bupivacaine, carbamazepine, cholesterol, clonazepam, dehydroepiandrosterone, dexamethasone, diazepam, estradiol, estriol, hydroxyprogesterone, methimazole, phenobarbital, prednisone, progesterone, ritodrine, scopolamine, testosterone
**Interfering:** prednisolone

## KEY WORDS
SPE

## REFERENCE
Dodds, H.M.; Maguire, D.J.; Mortimer, R.H.; Addison, R.S.; Cannell, G.R. High performance liquid chromatographic separation of cortisol, cortisone, and their 20-reduced metabolites in perfusion media. *J.Liq.Chromatogr.*, **1995**, *18*, 1809–1820

## SAMPLE
**Matrix:** saliva
**Sample preparation:** 0.5 mL Saliva + 0.5 mL water + 1 mL mobile phase A, filter, inject a 400 μL aliquot onto column A and elute to waste with mobile phase A, after 7 min elute the contents of column A onto column B with mobile phase A, after 8 min remove column A from the circuit and elute column B with mobile phase B, start the gradient, monitor the effluent from column B.

## HPLC VARIABLES
**Column:** A 100 × 4.6 Capcell pak MF [PCMF, silicone polymer-coated silica with diol and phenyl groups] (Shiseido); B Capcell pak CN (Shiseido)
**Mobile phase:** A MeCN:water 10:90 containing 2 mM trisodium citrate, adjusted to pH 6.5 with HCl; B Gradient. X was MeCN:water 10:90. Y was MeCN. X:Y from 100:0 to 72.3:27.7 over 5 min, maintain at 72.3:27.7 for 16 min, return to initial conditions over 1 min.
**Column temperature:** 40
**Flow rate:** A 0.5; B 0.5
**Injection volume:** 400
**Detector:** F ex 488 (10 mW Ar$^+$ laser) em 537 following post-column reaction. The column effluent mixed with concentrated sulfuric acid pumped at 0.75 mL/min and flowed through a 2.5 m × 0.25 mm i.d. Dyflon reaction coil at 105° to the detector.

## CHROMATOGRAM
**Retention time:** 29.4
**Limit of quantitation:** 0.5 nM

## KEY WORDS
column-switching; heart-cut; post-column reaction

## REFERENCE
Okumura, T.; Nakajima, Y.; Takamatsu, T.; Matsuoka, M. Column-switching high-performance liquid chromatographic system with a laser-induced fluorimetric detector for direct, automated assay of salivary cortisol. *J.Chromatogr.B*, **1995**, *670*, 11–20

## SAMPLE
**Matrix:** solutions

## HPLC VARIABLES
**Column:** 300 × 4.1 10 μm Versapack C18 (Alltech)
**Mobile phase:** MeCN:water 40:60
**Flow rate:** 1
**Injection volume:** 50
**Detector:** UV 254

## KEY WORDS
for hydrocortisone and hydrocortisone acetate

## REFERENCE
Michniak, B.B.; Player, M.R.; Sowell, J.W. Synthesis and *in vitro* transdermal penetration enhancing activity of lactam N-acetic acid esters. *J.Pharm.Sci.*, **1996**, *85*, 150–154

## SAMPLE
**Matrix:** solutions
**Sample preparation:** Add 3 mL of a chloroform solution to 300 μL EtOH, add 1 mL 3 mg/mL acenaphthene-5-sulfonyl hydrazine in EtOH:toluene 10:90, evaporate to dryness under reduced pressure at 60°, reconstitute with 200 μL mobile phase, inject an aliquot. (Preparation of acenaphthene-5-sulfonyl hydrazine is as follows. Dissolve 20 g acenaphthene in 100 g nitrobenzene, cool to 0°, add 9 mL chlorosulfonic acid dropwise with stirring, maintain the temperature below 5°, when the addition is complete allow the temperature to rise to 20° over 30 min, add 500 mL water. Remove the aqueous layer and neutralize it with solid sodium carbonate, heat and add NaCl until precipitation occurs, cool in an ice bath for 1 h, filter, heat at 140° to remove traces of water and nitrobenzene to give acenaphthene-5-sulfonic acid sodium salt as a pale yellow solid (mp >300°). Grind 10 g acenaphthene-5-sulfonic acid sodium salt with 3.5 g phosphorus pentachloride in a mortar for 3 min, add ice and water, extract with 100 mL ethyl acetate. Wash the ethyl acetate layer with 5% sodium bicarbonate and with water until neutral, dry over anhydrous sodium sulfate, evaporate the ethyl acetate under a stream of nitrogen, chromatograph on a 300 × 20 column of silica gel H with toluene to give acenaphthene-5-sulfonyl chloride (mp 98-101°) as the first yellow band to elute. Cool a solution of 1 g acenaphthene-5-sulfonyl chloride in 3 mL THF to 10° and pass nitrogen through the solution, add 400 μL 85% hydrazine hydrate dropwise with stirring (Caution! Hydrazine hydrate is a carcinogen!), maintain the temperature between 10° and 15°, stir for a further 15 min. Filter the upper THF layer through Celite, wash the Celite with 1 mL THF. Stir the filtrate vigorously and add two 10 mL portions of water, cool in a refrigerator for 1 h, filter the precipitate, wash with water, dry, recrystallize from EtOH to give acenaphthene-5-sulfonyl hydrazine (mp 132-4°).)

## HPLC VARIABLES
**Column:** 500 × 1 10 μm silica
**Mobile phase:** Toluene:dioxane 90:10 (Caution! Dioxane is a carcinogen!)
**Detector:** F ex 230 em 350

## OTHER SUBSTANCES
**Simultaneous:** fluocinolone acetonide

## KEY WORDS
derivatization; normal phase

## REFERENCE
Gifford, L.A.; Owusu-Daaku, F.T.K.; Stevens, A.J. Acenaphthene fluorescence derivatization reagents for use in high-performance liquid chromatography. *J.Chromatogr.A*, **1995**, *715*, 201–212

## SAMPLE
**Matrix:** solutions

## HPLC VARIABLES
**Column:** 250 × 4.6 5 μm Nucleosil phenyl
**Mobile phase:** Gradient. Carbon dioxide:MeOH from 98:2 to 78:22 over 40 min.
**Column temperature:** 50
**Flow rate:** 2
**Detector:** UV

## CHROMATOGRAM
**Retention time:** 12.1

## OTHER SUBSTANCES
**Simultaneous:** estradiol, estriol, hydroxyprogesterone, norethisterone, testosterone, other steroids

## KEY WORDS
SFC; 200 bar

## REFERENCE
Hanson, M. Aspects of retention behaviour of steroids in packed column supercritical fluid chromatography. *Chromatographia*, **1995**, *40*, 58–68

## SAMPLE
**Matrix:** solutions
**Sample preparation:** Inject an aliquot of a 1 μM solution in MeOH.

## HPLC VARIABLES
**Column:** 470 × 4.6 5 μm Spheri-5 RP-18
**Mobile phase:** MeOH:water 56:44
**Flow rate:** 0.5
**Injection volume:** 10
**Detector:** UV 240

## CHROMATOGRAM
**Retention time:** 34

## OTHER SUBSTANCES
**Simultaneous:** cortisone, dehydrocorticosterone, methylprednisolone, prednisone, tetrahydrocortisol, tetrahydrocortisone
**Interfering:** prednisolone

## REFERENCE
Lukulay, P.H.; McGuffin, V.L. Comparison of solvent modulation with premixed mobile phases for the separation of corticosteroids by liquid chromatography. *J.Liq.Chromatogr.*, **1995**, *18*, 4039–4062

## SAMPLE
**Matrix:** solutions
**Sample preparation:** Inject a 10 μL aliquot of a 100 ppm solution.

## HPLC VARIABLES
**Column:** 150 × 4.6 Develosil ODS-5
**Mobile phase:** Gradient. MeOH:water from 50:50 to 90:10 over 15 min.
**Flow rate:** 1
**Injection volume:** 10
**Detector:** MS, JEOL JMS-SX102A reversed geometry (BE), accelerating voltage +5 kV, air pressure chemical ionization APCI, nebulizer 290°, ion source chamber 400°, discharge electrode, skimmer 1 aperture 300 μm, skimmer 2 aperture 400 μm, no nebulizer gas

## CHROMATOGRAM
**Retention time:** 6

## OTHER SUBSTANCES
**Simultaneous:** corticosterone, cortisone, progesterone

## REFERENCE
Nojima, K.; Fujimaki, S.; Hertsens, R.C.; Morita, T. Application of liquid chromatography-atmospheric pressure chemical ionization mass spectrometry to a sector mass spectrometer. *J.Chromatogr.A*, **1995**, *712*, 17–19

## SAMPLE
**Matrix:** solutions

## HPLC VARIABLES
**Column:** 10 μm Alltech octadecylsilyl
**Mobile phase:** MeCN:water 45:55
**Flow rate:** 1.5
**Detector:** UV 242

## KEY WORDS
water

## REFERENCE
Phares, K.; Cho, M.; Johnson, K.; Swarbrick, J. Drug transport across nylon 610 films: Influence of synthesis variables. *Pharm.Res.*, **1995**, *12*, 248–256

## SAMPLE
**Matrix:** solutions

## HPLC VARIABLES
**Column:** 250 × 4.6 Zorbax RX
**Mobile phase:** Gradient. A was 10 mL concentrated orthophosphoric acid and 7 mL triethylamine in 1 L water. B was 10 mL concentrated orthophosphoric acid and 7 mL triethylamine in 200 mL water, make up to 1 L with MeCN. A:B from 100:0 to 0:100 over 30 min, maintain at 0:100 for 5 min.
**Column temperature:** 30
**Flow rate:** 2
**Detector:** UV 210

## OTHER SUBSTANCES
**Also analyzed:** acepromazine, acetaminophen, acetophenazine, albuterol, aminophylline, amitriptyline, amobarbital, amoxapine, amphetamine, amylocaine, antipyrine, aprobar-

bital, aspirin, atenolol, atropine, avermectin, barbital, benzocaine, benzoic acid, benzotropine, benzphetamine, berberine, bibucaine, bromazepan, brompheniramine, buprenorphine, buspirone, butabarbital, butacaine, butethal, caffeine, carbamazepine, carbromal, chloramphenicol, chlordiazepoxide, chloroquine, chlorothiazide, chloroxylenol, chlorphenesin, chlorpheniramine, chlorpromazine, chlorpropamide, chlortetracycline, cimetidine, cinchonidine, cinchonine, clenbuterol, clonazepam, clonixin, clorazepate, cocaine, codeine, colchicine, cortisone, coumarin, cyclazocine, cyclobenzaprine, cyclothiazide, cyheptamide, cymarin, danazol, danthron, dapsone, debrisoquine, desipramine, dexamethasone, dextromethorphan, dextropropoxyphene, diamorphine, diazepam, diclofenac, diethylpropion, diethylstilbestrol, diflunisal, digitoxin, digoxin, diltiazem, diphenhydramine, diphenoxylate, diprenorphine, dipyrone, disulfiram, dopamine, doxapram, doxepin, dronabinol, ephedrine, epinephrine, epinine, estradiol, estriol, estrone, ethacrynic acid, ethosuximide, etonitazene, etorphine, eugenol, famotidine, fenbendazole, fencamfamine, fenoprofen, fenproporex, fentanyl, flubendazole, flufenamic acid, flunitrazepam, 5-fluorouracil, fluoxymesterone, fluphenazine, furosemide, gentisic acid, gitoxigenin, glipizide, glunixin, glutethimide, glybenclamide, guaiacol, halazepam, haloperidol, hydrochlorothiazide, hydrocodone, hydromorphone, hydroxyquinoline, ibogaine, ibuprofen, iminostilbene, imipramine, indomethacin, isocarbostyril, isocarboxazid, isoniazid, isoproterenol, isoxsuprine, ivermectin, ketamine, ketoprofen, kynurenic acid, levorphanol, lidocaine, lorazepam, lormetazepam, loxapine, mazindol, mebendazole, meclizine, meclofenamic acid, medazepam, mefenamic acid, megestrol, mepacrine, meperidine, mephentermine, mephenytoin, mephesin, mephobarbital, mepivacaine, mescaline, mesoridazine, methadone, methamphetamine, methapyrilene, methaqualone, methazolamide, methocarbamol, methoxamine, methsuximide, methyl salicylate, methyldopa, methyldopamine, methylphenidate, methylprednisolone, methyltestosterone, methyprylon, metoprolol, mibolerone, morphine, nadolol, nalorphine, naloxone, naltrexone, naphazoline, naproxen, nefopam, niacinamide, nicotine, nicotinic acid, nifedipine, niflumic acid, nitrazepam, norepinephrine, nortriptyline, noscapine, nylidrin, oxazepam, oxycodone, oxymorphone, oxyphenbutazone, oxytetracycline, papaverine, pargyline, pemoline, pentazocine, pentobarbital, persantine, phenacetin, phenazocine, phenazopyridine, phencyclidine, phendimetrazine, phenelzine, pheniramine, phenobarbital, phenothiazine, phensuximide, phentermine, phenylbutazone, phenylephrine, phenylpropanolamine, piperocaine, prazepam, prednisolone, primidone, probenecid, progesterone, propiomazine, propranolol, propylparaben, pseudoephedrine, puromycin, pyrilamine, pyrithyldione, quazepam, quinaldic acid, quinidine, quinine, ranitidine, recinnamine, reserpine, resorcinol, saccharin, albuterol, salicylamide, salicylic acid, scopolamine, scopoletin, secobarbital, strychnine, sulfacetamide, sufadiazine, sulfadimethoxine, sulfaethidole, sulfamerazine, sulfamethazine, sulfamethoxizole, sulfanilamide, sulfapyridine, sulfasoxizole, sulindac, tamoxifen, temazepam, testosterone, tetracaine, tetracycline, tetramisole, thebaine, theobromine, theophylline, thiabendazole, thiamine, thiamylal, thiobarbituric acid, thioridazine, thiosalicylic acid, thiothixene, thymol, tolazamide, tolazoline, tobutamide, tolmetin, tranylcypromine, triamcinolone, tribenzylamine, trichloromethiazide, trifluoperazine, trihexyphenidyl, trimethoprim, tripelennamine, triproilidine, tropacocaine, tyramine, verapamil, vincamine, warfarin, yohimbine, zoxazolamine

## REFERENCE

Hill, D.W.; Kind, A.J. Reversed-phase solvent gradient HPLC retention indexes of drugs. *J.Anal.Toxicol.*, **1994**, *18*, 233–242

## SAMPLE

**Matrix:** solutions
**Sample preparation:** Prepare a 25 μg/mL solution in mobile phase, inject an aliquot.

## HPLC VARIABLES

**Column:** 250 × 4.6 Partisil 10 ODS-1
**Mobile phase:** MeOH:water 55:45
**Column temperature:** 40

**Flow rate:** 1.5
**Detector:** UV 240

## CHROMATOGRAM
**Retention time:** k′ 1.094

## OTHER SUBSTANCES
**Also analyzed:** androsterone (UV 210), cortexolone (UV 240), cortisone (UV 240), estradiol (UV 280), estrone (UV 280), ethinyl estradiol (UV 280), ethisterone (UV 240), hydroxy-progesterone (UV 240), lynestrenol (UV 210), medroxyprogesterone acetate (UV 240), medroxyprogesterone (UV 240), methandienone (UV 240), methylandrostenediol (UV 210), methylprednisolone acetate (UV 240), methylprednisolone (UV 240), methyltestos-terone (UV 240), nandrolone (UV 240), norethisterone (UV 240), prednisolone acetate (UV 240), prednisolone (UV 240), prednisone (UV 240), pregnenolone (UV 210), progesterone (UV 240), testosterone (UV 240)

## REFERENCE
Sadlej-Sosnowska, N. Structure retention relationship for steroid hormones. Functional groups as structural descriptors. *J.Liq.Chromatogr.*, **1994**, *17*, 2319–2330

## SAMPLE
**Matrix:** solutions
**Sample preparation:** Condition a Bond Elut C18 SPE cartridge with 4 mL water then 3 mL MeOH. Add aqueous steroid solution to the SPE cartridge, elute with MeOH, inject a 20 μL aliquot.

## HPLC VARIABLES
**Column:** 150 × 4 5 μm Nucleosil C18
**Mobile phase:** MeOH:water 70:30
**Flow rate:** 1
**Injection volume:** 20
**Detector:** UV 240

## CHROMATOGRAM
**Limit of detection:** 3.5 μg/mL

## OTHER SUBSTANCES
**Also analyzed:** prednisolone

## KEY WORDS
SPE; for hydrocortisone or hydrocortisone 21-acetate

## REFERENCE
Valenta, C.; Janout, H. Corticosteroid analysis by HPLC with increased sensitivity by use of precolumn concentration. *J.Liq.Chromatogr.*, **1994**, *17*, 1141–1146

## SAMPLE
**Matrix:** solutions
**Sample preparation:** Evaporate solution (eluate from preparative HPLC) to dryness under a stream of nitrogen, reconstitute with 10 μL 2 μg/mL 9-anthroylnitrile (Wako) in MeCN and 10 μL triethylamine:MeCN 30:70 under nitrogen, let stand at room temperature for 20 min, add 5 μL water, after 6 min add 50 μL 600 mM acetic acid in MeCN, evaporate to dryness under a stream of nitrogen at 37°, reconstitute with 90 μL MeOH: 0.4 N $NaH_2PO_4$ 60:40, add to a Cyclobond I silica-bonded β-cyclodextrin SPE cartridge (Astec), wash with 1 mL water, wash with 8 mL MeOH:water 25:75 containing 7.5 mM pH 7.0 phosphate buffer, elute with 1 mL MeOH, evaporate to dryness under a stream

of nitrogen, reconstitute with mobile phase, inject an aliquot onto column A and elute to waste with mobile phase, after the solvent front has passed through divert the effluent from column A onto column B, monitor the effluent from column B.

## HPLC VARIABLES
**Column:** A 30 × 2.1 silica (Brownlee); B 150 × 2 Hypersil
**Mobile phase:** Hexane:ethyl acetate 67:33 (half-saturated with water)
**Flow rate:** 0.5
**Detector:** F ex 305-395 em 430-470

## CHROMATOGRAM
**Retention time:** 5.48
**Limit of detection:** 9 pg

## OTHER SUBSTANCES
**Simultaneous:** cortisone, prednisolone

## KEY WORDS
derivatization; SPE; column-switching; normal phase

## REFERENCE
Haegele, A.D.; Wade, S.E. Ultrasensitive differential measurement of cortisol and cortisone in biological samples using fluorescent ester derivatives in normal phase HPLC. *J.Liq.Chromatogr.*, **1991**, *14*, 1133–1148

## SAMPLE
**Matrix:** solutions
**Sample preparation:** Sample + 400 µL 5 mM DBD-PZ + 70 mM diethylphosphorocyanidate in MeCN, react for 6 h, inject a 1 µL aliquot. (DBD-PZ prepared from 123 mg 4-(N, N-dimethylaminosulfonyl)-7-fluoro-2,1,3-benzoxadiazole in 20 mL MeCN added dropwise to 129 mg piperazine in 20 mL MeCN at room temperature, stir for 30 min, evaporate under reduced pressure, dissolve residue in 50 mL 5% HCl, extract three times with 20 mL ethyl acetate, discard ethyl acetate extracts, adjust pH of aqueous solution to 13-14 with 5% NaOH, extract five times with 50 mL ethyl acetate, combine extracts, wash with 20 mL water, dry over anhydrous sodium sulfate, evaporate under vacuum to give DBD-PZ as orange crystals, mp 121-2°.)

## HPLC VARIABLES
**Column:** 150 × 4.6 5 µm Inertsil ODS-2
**Mobile phase:** MeCN:water 45:55
**Column temperature:** 40
**Flow rate:** 1
**Injection volume:** 1
**Detector:** F ex 437 em 561

## CHROMATOGRAM
**Retention time:** 13 (hydrocortisone succinate)
**Limit of detection:** 14 fmol

## OTHER SUBSTANCES
**Simultaneous:** alprostadil, dinoprost, prednisolone succinate

## REFERENCE
Toyo'oka, T.; Ishibashi, M.; Takeda, Y.; Nakashima, K.; Akiyama, S.; Uzu, S.; Imai, K. Precolumn fluorescence tagging reagent for carboxylic acids in high-performance liquid chromatography: 4-substituted-7-aminoalkylamino-2,1,3-benzoxadiazoles. *J.Chromatogr.*, **1991**, *588*, 61–71

## SAMPLE
**Matrix:** solutions
**Sample preparation:** Dissolve in MeOH:water 1:1 at a concentration of 50 µg/mL, inject a 10 µL aliquot.

## HPLC VARIABLES
**Column:** 300 × 3.9 10 µm µBondapak C18
**Mobile phase:** MeOH:acetic acid:triethylamine:water 60:1.5:0.5:38
**Flow rate:** 1.5
**Injection volume:** 10
**Detector:** UV 240

## CHROMATOGRAM
**Retention time:** 8 (hydrocortisone acetate)

## OTHER SUBSTANCES
**Simultaneous:** methyltestosterone, norethindrone, prednisolone, prednisolone succinate, prednisone, progesterone

## REFERENCE
Roos, R.W.; Lau-Cam, C.A. General reversed-phase high-performance liquid chromatographic method for the separation of drugs using triethylamine as a competing base. *J.Chromatogr.*, **1986**, *370*, 403–418

## SAMPLE
**Matrix:** solutions
**Sample preparation:** Dissolve in MeOH:water 1:1 at a concentration of 50 µg/mL, inject a 10 µL aliquot.

## HPLC VARIABLES
**Column:** 300 × 3.9 10 µm µBondapak C18
**Mobile phase:** MeOH:acetic acid:triethylamine:water 60:1.5:0.5:38
**Flow rate:** 1.5
**Injection volume:** 10
**Detector:** UV

## CHROMATOGRAM
**Retention time:** $k'$ 1.27

## REFERENCE
Roos, R.W.; Lau-Cam, C.A. General reversed-phase high-performance liquid chromatographic method for the separation of drugs using triethylamine as a competing base. *J.Chromatogr.*, **1986**, *370*, 403–418

## SAMPLE
**Matrix:** solutions

## HPLC VARIABLES
**Column:** 250 × 4.6 5 µm SI-100 (Brownlee)
**Mobile phase:** Butyl chloride:THF:MeOH:glacial acetic acid 95:7:3.5:3 (Butyl chloride was 50% water saturated.)
**Injection volume:** 20
**Detector:** UV 254

## CHROMATOGRAM
**Retention time:** 16 (hydrocortisone), 10 (hydrocortisone acetate)

## OTHER SUBSTANCES
**Simultaneous:** 4-androstene-3,11,17-trione, cortisone, cortisone acetate

## KEY WORDS
normal phase

## REFERENCE
Kane, M.P.; Tsuji, K. Radiolytic degradation scheme for $^{60}$Co-irradiated corticosteroids. *J.Pharm.Sci.*, **1983**, *72*, 30–35

## SAMPLE
**Matrix:** solutions

## HPLC VARIABLES
**Column:** 500 × 1 C18 (Alltech)
**Mobile phase:** MeOH:water 65:35
**Flow rate:** 0.04
**Injection volume:** 0.5
**Detector:** UV 254; MS, Hewlett Packard 5985, home-made interface (details in paper)

## CHROMATOGRAM
**Retention time:** 19

## OTHER SUBSTANCES
**Simultaneous:** betamethasone

## KEY WORDS
microbore

## REFERENCE
Eckers, C.; Skrabalak, D.S.; Henion, J. On-line direct liquid introduction interface for micro-liquid chromatography/mass spectrometry: application to drug analysis. *Clin.Chem.*, **1982**, *28*, 1882–1886

## SAMPLE
**Matrix:** tissue
**Sample preparation:** Extract 70-125 mg tissue four times with 5 mL portions of ether: chloroform 80:20. Combine the organic layers and evaporate them to dryness under a stream of nitrogen, reconstitute the residue in 100 μL MeOH, inject an aliquot.

## HPLC VARIABLES
**Column:** 80 mm long 10 μm octadecylsilane radial compression (Radial-Pak) (Waters)
**Mobile phase:** Gradient. A was MeOH:water 50:50. B was MeOH. A:B from 100:0 to 70:30 over 20 min, to 0:100 over 20 min.
**Flow rate:** 2
**Detector:** UV 254

## CHROMATOGRAM
**Retention time:** 10

## OTHER SUBSTANCES
**Extracted:** androstenedione, deoxycortisol, 17-hydroxyprogesterone, testosterone

**Simultaneous:** estradiol, estriol, pregnenolone, progesterone, testosterone enanthate, testosterone propionate

## KEY WORDS
tumor

## REFERENCE
Kessler, M.J. Analysis of steroids from normal and tumor tissue by HPLC. *Clin.Chim.Acta*, **1982**, *125*, 21–30

## SAMPLE
**Matrix:** urine
**Sample preparation:** 10 mL Urine + 40 µL 25 µg/mL corticosterone, vortex briefly, add 1 mL 100 mM NaOH, vortex briefly, add 3 mL dichloromethane, rotate at 20 rpm for 45 min, centrifuge at 1000 g for 15 min, discard the aqueous layer, centrifuge at 1000 g for 10 min, discard the aqueous layer, add 150 mg NaCl, break up emulsion, centrifuge for 10 min. Remove the organic layer and evaporate it to dryness under a stream of nitrogen at 45°, reconstitute the residue in 150 µL MeOH, inject an aliquot.

## HPLC VARIABLES
**Column:** 150 × 3.9 4 µm Nova-Pak C18
**Mobile phase:** Gradient. MeOH:water from 30:70 to 44:56 over 6 min, maintain at 44:56 for 14 min, return to initial conditions over 3 min, re-equilibrate for 5 min.
**Flow rate:** 1
**Detector:** UV 246

## CHROMATOGRAM
**Retention time:** 13.6
**Internal standard:** corticosterone (17.8)

## OTHER SUBSTANCES
**Extracted:** cortisone

## REFERENCE
Lee, Y.S.; Lorenzo, B.J.; Koufis, T.; Reidenberg, M.M. Grapefruit juice and its flavonoids inhibit 11β-hydroxysteroid dehydrogenase. *Clin.Pharmacol.Ther.*, **1996**, *59*, 62–71

## SAMPLE
**Matrix:** urine
**Sample preparation:** Inject 500 µL urine onto column A with mobile phase A and elute to waste, after 5 min backflush the contents of column A onto column B with mobile phase B and start the gradient, monitor the effluent from column B.

## HPLC VARIABLES
**Column:** A 10 µm Ultrabiosep C18 SFCC (ISRP); B 250 × 4.6 5 µm Ultrabase C18 SFCC
**Mobile phase:** A water; B Gradient. MeCN:water 20:80 for 10 min, to 40:60 over 10 min, re-equilibrate with 20:80 for 5 min. (Re-equilibrate column A with mobile phase A for 5 min before the next injection.)
**Flow rate:** 1
**Injection volume:** 500
**Detector:** UV 242

## CHROMATOGRAM
**Retention time:** 22.77
**Limit of detection:** 3 ng/mL
**Limit of quantitation:** 10 ng/mL

## OTHER SUBSTANCES
**Extracted:** 6β-hydroxycortisol
**Simultaneous:** corticosterone, cortisone, deoxycorticosterone, 11-deoxycortisol, predniso-
lone, prednisone

## KEY WORDS
column-switching

## REFERENCE
Bidart, M.; Lesgards, G. Direct injection analysis of 6β-hydroxycortisol and cortisol in urine by HPLC-
UV with on-line IRSP column. *J.Liq.Chromatogr.*, **1995**, *18*, 725–738

## SAMPLE
**Matrix:** urine
**Sample preparation:** 3 mL Urine + 100 ng methylprednisolone + 0.25 g NaCl, adjust pH
to 9.0 with 0.5 g $Na_2HPO_4$, add 4 mL dichloromethane, vortex 1 min, centrifuge at 3700
g for 3 min. Remove organic phase and dry it over anhydrous sodium sulfate. Evaporate
a 3 mL aliquot to dryness under vacuum, reconstitute residue with 200 μL MeOH, inject
20 μL aliquot.

## HPLC VARIABLES
**Column:** 250 × 4.6 5 μm Hypersil 5-ODS
**Mobile phase:** MeCN:water 30:70
**Column temperature:** 30
**Flow rate:** 1
**Injection volume:** 20
**Detector:** UV 245

## CHROMATOGRAM
**Retention time:** 7
**Internal standard:** methylprednisolone (14)
**Limit of detection:** 51 pg

## OTHER SUBSTANCES
**Extracted:** cortisone
**Noninterfering:** corticosterone, deflazacort, deoxycorticosterone, fluorocortisone acetate,
21-hydroxydeflazacort, 11α-hydroxyprogesterone, prednisolone, prednisone, triamcinolone
acetonide
**Interfering:** fluorocortisone

## REFERENCE
Santos-Montes, A.; Gonzalo-Lumbreras, R.; Izquierdo-Hornillos, R. Simultaneous determination of cortisol and cortisone in urine by reversed-phase high-performance liquid chromatography. Clinical and
doping control applications. *J.Chromatogr.B*, **1995**, *673*, 27–33

## SAMPLE
**Matrix:** urine
**Sample preparation:** 3 mL Urine + 0.25 g NaCl, adjust pH to 9.0 with 0.5 g $Na_2HPO_4$,
add 4 mL dichloromethane, vortex 1 min, centrifuge at 3700 g for 3 min. Remove organic
phase and dry it over anhydrous sodium sulfate. Evaporate a 3 mL aliquot to dryness
under vacuum, reconstitute residue with 200 μL 5 μg/mL IS in MeOH, inject 20 μL
aliquot.

## HPLC VARIABLES
**Column:** 250 × 4.6 Hypersil ODS

**Mobile phase:** MeCN:water 32:68
**Column temperature:** 30
**Flow rate:** 1
**Injection volume:** 20
**Detector:** UV 245

## CHROMATOGRAM
**Retention time:** 6
**Internal standard:** methylprednisolone (9)

## OTHER SUBSTANCES
**Simultaneous:** betamethasone, corticosterone, cortisone, dexamethasone, fluorocortisone acetate, hydroxyprogesterone, triamcinolone, triamcinolone acetonide
**Interfering:** fluorocortisone, prednisolone, prednisone

## KEY WORDS
SPE also discussed

## REFERENCE
Santos-Montes, A.; Gonzalo-Lumbreras, R.; Gasco-Lopez, A.I.; Izquierdo-Hornillos, R. Solvent and solid-phase extraction of natural and synthetic corticoids in human urine. *J.Chromatogr.B*, **1994**, *652*, 83–89

## SAMPLE
**Matrix:** urine
**Sample preparation:** Equilibrate a Sephadex G-25M column with 100 mM pH 7.0 phosphate buffer. Condition a Bond-Elut C18 SPE cartridge with 1 mL MeCN, 4 mL acetone: water 20:80, and 4 mL water. 2 mL Urine + 500 μL 500 mM pH 5.0 acetate buffer + 50 μL 1 μg/mL fluorocortisone in MeOH + 160 μL 100000 Fishmann U/mL β-glucuronidase and 800000 Roy U/mL arylsulfatase (from Helix pomatia, Boehringer Mannheim), heat at 37° for 24 h, filter (0.45 μm), add to the Sephadex column, wash with three 2 mL portions of 100 mM pH 7.0 phosphate buffer, elute with four 2 mL portions of 100 mM pH 7.0 phosphate buffer. Add the eluate to the SPE cartridge, wash with 4 mL water, wash with 4 mL acetone:water 20:80, elute with 1 mL MeCN. Evaporate the eluate to dryness under a stream of nitrogen, reconstitute the residue in 100 μL MeOH, add 20 μL cupric acetate solution, let stand at room temperature for 1 h, add 100 μL reagent, heat at 60° for 40 min, cool, centrifuge briefly at 1000 g, inject a 100 μL aliquot of the supernatant. (Cupric acetate solution was 0.7 g cupric acetate in 10 mL water diluted to 100 mL with MeOH. Reagent was 7 mM 1,2-diamino-4,5-methylenedioxybenzene in water containing 200 mM β-mercaptoethanol and 250 mM sodium hydrosulfite, store in the dark at 4°, stable for at least 2 weeks. Prepare 1,2-diamino-4,5-methylenedioxybenzene as follows. Add 5 g 1,2-(methylenedioxy)-4-nitrobenzene to 37.5 mL concentrated nitric acid and 12.5 mL glacial acetic acid, pour the yellow-colored solution into water, recrystallize the 1,2-dinitro-4,5-methylenedioxybenzene from EtOH (Rec.Trav.Chim.Pays-Bas 1930, 49, 45). Dissolve 5 g 1,2-dinitro-4,5-methylenedioxybenzene in 200 mL benzene (Caution! Benzene is a carcinogen!), add 100 g 80 mesh iron powder, add 20 mL concentrated HCl in small portions over 1 h while heating the mixture under reflux. Reflux for 4 h, add 10 mL water, reflux for 2 h, cool, make alkaline with 2.6 M NaOH, extract three times with 200 mL portions of benzene. Combine the extracts, evaporate to dryness to give 1,2-diamino-4,5-methylenedioxybenzene, mix with 10 mL concentrated HCl, recrystallize from EtOH to give 1,2-diamino-4,5-methylenedioxybenzene dihydrochloride, mp 176-9° (Chem.Pharm.Bull. 1987, 35, 687).)

## HPLC VARIABLES
**Column:** 250 × 4.6 5 μm L-Column ODS (Chemicals Inspection and Testing Institute, Tokyo)

**Mobile phase:** MeOH:MeCN:500 mM ammonium acetate 50:10:40 (After each injection wash with MeOH:water 80:20 for 20 min, re-equilibrate for 20 min.)
**Flow rate:** 1
**Injection volume:** 100
**Detector:** F ex 350 em 390

## CHROMATOGRAM
**Retention time:** 38.5
**Internal standard:** fludrocortisone (35.6)
**Limit of detection:** 1.18 ng/mL

## OTHER SUBSTANCES
**Extracted:** aldosterone, tetrahydroaldosterone
**Noninterfering:** corticosterone, cortisone, hydroxycorticosteroids

## KEY WORDS
SPE; derivatization

## REFERENCE
Yoshitake, T.; Ishida, J.; Sonezaki, S.; Yamaguchi, M. High performance liquid chromatographic determination of 3α,5β-tetrahydroaldosterone and cortisol in human urine with fluorescence detection. *Biomed.Chromatogr.*, **1992**, *6*, 217–221

## SAMPLE
**Matrix:** urine
**Sample preparation:** 3 mL Urine + 1.5 μg betamethasone + 100 mg $K_2HPO_4$ + 500 mg anhydrous sodium sulfate + 5 mL diethyl ether, shake mechanically for 10 min, centrifuge at 2500 g for 5 min. Remove the organic layer and evaporate it to dryness under vacuum, reconstitute the residue in 200 μL MeOH, filter (0.45 μm), inject a 15 μL aliquot.

## HPLC VARIABLES
**Column:** 100 × 4.6 5 μm Hypersil ODS
**Mobile phase:** Gradient. MeCN:water from 4:96 to 30:70 over 10 min, to 45:55 over 5 min, to 50:50 over 3 min
**Column temperature:** 40
**Flow rate:** 1
**Injection volume:** 15
**Detector:** UV 246

## CHROMATOGRAM
**Retention time:** 11.32
**Internal standard:** betamethasone (12.83)
**Limit of detection:** 10 ng/mL

## OTHER SUBSTANCES
**Extracted:** corticosterone, cortisone, deoxycorticosterone, hydrocortisone, prednisolone, prednisone, triamcinolone, triamcinolone acetonide

## REFERENCE
Park, S.-J.; Kim, Y.-J.; Pyo, H.-S.; Park, J. Analysis of corticosteroids in urine by HPLC and thermospray LC/MS. *J.Anal.Toxicol.*, **1990**, *14*, 102–108

## SAMPLE
**Matrix:** urine
**Sample preparation:** Adjust pH of 2 mL urine to 6.5, add 100 μL 500 Fishman U/mL β-glucuronidase (from E. coli), add 200 μL 200 mM pH 6.5 phosphate buffer, add 1 drop

chloroform, mix well, heat at 37° for 24 h, add 20 μL 100 μg/mL betamethasone in MeOH, add 4 mL dichloromethane, shake for 3 min. Remove the organic layer and wash it with 500 μL 100 mM NaOH, wash with 500 μL water. Remove the organic layer and evaporate it to dryness at 80°, reconstitute the residue in 100 μL mobile phase, inject a 10 μL aliquot.

## HPLC VARIABLES
**Column:** 250 × 4.6 10 μm Finepak C18
**Mobile phase:** MeOH:water 50:50
**Column temperature:** 40
**Flow rate:** 0.8
**Injection volume:** 10
**Detector:** F ex 370 em 480 following post-column reaction. The column effluent mixed with 400 mM NaOH and reagent pumped at 0.5 mL/min and the mixture flowed through a 30 m × 0.5 mm ID PTFE coil at 95° and another coil immersed in water to the detector. (Reagent was 0.5% benzamidine hydrochloride in isopropanol:water 50:50.)

## CHROMATOGRAM
**Retention time:** 18
**Internal standard:** betamethasone (20)

## OTHER SUBSTANCES
**Simultaneous:** tetrahydrocortisol, tetrahydrocortisone, tetrahydro-11-deoxycortisol
**Noninterfering:** aldosterone, androsterone, corticosterone, dehydroepiandrosterone, 11-deoxycorticosterone, 16-hydroxydehydroepiandrosterone, progesterone

## KEY WORDS
post-column reaction

## REFERENCE
Seki, T.; Yamaguchi, Y. New fluorimetric determination of 17-hydroxycorticosteroids after high-performance liquid chromatography using post-column derivatization with benzamidine. *J.Chromatogr.*, **1984**, *305*, 188–193

◆ ◆ ◆

## ANNOTATED BIBLIOGRAPHY
Bast, G.E.; Kampffmeyer, H.G. Absorption and metabolism of hydrocortisone 21-butyrate, 21-hemisuccinate and hydrocortisone by skin of the rabbit ear during single-pass perfusion. *Xenobiotica*, **1994**, *24*, 1029–1042 [extracted cortisone, hydrocortisone, hydrocortisone 21-butyrate, hydrocortisone 21-hemisuccinate, hydrocortisone sulfate; perfusate; effusate; hydrocortisone acetate (IS); gradient]

Dolezalova, M. Routine high-performance liquid chromatographic determination of urinary unconjugated cortisol using solid-phase extraction and ultraviolet detection. *Clin.Chim.Acta*, **1994**, *231*, 129–137

Inoue, S.; Inokuma, M.; Harada, T.; Shibutani, Y.; Yoshitake, T.; Charles, B.; Ishida, J.; Yamaguchi, M. Simultaneous high-performance liquid chromatographic determination of 6β-hydroxycortisol and cortisol in urine with fluorescence detection and its application for estimating hepatic drug-metabolizing enzyme induction. *J.Chromatogr.B*, **1994**, *661*, 15–23 [extracted 6β-hydroxycortisol; urine; fluorescence detection; fludrocortisone (IS); gradient; LOD 0.95 ng/mL; human; monkey]

Lykkesfeldt, J.; Loft, S.; Poulsen, H.E. Simultaneous determination of urinary free cortisol and 6β-hydroxycortisol by high-performance liquid chromatography to measure human CYP3A activity. *J.Chromatogr.B*, **1994**, *660*, 23–29 [urine; extracted 6β-hydroxycortisol; dexamethasone (IS); SPE; gradient; LOQ 1 ng/mL]

Valvo, L.; Paris, A.; Savella, A.L.; Gallinella, B.; Ciranni Signoretti, E. General high-performance liquid chromatographic procedures for the rapid screening of natural and synthetic corticosteroids. *J.Pharm.Biomed.Anal.*, **1994**, *12*, 805–810 [gradient; reverse phase; normal phase; for hydrocortisone, hydrocortisone 21-acetate, hydrocortisone 21-hemisuccinate; also beclomethasone, beclomethasone 17,21-dipropionate, betamethasone, betamethasone 21-acetate, betamethasone 17,21-dipropionate, betamethasone 21-disodium phosphate, betamethasone 17-valerate, cortisone, cortisone 21-acetate, 11-deoxycorticosterone 21-acetate, dexamethasone, dexamethasone 21-acetate, dexamethasone 21-disodium phosphate, fluocinolone, fluocinolone acetonide, 9α-fluorohydrocortisone 21-acetate, 9α-fluorohydrocortisone, 9α-fluoroprednisolone, 9α-fluoroprednisolone 21-acetate, 6α-methylprednisolone, 6α-methylprednisolone 21-acetate, 6α-methylprednisolone 21-sodium succinate, prednisolone, prednisolone 21-acetate, prednisolone 21-disodium phosphate, prednisolone 21-pivalate, prednisolone 21-sodium succinate, prednisone, triamcinolone, triamcinolone acetonide]

Li, Y.-M.; Chen, L.-R.; Qu, Y. Use of micellar mobile phases and an HPLC column switching system for direct injection determination of urinary free cortisol. *J.Liq.Chromatogr.*, **1993**, *16*, 2583−2594 [column-switching; direct injection; urine; LOD 1.2 ng/mL]

Santos-Montes, A.; Gasco-Lopez, A.I.; Izquierdo-Hornillos, R. Optimization of the high-performance liquid chromatographic separation of a mixture of natural and synthetic corticosteroids. *J.Chromatogr.*, **1993**, *620*, 15−23 [simultaneous betamethasone, corticosterone, cortisone, deoxycorticosterone, dexamethasone, fluorocortisone, hydroxyprogesterone, methylprednisolone, prednisolone, prednisone, triamcinolone]

Teffera, Y.; Abramson, F.P.; McLean, M.; Vestal, M. Development of an isotope-selective high-performance liquid chromatography detector using chemical-reaction-interface mass spectrometry: application to deuterated cortisol metabolites in urine. *J.Chromatogr.*, **1993**, *620*, 89−96

Bhounsule, G.J.; Gorule, V.S.; Patil, G.V. Simultaneous determination of 5-bromosalicyl-4-chloranilide, bamipine lactate and hydrocortisone acetate by high-performance liquid chromatography. *Indian Drugs*, **1992**, *29*, 594−597

Nozaki, O.; Ohata, T.; Ohba, Y.; Moriyama, H.; Kato, Y. Determination of urinary free cortisol by high performance liquid chromatography with sulphuric acid-ethanol derivatization and column switching. *Biomed.Chromatogr.*, **1992**, *6*, 109−114 [column-switching; urine; derivatization]

Nozaki, O.; Ohata, T.; Ohba, Y.; Moriyama, H.; Kato, Y. Determination of serum cortisol by reversed-phase liquid chromatography using precolumn sulphuric acid-ethanol fluorescence derivatization and column switching. *J.Chromatogr.*, **1991**, *570*, 1−11

Qin, Y.; Liang, D.; Zeng, J.; Mao, W. [Determination of hydrocortisone and methylprednisolone in plasma by reversed-phase HPLC]. *Hua Hsi I Ko Ta Hsueh Hsueh Pao*, **1991**, *22*, 270−273

Shalaby, A.; Shahjahan, M. Improved high performance liquid chromatographic method for the determination of some corticosteroids. *J.Liq.Chromatogr.*, **1991**, *14*, 1267−1274 [formulations; ointment; lotion; tablets; injections; simultaneous dexamethasone, prednisolone]

Wade, S.E.; Haegele, A.D. Corticosteroid analysis by HPLC-UV facilitated by use of an injector-mounted extraction column. *J.Liq.Chromatogr.*, **1991**, *14*, 1257−1266 [rabbit; serum; LOD 300 pg; column-switching; column temp 50]

Wade, S.E.; Haegele, A.D. Differential measurement of cortisol and cortisone in human saliva by HPLC with UV detection. *J.Liq.Chromatogr.*, **1991**, *14*, 1813−1827 [saliva; UV detection; LOD 0.5 ng/mL; column-switching; column temp 50; SPE]

Wanwimolruk, S. Rapid high-performance liquid chromatographic analysis and stability study of hydrocortisone 17-butyrate in cream preparations. *Pharm.Res.*, **1991**, *8*, 547−549

Alvinerie, M.; Sutra, J.F.; Galtier, P.; Houin, G.; Toutain, P.L. Simultaneous measurement of prednisone, prednisolone and hydrocortisone in plasma by high performance liquid chromatography. *Ann Biol.Clin.(Paris)*, **1990**, *48*, 87−90

Esteban, N.V.; Yergey, A.L.; Liberato, D.J.; Loughlin, T.; Loriaux, D.L. Stable isotope dilution method using thermospray liquid chromatography/mass spectrometry for quantification of daily cortisol production in humans. *Biomed.Environ.Mass.Spectrom.*, **1988**, *15*, 603−608 [thermospray; LC-MS]

Sheikh, S.U.; Touchstone, J. HPLC of steroids in non-aqueous mobile phase at subambient temperature. *J.Liq.Chromatogr.*, **1987**, *10*, 2489−2496 [column temp -50; simultaneous cortisone, desoxycorticosterone, estradiol, estrone]

Das Gupta, V. Quantitation of glipizide and glyburide in tablets using high performance liquid chromatography. *J.Liq.Chromatogr.*, **1986**, *9*, 3607−3615 [simultaneous glipizide, glyburide; tablets; hydrocortisone is IS]

Derendorf, H.; Rohdewald, P.; Hochhaus, G.; Moellmann, H. HPLC determination of glucocorticoid alcohols, their phosphates and hydrocortisone in aqueous solutions and biological fluids. *J.Pharm.Biomed.Anal.*, **1986**, *4*, 197−206

Gutenberger, S.K.; Olson, D.P.; Kagel, R.A. Comparison of reversed phase high-performance liquid chromatography and competitive protein binding assay in the quantification of cortisol in bovine plasma. *J.Liq.Chromatogr.*, **1985**, *8*, 107–124 [plasma; cow; dexamethasone (IS)]

Carson, S.W.; Jusko, W.J. Simultaneous analysis of cortexolone and cortisol by high-performance liquid chromatography for use in the metyrapone test. *J.Chromatogr.*, **1984**, *306*, 345–350

Molokhia, A.M.; El-Hoofy, S.; Al-Rahman, S. A HPLC method for the determination of butaperazine in solutions, tablets, plasma and bile. *J.Liq.Chromatogr.*, **1984**, *7*, 1643–1649 [butaperazine; solutions; tablets; plasma; bile; hydrocortisone is IS; fluorescence detection]

Benjamin, E.J.; Conley, D.L. On-line HPLC method for clean-up and analysis of hydrocortisone and sulconazole nitrate in a cream. *Int.J.Pharm.*, **1983**, *13*, 205–217

Gandelman, M.S.; Birks, J.W. Liquid chromatographic detection of cardiac glycosides, saccharides and hydrocortisone based on the photoreduction of 2-tert-butylanthraquinone. *Anal.Chim.Acta*, **1983**, *155*, 159–171

Seki, T.; Yamaguchi, Y. New fluorimetric detection method of corticosteroids after high-performance liquid chromatography using post-column derivatization with glycinamide. *J.Liq.Chromatogr.*, **1983**, *6*, 1131–1138 [post-column reaction; derivatization; urine; also corticosterone, cortisone, deoxycortisol, prednisolone, tetrahydrocortisone, tetrahydrodeoxycortisol]

Lewbart, M.L.; Elverson, R.A. Determination of urinary free cortisol and cortisone by sequential thin layer and high-performance liquid chromatography. *J.Steroid Biochem.*, **1982**, *17*, 185–190

Ost, L.; Falk, O.; Lantto, O.; Bjorkhem, I. Simultaneous determination of prednisolone and cortisol in serum by HPLC and by isotope dilution–mass spectrometry. *Scand.J.Clin.Lab.Invest.*, **1982**, *42*, 181–187

Rego, A.; Nelson, B. Simultaneous determination of hydrocortisone and benzyl alcohol in pharmaceutical formulations by reversed-phase high-pressure liquid chromatography. *J.Pharm.Sci.*, **1982**, *71*, 1219–1223 [simultaneous benzaldehyde, benzoic acid, benzyl alcohol; phenethyl alcohol (IS); creams; gels; ointments; solutions]

Shihabi, Z.K.; Andrews, R.I.; Scaro, J. Liquid chromatographic assay of urinary free cortisol. *Clin.Chim.Acta*, **1982**, *124*, 75–83

Matsuzawa, T.; Sugimoto, N.; Ishiguro, I. A simple micromethod for determining human serum cortisol by high-pressure liquid chromatography using 0.1 ml serum. *Anal.Biochem.*, **1981**, *115*, 250–253

Okumura, T. Application of thin-layer chromatography to high-performance liquid chromatographic separation of steroidal hormones and cephalosporin antibiotics. *J.Liq.Chromatogr.*, **1981**, *4*, 1035–1064 [normal phase; also betamethasone, cephalexin, cephaloglycine, cephaloridine, cephalothin, cortisone, dexamethasone, hydrocortisone]

de Vries, C.P.; Lomecky-Janousek, M.; Popp-Snijders, C. Rapid quantitative assay of plasma 11-deoxycortisol and cortisol by high-performance liquid chromatography for use in the metyrapone test. *J.Chromatogr.*, **1980**, *183*, 87–91

Hansen, J.; Bundgaard, H. Studies on the stability of corticosteroids. III. Separation and quantitation of hydrocortisone and its degradation products by high-performance liquid chromatography. *Arch.Pharm.Chemi, Sci.Ed.*, **1980**, *8*, 91–99

Petersen, M.C.; Nation, R.L.; Ashley, J.J. Simultaneous determination of betamethasone, betamethasone acetate and hydrocortisone in biological fluids using high-performance liquid chromatography. *J.Chromatogr.*, **1980**, *183*, 131–139

# Ibuprofen

**Molecular formula:** $C_{13}H_{18}O_2$
**Molecular weight:** 206.3
**CAS Registry No.:** 15687-27-1, 58560-75-1 ($\pm$ mixture), 61054-06-6 (Al salt), 112017-99-9 (piconol)

## SAMPLE
**Matrix:** blood
**Sample preparation:** 100 µL Plasma + 50 µL 2 µg/mL flurbiprofen + 25 µL 2 M HCl, vortex for 15 s, add 2 mL isooctane:isopropanol 85:15, rotate for 5 min, centrifuge at 3000 rpm for 10 min. Remove the upper organic layer and evaporate it to dryness under a stream of nitrogen at 45°, reconstitute the residue in 25 µL 5 mg/mL 5-bromoacetyl acenaphthene in MeCN, add 10 µL 3% triethylamine in MeCN, vortex for 30 s, heat at 75° for 5 min, evaporate to dryness under reduced pressure, reconstitute with 25 µL MeCN, inject a 20 µL aliquot. (Prepare 5-bromoacetyl acenaphthene as follows. Add 43 g bromoacetylchloride to 43 g acenaphthene dissolved in 200 mL dichloroethane, cool to -5° in an ice/salt bath, stir vigorously and add 38 g aluminum chloride in small portions over 90 min, do not allow temperature to go above 3°, place under reduced pressure for 30 min, add an excess of crushed ice. Separate the dichloroethane layer and wash it with two 100 mL portions of dilute HCl, wash with 100 mL 5% sodium carbonate solution. Dry the organic layer over anhydrous magnesium sulfate, remove the solvent under reduced pressure, allow the oily residue to solidify, remove liquid by blotting with filter paper. Purify the solid by chromatography on a 300 × 20 column of 60-120 mesh silica gel, elute with toluene, unreacted acenaphthene elutes first followed by 5-bromoacetyl acenaphthene (mp 87-90°).)

## HPLC VARIABLES
**Column:** 250 × 4.6 5 µm Hypersil C18
**Mobile phase:** MeCN:water 90:10
**Flow rate:** 1
**Injection volume:** 20
**Detector:** F ex 250 em 450

## CHROMATOGRAM
**Internal standard:** flurbiprofen
**Limit of detection:** 2.5 pmole

## KEY WORDS
rat; plasma; protect from light; pharmacokinetics; derivatization

## REFERENCE
Gifford, L.A.; Owusu-Daaku, F.T.K.; Stevens, A.J. Acenaphthene fluorescence derivatization reagents for use in high-performance liquid chromatography. *J.Chromatogr.A*, **1995**, *715*, 201–212

## SAMPLE
**Matrix:** blood
**Sample preparation:** 500 µL Plasma + 125 µL 40 mM decanoic acid in MeCN, mix. Dialyze a 100 µL sample against 20 mM pH 7.0 phosphate buffer using a Gilson Cuprophane membrane (molecular mass cut-off 15 kDa). Continuously pump the buffer through the dialysis cell and through column A at 3 mL/min for 9.6 min, backflush the contents of column A onto column B with the mobile phase, monitor the effluent from column B. (After each injection flush plasma channel with 1 mL 0.05% Triton X-100, with 1 mL 1 mM HCl, and with 2 mL water. After each injection flush buffer channel with 3 mL 20 mM pH 7.0 phosphate buffer and condition column A with 1 mL 20 mM pH 7.0 phosphate buffer.)

## HPLC VARIABLES
**Column:** A 10 × 2 40 µm Bondesil C18 (Analytichem); B 250 × 3.1 5 µm C18 (RoSil Research Separation Laboratories)
**Mobile phase:** MeCN:MeOH:20 mM pH 3.2 phosphate buffer 50:10:40
**Flow rate:** 1
**Injection volume:** 100
**Detector:** UV 264

## CHROMATOGRAM
**Retention time:** 11
**Limit of detection:** 2 µg/mL

## OTHER SUBSTANCES
**Extracted:** fenoprofen (UV 272), flurbiprofen (UV 247), ketoprofen (UV 261), naproxen (UV 272)

## KEY WORDS
plasma; dialysis; column-switching

## REFERENCE
Herráez-Hernández, R.; Van de Merbel, N.C.; Brinkman, U.A.T. Determination of the total concentration of highly protein-bound drugs in plasma by on-line dialysis and column liquid chromatography: application to non-steroidal anti-inflammatory drugs. *J.Chromatogr.B*, **1995**, *666*, 127–137

## SAMPLE
**Matrix:** blood
**Sample preparation:** 1 mL Plasma + 100 µL 1 M HCl + 100 µL fenoprofen solution + 10 mL ether, stir for 10 min, centrifuge at 6000 rpm for 5 min, repeat extraction 3 more times. Combine the organic layers and evaporate them to dryness under reduced pressure (0.5 bar), reconstitute the residue in 1 mL MeOH, inject a 50 µL aliquot.

## HPLC VARIABLES
**Guard column:** 4 × 4 5 µm Licrospher 100 RP-18
**Column:** 125 × 4 5 µm Licrospher 100 RP-18
**Mobile phase:** MeCN:pH 4.8 sodium acetate buffer 40:60
**Injection volume:** 50
**Detector:** UV 223

## CHROMATOGRAM
**Internal standard:** fenoprofen
**Limit of detection:** LOD 5 ng/mL

## KEY WORDS
plasma; pharmacokinetics

## REFERENCE
Kleinbloesem, C.H.; Ouwerkerk, M.; Spitznagel, W.; Wilkinson, F.E.; Kaiser, R.R. Pharmacokinetics and bioavailability of percutaneous ibuprofen. *Arzneimittelforschung*, **1995**, *45*, 1117–1121

## SAMPLE
**Matrix:** blood
**Sample preparation:** 500 µL Plasma + 100 µL 2.5 µg/mL S-(+)-naproxen + 500 µL 600 mM sulfuric acid + 15 mL dichloromethane, mix for 20 min, centrifuge at 2000 g for 5 min. Remove the organic layer and evaporate it to dryness under a stream of nitrogen at room temperature, reconstitute the residue in 100 µL 50 mM triethylamine in MeCN +

50 μL 60 mM ethyl chloroformate in MeCN, vortex for 30 s, add 50 μL 100 mM L-leucinamide in MeOH:triethylamine 100:14, let stand for 2 min, add 50 μL water, inject a 10-50 μL aliquot of the reaction mixture.

## HPLC VARIABLES
**Column:** 250 × 4.6 5 μm Ultrabase C18 (Shandon)
**Mobile phase:** MeCN:60 mM $KH_2PO_4$:triethylamine 49:51:0.1
**Flow rate:** 1.8
**Injection volume:** 10-50
**Detector:** UV 225

## CHROMATOGRAM
**Retention time:** 5.5 (R-(−)), 5.8 (S-(+))
**Internal standard:** S-(+)-naproxen (2.5)
**Limit of detection:** 100 ng/mL

## OTHER SUBSTANCES
**Extracted:** flurbiprofen (UV 275), ketoprofen (UV 275)

## KEY WORDS
plasma; chiral; derivatization

## REFERENCE
Péhourcq, F.; Lagrange, F.; Labat, L.; Bannwarth, B. Simultaneous measurement of flurbiprofen, ibuprofen, and ketoprofen enantiomer concentrations in plasma using L-leucinamide as the chiral coupling component. *J.Liq.Chromatogr.*, **1995**, *18*, 3969–3979

## SAMPLE
**Matrix:** blood
**Sample preparation:** 500 μL Plasma + 20 μL 500 μg/mL phenylbutazone in MeOH + 1.5 mL MeOH, vortex, centrifuge for 15 min at 3000 g. Remove the supernatant and evaporate it to 500 μL using a vortex evaporator, inject a 20 μL aliquot.

## HPLC VARIABLES
**Guard column:** present but not specified
**Column:** 10 μm RP-8 (Alltech)
**Mobile phase:** MeOH:1% acetic acid 78:22
**Injection volume:** 20
**Detector:** UV 235

## CHROMATOGRAM
**Internal standard:** phenylbutazone
**Limit of quantitation:** 100 ng/mL

## OTHER SUBSTANCES
**Extracted:** ibudice

## KEY WORDS
dog; plasma; pharmacokinetics

## REFERENCE
Samara, E.; Avnir, D.; Ladkani, D.; Bialer, M. Pharmacokinetic analysis of diethylcarbonate prodrugs of ibuprofen and naproxen. *Biopharm.Drug Dispos.*, **1995**, *16*, 201–210

## SAMPLE
**Matrix:** blood

**Sample preparation:** Erythrocytes. 500 μL Erythrocytes + 5 μL 1 mg/mL indomethacin in MeOH + 900 μL water, shake for 5 min, sonicate for 5 min, let stand at room temperature for 5 min, add 400 μL 3 M HCl, shake for 5 min, add 6 mL dichloromethane, shake, centrifuge at 1930 g for 5 min. Remove 5 mL of the organic layer and evaporate it to dryness under a stream of nitrogen, reconstitute the residue in 100 μL mobile phase, inject a 10 μL aliquot. Plasma. 500 μL Plasma + 5 μL 1 mg/mL indomethacin in MeOH, acidify gradually with 900 μL 1 M HCl, shake, add 200 μL 3 M HCl, shake for 5 min, add 6 mL dichloromethane, shake, centrifuge at 1930 g for 5 min. Remove 5 mL of the organic layer and evaporate it to dryness under a stream of nitrogen, reconstitute the residue in 100 μL mobile phase, inject a 10 μL aliquot.

## HPLC VARIABLES
**Column:** 150 × 3.3 5 μm C18 glass column (Tessek)
**Mobile phase:** MeOH:water 66:30 adjusted to pH 3.0 with 5% perchloric acid
**Flow rate:** 1.3
**Injection volume:** 10
**Detector:** UV 222

## CHROMATOGRAM
**Retention time:** 9.9
**Internal standard:** indomethacin (7.8)
**Limit of detection:** 20 (plasma), 30 (erythrocytes) ng/mL

## OTHER SUBSTANCES
**Simultaneous:** diazepam, phenylanthranilic acid

## KEY WORDS
plasma; erythrocytes; rabbit; pharmacokinetics

## REFERENCE
Sochor, J.; Klimes, J.; Sedlácek, J.; Zahradnicek, M. Determination of ibuprofen in erythrocytes and plasma by high-performance liquid chromatography. *J.Pharm.Biomed.Anal.*, **1995**, *13*, 899–903

## SAMPLE
**Matrix:** blood
**Sample preparation:** 2 mL Whole blood or plasma + 2 mL buffer + 5 mL chloroform: isopropanol:n-heptane 60:14:26, shake gently horizontally for 10 min, centrifuge at 2800 g for 10 min. Remove the lower organic layer and evaporate it to dryness under vacuum at 45°, reconstitute the residue in 100 μL mobile phase, centrifuge at 2800 g for 5 min, inject a 50 μL aliquot of the supernatant. (Buffer was saturated ammonium chloride solution 25% diluted with water, adjusted to pH 9.5 with 25% ammonia solution.)

## HPLC VARIABLES
**Column:** 300 × 3.9 4 μm NovaPack C18
**Mobile phase:** MeOH:THF:buffer 65:5:30 (Buffer was 0.68 g/L (10 mM (sic)) $KH_2PO_4$ adjusted to pH 2.6 with concentrated orthophosphoric acid.) (At the end of each session wash the column with water for 1 h and MeOH for 1 h, re-equilibrate for 30 min.)
**Column temperature:** 30
**Flow rate:** 0.8
**Injection volume:** 50
**Detector:** UV 220

## CHROMATOGRAM
**Retention time:** 10.58
**Limit of detection:** <120 ng/mL

## OTHER SUBSTANCES

**Extracted:** acebutolol, acenocoumarol, acepromazine, aceprometazine, acetaminophen, aconitine, ajmaline, albuterol, alimemazine, alminoprofen, alprazolam, alprenolol, amisulpride, amitriptyline, amodiaquine, amoxapine, aspirin, astemizole, atenolol, benazepril, benperidol, benzocaine, benzoylecgonine, bepridil, betaxolol, bisoprolol, bromazepam, brompheniramine, bumadizone, bupivacaine, buprenorphine, buspirone, caffeine, carbamazepine, carbinoxamine, carpipramine, carteolol, celiprolol, cetirizine, chlordiazepoxide, chlormezanone, chlorophenacinone, chloroquine, chlorpheniramine, chlorpromazine, chlorpropamide, cibenzoline, cicletanine, clemastine, clobazam, clomipramine, clonazepam, clonidine, clorazepate, clozapine, cocaine, codeine, colchicine, cyamemazine, cyclizine, cycloguanil, cyproheptadine, cytarabine, dacarbazine, daunorubicin, debrisoquine, demexiptiline, desipramine, dextromethorphan, dextromoramide, dextropropoxyphene, diazepam, diazoxide, diclofenac, dihydralazine, diltiazem, diphenhydramine, dipyridamole, disopyramide, dosulepine, doxepin, doxylamine, droperidol, ephedrine, estazolam, etodolac, fenfluramine, fenoprofen, fentiazac, flecainide, flumazenil, flunitrazepam, fluoxetine, fluphenazine, flurbiprofen, fluvoxamine, glibenclamide, glibornuride, glipizide, glutethimide, haloperidol, histapyrrodine, hydroxychloroquine, hydroxyzine, imipramine, indomethacin, iproniazid, ketamine, ketoprofen, labetalol, levomepromazine, lidocaine, lisinopril, loperamide, loprazolam, loratadine, lorazepam, loxapine, maprotiline, medazepam, medifoxamine, mefenamic acid, mefenidramine, mefloquine, melphalan, meperidine, mephenesin, mephentermine, mepivacaine, metapramine, metformin, methadone, methaqualone, methocarbamol, methotrexate, metipranolol, metoclopramide, metoprolol, mexiletine, mianserine, midazolam, minoxidil, moclobemide, moperone, morphine, nadolol, nalbuphine, nalorphine, naloxone, naltrexone, naproxen, nialamide, nicardipine, nifedipine, niflumic acid, nimodipine, nitrazepam, nitrendipine, nizatidine, nomifensine, nortriptyline, omeprazole, opipramol, oxazepam, oxprenolol, penbutolol, penfluridol, pentazocine, phencyclidine, phenobarbital, phenol, phenylbutazone, pimozide, pindolol, pipamperone, piroxicam, prazepam, prazosin, prilocaine, procainamide, procarbazine, proguanil, promethazine, propafenone, propranolol, protriptyline, pyrimethamine, quinidine, quinine, quinupramine, ramipril, ranitidine, reserpine, ritodrine, secobarbital, sotalol, strychnine, sulfinpyrazole, sulindac, sulpride, sultopride, suriclone, temazepam, tenoxicam, terbutaline, terfenadine, tetracaine, tetrazepam, thiopental, thioproperazine, thioridazine, tianeptine, tiapride, tiaprofenic acid, ticlopidine, timolol, tioclomarol, tofisopam, tolbutamide, toloxatone, trazodone, triazolam, trifluoperazine, trifluperidol, trimipramine, triprolidine, tropatenine, verapamil, viloxazine, vinblastine, vincristine, vindesine, warfarin, yohimbine, zolpidem, zopiclone, zorubicine

**Interfering:** alpidem, chlorambucil, floctafenine, ibuprofen, lidoflazine

## KEY WORDS

whole blood; plasma

## REFERENCE

Tracqui, A.; Kintz, P.; Mangin, P. Systematic toxicological analysis using HPLC/DAD. *J.Forensic Sci.*, **1995**, *40*, 254–262

## SAMPLE

**Matrix:** blood

**Sample preparation:** 500 µL Plasma + 100 µL 200 µg/mL tridecanoic acid in ethylene chloride + 200 µL 600 mM sulfuric acid + 3 mL isooctane:isopropanol 95:5, extract. Remove the organic layer and evaporate it to dryness, reconstitute the residue in 1 mL 2.4 mg/mL 2-ethoxy-1-ethoxycarbonyl-1,2-dihydroquinoline (EEDQ) in ethylene chloride, add 5 mL reagent, reflux for 10 min, dilute with 10 mL ethylene chloride, wash with an equal volume of 200 mM NaOH, wash with an equal volume of 1 M HCl, wash with an equal volume of water. Remove the organic layer and dry it over sodium sulfate, evaporate to dryness, reconstitute the residue in mobile phase, inject an aliquot. (Prepare reagent by dissolving 5 mg p-nitrobenzylamine hydrochloride in 5 mL 200 mM NaOH, extract with 5 mL ethylene chloride, dry the organic layer over anhydrous sodium sulfate, use this solution as the reagent.)

## HPLC VARIABLES
**Column:** 100 × 4.6 3 μm (R)-(−)-(1-naphthyl)ethylurea (Prepare by pumping 2 g (R)-(−)-1-(1-naphthyl)ethyl isocyanate in 100 mL dichloromethane through a 100 × 4.6 3 μm aminopropyl-silanized silica column (Regis) at 2 mL/min (without detector), after 12.5 min recycle the mobile phase, after 2 h wash the column with 300 mL dichloromethane at 2 mL/min, wash with hexane:isopropanol 80:20 until a steady baseline is achieved.)
**Mobile phase:** Hexane:isopropanol 87.5:12.5
**Flow rate:** 1.5
**Detector:** UV 235

## CHROMATOGRAM
**Retention time:** 21 (S), 22.5 (R)
**Internal standard:** tridecanoic acid (19)
**Limit of quantitation:** 2.5 μg/mL

## KEY WORDS
derivatization; dog; chiral; plasma; pharmacokinetics

## REFERENCE
Ahn, H.-Y.; Shiu, G.K.; Trafton, W.F.; Doyle, T.D. Resolution of the enantiomers of ibuprofen; comparison study of diastereomeric method and chiral stationary phase method. *J.Chromatogr.B*, **1994**, *653*, 163–169

## SAMPLE
**Matrix:** blood
**Sample preparation:** 500 μL Plasma + 50 ng flurbiprofen + 150 μL 1 M phosphoric acid + 5 mL hexane:ether 80:20, extract. Remove the organic layer and evaporate it to dryness, reconstitute the residue in 200 μL mobile phase, inject an aliquot.

## HPLC VARIABLES
**Column:** 4 μm Nova Pak C18
**Mobile phase:** MeCN:water:acetic acid 59:40.5:0.5
**Flow rate:** 1.3
**Detector:** UV 233

## CHROMATOGRAM
**Internal standard:** flurbiprofen
**Limit of detection:** 400 ng/mL

## KEY WORDS
pharmacokinetics; plasma

## REFERENCE
al-Meshal, M.A.; El-Sayed, Y.M.; al-Balla, S.R.; Gouda, M.W. The effect of colestipol and cholestyramine on ibuprofen bioavailability in man. *Biopharm.Drug Dispos.*, **1994**, *15*, 463–471

## SAMPLE
**Matrix:** blood
**Sample preparation:** 10 μL Plasma is injected into MeCN pumped at 0.2 mL/min, precipitated proteins are removed by 0.5 and 0.2 μm filters in series, MeCN containing sample is switched into mobile phase allowed to pass onto analytical column. Next filter unit is switched out of circuit and back-flushed to waste with 100 mM sodium dodecyl sulfate at

2 mL/min, analytical column is eluted in normal fashion with mobile phase. Equilibrate filters with MeCN for 5 min before next injection.

## HPLC VARIABLES
**Guard column:** 20 mm Brownlee C18
**Column:** 250 × 4.6 5 μm Ultrasphere C18
**Mobile phase:** MeCN:buffer 70:30 (Buffer was 0.094% triethylamine in water adjusted to pH 3 with glacial acetic acid.)
**Flow rate:** 1
**Injection volume:** 10
**Detector:** UV 280

## CHROMATOGRAM
**Retention time:** 6.39

## KEY WORDS
plasma; dog

## REFERENCE
Asafu-Adjaye, E.B.; Su, S.Y.; Shiu, G.K. Switching-valve-filter technique for the direct injection and analysis of drugs in plasma using high-performance liquid chromatography. *J.Chromatogr.B*, **1994**, *652*, 35–42

## SAMPLE
**Matrix:** blood
**Sample preparation:** Inject sample onto column A and elute to waste with mobile phase A, after 2 min backflush the contents of column A onto column B with mobile phase B for 2 min, elute column B with mobile phase B, monitor effluent from column B.

## HPLC VARIABLES
**Column:** A 10 × 3 BioTrap Acid C18 (ChromTech); B 10 × 3 CT-sil C18 guard column + 100 × 4.6 5 μm CT-sil C18 (ChromTech)
**Mobile phase:** A 200 mM pH 2.1 phosphate buffer; B MeOH:82 mM pH 6.0 phosphate buffer 65:35
**Flow rate:** A 0.55; B 1
**Injection volume:** 10
**Detector:** F ex 225 em 535

## CHROMATOGRAM
**Retention time:** 5.8
**Limit of quantitation:** 520 ng/mL

## KEY WORDS
plasma; column-switching; direct injection

## REFERENCE
Hermansson, J.; Grahn, A. Determination of drugs by direct injection of plasma into a biocompatible extraction column based on a protein-entrapped hydrophobic phase. *J.Chromatogr.A*, **1994**, *660*, 119–129

## SAMPLE
**Matrix:** blood
**Sample preparation:** Place 100 μL 100 μM diclofenac in dichloromethane in the bottom of a tube and evaporate it to dryness under a stream of nitrogen, add 100 μL plasma, add 25 μL 1 M HCl, mix, add to a dry Chem Elut diatomaceous earth SPE cartridge (Varian), let stand for 5 min, elute with 6 mL n-hexane:diethyl ether:isopropanol 50:

50:1. Evaporate the eluate to dryness under a stream of nitrogen at 40°, reconstitute the residue in 200 μL 2 mM (−)-APMB in dichloromethane, add 100 μL 20 mM 2,2'-dipyridyl disulfide in dichloromethane, add 100 μL 20 mM triphenylphosphine in dichloromethane, mix, let stand at room temperature for 5 min. Evaporate to dryness under a stream of nitrogen at 40°, reconstitute the residue in 400 μL mobile phase, inject a 10 μL aliquot. ((−)-APMB is (−)-2-[4-(1-aminoethyl)phenyl]-6-methoxybenzoxazole. Synthesis is as follows. Hydrogenate 5-methoxy-2-nitrophenol in EtOH over platinum oxide to give 2-amino-5-methoxyphenol (J. Org. Chem. 1957, 22, 220). It should be possible to prepare ethyl 4-acetylbenzimidate hydrochloride ($CH_3COC_6H_4C(=NH)OC_2H_5.HCl$) by passing dry hydrogen chloride into a mixture of 4-acetylbenzonitrile and 1.2-1.5 equivalents EtOH in an inert solvent (e.g., benzene, chloroform, dioxane, ether, nitrobenzene (Caution! Benzene, chloroform, and dioxane are carcinogens!) at 0-5°, the benzimidate should crystallize from the mixture in 7-10 days (J. Chem. Soc. 1942, 103). Add a solution of 5.5 g 2-amino-5-methoxyphenol in 200 mL MeOH to 9 g ethyl 4-acetylbenzimidate hydrochloride, stir at 60-70° for 4 h, evaporate to dryness under reduced pressure, recrystallize from EtOH to give 4-(6-methoxy-2-benzoxazolyl)acetophenone as fine orange-yellow crystals (mp 167°) (J. Chromatogr. 1990, 532, 65). Add 7.0 g hydroxylamine hydrochloride and 8.2 g sodium acetate to 10.1 g 4-(6-methoxy-2-benzoxazolyl)acetophenone in 500 mL EtOH:water 95:5, reflux for 1 h, pour into ice-water, filter, recrystallize from EtOH:water 90:10 to give 4-(6-methoxy-2-benzoxazolyl)acetophenone oxime as faint reddish needles (mp 212°). Dissolve 4.7 g 4-(6-methoxy-2-benzoxazolyl)acetophenone oxime in 300 mL MeOH, add 3 g 10% palladium on charcoal, add 10.5 g ammonium formate, reflux for 30 min, filter, evaporate the filtrate to dryness under reduced pressure. Take up the residue in 100 mL 5% HCl and wash the aqueous phase with 100 mL ethyl acetate. Adjust the pH of the aqueous layer to 13-14 with 10% NaOH and extract with 200 mL ethyl acetate. Wash the organic layer with 100 mL water and dry it over anhydrous sodium sulfate, evaporate to dryness under reduced pressure to give racemic 2-[4-(1-aminoethyl)phenyl]-6-methoxybenzoxazole. Dissolve 3.6 g racemic 2-[4-(1-aminoethyl)phenyl]-6-methoxybenzoxazole in 50 mL EtOH and add 3.5 g (S)-(−)-α-methoxy-α-trifluoromethylphenylacetic acid, allow to stand overnight at 5°. Collect the precipitate and fractionally crystallize it from EtOH 4 times. Take up the final product in 5% NaOH and extract it with ethyl acetate, wash the organic layer with water, dry over anhydrous sodium sulfate, evaporate to dryness under reduced pressure, recrystallize from EtOH to give (−)-2-[4-(1-aminoethyl)phenyl]-6-methoxybenzoxazole as pale yellow crystals (mp 74°) (J. Chromatogr. 1993, 645, 75).)

## HPLC VARIABLES
**Column:** 150 × 4.6 5 μm TSK gel ODS-80TS (Tosoh)
**Mobile phase:** MeCN:water:acetic acid 70:30:0.1
**Column temperature:** 40
**Flow rate:** 1
**Injection volume:** 10
**Detector:** F ex 320 em 380

## CHROMATOGRAM
**Retention time:** 11.0 (S), 12.2 (R)
**Internal standard:** diclofenac (14.0)
**Limit of quantitation:** 200 ng/mL (S); 400 ng/mL (R)

## KEY WORDS
derivatization; rat; plasma; chiral; pharmacokinetics; SPE

## REFERENCE
Kondo, J.; Suzuki, N.; Naganuma, H.; Imaoka, T.; Kawasaki, T.; Nakanishi, A.; Kawahara, Y. Enantiospecific determination of ibuprofen in rat plasma using chiral fluorescence derivatization reagent, (−)-2-[4-(1- aminoethyl)phenyl]-6-methoxybenzoxazole. *Biomed.Chromatogr.*, **1994**, 8, 170−174

## SAMPLE
**Matrix:** blood

**Sample preparation:** Automated SPE by ASPEC system. Condition a C18 Clean-Up SPE cartridge (CEC 18111, Worldwide Monitoring) with 2 mL MeOH then 2 mL water. 1 mL Plasma + 1 mL water, vortex, add to the SPE cartridge, wash with 3 mL water, wash with 3 mL 750 mL/L methanol. Elute with three aliquots of 300 μL 0.1 M ammonium acetate in MeOH. Add 0.5 mL 0.5 M NaOH and 4 mL 50 mL/L isopropanol in heptane to eluate, mix thoroughly. Allow 5 min for phase separation. Remove upper heptane phase and add it to 300 μL 0.1 M phosphoric acid (pH 2.5), mix, separate, inject a 100 μL aliquot of the aqueous phase.

## HPLC VARIABLES
**Guard column:** LC-8-DB (Supelco)
**Column:** 150 × 4.6 LC-8-DB (Supelco)
**Mobile phase:** MeCN:buffer 35:65 (Buffer was 10 mL/L triethylamine in water adjusted to pH 5.5 with glacial acetic acid.)
**Flow rate:** 2
**Injection volume:** 100
**Detector:** UV 228

## CHROMATOGRAM
**Retention time:** 4.0

## OTHER SUBSTANCES
**Extracted:** acetazolamide, amitriptyline, chlordiazepoxide, chlorimipramine, chlorpromazine, dextromethorphan, diazepam, diphenhydramine, doxepin, encainide, fentanyl, flecainide, fluoxetine, flurazepam, haloperidol, hydroxyethylflurazepam, imipramine, lidocaine, maprotiline, methadone, mexiletine, midazolam, norchlorimipramine, nordoxepin, nordiazepam, norfluoxetine, nortriptyline, pentazocine, propoxyphene, propranolol, quinidine, temazepam, trazodone, trimipramine, verapamil
**Noninterfering:** acetaminophen, acetylmorphine, amiodarone, amobarbital, amphetamine, bendroflumethiazide, benzocaine, benzoylecgonine, benzthiazide, butalbital, carbamazepine, chlorothiazide, clonazepam, cocaine, codeine, cotinine, cyclosporine, cyclothiazide, desalkylflurazepam, diamorphine, dicumerol, ephedrine, ethacrynic acid, ethanol, ethchlorvynol, ethosuximide, furosemide, glutethimide, hydrochlorothiazide, hydrocodone, hydroflumethiazide, hydromorphone, lorazepam, mephentermine, meprobamate, methamphetamine, metharbital, methoxsalen, methoxyphenteramine, methsuximide, methylcyclothiazide, metoprolol, MHPG, monoacetylmorphine, morphine, normethsuximide, oxazepam, oxycodone, oxymorphone, pentobarbital, phencyclidine, phenteramine, phenylephrine, phenytoin, polythiazide, primidone, prochlorperazine, salicylic acid, sulfanilamide, THC-COOH, theophylline, thiazolam, thiopental, thioridazine, tocainide, trichloromethiazide, trifluoperazine, valproic acid, warfarin
**Interfering:** desipramine, methaqualone, norverapamil, promazine, propafenone, protriptyline

## KEY WORDS
plasma; SPE

## REFERENCE
Nichols, J.H.; Charlson, J.R.; Lawson, G.M. Automated HPLC assay of fluoxetine and norfluoxetine in serum. *Clin.Chem.*, **1994**, *40*, 1312–1316

## SAMPLE
**Matrix:** blood
**Sample preparation:** Condition a 1 mL 60 μm Separon SGX C18 SPE cartridge with 5 mL MeOH, 5 mL water, and 5 mL buffer. 250 μL Blood + 5 μL 500 μg/mL indomethacin in MeOH + 500 μL water, shake for 5 min, sonicate for 5 min, let stand at room temperature for 5 min, add 1 mL buffer, shake for 5 min, centrifuge at 1930 g for 10 min, add the supernatant to the SPE cartridge, wash with 5 mL buffer, wash with 10 mL

water, dry with vacuum for 5 min, elute with dichloromethane. Evaporate eluate to dryness under a stream of nitrogen, dissolve in 100 μL mobile phase, inject 10 μL aliquot. (Buffer was 66 mM KH$_2$PO$_4$ adjusted to pH 2.0 with phosphoric acid.)

## HPLC VARIABLES
**Column:** 150 × 3.3 5 μm Separon SGX C18 glass column
**Mobile phase:** MeOH water 220:100, adjusted to pH 3.0 with 5% perchloric acid
**Flow rate:** 1.3
**Injection volume:** 10
**Detector:** UV 222

## CHROMATOGRAM
**Retention time:** 9.9
**Internal standard:** indomethacin (7.8)
**Limit of detection:** 100 ng/mL
**Limit of quantitation:** 300 ng/mL

## KEY WORDS
SPE; rabbit; human

## REFERENCE
Sochor, J.; Klimes, J.; Zahradnícek, M.; Sedlácek, J. High-performance liquid chromatographic assay for ibuprofen in whole blood using solid-phase extraction. *J.Chromatogr.B*, **1994**, *654*, 282–286

## SAMPLE
**Matrix:** blood
**Sample preparation:** Condition a 100 mg Bond-Elut SPE cartridge with 4 mL MeOH, then 4 mL water, then 4 mL 10 mM phosphoric acid. 500 μL Plasma + 25 μL 80 μg/mL ibufenac in MeOH, vortex 30 s, stand at RT for 10 min, add 1 mL MeCN, vortex 30 s, centrifuge at 1200 g. Remove supernatant and add it to 8.5 mL 10 mM pH 2 phosphoric acid. Add this solution to the SPE cartridge, wash with 2 mL MeCN:10 mM phosphoric acid 20:80, centrifuge at 1800 g for 3 min to remove liquid completely, elute with 1 mL MeCN:10 mM phosphoric acid 1:1, centrifuge at 1800 g for 3 min to remove all of eluent. Evaporate eluent to dryness under a stream of nitrogen at 37°, reconstitute in 400 μL MeCN:MeOH:1% acetic acid (pH 3) 10:15:75, sonicate 3 min, vortex 1 min, inject 100 μL aliquot

## HPLC VARIABLES
**Guard column:** 15 × 3.2 7 μm Brownlee RP-18
**Column:** 150 × 4.6 5 μm Axxiom ODS
**Mobile phase:** MeOH:10 mM pH 2.2 trifluoroacetic acid 57:43
**Flow rate:** 1.2
**Injection volume:** 100
**Detector:** UV 225; UV 214

## CHROMATOGRAM
**Retention time:** 24.8
**Internal standard:** ibufenac (16.1)
**Limit of quantitation:** 500 ng/mL

## OTHER SUBSTANCES
**Extracted:** metabolites, glucuronides
**Noninterfering:** aspirin, acetaminophen, salicylic acid

## KEY WORDS
plasma; SPE

## REFERENCE

Castillo, M.; Smith, P.C. Direct determination of ibupropfen and ibuprofen acyl glucuronide in plasma by high-performance liquid chromatography using solid-phase extraction. *J.Chromatogr.*, **1993**, *614*, 109–116

## SAMPLE

**Matrix:** blood
**Sample preparation:** 500 μL Plasma + 50 μL 200 μg/mL fenoprofen in MeOH:water 1: 4 + 200 μL 1 M sulfuric acid + 3 mL isooctane:isopropanol 95:5, vortex 30 s, centrifuge at 1800 g for 5 min. Remove organic layer and evaporate it to dryness. Add 300 μL 50 mM triethylamine in MeCN and 50 μL 6 mM ethyl chloroformate in MeCN, wait 30 s, add 25 μL 0.1% (S)-naphthylethylamine in MeCN:triethylamine 98:2, after 3 min add 25 μL 2.5% ethanolamine in MeCN, inject 2-30 μL aliquot.

## HPLC VARIABLES

**Column:** 100 × 4.6 5 μm Partisil ODS 3 RAC
**Mobile phase:** MeCN:water:acetic acid:triethylamine 60:40:0.1:0.02, final pH 5.0 (After every third injection flush with MeCN for 6 min at 1.6 mL/min, equilibrate with mobile phase for 9 min.)
**Flow rate:** 1.2
**Injection volume:** 2-30
**Detector:** F ex 280 em 320

## CHROMATOGRAM

**Retention time:** 10.5 (S-(+)), 11.8 (R-(−))
**Internal standard:** fenoprofen (7.5 (S), 8.8 (R))
**Limit of detection:** 10 ng/mL

## KEY WORDS

plasma; chiral; also UV 232 (Clin. Chem. 1988, 34 ,493)

## REFERENCE

Lemko, C.H.; Caillé, G.; Foster, R.T. Stereospecific high-performance liquid chromatographic assay of ibuprofen: improved sensitivity and sample processing efficiency. *J.Chromatogr.*, **1993**, *619*, 330–335

## SAMPLE

**Matrix:** blood
**Sample preparation:** 1 mL Plasma + 500 μL MeCN, vortex 2 min, centrifuge at 2000 g for 4 min. Remove supernatant and saturate it with anhydrous ammonium sulfate. Centrifuge at 2000 g for 2 min, inject 50-100 μL aliquot of the supernatant.

## HPLC VARIABLES

**Column:** 150 × 4.6 3 μm Spherisorb ODS
**Mobile phase:** MeCN:pH 2.2 phosphoric acid 50:50
**Flow rate:** 1
**Injection volume:** 50-100
**Detector:** UV 220

## CHROMATOGRAM

**Retention time:** 13
**Limit of detection:** 25 ng/mL

## OTHER SUBSTANCES

**Noninterfering:** acetaminophen, aspirin, 5-azacytidine, baclofen, cimetidine, cyclophosphamide, cyclosporin A, famotidine, 5-fluorouracil, ranitidine, verapamil

## KEY WORDS
plasma

## REFERENCE
Rustum, A.M. Assay of ibuprofen in human plasma by rapid and sensitive reversed-phase high-performance liquid chromatography: application to a single dose pharmacokinetic study. *J.Chromatogr.Sci.*, **1991**, *29*, 16–20

## SAMPLE
**Matrix:** blood
**Sample preparation:** 1 mL Serum + 50 μL 1 M HCl, add 4 mL hexane:isopropanol 85:15, vortex for 30 s, centrifuge at 2000 g for 5 min, repeat extraction. Combine the organic layers and evaporate them to dryness under a stream of nitrogen at 40°, reconstitute the residue in 300 μL MeCN:20 mM pH 3.5 phosphate buffer 40:60 containing 5 mM tetrabutylammonium bromide, inject a 200 μL aliquot onto column A with mobile phase A, collect the eluate containing ibuprofen in a sample loop and inject it onto column B with mobile phase B. Collect the eluate containing ibuprofen in a sample loop and inject it onto column C with mobile phase C, monitor the effluent from column C.

## HPLC VARIABLES
**Column:** A 70 × 4.6 5 μm YMC ODS A type (Yakamura Chemical); B 70 × 4.6 5 μm YMC ODS A type (Yakamura Chemical); C 150 × 4.6 5 μm TSK gel ODS 80 TM (Tosoh)
**Mobile phase:** A MeCN:20 mM pH 3.5 phosphate buffer 40:60 containing 5 mM tetrabutylammonium bromide; B MeCN:20 mM pH 7 phosphate buffer 30:70 containing 5 mM tetrabutylammonium bromide; C MeCN:20 mM pH 7 phosphate buffer 30:70
**Column temperature:** 40
**Flow rate:** 1
**Injection volume:** 200
**Detector:** UV 221

## CHROMATOGRAM
**Retention time:** 41
**Limit of detection:** 0.5 ng/mL

## KEY WORDS
serum; column-switching; heart-cut

## REFERENCE
Yamashita, K.; Motohashi, M.; Yashiki, T. Column-switching techniques for high-performance liquid chromatography of ibuprofen and mefenamic acid in human serum with short-wavelength ultraviolet detection. *J.Chromatogr.*, **1991**, *570*, 329–338

## SAMPLE
**Matrix:** blood
**Sample preparation:** 500 μL Plasma + 200 μL 2 M HCl + 6 mL ice-cold hexane:diethyl ether 8:2, extract, centrifuge at 1500 g for 10 min. Remove 5 mL of organic layer and evaporate it to dryness under a stream of nitrogen. Dissolve in 250 μL isopropanol:water 2:8, inject a 40 μL aliquot.

## HPLC VARIABLES
**Column:** 100 × 4.0 AGP (EnantioPac)
**Mobile phase:** 20 mM pH 6.7 phosphate buffer containing 0.5% isopropanol and 5 mM dimethyloctylamine
**Column temperature:** 15
**Flow rate:** 0.5

**Injection volume:** 40
**Detector:** UV 220

## CHROMATOGRAM
**Retention time:** 14(R), 17(S)
**Limit of quantitation:** 100 ng/mL

## OTHER SUBSTANCES
**Simultaneous:** fenoprofen

## KEY WORDS
plasma

## REFERENCE
Menzel-Soglowek, S.; Geisslinger, G.; Brune, K. Stereoselective high-performance liquid chromatographic determination of ketoprofen, ibuprofen and fenoprofen in plasma using a chiral $\alpha_1$-acid glycoprotein column. *J.Chromatogr.*, **1990**, *532*, 295–303

## SAMPLE
**Matrix:** blood
**Sample preparation:** Dialyze (Spectrum Medical Industries, Inc. Spectrophor 2, 12000-14000 molecular weight cutoff) 3.5 mL plasma with 3.5 mL buffer at 37° with one 6 cm oscillation per s for 16-17 h. Remove a 2 mL aliquot of the buffer and add it to 500 μL 2 M sulfuric acid and 10 mL heptane:isopropanol 95:5, vortex for 1 min, centrifuge at 1000 g for 5 min. Remove the organic layer and evaporate it to dryness under a stream of nitrogen at 50°, reconstitute the residue in 100 μL 1% thionyl chloride in dichloromethane (freshly prepared), vortex briefly, heat at 70° in a tube securely sealed with a PTFE-lined cap for 1 h, let cool for 15 min, add 500 μL 1% S-(−)-1-phenylethylamine in dichloromethane (freshly prepared), vortex briefly, let stand at room temperature for 20 min, add 500 μL 2 M sulfuric acid, add 5 mL heptane, vortex for 1 min, centrifuge at 1000 g for 5 min. Remove the organic layer and evaporate it to dryness under a stream of nitrogen at 50°, reconstitute the residue in 100 μL mobile phase, vortex briefly, inject a 50-90 μL aliquot. (Buffer was isotonic pH 7.4 phosphate buffer prepared from 67 mM $NaH_2PO_4$, 67 mM $Na_2HPO_4$, and NaCl.)

## HPLC VARIABLES
**Column:** 250 × 4 5 μm Hibar Lichrosorb Si60
**Mobile phase:** Heptane:isopropanol 97.5:2.5
**Flow rate:** 2
**Injection volume:** 50-90
**Detector:** UV 216

## CHROMATOGRAM
**Retention time:** 3 (R-(−)), 7 (S-(+))

## KEY WORDS
plasma; normal phase; derivatization; chiral

## REFERENCE
Evans, A.M.; Nation, R.L.; Sansom, L.N.; Bochner, F.; Somogyi, A.A. Stereoselective plasma protein binding of ibuprofen enantiomers. *Eur.J.Clin.Pharmacol.*, **1989**, *36*, 283–290

## SAMPLE
**Matrix:** blood
**Sample preparation:** 500 μL Plasma + 100 μL 100 μg/mL IS in 10 mM NaOH + 200 μL 600 mM sulfuric acid + 3 mL isooctane:isopropanol 95:5, vortex for 30 s, centrifuge at

1800 g for 5 min. Remove the organic layer and evaporate it to dryness, reconstitute the residue in 100 μL 50 mM triethylamine in MeCN, add 50 μL 6 mM ethyl chloroformate in MeCN, after 30 s add 25 μL 1 mL/L (S)-(−)-1-(1-naphthyl)ethylamine in MeCN, let stand for 3 min, add 500 μL 250 mM HCl, add 2 mL chloroform, vortex for 15 s, centrifuge at 1800 g for 2 min. Remove the organic layer and evaporate it to dryness under reduced pressure, reconstitute the residue in 200 μL mobile phase, inject a 10-50 μL aliquot.

## HPLC VARIABLES
**Guard column:** 20 mm long 37-53 μm reversed-phase
**Column:** 100 × 4.6 5 μm Partisil 5 ODS-3
**Mobile phase:** MeCN:water:acetic acid:triethylamine 55:45:0.1:0.02, pH 4.9
**Flow rate:** 1
**Injection volume:** 10-50
**Detector:** UV 232

## CHROMATOGRAM
**Retention time:** 18.5 (S), 21.0 (R)
**Internal standard:** (±)-2-(4-benzoylphenyl)butyric acid (11, 13 (enantiomers))
**Limit of quantitation:** 100 ng/mL

## OTHER SUBSTANCES
**Simultaneous:** etodolac (enantiomers not resolved), flurbiprofen, ketoprofen, tiaprofenic acid (not derivatized)

## KEY WORDS
derivatization; plasma; chiral; pharmacokinetics

## REFERENCE
Mehvar, R.; Jamali, F.; Pasutto, F.M. Liquid-chromatographic assay of ibuprofen enantiomers in plasma. *Clin.Chem.*, **1988**, *34*, 493–496

## SAMPLE
**Matrix:** blood
**Sample preparation:** Condition a 1 mL Bond-Elut C8 SPE cartridge with 2 mL MeOH then 1 mL 10 mM HCl, do not allow it to dry completely. Sonicate 1 mL whole blood for 20-30 min then apply it to the SPE cartridge. Wash with 100 μL water, elute with three 500 μL portions of MeOH:MeCN:1% aqueous ammonium hydroxide 50:20:30, combine eluents and evaporate to dryness under a stream of nitrogen at 40°. Redissolve in 1 mL MeOH, inject a 20 μL aliquot.

## HPLC VARIABLES
**Column:** 250 × 4.5 5 μm Spherisorb ODS
**Mobile phase:** MeCN:MeOH:buffer 35:13:52 (Buffer was water adjusted to pH 3.2 with orthophosphoric acid.)
**Flow rate:** 1
**Injection volume:** 20
**Detector:** UV 250

## CHROMATOGRAM
**Retention time:** 8.5

## OTHER SUBSTANCES
**Simultaneous:** acetaminophen, fenoprofen, indomethacin, ketoprofen, naproxen, salicylic acid

## KEY WORDS
whole blood; SPE

## REFERENCE

Moore, C.M.; Tebbett, I.R. Rapid extraction of anti-inflammatory drugs in whole blood for HPLC analysis. *Forensic Sci.Int.*, **1987**, *34*, 155–158

## SAMPLE

**Matrix:** blood
**Sample preparation:** 25 μL Plasma + 50 μL 40 μg/mL mefenamic acid in MeCN, vortex 10 s, centrifuge at 11000 g for 2 min, inject a 20 μL aliquot

## HPLC VARIABLES

**Guard column:** 30 × 4.6 10 μm Brownlee RP-18
**Column:** 250 × 4.5 5 μm IBM octadecyl
**Mobile phase:** MeCN:MeOH:water:85% phosphoric acid 58:5:37:0.05
**Flow rate:** 1.8
**Injection volume:** 20
**Detector:** UV 196

## CHROMATOGRAM

**Retention time:** 6.8
**Internal standard:** mefenamic acid (9.8)
**Limit of detection:** 500 ng/mL

## OTHER SUBSTANCES

**Noninterfering:** acetaminophen, caffeine, carbamazepine, chloramphenicol, desipramine, digoxin, disopyramide, ethosuximide, gentamicin, imipramine, lidocaine, methotrexate, N-acetylprocainamide, phenobarbital, phenytoin, primidone, procainamide, quinidine, salicylic acid, theophylline, tobramycin, valproic acid

## KEY WORDS

plasma; rat

## REFERENCE

Shah, A.; Jung, D. Improved high-performance liquid chromatographic assay of ibuprofen in plasma. *J.Chromatogr.*, **1985**, *344*, 408–411

## SAMPLE

**Matrix:** blood
**Sample preparation:** 1 mL Plasma + 100 μL 100 μg/mL tolmetin + 0.5 mL 1 M HCl + 10 mL dichloromethane, shake 10 min, centrifuge at 1000 g for 5 min. Remove the organic phase and evaporate it to dryness under a stream of nitrogen at 40°. Reconstitute with 200 μL mobile phase, inject 10-30 μL aliquot.

## HPLC VARIABLES

**Column:** 250 × 4.5 Partisil 10 ODS-3
**Mobile phase:** MeOH:water:phosphoric acid 700:300:1
**Flow rate:** 2.5
**Injection volume:** 10-30
**Detector:** UV 220

## CHROMATOGRAM

**Retention time:** 7.7
**Internal standard:** tolmetin (3.2)
**Limit of quantitation:** 5 μg/mL

## KEY WORDS

plasma

## REFERENCE

Lockwood, G.F.; Wagner, J.G. High-performance liquid chromatographic determination of ibuprofen and its major metabolites in biological fluids. *J.Chromatogr.*, **1982**, *232*, 335–343

## SAMPLE

**Matrix:** blood, synovial fluid

**Sample preparation:** 0.5 mL Plasma or synovial fluid + 50 µL 300 µg/mL flurbiprofen + 200 µL 2 M HCl, vortex 15 s, add 5 mL hexane:diethyl ether 1:1, tumble 10 min on a rotary mixer, centrifuge at 10000 g for 5 min. Remove organic layer and evaporate it to dryness under vacuum centrifugation. Reconstitute residue in 300 µL MeOH + 200 µL water, sonicate 5 min, vortex 15 s, inject aliquot.

## HPLC VARIABLES

**Guard column:** 20 × 2 Perisorb RP18 30-40 µm pellicular

**Column:** 125 × 4.6 5 µm Spherisorb ODS 1

**Mobile phase:** MeOH:water 65:35, adjusted to pH 3.3 with phosphoric acid

**Flow rate:** 1

**Injection volume:** 20

**Detector:** UV 220

## CHROMATOGRAM

**Retention time:** 8

**Internal standard:** flurbiprofen (6.5)

**Limit of detection:** <5 µg/mL

## KEY WORDS

plasma

## REFERENCE

Blagbrough, I.S.; Daykin, M.M.; Doherty, M.; Pattrick, M.; Shaw, P.N. High-performance liquid chromatographic determination of naproxen, ibuprofen and diclofenac in plasma and synovial fluid in man. *J.Chromatogr.*, **1992**, *578*, 251–257

## SAMPLE

**Matrix:** blood, urine

**Sample preparation:** Plasma. 1 mL Plasma + 100 µL 1 M HCl + 100 µL 250 µg/mL flurbiprofen + 100 µL water + 6 mL ether:hexane 20:80, shake for 10 min, centrifuge at 900 g for 5 min. Remove 4 mL of the organic layer and evaporate it to dryness under a stream of nitrogen at 20°, reconstitute the residue in 100 µL 10 mM NaOH, sonicate for 3 min, add 50 µL 100 mM pH 7.0 phosphate buffer containing 0.1% dimethyloctylamine, sonicate for 3 min, inject a 5 µL aliquot. Urine. 1 mL Urine + 500 µL 1 M NaOH, let stand for 30 min, add 700 µL 1 M HCl, add 100 µL 250 µg/mL flurbiprofen, add 1 mL water, add 5 mL ether:hexane 20:80, shake for 10 min, centrifuge at 900 g for 5 min. Remove 4 mL of the organic layer and evaporate it to dryness under a stream of nitrogen at 20°, reconstitute the residue in 100 µL 10 mM NaOH, sonicate for 3 min, add 50 µL 100 mM pH 7.0 phosphate buffer containing 0.1% dimethyloctylamine, sonicate for 3 min, inject a 5 µL aliquot. (To measure unconjugated ibuprofen in urine proceed as for plasma.)

## HPLC VARIABLES

**Guard column:** 10 × 3 5 µm Chiral-AGP (ChromTech)

**Column:** 100 × 4 5 µm Chiral-AGP (ChromTech)

**Mobile phase:** Gradient. A was 10 mM pH 7.0 phosphate buffer containing 1 mM dimethyloctylamine. B was isopropanol:10 mM pH 7.0 phosphate buffer 50:50 containing 1 mM dimethyloctylamine. A:B 99.2:0.8 for 5 min, to 59:41 over 10 min, re-equilibrate for 10 min.

**Flow rate:** 0.9

**Injection volume:** 5
**Detector:** UV 220 for 7 min, then UV 245

## CHROMATOGRAM
**Retention time:** 3.1 (R), 5.3 (S)
**Internal standard:** flurbiprofen (9.5, 11.6)
**Limit of detection:** 100 ng/mL
**Limit of quantitation:** 250 ng/mL

## KEY WORDS
plasma; chiral

## REFERENCE
de Vries, J.X.; Schmitz-Kummer, E.; Siemon, D. The analysis of ibuprofen enantiomers in human plasma by high-performance liquid chromatography on an α1-acid glycoprotein chiral stationary phase. *J.Liq.Chromatogr.*, **1994**, *17*, 2127–2145

## SAMPLE
**Matrix:** blood, urine
**Sample preparation:** 100 μL Urine or rat plasma or 500 μL human plasma + 50 μL 25 μg/mL fenoprofen in MeOH:10 mM NaOH 10:90 + 200 μL 600 mM sulfuric acid + 3 mL isooctane:isopropanol 95:5, vortex for 30 s, centrifuge at 1800 g for 5 min. Remove the organic layer and evaporate it to dryness, reconstitute the residue in 200 μL 50 mM triethylamine in MeCN, add 50 μL 6 mM ethyl chloroformate in MeCN, vortex for 30 s, add 50 μL 500 mM R-(+)-α-phenylethylamine in MeCN:triethylamine 80:20, vortex briefly, let stand for 2 min, add 1 mL 250 mM HCl, add 3 mL chloroform, vortex for 30 s, centrifuge at 1800 g for 2 min. Remove the organic layer and evaporate it to dryness, reconstitute the residue in 200 μL mobile phase, inject a 10-150 μL aliquot.

## HPLC VARIABLES
**Guard column:** 20 × 4.6 37-53 μm reversed-phase
**Column:** 100 × 4.6 5 μm C18 (Phenomenex)
**Mobile phase:** MeCN:water:acetic acid:triethylamine 46.5:53.5:0.1:0.03, pH 4.9
**Flow rate:** 1.6
**Injection volume:** 10-150
**Detector:** UV 225

## CHROMATOGRAM
**Retention time:** 15.69 (R), 17.65 (S)
**Internal standard:** fenoprofen (11.70, 13.40 (enantiomers))
**Limit of quantitation:** 250 ng/mL

## KEY WORDS
derivatization; human; rat; plasma; chiral; pharmacokinetics

## REFERENCE
Wright, M.R.; Sattari, S.; Brocks, D.R.; Jamali, F. Improved high-performance liquid chromatographic assay method for the enantiomers of ibuprofen. *J.Chromatogr.*, **1992**, *583*, 259–265

## SAMPLE
**Matrix:** bulk
**Sample preparation:** 10 mg Compound + 10 mg 1-(3-dimethylaminopropyl)-3-ethylcarbodiimide hydrochloride + 2 drops 3,5-dimethylaniline + 1.5 mL dichloromethane, mix, after 30 min add 1 mL 1 M HCl, shake vigorously. Remove the lower organic layer and dry it over anhydrous magnesium sulfate, inject an aliquot.

## HPLC VARIABLES
**Column:** 250 × 4.6 5 μm D N-(3,5-dinitrobenzoyl)phenylglycine (Regis)
**Mobile phase:** Hexane:isopropanol 80:20
**Flow rate:** 2
**Injection volume:** 20
**Detector:** UV 254; UV 280

## CHROMATOGRAM
**Retention time:** k′ 1.23 (for first enantiomer)

## OTHER SUBSTANCES
**Also analyzed:** carprofen, cicloprofen, etodolac, fenoprofen, flurbiprofen, ketoprofen, naproxen, pirprofen, tiaprofenic acid

## KEY WORDS
derivatization; α = 1.30; chiral

## REFERENCE
Pirkle, W.H.; Murray, P.G. The separation of the enantiomers of a variety of non-steroidal anti-inflammatory drugs (NSAIDS) as their anilide derivatives using a chiral stationary phase. *J.Liq.Chromatogr.*, **1990**, *13*, 2123–2134

## SAMPLE
**Matrix:** solutions

## HPLC VARIABLES
**Column:** 250 × 4.6 5 μm Supelcosil LC-DP (A) or 250 × 4 5 μm LiChrospher 100 RP-8 (B)
**Mobile phase:** MeCN:0.025% phosphoric acid:buffer 25:10:5 (A) or 60:25:15 (B) (Buffer was 9 mL concentrated phosphoric acid and 10 mL triethylamine in 900 mL water, adjust pH to 3.4 with dilute phosphoric acid, make up to 1 L.)
**Flow rate:** 0.6
**Injection volume:** 25
**Detector:** UV 229

## CHROMATOGRAM
**Retention time:** 8.12 (A), 10.50 (B)

## OTHER SUBSTANCES
**Also analyzed:** acebutolol, acepromazine, acetaminophen, acetazolamide, acetophenazine, albuterol, alprazolam, amitriptyline, amobarbital, amoxapine, antipyrine, atenolol, atropine, azatadine, baclofen, benzocaine, bromocriptine, brompheniramine, brotizolam, bupivacaine, buspirone, butabarbital, butalbital, caffeine, carbamazepine, cetirizine, chlorcyclizine, chlordiazepoxide, chlormezanone, chloroquine, chlorpheniramine, chlorpromazine, chlorpropamide, chlorprothixene, chlorthalidone, chlorzoxazone, cimetidine, cisapride, clomipramine, clonazepam, clonidine, clozapine, cocaine, codeine, colchicine, cyclizine, cyclobenzaprine, dantrolene, desipramine, diazepam, diclofenac, diflunisal, diltiazem, diphenhydramine, diphenidol, diphenoxylate, dipyridamole, disopyramide, dobutamine, doxapram, doxepin, droperidol, encainide, ethidium bromide, ethopropazine, fenoprofen, fentanyl, flavoxate, fluoxetine, fluphenazine, flurazepam, flurbiprofen, fluvoxamine, furosemide, glutethimide, glyburide, guaifenesin, haloperidol, homatropine, hydralazine, hydrochlorothiazide, hydrocodone, hydromorphone, hydroxychloroquine, hydroxyzine, imipramine, indomethacin, ketoconazole, ketoprofen, ketorolac, labetalol, levorphanol, lidocaine, loratadine, lorazepam, lovastatin, loxapine, mazindol, mefenamic acid, meperidine, mephenytoin, mepivacaine, mesoridazine, metaproterenol, methadone, methdilazine, methocarbamol, methotrexate, methotrimeprazine, methoxamine, methyldopa, methylphenidate, metoclopramide, metolazone, metoprolol, metronidazole, mida-

zolam, moclobemide, morphine, nadolol, nalbuphine, naloxone, naphazoline, naproxen, nifedipine, nizatidine, norepinephrine, nortriptyline, oxazepam, oxycodone, oxymetazoline, paroxetine, pemoline, pentazocine, pentobarbital, pentoxifylline, perphenazine, pheniramine, phenobarbital, phenol, phenolphthalein, phentolamine, phenylbutazone, phenyltoloxamine, phenytoin, pimozide, pindolol, piroxicam, pramoxine, prazepam, prazosin, probenecid, procainamide, procaine, prochlorperazine, procyclidine, promazine, promethazine, propafenone, propantheline, propiomazine, propofol, propranolol, protriptyline, quazepam, quinidine, quinine, racemethorphan, ranitidine, remoxipride, risperidone, salicylic acid, scopolamine, secobarbital, sertraline, sotalol, spironolactone, sulfinpyrazone, sulindac, temazepam, terbutaline, terfenadine, tetracaine, theophylline, thiethylperazine, thiopental, thioridazine, thiothixene, timolol, tocainide, tolbutamide, tolmetin, trazodone, triamterene, triazolam, trifluoperazine, triflupromazine, trimeprazine, trimethoprim, trimipramine, verapamil, warfarin, xylometazoline, yohimbine, zopiclone

## KEY WORDS
some details of plasma extraction

## REFERENCE
Koves, E.M. Use of high-performance liquid chromatography-diode array detection in forensic toxicology. *J.Chromatogr.A*, **1995**, *692*, 103–119

## SAMPLE
**Matrix:** solutions

## HPLC VARIABLES
**Column:** 150 × 4.6 cellulose 3,5-dimethylphenylcarbamate/10-undecenoate bonded to allylsilica
**Mobile phase:** Heptane:isopropanol:trifluoroacetic acid 98:2:0.1
**Flow rate:** 1
**Injection volume:** 1000
**Detector:** UV 254

## CHROMATOGRAM
**Retention time:** $k'$ 1.19

## KEY WORDS
chiral; $\alpha$ = 1.09

## REFERENCE
Oliveros, L.; Lopez, P.; Minguillon, C.; Franco, P. Chiral chromatographic discrimination ability of a cellulose 3,5-dimethylphenylcarbamate/10-undecenoate mixed derivative fixed on several chromatographic matrices. *J.Liq.Chromatogr.*, **1995**, *18*, 1521–1532

## SAMPLE
**Matrix:** solutions

## HPLC VARIABLES
**Column:** 150 mm long 5 μm Microsorb-MV C18
**Mobile phase:** MeCN:10 mM pH 7 phosphate buffer 34:66
**Flow rate:** 1.5
**Detector:** UV 220

## REFERENCE
Phillips, C.A.; Michniak, B.B. Transdermal delivery of drugs with differing lipophilicities using azone analogs as dermal penetration enhancers. *J.Pharm.Sci.*, **1995**, *84*, 1427–1433

## SAMPLE
**Matrix:** solutions
**Sample preparation:** 1 mL 5 mM Ibuprofen in dichloromethane + 300 μL 1 mg/mL hydroxybenzotriazole in dichloromethane:pyridine 99:1 + 300 μL 11 mg/mL 1-ethyl-3-dimethylaminopropylcarbodiimide in dichloromethane + 300 μL 3.47 mg/mL 1-naphthylamine (Caution! 1-Naphthylamine in a carcinogen!) in dichloromethane, vortex, let stand for 1 h, evaporate to dryness under a stream of nitrogen, reconstitute with 5 mL MeOH, inject an aliquot.

## HPLC VARIABLES
**Column:** 150 × 2.1 Tolycellulose EXP B101 (tris(4-methylbenzoate)cellulose covalently bonded to 10 μm aminopropylsilica)
**Mobile phase:** MeOH:buffer 85:15 (Buffer was 14.05 g/L sodium perchlorate adjusted to pH 2.0.)
**Flow rate:** 0.21
**Injection volume:** 1
**Detector:** UV 230; UV 254

## CHROMATOGRAM
**Retention time:** k' 1.43 (first enantiomer)

## OTHER SUBSTANCES
**Also analyzed:** fenoprofen, flurbiprofen, ketoprofen, tiaprofenic acid

## KEY WORDS
derivatization; narrow-bore; chiral; α = 1.31; see also Biomed. Chromatogr. 1995, 9, 292

## REFERENCE
Van Overbeke, A.; Baeyens, W.; Van Der Weken, G.; Van de Voorde, I.; Dewaele, C. Comparative chromatographic study on the chiral separation of the 1-naphthylamine derivative of ketoprofen on cellulose-based columns of different sizes. *Biomed.Chromatogr.*, **1995**, *9*, 289–290

## SAMPLE
**Matrix:** solutions

## HPLC VARIABLES
**Column:** 250 × 4.6 Zorbax RX
**Mobile phase:** Gradient. A was 10 mL concentrated orthophosphoric acid and 7 mL triethylamine in 1 L water. B was 10 mL concentrated orthophosphoric acid and 7 mL triethylamine in 200 mL water, make up to 1 L with MeCN. A:B from 100:0 to 0:100 over 30 min, maintain at 0:100 for 5 min.
**Column temperature:** 30
**Flow rate:** 2
**Detector:** UV 210

## OTHER SUBSTANCES
**Also analyzed:** acepromazine, acetaminophen, acetophenazine, albuterol, aminophylline, amitriptyline, amobarbital, amoxapine, amphetamine, amylocaine, antipyrine, aprobarbital, aspirin, atenolol, atropine, avermectin, barbital, benzocaine, benzoic acid, benzotropine, benzphetamine, berberine, bibucaine, bromazepan, brompheniramine, buprenorphine, buspirone, butabarbital, butacaine, butethal, caffeine, carbamazepine, carbromal, chloramphenicol, chlordiazepoxide, chloroquine, chlorothiazide, chloroxylenol, chlorphenesin, chlorpheniramine, chlorpromazine, chlorpropamide, chlortetracycline, cimetidine, cinchonidine, cinchonine, clenbuterol, clonazepam, clonixin, clorazepate, cocaine, codeine, colchicine, cortisone, coumarin, cyclazocine, cyclobenzaprine, cyclothiazide, cyheptamide, cymarin, danazol, danthron, dapsone, debrisoquine, desipramine, dexamethasone, dex-

tromethorphan, dextropropoxyphene, diamorphine, diazepam, diclofenac, diethylpropion, diethylstilbestrol, diflunisal, digitoxin, digoxin, diltiazem, diphenhydramine, diphenoxylate, diprenorphine, dipyrone, disulfiram, dopamine, doxapram, doxepin, dronabinol, ephedrine, epinephrine, epinine, estradiol, estriol, estrone, ethacrynic acid, ethosuximide, etonitazene, etorphine, eugenol, famotidine, fenbendazole, fencamfamine, fenoprofen, fenproporex, fentanyl, flubendazole, flufenamic acid, flunitrazepam, 5-fluorouracil, fluoxymesterone, fluphenazine, furosemide, gentisic acid, gitoxigenin, glipizide, glunixin, glutethimide, glybenclamide, guaiacol, halazepam, haloperidol, hydrochlorothiazide, hydrocodone, hydrocortisone, hydromorphone, hydroxyquinoline, ibogaine, iminostilbene, imipramine, indomethacin, isocarbostyril, isocarboxazid, isoniazid, isoproterenol, isoxsuprine, ivermectin, ketamine, ketoprofen, kynurenic acid, levorphanol, lidocaine, lorazepam, lormetazepam, loxapine, mazindol, mebendazole, meclizine, meclofenamic acid, medazepam, mefenamic acid, megestrol, mepacrine, meperidine, mephentermine, mephenytoin, mephesin, mephobarbital, mepivacaine, mescaline, mesoridazine, methadone, methamphetamine, methapyrilene, methaqualone, methazolamide, methocarbamol, methoxamine, methsuximide, methyl salicylate, methyldopa, methyldopamine, methylphenidate, methylprednisolone, methyltestosterone, methyprylon, metoprolol, mibolerone, morphine, nadolol, nalorphine, naloxone, naltrexone, naphazoline, naproxen, nefopam, niacinamide, nicotine, nicotinic acid, nifedipine, niflumic acid, nitrazepam, norepinephrine, nortriptyline, noscapine, nylidrin, oxazepam, oxycodone, oxymorphone, oxyphenbutazone, oxytetracycline, papaverine, pargyline, pemoline, pentazocine, pentobarbital, persantine, phenacetin, phenazocine, phenazopyridine, phencyclidine, phendimetrazine, phenelzine, pheniramine, phenobarbital, phenothiazine, phensuximide, phentermine, phenylbutazone, phenylephrine, phenylpropanolamine, piperocaine, prazepam, prednisolone, primidone, probenecid, progesterone, propiomazine, propranolol, propylparaben, pseudoephedrine, puromycin, pyrilamine, pyrithyldione, quazepam, quinaldic acid, quinidine, quinine, ranitidine, recinnamine, reserpine, resorcinol, saccharin, albuterol, salicylamide, salicylic acid, scopolamine, scopoletin, secobarbital, strychnine, sulfacetamide, sulfadiazine, sulfadimethoxine, sulfaethidole, sulfamerazine, sulfamethazine, sulfamethoxizole, sulfanilamide, sulfapyridine, sulfasoxizole, sulindac, tamoxifen, temazepam, testosterone, tetracaine, tetracycline, tetramisole, thebaine, theobromine, theophylline, thiabendazole, thiamine, thiamylal, thiobarbituric acid, thioridazine, thiosalicylic acid, thiothixene, thymol, tolazamide, tolazoline, tobutamide, tolmetin, tranylcypromine, triamcinolone, tribenzylamine, trichloromethiazide, trifluoperazine, trihexyphenidyl, trimethoprim, tripelennamine, triprolidine, tropacocaine, tyramine, verapamil, vincamine, warfarin, yohimbine, zoxazolamine

## REFERENCE
Hill, D.W.; Kind, A.J. Reversed-phase solvent gradient HPLC retention indexes of drugs. *J.Anal.Toxicol.*, **1994**, *18*, 233–242

## SAMPLE
**Matrix:** solutions

## HPLC VARIABLES
**Column:** 250 × 4.6 10 μm Chiralpak AD (Daicel)
**Mobile phase:** Carbon dioxide:MeOH 96:4
**Column temperature:** 30
**Flow rate:** 2.5
**Detector:** UV 210

## CHROMATOGRAM
**Retention time:** 4.3, 5 (enantiomers)

## OTHER SUBSTANCES
**Simultaneous:** fenoprofen, flurbiprofen, ketoprofen, naproxen

## KEY WORDS
SFC; 250 bar; chiral

## REFERENCE
Kot, A.; Sandra, P.; Venema, A. Sub- and supercritical fluid chromatography on packed columns: A versatile tool for the enantioselective separation of basic and acidic drugs. *J.Chromatogr.Sci.*, **1994**, *32*, 439–448

## SAMPLE
**Matrix:** solutions
**Sample preparation:** Mix 1 mL 100 μg/mL compound in dichloromethane with 300 μL 100 μg/mL 1-hydroxybenzotriazole in dichloromethane:pyridine 99:1, 300 μL 1.1 mg/mL 1-ethyl-3-dimethylaminopropylcarbodiimide hydrochloride in dichloromethane, and 300 μL 300 μg/mL benzylamine in dichloromethane, vortex, let stand at room temperature for 1.5 h, evaporate to dryness under a stream of nitrogen, reconstitute the residue in 500 μL MeOH, inject an aliquot.

## HPLC VARIABLES
**Column:** 150 × 4.6 10 μm EXP B101 tris(4-methylbenzoate) cellulose on silica (Bio-Rad)
**Mobile phase:** MeOH:buffer 70:30 (Prepare buffer solution by dissolving 14.05 g sodium perchlorate in water, adjust pH to 2.0, make up to 1 L with water.)
**Flow rate:** 1
**Detector:** UV 230

## CHROMATOGRAM
**Retention time:** 7 (R), 9 (S)

## OTHER SUBSTANCES
**Also analyzed:** benoxaprofen (MeOH:buffer 80:20), carprofen, fenoprofen, flurbiprofen, ketoprofen, pirprofen, tiaprofenic acid

## KEY WORDS
derivatization; chiral

## REFERENCE
Van Overbeke, A.; Baeyens, W.; Van den Bossche, W.; Dewaele, C. Separation of 2-arylpropionic acids on a cellulose based chiral stationary phase by RP-HPLC. *J.Pharm.Biomed.Anal.*, **1994**, *12*, 901–909

## SAMPLE
**Matrix:** solutions
**Sample preparation:** Condition an Analytichem AASP propylbenzenesulfonic acid (SCX) SPE cartridge with isopropanol at 4 mL/min for 1.5 min and with mobile phase at 4 mL/min for 2 min. 100 μL Ibuprofen solution in dichloromethane + 100 μL 500 μg/mL 2-phenylpropionic acid in dichloromethane + 200 μL 100 mM triethylamine in dichloromethane + 100 μL 60 mM ethyl chloroformate, vortex for 15 s, let stand for 15 min, add 100 μL 500 mM p-anisidine in dichloromethane, vortex for 15 s, let stand for 5 min, add 600 μL isopropanol:hexane 10:90, vortex for 15 s, add a 25 μL aliquot to the SPE cartridge, elute the contents of the SPE cartridge onto the analytical column with mobile phase, after 2.43 min remove the SPE cartridge from the circuit, elute the analytical column with mobile phase and monitor the effluent from the column.

## HPLC VARIABLES
**Column:** 250 × 4.6 5 μm Rexchrom Regis Pirkle D-phenylglycine (Regis)
**Mobile phase:** Hexane:isopropanol 90:10

**Flow rate:** 2
**Injection volume:** 25
**Detector:** UV 254

## CHROMATOGRAM
**Retention time:** 10 (S), 12.5 (R)
**Internal standard:** 2-phenylpropionic acid (16 (S), 18 (R))
**Limit of detection:** 500 ng/mL

## KEY WORDS
chiral; derivatization; SPE

## REFERENCE
Nicoll-Griffith, D.; Scartozzi, M.; Chiem, N. Automated derivatization and high-performance liquid chromatographic analysis of ibuprofen enantiomers. *J.Chromatogr.A*, **1993**, *653*, 253–259

## SAMPLE
**Matrix:** solutions
**Sample preparation:** Dissolve the compound in 400 μL MeCN containing 5 mM DBD-PZ and 70 mM diethylphosphorocyanidate, let stand at room temperature for 6 h, inject a 1 μL aliquot. (DBD-PZ prepared from 123 mg 4-(N, N-dimethylaminosulfonyl)-7-fluoro-2,1,3-benzoxadiazole in 20 mL MeCN added dropwise to 129 mg piperazine in 20 mL MeCN at room temperature, stir for 30 min, evaporate under reduced pressure, dissolve residue in 50 mL 5% HCl, wash three times with 20 mL ethyl acetate, discard ethyl acetate washes, adjust pH of aqueous solution to 13-14 with 5% NaOH, extract five times with 50 mL ethyl acetate, combine extracts, wash with 20 mL water, dry over anhydrous sodium sulfate, evaporate under vacuum to give DBD-PZ as orange crystals, mp 121-2°.)

## HPLC VARIABLES
**Column:** 150 × 4.6 5 μm Inertsil ODS-2
**Mobile phase:** MeCN:water 65:35
**Column temperature:** 40
**Flow rate:** 1
**Injection volume:** 1
**Detector:** F ex 437 em 561

## CHROMATOGRAM
**Retention time:** 10
**Limit of detection:** 3.9 fmol

## OTHER SUBSTANCES
**Simultaneous:** indomethacin

## KEY WORDS
derivatization

## REFERENCE
Toyo'oka, T.; Ishibashi, M.; Takeda, Y.; Nakashima, K.; Akiyama, S.; Uzu, S.; Imai, K. Precolumn fluorescence tagging reagent for carboxylic acids in high-performance liquid chromatography: 4-substituted-7-aminoalkylamino-2,1,3-benzoxadiazoles. *J.Chromatogr.*, **1991**, *588*, 61–71

## SAMPLE
**Matrix:** solutions
**Sample preparation:** Inject a 50 μL aliquot of a solution in mobile phase.

## HPLC VARIABLES

**Column:** 100 × 4.6 5 μm Spheri-5 RP-8

**Mobile phase:** MeOH:buffer 30:70 (Prepare buffer by mixing 4 mM $Na_2HPO_4$ and 7 mM $KH_2PO_4$ to achieve pH 7.)

**Flow rate:** 1

**Injection volume:** 50

**Detector:** F ex 355 em 460 (408 nm cutoff filter) following post-column extraction. The column effluent mixed with 50 μg/mL reagent in mobile phase pumped at 0.5 mL/min and then with chloroform pumped at 1 mL/min and the mixture flowed through a 1.8 m × 0.3 mm ID knitted PTFE coil to a 50 μL membrane phase separator using a polyethylene-backed 0.5 μm Fluoropore membrane filter (design in paper). The organic phase flowed to the detector. (Synthesize the reagent, α-(3,4-dimethoxyphenyl)-4'-trimethylammoniumme-thylcinnamonitrile methosulfate, as follows. Stir 20 mmoles 3,4-dimethoxyphenylacetoni-trile and 20 mmoles p-toluamide in 50 mL EtOH at 50°, add 5 mL 50% aqueous KOH slowly, stir at 50° for 5 min, cool to room temperature, filter, dry the precipitate of α-(3,4-dimethoxyphenyl)-4'-methylcinnamonitrile. Dissolve 20 mmoles α-(3,4-dimethoxyphenyl)-4'-methylcinnamonitrile, 20 mmoles N-bromosuccinimide, and 20 mg benzoyl peroxide in 100 mL carbon tetrachloride (Caution! Carbon tetrachloride is a carcinogen!), reflux with stirring for 1.5 h, cool, filter, evaporate to dryness under reduced pressure, recrystallize from MeOH to give α-(3,4-dimethoxyphenyl)-4'-bromomethylcinnamonitrile. Vigorously stir 30 mmoles anhydrous dimethylamine in 100 mL dry benzene (Caution! Benzene is a carcinogen!) at 0°, very slowly add 10 mmoles α-(3,4-dimethoxyphenyl)-4'-bromomethylcinna-monitrile while stirring at 0°, stir at room temperature overnight, add 150 mL water, re-move the organic phase, extract the aqueous phase twice with 100 mL portions of diethyl ether, wash the organic layers with saturated NaCl solution, dry over anhydrous magne-sium sulfate, evaporate under reduced pressure to give α-(3,4-dimethoxyphenyl)-4'-dime-thylaminomethylcinnamonitrile (J.Chem.Eng.Data 1987, 32, 387). Reflux 10 mmoles α-(3,4-dimethoxyphenyl)-4'-dimethylaminomethylcinnamonitrile, 20 mmoles dimethyl sulfate (Caution! Dimethyl sulfate is a carcinogen and acutely toxic!), and 5 g potassium carbonate in 50 mL acetone for 1 h, cool to room temperature, filter, dry the precipitate under vacuum at room temperature overnight, recrystallize from chloroform containing 2-3 drops of 95% EtOH to give α-(3,4-dimethoxyphenyl)-4'-trimethylammoniummethylcinnamonitrile meth-osulfate (mp 212-215°). Protect solutions from light.)

## CHROMATOGRAM

**Retention time:** k' 4.1240

**Limit of detection:** 100 ng/mL

## OTHER SUBSTANCES

**Simultaneous:** ketoprofen, mefenamic acid, naproxen, probenecid, salicylic acid, valproic acid

## KEY WORDS

post-column extraction

## REFERENCE

Kim, M.; Stewart, J.T. HPLC post-column ion-pair extraction of acidic drugs using a substituted α-phenylcinnamonitrile quaternary ammonium salt as a new fluorescent ion-pair reagent. *J.Liq.Chromatogr.*, **1990**, *13*, 213–237

## SAMPLE

**Matrix:** solutions

## HPLC VARIABLES

**Column:** 250 × 4 OmniPac PAX-500 (Dionex)

**Mobile phase:** Gradient. A was MeCN:10 mM sodium carbonate 18:82. B was MeCN:50 mM sodium carbonate 33:67. A:B from 100:0 to 0:100 over 10 min.

**Flow rate:** 1
**Detector:** UV 254

## CHROMATOGRAM
**Retention time:** 4

## OTHER SUBSTANCES
**Simultaneous:** aspirin, carprofen, diflunisal, fenbufen, indomethacin, naproxen, tolmetin

## REFERENCE
Slingsby, R.W.; Rey, M. Determination of pharmaceuticals by multi-phase chromatography: Combined reversed phase and ion exchange in one column. *J.Liq.Chromatogr.*, **1990**, *13*, 107−134

## SAMPLE
**Matrix:** urine
**Sample preparation:** 100 μL Urine + 200 μL pH 5 citrate/NaOH buffer + 500 ng clofibric acid + 5 mL diethyl ether:dichloromethane 80:20, shake for 10 min, centrifuge. Remove the organic layer and evaporate it to dryness, add 200 μL toluene and evaporate it to remove traces of water. Reconstitute the residue in 500 μL dichloromethane, add 50 μL 1 mg/mL 1-hydroxybenzotriazole in dichloromethane:pyridine 99:1, add 50 μL 1 mg/mL 1-(3-dimethylaminopropyl)-3-ethylcarbodiimide hydrochloride in dichloromethane, add 50 μL 1 mg/mL FLOPA in dichloromethane, vortex, let stand at room temperature for 2 h, evaporate to dryness, reconstitute in mobile phase, inject a 10-20 μL aliquot. (To hydrolyze glucuronides add 100 μL 1 M NaOH to 100 μL urine, let stand for 1 h, add 100 μL 1 M HCl, proceed as above.) [FLOPA is the corresponding amine hydrochloride from (+)-(S)-flunoxaprofen. Synthesis is as follows (protect from light). 500 mg (+)-(S)-Flunoxaprofen in 50 mL dry toluene, slowly add 5 mL freshly distilled thionyl chloride, reflux for 1 h, evaporate to dryness under vacuum, dry the acyl chloride under vacuum over KOH for 2 days. Dissolve 0.5 mmoles acyl chloride in 5 mL acetone, add 0.6 mmoles sodium azide dissolved in ice water with stirring, stir at 0° for 30 min, add 10 mL ice-cold water, filter, dry solid in a desiccator under vacuum. Dissolve the solid in 1 mL toluene or dichloromethane (dried over 3 Ångström molecular sieve), reflux for 10 min, evaporate, store resulting isocyanate under vacuum over a desiccant. Dissolve 0.5 mmole isocyanate in 5 mL acetone, add 20 mL 8.5% phosphoric acid, heat to 80° for 1.5 h, adjust to pH 13, extract with diethyl ether:dichloromethane 4:1. Wash the organic layer twice with water, dry over anhydrous sodium sulfate, evaporate to dryness, dissolve in 1 mL toluene, evaporate to give crystals mp 91°. Dissolve in ether, add 0.5 M HCl in ether, filter, dissolve solid in a small volume of MeOH, precipitate with ether, dry FLOPA over phosphorus pentoxide under vacuum.]

## HPLC VARIABLES
**Column:** 250 × 4.6 5 μm Zorbax Sil
**Mobile phase:** n-Hexane:chloroform:EtOH 100:10:1.25
**Flow rate:** 2
**Injection volume:** 10-20
**Detector:** F ex 305 em 355

## CHROMATOGRAM
**Retention time:** 6.5 (R-(−)), 13.5 (R-(+))
**Internal standard:** clofibric acid (5)

## KEY WORDS
pharmacokinetics; chiral; derivatization; normal phase

## REFERENCE
Spahn, H.; Langguth, P. Chiral amines derived from 2-arylpropionic acids: novel reagents for the liquid chromatographic (LC) fluorescence assay of optically active carboxylic acid xenobiotics. *Pharm.Res.*, **1990**, *7*, 1262−1268

## SAMPLE
**Matrix:** urine
**Sample preparation:** Dilute urine 20-fold with 100 mM pH 2.0 phosphate buffer, extract twice with two volumes of diethyl ether, centrifuge at 5000 g for 5 min. Combine the organic layers and evaporate them to dryness under a stream of nitrogen below 30°. Reconstitute in 0.2-1 mL mobile phase, inject an aliquot.

## HPLC VARIABLES
**Guard column:** 4 × 4 5 μm LiChrospher 100 RP-18
**Column:** 250 × 4 5 μm LiChrospher CH-18
**Mobile phase:** MeOH:10 mM pH 6.0 phosphate buffer 80:20 containing 2.5 mM cethexonium bromide (Rinse with 100 mL MeOH:EtOH:water 50:25:25 at the end of the day.)
**Flow rate:** 1
**Injection volume:** 10
**Detector:** UV 278

## CHROMATOGRAM
**Retention time:** k′ 5.0

## OTHER SUBSTANCES
**Extracted:** glucuronides

## REFERENCE
Liu, H.-F.; Leroy, P.; Nicolas, A.; Magdalou, J.; Siest, G. Evaluation of a versatile reversed-phase high-performance liquid chromatographic system using cethexonium bromide as ion-pairing reagent for the analysis of glucuronic acid conjugates. *J.Chromatogr.*, **1989**, *493*, 137–147

## SAMPLE
**Matrix:** urine
**Sample preparation:** 1 mL Urine + 100 μL 1 mg/mL methylprednisolone + 1 mL 1.5 M HCl + 500 μL water + 10 mL dichloromethane, shake 20 min, centrifuge at 250 g for 3 min. Remove the organic phase and evaporate it to dryness under a stream of nitrogen at 40°. Reconstitute with 200 μL MeOH, inject 10-30 μL aliquot. (This assay determines free drug and metabolites. To determine total drug and metabolites (free plus conjugated) add 500 μL 1 M NaOH to 1 mL urine, let stand for 20 min at room temperature then proceed as above.)

## HPLC VARIABLES
**Column:** 250 × 4.5 Partisil 10 ODS-3
**Mobile phase:** Gradient. A was MeCN:water 28:72 containing 0.05% phosphoric acid and 0.05% acetone. B was MeCN:50 mM $KH_2PO_4$ 50:50. A:B 100:0 for 8 min then to 0:100 over 6 min
**Flow rate:** 2
**Injection volume:** 10-30
**Detector:** UV 220

## CHROMATOGRAM
**Retention time:** 21.8
**Internal standard:** methylprednisolone (13.6)
**Limit of quantitation:** 3 μg/mL

## OTHER SUBSTANCES
**Extracted:** metabolites

## REFERENCE
Lockwood, G.F.; Wagner, J.G. High-performance liquid chromatographic determination of ibuprofen and its major metabolites in biological fluids. *J.Chromatogr.*, **1982**, *232*, 335–343

◆——————————————————◆——————————————————◆

## ANNOTATED BIBLIOGRAPHY

Baeyens, W.R.G.; Van Der Weken, G.; Van Overbeke, A.; Zhang, X.R. A comparative study for the determination of ibuprofen in pharmaceutical preparations using different internal column diameters. *Biomed.Chromatogr.*, **1995**, *9*, 259–260

Haginaka, J.; Kanasugi, N. Enantioselectivity of bovine serum albumin-bonded columns produced with isolated protein fragments. *J.Chromatogr.A*, **1995**, *694*, 71–80 [chiral; also benzoin, clorazepate, fenoprofen, flurbiprofen, ibuprofen, ketoprofen, lorazepam, lormetazepam, oxazepam, pranoprofen, temazepam, warfarin]

Nakamura, K.; Fujima, H.; Kitagawa, H.; Wada, H.; Makino, K. Preparation and chromatographic characteristics of a chiral- recognizing perphenylated cyclodextrin column. *J.Chromatogr.A*, **1995**, *694*, 111–118 [chiral; also acetylpheneturide, alprenolol, arotinolol, atenolol, benzoin, biperiden, bunitrolol, chlormezanone, chlorphenesin, chlorpheniramine, eperisone, flavanone, oxprenolol, phenylethyl alcohol, phenylethylamine, pindolol, proglumide, propranolol, trihexyphenidyl]

Terfloth, G.J.; Pirkle, W.H.; Lynam, K.G.; Nicolas, E.C. Broadly applicable polysiloxane-based chiral stationary phase for high-performance liquid chromatography and supercritical fluid chromatography. *J.Chromatogr.A*, **1995**, *705*, 185–194 [chiral; HPLC; SFC; also carprofen, cicloprofen, etodolac, fenoprofen, flurbiprofen, naproxen, pirprofen, warfarin]

Naidong, W.; Lee, J.W. Development and validation of a liquid chromatographic method for the quantitation of ibuprofen enantiomers in human plasma. *J.Pharm.Biomed.Anal.*, **1994**, *12*, 551–556 [chiral; plasma; LOQ 1 µg/mL; column temp 20]

Shirley, M.A.; Guan, X.; Kaiser, D.G.; Halstead, G.W.; Baillie, T.A. Taurine conjugation of ibuprofen in humans and in rat liver in vitro. Relationship to metabolic chiral inversion. *J.Pharmacol.Exp.Ther.*, **1994**, *269*, 1166–1175 [human; urine; gradient; extracted metabolites; microsomal incubations; rat; liver]

Van den Mooter, G.; Samyn, C.; Kinget, R. The relation between swelling properties and enzymatic degradation of azo polymers designed for colon-specific drug delivery. *Pharm.Res.*, **1994**, *11*, 1737–1741 [naproxen (IS); rat; cecal content release medium]

Van Overbeke, A.; Baeyens, W.; Van den Bossche, W.; Dewaele, C. Enantiomeric separation of amide derivatives of some 2-arylpropionic acids by HPLC on a cellulose-based chiral stationary phase. *J.Pharm.Biomed.Anal.*, **1994**, *12*, 911–916 [chiral; derivatization; also, flurbiprofen, ketoprofen, tiaprofenic acid]

Jung, E.S.; Lee, H.S.; Rho, J.K.; Kwon, K.I. Simultaneous determination of ibuproxam and ibuprofen in human plasma by HPLC with column switching. *Chromatographia*, **1993**, *37*, 618–621 [plasma; column-switching; human; rat; pharmacokinetics; extracted metabolites, ibuproxam; N-phenylanthranilic acid (IS); LOD 100 ng/mL]

Szász, G.; Budvári-Bárany, Z.; Löre, A.; Radeczky, G.; Shalaby, A. HPLC of antiphlogistic acids on silica dynamically modified with cetylpyridinium chloride. *J.Liq.Chromatogr.*, **1993**, *16*, 2335–2345 [also diclofenac, fenoprofen, ketoprofen, naproxen, nicotinic acid, nifluminic acid, salicylic acid]

Pirkle, W.H.; Welch, C.J. An improved chiral stationary phase for the chromatographic separation of underivatized naproxen enantiomers. *J.Liq.Chromatogr.*, **1992**, *15*, 1947–1955 [chiral; also cicloprofen, fenoprofen, flurbiprofen, ketoprofen, naproxen, piprofen, tiaprofenic acid]

Nahata, M.C. Determination of ibuprofen in human plasma by high-performance liquid chromatography. *J.Liq.Chromatogr.*, **1991**, *14*, 187–192 [plasma; LOD 250 ng/mL; isobutylphenyl acetate(IS)]

George, R.D.; Contario, J.J. Quantitation of terfenadine, pseudoephedrine hydrochloride, and ibuprofen in a liquid animal dosing formulation using high performance liquid chromatography. *J.Liq.Chromatogr.*, **1988**, *11*, 475–488 [simultaneous degradation products, pseudoephedrine, terfenadine; formulations; stability-indicating; ion-exchange]

Karnes, H.T.; Rajasekharaiah, K.; Small, R.E.; Farthing, D. Automated solid phase extraction and HPLC analysis of ibuprofen in plasma. *J.Liq.Chromatogr.*, **1988**, *11*, 489–499 [plasma; SPE; flurbiprofen (IS); fluorescence detection]

Askholt, J.; Nielsen-Kudsk, F. Rapid HPLC-determination of ibuprofen and flurbiprofen in plasma for therapeutic drug control and pharmacokinetic applications. *Acta Pharmacol.Toxicol.(Copenh)*, **1986**, *59*, 382–386

Hermansson, J.; Erikson, M. Direct liquid chromatographic resolution of acidic drugs using a chiral α1-acid glycoprotein column (Enantiopac). *J.Liq.Chromatogr.*, **1986**, *9*, 621–639 [chiral; also bendroflumethiazide, disopyramide, ethotoin, hexobarbital, ketoprofen, naproxen, 2-phenoxypropionic acid, RAC 109]

Albert, K.S.; Raabe, A.; Garry, M.; Antal, E.J.; Gillespie, W.R. Determination of ibuprofen in capillary and venous plasma by high-performance liquid chromatography with ultraviolet detection. *J.Pharm.Sci.*, **1984**, *73*, 1487–1489

Aravind, M.K.; Miceli, J.N.; Kauffman, R.E. Determination of ibuprofen by high-performance liquid chromatography. *J.Chromatogr.*, **1984**, *308*, 350–353

Lee, E.J.; Williams, K.M.; Graham, G.G.; Day, R.O.; Champion, G.D. Liquid chromatographic determination and plasma concentration profile of optical isomers of ibuprofen in humans. *J.Pharm.Sci.*, **1984**, *73*, 1542–1544 [chiral]

Litowitz, H.; Olanoff, L.; Hoppel, C.L. Determination of ibuprofen in human plasma by high-performance liquid chromatography. *J.Chromatogr.*, **1984**, *311*, 443–448

Greenblatt, D.J.; Arendt, R.M.; Locniskar, A. Ibuprofen pharmacokinetics: use of liquid chromatography with radial compression separation. *Arzneimittelforschung*, **1983**, *33*, 1671–1673

Ali, A.; Kazmi, S.; Plakogiannis, F.M. High-pressure liquid chromatographic determination of ibuprofen in plasma. *J.Pharm.Sci.*, **1981**, *70*, 944–945

Kearns, G.L.; Wilson, J.T. Determination of ibuprofen in serum by high-performance liquid chromatography and application to ibuprofen disposition. *J.Chromatogr.*, **1981**, *226*, 183–190

Shimek, J.L.; Rao, N.G.; Khalil, S.K. High-pressure liquid chromatographic determination of ibuprofen in plasma. *J.Pharm.Sci.*, **1981**, *70*, 514–516

Snider, B.G.; Beaubien, L.J.; Sears, D.J.; Rahn, P.D. Determination of flurbiprofen and ibuprofen in dog serum with automated sample preparation. *J.Pharm.Sci.*, **1981**, *70*, 1347–1349

# Imipenem

**Molecular formula:** $C_{12}H_{17}N_3O_4S$
**Molecular weight:** 317.4
**CAS Registry No.:** 64221-86-9, 74431-23-5 (monohydrate)

## SAMPLE

**Matrix:** aqueous humor, blood
**Sample preparation:** 1 mL Plasma or aqueous humor + 1 mL solvent, remove 0.5 mL mixture and add it to 0.5 mL MeOH, mix for 15 min, centrifuge at 4° at 4000 g for 10 min, inject a 50 µL aliquot of the supernatant.

## HPLC VARIABLES

**Guard column:** 37-53 µm Pellicular ODS
**Column:** 250 × 4.6 Viosfer octadecyl (Violet)
**Mobile phase:** MeOH:100 mM pH 7.2 borate buffer 10:90
**Flow rate:** 1.5
**Injection volume:** 50
**Detector:** UV 313

## CHROMATOGRAM

**Retention time:** 5
**Limit of detection:** 400 (plasma), 150 (aqueous humor) ng/mL

## KEY WORDS

plasma

## REFERENCE

Carlucci, G.; Biordi, L.; Bologna, M. Human plasma and aqueous humor determination of imipenem by liquid chromatography with ultraviolet detection. *J.Liq.Chromatogr.*, **1993**, *16*, 2347–2358

## SAMPLE

**Matrix:** blood, tissue, urine
**Sample preparation:** Serum, plasma. Dilute serum or plasma 1:2 to 1:10 with buffer, centrifuge, inject a 20 µL aliquot of supernatant. Urine. Dilute urine 1:10 to 1:100 with buffer, centrifuge, inject a 20 µL aliquot of supernatant. Tissue (lung, gut). Cut tissue with a scalpel, homogenize with 1-3 mL buffer, centrifuge at 9600 g for 5 min three times, inject a 20 µL aliquot. Tissue (chondral). Cut tissue with a scalpel, homogenize with 3-6 mL buffer in an ice bath for 2-3 min, centrifuge at 9600 g for 5 min four or five times, inject a 100 µL aliquot. Pleural. Dilute human pleural samples with buffer, centrifuge, inject a 20 µL aliquot. (Buffer was 66.6 mM $K_2HPO_4$ adjusted to pH 7.40 with $KH_2PO_4$.)

## HPLC VARIABLES

**Column:** 200 × 4 5 µm Nucleosil C18
**Mobile phase:** 15 mM Phosphoric acid adjusted to pH 7.0 with tetrabutylammonium hydroxide
**Flow rate:** 1
**Injection volume:** 20-100
**Detector:** UV 313

## CHROMATOGRAM

**Retention time:** 4
**Limit of detection:** 300 ng/mL

## OTHER SUBSTANCES
**Simultaneous:** metabolites

## KEY WORDS
serum; plasma; lung; gut; pleural; chondral

## REFERENCE
Knöller, J.; König, W.; Schönfeld, W.; Bremm, K.D.; Köller, M. Application of high-performance liquid chromatography of some antibiotics in clinical microbiology. *J.Chromatogr.*, **1988**, *427*, 257–267

## SAMPLE
**Matrix:** blood, urine
**Sample preparation:** Add a stabilizing solution of 4-morpholineethanesulfonic acid buffer:ethylene glycol 1:1, prepare ultrafiltrate using an Amicon Centrifree micropartition system in a centrifuge, inject an aliquot.

## HPLC VARIABLES
**Column:** µBondapak C18
**Mobile phase:** MeOH:100 mM sodium acetate 2:98, pH adjusted to 6.0 with acetic acid
**Flow rate:** 1
**Detector:** UV 298

## CHROMATOGRAM
**Retention time:** 7.4
**Limit of detection:** 600 ng/mL

## KEY WORDS
plasma

## REFERENCE
Paradis, D.; Vallée, F.; Allard, S.; Bisson, C.; Daviau, N.; Drapeau, C.; Auger, F.; LeBel, M. Comparative study of pharmacokinetics and serum bactericidal activities of cefpirome, ceftazidime, ceftriaxone, imipenem, and ciprofloxacin. *Antimicrob.Agents Chemother.*, **1992**, *36*, 2085–2092

## SAMPLE
**Matrix:** enzyme incubations
**Sample preparation:** Add 2 volumes of MeOH, mix well, centrifuge at 3000 g for 15 min, inject a 20 µL aliquot of the supernatant.

## HPLC VARIABLES
**Column:** TSKgel ODS-80Tm (Tosoh)
**Mobile phase:** MeOH:100 mM pH 7.0 phosphate buffer 4:100
**Flow rate:** 0.75
**Injection volume:** 20
**Detector:** UV 290

## CHROMATOGRAM
**Limit of quantitation:** 1 µg/mL

## REFERENCE
Hikida, M.; Kawashima, K.; Yoshida, M.; Mitsuhashi, S. Inactivation of new carbapenem antibiotics by dehydropeptidase-I from porcine and human renal cortex. *J.Antimicrob.Chemother.*, **1992**, *30*, 129–134

## SAMPLE
**Matrix:** formulations
**Sample preparation:** Dilute 1:100 with mobile phase, inject a 30 μL aliquot.

## HPLC VARIABLES
**Column:** 200 × 4.6 RP-8 (Hewlett-Packard)
**Mobile phase:** MeCN:MeOH:buffer 0.4:0.5:99.1, adjusted to pH 7.00 with NaOH (Buffer was 4 mM 3-[N-morpholino]propanesulfonic acid (MOPS) containing 2 g/L sodium hexane sulfate.)
**Flow rate:** 1.8
**Injection volume:** 30
**Detector:** UV 250

## CHROMATOGRAM
**Retention time:** 5.5

## OTHER SUBSTANCES
**Extracted:** cilastatin

## KEY WORDS
injections; total parenteral nutrition; stability-indicating

## REFERENCE
Zaccardelli, D.S.; Krcmarik, C.S.; Wolk, R.; Khalidi, N. Stability of imipenem and cilastatin sodium in total parenteral nutrient solution. *J.Parenter.Enteral.Nutr.*, **1990**, *14*, 306–309

## SAMPLE
**Matrix:** solutions
**Sample preparation:** Inject a 10 μL aliquot of a 400 μg/mL solution.

## HPLC VARIABLES
**Column:** 150 × 4.6 Microsorb C8 80-315
**Mobile phase:** 1 mM $KH_2PO_4$ adjusted to pH 6.8 with 500 mM NaOH
**Flow rate:** 1.5
**Injection volume:** 10
**Detector:** UV 300

## REFERENCE
Connolly, M.; Debenedetti, P.G.; Tung, H.-H. Freeze crystallization of imipenem. *J.Pharm.Sci.*, **1996**, *85*, 174–177

## SAMPLE
**Matrix:** solutions
**Sample preparation:** Prepare a 2.5-5 μg/mL solution, inject a 20 μL aliquot.

## HPLC VARIABLES
**Column:** 80 × 4.6 3.65 μm Zorbax Rx-SIL (similar to Zorbax SB-C8 (Mac-Mod Analytical))
**Mobile phase:** MeCN:buffer 0.5:99.5 (Buffer was 0.1% acetic acid adjusted to pH 7 with ammonium hydroxide.)
**Flow rate:** 1
**Injection volume:** 20
**Detector:** UV 296

## CHROMATOGRAM
**Retention time:** k' 3.5

## REFERENCE
Kirkland, K.M.; McCombs, D.A.; Kirkland, J.J. Rapid, high-resolution high-performance liquid chromatographic analysis of antibiotics. *J.Chromatogr.A*, **1994**, *660*, 327–337

## SAMPLE
**Matrix:** solutions

## HPLC VARIABLES
**Column:** 150 × 4.6 PLRP-S styrene-divinylbenzene copolymer (Polymer Labs)
**Mobile phase:** Gradient. MeCN:20 mM pH 7.2 $KH_2PO_4$-NaOH buffer 3:97 to 7:93 in 22 min.
**Column temperature:** 40
**Flow rate:** 1.2
**Detector:** UV 295

## CHROMATOGRAM
**Retention time:** 7

## OTHER SUBSTANCES
**Simultaneous:** degradation products

## KEY WORDS
pH >4 buffer

## REFERENCE
Smith, G.B.; Dezeny, G.C.; Douglas, A.W. Stability and kinetics of degradation of imipenem in aqueous solution. *J.Pharm.Sci.*, **1990**, *79*, 732–740

## SAMPLE
**Matrix:** solutions

## HPLC VARIABLES
**Column:** 250 × 4.6 Partisil PXS 5/25 PAC
**Mobile phase:** Gradient. MeCN:water from 25:75 to 50:50 in 30 min.
**Flow rate:** 2
**Detector:** UV 320

## CHROMATOGRAM
**Retention time:** 6

## OTHER SUBSTANCES
**Simultaneous:** degradation products

## KEY WORDS
pH 4 buffer

## REFERENCE
Smith, G.B.; Dezeny, G.C.; Douglas, A.W. Stability and kinetics of degradation of imipenem in aqueous solution. *J.Pharm.Sci.*, **1990**, *79*, 732–740

## SAMPLE
**Matrix:** solutions
**Sample preparation:** Inject a 25 μL aliquot.

## HPLC VARIABLES
**Column:** 100 × 4 5 μm ODS-Hypersil
**Mobile phase:** MeCN:10 mM ammonium acetate 20:80
**Flow rate:** 2
**Injection volume:** 25
**Detector:** UV 300

## OTHER SUBSTANCES
**Also analyzed:** penicillin G (UV 227)

## REFERENCE
Eley, A.; Greenwood, D. Beta-lactamases of type culture strains of the *Bacteroides fragilis* group and of strains that hydrolse cefoxitin, latamoxef and imipenem. *J.Med.Microbiol.*, **1986**, *21*, 49–57

## ANNOTATED BIBLIOGRAPHY

Jenke, D.R. Drug binding by reservoirs in elastomeric infusion devices. *Pharm.Res.*, **1994**, *11*, 984–989 [formulations; saline; 5% dextrose; simultaneous cilastatin]

Ebey, W.J.; Boucher, B.A.; Pieper, J.A. A rapid HPLC method for determination of imipenem in plasma. *J.Liq.Chromatogr.*, **1988**, *11*, 3471–3481 [plasma; LOD 1000 ng/mL]

Krausse, R.; Ullmann, U. Determination of imipenem and cilastatin in serum and tissue by high-pressure liquid chromatography. *Infection*, **1986**, *14*, 243–245

Gravallese, D.A.; Musson, D.G.; Pauliukonis, L.T.; Bayne, W.F. Determination of imipenem (N-formimidoyl thienamycin) in human plasma and urine by high-performance liquid chromatography, comparison with microbiological methodology and stability. *J.Chromatogr.*, **1984**, *310*, 71–84 [plasma; urine; serotonin (IS); ultrafiltrate; LOD 300 ng/mL (plasma); LOD 1 μg/mL (urine); non-interfering cilastatin; pharmacokinetics]

Myers, C.M.; Blumer, J.L. Determination of imipenem and cilastatin in serum by high-pressure liquid chromatography. *Antimicrob.Agents Chemother.*, **1984**, *26*, 78–81

# Indapamide

**Molecular formula:** $C_{16}H_{16}ClN_3O_3S$
**Molecular weight:** 365.8
**CAS Registry No.:** 26807-65-8

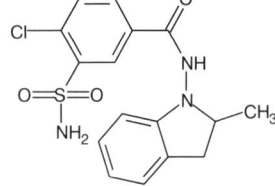

## SAMPLE
**Matrix:** blood
**Sample preparation:** 1 mL Whole blood + 1 mL 4 μg/mL glipizide in 50 mM pH 6.6 $KH_2PO_4$, vortex, add 6 mL diethyl ether, shake at 150 ± 20 oscillations/min on a reciprocating shaker for 15 min, centrifuge at 1500 g for 10 min. Remove the organic layer and evaporate it to dryness at 37° under a stream of nitrogen. Dissolve residue in 200 μL mobile phase, centrifuge at 1000 g for 3 min, inject a 40 μL aliquot of the supernatant.

## HPLC VARIABLES
**Column:** 150 × 4.6 5 μm Nucleosil C18
**Mobile phase:** MeCN:isopropanol:buffer 30:5:65 (Buffer was 80 mM ammonium acetate adjusted to pH 3.5 with concentrated HCl.)
**Flow rate:** 1
**Injection volume:** 40
**Detector:** UV 241

## CHROMATOGRAM
**Retention time:** 5.2
**Internal standard:** glipizide (5.9)
**Limit of quantitation:** 10 ng/mL

## OTHER SUBSTANCES
**Noninterfering:** acetaminophen, aspirin, caffeine, dextromethorphan, ibuprofen, nicotine, phenylpropanolamine, theophylline

## KEY WORDS
whole blood

## REFERENCE
Miller, R.B.; Dadgar, D.; Lalande, M. High-performance liquid chromatographic method for the determination of indapamide in human whole blood. *J.Chromatogr.*, **1993**, *614*, 293–298

## SAMPLE
**Matrix:** blood
**Sample preparation:** Condition a Sep-Pak C18 SPE cartridge with 5 mL MeOH and 5 mL 10 mM HCl. 2 mL Plasma + 2 mL 2 μg/mL IS in water + 10 mL 10 mM HCl, add to the SPE cartridge, wash with 4 mL 10 mM HCl, elute with 1 mL MeOH, inject a 20 μL aliquot of the eluate.

## HPLC VARIABLES
**Column:** 250 × 4 10 μm LiChrosorb SI 60 ODS
**Mobile phase:** MeCN:10 mM pH 3.5 sodium phosphate 30:70
**Flow rate:** 1
**Injection volume:** 20
**Detector:** UV 241

## CHROMATOGRAM
**Retention time:** 9

**Internal standard:** sulfadimethoxime (6.2)
**Limit of detection:** 10 ng/mL

## KEY WORDS
plasma; SPE; see Anal.Abs. 1986, 48, 12D80

## REFERENCE
Gaetani, E.; Laureri, C.F.; Vitto, M.; Borghi, L.; Elia, G.F.; Novarini, A. Determinazione dell'indapamide nel plasma mediante HPLC [Determination of indapamide in plasma by HPLC]. *Boll.Chim.Farm.*, **1986**, *125*, 35–37

## SAMPLE
**Matrix:** blood
**Sample preparation:** Keep tubes in crushed ice except when being processed throughout this procedure. 2 mL Plasma + 100 µL 10 µg/mL sulfanilamide in MeCN + 8 mL diethyl ether, vortex for 2 min, centrifuge at 4°. Remove ether layer and add it to 1 mL 100 mM NaOH, vortex for 2 min, centrifuge at 4°. Remove aqueous layer and add it to 1 mL 100 mM HCl and 500 µL 50 mM pH 7.4 sodium phosphate, add 8 mL ether, vortex for 2 min, centrifuge at 4°. Evaporate ether layer to dryness, reconstitute in 200 µL mobile phase, inject 50 µL aliquot.

## HPLC VARIABLES
**Column:** 250 × 4.6 5 µm Zorbax ODS
**Mobile phase:** MeCN:100 mM pH 3.6 sodium acetate buffer 43:57
**Column temperature:** 54
**Flow rate:** 1
**Injection volume:** 50
**Detector:** UV 241

## CHROMATOGRAM
**Retention time:** 6.3
**Internal standard:** sulfanilamide (5.3)
**Limit of detection:** 25 ng/mL

## KEY WORDS
plasma

## REFERENCE
Choi, R.L.; Rosenberg, M.; Grebow, P.E.; Huntley, T.E. High-performance liquid chromatographic analysis of indapamide (RHC 2555) in urine, plasma and blood. *J.Chromatogr.*, **1982**, *230*, 181–187

## SAMPLE
**Matrix:** bulk, formulations, urine
**Sample preparation:** Bulk, tablets. Weigh out amount containing 10 mg indapamide, dissolve in 10 mL 150 µg/mL sulfisoxazole in MeOH, make up to 100 mL with MeOH, inject an aliquot. Urine. 1 mL Urine + 1 mL 150 µg/mL sulfisoxazole in MeOH + 10 mL ethyl acetate, vortex, centrifuge. Remove the organic layer and evaporate it to dryness under a stream of nitrogen, reconstitute the residue in 1 mL mobile phase, inject a 10 µL aliquot. (If indapamide concentration is >100 ng/mL mix 1 mL urine, 1 mL 150 µg/mL sulfisoxazole in MeOH, and 8 mL mobile phase, vortex, inject a 10 µL aliquot.)

## HPLC VARIABLES
**Guard column:** 20 × 4 Bondapak C18 Corasil
**Column:** 300 × 4 10 µm µBondapak C18
**Mobile phase:** MeCN:buffer 35:65 (Buffer was water adjusted to pH 2.8 with 10% phosphoric acid.)

**Flow rate:** 2
**Injection volume:** 10
**Detector:** UV 254

## CHROMATOGRAM
**Retention time:** 7
**Internal standard:** sulfisoxazole (4)
**Limit of detection:** 25 ng/mL

## OTHER SUBSTANCES
**Extracted:** metabolites

## KEY WORDS
tablets

## REFERENCE
Pietta, P.; Calatroni, A.; Rava, A. High-performance liquid chromatographic assay for monitoring indapamide and its major metabolite in urine. *J.Chromatogr.*, **1982**, *228*, 377–381

## SAMPLE
**Matrix:** solutions

## HPLC VARIABLES
**Column:** Chiralcel OD-R
**Mobile phase:** MeCN:water 40:60
**Column temperature:** 40
**Flow rate:** 1
**Detector:** UV 254

## CHROMATOGRAM
**Retention time:** 11.3, 15.8 (enantiomers)

## KEY WORDS
chiral

## REFERENCE
*Application Guide for Chiral Column Selection, Second Edition, Chiral Technologies Inc., Exton PA,* **1995**, p. 39

## SAMPLE
**Matrix:** solutions

## HPLC VARIABLES
**Column:** 250 × 4.6 Chirex 3022 (Phenomenex)
**Mobile phase:** Hexane:1,2-dichloroethane:EtOH/trifluoroacetic acid 58:35:7 (EtOH/trifluoroacetic acid was premixed 20:1.)
**Flow rate:** 0.7-1
**Injection volume:** 20
**Detector:** UV 248

## KEY WORDS
chiral; $\alpha$ = 1.08

## REFERENCE
Cleveland, T. Pirkle-concept chiral stationary phases for the HPLC separation of pharmaceutical racemates. *J.Liq.Chromatogr.*, **1995**, *18*, 649–671

## SAMPLE
**Matrix:** solutions
**Sample preparation:** Inject a 20 μL aliquot.

## HPLC VARIABLES
**Column:** 250 × 4.6 5 μm Lichrosorb RP-18
**Mobile phase:** MeOH:buffer 50:50 (Buffer was 1% aqueous acetic acid containing 0.2% triethylamine.)
**Flow rate:** 1
**Injection volume:** 20
**Detector:** UV 250

## CHROMATOGRAM
**Retention time:** 11
**Limit of detection:** 500 ng/mL

## OTHER SUBSTANCES
**Simultaneous:** degradation products

## KEY WORDS
stability-indicating

## REFERENCE
Padval, M.V.; Bhargava, H.N. Liquid chromatographic determination of indapamide in the presence of its degradation products. *J.Pharm.Biomed.Anal.*, **1993**, *11*, 1033–1036

## ANNOTATED BIBLIOGRAPHY
Chen, D. [Determination of indapamide in human serum by high performance liquid chromatography]. *Chung Kuo I Hsueh Ko Hsueh Yuan Hsueh Pao*, **1990**, *12*, 286–289

# Insulin

**Molecular formula:** $C_{258}H_{383}N_{65}O_{77}S_6$ (human)

**Molecular weight:** 5807.6 (human)

**CAS Registry No.:** 9004-10-8 (injection), 8049-62-5 (zinc suspension), 11061-68-0 (human), 12584-58-6 (pig), 11070-73-8 (cow), 9004-14-2 (neutral insulin), 8049-62-5 (isophane insulin), 9004-17-5 (protamine zinc suspension)

## SAMPLE
**Matrix:** blood, tissue

**Sample preparation:** Pancreas. 1 g Tissue + 25 mL 6% trichloroacetic acid, homogenize, centrifuge, wash with 1 mL 5% trichloroacetic acid, extract precipitate twice at 37° by shaking for 2 h with 4 mL acid ethanol (A), adjust pH to 8.5-9 with concentrated ammonium hydroxide, inject 200 μL aliquot. (Acid ethanol (A) was 15 mL EtOH + 5 mL water + 3 mL concentrated HCl.) Plasma. 1 mL Plasma + 2 mL water + 7.5 mL cold acid ethanol (B), stand at 4° for 12 h, centrifuge at 2800 rpm at 4° for 20 min, adjust pH of supernatant to 8.3 with concentrated ammonium hydroxide, keep at 4° for 15 min, centrifuge at 2800 rpm at 4° for 20 min, adjust pH of supernatant to 5.3 with 4 M HCl, for each 1 mL add 25 μL 2 M ammonium acetate, readjust pH to 5.3, to each 10 mL slowly add 15 mL cold EtOH and 25 mL diethyl ether, keep at 4° for 12 h, centrifuge at 2800 rpm at 4° for 30 min. Remove precipitate and dry it under nitrogen gas. Dissolve in 100 mM pH 3.10 $NaH_2PO_4$, inject aliquot. (Acid ethanol (B) was 375 mL 95% EtOH and 7.5 mL concentrated HCl.)

## HPLC VARIABLES
**Column:** 250 × 4.5 Spherisorb S5 ODS2

**Mobile phase:** Gradient. A was MeCN. B was 100 mM pH 3.10 $NaH_2PO_4$. A:B from 0:100 to 28:72 over 14 min, to 28.8:71.2 over 8 min, to 29.2:71.8 over 10 min, to 39.2:61.8 over 5 min, to 60:40 over 4 min.

**Flow rate:** 1

**Injection volume:** 200

**Detector:** UV 220

## CHROMATOGRAM
**Retention time:** 33 (human), 35 (porcine)

## KEY WORDS
pancreas; serum; human; pig

## REFERENCE
Knip, M. Analysis of pancreatic peptide hormones by reversed-phase high-performance liquid chromatography. *Horm.Metab.Res.*, **1984**, *16*, 487–491

## SAMPLE
**Matrix:** bulk

**Sample preparation:** Dissolve in mobile phase at 100-200 μg/mL, inject 100 μL aliquot.

## HPLC VARIABLES
**Guard column:** 20 × 4.6 3 μm Supelcosil LC-18DB ODS

**Column:** 150 × 4.6 3 μm Supelcosil LC-18DB ODS

**Mobile phase:** Gradient. A was 500 mM sodium sulfate + 300 mM $NaH_2PO_4$ adjusted to pH 2.5 with perchloric acid. B was water. C was MeCN:water 60:40. A:B:C at 20:30:50 for 3 min then to 20:22.5:57.5 over 27 min.

**Column temperature:** 45

**Flow rate:** 1
**Injection volume:** 100
**Detector:** UV 210

## CHROMATOGRAM
**Retention time:** 18 (bovine), 19.5 (bovine MDA), 21 (human), 22 (porcine), 23 (human MDA), 24 (porcine MDA)

## KEY WORDS
cow; pig; human

## REFERENCE
Janssen, P.S.L.; van Nispen, J.W.; van Zeeland, M.J.M.; Melgers, P.A.T.A. Complementary information from isotachophoresis and high-performance liquid chromatography in peptide analysis. *J.Chromatogr.*, **1989**, *470*, 171–183

## SAMPLE
**Matrix:** formulations
**Sample preparation:** Dissolve 10 mg microspheres in 2 mL MeCN, centrifuge at 3000 rpm for 5 min, discard MeCN, repeat process three times, dissolve pellet in 5 mL pH 7.4 Tris buffer containing 0.1% trifluoroacetic acid, inject a 20 µL aliquot.

## HPLC VARIABLES
**Column:** 150 × 6 Asahipak ODP-50 6D (Asahi Chemical)
**Mobile phase:** MeCN:buffer 28:72 (Buffer was 0.3% ethanolamine adjusted to pH 2.0 with phosphoric acid.)
**Column temperature:** 40
**Flow rate:** 2
**Injection volume:** 20
**Detector:** UV 220

## CHROMATOGRAM
**Retention time:** 7

## OTHER SUBSTANCES
**Simultaneous:** degradation products

## KEY WORDS
microspheres; cow

## REFERENCE
Uchida, T.; Yagi, A.; Oda, Y.; Nakada, Y.; Goto, S. Instability of bovine insulin in poly(lactide-co-glycolide) (PLGA). *Chem.Pharm.Bull.*, **1996**, *44*, 235–236

## SAMPLE
**Matrix:** formulations
**Sample preparation:** 100 µL Injection + 50 µL 50 µg/mL benzoic acid in water + 250 µL water, vortex briefly, inject a 50 µL aliquot.

## HPLC VARIABLES
**Column:** 250 × 4.6 5 µm 300 Å silica for proteins and peptides (Vydac)
**Mobile phase:** MeCN:50 mM pH 2.4 $KH_2PO_4$ 25:75
**Flow rate:** 1
**Injection volume:** 50
**Detector:** UV 230

## CHROMATOGRAM
**Retention time:** 11.16
**Internal standard:** benzoic acid

## OTHER SUBSTANCES
**Simultaneous:** degradation products

## KEY WORDS
injections; stability-indicating

## REFERENCE
Hoyer, G.L.; Nolan, P.E., Jr.; LeDoux, J.H.; Moore, L.A. Selective stability-indicating high-performance liquid chromatographic assay for recombinant human regular insulin. *J.Chromatogr.A*, **1995**, *699*, 383–388

## SAMPLE
**Matrix:** formulations
**Sample preparation:** Prepare a 150 µg/mL solution in 10 mM HCl, inject a 20 µL aliquot.

## HPLC VARIABLES
**Column:** 250 × 4.6 5 µm Ultremex octadecylsilane
**Mobile phase:** MeCN:buffer 26:74 (Buffer was 200 mM sodium sulfate adjusted to pH 2.3.)
**Column temperature:** 40
**Flow rate:** 0.9
**Injection volume:** 20
**Detector:** UV 214

## CHROMATOGRAM
**Retention time:** 17.5

## KEY WORDS
injections

## REFERENCE
Lookabaugh, M.; Biswas, M.; Krull, I.S. Quantitation of insulin injection by high-performance liquid chromatography and high-performance capillary electrophoresis. *J.Chromatogr.*, **1991**, *549*, 357–366

## SAMPLE
**Matrix:** formulations

## HPLC VARIABLES
**Column:** 50 × 4.6 C4 wide pore (300 Å) (Supelco)
**Mobile phase:** Gradient. A was MeOH:isopropanol:water 4:1:95 containing 2.8 g/L NaCl. B was MeOH:isopropanol:water 60:10:30 containing 4.2 g/L NaCl. A:B from 45:55 to 30:70 over 5 min.
**Flow rate:** 1
**Injection volume:** 20
**Detector:** E, Bioanalytical systems Model LC-4B, dual glassy-carbon working electrode used in parallel mode, +0.65 V and +0.80 V (monitored), stainless steel auxiliary electrode, Ag/AgCl reference electrode following post-column reaction. The column effluent flowed at 0-5° through a 2 mL knitted coil of 0.5 mm i.d. PTFE tubing irradiated with a low pressure mercury lamp (Photronix Model 816) to the detector.

## CHROMATOGRAM
**Retention time:** 8

## OTHER SUBSTANCES
**Also analyzed:** b-lactoglobulin A, lysozyme, phenylalanine, ribonuclease A, tryptophan, tyramine

## KEY WORDS
post-column reaction

## REFERENCE
Dou, L.; Krull, I.S. Determination of aromatic and sulfur-containing amino acids, peptides, and proteins using high-performance liquid chromatography with photolytic electrochemical detection. *Anal.Chem.*, **1990**, *62*, 2599–2606

## SAMPLE
**Matrix:** solutions

## HPLC VARIABLES
**Column:** 150 × 4.6 Synchropak C4
**Mobile phase:** Gradient. A was 0.05% trifluoroacetic acid in water. B was 0.05% trifluoroacetic acid in MeCN. A:B from 74:26 to 38:62 over 15 min.
**Flow rate:** 1.5
**Detector:** UV 220

## CHROMATOGRAM
**Retention time:** 4.7

## REFERENCE
Ho, H.-O.; Hsiao, C.-C.; Sheu, M.-T. Preparation of microemulsions using polyglycerol fatty acid esters as surfactant for the delivery of protein drugs. *J.Pharm.Sci.*, **1996**, *85*, 138–143

## SAMPLE
**Matrix:** solutions

## HPLC VARIABLES
**Column:** 150 × 4.6 5 μm Zorbax 300 Å SB-C3
**Mobile phase:** Gradient. A was MeCN:water:trifluoroacetic acid 5:95:0.1. B was MeCN:water:trifluoroacetic acid 5:95:0.085. A:B from 85:15 to 47:53 over 20 min.
**Column temperature:** 35
**Flow rate:** 1
**Injection volume:** 10
**Detector:** UV 215

## CHROMATOGRAM
**Retention time:** 10

## OTHER SUBSTANCES
**Simultaneous:** angiotensin II, carbonic anhydrase, cytochrome C, leucine enkephalin, lysozyme, myoglobin, RNAase

## REFERENCE
Ricker, R.D.; Sandoval, L.A.; Permar, B.J.; Boyes, B.E. Improved reversed-phase high performance liquid chromatography columns for biopharmaceutical analysis. *J.Pharm.Biomed.Anal.*, **1996**, *14*, 93–105

## SAMPLE
**Matrix:** solutions

## HPLC VARIABLES
**Column:** 150 × 4.6 Develosil ODS-HG-5 (Nomura Chemical)
**Mobile phase:** MeCN:100 mM pH 9.0 phosphate buffer 26:74
**Column temperature:** 40
**Flow rate:** 0.8
**Detector:** UV 214

## CHROMATOGRAM
**Retention time:** 18

## OTHER SUBSTANCES
**Simultaneous:** degradation products

## REFERENCE
Yomota, C.; Yoshii, Y.; Takahata, T.; Okada, S. Separation of B-3 monodesamidoinsulin from human insulin by high-performance liquid chromatography under alkaline conditions. *J.Chromatogr.A*, **1996**, *721*, 89–96

## SAMPLE
**Matrix:** solutions

## HPLC VARIABLES
**Column:** 250 × 4.6 Protein & Peptide C18 (Vydac)
**Mobile phase:** MeCN:buffer 26:74 (Buffer was 28.4 g sodium sulfate and 2.7 mL phosphoric acid in 1 L water, pH adjusted to 2.3 with ethanolamine (if necessary).)
**Column temperature:** 40
**Flow rate:** 0.8
**Detector:** UV 214

## CHROMATOGRAM
**Retention time:** 15

## OTHER SUBSTANCES
**Simultaneous:** degradation products

## REFERENCE
Yomota, C.; Yoshii, Y.; Takahata, T.; Okada, S. Separation of B-3 monodesamidoinsulin from human insulin by high-performance liquid chromatography under alkaline conditions. *J.Chromatogr.A*, **1996**, *721*, 89–96

## SAMPLE
**Matrix:** solutions

## HPLC VARIABLES
**Column:** 150 × 4 Armsorb-Si-300 p (DM) (Armchrom, Yerevan, Armenia)
**Mobile phase:** Gradient. A was MeCN:1 M ammonium acetate 10:90. B was MeCN:1 M ammonium acetate 50:50. A:B from 76:24 to 66:34 over 40 min.
**Flow rate:** 0.8
**Detector:** UV 200

## CHROMATOGRAM
**Retention time:** 26

## OTHER SUBSTANCES
**Simultaneous:** proinsulin

## KEY WORDS
recombinant; comparison with capillary electrophoresis

## REFERENCE
Klyushnichenko, V.E.; Koulich, D.M.; Yakimov, S.A.; Maltsev, K.V.; Grishina, G.A.; Nazimov, I.V.; Wulf-son, A.N. Recombinant human insulin. III. High-performance liquid chromatography and high-performance capillary electrophoresis control in the analysis of step-by-step production of recombinant human insulin. *J.Chromatogr.A*, **1994**, *661*, 83–92

## SAMPLE
**Matrix:** solutions

## HPLC VARIABLES
**Column:** 600 × 7.5 TSK G 2000 SW (TOSOH)
**Mobile phase:** MeCN:buffer 5:95 (Buffer was 100 mM pH 7.0 phosphate buffer containing 200 mM sodium sulfate.)
**Flow rate:** 1
**Detector:** UV 200

## CHROMATOGRAM
**Retention time:** 19

## OTHER SUBSTANCES
**Simultaneous:** proinsulin

## KEY WORDS
recombinant; comparison with capillary electrophoresis; SEC

## REFERENCE
Klyushnichenko, V.E.; Koulich, D.M.; Yakimov, S.A.; Maltsev, K.V.; Grishina, G.A.; Nazimov, I.V.; Wulf-son, A.N. Recombinant human insulin. III. High-performance liquid chromatography and high-performance capillary electrophoresis control in the analysis of step-by-step production of recombinant human insulin. *J.Chromatogr.A*, **1994**, *661*, 83–92

## SAMPLE
**Matrix:** solutions
**Sample preparation:** Prepare a 1 mg/mL solution in MeCN:water 20:80, inject a 10 μL aliquot.

## HPLC VARIABLES
**Column:** Nucleosil 100-5 C18 RP
**Mobile phase:** Gradient. MeCN:buffer from 20:80:40:60 over 20 min. (Buffer was 83 mM phosphoric acid adjusted to pH 2.25 with triethylamine.)
**Injection volume:** 10
**Detector:** UV 215

## CHROMATOGRAM
**Retention time:** 15

## REFERENCE
Lenz, V.J.; Gattner, H.-G.; Leithäuser, M.; Brandenburg, D.; Wollmer, A.; Höcker, H. Proteolyses of a fluorogenic insulin derivative and native insulin in reversed micelles monitored by fluorescence emission, reversed-phase high-performance liquid chromatography, and capillary zone electrophoresis. *Anal.Biochem.*, **1994**, *221*, 85–93

**SAMPLE**
**Matrix:** solutions

**HPLC VARIABLES**
**Column:** 250 × 4.5 5 μm Kromasil C8 (Eka-Nobel)
**Mobile phase:** Gradient. A was MeCN:water 10:90 containing 0.1% trifluoroacetic acid. B
    was MeCN:water 90:10 containing 0.1% trifluoroacetic acid. A:B from 0:100 to 75:25
    over 8 min, to 25:75 over 12 min.
**Flow rate:** 2
**Detector:** UV 254

**CHROMATOGRAM**
**Retention time:** 10.7

**OTHER SUBSTANCES**
**Simultaneous:** angiotensin I, angiotensin II, bradykinin, leucin enkephalin, lysozyme, mel-
    ittin, methionine enkephalin, oxytocin

**REFERENCE**
*Bodman Product Guide, Bodman, Aston PA,* **1992**, p. 104

**SAMPLE**
**Matrix:** solutions

**HPLC VARIABLES**
**Column:** 150 × 4.1 10 μm PRP-3 (Hamilton)
**Mobile phase:** Gradient. A was 0.1% trifluoroacetic acid in 50 mM NaOH. B was 0.1%
    trifluoroacetic acid in MeCN. A:B from 100:0 to 40:60 over 30 min.
**Flow rate:** 2
**Detector:** UV 220

**CHROMATOGRAM**
**Retention time:** 14

**OTHER SUBSTANCES**
**Simultaneous:** cytochrome C, lysozyme, myoglobin, ribonuclease A, trypsin

**REFERENCE**
*Rainin Catalog 1991-2, Rainin Instrument Co., Woburn MA,* **1991**, p. 3.33

**SAMPLE**
**Matrix:** solutions

**HPLC VARIABLES**
**Column:** 250 × 4.6 PLRP-S 1000Å (Polymer Labs)
**Mobile phase:** Gradient. A was 0.1% trifluoroacetic acid in water. B was 0.1% trifluoroacetic
    acid in 95% MeCN. A:B from 80:20 to 40:60 over 22 min.
**Flow rate:** 1.5
**Detector:** UV 220

**CHROMATOGRAM**
**Retention time:** 5

## OTHER SUBSTANCES
**Simultaneous:** bovine serum albumin, cytochrome C, lysozyme, myoglobin, ovalbumin, ribonuclease

## REFERENCE
*Rainin Catalog 1991-2, Rainin Instrument Co., Woburn MA*, **1991**, p. 3.63

## SAMPLE
**Matrix:** solutions
**Sample preparation:** Dissolve 70 mg trinitrobenzenesulfonic acid in 1 mL 100 mM pH 8.2 sodium bicarbonate and immediately add an aliquot to 50 volumes of 10 mg/mL insulin in 100 mM pH 8.2 sodium bicarbonate, let stand in the dark at room temperature for 2 h, add to a 150 × 60 column of Sephadex G25 made up in 100 mM pH 8.2 sodium bicarbonate, collect the major colored band and lyophilize it. Reconstitute, inject an aliquot.

## HPLC VARIABLES
**Column:** μBondapak C18
**Mobile phase:** Gradient. MeCN:100 mM pH 3.6 sodium phosphate 25:75 for 10 min, to 45:55 over 1 h
**Flow rate:** 1
**Detector:** UV 280

## CHROMATOGRAM
**Retention time:** 50

## KEY WORDS
derivatization

## REFERENCE
Wallace, G.R.; McLeod, A.; Chain, B.M. Chromatographic analysis of the trinitrophenyl derivatives of insulin. *J.Chromatogr.*, **1988**, *427*, 239–246

## ANNOTATED BIBLIOGRAPHY
Lakhiari, H.; Legendre, E.; Muller, D.; Jozefonvicz, J. High-performance affinity chromatography of insulin on coated silica grafted with sialic acid. *J.Chromatogr.B*, **1995**, *664*, 163–173

Calvaruso, G.; Tesoriere, G.; Vento, R.; Giuliano, M.; Carabillò, M. High-performance liquid chromatographic method for the determination of insulin synthesis in biological systems. *J.Chromatogr.B*, **1994**, *660*, 259–264 [SEC; GPC; reverse phase]

Dimov, N.; Simeonov, S. Experimental models for optimization of insulin separation on reversed phase columns. *Biomed.Chromatogr.*, **1994**, *8*, 32–36 [gradient; cow; pig]

Kliushnichenko, V.E.; Iakimov, S.A.; Arutiunian, A.M.; Ivanov, A.E.; Mal'tsev, K.V.; Vul'fson, A.N. [Genetic engineering of human insulin. IV. Development and optimization of an analysis system using reversed phase high pressure liquid chromatography]. *Bioorg.Khim.*, **1994**, *20*, 1080–1088

Klyushnichenko, V.E.; Yakimov, S.A.; Arutyunyan, A.M.; Ivanov, A.E.; Maltsev, K.V.; Wulfson, A.N. Recombinant human insulin V. Optimization of the reversed-phase high-performance liquid chromatographic separation. *J.Chromatogr.B*, **1994**, *662*, 363–369 [gradient]

Ohkubo, T. High performance liquid chromatographic analysis of polypeptide hormones in transplanted rat islets. *Biomed.Chromatogr.*, **1994**, *8*, 301–305

Yamamoto, A.; Taniguchi, T.; Rikyuu, K.; Tsuji, T.; Fujita, T.; Murakami, M.; Muranishi, S. Effect of various protease inhibitors on the intestinal absorption and degradation of insulin in rats. *Pharm.Res.*, **1994**, *11*, 1496–1500 [rat; gradient]

Salem, I.I.; Bedmar, M.C.; Medina, M.M.; Cerezo, A. Insulin evaluation in pharmaceuticals: Variables in RP-HPLC and method validation. *J.Liq.Chromatogr.*, **1993**, *16*, 1183–1194 [formulations]

Cruz, N.; López, M.; Estrada, G.; Alvarado, X.; de Anda, R.; Balbás, P.; Gosset, G.; Bolivar, F. Preparative isolation of recombinant human insulin-A chain by ion exchange chromatography. *J.Liq.Chromatogr.*, **1992**, *15*, 2311–2324

Cruz, N.; Antonio, S.; de Anda, R.; Gosset, G.; Bolivar, F. Preparative isolation by high performance liquid chromatography of human insulin B chain produced in Escherichia coli. *J.Liq.Chromatogr.*, **1990**, *13*, 1517–1528

Caprioli, R.M.; DaGue, B.; Fan, T.; Moore, W.T. Microbore HPLC/mass spectrometry for the analysis of peptide mixtures using a continuous flow interface. *Biochem.Biophys.Res.Commun.*, **1987**, *146*, 291–299 [LC-MS; microbore; gradient; UV detection; cow; sheep; pig; horse]

Schrader, E.; Pfeiffer, E.F. The influence of motion and temperature upon the aggregational behaviour of soluble insulin formulations investigated by high performance liquid chromatography. *J.Liq.Chromatogr.*, **1985**, *8*, 1139–1157 [GPC; SEC; reverse-phase; pig; human]

Schrader, E.; Pfeiffer, E.F. HPLC-gel filtration of insulin during short and long time infusion by artificial delivery systems. *J.Liq.Chromatogr.*, **1985**, *8*, 1121–1137 [GPC; SEC]

Smith, H.W., Jr.; Atlins, L.M.; Binkley, D.A.; Richardson, W.G.; Miner, D.J. A universal HPLC determination of insulin potency. *J.Liq.Chromatogr.*, **1985**, *8*, 419–439 [cow; pig; human]

Ohta, M.; Tokunaga, H.; Kimura, T.; Satoh, H.; Kawamura, J. Analysis of insulins by high-performance liquid chromatography. III. Determination of insulin in various preparations. *Chem.Pharm.Bull.*, **1984**, *32*, 4641–4649

Ohta, M.; Tokunaga, H.; Kimura, T.; Yamaha, T. [Analysis of insulin by high-performance liquid chromatography. IV. Stability of insulin in hydrochloric acid]. *Yakugaku Zasshi*, **1984**, *104*, 1309–1313

Ohta, M.; Tokunaga, H.; Kimura, T.; Satoh, H.; Kawamura, J. [Analysis of insulin in preparations by high performance liquid chromatography]. *Yakugaku Zasshi*, **1982**, *102*, 1092–1094

Pocker, Y.; Biswas, S.B. A simple liquid chromatographic method for analysis of insulin and its derivatives. *J.Liq.Chromatogr.*, **1982**, *5*, 1–14 [SEC; GPC]

# Interferon

**Molecular formula:** $C_{860}H_{1353}N_{229}O_{255}S_9$
**Molecular weight:** 19269.1
**CAS Registry No.:** 76543-88-9 ($\alpha$A), 99210-65-8 ($\alpha$2B), 98059-61-1 (gamma)

## SAMPLE
**Matrix:** solutions
**Sample preparation:** Dilute PCR product solution 1:5 with water, inject an 80 µL aliquot.

## HPLC VARIABLES
**Column:** 35 × 4.6 2.5 µm TSK DEAE-NPR (Perkin-Elmer)
**Mobile phase:** Gradient. A was 25 mM pH 9.0 Tris-HCl buffer. B was 25 mM pH 9.0 Tris-HCl buffer containing 1 M NaCl. A:B from 70:30 to 45:55 over 30 s, to 35:65 over 2.5 min, to 0:100 over 30 s, maintain at 0:100 for 30 s, return to initial conditions over 30 s, re-equilibrate for 30 s.
**Flow rate:** 1
**Injection volume:** 80
**Detector:** UV 260

## CHROMATOGRAM
**Retention time:** 2.2

## REFERENCE
Zeillinger, R.; Schneeberger, C.; Speiser, P.; Kury, F. Rapid quantitative analysis of differential PCR products by high-performance liquid chromatography. *BioTechniques*, **1993**, *15*, 89–95

## SAMPLE
**Matrix:** solutions
**Sample preparation:** Inject an aliquot of a solution in glycerol:20 mM pH 7.2 sodium phosphate buffer 30:70.

## HPLC VARIABLES
**Column:** 150 × 3 Separon SGX C-18 glass column
**Mobile phase:** Gradient. A was 1 M pyridine adjusted to pH 5.0 with acetic acid. B was n-propanol. A:B 80:20 for 20 min, to 30:70 over 1 h.
**Flow rate:** 0.25
**Detector:** UV 280; bioassay

## CHROMATOGRAM
**Retention time:** 46

## REFERENCE
Aboagye-Mathiesen, G.; Toth, F.D.; Juhl, C.; Norskov-Lauritsen, N.; Petersen, P.M.; Ebbesen, P. Purification of human placental trophoblast interferon by two-dimensional high performance liquid chromatography. *Prep.Biochem.*, **1991**, *21*, 35–51

## SAMPLE
**Matrix:** solutions

## HPLC VARIABLES
**Column:** 600 × 7.8 10 µm Protein Pak J125 (Waters)

**Mobile phase:** Propylene glycol : buffer 25 : 75 (Buffer was 20 mM pH 7.0 sodium phosphate buffer containing 500 mM sodium sulfate and 0.04% Tween 20.)
**Flow rate:** 0.5
**Detector:** UV 214

## KEY WORDS
interferon-omega1; GPC

## REFERENCE
Adolf, G.R.; Maurer-Fogy, I.; Kalsner, I.; Cantell, K. Purification and characterization of natural human interferon omega 1. Two alternative cleavage sites for the signal peptidase. *J.Biol.Chem.*, **1990**, *265*, 9290–9295

## SAMPLE
**Matrix:** solutions

## HPLC VARIABLES
**Column:** 250 × 4.6 5 µm Bakerbond WP C18
**Mobile phase:** Gradient. A was 0.1% trifluoroacetic acid in water. B was 0.1% trifluoroacetic acid in MeCN. A:B 80:20 for 2 min, to 32:68 over 24 min, maintain at 32:68 for 10 min.
**Column temperature:** 30
**Flow rate:** 1
**Detector:** UV 214; UV 280

## CHROMATOGRAM
**Retention time:** 22

## KEY WORDS
interferon-omega1

## REFERENCE
Adolf, G.R.; Maurer-Fogy, I.; Kalsner, I.; Cantell, K. Purification and characterization of natural human interferon omega 1. Two alternative cleavage sites for the signal peptidase. *J.Biol.Chem.*, **1990**, *265*, 9290–9295

## SAMPLE
**Matrix:** solutions

## HPLC VARIABLES
**Column:** 300 × 4 Nucleosil 5C18
**Mobile phase:** Gradient. MeCN:0.1% trifluoroacetic acid from 35:65 to 55:45 over 40 min.
**Flow rate:** 0.9
**Detector:** UV 210

## CHROMATOGRAM
**Retention time:** 20 (Mf-1), 22 (Mf-2), 24 (Ms)

## REFERENCE
Nakagawa, S.; Honda, S.; Sugino, H.; Kusumoto, S.; Sasaoki, K.; Nishi, K.; Kakinuma, A. Characterization of three species of Escherichia coli-derived human leukocyte interferon A separated by reverse-phase high-performance liquid chromatography. *J.Interferon.Res.*, **1987**, *7*, 285–299

## SAMPLE
**Matrix:** solutions

## HPLC VARIABLES
**Guard column:** Whatman Copell ODS
**Column:** 250 × 10 Synchropak C18 RP-P
**Mobile phase:** Gradient. A was 0.025% trifluoroacetic acid in water. B was 0.025% trifluoroacetic acid in MeCN. A:B from 70:30 to 40:60 over 30 min.
**Flow rate:** 2
**Detector:** UV 220

## CHROMATOGRAM
**Retention time:** 26 (IFN-αA)

## OTHER SUBSTANCES
**Simultaneous:** other forms of interferon

## REFERENCE
Felix, A.M.; Heimer, E.P.; Tarnowski, S.J. Analysis of different forms of recombinant human interferons by high-performance liquid chromatography. *Methods Enzymol.*, **1986**, *119*, 242–248

## SAMPLE
**Matrix:** solutions

## HPLC VARIABLES
**Guard column:** Whatman Copell ODS
**Column:** 250 × 10 Synchropak C18 RP-P
**Mobile phase:** Gradient. A was 0.025% trifluoroacetic acid in water. B was 0.025% trifluoroacetic acid in MeCN. A:B from 70:30 to 40:60 over 30 min.
**Flow rate:** 2
**Detector:** UV 220

## CHROMATOGRAM
**Retention time:** 26 (IFN-αA)

## REFERENCE
Felix, A.M.; Heimer, E.P.; Lambros, T.J.; Swistok, J.; Tarnowski, S.J.; Wang, C.-T. Analysis of different forms of recombinant human leukocyte interferons and synthetic fragments by high-performance liquid chromatography. *J.Chromatogr.*, **1985**, *327*, 359–368

## SAMPLE
**Matrix:** solutions
**Sample preparation:** Centrifuge at 5000 g at 4° for 10 min, inject an aliquot.

## HPLC VARIABLES
**Column:** 50 × 5 10 μm Mono-S HPLC cation-exchange column (Pharmacia)
**Mobile phase:** Gradient. A was 10 mM pH 7.0 sodium phosphate in ethylene glycol:water 20:80. B was 10 mM pH 7.0 sodium phosphate + 400 mM NaCl in ethylene glycol:water 20:80. A:B 100:0 for 30 min then to 0:100 over 60 min.
**Flow rate:** 0.5
**Detector:** bioassay

## CHROMATOGRAM
**Retention time:** 70

## KEY WORDS
crude mixtures

## REFERENCE

Friedlander, J.; Fischer, D.G.; Rubinstein, M. Isolation of two discrete human interferon-gamma (immune) subtypes by high-performance liquid chromatography. *Anal.Biochem.*, **1984**, *137*, 115–119

## SAMPLE

**Matrix:** solutions

**Sample preparation:** Pump 125 mL of a solution of interferon in ethylene glycol:50 mM sodium phosphate buffer containing 1 M NaCl 50:50 onto the column (which was previously equilibrated with 1 M NaCl in ethylene glycol:water 50:50) then start the gradient.

## HPLC VARIABLES

**Column:** 300 × 4.6 10 μm Chromegabond octyl

**Mobile phase:** Gradient. A was pyridine:formic acid:isopropanol:n-butanol:water 8:8:20:3.3:60.7. B was pyridine:formic acid:isopropanol:n-butanol:water 8:8:25:20:39. A:B from 100:0 to 75:25 over 30 min, to 45:55 over 190 min, to 0:100 over 20 min, maintain at 0:100 for 1 h.

**Flow rate:** 0.45

**Injection volume:** 125000

**Detector:** F; bioassay

## CHROMATOGRAM

**Retention time:** 100

## REFERENCE

Friesen, H.J.; Stein, S.; Pestka, S. Purification of human fibroblast interferon by high-performance liquid chromatography. *Methods Enzymol.*, **1981**, *78*, 430–435

## SAMPLE

**Matrix:** solutions

## HPLC VARIABLES

**Column:** MN-cyanopropyl

**Mobile phase:** Gradient. Pyridine:formic acid:n-propanol:water from 8:8:0:84 to 8:8:40:44 over 1 h.

**Flow rate:** 0.3

**Detector:** F; bioassay

## CHROMATOGRAM

**Retention time:** 40

## REFERENCE

Friesen, H.J.; Stein, S.; Pestka, S. Purification of human fibroblast interferon by high-performance liquid chromatography. *Methods Enzymol.*, **1981**, *78*, 430–435

## ANNOTATED BIBLIOGRAPHY

Feng, W.; Geng, X. Studies on silica-bonded monoclonal antibody packing material for separation of recombinant interferon by high performance immunoaffinity chromatography. *Biomed.Chromatogr.*, **1993**, *7*, 317–320 [column was anti-interferon monoclonal antibody bonded to silica; gradient]

# Ipratropium Bromide

**Molecular formula:** $C_{20}H_{30}BrNO_3$
**Molecular weight:** 412.4
**CAS Registry No.:** 22254-24-6, 66985-17-9 (monohydrate)

## SAMPLE
**Matrix:** formulations
**Sample preparation:** Dilute with water to a concentration of 125 μg/mL, add a 300 μL aliquot to 250 μL 2 mg/mL mepivacaine hydrochloride in water, vortex, inject a 10 μL aliquot.

## HPLC VARIABLES
**Column:** 100 × 8 4 μm NovaPak C18 radial compression
**Mobile phase:** Gradient. A was 2.5 mM PIC B-8 Low UV (Waters) in THF:water 40:60. B was water. C was MeOH:water 50:50. A:B:C 50:50:0 for 7.7 min, to 60:15:25 over 5.3 min, re-equilibrate at initial conditions for 5 min.
**Flow rate:** 2
**Injection volume:** 10
**Detector:** UV 220

## CHROMATOGRAM
**Retention time:** 5.9
**Internal standard:** mepivacaine (8.2)

## OTHER SUBSTANCES
**Simultaneous:** albuterol, fenoterol, terbutaline

## KEY WORDS
nebulizer solutions; stability-indicating

## REFERENCE
Jacobson, G.A.; Peterson, G.M. High-performance liquid chromatographic assay for the simultaneous determination of ipratropium bromide, fenoterol, salbutamol and terbutaline in nebulizer solution. *J.Pharm.Biomed.Anal.*, **1994**, *12*, 825–832

# Ketoconazole

**Molecular formula:** $C_{26}H_{28}Cl_2N_4O_4$
**Molecular weight:** 531.4
**CAS Registry No.:** 65277-42-1

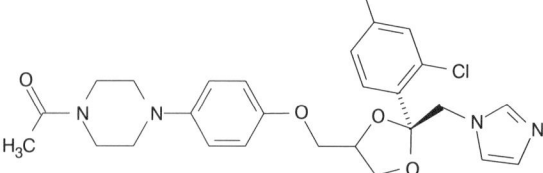

## SAMPLE
**Matrix:** blood
**Sample preparation:** 500 μL Plasma + 2 μg clotrimazole + hexane:isoamyl alcohol 98.5:1.5, vortex, centrifuge. Remove the organic layer and evaporate it to dryness, reconstitute the residue in 250 μL MeCN, inject an aliquot.

## HPLC VARIABLES
**Column:** 150 × 3.9 NovaPak C18
**Mobile phase:** MeCN:MeOH:50 mM phosphate buffer 40:5:55
**Detector:** UV 220

## CHROMATOGRAM
**Limit of detection:** 100-200 ng/mL

## KEY WORDS
plasma; pharmacokinetics

## REFERENCE
von Moltke, L.L.; Greenblatt, D.J.; Harmatz, J.S.; Duan, S.X.; Harrel, L.M.; Cotreau-Bibbo, M.M.; Pritchard, G.A.; Wright, C.E.; Shader, R.I. Triazolam biotransformation by human liver microsomes in vitro: Effects of metabolic inhibitors and clinical confirmation of a predicted interaction with ketoconazole. *J.Pharmacol.Exp.Ther.*, **1996**, *276*, 370–379

## SAMPLE
**Matrix:** blood
**Sample preparation:** 200 μL Serum + 500 μL MeCN, mix, centrifuge, inject a 200 μL aliquot of the supernatant.

## HPLC VARIABLES
**Guard column:** 15 mm long C18
**Column:** 250 × 4.6 5 μm C18 (Beckman)
**Mobile phase:** MeCN:50 mM pH 2.2 phosphoric acid 40:60
**Flow rate:** 2
**Injection volume:** 200
**Detector:** UV 207

## CHROMATOGRAM
**Limit of quantitation:** 20 ng/mL

## KEY WORDS
serum; pharmacokinetics

## REFERENCE
Chin, T.W.F.; Loeb, M.; Fong, I.W. Effects of an acidic beverage (Coca-Cola) on absorption of ketoconazole. *Antimicrob.Agents Chemother.*, **1995**, *39*, 1671–1675

## SAMPLE
**Matrix:** blood

**Sample preparation:** 200 μL Serum + 300 μL 5 μg/mL terconazole in MeCN, vortex, centrifuge, inject 75 μL supernatant.

## HPLC VARIABLES
**Column:** 300 × 4.5 μBondapak C18
**Mobile phase:** MeOH:buffer 60:40 (Buffer was 25 mM $KH_2PO_4$ + 4 mM heptanesulfonic acid, adjusted to pH 8.0 with 1 M NaOH.)
**Flow rate:** 1.8
**Injection volume:** 75
**Detector:** UV 226

## CHROMATOGRAM
**Internal standard:** terconazole
**Limit of quantitation:** 10 ng/mL

## KEY WORDS
serum

## REFERENCE
Carver, P.L.; Berardi, R.R.; Knapp, M.J.; Rider, J.M.; Kauffman, C.A.; Bradley, S.F.; Atassi, M. In vivo interaction of ketoconazole and sucralfate in healthy volunteers. *Antimicrob.Agents Chemother.*, **1994**, *38*, 326–329

## SAMPLE
**Matrix:** blood
**Sample preparation:** Condition a C18 SPE cartridge (Analytichem part 607303) by washing with 2 mL MeOH then 5 mL water. 1 mL Serum + 100 μL 3 mg/mL clotrimazole in MeOH + 200 μL ammonium hydroxide, add to cartridge, wash with 6 mL water, elute with 3 mL MeOH. Evaporate the residue to dryness under a stream of nitrogen at 45°, reconstitute in 1 mL mobile phase, inject 50 μL aliquot. (SPE preparation from J. Chromatogr. 1986, 377, 287.)

## HPLC VARIABLES
**Column:** 150 × 3.9 Novopak C18
**Mobile phase:** MeOH:MeCN:20 mM $KH_2PO_4$ 30:30:35, adjusted to pH 6.8
**Flow rate:** 2
**Injection volume:** 50
**Detector:** UV 254

## CHROMATOGRAM
**Retention time:** 4.3
**Internal standard:** clotrimazole (7.0)
**Limit of quantitation:** 200 ng/mL

## KEY WORDS
serum; SPE

## REFERENCE
Piscitelli, S.C.; Goss, T.F.; Wilton, J.H.; D'Andrea, D.T.; Goldstein, H.; Schentag, J.J. Effects of ranitidine and sucralfate on ketoconazole bioavailability. *Antimicrob.Agents Chemother.*, **1991**, *35*, 1765–1771

## SAMPLE
**Matrix:** blood
**Sample preparation:** Condition a 3 mL Bond-Elut C18 cartridge with 6 mL MeOH and 6 mL water. 1 mL Serum + 110 μL 50 mg/L terconazole in water + 250 μL 100 mM NaOH + 3 mL water, add to cartridge, wash with 9 mL water, wash with 200 μL MeOH, elute

with 1 mL MeOH. Evaporate eluent at 60°, resuspend in 200 μL mobile phase, centrifuge at 13000 g for 2 min, inject 20-40 μL.

## HPLC VARIABLES
**Guard column:** Chrompack C18
**Column:** 100 × 3 Hypersil ODS in a Chrompack glass cartridge
**Mobile phase:** MeCN:water 45:55 containing 500 μL/L diethylamine, pH adjusted to 8.0 with orthophosphoric acid
**Flow rate:** 0.6
**Injection volume:** 20-40
**Detector:** UV 254

## CHROMATOGRAM
**Retention time:** 5.0
**Internal standard:** terconazole (11.4)
**Limit of quantitation:** 50 ng/mL

## OTHER SUBSTANCES
**Noninterfering:** acetaminophen, acyclovir, allopurinol, amoxicillin, amphotericin B, ampicillin, aspirin, azlocillin, bendrofluazide, bumetanide, buprenorphine, carbenicillin, cefazolin, cefotaxime, cefoxitin, ceftazidime, cefuroxime, cephalexin, chlorambucil, chloramphenicol, chlordiazepoxide, chlorpheniramine, chlorpropamide, cyclophosphamide, cyclosporin, cytarabine, daunorubicin, dextropropoxyphene, dihydrocodeine, domperidone, flucytosine, furosemide, gentamicin, griseofulvin, melphalan, methotrexate, metochlopramide, metronidazole, miconazole, nabilone, netilmicin, nicotinamide, nitrazepam, penicillin G, piperacillin, prednisolone, procarbazine, prochlorperazine, riboflavin, rifampin, sulfamethoxazole, thioguanine, tobramycin, tolbutamide, trimethoprim
**Interfering:** diazepam

## KEY WORDS
serum

## REFERENCE
Turner, C.A.; Turner, A.; Warnock, D.W. High performance liquid chromatographic determination of ketoconazole in human serum. *J.Antimicrob.Chemother.*, **1986**, *18*, 757–763

## SAMPLE
**Matrix:** blood, tissue
**Sample preparation:** Condition a SPICE reversed-phase SPE cartridge by washing with 2 mL MeOH then 5 mL water. Tissue. 0.5 g Tissue + 100 μL 3 mg/mL clotrimazole in MeOH + 4 mL MeCN, homogenize for 2 min using a PTFE pestle in a tissue grinder, centrifuge at 1500 g for 15 min. Remove supernatant and evaporate it under a stream of nitrogen at 45°. Reconstitute residue in 1 mL 10 mM HCl, add 200 μL ammonium hydroxide to adjust pH to about 10.5, add to cartridge, wash with 6 mL water, elute with 3 mL MeOH. Evaporate the residue to dryness under a stream of nitrogen at 45°, reconstitute in 200 μL mobile phase, inject 50 μL aliquot. Plasma. 1 mL Plasma + 100 μL 3 mg/mL clotrimazole in MeOH + 200 μL ammonium hydroxide, add to cartridge, wash with 6 mL water, elute with 3 mL MeOH. Evaporate the residue to dryness under a stream of nitrogen at 45°, reconstitute in 1 mL mobile phase, inject 50 μL aliquot.

## HPLC VARIABLES
**Column:** 150 × 3.9 5 μm Novapak C18
**Mobile phase:** MeCN:MeOH:20 mM pH 6.8 KH$_2$PO$_4$/NaOH 30:35:35
**Flow rate:** 2
**Injection volume:** 50
**Detector:** UV 254

## CHROMATOGRAM
**Retention time:** 4.5
**Internal standard:** clotrimazole (6.8)
**Limit of detection:** 200 ng/mL (plasma), 400 ng/g (tissue)

## KEY WORDS
plasma; SPE; lung; liver; adrenal

## REFERENCE
Riley, C.M.; James, M.O. Determination of ketoconazole in the plasma, liver, lung and adrenal of the rat by high-performance liquid chromatography. *J.Chromatogr.*, **1986**, *377*, 287–294

## SAMPLE
**Matrix:** formulations
**Sample preparation:** Tablets. Powder tablets, weigh out amount equivalent to about 30 mg ketoconazole, add 100 mL MeOH, sonicate for 5 min, filter. Add a 2 mL aliquot of filtrate to 5 mL of 200 μg/mL clotrimazole in MeOH, make up to 25 mL with MeOH, inject 20 μL aliquot. Cream. Condition a 500 mg Bond-Elut diol cartridge with 6 mL dichloromethane. Weigh out cream equivalent to about 5 mg of drug, add 30 mL dichloromethane, sonicate for 3 min, make up to 100 mL with dichloromethane, filter. Add a 2 mL aliquot to the cartridge, wash with 2 mL dichloromethane:methanol 4:1, wash with 1 mL MeOH, elute with 3 mL MeOH:buffer 85:15. Add eluate to 1 mL 200 μg/mL clotrimazole in MeOH, make up to 5 mL with MeOH, inject 20 μL aliquot. (Buffer was 50 mM triethylamine adjusted to pH 7.0 with phosphoric acid.)

## HPLC VARIABLES
**Column:** 250 × 4.6 5 μm Spherisorb CN
**Mobile phase:** THF:buffer 30:70 (Buffer was 50 mM triethylamine adjusted to pH 3.0 with phosphoric acid.)
**Flow rate:** 1
**Injection volume:** 20
**Detector:** UV 230 [Enhanced sensitivity with photoreactor (Beam Boost model C6808 with 10 m × 0.3 mm reaction coil) followed by UV detection at 270 nm.]

## CHROMATOGRAM
**Retention time:** 7
**Internal standard:** clotrimazole (9.5)

## OTHER SUBSTANCES
**Simultaneous:** bifonazole, econazole, fenticonazole, isoconazole, miconazole, tioconazole

## KEY WORDS
tablets; creams; post-column reaction

## REFERENCE
Di Pietra, A.M.; Cavrini, V.; Andrisano, V.; Gatti, R. HPLC analysis of imidazole antimycotic drugs in pharmaceutical formulations. *J.Pharm.Biomed.Anal.*, **1992**, *10*, 873–879

## SAMPLE
**Matrix:** solutions

## HPLC VARIABLES
**Column:** 5 μm Deltabond CN (Keystone)
**Mobile phase:** Carbon dioxide:MeOH

**Flow rate:** 0.5 (CO2), 0.05 to 0.12 in 7 min (MeOH)
**Detector:** UV 254

## CHROMATOGRAM
**Retention time:** 7

## OTHER SUBSTANCES
**Simultaneous:** impurities

## KEY WORDS
outlet pressure 3600 psi; SFC; back-pressure regulator heated to 60°

## REFERENCE
Ashraf-Khorassani, M.; Levy, J.M. Addition of modifier in supercritical fluid chromatography using a microbore reciprocating pump. *Chromatographia*, **1995**, *40*, 78–84

## SAMPLE
**Matrix:** solutions

## HPLC VARIABLES
**Column:** 250 × 4.6 5 μm Supelcosil LC-DP (A) or 250 × 4 5 μm LiChrospher 100 RP-8 (B)
**Mobile phase:** MeCN:0.025% phosphoric acid:buffer 25:10:5 (A) or 60:25:15 (B) (Buffer was 9 mL concentrated phosphoric acid and 10 mL triethylamine in 900 mL water, adjust pH to 3.4 with dilute phosphoric acid, make up to 1 L.)
**Flow rate:** 0.6
**Injection volume:** 25
**Detector:** UV 229

## CHROMATOGRAM
**Retention time:** 11.30 (A), 5.92 (B)

## OTHER SUBSTANCES
**Also analyzed:** acebutolol, acepromazine, acetaminophen, acetazolamide, acetophenazine, albuterol, alprazolam, amitriptyline, amobarbital, amoxapine, antipyrine, atenolol, atropine, azatadine, baclofen, benzocaine, bromocriptine, brompheniramine, brotizolam, bupivacaine, buspirone, butabarbital, butalbital, caffeine, carbamazepine, cetirizine, chlorcyclizine, chlordiazepoxide, chlormezanone, chloroquine, chlorpheniramine, chlorpromazine, chlorpropamide, chlorprothixene, chlorthalidone, chlorzoxazone, cimetidine, cisapride, clomipramine, clonazepam, clonidine, clozapine, cocaine, codeine, colchicine, cyclizine, cyclobenzaprine, dantrolene, desipramine, diazepam, diclofenac, diflunisal, diltiazem, diphenhydramine, diphenidol, diphenoxylate, dipyridamole, disopyramide, dobutamine, doxapram, doxepin, droperidol, encainide, ethidium bromide, ethopropazine, fenoprofen, fentanyl, flavoxate, fluoxetine, fluphenazine, flurazepam, flurbiprofen, fluvoxamine, furosemide, glutethimide, glyburide, guaifenesin, haloperidol, homatropine, hydralazine, hydrochlorothiazide, hydrocodone, hydromorphone, hydroxychloroquine, hydroxyzine, ibuprofen, imipramine, indomethacin, ketoprofen, ketorolac, labetalol, levorphanol, lidocaine, loratadine, lorazepam, lovastatin, loxapine, mazindol, mefenamic acid, meperidine, mephenytoin, mepivacaine, mesoridazine, metaproterenol, methadone, methdilazine, methocarbamol, methotrexate, methotrimeprazine, methoxamine, methyldopa, methylphenidate, metoclopramide, metolazone, metoprolol, metronidazole, midazolam, moclobemide, morphine, nadolol, nalbuphine, naloxone, naphazoline, naproxen, nifedipine, nizatidine, norepinephrine, nortriptyline, oxazepam, oxycodone, oxymetazoline, paroxetine, pemoline, pentazocine, pentobarbital, pentoxifylline, perphenazine, pheniramine, phenobarbital, phenol, phenolphthalein, phentolamine, phenylbutazone, phenyltoloxamine, phenytoin, pimozide, pindolol, piroxicam, pramoxine, prazepam, prazosin, probenecid, procainamide, procaine, prochlorperazine, procyclidine, promazine, promethazine, propafenone, propantheline, propiomazine, propofol, propranolol, protripty-

line, quazepam, quinidine, quinine, racemethorphan, ranitidine, remoxipride, risperidone, salicylic acid, scopolamine, secobarbital, sertraline, sotalol, spironolactone, sulfinpyrazone, sulindac, temazepam, terbutaline, terfenadine, tetracaine, theophylline, thiethylperazine, thiopental, thioridazine, thiothixene, timolol, tocainide, tolbutamide, tolmetin, trazodone, triamterene, triazolam, trifluoperazine, triflupromazine, trimeprazine, trimethoprim, trimipramine, verapamil, warfarin, xylometazoline, yohimbine, zopiclone

## KEY WORDS
some details of plasma extraction

## REFERENCE
Koves, E.M. Use of high-performance liquid chromatography-diode array detection in forensic toxicology. *J.Chromatogr.A*, **1995**, *692*, 103–119

## SAMPLE
**Matrix:** tissue
**Sample preparation:** Skin sample extracted with 250 μL mobile phase, vortex 1 min, centrifuge at 8000 rpm for 10 min, inject 40 μL aliquot.

## HPLC VARIABLES
**Column:** 125 × 4.5 Whatman 5 μm reverse-phase C18
**Mobile phase:** MeCN:10 mM pH 6.0 $K_2HPO_4$ 65:35
**Flow rate:** 0.7
**Injection volume:** 40
**Detector:** UV 254

## CHROMATOGRAM
**Retention time:** 8.6
**Limit of detection:** 50 ng/mL

## KEY WORDS
skin

## REFERENCE
Pershing, L.K.; Corlett, J.; Jorgensen, C. In vivo pharmacokinetics and pharmacodynamics of topical ketoconazole and miconazole in human stratum corneum. *Antimicrob.Agents Chemother.*, **1994**, *38*, 90–95

## ANNOTATED BIBLIOGRAPHY

Hoffman, D.W.; Jones-King, K.L.; Ravaris, C.L.; Edkins, R.D. Electrochemical detection for high-performance liquid chromatography of ketoconazole in plasma and saliva. *Anal.Biochem.*, **1988**, *172*, 495–498

Badcock, N.R. Micro-determination of ketoconazole in plasma or serum by high-performance liquid chromatography. *J.Chromatogr.*, **1984**, *306*, 436–440

Pascucci, V.L.; Bennett, J.; Narang, P.K.; Chatterji, D.C. Quantitation of ketoconazole in biological fluids using high-performance liquid chromatography. *J.Pharm.Sci.*, **1983**, *72*, 1467–1469

# Ketoprofen

**Molecular formula:** $C_{16}H_{14}O_3$
**Molecular weight:** 254.3
**CAS Registry No.:** 22071-15-4

## SAMPLE
**Matrix:** bile, blood, perfusate
**Sample preparation:** Dilute bile with saline. Inject 200 μL plasma, 100 μL perfusate, or 100 μL diluted bile onto column A and elute to waste with mobile phase A, after 15 min backflush the contents of column A onto column B with mobile phase B, after 5 min remove column A from the circuit, elute column B with mobile phase B and monitor the effluent from column B. Re-equilibrate column A with mobile phase A for 5 min before the next injection. (See also J.Pharm.Sci. 1995, 84, 1327.)

## HPLC VARIABLES
**Column:** A 30 × 4.6 L-column (porous silica gel with internal surfaces coated with octadecyl groups and external surfaces coated with glycerylpropyl groups) (Chemical Inspection and Testing Institute, Tokyo); B 250 × 4.6 Sumichiral OA-2500S ((R)-N-(3,5-dinitrobenzoyl)-1-naphthylglycine bonded to aminopropyl silica) (Sumika, Osaka)
**Mobile phase:** A 20 mM pH 6.8 ammonium acetate buffer; B MeOH:buffer 95:5, pH 6.2 (Buffer was 1 M acetic acid:1 M ammonium acetate 20:80, pH 4.0. Dilute to 600 mM acetate before use.)
**Flow rate:** 1
**Injection volume:** 100-200
**Detector:** UV 262

## CHROMATOGRAM
**Retention time:** 27 (−), 29 (+)
**Limit of detection:** 20 ng/mL

## KEY WORDS
chiral; rat; column-switching; plasma

## REFERENCE
Yagi, M.; Shibukawa, A.; Nakagawa, T. Direct injection analysis of ketoprofen enantiomers in plasma using column-switching high-performance liquid chromatography system. *Chem.Pharm.Bull.*, **1990**, *38*, 2513−2517

## SAMPLE
**Matrix:** blood
**Sample preparation:** 1 mL Plasma + 100 μL tolmetin solution + 500 μL pH 1.8 phosphate buffer, extract with 1-butanol/MTBE. Remove the organic layer and add it to 500 μL pH 6.1 ammonium acetate buffer, mix, inject an aliquot of the aqueous layer.

## HPLC VARIABLES
**Column:** 150 × 4.6 5 μm Cosmosil C18
**Mobile phase:** MeCN:250 mM pH 5.0 ammonium acetate buffer 20:80
**Flow rate:** 1.8
**Detector:** UV 350

## CHROMATOGRAM
**Internal standard:** tolmetin (UV 258)
**Limit of quantitation:** 5 ng/mL

## KEY WORDS
plasma; pharmacokinetics

## REFERENCE
Shah, A.K.; Wei, G.; Lanman, R.C.; Bhargava, V.O.; Weir, S.J. Percutaneous absorption of ketoprofen from different anatomical sites in man. *Pharm.Res.*, **1996**, *13*, 168–172

## SAMPLE
**Matrix:** blood
**Sample preparation:** 500 μL Plasma + 500 μL MeCN, vortex for 10 s, centrifuge at 1500 g for 10 min. Remove 800 μL of the supernatant and add it to 5 mL dichloromethane, vortex for 30 s, centrifuge for 10 min, inject a 50 μL aliquot of the aqueous layer.

## HPLC VARIABLES
**Guard column:** 20 × 4.6 10 μm Spherisorb C8
**Column:** 250 × 4.6 5 μm Spherisorb ODS
**Mobile phase:** MeOH:buffer 40:60 (Buffer was 40 mM $Na_2HPO_4$ adjusted to pH 8 with orthophosphoric acid.)
**Column temperature:** 50
**Flow rate:** 1
**Injection volume:** 50
**Detector:** UV 360

## CHROMATOGRAM
**Retention time:** 8.6
**Internal standard:** ketoprofen

## OTHER SUBSTANCES
**Extracted:** piroxicam

## KEY WORDS
plasma; ketoprofen is IS

## REFERENCE
Edno, L.; Bressolle, F.; Combe, B.; Galtier, M. A reproducible and rapid HPLC assay for quantitation of piroxicam in plasma. *J.Pharm.Biomed.Anal.*, **1995**, *13*, 785–789

## SAMPLE
**Matrix:** blood
**Sample preparation:** 500 μL Plasma + 125 μL 40 mM decanoic acid in MeCN, mix. Dialyze a 100 μL sample against 20 mM pH 7.0 phosphate buffer using a Gilson Cuprophane membrane (molecular mass cut-off 15 kDa). Continuously pump the buffer through the dialysis cell and through column A at 3 mL/min for 9.6 min, backflush the contents of column A onto column B with the mobile phase, monitor the effluent from column B. (After each injection flush plasma channel with 1 mL 0.05% Triton X-100, with 1 mL 1 mM HCl, and with 2 mL water. After each injection flush buffer channel with 3 mL 20 mM pH 7.0 phosphate buffer and condition column A with 1 mL 20 mM pH 7.0 phosphate buffer.)

## HPLC VARIABLES
**Column:** A 10 × 2 40 μm Bondesil C18 (Analytichem); B 250 × 3.1 5 μm C18 (RoSil Research Separation Laboratories)
**Mobile phase:** MeCN:MeOH:20 mM pH 3.2 phosphate buffer 50:10:40
**Flow rate:** 1

**Injection volume:** 100
**Detector:** UV 261

## CHROMATOGRAM
**Retention time:** 5.5
**Limit of detection:** 100 ng/mL

## OTHER SUBSTANCES
**Extracted:** fenoprofen (UV 272), flurbiprofen (UV 247), ibuprofen (UV 264), naproxen (UV 272)

## KEY WORDS
plasma; dialysis; column-switching

## REFERENCE
Herráez-Hernández, R.; Van de Merbel, N.C.; Brinkman, U.A.T. Determination of the total concentration of highly protein-bound drugs in plasma by on-line dialysis and column liquid chromatography: application to non-steroidal anti-inflammatory drugs. *J.Chromatogr.B*, **1995**, *666*, 127–137

## SAMPLE
**Matrix:** blood
**Sample preparation:** 500 μL Plasma + 100 μL 50 μg/mL indomethacin + 500 μL 600 mM sulfuric acid + 15 mL dichloromethane, mix for 20 min, centrifuge at 2000 g for 5 min. Remove the organic layer and evaporate it to dryness under a stream of nitrogen at room temperature, reconstitute the residue in 100 μL 50 mM triethylamine in MeCN + 50 μL 60 mM ethyl chloroformate in MeCN, vortex for 30 s, add 50 μL 100 mM L-leucinamide in MeOH:triethylamine 100:14, let stand for 2 min, add 50 μL water, inject a 10-50 μL aliquot of the reaction mixture.

## HPLC VARIABLES
**Column:** 250 × 4.6 5 μm Ultrabase C18 (Shandon)
**Mobile phase:** MeCN:60 mM $KH_2PO_4$:triethylamine 49:51:0.1
**Flow rate:** 1.8
**Injection volume:** 10-50
**Detector:** UV 275

## CHROMATOGRAM
**Retention time:** 2.0 (R-(−)), 2.5 (S-(+))
**Internal standard:** indomethacin (5.3)
**Limit of detection:** 100 ng/mL

## OTHER SUBSTANCES
**Extracted:** flurbiprofen, ibuprofen (UV 225)

## KEY WORDS
plasma; chiral; derivatization

## REFERENCE
Péhourcq, F.; Lagrange, F.; Labat, L.; Bannwarth, B. Simultaneous measurement of flurbiprofen, ibuprofen, and ketoprofen enantiomer concentrations in plasma using L-leucinamide as the chiral coupling component. *J.Liq.Chromatogr.*, **1995**, *18*, 3969–3979

## SAMPLE
**Matrix:** blood

**Sample preparation:** 2 mL Whole blood or plasma + 2 mL buffer + 5 mL chloroform: isopropanol:n-heptane 60:14:26, shake gently horizontally for 10 min, centrifuge at 2800 g for 10 min. Remove the lower organic layer and evaporate it to dryness under vacuum at 45°, reconstitute the residue in 100 μL mobile phase, centrifuge at 2800 g for 5 min, inject a 50 μL aliquot of the supernatant. (Buffer was saturated ammonium chloride solution 25% diluted with water, adjusted to pH 9.5 with 25% ammonia solution.)

## HPLC VARIABLES
**Column:** 300 × 3.9 4 μm NovaPack C18
**Mobile phase:** MeOH:THF:buffer 65:5:30 (Buffer was 0.68 g/L (10 mM (sic)) KH$_2$PO$_4$ adjusted to pH 2.6 with concentrated orthophosphoric acid.) (At the end of each session wash the column with water for 1 h and MeOH for 1 h, re-equilibrate for 30 min.)
**Column temperature:** 30
**Flow rate:** 0.8
**Injection volume:** 50
**Detector:** UV 257

## CHROMATOGRAM
**Retention time:** 4.69
**Limit of detection:** <120 ng/mL

## OTHER SUBSTANCES
**Extracted:** acebutolol, acenocoumarol, acepromazine, aceprometazine, acetaminophen, aconitine, ajmaline, albuterol, alimemazine, alpidem, alprazolam, alprenolol, amisulpride, amitriptyline, amoxapine, aspirin, astemizole, atenolol, benazepril, benzocaine, benzoylecgonine, bepridil, betaxolol, bisoprolol, bromazepam, brompheniramine, bumadizone, bupivacaine, buprenorphine, buspirone, caffeine, carbamazepine, carbinoxamine, carpipramine, carteolol, celiprolol, cetirizine, chlorambucil, chlordiazepoxide, chlormezanone, chlorophenacinone, chlorpheniramine, chlorpromazine, chlorpropamide, cibenzoline, clemastine, clobazam, clomipramine, clonazepam, clonidine, clorazepate, clozapine, codeine, colchicine, cyamemazine, cyclizine, cycloguanil, cyproheptadine, cytarabine, dacarbazine, daunorubicin, debrisoquine, demexiptiline, desipramine, dextromethorphan, dextromoramide, dextropropoxyphene, diazepam, diazoxide, diclofenac, dihydralazine, diltiazem, diphenhydramine, dipyridamole, disopyramide, dosulepine, doxepin, ephedrine, estazolam, etodolac, fenfluramine, fenoprofen, fentiazac, flecainide, floctafenine, flumazenil, flunitrazepam, fluoxetine, fluphenazine, flurbiprofen, fluvoxamine, glibenclamide, glibornuride, glipizide, glutethimide, haloperidol, histapyrrodine, hydroxyzine, ibuprofen, imipramine, indomethacin, iproniazid, ketamine, levomepromazine, lidocaine, lidoflazine, lisinopril, loperamide, loprazolam, loratadine, lorazepam, loxapine, maprotiline, medazepam, medifoxamine, mefenamic acid, mefenidramine, mefloquine, melphalan, mephenesin, mephentermine, metapramine, metformin, methadone, methaqualone, methocarbamol, methotrexate, metipranolol, metoclopramide, metoprolol, mexiletine, mianserine, midazolam, minoxidil, moperone, morphine, nadolol, nalbuphine, nalorphine, naloxone, naltrexone, naproxen, nialamide, nicardipine, nifedipine, niflumic acid, nimodipine, nitrazepam, nitrendipine, nizatidine, nortriptyline, omeprazole, opipramol, oxazepam, oxprenolol, penbutolol, penfluridol, pentazocine, phencyclidine, phenobarbital, phenol, phenylbutazone, pimozide, pindolol, pipamperone, piroxicam, prazepam, prazosin, prilocaine, procainamide, procarbazine, proguanil, promethazine, propafenone, propranolol, protriptyline, pyrimethamine, quinidine, quinine, quinupramine, ramipril, ranitidine, reserpine, ritodrine, secobarbital, sotalol, strychnine, sulfinpyrazone, sulindac, sulpride, sultopride, suriclone, tenoxicam, terbutaline, terfenadine, tetracaine, tetrazepam, thiopental, thioproperazine, thioridazine, tianeptine, tiapride, tiaprofenic acid, ticlopidine, tioclomarol, tofisopam, tolbutamide, toloxatone, trazodone, triazolam, trifluoperazine, trifluperidol, trimipramine, triprolidine, tropatenine, verapamil, vinblastine, vincristine, vindesine, warfarin, yohimbine, zopiclone, zorubicine
**Interfering:** alminoprofen, amodiaquine, benperidol, chloroquine, cicletanine, cocaine, doxylamine, droperidol, hydroxychloroquine, ketoprofen, labetalol, meperidine, mepivacaine, moclobemide, nomifensine, temazepam, timolol, viloxazine, zolpidem

## KEY WORDS
whole blood; plasma

## REFERENCE
Tracqui, A.; Kintz, P.; Mangin, P. Systematic toxicological analysis using HPLC/DAD. *J.Forensic Sci.*, **1995**, *40*, 254–262

## SAMPLE
**Matrix:** blood
**Sample preparation:** 100 μL Plasma + 100 μL 20 μg/mL ibuprofen + 8 mL dichloromethane, shake for 20 min, centrifuge at 2500 rpm for 20 min. Remove 7 mL of the organic layer and evaporate it to dryness under nitrogen or at 60°. Dissolve residue in 200 μL mobile phase, inject a 20 μL aliquot.

## HPLC VARIABLES
**Column:** 150 × 6 Shimpack CLS-ODS (Shimadzu)
**Mobile phase:** MeCN:MeOH:0.5 mM phosphoric acid 30:30:40
**Column temperature:** 40
**Flow rate:** 1.5
**Injection volume:** 20
**Detector:** UV 258

## CHROMATOGRAM
**Internal standard:** ibuprofen

## KEY WORDS
plasma; rat

## REFERENCE
Lee, C.K.; Uchida, T.; Kitagawa, K.; Yagi, A.; Kim, N.-S.; Goto, S. Skin permeability of various drugs with different lipophilicity. *J.Pharm.Sci.*, **1994**, *83*, 562–565

## SAMPLE
**Matrix:** blood
**Sample preparation:** 1 mL Plasma + 50 μL 200 μg/mL S-naproxen in MeOH + 500 μL 2 M sulfuric acid + 8 mL n-hexane:ethyl acetate 90:10, mix gently at 30 rpm for 10 min, centrifuge at 1500 g for 10 min. Remove organic layer and evaporate it to dryness at 45° under a stream of nitrogen. Reconstitute in 100 μL 1.5% thionyl chloride in n-hexane (freshly prepared), heat at 75° for 1 h in a capped tube, cool to room temperature, add 500 μL 2% S-1-phenylethylamine in dichloromethane (freshly prepared), let stand for 15 min, add 500 μL 2 M sulfuric acid + 5 mL n-hexane, mix gently at 30 rpm for 10 min, centrifuge at 1500 g for 10 min. Remove organic layer and evaporate it to dryness at 45° under a stream of nitrogen. Reconstitute in 250 μL mobile phase, inject 200 μL aliquot.

## HPLC VARIABLES
**Column:** 250 × 4 5 μm SGE silica glass column
**Mobile phase:** n-Heptane:isopropanol 92:8
**Flow rate:** 1
**Injection volume:** 200
**Detector:** UV 254

## CHROMATOGRAM
**Retention time:** 5.2(R), 6.6(S)
**Internal standard:** S-naproxen (5.9)
**Limit of quantitation:** 150 ng/mL

## OTHER SUBSTANCES
**Simultaneous:** fenoprofen, ibuprofen, mefenamic acid, salicylic acid
**Noninterfering:** diazepam, digoxin, methylprednisolone, midazolam, nifedipine, penicilla-
mine, ranitidine, theophylline

## KEY WORDS
plasma; normal phase; derivatization; chiral

## REFERENCE
Hayball, P.J.; Nation, R.L.; Bochner, F.; Le Leu, R.K. Enantiospecific analysis of ketoprofen in plasma
by high-performance liquid chromatography. *J.Chromatogr.*, **1991**, *570*, 446–452

## SAMPLE
**Matrix:** blood
**Sample preparation:** 1 mL Plasma + 200 μL 2 M HCl + 6 mL ice-cold diethyl ether,
extract, centrifuge at 1500 g for 10 min. Remove 5 mL of organic layer and evaporate it
to dryness under a stream of nitrogen. Dissolve in 250 μL isopropanol:water 2:8, inject
a 40 μL aliquot.

## HPLC VARIABLES
**Column:** 100 × 4.0 AGP (EnantioPac)
**Mobile phase:** 20 mM pH 6.7 phosphate buffer containing 0.5% isopropanol and 5 mM
dimethyloctylamine
**Column temperature:** 15
**Flow rate:** 0.5
**Injection volume:** 40
**Detector:** UV 260

## CHROMATOGRAM
**Retention time:** 25(R), 32(S)
**Limit of quantitation:** 100 ng/mL

## KEY WORDS
plasma

## REFERENCE
Menzel-Soglowek, S.; Geisslinger, G.; Brune, K. Stereoselective high-performance liquid chromatographic determination of ketoprofen, ibuprofen and fenoprofen in plasma using a chiral $\alpha_1$-acid glycoprotein column. *J.Chromatogr.*, **1990**, *532*, 295–303

## SAMPLE
**Matrix:** blood
**Sample preparation:** Condition a 1 mL 100 mg C2 SPE cartridge (Analytichem) with 2
mL MeOH and 1 mL water. 100 μL Plasma + 20 μL MeOH + 500 μL 1 M HCl, vortex
for 15 s, add to the SPE cartridge, rinse out tube with 1 mL water, add rinse to the SPE
cartridge, elute with 1 mL mobile phase, vortex the eluate, inject a 10 μL aliquot.

## HPLC VARIABLES
**Guard column:** 30-40 μm pellicular Vydac Reversed-Phase
**Column:** 75 × 3.9 4 μm Nova-Pak phenyl
**Mobile phase:** MeOH:buffer 42:58 (Buffer was 100 mM $NaH_2PO_4$ adjusted to pH 7.0 with
50% aqueous NaOH.)
**Flow rate:** 1
**Injection volume:** 10
**Detector:** E, Bioanalytical Systems LC-4B, LC-17 thin-layer glassy carbon working elec-
trode +1.10 V, Ag/AgCl reference electrode following post-column reaction. The column

effluent flowed through an air-cooled 7.9 m × 0.3 mm ID PTFE coil irradiated by an SC3-9 ultraviolet lamp (UVP, Inc.) to the detector.

## CHROMATOGRAM
**Retention time:** 2.7
**Internal standard:** ketoprofen
**Limit of detection:** 23 ng/mL

## OTHER SUBSTANCES
**Extracted:** clofibric acid

## KEY WORDS
post-column reaction; plasma; SPE; ketoprofen is IS

## REFERENCE
Bachman, W.J.; Stewart, J.T. HPLC-photolysis-electrochemical detection in pharmaceutical analysis: Application to the determination of clofibric acid in human plasma. *J.Liq.Chromatogr.*, **1989**, *12*, 2947−2959

## SAMPLE
**Matrix:** blood
**Sample preparation:** 0.5-1 mL Plasma + 500 μL 4-8 μg/mL IS in water + 500 μL buffer + 4 mL dichloromethane:n-propanol 99:1, extract on a rotamixer, centrifuge at 1200 g. Remove the organic layer and evaporate it to dryness under a stream of air at 30°, add 5 drops toluene, evaporate to dryness under a stream of air at 30°, reconstitute the residue in 200 μL 50 mM triethylamine in MeCN, add 100 μL 60 mM ethyl chloroformate in MeCN, after 30 s add 100 μL 1 M l-leucinamide hydrochloride in MeOH containing 1 M triethylamine, after 2 min add 500 μL 250 mM HCl, extract with 4 mL ethyl acetate. Evaporate the organic layer to dryness under a stream of air at 30°, reconstitute the residue with 100 μL MeCN, add 400 μL 10 mM pH 6.5 phosphate buffer, inject a 60 μL aliquot. (Prepare buffer as follows. Neutralize a 1 M solution of tetrabutylammonium sulfate in water with NaOH, wash 5 times with dichloromethane, wash twice with heptane. Prepare a 100 mM pH 9.6 sodium carbonate buffer containing 0.5 M of the neutralized and washed tetrabutylammonium salt.)

## HPLC VARIABLES
**Guard column:** 30 × 4 Perisorb RP-18 (Merck)
**Column:** 250 × 4 7 μm LiChroCart RP-18 (Merck)
**Mobile phase:** MeCN:10 mM pH 6.5 phosphate buffer 38:62
**Flow rate:** 2
**Injection volume:** 10
**Detector:** UV 260

## CHROMATOGRAM
**Retention time:** 8.5 (−), 10 (+)
**Internal standard:** 2-(4-benzoylphenyl)butyric acid (12, 15 (enantiomers))
**Limit of quantitation:** 250 ng/mL

## KEY WORDS
plasma; derivatization; chiral; pharmacokinetics

## REFERENCE
Björkman, S. Determination of the enantiomers of ketoprofen in blood plasma by ion-pair extraction and high-performance liquid chromatography of leucinamide derivatives. *J.Chromatogr.*, **1987**, *414*, 465−471

## SAMPLE
**Matrix:** blood
**Sample preparation:** Condition a 1 mL Bond-Elut C8 SPE cartridge with 2 mL MeOH then 1 mL 10 mM HCl, do not allow it to dry completely. Sonicate 1 mL whole blood for 20-30 min then apply to cartridge. Wash with 100 μL water, elute with three 500 μL portions of MeOH:MeCN:1% aqueous ammonium hydroxide 50:20:30, combine eluents and evaporate to dryness under a stream of nitrogen at 40°. Redissolve in 1 mL MeOH, inject a 20 μL aliquot.

## HPLC VARIABLES
**Column:** 250 × 4.5 5 μm Spherisorb ODS
**Mobile phase:** MeCN:MeOH:buffer 35:13:52 (Buffer was water adjusted to pH 3.2 with orthophosphoric acid.)
**Flow rate:** 1
**Injection volume:** 20
**Detector:** UV 250

## CHROMATOGRAM
**Retention time:** 5

## OTHER SUBSTANCES
**Simultaneous:** acetaminophen, fenoprofen, ibuprofen, indomethacin, naproxen, salicylic acid

## KEY WORDS
whole blood; SPE

## REFERENCE
Moore, C.M.; Tebbett, I.R. Rapid extraction of anti-inflammatory drugs in whole blood for HPLC analysis. *Forensic Sci.Int.*, **1987**, *34*, 155–158

## SAMPLE
**Matrix:** blood
**Sample preparation:** 1 mL Plasma + 200 μL 1 M HCl + 4-5 mL ethyl acetate, vortex for 1.5-2 min, centrifuge at 400 g for 10 min. Remove the organic layer and evaporate it to dryness under reduced pressure at 40-50°, reconstitute the residue in 200 μL MeOH, inject a 20 μL aliquot.

## HPLC VARIABLES
**Column:** 100 × 5 10 μm spherical C18 radial compression (Waters)
**Mobile phase:** MeCN:water 45:55 containing 2.5 mL/L acetic acid
**Flow rate:** 1
**Injection volume:** 20
**Detector:** UV 280

## CHROMATOGRAM
**Retention time:** 5
**Internal standard:** ketoprofen

## OTHER SUBSTANCES
**Extracted:** oxaprozin
**Simultaneous:** acetaminophen, fenoprofen, flurbiprofen, indomethacin, phenylbutazone, salicylic acid
**Noninterfering:** ibuprofen, piroxicam

**KEY WORDS**
plasma; ketoprofen is IS

**REFERENCE**
Matlis, R.; Greenblatt, D.J. Rapid high-performance liquid chromatographic analysis of oxaprozin, a non-steroidal anti-inflammatory agent. *J.Chromatogr.*, **1984**, *310*, 445–449

**SAMPLE**
**Matrix:** blood, urine
**Sample preparation:** Plasma. 500 μL Plasma + 10 μg fenoprofen + 100 μL 600 mM sulfuric acid + 4 mL 2,2,4-trimethylpentane:isopropanol 95:5, vortex for 10 s, centrifuge at 1800 g for 5 min. Remove the organic layer and evaporate it to dryness under reduced pressure, reconstitute the residue in 180 μL mobile phase, vortex for 10 s, inject a 100 μL aliquot. Urine. 500 μL Urine + 20 μg fenoprofen + 100 μL 600 mM sulfuric acid + 4 mL 2,2,4-trimethylpentane:isopropanol 95:5, vortex for 10 s, centrifuge at 1800 g for 3 min. Remove the organic layer and add it to 3 mL water, vortex for 10 s, centrifuge for 3 min. Remove the aqueous phase and add it to 200 μL 600 mM sulfuric acid and 3 mL chloroform, vortex for 10 s, centrifuge for 3 min. Remove the organic phase and evaporate it to dryness, reconstitute the residue in 180 μL mobile phase, vortex for 10 s, inject a 100 μL aliquot.

**HPLC VARIABLES**
**Column:** 250 × 4.6 Chiralpak AD amylose carbamate (Chiral Technologies)
**Mobile phase:** Hexane:isopropanol:trifluoroacetic acid 80:19.9:0.1
**Flow rate:** 1
**Injection volume:** 100
**Detector:** UV 254

**CHROMATOGRAM**
**Retention time:** 7.0 (R-(−)), 8.3 (S-(+))
**Internal standard:** fenoprofen (5.3, 6.3 (enantiomers))
**Limit of quantitation:** 50 ng/mL (plasma); 200 ng/mL (urine)

**KEY WORDS**
plasma; chiral

**REFERENCE**
Carr, R.A.; Caillé, G.; Ngoc, A.H.; Foster, R.T. Stereospecific high-performance liquid chromatographic assay of ketoprofen in human plasma and urine. *J.Chromatogr.B*, **1995**, *668*, 175–181

**SAMPLE**
**Matrix:** blood, urine
**Sample preparation:** Add naproxen to plasma or urine, acidify with 1 M pH 2 phosphate buffer, extract with hexane:THF 80:20.

**HPLC VARIABLES**
**Guard column:** 20 × 4.6 Nucleosil OCS 10
**Column:** 150 × 6 YMC Pack A312 S5 120A ODS
**Mobile phase:** MeCN:50 mM phosphate buffer 19:83
**Column temperature:** 28
**Flow rate:** 2
**Injection volume:** 50
**Detector:** UV 262

**CHROMATOGRAM**
**Retention time:** 14.48 (plasma), 13.53 (urine)

**Internal standard:** naproxen (11.11 (plasma), 10.47 (urine))
**Limit of quantitation:** 20 ng/mL

## KEY WORDS
plasma; rat

## REFERENCE
Daffonchio, L.; Bestetti, A.; Clavenna, G.; Fedele, G.; Ferrari, M.P.; Omini, C. Effect of a new foam formulation of ketoprofen lysine salt in experimental models of inflammation and hyperalgesia. *Arzneimittelforschung*, **1995**, *45*, 590–594

## SAMPLE
**Matrix:** blood, urine
**Sample preparation:** Plasma. 100 μL Plasma + 100 μL 600 mM sulfuric acid + 4 mL isooctane:isopropanol 95:5, vortex 30 s, centrifuge at 1800 g for 5 min. Remove organic layer and add 4 mL water to it. Vortex for 30 s, centrifuge for 3 min. Remove organic layer and evaporate it to dryness on a Speed Vac concentrator. Reconstitute residue in 200 μL MeOH, vortex 30 s, add 100 μL 100 μg/mL indoprofen in water, inject a 20 μL aliquot. Urine. 100 μL Urine + 25 μL 1 M NaOH, add 125 μL 600 mM sulfuric acid, proceed as for plasma.

## HPLC VARIABLES
**Guard column:** 50 × 5 37-53 μm C18 material
**Column:** 100 × 4.6 5 μm Partisil 5 ODS-3
**Mobile phase:** MeCN:60 mM $KH_2PO_4$:triethylamine 25:75:0.1
**Flow rate:** 1
**Injection volume:** 20
**Detector:** UV 275

## CHROMATOGRAM
**Retention time:** 4.7
**Internal standard:** indoprofen (3.4)

## OTHER SUBSTANCES
**Simultaneous:** probenecid

## KEY WORDS
plasma; rat

## REFERENCE
Palylyk, E.L.; Jamali, F. Simultaneous determination of ketoprofen enantiomers and probenecid in plasma and urine by high-performance liquid chromatography. *J.Chromatogr.*, **1991**, *568*, 187–196

## SAMPLE
**Matrix:** blood, urine
**Sample preparation:** Plasma. 100 μL Plasma + 100 μL 100 μg/mL indoprofen in water + 100 μL 600 mM sulfuric acid + 5 mL isooctane:isopropanol 95:5, vortex 30 s, centrifuge at 1800 g for 5 min. Remove organic layer and add 5 mL water to it. Vortex for 30 s, centrifuge for 3 min. Remove organic layer and evaporate it to dryness on a Speed Vac concentrator. Reconstitute residue in 100 μL 50 mM triethylamine in MeCN, vortex 30 s, add 50 μL 60 mM ethyl chloroformate in MeCN, let stand 30 s, add 50 μL 1 M L-leucinamide hydrochloride and 1 M triethylamine in MeOH, let stand 2 min, add 50 μL water, inject 10-60 μL aliquots. Urine. 100 μL Urine + 25 μL 1 M NaOH, add 125 μL 600 mM sulfuric acid, proceed as for plasma.

## HPLC VARIABLES
**Guard column:** 50 × 5 37-53 μm C18 material
**Column:** 100 × 4.6 5 μm Partisil 5 ODS-3
**Mobile phase:** MeCN:60 mM $KH_2PO_4$:triethylamine 35:65:0.1
**Flow rate:** 1
**Injection volume:** 10-60
**Detector:** UV 275

## CHROMATOGRAM
**Retention time:** 10 (R), 12 (S)
**Internal standard:** indoprofen (6(R), 7(S))
**Limit of quantitation:** 500 ng/mL

## OTHER SUBSTANCES
**Simultaneous:** probenecid
**Also analyzed:** carprofen, cicloprofen, fenoprofen, flurbiprofen, indoprofen, pirprofen

## KEY WORDS
plasma; rat; chiral; derivatization

## REFERENCE
Palylyk, E.L.; Jamali, F. Simultaneous determination of ketoprofen enantiomers and probenecid in plasma and urine by high-performance liquid chromatography. *J.Chromatogr.*, **1991**, *568*, 187–196

## SAMPLE
**Matrix:** blood, urine
**Sample preparation:** 50 μL Plasma or urine + 10 μL 1 M NaOH, heat at 37° for 2 h, add 10 μL 1 M HCl, extract with 1 mL ethyl acetate. Remove the organic layer and evaporate it to dryness, reconstitute the residue in 100 μL 50 mM triethylamine in MeCN, add 50 μL 60 mM ethyl chloroformate in MeCN, let stand for 2 min, add 50 μL 1 M L-leucinamide in 1 M triethylamine in MeOH. Evaporate, take up the residue in 100 μL mobile phase, inject a 100 μL aliquot. (Hydrolysis of glucuronides may be omitted.)

## HPLC VARIABLES
**Column:** 250 × 4.6 5 μm Ultrasphere ODS
**Mobile phase:** MeCN:60 mM pH 6 potassium phosphate buffer 40:60
**Flow rate:** 1
**Injection volume:** 100
**Detector:** UV 272

## CHROMATOGRAM
**Retention time:** 13.6 (R), 16.0 (S)
**Internal standard:** ketoprofen

## OTHER SUBSTANCES
**Extracted:** fenoprofen, flunoxaprofen

## KEY WORDS
plasma; chiral; ketoprofen is IS

## REFERENCE
Volland, C.; Sun, H.; Benet, L.Z. Stereoselective analysis of fenoprofen and its metabolites. *J.Chromatogr.*, **1990**, *534*, 127–138

## SAMPLE
**Matrix:** blood, urine

**Sample preparation:** Plasma. 500 µL Plasma + 100 µL 100 µg/mL calcium fenoprofen in water + 100 µL 600 mM sulfuric acid + 3 mL isooctane:isopropanol 95:5, vortex for 30 s, centrifuge at 1800 RCF for 5 min. Remove the organic layer and add it to 3 mL water, vortex for 30 s, centrifuge for 3 min. Remove the aqueous layer and add it to 200 µL 600 mM sulfuric acid, add 3 mL chloroform, vortex for 30 s, centrifuge for 3 min. Remove the organic layer and evaporate it to dryness under reduced pressure, reconstitute with 100 µL 50 mM triethylamine in MeCN, add 50 µL 60 mM ethyl chloroformate in MeCN, after 30 s add 50 µL 1 M l-leucinamide hydrochloride in MeOH containing 1 M triethylamine, after 2 min add 50 µL water, inject a 10-40 µL aliquot. Urine. 100-500 µL Urine + 25-125 µL 1 M NaOH, mix, add a volume of 600 mM sulfuric acid equal to the volume of 1 M NaOH plus 100 µL, add 100 µL 100 µg/mL calcium fenoprofen in water, add 3 mL isooctane:isopropanol 95:5, vortex for 30 s, centrifuge at 1800 RCF for 5 min. Remove the organic layer and add it to 3 mL water, vortex for 30 s, centrifuge for 3 min. Remove the aqueous layer and add it to 200 µL 600 mM sulfuric acid, add 3 mL chloroform, vortex for 30 s, centrifuge for 3 min. Remove the organic layer and evaporate it to dryness under reduced pressure, reconstitute with 100 µL 50 mM triethylamine in MeCN, add 50 µL 60 mM ethyl chloroformate in MeCN, after 30 s add 50 µL 1 M l-leucinamide hydrochloride in MeOH containing 1 M triethylamine, after 2 min add 50 µL water, inject a 10-40 µL aliquot.

## HPLC VARIABLES
**Guard column:** 50 mm long 37-53 µm C18
**Column:** 100 mm long Partisil 5 ODS-3
**Mobile phase:** MeCN:60 mM $KH_2PO_4$:triethylamine 64:36:0.02
**Flow rate:** 1
**Injection volume:** 10-40
**Detector:** UV 275

## CHROMATOGRAM
**Retention time:** 9.8 (R-(−)), 11.3 (S-(+))
**Internal standard:** fenoprofen 17.7 (R-(−)), 19.9 (S-(+))
**Limit of quantitation:** 50 ng/mL

## OTHER SUBSTANCES
**Simultaneous:** flurbiprofen
**Interfering:** naproxen

## KEY WORDS
plasma; derivatization; chiral

## REFERENCE
Foster, R.T.; Jamali, F. High-performance liquid chromatographic assay of ketoprofen enantiomers in human plasma and urine. *J.Chromatogr.*, **1987**, *416*, 388−393

## SAMPLE
**Matrix:** blood, urine
**Sample preparation:** Plasma. 350 µL 2 µg/mL Naproxen in 10 mM pH 6.0 phosphate buffer containing 0.05% MeOH + 650 µL pH 6 phosphate buffer + 1 mL plasma + 0.5 mL 1 M pH 2 phosphate buffer + 10 mL diethyl ether, vortex 1 min, centrifuge at 2000 g for 3 min. Remove organic phase and evaporate it to dryness under a stream of nitrogen. Dissolve residue in 250 µL mobile phase, vortex for 15 s, inject aliquot. Urine. 350 µL 20 µg/mL Naproxen in 10 mM pH 6.0 phosphate buffer containing 0.5% MeOH + 650 µL pH 6 phosphate buffer + 1 mL urine + 1 mL 0.5 M pH 7 phosphate buffer + 10 mL diethyl ether, vortex 1 min, centrifuge at 2000 g for 3 min. Remove organic phase and evaporate it to dryness under a stream of nitrogen. Dissolve residue in 250 µL mobile phase, vortex for 15 s, inject aliquot.

## HPLC VARIABLES
**Guard column:** 40 × 3.2 30-44 μm Vydac reverse-phase
**Column:** 40 × 4.6 5 μm Spherisorb ODS
**Mobile phase:** MeCN:50 mM pH 7.0 phosphate buffer 6:94 to 8:92
**Flow rate:** 2
**Injection volume:** 5-200
**Detector:** UV 262

## CHROMATOGRAM
**Retention time:** 16
**Internal standard:** naproxen (10)
**Limit of detection:** 10 ng/mL

## OTHER SUBSTANCES
**Simultaneous:** metabolites, fenoprofen, probenecid, salicylic acid

## KEY WORDS
plasma

## REFERENCE
Upton, R.A.; Buskin, J.N.; Guentert, T.W.; Williams, R.L.; Riegelman, S. Convenient and sensitive high-performance liquid chromatography assay for ketoprofen, naproxen and other allied drugs in plasma or urine. *J.Chromatogr.*, **1980**, *190*, 119–128

## SAMPLE
**Matrix:** bulk
**Sample preparation:** 10 mg Compound + 10 mg 1-(3-dimethylaminopropyl)-3-ethylcarbodiimide hydrochloride + 2 drops 3,5-dimethylaniline + 1.5 mL dichloromethane, mix, after 30 min add 1 mL 1 M HCl, shake vigorously. Remove the lower organic layer and dry it over anhydrous magnesium sulfate, inject an aliquot.

## HPLC VARIABLES
**Column:** 250 × 4.6 5 μm D N-(3,5-dinitrobenzoyl)phenylglycine (Regis)
**Mobile phase:** Hexane:isopropanol 80:20
**Flow rate:** 2
**Injection volume:** 20
**Detector:** UV 254; UV 280

## CHROMATOGRAM
**Retention time:** k' 4.97 (for first enantiomer)

## OTHER SUBSTANCES
**Also analyzed:** carprofen, cicloprofen, etodolac, fenoprofen, flurbiprofen, ibuprofen, naproxen, pirprofen, tiaprofenic acid

## KEY WORDS
derivatization; α = 1.18; chiral

## REFERENCE
Pirkle, W.H.; Murray, P.G. The separation of the enantiomers of a variety of non-steroidal anti-inflammatory drugs (NSAIDS) as their anilide derivatives using a chiral stationary phase. *J.Liq.Chromatogr.*, **1990**, *13*, 2123–2134

## SAMPLE
**Matrix:** formulations

**Sample preparation:** Weigh out capsule contents equivalent to 50 mg ketoprofen, add 80 mL MeOH, sonicate for 10 min, make up to 100 mL with MeOH, filter. Dilute a 4 mL aliquot of the filtrate to 100 mL with mobile phase. Mix a 10 mL aliquot of the diluted solution with 10 mL 50 μg/mL ibuprofen in mobile phase, make up to 100 mL with mobile phase, inject a 50 μL aliquot.

## HPLC VARIABLES
**Column:** 100 × 4.6 5 μm Spheri-5 RP-8
**Mobile phase:** MeOH:buffer 30:70 (Prepare buffer by mixing 4 mM $Na_2HPO_4$ and 7 mM $KH_2PO_4$ to achieve pH 7.)
**Flow rate:** 1
**Injection volume:** 50
**Detector:** F ex 355 em 460 (408 nm cutoff filter) following post-column extraction. The column effluent mixed with 50 μg/mL reagent in mobile phase pumped at 0.5 mL/min and then with chloroform pumped at 1 mL/min and the mixture flowed through a 1.8 m × 0.3 mm ID knitted PTFE coil to a 50 μL membrane phase separator using a polyethylene-backed 0.5 μm Fluoropore membrane filter (design in paper). The organic phase flowed to the detector. (Synthesize the reagent, α-(3,4-dimethoxyphenyl)-4'-trimethylammoniummethylcinnamonitrile methosulfate, as follows. Stir 20 mmoles 3,4-dimethoxyphenylacetonitrile and 20 mmoles p-toluamide in 50 mL EtOH at 50°, add 5 mL 50% aqueous KOH slowly, stir at 50° for 5 min, cool to room temperature, filter, dry the precipitate of α-(3,4-dimethoxyphenyl)-4'-methylcinnamonitrile. Dissolve 20 mmoles α-(3,4-dimethoxyphenyl)-4'-methylcinnamonitrile, 20 mmoles N-bromosuccinimide, and 20 mg benzoyl peroxide in 100 mL carbon tetrachloride (Caution! Carbon tetrachloride is a carcinogen!), reflux with stirring for 1.5 h, cool, filter, evaporate to dryness under reduced pressure, recrystallize from MeOH to give α-(3,4-dimethoxyphenyl)-4'-bromomethylcinnamonitrile. Vigorously stir 30 mmoles anhydrous dimethylamine in 100 mL dry benzene (Caution! Benzene is a carcinogen!) at 0°, very slowly add 10 mmoles α-(3,4-dimethoxyphenyl)-4'-bromomethylcinnamonitrile while stirring at 0°, stir at room temperature overnight, add 150 mL water, remove the organic phase, extract the aqueous phase twice with 100 mL portions of diethyl ether, wash the organic layers with saturated NaCl solution, dry over anhydrous magnesium sulfate, evaporate under reduced pressure to give α-(3,4-dimethoxyphenyl)-4'-dimethylaminomethylcinnamonitrile (J.Chem.Eng.Data 1987, 32, 387). Reflux 10 mmoles α-(3,4-dimethoxyphenyl)-4'-dimethylaminomethylcinnamonitrile, 20 mmoles dimethyl sulfate (Caution! Dimethyl sulfate is a carcinogen and acutely toxic!), and 5 g potassium carbonate in 50 mL acetone for 1 h, cool to room temperature, filter, dry the precipitate under vacuum at room temperature overnight, recrystallize from chloroform containing 2-3 drops of 95% EtOH to give α-(3,4-dimethoxyphenyl)-4'-trimethylammoniummethylcinnamonitrile methosulfate (mp 212-215°). Protect solutions from light.)

## CHROMATOGRAM
**Retention time:** k' 1.5504
**Internal standard:** ibuprofen (k' 4.124)
**Limit of detection:** 20 ng/mL

## OTHER SUBSTANCES
**Simultaneous:** mefenamic acid, naproxen, probenecid, salicylic acid, valproic acid

## KEY WORDS
capsules; post-column extraction

## REFERENCE
Kim, M.; Stewart, J.T. HPLC post-column ion-pair extraction of acidic drugs using a substituted α-phenylcinnamonitrile quaternary ammonium salt as a new fluorescent ion-pair reagent. *J.Liq.Chromatogr.*, **1990**, *13*, 213–237

## SAMPLE
**Matrix:** solutions

## HPLC VARIABLES
**Column:** 125 × 3 Ecocart LiChrospher 100 RP-18
**Mobile phase:** Isopropanol:100 mM $KH_2PO_4$:formic acid 54:100:0.1
**Flow rate:** 0.6
**Detector:** UV 254

## CHROMATOGRAM
**Retention time:** 4.3
**Limit of quantitation:** 200-500 ng/mL

## OTHER SUBSTANCES
**Also analyzed:** acemetacin, diclofenac, flurbiprofen, indomethacin, lonazolac, naproxen, piroxicam, sulindac, tenoxicam

## REFERENCE
Baeyens, W.R.G.; Van Der Weken, G.; Van Overbeke, A.; Zhang, Z.D. Preliminary results on the LC-separation of non-steroidal anti-inflammatory agents in conventional and narrow-bore RP set-ups applying columns with different internal diameters. *Biomed.Chromatogr.*, **1995**, *9*, 261–262

## SAMPLE
**Matrix:** solutions

## HPLC VARIABLES
**Column:** 250 × 4.5 5 μm Ultrasphere ODS
**Mobile phase:** MeCN:10 mM tetrabutylammonium buffer 45:55
**Flow rate:** 1
**Detector:** UV 220

## CHROMATOGRAM
**Retention time:** 11.6

## OTHER SUBSTANCES
**Simultaneous:** fenoprofen

## REFERENCE
Bischer, A.; Iwaki, M.; Zia-Amirhosseini, P.; Benet, L.Z. Stereoselective reversible binding properties of the glucuronide conjugates of fenoprofen enantiomers to human serum albumin. *Drug Metab.Dispos.*, **1995**, *23*, 900–903

## SAMPLE
**Matrix:** solutions

## HPLC VARIABLES
**Column:** 250 × 4.6 5 μm Supelcosil LC-DP (A) or 250 × 4 5 μm LiChrospher 100 RP-8 (B)
**Mobile phase:** MeCN:0.025% phosphoric acid:buffer 25:10:5 (A) or 60:25:15 (B) (Buffer was 9 mL concentrated phosphoric acid and 10 mL triethylamine in 900 mL water, adjust pH to 3.4 with dilute phosphoric acid, make up to 1 L.)
**Flow rate:** 0.6
**Injection volume:** 25
**Detector:** UV 229

## CHROMATOGRAM
**Retention time:** 7.02 (A), 7.04 (B)

## OTHER SUBSTANCES
**Also analyzed:** acebutolol, acepromazine, acetaminophen, acetazolamide, acetophenazine, albuterol, alprazolam, amitriptyline, amobarbital, amoxapine, antipyrine, atenolol, atropine, azatadine, baclofen, benzocaine, bromocriptine, brompheniramine, brotizolam, bupivacaine, buspirone, butabarbital, butalbital, caffeine, carbamazepine, cetirizine, chlorcyclizine, chlordiazepoxide, chlormezanone, chloroquine, chlorpheniramine, chlorpromazine, chlorpropamide, chlorprothixene, chlorthalidone, chlorzoxazone, cimetidine, cisapride, clomipramine, clonazepam, clonidine, clozapine, cocaine, codeine, colchicine, cyclizine, cyclobenzaprine, dantrolene, desipramine, diazepam, diclofenac, diflunisal, diltiazem, diphenhydramine, diphenidol, diphenoxylate, dipyridamole, disopyramide, dobutamine, doxapram, doxepin, droperidol, encainide, ethidium bromide, ethopropazine, fenoprofen, fentanyl, flavoxate, fluoxetine, fluphenazine, flurazepam, flurbiprofen, fluvoxamine, furosemide, glutethimide, glyburide, guaifenesin, haloperidol, homatropine, hydralazine, hydrochlorothiazide, hydrocodone, hydromorphone, hydroxychloroquine, hydroxyzine, ibuprofen, imipramine, indomethacin, ketoconazole, ketorolac, labetalol, levorphanol, lidocaine, loratadine, lorazepam, lovastatin, loxapine, mazindol, mefenamic acid, meperidine, mephenytoin, mepivacaine, mesoridazine, metaproterenol, methadone, methdilazine, methocarbamol, methotrexate, methotrimeprazine, methoxamine, methyldopa, methylphenidate, metoclopramide, metolazone, metoprolol, metronidazole, midazolam, moclobemide, morphine, nadolol, nalbuphine, naloxone, naphazoline, naproxen, nifedipine, nizatidine, norepinephrine, nortriptyline, oxazepam, oxycodone, oxymetazoline, paroxetine, pemoline, pentazocine, pentobarbital, pentoxifylline, perphenazine, pheniramine, phenobarbital, phenol, phenolphthalein, phentolamine, phenylbutazone, phenyltoloxamine, phenytoin, pimozide, pindolol, piroxicam, pramoxine, prazepam, prazosin, probenecid, procainamide, procaine, prochlorperazine, procyclidine, promazine, promethazine, propafenone, propantheline, propiomazine, propofol, propranolol, protriptyline, quazepam, quinidine, quinine, racemethorphan, ranitidine, remoxipride, risperidone, salicylic acid, scopolamine, secobarbital, sertraline, sotalol, spironolactone, sulfinpyrazone, sulindac, temazepam, terbutaline, terfenadine, tetracaine, theophylline, thiethylperazine, thiopental, thioridazine, thiothixene, timolol, tocainide, tolbutamide, tolmetin, trazodone, triamterene, triazolam, trifluoperazine, triflupromazine, trimeprazine, trimethoprim, trimipramine, verapamil, warfarin, xylometazoline, yohimbine, zopiclone

## KEY WORDS
some details of plasma extraction

## REFERENCE
Koves, E.M. Use of high-performance liquid chromatography-diode array detection in forensic toxicology. *J.Chromatogr.A*, **1995**, *692*, 103–119

## SAMPLE
**Matrix:** solutions
**Sample preparation:** Prepare a 100 μM solution in buffer, inject a 20 μL aliquot.

## HPLC VARIABLES
**Column:** 100 × 4.6 column containing riboflavin binding proteins (Prepare as follows. Add riboflavin to saturate protein of egg yolk, homogenize with 3 volumes buffer, centrifuge, add the supernatant to a 500 × 30 column of DEAE-cellulose (Whatman) equilibrated with buffer, wash extensively with buffer to remove bound protein, elute riboflavin binding proteins (RFBP) with buffer containing 200 mM NaCl (RFBP has intense yellow color, absorption at 455 nm). Purify RFBP on a Sephadex G-100 column with 50 mM pH 7.5 Tris-HCl buffer as eluent, remove the bound riboflavin by extensive dialysis at pH 3.0. Add 4.5 g N, N-disuccinylimidyl carbonate to 3 g Nucleosil 5NH2 slurried in MeCN, filter, wash with MeCN, wash with 50 mM pH 7.5 phosphate buffer. Suspend 300 mg RFBP in 50 mM phosphate buffer, add the activated silica, mix gently for 2 h using a rotary evaporator, filter, wash with sterile water, wash with isopropanol:water 1:2, pack in a 100 × 4.6 column.) (Buffer was 100 mM pH 5.3 sodium acetate.)

**Mobile phase:** 50 mM pH 5.5 KH$_2$PO$_4$
**Flow rate:** 0.8
**Injection volume:** 20
**Detector:** UV

## CHROMATOGRAM
**Retention time:** k′ 2.56

## OTHER SUBSTANCES
**Simultaneous:** flurbiprofen, isradipine, nimodipine, suprofen

## KEY WORDS
chiral; α = 1.22

## REFERENCE
Massolini, G.; De Lorenzi, E.; Ponci, M.C.; Gandini, C.; Caccialanza, G.; Monaco, H.L. Egg yolk riboflavin binding protein as a new chiral stationary phase in high-performance liquid chromatography. *J.Chromatogr.A*, **1995**, *704*, 55–65

## SAMPLE
**Matrix:** solutions
**Sample preparation:** 1 mL 1.23 mg/mL ketoprofen in dichloromethane + 300 μL 1 mg/mL hydroxybenzotriazole in dichloromethane:pyridine 99:1 + 300 μL 11 mg/mL 1-ethyl-3-dimethylaminopropylcarbodiimide in dichloromethane + 300 μL 3.47 mg/mL 1-naphthylamine (Caution! 1-Naphthylamine in a carcinogen!) in dichloromethane, vortex, let stand for 1 h, evaporate to dryness under a stream of nitrogen, reconstitute with 5 mL MeOH, inject an aliquot.

## HPLC VARIABLES
**Column:** 150 × 2.1 Tolycellulose EXP B101 (tris(4-methylbenzoate)cellulose covalently bonded to 10 μm aminopropylsilica)
**Mobile phase:** MeOH:buffer 85:15 (Buffer was 14.05 g/L sodium perchlorate adjusted to pH 2.0.)
**Flow rate:** 0.21
**Injection volume:** 1
**Detector:** UV 220

## CHROMATOGRAM
**Retention time:** 5.9, 8.6 (enantiomers)
**Limit of detection:** 100 pg

## OTHER SUBSTANCES
**Also analyzed:** fenoprofen, flurbiprofen, ibuprofen, tiaprofenic acid

## KEY WORDS
derivatization; narrow-bore; chiral

## REFERENCE
Van Overbeke, A.; Baeyens, W.; Van Der Weken, G.; Van de Voorde, I.; Dewaele, C. Comparative chromatographic study on the chiral separation of the 1-naphthylamine derivative of ketoprofen on cellulose-based columns of different sizes. *Biomed.Chromatogr.*, **1995**, *9*, 289–290

## SAMPLE
**Matrix:** solutions
**Sample preparation:** Prepare a 500 μg/mL solution in isopropanol, inject a 20 μL aliquot.

## HPLC VARIABLES
**Column:** 250 × 4.6 Chiralcel OJ
**Mobile phase:** n-Hexane:isopropanol:acetic acid 80:20:0.5
**Flow rate:** 1
**Injection volume:** 20
**Detector:** UV 230

## CHROMATOGRAM
**Retention time:** 8, 10 (enantiomers)

## OTHER SUBSTANCES
**Simultaneous:** benoxaprofen, carprofen, ibuprofen, pirprofen, protizinic acid

## KEY WORDS
chiral

## REFERENCE
Van Overbeke, A.; Baeyens, W.; Dewaele, C. Comparative study on the enantiomeric separation of several non-steroidal anti-inflammatory drugs on two cellulose-based chiral stationary phases. *J.Liq.Chromatogr.*, **1995**, *18*, 2427–2443

## SAMPLE
**Matrix:** solutions

## HPLC VARIABLES
**Column:** 250 × 4.6 Zorbax RX
**Mobile phase:** Gradient. A was 10 mL concentrated orthophosphoric acid and 7 mL triethylamine in 1 L water. B was 10 mL concentrated orthophosphoric acid and 7 mL triethylamine in 200 mL water, make up to 1 L with MeCN. A:B from 100:0 to 0:100 over 30 min, maintain at 0:100 for 5 min.
**Column temperature:** 30
**Flow rate:** 2
**Detector:** UV 210

## OTHER SUBSTANCES
**Also analyzed:** acepromazine, acetaminophen, acetophenazine, albuterol, aminophylline, amitriptyline, amobarbital, amoxapine, amphetamine, amylocaine, antipyrine, aprobarbital, aspirin, atenolol, atropine, avermectin, barbital, benzocaine, benzoic acid, benzotropine, benzphetamine, berberine, bibucaine, bromazepan, brompheniramine, buprenorphine, buspirone, butabarbital, butacaine, butethal, caffeine, carbamazepine, carbromal, chloramphenicol, chlordiazepoxide, chloroquine, chlorothiazide, chloroxylenol, chlorphenesin, chlorpheniramine, chlorpromazine, chlorpropamide, chlortetracycline, cimetidine, cinchonidine, cinchonine, clenbuterol, clonazepam, clonixin, clorazepate, cocaine, codeine, colchicine, cortisone, coumarin, cyclazocine, cyclobenzaprine, cyclothiazide, cyheptamide, cymarin, danazol, danthron, dapsone, debrisoquine, desipramine, dexamethasone, dextromethorphan, dextropropoxyphene, diamorphine, diazepam, diclofenac, diethylpropion, diethylstilbestrol, diflunisal, digitoxin, digoxin, diltiazem, diphenhydramine, diphenoxylate, diprenorphine, dipyrone, disulfiram, dopamine, doxapram, doxepin, dronabinol, ephedrine, epinephrine, epinine, estradiol, estriol, estrone, ethacrynic acid, ethosuximide, etonitazene, etorphine, eugenol, famotidine, fenbendazole, fencamfamine, fenoprofen, fenproporex, fentanyl, flubendazole, flufenamic acid, flunitrazepam, 5-fluorouracil, fluoxymesterone, fluphenazine, furosemide, gentisic acid, gitoxigenin, glipizide, glunixin, glutethimide, glybenclamide, guaiacol, halazepam, haloperidol, hydrochlorothiazide, hydrocodone, hydrocortisone, hydromorphone, hydroxyquinoline, ibogaine, ibuprofen, iminostilbene, imipramine, indomethacin, isocarbostyril, isocarboxazid, isoniazid, isoproterenol, isoxsuprine, ivermectin, ketamine, kynurenic acid, levorphanol, lidocaine, loraze-

pam, lormetazepam, loxapine, mazindol, mebendazole, meclizine, meclofenamic acid, medazepam, mefenamic acid, megestrol, mepacrine, meperidine, mephentermine, mephenytoin, mephesin, mephobarbital, mepivacaine, mescaline, mesoridazine, methadone, methamphetamine, methapyrilene, methaqualone, methazolamide, methocarbamol, methoxamine, methsuximide, methyl salicylate, methyldopa, methyldopamine, methylphenidate, methylprednisolone, methyltestosterone, methyprylon, metoprolol, mibolerone, morphine, nadolol, nalorphine, naloxone, naltrexone, naphazoline, naproxen, nefopam, niacinamide, nicotine, nicotinic acid, nifedipine, niflumic acid, nitrazepam, norepinephrine, nortriptyline, noscapine, nylidrin, oxazepam, oxycodone, oxymorphone, oxyphenbutazone, oxytetracycline, papaverine, pargyline, pemoline, pentazocine, pentobarbital, persantine, phenacetin, phenazocine, phenazopyridine, phencyclidine, phendimetrazine, phenelzine, pheniramine, phenobarbital, phenothiazine, phensuximide, phentermine, phenylbutazone, phenylephrine, phenylpropanolamine, piperocaine, prazepam, prednisolone, primidone, probenecid, progesterone, propiomazine, propranolol, propylparaben, pseudoephedrine, puromycin, pyrilamine, pyrithyldione, quazepam, quinaldic acid, quinidine, quinine, ranitidine, recinnamine, reserpine, resorcinol, saccharin, albuterol, salicylamide, salicylic acid, scopolamine, scopoletin, secobarbital, strychnine, sulfacetamide, sufadiazine, sulfadimethoxine, sulfaethidole, sulfamerazine, sulfamethazine, sulfamethoxizole, sulfanilamide, sulfapyridine, sulfasoxizole, sulindac, tamoxifen, temazepam, testosterone, tetracaine, tetracycline, tetramisole, thebaine, theobromine, theophylline, thiabendazole, thiamine, thiamylal, thiobarbituric acid, thioridazine, thiosalicylic acid, thiothixene, thymol, tolazamide, tolazoline, tobutamide, tolmetin, tranylcypromine, triamcinolone, tribenzylamine, trichloromethiazide, trifluoperazine, trihexyphenidyl, trimethoprim, tripelennamine, triprolidine, tropacocaine, tyramine, verapamil, vincamine, warfarin, yohimbine, zoxazolamine

## REFERENCE

Hill, D.W.; Kind, A.J. Reversed-phase solvent gradient HPLC retention indexes of drugs. *J.Anal.Toxicol.*, **1994**, *18*, 233−242

## SAMPLE
**Matrix:** solutions

## HPLC VARIABLES
**Column:** 250 × 4.6 10 μm Chiralpak AD (Daicel)
**Mobile phase:** Carbon dioxide : MeOH 96 : 4
**Column temperature:** 30
**Flow rate:** 2.5
**Detector:** UV 210

## CHROMATOGRAM
**Retention time:** 13.5, 15 (enantiomers)

## OTHER SUBSTANCES
**Simultaneous:** fenoprofen, ibuprofen, flurbiprofen, naproxen

## KEY WORDS
SFC; 250 bar; chiral

## REFERENCE

Kot, A.; Sandra, P.; Venema, A. Sub- and supercritical fluid chromatography on packed columns: A versatile tool for the enantioselective separation of basic and acidic drugs. *J.Chromatogr.Sci.*, **1994**, *32*, 439−448

## SAMPLE
**Matrix:** solutions

**Sample preparation:** Mix 1 mL 100 μg/mL compound in dichloromethane with 300 μL 100 μg/mL 1-hydroxybenzotriazole in dichloromethane:pyridine 99:1, 300 μL 1.1 mg/mL 1-ethyl-3-dimethylaminopropylcarbodiimide hydrochloride in dichloromethane, and 300 μL 300 μg/mL benzylamine in dichloromethane, vortex, let stand at room temperature for 1.5 h, evaporate to dryness under a stream of nitrogen, reconstitute the residue in 500 μL MeOH, inject an aliquot.

## HPLC VARIABLES
**Column:** 150 × 4.6 10 μm EXP B101 tris(4-methylbenzoate) cellulose on silica (Bio-Rad)
**Mobile phase:** MeOH:buffer 70:30 (Prepare buffer solution by dissolving 14.05 g sodium perchlorate in water, adjust pH to 2.0, make up to 1 L with water.)
**Flow rate:** 1
**Detector:** UV 230

## CHROMATOGRAM
**Retention time:** k′ 4.06, k′ 5.69 (enantiomers)

## OTHER SUBSTANCES
**Also analyzed:** benoxaprofen (MeOH:buffer 80:20), carprofen, fenoprofen, flurbiprofen, ibuprofen, pirprofen, tiaprofenic acid

## KEY WORDS
derivatization; chiral

## REFERENCE
Van Overbeke, A.; Baeyens, W.; Van den Bossche, W.; Dewaele, C. Separation of 2-arylpropionic acids on a cellulose based chiral stationary phase by RP-HPLC. *J.Pharm.Biomed.Anal.*, **1994**, *12*, 901–909

◆━━━━━━━━━━━━━━━━━━◆━━━━━━━━━━━━━━━━━━◆

## ANNOTATED BIBLIOGRAPHY

Haginaka, J.; Kanasugi, N. Enantioselectivity of bovine serum albumin-bonded columns produced with isolated protein fragments. *J.Chromatogr.A*, **1995**, *694*, 71–80 [chiral; also benzoin, clorazepate, fenoprofen, flurbiprofen, ibuprofen, lorazepam, lormetazepam, oxazepam, pranoprofen, temazepam, warfarin]

Hyun, M.H.; Ryoo, J.-J.; Cho, Y.J.; Jin, J.S. Unusual examples of the liquid chromatographic resolution of racemates. Resolution of π-donor analytes on a π-donor chiral stationary phase. *J.Chromatogr.A*, **1995**, *692*, 91–96 [chiral; also alminoprofen, fenoprofen, flurbiprofen, ibuprofen, ketoprofen, naproxen]

Van Overbeke, A.; Baeyens, W.; Van den Bossche, W.; Dewaele, C. Enantiomeric separation of amide derivatives of some 2-arylpropionic acids by HPLC on a cellulose-based chiral stationary phase. *J.Pharm.Biomed.Anal.*, **1994**, *12*, 911–916 [chiral; derivatization; also flurbiprofen, ibuprofen, tiaprofenic acid]

Haginaka, J.; Murashima, T.; Fujima, H.; Wada, H. Direct injection assay of drug enantiomers in serum on ovomucoid-bonded silica materials by liquid chromatography. *J.Chromatogr.*, **1993**, *620*, 199–204 [direct injection; serum; chiral; also benzoin, chlorpheniramine, oxazepam]

Szász, G.; Budvári-Bárany, Z.; Löre, A.; Radeczky, G.; Shalaby, A. HPLC of antiphlogistic acids on silica dynamically modified with cetylpyridinium chloride. *J.Liq.Chromatogr.*, **1993**, *16*, 2335–2345 [also diclofenac, ibuprofen, ketoprofen, naproxen, nicotinic acid, nifluminic acid, salicylic acid]

Benoit, E.; Jaussaud, P.; Besse, S.; Videmann, B.; Courtot, D.; Delatour, P.; Bonnaire, Y. Identification of a benzhydrolic metabolite of ketoprofen in horses by gas chromatography-mass spectrometry and high-performance liquid chromatography. *J.Chromatogr.*, **1992**, *583*, 167–173 [plasma; SPE; gradient; extracted metabolites; LOQ 500 ng/mL; LOD 100 ng/mL; pharmacokinetics; horse]

Mannucci, C.; Bertini, J.; Cocchini, A.; Perico, A.; Salvagnini, F.; Triolo, A. High performance liquid chromatography simultaneous quantitation of ketoprofen and parabens in a commercial gel formulation. *J.Liq.Chromatogr.*, **1992**, *15*, 327–335 [gels; formulations; column temp 50; simultaneous butyl paraben, ethyl paraben, methyl paraben, propyl paraben]

Oda, Y.; Asakawa, N.; Yoshida, Y.; Sato, T. On-line determination and resolution of the enantiomers of ketoprofen in plasma using coupled achiral-chiral high-performance liquid chromatography. *J.Pharm.Biomed.Anal.*, **1992**, *10*, 81–87

Rainsford, K.D.; James, C.; Hunt, R.H.; Stetsko, P.I.; Rischke, J.A.; Karim, A.; Nicholson, P.A.; Smith, M.; Hantsbarger, G. Effects of misoprostol on the pharmacokinetics of indomethacin in human volunteers. *Clin.Pharmacol.Ther.*, **1992**, *51*, 415–421 [plasma; ketoprofen is IS; extracted indomethacin; pharmacokinetics]

Shibukawa, A.; Terakita, A.; He, J.Y.; Nakagawa, T. High-performance frontal analysis-high-performance liquid chromatographic system for stereoselective determination of unbound ketoprofen enantiomers in plasma after direct sample injection. *J.Pharm.Sci.*, **1992**, *81*, 710–715 [plasma; column-switching; column temp 15; chiral; LOD 1 nM; human; rat]

Wong, C.-Y.; Yeh, M.-K.; Wang, D.-P. High-performance liquid chromatographic determination of ketoprofen in pharmaceutical dosage forms and plasma. *J.Liq.Chromatogr.*, **1992**, *15*, 1215–1225 [plasma; formulations; isopropylphenazone (IS); rabbit; human; pharmacokinetics]

Corvetta, A.; Della Bitta, R.; Luchetti, M.M.; Pomponio, G.; Ciuffoletti, V. Tenoxicam and ketoprofen level monitoring with high performance liquid chromatography in patients affected by rheumatoid arthritis. *Clin.Exp.Rheumatol.*, **1991**, *9*, 143–148

Goto, S.; Kawata, M.; Suzuki, T.; Kim, N.-S.; Ito, C. Preparation and evaluation of Eudragit gels. I. Eudragit organogels containing drugs as rectal sustained-release preparations. *J.Pharm.Sci.*, **1991**, *80*, 958–961 [rabbit; plasma; ibuprofen (IS); pharmacokinetics]

Wanwimolruk, S.; Wanwimolruk, S.Z.; Zoest, A.R. Sensitive HPLC assay for ketoprofen in human plasma and its application to pharmacokinetic study. *J.Liq.Chromatogr.*, **1991**, *14*, 3685–3694 [plasma; pharmacokinetics; piroxicam (IS); microbore; LOD 50 ng/mL]

Schmitt, M.; Guentert, T.W. Biopharmaceutical evaluation of ketoprofen following intravenous, oral, and rectal administration in dogs. *J.Pharm.Sci.*, **1990**, *79*, 614–616 [plasma; dog; naproxen (IS); pharmacokinetics]

Chi, S.-C.; Jun, H.W. Quantitation of ketoprofen in isopropyl myristate by high performance liquid chromatography. *J.Liq.Chromatogr.*, **1989**, *12*, 2931–2945 [naproxen (IS)]

Lempiainen, M.; Makela, A.L. Determination of ketoprofen by high-performance liquid chromatography from serum and urine: clinical application in children with juvenile rheumatoid arthritis. *Int.J.Clin.Pharmacol.Res.*, **1987**, *7*, 265–271

Pietta, P.; Manera, E.; Ceva, P. Purity assay of ketoprofen by high-performance liquid chromatography. *J.Chromatogr.*, **1987**, *387*, 525–527

Hermansson, J.; Erikson, M. Direct liquid chromatographic resolution of acidic drugs using a chiral α1-acid glycoprotein column (Enantiopac). *J.Liq.Chromatogr.*, **1986**, *9*, 621–639 [chiral; also bendroflumethiazide, disopyramide, ethotoin, hexobarbital, ibuprofen, naproxen, 2-phenoxypropionic acid, RAC 109]

Royer, R.J.; Lapicque, F.; Netter, P.; Monot, C.; Bannwarth, B.; Cure, M.C. Estimation by high-performance liquid chromatography of ketoprofen in plasma. Application to the study of its protein binding. *Biomed.Pharmacother.*, **1986**, *40*, 100–105

Sallustio, B.C.; Abas, A.; Hayball, P.J.; Purdie, Y.J.; Meffin, P.J. Enantiospecific high-performance liquid chromatographic analysis of 2-phenylpropionic acid, ketoprofen and fenoprofen. *J.Chromatogr.*, **1986**, *374*, 329–337

Oka, K.; Aoshima, S.; Noguchi, M. Highly sensitive determination of ketoprofen in human serum and urine and its application to pharmacokinetic study. *J.Chromatogr.*, **1985**, *345*, 419–424

Kaye, C.M.; Sankey, M.G.; Holt, J.E. A high-pressure liquid chromatographic methods for the assay of ketoprofen in plasma and urine, and its application to determining the urinary excretion of free and conjugated ketoprofen following oral administrations of Orudis to man. *Br.J.Clin.Pharmacol.*, **1981**, *11*, 395–398

Bannier, A.; Brazier, J.L.; Ribon, B.; Quincy, C. Determination of ketoprofen in biological fluids by reversed-phase chromatography. *J.Pharm.Sci.*, **1980**, *69*, 763–765

# Ketorolac

**Molecular formula:** $C_{15}H_{13}NO_3$
**Molecular weight:** 255.3
**CAS Registry No.:** 74103-06-3, 74103-07-4
(tromethamine)

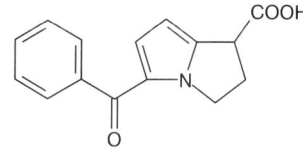

## SAMPLE
**Matrix:** blood
**Sample preparation:** Condition a 10 × 3 C18 SPE cartridge (Analytichem) with 2 mL MeOH, 2 mL water, and 2 mL 50 mM pH 3.5 sodium acetate at 2 mL/min. 550 μL Plasma + 550 μL 0.9% NaCl, vortex vigorously, add 25 μL 10 μg/mL ketoprofen in MeOH:water 10:90, add 1 mL to the SPE cartridge at 1 mL/min, wash with 1 mL 50 mM pH 3.5 sodium acetate at 1 mL/min, wash with 1.5 mL MeOH:0.1% acetic acid 20:80 at 1.5 mL/min, elute the contents of the cartridge onto the column with mobile phase.

## HPLC VARIABLES
**Guard column:** 15 × 3.2 7 μm Newguard RP-18
**Column:** 100 × 8 4 μm Nova-pak C18 radial pak
**Mobile phase:** Gradient. MeCN:0.1% acetic acid from 30:70 to 60:40 over 10 min, maintain at 60:40 for 2 min, to 100:0 over 3 min.
**Flow rate:** 2
**Detector:** UV 313 for 7.2 min then UV 258

## CHROMATOGRAM
**Retention time:** 6.7
**Internal standard:** ketoprofen (8.8)
**Limit of quantitation:** 5 ng/mL

## KEY WORDS
plasma; SPE

## REFERENCE
Solà, J.; Pruñonosa, J.; Colom, H.; Peraire, C.; Obach, R. Determination of ketorolac in human plasma by high-performance liquid chromatography after automated on-line solid-phase extraction. *J.Liq.Chrom.Rel.Technol.*, **1996**, *19*, 89–99

## SAMPLE
**Matrix:** blood
**Sample preparation:** Wash Amberlite XAD-2 polymeric adsorbent resin with water, acetone, and ethyl acetate and store it in ethyl acetate until use. 1 mL Blood + 100 μL 100 μg/mL cyclopentobarbital + 5 mL water + 1 g resin, vortex for 30 s or until homogeneous, let stand for 1 min, discard the supernatant, add 6 mL ethyl acetate, vortex for 1 min, centrifuge at 2000 rpm for 5 min. Remove the organic supernatant and evaporate it to dryness under a stream of air at 60°, reconstitute the residue in 100 μL MeCN, add 500 μL heptane, vortex for 5 s, centrifuge at 2000 rpm for 5 min, inject a 20 μL aliquot of the MeCN layer.

## HPLC VARIABLES
**Column:** 250 × 4.6 Lichrospher RP-8
**Mobile phase:** MeCN:buffer 36:64 (Buffer was 6.8 g $KH_2PO_4$ and 1 mL phosphoric acid in 1 L water, pH 3.)
**Flow rate:** 1.5

**Injection volume:** 20
**Detector:** UV 312

## CHROMATOGRAM
**Retention time:** 7.0
**Internal standard:** cyclopentobarbital (UV 190) (4.3)

## KEY WORDS
SPE

## REFERENCE
Logan, B.K.; Friel, P.N.; Peterson, K.L.; Predmore, D.B. Analysis of ketorolac in postmortem blood. *J.Anal.Toxicol.*, **1995**, *19*, 61−64

## SAMPLE
**Matrix:** blood
**Sample preparation:** 1 mL Plasma + 100 μL 5% zinc sulfate in water, vortex for 2 min, add 440 μL buffer, vortex for 1 min, centrifuge at 2000 g for 10 min, inject a 100 μL aliquot of the supernatant. (Buffer was 100 mM $NaH_2PO_4$ and 10 mM sodium lauryl sulfate, pH adjusted to 2.8 with phosphoric acid.)

## HPLC VARIABLES
**Column:** 250 × 4.6 5 μm Nucleosil C18
**Mobile phase:** MeCN:water 35:65 containing 10 mM $NaH_2PO_4$ and 1 mM sodium lauryl sulfate, pH adjusted to 2.8 with phosphoric acid
**Flow rate:** 1.5
**Injection volume:** 100
**Detector:** UV 355

## CHROMATOGRAM
**Retention time:** 10.3
**Internal standard:** ketorolac

## OTHER SUBSTANCES
**Extracted:** tenoxicam

## KEY WORDS
plasma; protect from light; ketorolac is IS

## REFERENCE
Mason, J.L.; Hobbs, G.J. Simple method for the analysis of tenoxicam in human plasma using high-performance liquid chromatography. *J.Chromatogr.B*, **1995**, *665*, 410−415

## SAMPLE
**Matrix:** blood
**Sample preparation:** 500 (Human) or 100 (rat) μL plasma + 50 μL 100 μg/mL naproxen in MeOH + 200 μL 600 mM sulfuric acid + 3 mL diethyl ether, vortex for 30 s, centrifuge at 2500 rpm for 5 min. Remove the organic layer and evaporate it to dryness under reduced pressure, reconstitute the residue in 200 μL mobile phase, inject a 10-120 μL aliquot.

## HPLC VARIABLES
**Guard column:** 50 mm long 5 μm silica (Phenomenex)
**Column:** 125 mm long Partisil 5 ODS 3 + 50 mm long chiral tert-leucine (Phenomenex)

**Mobile phase:** MeOH:ethyl acetate:isopropanol 50:50:2 containing 0.5 mM ammonium
  acetate
**Flow rate:** 0.8
**Injection volume:** 10-120
**Detector:** UV 313

## CHROMATOGRAM
**Retention time:** 12 (S), 13 (R)
**Internal standard:** naproxen (21)
**Limit of detection:** <10 ng/mL

## KEY WORDS
human; rat; plasma; pharmacokinetics; chiral; racemization does not occur in contrast to
  previous derivatization method (F. Jamali et al.; J.Liq.Chromatogr. 1989; 12; 1835)

## REFERENCE
Vakily, M.; Corrigan, B.; Jamali, F. The problem of racemization in the stereospecific assay and pharmacokinetic evaluation of ketorolac in human and rats. *Pharm.Res.*, **1995**, *12*, 1652–1657

## SAMPLE
**Matrix:** blood
**Sample preparation:** 1 mL Plasma + 100 ng sodium tolmetin + 200 μL 100 mM pH 4
  sodium acetate, extract twice with 5 mL diethyl ether. Combine the organic layers and
  evaporate them to dryness under a stream of nitrogen at 50°, reconstitute the residue in
  200 μL water, inject an 80 μL aliquot.

## HPLC VARIABLES
**Column:** 150 × 3.9 4 μm Novapak C18
**Mobile phase:** MeCN:1 mM pH 3 phosphoric acid 32:68
**Flow rate:** 1
**Injection volume:** 80
**Detector:** UV 313

## CHROMATOGRAM
**Retention time:** 7
**Internal standard:** tolmetin (11)

## KEY WORDS
plasma; pharmacokinetics

## REFERENCE
Flores-Murrieta, F.J.; Granados-Soto, V.; Castañeda-Hernández, G.; Herrera, J.E.; Hong, E. Comparative
  bioavailability of two oral formulations of ketorolac tromethamine: Dolac and Exodol. *Bio-*
  *pharm.Drug Dispos.*, **1994**, *15*, 129–136

## SAMPLE
**Matrix:** blood
**Sample preparation:** 500 μL Plasma + 50 μL 4 M phosphoric acid + 50 μL 200 μg/mL
  sodium S-(+)-naproxen in water + 400 μL hexane:pentan-2-ol 90:10, mix 4 times at 1000
  rpm for 30 s, centrifuge at 5000 rpm for 5 min, freeze in solid carbon dioxide for 5 min.
  Remove upper organic phase and add it to 150 μL 20 mM NaOH. Mix 4 times at 1200
  rpm for 30 s, centrifuge at 3000 rpm for 6 min, freeze in solid carbon dioxide for 5 min,
  discard upper organic phase, blot tube openings on clean filter paper. Allow to warm for
  5 min to above 0°, mix at 1200 rpm for 30 s, centrifuge at 5000 rpm for 5 min, inject a
  5-100 μL aliquot.

## HPLC VARIABLES
**Guard column:** 10 × 3 ChromTech diol guard column
**Column:** 100 × 4 ChromTech Chiral AGP-CSP
**Mobile phase:** Isopropanol:100 mM pH 5.5 $NaH_2PO_4$ (Use a 4 × 6 Waters Guard-Pak C18 between pump and injector.)
**Flow rate:** 0.9
**Injection volume:** 5-100
**Detector:** UV 325

## CHROMATOGRAM
**Retention time:** 3.3 (R), 4.8 (S)
**Internal standard:** sodium S-(+)-naproxen (6.4)
**Limit of detection:** 35 ng/mL

## OTHER SUBSTANCES
**Simultaneous:** metabolites

## KEY WORDS
plasma; recycle mobile phase; sheep

## REFERENCE
Mills, M.H.; Mather, L.E.; Gu, X.S.; Huang, J.L. Determination of ketorolac enantiomers in plasma using enantioselective liquid chromatography on an $\alpha_1$-acid glycoprotein chiral stationary phase and ultraviolet detection. *J.Chromatogr.B*, **1994**, *658*, 177−182

## SAMPLE
**Matrix:** blood
**Sample preparation:** 1 mL Serum + 100 μL 5% (w/v) zinc sulfate in water, vortex for 2 min, make up to 4 mL with MeOH, vortex for 2 min, centrifuge at 2000 g for 5 min, inject 100 μL of supernatant.

## HPLC VARIABLES
**Guard column:** 10 μm C18 Waters guard column
**Column:** 300 × 3.9 10 μm μBondapak C18
**Mobile phase:** MeCN:water 40:60 adjusted to pH 2.8 ± 0.1 with 85% orthophosphoric acid
**Flow rate:** 1.4
**Injection volume:** 100
**Detector:** UV 313

## CHROMATOGRAM
**Retention time:** 6
**Limit of detection:** 10 ng/mL

## KEY WORDS
serum

## REFERENCE
Chaudhary, R.S.; Gangwal, S.S.; Jindal, K.C.; Khanna, S. Reversed-phase high-performance liquid chromatography of ketorolac and its application to bioequivalence studies in human serum. *J.Chromatogr.*, **1993**, *614*, 180−184

## SAMPLE
**Matrix:** blood
**Sample preparation:** 1 mL Plasma + 50 μL 15 μg/mL (S)-ketoprofen in MeOH:water 1:4 + 75 μL 2 M sulfuric acid + 100 μL MeOH:water 1:4 + 8 mL hexane:ethyl acetate

80:20, rotary mix for 10 min, centrifuge at 2000 g for 10 min. Remove the organic layer and evaporate it to dryness at 55° under a stream of nitrogen. Reconstitute in 100 μL 1.5% thionyl chloride in n-hexane (freshly prepared),, vortex, heat at 80° for 30 min, cool to room temperature, add 500 μL reagent, vortex, let stand at room temperature for 10 min, evaporate to dryness under a stream of nitrogen, add 1 mL 2 M sulfuric acid, add 5 mL ethyl acetate, mix, centrifuge. Remove the organic layer and evaporate it to dryness at 55° under a stream of nitrogen. Reconstitute in 125 μL mobile phase, inject a 100 μL aliquot. (Reagent was 3% (S)-1-phenylethylamine in dry dichloromethane prepared within 30 min of use.)

## HPLC VARIABLES
**Column:** 100 × 8 4 μm Nova-Pak phenyl radially compressed bonded phase cartridge
**Mobile phase:** MeCN:20 mM sodium acetate buffer 50:50 containing 0.1% triethylamine, final pH 5.5
**Flow rate:** 2
**Injection volume:** 100
**Detector:** UV 310 for 8 min (derivatized ketorolac) then UV 254 (derivatized ketoprofen)

## CHROMATOGRAM
**Retention time:** 6.5 (S), 7.2 (R)
**Internal standard:** (S)-ketoprofen (8.6)
**Limit of quantitation:** 50 ng/mL

## KEY WORDS
plasma; derivatization; chiral

## REFERENCE
Hayball, P.J.; Tamblyn, J.G.; Holden, Y.; Wrobel, J. Stereoselective analysis of ketorolac in human plasma by high-performance liquid chromatography. *Chirality*, **1993**, *5*, 31–35

## SAMPLE
**Matrix:** blood
**Sample preparation:** 1 mL Plasma + 1 mL water:MeOH 9:1 + 500 μL 300 ng/mL m-hydroxyketorolac in water:MeOH 9:1 + 100 μL 0.5 M pH 3 sodium acetate + 5 mL diethyl ether, shake for 5 min, centrifuge at 3500 g for 5 min. Remove organic layer and add it to 3 mL hexane and 2 mL 100 mM NaOH. Shake for 5 min, centrifuge for 5 min, discard organic layer. Add 500 μL 2 M HCl to aqueous layer, add 8 mL diethyl ether, shake for 5 min, centrifuge at 3500 g for 5 min. Remove organic layer and evaporate it to dryness under a stream of nitrogen at 45°. Reconstitute residue in 100 μL mobile phase, vortex 15 s, inject a 30 μL aliquot.

## HPLC VARIABLES
**Guard column:** 70 × 2.1 30-38 μm Whatman HC Pellosil C18
**Column:** 250 × 4.6 5 μm Regis Spherisorb ODS
**Mobile phase:** MeCN:MeOH:20 mM pH 6 phosphate buffer containing 10 mM tetrabutyl ammonium phosphate 15:20:65 (Buffer was 88.9 parts 20 mM $KH_2PO_4$ + 11.1 parts 20 mM $Na_2HPO_4$.)
**Flow rate:** 0.8
**Injection volume:** 30
**Detector:** UV 313

## CHROMATOGRAM
**Retention time:** 7.2
**Internal standard:** m-hydroxyketorolac (8.3)
**Limit of quantitation:** 50 ng/mL

## OTHER SUBSTANCES
**Simultaneous:** metabolites

## KEY WORDS
plasma

## REFERENCE
Wu, A.T.; Massey, I.J. Simultaneous determination of ketorolac and its hydroxylated metabolite in plasma by high-performance liquid chromatography. *J.Chromatogr.*, **1990**, *534*, 241–246

## SAMPLE
**Matrix:** formulations
**Sample preparation:** Add 5 mL water to 200 mg powder, sonicate for 5 min, add 15 mL MeOH, sonicate for 10 min, centrifuge an aliquot at 2500 rpm for 10 min, dilute the supernatant with mobile phase, inject an aliquot.

## HPLC VARIABLES
**Column:** 250 × 4.6 Spherisorb ODS 1
**Mobile phase:** MeOH:water:acetic acid 59:40:1
**Flow rate:** 1.2
**Injection volume:** 25
**Detector:** UV 254

## CHROMATOGRAM
**Retention time:** 7

## OTHER SUBSTANCES
**Simultaneous:** degradation products

## KEY WORDS
powder

## REFERENCE
Brandl, M.; Magill, A.; Rudraraju, V.; Gordon, M.S. Approaches for improving the stability of ketorolac in powder blends. *J.Pharm.Sci.*, **1995**, *84*, 1151–1153

## SAMPLE
**Matrix:** formulations
**Sample preparation:** Add naproxen and dilute to a final concentration of about 20 µg/mL ketorolac tromethamine and 20 µg/mL naproxen.

## HPLC VARIABLES
**Column:** 250 × 4.6 5 µm Spherisorb ODS I
**Mobile phase:** MeOH:water:acetic acid 55:44:1
**Flow rate:** 1
**Injection volume:** 20
**Detector:** UV 254

## CHROMATOGRAM
**Retention time:** 10.37 (ketorolac tromethamine)
**Internal standard:** naproxen (14.93)

## OTHER SUBSTANCES
**Simultaneous:** degradation products

## KEY WORDS
infusion solutions; injections; stability-indicating

## REFERENCE
Floy, B.J.; Royko, C.G.; Fleitman, J.S. Compatibility of ketorolac tromethamine injection with common infusion fluids and administration sets. *Am.J.Hosp.Pharm.*, **1990**, *47*, 1097–1100

## SAMPLE
**Matrix:** perfusate

## HPLC VARIABLES
**Column:** 100 × 4 5 μm Chiral-AGP (Chrom-Tech)
**Mobile phase:** 60 mM pH 7.0 phosphate buffer
**Flow rate:** 1.3
**Detector:** UV (wavelength not specified)

## CHROMATOGRAM
**Retention time:** 2 (R), 3.2 (S)

## KEY WORDS
chiral

## REFERENCE
Roy, S.D.; Chatterjee, D.J.; Manoukian, E.; Divor, A. Permeability of pure enantiomers of ketorolac through human cadaver skin. *J.Pharm.Sci.*, **1995**, *84*, 987–990

## SAMPLE
**Matrix:** solutions
**Sample preparation:** Inject a 2 μL aliquot.

## HPLC VARIABLES
**Column:** 100 × 4.6 5 μm Chiral AGP (ChromTech)
**Mobile phase:** 40 mM pH 7.0 Potassium phosphate buffer
**Flow rate:** 0.6
**Injection volume:** 2
**Detector:** UV 323

## KEY WORDS
chiral

## REFERENCE
Brandl, M.; Conley, D.; Johnson, D. Racemization of ketorolac in aqueous solution. *J.Pharm.Sci.*, **1995**, *84*, 1045–1048

## SAMPLE
**Matrix:** solutions
**Sample preparation:** Acidify 5 mL solution with concentrated HCl, extract with two 5 mL portions of dichloromethane. Combine the organic layers and evaporate them to dryness, reconstitute the residue in 1 mL d-2-octanol:toluene:sulfuric acid 2:100:0.1, heat at 40°for 19 h, neutralize with 1 mL 20 mM sodium bicarbonate. Remove the organic layer and dry it over anhydrous sodium sulfate. Remove a 200 μL aliquot and evaporate it to dryness, reconstitute with 10 mL mobile phase, inject an aliquot.

## HPLC VARIABLES
**Column:** 250 × 4.6 5 μm Ultrasphere-Si

**Mobile phase:** Hexane:ethyl acetate 96:42
**Detector:** UV 325

## KEY WORDS
chiral; derivatization; normal phase

## REFERENCE
Brandl, M.; Conley, D.; Johnson, D. Racemization of ketorolac in aqueous solution. *J.Pharm.Sci.*, **1995**, *84*, 1045–1048

## SAMPLE
**Matrix:** solutions

## HPLC VARIABLES
**Column:** 250 × 4.6 Ultrasphere C8
**Mobile phase:** MeCN:water:acetic acid 45:55:0.2
**Flow rate:** 1
**Detector:** UV 314

## REFERENCE
Brandl, M.; Conley, D.; Johnson, D. Racemization of ketorolac in aqueous solution. *J.Pharm.Sci.*, **1995**, *84*, 1045–1048

## SAMPLE
**Matrix:** solutions

## HPLC VARIABLES
**Column:** 250 × 4.6 5 μm Supelcosil LC-DP (A) or 250 × 4 5 μm LiChrospher 100 RP-8 (B)
**Mobile phase:** MeCN:0.025% phosphoric acid:buffer 25:10:5 (A) or 60:25:15 (B) (Buffer was 9 mL concentrated phosphoric acid and 10 mL triethylamine in 900 mL water, adjust pH to 3.4 with dilute phosphoric acid, make up to 1 L.)
**Flow rate:** 0.6
**Injection volume:** 25
**Detector:** UV 229

## CHROMATOGRAM
**Retention time:** 6.32 (A), 5.55 (B)

## OTHER SUBSTANCES
**Also analyzed:** acebutolol, acepromazine, acetaminophen, acetazolamide, acetophenazine, albuterol, alprazolam, amitriptyline, amobarbital, amoxapine, antipyrine, atenolol, atropine, azatadine, baclofen, benzocaine, bromocriptine, brompheniramine, brotizolam, bupivacaine, buspirone, butabarbital, butalbital, caffeine, carbamazepine, cetirizine, chlorcyclizine, chlordiazepoxide, chlormezanone, chloroquine, chlorpheniramine, chlorpromazine, chlorpropamide, chlorprothixene, chlorthalidone, chlorzoxazone, cimetidine, cisapride, clomipramine, clonazepam, clonidine, clozapine, cocaine, codeine, colchicine, cyclizine, cyclobenzaprine, dantrolene, desipramine, diazepam, diclofenac, diflunisal, diltiazem, diphenhydramine, diphenidol, diphenoxylate, dipyridamole, disopyramide, dobutamine, doxapram, doxepin, droperidol, encainide, ethidium bromide, ethopropazine, fenoprofen, fentanyl, flavoxate, fluoxetine, fluphenazine, flurazepam, flurbiprofen, fluvoxamine, furosemide, glutethimide, glyburide, guaifenesin, haloperidol, homatropine, hydralazine, hydrochlorothiazide, hydrocodone, hydromorphone, hydroxychloroquine, hydroxyzine, ibuprofen, imipramine, indomethacin, ketoconazole, ketoprofen, labetalol, levorphanol, lidocaine, loratadine, lorazepam, lovastatin, loxapine, mazindol, mefenamic acid, meperidine, mephenytoin, mepivacaine, mesoridazine, metaproterenol, methadone, methdilazine, methocarbamol, methotrexate, methotrimeprazine, methoxamine, methyl-

dopa, methylphenidate, metoclopramide, metolazone, metoprolol, metronidazole, mida-zolam, moclobemide, morphine, nadolol, nalbuphine, naloxone, naphazoline, naproxen, nifedipine, nizatidine, norepinephrine, nortriptyline, oxazepam, oxycodone, oxymetazo-line, paroxetine, pemoline, pentazocine, pentobarbital, pentoxifylline, perphenazine, pheniramine, phenobarbital, phenol, phenolphthalein, phentolamine, phenylbutazone, phenyltoloxamine, phenytoin, pimozide, pindolol, piroxicam, pramoxine, prazepam, pra-zosin, probenecid, procainamide, procaine, prochlorperazine, procyclidine, promazine, pro-methazine, propafenone, propantheline, propiomazine, propofol, propranolol, protripty-line, quazepam, quinidine, quinine, racemethorphan, ranitidine, remoxipride, risperidone, salicylic acid, scopolamine, secobarbital, sertraline, sotalol, spironolactone, sulfinpyra-zone, sulindac, temazepam, terbutaline, terfenadine, tetracaine, theophylline, thiethyl-perazine, thiopental, thioridazine, thiothixene, timolol, tocainide, tolbutamide, tolmetin, trazodone, triamterene, triazolam, trifluoperazine, triflupromazine, trimeprazine, tri-methoprim, trimipramine, verapamil, warfarin, xylometazoline, yohimbine, zopiclone

## KEY WORDS
some details of plasma extraction

## REFERENCE
Koves, E.M. Use of high-performance liquid chromatography-diode array detection in forensic toxicology. *J.Chromatogr.A*, **1995**, *692*, 103–119

## SAMPLE
**Matrix:** solutions

## HPLC VARIABLES
**Column:** 250 × 4.5 5 μm Spherisorb ODS I
**Mobile phase:** MeCN:water:acetic acid 44:58:1, pH 3.0
**Flow rate:** 1
**Detector:** UV 314

## KEY WORDS
skin permeation; pharmacokinetics

## REFERENCE
Roy, S.D.; Manoukian, E.; Combs, D. Absorption of transdermally delivered ketorolac in humans. *J.Pharm.Sci.*, **1995**, *84*, 49–52

## SAMPLE
**Matrix:** solutions

## HPLC VARIABLES
**Column:** 250 × 4.5 5 μm Spherisorb ODS I
**Mobile phase:** MeCN:water:acetic acid 44:58:1, pH 3.0
**Flow rate:** 1
**Detector:** UV 314

## REFERENCE
Roy, S.D.; Manoukian, E. Transdermal delivery of ketorolac tromethamine: Permeation enhancement, device design, and pharmacokinetics in healthy humans. *J.Pharm.Sci.*, **1995**, *84*, 1190–1196

◆─────────────────◆─────────────────────────────────◆

## ANNOTATED BIBLIOGRAPHY
Flores-Murrieta, F.J.; Granados-Soto, V.; Hong, E. Determination of ketorolac in blood and plasma samples by high-performance liquid chromatography. *Boll.Chim.Farm.*, **1994**, *133*, 588–591

Hayball, P.J.; Holman, J.W.; Nation, R.L. Influence of octanoic acid on the reversible protein binding of ketorolac enantiomers to human serum albumin (HSA): comparative liquid chromatographic studies using a HSA chiral stationary phase. *J.Chromatogr.B*, **1994**, *662*, 128–133 [chiral]

Hayball, P.J.; Holman, J.W.; Nation, R.L.; Massy-Westropp, R.A.; Hamon, D.P. Marked enantioselective protein binding in humans of ketorolac in vitro: elucidation of enantiomer unbound fractions following facile synthesis and direct chiral HPLC resolution of tritium-labelled ketorolac. *Chirality*, **1994**, *6*, 642–648

Jones, D.J.; Bjorksten, A.R. Detection of ketorolac enantiomers in human plasma using enantioselective liquid chromatography. *J.Chromatogr.B*, **1994**, *661*, 165–167 [plasma; chiral; LOD 5 ng/mL; naproxen (IS)]

Liu, H.; Wehmeyer, K.R. Supercritical fluid extraction as a sample preparation technique for the direct isolation of drugs from plasma prior to analysis. *J.Chromatogr.B*, **1994**, *657*, 206–213 [SFE; extracted flavone; dog; plasma; p-fluoroketorolac (IS); LOD <25 ng/mL]

Roy, S.D.; Manoukian, E. Permeability of ketorolac acid and its ester analogs (prodrug) through human cadaver skin. *J.Pharm.Sci.*, **1994**, *83*, 1548–1553

Jamali, F.; Pasutto, F.M.; Lemko, C. HPLC of ketorolac enantiomers and application to pharmacokinetics in the rat. *J.Liq.Chromatogr.*, **1989**, *12*, 1835–1850 [chiral; pharmacokinetics; rat; plasma; ketoprofen (IS); racemization occurs-see M. Vakily et al., Pharm. Res. 1995, 12, 1652]

# Leuprolide

**Molecular formula:** $C_{59}H_{84}N_{16}O_{12}$
**Molecular weight:** 1209.4
**CAS Registry No.:** 53714-56-0 (leuprolide), 74381-53-6 (leuprolide acetate)

## SAMPLE
**Matrix:** formulations
**Sample preparation:** Mix sample with 2 mg/mL ethyl p-hydroxybenzoate in MeOH so as to give 100 µg/mL leuprolide acetate and 150 µg/mL ethyl p-hydroxybenzoate in saline, inject a 20 µL aliquot.

## HPLC VARIABLES
**Column:** 150 × 4-4.5 5 µm octadecylsilane (IBM, Bio-Rad, or Nucleosil)
**Mobile phase:** MeCN:buffer 23:77 Saturate mobile phase with silica by slurrying with Alltech Adsorbosil then filtering (0.4 µm). (Buffer was 87 mM $(NH_4)H_2PO_4$ adjusted to pH 6.5 with ammonium hydroxide.)
**Flow rate:** 2
**Injection volume:** 20
**Detector:** UV 220

## CHROMATOGRAM
**Retention time:** 19 (leuprolide acetate)
**Internal standard:** ethyl p-hydroxybenzoate (7)

## OTHER SUBSTANCES
**Simultaneous:** degradation products

## KEY WORDS
injections; stability-indicating

## REFERENCE
Sutherland, J.W.; Menon, G.N. HPLC of leuprolide acetate in injectable solutions. *J.Liq.Chromatogr.*, **1987**, *10*, 2281–2289

## SAMPLE
**Matrix:** solutions

## HPLC VARIABLES
**Column:** 200 × 3 Spherisorb S5ODS-2

**Mobile phase:** Gradient. A was 0.05% phosphoric acid containing 0.5% $(NH_4)_2SO_4$. B was MeCN. A:B from 82:18 to 64:36 over 25 min, maintain at 64:36 for 2.5 min, return to initial conditions over 1 min, re-equilibrate for 6.5 min. Alternatively, isocratic MeCN: 0.05% phosphoric acid containing 0.5% $(NH_4)_2SO_4$ 24:76.
**Flow rate:** 0.5
**Detector:** UV 210

## CHROMATOGRAM
**Retention time:** 20 (gradient), 13 (isocratic)

## OTHER SUBSTANCES
**Simultaneous:** buserelin, deslorelin, gonadorelin, goserelin, nafarelin

## KEY WORDS
comparison with capillary electrophoresis

## REFERENCE
Corran, P.H.; Sutcliffe, N. Identification of gonadorelin (LHRH) derivatives: comparison of reversed-phase high-performance liquid chromatography and micellar electrokinetic chromatography. *J.Chromatogr.*, **1993**, *636*, 87–94

# Levothyroxine

**Molecular formula:** $C_{15}H_{11}I_4NO_4$
**Molecular weight:** 775.9
**CAS Registry No.:** 55-03-8 (levothyroxine sodium),
25416-65-3 (levothyroxine sodium hydrate),
51-48-9 (levothyroxine)

## SAMPLE
**Matrix:** blood
**Sample preparation:** Equilibrate a Sep-Pak silica SPE cartridge with 5 mL ethyl acetate. 1 mL Serum + 3 mL 5% trichloroacetic acid + 4 mL ethyl acetate, vortex vigorously, centrifuge at 1500 g for 5 min. Remove organic layer and repeat extraction twice with 3 mL portions of ethyl acetate. Combine extracts, evaporate to about 1.5 mL, add to the SPE cartridge. Wash with 8 mL ethyl acetate, elute with 4 mL MeOH:ammonium hydroxide (90:10). Evaporate the eluent to dryness under nitrogen, reconstitute in 100 μL MeOH, inject.

## HPLC VARIABLES
**Column:** 250 × 4.6 5 μm Ultrasphere I.P.
**Mobile phase:** MeCN:buffer 35:65 (Buffer was 13.6 g sodium acetate, 1.0 g cupric sulfate pentahydrate, 0.92 g L-proline, and 0.34 g silver nitrate.)
**Flow rate:** 1.5
**Injection volume:** 100
**Detector:** E, Bioanalytical Systems Inc. TL-5 Kel-F glassy carbon thin-layer cell, LC-4 electronic controller, +0.78 V, 2-5 nA/V

## CHROMATOGRAM
**Retention time:** 8
**Limit of detection:** 3 ng/mL

## OTHER SUBSTANCES
**Extracted:** dextrothyroxine, triiodothyronine

## KEY WORDS
serum; SPE

## REFERENCE
Hay, I.D.; Annesley, T.M.; Jiang, N.S.; Gorman, C.A. Simultaneous determination of D- and L-thyroxine in human serum by liquid chromatography with electrochemical detection. *J.Chromatogr.*, **1981**, *226*, 383–390

## SAMPLE
**Matrix:** formulations
**Sample preparation:** Grind tablets containing about 1 mg levothyroxine, add 4.5 mL 0.5 mg/mL hydroxyprogesterone caproate in MeOH, add 20.5 mL 10 mM NaOH in MeOH:water 75:25, shake intermittently for 5 min, filter, discard first 5 mL filtrate, inject a 25 μL aliquot.

## HPLC VARIABLES
**Column:** 300 × 3.9 μBondapak CN

**Mobile phase:** MeCN:0.1% phosphoric acid in water 35:65
**Flow rate:** 3
**Injection volume:** 25
**Detector:** UV 225

## CHROMATOGRAM
**Retention time:** 5
**Internal standard:** hydroxyprogesterone caproate (8)

## KEY WORDS
tablets

## REFERENCE
Das Gupta, V.; Odom, C.; Bethea, C.; Plattenburg, J. Effect of excipients on the stability of levothyroxine sodium tablets. *J.Clin.Pharm.Ther.*, **1990**, *15*, 331–336

## SAMPLE
**Matrix:** formulations
**Sample preparation:** Weigh out powder equivalent to about 65 mg thyroid, add 5 mL enzyme solution, mix well, incubate at 37° for 28 h, agitate after 4-8 h and after 20-24 h, add 2 mL deactivating solution, mix well, centrifuge at 2000 rpm for 5-10 min, if necessary filter (0.45 µm). (The enzyme solution was about 150 protease units/mL of bacterial protease from Streptomyces griseus in 110 mM NaCL + 40 mM Tris buffer + 50 mM methimazole (pH adjusted to 8.4 ± 0.05 with 6 M HCl) reducing buffer. Deactivating solution was 1:100 phosphoric acid:MeCN.)

## HPLC VARIABLES
**Column:** 300 × 4 µBondapak C18
**Mobile phase:** MeCN:0.5% phosphoric acid in water 28:72
**Column temperature:** 34
**Flow rate:** 1.5
**Injection volume:** 200
**Detector:** UV 225

## CHROMATOGRAM
**Retention time:** 22

## OTHER SUBSTANCES
**Simultaneous:** liothyronine, L-3,3′,5′-triiodothyronine

## KEY WORDS
tablets; powders

## REFERENCE
Richheimer, S.L.; Jensen, C.B. Determination of liothyronine and levothyroxine in thyroid preparations by liquid chromatography. *J.Pharm.Sci.*, **1986**, *75*, 215–217

## SAMPLE
**Matrix:** formulations
**Sample preparation:** Powder tablets, weigh out amount equivalent to about 200 µg sodium levothyroxine, add 10 mL mobile phase, sonicate for 5 min, centrifuge. Filter (0.45 µm, 25 mm Acrodisc CR, Gelman) the supernatant, inject a 200 µL aliquot.

## HPLC VARIABLES
**Guard column:** 40 × 4 40 µm RP 201SC pellicular (Vydac)

**Column:** 300 × 4 μBondapak C18
**Mobile phase:** MeCN:buffer 60:40 (Buffer was pH 3.0 containing 5 mM 1-octanesulfonic acid and 5 mM tetramethylammonium chloride.)
**Flow rate:** 2
**Injection volume:** 200
**Detector:** UV 230

## CHROMATOGRAM
**Retention time:** 4.5

## OTHER SUBSTANCES
**Simultaneous:** liothyronine, 3,5-diiodo-L-thyronine

## KEY WORDS
tablets; stability-indicating

## REFERENCE
Richheimer, S.L.; Amer, T.M. Stability-indicating assay, dissolution, and content uniformity of sodium levothyroxine in tablets. *J.Pharm.Sci.*, **1983**, *72*, 1349–1351

## SAMPLE
**Matrix:** formulations
**Sample preparation:** Grind a tablet, add 50 μg 3,3′,5′-triiodothyronine, add 20 mL solvent A, stir for 10 min, add 40 mL solvent B, stir for 30 min, filter. Remove the upper layer and wash it six times with 15 mL portions of water saturated with butanol, evaporate under vacuum at 40-42°, reconstitute in 2.5 mL 3% ammonium hydroxide in MeOH, inject an aliquot (Anal.Lett. 1979, 12, 1201). (Prepare the solvents by mixing 1.8 L 1-butanol, 1.35 L water and 450 mL concentrated HCl, shake vigorously for 20 min, allow to separate. The lower layer was solvent A and the upper layer was solvent B.)

## HPLC VARIABLES
**Guard column:** 25 × 2.5 Co:Pel ODS
**Column:** 300 × 3.9 10 μm μBondapak C18
**Mobile phase:** MeOH:100 mM pH 5.0 ammonium acetate 50:50
**Flow rate:** 1.5
**Detector:** UV 254

## CHROMATOGRAM
**Retention time:** 19.5
**Internal standard:** 3,3′,5′-triiodothyronine

## OTHER SUBSTANCES
**Simultaneous:** liothyronine

## KEY WORDS
protect from light; tablets

## REFERENCE
Rapaka, R.S.; Knight, P.W.; Prasad, V.K. Reversed-phase high-performance liquid chromatographic analysis of liothyronine sodium and levothyroxine sodium in tablet formulations: preliminary studies on dissolution and content uniformity. *J.Pharm.Sci.*, **1981**, *70*, 131–134

## SAMPLE
**Matrix:** solutions

## HPLC VARIABLES
**Column:** 250 × 4.6 5 μm Phenomenex cyano-bonded silica
**Mobile phase:** MeCN:water:phosphoric acid 400:600:1
**Flow rate:** 1.5
**Detector:** UV 225

## CHROMATOGRAM
**Retention time:** 8.6

## OTHER SUBSTANCES
**Simultaneous:** degradation products, liothyronine

## REFERENCE
Won, C.M. Kinetics of degradation of levothyroxine in aqueous solution and in solid state. *Pharm.Res.*, **1992**, *9*, 131–137

## SAMPLE
**Matrix:** solutions
**Sample preparation:** Take up 1.5 mg levothyroxine in 200 μL 100 mM sodium bicarbonate and 400 μL reagent, stir in an ice bath for 30 min, evaporate to dryness below 30°, add 100 μL trifluoroacetic acid to the dry residue, let stand for 30 min at room temperature, add 2 mL 1 M sodium bicarbonate, centrifuge. Remove the precipitate and dissolve it in 600 μL MeOH:20 mM NaOH 50:50, inject a 15 μL aliquot. Reagent was 7 mg/mL BOC-L-Leu-SU (tert-butyloxy-L-leucine-N-hydroxysuccinimide ester) in MeOH, prepared immediately before use.)

## HPLC VARIABLES
**Column:** 150 × 3.2 7 μm LiChrosorb RP-18
**Mobile phase:** MeOH:water 60:40 containing 0.05% methanesulfonic acid
**Flow rate:** 1
**Injection volume:** 15
**Detector:** UV 230

## CHROMATOGRAM
**Retention time:** 9
**Limit of detection:** 0.05% of the D form

## OTHER SUBSTANCES
**Simultaneous:** impurities, dextrothyroxine

## KEY WORDS
derivatization; chiral

## REFERENCE
Lankmayr, E.P.; Budna, K.W.; Nachtmann, F. Separation of enantiomeric iodinated thyronines by liquid chromatography of diastereomers. *J.Chromatogr.*, **1980**, *198*, 471–479

## SAMPLE
**Matrix:** tissue
**Sample preparation:** 100 μL Thyroid tissue + 200 μL MeCN, mix, centrifuge. Remove a 100 μL aliquot of the supernatant and add it to 100 μL 4 nM dabsyl chloride in MeCN, heat at 70° for 10 min, add 400 μL MeOH:50 mM pH 7.0 phosphate buffer 50:50, inject a 20 μL aliquot.

## HPLC VARIABLES
**Column:** 150 × 4.6 5 μm Hypersil ODS

**Mobile phase:** Gradient. A was MeOH:25 mM pH 6.5 sodium acetate 56:44. B was MeOH. A:B from 80:20 to 35:65 over 15 min, maintain at 35:65 for 3 min, to 0:100 over 1 min, maintain at 0:100 for 2 min.
**Flow rate:** 1
**Injection volume:** 20
**Detector:** UV 436

## CHROMATOGRAM
**Retention time:** 17.5

## OTHER SUBSTANCES
**Extracted:** diiodothyronine (T2), liothyronine (T3)

## KEY WORDS
derivatization; thyroid

## REFERENCE
Jansen, E.H.J.M.; van den Berg, R.H.; Both-Miedema, R.; Doorn, L. Advantages and limitations of pre-column derivatization of amino acids with dabsyl chloride. *J.Chromatogr.*, **1991**, *553*, 123–133

◆━━━━━━━━━━━━━━━━━━━◆━━━━━━━━━━━━━━━━━━━◆

## ANNOTATED BIBLIOGRAPHY
Fish, L.H.; Schwartz, H.L.; Cavanaugh, J.; Steffes, M.W.; Bantle, J.P.; Oppenheimer, J.H. Replacement dose, metabolism, and bioavailability of levothyroxine in the treatment of hypothyroidism. Role of triiodothyronine in pituitary feedback in humans. *N.Engl.J.Med.*, **1987**, *316*, 764–770 [tablets]

Brower, J.F.; Toler, D.Y.; Reepmeyer, J.C. Determination of sodium levothyroxine in bulk, tablet, and injection formulations by high-performance liquid chromatography. *J.Pharm.Sci.*, **1984**, *73*, 1315–1317

Garnick, R.L.; Burt, G.F.; Long, D.A.; Bastian, J.W.; Aldred, J.P. High-performance liquid chromatographic assay for sodium levothyroxine in tablet formulations: content uniformity applications. *J.Pharm.Sci.*, **1984**, *73*, 75–77

Smith, D.J.; Biesemeyer, M.; Yaciw, C. The separation and determination of liothyronine and levothyroxine in tablets by reversed-phase high performance liquid chromatography. *J.Chromatogr.Sci.*, **1981**, *19*, 72–78

# Lisinopril

**Molecular formula:** $C_{21}H_{31}N_3O_5$
**Molecular weight:** 405.5
**CAS Registry No.:** 76547-98-3 (anhydrous), 83915-83-7 (dihydrate)

## SAMPLE
**Matrix:** blood
**Sample preparation:** 2 mL Whole blood or plasma + 2 mL buffer + 5 mL chloroform: isopropanol:n-heptane 60:14:26, shake gently horizontally for 10 min, centrifuge at 2800 g for 10 min. Remove the lower organic layer and evaporate it to dryness under vacuum at 45°, reconstitute the residue in 100 μL mobile phase, centrifuge at 2800 g for 5 min, inject a 50 μL aliquot of the supernatant. (Buffer was saturated ammonium chloride solution 25% diluted with water, adjusted to pH 9.5 with 25% ammonia solution.)

## HPLC VARIABLES
**Column:** 300 × 3.9 4 μm NovaPack C18
**Mobile phase:** MeOH:THF:buffer 65:5:30 (Buffer was 0.68 g/L (10 mM (sic)) $KH_2PO_4$ adjusted to pH 2.6 with concentrated orthophosphoric acid.) (At the end of each session wash the column with water for 1 h and MeOH for 1 h, re-equilibrate for 30 min.)
**Column temperature:** 30
**Flow rate:** 0.8
**Injection volume:** 50
**Detector:** UV 259

## CHROMATOGRAM
**Retention time:** 3.50
**Limit of detection:** <120 ng/mL

## OTHER SUBSTANCES
**Extracted:** acebutolol, acenocoumarol, acepromazine, aceprometazine, acetaminophen, aconitine, ajmaline, albuterol, alimemazine, alminoprofen, alpidem, alprazolam, alprenolol, amitriptyline, amodiaquine, amoxapine, astemizole, atenolol, benazepril, benperidol, benzoylecgonine, bepridil, betaxolol, bisoprolol, bromazepam, brompheniramine, bumadizone, bupivacaine, buprenorphine, buspirone, caffeine, carbamazepine, carbinoxamine, carpipramine, celiprolol, cetirizine, chlorambucil, chlordiazepoxide, chlorophenacinone, chloroquine, chlorpheniramine, chlorpromazine, chlorpropamide, cibenzoline, cicletanine, clemastine, clobazam, clomipramine, clonazepam, clonidine, clorazepate, clozapine, cocaine, colchicine, cyamemazine, cyclizine, cycloguanil, cyproheptadine, cytarabine, dacarbazine, daunorubicin, debrisoquine, demexiptiline, desipramine, dextromethorphan, dextromoramide, dextropropoxyphene, diazepam, diazoxide, diclofenac, dihydralazine, diltiazem, diphenhydramine, dipyridamole, disopyramide, dosulepin, doxepin, doxylamine, droperidol, ephedrine, estazolam, etodolac, fenfluramine, fenoprofen, fentiazac, flecainide, floctafenine, flumazenil, flunitrazepam, fluoxetine, fluphenazine, flurbiprofen, fluvoxamine, glibenclamide, glibornuride, glipizide, glutethimide, haloperidol, histapyrrodine, hydroxychloroquine, hydroxyzine, ibuprofen, imipramine, indomethacin, iproniazid, ketamine, ketoprofen, labetalol, levomepromazine, lidocaine, lidoflazine, loperamide, loprazolam, loratadine, lorazepam, loxapine, maprotiline, medazepam, medifoxamine, mefenamic acid, mefenidramine, mefloquine, melphalan, meperidine, mephentermine, mepivacaine, metapramine, methadone, methaqualone, methocarbamol, methotrexate, metipranolol, metoclopramide, metoprolol, mexiletine, mianserine, midazolam, minoxidil, moclobemide, moperone, morphine, nadolol, nalbuphine, naproxen, nialamide, nicardipine, nifedipine, niflumic acid, nimodipine, nitrazepam, nitrendipine, nomifensine, nortriptyline, omeprazole, opipramol, oxazepam, oxprenolol, penbutolol, penfluridol, pentazocine, phencyclidine, phenobarbital, phenylbutazone, pimozide, pindolol, pipamperone, piroxicam, prazepam, prazosin, prilocaine, procainamide, procarbazine, pro-

guanil, promethazine, propafenone, propranolol, protriptyline, pyrimethamine, quinidine, quinine, quinupramine, ramipril, ranitidine, reserpine, secobarbital, strychnine, sulfin-pyrazole, sulindac, sulpride, suriclone, temazepam, tenoxicam, terbutaline, terfenadine, tetracaine, tetrazepam, thiopental, thioproperazine, thioridazine, tianeptine, tiapride, tia-profenic acid, ticlopidine, timolol, tioclomarol, tofisopam, tolbutamide, toloxatone, trazo-done, triazolam, trifluoperazine, trifluperidol, trimipramine, triprolidine, tropatenine, ver-apamil, viloxazine, vinblastine, vincristine, vindesine, warfarin, yohimbine, zolpidem, zopiclone, zorubicine

**Interfering:** amisulpride, aspirin, benzocaine, carteolol, chlormezanone, codeine, mephe-nesin, metformin, nalorphine, naloxone, naltrexone, nizatidine, phenol, ritodrine, sotalol, sultopride

## KEY WORDS
whole blood; plasma

## REFERENCE
Tracqui, A.; Kintz, P.; Mangin, P. Systematic toxicological analysis using HPLC/DAD. *J.Forensic Sci.*, **1995**, *40*, 254–262

## SAMPLE
**Matrix:** solutions

## HPLC VARIABLES
**Column:** 250 × 4.6 Spherisorb 5 ODS-2
**Mobile phase:** n-Propanol:buffer 5:95 (Buffer was pH 3.0 phosphate buffer containing 0.4% triethylamine.)
**Flow rate:** 1
**Detector:** UV 240

## CHROMATOGRAM
**Retention time:** 8.2

## REFERENCE
Barbato, F.; Morrica, P.; Quaglia, F. Analysis of ACE inhibitor drugs by high performance liquid chro-matography. *Farmaco*, **1994**, *49*, 457–460

## SAMPLE
**Matrix:** solutions

## HPLC VARIABLES
**Column:** 300 × 3.9 µBondapak phenyl
**Mobile phase:** MeOH:water:85% phosphoric acid 60:40:0.05
**Column temperature:** 30-40
**Detector:** UV 215-220

## OTHER SUBSTANCES
**Also analyzed:** enalapril

## REFERENCE
Ranadive, S.A.; Chen, A.X.; Serajuddin, A.T. Relative lipophilicities and structural-pharmacological con-siderations of various angiotensin-converting enzyme (ACE) inhibitors. *Pharm.Res.*, **1992**, *9*, 1480–1486

## SAMPLE
**Matrix:** urine

**Sample preparation:** Condition a Sep-Pak C18 SPE cartridge with 10 mL MeOH, 10 mL water, and 20 mL 100 mM HCl. 1 mL Urine + 80 μL 8 μg/mL enalaprilat + 10 μL 6 M nitric acid, vortex for 30 s, add to the SPE cartridge, wash with 20 mL 100 mM HCl, elute with 3 mL MeCN:water 10:90, elute with 6 mL water. Combine the eluates and evaporate the MeCN under a stream of air at 65°, add 25 μL 6 M nitric acid, add this solution to the SPE cartridge, wash with 10 mL chloroform, elute with 6 mL MeOH. Evaporate the eluate to dryness under a stream of air at 65°, wash the residue with 1 mL MeCN, reconstitute with 500 μL MeOH:chloroform 10:90, vortex for 30 s. Put this solution in an another tube, evaporate to dryness, reconstitute with 100 μL mobile phase, inject a 10 μL aliquot.

## HPLC VARIABLES
**Column:** 300 × 3.9 10 μm μBondapak C18
**Mobile phase:** MeCN:MeOH:THF:15 mM pH 2.9 $KH_2PO_4$ 6:1:1:92
**Column temperature:** 40
**Flow rate:** 1.5
**Injection volume:** 10
**Detector:** UV 206

## CHROMATOGRAM
**Retention time:** 7.5
**Internal standard:** enalaprilat (10.5)
**Limit of quantitation:** 500 ng/mL

## KEY WORDS
SPE; pharmacokinetics

## REFERENCE
Wong, Y.-c.; Charles, B.G. Determination of the angiotensin-converting enzyme inhibitor lisinopril in urine using solid-phase extraction and reversed-phase high-performance liquid chromatography. *J.Chromatogr.B*, **1995**, *673*, 306–310

## ANNOTATED BIBLIOGRAPHY
Friedman, D.I.; Amidon, G.L. Intestinal absorption mechanism of dipeptide angiotensin converting enzyme inhibitors of the lysyl-proline type: lisinopril and SQ 29,852. *J.Pharm.Sci.*, **1989**, *78*, 995–998 [rat; perfusate]

# Loracarbef

**Molecular formula:** $C_{16}H_{16}ClN_3O_4$
**Molecular weight:** 349.8
**CAS Registry No.:** 121961-22-6 (monohydrate)

## SAMPLE
**Matrix:** blood, urine
**Sample preparation:** Plasma or serum. Condition a Sep-Pak C18 SPE cartridge with 2 mL MeOH then 2 mL water, do not allow to go dry. 500 μL Plasma or serum + 100 μL 100 μg/mL cephalexin in water + 50 μL 25% acetic acid, mix, add to SPE cartridge, wash with two 1 mL portions of water, elute with 3 mL MeOH. Evaporate eluate under nitrogen, add 200 μL mobile phase, vortex, inject a 25 μL aliquot. Urine. Dilute 100:1 (ratio may vary depending on concentration) with water, inject a 50 μL aliquot.

## HPLC VARIABLES
**Column:** 150 × 4.6 5 μm Supelcosil LC-18-DB
**Mobile phase:** MeOH:THF:buffer 16:4:80 (Buffer was 1 g sodium 1-heptanesulfonate + 15 mL triethylamine in 1 L water with the pH adjusted to 2.3 with concentrated phosphoric acid.)
**Column temperature:** 30
**Flow rate:** 1.4
**Injection volume:** 25-50
**Detector:** UV 265

## CHROMATOGRAM
**Retention time:** 7.5
**Internal standard:** cephalexin (9.2)
**Limit of quantitation:** 500 ng/mL

## OTHER SUBSTANCES
**Extracted:** cefaclor, hydroxyloracarbef
**Noninterfering:** acetaminophen, caffeine

## KEY WORDS
plasma; serum; SPE; pharmacokinetics

## REFERENCE
Kovach, P.M.; Lantz, R.J.; Brier, G. High-performance liquid chromatographic determination of loracarbef, a potential metabolite, cefaclor and cephalexin in human plasma, serum and urine. *J.Chromatogr.*, **1991**, *567*, 129–139

## SAMPLE
**Matrix:** bulk

## HPLC VARIABLES
**Column:** 250 × 4.6 5 μm C18 (YMC)
**Mobile phase:** Gradient. A was 6.9 g/L $(NH_4)H_2PO_4$ adjusted to pH 2.5 with phosphoric acid. B was MeCN:6.9 g/L $(NH_4)H_2PO_4$ adjusted to pH 2.5 with phosphoric acid 60:40. A:B from 100:0 to 0:100 over 30 min, maintain at 0:100 for 10 min.
**Flow rate:** 1
**Detector:** UV 220

## CHROMATOGRAM
**Retention time:** 18

## OTHER SUBSTANCES
**Simultaneous:** degradation products

## REFERENCE
Baertschi, S.W.; Dorman, D.E.; Spangle, L.A.; Collins, M.W.; Lorenz, L.J. Formation of fluorescent pyrazine derivatives via a novel degradation pathway of the carbacephalosporin loracarbef. *J.Pharm.Biomed.Anal.*, **1995**, *13*, 323–328

## SAMPLE
**Matrix:** solutions

## HPLC VARIABLES
**Column:** 250 × 4.6 YMC A303 ODS (YMC)
**Mobile phase:** Gradient. A was 2.4 g/L NaH$_2$PO$_4$ adjusted to pH 2.5 with phosphoric acid. B was MeCN:2.4 g/L NaH$_2$PO$_4$ adjusted to pH 2.5 with phosphoric acid 60:40. A:B from 0:100 to 100:0 over 30 min, maintain at 100:0 for 10 min.
**Flow rate:** 1
**Detector:** UV 210

## CHROMATOGRAM
**Retention time:** 15

## OTHER SUBSTANCES
**Simultaneous:** degradation products

## REFERENCE
Skibic, M.J.; Taylor, K.W.; Occolowitz, J.L.; Collins, M.W.; Paschal, J.W.; Lorenz, L.J.; Spangle, L.A.; Dorman, D.E.; Baertschi, S.W. Aqueous acidic degradation of the carbacephalosporin loracarbef. *J.Pharm.Sci.*, **1993**, *82*, 1010–1017

## SAMPLE
**Matrix:** solutions

## HPLC VARIABLES
**Column:** 250 × 4.4 Zorbax ODS
**Mobile phase:** MeCN:25 mM (NH$_4$)H$_2$PO$_4$ 10:90
**Flow rate:** 1
**Detector:** UV 254

## OTHER SUBSTANCES
**Also analyzed:** cefaclor

## REFERENCE
Pasini, C.E.; Indelicato, J.M. Pharmaceutical properties of loracarbef: the remarkable solution stability of an oral 1-carba-1-dethiacephalosporin antibiotic. *Pharm.Res.*, **1992**, *9*, 250–254

# Loratadine

**Molecular formula:** $C_{22}H_{23}ClN_2O_2$
**Molecular weight:** 382.9
**CAS Registry No.:** 79794-75-5

## SAMPLE
**Matrix:** blood
**Sample preparation:** 2 mL Whole blood or plasma + 2 mL buffer + 5 mL chloroform: isopropanol:n-heptane 60:14:26, shake gently horizontally for 10 min, centrifuge at 2800 g for 10 min. Remove the lower organic layer and evaporate it to dryness under vacuum at 45°, reconstitute the residue in 100 μL mobile phase, centrifuge at 2800 g for 5 min, inject a 50 μL aliquot of the supernatant. (Buffer was saturated ammonium chloride solution 25% diluted with water, adjusted to pH 9.5 with 25% ammonia solution.)

## HPLC VARIABLES
**Column:** 300 × 3.9 4 μm NovaPack C18
**Mobile phase:** MeOH:THF:buffer 65:5:30 (Buffer was 0.68 g/L (10 mM (sic)) $KH_2PO_4$ adjusted to pH 2.6 with concentrated orthophosphoric acid.) (At the end of each session wash the column with water for 1 h and MeOH for 1 h, re-equilibrate for 30 min.)
**Column temperature:** 30
**Flow rate:** 0.8
**Injection volume:** 50
**Detector:** UV 247

## CHROMATOGRAM
**Retention time:** 11.29
**Limit of detection:** <120 ng/mL

## OTHER SUBSTANCES
**Extracted:** acebutolol, acenocoumarol, acepromazine, aceprometazine, acetaminophen, aconitine, ajmaline, albuterol, alimemazine, alminoprofen, alprazolam, alprenolol, amisulpride, amitriptyline, amodiaquine, amoxapine, aspirin, astemizole, atenolol, benazepril, benperidol, benzocaine, benzoylecgonine, bepridil, betaxolol, bisoprolol, bromazepam, brompheniramine, bumadizone, bupivacaine, buprenorphine, buspirone, caffeine, carbamazepine, carbinoxamine, carpipramine, carteolol, celiprolol, cetirizine, chlorambucil, chlordiazepoxide, chlormezanone, chlorophenacinone, chloroquine, chlorpheniramine, chlorpropamide, cibenzoline, cicletanine, clemastine, clobazam, clomipramine, clonazepam, clonidine, clorazepate, clozapine, cocaine, codeine, colchicine, cyamemazine, cyclizine, cycloguanil, cyproheptadine, cytarabine, dacarbazine, daunorubicin, debrisoquine, demexiptiline, desipramine, dextromethorphan, dextromoramide, dextropropoxyphene, diazepam, diazoxide, diclofenac, dihydralazine, diltiazem, diphenhydramine, dipyridamole, disopyramide, dosulepine, doxepin, doxylamine, droperidol, ephedrine, estazolam, etodolac, fenfluramine, fenoprofen, fentiazac, flecainide, floctafenine, flumazenil, flunitrazepam, fluoxetine, fluphenazine, flurbiprofen, fluvoxamine, glibenclamide, glibornuride, glipizide, glutethimide, haloperidol, histapyrrodine, hydroxychloroquine, hydroxyzine, ibuprofen, imipramine, indomethacin, iproniazid, ketamine, ketoprofen, labetalol, levomepromazine, lidocaine, lidoflazine, lisinopril, loperamide, loprazolam, lorazepam, loxapine, maprotiline, medazepam, medifoxamine, mefenamic acid, mefenidramine, mefloquine, melphalan, meperidine, mephenesin, mephentermine, mepivacaine, metapramine, metformin, methadone, methaqualone, methocarbamol, methotrexate, metipranolol, metoclopramide, metoprolol, mexiletine, mianserine, midazolam, minoxidil, moclobemide,

moperone, morphine, nadolol, nalbuphine, nalorphine, naloxone, naltrexone, naproxen, nialamide, nicardipine, nifedipine, niflumic acid, nimodipine, nitrazepam, nitrendipine, nizatidine, nomifensine, nortriptyline, omeprazole, opipramol, oxazepam, oxprenolol, penbutolol, penfluridol, pentazocine, phencyclidine, phenobarbital, phenol, phenylbutazone, pimozide, pindolol, pipamperone, piroxicam, prazepam, prazosin, prilocaine, procainamide, procarbazine, proguanil, promethazine, propafenone, propranolol, protriptyline, pyrimethamine, quinidine, quinine, quinupramine, ramipril, ranitidine, reserpine, ritodrine, secobarbital, sotalol, strychnine, sulfinpyrazone, sulindac, sulpride, sultopride, suriclone, temazepam, tenoxicam, terbutaline, terfenadine, tetracaine, tetrazepam, thiopental, thioproperazine, thioridazine, tianeptine, tiapride, tiaprofenic acid, ticlopidine, timolol, tioclomarol, tofisopam, tolbutamide, toloxatone, trazodone, triazolam, trifluoperazine, trifluperidol, trimipramine, triprolidine, tropatenine, verapamil, viloxazine, vinblastine, vincristine, vindesine, warfarin, yohimbine, zolpidem, zopiclone, zorubicine
**Interfering:** alpidem, chlorpromazine

## KEY WORDS
whole blood; plasma

## REFERENCE
Tracqui, A.; Kintz, P.; Mangin, P. Systematic toxicological analysis using HPLC/DAD. *J.Forensic Sci.*, **1995**, *40*, 254–262

## SAMPLE
**Matrix:** blood
**Sample preparation:** 1 mL Plasma + 100 μL IS in 0.1% phosphoric acid + 100 μL buffer + 2 mL diethyl ether:n-hexane 75:25, shake for 10 min, centrifuge at 4000 rpm for 5 min, let stand at -27°for 2 h. Remove the organic layer and add it to 200 μL 12.5% phosphoric acid, shake for 10 min, centrifuge at 4000 rpm for 2 min, inject a 100 μL aliquot of the aqueous phase.

## HPLC VARIABLES
**Column:** 250 × 4.6 7 μm Nucleosil C18
**Mobile phase:** MeCN:water:$(NH_4)H_2PO_4$:orthophosphoric acid 110:150:1.5:8 (v/v/w/v)
**Flow rate:** 1.5
**Injection volume:** 100
**Detector:** F ex 290 em 460

## CHROMATOGRAM
**Retention time:** 5
**Internal standard:** propyl 4-(8-chloro-5,6-dihydro-11H-benzo-[5,6]-cyclohepta-[1,2-b]pyridin-11-ylidin)-1-piperidinecarboxylate (8.5)
**Limit of quantitation:** 0.5 ng/mL

## OTHER SUBSTANCES
**Noninterfering:** metabolites

## KEY WORDS
plasma; pharmacokinetics

## REFERENCE
Zhong, D.; Blume, H. HPLC-Bestimmung von Loratadin und seinen aktiven Metaboliten Descarboethoxyloratadin in Humanplasma [HPLC determination of loratadine and its active metabolite descarboethoxyloratadine in human plasma]. *Pharmazie*, **1994**, *49*, 736–739

## SAMPLE
**Matrix:** solutions

## HPLC VARIABLES
**Column:** 250 × 4.6 5 μm Supelcosil LC-DP (A) or 250 × 4 5 μm LiChrospher 100 RP-8 (B)
**Mobile phase:** MeCN : 0.025% phosphoric acid : buffer 25 : 10 : 5 (A) or 60 : 25 : 15 (B) (Buffer was 9 mL concentrated phosphoric acid and 10 mL triethylamine in 900 mL water, adjust pH to 3.4 with dilute phosphoric acid, make up to 1 L.)
**Flow rate:** 0.6
**Injection volume:** 25
**Detector:** UV 229

## CHROMATOGRAM
**Retention time:** 10.90 (A), 13.25 (B)

## OTHER SUBSTANCES
**Also analyzed:** acebutolol, acepromazine, acetaminophen, acetazolamide, acetophenazine, albuterol, alprazolam, amitriptyline, amobarbital, amoxapine, antipyrine, atenolol, atropine, azatadine, baclofen, benzocaine, bromocriptine, brompheniramine, brotizolam, bupivacaine, buspirone, butabarbital, butalbital, caffeine, carbamazepine, cetirizine, chlorcyclizine, chlordiazepoxide, chlormezanone, chloroquine, chlorpheniramine, chlorpromazine, chlorpropamide, chlorprothixene, chlorthalidone, chlorzoxazone, cimetidine, cisapride, clomipramine, clonazepam, clonidine, clozapine, cocaine, codeine, colchicine, cyclizine, cyclobenzaprine, dantrolene, desipramine, diazepam, diclofenac, diflunisal, diltiazem, diphenhydramine, diphenidol, diphenoxylate, dipyridamole, disopyramide, dobutamine, doxapram, doxepin, droperidol, encainide, ethidium bromide, ethopropazine, fenoprofen, fentanyl, flavoxate, fluoxetine, fluphenazine, flurazepam, flurbiprofen, fluvoxamine, furosemide, glutethimide, glyburide, guaifenesin, haloperidol, homatropine, hydralazine, hydrochlorothiazide, hydrocodone, hydromorphone, hydroxychloroquine, hydroxyzine, ibuprofen, imipramine, indomethacin, ketoconazole, ketoprofen, ketorolac, labetalol, levorphanol, lidocaine, lorazepam, lovastatin, loxapine, mazindol, mefenamic acid, meperidine, mephenytoin, mepivacaine, mesoridazine, metaproterenol, methadone, methdilazine, methocarbamol, methotrexate, methotrimeprazine, methoxamine, methyldopa, methylphenidate, metoclopramide, metolazone, metoprolol, metronidazole, midazolam, moclobemide, morphine, nadolol, nalbuphine, naloxone, naphazoline, naproxen, nifedipine, nizatidine, norepinephrine, nortriptyline, oxazepam, oxycodone, oxymetazoline, paroxetine, pemoline, pentazocine, pentobarbital, pentoxifylline, perphenazine, pheniramine, phenobarbital, phenol, phenolphthalein, phentolamine, phenylbutazone, phenyltoloxamine, phenytoin, pimozide, pindolol, piroxicam, pramoxine, prazepam, prazosin, probenecid, procainamide, procaine, prochlorperazine, procyclidine, promazine, promethazine, propafenone, propantheline, propiomazine, propofol, propranolol, protriptyline, quazepam, quinidine, quinine, racemethorphan, ranitidine, remoxipride, risperidone, salicylic acid, scopolamine, secobarbital, sertraline, sotalol, spironolactone, sulfinpyrazone, sulindac, temazepam, terbutaline, terfenadine, tetracaine, theophylline, thiethylperazine, thiopental, thioridazine, thiothixene, timolol, tocainide, tolbutamide, tolmetin, trazodone, triamterene, triazolam, trifluoperazine, triflupromazine, trimeprazine, trimethoprim, trimipramine, verapamil, warfarin, xylometazoline, yohimbine, zopiclone

## KEY WORDS
some details of plasma extraction

## REFERENCE
Koves, E.M. Use of high-performance liquid chromatography-diode array detection in forensic toxicology. *J.Chromatogr.A*, **1995**, *692*, 103–119

# Lorazepam

**Molecular formula:** $C_{15}H_{10}Cl_2N_2O_2$
**Molecular weight:** 321.2
**CAS Registry No.:** 846-49-1

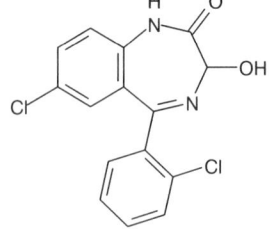

## SAMPLE
**Matrix:** blood
**Sample preparation:** 1-2 mL Plasma + 1 mL 1 M pH 10 bicarbonate buffer + 8 mL n-hexane:ethyl acetate 70:30, shake for 10 min, centrifuge. Remove the organic layer and evaporate it to dryness under a stream of nitrogen, reconstitute the residue in 200 μL mobile phase, inject an aliquot.

## HPLC VARIABLES
**Column:** 250 × 4.6 Ultrasphere ODS
**Mobile phase:** Gradient. A was MeCN:buffer 30:70. B was MeCN:buffer 70:30. A:B from 85:15 to 40:60 over 10 min (Waters curve no. 5), to 0:100 over 5 min (Waters curve no. 1), return to initial conditions over 10 min (Waters curve no. 1). (Buffer was 10 mM pH 3.35 $NaH_2PO_4$.)
**Flow rate:** 1
**Detector:** UV 229

## CHROMATOGRAM
**Internal standard:** lorazepam

## OTHER SUBSTANCES
**Extracted:** desmethyldiazepam, diazepam

## KEY WORDS
plasma; lorazepam is IS

## REFERENCE
Caraco, Y.; Tateishi, T.; Wood, A.J.J. Interethnic difference in omeprazole's inhibition of diazepam metabolism. *Clin.Pharmacol.Ther.*, **1995**, *58*, 62–72

## SAMPLE
**Matrix:** blood
**Sample preparation:** 2 mL Whole blood or plasma + 2 mL buffer + 5 mL chloroform:isopropanol:n-heptane 60:14:26, shake gently horizontally for 10 min, centrifuge at 2800 g for 10 min. Remove the lower organic layer and evaporate it to dryness under vacuum at 45°, reconstitute the residue in 100 μL mobile phase, centrifuge at 2800 g for 5 min, inject a 50 μL aliquot of the supernatant. (Buffer was saturated ammonium chloride solution 25% diluted with water, adjusted to pH 9.5 with 25% ammonia solution.)

## HPLC VARIABLES
**Column:** 300 × 3.9 4 μm NovaPack C18
**Mobile phase:** MeOH:THF:buffer 65:5:30 (Buffer was 0.68 g/L (10 mM (sic)) $KH_2PO_4$ adjusted to pH 2.6 with concentrated orthophosphoric acid.) (At the end of each session wash the column with water for 1 h and MeOH for 1 h, re-equilibrate for 30 min.)
**Column temperature:** 30
**Flow rate:** 0.8

**Injection volume:** 50
**Detector:** UV 230

## CHROMATOGRAM
**Retention time:** 4.19
**Limit of detection:** <120 ng/mL

## OTHER SUBSTANCES
**Extracted:** acebutolol, acenocoumarol, acepromazine, aceprometazine, acetaminophen, aconitine, ajmaline, albuterol, alimemazine, alminoprofen, alpidem, alprenolol, amisulpride, amitriptyline, amodiaquine, amoxapine, aspirin, astemizole, atenolol, benazepril, benperidol, benzocaine, benzoylecgonine, bepridil, betaxolol, bisoprolol, bromazepam, brompheniramine, bumadizone, bupivacaine, buprenorphine, buspirone, caffeine, carbamazepine, carbinoxamine, carpipramine, carteolol, celiprolol, cetirizine, chlorambucil, chlordiazepoxide, chlormezanone, chlorophenacinone, chloroquine, chlorpheniramine, chlorpromazine, chlorpropamide, cibenzoline, cicletanine, clemastine, clobazam, clomipramine, clonazepam, clonidine, clorazepate, clozapine, cocaine, codeine, colchicine, cyamemazine, cyclizine, cyproheptadine, cytarabine, dacarbazine, daunorubicin, debrisoquine, demexiptiline, desipramine, dextromethorphan, dextromoramide, dextropropoxyphene, diazepam, diazoxide, diclofenac, dihydralazine, diltiazem, diphenhydramine, dipyridamole, disopyramide, dosulepine, doxepin, doxylamine, droperidol, ephedrine, estazolam, etodolac, fenfluramine, fenoprofen, fentiazac, flecainide, floctafenine, flumazenil, flunitrazepam, fluoxetine, fluphenazine, flurbiprofen, fluvoxamine, glibenclamide, glibornuride, glipizide, glutethimide, haloperidol, histapyrrodine, hydroxychloroquine, hydroxyzine, ibuprofen, imipramine, indomethacin, iproniazid, ketoprofen, labetalol, levomepromazine, lidocaine, lidoflazine, lisinopril, loperamide, loprazolam, loratadine, loxapine, maprotiline, medazepam, medifoxamine, mefenamic acid, mefenidramine, mefloquine, melphalan, meperidine, mephenesin, mephentermine, mepivacaine, metapramine, metformin, methadone, methocarbamol, methotrexate, metipranolol, metoclopramide, mexiletine, mianserine, midazolam, minoxidil, moclobemide, moperone, morphine, nadolol, nalbuphine, nalorphine, naloxone, naltrexone, naproxen, nialamide, nicardipine, niflumic acid, nimodipine, nitrazepam, nitrendipine, nizatidine, nomifensine, nortriptyline, omeprazole, opipramol, oxazepam, oxprenolol, penbutolol, penfluridol, pentazocine, phencyclidine, phenobarbital, phenol, phenylbutazone, pimozide, pindolol, pipamperone, prazepam, prazosin, prilocaine, procainamide, procarbazine, proguanil, promethazine, propafenone, propranolol, protriptyline, pyrimethamine, quinidine, quinupramine, ramipril, ranitidine, reserpine, ritodrine, secobarbital, sotalol, strychnine, sulfinpyrazone, sulpride, sultopride, suriclone, temazepam, tenoxicam, terbutaline, terfenadine, tetracaine, tetrazepam, thiopental, thioproperazine, thioridazine, tianeptine, tiapride, tiaprofenic acid, ticlopidine, timolol, tioclomarol, tofisopam, tolbutamide, toloxatone, trazodone, triazolam, trifluoperazine, trifluperidol, trimipramine, triprolidine, tropatenine, verapamil, viloxazine, vinblastine, vincristine, vindesine, warfarin, yohimbine, zolpidem, zopiclone, zorubicine
**Interfering:** alprazolam, cycloguanil, ketamine, methaqualone, metoprolol, nifedipine, piroxicam, quinine, sulindac

## KEY WORDS
whole blood; plasma

## REFERENCE
Tracqui, A.; Kintz, P.; Mangin, P. Systematic toxicological analysis using HPLC/DAD. *J.Forensic Sci.*, **1995**, *40*, 254–262

## SAMPLE
**Matrix:** blood
**Sample preparation:** Make 1 mL serum alkaline with borate buffer, extract with cyclohexane:dichloromethane 60:40. Remove the organic layer and evaporate it to dryness, reconstitute the residue in mobile phase, inject an aliquot

## HPLC VARIABLES
**Column:** C18 DB (Supelco)
**Mobile phase:** MeCN:pH 2.5 phosphate buffer 37:63
**Detector:** UV 254

## OTHER SUBSTANCES
**Extracted:** bromazepam, clobazam, diazepam, fluvoxamine, oxazepam

## KEY WORDS
serum

## REFERENCE
Vandenberghe, H.; MacDonald, J.C. Analysis of fluvoxamine, clobazam and other benzodiazepines on the same HPLC system (Abstract 40). *Ther.Drug Monit.*, **1995**, *17*, 393

## SAMPLE
**Matrix:** blood
**Sample preparation:** Inject 100-200 µL plasma onto column A with mobile phase A and elute to waste, after 5 min backflush the contents of column A onto column B with mobile phase B, after 5 min remove column A from the circuit, elute column B with mobile phase B, monitor the effluent from column B. Wash column A with MeCN:water 60:40 at 1 mL/min for 6 min then re-equilibrate with pH 7.5 buffer for 10 min.

## HPLC VARIABLES
**Column:** A 45 × 4 12 µm TSK-gel G 3 PW (Tosohass); B 75 × 4.6 Ultrasphere ODS C18 3 µm
**Mobile phase:** A 50 mM pH 7.5 phosphate buffer; B Gradient. A was MeCN. B was 65 mM $KH_2PO_4$ + 1% diethylamine adjusted to pH 5.4 with phosphoric acid. A:B 22:78 for 5 min, to 25:75 over 10 min, to 60:40 over 15 min.
**Flow rate:** 1
**Injection volume:** 100-200
**Detector:** UV 230

## CHROMATOGRAM
**Retention time:** 22

## OTHER SUBSTANCES
**Extracted:** alprazolam, bromazepam, chlordiazepoxide, clobazam, clorazepate, clotiazepam, desmethylclobazam, desmethyldiazepam, diazepam, estazolam, flunitrazepam, loflazepate, medazepam, nitrazepam, oxazepam, prazepam, temazepam, tetrazepam, tofisopam, triazolam
**Noninterfering:** carbamazepine, ethosuximide, phenobarbital, phenytoin, primidone, valproic acid
**Interfering:** clonazepam

## KEY WORDS
plasma; column-switching

## REFERENCE
Lacroix, C.; Wojciechowski, F.; Danger, P. Monitoring of benzodiazepines (clobazam, diazepam and their main active metabolites) in human plasma by column-switching high-performance liquid chromatography. *J.Chromatogr.*, **1993**, *617*, 285–290

## SAMPLE
**Matrix:** blood

**Sample preparation:** 0.5 mL Plasma +10 μL 250 μg/mL 1-acetamidopyrene in MeOH + 200 μL 1 M ammonium sulfate + 800 μL cold MeCN, vortex for 30 s, store at -20° for at least 30 min, vortex, centrifuge at 1500 g for 30 min. Remove 400 μL of the upper organic layer and evaporate it under a stream of nitrogen. Reconstitute with 100 μL mobile phase, vortex for 30 s, inject a 75 μL aliquot.

## HPLC VARIABLES
**Column:** 300 × 3.9 10 μm μBondapak C18
**Mobile phase:** MeCN:buffer 47:53 (Buffer was 6.805 g potassium monophosphate in 1 L water, adjust pH to 6.00 with 10 M NaOH.)
**Flow rate:** 1
**Injection volume:** 75
**Detector:** UV 214

## CHROMATOGRAM
**Retention time:** 6.1
**Internal standard:** 1-acetamidopyrene (9.7)
**Limit of detection:** 0.781 ng/mL

## OTHER SUBSTANCES
**Simultaneous:** antipyrine, indocyanine green
**Noninterfering:** adenosine, albuterol, alphenal, aspirin, caffeine, carbamazepine, cefazolin, cephalexin, cephalothin, cimetidine, ciprofloxacin, claforan, desipramine, enoxacin, fleroxacin, furosemide, hydralazine, hydrochlorothiazide, minoxidil, norfloxacin, phenytoin, propafenone, sulindac, teicoplanin, theophylline, vancomycin
**Interfering:** some indocyanine green impurities

## KEY WORDS
plasma

## REFERENCE
Awni, W.M.; Bakker, L.J. Antipyrine, indocyanine green, and lorazepam determined in plasma by high-pressure liquid chromatography. *Clin.Chem.*, **1989**, *35*, 2124–2126

## SAMPLE
**Matrix:** blood
**Sample preparation:** 1 mL Plasma + 100 μL 1 μg/mL diazepam in MeOH + 2 mL phosphate buffer, vortex for 30 s, add 8 mL hexane:isoamyl alcohol 95:5, shake gently by hand for 10 min, vortex for 1 min, centrifuge at 400 g at 4° for 10 min. Remove organic layer and add it to 2 mL 6 M HCl, shake for 10 min, vortex for 1 min, centrifuge at 400 g for 10 min. Remove the aqueous layer and slowly add about 2 mL 6 M NaOH to it to achieve a pH greater than 7.0, add 2 mL phosphate buffer, vortex for 10 s, add 8 mL hexane:isoamyl alcohol 95:5, shake for 10 min, vortex for 1 min, centrifuge at 400 g for 10 min. Remove the organic layer and evaporate it to dryness at 40° with nitrogen, rinse residue from sides with hexane:isoamyl alcohol 95:5, again dry with nitrogen, take up residue in 40 μL MeOH, inject a 20-40 μL aliquot. (Phosphate buffer was 136.1 g $KH_2PO_4$ in 1 L water, adjusted to pH 7 with 1 M $K_2HPO_4$.)

## HPLC VARIABLES
**Column:** 300 × 3.9 10 μm μBondapak C18
**Mobile phase:** MeOH:MeCN:10 mM sodium acetate 40:12.5:47.5, at pH 4.6
**Flow rate:** 2.2
**Injection volume:** 20-40
**Detector:** UV 254

## CHROMATOGRAM
**Retention time:** 5.2

**Internal standard:** diazepam (9.2)
**Limit of quantitation:** 2.5 ng/mL

## OTHER SUBSTANCES
**Simultaneous:** midazolam, flunitrazepam, clonazepam, flurazepam, temazepam, nitrazepam
**Interfering:** oxazepam

## KEY WORDS
plasma

## REFERENCE
Egan, J.M.; Abernethy, D.R. Lorazepam analysis using liquid chromatography: improved sensitivity for single-dose pharmacokinetic studies. *J.Chromatogr.*, **1986**, *380*, 196–201

## SAMPLE
**Matrix:** blood
**Sample preparation:** 0.5 mL Plasma + 0.5 mL water + 0.5 mL 0.25 M NaOH, vortex, allow to stand at room temperature for 20 min, add 20 μL of 100 μg/mL phenacetin and 3 μg/mL flunitrazepam, vortex, add 5 mL diethyl ether, vortex for 30 s, centrifuge at 900 g for 5 min, freeze in acetone/dry ice for 5 min. Remove the supernatant and dry it under nitrogen. Reconstitute in 115 μL MeCN:0.1% pH 3 sodium phosphate buffer 30:70, inject an aliquot.

## HPLC VARIABLES
**Guard column:** 23 × 3.9 37-50 μm μBondapak phenyl
**Column:** 300 × 3.9 10 μm μBondapak phenyl
**Mobile phase:** Gradient. A was MeCN:0.1% pH 3 sodium phosphate buffer 5:95. B was MeCN:0.1% pH 3 sodium phosphate buffer 70:30. A:B 80:20 for 2.5 min, then to 45:55 over 20 min, then to 25:75 over 3 min, then to 80:20 over 3 min, equilibrate at 80:20 for 7 min.
**Column temperature:** 40
**Flow rate:** 2
**Injection volume:** 200
**Detector:** UV 229

## CHROMATOGRAM
**Retention time:** 15.11
**Internal standard:** flunitrazepam (17.90)
**Limit of quantitation:** 10.5 ng/mL

## OTHER SUBSTANCES
**Simultaneous:** antipyrine (at 254 nm)
**Noninterfering:** acetaminophen, allopurinol, indocyanine green, sulfamethoxazole, trimethoprim

## KEY WORDS
plasma

## REFERENCE
Riley, C.A.; Evans, W.E. Simultaneous analysis of antipyrine and lorazepam by high-performance liquid chromatography. *J.Chromatogr.*, **1986**, *382*, 199–205

## SAMPLE
**Matrix:** blood, CSF, gastric fluid, urine

**Sample preparation:** 200 μL Serum, urine, CSF, or gastric fluid + 300 μL reagent. Flush column A to waste with 500 μL 500 mM ammonium sulfate, inject sample onto column A, flush column A to waste with 500 μL 500 mM ammonium sulfate, backflush the contents of column A onto column B with mobile phase, monitor the effluent from column B. (Reagent was 8.05 M guanidine hydrochloride and 1.02 M ammonium sulfate in water.)

## HPLC VARIABLES
**Column:** A 40 μm preparative grade C18 (Analytichem); B 75 × 2.1 pellicular C18 (Whatman) + 250 × 4.6 5 μm C8 end-capped (Whatman)
**Mobile phase:** Gradient. A was 50 mM pH 4.5 $KH_2PO_4$. B was MeCN:isopropanol 80:20. A:B 90:10 for 1 min, to 30:70 over 20 min.
**Column temperature:** 50
**Flow rate:** 1.5
**Detector:** UV 220

## CHROMATOGRAM
**Retention time:** 12.57
**Internal standard:** heptanophenone (19)

## OTHER SUBSTANCES
**Extracted:** acetaminophen, allobarbital, azinphos, barbital, brallobarbitone, bromazepam, butethal, caffeine, carbamazepine, carbaryl, cephaloridine, chloramphenicol, chlordiazepoxide, chlorothiazide, chlorvinphos, clothiapine, cocaine, coomassie blue, desipramine, diazepam, diphenhydramine, dipipanone, ethylbromphos, flufenamic acid, formothion, griseofulvin, indomethacin, lidocaine, malathion, medazepam, midazolam, oxazepam, paraoxon, penicillin G, pentobarbital, prazepam, propoxyphene, prothiophos, quinine, salicylic acid, secobarbital, strychnine, sulfamethoxazole, theophylline, thiopental, thioridazine, trimethoprim

## KEY WORDS
serum; column-switching

## REFERENCE
Kruger, P.B.; Albrecht, C.F.De V.; Jaarsveld, P.P. Use of guanidine hydrochloride and ammonium sulfate in comprehensive in-line sorption enrichment of xenobiotics in biological fluids by high-performance liquid chromatography. *J.Chromatogr.*, **1993**, *612*, 191–198

## SAMPLE
**Matrix:** formulations
**Sample preparation:** Dilute with mobile phase, inject an aliquot.

## HPLC VARIABLES
**Column:** 250 × 4.6 5 μm cyano
**Mobile phase:** MeCN:100 mM $NaH_2PO_4$ 20:80 adjusted to pH 4.2 with phosphoric acid
**Flow rate:** 1.5
**Injection volume:** 20
**Detector:** UV 220

## CHROMATOGRAM
**Retention time:** 8.72

## OTHER SUBSTANCES
**Simultaneous:** granisetron (UV 300)

## KEY WORDS
stability-indicating; injections; saline

## REFERENCE

Mayron, D.; Gennaro, A.R. Stability and compatibility of granisetron hydrochloride in i.v. solutions and oral liquids and during simulated Y-site injection with selected drugs. *Am.J.Health-Syst.Pharm.*, **1996**, *53*, 294–304

## SAMPLE

**Matrix:** formulations
**Sample preparation:** Inject a 20 µL aliquot.

## HPLC VARIABLES

**Column:** 150 × 4.6 5 µm Spherisorb ODS-1
**Mobile phase:** MeOH:50 mM $(NH_4)H_2PO_4$ 50:50, adjusted to pH 7.22
**Flow rate:** 1.3 for 12 min then 2.1 for 14 min
**Injection volume:** 20
**Detector:** UV 240

## CHROMATOGRAM

**Retention time:** 9

## OTHER SUBSTANCES

**Simultaneous:** degradation products

## KEY WORDS

stability-indicating; 5% dextrose; saline; injections

## REFERENCE

Mancano, M.A.; Boullata, J.I.; Gelone, S.P.; Zitterman, R.E.; Borenstein, M.R. Availability of lorazepam after simulated administration from glass and polyvinyl chloride containers. *Am.J.Health-Syst.Pharm.*, **1995**, *52*, 2213–2216

## SAMPLE

**Matrix:** formulations
**Sample preparation:** Dilute with water.

## HPLC VARIABLES

**Column:** Waters C18 column (PN 86344)
**Mobile phase:** MeOH:50 mM $(NH_4)H_2PO_4$ 50:50, adjusted to pH 6.5 with ammonium hydroxide
**Flow rate:** 2
**Injection volume:** 20
**Detector:** UV 240

## CHROMATOGRAM

**Retention time:** 21.1

## KEY WORDS

saline; injections; stability-indicating

## REFERENCE

Stiles, M.L.; Allen, L.V., Jr.; Prince, S.J.; Holland, J.S. Stability of dexamethasone sodium phosphate, diphenhydramine hydrochloride, lorazepam, and metoclopramide hydrochloride in portable infusion-pump reservoirs. *Am.J.Hosp.Pharm.*, **1994**, *51*, 514–517

## SAMPLE
**Matrix:** formulations
**Sample preparation:** Inject an aliquot directly.

## HPLC VARIABLES
**Column:** 250 × 4.6 5 µm Vydac 208TP54 C8
**Mobile phase:** MeOH:50 mM $(NH_4)H_2PO_4$ 57:43, pH adjusted to 6.5 with ammonium hydroxide
**Flow rate:** 1
**Detector:** UV 254

## CHROMATOGRAM
**Retention time:** 6.1

## KEY WORDS
injections; saline; 5% dextrose; lactated Ringer's; stability-indicating

## REFERENCE
Trissel, L.A.; Pearson, S.D. Storage of lorazepam in three injectable solutions in polyvinyl chloride and polyolefin bags. *Am.J.Hosp.Pharm.*, **1994**, *51*, 368–372

## SAMPLE
**Matrix:** formulations
**Sample preparation:** Dilute with saline, inject a 10 µL aliquot.

## HPLC VARIABLES
**Column:** 250 × 4.6 Lichrosorb 10 RP 8
**Mobile phase:** MeOH:THF:water 50:5:50
**Flow rate:** 3
**Injection volume:** 10
**Detector:** UV 254

## CHROMATOGRAM
**Retention time:** 8

## OTHER SUBSTANCES
**Simultaneous:** diazepam, thiopental

## KEY WORDS
injections; saline

## REFERENCE
Martens, H.J.; de Goede, P.N.; van Loenen, A.C. Sorption of various drugs in polyvinyl chloride, glass, and polyethylene-lined infusion containers. *Am.J.Hosp.Pharm.*, **1990**, *47*, 369–373

## SAMPLE
**Matrix:** solutions

## HPLC VARIABLES
**Column:** 150 × 3.9 4 µm Nova pack C18
**Mobile phase:** MeOH:water 52:48
**Column temperature:** 48
**Flow rate:** 0.8

**Injection volume:** 20
**Detector:** UV 254

## CHROMATOGRAM
**Retention time:** 6

## OTHER SUBSTANCES
**Simultaneous:** bromazepam, chlordiazepoxide, clobazam, clorazepate, diazepam, flunitrazepam, nitrazepam, oxazepam, tofisopam

## REFERENCE
Guillaume, Y.; Guinchard, C. Thermodynamic behavior of mixed benzodiazepines by a new liquid chromatographic method. *Chromatographia*, **1995**, *40*, 193–196

## SAMPLE
**Matrix:** solutions

## HPLC VARIABLES
**Column:** 150 × 3.9 4 μm Nova pak C18
**Mobile phase:** MeCN:water 57:43
**Column temperature:** 44
**Flow rate:** 1.1
**Injection volume:** 20
**Detector:** UV 254

## CHROMATOGRAM
**Retention time:** 6.3

## OTHER SUBSTANCES
**Simultaneous:** bromazepam, chlordiazepoxide, clobazam, clorazepate, diazepam, flunitrazepam, nitrazepam, oxazepam, tofisopam

## REFERENCE
Guillaume, Y.; Guinchard, C. Marked difference between acetonitrile/water and methanol/water mobile phase systems on the thermodynamic behavior of benzodiazepines in reversed phase liquid chromatography. *Chromatographia*, **1995**, *41*, 84–87

## SAMPLE
**Matrix:** solutions

## HPLC VARIABLES
**Column:** 250 × 4.6 5 μm Supelcosil LC-DP (A) or 250 × 4 5 μm LiChrospher 100 RP-8 (B)
**Mobile phase:** MeCN:0.025% phosphoric acid:buffer 25:10:5 (A) or 60:25:15 (B) (Buffer was 9 mL concentrated phosphoric acid and 10 mL triethylamine in 900 mL water, adjust pH to 3.4 with dilute phosphoric acid, make up to 1 L.)
**Flow rate:** 0.6
**Injection volume:** 25
**Detector:** UV 229

## CHROMATOGRAM
**Retention time:** 6.14 (A), 5.78 (B)

## OTHER SUBSTANCES
**Also analyzed:** acebutolol, acepromazine, acetaminophen, acetazolamide, acetophenazine, albuterol, alprazolam, amitriptyline, amobarbital, amoxapine, antipyrine, atenolol, atro-

pine, azatadine, baclofen, benzocaine, bromocriptine, brompheniramine, brotizolam, bupivacaine, buspirone, butabarbital, butalbital, caffeine, carbamazepine, cetirizine, chlorcyclizine, chlordiazepoxide, chlormezanone, chloroquine, chlorpheniramine, chlorpromazine, chlorpropamide, chlorprothixene, chlorthalidone, chlorzoxazone, cimetidine, cisapride, clomipramine, clonazepam, clonidine, clozapine, cocaine, codeine, colchicine, cyclizine, cyclobenzaprine, dantrolene, desipramine, diazepam, diclofenac, diflunisal, diltiazem, diphenhydramine, diphenidol, diphenoxylate, dipyridamole, disopyramide, dobutamine, doxapram, doxepin, droperidol, encainide, ethidium bromide, ethopropazine, fenoprofen, fentanyl, flavoxate, fluoxetine, fluphenazine, flurazepam, flurbiprofen, fluvoxamine, furosemide, glutethimide, glyburide, guaifenesin, haloperidol, homatropine, hydralazine, hydrochlorothiazide, hydrocodone, hydromorphone, hydroxychloroquine, hydroxyzine, ibuprofen, imipramine, indomethacin, ketoconazole, ketoprofen, ketorolac, labetalol, levorphanol, lidocaine, loratadine, lovastatin, loxapine, mazindol, mefenamic acid, meperidine, mephenytoin, mepivacaine, mesoridazine, metaproterenol, methadone, methdilazine, methocarbamol, methotrexate, methotrimeprazine, methoxamine, methyldopa, methylphenidate, metoclopramide, metolazone, metoprolol, metronidazole, midazolam, moclobemide, morphine, nadolol, nalbuphine, naloxone, naphazoline, naproxen, nifedipine, nizatidine, norepinephrine, nortriptyline, oxazepam, oxycodone, oxymetazoline, paroxetine, pemoline, pentazocine, pentobarbital, pentoxifylline, perphenazine, pheniramine, phenobarbital, phenol, phenolphthalein, phentolamine, phenylbutazone, phenyltoloxamine, phenytoin, pimozide, pindolol, piroxicam, pramoxine, prazepam, prazosin, probenecid, procainamide, procaine, prochlorperazine, procyclidine, promazine, promethazine, propafenone, propantheline, propiomazine, propofol, propranolol, protriptyline, quazepam, quinidine, quinine, racemethorphan, ranitidine, remoxipride, risperidone, salicylic acid, scopolamine, secobarbital, sertraline, sotalol, spironolactone, sulfinpyrazone, sulindac, temazepam, terbutaline, terfenadine, tetracaine, theophylline, thiethylperazine, thiopental, thioridazine, thiothixene, timolol, tocainide, tolbutamide, tolmetin, trazodone, triamterene, triazolam, trifluoperazine, triflupromazine, trimeprazine, trimethoprim, trimipramine, verapamil, warfarin, xylometazoline, yohimbine, zopiclone

## KEY WORDS
some details of plasma extraction

## REFERENCE
Koves, E.M. Use of high-performance liquid chromatography-diode array detection in forensic toxicology. *J.Chromatogr.A*, **1995**, *692*, 103–119

## SAMPLE
**Matrix:** solutions
**Sample preparation:** Prepare a 100 μM solution in buffer, inject a 20 μL aliquot.

## HPLC VARIABLES
**Column:** 100 × 4.6 column containing riboflavin binding proteins (Prepare as follows. Add riboflavin to saturate protein of egg yolk, homogenize with 3 volumes buffer, centrifuge, add the supernatant to a 500 × 30 column of DEAE-cellulose (Whatman) equilibrated with buffer, wash extensively with buffer to remove bound protein, elute riboflavin binding proteins (RFBP) with buffer containing 200 mM NaCl (RFBP has intense yellow color, absorption at 455 nm). Purify RFBP on a Sephadex G-100 column with 50 mM pH 7.5 Tris-HCl buffer as eluent, remove the bound riboflavin by extensive dialysis at pH 3.0. Add 4.5 g N, N-disuccinylimidyl carbonate to 3 g Nucleosil 5NH2 slurried in MeCN, filter, wash with MeCN, wash with 50 mM pH 7.5 phosphate buffer. Suspend 300 mg RFBP in 50 mM phosphate buffer, add the activated silica, mix gently for 2 h using a rotary evaporator, filter, wash with sterile water, wash with isopropanol:water 1:2, pack in a 100 × 4.6 column.) (Buffer was 100 mM pH 5.3 sodium acetate.)
**Mobile phase:** EtOH:50 mM pH 5.5 $KH_2PO_4$ 5:95
**Flow rate:** 0.8
**Injection volume:** 20
**Detector:** UV

## CHROMATOGRAM
**Retention time:** k' 4.02

## OTHER SUBSTANCES
**Simultaneous:** bepridil, manidipine, nicardipine
**Interfering:** oxazepam

## KEY WORDS
chiral; $\alpha$ = 1.63

## REFERENCE
Massolini, G.; De Lorenzi, E.; Ponci, M.C.; Gandini, C.; Caccialanza, G.; Monaco, H.L. Egg yolk riboflavin binding protein as a new chiral stationary phase in high-performance liquid chromatography. *J.Chromatogr.A*, **1995**, *704*, 55–65

## SAMPLE
**Matrix:** solutions

## HPLC VARIABLES
**Column:** 150 × 4.6 cellulose 3,5-dimethylphenylcarbamate/10-undecenoate bonded to allylsilica
**Mobile phase:** Heptane:isopropanol 90:10
**Flow rate:** 1
**Injection volume:** 1000
**Detector:** UV 254

## CHROMATOGRAM
**Retention time:** k' 9.84

## KEY WORDS
chiral; $\alpha$ 1.64

## REFERENCE
Oliveros, L.; Lopez, P.; Minguillon, C.; Franco, P. Chiral chromatographic discrimination ability of a cellulose 3,5-dimethylphenylcarbamate/10-undecenoate mixed derivative fixed on several chromatographic matrices. *J.Liq.Chromatogr.*, **1995**, *18*, 1521–1532

## SAMPLE
**Matrix:** solutions
**Sample preparation:** Dissolve the compound, S-trolox methyl ether (Fluka), dicyclohexylcarbodiimide, and 4-dimethylaminopyridine in dichloromethane, stir at room temperature for 1 h, filter (0.45 μm), inject an aliquot.

## HPLC VARIABLES
**Column:** 300 × 0.32 5 μm LiChrosorb Diol
**Mobile phase:** Carbon dioxide:MeOH 91.5:8.5
**Column temperature:** 80
**Injection volume:** 0.2
**Detector:** UV 254

## CHROMATOGRAM
**Retention time:** 50.2 (second peak)

## KEY WORDS
derivatization; subcritical fluid chromatography; chiral; density of mobile phase 0.62 g/mL; resolution ($R_S$) 1.2

## REFERENCE

Almquist, S.R.; Petersson, P.; Walther, W.; Markides, K.E. Direct and indirect approaches to enantiomeric separation of benzodiazepines using micro column techniques. *J.Chromatogr.A*, **1994**, *679*,
139–146

## SAMPLE

**Matrix:** solutions
**Sample preparation:** Dilute in MeOH to a concentration of 10-80 mg/mL, inject an aliquot

## HPLC VARIABLES

**Column:** 150 × 3.9 5 μm Nova pak RP 18
**Mobile phase:** MeOH:water 50:50
**Column temperature:** 50
**Flow rate:** 0.82
**Injection volume:** 20
**Detector:** UV 254

## CHROMATOGRAM

**Retention time:** 8.5

## OTHER SUBSTANCES

**Simultaneous:** bromazepam, chlorazepate, chlordiazepoxide, clobazam, diazepam, flunitrazepam, nitrazepam, oxazepam, tofisopam

## KEY WORDS

conditions are optimized

## REFERENCE

Guillaume, Y.; Guinchard, C. Study and optimization of column efficiency in HPLC: Comparison of two
methods for separating ten benzodiazepines. *J.Liq.Chromatogr.*, **1994**, *17*, 1443–1459

## SAMPLE

**Matrix:** solutions

## HPLC VARIABLES

**Column:** 250 × 4.6 Zorbax RX
**Mobile phase:** Gradient. A was 10 mL concentrated orthophosphoric acid and 7 mL trie-
thylamine in 1 L water. B was 10 mL concentrated orthophosphoric acid and 7 mL trie-
thylamine in 200 mL water, make up to 1 L with MeCN. A:B from 100:0 to 0:100 over
30 min, maintain at 0:100 for 5 min.
**Column temperature:** 30
**Flow rate:** 2
**Detector:** UV 210

## OTHER SUBSTANCES

**Also analyzed:** acepromazine, acetaminophen, acetophenazine, albuterol, aminophylline,
amitriptyline, amobarbital, amoxapine, amphetamine, amylocaine, antipyrine, aprobarbital, aspirin, atenolol, atropine, avermectin, barbital, benzocaine, benzoic acid, benzotropine, benzphetamine, berberine, bibucaine, bromazepan, brompheniramine, buprenorphine, buspirone, butabarbital, butacaine, butethal, caffeine, carbamazepine, carbromal,
chloramphenicol, chlordiazepoxide, chloroquine, chlorothiazide, chloroxylenol, chlorphenesin, chlorpheniramine, chlorpromazine, chlorpropamide, chlortetracycline, cimetidine,
cinchonidine, cinchonine, clenbuterol, clonazepam, clonixin, clorazepate, cocaine, codeine,
colchicine, cortisone, coumarin, cyclazocine, cyclobenzaprine, cyclothiazide, cyheptamide,
cymarin, danazol, danthron, dapsone, debrisoquine, desipramine, dexamethasone, dex-

tromethorphan, dextropropoxyphene, diamorphine, diazepam, diclofenac, diethylpropion, diethylstilbestrol, diflunisal, digitoxin, digoxin, diltiazem, diphenhydramine, diphenoxylate, diprenorphine, dipyrone, disulfiram, dopamine, doxapram, doxepin, dronabinol, ephedrine, epinephrine, epinine, estradiol, estriol, estrone, ethacrynic acid, ethosuximide, etonitazene, etorphine, eugenol, famotidine, fenbendazole, fencamfamine, fenoprofen, fenproporex, fentanyl, flubendazole, flufenamic acid, flunitrazepam, 5-fluorouracil, fluoxymesterone, fluphenazine, furosemide, gentisic acid, gitoxigenin, glipizide, glunixin, glutethimide, glybenclamide, guaiacol, halazepam, haloperidol, hydrochlorothiazide, hydrocodone, hydrocortisone, hydromorphone, hydroxyquinoline, ibogaine, ibuprofen, iminostilbene, imipramine, indomethacin, isocarbostyril, isocarboxazid, isoniazid, isoproterenol, isoxsuprine, ivermectin, ketamine, ketoprofen, kynurenic acid, levorphanol, lidocaine, lormetazepam, loxapine, mazindol, mebendazole, meclizine, meclofenamic acid, medazepam, mefenamic acid, megestrol, mepacrine, meperidine, mephentermine, mephenytoin, mephesin, mephobarbital, mepivacaine, mescaline, mesoridazine, methadone, methamphetamine, methapyrilene, methaqualone, methazolamide, methocarbamol, methoxamine, methsuximide, methyl salicylate, methyldopa, methyldopamine, methylphenidate, methylprednisolone, methyltestosterone, methyprylon, metoprolol, mibolerone, morphine, nadolol, nalorphine, naloxone, naltrexone, naphazoline, naproxen, nefopam, niacinamide, nicotine, nicotinic acid, nifedipine, niflumic acid, nitrazepam, norepinephrine, nortriptyline, noscapine, nylidrin, oxazepam, oxycodone, oxymorphone, oxyphenbutazone, oxytetracycline, papaverine, pargyline, pemoline, pentazocine, pentobarbital, persantine, phenacetin, phenazocine, phenazopyridine, phencyclidine, phendimetrazine, phenelzine, pheniramine, phenobarbital, phenothiazine, phensuximide, phentermine, phenylbutazone, phenylephrine, phenylpropanolamine, piperocaine, prazepam, prednisolone, primidone, probenecid, progesterone, propiomazine, propranolol, propylparaben, pseudoephedrine, puromycin, pyrilamine, pyrithyldione, quazepam, quinaldic acid, quinidine, quinine, ranitidine, recinnamine, reserpine, resorcinol, saccharin, albuterol, salicylamide, salicylic acid, scopolamine, scopoletin, secobarbital, strychnine, sulfacetamide, sulfadiazine, sulfadimethoxine, sulfaethidole, sulfamerazine, sulfamethazine, sulfamethoxizole, sulfanilamide, sulfapyridine, sulfasoxizole, sulindac, tamoxifen, temazepam, testosterone, tetracaine, tetracycline, tetramisole, thebaine, theobromine, theophylline, thiabendazole, thiamine, thiamylal, thiobarbituric acid, thioridazine, thiosalicylic acid, thiothixene, thymol, tolazamide, tolazoline, tobutamide, tolmetin, tranylcypromine, triamcinolone, tribenzylamine, trichloromethiazide, trifluoperazine, trihexyphenidyl, trimethoprim, tripelennamine, triprolidine, tropacocaine, tyramine, verapamil, vincamine, warfarin, yohimbine, zoxazolamine

## REFERENCE

Hill, D.W.; Kind, A.J. Reversed-phase solvent gradient HPLC retention indexes of drugs. *J.Anal.Toxicol.*, **1994**, *18*, 233–242

## SAMPLE
**Matrix:** solutions

## HPLC VARIABLES
**Column:** 250 × 4.6 10 μm Chiralpak OD (Daicel)
**Mobile phase:** Carbon dioxide:MeCN:EtOH:diethylamine 69.5:15:15:0.5
**Column temperature:** 30
**Flow rate:** 2
**Detector:** UV 220

## CHROMATOGRAM
**Retention time:** 5.2, 7 (enantiomers)

## KEY WORDS
SFC; chiral; pressure 200 bar

## REFERENCE

Kot, A.; Sandra, P.; Venema, A. Sub- and supercritical fluid chromatography on packed columns: A versatile tool for the enantioselective separation of basic and acidic drugs. *J.Chromatogr.Sci.*, **1994**, *32*, 439–448

◆ ━━━━━━━━━━━━━━━━━━━━━ ◆ ━━━━━━━━━━━━━━━━━━━━━ ◆

## ANNOTATED BIBLIOGRAPHY

Haginaka, J.; Kanasugi, N. Enantioselectivity of bovine serum albumin-bonded columns produced with isolated protein fragments. *J.Chromatogr.A*, **1995**, *694*, 71–80 [chiral; also benzoin, clorazepate, fenoprofen, flurbiprofen, ibuprofen, ketoprofen, lormetazepam, oxazepam, pranoprofen, temazepam, warfarin]

Ficarra, R.; Ficarra, P.; Tommasini, S.; Carulli, M.; Costantino, D.; Calabrò, M.L. Chromatographic investigations of brotizolam. *Farmaco*, **1994**, *49*, 437–440 [simultaneous brotizolam; tablets; lorazepam is IS]

Herman, R.J.; Chaudhary, A.; Szakacs, C.B. Disposition of lorazepam in Gilbert's syndrome: Effects of fasting, feeding, and enterohepatic circulation. *J.Clin.Pharmacol.*, **1994**, *34*, 978–984 [plasma; urine]

Fujima, H.; Wada, H.; Miwa, T.; Haginaka, J. Chiral separation of lorazepam on ovomucoid-bonded columns: Peak coalescence due to racemization. *J.Liq.Chromatogr.*, **1993**, *16*, 879–891 [chiral; column temp 7]

Kondo, T.; Buss, D.C.; Routledge, P.A. A method for rapid determination of lorazepam by high-performance liquid chromatography. *Ther.Drug Monit.*, **1993**, *15*, 35–38

Gunawan, S.; Walton, N.Y.; Treiman, D.M. Analysis of lorazepam in rat brain using liquid/liquid and solid-phase extraction in combination with high performance liquid chromatography. *Biomed.Chromatogr.*, **1990**, *4*, 168–170 [SPE]

Noggle, F.T., Jr.; Clark, C.R.; DeRuiter, J. Liquid chromatographic separation of some common benzodiazepines and their metabolites. *J.Liq.Chromatogr.*, **1990**, *13*, 4005–4021 [also alprazolam, bromazepam, chlordiazepoxide, clobazam, clonazepam, clorazepate, demoxepam, diazepam, estazolam, fludiazepam, flunitrazepam, flurazepam, halazepam, lorazepam, midazolam, nitrazepam, nordiazepam, oxazepam, prazepam, temazepam, triazolam]

Gunawan, S.; Treiman, D.M. Determination of lorazepam in plasma of patients during status epilepticus by high-performance liquid chromatography. *Ther.Drug Monit.*, **1988**, *10*, 172–176

Pietrogrande, M.C.; Dondi, F.; Blo, G.; Borea, P.A.; Bighi, C. Retention behavior of benzodiazepines in normal-phase HPLC. Silica, cyano, and amino phases comparison. *J.Liq.Chromatogr.*, **1988**, *11*, 1313–1333 [also metabolites diazepam, medazepam, methyllorazepam, oxazepam, prazepam, temazepam]

Wong, S.H.Y.; McHugh, S.L.; Dolan, J.; Cohen, K.A. Tricyclic antidepressant analysis by reversed-phase liquid chromatography using phenyl columns. *J.Liq.Chromatogr.*, **1986**, *9*, 2511–2538 [also acetaminophen, amitriptyline, amobarbital, amoxapine, barbital, chlordiazepoxide, chlorpromazine, cimetidine, clomipramine, codeine, desipramine, desmethyldoxepin, diazepam, doxepin, fluphenazine, flurazepam, glutethimide, hydroxyamoxapine, imipramine, maprotiline, meperidine, metabolites, nortriptyline, oxazepam, pentobarbital, perphenazine, phenobarbital, phenytoin, propoxyphene, protriptyline, secobarbital, thioridazine, trazodone]

Pietrogrande, M.C.; Bighi, C.; Borea, P.A.; Barbaro, A.M.; Guerra, M.C.; Biagi, G.L. Relationship between log k' values of benzodiazepines and composition of the mobile phase. *J.Liq.Chromatogr.*, **1985**, *8*, 1711–1729 [also carbenicillin, chlordiazepoxide, diazepam, dicloxacillin, flurazepam, medazepam, methyl lorazepam, methyllorazepam, nitrazepam, oxazepam, prazepam, temazepam, testosterone]

Walmsley, L.M.; Chasseaud, L.F. High-performance liquid chromatographic determination of lorazepam in monkey plasma. *J.Chromatogr.*, **1981**, *226*, 155–163

# Lovastatin

**Molecular formula:** C₂₄H₃₆O₅

**Molecular weight:** 404.6

**CAS Registry No.:** 75330-75-5

## SAMPLE

**Matrix:** bile, blood

**Sample preparation:** Prepare a 1 mL Bond-Elut C2 SPE cartridge by washing with 1 mL MeOH and 1 mL buffer. Plasma. 500 µL Plasma + 500 µL buffer, keep in an ice water mixture, add 100 µL MeCN:water 60:40, add 50 µL 5 µg/mL methyl mevinolinic acid in MeCN:water 60:40, mix, add to the SPE cartridge, wash twice with 1 mL buffer, wash twice with 1 mL MeCN:buffer 20:80, elute with 400 µL MeCN:water 75:25, inject a 30 µL aliquot of eluant. Bile. Inject a 10 µL aliquot directly. (Buffer was 100 mM pH 7.2 potassium phosphate buffer.)

## HPLC VARIABLES

**Column:** 50 × 4.6 3 µm Sepralyte C18

**Mobile phase:** MeCN:50 mM ammonium phosphate + 10 mM phosphoric acid. Isocratic 50:50 (plasma). Gradient 20:80 to 75:25 over 10 min (bile).

**Flow rate:** 1.5

**Injection volume:** 30 (plasma), 10 (bile)

**Detector:** UV 238

## CHROMATOGRAM

**Retention time:** 4.95

**Internal standard:** methyl mevinolinic acid (3.50)

**Limit of detection:** 25 ng/mL

## OTHER SUBSTANCES

**Simultaneous:** mevinolinic acid

## KEY WORDS

plasma; human; rat; dog; SPE

## REFERENCE

Stubbs, R.J.; Schwartz, M.; Bayne, W.F. Determination of mevinolin and mevinolinic acid in plasma and bile by reversed-phase high-performance liquid chromatography. *J.Chromatogr.*, **1986**, *383*, 438–443

## SAMPLE

**Matrix:** bile, microsomal incubations

**Sample preparation:** Microsomal incubations. Extract 10 mL microsomal incubations with 30 mL ethyl acetate:acetone 2:1, centrifuge. Remove organic layer and dry it over anhydrous sodium sulfate, filter, evaporate under vacuum, reconstitute in n-propanol, inject an aliquot. Bile. Acidify to pH 3.5 with 20 mM formic acid, extract with toluene: n-propanol 85:15, extract with n-butanol saturated with water, combine extracts, evaporate under vacuum, reconstitute in n-propanol, inject an aliquot.

## HPLC VARIABLES

**Column:** 300 × 10 µBondapak ODS

**Mobile phase:** Gradient. MeCN:water from 30:70 to 90:10 over 30 min.
**Flow rate:** 3
**Detector:** UV 238

## CHROMATOGRAM
**Retention time:** 26

## OTHER SUBSTANCES
**Simultaneous:** metabolites

## REFERENCE
Vyas, K.P.; Kari, P.H.; Pitzenberger, S.M.; Halpin, R.A.; Ramjit, H.G.; Arison, B.; Murphy, J.S.; Hoffman, W.F.; Schwartz, M.S.; Ulm, E.H. Biotransformation of lovastatin. I. Structure elucidation of *in vitro* and *in vivo* metabolites in the rat and mouse. *Drug Metab.Dispos.*, **1990**, *18*, 203–211

## SAMPLE
**Matrix:** microsomal incubations
**Sample preparation:** 1 mL Microsomal incubation + 1 mL acetone + 2 mL ethyl acetate, extract, centrifuge. Remove the organic layer and dry it over anhydrous sodium sulfate, concentrate under a stream of nitrogen. Reconstitute in 200 µL n-propanol, inject 20-50 µL aliquot.

## HPLC VARIABLES
**Column:** 250 × 4.6 5 µm Ultrasphere ODS
**Mobile phase:** Gradient. MeCN:5 mM formic acid from 30:70 to 90:10 over 30 min.
**Flow rate:** 1
**Injection volume:** 20-50
**Detector:** UV 238

## CHROMATOGRAM
**Retention time:** 25

## OTHER SUBSTANCES
**Simultaneous:** metabolites

## REFERENCE
Vyas, K.P.; Kari, P.H.; Prakash, S.R.; Duggan, D.E. Biotransformation of lovastatin. II. In vitro metabolism by rat and mouse liver microsomes and involvement of cytochrome P-450 in dehydrogenation of lovastatin. *Drug Metab.Dispos.*, **1990**, *18*, 218–222

## SAMPLE
**Matrix:** solutions

## HPLC VARIABLES
**Column:** 250 × 4.6 5 µm Supelcosil LC-DP (A) or 250 × 4 5 µm LiChrospher 100 RP-8 (B)
**Mobile phase:** MeCN:0.025% phosphoric acid:buffer 25:10:5 (A) or 60:25:15 (B) (Buffer was 9 mL concentrated phosphoric acid and 10 mL triethylamine in 900 mL water, adjust pH to 3.4 with dilute phosphoric acid, make up to 1 L.)
**Flow rate:** 0.6
**Injection volume:** 25
**Detector:** UV 229

## CHROMATOGRAM
**Retention time:** 10.49 (A), 14.79 (B)

## OTHER SUBSTANCES
**Also analyzed:** acebutolol, acepromazine, acetaminophen, acetazolamide, acetophenazine, albuterol, alprazolam, amitriptyline, amobarbital, amoxapine, antipyrine, atenolol, atropine, azatadine, baclofen, benzocaine, bromocriptine, brompheniramine, brotizolam, bupivacaine, buspirone, butabarbital, butalbital, caffeine, carbamazepine, cetirizine, chlorcyclizine, chlordiazepoxide, chlormezanone, chloroquine, chlorpheniramine, chlorpromazine, chlorpropamide, chlorprothixene, chlorthalidone, chlorzoxazone, cimetidine, cisapride, clomipramine, clonazepam, clonidine, clozapine, cocaine, codeine, colchicine, cyclizine, cyclobenzaprine, dantrolene, desipramine, diazepam, diclofenac, diflunisal, diltiazem, diphenhydramine, diphenidol, diphenoxylate, dipyridamole, disopyramide, dobutamine, doxapram, doxepin, droperidol, encainide, ethidium bromide, ethopropazine, fenoprofen, fentanyl, flavoxate, fluoxetine, fluphenazine, flurazepam, flurbiprofen, fluvoxamine, furosemide, glutethimide, glyburide, guaifenesin, haloperidol, homatropine, hydralazine, hydrochlorothiazide, hydrocodone, hydromorphone, hydroxychloroquine, hydroxyzine, ibuprofen, imipramine, indomethacin, ketoconazole, ketoprofen, ketorolac, labetalol, levorphanol, lidocaine, loratadine, lorazepam, loxapine, mazindol, mefenamic acid, meperidine, mephenytoin, mepivacaine, mesoridazine, metaproterenol, methadone, methdilazine, methocarbamol, methotrexate, methotrimeprazine, methoxamine, methyldopa, methylphenidate, metoclopramide, metolazone, metoprolol, metronidazole, midazolam, moclobemide, morphine, nadolol, nalbuphine, naloxone, naphazoline, naproxen, nifedipine, nizatidine, norepinephrine, nortriptyline, oxazepam, oxycodone, oxymetazoline, paroxetine, pemoline, pentazocine, pentobarbital, pentoxifylline, perphenazine, pheniramine, phenobarbital, phenol, phenolphthalein, phentolamine, phenylbutazone, phenyltoloxamine, phenytoin, pimozide, pindolol, piroxicam, pramoxine, prazepam, prazosin, probenecid, procainamide, procaine, prochlorperazine, procyclidine, promazine, promethazine, propafenone, propantheline, propiomazine, propofol, propranolol, protriptyline, quazepam, quinidine, quinine, racemethorphan, ranitidine, remoxipride, risperidone, salicylic acid, scopolamine, secobarbital, sertraline, sotalol, spironolactone, sulfinpyrazone, sulindac, temazepam, terbutaline, terfenadine, tetracaine, theophylline, thiethylperazine, thiopental, thioridazine, thiothixene, timolol, tocainide, tolbutamide, tolmetin, trazodone, triamterene, triazolam, trifluoperazine, triflupromazine, trimeprazine, trimethoprim, trimipramine, verapamil, warfarin, xylometazoline, yohimbine, zopiclone

## KEY WORDS
some details of plasma extraction

## REFERENCE
Koves, E.M. Use of high-performance liquid chromatography-diode array detection in forensic toxicology. *J.Chromatogr.A*, **1995**, *692*, 103–119

## SAMPLE
**Matrix:** solutions
**Sample preparation:** Centrifuge at 2000 rpm, inject an aliquot.

## HPLC VARIABLES
**Column:** 150 × 4.6 5 µm Hypersil ODS
**Mobile phase:** MeCN:water:triethylamine:glacial acetic acid 500:500:1:1
**Column temperature:** 30
**Flow rate:** 2
**Detector:** UV 238

## CHROMATOGRAM
**Retention time:** 14.4 (4.4 hydroxy acid form)
**Limit of detection:** 10 ng/mL
**Limit of quantitation:** 25 ng/mL

## REFERENCE

Serajuddin, A.T.; Ranadive, S.A.; Mahoney, E.M. Relative lipophilicities, solubilities, and structure-pharmacological considerations of 3-hydroxy-3-methylglutaryl-coenzyme A (HMG-CoA) reductase inhibitors pravastatin, lovastatin, mevastatin, and simvastatin. *J.Pharm.Sci.*, **1991**, *80*, 830–834

## SAMPLE

**Matrix:** tissue
**Sample preparation:** Homogenize in 4 mL water, add 4 mL ice-cold acetone, centrifuge at 3000 rpm for 15 min, extract twice with an equal volume of hexanes. Evaporate extracts under vacuum, reconstitute in n-propanol, inject an aliquot.

## HPLC VARIABLES

**Column:** 300 × 7 Hamilton PRP-1
**Mobile phase:** Gradient. MeCN:5 mM formic acid from 30:70 to 90:10 over 30 min.
**Flow rate:** 3
**Detector:** UV 238

## CHROMATOGRAM

**Retention time:** 19

## OTHER SUBSTANCES

**Simultaneous:** metabolites

## KEY WORDS

liver

## REFERENCE

Vyas, K.P.; Kari, P.H.; Pitzenberger, S.M.; Halpin, R.A.; Ramjit, H.G.; Arison, B.; Murphy, J.S.; Hoffman, W.F.; Schwartz, M.S.; Ulm, E.H. Biotransformation of lovastatin. I. Structure elucidation of *in vitro* and *in vivo* metabolites in the rat and mouse. *Drug Metab.Dispos.*, **1990**, *18*, 203–211

# Medroxyprogesterone Acetate

**Molecular formula:** $C_{24}H_{34}O_4$
**Molecular weight:** 386.5
**CAS Registry No.:** 71-58-9 (medroxyprogesterone acetate),
520-85-4 (medroxyprogesterone)

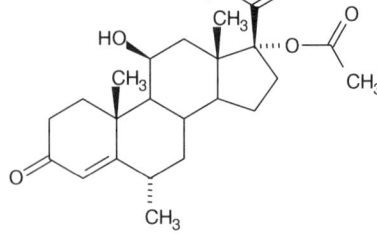

## SAMPLE
**Matrix:** blood
**Sample preparation:** Extract 0.1-1 mL plasma twice with 40 volumes of diethyl ether, evaporate the organic solvent, dissolve the residue in 100 μL MeOH:water 80:20, inject a 30 μL aliquot.

## HPLC VARIABLES
**Column:** 250 × 4.6 5 μm Beckman RP-ODS
**Mobile phase:** Gradient. MeCN:water from 40:60 to 70:30 over 20 min, then at 70:30 for 5 min.
**Injection volume:** 30
**Detector:** UV 238

## CHROMATOGRAM
**Retention time:** 20.5

## OTHER SUBSTANCES
**Extracted:** metabolites

## KEY WORDS
plasma

## REFERENCE
Sturm, G.; Haberlein, H.; Bauer, T.; Plaum, T.; Stalker, D.J. Mass spectrometric and high-performance liquid chromatographic studies of medroxyprogesterone acetate metabolites in human plasma. *J.Chromatogr.*, **1991**, *562*, 351–362

## SAMPLE
**Matrix:** blood
**Sample preparation:** 2 mL Plasma + 1 mL 0.5 μg/mL 16β-methylprogesterone in 200 mM pH 7.0 phosphate buffer, vortex for 3 s, add 7 mL hexane, mix on a rolling mixer for 30 min. Remove the hexane layer and evaporate it to dryness at 30° under nitrogen. Dissolve residue in 200 μL mobile phase, inject a 150 μL aliquot.

## HPLC VARIABLES
**Column:** 250 × 4.6 5 μm Spherisorb 5-ODS2
**Mobile phase:** MeOH: 20 mM pH 4 acetate buffer 79:21
**Flow rate:** 1.5
**Injection volume:** 150
**Detector:** UV 240

## CHROMATOGRAM
**Retention time:** 5.3
**Internal standard:** 16β-methylprogesterone (9.0)
**Limit of detection:** 10 ng/mL

## OTHER SUBSTANCES
**Simultaneous:** aldosterone, androstenedione, corticosterone, cortisone, estradiol, 17α-hydroxyprogesterone, progesterone, testosterone
**Noninterfering:** cholesterol
**Interfering:** lignocaine

## KEY WORDS
plasma

## REFERENCE
Read, J.; Mould, G.; Stevenson, D. Simple high-performance liquid chromatographic method for the determination of medroxyprogesterone acetate in human plasma. *J.Chromatogr.*, **1985**, *341*, 437–444

## SAMPLE
**Matrix:** formulations
**Sample preparation:** Grind tablets to a fine powder, weigh out an amount equivalent to about 10 mg medroxyprogesterone acetate, add 10 mL MeOH, sonicate for 5 min, dilute to 50 mL with MeOH. Dilute 25 mL of this solution to 50 mL with MeOH, filter (0.45 μm), inject a 20 μL aliquot.

## HPLC VARIABLES
**Column:** 150 × 3.9 5 μm Novapak C18
**Mobile phase:** MeOH:10 mM $(NH_4)_2HPO_4$ 80:20, adjust pH to 7.2 ± 0.1 with 85% phosphoric acid
**Flow rate:** 1
**Injection volume:** 20
**Detector:** UV 254

## CHROMATOGRAM
**Retention time:** 6.7

## KEY WORDS
tablets; stability-indicating

## REFERENCE
Fatmi, A.A.; Williams, G.V.; Hickson, E.A. Liquid chromatographic determination of medroxyprogesterone acetate in tablets. *J.Assoc.Off.Anal.Chem.*, **1988**, *71*, 528–530

## SAMPLE
**Matrix:** formulations
**Sample preparation:** Suspensions. Dilute 2 mL suspension to 1 L with EtOH, remove a 2.5 mL aliquot and add it to 1 mL 1 mg/mL hydrocortisone in EtOH. Dilute this mixture to 50 mL with EtOH, inject an aliquot. Tablets. Grind tablets to a fine powder, stir with 50 mL EtOH, make up to 100 mL with EtOH, filter (Whatman No. 1 paper), reject the first 20 mL of the filtrate. Mix 2 mL of the filtrate with 1 mL 1 mg/mL hydrocortisone in EtOH, make this mixture up to 50 mL with EtOH, inject an aliquot.

## HPLC VARIABLES
**Column:** 300 × 4 μBondapak CN
**Mobile phase:** MeOH:20 mM $KH_2PO_4$ 30:70
**Flow rate:** 2
**Injection volume:** 20
**Detector:** UV 254

## CHROMATOGRAM
**Retention time:** 7.5
**Internal standard:** hydrocortisone (2.5)

## OTHER SUBSTANCES
**Simultaneous:** benzyl benzoate, progesterone
**Noninterfering:** methylcellulose, mystristyl-gamma-picolinium chloride, polyethylene glycol 4000, thimerosal

## REFERENCE
Das Gupta, V. Quantitation of hydroxyprogesterone caproate, medroxyprogesterone acetate, and progesterone by reversed-phase high-pressure liquid chromatography. *J.Pharm.Sci.*, **1982**, *71*, 294–297

## SAMPLE
**Matrix:** solutions
**Sample preparation:** Prepare a 25 µg/mL solution in mobile phase, inject an aliquot.

## HPLC VARIABLES
**Column:** 250 × 4.6 Partisil 10 ODS-1
**Mobile phase:** MeOH:water 55:45
**Column temperature:** 40
**Flow rate:** 1.5
**Detector:** UV 240

## CHROMATOGRAM
**Retention time:** k′ 6.712 (medroxyprogesterone acetate), k′ 3.430 (medroxyprogesterone)

## OTHER SUBSTANCES
**Also analyzed:** androsterone (UV 210), cortexolone (UV 240), cortisone (UV 240), estradiol (UV 280), estrone (UV 280), ethinyl estradiol (UV 280), ethisterone (UV 240), hydrocortisone (UV 240), hydroxyprogesterone (UV 240), lynestrenol (UV 210), methandienone (UV 240), methylandrostenediol (UV 210), methylprednisolone acetate (UV 240), methylprednisolone (UV 240), methyltestosterone (UV 240), nandrolone (UV 240), norethisterone (UV 240), prednisolone acetate (UV 240), prednisolone (UV 240), prednisone (UV 240), pregnenolone (UV 210), progesterone (UV 240), testosterone (UV 240)

## REFERENCE
Sadlej-Sosnowska, N. Structure retention relationship for steroid hormones. Functional groups as structural descriptors. *J.Liq.Chromatogr.*, **1994**, *17*, 2319–2330

## SAMPLE
**Matrix:** solutions

## HPLC VARIABLES
**Column:** 150 × 4.6 5 µm Hypersil ODS
**Mobile phase:** MeOH:water 60:40
**Injection volume:** 250
**Detector:** UV

## CHROMATOGRAM
**Retention time:** 10 (for medroxyprogesterone)

## OTHER SUBSTANCES
**Simultaneous:** dienestrol, diethylstilbestrol, hexestrol, 17α-methyltestosterone, nandrolone, trenbolone, zeranol

## REFERENCE
Jansen, E.H.J.M.; Both-Miedema, R.; van den Berg, R.H. Application of optimization procedures for the separation of anabolic compounds by high-performance liquid chromatography. *J.Chromatogr.*, **1989**, *489*, 57–64

## SAMPLE
**Matrix:** solutions
**Sample preparation:** Dissolve in MeOH:water 1:1 at a concentration of 50 μg/mL, inject a 10 μL aliquot.

## HPLC VARIABLES
**Column:** 300 × 3.9 10 μm μBondapak C18
**Mobile phase:** MeOH:acetic acid:triethylamine:water 70:1.5:0.5:28
**Flow rate:** 1.5
**Injection volume:** 10
**Detector:** UV

## CHROMATOGRAM
**Retention time:** k′ 3.12

## REFERENCE
Roos, R.W.; Lau-Cam, C.A. General reversed-phase high-performance liquid chromatographic method for the separation of drugs using triethylamine as a competing base. *J.Chromatogr.*, **1986**, *370*, 403–418

## SAMPLE
**Matrix:** tissue
**Sample preparation:** Homogenize (Waring blender) tissue at full speed for 2 min, lyophilize, grind. Extract with supercritical carbon dioxide at 60° at 400 atmospheres with a 20 cm × 21 μm restrictor for 1 h, collect the extract in 1 mL MeOH cooled to 5°. Evaporate the MeOH to dryness under a stream of nitrogen, reconstitute the residue in 100 μL MeCN:MeOH:20 mM ammonium formate 15:15:70, inject an aliquot. Alternatively, vortex 5 g ground tissue with 10 mL 40 mM sodium acetate, adjust pH to 4.2-4.7 with glacial acetic acid, add 100 μL β-glucuronidase (Sigma), heat at 37° for 8 h, add 20 mL MeCN, vortex for 30 s, centrifuge at 5000 rpm for 20 min. Remove a 30 mL aliquot of the supernatant and add it to 8 mL hexane and 2 mL dichloromethane, rotate for 3 min, centrifuge at 2000 rpm for 2 min. Remove a 15 mL aliquot of the middle layer and evaporate it to dryness under a stream of nitrogen, reconstitute the residue in 1 mL dichloromethane, inject an aliquot.

## HPLC VARIABLES
**Column:** 50 × 4.6 5 μm Supelcosil
**Mobile phase:** Gradient. MeCN:MeOH:20 mM ammonium formate from 2.5:2.5:95 to 47.5:47.5:5 over 19 min.
**Flow rate:** 1
**Injection volume:** 20
**Detector:** UV 245; MS, Sciex TAGA 6000E tandem triple quadrupole, APCI

## CHROMATOGRAM
**Retention time:** 13.7
**Limit of detection:** 100 ppb

## OTHER SUBSTANCES
**Extracted:** dexamethasone, diethylstilbestrol, melengestrol acetate, trenbolone, triamcinolone acetonide, zeranol

## KEY WORDS
cow; muscle; liver; SFE

## REFERENCE
Huopalahti, R.P.; Henion, J.D. Application of supercritical fluid extraction and high performance liquid chromatography/mass spectrometry for the determination of some anabolic agents directly from bovine tissue samples. *J.Liq.Chrom.Rel.Technol.*, **1996**, *19*, 69–87

## SAMPLE
**Matrix:** urine
**Sample preparation:** 10 mL Urine + glucuronidase/sulfatase (Helix pomatia), incubate at 37° for 1 h, extract twice with 5 mL diethyl ether, add 225 μL water and evaporate ether under nitrogen, add 400 μL MeOH, inject a 250 μL aliquot of this mixture.

## HPLC VARIABLES
**Guard column:** 75 × 2.1 Corasil C18
**Column:** 150 × 4.6 5 μm Hypersil ODS
**Mobile phase:** MeOH:water 60:40
**Flow rate:** 2
**Injection volume:** 250
**Detector:** UV 240

## CHROMATOGRAM
**Retention time:** 14 (for medroxyprogesterone)
**Limit of detection:** about 6 ng/mL

## OTHER SUBSTANCES
**Simultaneous:** trans-diethylstilbestrol, 17α-methyltestosterone, nandrolone, 17β-trenbolone, zeranol

## KEY WORDS
cow

## REFERENCE
Jansen, E.H.; Both-Miedema, R.; van Blitterswijk, H.; Stephany, R.W. Separation and purification of several anabolics present in bovine urine by isocratic high-performance liquid chromatography. *J.Chromatogr.*, **1984**, *299*, 450–455

◆──────────────────────────◆──────────────────────────◆

## ANNOTATED BIBLIOGRAPHY
Mould, G.P.; Read, J.; Edwards, D.; Bye, A. A comparison of the high-performance liquid chromatography and RIA measurement of medroxyprogesterone acetate. *J.Pharm.Biomed.Anal.*, **1989**, 7, 119–122

Rees, H.D.; Bonsall, R.W.; Michael, R.P. Pre-optic and hypothalamic neurons accumulate [³H]medroxyprogesterone acetate in male cynomolgus monkeys. *Life Sci.*, **1986**, *39*, 1353–1359 [brain; tissue; UV detection]

Milano, G.; Carle, G.; Renee, N.; Boublil, J.L.; Namer, M. Determination of medroxyprogesterone acetate in plasma by high-performance liquid chromatography. *J.Chromatogr.*, **1982**, *232*, 413–417

# Methylphenidate

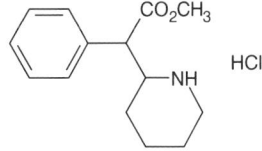

**Molecular formula:** $C_{14}H_{19}NO_2$
**Molecular weight:** 233.3
**CAS Registry No.:** 113-45-1 (methylphenidate), 298-59-9
(methylphenidate hydrochloride)

## SAMPLE
**Matrix:** blood
**Sample preparation:** 1 mL Plasma + 40 µL 1 µg/mL ethylphenidate + 1 mL 200 mM pH 9.1 carbonate buffer + 5 mL hexane:ethyl acetate 75:25, mix for 10 min, centrifuge at 1000 g for 5 min. Remove 4.5 mL of the organic layer and evaporate it to dryness under a stream of nitrogen at 55°, reconstitute the residue in 100 µL MeCN, inject a 75 µL aliquot.

## HPLC VARIABLES
**Column:** 150 × 4.6 5 µm Ultrasphere ODS
**Mobile phase:** MeCN:0.07% triethylamine 35:65, adjust pH to 3.4 with concentrated phosphoric acid.
**Flow rate:** 1.5
**Injection volume:** 75
**Detector:** UV 192

## CHROMATOGRAM
**Retention time:** 3.2
**Internal standard:** ethylphenidate (Prepare ethylphenidate by refluxing methylphenidate in acidic EtOH for 72 h.) (4.6)
**Limit of detection:** 2.5 ng/mL
**Limit of quantitation:** 5 ng/mL

## OTHER SUBSTANCES
**Noninterfering:** acetaminophen, caffeine, cimetidine, diazepam, diltiazem, doxepin, flurazepam, nifedipine, procainamide, quinidine, theophylline, verapamil

## KEY WORDS
plasma

## REFERENCE
Lalande, M.; Wilson, D.L.; McGilveray, I.J. HPLC determination of methylphenidate in human plasma. *J.Liq.Chromatogr.*, **1987**, *10*, 2257–2264

## SAMPLE
**Matrix:** blood, saliva, tissue, urine
**Sample preparation:** Homogenize (Polytron) tissue with 4 (whole brain) or 8 (brain striata) volumes of 100 mM pH 4.5 $NaH_2PO_4$ containing 0.5% NaF. Add 500 µL brain homogenate or 500 µL plasma, saliva, or urine containing 15 µL saturated NaF solution to 75 µL 150 µg/mL IS, add 50 µL 50% perchloric acid, mix vigorously for 10 s, let stand at room temperature for 10 min, add 1 mL water, mix briefly, centrifuge at 10° at 2500 (?) for 30 min. Remove the supernatant and add it to 750 µL saturated sodium carbonate solution, mix briefly, add 7.5 mL pentane:chloroform 95:5, rock gently for 10 min, centrifuge in a desk-top centrifuge for 2 min, freeze in dry ice/acetone for 2 min. Remove the organic layer and add it to 250 µL 100 mM HCl, mix vigorously for 10 s, centrifuge in a desk-top centrifuge for 1-2 min, freeze in dry ice/acetone for 3-5 min, discard the organic

layer. Allow the aqueous layer to thaw, remove any trace of organic solvent with a stream of nitrogen, inject a 75 μL aliquot of the aqueous layer.

## HPLC VARIABLES
**Guard column:** 15 × 3.2 7 μm Brownlee RP-8
**Column:** 250 × 4.6 5 μm Zorbax RX-C18
**Mobile phase:** MeCN:buffer 18:82 (Buffer was 100 mM K$_2$HPO$_4$ containing 0.5% triethylamine, adjusted to pH 2.7 with phosphoric acid.)
**Flow rate:** 2
**Injection volume:** 75
**Detector:** UV 235

## CHROMATOGRAM
**Retention time:** 4.6
**Internal standard:** 2β-carbomethoxy-3β-(4-chlorophenyl)tropane (RTI-31) (Research Biochemical International, Natick MA) (11.4)

## OTHER SUBSTANCES
**Extracted:** chlordiazepoxide, clozapine, cocaine, gepirone, pentazocine, pseudococaine
**Simultaneous:** acetaminophen, acetophenazine, amoxapine, amphetamine, atropine, benperidol, buspirone, caffeine, carbamazepine, chlorpheniramine, codeine, dextromethorphan, diazepam, diphenhydramine, flupenthixol, flurazepam, haloperidol, hydergine, hydrocodone, hydromorphone, lidocaine, loxapine, mepazine, meperidine, mesoridazine, methaquaalude, 3,4-methylenedioxyamphetamine, 3,4-methylenedioxyethylamphetamine, 3,4-methylenedioxymethamphetamine, morphine, norcocaine, oxazepam, pentobarbital, phenylpropanolamine, procainamide, procaine, propyl benzoylecgonine, quinidine, quinine, salicylic acid, secobarbital, theophylline, trazodone, 3-tropanyl-3,5-dichlorobenzoate, vancomycin, WIN 35428
**Noninterfering:** amitriptyline, benztropine methanesulfonate, butaperazine, butriptyline, carphenazine, chlorpromazine, clomipramine, cyclobenzaprine, dextropropoxyphene, dronabinol, ephedrine, ethchlorvynol, fluoxetine, fluphenazine, imipramine, meprobamate, methadone, methamphetamine, nicotine, norfluoxetine, nortriptyline, PCP, phenothiazine, pseudoephedrine

## KEY WORDS
rat; cow; plasma; brain

## REFERENCE
Bonate, P.L.; Davis, C.M.; Silverman, P.B.; Swann, A. Determination of cocaine in biological matrices using reversed phase HPLC: Application to plasma and brain tissue. *J.Liq.Chromatogr.*, **1995**, *18*, 3473–3494

## SAMPLE
**Matrix:** bulk
**Sample preparation:** Dissolve 100 mg bulk drug in 5 mL buffer, extract with 5 mL chloroform, repeat the extraction. Combine the organic layers and evaporate them to dryness under a stream of nitrogen at 35°, reconstitute the residue in 1 mL chloroform, inject a 20 μL aliquot. (Prepare buffer by adjusting the pH of 100 mM sodium bicarbonate to 10 with 100 mM NaOH.)

## HPLC VARIABLES
**Column:** 100 × 2.1 Sil-X (Perkin-Elmer)
**Mobile phase:** Chloroform:cyclohexane:EtOH:concentrated ammonium hydroxide 85:13.5:1.5:0.5
**Flow rate:** 0.4
**Injection volume:** 20
**Detector:** UV 254

## CHROMATOGRAM
**Retention time:** 20
**Limit of detection:** 0.1%

## KEY WORDS
normal phase; separation of diastereomers

## REFERENCE
Padmanabhan, G.R.; Fogel, J.; Mollica, J.A.; O'Connor, J.M.; Strusz, R. Application of high pressure liquid chromatography to the determination of diastereomer in methylphenidate hydrochloride. *J.Liq.Chromatogr.*, **1980**, *3*, 1079–1085

## SAMPLE
**Matrix:** solutions
**Sample preparation:** Inject a 10 µL aliquot.

## HPLC VARIABLES
**Column:** 150 × 4.6 Bio-Sil C18 HL 90-5 (Bio-Rad)
**Mobile phase:** Gradient. MeCN:80 mM phosphoric acid from 0:100 to 70:30 over 10 min.
**Flow rate:** 1
**Injection volume:** 10
**Detector:** UV 200

## CHROMATOGRAM
**Retention time:** 14

## OTHER SUBSTANCES
**Simultaneous:** amphetamine, cocaine, hydroxyamphetamine, nylidrin, oxymetazoline, phendimetrazine, tetrahydrozoline

## REFERENCE
*Life Science Research Products, Bio-Rad, Hercules CA,* **1995**, p. 76

## SAMPLE
**Matrix:** solutions

## HPLC VARIABLES
**Column:** 250 × 4.6 5 µm Supelcosil LC-DP (A) or 250 × 4 5 µm LiChrospher 100 RP-8 (B)
**Mobile phase:** MeCN:0.025% phosphoric acid:buffer 25:10:5 (A) or 60:25:15 (B) (Buffer was 9 mL concentrated phosphoric acid and 10 mL triethylamine in 900 mL water, adjust pH to 3.4 with dilute phosphoric acid, make up to 1 L.)
**Flow rate:** 0.6
**Injection volume:** 25
**Detector:** UV 229

## CHROMATOGRAM
**Retention time:** 8.61 (A), 4.65 (B)

## OTHER SUBSTANCES
**Also analyzed:** acebutolol, acepromazine, acetaminophen, acetazolamide, acetophenazine, albuterol, alprazolam, amitriptyline, amobarbital, amoxapine, antipyrine, atenolol, atropine, azatadine, baclofen, benzocaine, bromocriptine, brompheniramine, brotizolam, bupivacaine, buspirone, butabarbital, butalbital, caffeine, carbamazepine, cetirizine, chlorcyclizine, chlordiazepoxide, chlormezanone, chloroquine, chlorpheniramine, chlorpromazine, chlorpropamide, chlorprothixene, chlorthalidone, chlorzoxazone, cimeti-

dine, cisapride, clomipramine, clonazepam, clonidine, clozapine, cocaine, codeine, colchicine, cyclizine, cyclobenzaprine, dantrolene, desipramine, diazepam, diclofenac, diflunisal, diltiazem, diphenhydramine, diphenidol, diphenoxylate, dipyridamole, disopyramide, dobutamine, doxapram, doxepin, droperidol, encainide, ethidium bromide, ethopropazine, fenoprofen, fentanyl, flavoxate, fluoxetine, fluphenazine, flurazepam, flurbiprofen, fluvoxamine, furosemide, glutethimide, glyburide, guaifenesin, haloperidol, homatropine, hydralazine, hydrochlorothiazide, hydrocodone, hydromorphone, hydroxychloroquine, hydroxyzine, ibuprofen, imipramine, indomethacin, ketoconazole, ketoprofen, ketorolac, labetalol, levorphanol, lidocaine, loratadine, lorazepam, lovastatin, loxapine, mazindol, mefenamic acid, meperidine, mephenytoin, mepivacaine, mesoridazine, metaproterenol, metformin, methadone, methdilazine, methocarbamol, methotrexate, methotrimeprazine, methoxamine, methyldopa, metoclopramide, metolazone, metoprolol, metronidazole, midazolam, moclobemide, morphine, nadolol, nalbuphine, naloxone, naphazoline, naproxen, nifedipine, nizatidine, norepinephrine, nortriptyline, oxazepam, oxycodone, oxymetazoline, paroxetine, pemoline, pentazocine, pentobarbital, pentoxifylline, perphenazine, pheniramine, phenobarbital, phenol, phenolphthalein, phentolamine, phenylbutazone, phenyltoloxamine, phenytoin, pimozide, pindolol, piroxicam, pramoxine, prazepam, prazosin, probenecid, procainamide, procaine, prochlorperazine, procyclidine, promazine, promethazine, propafenone, propantheline, propiomazine, propofol, propranolol, protriptyline, quazepam, quinidine, quinine, racemethorphan, ranitidine, remoxipride, risperidone, salicylic acid, scopolamine, secobarbital, sertraline, sotalol, spironolactone, sulfinpyrazone, sulindac, temazepam, terbutaline, terfenadine, tetracaine, theophylline, thiethylperazine, thiopental, thioridazine, thiothixene, timolol, tocainide, tolbutamide, tolmetin, trazodone, triamterene, triazolam, trifluoperazine, triflupromazine, trimeprazine, trimethoprim, trimipramine, verapamil, warfarin, xylometazoline, yohimbine, zopiclone

## KEY WORDS
some details of plasma extraction

## REFERENCE
Koves, E.M. Use of high-performance liquid chromatography-diode array detection in forensic toxicology. *J.Chromatogr.A*, **1995**, *692*, 103–119

## SAMPLE
**Matrix:** solutions

## HPLC VARIABLES
**Column:** 250 × 4.6 Zorbax RX
**Mobile phase:** Gradient. A was 10 mL concentrated orthophosphoric acid and 7 mL triethylamine in 1 L water. B was 10 mL concentrated orthophosphoric acid and 7 mL triethylamine in 200 mL water, make up to 1 L with MeCN. A:B from 100:0 to 0:100 over 30 min, maintain at 0:100 for 5 min.
**Column temperature:** 30
**Flow rate:** 2
**Detector:** UV 210

## OTHER SUBSTANCES
**Also analyzed:** acepromazine, acetaminophen, acetophenazine, albuterol, aminophylline, amitriptyline, amobarbital, amoxapine, amphetamine, amylocaine, antipyrine, aprobarbital, aspirin, atenolol, atropine, avermectin, barbital, benzocaine, benzoic acid, benzotropine, benzphetamine, berberine, bibucaine, bromazepan, brompheniramine, buprenorphine, buspirone, butabarbital, butacaine, butethal, caffeine, carbamazepine, carbromal, chloramphenicol, chlordiazepoxide, chloroquine, chlorothiazide, chloroxylenol, chlorphenesin, chlorpheniramine, chlorpromazine, chlorpropamide, chlortetracycline, cimetidine, cinchonidine, cinchonine, clenbuterol, clonazepam, clonixin, clorazepate, cocaine, codeine, colchicine, cortisone, coumarin, cyclazocine, cyclobenzaprine, cyclothiazide, cyheptamide, cymarin, danazol, danthron, dapsone, debrisoquine, desipramine, dexamethasone, dex-

tromethorphan, dextropropoxyphene, diamorphine, diazepam, diclofenac, diethylpropion, diethylstilbestrol, diflunisal, digitoxin, digoxin, diltiazem, diphenhydramine, diphenoxylate, diprenorphine, dipyrone, disulfiram, dopamine, doxapram, doxepin, dronabinol, ephedrine, epinephrine, epinine, estradiol, estriol, estrone, ethacrynic acid, ethosuximide, etonitazene, etorphine, eugenol, famotidine, fenbendazole, fencamfamine, fenoprofen, fenproporex, fentanyl, flubendazole, flufenamic acid, flunitrazepam, 5-fluorouracil, fluoxymesterone, fluphenazine, furosemide, gentisic acid, gitoxigenin, glipizide, glunixin, glutethimide, glybenclamide, guaiacol, halazepam, haloperidol, hydrochlorothiazide, hydrocodone, hydrocortisone, hydromorphone, hydroxyquinoline, ibogaine, ibuprofen, iminostilbene, imipramine, indomethacin, isocarbostyril, isocarboxazid, isoniazid, isoproterenol, isoxsuprine, ivermectin, ketamine, ketoprofen, kynurenic acid, levorphanol, lidocaine, lorazepam, lormetazepam, loxapine, mazindol, mebendazole, meclizine, meclofenamic acid, medazepam, mefenamic acid, megestrol, mepacrine, meperidine, mephentermine, mephenytoin, mephesin, mephobarbital, mepivacaine, mescaline, mesoridazine, methadone, methamphetamine, methapyrilene, methaqualone, methazolamide, methocarbamol, methoxamine, methsuximide, methyl salicylate, methyldopa, methyldopamine, methylprednisolone, methyltestosterone, methyprylon, metoprolol, mibolerone, morphine, nadolol, nalorphine, naloxone, naltrexone, naphazoline, naproxen, nefopam, niacinamide, nicotine, nicotinic acid, nifedipine, niflumic acid, nitrazepam, norepinephrine, nortriptyline, noscapine, nylidrin, oxazepam, oxycodone, oxymorphone, oxyphenbutazone, oxytetracycline, papaverine, pargyline, pemoline, pentazocine, pentobarbital, persantine, phenacetin, phenazocine, phenazopyridine, phencyclidine, phendimetrazine, phenelzine, pheniramine, phenobarbital, phenothiazine, phensuximide, phentermine, phenylbutazone, phenylephrine, phenylpropanolamine, piperocaine, prazepam, prednisolone, primidone, probenecid, progesterone, propiomazine, propranolol, propylparaben, pseudoephedrine, puromycin, pyrilamine, pyrithyldione, quazepam, quinaldic acid, quinidine, quinine, ranitidine, recinnamine, reserpine, resorcinol, saccharin, albuterol, salicylamide, salicylic acid, scopolamine, scopoletin, secobarbital, strychnine, sulfacetamide, sufadiazine, sulfadimethoxine, sulfaethidole, sulfamerazine, sulfamethazine, sulfamethoxizole, sulfanilamide, sulfapyridine, sulfasoxizole, sulindac, tamoxifen, temazepam, testosterone, tetracaine, tetracycline, tetramisole, thebaine, theobromine, theophylline, thiabendazole, thiamine, thiamylal, thiobarbituric acid, thioridazine, thiosalicylic acid, thiothixene, thymol, tolazamide, tolazoline, tobutamide, tolmetin, tranylcypromine, triamcinolone, tribenzylamine, trichloromethiazide, trifluoperazine, trihexyphenidyl, trimethoprim, tripelennamine, triprolidine, tropacocaine, tyramine, verapamil, vincamine, warfarin, yohimbine, zoxazolamine

## REFERENCE

Hill, D.W.; Kind, A.J. Reversed-phase solvent gradient HPLC retention indexes of drugs. *J.Anal.Toxicol.*, **1994**, *18*, 233–242

## SAMPLE

**Matrix:** solutions

## HPLC VARIABLES

**Column:** 150 × 4.6 Supelcosil LC-ABZ
**Mobile phase:** MeCN:25 mM pH 6.9 potassium phosphate buffer 35:65
**Flow rate:** 1.5
**Injection volume:** 25
**Detector:** UV 254

## CHROMATOGRAM

**Retention time:** 3.495

## OTHER SUBSTANCES

**Also analyzed:** 6-acetylmorphine, amiloride, amphetamine, benzocaine, benzoylecgonine, caffeine, cocaine, codeine, doxylamine, fluoxetine, glutethimide, hexobarbital, hypoxanthine, levorphanol, LSD, meperidine, mephobarbital, methadone, methyprylon, N-norcodeine, oxazepam, oxycodone, phenylpropanolamine, prilocaine, procaine, terfenadine

## REFERENCE

Ascah, T.L. Improved separations of alkaloid drugs and other substances of abuse using Supelcosil LC-ABZ column. *Supelco Reporter*, **1993**, *12(3)*, 18–21

## SAMPLE
**Matrix:** solutions
**Sample preparation:** Dissolve in MeOH at a concentration of 1 mg/mL, inject a 20 μL aliquot.

## HPLC VARIABLES
**Column:** 250 × 5 Spherisorb S5W
**Mobile phase:** MeOH:buffer 90:10 (Buffer was 94 mL 35% ammonia and 21.5 mL 70% nitric acid in 884 mL water, adjust the pH to 10.1 with ammonia.)
**Flow rate:** 2
**Injection volume:** 20
**Detector:** UV 254

## CHROMATOGRAM
**Retention time:** 1.80

## OTHER SUBSTANCES
**Simultaneous:** acetylcodeine, amphetamine, benzphetamine, benzylmorphine, bromo-STP, buprenorphine, chlorphentermine, codeine, codeine-N-oxide, dextromoramide, dextropropoxyphene, diamorphine, diethylpropion, dihydrocodeine, dihydromorphine, dimethylamphetamine, dipipanone, ephedrine, epinephrine, ethoheptazine, ethylmorphine, etorphine, fencamfamin, fenfluramine, fentanyl, hydrocodone, 4-hydroxyamphetamine, levorphanol, mazindol, meperidine, mephentermine, mescaline, methadone, methamphetamine, methylenedioxyamphetamine, methylephedrine, monoacetylmorphine, morphine, morphine-3-glucuronide, morphine-N-oxide, naloxone, norcodeine, norlevorphanol, normetanephrine, normethadone, normorphine, norpethidine, norpseudoephedrine, noscapine, oxycodone, papaverine, pemoline, pentazocine, 2-phenethylamine, phenoperidine, phentermine, phenylephrine, phenylpropanolamine, pholcodeine, pipradol, piritramide, prolintane, pseudoephedrine, STP, thebacon, thebaine, trimethoxyamphetamine, tyramine
**Noninterfering:** dopamine, levodopa, methyldopa, methyldopate, norepinephrine
**Interfering:** caffeine, fenethyline, hydroxypethidine, levallorphan, nalorphine, norpipanone, phenazocine, phendimetrazine, phenelzine, tranylcypromine

## REFERENCE

Law, B.; Gill, R.; Moffat, A.C. High-performance liquid chromatography retention data for 84 basic drugs of forensic interest on a silica column using an aqueous methanol eluent. *J.Chromatogr.*, **1984**, *301*, 165–172

## SAMPLE
**Matrix:** solutions

## HPLC VARIABLES
**Column:** μBondapak C18 (A), or μBondapak alkyl phenyl (B), or μBondapak CN (C)
**Mobile phase:** MeOH:water:acetic acid 20:79:1 containing 5 mM methanesulfonic acid

## CHROMATOGRAM
**Retention time:** k′ 9.36 (A), k′ 8.95 (B), k′ 1.82 (C)

## OTHER SUBSTANCES
**Also analyzed:** acetaminophen, acetylcodeine, acetylmorphine, aminopyrene, aminopyrine, amobarbital, amphetamine, antipyrine, benzocaine, butabarbital, caffeine, cocaine, codeine, diamorphine, diazepam, diethylpropion, DMT, ephedrine, glutethimide, Lampa, lidocaine, LSD, MDA, mecloqualone, mescaline, methamphetamine, methapyrilene, methaqualone, methpyrilene, morphine, narcotine, papaverine, PCP, pentobarbital, phencyclidine, phendimetrazine, phenmetrazine, phenobarbital, phentermine, phenylpropanolamine, procaine, quinidine, quinine, secobarbital, strychnine, TCP, tetracaine, thebaine, theophylline

## REFERENCE
Lurie, I.S.; Demchuk, S.M. Optimization of a reverse phase ion-pair chromatographic separation for drugs of forensic interest. Part II—Factors effecting selectivity. *J.Liq.Chromatogr.*, **1981**, *4*, 357–374

## SAMPLE
**Matrix:** urine
**Sample preparation:** 2 mL Urine + 3 mL 5 M NaOH, vortex 30 s, add 12 mL diethyl ether, rotate for 5 min, centrifuge at 2500 rpm for 5 min. Remove the ether layer and evaporate it to dryness at 40° under a stream of nitrogen, reconstitute in 2 mL mobile phase, inject a 200 μL aliquot.

## HPLC VARIABLES
**Column:** 250 × 4.6 10 μm Alltech C18
**Mobile phase:** MeOH:water 50:50 containing 7 mL/L butylamine, adjusted to pH 3.2 with sulfuric acid
**Flow rate:** 1.8
**Injection volume:** 200
**Detector:** E, Bioanalytical Systems Model LC4B, dual glassy carbon working electrode cell half operated in the parallel mode + 1.0 V and +0.9 V, stainless steel auxiliary electrode cell half, Ag/AgCl reference electrode. The detector was preceded by a Photronix Model 816 UV irradiator which irradiated the mobile phase in a 9.144 m length of 0.5 mm i.d. × 1.6 mm o.d. PTFE tubing in a three-dimensional figure eight configuration. The irradiation apparatus was maintained at 0-5° using an ice bath.

## CHROMATOGRAM
**Retention time:** 4
**Limit of detection:** 750 ppb

## OTHER SUBSTANCES
**Simultaneous:** chlordiazepoxide, cocaine, nitrazepam, phenobarbital

## REFERENCE
Selavka, C.M.; Krull, I.S.; Lurie, I.S. Photolytic derivatization for improved LCEC determinations of pharmaceuticals in biological fluids. *J.Chromatogr.Sci.*, **1985**, *23*, 499–508

◆━━━━━━━━━━━━━━━━━━━━◆━━━━━━━━━━━━━━━━━━━━◆

## ANNOTATED BIBLIOGRAPHY
Schill, G.; Wainer, I.W.; Barkan, S.A. Chiral separation of cationic drugs on an α1-acid glycoprotein bonded stationary phase. *J.Liq.Chromatogr.*, **1986**, *9*, 641–666 [chiral; also atropine, bromdiphenhydramine, brompheniramine, bupivacaine, butorphanol, carbinoxamine, chlorpheniramine, clidinium, cocaine, cyclopentolate, dimethindene, diperidone, disopyramide, doxylamine, ephedrine, homatropine, labetalol, labetalol A, labetalol B, mepensolate, mepivacaine, methadone, methorphan, methylatropine, methylhomatropine, metoprolol, nadolol, nadolol A, nadolol B, oxprenolol, oxyphencyclimine, phenmetrazine, phenoxybenzamine, promethazine, pronethalol, propoxyphene, propranolol, pseudoephedrine, terbutaline, tocainide, tridihexethyl]

# Methylprednisolone

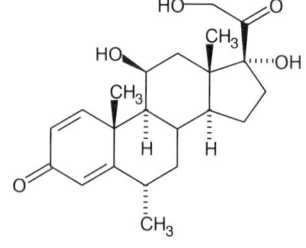

**Molecular formula:** $C_{22}H_{30}O_5$
**Molecular weight:** 374.5
**CAS Registry No.:** 83-43-2, 53-36-1 (acetate), 5015-36-1
(sodium phosphate), 2375-03-3 (sodium succinate), 2921-57-5
(hemisuccinate), 90350-40-6 (suleptanate)

## SAMPLE
**Matrix:** blood
**Sample preparation:** Extract 1 mL plasma containing dexamethasone with 12 mL dichloromethane. Remove the organic phase and wash it with 2 mL 100 mM NaOH, wash with 1 mL water, dry over 1 g anhydrous sodium sulfate. Evaporate to dryness under a stream of nitrogen, reconstitute with mobile phase, inject an aliquot.

## HPLC VARIABLES
**Column:** 150 × 4.6 3 μm Spherisorb silica
**Mobile phase:** Hexane:dichloromethane:EtOH:glacial acetic acid 26:69:3.4:2
**Flow rate:** 0.75
**Detector:** UV 254

## CHROMATOGRAM
**Retention time:** 7.4
**Internal standard:** dexamethasone (5.1)
**Limit of quantitation:** 10 ng/mL

## OTHER SUBSTANCES
**Extracted:** hydrocortisone, prednisolone

## KEY WORDS
plasma; pharmacokinetics; normal phase

## REFERENCE
Möllmann, H.; Hochhaus, G.; Rohatagi, S.; Barth, J.; Derendorf, H. Pharmacokinetic/pharmacodynamic evaluation of deflazacort in comparison to methylprednisolone and prednisolone. *Pharm.Res.*, **1995**, *12*, 1096–1100

## SAMPLE
**Matrix:** blood
**Sample preparation:** 1 mL Serum + 100 μL water containing 5 μg/mL 2,3-diaminonaphthalene and 3.5 μg/mL 18-hydroxy-11-deoxycorticosterone + 1 mL 250 mM NaOH + 7 mL diethyl ether, shake on a rotary shaker for 15 min, repeat extraction. Combine the organic layers and evaporate them to dryness under a stream of nitrogen at 30-40°, reconstitute the residue in 70 μL MeOH:100 mM perchloric acid 50:50, inject a 20 μL aliquot.

## HPLC VARIABLES
**Column:** 150 × 3.9 4 μm Nova-Pak C18
**Mobile phase:** Gradient. A was 58 mM $NaH_2PO_4$ containing 6 mM sodium heptanesulfonate, adjusted to pH 3.1 with concentrated phosphoric acid. B was MeCN:MeOH 85:15. A:B from 100:0 to 78:22 over 5 min, to 70:30 over 12 min, maintain at 70:30 for 4 min, to 65:35 over 9 min.
**Flow rate:** 1

**Injection volume:** 20
**Detector:** UV 245; UV 256; UV 343

## CHROMATOGRAM
**Retention time:** 18.70 (methylprednisolone), 21.36 (methylprednisolone acetate)
**Internal standard:** 2,3-diaminonaphthalene (10.71), 18-hydroxy-11-deoxycorticosterone (15.85)
**Limit of detection:** 1-10 ng/mL (245 nm)

## OTHER SUBSTANCES
**Extracted:** betamethasone, chloroquine, corticosterone, cortisone, dexamethasone, fluendrenolide, fluocinolone acetonide, fluorometholone, fluprednisolone, hydrocortisone, hydroxychloroquine, 17δ-hydroxyprogesterone, meprednisone, paramethasone, prednisolone, prednisone, progesterone, triamcinolone
**Noninterfering:** aspirin, ibuprofen, indomethacin, phenylbutazone, pregnenolone

## KEY WORDS
serum

## REFERENCE
Volin, P. Simple and specific reversed-phase liquid chromatographic method with diode-array detection for simultaneous determination of serum hydroxychloroquine, chloroquine and some corticosteroids. *J.Chromatogr.B*, **1995**, *666*, 347–353

## SAMPLE
**Matrix:** blood
**Sample preparation:** 2 mL Plasma + 500 ng fluorometholone + 5 mL hexane, shake horizontally at 60 cycles/min for 10 min, centrifuge at 1000 g for 5 min, discard the hexane layer, add 8 mL dichloromethane, shake horizontally at 60 cycles/min for 10 min, centrifuge at 1000 g for 5 min, repeat extraction with 8 mL dichloromethane. Combine the dichloromethane layers and add 300 mg anhydrous sodium sulfate, shake horizontally at 200 cycles/min for 5 min, centrifuge at 1000 g for 5 min. Remove the organic layer and evaporate it to dryness under a stream of nitrogen at 40°, reconstitute the residue in 200 μL mobile phase, add 2 mL hexane, vortex for 1 min, centrifuge at 1000 g for 5 min, discard the hexane layer, inject a 100 μL aliquot of the mobile phase layer.

## HPLC VARIABLES
**Guard column:** 30 × 4.6 5 μm Spheri-5 RP-18
**Column:** 250 × 4.6 5 μm Ultrasphere ODS
**Mobile phase:** MeCN:water:glacial acetic acid 33:62:5
**Flow rate:** 1
**Injection volume:** 100
**Detector:** UV 254

## CHROMATOGRAM
**Retention time:** 8.07 (methylprednisolone), 21.80 (methylprednisolone acetate)
**Internal standard:** fluorometholone (14.55)
**Limit of detection:** 1 ng/mL
**Limit of quantitation:** 2 ng/mL

## KEY WORDS
plasma; pharmacokinetics

## REFERENCE
Hopkins, N.K.; Wagner, C.M.; Brisson, J.; Addison, T.E. Validation of the simultaneous determination of methylprednisolone and methylprednisolone acetate in human plasma by high-performance liquid chromatography. *J.Chromatogr.*, **1992**, *577*, 87–93

## SAMPLE

**Matrix:** blood

**Sample preparation:** 1 mL Plasma + 100 µL 3 M sulfuric acid + 50 µL 6 µg/mL prednisone in MeCN:MeOH 50:50, mix, add 15 mL hexane:ethyl acetate 50:50, shake for 20 min, centrifuge, freeze at -70°. Remove the organic layer and add it to 1 mL 1 M nitric acid, shake, freeze. Remove the organic layer and dry it over anhydrous sodium sulfate, evaporate to dryness under a stream of nitrogen at 30°, reconstitute the residue in 100 µL mobile phase, inject an aliquot.

## HPLC VARIABLES

**Guard column:** 12.5 × 4 5 µm Zorbax SIL

**Column:** three 80 × 4 5 µm Zorbax SIL Reliance 5 columns in series

**Mobile phase:** Dichloromethane:hexane:EtOH:glacial acetic acid 69:26:2.3:1 (Pass the mobile phase through a 70 × 6 37-53 µm HC-Pellocil (Whatman) column.)

**Flow rate:** 2

**Detector:** UV 254

## CHROMATOGRAM

**Retention time:** 12 (methylprednisolone hemisuccinate), 19 (methylprednisolone)

**Internal standard:** prednisone (9)

**Limit of detection:** 10 ng/mL

## OTHER SUBSTANCES

**Extracted:** hydrocortisone

## KEY WORDS

plasma; pharmacokinetics; normal phase

## REFERENCE

Kong, A.-N.; Slaughter, R.L.; Jusko, W.J. Simultaneous analysis of methylprednisolone hemisuccinate, cortisol and methylprednisolone by normal-phase high-performance liquid chromatography in human plasma. *J.Chromatogr.*, **1988**, *432*, 308–314

## SAMPLE

**Matrix:** blood

**Sample preparation:** 1 mL Serum + 10 mL dichloromethane, shake for 1 h, centrifuge for 15 min. Discard the aqueous layer, wash the organic layer with 1 mL 100 mM NaOH and 1 mL water (vortex for 30 s and centrifuge for 15 min each time). Remove the organic layer and evaporate it to dryness under a stream of nitrogen at 40-45°, reconstitute the residue in 200 µL mobile phase, inject a 50 µL aliquot onto column A in series with column B and elute with mobile phase. After 2.1 min remove column A from the circuit and backflush it with mobile phase at 1 mL/min for 15 min, elute column B with mobile phase and monitor the effluent.

## HPLC VARIABLES

**Column:** A 30 mm long Spherisorb silica; B 250 × 4.6 6 µm Zorbax silica

**Mobile phase:** Butyl chloride:THF:MeOH:phosphoric acid 880:100:15:0.5 (Butyl chloride was 50% water-saturated.)

**Flow rate:** 2

**Injection volume:** 50

**Detector:** UV 236

## CHROMATOGRAM

**Retention time:** 24

**Internal standard:** 6α-methylprednisolone

## OTHER SUBSTANCES
**Extracted:** fluoxymesterone

## KEY WORDS
serum; normal phase; column-switching; 6α-methylprednisolone is IS

## REFERENCE
Capponi, V.J.; Cox, S.R.; Harrington, E.L.; Wright, C.E.; Antal, E.J.; Albert, K.S. Liquid chromatographic
    assay for fluoxymesterone in human serum with application to a preliminary bioavailability study.
    *J.Pharm.Sci.*, **1985**, *74*, 308–311

## SAMPLE
**Matrix:** blood
**Sample preparation:** 1 mL Plasma + 80 μL 3.125 μg/mL dexamethasone in MeOH, mix,
    add 15 mL dichloromethane, shake for 20 min, centrifuge. Remove organic phase and
    wash it with 1 mL 100 mM NaOH then with 1 mL water. Remove organic phase and dry
    it with 1 g anhydrous sodium sulfate. Evaporate to dryness at 45° under a stream of
    nitrogen, reconstitute in 200 μL mobile phase, inject.

## HPLC VARIABLES
**Guard column:** 70 × 6 37-53 μm Whatman HC-Pellocil
**Column:** 250 × 4.6 5-6 μm Zorbax SIL
**Mobile phase:** Hexane:dichloromethane:ethanol:acetic acid 26:69:3.4:1
**Flow rate:** 2
**Injection volume:** 200
**Detector:** UV 254

## CHROMATOGRAM
**Retention time:** 12
**Internal standard:** dexamethasone (8)
**Limit of detection:** 2 ng/mL
**Limit of quantitation:** 10 ng/mL

## OTHER SUBSTANCES
**Simultaneous:** beclomethasone, betamethasone, corticosterone, cortisone, fluocinonide, hy-
    drocortisone, methylprednisone, prednisolone, prednisone

## KEY WORDS
plasma; normal phase

## REFERENCE
Ebling, W.F.; Szefler, S.J.; Jusko, W.J. Analysis of cortisol, methylprednisolone, and methylprednisolone
    hemisuccinate. Absence of effects of troleandomycin on ester hydrolysis. *J.Chromatogr.*, **1984**, *305*,
    271–280

## SAMPLE
**Matrix:** blood
**Sample preparation:** 1 mL Plasma + 100 μL 2 μg/mL dexamethasone in EtOH:water
    10:90 + 100 μL 250 mM NaOH + 7 mL ether:dichloromethane 60:40, vortex for 30 s,
    centrifuge at 2000 rpm for 5 min. Remove 6 mL of the organic layer and evaporate it to
    dryness under a stream of air at 40°, reconstitute the residue in 100 μL
    dichloromethane:EtOH:water 95:4:1, inject a 50 μL aliquot.

## HPLC VARIABLES
**Column:** 250 × 4.5 5 μm Partisil silica

**Mobile phase:** Dichloromethane:EtOH:water 95:4:1
**Flow rate:** 1.5
**Injection volume:** 50
**Detector:** UV 239

## CHROMATOGRAM
**Retention time:** 18
**Internal standard:** dexamethasone (11.5)
**Limit of quantitation:** 25 ng/mL

## OTHER SUBSTANCES
**Extracted:** corticosterone, 11-deoxycortisol, hydrocortisone, 17-hydroxyprogesterone, prednisolone, prednisone, progesterone

## KEY WORDS
plasma; normal phase

## REFERENCE
Scott, N.R.; Chakraborty, J.; Marks, V. Determination of prednisolone, prednisone, and cortisol in human plasma by high-performance liquid chromatography. *Anal.Biochem.*, **1980**, *108*, 266–268

## SAMPLE
**Matrix:** formulations
**Sample preparation:** Inject a 20 μL aliquot.

## HPLC VARIABLES
**Column:** 250 × 4.6 5 μm Spherisorb CN
**Mobile phase:** MeCN:100 mM pH 4.5 $NaH_2PO_4$ 15:85
**Injection volume:** 20
**Detector:** UV 300

## CHROMATOGRAM
**Retention time:** 4.2
**Internal standard:** propylparaben (3.3)

## OTHER SUBSTANCES
**Simultaneous:** dexamethasone, granisetron

## KEY WORDS
injections; 5% dextrose; saline; water

## REFERENCE
Pinguet, F.; Rouanet, P.; Martel, P.; Fabbro, M.; Salabert, D.; Astre, C. Compatibility and stability of granisetron, dexamethasone, and methylprednisolone in injectable solutions. *J.Pharm.Sci.*, **1995**, *84*, 267–268

## SAMPLE
**Matrix:** formulations
**Sample preparation:** Prepare a solution in chloroform:n-butanol 95:5, inject a 5 μL aliquot.

## HPLC VARIABLES
**Guard column:** 5 μm Guard-Pak Resolve Si (dead volume 60-75 μL)
**Column:** 75 × 3.9 4 μm Nova-Pak silica
**Mobile phase:** Dichloromethane:EtOH 34:1

**Flow rate:** 0.7
**Injection volume:** 5
**Detector:** UV 240

## CHROMATOGRAM
**Retention time:** 10.5 (6α-methylprednisolone)

## OTHER SUBSTANCES
**Simultaneous:** betamethasone, cortisone, dexamethasone, hydrocortisone, prednisolone
**Interfering:** prednisone

## KEY WORDS
tablets; normal phase

## REFERENCE
Liu, K.-R.; Chen, S.-H.; Wu, S.-M.; Kou, H.-S.; Wu, H.-L. High-performance liquid chromatographic determination of betamethasone and dexamethasone. *J.Chromatogr.A*, **1994**, *676*, 455−460

## SAMPLE
**Matrix:** formulations
**Sample preparation:** Dilute 20-fold with mobile phase, inject a 10 μL aliquot.

## HPLC VARIABLES
**Column:** 300 × 3.9 10 μm μBondapak C18
**Mobile phase:** MeCN:50 mM pH 5.7 sodium acetate 33:67
**Flow rate:** 1
**Injection volume:** 10
**Detector:** UV 248

## CHROMATOGRAM
**Retention time:** 6.4 (methylprednisolone sodium succinate)

## KEY WORDS
water; injections; stability-indicating

## REFERENCE
Nahata, M.C.; Morosco, R.S.; Hipple, T.F. Stability of diluted methylprednisolone sodium succinate injection at two temperatures. *Am.J.Hosp.Pharm.*, **1994**, *51*, 2157−2159

## SAMPLE
**Matrix:** perfusate
**Sample preparation:** Condition a Sep-Pak C18 SPE cartridge with 10 mL MeOH and 10 mL water. Add 3 mL Perfusate to the SPE cartridge, wash three times with 10 mL aliquots of water, elute with 5 mL MeOH. Evaporate the eluate to dryness under a stream of nitrogen at 35°, reconstitute the residue in 100 μL mobile phase, inject a 50 μL aliquot.

## HPLC VARIABLES
**Guard column:** 15 × 3.2 Newguard RP-18
**Column:** two 250 × 4.6 Spheri-5 RP-18 columns in series
**Mobile phase:** MeOH:water 53:47
**Column temperature:** 40
**Flow rate:** 1.1
**Injection volume:** 50
**Detector:** UV 242

## CHROMATOGRAM
**Retention time:** 30
**Internal standard:** 6α-methylprednisolone

## OTHER SUBSTANCES
**Extracted:** metabolites, cortisone, dihydrocortisol, dihydrocortisone, hydrocortisone
**Simultaneous:** prednisolone
**Noninterfering:** acetaminophen, albuterol, betamethasone, bupivacaine, carbamazepine, cholesterol, clonazepam, dehydroepiandrosterone, dexamethasone, diazepam, estradiol, estriol, hydroxyprogesterone, methimazole, phenobarbital, prednisone, progesterone, ritodrine, scopolamine, testosterone

## KEY WORDS
SPE; methylprednisolone is IS

## REFERENCE
Dodds, H.M.; Maguire, D.J.; Mortimer, R.H.; Addison, R.S.; Cannell, G.R. High performance liquid chromatographic separation of cortisol, cortisone, and their 20-reduced metabolites in perfusion media. *J.Liq.Chromatogr.*, **1995**, *18*, 1809–1820

## SAMPLE
**Matrix:** solutions
**Sample preparation:** Inject an aliquot of a 1 μM solution in MeOH.

## HPLC VARIABLES
**Column:** 470 × 4.6 5 μm Spheri-5 RP-18
**Mobile phase:** MeOH:water 56:44
**Flow rate:** 0.5
**Injection volume:** 10
**Detector:** UV 240

## CHROMATOGRAM
**Retention time:** 55

## OTHER SUBSTANCES
**Simultaneous:** cortisone, dehydrocorticosterone, hydrocortisone, prednisolone, prednisone, tetrahydrocortisol, tetrahydrocortisone

## REFERENCE
Lukulay, P.H.; McGuffin, V.L. Comparison of solvent modulation with premixed mobile phases for the separation of corticosteroids by liquid chromatography. *J.Liq.Chromatogr.*, **1995**, *18*, 4039–4062

## SAMPLE
**Matrix:** solutions

## HPLC VARIABLES
**Column:** 250 × 4.6 Zorbax RX
**Mobile phase:** Gradient. A was 10 mL concentrated orthophosphoric acid and 7 mL triethylamine in 1 L water. B was 10 mL concentrated orthophosphoric acid and 7 mL triethylamine in 200 mL water, make up to 1 L with MeCN. A:B from 100:0 to 0:100 over 30 min, maintain at 0:100 for 5 min.
**Column temperature:** 30
**Flow rate:** 2
**Detector:** UV 210

## OTHER SUBSTANCES

**Also analyzed:** acepromazine, acetaminophen, acetophenazine, albuterol, aminophylline, amitriptyline, amobarbital, amoxapine, amphetamine, amylocaine, antipyrine, aprobarbital, aspirin, atenolol, atropine, avermectin, barbital, benzocaine, benzoic acid, benzotropine, benzphetamine, berberine, bibucaine, bromazepan, brompheniramine, buprenorphine, buspirone, butabarbital, butacaine, butethal, caffeine, carbamazepine, carbromal, chloramphenicol, chlordiazepoxide, chloroquine, chlorothiazide, chloroxylenol, chlorphenesin, chlorpheniramine, chlorpromazine, chlorpropamide, chlortetracycline, cimetidine, cinchonidine, cinchonine, clenbuterol, clonazepam, clonixin, clorazepate, cocaine, codeine, colchicine, cortisone, coumarin, cyclazocine, cyclobenzaprine, cyclothiazide, cyheptamide, cymarin, danazol, danthron, dapsone, debrisoquine, desipramine, dexamethasone, dextromethorphan, dextropropoxyphene, diamorphine, diazepam, diclofenac, diethylpropion, diethylstilbestrol, diflunisal, digitoxin, digoxin, diltiazem, diphenhydramine, diphenoxylate, diprenorphine, dipyrone, disulfiram, dopamine, doxapram, doxepin, dronabinol, ephedrine, epinephrine, epinine, estradiol, estriol, estrone, ethacrynic acid, ethosuximide, etonitazene, etorphine, eugenol, famotidine, fenbendazole, fencamfamine, fenoprofen, fenproporex, fentanyl, flubendazole, flufenamic acid, flunitrazepam, 5-fluorouracil, fluoxymesterone, fluphenazine, furosemide, gentisic acid, gitoxigenin, glipizide, glunixin, glutethimide, glybenclamide, guaiacol, halazepam, haloperidol, hydrochlorothiazide, hydrocodone, hydrocortisone, hydromorphone, hydroxyquinoline, ibogaine, ibuprofen, iminostilbene, imipramine, indomethacin, isocarbostyril, isocarboxazid, isoniazid, isoproterenol, isoxsuprine, ivermectin, ketamine, ketoprofen, kynurenic acid, levorphanol, lidocaine, lorazepam, lormetazepam, loxapine, mazindol, mebendazole, meclizine, meclofenamic acid, medazepam, mefenamic acid, megestrol, mepacrine, meperidine, mephentermine, mephenytoin, mephesin, mephobarbital, mepivacaine, mescaline, mesoridazine, methadone, methamphetamine, methapyrilene, methaqualone, methazolamide, methocarbamol, methoxamine, methsuximide, methyl salicylate, methyldopa, methyldopamine, methylphenidate, methyltestosterone, methyprylon, metoprolol, mibolerone, morphine, nadolol, nalorphine, naloxone, naltrexone, naphazoline, naproxen, nefopam, niacinamide, nicotine, nicotinic acid, nifedipine, niflumic acid, nitrazepam, norepinephrine, nortriptyline, noscapine, nylidrin, oxazepam, oxycodone, oxymorphone, oxyphenbutazone, oxytetracycline, papaverine, pargyline, pemoline, pentazocine, pentobarbital, persantine, phenacetin, phenazocine, phenazopyridine, phencyclidine, phendimetrazine, phenelzine, pheniramine, phenobarbital, phenothiazine, phensuximide, phentermine, phenylbutazone, phenylephrine, phenylpropanolamine, piperocaine, prazepam, prednisolone, primidone, probenecid, progesterone, propiomazine, propranolol, propylparaben, pseudoephedrine, puromycin, pyrilamine, pyrithyldione, quazepam, quinaldic acid, quinidine, quinine, ranitidine, recinnamine, reserpine, resorcinol, saccharin, albuterol, salicylamide, salicylic acid, scopolamine, scopoletin, secobarbital, strychnine, sulfacetamide, sufadiazine, sulfadimethoxine, sulfaethidole, sulfamerazine, sulfamethazine, sulfamethoxizole, sulfanilamide, sulfapyridine, sulfasoxizole, sulindac, tamoxifen, temazepam, testosterone, tetracaine, tetracycline, tetramisole, thebaine, theobromine, theophylline, thiabendazole, thiamine, thiamylal, thiobarbituric acid, thioridazine, thiosalicylic acid, thiothixene, thymol, tolazamide, tolazoline, tobutamide, tolmetin, tranylcypromine, triamcinolone, tribenzylamine, trichloromethiazide, trifluoperazine, trihexyphenidyl, trimethoprim, tripelennamine, triprolidine, tropacocaine, tyramine, verapamil, vincamine, warfarin, yohimbine, zoxazolamine

## REFERENCE

Hill, D.W.; Kind, A.J. Reversed-phase solvent gradient HPLC retention indexes of drugs. *J.Anal.Toxicol.*, **1994**, *18*, 233–242

## SAMPLE

**Matrix:** solutions
**Sample preparation:** Prepare a 25 µg/mL solution in mobile phase, inject an aliquot.

## HPLC VARIABLES

**Column:** 250 × 4.6 Partisil 10 ODS-1

**Mobile phase:** MeOH:water 55:45
**Column temperature:** 40
**Flow rate:** 1.5
**Detector:** UV 240

## CHROMATOGRAM
**Retention time:** k′ 1.509 (methylprednisolone), k′ 2.190 (methylprednisolone acetate)

## OTHER SUBSTANCES
**Also analyzed:** androsterone (UV 210), cortexolone (UV 240), cortisone (UV 240), estradiol (UV 280), estrone (UV 280), ethinyl estradiol (UV 280), ethisterone (UV 240), hydrocortisone (UV 240), hydroxyprogesterone (UV 240), lynestrenol (UV 210), medroxyprogesterone acetate (UV 240), medroxyprogesterone (UV 240), methandienone (UV 240), methylandrostenediol (UV 210), methyltestosterone (UV 240), nandrolone (UV 240), norethisterone (UV 240), prednisolone acetate (UV 240), prednisolone (UV 240), prednisone (UV 240), pregnenolone (UV 210), progesterone (UV 240), testosterone (UV 240)

## REFERENCE
Sadlej-Sosnowska, N. Structure retention relationship for steroid hormones. Functional groups as structural descriptors. *J.Liq.Chromatogr.*, **1994**, *17*, 2319–2330

## SAMPLE
**Matrix:** solutions
**Sample preparation:** Inject 20 μL aliquot of a MeOH solution.

## HPLC VARIABLES
**Column:** 250 × 4.6 5 μm Hypersil 5-ODS
**Mobile phase:** THF:water 23:77
**Column temperature:** 30
**Flow rate:** 1
**Injection volume:** 20
**Detector:** UV 245

## CHROMATOGRAM
**Retention time:** k′ 11.36

## OTHER SUBSTANCES
**Simultaneous:** metabolites, betamethasone, corticosterone, cortisone, deflazacort, deoxycorticosterone, dexamethasone, fludrocortisone, fludrocortisone acetate, fluorocortisone, fluorocortisone acetate, hydrocortisone, 21-hydroxydeflazacort, 11α-hydroxyprogesterone, prednisolone, prednisone, triamcinolone, triamcinolone acetonide

## REFERENCE
Santos-Montes, A.; Gonzalo-Lumbreras, R. Gasco-Lopez, A.I.; Izquierdo-Hornillos, R. Extraction and high-performance liquid chromatographic separation of deflazacort and its metabolite 21-hydroxy-deflazacort. Application to urine samples. *J.Chromatogr.B*, **1994**, *657*, 248–253

## SAMPLE
**Matrix:** solutions

## HPLC VARIABLES
**Column:** 250 × 4.6 5 μm SI-100 (Brownlee)
**Mobile phase:** Butyl chloride:THF:MeOH:glacial acetic acid 95:7:3.5:3 (Butyl chloride was 50% water saturated.)
**Injection volume:** 20
**Detector:** UV 254

## KEY WORDS
normal phase

## REFERENCE
Kane, M.P.; Tsuji, K. Radiolytic degradation scheme for $^{60}$Co-irradiated corticosteroids. *J.Pharm.Sci.*, **1983**, *72*, 30−35

## SAMPLE
**Matrix:** synovial fluid
**Sample preparation:** 100 μL Synovial fluid + 10 μL 10 μg/mL flumethasone in MeOH + 1 mL 100 mM NaOH + 10 mL dichloromethane, shake for 10 min, centrifuge at 8400 g for 10 min. Remove the organic layer and evaporate it to dryness under a stream of nitrogen at 40°, reconstitute the residue in 100 μL mobile phase, vortex, inject the whole amount.

## HPLC VARIABLES
**Column:** 100 × 8 10 μm Radial Pak B silica (Waters)
**Mobile phase:** Dichloromethane:MeOH:glacial acetic acid 96.8:2.4:0.8
**Flow rate:** 1.4
**Injection volume:** 100
**Detector:** UV 254

## CHROMATOGRAM
**Retention time:** 8 (methylprednisolone), 4 (methylprednisolone acetate)
**Internal standard:** flumethasone (6)
**Limit of detection:** 200 ng/mL

## KEY WORDS
normal phase; cow; pharmacokinetics

## REFERENCE
Alvinerie, M.; Toutain, P.L. Determination of methylprednisolone and methylprednisolone acetate in synovial fluid using high-performance liquid chromatography. *J.Chromatogr.*, **1984**, *309*, 385−390

## SAMPLE
**Matrix:** tissue
**Sample preparation:** Homogenize rat brain in 10 mL n-hexane:isopropanol 60:40 in a glass homogenizing tube with a PTFE pestle at 4° using 15 passes, transfer to an acid-washed glass tube, wash out homogenizing tube with three 2.5 mL aliquots of n-hexane:isopropanol 60:40, centrifuge with a table-top centrifuge, remove the supernatant, re-extract the pellet with 8 mL n-hexane:isopropanol 60:40. Combine the supernatants and filter (0.2 μm) them, evaporate to dryness under a stream of nitrogen, reconstitute the residue in n-hexane:isopropanol 60:40, inject an aliquot into the gradient elution system. Collect the methylprednisolone peak and rechromatograph isocratically.

## HPLC VARIABLES
**Column:** 250 × 4.6 5-6 μm Zorbax silica
**Mobile phase:** Gradient. A was n-hexane:isopropanol 60:40. B was n-hexane:isopropanol:water 56.7:37.8:5.5. A:B from 55:45 to 24:76 over 7 min, to 0:100 over 5 min, maintain at 0:100 for 22 min. (Methylprednisolone peak can be collected and rechromatographed isocratically at A:B 90:10 at 0.8 mL/min, retention time 7.17 min.)
**Column temperature:** 34
**Flow rate:** 1.8
**Detector:** UV 254

## CHROMATOGRAM
**Retention time:** 3.27
**Limit of detection:** 4.9 ng
**Limit of quantitation:** 9.8 ng

## KEY WORDS
rat; brain; normal phase

## REFERENCE
Murphy, E.J.; Slivka, A.P.; Rosenberger, T.A.; Horrocks, L.A. High-performance liquid chromatography separation and quantitation of methylprednisolone from rat brain. *Anal.Biochem.*, **1993**, *209*, 339–342

## SAMPLE
**Matrix:** urine
**Sample preparation:** 3 mL Urine + 0.25 g NaCl, adjust pH to 9.0 with 0.5 g $Na_2HPO_4$, add 4 mL dichloromethane, vortex 1 min, centrifuge at 3700 g for 3 min. Remove organic phase and dry it over anhydrous sodium sulfate. Evaporate a 3 mL aliquot to dryness under vacuum, reconstitute residue with 200 μL MeOH, inject 20 μL aliquot.

## HPLC VARIABLES
**Column:** 250 × 4.6 5 μm Hypersil 5-ODS
**Mobile phase:** MeCN:water 30:70
**Column temperature:** 30
**Flow rate:** 1
**Injection volume:** 20
**Detector:** UV 245

## CHROMATOGRAM
**Retention time:** 14
**Internal standard:** methylprednisolone

## OTHER SUBSTANCES
**Extracted:** cortisone, hydrocortisone
**Simultaneous:** fluorocortisone
**Noninterfering:** corticosterone, deflazacort, deoxycorticosterone, fluorocortisone acetate, 21-hydroxydeflazacort, 11α-hydroxyprogesterone, prednisolone, prednisone, triamcinolone acetonide

## KEY WORDS
methylprednisolone is IS

## REFERENCE
Santos-Montes, A.; Gonzalo-Lumbreras, R.; Izquierdo-Hornillos, R. Simultaneous determination of cortisol and cortisone in urine by reversed-phase high-performance liquid chromatography. Clinical and doping control applications. *J.Chromatogr.B*, **1995**, *673*, 27–33

## SAMPLE
**Matrix:** urine
**Sample preparation:** 3 mL Urine + 0.25 g NaCl, adjust pH to 9.0 with 0.5 g $Na_2HPO_4$, add 4 mL dichloromethane, vortex 1 min, centrifuge at 3700 g for 3 min. Remove organic phase and dry it over anhydrous sodium sulfate. Evaporate a 3 mL aliquot to dryness under vacuum, reconstitute residue with 200 μL 5 μg/mL IS in MeOH, inject 20 μL aliquot.

## HPLC VARIABLES
**Column:** 250 × 4.6 Hypersil ODS
**Mobile phase:** MeCN:water 32:68
**Column temperature:** 30
**Flow rate:** 1
**Injection volume:** 20
**Detector:** UV 245

## CHROMATOGRAM
**Retention time:** 9
**Internal standard:** prednisone (6.3)

## OTHER SUBSTANCES
**Simultaneous:** betamethasone, corticosterone, cortisone, dexamethasone, fluorocortisone, fluorocortisone acetate, hydrocortisone, hydroxyprogesterone, prednisolone, triamcinolone, triamcinolone acetonide

## KEY WORDS
SPE also discussed

## REFERENCE
Santos-Montes, A.; Gonzalo-Lumbreras, R.; Gasco-Lopez, A.I.; Izquierdo-Hornillos, R. Solvent and solid-phase extraction of natural and synthetic corticoids in human urine. *J.Chromatogr.B*, **1994**, *652*, 83–89

## SAMPLE
**Matrix:** urine
**Sample preparation:** 1 mL Urine + 1 mL 1.5 M HCl + 500 μL water + 10 mL dichloromethane, shake 20 min, centrifuge at 250 g for 3 min. Remove the organic phase and evaporate it to dryness under a stream of nitrogen at 40°. Reconstitute with 200 μL MeOH, inject 10-30 μL aliquot.

## HPLC VARIABLES
**Column:** 250 × 4.5 Partisil 10 ODS-3
**Mobile phase:** Gradient. A was MeCN:water 28:72 containing 0.05% phosphoric acid and 0.05% acetone. B was MeCN:50 mM $KH_2PO_4$ 50:50. A:B 100:0 for 8 min then to 0:100 over 6 min
**Flow rate:** 2
**Injection volume:** 10-30
**Detector:** UV 220

## CHROMATOGRAM
**Retention time:** 13.6
**Internal standard:** methylprednisolone

## OTHER SUBSTANCES
**Extracted:** ibuprofen, ibuprofen metabolites

## KEY WORDS
methylprednisolone is IS

## REFERENCE
Lockwood, G.F.; Wagner, J.G. High-performance liquid chromatographic determination of ibuprofen and its major metabolites in biological fluids. *J.Chromatogr.*, **1982**, *232*, 335–343

## ANNOTATED BIBLIOGRAPHY

Herman, B.D.; Sinclair, B.D.; Milton, N.; Nail, S.L. The effect of bulking agent on the solid-state stability of freeze-dried methylprednisolone sodium succinate. *Pharm.Res.*, **1994**, *11*, 1467–1473

Valvo, L.; Paris, A.; Savella, A.L.; Gallinella, B.; Ciranni Signoretti, E. General high-performance liquid chromatographic procedures for the rapid screening of natural and synthetic corticosteroids. *J.Pharm.Biomed.Anal.*, **1994**, *12*, 805–810 [gradient; reverse phase; normal phase; for 6α-methylprednisolone, 6α-methylprednisolone 21-acetate, 6α-methylprednisolone 21-sodium succinate; also beclomethasone, beclomethasone 17,21-dipropionate, betamethasone, betamethasone 21-acetate, betamethasone 17,21-dipropionate, betamethasone 21-disodium phosphate, betamethasone 17-valerate, cortisone, cortisone 21-acetate, 11-deoxycorticosterone 21-acetate, dexamethasone, dexamethasone 21-acetate, dexamethasone 21-disodium phosphate, fluocinolone, fluocinolone acetonide, 9α-fluorohydrocortisone 21-acetate, 9α-fluorohydrocortisone, 9α-fluoroprednisolone, 9α-fluoroprednisolone 21-acetate, hydrocortisone, hydrocortisone 21-acetate, hydrocortisone 21-hemisuccinate, prednisolone, prednisolone 21-acetate, prednisolone 21-disodium phosphate, prednisolone 21-pivalate, prednisolone 21-sodium succinate, prednisone, triamcinolone, triamcinolone acetonide]

Santos-Montes, A.; Gasco-Lopez, A.I.; Izquierdo-Hornillos, R. Optimization of the high-performance liquid chromatographic separation of a mixture of natural and synthetic corticosteroids. *J.Chromatogr.*, **1993**, *620*, 15–23 [simultaneous betamethasone, corticosterone, cortisone, deoxycorticosterone, dexamethasone, fluorocortisone, hydrocortisone, hydroxyprogesterone, prednisolone, prednisone, triamcinolone]

McGinley, P.A.; Braughler, J.M.; Hall, E.D. Determination of methylprednisolone in central nervous tissue and plasma using normal-phase high-performance liquid chromatography. *J.Chromatogr.*, **1982**, *230*, 29–35

# Metoprolol

**Molecular formula:** $C_{15}H_{25}NO_3$
**Molecular weight:** 267.4
**CAS Registry No.:** 37350-58-6, 56392-17-7 (tartrate), 119637-66-0
(fumarate), 98418-47-4 (succinate)

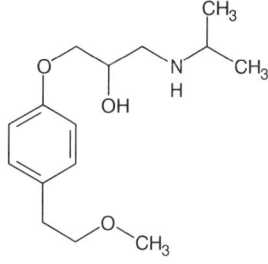

## SAMPLE
**Matrix:** blood
**Sample preparation:** Condition a Styrosorb cross-linked polystyrene (Biochrom, Moscow) or Sep-Pak C18 SPE cartridge with two 3 mL portions of MeOH and two 3 mL portions of water. Add 1 mL serum to the SPE cartridge, wash with two 3 mL portions of water, elute with 600 μL MeOH:diethylamine 99.7:0.3. Evaporate the eluate to dryness under a stream of air at 40°, reconstitute with 50 μL n-heptane:isopropanol:MeOH 83:13:4, inject a 20 μL aliquot onto a 250 × 4.6 10 μm Silasorb-NH2 (Elsico, Moscow) column and elute with n-heptane:isopropanol:MeOH 83:13:4 at 2.5 mL/min, collect the eluate containing metoprolol at about 4.1 min, evaporate to dryness, reconstitute with 30 μL mobile phase, inject a 10-20 μL aliquot.

## HPLC VARIABLES
**Guard column:** 50 × 4.6 10 μm Chiralcel OD
**Column:** 250 × 4.6 10 μm Chiralcel OD
**Mobile phase:** n-Heptane:isopropanol:MeOH 77:8:15
**Flow rate:** 1.3
**Injection volume:** 10-20
**Detector:** F ex 220 em 320 (cut-off filter)

## CHROMATOGRAM
**Retention time:** 3.5 (R), 4.6 (S)

## KEY WORDS
SPE; serum; silanize glassware; chiral

## REFERENCE
Rumiantsev, D.O.; Ivanova, T.V. Solid-phase extraction of Styrosorb cartridges as a sample pretreatment method in the stereoselective analysis of propranolol in human serum. *J.Chromatogr.B*, **1995**, *674*, 301–305

## SAMPLE
**Matrix:** blood
**Sample preparation:** 2 mL Whole blood or plasma + 2 mL buffer + 5 mL chloroform: isopropanol:n-heptane 60:14:26, shake gently horizontally for 10 min, centrifuge at 2800 g for 10 min. Remove the lower organic layer and evaporate it to dryness under vacuum at 45°, reconstitute the residue in 100 μL mobile phase, centrifuge at 2800 g for 5 min, inject a 50 μL aliquot of the supernatant. (Buffer was saturated ammonium chloride solution 25% diluted with water, adjusted to pH 9.5 with 25% ammonia solution.)

## HPLC VARIABLES
**Column:** 300 × 3.9 4 μm NovaPack C18
**Mobile phase:** MeOH:THF:buffer 65:5:30 (Buffer was 0.68 g/L (10 mM (sic)) $KH_2PO_4$ adjusted to pH 2.6 with concentrated orthophosphoric acid.) (At the end of each session wash the column with water for 1 h and MeOH for 1 h, re-equilibrate for 30 min.)

**Column temperature:** 30
**Flow rate:** 0.8
**Injection volume:** 50
**Detector:** UV 223

## CHROMATOGRAM
**Retention time:** 4.20
**Limit of detection:** <120 ng/mL

## OTHER SUBSTANCES
**Extracted:** acebutolol, acenocoumarol, acepromazine, aceprometazine, acetaminophen, aconitine, albuterol, alimemazine, alminoprofen, alpidem, alprenolol, amisulpride, amitriptyline, amodiaquine, amoxapine, aspirin, astemizole, atenolol, benazepril, benperidol, benzocaine, benzoylecgonine, bepridil, betaxolol, bisoprolol, bromazepam, brompheniramine, bumadizone, bupivacaine, buprenorphine, buspirone, caffeine, carbamazepine, carbinoxamine, carpipramine, carteolol, cetirizine, chlorambucil, chlordiazepoxide, chlormezanone, chlorophenacinone, chloroquine, chlorpheniramine, chlorpromazine, chlorpropamide, cibenzoline, cicletanine, clemastine, clomipramine, clonazepam, clonidine, clorazepate, clozapine, cocaine, codeine, colchicine, cyamemazine, cyclizine, cyproheptadine, cytarabine, dacarbazine, daunorubicin, demexiptiline, desipramine, dextromoramide, dextropropoxyphene, diazepam, diazoxide, diclofenac, dihydralazine, diltiazem, diphenhydramine, dipyridamole, dosulepine, doxepin, doxylamine, droperidol, ephedrine, estazolam, etodolac, fenfluramine, fenoprofen, fentiazac, flecainide, floctafenine, flumazenil, flunitrazepam, fluoxetine, fluphenazine, flurbiprofen, fluvoxamine, glibenclamide, glibornuride, glipizide, glutethimide, haloperidol, histapyrrodine, hydroxychloroquine, hydroxyzine, ibuprofen, imipramine, indomethacin, iproniazid, ketoprofen, labetalol, levomepromazine, lidocaine, lidoflazine, lisinopril, loperamide, loprazolam, loratadine, loxapine, maprotiline, medazepam, medifoxamine, mefenamic acid, mefenidramine, mefloquine, melphalan, meperidine, mephenesin, mepivacaine, metapramine, metformin, methadone, methocarbamol, methotrexate, metipranolol, metoclopramide, mexiletine, mianserine, midazolam, moclobemide, moperone, morphine, nadolol, nalbuphine, nalorphine, naloxone, naltrexone, naproxen, nialamide, nicardipine, niflumic acid, nimodipine, nitrendipine, nizatidine, nomifensine, nortriptyline, omeprazole, opipramol, oxprenolol, penbutolol, penfluridol, phencyclidine, phenobarbital, phenol, phenylbutazone, pimozide, pindolol, pipamperone, prazepam, prazosin, procainamide, procarbazine, proguanil, promethazine, propafenone, propranolol, protriptyline, pyrimethamine, quinupramine, ramipril, ranitidine, reserpine, ritodrine, secobarbital, sotalol, strychnine, sulfinpyrazone, sulpride, sultopride, suriclone, temazepam, tenoxicam, terbutaline, terfenadine, tetracaine, tetrazepam, thiopental, thioproperazine, thioridazine, tianeptine, tiapride, ticlopidine, timolol, tioclomarol, tolbutamide, toloxatone, trazodone, triazolam, trifluoperazine, trifluperidol, trimipramine, triprolidine, tropatenine, verapamil, viloxazine, vinblastine, vincristine, vindesine, warfarin, zolpidem, zorubicine
**Interfering:** ajmaline, alprazolam, celiprolol, clobazam, cycloguanil, debrisoquine, dextromethorphan, disopyramide, ketamine, lorazepam, mephentermine, methaqualone, metoprolol, minoxidil, nifedipine, nitrazepam, oxazepam, pentazocine, piroxicam, prilocaine, quinidine, quinine, sulindac, tiaprofenic acid, tofisopam, yohimbine, zopiclone

## KEY WORDS
whole blood; plasma

## REFERENCE
Tracqui, A.; Kintz, P.; Mangin, P. Systematic toxicological analysis using HPLC/DAD. *J.Forensic Sci.*, **1995**, *40*, 254–262

## SAMPLE
**Matrix:** blood
**Sample preparation:** 1 mL Plasma + 1 mL 5 μg/mL R-propranolol in water + 50 μL 100 mM NaOH + 4 mL chloroform, vortex for 30 s, centrifuge at 1800 g for 5 min. Remove

organic layer and evaporate it to dryness in vacuum, reconstitute in 200 μL 0.05% S-(+)-1-(1-naphthyl)ethyl isocyanate in chloroform, vortex for 30 s, inject a 75-150 μL aliquot.

## HPLC VARIABLES
**Column:** 250 mm long Whatman 5 μm silica
**Mobile phase:** Hexane:chloroform:MeOH 85:14:1
**Flow rate:** 2
**Injection volume:** 75-150
**Detector:** F ex 220 no emission filter

## CHROMATOGRAM
**Retention time:** 14.1 (R-(+)), 16.2 (S-(−))
**Internal standard:** R-propranolol (8.5)
**Limit of quantitation:** 5 ng/mL

## KEY WORDS
plasma; normal phase; derivatization; chiral

## REFERENCE
Bhatti, M.M.; Foster, R.T. Stereospecific high-performance liquid chromatographic assay of metoprolol. *J.Chromatogr.*, **1992**, *579*, 361–365

## SAMPLE
**Matrix:** blood
**Sample preparation:** 1 mL Serum + 100 μL 2 M NaOH + 4 mL dichloromethane, vortex for 10 s, centrifuge at 3000 rpm for 10 min. Remove the organic layer and evaporate it to dryness under a stream of nitrogen, reconstitute the residue in 100 μL mobile phase, inject an 80 μL aliquot.

## HPLC VARIABLES
**Guard column:** NewGuard C18 (Brownlee)
**Column:** 250 × 4.6 5 μm Dynamax Microsorb C18
**Mobile phase:** MeCN:0.1% triethylamine in water adjusted to pH 3.5 with 85% phosphoric acid 20:80
**Flow rate:** 1
**Injection volume:** 80
**Detector:** F ex 215

## CHROMATOGRAM
**Retention time:** 11.65
**Internal standard:** metoprolol

## OTHER SUBSTANCES
**Simultaneous:** pindolol

## KEY WORDS
serum; metoprolol is IS

## REFERENCE
Chmielowiec, D.; Schuster, D.; Gengo, F. Determination of pindolol in human serum by HPLC. *J.Chromatogr.Sci.*, **1991**, *29*, 37–39

## SAMPLE
**Matrix:** blood

**Sample preparation:** Condition an Analytichem 3 mL 200 mg SPE cartridge with 3 mL MeOH and 3 mL water. 1 mL Plasma + 30 µL 0.5 µg/mL d-propranolol in MeOH, vortex for 15 s, add to the SPE cartridge, wash with 3 mL water, wash with 1 mL MeOH:water 50:50, elute with two aliquots of 500 µL MeOH containing 0.1% triethylamine. Evaporate the eluate to dryness under nitrogen at 30°, reconstitute with 150 µL mobile phase, vortex for 30 s, inject a 100 µL aliquot.

## HPLC VARIABLES
**Guard column:** Chiracel OD (Daicel Chemical Industries)
**Column:** 250 × 4.6 Chiracel OD (Daicel Chemical Industries)
**Mobile phase:** Hexane:EtOH:N,N-diethylamine 95:5:0.1
**Flow rate:** 0.5
**Injection volume:** 100
**Detector:** F ex 220 em 320

## CHROMATOGRAM
**Retention time:** 19 (d), 22 (l)
**Internal standard:** d-propranolol (34)
**Limit of detection:** 4 ng/mL

## KEY WORDS
plasma; SPE; chiral

## REFERENCE
Herring, V.L.; Bastian, T.L.; Lalonde, R.L. Solid-phase extraction and direct high-performance liquid chromatographic determination of metoprolol enantiomers in plasma. *J.Chromatogr.*, **1991**, *567*, 221–227

## SAMPLE
**Matrix:** blood
**Sample preparation:** Condition a Bond-Elut C18 SPE cartridge with 2 mL MeCN and 1 mL MeCN:water 40:60. 995 µL Serum + 5 µL 20 µM oxprenolol in water, add 20 µL of this serum to 500 µL MeCN:water 40:60, vortex, add to SPE cartridge, wash with 2 mL MeCN:water 40:60, elute with 1 mL MeOH:water 1:1 containing 0.05% trifluoroacetic acid. Evaporate eluate to dryness, reconstitute in MeCN:water 20:80 containing 0.05% trifluoroacetic acid, inject.

## HPLC VARIABLES
**Column:** 150 × 4.6 5 µm TSK gel ODS 80Tm (Tosoh)
**Mobile phase:** MeCN:water 31:69 containing 0.05% trifluoroacetic acid
**Column temperature:** 40
**Injection volume:** 20
**Detector:** F ex 230 em 300

## CHROMATOGRAM
**Internal standard:** oxprenolol

## KEY WORDS
serum; SPE

## REFERENCE
Uzu, S.; Imai, K.; Nakashima, K.; Akiyama, S. Use of 4-(N,N-dimethylaminosulphonyl)-7-fluoro-2,1,3-benzoxadiazole as a labelling reagent for peroxyoxalate chemiluminescence detection and its application to the determination of the β-blocker metoprolol in serum by high-performance liquid chromatography. *Analyst*, **1991**, *116*, 1353–1357

## SAMPLE

**Matrix:** blood

**Sample preparation:** Condition a Bond-Elut C18 SPE cartridge with 2 mL MeCN and 1 mL MeCN:water 40:60. Add 20 µL serum to 500 µL MeCN:water 40:60, vortex, add to SPE cartridge, wash with 2 mL MeCN:water 40:60, elute with 1 mL MeOH:water 1:1 containing 0.05% trifluoroacetic acid. Evaporate eluate to dryness under reduced pressure at 40°, reconstitute with 50 µL 100 mM pH 9.0 borate buffer containing 2 mM EDTA, add 50 µL 50 mM DBD-F in MeCN, heat at 45°for 8 h, add 100 µL 100 mM acetic acid in MeCN:water 50:50, inject a 20 µL aliquot. (Synthesis of DBD-F is as follows. Dissolve 0.5 g magnesium sulfate heptahydrate and 6 g NaOH in 60 mL water, throughout the reaction keep the flask at about 20°with cold water cooling, add 15 mL 30% hydrogen peroxide, add 75 mL MeOH, add 12.1 g powdered benzoyl peroxide in one go, stir for 10 min, pour into 150 mL 20% sulfuric acid, extract three times with 50 mL portions of chloroform, determine peroxybenzoic acid concentration by iodometric titration (Tetrahedron 1967, 23, 3327). Slowly add 110 mL 1 M peroxybenzoic acid in chloroform to 7 g 2,6-difluoroaniline dissolved in 100 mL chloroform, stir at room temperature, when reaction is complete (iodometric titration) wash with 2% sodium thiosulfate, wash with 5% sodium carbonate, wash with water, dry over anhydrous sodium sulfate, evaporate to dryness under reduced pressure, recrystallize 2,6-difluoronitrosobenzene from EtOH (mp 108.5-109.5). Stir 8.5 g 2,6-difluoronitrosobenzene in 85 mL DMSO at room temperature and add a solution of 3.91 g sodium azide in 85 mL DMSO dropwise, let stand for about 1 h, add to a large volume of water, extract with ether, dry the extracts over anhydrous sodium sulfate, evaporate to dryness under reduced pressure and distil to give 4-fluoro-2,1,3-benzoxadiazole as a colorless oil (bp 83°/12 mm Hg) (J.Chem.Soc.(C) 1970, 1433). Add 11 mL chlorosulfonic acid dropwise to 3 g 4-fluoro-2,1,3-benzoxadiazole in 10 mL chloroform at 0-10° (use a calcium chloride drying tube), stir at room temperature for 1 h, reflux for 2 h, cool, slowly pour into ice water, remove the organic layer, extract the aqueous layer with chloroform, combine the organic layer, wash, dry over anhydrous magnesium sulfate, evaporate under reduced pressure, take up the residue in 5 mL benzene (Caution! Benzene is a carcinogen!), chromatograph on a 150 × 30 column of silica gel (100-200 mesh Kanto Chemical) with n-hexane:benzene 50:50, evaporate the appropriate fractions to give 4-(chlorosulfonyl)-7-fluoro-2,1,3-benzoxadiazole (CBD-F) as pale yellow needles (mp 64-66°) (Anal. Chem. 1984. 56, 2461). Stir 0.76 g CBD-F in 70 mL MeCN at 0-10°and add 1 g dimethylamine hydrochloride in 10 mL 100 mM pH 10 borax dropwise, adjust pH to 5 with 1 M HCl, concentrate to about 10 mL under reduced pressure, extract three times with 200 mL portions of diethyl ether, wash with water, dry over anhydrous magnesium sulfate, evaporate under reduced pressure, chromatograph on a 500 × 20 column of silica gel with chloroform, isolate the appropriate fraction and re-chromatograph on the same column with ethyl acetate:benzene 1:2 to give 4-(N, N-dimethylaminosulfonyl)-7-fluoro-2,1,3-benzoxadiazole (DBD-F) as white needles (mp 124-125°) (yield = 1% !). On a Merck no. 5714 60F$_{254}$ tlc plate eluted with chloroform DBD-F has Rf 0.32 and lies between two other reaction products (Analyst 1989, 114, 413). It is also reported that DBD-F can be purchased from Tokyo Kasei.)

## HPLC VARIABLES

**Column:** 250 × 4.6 5 µm TSK gel ODS 80Tm (Tosoh)

**Mobile phase:** MeCN:THF:50 mM pH 6.0 imidazole nitrate buffer 28:20:52

**Column temperature:** 40

**Flow rate:** 0.8

**Injection volume:** 20

**Detector:** Chemiluminescence following post-column reaction. The column effluent mixed with the reagent pumped at 1.4 mL/min and the mixture flowed to the detector. (Reagent was 0.25 mM bis[4-nitro-2-(3,6,9-trioxadecyloxycarbonyl)phenyl] oxalate (TDPO, Wako, Osaka) and 37.5 mM hydrogen peroxide in MeCN:ethyl acetate 50:50.)

## CHROMATOGRAM

**Retention time:** 9

**Limit of detection:** 0.8 ng/mL

## KEY WORDS
SPE; derivatization; post-column reaction; serum

## REFERENCE
Uzu, S.; Imai, K.; Nakashima, K.; Akiyama, S. Use of 4-(*N,N*-dimethylaminosulphonyl)-7-fluoro-2,1,3-benzoxadiazole as a labelling reagent for peroxyoxalate chemiluminescence detection and its application to the determination of the β-blocker metoprolol in serum by high-performance liquid chromatography. *Analyst*, **1991**, *116*, 1353–1357

## SAMPLE
**Matrix:** blood
**Sample preparation:** 1 mL Serum + 50 μL 1 μg/mL verapamil in MeOH + 500 μL 0.5 M HCl + 4 mL diethyl ether, vortex for 30 s, centrifuge at 2000 g for 5 min, discard organic layer, add 100 μL 2 M NaOH and 4 mL diethyl ether to the aqueous layer, vortex for 1 min, centrifuge at 2000 g for 10 min. Remove the organic layer and dry it at 35° under a stream of helium. Reconstitute in 150 μL mobile phase, inject a 25 μL aliquot.

## HPLC VARIABLES
**Column:** 250 × 4.6 Chiralcel OD (Diacel Chemical Industries)
**Mobile phase:** Hexane:isopropanol 90:10 which contained 10 mM octylamine
**Flow rate:** 1
**Injection volume:** 25
**Detector:** F ex 275 em 315

## CHROMATOGRAM
**Retention time:** 6.13 (R), 14.34 (S)
**Internal standard:** verapamil (9.80)
**Limit of detection:** 5 ng/mL

## OTHER SUBSTANCES
**Simultaneous:** procainamide
**Noninterfering:** diazepam, digoxin, furosemide, lidocaine, prochlorperazine, quinidine

## KEY WORDS
serum; chiral

## REFERENCE
Straka, R.J.; Johnson, K.A.; Marshall, P.S.; Remmel, R.P. Analysis of metoprolol enantiomers in human serum by liquid chromatography on a cellulose-based chiral stationary phase. *J.Chromatogr.*, **1990**, *530*, 83–93

## SAMPLE
**Matrix:** blood
**Sample preparation:** 1 mL Plasma + 1 mL 1 M pH 9.9 carbonate buffer + 6 mL water-saturated diethyl ether, agitate for 15 min, centrifuge at 4200 g for 5 min. Remove 5 mL of the organic layer and evaporate it to dryness under a stream of nitrogen at 35°, reconstitute the residue in 250 μL 142.4 mM triethylamine in dichloromethane, add 100 μL reagent, let stand for 15 min, evaporate to dryness under a stream of nitrogen at 35°, add 2 mL 100 mM NaOH, agitate for 10 min, add 6 mL ether, extract for 15 min, centrifuge. Remove 5 mL of the organic layer and evaporate it to dryness under a stream of nitrogen at 35°, cool in an ice bath, reconstitute the residue in 250 μL trifluoroacetic acid, let stand at 0° for 10 min, add 2 mL 2 M NaOH, extract with 6 mL ether. Remove a 5 mL aliquot of the organic layer and extract it with 100 μL 100 mM phosphoric acid, inject a 94 μL aliquot of the aqueous layer. (Purify triethylamine by drying it over NaOH pellets overnight, filter, add a volume of naphthylisocyanate equal to 2% of the volume of triethylamine, distill. Prepare reagent by dissolving 1 mmole N-tert-butoxycarbonyl-L-leucine (BOC-

L-Leu) in 3 mL dichloromethane, add 2 mL 250 mM dicyclohexylcarbodiimide in dichloromethane, let stand at 0° for 1 h, filter.)

## HPLC VARIABLES
**Column:** 100 × 3.2 10 μm μBondapak C18
**Mobile phase:** MeCN:100 mM pH 3.0 phosphate buffer 30:70
**Flow rate:** 0.5
**Injection volume:** 94
**Detector:** F ex 193 em (no cutoff filter)

## CHROMATOGRAM
**Retention time:** 5 (S), 6.5 (R)
**Limit of detection:** 0.5 ng/mL

## OTHER SUBSTANCES
**Also analyzed:** alprenolol

## KEY WORDS
pharmacokinetics; plasma; chiral; derivatization

## REFERENCE
Hermansson, J.; von Bahr, C. Determination of (R)- and (S)-alprenolol and (R)- and (S)-metoprolol as their diastereomeric derivatives in human plasma by reversed-phase liquid chromatography. *J.Chromatogr.*, **1982**, *227*, 113–127

## SAMPLE
**Matrix:** blood, urine
**Sample preparation:** Plasma. 1-2 mL Plasma + 100 μL 4 M NaOH + 100 μL water + 5 mL dichloromethane, shake by hand for 10 s, vortex vigorously for 3 min, centrifuge at 1500 g for 5 min. Remove the organic layer and evaporate it to dryness under a stream of air at 37°, reconstitute the residue in 100 μL mobile phase, filter (0.45 μm), inject a 40-60 μL aliquot. Urine. 0.1-1 mL Urine + 100 μL 4 M NaOH + 100 μL water + 5 mL dichloromethane, vortex vigorously for 1 min, centrifuge at 1500 g for 5 min. Remove the organic layer and add it to 1 mL 100 mM phosphoric acid, vortex for 1 min, centrifuge at 1500 g for 5 min, inject a 40 μL aliquot of the aqueous phase.

## HPLC VARIABLES
**Column:** 250 × 4 Wakosil 5C18 (Wako)
**Mobile phase:** MeCN:water:triethylamine 18:81:1, pH adjusted to 3.0 with phosphoric acid
**Flow rate:** 1
**Injection volume:** 40-60
**Detector:** UV 295

## CHROMATOGRAM
**Retention time:** 8.69
**Internal standard:** metoprolol tartrate

## OTHER SUBSTANCES
**Simultaneous:** acebutolol, N-acetylprocainamide, alprenolol, atenolol, bufetolol, bupranolol, carteolol, disopyramide, indenolol, lidocaine, nifedipine, pindolol, procainamide, propranolol, quinidine, timolol
**Noninterfering:** diltiazem, glycinexylidide, mexiletine, nicardipine, tocainide, verapamil
**Interfering:** nicainoprol

## KEY WORDS
plasma; metoprolol is IS

## REFERENCE

Kubota, K.; Nakamura, H.; Koyama, E.; Yamada, T.; Kikuchi, K.; Ishizaki, T. Simple and sensitive determination of timolol in human plasma and urine by high-performance liquid chromatography with ultraviolet detection. *J.Chromatogr.*, **1990**, *533*, 255–263

## SAMPLE

**Matrix:** blood, urine
**Sample preparation:** 500 μL Plasma or 100-200 μL urine + 100 (plasma) or 150 (urine) μL 10 (plasma) or 600 (urine) μg/mL pindolol in MeOH + 0.5 mL 1 M NaOH + 3 mL dichloromethane, vortex for 1 min, centrifuge at 1500 g for 10 min. Remove the organic layer and evaporate it to dryness under a stream of air at 35°, reconstitute the residue in 100 (plasma) or 200 (urine) μL mobile phase, inject a 20-30 μL aliquot.

## HPLC VARIABLES

**Column:** 250 × 4.6 5 μm Zorbax ODS
**Mobile phase:** MeCN:water:triethylamine 25:74:1, adjusted to pH 4.0 with phosphoric acid
**Flow rate:** 0.8
**Injection volume:** 20-30
**Detector:** F ex 230 em 300

## CHROMATOGRAM

**Retention time:** 8.6
**Internal standard:** pindolol (6.0)
**Limit of detection:** 2 ng/mL (plasma)

## OTHER SUBSTANCES

**Extracted:** metabolites

## KEY WORDS

plasma; pharmacokinetics

## REFERENCE

Horai, Y.; Ishizaki, T.; Kusaka, M.; Tsujimoto, G.; Hashimoto, K. Simultaneous determination of metoprolol and α-hydroxymetoprolol in human plasma and urine by liquid chromatography with a preliminary observation on metoprolol oxidation in Japanese subjects. *Ther.Drug Monit.*, **1988**, *10*, 428–433

## SAMPLE

**Matrix:** blood, urine
**Sample preparation:** Blood. Add MeCN to blood so that the ratio is 1:5. Remove a 1 mL aliquot and add 1 mL water, adjust pH to 9.8-10.2 with 7-10 drops 100 mM NaOH, add 2 mL benzene (Caution! Benzene is a carcinogen!), shake on an automatic shaker for 30 min, centrifuge, remove the organic phase and extract the aqueous layer again with 2 mL benzene for 20 min. Combine the organic layers and evaporate them under reduced pressure at 30°, dissolve the residue in 50 μL mobile phase, inject a 10-50 μL aliquot. Urine. 20-1000 μL Urine + 1 mL 200 mM pH 10.2 sodium borate buffer (Sörensen), adjust pH to 9.8-10.2 with 100 mM NaOH (if necessary), add 500 mg NaCl, extract with 4 mL benzene for 30 min, centrifuge. Remove the organic layer and evaporate it under reduced pressure at 30°, dissolve the residue in 50 μL mobile phase, inject a 10-50 μL aliquot.

## HPLC VARIABLES

**Column:** 250 × 4.6 10 μm μBondapak C18
**Mobile phase:** MeOH:water 48:52 containing 0.4% phosphoric acid and 0.2% heptanesulfonic acid
**Flow rate:** 2

**Injection volume:** 10-50
**Detector:** F ex 225 em 295

## CHROMATOGRAM
**Retention time:** 4.6
**Internal standard:** metoprolol

## OTHER SUBSTANCES
**Simultaneous:** dihydrolevobunolol

## KEY WORDS
human; dog; metoprolol is IS

## REFERENCE
Hengy, H.; Kölle, E.-U. Determination of levobunolol and dihydrolevobunolol in blood and urine by high-performance liquid chromatography using fluorescence detection. *J.Chromatogr.*, **1985**, *338*, 444–449

## SAMPLE
**Matrix:** bulk
**Sample preparation:** Dissolve 10 µmole compound (as free base or hydrochloride) in 500 µL MeCN, add 250 µL 5% sodium carbonate (for hydrochlorides only), add 500 µL 100 mM reagent in MeCN, vortex for 1 min, heat at 60°for 2 h, add 100 µmole L-proline, heat at 60°for 30 min. Remove a 100 µL aliquot and dilute it with mobile phase, neutralize with acetic acid, inject a 10 µL aliquot. (Prepare the reagent ((R, R)-N-(3,5-dinitroben-zoyl)-2-aminocyclohexylisothiocyanate) as follows. Add 0.7 mL carbon disulfide to 6 mL (1R,2R)-(−)-1,2-diaminocyclohexane, 12 mL water, and 12 mL EtOH, heat the oil bath to 80°, add 2.8 mL carbon disulfide dropwise (making sure that the product does not start to precipitate), when addition is complete reflux for 1 h, acidify with 500 µL 5 M HCl, reflux for 12 h, cool, filter, wash the solid with a little cold EtOH to give trans-4,5-tetra-methyleneimidazolidine-2-thione as a white fluffy solid (mp 148-150°) (Tetrahedron 1993, 49, 4419). Stir 7.97 g 3,5-dinitrobenzoyl chloride in 30 mL dichloroethane at 50°, add a solution of 6 g trans-4,5-tetramethyleneimidazolidine-2-thione in 120 mL dichloroethane containing a catalytic amount of 4-(dimethylamino)pyridine over 15 min, reflux for 2 h, remove the crystals of (R, R)-N-(3,5-dinitrobenzoyl)-2-aminocyclohexylisothiocyanate by filtration, evaporate the filtrate to dryness and dissolve the residue in 60 mL dichloroe-thane, reflux for 16 h to obtain more (R, R)-N-(3,5-dinitrobenzoyl)-2-aminocyclohexyliso-thiocyanate (mp >250°, $[\alpha]_{546}$ = -133° (c=1) in MeCN).)

## HPLC VARIABLES
**Column:** 125 × 4 5 µm Lichrospher 60 RP Select B
**Mobile phase:** MeCN:20 mM ammonium acetate 55:45
**Flow rate:** 1
**Injection volume:** 10
**Detector:** UV 254

## CHROMATOGRAM
**Retention time:** k′ 5.33, k′ 7.83 (enantiomers)

## OTHER SUBSTANCES
**Also analyzed:** acebutolol, alprenolol, atenolol, carazolol, carvedilol, formoterol, metham-phetamine, metipranolol, nifenanol, nitrilo atenolol, oxprenolol, pindolol, propranolol, xamoterol

## KEY WORDS
derivatization; chiral

## REFERENCE

Kleidernigg, O.P.; Posch, K.; Lindner, W. Synthesis and application of a new isothiocyanate as a chiral derivatizing agent for the indirect resolution of chiral amino alcohols and amines. *J.Chromatogr.A*, **1996**, *729*, 33–42

## SAMPLE

**Matrix:** bulk
**Sample preparation:** Prepare a 2 mg/mL solution, inject a 20 μL aliquot.

## HPLC VARIABLES

**Column:** 125 × 4 5 μm LiCHrospher RP-select B C8
**Mobile phase:** MeCN:buffer 17:83 (Buffer was 11.5 g ammonium dihydrogen phosphate and 10 mL 1 M phosphoric acid made up to 2 L, pH 3.2.)
**Flow rate:** 1
**Injection volume:** 20
**Detector:** UV 280

## CHROMATOGRAM

**Retention time:** 7

## OTHER SUBSTANCES

**Simultaneous:** impurities

## REFERENCE

Erickson, M.; Karlsson, K.-E.; Lamm, B.; Larsson, S.; Svensson, L.A.; Vessman, J. Identification of a new by-product detected in metoprolol tartrate. *J.Pharm.Biomed.Anal.*, **1995**, *13*, 567–574

## SAMPLE

**Matrix:** formulations
**Sample preparation:** Dissolve 2 mg tablet or capsule in 10 mL pH 10 solution, extract twice with 2 mL ether, combine extracts, filter, inject a 20 μL aliquot.

## HPLC VARIABLES

**Column:** 250 × 4.6 5 μm β-cyclodextrin bonded C18 (Advanced Separation Technologies)
**Mobile phase:** MeCN:MeOH:acetic acid:triethylamine 95:5:0.3:0.2
**Flow rate:** 1
**Injection volume:** 20
**Detector:** UV 254

## CHROMATOGRAM

**Retention time:** 12, 13 (enantiomers)

## OTHER SUBSTANCES

**Simultaneous:** atenolol, propranolol

## KEY WORDS

capsules; tablets; chiral

## REFERENCE

Tran, C.D.; Dotlich, M. Enantiomeric separation of beta-blockers by high performance liquid chromatography. *J.Chem.Educ.*, **1995**, *72*, 71–73

## SAMPLE

**Matrix:** formulations
**Sample preparation:** Take up in mobile phase, inject an aliquot.

## HPLC VARIABLES
**Column:** 250 × 4.6 10 μm LiChrosorb C2
**Mobile phase:** MeCN:buffer 35:65 (1 mL 100 mM HCl + 1200 mL water + 5.84 g NaCl, mix to dissolve, add 700 mL MeOH, make up to 2 L, apparent pH 4.5.)
**Flow rate:** 1.2
**Injection volume:** 20
**Detector:** UV 254

## CHROMATOGRAM
**Retention time:** 10.5

## OTHER SUBSTANCES
**Simultaneous:** atenolol, nadolol, alprenolol, acebutolol, oxprenolol, pindolol, practolol, propranolol, sotalol, timolol

## KEY WORDS
tablets

## REFERENCE
Patel, B.R.; Kirschbaum, J.J.; Poet, R.B. High-pressure liquid chromatography of nadolol and other beta-adrenergic blocking drugs. *J.Pharm.Sci.*, **1981**, *70*, 336–338

## SAMPLE
**Matrix:** saliva
**Sample preparation:** Condition a 100 mg 1 mL Bond-Elut C2 SPE cartridge with 1 mL MeOH, 1 mL water, and 1 mL pH 9.0 borate buffer. Centrifuge a cotton roll soaked with saliva at 1000 g for 5 min, remove the liquid supernatant. 1 mL Supernatant + 50 μL 10 μg/mL alprenolol, add to the SPE cartridge, wash with 500 μL water, wash with 500 μL MeCN, elute with two 500 μL portions of acidified MeOH. Evaporate the eluate to dryness under a stream of nitrogen at 60°, reconstitute the residue in 50 μL mobile phase, mix for 15 s, inject a 40 μL aliquot. (Acidified MeOH was 50 mL MeOH + 300 μL 96% acetic acid.)

## HPLC VARIABLES
**Guard column:** RCSS silica guard-pack (Waters)
**Column:** 250 × 4.6 Chiralcel OD-H
**Mobile phase:** n-Hexane:EtOH:diethylamine 91:8:1
**Flow rate:** 1
**Injection volume:** 40
**Detector:** F ex 225 em 320 cut-off filter

## CHROMATOGRAM
**Internal standard:** (S)-alprenolol

## KEY WORDS
SPE; chiral

## REFERENCE
Höld, K.M.; de Boer, D.; Zuidema, J.; Maes, R.A.A. Evaluation of the Salivette as sampling device for monitoring β-adrenoceptor blocking drugs in saliva. *J.Chromatogr.B*, **1995**, *663*, 103–110

## SAMPLE
**Matrix:** solutions
**Sample preparation:** Mix 20 μL of a 1 mM solution in MeOH or water with 50 μL pH 8 borate buffer and 50 μL 18 mM 2-(6-methoxy-2-naphthyl)-1-propyl chloroformate in ace-

tone, vortex, let stand at room temperature for 30 min, add 100 μL 10 mM trans-4-hydroxy-L-proline in water, mix, let stand for 2 min, add 2 mL dichloromethane, vortex for 30 s. Remove the organic layer and evaporate it to dryness under reduced pressure, reconstitute the residue in 100 μL mobile phase, inject an aliquot. (Prepare 2-(6-methoxy-2-naphthyl)-1-propyl chloroformate as follows. Stir 1.5 mmoles lithium aluminum hydride in THF, slowly add 2 mmoles (S)-naproxen in 20 mL anhydrous THF, reflux for 1 h, evaporate most of the solvent, cautiously add water with stirring, acidify with 6 N HCl, extract three times with diethyl ether. Combine the organic layers and dry them over anhydrous sodium sulfate, evaporate to dryness, chromatograph on silica gel with dichloromethane:MeOH 100:2 (flash chromatography), evaporate eluate to dryness, dry under vacuum over KOH to give 2-(6-methoxy-2-naphthyl)propanol as a white solid (mp 92-3°). Stir 0.5 mmoles 2-(6-methoxy-2-naphthyl)propanol and 0.5 mmoles triethylamine in 10 mL dry toluene at 0°, add 1 mL 20% phosgene in toluene (Caution! Phosgene is highly toxic, perform reaction in a chemical fume hood!) (Fluka), stir for 4 h, filter, evaporate to dryness under reduced pressure, dry under vacuum to give 2-(6-methoxy-2-naphthyl)-1-propyl chloroformate (mp 60°). Store under vacuum over phosphorus pentoxide at room temperature.)

## HPLC VARIABLES
**Column:** 250 × 4 5 μm Zorbax-SIL
**Mobile phase:** n-Hexane:isopropanol 100:1.5
**Flow rate:** 1.5
**Injection volume:** 100
**Detector:** UV 230; F ex 270 em 365

## CHROMATOGRAM
**Retention time:** k' 12.6 (S-(−)), k' 13.5 (R-(+))
**Limit of detection:** 0.2 ng/mL

## OTHER SUBSTANCES
**Simultaneous:** flecainide, propafenone, tocainide

## KEY WORDS
derivatization; chiral; normal phase

## REFERENCE
Büschges, R.; Linde, H.; Mutschler, E.; Spahn-Langguth, H. Chloroformates and isothiocyanates derived from 2-arylpropionic acids as chiral reagents: synthetic routes and chromatographic behaviour of the derivatives. *J.Chromatogr.A*, **1996**, *725*, 323–334

## SAMPLE
**Matrix:** solutions
**Sample preparation:** Mix 300 μL of a 30 μM solution in dichloromethane with 10 μL 20 mM 1-(6-methoxy-2-naphthyl)ethyl isothiocyanate in anhydrous dichloromethane and 50 μL 0.1% triethylamine in dichloromethane, vortex thoroughly, heat at 50° for 1.5 h, inject an aliquot. (Synthesize 1-(6-methoxy-2-naphthyl)ethyl isothiocyanate as follows (protect from light). Dissolve 500 mg (S)-(+)-naproxen in 50 mL dry toluene, slowly add 5 mL freshly distilled thionyl chloride, reflux for 1 h, evaporate to dryness under vacuum, dry the acyl chloride (mp 87.5°) under vacuum over KOH for 2 days. Dissolve 0.5 mmoles acyl chloride in 5 mL acetone, stir at 0°, add 0.6 mmoles sodium azide dissolved in ice water, stir at 0° for 30 min, add 10 mL ice-cold water, filter, dry solid in a desiccator under vacuum. Dissolve the solid in 1 mL toluene or dichloromethane (dried over 3 Å molecular sieve), reflux for 10 min, evaporate, store resulting isocyanate (mp 51°) under vacuum over a desiccant. Dissolve 0.5 mmole isocyanate in 5 mL acetone, add 20 mL 8.5% phosphoric acid, heat to 80° for 1.5 h, adjust to pH 13, extract with diethyl ether:dichloromethane 4:1. Wash the organic layer twice with water, dry over anhydrous sodium sulfate, evaporate to dryness, dissolve in 1 mL toluene, evaporate to give the amine from

naproxen as crystals (mp 53°) (Pharm.Res. 1990, 7, 1262). Dissolve 1 mmole 1,1-thiocarbonyldiimidazole in 15 mL ice-cold chloroform, stir at 0°, add dropwise 1 mmole of the amine dissolved in 10 mL chloroform, stir at room temperature for 1.5 h, evaporate to dryness, reconstitute with carbon tetrachloride (Caution! Carbon tetrachloride is a carcinogen!), filter, evaporate the filtrate to dryness, store the resulting oil in a desiccator, purify on a short silica gel column with dichloromethane:light petroleum 50:50 to give 1-(6-methoxy-2-naphthyl)ethyl isothiocyanate as a slightly yellow liquid (store in the freezer under argon).)

## HPLC VARIABLES
**Column:** 250 × 4 5 μm Zorbax ODS
**Mobile phase:** MeCN:water 55:45
**Flow rate:** 1
**Injection volume:** 100
**Detector:** UV 230; F ex 270 em 350

## CHROMATOGRAM
**Retention time:** k' 20.4 (S-(−)), 25.4 (R-(+))

## KEY WORDS
derivatization; chiral; F not much more sensitive than UV; α = 1.25

## REFERENCE
Büschges, R.; Linde, H.; Mutschler, E.; Spahn-Langguth, H. Chloroformates and isothiocyanates derived from 2-arylpropionic acids as chiral reagents: synthetic routes and chromatographic behaviour of the derivatives. *J.Chromatogr.A*, **1996**, *725*, 323−334

## SAMPLE
**Matrix:** solutions

## HPLC VARIABLES
**Column:** 250 × 4.6 CSP-4 (Prepare as follows. Add a solution of 1.07 g L-valyl-L-valyl-L-valine isopropylester (Bunseki Kagaku 1979, 28, 125) in 30 mL dry dioxane (Caution! Dioxane is a carcinogen!) dropwise to a mixture of 2.2 g 2,4,6-trichloro-1,3,5-triazine (cyanuric chloride) in 20 mL dry dioxane stirred at 0°, add 3 g anhydrous sodium carbonate at room temperature, stir, filter, evaporate to give a colorless solid. Dissolve 8.3 g of this solid in 30 mL dry dioxane, add 2 g N-(2-aminoethyl)-3-aminopropyltrimethoxysilane, add 1.5 g anhydrous sodium carbonate, reflux with stirring for 40 h, filter, add 3 g dried 10 μm LiChrosorb Si 100, reflux with slow stirring for 10 h, cool, filter. Wash the solid with dioxane, MeOH, and diethyl ether, dry under reduced pressure (J.Chromatogr. 1984, 292, 427).)
**Mobile phase:** Hexane:1,2-dichloroethane:EtOH:trifluoroacetic acid 62.5:35:0.625:0.25
**Detector:** UV

## CHROMATOGRAM
**Retention time:** k' 1.22 (first enantiomer)

## KEY WORDS
chiral; α = 1.05

## REFERENCE
Oi, N.; Kitahara, H.; Matsushita, Y.; Kisu, N. Enantiomer separation by gas and high-performance liquid chromatography with tripeptide derivatives as chiral stationary phases. *J.Chromatogr.A*, **1996**, *722*, 229−232

## SAMPLE
**Matrix:** solutions

## HPLC VARIABLES
**Guard column:** 10 × 3.2 5 μm Partisil ODS3
**Column:** 100 × 4.6 5 μm Partisil ODS3
**Mobile phase:** MeCN:buffer 25:75 (Buffer was 60 mM KH$_2$PO$_4$ adjusted to pH 3.0 with phosphoric acid.)
**Flow rate:** 0.6-1
**Injection volume:** 10-100
**Detector:** UV 270

## OTHER SUBSTANCES
**Also analyzed:** oxprenolol

## REFERENCE
Palm, K.; Luthman, K.; Ungell, A.-L.; Strandlund, G.; Artursson, P. Correlation of drug absorption with molecular surface properties. *J.Pharm.Sci.*, **1996**, *85*, 32–39

## SAMPLE
**Matrix:** solutions

## HPLC VARIABLES
**Column:** 250 × 4.6 Chirex 3022 (Phenomenex)
**Mobile phase:** Hexane:1,2-dichloroethane:EtOH/trifluoroacetic acid 60:35:5 (EtOH/trifluoroacetic acid was premixed 20:1.)
**Flow rate:** 0.7-1
**Injection volume:** 20
**Detector:** UV 276

## KEY WORDS
chiral; α = 1.08

## REFERENCE
Cleveland, T. Pirkle-concept chiral stationary phases for the HPLC separation of pharmaceutical racemates. *J.Liq.Chromatogr.*, **1995**, *18*, 649–671

## SAMPLE
**Matrix:** solutions
**Sample preparation:** Inject a 20 μL aliquot of a 1 mg/mL solution.

## HPLC VARIABLES
**Column:** 250 × 4.6 10 μm Chiralcel OD
**Mobile phase:** Hexane:isopropanol:diethylamine 80:20:0.1
**Flow rate:** 0.5
**Injection volume:** 20
**Detector:** UV 275

## CHROMATOGRAM
**Retention time:** k' 0.69, 2.02 (enantiomers)

## KEY WORDS
chiral

## REFERENCE
Ekelund, J.; van Arkens, A.; Bronnum-Hansen, K.; Fich, K.; Olsen, L.; Petersen, P.V. Chiral separations of β-blocking drug substances using chiral stationary phases. *J.Chromatogr.A*, **1995**, *708*, 253–261

## SAMPLE
**Matrix:** solutions
**Sample preparation:** Inject an aliquot of a 200 μM solution in MeOH.

## HPLC VARIABLES
**Column:** 100 × 4.7 7 μm Hypercarb (Shandon)
**Mobile phase:** MeOH containing 5 mM N-benzyloxycarbonylglycyl-L-proline and 4.5 mM NaOH
**Column temperature:** 17
**Injection volume:** 20
**Detector:** UV 270

## CHROMATOGRAM
**Retention time:** k′ 18 (first enantiomer)

## KEY WORDS
chiral; α = 1.09

## REFERENCE
Huynh, N.-H.; Karlsson, A.; Pettersson, C. Enantiomeric separation of basic drugs using N-benzyloxy-carbonylglyclyl-L-proline as counter ion in methanol. *J.Chromatogr.A*, **1995**, *705*, 275–287

## SAMPLE
**Matrix:** solutions

## HPLC VARIABLES
**Column:** 150 × 4.6 12 μm 1-myristoyl-2-[(13-carboxyl)-tridecoyl]-sn-3-glycerophosphocho-line chemically bonded to silica (Regis)
**Mobile phase:** MeCN:100 mM pH 7.0 phosphate buffer 20:80
**Flow rate:** 1
**Detector:** UV 254

## CHROMATOGRAM
**Retention time:** k′ 2.72

## OTHER SUBSTANCES
**Also analyzed:** acebutolol, alprenolol, antazoline, atenolol, betaxolol, bisoprolol, bopindolol, bupranolol, carteolol, celiprolol, chloropyramine, chlorpheniramine, cicloprolol, cimetidine, cinnarizine, cirazoline, clonidine, dilevalol, dimethindene, diphenhydramine, doxazosin, esmolol, famotidine, isothipendyl, ketotifen, metiamide, moxonidine, nadolol, naphazoline, nifenalol, nizatidine, oxprenolol, pheniramine, phentolamine, pindolol, pizotyline (pizotifen), practolol, prazosin, promethazine, propranolol, pyrilamine (mepyramine), ranitidine, roxatidine, sotalol, tiamenidine, timolol, tramazoline, tripelennamine, triprolidine, tymazoline, UK-14,304

## REFERENCE
Kaliszan, R.; Nasal, A.; Turowski, M. Binding site for basic drugs on $\alpha_1$-acid glycoprotein as revealed by chemometric analysis of biochromatographic data. *Biomed.Chromatogr.*, **1995**, *9*, 211–215

## SAMPLE
**Matrix:** solutions

## HPLC VARIABLES
**Column:** 250 × 4.6 5 μm Supelcosil LC-DP (A) or 250 × 4 5 μm LiChrospher 100 RP-8 (B)

**Mobile phase:** MeCN:0.025% phosphoric acid:buffer 25:10:5 (A) or 60:25:15 (B) (Buffer was 9 mL concentrated phosphoric acid and 10 mL triethylamine in 900 mL water, adjust pH to 3.4 with dilute phosphoric acid, make up to 1 L.)
**Flow rate:** 0.6
**Injection volume:** 25
**Detector:** UV 229

## CHROMATOGRAM
**Retention time:** 7.12 (A), 4.19 (B)

## OTHER SUBSTANCES
**Also analyzed:** acebutolol, acepromazine, acetaminophen, acetazolamide, acetophenazine, albuterol, alprazolam, amitriptyline, amobarbital, amoxapine, antipyrine, atenolol, atropine, azatadine, baclofen, benzocaine, bromocriptine, brompheniramine, brotizolam, bupivacaine, buspirone, butabarbital, butalbital, caffeine, carbamazepine, cetirizine, chlorcyclizine, chlordiazepoxide, chlormezanone, chloroquine, chlorpheniramine, chlorpromazine, chlorpropamide, chlorprothixene, chlorthalidone, chlorzoxazone, cimetidine, cisapride, clomipramine, clonazepam, clonidine, clozapine, cocaine, codeine, colchicine, cyclizine, cyclobenzaprine, dantrolene, desipramine, diazepam, diclofenac, diflunisal, diltiazem, diphenhydramine, diphenidol, diphenoxylate, dipyridamole, disopyramide, dobutamine, doxapram, doxepin, droperidol, encainide, ethidium bromide, ethopropazine, fenoprofen, fentanyl, flavoxate, fluoxetine, fluphenazine, flurazepam, flurbiprofen, fluvoxamine, furosemide, glutethimide, glyburide, guaifenesin, haloperidol, homatropine, hydralazine, hydrochlorothiazide, hydrocodone, hydromorphone, hydroxychloroquine, hydroxyzine, ibuprofen, imipramine, indomethacin, ketoconazole, ketoprofen, ketorolac, labetalol, levorphanol, lidocaine, loratadine, lorazepam, lovastatin, loxapine, mazindol, mefenamic acid, meperidine, mephenytoin, mepivacaine, mesoridazine, metaproterenol, metformin, methadone, methdilazine, methocarbamol, methotrexate, methotrimeprazine, methoxamine, methyldopa, methylphenidate, metoclopramide, metolazone, metronidazole, midazolam, moclobemide, morphine, nadolol, nalbuphine, naloxone, naphazoline, naproxen, nifedipine, nizatidine, norepinephrine, nortriptyline, oxazepam, oxycodone, oxymetazoline, paroxetine, pemoline, pentazocine, pentobarbital, pentoxifylline, perphenazine, pheniramine, phenobarbital, phenol, phenolphthalein, phentolamine, phenylbutazone, phenyltoloxamine, phenytoin, pimozide, pindolol, piroxicam, pramoxine, prazepam, prazosin, probenecid, procainamide, procaine, prochlorperazine, procyclidine, promazine, promethazine, propafenone, propantheline, propiomazine, propofol, propranolol, protriptyline, quazepam, quinidine, quinine, racemethorphan, ranitidine, remoxipride, risperidone, salicylic acid, scopolamine, secobarbital, sertraline, sotalol, spironolactone, sulfinpyrazone, sulindac, temazepam, terbutaline, terfenadine, tetracaine, theophylline, thiethylperazine, thiopental, thioridazine, thiothixene, timolol, tocainide, tolbutamide, tolmetin, trazodone, triamterene, triazolam, trifluoperazine, triflupromazine, trimeprazine, trimethoprim, trimipramine, verapamil, warfarin, xylometazoline, yohimbine, zopiclone

## KEY WORDS
some details of plasma extraction

## REFERENCE
Koves, E.M. Use of high-performance liquid chromatography-diode array detection in forensic toxicology. *J.Chromatogr.A*, **1995**, *692*, 103–119

## SAMPLE
**Matrix:** solutions

## HPLC VARIABLES
**Column:** 150 × 4.6 cellulose 3,5-dimethylphenylcarbamate/10-undecenoate bonded to allylsilica

**Mobile phase:** Heptane:isopropanol:diethylamine 80:20:0.1
**Flow rate:** 1
**Injection volume:** 1000
**Detector:** UV 254

## CHROMATOGRAM
**Retention time:** k′ 1.61

## KEY WORDS
chiral; α = 1.24

## REFERENCE
Oliveros, L.; Lopez, P.; Minguillon, C.; Franco, P. Chiral chromatographic discrimination ability of a cellulose 3,5-dimethylphenylcarbamate/10-undecenoate mixed derivative fixed on several chromatographic matrices. *J.Liq.Chromatogr.*, **1995**, *18*, 1521–1532

## SAMPLE
**Matrix:** solutions

## HPLC VARIABLES
**Column:** 300 × 3.9 5 μm Nova-Pak C18
**Mobile phase:** MeOH:buffer 30:70 (Buffer was pH 4.0 phosphate buffer (ionic strength = 0.1) containing 2.86 mM N, N-dimethyloctylamine, pH readjusted to 4.00 with 85% phosphoric acid.)
**Column temperature:** 30
**Flow rate:** 1
**Injection volume:** 100
**Detector:** UV 220

## CHROMATOGRAM
**Retention time:** k′ 2.81

## OTHER SUBSTANCES
**Also analyzed:** acebutolol, bunitrolol, carazolol, celiprolol, esmolol, mepindolol, timolol

## REFERENCE
Hamoir, T.; Verlinden, Y.; Massart, D.L. Reversed-phase liquid chromatography of β-adrenergic blocking drugs in the presence of a tailing suppressor. *J.Chromatogr.Sci.*, **1994**, *32*, 14–20

## SAMPLE
**Matrix:** solutions

## HPLC VARIABLES
**Column:** 250 × 4.6 Zorbax RX
**Mobile phase:** Gradient. A was 10 mL concentrated orthophosphoric acid and 7 mL triethylamine in 1 L water. B was 10 mL concentrated orthophosphoric acid and 7 mL triethylamine in 200 mL water, make up to 1 L with MeCN. A:B from 100:0 to 0:100 over 30 min, maintain at 0:100 for 5 min.
**Column temperature:** 30
**Flow rate:** 2
**Detector:** UV 210

## OTHER SUBSTANCES

**Also analyzed:** acepromazine, acetaminophen, acetophenazine, albuterol, aminophylline, amitriptyline, amobarbital, amoxapine, amphetamine, amylocaine, antipyrine, aprobarbital, aspirin, atenolol, atropine, avermectin, barbital, benzocaine, benzoic acid, benzotropine, benzphetamine, berberine, bibucaine, bromazepan, brompheniramine, buprenorphine, buspirone, butabarbital, butacaine, butethal, caffeine, carbamazepine, carbromal, chloramphenicol, chlordiazepoxide, chloroquine, chlorothiazide, chloroxylenol, chlorphenesin, chlorpheniramine, chlorpromazine, chlorpropamide, chlortetracycline, cimetidine, cinchonidine, cinchonine, clenbuterol, clonazepam, clonixin, clorazepate, cocaine, codeine, colchicine, cortisone, coumarin, cyclazocine, cyclobenzaprine, cyclothiazide, cyheptamide, cymarin, danazol, danthron, dapsone, debrisoquine, desipramine, dexamethasone, dextromethorphan, dextropropoxyphene, diamorphine, diazepam, diclofenac, diethylpropion, diethylstilbestrol, diflunisal, digitoxin, digoxin, diltiazem, diphenhydramine, diphenoxylate, diprenorphine, dipyrone, disulfiram, dopamine, doxapram, doxepin, dronabinol, ephedrine, epinephrine, epinine, estradiol, estriol, estrone, ethacrynic acid, ethosuximide, etonitazene, etorphine, eugenol, famotidine, fenbendazole, fencamfamine, fenoprofen, fenproporex, fentanyl, flubendazole, flufenamic acid, flunitrazepam, 5-fluorouracil, fluoxymesterone, fluphenazine, furosemide, gentisic acid, gitoxigenin, glipizide, glunixin, glutethimide, glybenclamide, guaiacol, halazepam, haloperidol, hydrochlorothiazide, hydrocodone, hydrocortisone, hydromorphone, hydroxyquinoline, ibogaine, ibuprofen, iminostilbene, imipramine, indomethacin, isocarbostyril, isocarboxazid, isoniazid, isoproterenol, isoxsuprine, ivermectin, ketamine, ketoprofen, kynurenic acid, levorphanol, lidocaine, lorazepam, lormetazepam, loxapine, mazindol, mebendazole, meclizine, meclofenamic acid, medazepam, mefenamic acid, megestrol, mepacrine, meperidine, mephentermine, mephenytoin, mephesin, mephobarbital, mepivacaine, mescaline, mesoridazine, methadone, methamphetamine, methapyrilene, methaqualone, methazolamide, methocarbamol, methoxamine, methsuximide, methyl salicylate, methyldopa, methyldopamine, methylphenidate, methylprednisolone, methyltestosterone, methyprylon, mibolerone, morphine, nadolol, nalorphine, naloxone, naltrexone, naphazoline, naproxen, nefopam, niacinamide, nicotine, nicotinic acid, nifedipine, niflumic acid, nitrazepam, norepinephrine, nortriptyline, noscapine, nylidrin, oxazepam, oxycodone, oxymorphone, oxyphenbutazone, oxytetracycline, papaverine, pargyline, pemoline, pentazocine, pentobarbital, persantine, phenacetin, phenazocine, phenazopyridine, phencyclidine, phendimetrazine, phenelzine, pheniramine, phenobarbital, phenothiazine, phensuximide, phentermine, phenylbutazone, phenylephrine, phenylpropanolamine, piperocaine, prazepam, prednisolone, primidone, probenecid, progesterone, propiomazine, propranolol, propylparaben, pseudoephedrine, puromycin, pyrilamine, pyrithyldione, quazepam, quinaldic acid, quinidine, quinine, ranitidine, recinnamine, reserpine, resorcinol, saccharin, albuterol, salicylamide, salicylic acid, scopolamine, scopoletin, secobarbital, strychnine, sulfacetamide, sufadiazine, sulfadimethoxine, sulfaethidole, sulfamerazine, sulfamethazine, sulfamethoxizole, sulfanilamide, sulfapyridine, sulfasoxizole, sulindac, tamoxifen, temazepam, testosterone, tetracaine, tetracycline, tetramisole, thebaine, theobromine, theophylline, thiabendazole, thiamine, thiamylal, thiobarbituric acid, thioridazine, thiosalicylic acid, thiothixene, thymol, tolazamide, tolazoline, tobutamide, tolmetin, tranylcypromine, triamcinolone, tribenzylamine, trichloromethiazide, trifluoperazine, trihexyphenidyl, trimethoprim, tripelennamine, triprolidine, tropacocaine, tyramine, verapamil, vincamine, warfarin, yohimbine, zoxazolamine

## REFERENCE

Hill, D.W.; Kind, A.J. Reversed-phase solvent gradient HPLC retention indexes of drugs. *J.Anal.Toxicol.*, **1994**, *18*, 233–242

## SAMPLE
**Matrix:** solutions

## HPLC VARIABLES
**Column:** 250 × 4.6 10 μm Chiralpak OD (Daicel)

**Mobile phase:** Carbon dioxide:MeOH:diethylamine 69.5:30:0.5
**Column temperature:** 30
**Flow rate:** 2
**Detector:** UV 223

## CHROMATOGRAM
**Retention time:** 2, 3.1 (enantiomers)

## KEY WORDS
SFC; chiral; pressure 200 bar

## REFERENCE
Kot, A.; Sandra, P.; Venema, A. Sub- and supercritical fluid chromatography on packed columns: A versatile tool for the enantioselective separation of basic and acidic drugs. *J.Chromatogr.Sci.*, **1994**, *32*, 439–448

## SAMPLE
**Matrix:** solutions
**Sample preparation:** Prepare a 25 µg/mL solution in MeCN:water 40:60, inject an aliquot.

## HPLC VARIABLES
**Column:** 250 × 4.6 3 µm silica (Phenomenex)
**Mobile phase:** MeCN:6.25 mM pH 3.0 phosphate buffer 40:60
**Flow rate:** 1
**Injection volume:** 50
**Detector:** UV 254

## CHROMATOGRAM
**Retention time:** 7.53

## OTHER SUBSTANCES
**Also analyzed:** atenolol, clonidine, diltiazem, nifedipine, prazosin, propranolol, verapamil

## REFERENCE
Simmons, B.R.; Stewart, J.T. HPLC separation of selected cardiovascular agents on underivatized silica using an aqueous organic mobile phase. *J.Liq.Chromatogr.*, **1994**, *17*, 2675–2690

## SAMPLE
**Matrix:** solutions
**Sample preparation:** Filter (0.22 µm), inject a 10 µL aliquot.

## HPLC VARIABLES
**Column:** 250 × 4.6 internal surface reversed-phase silica (Pinkerton) (Regis Chemical)
**Mobile phase:** Isopropanol:100 mM pH 6.8 $KH_2PO_4$ 10:90
**Flow rate:** 1
**Injection volume:** 10
**Detector:** UV 232-274 (wavelength of maximum absorption used)

## CHROMATOGRAM
**Retention time:** 26.0

## OTHER SUBSTANCES
**Simultaneous:** acebutolol, alprenolol, atenolol, carteolol, oxprenolol, pindolol

## REFERENCE

Ohshima, T.; Takagi, K.; Miyamoto, K.-I. High performance liquid chromatographic retention time of β-blockers as an index of pharmacological activity. *J.Liq.Chromatogr.*, **1993**, *16*, 3933–3939

## SAMPLE

**Matrix:** solutions
**Sample preparation:** Prepare a 10 µg/mL solution in MeOH, inject a 20 µL aliquot.

## HPLC VARIABLES

**Column:** 125 × 4.9 Spherisorb S5W silica
**Mobile phase:** MeOH containing 10 mM ammonium perchlorate and 1 mL/L 100 mM NaOH in MeOH, pH 6.7
**Flow rate:** 2
**Injection volume:** 20
**Detector:** E, LeCarbone, V25 glassy carbon electrode, + 1.2 V

## CHROMATOGRAM

**Retention time:** 2.1

## OTHER SUBSTANCES

**Also analyzed:** acebutolol, acepromazine, acetophenazine, N-acetylprocainamide, albuterol, alprenolol, amethocaine, amiodarone, amitriptyline, antazoline, atenolol, azacyclonal, bamethane, benactyzine, benperidol, benzethidine, benzocaine, benzoctamine, benzphetamine, benzquinamide, bromhexine, bromodiphenhydramine, bromperidol, brompheniramine, brompromazine, buclizine, bufotenine, bupivacaine, buprenorphine, butacaine, butethamate, chlorcyclizine, chlorpheniramine, chlorphenoxamine, chlorprenaline, chlorpromazine, chlorprothixene, cimetidine, cinchonidine, cinnarizine, clemastine, clomipramine, clonidine, cocaine, cyclazocine, cyclizine, cyclopentamine, cyproheptadine, deserpidine, desipramine, dextromoramide, dextropropoxyphene, dicyclomine, diethylcarbamazine, diethylpropion, diethylthiambutene, dihydroergotamine, dimethindene, dimethothiazine, diphenhydramine, diphenoxylate, dipipanone, diprenorphine, dipyridamole, disopyramide, dothiepin, doxapram, doxepin, doxylamine, droperidol, ephedrine, ergocornine, ergocristine, ergocristinine, ergocryptine, ergometrine, ergosine, ergosinine, ergotamine, ethopropazine, etorphine, etoxeridine, fenethazine, fenfluramine, fenoterol, fentanyl, flavoxate, fluopromazine, flupenthixol, fluphenazine, flurazepam, haloperidol, hydroxyzine, hyoscine, ibogaine, imipramine, indapamine, iprindole, isothipendyl, isoxsuprine, ketanserin, laudanosine, lidocaine, lofepramine, loxapine, maprotiline, mecamylamine, meclophenoxate, meclozine, medazepam, mephentermine, mepivacaine, meptazinol, mepyramine, mesoridazine, metaraminol, methadone, methamphetamine, methapyrilene, methdilazine, methotrimeprazine, methoxamine, methoxyphenamine, methoxypromazine, methylephedrine, methylergonovine, methysergide, metoclopramide, metopimazine, mianserin, morazone, nadolol, nalorphine, naloxone, naphazoline, nicotine, nifedipine, nomifensine, nortriptyline, noscapine, orphenadrine, oxeladin, oxprenolol, oxymetazolin, papaverine, pargyline, pecazine, penbutolol, pentazocine, penthienate, pericyazine, perphenazine, phenadoxone, phenampromide, phenazocine, phenbutrazate, phendimetrazine, phenelzine, phenglutarimide, phenindamine, pheniramine, phenmetrazine, phenomorphan, phenoperidine, phenothiazine, phenoxybenzamine, phentolamine, phenylephrine, phenyltoloxamine, physostigmine, piminodine, pimozide, pindolol, pipamazine, pipazethate, piperacetazine, piperidolate, pipradol, pirenzepine, piritramide, pizotifen, practolol, pramoxine, prazosin, prenylamine, prilocaine, primaquine, proadifen, procainamide, procaine, prochlorperazine, procyclidine, proheptazine, prolintane, promazine, promethazine, pronethalol, properidine, propiomazine, propranolol, prothipendyl, protriptyline, proxymetacaine, pseudoephedrine, pyrimethamine, quinidine, quinine, ranitidine, rescinnamine, sotalol, tacrine, terazosin, terbutaline, terfenadine, thenyldiamine, theophylline, thiethylperazine, thiopropazate, thioproperazine, thioridazine, thiothixene, thonzylamine, timolol, tocainide, tolpropamine, tolycaine, tranylcypromine, trazodone, trifluoperazine, trifluperidol, trimeperidine, trimeprazine, trimethobenzamide, trimethoprim, trimipramine, tripelennamine, triprolidine, tryptamine, verapamil, xylometazoline

## REFERENCE

Jane, I.; McKinnon, A.; Flanagan, R.J. High-performance liquid chromatographic analysis of basic drugs on silica columns using non-aqueous ionic eluents. II. Application of UV, fluorescence and electro-chemical oxidation detection. *J.Chromatogr.*, **1985**, *323*, 191–225

## SAMPLE

**Matrix:** urine

**Sample preparation:** 2 mL Urine + 30 µL 1 mg/mL (±)-toliprolol in MeOH + 1 mL 2 M potassium carbonate + 1 g NaCl, extract twice with 5 mL ethyl acetate. Remove the organic layers and dry them over anhydrous sodium sulfate, evaporate to dryness under a stream of nitrogen at 50°, reconstitute the residue in 100 µL 0.4% triethylamine in MeCN:MeOH 50:50, add 100 µL 1% S-(−)-menthyl chloroformate in MeCN (prepare weekly in MeCN dried over anhydrous sodium sulfate), let stand at room temperature for 1 h, evaporate to dryness under a stream of nitrogen, reconstitute with 300 µL MeOH, inject a 10 µL aliquot.

## HPLC VARIABLES

**Guard column:** 20 × 4 30 µm Hypersil HP ODS

**Column:** 250 × 4.6 Hypersil 5 C18

**Mobile phase:** Gradient. A was 13.8 g $NaH_2PO_4.H_2O$ and 1.59 g propylamine hydrochloride in 1 L water, pH adjusted to 3.2 with concentrated phosphoric acid. B was MeOH. A:B from 25:75 to 15:85 over 15 min, maintain at 15:85 for 5 min, to 10:90 over 5 min, maintain at 10:90 for 3 min.

**Flow rate:** 1

**Injection volume:** 10

**Detector:** F ex 223 em 340 (cut-off filter)

## CHROMATOGRAM

**Retention time:** 19.7 (−), 20.6 (+)

**Internal standard:** toliprolol (23.3 (−), 24.5 (+))

**Limit of detection:** 5 ng

## OTHER SUBSTANCES

**Extracted:** metabolites

## KEY WORDS

chiral; derivatization; pharmacokinetics

## REFERENCE

Li, F.; Cooper, S.F.; Côté, M. Determination of the enantiomers of metoprolol and its major acidic me-tabolite in human urine by high-performance liquid chromatography with fluorescence detection. *J.Chromatogr.B*, **1995**, *668*, 67–75

## SAMPLE

**Matrix:** urine

**Sample preparation:** 3 mL Urine + 500 µL 5 M NaOH + 1 g anhydrous sodium sulfate + 2 mL diethyl ether, shake mechanically for 15 min, centrifuge at 734 g for 5 min. Remove the organic layer and evaporate it to dryness under a stream of nitrogen at 60°, reconstitute the residue in 10 mL mobile phase, inject a 20 µL aliquot.

## HPLC VARIABLES

**Guard column:** µBondapak C18

**Column:** 300 × 3.9 10 µm µBondapak C18

**Mobile phase:** MeCN:water 40:60 containing 5 mM $KH_2PO_4$/$K_2HPO_4$ adjusted to pH 6.5 with phosphoric acid or 1 M KOH

**Column temperature:** 30 ± 0.2

**Flow rate:** 1.3
**Injection volume:** 20
**Detector:** E, EG&G Princeton Applied Research PAR Model 400, glassy carbon cell +1300 mV, d.c. mode, Ag/AgCl reference electrode (At the end of each day clean electrode with MeOH as mobile phase and potential -600 mV for 1 min and +1500 mV for 10 min, repeat 3 times. If necessary, wipe with a tissue wetted with water then a tissue wetted with MeOH.)

## CHROMATOGRAM
**Retention time:** 5.42
**Limit of quantitation:** 400 ppb

## OTHER SUBSTANCES
**Extracted:** alprenolol, nadolol, oxprenolol, timolol
**Simultaneous:** atenolol

## REFERENCE
Maguregui, M.I.; Alonso, R.M.; Jiménez, R.M. High-performance liquid chromatography with amperometric detection applied to the screening of β-blockers in human urine. *J.Chromatogr.B*, **1995**, **674**, 85–91

## SAMPLE
**Matrix:** urine
**Sample preparation:** 1 mL Urine + 10 mg β-glucuronidase/arylsulfatase (Helix pomatia, Sigma), heat at 37° overnight, add an equal volume of buffer, centrifuge at 2000 g for 5 min, inject an aliquot of the supernatant onto column A with mobile phase A and elute to waste. After 2.5 min backflush the contents of column A onto column B with mobile phase B, monitor the effluent from column B. For gradient elution, after 15 min re-equilibrate both columns for 12.5 min before the next injection. For isocratic elution, remove column A from the circuit after 1.25 min, re-equilibrate column A for 1.5 min. (Buffer was 200 mM boric acid adjusted to pH 9.5 with 5 M NaOH.)

## HPLC VARIABLES
**Column:** A 10 × 4.6 5 μm Spherisorb cyanopropyl; B 250 × 4.6 Capcell Pak C18 UG-120 (Shiseido)
**Mobile phase:** A water; B Gradient. MeCN:buffer from 3:97 to 30:70 over 30 min, to 40: 60 over 8 min (for screening) or isocratic 23:77 (Buffer was 3.4 mL/L phosphoric acid adjusted to pH 3.0 with 5 M NaOH.)
**Flow rate:** A 1.25; B 1
**Injection volume:** 100
**Detector:** UV 220

## CHROMATOGRAM
**Retention time:** 12 (gradient), 6 (isocratic)
**Limit of detection:** 250 ng/mL

## OTHER SUBSTANCES
**Extracted:** metabolites (gradient), acebutolol (gradient), alprenolol (gradient), amphetamine (gradient), atenolol (gradient), bopindolol (gradient), codeine (gradient), ephedrine (gradient), labetalol (gradient), morphine (gradient), nadolol (gradient), oxprenolol (gradient), pindolol (gradient), propranolol (gradient), timolol (gradient)

## KEY WORDS
column-switching

## REFERENCE

Saarinen, M.T.; Sirén, H.; Riekkola, M.-L. Screening and determination of β-blockers, narcotic analgesics and stimulants in urine by high-performance liquid chromatography with column switching. *J.Chromatogr.B*, **1995**, *664*, 341–346

---

## ANNOTATED BIBLIOGRAPHY

Ghosh, T.K.; Adir, J.; Xiang, S.-L.; Onyilofur, S. Transdermal delivery of metoprolol II: In-vitro skin permeation and bioavailability in hairless rats. *J.Pharm.Sci.*, **1995**, *84*, 158–160 [rat; plasma; pindolol (IS); fluorescence detection; LOD 5 ng/mL; pharmacokinetics]

Welch, C.J.; Perrin, S.R. Improved chiral stationary phase for β-blocker enantioseparations. *J.Chromatogr.A*, **1995**, *690*, 218–225 [chiral; also bufuralol]

Xie, H.G.; Zhou, H.H. Assay of metoprolol and alpha-hydroxymetoprolol in human urine by reversed-phase liquid chromatography with direct-injection. *Chung Kuo Yao Li Hsueh Pao*, **1995**, *16*, 32–35

Bailey, C.J.; Ruane, R.J.; Wilson, I.D. Packed-column supercritical fluid chromatography of β-blockers. *J.Chromatogr.Sci.*, **1994**, *23*, 426–429 [SFC; simultaneous alprenolol, atenolol, labetalol, metoprolol, oxprenolol, pindolol, practolol, propranolol, toliprolol, xamoterol]

Hermansson, J.; Grahn, A. Resolution of racemic drugs on a new chiral column based on silica-immobilized cellobiohydrolase. Characterization of the basic properties of the column. *J.Chromatogr.*, **1994**, *687*, 45–59 [chiral; also acebutolol, atenolol, betaxolol, bisoprolol, carbuterol, cathinone, cimetidine, dobutamine, dropropizine, epanolol, epinephrine, laudanosine, metanephrine, moprolol, norepinephrine, normetanephrine, octopamine, oxybutynine, pamatolol, practolol, prilocaine, propafenone, proxyphylline, sotalol, talinolol, tetrahydropapaveroline, tetramisole, timolol, tolamolol, toliprolol]

Armstrong, D.W.; Chen, S.; Chang, C.; Chang, S. A new approach for the direct resolution of racemic beta adrenergic blocking agents by HPLC. *J.Liq.Chromatogr.*, **1992**, *15*, 545–556 [chiral; also alprenolol, atenolol, cateolol, labetolol, nadolol, oxprenolol, pindolol, propranolol, timolol]

Leloux, M.S. Rapid chiral separation of metoprolol in plasma–application to the pharmacokinetics/pharmacodynamics of metoprolol enantiomers in the conscious goat. *Biomed.Chromatogr.*, **1992**, *6*, 99–105

Balmér, K.; Persson, A.; Lagerström, P.-O.; Persson, B.-A.; Schill, G. Liquid chromatographic separation of the enantiomers of metoprolol and its alpha-hydroxy metabolite on Chiralcel OD for determination in plasma and urine. *J.Chromatogr.*, **1991**, *553*, 391–397 [plasma; urine; human; dog; extracted metabolites; fluorescence detection; chiral; column temp 35; column temp 25; LOD 10 nM]

Shen, J.; Wanwimolruk, S.; Hung, C.T.; Zoest, A.R. Quantitative analysis of β-blockers in human plasma by reversed-phase ion-pair high-performance liquid chromatography using a microbore column. *J.Liq.Chromatogr.*, **1991**, *14*, 777–793 [plasma; microbore; also atenolol, labetalol, pindolol, propranolol; oxprenolol (IS); fluorescence detection; UV detection; LOD 1-10 ng/mL]

Persson, B.A.; Balmer, K.; Lagerstrom, P.O.; Schill, G. Enantioselective determination of metoprolol in plasma by liquid chromatography on a silica-bonded alpha 1-acid glycoprotein column. *J.Chromatogr.*, **1990**, *500*, 629–636

Bui, K.H.; French, S.B. Direct serum injection and analysis of drugs with aqueous mobile phases containing triethylammonium acetate. *J.Liq.Chromatogr.*, **1989**, *12*, 861–873 [direct injection; serum; plasma; dog; rat; fluorescence detection; UV detection; also antipyrine, atenolol, hexanophenone, naproxen, propranolol]

Rutledge, D.R.; Garrick, C. Rapid high-performance liquid chromatographic method for the measurement of the enantiomers of metoprolol in serum using a chiral stationary phase. *J.Chromatogr.*, **1989**, *497*, 181–190

Schuster, D.; Modi, M.W.; Lalka, D.; Gengo, F.M. Reversed-phase high-performance liquid chromatographic assay to quantitate diastereomeric derivatives of metoprolol enantiomers in plasma. *J.Chromatogr.*, **1988**, *433*, 318–325

Buhring, K.U.; Garbe, A. Determination of the new beta-blocker bisoprolol and of metoprolol, atenolol and propranolol in plasma and urine by high-performance liquid chromatography. *J.Chromatogr.*, **1986**, *382*, 215–224

Schill, G.; Wainer, I.W.; Barkan, S.A. Chiral separation of cationic drugs on an α1-acid glycoprotein bonded stationary phase. *J.Liq.Chromatogr.*, **1986**, *9*, 641–666 [chiral; also atropine, bromdiphenhydramine, brompheniramine, bupivacaine, butorphanol, carbinoxamine, chlorpheniramine, clidinium, cocaine, cyclopentolate, dimethindene, diperidone, disopyramide, doxylamine, ephedrine, homatropine, labetalol, labetalol A, labetalol B, mepensolate, mepivacaine, methadone, methorphan, methylatropine, methylhomatropine, methylphenidate, metoprolol, nadolol, nadolol A, nadolol B, oxprenolol, oxyphencyclimine, phenmetrazine, phenoxybenzamine, promethazine, pronethalol, propoxyphene, propranolol, pseudoephedrine, terbutaline, tocainide, tridihexethyl]

Harrison, P.M.; Tonkin, A.M.; McLean, A.J. Simple and rapid analysis of atenolol and metoprolol in plasma using solid-phase extraction and high-performance liquid chromatography. *J.Chromatogr.*, **1985**, *339*, 429–433 [SPE]

Lennard, M.S. Quantitative analysis of metoprolol and three of its metabolites in urine and liver microsomes by high-performance liquid chromatography. *J.Chromatogr.*, **1985**, *342*, 199–205

Gengo, F.M.; Ziemniak, M.A.; Kinkel, W.R.; McHugh, W.B. High-performance liquid chromatographic determination of metoprolol and alpha-hydroxymetoprolol concentrations in human serum, urine, and cerebrospinal fluid. *J.Pharm.Sci.*, **1984**, *73*, 961–963

Godbillon, J.; Duval, M. Determination of two metoprolol metabolites in human urine by high-performance liquid chromatography. *J.Chromatogr.*, **1984**, *309*, 198–202

Lecaillon, J.B.; Godbillon, J.; Abadie, F.; Gosset, G. Determination of metoprolol and its alpha-hydroxylated metabolite in human plasma by high-performance liquid chromatography. *J.Chromatogr.*, **1984**, *305*, 411–417

Johnston, G.D.; Nies, A.S.; Gal, J. Determination of metoprolol in human blood plasma using high-performance liquid chromatography. *J.Chromatogr.*, **1983**, *278*, 204–208

Lennard, M.S.; Silas, J.H. Rapid determination of metoprolol and alpha-hydroxymetoprolol in human plasma and urine by high-performance liquid chromatography. *J.Chromatogr.*, **1983**, *272*, 205–209

Mehta, A.C. High-performance liquid chromatographic determination of oxprenolol hydrochloride and metoprolol tartrate in tablets and injections. *Analyst*, **1982**, *107*, 1379–1382

Pautler, D.B.; Jusko, W.J. Determination of metoprolol and alpha-hydroxymetoprolol in plasma by high-performance liquid chromatography. *J.Chromatogr.*, **1982**, *228*, 215–222

Rosseel, M.T.; Belpaire, F.M.; Bekaert, I.; Bogaert, M.G. High-performance liquid chromatographic determination of metoprolol in plasma. *J.Pharm.Sci.*, **1982**, *71*, 114–115

Winkler, H.; Ried, W.; Lemmer, B. High-performance liquid chromatographic method for the quantitative analysis of the aryloxypropanolamines propranolol, metoprolol and atenolol in plasma and tissue. *J.Chromatogr.*, **1982**, *228*, 223–234

Lefebvre, M.A.; Girault, J.; Fourtillan, J.B. β-Blocking agents: Determination of biological levels using high performance liquid chromatography. *J.Liq.Chromatogr.*, **1981**, *4*, 483–500 [fluorescence detection; plasma; also acebutolol, atenolol, metoprolol, oxprenolol, pindolol, propranolol, sotalol, timolol]

# Midazolam

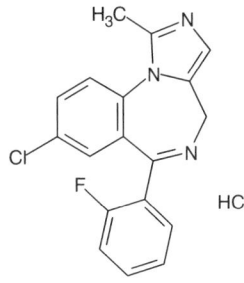

**Molecular formula:** $C_{18}H_{13}ClFN_3$
**Molecular weight:** 325.8
**CAS Registry No.:** 59467-70-8 (midazolam), 59467-96-8 (midazolam hydrochloride), 59467-94-6 (midazolam maleate)

## SAMPLE
**Matrix:** aqueous humor, blood, tissue, urine
**Sample preparation:** Homogenize tissue 1:2 (w/v). 1 mL Sample + 100 μL 10 μg/mL methaqualone + 1 mL ammonium chloride/ammonium hydroxide buffer (pH 9.2) + 3 mL n-butyl chloride, mix, centrifuge. Remove the organic layer and evaporate it to dryness under nitrogen at 45°, reconstitute the residue with 50 μL mobile phase, inject a 20 μL aliquot.

## HPLC VARIABLES
**Column:** 100 × 8 10 μm μBondapak C18
**Mobile phase:** MeCN:buffer 40:60, pH 3.3 (Buffer was 150 mL 100 mM $KH_2PO_4$ made up to 1 L, pH adjusted to 3.3 with 100 mM phosphoric acid.)
**Flow rate:** 2.5
**Injection volume:** 20
**Detector:** UV 220

## CHROMATOGRAM
**Retention time:** 3.93
**Internal standard:** methaqualone (6.42)

## KEY WORDS
plasma; liver; kidney

## REFERENCE
Ferslew, K.E.; Hagardorn, A.N.; McCormick, W.F. Postmortem determination of the biological distribution of sufentanil and midazolam after an acute intoxication. *J.Forensic Sci.*, **1989**, *34*, 249–257

## SAMPLE
**Matrix:** blood
**Sample preparation:** Condition a Sep-Pak C18 SPE cartridge with water, MeOH, and 100 mM ammonium acetate. 5 mL Plasma + 250 ng detomidine, add to the SPE cartridge, wash with 100 mM ammonium acetate, elute with MeOH:100 mM ammonium acetate 75:25. Evaporate the eluate to dryness under reduced pressure, reconstitute the residue in 200 μL mobile phase, inject a 50 μL aliquot.

## HPLC VARIABLES
**Column:** 150 × 4.6 5 μm Hitachi gel
**Mobile phase:** 3056
**Mobile phase:** MeOH:100 mM ammonium acetate 65:35
**Flow rate:** 1
**Injection volume:** 50
**Detector:** MS, Hitachi M-1000, APCI interface, drift voltage 21 V, nebulizer 260°, vaporizer 399°, multiplier voltage 1500 VF, m/z 326

## CHROMATOGRAM
**Retention time:** 10.5
**Internal standard:** detomidine (m/z 187) (6.5)
**Limit of quantitation:** 1-2 ng/mL

## OTHER SUBSTANCES
**Extracted:** atipamazole, medetomidine

## KEY WORDS
pig; plasma; pharmacokinetics; SPE

## REFERENCE
Kanazawa, H.; Nishimura, R.; Sasaki, N.; Takeuchi, A.; Takai, N.; Nagata, Y.; Matsushima, Y. Determination of medetomidine, atipamazole and midazolam by liquid chromatography-mass spectrometry. *Biomed.Chromatogr.*, **1995**, *9*, 188–191

## SAMPLE
**Matrix:** blood
**Sample preparation:** 1 mL Plasma + 300 µL 100 mM pH 9 borate buffer + 25 µL flurazepam in EtOH + 5 mL diethyl ether, mix at 60 rpm for 10 min, centrifuge at 15° at 1500 g for 5 min. Remove the organic layer and evaporate it to dryness under a stream of nitrogen at 45°, reconstitute the residue in 100 µL mobile phase, inject an aliquot.

## HPLC VARIABLES
**Column:** 150 × 4.6 5 µm Spherisorb CN
**Mobile phase:** MeOH:isopropanol 75:25 containing 0.015% perchloric acid
**Flow rate:** 1.5
**Detector:** UV 215

## CHROMATOGRAM
**Retention time:** 4.7
**Internal standard:** flurazepam (6.2)
**Limit of quantitation:** 2 ng/mL

## OTHER SUBSTANCES
**Extracted:** metabolites

## KEY WORDS
plasma

## REFERENCE
Lehmann, B.; Boulieu, R. Determination of midazolam and its unconjugated 1-hydroxy metabolite in human plasma by high-performance liquid chromatography. *J.Chromatogr.B*, **1995**, *674*, 138–142

## SAMPLE
**Matrix:** blood
**Sample preparation:** 600 µL Plasma + 600 µL IS solution, shake, centrifuge at 1500 g for 3 min, inject a 400 µL aliquot onto column A with mobile phase A, elute with mobile phase A for 4 min, backflush column A with mobile phase A for 1.5 min, backflush column A with mobile phase B for 4.5 min, backflush contents of column A onto column B with mobile phase C and start the gradient. After 3 min remove column A from circuit, monitor effluent from column B. (IS solution was 2.5 mL 200 µg/mL flurazepam in MeCN + 3.6 mL 2 M NaOH, add 200 mL MeCN, make up to 1 L with water.)

## HPLC VARIABLES
**Column:** A 17 × 4.6 37-50 μm Bondapak C18 Corasil; B 4 × 4 5 μm LiChrospher 60 RP-select B + 250 × 4 5 μm LiChrospher 60 RP-select B
**Mobile phase:** A 100 mM NaOH; B 2.7 g/L KH$_2$PO$_4$ adjusted to pH 8.0 with 2 M NaOH; C Gradient. I was 2.7 g/L KH$_2$PO$_4$ adjusted to pH 2.4 with 85% phosphoric acid. II was MeCN. I:II from 76:24 to 66:34 over 11 min.
**Flow rate:** A 1; B 1; C 1.5
**Injection volume:** 400
**Detector:** UV 230

## CHROMATOGRAM
**Retention time:** 20.5
**Internal standard:** flurazepam (21.5)
**Limit of detection:** 2 ng/mL
**Limit of quantitation:** 10 ng/mL

## OTHER SUBSTANCES
**Extracted:** metabolites

## KEY WORDS
plasma

## REFERENCE
Lauber, R.; Mosiman, M.; Bührer, M.; Zbinden, A.M. Automated determination of midazolam in human plasma by high-performance liquid chromatography using column switching. *J.Chromatogr.B*, **1994**, *654*, 69–75

## SAMPLE
**Matrix:** blood
**Sample preparation:** Condition a 12 mL 500 mg PrepSep C1 SPE cartridge with 3 mL MeOH and 3 mL water. 1 mL Plasma + 100 μL 3 μg/mL midazolam in MeOH, mix, add to SPE cartridge, wash with two 3 mL portions of water, wash with two 1 mL portions of MeOH:water 30:70, elute with two 1 mL portions of MeOH:50 mM pH 9.0 (NH$_4$)$_2$HPO$_4$ 90:10, evaporate the eluents under vacuum, dissolve the residue in 200 μL mobile phase, inject a 100 μL aliquot.

## HPLC VARIABLES
**Column:** 100 × 4.6 5 μm Spherisorb C8
**Mobile phase:** MeCN:MeOH:20 mM (NH$_4$)H$_2$PO$_4$ 5:35:60 containing 2 mL/L 200 mM tetrabutylammonium bromide, final pH adjusted to 4.10
**Column temperature:** 30
**Flow rate:** 1.5
**Injection volume:** 100
**Detector:** UV 254

## CHROMATOGRAM
**Retention time:** 10.2
**Internal standard:** clonazepam (12.4)
**Limit of quantitation:** 15 ng/mL

## OTHER SUBSTANCES
**Extracted:** metabolites

## KEY WORDS
plasma; SPE

## REFERENCE

Mastey, V.; Panneton, A.-C.; Donati, F.; Varin, F. Determination of midazolam and two of its metabolites in human plasma by high-performance liquid chromatography. *J.Chromatogr.B*, **1994**, *655*, 305–310

## SAMPLE

**Matrix:** blood

**Sample preparation:** Automated SPE by ASPEC system. Condition a C18 Clean-Up SPE cartridge (CEC 18111, Worldwide Monitoring) with 2 mL MeOH then 2 mL water. 1 mL Plasma + 1 mL 400 ng/mL protriptyline in water, vortex, add to column, wash with 3 mL water, wash with 3 mL 750 mL/L methanol. Elute with three aliquots of 300 μL 0.1 M ammonium acetate in MeOH. Add 0.5 mL 0.5 M NaOH and 4 mL 50 mL/L isopropanol in heptane to eluate, mix thoroughly. Allow 5 min for phase separation. Remove upper heptane phase and add it to 300 μL 0.1 M phosphoric acid (pH 2.5), mix, separate, inject a 100 μL aliquot of the aqueous phase.

## HPLC VARIABLES

**Guard column:** LC-8-DB (Supelco)

**Column:** 150 × 4.6 LC-8-DB (Supelco)

**Mobile phase:** MeCN:buffer 35:65 (Buffer was 10 mL/L triethylamine in water adjusted to pH 5.5 with glacial acetic acid.)

**Flow rate:** 2

**Injection volume:** 100

**Detector:** UV 228

## CHROMATOGRAM

**Retention time:** 7.6

**Internal standard:** protriptyline (4)

## OTHER SUBSTANCES

**Extracted:** acetazolamide, amitriptyline, chlordiazepoxide, chlorimipramine, chlorpromazine, desipramine, dextromethorphan, diazepam, diphenhydramine, doxepin, encainide, fentanyl, flecainide, fluoxetine, flurazepam, haloperidol, hydroxyethylflurazepam, ibuprofen, imipramine, lidocaine, maprotiline, methadone, methaqualone, mexiletine, norchlorimipramine, nordiazepam, nordoxepin, norfluoxetine, nortriptyline, norverapamil, pentazocine, promazine, propafenone, propoxyphene, propranolol, protriptyline, quinidine, temazepam, trazodone, trimipramine, verapamil

**Noninterfering:** acetaminophen, acetylmorphine, amiodarone, amobarbital, amphetamine, bendroflumethiazide, benzocaine, benzoylecgonine, benzthiazide, butalbital, carbamazepine, chlorothiazide, clonazepam, cocaine, codeine, cotinine, cyclosporine, cyclothiazide, desalkylflurazepam, diamorphine, dicumerol, ephedrine, ethacrynic acid, ethanol, ethchlorvynol, ethosuximide, furosemide, glutethimide, hydrochlorothiazide, hydrocodone, hydroflumethiazide, hydromorphone, lorazepam, mephentermine, meprobamate, methamphetamine, metharbital, methoxsalen, methoxyphenteramine, methsuximide, methylcyclothiazide, metoprolol, MHPG, monoacetylmorphine, morphine, normethsuximide, oxazepam, oxycodone, oxymorphone, pentobarbital, phencyclidine, phenteramine, phenylephrine, phenytoin, polythiazide, primidone, prochlorperazine, salicylic acid, sulfanilamide, THC-COOH, theophylline, thiazolam, thiopental, thioridazine, tocainide, trichloromethiazide, trifluoperazine, valproic acid, warfarin

## KEY WORDS

plasma; SPE

## REFERENCE

Nichols, J.H.; Charlson, J.R.; Lawson, G.M. Automated HPLC assay of fluoxetine and norfluoxetine in serum. *Clin.Chem.*, **1994**, *40*, 1312–1316

## SAMPLE
**Matrix:** blood
**Sample preparation:** Condition a 100 mg Bond-Elut C2 SPE cartridge with 1 volume MeOH and 1 volume 10 mM pH 8.0 phosphate buffer. 1 mL Plasma + 5 μg prazepam + 100 μL 1 M pH 8.0 potassium phosphate buffer, mix, add to the SPE cartridge, wash with 3 volumes of water, wash with 1 mL MeOH:water 30:70, wash with 1 mL water, elute with 1 mL MeOH:water 70:30, elute with 1 mL water. Evaporate the eluate to dryness, reconstitute with 200 μL mobile phase, inject an aliquot.

## HPLC VARIABLES
**Column:** 35 × 4.6 5 μm Ultrabase C18
**Mobile phase:** MeOH:water 60:40
**Flow rate:** 1
**Injection volume:** 20
**Detector:** UV 217

## CHROMATOGRAM
**Retention time:** 4
**Internal standard:** prazepam (8)
**Limit of detection:** 3 ng/mL
**Limit of quantitation:** 5 ng/mL

## KEY WORDS
plasma; SPE

## REFERENCE
Berrueta, L.A.; Gallo, B.; Vincente, F. Rapid determination of midazolam in plasma using SPE and HPLC. *Am.Lab.*, **1993**, *25 (Dec.)*, 20R–20T

## SAMPLE
**Matrix:** blood
**Sample preparation:** Condition a Sep-Pak C18 SPE cartridge with water, MeOH, and 100 mM ammonium acetate. Add 200 μL plasma to the SPE cartridge, wash with 100 mM ammonium acetate, elute with MeOH:100 mM ammonium acetate 3:1. Evaporate the eluate to dryness under reduced pressure, dissolve the residue in 200 μL mobile phase, inject a 20 μL aliquot.

## HPLC VARIABLES
**Column:** 150 × 4.6 Hitachi gel 3056 octadecylsilica
**Mobile phase:** MeOH:100 mM ammonium acetate 60:40
**Flow rate:** 1
**Injection volume:** 20
**Detector:** MS, Hitachi M1000, APCI, nebulizer 260°, vaporizer 399°

## CHROMATOGRAM
**Retention time:** 12.7
**Limit of detection:** 0.5-2.5 ng/mL

## OTHER SUBSTANCES
**Simultaneous:** atipamezole, atropine, butorphanol, flumazenil, ketamine, medetomidine, xylazine

## KEY WORDS
plasma; SPE; dog

## REFERENCE

Kanazawa, H.; Nagata, Y.; Matsushima, Y.; Takai, N.; Uchiyama, H.; Nishimura, R.; Takeuchi, A. Liquid chromatography-mass spectrometry for the determination of medetomidine and other anaesthetics in plasma. *J.Chromatogr.*, **1993**, *631*, 215–220

## SAMPLE

**Matrix:** blood
**Sample preparation:** Condition a 100 mg C18 Bond-Elut SPE cartridge with 2 mL MeOH and 2 mL water. 1 mL Plasma + 100 μL 10 μg/mL climazolam in MeOH + 1 mL MeCN:water 30:70, vortex for 10 s, centrifuge for 5 min at 4000 g, add to the SPE cartridge, wash with 2 mL MeCN:water 15:85, let dry for 3-4 min, elute with four 200 μL aliquots of MeOH. Evaporate the eluate under nitrogen, take up the residue in 100 μL MeOH, inject a 20 μL aliquot.

## HPLC VARIABLES

**Guard column:** 4 × 4 5 μm LiChrosorb 100 RP 18
**Column:** 125 × 4 5 μm LiChrospher 100 RP 18 endcapped
**Mobile phase:** MeCN:MeOH:THF:buffer 28:25:2:50 (Prepare a 1 M pH 5.6 phosphate buffer from 94.8 mL 1 M $KH_2PO_4$ + 5.2 mL 1 M $K_2HPO_4$. Dilute 10 mL of this buffer to 1 L to give the 10 mM pH 5.6 phosphate buffer used in the mobile phase.)
**Flow rate:** 1.3
**Injection volume:** 20
**Detector:** UV 254

## CHROMATOGRAM

**Retention time:** 7.48
**Internal standard:** climazolam (9.79)
**Limit of quantitation:** 50 ng/mL

## OTHER SUBSTANCES

**Extracted:** metabolites

## KEY WORDS

plasma; SPE

## REFERENCE

Sautou, V.; Chopineau, J.; Terrisse, M.P.; Bastide, P. Solid-phase extraction of midazolam and two of its metabolites from plasma for high-performance liquid chromatographic analysis. *J.Chromatogr.*, **1991**, *571*, 298–304

## SAMPLE

**Matrix:** blood
**Sample preparation:** 1 mL Plasma + 25 μL EtOH + 25 μL 3 μg/mL IS1 and 7.6 μg/mL IS2 in EtOH + 1 mL 100 mM $Na_2HPO_4$ adjusted to pH 10.5 with NaOH + 5 mL diethyl ether:dichloromethane 60:40, vortex for 30 s, centrifuge at 4° at 2000 g for 10 min. Remove the organic phase and add it to 1 mL 100 mM $Na_2HPO_4$ adjusted to pH 10.5 with NaOH, vortex for 30 s, centrifuge at 4° at 2000 g for 5 min. Remove the organic layer and evaporate it to dryness under a stream of nitrogen at 40°, reconstitute the residue in 50 μL mobile phase, inject a 1-15 μL aliquot.

## HPLC VARIABLES

**Column:** 100 × 4.6 3 μm CP-Microspher C18 (Chrompack)
**Mobile phase:** Gradient. A was MeOH:buffer 1:2. B was MeOH:water 80:20. A:B 93.8:6.2 for 5.5 min, to 60:40 over 0.15 min, maintain at 60:40 for 11.3 min, to 2.5:97.5 over 0.5 min, maintain at 2.5:97.5 for 3.5 min, return to initial conditions over 0.5 min (Buffer was 6 g/L $NaH_2PO_4$ and 1 mL/L triethylamine adjusted to pH 7.00 with NaOH.)

Column temperature: 40
Flow rate: 1.5
Injection volume: 1-15
Detector: UV 220

## CHROMATOGRAM
Retention time: 15.5
Internal standard: IS1 ethyl 7-chloro-5,6-dihydro-5-methyl-6-oxo-4H-imidazo[1,5-a][1,4] benzodiazepine-3-carboxylate (Ro 15-305) (6.4); IS2 climazolam (17.0)
Limit of detection: 1 ng/mL

## OTHER SUBSTANCES
Extracted: flumazenil

## KEY WORDS
plasma; pharmacokinetics

## REFERENCE
Vletter, A.A.; Burm, A.G.L.; Breimer, L.T.M.; Spierdijk, J. High-performance liquid chromatographic assay to determine midazolam and flumazenil simultaneously in human plasma. *J.Chromatogr.*, **1990**, *530*, 177–185

## SAMPLE
Matrix: blood, CSF, gastric fluid, urine
Sample preparation: 200 μL Serum, urine, CSF, or gastric fluid + 300 μL reagent. Flush column A to waste with 500 μL 500 mM ammonium sulfate, inject sample onto column A, flush column A to waste with 500 μL 500 mM ammonium sulfate, backflush the contents of column A onto column B with mobile phase, monitor the effluent from column B. (Reagent was 8.05 M guanidine hydrochloride and 1.02 M ammonium sulfate in water.)

## HPLC VARIABLES
Column: A 40 μm preparative grade C18 (Analytichem); B 75 × 2.1 pellicular C18 (Whatman) + 250 × 4.6 5 μm C8 end-capped (Whatman)
Mobile phase: Gradient. A was 50 mM pH 4.5 $KH_2PO_4$. B was MeCN:isopropanol 80:20. A:B 90:10 for 1 min, to 30:70 over 20 min.
Column temperature: 50
Flow rate: 1.5
Detector: UV 220

## CHROMATOGRAM
Retention time: 14.69
Internal standard: heptanophenone (19)

## OTHER SUBSTANCES
Extracted: acetaminophen, allobarbital, azinphos, barbital, brallobarbitone, bromazepam, butethal, caffeine, carbamazepine, carbaryl, cephaloridine, chloramphenicol, chlordiazepoxide, chlorothiazide, chlorvinphos, clothiapine, cocaine, coomassie blue, desipramine, diazepam, diphenhydramine, dipipanone, ethylbromphos, flufenamic acid, formothion, griseofulvin, indomethacin, lidocaine, lorazepam, malathion, medazepam, oxazepam, paraoxon, penicillin G, pentobarbital, prazepam, propoxyphene, prothiophos, quinine, salicylic acid, secobarbital, strychnine, sulfamethoxazole, theophylline, thiopental, thioridazine, trimethoprim

## KEY WORDS
serum; column-switching

## REFERENCE

Kruger, P.B.; Albrecht, C.F.De V.; Jaarsveld, P.P. Use of guanidine hydrochloride and ammonium sulfate in comprehensive in-line sorption enrichment of xenobiotics in biological fluids by high-performance liquid chromatography. *J.Chromatogr.*, **1993**, *612*, 191–198

## SAMPLE

**Matrix:** blood, urine
**Sample preparation:** Condition a C2 Bond-Elut SPE cartridge with 1 column volume methanol and 1 column volume buffer. Add 1 mL of urine buffered with pH 6 100 mM phosphate buffer or plasma buffered with pH 8 100 mM phosphate buffer to the SPE cartridge, wash with 3 column volumes of water, wash with 1 mL of MeOH:water 30:70, elute with 1 mL of MeOH:water 60:40. Evaporate the eluate to dryness and take up the residue in 200 μL mobile phase, inject an aliquot.

## HPLC VARIABLES

**Column:** 35 × 4.6 5 μm ultrabase C18 (Scharlau)
**Mobile phase:** MeOH:water 60:40
**Flow rate:** 1
**Injection volume:** 20
**Detector:** UV 217

## CHROMATOGRAM

**Retention time:** 4
**Internal standard:** prazepam (8)
**Limit of detection:** 93 ng/mL

## OTHER SUBSTANCES

**Also analyzed:** adinazolam, brotizolam, diazepam, nordazepam, oxazepam, temazepam

## KEY WORDS

plasma; SPE

## REFERENCE

Casas, M.; Berrueta, L.A.; Gallo, B.; Vicente, F. Solid-phase extraction of 1,4-benzodiazepines from biological fluids. *J.Pharm.Biomed.Anal.*, **1993**, *11*, 277–284

## SAMPLE

**Matrix:** blood, urine
**Sample preparation:** 500 μL Plasma or urine + 50 μL 4 μg/mL flurazepam in MeOH + 1 mL 100 mM pH 9 sodium phosphate buffer + 4 mL dichloromethane:diethyl ether 60:40, shake at 45 rpm for 15 min, centrifuge at 10° at 1870 g for 10 min. Remove the organic layer and evaporate it to dryness under a stream of nitrogen at 40°, reconstitute the residue in 80 μL MeOH, inject a 30 μL aliquot. (Deconjugate urine as follows. 250 μL Urine + 750 μL pH 5.4 acetate buffer + 500 U β-glucuronidase, heat at 37° for 18 h, add 20 μL 5 M NaOH, centrifuge, proceed as above using 5 mL dichloromethane:diethyl ether.)

## HPLC VARIABLES

**Guard column:** 30 × 4.6 30 μm C8
**Column:** 100 × 8 4 μm Nova Pak C18
**Mobile phase:** MeCN:40 mM sodium phosphate buffer 32:68 containing 1 mL/L triethylamine, final pH 7.2
**Flow rate:** 1.5
**Injection volume:** 30
**Detector:** UV 220

## CHROMATOGRAM
**Retention time:** 33.5
**Internal standard:** flurazepam (18.5)
**Limit of detection:** 10 ng/mL

## OTHER SUBSTANCES
**Extracted:** flumazenil, 1-hydroxymethylmidazolam, 4-hydroxymidazolam
**Noninterfering:** alfentanil, atropine, bupivacaine, lignocaine, neostigmine

## KEY WORDS
plasma

## REFERENCE
Chan, K.; Jones, R.D.M. Simultaneous determination of flumazenil, midazolam and metabolites in human biological fluids by liquid chromatography. *J.Chromatogr.*, **1993**, *619*, 154–160

## SAMPLE
**Matrix:** formulations

## HPLC VARIABLES
**Column:** 300 × 3.9 µBondapak C18
**Mobile phase:** MeCN:THF:dioxane:4.5 mM pH 7.0 potassium phosphate buffer 17.5:10:17.5:55 (Caution! Dioxane is a carcinogen!)
**Flow rate:** 1
**Detector:** UV 254

## CHROMATOGRAM
**Retention time:** 10.3

## OTHER SUBSTANCES
**Simultaneous:** benzyl alcohol

## KEY WORDS
injections; saline; stability-indicating

## REFERENCE
McMullin, S.T.; Burns Schaiff, R.A.; Dietzen, D.J. Stability of midazolam hydrochloride in polyvinyl chloride bags under fluorescent light. *Am.J.Health-Syst.Pharm.*, **1995**, *52*, 2018–2020

## SAMPLE
**Matrix:** formulations
**Sample preparation:** Inject an aliquot directly.

## HPLC VARIABLES
**Column:** 250 × 4.6 5 µm C18-modified HS silica (Vydac)
**Mobile phase:** MeOH:MeCN:THF:buffer 29.4:29.4:1.2:40 (Buffer was 6.1 mL $K_2HPO_4$ + 3.9 mL 1 M $KH_2PO_4$ made up to 1 L.)
**Injection volume:** 10
**Detector:** UV 220; UV 254

## CHROMATOGRAM
**Retention time:** 5.1

## OTHER SUBSTANCES
**Simultaneous:** degradation products, benzyl alcohol

## KEY WORDS
injections; stability-indicating; saline; 5% dextrose

## REFERENCE
Hagan, R.L.; Jacobs, L.F., III; Pimsler, M.; Merritt, G.J. Stability of midazolam hydrochloride in 5% dextrose injection or 0.9% sodium chloride injection over 30 days. *Am.J.Hosp.Pharm.*, **1993**, *50*, 2379–2381

## SAMPLE
**Matrix:** formulations
**Sample preparation:** Dilute with saline, inject a 10 μL aliquot.

## HPLC VARIABLES
**Column:** 300 × 3.9 μBondapak C18
**Mobile phase:** MeCN:100 mM $K_2HPO_4$ 33:67 adjusted to a final pH of 4.4 with phosphoric acid
**Flow rate:** 1.7
**Injection volume:** 10
**Detector:** UV 254

## CHROMATOGRAM
**Retention time:** 5.1

## OTHER SUBSTANCES
**Interfering:** promethazine

## KEY WORDS
injections; saline

## REFERENCE
Martens, H.J.; de Goede, P.N.; van Loenen, A.C. Sorption of various drugs in polyvinyl chloride, glass, and polyethylene-lined infusion containers. *Am.J.Hosp.Pharm.*, **1990**, *47*, 369–373

## SAMPLE
**Matrix:** microsomal incubations
**Sample preparation:** 200 μL Microsomal incubation + 200 μL cold MeOH:MeCN 35:21 + flunitrazepam in MeOH, centrifuge at 10000 rpm for 5 min, inject an aliquot of the supernatant.

## HPLC VARIABLES
**Guard column:** Zorbax RX-C18
**Column:** 250 × 4.6 Zorbax RX-C18
**Mobile phase:** MeCN:MeOH:10 mM pH 7.4 potassium phosphate buffer 21:35:44
**Flow rate:** 1
**Detector:** UV 220

## CHROMATOGRAM
**Retention time:** 17.5
**Internal standard:** flunitrazepam (7.5)

## KEY WORDS
human; liver

## REFERENCE

Ring, B.J.; Binkley, S.N.; Roskos, L.; Wrighton, S.A. Effect of fluoxetine, norfluoxetine, sertraline and desmethyl sertraline on human CYP3A catalyzed 1'-hydroxy midazolam formation *in vitro.* *J.Pharmacol.Exp.Ther.*, **1995**, *275*, 1131–1135

## SAMPLE

**Matrix:** microsomal incubations
**Sample preparation:** Add an equal volume of EtOH to the microsomal incubation, centrifuge, inject an aliquot of the supernatant.

## HPLC VARIABLES

**Guard column:** 4 × 4 5 μm RP18 Lichrospher 100
**Column:** 150 × 4.6 3 μm Supelcosil LC-8
**Mobile phase:** MeCN:MeOH:100 mM pH 6 ammonium acetate 17:26:57
**Column temperature:** 40
**Flow rate:** 2
**Detector:** UV 254

## KEY WORDS

monkey; mouse; rat; dog; human

## REFERENCE

Valles, B.; Schiller, C.D.; Coassolo, P.; De Sousa, G.; Wyss, R.; Jaeck, D.; Viger-Chougnet, A.; Rahmani, R. Metabolism of mofarotene in hepatocytes and liver microsomes from different species. Comparison with in vivo data and evaluation of the cytochrome P450 isoenzymes involved in human biotransformation. *Drug Metab.Dispos.*, **1995**, *23*, 1051–1057

## SAMPLE

**Matrix:** solutions

## HPLC VARIABLES

**Column:** 250 × 4.6 5 μm Supelcosil LC-DP (A) or 250 × 4 5 μm LiChrospher 100 RP-8 (B)
**Mobile phase:** MeCN:0.025% phosphoric acid:buffer 25:10:5 (A) or 60:25:15 (B) (Buffer was 9 mL concentrated phosphoric acid and 10 mL triethylamine in 900 mL water, adjust pH to 3.4 with dilute phosphoric acid, make up to 1 L.)
**Flow rate:** 0.6
**Injection volume:** 25
**Detector:** UV 229

## CHROMATOGRAM

**Retention time:** 10.21 (A), 6.30 (B)

## OTHER SUBSTANCES

**Also analyzed:** acebutolol, acepromazine, acetaminophen, acetazolamide, acetophenazine, albuterol, alprazolam, amitriptyline, amobarbital, amoxapine, antipyrine, atenolol, atropine, azatadine, baclofen, benzocaine, bromocriptine, brompheniramine, brotizolam, bupivacaine, buspirone, butabarbital, butalbital, caffeine, carbamazepine, cetirizine, chlorcyclizine, chlordiazepoxide, chlormezanone, chloroquine, chlorpheniramine, chlorpromazine, chlorpropamide, chlorprothixene, chlorthalidone, chlorzoxazone, cimetidine, cisapride, clomipramine, clonazepam, clonidine, clozapine, cocaine, codeine, colchicine, cyclizine, cyclobenzaprine, dantrolene, desipramine, diazepam, diclofenac, diflunisal, diltiazem, diphenhydramine, diphenidol, diphenoxylate, dipyridamole, disopyramide, dobutamine, doxapram, doxepin, droperidol, encainide, ethidium bromide, ethopropazine, fenoprofen, fentanyl, flavoxate, fluoxetine, fluphenazine, flurazepam, flurbiprofen, fluvoxamine, furosemide, glutethimide, glyburide, guaifenesin, haloperidol, homatropine, hydralazine, hydrochlorothiazide, hydrocodone, hydromorphone, hydroxychloroquine, hydroxyzine, ibuprofen, imipramine, indomethacin, ketoconazole, ketoprofen, ketorolac, labetalol, levorphanol, lidocaine, loratadine, lorazepam, lovastatin, loxapine, mazindol, mefenamic acid, meperidine, mephenytoin, mepivacaine, mesoridazine, metaproterenol, metformin, methadone, methdilazine, methocarbamol, methotrexate, methotrimeprazine, methoxamine, methyldopa, methylphenidate, metoclopramide, metolazone, metoprolol, metronidazole, moclobemide, morphine, nadolol, nalbuphine, naloxone, naphazoline, naproxen, nifedipine, nizatidine, norepinephrine, nortriptyline, oxazepam, oxycodone, oxymetazoline, paroxetine, pemoline, pentazocine, pentobarbital, pentoxifylline, perphenazine, pheniramine, phenobarbital, phenol, phenolphthalein, phentolamine, phenylbutazone, phenyltoloxamine, phenytoin, pimozide, pindolol, piroxicam, pramoxine, prazepam, prazosin, probenecid, procainamide, procaine, prochlorperazine, procyclidine, promazine, promethazine, propafenone, propantheline, propiomazine, propofol, propranolol, protriptyline, quazepam, quinidine, quinine, racemethorphan, ranitidine, remoxipride, risperidone, salicylic acid, scopolamine, secobarbital, sertraline, sotalol, spironolactone, sulfinpyrazone, sulindac, temazepam, terbutaline, terfenadine, tetracaine, theophylline, thiethylperazine, thiopental, thioridazine, thiothixene, timolol, tocainide, tolbutamide, tolmetin, trazodone, triamterene, triazolam, trifluoperazine, triflupromazine, trimeprazine, trimethoprim, trimipramine, verapamil, warfarin, xylometazoline, yohimbine, zopiclone

## KEY WORDS
some details of plasma extraction

## REFERENCE
Koves, E.M. Use of high-performance liquid chromatography-diode array detection in forensic toxicology. *J.Chromatogr.A*, **1995**, *692*, 103–119

## SAMPLE
**Matrix:** urine
**Sample preparation:** 2 mL Urine + 3 mL 500 mM NaOH, vortex for 30 s, add 12 mL diethyl ether, rotate for 5 min, centrifuge at 2500 rpm for 5 min. Remove the ether layer and evaporate it to dryness under a stream of nitrogen, reconstitute the residue in 2 mL mobile phase, inject an aliquot.

## HPLC VARIABLES
**Column:** 100 × 5 5 µm Waters Radial-Pak C18
**Mobile phase:** MeOH:200 mM NaCl 65:35
**Flow rate:** 1.2
**Injection volume:** 50
**Detector:** E, Bioanalytical Systems LC4B, glassy carbon working electrode operated in parallel mode, stainless steel auxiliary electrode, electrode potentials +1.0 V and 0.85 V, Ag/AgCl reference electrode following post-column reaction. The column effluent flowed through a 9.144 m × 0.5 mm i.d. figure eight coil of PTFE tubing in a UV irradiation unit maintained at 0-5° with an ice bath to the detector.

## CHROMATOGRAM
**Limit of detection:** 22 ng/mL

## OTHER SUBSTANCES
**Also analyzed:** benzophenone, clonazepam, diazepam, demoxepam, flurazepam

## KEY WORDS
post-column reaction

## REFERENCE
Selavka, C.M.; Krull, I.S.; Lurie, I.S. Photolytic derivatization for improved LCEC determinations of pharmaceuticals in biological fluids. *J.Chromatogr.Sci.*, **1985**, *23*, 499–508

## ANNOTATED BIBLIOGRAPHY

Wrighton, S.A.; Ring, B.J. Inhibition of human CYP3A catalyzed 1'-hydroxy midazolam formation by ketoconazole, nifedipine, erythromycin, cimetidine, and nizatidine. *Pharm.Res.*, **1994**, *11*, 921–924

Ha, H.R.; Rentsch, K.M.; Kneer, J.; Vonderschmitt, D.J. Determination of midazolam and its alpha-hydroxy metabolite in human plasma and urine by high-performance liquid chromatography. *Ther.Drug Monit.*, **1993**, *15*, 338–343

Noggle, F.T., Jr.; Clark, C.R.; DeRuiter, J. Liquid chromatographic separation of some common benzo-diazepines and their metabolites. *J.Liq.Chromatogr.*, **1990**, *13*, 4005–4021 [also metabolites, alpra-zolam, bromazepam, chlordiazepoxide, clobazam, clonazepam, clorazepate, demoxepam, diazepam, estazolam, fludiazepam, flunitrazepam, flurazepam, halazepam, lorazepam, nitrazepam, nordiaze-pam, oxazepam, prazepam, temazepam, triazolam]

Vasiliades, J.V.; Sahawneh, T. Midazolam determination by gas chromatography, liquid chromatography and gas chromatography-mass spectrometry. *J.Chromatogr.*, **1982**, *228*, 195–203

Vasiliades, J.; Sahawneh, T.H. Determination of midazolam by high-performance liquid chromatography. *J.Chromatogr.*, **1981**, *225*, 266–271

Vree, T.B.; Baars, A.M.; Booij, L.H.; Driessen, J.J. Simultaneous determination and pharmacokinetics of midazolam and its hydroxymetabolites in plasma and urine of man and dog by means of high-performance liquid chromatography. *Arzneimittelforschung*, **1981**, *31*, 2215–2219

# Mometasone Furoate

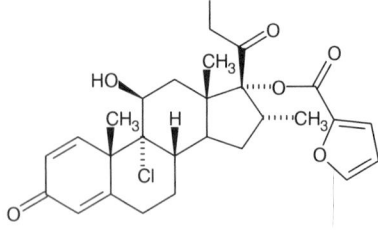

**Molecular formula:** $C_{27}H_{30}Cl_2O_6$
**Molecular weight:** 521.4
**CAS Registry No.:** 83919-23-7

## SAMPLE
**Matrix:** formulations
**Sample preparation:** Dissolve in MeCN:THF 94.3:5.7, inject an aliquot.

## HPLC VARIABLES
**Column:** 250 × 4.6 5 µm Suplex pKb-100 (Supelco)
**Mobile phase:** MeCN:THF:15 mM pH 4.1 acetate buffer 41.5:2.5:56
**Flow rate:** 2
**Detector:** UV 254

## CHROMATOGRAM
**Retention time:** 18

## OTHER SUBSTANCES
**Simultaneous:** degradation products, impurities, benzaldehyde, benzyl alcohol, clotrimazole, orthochlorophenyl-diphenylmethanol

## KEY WORDS
creams

## REFERENCE
Spangler, M. Isocratic reversed phase HPLC analysis of a pharmaceutical cream. *Supelco Reporter*, **1994**, *13(2)*, 12–13

# Mupirocin

**Molecular formula:** $C_{26}H_{44}O_9$
**Molecular weight:** 500.6
**CAS Registry No.:** 12650-69-0

## SAMPLE
**Matrix:** formulations
**Sample preparation:** 5 g Ointment + 100 mL MeCN + 300 mL 50 mM pH 6.3 phosphate buffer, mechanically stir for 30 min, make up to 500 mL with 50 mM pH 6.3 phosphate buffer, filter (0.2 μm), inject a 20 μL aliquot.

## HPLC VARIABLES
**Column:** 100 × 5 Waters radial-pak C18
**Mobile phase:** MeCN:50 mM pH 6.3 phosphate buffer 25:75
**Flow rate:** 2
**Injection volume:** 20
**Detector:** UV 229

## CHROMATOGRAM
**Retention time:** 5.5

## KEY WORDS
ointments

## REFERENCE
Jagota, N.K.; Stewart, J.T.; Warren, F.W.; John, P.M. Stability of mupirocin ointment (Bactroban) admixed with other proprietary dermatological products. *J.Clin.Pharm.Ther.*, **1992**, *17*, 181–184

## SAMPLE
**Matrix:** nutrient broth
**Sample preparation:** Filter, mix an aliquot of the filtrate with MeCN, centrifuge at 2500 g for 10 min, inject a 10 μL aliquot of the supernatant.

## HPLC VARIABLES
**Column:** 150 mm long MCH 5 Micro-Pak
**Mobile phase:** MeCN:water:orthophosphoric acid 30:70:1
**Flow rate:** 1
**Injection volume:** 10
**Detector:** UV 215

## CHROMATOGRAM
**Retention time:** 18.9

## OTHER SUBSTANCES
**Noninterfering:** metabolites

## REFERENCE
Cookson, B. Failure of mupirocin-resistant staphylococci to inactivate mupirocin. *Eur.J.Clin. Microbiol.Infect.Dis.*, **1989**, *8*, 1038–1040

## SAMPLE
**Matrix:** solutions

## HPLC VARIABLES
**Column:** μBondapak C18
**Mobile phase:** MeOH:50 mM pH 4.5 ammonium acetate 60:40
**Flow rate:** 1

## CHROMATOGRAM
**Retention time:** 15

## REFERENCE
Farmer, T.H.; Gilbart, J.; Elson, S.W. Biochemical basis of mupirocin resistance in strains of *Staphylococcus aureus*. *J.Antimicrob.Chemother.*, **1992**, *30*, 587–596

## SAMPLE
**Matrix:** solutions

## HPLC VARIABLES
**Column:** 300 × 3.9 μBondapak C18
**Mobile phase:** MeOH:50 mM pH 4.5 ammonium acetate 40:60
**Flow rate:** 2
**Injection volume:** 50
**Detector:** UV 235

## KEY WORDS
radiolabeled

## REFERENCE
Morecombe, D.J. High-efficiency preparative-scale reversed-phase high-performance liquid chromatographic purification of 14C-labelled antibiotics. *J.Chromatogr.*, **1987**, *389*, 389–395

# Nabumetone

**Molecular formula:** $C_{15}H_{16}O_2$
**Molecular weight:** 228.3
**CAS Registry No.:** 42924-53-8

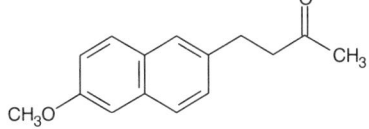

## SAMPLE
**Matrix:** blood
**Sample preparation:** 2 mL Plasma + 200 μL 200 μg/mL naproxen + 1 mL 500 mM pH 3 citrate buffer + 100 μL 1 M HCl + 5 mL ether, shake for 2 min, centrifuge. Remove the organic layer and evaporate it to dryness, reconstitute the residue in MeCN, inject an aliquot.

## HPLC VARIABLES
**Guard column:** Resolve C18 Guard-Pak (Waters)
**Column:** C18 Radial-Pak (Waters)
**Mobile phase:** MeCN:sodium acetate buffer 60:40
**Detector:** UV 280

## CHROMATOGRAM
**Internal standard:** naproxen

## OTHER SUBSTANCES
**Extracted:** metabolites

## KEY WORDS
plasma; pharmacokinetics

## REFERENCE
Brier, M.E.; Sloan, R.S.; Aronoff, G.R. Population pharmacokinetics of the active metabolite of nabumetone in renal dysfunction. *Clin.Pharmacol.Ther.*, **1995**, *57*, 622−627

## SAMPLE
**Matrix:** blood
**Sample preparation:** 0.5 mL Plasma + 2.5 μL 400 μg/mL 6-chloro-2-naphthylacetic acid in acetone, mix on a whirlmixer, add 6 mL n-hexane:ethyl acetate 50:50, add 0.7 mL 1.5 M HCl, shake mechanically for 30 min, centrifuge at 1500 g for 10 min. Remover organic layer and evaporate it to dryness at 37° under a stream of nitrogen. Dissolve residue in mobile phase, centrifuge, inject a 20 μL aliquot.

## HPLC VARIABLES
**Guard column:** Lichrosorb RP-8
**Column:** 250 × 4.6 5 μm Ultrasphere ODS
**Mobile phase:** MeOH:50 mM pH 3.0 sodium acetate 70:30
**Column temperature:** 40
**Flow rate:** 1
**Injection volume:** 20
**Detector:** F ex 284 em 320 (cut-off)

## CHROMATOGRAM
**Retention time:** 7
**Internal standard:** 6-chloro-2-naphthylacetic acid (6)
**Limit of detection:** 250 ng/mL

## OTHER SUBSTANCES
**Extracted:** metabolites
**Simultaneous:** acetaminophen, aspirin, naproxen, salicylic acid

## KEY WORDS
plasma

## REFERENCE
Ray, J.E.; Day, R.O. High-performance liquid chromatographic determination of a new anti-inflammatory
agent, nabumetone, and its major metabolite in plasma using fluorimetric detection. *J.Chromatogr.*,
**1984**, *336*, 234–238

## SAMPLE
**Matrix:** blood, synovial fluid
**Sample preparation:** 100 μL Plasma or synovial fluid + 200 μL 2 mg/mL IS in MeCN,
vortex thoroughly, centrifuge at 11000 rpm for 5 min, inject an 80 μL aliquot of the
supernatant. Tissue. Homogenize (Ultra-Turrax) 150 mg tissue with 600 (synovial membrane) or 1200 (fibrous capsule tissue) mg water for 1 min, vortex vigorously, sonicate for
30 min, vortex, centrifuge. Remove a 50 μL aliquot and add it to 100 μL 1 (synovial
membrane) or 0.33 (fibrous capsule tissue) mg/mL IS in MeCN, vortex thoroughly, centrifuge at 11000 rpm for 5 min, inject an 4 μL aliquot of the supernatant.

## HPLC VARIABLES
**Column:** 150 × 3.9 Resolve 5 C18 (Waters)
**Mobile phase:** MeCN:50 mM $NaH_2PO_4$ 40:60, pH 5.2 (For the active metabolite, 6-meth-
oxy-2-naphthylacetic acid, use MeCN:50 mM $NaH_2PO_4$ 30:70, pH 5.2.)
**Flow rate:** 1.6
**Injection volume:** 4-80
**Detector:** F ex 224 em 356

## CHROMATOGRAM
**Internal standard:** BRL 24333
**Limit of detection:** <21 ng/mL

## KEY WORDS
plasma; synovial membrane; fibrous capsule tissue

## REFERENCE
Miehlke, R.K.; Schneider, S.; Sörgel, F.; Muth, P.; Henschke, F.; Giersch, K.H.; Münzel, P. Penetration
of the active metabolite of nabumetone into synovial fluid and adherent tissue of patients undergoing
knee joint surgery. *Drugs*, **1990**, *40 Suppl 5*, 57–61

# Naproxen

**Molecular formula:** $C_{14}H_{14}O_3$
**Molecular weight:** 230.3
**CAS Registry No.:** 22204-53-1, 26159-34-2 (sodium salt)

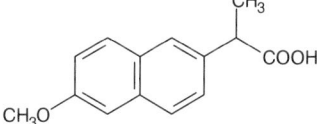

## SAMPLE
**Matrix:** aqueous humor
**Sample preparation:** 100 μL Aqueous humor + 500 μL MeCN, mix mechanically for 90 s, centrifuge at 3000 g for 20 min. Remove supernatant and dry it under nitrogen at room temperature. Dissolve residue in 50 μL mobile phase by swirl mixing for 1 min, centrifuge at 3000 g for 20 s. For concentrations of <20 ng/mL, reduce volume to 20-30 μL under nitrogen.

## HPLC VARIABLES
**Column:** 150 × 4.5 5 μm Ultrasphere octyl
**Mobile phase:** 505 mL MeCN containing 0.65 mL triethylamine + 495 mL 1.65% glacial acetic acid, apparent pH 4.35
**Column temperature:** 30
**Flow rate:** 1
**Injection volume:** 20
**Detector:** UV 280

## CHROMATOGRAM
**Retention time:** 3.89
**Internal standard:** naproxen

## OTHER SUBSTANCES
**Extracted:** diclofenac, flurbiprofen
**Simultaneous:** bacitracin, cortisone, diazepam, fluorometholone, hydrocortisone, imipramine, indomethacin, ketorolac, levobunolol, meclofenamic acid, metipranolol, neomycin, prednisolone, proraracaine, propranolol, salicylic acid, sulfacetamide, suprofen
**Noninterfering:** acebutolol, acetaminophen, acetazolamide, alprenolol, apraclonidine, atenolol, atropine, betamethasone, betaxolol, bupivacaine, caffeine, cyclopentolate, dexamethasone, diphenhydramine, erythromycin, haloperidol, lidocaine, phenylephrine, polymyxin B, procaine, scopolamine, timolol, tropicamide
**Interfering:** ketoprofen

## KEY WORDS
rabbit; human; naproxen is IS

## REFERENCE
Riegel, M.; Ellis, P.P. High-performance liquid chromatography assay for antiinflammatory agents diclofenac and flurbiprofen in ocular fluids. *J.Chromatogr.B*, **1994**, *654*, 140–145

## SAMPLE
**Matrix:** blood
**Sample preparation:** 500 μL Plasma + 500 μL 500 mM HCl, vortex for 1 min, add 5 mL ethyl acetate, extract for 20 min, centrifuge at 2500 rpm for 10 min. Remove the organic layer and evaporate it to dryness, reconstitute the residue in 200 μL MeCN, inject a 100 μL aliquot.

## HPLC VARIABLES
**Column:** 250 × 5 5 μm C18 (Machery & Nagel)

**Mobile phase:** MeCN:water:acetic acid 50:50:0.1
**Flow rate:** 1
**Injection volume:** 100
**Detector:** UV 280

## CHROMATOGRAM
**Internal standard:** naproxen

## OTHER SUBSTANCES
**Extracted:** diclofenac

## KEY WORDS
plasma; naproxen is IS

## REFERENCE
Ramakrishna, S.; Fadnavis, N.W.; Diwan, P.V. Comparative pharmacokinetic evaluation of compressed
suppositories of diclofenac sodium in humans. *Arzneimittelforschung*, **1996**, *46*, 175–177

## SAMPLE
**Matrix:** blood
**Sample preparation:** 500 μL Plasma + 125 μL 40 mM decanoic acid in MeCN, mix. Dialyze a 100 μL sample against 20 mM pH 7.0 phosphate buffer using a Gilson Cuprophane membrane (molecular mass cut-off 15 kDa). Continuously pump the buffer through the dialysis cell and through column A at 3 mL/min for 9.6 min, backflush the contents of column A onto column B with the mobile phase, monitor the effluent from column B. (After each injection flush plasma channel with 1 mL 0.05% Triton X-100, with 1 mL 1 mM HCl, and with 2 mL water. After each injection flush buffer channel with 3 mL 20 mM pH 7.0 phosphate buffer and condition column A with 1 mL 20 mM pH 7.0 phosphate buffer.)

## HPLC VARIABLES
**Column:** A 10 × 2 40 μm Bondesil C18 (Analytichem); B 250 × 3.1 5 μm C18 (RoSil Research Separation Laboratories)
**Mobile phase:** MeCN:MeOH:20 mM pH 3.2 phosphate buffer 50:10:40
**Flow rate:** 1
**Injection volume:** 100
**Detector:** UV 272

## CHROMATOGRAM
**Limit of detection:** 100 ng/mL

## OTHER SUBSTANCES
**Extracted:** fenoprofen (UV 272), flurbiprofen (UV 247), ibuprofen (UV 264), ketoprofen (UV 261)

## KEY WORDS
plasma; dialysis; column-switching

## REFERENCE
Herráez-Hernández, R.; Van de Merbel, N.C.; Brinkman, U.A.T. Determination of the total concentration of highly protein-bound drugs in plasma by on-line dialysis and column liquid chromatography: application to non-steroidal anti-inflammatory drugs. *J.Chromatogr.B*, **1995**, *666*, 127–137

## SAMPLE
**Matrix:** blood

**Sample preparation:** Add 5 μL 42% phosphoric acid to 100 μL plasma and freeze until required. 100 μL Acidified plasma + 10 μL 10 μg/mL homotryptophol in MeOH + 200 μL MeCN, vortex for 1 min, centrifuge, filter (0.45 μm), inject an aliquot of the supernatant.

## HPLC VARIABLES
**Column:** 250 × 4.6 5 μm Ultrasphere ODS
**Mobile phase:** MeCN:50 mM pH 6.0 ammonium acetate 25:75
**Flow rate:** 1
**Detector:** F ex 275 em 345

## CHROMATOGRAM
**Retention time:** 24.7
**Internal standard:** homotryptophol (22)
**Limit of detection:** 50 nM

## OTHER SUBSTANCES
**Extracted:** metabolites, glucuronides

## KEY WORDS
rat; plasma; pharmacokinetics

## REFERENCE
Iwaki, M.; Bischer, A.; Nguyen, A.C.; McDonagh, A.F.; Benet, L.Z. Stereoselective disposition of naproxen glucuronide in the rat. *Drug Metab.Dispos.*, **1995**, *23*, 1099–1103

## SAMPLE
**Matrix:** blood
**Sample preparation:** 1 mL Plasma + 100 μL 5% zinc sulfate in water, vortex for 2 min, add 3 mL MeOH, vortex for 2 min, add 440 μL buffer, vortex for 1 min, centrifuge at 27° at 2000 g for 10 min, inject a 100 μL aliquot of the supernatant. (Buffer was 100 mM NaH$_2$PO$_4$ containing 10 mM sodium lauryl sulfate, adjust pH to 2.8 with orthophosphoric acid, filter (0.45 μm).)

## HPLC VARIABLES
**Column:** 250 × 4.6 5 μm Nucleosil C18
**Mobile phase:** MeCN:water 35:65 containing 1 mM sodium lauryl sulfate and 10 mM NaH$_2$PO$_4$, pH adjusted to 2.8 with orthophosphoric acid
**Flow rate:** 1.5
**Injection volume:** 100
**Detector:** UV 280

## CHROMATOGRAM
**Retention time:** 5.8
**Internal standard:** naproxen

## OTHER SUBSTANCES
**Extracted:** diclofenac

## KEY WORDS
plasma; naproxen is IS

## REFERENCE
Mason, J.L.; Hobbs, G.J. A rapid high performance liquid chromatographic assay for the measurement of diclofenac in human plasma. *J.Liq.Chromatogr.*, **1995**, *18*, 2045–2058

## SAMPLE
**Matrix:** blood
**Sample preparation:** 500 μL Plasma + 20 μL 500 μg/mL phenylbutazone in MeOH + 1.5 mL MeOH, vortex, centrifuge for 15 min at 3000 g. Remove the supernatant and evaporate it to 500 μL using a vortex evaporator, inject a 20 μL aliquot.

## HPLC VARIABLES
**Guard column:** present but not specified
**Column:** 10 μm RP-8 (Alltech)
**Mobile phase:** MeOH:1% acetic acid 70:30
**Injection volume:** 20
**Detector:** UV 240

## CHROMATOGRAM
**Internal standard:** phenylbutazone
**Limit of quantitation:** 100 ng/mL

## OTHER SUBSTANCES
**Extracted:** napdice

## KEY WORDS
dog; plasma; pharmacokinetics

## REFERENCE
Samara, E.; Avnir, D.; Ladkani, D.; Bialer, M. Pharmacokinetic analysis of diethylcarbonate prodrugs of ibuprofen and naproxen. *Biopharm.Drug Dispos.*, **1995**, *16*, 201–210

## SAMPLE
**Matrix:** blood
**Sample preparation:** 200 μL Serum + 20 μL 1 mg/mL phenyl salicylate in mobile phase, vortex briefly, add 560 μL MeCN slowly with continuous vortexing, vortex for 30 s, centrifuge at 3500 rpm for 10 min, inject a 50 μL aliquot of the supernatant.

## HPLC VARIABLES
**Guard column:** 30 μm Perisorb C18
**Column:** 250 × 4.6 5 μm Partisil ODS-3 C18
**Mobile phase:** MeCN:40 mM pH 2.5 HCl buffer 47:53
**Flow rate:** 1.2
**Injection volume:** 50
**Detector:** F ex 230 em 370

## CHROMATOGRAM
**Retention time:** 5
**Internal standard:** phenyl salicylate (10)
**Limit of detection:** 1 ng/mL
**Limit of quantitation:** 2 ng/mL

## KEY WORDS
serum; dog; pharmacokinetics

## REFERENCE
Suh, H.; Jun, H.W.; Lu, G.W. Fluorometric high performance liquid chromatography for quantitation of naproxen in serum. *J.Liq.Chromatogr.*, **1995**, *18*, 3105–3115

## SAMPLE
**Matrix:** blood

**Sample preparation:** 2 mL Whole blood or plasma + 2 mL buffer + 5 mL chloroform: isopropanol:n-heptane 60:14:26, shake gently horizontally for 10 min, centrifuge at 2800 g for 10 min. Remove the lower organic layer and evaporate it to dryness under vacuum at 45°, reconstitute the residue in 100 μL mobile phase, centrifuge at 2800 g for 5 min, inject a 50 μL aliquot of the supernatant. (Buffer was saturated ammonium chloride solution 25% diluted with water, adjusted to pH 9.5 with 25% ammonia solution.)

## HPLC VARIABLES
**Column:** 300 × 3.9 4 μm NovaPack C18
**Mobile phase:** MeOH:THF:buffer 65:5:30 (Buffer was 0.68 g/L (10 mM (sic)) $KH_2PO_4$ adjusted to pH 2.6 with concentrated orthophosphoric acid.) (At the end of each session wash the column with water for 1 h and MeOH for 1 h, re-equilibrate for 30 min.)
**Column temperature:** 30
**Flow rate:** 0.8
**Injection volume:** 50
**Detector:** UV 230

## CHROMATOGRAM
**Retention time:** 5.34
**Limit of detection:** <120 ng/mL

## OTHER SUBSTANCES
**Extracted:** acebutolol, acenocoumarol, acepromazine, aceprometazine, acetaminophen, aconitine, ajmaline, albuterol, alimemazine, alminoprofen, alpidem, alprazolam, alprenolol, amisulpride, amitriptyline, amodiaquine, amoxapine, aspirin, astemizole, atenolol, benazepril, benperidol, benzocaine, benzoylecgonine, bepridil, betaxolol, bisoprolol, bromazepam, brompheniramine, bumadizone, caffeine, carbamazepine, carpipramine, carteolol, celiprolol, cetirizine, chlorambucil, chlordiazepoxide, chlormezanone, chlorophenacinone, chloroquine, chlorpheniramine, chlorpromazine, chlorpropamide, cibenzoline, cicletanine, clemastine, clobazam, clomipramine, clonazepam, clonidine, clozapine, cocaine, codeine, colchicine, cyamemazine, cyclizine, cycloguanil, cyproheptadine, cytarabine, dacarbazine, daunorubicin, debrisoquine, demexiptiline, desipramine, dextromethorphan, dextromoramide, dextropropoxyphene, diazepam, diazoxide, diclofenac, dihydralazine, diltiazem, diphenhydramine, dipyridamole, disopyramide, dosulepine, doxepin, doxylamine, droperidol, ephedrine, estazolam, etodolac, fenoprofen, fentiazac, floctafenine, flumazenil, flunitrazepam, fluoxetine, fluphenazine, flurbiprofen, fluvoxamine, glibenclamide, glibornuride, glipizide, glutethimide, haloperidol, histapyrrodine, hydroxychloroquine, hydroxyzine, ibuprofen, imipramine, indomethacin, iproniazid, ketamine, ketoprofen, labetalol, levomepromazine, lidocaine, lidoflazine, lisinopril, loperamide, loratadine, lorazepam, loxapine, maprotiline, medazepam, medifoxamine, mefenamic acid, mefenidramine, mefloquine, melphalan, meperidine, mephenesin, mephentermine, mepivacaine, metapramine, metformin, methadone, methaqualone, methocarbamol, methotrexate, metoclopramide, metoprolol, mexiletine, mianserine, minoxidil, moclobemide, moperone, morphine, nadolol, nalbuphine, nalorphine, naloxone, naltrexone, nialamide, nicardipine, nifedipine, niflumic acid, nimodipine, nitrazepam, nitrendipine, nizatidine, nomifensine, nortriptyline, omeprazole, opipramol, oxazepam, penbutolol, penfluridol, pentazocine, phenobarbital, phenol, phenylbutazone, pimozide, pindolol, pipamperone, piroxicam, prazepam, prazosin, prilocaine, procainamide, procarbazine, proguanil, promethazine, propafenone, propranolol, protriptyline, pyrimethamine, quinidine, quinine, quinupramine, ramipril, ranitidine, reserpine, ritodrine, secobarbital, sotalol, strychnine, sulfinpyrazole, sulindac, sulpride, sultopride, suriclone, temazepam, tenoxicam, terbutaline, terfenadine, tetracaine, tetrazepam, thioproperazine, thioridazine, tiapride, tiaprofenic acid, ticlopidine, timolol, tioclomarol, tofisopam, tolbutamide, toloxatone, trazodone, triazolam, trifluoperazine, trifluperidol, trimipramine, tropatenine, viloxazine, vinblastine, vincristine, vindesine, yohimbine, zolpidem, zopiclone, zorubicine
**Interfering:** bupivacaine, buprenorphine, buspirone, carbinoxamine, clorazepate, fenfluramine, flecainide, loprazolam, metipranolol, midazolam, oxprenolol, phencyclidine, thiopental, tianeptine, triprolidine, verapamil, warfarin

## KEY WORDS
whole blood; plasma

## REFERENCE
Tracqui, A.; Kintz, P.; Mangin, P. Systematic toxicological analysis using HPLC/DAD. *J.Forensic Sci.*, **1995**, *40*, 254–262

## SAMPLE
**Matrix:** blood
**Sample preparation:** 100 μL Plasma + 50 μL 1 M HCl + 1 mL 20 μg/mL indomethacin in chloroform, extract. Remove the organic layer and evaporate it to dryness under a stream of air, reconstitute the residue in 2 mL MeCN:water 50:50, inject a 50 μL aliquot.

## HPLC VARIABLES
**Column:** 300 × 3.9 10 μm μBondapak C18
**Mobile phase:** MeCN:water:orthophosphoric acid 50:49.5:0.5
**Flow rate:** 1
**Injection volume:** 50
**Detector:** UV 230

## CHROMATOGRAM
**Retention time:** 5.1
**Internal standard:** indomethacin (9.4)
**Limit of quantitation:** 1.5 μg/mL

## KEY WORDS
plasma; pharmacokinetics

## REFERENCE
Charles, B.G.; Mogg, G.A.G Comparative *in vitro* and *in vivo* bioavailability of naproxen from tablet and caplet formulations. *Biopharm.Drug Dispos.*, **1994**, *15*, 121–128

## SAMPLE
**Matrix:** blood
**Sample preparation:** Inject sample onto column A with mobile phase A and elute for 3 min. Backflush contents of column A onto column B with mobile phase B for 4 min and elute column B with mobile phase B and monitor eluant.

## HPLC VARIABLES
**Column:** A 10 × 3 BioTrap Acid C18 (ChromTech); B 10 × 3 CT-sil C18 guard column + 100 × 4.6 5 μm CT-sil C18 (ChromTech)
**Mobile phase:** A 200 mM pH 2.1 phosphate buffer; B MeOH:120 mM pH 3.0 phosphate buffer 65:35
**Flow rate:** A 0.55; B 1
**Injection volume:** 50
**Detector:** UV 328

## CHROMATOGRAM
**Retention time:** 6.5
**Limit of quantitation:** 1050 ng/mL

## KEY WORDS
plasma; column-switching; direct injection

## REFERENCE

Hermansson, J.; Grahn, A. Determination of drugs by direct injection of plasma into a biocompatible extraction column based on a protein-entrapped hydrophobic phase. *J.Chromatogr.A*, **1994**, *660*, 119−129

## SAMPLE

**Matrix:** blood
**Sample preparation:** 25 μL Plasma + 100 μL 10 μg/mL naproxen in MeCN, vortex 30 s, centrifuge at 11000-12300 g for 7 min. Remove supernatant and evaporate it under air at 55°. Dissolve residue in 50 μL mobile phase and inject a 20 μL aliquot.

## HPLC VARIABLES

**Column:** 300 × 3.9 10 μm μBondapak C18
**Mobile phase:** MeCN:80 mM pH 2.0 orthophosphoric acid 46:54
**Flow rate:** 1.1
**Injection volume:** 20
**Detector:** F ex 270 em 410

## CHROMATOGRAM

**Retention time:** 11.5
**Internal standard:** naproxen

## OTHER SUBSTANCES

**Extracted:** furosemide
**Noninterfering:** amikacin, amoxicillin, dexamethasone, gentamicin, indomethacin, morphine, phenobarbital, theophylline, vitamins

## KEY WORDS

plasma; microscale; neonatal; naproxen is IS

## REFERENCE

Sidhu, J.S.; Charles, B.G. Simple microscale high-performance liquid chromatographic method for determination of furosemide in neonatal plasma. *J.Chromatogr.*, **1993**, *612*, 161−165

## SAMPLE

**Matrix:** blood
**Sample preparation:** 500 μL Plasma + 200 μL 2 M HCl + 6 mL hexane:diethyl ether 8:2 (ice cold), extract, centrifuge at 1500 g for 5 min. Remove 5 mL of organic layer and evaporate it to dryness under a stream of nitrogen, redissolve in 500 μL 30 mM pH 7.5 phosphate buffer, inject 50-100 μL aliquot.

## HPLC VARIABLES

**Column:** 100 × 4 3 μm Nucleosil RP 8
**Mobile phase:** MeCN:water 40:60 , acidified with 1 mL 85% phosphoric acid
**Flow rate:** 1
**Injection volume:** 50-100
**Detector:** UV 246

## CHROMATOGRAM

**Retention time:** 4.2
**Internal standard:** S-naproxen
**Limit of quantitation:** 50 ng/mL

## OTHER SUBSTANCES

**Extracted:** flurbiprofen

## KEY WORDS
plasma; naproxen is IS

## REFERENCE
Giesslinger, G.; Menzel-Soglowek, S.; Schuster, O.; Brune, K. Stereoselective high-performance liquid chromatographic determination of flurbiprofen in human plasma. *J.Chromatogr.*, **1992**, *573*, 163–167

## SAMPLE
**Matrix:** blood
**Sample preparation:** 1 mL Plasma + 500 μL 2 M sulfuric acid + 8 mL n-hexane:ethyl acetate 90:10, mix gently at 30 rpm for 10 min, centrifuge at 1500 g for 10 min. Remove organic layer and evaporate it to dryness at 45° under a stream of nitrogen. Reconstitute in 100 μL 1.5% thionyl chloride in n-hexane (freshly prepared), heat at 75° for 1 h in a capped tube, cool to room temperature, add 500 μL 2% S-1-phenylethylamine in dichloromethane (freshly prepared), let stand for 15 min, add 500 μL 2 M sulfuric acid + 5 mL n-hexane, mix gently at 30 rpm for 10 min, centrifuge at 1500 g for 10 min. Remove organic layer and evaporate it to dryness at 45° under a stream of nitrogen. Reconstitute in 250 μL mobile phase, inject 200 μL aliquot.

## HPLC VARIABLES
**Column:** 250 × 4 5 μm SGE silica glass column
**Mobile phase:** n-Heptane:isopropanol 92:8
**Flow rate:** 1
**Injection volume:** 200
**Detector:** UV 254

## CHROMATOGRAM
**Retention time:** 5.9
**Internal standard:** S-naproxen
**Limit of quantitation:** 150 ng/mL

## OTHER SUBSTANCES
**Extracted:** ketoprofen
**Simultaneous:** fenoprofen, ibuprofen, mefenamic acid, salicylic acid
**Noninterfering:** diazepam, digoxin, methylprednisolone, midazolam, nifedipine, penicillamine, ranitidine, theophylline

## KEY WORDS
plasma; normal phase; naproxen is IS; derivatization; chiral

## REFERENCE
Hayball, P.J.; Nation, R.L.; Bochner, F.; Le Leu, R.K. Enantiospecific analysis of ketoprofen in plasma by high-performance liquid chromatography. *J.Chromatogr.*, **1991**, *570*, 446–452

## SAMPLE
**Matrix:** blood
**Sample preparation:** Condition a Sep-Pak C18 SPE cartridge with 2 mL MeOH, 5 mL water, and 1 mL buffer. 20-200 μL Plasma + 100 μL MeOH + 20 μL 50 μg/mL indomethacin in MeOH + 100 μL buffer + 100 μL water, vortex for 2 min, centrifuge at 1800 g for 10 min, apply supernatant to the SPE cartridge, wash with 5 mL water, elute with two 5 mL portions of MeOH, evaporate eluate and take up residue in 1 mL mobile phase. (Buffer was 250 mM ammonium phosphate buffer adjusted to pH 3.0 with orthophosphoric acid.

## HPLC VARIABLES
**Column:** 100 × 4.6 5 µm Brownlee RP18
**Mobile phase:** MeOH:25 mM pH 3.0 phosphate buffer 75:25 (Prepare buffer by diluting a 250 mM ammonium phosphate buffer adjusted to pH 3.0 with orthophosphoric acid.)
**Injection volume:** 20
**Detector:** E, ESA Coulochem Model 5100 A, +0.9 V

## CHROMATOGRAM
**Retention time:** 10.0
**Internal standard:** indomethacin (14.6)
**Limit of detection:** 15 ng/mL

## OTHER SUBSTANCES
**Extracted:** diflunisal, sulindac

## KEY WORDS
plasma; SPE

## REFERENCE
Kazemifard, A.G.; Moore, D.E. Liquid chromatography with amperometric detection for the determination of non-steroidal anti-inflammatory drugs in plasma. *J.Chromatogr.*, **1990**, *533*, 125–132

## SAMPLE
**Matrix:** blood
**Sample preparation:** Dog, rat. 100 µL Plasma + 100 µL 1 M HCl + 6 mL MTBE, vortex for 1 min, centrifuge at 1200 g for 2 min. Remove the organic layer and evaporate it to dryness under reduced pressure at 35°. Reconstitute in 1 mL 2% MeOH, vortex for 2 min, inject a 100 µL aliquot. Human. 500 µL Plasma + 200 µL 1 M HCl + 8 mL diethyl ether, vortex for 1 min, centrifuge at 1200 g for 2 min. Remove the organic layer and evaporate it to dryness under reduced pressure at 35°. Reconstitute in 2 mL 2% MeOH, vortex for 2 min, inject a 100 µL aliquot.

## HPLC VARIABLES
**Column:** 50 × 4.6 5 µm Spherisorb C8
**Mobile phase:** MeCN:THF:MeOH:2% acetic acid (pH 2.5) 5:18:5:72
**Flow rate:** 1.5
**Injection volume:** 100
**Detector:** UV 330

## CHROMATOGRAM
**Retention time:** 8
**Internal standard:** naproxen

## OTHER SUBSTANCES
**Extracted:** 5'-hydroxypiroxicam, piroxicam

## KEY WORDS
plasma; dog; rat; human; naproxen is IS

## REFERENCE
Gillilan, R.B.; Mason, W.D.; Fu, C.-H.J. Rapid analysis of piroxicam in dog, rat and human plasma by high-performance liquid chromatography. *J.Chromatogr.*, **1989**, *487*, 232–235

## SAMPLE
**Matrix:** blood

**Sample preparation:** 1 mL Plasma + 1 mL 100 mM HCl + 10 mL dichloromethane, extract. Dry organic layer at 50° under nitrogen, dissolve in 1 mL mobile phase, inject a 20 μL aliquot.

## HPLC VARIABLES
**Column:** 150 × 4.6 3 μm Alltech C8
**Mobile phase:** MeCN:80 mM phosphoric acid 35:65
**Flow rate:** 1
**Injection volume:** 20
**Detector:** F ex 235 em 405

## CHROMATOGRAM
**Retention time:** 3.5
**Internal standard:** naproxen

## OTHER SUBSTANCES
**Extracted:** bumetanide, furosemide

## KEY WORDS
plasma; naproxen is IS; horse

## REFERENCE
Singh, A.K.; McArdle, C.; Gordon, B.; Ashraf, M.; Granley, K. Simultaneous analysis of furosemide and bumetanide in horse plasma using high performance liquid chromatography. *Biomed.Chromatogr.*, **1989**, *3*, 262–265

## SAMPLE
**Matrix:** blood
**Sample preparation:** Condition a 1 mL Bond-Elut C8 SPE cartridge with 2 mL MeOH then 1 mL 10 mM HCl, do not allow it to dry completely. Sonicate 1 mL whole blood for 20-30 min then apply to the SPE cartridge. Wash with 100 μL water, elute with three 500 μL portions of MeOH:MeCN:1% aqueous ammonium hydroxide 50:20:30, combine eluents and evaporate to dryness under a stream of nitrogen at 40°. Redissolve in 1 mL MeOH, inject a 20 μL aliquot.

## HPLC VARIABLES
**Column:** 250 × 4.5 5 μm Spherisorb ODS
**Mobile phase:** MeCN:MeOH:buffer 35:13:52 (Buffer was water adjusted to pH 3.2 with orthophosphoric acid.)
**Flow rate:** 1
**Injection volume:** 20
**Detector:** UV 250

## CHROMATOGRAM
**Retention time:** 5

## OTHER SUBSTANCES
**Extracted:** acetaminophen, fenoprofen, ibuprofen, indomethacin, ketoprofen, salicylic acid

## KEY WORDS
whole blood; SPE

## REFERENCE
Moore, C.M.; Tebbett, I.R. Rapid extraction of anti-inflammatory drugs in whole blood for HPLC analysis. *Forensic Sci.Int.*, **1987**, *34*, 155–158

## SAMPLE
**Matrix:** blood
**Sample preparation:** Centrifuge serum at 15° at 203000 g for 20 h, remove 1 mL of the supernatant and add 250 μL 4 M HCl, add 5 mL distilled diethyl ether, rotate at 30 rpm for 15 min, centrifuge at 700 g for 5 min. Remove the organic layer and evaporate it to dryness under a stream of nitrogen at 40°, reconstitute the residue in 100 μL mobile phase, inject the whole amount.

## HPLC VARIABLES
**Column:** μBondapak C18 Radial Pak
**Mobile phase:** MeCN:100 mM $(NH_4)H_2PO_4$ 63:37
**Flow rate:** 1.5
**Injection volume:** 100
**Detector:** F ex 320 em 390 following post-column reaction. The column effluent mixed with 12% triethanolamine (for fluorescence enhancement) pumped at 0.5 mL/min and flowed through a 1 m reaction coil to the detector.

## CHROMATOGRAM
**Internal standard:** naproxen

## OTHER SUBSTANCES
**Extracted:** warfarin
**Simultaneous:** salicylic acid
**Noninterfering:** acetaminophen, carbamazepine, furosemide, hydrochlorothiazide, phenobarbital, phenytoin, spironolactone

## KEY WORDS
serum; post-column reaction; ultracentrifugate; naproxen is IS

## REFERENCE
Steyn, J.M.; van der Merwe, H.M.; de Kock, M.J. Reversed-phase high-performance liquid chromatographic method for the determination of warfarin from biological fluids in the low nanogram range. *J.Chromatogr.*, **1986**, *378*, 254–260

## SAMPLE
**Matrix:** blood
**Sample preparation:** 500 μL Plasma + 200 μL, mix on a whirlmixer, add 5 mL diethyl ether:n-hexane 1:1 + 700 μL 1.5 M HCl, shake 30 min, centrifuge at 1500 g for 15 min. Remove organic layer and evaporate it to dryness at 30° under a stream of dry air, take up residue in 1 mL MeOH, inject 10 μL aliquot.

## HPLC VARIABLES
**Column:** 150 × 4.6 5μm LiChrosorb RP-8
**Mobile phase:** MeOH:water 50:50 containing 10 mM tetramethylammonium hydrogen sulfate and 10 mM tris(hydroxymethyl)aminomethane (Tris)
**Column temperature:** 32
**Flow rate:** 1.4
**Injection volume:** 10
**Detector:** UV 254

## CHROMATOGRAM
**Retention time:** 2.55
**Internal standard:** naproxen
**Limit of detection:** <5 μg/mL

## OTHER SUBSTANCES
**Extracted:** diflunisal

## KEY WORDS
plasma; naproxen is IS

## REFERENCE
Van Loenhout, J.W.; Ketelaars, H.C.; Gribnau, F.W.; Van Ginneken, C.A.; Tan, Y. Rapid high-performance liquid chromatographic method for the quantitative determination of diflunisal in plasma. *J.Chromatogr.*, **1980**, *182*, 487–491

## SAMPLE
**Matrix:** blood, synovial fluid
**Sample preparation:** 0.5 mL Plasma or synovial fluid + 50 μL 10 mg/mL diphenylacetic acid + 200 μL 2 M HCl + 5 mL diethyl ether, tumble 10 min on a rotary mixer. Remove organic layer and evaporate it to dryness under vacuum centrifugation. Reconstitute residue in 500 μL mobile phase, inject aliquot.

## HPLC VARIABLES
**Guard column:** 20 × 2 Perisorb RP18 30-40 μm pellicular
**Column:** 125 × 4.6 5 μm Spherisorb ODS 1
**Mobile phase:** MeOH:Sorensen's phosphate buffer (pH 7) 37:63
**Flow rate:** 1
**Injection volume:** 20
**Detector:** UV 260

## CHROMATOGRAM
**Retention time:** 8
**Internal standard:** diphenylacetic acid (5)
**Limit of detection:** <5 μg/mL

## KEY WORDS
plasma

## REFERENCE
Blagbrough, I.S.; Daykin, M.M.; Doherty, M.; Pattrick, M.; Shaw, P.N. High-performance liquid chromatographic determination of naproxen, ibuprofen and diclofenac in plasma and synovial fluid in man. *J.Chromatogr.*, **1992**, *578*, 251–257

## SAMPLE
**Matrix:** blood, urine
**Sample preparation:** Acidify plasma or urine with 1 M pH 2 phosphate buffer, extract with hexane:THF 80:20.

## HPLC VARIABLES
**Guard column:** 20 × 4.6 Nucleosil OCS 10
**Column:** 150 × 6 YMC Pack A312 S5 120A ODS
**Mobile phase:** MeCN:50 mM phosphate buffer 19:83
**Column temperature:** 28
**Flow rate:** 2
**Injection volume:** 50
**Detector:** UV 262

## CHROMATOGRAM
**Retention time:** 11.11 (plasma), 10.47 (urine)
**Internal standard:** naproxen

## OTHER SUBSTANCES
**Extracted:** ketoprofen

## KEY WORDS
plasma; rat; naproxen is IS

## REFERENCE
Daffonchio, L.; Bestetti, A.; Clavenna, G.; Fedele, G.; Ferrari, M.P.; Omini, C. Effect of a new foam formulation of ketoprofen lysine salt in experimental models of inflammation and hyperalgesia. *Arzneimittelforschung*, **1995**, *45*, 590–594

## SAMPLE
**Matrix:** blood, urine
**Sample preparation:** Plasma. 1 mL Plasma + 1 mL 100 mM HCl + 10 mL dichloromethane, rotate for 10 min, centrifuge at 1500 g for 15 min. Remove the organic layer and evaporate it to dryness under a stream of nitrogen at 45°, reconstitute the residue in mobile phase, inject a 20 μL aliquot. Urine. 50 μL Urine + 1 mL mobile phase, inject a 20 μL aliquot.

## HPLC VARIABLES
**Column:** 75 × 4.6 3 μm Supelcosil LC-8
**Mobile phase:** MeCN:50 mM phosphoric acid 45:55
**Flow rate:** 1
**Injection volume:** 20
**Detector:** F ex 235 em 405; UV 235

## CHROMATOGRAM
**Retention time:** 3.2
**Limit of detection:** 20-50 ng/mL (F)

## OTHER SUBSTANCES
**Extracted:** ethacrynic acid, indomethacin, mefenamic acid, phenylbutazone, thiosalicylic acid
**Interfering:** flunixin (with UV detection not with F detection)

## KEY WORDS
plasma

## REFERENCE
Singh, A.K.; Jang, Y.; Mishra, U.; Granley, K. Simultaneous analysis of flunixin, naproxen, ethacrynic acid, indomethacin, phenylbutazone, mefenamic acid and thiosalicyclic acid in plasma and urine by high-performance liquid chromatography and gas chromatography-mass spectrometry. *J.Chromatogr.*, **1991**, *568*, 351–361

## SAMPLE
**Matrix:** blood, urine
**Sample preparation:** Plasma. 350 μL 10 μg/mL Ketoprofen in 10 mM pH 6.0 phosphate buffer containing 0.1% MeOH + 650 μL pH 6 phosphate buffer + 1 mL plasma or urine + 0.5 mL 1 M pH 2 (plasma) or 1 mL 0.5 M pH 7 (urine) phosphate buffer + 10 mL diethyl ether, vortex 1 min, centrifuge at 2000 g for 3 min. Remove organic phase and

evaporate it to dryness under a stream of nitrogen. Dissolve residue in 250 μL mobile phase, vortex for 15 s, inject aliquot.

## HPLC VARIABLES
**Guard column:** 40 × 3.2 30-44 μm Vydac reverse-phase
**Column:** 40 × 4.6 5 μm Spherisorb ODS
**Mobile phase:** MeCN:50 mM pH 7.0 phosphate buffer 6:94 to 8:92
**Flow rate:** 2
**Injection volume:** 5-200
**Detector:** UV 262

## CHROMATOGRAM
**Retention time:** 10
**Internal standard:** ketoprofen (16)
**Limit of detection:** 2 ng/mL

## OTHER SUBSTANCES
**Extracted:** metabolites, fenoprofen, probenecid, salicylic acid

## KEY WORDS
plasma

## REFERENCE
Upton, R.A.; Buskin, J.N.; Guentert, T.W.; Williams, R.L.; Riegelman, S. Convenient and sensitive high-performance liquid chromatography assay for ketoprofen, naproxen and other allied drugs in plasma or urine. *J.Chromatogr.*, **1980**, *190*, 119–128

## SAMPLE
**Matrix:** bulk
**Sample preparation:** 10 mg Compound + 10 mg 1-(3-dimethylaminopropyl)-3-ethylcarbodiimide hydrochloride + 2 drops 3,5-dimethylaniline + 1.5 mL dichloromethane, mix, after 30 min add 1 mL 1 M HCl, shake vigorously. Remove the lower organic layer and dry it over anhydrous magnesium sulfate, inject an aliquot.

## HPLC VARIABLES
**Column:** 250 × 4.6 5 μm D N-(3,5-dinitrobenzoyl)phenylglycine (Regis)
**Mobile phase:** Hexane:isopropanol 80:20
**Flow rate:** 2
**Injection volume:** 20
**Detector:** UV 254; UV 280

## CHROMATOGRAM
**Retention time:** $k'$ 9.10 (for first enantiomer)

## OTHER SUBSTANCES
**Also analyzed:** carprofen, cicloprofen, etodolac, fenoprofen, flurbiprofen, ibuprofen, ketoprofen, pirprofen, tiaprofenic acid

## KEY WORDS
derivatization; α = 1.17; chiral

## REFERENCE
Pirkle, W.H.; Murray, P.G. The separation of the enantiomers of a variety of non-steroidal anti-inflammatory drugs (NSAIDS) as their anilide derivatives using a chiral stationary phase. *J.Liq.Chromatogr.*, **1990**, *13*, 2123–2134

## SAMPLE
**Matrix:** dialysate
**Sample preparation:** Inject a 10 μL aliquot of dialysate (pH 7.4 isotonic phosphate buffer).

## HPLC VARIABLES
**Guard column:** 37-50 μm Corasil C18
**Column:** 100 × 4 5 μm Nucleosil C18
**Mobile phase:** MeCN:50 mM pH 3.0 phosphate buffer 48:52
**Flow rate:** 1.1
**Injection volume:** 10
**Detector:** F ex 262 em 356

## CHROMATOGRAM
**Retention time:** 2.5
**Internal standard:** naproxen

## OTHER SUBSTANCES
**Extracted:** flurbiprofen (F ex 258 em 310)

## KEY WORDS
mouse; rat; naproxen is IS

## REFERENCE
Evrard, P.A.; Deridder, G.; Verbeeck, R.K. Intravenous microdialysis in the mouse and the rat: Development and pharmacokinetic application of a new probe. *Pharm.Res.*, **1996**, *13*, 12−17

## SAMPLE
**Matrix:** formulations
**Sample preparation:** Dissolve in 100 mM pH 6-8 phosphate buffer (total ionic strength 0.5), inject an aliquot.

## HPLC VARIABLES
**Column:** 150 × 4.6 5 μm Supelcosil C18
**Mobile phase:** MeCN:50 mM phosphate buffer 25:75
**Flow rate:** 1
**Detector:** UV 222

## CHROMATOGRAM
**Retention time:** 4.2

## KEY WORDS
tablets

## REFERENCE
Chakrabarti, S.; Southard, M.Z. Control of poorly soluble drug dissolution in conditions simulating the gastrointestinal tract flow. 1. Effect of tablet geometry in buffered medium. *J.Pharm.Sci.*, **1996**, *85*, 313−319

## SAMPLE
**Matrix:** formulations
**Sample preparation:** Dissolve tablet in 10 mM HCl containing 90 mM KCl (pH 2.0), inject an aliquot.

## HPLC VARIABLES
**Column:** 50 mm long ODS Hypersil C18

**Mobile phase:** MeCN:50 mM pH 7.0 phosphate buffer 32:68 containing 5 mM tetrabutylammonium
**Flow rate:** 1
**Detector:** UV 220

## CHROMATOGRAM
**Retention time:** 6.0

## OTHER SUBSTANCES
**Simultaneous:** phenytoin

## KEY WORDS
tablets

## REFERENCE
Neervannan, S.; Dias, L.S.; Southard, M.Z.; Stella, V.J. A convective-diffusion model for dissolution of two non-interacting drug mixtures from co-compressed slabs under laminar hydrodynamic conditions. *Pharm.Res.*, **1994**, *11*, 1288–1295

## SAMPLE
**Matrix:** solutions
**Sample preparation:** Filter (0.45 µm), dilute the filtrate with mobile phase, inject an aliquot.

## HPLC VARIABLES
**Column:** 150 × 4.6 5 µm Hypersil ODS
**Mobile phase:** MeCN:10 mM pH 6 phosphate buffer 52:48
**Detector:** UV 250

## REFERENCE
Okimoto, K.; Rajewski, R.A.; Uekama, K.; Jona, J.A.; Stella, V.J. The interaction of charged and uncharged drugs with neutral (HP-β-CD) and anionically charged (SBE7-β-CD) β-cyclodextrins. *Pharm.Res.*, **1996**, *13*, 256–264

## SAMPLE
**Matrix:** solutions

## HPLC VARIABLES
**Column:** 125 × 3 Ecocart LiChrospher 100 RP-18
**Mobile phase:** Isopropanol:100 mM $KH_2PO_4$:formic acid 54:100:0.1
**Flow rate:** 0.6
**Detector:** UV 254

## CHROMATOGRAM
**Retention time:** 5.6
**Limit of quantitation:** 200-500 ng/mL

## OTHER SUBSTANCES
**Simultaneous:** acemetacin, diclofenac, flurbiprofen, indomethacin, ketoprofen, lonazolac, piroxicam, sulindac, tenoxicam

## REFERENCE
Baeyens, W.R.G.; Van Der Weken, G.; Van Overbeke, A.; Zhang, Z.D. Preliminary results on the LC-separation of non-steroidal anti-inflammatory agents in conventional and narrow-bore RP set-ups applying columns with different internal diameters. *Biomed.Chromatogr.*, **1995**, *9*, 261–262

## SAMPLE
**Matrix:** solutions

## HPLC VARIABLES
**Column:** 250 × 4.6 5 μm Supelcosil LC-DP (A) or 250 × 4 5 μm LiChrospher 100 RP-8 (B)
**Mobile phase:** MeCN:0.025% phosphoric acid:buffer 25:10:5 (A) or 60:25:15 (B) (Buffer was 9 mL concentrated phosphoric acid and 10 mL triethylamine in 900 mL water, adjust pH to 3.4 with dilute phosphoric acid, make up to 1 L.)
**Flow rate:** 0.6
**Injection volume:** 25
**Detector:** UV 229

## CHROMATOGRAM
**Retention time:** 6.97 (A), 7.18 (B)

## OTHER SUBSTANCES
**Also analyzed:** acebutolol, acepromazine, acetaminophen, acetazolamide, acetophenazine, albuterol, alprazolam, amitriptyline, amobarbital, amoxapine, antipyrine, atenolol, atropine, azatadine, baclofen, benzocaine, bromocriptine, brompheniramine, brotizolam, bupivacaine, buspirone, butabarbital, butalbital, caffeine, carbamazepine, cetirizine, chlorcyclizine, chlordiazepoxide, chlormezanone, chloroquine, chlorpheniramine, chlorpromazine, chlorpropamide, chlorprothixene, chlorthalidone, chlorzoxazone, cimetidine, cisapride, clomipramine, clonazepam, clonidine, clozapine, cocaine, codeine, colchicine, cyclizine, cyclobenzaprine, dantrolene, desipramine, diazepam, diclofenac, diflunisal, diltiazem, diphenhydramine, diphenidol, diphenoxylate, dipyridamole, disopyramide, dobutamine, doxapram, doxepin, droperidol, encainide, ethidium bromide, ethopropazine, fenoprofen, fentanyl, flavoxate, fluoxetine, fluphenazine, flurazepam, flurbiprofen, fluvoxamine, furosemide, glutethimide, glyburide, guaifenesin, haloperidol, homatropine, hydralazine, hydrochlorothiazide, hydrocodone, hydromorphone, hydroxychloroquine, hydroxyzine, ibuprofen, imipramine, indomethacin, ketoconazole, ketoprofen, ketorolac, labetalol, levorphanol, lidocaine, loratadine, lorazepam, lovastatin, loxapine, mazindol, mefenamic acid, meperidine, mephenytoin, mepivacaine, mesoridazine, metaproterenol, metformin, methadone, methdilazine, methocarbamol, methotrexate, methotrimeprazine, methoxamine, methyldopa, methylphenidate, metoclopramide, metolazone, metoprolol, metronidazole, midazolam, moclobemide, morphine, nadolol, nalbuphine, naloxone, naphazoline, nifedipine, nizatidine, norepinephrine, nortriptyline, oxazepam, oxycodone, oxymetazoline, paroxetine, pemoline, pentazocine, pentobarbital, pentoxifylline, perphenazine, pheniramine, phenobarbital, phenol, phenolphthalein, phentolamine, phenylbutazone, phenyltoloxamine, phenytoin, pimozide, pindolol, piroxicam, pramoxine, prazepam, prazosin, probenecid, procainamide, procaine, prochlorperazine, procyclidine, promazine, promethazine, propafenone, propantheline, propiomazine, propofol, propranolol, protriptyline, quazepam, quinidine, quinine, racemethorphan, ranitidine, remoxipride, risperidone, salicylic acid, scopolamine, secobarbital, sertraline, sotalol, spironolactone, sulfinpyrazone, sulindac, temazepam, terbutaline, terfenadine, tetracaine, theophylline, thiethylperazine, thiopental, thioridazine, thiothixene, timolol, tocainide, tolbutamide, tolmetin, trazodone, triamterene, triazolam, trifluoperazine, triflupromazine, trimeprazine, trimethoprim, trimipramine, verapamil, warfarin, xylometazoline, yohimbine, zopiclone

## KEY WORDS
some details of plasma extraction

## REFERENCE
Koves, E.M. Use of high-performance liquid chromatography-diode array detection in forensic toxicology. *J.Chromatogr.A*, **1995**, *692*, 103–119

## SAMPLE
**Matrix:** solutions

## HPLC VARIABLES
**Column:** 150 × 4.6 cellulose 3,5-dimethylphenylcarbamate/10-undecenoate bonded to allylsilica
**Mobile phase:** Heptane:isopropanol:trifluoroacetic acid 98:2:0.1
**Flow rate:** 1
**Injection volume:** 1000
**Detector:** UV 254

## CHROMATOGRAM
**Retention time:** k′ 4.48

## KEY WORDS
chiral; α = 1.16

## REFERENCE
Oliveros, L.; Lopez, P.; Minguillon, C.; Franco, P. Chiral chromatographic discrimination ability of a cellulose 3,5-dimethylphenylcarbamate/10-undecenoate mixed derivative fixed on several chromatographic matrices. *J.Liq.Chromatogr.*, **1995**, *18*, 1521–1532

## SAMPLE
**Matrix:** solutions

## HPLC VARIABLES
**Column:** 250 × 4.6 Zorbax RX
**Mobile phase:** Gradient. A was 10 mL concentrated orthophosphoric acid and 7 mL triethylamine in 1 L water. B was 10 mL concentrated orthophosphoric acid and 7 mL triethylamine in 200 mL water, make up to 1 L with MeCN. A:B from 100:0 to 0:100 over 30 min, maintain at 0:100 for 5 min.
**Column temperature:** 30
**Flow rate:** 2
**Detector:** UV 210

## OTHER SUBSTANCES
**Also analyzed:** acepromazine, acetaminophen, acetophenazine, albuterol, aminophylline, amitriptyline, amobarbital, amoxapine, amphetamine, amylocaine, antipyrine, aprobarbital, aspirin, atenolol, atropine, avermectin, barbital, benzocaine, benzoic acid, benzotropine, benzphetamine, berberine, bibucaine, bromazepan, brompheniramine, buprenorphine, buspirone, butabarbital, butacaine, butethal, caffeine, carbamazepine, carbromal, chloramphenicol, chlordiazepoxide, chloroquine, chlorothiazide, chloroxylenol, chlorphenesin, chlorpheniramine, chlorpromazine, chlorpropamide, chlortetracycline, cimetidine, cinchonidine, cinchonine, clenbuterol, clonazepam, clonixin, clorazepate, cocaine, codeine, colchicine, cortisone, coumarin, cyclazocine, cyclobenzaprine, cyclothiazide, cyheptamide, cymarin, danazol, danthron, dapsone, debrisoquine, desipramine, dexamethasone, dextromethorphan, dextropropoxyphene, diamorphine, diazepam, diclofenac, diethylpropion, diethylstilbestrol, diflunisal, digitoxin, digoxin, diltiazem, diphenhydramine, diphenoxylate, diprenorphine, dipyrone, disulfiram, dopamine, doxapram, doxepin, dronabinol, ephedrine, epinephrine, epinine, estradiol, estriol, estrone, ethacrynic acid, ethosuximide, etonitazene, etorphine, eugenol, famotidine, fenbendazole, fencamfamine, fenoprofen, fenproporex, fentanyl, flubendazole, flufenamic acid, flunitrazepam, 5-fluorouracil, fluoxymesterone, fluphenazine, furosemide, gentisic acid, gitoxigenin, glipizide, glunixin, glutethimide, glybenclamide, guaiacol, halazepam, haloperidol, hydrochlorothiazide, hydrocodone, hydrocortisone, hydromorphone, hydroxyquinoline, ibogaine, ibuprofen, iminostilbene, imipramine, indomethacin, isocarbostyril, isocarboxazid, isoniazid, isoproter-

enol, isoxsuprine, ivermectin, ketamine, ketoprofen, kynurenic acid, levorphanol, lidocaine, lorazepam, lormetazepam, loxapine, mazindol, mebendazole, meclizine,
meclofenamic acid, medazepam, mefenamic acid, megestrol, mepacrine, meperidine, mephentermine, mephenytoin, mephesin, mephobarbital, mepivacaine, mescaline, mesoridazine, methadone, methamphetamine, methapyrilene, methaqualone, methazolamide,
methocarbamol, methoxamine, methsuximide, methyl salicylate, methyldopa, methyldopamine, methylphenidate, methylprednisolone, methyltestosterone, methyprylon, metoprolol, mibolerone, morphine, nadolol, nalorphine, naloxone, naltrexone, naphazoline, nefopam, niacinamide, nicotine, nicotinic acid, nifedipine, niflumic acid, nitrazepam,
norepinephrine, nortriptyline, noscapine, nylidrin, oxazepam, oxycodone, oxymorphone,
oxyphenbutazone, oxytetracycline, papaverine, pargyline, pemoline, pentazocine, pentobarbital, persantine, phenacetin, phenazocine, phenazopyridine, phencyclidine, phendimetrazine, phenelzine, pheniramine, phenobarbital, phenothiazine, phensuximide, phentermine, phenylbutazone, phenylephrine, phenylpropanolamine, piperocaine, prazepam,
prednisolone, primidone, probenecid, progesterone, propiomazine, propranolol, propylparaben, pseudoephedrine, puromycin, pyrilamine, pyrithyldione, quazepam, quinaldic acid,
quinidine, quinine, ranitidine, recinnamine, reserpine, resorcinol, saccharin, albuterol,
salicylamide, salicylic acid, scopolamine, scopoletin, secobarbital, strychnine, sulfacetamide, sufadiazine, sulfadimethoxine, sulfaethidole, sulfamerazine, sulfamethazine, sulfamethoxizole, sulfanilamide, sulfapyridine, sulfasoxizole, sulindac, tamoxifen, temazepam,
testosterone, tetracaine, tetracycline, tetramisole, thebaine, theobromine, theophylline,
thiabendazole, thiamine, thiamylal, thiobarbituric acid, thioridazine, thiosalicylic acid,
thiothixene, thymol, tolazamide, tolazoline, tobutamide, tolmetin, tranylcypromine, triamcinolone, tribenzylamine, trichloromethiazide, trifluoperazine, trihexyphenidyl, trimethoprim, tripelennamine, triprolidine, tropacocaine, tyramine, verapamil, vincamine,
warfarin, yohimbine, zoxazolamine

## REFERENCE

Hill, D.W.; Kind, A.J. Reversed-phase solvent gradient HPLC retention indexes of drugs. *J.Anal.Toxicol.*,
**1994**, *18*, 233−242

## SAMPLE

**Matrix:** solutions

## HPLC VARIABLES

**Column:** 250 × 4.6 10 μm Chiralpak AD (Daicel)
**Mobile phase:** Carbon dioxide:MeOH 96:4
**Column temperature:** 30
**Flow rate:** 2.5
**Detector:** UV 210

## CHROMATOGRAM

**Retention time:** 18.5, 23 (enantiomers)

## OTHER SUBSTANCES

**Simultaneous:** fenoprofen, flurbiprofen, ibuprofen, ketoprofen

## KEY WORDS

SFC; 250 bar; chiral

## REFERENCE

Kot, A.; Sandra, P.; Venema, A. Sub- and supercritical fluid chromatography on packed columns: A
versatile tool for the enantioselective separation of basic and acidic drugs. *J.Chromatogr.Sci.*, **1994**,
*32*, 439−448

## SAMPLE
**Matrix:** solutions
**Sample preparation:** Inject an aliquot of a 50 μg/mL solution.

## HPLC VARIABLES
**Column:** 100 × 4.6 Chiral AGP CSP (ChromTech)
**Mobile phase:** Isopropanol:4 mM pH 7.0 phosphate buffer 0.5:99.5
**Flow rate:** 0.8
**Injection volume:** 20
**Detector:** UV 263

## CHROMATOGRAM
**Retention time:** 3 (R), 4.5. (S)

## REFERENCE
Kern, J.R. Chromatographic separation of the optical isomers of naproxen. *J.Chromatogr.*, **1991**, *543*, 355–306

## SAMPLE
**Matrix:** solutions
**Sample preparation:** Inject a 50 μL aliquot of a solution in mobile phase.

## HPLC VARIABLES
**Column:** 100 × 4.6 5 μm Spheri-5 RP-8
**Mobile phase:** MeOH:buffer 30:70 (Prepare buffer by mixing 4 mM $Na_2HPO_4$ and 7 mM $KH_2PO_4$ to achieve pH 7.)
**Flow rate:** 1
**Injection volume:** 50
**Detector:** F ex 355 em 460 (408 nm cutoff filter) following post-column extraction. The column effluent mixed with 50 μg/mL reagent in mobile phase pumped at 0.5 mL/min and then with chloroform pumped at 1 mL/min and the mixture flowed through a 1.8 m × 0.3 mm ID knitted PTFE coil to a 50 μL membrane phase separator using a polyethylene-backed 0.5 μm Fluoropore membrane filter (design in paper). The organic phase flowed to the detector. (Synthesize the reagent, α-(3,4-dimethoxyphenyl)-4'-trimethylammoniumme-thylcinnamonitrile methosulfate, as follows. Stir 20 mmoles 3,4-dimethoxyphenylacetoni-trile and 20 mmoles p-toluamide in 50 mL EtOH at 50°, add 5 mL 50% aqueous KOH slowly, stir at 50° for 5 min, cool to room temperature, filter, dry the precipitate of α-(3,4-dimethoxyphenyl)-4'-methylcinnamonitrile. Dissolve 20 mmoles α-(3,4-dimethoxyphenyl)-4'-methylcinnamonitrile, 20 mmoles N-bromosuccinimide, and 20 mg benzoyl peroxide in 100 mL carbon tetrachloride (Caution! Carbon tetrachloride is a carcinogen!), reflux with stirring for 1.5 h, cool, filter, evaporate to dryness under reduced pressure, recrystallize from MeOH to give α-(3,4-dimethoxyphenyl)-4'-bromomethylcinnamonitrile. Vigorously stir 30 mmoles anhydrous dimethylamine in 100 mL dry benzene (Caution! Benzene is a car-cinogen!) at 0°, very slowly add 10 mmoles α-(3,4-dimethoxyphenyl)-4'-bromomethylcinna-monitrile while stirring at 0°, stir at room temperature overnight, add 150 mL water, re-move the organic phase, extract the aqueous phase twice with 100 mL portions of diethyl ether, wash the organic layers with saturated NaCl solution, dry over anhydrous magne-sium sulfate, evaporate under reduced pressure to give α-(3,4-dimethoxyphenyl)-4'-dime-thylaminomethylcinnamonitrile (J.Chem.Eng.Data 1987, 32, 387). Reflux 10 mmoles α-(3,4-dimethoxyphenyl)-4'-dimethylaminomethylcinnamonitrile, 20 mmoles dimethyl sulfate (Caution! Dimethyl sulfate is a carcinogen and acutely toxic!), and 5 g potassium carbonate in 50 mL acetone for 1 h, cool to room temperature, filter, dry the precipitate under vacuum at room temperature overnight, recrystallize from chloroform containing 2-3 drops of 95% EtOH to give α-(3,4-dimethoxyphenyl)-4'-trimethylammoniummethylcinnamonitrile meth-osulfate (mp 212-215°). Protect solutions from light.)

## CHROMATOGRAM
**Retention time:** k′ 0.9690
**Limit of detection:** 100 ng/mL

## OTHER SUBSTANCES
**Simultaneous:** ibuprofen, ketoprofen, mefenamic acid, probenecid, salicylic acid, valproic acid

## KEY WORDS
post-column extraction

## REFERENCE
Kim, M.; Stewart, J.T. HPLC post-column ion-pair extraction of acidic drugs using a substituted α-phenylcinnamonitrile quaternary ammonium salt as a new fluorescent ion-pair reagent. *J.Liq.Chromatogr.*, **1990**, *13*, 213–237

## SAMPLE
**Matrix:** solutions

## HPLC VARIABLES
**Column:** 250 × 4 OmniPac PAX-500 (Dionex)
**Mobile phase:** Gradient. A was MeCN:10 mM sodium carbonate 18:82. B was MeCN:50 mM sodium carbonate 33:67. A:B from 100:0 to 0:100 over 10 min.
**Flow rate:** 1
**Detector:** UV 254

## CHROMATOGRAM
**Retention time:** 12

## OTHER SUBSTANCES
**Simultaneous:** aspirin, carprofen, diflunisal, fenbufen, ibuprofen, indomethacin, tolmetin

## REFERENCE
Slingsby, R.W.; Rey, M. Determination of pharmaceuticals by multi-phase chromatography: Combined reversed phase and ion exchange in one column. *J.Liq.Chromatogr.*, **1990**, *13*, 107–134

## ANNOTATED BIBLIOGRAPHY

Hyun, M.H.; Ryoo, J.-J.; Cho, Y.J.; Jin, J.S. Unusual examples of the liquid chromatographic resolution of racemates. Resolution of π-donor analytes on a π-donor chiral stationary phase. *J.Chromatogr.A*, **1995**, *692*, 91–96 [chiral; also alminoprofen, fenoprofen, flurbiprofen, ibuprofen, ketoprofen]

Terfloth, G.J.; Pirkle, W.H.; Lynam, K.G.; Nicolas, E.C. Broadly applicable polysiloxane-based chiral stationary phase for high-performance liquid chromatography and supercritical fluid chromatography. *J.Chromatogr.A*, **1995**, *705*, 185–194 [chiral; SFC; HPLC; also carprofen, cicloprofen, etodolac, fenoprofen, flurbiprofen, ibuprofen, pirprofen, warfarin]

Kempe, M.; Mosbach, K. Direct resolution of naproxen on a non-covalently molecularly imprinted chiral stationary phase. *J.Chromatogr.A*, **1994**, *664*, 276–279 [chiral]

Szász, G.; Budvári-Bárany, Z.; Löre, A.; Radeczky, G.; Shalaby, A. HPLC of antiphlogistic acids on silica dynamically modified with cetylpyridinium chloride. *J.Liq.Chromatogr.*, **1993**, *16*, 2335–2345 [also diclofenac, fenoprofen, ibuprofen, ketoprofen, nicotinic acid, nifluminic acid, salicylic acid]

Tsai, S.-W.; Wei, H.-J. Self-normalized analysis of lipase-catalyzed conversion of naproxen enantiomers. *J.Liq.Chromatogr.*, **1993**, *16*, 2993–3001 [chiral; normal phase; naphthalene (IS)]

Andersen, J.V.; Hansen, S.H. Simultaneous quantitative determination of naproxen, its metabolite 6-O-desmethylnaproxen and their five conjugates in plasma and urine samples by high-performance liquid chromatography on dynamically modified silica. *J.Chromatogr.*, **1992**, *577*, 325–333 [extracted metabolites; plasma; urine]

Pirkle, W.H.; Welch, C.J.; Lamm, B. Design, synthesis, and evaluation of an improved enantioselective naproxen selector. *J.Org.Chem.*, **1992**, *57*, 3854–3860 [chiral; also carprofen, cicloprofen, fenoprofen, flurbiprofen, ibuprofen, ketoprofen, pirprofen, tiaprofenic acid]

Pirkle, W.H.; Welch, C.J. An improved chiral stationary phase for the chromatographic separation of underivatized naproxen enantiomers. *J.Liq.Chromatogr.*, **1992**, *15*, 1947–1955 [chiral; also cicloprofen, fenoprofen, flurbiprofen, ibuprofen, ketoprofen, piprofen, tiaprofenic acid]

Vree, T.B.; Van den Biggelaar-Martea, M.; Verwey-Van Wissen, C.P.W.G.M. Determination of naproxen and its metabolite O-demethylnaproxen with their acyl glucuronides in human plasma and urine by means of direct gradient high-performance liquid chromatography. *J.Chromatogr.*, **1992**, *578*, 239–249 [plasma; urine; extracted metabolites; gradient; LOQ 1.5 μg/mL (plasma); LOQ 1 μg/mL (urine); pharmacokinetics]

Pirkle, W.H.; Welch, C.J. Chromatographic separation of underivatized naproxen enantiomers. *J.Liq.Chromatogr.*, **1991**, *14*, 3387–3396 [chiral]

Buszewski, B.; El Mouelhi, M.; Albert, K.; Bayer, E. Influence of the structure of chemically bonded C18 phases on HPLC separation of naproxen glucuronide diastereoisomers. *J.Liq.Chromatogr.*, **1990**, *13*, 505–524

Lafontaine, D.; Mailhot, C.; Vermeulen, M.; Bissonnette, B.; Lambert, C. Influence of chewable sucralfate or a standard meal on the bioavailability of naproxen. *Clin.Pharm.*, **1990**, *9*, 773–777 [pharmacokinetics; LOQ 1 μg/mL; diphenylacetic acid (IS)]

Wanwimolruk, S. A simple isocratic high-performance liquid chromatographic (HPLC) determination of naproxen in human plasma using a microbore column technique. *J.Liq.Chromatogr.*, **1990**, *13*, 1611–1625 [plasma; microbore; flurbiprofen (IS); fluorescence detection; LOD 100 ng/mL; interfering ephedrine; non-interfering acetaminophen, aspirin, caffeine, diazepam, diclofenac, indomethacin, lidocaine, lorazepam, metoprolol, phenylbutazone, phenytoin, pindolol, piroxicam, salicylic acid, salicyluric acid, sulfanilamide, sulfinpyrazole, theophylline, triazolam]

Bui, K.H.; French, S.B. Direct serum injection and analysis of drugs with aqueous mobile phases containing triethylammonium acetate. *J.Liq.Chromatogr.*, **1989**, *12*, 861–873 [direct injection; plasma; dog; rat; fluorescence detection; UV detection; also antipyrine, atenolol, hexanophenone, metoprolol, propranolol]

Pettersson, C.; Gioeli, C. Improved resolution of enantiomers of naproxen by the simultaneous use of a chiral stationary phase and a chiral additive in the mobile phase. *J.Chromatogr.*, **1988**, *435*, 225–228 [chiral]

Satterwhite, J.H.; Boudinot, F.D. High-performance liquid chromatographic determination of ketoprofen and naproxen in rat plasma. *J.Chromatogr.*, **1988**, *431*, 444–449 [ketoprofen (IS); LOD 5 ng/mL; pharmacokinetics]

Hermansson, J.; Erikson, M. Direct liquid chromatographic resolution of acidic drugs using a chiral α1-acid glycoprotein column (Enantiopac). *J.Liq.Chromatogr.*, **1986**, *9*, 621–639 [chiral; also bendroflumethiazide, disopyramide, ethotoin, hexobarbital, ibuprofen, ketoprofen, 2-phenoxypropionic acid, RAC 109]

Hylandskjaer, A.; Aarbakke, J. An improved microscale HPLC assay for naproxen plasma levels. *Acta Pharmacol.Toxicol.(Copenh.)*, **1983**, *52*, 78–80

Meinsma, D.A.; Radzik, D.M.; Kissinger, P.T. Determination of common analgesics in serum and urine by liquid chromatography/electrochemistry. *J.Liq.Chromatogr.*, **1983**, *6*, 2311–2335 [electrochemical detection; also acetaminophen, codeine, methyl salicylate, phenacetin, salicylic acid]

Shimek, J.L.; Rao, N.G.; Khalil, S.K. An isocratic high-pressure liquid chromatographic determination of naproxen and desmethylnaproxen in human plasma. *J.Pharm.Sci.*, **1982**, *71*, 436–439

van Loenhout, J.W.A.; van Ginneken, C.A.M.; Ketelaars, H.C.J.; Kimenai, P.M.; Tan, Y.; Gribnau, F.W.J. A high performance liquid chromatographic method for the quantitative determination of naproxen and des-methyl-naproxen in biological samples. *J.Liq.Chromatogr.*, **1982**, *5*, 549–561 [fluorescence detection; plasma; urine; extracted metabolites; column temp 35]

Broquaire, M.; Rovei, V.; Braithwaite, R. Quantitative determination of naproxen in plasma by a simple high-performance liquid chromatographic method. *J.Chromatogr.*, **1981**, *224*, 43–49

Burgoyne, R.F.; Brown, P.R.; Kaplan, S.R. The simultaneous assay of naproxen and salicylic acid in serum using high pressure liquid chromatography. *J.Liq.Chromatogr.*, **1980**, *3*, 101–111 [fluorescence detection]

# Neomycin

**Molecular formula:** $C_{23}H_{46}N_6O_{13}$ (neomycin B)
**Molecular weight:** 614.7 (neomycin B)
**CAS Registry No.:** 1404-04-2 (neomycin),
1406-04-8 (neomycin undecylenate),
1405-10-3 (neomycin sulfate),
1405-12-5 (neomycin palmitate)

## SAMPLE
**Matrix:** blood, urine
**Sample preparation:** Plasma. 1 mL Plasma + 100 μL 30% trichloroacetic acid in water, vortex, centrifuge at 4° at 3600 g for 30 min. Remove 450 μL of the supernatant and add it to 50 μL buffer, vortex, centrifuge at 4° at 3600 g for 30 min, inject a 5-25 μL aliquot of the supernatant. Urine. Centrifuge urine at 4° at 3600 g for 30 min. Remove a 180 μL aliquot and add it to 20 μL buffer, vortex, inject a 5-25 μL aliquot. (Buffer contained 110 mM sodium 1-pentanesulfonate and 70 mM acetic acid.)

## HPLC VARIABLES
**Guard column:** 20 × 4.6 5 μm Supelguard LC-8-DB (Supelco)
**Column:** 150 × 4.6 5 μm Supelcosil LC-8-DB
**Mobile phase:** MeOH:buffer 1.5:98.5 (Buffer contained 11 mM sodium 1-pentanesulfonate, 56 mM sodium sulfate, and 7 mM acetic acid.)
**Column temperature:** 32.5
**Injection volume:** 5-30
**Detector:** F ex 340 em 455 following post-column derivatization. The column effluent mixed with an o-phthalaldehyde reagent (Pierce) and flowed through a reaction coil (PCR 520, Applied Biosystems) at 33° to the detector.

## CHROMATOGRAM
**Retention time:** 16
**Limit of quantitation:** 250 ng/mL (plasma); 1000 ng/mL (urine)

## KEY WORDS
plasma; cow; pharmacokinetics; post-column reaction

## REFERENCE
Shaikh, B.; Jackson, J.; Guyer, G.; Ravis, W.R. Determination of neomycin in plasma and urine by high-performance liquid chromatography. Application to a preliminary pharmacokinetic study. *J.Chromatogr.*, **1991**, *571*, 189–198

## SAMPLE
**Matrix:** bulk
**Sample preparation:** Dissolve 20 mg neomycin sulfate powder in 100 mL 100 mM pH 8.0 sodium phosphate buffer. Remove a 10 mL aliquot and add it to 10 mL 40 mg/mL 2-naphthalenesulfonyl chloride in MeCN (prepare fresh daily), shake briefly, heat at 100-105° for 10 min, cool to room temperature, add 15 mL IS solution, shake vigorously for 10 min, centrifuge at <300 g for 3-5 min, inject a 50 μL aliquot of the lower organic layer. (Prepare IS solution by dissolving 2 mg prednisolone in a small amount of THF, make up to 100 mL with chloroform.)

## HPLC VARIABLES
**Column:** 125 × 4.6 5 μm P-E HS-5 silica (Perkin-Elmer)

**Mobile phase:** Chloroform:MeOH:acetic acid 95:2.3:2.5
**Flow rate:** 1.7
**Injection volume:** 50
**Detector:** UV 254

## CHROMATOGRAM
**Retention time:** 6.2 (neomycin B), 8.3 (neomycin C)
**Internal standard:** prednisolone (12.3)

## OTHER SUBSTANCES
**Simultaneous:** impurities, neamine

## KEY WORDS
derivatization; normal phase

## REFERENCE
Tsuji, K.; Jenkins, K.M. Derivatization of primary amines by 2-naphthalenesulfonyl chloride for high-performance liquid chromatographic assay of neomycin sulfate. *J.Chromatogr.*, **1986**, *369*, 105–115

## SAMPLE
**Matrix:** formulations
**Sample preparation:** 5 g Ointment + 3 mL MeOH, heat at 55° for 5 min, vortex twice for 20 s, centrifuge at 2000 g for 2 min, discard the supernatant, repeat the MeOH wash twice more, add 30 mL chloroform, heat at 55°, vortex for 15 s, add 10 mL MeOH:water 20:80, shake vigorously for 20 min, centrifuge at 1000 g for 3 min, remove the upper aqueous layer, repeat the MeOH/water extraction twice more. Combine the aqueous layers and make up to 50 mL with 20 mM pH 9.0 borate buffer. Remove a 10 mL aliquot and add it to 15 mL 150 mM 2,4-dinitrofluorobenzene in MeOH, heat at 100° for 45 min, cool, make up to 250 mL with mobile phase, inject an aliquot of the yellow organic layer.

## HPLC VARIABLES
**Column:** 250 × 4.6 5 μm LiChrosorb silica SI-100
**Mobile phase:** THF:chloroform:water:glacial acetic acid 39.2:59.8:0.8:0.2
**Flow rate:** 1
**Injection volume:** 40
**Detector:** UV 254

## CHROMATOGRAM
**Retention time:** 4 (neomycin C), 12 (neomycin B)

## OTHER SUBSTANCES
**Simultaneous:** hydrocortisone acetate
**Noninterfering:** cortisone acetate, fluorometholone, methylprednisolone, prednisolone acetate

## KEY WORDS
normal phase; ointment; derivatization

## REFERENCE
Binns, R.B.; Tsuji, K. High-performance liquid chromatographic analysis of neomycin in petrolatum-based ointments and in veterinary formulations. *J.Pharm.Sci.*, **1984**, *73*, 69–72

## SAMPLE
**Matrix:** formulations

**Sample preparation:** Powder. Dissolve 50 mg powder in 25 mL 20 mM pH 9.0 borate buffer, remove a 5 mL aliquot and add it to 15 mL 150 mM 2,4-dinitrofluorobenzene in MeOH (prepare fresh daily), heat at 100° for 45 min, cool, make up to 250 mL with mobile phase, discard the upper aqueous phase, inject a 20 μL aliquot of the lower organic phase. Ointment. Weigh out 5 g ointment, add 25 mL chloroform, heat at 60° until the ointment dissolved, shake vigorously for 15 min, centrifuge at 4000 g for 15 min, discard the chloroform, add 25 mL chloroform to the solid, centrifuge, discard the chloroform, add 10 mL 20 mM pH 9.0 borate buffer and 15 mL n-heptane to the solid, shake vigorously for 10 min, remove a 5 mL aliquot of the aqueous phase and add it to 15 mL 150 mM 2,4-dinitrofluorobenzene in MeOH (prepare fresh daily), heat at 100° for 45 min, cool, make up to 250 mL with mobile phase, discard the upper aqueous phase, inject a 20 μL aliquot of the lower organic phase.

## HPLC VARIABLES
**Column:** 250 × 4.6 5 μm LiChrosorb SI-100
**Mobile phase:** Chloroform:THF:water 60:39.2:0.8
**Flow rate:** 1
**Injection volume:** 20
**Detector:** UV 350

## CHROMATOGRAM
**Retention time:** 10 (neomycin C), 14 (neomycin B)
**Limit of detection:** 1 ng

## KEY WORDS
powder; ointment; normal phase; derivatization

## REFERENCE
Tsuji, K.; Goetz, J.F.; VanMeter, W.; Gusciora, K.A. Normal-phase high-performance liquid chromatographic determination of neomycin sulfate derivatized with 1-fluoro-2,4-dinitrobenzene. *J.Chromatogr.*, **1979**, *175*, 141–152

## SAMPLE
**Matrix:** milk
**Sample preparation:** Prepare a SPE column as follows. Shake Amberlite CG 50 resin in buffer, equilibrate for 2 h, fill a plugged Pasteur pipette to a height of 35 mm with the slurry, wash with water until the eluent is neutral. 40 mL Milk + 2 g NaCl, shake well, add a 10 mL aliquot to the SPE column, wash with 8 mL water, add 600 μL reagent, let stand for 2 min, elute with 3 mL MeOH:buffer 80:20, make up the eluate to 4 mL with MeOH, store at -8 to -10°, after 15 min inject a 10-20 μL aliquot. (Prepare buffer by dissolving 76 g potassium tetraborate in 400 mL water, adjusting pH to 11 with KOH, and making up to 500 mL with water. Prepare reagent by dissolving 100 mg o-phthalaldehyde in 10 mL MeOH and adding 200 μL 2-mercaptoethanol and 10 mL borate buffer. Prepare borate buffer by dissolving 3.1 g boric acid in 100 mL water and adjusting the pH to 10.5 with 50% KOH, make up to 125 mL with water.)

## HPLC VARIABLES
**Column:** 150 × 4.6 5 μm HISEP (Supelco)
**Mobile phase:** Gradient. A was 2 g/L tripotassium EDTA in MeOH:water 70:30. B was MeOH. A:B 100:0 to 40:60 over 15 min (LDC concave curve 4). (This curve holds the initial conditions for about 6 min.)
**Flow rate:** 1.7
**Injection volume:** 10-20
**Detector:** F ex 340 em KV418 filter

## CHROMATOGRAM
**Retention time:** 8.2, 19.8
**Limit of detection:** 50 ppb

## KEY WORDS
cow; derivatization; SPE

## REFERENCE
Agarwal, V.K. High performance liquid chromatographic determination of neomycin in milk using a HISEP column. *J.Liq.Chromatogr.*, **1990**, *13*, 2475–2487

## SAMPLE
**Matrix:** milk
**Sample preparation:** 1 mL Skim milk + 100 μL 20% trichloroacetic acid, vortex, centrifuge at 4° at 4000 rpm for 30 min. Remove a 180 μL aliquot and add it to 20 μL 100 mM sodium 1-pentanesulfonate containing 70 mM acetic acid, vortex, inject a 25 μL aliquot.

## HPLC VARIABLES
**Guard column:** 20 × 4.6 5 μm Supelguard LC-8-DB
**Column:** 150 × 4.6 5 μm Supelcosil LC-8-DB
**Mobile phase:** MeOH:buffer 1.5:98.5 (Buffer contained 10 mM pentanesulfonic acid, 56 mM sodium sulfate, and 7 mM acetic acid.)
**Column temperature:** 32.5
**Injection volume:** 25
**Detector:** F ex 340 em 455 following post-column reaction. The column effluent mixed with o-phthalaldehyde solution (Pierce) and flowed through a reaction coil at 33° to the detector.

## CHROMATOGRAM
**Retention time:** 20
**Limit of detection:** 150 ng/mL

## KEY WORDS
cow; derivatization; post-column reaction

## REFERENCE
Shaikh, B.; Jackson, J. Determination of neomycin in milk by reversed phase ion-pairing liquid chromatography. *J.Liq.Chromatogr.*, **1989**, *12*, 1497–1515

## SAMPLE
**Matrix:** perilymph
**Sample preparation:** Lyophilize 6 μL perilymph, reconstitute with 90 μL pyridine, add 10 μL benzoyl chloride, heat at 80° for 30 min, evaporate to dryness under a stream of nitrogen, add 1 mL MeOH, heat at 80° for 10 min, add 50 mg solid sodium carbonate, add 1 mL MeOH saturated with sodium carbonate, wash 3 times with 2 mL portions of n-hexane, add 1 mL water, remove any hexane which separates, extract with 3 mL chloroform. Wash the chloroform layer 3 times with 1 mL portions of MeOH:water 50:50, evaporate the chloroform layer to dryness, reconstitute with 15 μL chloroform, inject a 5 μL aliquot.

## HPLC VARIABLES
**Column:** 250 × 4.6 5-6 μm Zorbax SIL
**Mobile phase:** n-Hexane:THF 50:50
**Flow rate:** 2
**Injection volume:** 5
**Detector:** UV 230

## CHROMATOGRAM
**Retention time:** 8.11, 9.28, 10.08
**Limit of detection:** 10 ng

## OTHER SUBSTANCES
**Simultaneous:** dihydrostreptomycin, kanamycin, streptomycin

## KEY WORDS
derivatization; normal phase; guinea pig

## REFERENCE
Harada, T.; Iwamori, M.; Nagai, Y.; Nomura, Y. Analysis of aminoglycoside antibiotics as benzoyl derivatives by high-performance liquid chromatography and its application to the quantitation of neomycin in the perilymph. *J.Chromatogr.*, **1985**, *337*, 187–193

## SAMPLE
**Matrix:** solutions
**Sample preparation:** Inject a 20 μL aliquot of a solution in mobile phase.

## HPLC VARIABLES
**Column:** 150 × 4.6 5 μm Supelcosil LC-8
**Mobile phase:** THF:buffer 3:97 (Buffer was 7 mM acetic acid containing 5.6 mM sodium sulfate and 10 mM sodium pentanesulfonate.)
**Flow rate:** 1.75
**Injection volume:** 20
**Detector:** F ex 365 em 418 following post-column reaction. The column effluent mixed with the reagent pumped at 0.4 mL/min and the mixture flowed through a 50 × 4.6 column packed with 75 μm glass beads and a 3.05 m × 0.5 mm knitted PTFE coil at 40° to the detector. (The reagent was 400 mM boric acid containing 380 mM KOH, 6 mL/L 40% Brij-35, 4 mL/L mercaptoethanol, and 0.8 g/L orthophthalaldehyde.)

## CHROMATOGRAM
**Retention time:** 5

## KEY WORDS
post-column reaction

## REFERENCE
*Supelco Chromatography Products, Supelco, Inc., Bellefonte PA,* **1996**, p. A29

## SAMPLE
**Matrix:** solutions

## HPLC VARIABLES
**Column:** 150 × 4.5 5 μm Ultrasphere octyl
**Mobile phase:** MeCN:triethylamine:1.65% glacial acetic acid 505:0.65:495, pH 4.35
**Column temperature:** 30
**Flow rate:** 1
**Injection volume:** 20
**Detector:** UV 280

## CHROMATOGRAM
**Retention time:** 1.99
**Internal standard:** naproxen (3.89)

## OTHER SUBSTANCES
**Simultaneous:** bacitracin, cortisone acetate, diazepam, diclofenac, fluorometholone, flurbiprofen, hydrocortisone acetate, imipramine, indomethacin, ketoprofen, ketorolac tromethamine, levobunolol, meclofenamic acid, prednisolone acetate, proparacaine, salicylic acid, sulfacetamide, suprofen

**Noninterfering:** acebutolol, acetaminophen, acetazolamide, alprenolol, apraclonidine, atenolol, atropine, betamethasone, betaxolol, bupivacaine, caffeine, cyclopentolate, dexamethasone, diphenhydramine, erythromycin, haloperidol, lidocaine, phenylephrine, polymyxin B, procaine, scopolamine, timolol, tropicamide
**Interfering:** metipranolol, propranolol

## KEY WORDS
human; rabbit

## REFERENCE
Riegel, M.; Ellis, P.P. High-performance liquid chromatography assay for antiinflammatory agents diclofenac and flurbiprofen in ocular fluids. *J.Chromatogr.B*, **1994**, *654*, 140–145

## SAMPLE
**Matrix:** solutions
**Sample preparation:** 50 μL Buffered reaction mixture + 50 μL isopropanol + 50 μL reagent, heat at 60° for 10 min, centrifuge at 1000 g for 2 min, immediately inject a 50 μL aliquot of the supernatant. (Reagent was 80 mM o-phthalaldehyde and 250 mM thioglycolic acid in 1 M boric acid, pH adjusted to 10.4 with 40% KOH.)

## HPLC VARIABLES
**Column:** 100 × 5 Hypersil ODS
**Mobile phase:** MeOH:water:acetic acid 63.75:31.25:5 containing 5 g/L heptanesulfonic acid
**Flow rate:** 2
**Injection volume:** 50
**Detector:** UV 330

## CHROMATOGRAM
**Retention time:** 21

## KEY WORDS
reaction mixtures; derivatization

## REFERENCE
Lovering, A.M.; White, L.O.; Reeves, D.S. Identification of aminoglycoside-acetylating enzymes by high-pressure liquid chromatographic determination of their reaction products. *Antimicrob.Agents Chemother.*, **1984**, *26*, 10–12

## SAMPLE
**Matrix:** solutions

## HPLC VARIABLES
**Column:** 300 × 3.9 10 μm μBondapak C18
**Mobile phase:** MeCN:water 17:83 containing 16 mM sodium 1-hexanesulfonate and 20 mM Na$_3$PO$_4$, pH 3.5 (Connect a 250 × 4.6 column of Bondapak C18/Corasil or Co:Pell ODS between pump and injector. Flush column with MeOH:water 50:50 at the end of the day.)
**Column temperature:** 25
**Flow rate:** 1.5
**Injection volume:** 25
**Detector:** RI

## CHROMATOGRAM
**Retention time:** 8

## OTHER SUBSTANCES
**Simultaneous:** paromomycin

## REFERENCE
Whall, T.J. Determination of streptomycin sulfate and dihydrostreptomycin sulfate by high-performance liquid chromatography. *J.Chromatogr.*, **1981**, *219*, 89–100

## SAMPLE
**Matrix:** tissue
**Sample preparation:** Homogenize (Tissumizer) 1 g Ground tissue + 4 mL buffer at medium speed for 1 min, centrifuge at 3600 g for 20 min, remove the supernatant, re-homogenize pellet in 4 mL buffer for 10 min, centrifuge. Combine the supernatants, heat in a boiling water bath with occasional mixing for 5 min, centrifuge at 2000 g for 20 min, remove the supernatant, vortex the precipitate with 2 mL buffer for 30 s, centrifuge at 2000 g for 10 min. Combine the supernatants, acidify to pH 3.5-4 with 50-60 μL sulfuric acid, centrifuge at 2000 g for 10 min, inject an aliquot of the supernatant. (Buffer was 33.46 g $K_2HPO_4$ and 1.046 g $KH_2PO_4$ in 1 L water, pH 8.0.)

## HPLC VARIABLES
**Guard column:** 10 μm RP-18
**Column:** 150 × 4.6 5 μm Supelcosil LC-8-DB
**Mobile phase:** MeOH:buffer 1.5:98.5 (Buffer was 10 mM sodium 1-pentanesulfonate, 56 mM sodium sulfate, and 7 mM acetic acid.)
**Flow rate:** 1.5
**Detector:** F ex 340 em 455 following post-column reaction with derivatization reagent pumped at 0.9 mL/min. (Derivatization reagent was commercially available (Pierce) or prepared by adding 2.5 mL 2-mercaptoethanol and 2.5 mL Brij-35 to 850 mg o-phthalaldehyde in 10 mL MeOH, mix until decolorization is complete, add 1 L buffer, filter (0.45 μm), and refrigerate until used. Buffer was prepared by adjusting pH of 250 mM boric acid to 9.5 with 5 M KOH.)

## CHROMATOGRAM
**Retention time:** 22
**Limit of detection:** 3.5 ng

## OTHER SUBSTANCES
**Extracted:** paromomycin
**Simultaneous:** dihydrostreptomycin, streptomycin

## KEY WORDS
kidney; muscle; cow; pig; post-column reaction

## REFERENCE
Shaikh, B.; Allen, E.H.; Gridley, J.C. Determination of neomycin in animal tissues, using ion-pair liquid chromatography with fluorometric detection. *J.Assoc.Off.Anal.Chem.*, **1985**, *68*, 29–36

# Nifedipine

**Molecular formula:** $C_{17}H_{18}N_2O_6$
**Molecular weight:** 346.3
**CAS Registry No.:** 21829-25-4

## SAMPLE
**Matrix:** blood
**Sample preparation:** 2 mL Whole blood or plasma + 2 mL buffer + 5 mL chloroform: isopropanol:n-heptane 60:14:26, shake gently horizontally for 10 min, centrifuge at 2800 g for 10 min. Remove the lower organic layer and evaporate it to dryness under vacuum at 45°, reconstitute the residue in 100 μL mobile phase, centrifuge at 2800 g for 5 min, inject a 50 μL aliquot of the supernatant. (Buffer was saturated ammonium chloride solution 25% diluted with water, adjusted to pH 9.5 with 25% ammonia solution.)

## HPLC VARIABLES
**Column:** 300 × 3.9 4 μm NovaPack C18
**Mobile phase:** MeOH:THF:buffer 65:5:30 (Buffer was 0.68 g/L (10 mM (sic)) $KH_2PO_4$ adjusted to pH 2.6 with concentrated orthophosphoric acid.) (At the end of each session wash the column with water for 1 h and MeOH for 1 h, re-equilibrate for 30 min.)
**Column temperature:** 30
**Flow rate:** 0.8
**Injection volume:** 50
**Detector:** UV 236

## CHROMATOGRAM
**Retention time:** 4.21
**Limit of detection:** <120 ng/mL

## OTHER SUBSTANCES
**Extracted:** acebutolol, acenocoumarol, acepromazine, aceprometazine, acetaminophen, aconitine, albuterol, alimemazine, alminoprofen, alpidem, alprenolol, amisulpride, amitriptyline, amodiaquine, amoxapine, aspirin, astemizole, atenolol, benazepril, benperidol, benzocaine, benzoylecgonine, bepridil, betaxolol, bisoprolol, bromazepam, brompheniramine, bumadizone, bupivacaine, buprenorphine, buspirone, caffeine, carbamazepine, carbinoxamine, carpipramine, carteolol, cetirizine, chlorambucil, chlordiazepoxide, chlormezanone, chlorophenacinone, chloroquine, chlorpheniramine, chlorpromazine, chlorpropamide, cibenzoline, cicletanine, clemastine, clobazam, clomipramine, clonazepam, clonidine, clorazepate, clozapine, cocaine, codeine, colchicine, cyamemazine, cyclizine, cyproheptadine, cytarabine, dacarbazine, daunorubicin, demexiptiline, desipramine, dextromoramide, dextropropoxyphene, diazepam, diazoxide, diclofenac, dihydralazine, diltiazem, diphenhydramine, dipyridamole, dosulepine, doxepin, doxylamine, droperidol, ephedrine, estazolam, etodolac, fenfluramine, fenoprofen, fentiazac, flecainide, floctafenine, flumazenil, flunitrazepam, fluoxetine, fluphenazine, flurbiprofen, fluvoxamine, glibenclamine, glibornuride, glipizide, glutethimide, haloperidol, histapyrrodine, hydroxychloroquine, hydroxyzine, ibuprofen, imipramine, indomethacin, iproniazid, ketoprofen, labetalol, levomepromazine, lidocaine, lidoflazine, lisinopril, loperamide, loprazolam, loratadine, loxapine, maprotiline, medazepam, medifoxamine, mefenamic acid, mefenidramine, mefloquine, melphalan, meperidine, mephenesin, mepivacaine, metapramine, metformin, methadone, methocarbamol, methotrexate, metipranolol, metoclopramide, mexiletine, mianserine, midazolam, moclobemide, moperone, morphine, nadolol, nalbuphine, nalorphine, naloxone, naltrexone, naproxen, nialamide, nicardipine, niflumic acid, nimodipine, nitrendipine, nizatidine, nomifensine, nortriptyline, omeprazole, opipramol, oxprenolol, penbutolol, penfluridol, phencyclidine, phenobarbital, phenol, phenylbutazone,

pimozide, pindolol, pipamperone, prazepam, prazosin, procainide, procarbazine, proguanil, promethazine, propafenone, propranolol, protriptyline, pyrimethamine, quinupramine, ramipril, ranitidine, reserpine, ritodrine, secobarbital, sotalol, strychnine, sulfinpyrazole, sulpride, sultopride, suriclone, temazepam, tenoxicam, terbutaline, terfenadine,
tetracaine, tetrazepam, thiopental, thioproperazine, thioridazine, tianeptine, tiapride, ticlopidine, timolol, tioclomarol, tolbutamide, toloxatone, trazodone, triazolam, trifluoperazine, trifluperidol, trimipramine, triprolidine, tropatenine, verapamil, viloxazine, vinblastine, vincristine, vindesine, warfarin, zolpidem, zorubicine

**Interfering:** ajmaline, alprazolam, celiprolol, cycloguanil, debrisoquine, dextromethorphan,
disopyramide, ketamine, lorazepam, mephentermine, methaqualone, metoprolol, minoxidil, nitrazepam, oxazepam, pentazocine, piroxicam, prilocaine, quinidine, quinine, sulindac, tiaprofenic acid, tofisopam, yohimbine, zopiclone

## KEY WORDS
whole blood; plasma

## REFERENCE
Tracqui, A.; Kintz, P.; Mangin, P. Systematic toxicological analysis using HPLC/DAD. *J.Forensic Sci.*,
**1995**, *40*, 254–262

## SAMPLE
**Matrix:** blood
**Sample preparation:** 1 mL Plasma + 50 μL 10 μg/mL nisoldipine in MeOH + 100 μL 1
M NaOH, vortex for 3 s, add 5 mL 75:25 MTBE:isooctane, vortex for 30 s, centrifuge at
1800 g for 5 min. Transfer upper layer to a 100 × 13 mm glass tube, evaporate to dryness
without heating using a Speed Vac concentrator evaporator. Reconstitute in 200 μL mobile
phase, vortex for 15 s, inject a 150 μL aliquot.

## HPLC VARIABLES
**Column:** 100 × 8 4μm C8 Nova-Pak radial pack
**Mobile phase:** MeOH:water 65:35 adjusted to approx. pH 4 with 1% acetic acid and 0.03%
triethylamine
**Flow rate:** 1.1
**Injection volume:** 150
**Detector:** UV 350

## CHROMATOGRAM
**Retention time:** 6.5
**Internal standard:** nisoldipine (16)
**Limit of detection:** 1 ng/mL
**Limit of quantitation:** 5 ng/mL

## KEY WORDS
plasma; analysis conducted under sodium lamps

## REFERENCE
Grundy, J.S.; Kherani, R.; Foster, R.T. Sensitive high-performance liquid chromatographic assay for
nifedipine in human plasma utilizing ultraviolet detection. *J.Chromatogr.B*, **1994**, *654*, 146–151

## SAMPLE
**Matrix:** blood
**Sample preparation:** 1 mL Serum + 150 ng nitrazepam + 100 μL 1 M NaOH + 5 mL
hexane:dichloromethane 70:30, shake horizontally for 20 min, centrifuge for 10 min.
Remove the organic layer and evaporate it to dryness under a stream of nitrogen at 40°,
reconstitute the residue in 120 μL mobile phase, inject a 20 μL aliquot.

## HPLC VARIABLES
**Column:** 250 × 4.6 10 μm RP-18 (Kontron)
**Mobile phase:** MeOH:water:glacial acetic acid 70:30:1
**Flow rate:** 1.5
**Injection volume:** 20
**Detector:** UV 238

## CHROMATOGRAM
**Retention time:** 4.2
**Internal standard:** nitrazepam (3.8)
**Limit of detection:** 5 ng/mL
**Limit of quantitation:** 10 ng/mL

## KEY WORDS
serum; protect from light; pharmacokinetics

## REFERENCE
Jankowski, A.; Lamparczyk, H. Evaluation of chromatographic methods for the determination of nifedipine in human serum. *J.Chromatogr.A*, **1994**, *668*, 469–473

## SAMPLE
**Matrix:** blood
**Sample preparation:** Centrifuge plasma, inject a 650 μL aliquot onto column A and elute to waste with mobile phase A, after 45 s inject 650 μL MeOH:water 20:80, continue to elute column A to waste with mobile phase A, after 45 s backflush the contents of column A onto column B with mobile phase B, after 1 min remove column A from the circuit, elute column B with mobile phase B, monitor the effluent from column B. Re-equilibrate column A with mobile phase A for at least 45 s.

## HPLC VARIABLES
**Column:** A 10 × 4 25-40 μm LiChroprep RP-2; B 125 × 4 5 μm Hypersil ODS
**Mobile phase:** A 20 mM Ammonium carbamate; B MeCN:20 mM pH 4.0 acetate buffer 50:50
**Flow rate:** A 3; B 1
**Injection volume:** 650
**Detector:** UV 355

## CHROMATOGRAM
**Retention time:** 2.4
**Limit of detection:** 0.5 ng/mL

## KEY WORDS
plasma; pharmacokinetics; column-switching; protect from light

## REFERENCE
Nitsche, V.; Schütz, H.; Eichinger, A. Rapid high-performance liquid chromatographic determination of nifedipine in plasma with on-line precolumn solid-phase extraction. *J.Chromatogr.*, **1987**, *420*, 207–211

## SAMPLE
**Matrix:** blood
**Sample preparation:** 1 mL Serum + 50 μL 4 μg/mL nisoldipine in mobile phase, adjust to pH 9 with 1 M KOH, add 0.5 g KCl, add 3 mL diethyl ether, shake gently (Fisher Roto-Rack) for 10 min, centrifuge at 1000 rpm for 15 min. Separate and retain ether layer, extract aqueous layer as before. Evaporate all ether layers to dryness under dry nitrogen

at 20°. Reconstitute the residue in 50 μL mobile phase, inject a 20 μL aliquot. (Wrap tubes in Al foil.)

## HPLC VARIABLES
**Column:** 250 × 4 Brownlee 10 μm RP8
**Mobile phase:** MeOH:50 mM buffer 30:70 (Buffer was 12 g $NaH_2PO_4$ in 2 L water adjusted to pH 4 ± 0.5 with 80% phosphoric acid.)
**Flow rate:** 0.75
**Injection volume:** 20
**Detector:** UV 350

## CHROMATOGRAM
**Retention time:** 7
**Internal standard:** nisoldipine (13)
**Limit of detection:** 10 ng/mL

## OTHER SUBSTANCES
**Simultaneous:** photodegradation products of nifedipine, furosemide, hydrochlorothiazide, timolol, triamterene
**Noninterfering:** aspirin, acetaminophen, caffeine, chlorthalidone, diazepam, methyldopa, oxprenolol, propranolol

## KEY WORDS
serum

## REFERENCE
Snedden, W.; Fernandez, P.G.; Galway, B.A.; Kim, B.K. Specific HPLC assay for serum nifedipine. *Clin.Invest.Med.*, **1984**, 7, 173–178

## SAMPLE
**Matrix:** perfusate
**Sample preparation:** Add perfusate to an equal volume of nitrendipine in MeOH, vortex, centrifuge, inject an aliquot.

## HPLC VARIABLES
**Column:** 250 × 4.6 Nucleosil 5C18
**Mobile phase:** MeCN:water 40:60
**Detector:** UV 235

## CHROMATOGRAM
**Internal standard:** nitrendipine

## KEY WORDS
protect from light

## REFERENCE
Kobayashi, D.; Matsuzawa, T.; Sugibayashi, K.; Morimoto, Y.; Kobayashi, M.; Kimura, M. Feasibility of use of several cardiovascular agents in transdermal therapeutic systems with *l*-menthol-ethanol system on hairless rat and human skin. *Biol.Pharm.Bull.*, **1993**, *16*, 254–258

## SAMPLE
**Matrix:** solutions
**Sample preparation:** Inject a 5 μL aliquot.

## HPLC VARIABLES
**Column:** 33 × 4.6 3 μm Supelcosil LC-18-DB

**Mobile phase:** Gradient. MeCN:buffer from 10:90 to 50:50 over 10 min, maintain at 50:50 for 2 min. (Buffer was 25 mM $KH_2PO_4$ containing 0.02% triethylamine, pH 3.0.)
**Column temperature:** 25
**Flow rate:** 2
**Injection volume:** 5
**Detector:** UV 220

## CHROMATOGRAM
**Retention time:** 8.2

## OTHER SUBSTANCES
**Simultaneous:** digoxin, diltiazem, dipyridamole, dobutamine, flunarizine, lidoflazine, oxprenolol, pindolol, verapamil

## REFERENCE
*Supelco Chromatography Products, Supelco, Inc., Bellefonte PA,* **1996**, p. A27

## SAMPLE
**Matrix:** solutions

## HPLC VARIABLES
**Column:** 250 × 4.6 5 μm Supelcosil LC-DP (A) or 250 × 4 5 μm LiChrospher 100 RP-8 (B)
**Mobile phase:** MeCN:0.025% phosphoric acid:buffer 25:10:5 (A) or 60:25:15 (B) (Buffer was 9 mL concentrated phosphoric acid and 10 mL triethylamine in 900 mL water, adjust pH to 3.4 with dilute phosphoric acid, make up to 1 L.)
**Flow rate:** 0.6
**Injection volume:** 25
**Detector:** UV 229

## CHROMATOGRAM
**Retention time:** 7.42 (A), 7.91 (B)

## OTHER SUBSTANCES
**Also analyzed:** acebutolol, acepromazine, acetaminophen, acetazolamide, acetophenazine, albuterol, alprazolam, amitriptyline, amobarbital, amoxapine, antipyrine, atenolol, atropine, azatadine, baclofen, benzocaine, bromocriptine, brompheniramine, brotizolam, bupivacaine, buspirone, butabarbital, butalbital, caffeine, carbamazepine, cetirizine, chlorcyclizine, chlordiazepoxide, chlormezanone, chloroquine, chlorpheniramine, chlorpromazine, chlorpropamide, chlorprothixene, chlorthalidone, chlorzoxazone, cimetidine, cisapride, clomipramine, clonazepam, clonidine, clozapine, cocaine, codeine, colchicine, cyclizine, cyclobenzaprine, dantrolene, desipramine, diazepam, diclofenac, diflunisal, diltiazem, diphenhydramine, diphenidol, diphenoxylate, dipyridamole, disopyramide, dobutamine, doxapram, doxepin, droperidol, encainide, ethidium bromide, ethopropazine, fenoprofen, fentanyl, flavoxate, fluoxetine, fluphenazine, flurazepam, flurbiprofen, fluvoxamine, furosemide, glutethimide, glyburide, guaifenesin, haloperidol, homatropine, hydralazine, hydrochlorothiazide, hydrocodone, hydromorphone, hydroxychloroquine, hydroxyzine, ibuprofen, imipramine, indomethacin, ketoconazole, ketoprofen, ketorolac, labetalol, levorphanol, lidocaine, loratadine, lorazepam, lovastatin, loxapine, mazindol, mefenamic acid, meperidine, mephenytoin, mepivacaine, mesoridazine, metaproterenol, metformin, methadone, methdilazine, methocarbamol, methotrexate, methotrimeprazine, methoxamine, methyldopa, methylphenidate, metoclopramide, metolazone, metoprolol, metronidazole, midazolam, moclobemide, morphine, nadolol, nalbuphine, naloxone, naphazoline, naproxen, nizatidine, norepinephrine, nortriptyline, oxazepam, oxycodone, oxymetazoline, paroxetine, pemoline, pentazocine, pentobarbital, pentoxifylline, perphenazine, pheniramine, phenobarbital, phenol, phenolphthalein, phentolamine, phenylbutazone, phenyltoloxamine, phenytoin, pimozide, pindolol, piroxicam, pramoxine, prazepam, prazosin, probenecid, procainamide, procaine, prochlorperazine, procyclidine, promazine,

promethazine, propafenone, propantheline, propiomazine, propofol, propranolol, protriptyline, quazepam, quinidine, quinine, racemethorphan, ranitidine, remoxipride, risperidone, salicylic acid, scopolamine, secobarbital, sertraline, sotalol, spironolactone, sulfinpyrazone, sulindac, temazepam, terbutaline, terfenadine, tetracaine, theophylline, thiethylperazine, thiopental, thioridazine, thiothixene, timolol, tocainide, tolbutamide, tolmetin, trazodone, triamterene, triazolam, trifluoperazine, triflupromazine, trimeprazine, trimethoprim, trimipramine, verapamil, warfarin, xylometazoline, yohimbine, zopiclone

## KEY WORDS
some details of plasma extraction

## REFERENCE
Koves, E.M. Use of high-performance liquid chromatography-diode array detection in forensic toxicology. *J.Chromatogr.A*, **1995**, *692*, 103–119

## SAMPLE
**Matrix:** solutions

## HPLC VARIABLES
**Column:** 250 × 4.6 Zorbax RX
**Mobile phase:** Gradient. A was 10 mL concentrated orthophosphoric acid and 7 mL triethylamine in 1 L water. B was 10 mL concentrated orthophosphoric acid and 7 mL triethylamine in 200 mL water, make up to 1 L with MeCN. A:B from 100:0 to 0:100 over 30 min, maintain at 0:100 for 5 min.
**Column temperature:** 30
**Flow rate:** 2
**Detector:** UV 210

## OTHER SUBSTANCES
**Also analyzed:** acepromazine, acetaminophen, acetophenazine, albuterol, aminophylline, amitriptyline, amobarbital, amoxapine, amphetamine, amylocaine, antipyrine, aprobarbital, aspirin, atenolol, atropine, avermectin, barbital, benzocaine, benzoic acid, benzotropine, benzphetamine, berberine, bibucaine, bromazepan, brompheniramine, buprenorphine, buspirone, butabarbital, butacaine, butethal, caffeine, carbamazepine, carbromal, chloramphenicol, chlordiazepoxide, chloroquine, chlorothiazide, chloroxylenol, chlorphenesin, chlorpheniramine, chlorpromazine, chlorpropamide, chlortetracycline, cimetidine, cinchonidine, cinchonine, clenbuterol, clonazepam, clonixin, clorazepate, cocaine, codeine, colchicine, cortisone, coumarin, cyclazocine, cyclobenzaprine, cyclothiazide, cyheptamide, cymarin, danazol, danthron, dapsone, debrisoquine, desipramine, dexamethasone, dextromethorphan, dextropropoxyphene, diamorphine, diazepam, diclofenac, diethylpropion, diethylstilbestrol, diflunisal, digitoxin, digoxin, diltiazem, diphenhydramine, diphenoxylate, diprenorphine, dipyrone, disulfiram, dopamine, doxapram, doxepin, dronabinol, ephedrine, epinephrine, epinine, estradiol, estriol, estrone, ethacrynic acid, ethosuximide, etonitazene, etorphine, eugenol, famotidine, fenbendazole, fencamfamine, fenoprofen, fenproporex, fentanyl, flubendazole, flufenamic acid, flunitrazepam, 5-fluorouracil, fluoxymesterone, fluphenazine, furosemide, gentisic acid, gitoxigenin, glipizide, glunixin, glutethimide, glybenclamide, guaiacol, halazepam, haloperidol, hydrochlorothiazide, hydrocodone, hydrocortisone, hydromorphone, hydroxyquinoline, ibogaine, ibuprofen, iminostilbene, imipramine, indomethacin, isocarbostyril, isocarboxazid, isoniazid, isoproterenol, isoxsuprine, ivermectin, ketamine, ketoprofen, kynurenic acid, levorphanol, lidocaine, lorazepam, lormetazepam, loxapine, mazindol, mebendazole, meclizine, meclofenamic acid, medazepam, mefenamic acid, megestrol, mepacrine, meperidine, mephentermine, mephenytoin, mephesin, mephobarbital, mepivacaine, mescaline, mesoridazine, methadone, methamphetamine, methapyrilene, methaqualone, methazolamide, methocarbamol, methoxamine, methsuximide, methyl salicylate, methyldopa, methyldopamine, methylphenidate, methylprednisolone, methyltestosterone, methyprylon, meto-

prolol, mibolerone, morphine, nadolol, nalorphine, naloxone, naltrexone, naphazoline, naproxen, nefopam, niacinamide, nicotine, nicotinic acid, niflumic acid, nitrazepam, norepinephrine, nortriptyline, noscapine, nylidrin, oxazepam, oxycodone, oxymorphone, oxyphenbutazone, oxytetracycline, papaverine, pargyline, pemoline, pentazocine, pentobarbital, persantine, phenacetin, phenazocine, phenazopyridine, phencyclidine, phendimetrazine, phenelzine, pheniramine, phenobarbital, phenothiazine, phensuximide, phentermine, phenylbutazone, phenylephrine, phenylpropanolamine, piperocaine, prazepam, prednisolone, primidone, probenecid, progesterone, propiomazine, propranolol, propylparaben, pseudoephedrine, puromycin, pyrilamine, pyrithyldione, quazepam, quinaldic acid, quinidine, quinine, ranitidine, recinnamine, reserpine, resorcinol, saccharin, albuterol, salicylamide, salicylic acid, scopolamine, scopoletin, secobarbital, strychnine, sulfacetamide, sufadiazine, sulfadimethoxine, sulfaethidole, sulfamerazine, sulfamethazine, sulfamethoxizole, sulfanilamide, sulfapyridine, sulfasoxizole, sulindac, tamoxifen, temazepam, testosterone, tetracaine, tetracycline, tetramisole, thebaine, theobromine, theophylline, thiabendazole, thiamine, thiamylal, thiobarbituric acid, thioridazine, thiosalicylic acid, thiothixene, thymol, tolazamide, tolazoline, tobutamide, tolmetin, tranylcypromine, triamcinolone, tribenzylamine, trichloromethiazide, trifluoperazine, trihexyphenidyl, trimethoprim, tripelennamine, triprolidine, tropacocaine, tyramine, verapamil, vincamine, warfarin, yohimbine, zoxazolamine

## REFERENCE
Hill, D.W.; Kind, A.J. Reversed-phase solvent gradient HPLC retention indexes of drugs. *J.Anal.Toxicol.*, **1994**, *18*, 233–242

## SAMPLE
**Matrix:** solutions
**Sample preparation:** Prepare a solution in MeCN:water 40:60, inject an aliquot.

## HPLC VARIABLES
**Column:** 250 × 4.6 3 μm silica (Phenomenex)
**Mobile phase:** MeCN:6.25 mM pH 3.0 phosphate buffer 40:60
**Flow rate:** 1
**Injection volume:** 50
**Detector:** UV 254

## CHROMATOGRAM
**Retention time:** 3.24

## OTHER SUBSTANCES
**Also analyzed:** atenolol, clonidine, diltiazem, metoprolol, prazosin, propranolol, verapamil

## REFERENCE
Simmons, B.R.; Stewart, J.T. HPLC separation of selected cardiovascular agents on underivatized silica using an aqueous organic mobile phase. *J.Liq.Chromatogr.*, **1994**, *17*, 2675–2690

## SAMPLE
**Matrix:** solutions
**Sample preparation:** Centrifuge at 2000 g at 37° for 15 min, inject an aliquot.

## HPLC VARIABLES
**Column:** 100 × 8 5 μm C18 Novapak
**Mobile phase:** MeCN:10 mM pH 4.5 phosphate buffer 70:30
**Flow rate:** 2
**Detector:** UV 237

## OTHER SUBSTANCES
**Also analyzed:** felodipine, nicardipine, nimodipine, nitrendipine

## KEY WORDS
buffers

## REFERENCE
Diez, I.; Colom, H.; Moreno, J.; Obach, R.; Peraire, C.; Domenech, J. A comparative in vitro study of transdermal absorption of a series of calcium channel antagonists. *J.Pharm.Sci.*, **1991**, *80*, 931–934

## SAMPLE
**Matrix:** solutions
**Sample preparation:** Inject a 10 µL aliquot of a solution in MeCN.

## HPLC VARIABLES
**Column:** 100 × 4.6 CS-MP Spheri-5 cyano
**Mobile phase:** Gradient. MeCN:buffer from 10:90 to 40:60 over 10 min, re-equilibrate for 5 min. (Buffer was 50 mM $KH_2PO_4$ adjusted to pH 3 with phosphoric acid.)
**Flow rate:** 1.5
**Injection volume:** 10
**Detector:** UV 200; UV 272; UV 276; UV 280; UV 314

## CHROMATOGRAM
**Retention time:** 8.5

## OTHER SUBSTANCES
**Simultaneous:** degradation products, nitrendipine

## REFERENCE
Logan, B.K.; Patrick, K.S. Photodegradation of nifedipine relative to nitrendipine evaluated by liquid and gas chromatography. *J.Chromatogr.*, **1990**, *529*, 175–181

## SAMPLE
**Matrix:** solutions
**Sample preparation:** Prepare a 10 µg/mL solution in MeOH, inject a 20 µL aliquot.

## HPLC VARIABLES
**Column:** 125 × 4.9 Spherisorb S5W silica
**Mobile phase:** MeOH containing 10 mM ammonium perchlorate and 1 mL/L 100 mM NaOH in MeOH, pH 6.7
**Flow rate:** 2
**Injection volume:** 20
**Detector:** E, LeCarbone, V25 glassy carbon electrode, + 1.2 V

## CHROMATOGRAM
**Retention time:** 1.1

## OTHER SUBSTANCES
**Also analyzed:** acebutolol, acepromazine, acetophenazine, N-acetylprocainamide, albuterol, alprenolol, amethocaine, amiodarone, amitriptyline, antazoline, atenolol, azacyclonal, bamethane, benactyzine, benperidol, benzethidine, benzocaine, benzoctamine, benzphetamine, benzquinamide, bromhexine, bromodiphenhydramine, bromperidol, brompheniramine, brompromazine, buclizine, bufotenine, bupivacaine, buprenorphine, butacaine, butethamate, chlorcyclizine, chlorpheniramine, chlorphenoxamine, chlorprenaline, chlorpromazine, chlorprothixene, cimetidine, cinchonidine, cinnarizine, clemastine, clomipramine, clonidine, cocaine, cyclazocine, cyclizine, cyclopentamine, cyproheptadine, deserpidine, desipramine, dextromoramide, dextropropoxyphene, dicyclomine, diethylcarbamazine, diethylpropion, diethylthiambutene, dihydroergotamine, dimethindene, di-

methothiazine, diphenhydramine, diphenoxylate, dipipanone, diprenorphine, dipyridamole, disopyramide, dothiepin, doxapram, doxepin, doxylamine, droperidol, ephedrine, ergocornine, ergocristine, ergocristinine, ergocryptine, ergometrine, ergosine, ergosinine, ergotamine, ethopropazine, etorphine, etoxeridine, fenethazine, fenfluramine, fenoterol, fentanyl, flavoxate, fluopromazine, flupenthixol, fluphenazine, flurazepam, haloperidol, hydroxyzine, hyoscine, ibogaine, imipramine, indapamine, iprindole, isothipendyl, isoxsuprine, ketanserin, laudanosine, lidocaine, lofepramine, loxapine, maprotiline, mecamylamine, meclophenoxate, meclozine, medazepam, mephentermine, mepivacaine, meptazinol, mepyramine, mesoridazine, metaraminol, methadone, methamphetamine, methapyrilene, methdilazene, methotrimeprazine, methoxamine, methoxyphenamine, methoxypromazine, methylephedrine, methylergonovine, methysergide, metoclopramide, metopimazine, metoprolol, mianserin, morazone, nadolol, nalorphine, naloxone, naphazoline, nicotine, nomifensine, nortriptyline, noscapine, orphenadrine, oxeladin, oxprenolol, oxymetazolin, papaverine, pargyline, pecazine, penbutolol, pentazocine, penthienate, pericyazine, perphenazine, phenadoxone, phenampromide, phenazocine, phenbutrazate, phendimetrazine, phenelzine, phenglutarimide, phenindamine, pheniramine, phenmetrazine, phenomorphan, phenoperidine, phenothiazine, phenoxybenzamine, phentolamine, phenylephrine, phenyltoloxamine, physostigmine, piminodine, pimozide, pindolol, pipamazine, pipazethate, piperacetazine, piperidolate, pipradol, pirenzepine, piritramide, pizotifen, practolol, pramoxine, prazosin, prenylamine, prilocaine, primaquine, proadifen, procainamide, procaine, prochlorperazine, procyclidine, proheptazine, prolintane, promazine, promethazine, pronethalol, properidine, propiomazine, propranolol, prothipendyl, protriptyline, proxymetacaine, pseudoephedrine, pyrimethamine, quinidine, quinine, ranitidine, rescinnamine, sotalol, tacrine, terazosin, terbutaline, terfenadine, thenyldiamine, theophylline, thiethylperazine, thiopropazate, thioproperazine, thioridazine, thiothixene, thonzylamine, timolol, tocainide, tolpropamine, tolycaine, tranylcypromine, trazodone, trifluoperazine, trifluperidol, trimeperidine, trimeprazine, trimethobenzamide, trimethoprim, trimipramine, tripelennamine, triprolidine, tryptamine, verapamil, xylometazoline

## REFERENCE

Jane, I.; McKinnon, A.; Flanagan, R.J. High-performance liquid chromatographic analysis of basic drugs on silica columns using non-aqueous ionic eluents. II. Application of UV, fluorescence and electrochemical oxidation detection. *J.Chromatogr.*, **1985**, *323*, 191–225

## ANNOTATED BIBLIOGRAPHY

Grundy, J.S.; Kherani, R.; Foster, R.T. Photostability determination of commercially available nifedipine oral dosage formulations. *J.Pharm.Biomed.Anal.*, **1994**, *12*, 1529–1535 [photostability; simultaneous degradation products; capsules; tablets]

Hayase, N.; Itagaki, Y.-I.; Ogawa, S.; Akutsu, S.; Inagaki, S.-I.; Abiko, Y. Newly discovered photodegration products of nifedipine in hospital prescriptions. *J.Pharm.Sci.*, **1994**, *83*, 532–538 [simultaneous degradation products; column temp 30; tablets]

Kobayashi, D.; Matsuzawa, T.; Sugibayashi, K.; Morimoto, Y.; Kimura, M. Analysis of the combined effect of l-menthol and ethanol as skin permeation enhancers based on a two-layer skin model. *Pharm.Res.*, **1994**, *11*, 96–103 [also atenolol, morphine, naloxone, nitrendipine, vinpocetine]

Thongnopnua, P.; Viwatwongsa, K. Quantitative analysis of nifedipine in plasma by high-performance liquid chromatography. *J.Pharm.Biomed.Anal.*, **1994**, *12*, 119–125 [LOQ 7 ng/mL; butamben (IS)]

Erram, S.V.; Tipnis, H.P. Simultaneous determination of atenolol and nifedipine in solid dosage forms by RP-HPLC. *Indian Drugs*, **1992**, *29*, 436–438

Ohkubo, T.; Noro, H.; Sugawara, K. High-performance liquid chromatographic determination of nifedipine and a trace photodegradation product in hospital prescriptions. *J.Pharm.Biomed.Anal.*, **1992**, *10*, 67–70

Bammi, R.K.; Nayak, V.G.; Bhate, V.R.; Dhumal, S.N.; Purandare, S..M.; Dikshit, P.M.; Gaitonde, C.D. Analysis of nifedipine and related compounds in soft geletin capsules by liquid chromatography. *Drug Dev.Ind.Pharm.*, **1991**, *17*, 2239–2244

Beaulieu, N.; Curran, N.M.; Graham, S.J.; Sears, R.W.; Lovering, E.G. Validation of an HPLC method for nifedipine and its related substances in raw materials. *J.Liq.Chromatogr.*, **1991**, *14*, 1173–1183 [bulk; simultaneous impurities; LOQ 0.01-0.1%]

Jain, R.; Jain, C.L. Simultaneous microquantification of nifedipine and atenolol in multicomponent dosage forms using high performance liquid chromatography. *Microchem.J.*, **1991**, *44*, 187–192 [capsules; rugged]

Rau, H.L.; Aroor, A.R.; Rao, P.G. Simultaneous determination of nifedipine and atenolol by HPLC in combined dosage forms. *Indian Drugs*, **1991**, *28*, 283–284

Soons, P.A.; Schellens, J.H.M.; Roosemalen, M.C.M.; Breimer, D.D. Analysis of nifedipine and its pyridine metabolite dehydronifedipine in blood and plasma: review and improved high-performance liquid chromatographic methodology. *J.Pharm.Biomed.Anal.*, **1991**, *9*, 475–484

Telting-Diaz, M.; Kelly, M.T.; Hua, C.; Smyth, M.R. High-performance liquid chromatographic determination of nifedipine, nicardipine and pindolol using a carbon fiber flow-through amperometric detector. *J.Pharm.Biomed.Anal.*, **1991**, *9*, 889–893

Budvári-Bárány, Z.; Szász, G.; Radeczky, G.; Simonyi, I.; Shalaby, A. Some new data concerning the chromatographic purity test for nifedipine. *J.Liq.Chromatogr.*, **1990**, *13*, 3541–3551 [simultaneous impurities; bulk]

El-Sayed, A.A.; Ibrahim, M.M.K.; Omar, S.M. A high-performance liquid chromatographic method for studying the stability and dissolution of nifedipine from its marketed formulation in Egypt. *Egypt.J.Pharm.Sci.*, **1990**, *31*, 541–550

Fu, C.J.; Mason, W.D. A simplified method for determination of nifedipine in human plasma by high performance liquid chromatography. *Anal.Lett.*, **1989**, *22*, 2985–3002

Sheridan, M.E.; Clarke, G.S.; Robinson, M.L. Analysis of nifedipine in serum using solid-phase extraction and liquid chromatography. *J.Pharm.Biomed.Anal.*, **1989**, *7*, 519–522

Mascher, H.; Vergin, H. HPLC determination of nifedipine in plasma on normal phase. *Chromatographia*, **1988**, *25*, 919–922 [simultaneous degradation products; LOD 1 ng/mL]

Poetter, H.; Huelm, M. Assay of nifedipine and its by- and degradation products in the drug substance and dragees by liquid chromatography on formamide-saturated silica gel columns. *J.Pharm. Biomed.Anal.*, **1988**, *6*, 115–119

Huebert, N.D.; Spedding, M.; Haegele, K.D. Quantitative analysis of the dihydropyridines, 3-(2-furoyl)-5-methoxycarbonyl-2,6-dimethyl-4-(2-nitrophenyl)-1,4-dihydropyridine and nifedipine, by high-performance liquid chromatography with electrochemical detection. *J.Chromatogr.*, **1986**, *353*, 175–180

Snedden, W.; Fernandez, P.G.; Nath, C. High performance liquid chromatography analysis of nifedipine and some of its metabolites in hypertensive patients. *Can.J.Physiol.Pharmacol.*, **1986**, *64*, 290–296

Kleinbloesem, C.H.; van Harten, J.; van Brummelen, P.; Breimer, D.D. Liquid chromatographic determination of nifedipine in plasma and of its main metabolite in urine. *J.Chromatogr.*, **1984**, *308*, 209–216

Miyazaki, K.; Kohri, N.; Arita, T.; Shimono, H.; Katoh, K.; Nomura, A.; Yasuda, H. High-performance liquid chromatographic determination of nifedipine in plasma. *J.Chromatogr.*, **1984**, *310*, 219–222

Bach, P.R. Determination of nifedipine in serum or plasma by reversed-phase liquid chromatography. *Clin.Chem.*, **1983**, *29*, 1344–1348

Dokladalova, J.; Tykal, J.A.; Coco, S.J.; Durkee, P.E.; Quercia, G.T.; Korst, J.J. Occurrence and measurement of nifedipine and its nitropyridine derivatives in human blood plasma. *J.Chromatogr.*, **1982**, *231*, 451–458

Sadanaga, T.; Hikida, K.; Tameto, K.; Matsushima, Y.; Ohkura, Y. Determination of nifedipine in plasma by high-performance liquid chromatography. *Chem.Pharm.Bull.*, **1982**, *30*, 3807–3809

Pietta, P.; Rava, A.; Biondi, P. High-performance liquid chromatography of nifedipine, its metabolites and photochemical degradation products. *J.Chromatogr.*, **1981**, *210*, 516–521

# Nitrofurantoin

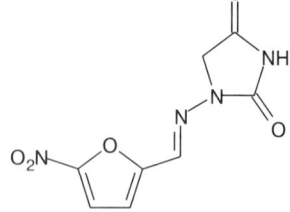

**Molecular formula:** $C_8H_6N_4O_5$
**Molecular weight:** 238.2
**CAS Registry No.:** 67-20-9, 54-87-5 (sodium salt), 17140-81-7
(monohydrate)

---

## SAMPLE
**Matrix:** blood, milk
**Sample preparation:** 1 mL Plasma or milk + furazolidone, extract with 5 mL dichloromethane in acidic medium (pH 3), mix, centrifuge. Remove organic phase and evaporate it to dryness at 50° under a stream of nitrogen. Take up residue in MeCN and inject an aliquot. (Protect from light during extraction procedure.)

---

## HPLC VARIABLES
**Column:** 75 × 4.6 Beckman XL 3 μm ODS
**Mobile phase:** MeCN:water 35:65
**Flow rate:** 1
**Detector:** UV 364

---

## CHROMATOGRAM
**Internal standard:** furazolidone
**Limit of detection:** 10 ng/mL

---

## KEY WORDS
plasma

---

## REFERENCE
Pons, G.; Rey, E.; Richard, M.O.; Vauzelle, F.; Francoual, C.; Moran, C.; d'Athis, P.; Badoual, J.; Olive, G. Nitrofurantoin excretion in human milk. *Dev.Pharmacol.Ther.*, **1990**, *14*, 148–152

---

## SAMPLE
**Matrix:** cell suspensions
**Sample preparation:** 200 μL Cell suspension + 200 μL 10% trichloroacetic acid + 100 μL 10 μg/mL furazolidone, centrifuge, inject a 75 μL aliquot of the supernatant.

---

## HPLC VARIABLES
**Column:** C18 (Waters)
**Mobile phase:** MeOH:water:glacial acetic acid 20:80:0.09, adjusted to pH 5 with NaOH
**Flow rate:** 1.5
**Injection volume:** 75
**Detector:** UV 370

---

## CHROMATOGRAM
**Retention time:** 6
**Internal standard:** furazolidone (8)

---

## REFERENCE
Minchin, R.F.; Ho, P.C.; Boyd, M.R. Reductive metabolism of nitrofurantoin by rat lung and liver in vitro. *Biochem.Pharmacol.*, **1986**, *35*, 575–580

---

## SAMPLE
**Matrix:** eggs, tissue

**Sample preparation:** Blend (Stomacher) 10 g homogenized tissue with 30 mL saline for 3 min, centrifuge at 2000 g, mix 20 mL of the supernatant with 2 mL 1% sodium azide. Dilute 10 mL homogenized egg with 10 mL saline, add 3 mL 10% sodium azide solution. Dialyze sample using a Cuprophan membrane (10000-15000 dalton cut-off) against water pumped at 0.36-1.44 mL/min for 3-9 min, pass the water through column A, flush the column with pure water for 8 min, backflush the contents of column A onto column B with mobile phase, after 5 min remove column A from the circuit, elute column B with mobile phase, monitor the effluent from column B. To increase sensitivity a number of sample batches can be dialyzed before the contents of column A are analyzed. (Caution! Sodium azide is carcinogenic, mutagenic, and highly toxic! Do not discharge to the sink!)

## HPLC VARIABLES
**Column:** A 60 × 4.6 37-50 μm Bondapak C18/Corasil; B 250 × 4.6 5 μm Hypersil ODS
**Mobile phase:** MeCN:100 mM pH 5 acetate buffer 20:80
**Flow rate:** 1
**Detector:** UV 365

## CHROMATOGRAM
**Retention time:** 8
**Limit of detection:** 1 ng/g (eggs); 2 ng/g (tissue)

## OTHER SUBSTANCES
**Extracted:** furaltadone, furazolidone, nitrofurazone

## KEY WORDS
protect from light; cow; muscle; dialysis

## REFERENCE
Aerts, M.M.; Beek, W.M.; Brinkman, U.A. On-line combination of dialysis and column-switching liquid chromatography as a fully automated sample preparation technique for biological samples. Determination of nitrofuran residues in edible products. *J.Chromatogr.*, **1990**, *500*, 453–468

## SAMPLE
**Matrix:** feed, formulations, milk
**Sample preparation:** Formulations. Dissolve formulation in DMF, filter, inject a 10 μL aliquot. Feeds. Stir 10 g finely ground feeds with 40 mL DMF for 30 min, centrifuge, filter, wash residues with DMF, dilute to 50 mL with DMF, inject a 10 μL aliquot. Milk. Lyophilize 200 mL milk, wash with 75 mL MeCN during 15 min. Extract residue with 15 mL DMF with stirring for 30 min, wash residue with a mixture of 25 mL MeCN+ 5 mL DMF. Combine all organic solutions and evaporate to dryness in vacuum. Treat residue with DMF, filter, dilute to 25 mL with DMF, filter before analysis, inject a 10 μL aliquot.

## HPLC VARIABLES
**Column:** 33 × 4.6 Perkin-Elmer Pecosphere 3x3 CR C18
**Mobile phase:** MeCN:100 mM pH 3.2 sodium acetate/acetic acid 10:90
**Flow rate:** 2
**Injection volume:** 10
**Detector:** UV 360

## CHROMATOGRAM
**Retention time:** 1.36
**Limit of detection:** 4.7 ng

## OTHER SUBSTANCES
**Simultaneous:** furaltadone, furazolidone

## REFERENCE

Galeano Díaz, T.; Lopez Martínez, L.; Martínez Galera, M.; Salinas, F. Rapid determination of nitrofurantoin, furazolidone and furaltadone in formulations, feed and milk by high performance liquid chromatography. *J.Liq.Chromatogr.*, **1994**, *17*, 457–475

## SAMPLE

**Matrix:** formulations

**Sample preparation:** Dissolve tablet in 10 mM HCl containing 90 mM KCl (pH 2.0), inject an aliquot.

## HPLC VARIABLES

**Column:** 50 mm long ODS Hypersil C18

**Mobile phase:** MeCN:50 mM pH 7.0 phosphate buffer 30:70

**Flow rate:** 1

**Detector:** UV 257

## CHROMATOGRAM

**Retention time:** 5.3

## OTHER SUBSTANCES

**Simultaneous:** hydrocortisone

## KEY WORDS

tablets

## REFERENCE

Neervannan, S.; Dias, L.S.; Southard, M.Z.; Stella, V.J. A convective-diffusion model for dissolution of two non-interacting drug mixtures from co-compressed slabs under laminar hydrodynamic conditions. *Pharm.Res.*, **1994**, *11*, 1288–1295

## SAMPLE

**Matrix:** formulations

**Sample preparation:** 5 mL Suspension + 20 mL water + 50 mL DMF, shake for 10 min, cool to room temperature, dilute to 100 mL with DMF, centrifuge an aliquot. 4 mL Supernatant + 15 mL 65 μg/mL acetanilide in mobile phase, filter (5 μm), discard first 2 mL, inject a 15 μL aliquot.

## HPLC VARIABLES

**Column:** 300 × 3.9 10 μm μBondapak C18

**Mobile phase:** MeCN:buffer 12:88 (Buffer was 6.8 g $KH_2PO_4$ in 500 mL water, add 30 mL 1 M NaOH, dilute to 1 L with water, pH was 7.0.)

**Flow rate:** 1.5

**Injection volume:** 15

**Detector:** UV 254

## CHROMATOGRAM

**Retention time:** 8

**Internal standard:** acetanilide

## KEY WORDS

suspensions

## REFERENCE

Juenge, E.C.; Kreienbaum, M.A.; Gurka, D.F. Assay of nitrofurantoin oral suspensions contaminated with 3-(5-nitrofurfurylideneamino)hydantoic acid. *J.Pharm.Sci.*, **1985**, *74*, 100–102

## SAMPLE
**Matrix:** solutions
**Sample preparation:** Dissolve in chloroform at a concentration of 1 µg/mL, inject an aliquot.

## HPLC VARIABLES
**Column:** 250 × 4 5 µm Lichrospher RP-18
**Mobile phase:** MeCN:10 mM sodium acetate 20:80, pH 5
**Column temperature:** 30
**Flow rate:** 1.6
**Injection volume:** 20
**Detector:** UV 365

## CHROMATOGRAM
**Retention time:** 9.8
**Limit of detection:** 22 ng/mL

## OTHER SUBSTANCES
**Simultaneous:** degradation products, carbadox, furaltadone, furazolidone, nitrofurazone

## REFERENCE
Kaniou, I.; Zachariadis, G.; Kalligas, G.; Tsoukali, H.; Stratis, J. Separation and determination of carbadox, nitrofurazone, nitrofurantoin, furazolidone, and furaltadone in their mixtures by thin layer and high performance liquid chromatography. *J.Liq.Chromatogr.*, **1994**, *17*, 1385–1398

## SAMPLE
**Matrix:** tissue
**Sample preparation:** Homogenize (Tekmar Tissuemizer SDT 1810 with SDT 182 EN shaft/blades) 5 g finely chopped muscle tissue and 20 mL MeCN at medium speed for 45 s, centrifuge at 1500 g for 5 min, add 15 mL MeCN-saturated hexane, shake for 30 s, discard the hexane layer, repeat the wash. Add 9 mL EtOH to the MeCN layer and evaporate under reduced pressure to 2-5 mL at 45°, add 2 mL EtOH, evaporate to 2 mL, add 2 mL EtOH, evaporate to dryness, add 1 mL mobile phase, sonicate for 5 min, centrifuge at 15400 g for 10 min, filter (0.45 µm) the supernatant, inject a 50 µL aliquot of the supernatant.

## HPLC VARIABLES
**Column:** 200 × 4.6 5 µm ODS Hypersil C18
**Mobile phase:** MeCN:1% acetic acid 25:75
**Column temperature:** 40
**Flow rate:** 1
**Injection volume:** 50
**Detector:** UV 375

## CHROMATOGRAM
**Retention time:** 3.5
**Limit of detection:** 1 ng/g
**Limit of quantitation:** 5 ng/g

## OTHER SUBSTANCES
**Extracted:** metabolites, furazolidone, nitrofurazone

## KEY WORDS
fish; muscle

## REFERENCE

Rupp, H.S.; Munns, R.K.; Long, A.R.; Plakas, S.M. Simultaneous determination of nitrofurazone, nitrofurantoin, and furazolidone in channel catfish (*Ictalurus punctatus*) muscle tissue by liquid chromatography. *J.AOAC Int.*, **1994**, *77*, 344–350

## ANNOTATED BIBLIOGRAPHY

Ebel, S.; Liedtke, R.; Missler, B. [Quantitative determination of nitrofurantoin in body fluids by direct injection HPLC]. *Arch.Pharm.(Weinheim).*, **1980**, *313*, 95–96

Jonen, H.G.; Oesch, F.; Platt, K.L. 4-Hydroxylation of nitrofurantoin in the rat. A 3-methylcholanthrene-inducible pathway of a relatively nontoxic compound. *Drug Metab.Dispos.*, **1980**, *8*, 446–451

# Nitroglycerin

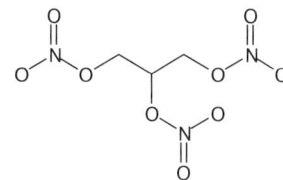

**Molecular formula:** $C_3H_5N_3O_9$
**Molecular weight:** 227.1
**CAS Registry No.:** 55-63-0

## SAMPLE
**Matrix:** blood
**Sample preparation:** 3 mL Plasma + 12 mL dichloromethane:ethyl acetate 1:1, shake mechanically at 250 cycles/min for 5 min, centrifuge at 550 g at 4° for 5 min. Remove the organic phase and evaporate it to about 20 μL under a stream of nitrogen at room temperature, inject.

## HPLC VARIABLES
**Column:** 250 × 4 10 μm Zorbax $NH_2$
**Mobile phase:** n-Hexane:MeOH 95:5
**Flow rate:** 5
**Injection volume:** 20
**Detector:** Thermal energy analyzer, Thermo Electron Corp. Model 502A, furnace temp 575°, argon 15 mL/min, oxygen 25 mL/min, MeOH/dry ice slush bath

## CHROMATOGRAM
**Retention time:** 5.0
**Internal standard:** nitroglycerin

## OTHER SUBSTANCES
**Extracted:** isosorbide dinitrate, isosorbide mononitrate

## KEY WORDS
plasma; nitroglycerin is IS

## REFERENCE
Maddock, J.; Lewis, P.A.; Woodward, A.; Massey, P.R.; Kennedy, S. Determination of isosorbide dinitrate and its mononitrate metabolites in human plasma by high-performance liquid chromatography-thermal energy analysis. *J.Chromatogr.*, **1983**, *272*, 129–136

## SAMPLE
**Matrix:** formulations
**Sample preparation:** Inject directly.

## HPLC VARIABLES
**Column:** 300 × 3.9 10 μm μBondapak C18
**Mobile phase:** MeCN:MeOH:water 6:6:13
**Column temperature:** 35
**Flow rate:** 2.8
**Detector:** UV 210

## CHROMATOGRAM
**Retention time:** 2.8

## KEY WORDS
injections; stability-indicating

## REFERENCE
Driver, P.S.; Jarvi, E.J.; Gratzer, P.L. Stability of nitroglycerin as nitroglycerin concentrate for injection stored in plastic syringes. *Am.J.Hosp.Pharm.*, **1993**, *50*, 2561–2563

## SAMPLE
**Matrix:** formulations

## HPLC VARIABLES
**Column:** C18
**Mobile phase:** MeOH:water 50:50
**Flow rate:** 1.3
**Detector:** UV 220

## CHROMATOGRAM
**Internal standard:** isosorbide dinitrate

## KEY WORDS
injections; 5% dextrose; stability-indicating

## REFERENCE
Pramar, Y.; Das Gupta, V.; Gardner, S.N.; Yau, B. Stabilities of dobutamine, dopamine, nitroglycerin and sodium nitroprusside in disposable plastic syringes. *J.Clin.Pharm.Ther.*, **1991**, *16*, 203–207

## SAMPLE
**Matrix:** formulations
**Sample preparation:** Dilute with saline, inject a 10 μL aliquot.

## HPLC VARIABLES
**Column:** 250 × 4.6 Lichrosorb 10 RP 8
**Mobile phase:** MeOH:water 50:50
**Flow rate:** 2
**Injection volume:** 10
**Detector:** UV 214

## CHROMATOGRAM
**Retention time:** 5.8

## OTHER SUBSTANCES
**Simultaneous:** isosorbide dinitrate

## KEY WORDS
injections; saline

## REFERENCE
Martens, H.J.; de Goede, P.N.; van Loenen, A.C. Sorption of various drugs in polyvinyl chloride, glass, and polyethylene-lined infusion containers. *Am.J.Hosp.Pharm.*, **1990**, *47*, 369–373

## SAMPLE
**Matrix:** formulations
**Sample preparation:** Weigh out an amount of finely powdered tablets or capsules equivalent to about 25 mg of drug. Add 50 mL buffer, shake for 30 min, add 10 mL MeOH, make up to 100 mL with buffer, filter (0.45 μm), inject a 20 μL aliquot. If the sample clumps when the buffer is added, agitate with a stirring rod and sonicate. (Buffer was MeOH:200 mM pH 4.7 ammonium acetate buffer:water 55:10:35.)

## HPLC VARIABLES
**Guard column:** 50 × 6.4 25-37 μm Whatman Co-Pell ODS
**Column:** 250 × 4.6 5 μm Ultrasphere ODS
**Mobile phase:** MeOH:200 mM pH 4.7 ammonium acetate buffer:water 55:10:35
**Flow rate:** 1
**Injection volume:** 20
**Detector:** UV 254

## CHROMATOGRAM
**Retention time:** 8
**Internal standard:** nitroglycerin

## OTHER SUBSTANCES
**Simultaneous:** isosorbide dinitrate, isosorbide mononitrate, pentaerythritol tetranitrate, saccharin

## KEY WORDS
tablets; capsules; nitroglycerin is IS

## REFERENCE
Carlson, M.; Thompson, R.D.; Snell, R.P. Determination of isosorbide dinitrate in pharmaceutical products by HPLC. *J.Chromatogr.Sci.*, **1988**, *26*, 574−578

## SAMPLE
**Matrix:** formulations

## HPLC VARIABLES
**Column:** 150 × 4.6 5 μm C8 (Altex)
**Mobile phase:** MeCN:water 40:60
**Flow rate:** 2.67

## KEY WORDS
tablets; mouth rinses

## REFERENCE
Noonan, P.K.; Benet, L.Z. Incomplete and delayed bioavailability of sublingual nitroglycerin. *Am.J.Cardiol.*, **1985**, *55*, 184−187

## SAMPLE
**Matrix:** formulations
**Sample preparation:** 1 mL Solution + 300 μL isosorbide dinitrate solution, mix, inject an aliquot.

## HPLC VARIABLES
**Column:** 300 × 3.9 10 μm μBondapak phenyl
**Mobile phase:** MeCN:THF:water 26:10:64
**Flow rate:** 2
**Detector:** UV 218

## CHROMATOGRAM
**Internal standard:** isosorbide dinitrate

## KEY WORDS
injections; 5% dextrose; saline

## REFERENCE
Thompson, M.; Smith, M.; Gragg, R.; Soliman, K.F. Stability of nitroglycerin and dobutamine in 5% dextrose and 0.9% sodium chloride injection. *Am.J.Hosp.Pharm.*, **1985**, *42*, 361–362

## SAMPLE
**Matrix:** formulations
**Sample preparation:** Powder tablets, weigh out a portion equivalent to 3 mg erythrityl tetranitrate, add to 10 mL 75 µg/mL nitroglycerin in MeOH, sonicate for 2 min, shake mechanically for 30 min, filter, inject an aliquot

## HPLC VARIABLES
**Guard column:** 40 × 4.6 µBondapak C18/Corasil
**Column:** 300 × 3.9 10 µm µBondapak C18
**Mobile phase:** MeOH:water 40:60
**Flow rate:** 1
**Injection volume:** 20
**Detector:** UV 220

## CHROMATOGRAM
**Retention time:** 14
**Internal standard:** nitroglycerin

## OTHER SUBSTANCES
**Simultaneous:** erythrityl tetranitrate, isosorbide dinitrate, pentaerythritol tetranitrate

## KEY WORDS
tablets; nitroglycerin is IS

## REFERENCE
Olsen, C.S.; Scroggins, H.S. High-performance liquid chromatographic determination of the nitrate esters isosorbide dinitrate, pentaerythritol tetranitrate, and erythrityl tetranitrate in various tablet forms. *J.Pharm.Sci.*, **1984**, *73*, 1303–1304

## SAMPLE
**Matrix:** formulations
**Sample preparation:** Dilute with water, inject an aliquot.

## HPLC VARIABLES
**Column:** 300 × 3.9 µBondapack phenyl
**Mobile phase:** MeCN:THF:water 26:10:64
**Flow rate:** 2
**Injection volume:** 10-40
**Detector:** UV 218

## CHROMATOGRAM
**Retention time:** 10
**Internal standard:** isosorbide dinitrate (6)

## KEY WORDS
injections; stability-indicating

## REFERENCE
Baaske, D.M.; Carter, J.E.; Amann, A.H. Rapid and accurate stability-indicating assay for nitroglycerin. *J.Pharm.Sci.*, **1979**, *68*, 481–483

## SAMPLE
**Matrix:** solutions
**Sample preparation:** Add 10 g 10% nitroglycerin solution in lactose to 125 mL MeOH, sonicate for 5 min, shake mechanically for 30 min, dilute to 200 mL with MeOH, let undissolved lactose settle, filter through paper. Dilute 5 mL to 250 mL with mobile phase, inject a 20 μL aliquot.

## HPLC VARIABLES
**Column:** 250 × 4.6 5 μm Ultrasphere ODS C18
**Mobile phase:** MeCN:water 65:35
**Flow rate:** 1
**Injection volume:** 20
**Detector:** UV 230

## CHROMATOGRAM
**Retention time:** 6
**Internal standard:** nitroglycerin

## OTHER SUBSTANCES
**Simultaneous:** pentaerythritol tetranitrate

## KEY WORDS
collaborative study; nitroglycerin is IS

## REFERENCE
Carlson, M. Liquid chromatographic determination of pentaerythritol tetranitrate in pharmaceuticals: collaborative study. *J.Assoc.Off.Anal.Chem.*, **1990**, *73*, 693–697

## SAMPLE
**Matrix:** solutions

## HPLC VARIABLES
**Column:** 150 × 4.6 5 μm Supelcosil LC-8
**Mobile phase:** MeOH:water 50:50
**Flow rate:** 1.5
**Detector:** UV 210

## CHROMATOGRAM
**Retention time:** 4.5-5

## REFERENCE
Roberts, M.E.; Mueller, K.R. Comparisons of in vitro nitroglycerin (TNG) flux across yucatan pig, hairless mouse, and human skins. *Pharm.Res.*, **1990**, *7*, 673–676

## SAMPLE
**Matrix:** solutions

## HPLC VARIABLES
**Column:** 33 × 4.6 3 μm C18 (Perkin-Elmer)
**Mobile phase:** MeCN:MeOH:water 24:24:52
**Flow rate:** 3.5
**Detector:** UV 210

## CHROMATOGRAM
**Retention time:** 0.6
**Limit of quantitation:** 1 µg/mL

## OTHER SUBSTANCES
**Simultaneous:** degradation products

## REFERENCE
Severin, G. Rapid high-performance liquid chromatographic procedure for nitroglycerin and its degradation products. *J.Chromatogr.*, **1985**, *320*, 445–449

◆————————————————◆————————————————◆

## ANNOTATED BIBLIOGRAPHY
Carlson, M.; Thompson, R.D. Determination of pentaerythritol tetranitrate in pharmaceuticals by high performance liquid chromatography. *J.Liq.Chromatogr.*, **1988**, *11*, 2603–2620 [simultaneous pentaerythritol tetranitrate; tablets; capsules; nitroglycerin is IS]

Torok, I.; Paal, T.; Koszegine Szalai, H.; Keseru, P. [High performance liquid chromatographic assay and determination of the even distribution of active ingredients in nitroglycerin and isosorbide dinitrate tablets]. *Acta Pharm.Hung.*, **1985**, *55*, 154–162

Baaske, D.M.; Karnatz, N.N.; Carter, J.E. High-performance liquid chromatographic assay for partially nitrated glycerins in nitroglycerin. *J.Pharm.Sci.*, **1983**, *72*, 194–196

Gelber, L.; Papas, A.N. Validation of high-performance liquid chromatographic methods for analysis of sustained-release preparations containing nitroglycerin, isosorbide dinitrate, or pentaerythritol tetranitrate. *J.Pharm.Sci.*, **1983**, *72*, 124–126

Yu, W.C.; Goff, E.U. Determination of vasodilators and their metabolites in plasma by liquid chromatography with a nitrosyl-specific detector. *Anal.Chem.*, **1983**, *55*, 29–32 [TEA detection; plasma; SPE; extracted isosorbide dinitrate, isosorbide mononitrate, pentaerythritol tetranitrate; LOD 0.1 ng]

# Nizatidine

**Molecular formula:** $C_{12}H_{21}N_5O_2S_2$
**Molecular weight:** 331.5
**CAS Registry No.:** 76963-41-2

## SAMPLE
**Matrix:** blood
**Sample preparation:** 2 mL Whole blood or plasma + 2 mL buffer + 5 mL chloroform: isopropanol:n-heptane 60:14:26, shake gently horizontally for 10 min, centrifuge at 2800 g for 10 min. Remove the lower organic layer and evaporate it to dryness under vacuum at 45°, reconstitute the residue in 100 μL mobile phase, centrifuge at 2800 g for 5 min, inject a 50 μL aliquot of the supernatant. (Buffer was saturated ammonium chloride solution 25% diluted with water, adjusted to pH 9.5 with 25% ammonia solution.)

## HPLC VARIABLES
**Column:** 300 × 3.9 4 μm NovaPack C18
**Mobile phase:** MeOH:THF:buffer 65:5:30 (Buffer was 0.68 g/L (10 mM (sic)) $KH_2PO_4$ adjusted to pH 2.6 with concentrated orthophosphoric acid.) (At the end of each session wash the column with water for 1 h and MeOH for 1 h, re-equilibrate for 30 min.)
**Column temperature:** 30
**Flow rate:** 0.8
**Injection volume:** 50
**Detector:** UV 322

## CHROMATOGRAM
**Retention time:** 3.52
**Limit of detection:** <120 ng/mL

## OTHER SUBSTANCES
**Extracted:** acebutolol, acenocoumarol, acepromazine, aceprometazine, acetaminophen, aconitine, ajmaline, albuterol, alimemazine, alminoprofen, alpidem, alprazolam, alprenolol, amitriptyline, amodiaquine, amoxapine, aspirin, astemizole, atenolol, benazepril, benperidol, benzoylecgonine, bepridil, betaxolol, bisoprolol, bromazepam, brompheniramine, bumadizone, bupivacaine, buprenorphine, buspirone, caffeine, carbamazepine, carbinoxamine, carpipramine, celiprolol, cetirizine, chlorambucil, chlordiazepoxide, chlormezanone, chlorophenacinone, chloroquine, chlorpheniramine, chlorpromazine, chlorpropamide, cibenzoline, cicletanine, clemastine, clobazam, clomipramine, clonazepam, clonidine, clorazepate, clozapine, cocaine, colchicine, cyamemazine, cyclizine, cycloguanil, cyproheptadine, cytarabine, dacarbazine, daunorubicin, debrisoquine, demexiptiline, desipramine, dextromethorphan, dextromoramide, dextropropoxyphene, diazepam, diazoxide, diclofenac, dihydralazine, diltiazem, diphenhydramine, dipyridamole, disopyramide, dosulepine, doxepin, doxylamine, droperidol, ephedrine, estazolam, etodolac, fenfluramine, fenoprofen, fentiazac, flecainide, floctafenine, flumazenil, flunitrazepam, fluoxetine, fluphenazine, flurbiprofen, fluvoxamine, glibenclamide, glibornuride, glipizide, glutethimide, haloperidol, histapyrrodine, hydroxychloroquine, hydroxyzine, ibuprofen, imipramine, indomethacin, iproniazid, ketamine, ketoprofen, labetalol, levomepromazine, lidocaine, lidoflazine, loperamide, loprazolam, loratadine, lorazepam, loxapine, maprotiline, medazepam, medifoxamine, mefenamic acid, mefenidramine, mefloquine, melphalan, meperidine, mephentermine, mepivacaine, metapramine, metformin, methadone, methaqualone, methocarbamol, methotrexate, metipranolol, metoclopramide, metoprolol, mexiletine, mianserine, midazolam, minoxidil, moclobemide, moperone, morphine, nadolol, nalbuphine, naproxen, nialamide, nicardipine, nifedipine, niflumic acid, nimodipine, nitrazepam, nitrendipine, nomifensine, nortriptyline, omeprazole, opipramol, oxazepam, oxprenolol, penbutolol, penfluridol, pentazocine, phencyclidine, phenobarbital, phenol, phenylbutazone, pimozide, pindolol, pipamperone, piroxicam, prazepam, prazosin, prilocaine,

procainamide, procarbazine, proguanil, promethazine, propafenone, propranolol, protriptyline, pyrimethamine, quinidine, quinine, quinupramine, ramipril, ranitidine, reserpine, secobarbital, strychnine, sulfinpyrazole, sulindac, sulpride, suriclone, temazepam, tenoxicam, terbutaline, terfenadine, tetracaine, tetrazepam, thiopental, thioproperazine, thioridazine, tianeptine, tiapride, tiaprofenic acid, ticlopidine, timolol, tioclomarol, tofisopam, tolbutamide, toloxatone, trazodone, triazolam, trifluoperazine, trifluperidol, trimipramine, triprolidine, tropatenine, verapamil, viloxazine, vinblastine, vincristine, vindesine, warfarin, yohimbine, zolpidem, zopiclone, zorubicine

**Interfering:** amisulpride, benzocaine, carteolol, codeine, lisinopril, mephenesin, nalorphine, naloxone, naltrexone, nizatidine, ritodrine, sotalol, sultopride

## KEY WORDS
whole blood; plasma

## REFERENCE
Tracqui, A.; Kintz, P.; Mangin, P. Systematic toxicological analysis using HPLC/DAD. *J.Forensic Sci.*, **1995**, *40*, 254–262

## SAMPLE
**Matrix:** blood, saliva, urine
**Sample preparation:** Plasma or saliva. 1 mL Plasma or saliva + 200 ng IS + 250 μL saturated sodium carbonate + 500 μL saturated solution of sodium sulfate in EtOH, extract into 5 mL dichloromethane. Evaporate the organic layer and dissolve the residue in 200 μL 100 mM HCl:mobile phase 1:1, inject an aliquot. Urine. 1 mL Urine + 2000 ng IS + 1 mL saturated sodium carbonate, extract into 5 mL dichloromethane. Evaporate the organic layer and dissolve the residue in 200 μL mobile phase, inject an aliquot.

## HPLC VARIABLES
**Guard column:** Brownlee Spheri-10 RP-8 guard column
**Column:** 250 × 4.6 6 μm Zorbax C8
**Mobile phase:** MeCN containing 1% triethylamine:20 mM ammonium acetate 18.5:81.5
**Flow rate:** 1
**Detector:** UV 313

## CHROMATOGRAM
**Internal standard:** N1-ethylnizatidine
**Limit of detection:** 10 ng/mL

## KEY WORDS
plasma

## REFERENCE
Knadler, M.P.; Bergstrom, R.F.; Callaghan, J.T.; Rubin, A. Nizatidine, an $H_2$-blocker. Its metabolism and disposition in man. *Drug Metab.Dispos.*, **1986**, *14*, 175–182

## SAMPLE
**Matrix:** blood, tissue
**Sample preparation:** Blood. 25 μL Plasma + 100 μL water + 100 μL 5 M NaOH + 5 mL dichloromethane, shake for 10 min, centrifuge at 1650 g for 10 min. Remove 4 mL of the organic layer and evaporate it to dryness under reduced pressure at room temperature, reconstitute the residue in 100 μL mobile phase, inject a 50 μL aliquot. (Hemolyze 25 μL whole blood with 200 μL water then proceed as above.) Tissue. Homogenize brain tissue with 100 μL water and 1 mL saline at 0° for 1 min, add 100 μL 500 mM NaOH, extract with 5 mL dichloromethane. Remove 3 mL of the organic layer and evaporate it to dryness, reconstitute the residue in 100 μL mobile phase, centrifuge at 10000 g, inject a 50 μL aliquot.

## HPLC VARIABLES
**Column:** 250 × 4 Senshu gel 7C18H (Senshu, Tokyo)
**Mobile phase:** MeCN:buffer 5:95. (Buffer was 5 mM $NaH_2PO_4$ containing 5 mM tetra-methylammonium chloride.)
**Column temperature:** 30
**Flow rate:** 2
**Injection volume:** 50
**Detector:** UV 320

## CHROMATOGRAM
**Internal standard:** nizatidine

## OTHER SUBSTANCES
**Extracted:** ranitidine

## KEY WORDS
nizatidine is IS; plasma; mouse; brain; whole blood

## REFERENCE
Shimokawa, M.; Yamamoto, K.; Kawakami, J.; Sawada, Y.; Iga, T. Effect of renal or hepatic dysfunction on neurotoxic convulsion induced by ranitidine in mice. *Pharm.Res.*, **1994**, *11*, 1519–1523

## SAMPLE
**Matrix:** formulations
**Sample preparation:** Dilute a 1 mL aliquot to a theoretical concentration of 30 µg/mL with MeCN:water 2:50, inject a 20 µL aliquot.

## HPLC VARIABLES
**Column:** 150 mm long 5 µm Waters Resolve C18
**Mobile phase:** MeOH:100 mM ammonium acetate:diethylamine 1280:2720:4, adjusted to a pH of 7.5
**Column temperature:** 30
**Flow rate:** 1
**Injection volume:** 20
**Detector:** UV 230

## CHROMATOGRAM
**Retention time:** 6.7

## OTHER SUBSTANCES
**Simultaneous:** nizatidine amide, nizatidine sulfoxide, phenol

## KEY WORDS
injections; stability-indicating

## REFERENCE
Raineri, D.L.; Cwik, M.J.; Rodvold, K.A.; Deyo, K.L.; Scaros, L.P.; Fischer, J.H. Stability of nizatidine in commonly used intravenous fluids and containers. *Am.J.Hosp.Pharm.*, **1988**, *45*, 1523–1529

## SAMPLE
**Matrix:** perfusate, urine
**Sample preparation:** Urine. Take 10 µL urine diluted 10 times with 10 mM pH 7.5 $Na_2HPO_4$ buffer, make up volume to 300 µL with 10 mM pH 7.5 $Na_2HPO_4$ buffer. Place solution on YM-10 ultrafiltration membrane with a cut-off of 10000, centrifuge at 4000 g for 20 min. Mix 180 µL filtrate with 20 µL MeOH, inject 50 µL. Perfusate. Take 10-100

μL perfusate, make up volume to 300 μL with 10 mM pH 7.5 Na₂HPO₄ buffer. Place solution on YM-10 ultrafiltration membrane with a cut-off of 10000, centrifuge at 4000 g for 20 min. Mix 180 μL filtrate with 20 μL MeOH, inject 50 μL.

## HPLC VARIABLES
**Guard column:** 75 × 2.1 5 μm LiChrosorb RP-18
**Column:** 150 × 4.6 5 μm LiChrosorb RP-18
**Mobile phase:** MeOH:10 mM pH 7.5 Na₂HPO₄ 30:70
**Column temperature:** 40
**Flow rate:** 1
**Injection volume:** 50
**Detector:** UV 228

## CHROMATOGRAM
**Retention time:** 10
**Internal standard:** nizatidine
**Limit of detection:** 50 ng/mL

## OTHER SUBSTANCES
**Extracted:** cimetidine

## KEY WORDS
nizatidine is IS

## REFERENCE
Boom, S.P.A.; Moons, M.M.; Russel, F.G.M. Renal tubular transport of cimetidine in the isolated perfused kidney of the rat. *Drug Metab.Dispos.*, **1994**, *22*, 148–153

## SAMPLE
**Matrix:** solutions

## HPLC VARIABLES
**Column:** 150 × 4.6 12 μm 1-myristoyl-2-[(13-carboxyl)-tridecoyl]-sn-3-glycerophosphocholine chemically bonded to silica (Regis)
**Mobile phase:** MeCN:100 mM pH 7.0 phosphate buffer 20:80
**Flow rate:** 1
**Detector:** UV 254

## CHROMATOGRAM
**Retention time:** k′ 0.43

## OTHER SUBSTANCES
**Also analyzed:** acebutolol, alprenolol, antazoline, atenolol, betaxolol, bisoprolol, bopindolol, bupranolol, carteolol, celiprolol, chloropyramine, chlorpheniramine, cicloprolol, cimetidine, cinnarizine, cirazoline, clonidine, dilevalol, dimethindene, diphenhydramine, doxazosin, esmolol, famotidine, isothipendyl, ketotifen, metiamide, metoprolol, moxonidine, nadolol, naphazoline, nifenalol, oxprenolol, pheniramine, phentolamine, pindolol, pizotyline (pizotifen), practolol, prazosin, promethazine, propranolol, pyrilamine (mepyramine), ranitidine, roxatidine, sotalol, tiamenidine, timolol, tramazoline, tripelennamine, triprolidine, tymazoline, UK-14,304

## REFERENCE
Kaliszan, R.; Nasal, A.; Turowski, M. Binding site for basic drugs on α₁-acid glycoprotein as revealed by chemometric analysis of biochromatographic data. *Biomed.Chromatogr.*, **1995**, *9*, 211–215

## SAMPLE

**Matrix:** solutions

## HPLC VARIABLES

**Column:** 250 × 4.6 5 μm Supelcosil LC-DP (A) or 250 × 4 5 μm LiChrospher 100 RP-8 (B)

**Mobile phase:** MeCN:0.025% phosphoric acid:buffer 25:10:5 (A) or 60:25:15 (B) (Buffer was 9 mL concentrated phosphoric acid and 10 mL triethylamine in 900 mL water, adjust pH to 3.4 with dilute phosphoric acid, make up to 1 L.)

**Flow rate:** 0.6

**Injection volume:** 25

**Detector:** UV 229

## CHROMATOGRAM

**Retention time:** 5.57 (A), 3.13 (B)

## OTHER SUBSTANCES

**Also analyzed:** acebutolol, acepromazine, acetaminophen, acetazolamide, acetophenazine, albuterol, alprazolam, amitriptyline, amobarbital, amoxapine, antipyrine, atenolol, atropine, azatadine, baclofen, benzocaine, bromocriptine, brompheniramine, brotizolam, bupivacaine, buspirone, butabarbital, butalbital, caffeine, carbamazepine, cetirizine, chlorcyclizine, chlordiazepoxide, chlormezanone, chloroquine, chlorpheniramine, chlorpromazine, chlorpropamide, chlorprothixene, chlorthalidone, chlorzoxazone, cimetidine, cisapride, clomipramine, clonazepam, clonidine, clozapine, cocaine, codeine, colchicine, cyclizine, cyclobenzaprine, dantrolene, desipramine, diazepam, diclofenac, diflunisal, diltiazem, diphenhydramine, diphenidol, diphenoxylate, dipyridamole, disopyramide, dobutamine, doxapram, doxepin, droperidol, encainide, ethidium bromide, ethopropazine, fenoprofen, fentanyl, flavoxate, fluoxetine, fluphenazine, flurazepam, flurbiprofen, fluvoxamine, furosemide, glutethimide, glyburide, guaifenesin, haloperidol, homatropine, hydralazine, hydrochlorothiazide, hydrocodone, hydromorphone, hydroxychloroquine, hydroxyzine, ibuprofen, imipramine, indomethacin, ketoconazole, ketoprofen, ketorolac, labetalol, levorphanol, lidocaine, loratadine, lorazepam, lovastatin, loxapine, mazindol, mefenamic acid, meperidine, mephenytoin, mepivacaine, mesoridazine, metaproterenol, metformin, methadone, methdilazine, methocarbamol, methotrexate, methotrimeprazine, methoxamine, methyldopa, methylphenidate, metoclopramide, metolazone, metoprolol, metronidazole, midazolam, moclobemide, morphine, nadolol, nalbuphine, naloxone, naphazoline, naproxen, nifedipine, norepinephrine, nortriptyline, oxazepam, oxycodone, oxymetazoline, paroxetine, pemoline, pentazocine, pentobarbital, pentoxifylline, perphenazine, pheniramine, phenobarbital, phenol, phenolphthalein, phentolamine, phenylbutazone, phenyltoloxamine, phenytoin, pimozide, pindolol, piroxicam, pramoxine, prazepam, prazosin, probenecid, procainamide, procaine, prochlorperazine, procyclidine, promazine, promethazine, propafenone, propantheline, propiomazine, propofol, propranolol, protriptyline, quazepam, quinidine, quinine, racemethorphan, ranitidine, remoxipride, risperidone, salicylic acid, scopolamine, secobarbital, sertraline, sotalol, spironolactone, sulfinpyrazone, sulindac, temazepam, terbutaline, terfenadine, tetracaine, theophylline, thiethylperazine, thiopental, thioridazine, thiothixene, timolol, tocainide, tolbutamide, tolmetin, trazodone, triamterene, triazolam, trifluoperazine, triflupromazine, trimeprazine, trimethoprim, trimipramine, verapamil, warfarin, xylometazoline, yohimbine, zopiclone

## KEY WORDS

some details of plasma extraction

## REFERENCE

Koves, E.M. Use of high-performance liquid chromatography-diode array detection in forensic toxicology. *J.Chromatogr.A*, **1995**, *692*, 103–119

# Norethindrone

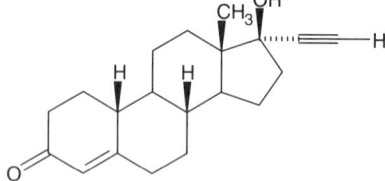

**Molecular formula:** C$_{20}$H$_{26}$O$_2$
**Molecular weight:** 298.4
**CAS Registry No.:** 68-22-4, 51-98-9 (acetate)

## SAMPLE
**Matrix:** blood
**Sample preparation:** Condition a Bond-Elut C18 SPE cartridge with MeOH and water. Add 500 µL plasma to the SPE cartridge, wash with 2 mL water, wash with 2 mL MeOH:water 20:80, elute with two 500 µL aliquots of MeOH. Evaporate the eluate to dryness under a stream of nitrogen, reconstitute the residue in 50 µL MeOH:water 20:80, inject a 20 µL aliquot.

## HPLC VARIABLES
**Column:** 100 × 1 5 µm Hypersil ODS
**Mobile phase:** MeCN:MeOH:water 25:25:50
**Flow rate:** 0.1
**Injection volume:** 20
**Detector:** UV

## CHROMATOGRAM
**Retention time:** 20
**Limit of detection:** 2 ng/mL

## OTHER SUBSTANCES
**Extracted:** androstenedione, 20α-hydroxy-4-pregnen-3-one, 17α-hydroxyprogesterone, progesterone, testosterone

## KEY WORDS
microbore; rat; plasma; SPE

## REFERENCE
Taylor, R.B.; Kendle, K.E.; Reid, R.G.; Hung, C.T. Chromatography of progesterone and its major metabolites in rat plasma using microbore high-performance liquid chromatography columns with conventional injection and detection systems. *J.Chromatogr.*, **1987**, *385*, 383−392

## SAMPLE
**Matrix:** blood
**Sample preparation:** 2 mL Plasma + 5 mL hexane:dichloromethane 1:1, shake for 10 min, centrifuge at 2000 rpm for 10 min. Remove 4 mL of the organic phase and evaporate it at 55° under nitrogen. Reconstitute residue in 300 µL mobile phase, inject a 150 µL aliquot.

## HPLC VARIABLES
**Column:** 250 × 4.6 5 µm LiChrosorb RP-8
**Mobile phase:** MeOH:MeCN:water 20:30:50
**Flow rate:** 2
**Injection volume:** 150
**Detector:** UV 254

## CHROMATOGRAM
**Retention time:** 8
**Limit of quantitation:** 2 ng/mL

## KEY WORDS
plasma

## REFERENCE
Loo, J.C.K.; Brien, R. Analysis of norethindrone in plasma by high-performance liquid chromatography. *J.Liq.Chromatogr.*, **1981**, *4*, 871–877

## SAMPLE
**Matrix:** bulk

## HPLC VARIABLES
**Column:** 250 × 4 10 μm LiChrosorb RP-18
**Mobile phase:** MeOH:water 70:30
**Flow rate:** 1
**Injection volume:** 25
**Detector:** UV 240

## CHROMATOGRAM
**Retention time:** 7

## OTHER SUBSTANCES
**Simultaneous:** ethinyl estradiol, impurities, norgestrel

## REFERENCE
Görög, S.; Herényi, B. Analysis of steroids. XXXVIII. The use of high-performance liquid chromatography with diode-array UV detection for estimating impurity profiles of steroid drugs. *J.Chromatogr.*, **1987**, *400*, 177–186

## SAMPLE
**Matrix:** food
**Sample preparation:** Dissolve apple extracts in THF, inject a 10 μL extract.

## HPLC VARIABLES
**Column:** 300 × 3.9 μBondapak C18
**Mobile phase:** MeOH:THF:water 11:26:63
**Flow rate:** 1.7
**Injection volume:** 10
**Detector:** UV 254

## CHROMATOGRAM
**Retention time:** 11.6

## OTHER SUBSTANCES
**Simultaneous:** degradation products, impurities

## KEY WORDS
apples; fruit; stability-indicating

## REFERENCE
Mayberry, D.O.; Kowblansky, M.; Lane, P.A.; Wray, P.E. Determination of norethindrone stability on Red Delicious apples. *J.Pharm.Sci.*, **1990**, *79*, 746–749

## SAMPLE
**Matrix:** formulations

**Sample preparation:** Dissolve 12 tablets in 600 mL water with stirring at 75 rpm, remove 3 mL sample, centrifuge at 3000 rpm for 15 min, inject a 250 µL aliquot.

## HPLC VARIABLES
**Column:** 100 × 4.6 3 µm Phenomenex IB-Sil 3 C18
**Mobile phase:** MeCN:water 40:60, pH 5.6
**Flow rate:** 1.2
**Injection volume:** 250
**Detector:** UV 200

## CHROMATOGRAM
**Retention time:** 7.7

## OTHER SUBSTANCES
**Simultaneous:** ethinyl estradiol

## KEY WORDS
tablets; modification of USP method

## REFERENCE
Dorantes, A.; Stavchansky, S. Modification of the U.S.P. dissolution method for the analysis of norethindrone and ethinyl estradiol tablets. *J.Pharm.Sci.*, **1994**, *83*, 379–381

## SAMPLE
**Matrix:** formulations
**Sample preparation:** Dissolve 6 tablets in 600 mL water:isopropanol 97:3, remove 5 mL samples, centrifuge at 1500 rpm for 10 min, inject a 50-200 µL aliquot.

## HPLC VARIABLES
**Column:** 300 × 3.9 10 µm Bondapak C18
**Mobile phase:** MeCN:water 55:45
**Flow rate:** 1
**Injection volume:** 50-200
**Detector:** UV 200

## OTHER SUBSTANCES
**Simultaneous:** mestranol

## KEY WORDS
tablets; modified USP method

## REFERENCE
Nguyen, H.T.; Shiu, G.K.; Worsley, W.N.; Skelly, J.P. Dissolution testing of norethindrone:ethinyl estradiol, norethindrone:mestranol, and norethindrone acetate:ethinyl estradiol combination tablets. *J.Pharm.Sci.*, **1990**, *79*, 163–167

## SAMPLE
**Matrix:** formulations
**Sample preparation:** Dissolve 6 tablets in 600 mL 100 mM HCl containing 0.02% sodium lauryl sulfate, remove 5 mL samples, centrifuge at 1500 rpm for 10 min, inject a 50-200 µL aliquot.

## HPLC VARIABLES
**Column:** 220 × 4.6 5 µm Spheri-5 C18
**Mobile phase:** MeCN:20 mM pH 6.0 phosphate buffer 35:65

**Flow rate:** 1.5
**Injection volume:** 50-200
**Detector:** UV 200

## CHROMATOGRAM
**Retention time:** 20.26

## OTHER SUBSTANCES
**Simultaneous:** ethinyl estradiol

## KEY WORDS
tablets; modified USP method

## REFERENCE
Nguyen, H.T.; Shiu, G.K.; Worsley, W.N.; Skelly, J.P. Dissolution testing of norethindrone:ethinyl estradiol, norethindrone:mestranol, and norethindrone acetate:ethinyl estradiol combination tablets. *J.Pharm.Sci.*, **1990**, *79*, 163–167

## SAMPLE
**Matrix:** formulations
**Sample preparation:** Dissolve 6 tablets in 600 mL 100 mM HCl containing 0.02% sodium lauryl sulfate, remove 5 mL samples, centrifuge at 1500 rpm for 10 min, inject a 50-200 µL aliquot.

## HPLC VARIABLES
**Column:** 220 × 4.6 5 µm Spheri-5 C18
**Mobile phase:** MeCN:20 mM pH 6.0 phosphate buffer 60:40
**Flow rate:** 1
**Injection volume:** 50-200
**Detector:** UV 240

## CHROMATOGRAM
**Retention time:** 12.73 (norethindrone acetate)

## KEY WORDS
tablets; modified USP procedure

## REFERENCE
Nguyen, H.T.; Shiu, G.K.; Worsley, W.N.; Skelly, J.P. Dissolution testing of norethindrone:ethinyl estradiol, norethindrone:mestranol, and norethindrone acetate:ethinyl estradiol combination tablets. *J.Pharm.Sci.*, **1990**, *79*, 163–167

## SAMPLE
**Matrix:** formulations
**Sample preparation:** Oils. 1 mL Sample + 25 mL MeOH:water 90:10, shake vigorously for 5 min, centrifuge, inject a 10 µL aliquot of the supernatant. Tablets. Grind a tablet to a fine powder, add 25 mL MeOH, sonicate for 5-10 min, filter (0.45 µm), discard first 5 mL of filtrate, inject a 10 µL aliquot of the remaining filtrate. Suspensions (aqueous). Make up 5 mL to 50 mL with MeOH, filter (0.45 µm), discard first 5 mL of filtrate, inject a 10 µL aliquot of the remaining filtrate.

## HPLC VARIABLES
**Column:** 250 × 4.6 5 µm Zorbax ODS
**Mobile phase:** MeOH:water 75:25
**Flow rate:** 1.5

**Injection volume:** 10
**Detector:** UV 240

## CHROMATOGRAM
**Retention time:** 4.6 (norethindrone), 8.6 (norethindrone acetone)
**Limit of detection:** 5 μg/mL

## OTHER SUBSTANCES
**Simultaneous:** aspirin, benzyl alcohol, benzyl benzoate, caffeine, calusterone, cortisone, dehydroepiandrosterone (UV 210), formebolone, mesterolone (UV 210), methandriol (UV 210), methandrostenolone, methenolone acetate, methyltestosterone, mibolerone, nandrolone, nandrolone acetate, nandrolone propionate, norethandrolone, norgestrel, oxymetholone, stanozolol, testolactone, testosterone, testosterone acetate, testosterone propionate, trenbolone acetate
**Interfering:** boldenone, ethisterone, fluoxymesterone, oxandrolone (UV 210)

## KEY WORDS
oils; tablets; suspensions

## REFERENCE
Walters, M.J.; Ayers, R.J.; Brown, D.J. Analysis of illegally distributed anabolic steroid products by liquid chromatography with identity confirmation by mass spectrometry or infrared spectrophotometry. *J.Assoc.Off.Anal.Chem.*, **1990**, *73*, 904–926

## SAMPLE
**Matrix:** formulations
**Sample preparation:** Dissolve 5 g cream containing 0.00925% fluocinonide, 0.00365% procinonide, and 0.0021% ciprocinonide in 2.5 mL THF, add norethindrone, dilute to 25 mL with MeOH, centrifuge, inject a 25 μL aliquot onto column A with mobile phase A and allow components to elute from column A to column B for 7 min. After 7 min remove column A from circuit, monitor effluent from column B. Back-flush column A with mobile phase B for 5 min, equilibrate column A with mobile phase A for 5 min before next injection.

## HPLC VARIABLES
**Column:** A 30 × 4.6 5 μm Spheri-5 ODS (Brownlee); B 70 × 2.1 Whatman Co:Pell ODS + 250 × 4.6 5 μm Ultrasphere C18
**Mobile phase:** A MeCN:THF:water 43:4:53; B MeOH:THF 75:25
**Flow rate:** A 1.5; B 1
**Injection volume:** 25
**Detector:** UV 260 for 22 min then UV 236

## CHROMATOGRAM
**Retention time:** 12
**Internal standard:** norethindrone

## OTHER SUBSTANCES
**Simultaneous:** ciprocinonide, fluocinolone acetonide, fluocinonide, procinonide

## KEY WORDS
creams; column-switching; norethindrone is IS

## REFERENCE
Conley, D.L.; Benjamin, E.J. Automated high-performance liquid chromatographic column switching technique for the on-line clean-up and analysis of drugs in topical cream formulations. *J.Chromatogr.*, **1983**, *257*, 337–344

## SAMPLE
**Matrix:** formulations
**Sample preparation:** 1 Tablet + 4 mL 50 mM $KH_2PO_4$, rotate 15 min, add 2 mL 1 μg/mL
o-phenylphenol in mobile phase, add 4 mL MeOH, rotate 15 min, centrifuge. Remove
supernatant, extract residue twice with 5 mL mobile phase (10 min rotation), combine
supernatants, inject 50 μL aliquot.

## HPLC VARIABLES
**Column:** 250 × 4.6 10 μm LiChrosorb RP8
**Mobile phase:** MeOH:50 mM $KH_2PO_4$ 3:2
**Flow rate:** 2
**Injection volume:** 50
**Detector:** UV 220

## CHROMATOGRAM
**Retention time:** 9 (norethindrone), 19.3 (norethindrone acetate)

## OTHER SUBSTANCES
**Simultaneous:** methyltestosterone, norgestrel
**Interfering:** ethinyl estradiol

## KEY WORDS
tablets; stability-indicating

## REFERENCE
Strusiak, S.H.; Hoogerheide, J.G.; Gardner, M.S. Determination of ethinyl estradiol in solid dosage forms
by high-performance liquid chromatography. *J.Pharm.Sci.*, **1982**, *71*, 636–640

## SAMPLE
**Matrix:** reaction mixtures
**Sample preparation:** If necessary, remove oxidizing power of solution by adding sodium
metabisulfite, inject a 20 μL aliquot.

## HPLC VARIABLES
**Guard column:** 15 × 4.6 5 μm Microsorb C8
**Column:** 250 × 4.6 5 μm Microsorb C8
**Mobile phase:** MeCN:100 mM $KH_2PO_4$ adjusted to pH 7 with 1 M KOH 50:50
**Flow rate:** 1
**Injection volume:** 20
**Detector:** UV 254

## CHROMATOGRAM
**Retention time:** 8.7
**Limit of detection:** 100 ng/mL

## REFERENCE
Lunn, G.; Rhodes, S.W.; Sansone, E.B.; Schmuff, N.R. Photolytic destruction and polymeric resin decontamination of aqueous solutions of pharmaceuticals. *J.Pharm.Sci.*, **1994**, *83*, 1289–1293

## SAMPLE
**Matrix:** solutions

## HPLC VARIABLES
**Column:** 50 × 4.6 5 μm Supelcosil LC-18
**Mobile phase:** MeOH:THF:water 10:20:70

**Flow rate:** 2
**Injection volume:** 20
**Detector:** UV 220

## CHROMATOGRAM
**Retention time:** 3 (norethindrone), 7.4 (norethindrone acetate)

## OTHER SUBSTANCES
**Simultaneous:** ethinyl estradiol, norethynodrel acetate, norgestrel

## REFERENCE
*Supelco Catalog, Supelco, Inc, Bellefonte PA,* **1996**, p. A130

## SAMPLE
**Matrix:** solutions
**Sample preparation:** Inject a 5 μL aliquot of a 10 μg/mL solution in MeOH.

## HPLC VARIABLES
**Column:** 75 × 4.6 3 μm Ultrasphere ODS
**Mobile phase:** MeCN:10 mM ammonium acetate buffer 45:55
**Flow rate:** 0.5
**Injection volume:** 5
**Detector:** UV 254

## CHROMATOGRAM
**Retention time:** 5.941

## OTHER SUBSTANCES
**Simultaneous:** boldenone, epimethandienone, epitestosterone, fluoxymesterone, 6β-hydroxymethandienone, methandienone, oxymetholone (UV 280), trenbolone

## REFERENCE
Barrón, D.; Pascual, J.A.; Segura, J.; Barbosa, J. Prediction of LC retention of steroids using solvatochromic parameters. *Chromatographia,* **1995**, *41,* 573–580

## SAMPLE
**Matrix:** solutions
**Sample preparation:** Inject a 20 μL aliquot of a solution in MeOH:water 50:50.

## HPLC VARIABLES
**Column:** 250 × 4 7 μm LichroCART RP-8 (Merck)
**Mobile phase:** MeCN:MeOH:water 32:37:31
**Flow rate:** 1
**Injection volume:** 20
**Detector:** UV 230

## CHROMATOGRAM
**Retention time:** 5

## OTHER SUBSTANCES
**Simultaneous:** fluoxymesterone, medrogestone, mestranol, progesterone, testosterone propionate

## REFERENCE
Gau, Y.S.; Sun, S.W.; Chem, R.R.-L. Optimization of high-performance liquid chromatographic separation for progestogenic, estrogenic, and androgenic steroids using factorial design. *J.Liq.Chromatogr.*, **1995**, *18*, 2373–2382

## SAMPLE
**Matrix:** solutions

## HPLC VARIABLES
**Column:** 250 × 4.6 10 μm Nucleosil C18
**Mobile phase:** MeCN:THF:water 12.9:22.4:64.7
**Flow rate:** 1
**Detector:** UV 240

## CHROMATOGRAM
**Retention time:** 8.5 (norethindrone), 15.5 (norethindrone acetate)

## OTHER SUBSTANCES
**Simultaneous:** estrone, ethinyl estradiol, mestranol, norgestrel

## REFERENCE
Gazdag, M.; Szepesi, G.; Szeleczki, E. Selection of high-performance liquid chromatographic methods in pharmaceutical analysis. I. Optimization for selectivity in reversed-phase chromatography. *J.Chromatogr.*, **1988**, *454*, 83–94

## SAMPLE
**Matrix:** solutions

## HPLC VARIABLES
**Column:** 250 × 4.6 5 μm LiChrosorb Si 60
**Mobile phase:** Hexane:dioxane:isopropanol 95:3:2
**Flow rate:** 1
**Detector:** UV 254

## CHROMATOGRAM
**Retention time:** 20 (norethindrone), 10 (norethindrone acetate)

## OTHER SUBSTANCES
**Simultaneous:** estrone, ethinyl estradiol, mestranol, norgestrel

## KEY WORDS
normal phase

## REFERENCE
Gazdag, M.; Szepesi, G.; Fábián-Varga, K. Selection of high-performance liquid chromatographic methods in pharmaceutical analysis. II. Optimization for selectivity in normal-phase systems. *J.Chromatogr.*, **1988**, *454*, 95–107

## SAMPLE
**Matrix:** solutions
**Sample preparation:** Dissolve in MeOH:water 1:1 at a concentration of 50 μg/mL, inject a 10 μL aliquot.

## HPLC VARIABLES
**Column:** 300 × 3.9 10 μm μBondapak C18

**Mobile phase:** MeOH:acetic acid:triethylamine:water 60:1.5:0.5:38
**Flow rate:** 1.5
**Injection volume:** 10
**Detector:** UV 240

## CHROMATOGRAM
**Retention time:** 12

## OTHER SUBSTANCES
**Simultaneous:** hydrocortisone acetate, methyltestosterone, prednisolone, prednisolone succinate, prednisone, progesterone

## REFERENCE
Roos, R.W.; Lau-Cam, C.A. General reversed-phase high-performance liquid chromatographic method for the separation of drugs using triethylamine as a competing base. *J.Chromatogr.*, **1986**, *370*, 403–418

## ANNOTATED BIBLIOGRAPHY
Tang, G.P.; Chen, Q.Q. [Column switching HPLC method for determination of in vivo release of norethindrone-alpha, beta-poly (3-hydroxypropyl)-DL-asparamide conjugate]. *Yao Hsueh Hsueh Pao*, **1994**, *29*, 301–305

Lee, G.J.-L.; Oyang, M.-H.; Bautista, J.; Kushinsky, S. Determination of ethinylestradiol and norethindrone in a single specimen of plasma by automated high-performance liquid chromatography and subsequent radioimmunoassay. *J.Liq.Chromatogr.*, **1987**, *10*, 2305–2318 [LOD 20-50 pg/mL]

Papas, A.N.; Marchese, S.M.; Delaney, M.F. Rapid determination of norethindrone and ethinylestradiol in oral contraceptive tablets by reversed-phase liquid chromatography. *LC Mag.*, **1985**, *3*, 354–358 [tablets; column temp 25; simultaneous ethinylestradiol; Chem.Abs., 102, 226114]

Swynnerton, N.F.; Fischer, J.B. Determination of ethynylestradiol and norethindrone in synthetic intestinal fluid and in timed-release oral formulations. *J.Liq.Chromatogr.*, **1980**, *3*, 1195–1204 [formulations; also ethinylestradiol]

# Norgestrel

**Molecular formula:** $C_{21}H_{28}O_2$
**Molecular weight:** 312.5
**CAS Registry No.:** 797-63-7, 797-64-8 ((-) form), 6533-00-2

## SAMPLE
**Matrix:** blood
**Sample preparation:** 1 mL Serum + 3 mL MTBE, vortex for 1 min, centrifuge at 1500 rpm for 5 min. Remove the organic layer and evaporate it to dryness under a stream of nitrogen at room temperature, reconstitute the residue in 100 μL MeOH, inject an aliquot.

## HPLC VARIABLES
**Guard column:** 18 mm long (Brownlee)
**Column:** 300 × 3.6 10 μm μBondapak C18
**Mobile phase:** MeOH:water 80:20
**Flow rate:** 1
**Detector:** UV 254; RIA

## CHROMATOGRAM
**Retention time:** 5.6

## OTHER SUBSTANCES
**Extracted:** metabolites, norgestimate

## KEY WORDS
serum

## REFERENCE
Wong, F.A.; Juzwin, S.J.; Tischio, N.S.; Flor, S.C. Determination of norgestimate in serum by automated high-performance liquid chromatography and subsequent radioimmunoassay. *J.Liq.Chromatogr.*, **1995**, *18*, 1851–1861

## SAMPLE
**Matrix:** bulk

## HPLC VARIABLES
**Column:** 250 × 4 10 μm LiChrosorb RP-18
**Mobile phase:** MeOH:water 70:30
**Flow rate:** 1
**Injection volume:** 25
**Detector:** UV 240

## CHROMATOGRAM
**Retention time:** 9

## OTHER SUBSTANCES
**Simultaneous:** impurities, norethindrone
**Interfering:** ethinyl estradiol

## REFERENCE
Görög, S.; Herényi, B. Analysis of steroids. XXXVIII. The use of high-performance liquid chromatography with diode-array UV detection for estimating impurity profiles of steroid drugs. *J.Chromatogr.*, **1987**, *400*, 177–186

## SAMPLE
**Matrix:** culture medium
**Sample preparation:** Extract culture medium twice with 2 volumes of ether, combine the extracts and evaporate them to dryness, reconstitute with MeOH, inject an aliquot.

## HPLC VARIABLES
**Column:** 300 × 3.9 Techopak 10 C18 (HPLC Technology)
**Mobile phase:** MeOH:water 70:30
**Flow rate:** 1.5
**Detector:** UV 240; Radioactivity

## CHROMATOGRAM
**Retention time:** 5

## OTHER SUBSTANCES
**Extracted:** metabolites, norgestimate

## KEY WORDS
tritium labeled

## REFERENCE
Wild, M.J.; Rudland, P.S.; Back, D.J. Metabolism of the oral contraceptive steroids ethynylestradiol and norgestimate by normal (Huma 7) and malignant (MCF-7 and ZR-75-1) human breast cells in culture. *J.Steroid Biochem.Mol.Biol.*, **1991**, *39*, 535–543

## SAMPLE
**Matrix:** formulations
**Sample preparation:** Centrifuge oil formulation at 30° at 2000 rpm for 30 min, filter (Whatman No. 1 paper), collect the last 4 mL of the filtrate. Dilute a 10 μL aliquot to 10 mL with MeCN:water 60:40 containing 0.3% Tween 80, remove a 2 mL aliquot and add it to 1 mL 3.33 μg/mL progesterone, vortex for 10 s, inject a 50 μL aliquot.

## HPLC VARIABLES
**Column:** 300 × 3.9 Novapak C18
**Mobile phase:** MeCN:water 60:40
**Flow rate:** 2
**Injection volume:** 50
**Detector:** UV 248

## CHROMATOGRAM
**Internal standard:** progesterone

## KEY WORDS
oils

## REFERENCE
Gao, Z.-H.; Shukla, A.J.; Johnson, J.R.; Crowley, W.R. Controlled release of a contraceptive steroid from biodegradable and injectable gel formulations: In vitro evaluation. *Pharm.Res.*, **1995**, *12*, 857–863

## SAMPLE
**Matrix:** formulations
**Sample preparation:** Oils. 1 mL Sample + 25 mL MeOH:water 90:10, shake vigorously for 5 min, centrifuge, inject a 10 μL aliquot of the supernatant. Tablets. Grind a tablet to a fine powder, add 25 mL MeOH, sonicate for 5-10 min, filter (0.45 μm), discard first 5 mL of filtrate, inject a 10 μL aliquot of the remaining filtrate. Suspensions (aqueous).

Make up 5 mL to 50 mL with MeOH, filter (0.45 μm), discard first 5 mL of filtrate, inject a 10 μL aliquot of the remaining filtrate.

## HPLC VARIABLES
**Column:** 250 × 4.6 5 μm Zorbax ODS
**Mobile phase:** MeOH:water 75:25
**Flow rate:** 1.5
**Injection volume:** 10
**Detector:** UV 240

## CHROMATOGRAM
**Retention time:** 5.9
**Limit of detection:** 5 μg/mL

## OTHER SUBSTANCES
**Simultaneous:** aspirin, benzyl alcohol, benzyl benzoate, boldenone, caffeine, calusterone, cortisone, dehydroepiandrosterone (UV 210), fluoxymesterone, formebolone, mesterolone (UV 210), methandriol (UV 210), methenolone acetate, methyltestosterone, mibolerone, nandrolone acetate, nandrolone propionate, norethandrolone, norethindrone, norethindrone acetate, oxandrolone (UV 210), oxymetholone, stanozolol, testolactone, testosterone acetate, testosterone propionate, trenbolone acetate
**Interfering:** ethisterone, methandrostenolone, nandrolone, testosterone

## KEY WORDS
oils; tablets; suspensions

## REFERENCE
Walters, M.J.; Ayers, R.J.; Brown, D.J. Analysis of illegally distributed anabolic steroid products by liquid chromatography with identity confirmation by mass spectrometry or infrared spectrophotometry. *J.Assoc.Off.Anal.Chem.*, **1990**, *73*, 904–926

## SAMPLE
**Matrix:** microsomal incubations, mucosal fluid
**Sample preparation:** Mucosal fluid. Extract 1 mL mucosal fluid twice with 5 mL diethyl ether, evaporate extracts to dryness, resuspend residue in 100 μL MeOH, inject an aliquot. Microsomal incubations. Extract 2.5 mL microsomal preparation with 5 mL diethyl ether, proceed as before.

## HPLC VARIABLES
**Guard column:** on-line guard column
**Column:** 100 × 8 μBondapak radial compression module
**Mobile phase:** MeOH:water 70:30
**Flow rate:** 1.5
**Injection volume:** 100
**Detector:** UV 240

## CHROMATOGRAM
**Retention time:** 6

## OTHER SUBSTANCES
**Simultaneous:** 3-ketonorgestimate, 17-deacetylnorgestimate, norgestimate

## REFERENCE
Madden, S.; Back, D.J. Metabolism of norgestimate by human gastrointestinal mucosa and liver microsomes in vitro. *J.Steroid Biochem.Mol.Biol.*, **1991**, *38*, 497–503

## SAMPLE
**Matrix:** solutions

## HPLC VARIABLES
**Column:** 50 × 4.6 5 μm Supelcosil LC-18
**Mobile phase:** MeOH:THF:water 10:20:70
**Flow rate:** 2
**Injection volume:** 20
**Detector:** UV 220

## CHROMATOGRAM
**Retention time:** 5.5

## OTHER SUBSTANCES
**Simultaneous:** ethinyl estradiol, norethindrone, norethindrone acetate, norethynodrel acetate

## REFERENCE
*Supelco Catalog, Supelco, Inc., Bellefonte PA,* **1996**, p. A130

## SAMPLE
**Matrix:** solutions
**Sample preparation:** Inject an aliquot of a 10 μg/mL solution.

## HPLC VARIABLES
**Column:** 150 × 4.6 Supelco ODS
**Mobile phase:** MeCN:water 25:75 containing 14 mM cyclodextrin
**Column temperature:** 0
**Flow rate:** 1-3
**Injection volume:** 20
**Detector:** UV 240

## CHROMATOGRAM
**Retention time:** 16 (D(+)), 18 (D(−))

## KEY WORDS
chiral

## REFERENCE
Lamparczyk, H.; Zarzycki, P.K.; Nowakowska, J. Effect of temperature on separation of norgestrel enantiomers by high-performance liquid chromatography. *J.Chromatogr.A*, **1994**, *668*, 413–417

## SAMPLE
**Matrix:** solutions
**Sample preparation:** Direct injection.

## HPLC VARIABLES
**Column:** 250 × 4.6 10 μm Partisil C-18 ODS-3
**Mobile phase:** MeCN:water 50:50
**Flow rate:** 2
**Detector:** UV 243

## CHROMATOGRAM
**Retention time:** 6.0

## KEY WORDS
see also J.Pharm.Sci. 1989, 78, 477

## REFERENCE
Catz, P.; Friend, D.R. In vitro evaluations of transdermal levonorgestrel. *Drug Des.Deliv.*, **1990**, *6*, 49–60

## SAMPLE
**Matrix:** solutions

## HPLC VARIABLES
**Column:** 250 × 4.6 10 μm Partisil ODS-3 C18
**Mobile phase:** MeCN:water 50:50
**Flow rate:** 2
**Detector:** UV 243

## CHROMATOGRAM
**Retention time:** 6.0

## REFERENCE
Friend, D.R.; Catz, P.; Heller, J.; Okagaki, M. Transdermal delivery of levonorgestrel IV: Evaluation of membranes. *J.Pharm.Sci.*, **1989**, *78*, 477–480

## SAMPLE
**Matrix:** solutions

## HPLC VARIABLES
**Guard column:** Chromsep reverse phase guard column (Chrompack)
**Column:** 100 × 3 5 μm Chromspher glass column
**Mobile phase:** MeCN:water 35:65
**Flow rate:** 0.4
**Detector:** UV 247

## CHROMATOGRAM
**Retention time:** 20

## OTHER SUBSTANCES
**Simultaneous:** nandrolone, trenbolone (at UV 340)

## REFERENCE
Haasnoot, W.; Schilt, R.; Hamers, A.R.; Huf, F.A.; Farjam, A.; Frei, R.W.; Brinkman, U.A. Determination of β-19-nortestosterone and its metabolite α-19-nortestosterone in biological samples at the sub parts per billion level by high-performance liquid chromatography with on-line immunoaffinity sample pretreatment. *J.Chromatogr.*, **1989**, *489*, 157–171

## SAMPLE
**Matrix:** solutions

## HPLC VARIABLES
**Column:** 250 × 4.6 10 μm Nucleosil C18
**Mobile phase:** MeCN:THF:water 12.9:22.4:64.7
**Flow rate:** 1
**Detector:** UV 240

## CHROMATOGRAM
**Retention time:** 13

## OTHER SUBSTANCES
**Simultaneous:** estrone, ethinyl estradiol, mestranol, norethindrone, norethindrone acetate

## REFERENCE
Gazdag, M.; Szepesi, G.; Szeleczki, E. Selection of high-performance liquid chromatographic methods in pharmaceutical analysis. I. Optimization for selectivity in reversed-phase chromatography. *J.Chromatogr.*, **1988**, *454*, 83–94

## SAMPLE
**Matrix:** solutions

## HPLC VARIABLES
**Column:** 250 × 4.6 5 μm LiChrosorb Si 60
**Mobile phase:** Hexane:dioxane:isopropanol 95:3:2
**Flow rate:** 1
**Detector:** UV 254

## CHROMATOGRAM
**Retention time:** 13

## OTHER SUBSTANCES
**Simultaneous:** estrone, ethinyl estradiol, mestranol, norethindrone, norethindrone acetate

## KEY WORDS
normal phase

## REFERENCE
Gazdag, M.; Szepesi, G.; Fábián-Varga, K. Selection of high-performance liquid chromatographic methods in pharmaceutical analysis. II. Optimization for selectivity in normal-phase systems. *J.Chromatogr.*, **1988**, *454*, 95–107

## SAMPLE
**Matrix:** solutions

## HPLC VARIABLES
**Column:** 250 × 4.6 μBondapak C18
**Mobile phase:** Dioxane:water 50:50 (Caution! Dioxane is a carcinogen!)
**Flow rate:** 1.4
**Detector:** UV 254

## OTHER SUBSTANCES
**Simultaneous:** norgestimate

## REFERENCE
Killinger, J.; Hahn, D.W.; Phillips, A.; Hetyei, N.S.; McGuire, J.L. The affinity of norgestimate for uterine progestogen receptors and its direct action on the uterus. *Contraception*, **1985**, *32*, 311–319

## SAMPLE
**Matrix:** tissue
**Sample preparation:** Incubate endometrial tissue with buffer, remove tissue, extract medium twice with 2 volumes of diethyl ether, evaporate to dryness, reconstitute in a small volume of MeOH, inject an aliquot.

## HPLC VARIABLES
**Column:** 300 × 3.9 Technopak 10 C18
**Mobile phase:** MeOH:water 70:30
**Flow rate:** 1.5
**Detector:** UV 240

## CHROMATOGRAM
**Retention time:** 6

## OTHER SUBSTANCES
**Simultaneous:** metabolites, norgestimate

## KEY WORDS
endometrial tissue

## REFERENCE
Wild, M.J.; Rudland, P.S.; Back, D.J. Metabolism of the oral contraceptive steroids ethynylestradiol, norgestimate and 3-ketodesogestrel by a human endometrial cancer cell line (HEC-1A) and endometrial tissue *in vitro. J.Steroid Biochem.Mol.Biol.*, **1993**, *45*, 407–420

# Nortriptyline

**Molecular formula:** $C_{19}H_{21}N$
**Molecular weight:** 263.4
**CAS Registry No.:** 72-69-5, 894-71-3 (HCl)

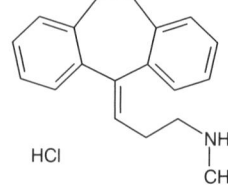

HCl

## SAMPLE

**Matrix:** blood
**Sample preparation:** 1 mL Serum + 500 μL 750 mM pH 10 sodium bicarbonate/carbonate buffer + 50 μL IS in EtOH:water 50:50 + 8 mL heptane:isoamyl alcohol 98:2, shake at 250 cycles/min for 5 min, centrifuge at 1500 g for 10 min, freeze in dry ice/EtOH. Remove the organic layer and add it to 150 μL 22 mM pH 2.5 $KH_2PO_4$/phosphoric acid buffer, shake at 250 cycles/min for 5 min, centrifuge at 1500 g for 10 min, freeze in dry ice/EtOH. Discard the organic layer, inject a 65 μL aliquot of the aqueous layer.

## HPLC VARIABLES

**Column:** 250 × 4.6 Supelco C18
**Mobile phase:** MeCN:buffer 45:55 (Buffer was 44 mM $KH_2PO_4$ containing 1.5 mL/L triethylamine, adjusted to pH 2.5 with phosphoric acid.)
**Flow rate:** 1.5
**Injection volume:** 65
**Detector:** UV 240

## CHROMATOGRAM

**Retention time:** 9.73
**Internal standard:** 1-(3-(dimethylamino)propyl)-1-(p-chlorophenyl)-1,3-dihydroisobenzofuran-5-carbonitrile (LU 10-202) (Lundbeck, Copenhagen) (8.33)

## OTHER SUBSTANCES

**Extracted:** metabolites, amitriptyline, citalopram
**Simultaneous:** chlorprothixene, clomipramine, clozapine, flupenthixol, haloperidol, levomepromazine, perphenazine, zuclopenthixol
**Noninterfering:** benzodiazepines
**Interfering:** desmethyllevomepromazine

## KEY WORDS

serum

## REFERENCE

Olesen, O.V.; Linnet, K. Simplified high-performance liquid chromatographic method for the determination of citalopram and desmethylcitalopram in serum without interference from commonly used psychotropic drugs and their metabolites. *J.Chromatogr.B*, **1996**, *675*, 83–88

## SAMPLE

**Matrix:** blood
**Sample preparation:** 1 mL Plasma + doxepin + NaOH + hexane:isoamyl alcohol 98:2, extract. Remove the organic phase and add it to 0.03% phosphoric acid, extract, inject an aliquot of the aqueous phase.

## HPLC VARIABLES

**Guard column:** C18
**Column:** 100 × 8 10 μm Resolve C8 (Waters)
**Mobile phase:** MeCN:MeOH:56 mM ammonium acetate:1 M ammonium hydroxide 100:10:4.5:2.6

**Flow rate:** 2.5
**Detector:** UV 220

## CHROMATOGRAM
**Retention time:** 31
**Internal standard:** doxepin (11.6)

## OTHER SUBSTANCES
**Extracted:** amitriptyline, fluoxetine, norfluoxetine

## KEY WORDS
plasma

## REFERENCE
el-Yazigi, A.; Chaleby, K.; Gad, A.; Raines, D.A. Steady-state kinetics of fluoxetine and amitriptyline in patients treated with a combination of these drugs as compared with those treated with amitriptyline alone. *J.Clin.Pharmacol.*, **1995**, *35*, 17−21

## SAMPLE
**Matrix:** blood
**Sample preparation:** 200 µL Plasma or serum + 100 µL 2 M pH 10.6 Tris buffer + 200 µL MTBE, vortex for 30 s, centrifuge at 9950 g for 3 min, inject a 50 µL aliquot of the organic layer.

## HPLC VARIABLES
**Guard column:** 10 × 4.6 SCX-10C5 (Hichrom)
**Column:** 150 × 4.6 Spherisorb S5 SCX (sulfopropyl-bonded silica) cation exchange
**Mobile phase:** 35 mM ammonium perchlorate in MeOH adjusted to an apparent pH of 6.7 with 100 mM NaOH in MeOH
**Flow rate:** 1.5
**Injection volume:** 50
**Detector:** UV 215

## CHROMATOGRAM
**Retention time:** 9
**Internal standard:** nortriptyline hydrochloride

## OTHER SUBSTANCES
**Extracted:** clozapine
**Simultaneous:** amitriptyline, chlorpromazine, clomipramine, dothiepin, doxepin, fluoxetine, fluphenazine, fluvoxamine, haloperidol, imipramine, maprotiline, mianserin, norclomipramine, nordothiepin, nordoxepin, norfluoxetine, nortriptyline, paroxetine, remoxipride, sulpride, thioridazine, trazodone
**Noninterfering:** carbamazepine, clonazepam, diazepam, flunitrazepam, lorazepam, nordiazepam, theophylline
**Interfering:** sertraline

## KEY WORDS
plasma; serum; nortriptyline is IS

## REFERENCE
McCarthy, P.T.; Hughes, S.; Paton, C. Measurement of clozapine and norclozapine in plasma/serum by high-performance liquid chromatography with ultraviolet detection. *Biomed.Chromatogr.*, **1995**, *9*, 36−41

## SAMPLE
**Matrix:** blood
**Sample preparation:** 2 mL Whole blood or plasma + 2 mL buffer + 5 mL chloroform: isopropanol:n-heptane 60:14:26, shake gently horizontally for 10 min, centrifuge at 2800 g for 10 min. Remove the lower organic layer and evaporate it to dryness under vacuum at 45°, reconstitute the residue in 100 μL mobile phase, centrifuge at 2800 g for 5 min, inject a 50 μL aliquot of the supernatant. (Buffer was saturated ammonium chloride solution 25% diluted with water, adjusted to pH 9.5 with 25% ammonia solution.)

## HPLC VARIABLES
**Column:** 300 × 3.9 4 μm NovaPack C18
**Mobile phase:** MeOH:THF:buffer 65:5:30 (Buffer was 0.68 g/L (10 mM (sic)) $KH_2PO_4$ adjusted to pH 2.6 with concentrated orthophosphoric acid.) (At the end of each session wash the column with water for 1 h and MeOH for 1 h, re-equilibrate for 30 min.)
**Column temperature:** 30
**Flow rate:** 0.8
**Injection volume:** 50
**Detector:** UV 239

## CHROMATOGRAM
**Retention time:** 9.28
**Limit of detection:** <120 ng/mL

## OTHER SUBSTANCES
**Extracted:** acebutolol, acenocoumarol, acepromazine, aceprometazine, acetaminophen, aconitine, ajmaline, albuterol, alimemazine, alminoprofen, alpidem, alprazolam, alprenolol, amisulpride, amodiaquine, amoxapine, aspirin, astemizole, atenolol, benazepril, benperidol, benzocaine, benzoylecgonine, bepridil, betaxolol, bisoprolol, bromazepam, brompheniramine, bumadizone, bupivacaine, buprenorphine, buspirone, caffeine, carbamazepine, carbinoxamine, carpipramine, carteolol, celiprolol, cetirizine, chlorambucil, chlordiazepoxide, chlormezanone, chlorophenacinone, chloroquine, chlorpheniramine, chlorpromazine, chlorpropamide, cibenzoline, cicletanine, clemastine, clobazam, clomipramine, clonazepam, clonidine, clorazepate, clozapine, cocaine, codeine, colchicine, cyamemazine, cyclizine, cycloguanil, cyproheptadine, cytarabine, dacarbazine, debrisoquine, demexiptiline, desipramine, dextromethorphan, dextromoramide, dextropropoxyphene, diazepam, diazoxide, dihydralazine, diltiazem, diphenhydramine, dipyridamole, disopyramide, dosulepine, doxepin, doxylamine, droperidol, ephedrine, estazolam, fenfluramine, fenoprofen, fentiazac, flecainide, floctafenine, flumazenil, flunitrazepam, fluphenazine, flurbiprofen, fluvoxamine, glibenclamide, glibornuride, glipizide, glutethimide, haloperidol, histapyrrodine, hydroxychloroquine, hydroxyzine, ibuprofen, imipramine, iproniazid, ketamine, ketoprofen, labetalol, levomepromazine, lidocaine, lidoflazine, lisinopril, loperamide, loprazolam, loratadine, lorazepam, loxapine, medazepam, medifoxamine, mefenamic acid, mefenidramine, melphalan, meperidine, mephenesin, mephentermine, mepivacaine, metapramine, metformin, methadone, methaqualone, methocarbamol, methotrexate, metipranolol, metoclopramide, metoprolol, mexiletine, mianserine, midazolam, minoxidil, moclobemide, moperone, morphine, nadolol, nalbuphine, nalorphine, naloxone, naltrexone, naproxen, nialamide, nicardipine, nifedipine, niflumic acid, nimodipine, nitrazepam, nitrendipine, nizatidine, nomifensine, omeprazole, opipramol, oxazepam, oxprenolol, penbutolol, penfluridol, pentazocine, phencyclidine, phenobarbital, phenol, phenylbutazone, pimozide, pindolol, pipamperone, piroxicam, prazepam, prazosin, prilocaine, procainamide, procarbazine, proguanil, promethazine, propafenone, propranolol, protriptyline, pyrimethamine, quinidine, quinine, quinupramine, ramipril, ranitidine, reserpine, ritodrine, secobarbital, sotalol, strychnine, sulfinpyrazole, sulindac, sulpride, sultopride, suriclone, temazepam, tenoxicam, terbutaline, terfenadine, tetracaine, tetrazepam, thiopental, thioproperazine, thioridazine, tianeptine, tiapride, tiaprofenic acid, ticlopidine, timolol, tofisopam, tolbutamide, toloxatone, trazodone, triazolam, trifluoperazine, trifluperidol, triprolidine, verapamil, viloxazine, vinblastine, vincristine, vindesine, warfarin, yohimbine, zolpidem, zopiclone, zorubicine

**Interfering:** amitriptyline, daunorubicin, diclofenac, etodolac, fluoxetine, indomethacin, maprotiline, mefloquine, tioclomarol, trimipramine, tropatenine

## KEY WORDS
whole blood; plasma

## REFERENCE
Tracqui, A.; Kintz, P.; Mangin, P. Systematic toxicological analysis using HPLC/DAD. *J.Forensic Sci.*, **1995**, *40*, 254–262

## SAMPLE
**Matrix:** blood
**Sample preparation:** 1 mL Serum + 500 μL 250 μg/mL protriptyline hydrochloride + 1 mL 500 mM NaOH + 4 mL toluene:n-hexane:isoamyl alcohol 77:22:3, mix for 10 min, centrifuge at 3000 rpm for 5 min. Remove the upper organic layer and evaporate it to dryness under a stream of air at 40°, reconstitute the residue in 200 μL MeOH, inject a 50 μL aliquot.

## HPLC VARIABLES
**Column:** 150 × 4.7 5 μm Supelcosil LC-PCN cyanopropyl
**Mobile phase:** MeCN:MeOH:10 mM pH 7.2 potassium phosphate buffer 60:15:25 (Prepare buffer by mixing 194 mL 1.36 g/L $KH_2PO_4$ with 274 mL 1.74 g/L $K_2HPO_4$.)
**Flow rate:** 2
**Injection volume:** 50
**Detector:** UV 254

## CHROMATOGRAM
**Retention time:** 6.95
**Internal standard:** protriptyline (8.1)

## OTHER SUBSTANCES
**Extracted:** amitriptyline, cyclobenzaprine
**Interfering:** norcycyclobenzaprine

## KEY WORDS
serum

## REFERENCE
Wong, E.C.C.; Koenig, J.; Turk, J. Potential interference of cyclobenzaprine and norcyclobenzaprine with HPLC measurement of amitriptyline and nortriptyline: resolution by GC-MS analysis. *J.Anal. Toxicol.*, **1995**, *19*, 218–224

## SAMPLE
**Matrix:** blood
**Sample preparation:** Automated SPE by ASPEC system. Condition a C18 Clean-Up SPE cartridge (CEC 18111, Worldwide Monitoring) with 2 mL MeOH then 2 mL water. 1 mL Plasma + 1 mL 400 ng/mL protriptyline in water, vortex, add to the SPE cartridge, wash with 3 mL water, wash with 3 mL 750 mL/L methanol. Elute with three aliquots of 300 μL 0.1 M ammonium acetate in MeOH. Add 0.5 mL 0.5 M NaOH and 4 mL 50 mL/L isopropanol in heptane to eluate, mix thoroughly. Allow 5 min for phase separation. Remove upper heptane phase and add it to 300 μL 0.1 M phosphoric acid (pH 2.5), mix, separate, inject a 100 μL aliquot of the aqueous phase.

## HPLC VARIABLES
**Guard column:** LC-8-DB (Supelco)
**Column:** 150 × 4.6 LC-8-DB (Supelco)

**Mobile phase:** MeCN:buffer 35:65 (Buffer was 10 mL/L triethylamine in water adjusted to pH 5.5 with glacial acetic acid.)
**Flow rate:** 2
**Injection volume:** 100
**Detector:** UV 228

## CHROMATOGRAM
**Retention time:** 4.8
**Internal standard:** protriptyline (4)

## OTHER SUBSTANCES
**Extracted:** acetazolamide, amitriptyline, chlordiazepoxide, chlorimipramine, chlorproma-zine, desipramine, dextromethorphan, diazepam, diphenhydramine, doxepin, encainide, fentanyl, flecainide, fluoxetine, flurazepam, haloperidol, hydroxyethylflurazepam, ibupro-fen, imipramine, lidocaine, maprotiline, methadone, methaqualone, mexiletine, midazo-lam, norchlorimipramine, nordiazepam, nordoxepin, norfluoxetine, norverapamil, pentaz-ocine, promazine, propafenone, propoxyphene, propranolol, protriptyline, quinidine, temazepam, trazodone, trimipramine, verapamil
**Noninterfering:** acetaminophen, acetylmorphine, amiodarone, amobarbital, amphetamine, bendroflumethiazide, benzocaine, benzoylecgonine, benzthiazide, butalbital, carbamaze-pine, chlorothiazide, clonazepam, cocaine, codeine, cotinine, cyclosporine, cyclothiazide, desalkylflurazepam, diamorphine, dicumerol, ephedrine, ethacrynic acid, ethanol, eth-chlorvynol, ethosuximide, furosemide, glutethimide, hydrochlorothiazide, hydrocodone, hydroflumethiazide, hydromorphone, lorazepam, mephentermine, meprobamate, meth-amphetamine, metharbital, methoxsalen, methoxyphenteramine, methsuximide, meth-ylcyclothiazide, metoprolol, MHPG, monoacetylmorphine, morphine, normethsuximide, oxazepam, oxycodone, oxymorphone, pentobarbital, phencyclidine, phenteramine, phen-ylephrine, phenytoin, polythiazide, primidone, prochlorperazine, salicylic acid, sulfanila-mide, THC-COOH, theophylline, thiazolam, thiopental, thioridazine, tocainide, trichlo-romethiazide, trifluoperazine, valproic acid, warfarin
**Interfering:** imipramine, maprotiline, verapamil

## KEY WORDS
plasma; SPE

## REFERENCE
Nichols, J.H.; Charlson, J.R.; Lawson, G.M. Automated HPLC assay of fluoxetine and norfluoxetine in serum. *Clin.Chem.*, **1994**, *40*, 1312–1316

## SAMPLE
**Matrix:** blood
**Sample preparation:** Condition a 1 mL BondElut C18 SPE cartridge with 1 mL 1 M HCl, 1 mL MeOH, 1 mL water, and 1 mL 1% potassium carbonate. 700 μL Serum + 50 μL 5 μg/mL trimipramine in 5% potassium bicarbonate + 700 μL MeCN, vortex, centrifuge at 1500 g for 5 min, add supernatant to SPE cartridge (at ca. 1 mL/min). Wash with 2 mL water and 1 mL MeCN, elute with 250 μL MeOH:35% perchloric acid 20:1 by gravity (10 min) then centrifuge for 20 s to remove rest of eluant, inject a 50 μL aliquot of the eluate.

## HPLC VARIABLES
**Guard column:** 15 mm 7 μm Brownlee RP-8
**Column:** 150 × 4.6 5 μm Ultrasphere Octyl
**Mobile phase:** MeCN:water 37.5:62.5 containing 0.5 g/L tetramethylammonium perchlo-rate and 0.5 mL/L 7% perchloric acid
**Flow rate:** 1.5
**Injection volume:** 50
**Detector:** UV 215

## CHROMATOGRAM
**Retention time:** 7.4
**Internal standard:** trimipramine (9.6)
**Limit of quantitation:** 5 ng/mL

## OTHER SUBSTANCES
**Extracted:** amitriptyline, clomipramine, desipramine, doxepin, fluoxetine, fluvoxamine, maprotiline, protriptyline
**Interfering:** imipramine

## KEY WORDS
serum; SPE

## REFERENCE
Gupta, R.N. An improved solid phase extraction procedure for the determination of antidepressants in serum by column liquid chromatography. *J.Liq.Chromatogr.*, **1993**, *16*, 2751–2765

## SAMPLE
**Matrix:** blood
**Sample preparation:** 500 μL Serum + 250 μL 100 mM lauryl sulfate, centrifuge at 2500 g for 8 min, inject a 250 μL aliquot of the supernatant onto column A with mobile phase A, elute with mobile phase A for 6 min, backflush contents of column A onto column B with mobile phase B for 4 min. Elute column B with mobile phase B for 6 min and conduct analysis. When not in use flush column A with mobile phase A. Every eight injections backflush column A with MeCN:water 70:30.

## HPLC VARIABLES
**Column:** A Guard-Pak 10 μm Resolve CN (Waters); B 150 × 3 7 μm Separon SGX CN (Tessek)
**Mobile phase:** A MeCN:water 3:97; B MeCN:buffer 26:74 (Buffer was 50 mM phosphoric acid, 50 mM ammonium phosphate, and 28 mM diethylamine, pH 2.55.)
**Flow rate:** 1
**Injection volume:** 250
**Detector:** UV 210

## CHROMATOGRAM
**Retention time:** 10
**Limit of detection:** 15-20 ng/mL

## OTHER SUBSTANCES
**Extracted:** amitriptyline

## KEY WORDS
serum; column-switching

## REFERENCE
Dolezalová, M. On-line solid-phase extraction and high-performance liquid chromatographic determination of nortriptyline and amitriptyline in serum. *J.Chromatogr.*, **1992**, *579*, 291–297

## SAMPLE
**Matrix:** blood
**Sample preparation:** Add 10 μL 20 μg/mL oxaprotiline in MeOH to 990 μL plasma or serum. Inject 100 μL plasma or serum onto column A with mobile phase A and elute to waste, after 15 min elute column A onto column B with mobile phase B for 2 min. Remove

column A from circuit and re-equilibrate it with mobile phase A for 5 min. Chromatograph on column B with mobile phase B.

## HPLC VARIABLES
**Column:** A 20 × 4.6 10 μm Hypersil MOS C8; B 20 × 4.6 5 μm Hypersil CPS CN + 250 × 4.6 5 μm Nucleosil 100 CN
**Mobile phase:** A MeOH:water 5:95; B MeOH:MeCN:10 mM pH 6.8 potassium phosphate buffer 188:578:235
**Flow rate:** 1.5
**Injection volume:** 100
**Detector:** UV 214

## CHROMATOGRAM
**Retention time:** 14.5
**Internal standard:** oxaprotiline (9.5)
**Limit of detection:** 20 ng/mL

## OTHER SUBSTANCES
**Extracted:** amitriptyline, clomipramine, clozapine, desipramine, doxepin, fluoxetine, fluvoxamine, imipramine, maprotiline, metoclopramide, norfluoxetine
**Noninterfering:** carbamazepine, chlordiazepoxide, clobazam, diazepam, flurazepam, fluspirilene, haloperidol, lorazepam, nitrazepam, nordiazepam, oxazepam, perazine, pimozide, spiroperidol, trifluperidol

## KEY WORDS
plasma; serum; column-switching

## REFERENCE
Härtter, S.; Wetzel, H.; Hiemke, C. Automated determination of fluvoxamine in plasma by column-switching high-performance liquid chromatography. *Clin.Chem.*, **1992**, *38*, 2082–2086

## SAMPLE
**Matrix:** blood
**Sample preparation:** For each 1 mL plasma or serum add 10 μL 14 μg/mL trimipramine in MeOH. Inject serum or plasma directly onto column A with mobile phase A, elute with mobile phase A to waste. After 15 min elute column A onto column B (foreflush) with mobile phase B. After 2 min remove column A from the circuit, elute column B with mobile phase B, monitor the effluent from column B. Re-equilibrate column A with mobile phase A.

## HPLC VARIABLES
**Column:** A 20 × 4.6 10 μm Hypersil MOS C8; B 20 × 4.6 5 μm Hypersil CPS CN + 250 × 4.6 5 μm Nucleosil 100 CN
**Mobile phase:** A MeOH:water 5:95; B MeCN:MeOH:buffer 578:188:235 (Buffer was 10 mM $K_2HPO_4$ adjusted to pH 6.8 with 85% phosphoric acid.)
**Flow rate:** 1.5
**Injection volume:** 100
**Detector:** UV 214

## CHROMATOGRAM
**Retention time:** 14.54
**Internal standard:** trimipramine (6.5)
**Limit of detection:** 5-10 ng/mL

## OTHER SUBSTANCES
**Extracted:** metabolites, amitriptyline, clomipramine, desipramine, doxepin, fluvoxamine, imipramine, maprotiline

**Noninterfering:** chlordiazepoxide, clobazam, clozapine, diazepam, flurazepam, fluspirilene, haloperidol, nitrazepam, oxazepam, perazine, pimozide, spiroperidol, trifluperidol

## KEY WORDS
plasma; serum; column-switching; LOD can be lowered to 1 ng/mL with 3 injections onto
column A before switching

## REFERENCE
Härtter, S.; Hiemke, C. Column switching and high-performance liquid chromatography in the analysis
of amitriptyline, nortriptyline and hydroxylated metabolites in human plasma or serum.
*J.Chromatogr.*, **1992**, *578*, 273–282

## SAMPLE
**Matrix:** blood
**Sample preparation:** 1 mL Plasma + 100 μL 1 M pH 7.6 K$_2$HPO$_4$, vortex for 5 s, add 6
mL ethyl acetate, vortex for 1 min, centrifuge at 1900 g for 5 min. Remove the organic
layer and evaporate it to dryness under a stream of nitrogen at 45°, reconstitute the
residue in 250 μL 40 mM sodium bicarbonate, add 20 μL 10 mg/mL dansyl chloride in
acetone, add 750 μL acetone, vortex for 30 s, let stand at room temperature at 22° for 15
min. Evaporate under a stream of nitrogen at 45°, reconstitute in 1 mL mobile phase,
vortex for 1 min, centrifuge at 1900 g for 10 min, inject a 100 μL aliquot of the
supernatant.

## HPLC VARIABLES
**Guard column:** 15 × 3.2 7 μm NewGuard RP-8 (Brownlee)
**Column:** 250 × 4.6 5 μm Supelcosil LC-18-DB
**Mobile phase:** MeCN:10 mM pH 7.2 potassium phosphate 85:15
**Flow rate:** 1.5
**Injection volume:** 100
**Detector:** F (wavelengths not given)

## CHROMATOGRAM
**Retention time:** 9.7
**Internal standard:** nortriptyline

## OTHER SUBSTANCES
**Extracted:** fluvoxamine
**Simultaneous:** desipramine

## KEY WORDS
plasma; derivatization; nortriptyline is IS

## REFERENCE
Pullen, R.H.; Fatmi, A.A. Determination of fluvoxamine in human plasma by high-performance liquid
chromatography with fluorescence detection. *J.Chromatogr.*, **1992**, *574*, 101–107

## SAMPLE
**Matrix:** blood
**Sample preparation:** 1 mL Plasma + 1 mL 0.6 M pH 9.8 carbonate buffer + 40 μL 5
μg/mL maprotiline in 10 mM HCl + 5 mL 200 g/L ethyl acetate in n-heptane, mix by
rocking for 10 min, centrifuge at 1500 g for 10 min. Remove organic layer and add it to
150 μL 100 mM HCl, mix 10 min, centrifuge at 1500 g for 10 min. Discard organic layer
and evaporate aqueous layer at 45° in a vacuum centrifuge for 1 h. Take up residue in
50 μL 1 M pH 10.3 carbonate buffer and 25 μL 10 mg/mL dansyl chloride in MeCN,
vortex, allow to react at room temperature for 45 min, evaporate at 45° in a vacuum

centrifuge for 20 min, reconstitute in 125 μL MeCN:water 75:25, vortex, centrifuge for 3-5 min, inject a 25-40 μL aliquot.

## HPLC VARIABLES
**Column:** 250 × 4.6 5 μm Supelcosil LC-18
**Mobile phase:** MeCN:25 mM KH$_2$PO$_4$ 75:25 containing 500 μL/L orthophosphoric acid and 600 μL/L n-butylamine
**Flow rate:** 2
**Injection volume:** 25-40
**Detector:** F ex 235 em 470 (cut-off)

## CHROMATOGRAM
**Retention time:** 14.18
**Internal standard:** maprotiline (12.8)

## OTHER SUBSTANCES
**Simultaneous:** amoxapine, clovoxamine, desipramine, fenfluramine, fluoxetine, fluvoxamine, norfluoxetine, propranolol, protriptyline, sertraline
**Noninterfering:** amitriptyline, atenolol, bupropion, carbamazepine, chlordiazepoxide, citalopram, clomipramine, clozapine, cyclobenzaprine, doxepin, imipramine, loxapine, metoprolol, mianserin, moclobemide, nomifensine, pindolol, thioridazine, tranylcypromine, trazodone, trimipramine

## KEY WORDS
plasma

## REFERENCE
Suckow, R.F.; Zhang, M.F.; Cooper, T.B. Sensitive and selective liquid-chromatographic assay of fluoxetine and norfluoxetine in plasma with fluorescence detection after precolumn derivatization. *Clin.Chem.*, **1992**, *38*, 1756–1761

## SAMPLE
**Matrix:** blood
**Sample preparation:** 1 mL Serum + 3 μL 20 ng/mL clobazam + 1 mL saturated sodium borate (adjusted to pH 11 with 6 M NaOH) + 5 mL n-hexane, mix 2 min, centrifuge at 3000 g for 10 min. Remove organic phase and evaporate to dryness under a stream of helium at 30°. Reconstitute in 20 μL mobile phase, inject a 10 μL aliquot.

## HPLC VARIABLES
**Guard column:** 20 mm long Pelliguard LC-8 40 μm (Supelco)
**Column:** 150 × 4.6 C8 5 μm (Supelco)
**Mobile phase:** MeCN:buffer 50:50 (Buffer was 1.2 mL butylamine in 1 L 10 mM NaH$_2$PO$_4$, pH adjusted to 3 with phosphoric acid.)
**Flow rate:** 1
**Injection volume:** 10
**Detector:** UV 254

## CHROMATOGRAM
**Retention time:** k' 2.460
**Internal standard:** clobazam (k' 1.344)
**Limit of detection:** 10 ng/mL

## OTHER SUBSTANCES
**Extracted:** amitriptyline, clomipramine, desipramine, imipramine
**Simultaneous:** alprazolam, clonazepam, diazepam, flunitrazepam, haloperidol, lorazepam, maprotiline, nitrazepam, triazolam

## KEY WORDS
serum

## REFERENCE
Segatti, M.P.; Nisi, G.; Grossi, F.; Mangiarotti, M.; Lucarelli, C. Rapid and simple high-performance liquid chromatographic determination of tricyclic antidepressants for routine and emergency serum analysis. *J.Chromatogr.*, **1991**, *536*, 319–325

## SAMPLE
**Matrix:** blood
**Sample preparation:** Inject 200 μL serum onto column A and elute with mobile phase A for 10 min then back-flush column A onto column B with mobile phase B for 4 min. Elute column B with mobile phase B and monitor the effluent. Remove column A from circuit and wash with MeCN:water 60:40 for 6 min then with mobile phase A for 10 min.

## HPLC VARIABLES
**Column:** A 40 × 4 TSKprecolumn PW (Tosoh); B 150 × 4 TSKgel ODS-80TM (Tosoh)
**Mobile phase:** A 50 mM pH 7.5 potassium phosphate; B MeCN:100 mM pH 2.7 potassium phosphate 32.5:67.5, containing 0.2 g/L sodium 1-heptanesulfonate
**Flow rate:** 1
**Injection volume:** 200
**Detector:** UV 210

## CHROMATOGRAM
**Retention time:** 15
**Limit of detection:** 10 ng/mL

## OTHER SUBSTANCES
**Extracted:** amitriptyline, amoxapine, clomipramine, doxepin, desipramine, maprotiline, trimipramine
**Interfering:** imipramine

## KEY WORDS
serum; column-switching; use gradient to determine metabolites

## REFERENCE
Matsumoto, K.; Kanba, S.; Kubo, H.; Yagi, G.; Iri, H.; Yuki, H. Automated determination of drugs in serum by column-switching high-performance liquid chromatography. IV. Separation of tricyclic and tetracyclic antidepressants and their metabolites. *Clin.Chem.*, **1989**, *35*, 453–456

## SAMPLE
**Matrix:** blood
**Sample preparation:** 1 mL Plasma + 50 μL MeOH:water 50:50 + 50 μL 10 μg/mL desmethyldoxepin in MeOH:water 50:50, mix, inject a 250 μL aliquot of this mixture onto column A and elute to waste with mobile phase A. After 1.5 min backflush the contents of column A onto column B with mobile phase B, monitor the effluent from column B.

## HPLC VARIABLES
**Column:** A 37-50 μm 10 × 1.5 Corasil RP C18; B 100 × 4 Techsphere 3CN (HPLC Technology)
**Mobile phase:** A water; B MeCN:50 mM acetate buffer 60:40, pH 7
**Flow rate:** A 0.8; B 0.9
**Injection volume:** 250
**Detector:** UV 215

## CHROMATOGRAM
**Retention time:** 10.0
**Internal standard:** desmethyldoxepin (8.5)
**Limit of detection:** 5-10 ng/mL

## OTHER SUBSTANCES
**Extracted:** metabolites, amitriptyline
**Simultaneous:** chloripramine, clomipramine, desipramine, doxepin, imipramine, protriptyline
**Noninterfering:** tranylcypromine
**Interfering:** cianopramine, trimipramine

## KEY WORDS
column-switching; plasma

## REFERENCE
Dadgar, D.; Power, A. Applications of column-switching technique in biopharmaceutical analysis. I. High-performance liquid chromatographic determination of amitriptyline and its metabolites in human plasma. *J.Chromatogr.*, **1987**, *416*, 99–109

## SAMPLE
**Matrix:** blood
**Sample preparation:** Condition a Bond Elut C-18 SPE cartridge twice with MeOH and twice with water. 500 µL Serum + 50 µL 1 µg/mL N-propionylprocainamide in 2.5 mM HCl, add to the SPE cartridge, wash with 2 volumes water, wash with 2 volumes 0.1 M acetic acid, wash with 1 volume MeOH:2.5 mM HCl 10:90. Add 200 µL 10 mM acetic acid and 5 mM diethylamine in MeOH to column, let stand 1 min, elute under vacuum, repeat, evaporate eluents to dryness under nitrogen at room temperature, reconstitute in 100 µL mobile phase, inject a 40 µL aliquot.

## HPLC VARIABLES
**Guard column:** Pelliguard LC-CN (Supelco)
**Column:** 150 × 4.6 5 µm Supelcosil LC-PCN
**Mobile phase:** MeCN:MeOH:10 mM pH 7.0 phosphate buffer 58:14:28
**Flow rate:** 1.2
**Injection volume:** 40
**Detector:** UV 254

## CHROMATOGRAM
**Retention time:** 14.0
**Internal standard:** N-propionylprocainamide (6)
**Limit of quantitation:** 25 ng/mL

## OTHER SUBSTANCES
**Extracted:** amitriptyline, desipramine, doxepin, imipramine, protriptyline, trimipramine
**Simultaneous:** atropine, butalbital, chlorpromazine, maprotiline, methadone, norpropoxyphene, phenylpropanolamine, procainamide, prochlorperazine, promethazine, propranolol, quinidine, trifluoperazine, trimeprazine
**Noninterfering:** acetaminophen, allopurinol, amikacin, amoxapine, amytal, bretylium, caffeine, carbamazepine, carisoprodol, chloramphenicol, chlordiazepoxide, chlorpropamide, clonazepam, codeine, diazepam, disopyramide, droperidol, ethinamate, ethinamate, ethosuximide, fluphenazine, flurazepam, furosemide, gentamicin, haloperidol, hydrochlorothiazide, hydroxyzine, ibuprofen, kanamycin, lidocaine, loxapine, meperidine, mephobarbital, meprobamate, methaqualone, methotrexate, morphine, nafcillin, naloxone, neomycin, perphenazine, phenacetin, phenobarbital, phenytoin, prazepam, primidone, procaine, propoxyphene, reserpine, salicylamide, salicylic acid, secobarbital, spironolactone, theophylline, thiopental, thioridazine, tobramycin, valproic acid, verapamil

## KEY WORDS
serum; SPE

## REFERENCE
Lin, W.-N.; Frade, P.D. Simultaneous quantitation of eight tricyclic antidepressants in serum by high-performance liquid chromatography. *Ther.Drug Monit.*, **1987**, *9*, 448–455

## SAMPLE
**Matrix:** blood
**Sample preparation:** 500 μL Serum + 250 μL di-iso-propyl ether:n-butyl alcohol 7:3 containing 800 ng/mL minaprine, centrifuge 2 min, shake, centrifuge 5 min, inject 50 μL aliquot of top organic layer.

## HPLC VARIABLES
**Guard column:** 30 × 4.6 5 μm Brownlee cyano spheri-5
**Column:** 250 × 4.6 5 μm Altex ultrasphere cyano
**Mobile phase:** MeCN:THF:water:2 M ammonium formate (pH 4.0) 700:100:195:5
**Column temperature:** 20
**Flow rate:** 1.5
**Injection volume:** 50
**Detector:** UV 240

## CHROMATOGRAM
**Retention time:** 6.5
**Internal standard:** minaprine (5.5)
**Limit of detection:** 20 ng/mL

## OTHER SUBSTANCES
**Simultaneous:** amitriptyline, diltiazem
**Also analyzed:** amiodarone, clomipramine, desipramine, haloperidol, imipramine, propafenone, verapamil

## KEY WORDS
serum

## REFERENCE
Mazzi, G. Simple and practical high-performance liquid chromatographic assay of some tricyclic drugs, haloperidol, diltiazem, verapamil, propafenone, and amiodarone. *Chromatographia*, **1987**, *24*, 313–316

## SAMPLE
**Matrix:** blood
**Sample preparation:** 1 mL Serum + 1 mL 450 mM NaOH + 5 mL hexane:isopropanol 95:5, shake for 5 min, centrifuge. Remove 4 mL of the organic layer and add it to 50 μL 200 mM HCl, shake for 2 min, centrifuge. Inject a 20 μL aliquot of the aqueous layer.

## HPLC VARIABLES
**Column:** 100 × 3 octyl CP-tm-Spher C8 glass column (Chrompack)
**Mobile phase:** MeCN:500 mM $NaH_2PO_4$ 35:65 adjusted to pH 2.2 with phosphoric acid
**Flow rate:** 0.8
**Injection volume:** 20
**Detector:** UV 210

## CHROMATOGRAM
**Retention time:** 4.5
**Limit of detection:** 10 ng/mL

## OTHER SUBSTANCES
**Simultaneous:** amitriptyline, doxepin

## KEY WORDS
serum

## REFERENCE
Van Damme, M.; Molle, L.; Abi Khalil, F. Useful sample handlings for reversed phase high performance liquid chromatography in emergency toxicology. *J.Toxicol.Clin.Toxicol.*, **1985**, *23*, 589–614

## SAMPLE
**Matrix:** blood
**Sample preparation:** 2 mL Serum or plasma + 100 μL 1 μg/mL IS in water + 0.5 mL water, vortex, extract with 10 mL toluene:isoamyl alcohol 99:1 for 10 min on a rotator, centrifuge for 5 min. Remove upper organic layer, evaporate under a stream of nitrogen at 37°, take up in 150 μL mobile phase, vortex for 2 min, add 0.5 mL hexane, vortex briefly, centrifuge for 5 min, discard upper hexane layer, inject a 100 μL aliquot of the lower layer.

## HPLC VARIABLES
**Column:** 250 × 4 Bio-Sil ODS-10 (Bio-Rad)
**Mobile phase:** MeCN:pH 4.5 50 mM phosphate buffer 30:70 (Buffer was 6.9 g $KH_2PO_4$ in 1 L adjusted to pH 4.5 with orthophosphoric acid.)
**Column temperature:** 45
**Flow rate:** 2.5
**Injection volume:** 100
**Detector:** UV 202

## CHROMATOGRAM
**Retention time:** 8.4
**Internal standard:** U-31485 (6.9)
**Limit of detection:** 1 ng/mL

## OTHER SUBSTANCES
**Extracted:** desipramine, protriptyline
**Noninterfering:** N-acetylprocainamide, amitriptyline, caffeine, chlordiazepoxide, chlorpromazine, diazepam, flurazepam, lorazepam, oxazepam, prazepam, procainamide, propranolol, thioridazine
**Interfering:** alprazolam, imipramine, triazolam

## KEY WORDS
serum; plasma

## REFERENCE
McCormick, S.R.; Nielsen, J.; Jatlow, P. Quantification of alprazolam in serum or plasma by liquid chromatography. *Clin.Chem.*, **1984**, *30*, 1652–1655

## SAMPLE
**Matrix:** blood
**Sample preparation:** 500 μL Plasma + 37 μL 2 μg/mL IS in MeOH + 500 μL pH 10 borate buffer + 1.5 mL hexane:isoamyl alcohol 95:5, shake for 10 min. Evaporate the organic layer to dryness under a stream of nitrogen, reconstitute in 100 μL MeOH, inject a 50 μL aliquot. (The borate buffer was prepared as follows. Prepare a solution of 61.8 g boric acid and 74.6 g KCl in 1 L water. Add 630 mL of this solution to 370 mL 106 g/L sodium carbonate solution. Adjust pH to 10.0 with 6 M NaOH and store at 35-37°.)

## HPLC VARIABLES
**Column:** 250 × 4.6 Zorbax Sil
**Mobile phase:** MeOH:ammonium hydroxide 998:2
**Flow rate:** 1.5
**Injection volume:** 50
**Detector:** UV 254

## CHROMATOGRAM
**Retention time:** 9
**Internal standard:** N-desmethylclomipramine hydrochloride (10)
**Limit of quantitation:** 20 ng/mL

## OTHER SUBSTANCES
**Extracted:** metabolites, amitriptyline, desipramine, 2-hydroxydesipramine, 2-hydroxyimipramine, imipramine
**Also analyzed:** clomipramine, desmethylclomipramine, desmethyldoxepin, doxepin, maprotiline, protriptyline
**Noninterfering:** chlordiazepoxide, diazepam, flurazepam, oxazepam, thioridazine

## KEY WORDS
plasma

## REFERENCE
Sutfin, T.A.; D'Ambrosio, R.; Jusko, W.J. Liquid-chromatographic determination of eight tri- and tetracyclic antidepressants and their major active metabolites. *Clin.Chem.*, **1984**, *30*, 471–474

## SAMPLE
**Matrix:** blood
**Sample preparation:** Condition a Bond-Elut C18 column with 2 volumes MeOH then 2 volumes water. Add 1 mL serum then 200 μL 700 ng/mL promazine in MeOH:0.1 M HCl 13:87 to each column, wash with 2 volumes water, wash with 2 volumes 0.1 M acetic acid, wash with MeOH/water, add 200 μL 10 mM ammonium acetate in MeOH, wait for 30 s, elute with vacuum, repeat elution process two more times. Combine eluates and evaporate them to dryness at 56-8° under compressed air. Reconstitute with 200 μL mobile phase, vortex 10 s, inject 75-100 μL aliquot. (MeOH/water was 500 mL MeOH:water 65:35 plus 25 μL concentrated HCl.)

## HPLC VARIABLES
**Column:** 250 × 4.6 5 μm Supelco
**Mobile phase:** 1 gal EtOH + 77 mL MeCN + 1.9 mL t-butylamine (EtOH:MeCN:t-butylamine 98:2:0.05)
**Flow rate:** 2
**Injection volume:** 75-100
**Detector:** UV 254

## CHROMATOGRAM
**Retention time:** 6.9
**Internal standard:** promazine (5.2)
**Limit of detection:** 4 ng/mL

## OTHER SUBSTANCES
**Extracted:** amitriptyline, desipramine, desmethyldoxepin, doxepin, imipramine, protriptyline
**Simultaneous:** N-acetylprocainamide, amoxapine, amphetamine, buprion, chlordiazepoxide, chlorimipramine, chlorpheniramine, chlorpromazine, cocaine, codeine, demoxepam, desmethylchlordiazepoxide, desmethyldisopyramide, dextropropoxyphene, diazepam, di-

sopyramide, fluphenazine, hydroxyamoxapine, 2-hydroxydesipramine, 2-hydroxyimipramine, 10-hydroxynortriptyline, iminostilbene, iprindole, loxepin, maprotiline, meperidine, methadone, mianserin, morphine, norzimeldine, oxapam, oxaprotiline, perphenazine, phenteramine, procainamide, prolixin, promethazine, propoxyphene, pyrilamine, quinidine, thioridazine, trifluoperazine, trifluopromazine, trimeprazine, trimipramine, zimeldine
**Noninterfering:** thiopropazine
**Interfering:** prochlorperazine

## KEY WORDS
serum; normal phase

## REFERENCE
Beierle, F.A.; Hubbard, R.W. Liquid chromatographic separation of antidepressant drugs: I. Tricyclics. *Ther.Drug Monit.*, **1983**, *5*, 279–292

## SAMPLE
**Matrix:** blood
**Sample preparation:** 2 mL Plasma + 100 μL 1 μg/mL loxapine in isopropanol:diethylamine 99.9:0.1 + 250 μL 25% potassium carbonate containing 0.1% diethylamine + 5 mL hexane:isoamyl alcohol 97:3, vortex for 30 s, centrifuge at 500 g for 3 min. Remove the organic layer and add it to 100 μL 250 mM HCl, vortex for 30 s, inject a 50 μL aliquot of the aqueous phase.

## HPLC VARIABLES
**Guard column:** 50 × 4.6 40 μm C8 (Supelco)
**Column:** 250 × 4.6 5 μm Supelcosil C8
**Mobile phase:** MeCN:water:diethylamine:85% phosphoric acid 53.3:45.1:1:0.4, pH adjusted to 7.2 with NaOH or phosphoric acid
**Flow rate:** 2
**Injection volume:** 50
**Detector:** UV 254

## CHROMATOGRAM
**Retention time:** k′ 3.45
**Internal standard:** loxapine (k′ 7.18)
**Limit of detection:** 2.5 ng/mL

## OTHER SUBSTANCES
**Extracted:** amitriptyline, chlordiazepoxide, chlorpromazine, desipramine, desmethldiazepam, desmethylchlordiazepoxide, desmethyldoxepin, diazepam, doxepin, fluphenazine, imipramine, oxazepam
**Noninterfering:** molindone, perphenazine, trifluoperazine
**Interfering:** haloperidol, thiothixene

## KEY WORDS
plasma

## REFERENCE
Kiel, J.S.; Abramson, R.K.; Morgan, S.L.; Voris, J.C. A rapid high performance liquid chromatographic method for the simultaneous measurement of six tricyclic antidepressants. *J.Liq.Chromatogr.*, **1983**, *6*, 2761–2773

## SAMPLE
**Matrix:** blood

**Sample preparation:** 10 mL Plasma or whole blood + 1 mL 1 M NaOH, extract twice with 10 mL hexane for 30 min. Remove the organic layers and evaporate them to dryness under a stream of nitrogen, reconstitute the residue in 1 mL 100 mM HCl, add 5 mL chloroform, vortex for 1 min, centrifuge. Remove a 4.5 mL aliquot of the organic layer and evaporate it to dryness, reconstitute the residue in 100 μL mobile phase, inject a 50 μL aliquot. (It is implied, but not explicitly stated in the paper, that this extraction procedure works for this compound.)

## HPLC VARIABLES
**Column:** 10 μm Micropak CN (Varian)
**Mobile phase:** MeCN:20 mM ammonium acetate 90:10
**Flow rate:** 2.5
**Injection volume:** 50
**Detector:** UV 254

## CHROMATOGRAM
**Retention time:** 10.7
**Limit of detection:** 10 ng/mL

## OTHER SUBSTANCES
**Simultaneous:** acetophenazine, amitriptyline, benztropine, butaperazine, carphenazine, chlorpromazine, fluphenazine, haloperidol, imipramine, mesoridazine, orphenadrine, piperacetazine, promazine, promethazine, thioridazine, thiothixene, trifluoperazine, triflupromazine, trihexyphenidyl, trimeprazine

## KEY WORDS
plasma; whole blood

## REFERENCE
Curry, S.H.; Brown, E.A.; Hu, O.Y.-P.; Perrin, J.H. Liquid chromatographic assay of phenothiazine, thioxanthene and butyrophenone neuroleptics and antihistamines in blood and plasma with conventional and radial compression columns and UV and electrochemical detection. *J.Chromatogr.*, **1982**, *231*, 361–376

## SAMPLE
**Matrix:** blood
**Sample preparation:** 1 mL Serum + 200 μL 10 μg/mL protriptyline in water + 200 μL 80 g/L NaHCO₃ + 5 mL hexane, vortex for 15 s, centrifuge for 5 min. Remove the hexane layer and evaporate it in a stream of nitrogen at 60°. Reconstitute in 100 μL mobile phase, vortex for 15 s, inject a 50 μL aliquot.

## HPLC VARIABLES
**Column:** 300 × 4 10 μm μBondapak CN
**Mobile phase:** MeCN:MeOH:5 mM phosphate buffer 60:15:25, adjusted to pH 7.0
**Flow rate:** 2
**Injection volume:** 50
**Detector:** UV 254

## CHROMATOGRAM
**Retention time:** 7.10
**Internal standard:** protriptyline (12.20)
**Limit of detection:** 10 ng/mL

## OTHER SUBSTANCES
**Extracted:** amitriptyline, chlorpromazine, desipramine, desmethyldoxepin, doxepin, imipramine, maprotiline, procainamide, propoxyphene, propranolol, thioridazine, trimipramine

**Noninterfering:** acetaminophen, caffeine, chlordiazepoxide, diazepam, methaqualone, salicylic acid, theophylline, trifluoperazine
**Interfering:** disopyramide

## KEY WORDS
serum

## REFERENCE
Koteel, P.; Mullins, R.E.; Gadsden, R.H. Sample preparation and liquid-chromatographic analysis for tricyclic antidepressants in serum. *Clin.Chem.*, **1982**, *28*, 462–466

## SAMPLE
**Matrix:** blood
**Sample preparation:** 2 mL Plasma + 1600 ng clomipramine in MeOH + 2 mL 1 M NaOH + 5 mL hexane:isoamyl alcohol 99:1, shake mechanically for 15 min, centrifuge at 1686 g for 5 min. Remove the organic phase and add it to 200 μL 0.05% orthophosphoric acid, shake for 15 min, centrifuge for 5 min, inject a 50 μL aliquot of the aqueous phase.

## HPLC VARIABLES
**Guard column:** μBondapak/Porasil
**Column:** μBondapak C18
**Mobile phase:** MeCN:buffer 40:60 (Buffer was 13.68 g $KH_2PO_4$ in 2 L water, adjusted to pH 4.7 with dilute KOH.)
**Column temperature:** 50
**Flow rate:** 2
**Injection volume:** 50
**Detector:** UV 254

## CHROMATOGRAM
**Retention time:** 4.5
**Internal standard:** clomipramine (7.5)
**Limit of detection:** 0.6 ng

## OTHER SUBSTANCES
**Extracted:** amitriptyline, desipramine, imipramine
**Simultaneous:** chlordiazepoxide, chlorpromazine, cimetidine, clomipramine, diazepam, doxepin, flurazepam, lorazepam, oxazepam, pentobarbital, perphenazine, phenobarbital, phenytoin, prochlorperazine, propoxyphene, secobarbital, thioridazine, trifluoperazine
**Noninterfering:** acetaminophen, codeine, meperidine

## KEY WORDS
plasma

## REFERENCE
Wong, S.H.Y.; McCauley, T. Reversed phase high-performance liquid chromatographic analysis of tricyclic antidepessants in plasma. *J.Liq.Chromatogr.*, **1981**, *4*, 849–862

## SAMPLE
**Matrix:** blood, dialysate
**Sample preparation:** Adjust pH of serum samples to 7.4 and dialyze 3 mL serum against 5 mL dialysis buffer at 37° in PTFE chambers for 4 h. Inject 1 mL mobile phase, 1 mL water, 2 mL dialysate, 1 mL water, and 1 mL MeCN:water 50:50 onto column A. Elute column A onto column B with mobile phase for 30 s then remove it from the circuit. Elute column B with mobile phase and monitor the effluent. (Dialysis buffer was 3.998 g $Na_2HPO_4.2H_2O$ + 0.775 g $NaH_2PO_4$ + 2.250 g NaCl + 0.055 g $Hg(NO_3)_2$ in 1 L water, pH was 7.4.)

## HPLC VARIABLES
**Column:** A 10 × 6 packed with 40 μm material from a Bond Elut cartridge (cat. no. 620303); B 100 × 4 3 μm Spherisorb ODS Superpac
**Mobile phase:** MeCN:85% phosphoric acid:triethylamine:water 49.55:0.225:0.225:50
**Flow rate:** 0.65
**Injection volume:** 1000-2000
**Detector:** UV 238

## CHROMATOGRAM
**Retention time:** 3.85
**Limit of detection:** 0.5 nM (5 mL sample)

## OTHER SUBSTANCES
**Extracted:** amitriptyline
**Simultaneous:** alprazolam, chlorpromazine, chlorprothixene, clomipramine, desclomipramine, desmethylimipramine, diazepam, flunitrazepam, fluphenazine, haloperidol, imipramine, levomepromazine, perphenazine, protriptyline, thioridazine, thioridazine sulfone, trimipramine, zimeldine, zuclopenthixol
**Noninterfering:** carbamazepine, clonazepam, lorazepam, nitrazepam, oxazepam, phenytoin
**Interfering:** maprotiline, promethazine, thioridazine sulfoxide

## KEY WORDS
serum; column-switching

## REFERENCE
Svensson, C.; Nyberg, G.; Mårtensson, E. High-performance liquid chromatographic quantitation of amitriptyline and nortriptyline in dialysate from plasma or serum using on-line solid-phase extraction. *J.Chromatogr.*, **1988**, *432*, 363–369

## SAMPLE
**Matrix:** blood, gastric contents, tissue, urine
**Sample preparation:** 1 mL Blood, urine, or gastric contents or 1 g tissue homogenate + 500 μL buffer + 8 mL n-hexane:ethyl acetate 70:30, mix on a rotary mixer for 10 min, centrifuge at 3000 g for 8 min. Remove the organic layer and evaporate it to dryness under a stream of nitrogen, reconstitute the residue in 100 μL 12.5 mM NaOH in MeOH:water 50:50, inject a 50 μL aliquot. (Buffer was 13.8 g potassium carbonate in 100 mL water, pH adjusted to 9.5 with concentrated HCl.)

## HPLC VARIABLES
**Guard column:** 4 × 4 30 μm LiChrocart Aluspher RP-select B (Merck)
**Column:** 125 × 4 5 μm Aluspher RP-select B (Merck)
**Mobile phase:** Gradient. A was 12.5 mM NaOH in MeOH. B was 12.5 mM NaOH in water. A:B 10:90 for 5 min, to 90:10 over 15 min, maintain at 90:10 for 5 min, return to initial conditions over 1 min, re-equilibrate for 5 min.
**Flow rate:** 1
**Injection volume:** 50
**Detector:** UV 230; UV 254

## CHROMATOGRAM
**Retention time:** 19

## OTHER SUBSTANCES
**Extracted:** alprenolol, amitriptyline, bromazepam, carbamazepine, chlordiazepoxide, chlorpromazine, clonazepam, desipramine, diazepam, flunitrazepam, haloperidol, nitrendipine, nordiazepam, pindolol, zolpidem

**Also analyzed:** acebutolol, acetaminophen, alprazolam, amphetamine, atenolol, betaxolol, brotizolam, caffeine, camazepam, captopril, chloroquine, clobazam, clomipramine, clothiapine, clotiazepam, cloxazolam, cocaine, codeine, diclofenac, dihydralazine, dihydrocodeine, dihydroergotamine, diphenhydramine, domperidone, doxepin, droperidol, ergotamine, ethyl loflazepate, fenethylline, fluoxetine, flupentixol, flurazepam, furosemide, gliclazide, hydrochlorothiazide, hydroxyzine, ibuprofen, imipramine, ketazolam, loprazolam, lorazepam, lormetazepam, maprotiline, medazepam, mepyramine, methadone, methaqualone, methyldopa, methylphenidate, metoclopramide, metoprolol, mexiletine, mianserin, midazolam, minoxidil, morphine, nadolol, nitrazepam, oxprenolol, papaverine, pentazocine, phenprocoumon, phenylbutazone, pipamperone, piritramide, practolol, prazepam, prazosin, promazine, promethazine, propoxyphene, propranolol, prothipendyl, quinine, sotalol, sulpride, thioridazine, trazodone, triazolam, trimipramine, tripelennamine, tyramine, verapamil, yohimbine

## REFERENCE

Lambert, W.E.; Meyer, E.; De Leenheer, A.P. Systematic toxicological analysis of basic drugs by gradient elution of an alumina-based HPLC packing material under alkaline conditions. *J.Anal.Toxicol.*, **1995**, *19*, 73–78

## SAMPLE

**Matrix:** blood, tissue

**Sample preparation:** Whole blood, serum. 1 mL Whole blood or serum + 1 µg cianopramine + 1 mL water, vortex, add 1 mL 200 mM sodium carbonate, vortex, add 6 mL hexane:1-butanol 95:5, gently agitate for 30 min, centrifuge at 2500 g for 5 min. Remove the organic layer and add it to 100 µL 0.2% phosphoric acid, agitate gently for 30 min, centrifuge for 5 min. Remove the organic layer and inject a 30 µL aliquot of the aqueous layer. Tissue. 0.5 mL Liver homogenate + 10 µg cianopramine + 500 µL 2% sodium tetraborate + 8 mL hexane:1-butanol 95:5, gently agitate for 30 min, centrifuge at 2500 g for 5 min. Remove the organic layer and add it to 400 µL 0.2% phosphoric acid, agitate gently for 30 min, centrifuge for 5 min. Remove the organic layer and inject a 30 µL aliquot of the aqueous layer.

## HPLC VARIABLES

**Guard column:** 15 × 3.2 7 µm RP-18 Newguard (Applied Biosystems)
**Column:** 100 × 4.6 5 µm Brownlee Spheri-5 RP-18
**Mobile phase:** MeCN:100 mM NaH$_2$PO$_4$:diethylamine 40:57.5:2.5
**Flow rate:** 2
**Injection volume:** 30
**Detector:** UV 220

## CHROMATOGRAM

**Retention time:** 7.27
**Internal standard:** cianopramine (8.93)
**Limit of detection:** 50 ng/mL

## OTHER SUBSTANCES

**Simultaneous:** amitriptyline, amoxapine, benztropine, brompheniramine, chlorpheniramine, chlorpromazine, clomipramine, cyproheptadine, desipramine, dothiepin, doxepin, fluoxetine, haloperidol, imipramine, loxapine, maprotiline, meperidine, mesoridazine, methadone, metoclopramide, mianserin, moclobemide, nomifensine, nordoxepin, norfluoxetine, norpropoxyphene, northiaden, pentobarbital, pheniramine, promethazine, propoxyphene, propranolol, protriptyline, quinidine, quinine, sulforidazine, thioridazine, thiothixene, tranylcypromine, trazodone, trihexyphenidyl, trimipramine, triprolidine

**Noninterfering:** dextromethorphan, norphetidine, phenoxybenzamine, prochlorperazine, trifluoperazine

**Interfering:** diphenhydramine

## KEY WORDS
serum; whole blood; liver

## REFERENCE
McIntyre, I.M.; King, C.V.; Skafidis, S.; Drummer, O.H. Dual ultraviolet wavelength high-performance liquid chromatographic method for the forensic or clinical analysis of seventeen antidepressants and some selected metabolites. *J.Chromatogr.*, **1993**, *621*, 215−223

## SAMPLE
**Matrix:** blood, tissue
**Sample preparation:** Serum. 1 mL Serum + 4 mL MeOH + 5 mL 2.5% perchloric acid, shake vigorously, centrifuge at 11000 g for 15 min. Add supernatant to 1 mL 4 M KOH, centrifuge. Add supernatant (9 mL) to 10 mL diethyl ether:ethyl acetate 85:15, shake for 15 min. Remove 8 mL of organic phase and evaporate it to dryness under a stream of nitrogen. Dissolve residue in 200 μL mobile phase buffer:MeOH 9:1, inject 100 μL aliquot. Tissue. 2 g Brain tissue + 10 mL 2.5% perchloric acid + 8 mL MeOH, homogenize, centrifuge at 11000 g for 15 min. Add supernatant to 4 mL 4 M KOH, centrifuge. Add supernatant to 20 mL diethyl ether:ethyl acetate 3:1, shake for 15 min. Remove 8 mL of organic phase and evaporate it to dryness under a stream of nitrogen. Dissolve residue in 200 μL mobile phase buffer:MeOH 9:1, inject 100 μL aliquot.

## HPLC VARIABLES
**Column:** 250 × 4.6 5 μm Cosmosil 5C18
**Mobile phase:** MeOH:THF:buffer 45:17:88 (Buffer was 1% triethylamine adjusted to pH 3.0 with phosphoric acid.)
**Column temperature:** 40
**Flow rate:** 1
**Injection volume:** 100
**Detector:** UV 254

## CHROMATOGRAM
**Retention time:** 14.2
**Internal standard:** nortriptyline
**Limit of detection:** 10 ng/g (tissue)

## OTHER SUBSTANCES
**Extracted:** desipramine, imipramine

## KEY WORDS
serum; rat; nortiptyline is IS

## REFERENCE
Sugita, S.; Kobayashi, A.; Suzuki, S.; Yoshida, T.; Nakazawa, K. High-performance liquid chromatographic determination of imipramine and its metabolites in rat brain. *J.Chromatogr.*, **1987**, *421*, 412−417

## SAMPLE
**Matrix:** hair
**Sample preparation:** Wash hair in water, rinse 3 times with MeOH, dry, weigh. 5-25 mg Washed hair + 1 mL 1 M NaOH, heat at 70° for 30 min, adjust pH to 9.5-10. 1 mL Extract + 1 μg protriptyline + 1 mL water + 1 mL 200 mM sodium carbonate buffer, mix, extract with hexane:butanol 95:5 for 20 min. Remove the organic layer and add it to 100 μL 0.2% orthophosphoric acid, mix for 20 min, inject a 30 μL aliquot of the aqueous layer.

## HPLC VARIABLES
**Guard column:** 15 × 3.2 7 μm Newguard RP-18

**Column:** 100 × 4.6 Spheri-5 RP-C18
**Mobile phase:** MeCN:buffer 40:60 (Buffer was 1.2 L 100 mM pH 7.0 $NaH_2PO_4$ + 30 mL diethylamine.)
**Flow rate:** 2
**Injection volume:** 30
**Detector:** UV 214

## CHROMATOGRAM
**Retention time:** 5
**Internal standard:** protriptyline (4)

## OTHER SUBSTANCES
**Extracted:** amitriptyline, clomipramine, desipramine, dothiepin, doxepin, haloperidol, imipramine, mianserin

## KEY WORDS
may be interferences

## REFERENCE
Couper, F.J.; McIntyre, I.M.; Drummer, O.H. Extraction of psychotropic drugs from human scalp hair. *J.Forensic Sci.*, **1995**, *40*, 83–86

## SAMPLE
**Matrix:** microsomal incubations
**Sample preparation:** 250 µL Microsomal incubation + 50 µL 1 M HCl, cool on ice, add desipramine, centrifuge at 16000 g for 5 min, inject an aliquot of the supernatant.

## HPLC VARIABLES
**Column:** 300 × 3.9 µBondapak C18
**Mobile phase:** MeCN:50 mM pH 6 potassium phosphate buffer 35:65
**Flow rate:** 1.4
**Detector:** UV 220

## CHROMATOGRAM
**Retention time:** 10.5
**Internal standard:** desipramine (9)
**Limit of detection:** LOD 50 pmole

## OTHER SUBSTANCES
**Extracted:** amitriptyline
**Noninterfering:** ketoconazole, α-naphthoflavone, quinidine

## KEY WORDS
human; liver

## REFERENCE
Schmider, J.; Greenblatt, D.J.; von Moltke, L.L.; Harmatz, J.S.; Shader, R.I. N-Demethylation of amitriptyline *in vitro*: Role of cytochrome P-450 3A (CYP3A) isoforms and effect of metabolic inhibitors. *J.Pharm.Exp.Ther.*, **1995**, *275*, 592–597

## SAMPLE
**Matrix:** solutions

## HPLC VARIABLES
**Column:** 250 × 4.6 5 µm Supelcosil LC-DP (A) or 250 × 4 5 µm LiChrospher 100 RP-8 (B)

**Mobile phase:** MeCN:0.025% phosphoric acid:buffer 25:10:5 (A) or 60:25:15 (B) (Buffer was 9 mL concentrated phosphoric acid and 10 mL triethylamine in 900 mL water, adjust pH to 3.4 with dilute phosphoric acid, make up to 1 L.)
**Flow rate:** 0.6
**Injection volume:** 25
**Detector:** UV 229

## CHROMATOGRAM
**Retention time:** 13.70 (A), 6.75 (B)

## OTHER SUBSTANCES
**Also analyzed:** acebutolol, acepromazine, acetaminophen, acetazolamide, acetophenazine, albuterol, alprazolam, amitriptyline, amobarbital, amoxapine, antipyrine, atenolol, atropine, azatadine, baclofen, benzocaine, bromocriptine, brompheniramine, brotizolam, bupivacaine, buspirone, butabarbital, butalbital, caffeine, carbamazepine, cetirizine, chlorcyclizine, chlordiazepoxide, chlormezanone, chloroquine, chlorpheniramine, chlorpromazine, chlorpropamide, chlorprothixene, chlorthalidone, chlorzoxazone, cimetidine, cisapride, clomipramine, clonazepam, clonidine, clozapine, cocaine, codeine, colchicine, cyclizine, cyclobenzaprine, dantrolene, desipramine, diazepam, diclofenac, diflunisal, diltiazem, diphenhydramine, diphenidol, diphenoxylate, dipyridamole, disopyramide, dobutamine, doxapram, doxepin, droperidol, encainide, ethidium bromide, ethopropazine, fenoprofen, fentanyl, flavoxate, fluoxetine, fluphenazine, flurazepam, flurbiprofen, fluvoxamine, furosemide, glutethimide, glyburide, guaifenesin, haloperidol, homatropine, hydralazine, hydrochlorothiazide, hydrocodone, hydromorphone, hydroxychloroquine, hydroxyzine, ibuprofen, imipramine, indomethacin, ketoconazole, ketoprofen, ketorolac, labetalol, levorphanol, lidocaine, loratadine, lorazepam, lovastatin, loxapine, mazindol, mefenamic acid, meperidine, mephenytoin, mepivacaine, mesoridazine, metaproterenol, metformin, methadone, methdilazine, methocarbamol, methotrexate, methotrimeprazine, methoxamine, methyldopa, methylphenidate, metoclopramide, metolazone, metoprolol, metronidazole, midazolam, moclobemide, morphine, nadolol, nalbuphine, naloxone, naphazoline, naproxen, nifedipine, nizatidine, norepinephrine, oxazepam, oxycodone, oxymetazoline, paroxetine, pemoline, pentazocine, pentobarbital, pentoxifylline, perphenazine, pheniramine, phenobarbital, phenol, phenolphthalein, phentolamine, phenylbutazone, phenyltoloxamine, phenytoin, pimozide, pindolol, piroxicam, pramoxine, prazepam, prazosin, probenecid, procainamide, procaine, prochlorperazine, procyclidine, promazine, promethazine, propafenone, propantheline, propiomazine, propofol, propranolol, protriptyline, quazepam, quinidine, quinine, racemethorphan, ranitidine, remoxipride, risperidone, salicylic acid, scopolamine, secobarbital, sertraline, sotalol, spironolactone, sulfinpyrazone, sulindac, temazepam, terbutaline, terfenadine, tetracaine, theophylline, thiethylperazine, thiopental, thioridazine, thiothixene, timolol, tocainide, tolbutamide, tolmetin, trazodone, triamterene, triazolam, trifluoperazine, triflupromazine, trimeprazine, trimethoprim, trimipramine, verapamil, warfarin, xylometazoline, yohimbine, zopiclone

## KEY WORDS
some details of plasma extraction

## REFERENCE
Koves, E.M. Use of high-performance liquid chromatography-diode array detection in forensic toxicology. *J.Chromatogr.A,* **1995,** *692,* 103–119

## SAMPLE
**Matrix:** solutions

## HPLC VARIABLES
**Guard column:** 30 × 2.1 Spheri-5 RP-8
**Column:** 220 × 2.1 Spheri-5 RP-8

**Mobile phase:** Gradient. A was 0.08% diethylamine and 0.09% phosphoric acid in water, pH 2.3. B was MeCN:water 90:10 containing 0.08% diethylamine and 0.09% phosphoric acid. A:B 95:5 for 2 min, to 0:100 over 15 min (?), maintain at 0:100 for 5 min.
**Column temperature:** 50
**Flow rate:** 0.5
**Detector:** UV 200

## CHROMATOGRAM
**Retention time:** 13.8

## OTHER SUBSTANCES
**Simultaneous:** amitriptyline, desipramine, desmethyldoxepin, doxepin, imipramine
**Also analyzed:** amphetamine, chlordiazepoxide, chlorpromazine, desalkylflurazepam, diazepam, diethylpropion, ephedrine, fenfluramine, flurazepam, mesoridazine, methamphetamine, norchlordiazepoxide, nordiazepam, oxazepam, phentermine, phenylpropanolamine, prazepam, promazine, thioridazine, thiothixene, trifluoperazine

## REFERENCE
*Rainin Catalog Ci-94, 1994-5, Rainin Instrument Co., Woburn MA,* **1994**, p. 7.24

## SAMPLE
**Matrix:** solutions
**Sample preparation:** Prepare a 1 mg/mL solution in MeOH, inject a 5 µL aliquot.

## HPLC VARIABLES
**Column:** 250 × 4.6 5 µm Lichrosphere cyanopropyl
**Mobile phase:** Carbon dioxide:MeOH:isopropylamine 90:10:0.05
**Column temperature:** 50
**Flow rate:** 3
**Injection volume:** 5
**Detector:** UV 220

## CHROMATOGRAM
**Retention time:** 4.03

## OTHER SUBSTANCES
**Simultaneous:** amitriptyline, benactyzine, buclizine, desipramine, hydroxyzine, imipramine, perphenazine, protriptyline, thioridazine

## KEY WORDS
SFC; pressure 200 bar

## REFERENCE
Berger, T.A.; Wilson, W.H. Separation of drugs by packed column supercritical fluid chromatography. 2. Antidepressants. *J.Pharm.Sci.*, **1994**, *83*, 287–290

## SAMPLE
**Matrix:** solutions

## HPLC VARIABLES
**Column:** 250 × 4.6 Zorbax RX
**Mobile phase:** Gradient. A was 10 mL concentrated orthophosphoric acid and 7 mL triethylamine in 1 L water. B was 10 mL concentrated orthophosphoric acid and 7 mL triethylamine in 200 mL water, make up to 1 L with MeCN. A:B from 100:0 to 0:100 over 30 min, maintain at 0:100 for 5 min.

**Column temperature:** 30
**Flow rate:** 2
**Detector:** UV 210

## OTHER SUBSTANCES

**Also analyzed:** acepromazine, acetaminophen, acetophenazine, albuterol, aminophylline, amitriptyline, amobarbital, amoxapine, amphetamine, amylocaine, antipyrine, aprobarbital, aspirin, atenolol, atropine, avermectin, barbital, benzocaine, benzoic acid, benzotropine, benzphetamine, berberine, bibucaine, bromazepan, brompheniramine, buprenorphine, buspirone, butabarbital, butacaine, butethal, caffeine, carbamazepine, carbromal, chloramphenicol, chlordiazepoxide, chloroquine, chlorothiazide, chloroxylenol, chlorphenesin, chlorpheniramine, chlorpromazine, chlorpropamide, chlortetracycline, cimetidine, cinchonidine, cinchonine, clenbuterol, clonazepam, clonixin, clorazepate, cocaine, codeine, colchicine, cortisone, coumarin, cyclazocine, cyclobenzaprine, cyclothiazide, cyheptamide, cymarin, danazol, danthron, dapsone, debrisoquine, desipramine, dexamethasone, dextromethorphan, dextropropoxyphene, diamorphine, diazepam, diclofenac, diethylpropion, diethylstilbestrol, diflunisal, digitoxin, digoxin, diltiazem, diphenhydramine, diphenoxylate, diprenorphine, dipyrone, disulfiram, dopamine, doxapram, doxepin, dronabinol, ephedrine, epinephrine, epinine, estradiol, estriol, estrone, ethacrynic acid, ethosuximide, etonitazene, etorphine, eugenol, famotidine, fenbendazole, fencamfamine, fenoprofen, fenproporex, fentanyl, flubendazole, flufenamic acid, flunitrazepam, 5-fluorouracil, fluoxymesterone, fluphenazine, furosemide, gentisic acid, gitoxigenin, glipizide, glunixin, glutethimide, glybenclamide, guaiacol, halazepam, haloperidol, hydrochlorothiazide, hydrocodone, hydrocortisone, hydromorphone, hydroxyquinoline, ibogaine, ibuprofen, iminostilbene, imipramine, indomethacin, isocarbostyril, isocarboxazid, isoniazid, isoproterenol, isoxsuprine, ivermectin, ketamine, ketoprofen, kynurenic acid, levorphanol, lidocaine, lorazepam, lormetazepam, loxapine, mazindol, mebendazole, meclizine, meclofenamic acid, medazepam, mefenamic acid, megestrol, mepacrine, meperidine, mephentermine, mephenytoin, mephesin, mephobarbital, mepivacaine, mescaline, mesoridazine, methadone, methamphetamine, methapyrilene, methaqualone, methazolamide, methocarbamol, methoxamine, methsuximide, methyl salicylate, methyldopa, methyldopamine, methylphenidate, methylprednisolone, methyltestosterone, methyprylon, metoprolol, mibolerone, morphine, nadolol, nalorphine, naloxone, naltrexone, naphazoline, naproxen, nefopam, niacinamide, nicotine, nicotinic acid, nifedipine, niflumic acid, nitrazepam, norepinephrine, noscapine, nylidrin, oxazepam, oxycodone, oxymorphone, oxyphenbutazone, oxytetracycline, papaverine, pargyline, pemoline, pentazocine, pentobarbital, persantine, phenacetin, phenazocine, phenazopyridine, phencyclidine, phendimetrazine, phenelzine, pheniramine, phenobarbital, phenothiazine, phensuximide, phentermine, phenylbutazone, phenylephrine, phenylpropanolamine, piperocaine, prazepam, prednisolone, primidone, probenecid, progesterone, propiomazine, propranolol, propylparaben, pseudoephedrine, puromycin, pyrilamine, pyrithyldione, quazepam, quinaldic acid, quinidine, quinine, ranitidine, recinnamine, reserpine, resorcinol, saccharin, albuterol, salicylamide, salicylic acid, scopolamine, scopoletin, secobarbital, strychnine, sulfacetamide, sufadiazine, sulfadimethoxine, sulfaethidole, sulfamerazine, sulfamethazine, sulfamethoxizole, sulfanilamide, sulfapyridine, sulfasoxizole, sulindac, tamoxifen, temazepam, testosterone, tetracaine, tetracycline, tetramisole, thebaine, theobromine, theophylline, thiabendazole, thiamine, thiamylal, thiobarbituric acid, thioridazine, thiosalicylic acid, thiothixene, thymol, tolazamide, tolazoline, tobutamide, tolmetin, tranylcypromine, triamcinolone, tribenzylamine, trichloromethiazide, trifluoperazine, trihexyphenidyl, trimethoprim, tripelennamine, triprolidine, tropacocaine, tyramine, verapamil, vincamine, warfarin, yohimbine, zoxazolamine

## REFERENCE

Hill, D.W.; Kind, A.J. Reversed-phase solvent gradient HPLC retention indexes of drugs. *J.Anal.Toxicol.*, **1994**, *18*, 233–242

## SAMPLE

**Matrix:** solutions

## HPLC VARIABLES
**Column:** 150 × 4.6 5 μm Adsorbosphere C18 (PEEK column) (retention times are longer and peaks broader with stainless steel column)
**Mobile phase:** MeCN:20 mM pH 3.2 KH$_2$PO$_4$ 23.4:76.6 containing 0.05% nonylamine
**Flow rate:** 1.2
**Detector:** UV 214

## CHROMATOGRAM
**Retention time:** 12

## OTHER SUBSTANCES
**Simultaneous:** amitriptyline, desmethyldoxepin, desipramine, doxepin, imipramine, loxapine, maprotiline, trazodone

## REFERENCE
*Alltech Chromatography Catalog 300, Alltech Associates, Inc., Deerfield IL*, **1993**, p. 440

## SAMPLE
**Matrix:** solutions
**Sample preparation:** Inject an aliquot of a solution in mobile phase.

## HPLC VARIABLES
**Column:** 250 × 4.6 Econosil C8
**Mobile phase:** MeCN:buffer 30:70 (Buffer was 20 mM KH$_2$PO$_4$ and 14 mM triethylamine adjusted to pH 3.0 with phosphoric acid.)
**Injection volume:** 20
**Detector:** UV 210

## CHROMATOGRAM
**Retention time:** 9.9
**Limit of quantitation:** < 1 μg/mL

## OTHER SUBSTANCES
**Simultaneous:** amitriptyline, amoxapine, carbamazepine, imipramine
**Also analyzed:** desipramine, cyclobenzaprine, doxepin, maprotiline, protriptyline

## KEY WORDS
UV spectra given

## REFERENCE
Ryan, T.W. Identification and quantification of tricyclic antidepressants by UV-photodiode array detection with multicomponent analysis. *J.Liq.Chromatogr.*, **1993**, *16*, 1545–1560

## SAMPLE
**Matrix:** solutions
**Sample preparation:** Dissolve in MeOH:water 1:1 at a concentration of 50 μg/mL, inject a 10 μL aliquot.

## HPLC VARIABLES
**Column:** 300 × 3.9 10 μm μBondapak C18
**Mobile phase:** MeOH:acetic acid:triethylamine:water 60:1.5:0.5:38
**Flow rate:** 1.5
**Injection volume:** 10
**Detector:** UV

## CHROMATOGRAM
**Retention time:** k' 2.60

## REFERENCE
Roos, R.W.; Lau-Cam, C.A. General reversed-phase high-performance liquid chromatographic method for the separation of drugs using triethylamine as a competing base. *J.Chromatogr.*, **1986**, *370*, 403–418

## SAMPLE
**Matrix:** solutions
**Sample preparation:** Prepare a 10 µg/mL solution in MeOH, inject a 20 µL aliquot.

## HPLC VARIABLES
**Column:** 125 × 4.9 Spherisorb S5W silica
**Mobile phase:** MeOH containing 10 mM ammonium perchlorate and 1 mL/L 100 mM NaOH in MeOH, pH 6.7
**Flow rate:** 2
**Injection volume:** 20
**Detector:** E, LeCarbone, V25 glassy carbon electrode, + 1.2 V

## CHROMATOGRAM
**Retention time:** 2.6

## OTHER SUBSTANCES
**Also analyzed:** acebutolol, acepromazine, acetophenazine, N-acetylprocainamide, albuterol, alprenolol, amethocaine, amiodarone, amitriptyline, antazoline, atenolol, azacyclonal, bamethane, benactyzine, benperidol, benzethidine, benzocaine, benzoctamine, benzphetamine, benzquinamide, bromhexine, bromodiphenhydramine, bromperidol, brompheniramine, brompromazine, buclizine, bufotenine, bupivacaine, buprenorphine, butacaine, butethamate, chlorcyclizine, chlorpheniramine, chlorphenoxamine, chlorprenaline, chlorpromazine, chlorprothixene, cimetidine, cinchonidine, cinnarizine, clemastine, clomipramine, clonidine, cocaine, cyclazocine, cyclizine, cyclopentamine, cyproheptadine, deserpidine, desipramine, dextromoramide, dextropropoxyphene, dicyclomine, diethylcarbamazine, diethylpropion, diethylthiambutene, dihydroergotamine, dimethindene, dimethothiazine, diphenhydramine, diphenoxylate, dipipanone, diprenorphine, dipyridamole, disopyramide, dothiepin, doxapram, doxepin, doxylamine, droperidol, ephedrine, ergocornine, ergocristine, ergocristinine, ergocryptine, ergometrine, ergosine, ergosinine, ergotamine, ethopropazine, etorphine, etoxeridine, fenethazine, fenfluramine, fenoterol, fentanyl, flavoxate, fluopromazine, flupenthixol, fluphenazine, flurazepam, haloperidol, hydroxyzine, hyoscine, ibogaine, imipramine, indapamine, iprindole, isothipendyl, isoxsuprine, ketanserin, laudanosine, lidocaine, lofepramine, loxapine, maprotiline, mecamylamine, meclophenoxate, meclozine, medazepam, mephentermine, mepivacaine, meptazinol, mepyramine, mesoridazine, metaraminol, methadone, methamphetamine, methapyrilene, methdilazene, methotrimeprazine, methoxamine, methoxyphenamine, methoxypromazine, methylephedrine, methylergonovine, methysergide, metoclopramide, metopimazine, metoprolol, mianserin, morazone, nadolol, nalorphine, naloxone, naphazoline, nicotine, nifedipine, nomifensine, nortriptyline, noscapine, orphenadrine, oxeladin, oxprenolol, oxymetazolin, papaverine, pargyline, pecazine, penbutolol, pentazocine, penthienate, pericyazine, perphenazine, phenadoxone, phenampromide, phenazocine, phenbutrazate, phendimetrazine, phenelzine, phenglutarimide, phenindamine, pheniramine, phenmetrazine, phenomorphan, phenoperidine, phenothiazine, phenoxybenzamine, phentolamine, phenylephrine, phenyltoloxamine, physostigmine, piminodine, pimozide, pindolol, pipamazine, pipazethate, piperacetazine, piperidolate, pipradol, pirenzepine, piritramide, pizotifen, practolol, pramoxine, prazosin, prenylamine, prilocaine, primaquine, proadifen, procainamide, procaine, prochlorperazine, procyclidine, proheptazine, prolintane, promazine, promethazine, pronethalol, properidine, propiomazine, propranolol, prothipendyl, protriptyline, proxymetacaine, pseudoephedrine, pyrimethamine, quinidine,

quinine, ranitidine, rescinnamine, sotalol, tacrine, terazosin, terbutaline, terfenadine, thenyldiamine, theophylline, thiethylperazine, thiopropazate, thioproperazine, thioridazine, thiothixene, thonzylamine, timolol, tocainide, tolpropamine, tolycaine, tranylcypromine, trazodone, trifluoperazine, trifluperidol, trimeperidine, trimeprazine, trimethobenzamide, trimethoprim, trimipramine, tripelennamine, triprolidine, tryptamine, verapamil, xylometazoline

## REFERENCE

Jane, I.; McKinnon, A.; Flanagan, R.J. High-performance liquid chromatographic analysis of basic drugs on silica columns using non-aqueous ionic eluents. II. Application of UV, fluorescence and electrochemical oxidation detection. *J.Chromatogr.*, **1985**, *323*, 191–225

## SAMPLE
**Matrix:** urine
**Sample preparation:** 1 mL Urine + 0.5 mL 1% trichloroacetic acid, centrifuge at 5200 g for 10 min, filter (0.2 μm), inject a 20 μL aliquot of the filtrate.

## HPLC VARIABLES
**Column:** 250 × 4 Lichrospher 5 μm 60 RP-select B
**Mobile phase:** Gradient. MeCN:50 mM pH 3.2 potassium phosphate buffer from 10:90 to 50:50 over 15 min.
**Flow rate:** 1.5
**Injection volume:** 20
**Detector:** UV 190-370

## CHROMATOGRAM
**Retention time:** 15

## OTHER SUBSTANCES
**Extracted:** benzoylecgonine, cocaine, diphenhydramine, ephedrine, lidocaine, morphine, nordiazepam, norpropoxyphene, phenylpropanolamine
**Also analyzed:** amitriptyline, amphetamine, codeine (different gradient), meperidine

## REFERENCE

Li, S.; Gemperline, P.J.; Briley, K.; Kazmierczak, S. Identification and quantitation of drugs of abuse in urine using the generalized rank annihilation method of curve resolution. *J.Chromatogr.B*, **1994**, *655*, 213–223

## SAMPLE
**Matrix:** urine
**Sample preparation:** 500 μL Urine + N-ethylnordiazepam + chlorpheniramine + 100 μL buffer, centrifuge at 11000 g for 30 s, inject a 500 μL aliquot onto column A with mobile phase A, after 0.6 min backflush column A with mobile phase A to waste for 1.6 min, elute column A with 250 μL mobile phase B, with 200 μL mobile phase C, and with 1.15 mL mobile phase D. Elute column A to waste until drugs start to emerge then elute onto column B. Elute column B to waste until drugs start to emerge, then elute onto column C. When all the drugs have emerged from column B, remove it from the circuit, elute column C with mobile phase D, monitor the effluent from column C. Flush column A with 7 mL mobile phase E, mobile phase D, and mobile phase A. Flush column B with 5 mL mobile phase E then with mobile phase D. (Buffer was 6 M ammonium acetate adjusted to pH 8.0 with 2 M KOH.)

## HPLC VARIABLES
**Column:** A 10 × 2.1 12-20 μm PRP-1 spherical poly(styrene-divinylbenzene) (Hamilton); B 10 × 3.2 11 μm Aminex A-28 (Bio-Rad); C 25 × 3.2 5 μm C8 (Phenomenex) + 150 × 4.6 5 μm silica (Macherey-Nagel)

**Mobile phase:** A 0.1% pH 8.0 potassium borate buffer; B 6 mM $KH_2PO_4$ containing 5 mM tetramethylammonium hydroxide, and 2 mM dimethyloctylamine, pH adjusted to 6.50 with phosphoric acid; C MeCN:buffer 40:60 (Buffer was 6 mM $KH_2PO_4$ containing 5 mM tetramethylammonium hydroxide, and 2 mM dimethyloctylamine, pH adjusted to 6.50 with phosphoric acid.); D MeCN:buffer 33:67 (Buffer was 6 mM $KH_2PO_4$ containing 5 mM tetramethylammonium hydroxide, and 2 mM dimethyloctylamine, pH adjusted to 6.50 with phosphoric acid.); E MeCN:buffer 70:30 (Buffer was 6 mM $KH_2PO_4$ containing 5 mM tetramethylammonium hydroxide, and 2 mM dimethyloctylamine, pH adjusted to 6.50 with phosphoric acid.)
**Column temperature:** 40 (B, C only)
**Flow rate:** A 5; B-E 1
**Injection volume:** 500
**Detector:** UV 210; UV 235

## CHROMATOGRAM
**Retention time:** k' 3.3
**Internal standard:** N-ethylnordiazepam (k' 2.1), chlorpheniramine (k' 5.9)
**Limit of detection:** 300 ng/mL

## OTHER SUBSTANCES
**Extracted:** amphetamine, benzoylecgonine, caffeine, cotinine, diazepam, ephedrine, lidocaine, nordiazepam, oxazepam, pentazocine, phenmetrazine, phenobarbital, phentermine, phenylpropanolamine, secobarbital, amitriptyline, codeine, flurazepam, hydrocodone, hydromorphone, imipramine, methadone, morphine
**Interfering:** desipramine, diphenhydramine, methamphetamine

## KEY WORDS
column-switching

## REFERENCE
Binder, S.R.; Regalia, M.; Biaggi-McEachern, M.; Mazhar, M. Automated liquid chromatographic analysis of drugs in urine by on-line sample cleanup and isocratic multi-column separation. *J.Chromatogr.*, **1989**, *473*, 325−341

## SAMPLE
**Matrix:** vitreous humor
**Sample preparation:** 600 μL Vitreous humor + 3 mL 0.1 M NaCl + 50 μL 4 μg/mL desmethylclomipramine in water, mix for a few s, add to a C18 SepPak attached to a 5 mL syringe, allow to flow through (10-15 min). Wash with 1 mL 0.1 M NaCl, wash with 1 mL water, wash with 3 mL reagent by gravity. Elute with 3 mL MeOH and push air through to remove as much as possible. Evaporate eluate in vacuum at 37°, vortex with 50 μL mobile phase for 1 min, inject 25 μL aliquot. (Reagent was isopropanol:n-heptane:1 M sulfuric acid 40:320:1.)

## HPLC VARIABLES
**Guard column:** 50 × 4.6 30 μm Permaphase ETH
**Column:** 250 × 4.6 5-6 μm Zorbax cyanopropyl
**Mobile phase:** MeCN:0.5 M acetic acid:n-butylamine 40:60:0.0022
**Flow rate:** 2.5
**Injection volume:** 25
**Detector:** UV 254

## CHROMATOGRAM
**Retention time:** 20.5
**Internal standard:** desmethylclomipramine
**Limit of detection:** 16.7 ng/mL

# 1028   Nortriptyline

## OTHER SUBSTANCES

**Simultaneous:** metabolites, amitriptyline, doxepin, imipramine

**Noninterfering:** acetaminophen, N-acetylprocainamide, amikacin, caffeine, carbamazepine, chloramphenicol, clonazepam, cyclosporine, diazepam, digoxin, disopyramide, ethosuximide, flurazepam, gentamicin, haloperidol, kanamycin, lidocaine, meprobamate,
methapyriline, methaqualone, methotrexate, methyprylon, netilmicin, pentazocine, pentobarbital, phenobarbital, phenytoin, prazepam, primidone, procainamide, propranolol,
quinidine, salicylic acid, secobarbital, streptomycin, theophylline, tobramycin, tocainide,
valproic acid, vancomycin

## REFERENCE

Evenson, M.A.; Engstrand, D.A. A SepPak HPLC method for tricyclic antidepressant drugs in human
vitreous humor. *J.Anal.Toxicol.*, **1989**, *13*, 322–325

◆─────────────────────────────◆─────────────────────────────◆

## ANNOTATED BIBLIOGRAPHY

Atta-Politou, J.; Tsarpalis, K.; Koutselinis, A. A modified simple and rapid reversed phase high performance liquid chromatographic method for quantification of amitriptyline and nortriptyline in plasma.
*J.Liq.Chromatogr.*, **1994**, *17*, 3969–3982 [extracted amitriptyline, clomipramine, doxepin, protriptyline; LOD 5 ng/mL; LOQ 10 ng/mL; interfering phenytoin; simultaneous chlordiazepoxide, chlorpheniramine, chlorpromazine, imipramine, maprotiline, phenobarbital, propxyphene, propranolol;
non-interfering alprazolam, artane, bromazepam, diazepam, haloperidol, lorazepam, nitrazepam,
oxazepam, pseudoephedrine, theophylline, triazolam]

Oshima, N.; Kotaki, H.; Sawada, Y.; Iga, T. Tissue distribution of amitriptyline after repeated administration in rats. *Drug Metab.Dispos.*, **1994**, *22*, 21–25 [rat; dialysate; plasma; tissue; liver; kidney;
lung; brain; muscle; heart; clomipramine (IS); column temp 35; pharmacokinetics]

Piperaki, S.; Parissi-Poulou, M.; Koupparis, M. A separation study of tricyclic antidepressant drugs by
HPLC with β-cyclodextrin bonded stationary phase. *J.Liq.Chromatogr.*, **1993**, *16*, 3487–3508 [also
amitriptyline, chloripramine, doxepin, imipramine, maprotiline, protriptyline]

Smith, C.S.; Abramson, R.K.; Morgan, S.L. An investigation of the metabolism of amitriptyline using
high performance liquid chromatography. *J.Liq.Chromatogr.*, **1986**, *9*, 1727–1745 [extracted amitriptyline; rat; tissue; liver]

Wong, S.H.Y.; McHugh, S.L.; Dolan, J.; Cohen, K.A. Tricyclic antidepressant analysis by reversed-phase
liquid chromatography using phenyl columns. *J.Liq.Chromatogr.*, **1986**, *9*, 2511–2538 [also acetaminophen, amitriptyline, amobarbital, amoxapine, barbital, chlordiazepoxide, chlorpromazine, cimetidine, clomipramine, codeine, desipramine, desmethyldoxepin, diazepam, doxepin, fluphenazine,
flurazepam, glutethimide, hydroxyamoxapine, imipramine, lorazepam, maprotiline, meperidine, oxazepam, pentobarbital, perphenazine, phenobarbital, phenytoin, propoxyphene, protriptyline, secobarbital, thioridazine, trazodone]

Suckow, R.F.; Cooper, T.B. Simultaneous determination of amitriptyline, nortriptyline and their respective isomeric 10-hydroxy metabolites in plasma by liquid chromatography. *J.Chromatogr.*, **1982**, *230*,
391–400

Jensen, K.M. Determination of amitriptyline-N-oxide, amitriptyline and nortriptyline in serum and
plasma by high-performance liquid chromatography. *J.Chromatogr.*, **1980**, *183*, 321–329

# Ofloxacin

**Molecular formula:** $C_{18}H_{20}FN_3O_4$
**Molecular weight:** 361.4
**CAS Registry No.:** 82419-36-1

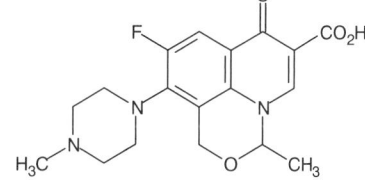

## SAMPLE
**Matrix:** blood
**Sample preparation:** 250 µL Serum + 50 µL pipemidic acid solution + 200 µL chloroform, extract. Extract the organic phase with 200 µL 100 mM NaOH, inject a 20 µL aliquot.

## HPLC VARIABLES
**Column:** C18
**Mobile phase:** MeCN:buffer 13:87 (Buffer was 10 mM $NaH_2PO_4$ and 5 mM tetrabutylammonium hydrogen sulfate, pH 2.7.)
**Flow rate:** 1.4
**Injection volume:** 20
**Detector:** F ex 282

## CHROMATOGRAM
**Internal standard:** pipemidic acid
**Limit of quantitation:** 50 ng/mL

## KEY WORDS
serum; pharmacokinetics

## REFERENCE
Klepser, M.E.; Patel, K.B.; Nicolau, D.P.; Quintiliani, R.; Nightingale, C.H. Comparison of the bactericidal activities of ofloxacin and ciprofloxacin alone and in combination with ceftazidime and piperacillin against clinical strains of *Pseudomonas aeruginosa. Antimicrob.Agents Chemother.*, **1995**, *39*, 2503–2510

## SAMPLE
**Matrix:** blood
**Sample preparation:** 1 mL Plasma + 50 µL 180 µg/mL enoxacin in MeOH, shake briefly, add 3 mL MeCN, shake at 1000 rpm for 30 s, centrifuge at 2600 g for 10 min. Remove 3 mL of the supernatant and evaporate it to dryness under a stream of nitrogen at 60°, reconstitute the residue in 100 µL 60 mM $KH_2PO_4$ adjusted to pH 2.6 with orthophosphoric acid, inject a 30 µL aliquot.

## HPLC VARIABLES
**Column:** 150 × 3 7 µm Separon SGX C18 (Tessek, Prague)
**Mobile phase:** THF:buffer 5.5:94.5 (Buffer was 60 mM $KH_2PO_4$ adjusted to pH 2.6 with orthophosphoric acid:triethylamine 97:3 containing 50 µg/mL sodium azide. Caution! Sodium azide is carcinogenic and toxic and must not be discharged to the plumbing system!)
**Flow rate:** 0.7
**Injection volume:** 30
**Detector:** F ex 282 em 450

## CHROMATOGRAM
**Retention time:** 6.7
**Internal standard:** enoxacin (5.0)
**Limit of detection:** 8 ng/mL
**Limit of quantitation:** 20 ng/mL

## KEY WORDS
plasma

## REFERENCE
Macek, J.; Ptácek, P. Determination of ofloxacin in human plasma using high-performance liquid chromatography and fluorescence detection. *J.Chromatogr.B*, **1995**, *673*, 316–319

## SAMPLE
**Matrix:** blood
**Sample preparation:** 500 µL Serum + 100 µL water + 1 mL MeCN, shake mechanically for 30 s, centrifuge at 10000 g for 5 min. Remove 1.5 mL of the supernatant and evaporate it to dryness under reduced pressure at 40°, reconstitute the residue in 300 µL mobile phase, inject a 25 µL aliquot.

## HPLC VARIABLES
**Guard column:** 30 × 4 30-40 µm Perisorb RP-18 (Merck)
**Column:** 125 × 4 5 µm Nucleosil 100 SA
**Mobile phase:** MeCN:buffer 75:25, adjusted to pH 3.82 with concentrated phosphoric acid, final sodium concentration 23 mM (Buffer was 6.74 mL concentrated phosphoric acid and 40 mL 2 M NaOH made up to 990 mL with water, adjust pH to 2.92 with concentrated phosphoric acid, make up to 1 L with water.)
**Flow rate:** 1.5
**Injection volume:** 25
**Detector:** F ex 295 em 525

## CHROMATOGRAM
**Retention time:** 8.0
**Internal standard:** ofloxacin

## OTHER SUBSTANCES
**Extracted:** sparfloxacin

## KEY WORDS
ofloxacin is IS; serum; protect from light

## REFERENCE
Borner, K.; Borner, E.; Lode, H. Determination of sparfloxacin in serum and urine by high-performance liquid chromatography. *J.Chromatogr.*, **1992**, *579*, 285–289

## SAMPLE
**Matrix:** blood
**Sample preparation:** 500 µL Serum + 500 µL 7% perchloric acid, vortex for 10 s, centrifuge at >700 g for 10 min, inject a 20 µL aliquot of the supernatant.

## HPLC VARIABLES
**Column:** 100 × 8 µBondapak C18 Radial-PAK
**Mobile phase:** MeOH:18 mM $KH_2PO_4$ containing 0.13 mM heptanesulfonic acid:concentrated phosphoric acid 30:70:0.1
**Injection volume:** 20
**Detector:** F ex 294 em 475

## CHROMATOGRAM
**Retention time:** 5.6

## KEY WORDS
serum

## REFERENCE

Griggs, D.J.; Wise, R. A simple isocratic high-pressure liquid chromatographic assay of quinolones in serum. *J.Antimicrob.Chemother.*, **1989**, *24*, 437−445

## SAMPLE

**Matrix:** blood

**Sample preparation:** 50 μL Plasma + 1 mL 100 mM pH 7.0 $K_2HPO_4$ adjusted to pH 7.0 with 85% orthophosphoric acid + 100 μL 300 μg/mL nalidixic acid in water + 3 mL dichloromethane:isoamyl alcohol 9:1, shake vigorously for 10 min, centrifuge at 2270 g for 10 min. Remove 2 mL of the organic phase and evaporate it to dryness under a stream of nitrogen at 40°. Reconstitute residue in 100 μL MeOH:50 mM NaOH 2:1, vortex, inject a 10 μL aliquot.

## HPLC VARIABLES

**Column:** 150 × 4.6 5 μm Chemcosorb 5-ODS-H

**Mobile phase:** MeOH:5 mM sodium lauryl sulfate 2:1, adjusted to pH 2.5 with 85% phosphoric acid (Better separation obtained at pH 2.35, J.Chromatogr. 1990, 530, 186.)

**Column temperature:** 40

**Flow rate:** 0.6

**Injection volume:** 10

**Detector:** UV 300

## CHROMATOGRAM

**Retention time:** 5.9

**Internal standard:** nalidixic acid (5.0)

**Limit of detection:** 150 ng/mL

## OTHER SUBSTANCES

**Extracted:** felbinac, fenbufen, norfloxacin

**Interfering:** enoxacin

## KEY WORDS

plasma; rat

## REFERENCE

Katagiri, Y.; Naora, K.; Ichikawa, N.; Hayashibara, M.; Iwamoto, K. Simultaneous determination of ofloxacin, fenbufen and felbinac in rat plasma by high-performance liquid chromatography. *J.Chromatogr.*, **1988**, *431*, 135−142

## SAMPLE

**Matrix:** blood

**Sample preparation:** 1 mL Plasma + 2 mL 100 mM pH 7.0 phosphate buffer + 100 μL 1 mg/mL flufenamic acid in phosphate buffer + 6 mL chloroform, agitate on a rotary mixer for 15 min, centrifuge at 4000 g at 4° for 5 min. Remove the organic phase and evaporate it to dryness under nitrogen at 37°. Suspend the residue in 100 μL 100 mM pH 7.0 phosphate buffer, vortex for 45 s, centrifuge at 9000 g for 2 min, inject a 20 μL aliquot of the supernatant.

## HPLC VARIABLES

**Column:** 150 × 4.5 5 μm Hypersil ODS

**Mobile phase:** MeCN:40 mM phosphoric acid 45:55, final pH 2.56-2.59

**Column temperature:** 17

**Flow rate:** 2

**Injection volume:** 20

**Detector:** F ex 370 em 400-700

## CHROMATOGRAM
**Retention time:** 3.83
**Internal standard:** flufenamic acid (2.67)
**Limit of quantitation:** 2.5 ng/mL

## KEY WORDS
plasma

## REFERENCE
Notarianni, L.J.; Jones, R.W. Method for the determination of ofloxacin, a quinolone carboxylic acid antimicrobial, by high-performance liquid chromatography. *J.Chromatogr.*, **1988**, *431*, 461–464

## SAMPLE
**Matrix:** blood, CSF
**Sample preparation:** Serum. 1 mL Serum + 500 μL 500 mM pH 7 phosphate buffer + 8 mL dichloromethane, rotate for 10 min at 20 rpm, centrifuge at 4000 rpm for 15 min. Remove the organic layer and evaporate it to dryness at 40° in a vortex evaporator. Dissolve in 1 mL mobile phase, inject a 100 μL aliquot. CSF. Inject a 20 μL aliquot directly.

## HPLC VARIABLES
**Column:** 250 × 4 5 μm Nucleosil C18
**Mobile phase:** MeCN:40 mM pH 2.2 phosphate buffer containing tetrabutylammonium chloride as the ion-pairing reagent 15:85
**Flow rate:** 1
**Injection volume:** 20 (CSF), 100 (serum)
**Detector:** F ex 295 em 475

## CHROMATOGRAM
**Limit of detection:** 90 ng/mL (serum); 70 ng/mL (CSF)

## KEY WORDS
serum

## REFERENCE
Bitar, N.; Claes, R.; Van der Auwera, P. Concentrations of ofloxacin in serum and cerebrospinal fluid of patients without meningitis receiving the drug intravenously and orally. *Antimicrob.Agents Chemother.*, **1989**, *33*, 1686–1690

## SAMPLE
**Matrix:** blood, intestinal efflux
**Sample preparation:** Intestinal efflux. Freeze intestinal efflux at -80°, lyophilize, reconstitute with 1 mL MeOH:100 mM phosphoric acid 50:50, centrifuge at 3000 rpm for 10 min, inject a 20 μL aliquot. Serum. Deproteinize serum with MeOH.

## HPLC VARIABLES
**Column:** 150 × 3.9 Novapack C18
**Mobile phase:** MeOH:buffer 25:75 (Buffer was 10 mM pH 3.0 potassium phosphate buffer containing 25 mM sodium heptanesulfonate (PIC B7) and 20 mM triethylamine.)
**Flow rate:** 1.5
**Injection volume:** 20
**Detector:** F ex 330 em 440

## CHROMATOGRAM
**Internal standard:** ofloxacin

## OTHER SUBSTANCES
**Extracted:** ciprofloxacin

## KEY WORDS
serum; rat; ofloxacin is IS

## REFERENCE
Rubinstein, E.; Dautrey, S.; Farinoti, R.; St.Julien, L.; Ramon, J.; Carbon, C. Intestinal elimination of sparfloxacin, fleroxacin, and ciprofloxacin in rats. *Antimicrob.Agents Chemother.*, **1995**, *39*, 99–102

## SAMPLE
**Matrix:** blood, saliva, sputum
**Sample preparation:** 200 μL Serum, sputum, or saliva + 200 μL 1.5 μg/mL pipemidic acid in 100 mM pH 7.0 phosphate buffer + 5 mL dichloromethane, shake for 15 min, centrifuge at 1600 g for 10 min. Remove 4 mL of organic phase and evaporate to dryness under a stream of nitrogen. Dissolve the residue in 200 μL mobile phase, sonicate for 10 min, inject.

## HPLC VARIABLES
**Column:** 150 × 6 5 μm YMC-Pack ODS-AM
**Mobile phase:** MeCN:10 mM phosphoric acid (pH 7.0) containing 20 mM tetrabutylammonium hydrogen sulfate 1:10
**Flow rate:** 1.3
**Injection volume:** 200
**Detector:** F ex 285 em 460

## CHROMATOGRAM
**Retention time:** 8
**Internal standard:** pipemidic acid (5)
**Limit of quantitation:** 10 ng/mL

## KEY WORDS
serum

## REFERENCE
Koizumi, F.; Ohnishi, A.; Takemura, H.; Okubo, S.; Kagami, T.; Tanaka, T. Effective monitoring of concentrations of ofloxacin in saliva of patients with chronic respiratory tract infections. *Antimicrob.Agents Chemother.*, **1994**, *38*, 1140–1143

## SAMPLE
**Matrix:** blood, tissue, urine
**Sample preparation:** Serum, plasma. Dilute serum or plasma 1:2 to 1:10 with 30 mM phosphoric acid, centrifuge, inject a 20 μL aliquot of supernatant. Urine. Dilute urine 1:10 to 1:100 with 30 mM phosphoric acid, centrifuge, inject a 20 μL aliquot of supernatant. Tissue (lung, gut). Cut tissue with a scalpel, homogenize with 1-3 mL buffer, centrifuge at 9600 g for 5 min three times, inject a 20 μL aliquot. Tissue (chondral). Cut tissue with a scalpel, homogenize with 3-6 mL buffer in an ice bath for 2-3 min, centrifuge at 9600 g for 5 min four or five times, inject a 100 μL aliquot. Pleural. Dilute human pleural samples with buffer, centrifuge, inject a 20 μL aliquot. (Buffer was 66.6 mM $K_2HPO_4$ adjusted to pH 7.40 with $KH_2PO_4$.)

## HPLC VARIABLES
**Column:** 200 × 4 5 μm Nucleosil C18
**Mobile phase:** MeOH:MeCN:buffer 13:7:80, adjusted to pH 3.0 with phosphoric acid (Buffer was 15 mM phosphoric acid adjusted to pH 3.0 with tetrabutylammonium hydroxide.)

**Flow rate:** 1
**Injection volume:** 20-100
**Detector:** F ex 278 em 446

## CHROMATOGRAM
**Limit of detection:** 10 ng/mL

## OTHER SUBSTANCES
**Simultaneous:** ciprofloxacin, norfloxacin

## KEY WORDS
serum; plasma; lung; gut; pleural; chondral

## REFERENCE
Knöller, J.; König, W.; Schönfeld, W.; Bremm, K.D.; Köller, M. Application of high-performance liquid chromatography of some antibiotics in clinical microbiology. *J.Chromatogr.*, **1988**, *427*, 257–267

## SAMPLE
**Matrix:** blood, urine
**Sample preparation:** 1 mL Plasma or serum or 200 μL urine + 1 mL pH 7 buffer + 5 mL dichloromethane, shake, centrifuge at 2500 g for 5 min. Remove 4 mL organic layer and evaporate it under nitrogen, dissolve the residue in 200 μL mobile phase.

## HPLC VARIABLES
**Guard column:** CT 30/6/4 Resolvosil-BSA-7 (Macherey-Nagel)
**Column:** 150 × 4 ET 150/8/4 Resolvosil-BSA-7 (Macherey-Nagel)
**Mobile phase:** Isopropanol:200 mM pH 8.0 phosphate buffer 3:97
**Flow rate:** 1
**Injection volume:** 100
**Detector:** F ex 298 em 458

## CHROMATOGRAM
**Retention time:** 5.2 (−), 7.5 (+)
**Limit of detection:** 3 ng/mL (plasma); 80 ng/mL (urine)

## KEY WORDS
plasma; serum; chiral

## REFERENCE
Lehr, K.-H.; Damm, P. Quantification of the enantiomers of ofloxacin in biological fluids by high-performance liquid chromatography. *J.Chromatogr.*, **1988**, *425*, 153–161

## SAMPLE
**Matrix:** blood, urine
**Sample preparation:** 500 μL Plasma or serum or 200 μL urine + 1 mL pH 7 buffer + 2 mL dichloromethane, shake, centrifuge at 2500 g for 5 min. Remove 1 mL organic layer and add it to 1 mL dichloromethane containing 20 μL diphenylphosphinyl chloride and 20 μL triethylamine, vortex for 10 s, add 500 μL L-leucinamide solution, shake for 10 min, extract with 200 μL 1 M HCl (500 μL 0.5 M HCl for urine samples), inject 100 μL of aqueous supernatant. (Prepare L-leucinamide solution by adding 10 mL dichloromethane and 1 mL 5 M NaOH to 500 mg L-leucinamide hydrochloride, shake, discard upper aqueous layer, store dichloromethane layer over anhydrous sodium sulfate.)

## HPLC VARIABLES
**Column:** 125 × 4.6 5 μm Nucleosil 120-5C18

**Mobile phase:** MeCN:200 mM phosphoric acid adjusted to pH 1.85 with tetraethylammonium hydroxide 20:80
**Column temperature:** 40
**Flow rate:** 1.5
**Injection volume:** 100
**Detector:** F ex 298 em 458

## CHROMATOGRAM
**Retention time:** 2.6 (−), 3.8 (+)
**Limit of detection:** 3 ng/mL (plasma); 80 ng/mL (urine)

## KEY WORDS
plasma; serum; derivatization; chiral

## REFERENCE
Lehr, K.-H.; Damm, P. Quantification of the enantiomers of ofloxacin in biological fluids by high-performance liquid chromatography. *J.Chromatogr.*, **1988**, *425*, 153−161

## SAMPLE
**Matrix:** blood, urine
**Sample preparation:** Plasma. 500 μL Plasma + 50 μL 200 μg/mL desmethylpefloxacine in 10 mM HCl + 1.5 mL 50 mM pH 9 borate buffer containing 50 mM KCl and 21 mM NaOH, vortex for 10 s, add 8 mL dichloromethane, shake for 5 min, centrifuge at 1500 g for 5 min. Remove 7 mL of the organic phase and add it to 200 μL 100 mM HCl, shake for 5 min, centrifuge at 1500 g, inject a 5 μL aliquot of the aqueous layer. Urine. Dilute with water 1:40, add 100 μg/mL desmethylpefloxacine, inject a 5 μL aliquot.

## HPLC VARIABLES
**Column:** 300 × 3.9 10 μm μBondapak C18
**Mobile phase:** MeCN:water 10:90 containing 1.36 g/L $KH_2PO_4$ and 40 mL/L 5 mM aqueous tetrabutylammonium phosphate, pH adjusted to 2.9 with formic acid
**Flow rate:** 0.7
**Injection volume:** 5
**Detector:** F ex 290 em 500

## CHROMATOGRAM
**Retention time:** 8.7
**Internal standard:** desmethylpefloxacine (10.2)
**Limit of detection:** 20 ng/mL

## KEY WORDS
plasma

## REFERENCE
Mignot, A.; Lefebvre, M.A.; Fourtillan, J.B. High-performance liquid chromatographic determination of ofloxacin in plasma and urine. *J.Chromatogr.*, **1988**, *430*, 192−197

## SAMPLE
**Matrix:** blood, vitreous humor
**Sample preparation:** Serum. 20 μL Serum + 130 μL mobile phase, mix, filter, inject a 100 μL aliquot of the filtrate. Vitreous humor. 15 μL Vitreous humor + 135 μL mobile phase, mix, filter, inject a 100 μL aliquot of the filtrate.

## HPLC VARIABLES
**Column:** 220 × 2.1 5 μm Nucleosil C18
**Mobile phase:** MeCN:100 mM pH 3.82 phosphoric acid 75:25

**Flow rate:** 0.2
**Injection volume:** 100
**Detector:** UV 240; UV 280; F ex 280 em 445

## CHROMATOGRAM
**Limit of detection:** 10 ng/mL

## KEY WORDS
rabbit; serum; pharmacokinetics

## REFERENCE
Perkins, R.J.; Liu, W.; Drusano, G.; Madu, A.; Mayers, M.; Madu, C.; Miller, M.H. Pharmacokinetics of ofloxacin in serum and vitreous humor of albino and pigmented rabbits. *Antimicrob.Agents Chemother.*, **1995**, *39*, 1493–1498

## SAMPLE
**Matrix:** bronchoalveolar lavage fluid
**Sample preparation:** 1 mL Bronchoalveolar lavage fluid + 20 μL 30 μg/mL IS + 1 mL 100 mM pH 7.0 phosphate buffer, vortex, add 5 mL chloroform, shake horizontally at 2 strokes/s for 10 min, centrifuge at 1500 g for 10 min. Remove the organic layer and evaporate it to dryness under a stream of nitrogen at 40°. Reconstitute residue in 400 μL mobile phase, inject a 20 μL aliquot.

## HPLC VARIABLES
**Column:** 150 × 4.6 5 μm TSK ODS-80TM (Tosoh)
**Mobile phase:** THF:0.5% $KH_2PO_4$ adjusted to pH 2.9 with orthophosphoric acid 1:16
**Flow rate:** 1
**Injection volume:** 20
**Detector:** F ex 290 em 460

## CHROMATOGRAM
**Retention time:** 12
**Internal standard:** 9-fluoro-2,3-dihydro-3-methyl-10-(1-imidazoyl)-7-oxo-7H-pyrido[1,2,3-de][1,4]benzoxazine-6-carboxylic acid (8)
**Limit of quantitation:** 0.5 ng/mL

## REFERENCE
Matsubayashi, K.; Une, T.; Osada, Y. Determination of ofloxacin in bronchoalveolar lavage fluid by high-performance liquid chromatography and fluorimetric detection. *J.Chromatogr.*, **1989**, *495*, 354–357

## SAMPLE
**Matrix:** cell lysate
**Sample preparation:** Sonicate lysed cells for 5 min, filter (0.2 μm PTFE), add timolol, inject an aliquot.

## HPLC VARIABLES
**Guard column:** 4 × 4 5 μm LiChrospher 100 RP-18
**Column:** 250 × 4 5 μm LiChrospher 100 RP-18
**Mobile phase:** MeCN:buffer 18:82 (Buffer was 0.5% triethylamine adjusted to pH 2.5 with 78% phosphoric acid.)
**Flow rate:** 1
**Injection volume:** 100
**Detector:** UV 293

## CHROMATOGRAM
**Retention time:** 5.40

**Internal standard:** timolol (7.70)
**Limit of detection:** 18 ng/mL

## REFERENCE
Fresta, M.; Spadaro, A.; Cerniglia, G.; Ropero, I.M.; Puglisi, G.; Furneri, P.M. Intracellular accumulation of ofloxacin-loaded liposomes in human synovial fibroblasts. *Antimicrob.Agents Chemother.*, **1995**, *39*, 1372–1375

## SAMPLE
**Matrix:** cells
**Sample preparation:** 100 μL Cell suspension + 100 μL cefoperazone solution + 100 μL Hanks balanced salt solution, sonicate 30 min, add 800 μL MeCN, centrifuge at 13000 g for 5 min, remove supernatant. Dry supernatant under air, dissolve in 100 μL mobile phase, inject 75 μL.

## HPLC VARIABLES
**Column:** μBondapak C18
**Mobile phase:** MeCN:5 mM pH 2.0 tetrabutylammonium hydrogen sulfate 10:90
**Flow rate:** 1
**Injection volume:** 75
**Detector:** UV 280

## CHROMATOGRAM
**Retention time:** 10.6
**Internal standard:** ciprofloxacin
**Limit of detection:** 100-1000 ng/mL

## REFERENCE
Darouiche, R.O.; Hamill, R.J. Antibiotic penetration of and bactericidal activity within endothelial cells. *Antimicrob.Agents Chemother.*, **1994**, *38*, 1059–1064

## SAMPLE
**Matrix:** cells
**Sample preparation:** Incubate cells in 2 mL 100 mM pH 3.0 glycine-HCl buffer for 2 h at room temperature, centrifuge at 5600 g for 5 min, inject an aliquot.

## HPLC VARIABLES
**Column:** Bondapak C18
**Mobile phase:** MeCN:25 mM phosphoric acid adjusted to pH 3.0 with tetrabutylammonium hydroxide 25:75
**Flow rate:** 1.5
**Detector:** F ex 340 em 425

## OTHER SUBSTANCES
**Also analyzed:** ciprofloxacin, fleroxacin, lomefloxacin, norfloxacin, temafloxacin

## REFERENCE
Pascual, A.; Garcia, I.; Conejo, M.C.; Perea, E.J. Fluorometric and high-performance liquid chromatographic measurement of quinolone uptake by human neutrophils. *Eur.J.Clin.Microbiol.Infect.Dis.*, **1991**, *10*, 969–971

## SAMPLE
**Matrix:** formulations
**Sample preparation:** Dissolve 40 mg freeze-dried nanoparticles in 25 mL acetone:MeOH 90:10 containing a few drops of 100 mM HCl, filter (0.45 μm), inject an aliquot.

## HPLC VARIABLES
**Column:** 250 × 4.6 5 μm Partisil 10-ODS-1 C18
**Mobile phase:** MeCN:10 mM $KH_2PO_4$:triethylamine 14:86:0.2
**Flow rate:** 1.5
**Injection volume:** 20
**Detector:** UV 292

## OTHER SUBSTANCES
**Simultaneous:** pefloxacin (UV 275)

## KEY WORDS
nanoparticles

## REFERENCE
Fresta, M.; Puglisi, G.; Giammona, G.; Cavallaro, G.; Micali, N.; Furneri, P.M. Pefloxacine mesilate and ofloxacin-loaded polyethylcyanoacrylate nanoparticles: Characterization of the colloidal drug carrier formulation. *J.Pharm.Sci.*, **1995**, *84*, 895–902

## SAMPLE
**Matrix:** hair
**Sample preparation:** Wash hair successively with 0.1% sodium dodecyl sulfate and water for 30 min, repeat twice, blot between 2 sheets of paper towel, allow to dry at room temperature. Take a 1 cm fragment of hair, add 500 μL 1 M NaOH, heat at 80° for 30 min, cool, add 500 μL 1 M HCl, add 1 mL 100 mM pH 4.6 potassium hydrogen citrate buffer, add 50 μL 1 μg/mL IS in water. Add the mixture to a Bond-Elut C8 SPE cartridge, elute with 2 mL THF:25 mM orthophosphoric acid 20:80, evaporate eluate to dryness in vacuum, dissolve residue in 150 μL mobile phase, vortex, inject a 60 μL aliquot.

## HPLC VARIABLES
**Column:** 150 × 4.6 Tosoh 5 μm TSKgel ODS-80Ts
**Mobile phase:** MeCN:25 mM orthophosphoric acid adjusted to pH 3.0 with 0.5 M tetra-n-butylamine hydroxide 5:95
**Column temperature:** 40
**Flow rate:** 1
**Injection volume:** 60
**Detector:** F ex 295 em 490

## CHROMATOGRAM
**Retention time:** 8.1
**Internal standard:** (R)-9-fluoro-2,3-dihydro-3-methyl-10-(4-ethyl-1-piperazinyl)-7-oxo-7H-pyrido[1,2,3-de][1,4]benzoxazine-6-carboxylic acid (DS-4632) (10.2) (determine at F ex 280 em 445)
**Limit of detection:** 0.2 ng/mL

## OTHER SUBSTANCES
**Extracted:** ciprofloxacin, norfloxacin (determine at F ex 280 em 445)

## KEY WORDS
SPE

## REFERENCE
Mizuno, A.; Uematsu, T.; Nakashima, M. Simultaneous determination of ofloxacin, norfloxacin and ciprofloxacin in human hair by high-performance liquid chromatography and fluorescence detection. *J.Chromatogr.B*, **1994**, *653*, 187–193

## SAMPLE
**Matrix:** hair
**Sample preparation:** Wash hair several times with 0.1% sodium dodecyl sulfate and with water. Add 500 μL 1 M NaOH to 5-10 cm hair, heat at 80° for 30 min, cool, add 500 μL 1 M HCl, add 1 mL 500 mM pH 7.0 phosphate buffer, add 100 μL 2.5 μg/mL IS, add 5 mL chloroform, shake mechanically for 20 min, centrifuge at 1670 g for 10 min. Remove the organic layer and evaporate it to dryness under a stream of nitrogen at 50°, reconstitute the residue in mobile phase, inject a 100 μL aliquot.

## HPLC VARIABLES
**Guard column:** 15 × 4.6 TSKguardgel ODS-120T (Tosoh)
**Column:** 250 × 4.6 5 μm TSKgel ODS-120T (Tosoh)
**Mobile phase:** MeCN:buffer 18:82 (Buffer was 6.8 g/L $KH_2PO_4$ adjusted to pH 2.6 with phosphoric acid.)
**Flow rate:** 0.8
**Injection volume:** 100
**Detector:** F ex 290 em 460

## CHROMATOGRAM
**Retention time:** 10
**Internal standard:** 9-fluoro-2,3-dihydro-3-methyl-10-(1-imidazolyl)-7-oxo-7H-pyrido[1,2,3-de][1,4]benzoxazine-6-carboxylic acid (DL-8357) (8)

## KEY WORDS
pharmacokinetics (Ther.Drug Monit. 1995, 17, 101)

## REFERENCE
Miyazawa, N.; Uematsu, T.; Mizuno, A.; Nagashima, S.; Nakashima, M. Ofloxacin in human hair determined by high performance liquid chromatography. *Forensic Sci.Int.*, **1991**, *51*, 65–77

## SAMPLE
**Matrix:** solutions
**Sample preparation:** Filter (0.45 μm) a solution in MeCN:water 10:90, inject an aliquot of the filtrate.

## HPLC VARIABLES
**Column:** 250 × 4 5 μm LiChrospher 100 RP-18
**Mobile phase:** MeCN:buffer 7:93 (Buffer was 25 mM phosphoric acid adjusted to pH 3.89 with 100 mM tetrabutylammonium hydroxide.)
**Flow rate:** 1
**Injection volume:** 10
**Detector:** UV 295

## CHROMATOGRAM
**Retention time:** 8.8

## OTHER SUBSTANCES
**Simultaneous:** ciprofloxacin (UV 280), enoxacin (UV 280), fleroxacin (UV 280), norfloxacin (UV 280), pipemidic acid (UV 280)

## REFERENCE
Barbosa, J.; Bergés, R.; Sanz-Nebot, V. Solvatochromic parameter values and pH in aqueous-organic mixtures used in liquid chromatography. Prediction of retention of a series of quinolones. *J.Chromatogr.A*, **1996**, *719*, 27–36

## SAMPLE
**Matrix:** solutions

## HPLC VARIABLES
**Guard column:** 20 mm long Supelguard LC-18S (Supelco)
**Column:** 250 × 4.6 Suplecosil LC-18S
**Mobile phase:** MeCN:buffer:water 10:3.5:86.5 (Buffer was 400 mM tetrabutylammonium hydroxide adjusted to pH 2.85.)
**Flow rate:** 1.8
**Detector:** UV 280

## REFERENCE
Sinko, P.J.; Hu, P. Determining intestinal metabolism and permeability for several compounds in rats. Implications on regional bioavailability in humans. *Pharm.Res.*, **1996**, *13*, 108–113

## SAMPLE
**Matrix:** solutions
**Sample preparation:** Prepare a 20 μg/mL solution in MeCN:water 10:90, filter (0.45 μm), inject an aliquot.

## HPLC VARIABLES
**Column:** 250 × 4 5 μm LiChrospher 100 RP-18
**Mobile phase:** MeCN:25 mM phosphoric acid 7:93, adjusted to pH 3.09 with 100 mM tetrabutylammonium hydroxide
**Flow rate:** 1
**Injection volume:** 10
**Detector:** UV 295

## CHROMATOGRAM
**Retention time:** 6.2

## OTHER SUBSTANCES
**Simultaneous:** ciprofloxacin (UV 280), norfloxacin (UV 280), pipemidic acid (UV 280)

## REFERENCE
Barbosa, J.; Bergés, R.; Sanz-Nebot, V. Linear solvation energy relationships in reversed-phase liquid chromatography. Prediction of retention of several quinolones. *J.Liq.Chromatogr.*, **1995**, *18*, 3445–3463

## SAMPLE
**Matrix:** solutions

## HPLC VARIABLES
**Column:** 125 × 4.6 3 μm ODS-Hypersil C18
**Mobile phase:** MeOH:THF:670 mM pH 3.0 phosphate buffer 20:0.8:79.2 plus 2 g/L tetrabutylammonium hydrogen sulfate and 2 mL/L 85% phosphoric acid
**Flow rate:** 1
**Injection volume:** 20
**Detector:** UV 286

## CHROMATOGRAM
**Retention time:** 6.38

## OTHER SUBSTANCES
**Simultaneous:** photodegradation products, fleroxacin
**Interfering:** ciprofloxacin

## REFERENCE
Tiefenbacher, E.-M.; Haen, E.; Przybilla, B.; Kurz, H. Photodegradation of some quinolones used as antimicrobial therapeutics. *J.Pharm.Sci.*, **1994**, *83*, 463–467

## SAMPLE
**Matrix:** tears
**Sample preparation:** Dry tear sample under nitrogen.

## HPLC VARIABLES
**Column:** 250 × 4.6 5 μm Ultrasphere ODS
**Mobile phase:** MeCN:phosphoric acid:sodium lauryl sulfate 40:1:0.2:58.8
**Flow rate:** 1.2
**Detector:** F ex 358 em 495

## CHROMATOGRAM
**Retention time:** 6.7
**Internal standard:** triamterene (8.1)
**Limit of detection:** 10 ng

## REFERENCE
Richman, J.; Zolezio, H.; Tang-Liu, D. Comparison of ofloxacin, gentamicin, and tobramycin concentrations in tears and in vitro MICs for 90% of test organisms. *Antimicrob.Agents Chemother.*, **1990**, *34*, 1602–1604

## SAMPLE
**Matrix:** tissue
**Sample preparation:** Condition a 500 mg Bond Elut C18 SPE cartridge with 5 mL MeOH and 10 mL water. Homogenize a 5 g tissue sample with 100 mL MeCN:0.2% metaphosphoric acid 30:70 at high spped, filter through ca. 2 mm Hyflo Super-Cel coated on a suction funnel. Evaporate filtrate under reduced pressure at 50° to about 30 mL. Apply remaining solution to the SPE cartridge, wash with 20 mL water, elute with 10 mL MeOH. Evaporate eluate to dryness under reduced pressure, dissolve in 1 mL mobile phase, inject a 10 μL aliquot.

## HPLC VARIABLES
**Column:** 150 × 4.6 Wakosil II 5C18 HG (Wako)
**Mobile phase:** MeCN:50 mM pH 2.4 phosphate buffer 20:80 containing 2.5 mM 1-heptanesulfonic acid
**Flow rate:** 0.6
**Injection volume:** 10
**Detector:** F ex 295 em 455

## CHROMATOGRAM
**Retention time:** 10
**Limit of detection:** 10 ng/g

## OTHER SUBSTANCES
**Extracted:** benofloxacin, danofloxacin, enrofloxacin

## KEY WORDS
chicken; SPE

## REFERENCE

Horie, M.; Saito, K.; Nose, N.; Nakazawa, H. Simultaneous determination of benofloxacin, danofloxacin, enrofloxacin and ofloxacin in chicken tissues by high-performance liquid chromatography. *J.Chromatogr.B*, **1994**, *653*, 69–76

## SAMPLE

**Matrix:** urine
**Sample preparation:** 1 mL Urine + 1 mL MeCN, centrifuge at 4000 g for 15 min, inject a 10 μL aliquot of the supernatant.

## HPLC VARIABLES

**Column:** 150 × 4.5 5 μm Hypersil ODS
**Mobile phase:** MeCN:40 mM phosphoric acid 45:55, final pH 2.56-2.59
**Column temperature:** 23
**Flow rate:** 2
**Injection volume:** 10
**Detector:** UV 254

## CHROMATOGRAM

**Retention time:** 4.63
**Limit of detection:** 1 μg/mL

## REFERENCE

Notarianni, L.J.; Jones, R.W. Method for the determination of ofloxacin, a quinolone carboxylic acid antimicrobial, by high-performance liquid chromatography. *J.Chromatogr.*, **1988**, *431*, 461–464

## ANNOTATED BIBLIOGRAPHY

Barbato, F.; Morrica, P.; Seccia, S.; Ventriglia, M. High performance liquid chromatographic analysis of quinolone antibacterial agents. *Farmaco*, **1994**, *49*, 407–410 [simultaneous cinoxacin, ciprofloxacin, flumequine, nalidixic acid, norfloxacin, oxolinic acid, pefloxacin, piromidic acid]

Fabre, D.; Bressolle, F.; Kinowski, J.M.; Bouvet, O.; Paganin, F.; Galtier, M. A reproducible, simple and sensitive HPLC assay for determination of ofloxacin in plasma and lung tissue. Application in pharmacokinetic studies. *J.Pharm.Biomed.Anal.*, **1994**, *12*, 1463–1469 [pharmacokinetics; LOQ 5 ng/mL; column temp 50; fluorescence detection]

Lehto, P.; Kivistö, K.T. Effect of sucralfate on absorption of norfloxacin and ofloxacin. *Antimicrob.Agents Chemother.*, **1994**, *38*, 248–251 [extracted norfloxacin; plasma; urine; LOQ 100 ng/mL; pharmacokinetics]

Lyon, D.J.; Cheung, S.W.; Chan, C.Y.; Cheng, A.F. Rapid HPLC assay of clinafloxacin, fleroxacin, levofloxacin, sparfloxacin and tosufloxacin. *J.Antimicrob.Chemother.*, **1994**, *34*, 446–448

Mueller, B.A.; Brierton, D.G.; Abel, S.R.; Bowman, L. Effect of feeding with Ensure on oral bioavailabilities of ofloxacin and ciprofloxacin. *Antimicrob.Agents Chemother.*, **1994**, *38*, 2101–2105 [extracted ciprofloxacin; plasma; ultrafiltration; fluorescence detection; A-57084 (IS); LOQ 9.5 ng/mL; pharmacokinetics]

Ohkubo, T.; Kudo, M.; Sugawara, K.; Sawada, Y. The determination of ofloxacin in human skin tissue by high-performance liquid chromatography and correlation between skin tissue concentration and serum level of ofloxacin. *J.Pharm.Pharmacol.*, **1994**, *46*, 522–524

Sánchez Navarro, A.; Martínez Cabarga, M.; Dominguez-Gil Hurlé, A. Oral absorption of ofloxacin administered together with aluminum. *Antimicrob.Agents Chemother.*, **1994**, *38*, 2510–2512 [plasma; ciprofloxacin (IS); LOD 50 ng/mL; pharmacokinetics]

Zeng, S.; Zhang, L.; Liu, Z.Q. [Quantification of the enantiomers of ofloxacin in human urine by RP-HPLC with chiral mobile phase additive]. *Yao Hsueh Hsueh Pao*, **1994**, *29*, 223–227

Israel, D.; Gillum, G.; Turik, M.; Harvey, K.; Ford, J.; Dalton, H.; Towle, M.; Echols, R.; Heller, A.H.; Polk, R. Pharmacokinetics and serum bactericidal titers of ciprofloxacin and ofloxacin following multiple oral doses in healthy volunteers. *Antimicrob.Agents Chemother.*, **1993**, *37*, 2193–2199 [pharmacokinetics; serum; extracted ciprofloxacin; norfloxacin (IS)]

Ohkubo, T.; Kudo, M.; Sugawara, K. Determination of ofloxacin in human serum by high-performance liquid chromatography with column switching. *J.Chromatogr.*, **1992**, *573*, 289–293

Budvári-Bárány, Z.; Szász, G.; Takács-Novák, K.; Hermecz, I.; Lore, A. The pH influence on the HPLC-retention of chemotherapeutic fluoroquinolone derivatives. *J.Liq.Chromatogr.*, **1991**, *14*, 3411–3424 [also amifloxacin, ciprofloxacin, lomefloxacin, nalidixic acid, norfloxacin, oxolinic acid, pefloxacin]

Le Coguic, A.; Bidault, R.; Farinotti, R.; Dauphin, A. Determination of ofloxacin in plasma and urine by liquid chromatography. *J.Chromatogr.*, **1988**, *434*, 320–323 [fluorescence detection; LOD 0.5 ng/mL]

Carlucci, G.; Guadagni, S.; Palumbo, G. Determination of ofloxacin, a new oxazine derivative in human serum, urine, and bile by high-performance liquid chromatography. *J.Liq.Chromatogr.*, **1986**, *9*, 2539–2547 [norfloxacin (IS)]

Groeneveld, A.J.; Brouwers, J.R. Quantitative determination of ofloxacin, ciprofloxacin, norfloxacin and pefloxacin in serum by high pressure liquid chromatography. *Pharm.Weekbl.[Sci].*, **1986**, *8*, 79–84

# Omeprazole

**Molecular formula:** C₁₇H₁₉N₃O₃S
**Molecular weight:** 345.4
**CAS Registry No.:** 73590-58-6

---

## SAMPLE
**Matrix:** blood
**Sample preparation:** 1 mL Serum + 1 µg IS + 5 mL dichloromethane, extract. Remove the organic layer and evaporate it to dryness under a stream of nitrogen at room temperature, reconstitute the residue in mobile phase, inject a 100 µL aliquot.

---

## HPLC VARIABLES
**Column:** 300 × 3.9 µBondapak C18
**Mobile phase:** MeOH:water 50:50 containing 1% triethylamine, pH adjusted to 7.4 with phosphoric acid
**Flow rate:** 2
**Injection volume:** 100
**Detector:** UV 302

---

## CHROMATOGRAM
**Retention time:** 5.3
**Internal standard:** 5-methyl-2-[[(4-methoxy-3,5-dimethyl-2-pyridinyl)methyl]sulfinyl]-1H-benzimidazole (H168/24) (8.4)
**Limit of quantitation:** 50 ng

---

## OTHER SUBSTANCES
**Extracted:** metabolites

---

## KEY WORDS
serum

---

## REFERENCE
Balian, J.D.; Sukhova, N.; Harris, J.W.; Hewett, J.; Pickle, L.; Goldstein, J.A.; Woosley, R.L.; Flockhart, D.A. The hydroxylation of omeprazole correlates with S-mephenytoin metabolism: A population study. *Clin.Pharmacol.Ther.*, **1995**, *57*, 662–669

---

## SAMPLE
**Matrix:** blood
**Sample preparation:** Condition a 100 mg C2 SPE cartridge (Varian) with 1 mL MeOH and 1 mL water. 1 mL Plasma + 50 µL 5 µg/mL IS, add to SPE cartridge, apply a vacuum so the sample moves through at 0.5 mL/min, wash with 1 mL water, dry under vacuum for 10 min, elute with two 1 mL portions of MeCN (do not allow to run dry). Evaporate the eluate to dryness under a stream of nitrogen at room temperature, reconstitute the residue in 50 µL MeOH, vortex for 2 min, add 200 µL pH 9.3 carbonate buffer, inject a 100-190 µL aliquot.

---

## HPLC VARIABLES
**Column:** 150 × 4 5 µm Resolvosil BSA-7 bovine serum albumin (Macherey-Nagel)
**Mobile phase:** n-Propanol:50 mM pH 7.0 ammonium phosphate buffer 0.75:99.25 (n-Propanol concentration should be optimized for each column over the range 0.05-1%.)
**Flow rate:** 1.5
**Injection volume:** 100-190
**Detector:** UV 302

## CHROMATOGRAM
**Retention time:** 13.88 (+), 18.27 (−)
**Internal standard:** 2-([(4-methoxy-2-pyridinyl)methyl]sulfinyl)-4,6-dimethyl-1H-benzimidazole (Astra, Hassle AB, Sweden) (9.73)
**Limit of quantitation:** 15 ng/mL

## OTHER SUBSTANCES
**Extracted:** metabolites

## KEY WORDS
plasma; chiral; protect from light; SPE

## REFERENCE
Cairns, A.M.; Chiou, R.H.-Y.; Rogers, J.D.; Demetriades, J.L. Enantioselective high-performance liquid chromatographic determination of omeprazole in human plasma. *J.Chromatogr.B*, **1995**, *666*, 323–328

## SAMPLE
**Matrix:** blood
**Sample preparation:** 500 µL Plasma + 150 µL water + 5 mL diethyl ether:dichloromethane 70:30, vortex for 5 min, centrifuge at 6° at 300-850 g for 10 min. Remove 4 mL of the organic layer and evaporate to dryness under reduced pressure at room temperature, reconstitute the residue in 500 µL buffer, refrigerate until injection, inject a 100 µL aliquot. (Buffer was MeCN:water 35:65 containing 1 mL/L n-octylamine and 5 mM N-acetohydroxamic acid, pH adjusted to 7.5 with 85% phosphoric acid.)

## HPLC VARIABLES
**Column:** 150 or 250 × 4.6 5 µm Hi-Chrom Reversible octadecylsilane (Regis)
**Mobile phase:** MeCN:water 35:65 containing 1 mL/L n-octylamine and 5 mM N-acetohydroxamic acid, pH adjusted to 7.0 with 85% phosphoric acid
**Column temperature:** 40-43
**Flow rate:** 1 for 15 min then 2.5
**Injection volume:** 100
**Detector:** UV 285

## CHROMATOGRAM
**Retention time:** 7.6
**Internal standard:** omeprazole

## OTHER SUBSTANCES
**Extracted:** lansoprazole

## KEY WORDS
plasma; omeprazole is IS; omeprazole has been determined using this assay, LOQ 10 ng/mL (Antimicrob. Agents Chemother. 1995, 39, 2078)

## REFERENCE
Karol, M.D.; Granneman, G.R.; Alexander, K. Determination of lansoprazole and five metabolites in plasma by high-performance liquid chromatography. *J.Chromatogr.B*, **1995**, *668*, 182–186

## SAMPLE
**Matrix:** blood
**Sample preparation:** 1 mL Plasma + 100 µL MeOH:75 mM pH 9.3 carbonate buffer 20:80 + 5 mL dichloromethane:MeCN 80:20, mix for 10 min, centrifuge at 2000 g for 5

min. Remove 4 mL of the organic layer and evaporate it to dryness under a stream of nitrogen, reconstitute the residue in 250 μL mobile phase, inject a 100 μL aliquot.

## HPLC VARIABLES
**Column:** 250 × 4.6 5 μm Nucleosil 5C18
**Mobile phase:** MeCN:20 mM pH 7.5 phosphate buffer 37:63
**Column temperature:** 40
**Flow rate:** 1
**Injection volume:** 100
**Detector:** UV 302

## CHROMATOGRAM
**Limit of quantitation:** 5 ng/mL

## KEY WORDS
rat; plasma; pharmacokinetics

## REFERENCE
Katashima, M.; Yamamoto, K.; Sugiura, M.; Sawada, Y.; Iga, T. Comparative pharmacokinetic/pharmacodynamic study of proton pump inhibitors, omeprazole and lansoprazole in rats. *Drug Metab.Dispos.*, **1995**, *23*, 718–723

## SAMPLE
**Matrix:** blood
**Sample preparation:** 2 mL Whole blood or plasma + 2 mL buffer + 5 mL chloroform: isopropanol:n-heptane 60:14:26, shake gently horizontally for 10 min, centrifuge at 2800 g for 10 min. Remove the lower organic layer and evaporate it to dryness under vacuum at 45°, reconstitute the residue in 100 μL mobile phase, centrifuge at 2800 g for 5 min, inject a 50 μL aliquot of the supernatant. (Buffer was saturated ammonium chloride solution 25% diluted with water, adjusted to pH 9.5 with 25% ammonia solution.)

## HPLC VARIABLES
**Column:** 300 × 3.9 4 μm NovaPack C18
**Mobile phase:** MeOH:THF:buffer 65:5:30 (Buffer was 0.68 g/L (10 mM (sic)) $KH_2PO_4$ adjusted to pH 2.6 with concentrated orthophosphoric acid.) (At the end of each session wash the column with water for 1 h and MeOH for 1 h, re-equilibrate for 30 min.)
**Column temperature:** 30
**Flow rate:** 0.8
**Injection volume:** 50
**Detector:** UV 303

## CHROMATOGRAM
**Retention time:** 3.72
**Limit of detection:** <120 ng/mL

## OTHER SUBSTANCES
**Extracted:** acenocoumarol, acepromazine, aceprometazine, acetaminophen, aconitine, ajmaline, albuterol, alimemazine, alminoprofen, alpidem, alprazolam, alprenolol, amisulpride, amitriptyline, amodiaquine, amoxapine, aspirin, astemizole, atenolol, benazepril, benperidol, benzocaine, benzoylecgonine, bepridil, betaxolol, bisoprolol, brompheniramine, bumadizone, bupivacaine, buprenorphine, buspirone, caffeine, carbinoxamine, carpipramine, carteolol, celiprolol, cetirizine, chlorambucil, chlordiazepoxide, chlormezanone, chlorophenacinone, chloroquine, chlorpheniramine, chlorpromazine, chlorpropamide, cibenzoline, cicletanine, clemastine, clobazam, clomipramine, clonazepam, clonidine, clorazepate, clozapine, cocaine, codeine, colchicine, cyamemazine, cyclizine, cycloguanil, cyproheptadine, cytarabine, dacarbazine, daunorubicin, debrisoquine, demexiptiline, desipramine, dextromethorphan, dextromoramide, dextropropoxyphene, diazepam, diazox-

ide, diclofenac, diltiazem, diphenhydramine, dipyridamole, disopyramide, dosulepine, doxepin, doxylamine, droperidol, ephedrine, estazolam, etodolac, fenfluramine, fenoprofen, fentiazac, flecainide, floctafenine, flumazenil, flunitrazepam, fluoxetine, fluphenazine, flurbiprofen, fluvoxamine, glibenclamide, glibornuride, glipizide, glutethimide, haloperidol, histapyrrodine, hydroxychloroquine, hydroxyzine, ibuprofen, imipramine, indomethacin, iproniazid, ketamine, ketoprofen, labetalol, levomepromazine, lidocaine, lidoflazine, lisinopril, loperamide, loprazolam, loratadine, lorazepam, loxapine, maprotiline, medazepam, medifoxamine, mefenamic acid, mefenidramine, mefloquine, melphalan, meperidine, mephenesin, mephentermine, mepivacaine, metapramine, metformin, methadone, methaqualone, methocarbamol, methotrexate, metipranolol, metoclopramide, metoprolol, mexiletine, mianserine, midazolam, minoxidil, moclobemide, moperone, morphine, nalorphine, naloxone, naltrexone, naproxen, nialamide, nicardipine, nifedipine, niflumic acid, nimodipine, nitrazepam, nitrendipine, nizatidine, nomifensine, nortriptyline, opipramol, oxazepam, oxprenolol, penbutolol, penfluridol, pentazocine, phencyclidine, phenobarbital, phenol, phenylbutazone, pimozide, pindolol, pipamperone, piroxicam, prazepam, prazosin, prilocaine, proguanil, promethazine, propafenone, propranolol, protriptyline, pyrimethamine, quinidine, quinine, quinupramine, ramipril, ranitidine, reserpine, ritodrine, secobarbital, sotalol, sulfinpyrazole, sulindac, sulpride, sultopride, suriclone, temazepam, tenoxicam, terbutaline, terfenadine, tetracaine, tetrazepam, thiopental, thioproperazine, thioridazine, tianeptine, tiapride, tiaprofenic acid, ticlopidine, timolol, tioclomarol, tofisopam, tolbutamide, toloxatone, trazodone, triazolam, trifluoperazine, trifluperidol, trimipramine, triprolidine, tropatenine, verapamil, viloxazine, vinblastine, vincristine, vindesine, warfarin, yohimbine, zolpidem, zopiclone, zorubicine
**Interfering:** acebutolol, bromazepam, carbamazepine, dihydralazine, nadolol, nalbuphine, procainamide, procarbazine, strychnine

## KEY WORDS
whole blood; plasma

## REFERENCE
Tracqui, A.; Kintz, P.; Mangin, P. Systematic toxicological analysis using HPLC/DAD. *J.Forensic Sci.*, **1995**, *40*, 254–262

## SAMPLE
**Matrix:** blood
**Sample preparation:** 1 mL Plasma + 500 ng IS + 6 mL dichloromethane, extract, centrifuge. Remove organic layer and evaporate it, dissolve the residue in 100 μL mobile phase, inject an aliquot.

## HPLC VARIABLES
**Column:** 120 mm long 5 μm Hypersil ODS
**Mobile phase:** MeOH:water:triethylamine 55:44:1, pH 7
**Flow rate:** 0.8
**Detector:** UV 302

## CHROMATOGRAM
**Retention time:** 7.2
**Internal standard:** 5-methyl-2-[[(4-methoxy-3,5-dimethyl-2-pyridinyl)methyl]sulfinyl]-1H-benzimidazole (H 168/24) (11.0)
**Limit of quantitation:** 10 ng/mL

## KEY WORDS
plasma

## REFERENCE
Treiber, G.; Walker, S.; Klotz, U. Omeprazole-induced increase in the absorption of bismuth from tripotassium dicitrato bismuthate. *Clin.Pharmacol.Ther.*, **1994**, *55*, 486–491

## SAMPLE
**Matrix:** blood
**Sample preparation:** Condition a Bond Elut C18 SPE cartridge with 2 mL MeOH and 2 mL water. Add 20 μL plasma + 40 ng nitrazepam to SPE cartridge, wash with 2 mL water, elute with 250 μL MeOH, inject a 60 μL aliquot of the eluate.

## HPLC VARIABLES
**Column:** 125 × 4 4 μm LiChro CART C18
**Mobile phase:** MeCN:20 mM pH 7.4 Na$_2$HPO$_4$ 35:65
**Flow rate:** 1
**Injection volume:** 60
**Detector:** UV 310

## CHROMATOGRAM
**Internal standard:** nitrazepam
**Limit of detection:** 25 ng/mL

## KEY WORDS
plasma; rat; SPE; pharmacokinetics

## REFERENCE
Watanabe, K.; Furuno, K.; Eto, K.; Oishi, R.; Gomita, Y. First-pass metabolism of omeprazole in rats. *J.Pharm.Sci.*, **1994**, *83*, 1131–1134

## SAMPLE
**Matrix:** blood
**Sample preparation:** 1 mL Plasma + 100 μL 1 M NaH$_2$PO$_4$ + 100 μL 50-100 μg/mL IS in MeOH:pH 9.3 carbonate buffer (I = 0.1) 20:80, mix, add 1 mL dichloromethane, shake for 10 min, centrifuge twice at 2500 g for 10 min, inject a 150 μL aliquot of the organic layer.

## HPLC VARIABLES
**Column:** 150 × 4.5 5 μm LiChrosorb Si 60
**Mobile phase:** Dichloromethane:MeOH containing 5% concentrated ammonium hydroxide 96.5:3.5
**Flow rate:** 1.5
**Injection volume:** 150
**Detector:** UV 302

## CHROMATOGRAM
**Retention time:** 4
**Internal standard:** 4,6-dimethyl-2-[[(4-methoxy-2-pyridinyl)methyl]sulfinyl]-1H-benzimidazole (H153/52) (5.5)
**Limit of detection:** 20 nM (plasma)

## OTHER SUBSTANCES
**Extracted:** metabolites

## KEY WORDS
plasma; normal phase

## REFERENCE
Lagerström, P.-O.; Persson, B.-A. Determination of omeprazole and metabolites in plasma and urine by liquid chromatography. *J.Chromatogr.*, **1984**, *309*, 347–356

## SAMPLE
**Matrix:** blood, urine
**Sample preparation:** Plasma. 1 mL Plasma + 100 μL 100 μg/mL phenacetin in MeOH + 5 mL dichloromethane + 250 mg NaCl + 500 μL 0.5 M pH 8.0 phosphate buffer, vortex for 10 min, centrifuge at 1610 g for 30 min. Remove the organic phase and evaporate it under vacuum at 40°, reconstitute the residue with 300 μL mobile phase, filter (0.45 μm), inject a 30 μL aliquot. Urine. 1 mL Urine + 100 μL 200 μg/mL phenacetin in MeOH + 5 mL dichloromethane + 250 mg NaCl + 500 μL 0.5 M pH 8.0 phosphate buffer, vortex for 10 min, centrifuge at 1610 g for 30 min. Remove the organic phase and evaporate it under vacuum at 40°, reconstitute the residue with 200 μL mobile phase, inject a 10-30 μL aliquot.

## HPLC VARIABLES
**Column:** 250 × 4.6 5 μm Capcell Pak C18 SG 120 (Shiseido)
**Mobile phase:** MeCN:50 mM pH 8.5 phosphate buffer 25:75
**Flow rate:** 0.8
**Injection volume:** 10-30
**Detector:** UV 302

## CHROMATOGRAM
**Retention time:** 16.8
**Internal standard:** phenacetin (13.6)
**Limit of quantitation:** 10 ng/mL

## OTHER SUBSTANCES
**Extracted:** metabolites
**Noninterfering:** cimetidine, ranitidine, famotidine

## KEY WORDS
plasma; pharmacokinetics

## REFERENCE
Kobayashi, K.; Chiba, K.; Sohn, D.-R.; Kato, Y.; Ishizaki, T. Simultaneous determination of omeprazole and its metabolites in plasma and urine by reversed-phase high-performance liquid chromatography with an alkaline-resistant polymer-coated C18 column. *J.Chromatogr.*, **1992**, *579*, 299–305

## SAMPLE
**Matrix:** blood, urine
**Sample preparation:** 1 mL Plasma or urine + 100 μL 1 M NaH$_2$PO$_4$ + 100 μL 50-100 μg/mL IS in MeOH:pH 9.3 carbonate buffer (I = 0.1) 20:80, mix, add 10 mL dichloromethane, shake for 10 min, centrifuge at 2500 g for 10 min. Remove 8 mL of the organic layer and evaporate it to dryness under a stream of nitrogen at room temperature, reconstitute the residue in 500 μL MeCN:pH 7.5 phosphate buffer (I=0.05), inject a 150 μL aliquot.

## HPLC VARIABLES
**Guard column:** 30 × 4.6 Spheri-5 RP-8
**Column:** 150 × 4.5 5 μm Polygosil C18 (Macherey-Nagel)
**Mobile phase:** MeCN:pH 7.5 phosphate buffer (I=0.05) 30:70
**Flow rate:** 1
**Injection volume:** 150
**Detector:** UV 302

## CHROMATOGRAM
**Retention time:** 7.5

**Internal standard:** 5-methyl-2-[[(4-methoxy-3,5-dimethyl-2-pyridinyl)methyl]sulfinyl]-1H-benzimidazole (H168/24) (10)
**Limit of quantitation:** 25 nM (plasma); 50 nM (urine)

## OTHER SUBSTANCES
**Extracted:** metabolites
**Noninterfering:** cimetidine, ranitidine

## KEY WORDS
plasma

## REFERENCE
Lagerström, P.-O.; Persson, B.-A. Determination of omeprazole and metabolites in plasma and urine by liquid chromatography. *J.Chromatogr.*, **1984**, *309*, 347–356

## SAMPLE
**Matrix:** microsomal incubations
**Sample preparation:** 1 mL Incubation mixture + 2 mL dichloromethane:butanol 99:1, cool on ice, add 100 µL 1 M $NaH_2PO_4$ + IS, vortex for 1 min, centrifuge at 1000 g for 5 min. Remove 1.5 mL of the organic phase and evaporate it to dryness under nitrogen, reconstitute in 150 µL mobile phase, inject a 50 µL aliquot.

## HPLC VARIABLES
**Guard column:** 15 × 3 7 µm Brownlee Aquapore silica
**Column:** 125 × 4 4 µm Superspher SI-60
**Mobile phase:** Dichloromethane:5% ammonium hydroxide in MeOH:isopropanol 191:8:1
**Flow rate:** 1.5
**Injection volume:** 50
**Detector:** UV 302

## CHROMATOGRAM
**Retention time:** 2.5
**Internal standard:** 4,6-Dimethyl-2-[[(4-methoxy-2-pyridinyl)methyl]sulfinyl]-1H-benzimidazole (3.3)

## OTHER SUBSTANCES
**Extracted:** metabolites

## KEY WORDS
normal phase

## REFERENCE
Andersson, T.; Lagerstrom, P.-O.; Miners, J.O.; Veronese, M.E.; Weidolf, L.; Birkett, D.J. High-performance liquid chromatographic assay for human liver microsomal omeprazole metabolism. *J.Chromatogr.*, **1993**, *619*, 291–297

## SAMPLE
**Matrix:** solutions
**Sample preparation:** Prepare a solution in EtOH, inject a 10-20 µL aliquot.

## HPLC VARIABLES
**Column:** 250 × 4.6 Chiralpak AD
**Mobile phase:** Hexane:EtOH 80:20
**Column temperature:** 35
**Flow rate:** 1

**Injection volume:** 10-20
**Detector:** UV 302

## CHROMATOGRAM
**Retention time:** k' 4.86 (of first (−) enantiomer)

## OTHER SUBSTANCES
**Simultaneous:** lansoprazole, pantoprazole, timoprazole

## KEY WORDS
chiral; α = 1.81

## REFERENCE
Balmér, K.; Persson, B.-A.; Lagerström, P.-O. Stereoselective effects in the separation of enantiomers of omeprazole and other substituted benzimidazoles on different chiral stationary phases. *J.Chromatogr.A*, **1994**, *660*, 269–273

## ANNOTATED BIBLIOGRAPHY
Yasuda, S.; Horai, Y.; Tomono, Y.; Nakai, H.; Yamato, C.; Manabe, K.; Kobayashi, K.; Chiba, K.; Ishizaki, T. Comparison of the kinetic disposition and metabolism of E3810, a new proton pump inhibitor, and omeprazole in relation to S-mephenytoin 4'-hydroxylation status. *Clin.Pharmacol.Ther.*, **1995**, *58*, 143–154 [LOQ 5 ng/mL; pharmacokinetics]

Kobayashi, K.; Chiba, K.; Tani, M.; Kuroiwa, Y.; Ishizaki, T. Development and preliminary application of a high-performance liquid chromatographic assay for omeprazole metabolism in human liver microsomes. *J.Pharm.Biomed.Anal.*, **1994**, *12*, 839–844

Arvidsson, T.; Collijn, E.; Tivert, A.M.; Rosen, L. Peak distortion in the column liquid chromatographic determination of omeprazole dissolved in borax buffer. *J.Chromatogr.*, **1991**, *586*, 271–276

Erlandsson, P.; Isaksson, R.; Lorentzon, P.; Lindberg, P. Resolution of the enantiomers of omeprazole and some of its analogues by liquid chromatography on a trisphenylcarbamoylcellulose-based stationary phase. The effect of the enantiomers of omeprazole on gastric glands. *J.Chromatogr.*, **1990**, *532*, 305–319

Amantea, M.A.; Narang, P.K. Improved procedure for quantitation of omeprazole and metabolites using reversed-phase high-performance liquid chromatography. *J.Chromatogr.*, **1988**, *426*, 216–222

Mihaly, G.W.; Prichard, P.J.; Smallwood, R.A.; Yeomans, N.D.; Louis, W.J. Simultaneous high-performance liquid chromatographic analysis of omeprazole and its sulphone and sulphide metabolites in human plasma and urine. *J.Chromatogr.*, **1983**, *278*, 311–319

# Ondansetron

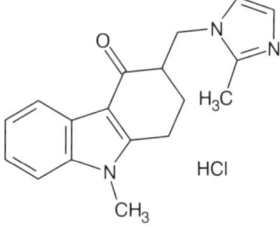

**Molecular formula:** $C_{18}H_{19}N_3O$
**Molecular weight:** 293.4
**CAS Registry No.:** 99614-02-5 (ondansetron), 116002-70-1
(ondansetron), 99614-01-4 (ondansetron hydrochloride dihydrate),
103639-04-9 (ondansetron hydrochloride dihydrate)

## SAMPLE
**Matrix:** beverages
**Sample preparation:** Juice, soft drinks. Inject a 10 μL aliquot. Tea. Cool 60 mL tea in an
ice bath to about 0° in 1 min. Remove a 10 mL aliquot, add 1 drop concentrated HCl, add
10 mL MeCN, mix, inject a 10 μL aliquot.

## HPLC VARIABLES
**Guard column:** 30 × 4.6 5 μm C18 (Phase Separations)
**Column:** 250 × 4.5 5 μm Spherisorb cyanopropyl
**Mobile phase:** MeCN:20 mM pH 5.4 phosphate buffer 60:40
**Flow rate:** 2
**Injection volume:** 10
**Detector:** UV 216

## CHROMATOGRAM
**Retention time:** 6.5

## KEY WORDS
tea; apple juice; soft drinks; stability indicating

## REFERENCE
Yamreudeewong, W.; Danthi, S.N.; Hill, R.A.; Fox, J.L. Stability of ondansetron hydrochloride injection
in various beverages. *Am.J.Health-Syst.Pharm.*, **1995**, *52*, 2011–2014

## SAMPLE
**Matrix:** blood
**Sample preparation:** Condition a 1 mL 100 mg cyanopropyl SPE cartridge (J & W Scientific) with 2 volumes of MeOH and 2 volumes of water, do not allow to dry. 1 mL Serum
+ 30 μL 10 μg/mL prazosin in EtOH, vortex for 10 s, add to SPE cartridge, wash with 1
volume MeOH:water 10:90, dry under full vacuum for 20 min, elute with four 250 μL
aliquots of MeOH:triethylamine 999:1. Evaporate the eluate to dryness under a stream
of nitrogen at room temperature, dissolve the residue in 250 μL mobile phase, inject a
100 μL aliquot.

## HPLC VARIABLES
**Guard column:** 500 × 4.6 10 μm Chiralcel OD precolumn (sic)
**Column:** 250 × 4.6 10 μm Chiralcel OD
**Mobile phase:** Hexane:95% EtOH:isopropanol:MeCN 65:25:10:1
**Flow rate:** 1
**Injection volume:** 100
**Detector:** UV 216

## CHROMATOGRAM
**Retention time:** 9.96 (R-(−)), 11.56 (S-(+))
**Internal standard:** prazosin (7.99)

**Limit of detection:** 2.5 ng/mL
**Limit of quantitation:** 10 ng/mL

## KEY WORDS
serum; SPE; chiral

## REFERENCE
Kelly, J.W.; He, L.; Stewart, J.T. High-performance liquid chromatographic separation of ondansetron enantiomers in serum using a cellulose-derivatized stationary phase and solid-phase extraction. *J.Chromatogr.*, **1993**, *622*, 291–295

## SAMPLE
**Matrix:** blood
**Sample preparation:** Condition a 100 mg Bond Elut CN SPE cartridge with two 1 mL portions of MeOH, two 1 mL portions of isopropanol:ammonia 99:1, and two 1 mL portions of water, do not allow to dry. Add 50 µL 0.5 M HCl then 1 mL plasma to the SPE cartridge, dry the cartridge with a vacuum, wash with two 1 mL portions of water, allow to dry, wash with two 1 mL portions of MeCN, allow to dry, elute with two 1 mL portions of isopropanol:ammonia 99:1, evaporate the eluate at 45° under vacuum, reconstitute in 200 µL mobile phase, sonicate for 5 min, vortex, inject a 100 µL aliquot.

## HPLC VARIABLES
**Column:** 100 × 4.6 3 µm Spherisorb S3W silica
**Mobile phase:** MeCN:buffer 40:60 (Buffer was 25 mM sodium acetate adjusted to pH 4.2 with glacial acetic acid.)
**Column temperature:** 35
**Flow rate:** 1
**Injection volume:** 100
**Detector:** UV 305

## CHROMATOGRAM
**Retention time:** 4-5
**Limit of quantitation:** 1 ng/mL

## OTHER SUBSTANCES
**Simultaneous:** metabolites, fluorouracil, melphalan
**Noninterfering:** cisplatin

## KEY WORDS
SPE; plasma

## REFERENCE
Colthup, P.V.; Felgate, C.C.; Palmer, J.L.; Scully, N.L. Determination of ondansetron pharmacokinetics in the young and elderly. *J.Pharm.Sci.*, **1991**, *80*, 868–871

## SAMPLE
**Matrix:** blood
**Sample preparation:** Condition a 100 mg Bondelut CN SPE cartridge with 2 mL MeOH, 2 mL isopropanol:ammonia 99:1, and 2 mL water. Mix 1 mL plasma with 50 µL 0.5 M HCl and add to the SPE cartridge, wash with 2 mL water, wash with 2 mL MeCN, elute with 2 mL isopropanol:ammonia 99:1, evaporate the eluate, dissolve the residue in 200 µL mobile phase, inject a 100 µL aliquot.

## HPLC VARIABLES
**Column:** 100 × 4.6 3 µm Spherisorb S3W silica
**Mobile phase:** MeCN:25 mM pH 4.2 sodium acetate 40:60

**Column temperature:** 35
**Injection volume:** 100
**Detector:** UV 305

## CHROMATOGRAM
**Limit of quantitation:** 1 ng/mL

## KEY WORDS
plasma; SPE; pharmacokinetics

## REFERENCE
Colthup, P.V.; Palmer, J.L. The determination in plasma and pharmacokinetics of ondansetron. *Eur.J.Cancer Clin.Oncol.*, **1989**, *25 Suppl 1*, S71–S74

## SAMPLE
**Matrix:** formulations
**Sample preparation:** Dilute with mobile phase, inject a 20 μL aliquot.

## HPLC VARIABLES
**Guard column:** ODS
**Column:** 250 × 4.6 5 μm Spherisorb ODS-1
**Mobile phase:** MeCN:100 mM pH 4.5 $KH_2PO_4$ 40:60
**Flow rate:** 1.2
**Injection volume:** 20
**Detector:** UV 210

## CHROMATOGRAM
**Retention time:** 11.6
**Limit of detection:** 70 ng/mL

## OTHER SUBSTANCES
**Simultaneous:** diphenhydramine, methyl paraben, propyl paraben

## KEY WORDS
injections; saline

## REFERENCE
Ye, L.; Stewart, J.T. HPLC determination of an ondansetron and diphenhydramine mixture in 0.9% sodium chloride injection. *J.Liq.Chromatogr.& Rel.Technol.*, **1996**, *19*, 711–718

## SAMPLE
**Matrix:** formulations
**Sample preparation:** Dilute with an equal volume of saline (if necessary), filter (0.2 μm), inject a 50 μL aliquot of the filtrate.

## HPLC VARIABLES
**Column:** 250 × 4.6 5 μm Accubond CN (J & W)
**Mobile phase:** MeCN:buffer 60:40 adjusted to pH 6.0 with 1 M NaOH (Buffer was 20 mM $KH_2PO_4$ containing 5 mM octanesulfonic acid.)
**Flow rate:** 1.5
**Injection volume:** 50
**Detector:** UV 305

## CHROMATOGRAM
**Retention time:** 8-9

## KEY WORDS
saline; injections; stability-indicating

## REFERENCE
Chung, K.C.; Moon, Y.S.K.; Chin, A.; Ulrich, R.W.; Gill, M.A. Compatibility of ondansetron hydrochloride and piperacillin sodium tazobactam sodium during simulated Y-site administration. *Am.J.Health-Syst.Pharm.*, **1995**, *52*, 1554–1556

## SAMPLE
**Matrix:** formulations
**Sample preparation:** Filter (0.22 μm), inject a 20 μL aliquot.

## HPLC VARIABLES
**Column:** 250 × 4.6 5 μm Ultrasphere reverse-phase
**Mobile phase:** MeCN:water 40:60
**Flow rate:** 1.5
**Injection volume:** 20
**Detector:** UV 200

## CHROMATOGRAM
**Retention time:** 3.3

## OTHER SUBSTANCES
**Simultaneous:** cyclophosphamide

## KEY WORDS
injections; saline; 5% dextrose

## REFERENCE
Fleming, R.A.; Olsen, D.J.; Savage, P.D.; Fox, J.L. Stability of ondansetron hydrochloride and cyclophosphamide in injectable solutions. *Am.J.Health-Syst.Pharm.*, **1995**, *52*, 514–516

## SAMPLE
**Matrix:** formulations
**Sample preparation:** Filter (0.22 μm), inject a 20 μL aliquot.

## HPLC VARIABLES
**Column:** 300 × 4.6 5 μm Spherisorb nitrile
**Mobile phase:** MeCN:20 mM pH 5.4 phosphate buffer 60:40
**Flow rate:** 1.2
**Injection volume:** 20
**Detector:** UV 216

## CHROMATOGRAM
**Retention time:** 9.3

## OTHER SUBSTANCES
**Noninterfering:** cyclophosphamide

## KEY WORDS
injections; saline; 5% dextrose; stability-indicating

## REFERENCE
Fleming, R.A.; Olsen, D.J.; Savage, P.D.; Fox, J.L. Stability of ondansetron hydrochloride and cyclophosphamide in injectable solutions. *Am.J.Health-Syst.Pharm.*, **1995**, *52*, 514–516

## SAMPLE
**Matrix:** formulations
**Sample preparation:** Dilute with 2 volumes water (if necessary), filter (0.45 μm), inject an aliquot.

## HPLC VARIABLES
**Column:** 220 × 4.6 10 μm Spheri-10
**Mobile phase:** MeCN:DMF:pH 6.5 sodium acetate 15:4:85
**Flow rate:** 1
**Detector:** UV 305

## KEY WORDS
total parental nutrient; injections; stability-indicating

## REFERENCE
Kirkham, J.C.; Rutherford, E.T.; Cunningham, G.N.; Daneshmand, K.A.; Falls, A.L. Stability of ondansetron hydrochloride in a total parenteral nutrient admixture. *Am.J.Health-Syst.Pharm.*, **1995**, *52*, 1557–1558

## SAMPLE
**Matrix:** formulations
**Sample preparation:** Inject a 20 μL aliquot.

## HPLC VARIABLES
**Column:** 100 × 4.6 Spheri-5 RP-8
**Mobile phase:** MeCN:buffer 45:55 (Buffer was 10 mM $KH_2PO_4$ adjusted to pH 4.0 with 1 M KOH.)
**Flow rate:** 1
**Injection volume:** 20
**Detector:** UV 254

## CHROMATOGRAM
**Retention time:** 5.9
**Limit of detection:** 90 ng/mL

## OTHER SUBSTANCES
**Simultaneous:** glycopyrrolate
**Noninterfering:** degradation products

## KEY WORDS
injections; saline

## REFERENCE
Venkateshwaran, T.G.; King, D.T.; Stewart, J.T. HPLC determination of ondansetron-atropine and ondansetron-glycopyrrolate mixtures in 0.9% sodium chloride injection. *J.Liq.Chromatogr.*, **1995**, *18*, 2647–2659

## SAMPLE
**Matrix:** formulations

## HPLC VARIABLES
**Column:** 250 × 4.6 5 μm Zorbax Rx-C8 base deactivated octylsilane (Mac-Mod Analytical)
**Mobile phase:** MeCN:10 mM $KH_2PO_4$ adjusted to pH 4.0 with 1 M KOH 23:77
**Flow rate:** 1

**Injection volume:** 20
**Detector:** UV 273

## CHROMATOGRAM
**Retention time:** 14.8
**Limit of detection:** 20 ng/mL

## OTHER SUBSTANCES
**Simultaneous:** metoclopramide

## KEY WORDS
saline; injections

## REFERENCE
Venkateshwaran, T.G.; King, D.T.; Stewart, J.T. HPLC determination of a metoclopramide and ondansetron mixture in 0.9% sodium chloride injection. *J.Liq.Chromatogr.*, **1995**, *18*, 117–126

## SAMPLE
**Matrix:** formulations
**Sample preparation:** 1 mL Sample + 50 μL 150 ng/mL cimetidine, inject a 10 μL aliquot.

## HPLC VARIABLES
**Column:** 250 × 4.8 Spherisorb S5CN
**Mobile phase:** MeCN:buffer 50:50 (Buffer was 20 mM $KH_2PO_4$ adjusted to pH 5.4 with 1 M NaOH.)
**Flow rate:** 1.5
**Injection volume:** 10
**Detector:** UV 216

## CHROMATOGRAM
**Retention time:** 9.5
**Internal standard:** cimetidine (4.5)

## KEY WORDS
injections; stability-indicating; 5% dextrose

## REFERENCE
Bosso, J.A.; Prince, R.A.; Fox, J.L. Compatibility of ondansetron hydrochloride with fluconazole, ceftazidime, aztreonam, and cefazolin sodium under simulated Y-site conditions. *Am.J.Hosp.Pharm.*, **1994**, *51*, 389–391

## SAMPLE
**Matrix:** formulations
**Sample preparation:** Dilute 1:2, inject a 50 μL aliquot.

## HPLC VARIABLES
**Column:** 250 × 4.6 5 μm Accubond CN (J & W)
**Mobile phase:** MeCN:20 mM $KH_2PO_4$ and 5 mM octanesulfonic acid 50:50, pH adjusted to 6.0 with 1 M NaOH
**Flow rate:** 1.5
**Injection volume:** 50
**Detector:** UV 305

## CHROMATOGRAM
**Retention time:** 8.5

## OTHER SUBSTANCES
**Noninterfering:** ranitidine, paclitaxel

## KEY WORDS
injections; 5% dextrose; stability-indicating

## REFERENCE
Burm, J.-P.; Jhee, S.S.; Chin, A.; Moon, Y.S.K.; Jeong, E.; Nii, L.; Fox, J.L.; Gill, M.A. Stability of paclitaxel with ondansetron hydrochloride or ranitidine hydrochloride during simulated Y-site administration. *Am.J.Hosp.Pharm.*, **1994**, *51*, 1201–1204

## SAMPLE
**Matrix:** formulations

## HPLC VARIABLES
**Column:** 300 × 3.9 10 μm μBondapak phenyl
**Mobile phase:** MeCN:buffer 50:50 (Buffer was 20 mM $KH_2PO_4$ adjusted to pH 5.4 with 1 M NaOH.)
**Flow rate:** 1
**Injection volume:** 20
**Detector:** UV 233

## CHROMATOGRAM
**Retention time:** 10.5
**Limit of detection:** 200 ng/mL

## OTHER SUBSTANCES
**Simultaneous:** degradation products, doxorubicin, methyl paraben, vincristine

## KEY WORDS
injections; saline

## REFERENCE
King, D.T.; Venkateshwaran, T.G.; Stewart, J.T. HPLC determination of a vincristine, doxorubicin, and ondansetron mixture in 0.9% sodium chloride injection. *J.Liq.Chromatogr.*, **1994**, *17*, 1399–1411

## SAMPLE
**Matrix:** formulations
**Sample preparation:** Dilute 1:10, inject a 10 μL aliquot.

## HPLC VARIABLES
**Column:** 300 × 4.6 5 μm Spherisorb CN
**Mobile phase:** MeCN:20 mM $KH_2PO_4$ 50:50, pH adjusted to 5.40 with 1 M NaOH
**Flow rate:** 1.5
**Injection volume:** 10
**Detector:** UV 216

## CHROMATOGRAM
**Retention time:** 8.8-10.0

## OTHER SUBSTANCES
**Simultaneous:** hydromorphone, morphine

## KEY WORDS
injections; saline; stability-indicating

## REFERENCE

Trissel, L.A.; Xu, Q.; Martinez, J.F.; Fox, J.L. Compatibility and stability of ondansetron hydrochloride with morphine sulfate and with hydromorphone hydrochloride in 0.9% sodium chloride injection at 4, 22, and 32°C. *Am.J.Hosp.Pharm.*, **1994**, *51*, 2138−2142

## SAMPLE
**Matrix:** formulations

## HPLC VARIABLES
**Column:** 220 × 4.6 5 μm silica (Brownlee)
**Mobile phase:** MeCN:6.25 mM NaH$_2$PO$_4$ adjusted to pH 3.0 with concentrated phosphoric acid 40:60
**Flow rate:** 1
**Injection volume:** 50
**Detector:** UV 216

## CHROMATOGRAM
**Retention time:** 8.6
**Limit of detection:** 8 ng/mL

## OTHER SUBSTANCES
**Simultaneous:** dacarbazine, doxorubicin
**Noninterfering:** degradation products

## KEY WORDS
injections; 5% dextrose

## REFERENCE

King, D.T.; Stewart, J.T. HPLC determination of dacarbazine, doxorubicin, and ondansetron mixture in 5% dextrose injection on underivatized silica with an aqueous-organic mobile phase. *J.Liq.Chromatogr.*, **1993**, *16*, 2309−2323

## SAMPLE
**Matrix:** microsomal incubations
**Sample preparation:** 750 μL Microsomal incubation + 1.5 mL MeCN, concentrate, inject an aliquot.

## HPLC VARIABLES
**Column:** 150 × 4.6 5 μm Hypersil C8 BDS
**Mobile phase:** MeCN:40 mM pH 4.5 ammonium acetate 20:80
**Flow rate:** 0.8
**Detector:** Radioactivity

## CHROMATOGRAM
**Retention time:** 26

## OTHER SUBSTANCES
**Extracted:** metabolites

## KEY WORDS
human; liver; [14]C labeled

## REFERENCE

Dixon, C.M.; Colthup, P.V.; Serabjit-Singh, C.J.; Kerr, B.M.; Boehlert, C.C.; Park, G.R.; Tarbit, M.H. Multiple forms of cytochrome P450 are involved in the metabolism of ondansetron in humans. *Drug Metab.Dispos.*, **1995**, *23*, 1225−1230

## SAMPLE
**Matrix:** solutions

## HPLC VARIABLES
**Column:** 250 × 4.6 10 μm Partisil ODS1
**Mobile phase:** MeOH:50 mM pH 3.0 phosphoric acid 40:60
**Column temperature:** 30
**Flow rate:** 1.5
**Detector:** Radioactivity

## KEY WORDS
$^{14}$C labeled

## REFERENCE
Collett, A.; Sims, E.; Walker, D.; He, Y.-L.; Ayrton, J.; Rowland, M.; Warhurst, G. Comparison of HT29-18-C$_1$ and Caco-2 cell lines as models for studying intestinal paracellular drug absorption. *Pharm.Res.*, **1996**, *13*, 216–221

## SAMPLE
**Matrix:** solutions
**Sample preparation:** Inject an aliquot of an aqueous solution.

## HPLC VARIABLES
**Column:** 250 × 4 Nucleosil C18
**Mobile phase:** MeOH:THF:buffer 30:5:65 (Buffer was 100 mM triethylamine adjusted to pH 3.0 with nitric acid.)
**Flow rate:** 0.8
**Injection volume:** 20
**Detector:** UV 254

## CHROMATOGRAM
**Retention time:** 7.40

## OTHER SUBSTANCES
**Simultaneous:** granisetron, tropisetron

## REFERENCE
Barbato, F.; Immacolata La Rotonda, M.; Quaglia, F. Retention behaviour of anti-emetic serotonin antagonists in reversed phase high performance liquid chromatography. *Farmaco*, **1995**, *50*, 875–880

## SAMPLE
**Matrix:** solutions

## HPLC VARIABLES
**Column:** 250 × 4.6 5 μm YMC-Pack ODS-A
**Mobile phase:** MeCN:MeOH:isopropanol:20 mM NaH$_2$PO$_4$ 50:15:5:30
**Flow rate:** 1
**Detector:** UV 216

## CHROMATOGRAM
**Retention time:** 3.7

## OTHER SUBSTANCES
**Simultaneous:** impurities

## REFERENCE

Ong, C.P.; Chow, K.K.; Ng, C.L.; Ong, F.M.; Lee, H.K.; Li, S.F.Y. Use of overlapping resolution mapping scheme for optimization of the high-performance liquid chromatographic separation of pharmaceuticals. *J.Chromatogr.A*, **1995**, *692*, 207–212

## ANNOTATED BIBLIOGRAPHY

Fischer, V.; Vickers, A.E.M.; Heitz, F.; Mahadevan, S.; Baldeck, J.-P.; Minery, P.; Tynes, R. The polymorphic cytochrome P-4502D6 is involved in the metabolism of both 5-hydroxytryptamine antagonists, tropisetron and ondansetron. *Drug Metab.Dispos.*, **1994**, *22*, 269–274 [microsomal incubations; human; liver; extracted metabolites; gradient; column temp 40; radioactivity detection]

Pompilio, F.M.; Fox, J.L.; Inagaki, K.; Burm, J.-P.; Jhee, S.; Gill, M.A. Stability of ranitidine hydrochloride with ondansetron hydrochloride or fluconazole during simulated Y-site administration. *Am.J.Hosp.Pharm.*, **1994**, *51*, 391–394 [stability-indicating; injections; 5% dextrose]

Williams, C.L.; Sanders, P.L.; Laizure, S.C.; Stevens, R.C.; Fox, J.L.; Hak, L.J. Stability of ondansetron hydrochloride in syrups compounded from tablets. *Am.J.Hosp.Pharm.*, **1994**, *51*, 806–809 [stability-indicating; syrup]

Graham, C.L.; Dukes, G.E.; Kao, C.-F.; Bertsch, J.M.; Hak, L.J. Stability of ondansetron in large-volume parenteral solutions. *Ann.Pharmacother.*, **1992**, *26*, 768–771 [injections; saline; 5% dextrose; lactated Ringer's solution; simultaneous degradation products]

# Oxaprozin

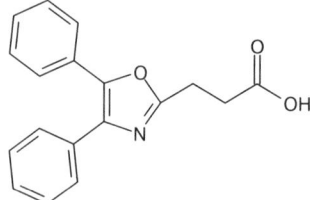

**Molecular formula:** $C_{18}H_{15}NO_3$
**Molecular weight:** 293.3
**CAS Registry No.:** 21256-18-8

## SAMPLE
**Matrix:** blood
**Sample preparation:** 1 mL Plasma + 50 μg ketoprofen + 200 μL 1 M HCl + 4-5 mL ethyl acetate, vortex for 1.5-2 min, centrifuge at 400 g for 10 min. Remove the organic layer and evaporate it to dryness under reduced pressure at 40-50°, reconstitute the residue in 200 μL MeOH, inject a 20 μL aliquot.

## HPLC VARIABLES
**Column:** 100 × 5 10 μm spherical C18 radial compression (Waters)
**Mobile phase:** MeCN:water 45:55 containing 2.5 mL/L acetic acid
**Flow rate:** 1
**Injection volume:** 20
**Detector:** UV 280

## CHROMATOGRAM
**Retention time:** 8
**Internal standard:** ketoprofen (5)
**Limit of detection:** 500 ng/mL

## OTHER SUBSTANCES
**Simultaneous:** acetaminophen, indomethacin, phenylbutazone, salicylic acid
**Noninterfering:** ibuprofen, piroxicam
**Interfering:** fenoprofen, flurbiprofen

## KEY WORDS
plasma; pharmacokinetics

## REFERENCE
Matlis, R.; Greenblatt, D.J. Rapid high-performance liquid chromatographic analysis of oxaprozin, a non-steroidal anti-inflammatory agent. *J.Chromatogr.*, **1984**, *310*, 445–449

## SAMPLE
**Matrix:** blood
**Sample preparation:** 500 μL Plasma + 2.5 mL water + 1 mL 20-60 μg/mL ketoprofen in 100 mM pH 4.8 acetate buffer, adjust to pH 2 with about 200 μL 1 M HCl, add 8 mL ether, shake mechanically for 10 min, centrifuge at 2000 rpm for 5 min. Remove 7 mL of the organic layer and evaporate it to dryness under reduced pressure at room temperature, reconstitute the residue in 500 μL mobile phase, inject a 10 μL aliquot.

## HPLC VARIABLES
**Column:** 300 × 4.6 10 μm Chromegabond C18 (E.S. Industries)
**Mobile phase:** MeCN:buffer 62:38 (Buffer was 50 mM $NaH_2PO_4$ adjusted to pH 3.9 with phosphoric acid.)
**Flow rate:** 1.5
**Injection volume:** 10
**Detector:** UV 290

## CHROMATOGRAM
**Retention time:** 5.5
**Internal standard:** ketoprofen (3.5)
**Limit of detection:** 500 ng/mL

## KEY WORDS
plasma; human; pharmacokinetics

## REFERENCE
McHugh, S.L.; Kirkman, S.K.; Knowles, J.A. Macro- and micromethods for high-performance liquid chromatographic analysis of oxaprozin in plasma. *J.Pharm.Sci.*, **1980**, *69*, 794–796

## SAMPLE
**Matrix:** blood, urine
**Sample preparation:** 50 μL Plasma or urine + 50 μL 100 mM pH 7 phosphate buffer + 50 μL 5-20 μg/mL ketoprofen in 100 mM pH 7 phosphate buffer + 200 μL ether, vortex for 5 s, centrifuge, repeat extraction twice. Combine the organic layers and evaporate them to dryness under a stream of nitrogen, reconstitute the residue in 50 μL mobile phase, inject a 20 μL aliquot.

## HPLC VARIABLES
**Guard column:** 50 × 2.1 37-45 μm Co:Pell ODS
**Column:** 300 × 4.6 10 μm Chromegabond C18 (E.S. Industries)
**Mobile phase:** MeCN:buffer 60:40 (Buffer was 50 mM $NaH_2PO_4$ adjusted to pH 3.9 with phosphoric acid.)
**Flow rate:** 1.5
**Injection volume:** 20
**Detector:** UV 280

## CHROMATOGRAM
**Retention time:** 4.5
**Internal standard:** ketoprofen (3.5)
**Limit of detection:** 100 ng/mL

## OTHER SUBSTANCES
**Extracted:** metabolites

## KEY WORDS
plasma; rat; pharmacokinetics

## REFERENCE
McHugh, S.L.; Kirkman, S.K.; Knowles, J.A. Macro- and micromethods for high-performance liquid chromatographic analysis of oxaprozin in plasma. *J.Pharm.Sci.*, **1980**, *69*, 794–796

## SAMPLE
**Matrix:** bulk
**Sample preparation:** Prepare a 400 μg/mL solution in mobile phase, inject a 10 μL aliquot.

## HPLC VARIABLES
**Column:** 300 × 3.9 μBondapak C18
**Mobile phase:** MeCN:MeOH:buffer 25:25:50, adjusted to pH 4.2 with 85% phosphoric acid (Buffer was 10 mM $KH_2PO_4$ containing 5 mM sodium 1-decanesulfonate.)
**Flow rate:** 1

**Injection volume:** 10
**Detector:** UV 254

## CHROMATOGRAM
**Retention time:** 19.8
**Limit of detection:** 54 ng/mL

## OTHER SUBSTANCES
**Simultaneous:** impurities

## KEY WORDS
stability-indicating

## REFERENCE
Ibrahim, F.B. Quantitative determination of oxaprozin and several of its related compounds by high-performance reversed-phase liquid chromatography. *J.Liq.Chromatogr.*, **1995**, *18*, 2621–2633

## SAMPLE
**Matrix:** solutions

## HPLC VARIABLES
**Column:** 100 × 2.1 C-8 (Brownlee)
**Mobile phase:** MeCN:buffer 28:72 (Buffer was 500 mM pH 4.6 sodium phosphate buffer containing 5 mM tetrabutylammonium phosphate.)
**Flow rate:** 0.5
**Detector:** UV 254

## CHROMATOGRAM
**Retention time:** 15.71

## OTHER SUBSTANCES
**Simultaneous:** metabolites, glucuronides

## REFERENCE
Wells, D.S.; Janssen, F.W.; Ruelius, H.W. Interactions between oxaprozin glucuronide and human serum albumin. *Xenobiotica*, **1987**, *17*, 1437–1449

# Oxycodone

**Molecular formula:** $C_{18}H_{21}NO_4$
**Molecular weight:** 315.4
**CAS Registry No.:** 76-42-6, 124-90-3 (HCl), 64336-55-6 (terephthalate)

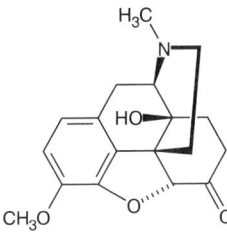

## SAMPLE
**Matrix:** bile, blood
**Sample preparation:** 0.5 mL Blood or bile + 10 (blood) or 15 (bile) μL 100 μg/mL nalorphine in MeOH + 300 μL 1.1 M pH 5.0 sodium acetate buffer + 3000-3500 U of Patella vulgata glucuronidase, incubate at 55° overnight, add 0.5 mL borate buffer to achieve a pH of 8.3-8.5. Add 8 mL chloroform:isopropanol 90:10, gently rotate for 30 min, centrifuge at 3500 rpm for 10 min, remove aqueous layer. Wash organic layer (twice for blood, three times for bile) with 3 mL 100 mM pH 9.9 sodium phosphate buffer with gentle rotation for 10 min and centrifugation each time. Add organic layer to 200 (blood) or 400 (bile) μL 0.2% phosphoric acid, gently rotate for 30 min, discard organic layer, inject 50 μL of the acid layer. (Borate buffer was 50 mM boric acid and 43 mM sodium tetraborate, adjusted to pH 9.8.)

## HPLC VARIABLES
**Guard column:** Nova-Pak phenyl guard column
**Column:** 150 × 3.9 5 μm Nova-Pak phenyl
**Mobile phase:** MeCN:10 mM pH 6.6 $NaH_2PO_4$ 10:90
**Flow rate:** 1.2
**Injection volume:** 50
**Detector:** UV 210; F ex 220 em 370 (cut-off)

## CHROMATOGRAM
**Retention time:** 21.4
**Internal standard:** nalorphine (23.5)

## OTHER SUBSTANCES
**Simultaneous:** codeine, dihydrocodeine, hydrocodone, 6-monoacetylmorphine, morphine
**Noninterfering:** acetylcodeine, amitriptyline, amphetamine, diamorphine, diazepam, dothiepin, doxepin, ephedrine, ephedrine, hydromorphone, mesoridazine, methadone, methamphetamine, 3-monoacetylmorphine, nordiazepam, norpropoxyphene, nortriptyline, oxazepam, propoxyphene, pseudoephedrine, quinidine, quinine, sulfamethoxazole, sulforidazine, thioridazine

## KEY WORDS
UV and F detection used together

## REFERENCE
Crump, K.L.; McIntyre, I.M.; Drummer, O.H. Simultaneous determination of morphine and codeine in blood and bile using dual ultraviolet and fluorescence high-performance liquid chromatography. *J.Anal.Toxicol.*, **1994**, *18*, 208−212

## SAMPLE
**Matrix:** blood
**Sample preparation:** 1 mL Plasma + 60 μL 1 μg/mL codeine in water + 100 μL 1 M phosphoric acid + 5 mL butyl chloride, mix for 1.5 min, centrifuge at 1500 g for 3 min, discard upper organic layer. To the aqueous layer add 500 μL pH 10 1 M carbonate buffer

and 5 mL butyl chloride, mix for 1.5 min, centrifuge at 1500 g for 3 min, remove organic layer and repeat extraction. Combine butyl chloride layers and evaporate them to dryness under a stream of air at 40°. Reconstitute the residue in 200 μL mobile phase, inject a 150 μL aliquot.

## HPLC VARIABLES
**Guard column:** Novapak C18 guard column
**Column:** 4 μm Novapak C18 in a Waters RCM 8 × 10 radial compression unit
**Mobile phase:** MeOH:MeCN:10 mM pH 7 phosphate buffer 230:20:1000, containing 40 mg/L cetyltrimethylammonium bromide (cetavlon)
**Flow rate:** 2
**Injection volume:** 150
**Detector:** E, Waters Model 460, working electrode 1.10 V

## CHROMATOGRAM
**Retention time:** 8.5
**Internal standard:** codeine (13.5)
**Limit of detection:** 5 ng/mL

## OTHER SUBSTANCES
**Noninterfering:** acetaminophen, amitriptyline, aspirin, atenolol, camazepam, carbamazepine, chlorimipramine, chlorthalidone, clonazepam, cortisone, desipramine, diazepam, halazepam, hydrochlorothiazide, hydrocortisone, imipramine, lorazepam, maprotiline, meperidine, methylphenobarbital, methylprednisolone, metoclopramide, midazolam, morphine, nalorphine, naloxone, nitrazepam, nortriptyline, oxazepam, oxprenolol, phenobarbital, phenytoin, pindolol, prazepam, prednisolone, prednisone, primidone, prochlorperazine, propranolol, salicylic acid, temazepam

## KEY WORDS
plasma

## REFERENCE
Smith, M.T.; Watt, J.A.; Mapp, G.P.; Cramond, T. Quantitation of oxycodone in human plasma using high-performance liquid chromatography with electrochemical detection. *Ther.Drug Monit.*, **1991**, *13*, 126–130

## SAMPLE
**Matrix:** blood
**Sample preparation:** 2 mL Plasma + 50 μL 2 μg/mL methadone in MeOH + 500 μL pH 9.6 carbonate buffer + 6 mL butyl chloride, shake on a mechanical shaker at 100 rpm for 15 min, centrifuge at 2000 g for 5 min. Remove the organic layer and add it to 3 mL 200 mM HCl. Shake for 15 min, centrifuge, remove aqueous layer. Add aqueous layer to 3 drops 60% NaOH and 6 mL butyl chloride, shake for 15 min, centrifuge. Remove organic layer and evaporate it to dryness at 50° under a stream of nitrogen, reconstitute residue in 60 μL mobile phase, inject a 40 μL aliquot.

## HPLC VARIABLES
**Column:** 10 μm Brownlee RP-8
**Mobile phase:** MeCN:MeOH:10 mM KH$_2$PO$_4$ 50:30:20
**Flow rate:** 1
**Injection volume:** 40
**Detector:** E, Bioanalytical Systems, glassy carbon working electrode 1.20 V, Ag/AgCl reference electrode

## CHROMATOGRAM
**Retention time:** 6.4

**Internal standard:** methadone (16)
**Limit of quantitation:** 2 ng/mL

## OTHER SUBSTANCES
**Simultaneous:** fentanyl, meperidine, phenoperidine

## KEY WORDS
plasma

## REFERENCE
Schneider, J.J.; Triggs, E.J.; Bourne, D.W.; Stephens, I.D.; Haviland, A.M. Determination of oxycodone in human plasma by high-performance liquid chromatography with electrochemical detection. *J.Chromatogr.*, **1984**, *308*, 359–362

## SAMPLE
**Matrix:** solutions
**Sample preparation:** Inject a 10 µL aliquot of a 50 µg/mL solution in water.

## HPLC VARIABLES
**Column:** 150 × 4.6 5 µm Supelcosil LC-ABZ
**Mobile phase:** MeCN:25 mM pH 7.5 potassium phosphate buffer 35:65
**Flow rate:** 1.5
**Injection volume:** 10
**Detector:** UV 254

## CHROMATOGRAM
**Retention time:** 3.6

## OTHER SUBSTANCES
**Simultaneous:** codeine, meperidine

## REFERENCE
*Supelco Chromatography Products, Supelco, Inc., Bellefonte PA*, **1996**, p. A23

## SAMPLE
**Matrix:** solutions

## HPLC VARIABLES
**Column:** 250 × 4.6 5 µm Supelcosil LC-DP (A) or 250 × 4 5 µm LiChrospher 100 RP-8 (B)
**Mobile phase:** MeCN:0.025% phosphoric acid:buffer 25:10:5 (A) or 60:25:15 (B) (Buffer was 9 mL concentrated phosphoric acid and 10 mL triethylamine in 900 mL water, adjust pH to 3.4 with dilute phosphoric acid, make up to 1 L.)
**Flow rate:** 0.6
**Injection volume:** 25
**Detector:** UV 229

## CHROMATOGRAM
**Retention time:** 6.51 (A), 5.82 (B)

## OTHER SUBSTANCES
**Also analyzed:** acebutolol, acepromazine, acetaminophen, acetazolamide, acetophenazine, albuterol, alprazolam, amitriptyline, amobarbital, amoxapine, antipyrine, atenolol, atropine, azatadine, baclofen, benzocaine, bromocriptine, brompheniramine, brotizolam, bupivacaine, buspirone, butabarbital, butalbital, caffeine, carbamazepine, cetirizine, chlorcyclizine, chlordiazepoxide, chlormezanone, chloroquine, chlorpheniramine,

chlorpromazine, chlorpropamide, chlorprothixene, chlorthalidone, chlorzoxazone, cimetidine, cisapride, clomipramine, clonazepam, clonidine, clozapine, cocaine, codeine, colchicine, cyclizine, cyclobenzaprine, dantrolene, desipramine, diazepam, diclofenac, diflunisal, diltiazem, diphenhydramine, diphenidol, diphenoxylate, dipyridamole, disopyramide, dobutamine, doxapram, doxepin, droperidol, encainide, ethidium bromide, ethopropazine, fenoprofen, fentanyl, flavoxate, fluoxetine, fluphenazine, flurazepam, flurbiprofen, fluvoxamine, furosemide, glutethimide, glyburide, guaifenesin, haloperidol, homatropine, hydralazine, hydrochlorothiazide, hydrocodone, hydromorphone, hydroxychloroquine, hydroxyzine, ibuprofen, imipramine, indomethacin, ketoconazole, ketoprofen, ketorolac, labetalol, levorphanol, lidocaine, loratadine, lorazepam, lovastatin, loxapine, mazindol, mefenamic acid, meperidine, mephenytoin, mepivacaine, mesoridazine, metaproterenol, metformin, methadone, methdilazine, methocarbamol, methotrexate, methotrimeprazine, methoxamine, methyldopa, methylphenidate, metoclopramide, metolazone, metoprolol, metronidazole, midazolam, moclobemide, morphine, nadolol, nalbuphine, naloxone, naphazoline, naproxen, nifedipine, nizatidine, norepinephrine, nortriptyline, oxazepam, oxymetazoline, paroxetine, pemoline, pentazocine, pentobarbital, pentoxifylline, perphenazine, pheniramine, phenobarbital, phenol, phenolphthalein, phentolamine, phenylbutazone, phenyltoloxamine, phenytoin, pimozide, pindolol, piroxicam, pramoxine, prazepam, prazosin, probenecid, procainamide, procaine, prochlorperazine, procyclidine, promazine, promethazine, propafenone, propantheline, propiomazine, propofol, propranolol, protriptyline, quazepam, quinidine, quinine, racemethorphan, ranitidine, remoxipride, risperidone, salicylic acid, scopolamine, secobarbital, sertraline, sotalol, spironolactone, sulfinpyrazone, sulindac, temazepam, terbutaline, terfenadine, tetracaine, theophylline, thiethylperazine, thiopental, thioridazine, thiothixene, timolol, tocainide, tolbutamide, tolmetin, trazodone, triamterene, triazolam, trifluoperazine, triflupromazine, trimeprazine, trimethoprim, trimipramine, verapamil, warfarin, xylometazoline, yohimbine, zopiclone

## KEY WORDS
some details of plasma extraction

## REFERENCE
Koves, E.M. Use of high-performance liquid chromatography-diode array detection in forensic toxicology. *J.Chromatogr.A*, **1995**, *692*, 103−119

## SAMPLE
**Matrix:** solutions

## HPLC VARIABLES
**Column:** 250 × 4.6 Zorbax RX
**Mobile phase:** Gradient. A was 10 mL concentrated orthophosphoric acid and 7 mL triethylamine in 1 L water. B was 10 mL concentrated orthophosphoric acid and 7 mL triethylamine in 200 mL water, make up to 1 L with MeCN. A:B from 100:0 to 0:100 over 30 min, maintain at 0:100 for 5 min.
**Column temperature:** 30
**Flow rate:** 2
**Detector:** UV 210

## OTHER SUBSTANCES
**Also analyzed:** acepromazine, acetaminophen, acetophenazine, albuterol, aminophylline, amitriptyline, amobarbital, amoxapine, amphetamine, amylocaine, antipyrine, aprobarbital, aspirin, atenolol, atropine, avermectin, barbital, benzocaine, benzoic acid, benzotropine, benzphetamine, berberine, bibucaine, bromazepan, brompheniramine, buprenorphine, buspirone, butabarbital, butacaine, butethal, caffeine, carbamazepine, carbromal, chloramphenicol, chlordiazepoxide, chloroquine, chlorothiazide, chloroxylenol, chlorphenesin, chlorpheniramine, chlorpromazine, chlorpropamide, chlortetracycline, cimetidine, cinchonidine, cinchonine, clenbuterol, clonazepam, clonixin, clorazepate, cocaine, codeine,

colchicine, cortisone, coumarin, cyclazocine, cyclobenzaprine, cyclothiazide, cyheptamide, cymarin, danazol, danthron, dapsone, debrisoquine, desipramine, dexamethasone, dextromethorphan, dextropropoxyphene, diamorphine, diazepam, diclofenac, diethylpropion, diethylstilbestrol, diflunisal, digitoxin, digoxin, diltiazem, diphenhydramine, diphenoxylate, diprenorphine, dipyrone, disulfiram, dopamine, doxapram, doxepin, dronabinol, ephedrine, epinephrine, epinine, estradiol, estriol, estrone, ethacrynic acid, ethosuximide, etonitazene, etorphine, eugenol, famotidine, fenbendazole, fencamfamine, fenoprofen, fenproporex, fentanyl, flubendazole, flufenamic acid, flunitrazepam, 5-fluorouracil, fluoxymesterone, fluphenazine, furosemide, gentisic acid, gitoxigenin, glipizide, glunixin, glutethimide, glybenclamide, guaiacol, halazepam, haloperidol, hydrochlorothiazide, hydrocodone, hydrocortisone, hydromorphone, hydroxyquinoline, ibogaine, ibuprofen, iminostilbene, imipramine, indomethacin, isocarbostyril, isocarboxazid, isoniazid, isoproterenol, isoxsuprine, ivermectin, ketamine, ketoprofen, kynurenic acid, levorphanol, lidocaine, lorazepam, lormetazepam, loxapine, mazindol, mebendazole, meclizine, meclofenamic acid, medazepam, mefenamic acid, megestrol, mepacrine, meperidine, mephentermine, mephenytoin, mephesin, mephobarbital, mepivacaine, mescaline, mesoridazine, methadone, methamphetamine, methapyrilene, methaqualone, methazolamide, methocarbamol, methoxamine, methsuximide, methyl salicylate, methyldopa, methyldopamine, methylphenidate, methylprednisolone, methyltestosterone, methyprylon, metoprolol, mibolerone, morphine, nadolol, nalorphine, naloxone, naltrexone, naphazoline, naproxen, nefopam, niacinamide, nicotine, nicotinic acid, nifedipine, niflumic acid, nitrazepam, norepinephrine, nortriptyline, noscapine, nylidrin, oxazepam, oxymorphone, oxyphenbutazone, oxytetracycline, papaverine, pargyline, pemoline, pentazocine, pentobarbital, persantine, phenacetin, phenazocine, phenazopyridine, phencyclidine, phendimetrazine, phenelzine, pheniramine, phenobarbital, phenothiazine, phensuximide, phentermine, phenylbutazone, phenylephrine, phenylpropanolamine, piperocaine, prazepam, prednisolone, primidone, probenecid, progesterone, propiomazine, propranolol, propylparaben, pseudoephedrine, puromycin, pyrilamine, pyrithyldione, quazepam, quinaldic acid, quinidine, quinine, ranitidine, recinnamine, reserpine, resorcinol, saccharin, albuterol, salicylamide, salicylic acid, scopolamine, scopoletin, secobarbital, strychnine, sulfacetamide, sufadiazine, sulfadimethoxine, sulfaethidole, sulfamerazine, sulfamethazine, sulfamethoxizole, sulfanilamide, sulfapyridine, sulfasoxizole, sulindac, tamoxifen, temazepam, testosterone, tetracaine, tetracycline, tetramisole, thebaine, theobromine, theophylline, thiabendazole, thiamine, thiamylal, thiobarbituric acid, thioridazine, thiosalicylic acid, thiothixene, thymol, tolazamide, tolazoline, tobutamide, tolmetin, tranylcypromine, triamcinolone, tribenzylamine, trichloromethiazide, trifluoperazine, trihexyphenidyl, trimethoprim, tripelennamine, triprolidine, tropacocaine, tyramine, verapamil, vincamine, warfarin, yohimbine, zoxazolamine

## REFERENCE

Hill, D.W.; Kind, A.J. Reversed-phase solvent gradient HPLC retention indexes of drugs. *J.Anal.Toxicol.*, **1994**, *18*, 233–242

## SAMPLE
**Matrix:** solutions

## HPLC VARIABLES
**Column:** 150 × 4.6 Supelcosil LC-ABZ
**Mobile phase:** MeCN:25 mM pH 6.9 potassium phosphate buffer 35:65
**Flow rate:** 1.5
**Injection volume:** 25
**Detector:** UV 254

## CHROMATOGRAM
**Retention time:** 2.670

## OTHER SUBSTANCES
**Also analyzed:** 6-acetylmorphine, amiloride, amphetamine, benzocaine, benzoylecgonine, caffeine, cocaine, codeine, doxylamine, fluoxetine, glutethimide, hexobarbital, hypoxanthine, levorphanol, LSD, meperidine, mephobarbital, methadone, methylphenidate, methyprylon, N-norcodeine, oxazepam, phenylpropanolamine, prilocaine, procaine, terfenadine

## REFERENCE
Ascah, T.L. Improved separations of alkaloid drugs and other substances of abuse using Supelcosil LC-ABZ column. *Supelco Reporter*, **1993**, *12(3)*, 18−21

## SAMPLE
**Matrix:** solutions

## HPLC VARIABLES
**Column:** 150 × 3.9 µBondapak C18
**Mobile phase:** MeCN:phosphate buffer 36:64, pH adjusted to 4.0
**Flow rate:** 1.5
**Injection volume:** 100
**Detector:** UV 254

## CHROMATOGRAM
**Retention time:** 2.5
**Limit of detection:** 50 ng/mL

## REFERENCE
Tien, J.H. Transdermal-controlled administration of oxycodone. *J.Pharm.Sci.*, **1991**, *80*, 741−743

## SAMPLE
**Matrix:** solutions
**Sample preparation:** Dissolve in MeOH at a concentration of 1 mg/mL, inject a 20 µL aliquot.

## HPLC VARIABLES
**Column:** 250 × 5 Spherisorb S5W
**Mobile phase:** MeOH:buffer 90:10 (Buffer was 94 mL 35% ammonia and 21.5 mL 70% nitric acid in 884 mL water, adjust the pH to 10.1 with ammonia.)
**Flow rate:** 2
**Injection volume:** 20
**Detector:** UV 254

## CHROMATOGRAM
**Retention time:** 2.44 (tails)

## OTHER SUBSTANCES
**Simultaneous:** benzphetamine, benzylmorphine, bromo-STP, buprenorphine, caffeine, codeine, codeine-N-oxide, dextromoramide, dextropropoxyphene, diamorphine, diethylpropion, dihydrocodeine, dihydromorphine, dimethylamphetamine, dipipanone, ephedrine, epinephrine, ethoheptazine, ethylmorphine, etorphine, fenethyline, fentanyl, hydrocodone, 4-hydroxyamphetamine, hydroxypethidine, levallorphan, levorphanol, mazindol, meperidine, mephentermine, mescaline, methadone, methamphetamine, methylephedrine, methylphenidate, morphine, morphine-3-glucuronide, morphine-N-oxide, nalorphine, naloxone, norcodeine, norlevorphanol, normetanephrine, normethadone, normorphine, norpethidine, norpipanone, noscapine, papaverine, pemoline, pentazocine, phenazocine, phendimetrazine, phenelzine, 2-phenethylamine, phenoperidine, phenylephrine, pholcodine, pipradol, piritramide, prolintane, pseudoephedrine, STP, tranylcypromine, trimethoxyamphetamine, tyramine

**Noninterfering:** dopamine, levodopa, methyldopa, methyldopate, norepinephrine
**Interfering:** acetylcodeine, amphetamine, chlorphentermine, fencamfamin, fenfluramine, methylenedioxyamphetamine, monoacetylmorphine, norpseudoephedrine, phentermine, phenylpropanolamine, thebacon, thebaine

## REFERENCE
Law, B.; Gill, R.; Moffat, A.C. High-performance liquid chromatography retention data for 84 basic drugs of forensic interest on a silica column using an aqueous methanol eluent. *J.Chromatogr.*, **1984**, *301*, 165–172

# Paclitaxel

**Molecular formula:** $C_{47}H_{51}NO_{14}$
**Molecular weight:** 853.9
**CAS Registry No.:** 33069-62-4

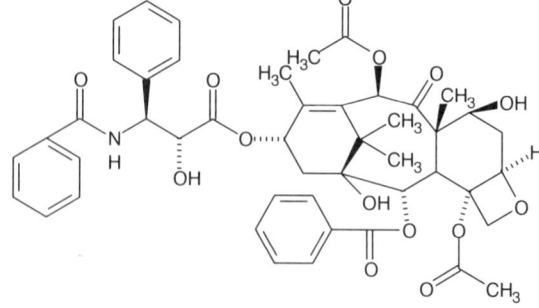

## SAMPLE
**Matrix:** blood
**Sample preparation:** 1 mL Plasma + 300 ng N-nitrosodiphenylamine (Caution! N-Nitrosodiphenylamine is a carcinogen!) + 400 µL 500 mM $Na_2HPO_4$, vortex for 5 s, add 5 mL ethyl acetate, mix for 1 min, centrifuge at 3000 rpm for 10 min. Remove the organic layer and evaporate it to dryness under a stream of nitrogen at room temperature, reconstitute the residue in 200 µL mobile phase, vortex for 30 s, sonicate for 30 s, inject a 100 µL aliquot.

## HPLC VARIABLES
**Guard column:** Guard-PAK C18 (Waters)
**Column:** 100 × 8 4 µm Nova Pak C18 radial compression
**Mobile phase:** MeCN:buffer 45.5:55.5 (Buffer was 1 mM $Na_2HPO_4$ adjusted to pH 5 with phosphoric acid.)
**Flow rate:** 4.5
**Injection volume:** 100
**Detector:** UV 227

## CHROMATOGRAM
**Retention time:** 5.26
**Internal standard:** N-nitrosodiphenylamine (6.45)
**Limit of quantitation:** 10 ng/mL

## OTHER SUBSTANCES
**Simultaneous:** carmustine, 5-fluorouracil, hydrocortisone, teniposide, thiotepa
**Noninterfering:** acetaminophen, aspirin, bleomycin, busulfan, carboplatin, cyclophosphamide, cytarabine, dacarbazine, hydroxyurea, ifosfamide, methotrexate, mitomycin, mitoxantrone, procarbazine

## KEY WORDS
plasma; pharmacokinetics

## REFERENCE
el-Yazigi, A.; Yusuf, A. Expedient liquid chromatographic assay for paclitaxel in plasma after its administration to cancer patients. *Ther.Drug Monit.*, **1995**, *17*, 511–515

## SAMPLE
**Matrix:** blood
**Sample preparation:** Condition a 1 mL cyano Bond Elut SPE cartridge with 2 mL MeOH and 2 mL 10 mM pH 5.0 ammonium acetate buffer. 1 mL Plasma + 1 mL 200 mM ammonium acetate, mix, add a 1 mL aliquot to the SPE cartridge, wash with 2 mL 10 mM pH 5.0 ammonium acetate, wash with 1 mL MeOH:10 mM pH 5.0 ammonium acetate 20:80, wash with 1 mL hexane, dry under vacuum for 1 min, elute with 2 mL MeCN:triethylamine 100:0.1, evaporate the eluate to dryness under a stream of nitrogen at 30°, reconstitute with 200 µL mobile phase, vortex for 30 s, inject a 50 µL aliquot.

## HPLC VARIABLES
**Guard column:** 4 × 4 5 µm LiChrospher RP-8
**Column:** 150 × 4.6 5 µm APEX-octyl (Jones Chromatography)
**Mobile phase:** MeCN:MeOH:20 mM pH 5.0 ammonium acetate buffer 40:10:50
**Injection volume:** 50
**Detector:** UV 227

## CHROMATOGRAM
**Retention time:** 10.1
**Limit of quantitation:** 10 ng/mL

## OTHER SUBSTANCES
**Extracted:** metabolites

## KEY WORDS
plasma; SPE

## REFERENCE
Huizing, M.T.; Sparreboom, A.; Rosing, H.; van Tellingen, O.; Pinedo, H.M.; Beijnen, J.H. Quantification of paclitaxel metabolites in human plasma by high-performance liquid chromatography. *J.Chromatogr.B*, **1995**, *674*, 261–268

## SAMPLE
**Matrix:** blood
**Sample preparation:** Plasma. 500 µL Plasma + 20 µL 10 µg/mL diethylstilbestrol in MeOH, vortex for 20 s, add 5 mL MTBE, vortex for 20 s, centrifuge at 170 g for 5 min, freeze in dry ice/acetone. Remove the organic layer and put it into a clean tube (twice), evaporate it to dryness under a stream of nitrogen at room temperature, reconstitute the residue in 200 µL MeCN:water 45:55, inject a 50 µL aliquot. Microsomal incubations. 250 µL Microsomal incubation + 5 mL MTBE, vortex for 20 s, centrifuge at 170 g for 5 min, freeze in dry ice/acetone. Remove the organic layer and put it into a clean tube (twice), evaporate it to dryness under a stream of nitrogen at room temperature, reconstitute the residue in 200 µL MeCN:water 45:55, inject a 50 µL aliquot.

## HPLC VARIABLES
**Guard column:** 30 × 3.2 5 µm Hypersil C8
**Column:** 150 × 3.2 5 µm Hypersil C8
**Mobile phase:** MeOH:30 mM pH 6 potassium phosphate buffer 58:42 (After 25 min wash column with MeOH:water 95:5 for 7 min, re-equilibrate for 13 min. Wash is not necessary for microsomal incubations.)
**Flow rate:** 0.4
**Injection volume:** 50
**Detector:** UV 227

## CHROMATOGRAM
**Retention time:** 22
**Internal standard:** diethylstilbestrol (17)
**Limit of detection:** 20 nM

## OTHER SUBSTANCES
**Extracted:** metabolites

## KEY WORDS
microsomal incubations; plasma; human; liver; pharmacokinetics

## REFERENCE

Sonnichsen, D.S.; Liu, Q.; Schuetz, E.G.; Schuetz, J.D.; Pappo, A.; Relling, M.V. Variability of human cytochrome P450 paclitaxel metabolism. *J.Pharm.Exp.Ther.*, **1995**, *275*, 566–575

## SAMPLE

**Matrix:** blood
**Sample preparation:** Dilute 50-500 µL mouse plasma to 500 µL with water, add 25 µL 40 µg/mL N-octylbenzamide in MeOH, mix, add 4 mL MTBE, vortex for 30 s, centrifuge at 500 g at 4° for 10 min. Remove 3 mL of the organic layer and evaporate it to dryness under a stream of nitrogen, reconstitute the residue in 100 µL MeOH, inject a 70 µL aliquot. (Full details for the synthesis of N-octylbenzamide are given in the paper.)

## HPLC VARIABLES

**Guard column:** 6.6 × 3 10 µm µBondapak C18
**Column:** 300 × 3.9 10 µm µBondapak C18
**Mobile phase:** MeOH:water 70:30
**Flow rate:** 2
**Injection volume:** 70
**Detector:** UV 227

## CHROMATOGRAM

**Retention time:** 7.35
**Internal standard:** N-octylbenzamide (13.00)
**Limit of quantitation:** 0.15 nM

## KEY WORDS

plasma; mouse

## REFERENCE

Sharma, A.; Conway, W.D.; Straubinger, R.M. Reversed-phase high-performance liquid chromatographic determination of taxol in mouse plasma. *J.Chromatogr.B*, **1994**, *655*, 315–319

## SAMPLE

**Matrix:** blood
**Sample preparation:** Condition a 1 mL Bond Elut cyano SPE cartridge with 2 mL MeOH and 2 mL 10 mM pH 5.0 ammonium acetate, do not allow to dry. 500 µL Plasma + 500 µL 200 mM pH 5.0 ammonium acetate, vortex for 20 s, add to the SPE cartridge, wash with 2 mL 10 mM pH 5.0 ammonium acetate, wash with 2 mL MeOH:10 mM pH 5.0 ammonium acetate 20:80, wash with 1 mL hexane, dry under vacuum for 1 min. Elute with two 1 mL volumes of 0.1% triethylamine in MeCN, evaporate the eluate to dryness at 30° under nitrogen, reconstitute the residue in 200 µL MeCN:MeOH:water 40:10:50 containing 10 mM ammonium acetate adjusted to pH 5.0 with glacial acetic acid, vortex for 30 s, inject a 100 µL aliquot.

## HPLC VARIABLES

**Column:** 250 × 4.6 5 µm octyl (Jones Chromatography)
**Mobile phase:** MeCN:MeOH:water:1 M ammonium acetate adjusted to pH 5.0 with glacial acetic acid 40:10:49:1
**Flow rate:** 1
**Injection volume:** 100
**Detector:** UV 227

## CHROMATOGRAM

**Retention time:** 10.6
**Limit of quantitation:** 10 ng/mL

## OTHER SUBSTANCES
**Simultaneous:** amsacrine, baccatin III, cephalomannine, daunorubicin, epipaclitaxel, paclitaxel C

## KEY WORDS
plasma; SPE; pharmacokinetics; methyl paclitaxel can be used as IS

## REFERENCE
Willey, T.A.; Bekos, E.J.; Gaver, R.C.; Duncan, G.F.; Tay, L.K.; Beijnen, J.H.; Farmen, R.H. High-performance liquid chromatographic procedure for the quantitative determination of paclitaxel (Taxol) in human plasma. *J.Chromatogr.*, **1993**, *621*, 231–238

## SAMPLE
**Matrix:** blood, culture medium, tissue
**Sample preparation:** Plasma, culture medium. 0.2-1 mL plasma or culture medium + 100 μL 8.7 μg/mL cephalomannine in MeOH + three volumes ethyl acetate, extract, repeat extraction. Combine extracts and centrifuge them at 2000 g for 5 min. Remove the supernatant and evaporate it to dryness under a stream of nitrogen, reconstitute the residue in 100-150 μL MeCN:water 37.5:62.5, inject a 10-60 μL aliquot onto column A with mobile phase A and elute to waste. After 8 min direct the effluent from column A onto column B. After another 7 min elute column B with mobile phase B, monitor the effluent from column B, re-equilibrate column A with mobile phase A. Tissue. 40 mg Dog bladder tissue + 100 μL 8.7 μg/mL cephalomannine in MeOH + 3-4 mL ethyl acetate or MeCN, homogenize (Tekmar) for 1 min, wash homogenizer probe with 3-4 mL same solvent for 15-20 s. Combine organic layers and centrifuge them at 2000 g for 5 min. Remove the supernatant and evaporate it to dryness under a stream of nitrogen, reconstitute the residue in 100-150 μL MeCN:water 37.5:62.5, inject a 10-60 μL aliquot onto column A with mobile phase A and elute to waste. After 8 min direct the effluent from column A onto column B. After another 7 min elute column B with mobile phase B, monitor the effluent from column B, re-equilibrate column A with mobile phase A.

## HPLC VARIABLES
**Column:** A Nova-Pak C8 guard column + 75 × 3.9 3 μm Nova-Pak C8; B 250 × 4.6 5 μm Bakerbond C18
**Mobile phase:** A MeCN:water 37.5:62.5; B MeCN:water 49:51
**Flow rate:** A 1; B 1.2
**Injection volume:** 10-60
**Detector:** UV 229

## CHROMATOGRAM
**Retention time:** 21.35
**Internal standard:** cephalomannine (20.11)
**Limit of detection:** 5 ng/mL (plasma); 5 ng/inj. (tissue)

## KEY WORDS
plasma; column-switching; dog; bladder; heart-cut

## REFERENCE
Song, D.; Au, J.L.-S. Isocratic high-performance liquid chromatographic assay of taxol in biological fluids and tissues using automated column switching. *J.Chromatogr.B*, **1995**, *663*, 337–344

## SAMPLE
**Matrix:** blood, feces, tissue, urine
**Sample preparation:** Condition a 1 mL 100 mg Bond Elut cyano SPE cartridge with 2 mL MeOH and 2 mL 10 mM pH 5.0 ammonium acetate buffer. Dilute urine with five volumes of human plasma. Homogenize (Biospec) tissue or feces in 5-10 volumes of 4% Boseral in

water at 4°. 200 (Feces, plasma), 200-1000 (tissue), or 1000 (urine) μL + 25 μL 20 μg/mL 2'-methylpaclitaxel in MeOH + 4 mL diethyl ether, mix vigorously for 5 min, centrifuge at 460 g for 5 min, freeze in EtOH/dry ice, repeat the extraction. Combine the organic layers and evaporate them to dryness under vacuum at 43°, reconstitute in 250 μL human plasma, add 300 μL 200 mM pH 5.0 ammonium acetate buffer, mix, add 500 μL to the SPE cartridge, wash with 2 mL 10 mM pH 5.0 ammonium acetate buffer, wash with 1 mL MeOH:10 mM pH 5.0 ammonium acetate buffer 20:80, elute with 500 μL MeCN: triethylamine 100:0.1. Evaporate the eluate to dryness under vacuum, reconstitute 200 μL in MeCN:MeOH:water 40:10:50, vortex for 30 s, inject a 50 μL aliquot.

## HPLC VARIABLES
**Column:** 150 × 4.6 5 μm APEX-octyl (Jones Chromatography)
**Mobile phase:** MeCN:MeOH:200 mM pH 5.0 ammonium acetate buffer 40:10:50
**Flow rate:** 1
**Injection volume:** 50
**Detector:** UV 227

## CHROMATOGRAM
**Retention time:** 10.6
**Internal standard:** 2'-methylpaclitaxel (15.3)
**Limit of detection:** 15 ng/mL (plasma)
**Limit of quantitation:** 25-125 ng/g

## OTHER SUBSTANCES
**Extracted:** metabolites

## KEY WORDS
plasma; mouse; SPE; brain; fat; muscle; breast; colon; appendix; intestine; stomach; liver; gall bladder; kidney; lung; spleen; heart; uterus; ovary; thymus; lymph nodes

## REFERENCE
Sparreboom, A.; van Tellingen, O.; Nooijen, W.J.; Beijnen, J.H. Determination of paclitaxel and metabolites in mouse plasma, tissues, urine and faeces by semi-automated reversed-phase high-performance liquid chromatography. *J.Chromatogr.B*, **1995**, *664*, 383–391

## SAMPLE
**Matrix:** blood, urine
**Sample preparation:** 800 μL Plasma or urine + 200 μL 7.5 μM N-cyclohexylbenzamide in MeCN, vortex, add 3 mL ethyl acetate, shake vigorously for 2 min, centrifuge at 150 g for 2 min. Remove the organic layer and evaporate it under nitrogen at room temperature, dissolve the residue in 100 μL MeCN, inject an aliquot.

## HPLC VARIABLES
**Guard column:** C-18 Guard-Pak
**Column:** 100 × 8 10 μm C18 Radial-Pak
**Mobile phase:** MeCN:water from 35:65 to 100:0 exponentially over 20 min (Waters exponential gradient curve no. 9), keep at 100:0 for 7 min, re-equilibrate at 35:65 for 5 min.
**Flow rate:** 2.5
**Detector:** UV 227

## CHROMATOGRAM
**Retention time:** 19.5
**Internal standard:** N-cyclohexylbenzamide (12.5)
**Limit of quantitation:** 30 nM

## KEY WORDS
plasma

## REFERENCE

Longnecker, S.M.; Donehower, R.C.; Cates, A.E.; Chen, T.-L.; Brundrett, R.B.; Grochow, L.B.; Ettinger, D.S.; Colvin, M. High-performance liquid chromatographic assay for taxol in human plasma and urine and pharmacokinetics in a phase I trial. *Cancer Treat.Rep.*, **1987**, *71*, 53–59

## SAMPLE

**Matrix:** cell cultures
**Sample preparation:** Filter (Miracloth, Calbiochem). Evaporate 1 mL of filtrate to dryness under vacuum, reconstitute in 200 μL MeOH:acetic acid 99.99:0.01, sonicate for 1 min, centrifuge at 16000 g for 15 min, filter (0.2 μm PVDF), inject an aliquot of the filtrate.

## HPLC VARIABLES

**Guard column:** Upchurch ODS
**Column:** 250 × 4.6 4 μm Curosil G (Phenomenex)
**Mobile phase:** MeCN:water 52.5:47.5
**Flow rate:** 1
**Detector:** UV 228

## CHROMATOGRAM

**Retention time:** 18

## OTHER SUBSTANCES

**Simultaneous:** impurities, related compounds

## REFERENCE

Ketchum, R.E.B.; Gibson, D.M. A novel method of isolating taxanes from cell suspension cultures of yew (*Taxus* spp.). *J.Liq.Chromatogr.*, **1995**, *18*, 1093–1111

## SAMPLE

**Matrix:** cell cultures
**Sample preparation:** Homogenize (Omni-mix) 50 mg bark or foliage in 1.5 mL MeOH for 2 min, sonicate for 5 min, centrifuge, filter (0.2 μm) the supernatant, inject an aliquot of the filtrate.

## HPLC VARIABLES

**Guard column:** phenyl (Rainin)
**Column:** 250 × 4.6 8 μm Dynamax 60 Å phenyl (Rainin)
**Mobile phase:** MeOH:MeCN:50 mM pH 4.4 acetate buffer 20:39:41
**Flow rate:** 1
**Detector:** UV 227

## CHROMATOGRAM

**Retention time:** 19

## OTHER SUBSTANCES

**Extracted:** cephalomannine

## REFERENCE

Wickremesinhe, E.R.M.; Arteca, R.N. Methodology for the identification and purification of taxol and cephalomannine from Taxus callus cultures. *J.Liq.Chromatogr.*, **1993**, *16*, 3263–3274

## SAMPLE

**Matrix:** formulations
**Sample preparation:** Dilute with mobile phase, inject an aliquot.

## HPLC VARIABLES
**Column:** 250 × 4.6 5 μm C18
**Mobile phase:** MeCN:water 40:60
**Flow rate:** 2.25
**Injection volume:** 20
**Detector:** UV 254

## CHROMATOGRAM
**Retention time:** 4.95

## KEY WORDS
stability-indicating; injections; saline

## REFERENCE
Mayron, D.; Gennaro, A.R. Stability and compatibility of granisetron hydrochloride in i.v. solutions and oral liquids and during simulated Y-site injection with selected drugs. *Am.J.Health-Syst.Pharm.*, **1996**, *53*, 294–304

## SAMPLE
**Matrix:** formulations
**Sample preparation:** Dilute 1:4, inject a 20 μL aliquot.

## HPLC VARIABLES
**Column:** 250 × 4.6 5 μm Adsorbosphere C18
**Mobile phase:** MeCN:12.5 mM ammonium phosphate 60:40, pH adjusted to 4.5 with 1 M HCl
**Flow rate:** 1
**Injection volume:** 20
**Detector:** UV 227

## CHROMATOGRAM
**Retention time:** 10.0

## OTHER SUBSTANCES
**Noninterfering:** ondansetron, ranitidine

## KEY WORDS
injections; 5% dextrose; stability-indicating

## REFERENCE
Burm, J.-P.; Jhee, S.S.; Chin, A.; Moon, Y.S.K.; Jeong, E.; Nii, L.; Fox, J.L.; Gill, M.A. Stability of paclitaxel with ondansetron hydrochloride or ranitidine hydrochloride during simulated Y-site administration. *Am.J.Hosp.Pharm.*, **1994**, *51*, 1201–1204

## SAMPLE
**Matrix:** formulations
**Sample preparation:** Filter (5 μm), dilute 1 mg/mL solutions tenfold with the same solvent, inject an aliquot.

## HPLC VARIABLES
**Column:** 250 × 4.6 5 μm Vydac C18
**Mobile phase:** MeCN:water 53:47
**Flow rate:** 1.5
**Detector:** UV 254

## CHROMATOGRAM
**Retention time:** 6.09

## KEY WORDS
injections; 5% dextrose; saline; stability-indicating

## REFERENCE
Xu, Q.; Trissel, L.A.; Martinez, J.F. Stability of paclitaxel in 5% dextrose injection or 0.9% sodium chloride injection at 4, 22, or 32°C. *Am.J.Hosp.Pharm.*, **1994**, *51*, 3058–3060

## SAMPLE
**Matrix:** microsomal incubations
**Sample preparation:** Extract microsomal incubation with 5 volumes ethyl acetate, evaporate the organic layer to dryness under a stream of nitrogen, reconstitute the residue in 500 μL mobile phase, inject an aliquot.

## HPLC VARIABLES
**Column:** 250 × 3.2 6 μm Curosil G (Phenomenex)
**Mobile phase:** MeCN:water 35:65
**Flow rate:** 0.6
**Detector:** UV 229

## CHROMATOGRAM
**Retention time:** 58

## OTHER SUBSTANCES
**Extracted:** metabolites

## KEY WORDS
rat; liver

## REFERENCE
Anderson, C.D.; Wang, J.; Kumar, G.N.; McMillan, J.M.; Walle, U.K.; Walle, T. Dexamethasone induction of taxol metabolism in the rat. *Drug Metab.Dispos.*, **1995**, *23*, 1286–1290

## SAMPLE
**Matrix:** plants
**Sample preparation:** Extract needles with MeOH, concentrate the extract under reduced pressure at <30°, partition concentrate with 0.8 volumes of water and 0.8 volumes of chloroform, repeat the extraction with 0.6 volumes chloroform then 0.4 volumes chloroform, concentrate under reduced pressure, dry under vacuum at 35-40°, dissolve in MeCN/water, inject an aliquot.

## HPLC VARIABLES
**Column:** 250 × 4.6 5 μm Partisil C8
**Mobile phase:** MeCN:MeOH:water 50:10:40
**Flow rate:** 0.5
**Detector:** UV 254

## KEY WORDS
needles; details of preparative HPLC also in paper

## REFERENCE
Rao, K.V.; Bhakuni, R.S.; Juchum, J.; Davies, R.M. A large scale process for paclitaxel and other taxanes from the needles of *Taxus x media Hicksii* and *Taxus floridana* using reverse phase column chromatography. *J.Liq.Chrom.Rel.Technol.*, **1996**, *19*, 427–447

## SAMPLE
**Matrix:** plants
**Sample preparation:** Shake 2 g ground plant material and 100 mL MeOH on a flat-bed orbital shaker for 12-16 h, allow to settle. Remove a 10 mL aliquot and add it to 10 mL 5% NaCl solution, wash twice with 10 mL portions of hexane, discard the hexane washes, extract four times with 10 mL portions of dichloromethane. Filter the dichloromethane extracts through 3 g anhydrous sodium sulfate, evaporate the filtrate to dryness, reconstitute with 2 mL dichloromethane, add to a 250 × 10 column dry packed with 1 g 149-250 μm Davsil silica (Alltech), wash with two 2 mL portions of dichloromethane, wash with two 5 mL portions of acetone:dichloromethane 4:96, elute with two 5 mL portions of acetone:dichloromethane 50:50. Combine the eluates and evaporate them to dryness under a stream of nitrogen at 35°, reconstitute the residue in 1 mL MeOH, inject a 10 μL aliquot.

## HPLC VARIABLES
**Column:** 150 × 4.6 5 μm Zorbax SB-C8
**Mobile phase:** Gradient. A was MeCN. B was MeOH:water 20:80. A:B 20:80 for 5 min, to 25:75 over 0.1 min, maintain at 25:75 for 8.9 min, to 35:65 over 10 min, maintain at 35:65 for 17 min, to 65:35 over 3 min, maintain at 65:35 for 11 min, return to initial conditions over 5 min, re-equilibrate for 20 min.
**Column temperature:** 35
**Flow rate:** 1
**Injection volume:** 10
**Detector:** UV 227

## CHROMATOGRAM
**Retention time:** 42.5
**Limit of detection:** 660 ng/g

## OTHER SUBSTANCES
**Extracted:** baccatin III, cephalomannine

## KEY WORDS
SPE; leaves; twigs

## REFERENCE
Lauren, D.R.; Jensen, D.J.; Douglas, J.A. Analysis of taxol, 10-deacetylbaccatin III and related compounds in *Taxus baccata*. *J.Chromatogr.A*, **1995**, *712*, 303–309

## SAMPLE
**Matrix:** plants
**Sample preparation:** Homogenize (Omni-mix) 50 mg bark or foliage in 1.5 mL MeOH for 2 min, sonicate for 5 min, centrifuge, filter (0.2 μm) the supernatant, inject an aliquot of the filtrate.

## HPLC VARIABLES
**Column:** 1000 × 4.7 7 μm porous graphitized carbon (Shandon)
**Mobile phase:** Dioxane:water 46:54 (Caution! Dioxane is a carcinogen!)
**Flow rate:** 0.8
**Injection volume:** 20
**Detector:** UV 228

## CHROMATOGRAM
**Retention time:** 18
**Limit of detection:** 30 ng/mL

## KEY WORDS
bark; foliage

## REFERENCE
Forgács, E. Use of porous graphitized carbon column for determining taxol in *Taxus baccata*. *Chromatographia*, **1994**, *39*, 740−742

## SAMPLE
**Matrix:** plants
**Sample preparation:** Grind yew needles to <3 mm in a blender. Weigh out 3-4 g, add 100 mL MeOH, shake for 16 h on a wrist-action shaker, filter (Whatman 1 or 2 paper), wash the solid with 25 mL MeOH. Evaporate the filtrate to dryness under reduced pressure at 40-43°, reconstitute the residue in 10 mL MeOH and 1 mL water. Condition a 47 mm Empore SPE extraction disk (3M Corp.) with 15 mL ethyl acetate, 15 mL MeOH and 15 mL water. Use a 47 mm polypropylene separator with 10 μm pore size (Gelman 61757) in front of the extraction disk. Pass 10 mL water and 7 mL crude extract through the disk, wash with 15 mL water, wash with 15 mL MeOH:water 20:80, 15 mL MeOH:water 45:55, elute with 20 mL MeOH:water 80:20, filter (2 μm) the eluate, inject a 10 μL aliquot.

## HPLC VARIABLES
**Guard column:** Taxsil guard cartridge (0335-CS) (MetaChem)
**Column:** 250 × 4.6 5 μm Taxsil (0335-250 × 046) (MetaChem)
**Mobile phase:** Gradient. A was MeCN. B was MeOH:water 30:70. A:B 41:59 for 15 min, to 65:35 over 5 min, maintain at 65:35 for 10 min, to 41:59 over 5 min, maintain at 41:59 for 5 min.
**Column temperature:** 32
**Flow rate:** 1
**Injection volume:** 10
**Detector:** UV 230

## CHROMATOGRAM
**Retention time:** 10

## KEY WORDS
SPE

## REFERENCE
Mattina, M.J.I.; MacEachern, G.J. Extraction, purification by solid-phase extraction and high-performance liquid chromatographic analysis of taxanes from ornamental Taxus needles. *J.Chromatogr.A*, **1994**, *679*, 269−275

## SAMPLE
**Matrix:** solutions

## HPLC VARIABLES
**Column:** 250 × 4.6 5 μm LiChrospher diol
**Mobile phase:** Gradient. MeOH:carbon dioxide 8:92 for 3 min, to 28:72 over 25 min, to 35:65 over 5.7 min, maintain at 35:65 for 4 min.
**Column temperature:** 30
**Flow rate:** 2
**Detector:** UV 227

## CHROMATOGRAM
**Retention time:** 15.77

## OTHER SUBSTANCES
**Simultaneous:** impurities, degradation products

## KEY WORDS
SFC; pressure 150 bar

## REFERENCE
Jagota, N.K.; Nair, J.B.; Frazer, R.; Klee, M.; Wang, M.Z. Supercritical fluid chromatography of paclitaxel. *J.Chromatogr.A*, **1996**, *721*, 315–322

## SAMPLE
**Matrix:** solutions
**Sample preparation:** Inject a 20 μL aliquot.

## HPLC VARIABLES
**Guard column:** C-18 Guard Pak
**Column:** 100 × 8 10 μm Resolve C-18 radial compression (Waters)
**Mobile phase:** Gradient. MeCN:water from 45:55 to 100:0 over 20 min (exponential gra-
dient), maintain at 100:0 for 5 min, re-equilibrate at initial conditions for 5 min.
**Column temperature:** 21
**Flow rate:** 2.5
**Injection volume:** 20
**Detector:** UV 227

## CHROMATOGRAM
**Retention time:** 9.33

## REFERENCE
Wenk, M.R.; Fahr, A.; Reszka, R.; Seelig, J. Paclitaxel partitioning into lipid bilayers. *J.Pharm.Sci.*,
**1996**, *85*, 228–231

## SAMPLE
**Matrix:** solutions

## HPLC VARIABLES
**Column:** 250 × 4.6 5 μm Partisil C8
**Mobile phase:** MeCN:MeOH:water 50:10:40
**Flow rate:** 0.5
**Detector:** UV 254

## CHROMATOGRAM
**Retention time:** 17

## OTHER SUBSTANCES
**Simultaneous:** impurities

## KEY WORDS
details of preparative chromatography

## REFERENCE
Rao, K.V.; Hanuman, J.B.; Alvarez, C.; Stoy, M.; Juchum, J.; Davies, R.M.; Baxley, R. A new large-scale
process for taxol and related taxanes from *Taxus brevifolia*. *Pharm.Res.*, **1995**, *12*, 1003–1010

## SAMPLE
**Matrix:** urine

**Sample preparation:** 19 mL Urine + 1 mL Cremophor EL:EtOH 50:50 (Sigma), mix. 500 µL Sample + 500 µL 200 mM pH 5.0 ammonium acetate buffer + 5 mL n-butyl chloride, rotate slowly for 10 min, centrifuge at 2500 g for 10 min, freeze in dry ice/acetone. Remove the organic layer and evaporate it to dryness under a stream of nitrogen at 40°, reconstitute the residue in 200 µL MeCN:MeOH:water 40:50:10 (40:10:50 ?), vortex for 30 s, inject a 50 µL aliquot.

## HPLC VARIABLES
**Guard column:** 4 × 4 5 µm LiChrospher RP-8
**Column:** 150 × 4.6 5 µm APEX-octyl (Jones Chromatography)
**Mobile phase:** MeCN:MeOH:20 mM pH 5.0 ammonium acetate buffer 40:10:50
**Flow rate:** 1
**Injection volume:** 50
**Detector:** UV 227

## CHROMATOGRAM
**Retention time:** 10
**Limit of quantitation:** 250 ng/mL

## REFERENCE
Huizing, M.T.; Rosing, H.; Koopman, F.; Keung, A.C.F.; Pinedo, H.M.; Beijnen, J.H. High-performance liquid chromatographic procedures for the quantitative determination of paclitaxel (Taxol) in human urine. *J.Chromatogr.B*, **1995**, *664*, 373–382

## SAMPLE
**Matrix:** urine
**Sample preparation:** Condition a 1 mL 100 mg Bond Elut cyano SPE cartridge with 2 mL MeOH and 2 mL 10 mM pH 5.0 ammonium acetate buffer. 19 mL Urine + 1 mL Cremophor EL:EtOH 50:50 (Sigma), mix. 500 µL Sample + 500 µL 200 mM pH 5.0 ammonium acetate buffer, vortex for 20 s, add to the SPE cartridge, wash with 2 mL 10 mM pH 5.0 ammonium acetate buffer, wash with 1 mL MeOH:10 mM pH 5.0 ammonium acetate buffer 20:80, wash with 1 mL hexane, dry under vacuum for 1 min, elute with 2 mL MeCN:triethylamine 100:0.1. Evaporate the eluate to dryness under a stream of nitrogen at 30°, reconstitute in 200 µL MeCN:MeOH:water 40:50:10 (40:10:50 ?), vortex for 30 s, inject a 50 µL aliquot.

## HPLC VARIABLES
**Guard column:** 4 × 4 5 µm LiChrospher RP-8
**Column:** 150 × 4.6 5 µm APEX-octyl (Jones Chromatography)
**Mobile phase:** MeCN:MeOH:20 mM pH 5.0 ammonium acetate buffer 40:10:50
**Flow rate:** 1
**Injection volume:** 50
**Detector:** UV 227

## CHROMATOGRAM
**Retention time:** 10
**Limit of detection:** 8 ng/mL
**Limit of quantitation:** 10 ng/mL

## KEY WORDS
SPE

## REFERENCE
Huizing, M.T.; Rosing, H.; Koopman, F.; Keung, A.C.F.; Pinedo, H.M.; Beijnen, J.H. High-performance liquid chromatographic procedures for the quantitative determination of paclitaxel (Taxol) in human urine. *J.Chromatogr.B*, **1995**, *664*, 373–382

◆————————————————————————————————————◆————————————————————————————————————◆

## ANNOTATED BIBLIOGRAPHY

Wu, D.-R.; Lohse, K.; Greenblatt, H.C. Preparative separation of taxol in normal- and reversed-phase operations. *J.Chromatogr.A*, **1995**, *702*, 233–241 [normal phase; reverse phase]

Alkan-Onyuksel, H.; Ramakrishnan, S.; Chai, H.-B.; Pezzuto, J.M. Mixed micellar formulation suitable for the parenteral administration of taxol. *Pharm.Res.*, **1994**, *11*, 206–212 [formulations]

Klecker, R.W.; Jamis-Dow, C.A.; Egorin, M.J.; Erkmen, K.; Parker, R.J.; Stevens, R.; Collins, J.M. Effect of cimetidine, probenecid, and ketoconazole on the distribution, biliary secretion, and metabolism of [3H]taxol in the Sprague-Dawley rat. *Drug Metab.Dispos.*, **1994**, *22*, 254–258 [rat; plasma; bile; urine; lung; spleen; liver; kidney; heart; muscle; brain; testes; fat; UV detection; radioactivity detection; tritium labeled]

Kopycki, W.J.; ElSohly, H.N.; McChesney, J.D. HPLC determination of taxol and related compounds in Taxus plant extracts. *J.Liq.Chromatogr.*, **1994**, *17*, 2569–2591 [gradient]

Kumar, G.N.; Walle, U.K.; Walle, T. Cytochrome P450 3A-mediated human liver microsomal taxol 6α-hydroxylation. *J.Pharmacol.Exp.Ther.*, **1994**, *268*, 1160–1165 [microsomal incubations; extracted metabolites]

Mase, H.; Hiraoka, M.; Suzuki, A.; Nakanomyo, H. [Determination of new anticancer drug, paclitaxel, in biological fluids by high performance liquid chromatography]. *Yakugaku.Zasshi.*, **1994**, *114*, 351–355

Bitsch, F.; Ma, W.; Macdonald, F.; Nieder, M.; Shackleton, C.H. Analysis of taxol and related diterpenoids from cell cultures by liquid chromatography-electrospray mass spectrometry. *J.Chromatogr.*, **1993**, *615*, 273–280 [LC-MS; electrospray; microbore; gradient; LOD 100 pg]

Castor, T.P.; Tyler, T.A. Determination of taxol in Taxus media needles in the presence of interfering components. *J.Liq.Chromatogr.*, **1993**, *16*, 723–731 [SPE; plants; simultaneous baccatin, cephalomannine]

Ketchum, E.B.; Gibson, D.M. Rapid isocratic reversed phase HPLC of taxanes on new columns developed specifically for taxol analysis. *J.Liq.Chromatogr.*, **1993**, *16*, 2519–2530 [simultaneous baccatin, cephalomannine; gradient; column temp 30]

Wheeler, N.C.; Jech, K.; Masters, S.; Brobst, S.W.; Alvarado, A.B.; Hoover, A.J.; Snader, K.M. Effects of genetic, epigenetic, and environmental factors on taxol content in Taxus brevifolia and related species. *J.Nat.Prod.*, **1992**, *55*, 432–440 [plants]

Waugh, W.N.; Trissel, L.A.; Stella, V.J. Stability, compatibility, and plasticizer extraction of taxol (NSC-125973) injection diluted in infusion solutions and stored in various containers. *Am.J.Hosp.Pharm.*, **1991**, *48*, 1520–1524 [stability-indicating; simultaneous degradation products; injections; 5% dextrose; saline]

Monsarrat, B.; Mariel, E.; Cros, S.; Garès, M.; Guénard, D.; Guéritte-Voegelein, F.; Wright, M. Taxol metabolism. Isolation and identification of three major metabolites of taxol in rat bile. *Drug Metab.Dispos.*, **1990**, *18*, 895–901 [rat; bile; extracted metabolites]

Witherup, K.M.; Look, S.A.; Stasko, M.W.; Ghiorzi, T.J.; Muschik, G.M.; Cragg, G.M. *Taxus* spp. needles contain amounts of taxol comparable to the bark of *Taxus brevifolia*: analysis and isolation. *J.Nat.Prod.*, **1990**, *53*, 1249–1255 [plants; analytical; preparative]

Witherup, K.M.; Look, S.A.; Stasko, M.W.; McCloud, T.G.; Issaq, H.J.; Muschik, G.M. High performance liquid chromatographic separation of taxol and related compounds from Taxus brevifolia. *J.Liq.Chromatogr.*, **1989**, *12*, 2117–2132 [plants; simultaneous baccatin, cephalomannine; gradient; isocratic]

Xu, L.X.; Liu, A.R. Determination of taxol in the extract of taxus chinensis by reversed phase HPLC. *Yao Hsueh Hsueh Pao*, **1989**, *24*, 552–555

# Penicillin V

**Molecular formula:** $C_{16}H_{18}N_2O_5S$
**Molecular weight:** 350.4
**CAS Registry No.:** 87-08-1, 132-98-9 (potassium salt),
5928-84-7 (benzathine), 63690-57-3 (benzathine
tetrahydrate), 6591-72-6 (hydrabamine)

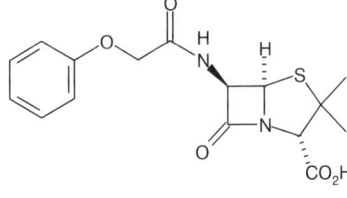

## SAMPLE
**Matrix:** blood
**Sample preparation:** Condition a Baker C18 SPE cartridge with 5 mL water and 5 mL 2% NaCl, do not allow to run dry. 2 mL Plasma + 30 mL water + 2 mL 170 mM sulfuric acid + 2 mL 5% sodium tungstate solution, vortex for 30 s, centrifuge at 2200 g for 10 min, filter supernatant (GF/B glass fiber filter), add 10 mL 20% NaCL, mix, add to the SPE cartridge at 3 mL/min, wash with 5 mL 2% NaCl, wash with 5 mL water, draw air through cartridge for 5 min, elute with 500 μL elution solution. Add 500 μL derivatization reagent to the eluate, vortex for 20 s, allow to react at 65° for 30 min, cool to room temperature, vortex, filter (0.45 μm), inject 50-100 μL aliquots. (Prepare derivatization reagent by dissolving 34.45 g 1,2,4-triazole in 150 mL water, add 25 mL 10 mM mercuric chloride solution, mix, adjust the pH to 9.0 ± 0.5 with 5 M NaOH, dilute to 250 mL with water. Prepare elution solution by mixing 60 mL MeCN and 5 mL buffer and making up to 100 mL with water. The buffer was 0.994 g $Na_2HPO_4$ + 1.794 g $NaH_2PO_4.H_2O$ in 100 mL water, pH 6.5.)

## HPLC VARIABLES
**Column:** 150 × 3.9 4 μm Nova-Pak C18
**Mobile phase:** MeCN:buffer 25:75 (Buffer contained 4.969 g $Na_2HPO_4$ + 8.969 g $NaH_2PO_4.H_2O$ + 2.482 g anhydrous sodium thiosulfate per liter.)
**Flow rate:** 1
**Injection volume:** 50-100
**Detector:** UV 325

## CHROMATOGRAM
**Retention time:** 5.8
**Internal standard:** penicillin V

## OTHER SUBSTANCES
**Extracted:** penicillin G

## KEY WORDS
plasma; cow; SPE; penicillin V is IS

## REFERENCE
Boison, J.O.; Korsrud, G.O.; MacNeil, J.D.; Keng, L.; Papich, M. Determination of penicillin G in bovine plasma by high-performance liquid chromatography after pre-column derivatization. *J.Chromatogr.*, **1992**, *576*, 315–320

## SAMPLE
**Matrix:** blood
**Sample preparation:** 500 μL Serum + 100 μL 50 μg/mL chloramphenicol in water + 1 mL 1 M pH 3.0 sodium acetate + 5 mL diethyl ether, shake for 20 min, centrifuge at 1200 g. Remove the organic layer and evaporate it to dryness at 40° under a stream of nitrogen. Reconstitute the residue in 100 μL water, inject a 50 μL aliquot.

## HPLC VARIABLES
**Column:** 300 × 3.9 10 μm μBondapak C18
**Mobile phase:** MeCN:10 mM potassium acetate buffer 20:80, pH 6.5
**Flow rate:** 1.6
**Injection volume:** 50
**Detector:** UV 215

## CHROMATOGRAM
**Retention time:** 3.10
**Internal standard:** chloramphenicol (5.10)
**Limit of detection:** 30 ng/mL

## OTHER SUBSTANCES
**Noninterfering:** amikacin, amiloride, amoxicillin, ampicillin, cephalexin, doxycycline, ethosuximide, gentamicin, hydrochlorothiazide, netilmicin, phenacetin, phenemal, phenytoin, primidone, sisomicin, tetracycline, tobramycin
**Interfering:** cloxacillin, penicillin G procaine

## KEY WORDS
serum

## REFERENCE
Lindberg, R.L.; Huupponen, R.K.; Huovinen, P. Rapid high-pressure liquid chromatographic method for analysis of phenoxymethylpenicillin in human serum. *Antimicrob.Agents Chemother.*, **1984**, *26*, 300–302

## SAMPLE
**Matrix:** blood, urine
**Sample preparation:** Plasma. 1 mL Plasma + 50 μL 100 μg/mL oxacillin in water + 20 μL 4% aqueous sodium dodecyl hydrogen sulfate solution, shake for 30 min, filter (Amicon MPS-1 micropartition system, YMT membrane) while centrifuging, adjust the pH of the ultrafiltrate to 6.3-6.5 with pH 4 citrate buffer, inject a 500 μL aliquot onto column A with mobile phase A and elute to waste, after 10 min elute the contents of column A onto column B with mobile phase B, elute with mobile phase B, monitor the effluent from column B. Urine. 5-100 μL Urine + 50 μL 100 μg/mL oxacillin in water, make up to 500 μL with water, inject onto column A with mobile phase A and elute to waste, after 10 min elute the contents of column A onto column B with mobile phase B, elute with mobile phase B, monitor the effluent from column B.

## HPLC VARIABLES
**Column:** A 50 × 4 Nucleosil 5-C18; B 250 × 5 Nucleosil 5-C18
**Mobile phase:** A MeCN:33 mM NaH$_2$PO$_4$ 5:95; B MeCN:33 mM NaH$_2$PO$_4$ 20:80
**Injection volume:** 500
**Detector:** UV 210

## CHROMATOGRAM
**Internal standard:** oxacillin
**Limit of quantitation:** 50 ng/mL

## KEY WORDS
plasma; column-switching; pharmacokinetics; ultrafiltrate

## REFERENCE
Lintz, W.; Hirsch, I.; Osterloh, G.; Schmidt-Böthelt, E.; Sous, H. Bioverfügbarkeit von Penicillin V in einer wäßrigen Zubereitungsform [Bioavailability of penicillin V in aqueous dosage forms]. *Arzneimittelforschung*, **1984**, *34*, 66–71

## SAMPLE

**Matrix:** cheese, milk, yogurt

**Sample preparation:** Condition a 6 mL 500 mg Bond Elut C18 SPE cartridge with 20 mL MeOH, 20 mL water, and 10 mL 2% NaCl, do not allow to go dry. 5 mL Milk (or 5 g yogurt or cottage cheese + 4 mL 1 M pH 6 phosphate buffer) + 25 mL water + 4 mL 170 mM sulfuric acid + 40 mL 5% sodium tungstate, vortex for 30 s, centrifuge at 1500 g for 10 min, remove the supernatant, add 10 mL 20% NaCl to the residue, vortex for 10 s, centrifuge. Combine the supernatants and add them to the SPE cartridge, wash with 10 mL 2% NaCl, wash with 10 mL water, elute with 1 mL MeCN:200 mM pH 6.5 sodium phosphate buffer:water 60:5:35. Add 1 mL reagent to the eluate, vortex for 10 s, heat at 65° for 30 min, cool to room temperature, vortex, filter (Acro 0.45 μm), inject a 50-100 μL aliquot of the filtrate. (Prepare reagent by dissolving 34.45 g 1,2,4-triazole in 150 mL water, add 25 mL 10 mM mercuric chloride solution, mix, adjust pH to 9.0 ± 0.5 with 5 M NaOH, make up to 250 mL with water.)

## HPLC VARIABLES

**Column:** 150 × 3.9 4 μm Nova-Pak C18

**Mobile phase:** MeCN:buffer 25:75 (Buffer was 4.696 g $Na_2HPO_4$, 8.969 g $NaH_2PO_4.H_2O$, and 2.482 g anhydrous sodium thiosulfate in 1 L water.)

**Flow rate:** 0.8

**Injection volume:** 50-100

**Detector:** UV 325

## CHROMATOGRAM

**Retention time:** 7

**Internal standard:** penicillin V

## OTHER SUBSTANCES

**Extracted:** penicillin G

## KEY WORDS

derivatization; cow; SPE; penicillin V is IS

## REFERENCE

Boison, J.O.K.; Keng, L.J.-Y.; MacNeil, J.D. Analysis of penicillin G in milk by liquid chromatography. *J.AOAC Int.*, **1994**, 77, 565–570

## SAMPLE

**Matrix:** cheese, milk, yogurt

**Sample preparation:** Condition a 6 mL 500 mg Bond Elut C18 SPE cartridge with 20 mL MeOH, 20 mL water, and 10 mL 2% NaCl, do not allow to go dry. 5 mL Milk (or 5 g yogurt or cottage cheese + 4 mL 1 M pH 6 phosphate buffer) + 25 mL water + 4 mL 170 mM sulfuric acid + 40 mL 5% sodium tungstate, vortex for 30 s, centrifuge at 1500 g for 10 min, remove the supernatant, add 10 mL 20% NaCl to the residue, vortex for 10 s, centrifuge. Combine the supernatants and add them to the SPE cartridge, wash with 10 mL 2% NaCl, wash with 10 mL water, elute with 750 μL MeCN:200 mM ammonium acetate:water 60:5:35, filter (Acro 0.45 μm), inject a 50-100 μL aliquot of the filtrate.

## HPLC VARIABLES

**Column:** 150 × 3.9 4 μm Nova-Pak C18

**Mobile phase:** Gradient. A was MeCN. B was MeCN:150 mM ammonium acetate 10:90. A:B 0:100 for 10 min, to 30:70 over 10 min, return to initial conditions over 10 min.

**Flow rate:** 0.9

**Injection volume:** 50-100

**Detector:** MS, VG Trio II, probe tip 255°, source 180°, thermospray/plasmaspray, m/z 335, m/z 160

## CHROMATOGRAM
**Internal standard:** penicillin V

## OTHER SUBSTANCES
**Extracted:** penicillin G

## KEY WORDS
cow; SPE; penicillin V is IS

## REFERENCE
Boison, J.O.K.; Keng, L.J.-Y.; MacNeil, J.D. Analysis of penicillin G in milk by liquid chromatography. *J.AOAC Int.*, **1994**, *77*, 565–570

## SAMPLE
**Matrix:** fermentation broth
**Sample preparation:** Adjust pH of fermentation broth to 7, centrifuge at 8000 g for 10 min, add MeCN, centrifuge, add dichloromethane to the supernatant, vortex for 10 s, shake for 15 min, centrifuge at 8000 g for 15 min. Add 1 mL of the aqueous layer to 100 μL reagent, heat at 50° for 50 min, cool in an ice bath, inject a 20 μL aliquot. (Prepare reagent by dissolving 4.125 g imidazole in 2.5 mL water, add 1 mL HCl, add 500 μL 110 mM mercury(II) chloride, add 1.5 mL HCl. Recrystallize imidazole twice from isopropanol.)

## HPLC VARIABLES
**Guard column:** 10 × 4 5 μm Spherisorb C18
**Column:** 20 × 4.6 5 μm Spherisorb C18 S5ODS2
**Mobile phase:** Gradient. MeCN:buffer from 16.5:83.5 to 31.5:68.5 over 17 min (Buffer was 10 mM $NaH_2PO_4$ containing 10 mM EDTA, adjusted to pH 6.5 with 2 M NaOH.)
**Flow rate:** 2
**Injection volume:** 20
**Detector:** UV 325

## CHROMATOGRAM
**Retention time:** 14.5
**Limit of detection:** 1 μg/mL

## OTHER SUBSTANCES
**Extracted:** methicillin, penicillin G, penicillin X

## KEY WORDS
derivatization

## REFERENCE
Rogers, M.E.; Adlard, M.W.; Saunders, G.; Holt, G. High-performance liquid chromatographic determination of penicillins following derivatization to mercury-stabilized penicillenic acids. *J.Liq. Chromatogr.*, **1983**, *6*, 2019–2031

## SAMPLE
**Matrix:** formulations
**Sample preparation:** Powder 20 tablets, dissolve a portion of the powder in water such that the concentration of penicillin V potassium is 0.6 mg/mL, mix well, filter. Mix 20 mL filtrate with 15 mL 0.4 mg/mL sulfadimethoxine in MeCN:water 50:50, dilute to 100 mL with water, mix well, inject a 20 μL aliquot.

## HPLC VARIABLES
**Guard column:** 40 mm long 30-50 μm Whatman Co:Pell ODS

**Column:** 300 × 3.9 10 μm μBondapak C18
**Mobile phase:** MeCN:MeOH:10 mM KH$_2$PO$_4$ 21:4:75
**Flow rate:** 1-1.5
**Injection volume:** 20
**Detector:** UV 225

## CHROMATOGRAM
**Internal standard:** sulfadimethoxine

## KEY WORDS
tablets; collaborative study

## REFERENCE
Mopper, B. Liquid chromatographic determination of penicillin V potassium in tablets: collaborative study. *J.Assoc.Off.Anal.Chem.*, **1989**, *72*, 883–884

## SAMPLE
**Matrix:** formulations
**Sample preparation:** Blend tablets and capsules with water in a high-speed blender for 5 min, filter, dilute with mobile phase, inject a 20 μL aliquot. Dilute oral suspensions and injections with mobile phase, inject a 20 μL aliquot.

## HPLC VARIABLES
**Guard column:** 70 mm long Co:Pell ODS
**Column:** 300 × 4.6 10 μm Chromegabond C18 (E.S. Industries)
**Mobile phase:** MeCN:MeOH:10 mM KH$_2$PO$_4$ 19:11:70
**Flow rate:** 1
**Injection volume:** 20
**Detector:** UV 225

## CHROMATOGRAM
**Retention time:** 8.4
**Limit of detection:** 1.02 μg/mL

## OTHER SUBSTANCES
**Simultaneous:** amoxicillin, ampicillin, cloxacillin, dicloxacillin, methicillin, nafcillin, oxacillin, penicillin G

## KEY WORDS
tablets; capsules; oral suspensions; injections

## REFERENCE
Briguglio, G.T.; Lau-Cam, C.A. Separation and identification of nine penicillins by reverse phase liquid chromatography. *J.Assoc.Off.Anal.Chem.*, **1984**, *67*, 228–231

## SAMPLE
**Matrix:** milk
**Sample preparation:** 10 mL Milk + 2 mL 200 mM tetraethylammonium chloride, stir, slowly add 38 mL MeCN over 30 s, let stand for 5 min, decant the supernatant through a plug of glass wool. 40 mL Filtrate + 1 mL water, evaporate under reduced pressure to 1-2 mL, make up to 4 mL with water, filter (0.45 μm polyvinylidene difluoride). Inject 2 mL into an LC system (150 × 4.6 5 μm Supelcosil LC-18; MeCN:10 mM KH$_2$PO$_4$ 0:100 for 3 min, to 40:60 over 27 min, to 0:100 over 1 min; 1 mL/min; UV 210 and 295), collect a 1.5 mL fraction at retention time for penicillin V (25 min), evaporate to 1 mL, inject a 200 μL aliquot.

## HPLC VARIABLES

**Column:** 150 × 4.6 5 µm Supelcosil LC-18-DB
**Mobile phase:** MeCN:buffer 33:67 (Buffer was 5 mM phosphoric acid and 5 mM $KH_2PO_4$.)
**Flow rate:** 1
**Injection volume:** 200
**Detector:** UV 210

## CHROMATOGRAM

**Limit of quantitation:** 2-5 ppb; cow

## OTHER SUBSTANCES

**Extracted:** amoxicillin, ampicillin, ceftiofur, cephapirin, cloxacillin, penicillin G

## REFERENCE

Moats, W.A.; Harik-Khan, R. Liquid chromatographic determination of β-lactam antibiotics in milk: A multiresidue approach. *J.AOAC Int.*, **1995**, *78*, 49–54

## SAMPLE

**Matrix:** milk
**Sample preparation:** Add 2 volumes MeCN to milk, stand 5 min, decant aqueous portion, suction filter, extract with an equal volume of dichloromethane:hexane 50:50, centrifuge aqueous phase at 3000 rpm for 10 min. Dilute 3:1 with 20 mM sodium acetate buffer and filter (0.2 µm nylon). Inject 50 µL onto column with mobile phase A, run mobile phase A for 30 min and elute to waste. After 30 min switch to mobile phase B and elute through detector.

## HPLC VARIABLES

**Column:** 100 × 8 Radial-Pak 10µm µBondapak C18
**Mobile phase:** A 20 mM sodium acetate buffer; B Gradient. MeCN:MeOH:20 mM sodium acetate buffer from 15:10:75 to 30:0:70 over 15 min and hold at 30:0:70
**Flow rate:** A 3; B 2
**Injection volume:** 50
**Detector:** E, Waters 464 pulsed electrochemical detector, thin layer cell, Ag/AgCl reference electrode, E1 = 1300 mV for 0.166 s, E2 = 1500 mV for 0.166 s, E3 = -200 mV for 0.333 s.

## CHROMATOGRAM

**Retention time:** 11.8
**Limit of detection:** 0.2 ppm

## OTHER SUBSTANCES

**Extracted:** ampicillin, cloxacillin, dicloxacillin, methicillin, nafcillin, oxacillin, penicillin G

## REFERENCE

Kirchmann, E.; Earley, R.L.; Welch, L.E. The electrochemical detection of penicillins in milk. *J.Liq.Chromatogr.*, **1994**, *17*, 1755–1772

## SAMPLE

**Matrix:** milk
**Sample preparation:** 50 g Milk + 2 drops penicillinase (Difco Laboratories), let stand 1 h at 37°, add 50 mL MeCN, shake vigorously for 1 min, centrifuge at 9000 g for 10 min, decant, add 5 g NaCl, swirl to dissolve, add 100 mL dichloromethane, shake for 1 min, centrifuge at 1000 g for 10 min. Remove top aqueous layer and extract organic layer with 25 mL 10% NaCl by shaking and centrifuging as before. Combine aqueous layers, add 1 mL 0.3% mercuric chloride in water, let stand 30 min, add 1 mL 2 M HCl, extract with three 50 mL portions of dichloromethane by shaking each portion for 1 min and centri-

fuging at 1000 g for 10 min, filter dichloromethane extracts through 30 g anhydrous sodium sulfate, evaporate to dryness under reduced pressure at 35°, if water remains add 5-10 mL MeOH to flask and complete evaporation. Dissolve residue in 1 mL 10% acetic acid, add 0.5 mL 0.08% dansyl hydrazine in 10% acetic acid, let stand 90 min to overnight in the dark, transfer reaction mixture to a separatory funnel with three 25 mL portions of dichloromethane, add 5 mL 2 M HCl, shake for 1 min, wash organic layer with 5 mL 5% NaHCO₃ solution, filter through 10-20 g anhydrous sodium sulfate. Extract acid aqueous layer again with 25 mL dichloromethane. Combine dichloromethane layers and evaporate to dryness at 35° under reduced pressure. Dissolve residue in 2 mL IS solution, inject a 20 μL aliquot. (Prepare IS solution by dissolving 10 μL benzaldehyde in 100 mL dichloromethane, evaporate 1 mL to dryness under reduced pressure, dissolve residue in 1 mL 10% acetic acid, add 0.5 mL 0.08% dansyl hydrazine in 10% acetic acid, let stand 90 min to overnight in the dark, transfer reaction mixture to a separatory funnel with three 25 mL portions of dichloromethane, add 5 mL 2 M HCl, shake for 1 min, wash organic layer with 5 mL 5% NaHCO₃ solution, filter through 10-20 g anhydrous sodium sulfate. Extract acid aqueous layer again with 25 mL dichloromethane. Combine dichloromethane layers and evaporate to dryness at 35° under reduced pressure. Dissolve residue in 100 mL MeCN then dilute an aliquot 1:4 with MeCN.)

## HPLC VARIABLES
**Column:** 250 × 4 10 μm Lichrosorb RP-18
**Mobile phase:** MeCN:water 58:42
**Flow rate:** 1
**Injection volume:** 20
**Detector:** F ex 254 em 500 (filter)

## CHROMATOGRAM
**Retention time:** 6.73
**Internal standard:** benzaldehyde (derivatized) (12.18)
**Limit of detection:** 5 ng/g

## OTHER SUBSTANCES
**Extracted:** cloxacillin, dicloxacillin, methicillin, nafcillin, penicillin G, phenethicillin
**Interfering:** oxacillin

## KEY WORDS
derivatization

## REFERENCE
Munns, R.K.; Shimoda, W.; Roybal, J.E.; Vieira, C. Multiresidue method for determination of eight neutral β-lactam penicillins in milk by fluorescence-liquid chromatography. *J.Assoc.Off.Anal.Chem.*, **1985**, *68*, 968–971

## SAMPLE
**Matrix:** milk
**Sample preparation:** Condition a Sep-Pak C18 SPE cartridge with 20 mL MeOH, 20 mL water, and 2 mL 2% NaCl. Pass through 30 g filtered (glass-wool plug) milk at 2 mL/min, wash with 5 mL water, wash with 10 mL MeOH:water:20% NaCl 10:80:10 containing 20 mM 18-crown-6, elute with 10 mL 15% (v/v) MeOH, inject a 100 μL aliquot.

## HPLC VARIABLES
**Guard column:** 50 × 2.1 Permaphase ETH (Du Pont)
**Column:** 150 × 4.3 LiChrosorb RP-18
**Mobile phase:** MeOH:water:0.2 M pH 4.0 phosphate buffer 25:65:10 containing 11 mM sodium 1-heptanesulfonate
**Column temperature:** 45
**Flow rate:** 1

**Injection volume:** 100
**Detector:** UV 210

## CHROMATOGRAM
**Retention time:** 18.5
**Limit of detection:** 30 ng/g

## OTHER SUBSTANCES
**Extracted:** ampicillin, penicillin G

## KEY WORDS
cow; SPE

## REFERENCE
Terada, H.; Sakabe, Y. Studies on residual antibacterials in foods. IV. Simultaneous determination of penicillin G, penicillin V and ampicillin in milk by high-performance liquid chromatography. *J.Chromatogr.*, **1985**, *348*, 379–387

## SAMPLE
**Matrix:** solutions
**Sample preparation:** Separate buffer containing drug from human serum albumin by centrifuging at 37° at 700 g for 3 min using a Micropartition System MPS-1 (Amicon) unit, inject a 10-20 μL aliquot of the ultrafiltrate.

## HPLC VARIABLES
**Guard column:** C18/Corasil (Waters)
**Column:** 300 × 3.9 μBondapak C18
**Mobile phase:** MeCN:10 mM ammonium acetate 25:75
**Flow rate:** 1.5
**Injection volume:** 10-20
**Detector:** UV 220

## REFERENCE
Terasaki, T.; Nouda, H.; Tsuji, A. Selective analysis of mutual displacement effects at the primary binding sites of phenoxymethylpenicillin and cephalothin bindings to human serum albumin. *J.Pharmacobiodyn.*, **1992**, *15*, 91–97

## SAMPLE
**Matrix:** solutions
**Sample preparation:** Prepare an aqueous solution, inject a 200 μL aliquot.

## HPLC VARIABLES
**Guard column:** present but not specified
**Column:** 150 × 4.6 4 μm Micropak SPC-18 C18
**Mobile phase:** Gradient. MeCN:10 mM orthophosphoric acid from 15:85 to 60:40 over 20 min
**Flow rate:** 1
**Injection volume:** 200
**Detector:** UV 220

## CHROMATOGRAM
**Retention time:** 15

## OTHER SUBSTANCES
**Simultaneous:** carbenicillin, cloxacillin, dicloxacillin, methicillin, nafcillin, penicillin G

## REFERENCE
Moats, W.A. Effect of the silica support of bonded reversed-phase columns on chromatography of some antibiotic compounds. *J.Chromatogr.*, **1986**, *366*, 69–78

## SAMPLE
**Matrix:** tissue
**Sample preparation:** Homogenize (Ultra-Turrax) 25 g tissue with 25 mL MeCN for 1 min, add 5 mL 500 mM pH 2.2 phosphate buffer while the homogenizer is still running, add 65 mL MeCN, homogenize for 1 min, centrifuge at 4000 g for 10 min. Remove the supernatant and add it to 7 g NaCl and 50 mL dichloromethane, shake for 2 min, allow to stand for 30 min. Remove the upper organic layer and add it to 5 g anhydrous sodium sulfate, shake for 30 s, filter through a cotton-wool plug, evaporate to about 4 mL under reduced pressure at 30°, add 3 mL dichloromethane, evaporate to about 4 mL, add 3 mL light petroleum, evaporate to about 0.5 mL, Suspend this residue with sonication in three 3 mL portions of light petroleum and place these fractions in a separate tube, rinse the original tube with 2 mL pH 7 phosphate buffer. Add the phosphate buffer rinse to the light petroleum extracts, vortex for 30 s, centrifuge, remove the aqueous layer. Extract the light petroleum layer with 2 mL pH 7 phosphate buffer and with two 1.5 mL portions of pH 7 phosphate buffer, combine all the aqueous phases, centrifuge, inject a 200 µL aliquot onto column A and elute to waste with mobile phase B, after 15 min elute to waste with mobile phase C at 2 mL/min, after 10 min elute the contents of column A onto column B with mobile phase D, after 2 min remove column A from the circuit, elute column B with mobile phase D, monitor the effluent from column B. (Wash column A with mobile phase A at 2 mL/min for 7 min, with mobile phase A at 1 mL/min for 5 min, with mobile phase B at 2 mL/min for 8 min, and with mobile phase B at 1 mL/min for 6 min.)

## HPLC VARIABLES
**Column:** A 4 × 4 5 µm LiChrospher 100 RP-18e; B 250 × 4 5 µm LiChrospher 100 RP-18e
**Mobile phase:** A MeCN:water 50:50; B 20 mM pH 7 phosphate buffer; C MeCN:20 mM pH 3 phosphate buffer 10:90; D MeCN:200 mM pH 3.0 phosphate buffer 35:65 containing 2 mM disodium EDTA
**Column temperature:** 35
**Flow rate:** 1 (except where indicated)
**Injection volume:** 200
**Detector:** E, Merck Model L3500, glassy carbon working electrode +0.65 V, stainless-steel auxiliary electrode, Ag/AgCl reference electrode following post-column reaction. The column effluent flowed through a 10 m × 0.3 mm ID woven PTFE coil illuminated by a UV 254 low-pressure mercury lamp to the detector.

## CHROMATOGRAM
**Retention time:** 6.1
**Limit of detection:** 1.4 ng

## OTHER SUBSTANCES
**Extracted:** cloxacillin, dicloxacillin, oxacillin, penicillin G

## KEY WORDS
post-column reaction; cow; muscle; column-switching

## REFERENCE
Lihl, S.; Rehorek, A.; Petz, M. High-performance liquid chromatographic determination of penicillins by means of automated solid-phase extraction and photochemical degradation with electrochemical detection. *J.Chromatogr.A*, **1996**, *729*, 229–235

## SAMPLE
**Matrix:** tissue

**Sample preparation:** Blend 15 g tissue with 45 mL (60 mL for liver and kidney) water in a 300 or 500 mL blender jar at half power (or less to control foaming) for 2 min. Add a 20 mL aliquot of homogenate to 40 mL MeCN, mix, after 5 min decant supernatant through a plug of glass wool, collect 30 mL. Shake vigorously 30 mL filtrate, 10 mL 200 mM phosphoric acid, and 20 mL dichloromethane, remove organic layer and extract aqueous layer with 10 mL dichloromethane (and 10 mL MeCN for liver and kidneys). Combine dichloromethane layers, add 15 mL MeCN, add 40 mL hexane, wash the mixture twice with 4 mL portions of water, extract the organic layer four times with 1 mL 10 mM pH 7 buffer. Combine extracts and add 0.1-0.2 mL tert-butanol, place in a rotary evaporator without heating at first. When the flask becomes cold warm to 50°, concentrate to less than 1 mL, adjust to a final volume of 1 mL, filter (Gelman Acrodisc LCPVDF), inject a 200 μL aliquot.

## HPLC VARIABLES
**Guard column:** Polymer Labs guard cartridge
**Column:** 150 × 4.6 5 μm 100 Å PLRP-S polystyrene-divinylbenzene (Polymer Labs)
**Mobile phase:** MeCN:buffer 18:82, after run was over flush at 35:65 for 5 min then re-equilibrate with 18:82 for 9 min. (Buffer was 10 mM pH 7 phosphate buffer prepared from 1.36 $KH_2PO_4$ and 2.84 g $Na_2HPO_4$ in 3 L water.)
**Flow rate:** 1
**Injection volume:** 200
**Detector:** UV 210

## CHROMATOGRAM
**Retention time:** 9-11
**Limit of detection:** 10 ng/g

## KEY WORDS
cow; pig

## REFERENCE
Moats, W.A. High-performance liquid chromatographic determination of penicillin G, penicillin V and cloxacillin in beef and pork tissues. *J.Chromatogr.*, **1992**, *593*, 15–20

## SAMPLE
**Matrix:** tissue
**Sample preparation:** Condition a 6 mL 500 mg Bond Elut C18 SPE cartridge with 20 mL MeOH, 20 mL water, and 10 mL 2% NaCl, do not allow to go dry. 5 g Tissue + 20 mL water, homogenize (Polytron, 20 mm probe), rinse probe with water so that total volume is 35 mL, shake mechanically for 5 min, add 5 mL 170 mM sulfuric acid, add 5 mL 5% sodium tungstate, vortex for 20 s, centrifuge at 2200 g for 10 min, remove the supernatant, add 15 mL water to the residue, shake for 5 min, centrifuge at 2200 g for 10 min. Combine the supernatants and filter (Whatman GF/B) them, add 10 mL 20% NaCl to the filtrate, mix thoroughly, add to the SPE cartridge at 3 mL/min, wash with 10 mL 2% NaCl, wash with 10 mL water, draw air through the cartridge for 5 min, immediately elute with 1 mL MeCN:200 mM pH 6.5 sodium phosphate buffer:water 60:5:35. Add 1 mL reagent to the eluate, vortex, heat at 65° for 30 min, cool rapidly to room temperature, vortex, filter (Acro 0.45 μm), inject a 80-100 μL aliquot of the filtrate. (Prepare reagent by dissolving 34.45 g 1,2,4-triazole in 150 mL water, add 25 mL 10 mM mercuric chloride solution, mix, adjust pH to 9.0 ± 0.5 with 5 M NaOH, make up to 250 mL with water.)

## HPLC VARIABLES
**Column:** 150 × 3.9 4 μm Nova-Pak C18
**Mobile phase:** MeCN:buffer 25:75 (Buffer was 4.969 g $Na_2HPO_4$, 8.969 g $NaH_2PO_4.H_2O$, and 2.482 g anhydrous sodium thiosulfate in 1 L water.)
**Flow rate:** 0.8

**Injection volume:** 80-100
**Detector:** UV 325

## CHROMATOGRAM
**Retention time:** 7.6
**Internal standard:** penicillin V

## OTHER SUBSTANCES
**Extracted:** penicillin G
**Simultaneous:** ampicillin, chloramphenicol

## KEY WORDS
muscle; liver; kidney; derivatization; cow; SPE; penicillin V is IS

## REFERENCE
Boison, J.O.; Salisbury, C.D.C.; Chan, W.; MacNeil, J.D. Determination of penicillin G residues in edible animal tissues by liquid chromatography. *J.Assoc.Off.Anal.Chem.*, **1991**, *74*, 497–501

## ANNOTATED BIBLIOGRAPHY
Blanchflower, W.J.; Hewitt, S.A.; Kennedy, G. Confirmatory assay for the simultaneous determination of five penicillins in muscle, kidney and milk using liquid chromatography-electrospray mass spectrometry. *Analyst*, **1994**, *119*, 2595–2601 [LC-MS; meat; LOD 5-100 ng/g; extracted cloxacillin, dicloxacillin, oxacillin, penicillin G; nafcillin (IS)]

Orford, C.D.; Perry, D.; Adlard, M.W. The determination of naturally produced penicillins and their biosynthetic precursors using pre-column derivatisation with dansylaziridine. *J.Liq.Chromatogr.*, **1991**, *14*, 2665–2684 [also penicillin G, penicillin K, penicillin X; LOD 1000 ng/mL; fluorescence detection]

Martín, J.; Méndez, R.; Negro, A. Effect of temperature on HPLC separations of penicillins. *J.Liq.Chromatogr.*, **1988**, *11*, 1707–1716 [also amoxicillin, ampicillin, cloxacillin, piperacillin, penicillin G; column temp 15-55]

Mopper, B. Liquid chromatographic determination of penicillin V potassium in tablets and powders for oral solution. *J.Assoc.Off.Anal.Chem.*, **1987**, *70*, 39–41

Möller, J.; Hiddessen, R.; Niehoff, J.; Schügerl, K. On-line high-performance liquid chromatography for monitoring fermentation processes for penicillin production. *Anal.Chim.Acta*, **1986**, *190*, 195–203 [simultaneous impurities]

Moats, W.A. Determination of penicillin G, penicillin V, and cloxacillin in milk by reversed-phase high-performance liquid chromatography. *J.Agric.Food Chem.*, **1983**, *31*, 880–883 [gradient; LOD 5 ppb; extracted cloxacillin, penicillin G]

LeBelle, M.J.; Lauriault, G.; Wilson, W.L. High performance liquid chromatographic analysis of penicillin V solid dosage forms. *J.Liq.Chromatogr.*, **1980**, *3*, 1573–1578

# Pentoxifylline

**Molecular formula:** $C_{13}H_{18}N_4O_3$
**Molecular weight:** 278.3
**CAS Registry No.:** 6493-05-6

## SAMPLE
**Matrix:** blood
**Sample preparation:** 100 μL Plasma + 100 μL 20 μg/mL tetracaine hydrochloride + 8 mL dichloromethane, shake for 20 min, centrifuge at 2500 rpm for 20 min. Remove 7 mL of the organic layer and evaporate it to dryness under nitrogen or at 60°. Dissolve residue in 200 μL mobile phase, inject a 20 μL aliquot.

## HPLC VARIABLES
**Column:** 150 × 6 Shimpack CLS-ODS (Shimadzu)
**Mobile phase:** MeCN:MeOH:0.5 mM phosphoric acid 20:12:68
**Column temperature:** 40
**Flow rate:** 1.5
**Injection volume:** 20
**Detector:** UV 273

## CHROMATOGRAM
**Internal standard:** tetracaine hydrochloride

## KEY WORDS
plasma; rat

## REFERENCE
Lee, C.K.; Uchida, T.; Kitagawa, K.; Yagi, A.; Kim, N.-S.; Goto, S. Skin permeability of various drugs with different lipophilicity. *J.Pharm.Sci.*, **1994**, *83*, 562–565

## SAMPLE
**Matrix:** blood
**Sample preparation:** Condition a 100 mg 1 mL Baker C18 SPE cartridge with 1 mL MeCN and 1 mL 60 mM pH 5.0 phosphate buffer. 500 μL Plasma + 150 ng 7-(2'-chloroethyl)theophylline + 100 μL 200 mM HCl + 400 μL 60 mM pH 5.0 phosphate buffer (final pH 5.3), centrifuge at 10000 g for 3 min. Add the supernatant to the SPE cartridge, wash twice with 1 mL 60 mM pH 5.0 phosphate buffer, dry, elute three times with 200 μL MeCN. Evaporate the eluate at 37° under a stream of air, reconstitute in 100 μL mobile phase, vortex, inject a 20 μL aliquot.

## HPLC VARIABLES
**Column:** 125 × 4 5 μm LiChrospher 100 RP-18
**Mobile phase:** MeCN:dioxan:water 6.5:6.5:87, acidified to pH 3.0 with glacial acetic acid (0.5% v/v)
**Flow rate:** 1
**Injection volume:** 20
**Detector:** UV 275

## CHROMATOGRAM
**Retention time:** 12.5

**Internal standard:** 7-(2'-chloroethyl)theophylline (11.2)
**Limit of detection:** 25 ng/mL

## OTHER SUBSTANCES
**Extracted:** metabolites
**Noninterfering:** caffeine, theobromine, theophylline

## KEY WORDS
plasma; SPE

## REFERENCE
Mancinelli, A.; Pace, S.; Marzo, A.; Martelli, E.A.; Passetti, G. Determination of pentoxifylline and its metabolites in human plasma by high-performance liquid chromatography with solid-phase extraction. *J.Chromatogr.*, **1992**, *575*, 101–107

## SAMPLE
**Matrix:** blood
**Sample preparation:** 100 μL Plasma + 20 μL 172 μM acetophenone + 300 μL MeOH, vortex, centrifuge at 1000 g for 5 min, inject a 50 μL aliquot of the supernatant.

## HPLC VARIABLES
**Guard column:** 5 μm Hisep (Supelco)
**Column:** 150 × 4.6 5 μm Hisep (Supelco)
**Mobile phase:** MeCN:water 10:90
**Injection volume:** 50
**Detector:** UV 280

## CHROMATOGRAM
**Retention time:** 3.3
**Internal standard:** acetophenone (4.5)
**Limit of detection:** 3.95 nM

## KEY WORDS
plasma; rabbit; pharmacokinetics

## REFERENCE
Lockemeyer, M.R.; Smith, C.V. Analysis of pentoxifylline in rabbit plasma using a Hisep high-performance liquid chromatography column. *J.Chromatogr.*, **1990**, *532*, 162–167

## SAMPLE
**Matrix:** blood
**Sample preparation:** 1 mL Plasma + 20 μL 0.5 mM IS in MeCN:water 22:78 + 100 μL 3 M HCl + 5 mL dichloromethane:chloroform 1:1, mix on a rotary mixer at 20 rpm for 10 min, centrifuge at 1000 g for 10 min. Remove the organic layer and evaporate it to dryness under a stream of nitrogen at room temperature. Dissolve the residue in 200 μL mobile phase, vortex, centrifuge at 1000 g for 1 min, inject a 20 μL aliquot.

## HPLC VARIABLES
**Column:** 150 × 4.6 5 μm Zorbax ODS
**Mobile phase:** MeCN:water:acetic acid 22:77.9:0.1
**Flow rate:** 0.75
**Injection volume:** 50
**Detector:** UV 274

## CHROMATOGRAM
**Retention time:** 8

**Internal standard:** 1-(6′-oxoheptyl)-3-dimethylxanthine (15)
**Limit of detection:** 53 nmol/L

## OTHER SUBSTANCES
**Extracted:** metabolites

## KEY WORDS
plasma; also urine using SPE

## REFERENCE
Lambert, W.E.; Yousouf, M.A.; Van Liedekerke, B.M.; De Roose, J.E.; De Leenheer, A.P. Simultaneous determination of pentoxifylline and three metabolites in biological fluids by liquid chromatography. *Clin.Chem.*, **1989**, *35*, 298–301

## SAMPLE
**Matrix:** blood
**Sample preparation:** Condition a 100 mg Baker cyanopropyl SPE cartridge with 2 mL MeOH and 2 mL water. 1 mL Plasma + 100 µL 5 µg/mL IS in water + 2 mL MeCN, mix, centrifuge at 600 g for 20 min. Remove the supernatant and evaporate it to dryness under a stream of nitrogen at 40°. Reconstitute the residue in 1 mL water, add it to the SPE cartridge, wash with 200 µL water, elute with 1 mL MeCN. Evaporate the eluate to dryness under a stream of nitrogen at 45°, reconstitute in 1 mL mobile phase, inject a 20 µL aliquot.

## HPLC VARIABLES
**Guard column:** 30 × 4 10 µm LiChrosorb cyanopropyl
**Column:** 125 × 4 5 µm LiChrosorb cyanopropyl
**Mobile phase:** MeCN:water 1:99
**Flow rate:** 1
**Injection volume:** 20
**Detector:** UV 280

## CHROMATOGRAM
**Retention time:** 8
**Internal standard:** 1-(6-oxoheptyl)-3,7-dimethylxanthine (9.5)
**Limit of detection:** 10 ng/mL

## OTHER SUBSTANCES
**Extracted:** metabolites
**Simultaneous:** caffeine, chlorthalidone, diclofenac, estriol, furosemide, phenacetin, phenobarbital, theobromine, theophylline, triamterene
**Noninterfering:** albuterol, amiloride, atenolol, mexiletine, tiapride

## KEY WORDS
plasma; SPE

## REFERENCE
Musch, G.; Hamoir, T.; Massart, D.L. Determination of pentoxifylline and its 5-hydroxy metabolite in human plasma by solid-phase extraction and high-performance liquid chromatography with ultraviolet detection. *J.Chromatogr.*, **1989**, *495*, 215–226

## SAMPLE
**Matrix:** blood
**Sample preparation:** 250 µL Whole blood + 4 mL MeCN, vortex for 30 s, centrifuge at 3000 g for 10 min. Remove a 3.7 mL aliquot of the supernatant and add it to 35 µL 2 µg/mL IS in MeOH, vortex for 10 s, filter (0.45 µm PTFE), evaporate the filtrate to dry-

ness under a stream of nitrogen at 47°, reconstitute the residue in 2 mL hexane, vortex for 15 s, add 300 μL water, vortex for 30 s, centrifuge at 1250 g for 15 min, inject a 250 μL aliquot of the aqueous phase.

## HPLC VARIABLES
**Column:** 250 × 4.6 5 μm LC-8-DB (Supelco)
**Mobile phase:** MeCN:water 24:76
**Flow rate:** 1
**Injection volume:** 250
**Detector:** UV 280

## CHROMATOGRAM
**Retention time:** 6.4
**Internal standard:** 3-isobutyl-1-methylxanthine (7.6)
**Limit of detection:** 5 ng/mL

## OTHER SUBSTANCES
**Extracted:** metabolites
**Simultaneous:** antipyrine, caffeine, 8-chlorotheophylline, dimethyluric acid, dyphylline, hypoxanthine, methyluric acid, methylxanthine, pentifylline, phenacetin, 8-phenyltheophylline, theobromine, uric acid, xanthine
**Interfering:** pentobarbital

## KEY WORDS
whole blood; human; rat

## REFERENCE
Grasela, D.M.; Rocci, M.L., Jr. High-performance liquid chromatographic analysis of pentoxifylline and 1-(5'-hydroxyhexyl)-3,7-dimethylxanthine in whole blood. *J.Chromatogr.*, **1987**, *419*, 368–374

## SAMPLE
**Matrix:** blood
**Sample preparation:** 250 μL Plasma + 20 μL 20 μg/mL phenacetin in mobile phase, vortex for 15 s, add 2.5 mL MeCN, vortex for 15 s, centrifuge at 1980 g for 10 min. Remove the supernatant and filter (PTFE) it, evaporate the filtrate to dryness under a stream of nitrogen at 70°, reconstitute the residue in 125 μL water, vortex, inject a 100 μL aliquot.

## HPLC VARIABLES
**Column:** 5 μm μBondapak C18
**Mobile phase:** MeCN:water 24:76
**Flow rate:** 1
**Injection volume:** 100
**Detector:** UV 280

## CHROMATOGRAM
**Retention time:** 8.0
**Internal standard:** phenacetin (14.7)
**Limit of quantitation:** 25 ng/mL

## OTHER SUBSTANCES
**Extracted:** metabolites
**Simultaneous:** aminopyrine, aspirin, caffeine, carbidopa, β-hydroxyethyltheophylline, β-hydroxypropyltheophylline, methicillin, methyldopa, nafcillin, nifedipine, penicillamine, penicillin V, penicillin G, phthalic acid, probenecid, salicylic acid, theobromine, theophylline

## KEY WORDS
plasma; rat; pharmacokinetics

## REFERENCE
Luke, D.R.; Rocci, M.L., Jr. Determination of pentoxifylline and a major metabolite, 3,7-dimethyl-1-(5'-hydroxyhexyl)xanthine, by high-performance liquid chromatography. *J.Chromatogr.*, **1986**, *374*, 191–195

## SAMPLE
**Matrix:** blood
**Sample preparation:** Inject a 500 μL aliquot of plasma onto column A and elute to waste with mobile phase A, after 4.5 min backflush the contents of column A onto column B with mobile phase B, monitor the effluent from column B.

## HPLC VARIABLES
**Column:** A 5 × 4.6 25-40 μm Lichroprep RP 2; B 60 × 4.6 7 μm Nucleosil phenyl
**Mobile phase:** A 5 mM pH 7.1 $Na_2HPO_4$; B MeCN:5 mM pH 6.7 $Na_2HPO_4$ 23:77
**Column temperature:** 45
**Flow rate:** A 1.8; B 0.5
**Injection volume:** 500
**Detector:** UV 273

## CHROMATOGRAM
**Retention time:** 5.1
**Limit of detection:** 5 ng/mL

## OTHER SUBSTANCES
**Extracted:** metabolites

## KEY WORDS
plasma; column-switching

## REFERENCE
von Stetten, O.; Arnold, P.; Aumann, M.; Guserle, R. Direct measurement of pentoxifylline and its hydroxymetabolite from plasma using HPLC with column switching technique. *Chromatographia*, **1984**, *19*, 415–417

## SAMPLE
**Matrix:** solutions
**Sample preparation:** Inject a 20 μL aliquot.

## HPLC VARIABLES
**Column:** 150 × 6 5 μm 3-aminopropylsilyl silica gel with the amino group derivatized with 1,8-naphthalic anhydride (Bunseki Kagaku 1993, 42, 817)
**Mobile phase:** MeOH:buffer 50:50 (Prepare buffer by dissolving 6.183 g boric acid and 1.461 g NaCl in 500 mL water, adjust pH to 6.4 with sodium borate solution.)
**Column temperature:** 30
**Flow rate:** 1
**Injection volume:** 20
**Detector:** UV 270

## CHROMATOGRAM
**Retention time:** 20
**Internal standard:** 1,3-dimethyl-7-(2-hydroxyethyl)xanthine (12)

## OTHER SUBSTANCES
**Simultaneous:** caffeine, hypoxanthine, propentofylline, theobromine, theophylline, uric acid, xanthine

## REFERENCE
Nakashima, K.; Inoue, K.; Mayahara, K.; Kuroda, N.; Hamachi, Y.; Akiyama, S. Use of 3-(1,8-naphthal-imido)propyl-modified silyl silica gel as a stationary phase for the high-performance liquid chromatographic separation of purine derivatives. *J.Chromatogr.A*, **1996**, *722*, 107–113

## SAMPLE
**Matrix:** solutions

## HPLC VARIABLES
**Column:** 250 × 4.6 5 μm Supelcosil LC-DP (A) or 250 × 4 5 μm LiChrospher 100 RP-8 (B)
**Mobile phase:** MeCN:0.025% phosphoric acid:buffer 25:10:5 (A) or 60:25:15 (B) (Buffer was 9 mL concentrated phosphoric acid and 10 mL triethylamine in 900 mL water, adjust pH to 3.4 with dilute phosphoric acid, make up to 1 L.)
**Flow rate:** 0.6
**Injection volume:** 25
**Detector:** UV 229

## CHROMATOGRAM
**Retention time:** 5.01 (A), 4.15 (B)

## OTHER SUBSTANCES
**Also analyzed:** acebutolol, acepromazine, acetaminophen, acetazolamide, acetophenazine, albuterol, alprazolam, amitriptyline, amobarbital, amoxapine, antipyrine, atenolol, atropine, azatadine, baclofen, benzocaine, bromocriptine, brompheniramine, brotizolam, bupivacaine, buspirone, butabarbital, butalbital, caffeine, carbamazepine, cetirizine, chlorcyclizine, chlordiazepoxide, chlormezanone, chloroquine, chlorpheniramine, chlorpromazine, chlorpropamide, chlorprothixene, chlorthalidone, chlorzoxazone, cimetidine, cisapride, clomipramine, clonazepam, clonidine, clozapine, cocaine, codeine, colchicine, cyclizine, cyclobenzaprine, dantrolene, desipramine, diazepam, diclofenac, diflunisal, diltiazem, diphenhydramine, diphenidol, diphenoxylate, dipyridamole, disopyramide, dobutamine, doxapram, doxepin, droperidol, encainide, ethidium bromide, ethopropazine, fenoprofen, fentanyl, flavoxate, fluoxetine, fluphenazine, flurazepam, flurbiprofen, fluvoxamine, furosemide, glutethimide, glyburide, guaifenesin, haloperidol, homatropine, hydralazine, hydrochlorothiazide, hydrocodone, hydromorphone, hydroxychloroquine, hydroxyzine, ibuprofen, imipramine, indomethacin, ketoconazole, ketoprofen, ketorolac, labetalol, levorphanol, lidocaine, loratadine, lorazepam, lovastatin, loxapine, mazindol, mefenamic acid, meperidine, mephenytoin, mepivacaine, mesoridazine, metaproterenol, metformin, methadone, methdilazine, methocarbamol, methotrexate, methotrimeprazine, methoxamine, methyldopa, methylphenidate, metoclopramide, metolazone, metoprolol, metronidazole, midazolam, moclobemide, morphine, nadolol, nalbuphine, naloxone, naphazoline, naproxen, nifedipine, nizatidine, norepinephrine, nortriptyline, oxazepam, oxycodone, oxymetazoline, paroxetine, pemoline, pentazocine, pentobarbital, perphenazine, pheniramine, phenobarbital, phenol, phenolphthalein, phentolamine, phenylbutazone, phenyltoloxamine, phenytoin, pimozide, pindolol, piroxicam, pramoxine, prazepam, prazosin, probenecid, procainamide, procaine, prochlorperazine, procyclidine, promazine, promethazine, propafenone, propantheline, propiomazine, propofol, propranolol, protriptyline, quazepam, quinidine, quinine, racemethorphan, ranitidine, remoxipride, risperidone, salicylic acid, scopolamine, secobarbital, sertraline, sotalol, spironolactone, sulfinpyrazone, sulindac, temazepam, terbutaline, terfenadine, tetracaine, theophylline, thiethylperazine, thiopental, thioridazine, thiothixene, timolol, tocainide, tolbutamide, tolmetin, trazodone, triamterene, triazolam, trifluoperazine, triflupromazine, trimeprazine, trimethoprim, trimipramine, verapamil, warfarin, xylometazoline, yohimbine, zopiclone

## KEY WORDS
some details of plasma extraction

## REFERENCE
Koves, E.M. Use of high-performance liquid chromatography-diode array detection in forensic toxicology. *J.Chromatogr.A*, **1995**, *692*, 103–119

## SAMPLE
**Matrix:** urine
**Sample preparation:** 1 mL Urine 100 μL 1 mg/mL IS in 10 mM NaOH, mix, add 6 mL dichloromethane, add 1 mL 1 M HCl, extract using a mechanical rotary inversion mixer at 20 rpm for 15 min, centrifuge at 2000 g for 5 min. Remove the organic phase and evaporate it to dryness under a stream of nitrogen at 40°. Reconstitute the residue in 2 mL mobile phase, inject an aliquot.

## HPLC VARIABLES
**Column:** 150 × 3 5 μm Spherisorb-ODS 1
**Mobile phase:** MeOH:20 mM orthophosphoric acid 1:2.5, adjusted to pH 4 with 6 M NaOH
**Flow rate:** 1
**Detector:** UV 274

## CHROMATOGRAM
**Retention time:** 18
**Internal standard:** 1-(3'-carboxypropyl)-3-methyl-7-propylxanthine (11.6)

## OTHER SUBSTANCES
**Extracted:** metabolites

## KEY WORDS
pentoxifylline not usually seen in urine

## REFERENCE
Bryce, T.A.; Burrows, J.L.; Jolley, K.W. Determination of 1-(3'-carboxypropyl)-3,7-dimethylxanthine and 1-(4'-carboxybutyl)-3,7-dimethylxanthine, two major metabolites of oxpentifylline, in urine by high-performance liquid chromatography. *J.Chromatogr.*, **1985**, *344*, 397–402

◆⎯⎯⎯⎯⎯⎯⎯⎯⎯⎯⎯⎯◆⎯⎯⎯⎯⎯⎯⎯⎯⎯⎯⎯⎯◆

## ANNOTATED BIBLIOGRAPHY
Ostrovska, V.; Pechova, A.; Kovacova, D.; Svobodova, X.; Kusala, S. [Determination of pentoxifylline and its major metabolites in human plasma using HPLC]. *Cesk.Farm.*, **1990**, *39*, 158–160

Garnier-Moiroux, A.; Poirier, J.M.; Cheymol, G.; Jaillon, P. High-performance liquid chromatographic determination of pentoxifylline and its hydroxy metabolite in human plasma. *J.Chromatogr.*, **1987**, *416*, 183–188

Rieck, W.; Platt, D. Determination of 3,7-dimethyl-1-(5-oxohexyl)-xanthine (pentoxifylline) and its 3,7-dimethyl-1-(5-hydroxyhexyl)-xanthine metabolite in the plasma of patients with multiple diseases using high-performance liquid chromatography. *J.Chromatogr.*, **1984**, *305*, 419–427

Smith, R.V.; Yang, S.K.; Davis, P.J.; Bauza, M.T. Determination of pentoxifylline and its major metabolites in microbial extracts by thin-layer and high-performance liquid chromatography. *J.Chromatogr.*, **1983**, *281*, 281–287

Chivers, D.A.; Birkett, D.J.; Miners, J.O. Simultaneous determination of pentoxifylline and its hydroxy metabolite in plasma by high-performance liquid chromatography. *J.Chromatogr.*, **1981**, *225*, 261–265

# Phenylpropanolamine

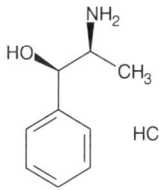

**Molecular formula:** C$_9$H$_{14}$ClNO
**Molecular weight:** 151.2
**CAS Registry No.:** 154-41-6 (phenylpropanolamine hydrochloride),
14838-15-4 (phenylpropanolamine)

## SAMPLE

**Matrix:** blood, urine
**Sample preparation:** Plasma. 1 mL Plasma + 50 μL 1 M NaOH + 4 mL diethyl ether, extract, repeat extraction. Combine organic layers and evaporate them to dryness at 40° under a stream of nitrogen. Dissolve the residue in 300 μL mobile phase A, inject a 200 μL aliquot. Urine. 100 μL Urine + 1 mL 100 mM NaOH + 4 mL diethyl ether, extract, repeat extraction. Combine organic layers and evaporate them to dryness at 40° under a stream of nitrogen. Dissolve the residue in 300 μL mobile phase A, inject a 100 μL aliquot. Inject onto column A and elute with mobile phase A, at an appropriate time (typically, 6.5 min) the effluent from column A was eluted onto column B using mobile phase A, when all compounds of interest had passed onto column B (typically, 8.7 min) elution of column B was continued with mobile phase B and the effluent from column B was monitored.

## HPLC VARIABLES

**Column:** A 70 × 4.6 5 μm A type YMC ODS (Yamamura); B 100 × 4.6 5 μm A type YMC ODS (Yamamura)
**Mobile phase:** A MeCN:buffer 5:95 containing 5 mM sodium butanesulfonate; B MeCN:buffer 5:95 (Buffer was 20 mM KH$_2$PO$_4$ adjusted to pH 3.5 with 10% orthophosphoric acid.)
**Column temperature:** 40
**Flow rate:** 1
**Injection volume:** 100-200
**Detector:** UV 205

## CHROMATOGRAM

**Retention time:** 13
**Limit of detection:** 0.4 ng/mL (plasma); 8 ng/mL (urine)

## KEY WORDS

plasma; column-switching; heart cut

## REFERENCE

Yamashita, K.; Motohashi, M.; Yashiki, T. High-performance liquid chromatographic determination of phenylpropanolamine in human plasma and urine, using column switching combined with ion-pair chromatography. *J.Chromatogr.*, **1990**, *527*, 103–114

## SAMPLE

**Matrix:** bulk
**Sample preparation:** Prepare a 10 mg/mL solution in 500 mM sodium bicarbonate solution, extract a 10 mL aliquot twice with 15 mL portions of dichloromethane. Combine the extracts and add 10 μL phenylisothiocyanate, evaporate to dryness under a stream of air, reconstitute with 10 mL MeOH, inject a 10 μL aliquot.

## HPLC VARIABLES

**Guard column:** 70 × 2.1 CO:PELL ODS
**Column:** 300 × 3.9 μBondapak C18

**Mobile phase:** MeOH:water:acetic acid 45:54:1
**Flow rate:** 2
**Injection volume:** 10
**Detector:** UV 254

## CHROMATOGRAM
**Retention time:** 10

## OTHER SUBSTANCES
**Simultaneous:** ephedrine, lidocaine, pseudoephedrine

## KEY WORDS
derivatization

## REFERENCE
Noggle, F.T., Jr.; Clark, C.R. Liquid chromatographic analysis of samples containing cocaine, local anesthetics, and other amines. *J.Assoc.Off.Anal.Chem.*, **1983**, *66*, 151–157

## SAMPLE
**Matrix:** formulations
**Sample preparation:** Grind ten tablets to a fine powder, add 15 mL MeCN:buffer 50:50, sonicate for 15 min, make up to 20 mL with MeCN:buffer 50:50, filter, dilute tenfold, inject a 10 μL aliquot. (Buffer was 50 mM sodium perchlorate adjusted to pH 3.0 with 70% perchloric acid.)

## HPLC VARIABLES
**Column:** 250 × 4 10 μm Lichrosorb RP-18
**Mobile phase:** Gradient. MeCN:buffer from 15:85 to 95:5 over 20 min, re-equilibrate at initial conditions for 10 min. (Buffer was 50 mM sodium perchlorate adjusted to pH 3.0 with 70% perchloric acid.)
**Column temperature:** 25
**Flow rate:** 2
**Injection volume:** 10
**Detector:** UV 254

## CHROMATOGRAM
**Retention time:** 2.5

## OTHER SUBSTANCES
**Simultaneous:** buzepide (UV 220), clocinizine

## KEY WORDS
tablets

## REFERENCE
Cavazzutti, G.; Gagliardi, L.; De Orsi, D.; Tonelli, D. Simultaneous determination of buzepide, phenyl-propanolamine, and clocinizine on pharmaceutical preparations by ion-pair reversed-phase HPLC. *J.Liq.Chromatogr.*, **1995**, *18*, 227–234

## SAMPLE
**Matrix:** formulations
**Sample preparation:** Finely powder half a tablet, add 9 mL mobile phase, sonicate for 20 min, make up to 10 mL with mobile phase, filter (Whatman type 40 and 0.2 μm Millipore), inject an aliquot of the filtrate.

## HPLC VARIABLES
**Column:** 250 × 4 5 µm LiChrospher 100 CN
**Mobile phase:** MeCN:THF:buffer 7:6:87 (Buffer was 0.8% acetic acid containing 5 mM sodium hexanesulfonate, 10 mM di-n-butylamine, and 0.12% phosphoric acid, pH 3.3.)
**Flow rate:** 1
**Injection volume:** 20
**Detector:** UV 260

## CHROMATOGRAM
**Retention time:** 3.5
**Limit of detection:** 15.1 µg/mL

## OTHER SUBSTANCES
**Simultaneous:** acetaminophen (UV 310), caffeine (UV 298), chlorpheniramine (UV 265), guaifenesin (glycerylguaicolate) (UV 284)

## KEY WORDS
tablets

## REFERENCE
Indrayanto, G.; Sunarto, A.; Adriani, Y. Simultaneous assay of phenylpropanolamine hydrochloride, caffeine, paracetamol, glycerylguaiacolate and chlorpheniramine in Silabat™ tablet using HPLC with diode array detection. *J.Pharm.Biomed.Anal.*, **1995**, *13*, 1555–1559

## SAMPLE
**Matrix:** formulations
**Sample preparation:** Dilute 10 mL to 50 mL with water, inject an aliquot.

## HPLC VARIABLES
**Column:** 250 × 4.6 Whatman 10 µm PXS SCX
**Mobile phase:** MeOH:100 mM $(NH_4)H_2PO_4$, apparent pH 6.2
**Column temperature:** 40
**Flow rate:** 2
**Injection volume:** 20
**Detector:** UV 263

## CHROMATOGRAM
**Retention time:** 3.12

## OTHER SUBSTANCES
**Simultaneous:** dextromethorphan

## KEY WORDS
liquid formulations; stability-indicating

## REFERENCE
Wilson, T.D.; Jump, W.G.; Neumann, W.C.; San Martin, T. Validation of improved methods for high-performance liquid chromatographic determination of phenylpropanolamine, dextromethorphan, guaifenesin and sodium benzoate in a cough-cold formulation. *J.Chromatogr.*, **1993**, *641*, 241–248

## SAMPLE
**Matrix:** formulations
**Sample preparation:** Tablets. One tablet + 50 mL MeOH, sonicate, make up to 100 mL with MeOH, centrifuge for 15 min. Remove 1 mL supernatant, make up to 10 mL with mobile phase, inject a 50 µL aliquot. Drops. Dilute drops with the mobile phase so that

the concentration of phenylpropanolamine hydrochloride is 50 μg/mL, inject a 50 μL aliquot.

## HPLC VARIABLES
**Column:** 100 × 4.6 Cyclobond I (Advanced Separation Technologies)
**Mobile phase:** MeOH:50 mM NaH$_2$PO$_4$ adjusted to pH 7.0 with 0.1 M NaOH 30:70
**Column temperature:** 35
**Flow rate:** 1.5
**Injection volume:** 50
**Detector:** UV 254

## CHROMATOGRAM
**Retention time:** 1.8

## OTHER SUBSTANCES
**Simultaneous:** pheniramine, pyrilamine (mepyramine)

## KEY WORDS
tablets; drops

## REFERENCE
el-Gizawy, S.M.; Ahmed, A. High-performance liquid chromatographic determination of mepyramine maleate, pheniramine maleate and phenylpropanolamine hydrochloride in tablets and drops. *Analyst*, **1987**, *112*, 867–869

## SAMPLE
**Matrix:** formulations
**Sample preparation:** Injections. Dilute 1.5 mL of a 20 mg/mL injection to 100 mL with water, remove a 10 mL aliquot and add it to 8 mL 2% phenylpropanolamine hydrochloride, make up to 100 mL with water, inject a 20 μL aliquot. Tablets. Grind tablets to a fine powder, weigh out amount equivalent to about 10 mg hydralazine, mix thoroughly with 2 mL 500 mM HCl, make up to 100 mL with water, shake for 2-3 min, filter, discard first 15 mL. 15 mL Filtrate + 4 mL 2% phenylpropanolamine hydrochloride, make up to 50 mL with water, inject a 20 μL aliquot.

## HPLC VARIABLES
**Column:** 300 × 3.9 μBondapak C18
**Mobile phase:** MeOH:15 mM KH$_2$PO$_4$:glacial acetic acid 2:97.9:0.1
**Flow rate:** 3
**Injection volume:** 20
**Detector:** UV 256

## CHROMATOGRAM
**Retention time:** 3
**Internal standard:** phenylpropanolamine

## OTHER SUBSTANCES
**Simultaneous:** hydralazine, hydrochlorothiazide

## KEY WORDS
injections; tablets; phenylpropanolamine is IS

## REFERENCE
Das Gupta, V. Quantitation of hydralazine hydrochloride in pharmaceutical dosage forms using high-performance liquid chromatography. *J.Liq.Chromatogr.*, **1985**, *8*, 2497–2509

## SAMPLE
**Matrix:** formulations
**Sample preparation:** Grind tablets to a fine powder and weigh out powder equivalent to about 80 mg metronidazole, add 90 mL water, heat to 60° with stirring, cool to room temperature, make up to 100 mL with water, filter, reject first 20 mL of filtrate. Mix 12.5 mL filtrate with 7 mL 10 mg/mL phenylpropanolamine hydrochloride in water, make up to 50 mL with water, inject a 20 μL aliquot.

## HPLC VARIABLES
**Column:** 300 × 4 10 μm μBondapak phenyl
**Mobile phase:** 20 mM $KH_2PO_4$, pH 4.2
**Flow rate:** 2.5
**Injection volume:** 20
**Detector:** UV 254

## CHROMATOGRAM
**Retention time:** 3.5
**Internal standard:** phenylpropanolamine hydrochloride

## OTHER SUBSTANCES
**Simultaneous:** metronidazole
**Noninterfering:** excipients, mannitol

## KEY WORDS
tablets; stability-indicating; phenylpropanolamine is IS

## REFERENCE
Das Gupta, V. Quantitation of metronidazole in pharmaceutical dosage forms using high-performance liquid chromatography. *J.Pharm.Sci.*, **1984**, *73*, 1331–1333

## SAMPLE
**Matrix:** formulations
**Sample preparation:** Leach 200 or 300 mg ground capsule or tablet with water or mobile phase and dilute to 50 mL, sonicate for 5 min, centrifuge at 2500 rpm for 5 min, inject an aliquot. Dilute 4-25 mL of liquid formulations to 250 mL with water, inject an aliquot.

## HPLC VARIABLES
**Column:** Partisil-10 C8
**Mobile phase:** MeOH:MeCN:water:PIC-B5 50:170:755:25 (PIC-B5 (Waters) is 200 mM sodium pentanesulfonate in glacial acetic acid.)
**Flow rate:** 2
**Injection volume:** 20
**Detector:** UV 254

## CHROMATOGRAM
**Retention time:** 4.5

## OTHER SUBSTANCES
**Simultaneous:** impurities, degradation products, benzoic acid, guaifenesin, phenylephrine

## KEY WORDS
tablets; capsules; liquid formulations; stability-indicating

## REFERENCE

Schieffer, G.W.; Smith, W.O.; Lubey, G.S.; Newby, D.G. Determination of the structure of a synthetic impurity in guaifenesin: modification of a high-performance liquid chromatographic method for phenylephrine hydrochloride, phenylpropanolamine hydrochloride, guaifenesin, and sodium benzoate in dosage forms. *J.Pharm.Sci.*, **1984**, *73*, 1856–1858

## SAMPLE

**Matrix:** formulations
**Sample preparation:** Capsules and Tablets. Leach 1 g of ground capsule or tablet with 250 mL 0.4 mg/mL 2,5-dihydroxybenzoic acid in water, sonicate for 10 min, centrifuge at 2500 rpm for 5 min, inject an aliquot. Liquid formulations. Dilute 4-25 mL of the formulation to 250 mL with 0.4 mg/mL 2,5-dihydroxybenzoic acid in water, inject an aliquot.

## HPLC VARIABLES

**Column:** 250 × 4.6 Partisil 10 C8
**Mobile phase:** MeOH:water:PIC-B5 300:675:25 (PIC-B5 (Waters) is 200 mM sodium pentanesulfonate in glacial acetic acid.)
**Flow rate:** 2
**Injection volume:** 20
**Detector:** UV 254

## CHROMATOGRAM

**Retention time:** 6
**Internal standard:** 2,5-dihydroxybenzoic acid (4.5)

## OTHER SUBSTANCES

**Simultaneous:** impurities, degradation products, guaifenesin, phenylephrine

## KEY WORDS

capsules; tablets; liquid formulations; stability-indicating

## REFERENCE

Schieffer, G.W.; Hughes, D.E. Simultaneous stability-indicating determination of phenylephrine hydrochloride, phenylpropanolamine hydrochloride, and guaifenesin in dosage forms by reversed-phase paired-ion high-performance liquid chromatography. *J.Pharm.Sci.*, **1983**, *72*, 55–59

## SAMPLE

**Matrix:** formulations
**Sample preparation:** Crush 10 tablets, add 250 mL 50 mM HCl in EtOH:water 50:50, heat for 15 min on a steam bath, shake mechanically for 2 h, filter (glass fiber GF/A, Whatman), inject a 30 µL aliquot of the filtrate.

## HPLC VARIABLES

**Column:** 250 × 4.6 10 µm Partisil-10-ODS
**Mobile phase:** MeCN:buffer 50:50 (Buffer was 2.85 mM ethylenediamine sulfate adjusted to pH 7.44 ± 0.02 with 1 M ammonium hydroxide.)
**Flow rate:** 3.8
**Injection volume:** 30
**Detector:** UV 216.5

## CHROMATOGRAM

**Retention time:** 4

## OTHER SUBSTANCES

**Simultaneous:** aposcopolamine, methscopolamine, pheniramine, pyrilamine, tropic acid

## KEY WORDS
tablets

## REFERENCE
Heidemann, D.R. High-pressure liquid chromatographic determination of methscopolamine nitrate, phenylpropanolamine hydrochloride, pyrilamine maleate, and pheniramine maleate in tablets. *J.Pharm.Sci.*, **1981**, *70*, 820–822

## SAMPLE
**Matrix:** solutions

## HPLC VARIABLES
**Guard column:** 30 × 2.1 Spheri-5 RP-8
**Column:** 220 × 2.1 Spheri-5 RP-8
**Mobile phase:** Gradient. A was 0.08% diethylamine and 0.09% phosphoric acid in water, pH 2.3. B was MeCN:water 90:10 containing 0.08% diethylamine and 0.09% phosphoric acid. A:B 95:5 for 2 min, to 0:100 over 15 min (?), maintain at 0:100 for 5 min.
**Column temperature:** 50
**Flow rate:** 0.5
**Detector:** UV 200

## CHROMATOGRAM
**Retention time:** 4.5

## OTHER SUBSTANCES
**Simultaneous:** amphetamine, diethylpropion, ephedrine, fenfluramine, methamphetamine, phentermine
**Also analyzed:** amitriptyline, chlordiazepoxide, chlorpromazine, desalkylflurazepam, desipramine, desmethyldoxepin, diazepam, doxepin, flurazepam, imipramine, mesoridazine, norchlordiazepoxide, nordiazepam, nortriptyline, oxazepam, prazepam, promazine, thioridazine, thiothixene, trifluoperazine

## REFERENCE
*Rainin Catalog Ci-94, 1994-5, Rainin Instrument Co., Woburn MA, p. 7.24*

## SAMPLE
**Matrix:** solutions

## HPLC VARIABLES
**Column:** 250 × 4.6 Zorbax RX
**Mobile phase:** Gradient. A was 10 mL concentrated orthophosphoric acid and 7 mL triethylamine in 1 L water. B was 10 mL concentrated orthophosphoric acid and 7 mL triethylamine in 200 mL water, make up to 1 L with MeCN. A:B from 100:0 to 0:100 over 30 min, maintain at 0:100 for 5 min.
**Column temperature:** 30
**Flow rate:** 2
**Detector:** UV 210

## OTHER SUBSTANCES
**Also analyzed:** acepromazine, acetaminophen, acetophenazine, albuterol, aminophylline, amitriptyline, amobarbital, amoxapine, amphetamine, amylocaine, antipyrine, aprobarbital, aspirin, atenolol, atropine, avermectin, barbital, benzocaine, benzoic acid, benzotropine, benzphetamine, berberine, bibucaine, bromazepan, brompheniramine, buprenorphine, buspirone, butabarbital, butacaine, butethal, caffeine, carbamazepine, carbromal, chloramphenicol, chlordiazepoxide, chloroquine, chlorothiazide, chloroxylenol, chlorphe-

nesin, chlorpheniramine, chlorpromazine, chlorpropamide, chlortetracycline, cimetidine, cinchonidine, cinchonine, clenbuterol, clonazepam, clonixin, clorazepate, cocaine, codeine, colchicine, cortisone, coumarin, cyclazocine, cyclobenzaprine, cyclothiazide, cyheptamide, cymarin, danazol, danthron, dapsone, debrisoquine, desipramine, dexamethasone, dextromethorphan, dextropropoxyphene, diamorphine, diazepam, diclofenac, diethylpropion, diethylstilbestrol, diflunisal, digitoxin, digoxin, diltiazem, diphenhydramine, diphenoxylate, diprenorphine, dipyrone, disulfiram, dopamine, doxapram, doxepin, dronabinol, ephedrine, epinephrine, epinine, estradiol, estriol, estrone, ethacrynic acid, ethosuximide, etonitazene, etorphine, eugenol, famotidine, fenbendazole, fencamfamine, fenoprofen, fenproporex, fentanyl, flubendazole, flufenamic acid, flunitrazepam, 5-fluorouracil, fluoxymesterone, fluphenazine, furosemide, gentisic acid, gitoxigenin, glipizide, glunixin, glutethimide, glybenclamide, guaiacol, halazepam, haloperidol, hydrochlorothiazide, hydrocodone, hydrocortisone, hydromorphone, hydroxyquinoline, ibogaine, ibuprofen, iminostilbene, imipramine, indomethacin, isocarbostyril, isocarboxazid, isoniazid, isoproterenol, isoxsuprine, ivermectin, ketamine, ketoprofen, kynurenic acid, levorphanol, lidocaine, lorazepam, lormetazepam, loxapine, mazindol, mebendazole, meclizine, meclofenamic acid, medazepam, mefenamic acid, megestrol, mepacrine, meperidine, mephentermine, mephenytoin, mephesin, mephobarbital, mepivacaine, mescaline, mesoridazine, methadone, methamphetamine, methapyrilene, methaqualone, methazolamide, methocarbamol, methoxamine, methsuximide, methyl salicylate, methyldopa, methyldopamine, methylphenidate, methylprednisolone, methyltestosterone, methyprylon, metoprolol, mibolerone, morphine, nadolol, nalorphine, naloxone, naltrexone, naphazoline, naproxen, nefopam, niacinamide, nicotine, nicotinic acid, nifedipine, niflumic acid, nitrazepam, norepinephrine, nortriptyline, noscapine, nylidrin, oxazepam, oxycodone, oxymorphone, oxyphenbutazone, oxytetracycline, papaverine, pargyline, pemoline, pentazocine, pentobarbital, persantine, phenacetin, phenazocine, phenazopyridine, phencyclidine, phendimetrazine, phenelzine, pheniramine, phenobarbital, phenothiazine, phensuximide, phentermine, phenylbutazone, phenylephrine, piperocaine, prazepam, prednisolone, primidone, probenecid, progesterone, propiomazine, propranolol, propylparaben, pseudoephedrine, puromycin, pyrilamine, pyrithyldione, quazepam, quinaldic acid, quinidine, quinine, ranitidine, recinnamine, reserpine, resorcinol, saccharin, albuterol, salicylamide, salicylic acid, scopolamine, scopoletin, secobarbital, strychnine, sulfacetamide, sulfadiazine, sulfadimethoxine, sulfaethidole, sulfamerazine, sulfamethazine, sulfamethoxizole, sulfanilamide, sulfapyridine, sulfasoxizole, sulindac, tamoxifen, temazepam, testosterone, tetracaine, tetracycline, tetramisole, thebaine, theobromine, theophylline, thiabendazole, thiamine, thiamylal, thiobarbituric acid, thioridazine, thiosalicylic acid, thiothixene, thymol, tolazamide, tolazoline, tobutamide, tolmetin, tranylcypromine, triamcinolone, tribenzylamine, trichloromethiazide, trifluoperazine, trihexyphenidyl, trimethoprim, tripelennamine, triprolidine, tropacocaine, tyramine, verapamil, vincamine, warfarin, yohimbine, zoxazolamine

**REFERENCE**
Hill, D.W.; Kind, A.J. Reversed-phase solvent-gradient HPLC retention indexes of drugs. *J.Anal.Toxicol.*, **1994**, *18*, 233–242

**SAMPLE**
**Matrix:** solutions

**HPLC VARIABLES**
**Column:** 150 × 4.6 Supelcosil LC-ABZ
**Mobile phase:** MeCN:25 mM pH 6.9 potassium phosphate buffer 35:65
**Flow rate:** 1.5
**Injection volume:** 25
**Detector:** UV 254

**CHROMATOGRAM**
**Retention time:** 1.312

## OTHER SUBSTANCES
**Also analyzed:** 6-acetylmorphine, amiloride, amphetamine, benzocaine, benzoylecgonine, caffeine, cocaine, codeine, doxylamine, fluoxetine, glutethimide, hexobarbital, hypoxanthine, levorphanol, LSD, meperidine, mephobarbital, methadone, methylphenidate, methyprylon, N-norcodeine, oxazepam, oxycodone, prilocaine, procaine, terfenadine

## REFERENCE
Ascah, T.L. Improved separations of alkaloid drugs and other substances of abuse using Supelcosil LC-ABZ column. *Supelco Reporter*, **1993**, *12(3)*, 18–21

## SAMPLE
**Matrix:** solutions
**Sample preparation:** Prepare a 0.5 mg/mL solution in water, inject a 5 µL aliquot.

## HPLC VARIABLES
**Column:** 250 × 4.6 Zorbax RX
**Mobile phase:** Gradient. A was 150 mM phosphoric acid and 50 mM triethylamine. B was MeCN:water 80:20 containing 150 mM phosphoric acid and 50 mM triethylamine. A:B 100:0 for 2.2 min then to 0:100 over 30 min.
**Column temperature:** 30
**Flow rate:** 2
**Injection volume:** 5
**Detector:** UV 210

## CHROMATOGRAM
**Retention time:** 4.9

## OTHER SUBSTANCES
**Simultaneous:** acetaminophen, aprobarbital, butabarbital, chlordiazepoxide, chloroxylenol, chlorpromazine, clenbuterol, cortisone, danazol, diflunisal, doxapram, estrone, fluoxymesterone, mefenamic acid, methyltestosterone, nicotine, oxazepam, phentermine, progesterone, sulfamethazine, sulfanilamide, testosterone, testosterone propionate, tranylcypromine, tripelennamine

## REFERENCE
Hill, D.W.; Kind, A.J. The effects of type B silica and triethylamine on the retention of drugs in silica based reverse phase high performance chromatography. *J.Liq.Chromatogr.*, **1993**, *16*, 3941–3964

## SAMPLE
**Matrix:** solutions
**Sample preparation:** Prepare a 1.5 mg/mL solution, inject a 10 µL aliquot.

## HPLC VARIABLES
**Guard column:** Supelguard LC-8-DB (Supelco)
**Column:** 50 × 4.6 Supelcosil LC-8-DB
**Mobile phase:** MeCN:buffer 10:90 containing 0.02% triethylamine (Buffer was $KH_2PO_4$ adjusted to pH 2.0 with phosphoric acid.)
**Column temperature:** 35
**Flow rate:** 2
**Injection volume:** 10
**Detector:** UV 254

## CHROMATOGRAM
**Retention time:** 1

## OTHER SUBSTANCES
**Simultaneous:** chlorpheniramine, methscopolamine, pseudoephedrine, triprolidine

## REFERENCE
*Supelco Catalog, Supelco, Inc., Bellefonte MA*, **1992**, p. 179

## SAMPLE
**Matrix:** solutions
**Sample preparation:** Dissolve in MeOH:water 1:1 at a concentration of 50 µg/mL, inject a 10 µL aliquot.

## HPLC VARIABLES
**Column:** 300 × 3.9 10 µm µBondapak C18
**Mobile phase:** MeOH:acetic acid:triethylamine:water 5:1.5:0.5:93
**Flow rate:** 1.5
**Injection volume:** 10
**Detector:** UV 254

## CHROMATOGRAM
**Retention time:** 7

## OTHER SUBSTANCES
**Simultaneous:** ephedrine, pseudoephedrine

## REFERENCE
Roos, R.W.; Lau-Cam, C.A. General reversed-phase high-performance liquid chromatographic method for the separation of drugs using triethylamine as a competing base. *J.Chromatogr.*, **1986**, *370*, 403–418

## SAMPLE
**Matrix:** solutions
**Sample preparation:** Dissolve in MeOH at a concentration of 1 mg/mL, inject a 20 µL aliquot.

## HPLC VARIABLES
**Column:** 250 × 5 Spherisorb S5W
**Mobile phase:** MeOH:buffer 90:10 (Buffer was 94 mL 35% ammonia and 21.5 mL 70% nitric acid in 884 mL water, adjust the pH to 10.1 with ammonia.)
**Flow rate:** 2
**Injection volume:** 20
**Detector:** UV 254

## CHROMATOGRAM
**Retention time:** 2.25

## OTHER SUBSTANCES
**Simultaneous:** amphetamine, benzphetamine, benzylmorphine, bromo-STP, buprenorphine, caffeine, codeine, codeine-N-oxide, dextromoramide, dextropropoxyphene, diethylpropion, dihydrocodeine, dihydromorphine, dimethylamphetamine, ephedrine, ethoheptazine, ethylmorphine, etorphine, fenethyline, fenfluramine, fentanyl, hydrocodone, 4-hydroxyamphetamine, hydroxypethidine, levallorphan, levorphanol, mazindol, mephentermine, mescaline, methadone, methamphetamine, methylenedioxyamphetamine, methylephedrine, methylphenidate, morphine, morphine-3-glucuronide, morphine-N-oxide, nalorphine, naloxone, norcodeine, norlevorphanol, normetanephrine, normethadone, normorphine, norpethidine, norpipanone, noscapine, papaverine, pemoline, phenazocine, phendimetrazine, phenelzine, 2-phenethylamine, phenoperidine, phenylephrine, pholcodeine, piritramide, prolintane, pseudoephedrine, STP, thebaine, tranylcypromine, trimethoxyamphetamine, tyramine

**Noninterfering:** dopamine, levodopa, methyldopa, methyldopate, norepinephrine
**Interfering:** acetylcodeine, chlorphentermine, diamorphine, dipipanone, epinephrine, fencamfamin, meperidine, monoacetylmorphine, norpseudoephedrine, oxycodone, pentazocine, phentermine, pipradol, thebacon

## REFERENCE

Law, B.; Gill, R.; Moffat, A.C. High-performance liquid chromatography retention data for 84 basic drugs of forensic interest on a silica column using an aqueous methanol eluent. *J.Chromatogr.*, **1984**, *301*, 165–172

## SAMPLE

**Matrix:** urine
**Sample preparation:** 1 mL Urine + 0.5 mL 1% trichloroacetic acid, centrifuge at 5200 g for 10 min, filter (0.2 μm), inject 20 μL aliquot.

## HPLC VARIABLES

**Column:** 250 × 4 Lichrospher 5 μm 60 RP-select B
**Mobile phase:** Gradient. MeCN:50 mM pH 3.2 potassium phosphate buffer from 10:90 to 50:50 over 15 min.
**Flow rate:** 1.5
**Injection volume:** 20
**Detector:** UV 190-370

## CHROMATOGRAM

**Retention time:** 4

## OTHER SUBSTANCES

**Extracted:** benzoylecgonine, cocaine, diphenhydramine, ephedrine, lidocaine, morphine, nordiazepam, norpropoxyphene, nortriptyline
**Also analyzed:** amitriptyline, amphetamine, codeine (different gradient), meperidine

## REFERENCE

Li, S.; Gemperline, P.J.; Briley, K.; Kazmierczak, S. Identification and quantitation of drugs of abuse in urine using the generalized rank annihilation method of curve resolution. *J.Chromatogr.B*, **1994**, *655*, 213–223

## SAMPLE

**Matrix:** urine
**Sample preparation:** 5 mL Urine + 25 μL mobile phase + 100 μL 10 M NaOH + 2 mL diethyl ether + 3 g sodium sulfate, shake for 20 min, centrifuge at 1200 g for 5 min. Remove the organic layer and evaporate it to dryness under a stream of nitrogen, reconstitute the residue in 100 μL mobile phase, inject a 20 μL aliquot.

## HPLC VARIABLES

**Column:** 125 × 4 5 μm LiChrospher 60 RP Select B
**Mobile phase:** 200 mM pH 5.5 phosphate buffer containing 150 mM triethylamine (Wash column with MeOH for 15 min and with water for 15 min at the end of each day.)
**Column temperature:** 40
**Flow rate:** 1.3
**Injection volume:** 20
**Detector:** UV 215

## CHROMATOGRAM

**Retention time:** 11.5
**Internal standard:** phenylpropanolamine
**Limit of detection:** 500 ng/mL

## OTHER SUBSTANCES

**Extracted:** ephedrine, ethylephedrine, N-methylephedrine, norephedrine, norpseudoephedrine, pseudoephedrine

**Noninterfering:** amfepramone, amphetamine, caffeine, chlorphentermine, cocaine, codeine, cropropamide, crotethamide, dimethylamphetamine, etamivan, fencamfamine, heptaminol, leptazol, lidocaine, meperidine, methoxamine, methylamphetamine, methylphenidate, nicotine, niketamide, phendimetrazine, phenmetrazine, pipradol, procaine, prolintane, strychnine

## KEY WORDS

phenylpropanolamine is IS

## REFERENCE

Imaz, C.; Carreras, D.; Navajas, R.; Rodriguez, C.; Rodriguez, A.F.; Maynar, J.; Cortes, R. Determination of ephedrines in urine by high-performance liquid chromatography. *J.Chromatogr.*, **1993**, *631*, 201–205

## SAMPLE

**Matrix:** urine

**Sample preparation:** 500 µL Urine + chlorpheniramine + 100 µL buffer, centrifuge at 11000 g for 30 s, inject a 500 µL aliquot onto column A with mobile phase A, after 0.6 min backflush column A with mobile phase A to waste for 1.6 min, elute column A with 250 µL mobile phase B, with 200 µL mobile phase C, and with 1.15 mL mobile phase D. Elute column A to waste until drugs start to emerge then elute onto column B. Elute column B to waste until drugs start to emerge, then elute onto column C. When all the drugs have emerged from column B, remove it from the circuit. Elute column C with mobile phase D, monitor the effluent from column C. Flush column A with 7 mL mobile phase E, mobile phase D, and mobile phase A. Flush column B with 5 mL mobile phase E then with mobile phase D. (Buffer was 6 M ammonium acetate adjusted to pH 8.0 with 2 M KOH.)

## HPLC VARIABLES

**Column:** A 10 × 2.1 12-20 µm PRP-1 spherical poly(styrene-divinylbenzene) (Hamilton); B 10 × 3.2 11 µm Aminex A-28 (Bio-Rad); C 25 × 3.2 5 µm C8 (Phenomenex) + 150 × 4.6 5 µm silica (Macherey-Nagel)

**Mobile phase:** A 0.1% pH 8.0 potassium borate buffer; B 6 mM $KH_2PO_4$ containing 5 mM tetramethylammonium hydroxide, and 2 mM dimethyloctylamine, pH adjusted to 6.50 with phosphoric acid; C MeCN:buffer 40:60 (Buffer was 6 mM $KH_2PO_4$ containing 5 mM tetramethylammonium hydroxide, and 2 mM dimethyloctylamine, pH adjusted to 6.50 with phosphoric acid.); D MeCN:buffer 33:67 (Buffer was 6 mM $KH_2PO_4$ containing 5 mM tetramethylammonium hydroxide, and 2 mM dimethyloctylamine, pH adjusted to 6.50 with phosphoric acid.); E MeCN:buffer 70:30 (Buffer was 6 mM $KH_2PO_4$ containing 5 mM tetramethylammonium hydroxide, and 2 mM dimethyloctylamine, pH adjusted to 6.50 with phosphoric acid.)

**Column temperature:** 40 (B, C)

**Flow rate:** A 5; B-E 1

**Injection volume:** 500

**Detector:** UV 210; UV 235

## CHROMATOGRAM

**Retention time:** k' 2.2

**Internal standard:** chlorpheniramine (k' 5.9)

**Limit of detection:** 300 ng/mL

## OTHER SUBSTANCES
**Extracted:** amitriptyline, benzoylecgonine, caffeine, codeine, cotinine, desipramine, diazepam, diphenhydramine, ephedrine, flurazepam, hydrocodone, hydromorphone, imipramine, lidocaine, methadone, methamphetamine, morphine, nordiazepam, nortriptyline, oxazepam, pentazocine, phenmetrazine, phenobarbital, secobarbital
**Interfering:** amphetamine, phentermine

## KEY WORDS
column-switching

## REFERENCE
Binder, S.R.; Regalia, M.; Biaggi-McEachern, M.; Mazhar, M. Automated liquid chromatographic analysis of drugs in urine by on-line sample cleanup and isocratic multi-column separation. *J.Chromatogr.*, **1989**, *473*, 325–341

## ANNOTATED BIBLIOGRAPHY

Doyle, T.D.; Brunner, C.A.; Vick, J.A. Enantiomeric analysis of phenylpropanolamine in plasma via resolution of dinitrophenylurea derivatives on a high performance liquid chromatographic chiral stationary phase. *Biomed.Chromatogr.*, **1991**, *5*, 43–46

Stockley, C.S.; Wing, L.M.; Miners, J.O. Stereospecific high-performance liquid chromatographic assay for the enantiomers of phenylpropanolamine in human plasma. *Ther.Drug Monit.*, **1991**, *13*, 332–338

Shi, R.J.Y.; Gee, W.L.; Williams, R.L.; Benet, L.Z.; Lin, E.T. Ion-pair liquid chromatographic analysis of phenylpropanolamine in plasma and urine by post-column derivatization with o-phthalaldehyde. *J.Liq.Chromatogr.*, **1985**, *8*, 1489–1500 [post-column reaction; fluorescence detection; LOD 2 ng/mL; α-methylbenzylamine (IS)]

Costanzo, S.J. Selection of mixed ion-pair modifiers for high-performance liquid chromatographic mobile phase. *J.Chromatogr.*, **1984**, *314*, 402–407 [simultaneous benzoic acid, guaifenesin, phenylephrine]

Das Gupta, V.; Heble, A.R. Quantitation of acetaminophen, chlorpheniramine maleate, dextromethorphan hydrobromide, and phenylpropanolamine hydrochloride in combination using high-performance liquid chromatography. *J.Pharm.Sci.*, **1984**, *73*, 1553–1556

Dye, D.; East, T.; Bayne, W.F. High-performance liquid chromatographic method for post-column, in-line derivatization with o-phthalaldehyde and fluorometric detection of phenylpropanolamine in human urine. *J.Chromatogr.*, **1984**, *284*, 457–461

Dowse, R.; Haigh, J.M.; Kanfer, I. Determination of phenylpropanolamine in serum and urine by high-performance liquid chromatography. *J.Pharm.Sci.*, **1983**, *72*, 1018–1020

Tan, H.S.; Salvador, G.C. Assay of mixtures of phenylpropanolamine hydrochloride and caffeine in appetite suppressant formulations by high-performance liquid chromatography. *J.Chromatogr.*, **1983**, *261*, 111–116

Lurie, I.S. Improved isocratic mobile phases for the reverse phase ion-pair chromatographic analysis of drugs of forensic interest. *J.Liq.Chromatogr.*, **1981**, *4*, 399–408 [also amobarbital, amphetamine, butabarbital, caffeine, diamorphine, ephedrine, methamphetamine, pentobarbital, phenobarbital, phentermine, secobarbital]

Lurie, I.S.; Demchuk, S.M. Optimization of a reverse phase ion-pair chromatographic separation for drugs of forensic interest. II. Factors effecting selectivity. *J.Liq.Chromatogr.*, **1981**, *4*, 357–374 [also acetaminophen, acetylcodeine, acetylmorphine, aminopyrene, aminopyrine, amobarbital, amphetamine, antipyrine, benzocaine, butabarbital, caffeine, cocaine, codeine, diamorphine, diazepam, diethylpropion, DMT, ephedrine, glutethimide, Lampa, lidocaine, LSD, MDA, mecloqualone, mescaline, methamphetamine, methapyrilene, methaqualone, methpyrilene, methylphenidate, morphine, narcotine, papaverine, PCP, pentobarbital, phencyclidine, phendimetrazine, phenmetrazine, phenobarbital, phentermine, procaine, quinidine, quinine, secobarbital, strychnine, TCP, tetracaine, thebaine, theophylline]

Mason, W.D.; Amick, E.N. High-pressure liquid chromatographic analysis of phenylpropanolamine in human plasma following derivatization with o-phthalaldehyde. *J.Pharm.Sci.*, **1981**, *70*, 707–709

Achari, R.G.; Jacob, J.T. A study of the retention behavior of some basic drug substances by ion-pair HPLC. *J.Liq.Chromatogr.*, **1980**, *3*, 81–92 [also N-acetylprocainamide, antazoline, atropine, caffeine, chlorpheniramine, codeine, ephedrine, epinephrine, naphazoline, papaverine, pheniramine, phenylephrine, phenylpropanolamine, procainamide, quinidine, scopolamine, xylocaine]

Muhammad, N.; Bodnar, J.A. Quantitative determination of guaifenesin, phenylpropanolamine hydrochloride, sodium benzoate & codeine phosphate in cough syrups by high-pressure liquid chromatography. *J.Liq.Chromatogr.*, **1980**, *3*, 113–122

# Phenytoin

**Molecular formula:** C₁₅H₁₂N₂O₂
**Molecular weight:** 252.3
**CAS Registry No.:** 57-41-0, 630-93-3 (sodium salt)

## SAMPLE
**Matrix:** blood
**Sample preparation:** Add two volumes of MeCN to the mouse serum, mix, centrifuge at 1500 g for 5 min, inject a 5 µL aliquot of the supernatant.

## HPLC VARIABLES
**Guard column:** Sentry (Waters)
**Column:** 150 × 4.6 Nova-Pak C18
**Mobile phase:** MeCN:MeOH:10 mM pH 7.0 phosphate buffer 50:30:110
**Column temperature:** 40
**Flow rate:** 0.5
**Injection volume:** 5
**Detector:** UV 214

## CHROMATOGRAM
**Retention time:** 9.5

## OTHER SUBSTANCES
**Extracted:** carbamazepine, carbamazepine-10,11-epoxide, phenobarbital, phenylethyl malonamide, primidone

## KEY WORDS
serum; mouse

## REFERENCE
Capparella, M.; Foster, W., III; Larrousse, M.; Phillips, D.J.; Pomfret, A.; Tuvim, Y. Characteristics and applications of a new high-performance liquid chromatography guard column. *J.Chromatogr.A*, **1995**, *691*, 141–150

## SAMPLE
**Matrix:** blood
**Sample preparation:** Plasma + methyl p-hydroxybenzoate in MeCN, centrifuge, mix the supernatant with an equal amount of water, inject an aliquot.

## HPLC VARIABLES
**Column:** ODS-1
**Mobile phase:** MeCN:MeOH:10 mM pH 5.75 potassium phosphate buffer 25:5:70
**Detector:** UV 210

## CHROMATOGRAM
**Internal standard:** methyl p-hydroxybenzoate
**Limit of quantitation:** 100 ng/mL

## KEY WORDS
plasma

## REFERENCE

Fleishaker, J.C.; Pearson, L.K.; Peters, G.R. Phenytoin causes a rapid increase in 6β-hydroxycortisol urinary excretion in humans–A putative measure of CYP3A induction. *J.Pharm.Sci.*, **1995**, *84*, 292–294

## SAMPLE
**Matrix:** blood
**Sample preparation:** Centrifuge serum at 15600 g for 3 min, inject a 400 μL aliquot of the supernatant.

## HPLC VARIABLES
**Guard column:** 5 μm GFF-II-S5-80 glycine-phenylalanine-phenylalanine internal surface bonded phase (Regis)
**Column:** 150 × 4.6 5 μm GFF-II-S5-80 glycine-phenylalanine-phenylalanine internal surface bonded phase (Regis)
**Mobile phase:** THF:12.5 mM pH 7.4 phosphate buffer 1:99
**Column temperature:** 37
**Flow rate:** 1
**Injection volume:** 400
**Detector:** UV 254

## CHROMATOGRAM
**Retention time:** 17.5 (free phenytoin), 14.5 (bound phenytoin)

## KEY WORDS
serum; direct injection

## REFERENCE

Gurley, B.J.; Marx, M.; Olsen, K. Phenytoin free fraction determination: comparison of an improved direct serum injection high-performance liquid chromatographic method to ultrafiltration coupled with fluorescence polarization immunoassay. *J.Chromatogr.B*, **1995**, *670*, 358–364

## SAMPLE
**Matrix:** blood
**Sample preparation:** Filter (0.22 μm), inject a 5 μL aliquot of the filtrate.

## HPLC VARIABLES
**Column:** 250 × 4 5 μm LiChrospher 100 Diol
**Mobile phase:** MeCN:50 mM pH 6.9 phosphate buffer 12:88
**Flow rate:** 0.6
**Injection volume:** 5
**Detector:** UV 254

## CHROMATOGRAM
**Retention time:** 7

## OTHER SUBSTANCES
**Extracted:** carbamazepine, phenobarbital

## KEY WORDS
serum; direct injection

## REFERENCE

Nimura, N.; Itoh, H.; Kinoshita, T. Diol-bonded silica gel as a restricted access packing forming a binary-layered phase for direct injection of serum for the determination of drugs. *J.Chromatogr.A*, **1995**, *689*, 203–210

## SAMPLE
**Matrix:** blood
**Sample preparation:** 50 μL Plasma + 100 μL MeCN, centrifuge, inject a 5 μL aliquot of the supernatant.

## HPLC VARIABLES
**Column:** 100 × 4.6 3.5 μm Zorbax SB
**Mobile phase:** MeCN:MeOH:10 mM pH 7.1 phosphate buffer 7:34:59
**Flow rate:** 1.5
**Injection volume:** 5
**Detector:** UV 220

## CHROMATOGRAM
**Limit of detection:** <1 μM

## OTHER SUBSTANCES
**Extracted:** carbamazepine, carbamazepine epoxide, hydroxycarbamazepine, lamotrigine, oxcarbazepine, phenobarbital
**Also analyzed:** ibuprofen, naproxen, trimethoprim

## KEY WORDS
plasma

## REFERENCE
Svensson, J.O. Simple HPLC method for determination of antiepileptic drugs in plasma (Abstract 102). *Ther.Drug Monit.*, **1995**, *17*, 408

## SAMPLE
**Matrix:** blood
**Sample preparation:** 100 μL Serum + 25 μL IS solution + 50 μL 200 mM HCl + 3 mL dichloromethane, vortex for 2 min, centrifuge at 1200 g for 5 min. Remove the organic layer and evaporate it to dryness under a stream of nitrogen at 40°, reconstitute the residue in 100 μL mobile phase, inject a 50 μL aliquot.

## HPLC VARIABLES
**Column:** 250 × 4.6 5 μm TSKgel ODS-80TM (Tosoh)
**Mobile phase:** MeCN:8 mM $NaH_2PO_4$ 35:65
**Flow rate:** 0.8
**Injection volume:** 50
**Detector:** UV 215

## CHROMATOGRAM
**Retention time:** 16
**Internal standard:** 5-(4-methylphenyl)-5-phenylhydantoin
**Limit of detection:** 5 ng/mL

## OTHER SUBSTANCES
**Extracted:** metabolites

## KEY WORDS
serum; rat; pharmacokinetics

## REFERENCE
Tanaka, E.; Ishikawa, A.; Misawa, S. Changes in the metabolism of three model substrates catalysed by different P450 isozymes when administered as a cocktail to the carbon tetrachloride-intoxicated rat. *Xenobiotica*, **1995**, *25*, 1111–1118

## SAMPLE
**Matrix:** blood
**Sample preparation:** 100 µL Serum + 50 µL 10 µg/mL IS + 100 µL 200 mM HCl + 3 mL dichloromethane, vortex for 2 min, centrifuge at 1200 g for 5 min. Remove the organic layer and evaporate it to dryness under a stream of nitrogen at 40°, reconstitute the residue in 100 µL mobile phase, inject a 50 µL aliquot.

## HPLC VARIABLES
**Column:** 250 × 4.6 5 µm TSKgel ODS 80TM (Tosoh)
**Mobile phase:** MeCN:8 mM pH 6 NaH$_2$PO$_4$ 35:65
**Flow rate:** 0.8
**Injection volume:** 50
**Detector:** UV 215

## CHROMATOGRAM
**Retention time:** 16
**Internal standard:** 5-(4-methylphenyl)-5-phenylhydantoin (24.7)
**Limit of quantitation:** 50 ng/mL

## OTHER SUBSTANCES
**Extracted:** metabolites
**Simultaneous:** carbamazepine, diazepam, ethosuximide, nitrazepam, oxazolam, phenobarbital, primidone

## KEY WORDS
rat; serum; pharmacokinetics

## REFERENCE
Tanaka, E.; Sakamoto, N.; Inubushi, M.; Misawa, S. Simultaneous determination of plasma phenytoin and its primary hydroxylated metabolites in carbon tetrachloride-intoxicated rats by high-performance liquid chromatography. *J.Chromatogr.B*, **1995**, *673*, 147–151

## SAMPLE
**Matrix:** blood
**Sample preparation:** 500 µL Serum + 600 µL allobarbital in 75 mM pH 6.8 phosphate buffer, add 200 units β-glucuronidase, heat at 37°for 30 min, add 1 mL of this solution to an Extrelut-1 SPE cartridge, let stand for 10 min, elute with 2.5 mL MTBE. Evaporate the eluate to dryness under a stream of nitrogen, reconstitute the residue in 50 µL MeOH:water 50:50, inject a 10 µL aliquot onto columns A and B in series and elute with mobile phase A. After 12 min elute column A with mobile phase B, continue to elute column B with mobile phase A. Carbamazepine diol, carbamazepine epoxide, phenytoin, and carbamazepine elute from column A and the enantiomers of 5-(p-hydroxyphenyl)-5-phenylhydantoin and mephobarbital, phenobarbital, zonisamide, and allobarbital elute from column B. Re-equilibrate columns A and B with mobile phase A for 5 min before the next injection.

## HPLC VARIABLES
**Column:** A 250 × 4 4 µm Superspher RP-18e (E. Merck); B 250 × 4 4 µm Superspher RP-18e (E. Merck)
**Mobile phase:** A MeOH:11.2 mM β-cyclodextrin in 20 mM KH$_2$PO$_4$ 5:95; B MeCN:20 mM KH$_2$PO$_4$ 16:84
**Flow rate:** 0.8
**Injection volume:** 10
**Detector:** UV 210 (A); UV 210 (B)

## CHROMATOGRAM
**Retention time:** 19 (column A)

**Internal standard:** allobarbital
**Limit of detection:** 3.9 ng/mL

## OTHER SUBSTANCES
**Extracted:** metabolites, carbamazepine, carbamazepine diol, carbamazepine epoxide, 5-(p-hydroxyphenyl)-5-phenylhydantoin, mephobarbital, phenobarbital, zonisamide

## KEY WORDS
serum; column-switching; SPE

## REFERENCE
Eto, S.; Noda, H.; Noda, A. Simultaneous determination of antiepileptic drugs and their metabolites, including chiral compounds, via β-cyclodextrin inclusion complexes by a column-switching chromatographic technique. *J.Chromatogr.B*, **1994**, *658*, 385–390

## SAMPLE
**Matrix:** blood
**Sample preparation:** Inject sample onto column A with mobile phase A and elute for 3 min. Backflush contents of column A onto column B with mobile phase B for 6 min and elute column B with mobile phase B and monitor eluant.

## HPLC VARIABLES
**Column:** A 10 × 3 BioTrap Acid C18 (ChromTech); B 10 × 3 CT-sil C18 guard column + 150 × 4.6 5 μm CT-sil C18 (ChromTech)
**Mobile phase:** A 82 mM pH 6.0 phosphate buffer; B MeCN : 82 mM pH 6.0 phosphate buffer 50:50
**Flow rate:** A 0.55; B 1
**Injection volume:** 50
**Detector:** UV 254

## CHROMATOGRAM
**Retention time:** 10.5
**Limit of quantitation:** 4.56 μg/mL

## OTHER SUBSTANCES
**Simultaneous:** carbamazepine

## KEY WORDS
plasma; column-switching; direct injection

## REFERENCE
Hermansson, J.; Grahn, A. Determination of drugs by direct injection of plasma into a biocompatible extraction column based on a protein-entrapped hydrophobic phase. *J.Chromatogr.A*, **1994**, *660*, 119–129

## SAMPLE
**Matrix:** blood
**Sample preparation:** Condition an Extrashot-Silica (diatomaceous earth) SPE cartridge (Kusano Scientific) with 200 μL EtOH and 200 μL dichloromethane, force out the remaining solvent with 500 μL air. Add 5 μL serum to the surface of the cartridge and pass 130 μL dichloromethane gently through the cartridge into the 100 μL sample loop.

## HPLC VARIABLES
**Column:** 125 × 4 5 μm LiChrosorb Si60
**Mobile phase:** n-Hexane : dichloromethane : EtOH : acetic acid 82.8 : 15 : 2 : 0.2

**Flow rate:** 1
**Injection volume:** 100
**Detector:** UV 240

## CHROMATOGRAM
**Retention time:** 7.9
**Limit of quantitation:** 5 μg/mL

## OTHER SUBSTANCES
**Extracted:** carbamazepine, phenobarbital

## KEY WORDS
serum; normal phase; SPE

## REFERENCE
Kouno, Y.; Ishikura, C.; Homma, M.; Oka, K. Simple and accurate high-performance liquid chromato-
graphic method for the measurement of three antiepileptics in therapeutic drug monitoring.
*J.Chromatogr.*, **1993**, *622*, 47–52

## SAMPLE
**Matrix:** blood
**Sample preparation:** 100 μL Serum + 200 μL 20 μg/mL butalbital in MeCN, vortex,
centrifuge 5 min, inject an aliquot of the supernatant.

## HPLC VARIABLES
**Column:** 125 × 4 LiChroSpher RP-8 5 μm
**Mobile phase:** MeCN:water:100 mM pH 7.0 phosphate buffer 20:75:5
**Column temperature:** 45
**Flow rate:** 2
**Injection volume:** 50
**Detector:** UV 212

## CHROMATOGRAM
**Retention time:** 10.6
**Internal standard:** butalbital (3.8)

## OTHER SUBSTANCES
**Extracted:** carbamazepine, phenobarbital
**Simultaneous:** 1

## KEY WORDS
serum

## REFERENCE
Hannak, D.; Haux, P.; Scharbert, F.; Kattermann, R. Liquid chromatographic analysis of phenobarbital,
phenytoin, and theophylline. *Wien.Klin.Wochenschr.Suppl.*, **1992**, *191*, 27–31

## SAMPLE
**Matrix:** blood
**Sample preparation:** Inject 20 μL serum onto column A with mobile phase A and elute to
waste, after 1.5 min backflush the contents of column A onto column B with mobile phase
B, after 1 min remove column A from the circuit, elute column B with mobile phase B,
monitor the effluent from column B. Re-equilibrate column A with mobile phase A.

## HPLC VARIABLES
**Column:** A 30 × 4.6 IRSP silica (for preparation see Anal. Chem. 1989, 61, 2445); B 150 × 4.6 5 μm Nucleosil C18
**Mobile phase:** A 14 mM $NaH_2PO_4$ containing 6 mM $Na_2HPO_4$; B MeCN:MeOH:14 mM $NaH_2PO_4$ containing 6 mM $Na_2HPO_4$ 15:20:65
**Flow rate:** 0.8
**Injection volume:** 20
**Detector:** UV 230

## CHROMATOGRAM
**Retention time:** 26
**Limit of detection:** 1 μg/mL

## OTHER SUBSTANCES
**Extracted:** carbamazepine, phenobarbital, primidone

## KEY WORDS
serum; column-switching

## REFERENCE
Haginaka, J.; Wakai, J.; Yasuda, H.; Kimura, Y. Determination of anticonvulsant drugs and methyl xanthine derivatives in serum by liquid chromatography with direct injection: column-switching method using a new internal-surface reversed-phase silica support as a precolumn. *J.Chromatogr.*, **1990**, *529*, 455–461

## SAMPLE
**Matrix:** blood
**Sample preparation:** 500 μL Plasma + 500 μL 1 M pH 5.0 sodium acetate buffer + 50 μL 210 μg/mL tolybarbital in MeOH, vortex for 15 s, add 4 mL dichloromethane:ethyl acetate 2:1, shake for 5 min, repeat extraction. Combine the organic layers and evaporate them to dryness under a stream of nitrogen at 40°. Reconstitute the residue in 200 μL mobile phase, inject a 20 μL aliquot.

## HPLC VARIABLES
**Column:** 250 × 4.6 5 μm Spherisorb Octyl C8
**Mobile phase:** MeOH:MeCN:THF:10 mM pH 6.5 ammonium phosphate buffer 16:11:7:66
**Column temperature:** 40
**Flow rate:** 1.5
**Injection volume:** 20
**Detector:** UV 215

## CHROMATOGRAM
**Retention time:** 11.6
**Internal standard:** tolybarbital (6.7)
**Limit of detection:** 500 ng/mL

## OTHER SUBSTANCES
**Extracted:** metabolites, carbamazepine, carbamazepine-10,11-epoxide, carbamazepinediol, cyheptamide, felbamate, 5-(p-hydroxyphenyl)-5-phenylhydantoin
**Also analyzed:** ethosuximide, ethotoin, lorazepam, phenobarbital, phenylethylmalonamide, primidone

## KEY WORDS
plasma

## REFERENCE

Remmel, R.P.; Miller, S.A.; Graves, N.M. Simultaneous assay of felbamate plus carbamazepine, phenytoin, and their metabolites by liquid chromatography with mobile phase optimization. *Ther.Drug Monit.*, **1990**, *12*, 90–96

## SAMPLE
**Matrix:** blood
**Sample preparation:** Prepare an SPE cartridge by plugging the end of a 1 mL disposable pipette tip with glass wool and adding about 100 mg Chromosorb P/NAW. Add 50 µL plasma then 50 µL 10 µg/mL tolylphenobarbital in 200 mM HCl to the SPE cartridge, let stand for 2 min, elute with 1 mL chloroform:isopropanol 6:1. Evaporate the eluate to dryness under a stream of nitrogen at 30°, reconstitute the residue in 100 µL mobile phase, inject a 15 µL aliquot.

## HPLC VARIABLES
**Column:** 150 × 4.6 5 µm Supelcosil-LC-8
**Mobile phase:** MeCN:water 20:80
**Flow rate:** 3.3
**Injection volume:** 15
**Detector:** UV 208

## CHROMATOGRAM
**Retention time:** 12.54
**Internal standard:** tolylphenobarbital (7.57)
**Limit of detection:** 50-100 ng/mL

## OTHER SUBSTANCES
**Extracted:** amobarbital, barbital, butabarbital, caffeine, carbamazepine, carbamazepine diol, carbamazepine epoxide, chloramphenicol, ethosuximide, glutethimide, mephenytoin, methaqualone, methyprylon, nirvanol, pentobarbital, phenacemide, phenobarbital, primidone, secobarbital, theophylline
**Noninterfering:** acetaminophen, N-acetylprocainamide, amikacin, amitriptyline, clonazepam, cyclosporine, desipramine, diazepam, digoxin, disopyramide, gentamicin, p-hydroxyphenobarbital, imipramine, lidocaine, methotrexate, netilmicin, nortriptyline, procainamide, quinidine, salicylic acid, sulfamethoxazole, tobramycin, trimethoprim, valproic acid, vancomycin

## KEY WORDS
plasma; SPE

## REFERENCE

Svinarov, D.A.; Dotchev, D.C. Simultaneous liquid-chromatographic determination of some bronchodilators, anticonvulsants, chloramphenicol, and hypnotic agents, with Chromosorb P columns used for sample preparation. *Clin.Chem.*, **1989**, *35*, 1615–1618

## SAMPLE
**Matrix:** blood
**Sample preparation:** 100 µL Serum + 100 µL buffer + 1.5 mL IS in 5% isopropanol in chloroform, vortex for 30 s, centrifuge. Remove the organic layer and evaporate it to dryness under a stream of air at room temperature, reconstitute the residue in 100 µL mobile phase, inject a 6-10 µL aliquot. (Buffer was 13.6 g $KH_2PO_4$ in 90 mL water, pH adjusted to 6.8 with about 3 mL 10 M NaOH, made up to 100 mL.)

## HPLC VARIABLES
**Guard column:** 20 × 4.6 Supelguard LC-1 (Supelco)
**Column:** 250 × 4.6 5 µm Supelcosil LC-1 (Supelco)

**Mobile phase:** MeOH:MeCN:buffer 17.5:17.5:65 (Buffer was 2.72 g $KH_2PO_4$ in 1.9 L water, pH adjusted to 6.3 with about 2 mL 1 M NaOH, made up to 2 L.)
**Flow rate:** 2
**Injection volume:** 6-10
**Detector:** UV 204

## CHROMATOGRAM
**Retention time:** 6.28
**Internal standard:** 5-ethyl-5-p-tolybarbituric acid (tolybarb) (4.80)

## OTHER SUBSTANCES
**Extracted:** acetaminophen, amobarbital, barbital, caffeine, carbamazepine, chloramphenicol, ethosuximide, mephobarbital, methsuximide, pentobarbital, phenobarbital, primidone, secobarbital, theophylline, thiopental
**Also analyzed:** acetanilide, N-acetylcysteine, N-acetylprocainamide, ampicillin, aspirin, butabarbital, butalbital, chlorpropamide, cimetidine, codeine, cyheptamide, diazoxide, diflunisal, diphylline, disopyramide, ethchlorvynol, gentisic acid, glutethimide, heptabarbital, hexobarbital, ibuprofen, indomethacin, ketoprofen, mefenamic acid, mephenytoin, methaqualone, methsuximide, methyl salicylate, methyprylon, morphine, naproxen, nirvanol, oxphenylbutazone, phenacetin, phensuximide, phenylbutazone, procainamide, salicylamide, salicylic acid, sulfamethoxazole, sulindac, tolmetin, trimethoprim, vancomycin
**Noninterfering:** amikacin, gentamicin, meprobamate, netilmicin, quinidine, tetracycline, tobramycin, valproic acid

## KEY WORDS
serum

## REFERENCE
Meatherall, R.; Ford, D. Isocratic liquid chromatographic determination of theophylline, acetaminophen, chloramphenicol, caffeine, anticonvulsants, and barbiturates in serum. *Ther.Drug Monit.*, **1988**, *10*, 101–115

## SAMPLE
**Matrix:** blood
**Sample preparation:** 100 μL Plasma + 200 μL 1 M HCl saturated with ammonium sulfate, vortex for 20 s, add 60 μL 10 μg/mL 4-methylprimidone in MeCN, vortex for 20 s, centrifuge at 2700 g for 5 min, inject a 5-10 μL aliquot of the MeCN layer.

## HPLC VARIABLES
**Column:** 250 × 4 5 μm LiChrosorb RP-18
**Mobile phase:** MeOH:THF:50 mM pH 5.9 phosphate buffer 44:1:55
**Column temperature:** 50
**Flow rate:** 1.1
**Injection volume:** 5-10
**Detector:** UV 210

## CHROMATOGRAM
**Retention time:** 8
**Internal standard:** 4-methylprimidone (5)
**Limit of detection:** 50 ng/mL

## OTHER SUBSTANCES
**Extracted:** carbamazepine, phenobarbital, primidone, valproic acid
**Simultaneous:** acetaminophen, caffeine, chloramphenicol, diazepam, ethosuximide, ethylphenylmalonamide, glutethimide, lidocaine, methylphenobarbital, pentobarbital, salicylic acid, theophylline

## KEY WORDS
plasma

## REFERENCE
Kushida, K.; Ishizaki, T. Concurrent determination of valproic acid with other antiepileptic drugs by high-performance liquid chromatography. *J.Chromatogr.*, **1985**, *338*, 131–139

## SAMPLE
**Matrix:** blood
**Sample preparation:** 200 μL Serum + 1 mL 5 μg/mL 5-tolyl-5-phenylhydantoin in MeCN, agitate for 3 min. Remove the supernatant and evaporate it to dryness, dissolve the residue in 1 mL mobile phase, inject a 50 μL aliquot.

## HPLC VARIABLES
**Guard column:** pellicular reversed phase (Chrompack 28653)
**Column:** 100 × 3 octyl CP-tm-Spher C8 glass column (Chrompack)
**Mobile phase:** MeCN:50 mM $NaH_2PO_4$ 30:70 adjusted to pH 2.2 with phosphoric acid
**Flow rate:** 0.9
**Injection volume:** 50
**Detector:** UV 210

## CHROMATOGRAM
**Retention time:** 6
**Internal standard:** 5-tolyl-5-phenylhydantoin (11)
**Limit of detection:** 1 μg/mL

## OTHER SUBSTANCES
**Extracted:** phenobarbital

## KEY WORDS
serum

## REFERENCE
Van Damme, M.; Molle, L.; Abi Khalil, F. Useful sample handlings for reversed phase high performance liquid chromatography in emergency toxicology. *J.Toxicol.Clin.Toxicol.*, **1985**, *23*, 589–614

## SAMPLE
**Matrix:** blood
**Sample preparation:** 50 μL Serum + 50 μL 10 μg/mL IS in MeCN, vortex for 10 s, centrifuge at 3000 g for 1 min, remove the supernatant and place it in another tube, centrifuge for 1 min, inject a 20 μL aliquot of the supernatant.

## HPLC VARIABLES
**Column:** 100 × 8 5 μm Nova Pak C18 Radial pak
**Mobile phase:** MeCN:MeOH:acetone:buffer 8:21:10:61, adjusted to pH 7.95 ± 0.02 with NaOH (Buffer was 1.36 g/L $KH_2PO_4$.)
**Flow rate:** 2.8
**Injection volume:** 20
**Detector:** UV 200

## CHROMATOGRAM
**Retention time:** 7.70
**Internal standard:** tolybarb (5-ethyl-5-(p-methylphenyl)barbituric acid) (4.89)
**Limit of detection:** 500 ng/mL

## OTHER SUBSTANCES

**Extracted:** metabolites, carbamazepine, ethosuximide, phenobarbital, primidone
**Simultaneous:** acetaminophen, N-acetylprocainamide, aspirin, ampicillin, caffeine, cephapirin, chloramphenicol, digoxin, disopyramide, hexobarbital, indomethacin, lidocaine, mephobarbital, methsuximide, nafcillin, pentobarbital, phenylethylmalonamide, procainamide, quinidine, salicylic acid, secobarbital, sulfamerazine, sulfamethazine, terbutaline, tetracycline, theobromine, theophylline
**Noninterfering:** acetazolamide, amikacin, cephalosporin C, gentamicin, propranolol, sulfadiazine, sulfamethoxazole, sulfisoxazole, tobramycin, valproic acid, verapamil

## KEY WORDS

serum

## REFERENCE

Ou, C.-N.; Rognerud, C.L. Simultaneous measurement of ethosuximide, primidone, phenobarbital, phenytoin, carbamazepine, and their bioactive metabolites by liquid chromatography. *Clin.Chem.*, **1984**, *30*, 1667–1670

## SAMPLE

**Matrix:** blood
**Sample preparation:** 500 μL Serum + 50 μL 7 μg/mL IS in water + 1 mL buffer, vortex for 10 s, add 5 mL n-hexane:ether:n-propanol 49:49:2, shake gently for 20 min, centrifuge at 1000 g for 5 min, repeat the extraction. Combine the organic layers and evaporate them to dryness under a stream of nitrogen at 50°, reconstitute the residue in 300 μL mobile phase, inject a 50-100 μL aliquot. (Buffer was 10 mM sodium acetate:10 mM acetic acid 88.5:11.5, pH 5.5.)

## HPLC VARIABLES

**Column:** $250 \times 4.6$ 5 μm Partisil 5 ODS-3
**Mobile phase:** MeCN:buffer 28:72 (Buffer was 300 μL 1 M $KH_2PO_4$ and 50 μL 900 mM phosphoric acid in 1.8 L water, pH 4.4.)
**Column temperature:** 50
**Flow rate:** 2.8
**Injection volume:** 50-100
**Detector:** UV 195

## CHROMATOGRAM

**Retention time:** 7.8
**Internal standard:** 5-(4-methylphenyl)-5-phenylhydantoin (11.5)

## OTHER SUBSTANCES

**Extracted:** ethosuximide, secobarbital
**Simultaneous:** mephobarbital, paramethadione, phenobarbital, primidone
**Noninterfering:** chlorazepate, clonazepam, diazepam, thioridazine, valproic acid
**Interfering:** carbamazepine

## KEY WORDS

serum

## REFERENCE

Levine, H.L.; Cohen, M.E.; Duffner, P.K.; Kustas, K.A.; Shen, D.D. An improved high-pressure liquid chromatographic assay for secobarbital in serum. *J.Pharm.Sci.*, **1982**, *71*, 1281–1283

## SAMPLE

**Matrix:** blood

**Sample preparation:** 400 µL Serum or plasma + 400 µL 10 µg/mL IS in acetone, vortex for 10 s, centrifuge at 4500-5000 g for 1 min, remove the supernatant to another tube, centrifuge for 30 s, inject a 5-7.5 µL aliquot.

## HPLC VARIABLES
**Column:** 300 × 3.9 µBondapak C18
**Mobile phase:** MeCN:MeOH:buffer 17:28:55, final pH 6.8-7.0 (Buffer was 400 µL 1 M KH$_2$PO$_4$ in 1 L water, pH adjusted to 6.0 with 900 mM phosphoric acid.)
**Column temperature:** 30
**Flow rate:** 0.7
**Injection volume:** 5-7.5
**Detector:** UV 195

## CHROMATOGRAM
**Retention time:** 16.1
**Internal standard:** tolybarb (5-ethyl-5-(p-methylphenyl)barbituric acid) (13.8)

## OTHER SUBSTANCES
**Extracted:** carbamazepine, N-desmethylmethsuximide, ethosuximide, phenobarbital, primidone
**Simultaneous:** acetaminophen, butalbital, caffeine, hexobarbital, methsuximide, phenacetin, phenylethylmalonamide, salicylic acid

## KEY WORDS
serum; plasma

## REFERENCE
Szabo, G.K.; Browne, T.R. Improved isocratic liquid-chromatographic simultaneous measurement of phenytoin, phenobarbital, primidone, carbamazepine, ethosuximide, and N-desmethylmethsuximide in serum. *Clin.Chem.*, **1982**, *28*, 100–104

## SAMPLE
**Matrix:** blood
**Sample preparation:** 200 µL Serum + 200 µL 50 µg/mL hexobarbital in MeCN + 25 µL glacial acetic acid, vortex for 10 s, centrifuge for 1 min, inject a 30-100 µL aliquot of the supernatant.

## HPLC VARIABLES
**Column:** µBondapak C18
**Mobile phase:** Gradient. MeCN:7.5 g/L NaH$_2$PO$_4$ adjusted to pH 3.2 with phosphoric acid 5:95 to 22:78 over 24 min, to 45:55 over 10 min, maintain at 45:55 for 5 min. Reequilibrate with 5:95 for 5 min.
**Column temperature:** 50
**Flow rate:** 3
**Injection volume:** 30-100
**Detector:** UV 210

## CHROMATOGRAM
**Retention time:** 24.0
**Internal standard:** hexobarbital (20.6)
**Limit of detection:** 200-2000 ng/mL

## OTHER SUBSTANCES
**Extracted:** acetaminophen, amobarbital, butabarbital, butalbital, chlordiazepoxide, diazepam, ethchlorvynol, flurazepam, glutethimide, methaqualone, methyprylon, nitrazepam, pentobarbital, phenobarbital, primidone, salicylic acid, secobarbital, theophylline

**Simultaneous:** amitriptyline, caffeine, clomipramine, codeine, desipramine, ethotoin, imipramine, lidocaine, mesantoin, methsuximide, nirvanol, nortriptyline, oxazepam, procainamide, phenylpropanolamine, propranolol, quinidine

## KEY WORDS
serum

## REFERENCE
Kabra, P.M.; Stafford, B.E.; Marton, L.J. Rapid method for screening toxic drugs in serum with liquid chromatography. *J.Anal.Toxicol.*, **1981**, *5*, 177–182

## SAMPLE
**Matrix:** blood
**Sample preparation:** 500 μL Plasma + 100 μL heptabarbital in MeOH + 500 μL 400 mM pH 7.0 sodium phosphate buffer + 10 mL ethyl acetate, extract. Evaporate the extract to dryness at 50°, reconstitute the residue in 20 μL MeOH, inject a 3 μL aliquot.

## HPLC VARIABLES
**Guard column:** 50 × 2.1 Whatman Co:Pell ODS
**Column:** 125 × 4.5 5 μm SAS Hypersil
**Mobile phase:** MeCN:buffer 20:80 (Buffer was 5 mM tetrabutylammonium hydroxide adjusted to pH 7.5 with phosphoric acid.)
**Flow rate:** 1.6
**Injection volume:** 3
**Detector:** UV 200

## CHROMATOGRAM
**Retention time:** 18.6
**Internal standard:** heptabarbital (9.8)
**Limit of quantitation:** 5.2 μM

## OTHER SUBSTANCES
**Extracted:** carbamazepine, ethosuximide, pheneturide, phenobarbital, primidone
**Simultaneous:** amobarbital, barbital, butabarbital, cyclobarbital, ethotoin, ethylphenacemide, glutethimide, methsuximide, pentobarbital, phenylethylmalonamide, secobarbital, sulfamethoxazole, sulthiame

## KEY WORDS
plasma; horse

## REFERENCE
Christofides, J.A.; Fry, D.E. Measurement of anticonvulsants in serum by reversed-phase ion-pair liquid chromatography. *Clin.Chem.*, **1980**, *26*, 499–501

## SAMPLE
**Matrix:** blood
**Sample preparation:** 200 μL Serum or plasma + 200 μL 20 μg/mL IS in MeOH:water 10:90 + 75 μL glacial acetic acid, vortex for 30 s, add 5 mL chloroform, shake for 5 min, centrifuge at 2000 rpm for 5 min. Remove the organic layer and evaporate it to dryness under a stream of nitrogen, reconstitute the residue in 200 μL mobile phase, inject a 40 μL aliquot.

## HPLC VARIABLES
**Guard column:** 30 × 2.1 Permaphase ETH (DuPont)
**Column:** 250 × 4.6 CLC 1 C8 (DuPont)

**Mobile phase:** MeCN:buffer 35:65 (Buffer was 20 mM $KH_2PO_4$ and 1 mM $K_2HPO_4$ adjusted to pH 5.6.)
**Column temperature:** 25
**Flow rate:** 2
**Injection volume:** 40
**Detector:** UV 220

## CHROMATOGRAM
**Retention time:** 6.3
**Internal standard:** alphenal (5-allyl-5-phenylbarbituric acid) (4.4)
**Limit of quantitation:** 500 ng/mL

## OTHER SUBSTANCES
**Extracted:** carbamazepine, ethosuximide, phenobarbital, primidone
**Simultaneous:** amobarbital, barbital, chlordiazepoxide, codeine, ethotoin, glutethimide, hexobarbital, hydrocortisone, mephenytoin, mephobarbital, metharbital, methsuximide, nitrazepam, phenacetin, phensuximide, secobarbital
**Noninterfering:** acetaminophen, acetazolamide, amphetamine, bilirubin, caffeine, diazepam, dimenhydrinate, meperidine, meprobamate, methamphetamine, methaqualone, methylphenidate, nicotine, propoxyphene, theophylline, valproic acid
**Interfering:** pentobarbital

## KEY WORDS
serum; plasma

## REFERENCE
Rydzewski, R.S.; Gadsden, R.H.; Phelps, C.A. Simultaneous rapid HPLC determination of anticonvulsant drugs in plasma and correlation with EMIT. *Ann.Clin.Lab.Sci.*, **1980**, *10*, 89–94

## SAMPLE
**Matrix:** blood, CSF
**Sample preparation:** 200 µL Serum, plasma, or CSF + 300 µL reagent. Flush column A to waste with 500 µL 500 mM ammonium sulfate, inject sample onto column A, flush column A to waste with 500 µL 500 mM ammonium sulfate, elute the contents of column A onto column B with mobile phase, monitor the effluent from column B. (Reagent was 8.05 M guanidine hydrochloride and 1.02 M ammonium sulfate in water.)

## HPLC VARIABLES
**Column:** A 30 × 2.1 40 µm preparative grade C18 (Analytichem); B 250 × 4.6 10 µm Partisil C8
**Mobile phase:** Gradient. A was 50 mM pH 4.5 $KH_2PO_4$. B was MeCN:isopropanol 80:20. A:B 90:10 for 1 min, to 30:70 over 15 min, maintain at 30:70 for 4 min.
**Column temperature:** 50
**Flow rate:** 1.5
**Detector:** UV 280 for 5 min then UV 254

## CHROMATOGRAM
**Retention time:** 11.90
**Internal standard:** heptanophenone (19.2)

## OTHER SUBSTANCES
**Extracted:** acetazolamide, ampicillin, bromazepam, caffeine, carbamazepine, chloramphenicol, chlorothiazide, diazepam, droperidol, ethionamide, furosemide, isoniazid, methadone, penicillin G, phenobarbital, prazepam, propoxyphene, pyrazinamide, rifampin, trimeprazine, trimethoprim

## KEY WORDS
serum; plasma; column-switching

## REFERENCE
Seifart, H.I.; Kruger, P.B.; Parkin, D.P.; van Jaarsveld, P.P.; Donald, P.R. Therapeutic monitoring of antituberculosis drugs by direct in-line extraction on a high-performance liquid chromatography system. *J.Chromatogr.*, **1993**, *619*, 285–290

## SAMPLE
**Matrix:** blood, dialysate
**Sample preparation:** Dialyze 400 μL plasma against 175 μL acceptor solution through a Cuprophane membrane (15 kDa cut-off) at 37° for 10 min, inject 500 μL acceptor solution (including the portion used for dialysis) onto column A at 0.71 mL/min, elute the contents of column A onto column B with mobile phase, remove column A from circuit and condition it with 1 mL acceptor solution, elute column B with mobile phase and monitor the effluent. Flush acceptor channel with 5 mL acceptor solution and plasma channel with 8 mL acceptor solution containing 25 μg/mL Triton X-100. (Acceptor solution contained 5.9 g NaCl, 4.1 g sodium acetate, 0.3 g KCl, and 1.65 g sodium citrate in 1 L water, adjusted to pH 7.4 with citric acid.)

## HPLC VARIABLES
**Column:** A 5 × 1.6 Hypersil ODS-2; B 100 × 3 5 μm Spherisorb ODS-2
**Mobile phase:** MeCN:THF:20 mM pH 6.0 phosphate buffer 22:6.5:71.5
**Column temperature:** 37
**Flow rate:** 0.6
**Detector:** UV 240

## CHROMATOGRAM
**Retention time:** 6
**Limit of detection:** 800 ng/mL

## OTHER SUBSTANCES
**Extracted:** carbamazepine, phenobarbital

## KEY WORDS
plasma; column-switching; dialysis

## REFERENCE
Johansen, K.; Krogh, M.; Andresen, A.T.; Christophersen, A.S.; Lehne, G.; Rasmussen, K.E. Automated analysis of free and total concentrations of three antiepileptic drugs in plasma with on-line dialysis and high-performance liquid chromatography. *J.Chromatogr.B*, **1995**, *669*, 281–288

## SAMPLE
**Matrix:** blood, dialysate
**Sample preparation:** Dialyze 400 μL plasma against 175 μL acceptor solution through a 4 cm² cuprophan (cellulose acetate, molecular mass cut-off 15000) membrane for 25 min at 20° and 10 min at 37°, inject 500 μL acceptor solution (including the dialysis sample) onto column A at 0.5 mL/min, elute column A onto column B with mobile phase, remove column A from circuit and condition it with 1 mL acceptor solution, elute column B with mobile phase and monitor the effluent. Flush acceptor channel with 5 mL acceptor solution and plasma channel with 8 mL acceptor solution containing 50 μg/mL Triton X-100. (Acceptor solution contained 5.9 g NaCl, 4.1 g sodium acetate, 0.3 g KCl, and 1.65 g sodium citrate in 1 L water, adjusted to pH 7.4 with citric acid.)

## HPLC VARIABLES
**Column:** A 10 × 2 36 μm polystyrene-divinylbenzene (Dyno Particles); B 100 × 3 5 μm Spherisorb ODS-2
**Mobile phase:** MeCN:THF:20 mM pH 5 phosphate buffer 25:7.5:67.5
**Column temperature:** 37
**Flow rate:** 0.7
**Detector:** UV 220

## CHROMATOGRAM
**Retention time:** 5
**Limit of detection:** 600 ng/mL

## OTHER SUBSTANCES
**Noninterfering:** carbamazepine, lamotrigine, oxcarbazepine, phenobarbital, valproic acid

## KEY WORDS
plasma; column-switching; dialysis

## REFERENCE
Andresen, A.T.; Rasmussen, K.E.; Rugstad, H.E. Automated determination of free phenytoin in human plasma with on-line equilibrium dialysis and column-switching high-performance liquid chromatography. *J.Chromatogr.*, **1993**, *621*, 189–198

## SAMPLE
**Matrix:** blood, saliva, urine
**Sample preparation:** Serum. 100 μL Serum + 200 μL MeCN, vortex for 10 s, centrifuge at 1500 g for 5 min, inject a 2 μL aliquot of the supernatant. Saliva. 250 μL Saliva + 50 μL MeCN, centrifuge at 1500 g for 5 min, inject a 2 μL aliquot of the supernatant. Urine. Condition a Sep-Pak SPE cartridge with 5 mL MeCN then 20 mL water. Add 2 mL urine to the cartridge, wash with 20 mL water, elute with 500 μL MeCN, inject 2 μL of the eluent.

## HPLC VARIABLES
**Guard column:** 20 × 2 3 μm ODS-Hypersil
**Column:** 250 × 2 3 μm ODS-Hypersil
**Mobile phase:** MeCN:MeOH:10 mM pH 7.0 phosphate buffer 50:30:110
**Column temperature:** 40
**Flow rate:** 0.2
**Injection volume:** 2
**Detector:** UV 200

## CHROMATOGRAM
**Retention time:** 11.6
**Limit of quantitation:** 780 ng/mL

## OTHER SUBSTANCES
**Simultaneous:** carbamazepine, carbamazepine-10,11-epoxide, clonazepam, dihydrodihydroxycarbamazepine, hexobarbital, p-hydroxyphenobarbital, 5-(m-hydroxyphenyl)-5-phenylhydantoin, 5-(p-hydroxyphenyl)-5-phenylhydantoin, nitrazepam, phenobarbital, phenylethylmaleimide, primidone
**Noninterfering:** chlordiazepoxide, cyheptamide, diazepam, lorazepam, nordiazepam, oxazepam, prazepam, temazepam

## KEY WORDS
serum; SPE

## REFERENCE

Liu, H.; Delgado, M.; Forman, L.J.; Eggers, C.M.; Montoya, J.L. Simultaneous determination of carbamazepine, phenytoin, phenobarbital, primidone and their principal metabolites by high-performance liquid chromatography with photodiode-array detection. *J.Chromatogr.*, **1993**, *616*, 105–115

## SAMPLE

**Matrix:** blood, tissue

**Sample preparation:** Tissue. Homogenize 20-200 mg brain tissue with 1 mL 1.5 μg/mL IS in 1% ammonium acetate buffer containing 1% sodium azide:MeCN 99:1, flush apparatus with 1 mL extraction buffer, add 1 mL acetone, shake for 5 min, centrifuge for 10 min. Add sample to an Extrelut-3 SPE cartridge (Kieselguhr), add 1 mL extraction buffer, wait for 10 min, elute with 15 mL extraction solvent. Evaporate the eluate, take up residue in 50 μL MeOH, add 50 μL water, inject a 10-25 μL aliquot. Serum. 100 μL Serum + 1 mL 1.5 μg/mL IS in 1% ammonium acetate buffer containing 1% sodium azide:MeCN 99:1, mix, add 1 mL extraction buffer, mix, add 1 mL acetone, shake for 5 min, centrifuge for 10 min. Add sample to an Extrelut-3 SPE cartridge (Kieselguhr), add 1 mL extraction buffer, wait for 10 min, elute with 15 mL extraction solvent. Evaporate the eluate, take up residue in 50 μL MeOH, add 50 μL water, inject a 10-25 μL aliquot. (Extraction buffer was 20 g $NaH_2PO_4.2H_2O$ + 4.5 g $Na_2HPO_4.2H_2O$ + 1.5 g sodium azide in 1 L water, pH 6. Extraction solvent was dichloromethane:isopropanol 97:3. Caution! Sodium azide is highly toxic!)

## HPLC VARIABLES

**Column:** 200 × 2.1 5 μm Hypersil ODS

**Mobile phase:** Gradient. A was MeCN:50 mM $(NH_4)H_2PO_4$ (pH 4.4) 10:90. B was MeCN: 50 mM $(NH_4)H_2PO_4$ (pH 4.4) 60:40. A:B from 85:15 to 55:45 over 9.5 min, keep at 55:45 for 0.5 min, return to 85:15 over 0.5 min.

**Column temperature:** 65

**Flow rate:** 0.3

**Injection volume:** 10-25

**Detector:** UV 207

## CHROMATOGRAM

**Retention time:** 10.18

**Internal standard:** 5-ethyl-5-(p-tolyl)barbituric acid (9.07)

**Limit of quantitation:** 250 ng/mL

## OTHER SUBSTANCES

**Extracted:** carbamazepine, carbamazepine-10,11-epoxide, N-desmethylmethsuximide, phenobarbital, primidone

## KEY WORDS

serum; SPE; brain

## REFERENCE

Juergens, U.; Rambeck, B. Sensitive analysis of antiepileptic drugs in very small portions of human brain by microbore HPLC. *J.Liq.Chromatogr.*, **1987**, *10*, 1847–1863

## SAMPLE

**Matrix:** blood, urine

**Sample preparation:** 100 μL Plasma or urine + 100 μL 20000 U/mL β-glucuronidase in 200 mM pH 4.9 sodium acetate buffer, heat at 37° for 20 h, add 20 μL 100 μg/mL mephenytoin, add 2 volumes of ethyl acetate, extract, centrifuge at 1000 g for 5 min, repeat extraction four more times. Combine the organic layers and evaporate them to dryness under a stream of nitrogen, reconstitute the residue in 20 μL MeOH, inject a 5 μL aliquot.

## HPLC VARIABLES
**Column:** 150 × 4.6 5 μm RP C18 (Beckman)
**Mobile phase:** Gradient. Isopropanol:water 20:80, to 25:75 after 7 min (step gradient).
**Flow rate:** 1.4
**Injection volume:** 5
**Detector:** UV 225

## CHROMATOGRAM
**Retention time:** 15
**Internal standard:** mephenytoin (8)
**Limit of detection:** 5 ng

## OTHER SUBSTANCES
**Extracted:** metabolites

## KEY WORDS
mouse; plasma

## REFERENCE
Lum, J.T.; Vassanji, N.A.; Wells, P.G. Analysis of the toxicologically relevant metabolites of phenytoin in biological samples by high-performance liquid chromatography. *J.Chromatogr.*, **1985**, *338*, 242–248

## SAMPLE
**Matrix:** cell suspensions
**Sample preparation:** Centrifuge cell suspensions, inject a 200 μL aliquot of the supernatant.

## HPLC VARIABLES
**Column:** 250 × 4.5 5 μm C18 (IBM)
**Mobile phase:** MeCN:5 mM pH 3.0 phosphate buffer 27:73
**Flow rate:** 1.2
**Injection volume:** 200
**Detector:** UV 210

## CHROMATOGRAM
**Retention time:** 19

## OTHER SUBSTANCES
**Extracted:** metabolites

## REFERENCE
Mays, D.C.; Pawluk, L.J.; Apseloff, G.; Davis, W.B.; She, Z.W.; Sagone, A.L.; Gerber, N. Metabolism of phenytoin and covalent binding of reactive intermediates in activated human neutrophils. *Biochem.Pharmacol.*, **1995**, *50*, 367–380

## SAMPLE
**Matrix:** enzyme preparations
**Sample preparation:** 200 μL Enzyme preparation + 50 μL 20% trichloroacetic acid, vortex, centrifuge at 14000 rpm for 12 min, inject an aliquot of the supernatant.

## HPLC VARIABLES
**Guard column:** 30 × 3.2 5 μm Hypersil C18
**Column:** 250 × 3.2 5 μm Hypersil C18
**Mobile phase:** MeOH:50 mM pH 6.5 ammonium phosphate 35:65

**Column temperature:** 40
**Flow rate:** 0.8
**Injection volume:** 100
**Detector:** UV 214

## OTHER SUBSTANCES
**Extracted:** fosphenytoin

## REFERENCE
TenHoor, C.N.; Stewart, B.H. Reconversion of fosphenytoin in the presence of intestinal alkaline phosphatase. *Pharm.Res.*, **1995**, *12*, 1806–1809

## SAMPLE
**Matrix:** formulations
**Sample preparation:** Dissolve tablet in 10 mM HCl containing 90 mM KCl (pH 2.0), inject an aliquot.

## HPLC VARIABLES
**Column:** 50 mm long ODS Hypersil C18
**Mobile phase:** MeCN:50 mM pH 7.0 phosphate buffer 32:68 containing 5 mM tetrabutylammonium
**Flow rate:** 1
**Detector:** UV 220

## CHROMATOGRAM
**Retention time:** 8.4

## OTHER SUBSTANCES
**Simultaneous:** naproxen

## KEY WORDS
tablets

## REFERENCE
Neervannan, S.; Dias, L.S.; Southard, M.Z.; Stella, V.J. A convective-diffusion model for dissolution of two non-interacting drug mixtures from co-compressed slabs under laminar hydrodynamic conditions. *Pharm.Res.*, **1994**, *11*, 1288–1295

## SAMPLE
**Matrix:** formulations
**Sample preparation:** Dissolve in water with rotating paddle, filter (0.45 μm), inject an aliquot.

## HPLC VARIABLES
**Column:** 150 × 4.6 Zorbax ODS
**Mobile phase:** MeCN:water 40:60
**Flow rate:** 1.5
**Detector:** UV 200

## KEY WORDS
oral dosage

## REFERENCE
Watanabe, A.; Hanawa, T.; Sugihara, M.; Yamamoto, K. Release profiles of phenytoin from new oral dosage form for the elderly. *Chem.Pharm.Bull.*, **1994**, *42*, 1642–1645

## SAMPLE
**Matrix:** microsomal incubations
**Sample preparation:** 2 mL Microsomal incubation + 25 µL 1 mM IS in EtOH + 5 mL
MTBE, rotate for 20 min. Remove the organic layer and evaporate it to dryness under a
stream of nitrogen at 60°, reconstitute the residue in 300 µL 0.05% orthophosphoric acid,
vortex, inject a 100 µL aliquot.

## HPLC VARIABLES
**Column:** Spherisorb 5 ODS
**Mobile phase:** MeCN:0.05% orthophosphoric acid 27:73
**Flow rate:** 1
**Injection volume:** 100
**Detector:** UV 214

## CHROMATOGRAM
**Retention time:** 24
**Internal standard:** 5-(4'-methylphenyl)-5-phenylhydantoin (42)

## OTHER SUBSTANCES
**Extracted:** metabolites

## KEY WORDS
rat; liver

## REFERENCE
Ashforth, E.I.L.; Carlile, D.J.; Chenery, R.; Houston, J.B. Prediction of *in vivo* disposition from *in vitro*
systems: Clearance of phenytoin and tolbutamide using rat hepatic microsomal and hepatocyte data.
*J.Pharm.Exp.Ther.*, **1995**, *274*, 761–766

## SAMPLE
**Matrix:** solutions

## HPLC VARIABLES
**Column:** 250 × 4.6 10 µm Spherisorb ODS-2
**Mobile phase:** MeCN:water 40:60
**Flow rate:** 0.8
**Detector:** UV 228

## CHROMATOGRAM
**Retention time:** 8

## REFERENCE
Mithani, S.D.; Bakatselou, V.; TenHoor, C.N.; Dressman, J.B. Estimation of the increase in solubility of
drugs as a function of bile salt concentration. *Pharm.Res.*, **1996**, *13*, 163–167

## SAMPLE
**Matrix:** solutions

## HPLC VARIABLES
**Column:** 250 × 4.6 5 µm Supelcosil LC-DP (A) or 250 × 4 5 µm LiChrospher 100 RP-8 (B)
**Mobile phase:** MeCN:0.025% phosphoric acid:buffer 25:10:5 (A) or 60:25:15 (B) (Buffer
was 9 mL concentrated phosphoric acid and 10 mL triethylamine in 900 mL water, adjust
pH to 3.4 with dilute phosphoric acid, make up to 1 L.)
**Flow rate:** 0.6

**Injection volume:** 25
**Detector:** UV 229

## CHROMATOGRAM
**Retention time:** 6.11 (A), 5.26 (B)

## OTHER SUBSTANCES
**Also analyzed:** acebutolol, acepromazine, acetaminophen, acetazolamide, acetophenazine, albuterol, alprazolam, amitriptyline, amobarbital, amoxapine, antipyrine, atenolol, atropine, azatadine, baclofen, benzocaine, bromocriptine, brompheniramine, brotizolam, bupivacaine, buspirone, butabarbital, butalbital, caffeine, carbamazepine, cetirizine, chlorcyclizine, chlordiazepoxide, chlormezanone, chloroquine, chlorpheniramine, chlorpromazine, chlorpropamide, chlorprothixene, chlorthalidone, chlorzoxazone, cimetidine, cisapride, clomipramine, clonazepam, clonidine, clozapine, cocaine, codeine, colchicine, cyclizine, cyclobenzaprine, dantrolene, desipramine, diazepam, diclofenac, diflunisal, diltiazem, diphenhydramine, diphenidol, diphenoxylate, dipyridamole, disopyramide, dobutamine, doxapram, doxepin, droperidol, encainide, ethidium bromide, ethopropazine, fenoprofen, fentanyl, flavoxate, fluoxetine, fluphenazine, flurazepam, flurbiprofen, fluvoxamine, furosemide, glutethimide, glyburide, guaifenesin, haloperidol, homatropine, hydralazine, hydrochlorothiazide, hydrocodone, hydromorphone, hydroxychloroquine, hydroxyzine, ibuprofen, imipramine, indomethacin, ketoconazole, ketoprofen, ketorolac, labetalol, levorphanol, lidocaine, loratadine, lorazepam, lovastatin, loxapine, mazindol, mefenamic acid, meperidine, mephenytoin, mepivacaine, mesoridazine, metaproterenol, metformin, methadone, methdilazine, methocarbamol, methotrexate, methotrimeprazine, methoxamine, methyldopa, methylphenidate, metoclopramide, metolazone, metoprolol, metronidazole, midazolam, moclobemide, morphine, nadolol, nalbuphine, naloxone, naphazoline, naproxen, nifedipine, nizatidine, norepinephrine, nortriptyline, oxazepam, oxycodone, oxymetazoline, paroxetine, pemoline, pentazocine, pentobarbital, pentoxifylline, perphenazine, pheniramine, phenobarbital, phenol, phenolphthalein, phentolamine, phenylbutazone, phenyltoloxamine, pimozide, pindolol, piroxicam, pramoxine, prazepam, prazosin, probenecid, procainamide, procaine, prochlorperazine, procyclidine, promazine, promethazine, propafenone, propantheline, propiomazine, propofol, propranolol, protriptyline, quazepam, quinidine, quinine, racemethorphan, ranitidine, remoxipride, risperidone, salicylic acid, scopolamine, secobarbital, sertraline, sotalol, spironolactone, sulfinpyrazone, sulindac, temazepam, terbutaline, terfenadine, tetracaine, theophylline, thiethylperazine, thiopental, thioridazine, thiothixene, timolol, tocainide, tolbutamide, tolmetin, trazodone, triamterene, triazolam, trifluoperazine, triflupromazine, trimeprazine, trimethoprim, trimipramine, verapamil, warfarin, xylometazoline, yohimbine, zopiclone

## KEY WORDS
some details of plasma extraction

## REFERENCE
Koves, E.M. Use of high-performance liquid chromatography-diode array detection in forensic toxicology. *J.Chromatogr.A*, **1995**, *692*, 103–119

## SAMPLE
**Matrix:** solutions

## HPLC VARIABLES
**Column:** 250 × 4 OmniPac PAX-500 (Dionex)
**Mobile phase:** Gradient. A was MeCN:5 mM sodium carbonate 9:81. B was MeCN:20 mM sodium carbonate 20:80. A:B from 100:0 to 0:100 over 10 min.
**Flow rate:** 1
**Detector:** UV 254

## CHROMATOGRAM
**Retention time:** 12.5

## OTHER SUBSTANCES
**Simultaneous:** allobarbital, amobarbital, barbital, barbituric acid, butabarbital, mephobarbital, methabarbital, methohexital, phenobarbital, secobarbital, thiamylal

## REFERENCE
Slingsby, R.W.; Rey, M. Determination of pharmaceuticals by multi-phase chromatography: Combined reversed phase and ion exchange in one column. *J.Liq.Chromatogr.*, **1990**, *13*, 107–134

## ANNOTATED BIBLIOGRAPHY

Kanda, T.; Kutsuna, H.; Ohtsu, Y.; Yamaguchi, M. Synthesis of polymer-coated mixed-functional packing materials for direct analysis of drug-containing serum and plasma by high-performance liquid chromatography. *J.Chromatogr.A*, **1994**, *672*, 51–57 [column temp 40; cow; human; extracted carbamazepine, phenobarbital; also chloramphenicol, indomethacin, theophylline, trimethoprim; direct injection]

Ramachandran, S.; Underhill, S.; Jones, S.R. Measurement of lamotrigine under conditions measuring phenobarbitone, phenytoin, and carbamazepine using reversed-phase high-performance liquid chromatography at dual wavelengths. *Ther.Drug Monit.*, **1994**, *16*, 75–82 [also carbamazepine, lamotrigine, phenobarbital; serum; hexabarbital (IS); LOD 200 ng/mL; non-interfering ethosuximide, oxcarbazepine, primidone, valproic acid]

Romanyshyn, L.A.; Wichmann, J.K.; Kucharczyk, N.; Shumaker, R.C.; Ward, D.; Sofia, R.D. Simultaneous determination of felbamate, primidone, phenobarbital, carbamazepine, two carbamazepine metabolites, phenytoin, and one phenytoin metabolite in human plasma by high-performance liquid chromatography. *Ther.Drug Monit.*, **1994**, *16*, 90–99 [column temp 40-50°; LOQ 195-391 ng/mL; non-interfering clonazepam, valproic acid; simultaneous acetaminophen, aspirin, brompheniramine, caffeine, chlorpheniramine, dextromethorphan, dimethadione, ethosuximide, ethotoin, ibuprofen, iministilbene, mephenytoin, mephobarbital, metharbital, methsuximide, paramethadione, phenacemide, phensuximide, phenylpropanolamine, theophylline, trimethadione; extracted metabolites, carbamazepine, felbamate, phenobarbital, primidone]

Smigol, V.; Svec, F.; Fréchet, J.M.J. Novel uniformly sized polymeric stationary phase with hydrophilized large pores for direct injection HPLC determination of drugs in biological fluids. *J.Liq.Chromatogr.*, **1994**, *17*, 891–911 [direct injection; cow; plasma; also aspirin, caffeine, carbamazepine, lidocaine, salicylic acid, theobromine, theophylline]

Sudo, Y.; Akiba, M.; Sakaki, T.; Takahata, Y. Glycerylalkylsilylated silica gels for direct injection analysis of drugs in serum by high-performance liquid chromatography. *J.Liq.Chromatogr.*, **1994**, *17*, 1743–1754 [extracted carbamazepine, phenobarbital]

Bahal, N.; Nahata, M.C. Determination of phenytoin and its major metabolite, 5-(p-hydroxyphenyl)-5-phenylhydantoin in urine by high-performance liquid chromatography. *J.Liq.Chromatogr.*, **1993**, *16*, 1135–1142

Herbranson, D.E.; Kriss-Danziger, P. Development and validation of a high performance liquid chromatographic (HPLC) method for the determination of phenytoin prodrug (fosphenytoin) in solutions, parenteral formulations, and active drug substance. *J.Liq.Chromatogr.*, **1993**, *16*, 1143–1161 [stability-indicating; LOD 100 ng/mL]

Soto-Otero, R.; Mendez-Alvarez, E.; Sierra-Marcuño, G. High-performance liquid chromatographic measurement of phenytoin, phenobarbital and their major metabolites in serum, brain tissue, and urine. *J.Liq.Chromatogr.*, **1988**, *11*, 3021–3040 [rat]

Vio, L.; Mamolo, M.G.; Furlan, G. Simultaneous determination of phenobarbital, phenytoin and methylphenobarbital in tablet formulations by HPLC. *Farmaco.[Prat].*, **1988**, *43*, 157–164

Asberg, A.; Haffner, F. Analysis of serum concentration of phenobarbital, phenytoin, carbamazepine and carbamazepine 10,11-epoxide by solvent-recycled liquid chromatography. *Scand.J.Clin.Lab.Invest.*, **1987**, *47*, 389–392

Soto-Otero, R.; Méndez-Alvarez, E.; Sierra-Marcuño, G. Simultaneous determination of ethosuximide, phenobarbital, phenytoin, and carbamazepine in brain tissue by HPLC. *J.Liq.Chromatogr.*, **1985**, *8*, 753–763

Gerson, B.; Bell, F.; Chan, S. Antiepileptic agents–primidone, phenobarbital, phenytoin, and carbamazepine by reversed-phase liquid chromatography. *Clin.Chem.*, **1984**, *30*, 105–108

Greenblatt, D.J.; Matlis, R.; Abernethy, D.R.; Oche, H.R. Improved liquid chromatographic analysis of phenytoin and salicylate using radial compression separation. *J.Chromatogr.*, **1983**, *275*, 450–457

Kabra, P.M.; Nelson, M.A.; Marton, L.J. Simultaneous very fast liquid-chromatographic analysis of ethosuximide, primidone, phenobarbital, phenytoin, and carbamazepine in serum. *Clin.Chem.*, **1983**, *29*, 473–476

Neels, H.M.; Totte, J.A.; Verkerk, R.M.; Vlietinck, A.J.; Scharpe, S.L. Simultaneous high performance liquid-chromatographic determination of carbamazepine, carbamazepine-10,11-epoxide, ethosuximide, phenobarbital, phenytoin, primidone and phenylethylmalonamide in plasma. *J.Clin. Chem.Clin.Biochem.*, **1983**, *21*, 295–299

Turnell, D.C.; Trevor, S.C.; Cooper, J.D. A rapid procedure for the simultaneous estimation of the anticonvulsant drugs, ethosuximide, phenobarbitone, phenytoin, and carbamazepine in serum using high-pressure liquid chromatography. *Ann Clin.Biochem.*, **1983**, *20 Pt 1*, 37–40

Kinberger, B.; Holmen, A. Analysis for carbamazepine and phenytoin in serum with a high-speed liquid chromatography system (Perkin-Elmer). *Clin.Chem.*, **1982**, *28*, 718–719

# Piroxicam

**Molecular formula:** $C_{15}H_{13}N_3O_4S$
**Molecular weight:** 331.4
**CAS Registry No.:** 36322-90-4, 87234-24-0 (cinnamic acid ester),
85056-47-9 (piroxicam olamine)

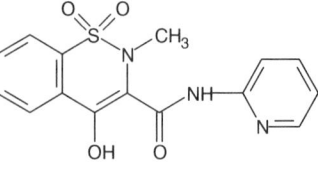

## SAMPLE
**Matrix:** bile, blood, urine
**Sample preparation:** Plasma. 500 µL Plasma + 1 mL 10 mM NaOH + 100 µL 100 µg/mL isoxicam in 10 mM NaOH + 250 µL 1 M HCl + 5 mL diethyl ether, shake for 6 min at 240 rpm on an orbital shaker, centrifuge at 4° at 900 g for 10 min. Remove the organic layer and evaporate it to dryness under a stream of nitrogen at 37°, reconstitute the residue in 250 µL MeCN:water 50:50, vortex for 1 min, inject a 50 µL aliquot. Urine, bile. 500 µL Urine or bile + 1 mL 10 mM NaOH + 100 µL 100 µg/mL isoxicam in 10 mM NaOH + 250 µL 1 M HCl + 5 mL diethyl ether, shake for 6 min at 240 rpm on an orbital shaker, centrifuge at 4° at 900 g for 10 min. Remove the organic layer and add it to 2 mL pH 4.9 citric acid-phosphate buffer, shake, centrifuge. Remove the organic layer and evaporate it to dryness under a stream of nitrogen at 37°, reconstitute the residue in 250 µL MeCN:water 50:50, vortex for 1 min, inject a 50 µL aliquot.

## HPLC VARIABLES
**Guard column:** Techsil C10 CN (HPLC Technology)
**Column:** 200 × 3.9 10 µm Techsil C10 CN (HPLC Technology)
**Mobile phase:** MeCN:water 22:78 (plasma, urine) or 10:90 (bile) containing 50 mM $NaH_2PO_4$, pH 3.5
**Flow rate:** 2.5
**Injection volume:** 50
**Detector:** UV 365

## CHROMATOGRAM
**Retention time:** 4 (plasma), 6.8 (bile)
**Internal standard:** isoxicam (8 (plasma), 17 (bile))
**Limit of detection:** 50 ng/mL

## OTHER SUBSTANCES
**Extracted:** metabolites

## KEY WORDS
plasma; pharmacokinetics; do not use plasticware

## REFERENCE
Milligan, P.A. Determination of piroxicam and its major metabolites in the plasma, urine and bile of humans by high-performance liquid chromatography. *J.Chromatogr.*, **1992**, *576*, 121–128

## SAMPLE
**Matrix:** blood
**Sample preparation:** 500 µL Plasma + 20 µL ketoprofen solution, mix, add 500 µL MeCN, vortex for 10 s, centrifuge at 1500 g for 10 min. Remove 800 µL of the supernatant and add it to 5 mL dichloromethane, vortex for 30 s, centrifuge for 10 min, inject a 50 µL aliquot of the aqueous layer.

## HPLC VARIABLES
**Guard column:** 20 × 4.6 10 µm Spherisorb C8

**Column:** 250 × 4.6 5 μm Spherisorb ODS
**Mobile phase:** MeOH:buffer 40:60 (Buffer was 40 mM Na$_2$HPO$_4$ adjusted to pH 8 with orthophosphoric acid.)
**Column temperature:** 50
**Flow rate:** 1
**Injection volume:** 50
**Detector:** UV 360

## CHROMATOGRAM
**Retention time:** 4.4
**Internal standard:** ketoprofen (8.6)
**Limit of detection:** 100 ng/mL
**Limit of quantitation:** 300 ng/mL

## KEY WORDS
plasma

## REFERENCE
Edno, L.; Bressolle, F.; Combe, B.; Galtier, M. A reproducible and rapid HPLC assay for quantitation of piroxicam in plasma. *J.Pharm.Biomed.Anal.*, **1995**, *13*, 785–789

## SAMPLE
**Matrix:** blood
**Sample preparation:** 1 mL Plasma + 100 μL 20 μg/mL isoxicam in MeOH + 200 μL 1 M HCl, vortex at slow speed for 30 s, add 10 mL dichloromethane, shake vigorously for 30 s, centrifuge at 2500 rpm for 5 min. Remove the organic phase and add it to 20 mg anhydrous sodium sulfate, filter, evaporate to dryness under a stream of nitrogen at 35°, reconstitute the residue in 200 μL MeOH, vortex for 30 s, centrifuge at 15000 g for 5 min, inject 20 μL of the supernatant.

## HPLC VARIABLES
**Column:** 250 × 4 5 μm LiChrospher 60 RP-Select B
**Mobile phase:** MeOH:water:acetic acid 48:45:7, pH 2.47
**Flow rate:** 1.1
**Injection volume:** 20
**Detector:** UV 340

## CHROMATOGRAM
**Retention time:** 6.1
**Internal standard:** isoxicam (9.75)
**Limit of quantitation:** 20 ng/mL

## OTHER SUBSTANCES
**Simultaneous:** droxicam

## KEY WORDS
protect from light; plasma; pharmacokinetics

## REFERENCE
Maya, M.T.; Pais, J.P.; Morais, J.A. A rapid method for the determination of piroxicam in plasma by high-performance liquid chromatography. *J.Pharm.Biomed.Anal.*, **1995**, *13*, 319–322

## SAMPLE
**Matrix:** blood
**Sample preparation:** 2 mL Whole blood or plasma + 2 mL buffer + 5 mL chloroform: isopropanol:n-heptane 60:14:26, shake gently horizontally for 10 min, centrifuge at 2800

g for 10 min. Remove the lower organic layer and evaporate it to dryness under vacuum at 45°, reconstitute the residue in 100 μL mobile phase, centrifuge at 2800 g for 5 min, inject a 50 μL aliquot of the supernatant. (Buffer was saturated ammonium chloride solution 25% diluted with water, adjusted to pH 9.5 with 25% ammonia solution.)

## HPLC VARIABLES
**Column:** 300 × 3.9 4 μm NovaPack C18
**Mobile phase:** MeOH:THF:buffer 65:5:30 (Buffer was 0.68 g/L (10 mM (sic)) $KH_2PO_4$ adjusted to pH 2.6 with concentrated orthophosphoric acid.) (At the end of each session wash the column with water for 1 h and MeOH for 1 h, re-equilibrate for 30 min.)
**Column temperature:** 30
**Flow rate:** 0.8
**Injection volume:** 50
**Detector:** UV 338

## CHROMATOGRAM
**Retention time:** 4.20
**Limit of detection:** <120 ng/mL

## OTHER SUBSTANCES
**Extracted:** acebutolol, acenocoumarol, acepromazine, aceprometazine, acetaminophen, aconitine, albuterol, alimemazine, alminoprofen, alpidem, alprenolol, amisulpride, amitriptyline, amodiaquine, amoxapine, aspirin, astemizole, atenolol, benazepril, benperidol, benzocaine, benzoylecgonine, bepridil, betaxolol, bisoprolol, bromazepam, brompheniramine, bumadizone, bupivacaine, buprenorphine, buspirone, caffeine, carbamazepine, carbinoxamine, carpipramine, carteolol, cetirizine, chlorambucil, chlordiazepoxide, chlormezanone, chlorophenacinone, chloroquine, chlorpheniramine, chlorpromazine, chlorpropamide, cibenzoline, cicletanine, clemastine, clomipramine, clonazepam, clonidine, clorazepate, clozapine, cocaine, codeine, colchicine, cyamemazine, cyclizine, cyproheptadine, cytarabine, dacarbazine, daunorubicin, demexiptiline, desipramine, dextromoramide, dextropropoxyphene, diazepam, diazoxide, diclofenac, dihydralazine, diltiazem, diphenhydramine, dipyridamole, dosulepine, doxepin, doxylamine, droperidol, ephedrine, estazolam, etodolac, fenfluramine, fenoprofen, fentiazac, flecainide, floctafenine, flumazenil, flunitrazepam, fluoxetine, fluphenazine, flurbiprofen, fluvoxamine, glibenclamide, glibornuride, glipizide, glutethimide, haloperidol, histapyrrodine, hydroxychloroquine, hydroxyzine, ibuprofen, imipramine, indomethacin, iproniazid, ketoprofen, labetalol, levomepromazine, lidocaine, lidoflazine, lisinopril, loperamide, loprazolam, loratadine, loxapine, maprotiline, medazepam, medifoxamine, mefenamic acid, mefenidramine, mefloquine, melphalan, meperidine, mephenesin, mepivacaine, metapramine, metformin, methadone, methocarbamol, methotrexate, metipranolol, metoclopramide, mexiletine, mianserine, midazolam, moclobemide, moperone, morphine, nadolol, nalbuphine, nalorphine, naloxone, naltrexone, naproxen, nialamide, nicardipine, niflumic acid, nimodipine, nitrendipine, nizatidine, nomifensine, nortriptyline, omeprazole, opipramol, oxprenolol, penbutolol, penfluridol, phencyclidine, phenobarbital, phenol, phenylbutazone, pimozide, pindolol, pipamperone, prazepam, prazosin, procainamide, procarbazine, proguanil, promethazine, propafenone, propranolol, protriptyline, pyrimethamine, quinupramine, ramipril, ranitidine, reserpine, ritodrine, secobarbital, sotalol, strychnine, sulfinpyrazole, sulpride, sultopride, suriclone, temazepam, tenoxicam, terbutaline, terfenadine, tetracaine, tetrazepam, thiopental, thioproperazine, thioridazine, tianeptine, tiapride, ticlopidine, timolol, tioclomarol, tolbutamide, toloxatone, trazodone, triazolam, trifluoperazine, trifluperidol, trimipramine, triprolidine, tropatenine, verapamil, viloxazine, vinblastine, vincristine, vindesine, warfarin, zolpidem, zorubicine
**Interfering:** ajmaline, alprazolam, celiprolol, clobazam, cycloguanil, debrisoquine, dextromethorphan, disopyramide, ketamine, lorazepam, mephentermine, methaqualone, metoprolol, minoxidil, nifedipine, nitrazepam, oxazepam, pentazocine, prilocaine, quinidine, quinine, sulindac, tiaprofenic acid, tofisopam, yohimbine, zopiclone

## KEY WORDS
whole blood; plasma

## REFERENCE
Tracqui, A.; Kintz, P.; Mangin, P. Systematic toxicological analysis using HPLC/DAD. *J.Forensic Sci.*, **1995**, *40*, 254–262

## SAMPLE
**Matrix:** blood
**Sample preparation:** Inject 20 µL plasma onto column A with mobile phase A, after 1 min switch mobile phase A to drain and leave column A to sit for 1 min. Elute the contents of column A onto column B with mobile phase B. After 1 min remove column A from circuit and re-equilibrate it with mobile phase A. Elute column B with mobile phase B and monitor the effluent.

## HPLC VARIABLES
**Column:** A 100 × 2 40 µm Bond Elut C2 (Chrompack) (Before use condition with MeCN, water, mobile phase A. Replace column after 40 injections.); B 20 × 4.6 Supelcosil LC18 DB + 150 × 4.6 5 µm Supelcosil LC18 DB
**Mobile phase:** A 150 mM $NaH_2PO_4$ adjusted to pH 3.5 with orthophosphoric acid; B MeCN:20 mM $Na_2HPO_4$ + 0.5 mM triethylamine adjusted to pH 3.1 with orthophosphoric acid 60:40
**Flow rate:** A 1; B 1
**Injection volume:** 20
**Detector:** UV 331

## CHROMATOGRAM
**Retention time:** 5.25
**Limit of detection:** 100 ng/mL
**Limit of quantitation:** 200 ng/mL

## KEY WORDS
plasma; column-switching

## REFERENCE
Saeed, K.; Becher, M. On-line solid-phase extraction of piroxicam prior to its determination by high-performance liquid chromatography. *J.Chromatogr.*, **1991**, *567*, 185–193

## SAMPLE
**Matrix:** blood
**Sample preparation:** Animal. 100 µL Plasma + 100 µL 1 M HCl + 100 µL 20 µg/mL naproxen in 2% MeOH + 6 mL MTBE, vortex for 1 min, centrifuge at 1200 g for 2 min. Remove the organic layer and evaporate it to dryness under reduced pressure at 35°. Reconstitute in 1 mL 2% MeOH, vortex for 2 min, inject a 100 µL aliquot. Human. 500 µL Plasma + 200 µL 1 M HCl + 100 µL 100 µg/mL naproxen in 2% MeOH + 8 mL diethyl ether, vortex for 1 min, centrifuge at 1200 g for 2 min. Remove the organic layer and evaporate it to dryness under reduced pressure at 35°. Reconstitute in 2 mL 2% MeOH, vortex for 2 min, inject a 100 µL aliquot.

## HPLC VARIABLES
**Column:** 50 × 4.6 5 µm Spherisorb C8
**Mobile phase:** MeCN:THF:MeOH:2% acetic acid (pH 2.5) 5:18:5:72
**Flow rate:** 1.5
**Injection volume:** 100
**Detector:** UV 330 (animal); UV 360 (human)

## CHROMATOGRAM
**Retention time:** 2.2
**Internal standard:** naproxen (8) (determined at UV 330 nm in each case)
**Limit of quantitation:** 250 ng/mL

## OTHER SUBSTANCES
**Extracted:** metabolites, 5′-hydroxypiroxicam

## KEY WORDS
plasma; dog; rat; human

## REFERENCE
Gillilan, R.B.; Mason, W.D.; Fu, C.-H.J. Rapid analysis of piroxicam in dog, rat and human plasma by high-performance liquid chromatography. *J.Chromatogr.*, **1989**, *487*, 232–235

## SAMPLE
**Matrix:** blood
**Sample preparation:** 50 μL Serum + 20 μL 100 mM pH 4.8 citrate buffer + 5 μL 30 μg/mL isoxicam in MeOH + 3 mL dichloromethane, shake for 45 s, centrifuge at 4000 rpm. Remove the organic layer and evaporate it to dryness under a stream of nitrogen at 40°, reconstitute the residue in 100 μL mobile phase, mix for 10 s, centrifuge, inject a 20 μL aliquot.

## HPLC VARIABLES
**Column:** 150 × 4.6 7 μm Separon SGX CN
**Mobile phase:** MeCN:10 mM phosphoric acid 70:30
**Flow rate:** 0.4
**Injection volume:** 20
**Detector:** UV 350

## CHROMATOGRAM
**Retention time:** 4.10
**Internal standard:** isoxicam (2.74)
**Limit of detection:** 40 ng/mL

## KEY WORDS
serum

## REFERENCE
Migulla, H.; Alken, R.G.; Hüller, H. Mikromethode zur Bestimmung der Piroxicamkonzentration im Serum [Micromethod for the determination of piroxicam concentration in serum]. *Pharmazie*, **1988**, *43*, 866–867

## SAMPLE
**Matrix:** blood, exudate
**Sample preparation:** 150 μL Plasma or exudate + 50 μL 5 μg/mL 6-methylpiroxicam + 200 μL 1 M pH 2 phosphate buffer, vortex, add 2 mL dichloromethane, rotate for 10 min, centrifuge. Remove the organic layer and evaporate it to dryness under a stream of nitrogen at 45°, reconstitute the residue in 120 μL mobile phase, inject a 100 μL aliquot.

## HPLC VARIABLES
**Column:** 200 × 4.6 Hypersil ODS
**Mobile phase:** MeOH:pH 8 phosphate buffer 45:55
**Flow rate:** 1

**Injection volume:** 100
**Detector:** UV 360

## CHROMATOGRAM
**Internal standard:** 6-methylpiroxicam
**Limit of detection:** 10 ng/mL

## KEY WORDS
plasma; rat; pharmacokinetics

## REFERENCE
Stevens, A.J.; Martin, S.W.; Brennan, B.S.; McLachlan, A.; Gifford, L.A.; Rowland, M.; Houston, J.B. Regional drug delivery II: Relationship between drug targeting index and pharmacokinetic parameters for three non-steroidal anti-inflammatory drugs using the rat air pouch model of inflammation. *Pharm.Res.*, **1995**, *12*, 1987–1996

## SAMPLE
**Matrix:** blood, tissue
**Sample preparation:** 1 mL Plasma or tissue + 50 μL 100 μg/mL isoxicam in MeOH + 700 mg potassium carbonate + 1 mL THF + 500 μL EtOH, vortex for 30 s, centrifuge at 2000 g for 5 min. Remove the supernatant and evaporate it to dryness under a stream of nitrogen at 70°. Reconstitute residue in 100 μL THF, vortex for 5 s, filter (0.5 μm), inject a 20 μL aliquot.

## HPLC VARIABLES
**Column:** 150 × 3.9 3 μm Novapak C18
**Mobile phase:** THF:water 45:55 with 1% acetic acid and 5 mM 1-heptanesulfonic acid (PIC B-7, Waters)
**Flow rate:** 0.7
**Injection volume:** 20
**Detector:** UV 313

## CHROMATOGRAM
**Retention time:** 4.2
**Internal standard:** isoxicam (8.0)
**Limit of detection:** 100 ng/mL

## KEY WORDS
plasma; rat; skin; muscle

## REFERENCE
Cerretani, D.; Micheli, L.; Fiaschi, A.I.; Giorgi, G. Rapid and sensitive determination of piroxicam in rat plasma, muscle and skin by high-performance liquid chromatography. *J.Chromatogr.*, **1993**, *614*, 103–108

## SAMPLE
**Matrix:** blood, urine
**Sample preparation:** Plasma. 500 μL Plasma + 200 μL 10% trichloroacetic acid + 100 μL 15 μg/mL naproxen in MeOH + 700 μL MeOH, vortex for 30 s, centrifuge at 2105 g for 10 min, inject a 100 μL aliquot of the supernatant. Urine. 1 mL Urine + 400 μL 1 M NaOH, vortex, let stand at room temperature for 2 h, add 300 μL MeOH, add 100 μL 15 μg/mL naproxen in MeOH, mix, inject a 100 μL aliquot.

## HPLC VARIABLES
**Guard column:** 100 × 2 Spherisorb pellicular revered-phase
**Column:** 250 × 4.5 5 μm Spherisorb C18

**Mobile phase:** MeCN : 100 mM sodium acetate 33 : 67, adjusted to pH 3.3 with glacial acetic acid
**Column temperature:** 40
**Flow rate:** 2.5
**Injection volume:** 100
**Detector:** UV 330

## CHROMATOGRAM
**Retention time:** 2.6
**Internal standard:** naproxen (3.2)
**Limit of detection:** 50 ng/mL

## OTHER SUBSTANCES
**Extracted:** metabolites

## KEY WORDS
plasma

## REFERENCE
Avgerinos, A.; Axarlis, S.; Dragatsis, J.; Karidas, T.; Malamataris, S. Extractionless high-performance liquid chromatographic method for the simultaneous determination of piroxicam and 5'-hydroxypiroxicam in human plasma and urine. *J.Chromatogr.B*, **1995**, *673*, 142–146

## SAMPLE
**Matrix:** blood, urine
**Sample preparation:** 500 µL Plasma or urine + 1 mL 100 mM NaOH + 100 µL 100 µg/mL isoxicam in 10 mM NaOH + 250 µL 1 M HCl + 5 mL diethyl ether, shake mechanically for 5 min, centrifuge at 1150 g at 4° for 5 min. Remove the ether layer and evaporate it to dryness at 35° under a stream of nitrogen. Reconstitute the residue in 250 µL 50 mM TRIS, inject a 50 µL aliquot.

## HPLC VARIABLES
**Column:** 150 × 3.9 10 µm µBondapak CN (plasma) or 300 × 3.9 10 µm µBondapak C18 (urine)
**Mobile phase:** MeCN : water 25 : 75 containing 50 mM $NaH_2PO_4$, final pH 3.2 (plasma) or THF : 5 mM sodium octylsulfonate buffer : glacial acetic acid 45 : 54 : 1 (urine)
**Flow rate:** 1.5
**Injection volume:** 50
**Detector:** UV 365

## CHROMATOGRAM
**Retention time:** 3 (plasma), 4 (urine)
**Internal standard:** isoxicam (6 (plasma), 7 (urine))
**Limit of detection:** 50 ng/mL (plasma); 100 ng/mL (urine)

## OTHER SUBSTANCES
**Extracted:** 5'-hydroxypiroxicam

## KEY WORDS
plasma

## REFERENCE
Richardson, C.J.; Ross, S.G.; Blocka, K.L.; Verbeeck, R.K. High-performance liquid chromatographic analysis of piroxicam and its major metabolite 5'-hydroxypiroxicam in human plasma and urine. *J.Chromatogr.*, **1986**, *382*, 382–388

## SAMPLE
**Matrix:** feed, formulations
**Sample preparation:** Capsules. Weigh out contents of a capsule (5-20 mg piroxicam), shake with 70 mL 10 mM HCl in MeOH for 30 min, make up to 100 mL with 10 mM HCl in MeOH, centrifuge at 2000 rpm for 10 min, dilute to 50 μg/mL with 10 mM HCl in MeOH before analysis, inject a 25 μL aliquot, use UV determination at 254 nm. Tablets (Formulation I). Disintegrate tablet with 5 mL water, shake with 70 mL 10 mM HCl in MeOH for 30 min, make up to 100 mL with 10 mM HCl in MeOH, centrifuge at 2000 rpm for 10 min, dilute to 50 μg/mL with 10 mM HCl in MeOH before analysis, inject a 25 μL aliquot, use UV determination at 254 nm. Tablets (Formulation II). Shake with 70 mL 10 mM NaOH in MeOH for 30 min, make up to 100 mL with MeOH:water 45:55, centrifuge at 2000 rpm for 10 min, dilute to 50 μg/mL with MeOH:water 45:55 before analysis, inject a 25 μL aliquot, use UV determination at 254 nm. Ointment. Dissolve 500 mg 10% ointment in 4 mL chloroform, warm if necessary, make up to 100 mL with 10 mM HCl in MeOH, mix, filter (Whatman GF/A), discard first few mL of filtrate, dilute to tenfold with 10 mM HCl in MeOH before analysis, inject a 25 μL aliquot, use UV detection at 340 nm. Suppositories. Weigh out, dissolve completely in 20 mL warm MeOH:isopropanol 1:1, add 60 mL mobile phase, shake for 20 min, make up to 100 mL with mobile phase, mix, filter (Whatman GF/A), dilute if necessary with mobile phase, inject a 25 μL aliquot, use UV determination at 254 nm. Ophthalmic suspensions. Shake until homogeneous, dilute a 5 mL aliquot to 100 mL with MeOH, dilute 10 fold with mobile phase, adjust pH to 4.0, inject a 20 μL aliquot, adjust mobile phase pH to 4.0. Rodent feed. Shake a 10 g sample with 25 mL MeCN:glacial acetic acid 90:10 for 30 min (or up to 6 h for aged samples), centrifuge at 1500 rpm for 10 min, inject a 50 μL aliquot, use UV determination at 365 nm, adjust mobile phase pH to 4.0.

## HPLC VARIABLES
**Column:** 300 × 3.9 10 μm Waters ODS, cat no. 27324
**Mobile phase:** MeOH:buffer 45:55 (Buffer was 400 mL 100 mM citric acid, add 200 mM $Na_2HPO_4.7H_2O$ until pH = 3 (about 102 mL), dilute to 1 L with water.)
**Flow rate:** 1.2
**Injection volume:** 25-50
**Detector:** UV 254; UV 340; UV 365

## CHROMATOGRAM
**Retention time:** 10

## OTHER SUBSTANCES
**Also analyzed:** impurities, degradation products, metabolites, piroxicam analogues

## KEY WORDS
capsules; tablets; ointments; suppositories; ophthalmic suspensions; stability-indicating

## REFERENCE
Richards, J.A.; Cole, D.A.; Hickey, R.J.; Sokol, L.S.-W. High performance liquid chromatography assay for piroxicam in pharmaceutical products. *J.Chromatogr.Sci.*, **1987**, *25*, 292–295

## SAMPLE
**Matrix:** solutions

## HPLC VARIABLES
**Column:** 125 × 3 Ecocart LiChrospher 100 RP-18
**Mobile phase:** Isopropanol:100 mM $KH_2PO_4$:formic acid 54:100:0.1
**Flow rate:** 0.6
**Detector:** UV 254

## CHROMATOGRAM
**Retention time:** 2.3
**Limit of quantitation:** 200-500 ng/mL

## OTHER SUBSTANCES
**Simultaneous:** acemetacin, diclofenac, flurbiprofen, indomethacin, ketoprofen, lonazolac, naproxen, sulindac, tenoxicam

## REFERENCE
Baeyens, W.R.G.; Van Der Weken, G.; Van Overbeke, A.; Zhang, Z.D. Preliminary results on the LC-separation of non-steroidal anti-inflammatory agents in conventional and narrow-bore RP set-ups applying columns with different internal diameters. *Biomed.Chromatogr.*, **1995**, *9*, 261–262

## SAMPLE
**Matrix:** solutions

## HPLC VARIABLES
**Column:** 250 × 4.6 5 μm Supelcosil LC-DP (A) or 250 × 4 5 μm LiChrospher 100 RP-8 (B)
**Mobile phase:** MeCN:0.025% phosphoric acid:buffer 25:10:5 (A) or 60:25:15 (B) (Buffer was 9 mL concentrated phosphoric acid and 10 mL triethylamine in 900 mL water, adjust pH to 3.4 with dilute phosphoric acid, make up to 1 L.)
**Flow rate:** 0.6
**Injection volume:** 25
**Detector:** UV 229

## CHROMATOGRAM
**Retention time:** 6.84 (A), 6.67 (B)

## OTHER SUBSTANCES
**Also analyzed:** acebutolol, acepromazine, acetaminophen, acetazolamide, acetophenazine, albuterol, alprazolam, amitriptyline, amobarbital, amoxapine, antipyrine, atenolol, atropine, azatadine, baclofen, benzocaine, bromocriptine, brompheniramine, brotizolam, bupivacaine, buspirone, butabarbital, butalbital, caffeine, carbamazepine, cetirizine, chlorcyclizine, chlordiazepoxide, chlormezanone, chloroquine, chlorpheniramine, chlorpromazine, chlorpropamide, chlorprothixene, chlorthalidone, chlorzoxazone, cimetidine, cisapride, clomipramine, clonazepam, clonidine, clozapine, cocaine, codeine, colchicine, cyclizine, cyclobenzaprine, dantrolene, desipramine, diazepam, diclofenac, diflunisal, diltiazem, diphenhydramine, diphenidol, diphenoxylate, dipyridamole, disopyramide, dobutamine, doxapram, doxepin, droperidol, encainide, ethidium bromide, ethopropazine, fenoprofen, fentanyl, flavoxate, fluoxetine, fluphenazine, flurazepam, flurbiprofen, fluvoxamine, furosemide, glutethimide, glyburide, guaifenesin, haloperidol, homatropine, hydralazine, hydrochlorothiazide, hydrocodone, hydromorphone, hydroxychloroquine, hydroxyzine, ibuprofen, imipramine, indomethacin, ketoconazole, ketoprofen, ketorolac, labetalol, levorphanol, lidocaine, loratadine, lorazepam, lovastatin, loxapine, mazindol, mefenamic acid, meperidine, mephenytoin, mepivacaine, mesoridazine, metaproterenol, metformin, methadone, methdilazine, methocarbamol, methotrexate, methotrimeprazine, methoxamine, methyldopa, methylphenidate, metoclopramide, metolazone, metoprolol, metronidazole, midazolam, moclobemide, morphine, nadolol, nalbuphine, naloxone, naphazoline, naproxen, nifedipine, nizatidine, norepinephrine, nortriptyline, oxazepam, oxycodone, oxymetazoline, paroxetine, pemoline, pentazocine, pentobarbital, pentoxifylline, perphenazine, pheniramine, phenobarbital, phenol, phenolphthalein, phentolamine, phenylbutazone, phenyltoloxamine, phenytoin, pimozide, pindolol, pramoxine, prazepam, prazosin, probenecid, procainamide, procaine, prochlorperazine, procyclidine, promazine, promethazine, propafenone, propantheline, propiomazine, propofol, propranolol, protriptyline, quazepam, quinidine, quinine, racemethorphan, ranitidine, remoxipride, risperidone, salicylic acid, scopolamine, secobarbital, sertraline, sotalol, spironolactone, sulfinpyrazone, sulindac, temazepam, terbutaline, terfenadine, tetracaine, theophylline,

thiethylperazine, thiopental, thioridazine, thiothixene, timolol, tocainide, tolbutamide, tolmetin, trazodone, triamterene, triazolam, trifluoperazine, triflupromazine, trimeprazine, trimethoprim, trimipramine, verapamil, warfarin, xylometazoline, yohimbine, zopiclone

## KEY WORDS
some details of plasma extraction

## REFERENCE
Koves, E.M. Use of high-performance liquid chromatography-diode array detection in forensic toxicology. *J.Chromatogr.A*, **1995**, *692*, 103–119

## ANNOTATED BIBLIOGRAPHY

Troconiz, J.I.; Lopez-Bustamante, L.G.; Fos, D. High-performance liquid chromatographic analysis of piroxicam and tenoxicam in plasma, blood and buffer solution. Application to pharmacokinetic studies in small laboratory animals. *Arzneimittelforschung*, **1993**, *43*, 679–681

Bartlett, A.; Costa, A.; Martínez, L.; Roser, R.; Sagarra, R.; Sánchez, J. The effect of antacid and ranitidine on droxicam pharmacokinetics. *J.Clin.Pharmacol.*, **1992**, *32*, 1115–1119 [plasma; isoxicam (IS); LOD 25 ng/mL]

Jiang, X.G.; Ge, S.D.; Wanga, X.J.; Xi, N.Z. [A HPLC method for determining piroxicam in body fluids]. *Chung Kuo Yao Li Hsueh Pao*, **1991**, *12*, 381–384

Wanwimolruk, S.; Wanwimolruk, S.Z.; Zoest, A.R. A simple and sensitive HPLC assay for piroxicam in plasma and its application to bioavailability study. *J.Liq.Chromatogr.*, **1991**, *14*, 2373–2381 [isoxicam (IS); LOD 50 ng/mL; microbore]

Sánchez, J.; Martínez, L.; García-Barbal, J.; Roser, R.; Bartlett, A.; Sagarra, R. The influence of gastric emptying on droxicam pharmacokinetics. *J.Clin.Pharmacol.*, **1989**, *29*, 739–745 [plasma; LOD 100 ng/mL]

Boudinot, F.D.; Ibrahim, S.S. High-performance liquid chromatographic assay for piroxicam in human plasma. *J.Chromatogr.*, **1988**, *430*, 424–428 [6'-methylpiroxicam (IS); LOD 2 ng/mL; pharmacokinetics]

Ge, S.D.; Cheng, Q.Y.; Wang, X.J.; Xi, N.Z. [HPLC determination of piroxicam contents in piroxicam suppositories]. *Yao Hsueh Hsueh Pao*, **1988**, *23*, 38–41

Ahmed, A.N.; El-Gizawy, S.M. Use of chemically-bonded cyclodextrin stationary phase for high performance liquid chromatographic determination of Feldene capsules. *J.Chromatogr.Sci.*, **1987**, *25*, 424–426 [column temp 35]

Macek, J.; Vacha, J. Rapid and sensitive method for determination of piroxicam in human plasma by high-performance liquid chromatography. *J.Chromatogr.*, **1987**, *420*, 445–449 [column temp 35; glass column; LOD 150 ng/mL; pharmacokinetics; simultaneous antipyrine, diclofenac, hydroxychloroquine, ibuprofen, isoxicam, plaquenil, sulfamethazine]

Chen, X.L.; Xi, N.Z.; Ge, S.D.; Sun, S.L. [HPLC determination of piroxicam in human serum and its pharmacokinetic parameters]. *Yao Hsueh Hsueh Pao*, **1986**, *21*, 692–697

Dixon, J.S.; Lowe, J.R.; Galloway, D.B. Rapid method for the determination of either piroxicam or tenoxicam in plasma using high-performance liquid chromatography. *J.Chromatogr.*, **1984**, *310*, 455–459

Fraser, A.D.; Woodbury, J.F. Liquid chromatographic determination of piroxicam in serum. *Ther.Drug Monit.*, **1983**, *5*, 239–242

Twomey, T.M.; Bartolucci, S.R.; Hobbs, D.C. Analysis of piroxicam in plasma by high-performance liquid chromatography. *J.Chromatogr.*, **1980**, *183*, 104–108

# Polymyxin

**Molecular formula:** $C_{56}H_{98}N_{16}O_{13}$ ($B_1$)
**Molecular weight:** 1203.5 ($B_1$)
**CAS Registry No.:** 1406-11-7 (polymyxin), 1404-26-8 (polymyxin B), 1405-20-5 (polymyxin B sulfate)

## SAMPLE
**Matrix:** formulations
**Sample preparation:** Sandwich cream or ointment between two layers of 200 mesh silica gel, extract with carbon dioxide:MeOH 95:5 at 300 atmospheres at 55° at 2 mL/min for 75 min (restrictor 300°), sonicate the SFE tube, frits, and silica gel with MeOH:100 mM HCl 25:75 containing 0.1% Tween 80 for 15 min, filter (0.2 μm), inject an aliquot of the filtrate. (SFE removes the hydrocarbon base of the cream or ointment leaving behind the insoluble polymyxin.)

## HPLC VARIABLES
**Column:** 250 × 4.6 5 μm Synchropak SCD
**Mobile phase:** MeCN:buffer 21.5:78.5 (Buffer was 100 mM $KH_2PO_4$ containing 0.1% trifluoroacetic acid.)
**Flow rate:** 1.5
**Injection volume:** 50
**Detector:** UV 215

## CHROMATOGRAM
**Limit of quantitation:** 0.016%

## KEY WORDS
SFE; cream; ointment

## REFERENCE
Moore, W.N.; Taylor, L.T. Analytical inverse supercritical fluid extraction of polar pharmaceutical compounds from cream and ointment matrices. *J.Pharm.Biomed.Anal.*, **1994**, *12*, 1227–1232

## SAMPLE
**Matrix:** solutions

## HPLC VARIABLES
**Guard column:** 45 × 4.7 Ultrasphere C18
**Column:** 250 × 4.7 Ultrasphere C18
**Mobile phase:** Gradient. A was 0.15% trifluoroacetic acid in water. B was 0.15% trifluoroacetic acid in MeCN. A:B from 100:0 to 50:50 over 25 min.
**Flow rate:** 2
**Detector:** UV 215

## CHROMATOGRAM
**Retention time:** 18.5

## OTHER SUBSTANCES
**Simultaneous:** degradation products

## REFERENCE
Danner, R.L.; Joiner, K.A.; Rubin, M.; Patterson, W.H.; Johnson, N.; Ayers, K.M.; Parrillo, J.E. Purification, toxicity, and antiendotoxin activity of polymyxin B nonapeptide. *Antimicrob.Agents Chemother.*, **1989**, *33*, 1428–1434

## SAMPLE
**Matrix:** solutions
**Sample preparation:** Inject a 10 μL aliquot of a 1 mg/mL solution.

## HPLC VARIABLES
**Column:** 250 × 4.6 5 μm Vydac TP C18
**Mobile phase:** Gradient. A was 0.1% trifluoroacetic acid in water. B was 0.075% trifluoroacetic acid in MeCN. A:B from 90:10 to 20:80 over 20 min.
**Flow rate:** 1.2
**Injection volume:** 10
**Detector:** UV 215

## CHROMATOGRAM
**Retention time:** 12.0

## OTHER SUBSTANCES
**Simultaneous:** degradation products.

## REFERENCE
Vaara, M. Analytical and preparative high-performance liquid chromatography of the papain-cleaved derivative of polymyxin B. *J.Chromatogr.*, **1988**, *441*, 423–430

## SAMPLE
**Matrix:** solutions
**Sample preparation:** Prepare a 400-500 μg/mL solution in water, inject a 20 μL aliquot.

## HPLC VARIABLES
**Column:** 150 × 4.6 5 μm Nucleosil 5 C18
**Mobile phase:** MeCN:buffer 23:77 (Buffer was 23 mM phosphoric acid containing 10 mM acetic acid and 50 mM sodium sulfate, adjust pH to 2.5 with triethylamine.)
**Flow rate:** 0.9
**Injection volume:** 20
**Detector:** UV 220

## CHROMATOGRAM
**Retention time:** 8 (polymyxin B2), 15 (polymyxin B1)

## OTHER SUBSTANCES
**Also analyzed:** colistin (polymyxin E)

## REFERENCE
Elverdam, I.; Larsen, P.; Lund, E. Isolation and characterization of three new polymyxins in polymyxins B and E by high-performance liquid chromatography. *J.Chromatogr.*, **1981**, *218*, 653–661

## SAMPLE
**Matrix:** solutions
**Sample preparation:** Prepare a 10-100 μg/mL solution in mobile phase, inject an aliquot.

## HPLC VARIABLES
**Column:** 250 × 4.6 5 μm Ultrasphere ion-pair
**Mobile phase:** MeCN:buffer 23:77 (Prepare mobile phase by mixing 230 mL MeCN, 700 mL water, and 38 g $Na_3PO_4$, adjust pH to 3.0 with phosphoric acid, make up to 1 L with water.)
**Column temperature:** 27
**Flow rate:** 1

**Injection volume:** 10
**Detector:** UV 185; UV 200

## CHROMATOGRAM
**Retention time:** 7 (B2), 8 (B3), 13 (B1)
**Limit of detection:** 30 ng (UV 185)

## OTHER SUBSTANCES
**Simultaneous:** colistin

## REFERENCE
Whall, T.J. High-performance liquid chromatography of polymyxin B sulfate and colistin sulfate. *J.Chromatogr.*, **1981**, *208*, 118–123

## SAMPLE
**Matrix:** solutions
**Sample preparation:** Inject a 5-50 µL aliquot of a 1 mg/mL solution in water.

## HPLC VARIABLES
**Column:** 200 × 4 Nucleosil 5C18
**Mobile phase:** MeCN:buffer 22.5:77.5 (Buffer was 5 mM pH 3.0 tartrate buffer containing 5 mM sodium 1-butanesulfonate and 50 mM sodium sulfate.)
**Flow rate:** 1
**Injection volume:** 5-50
**Detector:** UV 220

## CHROMATOGRAM
**Retention time:** 11 (polymyxin $B_2$), 13 (polymyxin $B_3$), 23.5 (polymyxin $B_1$)

## OTHER SUBSTANCES
**Simultaneous:** colistin, polymyxin C, polymyxin D, polymyxin E, polymyxin M, polymyxin S, polymyxin T

## REFERENCE
Terabe, S.; Konaka, R.; Shoji, J. Separation of polymyxins and octapeptins by high-performance liquid chromatography. *J.Chromatogr.*, **1979**, *173*, 313–320

## ANNOTATED BIBLIOGRAPHY
Kitamura-Matsunaga, H.; Kimura, Y.; Araki, T.; Izumiya, N. A sodium-containing polymyxin derived from polymyxin-complex during chromatography. *J.Antibiot.(Tokyo)*, **1984**, *37*, 1605–1610 [for colistin, polymyxin E; analytical; preparative]

# Pravastatin

**Molecular formula:** C$_{23}$H$_{36}$O$_7$
**Molecular weight:** 423.5
**CAS Registry No.:** 81131-70-6 (pravastatin sodium),
81093-37-0 (pravastatin)

## SAMPLE
**Matrix:** blood
**Sample preparation:** Condition an immobilized antibody column (preparation details in paper) with 4 mL water, 4 mL MeOH:water 80:20, and 4 mL PBS. Add 1 mL plasma to the column, let stand for 15 min, wash with two 4 mL portions of water, wash with 4 mL PBS, wash with two 4 mL portions of water, elute with 4 mL MeOH. Add 100 µL 100 ng/mL IS in water to the eluate, evaporate to dryness, reconstitute with 100 µL DMF, add 100 µL 47.6 µg/mL triethylamine in dioxane (Caution! Dioxane is a carcinogen!), add 100 µL 51.8 µL/mL diethyl phosphorocyanidate in dioxane, add 100 µL 100 µg/mL N-dansylethylenediamine (Molecular Probes, Inc., Eugene OR) in dioxane, stir for 10 s, let stand at room temperature for 10 min, inject a 10 µL aliquot onto column A and elute to waste with mobile phase A. After 7.5 min elute the contents of column A onto column B with mobile phase B. After 2.5 min remove column A from the circuit. Elute column B with mobile phase B and monitor the effluent from column B.

## HPLC VARIABLES
**Column:** A 150 × 4.6 Ultron 300-C4 (Shinwa Chemical Industries); B 150 × 6 Cosmosil 5C18-AR (Nacalai Tesque)
**Mobile phase:** A MeCN:10 mM pH 2.4 citric acid 30:70; B MeCN:5 mM pH 2.6 citric acid 50:50
**Flow rate:** 1
**Injection volume:** 10
**Detector:** F ex 325 (40 mW He-Cd laser); F ex 350 em 530; UV 239

## CHROMATOGRAM
**Retention time:** 17.3
**Internal standard:** R-416 (Sankyo) (19)
**Limit of quantitation:** 0.1 ng/mL (laser fluorescence)

## KEY WORDS
column-switching; dog; rat; laser; plasma; SPE; derivatization; pharmacokinetics

## REFERENCE
Dumousseaux, C.; Muramatsu, S.; Takasaki, W.; Takahagi, H. Highly sensitive and specific determination of pravastatin sodium in plasma by high-performance liquid chromatography with laser-induced fluorescence detection after immobilized antibody extraction. *J.Pharm.Sci.*, **1994**, *83*, 1630–1636

## SAMPLE
**Matrix:** blood
**Sample preparation:** Condition a 1 mL C2 Bond-Elut SPE cartridge with 2 mL water, 2 mL MeOH, and 2 mL water. 1 mL Plasma + 1 mL 100 mM pH 7.2 KH$_2$PO$_4$, add to the SPE cartridge, wash with 2 mL water, elute with 500 µL MeCN:water 75:25. Evaporate the eluate to dryness under reduced pressure, reconstitute with 80 µL IS solution, cen-

trifuge, inject a 50 μL aliquot. (Prepare the IS solution by mixing 2 mL 100 μg/mL simvastatin β-hydroxyacid in MeOH, 1 mL MeOH, and 1 mL 100 mM pH 5 KH₂PO₄.)

## HPLC VARIABLES
**Guard column:** 20 mm long Supelguard LC-18
**Column:** 50 × 4.6 3 μm Supelcosil LC-18
**Mobile phase:** MeCN:50 mM pH 3.5 ammonium phosphate 26:74
**Column temperature:** 50
**Flow rate:** 1.6
**Injection volume:** 50
**Detector:** UV 238

## CHROMATOGRAM
**Retention time:** 3.0
**Internal standard:** simvastatin β-hydroxyacid (4.8)
**Limit of detection:** 2 ng/mL

## KEY WORDS
pharmacokinetics; plasma; SPE

## REFERENCE
Iacona, I.; Regazzi, M.B.; Buggia, I.; Villani, P.; Fiorito, V.; Molinaro, M.; Guarnone, E. High-performance liquid chromatography determination of pravastatin in plasma. *Ther.Drug Monit.*, **1994**, *16*, 191–195

## SAMPLE
**Matrix:** blood, feces, urine
**Sample preparation:** Plasma. 2 mL Plasma + 4 mL MeCN, centrifuge at 700 g for 10 min, remove the supernatant and wash the precipitate twice with 2 mL MeCN:water 2:1. Combine the supernatants and evaporate them to dryness under vacuum, reconstitute the residue in 1 mL MeCN:water 2:1, centrifuge at 10000 g, remove the supernatant and add it to 500 μL water, centrifuge at 10000 g, inject a 1 mL aliquot. Urine. Centrifuge at 10000 g, inject an aliquot. Feces. 1 g Homogenized feces + 2 mL MeCN, sonicate for 5 min, shake in a wrist action shaker for 20 min, centrifuge at 700 g for 10 min. Remover the supernatant and wash the precipitate twice with 1 mL MeCN:water 2:1, combine the supernatants, inject a 500 μL aliquot.

## HPLC VARIABLES
**Guard column:** present but not specified
**Column:** 500 × 9.4 Partisil 10 ODS-3 C18
**Mobile phase:** Gradient. MeCN:10 mM pH 7.2 potassium phosphate buffer containing 5 mM tetrabutylammonium hydrogen sulphate at 25:75 for 20 min, then to 50:50 over 45 min, hold at 50:50 for 10 min.
**Flow rate:** 4
**Injection volume:** 500-1000
**Detector:** Collect fractions and measure radioactivity; UV 245

## CHROMATOGRAM
**Retention time:** 33

## OTHER SUBSTANCES
**Extracted:** metabolites

## KEY WORDS
plasma; semi-preparative; radiolabeled starting material

## REFERENCE

Everett, D.W.; Chando, T.J.; Didonato, G.C.; Singhvi, S.M.; Pan, H.Y.; Weinstein, S.H. Biotransformation of pravastatin sodium in humans. *Drug Metab.Dispos.*, **1991**, *19*, 740–748

## SAMPLE

**Matrix:** solutions
**Sample preparation:** Inject an aliquot of an aqueous solution.

## HPLC VARIABLES

**Column:** Cosmosil 5C18-AR (Nacalai Tesque)
**Mobile phase:** A MeCN:10 mM pH 2.4 citric acid 30:70
**Flow rate:** 1
**Detector:** UV 235

## REFERENCE

Dumousseaux, C.; Muramatsu, S.; Takasaki, W.; Takahagi, H. Highly sensitive and specific determination of pravastatin sodium in plasma by high-performance liquid chromatography with laser-induced fluorescence detection after immobilized antibody extraction. *J.Pharm.Sci.*, **1994**, *83*, 1630–1636

## SAMPLE

**Matrix:** solutions
**Sample preparation:** Centrifuge at 2000 rpm, inject an aliquot.

## HPLC VARIABLES

**Column:** 300 × 3.9 μBondapak
**Mobile phase:** MeOH:water:triethylamine:glacial acetic acid 500:500:1:1
**Column temperature:** 30
**Flow rate:** 1.3
**Detector:** UV 238

## CHROMATOGRAM

**Retention time:** 17.5 (10.1 hydroxy acid form)
**Limit of detection:** 10 ng/mL
**Limit of quantitation:** 25 ng/mL

## REFERENCE

Serajuddin, A.T.; Ranadive, S.A.; Mahoney, E.M. Relative lipophilicities, solubilities, and structure-pharmacological considerations of 3-hydroxy-3-methylglutaryl-coenzyme A (HMG-CoA) reductase inhibitors pravastatin, lovastatin, mevastatin, and simvastatin. *J.Pharm.Sci.*, **1991**, *80*, 830–834

# Prednisone

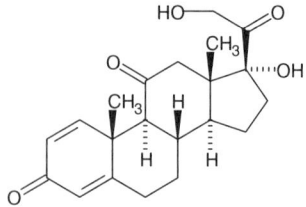

**Molecular formula:** $C_{21}H_{26}O_5$
**Molecular weight:** 358.4
**CAS Registry No.:** 53-03-2

## SAMPLE
**Matrix:** blood
**Sample preparation:** Condition an Empore C8 SPE extraction disc (3M Co.) by adding 500 µL MeOH and forcing through three drops, discard the remaining liquid, add water, force through three drops, discard the water. 300 µL Serum + 150 µL IS solution, let stand at room temperature for 10 min, add 800 µL saturated sodium borate solution, mix, centrifuge at 12400 g for 3 min (if necessary), add to SPE extraction disc, centrifuge at 100-120 g for 5 min, force through 200 µL water, force through 500 µL MeOH:water 18:82, elute with 50 µL MeCN then 150 µL water, mix the eluates, inject a 20 µL aliquot. (IS solution contained 0.5 mg/L fludrocortisone and 0.75 mg/L methylprednisolone in 400 mM HCl.) (The extraction disc permits use of lower volumes of eluate than a conventional SPE cartridge.)

## HPLC VARIABLES
**Guard column:** 20 × 2 30 µm Permaphase ETH (Du Pont)
**Column:** 250 × 2 Ultrasphere C18 or 250 × 4.6 Ultrasphere C18
**Mobile phase:** THF:water 20:80 (Use a 150 × 4.6 37-53 µm silica gel (Whatman) saturating column (held at 55°) between the pump and the injector.)
**Column temperature:** 55
**Flow rate:** 0.18 (250 × 2) or 0.8 (250 × 4.6)
**Injection volume:** 20
**Detector:** UV 254

## CHROMATOGRAM
**Retention time:** 10
**Internal standard:** fludrocortisone (15), methylprednisolone (20)
**Limit of detection:** 4 ng/mL

## OTHER SUBSTANCES
**Extracted:** corticosterone, cortisone, hydrocortisone, prednisolone
**Simultaneous:** aldosterone, androsteindione, beclomethasone, 11-deoxycorticosterone, 11-deoxycortisol, 21-deoxycortisone, dexamethasone, 17-hydroxyprogesterone, metyrapone, pregnenolone, progesterone, testosterone, triamcinolone

## KEY WORDS
serum; SPE; extraction disc

## REFERENCE
Lensmeyer, G.L.; Onsager, C.; Carlson, I.H.; Wiebe, D.A. Use of particle-loaded membranes to extract steroids for high-performance liquid chromatographic analyses. Improved analyte stability and detection. *J.Chromatogr.A*, **1995**, *691*, 239–246

## SAMPLE
**Matrix:** blood
**Sample preparation:** 1 mL Serum + 100 µL water containing 5 µg/mL 2,3-diaminonaphthalene and 3.5 µg/mL 18-hydroxy-11-deoxycorticosterone + 1 mL 250 mM NaOH + 7 mL diethyl ether, shake on a rotary shaker for 15 min, repeat extraction. Combine the

organic layers and evaporate them to dryness under a stream of nitrogen at 30-40°, reconstitute the residue in 70 μL MeOH:100 mM perchloric acid 50:50, inject a 20 μL aliquot.

## HPLC VARIABLES

**Column:** 150 × 3.9 4 μm Nova-Pak C18

**Mobile phase:** Gradient. A was 58 mM NaH$_2$PO$_4$ containing 6 mM sodium heptanesulfonate, adjusted to pH 3.1 with concentrated phosphoric acid. B was MeCN:MeOH 85:15. A:B from 100:0 to 83:17 over 5 min, to 75:25 over 12 min, to 70:30 over 4 min, to 65:35 over 9 min.

**Flow rate:** 1

**Injection volume:** 20

**Detector:** UV 245; UV 256; UV 343

## CHROMATOGRAM

**Retention time:** 14.15

**Internal standard:** 2,3-diaminonaphthalene (10.71), 18-hydroxy-11-deoxycorticosterone (15.85)

**Limit of detection:** 1-10 ng/mL (245 nm)

## OTHER SUBSTANCES

**Extracted:** betamethasone, chloroquine, corticosterone, cortisone, dexamethasone, fluendrenolide, fluocinolone acetonide, fluorometholone, fluprednisolone, hydrocortisone, hydroxychloroquine, 17δ-hydroxyprogesterone, meprednisone, methylprednisolone, methylprednisolone acetate, paramethasone, prednisolone, progesterone, triamcinolone

**Noninterfering:** aspirin, ibuprofen, indomethacin, phenylbutazone, pregnenolone

## KEY WORDS

serum

## REFERENCE

Volin, P. Simple and specific reversed-phase liquid chromatographic method with diode-array detection for simultaneous determination of serum hydroxychloroquine, chloroquine and some corticosteroids. *J.Chromatogr.B*, **1995**, *666*, 347−353

## SAMPLE

**Matrix:** blood

**Sample preparation:** 0.5-1 mL Plasma + 5 mL MeOH, vortex for 10 s, centrifuge at 2000 g for 10 min, inject a 15 μL aliquot of the supernatant.

## HPLC VARIABLES

**Guard column:** CN Guard-Pak

**Column:** 150 × 4.6 5 μm APEX octyl EC C8 (Jones Chromatography)

**Mobile phase:** Gradient. MeOH:buffer from 55:45 to 20:80 over 20 min. (Buffer was water acidified to pH 3.0 with trifluoroacetic acid.)

**Flow rate:** 1.5

**Injection volume:** 15

**Detector:** UV 254

## CHROMATOGRAM

**Retention time:** 5.5

**Internal standard:** prednisone

## OTHER SUBSTANCES

**Extracted:** novobiocin

## KEY WORDS
plasma; prednisone is IS

## REFERENCE
Chen, T.-L.; Kennedy, M.J.; Dunlap, V.M.; Colvin, O.M. Determination of plasma novobiocin levels by a reversed-phase high-performance liquid chromatographic assay. *J.Chromatogr.B*, **1994**, *652*, 109–113

## SAMPLE
**Matrix:** blood
**Sample preparation:** Prepare a Sep-Pak Plus Environmental C18 SPE cartridge by washing with 15 mL MeOH then 15 mL water. 1 mL Serum + 100 μL 3 μg/mL betamethasone in isopropanol:MeCN 1:1 + 100 μL isopropanol:acetonitrile 1:1, mix, add to the SPE cartridge, wash with 10 mL water, elute with 3 mL MeOH. Evaporate the eluate at 50° under a stream of nitrogen, reconstitute in 200 μL mobile phase A, inject a 20 μL aliquot.

## HPLC VARIABLES
**Guard column:** μBondapak C18 guard column
**Column:** 250 × 4.6 5 μm Hypersil ODS
**Mobile phase:** Gradient. A was isopropanol:50 mM pH 4.5 acetate buffer 10:90. B was isopropanol:50 mM pH 4.5 acetate buffer 30:70. A:B from 90:10 to 30:70 over 25 min, hold at 30:70 for 5 min, to 90:10 over 5 min, hold at 90:10 for 15 min before next injection.
**Column temperature:** 40
**Flow rate:** 1
**Injection volume:** 20
**Detector:** UV 254

## CHROMATOGRAM
**Retention time:** 23
**Internal standard:** betamethasone (33)
**Limit of quantitation:** 10 ng/mL

## OTHER SUBSTANCES
**Extracted:** metabolites, cortisone, hydrocortisone, prednisolone

## KEY WORDS
serum; SPE

## REFERENCE
Hirata, H.; Kasama, T.; Sawai, Y.; Fike, R.R. Simultaneous determination of deflazacort metabolites II and III, cortisol, cortisone, prednisolone and prednisone in human serum by reversed-phase high-performance liquid chromatography. *J.Chromatogr.B*, **1994**, *658*, 55–61

## SAMPLE
**Matrix:** blood
**Sample preparation:** 1 mL Plasma + 50 μL 4 μg/mL betamethasone in EtOH + 15 mL dichloromethane, shake horizontally for 15 min, centrifuge at 1500 g for 15 min. Remove the organic layer and wash it with 100 μL 100 mM NaOH then 1 mL water. Remove the aqueous phase and dry the organic phase over 1 g of anhydrous sodium sulfate. Evaporate the organic phase to dryness under a stream of nitrogen at not more than 37°, reconstitute in 200 μL mobile phase, inject a 175 μL aliquot.

## HPLC VARIABLES
**Guard column:** 20 × 2 30-38 μm HC Pellosil
**Column:** 250 × 4.6 5-6 μm Zorbax SIL

**Mobile phase:** Heptane:dichloromethane:glacial acetic acid:ethanol 350:600:10:35
**Flow rate:** 2
**Injection volume:** 175
**Detector:** UV 254

## CHROMATOGRAM
**Retention time:** 9
**Internal standard:** betamethasone (12)
**Limit of detection:** 5 ng/mL
**Limit of quantitation:** 10 ng/mL

## OTHER SUBSTANCES
**Extracted:** hydrocortisone, prednisolone
**Simultaneous:** hydrocortisone, prednisolone
**Noninterfering:** cyclosporin, ethinyl estradiol, ketoconazole, levonorgestrel, rapamycin, tacrolimus, tenidap, tetrahydrocortisone

## KEY WORDS
plasma; normal phase

## REFERENCE
Jusko, W.J.; Pyszczynski, N.A.; Bushway, M.S.; D'Ambrosio, R.; Mis, S.M. Fifteen years of operation of a high-performance liquid chromatographic assay for prednisolone, cortisol and prednisone in plasma. *J.Chromatogr.B*, **1994**, *658*, 47−54

## SAMPLE
**Matrix:** blood
**Sample preparation:** 1 mL Plasma + 100 µL 2 µg/mL 11-deoxy-17-hydroxycorticosterone in MeOH, mix for 5 s, add 7 mL dichloromethane, shake on a mechanical shaker for 5 min. Remove the organic phase and add it to 2 mL 100 mM HCl, shake, centrifuge for 5 min, wash the organic layer with 2 mL 200 mM NaOH, wash the organic layer with 2 mL water. Evaporate the organic layer to dryness, reconstitute the residue with 75 µL mobile phase, mix for 20 s, centrifuge at 10000 g for 2 min, inject 30 µL of the supernatant

## HPLC VARIABLES
**Column:** 250 × 4.6 5 µm Spherisorb C6
**Mobile phase:** MeOH:water 40:60
**Flow rate:** 1.4
**Injection volume:** 30
**Detector:** UV 254

## CHROMATOGRAM
**Retention time:** 11.8
**Internal standard:** 11-deoxy-17-hydroxycorticosterone (29.4)
**Limit of detection:** 5 ng/mL

## OTHER SUBSTANCES
**Extracted:** prednisolone
**Simultaneous:** acetaminophen, aldosterone, allopurinol, amitriptyline, caffeine, calcitriol, cephalothin, chlordiazepoxide, chlorothiazide, corticosterone, cortisone, dexamethasone, diazepam, ephedrine, ethinyl estradiol, furosemide, hydrocortisone, ibuprofen, imipramine, indomethacin, mechlorethamine, methylprednisone, metolazone, nandrolone, naproxen, phenacetin, phenobarbital, phenytoin, probenecid, progesterone, propranolol, sulfasalazine, testosterone, theophylline, vincristine

## KEY WORDS
plasma

## REFERENCE

Cheng, M.H.; Huang, W.Y.; Lipsey, A.I. Simultaneous liquid-chromatographic determination of prednisone and prednisolone in plasma. *Clin.Chem.*, **1988**, *34*, 1897–1899

## SAMPLE

**Matrix:** blood

**Sample preparation:** 2 mL Plasma + 40 μL 5 μg/mL dexamethasone in MeOH, vortex 30 s, add 5 mL dichloromethane:diethyl ether 50:50, vortex for 15 s, repeat extraction, combine organic layers and wash them with 4 mL 100 mM NaOH, centrifuge. Remove the organic layer and dry it over anhydrous sodium sulfate, evaporate to dryness under a stream of nitrogen at 40°, dissolve the residue in 150 μL dichloromethane, inject a 100 μL aliquot.

## HPLC VARIABLES

**Column:** 250 × 4.6 5 μm LiChrosorb Si-60

**Mobile phase:** Dichloromethane:water-saturated dichloromethane:THF:MeOH:glacial acetic acid 664.5:300:10:25:0.5

**Flow rate:** 0.8

**Injection volume:** 100

**Detector:** UV 254

## CHROMATOGRAM

**Retention time:** 17

**Internal standard:** dexamethasone (23.5)

**Limit of detection:** 5 ng/mL

## OTHER SUBSTANCES

**Extracted:** cortisone, hydrocortisone, prednisolone, prednisolone acetate

## KEY WORDS

plasma; normal phase; pig

## REFERENCE

Prasad, V.K.; Ho, B.; Haneke, C. Simultaneous determination of prednisolone acetate, prednisolone, prednisone, cortisone and hydrocortisone in swine plasma using solid-phase and liquid-liquid extraction techniques. *J.Chromatogr.*, **1986**, *378*, 305–316

## SAMPLE

**Matrix:** blood

**Sample preparation:** Condition a Bond-Elut C18 SPE cartridge with 2 mL MeCN, 2 mL acetone:water 2:98, and 4 mL water. Do not allow cartridge to run dry. 2 mL Plasma + 40 μL 5 μg/mL dexamethasone in MeOH, add to SPE cartridge, allow to sit for 15 min, wash twice with 2 mL water, wash twice with 2 mL acetone:water 2:98, pull a vacuum on the column for 15 min, elute with 1 mL MeCN under vacuum. Evaporate the eluate to dryness under a stream of nitrogen at 40°, dissolve the residue in 150 μL dichloromethane, inject a 100 μL aliquot.

## HPLC VARIABLES

**Column:** 250 × 4.6 5 μm LiChrosorb Si-60

**Mobile phase:** Dichloromethane:water-saturated dichloromethane:THF:MeOH:glacial acetic acid 664.5:300:10:25:0.5

**Flow rate:** 0.8

**Injection volume:** 100

**Detector:** UV 254

## CHROMATOGRAM
**Retention time:** 17
**Internal standard:** dexamethasone (23.5)
**Limit of detection:** 10 ng/mL

## OTHER SUBSTANCES
**Extracted:** cortisone, hydrocortisone, prednisolone, prednisolone acetate

## KEY WORDS
plasma; normal phase; pig; SPE

## REFERENCE
Prasad, V.K.; Ho, B.; Haneke, C. Simultaneous determination of prednisolone acetate, prednisolone, prednisone, cortisone and hydrocortisone in swine plasma using solid-phase and liquid-liquid extraction techniques. *J.Chromatogr.*, **1986**, *378*, 305–316

## SAMPLE
**Matrix:** blood
**Sample preparation:** 1 mL Plasma + 150 ng dexamethasone + 1 mL 100 mM NaOH + 10 mL ether:dichloromethane 60:40, shake for 10 min, centrifuge at 300 g for 5 min. Remove the organic layer and add it to 1 mL 100 mM HCl, shake for 5 min, centrifuge at 300 g for 5 min. Remove the organic layer and evaporate it to dryness under a stream of nitrogen, reconstitute the residue in 100 μL mobile phase, inject a 50 μL aliquot.

## HPLC VARIABLES
**Guard column:** μBondapak/Corasil (Waters)
**Column:** 300 × 3.9 10 μm μPorasil (Waters)
**Mobile phase:** Dichloromethane:glacial acetic acid 99:1 (Prepare dichloromethane as follows. Stir 500 mL dichloromethane, 30 mL EtOH, and 30 mL water for 1 h, use the lower organic layer.)
**Flow rate:** 2
**Injection volume:** 50
**Detector:** UV 254

## CHROMATOGRAM
**Retention time:** 4
**Internal standard:** dexamethasone (5)

## OTHER SUBSTANCES
**Extracted:** hydrocortisone, prednisolone

## KEY WORDS
plasma; normal phase

## REFERENCE
Hartley, R.; Brocklebank, J.T. Determination of prednisolone in plasma by high-performance liquid chromatography. *J.Chromatogr.*, **1982**, *232*, 406–412

## SAMPLE
**Matrix:** blood
**Sample preparation:** 1 mL Plasma + 5 mL water + 1 mL 2 μg/mL equilenin in MeOH + 50 μL 0.1 M NaOH to adjust pH to 10, vortex briefly after each addition, shake with 10 mL dichloromethane for 10 min, centrifuge at 2000 g for 10 min. Wash organic layer twice with 2 mL water, centrifuge 5 min, evaporate 8 mL of organic phase to dryness at

40° under a stream of nitrogen, reconstitute residue in 150 μL mobile phase, inject 25 μL aliquot.

## HPLC VARIABLES
**Column:** 300 × 4 10 μm μBondapak C18
**Mobile phase:** MeOH:buffer 60:40 (Buffer was 10 mL 200 mM acetic acid + 15 mL 200 mM sodium acetate made up to 1 L, pH 4.8.)
**Flow rate:** 1
**Injection volume:** 25
**Detector:** UV 254

## CHROMATOGRAM
**Retention time:** 4
**Internal standard:** equilenin (7.5)
**Limit of detection:** 5 ng/mL

## OTHER SUBSTANCES
**Extracted:** betamethasone, deoxycortisol, dexamethasone, hydrocortisone, prednisolone, triamcinolone

## KEY WORDS
Anal.Abs. 1982, 43, 4D182; plasma

## REFERENCE
Bouquet, S.; Brisson, A.M.; Gombert, J. Dosage du cortisol et du 11-désoxycortisol plasmatiques par chromatographie liquide haute performance [Cortisol and 11-desoxycortisol determination in blood by high performance liquid chromatography]. *Ann.Biol.Clin.(Paris)*, **1981**, *39*, 189–191

## SAMPLE
**Matrix:** blood
**Sample preparation:** 1 mL Plasma + 4 mL 59.5 ng/mL triamcinolone acetonide in dichloromethane, shake at high speed for 15 min, centrifuge at 2000 rpm for 10 min, remove aqueous layer, add 5 mL saturated sodium bicarbonate solution to the organic layer, shake at high speed for 5 min, centrifuge at 2000 rpm for 10 min, remove aqueous layer. Place organic layer in a pointed tube and evaporate to dryness at 45° under a stream of nitrogen. Reconstitute with 50 μL mobile phase, inject 20 μL aliquot.

## HPLC VARIABLES
**Column:** 10 μm Porasil
**Mobile phase:** Hexane:dichloromethane:ethanol:acetic acid 68.8:25:6:0.2
**Flow rate:** 2.5
**Injection volume:** 20
**Detector:** UV 254

## CHROMATOGRAM
**Retention time:** 4.5
**Internal standard:** triamcinolone acetonide (3)

## OTHER SUBSTANCES
**Extracted:** hydrocortisone, prednisolone

## KEY WORDS
plasma; normal phase

## REFERENCE

Agabeyoglu, I.T.; Wagner, J.G.; Kay, D.R. A sensitive high-pressure liquid chromatographic method for the determination of prednisone, prednisolone and hydrocortisone in plasma. *Res.Commun. Chem.Pathol.Pharmacol.*, **1980**, *28*, 163–176

## SAMPLE

**Matrix:** blood

**Sample preparation:** 1 mL Plasma + 100 µL 2 µg/mL dexamethasone in EtOH:water 10:90 + 100 µL 250 mM NaOH + 7 mL ether:dichloromethane 60:40, vortex for 30 s, centrifuge at 2000 rpm for 5 min. Remove 6 mL of the organic layer and evaporate it to dryness under a stream of air at 40°, reconstitute the residue in 100 µL dichloromethane:EtOH:water 95:4:1, inject a 50 µL aliquot.

## HPLC VARIABLES

**Column:** 250 × 4.5 5 µm Partisil silica

**Mobile phase:** Dichloromethane:EtOH:water 95:4:1

**Flow rate:** 1.5

**Injection volume:** 50

**Detector:** UV 239

## CHROMATOGRAM

**Retention time:** 7.7

**Internal standard:** dexamethasone (11.5)

**Limit of detection:** 25 ng/mL

## OTHER SUBSTANCES

**Extracted:** corticosterone, 11-deoxycortisol, hydrocortisone, 17-hydroxyprogesterone, 6α-methylprednisolone, prednisolone, progesterone

## KEY WORDS

plasma; normal phase

## REFERENCE

Scott, N.R.; Chakraborty, J.; Marks, V. Determination of prednisolone, prednisone, and cortisol in human plasma by high-performance liquid chromatography. *Anal.Biochem.*, **1980**, *108*, 266–268

## SAMPLE

**Matrix:** blood, urine

**Sample preparation:** Urine. Adjust pH of 5 mL urine to 7 with 500 mM pH 7 ammonium acetate buffer, extract twice with 7 mL portions of ethyl acetate. Combine the organic layers and evaporate them to dryness under a stream of nitrogen at 40°, reconstitute the residue in 200 µL MeOH, inject a 20 µL aliquot. Serum. Extract 2 mL serum twice with 4 mL portions of ethyl acetate. Combine the organic layers and evaporate them to dryness under a stream of nitrogen at 40°, reconstitute the residue in 200 µL MeOH, inject a 20 µL aliquot.

## HPLC VARIABLES

**Guard column:** 4 × 4 5 µm Lichrospher 100 RP-18

**Column:** 125 × 4 5 µm Lichrospher 100 RP-18

**Mobile phase:** Gradient. A was 10 mM ammonium acetate. B was MeOH. A:B from 50:50 to 0:100 over 7 min, maintain at 0:100 for 3 min.

**Flow rate:** 1

**Injection volume:** 20

**Detector:** MS, PE Sciex API III LC-MS-MS, heated nebulizer interface at 75 psi and 300°, auxiliary nitrogen 1.4 L/min, nitrogen curtain 1.2 L/min, negative mode ionization, discharge current -3 μA, collision gas argon with $570 \times 10^{12}$ molecules/sq.cm., dwell time 10 ms, step size 1 amu, m/z 327, SIM

## CHROMATOGRAM
**Retention time:** 5.5
**Internal standard:** prednisone

## OTHER SUBSTANCES
**Extracted:** triamcinolone acetonide (m/z 413)

## KEY WORDS
horse; serum; prednisone is IS

## REFERENCE
Koupai-Abyazani, M.R.; Yu, N.; Esaw, B.; Laviolette, B. Determination of triamcinolone acetonide in equine serum and urine by liquid chromatography-atmospheric pressure ionization mass spectrometry. *J.Anal.Toxicol.*, **1995**, *19*, 182–186

## SAMPLE
**Matrix:** solutions
**Sample preparation:** Inject an aliquot of a 1 μM solution in MeOH.

## HPLC VARIABLES
**Column:** $470 \times 4.6$ 5 μm Spheri-5 RP-18
**Mobile phase:** MeOH:water 56:44
**Flow rate:** 0.5
**Injection volume:** 10
**Detector:** UV 240

## CHROMATOGRAM
**Retention time:** 26

## OTHER SUBSTANCES
**Simultaneous:** cortisone, dehydrocorticosterone, hydrocortisone, methylprednisolone, prednisolone, tetrahydrocortisol, tetrahydrocortisone

## REFERENCE
Lukulay, P.H.; McGuffin, V.L. Comparison of solvent modulation with premixed mobile phases for the separation of corticosteroids by liquid chromatography. *J.Liq.Chromatogr.*, **1995**, *18*, 4039–4062

## SAMPLE
**Matrix:** solutions
**Sample preparation:** Prepare a 25 μg/mL solution in mobile phase, inject an aliquot.

## HPLC VARIABLES
**Column:** $250 \times 4.6$ Partisil 10 ODS-1
**Mobile phase:** MeOH:water 55:45
**Column temperature:** 40
**Flow rate:** 1.5
**Detector:** UV 240

## CHROMATOGRAM
**Retention time:** k′ 0.8779

## OTHER SUBSTANCES

**Also analyzed:** androsterone (UV 210), cortexolone, cortisone, estradiol (UV 280), estrone (UV 280), ethinyl estradiol (UV 280), ethisterone, hydrocortisone, hydroxyprogesterone, lynestrenol (UV 210), medroxyprogesterone acetate, medroxyprogesterone, methandienone, methylandrostenediol (UV 210), methylprednisolone acetate, methylprednisolone, methyltestosterone, nandrolone, norethisterone, prednisolone acetate, prednisolone, pregnenolone (UV 210), progesterone, testosterone

## REFERENCE

Sadlej-Sosnowska, N. Structure retention relationship for steroid hormones. Functional groups as structural descriptors. *J.Liq.Chromatogr.*, **1994**, *17*, 2319–2330

## SAMPLE

**Matrix:** solutions

## HPLC VARIABLES

**Column:** 150 × 6 5 μm Shim-pack CLC-ODS
**Mobile phase:** MeOH:THF:water 26:18:56
**Column temperature:** 48
**Flow rate:** 1
**Injection volume:** 20
**Detector:** UV 240

## CHROMATOGRAM

**Retention time:** 8.8 (prednisone acetate)

## OTHER SUBSTANCES

**Simultaneous:** androstenedione, corticosterone, cortisol, cortisone, 11-deoxycorticosterone, 11-deoxycortisol, dexamethasone acetate, estradiol, estriol, estrone, 17α-hydroxyprogesterone, progesterone, testosterone

## REFERENCE

Wei, J.Q.; Wei, J.L.; Zhou, X.T. Optimization of an isocratic reversed phase liquid chromatographic system for the separation of fourteen steroids using factorial design and computer simulation. *Biomed.Chromatogr.*, **1990**, *4*, 34–38

## SAMPLE

**Matrix:** solutions
**Sample preparation:** Dissolve in MeOH:water 1:1 at a concentration of 50 μg/mL, inject a 10 μL aliquot.

## HPLC VARIABLES

**Column:** 300 × 3.9 10 μm μBondapak C18
**Mobile phase:** MeOH:acetic acid:triethylamine:water 60:1.5:0.5:38
**Flow rate:** 1.5
**Injection volume:** 10
**Detector:** UV 240

## CHROMATOGRAM

**Retention time:** 4

## OTHER SUBSTANCES

**Simultaneous:** hydrocortisone acetate, methyltestosterone, norethindrone, prednisolone, prednisolone succinate, progesterone

## REFERENCE
Roos, R.W.; Lau-Cam, C.A. General reversed-phase high-performance liquid chromatographic method for the separation of drugs using triethylamine as a competing base. *J.Chromatogr.*, **1986**, *370*, 403–418

## SAMPLE
**Matrix:** solutions

## HPLC VARIABLES
**Column:** 250 × 4.6 5 μm RP-18 C18 (Brownlee)
**Mobile phase:** MeCN:water 40:60
**Injection volume:** 20
**Detector:** UV 254

## REFERENCE
Kane, M.P.; Tsuji, K. Radiolytic degradation scheme for $^{60}$Co-irradiated corticosteroids. *J.Pharm.Sci.*, **1983**, *72*, 30–35

## SAMPLE
**Matrix:** solutions, tissue
**Sample preparation:** Buffer. Condition a 3 mL Baker C18 SPE cartridge with 400 μL MeOH. Add 100-500 μL 0.1 M pH 4.5 acetate buffer containing steroids to SPE cartridge, wash with 1 mL water, elute with 1 mL MeOH. Evaporate the eluate to dryness under a stream of air, reconstitute in 40-200 μL dichloromethane:MeOH 98:2, inject a 40 μL aliquot. Skin tissue. Crush tissue with 3 mL MeOH:100 mM pH 4.5 acetate buffer 20:80 using a Polytron tissue homogenizer, wash homogenizer twice with the same solution, combine all solutions, add 10 mL dichloromethane, vortex for 1 min, centrifuge at 3000 rpm for 5 min. Filter the organic phase through a column of Celite 545, evaporate to dryness under a stream of air, reconstitute in 40-200 μL dichloromethane:MeOH 98:2, inject a 40 μL aliquot.

## HPLC VARIABLES
**Column:** 250 × 4 5 μm LiChrosorb Si-60
**Mobile phase:** n-Hexane:dichloromethane:MeOH:water 63.9:30:6:0.1
**Flow rate:** 1.2
**Injection volume:** 40-200
**Detector:** UV 240

## CHROMATOGRAM
**Retention time:** 5.3
**Internal standard:** prednisone

## OTHER SUBSTANCES
**Extracted:** betamethasone

## KEY WORDS
buffer; SPE; prednisone is IS; normal phase

## REFERENCE
Kubota, K.; Maibach, H.I. In vitro percutaneous permeation of betamethasone and betamethasone 17-valerate. *J.Pharm.Sci.*, **1993**, *82*, 1039–1045

## SAMPLE
**Matrix:** urine

**Sample preparation:** 3 mL Urine + 0.25 g NaCl, adjust pH to 9.0 with 0.5 g Na$_2$HPO$_4$, add 4 mL dichloromethane, vortex 1 min, centrifuge at 3700 g for 3 min. Remove organic phase and dry it over anhydrous sodium sulfate. Evaporate a 3 mL aliquot to dryness under vacuum, reconstitute residue with 200 μL 5 μg/mL IS in MeOH, inject 20 μL aliquot.

## HPLC VARIABLES
**Column:** 250 × 4.6 Hypersil ODS
**Mobile phase:** MeCN:water 32:68
**Column temperature:** 30
**Flow rate:** 1
**Injection volume:** 20
**Detector:** UV 245

## CHROMATOGRAM
**Retention time:** 6.3
**Internal standard:** methylprednisolone (9)

## OTHER SUBSTANCES
**Simultaneous:** betamethasone, corticosterone, cortisone, dexamethasone, fluorocortisone acetate, hydroxyprogesterone, prednisolone, triamcinolone, triamcinolone acetonide
**Interfering:** hydrocortisone, fluorocortisone

## KEY WORDS
SPE also discussed

## REFERENCE
Santos-Montes, A.; Gonzalo-Lumbreras, R.; Gasco-Lopez, A.I.; Izquierdo-Hornillos, R. Solvent and solid-phase extraction of natural and synthetic corticoids in human urine. *J.Chromatogr.B*, **1994**, *652*, 83–89

## SAMPLE
**Matrix:** urine
**Sample preparation:** Dilute, if necessary, 100 μL-1 mL urine to 1 mL with water, add 500 ng betamethasone, add to a Chem Elut high-surface-area diatomaceous earth extraction column, after 5 min elute with two 6 mL portions of ethyl acetate, combine the eluates and wash them twice with 1 mL 200 mM NaOH. Dry the organic layer over 1 g anhydrous sodium sulfate, evaporate to dryness at 30° under a stream of nitrogen, reconstitute the residue in 250 μL mobile phase, inject a 50 μL aliquot.

## HPLC VARIABLES
**Guard column:** 70 × 6 37-53 μm HC-Pellocil
**Column:** 250 × 4.6 5-6 μm Zorbax SIL
**Mobile phase:** Dichloromethane:glacial acetic acid:MeOH 91.3:7.5:1.2
**Flow rate:** 2
**Injection volume:** 50
**Detector:** UV 254

## CHROMATOGRAM
**Retention time:** 7
**Internal standard:** betamethasone (8.5)
**Limit of quantitation:** 50 ng/mL

## OTHER SUBSTANCES
**Extracted:** metabolites, hydrocortisone, 6β-hydroxycortisol, 6β-hydroxyprednisolone, 20α-hydroxyprednisolone, 20β-hydroxyprednisolone, 20β-hydroxyprednisone, prednisolone

## KEY WORDS
SPE

## REFERENCE
Garg, V.; Jusko, W.J. Simultaneous analysis of prednisone, prednisolone and their major hydroxylated metabolites in urine by high-performance liquid chromatography. *J.Chromatogr.*, **1991**, *567*, 39–47

## SAMPLE
**Matrix:** urine
**Sample preparation:** 3 mL Urine + 1.5 µg betamethasone + 100 mg $K_2HPO_4$ + 500 mg anhydrous sodium sulfate + 5 mL diethyl ether, shake mechanically for 10 min, centrifuge at 2500 g for 5 min. Remove the organic layer and evaporate it to dryness under vacuum, reconstitute the residue in 200 µL MeOH, filter (0.45 µm), inject a 15 µL aliquot.

## HPLC VARIABLES
**Column:** 60 × 4.6 3 µm Hypersil ODS
**Mobile phase:** Gradient. MeOH:150 mM ammonium acetate from 40:60 to 50:50 over 6 min, maintain at 50:50 for 1 min, to 60:40 over 3 min, maintain at 60:40 for 5 min.
**Flow rate:** 0.8
**Injection volume:** 15
**Detector:** MS, Hewlett-Packard HP 5988A, vaporizer probe 92° decreased to 89° over 6 min, decreased to 86° over 3 min, maintain at 86° for 5 min, ion source 276°, emission current 150 µA, electron energy 955 eV, positive ion mode, filament on

## CHROMATOGRAM
**Retention time:** 4.2
**Internal standard:** betamethasone (7)
**Limit of detection:** 1-5 ng (SIM)

## OTHER SUBSTANCES
**Extracted:** corticosterone, cortisone, deoxycorticosterone, hydrocortisone, 11α-hydroxyprogesterone, prednisolone, triamcinolone, triamcinolone acetonide

## REFERENCE
Park, S.-J.; Kim, Y.-J.; Pyo, H.-S.; Park, J. Analysis of corticosteroids in urine by HPLC and thermospray LC/MS. *J.Anal.Toxicol.*, **1990**, *14*, 102–108

## SAMPLE
**Matrix:** urine
**Sample preparation:** 3 mL Urine + 1.5 µg betamethasone + 100 mg $K_2HPO_4$ + 500 mg anhydrous sodium sulfate + 5 mL diethyl ether, shake mechanically for 10 min, centrifuge at 2500 g for 5 min. Remove the organic layer and evaporate it to dryness under vacuum, reconstitute the residue in 200 µL MeOH, filter (0.45 µm), inject a 15 µL aliquot.

## HPLC VARIABLES
**Column:** 100 × 4.6 5 µm Hypersil ODS
**Mobile phase:** Gradient. MeOH:water from 4:96 to 30:70 over 10 min, to 45:55 over 5 min, to 50:50 over 3 min.
**Column temperature:** 40
**Flow rate:** 1
**Injection volume:** 15
**Detector:** UV 246

## CHROMATOGRAM
**Retention time:** 3.5

**Internal standard:** betamethasone
**Limit of detection:** 10 ng/mL

## OTHER SUBSTANCES
**Extracted:** corticosterone, cortisone, deoxycorticosterone, hydrocortisone, 11α-hydroxypro-
gesterone, prednisolone, triamcinolone, triamcinolone acetonide

## REFERENCE
Park, S.-J.; Kim, Y.-J.; Pyo, H.-S.; Park, J. Analysis of corticosteroids in urine by HPLC and thermospray
LC/MS. *J.Anal.Toxicol.*, **1990**, *14*, 102–108

## SAMPLE
**Matrix:** urine
**Sample preparation:** Dilute 0.1-1 mL urine to 1 mL with water if necessary, add 50 μL
10 μg/mL dexamethasone in MeOH, vortex, add to a Chem-Elut cartridge (cat. no. 1003),
allow to stand for 5 min, elute twice with 5 mL portions of ethyl acetate (which are first
used to rinse the sample tube) at 5 min intervals, wash eluate twice with 1 mL 200 mM
NaOH, add 1 g anhydrous sodium sulfate, let stand for 30 min. Evaporate the organic
phase at 30°under a stream of nitrogen. Reconstitute the residue in 250 μL mobile phase,
inject a 50 μL aliquot.

## HPLC VARIABLES
**Column:** 250 × 4.6 5 μm Beckman Si column
**Mobile phase:** Dichloromethane:MeOH:THF:glacial acetic acid 96.9:2:1:10.1
**Flow rate:** 1.3
**Injection volume:** 50
**Detector:** UV 254

## CHROMATOGRAM
**Retention time:** 6.4
**Internal standard:** dexamethasone (8.0)
**Limit of detection:** 50 ng/mL

## OTHER SUBSTANCES
**Extracted:** 6β-hydroxyprednisolone, prednisolone

## KEY WORDS
normal phase;SPE

## REFERENCE
Teng, R.-L.; Benet, L.Z. Simultaneous measurement of prednisone, prednisolone and 6β-hydroxypred-
nisolone in urine by high-performance liquid chromatography using dexamethasone as the internal
standard. *J.Chromatogr.*, **1989**, *493*, 421–423

## ANNOTATED BIBLIOGRAPHY
Valvo, L.; Paris, A.; Savella, A.L.; Gallinella, B.; Ciranni Signoretti, E. General high-performance liquid
chromatographic procedures for the rapid screening of natural and synthetic corticosteroids.
*J.Pharm.Biomed.Anal.*, **1994**, *12*, 805–810 [gradient; reverse phase; normal phase; also beclomethasone, beclomethasone 17,21-dipropionate, betamethasone, betamethasone 21-acetate, betamethasone
17,21-dipropionate, betamethasone 21-disodium phosphate, betamethasone 17-valerate, cortisone,
cortisone 21-acetate, 11-deoxycorticosterone 21-acetate, dexamethasone, dexamethasone 21-acetate,
dexamethasone 21-disodium phosphate, fluocinolone, fluocinolone acetonide, 9α-fluorohydrocortisone
21-acetate, 9α-fluorohydrocortisone, 9α-fluoroprednisolone, 9α-fluoroprednisolone 21-acetate, hydrocortisone, hydrocortisone 21-acetate, hydrocortisone 21-hemisuccinate, 6α-methylprednisolone,
6α-methylprednisolone 21-acetate, 6α-methylprednisolone 21-sodium succinate, prednisolone,

prednisolone 21-acetate, prednisolone 21-disodium phosphate, prednisolone 21-pivalate, prednisolone 21-sodium succinate, triamcinolone, triamcinolone acetonide]

Santos-Montes, A.; Gasco-Lopez, A.I.; Izquierdo-Hornillos, R. Optimization of the high-performance liquid chromatographic separation of a mixture of natural and synthetic corticosteroids. *J.Chromatogr.*, **1993**, *620*, 15−23 [urine; column temp 30; extracted betamethasone, corticosterone, cortisone, deoxycorticosterone, dexamethasone, fluorocortisone, hydrocortisone, hydroxyprogesterone, methylprednisolone, prednisolone, triamcinolone]

Ulrich, B.; Frey, F.J.; Speck, R.F.; Frey, B.M. Pharmacokinetics/pharmacodynamics of ketoconazole-prednisolone interaction. *J.Pharmacol.Exp.Ther.*, **1992**, *260*, 487−490 [mouse; plasma; dexamethasone (IS); LOD 10 ng/mL; extracted prednisolone]

Cannell, G.R.; Mortimer, R.H.; Maguire, D.J.; Addison, R.S. Liquid chromatographic analysis of prednisolone, prednisone and their 20-reduced metabolites in perfusion media. *J.Chromatogr.*, **1991**, *563*, 341−347

McBride, J.H.; Rodgerson, D.O.; Park, S.S.; Reyes, A.F. Rapid liquid-chromatographic method for simultaneous determination of plasma prednisone, prednisolone, and cortisol in pediatric renal-transplant recipients. *Clin.Chem.*, **1991**, *37*, 643−646

Alvinerie, M.; Sutra, J.F.; Galtier, P.; Houin, G.; Toutain, P.L. Simultaneous measurement of prednisone, prednisolone and hydrocortisone in plasma by high performance liquid chromatography. *Ann.Biol.Clin.(Paris)*, **1990**, *48*, 87−90

Huber, F.; Wiedemann, M.; Heinrich, G.; Salama, Z.; Jaeger, H. Development of a high performance liquid chromatography method for the simultaneous measurement of prednisone and prednisolone. *Arzneimittelforschung*, **1990**, *40*, 926−931

Lasic, S.; Bobarevic, N.; Nikolin, B. Simultaneous determination of prednisone, prednisolone, cortisol and dexamethasone in plasma by high-performance liquid chromatography. *J.Pharm.Biomed.Anal.*, **1989**, *7*, 777−782

Frey, B.M.; Frey, F.J. Simultaneous measurement of prednisone, prednisolone and 6β-hydroxyprednisolone in urine by high-performance liquid chromatography coupled with a radioactivity detector. *J.Chromatogr.*, **1982**, *229*, 283−292

Ui, T.; Mitsunaga, M.; Tanaka, T.; Horiguchi, M. Determination of prednisone and prednisolone in human serum by high-performance liquid chromatography−especially on impaired conversion of corticosteroids in patients with chronic liver disease. *J.Chromatogr.*, **1982**, *239*, 711−716

Rocci, M.L., Jr.; Jusko, W.J. Analysis of prednisone, prednisolone and their 20β-hydroxylated metabolites by high-performance liquid chromatography. *J.Chromatogr.*, **1981**, *224*, 221−227 [perfusate; urine; 6β-hydroxycortisol (IS); triamcinolone (IS); LOD 10 ng/mL; extracted 20β-hydroxyprednisolone, 20β-hydroxyprednisone, prednisolone]

Rose, J.Q.; Jusko, W.J. Corticosteroid analysis in biological fluids by high-performance liquid chromatography. *J.Chromatogr.*, **1979**, *162*, 273−280 [plasma; urine; saliva; extracted hydrocortisone, prednisolone; dexamethasone (IS); LOD 5 ng/mL; pharmacokinetics; simultaneous beclomethasone, methylprednisolone, triamcinolone acetonide]

# Promethazine

**Molecular formula:** $C_{17}H_{20}N_2S$
**Molecular weight:** 284.4
**CAS Registry No.:** 60-87-7 (promethazine), 58-33-3 (promethazine hydrochloride), HCl
17693-51-5 (promethazine teoclate)

## SAMPLE
**Matrix:** blood
**Sample preparation:** 2 mL Whole blood or plasma + 2 mL buffer + 5 mL chloroform: isopropanol:n-heptane 60:14:26, shake gently horizontally for 10 min, centrifuge at 2800 g for 10 min. Remove the lower organic layer and evaporate it to dryness under vacuum at 45°, reconstitute the residue in 100 μL mobile phase, centrifuge at 2800 g for 5 min, inject a 50 μL aliquot of the supernatant. (Buffer was saturated ammonium chloride solution 25% diluted with water, adjusted to pH 9.5 with 25% ammonia solution.)

## HPLC VARIABLES
**Column:** 300 × 3.9 4 μm NovaPack C18
**Mobile phase:** MeOH:THF:buffer 65:5:30 (Buffer was 0.68 g/L (10 mM (sic)) $KH_2PO_4$ adjusted to pH 2.6 with concentrated orthophosphoric acid.) (At the end of each session wash the column with water for 1 h and MeOH for 1 h, re-equilibrate for 30 min.)
**Column temperature:** 30
**Flow rate:** 0.8
**Injection volume:** 50
**Detector:** UV 251

## CHROMATOGRAM
**Retention time:** 7.46
**Limit of detection:** <120 ng/mL

## OTHER SUBSTANCES
**Extracted:** acebutolol, acenocoumarol, acepromazine, aceprometazine, acetaminophen, aconitine, ajmaline, albuterol, alimemazine, alminoprofen, alpidem, alprazolam, alprenolol, amisulpride, amitriptyline, amodiaquine, aspirin, astemizole, atenolol, benazepril, benperidol, benzocaine, benzoylecgonine, bepridil, bisoprolol, bromazepam, brompheniramine, bumadizone, bupivacaine, buprenorphine, buspirone, caffeine, carbamazepine, carbinoxamine, carpipramine, carteolol, celiprolol, cetirizine, chlorambucil, chlordiazepoxide, chlormezanone, chlorophenacinone, chloroquine, chlorpheniramine, chlorpromazine, chlorpropamide, cibenzoline, cicletanine, clemastine, clobazam, clomipramine, clonazepam, clonidine, clorazepate, clozapine, cocaine, codeine, colchicine, cyamemazine, cyclizine, cycloguanil, cytarabine, dacarbazine, daunorubicin, debrisoquine, demexiptiline, desipramine, dextromethorphan, dextromoramide, diazepam, diazoxide, diclofenac, dihydralazine, diltiazem, diphenhydramine, dipyridamole, disopyramide, dosulepine, doxepin, doxylamine, droperidol, ephedrine, estazolam, etodolac, fenfluramine, fentiazac, flecainide, floctafenine, flumazenil, flunitrazepam, fluoxetine, fluphenazine, flurbiprofen, fluvoxamine, glibenclamide, glibornuride, glipizide, glutethimide, haloperidol, histapyrrodine, hydroxychloroquine, hydroxyzine, ibuprofen, imipramine, indomethacin, iproniazid, ketamine, ketoprofen, labetalol, levomepromazine, lidocaine, lidoflazine, lisinopril, loperamide, loprazolam, loratadine, lorazepam, maprotiline, medazepam, medifoxamine, mefenamic acid, mefenidramine, mefloquine, melphalan, meperidine, mephenesin, mephentermine, mepivacaine, metapramine, metformin, methaqualone, methocarbamol, methotrexate, metipranolol, metoclopramide, metoprolol, mexiletine, mianserine, midazolam, minoxidil, moclobemide, moperone, morphine, nadolol, nalbuphine, nalorphine, naloxone, naltrexone, naproxen, nialamide, nicardipine, nifedipine, niflumic acid, nimodipine, nitrazepam, nitrendipine, nizatidine, nomifensine, nortriptyline, omeprazole, ox-

azepam, oxprenolol, penbutolol, penfluridol, pentazocine, phencyclidine, phenobarbital, phenol, phenylbutazone, pimozide, pindolol, pipamperone, piroxicam, prazepam, prazosin, prilocaine, procainamide, procarbazine, proguanil, propranolol, protriptyline, pyrimethamine, quinidine, quinine, ramipril, ranitidine, reserpine, ritodrine, secobarbital, sotalol, strychnine, sulfinpyrazole, sulindac, sulpride, sultopride, suriclone, temazepam, tenoxicam, terbutaline, terfenadine, tetracaine, tetrazepam, thiopental, thioridazine, tianeptine, tiapride, tiaprofenic acid, ticlopidine, timolol, tioclomarol, tofisopam, tolbutamide, toloxatone, trazodone, triazolam, trifluoperazine, trifluperidol, trimipramine, triprolidine, tropatenine, verapamil, viloxazine, vinblastine, vincristine, vindesine, warfarin, yohimbine, zolpidem, zopiclone, zorubicine

**Interfering:** amoxapine, betaxolol, cyproheptadine, dextropropoxyphene, fenoprofen, loxapine, methadone, opipramol, promethazine, propafenone, quinupramine, thioproperazine

## KEY WORDS
whole blood; plasma

## REFERENCE
Tracqui, A.; Kintz, P.; Mangin, P. Systematic toxicological analysis using HPLC/DAD. *J.Forensic Sci.*, **1995**, *40*, 254–262

## SAMPLE
**Matrix:** blood
**Sample preparation:** 1 mL Serum + 2 mL water + 2 mL 2 M NaOH, vortex for 10 s, add 5 mL water-saturated n-heptane:isoamyl alcohol 99:1, shake gently for 20 min, centrifuge at 4° at 2800 g, remove organic layer and repeat the extraction. Combine the organic layers and evaporate them to dryness under reduced pressure. Dissolve the residue in 500 μL MeCN, inject a 30 μL aliquot (store at 5°).

## HPLC VARIABLES
**Column:** 250 × 4.6 5 μm Nucleosil 100 CN
**Mobile phase:** MeCN:pyridine:140 mM sodium acetate pH 3.1 698:2:300
**Flow rate:** 0.9
**Injection volume:** 100
**Detector:** E, Environmental Sciences Assoc. Coulochem II, Model 5011 detector cell, oxidative screen mode, screen electrode +0.25 V, sample electrode +0.5 V

## CHROMATOGRAM
**Retention time:** 9
**Limit of detection:** 0.1 ng/mL

## OTHER SUBSTANCES
**Extracted:** chlorprothixene
**Interfering:** methotrimeprazine (levomepromazine)

## KEY WORDS
serum; work under yellow light; recirculate mobile phase

## REFERENCE
Bagli, M.; Rao, M.L.; Höflich, G. Quantification of chlorprothixene, levomepromazine and promethazine in human serum using high-performance liquid chromatography with coulometric electrochemical detection. *J.Chromatogr.B*, **1994**, *657*, 141–148

## SAMPLE
**Matrix:** blood
**Sample preparation:** 1 mL Serum + 100 μL 100 ng/mL chlorpromazine, vortex at low speed for 10 s, add 1 mL 650 mM sodium carbonate, vortex, add 7 mL pentane:ethyl

acetate 50:50, shake vigorously for 15 min, centrifuge at 1110 g for 10 min. Remove the organic layer and evaporate it to dryness at 65° under nitrogen. Reconstitute the residue in 300 μL MeCN:MeOH:isopropanol:water: 1 M ammonium acetate pH 5.0 83:5:5: 6.65:0.35, sonicate for 5 min, vortex for 30 s, inject a 100 μL aliquot.

## HPLC VARIABLES
**Column:** 150 × 4.6 5 μm Burdick & Jackson CN
**Mobile phase:** MeCN:MeOH:isopropanol:water: 1 M ammonium acetate pH 7.2 83:5:5: 6.65:0.35
**Flow rate:** 1.5
**Injection volume:** 100
**Detector:** E, ESA Model 5100A Coulochem detector, Model 5011 analytical cell, detector 1 +0.50 V, detector 2 +0.70 V

## CHROMATOGRAM
**Retention time:** 7.7
**Internal standard:** chlorpromazine (10.2)
**Limit of quantitation:** 0.2 ng/mL

## KEY WORDS
serum

## REFERENCE
Fox, A.R.; McLoughlin, D.A. Rapid, sensitive high-performance liquid chromatographic method for the quantification of promethazine in human serum with electrochemical detection. *J.Chromatogr.*, **1993**, *631*, 255–259

## SAMPLE
**Matrix:** blood
**Sample preparation:** 10 mL Plasma or whole blood + 1 mL 1 M NaOH, extract twice with 10 mL hexane for 30 min. Remove the organic layers and evaporate them to dryness under a stream of nitrogen, reconstitute the residue in 1 mL 100 mM HCl, add 5 mL chloroform, vortex for 1 min, centrifuge. Remove a 4.5 mL aliquot of the organic layer and evaporate it to dryness, reconstitute the residue in 100 μL mobile phase, inject a 50 μL aliquot.

## HPLC VARIABLES
**Column:** 10 μm Micropak CN (Varian)
**Mobile phase:** MeCN:20 mM ammonium acetate 90:10
**Flow rate:** 2.5
**Injection volume:** 50
**Detector:** UV 254

## CHROMATOGRAM
**Retention time:** 5.7
**Limit of detection:** 10 ng/mL

## OTHER SUBSTANCES
**Extracted:** amitriptyline, butaperazine, carphenazine, chlorpromazine, promazine, thioridazine, trifluoperazine, trimeprazine
**Simultaneous:** acetophenazine, benztropine, haloperidol, imipramine, mesoridazine, nortriptyline, orphenadrine, piperacetazine, thiothixene, triflupromazine, trihexyphenidyl

## KEY WORDS
plasma; whole blood

## REFERENCE

McIntyre, I.M.; King, C.V.; Skafidis, S.; Drummer, O.H. Dual ultraviolet wavelength high-performance liquid chromatographic method for the forensic or clinical analysis of seventeen antidepressants and some selected metabolites. *J.Chromatogr.*, **1993**, *621*, 215–223

## SAMPLE

**Matrix:** blood
**Sample preparation:** Plasma. 1-5 mL Plasma + 1 mL 1 M NaOH, extract with mixed hexanes for 30 min, centrifuge. Remove a 9 mL aliquot of the hexane layer and evaporate it to dryness under a stream of nitrogen at 30°, dissolve residue in 100 μL mobile phase, inject a 50 μL aliquot. Whole blood. 10 mL Whole blood + 1 mL 1 M NaOH, extract with 15 mL mixed hexanes for 1 h. Remove an aliquot of the hexane layer and evaporate it to dryness, reconstitute the residue in 1 mL 100 mM HCl, extract with 5 mL chloroform by vortexing for 1 min, centrifuge. Remove a 4.5 mL aliquot of the chloroform layer, evaporate to dryness, dissolve in 10 μL mobile phase, inject an aliquot.

## HPLC VARIABLES

**Column:** 10 μm Micropak CN
**Mobile phase:** MeCN:5 mM ammonium acetate 90:10 (vary ammonium acetate concentration to achieve best separation)
**Flow rate:** 2.5
**Injection volume:** 10-50
**Detector:** UV 254; E, Bioanalytical Systems LC-4A, glassy carbon electrode +0.9 V, Ag/AgCl reference electrode

## CHROMATOGRAM

**Retention time:** 13.4
**Limit of detection:** 10 ng/mL (UV); 0.1 ng/mL (E)

## OTHER SUBSTANCES

**Extracted:** acetophenazine, amitriptyline, benztropine, butaperazine, carphenazine, chlorpromazine, fluphenazine, haloperidol, imipramine, mesoridazine, nortriptyline, piperacetazine, orphenadrine, promazine, thioridazine, thiothixene, trifluoperazine, triflupromazine, trihexyphenidyl, trimeprazine

## KEY WORDS

plasma; whole blood

## REFERENCE

Curry, S.H.; Brown, E.A.; Hu, O.Y.-P.; Perrin, J.H. Liquid chromatographic assay of phenothiazine, thioxanthene and butyrophenone neuroleptics and antihistamines in blood and plasma with conventional and radial compression columns and UV and electrochemical detection. *J.Chromatogr.*, **1982**, *231*, 361–376

## SAMPLE

**Matrix:** blood
**Sample preparation:** 2 mL Plasma or serum + 100 mg sodium carbonate, vortex for 3 s, add 10 mL 3 ng/mL triflupromazine in hexane:MeOH 99.7:0.3, shake for 15 min, centrifuge at 2000 rpm for 5 min. Remove 9.3 mL of the organic layer and evaporate it to dryness under a stream of air at 40°, reconstitute the residue in 50 μL ethyl acetate, add 25 μL trichloroethyl chloroformate, vortex, heat at 120° for 20 min, cool. Evaporate to dryness under a stream of air at 40°, reconstitute the residue in 100 μL MeOH, inject a 50 μL aliquot.

## HPLC VARIABLES

**Column:** 300 × 4 MCH-10 reversed-phase (Varian)

**Mobile phase:** MeOH:water 84:16
**Flow rate:** 2
**Injection volume:** 50
**Detector:** UV 254

## CHROMATOGRAM
**Retention time:** 4.2
**Internal standard:** triflupromazine (6)
**Limit of detection:** 1 ng/mL

## OTHER SUBSTANCES
**Simultaneous:** chlorpromazine, chlorprothixene, phenothiazine-5-oxide, promazine, trimeprazine
**Interfering:** phenothiazine

## KEY WORDS
plasma; serum; derivatization

## REFERENCE
Wallace, J.E.; Shimek, E.L., Jr.; Harris, S.C.; Stavchansky, S. Determination of promethazine in serum by liquid chromatography. *Clin.Chem.*, **1981**, *27*, 253–255

## SAMPLE
**Matrix:** blood, urine
**Sample preparation:** Urine. Condition a 3 mL Varian Certify SPE cartridge with two cartridge volumes of MeOH and two cartridge volumes of 100 mM pH 6.0 $KH_2PO_4$. 5 mL Urine + 50 µL 1 mg/mL chlorpromazine in water + 5 mL water + 1 mL phosphate buffer, vortex for 1 min, add to the SPE cartridge, rinse the tube with 1 mL phosphate buffer, add the rinse to the SPE cartridge, dry under vacuum for 2-5 min, wash with three 500 µL portions of MeOH:concentrated HCl 99:1 and two 500 µL portions of 1 M acetic acid (allow to dry after each wash), elute with six 1 mL portions of 2% ammonium hydroxide in EtOH. Evaporate the eluate to dryness under a stream of nitrogen at room temperature, reconstitute the residue in 1 mL mobile phase, vortex for 3 min, inject an aliquot. Serum. Condition a 3 mL Varian Certify SPE cartridge with two cartridge volumes of MeOH and two cartridge volumes of 100 mM pH 6.0 $KH_2PO_4$. 3 mL Serum + 50 µL 1 mg/mL chlorpromazine in water + 5 mL water + 1 mL phosphate buffer, vortex for 1 min, add to the SPE cartridge, rinse the tube with 1 mL phosphate buffer, add the rinse to the SPE cartridge, dry under vacuum for 2-5 min, wash with three 500 µL portions of MeOH:concentrated HCl 99:1 and two 500 µL portions of 1 M acetic acid (allow to dry after each wash), elute with six 1 mL portions of 2% ammonium hydroxide in EtOH. Evaporate the eluate to dryness under a stream of nitrogen at room temperature, reconstitute the residue in 300 µL mobile phase, vortex for 3 min, inject an aliquot. (To determine low levels of promethazine enantiomers the residue was reconstituted with 300 µL mobile phase containing 5 mg/mL enantiomer. The amount of promethazine in the serum was calculated by the increase in peak area over the control value.)

## HPLC VARIABLES
**Column:** 100 × 4.6 5 µm KK-CARNU (α-R-naphthyl)urea (YMC)
**Mobile phase:** Hexane:1,2-dichloroethane:EtOH:trifluoroacetic acid 40:15:10:0.1
**Flow rate:** 1
**Injection volume:** 20
**Detector:** F ex 254 em 280 (filter)

## CHROMATOGRAM
**Retention time:** 12.40 (R-(+)), 13.28 (S-(−))
**Internal standard:** chlorpromazine (19.13)
**Limit of detection:** 400 ng/mL

## KEY WORDS
chiral; serum; SPE

## REFERENCE
Ponder, G.W.; Stewart, J.T. A liquid chromatographic method for the determination of promethazine enantiomers in human urine and serum using solid-phase extraction and fluorescence detection. *J.Pharm.Biomed.Anal.*, **1995**, *13*, 1161−1166

## SAMPLE
**Matrix:** bulk
**Sample preparation:** Prepare a solution in the mobile phase.

## HPLC VARIABLES
**Column:** 250 × 4.6 Cyclobond I (Advanced Separation Technologies)
**Mobile phase:** MeOH:buffer 17:83 (Buffer was 2.5% triethylamine acetate, pH 5.50.)
**Flow rate:** 0.8
**Injection volume:** 20
**Detector:** UV 254

## CHROMATOGRAM
**Retention time:** 10.2

## OTHER SUBSTANCES
**Simultaneous:** isopromethazine

## REFERENCE
Piperaki, S.; Perakis, A.; Parissi-Poulou, M. Liquid chromatographic retention behaviour and separation of promethazine and isopromethazine on β-cyclodextrin bonded-phase column. *J.Chromatogr.A*, **1994**, *660*, 339−350

## SAMPLE
**Matrix:** formulations
**Sample preparation:** Dilute with saline, inject a 10 μL aliquot.

## HPLC VARIABLES
**Column:** 300 × 3.9 μBondapak C18
**Mobile phase:** MeCN:100 mM $K_2HPO_4$ 33:67 adjusted to a final pH of 4.4 with phosphoric acid
**Flow rate:** 1.7
**Injection volume:** 10
**Detector:** UV 254

## CHROMATOGRAM
**Retention time:** 5.5

## OTHER SUBSTANCES
**Interfering:** midazolam

## KEY WORDS
injections; saline

## REFERENCE
Martens, H.J.; de Goede, P.N.; van Loenen, A.C. Sorption of various drugs in polyvinyl chloride, glass, and polyethylene-lined infusion containers. *Am.J.Hosp.Pharm.*, **1990**, *47*, 369−373

## SAMPLE

**Matrix:** formulations

**Sample preparation:** Sonicate in mobile phase, dilute the clear supernatant to a concentration of 3-10 µg/mL, filter (0.5 µm), inject an aliquot.

## HPLC VARIABLES

**Column:** 250 × 2.6 10 µm Porasil silica gel (Perkin-Elmer)
**Mobile phase:** MeOH containing 0.02% $NH_4OH$:dichloromethane 20:80
**Flow rate:** 1
**Injection volume:** 20
**Detector:** UV 310

## CHROMATOGRAM

**Retention time:** 2

## OTHER SUBSTANCES

**Simultaneous:** aminophylline

## KEY WORDS

suppositories; normal phase

## REFERENCE

Kountourellis, J.E.; Raptouli, A.; Georgarakis, M. Simultaneous determination of aminophylline and promethazine in suppositories by high-performance liquid chromatography. *Pharmazie*, **1986**, *41*, 600–601

## SAMPLE

**Matrix:** formulations

**Sample preparation:** Tablets. Powder tablets, weigh out amount equivalent to about 10 mg, add 75 mL mobile phase, sonicate for 20 min, dilute to 100 mL with mobile phase, mix, filter (0.45 µm) (discard first 10 mL of filtrate), inject a 20 µL aliquot of the filtrate. Syrups, elixirs, injections. Measure out amount equivalent to about 10 mg, add 75 mL mobile phase, sonicate for 20 min, dilute to 100 mL with mobile phase, mix, inject a 20 µL aliquot.

## HPLC VARIABLES

**Column:** 300 × 3.9 10 µm µBondapak CN
**Mobile phase:** MeOH:3 mM ammonium acetate 90:10
**Flow rate:** 1.3
**Injection volume:** 20
**Detector:** UV 254

## CHROMATOGRAM

**Retention time:** 3.3

## OTHER SUBSTANCES

**Also analyzed:** chlorpheniramine, cyclizine, doxylamine, mesoridazine, pentazocine, protriptyline, pyrilamine, pyrimethamine, tripelennamine

## KEY WORDS

tablets; syrups; elixirs; injections

## REFERENCE

Walker, S.T. Liquid chromatographic determination of organic nitrogenous bases in dosage forms: a progress report. *J.Assoc.Off.Anal.Chem.*, **1985**, *68*, 539–542

## SAMPLE
**Matrix:** solutions

## HPLC VARIABLES
**Column:** 250 × 4.6 Chirex 3020 (Phenomenex)
**Mobile phase:** Hexane:1,2-dichloroethane:EtOH/trifluoroacetic acid 60:35:5 (EtOH/trifluoroacetic acid was premixed 20:1.)
**Flow rate:** 1
**Injection volume:** 20
**Detector:** UV 254

## CHROMATOGRAM
**Retention time:** 6.4, 6.8 (enantiomers)

## KEY WORDS
chiral

## REFERENCE
Cleveland, T. Pirkle-concept chiral stationary phases for the HPLC separation of pharmaceutical racemates. *J.Liq.Chromatogr.*, **1995**, *18*, 649–671

## SAMPLE
**Matrix:** solutions

## HPLC VARIABLES
**Column:** 100 × 4 5 μm CHIRAL-AGP (ChromTech)
**Mobile phase:** MeCN:59 mM pH 4.0 acetate buffer 1:99
**Flow rate:** 0.9
**Injection volume:** 20
**Detector:** UV 225

## CHROMATOGRAM
**Retention time:** k′ 7.68, 10.1 (enantiomers)

## KEY WORDS
chiral

## REFERENCE
Hermansson, J.; Grahn, A. Optimization of the separation of enantiomers of basic drugs. Retention mechanisms and dynamic modification of the chiral bonding properties on an $\alpha_1$-acid glycoprotein column. *J.Chromatogr.A*, **1995**, *694*, 57–69

## SAMPLE
**Matrix:** solutions
**Sample preparation:** Inject an aliquot of a 200 μM solution in MeOH.

## HPLC VARIABLES
**Column:** 100 × 4.7 7 μm Hypercarb (Shandon)
**Mobile phase:** MeOH containing 5 mM N-benzyloxycarbonylglycyl-L-proline and 4.5 mM NaOH
**Column temperature:** 17
**Injection volume:** 20
**Detector:** UV 270

## CHROMATOGRAM
**Retention time:** k′ 58 (first enantiomer)

## KEY WORDS
chiral; α = 1.23

## REFERENCE
Huynh, N.-H.; Karlsson, A.; Pettersson, C. Enantiomeric separation of basic drugs using N-benzyloxy-carbonylglyclyl-L-proline as counter ion in methanol. *J.Chromatogr.A*, **1995**, *705*, 275–287

## SAMPLE
**Matrix:** solutions

## HPLC VARIABLES
**Column:** 150 × 4.6 12 μm 1-myristoyl-2-[(13-carboxyl)-tridecoyl]-sn-3-glycerophosphocholine chemically bonded to silica (Regis)
**Mobile phase:** MeCN:100 mM pH 7.0 phosphate buffer 20:80
**Flow rate:** 1
**Detector:** UV 254

## CHROMATOGRAM
**Retention time:** k′ 32.21

## OTHER SUBSTANCES
**Also analyzed:** acebutolol, alprenolol, antazoline, atenolol, betaxolol, bisoprolol, bopindolol, bupranolol, carteolol, celiprolol, chloropyramine, chlorpheniramine, cicloprolol, cimetidine, cinnarizine, cirazoline, clonidine, dilevalol, dimethindene, diphenhydramine, doxazosin, esmolol, famotidine, isothipendyl, ketotifen, metiamide, metoprolol, moxonidine, nadolol, naphazoline, nifenalol, nizatidine, oxprenolol, pheniramine, phentolamine, pindolol, pizotyline (pizotifen), practolol, prazosin, promethazine, propranolol, pyrilamine (mepyramine), ranitidine, roxatidine, sotalol, tiamenidine, timolol, tramazoline, tripelennamine, triprolidine, tymazoline, UK-14,304

## REFERENCE
Kaliszan, R.; Nasal, A.; Turowski, M. Binding site for basic drugs on $\alpha_1$-acid glycoprotein as revealed by chemometric analysis of biochromatographic data. *Biomed.Chromatogr.*, **1995**, *9*, 211–215

## SAMPLE
**Matrix:** solutions

## HPLC VARIABLES
**Column:** 250 × 4.6 5 μm Supelcosil LC-DP (A) or 250 × 4 5 μm LiChrospher 100 RP-8 (B)
**Mobile phase:** MeCN:0.025% phosphoric acid:buffer 25:10:5 (A) or 60:25:15 (B) (Buffer was 9 mL concentrated phosphoric acid and 10 mL triethylamine in 900 mL water, adjust pH to 3.4 with dilute phosphoric acid, make up to 1 L.)
**Flow rate:** 0.6
**Injection volume:** 25
**Detector:** UV 229

## CHROMATOGRAM
**Retention time:** 13.20 (A), 6.40 (B)

## OTHER SUBSTANCES
**Also analyzed:** acebutolol, acepromazine, acetaminophen, acetazolamide, acetophenazine, albuterol, alprazolam, amitriptyline, amobarbital, amoxapine, antipyrine, atenolol, atro-

pine, azatadine, baclofen, benzocaine, bromocriptine, brompheniramine, brotizolam, bupivacaine, buspirone, butabarbital, butalbital, caffeine, carbamazepine, cetirizine, chlorcyclizine, chlordiazepoxide, chlormezanone, chloroquine, chlorpheniramine, chlorpromazine, chlorpropamide, chlorprothixene, chlorthalidone, chlorzoxazone, cimetidine, cisapride, clomipramine, clonazepam, clonidine, clozapine, cocaine, codeine, colchicine, cyclizine, cyclobenzaprine, dantrolene, desipramine, diazepam, diclofenac, diflunisal, diltiazem, diphenhydramine, diphenidol, diphenoxylate, dipyridamole, disopyramide, dobutamine, doxapram, doxepin, droperidol, encainide, ethidium bromide, ethopropazine, fenoprofen, fentanyl, flavoxate, fluoxetine, fluphenazine, flurazepam, flurbiprofen, fluvoxamine, furosemide, glutethimide, glyburide, guaifenesin, haloperidol, homatropine, hydralazine, hydrochlorothiazide, hydrocodone, hydromorphone, hydroxychloroquine, hydroxyzine, ibuprofen, imipramine, indomethacin, ketoconazole, ketoprofen, ketorolac, labetalol, levorphanol, lidocaine, loratadine, lorazepam, lovastatin, loxapine, mazindol, mefenamic acid, meperidine, mephenytoin, mepivacaine, mesoridazine, metaproterenol, metformin, methadone, methdilazine, methocarbamol, methotrexate, methotrimeprazine, methoxamine, methyldopa, methylphenidate, metoclopramide, metolazone, metoprolol, metronidazole, midazolam, moclobemide, morphine, nadolol, nalbuphine, naloxone, naphazoline, naproxen, nifedipine, nizatidine, norepinephrine, nortriptyline, oxazepam, oxycodone, oxymetazoline, paroxetine, pemoline, pentazocine, pentobarbital, pentoxifylline, perphenazine, pheniramine, phenobarbital, phenol, phenolphthalein, phentolamine, phenylbutazone, phenyltoloxamine, phenytoin, pimozide, pindolol, piroxicam, pramoxine, prazepam, prazosin, probenecid, procainamide, procaine, prochlorperazine, procyclidine, promazine, propafenone, propantheline, propiomazine, propofol, propranolol, protriptyline, quazepam, quinidine, quinine, racemethorphan, ranitidine, remoxipride, risperidone, salicylic acid, scopolamine, secobarbital, sertraline, sotalol, spironolactone, sulfinpyrazone, sulindac, temazepam, terbutaline, terfenadine, tetracaine, theophylline, thiethylperazine, thiopental, thioridazine, thiothixene, timolol, tocainide, tolbutamide, tolmetin, trazodone, triamterene, triazolam, trifluoperazine, triflupromazine, trimeprazine, trimethoprim, trimipramine, verapamil, warfarin, xylometazoline, yohimbine, zopiclone

## KEY WORDS
also some details of plasma extraction

## REFERENCE
Koves, E.M. Use of high-performance liquid chromatography-diode array detection in forensic toxicology. *J.Chromatogr.A*, **1995**, *692*, 103–119

## SAMPLE
**Matrix:** solutions
**Sample preparation:** Make up a 500 ng/mL solution in mobile phase, inject an aliquot.

## HPLC VARIABLES
**Column:** 250 × 4.6 5 µm Chiralcel-OJ
**Mobile phase:** Hexane:EtOH 50:50
**Flow rate:** 0.8
**Injection volume:** 100
**Detector:** F ex 254 em 280 (filter)

## CHROMATOGRAM
**Retention time:** 8.30, 9.93 (enantiomers)

## KEY WORDS
chiral

## REFERENCE
Ponder, G.W.; Butram, S.L.; Adams, A.G.; Ramanathan, C.S.; Stewart, J.T. Resolution of promethazine, ethopropazine, trimeprazine and trimipramine enantiomers on selected chiral stationary phases using high-performance liquid chromatography. *J.Chromatogr.A*, **1995**, *692*, 173–182

## SAMPLE
**Matrix:** solutions
**Sample preparation:** Prepare a 1 mg/mL solution in MeOH, inject a 5 μL aliquot.

## HPLC VARIABLES
**Column:** 250 × 4.6 5 μm Lichrosphere cyanopropyl
**Mobile phase:** Carbon dioxide:MeOH:isopropylamine 90:10:0.05
**Column temperature:** 50
**Flow rate:** 3
**Injection volume:** 5
**Detector:** UV 254

## CHROMATOGRAM
**Retention time:** 2.49

## OTHER SUBSTANCES
**Simultaneous:** acetophenazine, carphenazine, deserpidine, ethopropazine, methotrimeprazine, methotrimeprazine, perphenazine, promazine, reserpine, thiothixene, triflupromazine
**Interfering:** chlorprothixene, propiomazine

## KEY WORDS
SFC; pressure 200 bar

## REFERENCE
Berger, T.A.; Wilson, W.H. Separation of drugs by packed column supercritical fluid chromatography. 1. Phenothiazine antipsychotics. *J.Pharm.Sci.*, **1994**, *83*, 281–286

## SAMPLE
**Matrix:** solutions

## HPLC VARIABLES
**Column:** 250 × 4.6 amylose tris(3,5-dichlorophenylcarbamate)
**Mobile phase:** Hexane:isopropanol 98:2
**Flow rate:** 0.5
**Detector:** UV

## CHROMATOGRAM
**Retention time:** k′ 1.84 (of first (+) enantiomer)

## KEY WORDS
chiral; α = 1.24

## REFERENCE
Okamoto, Y.; Aburatani, R.; Hatano, K.; Hatada, K. Optical resolution of racemic drugs by chiral HPLC on cellulose and amylose tris(phenylcarbamate) derivatives. *J.Liq.Chromatogr.*, **1988**, *11*, 2147–2163

## SAMPLE
**Matrix:** solutions
**Sample preparation:** Dissolve in MeOH:water 1:1 at a concentration of 50 μg/mL, inject a 10 μL aliquot.

## HPLC VARIABLES
**Column:** 300 × 3.9 10 μm μBondapak C18

**Mobile phase:** MeOH:acetic acid:triethylamine:water 60:1.5:0.5:38
**Flow rate:** 1.5
**Injection volume:** 10
**Detector:** UV 254

## CHROMATOGRAM
**Retention time:** 11

## OTHER SUBSTANCES
**Simultaneous:** 8-chlorotheophylline, diphenhydramine, diphenylpyraline

## REFERENCE
Roos, R.W.; Lau-Cam, C.A. General reversed-phase high-performance liquid chromatographic method
for the separation of drugs using triethylamine as a competing base. *J.Chromatogr.*, **1986**, *370*,
403–418

## SAMPLE
**Matrix:** solutions
**Sample preparation:** Dissolve in MeOH:water 1:1 at a concentration of 50 µg/mL, inject
a 10 µL aliquot.

## HPLC VARIABLES
**Column:** 300 × 3.9 10 µm µBondapak C18
**Mobile phase:** MeOH:acetic acid:triethylamine:water 70:1.5:0.5:28
**Flow rate:** 1.5
**Injection volume:** 10
**Detector:** UV 254

## CHROMATOGRAM
**Retention time:** 5.5

## OTHER SUBSTANCES
**Simultaneous:** acetophenazine, butaperazine, chlorpromazine, mesoridazine, prochlorperazine, thiethylperazine, thioridazine

## REFERENCE
Roos, R.W.; Lau-Cam, C.A. General reversed-phase high-performance liquid chromatographic method
for the separation of drugs using triethylamine as a competing base. *J.Chromatogr.*, **1986**, *370*,
403–418

## SAMPLE
**Matrix:** solutions
**Sample preparation:** Prepare a 10 µg/mL solution in MeOH, inject a 20 µL aliquot.

## HPLC VARIABLES
**Column:** 125 × 4.9 Spherisorb S5W silica
**Mobile phase:** MeOH containing 10 mM ammonium perchlorate and 1 mL/L 100 mM
NaOH in MeOH, pH 6.7
**Flow rate:** 2
**Injection volume:** 20
**Detector:** E, LeCarbone, V25 glassy carbon electrode, + 1.2 V

## CHROMATOGRAM
**Retention time:** 5.4

## OTHER SUBSTANCES

**Also analyzed:** acebutolol, acepromazine, acetophenazine, N-acetylprocainamide, albu-terol, alprenolol, amethocaine, amiodarone, amitriptyline, antazoline, atenolol, azacy-clonal, bamethane, benactyzine, benperidol, benzethidine, benzocaine, benzoctamine, benzphetamine, benzquinamide, bromhexine, bromodiphenhydramine, bromperidol, brompheniramine, bromhromazine, buclizine, bufotenine, bupivacaine, buprenorphine, butacaine, butethamate, chlorcyclizine, chlorpheniramine, chlorphenoxamine, chlorpren-aline, chlorpromazine, chlorprothixene, cimetidine, cinchonidine, cinnarizine, clemastine, clomipramine, clonidine, cocaine, cyclazocine, cyclizine, cyclopentamine, cyproheptadine, deserpidine, desipramine, dextromoramide, dextropropoxyphene, dicyclomine, diethylcar-bamazine, diethylpropion, diethylthiambutene, dihydroergotamine, dimethindene, di-methothiazine, diphenhydramine, diphenoxylate, dipipanone, diprenorphine, dipyrida-mole, disopyramide, dothiepin, doxapram, doxepin, doxylamine, droperidol, ephedrine, ergocornine, ergocristine, ergocristinine, ergocryptine, ergometrine, ergosine, ergosinine, ergotamine, ethopropazine, etorphine, etoxeridine, fenethazine, fenfluramine, fenoterol, fentanyl, flavoxate, fluopromazine, flupenthixol, fluphenazine, flurazepam, haloperidol, hydroxyzine, hyoscine, ibogaine, imipramine, indapamine, iprindole, isothipendyl, isox-suprine, ketanserin, laudanosine, lidocaine, lofepramine, loxapine, maprotiline, mecamy-lamine, meclophenoxate, meclozine, medazepam, mephentermine, mepivacaine, mep-tazinol, mepyramine, mesoridazine, metaraminol, methadone, methamphetamine, methapyrilene, methdilazine, methotrimeprazine, methoxamine, methoxyphenamine, methoxypromazine, methylephedrine, methylergonovine, methysergide, metoclopramide, metopimazine, metoprolol, mianserin, morazone, nadolol, nalorphine, naloxone, napha-zoline, nicotine, nifedipine, nomifensine, nortriptyline, noscapine, orphenadrine, oxeladin, oxprenolol, oxymetazolin, papaverine, pargyline, pecazine, penbutolol, pentazocine, pen-thienate, pericyazine, perphenazine, phenadoxone, phenampromide, phenazocine, phen-butrazate, phendimetrazine, phenelzine, phenglutarimide, phenindamine, pheniramine, phenmetrazine, phenomorphan, phenoperidine, phenothiazine, phenoxybenzamine, phen-tolamine, phenylephrine, phenyltoloxamine, physostigmine, piminodine, pimozide, pin-dolol, pipamazine, pipazethate, piperacetazine, piperidolate, pipradol, pirenzepine, piri-tramide, pizotifen, practolol, pramoxine, prazosin, prenylamine, prilocaine, primaquine, proadifen, procainamide, procaine, prochlorperazine, procyclidine, proheptazine, prolin-tane, promazine, pronethalol, properidine, propiomazine, propranolol, prothipendyl, pro-triptyline, proxymetacaine, pseudoephedrine, pyrimethamine, quinidine, quinine, raniti-dine, rescinnamine, sotalol, tacrine, terazosin, terbutaline, terfenadine, thenyldiamine, theophylline, thiethylperazine, thiopropazate, thioproperazine, thioridazine, thiothixene, thonzylamine, timolol, tocainide, tolpropamine, tolycaine, tranylcypromine, trazodone, tri-fluoperazine, trifluperidol, trimeperidine, trimeprazine, trimethobenzamide, trimetho-prim, trimipramine, tripelennamine, triprolidine, tryptamine, verapamil, xylometazoline

## REFERENCE

Jane, I.; McKinnon, A.; Flanagan, R.J. High-performance liquid chromatographic analysis of basic drugs on silica columns using non-aqueous ionic eluents. II. Application of UV, fluorescence and electro-chemical oxidation detection. *J.Chromatogr.*, **1985**, *323*, 191–225

◆────────────────────◆────────────────────◆

## ANNOTATED BIBLIOGRAPHY

Radwanska, A.; Frackowiak, T.; Ibrahim, H.; Aubry, A.-F.; Kaliszan, R. Chromatographic modelling of interactions between melanin and phenothiazine and dibenzapine drugs. *Biomed.Chromatogr.*, **1995**, *9*, 233–237 [also acetopromazine, chlorpromazine, clomipramine, ethopromazine, fluphenazine, imip-ramine, perphenazine, prochlorperazine, promazine, propiomazine, thioridazine, trifluoperazine, tri-fluopromazine, trimeprazine]

Koytchev, R.; Alken, R.-G.; Kirkov, V.; Neshev, G.; Vagaday, M.; Kunter, U. Absolute Bioverfügbarkeit von Chlorpromazin, Promazin und Promethazin [Absolute bioavailability of chlorpromazine, prom-azine and promethazine]. *Arzneimittelforschung*, **1994**, *44*, 121–125 [column temp 40; LOD 1-2 ng/mL; serum; pharmacokinetics]

Tracqui, A.; Kintz, P.; Kreissig, P.; Mangin, P. Simple and rapid screening procedure for 27 neuroleptics using HPLC/DAD. *J.Liq.Chromatogr.*, **1992**, *15*, 1381–1396 [column temp 30; blood; urine; extracted acepromazine, aceprometazine, alimemazine, amisulpride, benperidol, chlorpromazine, cyamemazine, droperidol, fluanisone, haloperidol, levomepromazine, methotrimeprazine, metoclopramide, moperone, penfluridol, pimozide, pipamperone, pipotiazine, prochlorperazine, propericiazine, sulpride, sultopride, thioproperazine, thioridazine, tiapride, trifluoperazine, trifluperidol]

Kountourellis, J.E.; Markopoulou, C.K. A simultaneous analysis by high performance liquid chromatography of bamipine combined with tricyclic antidepressants and/or antipsychotics in dosage forms. *J.Liq.Chromatogr.*, **1991**, *14*, 2969–2977 [simultaneous bamipine, chlorprothixene, haloperidol, imipramine, prochlorperazine, thioridazine, trifluoperazine, trimeprazine, trimipramine]

Schill, G.; Wainer, I.W.; Barkan, S.A. Chiral separation of cationic drugs on an α1-acid glycoprotein bonded stationary phase. *J.Liq.Chromatogr.*, **1986**, *9*, 641–666 [also atropine, bromdiphenhydramine, brompheniramine, bupivacaine, butorphanol, carbinoxamine, chlorpheniramine, clidinium, cocaine, cyclopentolate, dimethindene, diperidone, disopyramide, doxylamine, ephedrine, homatropine, labetalol, labetalol A, labetalol B, mepensolate, mepivacaine, methadone, methorphan, methylatropine, methylhomatropine, methylphenidate, metoprolol, nadolol, nadolol A, nadolol B, oxprenolol, oxyphencyclimine, phenmetrazine, phenoxybenzamine, pronethalol, propoxyphene, propranolol, pseudoephedrine, terbutaline, tocainide, tridihexethyl]

Leelavathi, D.E.; Dressler, D.E.; Soffer, E.F.; Yachetti, S.D.; Knowles, J.A. Determination of promethazine in human plasma by automated high-performance liquid chromatography with electrochemical detection and by gas chromatography-mass spectrometry. *J.Chromatogr.*, **1985**, *339*, 105–115 [LOD 0.1 ng/mL; column-switching; thioridazine (IS)]

Allender, W.J.; Archer, A.W. Liquid chromatographic analysis of promethazine and its major metabolites in human postmortem material. *J.Forensic Sci.*, **1984**, *29*, 515–526

Stavchansky, S.; Wallace, J.; Chue, M.; Newburger, J. High pressure liquid chromatographic determination of promethazine hydrochloride in the presence of its thermal and photolytic degradation products: A stability indicating assay. *J.Liq.Chromatogr.*, **1983**, *6*, 1333–1344 [promazine (IS); column temp 45]

Patel, R.B.; Welling, P.G. High-pressure liquid chromatographic determination of promethazine plasma levels in the dog after oral, intramuscular, and intravenous dosage. *J.Pharm.Sci.*, **1982**, *71*, 529–532

Taylor, G.; Houston, J.B. Simultaneous determination of promethazine and two of its circulating metabolites by high-performance liquid chromatography. *J.Chromatogr.*, **1982**, *230*, 194–198 [imipramine (IS); LOD 0.2 ng/mL; whole blood; pharmacokinetics]

Brinkman, U.A.T.; Welling, P.L.M.; De Vries, G.; Scholten, A.H.M.T.; Frei, R.W. Liquid chromatography of demoxepam and phenothiazines using a post-column photochemical reactor and fluorescence detection. *J.Chromatogr.*, **1981**, *217*, 463–471 [post-column reaction; fluorescence detection; also demoxepam, chlorpromazine, methotrimeprazine, nedaltran]

Wallace, J.E.; Shimek, E.L.J.; Stavchansky, S.; Harris, S.C. Determination of promethazine and other phenothiazine compounds by liquid chromatography wih electrochemical detection. *Anal.Chem.*, **1981**, *53*, 960–962

DiGregorio, G.J.; Ruch, E. Human and whole blood and parotid saliva concentrations of oral and intramuscular promethazine. *J.Pharm.Sci.*, **1980**, *69*, 1457–1461 [promazine (IS); LOQ 2.5 ng/mL (whole blood); LOQ 0.5 ng/mL (saliva)]

# Propofol

**Molecular formula:** C$_{12}$H$_{18}$O
**Molecular weight:** 178.3
**CAS Registry No.:** 2078-54-8

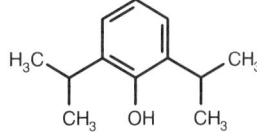

## SAMPLE
**Matrix:** blood
**Sample preparation:** 1 ml Whole blood + 20 μL thymol in MeOH + 1 mL 100 mM NaH$_2$PO$_4$ + 5 mL cyclohexane, shake at 200 rpm for 15 min, centrifuge at 1200 g for 5 min. Remove 4.5 mL of the organic layer and add it to 50 μL 1.875% tetraethylammonium hydroxide in EtOH, evaporate to dryness under a stream of nitrogen, reconstitute the residue in mobile phase, inject an aliquot.

## HPLC VARIABLES
**Column:** 250 × 4 C18
**Mobile phase:** MeCN:buffer 67:33 (Buffer was water adjusted to pH 4.0 with acetic acid.)
**Detector:** UV

## CHROMATOGRAM
**Retention time:** 13
**Internal standard:** thymol (5-methyl-2-isopropylphenol) (8)
**Limit of quantitation:** 500 ng/mL

## KEY WORDS
whole blood; pharmacokinetics

## REFERENCE
Dawidowicz, A.L.; Fijalkowska, A. Determination of propofol in blood by HPLC. Comparison of the extraction and precipitation methods. *J.Chromatogr.Sci.*, **1995**, *33*, 377–382

## SAMPLE
**Matrix:** blood
**Sample preparation:** Condition a 1 mL Supelclean LC-18 SPE cartridge (Supelco) with two 1 mL portions of MeCN and two 1 mL portions of water. 50 μL Whole blood + 25 ng thymol + 200 μL 100 mM KH$_2$PO$_4$, mix, add to the SPE cartridge, wash with two 1 mL portions of water, elute with MeCN. Discard the first 50 μL eluate and collect the next 600 μL, evaporate to about 200 μL under a stream of nitrogen, inject a 30 μL aliquot.

## HPLC VARIABLES
**Guard column:** Chrompack RP guard column
**Column:** 100 × 4.6 5 μm Spherisorb S5ODS2
**Mobile phase:** MeCN:water:85% orthophosphoric acid 46:54:0.1
**Flow rate:** 1
**Injection volume:** 30
**Detector:** F ex 276 em 310

## CHROMATOGRAM
**Internal standard:** thymol
**Limit of quantitation:** 50 ng/mL

## KEY WORDS
whole blood; SPE; pharmacokinetics

## REFERENCE

Lee, H.-S.; Khoo, Y.-M.; Chua, B.-C.; Ng, A.S.-B.; Tan, S.S.-W.; Chew, S.-L. Pharmacokinetics of propofol infusion in Asian patients undergoing coronary artery bypass grafting. *Ther.Drug Monit.*, **1995**, *17*, 336–341

## SAMPLE

**Matrix:** blood

**Sample preparation:** 500 μL Serum + 500 μL 20 μg/mL dibutyl phthalate in MeCN:65% perchloric acid 67:33, vortex for 1 min, centrifuge at 1150 g for 5 min, inject a 20 μL aliquot of the supernatant.

## HPLC VARIABLES

**Column:** 250 × 4.6 10 μm Spherisorb C18

**Mobile phase:** MeCN:water:acetic acid 67:33:0.04, pH 4.0

**Flow rate:** 1.5

**Injection volume:** 20

**Detector:** UV 270

## CHROMATOGRAM

**Internal standard:** dibutyl phthalate

**Limit of detection:** 100 ng/mL

## KEY WORDS

serum

## REFERENCE

Pavan, I.; Buglione, E.; Massiccio, M.; Gregoretti, C.; Burbi, L.; Berardino, M. Monitoring propofol serum levels by rapid and sensitive reversed-phase high-performance liquid chromatography during prolonged sedation in ICU patients. *J.Chromatogr.Sci.*, **1992**, *30*, 164–166

## SAMPLE

**Matrix:** blood, urine

**Sample preparation:** Inject a 10-400 μL aliquot of serum or urine onto column A with mobile phase A and elute to waste, after 2 min backflush the contents of column A onto column B with mobile phase B, elute with mobile phase B, monitor the effluent from column B.

## HPLC VARIABLES

**Column:** A 20 × 4 37-50 μm Bondapak C18/Corasil; B 125 × 4 5 μm LiChrosorb RP-18

**Mobile phase:** A 10 mM pH 7.88 KH$_2$PO$_4$; B MeCN:10 mM pH 7.88 KH$_2$PO$_4$ 70:30

**Flow rate:** A 1.5; B 1.3

**Injection volume:** 10-400

**Detector:** F ex 276 em 310

## CHROMATOGRAM

**Retention time:** 5

**Limit of quantitation:** 5 ng/mL

## KEY WORDS

serum; column-switching

## REFERENCE

Altmayer, P.; Büch, U.; Büch, H.P.; Larsen, R. Rapid and sensitive pre-column extraction high-performance liquid chromatographic assay for propofol in biological fluids. *J.Chromatogr.*, **1993**, *612*, 326–330

## SAMPLE
**Matrix:** formulations
**Sample preparation:** Filter (0.45 μm). Add an aliquot to an aliquot of 1 mg/mL thymol, make up to 3 mL with MeOH, inject a 50 μL aliquot.

## HPLC VARIABLES
**Column:** 250 × 4.6 5 μm Hypersil ODS C18
**Mobile phase:** MeCN:MeOH:water 55:10:35
**Flow rate:** 2
**Injection volume:** 50
**Detector:** UV 270

## CHROMATOGRAM
**Retention time:** 7.5
**Internal standard:** thymol (4.0)

## KEY WORDS
stability-indicating; parenteral nutrient solutions

## REFERENCE
Bhatt-Mehta, V.; Paglia, R.E.; Rosen, D.A. Stability of propofol with parenteral nutrient solutions during simulated Y-site injection. *Am.J.Health-Syst.Pharm.*, **1995**, *52*, 192–196

## SAMPLE
**Matrix:** solutions

## HPLC VARIABLES
**Column:** 250 × 4.6 5 μm Supelcosil LC-DP (A) or 250 × 4 5 μm LiChrospher 100 RP-8 (B)
**Mobile phase:** MeCN:0.025% phosphoric acid:buffer 25:10:5 (A) or 60:25:15 (B) (Buffer was 9 mL concentrated phosphoric acid and 10 mL triethylamine in 900 mL water, adjust pH to 3.4 with dilute phosphoric acid, make up to 1 L.)
**Flow rate:** 0.6
**Injection volume:** 25
**Detector:** UV 229

## CHROMATOGRAM
**Retention time:** 10.07 (A), 15.24 (B)

## OTHER SUBSTANCES
**Also analyzed:** acebutolol, acepromazine, acetaminophen, acetazolamide, acetophenazine, albuterol, alprazolam, amitriptyline, amobarbital, amoxapine, antipyrine, atenolol, atropine, azatadine, baclofen, benzocaine, bromocriptine, brompheniramine, brotizolam, bupivacaine, buspirone, butabarbital, butalbital, caffeine, carbamazepine, cetirizine, chlorcyclizine, chlordiazepoxide, chlormezanone, chloroquine, chlorpheniramine, chlorpromazine, chlorpropamide, chlorprothixene, chlorthalidone, chlorzoxazone, cimetidine, cisapride, clomipramine, clonazepam, clonidine, clozapine, cocaine, codeine, colchicine, cyclizine, cyclobenzaprine, dantrolene, desipramine, diazepam, diclofenac, diflunisal, diltiazem, diphenhydramine, diphenidol, diphenoxylate, dipyridamole, disopyramide, dobutamine, doxapram, doxepin, droperidol, encainide, ethidium bromide, ethopropazine, fenoprofen, fentanyl, flavoxate, fluoxetine, fluphenazine, flurazepam, flurbiprofen, fluvoxamine, furosemide, glutethimide, glyburide, guaifenesin, haloperidol, homatropine, hydralazine, hydrochlorothiazide, hydrocodone, hydromorphone, hydroxychloroquine, hydroxyzine, ibuprofen, imipramine, indomethacin, ketoconazole, ketoprofen, ketorolac, labetalol, levorphanol, lidocaine, loratadine, lorazepam, lovastatin, loxapine, mazindol, mefenamic acid, meperidine, mephenytoin, mepivacaine, mesoridazine, metaproterenol, metformin, methadone, methdilazine, methocarbamol, methotrexate, methotrimeprazine,

methoxamine, methyldopa, methylphenidate, metoclopramide, metolazone, metoprolol, metronidazole, midazolam, moclobemide, morphine, nadolol, nalbuphine, naloxone, naphazoline, naproxen, nifedipine, nizatidine, norepinephrine, nortriptyline, oxazepam, oxycodone, oxymetazoline, paroxetine, pemoline, pentazocine, pentobarbital, pentoxifylline, perphenazine, pheniramine, phenobarbital, phenol, phenolphthalein, phentolamine, phenylbutazone, phenyltoloxamine, phenytoin, pimozide, pindolol, piroxicam, pramoxine, prazepam, prazosin, probenecid, procainamide, procaine, prochlorperazine, procyclidine, promazine, promethazine, propafenone, propantheline, propiomazine, propranolol, protriptyline, quazepam, quinidine, quinine, racemethorphan, ranitidine, remoxipride, risperidone, salicylic acid, scopolamine, secobarbital, sertraline, sotalol, spironolactone, sulfinpyrazone, sulindac, temazepam, terbutaline, terfenadine, tetracaine, theophylline, thiethylperazine, thiopental, thioridazine, thiothixene, timolol, tocainide, tolbutamide, tolmetin, trazodone, triamterene, triazolam, trifluoperazine, triflupromazine, trimeprazine, trimethoprim, trimipramine, verapamil, warfarin, xylometazoline, yohimbine, zopiclone

## KEY WORDS
some details of plasma extraction

## REFERENCE
Koves, E.M. Use of high-performance liquid chromatography-diode array detection in forensic toxicology. *J.Chromatogr.A*, **1995**, *692*, 103–119

## ANNOTATED BIBLIOGRAPHY

Fan, S.Z.; Yu, H.Y.; Chen, Y.L.; Liu, C.C. Propofol concentration monitoring in plasma or whole blood by gas chromatography and high-performance liquid chromatography. *Anesth.Analg.*, **1995**, *81*, 175–178

Bailey, L.C.; Tang, K.T.; Rogozinski, B.A. The determination of 2,6-diisopropylphenol (propofol) in an oil in water emulsion dosage form by high-performance liquid chromatography and by second derivative UV spectroscopy. *J.Pharm.Biomed.Anal.*, **1991**, *9*, 501–506

Chan, K.; So, A.P. The measurement of propofol in human blood samples by liquid chromatography. *Methods Find.Exp.Clin.Pharmacol.*, **1990**, *12*, 135–139

Mazzi, G.; Schinella, M. Simple and practical high-performance liquid chromatographic assay of propofol in human blood by phenyl column chromatography with electrochemical detection. *J.Chromatogr.*, **1990**, *528*, 537–541

# Propoxyphene

**Molecular formula:** $C_{22}H_{29}NO_2$
**Molecular weight:** 339.5
**CAS Registry No.:** 469-62-5 (propoxyphene), 1639-60-7
(propoxyphene hydrochloride), 26570-10-5 (propoxyphene
napsylate monohydrate), 55557-30-7 (l-form propoxyphene
napsylate monohydrate), 2338-37-6 (l form propoxyphene),
17140-78-2 (l-form propoxyphene napsylate anhydrous)

## SAMPLE
**Matrix:** blood
**Sample preparation:** 2 mL Whole blood or plasma + 2 mL buffer + 5 mL chloroform:
isopropanol:n-heptane 60:14:26, shake gently horizontally for 10 min, centrifuge at 2800
g for 10 min. Remove the lower organic layer and evaporate it to dryness under vacuum
at 45°, reconstitute the residue in 100 μL mobile phase, centrifuge at 2800 g for 5 min,
inject a 50 μL aliquot of the supernatant. (Buffer was saturated ammonium chloride
solution 25% diluted with water, adjusted to pH 9.5 with 25% ammonia solution.)

## HPLC VARIABLES
**Column:** 300 × 3.9 4 μm NovaPack C18
**Mobile phase:** MeOH:THF:buffer 65:5:30 (Buffer was 0.68 g/L (10 mM (sic)) $KH_2PO_4$
adjusted to pH 2.6 with concentrated orthophosphoric acid.) (At the end of each session
wash the column with water for 1 h and MeOH for 1 h, re-equilibrate for 30 min.)
**Column temperature:** 30
**Flow rate:** 0.8
**Injection volume:** 50
**Detector:** UV 259

## CHROMATOGRAM
**Retention time:** 7.24
**Limit of detection:** <120 ng/mL

## OTHER SUBSTANCES
**Extracted:** acebutolol, acenocoumarol, acepromazine, aceprometazine, acetaminophen,
aconitine, ajmaline, albuterol, alimemazine, alminoprofen, alpidem, alprazolam, alprenolol, amisulpride, amitriptyline, amodiaquine, amoxapine, aspirin, astemizole, atenolol,
benazepril, benperidol, benzocaine, benzoylecgonine, bepridil, bisoprolol, bromazepam,
brompheniramine, bumadizone, bupivacaine, buprenorphine, buspirone, caffeine, carbamazepine, carbinoxamine, carpipramine, carteolol, celiprolol, cetirizine, chlorambucil,
chlordiazepoxide, chlormezanone, chlorophenacinone, chloroquine, chlorpheniramine,
chlorpromazine, chlorpropamide, cibenzoline, cicletanine, clemastine, clobazam, clomipramine, clonazepam, clonidine, clorazepate, clozapine, cocaine, codeine, colchicine, cyamemazine, cyclizine, cycloguanil, cyproheptadine, cytarabine, dacarbazine, daunorubicin,
debrisoquine, demexiptiline, desipramine, dextromethorphan, diazepam, diazoxide,
diclofenac, dihydralazine, diltiazem, diphenhydramine, dipyridamole, disopyramide, doxepin, doxylamine, droperidol, ephedrine, estazolam, etodolac, fenfluramine, fentiazac, flecainide, floctafenine, flumazenil, flunitrazepam, fluoxetine, fluphenazine, flurbiprofen,
fluvoxamine, glibenclamide, glibornuride, glipizide, glutethimide, haloperidol,
histapyrrodine, hydroxychloroquine, hydroxyzine, ibuprofen, imipramine, indomethacin,
iproniazid, ketamine, ketoprofen, labetalol, levomepromazine, lidocaine, lidoflazine, lisinopril, loperamide, loprazolam, loratadine, lorazepam, maprotiline, medazepam, medifoxamine, mefenamic acid, mefenidramine, mefloquine, melphalan, meperidine, mephenesin,
mephentermine, mepivacaine, metapramine, metformin, methaqualone, methocarbamol,
methotrexate, metipranolol, metoclopramide, metoprolol, mexiletine, mianserine, mida-

zolam, minoxidil, moclobemide, moperone, morphine, nadolol, nalbuphine, nalorphine, naloxone, naltrexone, naproxen, nialamide, nicardipine, nifedipine, niflumic acid, nimodipine, nitrazepam, nitrendipine, nizatidine, nomifensine, nortriptyline, omeprazole, opipramol, oxazepam, oxprenolol, penbutolol, penfluridol, pentazocine, phencyclidine, phenobarbital, phenol, phenylbutazone, pimozide, pindolol, pipamperone, piroxicam, prazepam, prazosin, prilocaine, procainamide, procarbazine, proguanil, propranolol, protriptyline, pyrimethamine, quinidine, quinine, quinupramine, ramipril, ranitidine, reserpine, ritodrine, secobarbital, sotalol, strychnine, sulfinpyrazole, sulindac, sulpride, sultopride, suriclone, temazepam, tenoxicam, terbutaline, terfenadine, tetracaine, tetrazepam, thiopental, thioridazine, tianeptine, tiapride, tiaprofenic acid, ticlopidine, timolol, tioclomarol, tofisopam, tolbutamide, toloxatone, trazodone, triazolam, trifluoperazine, trifluperidol, trimipramine, triprolidine, tropatenine, verapamil, viloxazine, vinblastine, vincristine, vindesine, warfarin, yohimbine, zolpidem, zopiclone, zorubicine
**Interfering:** betaxolol, dextromoramide, dextropropoxyphene, dosulepine, fenoprofen, loxapine, methadone, promethazine, propafenone, thioproperazine

## KEY WORDS
whole blood; plasma

## REFERENCE
Tracqui, A.; Kintz, P.; Mangin, P. Systematic toxicological analysis using HPLC/DAD. *J.Forensic Sci.*, **1995**, *40*, 254–262

## SAMPLE
**Matrix:** blood
**Sample preparation:** Condition a 3 mL Bond Elut Certify SPE cartridge with 2 mL MeOH and 2 mL 100 mM pH 6.0 phosphate buffer, do not allow to dry. 1 mL Blood + 6 mL 100 mM pH 6.0 phosphate buffer, vortex, sonicate, centrifuge, add the supernatant to the SPE cartridge, wash with water, wash with 1 mM pH 3.3 acetic acid, dry by suction, wash with 2 mL acetone:chloroform 50:50, elute with 3 mL ethyl acetate:ammonia 98:2. Evaporate the eluate under a stream of nitrogen at 40°, reconstitute the residue in 50 μL MeOH, inject a 10 μL aliquot.

## HPLC VARIABLES
**Column:** 150 × 3.9 4 μm Nova-Pack C18
**Mobile phase:** MeOH:50 mM ammonium acetate 75:25 (Mix column effluent with 50 mM ammonium acetate pumped at 0.5 mL/min.)
**Flow rate:** 0.6
**Injection volume:** 10
**Detector:** MS, Finnigan MAT TSQ 700 tandem quadrupole, MAT TSP-2 interface, thermospray, selective reaction monitoring m/z 340-266, collision offset -6 V, repeller 100 V, vaporizer 130°, source 200°, filament on 200 μA, argon 2.5 mTorr, multiplier 1500 V, dynode 15 kV, scan time 1.20 s, MSMSC factor 10

## CHROMATOGRAM
**Retention time:** 7.10 (dextropropoxyphene)
**Limit of detection:** 50 pg

## OTHER SUBSTANCES
**Extracted:** benperidol, dextromoramide, droperidol, haloperidol, methadone, penfluridol, pimozide, pipamperidone

## KEY WORDS
SPE; LC/MS

## REFERENCE

Verweij, A.M.A.; Hordijk, M.L.; Lipman, P.J.L. Quantitative liquid chromatographic thermospray-tandem mass spectrometric analysis of some analgesics and tranquilizers of the methadone, butyrophenone, or diphenylbutylpiperidine groups in whole blood. *J.Anal.Toxicol.*, **1995**, *19*, 65–68

## SAMPLE

**Matrix:** blood

**Sample preparation:** Automated SPE by ASPEC system. Condition a C18 Clean-Up SPE cartridge (CEC 18111, Worldwide Monitoring) with 2 mL MeOH then 2 mL water. 1 mL Plasma + 1 mL 400 ng/mL protriptyline in water, vortex, add to the SPE cartridge, wash with 3 mL water, wash with 3 mL 750 mL/L methanol. Elute with three aliquots of 300 μL 0.1 M ammonium acetate in MeOH. Add 0.5 mL 0.5 M NaOH and 4 mL 50 mL/L isopropanol in heptane to eluate, mix thoroughly. Allow 5 min for phase separation. Remove upper heptane phase and add it to 300 μL 0.1 M phosphoric acid (pH 2.5), mix, separate, inject a 100 μL aliquot of the aqueous phase.

## HPLC VARIABLES

**Guard column:** LC-8-DB (Supelco)

**Column:** 150 × 4.6 LC-8-DB (Supelco)

**Mobile phase:** MeCN:buffer 35:65 (Buffer was 10 mL/L triethylamine in water adjusted to pH 5.5 with glacial acetic acid.)

**Flow rate:** 2

**Injection volume:** 100

**Detector:** UV 228

## CHROMATOGRAM

**Retention time:** 5.2

**Internal standard:** protriptyline (4)

## OTHER SUBSTANCES

**Extracted:** acetazolamide, amitriptyline, chlordiazepoxide, chlorimipramine, chlorpromazine, desipramine, dextromethorphan, diazepam, diphenhydramine, doxepin, encainide, fentanyl, flecainide, fluoxetine, flurazepam, haloperidol, hydroxyethylflurazepam, ibuprofen, imipramine, lidocaine, methaqualone, mexiletine, midazolam, norchlorimipramine, nordoxepin, nortriptyline, norverapamil, pentazocine, promazine, propafenone, propranolol, protriptyline, quinidine, trazodone, trimipramine, verapamil

**Noninterfering:** acetaminophen, acetylmorphine, amiodarone, amobarbital, amphetamine, bendroflumethiazide, benzocaine, benzoylecgonine, benzthiazide, butalbital, carbamazepine, chlorothiazide, clonazepam, cocaine, codeine, cotinine, cyclosporine, cyclothiazide, desalkylflurazepam, diamorphine, dicumerol, ephedrine, ethacrynic acid, ethanol, ethchlorvynol, ethosuximide, furosemide, glutethimide, hydrochlorothiazide, hydrocodone, hydroflumethiazide, hydromorphone, lorazepam, mephentermine, meprobamate, methamphetamine, metharbital, methoxsalen, methoxyphenteramine, methsuximide, methylcyclothiazide, metoprolol, MHPG, monoacetylmorphine, morphine, normethsuximide, oxazepam, oxycodone, oxymorphone, pentobarbital, phencyclidine, phenteramine, phenylephrine, phenytoin, polythiazide, primidone, prochlorperazine, salicylic acid, sulfanilamide, THC-COOH, theophylline, thiazolam, thiopental, thioridazine, tocainide, trichloromethiazide, trifluoperazine, valproic acid, warfarin

**Interfering:** maprotiline, methadone, nordiazepam, norfluoxetine, temazepam

## KEY WORDS

plasma; SPE

## REFERENCE

Nichols, J.H.; Charlson, J.R.; Lawson, G.M. Automated HPLC assay of fluoxetine and norfluoxetine in serum. *Clin.Chem.*, **1994**, *40*, 1312–1316

## SAMPLE
**Matrix:** blood
**Sample preparation:** 2 mL Whole blood + 20 µL 100 µg/mL methadone in MeOH + 1 mL pH 11 Normex buffer (Carlo Erba), vortex for 20 s, add to a 3 mL Extrelut cartridge, wait for 10 min, elute with diethyl ether:dichloromethane 70:30. Add the eluate to 100 µL 10 mM HCl and evaporate it at 40° under a stream of nitrogen. Wash the residual acid solution with 3 mL diethyl ether, inject a 40 µL aliquot of the aqueous phase.

## HPLC VARIABLES
**Guard column:** 5 × 6 µBondapak C18 Guard-Pak
**Column:** 300 × 3.9 10 µm µBondapak C18
**Mobile phase:** MeCN:buffer 75:25 (Buffer was 0.1% acetic acid in 10 mM ammonium acetate.)
**Flow rate:** 1.5
**Injection volume:** 40
**Detector:** UV 215

## CHROMATOGRAM
**Retention time:** 13.42
**Internal standard:** methadone (16.40)
**Limit of detection:** 40 ng/mL

## OTHER SUBSTANCES
**Extracted:** dextromoramide, norpropoxyphene
**Simultaneous:** alimemazine, amitriptyline, amphetamine, bupivacaine, buprenorphine, clomipramine, cocaine, codeine, ethylmorphine, glafenine, meperidine, methamphetamine, methotrimeprazine, morphine, norcodeine, nortriptyline, pentazocine
**Noninterfering:** acetaminophen, aspirin, barbiturates, benzoylecgonine, caffeine, diazepam, flunitrazepam, lidocaine, lorazepam, nalorphine, naloxone, nitrazepam, nordiazepam, norflunitrazepam, normorphine, oxazepam, prazepam, triazolam, zolpidem
**Interfering:** cyamemazine, desipramine, imipramine

## KEY WORDS
whole blood

## REFERENCE
Rop, P.P.; Grimaldi, F.; Bresson, M.; Fornaris, M.; Viala, A. Simultaneous determination of dextromoramide, propoxyphene and norpropoxyphene in necropsic whole blood by liquid chromatography. *J.Chromatogr.*, **1993**, *615*, 357–364

## SAMPLE
**Matrix:** blood
**Sample preparation:** 1 mL Plasma + 20 µL 20 µM IS in water + 300 µL 100 mM pH 10.0 sodium carbonate buffer + 4 mL n-hexane:diethyl ether:n-butanol 70:25:5, rotate gently for 10 min, centrifuge at 3200 g at 2° for 10 min. Remove the organic layer and add it to 250 µL pH 2.0 phosphate buffer, vortex for 15 s, centrifuge, inject 175 µL of the aqueous layer.

## HPLC VARIABLES
**Column:** 100 × 4.6 YMC S-3 120A ODS (YMC)
**Mobile phase:** MeCN:buffer 39.5:60.5 + 0.2 mM N, N-dimethyloctylamine + 1 mM sodium decyl sulfate (Buffer was 50 mM $NaH_2PO_4$ adjusted to pH 2.0 with orthophosphoric acid.)
**Flow rate:** 1.3
**Injection volume:** 175
**Detector:** UV 210

## CHROMATOGRAM
**Retention time:** 6
**Internal standard:** 1-benzyl-3-methylethylamino-2-methyl-1-phenylpropyl propionate picrate (7.5)
**Limit of quantitation:** 2 nM

## OTHER SUBSTANCES
**Extracted:** norpropoxyphene

## KEY WORDS
plasma

## REFERENCE
Pettersson, K.J.; Nilsson, L.B. Determination of dextropropoxyphene and norpropoxyphene in plasma by high-performance liquid chromatography. *J.Chromatogr.*, **1992**, *581*, 161–164

## SAMPLE
**Matrix:** blood
**Sample preparation:** 1 mL Serum + 200 ng doxepin or desipramine + 500 μL 2% pH 9.5 sodium tetraborate + 9 mL freshly prepared hexane:isoamyl alcohol 99:1, shake vigorously for 5 min, centrifuge. Remove 8.5 mL of the organic phase and add it to 200 μL 50 mM HCl, shake well for 1 min, centrifuge, inject a 50 μL aliquot of the aqueous phase.

## HPLC VARIABLES
**Column:** 300 × 4 μBondapak phenyl
**Mobile phase:** MeCN:0.01% phosphoric acid containing 0.01% NaCl 35:65, final pH 2.8
**Flow rate:** 1.5
**Injection volume:** 50
**Detector:** UV 210

## CHROMATOGRAM
**Retention time:** 18.6
**Internal standard:** doxepin (12.2), desipramine (14.2)
**Limit of detection:** 10 ng/mL

## OTHER SUBSTANCES
**Extracted:** cocaine, dextromoramide, meperidine, methadone, normeperidine, norpropoxyphene, pentazocine
**Simultaneous:** amitriptyline, buprenorphine, chlorpromazine, codeine, desmethyldoxepin, diphenhydramine, ephedrine, imipramine, nortriptyline, oxazepam, oxycodone, pericyazine, pheniramine, propranolol, quinine, thiopropazate, thioridazine

## KEY WORDS
serum

## REFERENCE
Hackett, L.P.; Dusci, L.J.; Ilett, K.F. The analysis of several nonopiate narcotic analgesics and cocaine in serum using high-performance liquid chromatography. *J.Anal.Toxicol.*, **1987**, *11*, 269–271

## SAMPLE
**Matrix:** blood, CSF
**Sample preparation:** 200 μL Serum, plasma, or CSF + 300 μL reagent. Flush column A to waste with 500 μL 500 mM ammonium sulfate, inject sample onto column A, flush column A to waste with 500 μL 500 mM ammonium sulfate, elute the contents of column

A onto column B with mobile phase, monitor the effluent from column B. (Reagent was 8.05 M guanidine HCl and 1.02 M ammonium sulfate in water.)

## HPLC VARIABLES
**Column:** A 30 × 2.1 40 μm preparative grade C18 (Analytichem); B 250 × 4.6 10 μm Partisil C8
**Mobile phase:** Gradient. A was 50 mM pH 4.5 $KH_2PO_4$. B was MeCN:isopropanol 80:20. A:B 90:10 for 1 min, to 30:70 over 15 min, maintain at 30:70 for 4 min.
**Column temperature:** 50
**Flow rate:** 1.5
**Detector:** UV 280 for 5 min then UV 254

## CHROMATOGRAM
**Retention time:** 6.96
**Internal standard:** heptanophenone (19.2)

## OTHER SUBSTANCES
**Extracted:** acetazolamide, ampicillin, bromazepam, caffeine, carbamazepine, chloramphenicol, chlorothiazide, diazepam, droperidol, ethionamide, furosemide, isoniazid, methadone, penicillin G, phenobarbital, phenytoin, prazepam, pyrazinamide, rifampin, trimeprazine, trimethoprim

## KEY WORDS
serum; plasma; column-switching

## REFERENCE
Seifart, H.I.; Kruger, P.B.; Parkin, D.P.; van Jaarsveld, P.P.; Donald, P.R. Therapeutic monitoring of antituberculosis drugs by direct in-line extraction on a high-performance liquid chromatography system. *J.Chromatogr.*, **1993**, *619*, 285–290

## SAMPLE
**Matrix:** blood, CSF, gastric fluid, urine
**Sample preparation:** 200 μL Serum, urine, CSF, or gastric fluid + 300 μL reagent. Flush column A to waste with 500 μL 500 mM ammonium sulfate, inject sample onto column A, flush column A to waste with 500 μL 500 mM ammonium sulfate, backflush the contents of column A onto column B with mobile phase, monitor the effluent from column B. (Reagent was 8.05 M guanidine hydrochloride and 1.02 M ammonium sulfate in water.)

## HPLC VARIABLES
**Column:** A 40 μm preparative grade C18 (Analytichem); B 75 × 2.1 pellicular C18 (Whatman) + 250 × 4.6 5 μm C8 end-capped (Whatman)
**Mobile phase:** Gradient. A was 50 mM pH 4.5 $KH_2PO_4$. B was MeCN:isopropanol 80:20. A:B 90:10 for 1 min, to 30:70 over 20 min.
**Column temperature:** 50
**Flow rate:** 1.5
**Detector:** UV 220

## CHROMATOGRAM
**Retention time:** 6.90
**Internal standard:** heptanophenone (19)

## OTHER SUBSTANCES
**Extracted:** acetaminophen, allobarbital, azinphos, barbital, brallobarbitone, bromazepam, butethal, caffeine, carbamazepine, carbaryl, cephaloridine, chloramphenicol, chlordiazepoxide, chlorothiazide, chlorvinphos, clothiapine, cocaine, coomassie blue, desipramine, diazepam, diphenhydramine, dipipanone, ethylbromphos, flufenamic acid, formothion,

griseofulvin, indomethacin, lidocaine, lorazepam, malathion, medazepam, midazolam, oxazepam, paraoxon, penicillin G, pentobarbital, prazepam, prothiophos, quinine, salicylic acid, secobarbital, strychnine, sulfamethoxazole, theophylline, thiopental, thioridazine, trimethoprim

## KEY WORDS
serum; column-switching

## REFERENCE
Kruger, P.B.; Albrecht, C.F.De V.; Jaarsveld, P.P. Use of guanidine hydrochloride and ammonium sulfate in comprehensive in-line sorption enrichment of xenobiotics in biological fluids by high-performance liquid chromatography. *J.Chromatogr.*, **1993**, *612*, 191–198

## SAMPLE
**Matrix:** blood, milk
**Sample preparation:** Adjust 0.2-1 mL plasma or breast milk to 1 mL with blank biological fluid if necessary, add 50 µL 12 µg/mL IS in EtOH, add 500 µL 1 M pH 9.80 carbonate: bicarbonate buffer, add 10 mL n-butyl chloride, shake at high speed on a reciprocating shaker for 15 min, centrifuge at 2000 g for 5 min. Remove the organic phase and add it to 5 mL 200 mM HCl, shake at high speed on a reciprocating shaker for 15 min, centrifuge at 2000 g for 5 min. Remove the aqueous phase and wash it with 5 mL ether. Add 500 µL 4 M NaOH to the aqueous phase, add 10 mL chloroform, shake at high speed on a reciprocating shaker for 15 min, centrifuge at 2000 g for 5 min. Remove the organic phase and evaporate it to dryness under nitrogen at 55°, dissolve the residue in 200 µL mobile phase, inject a 100 µL aliquot.

## HPLC VARIABLES
**Column:** 300 × 3.9 µBondapak C18
**Mobile phase:** MeCN:2 mM sulfuric acid 50:50
**Flow rate:** 1.5
**Injection volume:** 100
**Detector:** UV 205

## CHROMATOGRAM
**Retention time:** 5.7
**Internal standard:** 2-diethylaminoethyl-2,2-diphenylvalerate hydrochloride (SKF 525-A) (8.8)
**Limit of detection:** 20 ng/mL

## OTHER SUBSTANCES
**Extracted:** norpropoxyphene

## KEY WORDS
plasma; breast

## REFERENCE
Kunka, R.L.; Yong, C.-L.; Ladik, C.F.; Bates, T.R. Liquid chromatographic determination of propoxyphene and norpropoxyphene in plasma and breast milk. *J.Pharm.Sci.*, **1985**, *74*, 103–104

## SAMPLE
**Matrix:** blood, tissue
**Sample preparation:** Blood. 1 mL Whole blood or serum + 1 µg cianopramine + 1 mL water, vortex, add 1 mL 200 mM sodium carbonate, vortex, add 6 mL hexane:1-butanol 95:5, gently agitate for 30 min, centrifuge at 2500 g for 5 min. Remove the organic layer and add it to 100 µL 0.2% phosphoric acid, agitate gently for 30 min, centrifuge for 5 min. Remove the organic layer and inject a 30 µL aliquot of the aqueous layer. Tissue.

0.5 mL Liver homogenate + 10 μg cianopramine + 500 μL 2% sodium tetraborate + 8 mL hexane:1-butanol 95:5, gently agitate for 30 min, centrifuge at 2500 g for 5 min. Remove the organic layer and add it to 400 μL 0.2% phosphoric acid, agitate gently for 30 min, centrifuge for 5 min. Remove the organic layer and inject a 30 μL aliquot of the aqueous layer.

## HPLC VARIABLES
**Guard column:** 15 × 3.2 7 μm RP-18 Newguard (Applied Biosystems)
**Column:** 100 × 4.6 5 μm Brownlee Spheri-5 RP-18
**Mobile phase:** MeCN:100 mM NaH$_2$PO$_4$:diethylamine 40:57.5:2.5
**Flow rate:** 2
**Injection volume:** 30
**Detector:** UV 220

## CHROMATOGRAM
**Retention time:** 12.76
**Internal standard:** cianopramine (8.93)

## OTHER SUBSTANCES
**Extracted:** amitriptyline, amoxapine, benztropine, brompheniramine, chlorpheniramine, chlorpromazine, clomipramine, cyproheptadine, desipramine, diphenhydramine, dothiepin, doxepin, fluoxetine, haloperidol, imipramine, loxapine, maprotiline, meperidine, methadone, metoclopramide, mianserin, moclobemide, nomifensine, nordoxepin, norfluoxetine, norpropoxyphene, northiaden, nortriptyline, pentobarbital, pheniramine, promethazine, propranolol, protriptyline, quinidine, quinine, sulforidazine, thioridazine, thiothixene, tranylcypromine, trazodone, trihexyphenidyl, trimipramine, triprolidine
**Noninterfering:** dextromethorphan, norphetidine, phenoxybenzamine, prochlorperazine, trifluoperazine
**Interfering:** mesoridazine

## KEY WORDS
whole blood; serum; liver

## REFERENCE
McIntyre, I.M.; King, C.V.; Skafidis, S.; Drummer, O.H. Dual ultraviolet wavelength high-performance liquid chromatographic method for the forensic or clinical analysis of seventeen antidepressants and some selected metabolites. *J.Chromatogr.*, **1993**, *621*, 215−223

## SAMPLE
**Matrix:** blood, tissue
**Sample preparation:** 2 mL Blood or 250 mg liver (homogenized with 3 parts water) or 500 mg brain (homogenized with 3 parts water) + 2 μg SKF-525-A + 1.5 mL pH 9.5 ammonium carbonate/ammonium hydroxide buffer + 10 mL hexane:isopropanol 99:1, rotate at 10 rpm for 10 min, centrifuge at 3500 rpm for 10 min. Remove the organic layer and add it to 2.5 mL 0.25 M sulfuric acid, rotate for 5 min, centrifuge at 1500 rpm for 5 min. Remove the aqueous layer and add concentrated ammonium hydroxide to make the pH 9.5, add 1.5 mL chloroform, vortex for 15 s, centrifuge at 1500 rpm for 5 min. Remove the organic layer and add 1 drop of 1% HCl in MeOH, evaporate to dryness at 50° under vacuum, reconstitute with 200 μL mobile phase, inject a 100 μL aliquot.

## HPLC VARIABLES
**Guard column:** 30 × 2.1 Whatman C18 pellicular
**Column:** 250 × 4.6 Spherisorb S-5-ODS
**Mobile phase:** MeCN:MeOH:buffer 48:4:48 (Buffer was 1980 mL water + 20 mL 85% phosphoric acid + 3.7 mL methanesulfonic acid adjusted to pH 3.0 with 5 M NaOH.)
**Column temperature:** 60
**Flow rate:** 2

**Injection volume:** 100
**Detector:** UV 220

## CHROMATOGRAM
**Retention time:** 6.0
**Internal standard:** SKF-525-A (11.2)
**Limit of quantitation:** 100 ng/mL

## OTHER SUBSTANCES
**Extracted:** N-desmethyldiazepam, diazepam, methadone, norpropoxyphene

## KEY WORDS
liver; brain

## REFERENCE
Rio, J.; Hodnett, N.; Bidanset, J.H. The determination of propoxyphene, norpropoxyphene, and methadone in postmortem blood and tissues by high-performance liquid chromatography. *J.Anal.Toxicol.*, **1987**, *11*, 222–224

## SAMPLE
**Matrix:** formulations
**Sample preparation:** Add one tablet to 10 mL MeOH and 80 mL dichloromethane, sonicate for 5 min, dilute to 100 mL with dichloromethane, dilute a 2 mL aliquot to 25 mL with mobile phase, inject a 10 μL aliquot.

## HPLC VARIABLES
**Guard column:** 20 × 4.6 5 μm Supelcosil
**Column:** 33 × 4.6 3 μm Supelcosil
**Mobile phase:** Dichloromethane:3.33% ammonium hydroxide in MeOH 98.5:1.5
**Flow rate:** 2
**Injection volume:** 10
**Detector:** UV 244

## CHROMATOGRAM
**Retention time:** 2
**Limit of detection:** 2500 ng/mL

## OTHER SUBSTANCES
**Simultaneous:** acetaminophen
**Noninterfering:** impurities, 4-aminophenol, p-hydroxyacetophenone, p-nitrophenol

## KEY WORDS
tablets; normal phase

## REFERENCE
Ascah, T.L.; Hunter, B.T. Simultaneous high-performance liquid chromatographic determination of propoxyphene and acetaminophen in pharmaceutical preparations. *J.Chromatogr.*, **1988**, *455*, 279–289

## SAMPLE
**Matrix:** solutions

## HPLC VARIABLES
**Column:** 250 × 4.6 Zorbax RX
**Mobile phase:** Gradient. A was 10 mL concentrated orthophosphoric acid and 7 mL triethylamine in 1 L water. B was 10 mL concentrated orthophosphoric acid and 7 mL trie-

thylamine in 200 mL water, make up to 1 L with MeCN. A:B from 100:0 to 0:100 over 30 min, maintain at 0:100 for 5 min.

**Column temperature:** 30
**Flow rate:** 2
**Detector:** UV 210

## OTHER SUBSTANCES

**Also analyzed:** acepromazine, acetaminophen, acetophenazine, albuterol, aminophylline, amitriptyline, amobarbital, amoxapine, amphetamine, amylocaine, antipyrine, aprobarbital, aspirin, atenolol, atropine, avermectin, barbital, benzocaine, benzoic acid, benzotropine, benzphetamine, berberine, bibucaine, bromazepan, brompheniramine, buprenorphine, buspirone, butabarbital, butacaine, butethal, caffeine, carbamazepine, carbromal, chloramphenicol, chlordiazepoxide, chloroquine, chlorothiazide, chloroxylenol, chlorphenesin, chlorpheniramine, chlorpromazine, chlorpropamide, chlortetracycline, cimetidine, cinchonidine, cinchonine, clenbuterol, clonazepam, clonixin, clorazepate, cocaine, codeine, colchicine, cortisone, coumarin, cyclazocine, cyclobenzaprine, cyclothiazide, cyheptamide, cymarin, danazol, danthron, dapsone, debrisoquine, desipramine, dexamethasone, dextromethorphan, diamorphine, diazepam, diclofenac, diethylpropion, diethylstilbestrol, diflunisal, digitoxin, digoxin, diltiazem, diphenhydramine, diphenoxylate, diprenorphine, dipyrone, disulfiram, dopamine, doxapram, doxepin, dronabinol, ephedrine, epinephrine, epinine, estradiol, estriol, estrone, ethacrynic acid, ethosuximide, etonitazene, etorphine, eugenol, famotidine, fenbendazole, fencamfamine, fenoprofen, fenproporex, fentanyl, flubendazole, flufenamic acid, flunitrazepam, 5-fluorouracil, fluoxymesterone, fluphenazine, furosemide, gentisic acid, gitoxigenin, glipizide, glunixin, glutethimide, glybenclamide, guaiacol, halazepam, haloperidol, hydrochlorothiazide, hydrocodone, hydrocortisone, hydromorphone, hydroxyquinoline, ibogaine, ibuprofen, iminostilbene, imipramine, indomethacin, isocarbostyril, isocarboxazid, isoniazid, isoproterenol, isoxsuprine, ivermectin, ketamine, ketoprofen, kynurenic acid, levorphanol, lidocaine, lorazepam, lormetazepam, loxapine, mazindol, mebendazole, meclizine, meclofenamic acid, medazepam, mefenamic acid, megestrol, mepacrine, meperidine, mephentermine, mephenytoin, mephesin, mephobarbital, mepivacaine, mescaline, mesoridazine, methadone, methamphetamine, methapyrilene, methaqualone, methazolamide, methocarbamol, methoxamine, methsuximide, methyl salicylate, methyldopa, methyldopamine, methylphenidate, methylprednisolone, methyltestosterone, methyprylon, metoprolol, mibolerone, morphine, nadolol, nalorphine, naloxone, naltrexone, naphazoline, naproxen, nefopam, niacinamide, nicotine, nicotinic acid, nifedipine, niflumic acid, nitrazepam, norepinephrine, nortriptyline, noscapine, nylidrin, oxazepam, oxycodone, oxymorphone, oxyphenbutazone, oxytetracycline, papaverine, pargyline, pemoline, pentazocine, pentobarbital, persantine, phenacetin, phenazocine, phenazopyridine, phencyclidine, phendimetrazine, phenelzine, pheniramine, phenobarbital, phenothiazine, phensuximide, phentermine, phenylbutazone, phenylephrine, phenylpropanolamine, piperocaine, prazepam, prednisolone, primidone, probenecid, progesterone, propiomazine, propranolol, propylparaben, pseudoephedrine, puromycin, pyrilamine, pyrithyldione, quazepam, quinaldic acid, quinidine, quinine, ranitidine, recinnamine, reserpine, resorcinol, saccharin, albuterol, salicylamide, salicylic acid, scopolamine, scopoletin, secobarbital, strychnine, sulfacetamide, sufadiazine, sulfadimethoxine, sulfaethidole, sulfamerazine, sulfamethazine, sulfamethoxizole, sulfanilamide, sulfapyridine, sulfasoxizole, sulindac, tamoxifen, temazepam, testosterone, tetracaine, tetracycline, tetramisole, thebaine, theobromine, theophylline, thiabendazole, thiamine, thiamylal, thiobarbituric acid, thioridazine, thiosalicylic acid, thiothixene, thymol, tolazamide, tolazoline, tobutamide, tolmetin, tranylcypromine, triamcinolone, tribenzylamine, trichloromethiazide, trifluoperazine, trihexyphenidyl, trimethoprim, tripelennamine, triprolidine, tropacocaine, tyramine, verapamil, vincamine, warfarin, yohimbine, zoxazolamine

## REFERENCE

Hill, D.W.; Kind, A.J. Reversed-phase solvent gradient HPLC retention indexes of drugs. *J.Anal.Toxicol.*, **1994**, *18*, 233–242

## SAMPLE
**Matrix:** solutions

## HPLC VARIABLES
**Column:** 300 × 3.9 µBondapak C18
**Mobile phase:** MeCN:10 mm $KH_2PO_4$ + 5 mM 1-decanesulfonic acid:heptylamine 50:50: 0.1, pH adjusted to 7.9 with 85% phosphoric acid
**Flow rate:** 2
**Injection volume:** 10
**Detector:** UV 220

## CHROMATOGRAM
**Retention time:** 13.6
**Internal standard:** papaverine (2.7)
**Limit of detection:** 50 ng/mL

## KEY WORDS
stability-indicating

## REFERENCE
Ibrahim, F.B. Simultaneous determination and separation of several barbiturates and analgesic products by ion-pair high-performance liquid chromatography. *J.Liq.Chromatogr.*, **1993**, *16*, 2835−2851

## SAMPLE
**Matrix:** solutions
**Sample preparation:** Prepare a 10 µg/mL solution in MeOH, inject a 20 µL aliquot.

## HPLC VARIABLES
**Column:** 125 × 4.9 Spherisorb S5W silica
**Mobile phase:** MeOH containing 10 mM ammonium perchlorate and 1 mL/L 100 mM NaOH in MeOH, pH 6.7
**Flow rate:** 2
**Injection volume:** 20
**Detector:** E, LeCarbone, V25 glassy carbon electrode, + 1.2 V

## CHROMATOGRAM
**Retention time:** 2.5

## OTHER SUBSTANCES
**Also analyzed:** acebutolol, acepromazine, acetophenazine, N-acetylprocainamide, albuterol, alprenolol, amethocaine, amiodarone, amitriptyline, antazoline, atenolol, azacyclonal, bamethane, benactyzine, benperidol, benzethidine, benzocaine, benzoctamine, benzphetamine, benzquinamide, bromhexine, bromodiphenhydramine, bromperidol, brompheniramine, brompromazine, buclizine, bufotenine, bupivacaine, buprenorphine, butacaine, butethamate, chlorcyclizine, chlorpheniramine, chlorphenoxamine, chlorprenaline, chlorpromazine, chlorprothixene, cimetidine, cinchonidine, cinnarizine, clemastine, clomipramine, clonidine, cocaine, cyclazocine, cyclizine, cyclopentamine, cyproheptadine, deserpidine, desipramine, dextromoramide, dicyclomine, diethylcarbamazine, diethylpropion, diethylthiambutene, dihydroergotamine, dimethindene, dimethothiazine, diphenhydramine, diphenoxylate, dipipanone, diprenorphine, dipyridamole, disopyramide, dothiepin, doxapram, doxepin, doxylamine, droperidol, ephedrine, ergocornine, ergocristine, ergocristinine, ergocryptine, ergometrine, ergosine, ergosinine, ergotamine, ethopropazine, etorphine, etoxeridine, fenethazine, fenfluramine, fenoterol, fentanyl, flavoxate, fluopromazine, flupenthixol, fluphenazine, flurazepam, haloperidol, hydroxyzine, hyoscine, ibogaine, imipramine, indapamine, iprindole, isothipendyl, isoxsuprine, ketanserin, laudanosine, lidocaine, lofepramine, loxapine, maprotiline, mecamylamine, meclophenoxate,

meclozine, medazepam, mephentermine, mepivacaine, meptazinol, mepyramine, mesoridazine, metaraminol, methadone, methamphetamine, methapyrilene, methdilazene, methotrimeprazine, methoxamine, methoxyphenamine, methoxypromazine, methylephedrine, methylergonovine, methysergide, metoclopramide, metopimazine, metoprolol, mianserin, morazone, nadolol, nalorphine, naloxone, naphazoline, nicotine, nifedipine, nomifensine, nortriptyline, noscapine, orphenadrine, oxeladin, oxprenolol, oxymetazolin, papaverine, pargyline, pecazine, penbutolol, pentazocine, penthienate, pericyazine, perphenazine, phenadoxone, phenampromide, phenazocine, phenbutrazate, phendimetrazine, phenelzine, phenglutarimide, phenindamine, pheniramine, phenmetrazine, phenomorphan, phenoperidine, phenothiazine, phenoxybenzamine, phentolamine, phenylephrine, phenyltoloxamine, physostigmine, piminodine, pimozide, pindolol, pipamazine, pipazethate, piperacetazine, piperidolate, pipradol, pirenzepine, piritramide, pizotifen, practolol, pramoxine, prazosin, prenylamine, prilocaine, primaquine, proadifen, procainamide, procaine, prochlorperazine, procyclidine, proheptazine, prolintane, promazine, promethazine, pronethalol, properidine, propiomazine, propranolol, prothipendyl, protriptyline, proxymetacaine, pseudoephedrine, pyrimethamine, quinidine, quinine, ranitidine, rescinnamine, sotalol, tacrine, terazosin, terbutaline, terfenadine, thenyldiamine, theophylline, thiethylperazine, thiopropazate, thioproperazine, thioridazine, thiothixene, thonzylamine, timolol, tocainide, tolpropamine, tolycaine, tranylcypromine, trazodone, trifluoperazine, trifluperidol, trimeperidine, trimeprazine, trimethobenzamide, trimethoprim, trimipramine, tripelennamine, triprolidine, tryptamine, verapamil, xylometazoline

## REFERENCE
Jane, I.; McKinnon, A.; Flanagan, R.J. High-performance liquid chromatographic analysis of basic drugs on silica columns using non-aqueous ionic eluents. II. Application of UV, fluorescence and electrochemical oxidation detection. *J.Chromatogr.*, **1985**, *323*, 191–225

## SAMPLE
**Matrix:** solutions
**Sample preparation:** Dissolve in MeOH at a concentration of 1 mg/mL, inject a 20 μL aliquot.

## HPLC VARIABLES
**Column:** 250 × 5 Spherisorb S5W
**Mobile phase:** MeOH:buffer 90:10 (Buffer was 94 mL 35% ammonia and 21.5 mL 70% nitric acid in 884 mL water, adjust the pH to 10.1 with ammonia.)
**Flow rate:** 2
**Injection volume:** 20
**Detector:** UV 254

## CHROMATOGRAM
**Retention time:** 1.57

## OTHER SUBSTANCES
**Simultaneous:** acetylcodeine, amphetamine, benzylmorphine, bromo-STP, chlorphentermine, codeine, codeine-N-oxide, diamorphine, dihydrocodeine, dihydromorphine, dimethylamphetamine, dipipanone, ephedrine, epinephrine, ethoheptazine, ethylmorphine, fencamfamin, fenfluramine, hydrocodone, 4-hydroxyamphetamine, hydroxypethidine, levallorphan, levorphanol, meperidine, mephentermine, mescaline, methadone, methamphetamine, methylenedioxyamphetamine, methylephedrine, methylphenidate, monoacetylmorphine, morphine, morphine-3-glucuronide, morphine-N-oxide, norcodeine, norlevorphanol, normetanephrine, normethadone, normorphine, norpethidine, norpipanone, norpseudoephedrine, oxycodone, pentazocine, phenelzine, 2-phenethylamine, phentermine, phenylephrine, phenylpropanolamine, pholcodine, pipradol, prolintane, pseudoephedrine, STP, thebacon, thebaine, trimethoxyamphetamine, tyramine
**Noninterfering:** dopamine, levodopa, methyldopa, methyldopate, norepinephrine

**Interfering:** benzphetamine, buprenorphine, caffeine, dextromoramide, diethylpropion, etorphine, fenethyline, fentanyl, mazindol, nalorphine, naloxone, noscapine, papaverine, pemoline, phenazocine, phendimetrazine, phenoperidine, piritramide, tranylcypromine

## REFERENCE

Law, B.; Gill, R.; Moffat, A.C. High-performance liquid chromatography retention data for 84 basic drugs of forensic interest on a silica column using an aqueous methanol eluent. *J.Chromatogr.*, **1984**, *301*, 165–172

◆ ◆ ◆

## ANNOTATED BIBLIOGRAPHY

Kinney, C.D.; Kelly, J.G. Liquid chromatographic determination of paracetamol and dextropropxyphene in plasma. *J.Chromatogr.*, **1987**, *419*, 433–437 [verapamil (IS); electrochemical detection; column temp 30; LOD 2 ng/mL; pharmacokinetics]

Lurie, I.S.; McGuiness, K. The quantitation of heroin and selected basic impurities via reversed phase HPLC. II. The analysis of adulterated samples. *J.Liq.Chromatogr.*, **1987**, *10*, 2189–2204 [UV detection; electrochemical detection; also acetaminophen, acetylcodeine, acetylmorphine, acetylprocaine, aminopyrene, amitriptyline, antipyrene, aspirin, barbital, benztropine, caffeine, cocaine, codeine, diamorphine, diazepam, diphenhydramine, dipyrone, ephedrine, ethylmorphine, lidocaine, meconin, methamphetamine, methapyrilene, methaqualone, monoacetylmorphine, morphine, nalorphine, niacinamide, noscapine, papaverine, phenacetin, phenmetrazine, phenobarbital, phenolphthalein, procaine, pyrilamine, quinidine, quinine, salicylamide, salicylic acid, secobarbital, strychnine, tartaric acid, tetracaine, thebaine, tripelennamine, tropacocaine, vitamin B3, vitamin B5]

Schill, G.; Wainer, I.W.; Barkan, S.A. Chiral separation of cationic drugs on an α1-acid glycoprotein bonded stationary phase. *J.Liq.Chromatogr.*, **1986**, *9*, 641–666 [chiral; also atropine, bromdiphenhydramine, brompheniramine, bupivacaine, butorphanol, carbinoxamine, chlorpheniramine, clidinium, cocaine, cyclopentolate, dimethindene, diperidone, disopyramide, doxylamine, ephedrine, homatropine, labetalol, labetalol A, labetalol B, mepensolate, mepivacaine, methadone, methorphan, methylatropine, methylhomatropine, methylphenidate, metoprolol, nadolol, nadolol A, nadolol B, oxprenolol, oxyphencyclimine, phenmetrazine, phenoxybenzamine, promethazine, pronethalol, propranolol, pseudoephedrine, terbutaline, tocainide, tridihexethyl]

Wong, S.H.Y.; McHugh, S.L.; Dolan, J.; Cohen, K.A. Tricyclic antidepressant analysis by reversed-phase liquid chromatography using phenyl columns. *J.Liq.Chromatogr.*, **1986**, *9*, 2511–2538 [also acetaminophen, amitriptyline, amobarbital, amoxapine, barbital, chlordiazepoxide, chlorpromazine, cimetidine, clomipramine, codeine, desipramine, desmethyldoxepin, diazepam, doxepin, fluphenazine, flurazepam, glutethimide, hydroxyamoxapine, imipramine, internal standard, lorazepam, maprotiline, meperidine, metabolites, nortriptyline, oxazepam, pentobarbital, perphenazine, phenobarbital, phenytoin, protriptyline, secobarbital, thioridazine, trazodone]

Angelo, H.R.; Kranz, T.; Strom, J.; Thisted, B.; Bredgaard, M. High-performance liquid chromatographic method for the determination of dextropropxyphene and nordextropropoxyphene in serum. *J.Chromatogr.*, **1985**, *345*, 413–418 [pyrroliphene (IS); LOD 100 nM; simultaneous clomipramine, clopenthixol, codeine, ketobemidone, levomepromazine, meperidine, methadone; non-interfering acetaminophen, amitriptyline, aprobarbital, barbital, carbamazepine, chlordiazepoxide, chlorpromazine, chlorprothixene, desimipramine, diazepam, disopyramide, flupenthixol, imipramine, lidocaine, methadone, morphine, nitrazepam, nortriptyline, oxazepam, phenobarbital, phenytoin, theophylline, thioridazine]

# Quinapril

**Molecular formula:** $C_{25}H_{30}N_2O_5$
**Molecular weight:** 438.5
**CAS Registry No.:** 85441-61-8 (quinapril),
90243-99-5 (quinapril hydrochloride monohydrate),
82586-55-8 (quinapril hydrochloride)

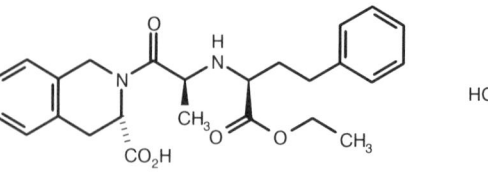

HCl

## SAMPLE
**Matrix:** blood, urine
**Sample preparation:** Condition a 100 mg 1 mL Bond-Elut C18 SPE cartridge with two 1 mL portions of MeOH, two 1 mL portions of water, and two 1 mL portions of 100 mM HCl. Condition a CBA-Bond-Elut SPE cartridge with MeOH and water. 1 mL Plasma + 1 mL buffer + 250 ng IS (or 50-500 μL urine + 500 ng IS), add to the C18 SPE cartridge, wash with two 1 mL portions of pH 3.4 water, wash with two 1 mL portions of distilled n-hexane, dry under vacuum for 20-30 min, elute with three 1 mL portions of chloroform:MeOH 2:1. Evaporate the eluate to dryness, reconstitute with 50 μL chloroform:MeOH 50:50 and 50 μL 2 mg/mL 9-anthryldiazomethane in MTBE, vortex for a few s, heat at 40° for 90 min, evaporate to dryness, reconstitute with two 100 μL portions of MeCN, add to the CBA SPE cartridge, wash with two 1 mL portions of MeCN, elute with three 1 mL portions of MeCN:triethylamine 99.8:0.2, evaporate the eluate to dryness under reduced pressure, reconstitute with 200 μL MeCN, inject a 20-50 μL aliquot. (Prepare 9-anthryldiazomethane as follows. Stir 8.8 g 9-anthraldehyde and 8.5 g 80% hydrazine hydrate (Caution! Hydrazine hydrate is a carcinogen!) in 150 mL EtOH at room temperature for 3 h, filter off the solid 9-anthraldehyde hydrazone and dry under vacuum (mp 124-6°) (Bull. Chem. Soc. Jpn. 1967, 40, 691). Dissolve 220 mg 9-anthraldehyde hydrazone in 100 mL anhydrous ether, add 800 mg activated manganese dioxide, follow the reaction by reverse-phase HPLC using MeCN at 0.4 mL/min and UV 254. At the end of the reaction filter off the manganese and wash it with 20 mL ether, evaporate the filtrate to obtain 9-anthryldiazomethane (mp 64-6°) (Anal.Biochem. 1980, 107, 116 and 1983, 132 456). Prepare activated manganese dioxide as follows. Stir a solution of 20 g potassium permanganate in 250 mL water at room temperature, add 10 g activated carbon (Nuchar C-190 or C-190N), stir for 16 h, filter (Buchner funnel), wash 4 times with 50 mL portions of water, dry in air, dry in an oven at 105-110° for 8-24 h (J.Org.Chem. 1970, 35, 3971).)

## HPLC VARIABLES
**Column:** 125 × 4.6 5 μm Spherisorb ODS II
**Mobile phase:** MeCN:MeOH:water 45:40:15 containing 0.24% ammonium perchlorate and 0.02% triethylamine
**Flow rate:** 1.6
**Injection volume:** 20-50
**Detector:** F ex 360 em 440

## CHROMATOGRAM
**Retention time:** 3.8
**Internal standard:** [2S-[1[R*(R*)]],2R*]-1-[2-[[(1-carboxy-3-phenyl)propyl]amino]-1-oxo-propyl]-octahydro-1H-indole-2-carboxylic acid (PD 110021, Warner-Lambert) (16)
**Limit of detection:** 5 ng/mL (plasma)
**Limit of quantitation:** 20 ng/mL (plasma); 100 ng/mL (urine)

## OTHER SUBSTANCES
**Extracted:** metabolites, quinaprilat
**Simultaneous:** enalapril, enalaprilat

## KEY WORDS
plasma; derivatization; pharmacokinetics; SPE

## REFERENCE
Hengy, H.; Most, M. Determination of the new ACE-inhibitor quinapril and its active metabolite quinaprilate in plasma and urine by high-performance liquid chromatography and pre-column labelling for fluorescent-detection. *J.Liq.Chromatogr.*, **1988**, *11*, 517−530

## SAMPLE
**Matrix:** perfusate
**Sample preparation:** Add taurocholic acid to perfusate, filter, inject an aliquot.

## HPLC VARIABLES
**Column:** 5 μm Ultrasphere C18
**Mobile phase:** MeOH:50 mM pH 7.4 sodium phosphate buffer 70:30
**Detector:** UV 220

## CHROMATOGRAM
**Retention time:** 5.5
**Internal standard:** taurocholic acid (7.5)

## REFERENCE
Hu, M.; Zheng, L.; Chen, J.; Liu, L.; Zhu, Y.; Dantzig, A.H.; Stratford, R.E., Jr. Mechanisms of transport of quinapril in Caco-2 cell monolayers: Comparison with cephalexin. *Pharm.Res.*, **1995**, *12*, 1120−1125

## SAMPLE
**Matrix:** perfusate, urine
**Sample preparation:** 100 μL Urine or perfusate + 200 μL MeCN, vortex for 5 s, sonicate for 5 min, centrifuge at 11150 g for 5 min, inject a 100 μL aliquot.

## HPLC VARIABLES
**Guard column:** 10 mm long 5 μm Econosil C8
**Column:** 250 × 4.6 5 μm Econosil C8
**Mobile phase:** MeCN:buffer 55:45 (Buffer was 0.01% triethylamine adjusted to pH 2.00 with phosphoric acid.)
**Flow rate:** 1
**Injection volume:** 100
**Detector:** UV 215; Radioactivity

## CHROMATOGRAM
**Retention time:** 3.4

## OTHER SUBSTANCES
**Extracted:** metabolites, quinaprilat

## KEY WORDS
tritium labelled; dog; pharmacokinetics

## REFERENCE
Kugler, A.R.; Olson, S.C.; Smith, D.E. Determination of quinapril and quinaprilat by high-performance liquid chromatography with radiochemical detection, coupled to liquid scintillation counting spectrometry. *J.Chromatogr.B*, **1995**, *666*, 360−367

## SAMPLE
**Matrix:** solutions

## HPLC VARIABLES
**Column:** 250 × 4.6 Spherisorb 5 ODS-2
**Mobile phase:** n-Propanol:buffer 20:80 (Buffer was pH 3.0 phosphate buffer containing 0.4% triethylamine.)
**Flow rate:** 1
**Detector:** UV 240

## CHROMATOGRAM
**Retention time:** 22

## OTHER SUBSTANCES
**Simultaneous:** benzepril, captopril, cilazapril, enalapril, ramipril

## REFERENCE
Barbato, F.; Morrica, P.; Quaglia, F. Analysis of ACE inhibitor drugs by high performance liquid chromatography. *Farmaco*, **1994**, *49*, 457–460

# Ramipril

**Molecular formula:** $C_{23}H_{32}N_2O_5$
**Molecular weight:** 416.5
**CAS Registry No.:** 87333-19-5

## SAMPLE
**Matrix:** blood
**Sample preparation:** 2 mL Whole blood or plasma + 2 mL buffer + 5 mL chloroform: isopropanol:n-heptane 60:14:26, shake gently horizontally for 10 min, centrifuge at 2800 g for 10 min. Remove the lower organic layer and evaporate it to dryness under vacuum at 45°, reconstitute the residue in 100 μL mobile phase, centrifuge at 2800 g for 5 min, inject a 50 μL aliquot of the supernatant. (Buffer was saturated ammonium chloride solution 25% diluted with water, adjusted to pH 9.5 with 25% ammonia solution.)

## HPLC VARIABLES
**Column:** 300 × 3.9 4 μm NovaPack C18
**Mobile phase:** MeOH:THF:buffer 65:5:30 (Buffer was 0.68 g/L (10 mM (sic)) $KH_2PO_4$ adjusted to pH 2.6 with concentrated orthophosphoric acid.) (At the end of each session wash the column with water for 1 h and MeOH for 1 h, re-equilibrate for 30 min.)
**Column temperature:** 30
**Flow rate:** 0.8
**Injection volume:** 50
**Detector:** UV 259

## CHROMATOGRAM
**Retention time:** 6.04
**Limit of detection:** <120 ng/mL

## OTHER SUBSTANCES
**Extracted:** acebutolol, acenocoumarol, acepromazine, acetaminophen, ajmaline, albuterol, alimemazine, alminoprofen, alpidem, alprazolam, amisulpride, amitriptyline, amodiaquine, amoxapine, aspirin, astemizole, atenolol, benazepril, benperidol, benzocaine, benzoylecgonine, bepridil, betaxolol, bromazepam, brompheniramine, bumadizone, bupivacaine, buprenorphine, buspirone, caffeine, carbamazepine, carbinoxamine, carpipramine, carteolol, celiprolol, cetirizine, chlorambucil, chlordiazepoxide, chlormezanone, chloroquine, chlorpheniramine, chlorpromazine, chlorpropamide, cibenzoline, cicletanine, clemastine, clobazam, clomipramine, clonazepam, clonidine, clorazepate, clozapine, cocaine, codeine, colchicine, cyamemazine, cyclizine, cycloguanil, cyproheptadine, cytarabine, dacarbazine, daunorubicin, debrisoquine, demexiptiline, desipramine, dextromethorphan, dextromoramide, dextropropoxyphene, diazoxide, diclofenac, dihydralazine, diphenhydramine, dipyridamole, disopyramide, dosulepine, doxylamine, droperidol, ephedrine, estazolam, etodolac, fenfluramine, fenoprofen, fentiazac, flecainide, floctafenine, flumazenil, flunitrazepam, fluoxetine, fluphenazine, flurbiprofen, fluvoxamine, glipizide, glutethimide, histapyrrodine, hydroxychloroquine, hydroxyzine, ibuprofen, imipramine, indomethacin, iproniazid, ketamine, ketoprofen, labetalol, levomepromazine, lidocaine, lidoflazine, lisinopril, loperamide, loprazolam, loratadine, lorazepam, loxapine, maprotiline, medazepam, medifoxamine, mefenamic acid, mefenidramine, mefloquine, melphalan, meperidine, mephenesin, mephentermine, mepivacaine, metapramine, metformin, methadone, methaqualone, methocarbamol, methotrexate, metipranolol, metoclopramide, metoprolol, mexiletine, midazolam, minoxidil, moclobemide, moperone, morphine, nadolol, nalbuphine, nalorphine, naloxone, naltrexone, naproxen, nialamide, nifedipine, niflumic acid, nimodipine, nitrazepam, nizatidine, nomifensine, nortriptyline, omeprazole, opipramol, oxazepam, oxprenolol, penbutolol, penfluridol, pentazocine, phencyclidine, phenobarbital, phenol, phenylbutazone, pimozide, pindolol, pipamperone, piroxicam, prazepam, prazosin, prilocaine, procainamide, procarbazine, proguanil, promethazine, propafenone, propran-

olol, protriptyline, pyrimethamine, quinidine, quinine, quinupramine, ranitidine, ritodrine, secobarbital, sotalol, strychnine, sulfinpyrazole, sulindac, sulpride, sultopride, suriclone, temazepam, tenoxicam, terbutaline, terfenadine, tetrazepam, thiopental, thioproperazine, thioridazine, tianeptine, tiapride, tiaprofenic acid, ticlopidine, timolol, tioclomarol, tofisopam, tolbutamide, toloxatone, trazodone, triazolam, trifluoperazine, trifluperidol, trimipramine, triprolidine, tropatenine, verapamil, viloxazine, vincristine, vindesine, warfarin, yohimbine, zolpidem, zopiclone, zorubicine

**Interfering:** aceprometazine, aconitine, alprenolol, bisoprolol, chlorophenacinone, diazepam, diltiazem, doxepin, glibenclamide, glibornuride, haloperidol, mianserine, nicardipine, nitrendipine, reserpine, tetracaine, vinblastine

## KEY WORDS
whole blood; plasma

## REFERENCE
Tracqui, A.; Kintz, P.; Mangin, P. Systematic toxicological analysis using HPLC/DAD. *J.Forensic Sci.*, **1995**, *40*, 254–262

## SAMPLE
**Matrix:** solutions
**Sample preparation:** Water, apple juice. Centrifuge at 350 g for 10 min, filter (0.45 μm) the supernatant, dilute the filtrate with 1-2 volumes of water, inject a 50 μL aliquot. Applesauce. 1 g Applesauce + 5 mL 100 mM HCl, vortex for 2 min, centrifuge at 350 g for 10 min, filter (0.45 μm) the supernatant, dilute the filtrate with 1-2 volumes of water, inject a 50 μL aliquot.

## HPLC VARIABLES
**Column:** C18 (J.T. Baker)
**Mobile phase:** MeCN:buffer 38.5:61.5, adjusted to pH 2.1 ± 0.1 with concentrated phosphoric acid (Buffer was 14 g sodium perchlorate monohydrate and 100 mL 500 mM phosphoric acid made up to 1 L with water, pH adjusted to 2.5 ± 0.1 with triethylamine.)
**Flow rate:** 1
**Injection volume:** 50
**Detector:** UV 210

## CHROMATOGRAM
**Retention time:** 12.1
**Limit of detection:** 100 ng/mL

## OTHER SUBSTANCES
**Simultaneous:** impurities, degradation products

## KEY WORDS
water; apple juice; applesauce; stability-indicating

## REFERENCE
Allen, L.V., Jr.; Stiles, M.L.; Prince, S.J.; McLaury, H.-J.; Sylvestri, M.F. Stability of ramipril in water, apple juice, and applesauce. *Am.J.Health-Syst.Pharm.*, **1995**, *52*, 2433–2436

## SAMPLE
**Matrix:** solutions

## HPLC VARIABLES
**Column:** 250 × 4.6 Spherisorb 5 ODS-2
**Mobile phase:** n-Propanol:buffer 20:80 (Buffer was pH 3.0 phosphate buffer containing 0.4% triethylamine.)

**Flow rate:** 1
**Detector:** UV 240

## CHROMATOGRAM
**Retention time:** 16

## OTHER SUBSTANCES
**Simultaneous:** benzepril, captopril, cilazapril, enalapril, quinapril

## REFERENCE
Barbato, F.; Morrica, P.; Quaglia, F. Analysis of ACE inhibitor drugs by high performance liquid chromatography. *Farmaco*, **1994**, *49*, 457–460

## SAMPLE
**Matrix:** solutions

## HPLC VARIABLES
**Column:** 300 × 3.9 μBondapak phenyl
**Mobile phase:** MeOH:water:85% phosphoric acid 55:45:0.05
**Detector:** UV 215-220

## REFERENCE
Ranadive, S.A.; Chen, A.X.; Serajuddin, A.T. Relative lipophilicities and structural-pharmacological considerations of various angiotensin-converting enzyme (ACE) inhibitors. *Pharm.Res.*, **1992**, *9*, 1480–1486

## SAMPLE
**Matrix:** solutions

## HPLC VARIABLES
**Column:** 5 μm Nucleosil SA
**Mobile phase:** MeCN:buffer 2:1 (Buffer was 0.15% $KH_2PO_4$ adjusted to pH 3.0 with phosphoric acid.)
**Flow rate:** 1
**Detector:** UV 210

## CHROMATOGRAM
**Retention time:** 13

## OTHER SUBSTANCES
**Simultaneous:** impurities, diastereomer (RS-SSS)

## REFERENCE
Ito, M.; Kuriki, T.; Goto, J.; Nambara, T. Separation of ramipril optical isomers by high-performance liquid chromatography. *J.Liq.Chromatogr.*, **1990**, *13*, 991–1000

## SAMPLE
**Matrix:** solutions

## HPLC VARIABLES
**Column:** Nucleosil 5 C18
**Mobile phase:** MeOH:buffer 50:50 (Buffer was 0.15% $KH_2PO_4$ adjusted to pH 2.4 with phosphoric acid.)

**Flow rate:** 1
**Detector:** UV 210

## CHROMATOGRAM
**Retention time:** k′ 10.60

## OTHER SUBSTANCES
**Simultaneous:** impurities, diastereomers (SS-RRR, RS-RRR, SR-SSS)

## REFERENCE
Ito, M.; Kuriki, T.; Goto, J.; Nambara, T. Separation of ramipril optical isomers by high-performance liquid chromatography. *J.Liq.Chromatogr.*, **1990**, *13*, 991−1000

# Ranitidine

**Molecular formula:** $C_{13}H_{22}N_4O_3S$
**Molecular weight:** 314.4
**CAS Registry No.:** 66357-35-5, 66357-59-3 (HCl)

## SAMPLE

**Matrix:** blood
**Sample preparation:** 1 mL Plasma + 25 μL 10 μg/mL procainamide in MeOH + 15 μL 6 M NaOH + 2 mL ethyl acetate:isopropanol 96:4, shake mechanically for 20 min, centrifuge at 3000 rpm for 10 min. Remove the organic layer and evaporate it to dryness under a stream of nitrogen at 40°, reconstitute the residue in 100 μL MeOH, inject a 75 μL aliquot.

## HPLC VARIABLES

**Column:** 150 × 3.9 10 μm μBondapak C18
**Mobile phase:** MeCN:10 mM pH 5.1 $KH_2PO_4$ 8:92
**Flow rate:** 2.5
**Injection volume:** 75
**Detector:** UV 330

## CHROMATOGRAM

**Retention time:** 5
**Internal standard:** procainamide (3)
**Limit of detection:** 20 ng/mL

## OTHER SUBSTANCES

**Simultaneous:** lidocaine
**Noninterfering:** brompheniramine, chlorpheniramine, cimetidine, diazepam, diclofenac, glyburide, ibuprofen, ketoprofen, metoclopramide, naproxen, phenylbutazone, verapamil

## KEY WORDS

plasma; pharmacokinetics

## REFERENCE

al-Khamis, K.I.; El-Sayed, Y.M.; Al-Rashood, K.A.; Bawazir, S.A. High-performance liquid chromatographic determination of ranitidine in human plasma. *J.Liq.Chromatogr.*, **1995**, *18*, 277−286

## SAMPLE

**Matrix:** blood
**Sample preparation:** 2 mL Whole blood or plasma + 2 mL buffer + 5 mL chloroform: isopropanol:n-heptane 60:14:26, shake gently horizontally for 10 min, centrifuge at 2800 g for 10 min. Remove the lower organic layer and evaporate it to dryness under vacuum at 45°, reconstitute the residue in 100 μL mobile phase, centrifuge at 2800 g for 5 min, inject a 50 μL aliquot of the supernatant. (Buffer was saturated ammonium chloride solution 25% diluted with water, adjusted to pH 9.5 with 25% ammonia solution.)

## HPLC VARIABLES

**Column:** 300 × 3.9 4 μm NovaPack C18
**Mobile phase:** MeOH:THF:buffer 65:5:30 (Buffer was 0.68 g/L (10 mM (sic)) $KH_2PO_4$ adjusted to pH 2.6 with concentrated orthophosphoric acid.) (At the end of each session wash the column with water for 1 h and MeOH for 1 h, re-equilibrate for 30 min.)
**Column temperature:** 30
**Flow rate:** 0.8

**Injection volume:** 50
**Detector:** UV 322

## CHROMATOGRAM
**Retention time:** 3.38
**Limit of detection:** <120 ng/mL

## OTHER SUBSTANCES
**Extracted:** acebutolol, acenocoumarol, acepromazine, aceprometazine, acetaminophen, aconitine, ajmaline, alimemazine, alminoprofen, alpidem, alprazolam, alprenolol, amitriptyline, amodiaquine, amoxapine, astemizole, benazepril, benperidol, benzocaine, benzoylecgonine, bepridil, betaxolol, bisoprolol, bromazepam, brompheniramine, bumadizone, bupivacaine, buprenorphine, buspirone, caffeine, carbamazepine, carbinoxamine, carpipramine, carteolol, celiprolol, cetirizine, chlorambucil, chlordiazepoxide, chlorophenacinone, chloroquine, chlorpheniramine, chlorpromazine, chlorpropamide, cibenzoline, cicletanine, clemastine, clobazam, clomipramine, clonazepam, clonidine, clorazepate, clozapine, cocaine, colchicine, cyamemazine, cyclizine, cycloguanil, cyproheptadine, cytarabine, dacarbazine, daunorubicin, debrisoquine, demexiptiline, desipramine, dextromethorphan, dextromoramide, dextropropoxyphene, diazepam, diazoxide, diclofenac, dihydralazine, diltiazem, diphenhydramine, dipyridamole, disopyramide, dosulepine, doxepin, doxylamine, droperidol, ephedrine, estazolam, etodolac, fenfluramine, fenoprofen, fentiazac, flecainide, floctafenine, flumazenil, flunitrazepam, fluoxetine, fluphenazine, flurbiprofen, fluvoxamine, glibenclamide, glibornuride, glipizide, glutethimide, haloperidol, histapyrrodine, hydroxychloroquine, hydroxyzine, ibuprofen, imipramine, indomethacin, iproniazid, ketamine, ketoprofen, labetalol, levomepromazine, lidocaine, lidoflazine, lisinopril, loperamide, loprazolam, loratadine, lorazepam, loxapine, maprotiline, medazepam, medifoxamine, mefenamic acid, mefenidramine, mefloquine, melphalan, meperidine, mephenesin, mephentermine, mepivacaine, metapramine, methadone, methaqualone, methocarbamol, methotrexate, metipranolol, metoclopramide, metoprolol, mexiletine, mianserine, midazolam, minoxidil, moclobemide, moperone, nadolol, nalbuphine, nalorphine, naloxone, naltrexone, naproxen, nialamide, nicardipine, nifedipine, niflumic acid, nimodipine, nitrazepam, nitrendipine, nizatidine, nomifensine, nortriptyline, omeprazole, opipramol, oxazepam, oxprenolol, penbutolol, penfluridol, pentazocine, phencyclidine, phenylbutazone, pimozide, pindolol, pipamperone, piroxicam, prazepam, prazosin, prilocaine, procainamide, procarbazine, proguanil, promethazine, propafenone, propranolol, protriptyline, pyrimethamine, quinidine, quinine, quinupramine, ramipril, reserpine, secobarbital, sotalol, strychnine, sulfinpyrazole, sulindac, sulpride, suriclone, temazepam, tenoxicam, terfenadine, tetracaine, tetrazepam, thiopental, thioproperazine, thioridazine, tianeptine, tiaprofenic acid, ticlopidine, timolol, tioclomarol, tofisopam, tolbutamide, trazodone, triazolam, trifluoperazine, trifluperidol, trimipramine, triprolidine, tropatenine, verapamil, viloxazine, vinblastine, vincristine, vindesine, warfarin, yohimbine, zolpidem, zopiclone, zorubicine
**Interfering:** albuterol, amisulpride, aspirin, atenolol, chlormezanone, codeine, metformin, morphine, phenobarbital, phenol, ritodrine, sultopride, terbutaline, tiapride, toloxatone

## KEY WORDS
whole blood; plasma

## REFERENCE
Tracqui, A.; Kintz, P.; Mangin, P. Systematic toxicological analysis using HPLC/DAD. *J.Forensic Sci.*, **1995**, *40*, 254–262

## SAMPLE
**Matrix:** blood
**Sample preparation:** Mix plasma with sodium carbonate solution and extract twice with 3 mL dichloromethane. Remove the organic layer and evaporate it to dryness, reconstitute the residue in 100 μL mobile phase, inject a 40 μL aliquot.

## HPLC VARIABLES
**Column:** 150 × 3.6 ODS Hypersil
**Mobile phase:** MeCN:isopropanol:pH 7 sodium acetate buffer:water 21:6:20:53
**Flow rate:** 0.4
**Injection volume:** 40
**Detector:** UV 280

## CHROMATOGRAM
**Limit of detection:** 20 ng/mL

## OTHER SUBSTANCES
**Noninterfering:** diclofenac

## KEY WORDS
plasma

## REFERENCE
Van Gelderen, M.E.M.; Olling, M.; Barends, D.M.; Meulenbelt, J.; Salomons, P.; Rauws, A.G. The bio-availability of diclofenac from enteric coated products in healthy volunteers with normal and artifi-cially decreased gastric acidity. *Biopharm.Drug Dispos.*, **1994**, *15*, 775–788

## SAMPLE
**Matrix:** blood
**Sample preparation:** Condition a 100 mg Bond Elut CN cartridge with 3 mL water, 3 mL MeOH, and 3 mL 50 mM pH 7.0 sodium phosphate buffer. Mix 4 volumes of plasma with 1 volume of 100 mM $NaH_2PO_4$. Add 0.5-1 mL plasma to the SPE cartridge, wash with 5 mL water, elute with 1 mL MeOH. Evaporate the eluate to dryness, reconstitute in 120 µL phosphate buffer, inject a 100 µL aliquot.

## HPLC VARIABLES
**Column:** 300 × 3.9 10 µm µBondapak C18
**Mobile phase:** MeOH:50 mM pH 7.0 sodium phosphate buffer 60:40
**Flow rate:** 0.8
**Injection volume:** 100
**Detector:** UV 315

## CHROMATOGRAM
**Retention time:** 8
**Limit of detection:** 1 ng/mL

## KEY WORDS
plasma; SPE; pharmacokinetics

## REFERENCE
Biermann, B.; Sommer, N.; Winne, D.; Schumm, F.; Maier, U.; Breyer-Pfaff, U. Renal clearance of pyr-idostigmine in myasthenic patients and volunteers under the influence of ranitidine and pirenzepine. *Xenobiotica*, **1993**, *23*, 1263–1275

## SAMPLE
**Matrix:** blood
**Sample preparation:** Condition a 100 mg LRC Bond Elut unendcapped cyanopropyl SPE cartridge with 1 mL MeOH and 1 mL water. 1 mL Serum or plasma + 250 µL 2 µg/mL procainamide in buffer, vortex, centrifuge at 1300 rpm for 10 min (plasma only), add 1 mL sample to the SPE cartridge, wash twice with 1 mL water, dry with nitrogen for 30 s, elute with two 250 µL aliquots of MeOH:buffer 60:40, inject 100 µL aliquot of eluate.

(Buffer was 50 mM $KH_2PO_4$ and $Na_2HPO_4$ 50:50 v/v adjusted to pH 6.0 with phosphoric acid.)

## HPLC VARIABLES
**Guard column:** 15 × 4.6 7 μm Brownlee RP-18
**Column:** 250 × 4.6 5 μm Spherisorb ODS-1
**Mobile phase:** MeOH:THF:buffer 65:5:30 (Buffer was 50 mM $KH_2PO_4$ and $Na_2HPO_4$ 50:50 v/v adjusted to pH 6.0 with phosphoric acid.)
**Column temperature:** 48
**Flow rate:** 1.25
**Injection volume:** 100
**Detector:** UV 320

## CHROMATOGRAM
**Retention time:** 4.5
**Internal standard:** procainamide (5.4)
**Limit of quantitation:** 10 ng/mL

## KEY WORDS
serum; plasma; robotic sample preparation; SPE

## REFERENCE
Lloyd, T.L.; Perschy, T.B.; Gooding, A.E.; Tomlinson, J.J. Robotic solid phase extraction and high performance liquid chromatographic analysis of ranitidine in serum or plasma. *Biomed.Chromatogr.*, **1992**, *6*, 311–316

## SAMPLE
**Matrix:** blood
**Sample preparation:** 1 mL Plasma + 200 μL MeCN + 50 mg $ZnSO_4$, vortex for 2 min, centrifuge at 2200 g for 3 min. Remove the supernatant, filter (0.45 μm), inject an aliquot. (Tubes were cleaned by sonication with acetone for 30 min, sonication twice with MeCN for 15 min, and drying at 100°. Stir 100 g $ZnSO_4$ crystals with 200 mL MeCN for 30 min, decant MeCN, repeat twice with fresh MeCN, dry crystals in a hood at room temperature.)

## HPLC VARIABLES
**Column:** 50 × 4.1 Alltech PLRP-S
**Mobile phase:** MeCN:5 mM $K_2HPO_4$ + 0.5 mM tetraethylammonium hydroxide 20:80, pH 11
**Flow rate:** 1
**Detector:** UV 315

## CHROMATOGRAM
**Retention time:** 7
**Limit of quantitation:** 5 ng/mL

## OTHER SUBSTANCES
**Simultaneous:** metabolites
**Noninterfering:** acetaminophen, ampicillin, cimetidine, creatinine, tetracycline

## KEY WORDS
plasma

## REFERENCE
Rustum, A.M. Rapid and sensitive HPLC determination of ranitidine in plasma. Application to pharmacokinetics study. *J.Liq.Chromatogr.*, **1988**, *11*, 2315–2335

## SAMPLE
**Matrix:** blood, milk
**Sample preparation:** 100 μL Serum or whole milk + 25 μL MeOH + 4 mL dichlorome-thane, vortex, centrifuge. Remove the organic layer and evaporate it to dryness under a stream of nitrogen, reconstitute the residue in 200 μL mobile phase, inject a 50 μL aliquot.

## HPLC VARIABLES
**Guard column:** 15 × 3.2 RP-18 (Applied Biosystems)
**Column:** 300 × 3.9 μBondapak C18
**Mobile phase:** MeCN:water:glacial acetic acid:triethylamine 15:85:0.15:0.02
**Flow rate:** 1
**Injection volume:** 50
**Detector:** UV 228

## CHROMATOGRAM
**Retention time:** 10.7
**Internal standard:** ranitidine

## OTHER SUBSTANCES
**Extracted:** cimetidine

## KEY WORDS
serum; whole milk; ranitidine is IS

## REFERENCE
Oo, C.Y.; Kuhn, R.J.; Desai, N.; McNamara, P.J. Active transport of cimetidine into human milk. *Clin.Pharmacol.Ther.*, **1995**, *58*, 548–555

## SAMPLE
**Matrix:** blood, tissue
**Sample preparation:** Blood. 25 μL Plasma + 100 μL 75 μM nizatidine + 100 μL 5 M NaOH + 5 mL dichloromethane, shake for 10 min, centrifuge at 1650 g for 10 min. Remove 4 mL of the organic layer and evaporate it to dryness under reduced pressure at room temperature, reconstitute the residue in 100 μL mobile phase, inject a 50 μL aliquot. (Hemolyze 25 μL whole blood with 200 μL water then proceed as above.) Tissue. Homogenize brain tissue with 100 μL 30 μM nizatidine and 1 mL saline at 0° for 1 min, add 100 μL 500 mM NaOH, extract with 5 mL dichloromethane. Remove 3 mL of the organic layer and evaporate it to dryness, reconstitute the residue in 100 μL mobile phase, centrifuge at 10000 g, inject a 50 μL aliquot.

## HPLC VARIABLES
**Column:** 250 × 4 Senshu gel 7C18H (Senshu, Tokyo)
**Mobile phase:** MeCN:buffer 5:95. (Buffer was 5 mM $NaH_2PO_4$ containing 5 mM tetra-methylammonium chloride.)
**Column temperature:** 30
**Flow rate:** 2
**Injection volume:** 50
**Detector:** UV 320

## CHROMATOGRAM
**Internal standard:** nizatidine
**Limit of detection:** 1 μM (blood); 0.5 nmole/g (brain)

## KEY WORDS
plasma; mouse; brain; whole blood

## REFERENCE

Shimokawa, M.; Yamamoto, K.; Kawakami, J.; Sawada, Y.; Iga, T. Effect of renal or hepatic dysfunction on neurotoxic convulsion induced by ranitidine in mice. *Pharm.Res.*, **1994**, *11*, 1519–1523

## SAMPLE

**Matrix:** blood, urine

**Sample preparation:** Plasma. 250 μL Plasma + 75 μL 20 μg/mL metoclopramide + 300 μL 2 M NaOH, vortex for 30 s, add 5 mL dichloromethane, mix for 2 min, centrifuge at 10000 rpm for 15 min. Remove the organic layer and evaporate to dryness under nitrogen. Reconstitute the residue in 150 μL mobile phase, inject a 50 μL aliquot. Urine. Dilute 100 μL urine + 75 μL 2 mg/mL metoclopramide to 10 mL with water, inject a 50 μL aliquot.

## HPLC VARIABLES

**Column:** 250 × 4.6 Spherisorb S 5 ODS II

**Mobile phase:** MeCN:MeOH:buffer 60:30:5, pH adjusted to 3.8 with glacial acetic acid (Buffer was 50 mM ammonium acetate containing 10 mM sodium octane sulfonate.)

**Flow rate:** 1.2

**Injection volume:** 50

**Detector:** UV 330

## CHROMATOGRAM

**Retention time:** 5

**Internal standard:** metoclopramide (8)

**Limit of detection:** 20 ng/mL

## OTHER SUBSTANCES

**Simultaneous:** acetaminophen, chlorodiazepam, chlorpromazine, desmethyldiazepam, diazepam, oxazepam, theobromine, theophylline

## KEY WORDS

plasma

## REFERENCE

Shiekh Salem, M.; Gharaibeh, A.M.; Alkaysi, H.N.; Badwan, A. High-performance liquid chromatographic analysis of ranitidine in plasma and urine. *J.Clin.Pharm.Ther.*, **1988**, *13*, 351–357

## SAMPLE

**Matrix:** blood, urine

**Sample preparation:** Plasma. 2 mL Plasma + 200 μL 5 M NaOH, vortex for a few seconds, add 10 mL dichloromethane, vortex for 10 min, centrifuge at 1200 g for 10 min. Remove the organic layer and evaporate it to dryness under nitrogen at 40°. Reconstitute in 100 μL water, inject a 20-50 μL aliquot. Urine. Dilute 1:50 with water, vortex for a few seconds, centrifuge at 1200 g for 5 min, extract the supernatant as described above.

## HPLC VARIABLES

**Guard column:** 50 × 4.6 Spherisorb CN

**Column:** 150 × 4.6 6 μm Zorbax CN

**Mobile phase:** MeCN:50 mM $KH_2PO_4$ 15:85 containing 5 mM octanesulfonic acid

**Flow rate:** 2

**Injection volume:** 20-50

**Detector:** UV 318

## CHROMATOGRAM

**Retention time:** 3.1

**Internal standard:** AH 20480 (4.2)
**Limit of detection:** 0.5 ng/mL

## KEY WORDS
plasma

## REFERENCE
Mullersman, G.; Derendorf, H. Rapid analysis of ranitidine in biological fluids and determination of its erythrocyte partitioning. *J.Chromatogr.*, **1986**, *381*, 385–391

## SAMPLE
**Matrix:** formulations
**Sample preparation:** 100 μL Injection + 50 μL 200 μg/mL caffeine in water + 850 μL water, vortex briefly, inject a 50 μL aliquot.

## HPLC VARIABLES
**Column:** 250 × 4.6 Ultrasphere-ODS C18
**Mobile phase:** MeCN:buffer 8:92 (Buffer was 10 mM $KH_2PO_4$ adjusted to pH 5.0 with phosphoric acid or NaOH.)
**Flow rate:** 2
**Injection volume:** 50
**Detector:** UV 262

## CHROMATOGRAM
**Retention time:** 6
**Internal standard:** caffeine (3)

## OTHER SUBSTANCES
**Simultaneous:** degradation products

## KEY WORDS
stability-indicating; injections

## REFERENCE
Hoyer, G.L.; LeDoux, J.; Nolan, P.E., Jr. A sensitive stability indicating assay for the $H_2$ blocker ranitidine. *J.Liq.Chromatogr.*, **1995**, *18*, 1239–1249

## SAMPLE
**Matrix:** formulations
**Sample preparation:** Dilute 1:4, inject a 20 μL aliquot.

## HPLC VARIABLES
**Column:** 250 × 4.6 5 μm Adsorbosphere C18
**Mobile phase:** MeOH:100 mM ammonium acetate 65:35
**Flow rate:** 1
**Injection volume:** 20
**Detector:** UV 322

## CHROMATOGRAM
**Retention time:** 8.0

## OTHER SUBSTANCES
**Noninterfering:** ondansetron, paclitaxel

## KEY WORDS
injections; 5% dextrose; stability-indicating

## REFERENCE
Burm, J.-P.; Jhee, S.S.; Chin, A.; Moon, Y.S.K.; Jeong, E.; Nii, L.; Fox, J.L.; Gill, M.A. Stability of paclitaxel with ondansetron hydrochloride or ranitidine hydrochloride during simulated Y-site administration. *Am.J.Hosp.Pharm.*, **1994**, *51*, 1201–1204

## SAMPLE
**Matrix:** formulations
**Sample preparation:** Dilute 4:1 with saline, inject a 20 µL aliquot.

## HPLC VARIABLES
**Column:** 250 × 4.6 5 µm Adsorbosphere
**Mobile phase:** MeOH:100 mM ammonium acetate 65:35, pH adjusted to 4.0 with 1 M NaOH
**Flow rate:** 1.2
**Injection volume:** 20
**Detector:** UV 322

## CHROMATOGRAM
**Retention time:** 7.5

## KEY WORDS
injections; saline; stability-indicating

## REFERENCE
Choi, J.-S.; Burm, J.-P.; Jhee, S.S.; Chin, A.; Ulrich, R.W.; Gill, M.A. Stability of piperacillin sodium-tazobactam sodium and ranitidine hydrochloride in 0.9% sodium chloride injection during simulated Y-site administration. *Am.J.Hosp.Pharm.*, **1994**, *51*, 2273–2276

## SAMPLE
**Matrix:** formulations
**Sample preparation:** Tablets. Grind tablets to a fine powder, weigh out a quantity equivalent to about 150 mg ranitidine hydrochloride, add 50 mL MeOH:water 1:1, sonicate for 10 min, dilute to 100 mL with MeOH:water 1:1, filter, dilute 5 mL filtrate to 100 mL with MeOH:water 1:1, inject a 20 µL aliquot. Injections, syrup. Measure out a volume equivalent to about 75 mg ranitidine hydrochloride, dilute to 50 mL with MeOH:water 1:1, mix, dilute 5 mL to 100 mL with MeOH:water 1:1, inject a 20 µL aliquot.

## HPLC VARIABLES
**Column:** 150 × 4.6 5 µm Microsorb-MV C18
**Mobile phase:** MeOH:10 mM $Na_2HPO_4$ adjusted to pH 7.0 with phosphoric acid 50:50
**Flow rate:** 1
**Injection volume:** 20
**Detector:** UV 320

## CHROMATOGRAM
**Retention time:** 3.44

## KEY WORDS
tablets; injections; syrups

## REFERENCE
Lau-Cam, C.A.; Rahman, M.; Roos, R.W. Rapid reversed phase high performance liquid chromatographic assay method for ranitidine hydrochloride in dosage forms. *J.Liq.Chromatogr.*, **1994**, *17*, 1089–1104

## SAMPLE
**Matrix:** formulations
**Sample preparation:** 100 μL Solution + 1.9 mL MeOH:water 20:80, inject a 50 μL aliquot.

## HPLC VARIABLES
**Guard column:** 5 μm Adsorbosphere C18
**Column:** 250 × 4.6 5 μm Adsorbosphere C18
**Mobile phase:** MeOH:100 mM ammonium acetate 65:35
**Flow rate:** 1
**Injection volume:** 50
**Detector:** UV 322

## CHROMATOGRAM
**Retention time:** 6.5

## KEY WORDS
stability-indicating; injections; saline

## REFERENCE
Inagaki, K.; Gill, M.A.; Okamoto, M.P.; Takagi, J. Stability of ranitidine hydrochloride with aztreonam, ceftazidime, or piperacillin sodium during simulated Y-site administration. *Am.J.Hosp.Pharm.*, **1992**, *49*, 2769–2772

## SAMPLE
**Matrix:** hepatocyte suspensions
**Sample preparation:** Add 2 volumes acetone to the hepatocyte suspension, centrifuge at 13000 g for 5 min. Remove the clear supernatant and evaporate it to dryness in a centrifugal evaporator, reconstitute the residue in 1 mL MeOH, sonicate, centrifuge at 13000 g for 5 min, inject an aliquot.

## HPLC VARIABLES
**Column:** 100 × 4.6 5 μm ODS 1 (Jones Chromatography)
**Mobile phase:** MeCN:MeOH:35 mM pH 7.0 phosphate buffer 5:46:49
**Column temperature:** 40
**Detector:** UV 320

## CHROMATOGRAM
**Retention time:** 6.2
**Limit of quantitation:** 10 ng

## OTHER SUBSTANCES
**Extracted:** metabolites

## KEY WORDS
dog; rat

## REFERENCE
Cross, D.M.; Bell, J.A.; Wilson, K. Kinetics of ranitidine metabolism in dog and rat isolated hepatocytes. *Xenobiotica*, **1995**, *25*, 367–375

## SAMPLE
**Matrix:** solutions

## HPLC VARIABLES
**Column:** 250 × 4.6 10 μm Partisil ODS1
**Mobile phase:** MeOH:50 mM pH 3.0 phosphoric acid 10:90
**Column temperature:** 30
**Flow rate:** 1.5
**Detector:** Radioactivity

## OTHER SUBSTANCES
**Also analyzed:** atenolol, cimetidine, hydrochlorothiazide

## KEY WORDS
[14]C labeled

## REFERENCE
Collett, A.; Sims, E.; Walker, D.; He, Y.-L.; Ayrton, J.; Rowland, M.; Warhurst, G. Comparison of HT29-18-$C_1$ and Caco-2 cell lines as models for studying intestinal paracellular drug absorption. *Pharm.Res.*, **1996**, *13*, 216–221

## SAMPLE
**Matrix:** solutions

## HPLC VARIABLES
**Column:** 150 × 4.6 12 μm 1-myristoyl-2-[(13-carboxyl)-tridecoyl]-sn-3-glycerophosphocholine chemically bonded to silica (Regis)
**Mobile phase:** MeCN:100 mM pH 7.0 phosphate buffer 20:80
**Flow rate:** 1
**Detector:** UV 254

## CHROMATOGRAM
**Retention time:** k' 0.96

## OTHER SUBSTANCES
**Also analyzed:** acebutolol, alprenolol, antazoline, atenolol, betaxolol, bisoprolol, bopindolol, bupranolol, carteolol, celiprolol, chloropyramine, chlorpheniramine, cicloprolol, cimetidine, cinnarizine, cirazoline, clonidine, dilevalol, dimethindene, diphenhydramine, doxazosin, esmolol, famotidine, isothipendyl, ketotifen, metiamide, metoprolol, moxonidine, nadolol, naphazoline, nifenalol, nizatidine, oxprenolol, pheniramine, phentolamine, pindolol, pizotyline (pizotifen), practolol, prazosin, promethazine, propranolol, pyrilamine (mepyramine), ranitidine, roxatidine, sotalol, tiamenidine, timolol, tramazoline, tripelennamine, triprolidine, tymazoline, UK-14,304

## REFERENCE
Kaliszan, R.; Nasal, A.; Turowski, M. Binding site for basic drugs on $\alpha_1$-acid glycoprotein as revealed by chemometric analysis of biochromatographic data. *Biomed.Chromatogr.*, **1995**, *9*, 211–215

## SAMPLE
**Matrix:** solutions

## HPLC VARIABLES
**Column:** 250 × 4.6 5 μm Supelcosil LC-DP (A) or 250 × 4 5 μm LiChrospher 100 RP-8 (B)
**Mobile phase:** MeCN:0.025% phosphoric acid:buffer 25:10:5 (A) or 60:25:15 (B) (Buffer was 9 mL concentrated phosphoric acid and 10 mL triethylamine in 900 mL water, adjust pH to 3.4 with dilute phosphoric acid, make up to 1 L.)
**Flow rate:** 0.6

**Injection volume:** 25
**Detector:** UV 229

## CHROMATOGRAM
**Retention time:** 5.88 (A), 3.27 (B)

## OTHER SUBSTANCES
**Also analyzed:** acebutolol, acepromazine, acetaminophen, acetazolamide, acetophenazine, albuterol, alprazolam, amitriptyline, amobarbital, amoxapine, antipyrine, atenolol, atropine, azatadine, baclofen, benzocaine, bromocriptine, brompheniramine, brotizolam, bupivacaine, buspirone, butabarbital, butalbital, caffeine, carbamazepine, cetirizine, chlorcyclizine, chlordiazepoxide, chlormezanone, chloroquine, chlorpheniramine, chlorpromazine, chlorpropamide, chlorprothixene, chlorthalidone, chlorzoxazone, cimetidine, cisapride, clomipramine, clonazepam, clonidine, clozapine, cocaine, codeine, colchicine, cyclizine, cyclobenzaprine, dantrolene, desipramine, diazepam, diclofenac, diflunisal, diltiazem, diphenhydramine, diphenidol, diphenoxylate, dipyridamole, disopyramide, dobutamine, doxapram, doxepin, droperidol, encainide, ethidium bromide, ethopropazine, fenoprofen, fentanyl, flavoxate, fluoxetine, fluphenazine, flurazepam, flurbiprofen, fluvoxamine, furosemide, glutethimide, glyburide, guaifenesin, haloperidol, homatropine, hydralazine, hydrochlorothiazide, hydrocodone, hydromorphone, hydroxychloroquine, hydroxyzine, ibuprofen, imipramine, indomethacin, ketoconazole, ketoprofen, ketorolac, labetalol, levorphanol, lidocaine, loratadine, lorazepam, lovastatin, loxapine, mazindol, mefenamic acid, meperidine, mephenytoin, mepivacaine, mesoridazine, metaproterenol, metformin, methadone, methdilazine, methocarbamol, methotrexate, methotrimeprazine, methoxamine, methyldopa, methylphenidate, metoclopramide, metolazone, metoprolol, metronidazole, midazolam, moclobemide, morphine, nadolol, nalbuphine, naloxone, naphazoline, naproxen, nifedipine, nizatidine, norepinephrine, nortriptyline, oxazepam, oxycodone, oxymetazoline, paroxetine, pemoline, pentazocine, pentobarbital, pentoxifylline, perphenazine, pheniramine, phenobarbital, phenol, phenolphthalein, phentolamine, phenylbutazone, phenyltoloxamine, phenytoin, pimozide, pindolol, piroxicam, pramoxine, prazepam, prazosin, probenecid, procainamide, procaine, prochlorperazine, procyclidine, promazine, promethazine, propafenone, propantheline, propiomazine, propofol, propranolol, protriptyline, quazepam, quinidine, quinine, racemethorphan, remoxipride, risperidone, salicylic acid, scopolamine, secobarbital, sertraline, sotalol, spironolactone, sulfinpyrazone, sulindac, temazepam, terbutaline, terfenadine, tetracaine, theophylline, thiethylperazine, thiopental, thioridazine, thiothixene, timolol, tocainide, tolbutamide, tolmetin, trazodone, triamterene, triazolam, trifluoperazine, triflupromazine, trimeprazine, trimethoprim, trimipramine, verapamil, warfarin, xylometazoline, yohimbine, zopiclone

## KEY WORDS
some details of plasma extraction

## REFERENCE
Koves, E.M. Use of high-performance liquid chromatography-diode array detection in forensic toxicology. *J.Chromatogr.A*, **1995**, *692*, 103–119

## SAMPLE
**Matrix:** solutions

## HPLC VARIABLES
**Column:** 250 × 4.6 Zorbax RX
**Mobile phase:** Gradient. A was 10 mL concentrated orthophosphoric acid and 7 mL triethylamine in 1 L water. B was 10 mL concentrated orthophosphoric acid and 7 mL triethylamine in 200 mL water, make up to 1 L with MeCN. A:B from 100:0 to 0:100 over 30 min, maintain at 0:100 for 5 min.
**Column temperature:** 30

**Flow rate:** 2
**Detector:** UV 210

## OTHER SUBSTANCES
**Also analyzed:** acepromazine, acetaminophen, acetophenazine, albuterol, aminophylline, amitriptyline, amobarbital, amoxapine, amphetamine, amylocaine, antipyrine, aprobarbital, aspirin, atenolol, atropine, avermectin, barbital, benzocaine, benzoic acid, benzotropine, benzphetamine, berberine, bibucaine, bromazepan, brompheniramine, buprenorphine, buspirone, butabarbital, butacaine, butethal, caffeine, carbamazepine, carbromal, chloramphenicol, chlordiazepoxide, chloroquine, chlorothiazide, chloroxylenol, chlorphenesin, chlorpheniramine, chlorpromazine, chlorpropamide, chlortetracycline, cimetidine, cinchonidine, cinchonine, clenbuterol, clonazepam, clonixin, clorazepate, cocaine, codeine, colchicine, cortisone, coumarin, cyclazocine, cyclobenzaprine, cyclothiazide, cyheptamide, cymarin, danazol, danthron, dapsone, debrisoquine, desipramine, dexamethasone, dextromethorphan, dextropropoxyphene, diamorphine, diazepam, diclofenac, diethylpropion, diethylstilbestrol, diflunisal, digitoxin, digoxin, diltiazem, diphenhydramine, diphenoxylate, diprenorphine, dipyrone, disulfiram, dopamine, doxapram, doxepin, dronabinol, ephedrine, epinephrine, epinine, estradiol, estriol, estrone, ethacrynic acid, ethosuximide, etonitazene, etorphine, eugenol, famotidine, fenbendazole, fencamfamine, fenoprofen, fenproporex, fentanyl, flubendazole, flufenamic acid, flunitrazepam, 5-fluorouracil, fluoxymesterone, fluphenazine, furosemide, gentisic acid, gitoxigenin, glipizide, glunixin, glutethimide, glybenclamide, guaiacol, halazepam, haloperidol, hydrochlorothiazide, hydrocodone, hydrocortisone, hydromorphone, hydroxyquinoline, ibogaine, ibuprofen, iminostilbene, imipramine, indomethacin, isocarbostyril, isocarboxazid, isoniazid, isoproterenol, isoxsuprine, ivermectin, ketamine, ketoprofen, kynurenic acid, levorphanol, lidocaine, lorazepam, lormetazepam, loxapine, mazindol, mebendazole, meclizine, meclofenamic acid, medazepam, mefenamic acid, megestrol, mepacrine, meperidine, mephentermine, mephenytoin, mephesin, mephobarbital, mepivacaine, mescaline, mesoridazine, methadone, methamphetamine, methapyrilene, methaqualone, methazolamide, methocarbamol, methoxamine, methsuximide, methyl salicylate, methyldopa, methyldopamine, methylphenidate, methylprednisolone, methyltestosterone, methyprylon, metoprolol, mibolerone, morphine, nadolol, nalorphine, naloxone, naltrexone, naphazoline, naproxen, nefopam, niacinamide, nicotine, nicotinic acid, nifedipine, niflumic acid, nitrazepam, norepinephrine, nortriptyline, noscapine, nylidrin, oxazepam, oxycodone, oxymorphone, oxyphenbutazone, oxytetracycline, papaverine, pargyline, pemoline, pentazocine, pentobarbital, persantine, phenacetin, phenazocine, phenazopyridine, phencyclidine, phendimetrazine, phenelzine, pheniramine, phenobarbital, phenothiazine, phensuximide, phentermine, phenylbutazone, phenylephrine, phenylpropanolamine, piperocaine, prazepam, prednisolone, primidone, probenecid, progesterone, propiomazine, propranolol, propylparaben, pseudoephedrine, puromycin, pyrilamine, pyrithyldione, quazepam, quinaldic acid, quinidine, quinine, recinnamine, reserpine, resorcinol, saccharin, albuterol, salicylamide, salicylic acid, scopolamine, scopoletin, secobarbital, strychnine, sulfacetamide, sulfadiazine, sulfadimethoxine, sulfaethidole, sulfamerazine, sulfamethazine, sulfamethoxizole, sulfanilamide, sulfapyridine, sulfasoxizole, sulindac, tamoxifen, temazepam, testosterone, tetracaine, tetracycline, tetramisole, thebaine, theobromine, theophylline, thiabendazole, thiamine, thiamylal, thiobarbituric acid, thioridazine, thiosalicylic acid, thiothixene, thymol, tolazamide, tolazoline, tobutamide, tolmetin, tranylcypromine, triamcinolone, tribenzylamine, trichloromethiazide, trifluoperazine, trihexyphenidyl, trimethoprim, tripelennamine, triprolidine, tropacocaine, tyramine, verapamil, vincamine, warfarin, yohimbine, zoxazolamine

## REFERENCE
Hill, D.W.; Kind, A.J. Reversed-phase solvent-gradient HPLC retention indexes of drugs. *J.Anal.Toxicol.*, **1994**, *18*, 233–242

## SAMPLE
**Matrix:** solutions
**Sample preparation:** Prepare a 10 µg/mL solution in MeOH, inject a 20 µL aliquot.

## HPLC VARIABLES
**Column:** 125 × 4.9 Spherisorb S5W silica
**Mobile phase:** MeOH containing 10 mM ammonium perchlorate and 1 mL/L 100 mM NaOH in MeOH, pH 6.7
**Flow rate:** 2
**Injection volume:** 20
**Detector:** E, LeCarbone, V25 glassy carbon electrode, + 1.2 V

## CHROMATOGRAM
**Retention time:** 3.1

## OTHER SUBSTANCES
**Also analyzed:** acebutolol, acepromazine, acetophenazine, N-acetylprocainamide, albuterol, alprenolol, amethocaine, amiodarone, amitriptyline, antazoline, atenolol, azacyclonal, bamethane, benactyzine, benperidol, benzethidine, benzocaine, benzoctamine, benzphetamine, benzquinamide, bromhexine, bromodiphenhydramine, bromperidol, brompheniramine, brompromazine, buclizine, bufotenine, bupivacaine, buprenorphine, butacaine, butethamate, chlorcyclizine, chlorpheniramine, chlorphenoxamine, chlorprenaline, chlorpromazine, chlorprothixene, cimetidine, cinchonidine, cinnarizine, clemastine, clomipramine, clonidine, cocaine, cyclazocine, cyclizine, cyclopentamine, cyproheptadine, deserpidine, desipramine, dextromoramide, dextropropoxyphene, dicyclomine, diethylcarbamazine, diethylpropion, diethylthiambutene, dihydroergotamine, dimethindene, dimethothiazine, diphenhydramine, diphenoxylate, dipipanone, diprenorphine, dipyridamole, disopyramide, dothiepin, doxapram, doxepin, doxylamine, droperidol, ephedrine, ergocornine, ergocristine, ergocristinine, ergocryptine, ergometrine, ergosine, ergosinine, ergotamine, ethopropazine, etorphine, etoxeridine, fenethazine, fenfluramine, fenoterol, fentanyl, flavoxate, fluopromazine, flupenthixol, fluphenazine, flurazepam, haloperidol, hydroxyzine, hyoscine, ibogaine, imipramine, indapamine, iprindole, isothipendyl, isoxsuprine, ketanserin, laudanosine, lidocaine, lofepramine, loxapine, maprotiline, mecamylamine, meclophenoxate, meclozine, medazepam, mephentermine, mepivacaine, meptazinol, mepyramine, mesoridazine, metaraminol, methadone, methamphetamine, methapyrilene, methdilazene, methotrimeprazine, methoxamine, methoxyphenamine, methoxypromazine, methylephedrine, methylergonovine, methysergide, metoclopramide, metopimazine, metoprolol, mianserin, morazone, nadolol, nalorphine, naloxone, naphazoline, nicotine, nifedipine, nomifensine, nortriptyline, noscapine, orphenadrine, oxeladin, oxprenolol, oxymetazolin, papaverine, pargyline, pecazine, penbutolol, pentazocine, penthienate, pericyazine, perphenazine, phenadoxone, phenampromide, phenazocine, phenbutrazate, phendimetrazine, phenelzine, phenglutarimide, phenindamine, pheniramine, phenmetrazine, phenomorphan, phenoperidine, phenothiazine, phenoxybenzamine, phentolamine, phenylephrine, phenyltoloxamine, physostigmine, piminodine, pimozide, pindolol, pipamazine, pipazethate, piperacetazine, piperidolate, pipradol, pirenzepine, piritramide, pizotifen, practolol, pramoxine, prazosin, prenylamine, prilocaine, primaquine, proadifen, procainamide, procaine, prochlorperazine, procyclidine, proheptazine, prolintane, promazine, promethazine, pronethalol, properidine, propiomazine, propranolol, prothipendyl, protriptyline, proxymetacaine, pseudoephedrine, pyrimethamine, quinidine, quinine, rescinnamine, sotalol, tacrine, terazosin, terbutaline, terfenadine, thenyldiamine, theophylline, thiethylperazine, thiopropazate, thioproperazine, thioridazine, thiothixene, thonzylamine, timolol, tocainide, tolpropamine, tolycaine, tranylcypromine, trazodone, trifluoperazine, trifluperidol, trimeperidine, trimeprazine, trimethobenzamide, trimethoprim, trimipramine, tripelennamine, triprolidine, tryptamine, verapamil, xylometazoline

## REFERENCE
Jane, I.; McKinnon, A.; Flanagan, R.J. High-performance liquid chromatographic analysis of basic drugs on silica columns using non-aqueous ionic eluents. II. Application of UV, fluorescence and electrochemical oxidation detection. *J.Chromatogr.*, **1985**, *323*, 191–225

## ANNOTATED BIBLIOGRAPHY

Pompilio, F.M.; Fox, J.L.; Inagaki, K.; Burm, J.-P.; Jhee, S.; Gill, M.A. Stability of ranitidine hydrochloride with ondansetron hydrochloride or fluconazole during simulated Y-site administration. *Am.J.Hosp.Pharm.*, **1994**, *51*, 391–394 [injections; 5% dextrose; stability-indicating]

Smith, M.S.; Oxford, J.; Evans, M.B. Improved method for the separation of ranitidine and its metabolites based on supercritical fluid chromatography. *J.Chromatogr.A*, **1994**, *683*, 402–406 [SFC]

Stiles, M.L.; Allen, L.V., Jr.; Prince, S. Stability of ranitidine hydrochloride during simulated home care use. *Am.J.Hosp.Pharm.*, **1994**, *51*, 1706–1707 [stability-indicating; injections; saline; 5% dextrose]

Suttle, A.B.; Brouwer, K.L.R. Bile flow but not enterohepatic recirculation influences the pharmacokinetics of ranitidine in the rat. *Drug Metab.Dispos.*, **1994**, *22*, 224–232 [serum; bile; pharmacokinetics; procaine (IS); LOQ 75 ng/mL (serum); LOQ 500 ng/mL (bile)]

Zhang, H.; Stewart, J.T. HPLC determination of norepinephrine bitartrate in 5% dextrose injection on underivatized silica with an aqueous-organic mobile phase. *J.Liq.Chromatogr.*, **1993**, *16*, 2861–2871 [injections; 5% dextrose]

Kaka, J.S. Rapid method for cimetidine and ranitidine determination in human and rat plasma by HPLC. *J.Liq.Chromatogr.*, **1988**, *11*, 3447–3456 [LOD 50-100 ng/mL]

Tracqui, A.; Kintz, P.; Mangin, P.; Lugnier, A.A.; Chaumont, A.J. A new rapid HPLC assay for the simultaneous determination of two histamine H2-receptor antagonists, cimetidine and ranitidine, in human plasma. *J.Toxicol.Clin.Exp.*, **1988**, *8*, 387–394

Lant, M.S.; Martin, L.E.; Oxford, J. Qualitative and quantitative analysis of ranitidine and its metabolites by high-performance liquid chromatography-mass spectrometry. *J.Chromatogr.*, **1985**, *323*, 143–152 [UV detection; LC-MS]

Boutagy, J.; More, D.G.; Munro, I.A.; Shenfield, G.M. Simultaneous analysis of cimetidine and ranitidine in human plasma by HPLC. *J.Liq.Chromatogr.*, **1984**, *7*, 1651–1664

Fleitman, J.; Torosian, G.; Perrin, J.H. Improved high-performance liquid chromatographic assay for cimetidine using ranitidine as an internal standard. *J.Chromatogr.*, **1982**, *229*, 255–258

Vandenberghe, H.M.; MacLeod, S.M.; Mahon, W.A.; Lebert, P.A.; Soldin, S.J. Analysis of ranitidine in serum by high performance liquid chromatography. *Ther.Drug Monit.*, **1980**, *2*, 379–384 [serum; plasma; metiamide (IS); LOD 10 ng/mL; extracted metabolites; non-interfering amitriptyline, carbamazepine, diazepam, ethosuximide, flurazepam, oxazepam, phenobarbital, phenytoin, primidone, theophylline]

# Retinoic Acid

**Molecular formula:** $C_{20}H_{28}O_2$
**Molecular weight:** 300.4
**CAS Registry No.:** 302-79-4 (tretinoin (all-trans)), 4759-48-2 (isotretinoin (13-cis))

## SAMPLE

**Matrix:** blood

**Sample preparation:** Condition a 100 mg methyl-C1 Accubond SPE cartridge (J&W) with three 1 mL portions of MeOH and three 1 mL portions of 1% ammonium acetate. 500 µL Plasma + 20 µL 1 µg/mL acitretin in MeCN containing 10 mM BHT + 1 mL 10 mM BHT in isopropanol, vortex, rotate for 15 min, centrifuge at 16000 g for 10 min. Remove the supernatant and add it to 11 mL 1% ammonium acetate, add to the SPE cartridge, wash with 1 mL 0.1% ammonium acetate, wash with 1 mL MeOH:0.1% ammonium acetate 50:50, dry under vacuum for 30 s, elute with 1.5 mL 10 mM BHT in MeCN. Add 10 µL pentafluorobenzyl bromide and 10 µL 10 mg/mL potassium carbonate in MeCN: water 50:50 to the eluate, vortex, let stand at room temperature for 1 h, evaporate to dryness under reduced pressure for 2 h, reconstitute with 20-100 µL 10 mM BHT in MeCN, inject a 20 µL aliquot.

## HPLC VARIABLES

**Column:** $150 \times 3.9$ Nova-Pak C18 + $75 \times 3.9$ Nova-Pak C18 (in series)
**Mobile phase:** Gradient. MeCN:buffer 80:20 for 10 min, to 90:10 (step gradient). (Buffer was 100 mM ammonium acetate adjusted to pH 5.0 with acetic acid.)
**Column temperature:** 40
**Injection volume:** 20
**Detector:** UV 369; MS Hewlett-Packard model 5988A, particle beam interface nebulizer 60°, helium 35 psi, m/z 299

## CHROMATOGRAM

**Retention time:** 26 (isotretinoin), 28.2 (tretinoin)
**Internal standard:** acitretin (m/z 325) (16)
**Limit of detection:** 0.05 ng/mL

## OTHER SUBSTANCES

**Extracted:** 9-cis-retinoic acid

## KEY WORDS

plasma; protect from light; derivatization; SPE

## REFERENCE

Lehman, P.A.; Franz, T.J. A sensitive high-pressure liquid chromatography/particle beam/mass spectrometry assay for the determination of *all-trans*-retinoic acid and 13-*cis*-retinoic acid in human plasma. *J.Pharm.Sci.*, **1996**, *85*, 287–290

## SAMPLE

**Matrix:** blood

**Sample preparation:** 10 μL Serum + 30 μL 200-500 ng/mL retinyl acetate in isopropanol:dichloroethane 2:1 + 5 μL glacial acetic acid, vortex for 30 s, centrifuge for 1 min, inject a 10-20 μL aliquot.

## HPLC VARIABLES
**Guard column:** C18 (Upchurch)
**Column:** 150 × 4.6 Ultracarb 5 ODS30 (Phenomenex)
**Mobile phase:** MeCN:dichloromethane:MeOH 85:12:3 containing 0.1% ammonium acetate (dissolve ammonium acetate in MeOH first)
**Flow rate:** 1
**Injection volume:** 10-20
**Detector:** UV 335

## CHROMATOGRAM
**Retention time:** 9 (tretinoin)
**Internal standard:** retinyl acetate (5.5)
**Limit of detection:** 10 ng/mL
**Limit of quantitation:** 50 ng/mL

## OTHER SUBSTANCES
**Extracted:** vitamin A

## KEY WORDS
protect from light; serum

## REFERENCE
Barua, A.B.; Kostic, D.; Barua, M.; Olson, J.A. Determination of retinol and retinoic acid in capillary blood by high-performance liquid chromatography. *J.Liq.Chromatogr.*, **1995**, *18*, 1459–1471

## SAMPLE
**Matrix:** blood
**Sample preparation:** Condition a Bakerbond SPE octadecyl SPE cartridge with 2 mL MeOH and 2 mL 1 M acetic acid. 1 mL Serum + 3 mL 1 M acetic acid + 50 μL 10 μM IS in DMSO, mix, add to the SPE cartridge, wash with 2 mL acetone:1 M acetic acid 50:50, dry under vacuum for 15 min, elute with 500 μL MeCN. Evaporate the eluate and take up the residue in 200 μL mobile phase, inject a 50 μL aliquot.

## HPLC VARIABLES
**Column:** 250 × 4 5 μm Lichrospher Si-60
**Mobile phase:** Hexane:dichloromethane:1,4-dioxane 78:18:4 containing 1% acetic acid
**Flow rate:** 0.8
**Injection volume:** 50
**Detector:** UV 360

## CHROMATOGRAM
**Retention time:** 6 (isotretinoin), 7 (tretinoin)
**Internal standard:** (all-E)-3-methyl-7-(5,6,7,8-tetrahydro-5,5,8,8-tetramethyl-2-naphthyl)-2,4,6-octanoic acid (Ro 13-6307) (7.5)
**Limit of detection:** 1.5 ng/mL

## OTHER SUBSTANCES
**Extracted:** metabolites

## KEY WORDS
protect from light; serum; SPE; normal phase; pharmacokinetics

## REFERENCE

Lefebvre, P.; Agadir, A.; Cornic, C.; Gourmel, B.; Hue, B.; Dreux, C.; Degos, L.; Chomienne, C. Simultaneous determination of all-*trans* and 13-*cis* retinoic acids and their 4-oxo metabolites by adsorption liquid chromatography after solid-phase extraction. *J.Chromatogr.B*, **1995**, *666*, 55–61

## SAMPLE

**Matrix:** blood

**Sample preparation:** 500 µL Plasma + 1 mL IS in EtOH, vortex for 30 s, add 5 mL water, add 7.5 mL n-hexane, add 300 µL 2 M HCl, rotate for 10 min, centrifuge at 1250 g for 8 min. Remove the organic layer and evaporate it at room temperature under a stream of nitrogen. Dissolve the residue in 100 µL mobile phase, inject a 50 µL aliquot.

## HPLC VARIABLES

**Column:** 150 × 4.6 5 µm Spherisorb S5W

**Mobile phase:** n-Hexane:isopropanol:acetic acid 200:0.7:0.135

**Flow rate:** 0.9

**Injection volume:** 50

**Detector:** UV 350

## CHROMATOGRAM

**Retention time:** 8 (isotretinoin), 11 (tretinoin)

**Internal standard:** Ro 15-1570 (22)

**Limit of detection:** 0.5 ng/mL

## OTHER SUBSTANCES

**Extracted:** retinol

## KEY WORDS

plasma; normal phase

## REFERENCE

Meyer, E.; Lambert, W.E.; De Leenheer, A.P. Simultaneous determination of endogenous retinoic acid isomers and retinol in human plasma by isocratic normal-phase HPLC with ultraviolet detection. *Clin.Chem.*, **1994**, *40*, 48–51

## SAMPLE

**Matrix:** blood

**Sample preparation:** 200 µL Serum + 2 mL diluting agent (amber tube), mix for 10 s, add 5 mL n-hexane, vortex for 30 s, centrifuge at 1600 g at 4° for 5 min. Remove the organic phase and concentrate to < 1 mL under vacuum below 30°, evaporate the rest of the solvent under nitrogen, dissolve the residue in 25 µL MeCN:MeOH 2:1, mix for 30 s, centrifuge at 8000 g for 1 min, inject a 20 µL aliquot. (Diluting agent was MeCN:100 mM ammonium acetate 25:75, pH adjusted to 5.5 with acetic acid.)

## HPLC VARIABLES

**Column:** 150 × 4.6 5 µm Chemcosorb 5-ODS-H

**Mobile phase:** MeCN:MeOH:100 mM ammonium acetate 46.7:23.3:30, pH adjusted to 7.0

**Column temperature:** 50

**Flow rate:** 1

**Injection volume:** 20

**Detector:** UV 340

## CHROMATOGRAM
**Retention time:** 14.9 (isotretinoin), 17.0 (tretinoin)
**Limit of quantitation:** 0.5 ng/mL

## OTHER SUBSTANCES
**Extracted:** vitamin A (retinol)

## KEY WORDS
serum

## REFERENCE
Takeda, N.; Yamamoto, A. Simultaneous determination of 13-cis- and all-trans-retinoic acids and retinol in human serum by high-performance liquid chromatography. *J.Chromatogr.B*, **1994**, *657*, 53–59

## SAMPLE
**Matrix:** blood
**Sample preparation:** 500 μL Plasma or serum + 100 μL 6 μg/mL 9-methylanthracene in MeOH + 1.5 mL MeCN + 100 μL 100 mM perchloric acid, flush headspace of vial with argon, vortex, centrifuge, inject a 50 μL aliquot of the supernatant. Sonicate serum, solvents, and mobile phase under vacuum before use. Use low-actinic glassware and yellow light.

## HPLC VARIABLES
**Column:** 250 × 4.6 5 μm Zorbax C18
**Mobile phase:** MeCN:0.5% acetic acid 85:15 containing 0.05% sodium hexanesulfonate
**Flow rate:** 2
**Injection volume:** 50
**Detector:** UV 365

## CHROMATOGRAM
**Retention time:** 6 (isotretinoin), 8 (tretinoin)
**Internal standard:** 9-methylanthracene (5)
**Limit of detection:** 12 ng/mL

## OTHER SUBSTANCES
**Simultaneous:** 4-oxo-13-cis-retinoic acid, retinol

## KEY WORDS
plasma; serum

## REFERENCE
Gadde, R.R.; Burton, F.W. Simple reversed-phase high-performance liquid chromatographic method for 13-cis-retinoic acid in serum. *J.Chromatogr.*, **1992**, *593*, 41–46

## SAMPLE
**Matrix:** blood
**Sample preparation:** 0.5-2 mL Plasma + 100 μL pH 7 phosphate buffer + 2 mL diethyl ether:ethyl acetate 50:50, vortex gently for 5 min, centrifuge at 2000 g for 10 min. Remove the organic layer and evaporate it to dryness under a stream of nitrogen, reconstitute the residue in 30-100 μL MeOH, inject a 25 μL aliquot.

## HPLC VARIABLES
**Column:** 250 × 4.6 5 μm Nucleosil C18
**Mobile phase:** MeOH:1% aqueous acetic acid 85:15
**Flow rate:** 1.5

**Injection volume:** 25
**Detector:** UV 350

## CHROMATOGRAM
**Retention time:** 12 (isotretinoin), 15 (tretinoin)
**Limit of detection:** 2 ng/mL

## OTHER SUBSTANCES
**Extracted:** acitretin, 13-cis-acitretin, etretinate, 4-oxo-13-cis-retinoic acid
**Noninterfering:** antidepressants, benzodiazepines, psoralen

## KEY WORDS
plasma; handle under yellow light

## REFERENCE
Bun, H.; al-Mallah, N.R.; Aubert, C.; Cano, J.P. High-performance liquid chromatography of aromatic retinoids and isotretinoin in biological fluids. *Methods Enzymol.*, **1990**, *189*, 167–172

## SAMPLE
**Matrix:** blood
**Sample preparation:** 500 μL Plasma + 750 μL 100 ng/mL acitretin in MeCN:9 mM NaOH 20:80, centrifuge at 1500 g for 3 min, inject a 500 μL aliquot onto column A with mobile phase A and elute for 7 min, elute column A in backflush mode with mobile phase A for 3 min, backflush contents of column A onto column B with mobile phase B and start the gradient for mobile phase B. At the end of the process flush the lines with component B of mobile phase B, re-equilibrate columns for 4 min. (Keep sample at 10° in the autosampler.)

## HPLC VARIABLES
**Column:** A 14 × 4.6 37-50 μm Bondapak C18 Corasil (column fitted with 3 μm sieves not glass fiber filters); B 30 × 4 5 μm Spherisorb ODS 1 + 125 × 4 5 μm Spherisorb ODS 1 + 125 × 4 5 μm Spherisorb ODS 1
**Mobile phase:** A MeCN:1% ammonium acetate 10:90; B Gradient. A was MeCN:water:10% ammonium acetate:acetic acid 600:400:4:30. B was MeCN:water:10% ammonium acetate:acetic acid 850:146:4:10. A:B 100:0 to 70:30 over 6 min, then to 0:100 over 5 min, stay at 0:100 for 11 min.
**Flow rate:** A 1.5; B 1
**Injection volume:** 500
**Detector:** UV 360

## CHROMATOGRAM
**Retention time:** 25 (isotretinoin), 27 (tretinoin)
**Internal standard:** acitretin (23)
**Limit of detection:** 0.5-1 ng/mL
**Limit of quantitation:** 2 ng/mL

## OTHER SUBSTANCES
**Extracted:** metabolites, isotretinoin, 4-oxoisotretinoin, 4-oxotretinoin

## KEY WORDS
plasma; column-switching

## REFERENCE
Wyss, R. Determination of retinoids in plasma by high-performance liquid chromatography and automated column switching. *Methods Enzymol.*, **1990**, *189*, 146–155

## SAMPLE
**Matrix:** blood
**Sample preparation:** 500 μL Plasma + 50 μL 5% perchloric acid, vortex for 30 s, add 500 μL ethyl acetate, whirl for 1 min, centrifuge at 13000 g for 1 min, inject a 50 μL aliquot of the organic layer.

## HPLC VARIABLES
**Guard column:** present but not specified
**Column:** 250 × 4.6 5 μm Ultrasphere ODS + 300 × 4 10 μm μBondapak in series
**Mobile phase:** MeCN:1% ammonium acetate 95:5
**Flow rate:** 2.5
**Injection volume:** 50
**Detector:** UV 340; UV 365

## CHROMATOGRAM
**Retention time:** 6.5 (isotretinoin), 8.2 (tretinoin)

## OTHER SUBSTANCES
**Extracted:** vitamin A

## KEY WORDS
protect from light; plasma

## REFERENCE
Peng, Y.-M.; Xu, M.-J.; Alberts, D.S. Analysis and stability of retinol in plasma. *J.Natl.Cancer Inst.*, **1987**, *78*, 95–99

## SAMPLE
**Matrix:** blood
**Sample preparation:** 500 μL Plasma + 100 μL 5% perchloric acid, mix rapidly, add 500 μL ethyl acetate, mix for 60-90 s, centrifuge at 13000 g for 5 min, inject an aliquot of the supernatant.

## HPLC VARIABLES
**Column:** Reversed-phase C18 (Waters or Bio-Rad)
**Mobile phase:** MeCN:1% ammonium acetate 75:25
**Flow rate:** 2.5
**Detector:** UV 340

## CHROMATOGRAM
**Retention time:** 3 (isotretinoin), 4 (tretinoin)
**Limit of detection:** 20 ng/mL

## OTHER SUBSTANCES
**Extracted:** vitamin A

## KEY WORDS
plasma; pharmacokinetics; protect from light

## REFERENCE
Davis, T.P.; Peng, Y.-M.; Goodman, G.E.; Alberts, D.S. HPLC, MS, and pharmacokinetics of melphalan, bisantrene and 13-cis retinoic acid. *J.Chromatogr.Sci.*, **1982**, *20*, 511–516

## SAMPLE
**Matrix:** blood, microsomal incubations

**Sample preparation:** 500 μL Plasma or 250 μL microsomal incubation + 25 μL 20 μg/mL IS in MeCN + 350 μL 1-butanol:MeCN 50:50, mix thoroughly, add 300 μL saturated K₂HPO₄, mix, centrifuge at 3000 g for 10 min, inject an aliquot of the organic layer.

## HPLC VARIABLES
**Column:** Adsorbosphere C18
**Mobile phase:** Gradient. MeCN:10 mM ammonium acetate from 50:50 to 95:5 over 10 min
**Flow rate:** 1.5
**Detector:** UV 365

## CHROMATOGRAM
**Internal standard:** Ro 23-4736
**Limit of detection:** 10 ng/mL

## OTHER SUBSTANCES
**Extracted:** metabolites

## KEY WORDS
plasma; protect from light; human; liver; for tretinoin; pharmacokinetics

## REFERENCE
Schwartz, E.L.; Hallam, S.; Gallagher, R.E.; Wiernik, P.H. Inhibition of all-*trans*-retinoic acid metabolism by fluconazole *in vitro* and in patients with acute promyelocytic leukemia. *Biochem.Pharmacol.*, **1995**, *50*, 923–928

## SAMPLE
**Matrix:** blood, tissue
**Sample preparation:** Plasma. Add 5 volumes of MeOH to plasma, cool to -20°, centrifuge at 16500 g for 5 min. Remove the supernatant and evaporate it to dryness under reduced pressure, reconstitute the residue in mobile phase, centrifuge at 16500 g for 5 min, inject a 10-250 μL aliquot. Liver. Homogenize (Kinematica-Polytron with an Aggregat PTA 10TS generator) rat liver with 5 volumes of MeOH at 20000 rpm for 20 s, let stand at -20° overnight, centrifuge at 0-4° at 3200 g for 10 min. Remove the supernatant and evaporate it to dryness under reduced pressure, reconstitute the residue in mobile phase, centrifuge at 16500 g for 5 min, inject a 10-250 μL aliquot.

## HPLC VARIABLES
**Guard column:** 20 × 4 5 μm C18 (Hewlett-Packard or Alltech)
**Column:** 250 × 4.6 5 μm Microsorb-MV
**Mobile phase:** MeCN:10 mM ammonium acetate:glacial acetic acid 80:20:1
**Column temperature:** 40
**Flow rate:** 1
**Injection volume:** 10-250
**Detector:** UV 348

## CHROMATOGRAM
**Retention time:** 13.6 (isotretinoin), 16.3 (tretinoin)

## OTHER SUBSTANCES
**Extracted:** metabolites

## KEY WORDS
rat; plasma; liver

## REFERENCE
Shirley, M.A.; Bennani, Y.L.; Boehm, M.F.; Breau, A.P.; Pathirana, C.; Ulm, E.H. Oxidative and reductive metabolism of 9-*cis*-retinoic acid in the rat. Identification of the 13,14-dihydro-9-*cis*-retinoic acid and its taurine conjugate. *Drug Metab.Dispos.*, **1996**, *24*, 293–302

## SAMPLE
**Matrix:** culture media
**Sample preparation:** 100 μL Culture media + 200 μL ice-cold EtOH, mix thoroughly, let stand for 15 min, centrifuge at 12000 g for 15 min, inject an aliquot of the supernatant.

## HPLC VARIABLES
**Guard column:** Whatman CO:PELL ODS guard column
**Column:** 100 × 8 5 μm Nova-Pak C18 (radial-packed)
**Mobile phase:** MeOH:100 mM pH 7.0 ammonium acetate 90:10
**Flow rate:** 1
**Detector:** UV 340

## CHROMATOGRAM
**Retention time:** 9.44 (tretinoin)

## OTHER SUBSTANCES
**Extracted:** acitretin, etretinate, isotretin, motretinid, retinal, vitamin A

## REFERENCE
Kochhar, D.M.; Penner, J.D.; Minutella, L.M. Biotransformation of etretinate and developmental toxicity of etretin and other aromatic retinoids in teratogenesis bioassays. *Drug Metab.Dispos.*, **1989**, *17*, 618–624

## SAMPLE
**Matrix:** solutions
**Sample preparation:** Inject an aliquot of a solution in the mobile phase.

## HPLC VARIABLES
**Column:** 250 × 4.6 5 μm Zorbax Rx-Sil
**Mobile phase:** Heptane:THF:acetic acid 96.5:3.5:0.015
**Flow rate:** 1.4
**Detector:** MS, Finnigan 4023, particle beam interface (Vestec universal interface model 700), electron impact mode 77 eV, scan m/z 200-350 (positive ion), multiplier voltage 1200 V, source 300°; UV 365

## CHROMATOGRAM
**Retention time:** 13 (isotretinoin), 16 (tretinoin)

## OTHER SUBSTANCES
**Simultaneous:** degradation products

## KEY WORDS
normal phase

## REFERENCE
Bempong, D.K.; Honigberg, I.L.; Meltzer, N.M. Normal phase LC-MS determination of retinoic acid degradation products. *J.Pharm.Biomed.Anal.*, **1995**, *13*, 285–291

## SAMPLE
**Matrix:** solutions

## HPLC VARIABLES
**Column:** 250 × 4.6 8 μm Unisphere-PBD (polybutadiene on alumina) (Biotage, Charlottesville, VA)
**Mobile phase:** MeOH:water 92:8
**Flow rate:** 1
**Detector:** UV 330

## CHROMATOGRAM
**Retention time:** 3.5 (tretinoin)

## OTHER SUBSTANCES
**Simultaneous:** retinol acetate, vitamin A, vitamin E

## REFERENCE
Jedrejewski, P.T.; Taylor, L.T. Comparison of silica-, alumina-, and polymer-based stationary phases for reversed-phase liquid chromatography. *J.Chromatogr.Sci.*, **1995**, *33*, 438–445

## SAMPLE
**Matrix:** solutions

## HPLC VARIABLES
**Column:** 250 × 4.6 5 μm Suplex-pKb-100 (Supelco)
**Mobile phase:** MeOH:acetic acid 99.95:0.05
**Flow rate:** 1
**Detector:** UV 350

## CHROMATOGRAM
**Retention time:** 13.5 (isotretinoin), 22 (tretinoin)

## OTHER SUBSTANCES
**Simultaneous:** other isomers

## KEY WORDS
photoisomerization solutions

## REFERENCE
Sundquist, A.R.; Stahl, W.; Steigel, A.; Sies, H. Separation of retinoic acid all-trans, mono-cis and poly-cis isomers by reversed-phase high-performance liquid chromatography. *J.Chromatogr.*, **1993**, *637*, 201–205

## SAMPLE
**Matrix:** tissue
**Sample preparation:** 100-120 mg Frog embryos + 100 μL isopropanol, sonicate on ice, vortex for 1 min, centrifuge at 4° at 4000 g for 20 min, inject an aliquot of the supernatant.

## HPLC VARIABLES
**Guard column:** 20 × 4 10 μm LiChrosorb RB 18
**Column:** 125 × 4.6 3 μm Spherisorb ODS II
**Mobile phase:** Gradient. MeOH:40 mM pH 7.3 ammonium acetate from 55:45 to 100:0 over 18 min.
**Flow rate:** 1.6
**Detector:** UV 354

## CHROMATOGRAM
**Retention time:** 12.5 (isotretinoin), 12.7 (9-cis-retinoic acid), 13.5 (tretinoin)

## OTHER SUBSTANCES
**Extracted:** metabolites, vitamin A

## KEY WORDS
frog; embryo

## REFERENCE
Creech Kraft, J.; Juchau, M.R. *Xenopus laevis*: A model system for the study of embryonic retinoid metabolism. III. Isomerization and metabolism of all-*trans*-retinoic acid and 9-*cis*-retinoic acid and their dysmorphogenic effects in embryos during neurulation. *Drug Metab.Dispos.*, **1995**, *23*, 1058–1071

## SAMPLE
**Matrix:** tissue
**Sample preparation:** 100-120 mg Tadpole embryos + 100 mL isopropanol, sonicate on ice, vortex for 1 min, centrifuge at 4000 g at 4° for 20 min, inject an aliquot of the supernatant.

## HPLC VARIABLES
**Guard column:** 20 × 4 10 μm LiChrosorb RB 18
**Column:** 125 × 4.6 3 μm Spherisorb ODS II
**Mobile phase:** Gradient. MeOH:40 mM pH 7.3 ammonium acetate from 55:45 to 100:0 over 20 min
**Flow rate:** 1.6
**Detector:** UV 354

## CHROMATOGRAM
**Retention time:** 11.8 (isotretinoin), 12.2 (tretinoin)

## OTHER SUBSTANCES
**Extracted:** metabolites, tretinoin, vitamin A

## KEY WORDS
handle under yellow light; tadpoles; embryos

## REFERENCE
Creech Kraft, J.; Kimelman, D.; Juchau, M.R. *Xenopus Laevis*: A model system for the study of embryonic retinoid metabolism. I. Embryonic metabolism of 9-*cis*- and all-*trans*-retinals and retinols and their corresponding acid forms. *Drug Metab.Dispos.*, **1995**, *23*, 72–82

◆ ─────────────────────── ◆ ─────────────────────── ◆

## ANNOTATED BIBLIOGRAPHY
Achkar, C.C.; Bentel, J.M.; Boylan, J.F.; Scher, H.I.; Gudas, L.J.; Miller, W.H., Jr. Differences in the pharmacokinetic properties of orally administered all-trans-retinoic acid and 9-cis-retinoic acid in the plasma of nude mice. *Drug Metab.Dispos.*, **1994**, *22*, 451–458 [gradient; LOD 0.5 ng/mL; extracted metabolites, isotretinoin, tretinoin, vitamin A]

Eckhoff, C.; Chari, S.; Kromka, M.; Staudner, H.; Juhasz, L.; Rudiger, H.; Agnish, N. Teratogenicity and transplacental pharmacokinetics of 13-cis-retinoic acid in rabbits. *Toxicol.Appl.Pharmacol.*, **1994**, *125*, 34–41 [plasma; acitretin (IS); SPE; column temp 35; tissue; extracted metabolites, isotretinoin, tretinoin]

Guiso, G.; Rambaldi, A.; Dimitrova, B.; Biondi, A.; Caccia, S. Determination of orally administered all-trans-retinoic acid in human plasma by high-performance liquid chromatography. *J.Chromatogr.B*, **1994**, *656*, 239–244 [all-trans-retinyl acetate (IS); extracted metabolites, isotretinoin, tretinoin; LOD 10 ng/mL; pharmacokinetics; non-interfering allopurinol, amikacin, aracytin, ceftazidime, ciprofloxacin, doxorubicin, fluconazole, prednisone]

Jiang, X.G.; Xi, N.Z. [A reversed-phase HPLC method for determining tretinoin]. *Chung Kuo Yao Li Hsueh Pao*, **1994**, *15*, 458–461

Sass, J.O.; Nau, H. Single-run analysis of isomers of retinoyl-β-D-glucuronide and retinoic acid by reversed-phase high-performance liquid chromatography. *J.Chromatogr.A*, **1994**, *685*, 182–188 [LOD 0.25 ng; column temp 60; gradient; simultaneous metabolites, isotretinoin, tretinoin]

Ranalder, U.B.; Lausecker, B.B.; Huselton, C. Micro liquid chromatography-mass spectrometry with direct liquid introduction used for separation and quantitation of all-*trans*- and 13-*cis*-retinoic acids and their 4-oxo metabolites in human plasma. *J.Chromatogr.*, **1993**, *617*, 129–135 [LC-MS; LOQ 0.3 ng/mL; extracted metabolites, isotretinoin, tretinoin]

Tan, X.; Meltzer, N.; Lindenbaum, S. Solid-state stability studies of 13-*cis*-retinoic acid and all-*trans*-retinoic acid using microcalorimetry and HPLC analysis. *Pharm.Res.*, **1992**, *9*, 1203–1208 [solutions; simultaneous degradation products, isotretinoin, tretinoin]

Bryan, P.D.; Honigberg, I.L.; Meltzer, N.M. Electrochemical detection of retinoids using normal phase HPLC. *J.Liq.Chromatogr.*, **1991**, *14*, 2287–2295 [LOD 1 ng; also acitretin, isotretinoin, tretinoin, vitamin A palmitate]

Dobie, A.K.; Yang, K.Y.; De, N.C. High-performance liquid chromatographic procedure for retinoic acid in ophthalmic solution. *J.Liq.Chromatogr.*, **1991**, *14*, 1219–1226

Creech Kraft, J.; Echoff, C.; Kuhnz, W.; Löfberg, B.; Nau, H. Automated determination of 13-cis- and all-trans-retinoic acid, their 4-oxo- metabolites and retinol in plasma, amniotic fluid and embryo by reversed-phase high-performance liquid chromatography with a precolumn switching technique. *J.Liq.Chromatogr.*, **1988**, *11*, 2051–2069 [gradient; LOD 2 ng/mL; mouse; extracted metabolites, isotretinoin, tretinoin, vitamin A]

Wyss, R.; Bucheli, F. Quantitative analysis of retinoids in biological fluids by high-performance liquid chromatography using column switching. I. Determination of isotretinoin and tretinoin and their 4-oxo metabolites in plasma. *J.Chromatogr.*, **1988**, *424*, 303–314 [plasma; LOQ 2 ng/mL; extracted metabolites, isotretinoin, tretinoin; etretin (IS)]

Furr, H.C.; Amédée-Manesme, O.; Olson, J.A. Gradient reversed-phase high-performance liquid chromatographic separation of naturally occurring retinoids. *J.Chromatogr.*, **1984**, *309*, 299–307 [gradient; rat; human; pig; liver; kidney; extracted retinyl esters, tretinoin, vitamin A]

Shelley, R.; Price, J.C.; Jun, H.W.; Cadwallader, D.E.; Capomacchia, A.C. Improved and rapid high-performance liquid chromatographic assay for 13-cis-retinoic acid or all-trans-retinoic acid. *J.Pharm.Sci.*, **1982**, *71*, 262–264 [rat; serum; extracted isotretinoin, retinol acetate, tretinoin, vitamin A; pharmacokinetics; LOQ 100 ng/mL]

Vane, F.M.; Stoltenborg, J.K.; Buggé, C.J. Determination of 13-cis-retinoic acid and its major metabolite, 4-oxo-13-cis-retinoic acid, in human blood by reversed-phase high-performance liquid chromatography. *J.Chromatogr.*, **1982**, *227*, 471–484 [gradient; whole blood; extracted metabolites, isotretinoin, tretinoin, vitamin A; LOQ 10 ng/mL; pharmacokinetics]

# Salmeterol

**Molecular formula:** $C_{25}H_{37}NO_4$
**Molecular weight:** 415.6
**CAS Registry No.:** 89365-50-4

## SAMPLE

**Matrix:** blood
**Sample preparation:** Condition a 100 mg C2 SPE cartridge with 1 mL MeOH and two 1 mL aliquots of water, do not allow to dry. Add 1 mL plasma to the SPE cartridge, wash with 1 mL water, allow to dry, wash with 1 mL MeCN, allow to dry, elute with 1.5 mL isopropanol. Evaporate the eluate to dryness under reduced pressure at 45°, reconstitute the residue in 100 μL mobile phase, sonicate for 5 min, vortex, inject a 75 μL aliquot.

## HPLC VARIABLES

**Column:** 150 × 4.6 5 μm PLRP-S poly(styrene-divinylbenzene) (Polymer Laboratories)
**Mobile phase:** MeCN:MeOH:water:1 M ammonium sulfate 5:60:30:5, adjusted to pH 4.0 with sulfuric acid
**Column temperature:** 40
**Flow rate:** 1
**Injection volume:** 75
**Detector:** F ex 230 em 305

## CHROMATOGRAM

**Retention time:** 7-8
**Limit of quantitation:** 1 ng/mL

## OTHER SUBSTANCES

**Extracted:** metabolites

## KEY WORDS

rat; dog; plasma; SPE; pharmacokinetics

## REFERENCE

Colthup, P.V.; Young, G.C.; Felgate, C.C. Determination of salmeterol in rat and dog plasma by high-performance liquid chromatography with fluorescence detection. *J.Pharm.Sci.*, **1993**, *82*, 323–325

# Scopolamine

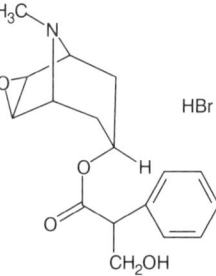

**Molecular formula:** $C_{17}H_{21}NO_4$

**Molecular weight:** 303.4

**CAS Registry No.:** 51-34-3 (scopolamine), 6533-68-2 (scopolamine hydrobromide trihydrate), 114-49-8 (scopolamine hydrobromide anhydrous)

## SAMPLE

**Matrix:** blood

**Sample preparation:** 1 mL Plasma + 100 μL MeOH, vortex briefly, add 50 μL 1 M ammonium hydroxide, mix, add 5 mL dichloromethane, shake horizontally for 5 min, centrifuge at 2500 rpm for 5 min. Remove the organic layer and evaporate it to dryness under a stream of nitrogen at 40°, reconstitute the residue in 100 μL mobile phase, inject a 20 μL aliquot.

## HPLC VARIABLES

**Guard column:** 10 × 2 5 μm BDS C18 (Keystone)

**Column:** 50 × 3 3 μm BDS C18 (Keystone)

**Mobile phase:** MeCN:MeOH:10 mM ammonium acetate 62.5:37.5:15

**Flow rate:** 0.5

**Injection volume:** 20

**Detector:** MS, Perkin Elmer Sciex API III-Plus triple quadrupole, APCI, nebulizer 400° and 80 psi, auxiliary nitrogen 1.2 L/min, curtain gas 1.2 L/min, interface 55°, collision gas argon, electron multiplier 3000 V, declustering potential 35 V, collision energy 35 eV

## CHROMATOGRAM

**Retention time:** 0.8

**Internal standard:** scopolamine

## OTHER SUBSTANCES

**Extracted:** hyoscyamine

## KEY WORDS

plasma; protect from light; scopolamine is IS

## REFERENCE

Xu, A.; Havel, J.; Linderholm, K.; Hulse, J. Development and validation of an LC/MS/MS method for the determination of L-hyoscyamine in human plasma. *J.Pharm.Biomed.Anal.*, **1996**, *14*, 33–42

## SAMPLE

**Matrix:** bulk, plants

**Sample preparation:** Place 0.5 g powdered crude drug in 25 mL mobile phase, reflux 30 min, cool, centrifuge at 1600 g, decant, wash residue twice with 10 mL portions of mobile phase, combine extracts and washings, make up to 50 mL with mobile phase, inject 10 μL aliquot.

## HPLC VARIABLES

**Column:** 150 × 4 5 μm TSK gel 120A ODS

**Mobile phase:** MeCN:67 mM pH 2.5 phosphate buffer 35:65 containing 17.5 mM sodium dodecylsulfate

**Column temperature:** 35

**Flow rate:** 1.5
**Injection volume:** 10
**Detector:** UV 210

## CHROMATOGRAM
**Retention time:** 10

## OTHER SUBSTANCES
**Simultaneous:** hyoscyamine

## REFERENCE
Oshima, T.; Sagara, K.; Tong, Y.Y.; Zhang, G.; Chen, Y.H. Application of ion-pair high performance liquid chromatography for analysis of hyoscyamine and scopolamine in solanaceous crude drugs. *Chem.Pharm.Bull.*, **1989**, *37*, 2456–2458

## SAMPLE
**Matrix:** formulations
**Sample preparation:** Tablets. Crush tablet, add 2 mL water, add 500 µL 0.1 M NaOH, if not basic to litmus add 0.1 M NaOH dropwise until it is. Extract with 10 mL chloroform then with three 5 mL portions of chloroform. Pass all extracts through a sodium sulfate column and then wash column with 2 mL chloroform. Evaporate all organic extracts with heating under a stream of air. Take up residue in 1 mL MeOH, inject a 100 µL aliquot. Injections. Evaporate 2 mL to dryness on a steam bath, take up residue in 5 mL water, add 500 µL 0.1 M NaOH, if not basic to litmus add 0.1 M NaOH dropwise until it is. Extract with 10 mL chloroform then with three 5 mL portions of chloroform. Pass all extracts through a sodium sulfate column and then wash column with 2 mL chloroform. Evaporate all organic extracts with heating under a stream of air. Take up residue in 2 mL MeOH, inject a 100 µL aliquot. (Sodium sulfate column was a 300 × 20 glass chromatography column containing 10 g anhydrous sodium sulfate washed with 10 mL chloroform before use.)

## HPLC VARIABLES
**Column:** 250 × 4.6 Alltech C18
**Mobile phase:** MeOH:20 g/L sodium pentanesulfonate 95:5
**Flow rate:** 1
**Injection volume:** 100
**Detector:** F ex 255 em 285

## CHROMATOGRAM
**Retention time:** 3.5

## OTHER SUBSTANCES
**Simultaneous:** atropine
**Noninterfering:** ergotamine, phenobarbital

## KEY WORDS
tablets; injections

## REFERENCE
Cieri, U.R. Determination of small quantities of atropine in commercial preparations by liquid chromatography with fluorescence detection. *J.Assoc.Off.Anal.Chem.*, **1985**, *68*, 1042–1045

## SAMPLE
**Matrix:** formulations
**Sample preparation:** Tablets, capsules. Powder tablets or remove contents of capsules, weigh out amount equivalent to about 600 µg hyoscyamine sulfate-atropine sulfate, add

25 mL 25 mM sulfuric acid, shake for 15 min, centrifuge at 3000 rpm for 5 min. Remove 5 mL of the supernatant and extract it twice with 30 mL portions of dichloromethane, discard the organic phase, add 2 mL buffer to the aqueous phase, extract with four 30 mL portions of dichloromethane, filter extracts through dichloromethane-rinsed glass wool, add 3 mL 2.25 μg/mL theophylline in dichloromethane, distill off the dichloromethane through a Snyder column by using a steam bath. When the volume reaches 10 mL, rinse the column with 1-2 mL dichloromethane, continue distillation to 0.5-1 mL, remove the column and rinse the concentrator tube-column junction with 1 mL dichloromethane, evaporate to 1 mL with a stream of air at 40°, add 100 μL 1% concentrated HCl in MeOH, mix, evaporate to dryness with a stream of air at 40°, rinse the sides of the concentrator tube with 500 μL MeOH, evaporate to dryness with a stream of air at 40°, dissolve the residue in 300 μL water, inject a 20 μL aliquot. Elixirs. Add an amount equivalent to about 600 μg hyoscyamine sulfate-atropine sulfate to a 150 mL beaker, warm at 40° with a current of air for 30 min to remove alcohol, cool, make up to 25 mL with water, remove 5 mL of this solution, add 2 mL 100 mM sulfuric acid, extract twice with 30 mL portions of dichloromethane, discard the organic phase, add 2 mL buffer to the aqueous phase, extract with four 30 mL portions of dichloromethane, filter extracts through dichloromethane-rinsed glass wool, add 3 mL 2.25 μg/mL theophylline in dichloromethane, distill off the dichloromethane through a Snyder column by using a steam bath. When the volume reaches 10 mL, rinse the column with 1-2 mL dichloromethane, continue distillation to 0.5-1 mL, remove the column and rinse the concentrator tube-column junction with 1 mL dichloromethane, evaporate to 1 mL with a stream of air at 40°, add 100 μL 1% concentrated HCl in MeOH, mix, evaporate to dryness with a stream of air at 40°, rinse the sides of the concentrator tube with 500 μL MeOH, evaporate to dryness with a stream of air at 40°, dissolve the residue in 300 μL water, inject a 20 μL aliquot. (Buffer was 5.3 g anhydrous sodium carbonate and 4.2 g sodium bicarbonate in 100 mL water, pH 9.4. Pass dichloromethane through 75 g basic aluminum oxide, Brockmann Activity Grade 1, store over 25 g alumina/4 L.)

## HPLC VARIABLES
**Column:** 250 × 4 5 μm Spherisorb ODS
**Mobile phase:** MeOH:buffer 250:525 (The 50 mM tetramethylammonium phosphate buffer was prepared from 500 mL water + 23 mL 20% tetramethylammonium hydroxide in MeOH + 10 mL concentrated phosphoric acid, adjust to pH 2.0 with concentrated phosphoric acid, make up to 1 L with water.
**Flow rate:** 0.8
**Injection volume:** 20
**Detector:** UV 220

## CHROMATOGRAM
**Retention time:** 5
**Internal standard:** theophylline (6.5)

## OTHER SUBSTANCES
**Simultaneous:** atropine, hyoscyamine, phenobarbital

## KEY WORDS
tablets; capsules; elixirs

## REFERENCE
Pennington, L.J.; Schmidt, W.F. Belladonna alkaloids and phenobarbital combination pharmaceuticals analysis. I: High-performance liquid chromatographic determinations of hyoscyamine-atropine and scopolamine. *J.Pharm.Sci.*, **1982**, *71*, 951–953

## SAMPLE
**Matrix:** plants

**Sample preparation:** 100 mg Freeze-dried powdered plant leaves + 10 mL mobile phase, heat at 40° for 15 min, filter, inject a 20 μL aliquot.

## HPLC VARIABLES
**Column:** 150 × 4 4 μm Novapack C18
**Mobile phase:** MeCN:water 12.5:87.5 with 0.3% phosphoric acid adjusted to pH 2.2 with triethylamine
**Flow rate:** 0.8
**Injection volume:** 20
**Detector:** UV 204

## CHROMATOGRAM
**Retention time:** 5.35
**Limit of detection:** 50 μg/g

## OTHER SUBSTANCES
**Simultaneous:** hyoscyamine, tropic acid

## KEY WORDS
freeze-dried plant leaves; leaves

## REFERENCE
Fliniaux, M.-A.; Manceau, F.; Jacquin-Dubreuil, A. Simultaneous analysis of l-hyoscyamine, l-scopolamine and dl- tropic acid in plant material by reversed phase high-performance liquid chromatography. *J.Chromatogr.*, **1993**, *644*, 193–197

## SAMPLE
**Matrix:** plants
**Sample preparation:** Extract 0.1 g dry plant material with 10 mL MeOH for 10 min under reflux, filter, inject aliquot.

## HPLC VARIABLES
**Guard column:** 40 × 4 10 μm Hypersil ODS
**Column:** 250 × 4 10 μm Hypersil ODS
**Mobile phase:** MeOH:water 45:55 containing 0.1% phosphoric acid adjusted to pH 7 with triethylamine
**Flow rate:** 1
**Detector:** UV 229

## CHROMATOGRAM
**Retention time:** 9.3

## OTHER SUBSTANCES
**Simultaneous:** hyoscyamine

## REFERENCE
Hagemann, K.; Piek, K.; Stöckigt, J.; Weiler, E.W. Monoclonal antibody-based enzyme immunoassay for the quantitative determination of the tropane alkaloid, scopolamine. *Planta Med.*, **1992**, *58*, 68–72

## SAMPLE
**Matrix:** plants
**Sample preparation:** Dissolve alkaloids in 1 mL MeOH, add 40 ng homatropine, inject aliquot.

## HPLC VARIABLES
**Column:** 150 × 4.1 5 μm Hamilton PRP-1

**Mobile phase:** MeCN:100 mM pH 10.4 ammonium acetate 30:70
**Flow rate:** 1
**Injection volume:** 20
**Detector:** MS thermospray, VG Trio-2, ion source 150°, vaporizer tip 170°, repeller electrode 150 V, m/z 304

## CHROMATOGRAM
**Internal standard:** homatropine (m/z 276)
**Limit of detection:** 2.5 ng/mL

## OTHER SUBSTANCES
**Simultaneous:** hyoscyamine

## REFERENCE
Auriola, S.; Martinsen, A.; Oksman-Caldentey, K.M.; Naaranlahti, T. Analysis of tropane alkaloids with thermospray high-performance liquid chromatography-mass spectrometry. *J.Chromatogr.*, **1991**, *562*, 737−744

## SAMPLE
**Matrix:** solutions

## HPLC VARIABLES
**Column:** 150 × 2 PRP-1 (Keystone)
**Mobile phase:** MeCN:buffer 34:66 (Buffer was 80 mM phosphate and 20 mM borate, pH 9.0.)
**Flow rate:** 0.15
**Injection volume:** 1
**Detector:** Chemiluminescence following post-column reaction. A 1 mM solution of Ru(2,2'-bipyridine)$_3^{++}$ in 50 mM sodium sulfate (continuously sparged with helium) was oxidized to Ru(2,2'-bipyridine)$_3^{+++}$ using a Princeton Applied Research Model 174A polarographic analyzer with a platinum gauze working electrode, a platinum wire auxiliary electrode, and a silver wire reference electrode. The Ru solution pumped at 0.3 mL/min mixed with the column effluent in the flow cell of the detector, a fluorescence detector with the light source removed.

## CHROMATOGRAM
**Retention time:** 4.5
**Limit of detection:** 0.1-1 µg/mL

## OTHER SUBSTANCES
**Simultaneous:** atropine

## KEY WORDS
post-column reaction

## REFERENCE
Holeman, J.A.; Danielson, N.D. Microbore liquid chromatography of tertiary amine anticholinergic pharmaceuticals with tris(2,2'-bipyridine)ruthenium(III) chemiluminescence detection. *J.Chromatogr. Sci.*, **1995**, *33*, 297−302

## SAMPLE
**Matrix:** solutions

## HPLC VARIABLES
**Column:** 250 × 4.6 5 µm Supelcosil LC-DP (A) or 250 × 4 5 µm LiChrospher 100 RP-8 (B)

**Mobile phase:** MeCN:0.025% phosphoric acid:buffer 25:10:5 (A) or 60:25:15 (B) (Buffer was 9 mL concentrated phosphoric acid and 10 mL triethylamine in 900 mL water, adjust pH to 3.4 with dilute phosphoric acid, make up to 1 L.)
**Flow rate:** 0.6
**Injection volume:** 25
**Detector:** UV 229

## CHROMATOGRAM
**Retention time:** 6.95 (A), 3.70 (B)

## OTHER SUBSTANCES
**Also analyzed:** acebutolol, acepromazine, acetaminophen, acetazolamide, acetophenazine, albuterol, alprazolam, amitriptyline, amobarbital, amoxapine, antipyrine, atenolol, atropine, azatadine, baclofen, benzocaine, bromocriptine, brompheniramine, brotizolam, bupivacaine, buspirone, butabarbital, butalbital, caffeine, carbamazepine, cetirizine, chlorcyclizine, chlordiazepoxide, chlormezanone, chloroquine, chlorpheniramine, chlorpromazine, chlorpropamide, chlorprothixene, chlorthalidone, chlorzoxazone, cimetidine, cisapride, clomipramine, clonazepam, clonidine, clozapine, cocaine, codeine, colchicine, cyclizine, cyclobenzaprine, dantrolene, desipramine, diazepam, diclofenac, diflunisal, diltiazem, diphenhydramine, diphenidol, diphenoxylate, dipyridamole, disopyramide, dobutamine, doxapram, doxepin, droperidol, encainide, ethidium bromide, ethopropazine, fenoprofen, fentanyl, flavoxate, fluoxetine, fluphenazine, flurazepam, flurbiprofen, fluvoxamine, furosemide, glutethimide, glyburide, guaifenesin, haloperidol, homatropine, hydralazine, hydrochlorothiazide, hydrocodone, hydromorphone, hydroxychloroquine, hydroxyzine, ibuprofen, imipramine, indomethacin, ketoconazole, ketoprofen, ketorolac, labetalol, levorphanol, lidocaine, loratadine, lorazepam, lovastatin, loxapine, mazindol, mefenamic acid, meperidine, mephenytoin, mepivacaine, mesoridazine, metaproterenol, metformin, methadone, methdilazine, methocarbamol, methotrexate, methotrimeprazine, methoxamine, methyldopa, methylphenidate, metoclopramide, metolazone, metoprolol, metronidazole, midazolam, moclobemide, morphine, nadolol, nalbuphine, naloxone, naphazoline, naproxen, nifedipine, nizatidine, norepinephrine, nortriptyline, oxazepam, oxycodone, oxymetazoline, paroxetine, pemoline, pentazocine, pentobarbital, pentoxifylline, perphenazine, pheniramine, phenobarbital, phenol, phenolphthalein, phentolamine, phenylbutazone, phenyltoloxamine, phenytoin, pimozide, pindolol, piroxicam, pramoxine, prazepam, prazosin, probenecid, procainamide, procaine, prochlorperazine, procyclidine, promazine, promethazine, propafenone, propantheline, propiomazine, propofol, propranolol, protriptyline, quazepam, quinidine, quinine, racemethorphan, ranitidine, remoxipride, risperidone, salicylic acid, secobarbital, sertraline, sotalol, spironolactone, sulfinpyrazone, sulindac, temazepam, terbutaline, terfenadine, tetracaine, theophylline, thiethylperazine, thiopental, thioridazine, thiothixene, timolol, tocainide, tolbutamide, tolmetin, trazodone, triamterene, triazolam, trifluoperazine, triflupromazine, trimeprazine, trimethoprim, trimipramine, verapamil, warfarin, xylometazoline, yohimbine, zopiclone

## KEY WORDS
some details of plasma extraction

## REFERENCE
Koves, E.M. Use of high-performance liquid chromatography-diode array detection in forensic toxicology. *J.Chromatogr.A*, **1995**, *692*, 103–119

## SAMPLE
**Matrix:** solutions
**Sample preparation:** Centrifuge.

## HPLC VARIABLES
**Column:** 250 × 4.6 5 µm Spherisorb S.50 DS1

**Mobile phase:** MeCN:buffer 50:50 (Buffer was 10 mM $KH_2PO_4$ + 0.0028% heptanesulfonic acid, pH 2.0.)
**Flow rate:** 1.5
**Injection volume:** 20
**Detector:** UV 274

## REFERENCE
Calpena, A.C.; Blanes, C.; Moreno, J.; Obach, R.; Domenech, J. A comparative in vitro study of transdermal absorption of antiemetics. *J.Pharm.Sci.*, **1994**, *83*, 29–33

## SAMPLE
**Matrix:** solutions

## HPLC VARIABLES
**Column:** 250 × 4.6 Zorbax RX
**Mobile phase:** Gradient. A was 10 mL concentrated orthophosphoric acid and 7 mL triethylamine in 1 L water. B was 10 mL concentrated orthophosphoric acid and 7 mL triethylamine in 200 mL water, make up to 1 L with MeCN. A:B from 100:0 to 0:100 over 30 min, maintain at 0:100 for 5 min.
**Column temperature:** 30
**Flow rate:** 2
**Detector:** UV 210

## OTHER SUBSTANCES
**Also analyzed:** acepromazine, acetaminophen, acetophenazine, albuterol, aminophylline, amitriptyline, amobarbital, amoxapine, amphetamine, amylocaine, antipyrine, aprobarbital, aspirin, atenolol, atropine, avermectin, barbital, benzocaine, benzoic acid, benzotropine, benzphetamine, berberine, bibucaine, bromazepan, brompheniramine, buprenorphine, buspirone, butabarbital, butacaine, butethal, caffeine, carbamazepine, carbromal, chloramphenicol, chlordiazepoxide, chloroquine, chlorothiazide, chloroxylenol, chlorphenesin, chlorpheniramine, chlorpromazine, chlorpropamide, chlortetracycline, cimetidine, cinchonidine, cinchonine, clenbuterol, clonazepam, clonixin, clorazepate, cocaine, codeine, colchicine, cortisone, coumarin, cyclazocine, cyclobenzaprine, cyclothiazide, cyheptamide, cymarin, danazol, danthron, dapsone, debrisoquine, desipramine, dexamethasone, dextromethorphan, dextropropoxyphene, diamorphine, diazepam, diclofenac, diethylpropion, diethylstilbestrol, diflunisal, digitoxin, digoxin, diltiazem, diphenhydramine, diphenoxylate, diprenorphine, dipyrone, disulfiram, dopamine, doxapram, doxepin, dronabinol, ephedrine, epinephrine, epinine, estradiol, estriol, estrone, ethacrynic acid, ethosuximide, etonitazene, etorphine, eugenol, famotidine, fenbendazole, fencamfamine, fenoprofen, fenproporex, fentanyl, flubendazole, flufenamic acid, flunitrazepam, 5-fluorouracil, fluoxymesterone, fluphenazine, furosemide, gentisic acid, gitoxigenin, glipizide, glunixin, glutethimide, glybenclamide, guaiacol, halazepam, haloperidol, hydrochlorothiazide, hydrocodone, hydrocortisone, hydromorphone, hydroxyquinoline, ibogaine, ibuprofen, iminostilbene, imipramine, indomethacin, isocarbostyril, isocarboxazid, isoniazid, isoproterenol, isoxsuprine, ivermectin, ketamine, ketoprofen, kynurenic acid, levorphanol, lidocaine, lorazepam, lormetazepam, loxapine, mazindol, mebendazole, meclizine, meclofenamic acid, medazepam, mefenamic acid, megestrol, mepacrine, meperidine, mephentermine, mephenytoin, mephesin, mephobarbital, mepivacaine, mescaline, mesoridazine, methadone, methamphetamine, methapyrilene, methaqualone, methazolamide, methocarbamol, methoxamine, methsuximide, methyl salicylate, methyldopa, methyldopamine, methylphenidate, methylprednisolone, methyltestosterone, methyprylon, metoprolol, mibolerone, morphine, nadolol, nalorphine, naloxone, naltrexone, naphazoline, naproxen, nefopam, niacinamide, nicotine, nicotinic acid, nifedipine, niflumic acid, nitrazepam, norepinephrine, nortriptyline, noscapine, nylidrin, oxazepam, oxycodone, oxymorphone, oxyphenbutazone, oxytetracycline, papaverine, pargyline, pemoline, pentazocine, pentobarbital, persantine, phenacetin, phenazocine, phenazopyridine, phencyclidine, phendimetrazine, phenelzine, pheniramine, phenobarbital, phenothiazine, phensuximide, phentermine, phenylbutazone, phenylephrine, phenylpropanolamine, piperocaine, praze-

pam, prednisolone, primidone, probenecid, progesterone, propiomazine, propranolol, propylparaben, pseudoephedrine, puromycin, pyrilamine, pyrithyldione, quazepam, quinaldic
acid, quinidine, quinine, ranitidine, recinnamine, reserpine, resorcinol, saccharin, albuterol, salicylamide, salicylic acid, scopoletin, secobarbital, strychnine, sulfacetamide, sufadiazine, sulfadimethoxine, sulfaethidole, sulfamerazine, sulfamethazine, sulfamethoxizole, sulfanilamide, sulfapyridine, sulfasoxizole, sulindac, tamoxifen, temazepam,
testosterone, tetracaine, tetracycline, tetramisole, thebaine, theobromine, theophylline,
thiabendazole, thiamine, thiamylal, thiobarbituric acid, thioridazine, thiosalicylic acid,
thiothixene, thymol, tolazamide, tolazoline, tobutamide, tolmetin, tranylcypromine, triamcinolone, tribenzylamine, trichloromethiazide, trifluoperazine, trihexyphenidyl, trimethoprim, tripelennamine, triprolidine, tropacocaine, tyramine, verapamil, vincamine,
warfarin, yohimbine, zoxazolamine

## REFERENCE

Hill, D.W.; Kind, A.J. Reversed-phase solvent-gradient HPLC retention indexes of drugs. *J.Anal.Toxicol.*,
**1994**, *18*, 233–242

## SAMPLE

**Matrix:** solutions
**Sample preparation:** Dissolve in MeOH or water to 0.1%.

## HPLC VARIABLES

**Column:** two 250 mm β-cyclodextrin bonded phase columns in series (Advanced Separation
Technologies)
**Mobile phase:** MeCN:1% pH 4.1 aqueous triethylammonium acetate 4:96
**Flow rate:** 0.5
**Injection volume:** 1
**Detector:** UV

## CHROMATOGRAM

**Retention time:** 31 (d-isomer), 34 (l-isomer)

## OTHER SUBSTANCES

**Simultaneous:** atropine, cocaine

## KEY WORDS

chiral

## REFERENCE

Armstrong, D.W.; Han, S.M.; Han, Y.I. Separation of optical isomers of scopolamine, cocaine, homatropine, and atropine. *Anal.Biochem.*, **1987**, *167*, 261–264

## SAMPLE

**Matrix:** solutions
**Sample preparation:** Dissolve in MeOH:water 1:1 at a concentration of 50 μg/mL, inject
a 10 μL aliquot.

## HPLC VARIABLES

**Column:** 300 × 3.9 10 μm μBondapak C18
**Mobile phase:** MeOH:acetic acid:triethylamine:water 15:1.5:0.5:83
**Flow rate:** 1.5
**Injection volume:** 10
**Detector:** UV 230

## CHROMATOGRAM

**Retention time:** 8

## OTHER SUBSTANCES
**Simultaneous:** atropine, atropine methyl, homatropine, methscopolamine, tropic acid

## REFERENCE
Roos, R.W.; Lau-Cam, C.A. General reversed-phase high-performance liquid chromatographic method for the separation of drugs using triethylamine as a competing base. *J.Chromatogr.*, **1986**, *370*, 403–418

## ANNOTATED BIBLIOGRAPHY
Theodoridis, G.; Papadoyannis, I.; Vasilikiotis, G.; Tsoukali-Papadopoulou, H. Reversed-phase high-performance liquid chromatography–photodiode-array analysis of alkaloid drugs of forensic interest. *J.Chromatogr.B*, **1995**, *668*, 253–263 [also amphetamine, bamifylline, caffeine, cocaine, codeine, diamorphine, ethylmorphine, flufenamic acid, hyoscyamine, methadone, morphine, nalorphine, norcodeine, papaverine, quinine, strychnine, theobromine, theophylline, tolfenamic acid]

Whelpton, R.; Hurst, P.R.; Metcalfe, R.F.; Saunders, S.A. Liquid chromatographic determination of hyoscine (scopolamine) in urine using solid phase extraction. *Biomed.Chromatogr.*, **1992**, *6*, 198–204

Achari, R.G.; Jacob, J.T. A study of the retention behavior of some basic drug substances by ion-pair HPLC. *J.Liq.Chromatogr.*, **1980**, *3*, 81–92 [also N-acetylprocainamide, antazoline, atropine, caffeine, chlorpheniramine, codeine, ephedrine, epinephrine, naphazoline, papaverine, pheniramine, phenylephrine, phenylpropanolamine, procainamide, quinidine, xylocaine]

# Sertraline

**Molecular formula:** $C_{17}H_{17}Cl_2N$
**Molecular weight:** 306.2
**CAS Registry No.:** 79617-96-2 (sertraline), 79559-97-0 (sertraline hydrochloride)

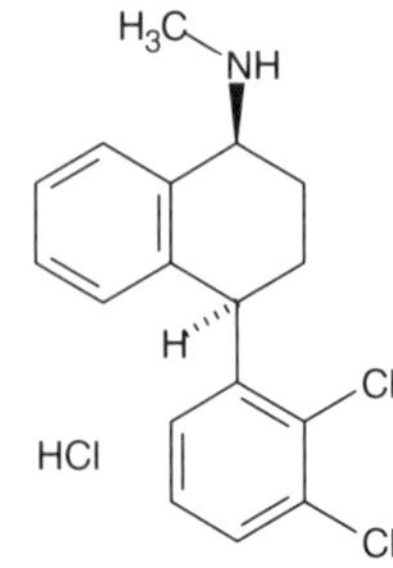


## SAMPLE
**Matrix:** blood
**Sample preparation:** Condition a 1 mL Bond Elut C18 SPE cartridge with 1 M HCl, twice with MeOH, and once with water. 500 μL Serum + 200 μL 500 ng/mL IS in 10 mg/mL potassium bicarbonate + 500 μL MeCN, centrifuge, add to the SPE cartridge, wash twice with water, wash once with MeCN, elute (by gravity) with 250 μL MeOH containing 25 mL/L 35% perchloric acid, inject a 20 μL aliquot of the eluate.

## HPLC VARIABLES
**Column:** 150 × 4.6 5 μm Ultrasphere ODS
**Mobile phase:** MeCN:water:7% perchloric acid 400:750:0.5 containing 0.5 g tetramethylammonium perchlorate
**Flow rate:** 1.8
**Injection volume:** 20
**Detector:** UV 205

## CHROMATOGRAM
**Retention time:** 11
**Internal standard:** CP-53,630-1 (8)
**Limit of quantitation:** 10 ng/mL

## OTHER SUBSTANCES
**Simultaneous:** desmethylsertraline

## KEY WORDS
serum; SPE

## REFERENCE

Gupta, R.N.; Dziurdzy, S.A. Therapeutic monitoring of sertraline. *Clin.Chem.*, **1994**, *40*, 498–499


## SAMPLE
**Matrix:** blood
**Sample preparation:** 1 mL Blood + 100 μL 1 μg/mL metycaine + 1 mL pH 9 saturated potassium borate buffer, mix, add 5 mL butyl chloride, extract. Remove the organic layer and add it to 1 mL 100 mM sulfuric acid, extract. Remove the aqueous layer and basify it with concentrated ammonium hydroxide, add 50 μL chloroform, extract. Remove the chloroform layer, evaporate to dryness under air at 60°, reconstitute in 100 μL MeOH, inject a 20 μL aliquot.

## HPLC VARIABLES
**Column:** Lichrospher RP-8
**Mobile phase:** MeCN:50 mM pH 3 phosphate buffer 45:55
**Flow rate:** 1.5

**Injection volume:** 20
**Detector:** UV 230

## CHROMATOGRAM
**Retention time:** 6.3
**Internal standard:** metycaine (4)
**Limit of quantitation:** 50 ng/mL

## OTHER SUBSTANCES
**Simultaneous:** desmethylsertraline, diphenhydramine

## REFERENCE
Logan, B.K.; Friel, P.N.; Case, G.A. Analysis of sertraline (Zoloft) and its major metabolite in postmortem specimens by gas and liquid chromatography. *J.Anal.Toxicol.*, **1994**, *18*, 139–142

## SAMPLE
**Matrix:** blood
**Sample preparation:** Condition a 200 mg 10 mL Clean Screen copolymeric ZSDAU020 SPE cartridge with 3 mL MeOH, 3 mL water, and 1 mL 100 mM pH 6 $KH_2PO_4$. 1 mL Serum + 2 mL 100 mM pH 6 $KH_2PO_4$ + 250 µL 2 µg/mL protriptyline, add to the SPE cartridge, wash with 1 mL water, wash with 1 mL 1 M acetic acid, wash with 3 mL MeOH, dry under vacuum for 5 min, elute with 3 mL dichloromethane:isopropanol:ammonium hydroxide 78:20:2. Add a drop of diethylamine to the eluate and evaporate it to dryness at 40° under a stream of nitrogen. Reconstitute the residue in 100 µL mobile phase, inject a 50 µL aliquot.

## HPLC VARIABLES
**Column:** 150 × 4.6 5 µm Supelco LC-8
**Mobile phase:** MeCN:MeOH:water:diethylamine 1000:800:1200:2, containing 1.5 g sodium pentanesulfonate, pH adjusted to 5.5-6.0 with phosphoric acid
**Flow rate:** 1.5
**Injection volume:** 50
**Detector:** UV 600 for 2 min then UV 214

## CHROMATOGRAM
**Retention time:** 9
**Internal standard:** protriptyline (6)
**Limit of detection:** 10 ng/mL

## OTHER SUBSTANCES
**Simultaneous:** desmethylsertraline

## KEY WORDS
serum; SPE

## REFERENCE
Rogowsky, D.; Marr, M.; Long, G.; Moore, C. Determination of sertraline and desmethylsertraline in human serum using copolymeric bonded-phase extraction, liquid chromatography and gas chromatography-mass spectrometry. *J.Chromatogr.B*, **1994**, *655*, 138–141

## SAMPLE
**Matrix:** blood
**Sample preparation:** 1 mL Plasma + 1 mL 0.6 M pH 9.8 carbonate buffer + 40 µL 5 µg/mL maprotiline in 10 mM HCl + 5 mL 200 g/L ethyl acetate in n-heptane, mix by rocking for 10 min, centrifuge at 1500 g for 10 min. Remove organic layer and add it to 150 µL 100 mM HCl, mix 10 min, centrifuge at 1500 g for 10 min. Discard organic layer

and evaporate aqueous layer at 45° in a vacuum centrifuge for 1 h. Take up residue in 50 μL 1 M pH 10.3 carbonate buffer and 25 μL 10 mg/mL dansyl chloride in MeCN, vortex, allow to react at room temperature for 45 min, evaporate at 45° in a vacuum centrifuge for 20 min, reconstitute in 125 μL MeCN:water 75:25, vortex, centrifuge for 3-5 min, inject a 25-40 μL aliquot.

## HPLC VARIABLES
**Column:** 250 × 4.6 5 μm Supelcosil LC-18
**Mobile phase:** MeCN:25 mM KH$_2$PO$_4$ 75:25 containing 500 μL/L orthophosphoric acid and 600 μL/L n-butylamine
**Flow rate:** 2
**Injection volume:** 25-40
**Detector:** F ex 235 em 470 (cut-off)

## CHROMATOGRAM
**Retention time:** 25.60
**Internal standard:** maprotiline (12.8)

## OTHER SUBSTANCES
**Simultaneous:** amoxapine, clovoxamine, desipramine, fenfluramine, fluoxetine, fluvoxamine, norfluoxetine, nortriptyline, propranolol, protriptyline
**Noninterfering:** amitriptyline, atenolol, bupropion, carbamazepine, chlordiazepoxide, citalopram, clomipramine, clozapine, cyclobenzaprine, doxepin, imipramine, loxapine, metoprolol, mianserin, moclobemide, nomifensine, pindolol, thioridazine, tranylcypromine, trazodone, trimipramine

## KEY WORDS
plasma

## REFERENCE
Suckow, R.F.; Zhang, M.F.; Cooper, T.B. Sensitive and selective liquid-chromatographic assay of fluoxetine and norfluoxetine in plasma with fluorescence detection after precolumn derivatization. *Clin.Chem.*, **1992**, *38*, 1756–1761

## SAMPLE
**Matrix:** solutions

## HPLC VARIABLES
**Column:** 250 × 4.6 5 μm Supelcosil LC-DP (A) or 250 × 4 5 μm LiChrospher 100 RP-8 (B)
**Mobile phase:** MeCN:0.025% phosphoric acid:buffer 25:10:5 (A) or 60:25:15 (B) (Buffer was 9 mL concentrated phosphoric acid and 10 mL triethylamine in 900 mL water, adjust pH to 3.4 with dilute phosphoric acid, make up to 1 L.)
**Flow rate:** 0.6
**Injection volume:** 25
**Detector:** UV 229

## CHROMATOGRAM
**Retention time:** 14.50 (A), 7.68 (B)

## OTHER SUBSTANCES
**Also analyzed:** acebutolol, acepromazine, acetaminophen, acetazolamide, acetophenazine, albuterol, alprazolam, amitriptyline, amobarbital, amoxapine, antipyrine, atenolol, atropine, azatadine, baclofen, benzocaine, bromocriptine, brompheniramine, brotizolam, bupivacaine, buspirone, butabarbital, butalbital, caffeine, carbamazepine, cetirizine, chlorcyclizine, chlordiazepoxide, chlormezanone, chloroquine, chlorpheniramine, chlorpromazine, chlorpropamide, chlorprothixene, chlorthalidone, chlorzoxazone, cimeti-

dine, cisapride, clomipramine, clonazepam, clonidine, clozapine, cocaine, codeine, colchicine, cyclizine, cyclobenzaprine, dantrolene, desipramine, diazepam, diclofenac, diflunisal, diltiazem, diphenhydramine, diphenidol, diphenoxylate, dipyridamole, disopyramide, dobutamine, doxapram, doxepin, droperidol, encainide, ethidium bromide, ethopropazine, fenoprofen, fentanyl, flavoxate, fluoxetine, fluphenazine, flurazepam, flurbiprofen, fluvoxamine, furosemide, glutethimide, glyburide, guaifenesin, haloperidol, homatropine, hydralazine, hydrochlorothiazide, hydrocodone, hydromorphone, hydroxychloroquine, hydroxyzine, ibuprofen, imipramine, indomethacin, ketoconazole, ketoprofen, ketorolac, labetalol, levorphanol, lidocaine, loratadine, lorazepam, lovastatin, loxapine, mazindol, mefenamic acid, meperidine, mephenytoin, mepivacaine, mesoridazine, metaproterenol, metformin, methadone, methdilazine, methocarbamol, methotrexate, methotrimeprazine, methoxamine, methyldopa, methylphenidate, metoclopramide, metolazone, metoprolol, metronidazole, midazolam, moclobemide, morphine, nadolol, nalbuphine, naloxone, naphazoline, naproxen, nifedipine, nizatidine, norepinephrine, nortriptyline, oxazepam, oxycodone, oxymetazoline, paroxetine, pemoline, pentazocine, pentobarbital, pentoxifylline, perphenazine, pheniramine, phenobarbital, phenol, phenolphthalein, phentolamine, phenylbutazone, phenyltoloxamine, phenytoin, pimozide, pindolol, piroxicam, pramoxine, prazepam, prazosin, probenecid, procainamide, procaine, prochlorperazine, procyclidine, promazine, promethazine, propafenone, propantheline, propiomazine, propofol, propranolol, protriptyline, quazepam, quinidine, quinine, racemethorphan, ranitidine, remoxipride, risperidone, salicylic acid, scopolamine, secobarbital, sotalol, spironolactone, sulfinpyrazone, sulindac, temazepam, terbutaline, terfenadine, tetracaine, theophylline, thiethylperazine, thiopental, thioridazine, thiothixene, timolol, tocainide, tolbutamide, tolmetin, trazodone, triamterene, triazolam, trifluoperazine, triflupromazine, trimeprazine, trimethoprim, trimipramine, verapamil, warfarin, xylometazoline, yohimbine, zopiclone

## KEY WORDS
some details of plasma extraction

## REFERENCE
Koves, E.M. Use of high-performance liquid chromatography-diode array detection in forensic toxicology. *J.Chromatogr.A*, **1995**, *692*, 103–119

## SAMPLE
**Matrix:** tissue
**Sample preparation:** Homogenize 1 mL liver in 5 mL water. Centrifuge 1 mL homogenate, add the supernatant to 1.2 µg clomipramine, add 75 µL MeOH, add 75 µL MeCN, add 100 µL 1 M HCl, vortex, centrifuge, inject an aliquot of the supernatant.

## HPLC VARIABLES
**Column:** 300 × 3.9 µBondapak C18
**Mobile phase:** MeCN:50 mM potassium phosphate buffer 35:65
**Flow rate:** 1.3
**Detector:** UV 213

## CHROMATOGRAM
**Internal standard:** clomipramine

## OTHER SUBSTANCES
**Extracted:** metabolites, desmethylsertraline
**Also analyzed:** fluoxetine

## KEY WORDS
mouse; liver

## REFERENCE
von Moltke, L.L.; Greenblatt, D.J.; Cotreau-Bibbo, M.M.; Duan, S.X.; Harmatz, J.S.; Shader, R.I. Inhibition of desipramine hydroxylation in vitro by serotonin-reuptake-inhibitor antidepressants and by quinidine and ketoconazole: A model system to predict drug interactions in vivo. *J.Pharmacol.Exp.Ther.*, **1994**, *268*, 1278–1283

## SAMPLE
**Matrix:** tissue
**Sample preparation:** 130 mg Brain tissue + 1.1 mL EtOH + 3.3 µg tetracaine hydrochloride, disrupt by sonication, centrifuge, inject a 20 µL aliquot.

## HPLC VARIABLES
**Column:** 300 × 4.1 10 µm Versapack C18
**Mobile phase:** MeCN:250 mM pH 2.7 potassium phosphate buffer 30:70
**Flow rate:** 2
**Injection volume:** 20
**Detector:** UV 235

## CHROMATOGRAM
**Retention time:** 12
**Internal standard:** tetracaine hydrochloride (4.9)
**Limit of detection:** 6.12 µg/g

## OTHER SUBSTANCES
**Extracted:** desmethylsertraline

## KEY WORDS
brain; mouse

## REFERENCE
Wiener, H.L.; Kramer, H.K.; Reith, M.E.A. Separation and determination of sertraline and its metabolite, desmethylsertraline, in mouse cerebral cortex by reversed-phase high-performance liquid chromatography. *J.Chromatogr.*, **1990**, *527*, 467–472

# Simvastatin

**Molecular formula:** $C_{25}H_{38}O_5$
**Molecular weight:** 418.6
**CAS Registry No.:** 79902-63-9

## SAMPLE
**Matrix:** bile, microsomal incubations, tissue
**Sample preparation:** Microsomal incubations. 1 mL Microsomal incubation + 1 mL acetone, extract with 2 mL ethyl acetate. Remove the organic layer and dry it over anhydrous sodium sulfate, evaporate to dryness under a stream of nitrogen, reconstitute the residue in 200 μL n-propanol, inject a 20 μL aliquot. Bile. Adjust pH of bile to 4. Extract 2 mL acidified bile with 10 mL MTBE:ethyl acetate 75:25. Remove the organic layer and dry it over anhydrous sodium sulfate, evaporate to dryness under a stream of nitrogen, reconstitute the residue in 200 μL n-propanol, inject a 20 μL aliquot. Tissue. Homogenize liver with 4 volumes of water. Extract a 500 μL aliquot with 500 μL MeCN. Evaporate the supernatant to dryness under a stream of nitrogen, reconstitute the residue in 200 μL n-propanol, inject a 20 μL aliquot.

## HPLC VARIABLES
**Column:** 250 × 4.6 5 μm HIBAR Lichrospher 100 CH-18
**Mobile phase:** Gradient. MeCN:5 mM formic acid from 30:70 to 90:10 over 30 min
**Flow rate:** 2
**Injection volume:** 20
**Detector:** UV 238

## CHROMATOGRAM
**Retention time:** 24

## OTHER SUBSTANCES
**Extracted:** metabolites

## KEY WORDS
liver; mouse; human; dog; rat

## REFERENCE
Vickers, S.; Duncan, C.A.; Vyas, K.P.; Kari, P.H.; Arison, B.; Prakash, S.R.; Ramjit, H.G.; Pitzenberger, S.M.; Stokker, G.; Duggan, D.E. *In vitro* and *in vivo* biotransformation of simvastatin, an inhibitor of HMG CoA reductase. *Drug Metab.Dispos.*, **1990**, *18*, 476–483

## SAMPLE
**Matrix:** blood
**Sample preparation:** 1 mL Plasma + 50 μL MeCN:water 60:40, shake at 150 cycles/min for 2 min, centrifuge at 1500 g for 3 min, remove the supernatant and extract the residue again with 400 μL MeCN, combine the supernatants, centrifuge, evaporate to dryness with a nitrogen stream under vacuum, reconstitute with 200 μL MeCN:water 25:75, filter (0.45 μm), inject a 20 μL aliquot.

## HPLC VARIABLES
**Guard column:** 20 mm long 40 μm pelliguard (Supelco)

**Column:** 250 × 4.6 5 μm ODS Hypersil
**Mobile phase:** MeCN:25 mM pH 4.5 NaH₂PO₄ 65:35
**Flow rate:** 1.5
**Injection volume:** 20
**Detector:** UV 238

## CHROMATOGRAM
**Retention time:** 7.2 (3.6 hydroxy acid form)
**Limit of detection:** 15 ng/mL
**Limit of quantitation:** 20 ng/mL

## KEY WORDS
plasma

## REFERENCE
Carlucci, G.; Mazzeo, P.; Biordi, L.; Bologna, M. Simultaneous determination of simvastatin and its hydroxy acid form in human plasma by high-performance liquid chromatography with UV detection. *J.Pharm.Biomed.Anal.*, **1992**, *10*, 693–697

## SAMPLE
**Matrix:** solutions
**Sample preparation:** Inject a 50 μL aliquot of a solution in MeOH:100 mM pH 5 KH₂PO₄ 75:25.

## HPLC VARIABLES
**Guard column:** 20 mm long Supelguard LC-18
**Column:** 50 × 4.6 3 μm Supelcosil LC-18
**Mobile phase:** MeCN:50 mM pH 3.5 ammonium phosphate 26:74
**Column temperature:** 50
**Flow rate:** 1.6
**Injection volume:** 50
**Detector:** UV 238

## CHROMATOGRAM
**Retention time:** 4.8 (simvastatin β-hydroxyacid)
**Internal standard:** simvastatin β-hydroxyacid

## OTHER SUBSTANCES
**Simultaneous:** pravastatin

## KEY WORDS
simvastatin β-hydroxyacid is IS

## REFERENCE
Iacona, I.; Regazzi, M.B.; Buggia, I.; Villani, P.; Fiorito, V.; Molinaro, M.; Guarnone, E. High-performance liquid chromatography determination of pravastatin in plasma. *Ther.Drug Monit.*, **1994**, *16*, 191–195

## SAMPLE
**Matrix:** solutions
**Sample preparation:** Centrifuge at 2000 rpm, inject an aliquot.

## HPLC VARIABLES
**Column:** 150 × 4.6 5 μm Hypersil ODS
**Mobile phase:** MeCN:water:triethylamine:glacial acetic acid 580:420:1:1

**Column temperature:** 30
**Flow rate:** 2
**Detector:** UV 238

## CHROMATOGRAM
**Retention time:** 10.7 (3.5 hydroxy acid form)
**Limit of detection:** 10 ng/mL
**Limit of quantitation:** 25 ng/mL

## REFERENCE
Serajuddin, A.T.; Ranadive, S.A.; Mahoney, E.M. Relative lipophilicities, solubilities, and structure-pharmacological considerations of 3-hydroxy-3-methylglutaryl-coenzyme A (HMG-CoA) reductase inhibitors pravastatin, lovastatin, mevastatin, and simvastatin. *J.Pharm.Sci.*, **1991**, *80*, 830–834

## ANNOTATED BIBLIOGRAPHY
Vickers, S.; Duncan, C.A.; Chen, I.W.; Rosegay, A.; Duggan, D.E. Metabolic disposition studies on simvastatin, a cholesterol-lowering prodrug. *Drug Metab.Dispos.*, **1990**, *18*, 138–145 [plasma; tissue; mouse; dog; human; SPE; gradient; liver; prostate; testes; adrenal; kidney; lung; spleen; forestom; glandstom]

# Somatropin

**Molecular formula:** $C_{990}H_{1528}N_{262}O_{300}S_7$
**Molecular weight:** 21500.0
**CAS Registry No.:** 9002-72-6, 12629-01-5 (human)

## SAMPLE
**Matrix:** formulations
**Sample preparation:** Prepare a 1 mg/mL solution in 12.5 mM ammonium bicarbonate, filter (0.45 µm), inject an aliquot.

## HPLC VARIABLES
**Guard column:** 40 × 6 TSK gel SWXL (Tosoh)
**Column:** 300 × 7.8 TSK gel G2000SWXL
**Mobile phase:** 50 mM pH 6.8 Phosphate buffer containing 300 mM NaCl and 0.05% sodium azide (Caution! Sodium azide is carcinogenic, highly toxic, and can form explosive compounds if discharged to the sewer!)
**Flow rate:** 0.8
**Injection volume:** 20
**Detector:** UV 280; RI

## CHROMATOGRAM
**Retention time:** 12.5

## OTHER SUBSTANCES
**Simultaneous:** impurities

## KEY WORDS
recombinant; GPC

## REFERENCE
Kawashige, M.; Sendo, T.; Otsubo, K.; Aoyama, T.; Oishi, R. Quality evaluation of commercial lyophilized human growth hormone preparations. *Biol.Pharm.Bull.*, **1995**, *18*, 1793−1796

## SAMPLE
**Matrix:** solutions

## HPLC VARIABLES
**Column:** Zorbax GF-250
**Mobile phase:** 50 mM ammonium bicarbonate
**Flow rate:** 0.6
**Detector:** UV 215

## CHROMATOGRAM
**Retention time:** 13.5, 14.5

## KEY WORDS
SEC

## REFERENCE
Katakam, M.; Bell, L.N.; Banga, A.K. Effect of surfactants on the physical stability of recombinant human growth hormone. *J.Pharm.Sci.*, **1995**, *84*, 713−716

## SAMPLE
**Matrix:** solutions

## HPLC VARIABLES
**Guard column:** 30 × 4.6 10 μm RP 300 Aquapore MPLC (Brownlee)
**Column:** 100 × 4.6 10 μm RP 300 Aquapore MPLC (Brownlee)
**Mobile phase:** Gradient. A was 0.1% trifluoroacetic acid in water. B was 0.1% trifluoroacetic acid in MeCN. A:B from 54:46 to 43:57 over 22 min.
**Detector:** UV 214

## KEY WORDS
recombinant; cow

## REFERENCE
Stevenson, C.L.; Hageman, M.J. Estimation of recombinant bovine somatropin solubility by excluded-volume interaction with polyethylene glycols. *Pharm.Res.*, **1995**, *12*, 1671–1676

## SAMPLE
**Matrix:** solutions

## HPLC VARIABLES
**Column:** 250 × 4.6 5 μm Vydac C4
**Mobile phase:** Gradient. MeCN:0.05% trifluoroacetic acid from 43:57 to 55:45 over 30 min.
**Column temperature:** 45
**Flow rate:** 0.5
**Injection volume:** 20
**Detector:** UV 220

## CHROMATOGRAM
**Retention time:** 28

## OTHER SUBSTANCES
**Simultaneous:** somatrem

## REFERENCE
Arcelloni, C.; Fermo, I.; Banfi, G.; Pontiroli, A.E.; Paroni, R. Capillary electrophoresis for protein analysis: separation of human growth hormone and human insulin molecular forms. *Anal.Biochem.*, **1993**, *212*, 160–167

## SAMPLE
**Matrix:** solutions

## HPLC VARIABLES
**Column:** 600 × 7.5 10 μm TSK G3000 SW (Pharmacia)
**Mobile phase:** 10 mM pH 7.2 $K_2HPO_4$ containing 300 mM NaCl
**Flow rate:** 1
**Injection volume:** 20
**Detector:** UV 254

## CHROMATOGRAM
**Retention time:** 17 (44K), 21 (22K)

## REFERENCE
Arcelloni, C.; Fermo, I.; Banfi, G.; Pontiroli, A.E.; Paroni, R. Capillary electrophoresis for protein analysis: separation of human growth hormone and human insulin molecular forms. *Anal.Biochem.*, **1993**, *212*, 160–167

## SAMPLE
**Matrix:** solutions

## HPLC VARIABLES
**Column:** 250 × 4.6 5 μm C4 (Vydac)
**Mobile phase:** Isopropanol:500 mM pH 6.5 KH$_2$PO$_4$ 29:71
**Column temperature:** 45
**Flow rate:** 1
**Detector:** F ex 295 em 348

## CHROMATOGRAM
**Retention time:** 25

## OTHER SUBSTANCES
**Simultaneous:** somatrem

## REFERENCE
Oroszlan, P.; Wicar, S.; Teshima, G.; Wu, S.L.; Hancock, W.S.; Karger, B.L. Conformational effects in the reversed-phase chromatographic behavior of recombinant human growth hormone (rhGH) and N-methionyl recombinant human growth hormone (Met-hGH). *Anal.Chem.*, **1992**, *64*, 1623–1631

## SAMPLE
**Matrix:** solutions
**Sample preparation:** Dissolve in 100 mM NaH$_2$PO$_4$ adjusted to pH 2.1 with orthophosphoric acid, inject a 100 μL aliquot.

## HPLC VARIABLES
**Column:** 250 × 4 Aquapore RP 300 (Kontron)
**Mobile phase:** Gradient. A was 100 mM NaH$_2$PO$_4$ adjusted to pH 2.1 with orthophosphoric acid. B was MeOH. A:B from 90:10 to 35:65 over 180 min.
**Flow rate:** 1
**Injection volume:** 100
**Detector:** UV 225

## CHROMATOGRAM
**Retention time:** 3

## OTHER SUBSTANCES
**Simultaneous:** adrenocorticotropin hormone and fragments, endorphins, lipotropic hormone and fragments, melanotropin, menotropins, prolactin

## REFERENCE
Richter, W.O.; Schwandt, P. Separation of neuropeptides by HPLC: evaluation of different supports, with analytical and preparative applications to human and porcine neurophysins, β-lipotropin, adrenocorticotropic hormone, and β-endorphin. *J.Neurochem.*, **1985**, *44*, 1697–1703

## SAMPLE
**Matrix:** solutions
**Sample preparation:** Prepare a 1 mg/mL solution in 1% ammonium bicarbonate, inject a 100 μL aliquot.

## HPLC VARIABLES
**Column:** 250 × 4.1 Synchropak RP-P (Synchrom)
**Mobile phase:** Gradient. A was 0.1% trifluoroacetic acid in water. B was 0.1% trifluoroacetic acid in n-propanol. A:B from 100:0 to 50:50 over 50 min.

**Flow rate:** 1
**Injection volume:** 100
**Detector:** UV 220

## CHROMATOGRAM
**Retention time:** 41

## REFERENCE
Kohr, W.J.; Keck, R.; Harkins, R.N. Characterization of intact and trypsin-digested biosynthetic human growth hormone by high-pressure liquid chromatography. *Anal.Biochem.*, **1982**, *122*, 348–359

◆━━━━━━━━━━━━━━━━━━━━◆━━━━━━━━━━━━━━━━━━━◆

## ANNOTATED BIBLIOGRAPHY
Chang, J.P.; Tucker, R.C.; Ghrist, B.F.; Coleman, M.R. Non-denaturing assay for the determination of the potency of recombinant bovine somatotropin by high-performance size-exclusion chromatography. *J.Chromatogr.A*, **1994**, *675*, 113–122

# Sucralfate

**Molecular formula:** $C_{12}H_{54}Al_{16}O_{75}S_8$
**Molecular weight:** 2086.7
**CAS Registry No.:** 54182-58-0

$R = SO_3[Al_2(OH)_x(H_2O)_y]$

## SAMPLE
**Matrix:** blood, feces, urine
**Sample preparation:** Urine. Centrifuge, inject an aliquot as described below. Feces. Homogenize with five times the volume of 0.01 M $(NH_4)_2SO_4$, re-extract the precipitate with 0.01 M $(NH_4)_2SO_4$, combine the supernatants, inject an aliquot as described below. Plasma. Dilute with two volumes water, apply to 20 × 4 25-40 μm LiChroprep-NH2 equilibrated with 0.01 M $(NH_4)_2SO_4$, wash with 10 mL 0.1 M $(NH_4)_2SO_4$, elute with 1 mL 1 M $(NH_4)_2SO_4$, dilute the eluate with 10 volumes of water, inject an aliquot as described below. Apply up to 5 mL sample to column A with mobile phase A and elute for 10 min then backflush contents of column A onto column B with mobile phase B. Elute column B with mobile phase B and monitor the output. Re-equilibrate column A with mobile phase A.

## HPLC VARIABLES
**Column:** A 40 × 4 25-40 μm LiChroprep NH2; B 250 × 4 10 μm LiChrosorb-NH2
**Mobile phase:** A 0.01 M $(NH_4)_2SO_4$; B 0.8 M $(NH_4)_2SO_4$
**Flow rate:** 1
**Injection volume:** 5000
**Detector:** Radioactivity

## CHROMATOGRAM
**Retention time:** 8-9 (after column switch)

## KEY WORDS
column-switching; radiolabeled compound; plasma

## REFERENCE
Steiner, K.; Bühring, K.U.; Faro, H.-P.; Garbe, A.; Nowak, H. Sucralfate: pharmacokinetics, metabolism and selective binding to experimental gastric and duodenal ulcers in animals. *Arzneimittelforschung*, **1982**, *32*, 512–518

# Sulbactam

**Molecular formula:** $C_8H_{11}NO_5S$
**Molecular weight:** 233.2
**CAS Registry No.:** 68373-14-8 (sulbactam), 83031-43-0 (sulbactam benzathine),
69388-79-0 (sulbactam pivoxil), 69388-84-7 (sulbactam sodium)

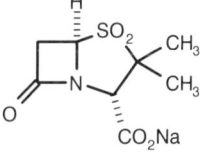

## SAMPLE
**Matrix:** blood
**Sample preparation:** 1 mL Serum + 400 µL 10 mg/mL zinc sulfate containing 350 µg/mL
  benzoic acid, vortex for 30 s, centrifuge at 5500 g for 5 min, inject a 20 µL aliquot of the
  supernatant.

## HPLC VARIABLES
**Column:** 150 × 4.3 5 µm Nova Pak
**Mobile phase:** MeOH:pH 6.30 phosphate buffer 5:95
**Column temperature:** 45
**Flow rate:** 2
**Injection volume:** 20
**Detector:** UV 225

## CHROMATOGRAM
**Retention time:** 1.35
**Internal standard:** benzoic acid (1.9)
**Limit of detection:** 10 µg/mL

## OTHER SUBSTANCES
**Extracted:** tazobactam

## KEY WORDS
serum

## REFERENCE
Guillaume, Y.; Peyrin, E.; Guinchard, C. Rapid determination of sulbactam and tazobactam in human
  serum by high-performance liquid chromatography. *J.Chromatogr.B*, **1995**, *665*, 363–371

## SAMPLE
**Matrix:** blood
**Sample preparation:** Filter (Amicon MPS-1 with YMT membrane) while centrifuging at
  1500 g for 10 min, inject a 20 µL aliquot of the ultrafiltrate.

## HPLC VARIABLES
**Guard column:** 30 × 4.6 10 µm Develosil ODS-10
**Column:** 150 × 4.6 5 µm Develosil ODS-5 (Nomura Chemicals)
**Mobile phase:** MeOH:buffer 1:1.5 (Prepare buffer by dissolving 1.791 g $Na_2HPO_4.12H_2O$
  and 0.780 g $NaH_2PO_4.2H_2O$ in 10 L water, add tetrabutylammonium bromide to a final
  concentration of 5 mM.)
**Flow rate:** 0.8
**Injection volume:** 20
**Detector:** UV 278 following post-column reaction. The column effluent mixed with MeOH:
  750 mM NaOH 1:1.5 pumped at 0.2 mL/min and this mixture flowed through a 1 m ×
  0.5 mm ID coil to the detector.

## CHROMATOGRAM
**Retention time:** 6

**Limit of detection:** 25 ng/mL
**Limit of quantitation:** 50 ng/mL

## OTHER SUBSTANCES
**Noninterfering:** ampicillin, cefoperazone, ticarcillin

## KEY WORDS
ultrafiltrate; plasma; post-column reaction

## REFERENCE
Haginaka, J.; Wakai, J.; Yasuda, H.; Uno, T.; Nakagawa, T. Improved high-performance liquid chromatographic assay of clavulanic acid and sulbactam by postcolumn alkaline degradation. *J.Liq.Chromatogr.*, **1985**, *8*, 2521–2534

## SAMPLE
**Matrix:** blood, CSF
**Sample preparation:** 1 mL Plasma or CSF + 200 µL hydrochloric acid R.P. + 100 µL 10 µg/mL pyrogallol in mobile phase + 6 mL diethyl ether, vortex for 1 min, centrifuge at 900 g at 4° for 10 min. Remove the organic layer and evaporate it to dryness under vacuum, reconstitute the residue in 100 µL mobile phase, inject a 50 µL aliquot.

## HPLC VARIABLES
**Column:** 300 × 4.6 10 µm µBondapak C18
**Mobile phase:** MeOH:water 20:80 with 0.5% glacial acetic acid
**Flow rate:** 0.75
**Injection volume:** 50
**Detector:** UV 225

## CHROMATOGRAM
**Retention time:** 8
**Internal standard:** pyrogallol (6)
**Limit of detection:** 500 ng/mL

## OTHER SUBSTANCES
**Simultaneous:** amoxicillin, ampicillin, cefoxitin

## KEY WORDS
plasma

## REFERENCE
Fredj, G.; Paillet, M.; Aussel, F.; Brouard, A.; Barreteau, H.; Divine, C.; Micoud, M. Determination of sulbactam in biological fluids by high-performance liquid chromatography. *J.Chromatogr.*, **1986**, *383*, 218–222

## SAMPLE
**Matrix:** blood, saliva, urine
**Sample preparation:** 1 mL Plasma or saliva or 20 µL urine + 20 µL 50 µg/mL salicylhydroxamic acid + 200 µg N-hydrochloric acid (?) + 4 mL redistilled diethyl ether, vortex for 30 s, centrifuge at 500 g for 5 min. Remove the organic layer and evaporate it to dryness under a stream of nitrogen at 35°, reconstitute the residue in 150 µL mobile phase, inject a 50 µL aliquot.

## HPLC VARIABLES
**Column:** 300 × 4.6 Magnusphere C18 (Magnus Scientific, Sandbach, England)

**Mobile phase:** MeOH : 50 mM ammonium phosphate : phosphoric acid 5 : 95 : 0.1 (plasma) or 0 : 99.9 : 0.1 (urine)
**Flow rate:** 1
**Injection volume:** 50
**Detector:** UV 225

## CHROMATOGRAM
**Retention time:** 7.5 (plasma), 16.5 (urine)
**Internal standard:** salicylhydroxamic acid (14.0 (plasma), 21.5 (urine))
**Limit of quantitation:** 200 ng/mL (plasma), 20000 ng/mL (urine)

## OTHER SUBSTANCES
**Noninterfering:** amoxicillin, carbenicillin, cefazolin, cefuroxime, cephalexin, cephaloridine, cephradine, cloxacillin, flucloxacillin, mecillinam, penicillin G, tetracycline

## KEY WORDS
plasma; pharmacokinetics

## REFERENCE
Rogers, H.J.; Bradbrook, I.D.; Morrison, P.J.; Spector, R.G.; Cox, D.A.; Lees, L.J. Pharmacokinetics and bioavailability of sultamicillin estimated by high performance liquid chromatography. *J.Antimicrob.Chemother.*, **1983**, *11*, 435–445

## SAMPLE
**Matrix:** blood, urine
**Sample preparation:** Serum. Filter using Molcut II (Millipore), inject a 50 μL aliquot of the ultrafiltrate. Urine. Dilute ten-fold with water, filter (Gelman acrylate copolymer 0.45 μm), inject a 20 μL aliquot of the filtrate.

## HPLC VARIABLES
**Guard column:** 4 × 4 5 μm LiChrospher RP-18(e)
**Column:** 250 × 4 5 μm LiChrospher RP-18(e)
**Mobile phase:** MeOH : 20 mM tetrabutylammonium bromide + 5 mM $Na_2HPO_4$ + 5 mM $NaH_2PO_4$ 1 : 1.75
**Flow rate:** 0.8
**Injection volume:** 20-50
**Detector:** UV 270 following post-column reaction. The column effluent mixed with 2 M NaOH and 0.05% sodium hypochlorite solution pumped at 0.1 mL/min in a 400 × 0.5 mm hollow fiber membrane reactor at 40°, this mixture flowed through a 1400 × 0.3 mm knitted open tubular reactor at 50° to the UV detector.

## CHROMATOGRAM
**Retention time:** 8
**Limit of detection:** 5 ng

## OTHER SUBSTANCES
**Extracted:** ampicillin

## KEY WORDS
serum; post-column reaction

## REFERENCE
Haginaka, J.; Nishimura, Y. Simultaneous determination of ampicillin and sulbactam by liquid chromatography: post-column reaction with sodium hydroxide and sodium hypochlorite using an active hollow-fiber membrane reactor. *J.Chromatogr.*, **1990**, *532*, 87–94

## SAMPLE
**Matrix:** blood, urine
**Sample preparation:** Serum. 1 mL Serum + 1 mL MeCN, shake at 0° for 15 min, centrifuge at 8000 g for 10 min. Add supernatant to 10 mL dichloromethane, shake at 0° for 15 min, centrifuge at 8000 g for 10 min, discard organic layer. Mix 200 μL aqueous layer with 400 μL reagent, after 40 min at 40° inject 40 μL aliquot. Urine. Dilute 10-fold, mix 200 μL with 400 μL reagent, after 40 min at 40° inject 40 μL aliquot. (Reagent was 3.45 g 1,2,4-triazole dissolved in 15 mL water, adjust pH to 6.0 with 4 M NaOH, make up to 25 mL.)

## HPLC VARIABLES
**Guard column:** 50 × 4.6 5μm Spherisorb C-18
**Column:** 250 × 4.6 5μm Spherisorb C-18
**Mobile phase:** MeCN:20 mM pH 7.0 phosphate buffer 0.2:99.8
**Flow rate:** 0.5
**Injection volume:** 40
**Detector:** UV 325

## CHROMATOGRAM
**Retention time:** 7
**Limit of detection:** 50 ng/mL

## OTHER SUBSTANCES
**Simultaneous:** ampicillin (with gradient to 25:75 over 25 min)

## KEY WORDS
serum; derivatization

## REFERENCE
Shah, A.J.; Adlard, M.W.; Stride, J.D. A sensitive assay for clavulanic acid and sulbactam in biological fluids by high-performance liquid chromatography and precolumn derivatization. *J.Pharm. Biomed.Anal.*, **1990**, *8*, 437−443

## SAMPLE
**Matrix:** blood, urine
**Sample preparation:** Dilute urine ten-fold with water. 50 μL Plasma or 100 μL diluted urine + 150 μL MeCN, vortex for 30 s, incubate at room temperature for 5 min, centrifuge at 1500 g for 10 min. Remove a 100 μL aliquot of the supernatant and add it to 100 μL triazole solution, heat at 50° for 15 min, cool to room temperature, centrifuge at 1500 g for 5 min, inject a 20 μL aliquot of the supernatant. (Triazole solution was 13.81 g 1,2,4-triazole in 60 mL water, pH adjusted to 10.0 ± 0.05 with saturated NaOH solution, made up to 100 mL with water.)

## HPLC VARIABLES
**Guard column:** 30 × 4.6 10 μm Develosil ODS-10
**Column:** 150 × 4.6 5 μm Develosil ODS-5
**Mobile phase:** MeCN:5 mM tetrabutylammonium bromide + 1 mM $Na_2HPO_4$ + 1 mM $NaH_2PO_4$ 25:75
**Column temperature:** 50
**Flow rate:** 1
**Injection volume:** 20
**Detector:** UV 326

## CHROMATOGRAM
**Retention time:** 7
**Limit of detection:** 200 ng/mL (plasma); 1 μg/mL (urine)

## OTHER SUBSTANCES
**Noninterfering:** degradation products, amoxicillin, ampicillin, cefoperazone, penicillin G

## KEY WORDS
plasma; human; rat; derivatization

## REFERENCE
Haginaka, J.; Wakai, J.; Yasuda, H.; Uno, T.; Nakagawa, T. High-performance liquid chromatographic assay of sulbactam using pre-column reaction with 1,2,4-triazole. *J.Chromatogr.*, **1985**, *341*, 115–122

## SAMPLE
**Matrix:** formulations
**Sample preparation:** Dilute 1:8 with water, combine a 100 μL aliquot of the diluted solution with 100 μL cimetidine solution and 200 μL water, inject a 20 μL aliquot.

## HPLC VARIABLES
**Column:** 3.9 × 300 μBondapak C18
**Mobile phase:** MeCN:MeOH:10 mM pH 2.6-2.7 phosphate buffer 7:14:79 containing 5 mM tetrabutylammonium hydrogen sulfate
**Flow rate:** 1
**Injection volume:** 20
**Detector:** UV 225

## CHROMATOGRAM
**Retention time:** 5.79
**Internal standard:** cimetidine (3.27)

## OTHER SUBSTANCES
**Simultaneous:** ampicillin, aztreonam

## KEY WORDS
stability-indicating; saline; injections

## REFERENCE
Belliveau, P.P.; Nightingale, C.H.; Quintiliani, R. Stability of aztreonam and ampicillin sodium-sulbactam sodium in 0.9% sodium chloride injection. *Am.J.Hosp.Pharm.*, **1994**, *51*, 901–904

## SAMPLE
**Matrix:** formulations
**Sample preparation:** Dilute injection with mobile phase, inject an aliquot.

## HPLC VARIABLES
**Column:** μBondapak C18
**Mobile phase:** MeCN:buffer 17.5:82.5 (Buffer was 5 mM tetrabutylammonium hydroxide adjusted to pH 5.0 with concentrated phosphoric acid.)
**Column temperature:** 25
**Flow rate:** 2
**Injection volume:** 10
**Detector:** UV 230

## CHROMATOGRAM
**Retention time:** 10.1

## OTHER SUBSTANCES
**Simultaneous:** ampicillin

## KEY WORDS
injections

## REFERENCE
Mushinsky, R.F.; Reynolds, M.L.; Nicholson, C.A.; Crider, L.L.; Forcier, G.A. Stability of sulbactam/ampicillin in diluents for parenteral administration. *Rev.Infect.Dis.*, **1986**, *8 Suppl 5*, S523–S527

## SAMPLE
**Matrix:** urine
**Sample preparation:** Dilute 10-fold with water, filter (0.45 μm acrylate copolymer), inject a 20 μL aliquot.

## HPLC VARIABLES
**Guard column:** 30 × 4.6 10 μm Develosil ODS-10
**Column:** 150 × 4.6 5 μm Develosil ODS-5 (Nomura Chemicals)
**Mobile phase:** MeOH:buffer 1:2 (Prepare buffer by dissolving 1.791 g $Na_2HPO_4.12H_2O$ and 0.780 g $NaH_2PO_4.2H_2O$ in 10 L water, add tetrabutylammonium bromide to a final concentration of 5 mM.)
**Flow rate:** 0.8
**Injection volume:** 20
**Detector:** UV 272 following post-column reaction. The column effluent mixed with MeOH: 750 mM NaOH 1:2 pumped at 0.2 mL/min and this mixture flowed through a 1 m × 0.5 mm ID coil to the detector.

## CHROMATOGRAM
**Retention time:** 9
**Limit of detection:** 500 ng/mL
**Limit of quantitation:** 1 μg/mL

## OTHER SUBSTANCES
**Noninterfering:** ampicillin, cefoperazone, ticarcillin

## KEY WORDS
post-column reaction

## REFERENCE
Haginaka, J.; Wakai, J.; Yasuda, H.; Uno, T.; Nakagawa, T. Improved high-performance liquid chromatographic assay of clavulanic acid and sulbactam by postcolumn alkaline degradation. *J.Liq.Chromatogr.*, **1985**, *8*, 2521–2534

## ANNOTATED BIBLIOGRAPHY
Bawdon, R.E.; Madsen, P.O. High-pressure liquid chromatographic assay of sulbactam in plasma, urine, and tissue. *Antimicrob.Agents Chemother.*, **1986**, *30*, 231–233

Haginaka, J.; Yasuda, H.; Uno, T.; Nakagawa, T. Sulbactam: alkaline degradation and determination by high-performance liquid chromatography. *Chem.Pharm.Bull.*, **1984**, *32*, 2752–2758

# Sulfamethoxazole

**Molecular formula:** $C_{10}H_{11}N_3O_3S$
**Molecular weight:** 253.3
**CAS Registry No.:** 723-46-6

## SAMPLE
**Matrix:** blood
**Sample preparation:** Condition a 1 mL Bakerbond C18 SPE cartridge with MeOH and 50 mM pH 5.5 citrate buffer. 500 μL Serum + 17 μL 600 μg/mL sulfamethazine in MeCN: water 20:80 + 500 μL 50 mM pH 5.5 citrate buffer, vortex, add to the SPE cartridge, wash twice with 50 mM pH 5.5 citrate buffer, air dry, elute with MeOH. Evaporate the eluate to dryness under a stream of nitrogen at 45°, reconstitute the residue in 250 μL mobile phase, inject a 10 μL aliquot.

## HPLC VARIABLES
**Guard column:** μBondapak C18 Guard-Pak
**Column:** 300 × 3.9 μBondapak C18
**Mobile phase:** MeCN:1% acetic acid 18:82
**Flow rate:** 2.5
**Injection volume:** 100
**Detector:** UV 240

## CHROMATOGRAM
**Retention time:** 8.93
**Internal standard:** sulfamethazine (4.19)

## OTHER SUBSTANCES
**Extracted:** trimethoprim

## KEY WORDS
serum; SPE

## REFERENCE
Moore, K.H.P.; Brouwer, K.L.R. High-performance liquid chromatographic evaluation of the effect of heat treatment on trimethoprim and sulfamethoxazole stability in serum. *Ther.Drug Monit.*, **1995**, *17*, 356–360

## SAMPLE
**Matrix:** blood
**Sample preparation:** 500 μL Plasma, whole blood, or red blood cells + 100 μL MeOH + 500 μL water + 100 μL buffer + 6 mL ethylene dichloride, shake on an orbital mixer for 20 min, centrifuge at 1000 g for 10 min. Remove the organic layer and evaporate it to dryness at 60° on a vortex evaporator, reconstitute the residue in 200 μL mobile phase, inject a 20 μL aliquot. (Buffer was prepared by adding 100 μL acetic acid to 9.9 mL phosphate buffer, pH 3.40.)

## HPLC VARIABLES
**Column:** 300 × 3.9 10 μm μBondapak C18

**Mobile phase:** MeCN:MeOH:1 M perchloric acid:water 30:9:0.8:95
**Flow rate:** 1.5
**Injection volume:** 20
**Detector:** UV 254

## CHROMATOGRAM
**Retention time:** 8
**Internal standard:** sulfamethoxazole

## OTHER SUBSTANCES
**Extracted:** sulfadoxine
**Simultaneous:** amodiaquine, chloroquine, dapsone, primaquine, pyrimethamine, quinidine, quinine, sulfalene

## KEY WORDS
plasma; whole blood; red blood cells; sulfamethoxazole is IS

## REFERENCE
Dua, V.K.; Sarin, R.; Sharma, V.P. Sulphadoxine concentrations in plasma, red blood cells and whole blood in healthy and *Plasmodium falciparum* malaria cases after treatment with Fansidar using high-performance liquid chromatography. *J.Pharm.Biomed.Anal.*, **1994**, *12*, 1317–1323

## SAMPLE
**Matrix:** blood
**Sample preparation:** Condition a 3 mL C18 Bond Elut SPE cartridge with 500 μL MeOH then 1 mL wash solution. 500 μL Serum + 500 μL wash solution, vortex for 3 s, add 25 μL 240 μg/mL p-nitrophenol in MeOH, vortex for 3 s, add to the SPE cartridge, wash with 1 mL wash solution, dry with vacuum applied for 30 s, elute with two 500 μL aliquots of MeOH:triethylamine 10:1. Combine the eluates and evaporate them to dryness under a stream of nitrogen, reconstitute in 200 μL mobile phase, inject a 50-100 μL aliquot. (Wash solution was 7 mL 3 M HCl + 1.6 g 1-octanesulfonic acid + 150 mL 100 mM disodium citrate made up to 800 mL with water, pH adjusted to 3.00 with 3 M HCl.)

## HPLC VARIABLES
**Guard column:** 20 × 2 40 μm Upchurch pellicular C18
**Column:** 300 × 3.9 10 μm μBondapak C18
**Mobile phase:** MeCN:buffer 24:76 with 0.8 g 1-octanesulfonic acid (Buffer was 7 mL 3 M HCl plus 150 mL 100 mM disodium citrate made up to 760 mL with water, pH adjusted to 3.00 with 3 M HCl.)
**Flow rate:** 1.5
**Injection volume:** 50-100
**Detector:** UV 230

## CHROMATOGRAM
**Retention time:** 6.8
**Internal standard:** p-nitrophenol (9.3)
**Limit of quantitation:** 250 ng

## OTHER SUBSTANCES
**Extracted:** $N^4$-acetylsulfamethoxazole, trimethoprim

## KEY WORDS
serum; SPE

## REFERENCE
Laizure, S.C.; Holden, C.L.; Stevens, R.C. Ion-paired high-performance liquid chromatographic separation of trimethoprim, sulfamethoxazole and $N^4$-acetylsulfamethoxazole with solid-phase extraction. *J.Chromatogr.*, **1990**, *528*, 235–242

## SAMPLE
**Matrix:** blood
**Sample preparation:** 100 µL Plasma + 200 µL MeOH + 100 µL 50 µg/mL sulfapyridine in MeOH + 1 mL 1 M pH 4.7 sodium acetate buffer + 8 mL dichloromethane, shake vigorously for 2 min, centrifuge at 1000 g at 10° for 10 min. Remove the organic layer and evaporate it to dryness at 50° under nitrogen, take up the residue in 250 µL mobile phase, inject a 20 µL aliquot.

## HPLC VARIABLES
**Guard column:** 75 × 4.6 Whatman CO:Pell
**Column:** 250 × 3.2 10 µm Spherisorb ODS
**Mobile phase:** MeOH:50 mM pH 7.4 phosphate buffer 20:80
**Flow rate:** 1.2
**Injection volume:** 20
**Detector:** UV 270

## CHROMATOGRAM
**Retention time:** 2.7
**Internal standard:** sulfapyridine (5.2)

## OTHER SUBSTANCES
**Extracted:** acetylsulfamethoxazole

## KEY WORDS
plasma

## REFERENCE
Astbury, C.; Dixon, J.S. Rapid method for the determination of either plasma sulfapyridine or sulfamethoxazole and their acetyl metabolites using high-performance liquid chromatography. *J.Chromatogr.*, **1987**, *414*, 223–227

## SAMPLE
**Matrix:** blood
**Sample preparation:** 1 mL Plasma + 100 µL 5 µg/mL sulfafurazole in MeOH:water 50: 50 + 200 µL 1 M pH 6.8 sodium phosphate buffer + 6 mL ethyl acetate, shake for 15 min, centrifuge at 600 g for 10 min. Remove the organic layer and evaporate it to dryness under a stream of nitrogen at 50°, reconstitute the residue in 175 µL MeOH:water 20: 80, vortex for 1 min, add 25 µL 40% trichloroacetic acid in 100 mM hydrochloric acid, vortex for 30 s, centrifuge at 8300 g for 5 min, inject a 100 µL aliquot of the supernatant.

## HPLC VARIABLES
**Column:** 250 × 4 5 µm LiChrosorb RP18
**Mobile phase:** MeCN:150 mM pH 4.85 ammonium phosphate buffer 12.6:87.4
**Flow rate:** 1.5
**Injection volume:** 100
**Detector:** UV 254

## CHROMATOGRAM
**Retention time:** 23
**Internal standard:** sulfafurazole (21)
**Limit of detection:** 80 ng/mL

## OTHER SUBSTANCES
**Extracted:** metabolites, N-acetylsulfamethoxazole, trimethoprim

## KEY WORDS
plasma; pharmacokinetics

## REFERENCE
Van der Steuijt, K.; Sonneveld, P. Concurrent analysis of methotrexate, trimethoprim, sulfamethoxazole and their major metabolites in plasma by high-performance liquid chromatography. *J.Chromatogr.*, **1987**, *422*, 328–333

## SAMPLE
**Matrix:** blood
**Sample preparation:** 1 mL Serum + 50 μL 50 μg/mL 4-dimethylaminobenzaldehyde in MeOH, vortex for 10 s, add 250 μL 60% perchloric acid, vortex for 10 s, centrifuge at 3000 rpm for 10 min, inject a 10-25 μL aliquot of the clear supernatant.

## HPLC VARIABLES
**Column:** 150 × 4.6 5 μm Hypersil ODS
**Mobile phase:** MeCN:buffer 25:75 (Buffer contained 50 mM $Na_2HPO_4$ adjusted to pH 2 with orthophosphoric acid.)
**Flow rate:** 2
**Injection volume:** 10-25
**Detector:** UV 254

## CHROMATOGRAM
**Retention time:** 3
**Internal standard:** 4-dimethylaminobenzaldehyde (7)
**Limit of quantitation:** 100 ng/mL

## KEY WORDS
serum

## REFERENCE
Hung, C.T.; Perrier, D.G. Determination of trimethoprim and sulfamethoxazole in serum by reversed-phase and ion pair HPLC. *J.Liq.Chromatogr.*, **1985**, *8*, 521–536

## SAMPLE
**Matrix:** blood, CSF, gastric fluid, urine
**Sample preparation:** 200 μL Serum, urine, CSF, or gastric fluid + 300 μL reagent. Flush column A to waste with 500 μL 500 mM ammonium sulfate, inject sample onto column A, flush column A to waste with 500 μL 500 mM ammonium sulfate, backflush the contents of column A onto column B with mobile phase, monitor the effluent from column B. (Reagent was 8.05 M guanidine hydrochloride and 1.02 M ammonium sulfate in water.)

## HPLC VARIABLES
**Column:** A 40 μm preparative grade C18 (Analytichem); B 75 × 2.1 pellicular C18 (Whatman) + 250 × 4.6 5 μm C8 end-capped (Whatman)
**Mobile phase:** Gradient. A was 50 mM pH 4.5 $KH_2PO_4$. B was MeCN:isopropanol 80:20. A:B 90:10 for 1 min, to 30:70 over 20 min.
**Column temperature:** 50
**Flow rate:** 1.5
**Detector:** UV 220

## CHROMATOGRAM
**Retention time:** 9.5
**Internal standard:** heptanophenone (19)

## OTHER SUBSTANCES
**Extracted:** acetaminophen, allobarbital, azinphos, barbital, brallobarbitone, bromazepam, butethal, caffeine, carbamazepine, carbaryl, cephaloridine, chloramphenicol, chlordiazepoxide, chlorothiazide, chlorvinphos, clothiapine, cocaine, coomassie blue, desipramine, diazepam, diphenhydramine, dipipanone, ethylbromphos, flufenamic acid, formothion, griseofulvin, indomethacin, lidocaine, lorazepam, malathion, medazepam, midazolam, oxazepam, paraoxon, penicillin G, pentobarbital, prazepam, propoxyphene, prothiophos, quinine, salicylic acid, secobarbital, strychnine, theophylline, thiopental, thioridazine, trimethoprim

## KEY WORDS
serum; column-switching

## REFERENCE
Kruger, P.B.; Albrecht, C.F.De V.; Jaarsveld, P.P. Use of guanidine hydrochloride and ammonium sulfate in comprehensive in-line sorption enrichment of xenobiotics in biological fluids by high-performance liquid chromatography. *J.Chromatogr.*, **1993**, *612*, 191–198

## SAMPLE
**Matrix:** blood, CSF, peritoneal fluid, synovial fluid, tissue, urine
**Sample preparation:** Condition a C18 SPE cartridge (J.T. Baker) with 1 mL MeOH, 1 mL water, and 1 mL 100 mM pH 4.5 acetate buffer. CSF, peritoneal fluid, serum, synovial fluid, tissue. Homogenize (TenBroeck tissue grinder) endometrial tissue with saline, centrifuge at 510 g, remove supernatant. Centrifuge fluids at 510 g. 1 mL Sample + 1 mL 100 mM acetate buffer, mix, add to the SPE cartridge, wash with 1 mL water, elute with 1 mL MeCN:MeOH:100 mM acetate buffer 12.5:12.5:75, add 12.5 ng ormetoprim, inject an aliquot. Urine. 1 mL Urine + 1 mL MeCN, centrifuge at 510 g for 10 min. Remove 25 μL of the supernatant and add it to 1 mL 100 mM acetate buffer, add to the SPE cartridge, wash with 1 mL water, elute with 1 mL MeCN:MeOH:100 mM acetate buffer 12.5:12.5:75, add 12.5 ng ormetoprim, inject an aliquot.

## HPLC VARIABLES
**Column:** C18 (Rainin)
**Mobile phase:** MeCN:MeOH:50 mM phosphate buffer 12.5:12.5:75, pH 3.0
**Flow rate:** 1.5
**Detector:** UV 280

## CHROMATOGRAM
**Retention time:** 11
**Internal standard:** ormetoprim (8)
**Limit of detection:** 50 ng/mL

## OTHER SUBSTANCES
**Extracted:** trimethoprim

## KEY WORDS
serum; endometrium; SPE; horse; pharmacokinetics

## REFERENCE
Brown, M.P.; Gronwall, R.; Castro, L. Pharmacokinetics and body fluid and endometrial concentrations of trimethoprim-sulfamethoxazole in mares. *Am.J.Vet.Res.*, **1988**, *49*, 918–922

## SAMPLE
**Matrix:** blood, CSF, urine
**Sample preparation:** 50 µL Serum, urine, or CSF + 50 µL 100 µg/mL antipyrine in MeOH, vortex for 15 s, centrifuge at 10000 g for 2 min, inject a 20 µL aliquot of the supernatant.

## HPLC VARIABLES
**Guard column:** 37-50 µm µBondapak C18
**Column:** 300 × 3.9 10 µm µBondapak C18
**Mobile phase:** MeOH:buffer 35:65 (Buffer was 97% 67 mM KH$_2$PO$_4$ + 3% 67 mM Na$_2$HPO$_4$ and the pH was adjusted to 3.5 with phosphoric acid.)
**Flow rate:** 1
**Injection volume:** 20
**Detector:** UV 225

## CHROMATOGRAM
**Retention time:** 8.0
**Internal standard:** antipyrine (11.0)
**Limit of detection:** 200 ng/mL

## OTHER SUBSTANCES
**Extracted:** N-acetylsulfamethoxazole, trimethoprim

## KEY WORDS
serum

## REFERENCE
Weber, A.; Opheim, K.E.; Siber, G.R.; Ericson, J.F.; Smith, A.L. High-performance liquid chromatography quantitation of trimethoprim, sulfamethoxazole, and N4-acetylsulfamethoxazole in body fluids. *J.Chromatogr.*, **1983**, *278*, 337–345

## SAMPLE
**Matrix:** blood, milk
**Sample preparation:** 1 mL Serum or milk + 4 mL MeCN, vortex, centrifuge at 1000 g for 15 min. Remove the supernatant and evaporate it to dryness under a stream of nitrogen at 40°, reconstitute the residue in 50 µL water, mix vigorously, add 1 mL MeCN, centrifuge at 1000 g for 10 min. Remove the upper layer and evaporate it to dryness, reconstitute the residue in 1 mL 10 ng/mL p-aminobenzoic acid in 0.01% trichloroacetic acid, centrifuge at 1000 g for 10 min. Remove a 500 µL aliquot of the clear layer and add it to 100 µL 1 mg/mL fluorescamine in acetone (prepared fresh each day), mix for 1 min, inject a 50 µL aliquot.

## HPLC VARIABLES
**Column:** 300 × 3.9 10 µm Nova-Pak C18
**Mobile phase:** MeCN:10 mM KH$_2$PO$_4$ 30:70
**Flow rate:** 1
**Injection volume:** 50
**Detector:** F ex 390 em 475

## CHROMATOGRAM
**Retention time:** 12.5
**Internal standard:** p-aminobenzoic acid (5.5)
**Limit of detection:** 0.1 ng/mL

## OTHER SUBSTANCES
**Extracted:** sulfadiazine, sulfadimethoxine, sulfamethazine, sulfamonomethoxine, sulfathiazole

## KEY WORDS
cow; serum; derivatization

## REFERENCE
Tsai, C.-E.; Kondo, F. Liquid chromatographic determination of fluorescent derivatives of six sulfonamides in bovine serum and milk. *J.AOAC Int.*, **1995**, *78*, 674–678

## SAMPLE
**Matrix:** blood, tissue
**Sample preparation:** 1 mL Serum or homogenized tissue + 4 mL MeCN, vortex, centrifuge at 1000 g for 15 min. Remove the supernatant and evaporate it to dryness under a stream of nitrogen at 40°, reconstitute the residue in 50 μL water, mix vigorously, add 1 mL MeCN, centrifuge at 1000 g for 15 min. Remove the upper layer and evaporate it to dryness, reconstitute the residue in 1 mL 10 ng/mL sulfadiazine in 0.01% trichloroacetic acid, shake, add 100 μL hexane, shake, centrifuge at 1000 g for 15 min. Remove a 500 μL aliquot of the clear aqueous layer and add it to 100 μL 1 mg/mL fluorescamine in MeCN (freshly prepared), shake by hand for 1 min, inject a 50 μL aliquot.

## HPLC VARIABLES
**Column:** 300 × 3.9 10 μm Nova-Pack C18
**Mobile phase:** MeCN:10 mM $KH_2PO_4$ 30:70
**Flow rate:** 1
**Injection volume:** 50
**Detector:** F ex 390 em 475

## CHROMATOGRAM
**Retention time:** 14.1
**Internal standard:** sulfadiazine (7.1)
**Limit of detection:** 0.1 ng/mL

## OTHER SUBSTANCES
**Extracted:** sulfadimethoxine, sulfamethazine, sulfamonomethoxine

## KEY WORDS
serum; pig; derivatization; kidney; muscle; liver

## REFERENCE
Tsai, C.-E.; Kondo, F. A sensitive high-performance liquid chromatographic method for detecting sulfonamide residues in swine serum and tissues after fluorescamine derivatization. *J.Liq.Chromatogr.*, **1995**, *18*, 965–976

## SAMPLE
**Matrix:** blood, urine
**Sample preparation:** Plasma. 100 μL Plasma + 100 μL MeCN, centrifuge at 3000 g for 5 min, inject 20 μL of the supernatant. Urine. 7 mL Urine + 50 μL diethylamine, centrifuge at 3000 g for 5 min, dilute the supernatant 1:9 with 200 mM pH 5.0 $K_2HPO_4$ buffer, inject a 20 μL aliquot.

## HPLC VARIABLES
**Guard column:** 75 × 3.1 Chrompack pellicular reversed phase (cat. no. 28653)
**Column:** 250 × 4.6 5 μm Spherisorb ODS

**Mobile phase:** Gradient. MeCN:DMF:buffer A:B:C. Over 30 min from A:B:C 1:3:96 to 20:10:70, then over 3 min to 1:3:96, keep at 1:3:96 for 2 min before next injection. (Buffer was 0.9 g 85% phosphoric acid and 0.225 g tetramethylammonium chloride in 1 L water, pH 2.1.)
**Flow rate:** 1.2
**Injection volume:** 20
**Detector:** UV 271

## CHROMATOGRAM
**Retention time:** 23
**Limit of detection:** 35 ng/mL (plasma)
**Limit of quantitation:** 100 ng/mL (plasma); 1500 ng/mL (urine)

## OTHER SUBSTANCES
**Extracted:** metabolites

## KEY WORDS
plasma

## REFERENCE
Vree, T.B.; van der Ven, A.J.A.M.; Verwey-van Wissen, C.P.W.G.M.; van Ewijk-Beneken Kolmer, E.W.J.; Swolfs, A.E.M.; van Galen, P.M.; Amatdjais-Groenen, H. Isolation, identification and determination of sulfamethoxazole and its known metabolites in human plasma and urine by high-performance liquid chromatography. *J.Chromatogr.B*, **1994**, *658*, 327–340

## SAMPLE
**Matrix:** cell suspensions
**Sample preparation:** Cool cell suspension in an ice bath, centrifuge at 800 g at 4° for 15 min, inject an aliquot of the supernatant.

## HPLC VARIABLES
**Column:** μBondapak C18
**Mobile phase:** MeCN:water 20:80
**Flow rate:** 2
**Detector:** UV 254

## CHROMATOGRAM
**Limit of detection:** 250 ng/mL

## OTHER SUBSTANCES
**Also analyzed:** sulfadiazine, sulfamerazine, sulfanilamide

## REFERENCE
Climax, J.; Lenehan, T.J.; Lambe, R.; Kenny, M.; Caffrey, E.; Darragh, A. Interaction of antimicrobial agents with human peripheral blood leucocytes: uptake and intracellular localization of certain sulphonamides and trimethoprims. *J.Antimicrob.Chemother.*, **1986**, *17*, 489–498

## SAMPLE
**Matrix:** eggs, food, milk
**Sample preparation:** Honey. Dissolve 1 g honey in 10 mL water, homogenize, filter (0.45 μm), inject a 50 μL aliquot. Milk, eggs. 5 mL Milk or 0.4 g lyophilized eggs + 10 mL trichloroacetic acid solution (so as to give a final trichloroacetic acid concentration of 3%), homogenize, centrifuge at 5000 rpm for 5 min. Re-extract the residue with 10 mL 3% trichloroacetic acid. Combine the aqueous phases and make up to 25 mL with trichloroacetic acid solution, inject a 50 μL aliquot.

## HPLC VARIABLES
**Column:** 150 × 4.6 5 μm Spherisorb ODS-2
**Mobile phase:** Gradient. MeCN:water 3:97 for 5 min, to 40:60 over 15 min, return to initial conditions over 1 min, re-equilibrate for 10 min. (Wash column with MeCN:ethyl acetate 5:95 at the end of each day.)
**Flow rate:** 1
**Injection volume:** 50
**Detector:** UV 260

## CHROMATOGRAM
**Retention time:** 16
**Limit of detection:** 40 ng/mL

## OTHER SUBSTANCES
**Extracted:** sulfadiazine, sulfaguanidine, sulfapyridine, sulfathiazole

## KEY WORDS
honey

## REFERENCE
Viñas, P.; López Erroz, C.; Hernández Canals, A.; Hernández Córdoba, M. Liquid chromatographic analysis of sulfonamides in foods. *Chromatographia*, **1995**, *40*, 382–386

## SAMPLE
**Matrix:** eggs, milk, tissue
**Sample preparation:** Milk. Homogenize 3 g milk and 500 μL 30% trichloroacetic acid, centrifuge at 5000 rpm for 5 min. Remove the aqueous phase and extract the residue with 4 mL 3% trichloroacetic acid. Combine the aqueous layers and make up to 10 mL with trichloroacetic acid, filter (0.45 μm), inject a 50 μL aliquot. Fish, eggs. Homogenize (Ultra-Turrax) 3 g fish or 4 g eggs with 4 mL 3% trichloroacetic acid, centrifuge at 5000 rpm for 5 min. Remove the aqueous phase and extract the residue with 4 mL 3% trichloroacetic acid. Combine the aqueous layers and make up to 10 mL with trichloroacetic acid, filter (0.45 μm), inject a 50 μL aliquot.

## HPLC VARIABLES
**Guard column:** 5 μm Spherisorb ODS-2
**Column:** 150 × 4.6 5 μm Spherisorb ODS-2
**Mobile phase:** Gradient. MeCN:water 3:97 for 5 min, to 40:60 over 15 min, return to initial conditions over 1 min, re-equilibrate for 10 min. (At the end of each day wash with MeCN:ethyl acetate 5:95.)
**Flow rate:** 0.5
**Injection volume:** 50
**Detector:** F ex 302 em 412 following post-column reaction. The column effluent mixed with reagent 1 pumped at 0.25 mL/min and with reagent 2 pumped at 0.25 mL/min and this mixture flowed through a 2.5 m × 0.8 mm i.d. PTFE coil at 40° to the detector. (Reagent 1 was 10 mM o-phthalaldehyde in EtOH:700 mM phosphoric acid 2:98. Reagent 2 was 20 mM β-mercaptoethanol in EtOH:700 mM phosphoric acid 2:98.)

## CHROMATOGRAM
**Retention time:** 26
**Limit of detection:** 11 ng/mL

## OTHER SUBSTANCES
**Extracted:** sulfadiazine, sulfaguanidine, sulfanilamide, sulfapyridine
**Noninterfering:** sulfathiazole

## KEY WORDS
post-column reaction

## REFERENCE
Viñas, P.; Erroz, C.L.; Campillo, N.; Hernández-Córdoba, M. Determination of sulphonamides in foods by liquid chromatography with postcolumn fluorescence derivatization. *J.Chromatogr.A*, **1996**, *726*, 125–131

## SAMPLE
**Matrix:** formulations
**Sample preparation:** Powder tablets or pills. Weigh out an amount of powdered tablets or pills or capsule contents, dissolve in 5 mL EtOH, dilute with 150 mM HCl containing 40 mM sodium dodecyl sulfate. Dilute suspensions or drops with 150 mM HCl containing 40 mM sodium dodecyl sulfate. Filter solutions if necessary. 10 mL Solution in 150 mM HCl containing 40 mM sodium dodecyl sulfate + 1 mL 100 mM sodium nitrite, let stand for 5 min, add 1 mL 300 mM sulfamic acid, let stand for 10 min, add 500 μL 30 mM N-(1-naphthyl)ethylenediamine dihydrochloride, make up to 25 mL with water, inject an aliquot.

## HPLC VARIABLES
**Guard column:** 35 × 4.6 C18 (Scharlau)
**Column:** 125 × 4.6 5 μm Spherisorb ODS-2 C18
**Mobile phase:** Pentanol:50 mM sodium dodecyl sulfate 2.4:97.6, pH adjusted to 7 with 100 mM phosphate buffer
**Flow rate:** 1
**Injection volume:** 20
**Detector:** UV 490

## CHROMATOGRAM
**Retention time:** 7.5
**Limit of detection:** 100 ng/mL

## OTHER SUBSTANCES
**Simultaneous:** sulfacetamide, sulfadiazine, sulfaguanidine, sulfamerazine, sulfamethizole, sulfanilamide, sulfathiazole
**Noninterfering:** benzocaine

## KEY WORDS
tablets; pills; capsules; suspensions; drops; derivatization

## REFERENCE
Garcia-Alvarez-Coque, M.C.; Simo-Alfonso, E.F.; Ramis-Ramos, G.; Esteve-Romero, J.S. High-performance micellar liquid chromatography determination of sulphonamides in pharmaceuticals after azodye precolumn derivatization. *J.Pharm.Biomed.Anal.*, **1995**, *13*, 237–245

## SAMPLE
**Matrix:** microsomal incubations
**Sample preparation:** Add a volume of 15% perchloric acid equal to 10% of the volume of the microsomal incubation, centrifuge, inject an aliquot of the supernatant.

## HPLC VARIABLES
**Guard column:** 30 × 4.6 5 μm Ultracarb ODS(30) C18 (Phenomenex)
**Column:** 150 × 4.6 5 μm Ultracarb ODS(30) C18 (Phenomenex)
**Mobile phase:** MeCN:water:glacial acetic acid:triethylamine 20:80:1:0.05
**Flow rate:** 1.5
**Detector:** UV 254

## CHROMATOGRAM
**Retention time:** 8.7

## OTHER SUBSTANCES
**Extracted:** metabolites

## REFERENCE
Cribb, A.E.; Spielberg, S.P.; Griffin, G.P. N⁴-Hydroxylation of sulfamethoxazole by cytochrome P450 of the cytochrome P4502C subfamily and reduction of sulfamethoxazole hydroxylamine in human and rat hepatic microsomes. *Drug Metab.Dispos.*, **1995**, *23*, 406–414

## SAMPLE
**Matrix:** milk
**Sample preparation:** 500 µL Milk + 2 g C18 material + 10 µL MeOH + 10 µL 12.5 µg/mL sulfamerazine in MeOH, let stand for 1 min, grind with a glass pestle until homogeneous, place in a 10 mL syringe barrel plugged with filter paper, place filter paper on top, compress to 4.5 mL with a plunger, restrict column outlet with a 100 µL pipette tip, wash with 8 mL hexane, remove excess hexane with positive pressure, elute with 8 mL dichloromethane, elute excess dichloromethane with positive pressure. Evaporate the eluate under a stream of nitrogen, dissolve the residue in 100 µL MeOH and 400 µL 17 mM orthophosphoric acid, sonicate for 5-10 min, centrifuge at 13600 g for 5 min, filter supernatant (0.45 µm), inject a 20 µL aliquot. (C18 material was Analytichem 40 µm 18% load endcapped. Add 22 g to a 50 mL syringe barrel wash with 2 column volumes of hexane, 2 volumes of dichloromethane, and 2 volumes of MeOH, vacuum aspirate until dry.)

## HPLC VARIABLES
**Column:** 75 × 4 3 µm Supelcosil LC-18
**Mobile phase:** MeCN:17 mM orthophosphoric acid 10:90
**Column temperature:** 45
**Flow rate:** 1 for 5 min then 2 for remainder of run
**Injection volume:** 20
**Detector:** UV 270

## CHROMATOGRAM
**Retention time:** 7
**Internal standard:** sulfamerazine (3)
**Limit of detection:** 62.5 ng/mL

## OTHER SUBSTANCES
**Extracted:** sulfadiazine, sulfadimethoxine, sulfamethazine, sulfanilamide, sulfathiazole, sulfisoxazole

## KEY WORDS
SPE; matrix solid-phase dispersion

## REFERENCE
Long, A.R.; Short, C.R.; Barker, S.A. Method for the isolation and liquid chromatographic determination of eight sulfonamides in milk. *J.Chromatogr.*, **1990**, *502*, 87–94

## SAMPLE
**Matrix:** reaction mixtures
**Sample preparation:** If necessary, remove oxidizing power of solution by adding sodium metabisulfite, inject a 20 µL aliquot.

## HPLC VARIABLES
**Guard column:** 15 × 4.6 5 μm Microsorb C8
**Column:** 250 × 4.6 5 μm Microsorb C8
**Mobile phase:** MeCN:5.5 mM sodium octanesulfonate + 20 mM trisodium citrate dihydrate adjusted to pH 3 with concentrated HCl 35:65
**Flow rate:** 1
**Injection volume:** 20
**Detector:** UV 270

## CHROMATOGRAM
**Retention time:** 6.6
**Limit of detection:** 200 ng/mL

## REFERENCE
Lunn, G.; Rhodes, S.W.; Sansone, E.B.; Schmuff, N.R. Photolytic destruction and polymeric resin decontamination of aqueous solutions of pharmaceuticals. *J.Pharm.Sci.*, **1994**, *83*, 1289–1293

## SAMPLE
**Matrix:** solutions

## HPLC VARIABLES
**Column:** 300 × 0.35 5 μm Vydac IDI-TP C18 TMS capped
**Mobile phase:** Gradient. MeCN:MeOH:buffer 0:0:100 at start of run, to 0:5:95 after injection (step gradient), to 0:8:92 over 7 min, to 6:0:94 (step gradient), maintain at 6:0:94 for 14 min, to 0:16:84 over 5 min, to 0:18:82 over 5 min, to 0:30:70 over 5 min. (Buffer was 1 mM pH 2.72 phosphate buffer.)
**Column temperature:** 30
**Flow rate:** 0.006
**Injection volume:** 1
**Detector:** UV 270

## CHROMATOGRAM
**Retention time:** 42

## OTHER SUBSTANCES
**Simultaneous:** diaveridine, phthalyl sulfathiazole, pyrimethamine, succinyl sulfathiazole, sulfabenzamide, sulfacetamide, sulfachloropyridazine, sulfadiazine, sulfadimethoxine, sulfaguanidine, sulfamerazine, sulfameter, sulfamethazine, sulfamethizole, sulfamethoxypyridazine, sulfamoxole, sulfanilamide, sulfanilic acid, sulfapyridine, sulfaquinoxaline, sulfathiazole, sulfisomidine, sulfisoxazole, trimethoprim

## KEY WORDS
capillary HPLC

## REFERENCE
Ricci, M.C.; Cross, R.F. High performance liquid chromatographic analyses of sulphonamides and dihydrofolate reductase inhibitors. II. Separations in acetonitrile modified solutions, ternary gradient studies & flow programming. *J.Liq.Chromatogr.& Rel.Technol.*, **1996**, *19*, 547–564

## SAMPLE
**Matrix:** solutions

## HPLC VARIABLES
**Column:** 300 × 0.35 5 μm Vydac IDI-TP C18 TMS capped

**Mobile phase:** Gradient. MeOH:1 mM pH 2.72 phosphate buffer 0:100 at start of run, to 10:90 after injection (step gradient), to 12:88 over 30 min, to 18:82 over 5 min, to 30: 70 over 5 min.
**Column temperature:** 30
**Flow rate:** 0.006
**Injection volume:** 1
**Detector:** UV 270

## CHROMATOGRAM
**Retention time:** 46

## OTHER SUBSTANCES
**Simultaneous:** diaveridine, phthalyl sulfathiazole, pyrimethamine, succinyl sulfathiazole, sulfabenzamide, sulfacetamide, sulfachloropyridazine, sulfadiazine, sulfadimethoxine, sulfaguanidine, sulfamerazine, sulfameter, sulfamethazine, sulfamethizole, sulfamethoxypyridazine, sulfamoxole, sulfanilamide, sulfanilic acid, sulfapyridine, sulfaquinoxaline, sulfathiazole, sulfisomidine, sulfisoxazole, trimethoprim

## KEY WORDS
capillary HPLC

## REFERENCE
Ricci, M.C.; Cross, R.F. High-performance liquid chromatographic analyses of sulphonamides and dihydrofolate reductase inhibitors. I. Separations in methanol-modified solutions. *J.Liq.Chrom. Rel.Technol.*, **1996**, *19*, 365–381

## SAMPLE
**Matrix:** solutions

## HPLC VARIABLES
**Column:** 150 × 4.6 5 μm Ultrasphere C18
**Mobile phase:** MeCN:water 27:73, adjusted to pH 3.5 with acetic acid
**Flow rate:** 2
**Detector:** UV 254

## CHROMATOGRAM
**Retention time:** 2.4

## OTHER SUBSTANCES
**Simultaneous:** metabolites

## REFERENCE
Nakamura, H.; Uetrecht, J.; Cribb, A.E.; Miller, M.A.; Zahid, N.; Hill, J.; Josephy, P.D.; Grant, D.M.; Spielberg, S.P. *In vitro* formation, disposition and toxicity of N-acetoxysulfamethoxazole, a potent mediator of sulfamethoxazole toxicity. *J.Pharmacol.Exp.Ther.*, **1995**, *274*, 1099–1104

## SAMPLE
**Matrix:** solutions
**Sample preparation:** Prepare a 2.5-5 μg/mL solution, inject a 20 μL aliquot.

## HPLC VARIABLES
**Column:** 80 × 4.6 3.65 μm Zorbax Rx-SIL (similar to Zorbax SB-C8 (Mac-Mod Analytical))
**Mobile phase:** MeCN:buffer 27:73 (Buffer was 0.1% trifluoroacetic acid adjusted to pH 3 with ammonium hydroxide.)
**Flow rate:** 1

**Injection volume:** 20
**Detector:** UV 267

## CHROMATOGRAM
**Retention time:** k′ 2.8

## REFERENCE
Kirkland, K.M.; McCombs, D.A.; Kirkland, J.J. Rapid, high-resolution high-performance liquid chromatographic analysis of antibiotics. *J.Chromatogr.A*, **1994**, *660*, 327–337

## SAMPLE
**Matrix:** solutions
**Sample preparation:** Prepare a solution in mobile phase, inject a 20 µL aliquot.

## HPLC VARIABLES
**Column:** 250 × 4.6 Nucleosil 5C18
**Mobile phase:** MeCN:10 mM pH 5.6 phosphate buffer 8:92
**Flow rate:** 1
**Injection volume:** 20
**Detector:** UV 270

## CHROMATOGRAM
**Retention time:** 44.0

## OTHER SUBSTANCES
**Simultaneous:** N-acetylsulfisomidine, sulfachloropyridazine, sulfadimethoxine, sulfadoxine, sulfamethazine (sulfadimidine), sulfamethoxypyridazine, sulfamonomethoxine, sulfisomidine, sulfisoxazole

## REFERENCE
Nishikawa, M.; Takahashi, Y.; Ishihara, Y. High performance liquid chromatographic determination of sulfisomidine and N4-acetylsulisomidine in swine tissues. *J.Liq.Chromatogr.*, **1993**, *16*, 4031–4047

## SAMPLE
**Matrix:** solutions
**Sample preparation:** Prepare a solution in MeOH:water 25:75, inject a 5 µL aliquot.

## HPLC VARIABLES
**Column:** 250 × 2.1 5 µm 201TP (Vydac)
**Mobile phase:** Gradient. A was MeCN containing 0.1% trifluoroacetic acid. B was water containing 0.1% trifluoroacetic acid. A:B from 5:95 to 40:60 over 20 min.
**Flow rate:** 0.2
**Injection volume:** 5
**Detector:** UV 270; MS, Sciex API III triple quadrupole, IonSpray interface

## CHROMATOGRAM
**Retention time:** 14.86

## OTHER SUBSTANCES
**Simultaneous:** phthalylsulfathiazole, succinylsulfathiazole, sulfabenzamide, sulfacetamide, sulfachloropyridazine, sulfadiazine, sulfadimethoxine, sulfaguanidine, sulfamerazine, sulfameter, sulfamethazine, sulfamethizole, sulfamethoxypyridazine, sulfamoxole, sulfanilamide, sulfapyridine, sulfaquinoxaline, sulfathiazole, sulfisomidine, sulfisoxazole

## REFERENCE

Pleasance, S.; Blay, P.; Quilliam, M.A.; O'Hara, G. Determination of sulfonamides by liquid chromatography, ultraviolet diode array detection and ion-spray tandem mass spectrometry with application to cultured salmon flesh. *J.Chromatogr.*, **1991**, *558*, 155–173

## SAMPLE
**Matrix:** solutions

## HPLC VARIABLES
**Column:** 250 × 4 OmniPac PCX-500 (Dionex)
**Mobile phase:** Gradient. A was MeCN:110 mM perchloric acid and 20 mM sodium acetate 27:73. B was MeCN:110 mM perchloric acid and 100 mM sodium acetate 50:50. A:B from 100:0 to 0:100 over 5 min, then re-equilibrate.
**Flow rate:** 1
**Detector:** UV 254

## CHROMATOGRAM
**Retention time:** 6.8

## OTHER SUBSTANCES
**Simultaneous:** sulfamerazine, sulfamethazine, sulfamethizole, sulfanilamide, sulfanilic acid, sulfathiazole, sulfisoxazole

## REFERENCE
Slingsby, R.W.; Rey, M. Determination of pharmaceuticals by multi-phase chromatography: Combined reversed phase and ion exchange in one column. *J.Liq.Chromatogr.*, **1990**, *13*, 107–134

## SAMPLE
**Matrix:** solutions
**Sample preparation:** Dissolve in MeOH:water 1:1 at a concentration of 50 µg/mL, inject a 10 µL aliquot.

## HPLC VARIABLES
**Column:** 300 × 3.9 10 µm µBondapak C18
**Mobile phase:** MeOH:acetic acid:triethylamine:water 20:1.5:0.5:78
**Flow rate:** 1.5
**Injection volume:** 10
**Detector:** UV 254

## CHROMATOGRAM
**Retention time:** 14

## OTHER SUBSTANCES
**Simultaneous:** sulfachlorpyridine, sulfadiazine, sulfamerazine, sulfamethazine, sulfamethizole, sulfanilamide, sulfanilic acid, sulfapyridine, sulfisoxazole

## REFERENCE
Roos, R.W.; Lau-Cam, C.A. General reversed-phase high-performance liquid chromatographic method for the separation of drugs using triethylamine as a competing base. *J.Chromatogr.*, **1986**, *370*, 403–418

## SAMPLE
**Matrix:** solutions
**Sample preparation:** Inject an aliquot of a solution in MeOH.

## HPLC VARIABLES
**Column:** 300 × 3.9 μBondapak C18
**Mobile phase:** MeCN:water:acetic acid 12.5:86.5:1
**Flow rate:** 1.6
**Injection volume:** 10
**Detector:** UV 254

## CHROMATOGRAM
**Retention time:** 16

## OTHER SUBSTANCES
**Simultaneous:** sulfabenzamide, sulfacetamide, sulfachlorpyridazine, sulfadiazine, sulfadimethoxine, sulfamerazine, sulfamethazine, sulfamethizole, sulfanilamide, sulfapyridine, sulfisoxazole

## REFERENCE
Roos, R.W. High pressure liquid chromatographic determination of sulfisoxazole in pharmaceuticals and separation patterns of other sulfonamides. *J.Assoc.Off.Anal.Chem.*, **1981**, *64*, 851–854

## SAMPLE
**Matrix:** tissue
**Sample preparation:** Condition a 3 mL 500 mg Sep-Pak SPE cartridge with 20 mL MeOH and 20 mL water. 5 g Homogenized tissue + 40 μL 20 μg/mL sulfaethoxypyridazine in water + 25 mL chloroform, shake mechanically for 2 min, centrifuge at 3000 g for 5 min, remove the supernatant and separate the layers. Add the aqueous layer to the residue and repeat the extraction. Combine the chloroform layers and add 10 mL 10% NaCl in 100 mM NaOH, shake vigorously for 1 min, remove the upper aqueous layer and centrifuge it at 1500 g for 10 min. Remove 8 mL of the upper aqueous layer and add it to 10 mL 1 M pH 6 NaH$_2$PO$_4$, vortex for 20 s, add to the SPE cartridge, wash with 20 mL water, elute with 1 mL MeCN. Evaporate the eluate to dryness under a stream of nitrogen at 50°, reconstitute in 2 mL mobile phase, vortex for 20 s, heat at 50° for 5 min, cool, filter (Gelman Acrodisc 0.45 μm), inject a 20-50 μL aliquot.

## HPLC VARIABLES
**Column:** 250 × 4.6 5 μm Spherisorb C18 ODS
**Mobile phase:** MeCN:10 mM pH 4.6 ammonium acetate 28:721.2
**Flow rate:** 1.2
**Injection volume:** 20-50
**Detector:** UV 265; MS, VG TRIO 2 quadrupole, ion source 189°, capillary jet 320°

## CHROMATOGRAM
**Retention time:** 11.7
**Internal standard:** sulfaethoxypyridazine (12.8)

## OTHER SUBSTANCES
**Extracted:** sulfachloropyridazine, sulfadiazine, sulfadimethoxine, sulfadoxine, sulfamerazine, sulfamethazine, sulfamethoxypyridazine, sulfathiazole

## KEY WORDS
cow; pig; muscle; liver; SPE

## REFERENCE
Boison, J.O.; Keng, L.J.-Y. Determination of sulfadimethoxine and sulfamethazine residues in animal tissues by liquid chromatography and thermospray mass spectrometry. *J.AOAC Int.*, **1995**, *78*, 651–658

## SAMPLE
**Matrix:** tissue
**Sample preparation:** Extract with supercritical carbon dioxide into a MeOH solution.

## HPLC VARIABLES
**Guard column:** 20 mm long Supelguard LC-18
**Column:** 150 × 4.6 5 μm Supelcosil LC-18
**Mobile phase:** MeOH:100 mM $KH_2PO_4$ adjusted to pH 4.5 with phosphoric acid 28:72
**Flow rate:** 0.5
**Detector:** UV 270

## CHROMATOGRAM
**Retention time:** 18

## OTHER SUBSTANCES
**Simultaneous:** $N^4$-acetylsulfamethoxazole, sulfamerazine, sulfamethazine, sulfamethizole, sulfamethoxypyridazine

## KEY WORDS
chicken; pig; liver; muscle; SFE

## REFERENCE
Cross, R.F.; Ezzell, J.L.; Richter, B.E. The supercritical fluid extraction of polar drugs (sulphonamides) from inert matrices and meat animal products. *J.Chromatogr.Sci.*, **1993**, *31*, 162–169

## SAMPLE
**Matrix:** tissue
**Sample preparation:** 1-3 g Tissue + 2 (liver) or 3 (muscle) mL 0.7% trichloroacetic acid in acetone, mix in Whirlmixer, sonicate for 10 min at 40°, add 2 mL 10 mM pH 6 $Na_2HPO_4$, sonicate for 5 min, add 100 μL 500 mM NaOH, add 9 (muscle) or 10 (liver) mL dichloromethane, mix thoroughly for 1 min, centrifuge at 2240 g for 5 min. Remove 6 mL of the organic layer and evaporate it to dryness at 40°under a stream of nitrogen. Dissolve the residue in 400 (muscle) or 800 (liver) μL MeCN:10 mM pH 2.8 phosphate buffer 20 :80, sonicate, wash with 1 mL hexane. Sonicate the aqueous phase for 1 min, centrifuge through a Spin-X filter tube, inject a 10 μL aliquot of the supernatant.

## HPLC VARIABLES
**Guard column:** 20 × 4.6 5 μm Supelcosil LC-18 DB
**Column:** 250 × 4.6 5 μm Supelcosil LC-18 DB
**Mobile phase:** MeCN:buffer 20:80 (tissue) with 0.1% triethylamine added (Buffer was 25 mM sodium phosphate and 20 mM sodium 1-hexanesulfonate, pH adjusted to 2.8 with 5 M phosphoric acid.)
**Flow rate:** 0.9
**Injection volume:** 10
**Detector:** UV 270

## CHROMATOGRAM
**Retention time:** 18
**Internal standard:** sulfamethoxazole

## OTHER SUBSTANCES
**Extracted:** sulfadiazine, trimethoprim

## KEY WORDS
fish; salmon; trout; sulfamethoxazole is IS; muscle; liver

## REFERENCE

Hormazabal, V.; Rogstad, A. Simultaneous determination of sulphadiazine and trimethoprim in plasma and tissues of cultured fish for residual and pharmacokinetic studies. *J.Chromatogr.*, **1992**, *583*, 201–207

## SAMPLE

**Matrix:** tissue

**Sample preparation:** Blend 3 g meat with 30 mL chloroform for 2 min in a Polytron homogenizer, shake for 10 min, centrifuge at 1600 g for 5 min, filter (5A filter paper). Add 5 mL filtrate to 1 mL 3 M HCl, shake for 10 min, centrifuge at 1600 g for 5 min. 250 μL Aqueous layer + 250 μL 3.5 M sodium acetate solution, vortex, add 100 μL 0.2% fluorescamine in acetone, vortex, let stand for 20 min at room temperature, inject a 10 μL aliquot.

## HPLC VARIABLES

**Column:** 150 × 4.6 5 μm Chemcosorb 5-ODS-H

**Mobile phase:** MeCN:2% acetic acid 5:3

**Column temperature:** 55

**Flow rate:** 1

**Injection volume:** 10

**Detector:** F ex 405 em 495

## CHROMATOGRAM

**Retention time:** 12

**Limit of detection:** 5 pg/g

## OTHER SUBSTANCES

**Extracted:** sulfadiazine, sulfadimethoxine, sulfamerazine, sulfamethazine (sulfadimidine), sulfamonomethoxine, sulfaquinoxaline, sulfisomidine

## KEY WORDS

cow; pig; chicken; ham; sausage; roast beef; derivatization

## REFERENCE

Takeda, N.; Akiyama, Y. Pre-column derivatization of sulfa drugs with fluorescamine and high-performance liquid chromatographic determination at their residual levels in meat and meat products. *J.Chromatogr.*, **1991**, *558*, 175–180

## SAMPLE

**Matrix:** tissue

**Sample preparation:** 500 mg Tissue + 2 g C18 material + 10 μL MeOH, let stand for 2 min, grind gently with a glass pestle until homogeneous, place in a 10 mL syringe barrel plugged with filter paper, place filter paper on top, compress to 4.5 mL with a plunger, restrict column outlet with a 100 μL pipette tip, wash with 8 mL hexane, remove excess hexane with positive pressure, elute with 8 mL dichloromethane, elute excess dichloromethane with positive pressure. Evaporate the eluate under a stream of nitrogen at 40°, dissolve the residue in 500 μL mobile phase, sonicate for 5-10 min, centrifuge at 17000 g for 5 min, filter supernatant (0.45 μm), inject a 25 μL aliquot. (C18 material was Analytichem 40 μm 18% load endcapped. Add 22 g to a 50 mL syringe barrel wash with 2 column volumes of hexane, 2 volumes of dichloromethane, and 2 volumes of MeOH, vacuum aspirate until dry.)

## HPLC VARIABLES

**Column:** 300 × 4 MicroPak ODS

**Mobile phase:** MeCN:17 mM orthophosphoric acid 35:65

**Column temperature:** 40

**Flow rate:** 1
**Injection volume:** 25
**Detector:** UV 270

## CHROMATOGRAM
**Retention time:** 6
**Internal standard:** sulfamethoxazole

## OTHER SUBSTANCES
**Extracted:** sulfadimethoxine

## KEY WORDS
muscle; fish; SPE; catfish; matrix solid-phase dispersion; sulfamethoxazole is IS

## REFERENCE
Long, A.R.; Hsieh, L.C.; Malbrough, M.S.; Short, C.R.; Barker, S.A. Matrix solid phase dispersion isolation and liquid chromatographic determination of sulfadimethoxine in catfish (Ictalurus punctatus) muscle tissue. *J.Assoc.Off.Anal.Chem.*, **1990**, *73*, 868–871

## SAMPLE
**Matrix:** urine
**Sample preparation:** 2 mL Urine + 10 mL 150 mM HCl containing 40 mM sodium dodecyl sulfate + 1 mL 100 mM sodium nitrite, let stand for 5 min, add 1 mL 300 mM sulfamic acid, let stand for 10 min, add 500 µL 30 mM N-(1-naphthyl)ethylenediamine dihydrochloride, make up to 25 mL with water, inject a 20 µL aliquot.

## HPLC VARIABLES
**Guard column:** 35 × 4.6 C18 (Scharlau)
**Column:** 125 × 4.6 5 µm Spherisorb ODS-2 C18
**Mobile phase:** Pentanol:50 mM sodium dodecyl sulfate 2.4:97.6, pH adjusted to 7 with 100 mM phosphate buffer
**Flow rate:** 1
**Injection volume:** 20
**Detector:** UV 490

## CHROMATOGRAM
**Retention time:** 8.5
**Limit of detection:** 100 ng/mL

## OTHER SUBSTANCES
**Extracted:** sulfamethizole, sulfathiazole
**Interfering:** sulfadiazine, sulfaguanidine

## KEY WORDS
derivatization

## REFERENCE
Simó-Alfonso, E.F.; Ramis-Ramos, G.; García-Alvarez-Coque, M.C.; Esteve-Romero, J.S. Determination of sulphonamides in human urine by azo dye precolumn derivatization and micellar liquid chromatography. *J.Chromatogr.B*, **1995**, *670*, 183–187

## ANNOTATED BIBLIOGRAPHY
Barker, S.A.; Long, A.R. Preparation of milk samples for immunoassay and liquid chromatographic screening using matrix solid-phase dispersion. *J.AOAC Int.*, **1994**, *77*, 848–854 [MSPD; infant for-

mula; also albendazole, chloramphenicol, chlorsulon, chlortetracycline, fenbendazole, furazolidone, mebendazole, oxfendazole, oxytetracycline, sulfadiazine, sulfadimethoxine, sulfamerazine, sulfamethazine, sulfanilamide, sulfathiazole, sulfisoxazole, tetracycline, thiabendazole; LOD 0.1-2 ng]

Bonazzi, D.; Andrisano, V.; Di Pietra, A.M.; Cavrini, V. Analysis of trimethoprim−sulfonamide drug combinations in dosage forms by UV spectroscopy and liquid chromatography (HPLC). *Farmaco*, **1994**, *49*, 381−386 [simultaneous sulfadiazine, sulfamethoxypyridazine, trimethoprim]

Lee, B.L.; Delahunty, T.; Safrin, S. The hydroxylamine of sulfamethoxazole and adverse reactions in patients with acquired immunodeficiency syndrome. *Clin.Pharmacol.Ther.*, **1994**, *56*, 184−189 [urine; extracted N-acetylsulfamethoxazole, sulfamethoxazole hydroxylamine; dinitrobenzylpyridine (IS); normal phase; LOD 5 μg/mL]

Mokry, M.; Klimes, J.; Zahradnicek, M. HPLC analysis of some sulfonamides in selected pharmaceutical formulations. *Pharmazie*, **1994**, *49*, 333−335 [tablets; injections; acetanilide (IS); simultaneous trimethoprim; also phenacetin, phenazone, phenobarbital, phthalylsulfathiazole, sulfadimidine, sulfamethoxydiazine, sulfamoxole, sulfisoxazole]

Shah, K.P.; Chang, M.; Riley, C.M. Automated analytical systems for drug development studies. II-A system for dissolution testing. *J.Pharm.Biomed.Anal.*, **1994**, *12*, 1519−1527 [tablets; also acetaminophen, trimethoprim]

Mengelers, M.J.B.; Polman, A.M.M.; Aerts, M.M.L.; Kuiper, H.A.; Van Miert, A.S.J.P.A.M. Determination of sulfadimethoxine, sulfamethoxazole, trimethoprim and their main metabolites in lung and edible tissues from pigs by multi-dimensional liquid chromatography. *J.Liq.Chromatogr.*, **1993**, *16*, 257−278 [muscle; kidney; liver; LOD 10-50 ng/g; column temp 30; column-switching]

Endoh, Y.S.; Takahashi, Y.; Nishikawa, M. HPLC determination of sulfonamides, their N4-acetyl metabolites and diaminopyrimidine coccidiostats in chicken tissues. *J.Liq.Chromatogr.*, **1992**, *15*, 2091−2110 [skin; plasma; muscle; liver; kidney; LOD 20-50 ng/g; also N-acetyldiaveridine, N-acetylsulfadiazine, N-acetylsulfadimethoxine, N-acetylsulfamethoxazole, N-acetylsulfamonomethoxine, N-acetylsulfaquinoxaline, diaveridine, ormetoprim, sulfadiazine, sulfadimethoxine, sulfamonomethoxine, sulfaquinoxaline, trimethoprim]

Avgerinos, A.; Athanasiou, G.; Malamataris, S. Rapid simultaneous determination of trimethoprim, sulfamethoxazole and acetylsulfamethoxazole in human plasma and urine by high-performance liquid chromatography. *J.Pharm.Biomed.Anal.*, **1991**, *9*, 507−510

Nelis, H.J.; Leger, F.; Sorgeloos, P.; De Leenheer, A.P. Liquid chromatographic determination of efficacy of incorporation of trimethoprim and sulfamethoxazole in brine shrimp (Artemia spp.) used for prophylactic chemotherapy of fish. *Antimicrob.Agents Chemother.*, **1991**, *35*, 2486−2489

Basci, N.E.; Bozkurt, A.; Kayaalp, S.O.; Isimer, A. Comparison of colorimetric and high-performance liquid chromatographic determination of sulfamethoxazole and acetylsulfamethoxazole. *J.Chromatogr.*, **1990**, *527*, 174−181 [serum; sulfamethazine (IS)]

DeAngelis, D.V.; Woolley, J.L.; Sigel, C.W. High-performance liquid chromatographic assay for the simultaneous measurement of trimethoprim and sulfamethoxazole in plasma or urine. *Ther.Drug Monit.*, **1990**, *12*, 382−392

Varoquaux, O.; Cordonnier, P.; Advenier, C.; Pays, M. Simultaneous HPLC determination of trimethoprim, sulfamethoxazole and its N4-acetyl metabolite in biological fluids. *Methodol.Surv. Biochem.Anal.*, **1990**, *20*, 123−130

Cross, R.F. Narrow-bore high-performance liquid chromatography separations of 22 sulfonamides. *J.Chromatogr.*, **1989**, *478*, 422−428 [simultaneous phthalyl sulfathiazole, succinyl sulfathiazole, sulfabenzamide, sulfacetamide, sulfachloropyridazine, sulfadiazine, sulfadimethoxine, sulfaguanidine, sulfamerazine, sulfameter, sulfamethazine, sulfamethizole, sulfamethoxypyridazine, sulfamoxole, sulfanilamide, sulfanilic acid, sulfapyridine, sulfaquinoxaline, sulfathiazole, sulfisomidine, sulfisoxazole]

Long, A.R.; Hsieh, L.C.; Malbrough, M.S.; Short, C.R.; Barker, S.A. A multiresidue method for the isolation and liquid chromatographic determination of seven sulfonamides in infant formula. *J.Liq.Chromatogr.*, **1989**, *12*, 1601−1612 [extracted sulfadiazine, sulfadimethoxine, sulfamerazine, sulfamethazine, sulfathiazole, sulfisoxazole; column temp 45; SPE]

Mengelers, M.J.B.; Oorsprong, M.B.M.; Kuiper, H.A.; Aerts, M.M..L.; Van Gogh, E.R.; Van Miert, A.S.J.P.A.M. Determination of sulfadimethoxine, sulfamethoxazole, trimethoprim and their main metabolites in porcine plasma by column switching HPLC. *J.Pharm.Biomed.Anal.*, **1989**, *7*, 1765−1776

Erdmann, G.R.; Canafax, D.M.; Giebink, G.S. High-performance liquid chromatographic analysis of trimethoprim and sulfamethoxazole in microliter volumes of chinchilla middle ear effusion and serum. *J.Chromatogr.*, **1988**, *433*, 187–195 [LOQ 500 ng/mL; LOD 250 ng/mL; cimetidine (IS); column temp 45°; extracted trimethoprim; pharmacokinetics]

Robinson, J.W. Normal phase liquid chromatographic determination of sulfamethoxazole in tablets: collaborative study. *J.Assoc.Off.Anal.Chem.*, **1985**, *68*, 88–91

Spreux-Varoquaux, O.; Chapalain, J.P.; Cordonnier, P.; Advenier, C.; Pays, M.; Lamine, L. Determination of trimethoprim, sulfamethoxazole and its N4-acetyl metabolite in biological fluids by high-performance liquid chromatography. *J.Chromatogr.*, **1983**, *274*, 187–199

Cairnes, D.A.; Evans, W.E. High-performance liquid chromatographic assay of methotrexate, 7-hydroxymethotrexate, 4-deoxy-4-amino-N10-methylpteroic acid and sulfamethoxazole in serum, urine and cerebrospinal fluid. *J.Chromatogr.*, **1982**, *231*, 103–110

Essers, L.; Korte, H. Comparison of the conventional methods and high-performance liquid chromatography for the determination of trimethoprim, sulfamethoxazole and its metabolite in serum. *Chemotherapy (Basel)*, **1982**, *28*, 247–252

Gochin, R.; Kanfer, I.; Haigh, J.M. Simultaneous determination of trimethoprim, sulfamethoxazole and N4-acetylsulfamethoxazole in serum and urine by high-performance liquid chromatography. *J.Chromatogr.*, **1981**, *223*, 139–145

Ascalone, V. Assay of trimethoprim, sulfamethoxazole and its N4-acetyl metabolite in biological fluids by high-pressure liquid chromatography. *J.High Resolut.Chromatogr.Chromatogr.Commun.*, **1980**, *3*, 261–264

Ferry, D.G.; McQueen, E.G.; Hearn, M.T.W. Sulfamethoxazole and trimethoprim estimation by high performance liquid chromatography. *Proc.Univ.Otago Med.Sch.*, **1978**, *56*, 46–48

# Sumatriptan

**Molecular formula:** $C_{14}H_{21}N_3O_2S$
**Molecular weight:** 295.4
**CAS Registry No.:** 103628-46-2 (sumatriptan), 103628-48-4 (sumatriptan succinate)

## SAMPLE
**Matrix:** blood
**Sample preparation:** Condition a Bond-Elut C2 SPE cartridge with two 1 mL portions of MeOH and two 1 mL portions of water. 1 mL Plasma + 100 µL 100 ng/mL IS in water, vortex briefly, add to the SPE cartridge at 2 mL/min, wash with 1 mL water, wash with two 1 mL portions of MeOH:water 30:70, elute with 1 mL MeOH:10 mM pH 5.0 ammonium acetate 60:40, evaporate the eluate to dryness under reduced pressure at 50°, reconstitute with 200 µL mobile phase, sonicate for 10 min, vortex, inject a 25 µL aliquot.

## HPLC VARIABLES
**Column:** 250 × 4.6 5 µm CN (Beckman)
**Mobile phase:** MeCN:MeOH:water:trifluoroacetic acid 36:6:58:0.1
**Flow rate:** 1.2
**Injection volume:** 25
**Detector:** MS, Sciex Model API III triple quadrupole, nebulizer probe 500°, nebulizing gas 80 psi, auxiliary flow 2 L/min, corona discharge needle +3 µA, 0.1143 mm orifice, orifice 40 V, collision gas argon

## CHROMATOGRAM
**Retention time:** 3
**Internal standard:** 3-[2-(diethylamino)ethyl)-N-methyl-1H-indole-5-methanesulfonamide (L-737,404, Merck) (3.35)
**Limit of quantitation:** 0.5 ng/mL

## KEY WORDS
plasma; SPE; pharmacokinetics

## REFERENCE
McLoughlin, D.A.; Olah, T.V.; Ellis, J.D.; Gilbert, J.D.; Halpin, R.A. Quantitation of the 5HT$_{1D}$ agonists MK-462 and sumatriptan in plasma by liquid chromatography-atmospheric pressure chemical ionization mass spectrometry. *J.Chromatogr.A*, **1996**, *726*, 115–124

## SAMPLE
**Matrix:** blood
**Sample preparation:** Condition an AASP C2 SPE cartridge (Jones Chromatography) with 1.8 mL MeOH and 1.8 mL water. Centrifuge plasma, remove a 1 mL aliquot and add it to 50 ng IS, vortex, add to the SPE cartridge, wash with 1.8 mL water, wash with 1.8 mL MeOH:water 30:70, elute the contents of the SPE cartridge onto the analytical column with the mobile phase.

## HPLC VARIABLES
**Guard column:** C18 Guard-Pak
**Column:** 50 × 4.6 3 μm Spherisorb ODS-2
**Mobile phase:** MeOH:100 mM ammonium acetate 60:40
**Flow rate:** 1
**Detector:** MS, thermospray, interface tip 148-152°, ion source 250°, m/z 296

## CHROMATOGRAM
**Retention time:** 1.2
**Internal standard:** tritium-labeled sumatriptan (on N-methyl carbon) (m/z 299)
**Limit of quantitation:** 2 ng/mL

## KEY WORDS
pharmacokinetics; SPE; plasma

## REFERENCE
Oxford, J.; Lant, M.S. Development and validation of a liquid chromatographic-mass spectrometric assay for the determination of sumatriptan in plasma. *J.Chromatogr.*, **1989**, *496*, 137–146

## SAMPLE
**Matrix:** blood, urine
**Sample preparation:** Plasma. 1 mL Plasma + 100 μL 4 M NaOH + 2.6 mL ethyl acetate:dichloromethane 80:20, agitate on a rotational shaker for 15 min, centrifuge at 1000 g for 4 min. Remove 2 mL of the organic phase and add it to 300 μL pH 7 phosphate buffer and 2 mL hexane, agitate on a rotational shaker for 15 min, centrifuge at 1000 g for 4 min, freeze in acetone/dry ice, discard the organic layer, thaw the aqueous layer and inject a 200 μL aliquot. Urine. Dilute 20-fold with pH 7 phosphate buffer, inject a 20 μL aliquot.

## HPLC VARIABLES
**Guard column:** 10 mm long 5 μm Spherisorb ODS-1
**Column:** 125 × 4.6 5 μm Spherisorb ODS-1
**Mobile phase:** MeOH:buffer 60:40 (Buffer was 5.25 g $Na_2HPO_4.2H_2O$ and 2.79 g $KH_2PO_4$ in 1 L water, pH 7.0.)
**Column temperature:** 40
**Flow rate:** 1
**Injection volume:** 20-200
**Detector:** E, ESA Coulochem model 5100A, 5011 analytical cell +0.55 V (conditioning cell) +0.8 V (analytical cell), guard cell +0.9 V; F ex 280 em 350 (see Pharm. Res. 1995, 12, 138)

## CHROMATOGRAM
**Limit of quantitation:** 1 ng/mL (plasma); 200 ng/mL (urine)

## KEY WORDS
plasma; dog; rat; human; pharmacokinetics

## REFERENCE
Andrew, P.D.; Birch, H.L.; Phillpot, D.A. Determination of sumatriptan succinate in plasma and urine by high-performance liquid chromatography with electrochemical detection. *J.Pharm.Sci.*, **1993**, *82*, 73–76

# Temazepam

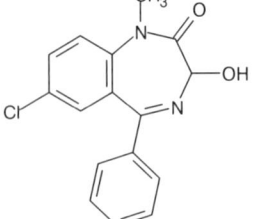

**Molecular formula:** $C_{16}H_{13}ClN_2O_2$
**Molecular weight:** 300.7
**CAS Registry No.:** 846-50-4

## SAMPLE
**Matrix:** blood
**Sample preparation:** 2 mL Whole blood or plasma + 2 mL buffer + 5 mL chloroform: isopropanol:n-heptane 60:14:26, shake gently horizontally for 10 min, centrifuge at 2800 g for 10 min. Remove the lower organic layer and evaporate it to dryness under vacuum at 45°, reconstitute the residue in 100 µL mobile phase, centrifuge at 2800 g for 5 min, inject a 50 µL aliquot of the supernatant. (Buffer was saturated ammonium chloride solution 25% diluted with water, adjusted to pH 9.5 with 25% ammonia solution.)

## HPLC VARIABLES
**Column:** 300 × 3.9 4 µm NovaPack C18
**Mobile phase:** MeOH:THF:buffer 65:5:30 (Buffer was 0.68 g/L (10 mM (sic)) $KH_2PO_4$ adjusted to pH 2.6 with concentrated orthophosphoric acid.) (At the end of each session wash the column with water for 1 h and MeOH for 1 h, re-equilibrate for 30 min.)
**Column temperature:** 30
**Flow rate:** 0.8
**Injection volume:** 50
**Detector:** UV 230

## CHROMATOGRAM
**Retention time:** 4.58
**Limit of detection:** <120 ng/mL

## OTHER SUBSTANCES
**Extracted:** acebutolol, acenocoumarol, acepromazine, aceprometazine, acetaminophen, aconitine, ajmaline, albuterol, alimemazine, alpidem, alprazolam, alprenolol, amisulpride, amitriptyline, amoxapine, aspirin, astemizole, atenolol, benazepril, benzocaine, benzoylecgonine, bepridil, betaxolol, bisoprolol, bromazepam, brompheniramine, bumadizone, bupivacaine, buprenorphine, buspirone, caffeine, carbamazepine, carbinoxamine, carpipramine, carteolol, celiprolol, cetirizine, chlorambucil, chlordiazepoxide, chlormezanone, chlorophenacinone, chlorpheniramine, chlorpromazine, chlorpropamide, cibenzoline, clemastine, clobazam, clomipramine, clonazepam, clonidine, clorazepate, clozapine, codeine, colchicine, cyamemazine, cyclizine, cycloguanil, cyproheptadine, cytarabine, dacarbazine, daunorubicin, debrisoquine, demexiptiline, desipramine, dextromethorphan, dextromoramide, dextropropoxyphene, diazepam, diazoxide, diclofenac, dihydralazine, diltiazem, diphenhydramine, dipyridamole, disopyramide, dosulepine, doxepin, ephedrine, estazolam, etodolac, fenfluramine, fenoprofen, fentiazac, flecainide, floctafenine, flumazenil, flunitrazepam, fluoxetine, fluphenazine, flurbiprofen, fluvoxamine, glibenclamide, glibornuride, glipizide, glutethimide, haloperidol, histapyrrodine, hydroxyzine, ibuprofen, imipramine, indomethacin, iproniazid, ketamine, levomepromazine, lidoflazine, lisinopril, loperamide, loprazolam, loratadine, lorazepam, loxapine, maprotiline, medazepam, medifoxamine, mefenamic acid, mefenidramine, mefloquine, melphalan, mephenesin, mephentermine, metapramine, metformin, methadone, methaqualone, methocarbamol, methotrexate, metipranolol, metoclopramide, metoprolol, mexiletine, mianserine, midazolam, minoxidil, moperone, morphine, nadolol, nalbuphine, nalorphine, naloxone, naltrexone, naproxen, nialamide, nicardipine, nifedipine, niflumic acid, nimodipine, nitrazepam, nitrendipine, nizatidine, nomifensine, nortriptyline, omeprazole, opipramol, oxazepam, oxprenolol, penbutolol, penfluridol, pentazocine, phencyclidine, phenobarbital, phenol, phenylbutazone, pimozide, pindolol, pipamperone, piroxicam, prazepam, prazosin,

prilocaine, procainamide, procarbazine, proguanil, promethazine, propafenone, propranolol, protriptyline, pyrimethamine, quinidine, quinine, quinupramine, ramipril, ranitidine, reserpine, ritodrine, sotalol, strychnine, sulfinpyrazole, sulindac, sulpride, sultopride, suriclone, tenoxicam, terbutaline, terfenadine, tetracaine, tetrazepam, thiopental, thioproperazine, thioridazine, tianeptine, tiapride, tiaprofenic acid, ticlopidine, tioclomarol, tofisopam, tolbutamide, toloxatone, trazodone, triazolam, trifluoperazine, trifluperidol, trimipramine, triprolidine, tropatenine, verapamil, vinblastine, vincristine, vindesine, warfarin, zopiclone, zorubicine

**Interfering:** alminoprofen, amodiaquine, benperidol, chloroquine, cicletanine, cocaine, doxylamine, droperidol, hydroxychloroquine, ketoprofen, labetalol, lidocaine, meperidine, mepivacaine, moclobemide, secobarbital, timolol, viloxazine, yohimbine, zolpidem

## KEY WORDS
whole blood; plasma

## REFERENCE
Tracqui, A.; Kintz, P.; Mangin, P. Systematic toxicological analysis using HPLC/DAD. *J.Forensic Sci.*, **1995**, *40*, 254–262

## SAMPLE
**Matrix:** blood
**Sample preparation:** Automated SPE by ASPEC system. Condition a C18 Clean-Up SPE cartridge (CEC 18111, Worldwide Monitoring) with 2 mL MeOH then 2 mL water. 1 mL Plasma + 1 mL 400 ng/mL protriptyline in water, vortex, add to the SPE cartridge, wash with 3 mL water, wash with 3 mL 750 mL/L methanol. Elute with three aliquots of 300 µL 0.1 M ammonium acetate in MeOH. Add 0.5 mL 0.5 M NaOH and 4 mL 50 mL/L isopropanol in heptane to eluate, mix thoroughly. Allow 5 min for phase separation. Remove upper heptane phase and add it to 300 µL 0.1 M phosphoric acid (pH 2.5), mix, separate, inject a 100 µL aliquot of the aqueous phase.

## HPLC VARIABLES
**Guard column:** LC-8-DB (Supelco)
**Column:** 150 × 4.6 LC-8-DB (Supelco)
**Mobile phase:** MeCN:buffer 35:65 (Buffer was 10 mL/L triethylamine in water adjusted to pH 5.5 with glacial acetic acid.)
**Flow rate:** 2
**Injection volume:** 100
**Detector:** UV 228

## CHROMATOGRAM
**Retention time:** 5.3
**Internal standard:** protriptyline (4)

## OTHER SUBSTANCES
**Extracted:** acetazolamide, amitriptyline, chlordiazepoxide, chlorimipramine, chlorpromazine, desipramine, dextromethorphan, diazepam, diphenhydramine, doxepin, encainide, fentanyl, flecainide, fluoxetine, flurazepam, haloperidol, hydroxyethylflurazepam, ibuprofen, imipramine, lidocaine, maprotiline, methadone, methaqualone, mexiletine, midazolam, norchlorimipramine, nordiazepam, nordoxepin, norfluoxetine, nortriptyline, norverapamil, pentazocine, promazine, propafenone, propoxyphene, propranolol, protriptyline, quinidine, trazodone, trimipramine, verapamil
**Noninterfering:** acetaminophen, acetylmorphine, amiodarone, amobarbital, amphetamine, bendroflumethiazide, benzocaine, benzoylecgonine, benzthiazide, butalbital, carbamazepine, chlorothiazide, clonazepam, cocaine, codeine, cotinine, cyclosporine, cyclothiazide, desalkylflurazepam, diamorphine, dicumerol, ephedrine, ethacrynic acid, ethanol, ethchlorvynol, ethosuximide, furosemide, glutethimide, hydrochlorothiazide, hydrocodone, hydroflumethiazide, hydromorphone, lorazepam, mephentermine, meprobamate, meth-

amphetamine, metharbital, methoxsalen, methoxyphenteramine, methsuximide, methylcyclothiazide, metoprolol, MHPG, monoacetylmorphine, morphine, normethsuximide, oxazepam, oxycodone, oxymorphone, pentobarbital, phencyclidine, phenteramine, phenylephrine, phenytoin, polythiazide, primidone, prochlorperazine, salicylic acid, sulfanilamide, THC-COOH, theophylline, thiazolam, thiopental, thioridazine, tocainide, trichloromethiazide, trifluoperazine, valproic acid, warfarin
**Interfering:** amitriptyline, maprotiline, methadone, nordiazepam, norfluoxetine, propoxyphene

## KEY WORDS
plasma; SPE

## REFERENCE
Nichols, J.H.; Charlson, J.R.; Lawson, G.M. Automated HPLC assay of fluoxetine and norfluoxetine in serum. *Clin.Chem.*, **1994**, *40*, 1312–1316

## SAMPLE
**Matrix:** blood
**Sample preparation:** Inject 100-200 μL plasma onto column A with mobile phase A and elute to waste, after 5 min backflush the contents of column A onto column B with mobile phase B, after 5 min remove column A from the circuit, elute column B with mobile phase B, monitor the effluent from column B. Wash column A with MeCN:water 60:40 at 1 mL/min for 6 min then re-equilibrate with pH 7.5 buffer for 10 min.

## HPLC VARIABLES
**Column:** A 45 × 4 12 μm TSK-gel G 3 PW (Tosohass); B 75 × 4.6 Ultrasphere ODS C18 3 μm
**Mobile phase:** A 50 mM pH 7.5 phosphate buffer; B Gradient. X was MeCN. Y was 65 mM $KH_2PO_4$ + 1% diethylamine adjusted to pH 5.4 with phosphoric acid. X:Y 22:78 for 5 min, to 25:75 over 10 min, to 60:40 over 15 min.
**Flow rate:** 1
**Injection volume:** 100-200
**Detector:** UV 230

## CHROMATOGRAM
**Retention time:** 23.5

## OTHER SUBSTANCES
**Extracted:** alprazolam, bromazepam, chlordiazepoxide, clobazam, clonazepam, clorazepate, clotiazepam, desmethylclobazam, desmethyldiazepam, diazepam, estazolam, flunitrazepam, loflazepate, lorazepam, medazepam, nitrazepam, oxazepam, prazepam, tetrazepam, tofisopam, triazolam
**Noninterfering:** carbamazepine, ethosuximide, phenobarbital, phenytoin, primidone, valproic acid

## KEY WORDS
plasma; column-switching

## REFERENCE
Lacroix, C.; Wojciechowski, F.; Danger, P. Monitoring of benzodiazepines (clobazam, diazepam and their main active metabolites) in human plasma by column-switching high-performance liquid chromatography. *J.Chromatogr.*, **1993**, *617*, 285–290

## SAMPLE
**Matrix:** blood

**Sample preparation:** Filter (0.5 μm) serum, inject 200 μL directly onto column A with mobile phase A, run with mobile phase A for 1.5 min then change to mobile phase B over 0.1 min, wash column A with mobile phase B for 10.5 min, backflush column A onto column B with mobile phase C for 7.5 min then switch column B out of circuit, elute column B with mobile phase C and monitor the eluant, re-equilibrate column A with mobile phase A for at least 5 min.

## HPLC VARIABLES
**Column:** A 15 × 3.2 5 μm Brownlee ODS; B 250 × 1 5 μm Adsorbosphere ODS
**Mobile phase:** A 10 mM sodium dodecyl sulfate; B water; C MeOH:water 65:35
**Flow rate:** A 1; B 1; C 0.06
**Injection volume:** 200
**Detector:** UV 242

## CHROMATOGRAM
**Retention time:** 26
**Limit of detection:** 30 ng/mL

## OTHER SUBSTANCES
**Simultaneous:** diazepam, nordiazepam, oxazepam

## KEY WORDS
serum; column-switching; microbore

## REFERENCE
Koenigbauer, M.J.; Curtis, M.A. Use of micellar mobile phases and microbore column switching for the assay of drugs in physiological fluids. *J.Chromatogr.*, **1988**, *427*, 277–285

## SAMPLE
**Matrix:** blood
**Sample preparation:** 1 mL Serum + 2 mL water + 50 μL 3.2 μg/mL estazolam in MeOH + 2 mL 100 mM NaOH, mix gently, add 8 mL diethyl ether, shake for 15 min, centrifuge at 2500 rpm for 5 min. Remove 4 mL of the organic layer and evaporate it to dryness under a stream of nitrogen at 40°, reconstitute the residue in 100 μL mobile phase, vortex for 30 s, inject a 50 μL aliquot.

## HPLC VARIABLES
**Column:** 50 × 4.6 Shim-pack FLC-C8 (Shimadzu)
**Mobile phase:** MeOH:buffer 53:47 (Buffer was 5 mM $Na_2HPO_4$ adjusted to pH 6.0 with phosphoric acid.)
**Flow rate:** 0.6
**Injection volume:** 50
**Detector:** UV 254

## CHROMATOGRAM
**Retention time:** 6
**Internal standard:** estazolam (4)
**Limit of detection:** 8 ng/mL

## OTHER SUBSTANCES
**Extracted:** diazepam, nordiazepam, oxazepam, triazolam
**Simultaneous:** bromazepam, flunitrazepam, nitrazepam, sulpride
**Noninterfering:** haloperidol, trihexyphenidyl
**Interfering:** clorazepate

## KEY WORDS
serum; pharmacokinetics

## REFERENCE

Tada, K.; Moroji, T.; Sekiguchi, R.; Motomura, H.; Noguchi, T. Liquid-chromatographic assay of diazepam and its major metabolites in serum, and application to pharmacokinetic study of high doses of diazepam in schizophrenics. *Clin.Chem.*, **1985**, *31*, 1712–1715

## SAMPLE

**Matrix:** blood, dialysate, urine

**Sample preparation:** Condition a 100 mg C18 Bond Elut SPE cartridge with 2 mL MeOH then 2 mL water. 1 mL Plasma or 1 mL urine or 10 mL dialysate + 1 mL MeCN:water 30:70, vortex for 10 s, centrifuge at 4000 g for 5 min, add to the SPE cartridge, wash with 2 mL MeCN:water 20:80, dry for 3 to 4 min, elute with four 200 µL aliquots of MeOH, evaporate combined eluates under nitrogen and take up residue in 100 µL MeOH, inject a 20 µL aliquot.

## HPLC VARIABLES

**Guard column:** 4 × 4 Lichrocart Lichrosorb RP 18-5

**Column:** 5 µm Lichrocart 125-4 Lichrospher 100 RP 18 endcapped

**Mobile phase:** MeCN:10 mM pH 5.6 buffer 40:60 (Prepare 1 M buffer from 94.8 mL 1 M $KH_2PO_4$ + 5.2 mL 1 M $K_2HPO_4$, dilute to 10 mM with water.)

**Flow rate:** 1.6

**Injection volume:** 20

**Detector:** UV 254

## CHROMATOGRAM

**Retention time:** 3.17

**Internal standard:** climazolam (5.62)

**Limit of quantitation:** 5 ng/mL

## OTHER SUBSTANCES

**Extracted:** oxazepam

## KEY WORDS

plasma; SPE

## REFERENCE

Chopineau, J.; Rivault, F.; Sautou, V.; Sommier, M.F. Determination of temazepam and its active metabolite, oxazepam in plasma, urine and dialysate using solid-phase extraction followed by high performance liquid chromatography. *J.Liq.Chromatogr.*, **1994**, *17*, 373–383

## SAMPLE

**Matrix:** blood, gastric contents, tissue, urine

**Sample preparation:** Tissue homogenates were 1:2 in water. 1 mL Sample + 1 mL saturated sodium borate buffer + 100 µL 20 µg/mL methyl clonazepam in water + 5 mL n-butyl chloride, rotate at 40 rpm for 30 min, centrifuge at 2500 rcf for 5 min. Remove the organic phase and evaporate it to dryness at 70° under a stream of air, reconstitute the residue in 300 µL mobile phase, vortex for 30 s, inject a 20 µL aliquot.

## HPLC VARIABLES

**Column:** 250 × 2.1 Analytichem ODS with an integral guard column

**Mobile phase:** MeCN:100 mM $KH_2PO_4$ 300:700, adjust pH to 3.00 with concentrated phosphoric acid

**Column temperature:** 60

**Flow rate:** 1.5

**Injection volume:** 20

**Detector:** UV 242

## CHROMATOGRAM
**Retention time:** 4.57
**Internal standard:** methyl clonazepam (5.36)
**Limit of detection:** 50 ng/mL

## OTHER SUBSTANCES
**Extracted:** diazepam, trazodone
**Also analyzed:** acetaminophen, alprazolam, amitriptyline, amoxapine, carbamazepine, chlordiazepoxide, chlorpromazine, chlorprothixene, clonazepam, demoxepam, desipramine, diphenhydramine, disopyramide, doxepin, ethotoin, flurazepam, glutethimide, haloperidol, haloperidol, imipramine, lidocaine, lorazepam, loxapine, maprotiline, mesantoin, mesoridazine, methaqualone, methotrimeprazine, nordiazepam, nortriptyline, oxazepam, pentazocine, perphenazine, phenacetin, phenobarbital, phenytoin, promazine, promethazine, propranolol, protriptyline, salicylic acid, thiothixene, trifluoperazine, triflupromazine, trimipramine
**Noninterfering:** chloral hydrate, codeine, ketamine, meperidine, methadone, methamphetamine, methyprylon, thioridazine

## KEY WORDS
serum; plasma; whole blood

## REFERENCE
Root, I.; Ohlson, G.B. Trazodone overdose: report of two cases. *J.Anal.Toxicol.*, **1984**, *8*, 91–94

## SAMPLE
**Matrix:** blood, milk
**Sample preparation:** 500 μL Plasma or milk + 25 μL 5 μg/mL flurazepam in water: MeCN 2.5:97.5 + 500 μL 67 mM pH 7.4 phosphate buffer + 7 mL diethyl ether, extract for 15 min (A). Remove ether layer and add it to 1 mL 1.5 M HCl, shake for 15 min. Freeze and discard ether phase. Basify aqueous phase with 1 mL 2 M NaOH, extract with 7 mL diethyl ether for 15 min. Evaporate ether at 37° under a stream of nitrogen and take up residue in mobile phase, inject an aliquot. (For plasma **only** ether at (A) can be evaporated at 37° under a stream of nitrogen, take up residue in mobile phase, inject an aliquot.)

## HPLC VARIABLES
**Guard column:** 25 × 4 5 μm LiChrospher 60 RP-select B
**Column:** 125 × 4 5 μm LiChrospher 60 RP-select B
**Mobile phase:** MeCN: 10 mM $KH_2PO_4$ 31:69, adjusted to pH 2.80 with phosphoric acid
**Column temperature:** 45
**Flow rate:** 2
**Injection volume:** 50
**Detector:** UV 241

## CHROMATOGRAM
**Retention time:** 6.3
**Internal standard:** flurazepam (3.0)

## OTHER SUBSTANCES
**Extracted:** diazepam, nordiazepam, oxazepam

## KEY WORDS
plasma; human; rabbit

## REFERENCE
Stebler, T.; Guentert, T.W. Determination of diazepam and nordazepam in milk and plasma in the presence of oxazepam and temazepam. *J.Chromatogr.*, **1991**, *564*, 330–337

## SAMPLE
**Matrix:** blood, tissue, urine
**Sample preparation:** 1 mL Blood, urine, or liver homogenate + 1 mL 1.15 M pH 6.4 phosphate buffer, add 25 µg prazepam, extract with 5 mL n-butyl chloride, centrifuge. Remove the organic layer and evaporate it in a vortex-evaporator. Reconstitute the residue in 100 µL MeOH, inject a 20 µL aliquot.

## HPLC VARIABLES
**Column:** 250 × 4.6 Partisil-10 ODS
**Mobile phase:** MeCN:1 mM pH 3.2 phosphate buffer 40:60
**Column temperature:** 50
**Flow rate:** 3
**Injection volume:** 20
**Detector:** UV 240

## CHROMATOGRAM
**Retention time:** 3.2
**Internal standard:** prazepam (7.2)

## KEY WORDS
liver

## REFERENCE
Martin, C.D.; Chan, S.C. Distribution of temazepam in body fluids and tissues in lethal overdose. *J.Anal.Toxicol.*, **1986**, *10*, 77–78

## SAMPLE
**Matrix:** blood, urine
**Sample preparation:** Wash a C2 Bond-Elut SPE cartridge with 1 column volume methanol and 1 column volume buffer. Add 1 mL of urine buffered with pH 6 100 mM phosphate buffer or plasma buffered with pH 8 100 mM phosphate buffer to the SPE cartridge, wash with 3 column volumes of water, wash with 1 mL of MeOH:water 30:70, elute with 1 mL of MeOH:water 60:40. Evaporate the eluate to dryness and take up the residue in 200 µL mobile phase, inject an aliquot.

## HPLC VARIABLES
**Column:** 35 × 4.6 5 µm ultrabase C18 (Scharlau)
**Mobile phase:** MeOH:water 60:40
**Flow rate:** 1.1
**Injection volume:** 20
**Detector:** UV 230

## CHROMATOGRAM
**Internal standard:** prazepam
**Limit of detection:** 88 ng/mL

## OTHER SUBSTANCES
**Also analyzed:** adinazolam, brotizolam, diazepam, midazolam, nordazepam, oxazepam

## KEY WORDS
plasma; SPE

## REFERENCE
Casas, M.; Berrueta, L.A.; Gallo, B.; Vicente, F. Solid-phase extraction of 1,4-benzodiazepines from biological fluids. *J.Pharm.Biomed.Anal.*, **1993**, *11*, 277–284

## SAMPLE
**Matrix:** blood, urine
**Sample preparation:** 1 mL Serum or urine + 50 μL 5 μg/mL nitrazepam in MeOH + 4 mL n-butyl chloride, vortex for 15 s, centrifuge at 9000 g for 5 min. Remove the organic layer and evaporate it to dryness under a stream of air at room temperature, reconstitute residue in 30 μL mobile phase, inject a 25 μL aliquot.

## HPLC VARIABLES
**Guard column:** 20 × 4.6 5 μm Supelguard LC-8
**Column:** 150 × 4.6 5 μm Supelcosil LC-8
**Mobile phase:** MeOH:water:buffer 55:25:20 (Buffer was 2.78 g $K_2HPO_4$ + 25.04 g $KH_2PO_4$ in 4 L water.)
**Flow rate:** 2.2
**Injection volume:** 25
**Detector:** UV 230

## CHROMATOGRAM
**Retention time:** 5.31
**Internal standard:** nitrazepam (3.19)
**Limit of detection:** 15.5 ng/mL
**Limit of quantitation:** 46.5 ng/mL

## OTHER SUBSTANCES
**Simultaneous:** alprazolam, chlordiazepoxide, clonazepam, N-desalkylflurazepam, N-desmethyldiazepam, diazepam, flurazepam, lorazepam, midazolam, oxazepam, triazolam

## KEY WORDS
serum

## REFERENCE
Kunsman, G.W.; Manno, J.E.; Przekop, M.A.; Manno, B.R.; Llorens, K.A.; Kunsman, C.M. Determination of temazepam and temazepam glucuronide by reversed-phase high-performance liquid chromatography. *J.Chromatogr.*, **1991**, *568*, 427–436

## SAMPLE
**Matrix:** formulations
**Sample preparation:** Weigh capsule contents, dissolve in 50 mL MeOH, add 5 mL 0.75 mg/mL sulfanilamide in MeOH, make up to 100 mL with MeOH, filter (0.45 μm), inject a 5 μL aliquot.

## HPLC VARIABLES
**Column:** 50 × 4.6 5 μm Supelcosil LC8DB C8
**Mobile phase:** MeOH:1% acetic acid 40:60
**Flow rate:** 1.5
**Injection volume:** 5
**Detector:** UV 254

## CHROMATOGRAM
**Retention time:** 2.8
**Internal standard:** sulfanilamide (0.8)

## OTHER SUBSTANCES
**Simultaneous:** degradation products

## KEY WORDS
capsules; stability-indicating

## REFERENCE

Fatmi, A.A.; Hickson, E.A. Determination of temazepam and related compounds in capsules by high-performance liquid chromatography. *J.Pharm.Sci.*, **1988**, *77*, 87–89

## SAMPLE

**Matrix:** formulations
**Sample preparation:** Dissolve contents of capsules in MeOH to give a final concentration of 0.1 mg/mL, filter (0.45 μm), inject a 20 μL aliquot.

## HPLC VARIABLES

**Column:** 250 × 4.6 LiChrosorb 10 RP-18
**Mobile phase:** MeOH:water 60:40
**Flow rate:** 3.5
**Injection volume:** 20
**Detector:** UV 258

## CHROMATOGRAM

**Retention time:** 4.5

## OTHER SUBSTANCES

**Simultaneous:** degradation products, diazepam

## KEY WORDS

capsules

## REFERENCE

Gordon, S.M.; Freeston, L.K.; Collins, A.J. Determination of temazepam and its major degradation products in soft gelatin capsules by isocratic reversed-phase high-performance liquid chromatography. *J.Chromatogr.*, **1986**, *368*, 180–183

## SAMPLE

**Matrix:** hair
**Sample preparation:** Wash hair in water, rinse 3 times with MeOH, dry, weigh. 5-25 mg Washed hair + 1 mL MeOH, heat at 55° for 18 h, adjust pH to 9.5-10. 1 mL Extract + 1 μg protriptyline + 1 mL water + 1 mL 200 mM sodium carbonate buffer, mix, extract with hexane:butanol 95:5 for 20 min. Remove the organic layer and add it to 100 μL 0.2% orthophosphoric acid, mix for 20 min, inject a 30 μL aliquot of the aqueous layer.

## HPLC VARIABLES

**Guard column:** 15 × 3.2 7 μm Newguard RP-18
**Column:** 100 × 4.6 Spheri-5 RP-C18
**Mobile phase:** MeCN:buffer 40:60 (Buffer was 1.2 L 100 mM pH 7.0 $NaH_2PO_4$ + 30 mL diethylamine.)
**Flow rate:** 2
**Injection volume:** 30
**Detector:** UV 214

## CHROMATOGRAM

**Internal standard:** protriptyline (4)

## OTHER SUBSTANCES

**Extracted:** amitriptyline, desipramine, diazepam, dothiepin, flunitrazepam, haloperidol, imipramine, nitrazepam, nortriptyline, oxazepam

## KEY WORDS

may be interferences

## REFERENCE

Couper, F.J.; McIntyre, I.M.; Drummer, O.H. Extraction of psychotropic drugs from human scalp hair. *J.Forensic Sci.*, **1995**, *40*, 83–86

## SAMPLE

**Matrix:** microsomal incubations
**Sample preparation:** 1 mL Microsomal incubation + 5 mL dichloromethane, extract, evaporate to dryness, reconstitute with MeOH, inject an aliquot.

## HPLC VARIABLES

**Column:** 150 × 4.6 Zorbax SB-C18
**Mobile phase:** MeCN:MeOH:water 10:40:50
**Flow rate:** 1
**Detector:** UV 232

## CHROMATOGRAM

**Internal standard:** 6β-hydroxyprogesterone

## OTHER SUBSTANCES

**Extracted:** diazepam

## KEY WORDS

rat; human; liver

## REFERENCE

Gelboin, H.V.; Krausz, K.W.; Goldfarb, I.; Buters, J.T.M.; Yang, S.K.; Gonzalez, F.J.; Korzekwa, K.R.; Shou, M. Inhibitory and non-inhibitory monoclonal antibodies to human cytochrome P450 3A3/4. *Biochem.Pharmacol.*, **1995**, *50*, 1841–1850

## SAMPLE

**Matrix:** solutions

## HPLC VARIABLES

**Column:** 250 × 4.6 5 μm Supelcosil LC-DP (A) or 250 × 4 5 μm LiChrospher 100 RP-8 (B)
**Mobile phase:** MeCN:0.025% phosphoric acid:buffer 25:10:5 (A) or 60:25:15 (B) (Buffer was 9 mL concentrated phosphoric acid and 10 mL triethylamine in 900 mL water, adjust pH to 3.4 with dilute phosphoric acid, make up to 1 L.)
**Flow rate:** 0.6
**Injection volume:** 25
**Detector:** UV 229

## CHROMATOGRAM

**Retention time:** 8.86 (A), 6.70 (B)

## OTHER SUBSTANCES

**Also analyzed:** acebutolol, acepromazine, acetaminophen, acetazolamide, acetophenazine, albuterol, alprazolam, amitriptyline, amobarbital, amoxapine, antipyrine, atenolol, atropine, azatadine, baclofen, benzocaine, bromocriptine, brompheniramine, brotizolam, bupivacaine, buspirone, butabarbital, butalbital, caffeine, carbamazepine, cetirizine, chlorcyclizine, chlordiazepoxide, chlormezanone, chloroquine, chlorpheniramine, chlorpromazine, chlorpropamide, chlorprothixene, chlorthalidone, chlorzoxazone, cimetidine, cisapride, clomipramine, clonazepam, clonidine, clozapine, cocaine, codeine, colchicine, cyclizine, cyclobenzaprine, dantrolene, desipramine, diazepam, diclofenac, diflunisal, diltiazem, diphenhydramine, diphenidol, diphenoxylate, dipyridamole, disopyramide, dobutamine, doxapram, doxepin, droperidol, encainide, ethidium bromide, ethopropazine, fenoprofen, fentanyl, flavoxate, fluoxetine, fluphenazine, flurazepam, flurbiprofen, fluvoxamine, furosemide, glutethimide, glyburide, guaifenesin, haloperidol, homatropine, hydralazine, hydrochlorothiazide, hydrocodone, hydromorphone, hydroxychloroquine, hydroxyzine, ibuprofen, imipramine, indomethacin, ketoconazole, ketoprofen, ketorolac, labetalol, levorphanol, lidocaine, loratadine, lorazepam, lovastatin, loxapine, mazindol, mefenamic acid, meperidine, mephenytoin, mepivacaine, mesoridazine, metaproterenol, metformin, methadone, methdilazine, methocarbamol, methotrexate, methotrimeprazine, methoxamine, methyldopa, methylphenidate, metoclopramide, metolazone, metoprolol, metronidazole, midazolam, moclobemide, morphine, nadolol, nalbuphine, naloxone, naphazoline, naproxen, nifedipine, nizatidine, norepinephrine, nortriptyline, oxazepam, oxycodone, oxymetazoline, paroxetine, pemoline, pentazocine, pentobarbital, pentoxifylline, perphenazine, pheniramine, phenobarbital, phenol, phenolphthalein, phentolamine, phenylbutazone, phenyltoloxamine, phenytoin, pimozide, pindolol, piroxicam, pramoxine, prazepam, prazosin, probenecid, procainamide, procaine, prochlorperazine, procyclidine, promazine, promethazine, propafenone, propantheline, propiomazine, propofol, propranolol, protriptyline, quazepam, quinidine, quinine, racemethorphan, ranitidine, remoxipride, risperidone, salicylic acid, scopolamine, secobarbital, sertraline, sotalol, spironolactone, sulfinpyrazone, sulindac, terbutaline, terfenadine, tetracaine, theophylline, thiethylperazine, thiopental, thioridazine, thiothixene, timolol, tocainide, tolbutamide, tolmetin, trazodone, triamterene, triazolam, trifluoperazine, triflupromazine, trimeprazine, trimethoprim, trimipramine, verapamil, warfarin, xylometazoline, yohimbine, zopiclone

## KEY WORDS
some details of plasma extraction

## REFERENCE
Koves, E.M. Use of high-performance liquid chromatography-diode array detection in forensic toxicology. *J.Chromatogr.A*, **1995**, *692*, 103–119

## SAMPLE
**Matrix:** solutions

## HPLC VARIABLES
**Column:** 150 × 4.6 cellulose 3,5-dimethylphenylcarbamate/10-undecenoate bonded to allylsilica
**Mobile phase:** MeCN:water 40:60
**Flow rate:** 1
**Injection volume:** 1000
**Detector:** UV 254

## CHROMATOGRAM
**Retention time:** k′ 5.92

## KEY WORDS
chiral; α = 1.17

## REFERENCE
Oliveros, L.; Lopez, P.; Minguillon, C.; Franco, P. Chiral chromatographic discrimination ability of a cellulose 3,5-dimethylphenylcarbamate/10-undecenoate mixed derivative fixed on several chromatographic matrices. *J.Liq.Chromatogr.*, **1995**, *18*, 1521–1532

## SAMPLE
**Matrix:** solutions
**Sample preparation:** Dissolve the compound, S-trolox methyl ether (Fluka), dicyclohexylcarbodiimide, and 4-dimethylaminopyridine in dichloromethane, stir at room temperature for 1 h, filter (0.45 μm), inject an aliquot.

## HPLC VARIABLES
**Column:** 300 × 0.32 5 µm LiChrosorb Diol
**Mobile phase:** Carbon dioxide:MeOH 91.5:8.5
**Column temperature:** 80
**Injection volume:** 0.2
**Detector:** UV 254

## CHROMATOGRAM
**Retention time:** 64.2 (second peak)

## KEY WORDS
derivatization; subcritical fluid chromatography; chiral; density of mobile phase 0.51 g/mL; resolution ($R_S$) 1.1

## REFERENCE
Almquist, S.R.; Petersson, P.; Walther, W.; Markides, K.E. Direct and indirect approaches to enantiomeric separation of benzodiazepines using micro column techniques. *J.Chromatogr.A*, **1994**, *679*, 139–146

## SAMPLE
**Matrix:** solutions

## HPLC VARIABLES
**Guard column:** 30 × 3.2 7 µm SI 100 ODS (not commercially available)
**Column:** 150 × 3.2 7 µm SI 100 ODS (not commercially available)
**Mobile phase:** MeCN:buffer 31.2:68.8 (Buffer was 6.66 g $KH_2PO_4$ and 4.8 g 85% phosphoric acid in 1 L water, pH 2.3.)
**Flow rate:** 0.5-1
**Detector:** UV 226; UV 308

## CHROMATOGRAM
**Retention time:** 8.6
**Internal standard:** 5-(4-methylphenyl)-5-phenylhydantoin (7.3)

## OTHER SUBSTANCES
**Also analyzed:** aspirin, caffeine, carbamazepine, chlordiazepoxide, chlorprothixene, clonazepam, diazepam, doxylamine, ethosuximide, furosemide, haloperidol, hydrochlorothiazide, methocarbamol, methotrimeprazine, nicotine, oxazepam, procaine, promazine, propafenone, propranolol, salicylamide, tetracaine, thiopental, triamterene, verapamil, zolpidem, zopiclone

## REFERENCE
Below, E.; Burrmann, M. Application of HPLC equipment with rapid scan detection to the identification of drugs in toxicological analysis. *J.Liq.Chromatogr.*, **1994**, *17*, 4131–4144

## SAMPLE
**Matrix:** solutions

## HPLC VARIABLES
**Column:** 250 × 4.6 Zorbax RX
**Mobile phase:** Gradient. A was 10 mL concentrated orthophosphoric acid and 7 mL triethylamine in 1 L water. B was 10 mL concentrated orthophosphoric acid and 7 mL triethylamine in 200 mL water, make up to 1 L with MeCN. A:B from 100:0 to 0:100 over 30 min, maintain at 0:100 for 5 min.
**Column temperature:** 30

**Flow rate:** 2
**Detector:** UV 210

## OTHER SUBSTANCES
**Also analyzed:** acepromazine, acetaminophen, acetophenazine, albuterol, aminophylline, amitriptyline, amobarbital, amoxapine, amphetamine, amylocaine, antipyrine, aprobarbital, aspirin, atenolol, atropine, avermectin, barbital, benzocaine, benzoic acid, benzotropine, benzphetamine, berberine, bibucaine, bromazepan, brompheniramine, buprenorphine, buspirone, butabarbital, butacaine, butethal, caffeine, carbamazepine, carbromal, chloramphenicol, chlordiazepoxide, chloroquine, chlorothiazide, chloroxylenol, chlorphenesin, chlorpheniramine, chlorpromazine, chlorpropamide, chlortetracycline, cimetidine, cinchonidine, cinchonine, clenbuterol, clonazepam, clonixin, clorazepate, cocaine, codeine, colchicine, cortisone, coumarin, cyclazocine, cyclobenzaprine, cyclothiazide, cyheptamide, cymarin, danazol, danthron, dapsone, debrisoquine, desipramine, dexamethasone, dextromethorphan, dextropropoxyphene, diamorphine, diazepam, diclofenac, diethylpropion, diethylstilbestrol, diflunisal, digitoxin, digoxin, diltiazem, diphenhydramine, diphenoxylate, diprenorphine, dipyrone, disulfiram, dopamine, doxapram, doxepin, dronabinol, ephedrine, epinephrine, epinine, estradiol, estriol, estrone, ethacrynic acid, ethosuximide, etonitazene, etorphine, eugenol, famotidine, fenbendazole, fencamfamine, fenoprofen, fenproporex, fentanyl, flubendazole, flufenamic acid, flunitrazepam, 5-fluorouracil, fluoxymesterone, fluphenazine, furosemide, gentisic acid, gitoxigenin, glipizide, glunixin, glutethimide, glybenclamide, guaiacol, halazepam, haloperidol, hydrochlorothiazide, hydrocodone, hydrocortisone, hydromorphone, hydroxyquinoline, ibogaine, ibuprofen, iminostilbene, imipramine, indomethacin, isocarbostyril, isocarboxazid, isoniazid, isoproterenol, isoxsuprine, ivermectin, ketamine, ketoprofen, kynurenic acid, levorphanol, lidocaine, lorazepam, lormetazepam, loxapine, mazindol, mebendazole, meclizine, meclofenamic acid, medazepam, mefenamic acid, megestrol, mepacrine, meperidine, mephentermine, mephenytoin, mephesin, mephobarbital, mepivacaine, mescaline, mesoridazine, methadone, methamphetamine, methapyrilene, methaqualone, methazolamide, methocarbamol, methoxamine, methsuximide, methyl salicylate, methyldopa, methyldopamine, methylphenidate, methylprednisolone, methyltestosterone, methyprylon, metoprolol, mibolerone, morphine, nadolol, nalorphine, naloxone, naltrexone, naphazoline, naproxen, nefopam, niacinamide, nicotine, nicotinic acid, nifedipine, niflumic acid, nitrazepam, norepinephrine, nortriptyline, noscapine, nylidrin, oxazepam, oxycodone, oxymorphone, oxyphenbutazone, oxytetracycline, papaverine, pargyline, pemoline, pentazocine, pentobarbital, persantine, phenacetin, phenazocine, phenazopyridine, phencyclidine, phendimetrazine, phenelzine, pheniramine, phenobarbital, phenothiazine, phensuximide, phentermine, phenylbutazone, phenylephrine, phenylpropanolamine, piperocaine, prazepam, prednisolone, primidone, probenecid, progesterone, propiomazine, propranolol, propylparaben, pseudoephedrine, puromycin, pyrilamine, pyrithyldione, quazepam, quinaldic acid, quinidine, quinine, ranitidine, recinnamine, reserpine, resorcinol, saccharin, albuterol, salicylamide, salicylic acid, scopolamine, scopoletin, secobarbital, strychnine, sulfacetamide, sufadiazine, sulfadimethoxine, sulfaethidole, sulfamerazine, sulfamethazine, sulfamethoxizole, sulfanilamide, sulfapyridine, sulfasoxizole, sulindac, tamoxifen, testosterone, tetracaine, tetracycline, tetramisole, thebaine, theobromine, theophylline, thiabendazole, thiamine, thiamylal, thiobarbituric acid, thioridazine, thiosalicylic acid, thiothixene, thymol, tolazamide, tolazoline, tobutamide, tolmetin, tranylcypromine, triamcinolone, tribenzylamine, trichloromethiazide, trifluoperazine, trihexyphenidyl, trimethoprim, tripelennamine, triprolidine, tropacocaine, tyramine, verapamil, vincamine, warfarin, yohimbine, zoxazolamine

## REFERENCE
Hill, D.W.; Kind, A.J. Reversed-phase solvent-gradient HPLC retention indexes of drugs. *J.Anal.Toxicol.*, **1994**, *18*, 233–242

## SAMPLE
**Matrix:** solutions
**Sample preparation:** Inject a 10 μL aliquot.

## HPLC VARIABLES
**Column:** 250 × 4.6 Vydac C18
**Mobile phase:** MeCN:20 mM pH 7.0 phosphate buffer 55:45
**Injection volume:** 10
**Detector:** UV 254

## CHROMATOGRAM
**Retention time:** 4
**Internal standard:** prazepam (7)

## OTHER SUBSTANCES
**Simultaneous:** degradation products

## REFERENCE
Yang, S.K. Acid-catalyzed ethanolysis of temazepam in anhydrous and aqueous ethanol solutions. *J.Pharm.Sci.*, **1994**, *83*, 898–902

## SAMPLE
**Matrix:** urine
**Sample preparation:** 1 mL Urine + 100 μL 5 mM pH 5.5 acetate buffer + 25 μL β-glucuronidase/arylsulfatase (0.235/0.065 U, Calbiochem), mix, heat at 37° for 16 h, add 50 μL 5-50 μg/mL prazepam in MeOH, add 1 mL saturated trisodium phosphate, add 3 mL dichloromethane, vortex for 2 min, centrifuge at 1610 g for 5 min. Remove a 2 mL aliquot of the organic layer and add it to 2 mL hexane and 2 mL 6 M HCl, vortex for 2 min, centrifuge at 1610 g for 5 min. Remove 1 mL of the aqueous phase and adjust pH to 6 with 1 mL 6 M NaOH and 1 mL saturated trisodium phosphate, add 3 mL dichloromethane, vortex for 2 min, centrifuge at 1610 g for 5 min. Remove the organic layer and evaporate it to dryness under a stream of nitrogen at 50°, reconstitute the residue in 150 μL mobile phase, inject a 60 μL aliquot.

## HPLC VARIABLES
**Column:** 250 × 4 5 μm LiChrospher 100 RP-18(e)
**Mobile phase:** MeOH:water:triethylamine 30:70:0.1, adjusted to pH 5.5 with phosphoric acid
**Flow rate:** 0.7
**Injection volume:** 60
**Detector:** UV 240

## CHROMATOGRAM
**Retention time:** 7.7
**Internal standard:** prazepam (17.0)
**Limit of detection:** 2 ng/mL

## OTHER SUBSTANCES
**Extracted:** desmethyldiazepam, diazepam, oxazepam
**Simultaneous:** amitriptyline, caffeine, carbamazepine, chlordiazepoxide, chlorpromazine, clonazepam, desipramine, flunitrazepam, flurazepam, haloperidol, imipramine, levomepromazine, mianserin, nitrazepam, nortriptyline, perphenazine, phenobarbital, phenytoin, sulpride, thioridazine, triazolam
**Interfering:** maprotiline

## REFERENCE
Chiba, K.; Horii, H.; Chiba, T.; Kato, Y.; Hirano, T.; Ishizaki, T. Development and preliminary application of high-performance liquid chromatographic assay of urinary metabolites of diazepam in humans. *J.Chromatogr.B*, **1995**, *668*, 77–84

## SAMPLE

**Matrix:** urine

**Sample preparation:** Heat 5 mL urine + 1 mL temazepam in MeOH with 1 mL β-glu-curonidase at 37° for 2.5 h, cool, adjust to pH 8.5 with saturated Na$_2$CO$_3$, extract with 10 mL dichloromethane. Evaporate, take up the residue in 200 μL mobile phase, inject an aliquot.

## HPLC VARIABLES

**Column:** 250 × 4.6 Brownlee 5 μm RP-8

**Mobile phase:** MeCN:10 mM KH$_2$PO$_4$:n-nonylamine 450:550:0.6, adjusted to pH 3.2 with phosphoric acid

**Flow rate:** 1.6

**Detector:** UV 225

## CHROMATOGRAM

**Retention time:** 7

**Internal standard:** temazepam

## OTHER SUBSTANCES

**Extracted:** alprazolam, triazolam

## KEY WORDS

temazepam is IS

## REFERENCE

Fraser, A.D. Urinary screening for alprazolam, triazolam, and their metabolites with the EMIT d.a.u. benzodiazepine metabolite assay. *J.Anal.Toxicol.*, **1987**, *11*, 263–266

◆————————————————◆————————————————◆

## ANNOTATED BIBLIOGRAPHY

Haginaka, J.; Kanasugi, N. Enantioselectivity of bovine serum albumin-bonded columns produced with isolated protein fragments. *J.Chromatogr.A*, **1995**, *694*, 71–80 [chiral; also benzoin, clorazepate, fenoprofen, flurbiprofen, ibuprofen, ketoprofen, lorazepam, lormetazepam, oxazepam, pranoprofen, warfarin]

Benhamou-Batut, F.; Demotes-Mainard, F.; Labat, L.; Vinçon, G.; Bannwarth, B. Determination of flunitrazepam by liquid chromatography. *J.Pharm.Biomed.Anal.*, **1994**, *12*, 931–936 [α-hydroxytriazolam (IS); extracted alprazolam, bromazepam, clobazam, clonazepam, diazepam, flunitrazepam, lorazepam, nitrazepam, nordiazepam, oxazepam, triazolam]

Yang, S.K. Base-catalysed rearrangement of temazepam. *J.Pharm.Biomed.Anal.*, **1994**, *12*, 209–219 [simultaneous degradation products]

Yang, T.J.; Pu, Q.L.; Yang, S.K. Hydrolysis of temazepam in simulated gastric fluid and its pharmacological consequence. *J.Pharm.Sci.*, **1994**, *83*, 1543–1547 [simultaneous degradation products]

Fernández, P.; Hermida, I.; Bermejo, A.M.; López-Rivadulla, M.; Cruz, A.; Concheiro, L. Simultaneous determination of diazepam and its metabolites in plasma by high-performance liquid chromatography. *J.Liq.Chromatogr.*, **1991**, *14*, 2587–2599 [extracted diazepam, nordiazepam, oxazepam; carbamazepine (IS); SPE]

Vree, T.B.; Baars, A.M.; Wuis, E.W. Direct high pressure liquid chromatographic analysis and preliminary pharmacokinetics of enantiomers of oxazepam and temazepam with their corresponding glucuronide conjugates. *Pharm.Weekbl.[Sci].*, **1991**, *13*, 83–90

Noggle, F.T., Jr.; Clark, C.R.; DeRuiter, J. Liquid chromatographic separation of some common benzodiazepines and their metabolites. *J.Liq.Chromatogr.*, **1990**, *13*, 4005–4021 [also alprazolam, bromazepam, chlordiazepoxide, clobazam, clonazepam, clorazepate, demoxepam, diazepam, estazolam, fludiazepam, flunitrazepam, flurazepam, halazepam, lorazepam, midazolam, nitrazepam, nordiazepam, oxazepam, prazepam, triazolam]

Bastos, M.L.A. Improvement of HPLC conditions for separation of diazepam and its metabolites in biological extracts. *J.Liq.Chromatogr.*, **1989**, *12*, 1919–1934 [liver; brain; bile; kidney; prazepam (IS); extracted demethyldiazepam, diazepam, oxazepam; gradient]

Pietrogrande, M.C.; Dondi, F.; Blo, G.; Borea, P.A.; Bighi, C. Retention behavior of benzodiazepines in normal-phase HPLC. Silica, cyano, and amino phases comparison. *J.Liq.Chromatogr.*, **1988**, *11*, 1313–1333 [also diazepam, lorazepam, medazepam, methyllorazepam, oxazepam, prazepam]

Blaschke, G. Chromatographic resolution of chiral drugs on polyamides and cellulose triacetate. *J.Liq.Chromatogr.*, **1986**, *9*, 341–368 [also camazepam, chlormezanon, chlorthalidone, ifosfamide, ketamine, methaqualone, mianserin, oxapadol, oxazepam, oxazolam, praziquantel, rolipram]

Pietrogrande, M.C.; Bighi, C.; Borea, P.A.; Barbaro, A.M.; Guerra, M.C.; Biagi, G.L. Relationship between log k' values of benzodiazepines and composition of the mobile phase. *J.Liq.Chromatogr.*, **1985**, *8*, 1711–1729 [also carbenicillin, chlordiazepoxide, diazepam, dicloxacillin, flurazepam, lorazepam, medazepam, methyllorazepam, nitrazepam, oxazepam, prazepam, testosterone]

Ho, P.C.; Triggs, E.J.; Heazlewood, V.; Bourne, D.W. Determination of nitrazepam and temazepam in plasma by high-performance liquid chromatography. *Ther.Drug Monit.*, **1983**, *5*, 303–307

# Terazosin

**Molecular formula:** $C_{19}H_{25}N_5O_4$
**Molecular weight:** 387.4
**CAS Registry No.:** 65390-64-7 (terazosin),
70024-40-7 (terazosin hydrochloride dihydrate),
63074-08-8 (terazosin hydrochloride)

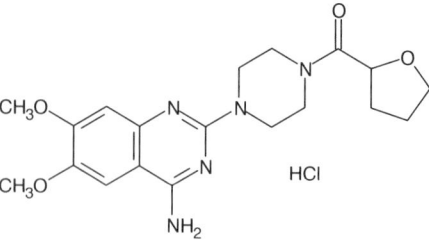

## SAMPLE
**Matrix:** blood, dialysate, urine
**Sample preparation:** 200 µL Serum, plasma, urine, or dialysate + 400 µL 200 ng/mL prazosin hydrochloride in MeCN at 5°, mix, centrifuge at 5° for 5-10 min. Remove the supernatant and evaporate it to dryness at 40-50° under a stream of air. Dissolve residue in 400 µL 100 mM mobile phase, inject a 50-100 µL aliquot.

## HPLC VARIABLES
**Column:** 300 × 3.9 5 µm Spherisorb ODS
**Mobile phase:** MeCN:THF:4-20 mM pH 5 sodium phosphate 25:6:69
**Flow rate:** 1.6
**Injection volume:** 50-100
**Detector:** F ex 250 em 370 (filter)

## CHROMATOGRAM
**Retention time:** 6
**Internal standard:** prazosin (9)
**Limit of quantitation:** 5 ng/mL (blood, dialysate), 50 ng/mL (urine)

## KEY WORDS
serum; plasma; human; dog

## REFERENCE
Patterson, S.E. Rapid and sensitive analysis of terazosin in plasma, peritoneal dialysis solution, and urine using high-performance liquid chromatography with fluorescence detection. *J.Chromatogr.*, **1984**, *311*, 206–212

## SAMPLE
**Matrix:** blood, dialysate, urine
**Sample preparation:** 1 mL Plasma, 1 mL dialysate, or 200 µL urine + 100 µL 1 M NaOH, mix, add 5 mL ethyl acetate:benzene 20:80 containing 2-50 ng/mL prazosin hydrochloride (Caution! Benzene is a carcinogen!), vortex for 5 min, centrifuge. Remove 4-4.5 mL of the organic phase and evaporate it to dryness at 40-50° under a stream of air. Dissolve residue in 300 µL 100 mM phosphoric acid or mobile phase, inject a 15-100 µL aliquot.

## HPLC VARIABLES
**Column:** 150 × 3.9 5 µm IBM C 1
**Mobile phase:** MeCN:THF:4-20 mM pH 7.0 sodium phosphate 22:6:72
**Flow rate:** 1
**Injection volume:** 15-100
**Detector:** F ex 250 em 370 (filter)

## CHROMATOGRAM
**Retention time:** 5
**Internal standard:** prazosin (8)
**Limit of quantitation:** 1 ng/mL (blood, dialysate), 10 ng/mL (urine)

## KEY WORDS

plasma; human; dog

## REFERENCE

Patterson, S.E. Rapid and sensitive analysis of terazosin in plasma, peritoneal dialysis solution, and urine using high-performance liquid chromatography with fluorescence detection. *J.Chromatogr.*, **1984**, *311*, 206–212

## SAMPLE

**Matrix:** blood, urine
**Sample preparation:** 50 µL Plasma or urine + 100 µL saturated NaCl + 50 µL 2 µg/mL dimethothiazine mesylate in water + 50 µL 4 M NaOH, vortex for 10 s, add 200 µL MTBE, vortex for 30 s, centrifuge at 9950 g for 2 min, inject a 110 µL aliquot of the organic phase.

## HPLC VARIABLES

**Column:** 250 × 5 5 µm Spherisorb S5W
**Mobile phase:** MeOH:10 mM ammonium perchlorate adjusted to pH 6.7 with 1 mL/L methanolic NaOH (0.1 M)
**Flow rate:** 2
**Injection volume:** 110
**Detector:** F ex 370 em 370-700

## CHROMATOGRAM

**Retention time:** 4
**Internal standard:** dimethothiazine mesylate (5)
**Limit of detection:** 1 µg/mL

## OTHER SUBSTANCES

**Extracted:** prazosin
**Simultaneous:** N-acetylprocainamide, ajmaline, chlorpromazine, desipramine, dipyridamole, doxazosin, flecainide, flurazepam, gallopamil, imipramine, ketanserin, metoprolol, mexiletine, mianserin, nadolol, nitrazepam, orphenadrine, oxprenolol, penbutolol, pindolol, prajmalium, procainamide, propranolol, protriptyline, pyrimethamine, quinidine, quinine, triamterene, trimipramine, verapamil
**Noninterfering:** amiodarone, atenolol, disopyramide, labetalol, lignocaine, lorcainide, methyldopa, nifedipine, prenalterol, propafenone, sotalol, timolol

## KEY WORDS

plasma

## REFERENCE

Bhamra, R.K.; Flanagan, R.J.; Holt, D.W. High-performance liquid chromatographic measurement of prazosin and terazosin in biological fluids. *J.Chromatogr.*, **1986**, *380*, 216–221

## SAMPLE

**Matrix:** bulk
**Sample preparation:** Dissolve 100 mg bulk material in 100 mL MeCN:50 mM pH 4.4 citrate buffer containing 10 mL 0.2 mg/mL 3,4-dimethoxybenzoic acid in MeOH, inject a 50 µL aliquot.

## HPLC VARIABLES

**Column:** 150 × 4.6 5 µm Zorbax Rx C8
**Mobile phase:** MeCN:isopropanol:50 mM pH 4.4 citrate buffer 175:50:1775
**Flow rate:** 2

**Injection volume:** 50
**Detector:** UV 254

## CHROMATOGRAM
**Retention time:** 27
**Internal standard:** 3,4-dimethoxybenzoic acid

## OTHER SUBSTANCES
**Simultaneous:** impurities

## KEY WORDS
rugged

## REFERENCE
Bauer, J.F.; Krogh, S.K.; Chang, Z.L.; Wong, C.F. Determination of minor impurities in terazosin hydrochloride by high-performance liquid chromatography. *J.Chromatogr.*, **1993**, *648*, 175–181

## SAMPLE
**Matrix:** solutions
**Sample preparation:** Prepare a 10 µg/mL solution in MeOH, inject a 20 µL aliquot.

## HPLC VARIABLES
**Column:** 125 × 4.9 Spherisorb S5W silica
**Mobile phase:** MeOH containing 10 mM ammonium perchlorate and 1 mL/L 100 mM NaOH in MeOH, pH 6.7
**Flow rate:** 2
**Injection volume:** 20
**Detector:** E, LeCarbone, V25 glassy carbon electrode, + 1.2 V

## CHROMATOGRAM
**Retention time:** 1.8

## OTHER SUBSTANCES
**Also analyzed:** acebutolol, acepromazine, acetophenazine, N-acetylprocainamide, albuterol, alprenolol, amethocaine, amiodarone, amitriptyline, antazoline, atenolol, azacyclonal, bamethane, benactyzine, benperidol, benzethidine, benzocaine, benzoctamine, benzphetamine, benzquinamide, bromhexine, bromodiphenhydramine, bromperidol, brompheniramine, brompromazine, buclizine, bufotenine, bupivacaine, buprenorphine, butacaine, butethamate, chlorcyclizine, chlorpheniramine, chlorphenoxamine, chlorprenaline, chlorpromazine, chlorprothixene, cimetidine, cinchonidine, cinnarizine, clemastine, clomipramine, clonidine, cocaine, cyclazocine, cyclizine, cyclopentamine, cyproheptadine, deserpidine, desipramine, dextromoramide, dextropropoxyphene, dicyclomine, diethylcarbamazine, diethylpropion, diethylthiambutene, dihydroergotamine, dimethindene, dimethothiazine, diphenhydramine, diphenoxylate, dipipanone, diprenorphine, dipyridamole, disopyramide, dothiepin, doxapram, doxepin, doxylamine, droperidol, ephedrine, ergocornine, ergocristine, ergocristinine, ergocryptine, ergometrine, ergosine, ergosinine, ergotamine, ethopropazine, etorphine, etoxeridine, fenethazine, fenfluramine, fenoterol, fentanyl, flavoxate, fluopromazine, flupenthixol, fluphenazine, flurazepam, haloperidol, hydroxyzine, hyoscine, ibogaine, imipramine, indapamine, iprindole, isothipendyl, isoxsuprine, ketanserin, laudanosine, lidocaine, lofepramine, loxapine, maprotiline, mecamylamine, meclophenoxate, meclozine, medazepam, mephentermine, mepivacaine, meptazinol, mepyramine, mesoridazine, metaraminol, methadone, methamphetamine, methapyrilene, methdilazine, methotrimeprazine, methoxamine, methoxyphenamine, methoxypromazine, methylephedrine, methylergonovine, methysergide, metoclopramide, metopimazine, metoprolol, mianserin, morazone, nadolol, nalorphine, naloxone, naphazoline, nicotine, nifedipine, nomifensine, nortriptyline, noscapine, orphenadrine, oxeladin, oxprenolol, oxymetazolin, papaverine, pargyline, pecazine, penbutolol, pentazocine, pen-

thienate, pericyazine, perphenazine, phenadoxone, phenampromide, phenazocine, phenbutrazate, phendimetrazine, phenelzine, phenglutarimide, phenindamine, pheniramine, phenmetrazine, phenomorphan, phenoperidine, phenothiazine, phenoxybenzamine, phentolamine, phenylephrine, phenyltoloxamine, physostigmine, piminodine, pimozide, pindolol, pipamazine, pipazethate, piperacetazine, piperidolate, pipradol, pirenzepine, piritramide, pizotifen, practolol, pramoxine, prazosin, prenylamine, prilocaine, primaquine, proadifen, procainamide, procaine, prochlorperazine, procyclidine, proheptazine, prolintane, promazine, promethazine, pronethalol, properidine, propiomazine, propranolol, prothipendyl, protriptyline, proxymetacaine, pseudoephedrine, pyrimethamine, quinidine, quinine, ranitidine, rescinnamine, sotalol, tacrine, terbutaline, terfenadine, thenyldiamine, theophylline, thiethylperazine, thiopropazate, thioproperazine, thioridazine, thiothixene, thonzylamine, timolol, tocainide, tolpropamine, tolycaine, tranylcypromine, trazodone, trifluoperazine, trifluperidol, trimeperidine, trimeprazine, trimethobenzamide, trimethoprim, trimipramine, tripelennamine, triprolidine, tryptamine, verapamil, xylometazoline

## REFERENCE

Jane, I.; McKinnon, A.; Flanagan, R.J. High-performance liquid chromatographic analysis of basic drugs on silica columns using non-aqueous ionic eluents. II. Application of UV, fluorescence and electrochemical oxidation detection. *J.Chromatogr.*, **1985**, *323*, 191–225

# Terbinafine

**Molecular formula:** $C_{21}H_{25}N$
**Molecular weight:** 291.4
**CAS Registry No.:** 91161-71-6

## SAMPLE
**Matrix:** blood
**Sample preparation:** 500 µL Plasma + 100 µL 5 µg/mL IS in water + 1 mL 200 mM pH 9 borate buffer + 8 mL n-hexane, shake horizontally at 200 rpm for 25 min, centrifuge at 2000 g for 10 min. Remove 7 mL of the supernatant and add it to 1 mL 500 mM sulfuric acid:isopropanol 85:15, shake for 15 min, centrifuge at 2000 g for 5 min, inject a 250 µL aliquot of the aqueous phase.

## HPLC VARIABLES
**Column:** 100 × 4.6 3 µm Pecospher C18
**Mobile phase:** MeCN:buffer 50:50 (Buffer was 3.35 mL triethylamine and 2.65 mL orthophosphoric acid in 2 L water.)
**Flow rate:** 1
**Injection volume:** 250
**Detector:** UV 224

## CHROMATOGRAM
**Retention time:** 9.9
**Internal standard:** 2-(1-naphthyl)-1-(3-phenylprop-2-enyl)piperidine hydrochloride (IW 85-190) (12.4)
**Limit of quantitation:** 2 ng/mL

## OTHER SUBSTANCES
**Extracted:** metabolites

## KEY WORDS
plasma; pharmacokinetics

## REFERENCE
Denouël, J.; Keller, H.P.; Schaub, P.; Delaborde, C.; Humbert, H. Determination of terbinafine and its desmethyl metabolite in human plasma by high-performance liquid chromatography. *J.Chromatogr.B*, **1995**, *663*, 353–359

## SAMPLE
**Matrix:** blood
**Sample preparation:** Add IS to plasma, adjust pH to 9, extract with n-hexane. Add the organic layer to 500 mM sulfuric acid:isopropanol 85:15, extract, inject an aliquot of the aqueous layer.

## HPLC VARIABLES
**Column:** 100 × 4.6 Spherisorb RP-18/5
**Mobile phase:** MeCN:10 mM $K_2HPO_4$:triethylamine 60:40:0.01
**Column temperature:** 50
**Flow rate:** 1
**Detector:** UV 224

## CHROMATOGRAM
**Internal standard:** 2-(1-naphthyl)-1-(3-phenylprop-2-enyl)piperidine hydrochloride (SDZ 85-190)
**Limit of quantitation:** 2 ng/mL

## OTHER SUBSTANCES
**Extracted:** metabolites

## KEY WORDS
plasma; pharmacokinetics

## REFERENCE
Kovarik, J.M.; Kirkesseli, S.; Humbert, H.; Grass, P.; Kutz, K. Dose-proportional pharmacokinetics of terbinafine and its N-demethylated metabolite in healthy volunteers. *Br.J.Dermatol.*, **1992**, *126 Suppl 39*, 8–13

## SAMPLE
**Matrix:** blood, milk
**Sample preparation:** Plasma. 100 μL Plasma + 300 μL 4 g/L ammonium acetate in MeOH, centrifuge, inject a 200 μL aliquot of the supernatant. Milk. 100 μL Milk + 100 μL 100 mM NaCl (?) + 800 μL MeCN, centrifuge, inject a 100 μL aliquot of the supernatant.

## HPLC VARIABLES
**Column:** 250 × 4.6 10 μm RP-8 (Merck)
**Mobile phase:** MeOH:10 mM pH 7 phosphate buffer 77:23 containing 4 g/L ammonium acetate
**Column temperature:** 35
**Flow rate:** 1.5
**Injection volume:** 100-200
**Detector:** E, Metrohm 656, glassy carbon electrode + 1.1 V

## CHROMATOGRAM
**Retention time:** 10.50 (plasma), 12.03 (milk)
**Limit of detection:** 50 ng/mL (plasma); 150 ng/mL (milk)

## OTHER SUBSTANCES
**Extracted:** metabolites

## KEY WORDS
plasma; human; rat; dog

## REFERENCE
Schatz, F.; Haberl, H. Analytical methods for the determination of terbinafine and its metabolites in human plasma, milk and urine. *Arzneimittelforschung*, **1989**, *39*, 527–532

## SAMPLE
**Matrix:** blood, urine
**Sample preparation:** Plasma. 750 μL Plasma + 50 μL 320 μg/mL IS + 25 μL 85% phosphoric acid + 750 μL EtOH:isopropanol 75:25, vortex for 10 s, let stand on crushed ice for 30 min, centrifuge at 1000 g for 15 min. Remove a 400 μL aliquot of the supernatant and add it to 400 μL 10 mM pH 5 phosphate buffer, vortex for 10 s, inject a 250 μL aliquot onto column A with mobile phase A and elute to waste. After 4 min backflush column A with mobile phase A to waste, after 1 min backflush the contents of column A onto column B with mobile phase B, after 10 min remove column A from the circuit, elute

column B with mobile phase B and monitor the effluent. (Recondition column A with mobile phase A for 30 min at 0.1 mL/min). Urine. 500 µL Urine + 50 µL 220 µg/mL IS + 200 µL 10000 IU/mL β-glucuronidase (bovine liver, Sigma) in 10 mL 10 mM pH 5 phosphate buffer, heat at 37° for 18 h, add 300 µL 10 mM pH 5 phosphate buffer, vortex for 10 s, centrifuge at 2500 g for 18 min, inject a 100-250 µL aliquot of the supernatant onto column A with mobile phase A and elute to waste. After 4 min backflush column A with mobile phase A to waste, after 1 min backflush the contents of column A onto column B with mobile phase B, after 10 min remove column A from the circuit, elute column B with mobile phase B and monitor the effluent. (Recondition column A with mobile phase A for 30 min at 0.1 mL/min).

## HPLC VARIABLES
**Column:** A 15 × 3.2 30-40 µm Perisorb RP-2 (Merck); B 220 × 4.6 5 µm Spheri-5 phenyl
**Mobile phase:** A 20 mM $KH_2PO_4$ containing 0.25% triethylamine, pH adjusted to 3.8 with 1 M phosphoric acid; B MeCN:buffer 55:45 (Buffer was 20 mM $KH_2PO_4$ containing 0.125% triethylamine, pH adjusted to 3.8 with 1 M phosphoric acid.)
**Column temperature:** 30-35
**Flow rate:** 1
**Injection volume:** 100-250
**Detector:** UV 224

## CHROMATOGRAM
**Retention time:** 34-36
**Internal standard:** 2-(1-naphthyl)-1-(3-phenylprop-2-enyl)piperidine hydrochloride (IW 85-190)
**Limit of quantitation:** 20 ng/mL (plasma); 100 ng/mL (urine)

## OTHER SUBSTANCES
**Extracted:** metabolites

## KEY WORDS
plasma; column-switching; pharmacokinetics

## REFERENCE
Zehender, H.; Denouël, J.; Roy, M.; Le Saux, L.; Schaub, P. Simultaneous determination of terbinafine (Lamisil) and five metabolites in human plasma and urine by high-performance liquid chromatography using on-line solid-phase extraction. *J.Chromatogr.B*, **1995**, *664*, 347–355

## SAMPLE
**Matrix:** nails
**Sample preparation:** 10-20 mg Nail clippings + 1 mL 1 M NaOH, let stand at room temperature for 48 h, adjust pH to 9 with 1 M HCl, adjust molarity to 0.2 with water, add 100 µL 10 mg/mL proteinase K (Type I fungal proteinase, P0390, Sigma), heat at 37° for 12-16 h, add 50 ng IS, vortex, add 3 mL hexane, vortex for 3 min, add 100 µL MeOH, centrifuge at 500 g for 10 min. Remove a 2.7 mL aliquot of the organic layer and add it to 200 µL 500 mM sulfuric acid:propanol 85:15, vortex for 3 min, centrifuge, inject a 100 µL aliquot of the aqueous layer.

## HPLC VARIABLES
**Guard column:** 20 × 2 5-20 µm Prepsil RPC8 (Apex)
**Column:** 250 mm long 5 µm Spherisorb C18
**Mobile phase:** MeCN:10 mM phosphate buffer:triethylamine 80:20:0.1, pH adjusted to 4.0 with 85% phosphoric acid
**Flow rate:** 2
**Injection volume:** 100
**Detector:** UV 224

## CHROMATOGRAM
**Retention time:** 7.5
**Internal standard:** 2-(1-naphthyl)-1-(3-phenylprop-2-enyl)piperidine hydrochloride (Sandoz 85-190) (8.4)

## OTHER SUBSTANCES
**Extracted:** metabolites

## REFERENCE
Dykes, P.J.; Thomas, R.; Finlay, A.Y. Determination of terbinafine in nail samples during systemic treatment for onychomycoses. *Br.J.Dermatol.*, **1990**, *123*, 481–486

# Terconazole

**Molecular formula:** $C_{26}H_{31}Cl_2N_5O_3$
**Molecular weight:** 532.5
**CAS Registry No.:** 67915-31-5

## SAMPLE
**Matrix:** blood
**Sample preparation:** Condition a 3 mL Bond-Elut C18 cartridge with 6 mL MeOH and 6 mL water. 1 mL Serum + 250 µL 100 mM NaOH + 3 mL water, add to cartridge, wash with 9 mL water, wash with 200 µL MeOH, elute with 1 mL MeOH. Evaporate eluent at 60°, resuspend in 200 µL mobile phase, centrifuge at 13000 g for 2 min, inject 20-40 µL.

## HPLC VARIABLES
**Guard column:** Chrompack C18
**Column:** 100 × 3 Hypersil ODS in a Chrompack glass cartridge
**Mobile phase:** MeCN:water 45:55 containing 500 µL/L diethylamine, pH adjusted to 8.0 with orthophosphoric acid
**Flow rate:** 0.6
**Injection volume:** 20-40
**Detector:** UV 254

## CHROMATOGRAM
**Retention time:** 11.4
**Internal standard:** terconazole

## OTHER SUBSTANCES
**Extracted:** diazepam, ketoconazole
**Noninterfering:** acetaminophen, acyclovir, allopurinol, amoxicillin, amphotericin B, ampicillin, aspirin, azlocillin, bendrofluazide, bumetanide, buprenorphine, carbenicillin, cefazolin, cefotaxime, cefoxitin, ceftazidime, cefuroxime, cephalexin, chlorambucil, chloramphenicol, chlordiazepoxide, chlorpheniramine, chlorpropamide, cyclophosphamide, cyclosporin, cytarabine, daunorubicin, dextropropoxyphene, dihydrocodeine, domperidone, flucytosine, furosemide, gentamicin, griseofulvin, melphalan, methotrexate, metoclopramide, metronidazole, miconazole, nabilone, netilmicin, nicotinamide, nitrazepam, penicillin G, piperacillin, prednisolone, procarbazine, prochlorperazine, riboflavin, rifampin, sulfamethoxazole, thioguanine, tobramycin, tolbutamide, trimethoprim

## KEY WORDS
serum; terconazole is IS

## REFERENCE
Turner, C.A.; Turner, A.; Warnock, D.W. High performance liquid chromatographic determination of ketoconazole in human serum. *J.Antimicrob.Chemother.*, **1986**, *18*, 757–763

## SAMPLE
**Matrix:** solutions

## HPLC VARIABLES
**Column:** 300 × 4.5 µBondapak C18
**Mobile phase:** MeOH:buffer 60:40 (Buffer was 25 mM $KH_2PO_4$ + 4 mM heptanesulfonic acid, adjusted to pH 8.0 with 1 M NaOH.)

**Flow rate:** 1.8
**Injection volume:** 75
**Detector:** UV 226

## OTHER SUBSTANCES
**Simultaneous:** ketoconazole

## REFERENCE
Carver, P.L.; Berardi, R.R.; Knapp, M.J.; Rider, J.M.; Kauffman, C.A.; Bradley, S.F.; Atassi, M. In vivo interaction of ketoconazole and sucralfate in healthy volunteers. *Antimicrob.Agents Chemother.*, **1994**, *38*, 326–329

# Terfenadine

**Molecular formula:** $C_{32}H_{41}NO_2$
**Molecular weight:** 471.7
**CAS Registry No.:** 50679-08-8

## SAMPLE
**Matrix:** blood
**Sample preparation:** 2 mL Whole blood or plasma + 2 mL buffer + 5 mL chloroform: isopropanol:n-heptane 60:14:26, shake gently horizontally for 10 min, centrifuge at 2800 g for 10 min. Remove the lower organic layer and evaporate it to dryness under vacuum at 45°, reconstitute the residue in 100 μL mobile phase, centrifuge at 2800 g for 5 min, inject a 50 μL aliquot of the supernatant. (Buffer was saturated ammonium chloride solution 25% diluted with water, adjusted to pH 9.5 with 25% ammonia solution.)

## HPLC VARIABLES
**Column:** 300 × 3.9 4 μm NovaPack C18
**Mobile phase:** MeOH:THF:buffer 65:5:30 (Buffer was 0.68 g/L (10 mM (sic)) $KH_2PO_4$ adjusted to pH 2.6 with concentrated orthophosphoric acid.) (At the end of each session wash the column with water for 1 h and MeOH for 1 h, re-equilibrate for 30 min.)
**Column temperature:** 30
**Flow rate:** 0.8
**Injection volume:** 50
**Detector:** UV 259

## CHROMATOGRAM
**Retention time:** 18.33
**Limit of detection:** <120 ng/mL

## OTHER SUBSTANCES
**Extracted:** acebutolol, acenocoumarol, acepromazine, aceprometazine, acetaminophen, aconitine, ajmaline, albuterol, alimemazine, alminoprofen, alpidem, alprazolam, alprenolol, amisulpride, amitriptyline, amodiaquine, amoxapine, aspirin, astemizole, atenolol, benazepril, benperidol, benzocaine, benzoylecgonine, betaxolol, bisoprolol, bromazepam, brompheniramine, bumadizone, bupivacaine, buprenorphine, buspirone, caffeine, carbamazepine, carbinoxamine, carpipramine, carteolol, celiprolol, cetirizine, chlorambucil, chlordiazepoxide, chlormezanone, chlorophenacinone, chloroquine, chlorpheniramine, chlorpromazine, chlorpropamide, cibenzoline, cicletanine, clemastine, clobazam, clomipramine, clonazepam, clonidine, clorazepate, clozapine, cocaine, codeine, colchicine, cyamemazine, cyclizine, cycloguanil, cyproheptadine, cytarabine, dacarbazine, daunorubicin, debrisoquine, demexiptiline, desipramine, dextromethorphan, dextromoramide, dextropropoxyphene, diazepam, diazoxide, diclofenac, dihydralazine, diltiazem, diphenhydramine, dipyridamole, disopyramide, dosulepine, doxepin, doxylamine, droperidol, ephedrine, estazolam, etodolac, fenfluramine, fenoprofen, fentiazac, flecainide, floctafenine, flumazenil, flunitrazepam, fluoxetine, fluphenazine, flurbiprofen, fluvoxamine, gliben-

clamide, glibornuride, glipizide, glutethimide, haloperidol, histapyrrodine, hydroxychloroquine, hydroxyzine, ibuprofen, imipramine, indomethacin, iproniazid, ketamine, ketoprofen, labetalol, levomepromazine, lidocaine, lidoflazine, lisinopril, loperamide, loprazolam, loratadine, lorazepam, loxapine, maprotiline, medazepam, medifoxamine, mefenamic acid, mefenidramine, mefloquine, melphalan, meperidine, mephenesin, mephentermine, mepivacaine, metapramine, metformin, methadone, methaqualone, methocarbamol, methotrexate, metipranolol, metoclopramide, metoprolol, mexiletine, mianserine, midazolam, minoxidil, moclobemide, moperone, morphine, nadolol, nalbuphine, nalorphine, naloxone, naltrexone, naproxen, nialamide, nicardipine, nifedipine, niflumic acid, nimodipine, nitrazepam, nitrendipine, nizatidine, nomifensine, nortriptyline, omeprazole, opipramol, oxazepam, oxprenolol, penbutolol, pentazocine, phencyclidine, phenobarbital, phenol, phenylbutazone, pimozide, pindolol, pipamperone, piroxicam, prazepam, prazosin, prilocaine, procainamide, procarbazine, proguanil, promethazine, propafenone, propranolol, protriptyline, pyrimethamine, quinidine, quinine, quinupramine, ramipril, ranitidine, reserpine, ritodrine, secobarbital, sotalol, strychnine, sulfinpyrazole, sulindac, sulpride, sultopride, suriclone, temazepam, tenoxicam, terbutaline, tetracaine, tetrazepam, thiopental, thioproperazine, thioridazine, tianeptine, tiapride, tiaprofenic acid, ticlopidine, timolol, tioclomarol, tofisopam, tolbutamide, toloxatone, trazodone, triazolam, trifluoperazine, trifluperidol, trimipramine, triprolidine, tropatenine, verapamil, viloxazine, vinblastine, vincristine, vindesine, warfarin, yohimbine, zolpidem, zopiclone, zorubicine
**Interfering:** bepridil, penfluridol

## KEY WORDS
whole blood; plasma

## REFERENCE
Tracqui, A.; Kintz, P.; Mangin, P. Systematic toxicological analysis using HPLC/DAD. *J.Forensic Sci.*, **1995**, *40*, 254–262

## SAMPLE
**Matrix:** blood
**Sample preparation:** 1 mL Plasma + 500 μL 100 mM HCl + 5 mL 50 ng/mL propranolol in MTBE:isopropanol 95:5, vortex for 20 s, centrifuge at 3000 g for 15 min, repeat the extraction. Combine the organic layers and evaporate them to dryness under a stream of nitrogen at 40°, dissolve the residue in 100 μL MeOH, inject a 40 μL aliquot.

## HPLC VARIABLES
**Column:** 150 × 4.5 5 μm Microsorb cyanopropylsilane
**Mobile phase:** MeCN:buffer 75:25 (Buffer was 1 mM sodium acetate adjusted to pH 4.0 with acetic acid.)
**Injection volume:** 40
**Detector:** F ex 230 em 300 (cut-off filter 270 nm)

## CHROMATOGRAM
**Retention time:** 15.1
**Internal standard:** propranolol (12.2)
**Limit of quantitation:** 2 ng/mL

## OTHER SUBSTANCES
**Extracted:** metabolites
**Noninterfering:** acetaminophen, aspirin, ibuprofen, pseudoephedrine, tricyclic antidepressants

## KEY WORDS
plasma

## REFERENCE
Surapaneni, S.; Khalil, S.K.W. A sensitive HPLC method for the determination of terfenadine and its metabolite in human plasma. *J.Liq.Chromatogr.*, **1994**, *17*, 2419–2428

## SAMPLE
**Matrix:** blood
**Sample preparation:** Condition a 3 mL 500 mg Analytichem SPE C18 cartridge with two 2 mL volumes of MeOH, 2 mL water, 1.5 mL 200 mM pH 4.0 acetate buffer. Add 1 mL 75 ng/mL IS in 200 mM pH 4.0 acetate buffer and 1 mL plasma to the SPE cartridge and blow through with nitrogen. Wash with 2 mL water, two 1 mL portions of MeOH:water 50:50, and 1 mL MeOH. Dry column by blowing nitrogen through at 8 mL/min for 5 min. Elute with two 500 µL portions of 50 mM triethylamine in MeOH. Evaporate the eluate to dryness at 55° under a stream of nitrogen and dissolve the residue in 200 µL mobile phase, inject a 150 µL aliquot.

## HPLC VARIABLES
**Column:** 250 × 4.6 5 µm Spherisorb S5CN cyano
**Mobile phase:** MeCN:MeOH:buffer 19:29:52 (Buffer was 12 mM, prepared from 0.5 g ammonium acetate and 5 mL acetic acid in 520 mL water.)
**Column temperature:** 35
**Flow rate:** 1.8
**Injection volume:** 150
**Detector:** F ex 230 em 280

## CHROMATOGRAM
**Retention time:** 13.5
**Internal standard:** MDL 26,042 (9.3)
**Limit of quantitation:** 10 ng/mL

## OTHER SUBSTANCES
**Extracted:** metabolites
**Simultaneous:** acetaminophen, ephedrine, ibufenac, ibuprofen, norpseudoephedrine, pseudoephedrine

## KEY WORDS
plasma; SPE

## REFERENCE
Coutant, J.E.; Westmark, P.A.; Nardella, P.A.; Walter, S.M.; Okerholm, R.A. Determination of terfenadine and terfenadine acid metabolite in plasma using solid-phase extraction and high-performance liquid chromatography with fluorescence detection. *J.Chromatogr.*, **1991**, *570*, 139–148

## SAMPLE
**Matrix:** blood
**Sample preparation:** Add 1 mL plasma to an Analytichem C18 SPE cartridge, elute with MeOH:200 mM pH 4 sodium acetate 95:5, evaporate eluate to dryness under reduced pressure, reconstitute in 200 µL mobile phase A, inject a 150 µL aliquot onto column A with mobile phase A, collect terfenadine fraction at 12 min, evaporate to dryness, dissolve residue in 100 µL mobile phase B, inject an 80 µL aliquot onto column B with mobile phase B, monitor effluent from column B.

## HPLC VARIABLES
**Column:** A 150 × 4.6 Spherisorb CN; B 150 × 4.6 5 µm Ultron ES-OVM ovomucoid (MAC-MOD Analytical)
**Mobile phase:** A MeOH:MeCN:12 mM pH 4.0 sodium acetate 35:13:52; B MeOH:25 mM pH 4.6 phosphate buffer 25:75

**Flow rate:** A 1.5; B 1
**Injection volume:** A 150; B 80
**Detector:** F ex 200 em 280

## CHROMATOGRAM
**Retention time:** 7 (S), 9 (R) (measured after injection on column B)

## OTHER SUBSTANCES
**Also analyzed:** metabolites

## KEY WORDS
plasma; SPE; rat; chiral

## REFERENCE
Zamani, K.; Conner, D.P.; Weems, H.B.; Yang, S.K.; Cantilena, L.R., Jr. Enantiomeric analysis of terfenadine in rat plasma by HPLC. *Chirality*, **1991**, *3*, 467−470

## SAMPLE
**Matrix:** microsomal incubations
**Sample preparation:** 2 mL Microsomal incubation + 1.6 mL cold MeCN + 400 μL 500
ng/mL IS in MeCN, centrifuge at 10000 g for 10 min. Add a 500 μL aliquot of the supernatant to 400 μL acetate buffer, inject a 200 μL aliquot.

## HPLC VARIABLES
**Column:** 250 × 4.6 5 μm Spherisorb CN
**Mobile phase:** MeCN:100 mM ammonium acetate buffer 40:60
**Column temperature:** 35
**Flow rate:** 1.3
**Injection volume:** 200
**Detector:** F ex 230 em 280

## CHROMATOGRAM
**Retention time:** 10
**Internal standard:** MDL 16,232 (Marion Merrell Dow) (11.5)

## OTHER SUBSTANCES
**Extracted:** metabolites, azacyclonol

## KEY WORDS
human; liver

## REFERENCE
Ling, K.-H.J.; Leeson, G.A.; Burmaster, S.D.; Hook, R.H.; Reith, M.K.; Cheng, L.K. Metabolism of terfenadine associated with CYP3A(4) activity in human hepatic microsomes. *Drug Metab.Dispos.*, **1995**,
*23*, 631−636

## SAMPLE
**Matrix:** microsomal incubations
**Sample preparation:** Add an equal volume of chilled MeOH to 0.25-1 mL microsomal
incubation, let stand on ice for 5 min, centrifuge at 2000 g for 10 min, inject a 50 μL
aliquot of the supernatant.

## HPLC VARIABLES
**Guard column:** 5 μm cyano guard column (YMC)
**Column:** 250 × 4.6 5 μm YMC-Pack cyano (YMC)

**Mobile phase:** Gradient. MeCN:10 mM pH 4.0 ammonium acetate 25:75 for 40 min, to 50:50 over 5 min
**Column temperature:** 35
**Flow rate:** 1
**Injection volume:** 50
**Detector:** UV 230; F ex 230 em 280; radioactivity

## CHROMATOGRAM
**Retention time:** 55

## OTHER SUBSTANCES
**Extracted:** metabolites

## KEY WORDS
human; liver; tritium labeled

## REFERENCE
Rodrigues, A.D.; Mulford, D.J.; Lee, R.D.; Surber, B.W.; Kukulka, M.J.; Ferrero, J.L.; Thomas, S.B.; Shet, M.S.; Estabrook, R.W. *In vitro* metabolism of terfenadine by a purified recombinant fusion protein containing cytochrome P4503A4 and NADPH-P450 reductase. Comparison to human liver microsomes and precision-cut liver tissue slices. *Drug Metab.Dispos.*, **1995**, *23*, 765–775

## SAMPLE
**Matrix:** solutions

## HPLC VARIABLES
**Column:** 250 × 4 5 μm LiChrospher 100 RP-8
**Mobile phase:** MeCN:0.025% phosphoric acid:buffer 60:25:15 (Buffer was 9 mL concentrated phosphoric acid and 10 mL triethylamine in 900 mL water, adjust pH to 3.4 with dilute phosphoric acid, make up to 1 L.)
**Flow rate:** 0.6
**Injection volume:** 25
**Detector:** UV 229

## CHROMATOGRAM
**Retention time:** 12.24

## OTHER SUBSTANCES
**Also analyzed:** acebutolol, acepromazine, acetaminophen, acetazolamide, acetophenazine, albuterol, alprazolam, amitriptyline, amobarbital, amoxapine, antipyrine, atenolol, atropine, azatadine, baclofen, benzocaine, bromocriptine, brompheniramine, brotizolam, bupivacaine, buspirone, butabarbital, butalbital, caffeine, carbamazepine, cetirizine, chlorcyclizine, chlordiazepoxide, chlormezanone, chloroquine, chlorpheniramine, chlorpromazine, chlorpropamide, chlorprothixene, chlorthalidone, chlorzoxazone, cimetidine, cisapride, clomipramine, clonazepam, clonidine, clozapine, cocaine, codeine, colchicine, cyclizine, cyclobenzaprine, dantrolene, desipramine, diazepam, diclofenac, diflunisal, diltiazem, diphenhydramine, diphenidol, diphenoxylate, dipyridamole, disopyramide, dobutamine, doxapram, doxepin, droperidol, encainide, ethidium bromide, ethopropazine, fenoprofen, fentanyl, flavoxate, fluoxetine, fluphenazine, flurazepam, flurbiprofen, fluvoxamine, furosemide, glutethimide, glyburide, guaifenesin, haloperidol, homatropine, hydralazine, hydrochlorothiazide, hydrocodone, hydromorphone, hydroxychloroquine, hydroxyzine, ibuprofen, imipramine, indomethacin, ketoconazole, ketoprofen, ketorolac, labetalol, levorphanol, lidocaine, loratadine, lorazepam, lovastatin, loxapine, mazindol, mefenamic acid, meperidine, mephenytoin, mepivacaine, mesoridazine, metaproterenol, metformin, methadone, methdilazine, methocarbamol, methotrexate, methotrimeprazine, methoxamine, methyldopa, methylphenidate, metoclopramide, metolazone, metoprolol, metronidazole, midazolam, moclobemide, morphine, nadolol, nalbuphine, naloxone, na-

phazoline, naproxen, nifedipine, nizatidine, norepinephrine, nortriptyline, oxazepam, ox-ycodone, oxymetazoline, paroxetine, pemoline, pentazocine, pentobarbital, pentoxifylline, perphenazine, pheniramine, phenobarbital, phenol, phenolphthalein, phentolamine, phenylbutazone, phenyltoloxamine, phenytoin, pimozide, pindolol, piroxicam, pramoxine, prazepam, prazosin, probenecid, procainamide, procaine, prochlorperazine, procyclidine, promazine, promethazine, propafenone, propantheline, propiomazine, propofol, propran-olol, protriptyline, quazepam, quinidine, quinine, racemethorphan, ranitidine, remoxi-pride, risperidone, salicylic acid, scopolamine, secobarbital, sertraline, sotalol, spironolac-tone, sulfinpyrazone, sulindac, temazepam, terbutaline, tetracaine, theophylline, thiethylperazine, thiopental, thioridazine, thiothixene, timolol, tocainide, tolbutamide, tolmetin, trazodone, triamterene, triazolam, trifluoperazine, triflupromazine, trimepra-zine, trimethoprim, trimipramine, verapamil, warfarin, xylometazoline, yohimbine, zopiclone

## KEY WORDS
some details of plasma extraction

## REFERENCE
Koves, E.M. Use of high-performance liquid chromatography-diode array detection in forensic toxicology. *J.Chromatogr.A*, **1995**, *692*, 103–119

## SAMPLE
**Matrix:** solutions

## HPLC VARIABLES
**Column:** 150 × 4.6 5 μm Ultron ES-OVM (Grom)
**Mobile phase:** Gradient. Isopropanol:100 mM pH 5.5 phosphate buffer 4:96 for 20 min, to 12:88 over 1 min, maintain at 12:88 for 19 min.
**Flow rate:** 1
**Detector:** UV 224; F ex 230 em 280

## CHROMATOGRAM
**Retention time:** 28 (R-(+)), 34 (S-(−))

## OTHER SUBSTANCES
**Simultaneous:** metabolites

## KEY WORDS
chiral

## REFERENCE
Terhechte, A.; Blaschke, G. Investigation of the stereoselective metabolism of the chiral H₁-antihista-minic drug terfenadine by high-performance liquid chromatography. *J.Chromatogr.A*, **1995**, *694*, 219–225

## SAMPLE
**Matrix:** solutions

## HPLC VARIABLES
**Column:** 150 × 4.6 Supelcosil LC-ABZ
**Mobile phase:** MeCN:25 mM pH 6.9 potassium phosphate buffer 35:65
**Flow rate:** 1.5
**Injection volume:** 25
**Detector:** UV 254

## CHROMATOGRAM
**Retention time:** 1.384

## OTHER SUBSTANCES
**Also analyzed:** 6-acetylmorphine, amiloride, amphetamine, benzocaine, benzoylecgonine, caffeine, cocaine, codeine, doxylamine, fluoxetine, glutethimide, hexobarbital, hypoxanthine, levorphanol, LSD, meperidine, mephobarbital, methadone, methylphenidate, methyprylon, N-norcodeine, oxazepam, oxycodone, phenylpropanolamine, prilocaine, procaine

## REFERENCE
Ascah, T.L. Improved separations of alkaloid drugs and other substances of abuse using Supelcosil LC-ABZ column. *Supelco Reporter*, **1993**, *12(3)*, 18–21

## SAMPLE
**Matrix:** solutions

## HPLC VARIABLES
**Column:** 150 × 4.6 5 μm MicroPak MCH-5
**Mobile phase:** MeCN:30 mM pH 3.0 phosphate buffer 70:30
**Flow rate:** 1.7
**Injection volume:** 20
**Detector:** UV 230

## REFERENCE
Fernández Otero, G.C.; Lucangioli, S.E.; Carducci, C.N. Adsorption of drugs in high-performance liquid chromatography injector loops. *J.Chromatogr.A*, **1993**, *654*, 87–91

## SAMPLE
**Matrix:** solutions

## HPLC VARIABLES
**Column:** 250 × 4.6 5 μm Cyclobond I β-cyclodextrin (Advanced Separations Technologies)
**Mobile phase:** Hexane:MeOH:EtOH 90:5:5
**Flow rate:** 2
**Detector:** UV 220

## CHROMATOGRAM
**Retention time:** 10.6 (S), 11.7 (R)

## KEY WORDS
chiral

## REFERENCE
Weems, H.; Zamani, K. Resolution of terfenadine enantiomers by beta-cyclodextrin chiral stationary phase high-performance liquid chromatography. *Chirality*, **1992**, *4*, 268–272

## SAMPLE
**Matrix:** solutions
**Sample preparation:** Prepare a 10 μg/mL solution in MeOH, inject a 20 μL aliquot.

## HPLC VARIABLES
**Column:** 125 × 4.9 Spherisorb S5W silica
**Mobile phase:** MeOH containing 10 mM ammonium perchlorate and 1 mL/L 100 mM NaOH in MeOH, pH 6.7

**Flow rate:** 2
**Injection volume:** 20
**Detector:** E, LeCarbone, V25 glassy carbon electrode, + 1.2 V

## CHROMATOGRAM
**Retention time:** 1.76

## OTHER SUBSTANCES
**Also analyzed:** acebutolol, acepromazine, acetophenazine, N-acetylprocainamide, albuterol, alprenolol, amethocaine, amiodarone, amitriptyline, antazoline, atenolol, azacyclonal, bamethane, benactyzine, benperidol, benzethidine, benzocaine, benzoctamine,
benzphetamine, benzquinamide, bromhexine, bromodiphenhydramine, bromperidol,
brompheniramine, bromopromazine, buclizine, bufotenine, bupivacaine, buprenorphine,
butacaine, butethamate, chlorcyclizine, chlorpheniramine, chlorphenoxamine, chlorprenaline, chlorpromazine, chlorprothixene, cimetidine, cinchonidine, cinnarizine, clemastine,
clomipramine, clonidine, cocaine, cyclazocine, cyclizine, cyclopentamine, cyproheptadine,
deserpidine, desipramine, dextromoramide, dextropropoxyphene, dicyclomine, diethylcarbamazine, diethylpropion, diethylthiambutene, dihydroergotamine, dimethindene, dimethothiazine, diphenhydramine, diphenoxylate, dipipanone, diprenorphine, dipyridamole, disopyramide, dothiepin, doxapram, doxepin, doxylamine, droperidol, ephedrine,
ergocornine, ergocristine, ergocristinine, ergocryptine, ergometrine, ergosine, ergosinine,
ergotamine, ethopropazine, etorphine, etoxeridine, fenethazine, fenfluramine, fenoterol,
fentanyl, flavoxate, fluopromazine, flupenthixol, fluphenazine, flurazepam, haloperidol,
hydroxyzine, hyoscine, ibogaine, imipramine, indapamine, iprindole, isothipendyl, isoxsuprine, ketanserin, laudanosine, lidocaine, lofepramine, loxapine, maprotiline, mecamylamine, meclophenoxate, meclozine, medazepam, mephentermine, mepivacaine, meptazinol, mepyramine, mesoridazine, metaraminol, methadone, methamphetamine,
methapyrilene, methdilazine, methotrimeprazine, methoxamine, methoxyphenamine,
methoxypromazine, methylephedrine, methylergonovine, methysergide, metoclopramide,
metopimazine, metoprolol, mianserin, morazone, nadolol, nalorphine, naloxone, naphazoline, nicotine, nifedipine, nomifensine, nortriptyline, noscapine, orphenadrine, oxeladin,
oxprenolol, oxymetazolin, papaverine, pargyline, pecazine, penbutolol, pentazocine, penthienate, pericyazine, perphenazine, phenadoxone, phenampromide, phenazocine, phenbutrazate, phendimetrazine, phenelzine, phenglutarimide, phenindamine, pheniramine,
phenmetrazine, phenomorphan, phenoperidine, phenothiazine, phenoxybenzamine, phentolamine, phenylephrine, phenyltoloxamine, physostigmine, piminodine, pimozide, pindolol, pipamazine, pipazethate, piperacetazine, piperidolate, pipradol, pirenzepine, piritramide, pizotifen, practolol, pramoxine, prazosin, prenylamine, prilocaine, primaquine,
proadifen, procainamide, procaine, prochlorperazine, procyclidine, proheptazine, prolintane, promazine, promethazine, pronethalol, properidine, propiomazine, propranolol, prothipendyl, protriptyline, proxymetacaine, pseudoephedrine, pyrimethamine, quinidine,
quinine, ranitidine, rescinnamine, sotalol, tacrine, terazosin, terbutaline, thenyldiamine,
theophylline, thiethylperazine, thiopropazate, thioproperazine, thioridazine, thiothixene,
thonzylamine, timolol, tocainide, tolpropamine, tolycaine, tranylcypromine, trazodone, trifluoperazine, trifluperidol, trimeperidine, trimeprazine, trimethobenzamide, trimethoprim, trimipramine, tripelennamine, triprolidine, tryptamine, verapamil, xylometazoline

## REFERENCE
Jane, I.; McKinnon, A.; Flanagan, R.J. High-performance liquid chromatographic analysis of basic drugs
on silica columns using non-aqueous ionic eluents. II. Application of UV, fluorescence and electrochemical oxidation detection. *J.Chromatogr.*, **1985**, *323*, 191–225

## SAMPLE
**Matrix:** urine
**Sample preparation:** 3 mL Urine + 3 mL 100 mM sodium bicarbonate, extract twice with
5 mL ethyl acetate. Combine extracts and evaporate them to dryness at 55° under a
stream of nitrogen. Dissolve the residue in 100 μL MeOH:water 20:80, inject a 10-30 μL
aliquot.

## HPLC VARIABLES
**Column:** 250 × 4.6 5 μm Cyclobond I
**Mobile phase:** MeOH:14 mM sodium perchlorate 75:25
**Flow rate:** 0.2
**Injection volume:** 10-30
**Detector:** UV 210; UV 254

## CHROMATOGRAM
**Retention time:** 43 ((+)-R), 47 ((−)-S)

## OTHER SUBSTANCES
**Simultaneous:** metabolites

## KEY WORDS
chiral

## REFERENCE
Chan, K.Y.; George, R.C.; Chen, T.-M.; Okerholm, R.A. Direct enantiomeric separation of terfenadine
and its major acid metabolite by high-performance liquid chromatography, and the lack of stereoselective terfenadine enantiomer biotransformation in man. *J.Chromatogr.*, **1991**, *571*, 291–297

◆——————————————◆——————————————◆

## ANNOTATED BIBLIOGRAPHY
Jurima-Romet, M.; Crawford, K.; Cyr, T.; Inaba, T. Terfenadine metabolism in human liver. In vitro
inhibition by macrolide antibiotics and azole antifungals. *Drug Metab.Dispos.*, **1994**, *22*, 849–857
[extracted metabolites; microsomal incubations; rat; column temp 35; LOQ 92 nM]

von Moltke, L.L.; Greenblatt, D.J.; Duan, S.X.; Harmatz, J.S.; Shader, R.I. In vitro prediction of the
terfenadine-ketoconazole pharmacokinetic interaction. *J.Clin.Pharmacol.*, **1994**, *34*, 1222–1227
[clomipramine (IS); microsomal incubations]

Chen, T.M.; Chan, K.Y.; Coutant, J.E.; Okerholm, R.A. Determination of the metabolites of terfenadine
in human urine by thermospray liquid chromatography-mass spectrometry. *J.Pharm.Biomed.Anal.*,
**1991**, *9*, 929–933 [LC-MS; UV detection]

George, R.D.; Contario, J.J. Quantitation of terfenadine, pseudoephedrine hydrochloride, and ibuprofen
in a liquid animal dosing formulation using high performance liquid chromatography.
*J.Liq.Chromatogr.*, **1988**, *11*, 475–488 [stability-indicating]

Gupta, S.K.; Gwilt, P.R.; Lim, J.K.; Waters, D.H. High-performance liquid chromatographic determination of terfenadine in commercial tablets. *J.Chromatogr.*, **1986**, *361*, 403–406

# Tetracycline

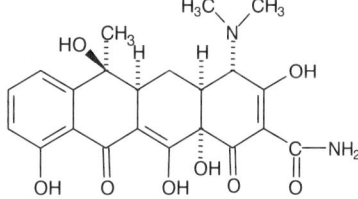

**Molecular formula:** $C_{22}H_{24}N_2O_8$
**Molecular weight:** 444.4
**CAS Registry No.:** 60-54-8, 6416-04-2 (trihydrate),
64-75-5 (HCl), 1336-20-5 (phosphate)

## SAMPLE
**Matrix:** blood
**Sample preparation:** 100 μL Serum + 200 μL 24% trichloroacetic acid in MeOH + 300 μL mobile phase buffer (A), vortex for 1 min, centrifuge at 2000 g for 15 min, inject 50 μL of the supernatant.

## HPLC VARIABLES
**Column:** 250 × 4.6 5 μm Capcell C18 type SG-120 (Shiseido)
**Mobile phase:** MeOH:buffer 60:40 (Buffer (A) was 100 mM pH 6.5 sodium acetate containing 35 mM calcium chloride and 25 mM disodium EDTA.)
**Column temperature:** 30 ± 0.2
**Flow rate:** 1
**Injection volume:** 50
**Detector:** F ex 390 em 512

## CHROMATOGRAM
**Retention time:** 8
**Limit of detection:** 30 ng/mL

## OTHER SUBSTANCES
**Also analyzed:** chlortetracycline, oxytetracycline

## KEY WORDS
serum

## REFERENCE
Iwaki, K.; Okumura, N.; Yamazaki, M. Rapid determination of tetracycline antibiotics in serum by reversed-phase high-performance liquid chromatography with fluorescence detection. *J.Chromatogr.*, **1993**, *619*, 319–323

## SAMPLE
**Matrix:** blood
**Sample preparation:** 250 μL Plasma + 50 μL 10 mM HCl + 1 mL buffer, vortex vigorously, add 6 mL ethyl acetate, add 500 μL isopropanol, shake for 2 min, centrifuge at 1500 g for 5 min. Remove the organic layer and evaporate it to dryness under reduced pressure at room temperature, reconstitute the residue in 300 μL mobile phase and 1 mL hexane, vortex, centrifuge at 9800 g for 5 min, inject a 100 μL aliquot of the lower aqueous layer. (Buffer was 3 M $NaH_2PO_4$ containing 1 M sodium sulfite, pH ca. 5.4.)

## HPLC VARIABLES
**Column:** 125 × 4 5 μm Lichrosorb RP 8
**Mobile phase:** MeCN:MeOH:10 mM oxalic acid 15:10:75
**Flow rate:** 1
**Injection volume:** 100
**Detector:** UV 357

## CHROMATOGRAM
**Retention time:** 4.5
**Internal standard:** tetracycline

## OTHER SUBSTANCES
**Extracted:** oxytetracycline

## KEY WORDS
plasma; cow; tetracycline is IS

## REFERENCE
Nelis, H.J.; Vandenbranden, J.; De Kruif, A.; Belpaire, F.; De Leenheer, A.P. Liquid chromatographic determination of oxytetracycline in bovine plasma by double-phase extraction. *J.Pharm.Sci.*, **1992**, *81*, 1216−1218

## SAMPLE
**Matrix:** blood
**Sample preparation:** 100 µL Plasma + 20 µL trifluoroacetic acid, mix 30 s in a whirl mixer, centrifuge at 5400 g for 5 min, inject supernatant (80 µL).

## HPLC VARIABLES
**Guard column:** 10 µm RP SAS-Hypersil
**Column:** 125 × 4.6 10 µm RP SAS-Hypersil
**Mobile phase:** MeCN:10 mM phosphoric acid 18:82
**Flow rate:** 2
**Injection volume:** 80
**Detector:** UV 365

## CHROMATOGRAM
**Retention time:** 2.1
**Limit of detection:** 80 ng/mL

## KEY WORDS
plasma

## REFERENCE
Krämer-Horaczynska, F. High-performance liquid chromatographic procedures for the quantitative analysis of 15 tetracycline derivatives in small blood samples. *J.Chromatogr.Sci.*, **1991**, *29*, 107−113

## SAMPLE
**Matrix:** blood, urine
**Sample preparation:** Serum. 500 µL Serum + 50 µL 6% ascorbic acid in water + 50 µL demeclocycline in MeOH/100 mM HCl + 1 mL buffer, mix for 30 s, add 6 mL ethyl acetate, rotate for 10 min, centrifuge at 3000 rpm for 6 min. Remove the organic layer and add it to 100 µL 0.2% ascorbic acid in MeOH, evaporate to dryness under vacuum while vortexing, reconstitute the residue in 200 µL mobile phase, mix, filter, keep in ice, inject a 20 µL aliquot. Urine. 100 µL Urine + 50 µL 6% ascorbic acid in water + 50 µL demeclocycline in MeOH/100 mM HCl + 400 µL buffer, mix for 30 s, add 3 mL ethyl acetate, rotate for 10 min, centrifuge at 3000 rpm for 6 min. Remove the organic layer and add it to 100 µL 0.2% ascorbic acid in MeOH, evaporate to dryness under vacuum while vortexing, reconstitute the residue in 200 µL mobile phase, mix, filter, keep in ice, inject a 20 µL aliquot. (Buffer was 27.6 g $NaH_2PO_4$ + 25.2 g sodium sulfite in 100 mL water, pH 6.1.)

## HPLC VARIABLES
**Column:** 100 × 2 5 µm Lichrosorb RP8

**Mobile phase:** MeCN:100 mM citric acid 24:76
**Flow rate:** 0.5
**Injection volume:** 20
**Detector:** UV 350

## CHROMATOGRAM
**Retention time:** 3
**Internal standard:** demeclocycline (4)
**Limit of detection:** 50 ng/mL

## OTHER SUBSTANCES
**Extracted:** chlortetracycline, doxycycline, methacycline, oxytetracycline

## KEY WORDS
serum

## REFERENCE
De Leenheer, A.P.; Nelis, H.J.C.F. Doxycycline determination in human serum and urine by high-performance liquid chromatography. *J.Pharm.Sci.*, **1979**, *68*, 999–1002

## SAMPLE
**Matrix:** bulk
**Sample preparation:** Prepare a 1 mg/mL solution, inject a 20 μL aliquot.

## HPLC VARIABLES
**Column:** 250 × 4.6 8 μm PLRP-S 100 Å poly(styrene-divinylbenzene) (Polymer Laboratories)
**Mobile phase:** t-Butanol:3.5% pH 9.0 K$_2$HPO$_4$:1% pH 9.0 tetrabutylammonium sulfate: 4% pH 9.0 sodium edetate:water 8.5:10:20:1:70.5 (w/v/v/v/v) (The solutions were adjusted to the required pH with 10% phosphoric acid and 8.5% NaOH.)
**Column temperature:** 60
**Flow rate:** 1
**Injection volume:** 20
**Detector:** UV 254

## CHROMATOGRAM
**Retention time:** 7.5

## OTHER SUBSTANCES
**Simultaneous:** impurities

## KEY WORDS
comparison of columns

## REFERENCE
Hendrix, C.; Roets, E.; Crommen, J.; de Beer, J.; Porqueras, E.; Van den Bossche, W.; Hoogmartens, J. Collaborative study of the analysis of tetracycline by liquid chromatography on poly(styrene-divinylbenzene). *J.Liq.Chromatogr.*, **1993**, *16*, 3321–3329

## SAMPLE
**Matrix:** bulk, formulations
**Sample preparation:** Bulk. Prepare a 10-100 μg/mL solution in buffer, inject an aliquot. Capsules, tablets. Prepare a 1 mg/mL solution of capsule contents or crushed tablets in buffer, sonicate for 10 min, filter (0.45 μm), dilute with buffer, inject an aliquot. (Buffer was 20 mM sodium perchlorate adjusted to pH 2.0 with perchloric acid.)

## HPLC VARIABLES
**Column:** 250 × 4.6 5 µm 100 Å PLRP-S polystyrene-divinylbenzene (Polymer Laboratories)
**Mobile phase:** MeCN:buffer 20:80 (Buffer was 20 mM sodium perchlorate adjusted to pH 2.0 with perchloric acid.)
**Flow rate:** 1
**Detector:** UV 280

## CHROMATOGRAM
**Retention time:** 27

## OTHER SUBSTANCES
**Simultaneous:** impurities, oxytetracycline

## KEY WORDS
capsules; tablets

## REFERENCE
Bryan, P.D.; Stewart, J.T. Chromatographic analysis of selected tetracyclines from dosage forms and bulk drug substance using polymeric columns with acidic mobile phases. *J.Pharm.Biomed.Anal.*, **1994**, *12*, 675–692

## SAMPLE
**Matrix:** bulk, formulations
**Sample preparation:** Bulk. Prepare a 10 mg/mL solution of tetracycline hydrochloride in water, inject a 20 µL aliquot. Prepare a 10 mg/mL solution of tetracycline in 100 mM HCl, inject a 20 µL aliquot. Formulations. Shake 500 mg capsule blend with 15 mL water and 1 mL concentrated ammonia until solid has dissolved, make up to 50 mL with pH 4.0 phosphate buffer, let stand for 10 min, filter, inject a 20 µL aliquot of the filtrate.

## HPLC VARIABLES
**Column:** 300 × 3.9 µBondapak phenyl
**Mobile phase:** Gradient. A was MeCN:water:phosphoric acid 240:1650:27, adjust pH to 2.2 with 45% KOH, make up to 2000 with water. B was MeCN:water:phosphoric acid 440:1500:27, adjust pH to 2.2 with 25% KOH, make up to 2000 with water. A:B 100:0 for 10 min then 0:100 for 5 min.
**Flow rate:** 2.6
**Injection volume:** 20
**Detector:** UV

## CHROMATOGRAM
**Retention time:** 7

## OTHER SUBSTANCES
**Simultaneous:** impurities, chlortetracycline

## REFERENCE
Muhammad, N.; Bodnar, J.A. Separation and quantitation of chlortetracycline, 4-epitetracycline, 4-epi-anhydrotetracycline, and anhydrotetracycline in tetracycline by high-performance liquid chromatography. *J.Pharm.Sci.*, **1980**, *69*, 928–930

## SAMPLE
**Matrix:** cell suspensions
**Sample preparation:** 100 µL Cell suspension + 100 µL cefoperazone solution + 100 µL Hanks balanced salt solution, sonicate 30 min, add 800 µL MeCN, centrifuge at 13000 g

for 5 min, remove supernatant. Dry supernatant under air, dissolve in 100 μL mobile phase, inject 75 μL.

## HPLC VARIABLES
**Column:** μBondapak C18
**Mobile phase:** MeCN:50 mM pH 3.1 KH$_2$PO$_4$ 45:55
**Flow rate:** 1
**Injection volume:** 75
**Detector:** UV 353

## CHROMATOGRAM
**Retention time:** 2.4
**Internal standard:** minocycline
**Limit of detection:** 100-1000 ng/mL

## REFERENCE
Darouiche, R.O.; Hamill, R.J. Antibiotic penetration of and bactericidal activity within endothelial cells. *Antimicrob.Agents Chemother.*, **1994**, *38*, 1059–1064

## SAMPLE
**Matrix:** food
**Sample preparation:** Prepare a 100 mg Baker 10 C18 cartridge by washing with MeOH, water, and 10 mL saturated aqueous Na$_2$EDTA. Dissolve 5 g honey in 20 mL 100 mM pH 4.0 Na$_2$EDTA-McIlvaine buffer, filter, apply to cartridge, wash with 20 mL water, air dry under vacuum for 5 min. Condition a Baker 10 COOH cartridge with ethyl acetate. Elute contents of C18 cartridge onto COOH cartridge with 50 mL ethyl acetate. Wash COOH cartridge with 10 mL MeOH, elute with 10 mL mobile phase, inject a 100 μL aliquot.

## HPLC VARIABLES
**Column:** 250 × 4.6 5 μm Bakerbond C8
**Mobile phase:** MeOH:MeCN:10 mM aqueous oxalic acid 1:1.5:3
**Flow rate:** 1
**Injection volume:** 100
**Detector:** UV 350

## CHROMATOGRAM
**Retention time:** 3.5
**Limit of detection:** 0.02 ppm

## OTHER SUBSTANCES
**Extracted:** chlortetracycline, doxycycline, oxytetracycline

## KEY WORDS
honey; SPE

## REFERENCE
Oka, H.; Ikai, Y.; Kawamura, N.; Uno, K.; Yamada, M.; Harada, K.; Uchiyama, M.; Asukabe, H.; Mori, Y.; Suzuki, M. Improvement of chemical analysis of antibiotics. IX. A simple method for residual tetracyclines analysis in honey using a tandem cartridge clean-up system. *J.Chromatogr.*, **1987**, *389*, 417–426

## SAMPLE
**Matrix:** food
**Sample preparation:** Condition a 500 mg Baker-10 C18 SPE cartridge with 10 mL MeOH, 10 mL water, and 10 mL saturated aqueous disodium EDTA. Condition a 500 mg Baker-10 COOH cartridge with MeOH:ethyl acetate 10:90. Dissolve 25 g honey in 50 mL 100

mM pH 4.0 disodium EDTA-McIlvaine buffer, filter. Add the filtrate to the C18 SPE cartridge, wash with 20 mL water, wash with 400 µL ethyl acetate, air dry under vacuum for 5 min, elute with 50 mL MeOH:ethyl acetate 10:90. Add a 5 mL aliquot to the COOH SPE cartridge, wash with 5 mL MeOH (?), elute with 10 mL mobile phase, inject a 100 µL aliquot.

## HPLC VARIABLES
**Column:** 75 × 4.6 3 µm Chemcosorb 3C8 (Chemco)
**Mobile phase:** MeCN:MeOH:10 mM aqueous oxalic acid 3:2:16, pH 3.0
**Flow rate:** 1
**Injection volume:** 100
**Detector:** UV 350

## CHROMATOGRAM
**Retention time:** 3
**Limit of detection:** 0.1 ppm

## OTHER SUBSTANCES
**Extracted:** chlortetracycline, demeclocycline (demethylchlortetracycline), doxycycline, methacycline, minocycline, oxytetracycline

## KEY WORDS
honey; SPE

## REFERENCE
Oka, H.; Ikai, Y.; Kawamura, N.; Uno, K.; Yamada, M.; Harada, K.; Suzuki, M. Improvement of chemical analysis of antibiotics. XII. Simultaneous analysis of seven tetracyclines in honey. *J.Chromatogr.*, **1987**, *400*, 253–261

---

## SAMPLE
**Matrix:** formulations
**Sample preparation:** Dissolve ointment in petroleum ether, add an equal volume of EtOH:water 70:30, dilute with MeOH to 100 µg/mL, inject a 10 µL aliquot.

## HPLC VARIABLES
**Column:** 300 × 3.9 10 µm LiChrosorb Si-60
**Mobile phase:** MeOH:water 5:95 containing 1.3 mM disodium citrate, 1 mM tetrabutylammonium bromide, 1.1 mM citric acid, and 8 mM EDTA.
**Flow rate:** 1
**Injection volume:** 10
**Detector:** UV 254

## CHROMATOGRAM
**Retention time:** $k'$ 1.56

## OTHER SUBSTANCES
**Simultaneous:** anhydrotetracycline, chlortetracycline, demeclocycline, doxycycline, epi-anhydrotetracycline, oxytetracycline, quatrimycin, rolitetracycline

## KEY WORDS
ointment

## REFERENCE
Lingeman, H.; van Munster, H.A.; Beynen, J.H.; Underberg, W.J.; Hulshoff, A. High-performance liquid chromatographic analysis of basic compounds on non-modified silica gel and aluminium oxide with aqueous solvent mixtures. *J.Chromatogr.*, **1986**, *352*, 261–274

## SAMPLE
**Matrix:** formulations
**Sample preparation:** Dissolve a capsule in 1 L 100 mM HCl, inject a 20 μL aliquot.

## HPLC VARIABLES
**Column:** 250 × 4.6 7 μm Zorbax TMS
**Mobile phase:** DMF:water 10:90 containing 5 mM sodium ethylenediaminetetraacetate, 100 mM citric acid, 20 mM sodium citrate, and 50 mM potassium nitrate.
**Flow rate:** 1.8
**Injection volume:** 20
**Detector:** UV 270

## CHROMATOGRAM
**Retention time:** 3.4

## OTHER SUBSTANCES
**Simultaneous:** sulfamethizole
**Noninterfering:** phenazopyridine

## KEY WORDS
capsules

## REFERENCE
Du Preez, J.L.; Botha, S.A.; Lötter, A.P. High-performance liquid chromatographic determination of phenazopyridine hydrochloride, tetracycline hydrochloride and sulphamethizole in combination. *J.Chromatogr.*, **1985**, *333*, 249–252

## SAMPLE
**Matrix:** milk
**Sample preparation:** Prepare a column by adding 1.5 mL of thoroughly mixed Chelating Sepharose Fast-Flow suspension in EtOH:water 20:80 (Pharmacia) to a 150 × 10 glass column, allow to drain, wash with three 2 mL portions of water, add 2 mL 10 mM copper(II) sulfate in water, wash with two 2 mL portions of water. Centrifuge 10 mL milk at 2100 g for 5 min, decant the skimmed milk, rinse the tube with two 1 mL portions of water. Add 10 mL pH 4.0 buffer to the milk and rinses, sonicate for 3 min, filter (Whatman 541 paper) the supernatant. Add the filtrate to the column, wash with 2 mL pH 4.0 buffer, wash with 2 mL water, wash with 2 mL MeOH, wash with 2 mL water, add 700 μL EDTA buffer to the column, elute with 3 mL EDTA buffer, add 20 μL 25 μg/mL demeclocycline hydrochloride in MeOH to the eluate, inject a 100 μL aliquot. (Prepare pH 4.0 buffer by adjusting 100 mM succinic acid to pH 4.0 with 10 M NaOH. Prepare EDTA buffer by dissolving 12.9 g citric acid monohydrate, 10.9 g $Na_2HPO_4$, 29.2 g NaCl, and 100 mmoles EDTA in 1 L water.)

## HPLC VARIABLES
**Guard column:** 5 × 3 PLRP-S (Polymer Laboratories)
**Column:** 250 × 4.6 5 μm 100 Å PLRP-S (Polymer Laboratories)
**Mobile phase:** MeCN:MeOH:buffer 15:10:60 (Buffer was 10 mM oxalic acid adjusted to pH 2.0 with 4 M HCl.)
**Flow rate:** 1
**Injection volume:** 100
**Detector:** F ex 406 em 515 following post-column reaction. The column effluent mixed with reagent pumped at 1 mL/min and the mixture flowed through a 600 μL reaction coil to the detector. (Reagent was 5% zirconyl chloride octahydrate in water.)

## CHROMATOGRAM
**Retention time:** 5.9

**Internal standard:** demeclocycline (8.3)
**Limit of detection:** 2 ng/mL

## OTHER SUBSTANCES
**Extracted:** chlortetracycline, oxytetracycline

## KEY WORDS
protect from light; cow; post-column reaction; derivatization; SPE

## REFERENCE
Croubels, S.; Van Peteghem, C.; Baeyens, W. Sensitive spectrofluorimetric determination of tetracycline residues in bovine milk. *Analyst*, **1994**, *119*, 2713–2716

## SAMPLE
**Matrix:** milk
**Sample preparation:** Prepare a column as follows. Swirl Chelating Sepharose Fast Flow resin (Pharmacia) in its bottle, add it to a polypropylene column to give a bed volume of 1.0-1.2 mL, wash 3 times with 2 mL portions of water, wash with 2 mL 10 mM copper sulfate, wash with two 2 mL portions of water. Centrifuge 5 mL milk at 10° at 1500 g for 15 min, remove the lower layer and add it to 10 mL succinate buffer, mix, centrifuge at 1500 g for 30 min, add the supernatant to the column. Wash with 2 mL succinate buffer, wash with 2 mL water, wash with 2 mL MeOH, wash with 2 mL water, wash with 700 µL citrate/phosphate buffer (be careful not to disturb bed), elute with 2.5 mL citrate/phosphate buffer (column is white and eluate is blue). Filter (Amicon Centricon 30, MW 30000 cut-off; pre-washed by centrifuging with 2 mL water) while centrifuging at 5000 g for 30-90 min, inject a 600 µL aliquot of the ultrafiltrate. (Prepare succinate buffer by dissolving 11.8 g succinic acid in 980 mL water, adjust pH to 4.0 with 10 M NaOH, make up to 1 L. Prepare the citrate/phosphate buffer by dissolving 12.9 g citric acid monohydrate, 10.9 g $Na_2HPO_4$, 37.2 g disodium EDTA dihydrate, and 29.2 g NaCl in 1 L water.)

## HPLC VARIABLES
**Column:** 150 × 4.6 5 µm PLRP-S (Polymer Labs)
**Mobile phase:** Gradient. MeCN:MeOH:10 mM oxalic acid 0:0:100 for 1 min, to 22:8:70 over 5 min, maintain at 22:8:70 for 11 min, return to initial conditions.
**Flow rate:** 1
**Injection volume:** 600
**Detector:** UV 355

## CHROMATOGRAM
**Retention time:** 13.3
**Limit of detection:** 0.52 ng/mL
**Limit of quantitation:** 1.10 ng/mL

## OTHER SUBSTANCES
**Extracted:** chlortetracycline, demeclocycline, doxyclycline, methacycline, minocycline, oxytetracycline
**Noninterfering:** chloramphenicol, gentian violet, hydromycin B, ivermectin, spectinomycin, sulfa drugs

## KEY WORDS
cow; SPE; ultrafiltrate

## REFERENCE
Carson, M.C. Simultaneous determination of multiple tetracycline residues in milk using metal chelate affinity chromatography. *J.AOAC Int.*, **1993**, *76*, 329–334

## SAMPLE

**Matrix:** milk

**Sample preparation:** 5 mL Milk + 1 mL 1 M HCl, mix, add 24 mL MeCN slowly with swirling over 30 s, let stand for 5 min, decant the clear supernatant through a plug of glass wool. 15 mL Filtrate + 15 mL dichloromethane + 30 mL hexane, mix, collect the aqueous layer. Extract the organic layer with 1 mL water. Combine the aqueous layers, make up to 4 mL with water, filter (13 mm, 0.45 μm, PVDF), inject a 1000 μL aliquot of the filtrate.

## HPLC VARIABLES

**Guard column:** 5 μm PLRP-S 100 Å polystyrene divinylbenzene (Polymer Laboratories)

**Column:** 150 × 4.6 5 μm PLRP-S 100 Å polystyrene divinylbenzene (Polymer Laboratories)

**Mobile phase:** Gradient. MeCN:buffer 20:80 for 3 min, to 38:62 over 22 min, maintain at 38:62 for 5 min, return to initial conditions for 1 min, re-equilibrate for 9 min. (Buffer was 3.94 g potassium oxalate + 3.61 g oxalic acid + 1.22 g sodium decanesulfonate in 1 L water, pH 2.30.)

**Flow rate:** 1

**Injection volume:** 1000

**Detector:** UV 365

## CHROMATOGRAM

**Retention time:** 18

**Limit of detection:** 5 ng/mL

## OTHER SUBSTANCES

**Extracted:** chlortetracycline, oxytetracycline

## REFERENCE

White, C.R.; Moats, W.A.; Kotula, K.L. Optimization of a liquid chromatographic method for determination of oxytetracycline, tetracycline, and chlortetracycline in milk. *J.AOAC Int.*, **1993**, *76*, 549– 554

## SAMPLE

**Matrix:** milk

**Sample preparation:** Place 22 g 40 μm, 18% load, end-capped bulk C18 material (Analytichem) in a 50 mL syringe barrel, wash with 2 column volumes hexane, dichloromethane, and MeOH, vacuum aspirate until dry. 2 g Bulk C18 material + 50 mg disodium EDTA + 50 mg oxalic acid + 500 μL milk + 10 μL MeOH, blend gently in a glass mortar and pestle for 30 s, place the mixture in a 10 mL plastic syringe barrel plugged with a piece of filter paper. Compress column volume to 4.5 mL, add a 100 μL pipette tip on the column outlet to restrict the flow. Wash with 8 mL hexane, remove excess hexane with positive pressure, elute with 8 mL MeCN:ethyl acetate 75:25. Evaporate the eluate to dryness under a stream of nitrogen at 40°, reconstitute the residue in 500 μL mobile phase, sonicate for 5-10 min, centrifuge at 17000 g for 5 min, filter the supernatant (0.45 μm), inject a 20 μL aliquot of the filtrate.

## HPLC VARIABLES

**Column:** 300 × 4 10 μm Micro Pak ODS

**Mobile phase:** MeCN:10 mM oxalic acid in water 30:70

**Flow rate:** 1

**Injection volume:** 20

**Detector:** UV 365

## CHROMATOGRAM

**Retention time:** 4.5

**Limit of detection:** 2 ng

## OTHER SUBSTANCES
**Extracted:** chlortetracycline, oxytetracycline

## KEY WORDS
cow; matrix solid-phase dispersion

## REFERENCE
Long, A.R.; Hsieh, L.C.; Malbrough, M.S.; Short, C.R.; Barker, S.A. Matrix solid-phase dispersion (MSPD) isolation and liquid chromatographic determination of oxytetracycline, tetracycline, and chlortetracycline in milk. *J.Assoc.Off.Anal.Chem.*, **1990**, *73*, 379–384

## SAMPLE
**Matrix:** milk
**Sample preparation:** 2 mL Milk + 4 mL buffer, filter (Amicon CF-25 ultrafiltration membrane) while centrifuging at 20° at 1000 g for 1 h, suspend solids in 2 mL buffer and repeat filtration for 40 min. Combine filtrates and inject a 500 μL aliquot as soon as possible. (Buffer (McIlvaine) was prepared by mixing 625 mL 28.41 g/L $Na_2HPO_4$ and 1 L 21.01 g/L citric acid monohydrate. The buffer was also 100 mM in disodium EDTA and the final pH was 4.0 ± 0.1.)

## HPLC VARIABLES
**Column:** 150 × 3.9 Novapak C18
**Mobile phase:** Gradient. MeCN:MeOH:10 mM oxalic acid 0:0:100 for 1 min, to 22:8:70 over 5 min, maintain at 22:8:70 for 10 min, re-equilibrate at 0:0:100 at 1.5 mL/min for 5 min and at 1 mL/min for 1 min. (Flush daily with 10 column volumes of water. Store column in MeOH:water 60:40, flush with water before use.)
**Column temperature:** 30
**Flow rate:** 1
**Injection volume:** 500
**Detector:** UV 360

## CHROMATOGRAM
**Retention time:** 10.8
**Limit of detection:** 11.6 ng/mL
**Limit of quantitation:** 31.3 ng/mL

## OTHER SUBSTANCES
**Extracted:** chlortetracycline, oxytetracycline

## KEY WORDS
cow; protect from light; ultrafiltrate

## REFERENCE
Thomas, M.H. Simultaneous determination of oxytetracycline, tetracycline, and chlortetracycline in milk by liquid chromatography. *J.Assoc.Off.Anal.Chem.*, **1989**, *72*, 564–567

## SAMPLE
**Matrix:** solutions

## HPLC VARIABLES
**Column:** 250 × 4.6 Zorbax RX
**Mobile phase:** Gradient. A was 10 mL concentrated orthophosphoric acid and 7 mL triethylamine in 1 L water. B was 10 mL concentrated orthophosphoric acid and 7 mL triethylamine in 200 mL water, make up to 1 L with MeCN. A:B from 100:0 to 0:100 over 30 min, maintain at 0:100 for 5 min.
**Column temperature:** 30

**Flow rate:** 2
**Detector:** UV 210

## OTHER SUBSTANCES
**Also analyzed:** acepromazine, acetaminophen, acetophenazine, albuterol, aminophylline, amitriptyline, amobarbital, amoxapine, amphetamine, amylocaine, antipyrine, aprobarbital, aspirin, atenolol, atropine, avermectin, barbital, benzocaine, benzoic acid, benzotropine, benzphetamine, berberine, bibucaine, bromazepan, brompheniramine, buprenorphine, buspirone, butabarbital, butacaine, butethal, caffeine, carbamazepine, carbromal, chloramphenicol, chlordiazepoxide, chloroquine, chlorothiazide, chloroxylenol, chlorphenesin, chlorpheniramine, chlorpromazine, chlorpropamide, chlortetracycline, cimetidine, cinchonidine, cinchonine, clenbuterol, clonazepam, clonixin, clorazepate, cocaine, codeine, colchicine, cortisone, coumarin, cyclazocine, cyclobenzaprine, cyclothiazide, cyheptamide, cymarin, danazol, danthron, dapsone, debrisoquine, desipramine, dexamethasone, dextromethorphan, dextropropoxyphene, diamorphine, diazepam, diclofenac, diethylpropion, diethylstilbestrol, diflunisal, digitoxin, digoxin, diltiazem, diphenhydramine, diphenoxylate, diprenorphine, dipyrone, disulfiram, dopamine, doxapram, doxepin, dronabinol, ephedrine, epinephrine, epinine, estradiol, estriol, estrone, ethacrynic acid, ethosuximide, etonitazene, etorphine, eugenol, famotidine, fenbendazole, fencamfamine, fenoprofen, fenproporex, fentanyl, flubendazole, flufenamic acid, flunitrazepam, 5-fluorouracil, fluoxymesterone, fluphenazine, furosemide, gentisic acid, gitoxigenin, glipizide, glunixin, glutethimide, glybenclamide, guaiacol, halazepam, haloperidol, hydrochlorothiazide, hydrocodone, hydrocortisone, hydromorphone, hydroxyquinoline, ibogaine, ibuprofen, iminostilbene, imipramine, indomethacin, isocarbostyril, isocarboxazid, isoniazid, isoproterenol, isoxsuprine, ivermectin, ketamine, ketoprofen, kynurenic acid, levorphanol, lidocaine, lorazepam, lormetazepam, loxapine, mazindol, mebendazole, meclizine, meclofenamic acid, medazepam, mefenamic acid, megestrol, mepacrine, meperidine, mephentermine, mephenytoin, mephesin, mephobarbital, mepivacaine, mescaline, mesoridazine, methadone, methamphetamine, methapyrilene, methaqualone, methazolamide, methocarbamol, methoxamine, methsuximide, methyl salicylate, methyldopa, methyldopamine, methylphenidate, methylprednisolone, methyltestosterone, methyprylon, metoprolol, mibolerone, morphine, nadolol, nalorphine, naloxone, naltrexone, naphazoline, naproxen, nefopam, niacinamide, nicotine, nicotinic acid, nifedipine, niflumic acid, nitrazepam, norepinephrine, nortriptyline, noscapine, nylidrin, oxazepam, oxycodone, oxymorphone, oxyphenbutazone, oxytetracycline, papaverine, pargyline, pemoline, pentazocine, pentobarbital, persantine, phenacetin, phenazocine, phenazopyridine, phencyclidine, phendimetrazine, phenelzine, pheniramine, phenobarbital, phenothiazine, phensuximide, phentermine, phenylbutazone, phenylephrine, phenylpropanolamine, piperocaine, prazepam, prednisolone, primidone, probenecid, progesterone, propiomazine, propranolol, propylparaben, pseudoephedrine, puromycin, pyrilamine, pyrithyldione, quazepam, quinaldic acid, quinidine, quinine, ranitidine, recinnamine, reserpine, resorcinol, saccharin, albuterol, salicylamide, salicylic acid, scopolamine, scopoletin, secobarbital, strychnine, sulfacetamide, sufadiazine, sulfadimethoxine, sulfaethidole, sulfamerazine, sulfamethazine, sulfamethoxizole, sulfanilamide, sulfapyridine, sulfasoxizole, sulindac, tamoxifen, temazepam, testosterone, tetracaine, tetramisole, thebaine, theobromine, theophylline, thiabendazole, thiamine, thiamylal, thiobarbituric acid, thioridazine, thiosalicylic acid, thiothixene, thymol, tolazamide, tolazoline, tobutamide, tolmetin, tranylcypromine, triamcinolone, tribenzylamine, trichloromethiazide, trifluoperazine, trihexyphenidyl, trimethoprim, tripelennamine, triprolidine, tropacocaine, tyramine, verapamil, vincamine, warfarin, yohimbine, zoxazolamine

## REFERENCE
Hill, D.W.; Kind, A.J. Reversed-phase solvent-gradient HPLC retention indexes of drugs. *J.Anal.Toxicol.*, **1994**, *18*, 233–242

## SAMPLE
**Matrix:** solutions
**Sample preparation:** Inject an aliquot of a 1 μg/mL solution in 10 mM HCl.

## HPLC VARIABLES

**Guard column:** present but not specified

**Column:** 150 × 4.6 5 μm PLRP-S styrene-divinylbenzene copolymer (Polymer Laboratories)

**Mobile phase:** Gradient. MeCN:50 mM pH 2.0 oxalate buffer 15:85, for 3 min to 60:40 over 17 min, maintain at 60:40 for 5 min, return to initial conditions over 1 min, re-equilibrate for 9 min. (After use flush with water for 10 min, store in MeCN:water 60:40.)

**Flow rate:** 1

**Detector:** UV 355

## CHROMATOGRAM

**Retention time:** 10

## OTHER SUBSTANCES

**Simultaneous:** chlortetracycline, oxytetracycline

## REFERENCE

White, C.R.; Moats, W.A.; Kotula, K.L. Comparative study of high performance liquid chromatographic methods for the determination of tetracycline antibiotics. *J.Liq.Chromatogr.*, **1993**, *16*, 2873–2890

## SAMPLE

**Matrix:** solutions

**Sample preparation:** Prepare a solution in 10 mM HCl, inject a 200 μL aliquot.

## HPLC VARIABLES

**Guard column:** present but not specified

**Column:** 150 × 4.6 5 μm PLRP-S styrene-divinyl benzene copolymer (Polymer Laboratories)

**Mobile phase:** Gradient. MeCN:10 mM orthophosphoric acid from 15:85 to 60:40 over 20 min

**Flow rate:** 1

**Injection volume:** 200

**Detector:** UV 355

## CHROMATOGRAM

**Retention time:** 8.5

## OTHER SUBSTANCES

**Simultaneous:** chlortetracycline, oxytetracycline

## REFERENCE

Moats, W.A. Effect of the silica support of bonded reversed-phase columns on chromatography of some antibiotic compounds. *J.Chromatogr.*, **1986**, *366*, 69–78

## SAMPLE

**Matrix:** tissue

**Sample preparation:** Condition a 500 mg Bond Elut cyclohexyl (CH) SPE cartridge with 10 mL MeOH and 10 mL water. Powder (domestic food blender) frozen kidney or muscle. Homogenize (Silverson Machines) 5 g powdered tissue and 45 mL 100 mM glycine in 1 M HCl for 1 min, add 5 g ammonium sulfate, shake for 30 s, let stand for 10 min, centrifuge at 2000 rpm for 15 min, filter (glass wool) the supernatant, repeat the extraction with 50 mL 100 mM glycine in 1 M HCl. Combine the filtrates and centrifuge an aliquot at 2200 rpm for 10 min, add a 20 mL aliquot of the supernatant to the SPE cartridge, wash with 10 mL water, elute with 7 mL MeOH. Evaporate the eluate to dryness under a stream of nitrogen at 65°, reconstitute the residue in 500 μL MeCN:20 mM oxalic acid 20:80, inject a 50 μL aliquot.

## HPLC VARIABLES

**Guard column:** Chromspher C8 (Chrompack)
**Column:** 200 × 3 5 μm Chromspher C8 glass column (Chrompack)
**Mobile phase:** Gradient. A was MeCN. B was MeCN:20 mM oxalic acid 10:90. A:B from 10:90 to 20:80 over 2 min, maintain at 20:80 for 8 min, to 25:75 over 1 min, maintain at 25:75 for 9 min, return to initial conditions over 5 min, re-equilibrate for 10 min.
**Flow rate:** 0.4
**Injection volume:** 50
**Detector:** F ex 390 em 490 following post-column reaction. The column effluent mixed with 750 mM aluminum chloride (degas by sonication, store in a brown bottle) pumped at 0.6 mL/min and flowed through a 13.7 m × 0.3 mm i.d. PTFE coil at 60° to the detector.

## CHROMATOGRAM

**Retention time:** 13.7
**Limit of detection:** 20 ng/g (muscle); 230 ng/g (kidney)

## OTHER SUBSTANCES

**Extracted:** chlortetracycline, oxytetracycline

## KEY WORDS

pig; cow; poultry; kidney; muscle; SPE; post-column reaction

## REFERENCE

McCracken, R.J.; Blanchflower, W.J.; Haggan, S.A.; Kennedy, D.G. Simultaneous determination of oxytetracycline, tetracycline and chlortetracycline in animal tissues using liquid chromatography, post-column derivatization with aluminium, and fluorescence detection. *Analyst*, **1995**, *120*, 1763–1766

## SAMPLE

**Matrix:** tissue
**Sample preparation:** Condition a 500 mg Separcol SI C18 SPE cartridge (Anapron) with 2 mL MeOH and 4 mL buffer. Homogenize 5 g muscle with 20 mL buffer and 3 mL n-hexane:dichloromethane 1:3 at 4°, centrifuge at 2400 g at 4° for 30 min, remove the supernatant, repeat homogenization with 10 mL buffer. Combine the supernatants, slowly add with constant stirring a volume of 1 g/mL trichloroacetic acid in water equal to 10% of the supernatant volume, stir for another min, keep in ice for 15 min, filter through paper, add to the SPE cartridge at no more than 10 mL/min, wash with 2 mL water, elute with 4 mL 10 mM oxalic acid in MeOH, inject a 10 μL aliquot. (Buffer was 15 g Na$_2$HPO$_4$.2H$_2$O + 13 g citric acid monohydrate + 3.72 g EDTA in 1 L water, pH 4.)

## HPLC VARIABLES

**Guard column:** 5 μm LiChrospher 100 RP-18 guard column
**Column:** 250 × 4 5 μm HP Spherisorb ODS 2
**Mobile phase:** MeOH:MeCN:10 mM aqueous oxalic acid 20:35:45
**Flow rate:** 1
**Injection volume:** 10
**Detector:** UV 360

## CHROMATOGRAM

**Retention time:** 6.5
**Limit of detection:** 50 ng/g

## OTHER SUBSTANCES

**Extracted:** chlortetracycline, oxytetracycline

## KEY WORDS

muscle; cow; pig; SPE

## REFERENCE

Sokol, J.; Matisova, E. Determination of tetracycline antibiotics in animal tissues of food-producing animals by high-performance liquid chromatography using solid-phase extraction. *J.Chromatogr.A*, **1994**, *669*, 75–80

## SAMPLE

**Matrix:** tissue

**Sample preparation:** Prepare an affinity column by filling a 10 mL column with 5 mL chelating Sepharose, allow to settle, wash with 20 mL 0.5% copper(II) sulfate solution, eliminate air bubbles by agitation, wash with 15 mL 50 mM pH 4 succinate buffer, do not allow to dry. Condition an Analytichem Bond Elut C18 SPE cartridge with 10 mL MeOH and 10 mL water, do not allow to dry. Homogenize 4 g minced kidney with 40 mL 50 mM pH 4 succinate buffer, sonicate for 10 min, centrifuge at 9000 rpm for 10 min, filter the supernatant through paper, repeat the extraction. Combine the supernatants and pass them through the affinity column at 5-7 mL/min, wash with 10 mL water, wash with 30 mL MeOH, wash with 20 mL water, elute with 50 mL 50 mM pH 4 succinate buffer containing 3.7% Titriplex III (ethylenedinitrilotetraacetic acid, disodium salt dihydrate). Add the eluate to the SPE cartridge at 5-7 mL/min, wash with 10 mL water, dry with air aspiration for 10 min, elute with 5 mL MeOH:MeCN 1:1, evaporate the eluate at 40° under a stream of nitrogen, dissolve the residue in 500 µL mobile phase, inject an aliquot. Protect from light through process. (The affinity columns may be re-used up to 15 times by washing with 20 mL water then 20 mL EtOH:water 20:80 then conditioning as described above.)

## HPLC VARIABLES

**Guard column:** Perisorb RP-8

**Column:** two 300 × 100 5 µm Chromspher C8 columns (cat. no. 28262) in series

**Mobile phase:** MeCN:10 mM pH 2 oxalic acid 20:80

**Flow rate:** 0.8

**Detector:** UV 365

## CHROMATOGRAM

**Retention time:** 6.5

**Limit of quantitation:** 10 ng/g

## OTHER SUBSTANCES

**Extracted:** chlortetracycline, demethylchlortetracycline, doxycycline, methacycline, oxytetracycline

## KEY WORDS

kidney; SPE

## REFERENCE

Degroodt, J.M.; Wyhowski de Bukanski, B.; Srebrnik, S. Multiresidue analysis of tetracyclines in kidney by HPLC and photodiode array detection. *J.Liq.Chromatogr.*, **1993**, *16*, 3515–3529

◆――――――――――――――◆――――――――――――――◆

## ANNOTATED BIBLIOGRAPHY

Croubles, S.; Baeyens, W.; Van Peteghem, C. Evaluation of a narrow-bore HPLC column for trace level analysis of tetracyclines–a comparison with a conventional column. *Biomed.Chromatogr.*, **1995**, *9*, 251–253 [simultaneous chlortetracycline, demeclocycline, oxytetracycline]

Barker, S.A.; Long, A.R. Preparation of milk samples for immunoassay and liquid chromatographic screening using matrix solid-phase dispersion. *J.AOAC Int.*, **1994**, *77*, 848–854 [MSPD; infant formula; also albendazole, chloramphenicol, chlorsulon, chlortetracycline, fenbendazole, furazolidone, mebendazole, oxfendazole, oxytetracycline, sulfadiazine, sulfadimethoxine, sulfamerazine, sulfamethazine, sulfamethoxazole, sulfanilamide, sulfathiazole, sulfisoxazole, thiabendazole; LOD 0.1-2 ng]

Muritu, J.W.; Kibwage, I.O.; Maitai, C.K.; Hoogmartens, J. Evaluation of tetracycline raw materials and finished products found on the Kenyan market. *J.Pharm.Biomed.Anal.*, **1994**, *12*, 1483–1488 [bulk; column temp 60; capsules; ointment; simultaneous impurities]

Walsh, J.R.; Walker, L.V.; Webber, J.J. Determination of tetracyclines in bovine and porcine muscle by high-performance liquid chromatography using solid-phase extraction. *J.Chromatogr.*, **1992**, *596*, 211–216

Ray, A.; Newton, V. Use of high-performance liquid chromatography to monitor stability of tetracycline and chlortetracycline in susceptibility determinations. *Antimicrob.Agents Chemother.*, **1991**, *35*, 1264–1266

Khan, N.H.; Wera, P.; Roets, E.; Hoogmartens, J. Quantitative analysis of tetracycline by high performance liquid chromatography on polystyrene-divinylbenzene packing materials. *J.Liq.Chromatogr.*, **1990**, *13*, 1351–1374 [column temp 60; simultaneous impurities, acetyldecarboxamidotetracycline, anhydrotetracycline, epianhydrotetracycline, epitetracycline; bulk]

LeDuc, B.W.; Greenblatt, D.J.; Friedman, H. Automated high-performance liquid chromatographic analysis of tetracycline in urine. *J.Chromatogr.*, **1989**, *490*, 474–477

Ray, A.; Harris, R. High-performance liquid chromatography as an alternative to microbiological measurements in the assay of tetracyclines. *J.Chromatogr.*, **1989**, *467*, 430–435

Martinez, E.E.; Shimoda, W. Liquid chromatographic determination of tetracycline residues in animal feeds. *J.Assoc.Off.Anal.Chem.*, **1988**, *71*, 477–480

Grobben-Verpoorten, A.; Dihuidi, K.; Roets, E.; Hoogmartens, J.; Vanderhaeghe, H. Determination of the stability of tetracycline suspensions by high performance liquid chromatography. *Pharm. Weekbl.[Sci].*, **1985**, *7*, 104–108

Oka, H.; Matsumoto, H.; Uno, K.; Harada, K.; Kadowaki, S.; Suzuki, M. Improvement of chemical analysis of antibiotics. VIII. Application of prepacked C18 cartridge for the analysis of tetracycline residues in animal liver. *J.Chromatogr.*, **1985**, *325*, 265–274

Vancaillie, R.; Dihuidi, K.; Roets, E.; Hoogmartens, J.; Vanderhaeghe, H. Determination by high performance liquid chromatography of the stability of alcoholic solutions for topical use containing tetracycline and 4-epitetracycline. *J.Pharm.Belg.*, **1985**, *40*, 168–172

Oka, H.; Suzuki, M. Improvement of chemical analysis of antibiotics. VII. Comparison of analytical methods for determination of impurities in tetracycline pharmaceutical preparations. *J.Chromatogr.*, **1984**, *314*, 303–311

Onji, Y.; Uno, M.; Tanigawa, K. Liquid chromatographic determination of tetracycline residues in meat and fish. *J.Assoc.Off.Anal.Chem.*, **1984**, *67*, 1135–1137

Dihuidi, K.; Roets, E.; Hoogmartens, J.; Vanderhaeghe, H. Influence of temperature on the stability of solid tetracycline hydrochloride, measured by high-performance liquid chromatography. *J.Chromatogr.*, **1982**, *246*, 350–355 [bulk; gradient; column temp 30; simultaneous impurities]

Hermansson, J. Rapid determination of tetracycline and lumecycline in human plasma and urine using high-performance liquid chromatography. *J.Chromatogr.*, **1982**, *232*, 385–393

Charles, B.G.; Cole, J.J.; Ravenscroft, P.J. Method for rapid determination of urinary tetracycline by high-performance liquid chromatography. *J.Chromatogr.*, **1981**, *222*, 152–155

# Theophylline

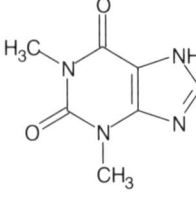

**Molecular formula:** $C_7H_8N_4O_2$
**Molecular weight:** 180.2
**CAS Registry No.:** 58-55-9, 32156-80-2 (diethanolamine), 573-41-1 (ethanolamine), 5600-19-1(isopropanolamine), 8002-89-9 (sodium acetate), 8000-10-1 (sodium glycinate), 5967-84-0 (monohydrate)

## SAMPLE
**Matrix:** bile, blood, tissue, urine
**Sample preparation:** Tissue. Macerate 2 g tissue with 5 mL water, add 15 mL MeCN, add 0.1 mL 1 mg/mL 2-acetamidophenol in ethanol, shake, centrifuge at 5200 g. Transfer supernatant to 50 mL tube containing 8 mL diethyl ether + 12 mL dichloromethane + 1 mL citrate buffer, vortex, proceed as in (A). Bile. 2 mL Bile + 3 mL water, add 15 mL MeCN, add 0.1 mL 1 mg/mL 2-acetamidophenol in ethanol, vortex, add 8 mL diethyl ether + 12 mL dichloromethane, vortex, centrifuge at 5200 g, proceed as in (A). Blood, urine. 5 mL Blood or urine + 15 mL acetone + 0.1 mL 1 mg/mL 2-acetamidophenol in ethanol, vortex for a few s, add 8 mL diethyl ether, vortex, centrifuge 5200 g. Transfer supernatant to 50 mL tube, add 12 mL dichloromethane, add 1 mL citrate buffer, vortex, proceed as in (A). (A). Discard lower, aqueous layer. Filter the organic layer through 3 g Florisil + 8 g anhydrous sodium sulfate and wash through with 15 mL diethyl ether. Evaporate filtrate to dryness under a stream of air at 40°. Reconstitute in 5 mL MeCN:0.1 N $NaH_2PO_4$ 60:40 + 3 mL hexane, vortex. Remove and discard upper hexane layer, add 8 mL 20% (v/v) isopropanol in chloroform to the aqueous layer, vortex. Remove and discard the upper aqueous layer and evaporate lower layer. Reconstitute residue in MeCN:water 10:90, inject an aliquot. (Citrate buffer was saturated sodium citrate containing enough sodium tunstate (sic) to bring pH to 8. To each 1 L of diethyl ether 10 mL of water and 1 mL of citrate buffer are added.)

## HPLC VARIABLES
**Column:** 100 × 4.6 C18 microbore
**Mobile phase:** MeCN:dilute phosphoric acid (1 mL 85% phosphoric acid in 140 mL water) 7:93
**Column temperature:** 40
**Flow rate:** 0.3
**Detector:** UV 271

## CHROMATOGRAM
**Retention time:** 3.2
**Internal standard:** 2-acetamidophenol (4.9)
**Limit of detection:** 50 (blood, urine); 200 (bile, tissue) ng/mL

## OTHER SUBSTANCES
**Extracted:** acetaminophen

## KEY WORDS
liver; whole blood

## REFERENCE
Mathis, D.F.; Budd, R.D. Extraction of acetaminophen and theophylline from post-mortem tissues and urine for high-performance liquid chromatographic analysis. *J.Chromatogr.*, **1988**, *439*, 466–469

## SAMPLE
**Matrix:** blood

**Sample preparation:** 100 μL Plasma + 10 μL 40 μg/mL β-hydroxyethyltheophylline + 10 μL perchloric acid, vortex for 10 s, centrifuge for 5 min, inject a 20 μL aliquot of the supernatant.

## HPLC VARIABLES
**Column:** 5 μm C18 Radpak (Waters)
**Mobile phase:** MeCN:MeOH:10 mM $KH_2PO_4$:triethylamine 5:7:88:0.02, pH 4.5
**Flow rate:** 3
**Injection volume:** 20
**Detector:** UV 273

## CHROMATOGRAM
**Retention time:** 5.5
**Internal standard:** β-hydroxyethyltheophylline (7.5)
**Limit of quantitation:** 625 ng/mL

## KEY WORDS
plasma; rat; pharmacokinetics

## REFERENCE
Angus, P.W.; Ng, C.Y.; Ghabrial, H.; Morgan, D.J.; Smallwood, R.A. Effects of chronic left ventricular failure on hepatic oxygenation and theophylline elimination in the rat. *Drug Metab.Dispos.*, **1995**, *23*, 485–489

## SAMPLE
**Matrix:** blood
**Sample preparation:** 1 mL Plasma + 10 μL 2 mg/mL IS in MeOH, mix, add 2 mL dichloromethane:diethyl ether 80:20, vortex for 15 s, centrifuge at 1500 g for 5 min, remove a 1.7 mL aliquot of the organic phase, repeat the extraction twice more with 2 mL portions of dichloromethane:diethyl ether 80:20. Combine the organic layers and evaporate them to dryness under a stream of nitrogen at room temperature, reconstitute the residue in 200 μL mobile phase, inject a 20 μL aliquot.

## HPLC VARIABLES
**Guard column:** 20 × 4.6 40 μm Pelliguard C18 (Supelco)
**Column:** 250 × 4.6 5 μm Viosfer C18 (Violet, Rome)
**Mobile phase:** MeCN:buffer 15:85 (Buffer was 100 mM phosphate buffer containing 5 mM tetrabutylammonium hydrogen sulfate, adjusted to pH 3 with orthophosphoric acid.)
**Flow rate:** 1.3
**Injection volume:** 20
**Detector:** UV 280

## CHROMATOGRAM
**Retention time:** 4.8
**Internal standard:** 2-[4-(2'-furoyl)phenyl]propionic acid (7.3)
**Limit of detection:** 20 ng/mL

## OTHER SUBSTANCES
**Extracted:** rufloxacin

## KEY WORDS
plasma

## REFERENCE
Carlucci, G.; Mazzeo, P.; Palumbo, G. Simultaneous determination of rufloxacin and theophylline by high-performance liquid chromatography in human plasma. *Analyst*, **1995**, *120*, 2493–2495

## SAMPLE
**Matrix:** blood
**Sample preparation:** Add 200 μL whole blood to 400 μL MeCN while mixing rapidly for 5 s, centrifuge at 1500 rpm for 2 min, mix 100 μL supernatant with 1 mL water, inject a 50 μL aliquot.

## HPLC VARIABLES
**Column:** 50 × 4.6 Little Champ reverse phase (Regis)
**Mobile phase:** MeCN:10 mM pH 3 sodium phosphate buffer 4:96
**Flow rate:** 1
**Injection volume:** 50
**Detector:** UV 273

## CHROMATOGRAM
**Limit of quantitation:** 500 nM

## KEY WORDS
rat; whole blood

## REFERENCE
Hoffman, D.J.; Seifert, T.; Borre, A.; Nellans, H.N. Method to estimate the rate and extent of intestinal absorption in conscious rats using an absorption probe and portal blood sampling. *Pharm.Res.*, **1995**, *12*, 889–894

## SAMPLE
**Matrix:** blood
**Sample preparation:** 100 μL Sample + 300 μL 7-(2-hydroxyethyl)theophylline in chloroform:isopropanol 50:50, extract. Remove 200 μL of the organic layer and evaporate it to dryness, reconstitute the residue in mobile phase, inject a 20 μL aliquot.

## HPLC VARIABLES
**Column:** Bio-Sil ODS-5S
**Mobile phase:** MeCN:THF:10 mM pH 4.75 sodium acetate 2:1:97
**Detector:** UV 273

## CHROMATOGRAM
**Internal standard:** 7-(2-hydroxyethyl)theophylline
**Limit of detection:** 500 ng/mL

## OTHER SUBSTANCES
**Extracted:** paraxanthine
**Noninterfering:** cephalosporins

## REFERENCE
Leonard, H.; Campbell, J.; Malliaros, D.; Berg, M.; Houser, S.; McLaughlin, L. Comparison of a theophylline HPLC reference method with an automated chemiluminescent immunoassay (Abstract 122). *Ther.Drug Monit.*, **1995**, *17*, 413

## SAMPLE
**Matrix:** blood
**Sample preparation:** Extract 250 μL plasma with chloroform:isopropanol 85:15. Remove the organic layer and evaporate it to dryness, reconstitute the residue in mobile phase, inject an aliquot.

## HPLC VARIABLES
**Column:** 100 × 4.6 microsphere C18 (Chrompack)
**Mobile phase:** MeCN:55 mM pH 4.0 sodium acetate buffer 6:94
**Flow rate:** 1.2
**Detector:** UV 278

## CHROMATOGRAM
**Internal standard:** theophylline

## OTHER SUBSTANCES
**Extracted:** caffeine

## KEY WORDS
plasma; pig; theophylline is IS

## REFERENCE
Monshouwer, M.; Witkamp, R.F.; Nijmeijer, S.M.; Pijpers, A.; Verheijden, J.H.M.; Van Miert, A.S.J.P.A.M. Selective effects of a bacterial infection (*Actinobacillus pleuropneumoniae*) on the hepatic clearance of caffeine, antipyrine, paracetamol, and indocyanine green in the pig. *Xenobiotica*, **1995**, *25*, 491–499

## SAMPLE
**Matrix:** blood
**Sample preparation:** 100 μL Plasma + 50 μL 20 μg/mL β-hydroxyethyltheophylline in MeOH + 200 μL 67 mM pH 8 phosphate buffer + 5 mL chloroform, shake for 10 min, centrifuge at 850 g for 5 min. Remove 4 mL of the organic layer and evaporate it to dryness under a stream of nitrogen at 40°, reconstitute the residue in 100 μL mobile phase, inject a 20 μL aliquot.

## HPLC VARIABLES
**Column:** reverse-phase
**Mobile phase:** MeCN:20 mM pH 5 acetate buffer 7:93
**Flow rate:** 1.5
**Injection volume:** 20
**Detector:** UV 275

## CHROMATOGRAM
**Internal standard:** β-hydroxyethyltheophylline

## KEY WORDS
plasma; rat; pharmacokinetics

## REFERENCE
Nagai, N.; Furuhata, M.; Ogata, H. Drug interactions between theophylline and H₂-antagonists, roxatidine acetate hydrochloride and cimetidine: Pharmacokinetic analysis in rats in vivo. *Biol.Pharm.Bull.*, **1995**, *18*, 1610–1613

## SAMPLE
**Matrix:** blood
**Sample preparation:** 100 μL Plasma + 10 μL 40 μg/mL β-hydroxyethyltheophylline + 10 μL perchloric acid, vortex for 10 s, centrifuge for 5 min, inject a 20 μL aliquot of the supernatant.

## HPLC VARIABLES
**Column:** 5 μm Radpak C18 (Waters)

**Mobile phase:** MeCN:MeOH:buffer 5:7:88 (Buffer was 10 mM $KH_2PO_4$ containing 0.02% triethylamine, pH 4.5.)
**Injection volume:** 20
**Detector:** UV 273

## CHROMATOGRAM
**Retention time:** 5.5
**Internal standard:** β-hydroxyethyltheophylline (7.5)
**Limit of quantitation:** 625 ng/mL

## KEY WORDS
rat; plasma

## REFERENCE
Ng, C.Y.; Angus, P.W.; Ghabrial, H.; Chou, S.T.; Arnolda, L.; Morgan, D.J.; Smallwood, R.A. Right heart failure impairs hepatic oxygenation and theophylline clearance in rats. *J.Pharm.Exp.Ther.*, **1995**, *273*, 1332–1336

## SAMPLE
**Matrix:** blood
**Sample preparation:** Filter (0.22 μm), inject a 20 μL aliquot of the filtrate.

## HPLC VARIABLES
**Column:** 250 × 4 5 μm LiChrospher 100 Diol
**Mobile phase:** MeCN:50 mM pH 6.9 phosphate buffer 1.8:98.2
**Flow rate:** 0.6
**Injection volume:** 20
**Detector:** UV 254

## CHROMATOGRAM
**Retention time:** 5.5

## OTHER SUBSTANCES
**Extracted:** caffeine

## KEY WORDS
serum; direct injection

## REFERENCE
Nimura, N.; Itoh, H.; Kinoshita, T. Diol-bonded silica gel as a restricted access packing forming a binary-layered phase for direct injection of serum for the determination of drugs. *J.Chromatogr.A*, **1995**, *689*, 203–210

## SAMPLE
**Matrix:** blood
**Sample preparation:** 25 μL Serum + 100 μL 250 ng/mL β-hydroxyethyltheophylline in water + 200 μL 200 mM pH 6.0 phosphate buffer + 3 mL dichloromethane, shake at 120 oscillations/min for 20 min, centrifuge at 174 g. Remove the organic layer and evaporate it to dryness under a stream of nitrogen at 60°, reconstitute the residue in 230 μL water, heat at 90° for 6 min, vortex, inject a 180 μL aliquot. (No details given for milk extraction.)

## HPLC VARIABLES
**Guard column:** Corasil Bondapak C18
**Column:** 5 μm radial-compression C18 (Waters)
**Mobile phase:** MeOH:THF:10 mM $KH_2PO_4$ 9:1:90, adjusted to pH 3.5

**Flow rate:** 1.2
**Injection volume:** 180
**Detector:** UV 214

## CHROMATOGRAM
**Retention time:** 9.4
**Internal standard:** β-hydroxyethyltheophylline (10.9)

## OTHER SUBSTANCES
**Extracted:** caffeine, paraxanthine, theobromine

## KEY WORDS
serum; pharmacokinetics

## REFERENCE
Oo, C.Y.; Burgio, D.E.; Kuhn, R.C.; Desai, N.; McNamara, P.J. Pharmacokinetics of caffeine and its
metabolites in lactation: Prediction of milk to serum concentration ratios. *Pharm.Res.*, **1995**, *12*,
313–316

## SAMPLE
**Matrix:** blood
**Sample preparation:** 200 µL Serum + 10 µL 500 ng/mL theophylline in water + 50 µL
isoamyl alcohol, vortex for 30 s, add 2 mL chloroform, vortex for 1 min, centrifuge at 1000
rpm for 5 min. Remove the organic layer and evaporate it to dryness under a stream of
nitrogen at 50°, reconstitute the residue in 100 µL mobile phase, inject a 50 µL aliquot.

## HPLC VARIABLES
**Column:** 150 × 3.9 5 µm Novapak C18
**Mobile phase:** MeCN:0.2% acetic acid 7.5:92.5 (After 4 min increase flow to 1.8 mL/min
over 1 min, maintain at 1.8 mL/min for 2 min.)
**Flow rate:** 0.8
**Injection volume:** 50
**Detector:** UV 270

## CHROMATOGRAM
**Retention time:** 3.4
**Internal standard:** zidovudine (6.0)
**Limit of detection:** 20 ng/mL

## KEY WORDS
rat; serum; pharmacokinetics

## REFERENCE
Radwan, M.A. HPLC assay of theophylline and zidovudine in rat serum. *J.Liq.Chromatogr.*, **1995**, *18*,
3301–3309

## SAMPLE
**Matrix:** blood
**Sample preparation:** Wash PCPure cartridge containing 0.4 g hydroxyapatite with 10 mL
MeCN and remove MeCN by evaporation. 75 µL Plasma + 25 µL of 50 µg/mL IS in 0.5%
aqueous MeCN injected onto PCPure cartridge, elute with MeCN:water 10:90. Use first
600 µL of eluate, inject 20 µL aliquots.

## HPLC VARIABLES
**Column:** 150 × 4.6 5 µm Inertsil ODS-2

**Mobile phase:** MeCN:water 5:95
**Column temperature:** 40
**Flow rate:** 1
**Injection volume:** 20
**Detector:** UV 280

## CHROMATOGRAM
**Retention time:** 7
**Internal standard:** 7-(2-hydroxyethyl)theophylline (10)
**Limit of detection:** 2.9 ng

## OTHER SUBSTANCES
**Simultaneous:** caffeine

## KEY WORDS
plasma

## REFERENCE
Iwase, H.; Gondo, K.; Koike, T.; Ono, I. Novel precolumn deproteinization method using a hydroxyapatite cartridge for the determination of theophylline and diazepam in human plasma by high-performance liquid chromatography with ultraviolet detection. *J.Chromatogr.B*, **1994**, *655*, 73–81

## SAMPLE
**Matrix:** blood
**Sample preparation:** 100 μL Plasma + 100 μL 20 μg/mL caffeine + 8 mL dichloromethane, shake for 20 min, centrifuge at 2500 rpm for 20 min. Remove 7 mL of the organic layer and evaporate it to dryness under nitrogen or at 60°. Dissolve residue in 200 μL mobile phase, inject a 20 μL aliquot.

## HPLC VARIABLES
**Column:** 150 × 6 Shimpack CLS-ODS (Shimadzu)
**Mobile phase:** MeCN:0.5 mM phosphoric acid 10:90
**Column temperature:** 40
**Flow rate:** 1.5
**Injection volume:** 20
**Detector:** UV 273

## CHROMATOGRAM
**Internal standard:** caffeine

## KEY WORDS
plasma; rat

## REFERENCE
Lee, C.K.; Uchida, T.; Kitagawa, K.; Yagi, A.; Kim, N.-S.; Goto, S. Skin permeability of various drugs with different lipophilicity. *J.Pharm.Sci.*, **1994**, *83*, 562–565

## SAMPLE
**Matrix:** blood
**Sample preparation:** 100 μL Plasma + 100 μL pH 7.4 phosphate buffer + 50 μL 20 μg/mL β-hydroxypropyltheophylline in pH 7.4 phosphate buffer + 5 mL chloroform:isopropanol 95:5, shake on a rotary mixer for 15 min, centrifuge at 800 g for 5 min. Evaporate organic layer under nitrogen at 45°, sonicate residue with 100 μL mobile phase, inject 25 μL aliquot.

## HPLC VARIABLES
**Guard column:** 10 × 4.9 Spherisorb ODS
**Column:** 250 × 4.9 Spherisorb S5 ODS2
**Mobile phase:** MeCN:buffer 15:85 adjusted to pH 3.0 with 85% phosphoric acid immediately before use (Buffer was 4.54 g $KH_2PO_4$ + 5.94 g $Na_2HPO_4.2H_2O$ + 1.49 g tetrabutylammonium hydrogen sulfate per L.)
**Flow rate:** 1.3
**Injection volume:** 25
**Detector:** UV 280

## CHROMATOGRAM
**Retention time:** 3.3
**Internal standard:** β-hydroxypropyltheophylline
**Limit of detection:** 500 ng/mL

## OTHER SUBSTANCES
**Extracted:** ciprofloxacin, enoxacin, norfloxacin

## KEY WORDS
plasma; rat

## REFERENCE
Davis, J.D.; Aarons, L.; Houston, J.B. Simultaneous assay of fluoroquinolones and theophylline in plasma by high-performance liquid chromatography. *J.Chromatogr.*, **1993**, *621*, 105–109

## SAMPLE
**Matrix:** blood
**Sample preparation:** 200 μL Serum + 200 μL 20 μg/mL antipyrine in mobile phase, filter (Millipore Millex 0.45 μm), inject a 25 μL aliquot.

## HPLC VARIABLES
**Guard column:** 40 × 4 C18 Corasil II
**Column:** 300 × 4 10 μm μBondapak phenyl
**Mobile phase:** Propanol:6 mM $C_{12}$ DAPS (Fluka) 3:97 ($C_{12}$ DAPS is 3-(dimethyldodecylammonio) propanesulfonate.)
**Injection volume:** 25
**Detector:** UV 273

## CHROMATOGRAM
**Retention time:** 5
**Internal standard:** antipyrine (8)
**Limit of detection:** 500 ng/mL

## OTHER SUBSTANCES
**Simultaneous:** albendazole, albendazole sulfoxide, aminophylline, amyleine, caffeine, flubendazole, β-hydroxytheophylline, mercaptopurine, nimorazole, procaine, theobromine
**Interfering:** dipropyline, metronidazole, tinidazole

## KEY WORDS
serum; micellar chromatography

## REFERENCE
Habel, D.; Guermouche, S.; Guermouche, M.H. Direct determination of theophylline in human serum by high-performance liquid chromatography using zwitterionic micellar mobile phase. Comparison with an enzyme multiplied immunoassa ytechnique. *Analyst*, **1993**, *118*, 1511–1513

## SAMPLE
**Matrix:** blood
**Sample preparation:** Dilute serum with an equal volume 7.5 μg/mL theobromine, inject a 20 μL aliquot directly.

## HPLC VARIABLES
**Column:** 150 × 4.6 ChromSpher 5 BioMatrix (Chrompack)
**Mobile phase:** MeCN:water 5:95
**Flow rate:** 1
**Injection volume:** 20
**Detector:** UV 280

## CHROMATOGRAM
**Retention time:** 2.4
**Internal standard:** theobromine (3.3)

## OTHER SUBSTANCES
**Simultaneous:** caffeine

## KEY WORDS
serum

## REFERENCE
Helmsing, P.J.; Huisman, R.; van der Weele, A. HPLC determination of caffeine and theophylline by direct serum injection. *Clin.Chem.*, **1993**, *39*, 1348–1349

## SAMPLE
**Matrix:** blood
**Sample preparation:** 10 μL Plasma + 300 μL 100 mM pH 6.0 $KH_2PO_4$ buffer + 100 μL 10 μg/mL diprophylline + 2 mL chloroform:isopropanol 50:50, vortex for 30 s, centrifuge at 2000 rpm for 10 min. Remove the organic layer and evaporate it to dryness under a stream of air at 40°, reconstitute the residue in 100 μL mobile phase, inject a 25 μL aliquot.

## HPLC VARIABLES
**Column:** NovaPak C18 radial compression
**Mobile phase:** MeOH:MeCN:10 mM $KH_2PO_4$ 9:2.5:90
**Flow rate:** 2
**Injection volume:** 25
**Detector:** E, ESA Coulochem Model 5100 A, Model 5010 analytical cell, first (screen) electrode 0.68 V, second (measuring) electrode 0.98 V, Model 5020 guard cell (before injector) 1.0 V; UV 270

## CHROMATOGRAM
**Retention time:** 8 (electrochemical detection)
**Internal standard:** diprophylline (UV detection) (10)
**Limit of detection:** 200 ng/mL

## OTHER SUBSTANCES
**Extracted:** caffeine, 3-methylxanthine, theobromine

## KEY WORDS
plasma

## REFERENCE
Augustijns, P.; Verbeke, N. A microassay method for the determination of theophylline in biological samples using HPLC with electrochemical detection. *J.Liq.Chromatogr.*, **1992**, *15*, 1303–1313

## SAMPLE
**Matrix:** blood
**Sample preparation:** 100 μL Serum + 100 μL 20 μg/mL hydroxyethyltheophylline in 2 M perchloric acid, vortex, centrifuge 5 min, inject 50 μL aliquot of supernatant.

## HPLC VARIABLES
**Column:** 125 × 4 LiChroSpher RP-8 5 μm
**Mobile phase:** MeOH:buffer 15:85 (Buffer was 5 mL 2 M sodium acetate + 845 mL water, pH adjusted to 4.0 with acetic acid.)
**Column temperature:** 45
**Flow rate:** 1.5
**Injection volume:** 50
**Detector:** UV 282

## CHROMATOGRAM
**Retention time:** 4.6
**Internal standard:** hydroxyethyltheophylline (5.6)

## OTHER SUBSTANCES
**Simultaneous:** caffeine, chloramphenicol

## KEY WORDS
serum

## REFERENCE
Hannak, D.; Haux, P.; Scharbert, F.; Kattermann, R. Liquid chromatographic analysis of phenobarbital, phenytoin, and theophylline. *Wien.Klin.Wochenschr.Suppl.*, **1992**, *191*, 27–31

## SAMPLE
**Matrix:** blood
**Sample preparation:** 500 μL Plasma + 50 μL 3 μg/mL 8-chlorotheophylline in mobile phase + 100 μL 1 M HCl + 3 mL dichloromethane, vortex for 2 min, centrifuge at 1200 g for 5 min. Remove the organic layer and evaporate it to dryness under a stream of nitrogen at 40°, reconstitute the residue in 100 μL mobile phase, inject a 20 μL aliquot.

## HPLC VARIABLES
**Column:** 150 × 4.6 5 μm TSK gel ODS-80TM (Tosoh)
**Mobile phase:** MeOH:100 mM $NaH_2PO_4$ 30:70
**Flow rate:** 0.8
**Injection volume:** 20
**Detector:** UV 274

## CHROMATOGRAM
**Retention time:** 5.9
**Internal standard:** 8-chlorotheophylline (11.7)
**Limit of quantitation:** 10 ng/mL

## OTHER SUBSTANCES
**Extracted:** metabolites, caffeine, paraxanthine, theobromine
**Noninterfering:** acetaminophen, aspirin, chlorpheniramine, ethenzamide, hydrocortisone, phenacetin, phenobarbital, phenylbutazone, phenytoin, prednisolone, prednisone, salicylic acid, trimethadione, vitamin B1

## KEY WORDS
plasma

## REFERENCE
Tanaka, E. Simultaneous determination of caffeine and its primary demethylated metabolites in human plasma by high-performance liquid chromatography. *J.Chromatogr.*, **1992**, *575*, 311–314

## SAMPLE
**Matrix:** blood
**Sample preparation:** Inject 20 μL serum onto column A with mobile phase A and elute to waste, after 1 min backflush the contents of column A onto column B with mobile phase B, after 1 min remove column A from the circuit, elute column B with mobile phase B, monitor the effluent from column B. Re-equilibrate column A with mobile phase A.

## HPLC VARIABLES
**Column:** A 30 × 4.6 IRSP silica (for preparation see Anal. Chem. 1989, 61, 2445); B 150 × 4.6 TSK gel ODS-80TM
**Mobile phase:** A 20 mM $NaH_2PO_4$; B MeCN:100 mM $NaH_2PO_4$ 10:90
**Flow rate:** A 0.8; B 1
**Injection volume:** 20
**Detector:** UV 275

## CHROMATOGRAM
**Retention time:** 7.5
**Limit of detection:** 50 ng/mL

## OTHER SUBSTANCES
**Extracted:** caffeine, theobromine

## KEY WORDS
serum; column-switching

## REFERENCE
Haginaka, J.; Wakai, J.; Yasuda, H.; Kimura, Y. Determination of anticonvulsant drugs and methyl xanthine derivatives in serum by liquid chromatography with direct injection: column-switching method using a new internal-surface reversed-phase silica support as a precolumn. *J.Chromatogr.*, **1990**, *529*, 455–461

## SAMPLE
**Matrix:** blood
**Sample preparation:** 100 μL Plasma or serum + 35 μL 22 μg/mL 3-ethylxanthine in 20 mM pH 4.0 acetate buffer, add to a Celute-MX SPE cartridge (Jones Chromatography), let stand for 10 min, elute with two portions of isopropanol:dichloromethane 10:90, evaporate the eluate to dryness under a stream of nitrogen below 37°, reconstitute the residue in 100 μL mobile phase B, centrifuge at 4400 rpm in a refrigerated centrifuge for 5 min, inject a 20 μL aliquot.

## HPLC VARIABLES
**Column:** 250 × 4.5 3 μm ODS Apex I (Jones Chromatography)
**Mobile phase:** Gradient. A was MeCN:THF:10 mM pH 4.0 acetate buffer 25:2:73. B was THF:10 mM pH 4.0 acetate buffer 0.01:99.99. From A:B 0:100 increasing at 2.1% A/min.
**Column temperature:** 50
**Flow rate:** 0.8
**Injection volume:** 20
**Detector:** UV 273

## CHROMATOGRAM
**Retention time:** 15.60
**Internal standard:** 3-ethylxanthine (13.64)

## OTHER SUBSTANCES
**Extracted:** acetaminophen, caffeine, dimethyluric acids, methylxanthines, paraxanthine, theobromine, trimethyluric acid

## KEY WORDS
plasma; serum; SPE

## REFERENCE
Leakey, T.E. Simultaneous analysis of theophylline, caffeine and eight of their metabolic products in human plasma by gradient high-performance liquid chromatography. *J.Chromatogr.*, **1990**, *507*, 199–220

## SAMPLE
**Matrix:** blood
**Sample preparation:** Centrifuge, filter (0.45 μm), inject an aliquot.

## HPLC VARIABLES
**Column:** 150 × 4.6 5 μm internal-surface, reversed-phase, Pinkerton-type, silica derivatized with glycine-phenylalanine-phenylalanine (Regis) (periodically reverse the column)
**Mobile phase:** 100 mM pH 6.8 phosphate buffer
**Flow rate:** 0.3
**Injection volume:** 10
**Detector:** UV 275

## CHROMATOGRAM
**Retention time:** 10.88
**Limit of detection:** <1000 ng/mL

## OTHER SUBSTANCES
**Extracted:** caffeine, doxofylline, dyphylline
**Noninterfering:** amitriptyline, amphetamine, atropine, benzoylecgonine, benztropine, caffeine, carbamazepine, carisoprodol, chlorpheniramine, chlorpromazine, chlorprothixene, cimetidine, cocaine, codeine, dextromethorphan, diazepam, diphenhydramine, diphenoxilate, disopyramide, doxepin, doxylamine, emetine, erythromycin, flurazepam, glutethimide, hydrocortisone, hydromorphone, hydroxyzine, imipramine, lidocaine, loxapine, meperidine, meprobamate, methadone, methamphetamine, methapyrilene, methaqualone, methocarbamol, methylphenidate, nicotine, nordiazepam, nortriptyline, orphenadrine, papaverine, pentazocine, phenacetin, phencyclidine, phenmetrazine, phenolphthalein, phentermine, phenylpropanolamine, phenytoin, prazepam, procainamide, procaine, propoxyphene, propranolol, protriptyline, pseudoephedrine, pyrilamine, quinine, salicylamide, spironolactone, strychnine, terpin hydrate, thioridazine, thiothixene, triamterene, trifluoperazine, triflupromazine, trihexyphenidyl, trimeprazine, trimethobenzamide, trimethoprim, tripelennamine
**Interfering:** acetaminophen

## KEY WORDS
serum; plasma; direct injection

## REFERENCE
Tagliaro, F.; Dorizzi, R.; Frigerio, A.; Marigo, M. Non-extraction HPLC method for simultaneous measurement of dyphylline and doxofylline in serum. *Clin.Chem.*, **1990**, *36*, 113–115

## SAMPLE
**Matrix:** blood
**Sample preparation:** Prepare an SPE cartridge by plugging the end of a 1 mL disposable pipette tip with glass wool and adding about 100 mg Chromosorb P/NAW. Add 50 μL plasma then 50 μL 10 μg/mL tolylphenobarbital in 200 mM HCl to the SPE cartridge, let stand for 2 min, elute with 1 mL chloroform:isopropanol 6:1. Evaporate the eluate to dryness under a stream of nitrogen at 30°, reconstitute the residue in 100 μL mobile phase, inject a 15 μL aliquot.

## HPLC VARIABLES
**Column:** 150 × 4.6 5 μm Supelcosil-LC-8
**Mobile phase:** MeCN:water 20:80
**Flow rate:** 3.3
**Injection volume:** 15
**Detector:** UV 208

## CHROMATOGRAM
**Retention time:** 0.77
**Internal standard:** tolylphenobarbital (7.57)
**Limit of detection:** 50-100 ng/mL

## OTHER SUBSTANCES
**Extracted:** amobarbital, barbital, butabarbital, caffeine, carbamazepine epoxide, carbamazepine, carbamazepinediol, chloramphenicol, ethosuximide, glutethimide, mephenytoin, methaqualone, methyprylon, nirvanol, pentobarbital, phenacemide, phenobarbital, phenytoin, primidone, secobarbital
**Noninterfering:** acetaminophen, N-acetylprocainamide, amikacin, amitriptyline, clonazepam, cyclosporine, desipramine, diazepam, digoxin, disopyramide, gentamicin, p-hydroxyphenobarbital, imipramine, lidocaine, methotrexate, netilmicin, nortriptyline, procainamide, quinidine, salicylic acid, sulfamethoxazole, tobramycin, trimethoprim, valproic acid, vancomycin

## KEY WORDS
plasma; SPE

## REFERENCE
Svinarov, D.A.; Dotchev, D.C. Simultaneous liquid-chromatographic determination of some bronchodilators, anticonvulsants, chloramphenicol, and hypnotic agents, with Chromosorb P columns used for sample preparation. *Clin.Chem.*, **1989**, *35*, 1615–1618

## SAMPLE
**Matrix:** blood
**Sample preparation:** 100 μL Serum + 100 μL buffer + 1.5 mL IS in 5% isopropanol in chloroform, vortex for 30 s, centrifuge. Remove the organic layer and evaporate it to dryness under a stream of air at room temperature, reconstitute the residue in 100 μL mobile phase, inject a 6-10 μL aliquot. (Buffer was 13.6 g $KH_2PO_4$ in 90 mL water, pH adjusted to 6.8 with about 3 mL 10 M NaOH, made up to 100 mL.)

## HPLC VARIABLES
**Guard column:** 20 × 4.6 Supelguard LC-1 (Supelco)
**Column:** 250 × 4.6 5 μm Supelcosil LC-1 (Supelco)
**Mobile phase:** MeOH:MeCN:buffer 17.5:17.5:65 (Buffer was 2.72 g $KH_2PO_4$ in 1.9 L water, pH adjusted to 6.3 with about 2 mL 1 M NaOH, made up to 2 L.)
**Flow rate:** 2
**Injection volume:** 6-10
**Detector:** UV 273

## CHROMATOGRAM
**Retention time:** 1.95
**Internal standard:** 3-isobutyl-1-methylxanthine (3.15)

## OTHER SUBSTANCES
**Extracted:** acetaminophen, amobarbital, barbital, caffeine, carbamazepine, chloramphenicol, ethosuximide, mephobarbital, methsuximide, pentobarbital, phenobarbital, phenytoin, primidone, secobarbital, thiopental
**Also analyzed:** acetanilide, N-acetylcysteine, N-acetylprocainamide, ampicillin, aspirin, butabarbital, butalbital, chlorpropamide, cimetidine, codeine, cyheptamide, diazoxide, diflunisal, diphylline, disopyramide, ethchlorvynol, gentisic acid, glutethimide, heptabarbital, hexobarbital, ibuprofen, indomethacin, ketoprofen, mefenamic acid, mephenytoin, methaqualone, methsuximide, methyl salicylate, methyprylon, morphine, naproxen, nirvanol, oxphenylbutazone, phenacetin, phensuximide, phenylbutazone, procainamide, salicylamide, salicylic acid, sulfamethoxazole, sulindac, tolmetin, trimethoprim, vancomycin
**Noninterfering:** amikacin, gentamicin, meprobamate, netilmicin, quinidine, tetracycline, tobramycin, valproic acid

## KEY WORDS
serum

## REFERENCE
Meatherall, R.; Ford, D. Isocratic liquid chromatographic determination of theophylline, acetaminophen, chloramphenicol, caffeine, anticonvulsants, and barbiturates in serum. *Ther.Drug Monit.*, **1988**, *10*, 101–115

## SAMPLE
**Matrix:** blood
**Sample preparation:** 200 μL Serum + 400 μL 100 μg/mL 8-chlorotheophylline in MeCN, vortex for 10 s, centrifuge at 15000 rpm for 3 min, inject a 20 μL aliquot of the supernatant.

## HPLC VARIABLES
**Column:** 250 × 4.6 5 μm Ultrasphere ODS
**Mobile phase:** MeCN:water:glacial acetic acid 4:84:12 containing 4.84 g/L Trizma, pH 2.3
**Flow rate:** 2
**Injection volume:** 20
**Detector:** UV 254

## CHROMATOGRAM
**Retention time:** 3.00
**Internal standard:** 8-chlorotheophylline (5.29)
**Limit of quantitation:** 2 μg/mL

## OTHER SUBSTANCES
**Extracted:** acetaminophen, salicylic acid
**Simultaneous:** caffeine, cefazolin, cimetidine, ergotamine, glutethimide, heparin, methamphetamine, propranolol, sulfamethoxazole, theobromine, tobutamide, trimethoprim
**Noninterfering:** amitriptyline, amobarbital, ampicillin, butabarbital, butalbital, celbenine, chlordiazepoxide, chlorpromazine, clorazepate, desipramine, diazepam, doxepin, ethchlorvynol, fluphenazine, hydroxyzine, ibuprofen, imipramine, isoniazid, lidocaine, mephobarbital, mesoridazine, methaqualone, methyluric acid, naprotyline, nordiazepam, nortriptyline, oxazepam, pentobarbital, perphenazine, phenelzine, phenmetrazine, phenobarbital, phenylbutazone, phenytoin, prednisolone, prednisone, procainamide, prochlorperazine, promazine, promethazine, propoxyphene, protriptyline, pyrilamine, secobarbital, thioridazine, thiothixene, timolol, trazodone, triazolam, trifluoperazine

## KEY WORDS
serum

## REFERENCE
Osterloh, J.; Yu, S. Simultaneous ion-pair and partition liquid chromatography of acetaminophen, theophylline and salicylate with application to 500 toxicologic specimens. *Clin.Chim.Acta*, **1988**, *175*, 239–248

## SAMPLE
**Matrix:** blood
**Sample preparation:** 100 μL Serum + 1 mL reagent, vortex for 10 min, centrifuge for 2-3 min. Remove the lower organic layer and evaporate it to dryness under a stream of nitrogen, reconstitute the residue in 200 μL mobile phase, inject a 40 μL aliquot. (Reagent was 1.5 mg β-hydroxyethyltheophylline and 25 μL glacial acetic acid in 100 mL chloroform:isopropanol 95:5.)

## HPLC VARIABLES
**Guard column:** 30 mm long 5 μm Ultrasphere ion-pair
**Column:** 150 × 4.6 5 μm Ultrasphere ion-pair
**Mobile phase:** MeCN:MeOH:water:1 M tetra-n-butylammonium hydroxide 3.5:3.5:91:2 containing 1.82 g Trizma base (Tris, tris(hydroxymethyl)aminomethane), pH adjusted to 7.50 ± 0.03 with concentrated HCl
**Flow rate:** 1.2
**Injection volume:** 40
**Detector:** UV 280

## CHROMATOGRAM
**Retention time:** k' 5.4
**Internal standard:** β-hydroxyethyltheophylline (k' 4.2)

## OTHER SUBSTANCES
**Extracted:** caffeine, 1,7-dimethyl xanthine, theobromine
**Simultaneous:** acetaminophen, acetazolamide, allopurinol, dimethylurea, dyphylline, 3-methylxanthine, oxypurinol, procainamide, sulfadiazine, sulfamethazine, uric acid
**Noninterfering:** ampicillin, cefazolin, cephalothin, cephapirin, chlorotheophylline, 1,3-dimethyluric acid, gentamicin, lidocaine, methicillin, methylurea, 3-methyluric acid, quinidine, sulfamerazine, 1,3,7-trimethyluric acid

## KEY WORDS
serum

## REFERENCE
Lauff, J.J. Ion-pair high-performance liquid chromatographic procedure for the quantitative analysis of theophylline in serum samples. *J.Chromatogr.*, **1987**, *417*, 99–109

## SAMPLE
**Matrix:** blood
**Sample preparation:** 200 μL Serum + 200 μL 4 μg/mL 8-chlorotheophylline in MeCN, mix, centrifuge, evaporate the supernatant to dryness, reconstitute in 400 μL mobile phase, inject a 20 μL aliquot.

## HPLC VARIABLES
**Column:** 100 × 3 8 μm octadecyl CP-tm-Spher C18 glass column (Chrompack)
**Mobile phase:** MeCN:20 mM sodium acetate 20:80, adjusted to pH 4.4 with phosphoric acid

**Flow rate:** 0.8
**Injection volume:** 20
**Detector:** UV 273

## CHROMATOGRAM
**Retention time:** 3.8
**Internal standard:** 8-chlorotheophylline (8.8)
**Limit of detection:** 1 μg/mL

## OTHER SUBSTANCES
**Simultaneous:** caffeine

## KEY WORDS
serum

## REFERENCE
Van Damme, M.; Molle, L.; Abi Khalil, F. Useful sample handlings for reversed phase high performance liquid chromatography in emergency toxicology. *J.Toxicol.Clin.Toxicol.*, **1985**, *23*, 589–614

## SAMPLE
**Matrix:** blood
**Sample preparation:** 500 μL Serum or plasma + 200 μL 100 mM pH 7.0 phosphate buffer + 3 mL 0.5 μg/mL 8-chlorotheophylline in isopropanol, stir at 40000 rpm for 5 s using dental micromotor with a PTFE mixing head, centrifuge at 3500 g for 2 min. Remove the supernatant and evaporate it to dryness under a stream of nitrogen at 60°, reconstitute the residue in 50 μL MeOH, inject a 10 μL aliquot.

## HPLC VARIABLES
**Guard column:** 50 × 3.2 30-38 μm Co:Pell ODS
**Column:** 250 × 4.6 5 μm Ultrasphere ODS
**Mobile phase:** MeCN:MeOH:10 mM pH 5.2 sodium acetate buffer 6:3:91
**Column temperature:** 40
**Flow rate:** 1.5
**Injection volume:** 10
**Detector:** UV 274

## CHROMATOGRAM
**Retention time:** 5.95
**Internal standard:** 8-chlorotheophylline (8.85)
**Limit of detection:** 100 ng/mL

## OTHER SUBSTANCES
**Extracted:** caffeine, dyphylline, paraxanthine, proxyphylline
**Simultaneous:** cefoxitin
**Noninterfering:** carbenicillin, cefoperazone, cephacetril, heparin, penicillin G, phenytoin, phenobarbital

## KEY WORDS
serum; plasma; pharmacokinetics

## REFERENCE
Wenk, M.; Eggs, B.; Follath, F. Simultaneous determination of diprophylline, proxyphylline and theophylline in serum by reversed-phase high-performance liquid chromatography. *J.Chromatogr.*, **1983**, *276*, 341–348

## SAMPLE
**Matrix:** blood
**Sample preparation:** 50 µL Serum + 50 µL 15 µg/mL β-hydroxyethyltheophylline in MeCN + 2 mL chloroform:isopropanol 95:5, mix for 30 s, centrifuge at 3000 g for 3 min. Remove the organic layer and evaporate it to dryness under a stream of nitrogen, reconstitute the residue in 50 µL MeOH, inject a 20 µL aliquot.

## HPLC VARIABLES
**Column:** µBondapak C18
**Mobile phase:** MeCN:buffer 9.75:90.25 (Buffer was 100 mM KH₂PO₄ adjusted to pH 4.0 with phosphoric acid.) (At the end of each day clean with water for 20 min and MeOH for 30 min.)
**Flow rate:** 2
**Injection volume:** 20
**Detector:** UV 254

## CHROMATOGRAM
**Retention time:** 4.8
**Internal standard:** β-hydroxyethyltheophylline (5.8)
**Limit of detection:** 500 ng/mL

## OTHER SUBSTANCES
**Extracted:** caffeine
**Simultaneous:** acetaminophen, N-acetylprocainamide, aspirin, procainamide, salicylic acid
**Noninterfering:** benzoic acid
**Interfering:** dyphylline (separated with MeCN:buffer 8:92), ampicillin

## KEY WORDS
serum

## REFERENCE
Ou, C.-N.; Frawley, V.L. Theophylline, dyphylline, caffeine, acetaminophen, salicylate, acetylsalicylate, procainamide, and N-acetylprocainamide determined in serum with a single liquid-chromatographic assay. *Clin.Chem.*, **1982**, *28*, 2157–2160

## SAMPLE
**Matrix:** blood
**Sample preparation:** 250 µL Serum + β-hydroxypropyltheophylline, vortex, add 1.5 g anhydrous sodium sulfite, add 2.5 mL chloroform:MeOH 90:10, shake vigorously for 30 s (work quickly to avoid forming a cake of sodium sulfite), centrifuge at 1000 g for 5 min. Remove the organic layer, filter (Whatman No. 1 paper pre-wetted with chloroform), rinse filter with 500 µL chloroform, evaporate the filtrate under a stream of nitrogen at 40°. Take up the residue in 100 µL dichloroethane and add it to 100 µL 100 mM ammonium carbonate, vortex for 10 s, centrifuge at 1000 g for 5 min, inject a 10 µL aliquot of the aqueous layer.

## HPLC VARIABLES
**Column:** 250 × 4 10 µm LiChrosorb RP-8
**Mobile phase:** MeOH:buffer 25:75 (Buffer was 1.5 mL 1 M KH₂PO₄ in 750 mL water, adjust to pH 3.0 with 900 mM perchloric acid.)
**Flow rate:** 2
**Injection volume:** 10
**Detector:** UV 275

## CHROMATOGRAM
**Retention time:** 4.5
**Internal standard:** β-hydroxypropyltheophylline (3.7)

## OTHER SUBSTANCES
**Simultaneous:** caffeine, dyphylline, theobromine
**Noninterfering:** acetaminophen, acetazolamide, albuterol, amitriptyline, amobarbital, carbamazepine, diazoxide, disopyramide, ethosuximide, isoproterenol, nitrazepam, nortriptyline, oxazepam, pentobarbital, phenobarbital, phenytoin, primidone, procainamide, salicylic acid, sulthiame

## KEY WORDS
serum

## REFERENCE
Paterson, N. High-performance liquid chromatographic method for the determination of diprophylline in human serum. *J.Chromatogr.*, **1982**, *232*, 450–455

## SAMPLE
**Matrix:** blood
**Sample preparation:** 200 μL Serum + 200 μL 50 μg/mL hexobarbital in MeCN + 25 μL glacial acetic acid, vortex for 10 s, centrifuge for 1 min, inject a 30-100 μL aliquot of the supernatant.

## HPLC VARIABLES
**Column:** μBondapak C18
**Mobile phase:** Gradient. MeCN:7.5 g/L $NaH_2PO_4$ adjusted to pH 3.2 with phosphoric acid 5:95 to 22:78 over 24 min, to 45:55 over 10 min, maintain at 45:55 for 5 min. Reequilibrate with 5:95 for 5 min.
**Column temperature:** 50
**Flow rate:** 3
**Injection volume:** 30-100
**Detector:** UV 210

## CHROMATOGRAM
**Retention time:** 4.3
**Internal standard:** hexobarbital (20.6)
**Limit of detection:** LOD 200-2000 ng/mL

## OTHER SUBSTANCES
**Extracted:** acetaminophen, amobarbital, butabarbital, butalbital, chlordiazepoxide, diazepam, ethchlorvynol, flurazepam, glutethimide, methaqualone, methyprylon, nitrazepam, pentobarbital, phenobarbital, phenytoin, primidone, salicylic acid, secobarbital
**Simultaneous:** amitriptyline, caffeine, clomipramine, codeine, desipramine, ethotoin, imipramine, lidocaine, mesantoin, methsuximide, nirvanol, nortriptyline, oxazepam, procainamide, phenylpropanolamine, propranolol, quinidine

## KEY WORDS
serum

## REFERENCE
Kabra, P.M.; Stafford, B.E.; Marton, L.J. Rapid method for screening toxic drugs in serum with liquid chromatography. *J.Anal.Toxicol.*, **1981**, *5*, 177–182

## SAMPLE
**Matrix:** blood

**Sample preparation:** 1 mL Plasma + 100 μL 0.2 mg/mL β-hydroxyethyltheophylline in buffer + 100 μL 40% aqueous trichloroacetic acid, vortex for 30 s, let stand for 5 min, centrifuge at 2000 g for 15 min, inject a 25 μL aliquot of the supernatant. (Buffer was 10 mM sodium acetate adjusted to pH 4.0 with glacial acetic acid.)

## HPLC VARIABLES
**Column:** 10 μm μBondapak C18
**Mobile phase:** MeCN:buffer 6:94 (Buffer was 10 mM sodium acetate adjusted to pH 4.0 with glacial acetic acid.)
**Column temperature:** 40
**Flow rate:** 2
**Injection volume:** 25
**Detector:** UV 274

## CHROMATOGRAM
**Retention time:** 6.5
**Internal standard:** β-hydroxyethyltheophylline (9)
**Limit of detection:** 1 μg/mL

## OTHER SUBSTANCES
**Extracted:** caffeine, dyphylline, theobromine

## KEY WORDS
plasma

## REFERENCE
Valia, K.H.; Hartman, C.A.; Kucharczyk, N.; Sofia, R.D. Simultaneous determination of dyphylline and theophylline in human plasma by high-performance liquid chromatography. *J.Chromatogr.*, **1980**, *221*, 170–175

## SAMPLE
**Matrix:** blood, CSF, gastric fluid, urine
**Sample preparation:** 200 μL Serum, urine, CSF, or gastric fluid + 300 μL reagent. Flush column A to waste with 500 μL 500 mM ammonium sulfate, inject sample onto column A, flush column A to waste with 500 μL 500 mM ammonium sulfate, backflush the contents of column A onto column B with mobile phase, monitor the effluent from column B. (Reagent was 8.05 M guanidine hydrochloride and 1.02 M ammonium sulfate in water.)

## HPLC VARIABLES
**Column:** A 40 μm preparative grade C18 (Analytichem); B 75 × 2.1 pellicular C18 (Whatman) + 250 × 4.6 5 μm C8 end-capped (Whatman)
**Mobile phase:** Gradient. A was 50 mM pH 4.5 $KH_2PO_4$. B was MeCN:isopropanol 80:20. A:B 90:10 for 1 min, to 30:70 over 20 min.
**Column temperature:** 50
**Flow rate:** 1.5
**Detector:** UV 220

## CHROMATOGRAM
**Retention time:** 5
**Internal standard:** heptanophenone (19)

## OTHER SUBSTANCES
**Extracted:** acetaminophen, allobarbital, azinphos, barbital, brallobarbitone, bromazepam, butethal, caffeine, carbamazepine, carbaryl, cephaloridine, chloramphenicol, chlordiazepoxide, chlorothiazide, chlorvinphos, clothiapine, cocaine, coomassie blue, desipramine, diazepam, diphenhydramine, dipipanone, ethylbromphos, flufenamic acid, formothion,

griseofulvin, indomethacin, lidocaine, lorazepam, malathion, medazepam, midazolam, oxazepam, paraoxon, penicillin G, pentobarbital, prazepam, propoxyphene, prothiophos, quinine, salicylic acid, secobarbital, strychnine, sulfamethoxazole, thiopental, thioridazine, trimethoprim

## KEY WORDS
serum; column-switching

## REFERENCE
Kruger, P.B.; Albrecht, C.F.De V.; Jaarsveld, P.P. Use of guanidine hydrochloride and ammonium sulfate in comprehensive in-line sorption enrichment of xenobiotics in biological fluids by high-performance liquid chromatography. *J.Chromatogr.*, **1993**, *612*, 191–198

## SAMPLE
**Matrix:** blood, gastric juice, pancreatic juice
**Sample preparation:** 500 μL Serum, pancreatic juice, or gastric juice + 25 μL 200 μg/mL β-hydroxyethyltheophylline in water + 500 μL MeCN, centrifuge at 6000 rpm for 10 min, remove supernatant and add it to 1.8 mL chloroform, vortex, centrifuge at 6000 rpm for 10 min. Remove the lower organic phase and evaporate it to dryness at 60° under a stream of air, dissolve the residue in 500 μL mobile phase, inject a 50 μL aliquot.

## HPLC VARIABLES
**Guard column:** C18 LichroCART
**Column:** 250 × 4.6 7 μm Lichrosorb C18
**Mobile phase:** THF:MeOH:10 mM pH 3.5 $KH_2PO_4$ 1:20:79
**Flow rate:** 0.8
**Injection volume:** 50
**Detector:** UV 280

## CHROMATOGRAM
**Retention time:** 7.03
**Internal standard:** β-hydroxyethyltheophylline (7.48)
**Limit of quantitation:** 250 ng/mL

## OTHER SUBSTANCES
**Extracted:** caffeine, 1-methyluric acid, 1-methylxanthine, paraxanthine, theobromine

## KEY WORDS
serum; dog

## REFERENCE
Casoli, P.; Vérine, H. High performance liquid chromatographic determination of methylxanthines in canine serum, gastric and pancreatic juices. *Biomed.Chromatogr.*, **1990**, *4*, 209–213

## SAMPLE
**Matrix:** blood, saliva
**Sample preparation:** 500 μL Plasma or saliva + 50 μL 50 ng/mL difloxacin, vortex briefly, add 500 μL 100 mM pH 7.4 phosphate buffer, add 4 mL dichloromethane, add 1 mL isopropanol, vortex for 30 s, shake gently for 30 min, centrifuge at 1500 g for 20 min. Remove the lower organic layer and evaporate it to dryness under a stream of nitrogen at 45°, reconstitute the residue in 500 μL mobile phase, inject a 50-200 μL aliquot.

## HPLC VARIABLES
**Guard column:** μBondapak C18 Guard-Pak
**Column:** 300 × 3.9 10 μm μBondapak C18

**Mobile phase:** MeOH:buffer 35:100 (Buffer was 5.44 g $KH_2PO_4$ and 4 mL tetrabutylam-
monium hydroxide in 1 L water, adjust pH to 2.5 with 85% phosphoric acid.)
**Flow rate:** 2
**Injection volume:** 50-200
**Detector:** UV 268

## CHROMATOGRAM
**Retention time:** 4.3
**Internal standard:** difloxacin (8.8)
**Limit of detection:** 100 ng/mL

## OTHER SUBSTANCES
**Extracted:** ciprofloxacin, enoxacin
**Simultaneous:** caffeine
**Noninterfering:** 1,3-dimethyluric acid, hypoxanthine, 1-methyluric acid, 1-methylxan-
thine, 3-methylxanthine, 7-methylxanthine, theobromine
**Interfering:** 1,7-dimethylxanthine

## KEY WORDS
plasma; pharmacokinetics

## REFERENCE
Zhai, S.; Korrapati, M.R.; Wei, X.; Muppalla, S.; Vestal, R.E. Simultaneous determination of theophylline, enoxacin and ciprofloxacin in human plasma and saliva by high-performance liquid chromatography. *J.Chromatogr.B*, **1995**, *669*, 372–376

## SAMPLE
**Matrix:** blood, saliva
**Sample preparation:** 100 µL Serum, plasma, or saliva + 100 µL 15 µg/mL 8-chlorotheo-
phylline in 6% perchloric acid, vortex for 10 s, centrifuge at 10800 g for 10 min, inject 20
µL of the clear supernatant.

## HPLC VARIABLES
**Guard column:** Whatman Co:Pell ODS
**Column:** 300 × 3.9 10 µm Bondex ODS (Phenomenex)
**Mobile phase:** THF:buffer 6:94 (Buffer was pH 4.0 10 mM, prepared from 41 mL 200 mM
acetic acid and 9 mL 200 mM sodium acetate made up to 1 L with water, pH adjusted to
4.0 with glacial acetic acid or 100 mM NaOH if necessary.)
**Flow rate:** 1
**Injection volume:** 20
**Detector:** UV 273

## CHROMATOGRAM
**Retention time:** 4.7
**Internal standard:** 8-chlorotheophylline (11)
**Limit of quantitation:** 70 ng/mL

## OTHER SUBSTANCES
**Simultaneous:** caffeine, paraxanthine, theobromine

## KEY WORDS
serum; plasma

## REFERENCE
Blanchard, J.; Harvey, S.; Morgan, W.J. A rapid and specific high-performance liquid chromatographic
assay for theophylline in biological fluids. *J.Chromatogr.Sci.*, **1990**, *28*, 303–306

## SAMPLE
**Matrix:** blood, tissue
**Sample preparation:** Tissue. Homogenize brain in twenty volumes of water, centrifuge at 1500 g, freeze at -20° for 2 h, centrifuge. 200 μL Supernatant + 100 μL pH 7.2 phosphate buffer, mix, add 6 mL dichloromethane:propanol 95:5, mix for 2 min, centrifuge at 1500 g for 10 min. Remove the organic layer and evaporate it to dryness under a stream of nitrogen at 35°, reconstitute the residue in 200 μL mobile phase, vortex for 1 min, inject a 25 μL aliquot. Serum. 200 μL Serum + 100 μL pH 7.2 phosphate buffer, mix, add 6 mL dichloromethane:propanol 95:5, mix for 2 min, centrifuge at 1500 g for 10 min. Remove the organic layer and evaporate it to dryness under a stream of nitrogen at 35°, reconstitute the residue in 200 μL mobile phase, vortex for 1 min, inject a 25 μL aliquot.

## HPLC VARIABLES
**Guard column:** 10 μm μBondapak C18
**Column:** 250 × 4.6 10 μm μBondapak C18
**Mobile phase:** THF:10 mM Na₂HPO₄ 3:97, pH adjusted to 6.5 with phosphoric acid
**Flow rate:** 2.5
**Injection volume:** 25
**Detector:** UV 273

## CHROMATOGRAM
**Retention time:** 5
**Limit of detection:** 62.5 ng/g (tissue); 62.5 ng/mL (serum)

## OTHER SUBSTANCES
**Extracted:** caffeine, paraxanthine, theobromine

## KEY WORDS
serum; rat; brain

## REFERENCE
Parra, P.; Limon, A.; Ferre, S.; Guix, T.; Jane, F. High-performance liquid chromatographic separation of caffeine, theophylline, theobromine and paraxanthine in rat brain and serum. *J.Chromatogr.*, **1991**, *570*, 185–190

## SAMPLE
**Matrix:** blood, urine
**Sample preparation:** Plasma. 500 μL Plasma + 25 μL 300 μM IS in EtOH:water 30:70 + 75 μL 1 M HCl, mix, add 5 mL ethyl acetate:isopropanol 90:10, shake for 10 min, centrifuge at 1000 g for 10 min, freeze at -30°. Remove the organic layer and evaporate it to dryness under a stream of nitrogen at 55°, reconstitute the residue in 300 μL mobile phase, sonicate until dissolved, centrifuge at 13000 g for 15 min, inject a 40 μL aliquot. Urine. Acidify 10 mL urine with 300 μL 1 M HCl. 40 μL Acidified urine + 50 μL 300 μM IS in EtOH:water 30:70, mix, make up to 400 μL with 10 mM pH 4.0 acetate buffer, add 5 mL ethyl acetate:isopropanol 93:7, shake for 10 min, centrifuge at 1000 g for 10 min, freeze at -30°. Remove the organic layer and evaporate it to dryness under a stream of nitrogen at 55°, reconstitute the residue in 300 μL mobile phase, vortex for 10 s, inject a 40 μL aliquot.

## HPLC VARIABLES
**Column:** 250 × 4.6 5 μm Ultrasphere ODS
**Mobile phase:** MeOH:10 mM pH 4.0 acetate buffer 9:91 (plasma) or 7:93 (urine)
**Column temperature:** 30
**Flow rate:** 1 for 11 min, 1.5 for 6 min, 2.5 for 13 min
**Injection volume:** 40
**Detector:** UV 273

## CHROMATOGRAM
**Retention time:** 20.2 (plasma), 20.9 (urine)
**Internal standard:** β-hydroxyethyltheophylline (24.2 (plasma), 28.4 (urine))
**Limit of detection:** 200 nM (plasma); 2 μM (urine)

## OTHER SUBSTANCES
**Extracted:** metabolites, 1,3-dimethyluric acid, 1-methyluric acid, 3-methylxanthine

## KEY WORDS
plasma

## REFERENCE
Rasmussen, B.B.; Brosen, K. Determination of theophylline and its metabolites in human urine and plasma by high-performance liquid chromatography. *J.Chromatogr.B*, **1996**, *676*, 169–174

## SAMPLE
**Matrix:** formulations
**Sample preparation:** Inject a 20 μL aliquot.

## HPLC VARIABLES
**Guard column:** 10 μm Dynamax C18
**Column:** 250 × 4.6 10 μm Dynamax C18
**Mobile phase:** MeCN:6.5 mM tetraheptylammonium bromide:100 mM pH 7.0 phosphate buffer:100 mM pH 5.0 citrate buffer 40:55.2:4.4:0.4
**Flow rate:** 1
**Injection volume:** 20
**Detector:** UV 271

## CHROMATOGRAM
**Retention time:** 3.0

## OTHER SUBSTANCES
**Simultaneous:** degradation products, ceftriaxone

## KEY WORDS
injections; use low actinic glassware; stability-indicating; saline; 5% dextrose

## REFERENCE
Parrish, M.A.; Bailey, L.C.; Medwick, T. Stability of ceftriaxone sodium and aminophylline or theophylline in intravenous mixtures. *Am.J.Hosp.Pharm.*, **1994**, *51*, 92–94

## SAMPLE
**Matrix:** formulations
**Sample preparation:** Dilute to 0.8-13 μg/mL, inject an aliquot.

## HPLC VARIABLES
**Column:** 150 × 4.6 5 μm Ultrasphere C18
**Mobile phase:** MeCN:100 mM pH 3.4 acetate buffer 1:10 (Buffer was 50 mL 100 mM sodium acetate diluted to 1 L with 100 mM acetic acid.)
**Flow rate:** 2
**Injection volume:** 20
**Detector:** UV 254

## CHROMATOGRAM
**Internal standard:** 5-methylresorcinol

## OTHER SUBSTANCES
**Simultaneous:** aminophylline, cefuroxime

## KEY WORDS
injections; 5% dextrose

## REFERENCE
Stewart, J.T.; Warren, F.W.; Johnson, S.M. Stability of cefuroxime sodium and aminophylline or theophylline. *Am.J.Hosp.Pharm.*, **1994**, *51*, 809–811

## SAMPLE
**Matrix:** formulations
**Sample preparation:** Shake, remove 2 mL of oral suspension, dilute to 40 mL with water,
vortex 1 min, centrifuge at 2000 rpm for 10 min. Dilute a 100 μL aliquot of supernatant
with 100 μL of 100 μg/mL theophylline and add 800 μL water, inject 20 μL aliquot.

## HPLC VARIABLES
**Column:** 300 mm 10 μm Waters reversed-phase C18
**Mobile phase:** MeCN:50 mM sodium acetate buffer 8:92, adjusted to pH 6.5
**Flow rate:** 1.5
**Injection volume:** 20
**Detector:** UV 254

## CHROMATOGRAM
**Retention time:** 4.3
**Internal standard:** theophylline

## OTHER SUBSTANCES
**Simultaneous:** famotidine

## KEY WORDS
stability-indicating; oral suspensions; theophylline is IS

## REFERENCE
Quercia, R.A.; Jay, G.T.; Fan, C.; Chow, M.S. Stability of famotidine in an extemporaneously prepared
oral liquid. *Am.J.Hosp.Pharm.*, **1993**, *50*, 691–693

## SAMPLE
**Matrix:** formulations
**Sample preparation:** Tablets. Weigh out powdered tablets containing 100 mg aminophylline, add 50 mL water, sonicate for 15 min, make up to 100 mL with water, mix, filter.
Remove a 5 mL aliquot of the filtrate and add it to 10 mL 5 mg/mL dansyl chloride in
acetone and 5 mL buffer, mix gently, let stand in the dark for 12 h, make up to 50 mL
with acetone:water 50:50, mix, inject an aliquot. Injections, oral liquids. Measure out an
amount containing 100 mg aminophylline, make up to 100 mL with water, mix. Remove
a 5 mL aliquot and add it to 10 mL 5 mg/mL dansyl chloride in acetone and 5 mL buffer,
mix gently, let stand in the dark for 12 h, make up to 50 mL with acetone:water 50:50,
mix, inject an aliquot. (Prepare buffer by dissolving 550 mg anhydrous sodium carbonate
in 300 mL water, add 300 mL acetone, mix.)

## HPLC VARIABLES
**Guard column:** 70 × 2.1 Co:Pell ODS
**Column:** 300 × 3.9 10 μm μBondapak C18
**Mobile phase:** MeOH:water:acetic acid:triethylamine 60:38:1.5:0.5 (A) or 65:33:1.5:0.5
(B)

**Flow rate:** 1.5
**Injection volume:** 20
**Detector:** UV 254

## CHROMATOGRAM
**Retention time:** 6.0 (mobile phase B), 7.45 (mobile phase A)

## OTHER SUBSTANCES
**Simultaneous:** ethylenediamine

## KEY WORDS
derivatization; tablets; injections; oral solutions

## REFERENCE
Lau-Cam, C.A.; Roos, R.W. Simultaneous high performance liquid chromatographic determination of theophylline and ethylenediamine in aminophylline dosage forms as their dansyl derivatives. *J.Liq.Chromatogr.*, **1991**, *14*, 1939–1956

## SAMPLE
**Matrix:** microsomal incubations
**Sample preparation:** 500 μL Microsomal incubation + 400 μL 2% zinc sulfate, centrifuge at 5000 rpm for 5 min. Remove the supernatant and add it to 30 mg ammonium sulfate and 10 μL 10 μg/mL 1,7-dimethyluric acid, extract with dichloromethane:isopropanol 80:20. Remove the organic layer and evaporate it to dryness, reconstitute, inject an aliquot.

## HPLC VARIABLES
**Guard column:** C18
**Column:** 250 × 4.6 Spherex 5 C18 (Phenomenex)
**Mobile phase:** Gradient. MeCN:25 μM pH 4.5 acetate buffer 2:98 for 20 min, 5:95 for 12 min, 9:91 for 8 min, re-equilibrate at initial conditions for 10 min.
**Flow rate:** 1
**Detector:** UV 276

## CHROMATOGRAM
**Retention time:** 35.5
**Internal standard:** 1,7-dimethyluric acid (33.6)

## OTHER SUBSTANCES
**Extracted:** metabolites, 1,3-dimethyluric acid, 1-methylxanthine, 3-methylxanthine

## KEY WORDS
human; liver

## REFERENCE
Tjia, J.F.; Colbert, J.; Back, D.J. Theophylline metabolism in human liver microsomes: Inhibition studies. *J.Pharmacol.Exp.Ther.*, **1996**, *276*, 912–917

## SAMPLE
**Matrix:** microsomal incubations
**Sample preparation:** 275 μL Microsomal incubation + 50 μL 30% perchloric acid, centrifuge at 2000 g for 10 min, inject a 60 μL aliquot of the supernatant.

## HPLC VARIABLES
**Guard column:** 20 × 4.6 40 μm Supelcosil LC-18

**Column:** 150 × 4.6 5 μm Supelcosil LC-18
**Mobile phase:** MeCN:THF:20 mM pH 3.5 sodium perchlorate 0.5:0.5:99
**Flow rate:** 1.5
**Injection volume:** 60
**Detector:** UV 280

## CHROMATOGRAM
**Retention time:** 14

## OTHER SUBSTANCES
**Extracted:** caffeine, paraxanthine, theobromine, trimethylurate

## KEY WORDS
monkey; liver

## REFERENCE
Bullock, P.; Pearce, R.; Draper, A.; Podval, J.; Bracken, W.; Veltman, J.; Thomas, P.; Parkinson, A. Induction of liver microsomal cytochrome P450 in cynomolgus monkeys. *Drug Metab.Dispos.*, **1995**, *23*, 736–748

## SAMPLE
**Matrix:** milk
**Sample preparation:** Centrifuge at 12800 g for 10 min, remove top layer of fat with a spatula. 450 μL Supernatant + 50 μL water + 150 μL 21.65 μg/mL proxyphylline in 6% perchloric acid, vortex for 5 s, cool on ice for 10-15 min, centrifuge at 12800 g for 10 min, inject a 50 μL aliquot of the supernatant.

## HPLC VARIABLES
**Guard column:** Co:Pell ODS glass beads (Whatman)
**Column:** 250 × 4.6 5 μm Ultrasphere ODS
**Mobile phase:** Gradient. A was 10 mM sodium acetate + 5 mM tetrabutylammonium hydrogen sulfate, adjusted to pH 4.9 with 2 M NaOH. B was MeOH:water 50:50 containing 10 mM sodium acetate + 5 mM tetrabutylammonium hydrogen sulfate, adjusted to pH 4.8 with glacial acetic acid. A:B 100:0 for 7.5 min, then to 85:15 over 7.5 min, then to 70:30 over 10 min, then to 68:32 over 4 min, then to 100:0 over 3 min.
**Flow rate:** 1.5
**Injection volume:** 50
**Detector:** UV 272

## CHROMATOGRAM
**Retention time:** 22.5
**Internal standard:** proxyphylline (25)

## OTHER SUBSTANCES
**Simultaneous:** caffeine, paraxanthine, theobromine

## REFERENCE
Blanchard, J.; Weber, C.W.; Shearer, L.E. HPLC analysis of methylxanthines in human breast milk. *J.Chromatogr.Sci.*, **1990**, *28*, 640–642

## SAMPLE
**Matrix:** perfusate
**Sample preparation:** Inject a 100 μL aliquot directly.

## HPLC VARIABLES
**Column:** 250 × 4.6 octyl Spherisorb S5-ODS

**Mobile phase:** MeCN:100 mM KH$_2$PO$_4$ 10:90
**Column temperature:** 40
**Flow rate:** 1
**Injection volume:** 100
**Detector:** UV 272

---

## REFERENCE

Michael-Baruch, E.; Shiri, Y.; Cohen, S. Alkali halide-assisted penetration of neostigmine across excised human skin: A combination of structured water disruption and a Donnan-like effect. *J.Pharm.Sci.*, **1994**, *83*, 1071–1076

---

## SAMPLE

**Matrix:** saliva
**Sample preparation:** Centrifuge saliva at 10000 rpm for 2 min. 1 mL Supernatant + 50 μL 80 μg/mL β-hydroxyethyltheophylline + 100 μL 100 mM HCl + 7 mL chloroform: isopropanol 95:5, shake for 10 min, centrifuge at 3000 rpm for 5 min. Remove 5 mL of the organic layer and evaporate it to dryness under a stream of nitrogen at 60°, reconstitute the residue in 200 μL mobile phase, filter, inject a 10 μL aliquot of the filtrate.

---

## HPLC VARIABLES

**Column:** 150 × 4.6 ODS 120T (Tosoh)
**Mobile phase:** MeCN:10 mM pH 4.0 acetate buffer 9:91
**Flow rate:** 1
**Injection volume:** 10
**Detector:** UV 270

---

## CHROMATOGRAM

**Internal standard:** β-hydroxyethyltheophylline

---

## KEY WORDS

pharmacokinetics

---

## REFERENCE

Kanke, M.; Katayama, H.; Nakamura, M. Application of curdlan to controlled drug delivery. II. In vitro and in vivo drug release studies of theophylline-containing curdlan tablets. *Biol.Pharm.Bull.*, **1995**, *18*, 1104–1108

---

## SAMPLE

**Matrix:** solutions
**Sample preparation:** Inject a 20 μL aliquot.

---

## HPLC VARIABLES

**Column:** 150 × 6 5 μm 3-aminopropylsilyl silica gel with the amino group derivatized with 1,8-naphthalic anhydride (Bunseki Kagaku 1993, 42, 817)
**Mobile phase:** MeOH:buffer 50:50 (Prepare buffer by dissolving 6.183 g boric acid and 1.461 g NaCl in 500 mL water, adjust pH to 6.4 with sodium borate solution.)
**Column temperature:** 30
**Flow rate:** 1
**Injection volume:** 20
**Detector:** UV 270

---

## CHROMATOGRAM

**Retention time:** 13.5
**Internal standard:** 1,3-dimethyl-7-(2-hydroxyethyl)xanthine (12)

## OTHER SUBSTANCES
**Simultaneous:** caffeine, hypoxanthine, pentoxifylline, propentofylline, theobromine, uric acid, xanthine

## REFERENCE
Nakashima, K.; Inoue, K.; Mayahara, K.; Kuroda, N.; Hamachi, Y.; Akiyama, S. Use of 3-(1,8-naphthalimido)propyl-modified silyl silica gel as a stationary phase for the high-performance liquid chromatographic separation of purine derivatives. *J.Chromatogr.A*, **1996**, *722*, 107–113

## SAMPLE
**Matrix:** solutions

## HPLC VARIABLES
**Column:** 250 × 4.6 5 μm Supelcosil LC-DP (A) or 250 × 4 5 μm LiChrospher 100 RP-8 (B)
**Mobile phase:** MeCN:0.025% phosphoric acid:buffer 25:10:5 (A) or 60:25:15 (B) (Buffer was 9 mL concentrated phosphoric acid and 10 mL triethylamine in 900 mL water, adjust pH to 3.4 with dilute phosphoric acid, make up to 1 L.)
**Flow rate:** 0.6
**Injection volume:** 25
**Detector:** UV 229

## CHROMATOGRAM
**Retention time:** 4.52 (A), 3.45 (B)

## OTHER SUBSTANCES
**Also analyzed:** acebutolol, acepromazine, acetaminophen, acetazolamide, acetophenazine, albuterol, alprazolam, amitriptyline, amobarbital, amoxapine, antipyrine, atenolol, atropine, azatadine, baclofen, benzocaine, bromocriptine, brompheniramine, brotizolam, bupivacaine, buspirone, butabarbital, butalbital, caffeine, carbamazepine, cetirizine, chlorcyclizine, chlordiazepoxide, chlormezanone, chloroquine, chlorpheniramine, chlorpromazine, chlorpropamide, chlorprothixene, chlorthalidone, chlorzoxazone, cimetidine, cisapride, clomipramine, clonazepam, clonidine, clozapine, cocaine, codeine, colchicine, cyclizine, cyclobenzaprine, dantrolene, desipramine, diazepam, diclofenac, diflunisal, diltiazem, diphenhydramine, diphenidol, diphenoxylate, dipyridamole, disopyramide, dobutamine, doxapram, doxepin, droperidol, encainide, ethidium bromide, ethopropazine, fenoprofen, fentanyl, flavoxate, fluoxetine, fluphenazine, flurazepam, flurbiprofen, fluvoxamine, furosemide, glutethimide, glyburide, guaifenesin, haloperidol, homatropine, hydralazine, hydrochlorothiazide, hydrocodone, hydromorphone, hydroxychloroquine, hydroxyzine, ibuprofen, imipramine, indomethacin, ketoconazole, ketoprofen, ketorolac, labetalol, levorphanol, lidocaine, loratadine, lorazepam, lovastatin, loxapine, mazindol, mefenamic acid, meperidine, mephenytoin, mepivacaine, mesoridazine, metaproterenol, metformin, methadone, methdilazine, methocarbamol, methotrexate, methotrimeprazine, methoxamine, methyldopa, methylphenidate, metoclopramide, metolazone, metoprolol, metronidazole, midazolam, moclobemide, morphine, nadolol, nalbuphine, naloxone, naphazoline, naproxen, nifedipine, nizatidine, norepinephrine, nortriptyline, oxazepam, oxycodone, oxymetazoline, paroxetine, pemoline, pentazocine, pentobarbital, pentoxifylline, perphenazine, pheniramine, phenobarbital, phenol, phenolphthalein, phentolamine, phenylbutazone, phenyltoloxamine, phenytoin, pimozide, pindolol, piroxicam, pramoxine, prazepam, prazosin, probenecid, procainamide, procaine, prochlorperazine, procyclidine, promazine, promethazine, propafenone, propantheline, propiomazine, propofol, propranolol, protriptyline, quazepam, quinidine, quinine, racemethorphan, ranitidine, remoxipride, risperidone, salicylic acid, scopolamine, secobarbital, sertraline, sotalol, spironolactone, sulfinpyrazone, sulindac, temazepam, terbutaline, terfenadine, tetracaine, thiethylperazine, thiopental, thioridazine, thiothixene, timolol, tocainide, tolbutamide, tolmetin, trazodone, triamterene, triazolam, trifluoperazine, triflupromazine, trimeprazine, trimethoprim, trimipramine, verapamil, warfarin, xylometazoline, yohimbine, zopiclone

## KEY WORDS
some details of plasma extraction

## REFERENCE
Koves, E.M. Use of high-performance liquid chromatography-diode array detection in forensic toxicology. *J.Chromatogr.A*, **1995**, *692*, 103–119

## SAMPLE
**Matrix:** solutions
**Sample preparation:** Inject a 50-200 μL aliquot of a solution in pH 7.4 Tyrode's buffer.

## HPLC VARIABLES
**Column:** 125 × 4 5 μm LiChrospher 60 RP-select B
**Mobile phase:** MeCN:10 mM pH 4 sodium acetate 12:88
**Flow rate:** 0.6
**Injection volume:** 50-200
**Detector:** UV 270

## OTHER SUBSTANCES
**Also analyzed:** cefazolin

## KEY WORDS
buffer

## REFERENCE
Saitoh, H.; Aungst, B.J. Possible involvement of multiple P-glycoprotein-mediated efflux systems in the transport of verapamil and other organic cations across rat intestine. *Pharm.Res.*, **1995**, *12*, 1304–1310

## SAMPLE
**Matrix:** solutions

## HPLC VARIABLES
**Column:** 250 × 4.6 Zorbax RX
**Mobile phase:** Gradient. A was 10 mL concentrated orthophosphoric acid and 7 mL triethylamine in 1 L water. B was 10 mL concentrated orthophosphoric acid and 7 mL triethylamine in 200 mL water, make up to 1 L with MeCN. A:B from 100:0 to 0:100 over 30 min, maintain at 0:100 for 5 min.
**Column temperature:** 30
**Flow rate:** 2
**Detector:** UV 210

## OTHER SUBSTANCES
**Also analyzed:** acepromazine, acetaminophen, acetophenazine, albuterol, aminophylline, amitriptyline, amobarbital, amoxapine, amphetamine, amylocaine, antipyrine, aprobarbital, aspirin, atenolol, atropine, avermectin, barbital, benzocaine, benzoic acid, benzotropine, benzphetamine, berberine, bibucaine, bromazepan, brompheniramine, buprenorphine, buspirone, butabarbital, butacaine, butethal, caffeine, carbamazepine, carbromal, chloramphenicol, chlordiazepoxide, chloroquine, chlorothiazide, chloroxylenol, chlorphenesin, chlorpheniramine, chlorpromazine, chlorpropamide, chlortetracycline, cimetidine, cinchonidine, cinchonine, clenbuterol, clonazepam, clonixin, clorazepate, cocaine, codeine, colchicine, cortisone, coumarin, cyclazocine, cyclobenzaprine, cyclothiazide, cyheptamide, cymarin, danazol, danthron, dapsone, debrisoquine, desipramine, dexamethasone, dextromethorphan, dextropropoxyphene, diamorphine, diazepam, diclofenac, diethylpropion, diethylstilbestrol, diflunisal, digitoxin, digoxin, diltiazem, diphenhydramine, diphenoxy-

late, diprenorphine, dipyrone, disulfiram, dopamine, doxapram, doxepin, dronabinol, ephedrine, epinephrine, epinine, estradiol, estriol, estrone, ethacrynic acid, ethosuximide, etonitazene, etorphine, eugenol, famotidine, fenbendazole, fencamfamine, fenoprofen, fenproporex, fentanyl, flubendazole, flufenamic acid, flunitrazepam, 5-fluorouracil, fluoxymesterone, fluphenazine, furosemide, gentisic acid, gitoxigenin, glipizide, glunixin, glutethimide, glybenclamide, guaiacol, halazepam, haloperidol, hydrochlorothiazide, hydrocodone, hydrocortisone, hydromorphone, hydroxyquinoline, ibogaine, ibuprofen, iminostilbene, imipramine, indomethacin, isocarbostyril, isocarboxazid, isoniazid, isoproterenol, isoxsuprine, ivermectin, ketamine, ketoprofen, kynurenic acid, levorphanol, lidocaine, lorazepam, lormetazepam, loxapine, mazindol, mebendazole, meclizine, meclofenamic acid, medazepam, mefenamic acid, megestrol, mepacrine, meperidine, mephentermine, mephenytoin, mephesin, mephobarbital, mepivacaine, mescaline, mesoridazine, methadone, methamphetamine, methapyrilene, methaqualone, methazolamide, methocarbamol, methoxamine, methsuximide, methyl salicylate, methyldopa, methyldopamine, methylphenidate, methylprednisolone, methyltestosterone, methyprylon, metoprolol, mibolerone, morphine, nadolol, nalorphine, naloxone, naltrexone, naphazoline, naproxen, nefopam, niacinamide, nicotine, nicotinic acid, nifedipine, niflumic acid, nitrazepam, norepinephrine, nortriptyline, noscapine, nylidrin, oxazepam, oxycodone, oxymorphone, oxyphenbutazone, oxytetracycline, papaverine, pargyline, pemoline, pentazocine, pentobarbital, persantine, phenacetin, phenazocine, phenazopyridine, phencyclidine, phendimetrazine, phenelzine, pheniramine, phenobarbital, phenothiazine, phensuximide, phentermine, phenylbutazone, phenylephrine, phenylpropanolamine, piperocaine, prazepam, prednisolone, primidone, probenecid, progesterone, propiomazine, propranolol, propylparaben, pseudoephedrine, puromycin, pyrilamine, pyrithyldione, quazepam, quinaldic acid, quinidine, quinine, ranitidine, recinnamine, reserpine, resorcinol, saccharin, albuterol, salicylamide, salicylic acid, scopolamine, scopoletin, secobarbital, strychnine, sulfacetamide, sufadiazine, sulfadimethoxine, sulfaethidole, sulfamerazine, sulfamethazine, sulfamethoxizole, sulfanilamide, sulfapyridine, sulfasoxizole, sulindac, tamoxifen, temazepam, testosterone, tetracaine, tetracycline, tetramisole, thebaine, theobromine, thiabendazole, thiamine, thiamylal, thiobarbituric acid, thioridazine, thiosalicylic acid, thiothixene, thymol, tolazamide, tolazoline, tobutamide, tolmetin, tranylcypromine, triamcinolone, tribenzylamine, trichloromethiazide, trifluoperazine, trihexyphenidyl, trimethoprim, tripelennamine, triprolidine, tropacocaine, tyramine, verapamil, vincamine, warfarin, yohimbine, zoxazolamine

## REFERENCE

Hill, D.W.; Kind, A.J. Reversed-phase solvent-gradient HPLC retention indexes of drugs. *J.Anal.Toxicol.*, **1994**, *18*, 233–242

## SAMPLE

**Matrix:** solutions
**Sample preparation:** Dissolve in MeOH:water 1:1 at a concentration of 50 µg/mL, inject a 10 µL aliquot.

## HPLC VARIABLES

**Column:** 300 × 3.9 10 µm µBondapak C18
**Mobile phase:** MeOH:acetic acid:triethylamine:water 25:1.5:0.5:73
**Flow rate:** 1.5
**Injection volume:** 10
**Detector:** UV 254

## CHROMATOGRAM

**Retention time:** 7

## OTHER SUBSTANCES

**Simultaneous:** caffeine, 8-chlorotheophylline, diphylline, theobromine

## REFERENCE
Roos, R.W.; Lau-Cam, C.A. General reversed-phase high-performance liquid chromatographic method for the separation of drugs using triethylamine as a competing base. *J.Chromatogr.*, **1986**, *370*, 403–418

## SAMPLE
**Matrix:** solutions
**Sample preparation:** Prepare a 10 µg/mL solution in MeOH, inject a 20 µL aliquot.

## HPLC VARIABLES
**Column:** 125 × 4.9 Spherisorb S5W silica
**Mobile phase:** MeOH containing 10 mM ammonium perchlorate and 1 mL/L 100 mM NaOH in MeOH, pH 6.7
**Flow rate:** 2
**Injection volume:** 20
**Detector:** E, LeCarbone, V25 glassy carbon electrode, + 1.2 V

## CHROMATOGRAM
**Retention time:** 1.04

## OTHER SUBSTANCES
**Also analyzed:** acebutolol, acepromazine, acetophenazine, N-acetylprocainamide, albuterol, alprenolol, amethocaine, amiodarone, amitriptyline, antazoline, atenolol, azacyclonal, bamethane, benactyzine, benperidol, benzethidine, benzocaine, benzoctamine, benzphetamine, benzquinamide, bromhexine, bromodiphenhydramine, bromperidol, brompheniramine, brompromazine, buclizine, bufotenine, bupivacaine, buprenorphine, butacaine, butethamate, chlorcyclizine, chlorpheniramine, chlorphenoxamine, chlorprenaline, chlorpromazine, chlorprothixene, cimetidine, cinchonidine, cinnarizine, clemastine, clomipramine, clonidine, cocaine, cyclazocine, cyclizine, cyclopentamine, cyproheptadine, deserpidine, desipramine, dextromoramide, dextropropoxyphene, dicyclomine, diethylcarbamazine, diethylpropion, diethylthiambutene, dihydroergotamine, dimethindene, dimethothiazine, diphenhydramine, diphenoxylate, dipipanone, diprenorphine, dipyridamole, disopyramide, dothiepin, doxapram, doxepin, doxylamine, droperidol, ephedrine, ergocornine, ergocristine, ergocristinine, ergocryptine, ergometrine, ergosine, ergosinine, ergotamine, ethopropazine, etorphine, etoxeridine, fenethazine, fenfluramine, fenoterol, fentanyl, flavoxate, fluopromazine, flupenthixol, fluphenazine, flurazepam, haloperidol, hydroxyzine, hyoscine, ibogaine, imipramine, indapamine, iprindole, isothipendyl, isoxsuprine, ketanserin, laudanosine, lidocaine, lofepramine, loxapine, maprotiline, mecamylamine, meclophenoxate, meclozine, medazepam, mephentermine, mepivacaine, meptazinol, mepyramine, mesoridazine, metaraminol, methadone, methamphetamine, methapyrilene, methdilazine, methotrimeprazine, methoxamine, methoxyphenamine, methoxypromazine, methylephedrine, methylergonovine, methysergide, metoclopramide, metopimazine, metoprolol, mianserin, morazone, nadolol, nalorphine, naloxone, naphazoline, nicotine, nifedipine, nomifensine, nortriptyline, noscapine, orphenadrine, oxeladin, oxprenolol, oxymetazolin, papaverine, pargyline, pecazine, penbutolol, pentazocine, penthienate, pericyazine, perphenazine, phenadoxone, phenampromide, phenazocine, phenbutrazate, phendimetrazine, phenelzine, phenglutarimide, phenindamine, pheniramine, phenmetrazine, phenomorphan, phenoperidine, phenothiazine, phenoxybenzamine, phentolamine, phenylephrine, phenyltoloxamine, physostigmine, piminodine, pimozide, pindolol, pipamazine, pipazethate, piperacetazine, piperidolate, pipradol, pirenzepine, piritramide, pizotifen, practolol, pramoxine, prazosin, prenylamine, prilocaine, primaquine, proadifen, procainamide, procaine, prochlorperazine, procyclidine, proheptazine, prolintane, promazine, promethazine, pronethalol, properidine, propiomazine, propranolol, prothipendyl, protriptyline, proxymetacaine, pseudoephedrine, pyrimethamine, quinidine, quinine, ranitidine, rescinnamine, sotalol, tacrine, terazosin, terbutaline, terfenadine, thenyldiamine, thiethylperazine, thiopropazate, thioproperazine, thioridazine, thiothixene, thonzylamine, timolol, tocainide, tolpropamine, tolycaine, tranylcypromine, trazo-

done, trifluoperazine, trifluperidol, trimeperidine, trimeprazine, trimethobenzamide, trimethoprim, trimipramine, tripelennamine, triprolidine, tryptamine, verapamil, xylometazoline

## REFERENCE
Jane, I.; McKinnon, A.; Flanagan, R.J. High-performance liquid chromatographic analysis of basic drugs on silica columns using non-aqueous ionic eluents. II. Application of UV, fluorescence and electrochemical oxidation detection. *J.Chromatogr.*, **1985**, *323*, 191–225

## SAMPLE
**Matrix:** solutions
**Sample preparation:** Dissolve in MeOH, inject a 1 µL aliquot.

## HPLC VARIABLES
**Column:** 150 × 1 3 µm Hitachi-Gel 3011 porous polymer (Hitachi)
**Mobile phase:** MeOH:ammonia 99:1
**Flow rate:** 0.03
**Injection volume:** 1
**Detector:** UV 254

## CHROMATOGRAM
**Retention time:** 3.56

## OTHER SUBSTANCES
**Also analyzed:** acetaminophen, aspirin, bucetin (3-hydroxy-p-butyrophenetidine), caffeine, dipyrone (sulpyrin), ethenzamide (o-ethoxybenzamide), mefenamic acid, phenacetin, salicylamide, salicylic acid, theobromine

## KEY WORDS
semi-micro; porous polymer

## REFERENCE
Matsushima, Y.; Nagata, Y.; Niyomura, M.; Takakusagi, K.; Takai, N. Analysis of antipyretics by semimicro liquid chromatography. *J.Chromatogr.*, **1985**, *332*, 269–273

## SAMPLE
**Matrix:** urine
**Sample preparation:** Dilute urine 10-fold with 5 µg/mL β-hydroxyethyltheophylline in water, mix, centrifuge at 14000 rpm for 2 min, inject a 25 µL aliquot of the supernatant.

## HPLC VARIABLES
**Guard column:** 40 × 2.5 10 µm Lichrosorb RP-2
**Column:** 150 × 4.6 5 µm Ultrasphere-ODS
**Mobile phase:** MeCN:THF:10 mM sodium acetate 3:0.1:96.9 containing 5 mM tetrabutylammonium hydrogen sulfate, pH 4.7
**Flow rate:** 1.5
**Injection volume:** 25
**Detector:** UV 280

## CHROMATOGRAM
**Retention time:** 9
**Internal standard:** β-hydroxyethyltheophylline (11)
**Limit of detection:** 1 µg/mL

## OTHER SUBSTANCES
**Extracted:** metabolites, caffeine, 1,3-dimethyluric acid, 1-methyluric acid, 3-methylxanthine

## REFERENCE
Tajerzadeh, H.; Dadashzadeh, S. An isocratic high-performance liquid chromatographic system for simultaneous determination of theophylline and its major metabolites in human urine. *J.Pharm.Biomed.Anal.*, **1995**, *13*, 1507–1512

## ANNOTATED BIBLIOGRAPHY

Hieda, Y.; Kashimura, S.; Hara, K.; Kageura, M. Highly sensitive and rapid determination of theophylline, theobromine and caffeine in human plasma and urine by gradient capillary high-performance liquid chromatography-frit-fast atom bombardment mass spectrometry. *J.Chromatogr.B*, **1995**, *667*, 241–246 [LC-MS; LOD 5 ng/mL]

Theodoridis, G.; Papadoyannis, I.; Vasilikiotis, G.; Tsoukali-Papadopoulou, H. Reversed-phase high-performance liquid chromatography–photodiode-array analysis of alkaloid drugs of forensic interest. *J.Chromatogr.B*, **1995**, *668*, 253–263 [also amphetamine, bamifylline, caffeine, cocaine, codeine, diamorphine, ethylmorphine, flufenamic acid, hyoscyamine, methadone, morphine, nalorphine, norcodeine, papaverine, quinine, scopolamine, strychnine, theobromine, tolfenamic acid]

Van den Mooter, G.; Samyn, C.; Kinget, R. *In vivo* evaluation of a colon-specific drug delivery system: An absorption study of theophylline from capsules coated with azo polymers in rats. *Pharm.Res.*, **1995**, *12*, 244–247 [plasma; LOD 250 ng/mL]

Hieda, Y.; Kashimura, S.; Hara, K.; Kageura, M. [Development of the method for analysis of theophylline and its related compounds using capillary high-performance liquid chromatography/fast atom bombardment-mass spectrometry]. *Nippon Hoigaku Zasshi*, **1994**, *48*, 253–262

Kanda, T.; Kutsuna, H.; Ohtsu, Y.; Yamaguchi, M. Synthesis of polymer-coated mixed-functional packing materials for direct analysis of drug-containing serum and plasma by high-performance liquid chromatography. *J.Chromatogr.A*, **1994**, *672*, 51–57 [direct injection; column temp 40; also carbamazepine, chloramphenicol, indomethacin, phenobarbital, phenytoin, trimethoprim]

Sarkar, M.A.; Jackson, B.J. Theophylline N-demethylations as probes for P4501A1 and P4501A2. *Drug Metab.Dispos.*, **1994**, *22*, 827–834 [theobromine (IS); LOD 50 ng/mL]

Smigol, V.; Svec, F.; Fréchet, J.M.J. Novel uniformly sized polymeric stationary phase with hydrophilized large pores for direct injection HPLC determination of drugs in biological fluids. *J.Liq.Chromatogr.*, **1994**, *17*, 891–911 [also aspirin, caffeine, carbamazepine, lidocaine, phenytoin, salicylic acid, theobromine]

Zhang, H.; Stewart, J.T. Determination of a cefuroxime and aminophylline/theophylline mixture by high-performance liquid chromatography. *J.Liq.Chromatogr.*, **1994**, *17*, 1327–1335 [simultaneous cefuroxime, orcinol]

Chow, A.T.; Meek, P.D.; Jusko, W.J. High-pressure liquid chromatographic assay of theophylline in dog feces following oral administration of sustained-release products. *J.Pharm.Sci.*, **1993**, *82*, 956–958

Abdel-Hay, M.H.; el-Din, M.S.; Abuirjeie, M.A. Simultaneous determination of theophylline and guaiphenesin by third-derivative ultraviolet spectrophotometry and high-performance liquid chromatography. *Analyst*, **1992**, *117*, 157–160

Abuirjeie, M.A.; el-Din, M.S.; Mahmoud, I.I. Determination of theobromine, theophylline and caffeine in various food products using derivative UV-spectrophotometric techniques and high-performance liquid chromatography. *J.Liq.Chromatogr.*, **1992**, *15*, 101–125 [ethyl paraben (IS)]

Mazzei, M.; Sottofattori, E.; Balbi, A.; Bottino, G.B. HPLC analysis of theophylline: bioequivalence study of two sustained-release formulations at steady state. *Farmaco*, **1992**, *47*, 769–777

Papadoyannis, I.; Georgarakis, M.; Samanidou, V.; Theodoridis, G. High-performance liquid chromatographic analysis of theophylline in the presence of caffeine in blood serum and pharmaceutical formulations. *J.Liq.Chromatogr.*, **1991**, *14*, 1587–1603

El-Sayed, Y.M.; Islam, S.I. High-performance liquid chromatographic analysis of theophylline in serum and its use in therapeutic drug monitoring. *J.Clin.Pharm.Ther.*, **1989**, *14*, 35–43

Lam, S.; Malikin, G. An improved micro-scale protein precipitation procedure for HPLC assay of therapeutic drugs in serum. *J.Liq.Chromatogr.*, **1989**, *12*, 1851–1872 [also acetaminophen, amiodarone, aspirin, caffeine, chloramphenicol, flecainide, pentobarbital, procainamide, pyrimethamine, quinidine, tocainide, trazodone; fluorescence detection; UV detection]

Thomas, J. [Analysis of theophylline and its sodium (or potassium) anisates by ultrafast high-performance liquid chromatography in different pharmaceutical preparations]. *J.Chromatogr.*, **1989**, *479*, 430–436

Hotchkiss, S.A.; Caldwell, J. High-performance liquid chromatographic assay for theophylline and its major metabolites in human urine. *J.Chromatogr.*, **1987**, *423*, 179–188

Kester, M.B.; Saccar, C.L.; Mansmann, H.C., Jr. A new simplified microassay for the quantitation of theophylline in serum by high-performance liquid chromatography. *J.Liq.Chromatogr.*, **1987**, *10*, 957–975 [LOD 100 ng/mL; extracted metabolites, caffeine, dimethyluric acid, dimethylxanthine, methylxanthine]

Kester, M.B.; Saccar, C.L.; Mansmann, H.C., Jr. Microassay for the simultaneous determination of theophylline and dyphylline in serum by high-performance liquid chromatography. *J.Chromatogr.*, **1987**, *416*, 91–97

Kester, M.B.; Saccar, C.L.; Rocci, M.L.J.; Mansmann, H.C., Jr. A new simplified assay for the quantitation of theophylline and its major metabolites in urine by high-performance liquid chromatography. *J.Pharm.Sci.*, **1987**, *76*, 238–241

Wong, S.H.Y.; Marzouk, N.; Aziz, O.; Sheeran, S. Microbore liquid chromatography for therapeutic drug monitoring and toxicology: Clinical analyses of theophylline, caffeine, procainamide, and N-acetyl procainamide. *J.Liq.Chromatogr.*, **1987**, *10*, 491–506

Alvi, S.U.; Castro, F. A simultaneous assay of theophylline, ephedrine hydrochloride, and phenobarbital in suspensions and tablets formulations by high performance liquid chromatography. *J.Liq.Chromatogr.*, **1986**, *9*, 2269–2279 [guaifenesin (IS)]

Wong, S.H.Y.; Marzouk, N.; McHugh, S.L.; Cazes, E. Simultaneous determination of theophylline and caffeine by reversed phase liquid chromatography using phenyl column. *J.Liq.Chromatogr.*, **1985**, *8*, 1797–1816 [hydroxyethyltheophylline (IS); also acetaminophen, cimetidine, codeine, dimethylxanthine, meperidine, pentobarbital, phenobarbital, secobarbital]

Ong, H.; Marleau, S. A bidimensional HPLC system for direct determination of theophylline in serum. *J.Liq.Chromatogr.*, **1984**, *7*, 779–791 [column-switching]

St-Pierre, M.V.; Tesoro, A.; Spino, M.; MacLeod, S.M. An HPLC method for the determination of theophylline and its metabolites in serum and urine. *J.Liq.Chromatogr.*, **1984**, *7*, 1593–1608 [extracted dimethyluric acid, methyluric acid, methylxanthine]

Klassen, R.; Stavric, B. HPLC separation of theophylline, paraxanthine, theobromine, caffeine and other caffeine metabolites in biological fluids. *J.Liq.Chromatogr.*, **1983**, *6*, 895–906 [plasma; milk; urine; saliva]

Ou, C.N.; Frawley, V.L. Concurrent measurement of theophylline and caffeine in neonates by an interference-free liquid-chromatographic method. *Clin.Chem.*, **1983**, *29*, 1934–1936

Agostini, O.; Chiari, A.; Ciofi Baffoni, D. Simultaneous high-performance liquid chromatographic determination of salbutamol sulphate, theophylline, and saccharin in a hydroalcoholic formulation. *Boll.Chim.Farm.*, **1982**, *121*, 612–618

Chen, T.M.; Chafetz, L. High-pressure liquid chromatographic assay of theophylline, ephedrine hydrochloride, and phenobarbital tablets. *J.Pharm.Sci.*, **1981**, *70*, 804–806

Lurie, I.S.; Demchuk, S.M. Optimization of a reverse phase ion-pair chromatographic separation for drugs of forensic interest. II. Factors effecting selectivity. *J.Liq.Chromatogr.*, **1981**, *4*, 357–374 [also acetaminophen, acetylcodeine, acetylmorphine, aminopyrene, aminopyrine, amobarbital, amphetamine, antipyrine, benzocaine, butabarbital, caffeine, cocaine, codeine, diamorphine, diazepam, diethylpropion, DMT, ephedrine, glutethimide, Lampa, lidocaine, LSD, MDA, mecloqualone, mescaline, methamphetamine, methapyrilene, methaqualone, methpyrilene, methylphenidate, morphine, narcotine, papaverine, PCP, pentobarbital, phencyclidine, phendimetrazine, phenmetrazine, phenobarbital, phentermine, phenylpropanolamine, procaine, quinidine, quinine, secobarbital, strychnine, TCP, tetracaine, thebaine]

Quattrone, A.J.; Putnam, R.S. A single liquid-chromatographic procedure for therapeutic monitoring of theophylline, acetaminophen, or ethosuximide. *Clin.Chem.*, **1981**, *27*, 129–132

Sommadossi, J.P.; Aubert, C.; Cano, J.P.; Durand, A.; Viala, A. Determination of theophylline in plasma by high performance liquid chromatography. *J.Liq.Chromatogr.*, **1981**, *4*, 97–107 [extracted caffeine, theobromine; normal phase]

Tan, H.S.; Booncong, P.C.; Fine, S.L. Simultaneous high-performance liquid chromatographic determination of theophylline, ephedrine hydrochloride, and phenobarbital in tablets. *J.Pharm.Sci.*, **1981**, *70*, 783–785

Van Aerde, P.; Moerman, E.; Van Severen, R.; Braeckman, P. Determination of plasma theophylline by straight-phase high-performance liquid chromatography: elimination of interfering caffeine metabolites. *J.Chromatogr.*, **1981**, *222*, 467–474

Weidner, N.; Dietzler, D.N.; Ladenson, J.H.; Kessler, G.; Larson, L.; Smith, C.H.; James, T.; McDonald, J.M. A clinically applicable high-pressure liquid chromatographic method for measurement of serum theophylline, with detailed evaluation of interferences. *Am.J.Clin.Pathol.*, **1980**, *73*, 79–86

# Timolol

**Molecular formula:** $C_{13}H_{24}N_4O_3S$
**Molecular weight:** 316.4
**CAS Registry No.:** 26839-75-8, 91524-16-2 (hemihydrate),
26921-17-5 (maleate)

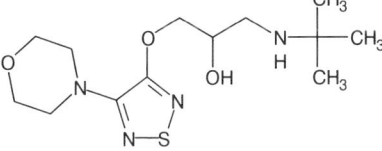

## SAMPLE
**Matrix:** aqueous humor
**Sample preparation:** 150 μL Aqueous humor + 50 μL 5 μg/mL aniline in water + 100 μL 400 mM NaOH, swirl-mix, add 3 mL isopropanol, swirl-mix for 2 min, centrifuge at 300 g for 5 min. Remove the organic layer and add it to 50 μL 100 mM HCl, swirl-mix for 2 min, centrifuge, inject a 20 μL aliquot of the acid layer.

## HPLC VARIABLES
**Column:** 150 × 4.5 5 μm Ultrasphere ODS
**Mobile phase:** MeCN:triethylamine:water:glacial acetic acid 395:5:1585:15
**Flow rate:** 1
**Injection volume:** 20
**Detector:** UV 280

## CHROMATOGRAM
**Retention time:** 4.5
**Internal standard:** aniline (3.5)
**Limit of quantitation:** 50 ng/mL

## OTHER SUBSTANCES
**Simultaneous:** alprenolol, betaxolol, caffeine, haloperidol, levobunolol, metoprolol, oxprenolol, pindolol, propranolol
**Noninterfering:** acetazolamide, atenolol, carbachol, dexamethasone, dipivefrin, epinephrine, gramicidin D, methazolamide, neomycin, pilocarpine, polymyxin B, sulfacetamide, theophylline
**Interfering:** acebutolol

## KEY WORDS
rabbit

## REFERENCE
Wu, P.-Y.; Riegel, M.; Ellis, P.P. High-performance liquid chromatographic assay for timolol in the aqueous humor of the eye. *J.Chromatogr.*, **1989**, *494*, 368–375

## SAMPLE
**Matrix:** blood
**Sample preparation:** 200 μL Plasma + 50 μL 50 mM pH 7.4 phosphate buffer + 500 μL 2% zinc sulfate in MeOH:water 50:50, mix, centrifuge at 13000 rpm for 5 min, inject an aliquot.

## HPLC VARIABLES
**Guard column:** 40 × 4.6 SynChropak bulk support (Knauer)
**Column:** 120 × 4.6 5 μm Spherisorb ODS1 C18
**Mobile phase:** MeCN:MeOH:pH 4.5 acetate buffer (ratio not given)
**Flow rate:** 1
**Detector:** UV 294

## CHROMATOGRAM
**Retention time:** 4.66

## OTHER SUBSTANCES
**Extracted:** cyclopropane carboxylic acid ester prodrug

## KEY WORDS
plasma

## REFERENCE
Hovgaard, L.; Brondsted, H.; Buur, A.; Bundgaard, H. Drug delivery studies in Caco-2 monolayers. Synthesis, hydrolysis, and transport of O-cyclopropane carboxylic acid ester prodrugs of various β-blocking agents. *Pharm.Res.*, **1995**, *12*, 387−392

## SAMPLE
**Matrix:** blood
**Sample preparation:** 2 mL Whole blood or plasma + 2 mL buffer + 5 mL chloroform: isopropanol:n-heptane 60:14:26, shake gently horizontally for 10 min, centrifuge at 2800 g for 10 min. Remove the lower organic layer and evaporate it to dryness under vacuum at 45°, reconstitute the residue in 100 μL mobile phase, centrifuge at 2800 g for 5 min, inject a 50 μL aliquot of the supernatant. (Buffer was saturated ammonium chloride solution 25% diluted with water, adjusted to pH 9.5 with 25% ammonia solution.)

## HPLC VARIABLES
**Column:** 300 × 3.9 4 μm NovaPack C18
**Mobile phase:** MeOH:THF:buffer 65:5:30 (Buffer was 0.68 g/L (10 mM (sic)) $KH_2PO_4$ adjusted to pH 2.6 with concentrated orthophosphoric acid.) (At the end of each session wash the column with water for 1 h and MeOH for 1 h, re-equilibrate for 30 min.)
**Column temperature:** 30
**Flow rate:** 0.8
**Injection volume:** 50
**Detector:** UV 299

## CHROMATOGRAM
**Retention time:** 4.76
**Limit of detection:** <120 ng/mL

## OTHER SUBSTANCES
**Extracted:** acebutolol, acepromazine, aceprometazine, acetaminophen, aconitine, ajmaline, albuterol, alimemazine, alpidem, alprazolam, alprenolol, amisulpride, amitriptyline, amoxapine, aspirin, astemizole, atenolol, benzocaine, benzoylecgonine, bepridil, betaxolol, bisoprolol, bromazepam, brompheniramine, bumadizone, bupivacaine, buprenorphine, buspirone, caffeine, carbamazepine, carbinoxamine, carpipramine, carteolol, celiprolol, cetirizine, chlorambucil, chlordiazepoxide, chlormezanone, chlorophenacinone, chlorpheniramine, chlorpromazine, chlorpropamide, cibenzoline, clemastine, clobazam, clomipramine, clonazepam, clonidine, clorazepate, clozapine, codeine, colchicine, cyamemazine, cyclizine, cycloguanil, cyproheptadine, cytarabine, dacarbazine, daunorubicin, debrisoquine, demexiptiline, desipramine, dextromethorphan, dextromoramide, dextropropoxyphene, diazepam, diazoxide, diclofenac, dihydralazine, diltiazem, diphenhydramine, disopyramide, dosulepine, doxepin, ephedrine, estazolam, etodolac, fenfluramine, fenoprofen, fentiazac, flecainide, floctafenine, flumazenil, flunitrazepam, fluoxetine, fluphenazine, flurbiprofen, fluvoxamine, glibenclamide, glibornuride, glipizide, glutethimide, haloperidol, histapyrrodine, hydroxyzine, ibuprofen, imipramine, indomethacin, iproniazid, ketamine, levomepromazine, lidocaine, lidoflazine, lisinopril, loperamide, loprazolam, loratadine, lorazepam, loxapine, maprotiline, medazepam, medifoxamine, mefenamic acid, mefenidramine, mefloquine, melphalan, mephenesin, mephentermine, metapra-

mine, metformin, methadone, methaqualone, methocarbamol, methotrexate, metipranolol, metoclopramide, metoprolol, mianserine, midazolam, minoxidil, moperone, morphine,
nadolol, nalbuphine, nalorphine, naloxone, naltrexone, naproxen, nialamide, nicardipine,
nifedipine, niflumic acid, nimodipine, nitrazepam, nitrendipine, nizatidine, nortriptyline,
omeprazole, opipramol, oxazepam, oxprenolol, penbutolol, penfluridol, pentazocine, phencyclidine, phenobarbital, phenol, phenylbutazone, pimozide, pindolol, piroxicam, prazepam, prazosin, prilocaine, procainamide, procarbazine, proguanil, promethazine, propafenone, propranolol, protriptyline, quinidine, quinine, quinupramine, ramipril, ranitidine,
reserpine, ritodrine, sotalol, strychnine, sulfinpyrazole, sulindac, sulpride, sultopride, suriclone, tenoxicam, terbutaline, terfenadine, tetracaine, tetrazepam, thiopental, thioproperazine, thioridazine, tianeptine, tiapride, tiaprofenic acid, tioclomarol, tofisopam, tolbutamide, toloxatone, triazolam, trifluoperazine, trifluperidol, trimipramine, triprolidine,
tropatenine, verapamil, vinblastine, warfarin, yohimbine, zopiclone, zorubicine
**Interfering:** acenocoumarol, alminoprofen, amodiaquine, benazepril, benperidol, chloroquine, cicletanine, cocaine, dipyridamole, doxylamine, droperidol, hydroxychloroquine, ketoprofen, labetalol, meperidine, mepivacaine, mexiletine, moclobemide, nomifensine, pipamperone, pyrimethamine, secobarbital, temazepam, ticlopidine, trazodone, viloxazine,
vincristine, vindesine, zolpidem

## KEY WORDS
whole blood; plasma

## REFERENCE
Tracqui, A.; Kintz, P.; Mangin, P. Systematic toxicological analysis using HPLC/DAD. *J.Forensic Sci.*,
**1995**, *40*, 254–262

## SAMPLE
**Matrix:** blood
**Sample preparation:** 1 mL Serum + 100 μL MeOH + 500 μL 500 mM NaOH, vortex for
10 s, add 5 mL dichloromethane:ether 2:3, shake gently for 5 min, centrifuge. Remove
the organic layer and evaporate it to dryness at 37° under a stream of nitrogen. Dissolve
the residue in 100 μL 100 mM HCl, add 500 μL ether, vortex briefly, centrifuge at 3000
rpm for 5 min, inject a 10-25 μL aliquot of the aqueous phase.

## HPLC VARIABLES
**Column:** 90 × 4.6 5 μm Hypersil ODS
**Mobile phase:** MeCN:buffer 35:65 (Buffer contained 30 mM sodium lauryl sulfate and 50
mM $Na_2HPO_4$ adjusted to pH 2 with orthophosphoric acid.)
**Flow rate:** 2
**Injection volume:** 10-25
**Detector:** UV 229

## CHROMATOGRAM
**Retention time:** 7
**Internal standard:** timolol maleate

## OTHER SUBSTANCES
**Extracted:** trimethoprim

## KEY WORDS
serum; timolol is IS

## REFERENCE
Hung, C.T.; Perrier, D.G. Determination of trimethoprim and sulfamethoxazole in serum by reversed-
phase and ion pair HPLC. *J.Liq.Chromatogr,* **1985**, *8*, 521–536

## SAMPLE

**Matrix:** blood, urine

**Sample preparation:** Plasma. 1-2 mL Plasma + 100 μL 4 M NaOH + 100 μL 100 μg/mL metoprolol tartrate in water + 5 mL dichloromethane, shake by hand for 10 s, vortex vigorously for 3 min, centrifuge at 1500 g for 5 min. Remove the organic layer and evaporate it to dryness under a stream of air at 37°, reconstitute the residue in 100 μL mobile phase, filter (0.45 μm), inject a 40-60 μL aliquot. Urine. 0.1-1 mL Urine + 100 μL 4 M NaOH + 100 μL 1 mg/mL metoprolol tartrate in water + 5 mL dichloromethane, vortex vigorously for 1 min, centrifuge at 1500 g for 5 min. Remove the organic layer and add it to 1 mL 100 mM phosphoric acid, vortex for 1 min, centrifuge at 1500 g for 5 min, inject a 40 μL aliquot of the aqueous phase.

## HPLC VARIABLES

**Column:** 250 × 4 Wakosil 5C18 (Wako)

**Mobile phase:** MeCN:water:triethylamine 18:81:1, pH adjusted to 3.0 with phosphoric acid

**Flow rate:** 1

**Injection volume:** 40-60

**Detector:** UV 295

## CHROMATOGRAM

**Retention time:** 7.36

**Internal standard:** metoprolol tartrate (8.69)

**Limit of detection:** 0.5 ng/mL

## OTHER SUBSTANCES

**Simultaneous:** N-acetylprocainamide, alprenolol, atenolol, bufetolol, bupranolol, carteolol, disopyramide, indenolol, lidocaine, nicainoprol, nifedipine, pindolol, procainamide, propranolol, quinidine

**Noninterfering:** diltiazem, glycinexylidide, mexiletine, nicardipine, tocainide, verapamil

**Interfering:** acebutolol

## KEY WORDS

plasma

## REFERENCE

Kubota, K.; Nakamura, H.; Koyama, E.; Yamada, T.; Kikuchi, K.; Ishizaki, T. Simple and sensitive determination of timolol in human plasma and urine by high-performance liquid chromatography with ultraviolet detection. *J.Chromatogr.*, **1990**, *533*, 255-263

## SAMPLE

**Matrix:** blood, urine

**Sample preparation:** Plasma. 1 mL Plasma + 100 μL 4 M NaOH + 0.25 μg phenacetin + 5 mL MTBE, shake gently for 10 min, centrifuge at 900 g for 5 min. Remove the organic layer and evaporate it to dryness at 40° under reduced pressure, reconstitute the residue in 100 μL mobile phase, inject a 50-100 μL aliquot. Urine. 1 mL Urine diluted 10-fold with water + 100 μL 4 M NaOH + 1 μg phenacetin + 5 mL MTBE, shake gently for 10 min, centrifuge at 900 g for 5 min. Remove the organic layer and evaporate it to dryness at 40° under reduced pressure, reconstitute the residue in 100 μL mobile phase, inject a 50-100 μL aliquot.

## HPLC VARIABLES

**Guard column:** 50 × 5 40 μm Waters reverse-phase guard column material

**Column:** 100 × 5 5 μm Hypersil 5-ODS

**Mobile phase:** MeCN:water 13:87 containing 1% triethylamine, adjusted to pH 3 with orthophosphoric acid

**Flow rate:** 2
**Injection volume:** 50-100
**Detector:** UV 295

## CHROMATOGRAM
**Retention time:** 3.7
**Internal standard:** phenacetin (8.6)
**Limit of quantitation:** 2 ng/mL

## OTHER SUBSTANCES
**Noninterfering:** acebutolol, acetaminophen, atenolol, canrenone, disopyramide, hydralazine, labetalol, lignocaine, nadolol, nifedipine, norverapamil, propranolol, sotalol, verapamil, warfarin

## KEY WORDS
plasma

## REFERENCE
Lennard, M.S.; Parkin, S. Timolol determination in plasma and urine by high-performance liquid chromatography with ultraviolet detection. *J.Chromatogr.*, **1985**, *338*, 249–252

## SAMPLE
**Matrix:** cell lysate
**Sample preparation:** Sonicate lysed cells for 5 min, filter (0.2 μm PTFE), inject an aliquot.

## HPLC VARIABLES
**Guard column:** 4 × 4 5 μm LiChrospher 100 RP-18
**Column:** 250 × 4 5 μm LiChrospher 100 RP-18
**Mobile phase:** MeCN:buffer 18:82 (Buffer was 0.5% triethylamine adjusted to pH 2.5 with
78% phosphoric acid.)
**Flow rate:** 1
**Injection volume:** 100
**Detector:** UV 293

## CHROMATOGRAM
**Retention time:** 7.70
**Internal standard:** timolol

## OTHER SUBSTANCES
**Extracted:** ofloxacin

## KEY WORDS
timolol is IS

## REFERENCE
Fresta, M.; Spadaro, A.; Cerniglia, G.; Ropero, I.M.; Puglisi, G.; Furneri, P.M. Intracellular accumulation
of ofloxacin-loaded liposomes in human synovial fibroblasts. *Antimicrob.Agents Chemother.*, **1995**, *39*,
1372–1375

## SAMPLE
**Matrix:** formulations
**Sample preparation:** Prepare a solution in mobile phase, inject an aliquot.

## HPLC VARIABLES
**Column:** 250 × 4.6 10 μm LiChrosorb C2

**Mobile phase:** MeCN:buffer 35:65 (1 mL 100 mM HCl + 1200 mL water + 5.84 g NaCl, mix to dissolve, add 700 mL MeOH, make up to 2 L, apparent pH 4.5.)
**Flow rate:** 1.2
**Injection volume:** 20
**Detector:** UV 254

## CHROMATOGRAM
**Retention time:** 8.2

## OTHER SUBSTANCES
**Simultaneous:** alprenolol, atenolol, metoprolol, nadolol, oxprenolol, pindolol, practolol, propranolol
**Interfering:** acebutolol, sotalol

## KEY WORDS
tablets

## REFERENCE
Patel, B.R.; Kirschbaum, J.J.; Poet, R.B. High-pressure liquid chromatography of nadolol and other β-adrenergic blocking drugs. *J.Pharm.Sci.*, **1981**, *70*, 336–338

## SAMPLE
**Matrix:** perfusate
**Sample preparation:** 50 μL Perfusate + 50 μL pH 7.4 phosphate-buffered saline or 100 mM HCl + 100 μL 500 μg/mL methyl p-hydroxybenzoate in MeOH, centrifuge at 12000 g for 10 min, inject a 50 μL aliquot of the supernatant.

## HPLC VARIABLES
**Column:** 150 × 4.6 Cosmosil 5C18-P (Nacalai Tesque)
**Mobile phase:** MeOH:50 mM $NaH_2PO_4$ 40:60
**Flow rate:** 1
**Injection volume:** 50
**Detector:** UV 290

## CHROMATOGRAM
**Internal standard:** methyl p-hydroxybenzoate

## KEY WORDS
rabbit

## REFERENCE
Sasaki, H.; Igarashi, Y.; Nagano, T.; Nishida, K.; Nakamura, J. Different effects of absorption promoters on corneal and conjunctival penetration of ophthalmic beta-blockers. *Pharm.Res.*, **1995**, *12*, 1146–1150

## SAMPLE
**Matrix:** solutions
**Sample preparation:** Inject a 20 μL aliquot of a 1 mg/mL solution.

## HPLC VARIABLES
**Column:** 250 × 4.6 10 μm Chiralcel OD
**Mobile phase:** Hexane:isopropanol:diethylamine 90:10:0.1
**Flow rate:** 0.5
**Injection volume:** 20
**Detector:** UV 300

## CHROMATOGRAM
**Retention time:** k' 0.85, 1.07 (enantiomers)

## KEY WORDS
chiral

## REFERENCE
Ekelund, J.; van Arkens, A.; Bronnum-Hansen, K.; Fich, K.; Olsen, L.; Petersen, P.V. Chiral separations of β-blocking drug substances using chiral stationary phases. *J.Chromatogr.A*, **1995**, *708*, 253–261

## SAMPLE
**Matrix:** solutions

## HPLC VARIABLES
**Column:** 150 × 4.6 12 μm 1-myristoyl-2-[(13-carboxyl)-tridecoyl]-sn-3-glycerophosphocholine chemically bonded to silica (Regis)
**Mobile phase:** MeCN:100 mM pH 7.0 phosphate buffer 20:80
**Flow rate:** 1
**Detector:** UV 254

## CHROMATOGRAM
**Retention time:** k' 2.43

## OTHER SUBSTANCES
**Also analyzed:** acebutolol, alprenolol, antazoline, atenolol, betaxolol, bisoprolol, bopindolol, bupranolol, carteolol, celiprolol, chloropyramine, chlorpheniramine, cicloprolol, cimetidine, cinnarizine, cirazoline, clonidine, dilevalol, dimethindene, diphenhydramine, doxazosin, esmolol, famotidine, isothipendyl, ketotifen, metiamide, metoprolol, moxonidine, nadolol, naphazoline, nifenalol, nizatidine, oxprenolol, pheniramine, phentolamine, pindolol, pizotyline (pizotifen), practolol, prazosin, promethazine, propranolol, pyrilamine (mepyramine), ranitidine, roxatidine, sotalol, tiamenidine, tramazoline, tripelennamine, triprolidine, tymazoline, UK-14,304

## REFERENCE
Kaliszan, R.; Nasal, A.; Turowski, M. Binding site for basic drugs on $\alpha_1$-acid glycoprotein as revealed by chemometric analysis of biochromatographic data. *Biomed.Chromatogr.*, **1995**, *9*, 211–215

## SAMPLE
**Matrix:** solutions

## HPLC VARIABLES
**Column:** 250 × 4.6 5 μm Supelcosil LC-DP (A) or 250 × 4 5 μm LiChrospher 100 RP-8 (B)
**Mobile phase:** MeCN:0.025% phosphoric acid:buffer 25:10:5 (A) or 60:25:15 (B) (Buffer was 9 mL concentrated phosphoric acid and 10 mL triethylamine in 900 mL water, adjust pH to 3.4 with dilute phosphoric acid, make up to 1 L.)
**Flow rate:** 0.6
**Injection volume:** 25
**Detector:** UV 229

## CHROMATOGRAM
**Retention time:** 7.04 (A), 3.94 (B)

## OTHER SUBSTANCES
**Also analyzed:** acebutolol, acepromazine, acetaminophen, acetazolamide, acetophenazine, albuterol, alprazolam, amitriptyline, amobarbital, amoxapine, antipyrine, atenolol, atro-

pine, azatadine, baclofen, benzocaine, bromocriptine, brompheniramine, brotizolam, bupivacaine, buspirone, butabarbital, butalbital, caffeine, carbamazepine, cetirizine, chlorcyclizine, chlordiazepoxide, chlormezanone, chloroquine, chlorpheniramine, chlorpromazine, chlorpropamide, chlorprothixene, chlorthalidone, chlorzoxazone, cimetidine, cisapride, clomipramine, clonazepam, clonidine, clozapine, cocaine, codeine, colchicine, cyclizine, cyclobenzaprine, dantrolene, desipramine, diazepam, diclofenac, diflunisal, diltiazem, diphenhydramine, diphenidol, diphenoxylate, dipyridamole, disopyramide, dobutamine, doxapram, doxepin, droperidol, encainide, ethidium bromide, ethopropazine, fenoprofen, fentanyl, flavoxate, fluoxetine, fluphenazine, flurazepam, flurbiprofen, fluvoxamine, furosemide, glutethimide, glyburide, guaifenesin, haloperidol, homatropine, hydralazine, hydrochlorothiazide, hydrocodone, hydromorphone, hydroxychloroquine, hydroxyzine, ibuprofen, imipramine, indomethacin, ketoconazole, ketoprofen, ketorolac, labetalol, levorphanol, lidocaine, loratadine, lorazepam, lovastatin, loxapine, mazindol, mefenamic acid, meperidine, mephenytoin, mepivacaine, mesoridazine, metaproterenol, metformin, methadone, methdilazine, methocarbamol, methotrexate, methotrimeprazine, methoxamine, methyldopa, methylphenidate, metoclopramide, metolazone, metoprolol, metronidazole, midazolam, moclobemide, morphine, nadolol, nalbuphine, naloxone, naphazoline, naproxen, nifedipine, nizatidine, norepinephrine, nortriptyline, oxazepam, oxycodone, oxymetazoline, paroxetine, pemoline, pentazocine, pentobarbital, pentoxifylline, perphenazine, pheniramine, phenobarbital, phenol, phenolphthalein, phentolamine, phenylbutazone, phenyltoloxamine, phenytoin, pimozide, pindolol, piroxicam, pramoxine, prazepam, prazosin, probenecid, procainamide, procaine, prochlorperazine, procyclidine, promazine, promethazine, propafenone, propantheline, propiomazine, propofol, propranolol, protriptyline, quazepam, quinidine, quinine, racemethorphan, ranitidine, remoxipride, risperidone, salicylic acid, scopolamine, secobarbital, sertraline, sotalol, spironolactone, sulfinpyrazone, sulindac, temazepam, terbutaline, terfenadine, tetracaine, theophylline, thiethylperazine, thiopental, thioridazine, thiothixene, tocainide, tolbutamide, tolmetin, trazodone, triamterene, triazolam, trifluoperazine, triflupromazine, trimeprazine, trimethoprim, trimipramine, verapamil, warfarin, xylometazoline, yohimbine, zopiclone

## KEY WORDS
some details of plasma extraction

## REFERENCE
Koves, E.M. Use of high-performance liquid chromatography-diode array detection in forensic toxicology. *J.Chromatogr.A*, **1995**, *692*, 103–119

## SAMPLE
**Matrix:** solutions

## HPLC VARIABLES
**Column:** 300 × 3.9 5 μm Nova-Pak C18
**Mobile phase:** MeOH:buffer 30:70 (Buffer was pH 4.0 phosphate buffer (ionic strength = 0.1) containing 2.86 mM N, N-dimethyloctylamine, pH readjusted to 4.00 with 85% phosphoric acid.)
**Column temperature:** 30
**Flow rate:** 1
**Injection volume:** 100
**Detector:** UV 220

## CHROMATOGRAM
**Retention time:** k' 2.77

## OTHER SUBSTANCES
**Also analyzed:** acebutolol, bunitrolol, carazolol, celiprolol, esmolol, mepindolol, metoprolol

## REFERENCE

Hamoir, T.; Verlinden, Y.; Massart, D.L. Reversed-phase liquid chromatography of β-adrenergic blocking drugs in the presence of a tailing suppressor. *J.Chromatogr.Sci.*, **1994**, *32*, 14–20

## SAMPLE

**Matrix:** solutions
**Sample preparation:** 50 μL Solution + 50 μL pH 7.4 PBS + 100 μL 500 μg/mL methyl p-hydroxybenzoate in MeOH, centrifuge at 12000 g for 10 min, inject a 50 μL aliquot.

## HPLC VARIABLES

**Column:** 150 × 4.6 Cosmosil 5C18-P (Nacalai Tesque)
**Mobile phase:** MeOH:50 mM $NaH_2PO_4$ 40:60
**Flow rate:** 1
**Injection volume:** 50
**Detector:** UV 290

## CHROMATOGRAM

**Internal standard:** methyl p-hydroxybenzoate

## KEY WORDS

buffers; Earle's balanced salt solution

## REFERENCE

Sasaki, H.; Igarishi, Y.; Nishida, K.; Nakamura, J. Intestinal permeability of ophthalmic β-blockers for predicting ocular permeability. *J.Pharm.Sci.*, **1994**, *83*, 1335–1338

## SAMPLE

**Matrix:** solutions

## HPLC VARIABLES

**Column:** 250 × 4.6 10 μm Chiralcel OD (Daicel)
**Mobile phase:** Hexane:isopropanol:diethylamine 95:5:0.4
**Column temperature:** 5
**Flow rate:** 0.7
**Detector:** UV 224

## CHROMATOGRAM

**Retention time:** 16.26 ((R)-(+)), 25.40 ((S)-(−))

## KEY WORDS

chiral

## REFERENCE

Aboul-Enein, H.Y.; Islam, M.R. Direct separation and optimization of timolol enantiomers on a cellulose tris-3,5-dimethylphenylcarbamate high-performance liquid chromatographic chiral stationary phase. *J.Chromatogr.*, **1990**, *511*, 109–114

## SAMPLE

**Matrix:** solutions
**Sample preparation:** Prepare a 10 μg/mL solution in MeOH, inject a 20 μL aliquot.

## HPLC VARIABLES

**Column:** 125 × 4.9 Spherisorb S5W silica

**Mobile phase:** MeOH containing 10 mM ammonium perchlorate and 1 mL/L 100 mM NaOH in MeOH, pH 6.7
**Flow rate:** 2
**Injection volume:** 20
**Detector:** E, LeCarbone, V25 glassy carbon electrode, + 1.2 V

## CHROMATOGRAM
**Retention time:** 1.94

## OTHER SUBSTANCES
**Also analyzed:** acebutolol, acepromazine, acetophenazine, N-acetylprocainamide, albuterol, alprenolol, amethocaine, amiodarone, amitriptyline, antazoline, atenolol, azacyclonal, bamethane, benactyzine, benperidol, benzethidine, benzocaine, benzoctamine, benzphetamine, benzquinamide, bromhexine, bromodiphenhydramine, bromperidol, brompheniramine, brompromazine, buclizine, bufotenine, bupivacaine, buprenorphine, butacaine, butethamate, chlorcyclizine, chlorpheniramine, chlorphenoxamine, chlorprenaline, chlorpromazine, chlorprothixene, cimetidine, cinchonidine, cinnarizine, clemastine, clomipramine, clonidine, cocaine, cyclazocine, cyclizine, cyclopentamine, cyproheptadine, deserpidine, desipramine, dextromoramide, dextropropoxyphene, dicyclomine, diethylcarbamazine, diethylpropion, diethylthiambutene, dihydroergotamine, dimethindene, dimethothiazine, diphenhydramine, diphenoxylate, dipipanone, diprenorphine, dipyridamole, disopyramide, dothiepin, doxapram, doxepin, doxylamine, droperidol, ephedrine, ergocornine, ergocristine, ergocristinine, ergocryptine, ergometrine, ergosine, ergosinine, ergotamine, ethopropazine, etorphine, etoxeridine, fenethazine, fenfluramine, fenoterol, fentanyl, flavoxate, fluopromazine, flupenthixol, fluphenazine, flurazepam, haloperidol, hydroxyzine, hyoscine, ibogaine, imipramine, indapamine, iprindole, isothipendyl, isoxsuprine, ketanserin, laudanosine, lidocaine, lofepramine, loxapine, maprotiline, mecamylamine, meclophenoxate, meclozine, medazepam, mephentermine, mepivacaine, meptazinol, mepyramine, mesoridazine, metaraminol, methadone, methamphetamine, methapyrilene, methdilazine, methotrimeprazine, methoxamine, methoxyphenamine, methoxypromazine, methylephedrine, methylergonovine, methysergide, metoclopramide, metopimazine, metoprolol, mianserin, morazone, nadolol, nalorphine, naloxone, naphazoline, nicotine, nifedipine, nomifensine, nortriptyline, noscapine, orphenadrine, oxeladin, oxprenolol, oxymetazolin, papaverine, pargyline, pecazine, penbutolol, pentazocine, penthienate, pericyazine, perphenazine, phenadoxone, phenampromide, phenazocine, phenbutrazate, phendimetrazine, phenelzine, phenglutarimide, phenindamine, pheniramine, phenmetrazine, phenomorphan, phenoperidine, phenothiazine, phenoxybenzamine, phentolamine, phenylephrine, phenyltoloxamine, physostigmine, piminodine, pimozide, pindolol, pipamazine, pipazethate, piperacetazine, piperidolate, pipradol, pirenzepine, piritramide, pizotifen, practolol, pramoxine, prazosin, prenylamine, prilocaine, primaquine, proadifen, procainamide, procaine, prochlorperazine, procyclidine, proheptazine, prolintane, promazine, promethazine, pronethalol, properidine, propiomazine, propranolol, prothipendyl, protriptyline, proxymetacaine, pseudoephedrine, pyrimethamine, quinidine, quinine, ranitidine, rescinnamine, sotalol, tacrine, terazosin, terbutaline, terfenadine, thenyldiamine, theophylline, thiethylperazine, thiopropazate, thioproperazine, thioridazine, thiothixene, thonzylamine, tocainide, tolpropamine, tolycaine, tranylcypromine, trazodone, trifluoperazine, trifluperidol, trimeperidine, trimeprazine, trimethobenzamide, trimethoprim, trimipramine, tripelennamine, triprolidine, tryptamine, verapamil, xylometazoline

## REFERENCE
Jane, I.; McKinnon, A.; Flanagan, R.J. High-performance liquid chromatographic analysis of basic drugs on silica columns using non-aqueous ionic eluents. II. Application of UV, fluorescence and electrochemical oxidation detection. *J.Chromatogr.*, **1985**, *323*, 191–225

## SAMPLE
**Matrix:** urine

**Sample preparation:** 3 mL Urine + 500 μL 5 M NaOH + 1 g anhydrous sodium sulfate + 2 mL diethyl ether, shake mechanically for 15 min, centrifuge at 734 g for 5 min. Remove the organic layer and evaporate it to dryness under a stream of nitrogen at 60°, reconstitute the residue in 10 mL mobile phase, inject a 20 μL aliquot.

## HPLC VARIABLES
**Guard column:** μBondapak C18
**Column:** 300 × 3.9 10 μm μBondapak C18
**Mobile phase:** MeCN:water 40:60 containing 5 mM $KH_2PO_4/K_2HPO_4$ adjusted to pH 6.5 with phosphoric acid or 1 M KOH
**Column temperature:** 30 ± 0.2
**Flow rate:** 1.3
**Injection volume:** 20
**Detector:** E, EG&G Princeton Applied Research PAR Model 400, glassy carbon cell +1300 mV, d.c. mode, Ag/AgCl reference electrode (At the end of each day clean electrode with MeOH as mobile phase and potential -600 mV for 1 min and +1500 mV for 10 min, repeat 3 times. If necessary, wipe with a tissue wetted with water then a tissue wetted with MeOH.)

## CHROMATOGRAM
**Retention time:** 4.88
**Limit of quantitation:** 15 ppb

## OTHER SUBSTANCES
**Extracted:** alprenolol, metoprolol, nadolol, oxprenolol
**Simultaneous:** atenolol

## REFERENCE
Maguregui, M.I.; Alonso, R.M.; Jiménez, R.M. High-performance liquid chromatography with amperometric detection applied to the screening of β-blockers in human urine. *J.Chromatogr.B*, **1995**, *674*, 85–91

## SAMPLE
**Matrix:** urine
**Sample preparation:** 1 mL Urine + 10 mg β-glucuronidase/arylsulfatase (Helix pomatia, Sigma), heat at 37° overnight, add an equal volume of buffer, centrifuge at 2000 g for 5 min, inject an aliquot of the supernatant onto column A with mobile phase A and elute to waste. After 2.5 min backflush the contents of column A onto column B with mobile phase B, monitor the effluent from column B. Re-equilibrate both columns for 12.5 min before the next injection. (Buffer was 200 mM boric acid adjusted to pH 9.5 with 5 M NaOH.)

## HPLC VARIABLES
**Column:** A 10 × 4.6 5 μm Spherisorb cyanopropyl; B 250 × 4.6 Capcell Pak C18 UG-120 (Shiseido)
**Mobile phase:** A water; B Gradient. MeCN:buffer from 3:97 to 30:70 over 30 min, to 40:60 over 8 min (Buffer was 3.4 mL/L phosphoric acid adjusted to pH 3.0 with 5 M NaOH.)
**Flow rate:** A 1.25; B 1
**Injection volume:** 100
**Detector:** UV 300

## CHROMATOGRAM
**Retention time:** 11.5
**Limit of detection:** 250 ng/mL

## OTHER SUBSTANCES
**Extracted:** alprenolol, amphetamine, atenolol, bopindolol, codeine, ephedrine, labetalol, metoprolol, morphine, nadolol, oxprenolol, pindolol, propranolol
**Interfering:** acebutolol

## KEY WORDS
column-switching

## REFERENCE
Saarinen, M.T.; Sirén, H.; Riekkola, M.-L. Screening and determination of β-blockers, narcotic analgesics and stimulants in urine by high-performance liquid chromatography with column switching. *J.Chromatogr.B*, **1995**, *664*, 341–346

## ANNOTATED BIBLIOGRAPHY
He, H.; Edeki, T.I.; Wood, A.J.J. Determination of low plasma timolol concentrations following topical application of timolol eye drops in humans by high-performance liquid chromatography with electrochemical detection. *J.Chromatogr.B*, **1994**, *661*, 351–356 [LOD 0.1072 ng/mL]

Hermansson, J.; Grahn, A. Resolution of racemic drugs on a new chiral column based on silica-immobilized cellobiohydrolase. Characterization of the basic properties of the column. *J.Chromatogr.*, **1994**, *687*, 45–59 [chiral; also acebutolol, atenolol, betaxolol, bisoprolol, carbuterol, cathinone, cimetidine, dobutamine, dopropizine, epanolol, epinephrine, laudanosine, metanephrine, metoprolol, moprolol, norepinephrine, normetanephrine, octopamine, oxybutynine, pamatolol, practolol, prilocaine, propafenone, proxyphylline, sotalol, talinolol, tetrahydropapaveroline, tetramisole, tolamolol, toliprolol]

Armstrong, D.W.; Chen, S.; Chang, C.; Chang, S. A new approach for the direct resolution of racemic beta adrenergic blocking agents by HPLC. *J.Liq.Chromatogr.*, **1992**, *15*, 545–556 [chiral; also alprenolol, atenolol, cateolol, labetolol, metoprolol, nadolol, oxprenolol, pindolol, propranolol]

Mazzo, D.J.; Snyder, P.A. High-performance liquid chromatography of timolol and potential degradates on dynamically modified silica. *J.Chromatogr.*, **1988**, *438*, 85–92 [simultaneous degradation products; column temp 35; transdermal patches; ophthalmic solutions; tablets]

Gregg, M.R.; Jack, D.B. Determination of timolol in plasma and breast milk using high-performance liquid chromatography with electrochemical detection. *J.Chromatogr.*, **1984**, *305*, 244–249

Mazzo, D.J. Simultaneous determination of maleic acid and timolol by high-performance liquid chromatography. *J.Chromatogr.*, **1984**, *299*, 503–507

Lefebvre, M.A.; Girault, J.; Fourtillan, J.B. β-Blocking agents: Determination of biological levels using high performance liquid chromatography. *J.Liq.Chromatogr.*, **1981**, *4*, 483–500 [fluorescence detection; plasma; also acebutolol, atenolol, metoprolol, oxprenolol, pindolol, propranolol, sotalol]

# Tramadol

**Molecular formula:** $C_{16}H_{25}NO_2$
**Molecular weight:** 263.4
**CAS Registry No.:** 27203-92-5 (tramadol), 22204-88-2 (tramadol hydrochloride)

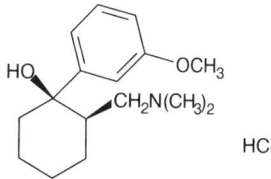

## SAMPLE
**Matrix:** solutions
**Sample preparation:** Dissolve in mobile phase.

## HPLC VARIABLES
**Guard column:** 15 × 3.2 7 μm Applied Biosystems pre-column
**Column:** 100 × 2 10 μm μPorasil
**Mobile phase:** MeCN:5 mM pH 3.75 sodium acetate 80:20
**Flow rate:** 1
**Injection volume:** 200
**Detector:** UV 214

## CHROMATOGRAM
**Retention time:** 11.0
**Limit of detection:** 6.3 ng/mL

## OTHER SUBSTANCES
**Simultaneous:** buprenorphine, butorphanol, codeine, ethylmorphine, fentanyl, morphine, nalbuphine
**Noninterfering:** atropine, diazepam, neostigmine, pancuronium, succinylcholine, thiopental
**Interfering:** meperidine

## REFERENCE
Ho, S.-T.; Wang, J.-J.; Ho, W.; Hu, O.Y.-P. Determination of buprenorphine by high-performance liquid chromatography with fluorescence detection: application to human and rabbit pharmacokinetic studies. *J.Chromatogr.*, **1991**, *570*, 339–350

## SAMPLE
**Matrix:** urine
**Sample preparation:** Adjust pH of 1 mL urine to >10 with 25% aqueous ammonia, add 4 mL n-hexane, shake for 15 min, centrifuge at 3000 g for 15 min. Remove the organic layer and evaporate it to dryness under a stream of nitrogen, reconstitute the residue in 100 μL mobile phase, inject a 50 μL aliquot.

## HPLC VARIABLES
**Guard column:** 25 × 4.6 10 μm Chiralpak AD amylose tris-3,5-dimethylphenyl carbamate
**Column:** 250 × 4.6 10 μm Chiralpak AD amylose tris-3,5-dimethylphenyl carbamate
**Mobile phase:** n-Hexane:isopropanol:diethylamine 97.5:2.5:0.01
**Flow rate:** 1
**Injection volume:** 50
**Detector:** UV 270

## CHROMATOGRAM
**Retention time:** 10 (+), 15 (−)

## OTHER SUBSTANCES
**Extracted:** metabolites

## KEY WORDS
chiral; pharmacokinetics

## REFERENCE
Elsing, B.; Blaschke, G. Achiral and chiral high-performance liquid chromatographic determination of tramadol and its major metabolites in urine after oral administration of racemic tramadol. *J.Chromatogr.*, **1993**, *612*, 223–230

## SAMPLE
**Matrix:** urine
**Sample preparation:** Adjust pH of 1 mL urine to 9 with 50-100 µL 50 mM pH 10.5 Tris buffer, add 50 µL 50 µg/mL IS in water, add 4 mL n-hexane:ethyl acetate 80:20, shake for 15 min, centrifuge at 3000 g for 15 min, repeat extraction. Combine the organic layers and evaporate them to dryness under a stream of nitrogen, reconstitute the residue in 100 µL mobile phase, inject a 50 µL aliquot.

## HPLC VARIABLES
**Guard column:** 25 × 4 5 µm RP Select B (Merck)
**Column:** 250 × 4 5 µm RP Select B (Merck)
**Mobile phase:** MeOH:50 mM pH 3.0 phosphate buffer 30:70
**Flow rate:** 1
**Injection volume:** 50
**Detector:** UV 270

## CHROMATOGRAM
**Retention time:** 12.8
**Internal standard:** 1-(m-ethoxyphenyl)-2-(dimethylaminomethyl)cyclohexan-1-ol (24.9)
**Limit of detection:** 80 ng/mL

## OTHER SUBSTANCES
**Extracted:** metabolites

## KEY WORDS
pharmacokinetics

## REFERENCE
Elsing, B.; Blaschke, G. Achiral and chiral high-performance liquid chromatographic determination of tramadol and its major metabolites in urine after oral administration of racemic tramadol. *J.Chromatogr.*, **1993**, *612*, 223–230

# Triamcinolone

**Molecular formula:** $C_{21}H_{27}FO_6$
**Molecular weight:** 394.4
**CAS Registry No.:** 124-94-7, 76-25-5 (acetonide), 1997-15-5
(acetonide disodium phosphate), 31002-79-6 (benetonide),
5611-51-8 (hexacetonide), 67-78-7 (diacetate),
4989-94-0 (furetonide), 989-96-8 (21-dihydrogen phosphate)

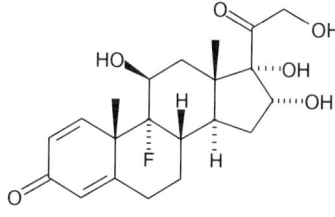

## SAMPLE
**Matrix:** blood
**Sample preparation:** Extract plasma with 12 mL dichloromethane, wash the organic layer
with 2 mL 100 mM NaOH and 1 mL water, dry the organic layer over 1 g anhydrous
sodium sulfate. Evaporate the organic layer to dryness under a stream of nitrogen, reconstitute with mobile phase, inject an aliquot.

## HPLC VARIABLES
**Column:** 150 × 4.6 3 μm Spherisorb silica
**Mobile phase:** Hexane:dichloromethane:EtOH:glacial acetic acid 26:69:3.4:2
**Flow rate:** 0.75
**Detector:** UV 254

## CHROMATOGRAM
**Retention time:** 6.0 (triamcinolone acetonide)
**Internal standard:** methylprednisolone (13.4)

## OTHER SUBSTANCES
**Extracted:** hydrocortisone
**Noninterfering:** cortisone

## KEY WORDS
plasma; normal phase

## REFERENCE
Rohatagi, S.; Hochhaus, G.; Möllmann, J.; Barth, J.; Galia, E.; Erdmann, M.; Sourgens, H.; Derendorf,
H. Pharmacokinetic and pharmacodynamic evaluation of triamcinolone acetonide after intravenous,
oral, and inhaled administration. *J.Clin.Pharmacol.*, **1995**, *35*, 1187–1193

## SAMPLE
**Matrix:** blood
**Sample preparation:** 1 mL Serum + 100 μL water containing 5 μg/mL 2,3-diaminona-
phthalene and 3.5 μg/mL 18-hydroxy-11-deoxycorticosterone + 1 mL 250 mM NaOH + 7
mL diethyl ether, shake on a rotary shaker for 15 min, repeat extraction. Combine the
organic layers and evaporate them to dryness under a stream of nitrogen at 30-40°, reconstitute the residue in 70 μL MeOH:100 mM perchloric acid 50:50, inject a 20 μL
aliquot.

## HPLC VARIABLES
**Column:** 150 × 3.9 4 μm Nova-Pak C18
**Mobile phase:** Gradient. A was 58 mM $NaH_2PO_4$ containing 6 mM sodium heptanesulfon-
ate, adjusted to pH 3.1 with concentrated phosphoric acid. B was MeCN:MeOH 85:15.
A:B from 100:0 to 78:22 over 5 min, to 70:30 over 12 min, maintain at 70:30 for 4 min,
to 65:35 over 9 min.

**Flow rate:** 1
**Injection volume:** 20
**Detector:** UV 245; UV 256; UV 343

## CHROMATOGRAM
**Retention time:** 9.59 (triamcinolone)
**Internal standard:** 2,3-diaminonaphthalene (10.71), 18-hydroxy-11-deoxycorticosterone (15.85)
**Limit of detection:** 1-10 ng/mL (245 nm)

## OTHER SUBSTANCES
**Extracted:** betamethasone, chloroquine, corticosterone, cortisone, dexamethasone, fluendrenolide, fluocinolone acetonide, fluorometholone, fluprednisolone, hydrocortisone, hydroxychloroquine, 17δ-hydroxyprogesterone, meprednisone, methylprednisolone, methylprednisolone acetate, paramethasone, prednisolone, prednisone, progesterone
**Noninterfering:** aspirin, ibuprofen, indomethacin, phenylbutazone, pregnenolone

## KEY WORDS
serum

## REFERENCE
Volin, P. Simple and specific reversed-phase liquid chromatographic method with diode-array detection for simultaneous determination of serum hydroxychloroquine, chloroquine and some corticosteroids. *J.Chromatogr.B*, **1995**, *666*, 347–353

## SAMPLE
**Matrix:** blood
**Sample preparation:** 100 µL Plasma + 10 µL 1 µg/mL fluocinolone acetonide in water + 1.2 mL dichloromethane, shake 1 min, repeat extraction. Combine organic layers and evaporate a 2 mL aliquot under reduced pressure below 40°. Dissolve the residue in 100 µL MeCN, add 10 µL reagent 1, add 10 µL reagent 2, heat at 70° for 20 min, cool to room temperature, add 100 µL water, add 200 µL MeOH:water 1:1, apply to a Sep-Pak C18 cartridge, wash vial into cartridge with 2 mL MeOH:water 1:1, wash cartridge with 40 mL MeOH:water 1:1, elute with 5 mL MeOH. Concentrate eluate to 500 µL by evaporation under reduced pressure below 40°, inject 20 µL aliquot. (Reagent 1 was prepared by dissolving 30 mg 2-(4-carboxyphenyl)-5,6-dimethylbenzimidazole in 3 mL pyridine, adding 700 mg 4-piperidinopyridine, and diluting to 10 mL with acetonitrile. Reagent 2 was 70 mg/mL 1-isopropyl-3-(3-dimethylaminopropyl)carbodiimide perchlorate in MeCN.)

## HPLC VARIABLES
**Guard column:** 50 × 4.6 7 µm Zorbax ODS
**Column:** 250 × 4.6 7 µm Zorbax ODS
**Mobile phase:** MeOH:water 75:25 containing 5 mM tetramethylammonium hydrogen sulfate
**Flow rate:** 0.4
**Injection volume:** 20
**Detector:** F ex 334 em 418

## CHROMATOGRAM
**Retention time:** 43.7 (triamcinolone), 49.4 (triamcinolone acetonide)
**Internal standard:** fluocinolone acetonide (40.7)
**Limit of detection:** 0.6 pg/mL; 3 ng/mL (acetonide)

## OTHER SUBSTANCES
**Extracted:** aldosterone, corticosterone, cortisone, dexamethasone, hydrocortisone

## KEY WORDS
plasma; derivatization

## REFERENCE
Katayama, M.; Masuda, Y.; Taniguchi, H. Determination of corticosteroids in plasma by high-performance liquid chromatography after pre-column derivatization with 2-(4-carboxyphenyl)-5,6-dimethylbenzimidazole. *J.Chromatogr.*, **1993**, *612*, 33–39

## SAMPLE
**Matrix:** blood
**Sample preparation:** Add 1 mL serum to a Sep Pak C18 SPE cartridge, wash with 4 mL water, elute with 4 mL MeOH, evaporate to dryness under vacuum, reconstitute in 50 µL MeCN:water 30:70, inject whole sample.

## HPLC VARIABLES
**Column:** 250 × 4.6 5 µm Ultrasphere ODS
**Mobile phase:** MeCN:water 30:70
**Flow rate:** 1
**Injection volume:** 50
**Detector:** enzyme immunoassay of fractions

## CHROMATOGRAM
**Retention time:** 6 (triamcinolone)
**Limit of detection:** 0.3 pg

## OTHER SUBSTANCES
**Extracted:** betamethasone, dexamethasone, flumethasone
**Noninterfering:** endogenous steroids

## KEY WORDS
serum; SPE; horse

## REFERENCE
Friedrich, A.; Schulz, R.; Meyer, H.H. Use of enzyme immunoassay and reverse-phase high-performance liquid chromatography to detect and confirm identity of dexamethasone in equine blood. *Am.J.Vet.Res.*, **1992**, *53*, 2213–2220

## SAMPLE
**Matrix:** blood
**Sample preparation:** 1 mL Plasma + 5 mL water + 1 mL 2 µg/mL equilenin in MeOH + 50 µL 0.1 M NaOH to adjust pH to 10, vortex briefly after each addition, shake with 10 mL dichloromethane for 10 min, centrifuge at 2000 g for 10 min. Wash organic layer twice with 2 mL water, centrifuge 5 min, evaporate 8 mL of organic phase to dryness at 40° under a stream of nitrogen, reconstitute residue in 150 µL mobile phase, inject a 25 µL aliquot.

## HPLC VARIABLES
**Column:** 300 × 4 10 µm µBondapak C18
**Mobile phase:** MeOH:buffer 60:40 (Buffer was 10 mL 200 mM acetic acid + 15 mL 200 mM sodium acetate made up to 1 L, pH 4.8.)
**Flow rate:** 1
**Injection volume:** 25
**Detector:** UV 254

## CHROMATOGRAM
**Retention time:** 3.5 (triamcinolone)

**Internal standard:** equilenin (7.5)
**Limit of detection:** 5 ng/mL

## OTHER SUBSTANCES
**Extracted:** betamethasone, deoxycortisol, dexamethasone, hydrocortisone, prednisolone, prednisone

## KEY WORDS
Anal. Abs. 1982, 43, 4D182; plasma

## REFERENCE
Bouquet, S.; Brisson, A.M.; Gombert, J. Dosage du cortisol et du 11-désoxycortisol plasmatiques par chromatographie liquide haute performance [Cortisol and 11-desoxycortisol determination in blood by high performance liquid chromatography]. *Ann.Biol.Clin.(Paris)*, **1981**, *39*, 189–191

## SAMPLE
**Matrix:** blood, tissue, urine
**Sample preparation:** Urine. 1 mL Urine + 1 mL MeOH:EtOH 50:50, centrifuge at 4000 g for 10 min. Remove the supernatant and evaporate to about 200 µL under a stream of nitrogen at 37°, inject a 5-20 µL aliquot. Plasma. Mix plasma with an equal volume of MeOH:EtOH 50:50, let stand at -20° for 30 min or overnight. Remove supernatant and wash precipitate twice with equal volumes of MeOH:EtOH 50:50. Combine the organic layers and evaporate them to dryness under a stream of nitrogen at 37°, reconstitute the residue in 200 µL MeOH:water 65:35, inject a 5-20 µL aliquot. Tissue. Homogenize (Polytron) fetal tissue in 10-15 mL MeOH:dimethoxymethane 50:50 for 1 min or until breakup was complete, shake at 37° overnight, centrifuge at 4000 g for 5 min. Filter (Whatman No. 1 filter paper) supernatant. Resuspend precipitate in MeOH:dimethoxymethane 50:50, filter, wash precipitate with MeOH. Combine filtrates, evaporate to dryness under nitrogen, resuspend residue in up to 500 µL MeOH:water 65:35, centrifuge, inject a 5-20 µL aliquot of the supernatant.

## HPLC VARIABLES
**Guard column:** 70 × 6 35-50 µm Bondapak C18 Corasil
**Column:** 250 × 10 5 µm LiChrosorb RP-18
**Mobile phase:** Gradient. MeOH:10 mM pH 6.9 ammonium acetate from 10:90 to 100:0 over 50 min (Waters No. 5 convex gradient).
**Flow rate:** 1.5
**Injection volume:** 5-20
**Detector:** UV 254

## CHROMATOGRAM
**Retention time:** 25.95 (triamcinolone), 33.99 (triamcinolone acetonide)

## OTHER SUBSTANCES
**Extracted:** metabolites, cortexolone, cortisol glucuronide, cortisone, hydrocortisone, 6β-hydroxycortisol

## KEY WORDS
plasma; monkey

## REFERENCE
Althaus, Z.R.; Rowland, J.M.; Freeman, J.P.; Slikker, W., Jr. Separation of some natural and synthetic corticosteroids in biological fluids and tissues by high-performance liquid chromatography. *J.Chromatogr.*, **1982**, *227*, 11–23

## SAMPLE
**Matrix:** blood, urine

**Sample preparation:** Urine. 5 mL Urine + 100 ng prednisone, adjust pH to 7 with 500 mM pH 7 ammonium acetate buffer, extract twice with 7 mL portions of ethyl acetate. Combine the organic layers and evaporate them to dryness under a stream of nitrogen at 40°, reconstitute the residue in 200 µL MeOH, inject a 20 µL aliquot. Serum. 2 mL Serum + 20 ng prednisone, extract twice with 4 mL portions of ethyl acetate. Combine the organic layers and evaporate them to dryness under a stream of nitrogen at 40°, reconstitute the residue in 200 µL MeOH, inject a 20 µL aliquot.

## HPLC VARIABLES
**Guard column:** 4 × 4 5 µm Lichrospher 100 RP-18
**Column:** 125 × 4 5 µm Lichrospher 100 RP-18
**Mobile phase:** Gradient. A was 10 mM ammonium acetate. B was MeOH. A:B from 50:50 to 0:100 over 7 min, maintain at 0:100 for 3 min.
**Flow rate:** 1
**Injection volume:** 20
**Detector:** MS, PE Sciex API III LC-MS-MS, heated nebulizer interface at 75 psi and 300°, auxiliary nitrogen 1.4 L/min, nitrogen curtain 1.2 L/min, negative mode ionization, discharge current -3 µA, collision gas argon with $570 \times 10^{12}$ molecules/sq.cm., dwell time 10 ms, step size 1 amu, m/z 413, SIM

## CHROMATOGRAM
**Retention time:** 8 (triamcinolone acetonide)
**Internal standard:** prednisone (m/z 327) (5.5)
**Limit of detection:** 0.1 ng/mL

## KEY WORDS
horse; serum; pharmacokinetics

## REFERENCE
Koupai-Abyazani, M.R.; Yu, N.; Esaw, B.; Laviolette, B. Determination of triamcinolone acetonide in equine serum and urine by liquid chromatography-atmospheric pressure ionization mass spectrometry. *J.Anal.Toxicol.*, **1995**, *19*, 182–186

## SAMPLE
**Matrix:** bulk
**Sample preparation:** Prepare a 2.5 mg/mL solution in MeOH, inject a 20 µL aliquot.

## HPLC VARIABLES
**Column:** 250 × 4.6 5 µm Lichrosorb Si-100
**Mobile phase:** Chloroform (stabilized with amylene):MeOH:water 978.6:20:1.4
**Flow rate:** 1
**Injection volume:** 20
**Detector:** UV 254

## CHROMATOGRAM
**Retention time:** 6.4 (triamcinolone acetonide)

## OTHER SUBSTANCES
**Simultaneous:** impurities

## KEY WORDS
normal phase

## REFERENCE
Cavina, G.; Alimenti, R.; Gallinella, B.; Valvo, L. The identification of related substances in triamcinolone acetonide by means of high-performance liquid chromatography with diode array detector and mass spectrometry. *J.Pharm.Biomed.Anal.*, **1992**, *10*, 685–692

## SAMPLE
**Matrix:** bulk
**Sample preparation:** Prepare a 2.5 mg/mL solution in MeOH, inject a 20 μL aliquot.

## HPLC VARIABLES
**Column:** 250 × 4.6 5 μm Spherisorb ODS
**Mobile phase:** MeCN:water 36:64
**Flow rate:** 1
**Injection volume:** 20
**Detector:** UV 240

## CHROMATOGRAM
**Retention time:** 6.4 (triamcinolone acetonide)

## OTHER SUBSTANCES
**Simultaneous:** impurities

## REFERENCE
Cavina, G.; Alimenti, R.; Gallinella, B.; Valvo, L. The identification of related substances in triamcinolone acetonide by means of high-performance liquid chromatography with diode array detector and
mass spectrometry. *J.Pharm.Biomed.Anal.*, **1992**, *10*, 685–692

## SAMPLE
**Matrix:** formulations
**Sample preparation:** Weigh out amount containing 10 mg triamcinolone acetonide, make
up to 50 mL with mobile phase. Remove a 2 mL aliquot and dilute it to 10 mL with mobile
phase, filter, inject a 20 μL aliquot of the filtrate.

## HPLC VARIABLES
**Column:** 125 × 4 5 μm LiChrospher 100 RP-18
**Mobile phase:** MeOH:water:96% acetic acid 55:44:1, pH 3.0
**Flow rate:** 1
**Injection volume:** 20
**Detector:** UV 254

## CHROMATOGRAM
**Retention time:** 5.02 (triamcinolone acetonide)

## OTHER SUBSTANCES
**Simultaneous:** salicylic acid

## KEY WORDS
topical solution

## REFERENCE
Kedor-Hackmann, E.R.M.; Gianotto, E.A.S.; Santoro, M.I.R.M. Determination of triamcinolone acetonide
and salicylic acid in pharmaceutical formulations by high performance liquid chromatography. *Pharmazie*, **1996**, *51*, 63

## SAMPLE
**Matrix:** formulations
**Sample preparation:** Dermatological patch (2 cm × 2 cm) + 2 mL hexane, shake mechanically for 10 min, add 8 mL mobile phase, mix thoroughly, centrifuge at 2500 rpm
for 10 min, remove 1 mL of lower phase, inject a 10 μL aliquot of this solution.

## HPLC VARIABLES
**Column:** 150 × 4.6 5 μm Spherisorb C8
**Mobile phase:** MeOH:water 70:30
**Flow rate:** 0.5
**Injection volume:** 10
**Detector:** UV 240

## KEY WORDS
Chem.Abs., 114, 129259; dermatological patches; stability-indicating; for triamcinolone acetonide

## REFERENCE
Edwardson, P.A.D.; Gardner, R.S. Problems associated with the extraction and analysis of triamcinolone acetonide in dermatological patches. *J.Pharm.Biomed.Anal.*, **1990**, *8*, 935–938

## SAMPLE
**Matrix:** formulations, solutions
**Sample preparation:** Ointment. 1 g Ointment + 5 mL MeOH + 5 mL water + 800 μL 1 mg/mL hydrocortisone in EtOH, stir until a clear solution forms, make up to 25 mL with water, inject a 20 μL aliquot. Solutions. 8 mL Solution + 800 μL 1 mg/mL hydrocortisone in EtOH + 5 mL MeOH, make up to 25 mL with water, inject a 20 μL aliquot.

## HPLC VARIABLES
**Column:** 300 × 4 μBondapak C18
**Mobile phase:** MeCN:200 mM $KH_2PO_4$ 32:68, pH 4.2
**Flow rate:** 3
**Injection volume:** 20
**Detector:** UV 254

## CHROMATOGRAM
**Retention time:** 8 (triamcinolone acetonide)
**Internal standard:** hydrocortisone (4)

## OTHER SUBSTANCES
**Simultaneous:** degradation products

## KEY WORDS
ointment; stability-indicating

## REFERENCE
Das Gupta, V. Stability of triamcinolone acetonide solutions as determined by high-performance liquid chromatography. *J.Pharm.Sci.*, **1983**, *72*, 1453–1456

## SAMPLE
**Matrix:** solutions

## HPLC VARIABLES
**Column:** 150 mm long 5 μm Microsorb-MV C18
**Mobile phase:** MeCN:water 60:40
**Flow rate:** 1
**Detector:** UV 240

## KEY WORDS
for triamcinolone acetonide

## REFERENCE

Phillips, C.A.; Michniak, B.B. Transdermal delivery of drugs with differing lipophilicities using azone analogs as dermal penetration enhancers. *J.Pharm.Sci.*, **1995**, *84*, 1427–1433

## SAMPLE
**Matrix:** solutions

## HPLC VARIABLES
**Column:** 250 × 4.6 Zorbax RX
**Mobile phase:** Gradient. A was 10 mL concentrated orthophosphoric acid and 7 mL triethylamine in 1 L water. B was 10 mL concentrated orthophosphoric acid and 7 mL triethylamine in 200 mL water, make up to 1 L with MeCN. A:B from 100:0 to 0:100 over 30 min, maintain at 0:100 for 5 min.
**Column temperature:** 30
**Flow rate:** 2
**Detector:** UV 210

## OTHER SUBSTANCES
**Also analyzed:** acepromazine, acetaminophen, acetophenazine, albuterol, aminophylline, amitriptyline, amobarbital, amoxapine, amphetamine, amylocaine, antipyrine, aprobarbital, aspirin, atenolol, atropine, avermectin, barbital, benzocaine, benzoic acid, benzotropine, benzphetamine, berberine, bibucaine, bromazepan, brompheniramine, buprenorphine, buspirone, butabarbital, butacaine, butethal, caffeine, carbamazepine, carbromal, chloramphenicol, chlordiazepoxide, chloroquine, chlorothiazide, chloroxylenol, chlorphenesin, chlorpheniramine, chlorpromazine, chlorpropamide, chlortetracycline, cimetidine, cinchonidine, cinchonine, clenbuterol, clonazepam, clonixin, clorazepate, cocaine, codeine, colchicine, cortisone, coumarin, cyclazocine, cyclobenzaprine, cyclothiazide, cyheptamide, cymarin, danazol, danthron, dapsone, debrisoquine, desipramine, dexamethasone, dextromethorphan, dextropropoxyphene, diamorphine, diazepam, diclofenac, diethylpropion, diethylstilbestrol, diflunisal, digitoxin, digoxin, diltiazem, diphenhydramine, diphenoxylate, diprenorphine, dipyrone, disulfiram, dopamine, doxapram, doxepin, dronabinol, ephedrine, epinephrine, epinine, estradiol, estriol, estrone, ethacrynic acid, ethosuximide, etonitazene, etorphine, eugenol, famotidine, fenbendazole, fencamfamine, fenoprofen, fenproporex, fentanyl, flubendazole, flufenamic acid, flunitrazepam, 5-fluorouracil, fluoxymesterone, fluphenazine, furosemide, gentisic acid, gitoxigenin, glipizide, glunixin, glutethimide, glybenclamide, guaiacol, halazepam, haloperidol, hydrochlorothiazide, hydrocodone, hydrocortisone, hydromorphone, hydroxyquinoline, ibogaine, ibuprofen, iminostilbene, imipramine, indomethacin, isocarbostyril, isocarboxazid, isoniazid, isoproterenol, isoxsuprine, ivermectin, ketamine, ketoprofen, kynurenic acid, levorphanol, lidocaine, lorazepam, lormetazepam, loxapine, mazindol, mebendazole, meclizine, meclofenamic acid, medazepam, mefenamic acid, megestrol, mepacrine, meperidine, mephentermine, mephenytoin, mephesin, mephobarbital, mepivacaine, mescaline, mesoridazine, methadone, methamphetamine, methapyrilene, methaqualone, methazolamide, methocarbamol, methoxamine, methsuximide, methyl salicylate, methyldopa, methyldopamine, methylphenidate, methylprednisolone, methyltestosterone, methyprylon, metoprolol, mibolerone, morphine, nadolol, nalorphine, naloxone, naltrexone, naphazoline, naproxen, nefopam, niacinamide, nicotine, nicotinic acid, nifedipine, niflumic acid, nitrazepam, norepinephrine, nortriptyline, noscapine, nylidrin, oxazepam, oxycodone, oxymorphone, oxyphenbutazone, oxytetracycline, papaverine, pargyline, pemoline, pentazocine, pentobarbital, persantine, phenacetin, phenazocine, phenazopyridine, phencyclidine, phendimetrazine, phenelzine, pheniramine, phenobarbital, phenothiazine, phensuximide, phentermine, phenylbutazone, phenylephrine, phenylpropanolamine, piperocaine, prazepam, prednisolone, primidone, probenecid, progesterone, propiomazine, propranolol, propylparaben, pseudoephedrine, puromycin, pyrilamine, pyrithyldione, quazepam, quinaldic acid, quinidine, quinine, ranitidine, recinnamine, reserpine, resorcinol, saccharin, albuterol, salicylamide, salicylic acid, scopolamine, scopoletin, secobarbital, strychnine, sulfacetamide, sufadiazine, sulfadimethoxine, sulfaethidole, sulfamerazine,

sulfamethazine, sulfamethoxizole, sulfanilamide, sulfapyridine, sulfasoxizole, sulindac, tamoxifen, temazepam, testosterone, tetracaine, tetracycline, tetramisole, thebaine, theobromine, theophylline, thiabendazole, thiamine, thiamylal, thiobarbituric acid, thioridazine, thiosalicylic acid, thiothixene, thymol, tolazamide, tolazoline, tobutamide, tolmetin, tranylcypromine, tribenzylamine, trichloromethiazide, trifluoperazine, trihexyphenidyl, trimethoprim, tripelennamine, triprolidine, tropacocaine, tyramine, verapamil, vincamine, warfarin, yohimbine, zoxazolamine

## KEY WORDS
for triamcinolone

## REFERENCE
Hill, D.W.; Kind, A.J. Reversed-phase solvent-gradient HPLC retention indexes of drugs. *J.Anal.Toxicol.*, **1994**, *18*, 233–242

## SAMPLE
**Matrix:** solutions
**Sample preparation:** Inject a 10-60 μL aliquot.

## HPLC VARIABLES
**Column:** 150 × 2 5 μm Spherisorb ODS1
**Mobile phase:** MeCN:water 50:50
**Flow rate:** 0.16-0.30
**Injection volume:** 10-60
**Detector:** UV 240

## CHROMATOGRAM
**Retention time:** 17 (triamcinolone acetonide)

## OTHER SUBSTANCES
**Simultaneous:** halcinonide

## KEY WORDS
Chem.Abs., 114, 136153

## REFERENCE
Gardner, R.S.; Walker, M.; Hollingsbee, D.A. A sensitive high-performance liquid chromatographic method for the assessment of percutaneous absorption of topical corticosteroids. *J.Pharm. Biomed.Anal.*, **1990**, *8*, 1083–1085

## SAMPLE
**Matrix:** solutions

## HPLC VARIABLES
**Column:** Phenyl (Waters) or 12% ODS (Whatman)
**Mobile phase:** MeCN:water 30:70
**Flow rate:** 2
**Detector:** UV 254

## CHROMATOGRAM
**Retention time:** 10.1 (phenyl) or 15.0 (ODS) (triamcinolone acetonide)

## OTHER SUBSTANCES
**Simultaneous:** fluoxymesterone

## KEY WORDS
Chem.Abs., 98, 166972; Proceedings of the Symposium on the Analysis of Steroids, Eger, Hungary, 1981

## REFERENCE
Kirschbaum, J.; Clay, R.; Poet, R. HPLC steroid analyses: generic column description and variable selectivity. *Anal.Chem.Symp.Ser. (Adv.Steroid Anal.)*, **1982**, *10*, 361–366

## SAMPLE
**Matrix:** solutions

## HPLC VARIABLES
**Column:** μBondapak ODS
**Mobile phase:** MeCN:water 30:70
**Flow rate:** 2
**Detector:** UV 254

## CHROMATOGRAM
**Retention time:** 12 (triamcinolone acetonide)
**Internal standard:** fluoxymesterone (10)

## REFERENCE
Kirschbaum, J. High-pressure liquid chromatography of triamcinolone acetonide: effect of different octadecylsilane columns on mobility. *J.Pharm.Sci.*, **1980**, *69*, 481–482

## SAMPLE
**Matrix:** tissue
**Sample preparation:** Homogenize (Waring blender) tissue at full speed for 2 min, lyophilize, grind. Extract with supercritical carbon dioxide at 60° at 400 atmospheres with a 20 cm × 21 μm restrictor for 1 h, collect the extract in 1 mL MeOH cooled to 5°. Evaporate the MeOH to dryness under a stream of nitrogen, reconstitute the residue in 100 μL MeCN:MeOH:20 mM ammonium formate 15:15:70, inject an aliquot. Alternatively, vortex 5 g ground tissue with 10 mL 40 mM sodium acetate, adjust pH to 4.2-4.7 with glacial acetic acid, add 100 μL β-glucuronidase (Sigma), heat at 37° for 8 h, add 20 mL MeCN, vortex for 30 s, centrifuge at 5000 rpm for 20 min. Remove a 30 mL aliquot of the supernatant and add it to 8 mL hexane and 2 mL dichloromethane, rotate for 3 min, centrifuge at 2000 rpm for 2 min. Remove a 15 mL aliquot of the middle layer and evaporate it to dryness under a stream of nitrogen, reconstitute the residue in 1 mL dichloromethane, inject an aliquot.

## HPLC VARIABLES
**Column:** 50 × 4.6 5 μm Supelcosil
**Mobile phase:** Gradient. MeCN:MeOH:20 mM ammonium formate from 2.5:2.5:95 to 47.5:47.5:5 over 19 min.
**Flow rate:** 1
**Injection volume:** 20
**Detector:** UV 245; MS, Sciex TAGA 6000E tandem triple quadrupole, APCI

## CHROMATOGRAM
**Retention time:** 10 (triamcinolone acetonide)
**Limit of detection:** 100 ppb

## OTHER SUBSTANCES
**Extracted:** dexamethasone, diethylstilbestrol, medroxyprogesterone, melengestrol acetate, trenbolone, zeranol

## KEY WORDS
cow; muscle; liver; SFE

## REFERENCE
Huopalahti, R.P.; Henion, J.D. Application of supercritical fluid extraction and high performance liquid chromatography/mass spectrometry for the determination of some anabolic agents directly from bovine tissue samples. *J.Liq.Chrom.Rel.Technol.*, **1996**, *19*, 69–87

## SAMPLE
**Matrix:** urine
**Sample preparation:** 3 mL Urine + 0.25 g NaCl, adjust pH to 9.0 with 0.5 g $Na_2HPO_4$, add 4 mL dichloromethane, vortex 1 min, centrifuge at 3700 g for 3 min. Remove organic phase and dry it over anhydrous sodium sulfate. Evaporate a 3 mL aliquot to dryness under vacuum, reconstitute residue with 200 μL 5 μg/mL IS in MeOH, inject 20 μL aliquot.

## HPLC VARIABLES
**Column:** 250 × 4.6 Hypersil ODS
**Mobile phase:** MeCN:water 32:68
**Column temperature:** 30
**Flow rate:** 1
**Injection volume:** 20
**Detector:** UV 245

## CHROMATOGRAM
**Retention time:** 2.9 (triamcinolone), 14 (triamcinolone acetonide)
**Internal standard:** methylprednisolone (9)

## OTHER SUBSTANCES
**Extracted:** betamethasone, corticosterone, cortisone, dexamethasone, fluorocortisone acetate, fluorocortisone, hydrocortisone, hydroxyprogesterone, prednisolone, prednisone

## KEY WORDS
SPE also discussed

## REFERENCE
Santos-Montes, A.; Gonzalo-Lumbreras, R.; Gasco-Lopez, A.I.; Izquierdo-Hornillos, R. Solvent and solid-phase extraction of natural and synthetic corticoids in human urine. *J.Chromatogr.B*, **1994**, *652*, 83–89

## SAMPLE
**Matrix:** urine
**Sample preparation:** 3 mL Urine + 1.5 μg betamethasone + 100 mg $K_2HPO_4$ + 500 mg anhydrous sodium sulfate + 5 mL diethyl ether, shake mechanically for 10 min, centrifuge at 2500 g for 5 min. Remove the organic layer and evaporate it to dryness under vacuum, reconstitute the residue in 200 μL MeOH, filter (0.45 μm), inject a 15 μL aliquot.

## HPLC VARIABLES
**Column:** 60 × 4.6 3 μm Hypersil ODS
**Mobile phase:** Gradient. MeOH:150 mM ammonium acetate from 40:60 to 50:50 over 6 min, maintain at 50:50 for 1 min, to 60:40 over 3 min, maintain at 60:40 for 5 min
**Flow rate:** 0.8
**Injection volume:** 15
**Detector:** MS, Hewlett-Packard HP 5988A, vaporizer probe 92° decreased to 89° over 6 min, decreased to 86° over 3 min, maintain at 86° for 5 min, ion source 276°, emission current 150 μA, electron energy 955 eV, positive ion mode, filament on

## CHROMATOGRAM
**Retention time:** 2.7 (triamcinolone), 7.5 (triamcinolone acetonide)
**Internal standard:** betamethasone (7)
**Limit of detection:** 1-5 ng (SIM)

## OTHER SUBSTANCES
**Extracted:** corticosterone, cortisone, deoxycorticosterone, hydrocortisone, 11α-hydroxypro-gesterone, prednisolone, prednisone

## REFERENCE
Park, S.-J.; Kim, Y.-J.; Pyo, H.-S.; Park, J. Analysis of corticosteroids in urine by HPLC and thermospray LC/MS. *J.Anal.Toxicol.*, **1990**, *14*, 102–108

## SAMPLE
**Matrix:** urine
**Sample preparation:** 3 mL Urine + 1.5 µg betamethasone + 100 mg K$_2$HPO$_4$ + 500 mg anhydrous sodium sulfate + 5 mL diethyl ether, shake mechanically for 10 min, centrifuge at 2500 g for 5 min. Remove the organic layer and evaporate it to dryness under vacuum, reconstitute the residue in 200 µL MeOH, filter (0.45 µm), inject a 15 µL aliquot.

## HPLC VARIABLES
**Column:** 100 × 4.6 5 µm Hypersil ODS
**Mobile phase:** Gradient. MeCN:water from 4:96 to 30:70 over 10 min, to 45:55 over 5 min, to 50:50 over 3 min
**Column temperature:** 40
**Flow rate:** 1
**Injection volume:** 15
**Detector:** UV 246

## CHROMATOGRAM
**Retention time:** 9.19 (triamcinolone), 13.82 (triamcinolone acetonide)
**Internal standard:** betamethasone
**Limit of detection:** 10 ng/mL

## OTHER SUBSTANCES
**Extracted:** corticosterone, cortisone, deoxycorticosterone, hydrocortisone, 11α-hydroxypro-gesterone, prednisolone, prednisone

## REFERENCE
Park, S.-J.; Kim, Y.-J.; Pyo, H.-S.; Park, J. Analysis of corticosteroids in urine by HPLC and thermospray LC/MS. *J.Anal.Toxicol.*, **1990**, *14*, 102–108

## ANNOTATED BIBLIOGRAPHY
Valvo, L.; Paris, A.; Savella, A.L.; Gallinella, B.; Ciranni Signoretti, E. General high-performance liquid chromatographic procedures for the rapid screening of natural and synthetic corticosteroids. *J.Pharm.Biomed.Anal.*, **1994**, *12*, 805–810 [gradient; reverse phase; normal phase; also beclomethasone, beclomethasone 17,21-dipropionate, betamethasone, betamethasone 21-acetate, betamethasone 17,21-dipropionate, betamethasone 21-disodium phosphate, betamethasone 17-valerate, cortisone, cortisone 21-acetate, 11-deoxycorticosterone 21-acetate, dexamethasone, dexamethasone 21-acetate, dexamethasone 21-disodium phosphate, fluocinolone, fluocinolone acetonide, 9α-fluorohydrocortisone 21-acetate, 9α-fluorohydrocortisone, 9α-fluoroprednisolone, 9α-fluoroprednisolone 21-acetate, hydrocortisone, hydrocortisone 21-acetate, hydrocortisone 21-hemisuccinate, 6α-methylprednisolone, 6α-methylprednisolone 21-acetate, 6α-methylprednisolone 21-sodium succinate, prednisolone, prednisolone 21-acetate, prednisolone 21-disodium phosphate, prednisolone 21-pivalate, prednisolone 21-sodium succinate, prednisone; for triamcinolone and triamcinolone acetonide]

Santos-Montes, A.; Gasco-Lopez, A.I.; Izquierdo-Hornillos, R. Optimization of the high-performance liquid chromatographic separation of a mixture of natural and synthetic corticosteroids. *J.Chromatogr.*, **1993**, *620*, 15–23 [simultaneous betamethasone, corticosterone, cortisone, deoxycorticosterone, dexamethasone, fluorocortisone, hydrocortisone, hydroxyprogesterone, methylprednisolone, prednisolone, prednisone; for triamcinolone]

Rowland, J.M.; Althaus, Z.R.; Slikker, W., Jr.; Hendrickx, A.G. Comparative distribution and metabolism of triamcinolone acetonide and cortisol in the rat embryomaternal unit. *Teratology*, **1983**, *27*, 333–341 [rat; plasma; tissue; gradient; for triamcinolone and triamcinolone acetonide; extracted cortisone, hydrocortisone]

Schöneshöfer, M.; Kage, A.; Weber, B. New "on-line" sample-pretreatment procedure for routine liquid-chromatographic assay of low-concentration compounds in body fluids, illustrated by triamcinolone assay. *Clin.Chem.*, **1983**, *29*, 1367–1371 [column-switching; LOD 5 ng/mL; urine; for triamcinolone]

Gordon, G.; Wood, P.R. Use of reversed-phase high-performance liquid chromatography in the analysis of products containing halcinonide and triamcinolone acetonide. *Proc.Anal.Div.Chem.Soc.*, **1977**, *14*, 30–32

Gordon, G.; Wood, P.R. Determination of triamcinolone acetonide in cream and suspension formulations by high-performance liquid chromatography. *Analyst*, **1976**, *101*, 876–882 [simultaneous excipients, prednisolone; for triamcinolone acetonide]

# Triamterene

**Molecular formula:** C$_{12}$H$_{11}$N$_7$
**Molecular weight:** 253.3
**CAS Registry No.:** 396-01-0

## SAMPLE
**Matrix:** bile, blood, urine
**Sample preparation:** 1 mL Plasma, whole blood, bile, or urine + 100 μL MeOH + 1 (plasma, blood, urine) or 0.5 (bile) mL 100 mM sodium bicarbonate + 6 mL diethyl ether, shake horizontally for 10 min, centrifuge at 2500 g at 10° for 10 (plasma, blood, urine) or 20 (bile) min, freeze in dry ice/acetone. Remove the organic layer and evaporate it to dryness, reconstitute the residue in 500 μL mobile phase, inject a 20 μL aliquot. (Hydrolyze conjugates as follows. 500 μL Plasma, whole blood, bile, or urine + 500 μL pH 5 buffer + 50 μL glusulase, heat at 50° for 24 h, add 50 μL 1 M NaOH, add 100 μL 1 μg/mL triamterene in MeOH, add 1 mL 100 mM sodium bicarbonate, proceed as above. In order to determine only the sulfate add 60 μL neutralized 100 mM 1,4-saccharolactone solution to inhibit β-glucuronidase.)

## HPLC VARIABLES
**Guard column:** 25 × 2.3 5 μm PRP-1 (Hamilton)
**Column:** 150 × 4.1 5 μm PRP-1 (Hamilton)
**Mobile phase:** MeOH:buffer 65:35 (Buffer was 1 g sodium carbonate and 2.902 g sodium bicarbonate in 1 L water, pH 9.8.)
**Flow rate:** 0.5
**Injection volume:** 20
**Detector:** F ex 330 em 420

## CHROMATOGRAM
**Retention time:** 10
**Internal standard:** triamterene

## OTHER SUBSTANCES
**Extracted:** dilevalol
**Noninterfering:** acebutolol, brefanolol, carteolol, carvedilol, enalapril, furosemide, hydrochlorothiazide, isosorbide dinitrate, metoprolol, nifedipine, piretanide, propranolol, verapamil, xamoterol, xipamide

## KEY WORDS
plasma; whole blood; triamterene is IS; pharmacokinetics

## REFERENCE
Neubeck, M.; Becker, C.; Henke, D.; Rösch, W.; Spahn-Langguth, H.; Mutschler, E. Pharmacokinetics of dilevalol and its conjugates in man. Assay method for plasma, blood, urine and bile samples and preliminary pharmacokinetic studies. *Arzneimittelforschung*, **1993**, *43*, 953–957

## SAMPLE
**Matrix:** blood
**Sample preparation:** 100 μL Plasma or whole blood + 400 μL 3.5 μg/mL furosemide in MeCN, vortex for 1 min, centrifuge at 3200 rpm for 10 min. Remove the supernatant and evaporate it to 200 μL under a stream of nitrogen, inject an aliquot.

## HPLC VARIABLES
**Column:** 300 × 4 10 μm Micro Pak MCH reverse phase

**Mobile phase:** MeCN:0.02% phosphoric acid 30:70 adjusted to pH 4.0 with 500 mM NaOH
**Flow rate:** 2
**Detector:** F ex 365 em 440

## CHROMATOGRAM
**Retention time:** 5.7
**Internal standard:** furosemide (8.1)
**Limit of quantitation:** 1 ng/mL

## KEY WORDS
whole blood; plasma; pharmacokinetics

## REFERENCE
Sörgel, F.; Lin, E.T.; Hasegawa, J.; Benet, L.Z. Liquid chromatographic analysis of triamterene and its major metabolite, hydroxytriamterene sulfate, in blood, plasma, and urine. *J.Pharm.Sci.*, **1984**, *73*, 831–833

## SAMPLE
**Matrix:** blood, urine
**Sample preparation:** Plasma. Dilute 200 μL plasma with 600 μL water, inject a 20 μL aliquot. Urine. Dilute 10 μL urine with 2 mL water, inject a 20 μL aliquot.

## HPLC VARIABLES
**Column:** 125 × 4 5 μm Spherisorb-amino
**Mobile phase:** Buffer containing 10 mM perchloric acid, 2 mM triethylamine, and 100 mM ammonium acetate
**Flow rate:** 3
**Injection volume:** 20
**Detector:** F ex 360 em 436

## CHROMATOGRAM
**Retention time:** 0.5
**Limit of detection:** 1 ng/mL

## OTHER SUBSTANCES
**Simultaneous:** hydroxytriamterene sulfate
**Noninterfering:** bendroflumethiazide, butizide, chlorthalidone, furosemide, hydrochlorothiazide, hydroxytriamterene, nifedipine

## KEY WORDS
plasma

## REFERENCE
Mascher, H.; Wasilewski, M. Simple and fast HPLC determination of triamterene and hydroxytriamterenesulphate in plasma and urine. *J.Liq.Chromatogr.*, **1994**, *17*, 1577–1585

## SAMPLE
**Matrix:** blood, urine
**Sample preparation:** Plasma. 150 μL Plasma + 10 μL 70% perchloric acid, shake for 15 min, centrifuge at 900 g for 10 min, inject 30 μL of the supernatant. Urine. Dilute urine with 2 volumes of MeOH, centrifuge at 900 g for 10 min, inject a 5 μL aliquot.

## HPLC VARIABLES
**Column:** 150 × 4.6 5 μm Nova-Pak C18

**Mobile phase:** MeCN:MeOH:buffer 14:8:70 (Buffer obtained from 6.75 mL 89% phosphoric acid in 900 mL water, pH adjusted to 2.8 with triethylamine, volume made up to 1 L.)
**Flow rate:** 0.8
**Injection volume:** 5-30
**Detector:** F ex 340-380 em 400-600

## CHROMATOGRAM
**Retention time:** 3.3
**Limit of quantitation:** 1 ng/mL

## OTHER SUBSTANCES
**Simultaneous:** metabolites

## KEY WORDS
plasma

## REFERENCE
Swart, K.J.; Botha, H. Rapid method for the determination of the diuretic triamterene and its metabolites in plasma and urine by high-performance liquid chromatography. *J.Chromatogr.*, **1987**, *413*, 315−319

## SAMPLE
**Matrix:** blood, urine
**Sample preparation:** Dilute 100 μL urine with 1 mL water. 1 mL Plasma or diluted urine + 40 μL 1 μg/mL p-methoxytriamterene in MeOH + 200 mg sodium bicarbonate + 5 mL ethyl acetate, shake for 1 min, centrifuge for 10 min. Remove the organic layer and evaporate it to dryness under a stream of nitrogen at 37°, wash sides of tube with ethyl acetate and again evaporate to dryness, dissolve residue in 20-50 μL MeOH, inject a 5-12 μL aliquot.

## HPLC VARIABLES
**Column:** 300 × 4 10 μm μBondapak C18
**Mobile phase:** MeOH:0.1% $KH_2PO_4$ 45:55, final pH adjusted to 3.8 with phosphoric acid
**Flow rate:** 2
**Injection volume:** 5-12
**Detector:** F ex 365 em 440; UV 230

## CHROMATOGRAM
**Retention time:** 2.6
**Internal standard:** p-methoxytriamterene (3.1)
**Limit of detection:** 1 ng/mL (F); 10 ng/mL (UV)

## OTHER SUBSTANCES
**Noninterfering:** metabolites

## KEY WORDS
plasma

## REFERENCE
Brodie, R.R.; Chasseaud, L.F.; Taylor, T.; Walmseley, L.M. Determination of the diuretic triamterene in the plasma and urine of humans by high-performance liquid chromatography. *J.Chromatogr.*, **1979**, *164*, 527−533

## SAMPLE
**Matrix:** formulations, urine

**Sample preparation:** Tablets. Pulverize tablets, add MeOH, shake for 30 min, sonicate for 5 min, filter (Albet 242 paper), wash solid with MeOH, make up filtrate to 50 mL with MeOH, inject a 20 μL aliquot. Urine. Adjust pH of 2 mL urine to 10.0 with 2 M KOH, add 1.5 mg NaCl, add 4 mL ethyl acetate, shake for 10 min, centrifuge at 2500 rpm for 5 min. Remove the organic layer and evaporate it to dryness under a stream of nitrogen at 40°, reconstitute the residue in 2 mL mobile phase, sonicate, inject a 20 μL aliquot.

## HPLC VARIABLES
**Guard column:** μBondapak C18
**Column:** 300 × 3.9 10 μm μBondapak C18
**Mobile phase:** MeCN:water 30:70 containing 5 mM $KH_2PO_4$/$K_2HPO_4$, pH adjusted to 5.5
**Flow rate:** 1
**Injection volume:** 20
**Detector:** E, EG&G Princeton Applied Research PAR Model 400, glassy carbon working electrode +1300 mV, Ag/AgCl reference electrode (At the end of each day clean electrode with mobile phase of MeOH at 1.5 mL/min, -800 mV for 2 min then +1600 mV for 5 min.)

## CHROMATOGRAM
**Retention time:** 5.01
**Limit of detection:** 0.1 ng/mL

## OTHER SUBSTANCES
**Extracted:** furosemide

## KEY WORDS
tablets; pharmacokinetics

## REFERENCE
Barroso, M.B.; Alonso, R.M.; Jiménez, R.M. Simultaneous determination of the diuretics triamterene and furosemide in pharmaceutical formulations and urine by HPLC-EC. *J.Liq.Chrom.Rel.Technol.*, **1996**, *19*, 231–246

## SAMPLE
**Matrix:** solutions

## HPLC VARIABLES
**Column:** 250 × 4.6 5 μm Supelcosil LC-DP (A) or 250 × 4 5 μm LiChrospher 100 RP-8 (B)
**Mobile phase:** MeCN:0.025% phosphoric acid:buffer 25:10:5 (A) or 60:25:15 (B) (Buffer was 9 mL concentrated phosphoric acid and 10 mL triethylamine in 900 mL water, adjust pH to 3.4 with dilute phosphoric acid, make up to 1 L.)
**Flow rate:** 0.6
**Injection volume:** 25
**Detector:** UV 229

## CHROMATOGRAM
**Retention time:** 6.73 (A), 3.90 (B)

## OTHER SUBSTANCES
**Also analyzed:** acebutolol, acepromazine, acetaminophen, acetazolamide, acetophenazine, albuterol, alprazolam, amitriptyline, amobarbital, amoxapine, antipyrine, atenolol, atropine, azatadine, baclofen, benzocaine, bromocriptine, brompheniramine, brotizolam, bupivacaine, buspirone, butabarbital, butalbital, caffeine, carbamazepine, cetirizine, chlorcyclizine, chlordiazepoxide, chlormezanone, chloroquine, chlorpheniramine, chlorpromazine, chlorpropamide, chlorprothixene, chlorthalidone, chlorzoxazone, cimetidine, cisapride, clomipramine, clonazepam, clonidine, clozapine, cocaine, codeine, colchicine, cyclizine, cyclobenzaprine, dantrolene, desipramine, diazepam, diclofenac, diflunisal,

diltiazem, diphenhydramine, diphenidol, diphenoxylate, dipyridamole, disopyramide, dobutamine, doxapram, doxepin, droperidol, encainide, ethidium bromide, ethopropazine, fenoprofen, fentanyl, flavoxate, fluoxetine, fluphenazine, flurazepam, flurbiprofen, fluvoxamine, furosemide, glutethimide, glyburide, guaifenesin, haloperidol, homatropine, hydralazine, hydrochlorothiazide, hydrocodone, hydromorphone, hydroxychloroquine, hydroxyzine, ibuprofen, imipramine, indomethacin, ketoconazole, ketoprofen, ketorolac, labetalol, levorphanol, lidocaine, loratadine, lorazepam, lovastatin, loxapine, mazindol, mefenamic acid, meperidine, mephenytoin, mepivacaine, mesoridazine, metaproterenol, metformin, methadone, methdilazine, methocarbamol, methotrexate, methotrimeprazine, methoxamine, methyldopa, methylphenidate, metoclopramide, metolazone, metoprolol, metronidazole, midazolam, moclobemide, morphine, nadolol, nalbuphine, naloxone, naphazoline, naproxen, nifedipine, nizatidine, norepinephrine, nortriptyline, oxazepam, oxycodone, oxymetazoline, paroxetine, pemoline, pentazocine, pentobarbital, pentoxifylline, perphenazine, pheniramine, phenobarbital, phenol, phenolphthalein, phentolamine, phenylbutazone, phenyltoloxamine, phenytoin, pimozide, pindolol, piroxicam, pramoxine, prazepam, prazosin, probenecid, procainamide, procaine, prochlorperazine, procyclidine, promazine, promethazine, propafenone, propantheline, propiomazine, propofol, propranolol, protriptyline, quazepam, quinidine, quinine, racemethorphan, ranitidine, remoxipride, risperidone, salicylic acid, scopolamine, secobarbital, sertraline, sotalol, spironolactone, sulfinpyrazone, sulindac, temazepam, terbutaline, terfenadine, tetracaine, theophylline, thiethylperazine, thiopental, thioridazine, thiothixene, timolol, tocainide, tolbutamide, tolmetin, trazodone, triazolam, trifluoperazine, triflupromazine, trimeprazine, trimethoprim, trimipramine, verapamil, warfarin, xylometazoline, yohimbine, zopiclone

## KEY WORDS
some details of plasma extraction

## REFERENCE
Koves, E.M. Use of high-performance liquid chromatography-diode array detection in forensic toxicology. *J.Chromatogr.A*, **1995**, *692*, 103–119

## SAMPLE
**Matrix:** solutions
**Sample preparation:** Inject a 10 µL aliquot of an aqueous solution.

## HPLC VARIABLES
**Column:** 250 × 4 5 µm Lichrosphere CN
**Mobile phase:** 2% acetic acid
**Flow rate:** 1
**Injection volume:** 10
**Detector:** UV 320

## CHROMATOGRAM
**Retention time:** 12.8

## OTHER SUBSTANCES
**Simultaneous:** caffeic acid

## REFERENCE
Uang, Y.-S.; Kang, F.-L.; Hsu, K.-Y. Determination of caffeic acid in rabbit plasma by high-performance liquid chromatography. *J.Chromatogr.B*, **1995**, *673*, 43–49

## SAMPLE
**Matrix:** solutions

## HPLC VARIABLES

**Guard column:** 30 × 3.2 7 μm SI 100 ODS (not commercially available)
**Column:** 150 × 3.2 7 μm SI 100 ODS (not commercially available)
**Mobile phase:** MeCN:buffer 31.2:68.8 (Buffer was 6.66 g $KH_2PO_4$ and 4.8 g 85% phosphoric acid in 1 L water, pH 2.3.)
**Flow rate:** 0.5-1
**Detector:** UV 211; UV 245

## CHROMATOGRAM

**Retention time:** 1.6
**Internal standard:** 5-(4-methylphenyl)-5-phenylhydantoin (7.3)

## OTHER SUBSTANCES

**Also analyzed:** aspirin, caffeine, carbamazepine, chlordiazepoxide, chlorprothixene, clonazepam, diazepam, doxylamine, ethosuximide, furosemide, haloperidol, hydrochlorothiazide, methocarbamol, methotrimeprazine, nicotine, oxazepam, procaine, promazine, propafenone, propranolol, salicylamide, temazepam, tetracaine, thiopental, verapamil, zolpidem, zopiclone

## REFERENCE

Below, E.; Burrmann, M. Application of HPLC equipment with rapid scan detection to the identification of drugs in toxicological analysis. *J.Liq.Chromatogr.*, **1994**, *17*, 4131–4144

## SAMPLE

**Matrix:** urine
**Sample preparation:** Inject an aliquot of urine directly onto column A with mobile phase A, after 1 min backflush the contents of column A onto column B with mobile phase B.

## HPLC VARIABLES

**Column:** A 20 × 2.1 Hypersil ODS-C18 30 μm; B 250 × 4 Hypersil ODS-C18 5 μm
**Mobile phase:** A Water; B Gradient. MeCN:buffer 15:85 for 1.5 min then to 80:20 over 8 min. Keep at 80:20 for 2.5 min then re-equilibrate with 15:85. (Buffer was 50 mM $NaH_2PO_4$ + 1.4 mL propylamine hydrochloride per liter adjusted to pH 3 with concentrated phosphoric acid.)
**Flow rate:** 1
**Injection volume:** 50
**Detector:** UV 230

## CHROMATOGRAM

**Retention time:** 6.9
**Limit of detection:** 7 ng/mL

## OTHER SUBSTANCES

**Extracted:** acetazolamide, amiloride, bendroflumethiazide, bumetanide, chlorthalidone, cyclothiazide, ethacrynic acid, furosemide, hydrochlorothiazide, probenecid, spironolactone

## KEY WORDS

column-switching

## REFERENCE

Campíns-Falco, P.; Herráez-Hernández, R.; Sevillano-Cabeza, A. Column-switching techniques for screening of diuretics and probenecid in urine samples. *Anal.Chem.*, **1994**, *66*, 244–248

## SAMPLE

**Matrix:** urine

**Sample preparation:** Inject 5 μL urine onto column A in mobile phase A, after 1 min elute the contents of column A onto column B with mobile phase B

## HPLC VARIABLES
**Column:** A 20 × 2.1 30 μm Hypersil ODS-C18; B 125 × 4 5 μm HP-LiChrospher 100 RP 18
**Mobile phase:** A water; B Gradient. A was MeCN. B was 3.45 g $NaH_2PO_4.H_2O$ + 700 μL propylamine hydrochloride in 500 mL water, pH adjusted to 3 with concentrated phosphoric acid. A:B from 0:100 to 50:50 after 4 min then hold at 50:50.
**Flow rate:** 1
**Injection volume:** 5
**Detector:** F ex 230 em 430

## CHROMATOGRAM
**Retention time:** 5.27
**Limit of detection:** 5 ng/mL

## KEY WORDS
column-switching

## REFERENCE
Campíns-Falcó, P.; Herráez-Hernández, R.; Sevillano-Cabeza, A. Determination of triamterene in urine by HPLC using fluorescence detection and column-switching. *Chromatographia*, **1994**, *38*, 29–34

## SAMPLE
**Matrix:** urine
**Sample preparation:** Condition a 1 mL 100 mg Bond-Elut C8 SPE cartridge with 500 μL MeOH and 500 μL water. 2 mL urine + 300 μL MeOH, add to the SPE cartridge, wash with 500 μL water, elute with 500 μL MeOH, filter (0.45 μm) the eluate, inject a 5 μL aliquot.

## HPLC VARIABLES
**Column:** 125 × 4 5 μm HP-LiChrospher 100 RP 18
**Mobile phase:** Gradient. MeCN:buffer from 0:100 to 30:70 over 5 min, maintain at 30:70. (Buffer was 3.45 g $NaH_2PO_4.H_2O$ + 700 μL propylamine hydrochloride in 500 mL water, adjust pH to 3 with concentrated phosphoric acid.)
**Flow rate:** 1
**Injection volume:** 5
**Detector:** UV 230

## CHROMATOGRAM
**Retention time:** 3.8
**Internal standard:** triamterene

## OTHER SUBSTANCES
**Extracted:** chlorthalidone
**Simultaneous:** atenolol, oprenolol, reserpine, spironolactone

## KEY WORDS
SPE; triamterene is IS

## REFERENCE
Campíns-Falcó, P.; Herráez-Hernández, R.; Sevillano-Cabeza, A. Simple and sensitive reversed-phase liquid chromatographic assay for analysis of chlorthalidone in urine. *J.Liq.Chromatogr.*, **1993**, *16*, 2571–2581

## SAMPLE
**Matrix:** urine
**Sample preparation:** Buffer urine to 4.9 by mixing with an equal volume of pH 4.9 200 mM sodium phosphate buffer. Inject a 40 μL aliquot onto column A with mobile phase A after 3 min backflush the contents of column A onto column B with mobile phase B and start the gradient. At the end of the run re-equilibrate for 10 min.

## HPLC VARIABLES
**Column:** A 20 × 4 5 μm Hypersil octadecylsilica ODS; B 200 × 4.6 5 μm Shiseido SG-120 polymer-based C18
**Mobile phase:** A water; B Gradient. MeCN:buffer from 7:93 to 15:85 over 3.5 min, to 50:50 over 8.5 min, maintain at 50:50 for 11 min (Buffer was 6.9 g $NaH_2PO_4.H_2O$ in 1 L water, pH adjusted to 3.1 with phosphoric acid.)
**Flow rate:** 1
**Injection volume:** 40
**Detector:** UV 360

## CHROMATOGRAM
**Retention time:** 11.6
**Limit of detection:** 500 ng/mL

## OTHER SUBSTANCES
**Extracted:** acetazolamide, amiloride, bendroflumethiazide, benzthiazide, bumetanide, caffeine, carbamazepine, chlorothiazide, chlorthalidone, clopamide, dichlorfenamide, ethacrynic acid, furosemide, hydrochlorothiazide, metyrapone, probenecid, spironolactone, trichlormethiazide

## KEY WORDS
column-switching; optimum detection wavelengths vary for each drug

## REFERENCE
Saarinen, M.; Sirén, H.; Riekkola, M.-L. A column switching technique for the screening of diuretics in urine by high performance liquid chromatography. *J.Liq.Chromatogr.*, **1993**, *16*, 4063–4078

## SAMPLE
**Matrix:** urine
**Sample preparation:** 5 mL Urine + 50 μL 100 μg/mL 7-propyltheophylline in MeOH + 200 μL ammonium chloride buffer + 2 g NaCl, extract with 6 mL ethyl acetate by rocking at 40 movements/min for 20 min and centrifuging at 800 g for 5 min, repeat extraction, combine organic layers, evaporate to dryness at 40° under a stream of nitrogen. Reconstitute in 200 μL MeCN:water 15:85 and inject 20 μL aliquots. (Ammonium chloride buffer was 28 g ammonium chloride in 100 mL water with the pH adjusted to 9.5 with concentrated ammonia solution.)

## HPLC VARIABLES
**Column:** 75 × 4.6 3 μm Ultrasphere ODS
**Mobile phase:** Gradient. MeCN:buffer from 10:90 to 15:85 over 2 min, to 55:45 over 3 min, to 60:40 over 3 min, maintain at 60:40 for 1 min, to 10:90 over 1 min, equilibrate at 10:90 for 2 min. (Buffer was 100 mM ammonium acetate adjusted to pH 3 with concentrated phosphoric acid.)
**Flow rate:** 1
**Injection volume:** 20
**Detector:** UV 270

## CHROMATOGRAM
**Retention time:** 3.7
**Internal standard:** 7-propyltheophylline (4.5)

## OTHER SUBSTANCES
**Simultaneous:** acetazolamide, amiloride, bendroflumethiazide, benzthiazide, bumetanide, buthiazide, caffeine, canrenone, chlorthalidone, clopamide, cyclothiazide, diclofenamide, ethacrynic acid, furosemide, hydrochlorothiazide, mesocarb, morazone, piretanide, polythiazide, probenecid, spironolactone, torsemide, xipamide

## REFERENCE
Ventura, R.; Nadal, T.; Alcalde, P.; Pascual, J.A.; Segura, J. Fast screening method for diuretics, probenecid and other compounds of doping interest. *J.Chromatogr.A*, **1993**, *655*, 233–242

## SAMPLE
**Matrix:** urine
**Sample preparation:** Make 5 mL urine alkaline (pH 9-10), add 2 g NaCl, extract twice with 6 mL ethyl acetate. Combine the organic layers and evaporate them to dryness under a stream of nitrogen, reconstitute the residue in 200 μL MeCN/water, inject a 10-20 μL aliquot.

## HPLC VARIABLES
**Column:** 100 × 4 5 μm SGE 100 GL-4 C18P (Scientific Glass Engineering)
**Mobile phase:** MeCN:MeOH:water:trifluoroacetic acid 4.5:10.5:85:0.5
**Flow rate:** 0.8 or 1
**Injection volume:** 10-20
**Detector:** MS, ZAB2-SEQ (VG), PSP source coupled to LC, source 250°, probe 240-260°, scan m/z 200-550; UV 270

## CHROMATOGRAM
**Retention time:** 4.0
**Limit of detection:** 50 ng (by MS)

## OTHER SUBSTANCES
**Extracted:** amiloride, bendroflumethiazide, benzthiazide, chlorthalidone, furosemide

## REFERENCE
Ventura, R.; Fraisse, D.; Becchi, M.; Paisse, O.; Segura, J. Approach to the analysis of diuretics and masking agents by high-performance liquid chromatography-mass spectrometry in doping control. *J.Chromatogr.*, **1991**, *562*, 723–736

## SAMPLE
**Matrix:** urine
**Sample preparation:** 2 mL Urine + 0.5 g solid buffer I (pH 5-5.5), vortex 15 s, add 4 mL ethyl acetate, agitate for 10 min, centrifuge at 600 g for 5 min. Remove organic layer and vortex it with 2 mL 5% aqueous lead acetate for 10 s, centrifuge at 600 g for 5 min, remove and keep organic phase. 2 mL Urine + 0.5 g solid buffer II (pH 9-9.5), vortex 15 s, add 4 mL ethyl acetate, agitate for 10 min, centrifuge at 600 g for 5 min. Remove organic layer and combine it with previous organic layer. Evaporate to dryness at 50° under a stream of nitrogen, reconstitute in 300 μL 50 μg/mL β-hydroxyethyltheophylline in MeOH, inject 5 μL aliquot. (Solid buffer I was $KH_2PO_4$:$Na_2HPO_4$ 99:1, solid buffer II was $NaHCO_3$:$K_2CO_3$ 3:2.)

## HPLC VARIABLES
**Column:** 250 × 4.6 5 μm HP Hypersil ODS (A) or HP LiChrosorb RP-18 (B)

**Mobile phase:** Gradient. MeCN:buffer from 15:85 at 2 min to 80:20 at 20 min (Buffer was 50 mM $NaH_2PO_4$ containing 16 mM propylamine hydrochloride, adjusted to pH 3 with concentrated phosphoric acid.)
**Flow rate:** 1
**Injection volume:** 5
**Detector:** UV 230;UV 275

## CHROMATOGRAM
**Retention time:** 6.8 (A), 7.6 (B)
**Internal standard:** β-hydroxyethyltheophylline (3.7 (A), 4.4 (B))
**Limit of detection:** 500 ng/mL

## OTHER SUBSTANCES
**Extracted:** acetazolamide, amiloride, bendroflumethiazide, benzthiazide, bumetanide, canrenone, chlorothiazide, chlorthalidone, cyclothiazide, dichlorphenamide, ethacrynic acid, furosemide, hydrochlorothiazide, hydroflumethiazide, methyclothiazide, metolazone, polythiazide, probenecid, quinethazone, spironolactone, trichloromethiazide
**Noninterfering:** acetaminophen, aspirin, caffeine, diflunisal, fenoprofen, ibuprofen, indomethacin, methocarbamol, naproxen, phenylbutazone, sulindac, tetracycline, theobromine, theophylline, tolmetin, trimethoprim, verapamil
**Interfering:** flumethiazide

## REFERENCE
Cooper, S.F.; Massé, R.; Dugal, R. Comprehensive screening procedure for diuretics in urine by high-performance liquid chromatography. *J.Chromatogr.*, **1989**, *489*, 65–88

## SAMPLE
**Matrix:** urine
**Sample preparation:** 20-100 μL Urine + 1 mL 500 μg/mL hydroflumethiazide in MeCN, vortex, centrifuge. Remove the supernatant, inject a 2 μL aliquot.

## HPLC VARIABLES
**Column:** 300 × 4 10 μm Micro Pak MCH reverse phase
**Mobile phase:** MeCN:0.02% phosphoric acid 13:87 adjusted to pH 5.3 with NaOH
**Flow rate:** 2
**Injection volume:** 2
**Detector:** F ex 365 em 440

## CHROMATOGRAM
**Retention time:** 8.6
**Internal standard:** hydroflumethiazide (7.0)
**Limit of quantitation:** 40 ng/mL

## OTHER SUBSTANCES
**Extracted:** metabolites

## KEY WORDS
pharmacokinetics

## REFERENCE
Sörgel, F.; Lin, E.T.; Hasegawa, J.; Benet, L.Z. Liquid chromatographic analysis of triamterene and its major metabolite, hydroxytriamterene sulfate, in blood, plasma, and urine. *J.Pharm.Sci.*, **1984**, *73*, 831–833

◆━━━━━━━━━━━━━━━━━━━━━━━━━━◆━━━━━━━━━━━━━━━━━━━━━━━━━━━◆

## ANNOTATED BIBLIOGRAPHY

Oertel, R.; Richter, K.; Dobrev, D.; Berndt, A.; Gramatté, T. Eine empfindliche HPLC-Methode zur Bestimmung von Triamteren im Serum [A sensitive HPLC-method for determination of triamterene in serum]. *Pharmazie*, **1994**, *49*, 700–702 [LOQ 2 ng/mL]

Bonet-Domingo, E.; Medina-Hernandez, M.J.; Garcia-Alvarez-Coque, M.C. A micellar liquid chromatographic procedure for the determination of amiloride, bendroflumethiazide, chlorthalidone, spironolactone and triamterene in pharmaceuticals. *J.Pharm.Biomed.Anal.*, **1993**, *11*, 711–716

Herráez-Hernández, R.; Campíns-Falcó, P.; Sevillano-Cabeza, A. Improved screening procedure for diuretics. *J.Liq.Chromatogr.*, **1992**, *15*, 2205–2224 [LOD 10-1000 ng/mL; gradient; urine; hydroxymethyltheophylline (IS); extracted acetazolamide, amiloride, bendroflumethiazide, bumetanide, chlorthalidone, cyclothiazide, ethacrynic acid, furosemide, hydrochlorothiazide, probenecid, spironolactone]

Campíns-Falcó, P.; Herráez-Hernández, R.; Sevillano-Cabeza, A. Solid-phase extraction techniques for assay of diuretics in human urine samples. *J.Liq.Chromatogr.*, **1991**, *14*, 3575–3590 [SPE; hydroxyethyltheophylline (IS); extracted acetazolamide, amiloride, bendroflumethiazide, bumetanide, chlorthalidone, ciclothiazide, ethacrynic acid, furosemide, hydrochlorothiazide, probenecid, spironolactone]

Shah, V.P.; Walker, M.A.; Prasad, V.K. Application of flow programming in the analysis of drugs and their metabolites in biological fluids. *J.Liq.Chromatogr.*, **1983**, *6*, 1949–1954 [urine; plasma; also chlorothiazide, hydrochlorothiazide]

Hasegawa, J.; Lin, E.T.; Williams, R.L.; Sörgel, F.; Benet, L.Z. Pharmacokinetics of triamterene and its metabolite in man. *J.Pharmacokinet.Biopharm.*, **1982**, *10*, 507–523 [plasma; urine; fluorescence detection; furosemide (IS); SPE]

Menon, G.N.; White, L.B. Simultaneous determination of hydrochlorothiazide and triamterene in capsule formulations by high-performance liquid chromatography. *J.Pharm.Sci.*, **1981**, *70*, 1083–1085 [m-hydroxyacetophenone (IS); stability-indicating; simultaneous degradation products]

Yakatan, G.J.; Cruz, J.E. High-performance liquid chromatographic analysis of triamterene and p-hydroxytriamterene in plasma. *J.Pharm.Sci.*, **1981**, *70*, 949–951 [extracted metabolites; LOD 20 ng/mL]

# Trimethoprim

**Molecular formula:** $C_{14}H_{18}N_4O_3$
**Molecular weight:** 290.3
**CAS Registry No.:** 738-70-5, 56585-33-2 (sulfate)

## SAMPLE
**Matrix:** blood
**Sample preparation:** 50 µL Serum + 75 µL acetone:10% trichloroacetic acid 1:2, vortex for 5 s, centrifuge for 4 min. Remove 62.5 µL of the supernatant and add it to 62.5 µL 50 mM $KH_2PO_4$, add 250 µL diethyl ether, vortex for 10 s, centrifuge for 5 min, filter (0.45 µm) the lower aqueous layer, inject a 20 µL aliquot of the filtrate.

## HPLC VARIABLES
**Column:** 75 × 4.6 TSK gel ODS-80TM (Tosoh)
**Mobile phase:** MeCN:50 mM pH 6.0 $KH_2PO_4$ 8:92
**Flow rate:** 1
**Injection volume:** 20
**Detector:** UV 235

## CHROMATOGRAM
**Retention time:** 30
**Internal standard:** trimethoprim

## OTHER SUBSTANCES
**Extracted:** vancomycin

## KEY WORDS
serum; trimethoprim is IS

## REFERENCE
Morishige, H.; Shuto, H.; Ieiri, I.; Otsubo, K.; Oishi, R. Instability of standard calibrators may be involved in overestimating vancomycin concentrations determined by fluorescence polarization immunoassay. *Ther.Drug Monit.*, **1996**, *18*, 80–85

## SAMPLE
**Matrix:** blood
**Sample preparation:** Condition a 1 mL Bakerbond C18 SPE cartridge with MeOH and 50 mM pH 5.5 citrate buffer. 500 µL Serum + 17 µL 600 µg/mL sulfamethazine in MeCN:water 20:80 + 500 µL 50 mM pH 5.5 citrate buffer, vortex, add to the SPE cartridge, wash twice with 50 mM pH 5.5 citrate buffer, air dry, elute with MeOH. Evaporate the eluate to dryness under a stream of nitrogen at 45°, reconstitute the residue in 250 µL mobile phase, inject a 10 µL aliquot.

## HPLC VARIABLES
**Guard column:** µBondapak C18 Guard-Pak
**Column:** 300 × 3.9 µBondapak C18
**Mobile phase:** MeCN:1% acetic acid 18:82
**Flow rate:** 2.5
**Injection volume:** 100
**Detector:** UV 240

## CHROMATOGRAM
**Retention time:** 3.14
**Internal standard:** sulfamethazine (4.19)

## OTHER SUBSTANCES
**Extracted:** sulfamethoxazole

## KEY WORDS
serum; SPE

## REFERENCE
Moore, K.H.P.; Brouwer, K.L.R. High-performance liquid chromatographic evaluation of the effect of heat treatment on trimethoprim and sulfamethoxazole stability in serum. *Ther.Drug Monit.*, **1995**, *17*, 356–360

## SAMPLE
**Matrix:** blood
**Sample preparation:** Condition a 3 mL 500 mg Sep-Pak with 10 mL MeOH and 10 mL water at 3 mL/min (do not allow to dry). 5 mL Serum + 50 μL 10 μg/mL ormetoprim in water, vortex for 15 s, allow to stand for 5 min, add 2 mL pH 11 NaH$_2$PO$_4$ (pH adjusted with 5 M NaOH), vortex for 10 s, add 30 mL dichloromethane, shake for 10 min, extract aqueous phase again with 20 mL dichloromethane. Combine organic layers and shake them with 10 mL 150 mM sulfuric acid. Remove 8 mL of the aqueous layer and add it to 10 mL phosphate buffer (adjusted to pH 6 with 1 M NaOH), add this to the SPE cartridge, wash with 10 mL water, draw air through the cartridge for 1 min, elute with 1 mL MeOH, evaporate the eluate at 50° under a stream of nitrogen, reconstitute the residue in 1 mL mobile phase, filter (0.45 μm), inject a 50 μL aliquot.

## HPLC VARIABLES
**Column:** 250 × 4 5 μm Inertsil C8
**Mobile phase:** MeCN:pH 3 ammonium formate-trifluoroacetic acid 20:80
**Flow rate:** 1
**Injection volume:** 50
**Detector:** MS, VG Trio 2 with a thermospray-plasmaspray LC interface, capillary probe tip at 300°

## CHROMATOGRAM
**Internal standard:** ormetoprim

## KEY WORDS
serum; cow; SPE; thermospray; plasmaspray

## REFERENCE
Nachilobe, P.; Boison, J.O.; Cassidy, R.M.; Fesser, A.C.E. Determination of trimethoprim in bovine serum by high-performance liquid chromatography with confirmation by thermospray liquid chromatography-mass spectrometry. *J.Chromatogr.*, **1993**, *616*, 243–252

## SAMPLE
**Matrix:** blood
**Sample preparation:** Condition a 3 mL 500 mg Sep-Pak SPE cartridge with 10 mL MeOH and 10 mL water at 3 mL/min (do not allow to dry). 5 mL Serum + 50 μL 10 μg/mL ormetoprim in water, vortex for 15 s, allow to stand for 5 min, add 2 mL pH 11 NaH$_2$PO$_4$ (pH adjusted with 5 M NaOH), vortex for 10 s, add 30 mL dichloromethane, shake for 10 min, extract aqueous phase again with 20 mL dichloromethane. Combine organic layers and shake them with 10 mL 150 mM sulfuric acid. Remove 8 mL of the aqueous layer

and add it to 10 mL phosphate buffer (adjusted to pH 6 with 1 M NaOH), add this to the SPE cartridge, wash with 10 mL water, draw air through the cartridge for 1 min, elute with 1 mL MeOH, evaporate the eluate at 50° under a stream of nitrogen, reconstitute the residue in 1 mL mobile phase, filter (0.45 μm), inject a 50 μL aliquot.

## HPLC VARIABLES
**Guard column:** C18 pre-column filter
**Column:** 250 × 4.6 5 μm Spherisorb ODS(2)
**Mobile phase:** MeCN:MeOH:buffer 150:50:800 (Buffer was 2.5 mL triethylamine in 900 mL water, add 5 mL glacial acetic acid, make up to 1 L with water.)
**Flow rate:** 1.5
**Injection volume:** 50
**Detector:** UV 225

## CHROMATOGRAM
**Retention time:** 7.5
**Internal standard:** ormetoprim (10)
**Limit of detection:** 5 ng/mL

## KEY WORDS
serum; cow; SPE

## REFERENCE
Nachilobe, P.; Boison, J.O.; Cassidy, R.M.; Fesser, A.C.E. Determination of trimethoprim in bovine serum by high-performance liquid chromatography with confirmation by thermospray liquid chromatography-mass spectrometry. *J.Chromatogr.*, **1993**, *616*, 243–252

## SAMPLE
**Matrix:** blood
**Sample preparation:** Condition a 3 mL C18 Bond Elut SPE cartridge with 500 μL MeOH then 1 mL wash solution. 500 μL Serum + 500 μL wash solution, vortex for 3 s, add 25 μL 240 μg/mL p-nitrophenol in MeOH, vortex for 3 s, add to the SPE cartridge, wash with 1 mL wash solution, dry with vacuum applied for 30 s, elute with two 500 μL aliquots of MeOH:triethylamine 10:1. Combine the eluates and evaporate them to dryness under a stream of nitrogen, reconstitute in 200 μL mobile phase, inject a 50-100 μL aliquot. (Wash solution was 7 mL 3 M HCl + 1.6 g 1-octanesulfonic acid + 150 mL 100 mM disodium citrate made up to 800 mL with water, pH adjusted to 3.00 with 3 M HCl.)

## HPLC VARIABLES
**Guard column:** 20 × 2 40 μm Upchurch pellicular C18
**Column:** 300 × 3.9 10 μm μBondapak C18
**Mobile phase:** MeCN:buffer 24:76 with 0.8 g/L 1-octanesulfonic acid (Buffer was 7 mL 3 M HCl plus 150 mL 100 mM disodium citrate made up to 760 mL with water, pH adjusted to 3.00 with 3 M HCl.)
**Flow rate:** 1.5
**Injection volume:** 50-100
**Detector:** UV 230

## CHROMATOGRAM
**Retention time:** 5.6
**Internal standard:** p-nitrophenol (9.3)
**Limit of quantitation:** 25 ng

## OTHER SUBSTANCES
**Simultaneous:** $N^4$-acetylsulfamethoxazole, sulfamethoxazole

## KEY WORDS
serum; SPE

## REFERENCE
Laizure, S.C.; Holden, C.L.; Stevens, R.C. Ion-paired high-performance liquid chromatographic separation of trimethoprim, sulfamethoxazole and N⁴-acetylsulfamethoxazole with solid-phase extraction. *J.Chromatogr.*, **1990**, *528*, 235–242

## SAMPLE
**Matrix:** blood
**Sample preparation:** 1 mL Plasma + 100 μL 5 μg/mL sulfafurazole in MeOH:water 50: 50 + 200 μL 1 M pH 6.8 sodium phosphate buffer + 6 mL ethyl acetate, shake for 15 min, centrifuge at 600 g for 10 min. Remove the organic layer and evaporate it to dryness under a stream of nitrogen at 50°, reconstitute the residue in 175 μL MeOH:water 20: 80, vortex for 1 min, add 25 μL 40% trichloroacetic acid in 100 mM hydrochloric acid, vortex for 30 s, centrifuge at 8300 g for 5 min, inject a 100 μL aliquot of the supernatant.

## HPLC VARIABLES
**Column:** 250 × 4 5 μm LiChrosorb RP18
**Mobile phase:** MeCN:150 mM pH 4.85 ammonium phosphate buffer 12.6:87.4
**Flow rate:** 1.5
**Injection volume:** 100
**Detector:** UV 254

## CHROMATOGRAM
**Retention time:** 10
**Internal standard:** sulfafurazole (21)
**Limit of detection:** 160 ng/mL

## OTHER SUBSTANCES
**Extracted:** N-acetylsulfamethoxazole, sulfamethoxazole

## KEY WORDS
plasma; pharmacokinetics

## REFERENCE
Van der Steuijt, K.; Sonneveld, P. Concurrent analysis of methotrexate, trimethoprim, sulfamethoxazole and their major metabolites in plasma by high-performance liquid chromatography. *J.Chromatogr.*, **1987**, *422*, 328–333

## SAMPLE
**Matrix:** blood
**Sample preparation:** 1 mL Serum + 100 μL 40 μg/mL timolol maleate in MeOH + 500 μL 500 mM NaOH, vortex for 10 s, add 5 mL dichloromethane:ether 2:3, shake gently for 5 min, centrifuge. Remove the organic layer and evaporate it to dryness at 37° under a stream of nitrogen. Dissolve the residue in 100 μL 100 mM HCl, add 500 μL ether, vortex briefly, centrifuge at 3000 rpm for 5 min, inject a 10-25 μL aliquot of the aqueous phase.

## HPLC VARIABLES
**Column:** 90 × 4.6 5 μm Hypersil ODS
**Mobile phase:** MeCN:buffer 35:65 (Buffer contained 30 mM sodium lauryl sulfate and 50 mM Na₂HPO₄ adjusted to pH 2 with orthophosphoric acid.)
**Flow rate:** 2

**Injection volume:** 10-25
**Detector:** UV 229

## CHROMATOGRAM
**Retention time:** 3
**Internal standard:** timolol maleate (7)
**Limit of quantitation:** 10 ng/mL

## KEY WORDS
serum

## REFERENCE
Hung, C.T.; Perrier, D.G. Determination of trimethoprim and sulfamethoxazole in serum by reversed-phase and ion pair HPLC. *J.Liq.Chromatogr.*, **1985**, *8*, 521–536

## SAMPLE
**Matrix:** blood
**Sample preparation:** 250 µL Plasma + 50 µL 100 mM NaOH + 1.5 mL ethyl acetate, vortex for 15 s, centrifuge at 18000 g for 1 min. Remove 1 mL of the organic phase and evaporate it to dryness at 40° under a stream of nitrogen, dissolve the residue in 250 µL 70 ng/mL sulfamethoxazole in mobile phase, inject a 10-100 µL aliquot.

## HPLC VARIABLES
**Guard column:** 40 × 4.6 5 µm Nucleosil C18
**Column:** 120 × 4.6 5 µm Nucleosil C18
**Mobile phase:** MeOH:70 mM pH 4.75 $KH_2PO_4$ 25:75
**Flow rate:** 1.5
**Injection volume:** 10-100
**Detector:** E, Metrohm Model 656 electrochemical and Model 641 VA detectors, working and auxiliary electrodes glassy carbon, reference electrode Ag/AgCl, 1200 mV; UV 280

## CHROMATOGRAM
**Retention time:** 6
**Internal standard:** sulfamethoxazole (7.5)
**Limit of detection:** 10 ng/mL (E); 100 ng/mL (UV)

## KEY WORDS
plasma; pig; human

## REFERENCE
Nordholm, L.; Dalgaard, L. Assay of trimethoprim in plasma and urine by high-performance liquid chromatography using electrochemical detection. *J.Chromatogr.*, **1982**, *233*, 427–431

## SAMPLE
**Matrix:** blood, CSF
**Sample preparation:** 200 µL Serum, plasma, or CSF + 300 µL reagent. Flush column A to waste with 500 µL 500 mM ammonium sulfate, inject sample onto column A, flush column A to waste with 500 µL 500 mM ammonium sulfate, elute the contents of column A onto column B with mobile phase, monitor the effluent from column B. (Reagent was 8.05 M guanidine hydrochloride and 1.02 M ammonium sulfate in water.)

## HPLC VARIABLES
**Column:** A 30 × 2.1 40 µm preparative grade C18 (Analytichem); B 250 × 4.6 10 µm Partisil C8
**Mobile phase:** Gradient. A was 50 mM pH 4.5 $KH_2PO_4$. B was MeCN:isopropanol 80:20. A:B 90:10 for 1 min, to 30:70 over 15 min, maintain at 30:70 for 4 min.

**Column temperature:** 50
**Flow rate:** 1.5
**Detector:** UV 280 for 5 min then UV 254

## CHROMATOGRAM
**Retention time:** 6.75
**Internal standard:** heptanophenone (19.2)

## OTHER SUBSTANCES
**Extracted:** acetazolamide, ampicillin, bromazepam, caffeine, carbamazepine, chloramphenicol, chlorothiazide, diazepam, droperidol, ethionamide, furosemide, isoniazid, methadone,
penicillin G, phenobarbital, phenytoin, prazepam, propoxyphene, pyrazinamide, rifampin,
trimeprazine

## KEY WORDS
serum; plasma; column-switching

## REFERENCE
Seifart, H.I.; Kruger, P.B.; Parkin, D.P.; van Jaarsveld, P.P.; Donald, P.R. Therapeutic monitoring of
antituberculosis drugs by direct in-line extraction on a high-performance liquid chromatography system. *J.Chromatogr.*, **1993**, *619*, 285–290

## SAMPLE
**Matrix:** blood, CSF, gastric fluid, urine
**Sample preparation:** 200 µL Serum, urine, CSF, or gastric fluid + 300 µL reagent. Flush
column A to waste with 500 µL 500 mM ammonium sulfate, inject sample onto column
A, flush column A to waste with 500 µL 500 mM ammonium sulfate, backflush the con-
tents of column A onto column B with mobile phase, monitor the effluent from column B.
(Reagent was 8.05 M guanidine hydrochloride and 1.02 M ammonium sulfate in water.)

## HPLC VARIABLES
**Column:** A 40 µm preparative grade C18 (Analytichem); B 75 × 2.1 pellicular C18 (What-
man) + 250 × 4.6 5 µm C8 end-capped (Whatman)
**Mobile phase:** Gradient. A was 50 mM pH 4.5 $KH_2PO_4$. B was MeCN:isopropanol 80:20.
A:B 90:10 for 1 min, to 30:70 over 20 min.
**Column temperature:** 50
**Flow rate:** 1.5
**Detector:** UV 220

## CHROMATOGRAM
**Retention time:** 6.5
**Internal standard:** heptanophenone (19)

## OTHER SUBSTANCES
**Extracted:** acetaminophen, allobarbital, azinphos, barbital, brallobarbitone, bromazepam,
butethal, caffeine, carbamazepine, carbaryl, cephaloridine, chloramphenicol, chlordiazepoxide, chlorothiazide, chlorvinphos, clothiapine, cocaine, coomassie blue, desipramine,
diazepam, diphenhydramine, dipipanone, ethylbromphos, flufenamic acid, formothion,
griseofulvin, indomethacin, lidocaine, lorazepam, malathion, medazepam, midazolam, oxazepam, paraoxon, penicillin G, pentobarbital, prazepam, propoxyphene, prothiophos, quinine, salicylic acid, secobarbital, strychnine, sulfamethoxazole, theophylline, thiopental,
thioridazine

## KEY WORDS
serum; column-switching

## REFERENCE

Kruger, P.B.; Albrecht, C.F.De V.; Jaarsveld, P.P. Use of guanidine hydrochloride and ammonium sulfate in comprehensive in-line sorption enrichment of xenobiotics in biological fluids by high-performance liquid chromatography. *J.Chromatogr.*, **1993**, *612*, 191–198

## SAMPLE

**Matrix:** blood, CSF, peritoneal fluid, synovial fluid, tissue, urine

**Sample preparation:** Condition a C18 SPE cartridge (J.T. Baker) with 1 mL MeOH, 1 mL water, and 1 mL 100 mM pH 4.5 acetate buffer. CSF, peritoneal fluid, serum, synovial fluid, tissue. Homogenize (TenBroeck tissue grinder) endometrial tissue with saline, centrifuge at 510 g, remove supernatant. Centrifuge fluids at 510 g. 1 mL Sample + 1 mL 100 mM acetate buffer, mix, add to the SPE cartridge, wash with 1 mL water, elute with 1 mL MeCN:MeOH:100 mM acetate buffer 12.5:12.5:75, add 12.5 ng ormetoprim, inject an aliquot. Urine. 1 mL Urine + 1 mL MeCN, centrifuge at 510 g for 10 min. Remove 25 μL of the supernatant and add it to 1 mL 100 mM acetate buffer, add to the SPE cartridge, wash with 1 mL water, elute with 1 mL MeCN:MeOH:100 mM acetate buffer 12.5:12.5:75, add 12.5 ng ormetoprim, inject an aliquot.

## HPLC VARIABLES

**Column:** C18 (Rainin)

**Mobile phase:** MeCN:MeOH:50 mM phosphate buffer 12.5:12.5:75, pH 3.0

**Flow rate:** 1.5

**Detector:** UV 280

## CHROMATOGRAM

**Retention time:** 6

**Internal standard:** ormetoprim (8)

**Limit of detection:** 15 ng/mL

## OTHER SUBSTANCES

**Extracted:** sulfamethoxazole

## KEY WORDS

serum; endometrium; SPE; horse; pharmacokinetics

## REFERENCE

Brown, M.P.; Gronwall, R.; Castro, L. Pharmacokinetics and body fluid and endometrial concentrations of trimethoprim-sulfamethoxazole in mares. *Am.J.Vet.Res.*, **1988**, *49*, 918–922

## SAMPLE

**Matrix:** blood, CSF, urine

**Sample preparation:** 50 μL Serum, urine, or CSF + 50 μL 100 μg/mL antipyrine in MeOH, vortex for 15 s, centrifuge at 10000 g for 2 min, inject a 20 μL aliquot of the supernatant.

## HPLC VARIABLES

**Guard column:** 37-50 μm μBondapak C18

**Column:** 300 × 3.9 10 μm μBondapak C18

**Mobile phase:** MeOH:buffer 35:65 (Buffer was 97% 67 mM $KH_2PO_4$ + 3% 67 mM $Na_2HPO_4$ and the pH was adjusted to 3.5 with phosphoric acid.)

**Flow rate:** 1

**Injection volume:** 20

**Detector:** UV 225

## CHROMATOGRAM

**Retention time:** 4.6

**Internal standard:** antipyrine (11.0)
**Limit of detection:** 50 ng/mL

## OTHER SUBSTANCES
**Extracted:** N-acetylsulfamethoxazole, sulfamethoxazole

## KEY WORDS
serum

## REFERENCE
Weber, A.; Opheim, K.E.; Siber, G.R.; Ericson, J.F.; Smith, A.L. High-performance liquid chromatography quantitation of trimethoprim, sulfamethoxazole, and N4-acetylsulfamethoxazole in body fluids. *J.Chromatogr.*, **1983**, *278*, 337–345

## SAMPLE
**Matrix:** blood, tissue
**Sample preparation:** Plasma. 500 μL Plasma + 100 μL 30 μg/mL sulfamethazine (sulfadimidine) in EtOH + 150 μL 3% trichloroacetic acid in EtOH + 100 μL EtOH, vortex, freeze at -20° for 5 min, centrifuge, freeze at -20° for 10 min, centrifuge through a Spin-X filter tube, inject a 10 μL aliquot of the supernatant. Tissue. 1-3 g Tissue + 3 (muscle) or 6 (liver) μL 1 mg/mL sulfamethoxazole in EtOH + 2 (liver) or 3 (muscle) mL 0.7% trichloroacetic acid in acetone, mix in Whirlmixer, sonicate for 10 min at 40°, add 2 mL 10 mM pH 6 $Na_2HPO_4$, sonicate for 5 min, add 100 μL 500 mM NaOH, add 9 (muscle) or 10 (liver) mL dichloromethane, mix thoroughly for 1 min, centrifuge at 2240 g for 5 min. Remove 6 mL of the organic layer and evaporate it to dryness at 40°under a stream of nitrogen. Dissolve the residue in 400 (muscle) or 800 (liver) μL MeCN:10 mM pH 2.8 phosphate buffer 20:80, sonicate, wash with 1 mL hexane. Sonicate the aqueous phase for 1 min, centrifuge through a Spin-X filter tube, inject a 10 μL aliquot of the supernatant.

## HPLC VARIABLES
**Guard column:** 20 × 4.6 5 μm Supelcosil LC-18 DB
**Column:** 250 × 4.6 5 μm Supelcosil LC-18 DB
**Mobile phase:** MeCN:buffer 23:77 (plasma) or 20:80 (tissue) with 0.1% triethylamine added (Buffer was 25 mM sodium phosphate and 20 mM sodium 1-hexanesulfonate, pH adjusted to 2.8 with 5 M phosphoric acid.)
**Flow rate:** 0.9
**Injection volume:** 10
**Detector:** UV 270

## CHROMATOGRAM
**Retention time:** 9 (plasma), 14 (tissue)
**Internal standard:** sulfamethazine (sulfadimidine) (8), sulfamethoxazole (18)
**Limit of quantitation:** 80 ng/g (muscle); 250 mg/mL (plasma)

## OTHER SUBSTANCES
**Simultaneous:** sulfadiazine

## KEY WORDS
plasma; fish; salmon; trout; muscle; liver

## REFERENCE
Hormazabal, V.; Rogstad, A. Simultaneous determination of sulphadiazine and trimethoprim in plasma and tissues of cultured fish for residual and pharmacokinetic studies. *J.Chromatogr.*, **1992**, *583*, 201–207

## SAMPLE
**Matrix:** blood, urine
**Sample preparation:** Blood. 320 μL Whole blood + 450 μL urea (1:1) + 30 μL 2 μg/mL sulfamethazine, shake mechanically for 15 min, filter (Amicon Micropartition System MPS-1) while centrifuging at 4° at 4000 rpm for at least 3 h, inject a 300 μL aliquot of the ultrafiltrate. Urine. Dilute urine if necessary. 320 μL Urine + 30 μL 2 μg/mL sulfamethazine, shake mechanically for 15 min, filter (Amicon Micropartition System MPS-1) while centrifuging at 4° at 4000 rpm for at least 3 h, inject a 300 μL aliquot of the ultrafiltrate.

## HPLC VARIABLES
**Column:** 150 mm long 5 μm Spherisorb ODS 2
**Mobile phase:** MeOH:MeCN:10 mM pH 5.0 sodium acetate buffer containing 4 mM triethylamine 14:14:72
**Flow rate:** 0.8
**Injection volume:** 300
**Detector:** UV 275

## CHROMATOGRAM
**Internal standard:** sulfamethazine
**Limit of quantitation:** 200 nM

## KEY WORDS
fish; whole blood; trout; rainbow trout; pharmacokinetics

## REFERENCE
Tan, W.P.; Wall, R.A. Disposition kinetics of trimethoprim in rainbow trout (*Oncorhynchus mykiss*). *Xenobiotica*, **1995**, *25*, 315–329

## SAMPLE
**Matrix:** blood, urine
**Sample preparation:** Dilute 1 mL urine with 5 or 10 mL water. 1 mL Serum, plasma, or diluted urine + 200 μL 1 M pH 6.8 KH$_2$PO$_4$ buffer + 6 mL ethyl acetate, vortex for 3 min, centrifuge at 2400 g for 5 min. Remove 5 mL of the organic layer and evaporate it to dryness under a stream of nitrogen at 50°, reconstitute the residue in 200 μL 30 μg/mL sulfadimethoxine in mobile phase, inject an aliquot.

## HPLC VARIABLES
**Guard column:** 25-40 μm LiChroprep Si 60 (Merck)
**Column:** 250 × 4 10 μm LiChrosorb Si 60
**Mobile phase:** Dichloromethane:MeOH:ammonia 80:19:1
**Flow rate:** 1.5
**Injection volume:** 20
**Detector:** UV 289

## CHROMATOGRAM
**Retention time:** 2.6
**Internal standard:** sulfadimethoxine (3.7)
**Limit of detection:** 30 ng/mL

## OTHER SUBSTANCES
**Extracted:** N-acetylsulfadiazine, sulfadiazine

## KEY WORDS
normal phase; serum; plasma; pharmacokinetics

## REFERENCE

Ascalone, V. Assay of trimethoprim, sulfadiazine and its N4-acetyl metabolite in biological fluids by normal-phase high-performance liquid chromatography. *J.Chromatogr.*, **1981**, *224*, 59–66

## SAMPLE

**Matrix:** cell cultures

**Sample preparation:** Condition a cyclohexyl-bonded silica Bond-elut SPE cartridge with 2 mL MeOH and 2 mL water. Centrifuge cell cultures at 6000 g at 4° for 15 min, add 100 μL supernatant and 100 μL 2 μg/mL sulfadiazine to the SPE cartridge, wash with 1 mL water, elute with 1.5 mL MeOH. Evaporate the eluate to dryness under a stream of air at 60°, reconstitute the residue in 100 μL water, vortex, inject a 20 μL aliquot.

## HPLC VARIABLES

**Column:** 100 × 4.6 5 μm ODS Hypersil

**Mobile phase:** MeOH:10 mM pH 2.5 phosphate buffer 5:95 containing 40 mM tetrabutylammonium bromide

**Flow rate:** 2

**Injection volume:** 20

**Detector:** UV 254

## CHROMATOGRAM

**Retention time:** 3

**Internal standard:** sulfadiazine (6)

**Limit of detection:** 100 ng/mL

## OTHER SUBSTANCES

**Extracted:** p-aminobenzoic acid, dibromopropamidine isethionate, sulfamerazine

## KEY WORDS

SPE

## REFERENCE

Taylor, R.B.; Richards, R.M.E.; Xing, D.K.-I. Determination of antibacterial agents in microbiological cultures by high-performance liquid chromatography. *Analyst*, **1990**, *115*, 797–799

## SAMPLE

**Matrix:** cell suspensions

**Sample preparation:** Cool cell suspension in an ice bath, centrifuge at 800 g at 4° for 15 min, inject an aliquot of the supernatant.

## HPLC VARIABLES

**Column:** μBondapak C18

**Mobile phase:** MeCN:MeOH:water 10:30:60 containing 10 mM K₂HPO₄

**Flow rate:** 2

**Detector:** UV 280

## CHROMATOGRAM

**Limit of detection:** 250 ng/mL

## OTHER SUBSTANCES

**Also analyzed:** brodimoprim

## REFERENCE

Climax, J.; Lenehan, T.J.; Lambe, R.; Kenny, M.; Caffrey, E.; Darragh, A. Interaction of antimicrobial agents with human peripheral blood leucocytes: uptake and intracellular localization of certain sulphonamides and trimethoprims. *J.Antimicrob.Chemother.*, **1986**, *17*, 489–498

## SAMPLE
**Matrix:** feed
**Sample preparation:** Weigh out 1 g ground feed, add 1 mL 1 mg/mL sulfamethazine in water, add 3 mL trichloroacetic acid solution, mix well, sonicate at 40° for 10 min, make up to 500 mL with MeCN:10 mM $Na_2HPO_4$ adjusted to pH 3 with phosphoric acid 20:80, mix well, filter a 500 μL aliquot (Costar spin-X (low type) 0.22 μm cellulose acetate) with centrifuging for 1 min, inject a 10 μL aliquot of the filtrate. (Prepare trichloroacetic acid solution by mixing 87 g trichloroacetic acid with 13 g water, add 0.7 mL of this solution to 99.3 mL acetone.)

## HPLC VARIABLES
**Guard column:** 20 × 4.6 5 μm Supelcosil-LC-18-DB
**Column:** 250 × 4.6 5 μm Supelcosil-LC-18-DB
**Mobile phase:** MeCN containing 0.1% triethylamine:10 mM pH 2.8 $Na_2HPO_4$ 21:79
**Flow rate:** 0.9
**Injection volume:** 10
**Detector:** UV 270

## CHROMATOGRAM
**Retention time:** 7
**Internal standard:** sulfamethazine (sulfadimidine) (8)
**Limit of quantitation:** 250 μg/g (muscle); 250 mg/mL (plasma)

## OTHER SUBSTANCES
**Extracted:** sulfadiazine

## REFERENCE
Hormazabal, V.; Steffanak, I.; Yndestad, M. Simultaneous extraction and determination of sulfadiazine and trimethoprim in medicated fish feed by high-performance liquid chromatography. *J.Chromatogr.*, **1993**, *648*, 183–186

## SAMPLE
**Matrix:** formulations
**Sample preparation:** Powder tablets, add 40 mg trimethoprim, dissolve in 70 mL MeOH, filter (paper), wash filter with MeOH, make up filtrate to 100 mL with MeOH. Dilute a 1 mL aliquot to 10 mL, inject a 10 μL aliquot.

## HPLC VARIABLES
**Column:** 250 × 4 10 μm Nucleosil C18
**Mobile phase:** MeOH:MeCN:water:triethylamine 20:20:60:0.15, pH adjusted to 3.0 with phosphoric acid
**Flow rate:** 1.5
**Injection volume:** 10
**Detector:** UV 240

## CHROMATOGRAM
**Retention time:** 3
**Internal standard:** trimethoprim

## OTHER SUBSTANCES
**Simultaneous:** nalidixic acid, phenazopyridine

## KEY WORDS
tablets; trimethoprim is IS

## REFERENCE

Sane, R.T.; Ghadge, J.K.; Jani, A.B.; Vaidya, A.J.; Kotwal, S.S. Simultaneous high-performance liquid chromatographic determination of haloperidol with propantheline bromide, nalidixic acid with phenazopyridine hydrochloride, and dipyridamole with aspirin in combined dosage (forms). *Indian Drugs*, **1992**, *29*, 240–244

## SAMPLE

**Matrix:** formulations
**Sample preparation:** Tablets. Powder tablets, weigh out an amount equivalent to about 100 mg trimethoprim, suspend in MeOH, sonicate for 2 min, filter, dilute with MeOH to 250 µg/mL. Suspensions. Dilute with MeOH to 250 µg/mL, sonicate, filter.

## HPLC VARIABLES

**Column:** 250 × 4.6 Partisil 10 ODS-3
**Mobile phase:** MeCN:water 25:75 containing 1% ammonium acetate, pH 6.90 ± 0.1
**Flow rate:** 1
**Injection volume:** 20
**Detector:** UV 254

## CHROMATOGRAM

**Retention time:** 9

## OTHER SUBSTANCES

**Simultaneous:** degradation products, methyl p-hydroxybenzoate, sulfamethoxazole
**Noninterfering:** n-propyl p-hydroxybenzoate

## KEY WORDS

tablets; suspensions; stability-indicating

## REFERENCE

Bergh, J.J.; Breytenbach, J.C.; Du Preez, J.L. High-performance liquid chromatographic analysis of trimethoprim in the presence of its degradation products. *J.Chromatogr.*, **1990**, *513*, 392–396

## SAMPLE

**Matrix:** formulations
**Sample preparation:** Tablets. Crush tablets, weigh out amount equivalent to about 40 mg trimethoprim, dissolve in EtOH, filter, make up to 200 mL with EtOH. Suspensions. Dilute 5 mL suspension to 100 mL with EtOH, filter.

## HPLC VARIABLES

**Column:** 250 × 4.6 7 µm Zorbax TMS
**Mobile phase:** MeCN:1-propanol:MeOH:THF:acetic acid:water 5:20:15:25:1:34
**Flow rate:** 2
**Injection volume:** 20
**Detector:** UV 271

## CHROMATOGRAM

**Retention time:** 4.37

## OTHER SUBSTANCES

**Simultaneous:** degradation products, methyl p-hydroxybenzoate, propyl p-hydroxybenzoate, sulfamethoxazole

## KEY WORDS

tablets; suspensions; stability-indicating

## REFERENCE
Bergh, J.J.; Breytenbach, J.C. Stability-indicating high-performance liquid chromatographic analysis of trimethoprim in pharmaceuticals. *J.Chromatogr.*, **1987**, *387*, 528–531

## SAMPLE
**Matrix:** reaction mixtures
**Sample preparation:** If necessary, remove oxidizing power of solution by adding sodium metabisulfite, inject a 20 µL aliquot.

## HPLC VARIABLES
**Guard column:** 15 × 4.6 5 µm Microsorb C8
**Column:** 250 × 4.6 5 µm Microsorb C8
**Mobile phase:** MeCN:5.5 mM sodium octanesulfonate + 20 mM trisodium citrate dihydrate adjusted to pH 3 with concentrated HCl 30:70
**Flow rate:** 1
**Injection volume:** 20
**Detector:** UV 230

## CHROMATOGRAM
**Retention time:** 6.7
**Limit of detection:** 200 ng/mL

## REFERENCE
Lunn, G.; Rhodes, S.W.; Sansone, E.B.; Schmuff, N.R. Photolytic destruction and polymeric resin decontamination of aqueous solutions of pharmaceuticals. *J.Pharm.Sci.*, **1994**, *83*, 1289–1293

## SAMPLE
**Matrix:** solutions

## HPLC VARIABLES
**Column:** 300 × 0.35 5 µm Vydac IDI-TP C18 TMS capped
**Mobile phase:** Gradient. MeCN:MeOH:buffer 0:0:100 at start of run, to 0:5:95 after injection (step gradient), to 0:8:92 over 7 min, to 6:0:94 (step gradient), maintain at 6:0:94 for 14 min, to 0:16:84 over 5 min, to 0:18:82 over 5 min, to 0:30:70 over 5 min. (Buffer was 1 mM pH 2.72 phosphate buffer.)
**Column temperature:** 30
**Flow rate:** 0.006
**Injection volume:** 1
**Detector:** UV 270

## CHROMATOGRAM
**Retention time:** 27.5

## OTHER SUBSTANCES
**Simultaneous:** diaveridine, phthalyl sulfathiazole, pyrimethamine, succinyl sulfathiazole, sulfabenzamide, sulfacetamide, sulfachloropyridazine, sulfadiazine, sulfadimethoxine, sulfaguanidine, sulfamerazine, sulfameter, sulfamethazine, sulfamethizole, sulfamethoxazole, sulfamethoxypyridazine, sulfamoxole, sulfanilamide, sulfanilic acid, sulfapyridine, sulfaquinoxaline, sulfathiazole, sulfisomidine, sulfisoxazole

## KEY WORDS
capillary HPLC

## REFERENCE

Ricci, M.C.; Cross, R.F. High performance liquid chromatographic analyses of sulphonamides and dih-ydrofolate reductase inhibitors. II. Separations in acetonitrile modified solutions, ternary gradient studies & flow programming. *J.Liq.Chromatogr.& Rel.Technol.*, **1996**, *19*, 547−564

## SAMPLE
**Matrix:** solutions

## HPLC VARIABLES
**Column:** 300 × 0.35 5 μm Vydac IDI-TP C18 TMS capped
**Mobile phase:** Gradient. MeOH:buffer 0:100 at start of run, to 10:90 after injection (step gradient), to 12:88 over 30 min, to 18:82 over 5 min, to 30:70 over 5 min. (Buffer was 1 mM pH 2.72 phosphate buffer.)
**Column temperature:** 30
**Flow rate:** 0.006
**Injection volume:** 1
**Detector:** UV 270

## CHROMATOGRAM
**Retention time:** 31

## OTHER SUBSTANCES
**Simultaneous:** diaveridine, phthalyl sulfathiazole, pyrimethamine, succinyl sulfathiazole, sulfabenzamide, sulfacetamide, sulfachloropyridazine, sulfadiazine, sulfadimethoxine, sulfaguanidine, sulfamerazine, sulfameter, sulfamethazine, sulfamethizole, sulfamethoxazole, sulfamethoxypyridazine, sulfamoxole, sulfanilamide, sulfanilic acid, sulfapyridine, sulfaquinoxaline, sulfathiazole, sulfisomidine, sulfisoxazole

## KEY WORDS
capillary HPLC

## REFERENCE

Ricci, M.C.; Cross, R.F. High-performance liquid chromatographic analyses of sulphonamides and dih-ydrofolate reductase inhibitors. I. Separations in methanol-modified solutions. *J.Liq. Chrom.Rel.Technol.*, **1996**, *19*, 365−381

## SAMPLE
**Matrix:** solutions

## HPLC VARIABLES
**Column:** 250 × 4.6 5 μm Supelcosil LC-DP (A) or 250 × 4 5 μm LiChrospher 100 RP-8 (B)
**Mobile phase:** MeCN:0.025% phosphoric acid:buffer 25:10:5 (A) or 60:25:15 (B) (Buffer was 9 mL concentrated phosphoric acid and 10 mL triethylamine in 900 mL water, adjust pH to 3.4 with dilute phosphoric acid, make up to 1 L.)
**Flow rate:** 0.6
**Injection volume:** 25
**Detector:** UV 229

## CHROMATOGRAM
**Retention time:** 6.67 (A), 3.65 (B)

## OTHER SUBSTANCES
**Also analyzed:** acebutolol, acepromazine, acetaminophen, acetazolamide, acetophenazine, albuterol, alprazolam, amitriptyline, amobarbital, amoxapine, antipyrine, atenolol, atropine, azatadine, baclofen, benzocaine, bromocriptine, brompheniramine, brotizolam,

bupivacaine, buspirone, butabarbital, butalbital, caffeine, carbamazepine, cetirizine, chlorcyclizine, chlordiazepoxide, chlormezanone, chloroquine, chlorpheniramine, chlorpromazine, chlorpropamide, chlorprothixene, chlorthalidone, chlorzoxazone, cimetidine, cisapride, clomipramine, clonazepam, clonidine, clozapine, cocaine, codeine, colchicine, cyclizine, cyclobenzaprine, dantrolene, desipramine, diazepam, diclofenac, diflunisal, diltiazem, diphenhydramine, diphenidol, diphenoxylate, dipyridamole, disopyramide, dobutamine, doxapram, doxepin, droperidol, encainide, ethidium bromide, ethopropazine, fenoprofen, fentanyl, flavoxate, fluoxetine, fluphenazine, flurazepam, flurbiprofen, fluvoxamine, furosemide, glutethimide, glyburide, guaifenesin, haloperidol, homatropine, hydralazine, hydrochlorothiazide, hydrocodone, hydromorphone, hydroxychloroquine, hydroxyzine, ibuprofen, imipramine, indomethacin, ketoconazole, ketoprofen, ketorolac, labetalol, levorphanol, lidocaine, loratadine, lorazepam, lovastatin, loxapine, mazindol, mefenamic acid, meperidine, mephenytoin, mepivacaine, mesoridazine, metaproterenol, metformin, methadone, methdilazine, methocarbamol, methotrexate, methotrimeprazine, methoxamine, methyldopa, methylphenidate, metoclopramide, metolazone, metoprolol, metronidazole, midazolam, moclobemide, morphine, nadolol, nalbuphine, naloxone, naphazoline, naproxen, nifedipine, nizatidine, norepinephrine, nortriptyline, oxazepam, oxycodone, oxymetazoline, paroxetine, pemoline, pentazocine, pentobarbital, pentoxifylline, perphenazine, pheniramine, phenobarbital, phenol, phenolphthalein, phentolamine, phenylbutazone, phenyltoloxamine, phenytoin, pimozide, pindolol, piroxicam, pramoxine, prazepam, prazosin, probenecid, procainamide, procaine, prochlorperazine, procyclidine, promazine, promethazine, propafenone, propantheline, propiomazine, propofol, propranolol, protriptyline, quazepam, quinidine, quinine, racemethorphan, ranitidine, remoxipride, risperidone, salicylic acid, scopolamine, secobarbital, sertraline, sotalol, spironolactone, sulfinpyrazone, sulindac, temazepam, terbutaline, terfenadine, tetracaine, theophylline, thiethylperazine, thiopental, thioridazine, thiothixene, timolol, tocainide, tolbutamide, tolmetin, trazodone, triamterene, triazolam, trifluoperazine, triflupromazine, trimeprazine, trimipramine, verapamil, warfarin, xylometazoline, yohimbine, zopiclone

## KEY WORDS
some details of plasma extraction

## REFERENCE
Koves, E.M. Use of high-performance liquid chromatography-diode array detection in forensic toxicology. *J.Chromatogr.A*, **1995**, *692*, 103–119

## SAMPLE
**Matrix:** solutions

## HPLC VARIABLES
**Column:** 250 × 4.6 Zorbax RX
**Mobile phase:** Gradient. A was 10 mL concentrated orthophosphoric acid and 7 mL triethylamine in 1 L water. B was 10 mL concentrated orthophosphoric acid and 7 mL triethylamine in 200 mL water, make up to 1 L with MeCN. A:B from 100:0 to 0:100 over 30 min, maintain at 0:100 for 5 min.
**Column temperature:** 30
**Flow rate:** 2
**Detector:** UV 210

## OTHER SUBSTANCES
**Also analyzed:** acepromazine, acetaminophen, acetophenazine, albuterol, aminophylline, amitriptyline, amobarbital, amoxapine, amphetamine, amylocaine, antipyrine, aprobarbital, aspirin, atenolol, atropine, avermectin, barbital, benzocaine, benzoic acid, benzotropine, benzphetamine, berberine, bibucaine, bromazepan, brompheniramine, buprenorphine, buspirone, butabarbital, butacaine, butethal, caffeine, carbamazepine, carbromal, chloramphenicol, chlordiazepoxide, chloroquine, chlorothiazide, chloroxylenol, chlorphe-

nesin, chlorpheniramine, chlorpromazine, chlorpropamide, chlortetracycline, cimetidine, cinchonidine, cinchonine, clenbuterol, clonazepam, clonixin, clorazepate, cocaine, codeine, colchicine, cortisone, coumarin, cyclazocine, cyclobenzaprine, cyclothiazide, cyheptamide, cymarin, danazol, danthron, dapsone, debrisoquine, desipramine, dexamethasone, dextromethorphan, dextropropoxyphene, diamorphine, diazepam, diclofenac, diethylpropion, diethylstilbestrol, diflunisal, digitoxin, digoxin, diltiazem, diphenhydramine, diphenoxylate, diprenorphine, dipyrone, disulfiram, dopamine, doxapram, doxepin, dronabinol, ephedrine, epinephrine, epinine, estradiol, estriol, estrone, ethacrynic acid, ethosuximide, etonitazene, etorphine, eugenol, famotidine, fenbendazole, fencamfamine, fenoprofen, fenproporex, fentanyl, flubendazole, flufenamic acid, flunitrazepam, 5-fluorouracil, fluoxymesterone, fluphenazine, furosemide, gentisic acid, gitoxigenin, glipizide, glunixin, glutethimide, glybenclamide, guaiacol, halazepam, haloperidol, hydrochlorothiazide, hydrocodone, hydrocortisone, hydromorphone, hydroxyquinoline, ibogaine, ibuprofen, iminostilbene, imipramine, indomethacin, isocarbostyril, isocarboxazid, isoniazid, isoproterenol, isoxsuprine, ivermectin, ketamine, ketoprofen, kynurenic acid, levorphanol, lidocaine, lorazepam, lormetazepam, loxapine, mazindol, mebendazole, meclizine, meclofenamic acid, medazepam, mefenamic acid, megestrol, mepacrine, meperidine, mephentermine, mephenytoin, mephesin, mephobarbital, mepivacaine, mescaline, mesoridazine, methadone, methamphetamine, methapyrilene, methaqualone, methazolamide, methocarbamol, methoxamine, methsuximide, methyl salicylate, methyldopa, methyldopamine, methylphenidate, methylprednisolone, methyltestosterone, methyprylon, metoprolol, mibolerone, morphine, nadolol, nalorphine, naloxone, naltrexone, naphazoline, naproxen, nefopam, niacinamide, nicotine, nicotinic acid, nifedipine, niflumic acid, nitrazepam, norepinephrine, nortriptyline, noscapine, nylidrin, oxazepam, oxycodone, oxymorphone, oxyphenbutazone, oxytetracycline, papaverine, pargyline, pemoline, pentazocine, pentobarbital, persantine, phenacetin, phenazocine, phenazopyridine, phencyclidine, phendimetrazine, phenelzine, pheniramine, phenobarbital, phenothiazine, phensuximide, phentermine, phenylbutazone, phenylephrine, phenylpropanolamine, piperocaine, prazepam, prednisolone, primidone, probenecid, progesterone, propiomazine, propranolol, propylparaben, pseudoephedrine, puromycin, pyrilamine, pyrithyldione, quazepam, quinaldic acid, quinidine, quinine, ranitidine, recinnamine, reserpine, resorcinol, saccharin, albuterol, salicylamide, salicylic acid, scopolamine, scopoletin, secobarbital, strychnine, sulfacetamide, sufadiazine, sulfadimethoxine, sulfaethidole, sulfamerazine, sulfamethazine, sulfamethoxizole, sulfanilamide, sulfapyridine, sulfasoxizole, sulindac, tamoxifen, temazepam, testosterone, tetracaine, tetracycline, tetramisole, thebaine, theobromine, theophylline, thiabendazole, thiamine, thiamylal, thiobarbituric acid, thioridazine, thiosalicylic acid, thiothixene, thymol, tolazamide, tolazoline, tobutamide, tolmetin, tranylcypromine, triamcinolone, tribenzylamine, trichloromethiazide, trifluoperazine, trihexyphenidyl, trimethoprim, tripelennamine, triprolidine, tropacocaine, tyramine, verapamil, vincamine, warfarin, yohimbine, zoxazolamine

## REFERENCE
Hill, D.W.; Kind, A.J. Reversed-phase solvent-gradient HPLC retention indexes of drugs. *J.Anal.Toxicol.*, **1994**, *18*, 233–242

## SAMPLE
**Matrix:** solutions
**Sample preparation:** Prepare a 2.5-5 µg/mL solution, inject a 20 µL aliquot.

## HPLC VARIABLES
**Column:** 80 × 4.6 3.65 µm Zorbax Rx-SIL (similar to Zorbax SB-C8 (Mac-Mod Analytical))
**Mobile phase:** MeCN:buffer 16:84 (Buffer was 0.1% trifluoroacetic acid adjusted to pH 3 with ammonium hydroxide.)
**Flow rate:** 1
**Injection volume:** 20
**Detector:** UV 220

## CHROMATOGRAM
**Retention time:** k′ 3.0

## REFERENCE
Kirkland, K.M.; McCombs, D.A.; Kirkland, J.J. Rapid, high-resolution high-performance liquid chromatographic analysis of antibiotics. *J.Chromatogr.A*, **1994**, *660*, 327–337

## SAMPLE
**Matrix:** solutions
**Sample preparation:** Centrifuge and filter cell solutions (0.22 μm), inject an aliquot.

## HPLC VARIABLES
**Guard column:** Guard-PAK C18 (Waters)
**Column:** 150 × 3.9 5 μm NOVA PAK C18
**Mobile phase:** MeOH:50 mM pH 6.0 $KH_2PO_4$ 35:65
**Flow rate:** 0.6
**Detector:** UV 254

## CHROMATOGRAM
**Retention time:** 5.8

## REFERENCE
Koga, H. High-performance liquid chromatography measurement of antimicrobial concentrations in polymorphonuclear leukocytes. *Antimicrob.Agents Chemother.*, **1987**, *31*, 1904–1908

## SAMPLE
**Matrix:** solutions
**Sample preparation:** Prepare a 10 μg/mL solution in MeOH, inject a 20 μL aliquot.

## HPLC VARIABLES
**Column:** 125 × 4.9 Spherisorb S5W silica
**Mobile phase:** MeOH containing 10 mM ammonium perchlorate and 1 mL/L 100 mM NaOH in MeOH, pH 6.7
**Flow rate:** 2
**Injection volume:** 20
**Detector:** E, LeCarbone, V25 glassy carbon electrode, + 1.2 V

## CHROMATOGRAM
**Retention time:** 1.94

## OTHER SUBSTANCES
**Also analyzed:** acebutolol, acepromazine, acetophenazine, N-acetylprocainamide, albuterol, alprenolol, amethocaine, amiodarone, amitriptyline, antazoline, atenolol, azacyclonal, bamethane, benactyzine, benperidol, benzethidine, benzocaine, benzoctamine, benzphetamine, benzquinamide, bromhexine, bromodiphenhydramine, bromperidol, brompheniramine, brompromazine, buclizine, bufotenine, bupivacaine, buprenorphine, butacaine, butethamate, chlorcyclizine, chlorpheniramine, chlorphenoxamine, chlorprenaline, chlorpromazine, chlorprothixene, cimetidine, cinchonidine, cinnarizine, clemastine, clomipramine, clonidine, cocaine, cyclazocine, cyclizine, cyclopentamine, cyproheptadine, deserpidine, desipramine, dextromoramide, dextropropoxyphene, dicyclomine, diethylcarbamazine, diethylpropion, diethylthiambutene, dihydroergotamine, dimethindene, dimethothiazine, diphenhydramine, diphenoxylate, dipipanone, diprenorphine, dipyridamole, disopyramide, dothiepin, doxapram, doxepin, doxylamine, droperidol, ephedrine, ergocornine, ergocristine, ergocristinine, ergocryptine, ergometrine, ergosine, ergosinine, ergotamine, ethopropazine, etorphine, etoxeridine, fenethazine, fenfluramine, fenoterol,

fentanyl, flavoxate, fluopromazine, flupenthixol, fluphenazine, flurazepam, haloperidol, hydroxyzine, hyoscine, ibogaine, imipramine, indapamine, iprindole, isothipendyl, isoxsuprine, ketanserin, laudanosine, lidocaine, lofepramine, loxapine, maprotiline, mecamylamine, meclophenoxate, meclozine, medazepam, mephentermine, mepivacaine, meptazinol, mepyramine, mesoridazine, metaraminol, methadone, methamphetamine, methapyrilene, methdilazene, methotrimeprazine, methoxamine, methoxyphenamine, methoxypromazine, methylephedrine, methylergonovine, methysergide, metoclopramide, metopimazine, metoprolol, mianserin, morazone, nadolol, nalorphine, naloxone, naphazoline, nicotine, nifedipine, nomifensine, nortriptyline, noscapine, orphenadrine, oxeladin, oxprenolol, oxymetazolin, papaverine, pargyline, pecazine, penbutolol, pentazocine, penthienate, pericyazine, perphenazine, phenadoxone, phenampromide, phenazocine, phenbutrazate, phendimetrazine, phenelzine, phenglutarimide, phenindamine, pheniramine, phenmetrazine, phenomorphan, phenoperidine, phenothiazine, phenoxybenzamine, phentolamine, phenylephrine, phenyltoloxamine, physostigmine, piminodine, pimozide, pindolol, pipamazine, pipazethate, piperacetazine, piperidolate, pipradol, pirenzepine, piritramide, pizotifen, practolol, pramoxine, prazosin, prenylamine, prilocaine, primaquine, proadifen, procainamide, procaine, prochlorperazine, procyclidine, proheptazine, prolintane, promazine, promethazine, pronethalol, properidine, propiomazine, propranolol, prothipendyl, protriptyline, proxymetacaine, pseudoephedrine, pyrimethamine, quinidine, quinine, ranitidine, rescinnamine, sotalol, tacrine, terazosin, terbutaline, terfenadine, thenyldiamine, theophylline, thiethylperazine, thiopropazate, thioproperazine, thioridazine, thiothixene, thonzylamine, timolol, tocainide, tolpropamine, tolycaine, tranylcypromine, trazodone, trifluoperazine, trifluperidol, trimeperidine, trimeprazine, trimethobenzamide, trimipramine, tripelennamine, triprolidine, tryptamine, verapamil, xylometazoline

## REFERENCE
Jane, I.; McKinnon, A.; Flanagan, R.J. High-performance liquid chromatographic analysis of basic drugs on silica columns using non-aqueous ionic eluents. II. Application of UV, fluorescence and electrochemical oxidation detection. *J.Chromatogr.*, **1985**, *323*, 191–225

## SAMPLE
**Matrix:** urine
**Sample preparation:** Condition a 1 mL 100 mg Bond Elut CN SPE cartridge with 3 mL MeCN and 3 mL water. 1 mL Urine + 1.1 mL water, add to the SPE cartridge, wash with 3 mL water, elute with 1 mL MeCN:water 15:85 containing 0.75% triethylamine and 0.375% phosphoric acid (85%), inject a 200 µL aliquot.

## HPLC VARIABLES
**Guard column:** 15 × 3.2 7 µm Aquapore C18 (Brownlee)
**Column:** 100 × 4.6 5 µm Hypersil ODS C18
**Mobile phase:** Gradient. MeCN:buffer from 9:91 to 35:65, re-equilibrate at initial conditions for 5 min. (Buffer was 0.16% triethylamine containing 0.08% orthophosphoric acid (85%), pH 4.2.)
**Flow rate:** 1.5
**Injection volume:** 200
**Detector:** UV 241

## CHROMATOGRAM
**Retention time:** 5.8
**Internal standard:** trimethoprim

## OTHER SUBSTANCES
**Extracted:** trimetrexate

## KEY WORDS
SPE; trimethoprim is IS

# REFERENCE

Tinsley, P.W.; LaCreta, F.P. Improved chromatographic method for the determination of trimetrexate in urine. *J.Chromatogr.*, **1990**, *529*, 468–472

◆━━━━━━━━━━━━━━━━━━━━━━━━━◆━━━━━━━━━━━━━━━━━━━━━━━━━◆

## ANNOTATED BIBLIOGRAPHY

Bonazzi, D.; Andrisano, V.; Di Pietra, A.M.; Cavrini, V. Analysis of trimethoprim-sulfonamide drug combinations in dosage forms by UV spectroscopy and liquid chromatography (HPLC). *Farmaco*, **1994**, *49*, 381–386 [simultaneous sulfadiazine, sulfamethoxazole, sulfamethoxypyridazine]

Kanda, T.; Kutsuna, H.; Ohtsu, Y.; Yamaguchi, M. Synthesis of polymer-coated mixed-functional packing materials for direct analysis of drug-containing serum and plasma by high-performance liquid chromatography. *J.Chromatogr.A*, **1994**, *672*, 51–57 [serum; plasma; direct injection; column temp 40; also carbamazepine, chloramphenicol, indomethacin, phenobarbital, phenytoin, theophylline]

Mokry, M.; Klimes, J.; Zahradnicek, M. HPLC analysis of some sulfonamides in selected pharmaceutical formulations. *Pharmazie*, **1994**, *49*, 333–335 [tablets; injections; acetanilide (IS); simultaneous sulfamethoxazole; also phenacetin, phenazone, phenobarbital, phthalylsulfathiazole, sulfadimidine, sulfamethoxydiazine, sulfamoxole, sulfisoxazole]

Shah, K.P.; Chang, M.; Riley, C.M. Automated analytical systems for drug development studies. II-A system for dissolution testing. *J.Pharm.Biomed.Anal.*, **1994**, *12*, 1519–1527 [tablets; also acetaminophen, sulfamethoxazole]

Mengelers, M.J.B.; Polman, A.M.M.; Aerts, M.M.L.; Kuiper, H.A.; Van Miert, A.S.J.P.A.M. Determination of sulfadimethoxine, sulfamethoxazole, trimethoprim and their main metabolites in lung and edible tissues from pigs by multi-dimensional liquid chromatography. *J.Liq.Chromatogr.*, **1993**, *16*, 257–278 [lung; muscle; kidney; liver; LOD 10-50 ng/g; column temp 30; column-switching]

Abounassif, M.A.; Hagga, M.E.M.; Gad-Kariem, E.A.; Al-Awadi, M.E. Simultaneous quantitation of sulfametrole and trimethoprim mixture by high-performance liquid chromatography and UV-spectrophotometry using least squares. *Acta Pharm.Fenn.*, **1992**, *101*, 51–56

Endoh, Y.S.; Takahashi, Y.; Nishikawa, M. HPLC determination of sulfonamides, their N4-acetyl metabolites and diaminopyrimidine coccidiostats in chicken tissues. *J.Liq.Chromatogr.*, **1992**, *15*, 2091–2110 [skin; plasma; muscle; liver; kidney; LOD 20-50 ng/g; also N-acetyldiaveridine, N-acetylsulfadiazine, N-acetylsulfadimethoxine, N-acetylsulfamethoxazole, N-acetylsulfamonomethoxine, N-acetylsulfaquinoxaline, diaveridine, ormethoprim, sulfadiazine, sulfadimethoxine, sulfamethoxazole, sulfamonomethoxine, sulfaquinoxaline]

Van't Klooster, G.A.E.; Kolker, H.J.; Woutersen-Van Nijnanten, F.M.A.; Noordhoek, J.; Van Miert, A.S.J.P.A.M. Determination of trimethoprim and its oxidative metabolites in cell culture media and microsomal incubation mixtures by high-performance liquid chromatography. *J.Chromatogr.*, **1992**, *579*, 355–360 [rat; liver; diaveridine (IS); LOD 150 ng/mL]

Avgerinos, A.; Athanasiou, G.; Malamataris, S. Rapid simultaneous determination of trimethoprim, sulfamethoxazole and acetylsulfamethoxazole in human plasma and urine by high-performance liquid chromatography. *J.Pharm.Biomed.Anal.*, **1991**, *9*, 507–510

DeAngelis, D.V.; Woolley, J.L.; Sigel, C.W. High-performance liquid chromatographic assay for the simultaneous measurement of trimethoprim and sulfamethoxazole in plasma or urine. *Ther.Drug Monit.*, **1990**, *12*, 382–392

McNally, V.; Lenehan, T.; Kelly, M.T.; Smyth, M.R. High-performance liquid chromatographic determination of trimethoprim and sulfadiazine in medicated fish feedstock. *Anal.Lett.*, **1990**, *23*, 2215–2231

Varoquaux, O.; Cordonnier, P.; Advenier, C.; Pays, M. Simultaneous HPLC determination of trimethoprim, sulfamethoxazole and its N4-acetyl metabolite in biological fluids. *Methodol.Surv. Biochem.Anal.*, **1990**, *20*, 123–130

Meatherall, R. High-performance liquid chromatographic determination of trimethoprim in serum. *Ther.Drug Monit.*, **1989**, *11*, 79–83

Mengelers, M.J.B.; Oorsprong, M.B.M.; Kuiper, H.A.; Aerts, M.M..L.; Van Gogh, E.R.; Van Miert, A.S.J.P.A.M. Determination of sulfadimethoxine, sulfamethoxazole, trimethoprim and their main metabolites in porcine plasma by column switching HPLC. *J.Pharm.Biomed.Anal.*, **1989**, *7*, 1765–1776

Tu, Y.-H.; Allen, L.V., Jr.; Fiorica, V.M.; Albers, D.D. Pharmacokinetics of trimethoprim in the rat. *J.Pharm.Sci.*, **1989**, *78*, 556–560 [plasma; brain; heart; lung; liver; spleen; kidney; prostate; testicles; seminal vesicles; LOD 100 ng/mL; chlorphenesin carbamate; pharmacokinetics]

Tu, Y.H.; Wang, D.-P.; Allen, L.V., Jr. Stability of a nonaqueous trimethoprim preparation. *Am.J.Hosp.Pharm.*, **1989**, *46*, 301–304 [propylene glycol; dimethylacetamide; LOQ 1 µg/mL]

Chen, G.; Tu, Y.H.; Allen, L.V., Jr.; Wang, D.P. Determination of trimethoprim in rat blood, plasma, prostate gland and seminal vesicles by high-performance liquid chromatography. *Int.J.Pharm.*, **1988**, *46*, 89–93

Erdmann, G.R.; Canafax, D.M.; Giebink, G.S. High-performance liquid chromatographic analysis of trimethoprim and sulfamethoxazole in microliter volumes of chinchilla middle ear effusion and serum. *J.Chromatogr.*, **1988**, *433*, 187–195 [LOQ 100 ng/mL; LOD 50 ng/mL; column temp 45; pharmacokinetics]

Svirbely, J.E.; Pesce, A.J. Trimethoprim analysis by LC. *J.Liq.Chromatogr.*, **1988**, *11*, 1075–1085 [LOD 50 ng/mL; serum; plasma; dialysate; urine; CSF; SPE; interfering atenolol, procainamide, oxazepam; non-interfering acetaminophen, amitriptyline, ampicillin, caffeine, cefitoxin, chlordiazepoxide, chlorpheniramine, chlorpromazine, cimetidine, clonidine, desipramine, diazepam, digoxin, diphenhydramine, doxepin, doxylamine, erythromycin, flurazepam, imipramine, lidocaine, minoxidil, nadolol, nortriptyline, prazosin, propranolol, quinidine, theophylline, thioridazine]

Hoppu, K.; Arjomaa, P. Determination of trimethoprim in pediatric samples by high-performance liquid chromatography. *Clin.Chim.Acta*, **1987**, *163*, 81–86

Svirbely, J.E.; Pesce, A.J. A high performance liquid chromatography method for trimethoprim utilizing solid-phase column extraction. *Ther.Drug Monit.*, **1987**, *9*, 216–220

Torel, J.; Cillard, J.; Cillard, P.; Vie, M. Simultaneous analysis of three antimicrobial agents in feed premixes by reversed-phase high-performance liquid chromatography. *J.Chromatogr.*, **1985**, *323*, 447–450 [simultaneous sulfamethazine, sulfamethoxypyridazine; LOD 80 ng]

Nordholm, L.; Dalgaard, L. Determination of trimethoprim metabolites including conjugates in urine using high-performance liquid chromatography with combined ultraviolet and electrochemical detection. *J.Chromatogr.*, **1984**, *305*, 391–399 [pig]

Spreux-Varoquaux, O.; Chapalain, J.P.; Cordonnier, P.; Advenier, C.; Pays, M.; Lamine, L. Determination of trimethoprim, sulfamethoxazole and its N4-acetyl metabolite in biological fluids by high-performance liquid chromatography. *J.Chromatogr.*, **1983**, *274*, 187–199 [LOD 15 ng/mL; plasma; urine; normal phase; gradient; simultaneous theophylline; non-interfering caffeine; pharmacokinetics]

Gochin, R.; Kanfer, I.; Haigh, J.M. Simultaneous determination of trimethoprim, sulfamethoxazole and N⁴-acetylsulfamethoxazole in serum and urine by high-performance liquid chromatography. *J.Chromatogr.*, **1981**, *223*, 139–145 [LOQ 100 ng/mL; column temp 30; sulfafurazole (IS); extracted aspirin, caffeine]

Ascalone, V. Assay of trimethoprim, sulfamethoxazole and its N4-acetyl metabolite in biological fluids by high-pressure liquid chromatography. *J.High Resolut.Chromatogr.Chromatogr.Commun.*, **1980**, *3*, 261–264

Weinfeld, R.E.; Macasieb, T.C. Determination of trimethoprim in biological fluids by high-performance liquid chromatography. *J.Chromatogr.*, **1979**, *164*, 73–84

Ferry, D.G.; McQueen, E.G.; Hearn, M.T.W. Sulfamethoxazole and trimethoprim estimation by high performance liquid chromatography. *Proc.Univ.Otago Med.Sch.*, **1978**, *56*, 46–48

# Valproic Acid

**Molecular formula:** $C_8H_{16}O_2$
**Molecular weight:** 144.2
**CAS Registry No.:** 99-66-1, 1069-66-5 (sodium salt), 76584-70-8 (semisodium salt)

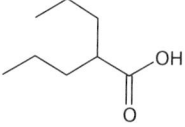

## SAMPLE
**Matrix:** blood
**Sample preparation:** Prepare ultrafiltrate from serum with an Amicon Centifree unit by centrifuging at 700 g for 10 min. 25 µL Ultrafiltrate + 475 µL 10 µg/mL undecylenic acid in MeCN, centrifuge. Remove 50 µL supernatant, add 100 µL 18-crown-6 solution, add 50 µL 1 mg/mL 4-bromomethyl-7-methoxycoumarin in MeCN, let stand in the dark at 65° for 30 min, inject a 5 µL aliquot. (Prepare 18-crown-6 solution by dissolving 100 mg potassium carbonate in 50 µL water, add 5 mL 20 mM 18-crown-6 in MeCN, sonicate for 30 min, add 5 mL MeCN.)

## HPLC VARIABLES
**Column:** 100 × 2.1 5 µm HP Hypersil-ODS
**Mobile phase:** MeOH:water 80:20
**Column temperature:** 40
**Flow rate:** 0.3
**Injection volume:** 5
**Detector:** F ex 322 em 695

## CHROMATOGRAM
**Retention time:** 2.5
**Internal standard:** undecylenic acid (4.5)
**Limit of quantitation:** 6.25 µg/mL

## OTHER SUBSTANCES
**Noninterfering:** carbamazepine, phenobarbital, phenytoin

## KEY WORDS
serum; derivatization

## REFERENCE
Liu, H.; Forman, L.J.; Montoya, J.; Eggers, C.; Barham, C.; Delgado, M. Determination of valproic acid by high-performance liquid chromatography with photodiode-array and fluorescence detection. *J.Chromatogr.*, **1992**, *576*, 163–169

## SAMPLE
**Matrix:** blood
**Sample preparation:** Filter (Amicon Centifree) serum at 550 g for 15 min. Remove a 25 µL aliquot of the ultrafiltrate and add it to 475 µL 1 µg/mL undecylenic acid in MeCN, centrifuge. Remove a 50 µL aliquot of the supernatant and add it to 100 µL 18-crown-6 suspension and 50 µL 1 mg/mL 4-bromomethyl-7-methoxycoumarin in MeCN, heat at 65° in the dark for 30 min, inject a 5 µL aliquot. (Prepare the 18-crown-6 suspension by dissolving 100 mg potassium carbonate in 50 µL water then adding this mixture to 5 mL 20 mM 18-crown-6 in MeCN, sonicate for 30 min, add 5 mL MeCN.)

## HPLC VARIABLES
**Column:** 100 × 2.1 5 µm HP Hypersil-ODS
**Mobile phase:** MeOH:water 80:20
**Column temperature:** 40
**Flow rate:** 0.3

**Injection volume:** 5
**Detector:** UV 200; UV 320; F ex 322 em 695

## CHROMATOGRAM
**Retention time:** 2.5
**Internal standard:** undecylenic acid (4.4)
**Limit of detection:** 1.25 ng/mL (S/N 12, fluorescence)

## OTHER SUBSTANCES
**Noninterfering:** acetaminophen, carbamazepine, chlordiazepoxide, clonazepam, desmethyldiazepam, diazepam, digoxin, disopyramide, ethosuximide, gentamicin, lidocaine, lorazepam, methotrexate, nitrazepam, oxazepam, phenobarbital, phenytoin, prazepam,
primidone, procainamide, quinidine, temazepam, theophylline, vancomycin

## KEY WORDS
serum; derivatization

## REFERENCE
Liu, H.; Montoya, J.L.; Forman, L.J.; Eggers, C.M.; Barham, C.F.; Delgado, M. Determination of free
valproic acid: evaluation of the Centrifree system and comparison between high-performance liquid
chromatography and enzyme immunoassay. *Ther.Drug Monit.*, **1992**, *14*, 513–521

## SAMPLE
**Matrix:** blood
**Sample preparation:** 200 μL Plasma + 500 μL water, shake, add 300 μL 25 mM perchloric
acid, add 3 mL cyclohexane, vortex for 3 min, centrifuge at 3500 rpm for 5 min, repeat
extraction twice more. Combine the organic layers and dry them over anhydrous sodium
sulfate, add 1 μmole sodium methoxide in MeOH, shake for 1 min, evaporate to dryness
under a stream of nitrogen at room temperature, reconstitute with 200 μL MeCN, add
100 μL buffer, add 200 μL reagent, heat at 70° for 40 min, inject an aliquot. (Prepare the
buffer by dissolving 3.8 g $KH_2PO_4$ and 5.96 g $Na_2HPO_4$ in 200 mL water, pH 7.4. The
reagent was 17 mg/mL 2-naphthacyl bromide in MeCN containing 1 mg/mL dicyclohex-
ane-18-crown-6.)

## HPLC VARIABLES
**Column:** 125 mm long 3 μm HS C18 (Perkin-Elmer)
**Mobile phase:** MeOH:water 77:23
**Flow rate:** 1
**Injection volume:** 6
**Detector:** UV 280

## CHROMATOGRAM
**Retention time:** 5.5
**Limit of detection:** 3.47 μg/mL

## KEY WORDS
plasma; derivatization; comparison with TLC

## REFERENCE
Corti, P.; Cenni, A.; Corbini, G.; Dreassi, E.; Murratzu, C.; Caricchia, A.M. Thin-layer chromatography
and densitometry in drug assay: comparison of methods for monitoring valproic acid in plasma.
*J.Pharm.Biomed.Anal.*, **1990**, *8*, 431–436

## SAMPLE
**Matrix:** blood

**Sample preparation:** Perform all operations with the exclusion of light. Evaporate 240 μL derivatization solution into a vial, add 400 μL 50 mM pH 7.0 phosphate buffer, add 100 μL plasma, add 10 μL 46 μg/mL undecylenic acid in MeCN, vortex for 5 s, heat at 70° for 40 min, add 500 μL MeCN, centrifuge at 3000 g for 5 min, inject a 20 μL aliquot. (Derivatization solution was 1.65 g Arkopal N-130 (a non-ionic surfactant, nonylphenol/13 unit chain polyoxyethylene) + 650 mg tetrahexylammonium bromide + 60 mg 4-bromo-methyl-7-methoxycoumarin in 20 mL acetone.)

## HPLC VARIABLES
**Guard column:** 10 × 3 5-20 μm LiChroprep RP-8
**Column:** 100 × 3 5 μm Chromspher C18
**Mobile phase:** Gradient. MeOH:water 80:20 for 3 min, then to 100:0 over 6 min, then held at 100:0 for 4 min.
**Injection volume:** 20
**Detector:** F ex 330 em 395

## CHROMATOGRAM
**Retention time:** 4
**Internal standard:** undecylenic acid (7)
**Limit of detection:** 1 μg/mL

## OTHER SUBSTANCES
**Extracted:** metabolites
**Noninterfering:** carbamazepine, phenobarbital, phenytoin

## KEY WORDS
plasma; derivatization

## REFERENCE
van der Horst, F.A.; Eikelboom, G.G.; Holthuis, J.J. High-performance liquid chromatographic determination of valproic acid in plasma using a micelle-mediated pre-column derivatization. *J.Chromatogr.*, **1988**, *456*, 191–199

## SAMPLE
**Matrix:** blood
**Sample preparation:** 250 μL Serum + 500 μL 250 ng/mL diazepam in MeCN, vortex for 15 s, centrifuge at 1400 g for 10 min, inject a 25 μL aliquot of the supernatant.

## HPLC VARIABLES
**Column:** 150 × 4.6 5 μm Zorbax ODS
**Mobile phase:** Gradient. MeCN:buffer 37:63 for 9 min, to 60:40, maintain at 60:40 for 1.5 min, re-equilibrate at initial conditions for 3 min. (Buffer was 10 mM $NaH_2PO_4$ adjusted to pH 2.3 with phosphoric acid.)
**Column temperature:** 40
**Flow rate:** 2.5
**Injection volume:** 25
**Detector:** UV 210

## CHROMATOGRAM
**Retention time:** 5.1
**Internal standard:** diazepam (7.7)
**Limit of detection:** 1.5 μg/mL

## OTHER SUBSTANCES
**Noninterfering:** carbamazepine, ethosuximide, phenobarbital, phenytoin, primidone

## KEY WORDS
serum; pharmacokinetics

## REFERENCE
Lovett, L.J.; Nygard, G.A.; Erdmann, G.R.; Burley, C.Z.; Wahba Kahlil, S.K. HPLC determination of valproic acid in human serum using ultraviolet detection. *J.Liq.Chromatogr.*, **1987**, *10*, 687–699

## SAMPLE
**Matrix:** blood
**Sample preparation:** 100 μL Plasma + 200 μL 1 M HCl saturated with ammonium sulfate, vortex for 20 s, add 60 μL 10 μg/mL 4-methylprimidone in MeCN, vortex for 20 s, centrifuge at 2700 g for 5 min, inject a 5-10 μL aliquot of the MeCN layer.

## HPLC VARIABLES
**Column:** 250 × 4 5 μm LiChrosorb RP-18
**Mobile phase:** MeOH:THF:50 mM pH 5.9 phosphate buffer 44:1:55
**Column temperature:** 50
**Flow rate:** 1.1
**Injection volume:** 5-10
**Detector:** UV 210

## CHROMATOGRAM
**Retention time:** 6.5
**Internal standard:** 4-methylprimidone (5)
**Limit of detection:** 2.6 μg/mL

## OTHER SUBSTANCES
**Extracted:** carbamazepine, phenobarbital, phenytoin, primidone
**Simultaneous:** acetaminophen, caffeine, chloramphenicol, diazepam, ethosuximide, ethylphenylmalonamide, glutethimide, lidocaine, methylphenobarbital, pentobarbital, salicylic acid, theophylline

## KEY WORDS
plasma

## REFERENCE
Kushida, K.; Ishizaki, T. Concurrent determination of valproic acid with other antiepileptic drugs by high-performance liquid chromatography. *J.Chromatogr.*, **1985**, *338*, 131–139

## SAMPLE
**Matrix:** blood
**Sample preparation:** 500 μL Serum + 100 μL 500 μg/mL hexanoic acid + 500 μL pH 2.2 phosphate buffer + 5 mL hexane, shake for 5 min, centrifuge for 1 min. Remove 4 mL of the organic layer and add it to 50 μL 200 mM NaOH, extract for 2 min, centrifuge for 1 min, inject a 20 μL aliquot of the aqueous phase.

## HPLC VARIABLES
**Column:** 100 × 3 octyl CP-tm-Spher C8 glass column (Chrompack)
**Mobile phase:** MeCN:500 mM NaH$_2$PO$_4$ 35:65 adjusted to pH 2.2 with phosphoric acid
**Flow rate:** 0.9
**Injection volume:** 20
**Detector:** UV 210

## CHROMATOGRAM
**Retention time:** 7.3

**Internal standard:** hexanoic acid (3)
**Limit of detection:** 5 μg/mL

## OTHER SUBSTANCES
**Simultaneous:** metabolites

## KEY WORDS
serum

## REFERENCE
Van Damme, M.; Molle, L.; Abi Khalil, F. Useful sample handlings for reversed phase high performance liquid chromatography in emergency toxicology. *J.Toxicol.Clin.Toxicol.*, **1985**, *23*, 589–614

## SAMPLE
**Matrix:** blood
**Sample preparation:** 50 μL Serum + 1 mL 4 μg/mL IS in MeCN, vortex for 10 s, centrifuge. Remove an 800 μL aliquot of the supernatant and add it to 200 μL 3 mg/mL α-bromoacetophenone in MeCN and 100 μL triethylamine, heat in an open tube at 80° for 30 min, cool, inject a 10 μL aliquot.

## HPLC VARIABLES
**Column:** 250 × 4.6 5 μm Finepak Sil C$_{18.5}$ (Japan Spectroscopic, Tokyo)
**Mobile phase:** MeCN:water 60:40
**Column temperature:** 30
**Flow rate:** 1
**Injection volume:** 10
**Detector:** UV 245

## CHROMATOGRAM
**Retention time:** 21
**Internal standard:** cyclohexane carboxylic acid (13)
**Limit of quantitation:** 500 ng/mL

## OTHER SUBSTANCES
**Extracted:** hexobarbital, phenobarbital
**Noninterfering:** carbamazepine, phenytoin

## KEY WORDS
serum; derivatization

## REFERENCE
Nakamura, M.; Kondo, K.; Nishioka, R.; Kawai, S. Improved procedure for the high-performance liquid chromatographic determination of valproic acid in serum as its phenacyl ester. *J.Chromatogr.*, **1984**, *310*, 450–454

## SAMPLE
**Matrix:** blood
**Sample preparation:** 100 μL Serum + 25 μL buffer + 250 μL 167 μg/mL nonanoic acid in MeCN, vortex for 10 s, centrifuge for 5 min. Remove a 200 μL aliquot of the supernatant and add it to 50 μL 20 mg/mL 4-bromophenacyl bromide in MeCN containing 500 μg/mL dicyclohexane-18-crown-6, heat at 70° for 15 min, cool, inject a 5 μL aliquot. (Prepare buffer by dissolving 19.04 g KH$_2$PO$_4$ and 37.4 g Na$_2$HPO$_4$.2H$_2$O in 1 L water, pH 7.0.)

## HPLC VARIABLES
**Column:** 100 × 5 5 μm Hypersil ODS

**Mobile phase:** MeCN:3 mM KH$_2$PO$_4$ 70:30
**Flow rate:** 2
**Injection volume:** 5
**Detector:** UV 254

## CHROMATOGRAM
**Retention time:** 4
**Internal standard:** nonanoic acid (6)
**Limit of detection:** 60 μM

## OTHER SUBSTANCES
**Noninterfering:** acetaminophen, clonazepam, diazepam, phenobarbital, phenytoin, primidone, salicylic acid, theophylline

## KEY WORDS
horse; serum; derivatization

## REFERENCE
Moody, J.P.; Allan, S.M. Measurement of valproic acid in serum as the 4-bromophenacyl ester by high performance liquid chromatography. *Clin.Chim.Acta*, **1983**, *127*, 263–269

## SAMPLE
**Matrix:** blood
**Sample preparation:** 250 μL Serum + 100 μL 2.5 mM heptanoic acid + 300 μL 2 M sulfuric acid + 1 mL petroleum ether (40-60°), shake for 5 min, centrifuge. Remove a 900 μL aliquot of the organic layer and add it to 10 μL tributylamine, evaporate to dryness at 70° in about 10 min, reconstitute the residue in 400 μL MeCN, add 50 μL 10 mg/mL p-bromophenacyl bromide in MeOH, heat at 70° for 10 min, inject a 20 μL aliquot.

## HPLC VARIABLES
**Column:** 250 × 4 10 μm Lichrosorb RP-2
**Mobile phase:** MeOH:50 mM pH 2.3-2.5 phosphate buffer 75:25
**Column temperature:** 50
**Flow rate:** 2
**Injection volume:** 20
**Detector:** UV 260

## CHROMATOGRAM
**Retention time:** 3.14
**Internal standard:** heptanoic acid (2.74)

## OTHER SUBSTANCES
**Noninterfering:** carbamazepine, clonazepam, ethosuximide, phenobarbital, phenytoin, primidone

## KEY WORDS
serum; derivatization

## REFERENCE
Ehrenthal, W.; Rochel, M. Kontrolle der Valproinsäure-Therapie durch Serumspiegelbestimmungen mit HPLC oder EMIT und durch gleichzeltige Bestimmung klinisch-chemischer Parameter [Drug monitoring for valproic acid with HPLC or EMIT and concomitant measurements of clinical-chemical parameters]. *Arzneimittelforschung*, **1982**, *32*, 449–452

## SAMPLE
**Matrix:** blood

**Sample preparation:** 1 mL Serum + 50 μg IS + 100 μL concentrated phosphoric acid + 2 mL dichloromethane, shake for 30 min, centrifuge at 2500 rpm for 10 min. Remove a 500 μL aliquot of the organic layer and add it to 100 μL reagent and 0.5 mg sodium bicarbonate, heat at 75° for 30 min, evaporate to 0.5 mL under a stream of nitrogen, make up to 1 mL with MeCN, inject an aliquot. (Reagent was 20 mM p-bromophenacyl bromide in MeCN containing 1 mM 18-crown-6, prepared by diluting Phenacyl-8 (Pierce) 5 times with MeCN.)

## HPLC VARIABLES
**Column:** C18
**Mobile phase:** MeCN:water 65:35
**Detector:** UV 254

## CHROMATOGRAM
**Retention time:** 10
**Internal standard:** cyclohexane carboxylic acid (6.5)
**Limit of quantitation:** 5 μg/mL

## KEY WORDS
serum; derivatization

## REFERENCE
Kline, W.F.; Enagonio, D.P.; Reeder, D.J.; May, W.E. Liquid chromatographic determination of valproic acid in human serum. *J.Liq.Chromatogr.*, **1982**, *5*, 1697–1709

## SAMPLE
**Matrix:** blood, dialysate
**Sample preparation:** Blood. 20 μL Plasma or whole blood + 1 mL 1 μg/mL nonanoic acid in MeCN, centrifuge. Remove 100 μL supernatant, add 100 μL suspension, 15 min before injection add 200 μL 0.5 mg/mL bromomethylmethoxycoumarin, mix, inject. Dialysate. Lyophilize, add 40 μL suspension (prepared with twice the amount of crown ether), add 40 μL 0.5 mg/mL bromomethylmethoxycoumarin, mix, inject after 15 min. (Suspension was 100 mg potassium carbonate + 50 μL water + 5 mL 20 mM 18-crown-6 in MeCN, sonicate for 1 h, add 5 mL MeCN, separate the suspension from the precipitated potassium carbonate.)

## HPLC VARIABLES
**Column:** 200 × 3 5 μm Chromspher ODS
**Mobile phase:** MeCN:2.5 M formic acid 75:25
**Flow rate:** 0.4
**Detector:** F ex 325 em 398 (cut-off filter)

## CHROMATOGRAM
**Retention time:** 8, 11 (double peak caused by impurity in derivatizing reagent)
**Internal standard:** nonanoic acid (12, 16)
**Limit of quantitation:** 1 μg/mL

## KEY WORDS
plasma; whole blood; brain; derivatization; human; rat

## REFERENCE
Wolf, J.H.; Veenma-van der Duin, L.; Korf, J. Automated analysis procedure for valproic acid in blood, serum and brain dialysate by high-performance liquid chromatography with bromomethylmethoxycoumarin as fluorescent label. *J.Chromatogr.*, **1989**, *487*, 496–502

## SAMPLE
**Matrix:** formulations

**Sample preparation:** Condition a Tech Elut C18 SPE cartridge with 3 mL MeOH and 2 mL water. Weigh out powdered tablets containing 40 mg sodium valproate, add 100 mL water, stir for 10 min, filter (paper), acidify the filtrate with phosphoric acid, add a 2 mL aliquot to the SPE cartridge, wash with two 2 mL portions of water, dry under vacuum, elute with 3 mL MeCN, dilute the eluate to 5 mL with MeCN. Mix 500 μL of this solution with 300 μL 178 μg/mL 2-bromoacetyl-6-methoxynaphthalene in MeCN, add 50 μL 3% triethylamine in MeCN, heat at 70° for 30 min, cool, add 50 μL 60.84 μg/mL IS in MeCN, inject a 50 μL aliquot. (Prepare 2-bromoacetyl-6-methoxynaphthalene by stirring equi-molar amounts of 2-acetyl-6-methoxynaphthalene (Janssen Chimica, Belgium) and phen-yltrimethylammonium tribromide in THF at room temperature for 3 h (Phosphorus and Sulfur 1985, 25, 357), purify by column chromatography on silica gel with chloroform: petroleum ether 50:50 (mp 109-112°).)

## HPLC VARIABLES
**Column:** 150 × 4.6 5 μm Hypersil RP-8
**Mobile phase:** MeCN:water 68:32
**Flow rate:** 1
**Injection volume:** 50
**Detector:** F ex 300 em 460

## CHROMATOGRAM
**Retention time:** 7
**Internal standard:** 6-methoxynaphthacyl ester of nonanoic acid (Dissolve 2 mmole nona-noic acid and 1 mmole 2-bromoacetyl-6-methoxynaphthalene in 10 mL anhydrous MeCN, add 500 μL triethylamine, heat to 60° for 30 min, cool, add 30 mL water, extract three times with 10 mL portions of diethyl ether. Combine the organic layers and wash them with 5% sodium bicarbonate solution, wash three times with 10 mL portions of water, dry over anhydrous sodium sulfate, evaporate to dryness under reduced pressure, recrys-tallize from water to give 6-methoxynaphthacyl ester of nonanoic acid (mp 66-8°).) (10)
**Limit of detection:** 2 pmole

## OTHER SUBSTANCES
**Simultaneous:** decanoic acid, dodecanoic acid, heptanoic acid, undecanoic acid
**Interfering:** octanoic acid

## KEY WORDS
derivatization; SPE; tablets

## REFERENCE
Gatti, R.; Cavrini, V.; Roveri, P. 2-Bromoacetyl-6-methoxynaphthalene: A useful fluorescent labelling reagent for HPLC analysis of carboxylic acids. *Chromatographia*, **1992**, *33*, 13−18

## SAMPLE
**Matrix:** formulations
**Sample preparation:** Weigh out capsule contents equivalent to 250 mg valproic acid, add 80 mL MeOH, sonicate for 10 min, make up to 100 mL with MeOH, filter, allow to settle for 30 min. Dilute a 4 mL aliquot to 100 mL with mobile phase. Mix a 20 mL aliquot of the diluted solution with 10 mL 50 μg/mL ibuprofen in mobile phase, make up to 100 mL with mobile phase, inject a 50 μL aliquot.

## HPLC VARIABLES
**Column:** 100 × 4.6 5 μm Spheri-5 RP-8
**Mobile phase:** MeOH:buffer 30:70 (Prepare buffer by mixing 4 mM $Na_2HPO_4$ and 7 mM $KH_2PO_4$ to achieve pH 7.)
**Flow rate:** 1
**Injection volume:** 50

**Detector:** F ex 355 em 460 (408 nm cutoff filter) following post-column extraction. The column effluent mixed with 50 μg/mL reagent in mobile phase pumped at 0.5 mL/min and then with chloroform pumped at 1 mL/min and the mixture flowed through a 1.8 m × 0.3 mm ID knitted PTFE coil to a 50 μL membrane phase separator using a polyethylene-backed 0.5 μm Fluoropore membrane filter (design in paper). The organic phase flowed to the detector. (Synthesize the reagent, α-(3,4-dimethoxyphenyl)-4′-trimethylammoniummethylcinnamonitrile methosulfate, as follows. Stir 20 mmoles 3,4-dimethoxyphenylacetonitrile and 20 mmoles p-toluamide in 50 mL EtOH at 50°, add 5 mL 50% aqueous KOH slowly, stir at 50° for 5 min, cool to room temperature, filter, dry the precipitate of α-(3,4-dimethoxyphenyl)-4′-methylcinnamonitrile. Dissolve 20 mmoles α-(3,4-dimethoxyphenyl)-4′-methylcinnamonitrile, 20 mmoles N-bromosuccinimide, and 20 mg benzoyl peroxide in 100 mL carbon tetrachloride (Caution! Carbon tetrachloride is a carcinogen!), reflux with stirring for 1.5 h , cool, filter, evaporate to dryness under reduced pressure, recrystallize from MeOH to give α-(3,4-dimethoxyphenyl)-4′-bromomethylcinnamonitrile. Vigorously stir 30 mmoles anhydrous dimethylamine in 100 mL dry benzene (Caution! Benzene is a carcinogen!) at 0°, very slowly add 10 mmoles α-(3,4-dimethoxyphenyl)-4′-bromomethylcinnamonitrile while stirring at 0°, stir at room temperature overnight, add 150 mL water, remove the organic phase, extract the aqueous phase twice with 100 mL portions of diethyl ether, wash the organic layers with saturated NaCl solution, dry over anhydrous magnesium sulfate, evaporate under reduced pressure to give α-(3,4-dimethoxyphenyl)-4′-dimethylaminomethylcinnamonitrile (J.Chem.Eng.Data 1987, 32, 387). Reflux 10 mmoles α-(3,4-dimethoxyphenyl)-4′-dimethylaminomethylcinnamonitrile, 20 mmoles dimethyl sulfate (Caution! Dimethyl sulfate is a carcinogen and acutely toxic!), and 5 g potassium carbonate in 50 mL acetone for 1 h, cool to room temperature, filter, dry the precipitate under vacuum at room temperature overnight, recrystallize from chloroform containing 2-3 drops of 95% EtOH to give α-(3,4-dimethoxyphenyl)-4′-trimethylammoniummethylcinnamonitrile methosulfate (mp 212-215°). Protect solutions from light.)

## CHROMATOGRAM
**Retention time:** k′ 0.3256
**Internal standard:** ibuprofen (k′ 4.124)
**Limit of detection:** 100 ng/mL

## OTHER SUBSTANCES
**Simultaneous:** ketoprofen, mefenamic acid, naproxen, probenecid, salicylic acid

## KEY WORDS
capsules; post-column extraction

## REFERENCE
Kim, M.; Stewart, J.T. HPLC post-column ion-pair extraction of acidic drugs using a substituted α-phenylcinnamonitrile quaternary ammonium salt as a new fluorescent ion-pair reagent. *J.Liq.Chromatogr.*, **1990**, *13*, 213–237

# Vancomycin

**Molecular formula:** $C_{66}H_{75}Cl_2N_9O_{24}$
**Molecular weight:** 1449.2
**CAS Registry No.:** 1404-90-6 (vancomycin), 1404-93-9 (vancomycin hydrochloride)

## SAMPLE
**Matrix:** blood
**Sample preparation:** 50 µL Serum + 9 µL 125 µg/mL trimethoprim + 75 µL acetone : 10% trichloroacetic acid 1:2, vortex for 5 s, centrifuge for 4 min. Remove 62.5 µL of the supernatant and add it to 62.5 µL 50 mM $KH_2PO_4$, add 250 µL diethyl ether, vortex for 10 s, centrifuge for 5 min, filter (0.45 µm) the lower aqueous layer, inject a 20 µL aliquot of the filtrate.

## HPLC VARIABLES
**Column:** 75 × 4.6 TSK gel ODS-80TM (Tosoh)
**Mobile phase:** MeCN:50 mM pH 6.0 $KH_2PO_4$ 8:92
**Flow rate:** 1
**Injection volume:** 20
**Detector:** UV 235

## CHROMATOGRAM
**Retention time:** 12
**Internal standard:** trimethoprim (30)
**Limit of detection:** 500 ng/mL

## OTHER SUBSTANCES
**Extracted:** degradation products

## KEY WORDS
serum; comparison with fluorescence polarization immunoassay

## REFERENCE

Morishige, H.; Shuto, H.; Ieiri, I.; Otsubo, K.; Oishi, R. Instability of standard calibrators may be involved in overestimating vancomycin concentrations determined by fluorescence polarization immunoassay. *Ther.Drug Monit.*, **1996**, *18*, 80–85

## SAMPLE

**Matrix:** blood
**Sample preparation:** 50 μL Serum or plasma + 25 μL 25 μg/mL ristocetin in water + 10 μL 15% perchloric acid, vortex for 5 min, centrifuge at 8° at 13000 g for 5 min, inject a 50 μL aliquot of the aqueous layer.

## HPLC VARIABLES

**Guard column:** amino (Burdick and Jackson)
**Column:** 250 × 4.6 5 μm Microsorb-MV NH2 or 250 × 4.6 5 μm CA AM5 amino propyl (Burdick and Jackson)
**Mobile phase:** MeCN:20 mM $NaH_2PO_4$:20 mM $Na_2HPO_4$ 62:14.25:23.75
**Flow rate:** 2
**Injection volume:** 50
**Detector:** UV 225

## CHROMATOGRAM

**Retention time:** 5.5
**Internal standard:** ristocetin (8.9)
**Limit of detection:** 320 ng/mL

## OTHER SUBSTANCES

**Noninterfering:** acetaminophen, N-acetylprocainamide, amikacin, amitriptyline, amlodipine, carbamazepine, cefotaxime, ceftazidime, chloramphenicol, ciprofloxacin, cisapride, clindamycin, clonidine, codeine, cyclosporine, digoxin, diphenhydramine, disopyramide, ethosuximide, fluconazole, gentamicin, gentamicin, heparin, labetalol, levothyroxine, lidocaine, lithium, methotrexate, metronidazole, minoxidil, nafcillin, nifedipine, phenobarbital, phenobarbital, phenytoin, phenytoin, primidone, procainamide, propranolol, quinidine, ranitidine, salicylic acid, theophylline, tobramycin, tobramycin, valproic acid, warfarin

## KEY WORDS

serum; plasma

## REFERENCE

Li, L.; Miles, M.V.; Hall, W.; Carson, S.W. An improved micromethod for vancomycin determination by high-performance liquid chromatography. *Ther.Drug Monit.*, **1995**, *17*, 366–370

## SAMPLE

**Matrix:** blood
**Sample preparation:** 1 mL Plasma + 50 μL 60% perchloric acid, mix, centrifuge. Wash the supernatant with 1 mL dichloromethane, inject a 50 μL aliquot of the aqueous phase.

## HPLC VARIABLES

**Column:** 150 × 4.6 5 μm Nucleosil RP-18
**Mobile phase:** MeCN:5 mM pH 2.8 $KH_2PO_4$ 10:90
**Column temperature:** 30
**Flow rate:** 1
**Injection volume:** 50
**Detector:** UV 229

## CHROMATOGRAM
**Retention time:** 4.8
**Limit of detection:** 200 ng/mL
**Limit of quantitation:** 1 μg/mL

## OTHER SUBSTANCES
**Simultaneous:** acetaminophen, aspirin, caffeine

## KEY WORDS
plasma

## REFERENCE
Luksa, J.; Marusic, A. Rapid high-performance liquid chromatographic determination of vancomycin in human plasma. *J.Chromatogr.B*, **1995**, *667*, 277–281

## SAMPLE
**Matrix:** blood
**Sample preparation:** 100 μL Plasma + 100 μL cold 40% trichloroacetic acid, mix for 2 min, centrifuge at 1000 g for 10 min. Remove the supernatant and add it to 1 mL diethyl ether, shake, discard the ether layer. Remove 50 μL of the aqueous layer and add it to 10 μL 10 ppm phenacetin, mix, inject a 20 μL aliquot.

## HPLC VARIABLES
**Column:** 300 × 3.9 10 μm μBondapak C18
**Mobile phase:** MeCN:MeOH:50 mM Pic-B7 8:17:75
**Flow rate:** 1
**Injection volume:** 20
**Detector:** UV 280

## CHROMATOGRAM
**Retention time:** 7.2
**Internal standard:** phenacetin (10.8)

## KEY WORDS
plasma; comparison with fluorescence polarization immunoassay

## REFERENCE
Najjar, T.A.; Al-Dhuwailie,.A.; Tekle, A. Comparison of high-performance liquid chromatography with fluorescence polarization immunoassay for the analysis of vancomycin in patients with chronic renal failure. *J.Chromatogr.B*, **1995**, *672*, 295–299

## SAMPLE
**Matrix:** cell suspensions
**Sample preparation:** 100 μL Cell suspension + 100 μL cefazolin solution + 100 μL Hanks balanced salt solution, sonicate 30 min, add 800 μL MeCN, centrifuge at 13000 g for 5 min, remove supernatant. Dry supernatant under air, dissolve in 100 μL mobile phase, inject 75 μL.

## HPLC VARIABLES
**Column:** μBondapak C18
**Mobile phase:** MeCN:50 mM pH 5.09 $KH_2PO_4$ 10:90
**Flow rate:** 1
**Injection volume:** 75
**Detector:** UV 254

## CHROMATOGRAM
**Retention time:** 14
**Internal standard:** cefazolin
**Limit of detection:** 100-1000 ng/mL

## REFERENCE
Darouiche, R.O.; Hamill, R.J. Antibiotic penetration of and bactericidal activity within endothelial cells. *Antimicrob.Agents Chemother.*, **1994**, *38*, 1059–1064

## SAMPLE
**Matrix:** formulations

## HPLC VARIABLES
**Column:** 250 × 4.6 5 μm Bakerbond phenylethyl
**Mobile phase:** MeCN:THF:buffer 7:1:92 (Buffer was 4 mL triethylamine in 2 L water, pH adjusted to 3.2 with phosphoric acid.)
**Flow rate:** 1
**Detector:** UV 280

## CHROMATOGRAM
**Retention time:** 3.03

## KEY WORDS
injections; saline; water; stability-indicating

## REFERENCE
Stiles, M.L.; Allen, L.V., Jr.; Prince, S.J. Stability of various antibiotics kept in an insulated pouch during administration via portable infusion pump. *Am.J.Health-Syst.Pharm.*, **1995**, *52*, 70–74

## SAMPLE
**Matrix:** formulations
**Sample preparation:** Dilute with mobile phase, inject a 20 μL aliquot.

## HPLC VARIABLES
**Column:** 300 × 4.6 5 μm Hypersil ODS
**Mobile phase:** MeCN:THF:buffer 8.3:1.4:90 (Buffer was 20 mL 1 M triethylamine in 900 mL water, adjusted to pH 3.2 with 1 M NaOH.)
**Flow rate:** 1.5
**Injection volume:** 20
**Detector:** UV 216

## CHROMATOGRAM
**Retention time:** 6.4

## OTHER SUBSTANCES
**Simultaneous:** aztreonam

## KEY WORDS
stability-indicating; injections; 5% dextrose; saline

## REFERENCE
Trissel, L.A.; Xu, Q.A.; Martinez, J.F. Compatibility and stability of aztreonam and vancomycin hydrochloride. *Am.J.Health-Syst.Pharm.*, **1995**, *52*, 2560–2564

## SAMPLE
**Matrix:** formulations
**Sample preparation:** Inject a 50 μL aliquot of the dialysis fluid.

## HPLC VARIABLES
**Column:** 250 × 3.6 5 μm Spherisorb ODS2
**Mobile phase:** MeCN:200 mM ammonium acetate:water 9:10:81 adjusted to pH 5.4 with glacial acetic acid
**Flow rate:** 1.5
**Injection volume:** 50
**Detector:** UV 214

## CHROMATOGRAM
**Retention time:** 9

## KEY WORDS
dialysis solutions; stability-indicating

## REFERENCE
Mawhinney, W.M.; Adair, C.G.; Gorman, S.P.; McClurg, B. Stability of vancomycin hydrochloride in peritoneal dialysis solution. *Am.J.Hosp.Pharm.*, **1992**, *49*, 137–139

## SAMPLE
**Matrix:** solutions
**Sample preparation:** Inject a 6-10 μL aliquot.

## HPLC VARIABLES
**Guard column:** 20 × 4.6 Supelguard LC-1 (Supelco)
**Column:** 250 × 4.6 5 μm Supelcosil LC-1 (Supelco)
**Mobile phase:** MeOH:MeCN:buffer 17.5:17.5:65 (Buffer was 2.72 g $KH_2PO_4$ in 1.9 L water, pH adjusted to 6.3 with about 2 mL 1 M NaOH, made up to 2 L.)
**Flow rate:** 2
**Injection volume:** 6-10
**Detector:** UV 204

## CHROMATOGRAM
**Retention time:** 1.83
**Internal standard:** 5-ethyl-5-p-tolybarbituric acid (tolybarb) (4.80)

## OTHER SUBSTANCES
**Simultaneous:** acetaminophen, acetanilide, N-acetylcysteine, N-acetylprocainamide, amobarbital, aspirin, barbital, butabarbital, butalbital, caffeine, carbamazepine, cimetidine, codeine, cyheptamide, diazoxide, diflunisal, disopyramide, ethchlorvynol, ethosuximide, gentisic acid, glutethimide, heptabarbital, hexobarbital, ibuprofen, indomethacin, ketoprofen, mefenamic acid, mephenytoin, mephobarbital, methaqualone, methsuximide, methyl salicylate, methyprylon, morphine, naproxen, nirvanol, oxphenylbutazone, pentobarbital, phenacetin, phenobarbital, phensuximide, phenylbutazone, phenytoin, primidone, procainamide, salicylamide, secobarbital, sulindac, thiopental, tolmetin, trimethoprim
**Noninterfering:** amikacin, gentamicin, meprobamate, netilmicin, quinidine, tetracycline, tobramycin, valproic acid
**Interfering:** ampicillin, chloramphenicol, chlorpropamide, diphylline, salicylic acid, sulfamethoxazole, theophylline

## REFERENCE

Meatherall, R.; Ford, D. Isocratic liquid chromatographic determination of theophylline, acetaminophen, chloramphenicol, caffeine, anticonvulsants, and barbiturates in serum. *Ther.Drug Monit.*, **1988**, *10*, 101–115

## SAMPLE

**Matrix:** solutions
**Sample preparation:** Inject a 25 μL aliquot.

## HPLC VARIABLES

**Guard column:** RCSS Guard-Pak (Waters)
**Column:** 100 × 8 C18 Radial Pak (Waters)
**Mobile phase:** MeOH:0.75% acetic acid 30:70, pH adjusted to 5.5 with triethylamine
**Flow rate:** 3
**Injection volume:** 25
**Detector:** UV 254

## CHROMATOGRAM

**Retention time:** 1.1

## OTHER SUBSTANCES

**Simultaneous:** acetaminophen, N-acetylprocainamide, cefaclor, cefamandole, cefazolin, cefotaxime, cefoxitin, cephalexin, cephalothin, cephapirin, chloramphenicol, cimetidine, miconazole, moxalactam, procainamide, sulfamethoxazole, theophylline, tobramycin

## REFERENCE

Danzer, L.A. Liquid-chromatographic determination of cephalosporins and chloramphenicol in serum. *Clin.Chem.*, **1983**, *29*, 856–858

## ANNOTATED BIBLIOGRAPHY

Demotes-Mainard, F.; Labat, L.; Vinçon, G.; Bannwarth, B. Column-switching high-performance liquid chromatographic determination of vancomycin in serum. *Ther.Drug Monit.*, **1994**, *16*, 293–297 [cefaloridine (IS); LOD 1 μg/mL; non-interfering cefoperazone, cefotiam; simultaneous cefonicid, ceforanide, cefotaxime, cefotetan, cefpiramide, cefsulodin, ceftazidime, ceftriaxone, desacetylcefotaxime]

Jenke, D.R. Drug binding by reservoirs in elastomeric infusion devices. *Pharm.Res.*, **1994**, *11*, 984–989 [injections; saline; 5% dextrose]

Hu, M.W.; Anne, L.; Forni, T.; Gottwald, K. Measurement of vancomycin in renally impaired patient samples using a new high-performance liquid chromatography method with vitamin B12 internal standard: comparison of high-performance liquid chromatography, emit, and fluorescence polarization immunoassay methods. *Ther.Drug Monit.*, **1990**, *12*, 562–569

Inman, E.L.; Clemens, R.L.; Olsen, B.A. Determination of EDTA in vancomycin by liquid chromatography with absorbance ratioing for peak identification. *J.Pharm.Biomed.Anal.*, **1990**, *8*, 513–520

Hosotsubo, H. Rapid and specific method for the determination of vancomycin in plasma by high-performance liquid chromatography on an aminopropyl column. *J.Chromatogr.*, **1989**, *487*, 421–427

Bauchet, J.; Pussard, E.; Garaud, J.J. Determination of vancomycin in serum and tissues by column liquid chromatography using solid-phase extraction. *J.Chromatogr.*, **1987**, *414*, 472–476

Greene, S.V.; Abdalla, T.; Morgan, S.L.; Bryan, C.S. High-performance liquid chromatographic analysis of vancomycin in plasma, bone, atrial appendage tissue and pericardial fluid. *J.Chromatogr.*, **1987**, *417*, 121–128

Rosenthal, A.F.; Sarfati, I.; A'Zary, E. Simplified liquid-chromatographic determination of vancomycin. *Clin.Chem.*, **1986**, *32*, 1016–1019

Jehl, F.; Gallion, C.; Thierry, R.C.; Monteil, H. Determination of vancomycin in human serum by high-pressure liquid chromatography. *Antimicrob.Agents Chemother.*, **1985**, *27*, 503–507

Hoagland, R.J.; Sherwin, J.E.; Phillips, J.M., Jr. Vancomycin: a rapid HPLC assay for a potent antibiotic. *J.Anal.Toxicol.*, **1984**, *8*, 75−77

McClain, J.B.; Bongiovanni, R.; Brown, S. Vancomycin quantitation by high-performance liquid chromatography in human serum. *J.Chromatogr.*, **1982**, *231*, 463−466

# Venlafaxine

**Molecular formula:** $C_{17}H_{27}NO_2$
**Molecular weight:** 277.4
**CAS Registry No.:** 93413-69-5 (venlafaxine),
99300-78-4 (venlafaxine hydrochloride)

HCl

---

## SAMPLE
**Matrix:** blood
**Sample preparation:** 1 mL Plasma + 50 μL 22.5 μg/mL IS in water + 200 μL saturated sodium borate, vortex, add 5 mL isopropyl ether (Caution! Isopropyl ether readily forms explosive peroxides!), shake for 15 min, centrifuge at 2500 rpm for 10 min. Remove a 4.5 mL aliquot of the organic phase and add it to 400 μL 10 mM HCl, shake for 15 min, centrifuge at 2500 rpm for 10 min. Discard the organic phase and add 1 mL saturated sodium borate solution to the aqueous layer, vortex, add 2 mL isopropyl ether, shake for 15 min, centrifuge at 2500 rpm for 5 min. Remove the organic layer and evaporate it to dryness under a stream of nitrogen, add 5-10 mg anhydrous sodium carbonate powder to the residue, add 200 μL 150 μg/mL naproxen chloride in dichloromethane, vortex, let stand in the dark for 20 h, evaporate to dryness under a stream of nitrogen, reconstitute with 1 mL water, vortex, add 2 mL chloroform, shake for 10 min, centrifuge at 2500 rpm for 5 min. Remove the chloroform layer and add it to 2 mL water, shake for 10 min, centrifuge at 2500 rpm for 5 min. Remove the organic layer and evaporate it to dryness under a stream of nitrogen, reconstitute the residue in 400 μL initial mobile phase, inject an 11 μL aliquot. (Prepare naproxen chloride as follows. React 8.1 g naproxen in 40 mL dichloromethane with 25 g oxalyl chloride at room temperature for 22 h, evaporate to dryness under reduced pressure at 42°, reconstitute the residue with 35 mL dichloromethane, slowly add 1.12 L n-hexane, chill to precipitate crystals, decant solvent carefully and rapidly, dry crystals rapidly under nitrogen, dry crystals of naproxen chloride (mp 95-7°) over calcium chloride in a vacuum desiccator.)

---

## HPLC VARIABLES
**Guard column:** 20 × 4.6 Supelguard LC-8-DB (Supelco)
**Column:** 150 × 4.6 5 μm Supelcosil LC-8-DB
**Mobile phase:** Gradient. MeCN:buffer from 50:50 to 54:46 over 30 min, to 55:45 over 6 min, to 57:43 over 14 min, maintain at 57:43 for 3 min, return to initial conditions over 3 min. (After 30 min increase flow rate to 2 mL/min over 6 min, maintain at 2 mL/min for 17 min, return to initial conditions over 3 min. Buffer was 100 mM $KH_2PO_4$ adjusted to pH 3.0 with 85% orthophosphoric acid, add 0.07% triethylamine to achieve a final pH of 3.25.)
**Flow rate:** 0.8
**Injection volume:** 11
**Detector:** UV 229

---

## CHROMATOGRAM
**Retention time:** 16.53 (R), 18.03 (S)
**Internal standard:** 1-[2-(dimethylamino)-1-(2-chlorophenyl)ethyl]cyclohexanol(Wy-45,818) (19.74, 22.38 (enantiomers))
**Limit of quantitation:** 50 ng/mL; 25 ng/mL (22μL inj)

---

## OTHER SUBSTANCES
**Extracted:** metabolites

---

## KEY WORDS
chiral; dog; rat; human; plasma; derivatization; pharmacokinetics

## REFERENCE

Wang, C.P.; Howell, S.R.; Scatina, J.; Sisenwine, S.F. The disposition of venlafaxine enantiomers in dogs, rats, and humans receiving venlafaxine. *Chirality*, **1992**, *4*, 84–90

## SAMPLE

**Matrix:** blood, urine

**Sample preparation:** Plasma. 1 mL Plasma + 50 μL 15 μg/mL IS in water + 200 μL saturated sodium borate + 5 mL purified isopropyl ether (Caution! Isopropyl ether readily forms explosive peroxides!), shake at 100 strokes/min for 15 min, centrifuge at 2500 rpm for 10 min. Remove a 4.5 mL aliquot of the organic layer and add it to 300 μL 10 mM HCl, shake at 150 strokes/min for 15 min, centrifuge at 2500 rpm for 10 min, discard the organic phase, vortex the aqueous layer under vacuum for 30 min (to remove traces of isopropyl ether), inject a 50 μL aliquot of the aqueous layer. Urine. 100 μL Urine + 1 mL 750 ng/mL IS in water + 200 μL saturated sodium borate + 5 mL purified isopropyl ether, shake at 100 strokes/min for 15 min, centrifuge at 2500 rpm for 10 min. Remove a 4.5 mL aliquot of the organic layer and add it to 300 μL 10 mM HCl, shake at 150 strokes/min for 15 min, centrifuge at 2500 rpm for 10 min, discard the organic phase, vortex the aqueous layer under vacuum for 30 min (to remove traces of isopropyl ether), inject a 50 μL aliquot of the aqueous layer. (Purify isopropyl ether each day by passing it through a column of neutral alumina (Brockman Activity 1, Fisher).)

## HPLC VARIABLES

**Column:** 150 × 4.6 5 μm Supelcosil LC-8DB

**Mobile phase:** MeCN:100 mM pH 4.4 ammonium phosphate 25.5:74.5 (plasma) or 20:80 (urine)

**Flow rate:** 1

**Injection volume:** 50

**Detector:** UV 229

## CHROMATOGRAM

**Retention time:** 10.1 (plasma), 12.2 (urine)

**Internal standard:** 1-[2-(dimethylamino)-1-(2-chlorophenyl)ethyl]cyclohexanol(Wy-45,818) (14.8 (plasma), 18.1 (urine))

**Limit of quantitation:** 10 ng/mL

## OTHER SUBSTANCES

**Extracted:** metabolites

## KEY WORDS

rat; dog; human; mouse; plasma; pharmacokinetics

## REFERENCE

Hicks, D.R.; Wolaniuk, D.; Russell, A.; Cavanaugh, N.; Kraml, M. A high-performance liquid chromatographic method for the simultaneous determination of venlafaxine and *O*-desmethylvenlafaxine in biological fluids. *Ther.Drug Monit.*, **1994**, *16*, 100–107

## SAMPLE

**Matrix:** urine

**Sample preparation:** 1 mL Urine + 1 mL pH 4.6 sodium acetate buffer + 200 μL Glusulase, heat at 37° overnight, add MeCN, filter. Evaporate to dryness, reconstitute with 100 μM $(NH_4)H_2PO_4$, inject an aliquot.

## HPLC VARIABLES

**Guard column:** 30 × 4.6 10 μm Spheri-10

**Column:** 250 × 4.6 5 μm LC-18-DB (Supelco)

**Mobile phase:** Gradient. MeCN:100 mM $(NH_4)H_2PO_4$ from 10:90 to 40:60 over 25 min.

**Flow rate:** 1
**Detector:** UV 229

## OTHER SUBSTANCES
**Extracted:** metabolites

## KEY WORDS
mouse; human

## REFERENCE
Howell, S.R.; Husbands, G.E.M.; Scatina, J.A.; Sisenwine, S.F. Metabolic disposition of [14]C-venlafaxine in mouse, rat, dog, rhesus monkey and man. *Xenobiotica*, **1993**, *23*, 349–359

# Verapamil

**Molecular formula:** $C_{27}H_{38}N_2O_4$
**Molecular weight:** 454.6
**CAS Registry No.:** 52-53-9 (verapamil),
152-11-4 (verapamil hydrochloride)

## SAMPLE
**Matrix:** blood
**Sample preparation:** 1 mL Plasma + 250 μL water + 5 mL MTBE, shake for 10 min, centrifuge at 1800 g for 10 min. Remove a 4 mL aliquot of the organic layer and add it to 150 μL 17 mM phosphoric acid, shake for 3 min, centrifuge at 2800 g for 10 min, inject a 50 μL aliquot of the aqueous layer.

## HPLC VARIABLES
**Column:** 250 × 4 10 μm Spherisorb ODS 2
**Mobile phase:** MeCN:buffer 60:40 (Buffer was 1.15 g/L $(NH_4)H_2PO_4$ containing 0.6 mL/L triethylamine, adjusted to pH 3.7 with phosphoric acid.)
**Flow rate:** 1
**Injection volume:** 50
**Detector:** UV 238

## CHROMATOGRAM
**Retention time:** 13.6
**Internal standard:** verapamil

## OTHER SUBSTANCES
**Extracted:** diltiazem

## KEY WORDS
plasma; verapamil is IS

## REFERENCE
Coors, C.; Schulz, H.-G.; Stache, F. Development and validation of a bioanalytical method for the quantification of diltiazem and desacetyldiltiazem in plasma by capillary zone electrophoresis. *J.Chromatogr.A*, **1995**, *717*, 235−243

## SAMPLE
**Matrix:** blood
**Sample preparation:** Add 200 μL whole blood to 400 μL MeCN while mixing rapidly for 5 s, centrifuge at 1500 rpm for 2 min, mix 200 μL supernatant with 700 μL 20 mM tetramethylammonium perchlorate containing 0.1% trifluoroacetic acid, inject a 50 μL aliquot.

## HPLC VARIABLES
**Column:** 50 × 4.6 Little Champ reverse phase (Regis)
**Mobile phase:** MeCN:MeOH:buffer 37:5:58 (Buffer was 20 mM tetramethylammonium perchlorate containing 0.1% trifluoroacetic acid.)
**Flow rate:** 1.4
**Injection volume:** 50
**Detector:** UV 228

## CHROMATOGRAM
**Limit of quantitation:** 100 nM

## KEY WORDS
rat; whole blood

## REFERENCE
Hoffman, D.J.; Seifert, T.; Borre, A.; Nellans, H.N. Method to estimate the rate and extent of intestinal absorption in conscious rats using an absorption probe and portal blood sampling. *Pharm.Res.*, **1995**, *12*, 889–894

## SAMPLE
**Matrix:** blood
**Sample preparation:** 1 mL Plasma + 50 μL 2.5 μg/mL (+)-glaucine + 100 μL water + 200 μL 2 M NaOH, mix, add 6 mL hexane: sec-butanol 998:2, vortex for 10 min, centrifuge at 1500 g for 10 min, freeze in dry ice/acetone. Remove the hexane layer, add 50 μL acetic anhydride, mix, evaporate to dryness under a stream of nitrogen at room temperature, reconstitute the residue in 200 μL MeCN:10 mM pH 4.8 potassium phosphate buffer 10:90, inject a 100 μL aliquot.

## HPLC VARIABLES
**Guard column:** 10 × 3 Chiral-AGP (ASTEC, Whippany, NJ)
**Column:** 150 × 4 Chiral-AGP (ASTEC, Whippany, NJ)
**Mobile phase:** MeCN:10 mM pH 6.65 potassium phosphate buffer
**Flow rate:** 0.9
**Injection volume:** 100
**Detector:** F ex 227 em 308

## CHROMATOGRAM
**Retention time:** 27 (R), 35 (S)
**Internal standard:** (+)-glaucine (20)
**Limit of quantitation:** 3.2 ng/mL

## OTHER SUBSTANCES
**Extracted:** metabolites, norverapamil

## KEY WORDS
plasma; chiral

## REFERENCE
Stagni, G.; Gillespie, W.R. Simultaneous analysis of verapamil and norverapamil enantiomers in human plasma by high-performance liquid chromatography. *J.Chromatogr.B*, **1995**, *667*, 349–354

## SAMPLE
**Matrix:** blood
**Sample preparation:** 2 mL Whole blood or plasma + 2 mL buffer + 5 mL chloroform: isopropanol:n-heptane 60:14:26, shake gently horizontally for 10 min, centrifuge at 2800 g for 10 min. Remove the lower organic layer and evaporate it to dryness under vacuum at 45°, reconstitute the residue in 100 μL mobile phase, centrifuge at 2800 g for 5 min, inject a 50 μL aliquot of the supernatant. (Buffer was saturated ammonium chloride solution 25% diluted with water, adjusted to pH 9.5 with 25% ammonia solution.)

## HPLC VARIABLES
**Column:** 300 × 3.9 4 μm NovaPack C18
**Mobile phase:** MeOH:THF:buffer 65:5:30 (Buffer was 0.68 g/L (10 mM (sic)) $KH_2PO_4$ adjusted to pH 2.6 with concentrated orthophosphoric acid.) (At the end of each session wash the column with water for 1 h and MeOH for 1 h, re-equilibrate for 30 min.)
**Column temperature:** 30

**Flow rate:** 0.8
**Injection volume:** 50
**Detector:** UV 230

## CHROMATOGRAM
**Retention time:** 5.38
**Limit of detection:** <120 ng/mL

## OTHER SUBSTANCES
**Extracted:** acebutolol, acenocoumarol, acepromazine, aceprometazine, acetaminophen, aconitine, ajmaline, albuterol, alimemazine, alminoprofen, alpidem, alprazolam, alprenolol, amisulpride, amitriptyline, amodiaquine, amoxapine, aspirin, astemizole, atenolol, benazepril, benperidol, benzocaine, benzoylecgonine, bepridil, betaxolol, bisoprolol, bromazepam, brompheniramine, bumadizone, caffeine, carbamazepine, carpipramine, carteolol, celiprolol, chlorambucil, chlormezanone, chlorophenacinone, chloroquine, chlorpromazine, chlorpropamide, cibenzoline, cicletanine, clemastine, clobazam, clomipramine, clonazepam, clonidine, clozapine, cocaine, codeine, colchicine, cyamemazine, cyclizine, cycloguanil, cyproheptadine, cytarabine, dacarbazine, daunorubicin, debrisoquine, demexiptiline, desipramine, dextromethorphan, dextromoramide, dextropropoxyphene, diazepam, diazoxide, diclofenac, dihydralazine, diltiazem, diphenhydramine, dipyridamole, disopyramide, dosulepine, doxepin, doxylamine, droperidol, ephedrine, estazolam, etodolac, fenoprofen, fentiazac, floctafenine, flumazenil, flunitrazepam, fluoxetine, fluphenazine, flurbiprofen, fluvoxamine, glibenclamide, glibornuride, glipizide, glutethimide, haloperidol, histapyrrodine, hydroxychloroquine, hydroxyzine, ibuprofen, imipramine, indomethacin, iproniazid, ketamine, ketoprofen, labetalol, levomepromazine, lidocaine, lidoflazine, lisinopril, loperamide, loratadine, lorazepam, loxapine, maprotiline, medazepam, medifoxamine, mefenamic acid, mefenidramine, mefloquine, melphalan, meperidine, mephenesin, mephentermine, mepivacaine, metformin, methadone, methaqualone, methocarbamol, methotrexate, metoclopramide, metoprolol, mexiletine, mianserine, minoxidil, moclobemide, morphine, nadolol, nalbuphine, nalorphine, naloxone, naltrexone, nialamide, nicardipine, nifedipine, niflumic acid, nimodipine, nitrazepam, nitrendipine, nizatidine, nomifensine, nortriptyline, omeprazole, opipramol, oxazepam, penbutolol, penfluridol, pentazocine, phenobarbital, phenol, phenylbutazone, pimozide, pindolol, pipamperone, piroxicam, prazepam, prazosin, prilocaine, procainamide, procarbazine, proguanil, promethazine, propafenone, propranolol, protriptyline, pyrimethamine, quinidine, quinine, quinupramine, ramipril, ranitidine, reserpine, ritodrine, secobarbital, sotalol, strychnine, sulfinpyrazole, sulindac, sulpride, sultopride, suriclone, temazepam, tenoxicam, terbutaline, terfenadine, tetracaine, tetrazepam, thioproperazine, thioridazine, tiapride, tiaprofenic acid, ticlopidine, timolol, tioclomarol, tofisopam, tolbutamide, toloxatone, trazodone, triazolam, trifluoperazine, trifluperidol, trimipramine, tropatenine, viloxazine, vinblastine, vincristine, vindesine, yohimbine, zolpidem, zopiclone, zorubicine
**Interfering:** bupivacaine, buprenorphine, buspirone, carbinoxamine, cetirizine, chlordiazepoxide, chlorpheniramine, clorazepate, fenfluramine, flecainide, loprazolam, metapramine, metipranolol, midazolam, moperone, naproxen, oxprenolol, phencyclidine, thiopental, tianeptine, triprolidine, warfarin

## KEY WORDS
whole blood; plasma

## REFERENCE
Tracqui, A.; Kintz, P.; Mangin, P. Systematic toxicological analysis using HPLC/DAD. *J.Forensic Sci.*, **1995**, *40*, 254–262

## SAMPLE
**Matrix:** blood
**Sample preparation:** 1 mL Serum + 20 µL 10 µg/mL IS in water + 500 µL 1 M NaOH + 5 mL diethyl ether, shake horizontally for 20 min, centrifuge at 1000 g for 10 min.

Remove the organic layer and add it to 14 mL 50 mM sulfuric acid, shake for 20 min, centrifuge at 400 g, discard the ether layer, remove traces of ether from the aqueous layer at 20°, inject a 100 μL aliquot of the aqueous layer.

## HPLC VARIABLES
**Column:** 150 × 3 5 μm CGC SGX18 (Tessek)
**Mobile phase:** MeOH:water:concentrated sulfuric acid 97:3:0.005
**Flow rate:** 0.4
**Injection volume:** 100

## CHROMATOGRAM
**Internal standard:** D-517-HCL
**Limit of detection:** 5 ng/mL

## OTHER SUBSTANCES
**Extracted:** norverapamil

## KEY WORDS
serum; pharmacokinetics

## REFERENCE
Vlcek, J.; Macek, K.; Hulek, P.; Brátová, M.; Fendrich, Z. Pharmacokinetic parameters of verapamil and its active metabolite norverapamil in patients with hepatopathy. *Arzneimittelforschung*, **1995**, *45*, 146–149

## SAMPLE
**Matrix:** blood
**Sample preparation:** 100 μL Plasma + 375 μL MeCN + 25 μL 1 mg/mL 5,6-benzoquinoline in mobile phase, vortex for 13 min, centrifuge at 5000 rpm for 5 min, inject an aliquot of the supernatant.

## HPLC VARIABLES
**Column:** 250 × 4 5 μm Ultracarb C20 ODS (Phenomenex)
**Mobile phase:** MeCN:0.07% orthophosphoric acid 33:67
**Flow rate:** 1
**Injection volume:** 5
**Detector:** F ex 280 em 320

## CHROMATOGRAM
**Retention time:** 10
**Internal standard:** 5,6-benzoquinoline (13)
**Limit of detection:** 3 ng/mL

## KEY WORDS
plasma; rabbit; pharmacokinetics

## REFERENCE
Yazan, Y.; Bozan, B. Rapid analysis of verapamil in plasma by reversed phase HPLC. *Pharmazie*, **1995**, *50*, 117–119

## SAMPLE
**Matrix:** blood
**Sample preparation:** Condition a 1 mL 50 mg Bond Elut 40 μm CN endcapped SPE cartridge with 1 mL MeOH then 1 mL buffer. 1.5 mL Plasma + 30 μL 5 μg/mL gallopamil in water, mix, add 1 mL to the SPE cartridge, wash with 1 mL buffer, elute with 240 μL

MeOH:2-aminoheptane 99.8:0.2, elute with 410 μL pH 3.0 acetate buffer (from mobile phase), mix eluates, inject a 250 μL aliquot. (Buffer was 250 mL 100 mM KH$_2$PO$_4$ + 195.5 mL 100 mM NaOH, make up to 1 L, pH 7.4.)

## HPLC VARIABLES
**Guard column:** 4 × 4 LiChroCART 5 μm LiChrospher 100 RP-18
**Column:** 250 × 4 LiChroCART 4 μm Superspher 100 RP-18
**Mobile phase:** MeCN:2-aminoheptane:buffer 30:0.5:70 (Buffer was 33 mL glacial acetic acid in water, pH adjusted to 3.0 with 0.01 M NaOH.)
**Column temperature:** 35
**Flow rate:** 1.1
**Injection volume:** 250
**Detector:** F ex 275 em 310

## CHROMATOGRAM
**Retention time:** 11
**Internal standard:** gallopamil (13)
**Limit of detection:** 1 ng/mL

## OTHER SUBSTANCES
**Extracted:** norverapamil

## KEY WORDS
plasma; SPE; rugged

## REFERENCE
Hubert, P.; Crommen, J. HPLC determination of verapamil and norverapamil in plasma using automated solid phase extraction for sample preparation and fluorometric detection. *J.Liq.Chromatogr.*, **1994**, *17*, 2147–2170

## SAMPLE
**Matrix:** blood
**Sample preparation:** Automated SPE by ASPEC system. Condition a C18 Clean-Up SPE cartridge (CEC 18111, Worldwide Monitoring) with 2 mL MeOH then 2 mL water. 1 mL Plasma + 1 mL 400 ng/mL protriptyline in water, vortex, add to the SPE cartridge, wash with 3 mL water, wash with 3 mL 750 mL/L methanol. Elute with three aliquots of 300 μL 0.1 M ammonium acetate in MeOH. Add 0.5 mL 0.5 M NaOH and 4 mL 50 mL/L isopropanol in heptane to eluate, mix thoroughly. Allow 5 min for phase separation. Remove upper heptane phase and add it to 300 μL 0.1 M phosphoric acid (pH 2.5), mix, separate, inject a 100 μL aliquot of the aqueous phase.

## HPLC VARIABLES
**Guard column:** LC-8-DB (Supelco)
**Column:** 150 × 4.6 LC-8-DB (Supelco)
**Mobile phase:** MeCN:buffer 35:65 (Buffer was 10 mL/L triethylamine in water adjusted to pH 5.5 with glacial acetic acid.)
**Flow rate:** 2
**Injection volume:** 100
**Detector:** UV 228

## CHROMATOGRAM
**Retention time:** 4.8
**Internal standard:** protriptyline (4)

## OTHER SUBSTANCES
**Extracted:** acetazolamide, amitriptyline, chlordiazepoxide, chlorimipramine, chlorpromazine, desipramine, dextromethorphan, diazepam, diphenhydramine, doxepin, encainide,
fentanyl, flecainide, fluoxetine, flurazepam, haloperidol, hydroxyethylflurazepam, ibuprofen, imipramine, lidocaine, maprotiline, methadone, methaqualone, mexiletine, midazolam, norchlorimipramine, nordiazepam, nordoxepin, norfluoxetine, norverapamil, pentazocine, promazine, propafenone, propoxyphene, propranolol, protriptyline, quinidine,
temazepam, trazodone, trimipramine, verapamil
**Noninterfering:** acetaminophen, acetylmorphine, amiodarone, amobarbital, amphetamine,
bendroflumethiazide, benzocaine, benzoylecgonine, benzthiazide, butalbital, carbamazepine, chlorothiazide, clonazepam, cocaine, codeine, cotinine, cyclosporine, cyclothiazide,
desalkylflurazepam, diamorphine, dicumerol, ephedrine, ethacrynic acid, ethanol, ethchlorvynol, ethosuximide, furosemide, glutethimide, hydrochlorothiazide, hydrocodone,
hydroflumethiazide, hydromorphone, lorazepam, mephentermine, meprobamate, methamphetamine, metharbital, methoxsalen, methoxyphenteramine, methsuximide, methylcyclothiazide, metoprolol, MHPG, monoacetylmorphine, morphine, normethsuximide,
oxazepam, oxycodone, oxymorphone, pentobarbital, phencyclidine, phenteramine, phenylephrine, phenytoin, polythiazide, primidone, prochlorperazine, salicylic acid, sulfanilamide, THC-COOH, theophylline, thiazolam, thiopental, thioridazine, tocainide, trichloromethiazide, trifluoperazine, valproic acid, warfarin
**Interfering:** imipramine, maprotiline, nortriptyline

## KEY WORDS
plasma; SPE

## REFERENCE
Nichols, J.H.; Charlson, J.R.; Lawson, G.M. Automated HPLC assay of fluoxetine and norfluoxetine in
serum. *Clin.Chem.*, **1994**, *40*, 1312–1316

## SAMPLE
**Matrix:** blood
**Sample preparation:** 1 mL Plasma + 40 µL 20 µg/mL deacetyldiltiazem in water + 1 mL
pH 10 borate buffer + 0.5 g NaCl, vortex for 30 s, add 6 mL hexane:isopropanol 95:5,
vortex for 30 s, shake for 10 min, centrifuge at 700 g for 10 min. Remove the organic
layer and add it to 200 µL 5 mM sulfuric acid, vortex for 30 s, shake for 10 min, centrifuge
at 700 g for 5 min, inject a 40 µL aliquot of the aqueous layer.

## HPLC VARIABLES
**Column:** 300 × 3.9 10 µm µBondapak C18
**Mobile phase:** MeCN:MeOH:buffer 22.5:22.5:55 (Buffer was 7.5 g sodium acetate trihy-
drate + 0.6 g 1-heptanesulfonic acid in 1 L water, adjust pH to 4.5 with acetic acid.)
**Flow rate:** 1.5
**Injection volume:** 40
**Detector:** UV 237

## CHROMATOGRAM
**Retention time:** 15.5
**Internal standard:** deacetyldiltiazem (7)
**Limit of detection:** 4 ng/mL

## OTHER SUBSTANCES
**Simultaneous:** celiprolol, diltiazem, propranolol
**Noninterfering:** atenolol, aspirin, caffeine, ibuprofen, lidocaine, metoprolol, nifedipine
**Interfering:** desipramine

## KEY WORDS
plasma

## REFERENCE
Rutledge, D.R.; Abadi, A.H.; Lopez, L.M. Simultaneous determination of verapamil and celiprolol in human plasma. *J.Chromatogr.Sci.*, **1994**, *32*, 153–156

## SAMPLE
**Matrix:** blood
**Sample preparation:** 1 mL Plasma + 1 mL 0.9% NaCl + 120 μL 2 M NaOH + 5 mL n-hexane, vortex for 30 s, mix for 10 min, centrifuge at 2500 g for 20 min, repeat extraction with 4 mL n-hexane. Combine the organic phases and evaporate them to dryness, rinse walls with 1 mL n-hexane, evaporate to dryness, dissolve residue in 50 μL isopropanol, inject a 25 μL aliquot.

## HPLC VARIABLES
**Guard column:** 50 × 4.6 Chiralpak AD (Baker)
**Column:** 250 × 4.6 Chiralpak AD (Baker)
**Mobile phase:** n-Hexane:isopropanol 90:10 with 0.1% diethylamine
**Flow rate:** 1
**Injection volume:** 25
**Detector:** F ex 223 no emission filter

## CHROMATOGRAM
**Retention time:** 10 (S-(−)), 12 (R-(+))
**Limit of quantitation:** 20 ng/mL

## OTHER SUBSTANCES
**Extracted:** norverapamil
**Also analyzed:** gallopamil

## KEY WORDS
plasma; chiral

## REFERENCE
Fieger, H.; Blaschke, G. Direct determination of the enantiomeric ratio of verapamil, its major metabolite norverapamil, and gallopamil in plasma by chiral high-performance liquid chromatography. *J.Chromatogr.*, **1992**, *575*, 255–260

## SAMPLE
**Matrix:** blood
**Sample preparation:** 1 mL plasma + 50 μL 400 ng/mL (+)-glaucine + 100 μL 2 M NaOH + 1 mL pH 7.0 sodium phosphate buffer (ionic strength 1) + 6 mL heptane, vortex for 1 min, centrifuge at 1500 g for 10 min, freeze in a dry ice bath. Decant heptane layer and evaporate it to dryness in a vacuum centrifuge at 60°, reconstitute the residue in 250 μL mobile phase, inject a 100 μL aliquot.

## HPLC VARIABLES
**Guard column:** 50 × 4 5 μm LiChrocart DIOL (column temp ambient)
**Column:** 250 × 4.6 10 μm Chiralpak AD (column temp 30°)
**Mobile phase:** Hexane:isopropanol:ethanol 85:7.5:7.5 containing 1% triethylamine
**Column temperature:** 30
**Flow rate:** 1
**Injection volume:** 100
**Detector:** F ex 272 em 317

## CHROMATOGRAM
**Retention time:** 7 (S), 8 (R)

**Internal standard:** (+)-glaucine (9.5)
**Limit of quantitation:** 2.5 ng/mL

## OTHER SUBSTANCES
**Extracted:** metabolites, norverapamil

## KEY WORDS
plasma; chiral

## REFERENCE
Shibukawa, A.; Wainer, I.W. Simultaneous direct determination of the enantiomers of verapamil and norverapamil in plasma using a derivatized amylose high-performance liquid chromatographic chiral stationary phase. *J.Chromatogr.*, **1992**, *574*, 85–92

## SAMPLE
**Matrix:** blood
**Sample preparation:** Add n-propyl p-aminobenzoate to plasma, deproteinize 1 mL plasma with 1 mL MeCN, inject an aliquot onto column A with mobile phase A, monitor effluent from column A with detector A, allow eluate containing verapamil to fill a 2 mL sample loop for 2 min, elute contents of sample loop onto column B with mobile phase B and at the same time allow pure mobile phase B to flow onto column B (ratio of column loop flow:pure mobile phase B flow 1:9), elute contents of column B onto column C with mobile phase C, monitor effluent from column C with detector B.

## HPLC VARIABLES
**Column:** A 150 × 4.6 Inertsil ODS-2; B 10 × 4 Ultron ES-OVMG; C 150 × 4.6 Ultron ES-OVM
**Mobile phase:** A MeCN:water 30:70 containing 5 mM sodium 1-pentanesulfonate, adjust pH to 3.0 with phosphoric acid; B 5 mM pH 7.5 phosphate buffer; C THF:EtOH:water 1:8:91 containing 20 mM $KH_2PO_4$
**Flow rate:** A 1; B 4; C 1
**Detector:** UV 230 (A); UV 230 (B)

## CHROMATOGRAM
**Retention time:** 14 (from column A), 23.5 (l enantiomer from column C), 25 (d enantiomer from column C)
**Internal standard:** n-propyl p-aminobenzoate (17, from column A)
**Limit of detection:** 1 ng/mL

## KEY WORDS
plasma; column-switching; chiral

## REFERENCE
Oda, Y.; Asakawa, N.; Kajima, T.; Yoshida, Y.; Sato, T. On-line determination and resolution of verapamil enantiomers by high-performance liquid chromatography with column switching. *J.Chromatogr.*, **1991**, *541*, 411–418

## SAMPLE
**Matrix:** blood
**Sample preparation:** 1 mL Plasma + 250 μL MeCN + 20 mg $MgSO_4$ + 10 mg $ZnSO_4$, vortex for 3 min, centrifuge at 2000 g for 6 min. Remove the supernatant, filter (0.45 μm), inject a 100 μL aliquot. (Tubes were cleaned by sonication with acetone for 30 min, sonication twice with MeCN for 15 min, and drying at 100°. Stir 100 g $ZnSO_4$ crystals with 200 mL MeCN for 30 min, decant MeCN, repeat twice with fresh MeCN, dry crystals in a hood at room temperature.)

## HPLC VARIABLES
**Column:** 80 × 2 5 μm Spherisorb C8
**Mobile phase:** MeCN:10 mM KH$_2$PO$_4$ 60:40, pH adjusted to 7.1 with 100 mM NaOH
**Flow rate:** 0.8
**Injection volume:** 100
**Detector:** UV 220

## CHROMATOGRAM
**Retention time:** 8
**Limit of quantitation:** 30 ng/mL

## KEY WORDS
plasma

## REFERENCE
Rustum, A.M. Measurement of verapamil in human plasma by reversed-phase high-performance liquid chromatography using a short octyl column. *J.Chromatogr.*, **1990**, *528*, 480−486

## SAMPLE
**Matrix:** blood
**Sample preparation:** 500 μL Serum + 250 μL diisopropyl ether:n-butyl alcohol 7:3 containing 400 ng/mL minaprine (Caution! Diisopropyl ether readily forms explosive peroxides!), centrifuge 2 min, shake, centrifuge 5 min, inject 50 μL aliquot of top organic layer.

## HPLC VARIABLES
**Guard column:** 30 × 4.6 5 μm Brownlee cyano spheri-5
**Column:** 250 × 4.6 5 μm Altex ultrasphere cyano
**Mobile phase:** MeCN:THF:water:2 M ammonium formate (pH 4.0) 700:100:195:5
**Column temperature:** 20
**Flow rate:** 1.5
**Injection volume:** 50
**Detector:** UV 235

## CHROMATOGRAM
**Retention time:** 6
**Internal standard:** minaprine (5.5)
**Limit of detection:** 20 ng/mL

## OTHER SUBSTANCES
**Also analyzed:** amiodarone, amitriptyline, clomipramine, desipramine, diltiazem, haloperidol, imipramine, nortriptyline, propafenone

## KEY WORDS
serum

## REFERENCE
Mazzi, G. Simple and practical high-performance liquid chromatographic assay of some tricyclic drugs, haloperidol, diltiazem, verapamil, propafenone, and amiodarone. *Chromatographia*, **1987**, *24*, 313−316

## SAMPLE
**Matrix:** formulations
**Sample preparation:** Dilute a 1 mL sample to 10 mL with mobile phase, inject an aliquot.

## HPLC VARIABLES
**Column:** 150 × 4.6 5 μm Spherisorb Phenyl

**Mobile phase:** MeCN:water:100 mM tetrabutylammonium hydrogen sulfate:500 mM KH$_2$PO$_4$ 15:50:25:10, pH 5.1
**Flow rate:** 2
**Injection volume:** 20
**Detector:** UV 268

## CHROMATOGRAM
**Retention time:** 10

## OTHER SUBSTANCES
**Simultaneous:** amrinone

## KEY WORDS
injections; stability-indicating; 5% dextrose; 0.45% NaCl

## REFERENCE
Riley, C.M.; Junkin, P. Stability of amrinone and digoxin, procainamide hydrochloride, propranolol hydrochloride, sodium bicarbonate, potassium chloride, or verapamil hydrochloride in intravenous admixtures. *Am.J.Hosp.Pharm.*, **1991**, *48*, 1245–1252

## SAMPLE
**Matrix:** formulations
**Sample preparation:** 1 mL Sample + 9 mL mobile phase, inject an aliquot.

## HPLC VARIABLES
**Column:** 150 × 4.6 5 μm Spherisorb phenyl
**Mobile phase:** MeCN:500 mM KH$_2$PO$_4$:100 mM tetrabutylammonium hydrogen sulfate: water 25:10:25:40 adjusted to pH 5.4 with 10 M NaOH
**Flow rate:** 2
**Injection volume:** 20
**Detector:** UV 268

## CHROMATOGRAM
**Retention time:** 5

## OTHER SUBSTANCES
**Simultaneous:** milrinone

## KEY WORDS
stability-indicating; 5% dextrose; injections

## REFERENCE
Riley, C.M. Stability of milrinone and digoxin, furosemide, procainamide hydrochloride, propranolol hydrochloride, quinidine gluconate, or verapamil hydrochloride in 5% dextrose injection. *Am.J.Hosp.Pharm.*, **1988**, *45*, 2079–2091

## SAMPLE
**Matrix:** reaction mixtures
**Sample preparation:** If necessary, remove oxidizing power of solution by adding sodium metabisulfite, inject a 20 μL aliquot.

## HPLC VARIABLES
**Guard column:** 15 × 4.6 5 μm Microsorb C8
**Column:** 250 × 4.6 5 μm Microsorb C8

**Mobile phase:** MeCN:5.5 mM sodium octanesulfonate + 20 mM trisodium citrate dihydrate adjusted to pH 3 with concentrated HCl 65:35
**Flow rate:** 1
**Injection volume:** 20
**Detector:** UV 230

## CHROMATOGRAM
**Retention time:** 7.1
**Limit of detection:** 1 µg/mL

## REFERENCE
Lunn, G.; Rhodes, S.W.; Sansone, E.B.; Schmuff, N.R. Photolytic destruction and polymeric resin decontamination of aqueous solutions of pharmaceuticals. *J.Pharm.Sci.*, **1994**, *83*, 1289–1293

## SAMPLE
**Matrix:** solutions

## HPLC VARIABLES
**Column:** 250 × 4.6 Chirex 3022 (Phenomenex)
**Mobile phase:** Hexane:1,2-dichloroethane:EtOH/trifluoroacetic acid 60:35:5 (EtOH/trifluoroacetic acid was premixed 20:1.)
**Flow rate:** 0.7-1
**Injection volume:** 20
**Detector:** UV 280

## KEY WORDS
chiral; $\alpha = 1.11$

## REFERENCE
Cleveland, T. Pirkle-concept chiral stationary phases for the HPLC separation of pharmaceutical racemates. *J.Liq.Chromatogr.*, **1995**, *18*, 649–671

## SAMPLE
**Matrix:** solutions

## HPLC VARIABLES
**Column:** 250 × 4.6 5 µm Supelcosil LC-DP (A) or 250 × 4 5 µm LiChrospher 100 RP-8 (B)
**Mobile phase:** MeCN:0.025% phosphoric acid:buffer 25:10:5 (A) or 60:25:15 (B) (Buffer was 9 mL concentrated phosphoric acid and 10 mL triethylamine in 900 mL water, adjust pH to 3.4 with dilute phosphoric acid, make up to 1 L.)
**Flow rate:** 0.6
**Injection volume:** 25
**Detector:** UV 229

## CHROMATOGRAM
**Retention time:** 13.28 (A), 6.96 (B)

## OTHER SUBSTANCES
**Also analyzed:** metabolites, acebutolol, acepromazine, acetaminophen, acetazolamide, acetophenazine, albuterol, alprazolam, amitriptyline, amobarbital, amoxapine, antipyrine, atenolol, atropine, azatadine, baclofen, benzocaine, bromocriptine, brompheniramine, brotizolam, bupivacaine, buspirone, butabarbital, butalbital, caffeine, carbamazepine, cetirizine, chlorcyclizine, chlordiazepoxide, chlormezanone, chloroquine, chlorpheniramine, chlorpromazine, chlorpropamide, chlorprothixene, chlorthalidone, chlorzoxazone, cimetidine, cisapride, clomipramine, clonazepam, clonidine, clozapine, cocaine, codeine, colchicine, cyclizine, cyclobenzaprine, dantrolene, desipramine, diazepam, diclofenac, diflunisal,

diltiazem, diphenhydramine, diphenidol, diphenoxylate, dipyridamole, disopyramide, dobutamine, doxapram, doxepin, droperidol, encainide, ethidium bromide, ethopropazine, fenoprofen, fentanyl, flavoxate, fluoxetine, fluphenazine, flurazepam, flurbiprofen, fluvoxamine, furosemide, glutethimide, glyburide, guaifenesin, haloperidol, homatropine, hydralazine, hydrochlorothiazide, hydrocodone, hydromorphone, hydroxychloroquine, hydroxyzine, ibuprofen, imipramine, indomethacin, ketoconazole, ketoprofen, ketorolac, labetalol, levorphanol, lidocaine, loratadine, lorazepam, lovastatin, loxapine, mazindol, mefenamic acid, meperidine, mephenytoin, mepivacaine, mesoridazine, metaproterenol, metformin, methadone, methdilazine, methocarbamol, methotrexate, methotrimeprazine, methoxamine, methyldopa, methylphenidate, metoclopramide, metolazone, metoprolol, metronidazole, midazolam, moclobemide, morphine, nadolol, nalbuphine, naloxone, naphazoline, naproxen, nifedipine, nizatidine, norepinephrine, nortriptyline, oxazepam, oxycodone, oxymetazoline, paroxetine, pemoline, pentazocine, pentobarbital, pentoxifylline, perphenazine, pheniramine, phenobarbital, phenol, phenolphthalein, phentolamine, phenylbutazone, phenyltoloxamine, phenytoin, pimozide, pindolol, piroxicam, pramoxine, prazepam, prazosin, probenecid, procainamide, procaine, prochlorperazine, procyclidine, promazine, promethazine, propafenone, propantheline, propiomazine, propofol, propranolol, protriptyline, quazepam, quinidine, quinine, racemethorphan, ranitidine, remoxipride, risperidone, salicylic acid, scopolamine, secobarbital, sertraline, sotalol, spironolactone, sulfinpyrazone, sulindac, temazepam, terbutaline, terfenadine, tetracaine, theophylline, thiethylperazine, thiopental, thioridazine, thiothixene, timolol, tocainide, tolbutamide, tolmetin, trazodone, triamterene, triazolam, trifluoperazine, triflupromazine, trimeprazine, trimethoprim, trimipramine, warfarin, xylometazoline, yohimbine, zopiclone

## KEY WORDS
some details of plasma extraction

## REFERENCE
Koves, E.M. Use of high-performance liquid chromatography-diode array detection in forensic toxicology. *J.Chromatogr.A*, **1995**, *692*, 103–119

## SAMPLE
**Matrix:** solutions
**Sample preparation:** Isolate verapamil using achiral preparative HPLC, freeze dry an appropriate aliquot of the mobile phase, inject an aliquot into this system.

## HPLC VARIABLES
**Column:** 150 × 4.6 OVM ovomucoid protein (Mac-Mod Analytical)
**Mobile phase:** EtOH:buffer 10:90 (Buffer was 13 mM $KH_2PO_4$ adjusted to pH 7.0 with phosphoric acid.)
**Flow rate:** 1
**Detector:** F ex 203 em 270 (cut-off)

## CHROMATOGRAM
**Retention time:** 12.6 (S), 16.2 (R)

## OTHER SUBSTANCES
**Also analyzed:** metabolites

## KEY WORDS
chiral

## REFERENCE
Lankford, S.M.; Bai, S.A. Determination of the stereochemical composition of the major metabolites of verapamil in dog urine with enantioselective liquid chromatographic techniques. *J.Chromatogr.B*, **1995**, *663*, 91–101

## SAMPLE
**Matrix:** solutions
**Sample preparation:** Inject a 50-200 μL aliquot of a solution in pH 7.4 Tyrode's buffer.

## HPLC VARIABLES
**Column:** 150 × 3.9 4 μm Nova-Pak C-18
**Mobile phase:** MeCN:50 mM phosphoric acid:triethylamine 40:60:0.1
**Column temperature:** 35
**Flow rate:** 0.6
**Injection volume:** 50-200
**Detector:** UV 230

## OTHER SUBSTANCES
**Also analyzed:** chlorpromazine, propantheline

## KEY WORDS
buffer

## REFERENCE
Saitoh, H.; Aungst, B.J. Possible involvement of multiple P-glycoprotein-mediated efflux systems in the transport of verapamil and other organic cations across rat intestine. *Pharm.Res.*, **1995**, *12*, 1304–1310

## SAMPLE
**Matrix:** solutions

## HPLC VARIABLES
**Column:** 150 × 3.9 Nova-Pak C18
**Mobile phase:** MeCN:buffer 35:65 containing 0.1% triethylamine (Buffer was 40 mM pH 4.0 phosphate buffer.)
**Detector:** UV 230

## REFERENCE
Surakitbanharn, Y.; McCandless, R.; Krzyzaniak, J.F.; Dannenfelser, R.-M.; Yalkowsky, S.H. Self-association of dexverapamil in aqueous solution. *J.Pharm.Sci.*, **1995**, *84*, 720–723

## SAMPLE
**Matrix:** solutions

## HPLC VARIABLES
**Guard column:** 30 × 3.2 7 μm SI 100 ODS (not commercially available)
**Column:** 150 × 3.2 7 μm SI 100 ODS (not commercially available)
**Mobile phase:** MeCN:buffer 31.2:68.8 (Buffer was 6.66 g $KH_2PO_4$ and 4.8 g 85% phosphoric acid in 1 L water, pH 2.3.)
**Flow rate:** 0.5-1
**Detector:** UV 225; UV 274

## CHROMATOGRAM
**Retention time:** 4.1
**Internal standard:** 5-(4-methylphenyl)-5-phenylhydantoin (7.3)

## OTHER SUBSTANCES

**Also analyzed:** aspirin, caffeine, carbamazepine, chlordiazepoxide, chlorprothixene, clonazepam, diazepam, doxylamine, ethosuximide, furosemide, haloperidol, hydrochlorothiazide, methocarbamol, methotrimeprazine, nicotine, oxazepam, procaine, promazine, propafenone, propranolol, salicylamide, temazepam, tetracaine, thiopental, triamterene, zolpidem, zopiclone

## REFERENCE

Below, E.; Burrmann, M. Application of HPLC equipment with rapid scan detection to the identification of drugs in toxicological analysis. *J.Liq.Chromatogr.*, **1994**, *17*, 4131–4144

## SAMPLE

**Matrix:** solutions

## HPLC VARIABLES

**Column:** 250 × 4.6 Zorbax RX
**Mobile phase:** Gradient. A was 10 mL concentrated orthophosphoric acid and 7 mL triethylamine in 1 L water. B was 10 mL concentrated orthophosphoric acid and 7 mL triethylamine in 200 mL water, make up to 1 L with MeCN. A:B from 100:0 to 0:100 over 30 min, maintain at 0:100 for 5 min.
**Column temperature:** 30
**Flow rate:** 2
**Detector:** UV 210

## OTHER SUBSTANCES

**Also analyzed:** acepromazine, acetaminophen, acetophenazine, albuterol, aminophylline, amitriptyline, amobarbital, amoxapine, amphetamine, amylocaine, antipyrine, aprobarbital, aspirin, atenolol, atropine, avermectin, barbital, benzocaine, benzoic acid, benzotropine, benzphetamine, berberine, bibucaine, bromazepan, brompheniramine, buprenorphine, buspirone, butabarbital, butacaine, butethal, caffeine, carbamazepine, carbromal, chloramphenicol, chlordiazepoxide, chloroquine, chlorothiazide, chloroxylenol, chlorphenesin, chlorpheniramine, chlorpromazine, chlorpropamide, chlortetracycline, cimetidine, cinchonidine, cinchonine, clenbuterol, clonazepam, clonixin, clorazepate, cocaine, codeine, colchicine, cortisone, coumarin, cyclazocine, cyclobenzaprine, cyclothiazide, cyheptamide, cymarin, danazol, danthron, dapsone, debrisoquine, desipramine, dexamethasone, dextromethorphan, dextropropoxyphene, diamorphine, diazepam, diclofenac, diethylpropion, diethylstilbestrol, diflunisal, digitoxin, digoxin, diltiazem, diphenhydramine, diphenoxylate, diprenorphine, dipyrone, disulfiram, dopamine, doxapram, doxepin, dronabinol, ephedrine, epinephrine, epinine, estradiol, estriol, estrone, ethacrynic acid, ethosuximide, etonitazene, etorphine, eugenol, famotidine, fenbendazole, fencamfamine, fenoprofen, fenproporex, fentanyl, flubendazole, flufenamic acid, flunitrazepam, 5-fluorouracil, fluoxymesterone, fluphenazine, furosemide, gentisic acid, gitoxigenin, glipizide, glunixin, glutethimide, glybenclamide, guaiacol, halazepam, haloperidol, hydrochlorothiazide, hydrocodone, hydrocortisone, hydromorphone, hydroxyquinoline, ibogaine, ibuprofen, iminostilbene, imipramine, indomethacin, isocarbostyril, isocarboxazid, isoniazid, isoproterenol, isoxsuprine, ivermectin, ketamine, ketoprofen, kynurenic acid, levorphanol, lidocaine, lorazepam, lormetazepam, loxapine, mazindol, mebendazole, meclizine, meclofenamic acid, medazepam, mefenamic acid, megestrol, mepacrine, meperidine, mephentermine, mephenytoin, mephesin, mephobarbital, mepivacaine, mescaline, mesoridazine, methadone, methamphetamine, methapyrilene, methaqualone, methazolamide, methocarbamol, methoxamine, methsuximide, methyl salicylate, methyldopa, methyldopamine, methylphenidate, methylprednisolone, methyltestosterone, methyprylon, metoprolol, mibolerone, morphine, nadolol, nalorphine, naloxone, naltrexone, naphazoline, naproxen, nefopam, niacinamide, nicotine, nicotinic acid, nifedipine, niflumic acid, nitrazepam, norepinephrine, nortriptyline, noscapine, nylidrin, oxazepam, oxycodone, oxymorphone, oxyphenbutazone, oxytetracycline, papaverine, pargyline, pemoline, pentazocine, pentobarbital, persantine, phenacetin, phenazocine, phenazopyridine, phencyclidine, phendimetrazine, phenelzine, pheniramine, phenobarbital, phenothiazine, phensuximide, phentermine, phenylbutazone, phenylephrine, phenylpropanolamine, piperocaine, prazepam, prednisolone, primidone, probenecid, progesterone, propiomazine, propranolol, propylparaben, pseudoephedrine, puromycin, pyrilamine, pyrithyldione, quazepam, quinaldic acid, quinidine, quinine, ranitidine, recinnamine, reserpine, resorcinol, saccharin, albuterol, salicylamide, salicylic acid, scopolamine, scopoletin, secobarbital, strychnine, sulfacetamide, sufadiazine, sulfadimethoxine, sulfaethidole, sulfamerazine, sulfamethazine, sulfamethoxizole, sulfanilamide, sulfapyridine, sulfasoxizole, sulindac, tamoxifen, temazepam, testosterone, tetracaine, tetracycline, tetramisole, thebaine, theobromine, theophylline, thiabendazole, thiamine, thiamylal, thiobarbituric acid, thioridazine, thiosalicylic acid, thiothixene, thymol, tolazamide, tolazoline, tobutamide, tolmetin, tranylcypromine, triamcinolone, tribenzylamine, trichloromethiazide, trifluoperazine, trihexyphenidyl, trimethoprim, tripelennamine, triprolidine, tropacocaine, tyramine, vincamine, warfarin, yohimbine, zoxazolamine

## REFERENCE

Hill, D.W.; Kind, A.J. Reversed-phase solvent-gradient HPLC retention indexes of drugs. *J.Anal.Toxicol.*, **1994**, *18*, 233–242

## SAMPLE

**Matrix:** solutions
**Sample preparation:** Prepare a 50 μg/mL solution in MeCN:water 40:60, inject an aliquot.

## HPLC VARIABLES

**Column:** 250 × 4.6 3 μm silica (Phenomenex)
**Mobile phase:** MeCN:6.25 mM pH 3.0 phosphate buffer 40:60
**Flow rate:** 1
**Injection volume:** 50
**Detector:** UV 254

## CHROMATOGRAM

**Retention time:** 10.29

## OTHER SUBSTANCES

**Also analyzed:** atenolol, clonidine, diltiazem, metoprolol, nifedipine, prazosin, propranolol

## REFERENCE

Simmons, B.R.; Stewart, J.T. HPLC separation of selected cardiovascular agents on underivatized silica using an aqueous organic mobile phase. *J.Liq.Chromatogr.*, **1994**, *17*, 2675–2690

## SAMPLE

**Matrix:** solutions

## HPLC VARIABLES

**Column:** 250 × 4.6 cellulose tris(3,5-difluorophenylcarbamate)
**Mobile phase:** Hexane:isopropanol:diethylamine 80:20:0.1
**Flow rate:** 0.5
**Detector:** UV

## CHROMATOGRAM

**Retention time:** 56 (+), 66 (−)

## KEY WORDS

chiral

## REFERENCE

Okamoto, Y.; Aburatani, R.; Hatano, K.; Hatada, K. Optical resolution of racemic drugs by chiral HPLC on cellulose and amylose tris(phenylcarbamate) derivatives. *J.Liq.Chromatogr.*, **1988**, *11*, 2147–2163

## SAMPLE

**Matrix:** solutions
**Sample preparation:** Prepare a 10 μg/mL solution in MeOH, inject a 20 μL aliquot.

## HPLC VARIABLES

**Column:** 125 × 4.9 Spherisorb S5W silica
**Mobile phase:** MeOH containing 10 mM ammonium perchlorate and 1 mL/L 100 mM NaOH in MeOH, pH 6.7
**Flow rate:** 2
**Injection volume:** 20
**Detector:** E, LeCarbone, V25 glassy carbon electrode, + 1.2 V

## CHROMATOGRAM

**Retention time:** 2.7

## OTHER SUBSTANCES

**Also analyzed:** acebutolol, acepromazine, acetophenazine, N-acetylprocainamide, albuterol, alprenolol, amethocaine, amiodarone, amitriptyline, antazoline, atenolol, azacyclonal, bamethane, benactyzine, benperidol, benzethidine, benzocaine, benzoctamine, benzphetamine, benzquinamide, bromhexine, bromodiphenhydramine, bromperidol, brompheniramine, brompromazine, buclizine, bufotenine, bupivacaine, buprenorphine, butacaine, butethamate, chlorcyclizine, chlorpheniramine, chlorphenoxamine, chlorprenaline, chlorpromazine, chlorprothixene, cimetidine, cinchonidine, cinnarizine, clemastine, clomipramine, clonidine, cocaine, cyclazocine, cyclizine, cyclopentamine, cyproheptadine, deserpidine, desipramine, dextromoramide, dextropropoxyphene, dicyclomine, diethylcarbamazine, diethylpropion, diethylthiambutene, dihydroergotamine, dimethindene, dimethothiazine, diphenhydramine, diphenoxylate, dipipanone, diprenorphine, dipyridamole, disopyramide, dothiepin, doxapram, doxepin, doxylamine, droperidol, ephedrine, ergocornine, ergocristine, ergocristinine, ergocryptine, ergometrine, ergosine, ergosinine, ergotamine, ethopropazine, etorphine, etoxeridine, fenethazine, fenfluramine, fenoterol, fentanyl, flavoxate, fluopromazine, flupenthixol, fluphenazine, flurazepam, haloperidol, hydroxyzine, hyoscine, ibogaine, imipramine, indapamine, iprindole, isothipendyl, isoxsuprine, ketanserin, laudanosine, lidocaine, lofepramine, loxapine, maprotiline, mecamylamine, meclophenoxate, meclozine, medazepam, mephentermine, mepivacaine, meptazinol, mepyramine, mesoridazine, metaraminol, methadone, methamphetamine, methapyrilene, methdilazene, methotrimeprazine, methoxamine, methoxyphenamine, methoxypromazine, methylephedrine, methylergonovine, methysergide, metoclopramide, metopimazine, metoprolol, mianserin, morazone, nadolol, nalorphine, naloxone, naphazoline, nicotine, nifedipine, nomifensine, nortriptyline, noscapine, orphenadrine, oxeladin, oxprenolol, oxymetazolin, papaverine, pargyline, pecazine, penbutolol, pentazocine, penthienate, pericyazine, perphenazine, phenadoxone, phenampromide, phenazocine, phenbutrazate, phendimetrazine, phenelzine, phenglutarimide, phenindamine, pheniramine, phenmetrazine, phenomorphan, phenoperidine, phenothiazine, phenoxybenzamine, phentolamine, phenylephrine, phenyltoloxamine, physostigmine, piminodine, pimozide, pindolol, pipamazine, pipazethate, piperacetazine, piperidolate, pipradol, pirenzepine, piritramide, pizotifen, practolol, pramoxine, prazosin, prenylamine, prilocaine, primaquine, proadifen, procainamide, procaine, prochlorperazine, procyclidine, proheptazine, prolintane, promazine, promethazine, pronethalol, properidine, propiomazine, propranolol, prothipendyl, protriptyline, proxymetacaine, pseudoephedrine, pyrimethamine, quinidine, quinine, ranitidine, rescinnamine, sotalol, tacrine, terazosin, terbutaline, terfenadine, thenyldiamine, theophylline, thiethylperazine, thiopropazate, thioproperazine, thioridazine, thiothixene, thonzylamine, timolol, tocainide, tolpropamine, tolycaine, tranylcypromine, trazodone, trifluoperazine, trifluperidol, trimeperidine, trimeprazine, trimethobenzamide, trimethoprim, trimipramine, tripelennamine, triprolidine, tryptamine, xylometazoline

## REFERENCE
Jane, I.; McKinnon, A.; Flanagan, R.J. High-performance liquid chromatographic analysis of basic drugs on silica columns using non-aqueous ionic eluents. II. Application of UV, fluorescence and electro-chemical oxidation detection. *J.Chromatogr.*, **1985**, *323*, 191–225

## SAMPLE
**Matrix:** urine
**Sample preparation:** Adjust pH of 2 mL urine to 11 with 5 M NaOH, extract twice with 10 mL pentane:dichloromethane 70:30. Combine the organic layers and evaporate to dryness under a stream of nitrogen, reconstitute the residue in 200 µL MeOH, inject a 100 µL aliquot. (Deconjugate 2 mL urine with 500 µL 100 mM pH 4.5 sodium acetate buffer and 100 µL Glusulase (90000 U/mL β-glucuronidase and 1000 U/mL sulfatase, Du Pont), heat at 37° for 18 h, proceed as before.)

## HPLC VARIABLES
**Column:** 250 × 10 10 µm C18 (Alltech)
**Mobile phase:** Gradient. A was MeCN. B was 50 mM ammonium acetate adjusted to pH 4.5 with acetic acid. A:B 0:100 to 70:30 over 70 min.
**Flow rate:** 3
**Injection volume:** 100
**Detector:** F ex 280 em 340 (cut-off)

## CHROMATOGRAM
**Retention time:** 66

## OTHER SUBSTANCES
**Extracted:** metabolites

## KEY WORDS
dog; preparative

## REFERENCE
Lankford, S.M.; Bai, S.A. Determination of the stereochemical composition of the major metabolites of verapamil in dog urine with enantioselective liquid chromatographic techniques. *J.Chromatogr.B*, **1995**, *663*, 91–101

## ANNOTATED BIBLIOGRAPHY
Romanová, D.; Brandsteterová, E.; Králiková, D.; Bozeková, L.; Kriska, M. Determination of verapamil and its metabolites in plasma using HPLC. *Pharmazie*, **1994**, *49*, 779–780 [plasma; serum; SPE; LOQ 50 ng/mL; fluorescence detection]

Miller, L.; Bergeron, R. Analytical and preparative resolution of enantiomers of verapamil and norverapamil using a cellulose-based chiral stationary phase in the reversed-phase mode. *J.Chromatogr.*, **1993**, *648*, 381–388 [chiral]

Muscara, M.N.; de-Nucci, G. Measurement of plasma verapamil levels by high-performance liquid chromatography. *Braz.J.Med.Biol.Res.*, **1993**, *26*, 753–763

Rasymas, A.K.; Boudoulas, H.; MacKichan, J. Determination of verapamil enantiomers in serum following racemate administration using HPLC. *J.Liq.Chromatogr.*, **1992**, *15*, 3013–3029 [chiral; achiral; fluorescence detection; extracted metabolites; imipramine (IS); LOD 0.2 ng/mL; pharmacokinetics]

Köppel, C.; Wagemann, A. Plasma level monitoring of D, L-verapamil and three of its metabolites by reversed-phase high-performance liquid chromatography. *J.Chromatogr.*, **1991**, *570*, 229–234 [LOD 5 ng/mL; fluorescence detection]

Lacroix, P.M.; Graham, S.J.; Lovering, E.G. High-performance liquid chromatographic method for the assay of verapamil hydrochloride and related compounds in raw material. *J.Pharm.Biomed.Anal.*, **1991**, *9*, 817–822

Oda, Y.; Asakawa, N.; Kajima, T.; Yoshida, Y.; Sato, T. Column-switching high-performance liquid chromatography for on-line simultaneous determination and resolution of enantiomers of verapamil and its metabolites in plasma. *Pharm.Res.*, **1991**, *8*, 997–1001

Chu, Y.-Q.; Wainer, I.W. Determination of the enantiomers of verapamil and norverapamil in serum using coupled achiral-chiral high-performance liquid chromatography. *J.Chromatogr.*, **1989**, *497*, 191–200 [plasma; chiral; column-switching; heart-cut; LOQ 25 ng/mL]

Salama, Z.B.; Dilger, C.; Czogalla, W.; Otto, R.; Jaeger, H. Quantitative determination of verapamil and metabolites in human serum by high-performance liquid chromatography and its application to biopharmaceutic investigations. *Arzneimittelforschung*, **1989**, *39*, 210–215

Bremseth, D.L.; Lima, J.J.; MacKichan, J.J. Specific HPLC method for the separation of verapamil and four major metabolites after oral dosing. *J.Liq.Chromatogr.*, **1988**, *11*, 2731–2749 [fluorescence detection; imipramine (IS); LOD 1 ng; plasma; serum; urine]

Pieper, J.A.; Rutledge, D.R. Determination of verapamil and its primary metabolites in serum by ion-pair adsorption high-performance liquid chromatography. *J.Chromatogr.Sci.*, **1988**, *26*, 473–477 [fluorescence detection; normal phase; LOD 0.22 ng; extracted metabolites, N-acetylprocainamide, procainamide, propranolol, quinidine; non-interfering digoxin, diltiazem, hydrochlorothiazide, lidocaine, phenobarbital, phenytoin, theophylline]

Johnson, S.M.; Khalil, S.K.W. An HPLC method for the determination of verapamil and norverapamil in human plasma. *J.Liq.Chromatogr.*, **1987**, *10*, 1187–1201 [trimipramine (IS); LOD 2 ng/mL; column temp 40; fluorescence detection]

Tsilifonis, D.C.; Wilk, K.; Reisch, R., Jr.; Daly, R.E. High performance liquid chromatographic assay of verapamil hydrochloride in dosage forms. *J.Liq.Chromatogr.*, **1985**, *8*, 499–511 [stability-indicating]

Lim, C.K.; Rideout, J.M.; Sheldon, J.W.S. Determination of verapamil and norverapamil in serum by high-performance liquid chromatography. *J.Liq.Chromatogr.*, **1983**, *6*, 887–893 [fluorescence detection]

Piotrovskii, V.K.; Rumiantsev, D.O.; Metelitsa, V.I. Ion-exchange high-performance liquid chromatography in drug assay in biological fluids. II. Verapamil. *J.Chromatogr.*, **1983**, *275*, 195–200

Cole, S.C.; Flanagan, R.J.; Johnston, A.; Holt, D.W. Rapid high-performance liquid chromatographic method for the measurement of verapamil and norverapamil in blood plasma or serum. *J.Chromatogr.*, **1981**, *218*, 621–629

Kuwada, M.; Tateyama, T.; Tsutsumi, J. Simultaneous determination of verapamil and its seven metabolites by high-performance liquid chromatography. *J.Chromatogr.*, **1981**, *222*, 507–511

Thomson, B.M.; Pannell, L.K. The analysis of verapamil in postmortem specimens by HPLC and GC. *J.Anal.Toxicol.*, **1981**, *5*, 105–109

Watson, E.; Kapur, P.A. High-performance liquid chromatographic determination of verapamil in plasma by fluorescence detection. *J.Pharm.Sci.*, **1981**, *70*, 800–801

Harapat, S.R.; Kates, R.E. High-performance liquid chromatographic analysis of verapamil. II. Simultaneous quantitation of verapamil and its active metabolite, norverapamil. *J.Chromatogr.*, **1980**, *181*, 484–489

Jaouni, T.M.; Leon, M.B.; Rosing, D.R.; Fales, H.M. Analysis of verapamil in plasma by liquid chromatography. *J.Chromatogr.*, **1980**, *182*, 473–477

Todd, G.D.; Bourne, D.W.; McAllister, R.G., Jr. Measurement of verapamil concentrations in plasma by gas chromatography and high pressure liquid chromatography. *Ther.Drug Monit.*, **1980**, *2*, 411–416

# Warfarin

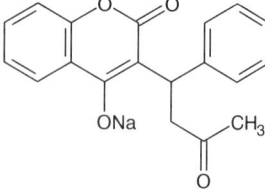

**Molecular formula:** $C_{19}H_{16}O_4$

**Molecular weight:** 308.3

**CAS Registry No.:** 81-81-2 (warfarin), 129-06-6 (warfarin sodium), 2610-86-8 (warfarin potassium)

## SAMPLE
**Matrix:** blood

**Sample preparation:** 1 mL Plasma + p-chlorowarfarin + 400 μL 1 M sulfuric acid + 12 mL ethyl acetate, shake for 30 min, centrifuge at 1500 rpm for 10 min. Remove 10 mL of the organic layer and evaporate it to dryness under a stream of nitrogen, reconstitute the residue in 200 μL MeCN:water 57:43, inject an aliquot.

## HPLC VARIABLES
**Column:** 250 × 4.6 Zorbax C-8

**Mobile phase:** MeCN:water:glacial acetic acid 57:52.6:0.4

**Flow rate:** 1

**Detector:** F ex 310 em 370 following post-column reaction. The column effluent mixed with 12% triethanolamine (for fluorescence enhancement) pumped at 0.3 mL/min and the mixture flowed through a 2 mL reaction coil to the detector.

## CHROMATOGRAM
**Retention time:** 9

**Internal standard:** p-chlorowarfarin (12)

**Limit of quantitation:** 6.3 ng/mL

## KEY WORDS
plasma; post-column reaction; pharmacokinetics

## REFERENCE
King, S.-Y.P.; Joslin, M.A.; Raudibaugh, K.; Pieniaszek, H.J., Jr.; Benedek, I.H. Dose-dependent pharmacokinetics of warfarin in healthy volunteers. *Pharm.Res.*, **1995**, *12*, 1874–1877

## SAMPLE
**Matrix:** blood

**Sample preparation:** 2 mL Whole blood or plasma + 2 mL buffer + 5 mL chloroform:isopropanol:n-heptane 60:14:26, shake gently horizontally for 10 min, centrifuge at 2800 g for 10 min. Remove the lower organic layer and evaporate it to dryness under vacuum at 45°, reconstitute the residue in 100 μL mobile phase, centrifuge at 2800 g for 5 min, inject a 50 μL aliquot of the supernatant. (Buffer was saturated ammonium chloride solution 25% diluted with water, adjusted to pH 9.5 with 25% ammonia solution.)

## HPLC VARIABLES
**Column:** 300 × 3.9 4 μm NovaPack C18

**Mobile phase:** MeOH:THF:buffer 65:5:30 (Buffer was 0.68 g/L (10 mM (sic)) $KH_2PO_4$ adjusted to pH 2.6 with concentrated orthophosphoric acid.) (At the end of each session wash the column with water for 1 h and MeOH for 1 h, re-equilibrate for 30 min.)

**Column temperature:** 30

**Flow rate:** 0.8

**Injection volume:** 50

**Detector:** UV 283

## CHROMATOGRAM
**Retention time:** 5.18

**Limit of detection:** <120 ng/mL

## OTHER SUBSTANCES

**Extracted:** acebutolol, acepromazine, aceprometazine, acetaminophen, aconitine, ajmaline, albuterol, alimemazine, alminoprofen, alpidem, alprazolam, alprenolol, amisulpride, amitriptyline, amodiaquine, amoxapine, aspirin, astemizole, atenolol, benperidol, benzocaine, benzoylecgonine, bepridil, betaxolol, bisoprolol, bromazepam, brompheniramine, bumadizone, caffeine, carbamazepine, carpipramine, carteolol, celiprolol, cetirizine, chlorambucil, chlormezanone, chlorophenacinone, chloroquine, chlorpheniramine, chlorpromazine, chlorpropamide, cibenzoline, cicletanine, clemastine, clobazam, clomipramine, clonazepam, clonidine, clozapine, cocaine, codeine, colchicine, cyamemazine, cyclizine, cycloguanil, cyproheptadine, cytarabine, dacarbazine, daunorubicin, debrisoquine, demexiptiline, desipramine, dextromethorphan, dextromoramide, dextropropoxyphene, diazepam, diazoxide, diclofenac, dihydralazine, diltiazem, diphenhydramine, disopyramide, dosulepine, doxepin, doxylamine, droperidol, ephedrine, estazolam, etodolac, fenoprofen, fentiazac, floctafenine, flumazenil, flunitrazepam, fluoxetine, fluphenazine, flurbiprofen, fluvoxamine, glibenclamide, glibornuride, glipizide, glutethimide, haloperidol, histapyrrodine, hydroxychloroquine, hydroxyzine, ibuprofen, imipramine, indomethacin, iproniazid, ketamine, ketoprofen, labetalol, levomepromazine, lidocaine, lidoflazine, lisinopril, loperamide, loprazolam, loratadine, lorazepam, loxapine, maprotiline, medazepam, medifoxamine, mefenamic acid, mefenidramine, mefloquine, melphalan, meperidine, mephenesin, mephentermine, mepivacaine, metformin, methadone, methaqualone, methocarbamol, methotrexate, metoclopramide, metoprolol, mianserine, minoxidil, moclobemide, moperone, morphine, nadolol, nalbuphine, nalorphine, naloxone, naltrexone, nialamide, nicardipine, nifedipine, niflumic acid, nimodipine, nitrazepam, nitrendipine, nizatidine, nomifensine, nortriptyline, omeprazole, opipramol, oxazepam, penbutolol, penfluridol, pentazocine, phenobarbital, phenol, phenylbutazone, pimozide, pindolol, piroxicam, prazepam, prazosin, prilocaine, procainamide, procarbazine, proguanil, promethazine, propafenone, propranolol, protriptyline, quinidine, quinine, quinupramine, ramipril, ranitidine, reserpine, ritodrine, secobarbital, sotalol, strychnine, sulfinpyrazone, sulindac, sulpride, sultopride, suriclone, temazepam, tenoxicam, terbutaline, terfenadine, tetracaine, tetrazepam, thioproperazine, thioridazine, tiapride, tiaprofenic acid, timolol, tioclomarol, tofisopam, tolbutamide, toloxatone, triazolam, trifluoperazine, trifluperidol, trimipramine, tropatenine, viloxazine, vinblastine, yohimbine, zolpidem, zopiclone, zorubicine

**Interfering:** acenocoumarol, benazepril, bupivacaine, buprenorphine, buspirone, carbinoxamine, chlordiazepoxide, clorazepate, dipyridamole, fenfluramine, flecainide, metapramine, metipranolol, mexiletine, midazolam, naproxen, oxprenolol, phencyclidine, pipamperone, pyrimethamine, thiopental, tianeptine, ticlopidine, trazodone, triprolidine, verapamil, vincristine, vindesine

## KEY WORDS
whole blood; plasma

## REFERENCE

Tracqui, A.; Kintz, P.; Mangin, P. Systematic toxicological analysis using HPLC/DAD. *J.Forensic Sci.*, **1995**, *40*, 254–262

## SAMPLE
**Matrix:** blood

**Sample preparation:** 1 mL Plasma + 100 µL 1 M HCl + 5 mL dichloromethane, stir for 10 min, centrifuge. Remove the organic layer and evaporate it to dryness under a stream of nitrogen at 40°, add 100 µL 1 M (−)-1-menthylchloroformate in dichloromethane, add 20 µL triethylamine, heat at 30° for 20 min, centrifuge, wash the supernatant with 3 mL 1 M HCl, evaporate to dryness, reconstitute the residue in 100 µL mobile phase, inject a 50 µL aliquot.

## HPLC VARIABLES
**Column:** 150 × 4.6 5 µm Nucleosil silica

**Mobile phase:** Heptane:ethyl acetate 93:7 (water concentration 51 ppm)
**Flow rate:** 1
**Injection volume:** 50
**Detector:** UV 310

## CHROMATOGRAM
**Retention time:** 10 (S), 11.5 (R)

## KEY WORDS
derivatization; plasma; chiral; normal phase

## REFERENCE
Aycard, M.; Letellier, S.; Maupas, B.; Guyon, F. Determination of (R) and (S) warfarin in plasma by high performance liquid chromatography using precolumn derivatization. *J.Liq.Chromatogr.*, **1992**, *15*, 2175–2182

## SAMPLE
**Matrix:** blood
**Sample preparation:** 1 mL Plasma + 50 µL 25 µg/mL IS + 100 µL 1 M HCl + 5 mL dichloromethane, stir for 10 min, centrifuge. Remove the organic layer and evaporate it to dryness under a stream of nitrogen at 40°, reconstitute the residue in 100 µL mobile phase, inject a 50 µL aliquot.

## HPLC VARIABLES
**Column:** 150 × 4.6 5 µm Spherisorb ODS 2
**Mobile phase:** MeCN:MeOH:50 mM pH 4 acetate buffer 11:44:45
**Flow rate:** 1.5
**Injection volume:** 50
**Detector:** UV 310

## CHROMATOGRAM
**Internal standard:** 8-chlorowarfarin
**Limit of detection:** 15 ng/mL

## KEY WORDS
plasma

## REFERENCE
Aycard, M.; Letellier, S.; Maupas, B.; Guyon, F. Determination of (R) and (S) warfarin in plasma by high performance liquid chromatography using precolumn derivatization. *J.Liq.Chromatogr.*, **1992**, *15*, 2175–2182

## SAMPLE
**Matrix:** blood
**Sample preparation:** Prepare a chromatographic column of 250 mg silica gel H in a Pasteur pipette, wash with three 1 mL portions of diethyl ether and three 1 mL portions of hexane:ether 4:1. Dry 40-500 µL 1 µg/mL (+)-p-chlorowarfarin in MeOH in a tube under a stream of nitrogen, add 1 mL citrated plasma, add 100 µL 1 M HCl, add 5 mL diethyl ether, vortex for 30 s, centrifuge at 1000 g for 5 min. Remove the ether layer and evaporate it at 40° under a stream of nitrogen, reconstitute in 1 mL hexane:ether 4:1, add to chromatographic column, wash with two 1 mL portions of hexane:ether 4:1, elute with 1 mL ether. Discard first 200 µL eluate, evaporate remainder at 40°, add 15 µL 50 mg/mL HCA in MeCN:water 25:1, add 10 µL 100 mg/mL N, N'-dicyclohexylcarbodiimide in MeCN:water 25:1, let stand for 10 min, inject a 20 µL aliquot within 4 h. (HCA was (−)-(1S,2R,4R)-endo-1,4,5,6,7,7-hexachlorobicyclo[2.2.1]hept-5-ene-2-carboxylic acid. Diethyl ether was freshly distilled before use.)

## HPLC VARIABLES
**Guard column:** 15 × 4 10 μm LiChrosorb C18
**Column:** 250 × 4 10 μm LiChrosorb C18
**Mobile phase:** MeCN:water 80:20
**Flow rate:** 1.6
**Injection volume:** 20
**Detector:** F ex 313 em 370 (cut-off filter) following post-column reaction. The column effluent mixed with 200 mM NaOH pumped at 0.5 mL/min and the mixture flowed through a 1 m reaction coil to the detector.

## CHROMATOGRAM
**Retention time:** 6 (S), 7 (R)
**Internal standard:** (+)-p-chlorowarfarin (9)
**Limit of detection:** 5 ng/mL

## KEY WORDS
plasma; SPE; derivatization; chiral; post-column reaction

## REFERENCE
Carter, S.R.; Duke, C.C.; Cutler, D.J.; Holder, G.M. Sensitive stereospecific assay of warfarin in plasma: reversed-phase high-performance liquid chromatographic separation using diastereoisomeric esters of (−)-(1S,2R,4R)-endo-1,4,5,6,7,7- hexachlorobicyclo[2.2.1]hept-5-ene-2-carboxylic acid. *J.Chromatogr.*, **1992**, *574*, 77−83

## SAMPLE
**Matrix:** blood
**Sample preparation:** Acidify 1 mL plasma to pH <2 with 100 μL 2 M HCl, extract with 5 mL diethyl ether:hexane 50:50. Remove the organic layer and evaporate it to dryness, reconstitute the residue in 20 μL 100 mg/mL carbobenzyloxy-L-proline in MeCN, 20 μL 1 mg/mL imidazole in MeCN, and 20 μL 100 mg/mL dicyclohexylcarbodiimide in MeCN, vortex for 10 s, let stand for 5-16 h, make up to 100 μL with MeCN, inject a 20 μL aliquot (J.Pharm.Sci. 1983, 72, 921; J.Clin.Pharmacol. 1995, 35, 1008).

## HPLC VARIABLES
**Column:** 100 × 4.6 3 μm Microspher C18 (Chrompack)
**Mobile phase:** MeCN:isopropanol:pH 6.6 phosphate buffer (I=0.017) 36:12:52
**Flow rate:** 1.5
**Injection volume:** 20
**Detector:** UV 313

## CHROMATOGRAM
**Retention time:** 14 (R), 17 (S)
**Internal standard:** p-chlorowarfarin (26)
**Limit of detection:** 60 ng/mL

## KEY WORDS
plasma; chiral; derivatization; pharmacokinetics

## REFERENCE
Sutfin, T.; Balmer, K.; Boström, H.; Eriksson, S.; Höglund, P.; Paulsen, O. Stereoselective interaction of omeprazole with warfarin in healthy men. *Ther.Drug Monit.*, **1989**, *11*, 176−184

## SAMPLE
**Matrix:** blood
**Sample preparation:** Centrifuge serum at 15° at 203000 g for 20 h, remove 1 mL of the supernatant and add it to 10 μL 1.5 μg/mL naproxen in 100 mM NaOH, add 250 μL 4

M HCl, add 5 mL distilled diethyl ether, rotate at 30 rpm for 15 min, centrifuge at 700 g for 5 min. Remove the organic layer and evaporate it to dryness under a stream of nitrogen at 40°, reconstitute the residue in 100 μL mobile phase, inject the whole amount.

## HPLC VARIABLES
**Column:** μBondapak C18 Radial Pak
**Mobile phase:** MeCN:100 mM $(NH_4)H_2PO_4$ 63:37
**Flow rate:** 1.5
**Injection volume:** 100
**Detector:** F ex 320 em 390 following post-column reaction. The column effluent mixed with 12% triethanolamine (for fluorescence enhancement) pumped at 0.5 mL/min and the mixture flowed through a 1 m reaction coil to the detector.

## CHROMATOGRAM
**Internal standard:** naproxen
**Limit of detection:** 0.3 ng/mL

## OTHER SUBSTANCES
**Simultaneous:** salicylic acid
**Noninterfering:** acetaminophen, carbamazepine, furosemide, hydrochlorothiazide, phenobarbital, phenytoin, spironolactone

## KEY WORDS
serum; post-column reaction; ultracentrifugate

## REFERENCE
Steyn, J.M.; van der Merwe, H.M.; de Kock, M.J. Reversed-phase high-performance liquid chromatographic method for the determination of warfarin from biological fluids in the low nanogram range. *J.Chromatogr.*, **1986**, *378*, 254–260

## SAMPLE
**Matrix:** blood
**Sample preparation:** 200 μL Plasma + 100 μL 8.46 μg/mL IS in water + 1 mL 100 mM potassium carbonate + 4 mL ether, shake for 3 min, centrifuge at 3000 rpm for 5 min, discard the organic layer. Acidify the aqueous layer with 1.5 mL 1 M HCl, add 6 mL ether, shake for 3 min, centrifuge at 3000 rpm for 3 min, freeze by immersion in liquid nitrogen for 40-60 s. Remove the organic layer and evaporate it to dryness under a stream of nitrogen at 45°, reconstitute the residue in 10 μL 100 mg/mL carbobenzyloxy-L-proline in MeCN, 10 μL 1 mg/mL imidazole in MeCN, and 10 μL 100 mg/mL dicyclohexylcarbodiimide in MeCN, vortex for 10 s, let stand for 2 h, centrifuge at 3000 rpm for 5 min, inject a 3-10 μL aliquot.

## HPLC VARIABLES
**Column:** 250 × 5 5 μm Spherisorb Si
**Mobile phase:** Hexane:ethyl acetate:MeOH:acetic acid 74.75:25:0.25:0.4
**Flow rate:** 1
**Injection volume:** 3-10
**Detector:** UV 313

## CHROMATOGRAM
**Retention time:** 14 (SS), 17 (RS)
**Internal standard:** 4'-fluorowarfarin (15, 19 (enantiomers))
**Limit of detection:** 160 ng (S); 96 ng (R)

## KEY WORDS
plasma; chiral; derivatization; normal phase

## REFERENCE

Banfield, C.; Rowland, M. Stereospecific high-performance liquid chromatographic analysis of warfarin in plasma. *J.Pharm.Sci.*, **1983**, *72*, 921–924

## SAMPLE

**Matrix:** cell suspensions

**Sample preparation:** Filter microbial cell suspensions. 4 mL Filtrate + 200 μL water + 100 μL 4 μg/mL 4'-hydroxywarfarin alcohols (potassium salt) in water + 4 mL cyclohexane:dichloromethane 90:10, extract twice, discard organic layers. Extract aqueous layer with 5 mL ethyl acetate. Extract the organic phase with 2 mL 100 mM KOH. Acidify the aqueous phase with 1 mL 5 M HCl, extract with 5 mL diethyl ether. Evaporate the ether layer to dryness under a stream of nitrogen, reconstitute the residue in 250 μL mobile phase, inject a 50-200 μL aliquot.

## HPLC VARIABLES

**Column:** 250 × 4.5 5 μm Beckman ODS C18

**Mobile phase:** MeOH:THF:1 M aqueous tetrabutylammonium phosphate:5 mM ammonium phosphate adjusted to pH 7.5 with ammonium hydroxide 30:7:1:62

**Flow rate:** 1

**Injection volume:** 50-200

**Detector:** F ex 290 emission filter No. 389

## CHROMATOGRAM

**Retention time:** 28

**Internal standard:** 4'-hydroxywarfarin alcohols (4, 6)

**Limit of detection:** 1 ng/mL

## OTHER SUBSTANCES

**Extracted:** metabolites

## REFERENCE

Wong, Y.W.; Davis, P.J. Analysis of warfarin and its metabolites by reversed-phase ion-pair liquid chromatography with fluorescence detection. *J.Chromatogr.*, **1989**, *469*, 281–291

## SAMPLE

**Matrix:** formulations

**Sample preparation:** Dilute with saline, inject a 50 μL aliquot.

## HPLC VARIABLES

**Column:** 100 × 8 Novapak C18 Radial-Pak

**Mobile phase:** MeCN:1.5% acetic acid 31:69 adjusted to pH 4.2 with 1 M NaOH

**Flow rate:** 4.2

**Injection volume:** 50

**Detector:** UV 254

## CHROMATOGRAM

**Retention time:** 14

## KEY WORDS

injections; saline

## REFERENCE

Martens, H.J.; de Goede, P.N.; van Loenen, A.C. Sorption of various drugs in polyvinyl chloride, glass, and polyethylene-lined infusion containers. *Am.J.Hosp.Pharm.*, **1990**, *47*, 369–373

## SAMPLE
**Matrix:** formulations
**Sample preparation:** Powder tablets, dissolve in mobile phase, sonicate, dilute with mobile phase to give a 200 μg/mL solution, inject an aliquot.

## HPLC VARIABLES
**Column:** 300 × 3.9 10 μm μBondapak C18
**Mobile phase:** THF:MeOH:water:acetic acid 35:10:65:0.1
**Flow rate:** 1.5
**Injection volume:** 20
**Detector:** UV 311

## OTHER SUBSTANCES
**Also analyzed:** dicumarol, phenprocoumon

## KEY WORDS
tablets

## REFERENCE
Moore, E.S. Liquid chromatographic determination of coumarin anticoagulants in tablets: collaborative study. *J.Assoc.Off.Anal.Chem.*, **1987**, *70*, 834–836

## SAMPLE
**Matrix:** microsomal incubations
**Sample preparation:** 500 μL Microsomal incubation + 10 μL 70% perchloric acid + 10 ng 7-ethoxycoumarin, centrifuge at 4° at 3000 g for 5 min, inject an aliquot of the supernatant.

## HPLC VARIABLES
**Guard column:** 11 × 4 10 μm Nucleosil C8
**Column:** 125 × 4 5 μm Nucleosil C18
**Mobile phase:** MeCN:0.5% phosphoric acid 36:62
**Flow rate:** 1.3
**Injection volume:** 100
**Detector:** UV 205

## CHROMATOGRAM
**Retention time:** 11.7
**Internal standard:** 7-ethoxycoumarin (7)

## OTHER SUBSTANCES
**Extracted:** metabolites

## KEY WORDS
human; liver

## REFERENCE
Lang, D.; Böcker, R. Highly sensitive and specific high-performance liquid chromatographic analysis of 7-hydroxywarfarin, a marker for human cytochrome P-4502C9 activity. *J.Chromatogr.B*, **1995**, *672*, 305–309

## SAMPLE
**Matrix:** solutions
**Sample preparation:** Filter (0.45 μm), dilute the filtrate with mobile phase, inject an aliquot.

## HPLC VARIABLES
**Column:** 150 × 4.6 5 μm Hypersil ODS
**Mobile phase:** MeCN:10 mM pH 4.7 acetate buffer 50:50
**Detector:** UV 214

## REFERENCE
Okimoto, K.; Rajewski, R.A.; Uekama, K.; Jona, J.A.; Stella, V.J. The interaction of charged and un-
charged drugs with neutral (HP-β-CD) and anionically charged (SBE7-β-CD) β-cyclodextrins.
*Pharm.Res.*, **1996**, *13*, 256–264

## SAMPLE
**Matrix:** solutions

## HPLC VARIABLES
**Column:** 250 × 4.6 Chirex 3022 (Phenomenex)
**Mobile phase:** Hexane:1,2-dichloroethane:EtOH/trifluoroacetic acid 88:10:2 (EtOH/
trifluoroacetic acid was premixed 20:1.)
**Flow rate:** 0.7-1
**Injection volume:** 20
**Detector:** UV 282

## KEY WORDS
chiral; α = 1.03

## REFERENCE
Cleveland, T. Pirkle-concept chiral stationary phases for the HPLC separation of pharmaceutical racemates. *J.Liq.Chromatogr.*, **1995**, *18*, 649–671

## SAMPLE
**Matrix:** solutions

## HPLC VARIABLES
**Column:** 250 × 4.6 5 μm Supelcosil LC-DP (A) or 250 × 4 5 μm LiChrospher 100 RP-8 (B)
**Mobile phase:** MeCN:0.025% phosphoric acid:buffer 25:10:5 (A) or 60:25:15 (B) (Buffer
was 9 mL concentrated phosphoric acid and 10 mL triethylamine in 900 mL water, adjust
pH to 3.4 with dilute phosphoric acid, make up to 1 L.)
**Flow rate:** 0.6
**Injection volume:** 25
**Detector:** UV 229

## CHROMATOGRAM
**Retention time:** 8.01 (A), 8.29 (B)

## OTHER SUBSTANCES
**Also analyzed:** acebutolol, acepromazine, acetaminophen, acetazolamide, acetophenazine,
albuterol, alprazolam, amitriptyline, amobarbital, amoxapine, antipyrine, atenolol, atropine, azatadine, baclofen, benzocaine, bromocriptine, brompheniramine, brotizolam,
bupivacaine, buspirone, butabarbital, butalbital, caffeine, carbamazepine, cetirizine,
chlorcyclizine, chlordiazepoxide, chlormezanone, chloroquine, chlorpheniramine,
chlorpromazine, chlorpropamide, chlorprothixene, chlorthalidone, chlorzoxazone, cimetidine, cisapride, clomipramine, clonazepam, clonidine, clozapine, cocaine, codeine, colchicine, cyclizine, cyclobenzaprine, dantrolene, desipramine, diazepam, diclofenac, diflunisal,
diltiazem, diphenhydramine, diphenidol, diphenoxylate, dipyridamole, disopyramide, dobutamine, doxapram, doxepin, droperidol, encainide, ethidium bromide, ethopropazine,
fenoprofen, fentanyl, flavoxate, fluoxetine, fluphenazine, flurazepam, flurbiprofen, fluvox-

amine, furosemide, glutethimide, glyburide, guaifenesin, haloperidol, homatropine, hydralazine, hydrochlorothiazide, hydrocodone, hydromorphone, hydroxychloroquine, hydroxyzine, ibuprofen, imipramine, indomethacin, ketoconazole, ketoprofen, ketorolac, labetalol, levorphanol, lidocaine, loratadine, lorazepam, lovastatin, loxapine, mazindol, mefenamic acid, meperidine, mephenytoin, mepivacaine, mesoridazine, metaproterenol, metformin, methadone, methdilazine, methocarbamol, methotrexate, methotrimeprazine, methoxamine, methyldopa, methylphenidate, metoclopramide, metolazone, metoprolol, metronidazole, midazolam, moclobemide, morphine, nadolol, nalbuphine, naloxone, naphazoline, naproxen, nifedipine, nizatidine, norepinephrine, nortriptyline, oxazepam, oxycodone, oxymetazoline, paroxetine, pemoline, pentazocine, pentobarbital, pentoxifylline, perphenazine, pheniramine, phenobarbital, phenol, phenolphthalein, phentolamine, phenylbutazone, phenyltoloxamine, phenytoin, pimozide, pindolol, piroxicam, pramoxine, prazepam, prazosin, probenecid, procainamide, procaine, prochlorperazine, procyclidine, promazine, promethazine, propafenone, propantheline, propiomazine, propofol, propranolol, protriptyline, quazepam, quinidine, quinine, racemethorphan, ranitidine, remoxipride, risperidone, salicylic acid, scopolamine, secobarbital, sertraline, sotalol, spironolactone, sulfinpyrazone, sulindac, temazepam, terbutaline, terfenadine, tetracaine, theophylline, thiethylperazine, thiopental, thioridazine, thiothixene, timolol, tocainide, tolbutamide, tolmetin, trazodone, triamterene, triazolam, trifluoperazine, triflupromazine, trimeprazine, trimethoprim, trimipramine, verapamil, xylometazoline, yohimbine, zopiclone

## KEY WORDS
some details of plasma extraction

## REFERENCE
Koves, E.M. Use of high-performance liquid chromatography-diode array detection in forensic toxicology. *J.Chromatogr.A,* **1995,** *692,* 103–119

## SAMPLE
**Matrix:** solutions

## HPLC VARIABLES
**Column:** 100 × 4 5 μm Chiral-AGP (ChromTech)
**Mobile phase:** Isopropanol:20 mM pH 6.0 potassium phosphate buffer 15:85
**Flow rate:** 1
**Detector:** UV 205

## KEY WORDS
chiral

## REFERENCE
Lang, D.; Böcker, R. Highly sensitive and specific high-performance liquid chromatographic analysis of 7-hydroxywarfarin, a marker for human cytochrome P-4502C9 activity. *J.Chromatogr.B,* **1995,** *672,* 305–309

## SAMPLE
**Matrix:** solutions
**Sample preparation:** Prepare a 100 μM solution in buffer, inject a 20 μL aliquot. (Buffer was 100 mM pH 5.3 sodium acetate.)

## HPLC VARIABLES
**Column:** 100 × 4.6 column containing riboflavin binding proteins (Prepare as follows. Add riboflavin to saturate protein of egg yolk, homogenize with 3 volumes buffer, centrifuge, add the supernatant to a 500 × 30 column of DEAE-cellulose (Whatman) equilibrated with buffer, wash extensively with buffer to remove bound protein, elute riboflavin bind-

ing proteins (RFBP) with buffer containing 200 mM NaCl (RFBP has intense yellow color, absorption at 455 nm). Purify RFBP on a Sephadex G-100 column with 50 mM pH 7.5 Tris-HCl buffer as eluent, remove the bound riboflavin by extensive dialysis at pH 3.0. Add 4.5 g N, N-disuccinylimidyl carbonate to 3 g Nucleosil 5NH2 slurried in MeCN, filter, wash with MeCN, wash with 50 mM pH 7.5 phosphate buffer. Suspend 300 mg RFBP in 50 mM phosphate buffer, add the activated silica, mix gently for 2 h using a rotary evaporator, filter, wash with sterile water, wash with isopropanol:water 1:2, pack in a 100 × 4.6 column. (Buffer was 100 mM pH 5.3 sodium acetate.))

**Mobile phase:** EtOH:50 mM pH 4.6 $KH_2PO_4$ 10:90
**Flow rate:** 0.8
**Injection volume:** 20
**Detector:** UV

## CHROMATOGRAM
**Retention time:** k′ 4.67

## KEY WORDS
chiral; α = 1.39

## REFERENCE
Massolini, G.; De Lorenzi, E.; Ponci, M.C.; Gandini, C.; Caccialanza, G.; Monaco, H.L. Egg yolk riboflavin binding protein as a new chiral stationary phase in high-performance liquid chromatography. *J.Chromatogr.A*, **1995**, *704*, 55–65

## SAMPLE
**Matrix:** solutions

## HPLC VARIABLES
**Column:** 150 × 4.6 cellulose 3,5-dimethylphenylcarbamate/10-undecenoate bonded to allylsilica
**Mobile phase:** Heptane:isopropanol 90:10
**Flow rate:** 1
**Injection volume:** 1000
**Detector:** UV 254

## CHROMATOGRAM
**Retention time:** k′ 4.85

## KEY WORDS
chiral; α = 1.81

## REFERENCE
Oliveros, L.; Lopez, P.; Minguillon, C.; Franco, P. Chiral chromatographic discrimination ability of a cellulose 3,5-dimethylphenylcarbamate/10-undecenoate mixed derivative fixed on several chromatographic matrices. *J.Liq.Chromatogr.*, **1995**, *18*, 1521–1532

## SAMPLE
**Matrix:** solutions

## HPLC VARIABLES
**Column:** 250 × 4.6 Zorbax RX
**Mobile phase:** Gradient. A was 10 mL concentrated orthophosphoric acid and 7 mL triethylamine in 1 L water. B was 10 mL concentrated orthophosphoric acid and 7 mL triethylamine in 200 mL water, make up to 1 L with MeCN. A:B from 100:0 to 0:100 over 30 min, maintain at 0:100 for 5 min.
**Column temperature:** 30

**Flow rate:** 2
**Detector:** UV 210

## OTHER SUBSTANCES

**Also analyzed:** acepromazine, acetaminophen, acetophenazine, albuterol, aminophylline, amitriptyline, amobarbital, amoxapine, amphetamine, amylocaine, antipyrine, aprobarbital, aspirin, atenolol, atropine, avermectin, barbital, benzocaine, benzoic acid, benzotropine, benzphetamine, berberine, bibucaine, bromazepan, brompheniramine, buprenorphine, buspirone, butabarbital, butacaine, butethal, caffeine, carbamazepine, carbromal, chloramphenicol, chlordiazepoxide, chloroquine, chlorothiazide, chloroxylenol, chlorphenesin, chlorpheniramine, chlorpromazine, chlorpropamide, chlortetracycline, cimetidine, cinchonidine, cinchonine, clenbuterol, clonazepam, clonixin, clorazepate, cocaine, codeine, colchicine, cortisone, coumarin, cyclazocine, cyclobenzaprine, cyclothiazide, cyheptamide, cymarin, danazol, danthron, dapsone, debrisoquine, desipramine, dexamethasone, dextromethorphan, dextropropoxyphene, diamorphine, diazepam, diclofenac, diethylpropion, diethylstilbestrol, diflunisal, digitoxin, digoxin, diltiazem, diphenhydramine, diphenoxylate, diprenorphine, dipyrone, disulfiram, dopamine, doxapram, doxepin, dronabinol, ephedrine, epinephrine, epinine, estradiol, estriol, estrone, ethacrynic acid, ethosuximide, etonitazene, etorphine, eugenol, famotidine, fenbendazole, fencamfamine, fenoprofen, fenproporex, fentanyl, flubendazole, flufenamic acid, flunitrazepam, 5-fluorouracil, fluoxymesterone, fluphenazine, furosemide, gentisic acid, gitoxigenin, glipizide, glunixin, glutethimide, glybenclamide, guaiacol, halazepam, haloperidol, hydrochlorothiazide, hydrocodone, hydrocortisone, hydromorphone, hydroxyquinoline, ibogaine, ibuprofen, iminostilbene, imipramine, indomethacin, isocarbostyril, isocarboxazid, isoniazid, isoproterenol, isoxsuprine, ivermectin, ketamine, ketoprofen, kynurenic acid, levorphanol, lidocaine, lorazepam, lormetazepam, loxapine, mazindol, mebendazole, meclizine, meclofenamic acid, medazepam, mefenamic acid, megestrol, mepacrine, meperidine, mephentermine, mephenytoin, mephesin, mephobarbital, mepivacaine, mescaline, mesoridazine, methadone, methamphetamine, methapyrilene, methaqualone, methazolamide, methocarbamol, methoxamine, methsuximide, methyl salicylate, methyldopa, methyldopamine, methylphenidate, methylprednisolone, methyltestosterone, methyprylon, metoprolol, mibolerone, morphine, nadolol, nalorphine, naloxone, naltrexone, naphazoline, naproxen, nefopam, niacinamide, nicotine, nicotinic acid, nifedipine, niflumic acid, nitrazepam, norepinephrine, nortriptyline, noscapine, nylidrin, oxazepam, oxycodone, oxymorphone, oxyphenbutazone, oxytetracycline, papaverine, pargyline, pemoline, pentazocine, pentobarbital, persantine, phenacetin, phenazocine, phenazopyridine, phencyclidine, phendimetrazine, phenelzine, pheniramine, phenobarbital, phenothiazine, phensuximide, phentermine, phenylbutazone, phenylephrine, phenylpropanolamine, piperocaine, prazepam, prednisolone, primidone, probenecid, progesterone, propiomazine, propranolol, propylparaben, pseudoephedrine, puromycin, pyrilamine, pyrithyldione, quazepam, quinaldic acid, quinidine, quinine, ranitidine, recinnamine, reserpine, resorcinol, saccharin, albuterol, salicylamide, salicylic acid, scopolamine, scopoletin, secobarbital, strychnine, sulfacetamide, sufadiazine, sulfadimethoxine, sulfaethidole, sulfamerazine, sulfamethazine, sulfamethoxizole, sulfanilamide, sulfapyridine, sulfasoxizole, sulindac, tamoxifen, temazepam, testosterone, tetracaine, tetracycline, tetramisole, thebaine, theobromine, theophylline, thiabendazole, thiamine, thiamylal, thiobarbituric acid, thioridazine, thiosalicylic acid, thiothixene, thymol, tolazamide, tolazoline, tobutamide, tolmetin, tranylcypromine, triamcinolone, tribenzylamine, trichloromethiazide, trifluoperazine, trihexyphenidyl, trimethoprim, tripelennamine, triprolidine, tropacocaine, tyramine, verapamil, vincamine, yohimbine, zoxazolamine

## REFERENCE

Hill, D.W.; Kind, A.J. Reversed-phase solvent-gradient HPLC retention indexes of drugs. *J.Anal.Toxicol.*, **1994**, *18*, 233–242

## SAMPLE

**Matrix:** solutions
**Sample preparation:** Dissolve in MeOH, inject a 1-2 μL aliquot.

## HPLC VARIABLES
**Guard column:** $10 \times 3$ 5 μm Chiral-AGP (ChromTech)
**Column:** $100 \times 4$ 5 μm Chiral-AGP (ChromTech)
**Mobile phase:** Gradient. A was 10 mM pH 7.0 phosphate buffer containing 1 mM N, N-dimethyloctylamine. B was 1 mM N, N-dimethyloctylamine in isopropanol. A:B from 100:0 to 80:20 over 10 min, stay at 80:20 for 15 min, equilibrate at 100:0 for 10 min.
**Flow rate:** 0.9
**Injection volume:** 1-2
**Detector:** F ex 292 em 380

## CHROMATOGRAM
**Retention time:** 9.4 (S), 10.4 (R)

## OTHER SUBSTANCES
**Simultaneous:** acenocoumarol, phenprocoumon, warfarin alcohol

## KEY WORDS
chiral

## REFERENCE
de Vries, J.X.; Schmitz-Kummer, E. Direct column liquid chromatographic enantiomer separation of the coumarin anticoagulants phenprocoumon, warfarin, acenocoumarol and metabolites on an α1-acid glycoprotein chiral stationary phase. *J.Chromatogr.*, **1993**, *644*, 315–320

## SAMPLE
**Matrix:** solutions
**Sample preparation:** Condition a 3 mL 200 mg Bond-Elut C18 SPE cartridge with 5 mL MeCN then 5 mL water (pH adjusted to 4.3 with acetic acid), do not allow to dry. Acidify water to pH 4.3 with acetic acid, run 1 L through the SPE cartridge with a vacuum, wash with 20 mL MeCN:water adjusted to pH 4.3 with acetic acid 20:80, elute with 1 mL MeCN:40 mM pH 7.4 phosphate buffer 1:1, inject a 30 μL aliquot.

## HPLC VARIABLES
**Column:** $150 \times 3.9$ 4 μm Nova-Pak C18
**Mobile phase:** MeOH:water:glacial acetic acid 62:38:1
**Flow rate:** 1
**Injection volume:** 30
**Detector:** UV 282; UV 306

## CHROMATOGRAM
**Retention time:** 5
**Limit of detection:** 0.02 ppb

## KEY WORDS
SPE; water

## REFERENCE
Dalbacke, J.; Dahlquist, I.; Persson, C. Determination of warfarin in drinking water by high-performance liquid chromatography after solid-phase extraction. *J.Chromatogr.*, **1990**, *507*, 381–387

## SAMPLE
**Matrix:** solutions

## HPLC VARIABLES
**Column:** $250 \times 4.6$ cellulose tris(3,5-dimethylphenylcarbamate)

**Mobile phase:** Hexane:isopropanol:formic acid 80:20:1
**Flow rate:** 0.5
**Detector:** UV

## CHROMATOGRAM
**Retention time:** 17 (−), 36 (+)

## KEY WORDS
chiral

## REFERENCE
Okamoto, Y.; Aburatani, R.; Hatano, K.; Hatada, K. Optical resolution of racemic drugs by chiral HPLC on cellulose and amylose tris(phenylcarbamate) derivatives. *J.Liq.Chromatogr.*, **1988**, *11*, 2147−2163

◆─────────────────────────◆─────────────────────────◆

## ANNOTATED BIBLIOGRAPHY

Haginaka, J.; Kanasugi, N. Enantioselectivity of bovine serum albumin-bonded columns produced with isolated protein fragments. *J.Chromatogr.A*, **1995**, *694*, 71−80 [chiral; also benzoin, clorazepate, fenoprofen, flurbiprofen, ibuprofen, ketoprofen, lorazepam, lormetazepam, oxazepam, pranoprofen, temazepam]

Terfloth, G.J.; Pirkle, W.H.; Lynam, K.G.; Nicolas, E.C. Broadly applicable polysiloxane-based chiral stationary phase for high-performance liquid chromatography and supercritical fluid chromatography. *J.Chromatogr.A*, **1995**, *705*, 185−194 [SFC; HPLC; also carprofen, cicloprofen, etodolac, fenoprofen, flurbiprofen, ibuprofen, naproxen, pirprofen]

Cai, W.M.; Hatton, J.; Pettigrew, L.C.; Dempsey, R.J.; Chandler, M.H.H. A simplified high-performance liquid chromatographic method for direct determination of warfarin enantiomers and their protein binding in stroke patients. *Ther.Drug Monit.*, **1994**, *16*, 509−512 [chiral; fluorescence detection; phenprocoumon (IS); LOD 8 ng/mL; post-column reaction detection; derivatization]

Ermer, J.C.; Hicks, D.R.; Wheeler, S.C.; Kraml, M.; Jusko, W.J. Concomitant etodolac affects neither the unbound clearance nor the pharmacologic effect of warfarin. *Clin.Pharmacol.Ther.*, **1994**, *55*, 305−316 [serum; nitrazepam (IS); SPE; LOD 15 ng/mL; pharmacokinetics]

Loun, B.; Hage, D.S. Chiral separation mechanisms in protein-based HPLC columns. 1. Thermodynamic studies of (R)- and (S)-warfarin binding to immobilized human serum albumin. *Anal.Chem.*, **1994**, *66*, 3814−3822 [column temp 37]

Wang, J.-P.; Unadkat, J.D.; McNamara, S.; O'Sullivan, T.A.; Smith, A.L.; Trager, W.F.; Ramsey, B. Disposition of drugs in cystic fibrosis VI. In vivo activity of cytochrome P450 isoforms involved in the metabolism of (R)-warfarin (including P450 3A4) is not enhanced in cystic fibrosis. *Clin.Pharmacol.Ther.*, **1994**, *55*, 528−534 [plasma; fluorowarfarin (IS); LOQ 50 ng/mL; pharmacokinetics]

Naidong, W.; Lee, J.W. Development and validation of a high-performance liquid chromatographic method for the quantitation of warfarin enantiomers in human plasma. *J.Pharm.Biomed.Anal.*, **1993**, *11*, 785−792

Shibukawa, A.; Nagao, M.; Kuroda, Y.; Nakagawa, T. Stereoselective determination of free warfarin concentration in protein binding equilibrium using direct sample injection and an on-line liquid chromatographic system. *Anal.Chem.*, **1990**, *62*, 712−716 [chiral; column-switching; column temp 37]

de Vries, J.X.; Volker, U. Separation of the enantiomers of phenprocoumon and warfarin by high-performance liquid chromatography using a chiral stationary phase. Determination of the enantiomeric ratio of phenprocoumon in human plasma and urine. *J.Chromatogr.*, **1989**, *493*, 149−156

Chan, K.; Woo, K.S. Determination of warfarin in human plasma by high performance liquid chromatography. *Methods Find.Exp.Clin.Pharmacol.*, **1988**, *10*, 699−703

Chu, Y.Q.; Wainer, I.W. The measurement of warfarin enantiomers in serum using coupled achiral/chiral, high-performance liquid chromatography (HPLC). *Pharm.Res.*, **1988**, *5*, 680−683

Wang, J.; Bonakdar, M. Sensitive measurements of warfarin by high-performance liquid chromatography with electrochemical detection. *J.Chromatogr.*, **1987**, *415*, 432−437

Ueland, P.M.; Kvalheim, G.; Lnning, P.E.; Kvinnsland, S. Determination of warfarin in human plasma by high performance liquid chromatography and photodiode array detector. *Ther.Drug Monit.*, **1985**, *7*, 329–335

Banfield, C.; Rowland, M. Stereospecific fluorescence high-performance liquid chromatographic analysis of warfarin and its metabolites in plasma and urine. *J.Pharm.Sci.*, **1984**, *73*, 1392–1396

Jeyaraj, G.L.; Porter, W.R. New method for the resolution of racemic warfarin and its analogues using low-pressure liquid chromatography. *J.Chromatogr.*, **1984**, *315*, 378–383

Perez, R.L. Simultaneous determination of warfarin, sulphaquinoxaline and fenitrothion in wheat-based rodenticide baits by high pressure liquid chromatography. *J.Liq.Chromatogr.*, **1983**, *6*, 353–365

Tasker, R.A.; Nakatsu, K. Rapid, reliable and sensitive assay for warfarin using normal-phase high-performance liquid chromatography. *J.Chromatogr.*, **1982**, *228*, 346–349

Lee, S.H.; Field, L.R.; Howald, W.N.; Trager, W.F. High-performance liquid chromatographic separation and fluorescence detection of warfarin and its metabolites by postcolumn acid/base manipulation. *Anal.Chem.*, **1981**, *53*, 467–471

Robinson, C.A.; Mungall, D.; Poon, M.C. Quantitation of plasma warfarin concentrations by high performance liquid chromatography. *Ther.Drug Monit.*, **1981**, *3*, 287–290

Trujillo, W.A. Determination of warfarin and sulfaquinoxaline in rodenticide concentrates by HPLC. *J.Liq.Chromatogr.*, **1980**, *3*, 1219–1226

# Zidovudine

**Molecular formula:** C₁₀H₁₃N₅O₄
**Molecular weight:** 267.2
**CAS Registry No.:** 30516-87-1

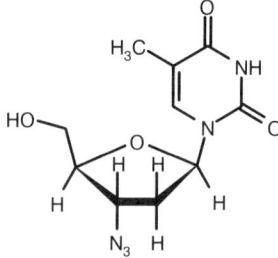

## SAMPLE
**Matrix:** bile, blood, tissue
**Sample preparation:** Condition a Baker C18 SPE cartridge with three 1 mL portions of MeOH, three 1 mL portions of water, and three 1 mL portions of buffer. Add a 200 μL aliquot of plasma, bile, or liver homogenate containing the IS to 200 μL buffer and add this mixture to the SPE cartridge, wash with two 200 μL portions of buffer, allow cartridge to dry, elute with two 100 μL portions of MeOH, combine the eluates, add an equal volume of water, inject a 40 μL aliquot. (Buffer was 200 mM pH 7.5 sodium phosphate containing 8 mM tetrabutylammonium sulfate.)

## HPLC VARIABLES
**Guard column:** μBondapak C18 Guard-pak
**Column:** 150 × 3.9 4 μm Novapak C18
**Mobile phase:** MeCN:200 mM pH 7.5 sodium phosphate + 8 mM tetrabutylammonium sulfate 5:95 (Wash column overnight with MeOH.)
**Flow rate:** 1.5
**Injection volume:** 20-50
**Detector:** UV 254

## CHROMATOGRAM
**Retention time:** 6.5
**Internal standard:** azidoddI

## OTHER SUBSTANCES
**Extracted:** metabolites
**Noninterfering:** ddC, didanosine

## KEY WORDS
plasma; liver; human; cat; rat; SPE

## REFERENCE
Molema, G.; Jansen, R.W.; Visser, J.; Meijer, D.K.F. Simultaneous analysis of azidothymidine and its mono-, di- and triphosphate derivatives in biological fluids, tissue and cultured cells by a rapid high-performance liquid chromatographic method. *J.Chromatogr.*, **1992**, *579*, 107–114

## SAMPLE
**Matrix:** blood
**Sample preparation:** Condition a 3 mL 200 mg Bond Elut C18 SPE cartridge with 3 mL MeOH and 2 mL buffer, do not allow to go dry. 1 mL Plasma + 25 μL 40 μg/mL 7-ethyltheophylline in water, vortex, add to the SPE cartridge at not more than 0.5 mL/min, wash with 1 mL buffer, air dry for 3 min, elute with 2 mL MeOH. Evaporate the eluate to dryness under a stream of nitrogen at 50°, reconstitute the residue in 100 μL MeCN: 5% acetic acid 10:90, vortex vigorously, centrifuge at 10000 g for 5 min, inject a 20 μL aliquot of the supernatant. (Prepare buffer by diluting 1.44 mL concentrated phosphoric acid to 1 L with water and adjusting the pH to 6.55 with concentrated ammonia.)

## HPLC VARIABLES
**Column:** 75 × 4.6 3 μm Ultrasphere ODS

**Mobile phase:** Gradient. A was 1.44 mL concentrated phosphoric acid and 4 mL n-octylamine in 1 L water, pH adjusted to 6.55 with concentrated ammonia. B was MeCN. A:B from 95:5 to 70:30 over 7 min, to 20:80 over 1.5 min, return to initial conditions over 1 min, re-equilibrate for 2.5 min.
**Flow rate:** 1
**Injection volume:** 20
**Detector:** UV 266

## CHROMATOGRAM
**Retention time:** 3.6
**Internal standard:** 7-ethyltheophylline (4.2)
**Limit of detection:** 7 ng/mL
**Limit of quantitation:** 22 ng/mL

## OTHER SUBSTANCES
**Extracted:** metabolites, caffeine

## KEY WORDS
plasma; SPE

## REFERENCE
Nadal, T.; Ortuño, J.; Pascual, J.A. Rapid and sensitive determination of zidovudine and zidovudine glucuronide in human plasma by ion-pair high-performance liquid chromatography. *J.Chromatogr.A*, **1996**, *721*, 127–137

## SAMPLE
**Matrix:** blood
**Sample preparation:** Filter (Millipore Ultrafree-MC, 10000 molecular mass limit) 250 µL serum while centrifuging at 17000 g for 1.5 h, inject a 50 µL aliquot of the clear ultrafiltrate.

## HPLC VARIABLES
**Column:** 150 × 3.9 4 µm Nova-Pak phenyl
**Mobile phase:** Isopropanol:20 mM pH 5 sodium citrate 2.5:97.5
**Flow rate:** 1
**Injection volume:** 50
**Detector:** UV 250

## CHROMATOGRAM
**Retention time:** 22.4

## OTHER SUBSTANCES
**Extracted:** didanosine, zalcitabine

## KEY WORDS
serum; ultrafiltrate

## REFERENCE
Rosell-Rovira, M.L.; Pou-Clavé, L.; Lopez-Galera, R.; Pascual-Mostaza, C. Determination of free serum didanosine by ultrafiltration and high-performance liquid chromatography. *J.Chromatogr.B*, **1996**, *675*, 89–92

## SAMPLE
**Matrix:** blood

**Sample preparation:** 100 µL Serum + 50 µL 5 µg/mL IS + 50 µL 2 M perchloric acid, mix, centrifuge at 9000 rpm for 10 min, inject an aliquot of the supernatant.

## HPLC VARIABLES
**Column:** 150 × 4.6 5 µm Hypersil ODS
**Mobile phase:** MeCN:40 mM pH 7.0 sodium acetate 12:88
**Flow rate:** 2
**Detector:** UV 263

## CHROMATOGRAM
**Internal standard:** 3′-azido-2′,3′-dideoxy-5-ethyluridine (CS-85)
**Limit of quantitation:** 50 ng/mL

## KEY WORDS
mouse; serum; pharmacokinetics

## REFERENCE
Manouilov, K.K.; White, C.A.; Boudinot, F.D.; Fedorov, I.I.; Chu, C.K. Lymphatic distribution of 3′-azido-3′-deoxythymidine and 3′-azido-2′,3′-dideoxyuridine in mice. *Drug Metab.Dispos.*, **1995**, *23*, 655–658

## SAMPLE
**Matrix:** blood
**Sample preparation:** 200 µL Plasma + 400 µL ethyl acetate, vortex for 30 s. Remove a 200 µL aliquot of the supernatant and evaporate it to dryness under a stream of nitrogen at 50°, reconstitute the residue in 200 µL mobile phase, vortex for 10 s, inject a 10 µL aliquot.

## HPLC VARIABLES
**Guard column:** 4 × 4 5 µm Lichrospher 60 RP-select B
**Column:** 125 × 4 5 µm Lichrospher 60 RP-select B
**Mobile phase:** MeOH:pH 7.0 phosphate buffer 20:80
**Flow rate:** 1
**Injection volume:** 10
**Detector:** UV 265

## KEY WORDS
plasma; rabbit; pharmacokinetics

## REFERENCE
Mirchandani, H.L.; Chien, Y.W. Intestinal absorption of dideoxynucleosides: Characterization using a multiloop in situ technique. *J.Pharm.Sci.*, **1995**, *84*, 44–48

## SAMPLE
**Matrix:** blood
**Sample preparation:** Heat blood at 56-58° for 30 min, add 500 µL serum to a lipophilic type W SPE cartridge (DuPont), wash with phosphate-buffered saline, elute with MeOH. Evaporate the eluate and reconstitute the residue with MeCN:water 15:85, inject an aliquot.

## HPLC VARIABLES
**Column:** 5 µm Resolve C18 (Waters)
**Mobile phase:** MeCN:25 mM pH 2.20 phosphate buffer 15:85
**Flow rate:** 1
**Detector:** UV 266; UV 254

## CHROMATOGRAM
**Limit of detection:** 40 nM

## OTHER SUBSTANCES
**Extracted:** metabolites, glucuronylzidovudine

## KEY WORDS
serum; pharmacokinetics; SPE

## REFERENCE
Moore, K.H.P.; Raasch, R.H.; Brouwer, K.L.R.; Opheim, K.; Cheeseman, S.H.; Eyster, E.; Lemon, S.M.; van der Horst, C.M. Pharmacokinetics and bioavailability of zidovudine and its glucuronidated metabolite in patients with human immunodeficiency virus infection and hepatic disease (AIDS Clinical Trials Group Protocol 062). *Antimicrob.Agents Chemother.*, **1995**, *39*, 2732–2737

## SAMPLE
**Matrix:** blood
**Sample preparation:** 200 μL Serum + 10 μL 500 ng/mL theophylline in water + 50 μL isoamyl alcohol, vortex for 30 s, add 2 mL chloroform, vortex for 1 min, centrifuge at 1000 rpm for 5 min. Remove the organic layer and evaporate it to dryness under a stream of nitrogen at 50°, reconstitute the residue in 100 μL mobile phase, inject a 50 μL aliquot.

## HPLC VARIABLES
**Column:** 150 × 3.9 5 μm Novapak C18
**Mobile phase:** MeCN:0.2% acetic acid 7.5:92.5 (After 4 min increase flow rate to 1.8 mL/min over 1 min, maintain at 1.8 mL/min for 2 min.)
**Flow rate:** 0.8
**Injection volume:** 50
**Detector:** UV 270

## CHROMATOGRAM
**Retention time:** 6.0
**Internal standard:** theophylline (3.4)
**Limit of detection:** 20 ng/mL

## KEY WORDS
rat; serum; pharmacokinetics

## REFERENCE
Radwan, M.A. HPLC assay of theophylline and zidovudine in rat serum. *J.Liq.Chromatogr.*, **1995**, *18*, 3301–3309

## SAMPLE
**Matrix:** blood
**Sample preparation:** Plasma. 500 μL Plasma + 500 μL 35% perchloric acid, vortex for 1 min, centrifuge at 4000 g at 4° for 5 min. Remove a 200 μL supernatant and add it to 800 μL 250 ng/mL IS. Urine. Dilute urine 1:50 with water, add 200 μL to 800 μL 250 ng/mL IS, inject a 200 μL aliquot. Purify on column A for 2 min then pass onto column B (?).

## HPLC VARIABLES
**Column:** A 30 × 3.9 40 μm Perisorb RP8; B 20 × 2 40 μm Bondapak C18 endcapped + 300 × 3 4 μm Novapak RP18 endcapped
**Mobile phase:** MeOH:THF:25 mM pH 3.1 phosphate buffer 3.7:2.8:93.5
**Flow rate:** A 0.7; B 1

**Injection volume:** 200
**Detector:** UV 270

## CHROMATOGRAM
**Retention time:** 16.8
**Internal standard:** 2-O-isopropylidene uridine (14)
**Limit of quantitation:** 20 ng/mL

## OTHER SUBSTANCES
**Extracted:** metabolites

## KEY WORDS
plasma

## REFERENCE
Schrive, I.; Plasse, J.C. Quantification of zidovudine and one of its metabolites in plasma and urine by solid-phase extraction and high-performance liquid chromatography. *J.Chromatogr.B*, **1994**, *657*, 233–237

## SAMPLE
**Matrix:** blood
**Sample preparation:** 200 µL Plasma + 50 µL 20% w/v trichloroacetic acid, mix, heat at 60° for 1 h, centrifuge at 2500 g for 2 min, inject an aliquot of the supernatant.

## HPLC VARIABLES
**Column:** 150 × 4 5 µm reversed-phase C18 (FSA Laboratory Supplies)
**Mobile phase:** MeCN:20 mM pH 2.7 KH$_2$PO$_4$ 15:85
**Flow rate:** 0.4
**Detector:** UV 267

## CHROMATOGRAM
**Retention time:** 16.8
**Internal standard:** β-hydroxyethyltheophylline (11)
**Limit of quantitation:** 430 nM

## OTHER SUBSTANCES
**Extracted:** metabolites
**Simultaneous:** acetaminophen
**Noninterfering:** aspirin, fluconazole, ketoconazole, sulfamethoxazole, trimethoprim

## KEY WORDS
plasma

## REFERENCE
Kamali, F.; Rawlins, M.D. Simple and rapid assay for zidovudine and zidovudine glucuronide in plasma using high-performance liquid chromatography. *J.Chromatogr.*, **1990**, *530*, 474–479

## SAMPLE
**Matrix:** blood, cell suspensions, perfusate
**Sample preparation:** Centrifuge cellular suspensions at 17000 g for 5 min, inject a 25 µL aliquot. Centrifuge perfusate at 17000 g for 5 min, inject a 50 µL aliquot. Dilute 1 mL plasma with 1 mL saturated ammonium sulfate, vortex for 30 s, centrifuge at 3000 g for 2 min, inject a 50 µL aliquot of the supernatant.

## HPLC VARIABLES
**Column:** 250 × 4 5 µm Phenyl Hypersil NC-04

**Mobile phase:** MeOH:1.4 g/L sodium acetate 20:80, adjusted to pH 6.55
**Flow rate:** 1
**Injection volume:** 25-50
**Detector:** UV 267

## CHROMATOGRAM
**Retention time:** 8
**Limit of detection:** 50 ng/mL

## KEY WORDS
plasma

## REFERENCE
Frijus-Plessen, N.; Michaelis, H.C.; Foth, H.; Kahl, G.F. Determination of 3'-azido-3'-deoxythymidine, 2',3'-dideoxycytidine, 3'-fluoro-3'-deoxythymidine and 2',3'-dideoxyinosine in biological samples by high-performance liquid chromatography. *J.Chromatogr.*, **1990**, *534*, 101–107

## SAMPLE
**Matrix:** blood, CSF
**Sample preparation:** Mix 100 µL plasma or 50 µL CSF with an equal volume of 5 µg/mL 7-β-hydroxypropyltheophylline in MeCN, centrifuge at 14000 rpm, inject a 20 µL aliquot of the supernatant.

## HPLC VARIABLES
**Column:** 250 × 4 LiChrospher RP-18e
**Mobile phase:** MeCN:water:acetic acid 15:84.9:0.1
**Flow rate:** 1
**Injection volume:** 20
**Detector:** UV 265

## CHROMATOGRAM
**Internal standard:** 7-β-hydroxypropyltheophylline

## KEY WORDS
rat; plasma; pharmacokinetics

## REFERENCE
Seki, T.; Sato, N.; Hasegawa, T.; Kawaguchi, T.; Juni, K. Nasal absorption of zidovudine and its transport to cerebrospinal fluid in rats. *Biol.Pharm.Bull.*, **1994**, *17*, 1135–1137

## SAMPLE
**Matrix:** blood, milk, tissue
**Sample preparation:** Tissue. Homogenize 1 g tissue with 10 mL water. Add 2 mL MeOH to 1 mL tissue homogenate, mix, incubate at 4° for 1 h, centrifuge at 4°, put supernatant in a fresh tube, centrifuge, add p-nitrophenol to the supernatant, inject a 10-50 µL aliquot. Serum. 1 mL Serum + 2 mL MeOH, mix, incubate at 4° for 1 h, centrifuge at 4°, add p-nitrophenol to the supernatant, inject a 10-50 µL aliquot. Milk. 1 mL Milk + 4 mL MeOH, mix, incubate at 4° for 1 h, centrifuge at 4°, add p-nitrophenol to the supernatant, inject a 10-50 µL aliquot.

## HPLC VARIABLES
**Guard column:** 10 µm C18 Guard-Pak
**Column:** 100 × 5 5 µm Radial-Pak NovA C18
**Mobile phase:** MeCN:100 mM ammonium acetate 6:94, adjusted to pH 4.5 with glacial acetic acid
**Flow rate:** 1

**Injection volume:** 10-50
**Detector:** UV 280

## CHROMATOGRAM
**Retention time:** 20
**Internal standard:** p-nitrophenol (37)
**Limit of detection:** 150 ng/g

## OTHER SUBSTANCES
**Extracted:** metabolites

## KEY WORDS
serum; mouse; brain; embryonic tissue; spleen; liver

## REFERENCE
Ruprecht, R.M.; Sharpe, A.H.; Jaenisch, R.; Trites, D. Analysis of 3'-azido-3'-deoxythymidine levels in tissues and milk by isocratic high-performance liquid chromatography. *J.Chromatogr.*, **1990**, *528*, 371–383

## SAMPLE
**Matrix:** blood, tissue
**Sample preparation:** Homogenize (Polytron Model PT-1200C) 1 mL whole blood or whole rat brain with 1 mL water for 3-5 min, add 4 mL MeCN, vortex, add 1 mL concentrated saline, allow to settle at -5° for 1 h. Remove the organic layer, filter, dilute 1:4 with 50 mM ammonium acetate, inject an aliquot.

## HPLC VARIABLES
**Column:** 250 × 4.6 5 μm Adsorbosphere C18
**Mobile phase:** MeCN:50 mM ammonium acetate:water 20:55:25 adjusted to pH 5.5 with glacial acetic acid
**Flow rate:** 1
**Detector:** UV 266

## CHROMATOGRAM
**Retention time:** 5.41
**Limit of detection:** 10-50 ng/mL
**Limit of quantitation:** 30-180 ng/mL

## OTHER SUBSTANCES
**Extracted:** metabolites

## KEY WORDS
whole blood; rat; brain; pharmacokinetics

## REFERENCE
Brewster, M.E.; Andersom, W.R.; Helton, D.O.; Bodor, N.; Pop, E. Dose-dependent brain delivery of zidovudine through the use of a zidovudine chemical delivery system. *Pharm.Res.*, **1995**, *12*, 796–798

## SAMPLE
**Matrix:** cell media
**Sample preparation:** Inject a 150 μL aliquot of intracellular or extracellular media.

## HPLC VARIABLES
**Column:** 250 × 4.6 5 μm Partisil ODS3

**Mobile phase:** Gradient. MeCN:50 mM pH 6.5 phosphate buffer from 5:95 to 90:10 over 32 min
**Flow rate:** 1
**Injection volume:** 150
**Detector:** Radioactivity

## CHROMATOGRAM
**Retention time:** 24

## OTHER SUBSTANCES
**Extracted:** metabolites

## KEY WORDS
liver; monkey; dog; rat; human

## REFERENCE
Nicolas, F.; De Sousa, G.; Thomas, P.; Placidi, M.; Lorenzon, G.; Rahmani, R. Comparative metabolism of 3'-azido-3'-deoxythymidine in cultured hepatocytes from rats, dogs, monkeys, and humans. *Drug Metab.Dispos.*, **1995**, *23*, 308–313

## SAMPLE
**Matrix:** cell suspensions
**Sample preparation:** 1 mL Cell suspension + 500 µL ice-cold MeCN + 500 µL water, centrifuge. Remove the supernatant and evaporate it to dryness, reconstitute the residue in 200 µL water, inject an aliquot.

## HPLC VARIABLES
**Column:** Partisil-10 SAX
**Mobile phase:** Gradient. 10 mM pH 3.6 ammonium phosphate:600 mM pH 3.8 ammonium phosphate 100:0 for 15 min then a convex gradient to 0:100 over 10 min then stay at 0:100 for 30 min.
**Flow rate:** 2
**Detector:** UV 254

## CHROMATOGRAM
**Retention time:** 6.5

## OTHER SUBSTANCES
**Extracted:** metabolites

## REFERENCE
Mukherji, E.; Au, J.L.-S.; Mathes, L.E. Differential antiviral activities and intracellular metabolism of 3'-azido-3'-deoxythymidine and 2',3'-dideoxyinosine in human cells. *Antimicrob.Agents Chemother.*, **1994**, *38*, 1573–1579

## SAMPLE
**Matrix:** dialysate
**Sample preparation:** Vortex dialysate and IS, inject a 5 µL aliquot.

## HPLC VARIABLES
**Column:** 200 × 2.1 5 µm Hypersil ODS
**Mobile phase:** MeCN:10 mM (NH$_4$)H$_2$PO$_4$ 15:85
**Flow rate:** 0.2
**Injection volume:** 5
**Detector:** UV 266

## CHROMATOGRAM
**Internal standard:** β-hydroxypropyltheophylline

## KEY WORDS
rabbit; pharmacokinetics

## REFERENCE
Wang, Y.; Sawchuk, R.J. Zidovudine transport in the rabbit brain during intravenous and intracerebroventricular infusion. *J.Pharm.Sci.*, **1995**, *84*, 871−876

## SAMPLE
**Matrix:** intestinal mucosal homogenate
**Sample preparation:** Homogenate mixture + 100 μL 250 mM NaCN, mix, centrifuge at 4° at 34000 g for 10 min, filter (0.45 μm) the supernatant, inject an aliquot of the filtrate.

## HPLC VARIABLES
**Column:** 150 × 3.9 Nova-Pak C18
**Mobile phase:** MeOH:100 mM potassium phosphate 25:75
**Flow rate:** 1
**Detector:** UV 254

## KEY WORDS
rat

## REFERENCE
Sinko, P.J.; Hu, P. Determining intestinal metabolism and permeability for several compounds in rats. Implications on regional bioavailability in humans. *Pharm.Res.*, **1996**, *13*, 108−113

## SAMPLE
**Matrix:** solutions
**Sample preparation:** Inject a 15 μL aliquot.

## HPLC VARIABLES
**Column:** 200 × 4.6 5 μm HP Hypersil ODS
**Mobile phase:** MeCN:20 mM pH 7.0 $Na_2HPO_4$ 20:80
**Column temperature:** 37
**Flow rate:** 1
**Injection volume:** 15
**Detector:** UV 265

## CHROMATOGRAM
**Retention time:** 4.18

## REFERENCE
Kim, D.-D.; Chien, Y.W. Transdermal delivery of dideoxynucleoside-type Anti-HIV drugs. 2. The effect of vehicle and enhancer on skin permeation. *J.Pharm.Sci.*, **1996**, *85*, 214−219

## SAMPLE
**Matrix:** solutions
**Sample preparation:** Prepare a 1 mg/mL solution in water, inject a 10 μL aliquot.

## HPLC VARIABLES
**Column:** 100 × 4.6 7 μm Hypercarb (Shandon)
**Mobile phase:** THF:water:trifluoroacetic acid 30:69.5:0.5

**Flow rate:** 1
**Injection volume:** 10
**Detector:** UV 280

## CHROMATOGRAM
**Retention time:** 10

## OTHER SUBSTANCES
**Simultaneous:** metabolites, zidovudine glucuronide

## REFERENCE
Ayrton, J.; Evans, M.B.; Harris, A.J.; Plumb, R.S. Porous graphitic carbon shows promise for the rapid chromatographic analysis of polar drug metabolites. *J.Chromatogr.B*, **1995**, *667*, 173–178

## SAMPLE
**Matrix:** solutions
**Sample preparation:** Inject a 15 μL aliquot.

## HPLC VARIABLES
**Column:** 200 × 4.6 5 μm HP Hypersil ODS
**Mobile phase:** MeCN:20 mM pH 7.0 $Na_2HPO_4$ 20:80
**Column temperature:** 37
**Flow rate:** 1
**Injection volume:** 15
**Detector:** UV 265

## CHROMATOGRAM
**Retention time:** 4.18

## REFERENCE
Kim, D.-D.; Chien, Y.W. Transdermal delivery of dideoxynucleoside-type anti-HIV drugs. 1. Stability studies for hairless rat skin permeation. *J.Pharm.Sci.*, **1995**, *84*, 1061–1066

## SAMPLE
**Matrix:** tissue
**Sample preparation:** 50 mg Lymph nodes + 50 μL 10 μg/mL IS, homogenize with 6 volumes of pH 7.4 phosphate buffer, add 3 volumes of MeCN, mix, centrifuge at 9000 rpm for 10 min. Remove the supernatant and evaporate it to dryness under a stream of nitrogen at 22°, reconstitute the residue in 150 μL mobile phase, inject a 100 μL aliquot.

## HPLC VARIABLES
**Column:** 150 × 4.6 5 μm Hypersil ODS
**Mobile phase:** MeCN:buffer 10:90 (Buffer was 0.3% acetic acid adjusted to pH 4.0 with triethylamine.)
**Flow rate:** 1.5
**Injection volume:** 100
**Detector:** UV 263

## CHROMATOGRAM
**Internal standard:** 3'-azido-2',3'-dideoxy-5-ethyluridine (CS-85)
**Limit of quantitation:** 50 ng/g

## KEY WORDS
mouse; lymph nodes; pharmacokinetics

## REFERENCE

Manouilov, K.K.; White, C.A.; Boudinot, F.D.; Fedorov, I.I.; Chu, C.K. Lymphatic distribution of 3′-azido-3′-deoxythymidine and 3′-azido-2′,3′-dideoxyuridine in mice. *Drug Metab.Dispos.*, **1995**, *23*, 655–658

## ANNOTATED BIBLIOGRAPHY

Almudaris, A.; Ashton, D.S.; Ray, A.; Valko, K. Trace analysis of impurities in 3′-azido-3′-deoxythymidine by reversed-phase high-performance liquid chromatography and thermospray mass spectrometry. *J.Chromatogr.A*, **1995**, *689*, 31–38 [LC-MS]

Bareggi, S.R.; Cinque, P.; Mazzei, M.; D′Arminio, A.; Ruggieri, A.; Pirola, R.; Nicolin, A.; Lazzarin, A. Pharmacokinetics of zidovudine in HIV-positive patients with liver disease. *J.Clin.Pharmacol.*, **1994**, *34*, 782–786 [plasma; extracted metabolites; LOD 10 ng/mL]

Nebinger, P.; Koel, M. Determination of serum zidovudine by ultrafiltration and high-performance liquid chromatography. *J.Pharm.Biomed.Anal.*, **1994**, *12*, 141–143 [ultrafiltrate; LOD 50 ng/mL; adenine (IS)]

Good, S.S.; Reynolds, D.J.; de Miranda, P. Simultaneous quantification of zidovudine and its glucuronide in serum by high-performance liquid chromatography. *J.Chromatogr.*, **1988**, *431*, 123–133

Unadkat, J.D.; Crosby, S.S.; Wang, J.P.; Hertel, C.C. Simple and rapid high-performance liquid chromatographic assay for zidovudine (azidothymidine) in plasma and urine. *J.Chromatogr.*, **1988**, *430*, 420–423

# Zolpidem

**Molecular formula:** $C_{19}H_{21}N_3O$
**Molecular weight:** 307.4
**CAS Registry No.:** 82626-48-0 (zolpidem), 99294-93-6 (zolpidem (+)-tartrate (2:1))

## SAMPLE
**Matrix:** blood
**Sample preparation:** 2 mL Whole blood or plasma + 2 mL buffer + 5 mL chloroform: isopropanol:n-heptane 60:14:26, shake gently horizontally for 10 min, centrifuge at 2800 g for 10 min. Remove the lower organic layer and evaporate it to dryness under vacuum at 45°, reconstitute the residue in 100 μL mobile phase, centrifuge at 2800 g for 5 min, inject a 50 μL aliquot of the supernatant. (Buffer was saturated ammonium chloride solution 25% diluted with water, adjusted to pH 9.5 with 25% ammonia solution.)

## HPLC VARIABLES
**Column:** 300 × 3.9 4 μm NovaPack C18
**Mobile phase:** MeOH:THF:buffer 65:5:30 (Buffer was 0.68 g/L (10 mM (sic)) $KH_2PO_4$ adjusted to pH 2.6 with concentrated orthophosphoric acid.) (At the end of each session wash the column with water for 1 h and MeOH for 1 h, re-equilibrate for 30 min.)
**Column temperature:** 30
**Flow rate:** 0.8
**Injection volume:** 50
**Detector:** UV 239

## CHROMATOGRAM
**Retention time:** 4.66
**Limit of detection:** <120 ng/mL

## OTHER SUBSTANCES
**Extracted:** acebutolol, acepromazine, aceprometazine, acetaminophen, aconitine, albuterol, alimemazine, alpidem, alprazolam, alprenolol, amisulpride, amitriptyline, amoxapine, aspirin, astemizole, atenolol, benazepril, benzocaine, benzoylecgonine, bepridil, betaxolol, bisoprolol, bromazepam, brompheniramine, bumadizone, bupivacaine, buprenorphine, buspirone, caffeine, carbamazepine, carbinoxamine, carpipramine, carteolol, cetirizine, chlorambucil, chlordiazepoxide, chlormezanone, chlorophenacinone, chlorpheniramine, chlorpromazine, chlorpropamide, cibenzoline, clemastine, clobazam, clomipramine, clonazepam, clonidine, clorazepate, clozapine, codeine, colchicine, cyamemazine, cyclizine, cycloguanil, cyproheptadine, cytarabine, dacarbazine, daunorubicin, debrisoquine, demexiptiline, desipramine, dextromethorphan, dextromoramide, dextropropoxyphene, diazepam, diazoxide, diclofenac, dihydralazine, diltiazem, diphenhydramine, dipyridamole, disopyramide, dosulepine, doxepin, ephedrine, estazolam, etodolac, fenfluramine, fenoprofen, fentiazac, flecainide, floctafenine, flumazenil, flunitrazepam, fluoxetine, fluphenazine, flurbiprofen, fluvoxamine, glibenclamide, glibornuride, glipizide, glutethimide, haloperidol, histapyrrodine, hydroxyzine, ibuprofen, imipramine, indomethacin, iproniazid, ketamine, levomepromazine, lidoflazine, lisinopril, loperamide, loprazolam, lorata-

dine, lorazepam, loxapine, maprotiline, medazepam, medifoxamine, mefenamic acid, mefenidramine, mefloquine, melphalan, mephenesin, mephentermine, metapramine, metformin, methadone, methaqualone, methocarbamol, methotrexate, metipranolol, metoclopramide, metoprolol, mianserine, midazolam, minoxidil, moperone, morphine, nadolol, nalbuphine, nalorphine, naloxone, naltrexone, naproxen, nialamide, nicardipine, nifedipine, niflumic acid, nimodipine, nitrazepam, nitrendipine, nizatidine, nortriptyline, omeprazole, opipramol, oxazepam, oxprenolol, penbutolol, penfluridol, pentazocine, phencyclidine, phenobarbital, phenol, phenylbutazone, pimozide, pindolol, pipamperone, piroxicam, prazepam, prazosin, prilocaine, procainamide, procarbazine, proguanil, promethazine, propafenone, propranolol, protriptyline, pyrimethamine, quinidine, quinine, quinupramine, ramipril, ranitidine, reserpine, ritodrine, sotalol, strychnine, sulfinpyrazole, sulindac, sulpride, sultopride, suriclone, tenoxicam, terbutaline, terfenadine, tetracaine, tetrazepam, thiopental, thioproperazine, thioridazine, tianeptine, tiapride, tiaprofenic acid, tioclomarol, tofisopam, tolbutamide, toloxatone, trazodone, triazolam, trifluoperazine, trifluperidol, trimipramine, triprolidine, tropatenine, verapamil, vinblastine, vincristine, warfarin, zopiclone, zorubicine

**Interfering:** acenocoumarol, ajmaline, alminoprofen, amodiaquine, benperidol, celiprolol, chloroquine, cicletanine, cocaine, doxylamine, droperidol, hydroxychloroquine, ketoprofen, labetalol, lidocaine, meperidine, mepivacaine, mexiletine, moclobemide, nomifensine, secobarbital, temazepam, ticlopidine, timolol, viloxazine, vindesine, yohimbine

## KEY WORDS
whole blood; plasma

## REFERENCE
Tracqui, A.; Kintz, P.; Mangin, P. Systematic toxicological analysis using HPLC/DAD. *J.Forensic Sci.*, **1995**, *40*, 254–262

## SAMPLE
**Matrix:** blood
**Sample preparation:** 2 mL Plasma + 2 mL buffer + 5 mL chloroform:isopropanol:n-heptane 60:14:26, shake gently on a horizontal agitator for 10 min, centrifuge at 2800 g for 10 min. Remove the organic layer and evaporate it to dryness under vacuum at 45°, reconstitute the residue in 100 μL mobile phase, inject a 50 μL aliquot. (Buffer was saturated ammonium chloride, diluted 25% with water, adjusted to pH 9.5 with 25% diluted ammonia solution.)

## HPLC VARIABLES
**Column:** 300 × 3.9 4 μm Nova-Pak C18
**Mobile phase:** MeOH:THF:buffer 65:5:30 (Buffer was 10 mM $KH_2PO_4$ adjusted to pH 2.6 with orthophosphoric acid. At the end of the day wash column with water at 0.8 mL/min for 1 h and MeOH at 0.8 mL/min for 1 h.)
**Column temperature:** 30
**Flow rate:** 0.8
**Injection volume:** 200
**Detector:** UV 207

## CHROMATOGRAM
**Retention time:** 4.66
**Limit of detection:** 23.1 ng/mL

## OTHER SUBSTANCES
**Extracted:** alpidem, suriclone, zopiclone
**Simultaneous:** nimodipine, p-nitrophenol, pyrimethamine, sultopride, tiaprofenic acid, vincristine
**Interfering:** ketotifen

## KEY WORDS
plasma

## REFERENCE
Tracqui, A.; Kintz, P.; Mangin, P. High-performance liquid chromatographic assay with diode-array detection for toxicological screening of zopiclone, zolpidem, suriclone and alpidem in human plasma. *J.Chromatogr.*, **1993**, *616*, 95–103

## SAMPLE
**Matrix:** blood
**Sample preparation:** 1 mL Serum or plasma + 1 mL saturated $Na_2HPO_4$ + 100 μL 10 μg/mL IS + 5 mL diisopropyl ether:isopropanol 95:5 (Caution! Diisopropyl ether readily forms explosive peroxides!), shake for 5 min, centrifuge. Remove the organic layer and evaporate it to dryness at 35°, reconstitute the residue in 50 μL mobile phase, vortex, inject a 20 μL aliquot.

## HPLC VARIABLES
**Column:** 100 × 3 Chrompack CP spher C8
**Mobile phase:** MeCN:50 mM $NaH_2PO_4$ 70:30 adjusted to pH 2.2 with 85% orthophosphoric acid
**Flow rate:** 0.8
**Injection volume:** 20
**Detector:** UV 210

## CHROMATOGRAM
**Retention time:** 2.6
**Internal standard:** N-6-dimethyl-2-(4-methylphenyl)-N-propylimidazo[1,2-α]pyridine-3-acetamide methanesulfonate (Synthelabo France) (6.4)

## OTHER SUBSTANCES
**Extracted:** flumazenil, prothipendyl

## KEY WORDS
serum; plasma; pharmacokinetics

## REFERENCE
Debailleul, G.; Khalil, F.A.; Lheureux, P. HPLC quantification of zolpidem and prothipendyl in a voluntary intoxication. *J.Anal.Toxicol.*, **1991**, *15*, 35–37

## SAMPLE
**Matrix:** blood, urine
**Sample preparation:** Dilute urine 10-fold with water. 1 mL Plasma or diluted urine + 20 μL 1.5 μg/mL IS in MeOH, vortex, centrifuge at 11000 g for 3 min, inject 100 (plasma) or 50 (diluted urine) μL supernatant onto column A and elute to waste with mobile phase A, after 2 min elute the contents of column A onto column B with mobile phase B, after 1.5 min remove column B from the circuit, elute column B with mobile phase B, monitor the effluent from column B. Backflush column A with MeCN:water 50:50 and MeOH:water 50:50 then forward flush with water.

## HPLC VARIABLES
**Column:** A 75 × 2.1 30-40 μm Perisorb C18 (Merck); B 20 × 4.6 40 μm Pelliguard (Supelco) + 150 × 4.6 5 μm Supelcosil LC 18-DB
**Mobile phase:** A water; B MeCN:MeOH:buffer 60:0.75:40 (Buffer was 50 mM $KH_2PO_4$ adjusted to pH 6.0 with 1 M KOH.)
**Flow rate:** A 2; B 1

**Injection volume:** 80-200
**Detector:** F ex 254 em 390

## CHROMATOGRAM
**Retention time:** 15.8
**Internal standard:** N-6-dimethyl-2-(4-methylphenyl)-N-propylimidazo[1,2-α]pyridine-3-acetamide (SL 83.0725, Synthélabo, France) (13.1)
**Limit of detection:** 0.2 ng/mL (plasma); 1 ng/mL (urine)
**Limit of quantitation:** 1 ng/mL (plasma)

## OTHER SUBSTANCES
**Extracted:** metabolites
**Simultaneous:** trazodone
**Noninterfering:** diazepam, flunitrazepam, lorazepam, nordiazepam, oxazepam, triazolam

## KEY WORDS
plasma; column-switching; pharmacokinetics

## REFERENCE
Ascalone, V.; Flaminio, L.; Guinebault, P.; Thénot, J.P.; Morselli, P.L. Determination of zolpidem, a new sleep-inducing agent, and its metabolites in biological fluids: pharmacokinetics, drug metabolism and overdosing investigations in humans. *J.Chromatogr.*, **1992**, *581*, 237–250

## SAMPLE
**Matrix:** blood, gastric contents, tissue, urine
**Sample preparation:** 1 mL Blood, urine, or gastric contents or 1 g tissue homogenate + 500 μL buffer + 8 mL n-hexane:ethyl acetate 70:30, mix on a rotary mixer for 10 min, centrifuge at 3000 g for 8 min. Remove the organic layer and evaporate it to dryness under a stream of nitrogen, reconstitute the residue in 100 μL 12.5 mM NaOH in MeOH:water 50:50, inject a 50 μL aliquot. (Buffer was 13.8 g potassium carbonate in 100 mL water, pH adjusted to 9.5 with concentrated HCl.)

## HPLC VARIABLES
**Guard column:** 4 × 4 30 μm LiChrocart Aluspher RP-select B (Merck)
**Column:** 125 × 4 5 μm Aluspher RP-select B (Merck)
**Mobile phase:** Gradient. A was 12.5 mM NaOH in MeOH. B was 12.5 mM NaOH in water. A:B 10:90 for 5 min, to 90:10 over 15 min, maintain at 90:10 for 5 min, return to initial conditions over 1 min, re-equilibrate for 5 min.
**Flow rate:** 1
**Injection volume:** 50
**Detector:** UV 230; UV 254

## CHROMATOGRAM
**Retention time:** 13

## OTHER SUBSTANCES
**Extracted:** alprenolol, amitriptyline, bromazepam, carbamazepine, chlordiazepoxide, chlorpromazine, clonazepam, desipramine, diazepam, flunitrazepam, haloperidol, nitrendipine, nordiazepam, nortriptyline, pindolol
**Also analyzed:** acebutolol, acetaminophen, alprazolam, amphetamine, atenolol, betaxolol, brotizolam, caffeine, camazepam, captopril, chloroquine, clobazam, clomipramine, clothiapine, clotiazepam, cloxazolam, cocaine, codeine, diclofenac, dihydralazine, dihydrocodeine, dihydroergotamine, diphenhydramine, domperidone, doxepin, droperidol, ergotamine, ethyl loflazepate, fenethylline, fluoxetine, flupentixol, flurazepam, furosemide, gliclazide, hydrochlorothiazide, hydroxyzine, ibuprofen, imipramine, ketazolam, loprazolam, lorazepam, lormetazepam, maprotiline, medazepam, mepyramine, methadone, methaqualone, methyldopa, methylphenidate, metoclopramide, metoprolol, mexiletine,

mianserin, midazolam, minoxidil, morphine, nadolol, nitrazepam, oxprenolol, papaverine, pentazocine, phenprocoumon, phenylbutazone, pipamperone, piritramide, practolol, prazepam, prazosin, promazine, promethazine, propoxyphene, propranolol, prothipendyl, quinine, sotalol, sulpride, thioridazine, trazodone, triazolam, trimipramine, tripelennamine, tyramine, verapamil, yohimbine

## REFERENCE

Lambert, W.E.; Meyer, E.; De Leenheer, A.P. Systematic toxicological analysis of basic drugs by gradient elution of an alumina-based HPLC packing material under alkaline conditions. *J.Anal.Toxicol.*, **1995**, *19*, 73–78

## SAMPLE
**Matrix:** hepatocyte suspensions
**Sample preparation:** 150 μL Hepatocyte suspension + 150 μL MeCN, centrifuge at 5000 g, add 250 ng IS, inject a 150 μL aliquot of the supernatant.

## HPLC VARIABLES
**Column:** 125 × 4 5 μm Lichrosphere 60
**Mobile phase:** MeCN:MeOH:20 mM pH 4.5 potassium phosphate buffer 4:30:66
**Flow rate:** 0.9
**Injection volume:** 150
**Detector:** F ex 254 em 390

## CHROMATOGRAM
**Retention time:** 20.84
**Internal standard:** SL 870105 (Synthélabo-Recherche, Paris) (22.56)

## OTHER SUBSTANCES
**Extracted:** metabolites

## KEY WORDS
human; liver

## REFERENCE

Pichard, L.; Gillet, G.; Bonfils, C.; Domergue, J.; Thénot, J.-P.; Maurel, P. Oxidative metabolism of zolpidem by human liver cytochrome P450S. *Drug Metab.Dispos.*, **1995**, *23*, 1253–1262

## SAMPLE
**Matrix:** microsomal incubations
**Sample preparation:** 500 μL Microsomal incubation + 500 μL MeCN, centrifuge at 5000 g, add IS to the supernatant at a concentration of 6.7 μg/mL, inject a 150 μL aliquot.

## HPLC VARIABLES
**Column:** 250 × 4.6 5 μm Interchrom C6 (Interchim, Montluçon, France)
**Mobile phase:** MeCN:25 mM pH 6.0 potassium phosphate buffer 30:65
**Flow rate:** 1
**Injection volume:** 150
**Detector:** F ex 254 em 390

## CHROMATOGRAM
**Retention time:** 18.90
**Internal standard:** N-dealkylated alpidem (13.20)

## OTHER SUBSTANCES
**Extracted:** metabolites

## KEY WORDS
human; liver

## REFERENCE
Pichard, L.; Gillet, G.; Bonfils, C.; Domergue, J.; Thénot, J.-P.; Maurel, P. Oxidative metabolism of zolpidem by human liver cytochrome P450S. *Drug Metab.Dispos.*, **1995**, *23*, 1253–1262

## SAMPLE
**Matrix:** solutions

## HPLC VARIABLES
**Guard column:** 30 × 3.2 7 μm SI 100 ODS (not commercially available)
**Column:** 150 × 3.2 7 μm SI 100 ODS (not commercially available)
**Mobile phase:** MeCN:buffer 31.2:68.8 (Buffer was 6.66 g $KH_2PO_4$ and 4.8 g 85% phosphoric acid in 1 L water, pH 2.3.)
**Flow rate:** 0.5-1
**Detector:** UV 203; UV 233; UV 292

## CHROMATOGRAM
**Retention time:** 1.8
**Internal standard:** 5-(4-methylphenyl)-5-phenylhydantoin (7.3)

## OTHER SUBSTANCES
**Also analyzed:** aspirin, caffeine, carbamazepine, chlordiazepoxide, chlorprothixene, clonazepam, diazepam, doxylamine, ethosuximide, furosemide, haloperidol, hydrochlorothiazide, methocarbamol, methotrimeprazine, nicotine, oxazepam, procaine, promazine, propafenone, propranolol, salicylamide, temazepam, tetracaine, thiopental, triamterene, verapamil, zopiclone

## REFERENCE
Below, E.; Burrmann, M. Application of HPLC equipment with rapid scan detection to the identification of drugs in toxicological analysis. *J.Liq.Chromatogr.*, **1994**, *17*, 4131–4144

## ANNOTATED BIBLIOGRAPHY
Vajta, S.; De Maack, F.; Devant, G.; Lesieur, M. Thermospray liquid chromatography tandem mass spectrometry: application to the elucidation of zolpidem metabolism. *Biomed.Environ. Mass.Spectrom.*, **1988**, *15*, 223–228

# NAME INDEX

| NAME | MONOGRAPH |
|------|-----------|
| 2-G | Guaifenesin |
| 27-400 | Cyclosporine |
| 38489 | Nortriptyline |
| 4311/b Ciba | Methylphenidate |
| 44089 | Valproic Acid |
| 640/359 | Cefuroxime |
| 66873 | Cephalexin |
| A 43818 | Leuprolide |
| A 56268 | Clarithromycin |
| Aarane | Cromolyn |
| Aararre | Cromolyn |
| AB08 | Doxycycline |
| Abacin | Sulfamethoxazole |
| Abacin | Trimethoprim |
| Abbocillin V | Penicillin V |
| Abboticine (erythromycin stearate) | Erythromycin |
| Abomecatin | Erythromycin |
| Abbott-43818 | Leuprolide |
| Abbott 44090 | Valproic Acid |
| Abbott 45975 | Terazosin |
| Abbott-50711 | Valproic Acid |
| Abbott-56268 | Clarithromycin |
| Abensanil | Acetaminophen |
| Aberel (tretinoin) | Retinoic Acid |
| Abricycline | Tetracycline |
| Acamol | Acetaminophen |
| Accupril | Quinapril |
| Accuprin | Quinapril |
| Accupro | Quinapril |
| Accurbron | Theophylline |
| Accuretic | Hydrochlorothiazide |
| Accuretic | Quinapril |
| Accutane (isotretinoin) | Retinoic Acid |
| Acediur | Captopril |
| Acenalin | Cisapride |
| Acenterine | Aspirin |

| NAME | MONOGRAPH |
|------|-----------|
| Atrophate | Atropine |
| Atropisol | Atropine |
| atroscine | Scopolamine |
| Atrosed | Atropine |
| Atrovent | Ipratropium Bromide |
| A/T/S | Erythromycin |
| Atumin | Dicyclomine |
| Augmentin | Amoxicillin |
| Augmentin | Clavulanic Acid |
| Ausocef | Cephalexin |
| Austrapen | Ampicillin |
| Austyn | Theophylline |
| Avantyl | Nortriptyline |
| Aventyl | Nortriptyline |
| Aventyl Hydrochloride | Nortriptyline |
| Avomine | Promethazine |
| AX 250 | Amoxicillin |
| Axer Alfa | Naproxen |
| Axid | Nizatidine |
| Axoren | Buspirone |
| Axoril (cefuroxime axetil) | Cefuroxime |
| Ay 6108 | Ampicillin |
| AY24236 | Etodolac |
| Aygestin (norethindrone acetate) | Norethindrone |
| Azamune | Azathioprine |
| Azanin | Azathioprine |
| Azantac | Ranitidine |
| azido-3'-deoxythymidine 3'-azidothymidine | Zidovudine Zidovudine |
| Azitrocin | Azithromycin |
| Azmacort (triamcinolone acetonide) | Triamcinolone |
| Azo-Gantanol | Sulfamethoxazole |
| Azoran | Azathioprine |
| azothioprine | Azathioprine |
| Azotrex | Tetracycline |
| AZT | Zidovudine |
| Aztec | Zidovudine |
| Azudoxat | Doxycycline |
| Azuglucon | Glyburide |
| Azupentat | Pentoxifylline |
| Bactoderm | Mupirocin |
| Bactramin | Sulfamethoxazole |
| Bactramin | Trimethoprim |
| Bactrim | Trimethoprim |
| Bactrim | Sulfamethoxazole |
| Bactroban | Mupirocin |
| Bactromin | Trimethoprim |
| Bactromin | Sulfamethoxazole |
| Bagren | Bromocriptine |
| Bajaten | Indapamide |
| Baktar | Sulfamethoxazole |
| Baktar | Trimethoprim |

| NAME | MONOGRAPH |
|---|---|
| Benzoestrofol (estradiol 3-benzoate) | Estradiol |
| Benzofoline (estradiol 3-benzoate) | Estradiol |
| [2a*R*-[2aα,4β,4aβ,6β,9(α*R**,β*S**),11α,12α,12bα]]-β-(benzoylamino)-α-hydroxybenzene-propanoic acid 6,12b-bis(acetyloxy)-12-(benzoyloxy)-2a,3,4,4a,5,6,9,10,11,12,12a,12b-dodecahydro-4,11-dihydroxy-4a,8,13,13-tetramethyl-5-oxo-7,11-methano-1*H*-cyclodeca[3,4]benz[1,2-*b*]-oxet-9-yl ester | Paclitaxel |
| (±)-5-benzoyl-2,3-dihydro-1*H*-pyrrolizine-1-carboxylic acid | Ketorolac |
| 5-benzoyl-1,2-dihydro-3*H*-pyrrolo[1,2-*a*]pyrrole-1-carboxylic acid | Ketorolac |
| *m*-benzoylhydratropic acid | Ketoprofen |
| 3-benzoyl-α-methylbenzeneacetic acid | Ketoprofen |
| 2-(3-benzoylphenyl)propionic acid | Ketoprofen |
| Benztrone (estradiol 3-benzoate) | Estradiol |
| Berkatens | Verapamil |
| Berkfuran | Nitrofurantoin |
| Berkfurin | Nitrofurantoin |
| Berofor Alpha 2 (Alfa-2c) | Interferon |
| Beromycin | Penicillin V |
| Beromycin 400 | Penicillin V |
| Beronald | Furosemide |
| Bespar | Buspirone |
| Betabactyl | Clavulanic Acid |
| beta-Corlan | Betamethasone |
| betadexamethasone | Betamethasone |
| Betadival | Betamethasone |
| Betafluorene (betamethasone 21-acetate) | Betamethasone |
| Betaloc | Metoprolol |
| Betamaze | Sulbactam |
| Betamox | Amoxicillin |
| Betapen VK | Penicillin V |
| Betaseron (Beta-1b) | Interferon |
| Betasolon | Betamethasone |
| Betasone (betamethasone 17,21-dipropionate) | Betamethasone |
| Betatrex (betamethasone 17-valerate) | Betamethasone |
| BetaVal (betamethasone 17-valerate) | Betamethasone |
| Betavet Soluspan | Betamethasone |
| Bethacil | Sulbactam |
| Betim | Timolol |
| Betnelan | Betamethasone |
| Betnesol (betamethasone 21-phosphate disodium salt) | Betamethasone |
| Betnesol-V (betamethasone 17-valerate) | Betamethasone |
| Betneval (betamethasone 17-valerate) | Betamethasone |
| Betnovate (betamethasone 17-valerate) | Betamethasone |
| Betsovet (betamethasone 21-adamantoate) | Betamethasone |
| Bextasol (betamethasone 17-valerate) | Betamethasone |
| Bialzepam | Diazepam |

| NAME | MONOGRAPH |
|---|---|
| Bristamycin (erythromycin stearate) | Erythromycin |
| Britacil | Ampicillin |
| Britiazem | Diltiazem |
| BRL 1341 | Ampicillin |
| BRL 2333 | Amoxicillin |
| BRL-4910A | Mupirocin |
| BRL 14151 | Clavulanic Acid |
| BRL 14151K | Clavulanic Acid |
| BRL 14777 | Nabumetone |
| 2-bromoergocryptine | Bromocriptine |
| 2-bromo-α-ergokryptin | Bromocriptine |
| (5'α)-2-bromo-12'-hydroxy-2'-(1-methylethyl)-5'-(2-methylpropyl)ergotaman-3',6',18-trione | Bromocriptine |
| Bronchoretard | Theophylline |
| Broncovaleas | Albuterol |
| Brondecon | Guaifenesin |
| Bronkodyl | Theophylline |
| Bronkotabs | Theophylline |
| Brufanic | Ibuprofen |
| Brufen | Ibuprofen |
| Brufort | Ibuprofen |
| Brumetidina | Cimetidine |
| Bruxicam | Piroxicam |
| Bruzem | Diltiazem |
| Buburone | Ibuprofen |
| Buccalsone (hydrocortisone 21-sodium succinate) | Hydrocortisone |
| Bufferin | Aspirin |
| Buspar | Buspirone |
| Buspimen | Buspirone |
| Buspinol | Buspirone |
| buspirone | Buspirone |
| Buspisal | Buspirone |
| 2-(*tert*-butylamino)-1-(4-hydroxy-3-hydroxymethylphenyl)ethanol | Albuterol |
| S-(—)-3-(3-*tert*-butylamino-2-hydroxypropoxy)-4-morpholino-1,2,5-thiadiazole | Timolol |
| α¹-[(*tert*-butylamino)methyl]-4-hydroxy-*m*-xylene-α,α'-diol | Albuterol |
| Butylenin | Ibuprofen |
| N-*tert*-butyl-3-oxo-4-aza-5α-androst-1-ene-17β-carboxamide | Finasteride |
| 1-(*p-tert*-butylphenyl)-4-[4'-(α-hydroxydi-phenylmethyl)-1'-piperidyl]butanol | Terfenadine |
| BW 56-72 | Trimethoprim |
| BW 57-322 | Azathioprine |
| BW 72U | Trimethoprim |
| BW 248U | Acyclovir |
| BW A509U | Zidovudine |
| Bykomycin | Neomycin |
| Cabermox | Amoxicillin |
| Calan | Verapamil |

| NAME | MONOGRAPH |
|------|-----------|
| [o-[(2,6-dichloroanilino)phenyl]acetic acid sodium salt | Diclofenac |
| 9,21-dichloro-11β,17-dihydroxy-16α-methyl-pregna-1,4-diene-3,20-dione 17-(2-furoate) | Mometasone Furoate |
| (11β,16α)-9,21-dichloro-17-[(2-furanylcarbonyl)oxy]-11-hydroxy-16-methylpregna-1,4-diene-3,20-dione | Mometasone Furoate |
| 2-[(2,6-dichlorophenyl)amino]benzeneacetic acid monosodium salt | Diclofenac |
| (1S-cis)-4-(3,4-dichlorophenyl)-1,2,3,4-tetrahydro-N-methyl-1-naphthalenamine | Sertraline |
| (1S,4S)-4-(3,4-dichlorophenyl)-1,2,3,4-tetrahydro-N-methyl-1-naphthylamine | Sertraline |
| cis-1-[4-[[2-(2,4-dichlorophenyl)-2-(1H-1,2,4-tria-zol-1-ylmethyl)-1,3-dioxolan-4-yl]methoxy]phenyl]-4-(1-methylethyl)piperazine | Terconazole |
| Dichlorosal | Hydrochlorothiazide |
| Dichlotride | Hydrochlorothiazide |
| Dichronic | Diclofenac |
| Diclobenin | Diclofenac |
| diclofenac | Diclofenac |
| Diclo-Phlogont | Diclofenac |
| Diclo-Puren | Diclofenac |
| Diclord | Diclofenac |
| Dicloreum | Diclofenac |
| Diclotride | Hydrochlorothiazide |
| Dicodid | Hydrocodone |
| Dicodrine | Hydrocodone |
| Dicromil | Desogestrel |
| Dicromil | Ethinyl Estradiol |
| Dicurin Procaine | Theophylline |
| dicyclomine | Dicyclomine |
| dicycloverin | Dicyclomine |
| (5α,6α)-7,8-didehydro-4,5-epoxy-3-methoxy-17-methylmorphinan-6-ol | Codeine |
| [2R-(2R*,3S*,4R*,5R*,8R*,10R*,11R*,12S*,13S*,14R*)]-13-[(2,6-dideoxy-3-C-methyl-3-O-methyl-α-L-ribo-hexopyranosyl)oxy]-2-ethyl-3,4,10-trihydroxy-3,5,6,8,10,12,14-heptamethyl-11-[[3,4,6-trideoxy-3-(dimethylamino)-β-D-xylo-hexopyranosyl]oxy]-1-oxa-6-azacyclopentadecan-15-one | Azithromycin |
| (3R*,4S*,5S*,6R*,7R*,9R*,11R*,12R*,13S*,14R*)-4-[(2,6-dideoxy-3-C-methyl-3-O-methyl-α-L-ribo-hexopyranosyl)oxy]-14-ethyl-7,12,13-trihydroxy-3,5,7,9,11,13-hexamethyl-6-[[3,4,6-trideoxy-3-(diethylamino)-β-D-xylo-hexopyranosyl]oxy]oxacyclotetra-decane-2,10-dione | Erythromycin |

| NAME | MONOGRAPH |
|---|---|
| EF-Cortelan Soluble (hydrocortisone 21-sodium succinate) | Hydrocortisone |
| Efcortelin | Hydrocortisone |
| Efcortesol (hydrocortisone 21-phosphate disodium salt) | Hydrocortisone |
| Eferox | Levothyroxine |
| Effederm (tretinoin) | Retinoic Acid |
| Effekton | Diclofenac |
| Effexor | Venlafaxine |
| Efpenix | Amoxicillin |
| Eismycin | Mupirocin |
| Ekko | Phenytoin |
| Ekomine | Atropine |
| Elavil | Amitriptyline |
| Elazor | Fluconazole |
| Eldecort | Hydrocortisone |
| Elisor | Pravastatin |
| Elixicon | Theophylline |
| Elixophyllin | Theophylline |
| Elobact (cefuroxime axetil) | Cefuroxime |
| Elocon | Mometasone Furoate |
| Eltrianyl | Sulfamethoxazole |
| Eltrianyl | Trimethoprim |
| Eltroxin | Levothyroxine |
| Emgel | Erythromycin |
| Emodin | Ibuprofen |
| Emotival | Lorazepam |
| Empecid | Clotrimazole |
| Empirin | Aspirin |
| Empirin | Codeine |
| Empracet | Acetaminophen |
| EMU | Erythromycin |
| E-Mycin | Erythromycin |
| E-Mycin E (erythromycin ethylsuccinate) | Erythromycin |
| Enacard | Enalapril |
| enalapril | Enalapril |
| Enaloc | Enalapril |
| Enantone | Leuprolide |
| Enap | Enalapril |
| Enapren | Enalapril |
| Encaprin | Aspirin |
| Encorton | Prednisone |
| Endecon | Acetaminophen |
| Endecon | Phenylpropanolamine |
| Endep | Amitriptyline |
| Etrafon | Amitriptyline |
| Endomixin | Neomycin |
| Endural | Furosemide |
| Endydol | Aspirin |
| Enelfa | Acetaminophen |
| Eneril | Acetaminophen |
| Englate | Theophylline |

| NAME | MONOGRAPH |
|---|---|
| (—)-6-[2-amino-2-(*p*-hydroxyphenyl)acetamido]-3,3-dimethyl-7-oxo-4-thia-1-azabicyclo[3.2.O]heptane-2-carboxylic acid | Amoxicillin |
| (±)-4-hydroxy-α$^1$-[[[6-(4-phenylbutoxy)hexyl]amino]methyl]-1,3-benzenedimethanol | Salmeterol |
| (±)-4-hydroxy-α′-[[[6-(4-phenylbutoxy)hexyl]amino]methyl]-*m*-xylene-α,′α-diol | Salmeterol |
| 5,5′-[(2-hydroxy-1,3-propanediyl)bis(oxy)]bis[4-oxo-4*H*-1-benzopyran-2-carboxylic acid] | Cromolyn |
| 5,5′-(2-hydroxytrimethylenedioxy)bis(4-oxochromene-2-carboxylic acid) | Cromolyn |
| 5,5′-[(2-hydroxytrimethylene)dioxy]bis(4-oxo-4*H*-1-benzopyran-2-carboxylic acid) | Cromolyn |
| Hydrozide | Hydrochlorothiazide |
| hyoscine | Scopolamine |
| hyoscyamine | Atropine |
| Hypertil | Captopril |
| Hy-Phen | Acetaminophen |
| Hy-Phen | Hydrocodone |
| Hypnovel | Midazolam |
| hypophyseal growth hormone | Somatropin |
| Hypothiazide | Hydrochlorothiazide |
| Hypurin | Insulin |
| Hysron | Medroxyprogesterone Acetate |
| Hytone | Hydrocortisone |
| Hytracin | Terazosin |
| Hytrin | Terazosin |
| Hytrinex | Terazosin |
| Ibiamox | Amoxicillin |
| Ibilex | Cephalexin |
| Ibinolo | Atenolol |
| Ibu-Attritin | Ibuprofen |
| Ibumetin | Ibuprofen |
| Ibuprocin | Ibuprofen |
| Ibu-slo | Ibuprofen |
| Ibutad | Ibuprofen |
| Ibutid | Ibuprofen |
| Ibutop | Ibuprofen |
| ICI 35868 | Propofol |
| ICI 66082 | Atenolol |
| ICI 118630 | Goserelin |
| Icipen | Penicillin V |
| Ifada | Famotidine |
| IFN-α$^2$ | Interferon |
| IFN-αA | Interferon |
| Iktorivil | Clonazepam |
| Iletin II | Insulin |
| Ilosone (erythromycin estolate) | Erythromycin |
| Ilotycin | Erythromycin |
| Ilotycin Gluheptate (erythromycin glucoheptonate) | Erythromycin |
| Imex | Tetracycline |

| NAME | MONOGRAPH |
|---|---|
| leuprolide | Leuprolide |
| leuprorelin | Leuprolide |
| Levanxene | Temazepam |
| Levanxol | Temazepam |
| Levaxin | Levothyroxine |
| Levium | Diazepam |
| Levius | Aspirin |
| Levlen | Ethinyl Estradiol |
| Levlen | Norgestrel |
| levofloxacin | Ofloxacin |
| levonorgestrel | Norgestrel |
| levopropoxyphene napsylate | Propoxyphene |
| Levothroid | Levothyroxine |
| Levothyrox | Levothyroxine |
| levothyroxine | Levothyroxine |
| Levsin | Atropine |
| Levsinex | Atropine |
| Lexibiotico | Cephalexin |
| Libanil | Glyburide |
| Lidaform-HC (hydrortisone acetate) | Hydrocortisone |
| Lidamantle-HC (hydrortisone acetate) | Hydrocortisone |
| Lidaprim | Trimethoprim |
| Lifeampil | Ampicillin |
| Limbitrol | Amitriptyline |
| Linaris | Sulfamethoxazole |
| Linaris | Trimethoprim |
| Linoral | Ethinyl Estradiol |
| Liomycin | Doxycycline |
| Liponorm | Simvastatin |
| Lipostat | Pravastatin |
| Lipozid | Gemfibrozil |
| Liptan | Ibuprofen |
| Lipur | Gemfibrozil |
| Liquamycin | Tetracycline |
| Liquiprin | Acetaminophen |
| Lisacort | Prednisone |
| Lisaglucon | Glyburide |
| Lisino | Loratadine |
| Litraderm (hydrocortisone acetate) | Hydrocortisone |
| Liviatin | Doxycycline |
| Lixidol | Ketorolac |
| Llonexina | Cephalexin |
| Lobufen | Ibuprofen |
| Locoid (hydrocortisone 17-butyrate) | Hydrocortisone |
| Lodalès | Simvastatin |
| Lodine | Etodolac |
| Loestrin | Ethinyl Estradiol |
| Loestrin (norethindrone acetate) | Norethindrone |
| Loftan | Albuterol |
| Logical | Valproic Acid |
| Logynon | Ethinyl Estradiol |
| Logynon | Norgestrel |

| NAME | MONOGRAPH |
|---|---|
| Mevacor | Lovastatin |
| Mevalotin | Pravastatin |
| Mevinacor | Lovastatin |
| mevinolin | Lovastatin |
| Mevlor | Lovastatin |
| Micrainin | Aspirin |
| Microgynon | Ethinyl Estradiol |
| Microgynon | Norgestrel |
| Microlut | Norgestrel |
| Micronase | Glyburide |
| Micronett | Norethindrone |
| Micronor | Norethindrone |
| Micronovum | Norethindrone |
| Microtrim | Trimethoprim |
| Microtrim | Sulfamethoxazole |
| Microval | Norgestrel |
| midazolam | Midazolam |
| Midol | Acetaminophen |
| Midol 200 | Ibuprofen |
| Midoxin | Doxycycline |
| Miketorin | Amitriptyline |
| Mildison | Hydrocortisone |
| Milligynon (norethindrone acetate) | Norethindrone |
| Millisrol | Nitroglycerin |
| Millithrol | Nitroglycerin |
| Milprem | Estrogens, Conjugated |
| Miltax | Ketoprofen |
| Milvane | Ethinyl Estradiol |
| Mindiab | Glipizide |
| Minetoin | Phenytoin |
| Mini-Pe | Norethindrone |
| Minidiab | Glipizide |
| Miniluteolas | Ethinyl Estradiol |
| Miniluteolas | Ethynodiol Diacetate |
| Minilyn | Ethinyl Estradiol |
| "mini-pill'' | Norethindrone |
| Minitran | Nitroglycerin |
| Minovlar (norethindrone acetate) | Norethindrone |
| Minulet | Ethinyl Estradiol |
| Miocaina | Guaifenesin |
| Miolisodal | Carisoprodol |
| Mioril | Carisoprodol |
| Miranax | Naproxen |
| Mirfat | Furosemide |
| MJ 9022-1 | Buspirone |
| MJF 11567-3 | Cefadroxil |
| MK-130 | Cyclobenzaprine |
| MK-208 | Famotidine |
| MK-421 | Enalapril |
| MK-521 | Lisinopril |
| MK-733 | Simvastatin |
| MK-787 | Imipenem |

| NAME | MONOGRAPH |
|------|-----------|
| Nycopren | Naproxen |
| Nysconitrine | Nitroglycerin |
| Nystaform-HC | Hydrocortisone |
| Obestat | Phenylpropanolamine |
| Odemase | Furosemide |
| Oedemex | Furosemide |
| Oestergon | Estradiol |
| oestradiol | Estradiol |
| Oestroform (estradiol 3-benzoate) | Estradiol |
| Oestrogel | Estradiol |
| Oflocet | Ofloxacin |
| Oflocin | Ofloxacin |
| Oflox | Ofloxacin |
| ofloxacine | Ofloxacin |
| Ohlexin | Cephalexin |
| Oliprevin | Pravastatin |
| Olivin | Enalapril |
| Omcilon | Triamcinolone |
| Omcilon-A (triamcinolone acetonide) | Triamcinolone |
| Omegamycin | Tetracycline |
| Omepral | Omeprazole |
| Omeprazen | Omeprazole |
| Omnipen | Ampicillin |
| Omnipen-N | Ampicillin |
| Omsat | Trimethoprim |
| Omsat | Sulfamethoxazole |
| ondansetron | Ondansetron |
| Ophthocort | Hydrocortisone |
| Ophthocort | Polymyxin |
| Opticrom | Cromolyn |
| Opticron | Cromolyn |
| Optisulin long | Insulin |
| Optium | Amoxicillin |
| Opturem | Ibuprofen |
| Oracef | Cephalexin |
| Oracéfal | Cefadroxil |
| Oracil-VK | Penicillin V |
| Oracilline | Penicillin V |
| Oracocin | Cephalexin |
| Oracon | Ethinyl Estradiol |
| Oradiol | Ethinyl Estradiol |
| Orafuran | Nitrofurantoin |
| Oragest | Medroxyprogesterone Acetate |
| Orapen V-K | Penicillin V |
| Oraprim | Trimethoprim |
| Oraprim | Sulfamethoxazole |
| Orasone | Prednisone |
| Oratren | Penicillin V |
| Oraxim (cefuroxime axetil) | Cefuroxime |
| Orbicilina | Ampicillin |
| Oresol | Guaifenesin |
| Oreson | Guaifenesin |

| NAME | MONOGRAPH |
|------|-----------|
| Ovcon | Norethindrone |
| Ovex B (estradiol 3-benzoate) | Estradiol |
| Ovin | Ethinyl Estradiol |
| Oviol | Desogestrel |
| Oviol | Ethinyl Estradiol |
| Ovocyclin | Estradiol |
| Ovocyclin Benzoate (estradiol 3-benzoate) | Estradiol |
| Ovocyclin dipropionate (estradiol dipropionate) | Estradiol |
| Ovocyclin M (estradiol 3-benzoate) | Estradiol |
| Ovocyclin-MB (estradiol 3-benzoate) | Estradiol |
| Ovocyclin-P (estradiol dipropionate) | Estradiol |
| Ovocylin | Estradiol |
| Ovoresta | Ethinyl Estradiol |
| Ovral | Ethinyl Estradiol |
| Ovral | Norgestrel |
| Ovran | Ethinyl Estradiol |
| Ovran | Norgestrel |
| Ovranette | Ethinyl Estradiol |
| Ovranette | Norgestrel |
| Ovrette | Norgestrel |
| Ovulen | Ethynodiol Diacetate |
| Ovysmen | Ethinyl Estradiol |
| Ovysmen | Norethindrone |
| Oxaldin | Ofloxacin |
| Oxapro | Oxaprozin |
| Oxcord | Nifedipine |
| Oxikon | Oxycodone |
| 1-(5-oxohexyl)-3,7-dimethylxanthine | Pentoxifylline |
| 1-(5-oxohexyl)theobromine | Pentoxifylline |
| oxpentifylline | Pentoxifylline |
| Oxycon | Oxycodone |
| oxydiazepam | Temazepam |
| Oxyphyllin | Theophylline |
| P 5O | Ampicillin |
| rt-PA | Alteplase |
| t-PA | Alteplase |
| Pabracort (hydrocortisone acetate) | Hydrocortisone |
| P-A-C Analgesic Tablets | Aspirin |
| Pacemol | Acetaminophen |
| Paceum | Diazepam |
| Pacitran | Diazepam |
| Paediathrocin (erythromycin ethylsuccinate) | Erythromycin |
| Paidomal | Theophylline |
| Paldesic | Acetaminophen |
| Paldomycin | Doxycycline |
| Palonyl | Ethinyl Estradiol |
| Pamelor | Nortriptyline |
| Pamocil | Amoxicillin |
| Panacef | Cefaclor |
| Panadol | Acetaminophen |
| Panaleve | Acetaminophen |
| Panalog (triamcinolone acetonide) | Triamcinolone |

| NAME | MONOGRAPH |
|---|---|
| Panasorb | Acetaminophen |
| Pancodine | Oxycodone |
| Percocet | Oxycodone |
| Pandel (hydrocortisone buteprate) | Hydrocortisone |
| Panets | Acetaminophen |
| Panex | Acetaminophen |
| Panfungol | Ketoconazole |
| Panlomyc | Terconazole |
| Panmycin | Tetracycline |
| Panmycin P | Tetracycline |
| Panodil | Acetaminophen |
| Panofen | Acetaminophen |
| Panophylline | Theophylline |
| Panoral | Cefaclor |
| Pantomicina (erythromycin stearate) | Erythromycin |
| Panurin | Hydrochlorothiazide |
| Panwarfin | Warfarin |
| Panzid | Ceftazidime |
| paracetamol | Acetaminophen |
| Paracort | Prednisone |
| Paralatin | Carboplatin |
| Paralergin | Astemizole |
| Paraspen | Acetaminophen |
| Parbetan (betamethasone 17-benzoate) | Betamethasone |
| Parelan | Acetaminophen |
| Parfuran | Nitrofurantoin |
| Parizac | Omeprazole |
| Parlodel | Bromocriptine |
| Parmol | Acetaminophen |
| Partrex | Tetracycline |
| Pasetocin | Amoxicillin |
| Pasolind | Acetaminophen |
| Pasolind N | Acetaminophen |
| Patrovina | Adiphenine |
| Paxate | Diazepam |
| Paxel | Diazepam |
| Paxofen | Ibuprofen |
| PCE | Erythromycin |
| PD 109452-2 | Quinapril |
| Pediamycin (erythromycin ethylsuccinate) | Erythromycin |
| Pediaprofen | Ibuprofen |
| Pediazole (erythromycin ethylsuccinate) | Erythromycin |
| Pediotic Suspension | Neomycin |
| Pediotic Suspension | Hydrocortisone |
| Pediotic Suspension | Polymyxin |
| Pedipen | Penicillin V |
| Pedisafe | Clotrimazole |
| Pelanin benzoate (estradiol 3-benzoate) | Estradiol |
| Pelanin Depot (estradiol 17-valerate) | Estradiol |
| PELS (erythromycin estolate) | Erythromycin |
| pemophyllin | Theophylline |
| Pen A | Ampicillin |

| NAME | MONOGRAPH |
|---|---|
| Penagen | Penicillin V |
| Penamox | Amoxicillin |
| Pen A/N | Ampicillin |
| Penapar VK | Penicillin V |
| Penavlon V | Penicillin V |
| Penbristol | Ampicillin |
| Penbritin | Ampicillin |
| Penbritin-S | Ampicillin |
| Penbrock | Ampicillin |
| Pencompren | Penicillin V |
| Penecort | Hydrocortisone |
| Penialmen | Ampicillin |
| Penicals | Penicillin V |
| penicillanic acid sulfone | Sulbactam |
| penicillanic acid 1,1-dioxide | Sulbactam |
| penicillin phenoxymethyl | Penicillin V |
| Pénicline | Ampicillin |
| Penimox | Amoxicillin |
| Penntuss | Codeine |
| Pen-Oral | Penicillin V |
| Penstabil | Ampicillin |
| Pensyn | Ampicillin |
| Pentrex | Ampicillin |
| Pentrexyl | Ampicillin |
| Pen-Vee | Penicillin V |
| Pen-Vee K | Penicillin V |
| Penvikal | Penicillin V |
| Pepcid | Famotidine |
| Pepcidina | Famotidine |
| Pepcidine | Famotidine |
| Pepdine | Famotidine |
| Pepdul | Famotidine |
| Peptan | Famotidine |
| Pepticum | Omeprazole |
| Peptol | Cimetidine |
| Percocet | Acetaminophen |
| Percodan | Aspirin |
| Percodan | Oxycodone |
| Percodan Demi | Oxycodone |
| Percogesic | Codeine |
| Percutol | Nitroglycerin |
| Perdorm | Temazepam |
| Perglottal | Nitroglycerin |
| Perlatanol | Estradiol |
| Perlinganit | Nitroglycerin |
| Perlutex | Medroxyprogesterone Acetate |
| Perovex | Ethinyl Estradiol |
| Persistin | Aspirin |
| Pfizer-E (erythromycin stearate) | Erythromycin |
| Pfizerpen VK | Penicillin V |
| Phaeva | Ethinyl Estradiol |
| Phenaphen | Acetaminophen |

| NAME | MONOGRAPH |
|------|-----------|
| Poviral | Acyclovir |
| Praeciglucon | Glyburide |
| Pramace | Ramipril |
| Pranoxen | Naproxen |
| Pravachol | Pravastatin |
| Pravaselect | Pravastatin |
| pravastatin | Pravastatin |
| Pravidel | Bromocriptine |
| Precortal | Prednisone |
| Prednilonga | Prednisone |
| 1,4-pregnadiene-17α,21-diol-3,11,20-trione | Prednisone |
| 4-pregnene-11β,17α,21 triol-3,20-dione | Hydrocortisone |
| Prelis | Metoprolol |
| Premarin | Estrogens, Conjugated |
| Prenormine | Atenolol |
| Prepcort | Hydrocortisone |
| Prepulsid | Cisapride |
| Pres | Enalapril |
| Pressural | Indapamide |
| Pretor | Cefotaxime |
| Prexan | Naproxen |
| Prilosec | Omeprazole |
| Primatene Tablets | Theophylline |
| Primaxin | Cilastatin |
| Primaxin | Imipenem |
| Primeral | Naproxen |
| Primofenac | Diclofenac |
| Primofol (estradiol 17-valerate) | Estradiol |
| Primogyn | Ethinyl Estradiol |
| Primogyn B (estradiol 3-benzoate) | Estradiol |
| Primogyn C | Ethinyl Estradiol |
| Primogyn I (estradiol 3-benzoate) | Estradiol |
| Primogyn M | Ethinyl Estradiol |
| Primolut N | Norethindrone |
| Primolut-Nor (norethindrone acetate) | Norethindrone |
| Primosiston (norethindrone acetate) | Norethindrone |
| Princillin | Ampicillin |
| Principen | Ampicillin |
| Principen/N | Ampicillin |
| Prinil | Lisinopril |
| Prinivil | Lisinopril |
| Prinzide | Lisinopril |
| Prinzide | Hydrochlorothiazide |
| proazamine | Promethazine |
| Procardia | Nifedipine |
| Procef | Cefprozil |
| Procrit | Epoetin |
| Proctocort | Hydrocortisone |
| Proctofoam-HC (hydrortisone acetate) | Hydrocortisone |
| Procyclomin | Dicyclomine |
| Prodasone | Medroxyprogesterone Acetate |
| Pro Dorm | Lorazepam |

| NAME | MONOGRAPH |
|---|---|
| Triamonide 40 (triamcinolone acetonide) | Triamcinolone |
| Triatec | Ramipril |
| Triavil | Amitriptyline |
| Tribrissen | Trimethoprim |
| Tricinolon (triamcinolone acetonide) | Triamcinolone |
| Tricortale | Triamcinolone |
| Tridil | Nitroglycerin |
| Tri-Ervonum | Ethinyl Estradiol |
| Triflucan | Fluconazole |
| α,α,α-trifluoro-2-methyl-4'-nitro-*m*-propiono-toluidine | Flutamide |
| Trigger | Ranitidine |
| Triglobe | Trimethoprim |
| Trigonyl | Sulfamethoxazole |
| Trigonyl | Trimethoprim |
| 11,17,21-trihydroxy-6-methyl-1,4-pregnadiene-3,20-dione | Methylprednisolone |
| 11,17,21-trihydroxypregn-4-ene-3,20-dione | Hydrocortisone |
| Tri-Levlen | Norgestrel |
| Tri-Levlen | Ethinyl Estradiol |
| Triludan | Terfenadine |
| Trimanyl | Trimethoprim |
| Trimesulf | Sulfamethoxazole |
| Trimesulf | Trimethoprim |
| 5-[(3,4,5-trimethoxyphenyl)methyl]-2,4-pyrimidinediamine | Trimethoprim |
| *N*,*N*,6-trimethyl-2-(4-methylphenyl)imidazo[1,2-*a*]pyridine-3-acetamide | Zolpidem |
| *N*,*N*,α-trimethyl-10*H*-phenothiazine-10-ethanamine | Promethazine |
| N,*N*,6-trimethyl-2-*p*-tolylimidazo[1,2-*a*]pyridine-3-acetamide | Zolpidem |
| Trimforte | Trimethoprim |
| Trimforte | Sulfamethoxazole |
| Tri-Minulet | Ethinyl Estradiol |
| Trimogal | Trimethoprim |
| Trimopan | Trimethoprim |
| Trimox | Amoxicillin |
| Trimpex | Trimethoprim |
| Trimysten | Clotrimazole |
| Trinalgon | Nitroglycerin |
| Trind | Phenylpropanolamine |
| trinitrin | Nitroglycerin |
| trinitroglycerol | Nitroglycerin |
| Trinitrosan | Nitroglycerin |
| Trinordiol | Ethinyl Estradiol |
| Trinordiol | Norgestrel |
| Tri-Norinyl | Ethinyl Estradiol |
| Tri-Norinyl | Norethindrone |
| Trinovum | Ethinyl Estradiol |
| Trinovum | Norethindrone |
| Triphacyclin | Tetracycline |

# MOLECULAR FORMULA INDEX

In general, molecular formulae are given for the simplest form of the compound. Thus, although a particular drug might be used as the hydrochloride, the molecular formula will refer to the free base.

| MOLECULAR FORMULA | COMPOUND |
|---|---|
| $C_3H_5N_3O_9$ | Nitroglycerin |
| $C_6H_{12}N_2O_4Pt$ | Carboplatin |
| $C_7H_8ClN_3O_4S_2$ | Hydrochlorothiazide |
| $C_7H_8N_4O_2$ | Theophylline |
| $C_8H_{11}NO_5S$ | Sulbactam |
| $C_8H_{11}N_5O_3$ | Acyclovir |
| $C_8H_{15}N_7O_2S_3$ | Famotidine |
| $C_8H_{16}O_2$ | Valproic acid |
| $C_8H_6N_4O_5$ | Nitrofurantoin |
| $C_8H_9NO_2$ | Acetaminophen |
| $C_8H_9NO_5$ | Clavulanic acid |
| $C_9H_7N_7O_2S$ | Azathioprine |
| $C_9H_8O_4$ | Aspirin |
| $C_9H_{14}ClNO$ | Phenylpropanolamine |
| $C_9H_{15}NO_3S$ | Captopril |
| $C_{10}H_{11}N_3O_3S$ | Sulfamethoxazole |
| $C_{10}H_{13}N_5O_4$ | Zidovudine |
| $C_{10}H_{14}O_4$ | Guaifenesin |
| $C_{10}H_{16}N_6S$ | Cimetidine |
| $C_{11}H_{11}F_3N_2O_3$ | Flutamide |
| $C_{11}H_{16}N_2O_3$ | Butalbital |
| $C_{12}H_{11}ClN_2O_5S$ | Furosemide |
| $C_{12}H_{11}N_7$ | Triamterene |
| $C_{12}H_{17}N_3O_4S$ | Imipenem |
| $C_{12}H_{18}O$ | Propofol |
| $C_{12}H_{21}N_5O_2S_2$ | Nizatidine |
| $C_{12}H_{24}N_2O_4$ | Carisoprodol |
| $C_{12}H_{54}Al_{16}O_{75}S_8$ | Sucralfate |
| $C_{13}H_{12}F_2N_6O$ | Fluconazole |
| $C_{13}H_{18}N_4O_3$ | Pentoxifylline |
| $C_{13}H_{18}O_2$ | Ibuprofen |
| $C_{13}H_{21}NO_3$ | Albuterol |

| MOLECULAR FORMULA | COMPOUND |
|---|---|
| $C_{13}H_{22}N_4O_3S$ | Ranitidine |
| $C_{13}H_{24}N_4O_3S$ | Timolol |
| $C_{14}H_{11}Cl_2NO_2$ | Diclofenac |
| $C_{14}H_{14}O_3$ | Naproxen |
| $C_{14}H_{18}N_4O_3$ | Trimethoprim |
| $C_{14}H_{19}NO_2$ | Methylphenidate |
| $C_{14}H_{21}N_3O_2S$ | Sumatriptan |
| $C_{14}H_{22}N_2O_3$ | Atenolol |
| $C_{15}H_{10}ClN_3O_3$ | Clonazepam |
| $C_{15}H_{10}Cl_2N_2O_2$ | Lorazepam |
| $C_{15}H_{11}I_4NO_4$ | Levothyroxine |
| $C_{15}H_{12}N_2O$ | Carbamazepine |
| $C_{15}H_{12}N_2O_2$ | Phenytoin |
| $C_{15}H_{13}NO_3$ | Ketorolac |
| $C_{15}H_{13}N_3O_4S$ | Piroxicam |
| $C_{15}H_{14}ClN_3O_4S$ | Cefaclor |
| $C_{15}H_{16}O_2$ | Nabumetone |
| $C_{15}H_{22}O_3$ | Gemfibrozil |
| $C_{15}H_{25}NO_3$ | Metoprolol |
| $C_{16}H_{13}ClN_2O$ | Diazepam |
| $C_{16}H_{13}ClN_2O_2$ | Temazepam |
| $C_{16}H_{14}O_3$ | Ketoprofen |
| $C_{16}H_{15}N_5O_7S_2$ | Cefixime |
| $C_{16}H_{16}ClN_3O_3S$ | Indapamide |
| $C_{16}H_{16}ClN_3O_4$ | Loracarbef |
| $C_{16}H_{16}N_4O_8S$ | Cefuroxime |
| $C_{16}H_{17}N_3O_4S$ | Cephalexin |
| $C_{16}H_{17}N_3O_5S$ | Cefadroxil |
| $C_{16}H_{17}N_5O_7S_2$ | Cefotaxime |
| $C_{16}H_{18}N_2O_5S$ | Penicillin V |
| $C_{16}H_{19}N_3O_4S$ | Ampicillin |
| $C_{16}H_{19}N_3O_5S$ | Amoxicillin |
| $C_{16}H_{25}NO_2$ | Tramadol |
| $C_{16}H_{26}N_2O_5S$ | Cilastatin |
| $C_{17}H_{13}ClN_4$ | Alprazolam |
| $C_{17}H_{17}Cl_2N$ | Sertraline |
| $C_{17}H_{18}FN_3O_3$ | Ciprofloxacin |
| $C_{17}H_{18}F_3NO$ | Fluoxetine |
| $C_{17}H_{18}N_2O_6$ | Nifedipine |
| $C_{17}H_{19}N_3O_3S$ | Omeprazole |
| $C_{17}H_{20}N_2S$ | Promethazine |
| $C_{17}H_{21}NO_3$ | Etodolac |
| $C_{17}H_{21}NO_4$ | Scopolamine |
| $C_{17}H_{23}NO_3$ | Atropine |
| $C_{17}H_{27}NO_2$ | Venlafaxine |
| $C_{18}H_{13}ClFN_3$ | Midazolam |
| $C_{18}H_{15}NO_3$ | Oxaprozin |
| $C_{18}H_{18}N_8O_7S_3$ | Ceftriaxone |
| $C_{18}H_{18}O_2$ | Estrogens, conjugated (equilenin) |
| $C_{18}H_{19}ClN_4$ | Clozapine |
| $C_{18}H_{19}N_3O$ | Ondansetron |
| $C_{18}H_{19}N_3O_5S$ | Cefprozil |

| MOLECULAR FORMULA | COMPOUND |
|---|---|
| $C_{18}H_{20}FN_3O_4$ | Ofloxacin |
| $C_{18}H_{20}O_2$ | Estrogens, conjugated (equilin) |
| $C_{18}H_{21}NO_3$ | Codeine |
| $C_{18}H_{21}NO_3$ | Hydrocodone |
| $C_{18}H_{21}NO_4$ | Oxycodone |
| $C_{18}H_{22}O_2$ | Estrogens, conjugated (estrone) |
| $C_{18}H_{22}O_2$ | Estrogens, conjugated (17α-dihydroequilin) |
| $C_{18}H_{24}O_2$ | Estradiol; Estrogens, conjugated |
| $C_{18}H_{33}ClN_2O_5S$ | Clindamycin |
| $C_{19}H_{16}O_4$ | Warfarin |
| $C_{19}H_{21}N$ | Nortriptyline |
| $C_{19}H_{21}N_3O$ | Zolpidem |
| $C_{19}H_{25}N_5O_4$ | Terazosin |
| $C_{19}H_{35}NO_2$ | Dicyclomine |
| $C_{20}H_{21}N$ | Cyclobenzaprine |
| $C_{20}H_{23}N$ | Amitriptyline |
| $C_{20}H_{24}O_2$ | Ethinyl estradiol |
| $C_{20}H_{25}ClN_2O_5$ | Amlodipine |
| $C_{20}H_{25}NO_2$ | Adiphenine |
| $C_{20}H_{26}O_2$ | Norethindrone |
| $C_{20}H_{28}N_2O_5$ | Enalapril |
| $C_{20}H_{28}O_2$ | Retinoic acid |
| $C_{20}H_{30}BrNO_3$ | Ipratropium bromide |
| $C_{21}H_{25}N$ | Terbinafine |
| $C_{21}H_{26}O_5$ | Prednisone |
| $C_{21}H_{27}FO_6$ | Triamcinolone |
| $C_{21}H_{27}N_5O_4S$ | Glipizide |
| $C_{21}H_{28}O_2$ | Norgestrel |
| $C_{21}H_{30}O_5$ | Hydrocortisone |
| $C_{21}H_{31}N_3O_5$ | Lisinopril |
| $C_{21}H_{31}N_5O_2$ | Buspirone |
| $C_{22}H_{17}ClN_2$ | Clotrimazole |
| $C_{22}H_{22}N_6O_7S_2$ | Ceftazidime |
| $C_{22}H_{23}ClN_2O_2$ | Loratadine |
| $C_{22}H_{24}N_2O_8$ | Tetracycline |
| $C_{22}H_{24}N_2O_8$ | Doxycycline |
| $C_{22}H_{26}N_2O_4S$ | Diltiazem |
| $C_{22}H_{29}FO_5$ | Betamethasone |
| $C_{22}H_{29}NO_2$ | Propoxyphene |
| $C_{22}H_{30}O$ | Desogestrel |
| $C_{22}H_{30}O_5$ | Methylprednisolone |
| $C_{23}H_{16}O_{11}$ | Cromolyn |
| $C_{23}H_{25}N_5O_5$ | Doxazosin |
| $C_{23}H_{28}ClN_3O_5S$ | Glyburide |
| $C_{23}H_{29}ClFN_3O_4$ | Cisapride |
| $C_{23}H_{32}N_2O_5$ | Ramipril |
| $C_{23}H_{36}O_7$ | Pravastatin |
| $C_{23}H_{36}N_2O_2$ | Finasteride |
| $C_{23}H_{46}N_6O_{13}$ | Neomycin B |
| $C_{24}H_{26}FNO_4$ | Fluvastatin |
| $C_{24}H_{28}N_2O_5$ | Benazepril |
| $C_{24}H_{32}O_4$ | Ethynodiol diacetate |

# CAS REGISTRY NUMBER INDEX

CAS REGISTRY
NUMBER                COMPOUND

| | |
|---|---|
| 5015-36-1 | Methylprednisolone sodium phosphate |
| 5534-05-4 | Betamethasone acibutate |
| 5534-09-8 | Beclomethasone dipropionate |
| 5593-20-4 | Betamethasone dipropionate |
| 5600-19-1 | Theophylline isopropanolamine |
| 5611-51-8 | Triamcinolone hexacetonide |
| 5786-21-0 | Clozapine |
| 5913-71-3 | Codeine acetate |
| 5928-84-7 | Penicillin V benzathine |
| 5967-84-0 | Theophylline monohydrate |
| 6000-74-4 | Hydrocortisone sodium phosphate |
| 6020-73-1 | Codeine salicylate |
| 6069-47-8 | Codeine monohydrate |
| 6202-23-9 | Cyclobenzaprine hydrochloride |
| 6416-04-2 | Tetracycline trihydrate |
| 6493-05-6 | Pentoxifylline |
| 6533-00-2 | Norgestrel |
| 6533-68-2 | Scopolamine hydrobromide trihydrate |
| 6591-72-6 | Penicillin V hydrabamine |
| 6835-16-1 | Atropine (hyoscyamine sulfate dihydrate) |
| 6854-40-6 | Codeine sulfate trihydrate |
| 7177-48-2 | Ampicillin trihydrate |
| 8000-10-1 | Theophylline sodium glycinate |
| 8002-89-9 | Theophylline sodium acetate |
| 8049-62-5 | Insulin (zinc suspension) |
| 9002-72-6 | Somatropin |
| 9004-10-8 | Insulin (injection) |
| 9004-14-2 | Insulin (neutral insulin) |
| 9004-17-5 | Insulin (protamine zinc suspension) |
| 10238-21-8 | Glyburide |
| 11061-68-0 | Insulin (human) |
| 11070-73-8 | Insulin (cow) |
| 12584-58-6 | Insulin (pig) |
| 12629-01-5 | Somatropin (human) |
| 12650-69-0 | Mupirocin |
| 13311-84-7 | Flutamide |
| 13609-67-1 | Hydrocortisone butyrate |
| 14838-15-4 | Phenylpropanolamine |
| 15307-79-6 | Diclofenac sodium |
| 15686-71-2 | Cephalexin |
| 15687-27-1 | Ibuprofen |
| 15826-37-6 | Cromolyn sodium |
| 16110-51-3 | Cromolyn |
| 17086-28-1 | Doxycycline monohydrate |
| 17140-78-2 | Propoxyphene napsylate anhydrous (l-form) |
| 17140-81-7 | Nitrofurantoin monohydrate |
| 17693-51-5 | Promethazine teoclate |
| 18323-44-9 | Clindamycin |
| 18559-94-9 | Albuterol |
| 20830-75-5 | Digoxin |
| 21256-18-8 | Oxaprozin |

# CROSS REFERENCE TO
## *THE MERCK INDEX*

For each drug the relevant abstract number in the 12th edition of *The Merck Index* (Budavari, S., Ed., *The Merck Index*, 12th edition, Merck & Co. Inc.: Whitehouse Station, NJ, 1996) is given. Much useful information, such as melting point, solubility, optical rotation, and references to reviews, can be found in *The Merck Index*.

| | |
|---|---|
| acetaminophen | 45 |
| acyclovir | 148 |
| adiphenine | 160 |
| albuterol | 217 |
| alprazolam | 320 |
| alteplase | 9608 |
| amitriptyline | 511 |
| amlodipine | 516 |
| amoxicillin | 617 |
| ampicillin | 628 |
| aspirin | 886 |
| astemizole | 891 |
| atenolol | 892 |
| atropine | 907, 4907 |
| azathioprine | 935 |
| azithromycin | 946 |
| beclomethasone | 1047 |
| benazepril | 1058 |
| betamethasone | 1226 |
| bromocriptine | 1437 |
| buspirone | 1528 |
| butalbital | 1536 |
| captopril | 1817 |
| carbamazepine | 1826 |
| carboplatin | 1870 |
| carisoprodol | 1889 |
| cefaclor | 1962 |
| cefadroxil | 1963 |
| cefixime | 1975 |
| cefotaxime | 1983 |

# CROSS REFERENCE TO
# *THE ORGANIC CHEMISTRY*
# *OF DRUG SYNTHESIS*

For each drug the relevant sections in the series *The Organic Chemistry of Drug Synthesis* by Lednicer *et al.* are identified. The volume number is followed by the page number. This series gives valuable information about the syntheses of various drugs, and this may be helpful in determining impurities, understanding degradation reactions, and so on. The volumes in this series are:

Lednicer, D.; Mitscher, L.A. *The Organic Chemistry of Drug Synthesis*, John Wiley & Sons, Inc.: New York, 1977.

Lednicer, D.; Mitscher, L.A. *The Organic Chemistry of Drug Synthesis*, Volume 2, John Wiley & Sons, Inc.: New York, 1980.

Lednicer, D.; Mitscher, L.A. *The Organic Chemistry of Drug Synthesis*, Volume 3, John Wiley & Sons, Inc.: New York, 1984.

Lednicer, D.; Mitscher, L.A.; Georg, G.I. *The Organic Chemistry of Drug Synthesis*, Volume 4, John Wiley & Sons, Inc.: New York, 1990.

Lednicer, D. *The Organic Chemistry of Drug Synthesis*, Volume 5, John Wiley & Sons, Inc.: New York, 1995.

| | |
|---|---|
| acetaminophen | 1 111 |
| acyclovir | 3 229; 4 31; 4 116; 4 165 |
| adiphenine | 1 81 |
| albuterol | 2 43 |
| alprazolam | 3 197 |
| amitriptyline | 1 151, 404 |
| amlodipine | 4 108 |
| amoxicillin | 4 179, 180 |
| ampicillin | 1 413; 2 437; 4 179 |
| astemizole | 3 177 |
| atenolol | 2 109 |
| atropine | 1 35, 71, 93; 2 71 |
| azathiprine | 2 464 |
| benazepril | 5 135 |
| betamethasone | 1 198 |
| buspirone | 2 300; 4 119 |